Complete Solutions Guide

PRECALCULUS

FOURTH EDITION

Larson / Hostetler

Dianna L. Zook
Indiana University
Purdue University at Fort Wayne, Indiana

Emily J. Keaton

HOUGHTON MIFFLIN COMPANY Boston New York

Sponsoring Editor: Christine B. Hoag
Senior Associate Editor: Maureen Brooks
Managing Editor: Catherine B. Cantin
Senior Project Editor: Karen Carter
Associate Project Editor: Rachel D'Angelo Wimberly
Editorial Assistant: Caroline Lipscomb
Production Supervisor: Lisa Merrill
Art Supervisor: Gary Crespo
Marketing Associate: Ros Kane
Marketing Assistant: Kate Burden Thomas

Printed in the United States of America.

International Standard Book Number: 0-669-41744-0

123456789-PO 01 00 99 98 97

PREFACE

This *Complete Solutions Guide* is a supplement to the textbook *Precalculus,* Fourth Edition, by Roland E. Larson and Robert P. Hostetler. All references to chapters, theorems, and definitions apply to this text.

Solutions to the exercises in the text are given, with all essential algebraic steps included, in three parts. Part I of this guide contains solutions to the odd-numbered exercises. Part II contains solutions to the Chapter Tests and Cumulative Tests. Part III contains solutions to the even-numbered text exercises and Focus on Concepts. Part IV contains solutions to Chapter Projects. We have made every effort to see that the solutions are correct. However, we would appreciate hearing about any errors or other suggestions for improvement.

We would like to thank our husbands, Edward Schlindwein and Mark Keaton, for their support during this project.

Dianna L. Zook
Emily J. Keaton

CONTENTS

PART I Solutions to Odd-Numbered Exercises

CHAPTER P
Prerequisites

CHAPTER P
Prerequisites

Section P.1 Real Numbers

- You should know the following sets.
  - (a) The set of real numbers includes the rational numbers and the irrational numbers.
  - (b) The set of rational numbers includes all real numbers that can be written as the ratio p/q of two integers, where $q \neq 0$.
  - (c) The set of irrational numbers includes all real numbers which are not rational.
  - (d) The set of integers: $\{\ldots, -3, -2, -1, 0, 1, 2, 3, \ldots\}$
  - (e) The set of whole numbers: $\{0, 1, 2, 3, 4, \ldots\}$
  - (f) The set of natural numbers: $\{1, 2, 3, 4, \ldots\}$
- The real number line is used to represent the real numbers.
- Know the inequality symbols.
  - (a) $a < b$ means a is less than b.
  - (b) $a \leq b$ means a is less than or equal to b.
  - (c) $a > b$ means a is greater than b.
  - (d) $a \geq b$ means a is greater than or equal to b.
- You should know that

$$|a| = \begin{cases} a, & \text{if } a \geq 0 \\ -a, & \text{if } a < 0. \end{cases}$$

- Know the properties of absolute value.
  - (a) $|a| \geq 0$ (b) $|-a| = |a|$ (c) $|ab| = |a|\,|b|$ (d) $\left|\dfrac{a}{b}\right| = \dfrac{|a|}{|b|}$
- The distance between a and b on the real line is $|b - a| = |a - b|$.
- You should be able to identify the terms in an algebraic expression.
- You should know and be able to use the basic rules of algebra.
- Commutative Property
  - (a) Addition: $a + b = b + a$ (b) Multiplication: $a \cdot b = b \cdot a$
- Associative Property
  - (a) Addition: $(a + b) + c = a + (b + c)$ (b) Multiplication: $(ab)c = a(bc)$
- Identity Property
  - (a) Addition: 0 is the identity; $a + 0 = 0 + a = a$.
  - (b) Multiplication: 1 is the identity; $a \cdot 1 = 1 \cdot a = a$.
- Inverse Property
  - (a) Addition: $-a$ is the inverse of a; $a + (-a) = -a + a = 0$.
  - (b) Multiplication: $1/a$ is the inverse of a, $a \neq 0$; $a(1/a) = (1/a)a = 1$.
- Distributive Property
  - (a) Left: $a(b + c) = ab + ac$ (b) Right: $(a + b)c = ac + bc$

continued

- **Properties of Negatives**
  - (a) $(-1)a = -a$
  - (b) $-(-a) = a$
  - (c) $(-a)b = a(-b) = -ab$
  - (d) $(-a)(-b) = ab$
  - (e) $-(a + b) = (-a) + (-b) = -a - b$
- **Properties of Zero**
  - (a) $a \pm 0 = a$
  - (b) $a \cdot 0 = 0$
  - (c) $0 \div a = 0/a = 0, a \neq 0$
  - (d) If $ab = 0$, then $a = 0$ or $b = 0$.
  - (e) $a/0$ is undefined.
- **Properties of Fractions** $(b \neq 0, d \neq 0)$
  - (a) Equivalent Fractions: $a/b = c/d$ if and only if $ad = bc$.
  - (b) Rule of Signs: $-a/b = a/-b = -(a/b)$ and $-a/-b = a/b$
  - (c) Equivalent Fractions: $a/b = ac/bc, c \neq 0$
  - (d) Addition and Subtraction
    1. Like Denominators: $(a/b) \pm (c/b) = (a \pm c)/b$
    2. Unlike Denominators: $(a/b) \pm (c/d) = (ad \pm bc)/bd$
  - (e) Multiplication: $(a/b) \cdot (c/d) = ac/bd$
  - (f) Division: $(a/b) \div (c/d) = (a/b) \cdot (d/c) = ad/bc$ if $c \neq 0$.
- **Properties of Equality**
  - (a) If $a = b$, then $a + c = b + c$.
  - (b) If $a = b$, then $ac = bc$.
  - (c) If $a + c = b + c$, then $a = b$.
  - (d) If $ac = bc$ and $c \neq 0$, then $a = b$.

Solutions to Odd-Numbered Exercises

1. $-9, -\frac{7}{2}, 5, \frac{2}{3}, \sqrt{2}, 0, 1$

(a) Natural numbers: 5, 1

(b) Integers: $-9, 5, 0, 1$

(c) Rational numbers: $-9, -\frac{7}{2}, 5, \frac{2}{3}, 0, 1$

(d) Irrational numbers: $\sqrt{2}$

3. 2.01, 0.666 . . . , −13, 0.010110111 . . .

(a) Natural numbers: none

(b) Integers: −13

(c) Rational numbers: 2.01, 0.666 . . . , −13

(d) Irrational numbers: 0.010110111 . . .

5. $-\pi, -\frac{1}{3}, \frac{6}{3}, \frac{1}{2}\sqrt{2}, -7.5$

(a) Natural numbers: $\frac{6}{3}$ (since it equals 2)

(b) Integers: $\frac{6}{3}$

(c) Rational numbers: $-\frac{1}{3}, \frac{6}{3}, -7.5$

(d) Irrational numbers: $-\pi, \frac{1}{2}\sqrt{2}$

7. $\frac{5}{8} = 0.625$

9. $\frac{41}{333} = 0.\overline{123}$

11. $-1 < 2.5$

13. $\frac{3}{2} < 7$

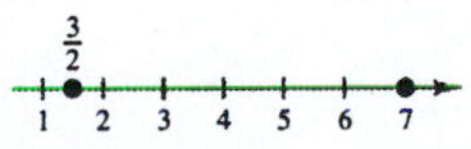

15. $-4 > -8$

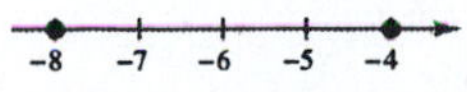

17. $\frac{5}{6} > \frac{2}{3}$

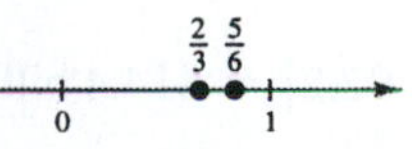

19. The inequality $x \le 5$ is the set of all real numbers less than or equal to 5. The interval is unbounded.

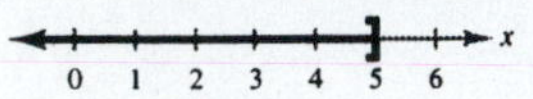

21. The inequality $x < 0$ is the set of all negative real numbers. The interval is unbounded.

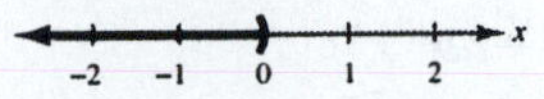

23. The inequality $x \ge 4$ is the set of all real numbers greater than or equal to 4. The interval is unbounded.

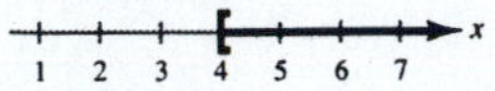

25. The inequality $-2 < x < 2$ is the set of all real numbers greater than -2 and less than 2. The interval is bounded.

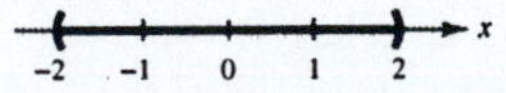

27. The inequality $-1 \le x < 0$ is the set of all negative real numbers greater than or equal to -1. The interval is bounded.

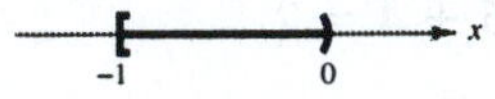

29. $\frac{127}{90} \approx 1.41111, \frac{584}{413} \approx 1.41404, \frac{7071}{5000} \approx 1.41420, \sqrt{2} \approx 1.41421, \frac{47}{33} \approx 1.42424$

31. $x < 0$

33. $y \ge 0$

35. $A \ge 30$

37. $|-10| = -(-10) = 10$

39. $|3 - \pi| = -(3 - \pi) = \pi - 3 \approx 0.1416$

41. $\frac{-5}{|-5|} = \frac{-5}{-(-5)} = \frac{-5}{5} = -1$

43. $-3|-3| = -3[-(-3)] = -9$

45. $-|16.25| + 20 = -16.25 + 20 = 3.75$

47. $|-3| > -|-3|$ since $3 > -3$.

49. $-5 = -|5|$ since $-5 = -5$.

51. $-|-2| = -|2|$ since $-2 = -2$.

53. $d(-1, 3) = |3 - (-1)| = |3 + 1| = 4$

55. $d(-\frac{5}{2}, 0) = \left|0 - (-\frac{5}{2})\right| = \frac{5}{2}$

57. $d(126, 75) = |75 - 126| = 51$

59. $d(\frac{16}{5}, \frac{112}{75}) = \left|\frac{112}{75} - \frac{16}{5}\right| = \frac{128}{75}$

61. $d(x, 5) = |x - 5|$ and $d(x, 5) \le 3$, thus $|x - 5| \le 3$.

63. $d(7, 18) = |7 - 18| = 11$ miles

65. $d(y, 0) = |y - 0| = |y|$ and $d(y, 0) \ge 6$, thus $|y| \ge 6$.

67.

Budgeted Expense, b	*Actual Expense, a*	$\lvert a - b\rvert$	$0.05b$
\$112,700	\$113,356	\$656	\$5635

The actual expense difference is greater than \$500 (but is less than 5% of the budget) so it does not pass the test.

69.

Budgeted Expense, b	*Actual Expense, a*	$\lvert a - b\rvert$	$0.05b$
\$37,640	\$37,335	\$305	\$1882

Since \$305 < \$500 and \$305 < \$1882, it passes the "budget variance test."

71. $|77.8 - 92.2| = \$14.4$ billion deficit for 1960

73. $|1031.3 - 1252.7| = \$221.4$ billion deficit for 1990

75. (a) $|u + v| \neq |u| + |v|$ if u is positive and v is negative or vice versa.

(b) $|u + v| \leq |u| + |v|$

They are equal when u and v have the same sign. If they differ in sign, $|u + v|$ is less than $|u| + |v|$.

77. $7x + 4$

Terms: $7x, 4$

79. $4x^3 + x - 5$

Terms: $4x^3, x, -5$

81. $4x - 6$

(a) $4(-1) - 6 = -4 - 6 = -10$

(b) $4(0) - 6 = 0 - 6 = -6$

83. $x^2 - 3x + 4$

(a) $(-2)^2 - 3(-2) + 4 = 4 + 6 + 4 = 14$

(b) $(2)^2 - 3(2) + 4 = 4 - 6 + 4 = 2$

85. $\dfrac{x + 1}{x - 1}$

(a) $\dfrac{1 + 1}{1 - 1} = \dfrac{2}{0}$ is undefined.

You cannot divide by zero.

(b) $\dfrac{-1 + 1}{-1 - 1} = \dfrac{0}{-2} = 0$

87. $x + 9 = 9 + x$

Commutative (addition)

89. $\dfrac{1}{(h + 6)}(h + 6) = 1, h \neq -6$

Inverse (multiplication)

91. $2(x + 3) = 2x + 6$

Distributive Property

93. $1 \cdot (1 + x) = 1 + x$

Identity (multiplication)

95. $x(3y) = (x \cdot 3)y$ Associative (multiplication)

$= (3x)y$ Commutative (multiplication)

97. $\dfrac{81 - (90 - 9)}{5} = \dfrac{81 - 81}{5} = \dfrac{0}{5} = 0$

99. $\dfrac{8 - 8}{-9 + (6 + 3)} = \dfrac{0}{-9 + 9} = \dfrac{0}{0}$ which is undefined.

101. $(4 - 7)(-2) = (-3)(-2) = 6$

103. $\frac{3}{16} + \frac{5}{16} = \frac{8}{16} = \frac{1}{2}$

105. $\dfrac{5}{8} - \dfrac{5}{12} + \dfrac{1}{6} = \dfrac{15}{24} - \dfrac{10}{24} + \dfrac{4}{24} = \dfrac{9}{24} = \dfrac{3}{8}$

107. $\dfrac{\not{4}}{5} \cdot \dfrac{1}{2} \cdot \dfrac{3}{\not{4}} \cdot = \dfrac{3}{10}$

109. $12 \div \frac{1}{4} = 12 \cdot \frac{4}{1} = 12 \cdot 4 = 48$

111. $-3 + \frac{3}{7} \approx -2.57$

113. $\dfrac{11.46 - 5.37}{3.91} \approx 1.56$

115.

n	1	0.5	0.01	0.0001	0.000001
$5/n$	5	10	500	50,000	5,000,000

117.

n	1	10	100	10,000	100,000
$5/n$	5	0.5	0.05	0.0005	0.00005

Section P.2 Exponents and Radicals

- You should know the properties of exponents.
  - (a) $a^1 = a$
  - (b) $a^0 = 1, a \neq 0$
  - (c) $a^m a^n = a^{m+n}$
  - (d) $a^m/a^n = a^{m-n}, a \neq 0$
  - (e) $a^{-n} = 1/a^n, a \neq 0$
  - (f) $(a^m)^n = a^{mn}$
  - (g) $(ab)^n = a^n b^n$
  - (h) $(a/b)^n = a^n/b^n, b \neq 0$
  - (i) $(a/b)^{-n} = (b/a)^n, a \neq 0, b \neq 0$
  - (j) $|a^2| = |a|^2 = a^2$
- You should be able to write numbers in scientific notation, $c \times 10^n$, where $1 \leq c < 10$ and n is an integer.
- You should be able to use your calculator to evaluate expressions involving exponents.
- You should know the properties of radicals.
  - (a) $\sqrt[n]{a^m} = (\sqrt[n]{a})^m$
  - (b) $\sqrt[n]{a} \cdot \sqrt[n]{b} = \sqrt[n]{ab}$
  - (c) $\dfrac{\sqrt[n]{a}}{\sqrt[n]{b}} = \sqrt[n]{\dfrac{a}{b}}$
  - (d) $\sqrt[m]{\sqrt[n]{a}} = \sqrt[mn]{a}$
  - (e) $(\sqrt[n]{a})^n = a$
  - (f) For n even, $\sqrt[n]{a^n} = |a|$.
    For n odd, $\sqrt[n]{a^n} = a$.
  - (g) $a^{1/n} = \sqrt[n]{a}$
  - (h) $a^{m/n} = (\sqrt[n]{a})^m = \sqrt[n]{a^m}, a \geq 0$
- You should be able to simplify radicals.
  - (a) All possible factors have been removed from the radical sign.
  - (b) All fractions have radical-free denominators.
  - (c) The index for the radical has been reduced as far as possible
- You should be able to use your calculator to evaluate radicals.

Solutions to Odd-Numbered Exercises

1. $-0.4^6 = -(0.4 \times 0.4 \times 0.4 \times 0.4 \times 0.4 \times 0.4)$

3. $(-10)(-10)(-10)(-10)(-10) = (-10)^5$

5. (a) $4^2 \cdot 3 = 16 \cdot 3 = 48$

(b) $3 \cdot 3^3 = 3^4 = 81$

7. (a) $(3^3)^2 = 3^6 = 729$

(b) $-3^2 = -9$

9. (a) $\dfrac{3}{3^{-4}} = 3^{1+4} = 3^5 = 243$

(b) $24(-2)^{-5} = \dfrac{24}{(-2)^5} = \dfrac{24}{-32} = -\dfrac{3}{4}$

11. $(-4)^3(5^2) = (-64)(25) = -1600$

13. $\dfrac{3^6}{7^3} = \dfrac{729}{343} \approx 2.125$

15. When $x = 2$, $-3x^3 = -3(2)^3 = -24$.

17. When $x = 10$, $6x^0 - (6x)^0 = 6(10)^0 - (60)^0 = 6(1) - 1 = 5$.

19. (a) $(-5z)^3 = (-5)^3z^3 = -125z^3$

(b) $5x^4(x^2) = 5x^{4+2} = 5x^6$

21. (a) $6y^2(2y^4)^2 = 6y^2 2^2 y^8 = 6 \cdot 4y^{2+8} = 24y^{10}$

(b) $\dfrac{3x^5}{x^3} = 3x^{5-3} = 3x^2$

23. (a) $\dfrac{7x^2}{x^3} = 7x^{2-3} = 7x^{-1} = \dfrac{7}{x}$

(b) $\dfrac{12(x+y)^3}{9(x+y)} = \dfrac{4}{3}(x+y)^{3-1} = \dfrac{4}{3}(x+y)^2$

25. (a) $(x+5)^0 = 1, x \neq -5$

(b) $(2x^2)^{-2} = \dfrac{1}{(2x^2)^2} = \dfrac{1}{4x^4}$

27. (a) $(-2x^2)^3(4x^3)^{-1} = \dfrac{-8x^6}{4x^3} = -2x^3$

(b) $\left(\dfrac{x}{10}\right)^{-1} = \dfrac{10}{x}$

29. (a) $(4a^{-2}b^3)^{-3} = 4^{-3}a^6b^{-9} = \dfrac{a^6}{4^3b^9} = \dfrac{a^6}{64b^9}$

(b) $\left(\dfrac{5x^2}{y^{-2}}\right)^{-4} = (5x^2y^2)^{-4} = \dfrac{1}{(5x^2y^2)^4} = \dfrac{1}{625x^8y^8}$

31. (a) $3^n \cdot 3^{2n} = 3^{n+2n} = 3^{3n}$

(b) $\left(\dfrac{a^{-2}}{b^{-2}}\right)\left(\dfrac{b}{a}\right)^3 = \left(\dfrac{b^2}{a^2}\right)\left(\dfrac{b^3}{a^3}\right) = \dfrac{b^5}{a^5}$

Radical Form	*Rational Exponent Form*
33. $\sqrt{9} = 3$ Given	$9^{1/2} = 3$ Answer
35. $\sqrt[5]{32} = 2$ Answer	$32^{1/5} = 2$ Given
37. $\sqrt{196} = 14$ Answer	$196^{1/2} = 14$ Given
39. $\sqrt[3]{-216} = -6$ Given	$(-216)^{1/3} = -6$ Answer
41. $\sqrt[3]{27^2} = (\sqrt[3]{27})^2 = 9$ Answer	$27^{2/3} = 9$ Given
43. $\sqrt[4]{81^3} = 27$ Given	$81^{3/4} = 27$ Answer

45. (a) $\sqrt{9} = 3$

(b) $\sqrt[3]{8} = 2$

47. (a) $-\sqrt[3]{-27} = -(-3) = 3$

(b) $\dfrac{4}{\sqrt{64}} = \dfrac{4}{8} = \dfrac{1}{2}$

49. (a) $(\sqrt[3]{-125})^3 = -125$

(b) $27^{1/3} = \sqrt[3]{27} = 3$

51. (a) $32^{-3/5} = \dfrac{1}{32^{3/5}} = \dfrac{1}{(\sqrt[5]{32})^3} = \dfrac{1}{(2)^3} = \dfrac{1}{8}$

(b) $\left(\dfrac{16}{81}\right)^{-3/4} = \left(\dfrac{81}{16}\right)^{3/4} = \left(\sqrt[4]{\dfrac{81}{16}}\right)^3 = \left(\dfrac{3}{2}\right)^3 = \dfrac{27}{8}$

53. (a) $\left(-\frac{1}{64}\right)^{-1/3} = (-64)^{1/3} = \sqrt[3]{-64} = -4$

(b) $\left(\frac{1}{\sqrt{32}}\right)^{-2/5} = (\sqrt{32})^{2/5} = \sqrt[5]{(\sqrt{32})^2} = \sqrt[5]{32} = 2$

55. (a) $\sqrt{57} \approx 7.550$

(b) $\sqrt[5]{-27^3} = (-27)^{3/5} \approx -7.225$

57. (a) $(1.2^{-2})\sqrt{75} + 3\sqrt{8} \approx 14.499$

(b) $\dfrac{-3 + \sqrt{21}}{3} \approx 0.528$

59. (a) $\sqrt{8} = \sqrt{4 \cdot 2} = \sqrt{4}\sqrt{2} = 2\sqrt{2}$

(b) $\sqrt[3]{24} = \sqrt[3]{8 \cdot 3} = \sqrt[3]{8}\sqrt[3]{3} = 2\sqrt[3]{3}$

61. (a) $\sqrt{72x^3} = \sqrt{36x^2 \cdot 2x} = 6x\sqrt{2x}$

(b) $\sqrt{\dfrac{18^2}{z^3}} = \dfrac{\sqrt{18^2}}{\sqrt{z^2 \cdot z}} = \dfrac{18}{z\sqrt{z}}$

63. (a) $\sqrt[3]{16x^5} = \sqrt[3]{8x^3 \cdot 2x^2} = 2x\sqrt[3]{2x^2}$

(b) $\sqrt{75x^2y^{-4}} = \sqrt{\dfrac{75x^2}{y^4}} = \dfrac{\sqrt{25x^2 \cdot 3}}{\sqrt{y^4}} = \dfrac{5|x|\sqrt{3}}{y^2}$

65. $5^{4/3} \cdot 5^{8/3} = 5^{12/3} = 5^4 = 625$

67. $\dfrac{(2x^2)^{3/2}}{2^{1/2}x^4} = \dfrac{2^{3/2}(x^2)^{3/2}}{2^{1/2}x^4} = \dfrac{2^{3/2}x^3}{2^{1/2}x^4} = 2^{3/2-1/2}x^{3-4} = 2^1x^{-1} = \dfrac{2}{x}$

69. $\dfrac{x^{-3} \cdot x^{1/2}}{x^{3/2} \cdot x^{-1}} = \dfrac{x^{1/2} \cdot x^1}{x^{3/2} \cdot x^3} = x^{1/2+1-3/2-3} = x^{-3} = \dfrac{1}{x^3}, x > 0$

71. (a) $\dfrac{1}{\sqrt{3}} = \dfrac{1}{\sqrt{3}} \cdot \dfrac{\sqrt{3}}{\sqrt{3}} = \dfrac{\sqrt{3}}{3}$

(b) $\dfrac{8}{\sqrt[3]{2}} = \dfrac{8}{\sqrt[3]{2}} \cdot \dfrac{\sqrt[3]{4}}{\sqrt[3]{4}} = \dfrac{8\sqrt[3]{4}}{2} = 4\sqrt[3]{4}$

73. (a) $\dfrac{2x}{5 - \sqrt{3}} = \dfrac{2x}{5 - \sqrt{3}} \cdot \dfrac{5 + \sqrt{3}}{5 + \sqrt{3}}$

$= \dfrac{2x(5 + \sqrt{3})}{25 - 3} = \dfrac{x(5 + \sqrt{3})}{11}$

(b) $\dfrac{3}{\sqrt{5} + \sqrt{6}} \cdot \dfrac{\sqrt{5} - \sqrt{6}}{\sqrt{5} - \sqrt{6}} = \dfrac{3(\sqrt{5} - \sqrt{6})}{5 - 6}$

$= -3(\sqrt{5} - \sqrt{6})$

$= 3(\sqrt{6} - \sqrt{5})$

75. (a) $\dfrac{\sqrt{8}}{2} = \dfrac{\sqrt{4 \cdot 2}}{2} = \dfrac{2\sqrt{2}}{2} = \dfrac{\sqrt{2}}{1} \cdot \dfrac{\sqrt{2}}{\sqrt{2}} = \dfrac{2}{\sqrt{2}}$

(b) $\sqrt[3]{\dfrac{9}{25}} = \dfrac{\sqrt[3]{9}}{\sqrt[3]{25}} \cdot \dfrac{\sqrt[3]{3}}{\sqrt[3]{3}} = \dfrac{\sqrt[3]{27}}{\sqrt[3]{75}} = \dfrac{3}{\sqrt[3]{75}}$

77. (a) $\dfrac{\sqrt{5} + \sqrt{3}}{3} = \dfrac{\sqrt{5} + \sqrt{3}}{3} \cdot \dfrac{\sqrt{5} - \sqrt{3}}{\sqrt{5} - \sqrt{3}} = \dfrac{5 - 3}{3(\sqrt{5} - \sqrt{3})}$

$= \dfrac{2}{3(\sqrt{5} - \sqrt{3})}$

(b) $\dfrac{\sqrt{7} - 3}{4} = \dfrac{\sqrt{7} - 3}{4} \cdot \dfrac{\sqrt{7} + 3}{\sqrt{7} + 3} = \dfrac{7 - 9}{4(\sqrt{7} + 3)}$

$= \dfrac{-2}{4(\sqrt{7} + 3)}$

$= -\dfrac{1}{2(\sqrt{7} + 3)}$

79. (a) $\sqrt[4]{3^2} = 3^{2/4} = 3^{1/2} = \sqrt{3}$

(b) $\sqrt[6]{(x + 1)^4} = (x + 1)^{4/6} = (x + 1)^{2/3} = \sqrt[3]{(x + 1)^2}$

81. (a) $\sqrt{\sqrt{32}} = (32^{1/2})^{1/2} = 32^{1/4} = \sqrt[4]{32} = \sqrt[4]{16 \cdot 2} = 2\sqrt[4]{2}$

(b) $\sqrt{\sqrt[4]{2x}} = ((2x)^{1/4})^{1/2} = (2x)^{1/8} = \sqrt[8]{2x}$

83. (a) $2\sqrt{50} + 12\sqrt{8} = 2\sqrt{25 \cdot 2} + 12\sqrt{4 \cdot 2}$

$= 2(5\sqrt{2}) + 12(2\sqrt{2}) = 10\sqrt{2} + 24\sqrt{2} = 34\sqrt{2}$

(b) $10\sqrt{32} - 6\sqrt{18} = 10\sqrt{16 \cdot 2} - 6\sqrt{9 \cdot 2}$

$= 10(4\sqrt{2}) - 6(3\sqrt{2}) = 40\sqrt{2} - 18\sqrt{2} = 22\sqrt{2}$

85. (a) $5\sqrt{x} - 3\sqrt{x} = 2\sqrt{x}$

(b) $-2\sqrt{9y} + 10\sqrt{y} = -2(3\sqrt{y}) + 10\sqrt{y} = -6\sqrt{y} + 10\sqrt{y} = 4\sqrt{y}$

87. $\sqrt{5} + \sqrt{3} \approx 3.968$ and $\sqrt{5 + 3} = \sqrt{8} \approx 2.828$
Thus, $\sqrt{5} + \sqrt{3} > \sqrt{5 + 3}$.

89. $\sqrt{3^2 + 2^2} = \sqrt{9 + 4} = \sqrt{13} \approx 3.606$
Thus, $5 > \sqrt{3^2 + 2^2}$.

91. $57{,}500{,}000 = 5.75 \times 10^7$ square miles

93. $0.0000899 = 8.99 \times 10^{-5}$

95. $5.24 \times 10^8 = 524{,}000{,}000$ servings

97. $4.8 \times 10^{-10} = 0.00000000048$ electrostatic units

99. (a) $750\left(1 + \dfrac{0.11}{365}\right)^{800} \approx 954.448$

(b) $\dfrac{67{,}000{,}000 + 93{,}000{,}000}{0.0052} = 30{,}769{,}230{,}769.2 \approx 3.077 \times 10^{10}$

101. (a) $\sqrt{4.5 \times 10^9} \approx 67{,}082.039$

(b) $\sqrt[3]{6.3 \times 10^4} \approx 39.791$

103. When any positive integer is squared, the units digit is 0, 1, 4, 5, 6, or 9. Therefore, $\sqrt{5233}$ is not an integer.

105. $T = 2\pi\sqrt{\dfrac{2}{32}} = 2\pi\sqrt{\dfrac{1}{16}} = 2\pi\left(\dfrac{1}{4}\right) = \dfrac{\pi}{2} \approx 1.57$ seconds

107. $r = 1 - \left(\dfrac{3225}{12{,}000}\right)^{1/4} \approx 0.280$ or 28%

109. $\text{Time} = \dfrac{\text{Distance}}{\text{Rate}} = \dfrac{93{,}000{,}000 \text{ miles}}{11{,}160{,}000 \text{ miles per minute}} = \dfrac{25}{3}$ minutes

Section P.3 Polynomials and Factoring

- Given a polynomial in x, $a_nx^n + a_{n-1}x^{n-1} + \ldots + a_1x + a_0$, where $a_n \neq 0$, and n is a nonnegative integer, you should be able to identify the following.
  - (a) Degree: n
  - (b) Terms: $a_nx^n, a_{n-1}x^{n-1}, \ldots, a_1x, a_0$
  - (c) Coefficients: $a_n, a_{n-1}, \ldots, a_1, a_0$
  - (d) Leading coefficient: a_n
  - (e) Constant term: a
- You should be able to add and subtract polynomials.
- You should be able to multiply polynomials by the Distributive Properties
- You should know the special binomial products.
  - (a) $(u + v)(u - v) = u^2 - v^2$
  - (b) $(u \pm v)^2 = u^2 \pm 2uv + v^2$
  - (c) $(u \pm v)^3 = u^3 \pm 3u^2v + 3uv^2 \pm v^3$
- You should be able to factor out all common factors, the first step in factoring.
- You should be able to factor the following special polynomial forms.
  - (a) $u^2 - v^2 = (u + v)(u - v)$
  - (b) $u^2 \pm 2uv + v^2 = (u \pm v)^2$
  - (c) $u^3 \pm v^3 = (u \pm v)(u^2 \mp uv + v^2)$
- You should be able to factor by grouping.
- You should be able to factor some trinomials by grouping.

Solutions to Odd-Numbered Exercises

1. $(6x + 5) - (8x + 15) = 6x + 5 - 8x - 15$
$= (6x - 8x) + (5 - 15)$
$= -2x - 10$

3. $-(x^3 - 2) + (4x^3 - 2x) = -x^3 + 2 + 4x^3 - 2x$
$= (4x^3 - x^3) - 2x + 2$
$= 3x^3 - 2x + 2$

5. $(15x^2 - 6) - (-8x^3 - 14x^2 - 17) = 15x^2 - 6 + 8x^3 + 14x^2 + 17$
$= 8x^3 + (15x^2 + 14x^2) + (-6 + 17)$
$= 8x^3 + 29x^2 + 11$

7. $5z - [3z - (10z + 8)] = 5z - (3z - 10z - 8)$
$= 5z - 3z + 10z + 8$
$= (5z - 3z + 10z) + 8$
$= 12z + 8$

9. $3x(x^2 - 2x + 1) = 3x(x^2) + 3x(-2x) + 3x(1)$
$= 3x^3 - 6x^2 + 3x$

11. $-5z(3z - 1) = -5z(3z) + (-5z)(-1)$
$= -15z^2 + 5z$

13. $(1 - x^3)(4x) = 1(4x) - x^3(4x) = 4x - 4x^4 = -4x^4 + 4x$

15. Add:
$$\begin{array}{r} 7x^3 - 2x^2 + 8 \\ -3x^3 \qquad\quad - 4 \\ \hline 4x^3 - 2x^2 + 4 \end{array}$$

17. Subtract:
$$\begin{array}{r} 5x^2 - 3x + 8 \\ - \qquad (x - 3) \\ \hline 5x^2 - 4x + 11 \end{array}$$

19. Multiply:
$$\begin{array}{r} -6x^2 + 15x - 4 \\ 5x + 3 \\ \hline -30x^3 + 75x^2 - 20x \qquad \\ - 18x^2 + 45x - 12 \\ \hline -30x^3 + 57x^2 + 25x - 12 \end{array}$$

21. Multiply:
$$\begin{array}{r} x^2 - x - 4 \qquad\quad \\ x^2 + 9 \qquad\quad \\ \hline x^4 - x^3 - 4x^2 \qquad\qquad \\ 9x^2 - 9x - 36 \\ \hline x^4 - x^3 + 5x^2 - 9x - 36 \end{array}$$

23. Multiply:
$$\begin{array}{l} x^2 - x + 1 \\ x^2 + x + 1 \\ \hline x^4 - x^3 + x^2 \\ \qquad x^3 - x^2 + x \\ \qquad\qquad x^2 - x + 1 \\ \hline x^4 - 0x^3 + x^2 + 0x + 1 = x^4 + x^2 + 1 \end{array}$$

25. $(x + 3)(x + 4) = x^2 + 4x + 3x + 12$ FOIL
$= x^2 + 7x + 12$

27. $(3x - 5)(2x + 1) = 6x^2 + 3x - 10x - 5$ FOIL
$= 6x^2 - 7x - 5$

29. $(2x + 3)^2 = (2x)^2 + 2(2x)(3) + 3^2$
$= 4x^2 + 12x + 9$

31. $(2x - 5y)^2 = 4x^2 - 2(5y)(2x) + 25y^2$
$= 4x^2 - 20xy + 25y^2$

33. $[(x - 3) + y]^2 = (x - 3)^2 + 2y(x - 3) + y^2$
$= x^2 - 6x + 9 + 2xy - 6y + y^2$
$= x^2 + 2xy + y^2 - 6x - 6y + 9$

35. $(x + 10)(x - 10) = x^2 - 100$

37. $(x + 2y)(x - 2y) = x^2 - (2y)^2 = x^2 - 4y^2$

39. $[(m - 3) + n][(m - 3) - n] = (m - 3)^2 - n^2$
$= m^2 - 6m + 9 - n^2$
$= m^2 - n^2 - 6m + 9$

41. $(2r^2 - 5)(2r^2 + 5) = (2r)^2 - 5^2 = 4r^4 - 25$

43. $(x + 1)^3 = x^3 + 3x^2(1) + 3x(1^2) + 1^3$
$= x^3 + 3x^2 + 3x + 1$

45. $(2x - y)^3 = (2x)^3 - 3(2x)^2y + 3(2x)y^2 - y^3$
$= 8x^3 - 12x^2y + 6xy^2 - y^3$

47. $(4x^3 - 3)^2 = (4x^3)^2 - 2(4x^3)(3) + (3)^2$
$= 16x^6 - 24x^3 + 9$

49. $5x(x + 1) - 3x(x + 1) = 2x(x + 1)$
$= 2x^2 + 2x$

51. $(u + 2)(u - 2)(u^2 + 4) = (u^2 - 4)(u^2 + 4)$
$= u^4 - 16$

53. No. $(x^2 + 1) + (-x^2 + 3) = 4$ which is not a second degree polynomial.

55. (a) $(x - 1)(x + 1) = x^2 - 1$
(b) $(x - 1)(x^2 + x + 1) = x^3 + x^2 + x - x^2 - x - 1 = x^3 - 1$
(c) $(x - 1)(x^3 + x^2 + x + 1) = x^4 + x^3 + x^2 + x - x^3 - x^2 - x - 1$
$= x^4 - 1$

From this pattern we have $(x - 1)(x^4 + x^3 + x^2 + x + 1) = x^5 - 1$.

57. (a) $500(1 + r)^2 = 500(r + 1)^2 = 500(r^2 + 2r + 1)$
$= 500r^2 + 1000r + 500$

(b)

r	$5\frac{1}{2}\%$	7%	8%	$8\frac{1}{2}\%$	9%
$500(1 + r)^2$	\$556.51	\$572.45	\$583.20	\$588.61	\$594.05

(c) As r increases, the amount increases.

59. $V = l \cdot w \cdot h$

$= \left(\frac{45 - 3x}{2}\right)(15 - 2x)(x)$

$= \frac{3}{2}x(x - 15)(2x - 15)$

When $x = 3$: $V = (18)(9)(3) = 486$ cubic cm

When $x = 5$: $V = (15)(5)(5) = 375$ cubic cm

When $x = 7$: $V = (12)(1)(7) = 84$ cubic cm

61. (a) $T = R + B = 1.1x + (0.14x^2 - 4.43x + 58.40)$

$= 0.14x^2 - 3.33x + 58.40$

(b)

x mi/hr	30	40	55
T feet	84.50	149.20	298.75

(c) As the speed increases, the total stopping distance increases.

63. $(x + 1)(x + 4) = x(x + 4) + 1(x + 4)$

$= x^2 + 4x + x + 4$

This illustrates the Distributive Property.

65. $3x + 6 = 3(x + 2)$

67. $2x^3 - 6x = 2x(x^2 - 3)$

69. $x^2 - 36 = x^2 - 6^2 = (x + 6)(x - 6)$

71. $16y^2 - 9 = (4y + 3)(4y - 3)$

73. $(x - 1)^2 - 4 = [(x - 1) + 2][(x - 1) - 2]$

$= (x + 1)(x - 3)$

75. $x^2 - 4x + 4 = x^2 - 2(2)x + 2^2 = (x - 2)^2$

77. $4t^2 + 4t + 1 = (2t)^2 + 2(2t)(1) + 1^2$

$= (2t + 1)^2$

79. $x^2 + x - 2 = (x + 2)(x - 1)$

81. $s^2 - 5s + 6 = (s - 3)(s - 2)$

83. $20 - y - y^2 = (5 + y)(4 - y)$ or $-(y + 5)(y - 4)$

85. $3x^2 - 5x + 2 = (3x - 2)(x - 1)$

87. $5x^2 + 26x + 5 = (5x + 1)(x + 5)$

89. $x^3 - 8 = x^3 - 2^3 = (x - 2)(x^2 + 2x + 4)$

91. $y^3 + 64 = y^3 + 4^3 = (y + 4)(y^2 - 4y + 16)$

93. $x^3 - x^2 + 2x - 2 = x^2(x - 1) + 2(x - 1)$

$= (x - 1)(x^2 + 2)$

95. $2x^3 - x^2 - 6x + 3 = x^2(2x - 1) - 3(2x - 1)$

$= (2x - 1)(x^2 - 3)$

97. $x^3 - 9x = x(x^2 - 9) = x(x + 3)(x - 3)$

99. $x^3 - 4x^2 = x^2(x - 4)$

101. $x^2 - 2x + 1 = (x - 1)^2$

103. $1 - 4x + 4x^2 = (1 - 2x)^2$

105. $2x^2 + 4x - 2x^3 = -2x(-x - 2 + x^2)$
$= -2x(x^2 - x - 2)$
$= -2x(x + 1)(x - 2)$

107. $9x^2 + 10x + 1 = (9x + 1)(x + 1)$

109. $3x^3 + x^2 + 15x + 5 = x^2(3x + 1) + 5(3x + 1)$
$= (3x + 1)(x^2 + 5)$

111. $x^4 - 4x^3 + x^2 - 4x = x(x^3 - 4x^2 + x - 4)$
$= x[x^2(x - 4) + (x - 4)]$
$= x(x - 4)(x^2 + 1)$

113. $25 - (z + 5)^2 = [5 + (z + 5)][5 - (z + 5)]$
$= -z(z + 10)$

115. $(x^2 + 1)^2 - 4x^2 = [(x^2 + 1) + 2x][(x^2 + 1) - 2x]$
$= (x^2 + 2x + 1)(x^2 - 2x + 1)$
$= (x + 1)^2(x - 1)^2$

117. $2t^3 - 16 = 2(t^3 - 8) = 2(t - 2)(t^2 + 2t + 4)$

119. $4x(2x - 1) + (2x - 1)^2 = (2x - 1)[4x + (2x - 1)]$
$= (2x - 1)(6x - 1)$

121. $2(x + 1)(x - 3)^2 - 3(x + 1)^2(x - 3) = (x + 1)(x - 3)[2(x - 3) - 3(x + 1)]$
$= (x + 1)(x - 3)[2x - 6 - 3x - 3]$
$= (x + 1)(x - 3)(-x - 9)$
$= -(x + 1)(x - 3)(x + 9)$

123. $7x(2)(x^2 + 1)(2x) - (x^2 + 1)^2(7) = 7(x^2 + 1)[4x^2 - (x^2 + 1)]$
$= 7(x^2 + 1)(3x^2 - 1)$

125. $2x(x - 5)^4 - x^2(4)(x - 5)^3 = 2x(x - 5)^3[(x - 5) - 2x]$
$= 2x(x - 5)^3(-x - 5)$
$= -2x(x - 5)^3(x + 5)$

127. $\frac{x^2}{2}(x^2 + 1)^4 - (x^2 + 1)^5 = (x^2 + 1)^4\left[\frac{x^2}{2} - (x^2 + 1)\right]$
$= (x^2 + 1)^4\left(-\frac{x^2}{2} - 1\right)$
$= -(x^2 + 1)^4\left(\frac{x^2}{2} + 1\right)$

129. $a^2 - b^2 = (a + b)(a - b)$
Matches model (b).

131. $a^2 + 2a + 1 = (a + 1)^2$
Matches model (a).

133. $3x^2 + 7x + 2 = (3x + 1)(x + 2)$

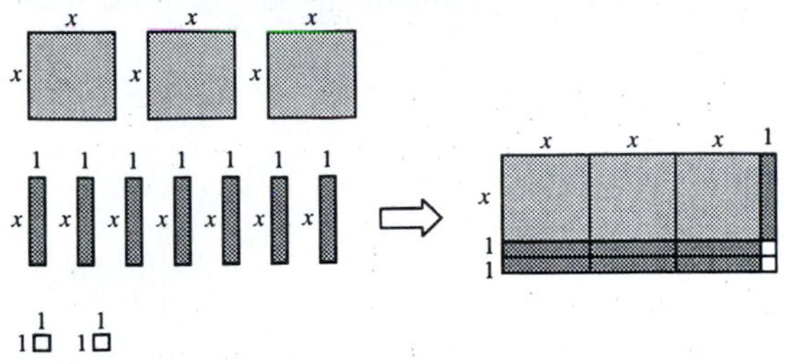

135. $2x^2 + 7x + 3 = (2x + 1)(x + 3)$

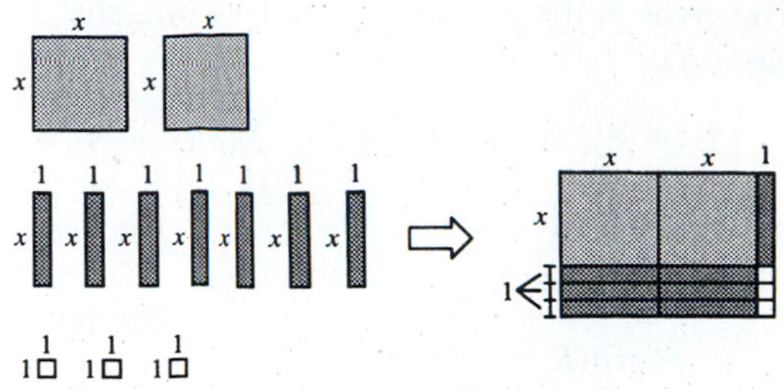

137. $A = \pi(r + 2)^2 - \pi r^2$

$= \pi[(r + 2)^2 - r^2]$

$= \pi[r^2 + 4r + 4 - r^2]$

$= \pi(4r + 4)$

$= 4\pi(r + 1)$

139. $A = 8(18) - 4x^2$

$= 4(36 - x^2)$

$= 4(6 - x)(6 + x)$

141. For $x^2 + bx - 15$ to be factorable, b must equal $m + n$ where $mn = -15$.

Factors of -15	Sum of factors
$(15)(-1)$	$15 + (-1) = 14$
$(-15)(1)$	$-15 + 1 = -14$
$(3)(-5)$	$3 + (-5) = -2$
$(-3)(5)$	$-3 + 5 = 2$

The possible b values are 14, -14, -2, or 2.

143. For $2x^2 + 5x + c$ to be factorable, the factors of $2c$ must add up to 5.

Possible c values	$2c$	Factors of $2c$ that add up to 5
2	4	$(1)(4) = 4$ and $1 + 4 = 5$
3	6	$(2)(3) = 6$ and $2 + 3 = 5$
-3	-6	$(6)(-1) = -6$ and $6 + (-1) = 5$
-7	-14	$(7)(-2) = -14$ and $7 + (-2) = 5$
-12	-24	$(8)(-3) = -24$ and $8 + (-3) = 5$

These are a few possible c values. There are *many* correct answers.

If $c = 2$: $2x^2 + 5x + 2 = (2x + 1)(x + 2)$
If $c = 3$: $2x^2 + 5x + 3 = (2x + 3)(x + 1)$
If $c = -3$: $2x^2 + 5x - 3 = (2x - 1)(x + 3)$
If $c = -7$: $2x^2 + 5x - 7 = (2x + 7)(x - 1)$
If $c = -12$: $2x^2 + 5x - 12 = (2x - 3)(x + 4)$

145. $9x^2 - 9x - 54 = 9(x^2 - x - 6) = 9(x + 2)(x - 3)$

The error in the problem in the book was that 3 was factored out of the first binomial but not out of the second binomial.

$$(3x + 6)(3x - 9) = 3(x + 2)(3)(x - 3) = 9(x + 2)(x - 3)$$

147. (a) $V = \pi R^2 h - \pi r^2 h$
$= \pi h(R^2 - r^2)$
$= \pi h(R - r)(R + r)$

(b) The average radius is $\dfrac{R + r}{2}$.

The thickness of the shell is $R - r$.

$$V = \pi h(R - r)(R + r) = 2\pi\left(\frac{R + r}{2}\right)(R - r)h$$

Section P.4 Fractional Expressions

- You should be able to find the domain of a fractional expression.
- You should know that a rational expression is the quotient of two polynomials.
- You should be able to simplify rational expressions by reducing them to lowest terms. This may involve factoring both the numerator and the denominator.
- You should be able to add, subtract, multiply, and divide rational expressions.
- You should be able to simplify compound fractions.

Solutions to Odd-Numbered Exercises

1. The domain of the polynomial $3x^2 - 4x + 7$ is the set of all real numbers.

3. The domain of the polynomial $4x^3 + 3, x \geq 0$ is the set of non-negative real numbers, since the polynomial is restricted to that set.

5. The domain of $\dfrac{1}{x - 2}$ is the set of all real numbers x such that $x \neq 2$.

7. The domain of $\dfrac{x - 1}{x(x - 4)}$ is the set of all real numbers x such that $x \neq 0$ and $x \neq 4$.

9. The domain of $\sqrt{x + 1}$ is the set of all real numbers x such that $x \geq -1$.

11. $\dfrac{5}{2x} = \dfrac{5(3x)}{(2x)(3x)} = \dfrac{5(3x)}{6x^2}, \quad x \neq 0$

The missing factor is $3x, x \neq 0$.

13. $\dfrac{x + 1}{x} = \dfrac{(x + 1)(x - 2)}{x(x - 2)}, \quad x \neq 2$

The missing factor is $x - 2, x \neq 2$.

15. $\dfrac{3x}{x - 3} = \dfrac{3x(x)}{(x - 3)(x)} = \dfrac{3x^2}{x^2 - 3x}, \quad x \neq 0$

The missing factor is $x, x \neq 0$.

17. $\dfrac{15x^2}{10x} = \dfrac{5x(3x)}{5x(2)} = \dfrac{3x}{2}, \quad x \neq 0$

19. $\dfrac{3xy}{xy + x} = \dfrac{x(3y)}{x(y + 1)} = \dfrac{3y}{y + 1}, \quad x \neq 0$

21. $\dfrac{x - 5}{10 - 2x} = \dfrac{x - 5}{-2(x - 5)} = -\dfrac{1}{2}, \quad x \neq 5$

23. $\dfrac{x^3 + 5x^2 + 6x}{x^2 - 4} = \dfrac{x(x + 2)(x + 3)}{(x + 2)(x - 2)} = \dfrac{x(x + 3)}{x - 2}, \quad x \neq -2$

25. $\dfrac{y^2 - 7y + 12}{y^2 + 3y - 18} = \dfrac{(y-3)(y-4)}{(y+6)(y-3)} = \dfrac{y-4}{y+6}, \quad x \neq 3$

27. $\dfrac{2 - x + 2x^2 - x^3}{x - 2} = \dfrac{(2-x) + x^2(2-x)}{-(2-x)} = \dfrac{(2-x)(1+x^2)}{-(2-x)} = -(1+x^2), \quad x \neq 2$

29. $\dfrac{z^3 - 8}{z^2 + 2z + 4} = \dfrac{(z-2)(z^2+2z+4)}{z^2+2z+4} = z - 2$

31.

x	0	1	2	3	4	5	6
$\dfrac{x^2 - 2x - 3}{x - 3}$	1	2	3	undef.	5	6	7
$x + 1$	1	2	3	4	5	6	7

The expressions are equivalent except at $x = 3$.

33. $\dfrac{5x^3}{2x^3 + 4} = \dfrac{5x^3}{2(x^3 + 2)}$. There are no common factors so this expression is in reduced form. In this case factors of terms were incorrectly cancelled.

35. $\dfrac{\pi r^2}{(2r)^2} = \dfrac{\pi r^2}{4r^2} = \dfrac{\pi}{4}, \quad r \neq 0$

37. $\dfrac{5}{x-1} \cdot \dfrac{x-1}{25(x-2)} = \dfrac{1}{5(x-2)}, \quad x \neq 1$

39. $\dfrac{(x+5)(x-3)}{x+2} \cdot \dfrac{1}{(x+5)(x+2)} = \dfrac{x-3}{(x+2)^2}, \quad x \neq -5$

41. $\dfrac{r}{r-1} \cdot \dfrac{r^2-1}{r^2} = \dfrac{r(r+1)(r-1)}{r^2(r-1)} = \dfrac{r+1}{r}, \quad r \neq 1$

43. $\dfrac{t^2 - t - 6}{t^2 + 6t + 9} \cdot \dfrac{t+3}{t^2 - 4} = \dfrac{(t-3)(t+2)(t+3)}{(t+3)^2(t+2)(t-2)} = \dfrac{t-3}{(t+3)(t-2)}, \quad t \neq -2$

45. $\dfrac{x^2 + xy - 2y^2}{x^3 + x^2y} \cdot \dfrac{x}{x^2 + 3xy + 2y^2} = \dfrac{(x+2y)(x-y)}{x^2(x+y)} \cdot \dfrac{x}{(x+2y)(x+y)}$

$$= \frac{x-y}{x(x+y)^2}, \quad x \neq -2y$$

47. $\dfrac{3(x+y)}{4} \div \dfrac{x+y}{2} = \dfrac{3(x+y)}{4} \cdot \dfrac{2}{x+y} = \dfrac{3}{2}, \quad x \neq -y$

49. $\dfrac{\left[\dfrac{x^2}{(x+1)^2}\right]}{\left[\dfrac{x}{(x+1)^3}\right]} = \dfrac{x^2}{(x+1)^2} \cdot \dfrac{(x+1)^3}{x} = x(x+1), \quad x \neq -1, 0$

51. $\dfrac{5}{x-1} + \dfrac{x}{x-1} = \dfrac{5+x}{x-1} = \dfrac{x+5}{x-1}$

53. $6 - \dfrac{5}{x+3} = \dfrac{6(x+3)}{(x+3)} - \dfrac{5}{x+3} = \dfrac{6(x+3) - 5}{x+3} = \dfrac{6x+13}{x+3}$

55. $\dfrac{3}{x-2} + \dfrac{5}{2-x} = \dfrac{3}{x-2} - \dfrac{5}{x-2} = -\dfrac{2}{x-2}$

57. $\dfrac{2}{x^2-4} - \dfrac{1}{x^2-3x+2} = \dfrac{2}{(x+2)(x-2)} - \dfrac{1}{(x-1)(x-2)}$

$= \dfrac{2(x-1)-(x+2)}{(x+2)(x-2)(x-1)} = \dfrac{x-4}{(x+2)(x-2)(x-1)}$

59. $\dfrac{1}{x^2-x-2} - \dfrac{x}{x^2-5x+6} = \dfrac{1}{(x-2)(x+1)} - \dfrac{x}{(x-2)(x-3)}$

$= \dfrac{(x-3)-x(x+1)}{(x+1)(x-2)(x-3)}$

$= \dfrac{-x^2-3}{(x+1)(x-2)(x-3)} = -\dfrac{x^2+3}{(x+1)(x-2)(x-3)}$

61. $-\dfrac{1}{x} + \dfrac{2}{x^2+1} + \dfrac{1}{x^3+x} = \dfrac{-(x^2+1)}{x(x^2+1)} + \dfrac{2x}{x(x^2+1)} + \dfrac{1}{x(x^2+1)}$

$= \dfrac{-x^2-1+2x+1}{x(x^2+1)} = \dfrac{-x^2+2x}{x(x^2+1)} = \dfrac{-x(x-2)}{x(x^2+1)}$

$= -\dfrac{x-2}{x^2+1} = \dfrac{2-x}{x^2+1}, \quad x \neq 0$

63. $x^2(x^2+1)^{-5} - (x^2+1)^{-4} = (x^2+1)^{-5}[x^2-(x^2+1)]$

$= -\dfrac{1}{(x^2+1)^5}$

65. $\dfrac{x+4}{x+2} - \dfrac{3x-8}{x+2} = \dfrac{(x+4)-(3x-8)}{x-2}$

$= \dfrac{x+4-3x+8}{x-2}$

$= \dfrac{-2x+12}{x-2}$

$= \dfrac{-2(x-6)}{x-2}$

The error was an incorrect subtraction in the numerator.

67. $\dfrac{\left(\dfrac{x}{2}-1\right)}{(x-2)} = \dfrac{\left(\dfrac{x}{2}-\dfrac{2}{2}\right)}{\left(\dfrac{x-2}{1}\right)}$

$= \dfrac{x-2}{2} \cdot \dfrac{1}{x-2}$

$= \dfrac{1}{2}, \quad x \neq 2$

69. $\dfrac{\left(\dfrac{1}{x}-\dfrac{1}{x+1}\right)}{\left(\dfrac{1}{x+1}\right)} = \dfrac{\dfrac{(x+1)-x}{x(x+1)}}{\dfrac{1}{x+1}} = \dfrac{1}{x(x+1)} \cdot \dfrac{x+1}{1} = \dfrac{1}{x}, \quad x \neq -1$

71. $$\frac{\left(\frac{x+3}{x-3}\right)^2}{\frac{1}{x+3}+\frac{1}{x-3}} = \frac{\frac{(x+3)^2}{(x-3)^2}}{\frac{(x-3)+(x+3)}{(x+3)(x-3)}}$$

$$= \frac{(x+3)^2}{(x-3)^2} \cdot \frac{(x+3)(x-3)}{2x} = \frac{(x+3)^3}{2x(x-3)}, \quad x \neq -3$$

73. $$\frac{\left[\frac{1}{(x+h)^2} - \frac{1}{x^2}\right]}{h} = \frac{\left[\frac{1}{(x+h)^2} - \frac{1}{x^2}\right]}{h} \cdot \frac{x^2(x+h)^2}{x^2(x+h)^2}$$

$$= \frac{x^2 - (x+h)^2}{hx^2(x+h)^2}$$

$$= \frac{x^2 - (x^2 + 2xh + h^2)}{hx^2(x+h)^2}$$

$$= \frac{-h(2x+h)}{hx^2(x+h)^2}$$

$$= -\frac{2x+h}{x^2(x+h)^2}, \quad h \neq 0$$

75. $$\frac{\left(\sqrt{x} - \frac{1}{2\sqrt{x}}\right)}{\sqrt{x}} = \frac{\left(\sqrt{x} - \frac{1}{2\sqrt{x}}\right)}{\sqrt{x}} \cdot \frac{2\sqrt{x}}{2\sqrt{x}} = \frac{2x-1}{2}, \quad x > 0$$

77. $$\frac{\frac{t^2}{\sqrt{t^2+1}} - \sqrt{t^2+1}}{t^2} = \frac{\left[\frac{t^2}{\sqrt{t^2+1}} - \sqrt{t^2+1}\right]}{t^2} \cdot \frac{\sqrt{t^2+1}}{\sqrt{t^2+1}}$$

$$= \frac{t^2 - (t^2+1)}{t^2\sqrt{t^2+1}} = -\frac{1}{t^2\sqrt{t^2+1}}$$

79. $$\frac{x(x+1)^{-3/4} - (x+1)^{1/4}}{x^2} = \frac{x(x+1)^{-3/4} - (x+1)^{1/4}}{x^2} \cdot \frac{(x+1)^{3/4}}{(x+1)^{3/4}}$$

$$= \frac{x(x+1)^0 - (x+1)^1}{x^2(x+1)^{3/4}}$$

$$= \frac{x - x - 1}{x^2(x+1)^{3/4}}$$

$$= -\frac{1}{x^2(x+1)^{3/4}}$$

81. $$\frac{\sqrt{x+2} - \sqrt{x}}{2} = \frac{\sqrt{x+2} - \sqrt{x}}{2} \cdot \frac{\sqrt{x+2} + \sqrt{x}}{\sqrt{x+2} + \sqrt{x}}$$

$$= \frac{(x+2) - x}{2\left(\sqrt{x+2} + \sqrt{x}\right)} = \frac{2}{2\left(\sqrt{x+2} + \sqrt{x}\right)}$$

$$= \frac{1}{\sqrt{x+2} + \sqrt{x}}$$

83. (a) $\frac{1}{16}$ minute

(b) $x\left(\frac{1}{16}\right) = \frac{x}{16}$ minutes

(c) $\frac{60}{16} = \frac{15}{4}$ minutes

85. Average $= \dfrac{\left(\frac{x}{3} + \frac{2x}{5}\right)}{2} = \dfrac{\left(\frac{x}{3} + \frac{2x}{5}\right)}{2} \cdot \dfrac{15}{15} = \dfrac{5x + 6x}{30} = \dfrac{11x}{30}$

87. (a) $r = \dfrac{\left(\dfrac{24[48(400) - 15{,}000]}{48}\right)}{\left[15{,}000 + \dfrac{48(400)}{12}\right]} \approx 0.1265 = 12.65\%$

(b) $r = \dfrac{\left[\dfrac{24(NM - P)}{N}\right]}{\left(P + \dfrac{NM}{12}\right)} = \dfrac{24(NM - P)}{N} \cdot \dfrac{12}{12P + NM} = \dfrac{288(NM - P)}{N(12P + NM)}$

$r = \dfrac{288[48(400) - 15{,}000]}{48[12(15{,}000) + 48(400)]} \approx 0.1265 = 12.65\%$

89. $T = 10\left(\dfrac{4t^2 + 16t + 75}{t^2 + 4t + 10}\right)$

(a)

t	0	1	2	3	4	5
T	75°	63.3°	55.9°	51.3°	48.3°	46.4°

(b)

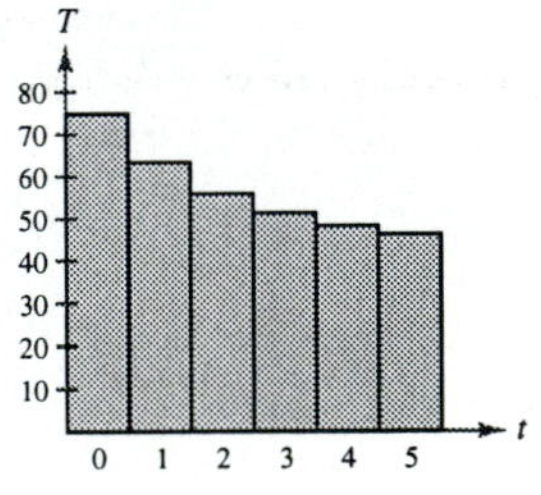

91. $\dfrac{x\left(\frac{x}{2}\right)}{x(2x + 1)} = \dfrac{\frac{x}{2}}{2x + 1} \cdot \dfrac{2}{2} = \dfrac{x}{2(2x + 1)}$

Section P.5 Solving Equations

- You should know how to solve linear equations.
  $ax + b = 0$
- An identity is an equation whose solution consists of every real number in its domain.
- To solve an equation you can:
  (a) Add or subtract the same quantity from both sides.
  (b) Multiply or divide both sides by the same nonzero quantity.
- To solve an equation that can be simplified to a linear equation:
  (a) Remove all symbols of grouping and all fractions.
  (b) Combine like terms.
  (c) Solve by algebra.
  (d) Check the answer.
- A "solution" that does not satisfy the original equation is called an extraneous solution.
- You should be able to solve a quadratic equation by factoring, if possible.
- You should be able to solve a quadratic equation of the form $u^2 = d$ by extracting square roots.
- You should be able to solve a quadratic equation by completing the square.
- You should know and be able to use the Quadratic Formula: For $ax^2 + bx + c = 0, a \neq 0,$

  $$x = \frac{-b \pm \sqrt{b^2 - 4ac}}{2a}.$$

- You should be able to solve polynomials of higher degree by factoring.
- For equations involving radicals or fractional powers, raise both sides to the same power.
- For equations with fractions, multiply both sides by the least common denominator to clear the fractions.
- For equations involving absolute value, remember that the expression inside the absolute value can be positive or negative.
- Always check for extraneous solutions.

Solutions to Odd-Numbered Exercises

1. $5x - 3 = 3x + 5$

(a) $5(0) - 3 \stackrel{?}{=} 3(0) + 5$

$-3 \neq 5$

$x = 0$ *is not* a solution.

(b) $5(-5) - 3 \stackrel{?}{=} 3(-5) + 5$

$-28 \neq -10$

$x = -5$ *is not* a solution.

(c) $5(4) - 3 \stackrel{?}{=} 3(4) + 5$

$17 = 17$

$x = 4$ *is* a solution.

(d) $5(10) - 3 \stackrel{?}{=} 3(10) + 5$

$47 \neq 35$

$x = 10$ *is not* a solution.

3. $3x^2 + 2x - 5 = 2x^2 - 2$

(a) $3(-3) + 2(-3) - 5 \stackrel{?}{=} 2(-3)^2 - 2$

$$16 = 16$$

$x = -3$ *is* a solution.

(b) $3(1)^2 + 2(1) - 5 \stackrel{?}{=} 2(1)^2 - 2$

$$0 = 0$$

$x = 1$ *is* a solution.

(c) $3(4)^2 + 2(4) - 5 \stackrel{?}{=} 2(4)^2 - 2$

$$51 \neq 30$$

$x = 4$ *is not* a solution.

(d) $3(-5)^2 + 2(-5) - 5 \stackrel{?}{=} 2(-5)^2 - 2$

$$60 \neq 48$$

$x = -5$ *is not* a solution.

5. $\dfrac{5}{2x} - \dfrac{4}{x} = 3$

(a) $\dfrac{5}{2(-1/2)} - \dfrac{5}{(-1/2)} \stackrel{?}{=} 3$

$$3 = 3$$

$x = -\frac{1}{2}$ *is* a solution.

(b) $\dfrac{5}{2(4)} - \dfrac{4}{4} \stackrel{?}{=} 3$

$$-\frac{3}{8} \neq 3$$

$x = 4$ *is not* a solution.

(c) $\dfrac{5}{2(0)} - \dfrac{4}{0}$ is undefined.

$x = 0$ *is not* a solution.

(d) $\dfrac{5}{2(1/4)} - \dfrac{4}{1/4} \stackrel{?}{=} 3$

$$-6 \neq 3$$

$x = \frac{1}{4}$ *is not* a solution.

7. $2(x - 1) = 2x - 2$ is an *identity* by the Distributive Property. It is true for all real values of x.

9. $-6(x - 3) + 5 = -2x + 10$ is *conditional.* There are real values of x for which the equation is not true.

11. $x^2 - 8x + 5 = (x - 4)^2 - 11$ is an *identity* since $(x - 4)^2 - 11 = x^2 - 8x + 16 - 11 = x^2 - 8x + 5$.

13. (a) Equivalent equations are derived from the substitution principle and simplification techniques. They have the same solution(s).

$2x + 3 = 8$ and $2x = 5$ are equivalent equations.

(b) Equivalent equations are produced by removing symbols of grouping, adding or subtracting the same quantity from both sides of the equation, multiplying or dividing both sides of the equation by the same nonzero quantity, or by interchanging the two sides of the equation.

15.
$$\begin{aligned} 2(x + 5) - 7 &= 3(x - 2) \\ 2x + 10 - 7 &= 3x - 6 \\ 2x + 3 &= 3x - 6 \\ -x &= -9 \\ x &= 9 \end{aligned}$$

17.
$$\begin{aligned} \frac{5x}{4} + \frac{1}{2} &= x - \frac{1}{2} \\ 4\left(\frac{5x}{4}\right) + 4\left(\frac{1}{2}\right) &= 4(x) - 4\left(\frac{1}{2}\right) \\ 5x + 2 &= 4x - 2 \\ x &= -4 \end{aligned}$$

19. $$\begin{aligned} 0.25x + 0.75(10 - x) &= 3 \\ 4(0.25x) + 4(0.75)(10 - x) &= 4(3) \\ x + 3(10 - x) &= 12 \\ x + 30 - 3x &= 12 \\ -2x &= -18 \\ x &= 9 \end{aligned}$$

21. $$\begin{aligned} x + 8 &= 2(x - 2) - x \\ x + 8 &= 2x - 4 - x \\ x + 8 &= x - 4 \\ 8 &= -4 \end{aligned}$$

Contradiction: no solution

23. $$\begin{aligned} \frac{100 - 4u}{3} &= \frac{5u + 6}{4} + 6 \\ 12\left(\frac{100 - 4u}{3}\right) &= 12\left(\frac{5u + 6}{4}\right) + 12(6) \\ 4(100 - 4u) &= 3(5u + 6) + 72 \\ 400 - 16u &= 15u + 18 + 72 \\ -31u &= -310 \\ u &= 10 \end{aligned}$$

25. $$\begin{aligned} \frac{5x - 4}{5x + 4} &= \frac{2}{3} \\ 3(5x - 4) &= 2(5x + 4) \\ 15x - 12 &= 10x + 8 \\ 5x &= 20 \\ x &= 4 \end{aligned}$$

27. $$\begin{aligned} 10 - \frac{13}{x} &= 4 + \frac{5}{x} \\ \frac{10x - 13}{x} &= \frac{4x + 5}{x} \\ 10x - 13 &= 4x + 5 \\ 6x &= 18 \\ x &= 3 \end{aligned}$$

29. $$\begin{aligned} \frac{1}{x - 3} + \frac{1}{x + 3} &= \frac{10}{x^2 - 9} \\ \frac{(x + 3) + (x - 3)}{x^2 - 9} &= \frac{10}{x^2 - 9} \\ 2x &= 10 \\ x &= 5 \end{aligned}$$

31. $$\begin{aligned} \frac{x}{x + 4} + \frac{4}{x + 4} + 2 &= 0 \\ \frac{x + 4}{x + 4} + 2 &= 0 \\ 1 + 2 &= 0 \\ 3 &= 0 \end{aligned}$$

Contradiction : no solution

33. $$\begin{aligned} \frac{7}{2x + 1} - \frac{8x}{2x - 1} &= -4 \\ 7(2x - 1) - 8x(2x + 1) &= -4(2x + 1)(2x - 1) \\ 14x - 7 - 16x^2 - 8x &= -16x^2 + 4 \\ 6x &= 11 \\ x &= \frac{11}{6} \end{aligned}$$

35. $$\begin{aligned} (x + 2)^2 + 5 &= (x + 3)^2 \\ x^2 + 4x + 4 + 5 &= x^2 + 6x + 9 \\ 4x + 9 &= 6x + 9 \\ -2x &= 0 \\ x &= 0 \end{aligned}$$

37. $$\begin{aligned} (x + 2)^2 - x^2 &= 4(x + 1) \\ x^2 + 4x + 4 - x^2 &= 4x + 4 \\ 4 &= 4 \end{aligned}$$

The equation is an identity; every real number is a solution.

39. $4 - 2(x - 2b) = ax + 3$

$4 - 2x + 4b = ax + 3$

$1 + 4b = ax + 2x$

$1 + 4b = x(a + 2)$

$\dfrac{1 + 4b}{a + 2} = x, \; a \neq -2$

41. (a)

x	-1	0	1	2	3	4
$3.2x - 5.8$	-9	-5.8	-2.6	0.6	3.8	7

(b) Since the sign changes from negative at 1 to positive at 2, the root is somewhere between 1 and 2. $1 < x < 2$

(c)

x	1.5	1.6	1.7	1.8	1.9	2
$3.2x - 5.8$	-1	-0.68	-0.36	-0.04	0.28	0.6

(d) Since the sign changes from negative at 1.8 to positive at 1.9, the root is somewhere between 1.8 and 1.9. $1.8 < x < 1.9$.

To improve accuracy, evaluate the expression in this interval and determine where the sign changes.

43. $6x^2 + 3x = 0$

$3x(2x + 1) = 0$

$3x = 0$ or $2x + 1 = 0$

$x = 0$ or $x = -\frac{1}{2}$

45. $x^2 - 2x - 8 = 0$

$(x - 4)(x + 2) = 0$

$x - 4 = 0$ or $x + 2 = 0$

$x = 4$ or $x = -2$

47. $3 + 5x - 2x^2 = 0$

$(3 - x)(1 + 2x) = 0$

$3 - x = 0$ or $1 + 2x = 0$

$x = 3$ or $x = -\frac{1}{2}$

49. $2x^2 = 19x + 33$

$2x^2 - 19x - 33 = 0$

$(2x + 3)(x - 11) = 0$

$2x + 3 = 0$ or $x - 11 = 0$

$x = -\frac{3}{2}$ or $x = 11$

51. $2x^4 - 18x^2 = 0$

$2x^2(x^2 - 9) = 0$

$2x^2(x + 3)(x - 3) = 0$

$2x^2 = 0$ or $x + 3 = 0$ or $x - 3 = 0$

$x = 0$ or $x = -3$ or $x = 3$

53. $x^3 - 2x^2 - 3x = 0$

$x(x^2 - 2x - 3) = 0$

$x(x + 1)(x - 3) = 0$

$x = 0$ or $x + 1 = 0$ or $x - 3 = 0$

$x = 0$ or $x = -1$ or $x = 3$

55. $2x^4 - 15x^3 + 18x^2 = 0$

$x^2(2x^2 - 15x + 18) = 0$

$x^2(2x - 3)(x - 6) = 0$

$x^2 = 0$ or $2x - 3 = 0$ or $x - 6 = 0$

$x = 0$ or $x = \frac{3}{2}$ or $x = 6$

57. $x^2 = 16$

$x = \pm 4$

$= \pm 4.00$

59. $3x^2 = 36$

$x^2 = 12$

$x = \pm 2\sqrt{3}$

$\approx \pm 3.46$

61. $(x - 12)^2 = 18$

$x - 12 = \pm 3\sqrt{2}$

$x = 12 \pm 3\sqrt{2}$

$x \approx 16.24$ or ≈ 7.76

63. $(x + 2)^2 = 12$

$x + 2 = \pm 2\sqrt{3}$

$x = -2 \pm 2\sqrt{3}$

$x \approx 1.46$ or $x \approx -5.46$

65. $x^2 - 2x = 0$

$x^2 - 2x + 1^2 = 0 + 1$

$x^2 - 2x + 1 = 1$

$(x - 1)^2 = 1$

$x - 1 = \pm\sqrt{1}$

$x = 1 \pm 1$

$x = 0$ or $x = 2$

67. $x^2 + 6x + 2 = 0$

$x^2 + 6x = -2$

$x^2 + 6x + 3^2 = -2 + 3^2$

$(x + 3)^2 = 7$

$x + 3 = \pm\sqrt{7}$

$x = -3 \pm \sqrt{7}$

69. $8 + 4x - x^2 = 0$

$-x^2 + 4x + 8 = 0$

$x^2 - 4x - 8 = 0$

$x^2 - 4x = 8$

$x^2 - 4x + 2^2 = 8 + 2^2$

$(x - 2)^2 = 12$

$x - 2 = \pm\sqrt{12}$

$x = 2 \pm 2\sqrt{3}$

71. $2x^2 + x - 1 = 0$

$$x = \frac{-b \pm \sqrt{b^2 - 4ac}}{2a}$$

$$= \frac{-1 \pm \sqrt{1^2 - 4(2)(-1)}}{2(2)}$$

$$= \frac{-1 \pm 3}{4} = \frac{1}{2}, -1$$

73. $x^2 + 8x - 4 = 0$

$$x = \frac{-b \pm \sqrt{b^2 - 4ac}}{2a}$$

$$= \frac{-8 \pm \sqrt{8^2 - 4(1)(-4)}}{2(1)}$$

$$= \frac{-8 \pm 4\sqrt{5}}{2}$$

$$= -4 \pm 2\sqrt{5}$$

75. $12x - 9x^2 = -3$

$-9x^2 + 12x + 3 = 0$

$$x = \frac{-b \pm \sqrt{b^2 - 4ac}}{2a}$$

$$= \frac{-12 \pm \sqrt{12^2 - 4(-9)(3)}}{2(-9)}$$

$$= \frac{-12 \pm 6\sqrt{7}}{-18} = \frac{2}{3} \pm \frac{\sqrt{7}}{3}$$

77. $3x + x^2 - 1 = 0$

$x^2 + 3x - 1 = 0$

$$x = \frac{-b \pm \sqrt{b^2 - 4ac}}{2a}$$

$$= \frac{-3 \pm \sqrt{3^2 - 4(1)(-1)}}{2(1)}$$

$$= \frac{-3 \pm \sqrt{13}}{2}$$

$$= -\frac{3}{2} \pm \frac{\sqrt{13}}{2}$$

79.

$$28x - 49x^2 = 4$$
$$-49x^2 + 28x - 4 = 0$$
$$x = \frac{-b \pm \sqrt{b^2 - 4ac}}{2a}$$
$$= \frac{-28 \pm \sqrt{28^2 - 4(-49)(-4)}}{2(-49)}$$
$$= \frac{-28 \pm 0}{-98} = \frac{2}{7}$$

81.

$$8t = 5 + 2t^2$$
$$-2t^2 + 8t - 5 = 0$$
$$t = \frac{-b \pm \sqrt{b^2 - 4ac}}{2a}$$
$$= \frac{-8 \pm \sqrt{8^2 - 4(-2)(-5)}}{2(-2)}$$
$$= \frac{-8 \pm 2\sqrt{6}}{-4} = 2 \pm \frac{\sqrt{6}}{2}$$

83. False. The product must equal zero to use the Zero-Factor Property.

85. (a) $ax^2 + bx = 0$

$x(ax + b) = 0$

$x = 0 \quad \text{or} \quad x = -\frac{b}{a}$

(b) $ax^2 - ax = 0$

$ax(x - 1) = 0$

$x = 0 \quad \text{or} \quad x = 1$

87.

$$x^4 - 4x^2 + 3 = 0$$
$$(x^2 - 3)(x^2 - 1) = 0$$
$$(x + \sqrt{3})(x - \sqrt{3})(x + 1)(x - 1) = 0$$
$$x + \sqrt{3} = 0 \Longrightarrow x = -\sqrt{3}$$
$$x - \sqrt{3} = 0 \Longrightarrow x = \sqrt{3}$$
$$x + 1 = 0 \Longrightarrow x = -1$$
$$x - 1 = 0 \Longrightarrow x = 1$$

89.

$$\frac{1}{t^2} + \frac{8}{t} + 15 = 0$$
$$1 + 8t + 15t^2 = 0$$
$$(1 + 3t)(1 + 5t) = 0$$
$$1 + 3t = 0 \Longrightarrow t = -\frac{1}{3}$$
$$1 + 5t = 0 \Longrightarrow t = -\frac{1}{5}$$

91.

$$2x + 9\sqrt{x} - 5 = 0$$
$$(2\sqrt{x} - 1)(\sqrt{x} + 5) = 0$$
$$\sqrt{x} = \frac{1}{2} \Longrightarrow x = \frac{1}{4}$$

($\sqrt{x} = -5$ is not a solution.)

93.

$$3x^{1/3} + 2x^{2/3} = 5$$
$$2x^{2/3} + 3x^{1/3} - 5 = 0$$
$$(2x^{1/3} + 5)(x^{1/3} - 1) = 0$$
$$2x^{1/3} + 5 = 0 \Longrightarrow x^{1/3} = -\tfrac{5}{2} \Longrightarrow x = \left(-\tfrac{5}{2}\right)^3 = -\tfrac{125}{8}$$
$$x^{1/3} - 1 = 0 \Longrightarrow x^{1/3} = 1 \Longrightarrow x = (1)^3 = 1$$

95. $\sqrt{x-10}-4=0$

$$\sqrt{x-10}=4$$
$$x-10=16$$
$$x=26$$

97. $\sqrt[3]{2x+5}+3=0$

$$\sqrt[3]{2x+5}=-3$$
$$2x+5=-27$$
$$2x=-32$$
$$x=-16$$

99. $x=\sqrt{11x-30}$

$$x^2=11x-30$$
$$x^2-11x+30=0$$
$$(x-5)(x-6)=0$$
$$x-5=0 \Rightarrow x=5$$
$$x-6=0 \Rightarrow x=6$$

101. $\sqrt{x+1}-3x=1$

$$\sqrt{x+1}=3x+1$$
$$x+1=9x^2+6x+1$$
$$0=9x^2+5x$$
$$0=x(9x+5)$$
$$x=0$$
$$9x+5=0 \Rightarrow x=-\tfrac{5}{9}, \text{ extraneous}$$

103. $\sqrt{x}-\sqrt{x-5}=1$

$$\sqrt{x}=1+\sqrt{x-5}$$
$$(\sqrt{x})^2=(1+\sqrt{x-5})^2$$
$$x=1+2\sqrt{x-5}+x-5$$
$$4=2\sqrt{x-5}$$
$$2=\sqrt{x-5}$$
$$4=x-5$$
$$9=x$$

105. $2\sqrt{x+1}-\sqrt{2x+3}=1$

$$2\sqrt{x+1}=\sqrt{2x+3}+1$$
$$(2\sqrt{x+1})^2=(\sqrt{2x+3}+1)^2$$
$$4(x+1)=2x+3+2\sqrt{2x+3}+1$$
$$4x+4=2x+4+2\sqrt{2x+3}$$
$$2x=2\sqrt{2x+3}$$
$$x=\sqrt{2x+3}$$
$$x^2=2x+3$$
$$x^2-2x-3=0$$
$$(x-3)(x+1)=0$$
$$x-3=0 \Rightarrow x=3$$
$$x+1=0 \Rightarrow x=-1, \text{ extraneous}$$

107. $(x-5)^{2/3}=16$

$$x-5=\pm 16^{3/2}$$
$$x-5=\pm 64$$
$$x=69,\ -59$$

109. $37.55=40-\sqrt{0.01x+1}$

$$\sqrt{0.01x+1}=2.45$$
$$0.01x+1=6.0025$$
$$0.01x=5.0025$$
$$x=500.25$$

Rounding x to the nearest whole unit yields $x \approx 500$ units.

111.
$$S = \pi r\sqrt{r^2 + h^2}$$
$$S^2 = \pi^2 r^2(r^2 + h^2)$$
$$S^2 = \pi^2 r^4 + \pi^2 r^2 h^2$$
$$\frac{S^2 - \pi^2 r^4}{\pi^2 r^2} = h^2$$
$$h = \frac{\sqrt{S^2 - \pi^2 r^4}}{\pi r}$$

113. $20 + \sqrt{20 - a} = b$
$$\sqrt{20 - a} = b - 20$$
$$20 - a = b^2 - 40b + 400$$
$$-a = b^2 - 40b + 380$$
$$a = -b^2 + 40b - 380$$

This formula gives the relationship between a and b. From the original equation we know that $a \le 20$ and $b \ge 20$. Choose a b value, where $b \ge 20$ and then solve for a, keeping in mind that $a \le 20$.

Some possibilities are:

$b = 20,\ a = 20$

$b = 21,\ a = 19$

$b = 22,\ a = 16$

$b = 23,\ a = 11$

$b = 24,\ a = 4$ ← (This is the one given in your textbook.)

$b = 25,\ a = -5$

115. $|x + 1| = 2$

$x + 1 = 2 \Rightarrow x = 1$

$-(x + 1) = 2 \Rightarrow x = -3$

117. $|2x - 1| = 5$

$2x - 1 = 5 \Rightarrow x = 3$

$-(2x - 1) = 5 \Rightarrow x = -2$

119. $|x^2 + 6x| = 3x + 18$

$x^2 + 6x = 3x + 18$

$x^2 + 3x - 18 = 0$

$(x + 6)(x - 3) = 0$

$x + 6 = 0 \Rightarrow x = -6$

$x - 3 = 0 \Rightarrow x = 3$

or

$-(x^2 + 6x) = 3x + 18$

$-x^2 - 6x = 3x + 18$

$0 = x^2 + 9x + 18$

$0 = (x + 3)(x + 6)$

$x + 3 = 0 \Rightarrow x = -3$

$x + 6 = 0 \Rightarrow x = -6$

The solutions are $x = -6$ and $x = \pm 3$.

121. $[x - (-3)](x - 5) = 0$

$(x + 3)(x - 5) = 0$

$x^2 - 2x - 15 = 0$

This equation has $x = -3$ and $x = 5$ as solutions, as does any nonzero multiple of this equation.

123.

$$\begin{aligned}
4x + 3y &= 100 \\
3y &= 100 - 4x \\
y &= \tfrac{1}{3}(100 - 4x) \\
\text{Area} &= \text{length} \cdot \text{width} \\
350 &= (2x)(y) \\
350 &= 2x\left[\tfrac{1}{3}(100 - 4x)\right] \\
350 &= \tfrac{2}{3}x(100 - 4x) \\
1050 &= 2x(100 - 4x) \\
1050 &= 200x - 8x^2
\end{aligned}$$

$$8x^2 - 200x + 1050 = 0$$

$$2(4x^2 - 100x + 525) = 0$$

$$2(2x - 35)(2x - 15) = 0$$

$$2x - 35 = 0 \quad \text{or} \quad 2x - 15 = 0$$

$$x = \tfrac{35}{2} \qquad\qquad x = \tfrac{15}{2}$$

$$y = 10 \qquad\qquad y = \tfrac{70}{3}$$

There are two different possible dimensions:

$x = \frac{35}{2} = 17.5$ meters and $y = 10$ meters

or $x = \frac{15}{2} = 7.5$ meters and $y = \frac{70}{3} = 23\frac{1}{3}$ meters

Section P.6 Solving Inequalities

- You should know the properties of inequalities.
  - (a) Transitive: $a < b$ and $b < c$ implies $a < c$.
  - (b) Addition: $a < b$ and $c < d$ implies $a + c < b + d$.
  - (c) Adding or Subtracting a Constant: $a \pm c < b \pm c$ if $a < b$.
  - (d) Multiplying or Dividing a Constant: For $a < b$,
    1. If $c > 0$, then $ac < bc$ and $\dfrac{a}{c} < \dfrac{b}{c}$.
    2. If $c < 0$,then $ac > bc$ and $\dfrac{a}{c} > \dfrac{b}{c}$.
- You should be able to solve absolute value or rational inequalities.
  - (a) $|x| < a$ if and only if $-a < x < a$.
  - (b) $|x| > a$ if and only if $x < -a$ or $x > a$.
- You should be able to solve polynomial inequalities.
  - (a) Find the critical number.
    1. Values that make the expression zero
    2. Values that make the expression undefined
  - (b) Test one value in each interval on the real number line resulting from the critical numbers.
  - (c) Determine the solution intervals.

Solutions to Odd-Numbered Exercises

1. Interval: $[-1, 3]$

Inequality: $-1 \le x \le 3$

The interval is bounded.

3. Interval: $(10, \infty)$

Inequality: $x > 10$ or $10 < x < \infty$

The interval is unbounded.

5. $x < 3$

Matches (c).

7. $-3 < x \le 4$

Matches (f).

9. $|x| < 3 \Rightarrow -3 < x < 3$

Matches (g).

11. $|x - 4| > 2 \Rightarrow x - 4 < -2$ or $x - 4 > 2$

$x < 2$ or $x > 6$

Matches (b).

13. (a) $x = 3$

$5(3) - 12 \overset{?}{>} 0$

$3 > 0$

Yes, $x = 3$ is a solution.

(b) $x = -3$

$5(-3) - 12 \overset{?}{>} 0$

$-27 \not> 0$

No, $x = -3$ is not a solution.

(c) $x = \frac{5}{2}$

$5\left(\frac{5}{2}\right) - 12 \overset{?}{>} 0$

$\frac{1}{2} > 0$

Yes, $x = \frac{5}{2}$ is a solution.

(d) $x = \frac{3}{2}$

$5\left(\frac{3}{2}\right) - 12 \overset{?}{>} 0$

$-\frac{9}{2} \not> 0$

No, $x = \frac{3}{2}$ is not a solution.

15. (a) $x = 4$

$0 \overset{?}{<} \dfrac{4 - 2}{4} \overset{?}{<} 2$

$0 < \dfrac{1}{2} < 2$

Yes, $x = 4$ is a solution.

(b) $x = 10$

$0 \overset{?}{<} \dfrac{10 - 2}{4} \overset{?}{<} 2$

$0 < 2 \not< 2$

No, $x = 10$ is not a solution.

(c) $x = 0$

$0 \overset{?}{<} \dfrac{0 - 2}{4} \overset{?}{<} 2$

$0 \not< -\dfrac{1}{2} < 2$

No, $x = 0$ is not a solution.

(d) $x = \dfrac{7}{2}$

$0 \overset{?}{<} \dfrac{(7/2) - 2}{4} \overset{?}{<} 2$

$0 < \dfrac{3}{8} < 2$

Yes, $x = \frac{7}{2}$ is a solution.

17. $x^2 - 3 < 0$

(a) $x = 3$

$(3)^2 - 3 \overset{?}{<} 0$

$6 \not< 0$

No, $x = 3$ is not a solution.

(b) $x = 0$

$(0)^2 - 3 \overset{?}{<} 0$

$-3 < 0$

Yes, $x = 0$ is a solution.

(c) $x = \frac{3}{2}$

$\left(\frac{3}{2}\right)^2 - 3 \overset{?}{<} 0$

$-\frac{3}{4} < 0$

Yes, $x = \frac{3}{2}$ is a solution.

(d) $x = -5$

$(-5)^2 - 3 \overset{?}{<} 0$

$22 \not< 0$

No, $x = -5$ is not a solution.

19. $\dfrac{x+2}{x-4} \geq 3$

(a) $x = 5$

$\dfrac{5+2}{5-4} \overset{?}{\geq} 3$

$7 \geq 3$

Yes, $x = 5$ is a solution.

(b) $x = 4$

$\dfrac{4+2}{4-4} \overset{?}{\geq} 3$

$\dfrac{6}{0}$ is undefined.

No, $x = 4$ is not a solution.

(c) $x = -\dfrac{9}{2}$

$\dfrac{-\frac{9}{2}+2}{-\frac{9}{2}-4} \overset{?}{\geq} 3$

$\dfrac{5}{17} \ngeq 3$

No, $x = -\frac{9}{2}$ is not a solution.

(d) $x = \dfrac{9}{2}$

$\dfrac{\frac{9}{2}+2}{\frac{9}{2}-4} \overset{?}{\geq} 3$

$13 \geq 3$

Yes, $x = \frac{9}{2}$ is a solution.

21. $4x < 12$

$\frac{1}{4}(4x) < \frac{1}{4}(12)$

$x < 3$

23. $-10x < 40$

$-\frac{1}{10}(-10) > -\frac{1}{10}(40)$

$x > -4$

25. $x - 5 \geq 7$

$x \geq 12$

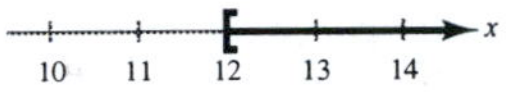

27. $4(x+1) < 2x + 3$

$4x + 4 < 2x + 3$

$2x < -1$

$x < -\frac{1}{2}$

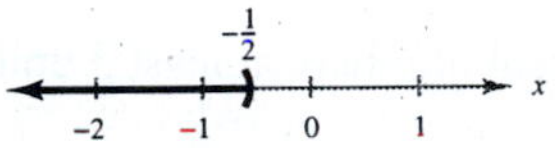

29. $4 - 2x < 3$

$-2x < -1$

$x > \frac{1}{2}$

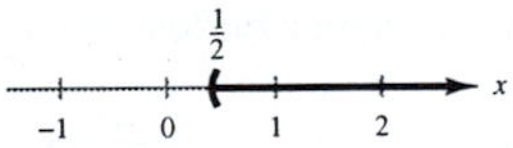

31. $1 < 2x + 3 < 9$

$-2 < 2x < 6$

$-1 < x < 3$

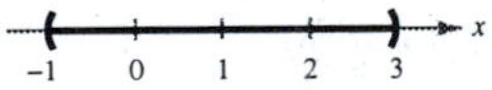

33. $-4 < \dfrac{2x-3}{3} < 4$

$-12 < 2x - 3 < 12$

$-9 < 2x < 15$

$-\dfrac{9}{2} < x < \dfrac{15}{2}$

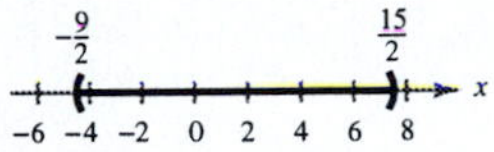

35. $\dfrac{3}{4} > x + 1 > \dfrac{1}{4}$

$-\dfrac{1}{4} > x > -\dfrac{3}{4}$

$-\dfrac{3}{4} < x < -\dfrac{1}{4}$

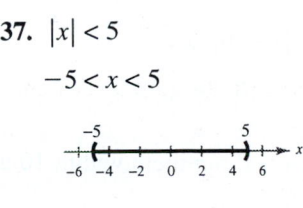

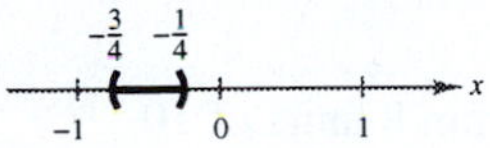

37. $|x| < 5$

$-5 < x < 5$

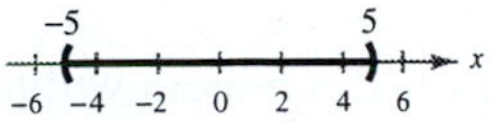

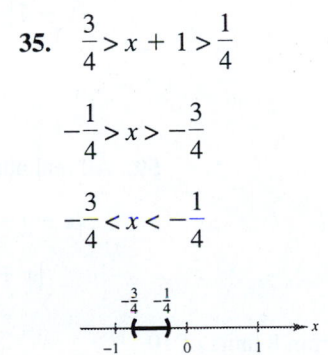

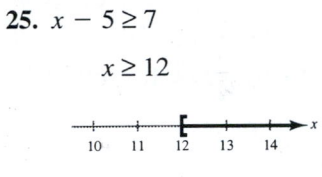

39. $\left|\frac{x}{2}\right| > 3$

$\frac{x}{2} < -3$ or $\frac{x}{2} > 3$

$x < -6$ $\quad$ $x > 6$

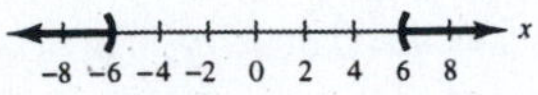

41. $|x - 20| \le 4$

$-4 \le x - 20 \le 4$

$16 \le x \le 24$

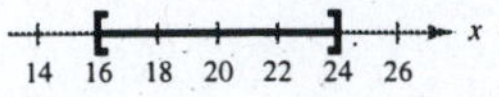

43. $|x - 20| \ge 4$

$x - 20 \le -4$ or $x - 20 \ge 4$

$x \le 16$ $\quad$ $x \ge 24$

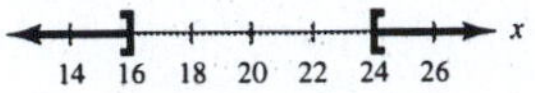

45. $\left|\frac{x - 3}{2}\right| \ge 5$

$\frac{x - 3}{2} \le -5$ or $\frac{x - 3}{2} \ge 5$

$x - 3 \le -10$ $\quad$ $x - 3 \ge 10$

$x \le -7$ $\quad$ $x \ge 13$

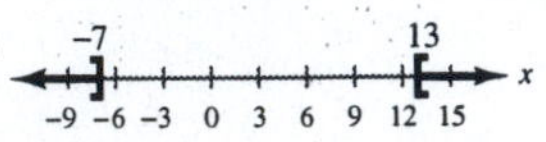

47. $|9 - 2x| - 2 < -1$

$|9 - 2x| < 1$

$-1 < 9 - 2x < 1$

$-10 < -2x < -8$

$5 > x > 4$

$4 < x < 5$

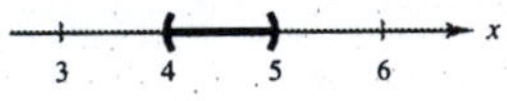

49. $2|x + 10| \ge 9$

$|x + 10| \ge \frac{9}{2}$

$x + 10 \le -\frac{9}{2}$ or $x + 10 \ge \frac{9}{2}$

$x \le -\frac{29}{2}$ $\quad$ $x \ge -\frac{11}{2}$

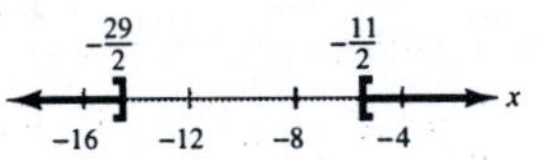

51. $|x - 5| < 0$

No solution. The absolute value of a number can never be less than zero.

53. The midpoint of the interval $[-3, 3]$ is 0. The interval represents all real numbers x no more than 3 units from 0.

$|x - 0| \le 3$

$|x| \le 3$

55. The graph shows all real numbers at least 3 units from 7.

$|x - 7| \ge 3$

57. All real numbers within 10 units of 12.

$|x - 12| < 10$

59. All real numbers more than 5 units from -3.

$|x - (-3)| > 5$

$|x + 3| > 5$

61. $|x - 10| < 8$ represents all real numbers within 8 units of 10.

63. $x^2 \le 9$

$x^2 - 9 \le 0$

$(x + 3)(x - 3) \le 0$

Critical numbers: $x = \pm 3$

Test intervals: $(-\infty, -3), (-3, 3), (3, \infty)$

Test: Is $(x + 3)(x - 3) \le 0$?

Solution set: $[-3, 3]$

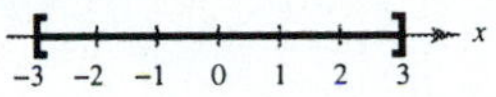

65. $x^2 > 4$

$x^2 - 4 > 0$

$(x + 2)(x - 2) > 0$

Critical numbers: $x = \pm 2$

Test intervals: $(-\infty, -2), (-2, 2), (2, \infty)$

Test: Is $x^2 - 4 > 0$?

Solution set: $(-\infty, -2) \cup (2, \infty)$

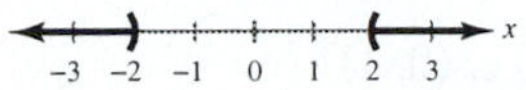

67. $(x + 2)^2 < 25$

$x^2 + 4x + 4 < 25$

$x^2 + 4x - 21 < 0$

$(x + 7)(x - 3) < 0$

Critical numbers: $x = -7, x = 3$

Test intervals: $(-\infty, -7), (-7, 3), (3, \infty)$

Test: Is $(x + 7)(x - 3) < 0$?

Solution set: $(-7, 3)$

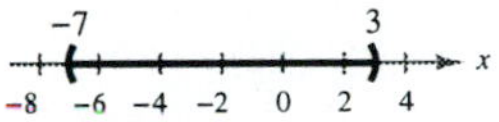

69. $x^2 + 4x + 4 \ge 9$

$x^2 + 4x - 5 \ge 0$

$(x + 5)(x - 1) \ge 0$

Critical numbers: $x = -5, x = 1$

Test intervals: $(-\infty, -5), (-5, 1), (1, \infty)$

Test: Is $(x + 5)(x - 1) \ge 0$?

Solution set: $(-\infty, -5] \cup [1, \infty)$

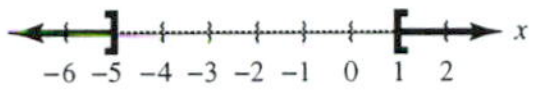

71. $3(x - 1)(x + 1) > 0$

Critical numbers: $x = \pm 1$

Test intervals: $(-\infty, -1), (-1, 1), (1, \infty)$

Test: Is $3(x - 1)(x + 1) > 0$?

Solution set: $(-\infty, -1) \cup (1, \infty)$

Solution set: $(-3, 2)$

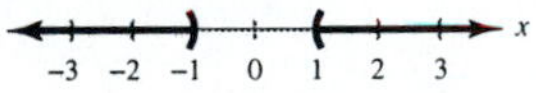

73. $x^2 + 2x - 3 < 0$

$(x + 3)(x - 1) < 0$

Critical numbers: $x = -3, x = 1$

Test intervals: $(-\infty, -3), (-3, 1), (1, \infty)$

Test: Is $(x + 3)(x - 1) < 0$?

Solution set: $(-3, 1)$

75. $4x^3 - 6x^2 < 0$

$2x^2(2x - 3) < 0$

Critical numbers: $x = 0, x = \frac{3}{2}$

Test intervals: $(-\infty, 0), \left(0, \frac{3}{2}\right), \left(\frac{3}{2}, \infty\right)$

Test: Is $2x^2(2x - 3) < 0$?

Solution set: $(-\infty, 0) \cup \left(0, \frac{3}{2}\right)$

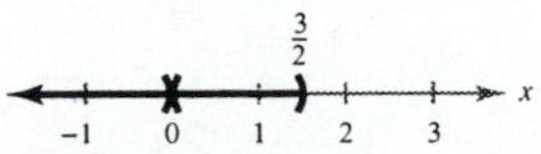

77. $(x - 1)^2(x + 2)^3 \ge 0$

Critical numbers: $x = 1, x = -2$

Test intervals: $(-\infty, -2), (-2, 1), (1, \infty)$

Test: Is $(x - 1)^2(x + 3)^3 \ge 0$?

Solution set: $[-2, \infty)$

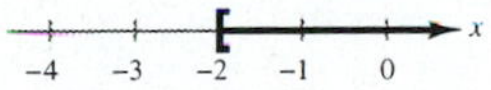

79. $\dfrac{1}{x} - x > 0$

$\dfrac{1 - x^2}{x} > 0$

Critical numbers: $x = 0, x = \pm 1$

Test intervals: $(-\infty, -1), (-1, 0), (0, 1), (1, \infty)$

Test: Is $\dfrac{1 - x^2}{x} > 0$?

Solution set: $(-\infty, -1) \cup (0, 1)$

81. $\dfrac{x + 6}{x + 1} - 2 < 0$

$\dfrac{x + 6 - 2(x + 1)}{x + 1} < 0$

$\dfrac{4 - x}{x + 1} < 0$

Critical numbers: $x = -1, x = 4$

Test intervals: $(-\infty, -1), (-1, 4), (4, \infty)$

Test: Is $\dfrac{4 - x}{x + 1} < 0$?

Solution set: $(-\infty, -1) \cup (4, \infty)$

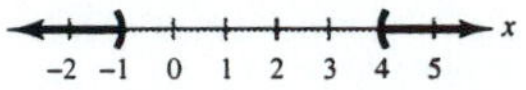

83. $\dfrac{4}{x + 5} > \dfrac{1}{2x + 3}$

$\dfrac{4}{x + 5} - \dfrac{1}{2x + 3} > 0$

$\dfrac{4(2x + 3) - (x + 5)}{(x + 5)(2x + 3)} > 0$

$\dfrac{7x + 7}{(x + 5)(2x + 3)} > 0$

Critical numbers: $x = -1, x = -5, x = -\dfrac{3}{2}$

Test intervals: $(-\infty, -5), \left(-5, -\dfrac{3}{2}\right), \left(-\dfrac{3}{2}, -1\right), (-1, \infty)$

Test: Is $\dfrac{7(x + 1)}{(x + 5)(2x + 3)} > 0$?

Solution set: $\left(-5, -\dfrac{3}{2}\right) \cup (-1, \infty)$

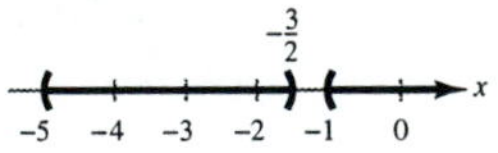

85. $x - 5 \geq 0$

$x \geq 5$

Domain: $[5, \infty)$

87. $4 - x^2 \ge 0$

$(2 + x)(2 - x) \ge 0$

Critical numbers: $x = \pm 2$

Test intervals: $(-\infty, -2), (-2, 2), (2, \infty)$

Test: Is $4 - x^2 \ge 0$?

Domain: $[-2, 2]$

89. $x^2 - 7x + 12 \ge 0$

$(x - 3)(x - 4) \ge 0$

Critical numbers: $x = 3, x = 4$

Test intervals: $(-\infty, 3), (3, 4), (4, \infty)$

Test: Is $(x - 3)(x - 4) \ge 0$?

Domain: $(-\infty, 3] \cup [4, \infty)$

91. $1250 < 1000[1 + r(2)]$

$1.250 < 1 + 2r$

$0.250 < 2r$

$0.125 < r$

$r > 12.5\%$

93. $R > C$

$115.95x > 95x + 750$

$20.95x > 750$

$x > 35.7995$

$x \ge 36$ units

95. $|h - 50| \le 30$

$-30 \le h - 50 \le 30$

$20 \le h \le 80$

Minimum: $h = 20$

Maximum: $h = 80$

97. $7.34 + 0.41t + 0.002t^2 \ge 25$

$0.002t^2 + 0.41t - 17.66 \ge 0$

By the Quadratic Formula, the critical numbers are $t \approx -241.55$ and $t \approx 36.55$. The only one that makes sense here is $t \approx 36.55$. This corresponds to the year 1996.

99. $|ax - b| \le c \implies$ c must be greater than or equal to zero.

$-c \le ax - b \le c$

$b - c \le ax \le b + c$

Let $a = 1$, then $b - c = 0$ and $b + c = 10$. This is true when $b = c = 5$. One set of values is : $a = 1,\ b = 5,\ c = 5$

(Note: This solution is not unique. Any positive multiple of these values will also work, such as:

$a = 2,\ b = c = 10$

$a = 3,\ b = c = 15$).

101. $x(50 - 0.0002x) - (12x + 150{,}000) \ge 1{,}650{,}000$

$-0.0002x^2 + 38x - 1{,}800{,}000 \ge 0$

By the Quadratic Formula, the critical numbers are $x = 90{,}000$ and $x = 100{,}000$.

The profit is at least \$1,650,000 when $90{,}000 \le x \le 100{,}000$ units

103.

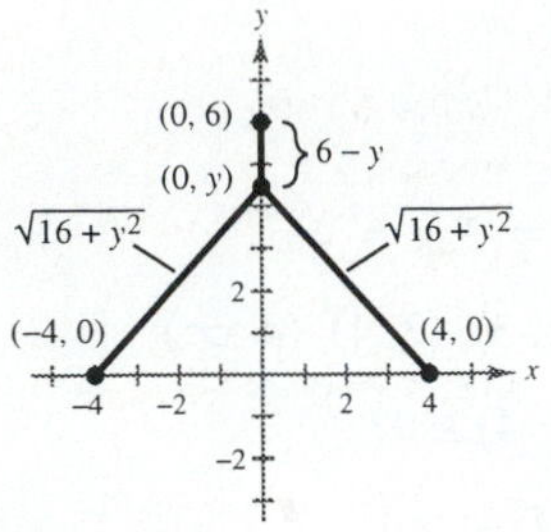

(a) $L = (6 - y) + 2\sqrt{16 + y^2}$

(b) $0 \le y \le 6$

When $y = 0$: $L = 14$

When $y = 6$: $L = 2\sqrt{52} = 4\sqrt{13} \approx 14.4$

(c)

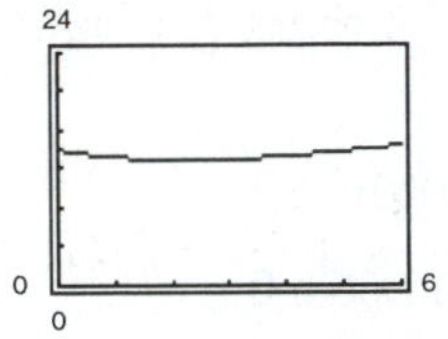

(d) $6 - y + 2\sqrt{16 + y^2} < 13$

$2\sqrt{16 + y^2} - y - 7 < 0$

To find the critical numbers, set:

$$2\sqrt{16 + y^2} - y - 7 = 0$$

$$2\sqrt{16 + y^2} = y + 7$$

$$4(16 + y^2) = y^2 + 14y + 49$$

$$64 + 4y^2 = y^2 + 14y + 49$$

$$3y^2 - 14y + 15 = 0$$

$$(3y - 5)(y - 3) = 0$$

$$y = \tfrac{5}{3},\ y = 3$$

By testing, the solution set is $\frac{5}{3} < y < 3$.

Section P.7 Errors and the Algebra of Calculus

- You should be able to recognize and avoid the common algebraic errors involving parentheses, fractions, exponents, radicals, and cancellation.
- You should be able to "unsimplify" algebraic expressions by the following methods.
  (a) Unusual Factoring
  (b) Inserting Factors or Terms
  (c) Rewriting with Negative Exponents
  (d) Writing a Fraction as a Sum of Terms

Solutions to Odd-Numbered Exercises

1. $2x - (3y + 4) = 2x - 3y - 4$

Distribute the minus sign.

3. $5z + 3(x - 2) = 5z + 3x - 6$

Use the Distributive Property.

5. $$-\frac{x-3}{x-1} = \frac{-(x-3)}{x-1} = \frac{3-x}{x-1}$$

Only the numerator is multiplied by (-1).

7. $$a\left(\frac{x}{y}\right) = \frac{a}{1} \cdot \frac{x}{y} = \frac{ax}{y}$$

Only the numerator is multiplied by a.

9. $(4x)^2 = 4^2x^2 = 16x^2$

Square both factors.

11. $\sqrt{x+9}$ does not simplify.

Do not apply the radical to the terms.

13. $\dfrac{6x+y}{6x-y}$ does not simplify.

Reduce common factors, not common factors of terms.

15. $\dfrac{1}{x+y^{-1}} = \dfrac{1}{x+(1/y)} \cdot \dfrac{y}{y} = \dfrac{y}{xy+1}$

The negative exponent is on a term of the denominator, not a factor.

17. $x(2x+1)^2 = x(4x^2+4x+1)$

Exponents are applied before multiplying.

19. $\sqrt[3]{x^3+7x^2} = \sqrt[3]{x^2(x+7)} = \sqrt[3]{x^2}\,\sqrt[3]{x+7}$

Radicals apply to every factor of the radicand.

21. $\dfrac{3}{x}+\dfrac{4}{y} = \dfrac{3}{x}\cdot\dfrac{y}{y}+\dfrac{4}{y}\cdot\dfrac{x}{x} = \dfrac{3y+4x}{xy}$

To add fractions, they must have a common denominator.

23. $\dfrac{1}{2y} = \dfrac{1}{2}\cdot\dfrac{1}{y}$

Use the definition for multiplying fractions.

25. $\dfrac{3x+2}{5} = \dfrac{1}{5}(3x+2)$

The required factor is $3x+2$.

27. $\dfrac{2}{3}x^2+\dfrac{1}{3}x+5 = \dfrac{2}{3}x^2+\dfrac{1}{3}x+\dfrac{15}{3} = \dfrac{1}{3}(2x^2+x+15)$

The required factor is $2x^2+x+15$.

29. $\frac{1}{3}x^3+5 = \frac{1}{3}x^3+\frac{15}{3} = \frac{1}{3}(x^3+15)$

The required factor is $\frac{1}{3}$.

31.
$$x(2x^2+15) = \frac{2x}{2}(2x^2+15) = \left(\frac{1}{2}\right)(2x)(2x^2+15)$$
$$= \left(\frac{1}{2}\right)(2x^2+15)(2x)$$

The required factor is $\frac{1}{2}$.

33.
$$x(1-2x^2)^3 = \frac{-4x}{-4}(1-2x^2)^3 = \left(-\frac{1}{4}\right)(-4x)(1-2x^2)^3$$
$$= \left(-\frac{1}{4}\right)(1-2x^2)^3(-4x)$$

The required factor is $-\frac{1}{4}$.

35.
$$\frac{1}{\sqrt{x}(1+\sqrt{x})^2} = \frac{1}{\sqrt{x}}\cdot\frac{1}{(1+\sqrt{x})^2} = (2)\left(\frac{1}{2\sqrt{x}}\right)\frac{1}{(1+\sqrt{x})^2}$$
$$= (2)\frac{1}{(1+\sqrt{x})^2}\left(\frac{1}{2\sqrt{x}}\right)$$

The required factor is 2.

37.
$$\frac{x+1}{(x^2+2x-3)^2} = \frac{1}{2}\cdot\frac{2(x+1)}{(x^2+2x-3)^2} = \left(\frac{1}{2}\right)\left(\frac{1}{(x^2+2x-3)^2}\right)(2x+2)$$

The required factor is $\frac{1}{2}$.

39. $\dfrac{3}{x} + \dfrac{5}{2x^2} - \dfrac{3}{2}x = \dfrac{6x}{2x^2} + \dfrac{5}{2x^2} - \dfrac{3x^3}{2x^2} = \left(\dfrac{1}{2x^2}\right)(6x + 5 - 3x^3)$

The required factor is $\dfrac{1}{2x^2}$.

41. $\dfrac{9x^2}{25} + \dfrac{16y^2}{49} = \dfrac{9}{25} \cdot \dfrac{x^2}{1} + \dfrac{16}{49} \cdot \dfrac{y^2}{1}$

$$= \frac{1}{25/9} \cdot \frac{x^2}{1} + \frac{1}{49/16} \cdot \frac{y^2}{1}$$

$$= \frac{x^2}{(25/9)} + \frac{y^2}{(49/16)}$$

The required factors are $\frac{25}{9}$ and $\frac{49}{16}$.

43. $\dfrac{x^2}{1/12} - \dfrac{y^2}{2/3} = x^2\left(\dfrac{12}{1}\right) - y^2\left(\dfrac{3}{2}\right) = \dfrac{12x^2}{1} - \dfrac{3y^2}{2}$

The required factors are 1 and 2.

45. $\sqrt{x} + \left(\sqrt{x}\right)^3 = \sqrt{x}\left(1 + \left(\sqrt{x}\right)^2\right) = \sqrt{x}\,(1 + x)$

The required factor is $1 + x$.

47. $3(2x + 1)x^{1/2} + 4x^{3/2} = x^{1/2}[3(2x + 1) + 4x]$

$$= x^{1/2}(6x + 3 + 4x)$$

$$= x^{1/2}(10x + 3)$$

The required factor is $10x + 3$.

49. $\dfrac{x^2}{\sqrt{x^2 + 1}} - \sqrt{x^2 + 1} = \dfrac{x^2}{\sqrt{x^2 + 1}} - \dfrac{\sqrt{x^2 + 1}}{1} \cdot \dfrac{\sqrt{x^2 + 1}}{\sqrt{x^2 + 1}}$

$$= \frac{x^2 - (x^2 + 1)}{\sqrt{x^2 + 1}} = \frac{-1}{\sqrt{x^2 + 1}}$$

$$= \frac{1}{\sqrt{x^2 + 1}}(-1)$$

The required factor is -1.

51. $\frac{1}{10}(2x + 1)^{5/2} - \frac{1}{6}(2x + 1)^{3/2} = \frac{3}{30}(2x + 1)^{3/2}(2x + 1)^1 - \frac{5}{30}(2x + 1)^{3/2}$

$$= \tfrac{1}{30}(2x + 1)^{3/2}[3(2x + 1) - 5]$$

$$= \tfrac{1}{30}(2x + 1)^{3/2}(6x - 2)$$

$$= \tfrac{1}{30}(2x + 1)^{3/2}2(3x - 1)$$

$$= \tfrac{1}{15}(2x + 1)^{3/2}(3x - 1)$$

The required factor is $3x - 1$.

53. $\dfrac{16 - 5x - x^2}{x} = \dfrac{16}{x} - \dfrac{5x}{x} - \dfrac{x^2}{x} = \dfrac{16}{x} - 5 - x$

55. $\dfrac{4x^3 - 7x^2 + 1}{x^{1/3}} = \dfrac{4x^3}{x^{1/3}} - \dfrac{7x^2}{x^{1/3}} + \dfrac{1}{x^{1/3}}$

$= 4x^{3-1/3} - 7x^{2-1/3} + \dfrac{1}{x^{1/3}}$

$= 4x^{8/3} - 7x^{5/3} + \dfrac{1}{x^{1/3}}$

57. $\dfrac{3 - 5x^2 - x^4}{\sqrt{x}} = \dfrac{3}{\sqrt{x}} - \dfrac{5x^2}{\sqrt{x}} - \dfrac{x^4}{\sqrt{x}}$

$= \dfrac{3}{\sqrt{x}} - 5x^{2-1/2} - x^{4-1/2}$

$= \dfrac{3}{\sqrt{x}} - 5x^{3/2} - x^{7/2}$

59. $\dfrac{-2(x^2 - 3)^{-3}(2x)(x + 1)^3 - 3(x + 1)^2(x^2 - 3)^{-2}}{[(x + 1)^3]^2} = \dfrac{(x^2 - 3)^{-3}(x + 1)^2[-4x(x + 1) - 3(x^2 - 3)]}{(x + 1)^6}$

$= \dfrac{-4x^2 - 4x - 3x^2 + 9}{(x^2 - 3)^3(x + 1)^4}$

$= \dfrac{-7x^2 - 4x + 9}{(x^2 - 3)^3(x + 1)^4}$

61. $\dfrac{(6x + 1)^3(27x^2 + 2) - (9x^3 + 2x)(3)(6x + 1)^2(6)}{[(6x + 1)^3]^2} = \dfrac{(6x + 1)^2[(6x + 1)(27x^2 + 2) - 18(9x^3 + 2x)]}{(6x + 1)^6}$

$= \dfrac{162x^3 + 12x + 27x^2 + 2 - 162x^3 - 36x}{(6x + 1)^4}$

$= \dfrac{27x^2 - 24x + 2}{(6x + 1)^4}$

63. $\dfrac{(x + 2)^{3/4}(x + 3)^{-2/3} - (x + 3)^{1/3}(x + 2)^{-1/4}}{[(x + 2)^{3/4}]^2} = \dfrac{(x + 2)^{-1/4}(x + 3)^{-2/3}[(x + 2) - (x + 3)]}{(x + 2)^{6/4}}$

$= \dfrac{x + 2 - x - 3}{(x + 2)^{1/4}(x + 3)^{2/3}(x + 2)^{6/4}}$

$= -\dfrac{1}{(x + 3)^{2/3}(x + 2)^{7/4}}$

65. $\dfrac{2(3x - 1)^{1/3} - (2x + 1)(\frac{1}{3})(3x - 1)^{-2/3}(3)}{(3x - 1)^{2/3}} = \dfrac{(3x - 1)^{-2/3}[2(3x - 1) - (2x + 1)]}{(3x - 1)^{2/3}}$

$= \dfrac{6x - 2 - 2x - 1}{(3x - 1)^{2/3}(3x - 1)^{2/3}}$

$= \dfrac{4x - 3}{(3x - 1)^{4/3}}$

67. (a) $y_1 = x^2\left(\frac{1}{3}\right)(x^2 + 1)^{-2/3}(2x) + (x^2 + 1)^{1/3}(2x)$

$$= 2x(x^2 + 1)^{-2/3}\left[\frac{x^2}{3} + (x^2 + 1)\right]$$

$$= 2x(x^2 + 1)^{-2/3}\left(\frac{4x^2}{3} + \frac{3}{3}\right)$$

$$= \frac{2x}{(x^2 + 1)^{2/3}} \cdot \frac{4x^2 + 3}{3}$$

$$= \frac{2x(4x^2 + 3)}{3(x^2 + 1)^{2/3}}$$

$$= y_2$$

(b)

x	-2	-1	$-\frac{1}{2}$	0	1	2	$\frac{5}{2}$
y_1	-8.7	-2.9	-1.1	0	2.9	8.7	12.5
y_2	-8.7	-2.9	-1.1	0	2.9	8.7	12.5

69. $y_1 = 2x\sqrt{1 - x^2} - \dfrac{x^3}{\sqrt{1 - x^2}}$ $y_2 = \dfrac{2 - 3x^2}{\sqrt{1 - x^2}}$

When $x = 0$, $y_1 = 0$. When $x = 0$, $y_2 = 1$.

Thus, $y_1 \neq y_2$.

$$y_1 = \frac{2x\sqrt{1 - x^2}}{1} - \frac{x^3}{\sqrt{1 - x^2}} = \frac{2x\sqrt{1 - x^2}}{1} \cdot \frac{\sqrt{1 - x^2}}{\sqrt{1 - x^2}} - \frac{x^3}{\sqrt{1 - x^2}}$$

$$= \frac{2x(1 - x^2) - x^3}{\sqrt{1 - x^2}} = \frac{2x - 2x^3 - x^3}{\sqrt{1 - x^2}}$$

$$= \frac{2x - 3x^3}{\sqrt{1 - x^2}}$$

Let $y_2 = \dfrac{2x - 3x^3}{\sqrt{1 - x^2}}$. Then $y_1 = y_2$.

Section P.8 Graphical Representation of Data

- You should be able to plot points.
- You should know that the distance between (x_1, y_1) and (x_2, y_2) in the plane is

$$d = \sqrt{(x_2 - x_1)^2 + (y_2 - y_1)^2}.$$

- You should know that the midpoint of the line segment joining (x_1, y_1) and (x_2, y_2) is

$$\left(\frac{x_1 + x_2}{2}, \frac{y_1 + y_2}{2}\right).$$

Solutions to Odd-Numbered Exercises

1.

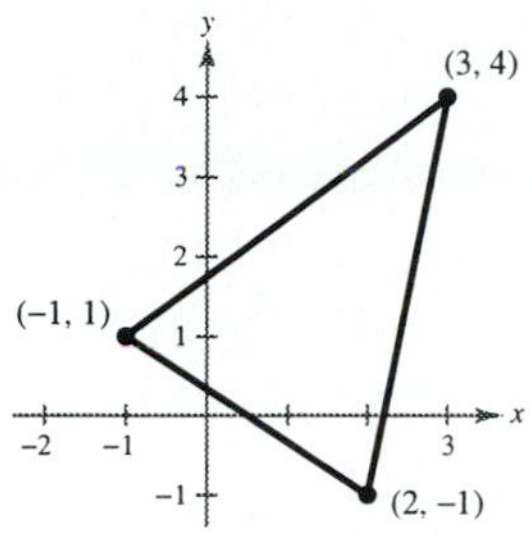

3.

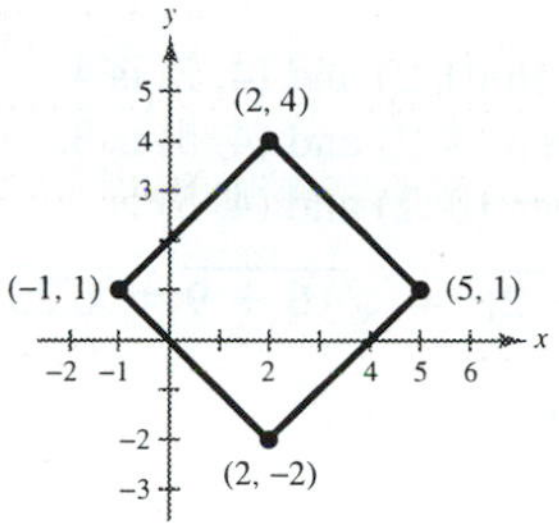

5. A: (2, 6), B: (−6, −2), C: (4, −4), D: (−3, 2)

7. (−3, 4)

9. (−5, −5)

11. On the x-axis, $y = 0$.
On the y-axis, $x = 0$.

13. $x > 0$ and $y < 0$ in Quadrant IV.

15. $x = -4$ and $y > 0$ in Quadrant II.

17. $y < -5$ in Quadrants III and IV.

19. $(x, -y)$ is in the second Quadrant means that (x, y) is in Quadrant III.

21. (x, y), $xy > 0$ means x and y have the same signs. This occurs in Quadrants I and III.

23. $(-2 + 2, -4 + 5) = (0, 1)$

$(2 + 2, -3 + 5) = (4, 2)$

$(-1 + 2, -1 + 5) = (1, 4)$

25.

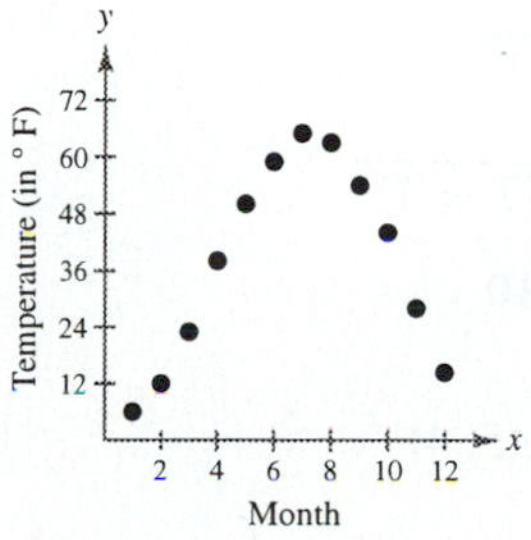

27. $y = 2 - \frac{1}{2}x$

x	-2	-1	$-\frac{1}{2}$	0	$\frac{1}{2}$	1	2
y	3	$\frac{5}{2}$	$\frac{9}{4}$	2	$\frac{7}{4}$	$\frac{3}{2}$	1

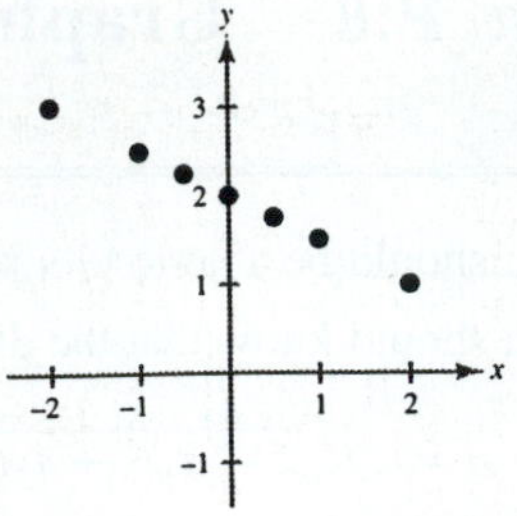

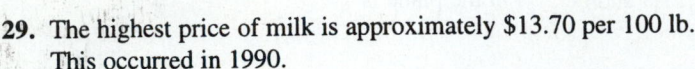

29. The highest price of milk is approximately \$13.70 per 100 lb. This occurred in 1990.

31. $\left(\dfrac{1200}{1995} \div \dfrac{51}{1967}\right)100 \approx 2300\%$

33. The minimum wage increased most rapidly in the 1970s.

35. The point (65, 83) represents an entrance exam score of 65.

37. $d = |5 - (-3)| = 8$

39. $d = |2 - (-3)| = 5$

41. (a) The distance between (0, 2) and (4, 2) is 4.
The distance between (4, 2) and (4, 5) is 3.
The distance between (0, 2) and (4, 5) is

$$\sqrt{(4-0)^2 + (5-2)^2} = \sqrt{16+9} = \sqrt{25} = 5.$$

(b) $4^2 + 3^2 = 16 + 9 = 25 = 5^2$

43. (a) The distance between (−1, 1) and (9, 1) is 10.
The distance between (9, 1) and (9, 4) is 3.
The distance between (−1, 1) and (9, 4) is

$$\sqrt{(9-(-1))^2 + (4-1)^2} = \sqrt{100+9} = \sqrt{109}.$$

(b) $10^2 + 3^2 = 109 = \left(\sqrt{109}\right)^2$

45. (a)

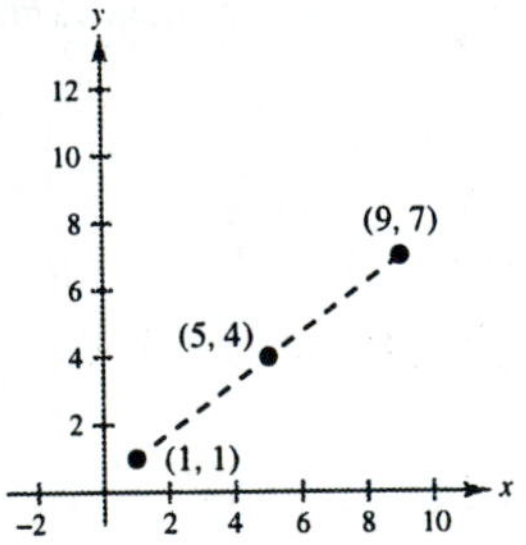

(b) $d = \sqrt{(9-1)^2 + (7-1)^2}$

$= \sqrt{64 + 36} = 10$

(c) $\left(\dfrac{9+1}{2}, \dfrac{7+1}{2}\right) = (5, 4)$

47. (a)

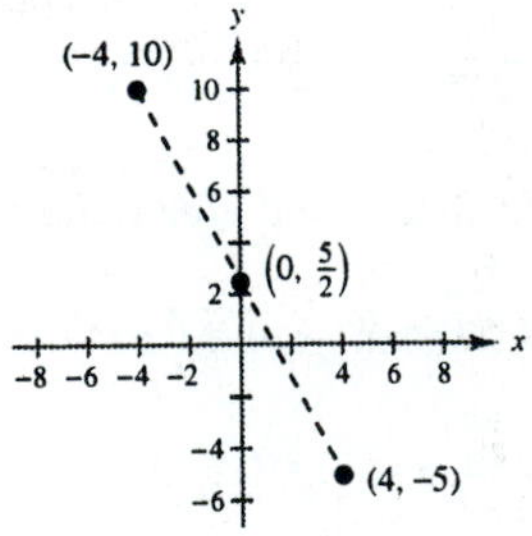

(b) $d = \sqrt{(4+4)^2 + (-5-10)^2}$

$= \sqrt{64 + 225} = 17$

(c) $\left(\dfrac{4-4}{2}, \dfrac{-5+10}{2}\right) = \left(0, \dfrac{5}{2}\right)$

49. (a)

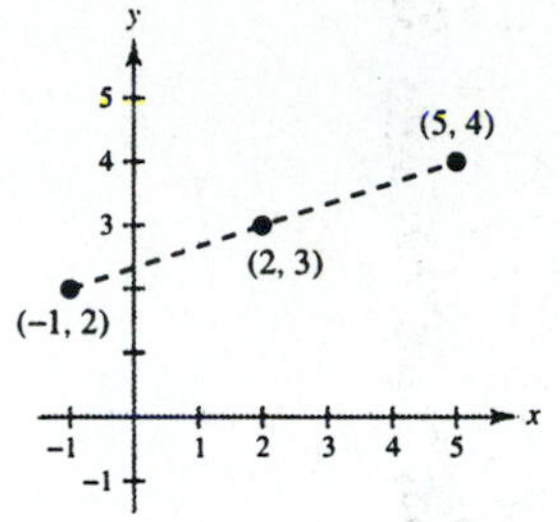

(b) $d = \sqrt{(5+1)^2 + (4-2)^2}$

$= \sqrt{36 + 4} = 2\sqrt{10}$

(c) $\left(\dfrac{-1+5}{2}, \dfrac{2+4}{2}\right) = (2, 3)$

51. (a)

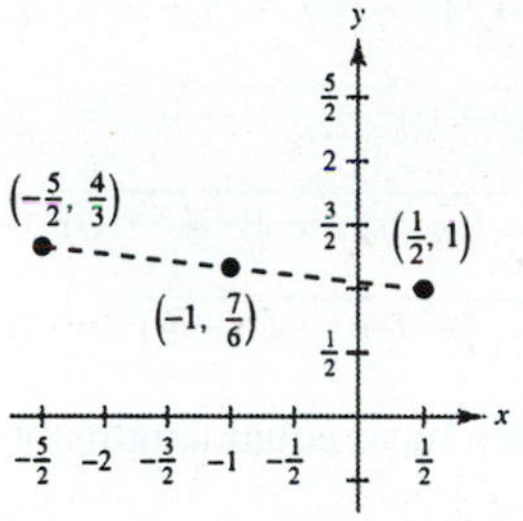

(b) $d = \sqrt{\left(\dfrac{1}{2} + \dfrac{5}{2}\right)^2 + \left(1 - \dfrac{4}{3}\right)^2}$

$d = \sqrt{9 + \dfrac{1}{9}} = \dfrac{\sqrt{82}}{3}$

(c) $\left(\dfrac{-\frac{5}{2} + \frac{1}{2}}{2}, \dfrac{\frac{4}{3} + 1}{2}\right) = \left(-1, \dfrac{7}{6}\right)$

53. (a)

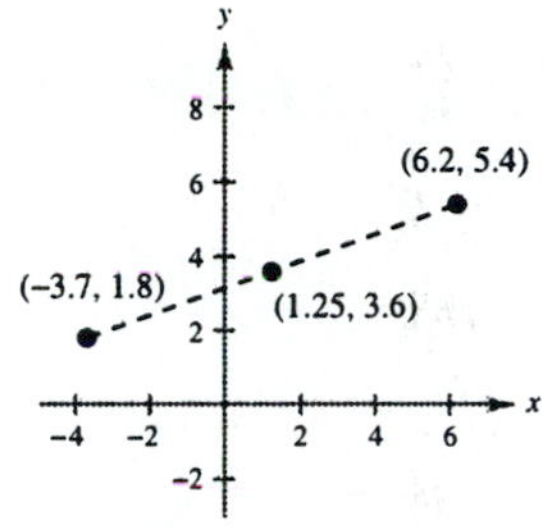

(b) $d = \sqrt{(6.2 + 3.7)^2 + (5.4 - 1.8)^2}$

$= \sqrt{98.01 + 12.96}$

$= \sqrt{110.97}$

(c) $\left(\dfrac{6.2 - 3.7}{2}, \dfrac{5.4 + 1.8}{2}\right) = (1.25, 3.6)$

55. (a)

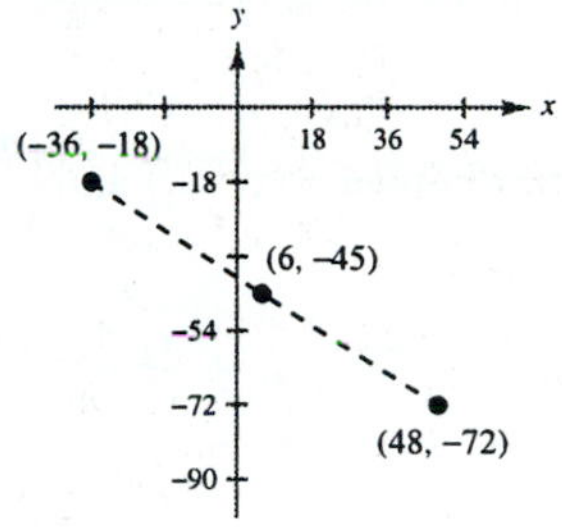

(b) $d = \sqrt{(48 + 36)^2 + (-72 + 18)^2}$

$= \sqrt{7056 + 2916}$

$= \sqrt{9972} = 6\sqrt{277}$

(c) $\left(\dfrac{-36 + 48}{2}, \dfrac{-18 - 72}{2}\right) = (6, -45)$

57. $\left(\dfrac{1991 + 1995}{2}, \dfrac{\$520{,}000 + \$740{,}000}{2}\right) = (1993, \$630{,}000)$

In 1993 the sales were $630,000.

59. $d_1 = \sqrt{(4-2)^2 + (0-1)^2} = \sqrt{5}$

$d_2 = \sqrt{(4+1)^2 + (0+5)^2} = \sqrt{50}$

$d_3 = \sqrt{(2+1)^2 + (1+5)^2} = \sqrt{45}$

$\left(\sqrt{5}\right)^2 + \left(\sqrt{45}\right)^2 = \left(\sqrt{50}\right)^2$

61. $d_1 = \sqrt{(0-1)^2 + (0-2)^2} = \sqrt{5}$

$d_2 = \sqrt{(0-2)^2 + (0-1)^2} = \sqrt{5}$

$d_3 = \sqrt{(3-1)^2 + (3-2)^2} = \sqrt{5}$

$d_4 = \sqrt{(3-2)^2 + (3-1)^2} = \sqrt{5}$

$d_1 = d_2 = d_3 = d_4$

63. $d_1 = \sqrt{(0-2)^2 + (9-5)^2} = \sqrt{4+16} = \sqrt{20} = 2\sqrt{5}$

$d_2 = \sqrt{(-2-0)^2 + (0-9)^2} = \sqrt{4+81} = \sqrt{85}$

$d_3 = \sqrt{(0-(-2))^2 + (-4-0)^2} = \sqrt{4+16} = \sqrt{20} = 2\sqrt{5}$

$d_4 = \sqrt{(0-2)^2 + (-4-5)^2} = \sqrt{4+81} = \sqrt{85}$

Opposite sides have equal lengths of $2\sqrt{5}$ and $\sqrt{85}$.

65. Since $x_m = \dfrac{x_1 + x_2}{2}$ and $y_m = \dfrac{y_1 + y_2}{2}$ we have:

$$2x_m = x_1 + x_2 \qquad 2y_m = y_1 + y_2$$

$$2x_m - x_1 = x_2 \qquad 2y_m - y_1 = y_2$$

Thus, $(x_2, y_2) = (2x_m - x_1, 2y_m - y_1)$.

67. The midpoint of the given line segment is $\left(\dfrac{x_1 + x_2}{2}, \dfrac{y_1 + y_2}{2}\right)$.

The midpoint between (x_1, y_1) and $\left(\dfrac{x_1 + x_2}{2}, \dfrac{y_1 + y_2}{2}\right)$ is

$$\left(\frac{x_1 + \frac{x_1 + x_2}{2}}{2}, \frac{y_1 + \frac{y_1 + y_2}{2}}{2}\right) = \left(\frac{3x_1 + x_2}{4}, \frac{3y_1 + y_2}{4}\right).$$

The midpoint between $\left(\dfrac{x_1 + x_2}{2}, \dfrac{y_1 + y_2}{2}\right)$ and (x_2, y_2) is

$$\left(\frac{\frac{x_1 + x_2}{2} + x_2}{2}, \frac{\frac{y_1 + y_2}{2} + y_2}{2}\right) = \left(\frac{x_1 + 3x_2}{4}, \frac{y_1 + 3y_2}{4}\right).$$

Thus, the three points are

$$\left(\frac{3x_1 + x_2}{4}, \frac{3y_1 + y_2}{4}\right), \left(\frac{x_1 + x_2}{2}, \frac{y_1 + y_2}{2}\right), \text{ and } \left(\frac{x_1 + 3x_2}{4}, \frac{y_1 + 3y_2}{4}\right).$$

69. $d = \sqrt{(45-10)^2 + (40-15)^2} = \sqrt{35^2 + 25^2} = \sqrt{1850} = 5\sqrt{74} \approx 43$ yards

71.

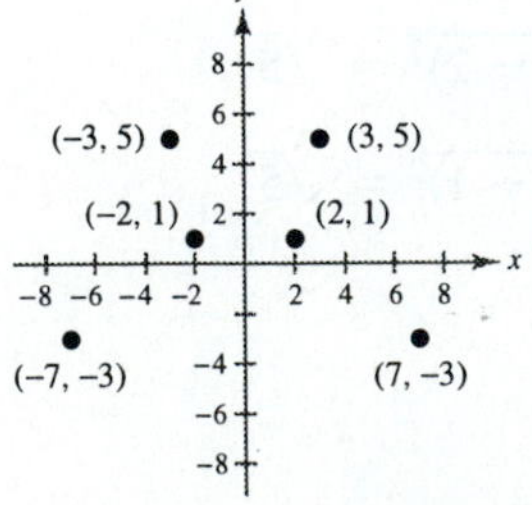

The points are reflected through the y-axis.

73. (a) It appears that the number of artists elected alternates between 7 and 8 per year in the 1990s. If this pattern continues, 8 would be elected in 1996.

(b) Since 1986 and 1987 were the first two years that artists were elected, there was a larger number of artists chosen.

❑ Review Exercises for Chapter P

Solutions to Odd-Numbered Exercises

1. $\{11, -14, -\frac{8}{9}, \frac{5}{2}, \sqrt{6}, 0.4\}$

(a) Natural numbers: 11

(b) Integers: 11, -14

(c) Rational numbers: 11, -14, $-\frac{8}{9}, \frac{5}{2}$, 0.4

(d) Irrational numbers: $\sqrt{6}$

3. (a) $\frac{5}{6} = 0.8\overline{3}$

(b) $\frac{7}{8} = 0.875$

5. $x \le 7$

The set consists of all real numbers less than or equal to 7.

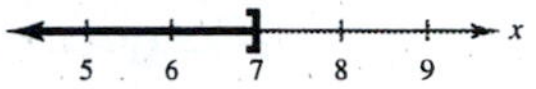

7. $d(x, 7) = |x - 7|$ and $d(x, 7) \ge 4$, thus $|x - 7| \ge 4$.

9. $d(y, -30) = |y - (-30)| = |y + 30|$ and $d(y, -30) < 5$, thus $|y + 30| < 5$.

11. $2x + (3x - 10) = (2x + 3x) - 10$

Illustrates the Associative Property of Addition

13. $0 + (a - 5) = a - 5$

Illustrates the Additive Identity Property

15. (a) $(-2z)^3 = (-2)^3z^3 = -8z^3$

(b) $(a^2b^4)(3ab^{-2}) = 3a^{2+1}b^{4+(-2)} = 3a^3b^2$

17. (a) $\dfrac{6^2u^3v^{-3}}{12u^{-2}v} = \dfrac{36u^{3-(-2)}v^{-3-1}}{12} = 3u^5v^{-4} = \dfrac{3u^5}{v^4}$

(b) $\dfrac{3^{-4}m^{-1}n^{-3}}{9^{-2}mn^{-3}} = \dfrac{9^2n^3}{3^4mmn^3} = \dfrac{81}{81m^2} = \dfrac{1}{m^2} = m^{-2}$

19. $30{,}296{,}000{,}000 = 3.0296 \times 10^{10}$

21. $4.833 \times 10^8 = 483{,}300{,}000$

23. (a) $1800(1 + 0.08)^{24} \approx 11{,}414.125$

(b) $0.0024\ (7{,}658{,}400) = 18{,}380.160$

25. Radical form: $\sqrt{16} = 4$

Rational exponent form: $16^{1/2} = 4$

27. (a) $\sqrt{4x^4} = 2x^2$

(b) $\sqrt{\dfrac{18u^2}{b^3}} = \sqrt{\dfrac{9u^2}{b^2} \cdot \dfrac{2}{b}} = \dfrac{3|u|}{b}\sqrt{\dfrac{2}{b}}$

29. $\dfrac{1}{2-\sqrt{3}} = \dfrac{1}{2-\sqrt{3}} \cdot \dfrac{2+\sqrt{3}}{2+\sqrt{3}} = \dfrac{2+\sqrt{3}}{4-3} = \dfrac{2+\sqrt{3}}{1} = 2+\sqrt{3}$

31. $\sqrt{50} - \sqrt{18} = \sqrt{25 \cdot 2} - \sqrt{9 \cdot 2} = 5\sqrt{2} - 3\sqrt{2} = 2\sqrt{2}$

33. $A = wh = 8\sqrt{3}\,\sqrt{24^2 - (8\sqrt{3})^2} = 8\sqrt{3}\,\sqrt{384}$

$= 8\sqrt{3}(8\sqrt{6}) = 64\sqrt{18} = 64(3\sqrt{2}) = 192\sqrt{2}$

35. $10(4 \cdot 7) = 10(28) = 280$

The error in $40 \cdot 70$ is an improper use of the Distributive Property.

37. $4\left(\dfrac{3}{7}\right) = \dfrac{4}{1}\left(\dfrac{3}{7}\right) = \dfrac{12}{7}$

Only the numerator is multiplied by 4.

39. $\dfrac{x-1}{1-x} = \dfrac{x-1}{-(x-1)} = -1,\ x \neq 1$

The error is an improper cancellation.

41. $(-x)^6 = x^6$

The exponent is to be applied to the whole quantity inside the parentheses.

43. $\sqrt{3^2 + 4^2} = \sqrt{9 + 16} = \sqrt{25} = 5$

Do not apply radicals term–by–term.

45. $\sqrt{10x} = \sqrt{10}\sqrt{x}$

Radicals apply to each factor.

47. $-(3x^2 + 2x) + (1 - 5x) = -3x^2 - 2x + 1 - 5x$

$= -3x^2 - 7x + 1$

49. $(2x - 3)^2 = (2x)^2 - 2(2x)(3) + 3^2$

$= 4x^2 - 12x + 9$

51. $(x^3 - 3x)(2x^2 + 3x + 5) = x^3(2x^2 + 3x + 5) - 3x(2x^2 + 3x + 5)$

$= 2x^5 + 3x^4 + 5x^3 - 6x^3 - 9x^2 - 15x$

$= 2x^5 + 3x^4 - x^3 - 9x^2 - 15x$

53. $x^3 - x = x(x^2 - 1) = x(x + 1)(x - 1)$

55. $2x^2 + 21x + 10 = (2x + 1)(x + 10)$

57. $x^3 - x^2 + 2x - 2 = x^2(x - 1) + 2(x - 1)$

$= (x - 1)(x^2 + 2)$

59. (a)

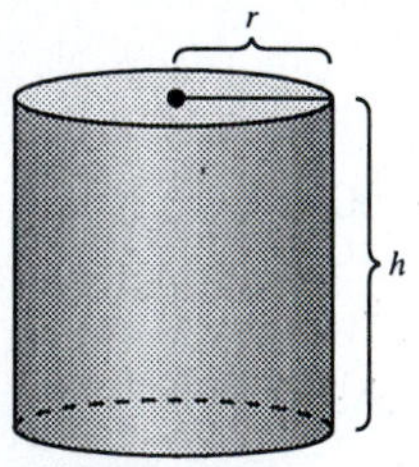

The surface area is the sum of the area of the side, $2\pi rh$, and the areas of the top and bottom which are each πr^2.

$S = 2\pi rh + \pi r^2 + \pi r^2 = 2\pi rh + 2\pi r^2$

(b) $S = 2\pi rh + 2\pi r^2 = 2\pi r(r + h)$

61. $\frac{2}{3}x^4 - \frac{3}{8}x^3 + \frac{5}{6}x^2 = \frac{16}{24}x^4 - \frac{9}{24}x^3 + \frac{20}{24}x^2$

$= \frac{1}{24}x^2(16x^2 - 9x + 20)$

The missing factor is $16x^2 - 9x + 20$.

63. $\dfrac{x^2-4}{x^4-2x^2-8}\cdot\dfrac{x^2+2}{x^2}=\dfrac{(x+2)(x-2)}{(x^2+2)(x^2-4)}\cdot\dfrac{x^2+2}{x^2}$

$=\dfrac{(x+2)(x-2)}{(x^2+2)(x+2)(x-2)}\cdot\dfrac{x^2+2}{x^2}$

$=\dfrac{1}{x^2}$

65. $2x+\dfrac{3}{2(x-4)}-\dfrac{1}{2(x+2)}=\dfrac{2x(2)(x-4)(x+2)+3(x+2)-(x-4)}{2(x-4)(x+2)}$

$=\dfrac{4x(x^2-2x-8)+3x+6-x+4}{2(x-4)(x+2)}$

$=\dfrac{4x^3-8x^2-32x+2x+10}{2(x-4)(x+2)}$

$=\dfrac{4x^3-8x^2-30x+10}{2(x-4)(x+2)}$

$=\dfrac{2(2x^3-4x^2-15x+5)}{2(x-4)(x+2)}$

$=\dfrac{2x^3-4x^2-15x+5}{(x-4)(x+2)}$

67. $\dfrac{1}{x-1}+\dfrac{1-x}{x^2+x+1}=\dfrac{(x^2+x+1)+(1-x)(x-1)}{(x-1)(x^2+x+1)}$

$=\dfrac{x^2+x+1-x^2+2x-1}{(x-1)(x^2+x+1)}$

$=\dfrac{3x}{(x-1)(x^2+x+1)}$

$=\dfrac{3x}{x^3-1}$

69. $\dfrac{\left[\dfrac{3a}{(a^2/x)-1}\right]}{\left(\dfrac{a}{x}-1\right)}=\dfrac{\dfrac{3ax}{a^2-x}}{\dfrac{a-x}{x}}$

$=\dfrac{3ax}{a^2-x}\cdot\dfrac{x}{a-x}$

$=\dfrac{3ax^2}{(a^2-x)(a-x)}$

$=\dfrac{3ax^2}{(a+x)(a-x)(a-x)}$

$=\dfrac{3ax^2}{(a+x)(a-x)^2}$

71. $3x-2(x+5)=10$

$3x-2x-10=10$

$x=20$

73. $4(x + 3) - 3 = 2(4 - 3x) - 4$

$$4x + 12 - 3 = 8 - 6x - 4$$
$$4x + 9 = -6x + 4$$
$$10x = -5$$
$$x = -\frac{1}{2}$$

75. $3\left(1 - \frac{1}{5t}\right) = 0$

$$1 - \frac{1}{5t} = 0$$
$$1 = \frac{1}{5t}$$
$$5t = 1$$
$$t = \frac{1}{5}$$

77. $6x = 3x^2$

$$0 = 3x^2 - 6x$$
$$0 = 3x(x - 2)$$
$$3x = 0 \implies x = 0$$
$$x - 2 = 0 \implies x = 2$$

79. $(x + 4)^2 = 18$

$$x + 4 = \pm\sqrt{18}$$
$$x = -4 \pm 3\sqrt{2}$$

81. $x^2 - 12x + 30 = 0$

$$x^2 - 12x = -30$$
$$x^2 - 12x + 36 = -30 + 36$$
$$(x - 6)^2 = 6$$
$$x - 6 = \pm\sqrt{6}$$
$$x = 6 \pm \sqrt{6}$$

83. $5x^4 - 12x^3 = 0$

$$x^3(5x - 12) = 0$$
$$x^3 = 0 \quad \text{or} \quad 5x - 12 = 0$$
$$x = 0 \quad \text{or} \quad x = \tfrac{12}{5}$$

85. $\dfrac{4}{(x - 4)^2} = 1$

$$4 = (x - 4)^2$$
$$\pm 2 = x - 4$$
$$4 \pm 2 = x$$
$$x = 6 \quad \text{or} \quad x = 2$$

87. $\sqrt{x + 4} = 3$

$$\left(\sqrt{x + 4}\right)^2 = (3)^2$$
$$x + 4 = 9$$
$$x = 5$$

89. $2\sqrt{x} - 5 = 0$

$$2\sqrt{x} = 5$$
$$4x = 25$$
$$x = \tfrac{25}{4}$$

91. $\sqrt{2x + 3} + \sqrt{x - 2} = 2$

$$\left(\sqrt{2x + 3}\right)^2 = \left(2 - \sqrt{x - 2}\right)^2$$
$$2x + 3 = 4 - 4\sqrt{x - 2} + x - 2$$
$$x + 1 = -4\sqrt{x - 2}$$
$$(x + 1)^2 = \left(-4\sqrt{x - 2}\right)^2$$
$$x^2 + 2x + 1 = 16(x - 2)$$
$$x^2 - 14x + 33 = 0$$
$$(x - 3)(x - 11) = 0$$

$x = 3$, extraneous or $x = 11$, extraneous

No solution

93. $(x - 1)^{2/3} - 25 = 0$

$$(x - 1)^{2/3} = 25$$
$$(x - 1)^2 = 25^3$$
$$x - 1 = \pm\sqrt{25^3}$$
$$x = 1 \pm 125$$
$$x = 126 \quad \text{or} \quad x = -124$$

95. $(x + 4)^{1/2} + 5x(x + 4)^{3/2} = 0$

$$(x + 4)^{1/2}[1 + 5x(x + 4)] = 0$$
$$(x + 4)^{1/2}(5x^2 + 20x + 1) = 0$$
$$(x + 4)^{1/2} = 0 \quad \text{OR} \quad 5x^2 + 20x + 1 = 0$$
$$x = -4 \qquad\qquad x = \frac{-20 \pm \sqrt{400 - 20}}{10}$$
$$x = \frac{-20 \pm 2\sqrt{95}}{10}$$
$$x = -2 \pm \frac{\sqrt{95}}{5}$$

97. $|x - 5| = 10$

$$x - 5 = -10 \quad \text{or} \quad x - 5 = 10$$
$$x = -5 \qquad\qquad x = 15$$

99. $|x^2 - 3| = 2x$

$$x^2 - 3 = 2x \quad \text{or} \quad x^2 - 3 = -2x$$
$$x^2 - 2x - 3 = 0 \qquad x^2 + 2x - 3 = 0$$
$$(x - 3)(x + 1) = 0 \qquad (x + 3)(x - 1) = 0$$
$$x = 3 \text{ or } x = -1 \qquad x = -3 \text{ or } x = 1$$

The only solutions to the original equation are $x = 3$ or $x = 1$. ($x = -3$ and $x = -1$ are extraneous.)

101. $V = \frac{1}{3}\pi r^2 h$

$$3V = \pi r^2 h$$
$$\frac{3V}{\pi h} = r^2$$
$$r = \sqrt{\frac{3V}{\pi h}}$$

Since r represents the radius of a cone, r is positive only.

103. $L = \frac{k}{3\pi r^2 p}$

$$3\pi r^2 pL = k$$
$$p = \frac{k}{3\pi r^2 L}$$

105. $x^2 - 2x \geq 3$

$x^2 - 2x - 3 \geq 0$

$(x + 1)(x - 3) \geq 0$

Critical numbers: $x = -1, x = 3$

Test intervals: $(-\infty, -1), (-1, 3), (3, \infty)$

Test: Is $(x + 1)(x - 3) \geq 0$?

Solution set: $(-\infty, -1] \cup [3, \infty)$

107. $\dfrac{x - 5}{3 - x} < 0$

Critical numbers: $x = 5, x = 3$

Test intervals: $(-\infty, 3), (3, 5), (5, \infty)$

Test: Is $\dfrac{x - 5}{3 - x} < 0$?

Solution set: $(-\infty, 3) \cup (5, \infty)$

109. $|x - 2| < 1$

$-1 < x - 2 < 1$

$1 < x < 3,$

which can be written as $(1, 3)$.

111. $\left|x - \dfrac{3}{2}\right| \geq \dfrac{3}{2}$

$x - \dfrac{3}{2} \leq -\dfrac{3}{2}$ or $x - \dfrac{3}{2} \geq \dfrac{3}{2}$

$x \leq 0$ or $x \geq 3,$

which can be written as $(-\infty, 0] \cup [3, \infty)$.

113. $\dfrac{x}{5} - 6 \leq -\dfrac{x}{2} + 6$

$10\left(\dfrac{x}{5} - 6\right) \leq 10\left(-\dfrac{x}{2} + 6\right)$

$2x - 60 \leq -5x + 60$

$7x \leq 120$

$x \leq \dfrac{120}{7}$

115. $(x - 4)|x| > 0$

Critical numbers: $x = 4, x = 0$

Test intervals: $(-\infty, 0), (0, 4), (4, \infty)$

Test: Is $(x - 4)|x| > 0$?

Solution set: $(4, \infty)$

117. $2x - 10 \geq 0$

$2x \geq 10$

$x \geq 5$

Domain: $[5, \infty)$

119.

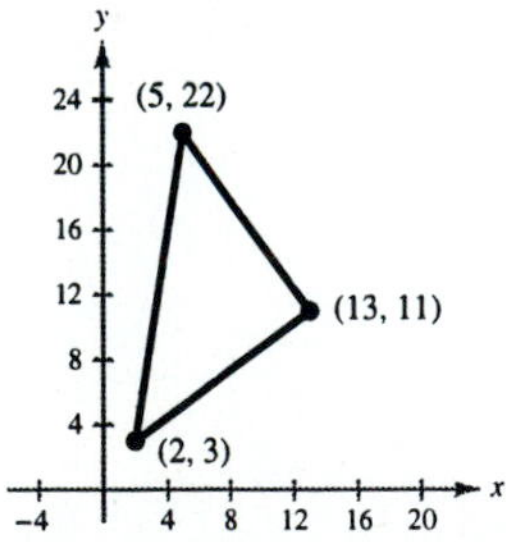

$d_1 = \sqrt{(13 - 5)^2 + (11 - 22)^2} = \sqrt{8^2 + (-11)^2} = \sqrt{64 + 121} = \sqrt{185}$

$d_2 = \sqrt{(2 - 13)^2 + (3 - 11)^2} = \sqrt{(-11)^2 + (-8)^2} = \sqrt{121 + 64} = \sqrt{185}$

$d_3 = \sqrt{(2 - 5)^2 + (3 - 22)^2} = \sqrt{(-3)^2 + (-19)^2} = \sqrt{9 + 361} = \sqrt{370}$

$d_1^2 + d_2^2 = 185 + 185 = 370 = d_3^2$

The points form a right triangle.

121. $x > 0$ and $y = -2$ is in Quadrant IV

123. $(-x, y)$ is in Quadrant III $\Longrightarrow$ (x, y) is in Quadrant IV.

125. (a)

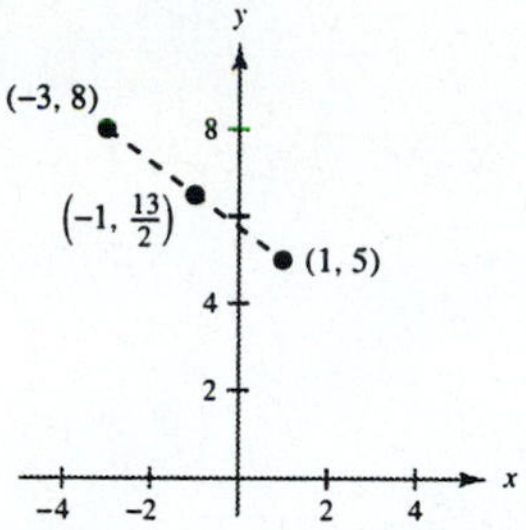

(b) $d = \sqrt{(1-(-3))^2 + (5-8)^2}$

$= \sqrt{4^2 + (-3)^2}$

$= \sqrt{16+9}$

$= \sqrt{25}$

$= 5$

(c) $\left(\frac{-3+1}{2}, \frac{8+5}{2}\right) = \left(-1, \frac{13}{2}\right)$

127. Each corner has a length of $2\sqrt{2}$ by the Pythagorean Theorem.

$P = 2(8\frac{1}{2} - 4) + 2(11 - 4) + 4(2\sqrt{2})$

$= 9 + 14 + 8\sqrt{2}$

$= 23 + 8\sqrt{2} \approx 34.3$ inches

129. (a) $2m + 2n = 2(m + n)$ which is even.

(b) $(2m + 1) + (2n + 1) = 2m + 2n + 2 = 2(m + n + 1)$ which is even.

(c) $(2n)(m) = 2nm$ which is even.

131. Original Price: $340 + 85 = 425$

Percent Discount: $85/425 = 0.20 = 20\%$

133. $2\pi(r + 1) - 2\pi r = 2\pi r + 2\pi - 2\pi r$

$= 2\pi$ meters longer

135.

$$\frac{90{,}000}{x} = \frac{90{,}000}{x+3} + 2500$$

$$90{,}000(x + 3) = 90{,}000x + 2500x(x + 3)$$

$$90{,}000x + 270{,}000 = 90{,}000x + 2500x^2 + 7500x$$

$$0 = 2500x^2 + 7500x - 270{,}000$$

$$0 = 2500(x^2 + 3x - 108)$$

$$0 = 2500(x + 12)(x - 9)$$

$$x = -12, \text{ extraneous}$$

$$x = 9$$

There are 9 people in the original group.

❑ Practice Test for Chapter P

1. Evaluate $\dfrac{|-42| - 20}{15 - |-4|}$.

2. Simplify $\dfrac{x}{z} - \dfrac{z}{y}$.

3. The distance between x and 7 is no more than 4. Use absolute value notation to describe this expression.

4. Evaluate $10(-x)^3$ for $x = 5$.

5. Simplify $(-4x^3)(-2x^{-5})\left(\frac{1}{16}x\right)$.

6. Change 0.0000412 to scientific notation.

7. Evaluate $125^{2/3}$.

8. Simplify $\sqrt[4]{64x^7y^9}$.

9. Rationalize the denominator and simplify $\dfrac{6}{\sqrt{12}}$.

10. Simplify $3\sqrt{80} - 7\sqrt{500}$.

11. Simplify $(8x^4 - 9x^2 + 2x - 1) - (3x^3 + 5x + 4)$.

12. Multiply $(x - 3)(x^2 + x - 7)$.

13. Multiply $[(x - 2) - y]^2$.

14. Factor $16x^4 - 1$.

15. Factor $6x^2 + 5x - 4$.

16. Factor $x^3 - 64$.

17. Combine and simplify $-\dfrac{3}{x} + \dfrac{x}{x^2 + 2}$.

18. Combine and simplify $\dfrac{x - 3}{4x} \div \dfrac{x^2 - 9}{x^2}$.

19. Simplify $\dfrac{1 - (1/x)}{1 - \dfrac{1}{1 - (1/x)}}$.

20. (a) Plot the points $(-3, 7)$ and $(5, -1)$,
    (b) find the distance between the points, and
    (c) find the midpoint of the line segment joining the points.

21. Solve $x^2 - 2x - 35 = 0$
    (a) by factoring.
    (b) by completing the square.
    (c) by the Quadratic Formula.

22. Solve $x^5 - 5x^3 + 4x = 0$ by factoring.

23. Solve $x = 2\sqrt{x + 3}$.

24. Solve $x^2 - 16 \le 0$.

25. Solve $\left|\dfrac{4 - x}{3}\right| > 2$.

CHAPTER 1
Functions and Their Graphs

CHAPTER 1
Functions and Their Graphs

Section 1.1 Graphs and Graphing Utilities

- You should be able to use the point-plotting method of graphing.
- You should be able to find x- and y-intercepts.
  (a) To find the x-intercepts, let $y = 0$ and solve for x.
  (b) To find the y-intercepts, let $x = 0$ and solve for y.
- You should be able to test for symmetry.
  (a) To test for x-axis symmetry, replace y with $-y$.
  (b) To test for y-axis symmetry, replace x with $-x$.
  (c) To test for origin symmetry, replace x with $-x$ and y with $-y$.
- You should know the standard equation of a circle with center (h, k) and radius r:
  $(x - h)^2 + (y - k)^2 = r^2$

Solutions to Odd-Numbered Exercises

1. $y = \sqrt{x + 4}$

(a) $(0, 2)$: $2 \stackrel{?}{=} \sqrt{0 + 4}$

$2 = 2$ 3

Yes, the point *is* on the graph.

(b) $(5, 3)$: $3 \stackrel{?}{=} \sqrt{5 + 4}$

$3 = \sqrt{9}$ 3

Yes, the point *is* on the graph.

3. $y = 4 - |x - 2|$

(a) $(1, 5)$: $5 \stackrel{?}{=} 4 - |1 - 2|$

$5 \neq 4 - 1$

No, the point *is not* on the graph.

(b) $(6, 0)$: $0 \stackrel{?}{=} 4 - |6 - 2|$

$0 = 4 - 4$ 3

Yes, the point *is* on the graph.

5. $2x - y - 3 = 0$

(a) $(1, 2)$: $2(1) - (2) - 3 \stackrel{?}{=} 0$

$-3 \neq 0$

No, the point *is not* on the graph.

(b) $(1, -1)$: $2(1) - (-1) - 3 \stackrel{?}{=} 0$

$2 + 1 - 3 = 0$ 3

Yes, the point *is* on the graph.

7. $x^2y - x^2 + 4y = 0$

(a) $\left(1, \frac{1}{5}\right)$: $(1)^2\left(\frac{1}{5}\right) - (1)^2 + 4\left(\frac{1}{5}\right) \stackrel{?}{=} 0$

$\frac{1}{5} - 1 + \frac{4}{5} = 0$ 3

Yes, the point *is* on the graph.

(b) $\left(2, \frac{1}{2}\right)$: $(2)^2\left(\frac{1}{2}\right) - (2)^2 + 4\left(\frac{1}{2}\right) \stackrel{?}{=} 0$

$2 - 4 + 2 = 0$ 3

Yes, the point *is* on the graph.

9. $y = -2x + 3$

x	-1	0	1	$\frac{3}{2}$	2
y	5	3	1	0	-1

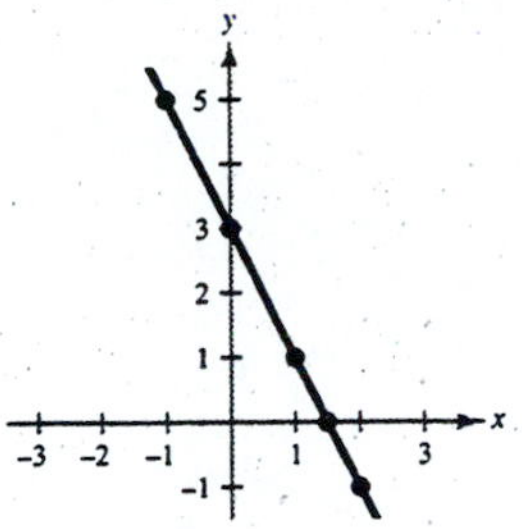

11. $y = x^2 - 2x$

x	-1	0	1	2	3
y	3	0	-1	0	3

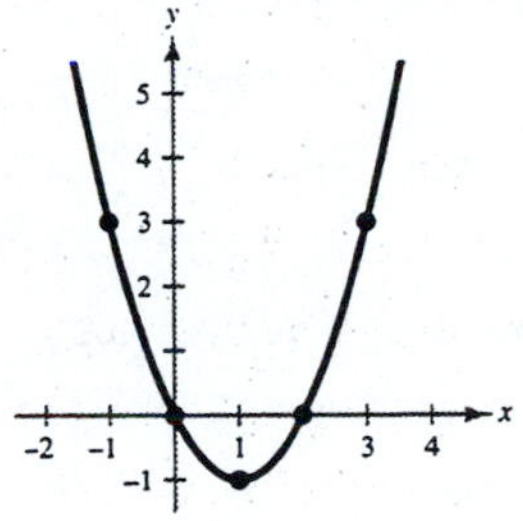

13.

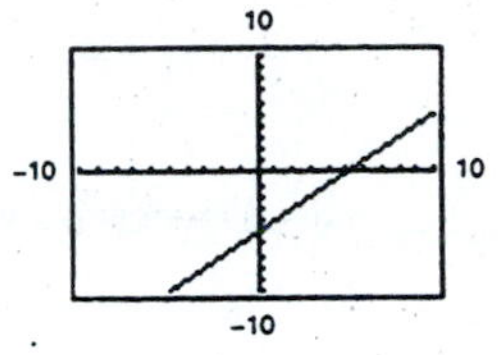

$y = x - 5$

Intercepts: $(5, 0), (0, -5)$

15.

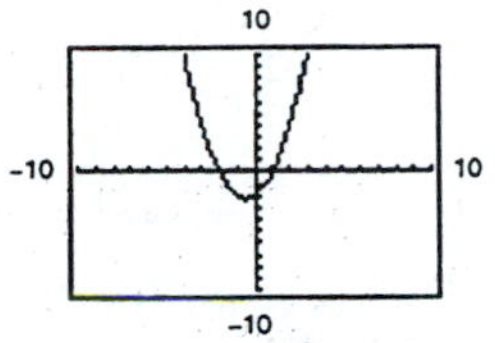

$y = x^2 + x - 2$

Intercepts: $(1, 0), (-2, 0), (0, -2)$

17.

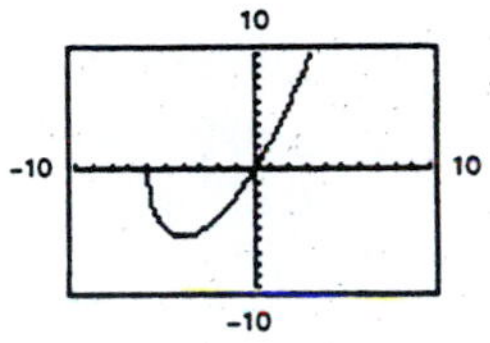

$y = x\sqrt{x + 6}$

Intercepts: $(0, 0), (-6, 0)$

19.

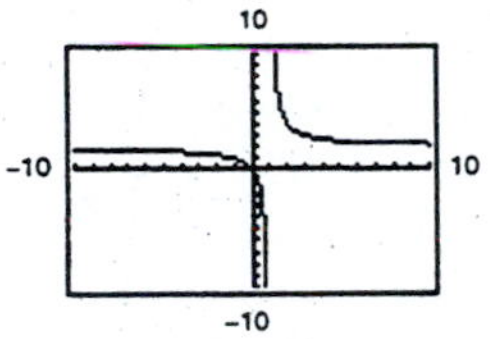

$y = \dfrac{2x}{x - 1}$

Intercept: $(0, 0)$

21. $(-x)^2 - y = 0 \implies x^2 - y = 0$

y-axis symmetry

23. $x - (-y)^2 = 0 \implies x - y^2 = 0$

x-axis symmetry

25. $-y = (-x)^3 \implies y = x^3$

Origin symmetry

27. $-y = \dfrac{-x}{(-x)^2 + 1} \implies y = \dfrac{x}{x^2 + 1}$

Origin symmetry

29. y-axis symmetry

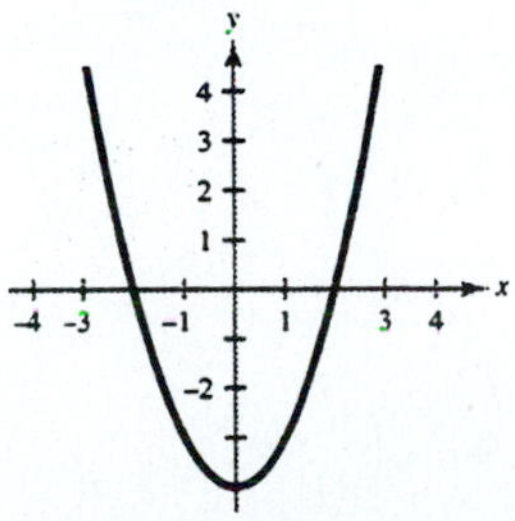

31. Origin symmetry

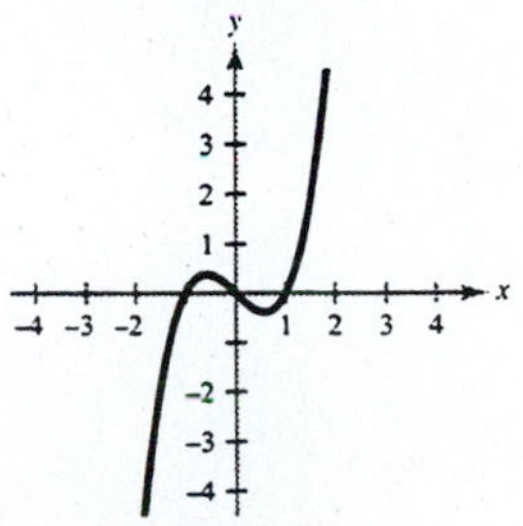

33. $y = 1 - x$ has intercepts $(1, 0)$ and $(0, 1)$. Matches graph (c).

35. $y = \sqrt{9 - x^2}$ has intercepts $(\pm 3, 0)$ and $(0, 3)$. Matches graph (f).

37. $y = x^3 - x + 1$ has a y-intercept of $(0, 1)$ and the points $(1, 1)$ and $(-2, -5)$ are on the graph. Matches graph (b).

39. $y = -3x + 2$
No symmetry

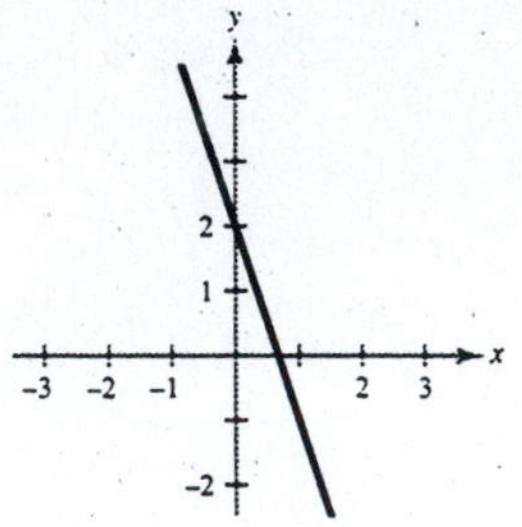

41. $y = 1 - x^2$
y-axis symmetry

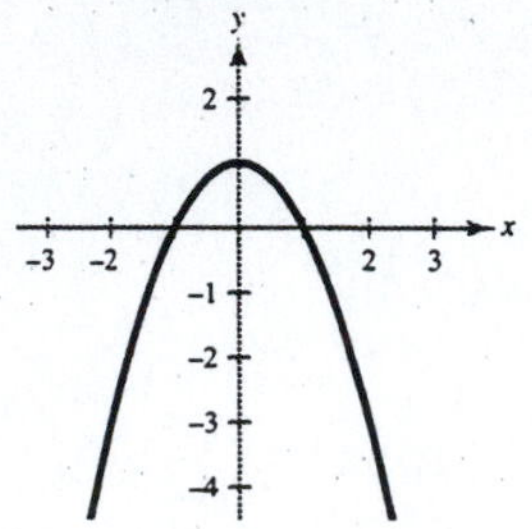

43. $y = x^2 - 3x$
No symmetry

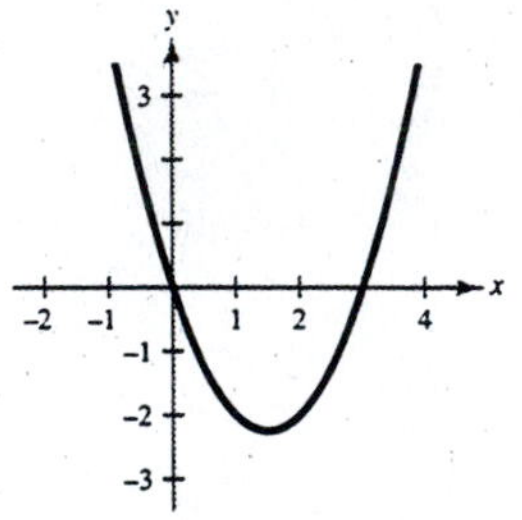

45. $y = x^3 + 2$
No symmetry

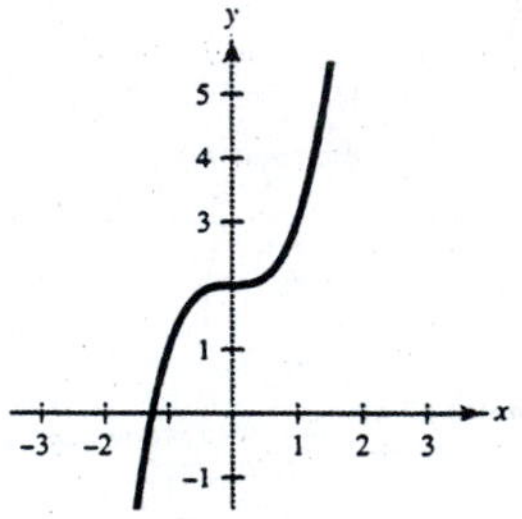

47. $y = \sqrt{x - 3}$
No symmetry
Domain: $x \geq 3$

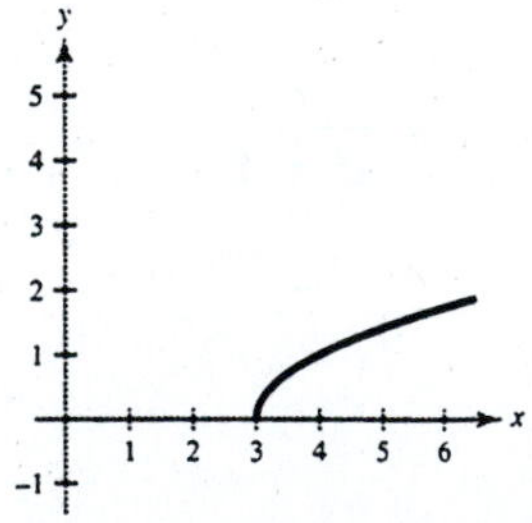

49. $y = |x - 2|$

No symmetry

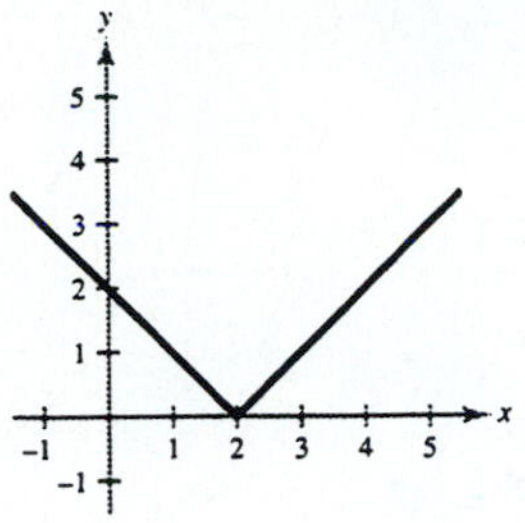

51. $x = y^2 - 1$

x-axis symmetry

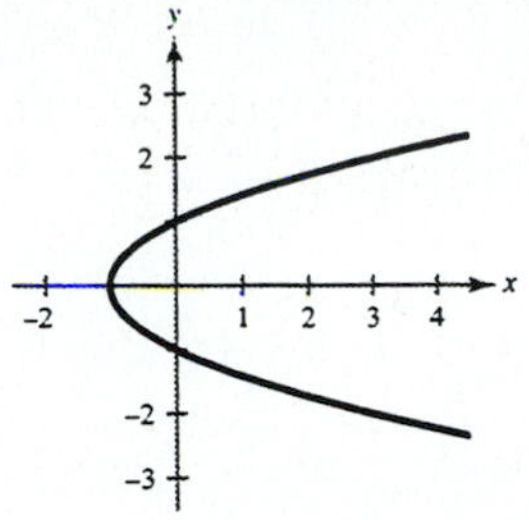

53. $y = 3 - \frac{1}{2}x$

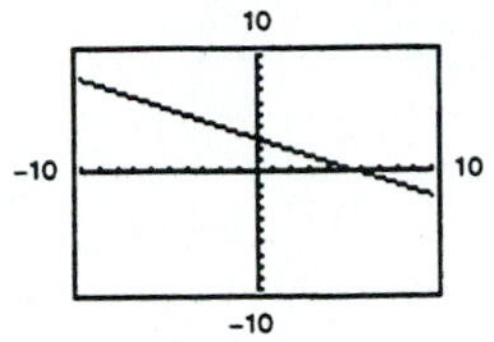

Intercepts: (6,0), (0, 3)

55. $y = x^2 - 4x + 3$

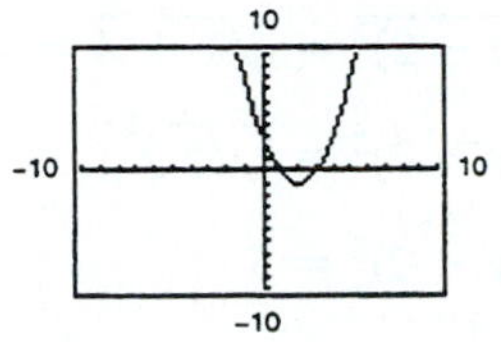

Intercepts: (3, 0), (1, 0), (0, 3)

57. $y = x(x - 2)^2$

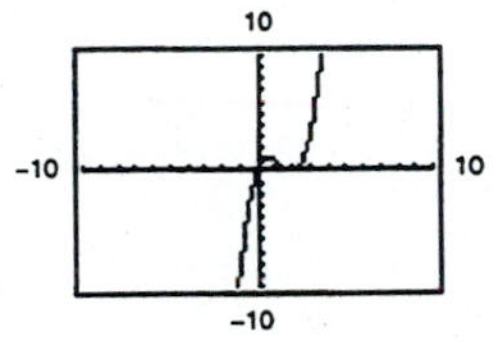

Intercepts: (0, 0), (2, 0)

59. $y = \sqrt[3]{x}$

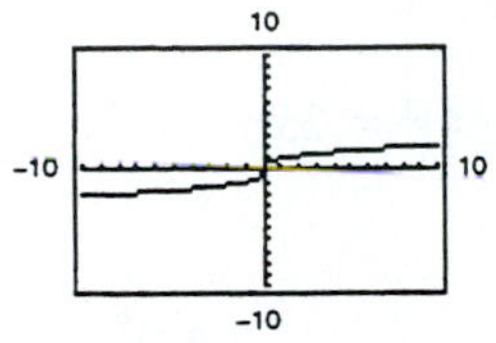

Intercepts: (0, 0)

61. $y = \frac{5}{2}x + 5$

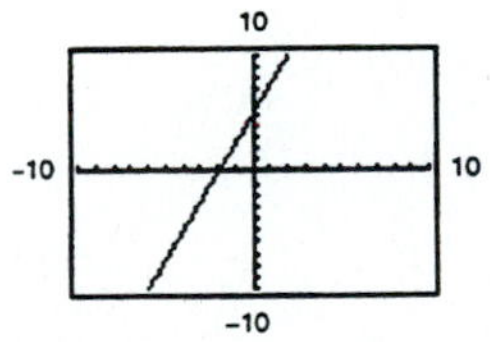

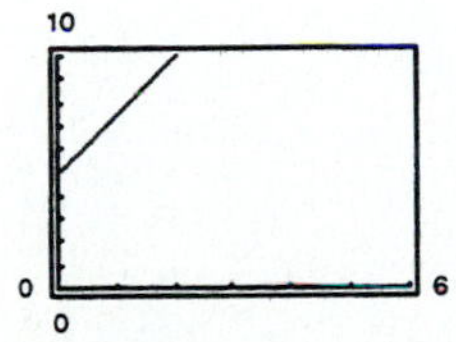

The standard setting gives a more complete graph.

63. $y = -x^2 + 10x - 5$

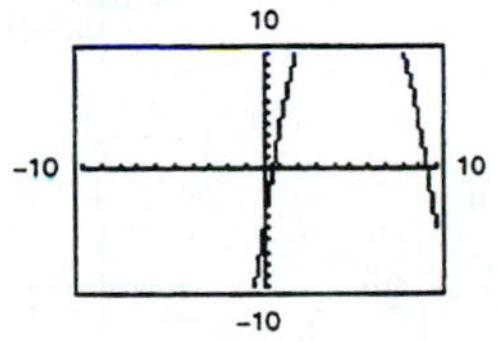

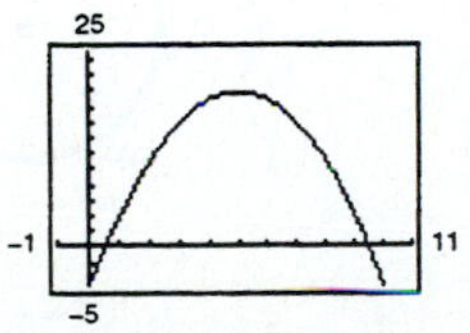

The specified setting gives a more complete graph.

65. $y = 4x^2 - 25$

Range/Window

Xmin = -5
Xmax = 5
Xscl = 1
Ymin = -30
Ymax = 30
Yscl = 10

67. $y = |x| + |x + 10|$

Range/Window

Xmin = -10
Xmax = 20
Xscl = 5
Ymin = -5
Ymax = 30
Yscl = 5

69. $x^2 + y^2 = 3^2$

$x^2 + y^2 = 9$

71. $(x - 2)^2 + [y - (-1)]^2 = 4^2$

$(x - 2)^2 + (y + 1)^2 = 16$

73. $r = \sqrt{(0 - (-1))^2 + (0 - 2)^2} = \sqrt{1 + 4} = \sqrt{5}$

$[x - (-1)]^2 + (y - 2)^2 = (\sqrt{5})^2$

$(x + 1)^2 + (y - 2)^2 = 5$

75. $r = \frac{1}{2}\sqrt{(6 - 0)^2 + (8 - 0)^2} = 5$

$\text{Center} = \left(\frac{0 + 6}{2}, \frac{0 + 8}{2}\right) = (3, 4)$

$(x - 3)^2 + (y - 4)^2 = 25$

77. Center: $(-2, -3)$; radius: 2

$[x - (-2)]^2 + [y - (-3)]^2 = 2^2$

$(x + 2)^2 + (y + 3)^2 = 4$

79. $x^2 + y^2 = 4$

Center: (0, 0); radius: 2

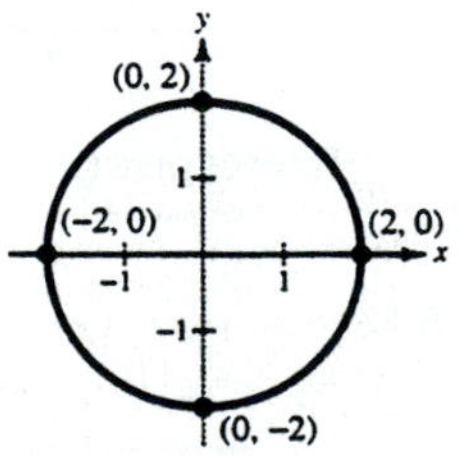

81. $(x - 1)^2 + (y + 3)^2 = 4$

$(x - 1)^2 + [y - (-3)]^2 = 2^2$

Center: $(1, -3)$; radius: 2

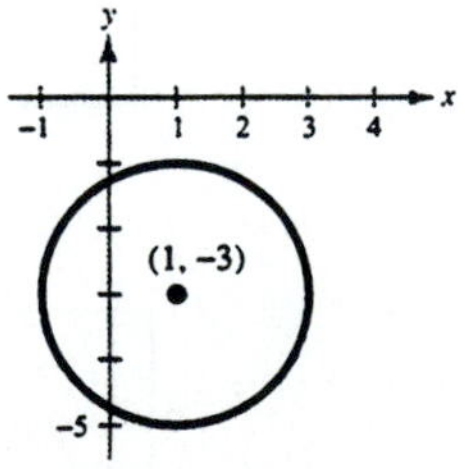

83. $\left(x - \frac{1}{2}\right)^2 + \left(y - \frac{1}{2}\right)^2 = \frac{9}{4}$

Center: $\left(\frac{1}{2}, \frac{1}{2}\right)$; radius: $\frac{3}{2}$

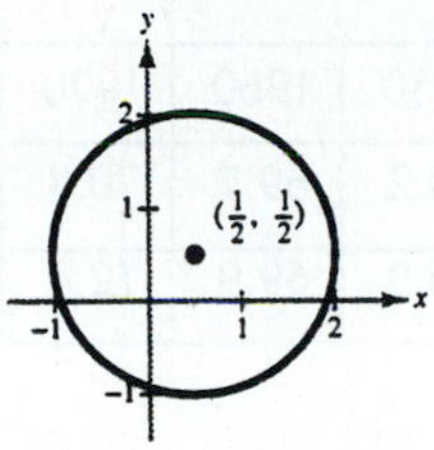

85. $y_1 = \sqrt{9 - x^2}$

$y_2 = -\sqrt{9 - x^2}$

A circle is bounded by their graphs.

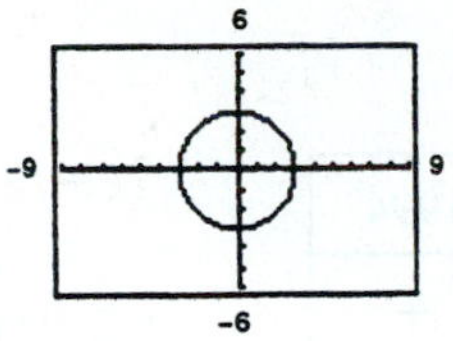

87. $y_1 = \frac{1}{4}(x^2 - 8)$

$y_2 = \frac{1}{4}x^2 - 2$

The graphs are identical. The Distributive Property is illustrated.

89. $y_1 = \frac{1}{5}[10(x^2 - 1)]$

$y_2 = 2(x^2 - 1)$

The graphs are identical. The Associative Property of Multiplication is illustrated.

91. $y = 225{,}000 - 20{,}000t, \ 0 \le t \le 8$

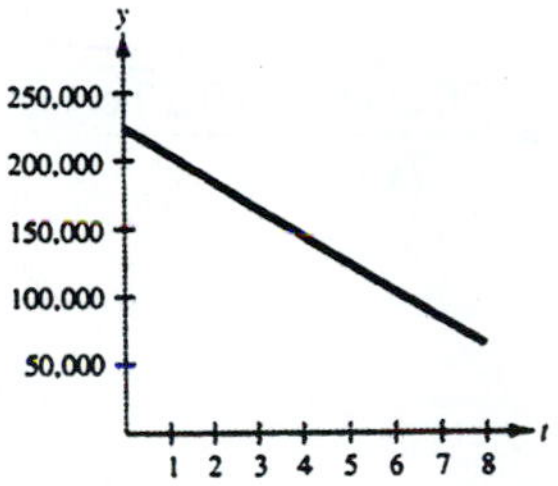

93. Most likely you would need to change the viewing window. For example, let $y_1 = x^2 + 12$. This graph would not show up on the standard window. Change the range/window to the following setting and try again.

Xmin = -5
Xmax = 5
Xscl = 1
Ymin = -5
Ymax = 40
Yscl = 5

95. (a)

Year	1920	1930	1940	1950	1960	1970	1980	1990
Life Expectancy	54.1	59.7	62.9	68.2	69.7	70.8	73.7	75.4
Model	52.8	58.7	63.3	66.9	69.9	72.4	74.6	76.4

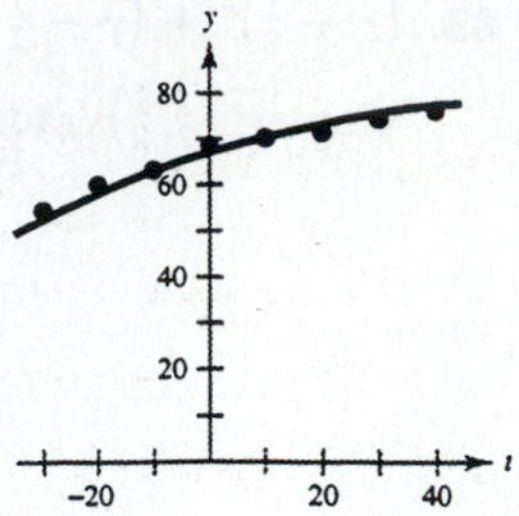

(b) When $t = 48$, $y \approx 77.7$ years.

(c) When $t = 50$, $y \approx 78.0$ years.

97. $y = 0.086t + 0.872,\ 0 \le t \le 4$

Year	1990	1991	1992	1993	1994
t	0	1	2	3	4
y	0.872	0.958	1.044	1.130	1.216

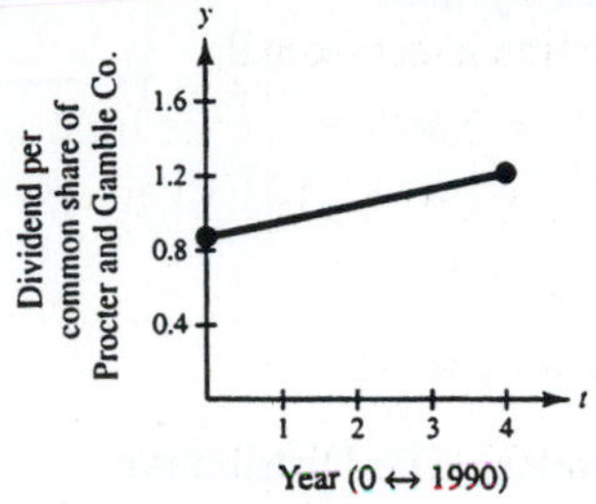

99. $9x^5 + 4x^3 - 7$

Terms: $9x^5, 4x^3, -7$

101. $\dfrac{1}{3 \cdot 4^{-1}} = \dfrac{4}{3} \neq 3 \cdot 4$ False

103. $\sqrt{18x} - \sqrt{2x} = 3\sqrt{2x} - \sqrt{2x} = 2\sqrt{2x}$

105. $\dfrac{70}{\sqrt{7x}} = \dfrac{70}{\sqrt{7x}} \cdot \dfrac{\sqrt{7x}}{\sqrt{7x}} = \dfrac{70\sqrt{7x}}{7x} = \dfrac{10\sqrt{7x}}{x}$

107. $\sqrt[6]{t^2} = t^{2/6} = |t|^{1/3} = \sqrt[3]{|t|}$

Section 1.2 Lines in the Plane and Slope

You should know the following important facts about lines.

- The graph of $y = mx + b$ is a straight line. It is called a linear equation.
- The slope of the line through (x_1, y_1) and (x_2, y_2) is
$$m = \frac{y_2 - y_1}{x_2 - x_1}.$$
- (a) If $m > 0$, the line rises from left to right.
  (b) If $m = 0$, the line is horizontal.
  (c) If $m < 0$, the line falls from left to right.
  (d) If m is undefined, the line is vertical.
- Equations of Lines
  (a) Slope-Intercept: $y = mx + b$
  (b) Point-Slope: $y - y_1 = m(x - x_1)$
  (c) Two-Point: $y - y_1 = \frac{y_2 - y_1}{x_2 - x_1}(x - x_1)$
  (d) General: $Ax + By + C = 0$
  (e) Vertical: $x = a$
  (f) Horizontal: $y = b$
- Given two distinct nonvertical lines
$$L_1: y = m_1x + b_1 \quad \text{and} \quad L_2: y = m_2x + b_2$$
  (a) L_1 is parallel to L_2 if and only if $m_1 = m_2$ and $b_1 \neq b_2$.
  (b) L_1 is perpendicular to L_2 if and only if $m_1 = -1/m_2$.

Solutions to Odd-Numbered Exercises

1. (a) $m = \frac{2}{3}$. Since the slope is positive, the line rises. Matches L_2.

(b) m is undefined. The line is vertical. Matches L_3.

(c) $m = -2$. The line falls. Matches L_1.

3.

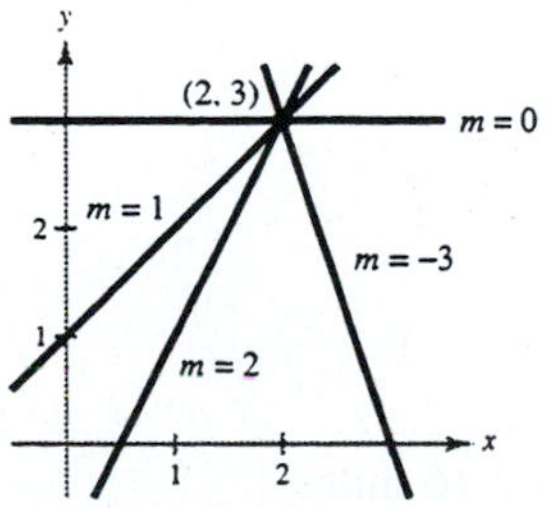

5. Slope $= \frac{\text{rise}}{\text{run}} = \frac{8}{5}$

7. Slope $= \frac{\text{rise}}{\text{run}} = \frac{0}{1} = 0$

9. Slope $= \frac{\text{rise}}{\text{run}} = \frac{-8}{2} = -4$

11.

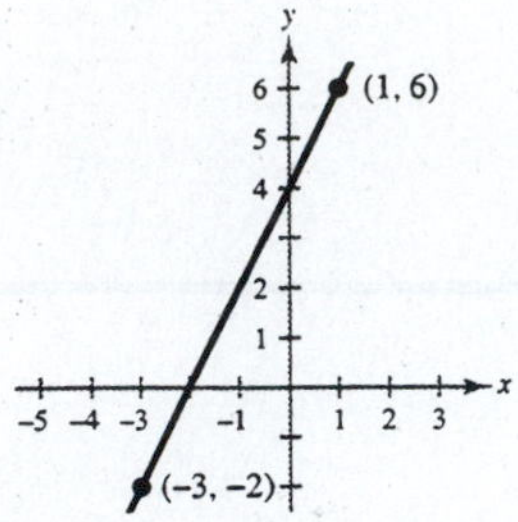

slope $= \dfrac{6+2}{1+3} = 2$

13.

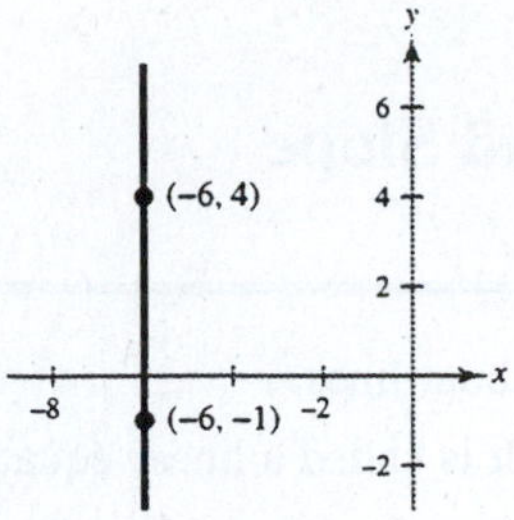

Slope is undefined.

15.

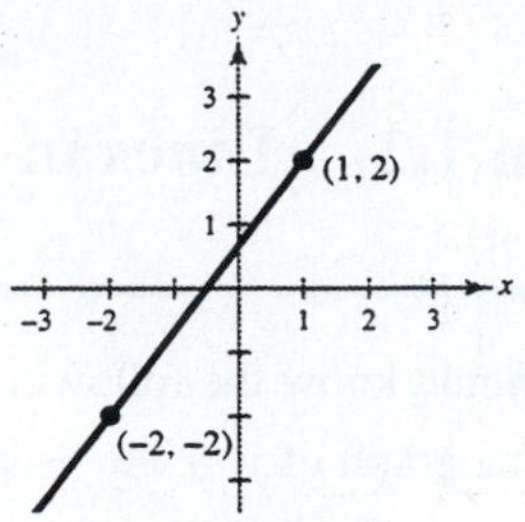

slope $= \dfrac{2+2}{1+2} = \dfrac{4}{3}$

17. Since $m = 0$, y does not change. Three points are $(0, 1)$, $(3, 1)$, and $(-1, 1)$.

19. Since $m = 1$, y increases by 1 for every one unit increase in x. Three points are $(6, -5)$, $(7, -4)$, and $(8, -3)$.

21. Since m is undefined, x does not change. Three points are $(-8, 0)$, $(-8, 2)$, and $(-8, 3)$.

23. Slope of L_1: $m = \dfrac{9+1}{5-0} = 2$

Slope of L_2: $m = \dfrac{1-3}{4-0} = -\dfrac{1}{2}$

L_1 and L_2 are perpendicular.

25. Slope of L_1: $m = \dfrac{0-6}{-6-3} = \dfrac{2}{3}$

Slope of L_2: $m = \dfrac{\frac{7}{3}+1}{5-0} = \dfrac{2}{3}$

L_1 and L_2 are parallel.

27. Yes, any pair of points on a line can be used to calculate the slope of the line. The rate of change remains the same on a line.

29. (a) $m = 135$. The sales are increasing 135 units per year.

(b) $m = 0$. There is no change in sales.

(c) $m = -40$. The sales are decreasing 40 units per year.

31. (a) The slope is negative and steep in 1989 to 1990.

(b) The slope is positive and steep in 1988 to 1989.

33. Slope $= \dfrac{\text{rise}}{\text{run}}$

$-\dfrac{12}{100} = -\dfrac{2000}{y}$

$-12y = -200{,}000$

$y = 16{,}666\frac{2}{3}$ feet ≈ 3.16 miles

35. $5x - y + 3 = 0$

$y = 5x + 3$

Slope: $m = 5$

y-intercept: $(0, 3)$

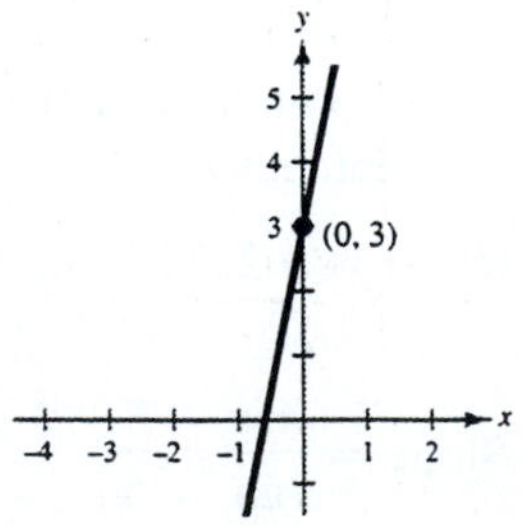

37. $5x - 2 = 0$

$x = \frac{2}{5}$

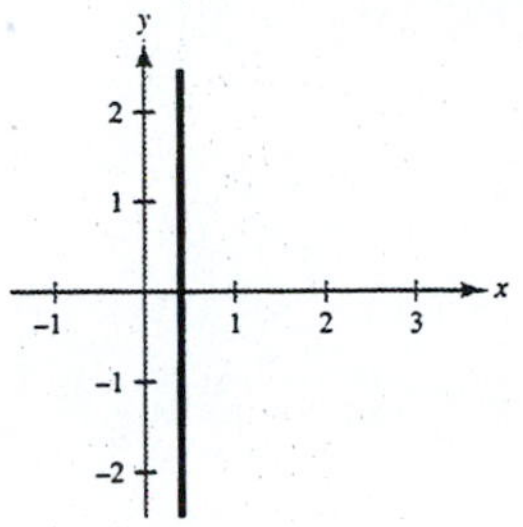

Slope: undefined

No y-intercept

39. $7x + 6y - 30 = 0$

$y = -\frac{7}{6}x + 5$

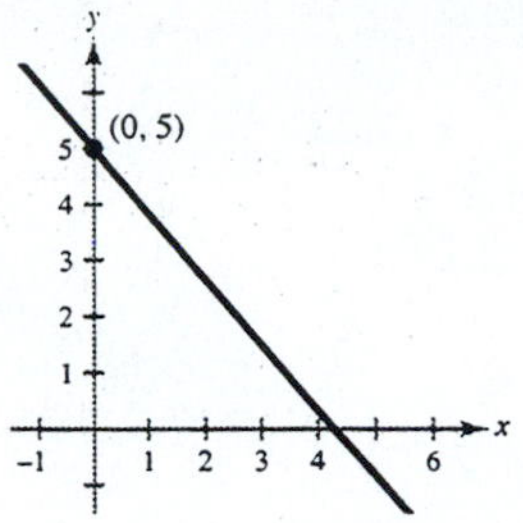

Slope: $m = -\frac{7}{6}$

y-intercept: $(0, 5)$

41. $y + 1 = \dfrac{5 + 1}{-5 - 5}(x - 5)$

$y = -\dfrac{3}{5}(x - 5) - 1$

$y = -\dfrac{3}{5}x + 2 \implies 3x + 5y - 10 = 0$

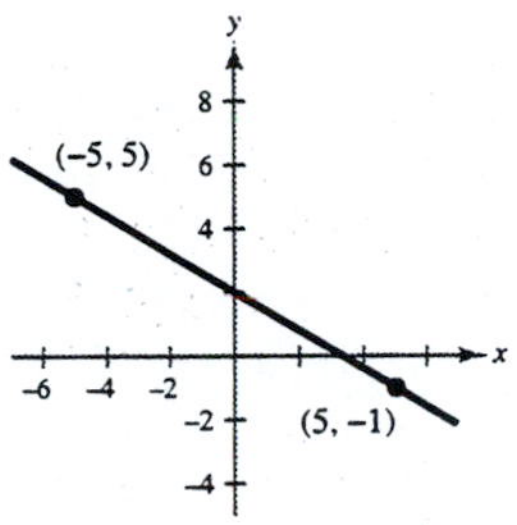

43. $y - \dfrac{1}{2} = \dfrac{\frac{5}{4} - \frac{1}{2}}{\frac{1}{2} - 2}(x - 2)$

$y = -\dfrac{1}{2}(x - 2) + \dfrac{1}{2}$

$y = -\dfrac{1}{2}x + \dfrac{3}{2} \implies x + 2y - 3 = 0$

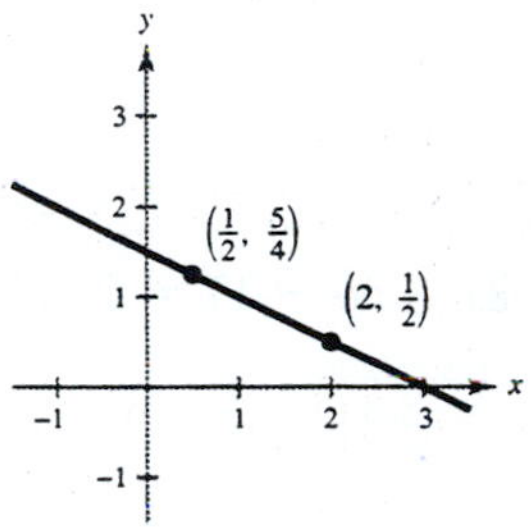

45. Since both points have $x = -8$, the slope is undefined.

$x = -8 \implies x + 8 = 0$

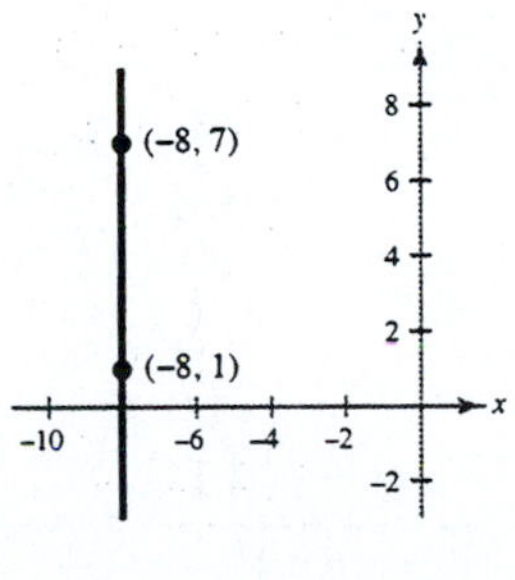

47. $y - 0.6 = \dfrac{-0.6 - 0.6}{-2 - 1}(x - 1)$

$y = 0.4(x - 1) + 0.6$

$y = 0.4x + 0.2 \implies 2x - 5y + 1 = 0$

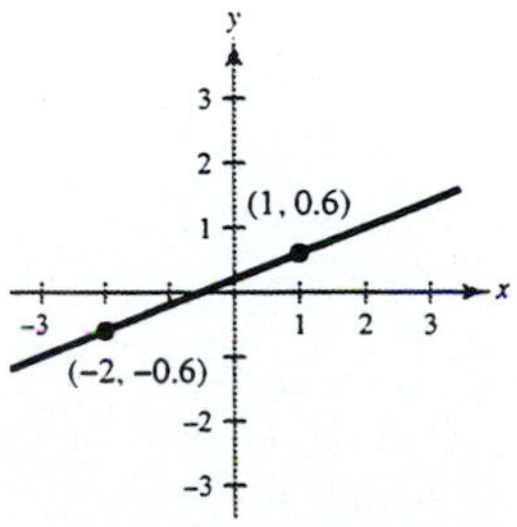

49. $y + 2 = 3(x - 0)$

$y = 3x - 2 \implies 3x - y - 2 = 0$

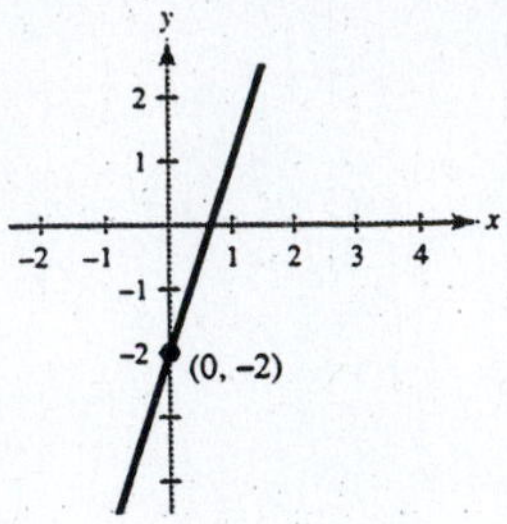

51. $y - 6 = -2(x + 3)$

$y = -2x \implies 2x + y = 0$

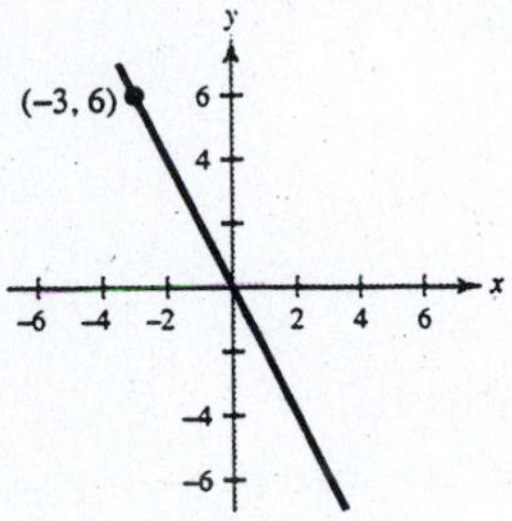

53. $y - 0 = -\frac{1}{3}(x - 4)$

$y = -\frac{1}{3}x + \frac{4}{3} \implies x + 3y - 4 = 0$

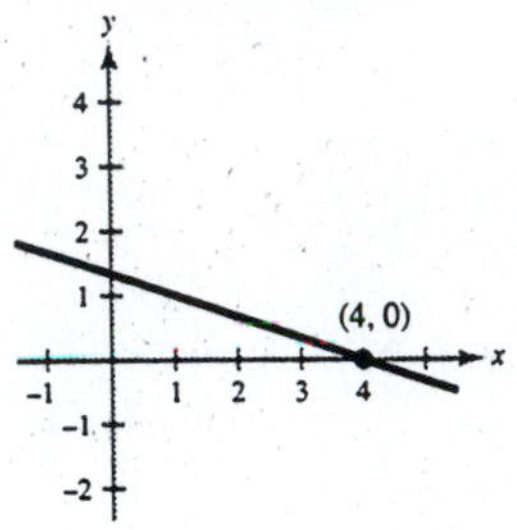

55. $x = 6$

$x - 6 = 0$

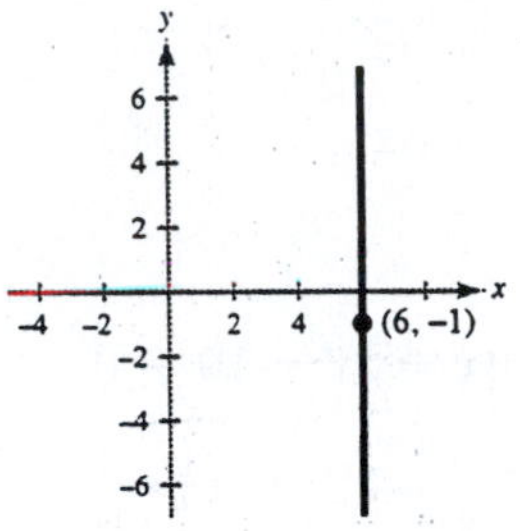

57. $y - \frac{5}{2} = \frac{4}{3}(x - 4)$

$y = \frac{4}{3}x - \frac{17}{6} \implies 8x - 6y - 17 = 0$

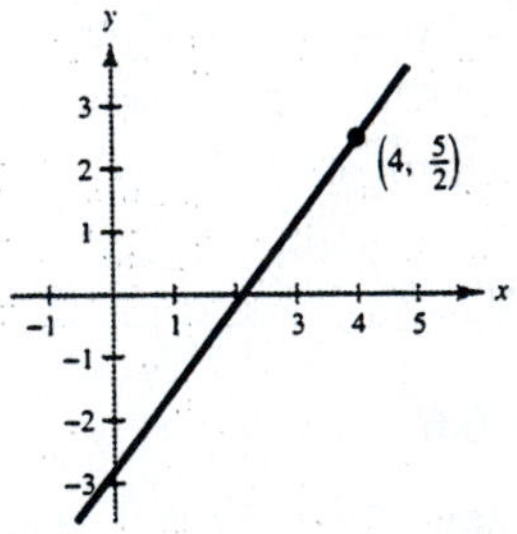

59. $\dfrac{x}{2} + \dfrac{y}{3} = 1$

$3x + 2y - 6 = 0$

61. $\dfrac{x}{-1/6} + \dfrac{y}{-2/3} = 1$

$6x + \dfrac{3}{2}y = -1$

$12x + 3y + 2 = 0$

63. $\dfrac{x}{a} + \dfrac{y}{a} = 1, \ a \neq 0$

$x + y = a$

$1 + 2 = a$

$3 = a$

$x + y = 3$

$x + y - 3 = 0$

65. $4x - 2y = 3$

$y = 2x - \frac{3}{2}$

slope: $m = 2$

(a) $y - 1 = 2(x - 2)$

$y = 2x - 3 \implies 2x - y - 3 = 0$

(b) $y - 1 = -\frac{1}{2}(x - 2)$

$y = -\frac{1}{2}x + 2 \implies x + 2y - 4 = 0$

67. $3x + 4y = 7$

$y = -\frac{3}{4}x + \frac{7}{4}$

slope: $m = -\frac{3}{4}$

(a) $y - 4 = -\frac{3}{4}(x + 6)$

$y = -\frac{3}{4}x - \frac{1}{2} \implies 3x + 4y + 2 = 0$

(b) $y - 4 = \frac{4}{3}(x + 6)$

$y = \frac{4}{3}x + 12 \implies 4x - 3y + 36 = 0$

69. $y = -3$

slope: $m = 0$

(a) $y = 0$

(b) $x = -1 \implies x + 1 = 0$

71. $L_1: y = \frac{1}{3}x - 2$

$L_2: y = \frac{1}{3}x + 3$

The lines are parallel.

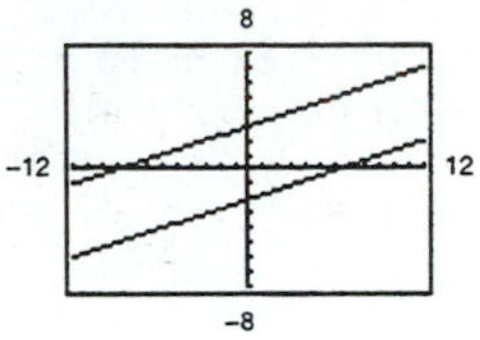

73. $L_1: y = \frac{1}{2}x - 3$

$L_2: y = -\frac{1}{2}x + 1$

Neither parallel nor perpendicular

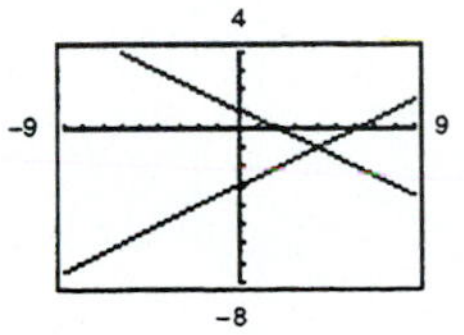

75. $L_1: y = \frac{2}{3}x - 3$

$L_2: y = -\frac{3}{2}x + 2$

The lines are perpendicular.

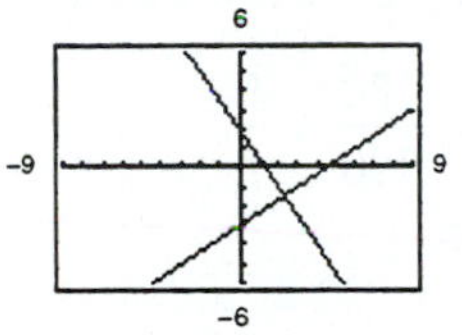

77. $y = 0.5x - 3$

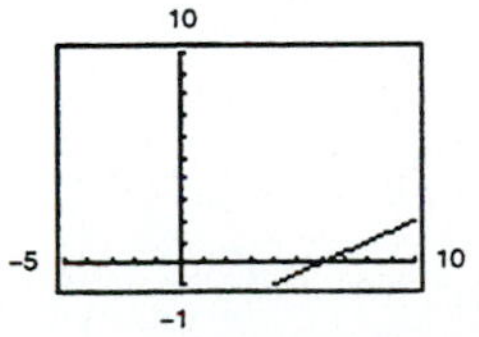

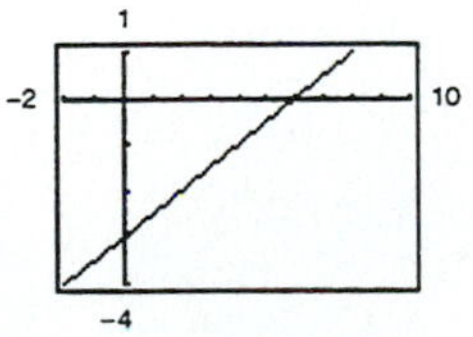

The second setting shows the x- and y-intercepts more clearly.

79. (a) $y = 2x$ (b) $y = -2x$ (c) $y = \frac{1}{2}x$

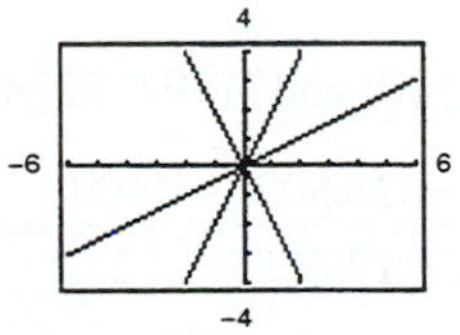

(b) and (c) are perpendicular.

81. (a) $y = -\frac{1}{2}x$ (b) $y = -\frac{1}{2}x + 3$ (c) $y = 2x - 4$

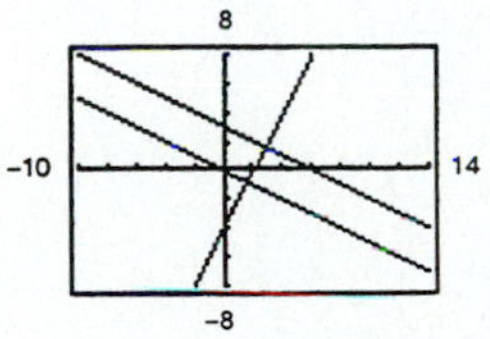

(a) and (b) are parallel.

(c) is perpendicular to (a) and (b).

83. $(6, 2540)$, $m = 125$

$V - 2540 = 125(t - 6)$

$V - 2540 = 125t - 750$

$V = 125t + 1790$

85. The slope is $m = -20$. This represents the decrease in the amount of the loan each week. Matches graph (b).

87. The slope is $m = 0.32$. This represents the increase in travel cost for each mile driven. Matches graph (a).

89. Set the distance between $(4, -1)$ and (x, y) equal to the distance between $(-2, 3)$ and (x, y).

$$\sqrt{(x-4)^2 + [y-(-1)]^2} = \sqrt{[x-(-2)]^2 + (y-3)^2}$$
$$(x-4)^2 + (y+1)^2 = (x+2)^2 + (y-3)^2$$
$$x^2 - 8x + 16 + y^2 + 2y + 1 = x^2 + 4x + 4 + y^2 - 6y + 9$$
$$-8x + 2y + 17 = 4x - 6y + 13$$
$$0 = 12x - 8y - 4$$
$$0 = 4(3x - 2y - 1)$$
$$0 = 3x - 2y - 1$$

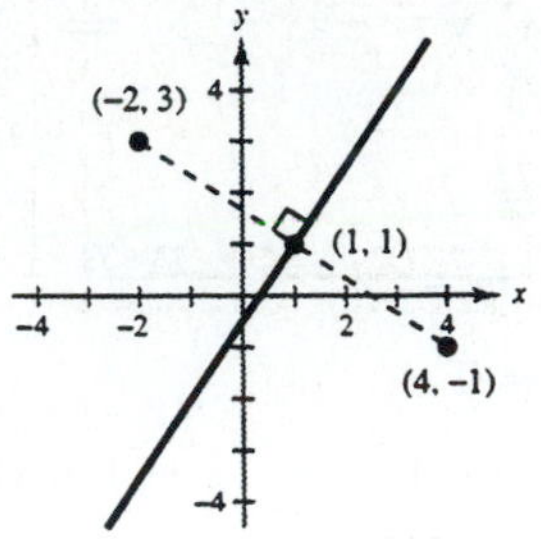

This line is the perpendicular bisector of the line segment connecting $(4, -1)$ and $(-2, 3)$.

91. Using the points $(0, 32)$ and $(100, 212)$, we have

$$m = \frac{212 - 32}{100 - 0} = \frac{180}{100} = \frac{9}{5}$$
$$F - 32 = \frac{9}{5}(C - 0)$$
$$F = \frac{9}{5}C + 32.$$

93. Using the points $(1995, 28{,}500)$ and $(1997, 32{,}900)$, we have

$$m = \frac{32{,}900 - 28{,}500}{1997 - 1995} = \frac{4400}{2} = 2200$$
$$S - 28{,}500 = 2200(t - 1995)$$
$$S = 2200t - 4{,}360{,}500$$

When $t = 2000$ we have $S = 2200(2000) - 4{,}360{,}500$ or \$39,500.

95. Using the points $(0, 875)$ and $(5, 0)$, where the first coordinate represents the year t and the second coordinate represents the value V, we have

$$m = \frac{0 - 875}{5 - 0} = -175$$
$$V = -175t + 875,\ 0 \le t \le 5.$$

97. Sale price = List price − 15% of the list price

$$S = L - 0.15L$$
$$S = 0.85L$$

99. (a) $C = 36{,}500 + 5.25t + 11.50t$
$= 16.75t + 36{,}500$

(b) $R = 27t$

(c) $P = R - C$
$= 27t - (16.75t + 36{,}500)$
$= 10.25t - 36{,}500$

(d) $0 = 10.25t - 36{,}500$
$36{,}500 = 10.25t$
$t \approx 3561$ hours

101. (a)

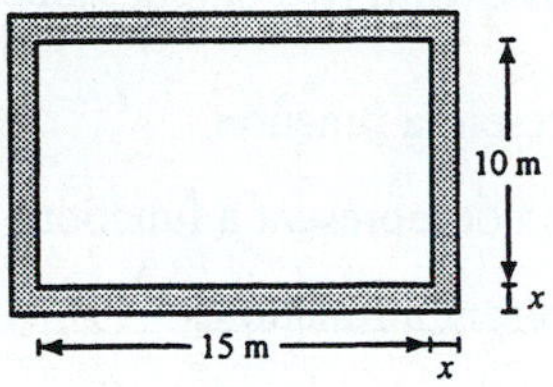

(b) $y = 2(15 + 2x) + 2(10 + 2x)$
$= 8x + 50$

(c)

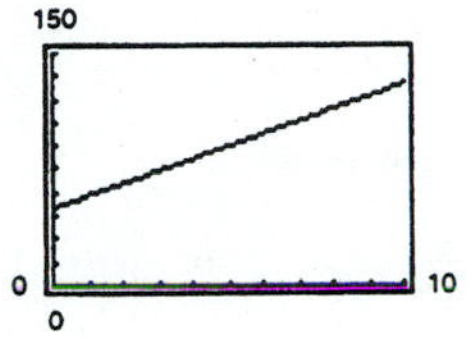

(d) Since $m = 8$, each 1 meter increase in x will increase y by 8 meters.

103. $C = 120 + 0.26x$

105. Two approximate points on this line are (6, 710) and (10, 1075).

$$m = \frac{1075 - 710}{10 - 6} \approx 91$$

$$y - 710 = 91(x - 6)$$
$$y = 91x + 164$$

This answer may vary depending on the points used.

107. $y = 8 - 3x$ is a linear equation with slope $m = -3$. Matches graph (d).

109. $y = \frac{1}{2}x^2 + 2x + 1$ is a quadratic equation. Its graph is a parabola. Matches graph (a).

Section 1.3 Functions

- Given a set or an equation, you should be able to determine if it represents a function.
- Given a function, you should be able to do the following.
  (a) Find the domain.
  (b) Evaluate it at specific values.

Solutions to Odd-Numbered Exercises

1. Yes, it does represent a function. Each domain value is matched with only one range value.

3. No, it does not represent a function. The domain values are each matched with three range values.

5. Yes, it does represent a function. Each input value is matched with only one output value.

7. No, it does not represent a function. The input values of 10 and 7 are each matched with two output values.

9. (a) Each element of A is matched with exactly one element of B, so it does represent a function.

(b) The element 1 in A is matched with two elements, -2 and 1 of B, so it does not represent a function.

(c) Each element of A is matched with exactly one element of B, so it does represent a function.

(d) The element 2 in A is not matched with an element of B, so it does not represent a function.

11. Each is a function. For each year there corresponds one and only one circulation.

13. $x^2 + y^2 = 4 \implies y = \pm\sqrt{4 - x^2}$

No, y *is not* a function of x.

15. $x^2 + y = 4 \implies y = 4 - x^2$

Yes, y *is* a function of x.

17. $2x + 3y = 4 \implies y = \frac{1}{3}(4 - 2x)$

Yes, y *is* a function of x.

19. $y^2 = x^2 - 1 \implies y = \pm\sqrt{x^2 - 1}$

No, y *is not* a function of x.

21. $y = |4 - x|$

Yes, y is a function of x.

23. $f(s) = \dfrac{1}{s + 1}$

(a) $f(4) = \dfrac{1}{(4) + 1} = \dfrac{1}{5}$

(b) $f(0) = \dfrac{1}{(0) + 1} = 1$

(c) $f(4x) = \dfrac{1}{(4x) + 1} = \dfrac{1}{4x + 1}$

(d) $f(x + c) = \dfrac{1}{(x + c) + 1} = \dfrac{1}{x + c + 1}$

25. $f(x) = 2x - 3$

(a) $f(1) = 2(1) - 3 = -1$

(b) $f(-3) = 2(-3) - 3 = -9$

(c) $f(x - 1) = 2(x - 1) - 3 = 2x - 5$

27. $h(t) = t^2 - 2t$

(a) $h(2) = 2^2 - 2(2) = 0$

(b) $h(1.5) = (1.5)^2 - 2(1.5) = -0.75$

(c) $h(x + 2) = (x + 2)^2 - 2(x + 2) = x^2 + 2x$

29. $f(y) = 3 - \sqrt{y}$

(a) $f(4) = 3 - \sqrt{4} = 1$

(b) $f(0.25) = 3 - \sqrt{0.25} = 2.5$

(c) $f(4x^2) = 3 - \sqrt{4x^2} = 3 - 2|x|$

31. $q(x) = \dfrac{1}{x^2 - 9}$

(a) $q(0) = \dfrac{1}{0^2 - 9} = -\dfrac{1}{9}$

(b) $q(3) = \dfrac{1}{3^2 - 9}$ is undefined.

(c) $q(y + 3) = \dfrac{1}{(y + 3)^2 - 9} = \dfrac{1}{y^2 + 6y}$

33. $f(x) = \dfrac{|x|}{x}$

(a) $f(2) = \dfrac{|2|}{2} = 1$

(b) $f(-2) = \dfrac{|-2|}{-2} = -1$

(c) $f(x - 1) = \dfrac{|x - 1|}{x - 1}$

35. $f(x) = \begin{cases} 2x + 1, & x < 0 \\ 2x + 2, & x \geq 0 \end{cases}$

(a) $f(-1) = 2(-1) + 1 = -1$

(b) $f(0) = 2(0) + 2 = 2$

(c) $f(2) = 2(2) + 2 = 6$

37. $f(x) = x^2 - 3$

x	-2	-1	0	1	2
$f(x)$	1	-2	-3	-2	1

39. $h(t) = \frac{1}{2}|t + 3|$

t	-5	-4	-3	-2	-1
$h(t)$	1	$\frac{1}{2}$	0	$\frac{1}{2}$	1

41. $f(x) = \begin{cases} -\frac{1}{2}x + 4, & x \leq 0 \\ (x - 2)^2, & x > 0 \end{cases}$

x	-2	-1	0	1	2
$f(x)$	5	$\frac{9}{2}$	4	1	0

43.
$$15 - 3x = 0$$
$$3x = 15$$
$$x = 5$$

45.
$$x^2 - 9 = 0$$
$$x^2 = 9$$
$$x = \pm 3$$

47.
$$f(x) = g(x)$$
$$x^2 = x + 2$$
$$x^2 - x - 2 = 0$$
$$(x + 1)(x - 2) = 0$$
$$x = -1 \text{ or } x = 2$$

49.
$$f(x) = g(x)$$
$$\sqrt{3x} + 1 = x + 1$$
$$\sqrt{3x} = x$$
$$3x = x^2$$
$$0 = x^2 - 3x$$
$$0 = x(x - 3)$$
$$x = 0 \text{ or } x = 3$$

51. $f(x) = 5x^2 + 2x - 1$

Since $f(x)$ is a polynomial, the domain is all real numbers x.

53. $h(t) = \dfrac{4}{t}$

Domain: All real numbers except $t = 0$

55. $g(y) = \sqrt{y - 10}$

Domain: $y - 10 \geq 0$

$y \geq 10$

57. $f(x) = \sqrt[4]{1 - x^2}$

Domain: $1 - x^2 \geq 0$

$-x^2 \geq -1$

$x^2 \leq 1$

$x^2 - 1 \leq 0$

$-1 \leq x \leq 1$ (See Section 1.8.)

59. $g(x) = \dfrac{1}{x} - \dfrac{1}{x + 2}$

Domain: All real numbers except $x = 0,\ x = -2$

61. $f(x) = x^2$

$\{(-2, 4), (-1, 1), (0, 0), (1, 1), (2, 4)\}$

63. $f(x) = \sqrt{x + 2}$

$\{(-2, 0), (-1, 1), (0, \sqrt{2}), (1, \sqrt{3}), (2, 2)\}$

65. The domain is the set of inputs of the function and the range is the set of corresponding outputs.

67. By plotting the points, we have a parabola, so $g(x) = cx^2$. Since $(-4, -32)$ is on the graph, we have $-32 = c(-4)^2 \implies c = -2$. Thus, $g(x) = -2x^2$.

69. Since the function is undefined at 0, we have $r(x) = c/x$. Since $(-8, -4)$ is on the graph, we have $-4 = c/-8 \implies c = 32$. Thus, $r(x) = 32/x$.

71.

$$\begin{aligned} f(x) &= x^2 - x + 1 \\ f(2 + h) &= (2 + h)^2 - (2 + h) + 1 \\ &= 4 + 4h + h^2 - 2 - h + 1 \\ &= h^2 + 3h + 3 \\ f(2) &= (2)^2 - 2 + 1 = 3 \\ f(2 + h) - f(2) &= h^2 + 3h \\ \frac{f(2 + h) - f(2)}{h} &= h + 3,\ h \neq 0 \end{aligned}$$

73. $f(x) = x^3$

$$f(x + c) = (x + c)^3 = x^3 + 3x^2c + 3xc^2 + c^3$$

$$\begin{aligned} \frac{f(x + c) - f(x)}{c} &= \frac{(x^3 + 3x^2c + 3xc^2 + c^3) - x^3}{c} \\ &= \frac{c(3x^2 + 3xc + c^2)}{c} \\ &= 3x^2 + 3xc + c^2,\ \ c \neq 0 \end{aligned}$$

75. $g(x) = 3x - 1$

$$\frac{g(x) - g(3)}{x - 3} = \frac{(3x - 1) - 8}{x - 3} = \frac{3x - 9}{x - 3} = \frac{3(x - 3)}{x - 3} = 3,\ x \neq 3$$

77. $A = \pi r^2,\ \ C = 2\pi r$

$$r = \frac{C}{2\pi}$$

$$A = \pi\left(\frac{C}{2\pi}\right)^2 = \frac{C^2}{4\pi}$$

79. (a)

Height, x	Width	Volume, V
1	$24 - 2(1)$	$1[24 - 2(1)]^2 = 484$
2	$24 - 2(2)$	$2[24 - 2(2)]^2 = 800$
3	$24 - 2(3)$	$3[24 - 2(3)]^2 = 972$
4	$24 - 2(4)$	$4[24 - 2(4)]^2 = 1024$
5	$24 - 2(5)$	$5[24 - 2(5)]^2 = 980$
6	$24 - 2(6)$	$6[24 - 2(6)]^2 = 864$

The volume is maximum when $x = 4$.

(b)

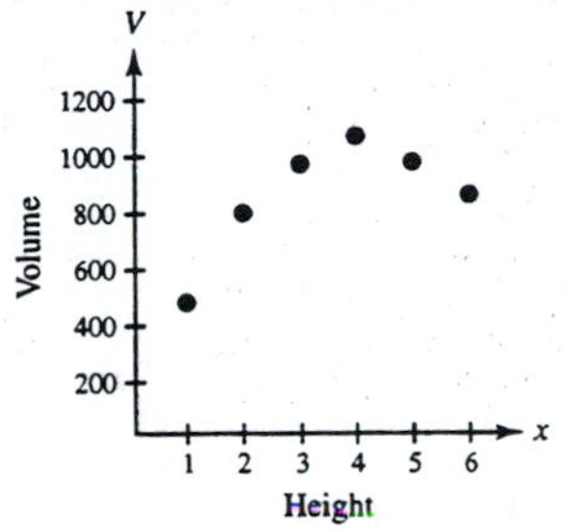

V is a function of x.

(c) $V = x(24 - 2x)^2$

Domain: $0 < x < 12$

81. $A = \frac{1}{2}bh = \frac{1}{2}xy$

Since $(0, y)$, $(2, 1)$, and $(x, 0)$ all lie on the same line, the slopes between any pair are equal.

$$\frac{1 - y}{2 - 0} = \frac{0 - 1}{x - 2}$$

$$\frac{1 - y}{2} = \frac{-1}{x - 2}$$

$$y = \frac{2}{x - 2} + 1$$

$$y = \frac{x}{x - 2}$$

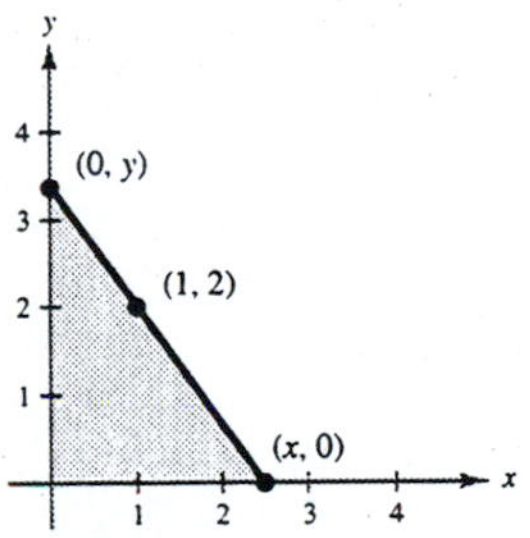

Therefore,

$$A = \frac{1}{2}x\left(\frac{x}{x - 2}\right) = \frac{x^2}{2(x - 2)}.$$

The domain of A includes x-values such that $x^2/[2(x - 2)] > 0$. Using methods of Section 1.8 we find that the domain is $x > 2$.

83. $V = l \cdot w \cdot h = x \cdot y \cdot x = x^2y$ where $4x + y = 108$.

Thus, $y = 108 - 4x$ and $V = x^2(108 - 4x) = 108x^2 - 4x^3$ where $0 < x < 27$.

85. (a) Cost = variable costs + fixed costs

$$C = 12.30x + 98{,}000$$

(b) Revenue = price per unit × number of units

$$R = 17.98x$$

(c) Profit = Revenue − Cost

$$P = 17.98x - (12.30x + 98{,}000)$$

$$P = 5.68x - 98{,}000$$

87. (a) $R = n(\text{rate}) = n[8.00 - 0.05(n - 80)],\ n \geq 80$

$$R = 12.00n - 0.05n^2 = 12n - \frac{n^2}{20} = \frac{240n - n^2}{20},\quad n \geq 80$$

(b)

n	90	100	110	120	130	140	150
$R(n)$	\$675	\$700	\$715	\$720	\$715	\$700	\$675

The revenue is maximum when 120 people take the trip.

89. (a)

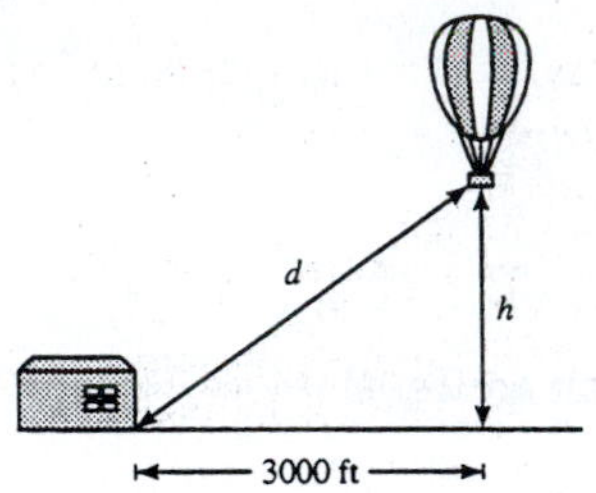

(b) $(3000)^2 + h^2 = d^2$

$$h = \sqrt{d^2 - (3000)^2}$$

Domain: $[3000, \infty)$
(since both $d \geq 0$ and $d^2 - (3000)^2 \geq 0$)

91. $\dfrac{t}{3} + \dfrac{t}{5} = 1$

$$15\left(\frac{t}{3} + \frac{t}{5}\right) = 15(1)$$

$$5t + 3t = 15$$

$$8t = 15$$

$$t = \frac{15}{8}$$

93. $\dfrac{3}{x(x+1)} - \dfrac{4}{x} = \dfrac{1}{x+1}$

$$x(x+1)\left[\frac{3}{x(x+1)} - \frac{4}{x}\right] = x(x+1)\left(\frac{1}{x+1}\right)$$

$$3 - 4(x+1) = x$$

$$3 - 4x - 4 = x$$

$$-1 = 5x$$

$$-\frac{1}{5} = x$$

69. $P = R - C = xp - C = x(100 - 0.0001x) - (350{,}000 + 30x)$

$= -0.0001x^2 + 70x - 350{,}000, \; 0 \le x$

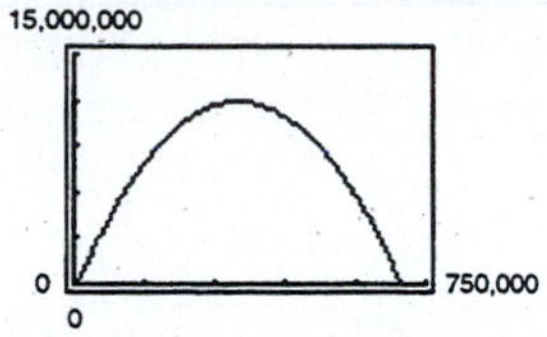

This function is maximized when $x = 350{,}000$ units.

71. $h = \text{top} - \text{bottom}$

$= (-x^2 + 4x - 1) - 2$

$= -x^2 + 4x - 3$

73. $h = \text{top} - \text{bottom}$

$= (4x - x^2) - 2x$

$= 2x - x^2$

75. $L = \text{right} - \text{left}$

$= \frac{1}{2}y^2 - 0$

$= \frac{1}{2}y^2$

77. $L = \text{right} - \text{left}$

$= 4 - y^2$

79. $y = -87.49 + 16.28t - 4.82t^2 - 1.20t^3$

(a) Domain: $-4 \le t \le 3$

(b)

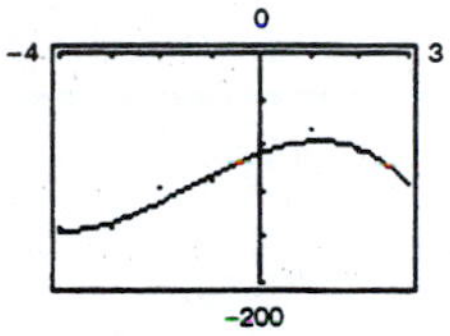

(c) Most accurate in 1986

Least accurate in 1990

(d) The balance would continue to decrease.

81. (a) For average salaries of college professors, a scale of \$10,000 would be appropriate.

(b) For the population of the United States, use a scale of 50,000,000.

(c) For the percent of the civilian workforce that is unemployed, use a scale of 1%.

83. $f(x) = a_{2n+1}x^{2n+1} + a_{2n-1}x^{2n-1} + \cdots + a_3x^3 + a_1x$

$f(-x) = a_{2n+1}(-x)^{2n+1} + a_{2n-1}(-x)^{2n-1} + \cdots + a_3(-x)^3 + a_1(-x)$

$= -a_{2n+1}x^{2n+1} - a_{2n-1}x^{2n-1} - \cdots - a_3x^3 - a_1x = -f(x)$

Therefore, $f(x)$ is odd.

85. $x^2 - 10x = 0$

$x(x - 10) = 0$

$x = 0$ or $x = 10$

87. $x^3 + x = 0$

$x(x^2 + 1) = 0$

$x = 0$ or $x^2 + 1 = 0$

$x^2 = -1$

$x = \pm\sqrt{-1} = \pm i$

Section 1.5 Translations and Combinations

- You should know the basic types of transformations.
  Let $y = f(x)$ and let c be a positive real number.
  1. $h(x) = f(x) + c$ Vertical shift c units upward
  2. $h(x) = f(x) - c$ Vertical shift c units downward
  3. $h(x) = f(x - c)$ Horizontal shift c units to the right
  4. $h(x) = f(x + c)$ Horizontal shift c units to the left
  5. $h(x) = -f(x)$ Reflection in the x-axis
  6. $h(x) = f(-x)$ Reflection in the y-axis
  7. $h(x) = cf(x), c > 1$ Vertical stretch
  8. $h(x) = cf(x), 0 < c < 1$ Vertical shrink
- Given two functions, f and g, you should be able to form the following functions (if defined):
  1. Sum: $(f + g)(x) = f(x) + g(x)$
  2. Difference: $(f - g)(x) = f(x) - g(x)$
  3. Product: $(fg)(x) = f(x)g(x)$
  4. Quotient: $(f/g)(x) = f(x)/g(x), g(x) \neq 0$
  5. Composition of f with g: $(f \circ g)(x) = f(g(x))$
  6. Composition of g with f: $(g \circ f)(x) = g(f(x))$

Solutions to Odd-Numbered Exercises

1. (a) $f(x) = x^3 + c$

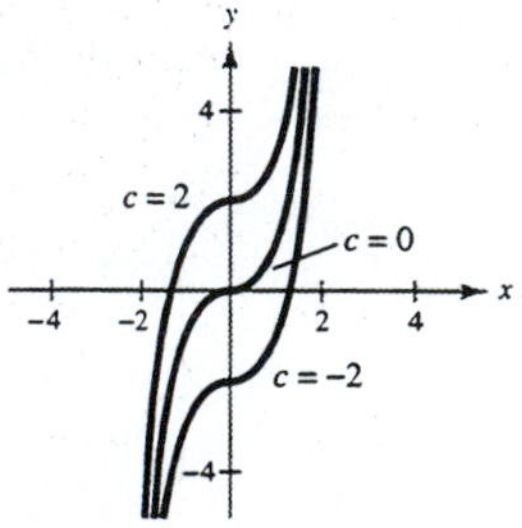

(b) $f(x) = (x - c)^3$

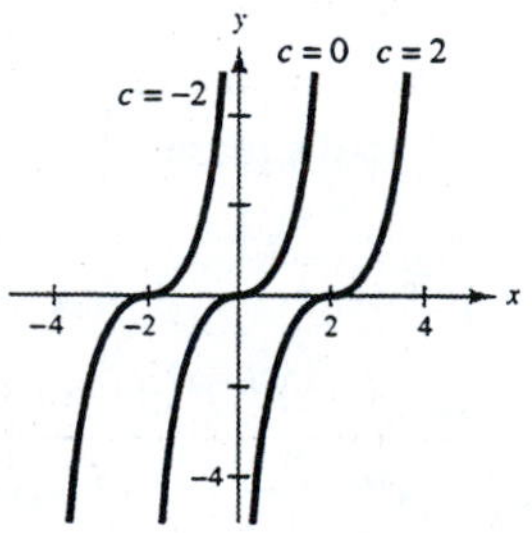

3. (a) $f(x) = |x + c|$

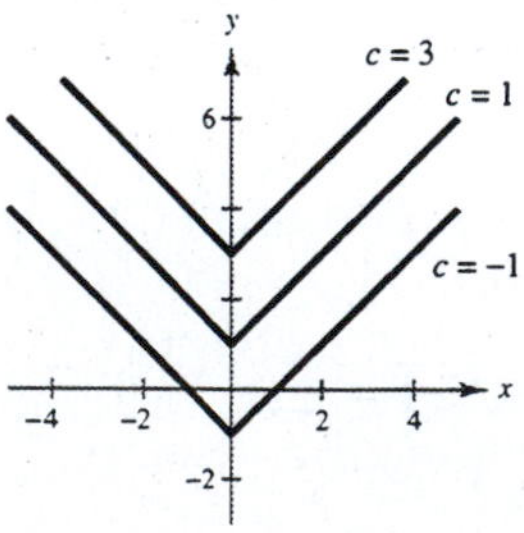

(b) $f(x) = |x - c|$

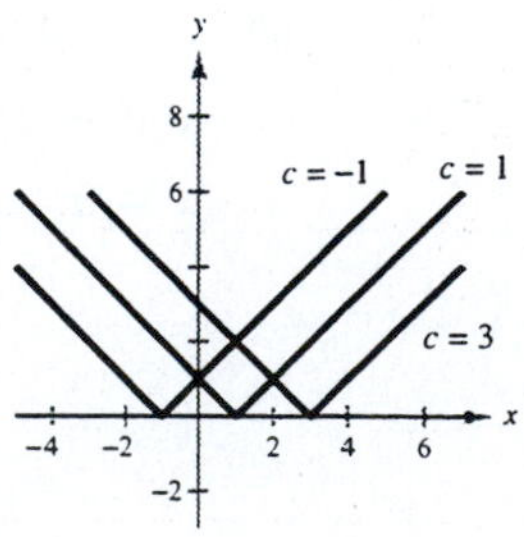

(c) $f(x) = |x + 4| + c$

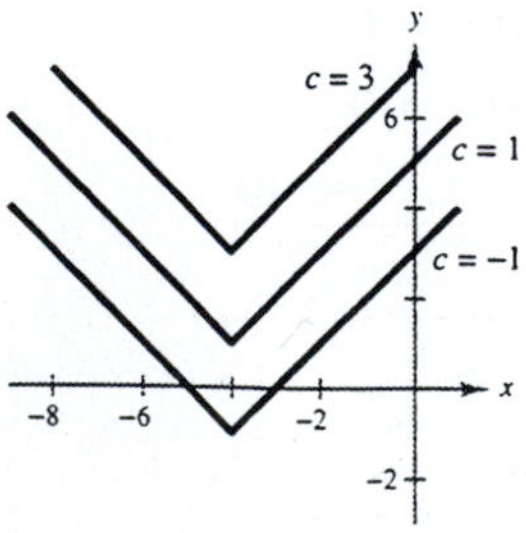

5. (a) $y = f(x) + 2$

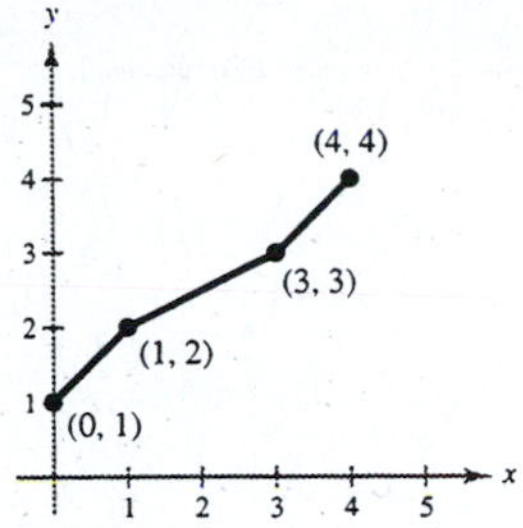

(b) $y = -f(x)$

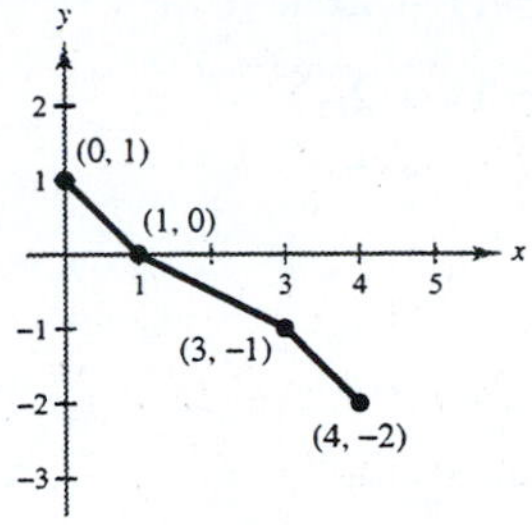

(c) $y = f(x - 2)$

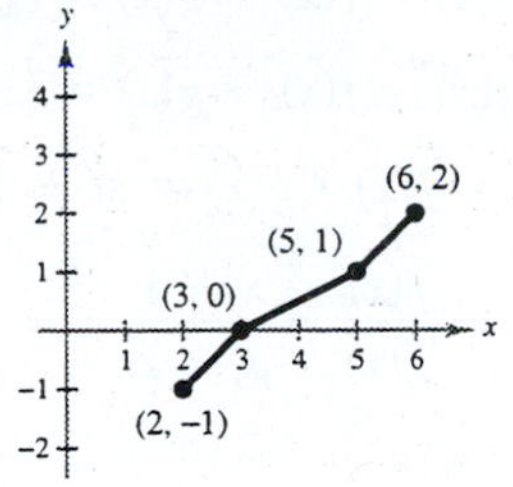

(d) $y = f(x + 3)$

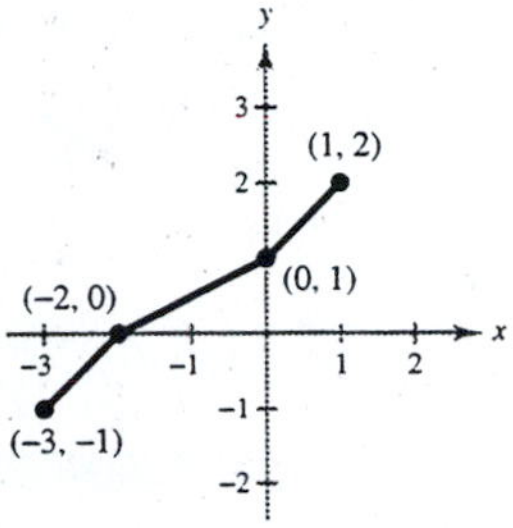

(e) $y = f(2x)$

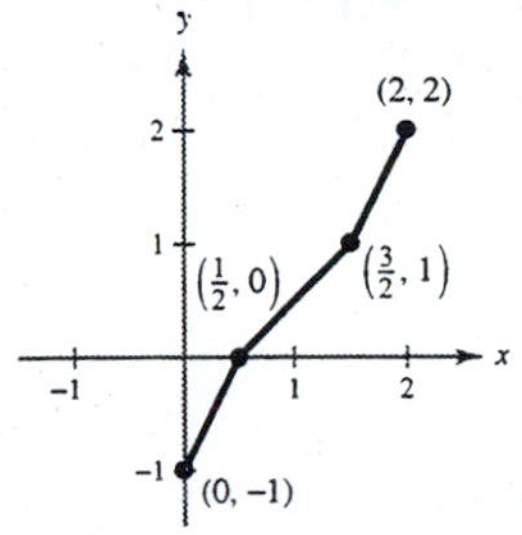

(f) $y = f(-x)$

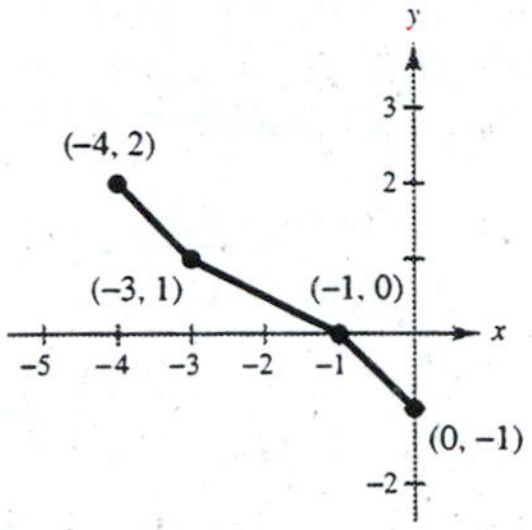

7. (a) Vertical shift one unit downward

$$y = x^2 - 1$$

(b) Vertical shift one unit upward, horizontal shift one unit to the left, and a reflection in the x-axis

$$y = 1 - (x + 1)^2$$

9. Horizontal shift two units to the right of $y = x^3$

$$y = (x - 2)^3$$

11. Reflection in the x-axis of $y = x^2$

$$y = -x^2$$

13. Reflection in the x-axis and a vertical shift one unit upward of $y = \sqrt{x}$

$$y = 1 - \sqrt{x}$$

15.

x	0	1	2	3
$f(x)$	2	3	1	2
$g(x)$	-1	0	$\frac{1}{2}$	0
$h(x) = f(x) + g(x)$	1	3	$\frac{3}{2}$	2

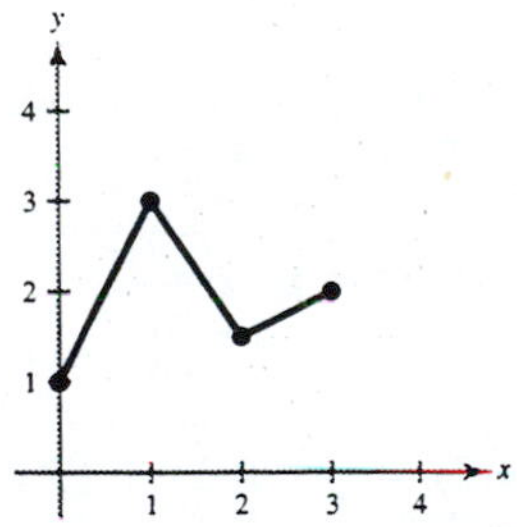

17.

x	-2	-1	0	1	2	3
$f(x)$	2	1	0	1	2	3
$g(x)$	4	3	2	1	0	1
$h(x) = f(x) + g(x)$	6	4	2	2	2	4

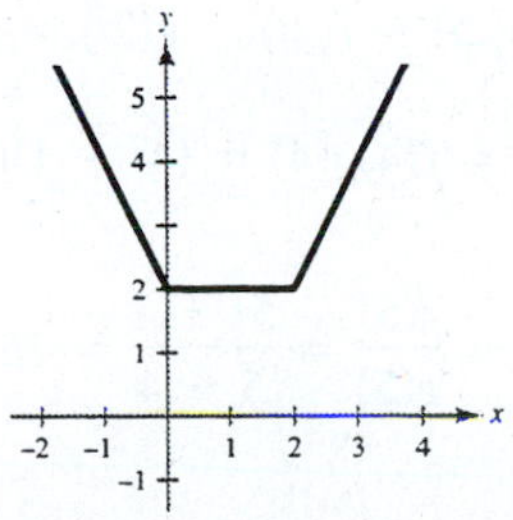

19. $f(x) = x + 1, g(x) = x - 1$

$(f + g)(x) = f(x) + g(x) = (x + 1) + (x - 1) = 2x$

$(f - g)(x) = f(x) - g(x) = (x + 1) - (x - 1) = 2$

$(fg)(x) = f(x) \cdot g(x) = (x + 1)(x - 1) = x^2 - 1$

$\left(\frac{f}{g}\right)(x) = \frac{f(x)}{g(x)} = \frac{x + 1}{x - 1}, \ x \neq 1$

21. $f(x) = x^2, g(x) = 1 - x$

$(f + g)(x) = f(x) + g(x) = x^2 + (1 - x) = x^2 - x + 1$

$(f - g)(x) = f(x) - g(x) = x^2 - (1 - x) = x^2 + x - 1$

$(fg)(x) = f(x) \cdot g(x) = x^2(1 - x) = x^2 - x^3$

$\left(\frac{f}{g}\right)(x) = \frac{f(x)}{g(x)} = \frac{x^2}{1 - x}, x \neq 1$

23. $f(x) = x^2 + 5, g(x) = \sqrt{1 - x}$

$(f + g)(x) = f(x) + g(x) = (x^2 + 5) + \sqrt{1 - x}$

$(f - g)(x) = f(x) - g(x) = (x^2 + 5) - \sqrt{1 - x}$

$(fg)(x) = f(x) \cdot g(x) = (x^2 + 5)\sqrt{1 - x}$

$\left(\frac{f}{g}\right)(x) = \frac{f(x)}{g(x)} = \frac{x^2 + 5}{\sqrt{1 - x}}, \ x < 1$

25. $f(x) = \frac{1}{x}, g(x) = \frac{1}{x^2}$

$(f + g)(x) = f(x) + g(x) = \frac{1}{x} + \frac{1}{x^2} = \frac{x + 1}{x^2}$

$(f - g)(x) = f(x) - g(x) = \frac{1}{x} - \frac{1}{x^2} = \frac{x - 1}{x^2}$

$(fg)(x) = f(x) \cdot g(x) = \frac{1}{x}\left(\frac{1}{x^2}\right) = \frac{1}{x^3}$

$\left(\frac{f}{g}\right)(x) = \frac{f(x)}{g(x)} = \frac{1/x}{1/x^2} = \frac{x^2}{x} = x, \ x \neq 0$

27. $(f + g)(3) = f(3) + g(3) = (3^2 + 1) + (3 - 4) = 9$

29. $(f - g)(0) = f(0) - g(0) = [0^2 + 1] - (0 - 4) = 5$

31. $(f - g)(2t) = f(2t) - g(2t) = [(2t)^2 + 1] - (2t - 4) = 4t^2 - 2t + 5$

33. $(fg)(4) = f(4)g(4) = (4^2 + 1)(4 - 4) = 0$

35. $\left(\frac{f}{g}\right)(5) = \frac{f(5)}{g(5)} = \frac{5^2 + 1}{5 - 4} = 26$

37. $\left(\frac{f}{g}\right)(-1) - g(3) = \frac{f(-1)}{g(-1)} - g(3)$

$$= \frac{(-1)^2 + 1}{-1 - 4} - (3 - 4)$$

$$= -\frac{2}{5} + 1 = \frac{3}{5}$$

39. $f(x) = \frac{1}{2}x$, $g(x) = x - 1$, $(f + g)(x) = \frac{3}{2}x - 1$

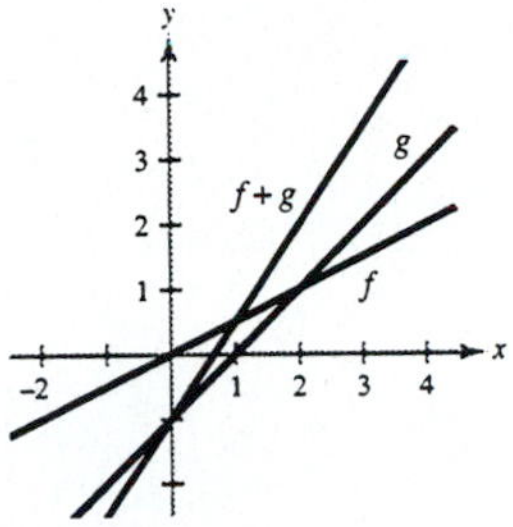

41. $f(x) = x^2$, $g(x) = -2x$, $(f + g)(x) = x^2 - 2x$

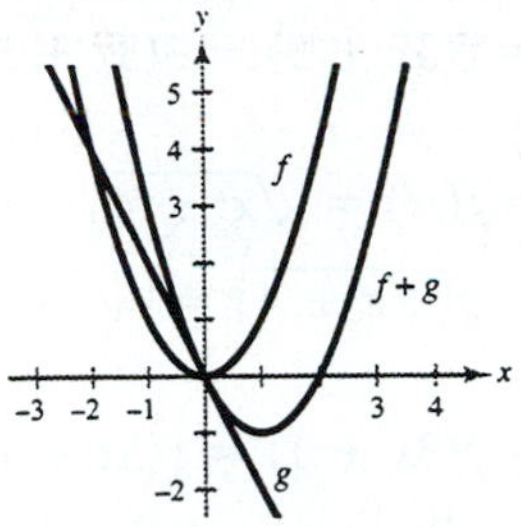

43. $f(x) = 3x$, $g(x) = -\frac{x^3}{10}$, $(f + g)(x) = 3x - \frac{x^3}{10}$

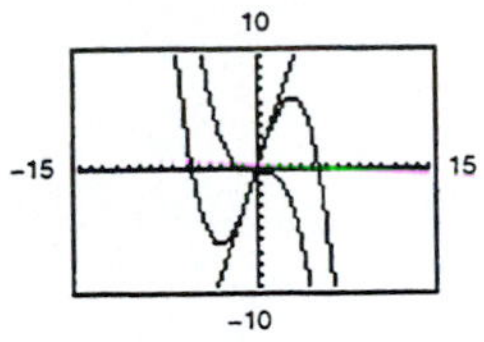

For $0 \le x \le 2$, $f(x)$ contributes most to the magnitude.
For $x > 6$, $g(x)$ contributes most to the magnitude.

45. $T(x) = R(x) + B(x) = \frac{3}{4}x + \frac{1}{15}x^2$

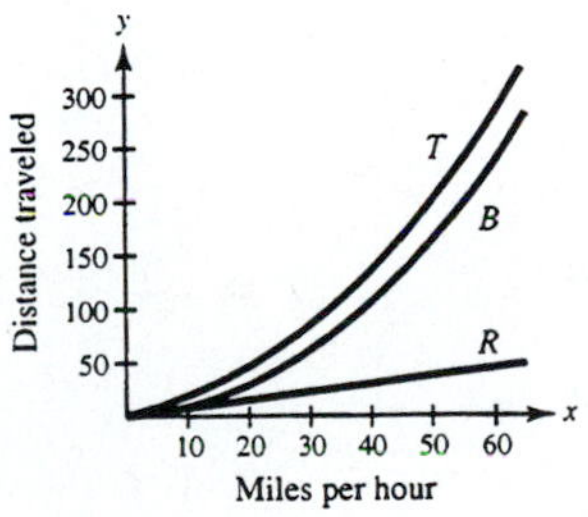

47.

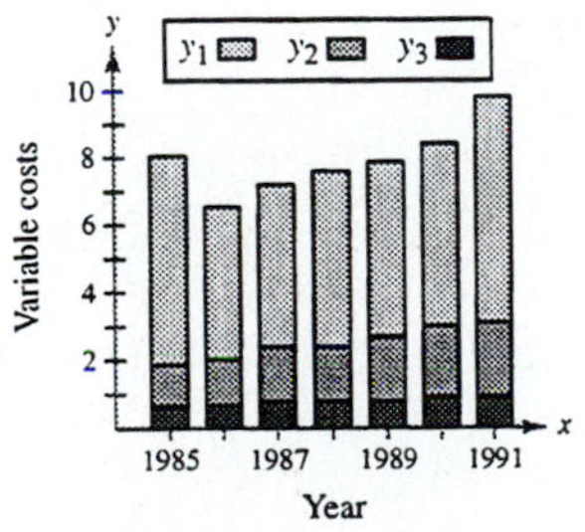

49. (a) T is a function of t since for each time t there corresponds one and only one temperature T.

(b) $T(4) = 60°$
$T(15) = 72°$

(c) $H(t) = T(t - 1)$; All the temperature changes would be one hour later.

(d) $H(t) = T(t) - 1$; The temperature would be decreased by one degree.

51. $f(x) = x^2, g(x) = x - 1$

(a) $(f \circ g)(x) = f(g(x)) = f(x - 1) = (x - 1)^2$

(b) $(g \circ f)(x) = g(f(x)) = g(x^2) = x^2 - 1$

(c) $(f \circ f)(x) = f(f(x)) = f(x^2) = (x^2)^2 = x^4$

53. $f(x) = 3x + 5, g(x) = 5 - x$

(a) $(f \circ g)(x) = f(g(x)) = f(5 - x) = 3(5 - x) + 5 = 20 - 3x$

(b) $(g \circ f)(x) = g(f(x)) = g(3x + 5) = 5 - (3x + 5) = -3x$

(c) $(f \circ f)(x) = f(f(x)) = f(3x + 5) = 3(3x + 5) + 5 = 9x + 20$

55. (a) $(f \circ g)(x) = f(g(x)) = f(x^2) = \sqrt{x^2 + 4}$

(b) $(g \circ f)(x) = g(f(x)) = g\left(\sqrt{x + 4}\right) = \left(\sqrt{x + 4}\right)^2 = x + 4, \ x \geq 4$

57. (a) $(f \circ g)(x) = f(g(x)) = f(3x + 1) = \frac{1}{3}(3x + 1) - 3 = x - \frac{8}{3}$

(b) $(g \circ f)(x) = g(f(x)) = g\left(\frac{1}{3}x - 3\right) = 3\left(\frac{1}{3}x - 3\right) + 1 = x - 8$

59. (a) $(f \circ g)(x) = f(g(x)) = f\left(\sqrt{x}\right) = (x^{1/2})^{1/2} = x^{1/4} = \sqrt[4]{x}$

(b) Since $f(x) = g(x)$, $(g \circ f)(x) = (f \circ g)(x) = \sqrt[4]{x}$.

61. (a) $(f \circ g)(x) = f(g(x)) = f(x + 6) = |x + 6|$

(b) $(g \circ f)(x) = g(f(x)) = g(|x|) = |x| + 6$

63. (a) $(f + g)(3) = f(3) + g(3) = 2 + 1 = 3$

(b) $\left(\frac{f}{g}\right)(2) = \frac{f(2)}{g(2)} = \frac{0}{2} = 0$

65. (a) $(f \circ g)(2) = f(g(2)) = f(2) = 0$

(b) $(g \circ f)(2) = g(f(2)) = g(0) = 4$

67. $g(x) = f(x) + 2$

Vertical shift 2 units upward

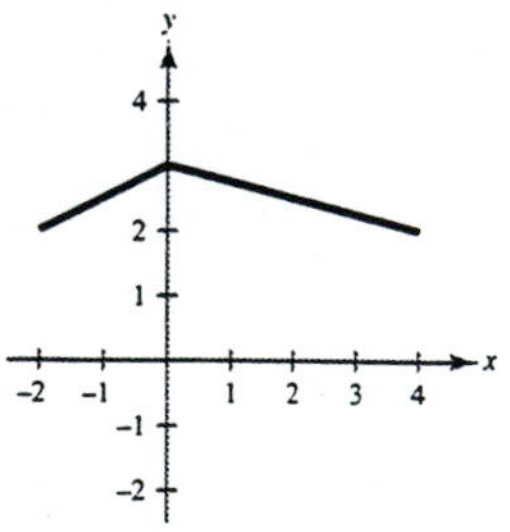

69. $g(x) = f(-x)$

Reflection in the y-axis

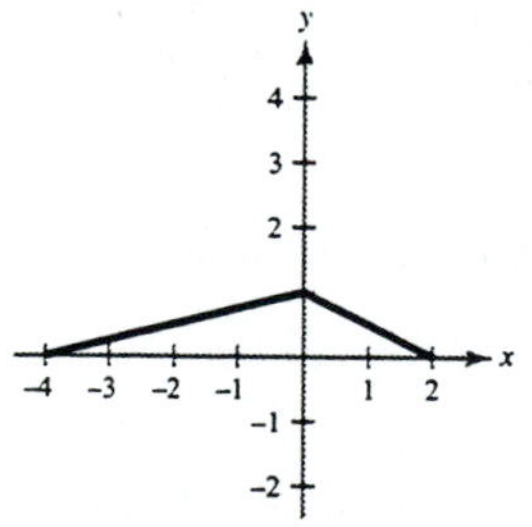

71. Let $f(x) = x^2$ and $g(x) = 2x + 1$, then $(f \circ g)(x) = h(x)$. This is not a unique solution. For example, if $f(x) = (x + 1)^2$ and $g(x) = 2x$, then $(f \circ g)(x) = h(x)$ as well.

73. Let $f(x) = \sqrt[3]{x}$ and $g(x) = x^2 - 4$, then $(f \circ g)(x) = h(x)$.
This answer is not unique. Other possibilities may be:

$$f(x) = \sqrt[3]{x - 4} \text{ and } g(x) = x^2$$
$$\text{or } f(x) = \sqrt[3]{-x} \text{ and } g(x) = 4 - x^2$$
$$\text{or } f(x) = \sqrt[9]{x} \text{ and } g(x) = (4 - x^2)^3$$

75. Let $f(x) = 1/x$ and $g(x) = x + 2$, then $(f \circ g)(x) = h(x)$. Again, this is not a unique solution. Other possibilities may be:

$$f(x) = \frac{1}{x + 2} \text{ and } g(x) = x$$

$$\text{or } f(x) = \frac{1}{x + 1} \text{ and } g(x) = x + 1$$

$$\text{or } f(x) = \frac{1}{x^2 + 2} \text{ and } g(x) = \sqrt{x}$$

77. (a) The domain of $f(x) = \sqrt{x}$ is $x \geq 0$.

(b) The domain of $g(x) = x^2 + 1$ is all real numbers.

(c) $(f \circ g)(x) = f(g(x)) = f(x^2 + 1) = \sqrt{x^2 + 1}$

The domain of $f \circ g$ is all real numbers.

79. (a) The domain of $f(x) = \dfrac{3}{x^2 - 1}$ is all real numbers except $x = \pm 1$.

(b) The domain of $g(x) = x + 1$ is all real numbers.

(c) $(f \circ g)(x) = f(g(x)) = f(x + 1) = \dfrac{3}{(x + 1)^2 - 1} = \dfrac{3}{x^2 + 2x} = \dfrac{3}{x(x + 2)}$

This domain of $f \circ g$ is all real numbers except $x = 0$ and $x = -2$.

81. $f(x) = 3x - 4$

$$\begin{aligned}\frac{f(x + h) - f(x)}{h} &= \frac{[3(x + h) - 4] - (3x - 4)}{h} \\ &= \frac{3x + 3h - 4 - 3x + 4}{h} \\ &= \frac{3h}{h} \\ &= 3\end{aligned}$$

83. $f(x) = \frac{4}{x}$

$$\frac{f(x+h)-f(x)}{h} = \frac{\frac{4}{x+h}-\frac{4}{x}}{h} = \frac{\frac{4x-4(x+h)}{x(x+h)}}{\frac{h}{1}}$$

$$= \frac{4x-4x-4h}{x(x+h)} \cdot \frac{1}{h}$$

$$= \frac{-4\cancel{h}}{x(x+h)} \cdot \frac{1}{\cancel{h}}$$

$$= \frac{-4}{x(x+h)}$$

85. (a) $r(x) = \frac{x}{2}$

(b) $A(r) = \pi r^2$

(c) $(A \circ r)(x) = A(r(x)) = A\left(\frac{x}{2}\right) = \pi\left(\frac{x}{2}\right)^2$

$(A \circ r)(x)$ represents the area of the circular base of the tank on the square foundation with side length y.

87. $(C \circ x)(t) = C(x(t))$

$= 60(50t) + 750$

$= 3000t + 750$

$(C \circ x)(t)$ represents the cost after t production hours.

89. (a) $R = p - 1200$

(b) $S = p - 0.08p = 0.92p$

(c) $(R \circ S)(p) = R(S(p)) = R(0.92p) = 0.92p - 1200$

$(S \circ R)(p) = S(R(p)) = S(p - 1200) = 0.92(p - 1200)$

$R \circ S$ represents taking a discount of 8% of the retail price and then receiving a \$1200 rebate. $S \circ R$ represents taking the \$1200 rebate first and then receiving an 8% discount on the difference.

(d) $(R \circ S)(18{,}400) = 0.92(18{,}400) - 1200 = \$15{,}728$

$(S \circ R)(18{,}400) = 0.92(18{,}400 - 1200) = \$15{,}824$

$R \circ S$ is a better deal. $S \circ R$ takes an 8% discount on a smaller amount.

91. Let $f(x)$ be an odd function, $g(x)$ be an even function and define $h(x) = f(x)g(x)$. Then

$h(-x) = f(-x)g(-x)$

$= [-f(x)]g(x)$ Since f is odd and g is even.

$= -f(x)g(x)$

$= -h(x)$

Thus, h is odd.

Section 1.6 Inverse Functions

- Two functions f and g are inverses of each other if $f(g(x)) = x$ for every x in the domain of g and $g(f(x)) = x$ for every x in the domain of f.
- Be able to find the inverse of a function, if it exists.
  1. Replace $f(x)$ with y.
  2. Interchange x and y.
  3. Solve for y. If this equation represents y as a function of x, then you have found $f^{-1}(x)$. If this equation does not represent y as a function of x, then f does not have an inverse function.
- A function f has an inverse function if and only if no **horizontal** line crosses the graph of f at more than one point.

Solutions to Odd-Numbered Exercises

1. The inverse is a line through $(-1, 0)$.
Matches graph (c).

3. The inverse is half a parabola starting at $(1, 0)$.
Matches graph (a).

5. $f^{-1}(x) = \frac{x}{8} = \frac{1}{8}x$

$$f(f^{-1}(x)) = f\left(\frac{x}{8}\right) = 8\left(\frac{x}{8}\right) = x$$

$$f^{-1}(f(x)) = f^{-1}(8x) = \frac{8x}{8} = x$$

7. $f^{-1}(x) = x - 10$

$$f(f^{-1}(x)) = f(x - 10) = (x - 10) + 10 = x$$

$$f^{-1}(f(x)) = f^{-1}(x + 10) = (x + 10) - 10 = x$$

9. $f^{-1}(x) = x^3$

$$f(f^{-1}(x)) = f(x^3) = \sqrt[3]{x^3} = x$$

$$f^{-1}(f(x)) = f^{-1}(\sqrt[3]{x}) = (\sqrt[3]{x})^3 = x$$

11. (a) $f(g(x)) = f\left(\frac{x}{2}\right) = 2\left(\frac{x}{2}\right) = x$

$$g(f(x)) = g(2x) = \frac{2x}{2} = x$$

(b)

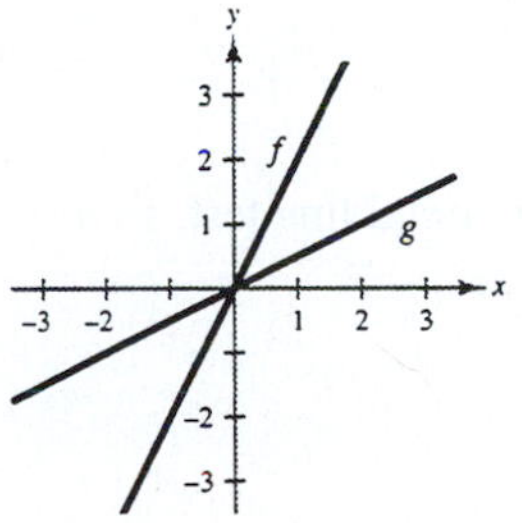

13. (a) $f(g(x)) = f\left(\frac{x-1}{5}\right) = 5\left(\frac{x-1}{5}\right) + 1 = x$

$$g(f(x)) = g(5x + 1) = \frac{(5x + 1) - 1}{5} = x$$

(b)

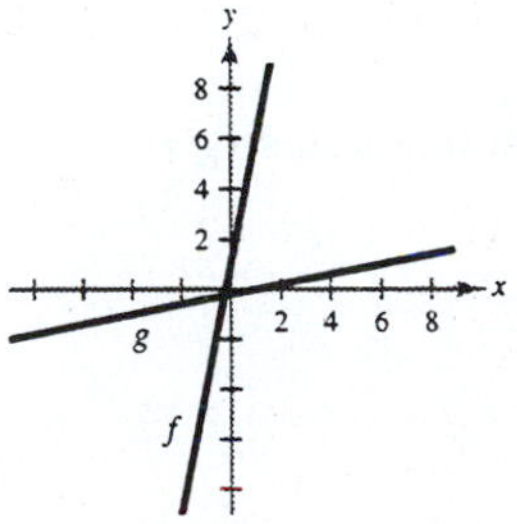

15. (a) $f(g(x)) = f\left(\sqrt[3]{x}\right) = \left(\sqrt[3]{x}\right)^3 = x$

$g(f(x)) = g(x^3) = \sqrt[3]{x^3} = x$

(b)

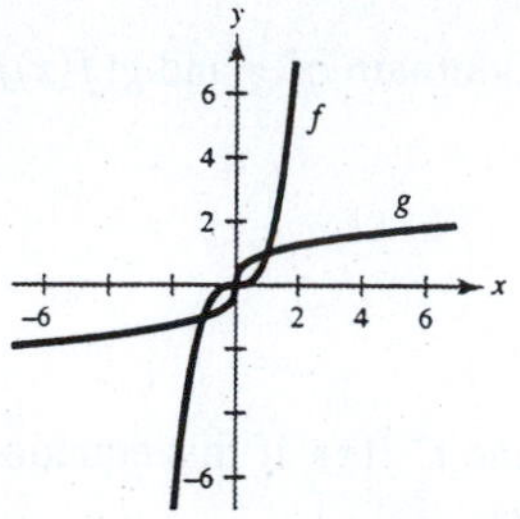

17. (a) $f(g(x)) = f(x^2 + 4),\ x \geq 0$

$= \sqrt{(x^2 + 4) - 4} = x$

$g(f(x)) = g\left(\sqrt{x - 4}\right)$

$= \left(\sqrt{x - 4}\right)^2 + 4 = x$

(b)

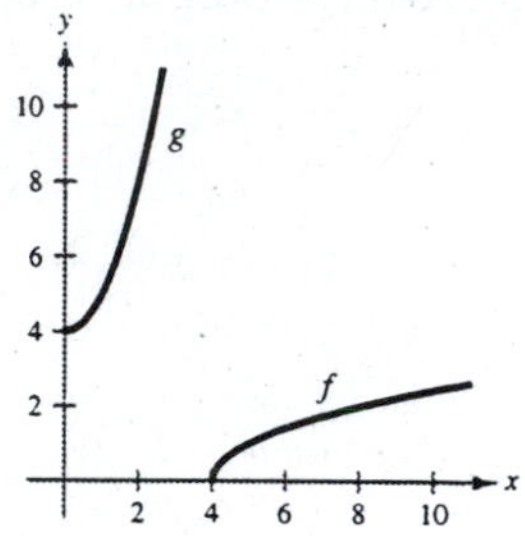

19. (a) $f(g(x)) = f\left(\sqrt{9 - x}\right),\ x \leq 9$

$= 9 - \left(\sqrt{9 - x}\right)^2 = x$

$g(f(x)) = g(9 - x^2),\ x \geq 0$

$= \sqrt{9 - (9 - x^2)} = x$

(b)

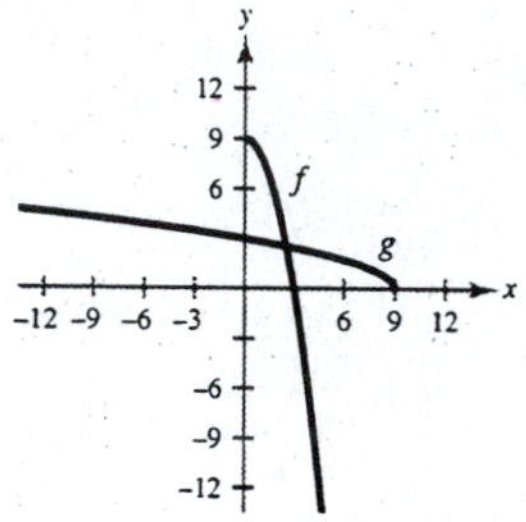

21. No, $\{(-2, -1), (1, 0), (2, 1), (1, 2), (-2, 3), (-6, 4)\}$ does not represent a function.

23. Since no horizontal line crosses the graph of f at more than one point, f **has** an inverse.

25. Since some horizontal lines cross the graph of f twice, f does **not** have an inverse.

27. $g(x) = \dfrac{4 - x}{6}$

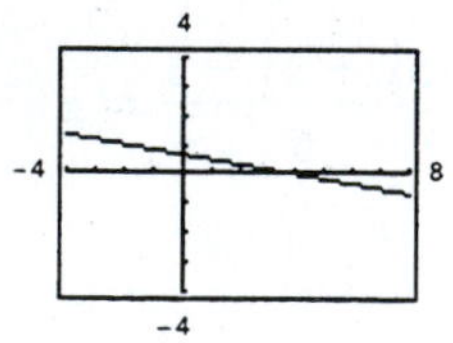

g passes the horizontal line test, so g **has** an inverse.

29. $h(x) = |x + 4| - |x - 4|$

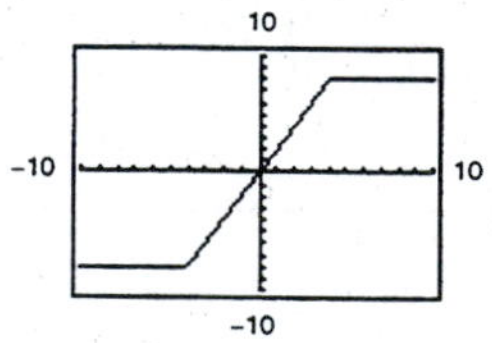

h does not pass the horizontal line test, so h does **not** have an inverse.

31. $f(x) = -2x\sqrt{16 - x^2}$

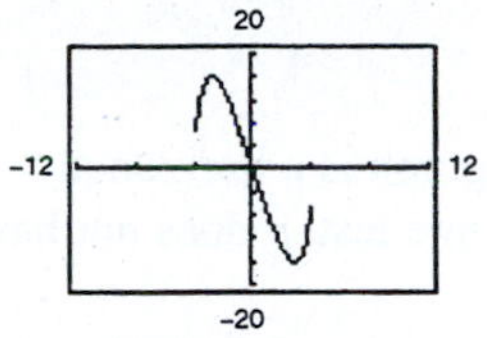

f does not pass the horizontal line test, so f does **not** have an inverse.

33. $f(x) = 2x - 3$

$y = 2x - 3$

$x = 2y - 3$

$y = \dfrac{x + 3}{2}$

$f^{-1}(x) = \dfrac{x + 3}{2}$

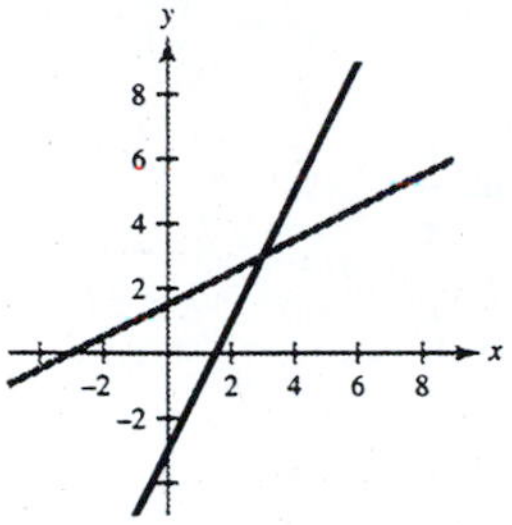

35. $f(x) = x^5$

$y = x^5$

$x = y^5$

$y = \sqrt[5]{x}$

$f^{-1}(x) = \sqrt[5]{x}$

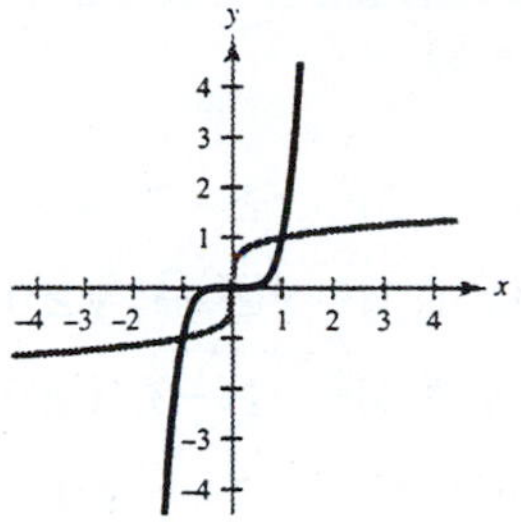

37. $f(x) = \sqrt{x}$

$y = \sqrt{x}$

$x = \sqrt{y}$

$y = x^2$

$f^{-1}(x) = x^2, \; x \geq 0$

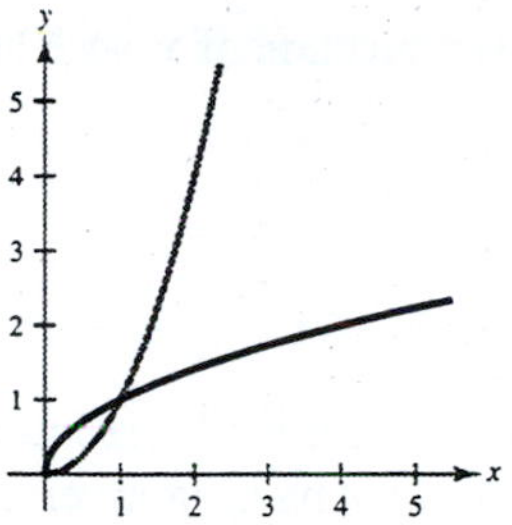

39. $f(x) = \sqrt{4 - x^2}, \; 0 \leq x \leq 2$

$y = \sqrt{4 - x^2}$

$x = \sqrt{4 - y^2}$

$f^{-1}(x) = \sqrt{4 - x^2}, \; 0 \leq x \leq 2$

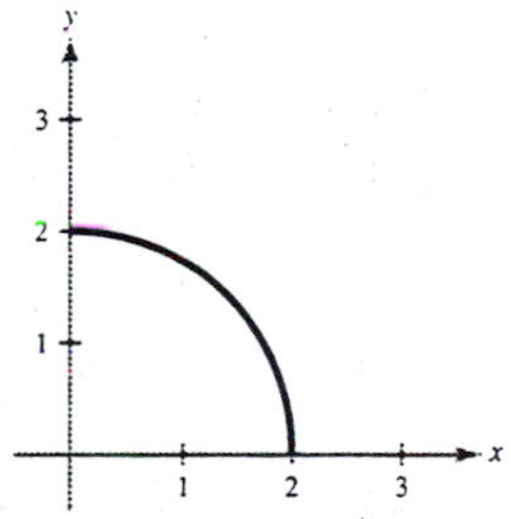

41. $f(x) = \sqrt[3]{x - 1}$

$y = \sqrt[3]{x - 1}$

$x = \sqrt[3]{y - 1}$

$x^3 = y - 1$

$y = x^3 + 1$

$f^{-1}(x) = x^3 + 1$

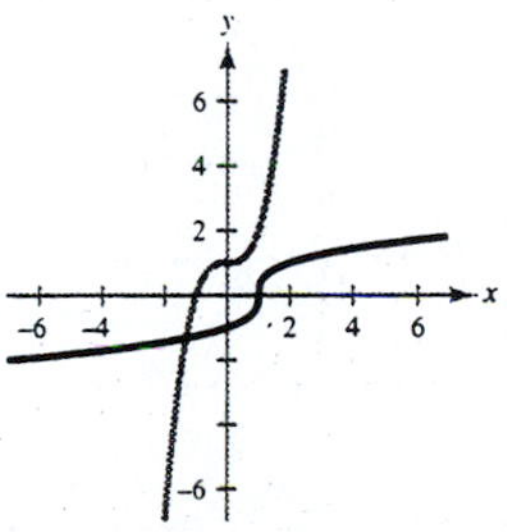

43. $f(x) = x^4$

$y = x^4$

$x = y^4$

$y = \pm\sqrt[4]{x}$

This does not represent y as a function of x. f does not have an inverse.

45. $g(x) = \frac{x}{8}$

$y = \frac{x}{8}$

$x = \frac{y}{8}$

$y = 8x$

This is a function of x, so g has an inverse.
$g^{-1}(x) = 8x$

47. $p(x) = -4$

$y = -4$

Since $y = -4$ for all x, the graph is a horizontal line and fails the horizontal line test. p does not have an inverse.

49. $f(x) = (x + 3)^2,\ x \geq -3 \implies y \geq 0$

$y = (x + 3)^2,\ x \geq -3,\ y \geq 0$

$x = (y + 3)^2,\ y \geq -3,\ x \geq 0$

$\sqrt{x} = y + 3,\ y \geq -3,\ x \geq 0$

$y = \sqrt{x} - 3,\ x \geq 0,\ y \geq -3$

This is a function of x, so f has an inverse.

$f^{-1}(x) = \sqrt{x} - 3,\ x \geq 0$

51. $h(x) = \frac{1}{x}$

$y = \frac{1}{x}$

$xy = 1$

$y = \frac{1}{x}$

This is a function of x, so h has an inverse.
$h^{-1}(x) = \frac{1}{x}$

53. $f(x) = \sqrt{2x + 3} \implies x \geq -\frac{3}{2},\ y \geq 0$

$y = \sqrt{2x + 3},\ x \geq -\frac{3}{2},\ y \geq 0$

$x = \sqrt{2y + 3},\ y \geq -\frac{3}{2},\ x \geq 0$

$x^2 = 2y + 3,\ x \geq 0,\ y \geq -\frac{3}{2}$

$y = \frac{x^2 - 3}{2},\ x \geq 0,\ y \geq -\frac{3}{2}$

This is a function of x, so f has an inverse.

$f^{-1}(x) = \frac{x^2 - 3}{2},\ x \geq 0$

55. $g(x) = x^2 - x^4$

The graph fails the horizontal line test, so g does not have an inverse.

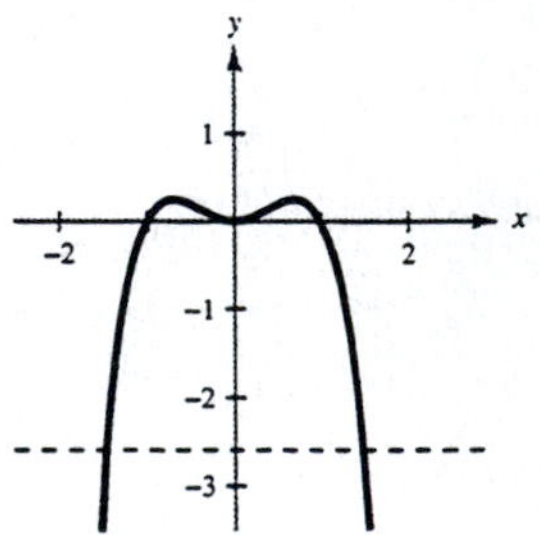

57. $f(x) = 25 - x^2,\ x \leq 0 \implies y \leq 25$

$y = 25 - x^2,\ x \leq 0,\ y \leq 25$

$x = 25 - y^2,\ y \leq 0,\ x \leq 25$

$y^2 = 25 - x,\ x \leq 25,\ y \leq 0$

$y = -\sqrt{25 - x},\ x \leq 25,\ y \leq 0$

This is a function of x, so f has an inverse.

$f^{-1}(x) = -\sqrt{25 - x},\ x \leq 25$

59. If we let $f(x) = (x-2)^2,\ x \ge 2$, then f has an inverse. [Note: we could also let $x \le 2$.]

$$f(x) = (x-2)^2,\ x \ge 2 \ \Rightarrow\ y \ge 0$$
$$y = (x-2)^2,\ x \ge 2,\ y \ge 0$$
$$x = (y-2)^2,\ x \ge 0,\ y \ge 2$$
$$\sqrt{x} = y - 2,\ x \ge 0,\ y \ge 2$$
$$\sqrt{x} + 2 = y,\ x \ge 0,\ y \ge 2$$

Thus, $f^{-1}(x) = \sqrt{x} + 2,\ x \ge 0$.

61. If we let $f(x) = |x+2|,\ x \ge -2$, then f has an inverse. [Note: we could also let $x \le -2$.]

$$f(x) = |x+2|,\ x \ge -2$$
$$f(x) = x + 2 \text{ when } x \ge -2.$$
$$y = x + 2,\ x \ge -2,\ y \ge 0$$
$$x = y + 2,\ x \ge 0,\ y \ge -2$$
$$x - 2 = y,\ x \ge 0,\ y \ge -2$$

Thus, $f^{-1}(x) = x - 2,\ x \ge 0$.

63.

x	$f(x)$
-2	-4
-1	-2
1	2
3	3

x	$f^{-1}(x)$
-4	-2
-2	-1
2	1
3	3

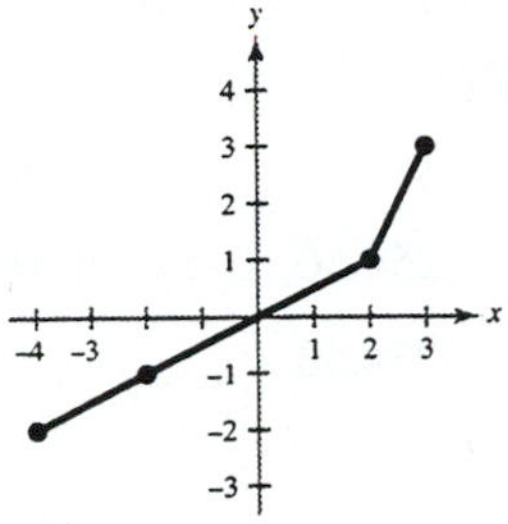

65. False, $f(x) = x^2$ is even and does not have an inverse.

67. True

In Exercises 69, 71, and 73, $f(x) = \frac{1}{8}x - 3,\ f^{-1}(x) = 8(x+3),\ g(x) = x^3,\ g^{-1}(x) = \sqrt[3]{x}$.

69. $(f^{-1} \circ g^{-1})(1) = f^{-1}(g^{-1}(1)) = f^{-1}(\sqrt[3]{1}) = 8(\sqrt[3]{1} + 3) = 32$

71. $(f^{-1} \circ f^{-1})(6) = f^{-1}(f^{-1}(6)) = f^{-1}(8[6+3]) = 8[8(6+3)+3] = 600$

73. $(f \circ g)(x) = f(g(x)) = f(x^3) = \frac{1}{8}x^3 - 3$

$$y = \tfrac{1}{8}x^3 - 3$$
$$x = \tfrac{1}{8}y^3 - 3$$
$$x + 3 = \tfrac{1}{8}y^3$$
$$8(x+3) = y^3$$
$$\sqrt[3]{8(x+3)} = y$$
$$(f \circ g)^{-1}(x) = 2\sqrt[3]{x+3}$$

In Exercises 75 and 77, $f(x) = x + 4,\ f^{-1}(x) = x - 4,\ g(x) = 2x - 5,\ g^{-1}(x) = \dfrac{x+5}{2}$.

75. $(g^{-1} \circ f^{-1})(x) = g^{-1}(f^{-1}(x)) = g^{-1}(x-4) = \dfrac{(x-4)+5}{2} = \dfrac{x+1}{2}$

77. $(f \circ g)(x) = f(g(x)) = f(2x-5) = (2x-5) + 4 = 2x - 1$

$$(f \circ g)^{-1}(x) = \frac{x+1}{2}$$

Note: Comparing Exercises 75 and 77, we see that $(f \circ g)^{-1}(x) = (g^{-1} \circ f^{-1})(x)$.

79. (a)
$$y = 8 + 0.75x$$
$$x = 8 + 0.75y$$
$$x - 8 = 0.75y$$
$$\frac{x-8}{0.75} = y$$
$$f^{-1}(x) = \frac{x-8}{0.75}$$

(b) x = hourly wage

y = number of units produced

(c) $y = \dfrac{22.25 - 8}{0.75} = 19$ units

81. (a)
$$y = 0.03x^2 + 254.50, \; 0 < x < 100$$
$$x = 0.03y^2 + 254.50$$
$$x - 254.50 = 0.03y^2$$
$$\frac{x - 254.50}{0.03} = y^2$$
$$\sqrt{\frac{x - 254.50}{0.03}} = y, \; 254.5 < x < 545.5$$
$$f^{-1}(x) = \sqrt{\frac{x - 254.50}{0.03}}$$

x = temperature in degrees Fahrenheit

y = percent load for a diesel engine

(b)

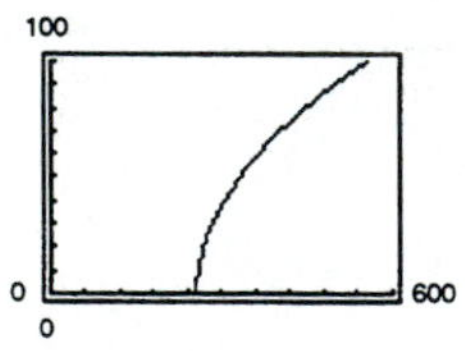

(c)
$$0.03x^2 + 254.50 < 500$$
$$0.03x^2 < 245.5$$
$$x^2 < 8183\tfrac{1}{3}$$
$$x < 90.46$$

Thus, $0 < x < 90.46$.

83. (a) Yes, since no y-value is paired with two different x-values, f^{-1} does exist.

(b) f^{-1} yields the year for a given average fuel consumption.

(c) $f^{-1}(19.95) = 8$

85. $x^2 = 64$
$$x = \pm\sqrt{64} = \pm 8$$

87. $4x^2 - 12x + 9 = 0$
$$(2x - 3)^2 = 0$$
$$2x - 3 = 0$$
$$x = \tfrac{3}{2}$$

89. $x^2 - 6x + 4 = 0$

$x^2 - 6x = -4$

$x^2 - 6x + 9 = -4 + 9$

$(x - 3)^2 = 5$

$x - 3 = \pm\sqrt{5}$

$x = 3 \pm \sqrt{5}$

91. $50 + 5x = 3x^2$

$0 = 3x^2 - 5x - 50$

$0 = (3x + 10)(x - 5)$

$3x + 10 = 0 \implies x = -\frac{10}{3}$

$x - 5 = 0 \implies x = 5$

93. Let $2n$ = first positive even integer. Then $2n + 2$ = next positive even integer.

$2n(2n + 2) = 288$

$4n^2 + 4n - 288 = 0$

$4(n^2 + n - 72) = 0$

$4(n + 9)(n - 8) = 0$

$n + 9 = 0 \implies n = -9$ Not a solution since the integers are positive.

$n - 8 = 0 \implies n = 8$

Thus, $2n = 16$ and $2n + 2 = 18$.

95.

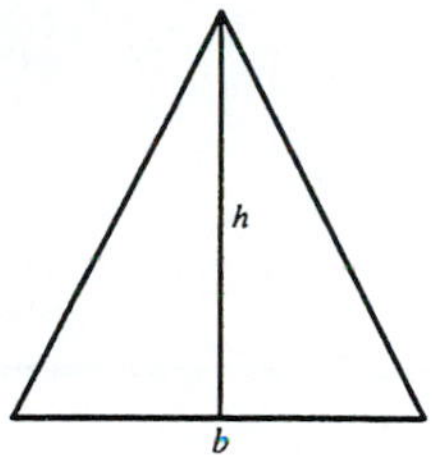

Given $b = h$ and $A = 10$ sq ft:

$A = \frac{1}{2}bh$

$10 = \frac{1}{2}bb$

$20 = b^2$

$\sqrt{20} = b$

$2\sqrt{5} = b$

Thus, $b = h = 2\sqrt{5}$ feet.

Section 1.7 Mathematical Modeling

You should know the following the following terms and formulas.

- Direct Variation (varies directly, directly proportional)
  - (a) $y = mx$
  - (b) $y = kx^n$ (as nth power)
- Inverse Variation (varies inversely, inversely proportional)
  - (a) $y = k/x$
  - (b) $y = k/(x^n)$ (as nth power)
- Joint Variation (varies jointly, jointly proportional)
  - (a) $z = kxy$
  - (b) $z = kx^n y^m$ (as nth power of x and mth power of y)
- k is called the constant of proportionality.
- Least Squares Regression Line $y = ax + b$ where

$$a = \frac{n\sum_{i=1}^{n} x_i y_i - \sum_{i=1}^{n} x_i \sum_{i=1}^{n} y_i}{n\sum_{i=1}^{n} x_i^2 - \left(\sum_{i=1}^{n} x_i\right)^2}$$

$$b = \frac{1}{n}\left(\sum_{i=1}^{n} y_i - a\sum_{i=1}^{n} x_i\right).$$

Solutions to Odd-Numbered Exercises

1. $y = 113{,}336.2 + 1265t,\ 7 \le t \le 13$

$t = 0$ represents 1980.

Year	1987	1988	1989	1990	1991	1992	1993
Actual Number	121,602	123,378	125,557	126,424	126,867	128,548	129,525
Model	122,191	123,456	124,721	125,986	127,251	128,516	129,781

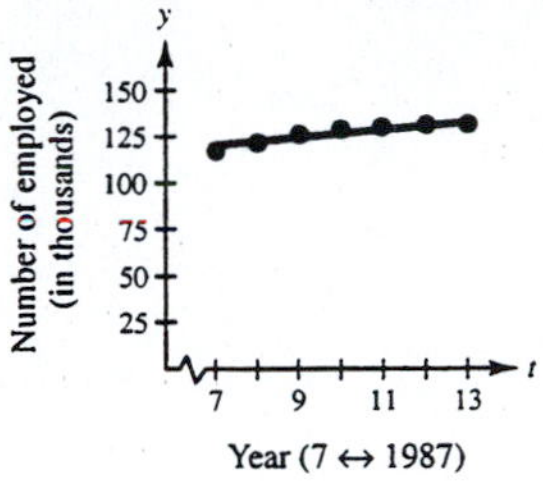

The model is a "good fit" for the actual data.

3. The graph appears to represent $y = 4/x$ which is an inverse variation.

5.

x	2	4	6	8	10
$y = x^2$	4	16	36	64	100

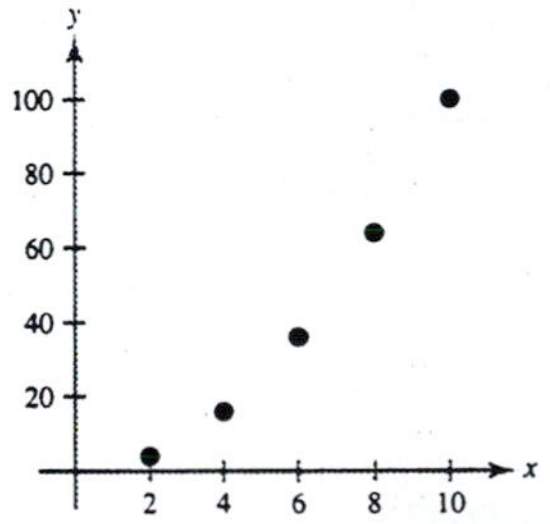

7.

x	2	4	6	8	10
$y = \frac{1}{2}x^2$	2	8	18	32	50

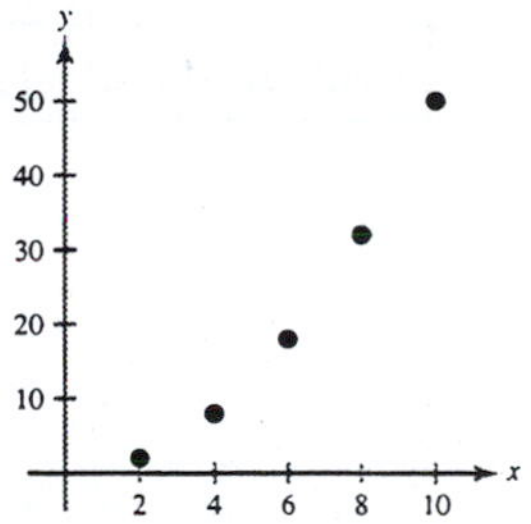

9.

x	2	4	6	8	10
$y = \dfrac{2}{x^2}$	$\frac{1}{2}$	$\frac{1}{8}$	$\frac{1}{18}$	$\frac{1}{32}$	$\frac{1}{50}$

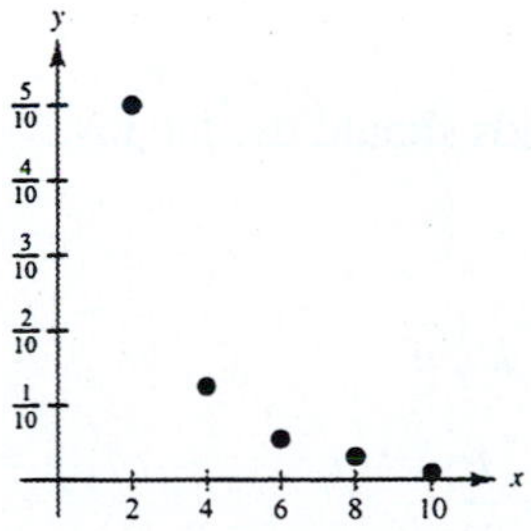

11.

x	2	4	6	8	10
$y = \dfrac{10}{x^2}$	$\frac{5}{2}$	$\frac{5}{8}$	$\frac{5}{18}$	$\frac{5}{32}$	$\frac{1}{10}$

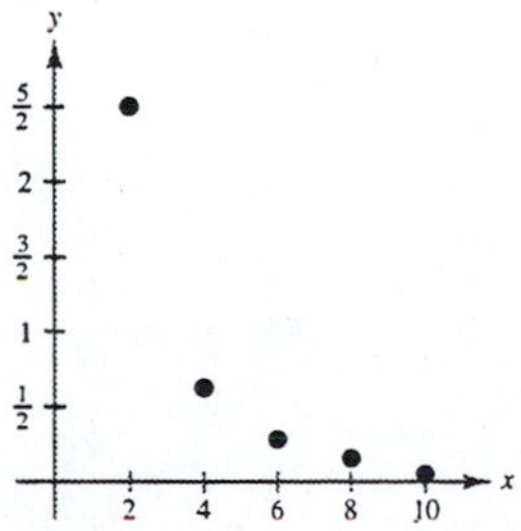

13. The chart represents the equation $y = \dfrac{5}{x}$.

15.
$$y = kx$$
$$-7 = k(10)$$
$$-\tfrac{7}{10} = k$$
$$y = -\tfrac{7}{10}x$$

This equation checks with the other points given in the chart.

17.
$$y = mx$$
$$12 = m(5)$$
$$\tfrac{12}{5} = m$$
$$y = \tfrac{12}{5}x$$

19.
$$y = mx$$
$$2050 = m(10)$$
$$205 = m$$
$$y = 205x$$

21.
$$I = kP$$
$$187.59 = k(2500)$$
$$0.075 = k$$
$$I = 0.075P$$

23.
$$y = kx$$
$$1840 = k(50{,}000)$$
$$0.0368 = k$$
$$y = 0.0368x$$

When $x = 85{,}000$ we have
$$y = 0.0368(85{,}000) = \$3128.$$

25.
$$y = kx$$
$$33 = k(13)$$
$$\tfrac{33}{13} = k$$
$$y = \tfrac{33}{13}x$$

Inches	5	10	20	25	30
Centimeters	12.7	25.4	50.8	63.5	76.2

27.
$$d = kF$$
$$0.15 = k(265)$$
$$\tfrac{3}{5300} = k$$
$$d = \tfrac{3}{5300}F$$

(a) $d = \tfrac{3}{5300}(90) \approx 0.05$ meter

(b) $0.1 = \tfrac{3}{5300}F$
$$\tfrac{530}{3} = F$$
$$F = 176\tfrac{2}{3} \text{ newtons}$$

29.
$$d = kF$$
$$1.9 = k(25) \implies k = 0.076$$
$$d = 0.076F$$

When the distance compressed is 3 inches, we have
$$3 = 0.076F$$
$$F \approx 39.47.$$

No child over 39.47 pounds should use the toy.

31. $A = kr^2$

33. $y = \dfrac{k}{x^2}$

35. $z = k\sqrt[3]{u}$

37. $z = kuv$

39. $F = \dfrac{kg}{r^2}$

41. $P = \dfrac{k}{V}$

43. $F = \dfrac{km_1m_2}{r^2}$

45. $A = \tfrac{1}{2}bh$

The area of a triangle is jointly proportional to the magnitude of the base and the height.

47. $V = \frac{4}{3}\pi r^3$

The volume of a sphere varies directly as the cube of its radius.

49. $r = \frac{d}{t}$

Average speed is directly proportional to the distance and inversely proportional to the time.

51.
$$A = kr^2$$
$$9\pi = k(3)^2$$
$$\pi = k$$
$$A = \pi r^2$$

53.
$$y = \frac{k}{x}$$
$$3 = \frac{k}{25}$$
$$75 = k$$
$$y = \frac{75}{x}$$

55.
$$h = \frac{k}{t^3}$$
$$\frac{3}{16} = \frac{k}{(4)^3}$$
$$\frac{3}{16} = \frac{k}{64}$$
$$k = 12$$
$$h = \frac{12}{t^3}$$

57.
$$z = kxy$$
$$64 = k(4)(8)$$
$$2 = k$$
$$z = 2xy$$

59.
$$F = krs^3$$
$$4158 = k(11)(3)^3$$
$$k = 14$$
$$F = 14rs^3$$

61.
$$z = \frac{kx^2}{y}$$
$$6 = \frac{k(6)^2}{4}$$
$$\frac{24}{36} = k$$
$$\frac{2}{3} = k$$
$$z = \frac{(2/3)x^2}{y} = \frac{2x^2}{3y}$$

63.
$$S = \frac{kL}{L - S}$$
$$4 = \frac{k(6)}{6 - 4}$$
$$4 = 3k$$
$$k = \frac{4}{3}$$
$$S = \frac{4/3L}{L - S} = \frac{4L}{3(L - S)}$$

65.
$$d = kv^2$$
$$0.02 = k\left(\frac{1}{4}\right)^2$$
$$k = 0.32$$
$$d = 0.32v^2$$
$$0.12 = 0.32v^2$$
$$v^2 = \frac{0.12}{0.32} = \frac{3}{8}$$
$$v = \frac{\sqrt{3}}{2\sqrt{2}} = \frac{\sqrt{6}}{4} \approx 0.61 \text{ mi/hr}$$

67.
$$r = \frac{kl}{A},\ A = \pi r^2 = \frac{\pi d^2}{4}$$
$$r = \frac{4kl}{\pi d^2}$$
$$66.17 = \frac{4(1000)k}{\pi(0.0126/12)^2}$$
$$k \approx 5.73 \times 10^{-8}$$
$$r = \frac{4(5.73 \times 10^{-8})l}{\pi(0.0126/12)^2}$$
$$33.5 = \frac{4(5.73 \times 10^{-8})l}{\pi(0.0126/12)^2}$$
$$\frac{33.5\pi(0.0126/12)^2}{4(5.73 \times 10^{-8})} = l$$
$$l \approx 506 \text{ feet}$$

69. $S = kt^2$

$144 = k(3)^2$

$16 = k$

$S = 16t^2$

$S = 16(5)^2 = 400$ feet

71. $P = kA = k(\pi r^2) = k\pi\left(\dfrac{d}{2}\right)^2$

$8.78 = k\pi\left(\dfrac{9}{2}\right)^2$

$\dfrac{4(8.78)}{81\pi} = k$

$k \approx 0.138$

However, we do not obtain \$11.78 when $d = 12$ inches.

$P = 0.138\pi\left(\dfrac{12}{2}\right)^2 \approx \$15.61$

Instead, $k = \dfrac{11.78}{36\pi} \approx 0.104$.

For the 15 inch pizza, we have $k = \dfrac{4(14.18)}{225\pi} \approx 0.080$.

The price is not directly proportional to the surface area. The best buy is the 15-inch pizza.

73. $v = \dfrac{k}{A}$

(a) $v = \dfrac{k}{0.75A} = \dfrac{4}{3}\left(\dfrac{k}{A}\right)$

The velocity is four-thirds the original.

(b) $v = \dfrac{k}{(1 + \frac{1}{3})A} = \dfrac{k}{\frac{4}{3}A} = \dfrac{3}{4}\left(\dfrac{k}{A}\right)$

The velocity is decreased by one-fourth.

75. (a)

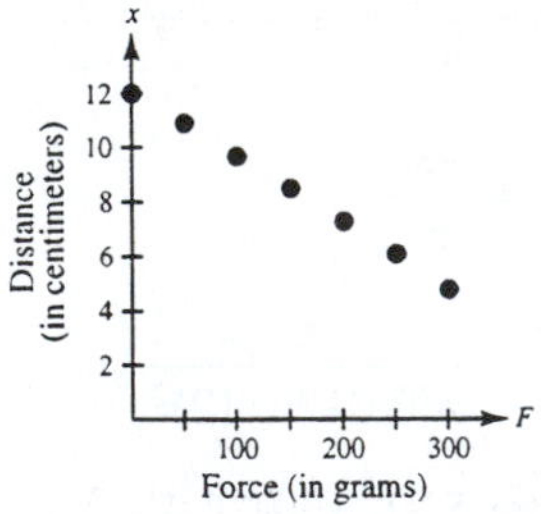

(b) It appears to fit Hooke's Law.

$k \approx \left|\dfrac{12.00 - 4.84}{0 - 300}\right| \approx |-0.0238| = 0.0238$

(c) $x = kF$

$9 = 0.0238F$

$F \approx 378$ grams

77. $y = \dfrac{262.76}{x^{2.12}}$

(a)

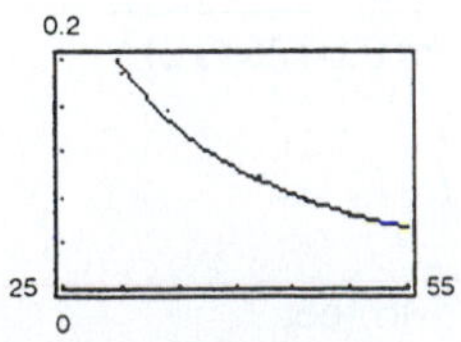

(b) $y = \dfrac{262.76}{(25)^{2.12}} \approx 0.2857$ microwatts per square centimeter

79. The data shown could be represented by a linear model which would be a good fit.

81. The points do not follow a linear pattern. A linear model would not be a good approximation.

83.

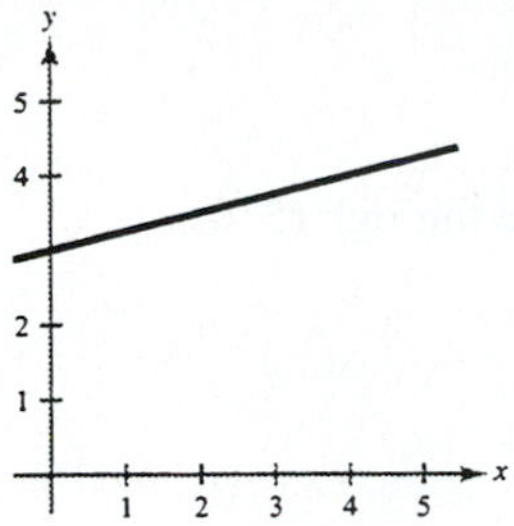

Using the points (0, 3) and (4, 4), we have $y = \frac{1}{4}x + 3$.

85.

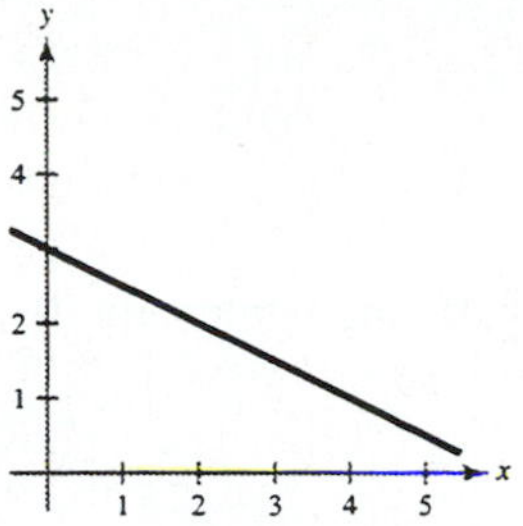

Using the points (2, 2) and (4, 1), we have $y = -\frac{1}{2}x + 3$.

87. (a) (7, 203), (8, 239), (9, 295), (10, 352), (11, 415), and (12, 488)

$$a = \frac{6(19929) - (57)(1992)}{6(559) - (57)^2} \approx 57.4$$

$$b = \frac{1}{6}[1992 - a(57)] \approx -213.6$$

$$y \approx 57.4t - 213.6$$

(b)

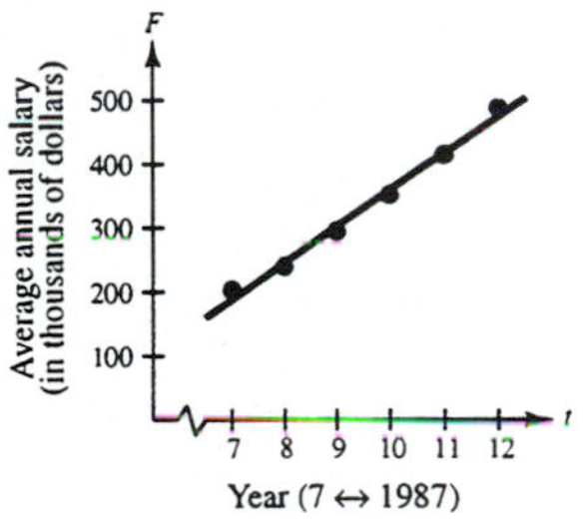

(c)

Year	t	y
1994	14	590.0
1995	15	647.4
1996	16	704.8

89. (a) Using the linear regression capabilities of a calculator yields $y \approx 1.145t + 124.425$.

(b)

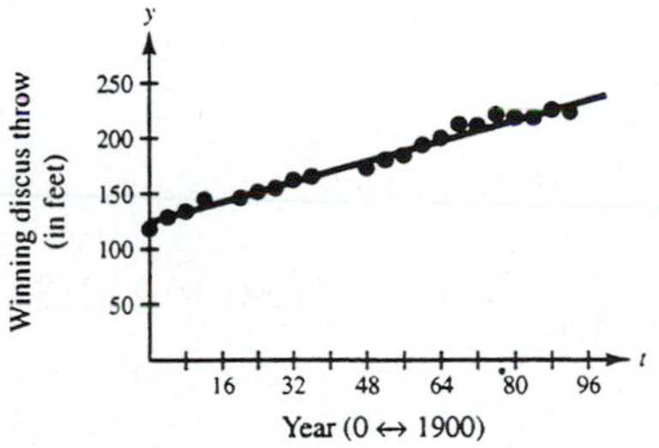

(c) Using the model with $t = 96$ yields 234.345 feet.

91. (a) Using the linear regression capabilities of a calculator yields $y \approx -0.74x + 106$.

(b)

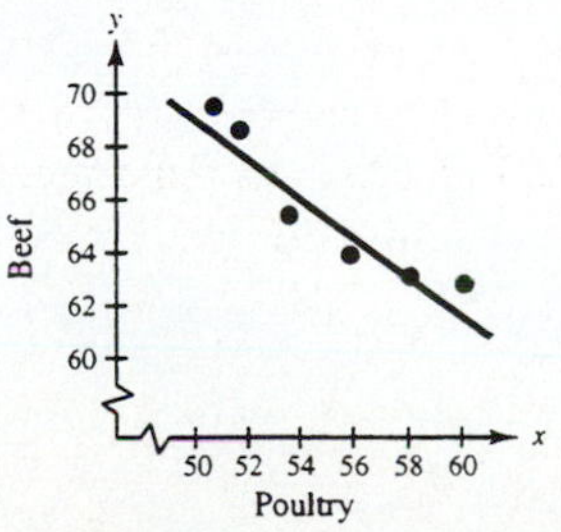

(c) Using the model with $x = 62$ yields 60.12 pounds.

(d) For each 1 pound increase in per capita consumption of poultry, the per capita consumption of beef decreases by an average of 0.74 pounds.

❑ Review Exercises for Chapter 1

Solutions to Odd-Numbered Exercises

1. $y - 2x - 3 = 0$

$y = 2x + 3$

Line with x-intercept $\left(-\frac{3}{2}, 0\right)$ and y-intercept $(0, 3)$

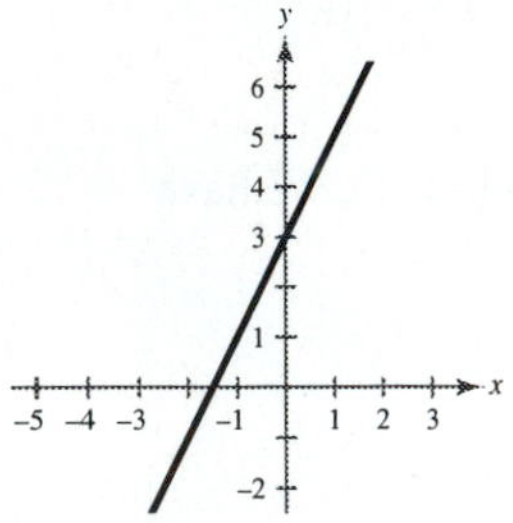

3. $x - 5 = 0$

$x = 5$ is a vertical line through $(5, 0)$.

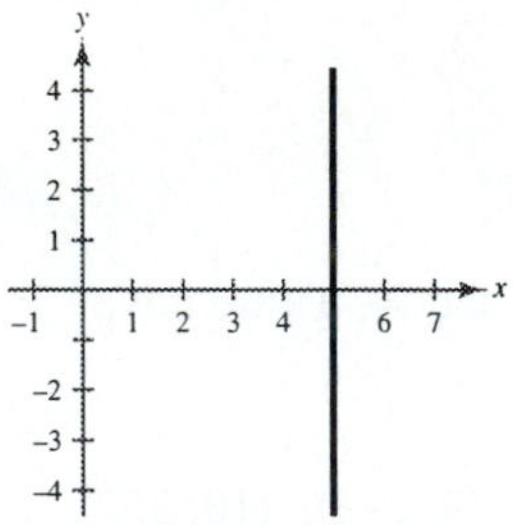

5. $y = \sqrt{5 - x}$

Domain: $(-\infty, 5]$

x	5	4	1	−4
y	0	1	2	3

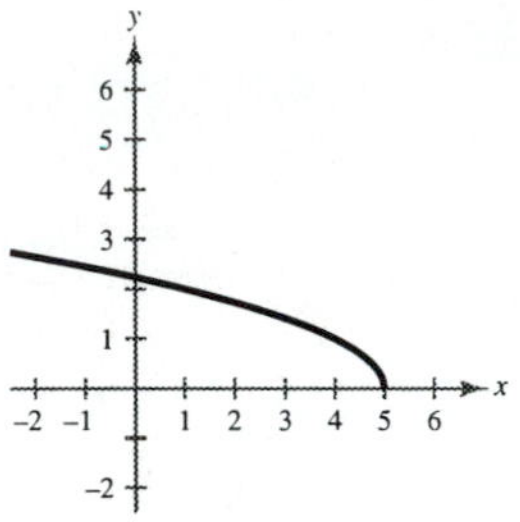

7. $y + 2x^2 = 0$

$y = -2x^2$ is a parabola

x	0	±1	±2
y	0	−2	−8

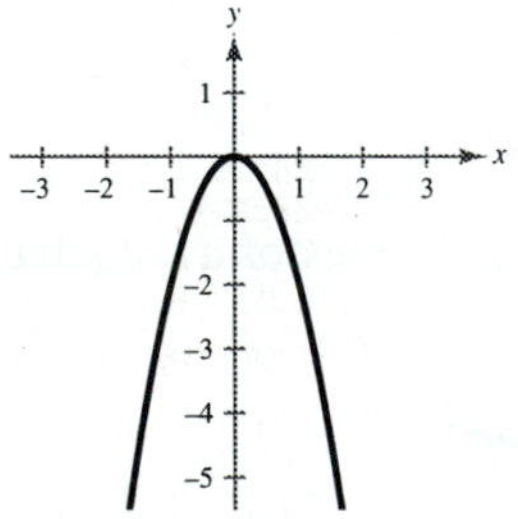

9. $y = \sqrt{25 - x^2}$

Domain: $-5 \le x \le 5$

x	0	±3	±4	±5
y	5	4	3	0

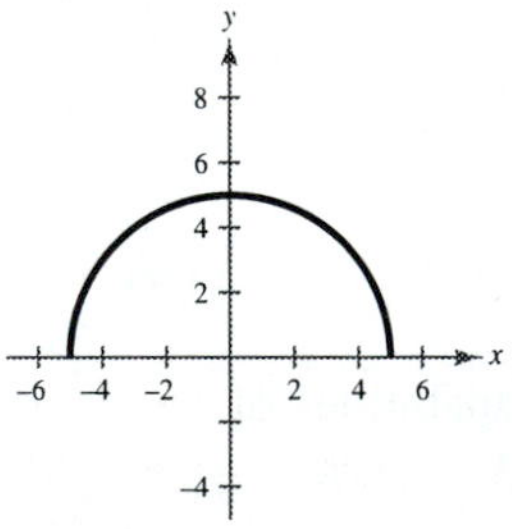

11. $y = \frac{1}{4}(x + 1)^3$

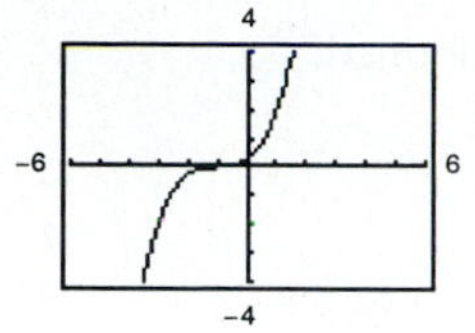

Intercepts: $(-1, 0), \left(0, \frac{1}{4}\right)$

13. $y = \frac{1}{4}x^4 - 2x^2$

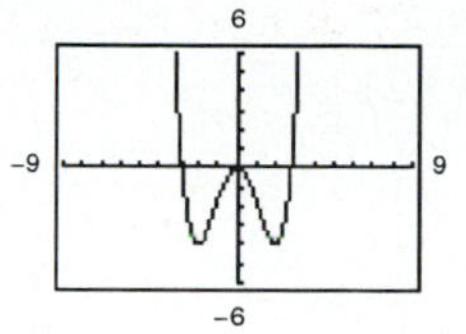

Intercepts: $(0, 0), \left(\pm 2\sqrt{2}, 0\right)$

15. $y = x\sqrt{9 - x^2}$

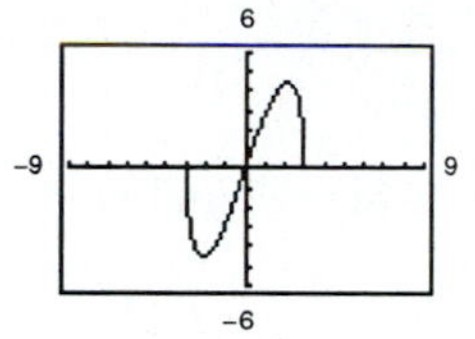

Intercepts: $(0, 0), (\pm 3, 0)$

17. $y = |x - 4| - 4$

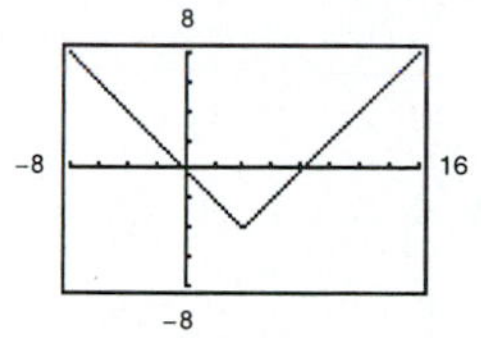

Intercepts: $(0, 0), (8, 0)$

19. $y^2 = 25 - x^2$

$y = \pm\sqrt{25 - x^2}$

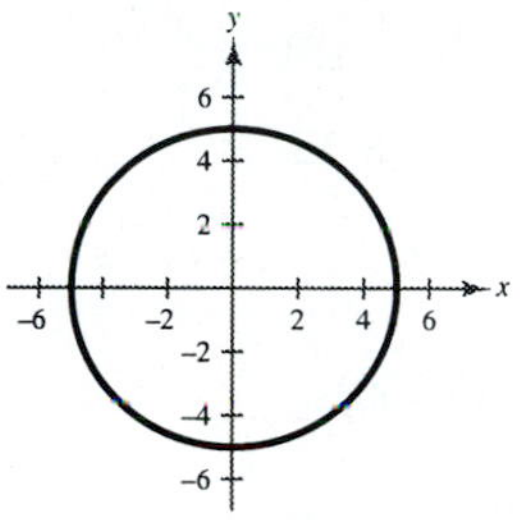

21. $y = 10x^3 - 21x^2$

Xmin = -2
Xmax = 3
Xscl = 1
Ymin = -20
Ymax = 15
Yscl = 5

23. $(x - 3)^2 + (y + 1)^2 = 9$

$(x - 3)^2 + [y - (-1)]^2 = 3^2$

Center: $(3, -1)$

Radius: $r = 3$

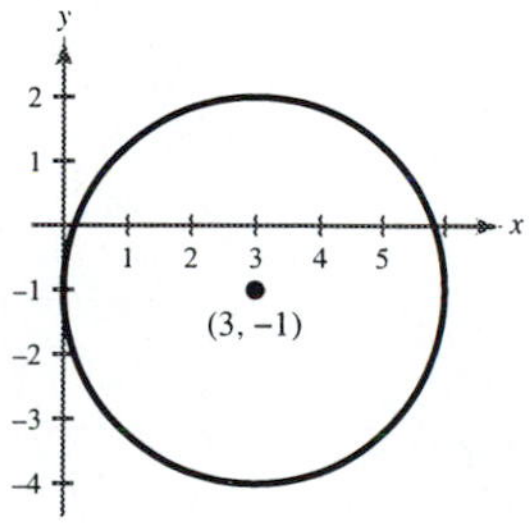

25. (a) $m = \frac{3}{2} > 0 \implies$ The line rises. Matches L_2.

(b) $m = 0 \implies$ The line is horizontal. Matches L_3.

(c) $m = -3 < 0 \implies$ The line falls. Matches L_1.

27. $(-4.5, 6)$, $(2.1, 3)$

$$m = \frac{3 - 6}{2.1 - (-4.5)} = \frac{-3}{6.6} = -\frac{30}{66} = -\frac{5}{11}$$

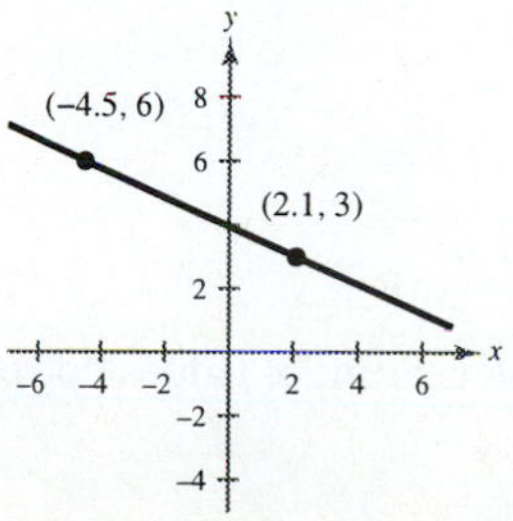

29. $(-2, 5), (0, t), (1, 1)$ are collinear.

$$\frac{t-5}{0-(-2)} = \frac{1-5}{1-(-2)}$$

$$\frac{t-5}{2} = \frac{-4}{3}$$

$$3(t-5) = -8$$

$$3t - 15 = -8$$

$$3t = 7$$

$$t = \frac{7}{3}$$

31. $(0, 0), (0, 10)$

$$m = \frac{10-0}{0-0} = \frac{10}{0} \text{ undefined.}$$

The line is vertical.

$x = 0$

33. $y - (-5) = \frac{3}{2}(x - 0)$

$$y + 5 = \frac{3}{2}x$$

$$y = \frac{3}{2}x - 5 \text{ or } 0 = 3x - 2y - 10$$

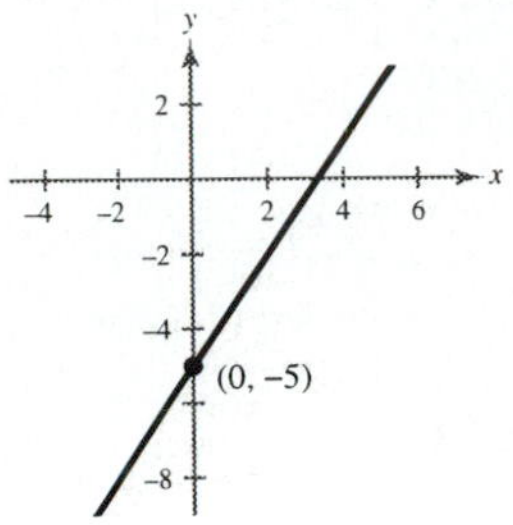

35. $5x - 4y = 8 \Rightarrow y = \frac{5}{4}x - 2$ and $m = \frac{5}{4}$

(a) Parallel slope: $m = \frac{5}{4}$

$$y - (-2) = \frac{5}{4}(x - 3)$$

$$4y + 8 = 5x - 15$$

$$0 = 5x - 4y - 23$$

(b) Perpendicular slope: $m = -\frac{4}{5}$

$$y - (-2) = -\frac{4}{5}(x - 3)$$

$$5y + 10 = -4x + 12$$

$$4x + 5y - 2 = 0$$

37. $(6, 12{,}500)$ $m = 850$

$$y - 12{,}500 = 850(t - 6)$$

$$y = 850t - 5100 + 12{,}500$$

$$y = 850t + 7400$$

39. $(2,\ 160{,}000), (3,\ 185{,}000)$

$$m = \frac{185{,}000 - 160{,}000}{3 - 2} = 25{,}000$$

$$S - 160{,}000 = 25{,}000(t - 2)$$

$$S = 25{,}000t + 110{,}000$$

For the fourth quarter let $t = 4$. Then we have

$$S = 25{,}000(4) + 110{,}000 = \$210{,}000.$$

41. $A = \{10, 20, 30, 40\}$ and $B = \{0, 2, 4, 6\}$

(a) 20 is matched with two elements in the range so it is not a function.

(b) function

(c) function

(d) 30 is not matched with any element of B so it is not a function.

43. $16x - y^4 = 0$

$$y^4 = 16x$$

$$y = \pm 2\sqrt[4]{x}$$

y is **not** a function of x. Some x-values correspond to two y-values.

45. $y = \sqrt{1-x}$

Each x-value, $x \le 1$, corresponds to only one y-value so y **is** a function of x.

47. $f(x) = x^2 + 1$

(a) $f(2) = (2)^2 + 1 = 5$

(b) $f(-4) = (-4)^2 + 1 = 17$

(c) $f(t^2) = (t^2)^2 + 1 = t^4 + 1$

(d) $-f(x) = -(x^2 + 1) = -x^2 - 1$

49. $g(s) = \dfrac{5}{3s - 9} = \dfrac{5}{3(s - 3)}$

Domain: All real numbers except $s = 3$.

51. $f(x) = \sqrt{x^2 + 8x}$

Domain: $x^2 + 8x \geq 0$

$x(x + 8) \geq 0$

Critical numbers: $x = 0, -8$

Test intervals: $(-\infty, -8), (-8, 0), (0, \infty)$

Solution set: $(-\infty, -8] \cup [0, \infty)$

53. $f(x) = (x^2 - 4)^2$

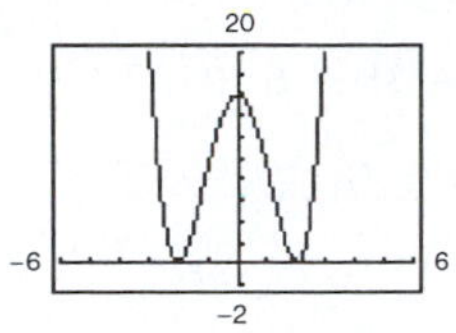

(a) Increasing on $(-2, 0)$ and $(2, \infty)$

Decreasing on $(-\infty, -2)$ and $(0, 2)$

(b) The graph has y-axis symmetry so the function is even.

55. $g(x) = |x + 2| - |x - 2|$

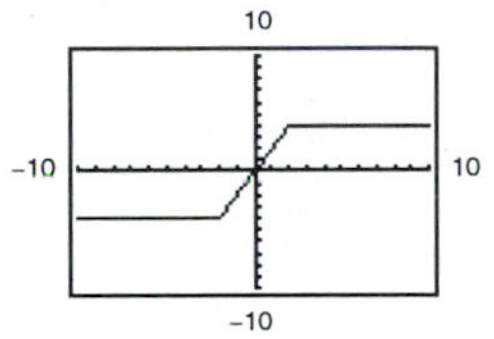

(a) Increasing on $(-2, 2)$

Constant on $(-\infty, -2] \cup [2, \infty)$

(b) The graph has origin symmetry so the function is odd

57. $v(t) = -32t + 48$

(a) $v(1) = 16$ ft/sec

(b) $0 = -32t + 48$

$t = \frac{48}{32} = 1.5$ sec

(c) $v(2) = -16$ ft/sec

59. $(f - g)(4) = f(4) - g(4)$

$= [3 - 2(4)] - \sqrt{4}$

$= -5 - 2$

$= -7$

61. $(h \circ g)(7) = h(g(7))$

$= h(\sqrt{7})$

$= 3(\sqrt{7})^2 + 2$

$= 23$

63. (a)
$$f(x) = \tfrac{1}{2}x - 3$$
$$y = \tfrac{1}{2}x - 3$$
$$x = \tfrac{1}{2}y - 3$$
$$x + 3 = \tfrac{1}{2}y$$
$$2(x + 3) = y$$
$$f^{-1}(x) = 2x + 6$$

(b)

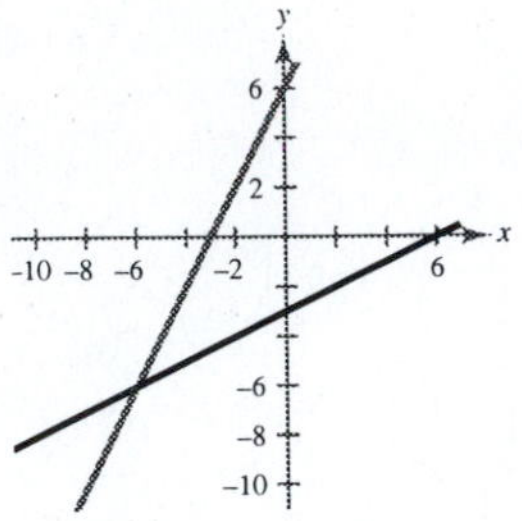

(c)
$$f^{-1}(f(x)) = f^{-1}\left(\tfrac{1}{2}x - 3\right) = 2\left(\tfrac{1}{2}x - 3\right) + 6 = x - 6 + 6 = x$$
$$f(f^{-1}(x)) = f(2x + 6) = \tfrac{1}{2}(2x + 6) - 3 = x + 3 - 3 = x$$

65. (a)
$$f(x) = \sqrt{x + 1}$$
$$y = \sqrt{x + 1}$$
$$x = \sqrt{y + 1}$$
$$x^2 = y + 1$$
$$x^2 - 1 = y$$
$$f^{-1}(x) = x^2 - 1, \ x \geq 0$$

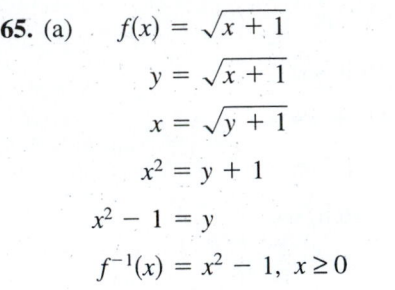

Note: The inverse must have a restricted domain.

(b)

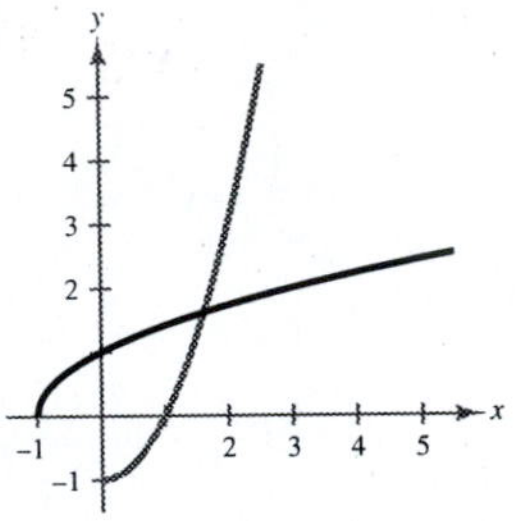

(c)
$$f^{-1}(f(x)) = f^{-1}\left(\sqrt{x + 1}\right) = \left(\sqrt{x + 1}\right)^2 - 1 = x + 1 - 1 = x$$
$$f(f^{-1}(x)) = f(x^2 - 1) = \sqrt{(x^2 - 1) + 1} = \sqrt{x^2} = x \text{ for } x \geq 0.$$

67.
$$F = cx\sqrt{y}$$
$$6 = c(9)\sqrt{4}$$
$$6 = 18c$$
$$\tfrac{1}{3} = c$$
$$F = \tfrac{1}{3}x\sqrt{y}$$

69.
$$z = \frac{cx^2}{y}$$
$$16 = \frac{c(5)^2}{2}$$
$$32 = 25c$$
$$\frac{32}{25} = c$$
$$z = \frac{32x^2}{25y}$$

71. (0, 13.52), (5, 13.52), (10, 15.46), (15, 18.20), (20, 21.02)

(a) $y = 0.3936 + 12.408$

(b)

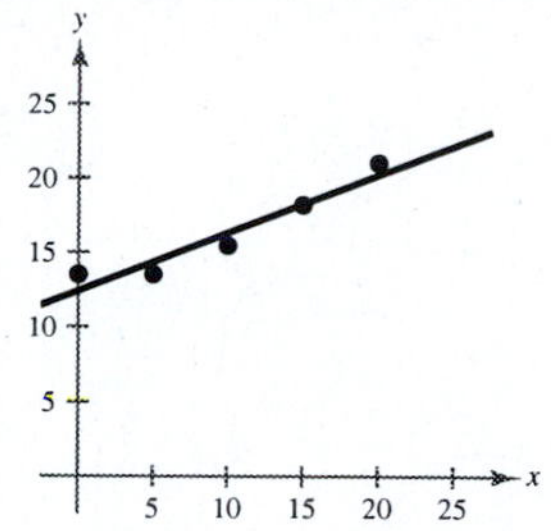

(c) The slope represents the increase per year in miles per gallon

(d) When $t = 30$, $y \approx 24.22$ miles per gallon

❑ Practice Test for Chapter 1

1. Graph $y = \sqrt{7 - x}$

2. Find the domain $y = \sqrt{25 - x^2}$

3. Write the standard equation of the circle with center $(-3, 5)$ and radius 6.

4. Find the equation of the line through $(2, 4)$ and $(3, -1)$.

5. Find the equation of the line with slope $m = 4/3$ and y-intercept $b = -3$.

6. Find the equation of the line through $(4, 1)$ perpendicular to the line $2x + 3y = 0$.

7. If it costs a company \$32 to produce 5 units of a product and \$44 to produce 9 units, how much does it cost to produce 20 units? (Assume that the cost function is linear.)

8. Given $f(x) = x^2 - 2x + 1$, find $f(x - 3)$.

9. Given $f(x) = 4x - 11$, find $\dfrac{f(x) - f(3)}{x - 3}$

10. Find the domain and range of $f(x) = \sqrt{36 - x^2}$.

11. Which equations determine y as a function of x?

(a) $6x - 5y + 4 = 0$

(b) $x^2 + y^2 = 9$

(c) $y^3 = x^2 + 6$

12. Sketch the graph of $f(x) = x^2 - 5$.

13. Sketch the graph of $f(x) = |x + 3|$.

14. Sketch the graph of $f(x) = \begin{cases} 2x + 1 & \text{if } x \geq 0, \\ x^2 - x & \text{if } x < 0. \end{cases}$

15. Use the graph of $f(x) = |x|$ to graph the following:

(a) $f(x + 2)$

(b) $-f(x) + 2$

16. Given $f(x) = 3x + 7$ and $g(x) = 2x^2 - 5$, find the following:

(a) $(g - f)(x)$

(b) $(fg)(x)$

17. Given $f(x) = x^2 - 2x + 16$ and $g(x) = 2x + 3$, find $f(g(x))$.

18. Given $f(x) = x^3 + 7$, find $f^{-1}(x)$.

19. Which of the following functions have inverses?

(a) $f(x) = |x - 6|$

(b) $f(x) = ax + b,\ a \neq 0$

(c) $f(x) = x^3 - 19$

20. Given $f(x) = \sqrt{\dfrac{3 - x}{x}}$, $0 < x \le 3$, find $f^{-1}(x)$.

Exercises 21–23, true or false?

21. $y = 3x + 7$ and $y = \frac{1}{3}x - 4$ are perpendicular.

22. $(f \circ g)^{-1} = g^{-1} \circ f^{-1}$

23. If a function has an inverse, then it must pass both the vertical line test and the horizontal line test.

24. If z varies directly as the cube of x and inversely as the square root of y, and $z = -1$ when $x = -1$ and $y = 25$, find z in terms of x and y.

25. Find the least square regression line for the data.

x	-2	-1	0	1	2	3
y	1	2.4	3	3.1	4	4.7

CHAPTER 2
Polynomial and Rational Functions

CHAPTER 2
Polynomial and Rational Functions

Section 2.1 Quadratic Functions

You should know the following facts about parabolas.

- $f(x) = ax^2 + bx + c,\ a \neq 0$, is a quadratic function, and its graph is a parabola.
- If $a > 0$, the parabola opens upward and the vertex is the minimum point. If $a < 0$, the parabola opens downward and the vertex is the maximum point.
- The vertex is $(-b/2a, f(-b/2a))$.
- To find the x-intercepts (if any), solve

  $$ax^2 + bx + c = 0.$$
- The standard form of the equation of a parabola is

  $$f(x) = a(x - h)^2 + k$$

  where $a \neq 0$.

  (a) The vertex is (h, k).

  (b) The axis is the vertical line $x = h$.

Solutions to Odd-Numbered Exercises

1. $f(x) = (x - 2)^2$ opens upward and has vertex $(2, 0)$. Matches graph (g).

3. $f(x) = x^2 - 2$ opens upward and has vertex $(0, -2)$. Matches graph (b).

5. $f(x) = 4 - (x - 2)^2 = -(x - 2)^2 + 4$ opens downward and has vertex $(2, 4)$. Matches graph (f).

7. $f(x) = x^2 + 3$ opens upward and has vertex $(0, 3)$. Matches graph (e).

9. (a) $y = \frac{1}{2}x^2$

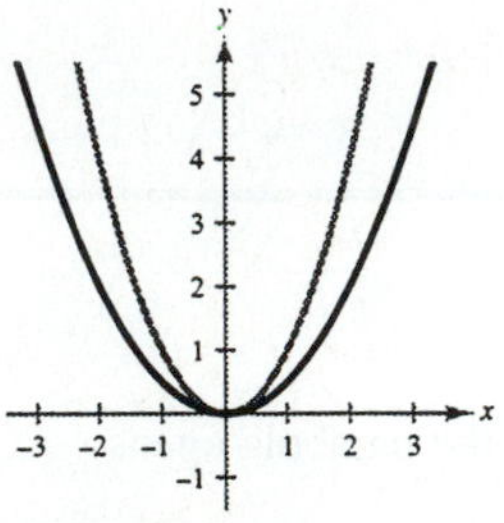

Vertical shrink

(b) $y = -\frac{1}{8}x^2$

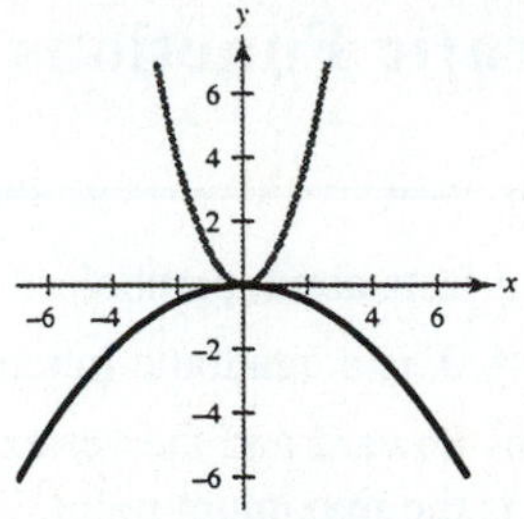

Vertical shrink and reflection in the x-axis

(c) $y = \frac{3}{2}x^2$

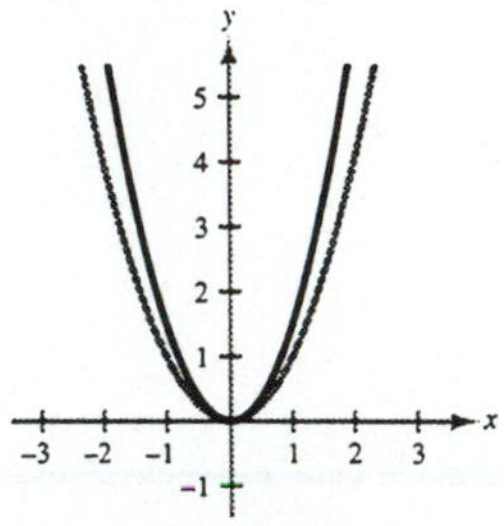

Vertical stretch

(d) $y = -3x^2$

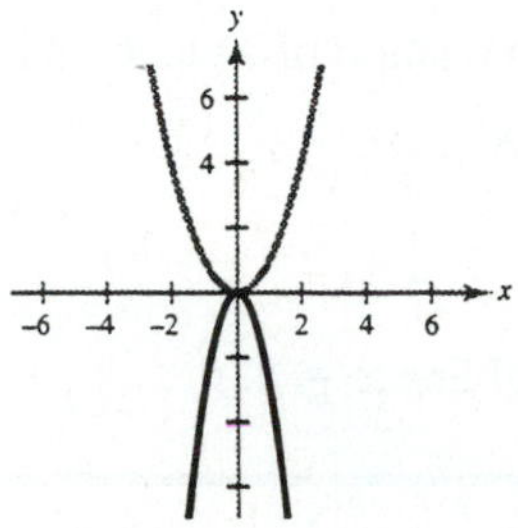

Vertical stretch and reflection in the x-axis

11. (a) $y = (x - 1)^2$

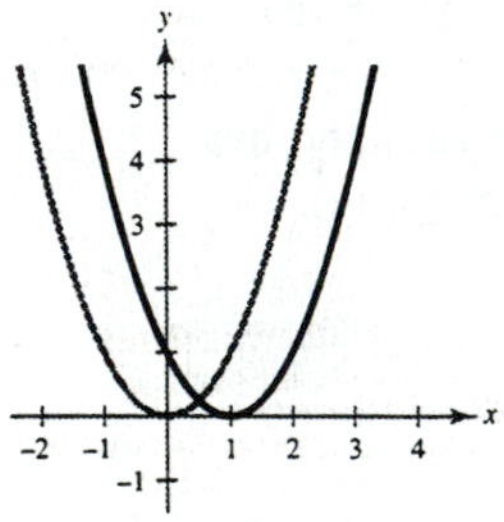

Horizontal translation one unit to the right

(b) $y = (x + 1)^2$

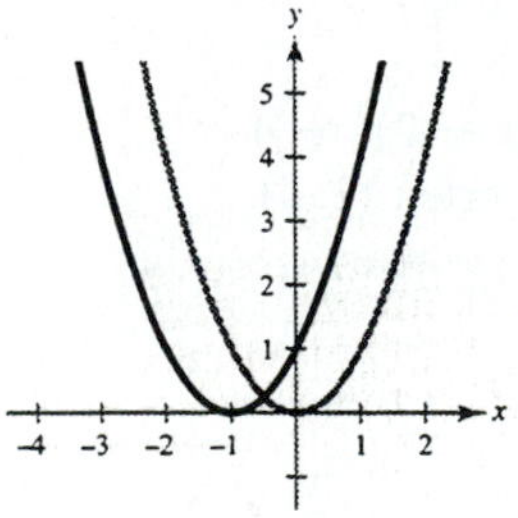

Horizontal translation one unit to the left

(c) $y = (x - 3)^2$

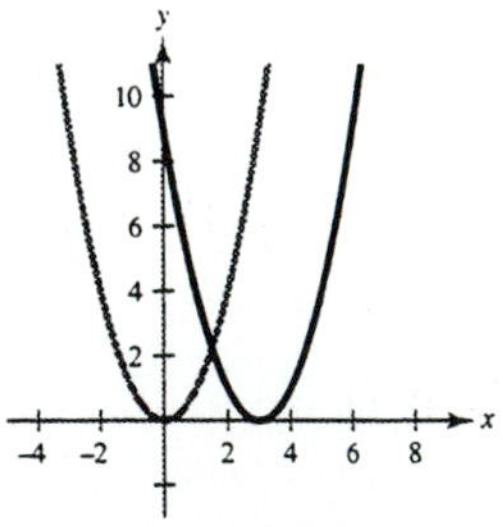

Horizontal translation three units to the right

(d) $y = (x + 3)^2$

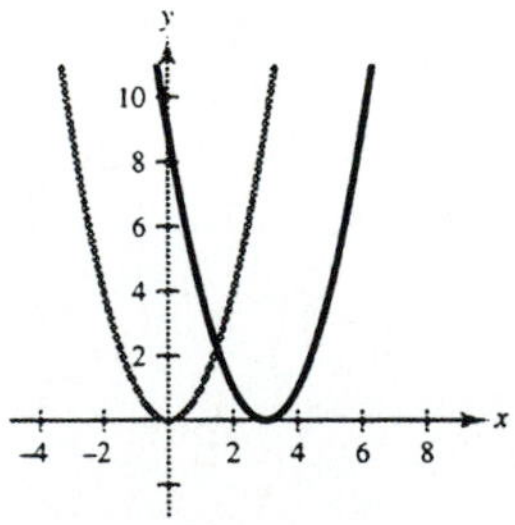

Horizontal translation three units to the left

13. $f(x) = x^2 - 5$

Vertex: $(0, -5)$

Intercepts: $\left(-\sqrt{5}, 0\right)$, $(0, -5)$, $\left(\sqrt{5}, 0\right)$

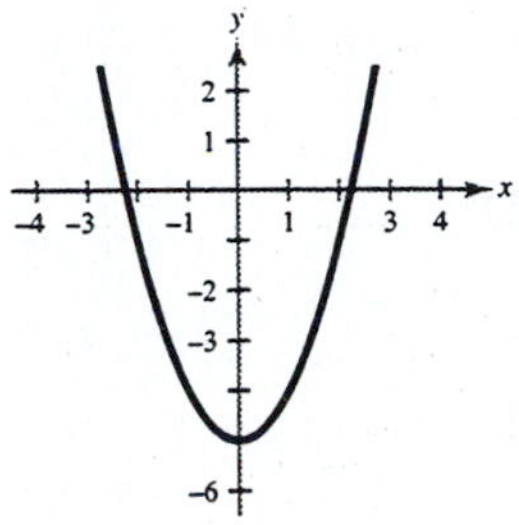

15. $f(x) = 16 - x^2$

Vertex: $(0, 16)$

Intercepts: $(-4, 0)$, $(0, 16)$, $(4, 0)$

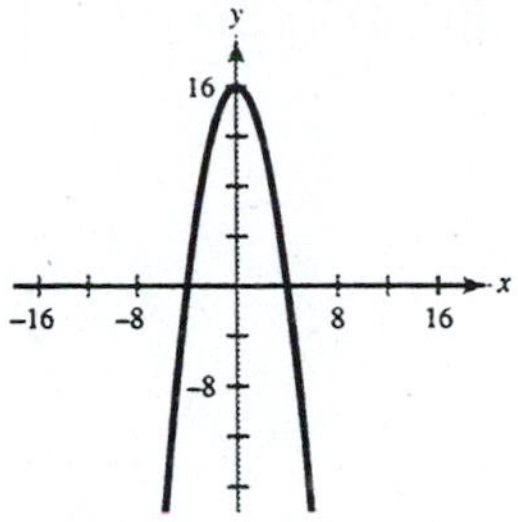

17. $f(x) = (x + 5)^2 - 6$

Vertex: $(-5, -6)$

Intercepts: $\left(-5 - \sqrt{6}, 0\right)$, $\left(-5 + \sqrt{6}, 0\right)$, $(0, 19)$

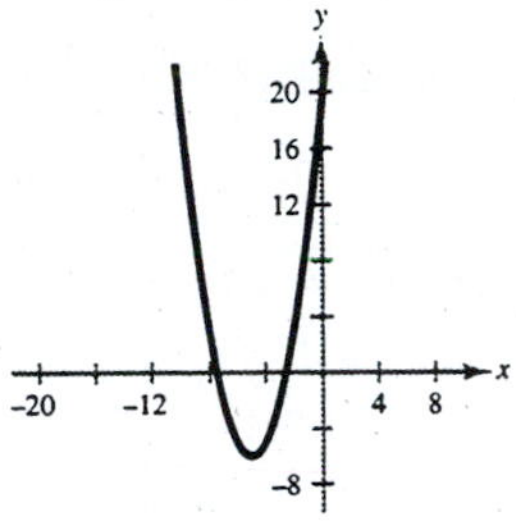

19. $h(x) = x^2 - 8x + 16 = (x - 4)^2$

Vertex: $(4, 0)$

Intercepts: $(0, 16)$, $(4, 0)$

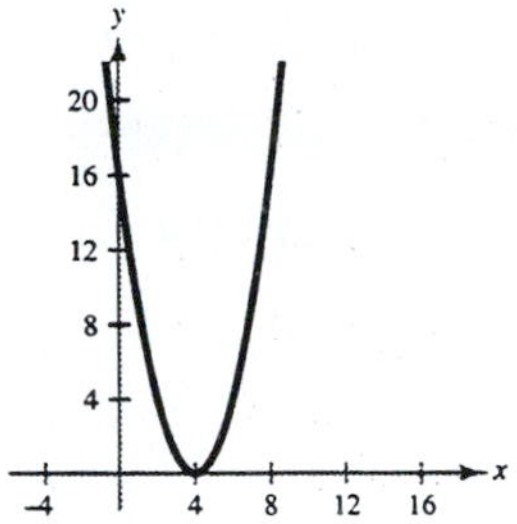

21. $f(x) = x^2 - x + \frac{5}{4} = \left(x - \frac{1}{2}\right)^2 + 1$

Vertex: $\left(\frac{1}{2}, 1\right)$

Intercept: $\left(0, \frac{5}{4}\right)$

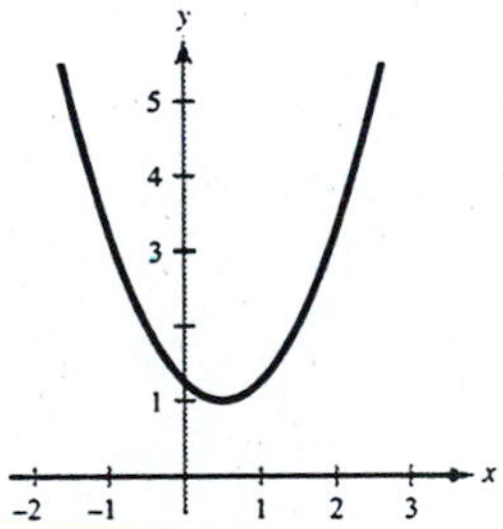

23. $f(x) = -x^2 + 2x + 5 = -(x - 1)^2 + 6$

Vertex: $(1, 6)$

Intercepts: $\left(1 - \sqrt{6}, 0\right)$, $(0, 5)$, $\left(1 + \sqrt{6}, 0\right)$

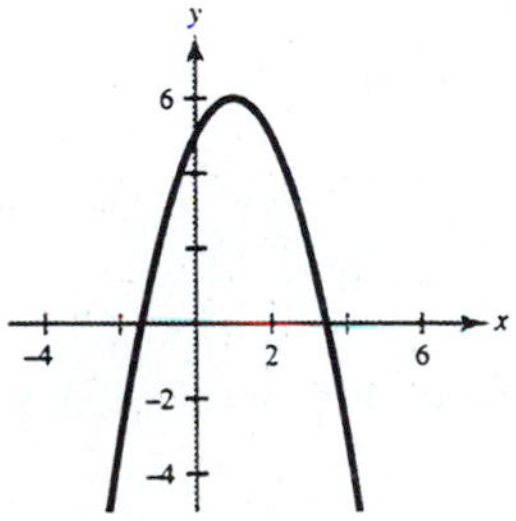

25. $h(x) = 4x^2 - 4x + 21 = 4\left(x - \frac{1}{2}\right)^2 + 20$

Vertex: $\left(\frac{1}{2}, 20\right)$

Intercept: $(0, 21)$

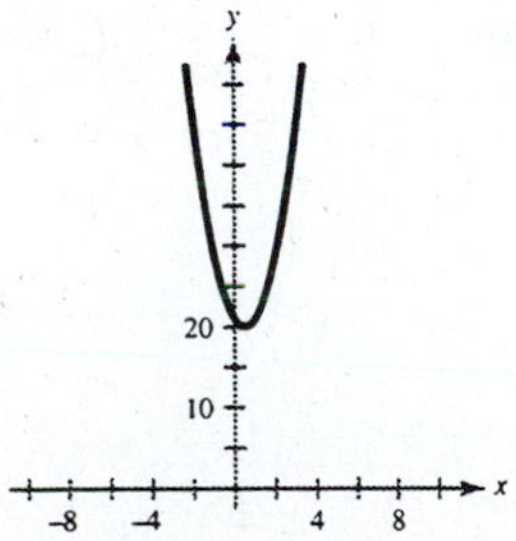

27. $f(x) = -(x^2 + 2x - 3) = -(x + 1)^2 + 4$

Vertex: $(-1, 4)$

Intercepts: $(-3, 0)$, $(0, 3)$, $(1, 0)$

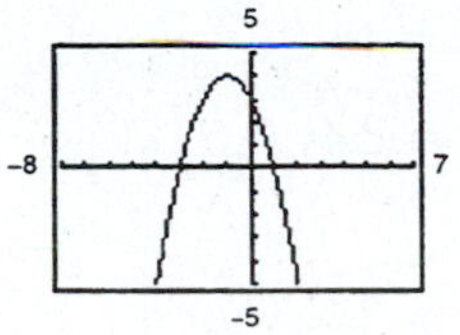

29. $f(x) = 2x^2 - 16x + 31$

$= 2(x - 4)^2 - 1$

Vertex: $(4, -1)$

Intercepts: $\left(4 \pm \frac{1}{2}\sqrt{2}, 0\right)$, $(0, 31)$

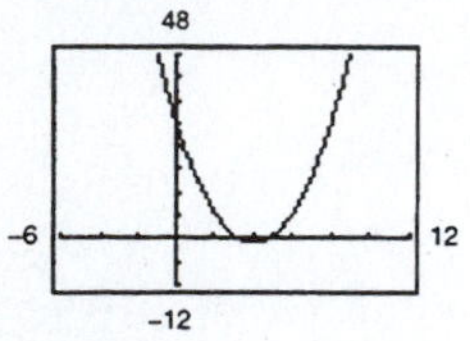

31. $(1, 0)$ is the vertex.

$f(x) = a(x - 1)^2 + 0 = a(x - 1)^2$

Since the graph passes through the point $(0, 1)$ we have:

$1 = a(0 - 1)^2$

$1 = a$

$f(x) = 1(x - 1)^2 = (x - 1)^2$

33. $(-1, 4)$ is the vertex.

$f(x) = a(x + 1)^2 + 4$

Since the graph passes through the point $(1, 0)$, we have:

$0 = a(1 + 1)^2 + 4$

$-4 = 4a$

$-1 = a$

$f(x) = -1(x + 1)^2 + 4 = -(x + 1)^2 + 4.$

35. $(-2, 2)$ is the vertex.

$f(x) = a(x + 2)^2 + 2$

Since the graph passes through the point $(-1, 0)$, we have:

$0 = a(-1 + 2)^2 + 2$

$-2 = a$

$f(x) = -2(x + 2)^2 + 2$

37. $(-2, 5)$ is the vertex.

$f(x) = a(x + 2)^2 + 5$

Since the graph passes through the point $(0, 9)$, we have:

$9 = a(0 + 2)^2 + 5$

$4 = 4a$

$1 = a$

$f(x) = 1(x + 2)^2 + 5 = (x + 2)^2 + 5$

39. $(3, 4)$ is the vertex.

$f(x) = a(x - 3)^2 + 4$

Since the graph passes through the point $(1, 2)$, we have:

$2 = a(1 - 3)^2 + 4$

$-2 = 4a$

$-\frac{1}{2} = a$

$f(x) = -\frac{1}{2}(x - 3)^2 + 4$

41. $(5, 12)$ is the vertex.

$f(x) = a(x - 5)^2 + 12$

Since the graph passes through the point $(7, 15)$, we have:

$15 = a(7 - 5)^2 + 12$

$3 = 4a \implies a = \frac{3}{4}$

$f(x) = \frac{3}{4}(x - 5)^2 + 12$

43. $y = x^2 - 16$

x-intercepts: $(\pm 4, 0)$

$0 = x^2 - 16$

$x^2 = 16$

$x = \pm 4$

45. $y = x^2 - 4x - 5$

x-intercepts: $(5, 0)$, $(-1, 0)$

$0 = x^2 - 4x - 5$

$0 = (x - 5)(x + 1)$

$x = 5$ or $x = -1$

47. $y = x^2 - 4x$

$0 = x^2 - 4x$

$0 = x(x - 4)$

$x = 0$ or $x = 4$

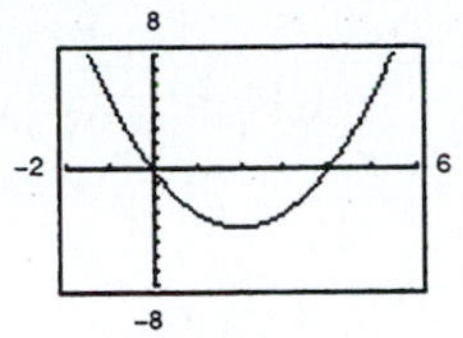

x-intercepts: $(0, 0)$, $(4, 0)$

49. $y = 2x^2 - 7x - 30$

$0 = 2x^2 - 7x - 30$

$0 = (2x + 5)(x - 6)$

$x = -\frac{5}{2}$ or $x = 6$

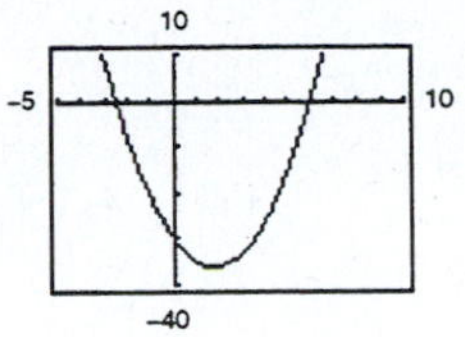

x-intercepts: $\left(-\frac{5}{2}, 0\right)$, $(6, 0)$

51. $f(x) = [x - (-1)](x - 3)$ opens upward

$= (x + 1)(x - 3)$

$= x^2 - 2x - 3$

$f(x) = -[x - (-1)](x - 3)$ opens downward

$= -(x + 1)(x - 3)$

$= -(x^2 - 2x - 3)$

$= -x^2 + 2x + 3$

Note: $f(x) = a(x + 1)(x - 3)$ has x-intercepts $(-1, 0)$ and $(3, 0)$ for all real numbers $a \neq 0$.

53. $f(x) = (x - 0)(x - 10)$ opens upward

$= x^2 - 10x$

$f(x) = -(x - 0)(x - 10)$ opens downward

$= -x^2 + 10x$

Note: $f(x) = a(x - 0)(x - 10) = ax(x - 10)$ has x-intercepts $(0, 0)$ and $(10, 0)$ for all real numbers $a \neq 0$.

55. $f(x) = [x - (-3)]\left[x - \left(-\frac{1}{2}\right)\right](2)$ opens upward

$= (x + 3)\left(x + \frac{1}{2}\right)(2)$

$= (x + 3)(2x + 1)$

$= 2x^2 + 7x + 3$

$f(x) = -(2x^2 + 7x + 3)$ opens downward

$= -2x^2 - 7x - 3$

Note: $f(x) = a(x + 3)(2x + 1)$ has x-intercepts $(-3, 0)$ and $\left(-\frac{1}{2}, 0\right)$ for all real numbers $a \neq 0$.

57. Let x = the first number and y = the second number. Then the sum is

$$x + y = 110 \Rightarrow y = 110 - x.$$

The product is $P(x) = xy = x(110 - x) = 110x - x^2$.

$$\begin{aligned} P(x) &= -x^2 + 110x \\ &= -(x^2 - 110x + 3025 - 3025) \\ &= -[(x - 55)^2 - 3025] \\ &= -(x - 55)^2 + 3025 \end{aligned}$$

The maximum value of the product occurs at the vertex of $P(x)$ and is 3025. This happens when $x = y = 55$.

59. Let x = the first number and y = the second number. Then the sum is

$$x + 2y = 24 \Rightarrow y = \frac{24 - x}{2}.$$

The product is $P(x) = xy = x\left(\frac{24 - x}{2}\right)$.

$$\begin{aligned} P(x) &= \frac{1}{2}(-x^2 + 24x) \\ &= -\frac{1}{2}(x^2 - 24x + 144 - 144) \\ &= -\frac{1}{2}[(x - 12)^2 - 144] = -\frac{1}{2}(x - 12)^2 + 72 \end{aligned}$$

The maximum value of the product occurs at the vertex of $P(x)$ and is 72. This happens when $x = 12$ and $y = (24 - 12)/2 = 6$. Thus, the numbers are 12 and 6.

61.

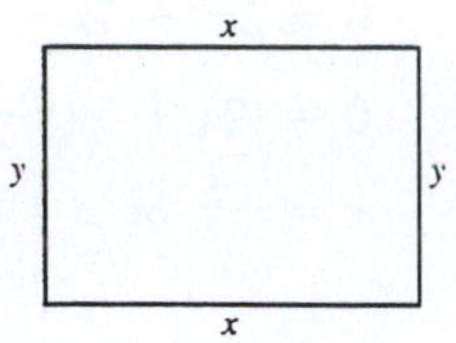

$2x + 2y = 100$

$y = 50 - x$

(a) $A(x) = xy = x(50 - x)$

Domain: $0 < x < 50$

(b)

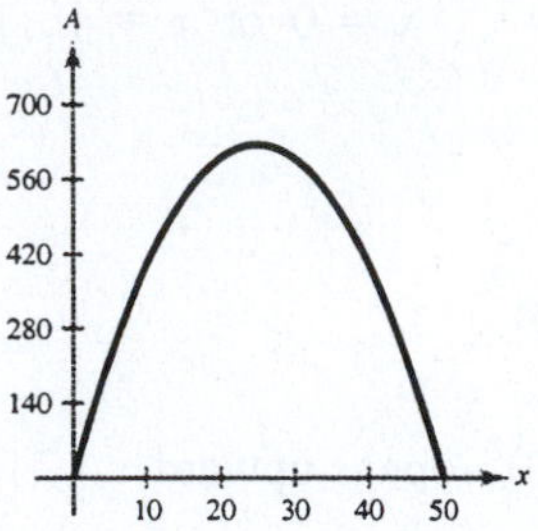

(c) The area is maximum (625 square feet) when $x = y = 25$. The rectangle has dimensions 25 ft × 25 ft.

63. (a) $4x + 3y = 200 \implies y = \frac{1}{3}(200 - 4x)$

x	y	Area
2	$\frac{1}{3}[200 - 4(2)]$	$2xy = 256$
4	$\frac{1}{3}[200 - 4(4)]$	$2xy \approx 491$
6	$\frac{1}{3}[200 - 4(6)]$	$2xy = 704$
8	$\frac{1}{3}[200 - 4(8)]$	$2xy = 896$
10	$\frac{1}{3}[200 - 4(10)]$	$2xy \approx 1067$
12	$\frac{1}{3}[200 - 4(12)]$	$2xy = 1216$

(b)

x	y	Area
20	$\frac{1}{3}[200 - 4(20)]$	$2xy = 1600$
22	$\frac{1}{3}[200 - 4(22)]$	$2xy \approx 1643$
24	$\frac{1}{3}[200 - 4(24)]$	$2xy = 1664$
26	$\frac{1}{3}[200 - 4(26)]$	$2xy = 1664$
28	$\frac{1}{3}[200 - 4(28)]$	$2xy \approx 1643$
30	$\frac{1}{3}[200 - 4(30)]$	$2xy = 1600$

(c) $A = 2xy = 2x\left(\frac{200 - 4x}{3}\right) = \frac{2x(4)(50 - x)}{3}$

$= \frac{8x(50 - x)}{3}$

(d)

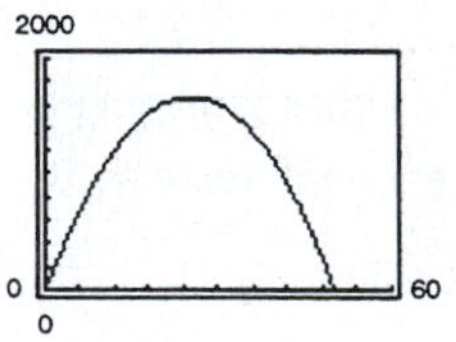

This area is maximum when $x = 25$ feet and $y = \frac{100}{3} = 33\frac{1}{3}$ feet.

(e) $A = \frac{8}{3}x(50 - x)$

$= -\frac{8}{3}(x^2 - 50x)$

$= -\frac{8}{3}(x^2 - 50x + 625 - 625)$

$= -\frac{8}{3}[(x - 25)^2 - 625]$

$= -\frac{8}{3}(x - 25)^2 + \frac{5000}{3}$

The maximum area occurs at the vertex and is 5000/3 square feet. This happens when $x = 25$ feet and $y = (200 - 4(25))/3 = 100/3$ feet. The dimensions are $2x = 50$ feet by $33\frac{1}{3}$ feet.

65. $R = 900x - 0.1x^2 = -0.1x^2 + 900x$

The vertex occurs at $x = -\frac{b}{2a} = -\frac{900}{2(-0.1)} = 4500$. The revenue is maximum when $x = 4500$ units.

67. $C = 800 - 10x + 0.25x^2 = 0.25x^2 - 10x + 800$

The vertex occurs at $x = -\dfrac{b}{2a} = -\dfrac{-10}{2(0.25)} = 20$. The cost is minimum when $x = 20$ fixtures.

69. $P = -0.0002x^2 + 140x - 250{,}000$

The vertex occurs at $x = -\dfrac{b}{2a} = -\dfrac{140}{2(-0.0002)} = 350{,}000$.

The profit is maximum when $x = 350{,}000$ units.

71. $y = -\dfrac{1}{12}x^2 + 2x + 4$

(a) When $x = 0$, $y = 4$ feet.

(b) The vertex occurs at $x = -\dfrac{b}{2a} = -\dfrac{2}{2\left(-\frac{1}{12}\right)} = 12$. The maximum height is

$$y = -\frac{1}{12}(12)^2 + 2(12) + 4 = 16 \text{ feet.}$$

(c) When the ball strikes the ground, $y = 0$.

$$0 = -\frac{1}{12}x^2 + 2x + 4$$

$$0 = x^2 - 24x - 48 \quad \text{Multiply both sides by } -12.$$

$$x = \frac{-(-24) \pm \sqrt{(-24)^2 - 4(1)(-48)}}{2(1)}$$

$$= \frac{24 \pm \sqrt{768}}{2} = \frac{24 \pm 16\sqrt{3}}{2} = 12 \pm 8\sqrt{3}$$

Using the positive value for x, we have $x = 12 + 8\sqrt{3} \approx 25.86$ feet.

73. $V = 0.77x^2 - 1.32x - 9.31, \; 5 \le x \le 40$

(a)

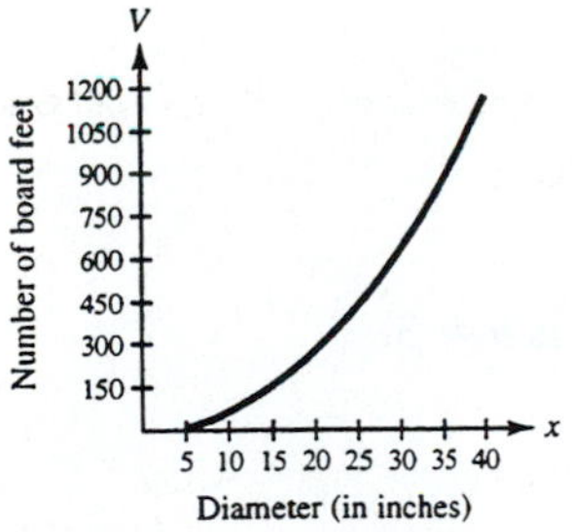

(b) $V(16) = 166.69$ board feet

(c) $500 = 0.77x^2 - 1.32x - 9.31$

$0 = 0.77x^2 - 1.32x - 509.31$

Using the Quadratic Formula and selecting the positive value for x, we have $x \approx 26.6$ inches in diameter.

75. $C = 4024.5 + 51.4t - 3.1t^2, -10 \le t \le 30$

(a)

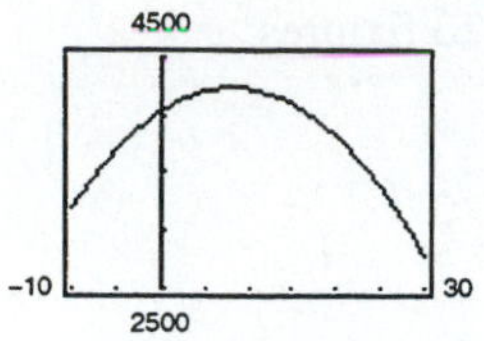

(b) $-\frac{b}{2a} = \frac{-51.4}{2(-3.1)} \approx 8.29$

The vertex occurs when $y \approx 4238$ which is the maximum average annual consumption. The warnings may not have had an immediate effect, but over time they and other findings about the health risks of cigarettes have had an effect.

(c) $C(0) = 4024.5$; annually $= \frac{116{,}530{,}000(4024.5)}{48{,}500{,}000} \approx 9670$; daily $= \frac{9670}{366} \approx 26$

77. If $f(x) = ax^2 + bx + c$ has two real zeros, then by the Quadratic Formula they are

$$x = \frac{-b \pm \sqrt{b^2 - 4ac}}{2a}.$$

The average of the zeros of f is

$$\frac{\frac{-b - \sqrt{b^2 - 4ac}}{2a} + \frac{-b + \sqrt{b^2 - 4ac}}{2a}}{2} = \frac{\frac{-2b}{2a}}{2} = -\frac{b}{2a}.$$

This is the x-coordinate of the vertex of the graph.

79. $(-4, 3)$ and $(2, 1)$

$$m = \frac{1 - 3}{2 - (-4)} = \frac{-2}{6} = -\frac{1}{3}$$

$$y - 1 = -\frac{1}{3}(x - 2)$$

$$3y - 3 = -x + 2$$

$$x + 3y - 5 = 0$$

81. $4x + 5y = 10 \implies y = -\frac{4}{5}x + 2$ and $m = -\frac{4}{5}$

The slope of the perpendicular line through $(0, 3)$ is $m = \frac{5}{4}$ and the y-intercept is $b = 3$.

$$y = \frac{5}{4}x + 3$$

$$0 = 5x - 4y + 12$$

Section 2.2 Polynomial Functions of Higher Degree

- You should know the following basic principles about polynomials.
- $f(x) = a_nx^n + a_{n-1}x^{n-1} + \cdots + a_2x^2 + a_1x + a_0$ is a polynomial function of degree n.
- If f is of odd degree and
  (a) $a_n > 0$, then
    1. $f(x) \to \infty$ as $x \to \infty$.
    2. $f(x) \to -\infty$ as $x \to -\infty$.
  (b) $a_n < 0$, then
    1. $f(x) \to -\infty$ as $x \to \infty$.
    2. $f(x) \to \infty$ as $x \to -\infty$.
- If f is of even degree and
  (a) $a_n > 0$, then
    1. $f(x) \to \infty$ as $x \to \infty$.
    2. $f(x) \to \infty$ as $x \to -\infty$.
  (b) $a_n < 0$, then
    1. $f(x) \to -\infty$ as $x \to \infty$.
    2. $f(x) \to -\infty$ as $x \to -\infty$.
- The following are equivalent for a polynomial function.
  (a) $x = a$ is a zero of a function.
  (b) $x = a$ is a solution of the polynomial equation $f(x) = 0$.
  (c) $(x - a)$ is a factor of the polynomial.
  (d) $(a, 0)$ is an x-intercept of the graph of f.
- A polynomial of degree n has at most n distinct zeros.
- If f is a polynomial function such that $a < b$ and $f(a) \neq f(b)$, then f takes on every value between $f(a)$ and $f(b)$ in the interval $[a, b]$.
- If you can find a value where a polynomial is positive and another value where it is negative, then there is at least one real zero between the values.

Solutions to Odd-Numbered Exercises

1. $f(x) = -2x + 3$ is a line with y-intercept $(0, 3)$. Matches graph (c).

3. $f(x) = -2x^2 - 5x$ is a parabola with x-intercepts $(0, 0)$ and $\left(-\frac{5}{2}, 0\right)$ and opens downward. Matches graph (h).

5. $f(x) = -\frac{1}{4}x^4 + 3x^2$ has intercepts $(0, 0)$ and $\left(\pm 2\sqrt{3}, 0\right)$. Matches graph (a).

7. $f(x) = x^4 + 2x^3$ has intercepts $(0, 0)$ and $(-2, 0)$. Matches graph (d).

9. $y = x^3$

(a) $f(x) = (x - 2)^3$

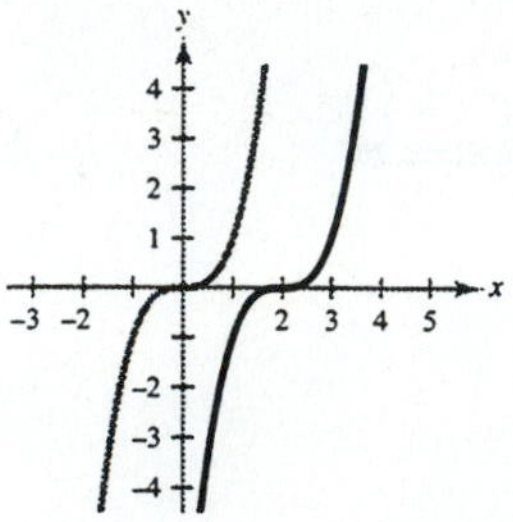

Horizontal shift two units to the right

(b) $f(x) = x^3 - 2$

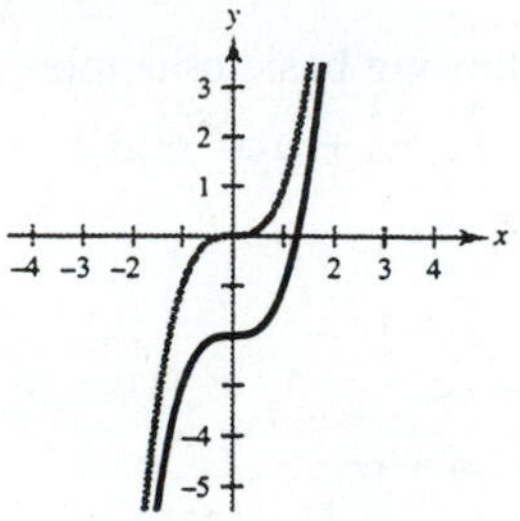

Vertical shift two units downward

(c) $f(x) = -\frac{1}{2}x^3$

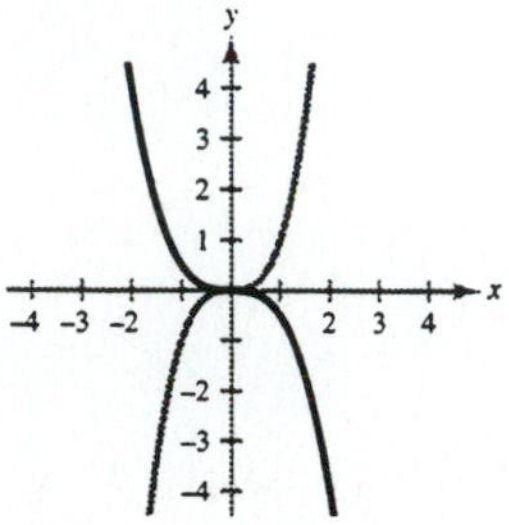

Reflection in the x-axis and a vertical shrink

(d) $f(x) = (x - 2)^3 - 2$

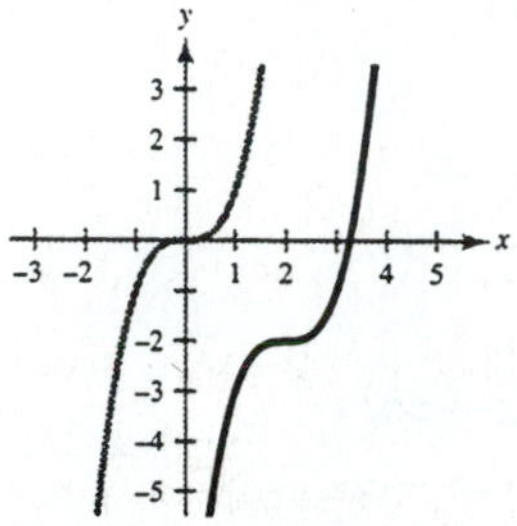

Horizontal shift two units to the right and a vertical shift two units downward

11. $y = x^4$

(a) $f(x) = (x + 3)^4$

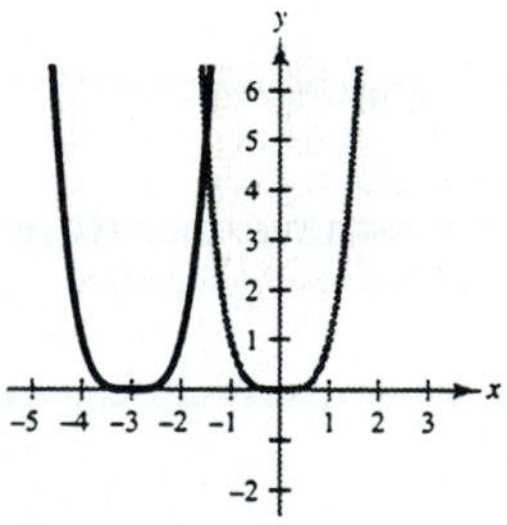

Horizontal shift three units to the left

(b) $f(x) = x^4 - 3$

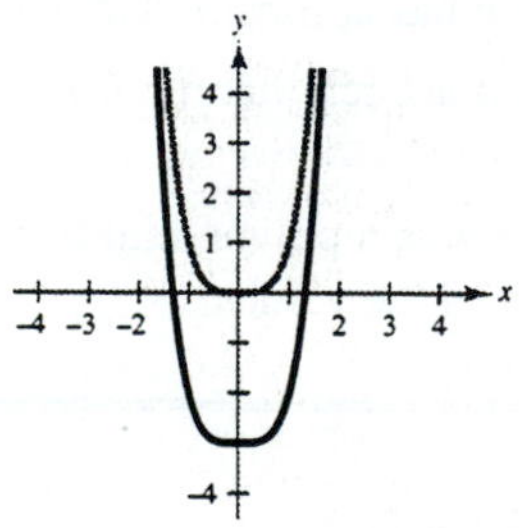

Vertical shift three units downward

(c) $f(x) = 4 - x^4$

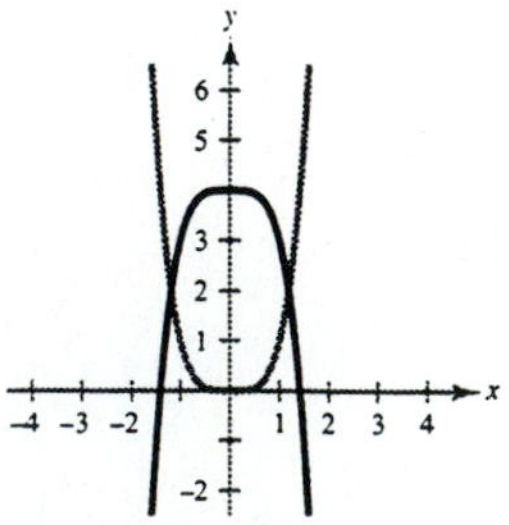

Reflection in the x-axis and then a vertical shift four units upward

(d) $f(x) = \frac{1}{2}(x - 1)^4$

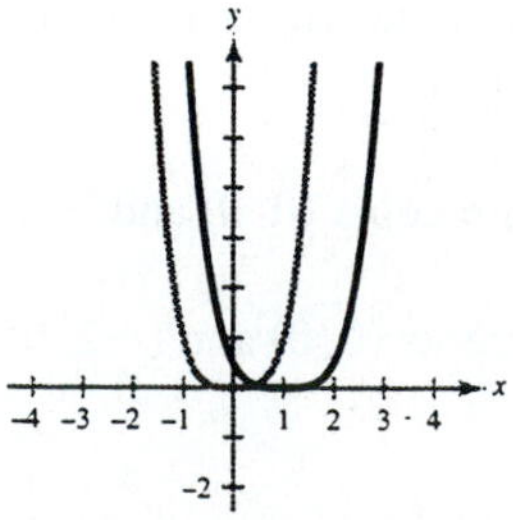

Horizontal shift one unit to the right and a vertical shrink

13. $f(x) = \frac{1}{3}x^3 + 5x$

Degree: 3

Leading coefficient: $\frac{1}{3}$

The degree is odd and the leading coefficient is positive. The graph falls to the left and rises to the right.

15. $g(x) = 5 - \frac{7}{2}x - 3x^2$

Degree: 2

Leading coefficient: -3

The degree is even and the leading coefficient is negative. The graph falls to the left and right.

17. $f(x) = -2.1x^5 + 4x^3 - 2$

Degree: 5

Leading coefficient: -2.1

The degree is odd and the leading coefficient is negative. The graph rises to the left and falls to the right.

19. $f(x) = 6 - 2x + 4x^2 - 5x^3$

Degree: 3

Leading coefficient: -5

The degree is odd and the leading coefficient is negative. The graph rises to the left and falls to the right.

21. $h(t) = -\frac{2}{3}(t^2 - 5t + 3)$

Degree: 2

Leading coefficient: $-\frac{2}{3}$

The degree is even and the leading coefficient is negative. The graph falls to the left and right.

23. $f(x) = 3x^3 - 9x + 1;\ g(x) = 3x^3$

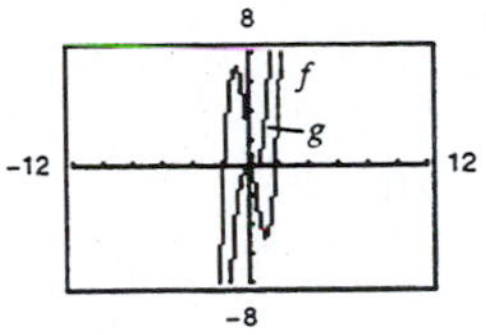

25. $f(x) = -(x^4 - 4x^3 + 16x);\ g(x) = -x^4$

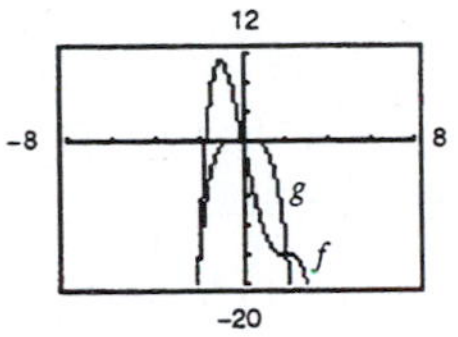

27. $f(x) = x^2 - 25$

$0 = (x + 5)(x - 5)$

$x = \pm 5$

29. $h(t) = t^2 - 6t + 9$

$0 = (t - 3)^2$

$t = 3$

31. $f(x) = x^2 + x - 2$

$0 = (x + 2)(x - 1)$

$x = -2, 1$

33. $f(x) = 3x^2 - 12x + 3$

$0 = 3(x^2 - 4x + 1)$

$$x = \frac{4 \pm \sqrt{16 - 4}}{2} = 2 \pm \sqrt{3}$$

35. $f(t) = t^3 - 4t^2 + 4t$

$0 = t(t - 2)^2$

$t = 0, 2$

37. $g(t) = \frac{1}{2}t^4 - \frac{1}{2}$

$0 = \frac{1}{2}(t + 1)(t - 1)(t^2 + 1)$

$t = \pm 1$

39. $f(x) = 2x^4 - 2x^2 - 40$

$0 = 2(x^2 + 4)(x + \sqrt{5})(x - \sqrt{5})$

$x = \pm\sqrt{5}$

41. $f(x) = 5x^4 + 15x^2 + 10$

$0 = 5(x^4 + 3x^2 + 2)$

$0 = 5(x^2 + 2)(x^2 + 1)$

No real zeros

43. $y = 4x^3 - 20x^2 + 25x$

$0 = 4x^3 - 20x^2 + 25x$

$0 = x(2x - 5)^2$

$x = 0$ or $x = \frac{5}{2}$

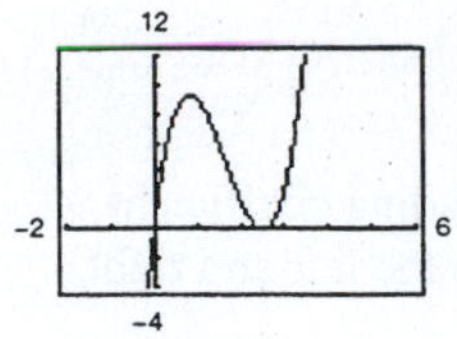

x-intercepts: $(0, 0), (\frac{5}{2}, 0)$

45. $y = x^5 - 5x^3 + 4x$

$0 = x^5 - 5x^3 + 4x$

$0 = x(x^2 - 1)(x^2 - 4)$

$0 = x(x + 1)(x - 1)(x + 2)(x - 2)$

$x = 0, \pm 1, \pm 2$

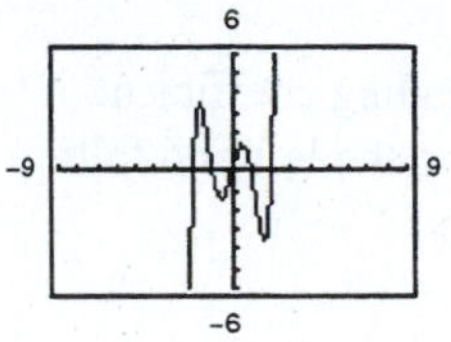

x-intercepts: $(0, 0), (\pm 1, 0), (\pm 2, 0)$

47. $f(x) = (x - 0)(x - 10)$

$f(x) = x^2 - 10x$

Note: $f(x) = a(x - 0)(x - 10) = ax(x - 10)$ has zeros 0 and 10 for all real numbers $a \neq 0$.

49. $f(x) = (x - 2)(x - (-6))$

$= (x - 2)(x + 6)$

$= x^2 + 4x - 12$

Note: $f(x) = a(x - 2)(x + 6)$ has zeros 2 and -6 for all real numbers $a \neq 0$.

51. $f(x) = (x - 0)(x - (-2))(x - (-3))$

$= x(x + 2)(x + 3)$

$= x^3 + 5x^2 + 6x$

Note: $f(x) = ax(x + 2)(x + 3)$ has zeros $0, -2, -3$ for all real numbers $a \neq 0$.

53. $f(x) = (x - 4)(x + 3)(x - 3)(x - 0)$

$= (x - 4)(x^2 - 9)x$

$= x^4 - 4x^3 - 9x^2 + 36x$

Note: $f(x) = a(x^4 - 4x^3 - 9x^2 + 36x)$ has these zeros for all real numbers $a \neq 0$.

55. $f(x) = \left[x - \left(1 + \sqrt{3}\right)\right]\left[x - \left(1 - \sqrt{3}\right)\right]$

$= \left[(x - 1) - \sqrt{3}\right]\left[(x - 1) + \sqrt{3}\right]$

$= (x - 1)^2 - \left(\sqrt{3}\right)^2$

$= x^2 - 2x + 1 - 3$

$= x^2 - 2x - 2$

Note: $f(x) = a(x^2 - 2x - 2)$ has these zeros for all real numbers $a \neq 0$.

57. $f(x) = x^3 - 3x^2 + 3$

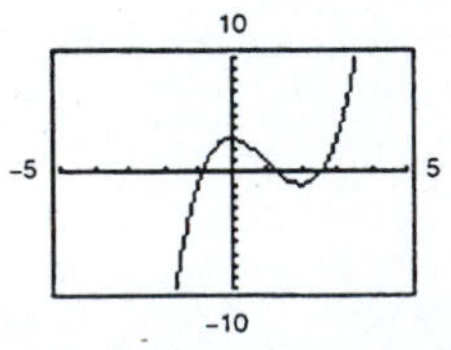

The functions has three zeros. They are in the intervals $(-1, 0)$, $(1, 2)$ and $(2, 3)$.

59. $g(x) = 3x^4 + 4x^3 - 3$

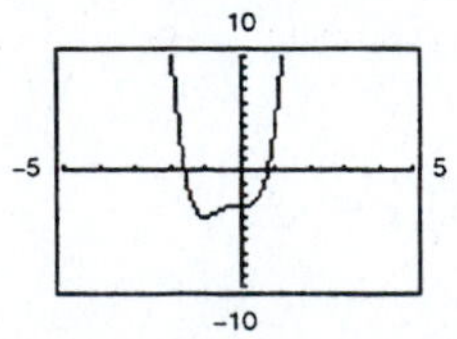

The function has two zeros. They are in the intervals $(-2, -1)$ and $(0, 1)$.

61. $f(x) = -\frac{3}{2}$

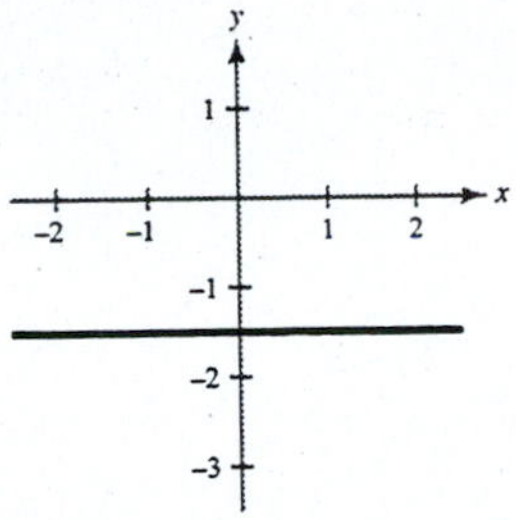

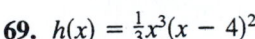

Horizontal line

63. $f(t) = \frac{1}{4}(t^2 - 2t + 15)$

$= \frac{1}{4}(t - 1)^2 + \frac{7}{2}$

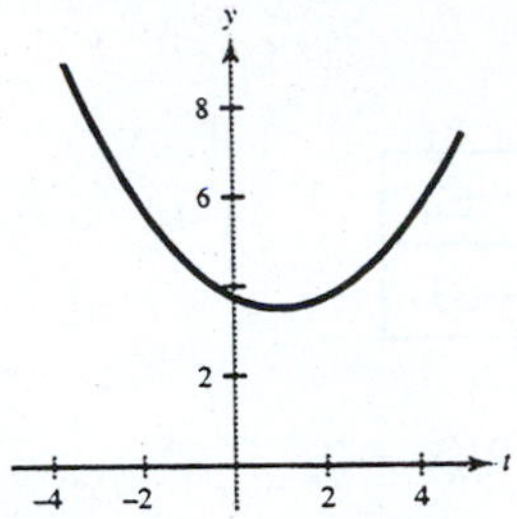

Parabola; opens upward

Vertex: $\left(1, \frac{7}{2}\right)$

65. $f(x) = x^3 - 3x^2 = x^2(x - 3)$

Zeros: 0 and 3

Right: Moves up

Left: Moves down

x	0	1	2	3	-1
$f(x)$	0	-2	-4	0	-4

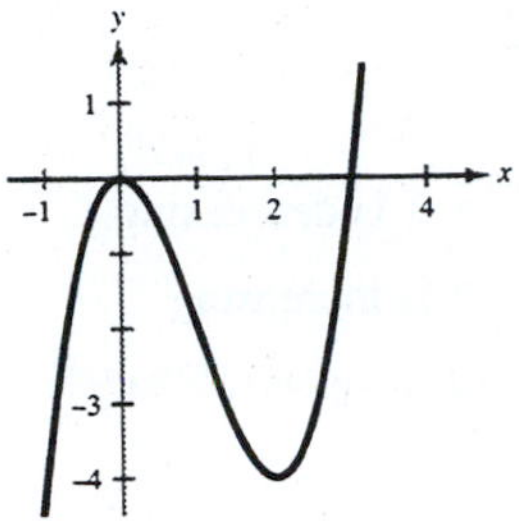

67. $g(t) = -\frac{1}{4}(t - 2)^2(t + 2)^2$

Zeros: 2 and -2

Right: Moves down

Left: Moves down

t	-3	-2	-1	0	1	2	3
$g(t)$	$-\frac{25}{4}$	0	$-\frac{9}{4}$	-4	$-\frac{9}{4}$	0	$-\frac{25}{4}$

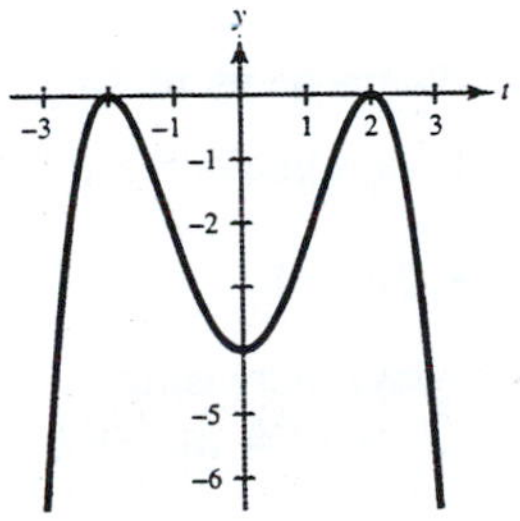

69. $h(x) = \frac{1}{3}x^3(x - 4)^2$

Zeros: 0 and 4

Right: Moves up

Left: Moves down

x	-1	0	1	2	3	4	5
$h(x)$	$-\frac{25}{3}$	0	3	$\frac{32}{3}$	9	0	$\frac{125}{3}$

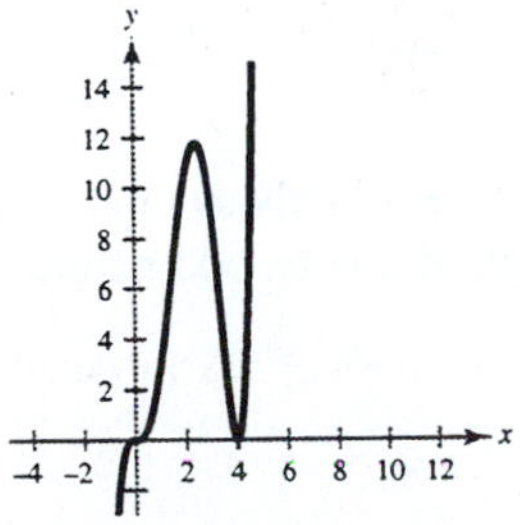

71. $f(x) = 1 - x^6$

x-intercepts: $(\pm 1, 0)$

y-intercept: $(0, 1)$

x	0	± 1	± 2
$f(x)$	1	0	-63

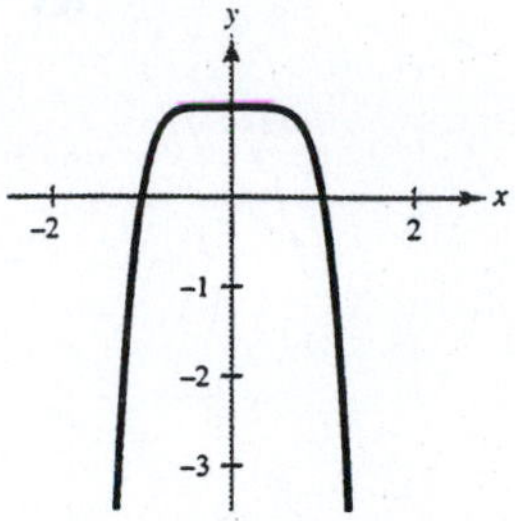

73. $f(x) = x^3 - 4x = x(x + 2)(x - 2)$

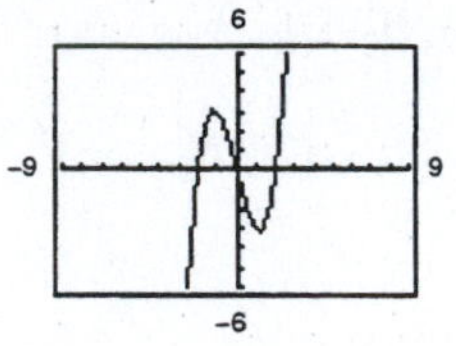

75. $g(x) = \frac{1}{5}(x + 1)^2(x - 3)(2x - 9)$

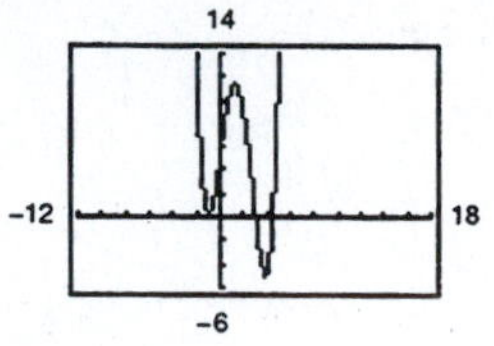

77. (a) $y_1 = -\frac{1}{3}(x - 2)^5 + 1$ is decreasing.

$y_2 = \frac{3}{5}(x + 2)^5 - 3$ is increasing.

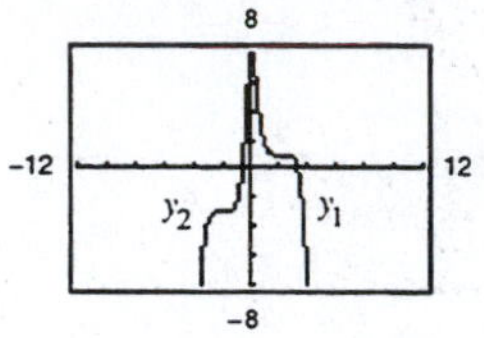

(b) If $a > 0$, $g(x)$ will always be increasing.

If $a < 0$, $g(x)$ will always be decreasing.

(c) $H(x) = x^5 - 3x^3 + 2x + 1$

Since $H(x)$ is not always increasing or always decreasing, $H(x) \neq a(x - h)^5 + k$.

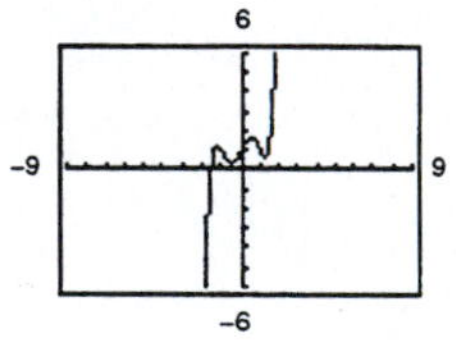

(d) $g(x) = a(x - h)^n + k$ is always increasing or always decreasing if n is an odd natural number.

79. $f(x) = x^4$; $f(x)$ is even.

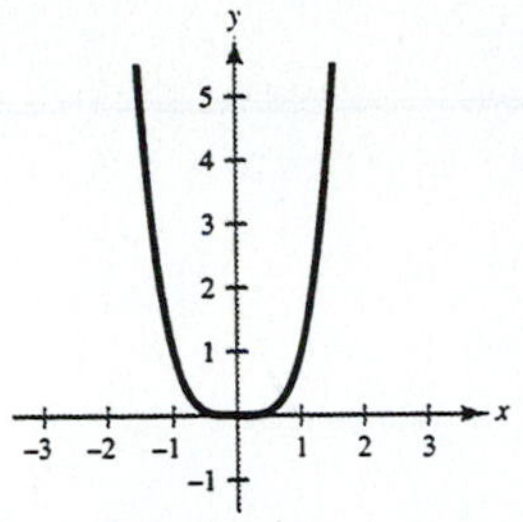

(a) $g(x) = f(x) + 2$

Vertical shift two units upward

$g(-x) = f(-x) + 2$

$= f(x) + 2$

$= g(x)$

Even

(b) $g(x) = f(x + 2)$

Horizontal shift two units to the left

Neither odd nor even

(c) $g(x) = f(-x) = (-x)^4 = x^4$

Reflection in the y-axis. The graph looks the same.
Even

(d) $g(x) = -f(x) = -x^4$

Reflection in the x-axis
Even

(e) $g(x) = f(\frac{1}{2}x) = \frac{1}{16}x^4$

Horizontal stretch

Even

(f) $g(x) = \frac{1}{2}f(x) = \frac{1}{2}x^4$

Vertical shrink

Even

(g) $g(x) = f(x^{3/4}) = (x^{3/4})^4 = x^3$

Odd

(h) $g(x) = (f \circ f)(x) = f(f(x)) = f(x^4) = (x^4)^4 = x^{16}$

Even

81. The point of diminishing returns (where the graph changes from curving upward to curving downward) occurs when $x = 200$. The point is (200, 160) which corresponds to spending $2,000,000 on advertising to obtain a revenue of $160 million.

83. (a) $(f \circ g)(3) = f(g(3)) = f(1) = 0$

(b) The graph does not pass the horizontal line test.

(c) Since $g(3) = 1,\ g^{-1}(1) = 3$.

(d) $(g^{-1} \circ f)(0) = g^{-1}(f(0)) = g^{-1}(0) = 1$, since $g(1) = 0$.

85. $3x^2 - y = 4$

$y = 3x^2 - 4$

Yes, y is a function of x. Each x-value corresponds to only one y-value.

87. $g(x) = x^2 - 1$

$$\frac{g(x) - g(4)}{x - 4} = \frac{(x^2 - 1) - 15}{x - 4}$$

$$= \frac{x^2 - 16}{x - 4}$$

$$= \frac{(x + 4)(x - 4)}{x - 4}$$

$$= x + 4, \quad x \neq 4$$

Section 2.3 Polynomial and Synthetic Division

You should know the following basic techniques and principles of polynomial division.

- The Division Algorithm (Long Division of Polynomials)
- Synthetic Division
- $f(k)$ is equal to the remainder of $f(x)$ divided by $(x - k)$.
- $f(k) = 0$ if and only if $(x - k)$ is a factor of $f(x)$.

Solutions to Odd-Numbered Exercises

1. $$\begin{aligned} y_2 &= 4 + \frac{4}{x-1} \\ &= \frac{4(x-1)+4}{x-1} \\ &= \frac{4x-4+4}{x-1} \\ &= \frac{4x}{x-1} \\ &= y_1 \end{aligned}$$

3. $$\begin{aligned} y_2 &= x - 2 + \frac{4}{x+2} \\ &= \frac{(x-2)(x+2)+4}{x+2} \\ &= \frac{x^2-4+4}{x+2} \\ &= \frac{x^2}{x+2} \\ &= y_1 \end{aligned}$$

5. $$\begin{aligned} y_2 &= x^3 - 4x + \frac{4x}{x^2+1} \\ &= \frac{(x^3-4x)(x^2+1)+4x}{x^2+1} \\ &= \frac{x^5+x^3-4x^3-4x+4x}{x^2+1} \\ &= \frac{x^5-3x^3}{x^2+1} \end{aligned}$$

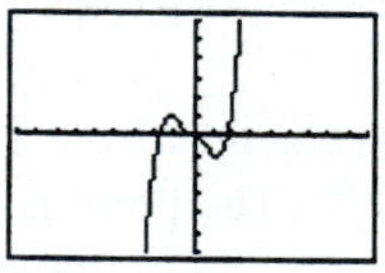

7. $$\begin{array}{r} 2x + 4 \\ x+3 \overline{\smash{)}\, 2x^2 + 10x + 12} \\ -(2x^2 + 6x) \\ \hline 4x + 12 \\ -(4x + 12) \\ \hline 0 \end{array}$$

$$\frac{2x^2 + 10x + 12}{x+3} = 2x + 4$$

9. $$\begin{array}{r} x^2 - 3x + 1 \\ 4x+5 \overline{\smash{)}\, 4x^3 - 7x^2 - 11x + 5} \\ -(4x^3 + 5x^2) \\ \hline -12x^2 - 11x \\ -(-12x^2 - 15x) \\ \hline 4x + 5 \\ -(4x + 5) \\ \hline 0 \end{array}$$

$$\frac{4x^3 - 7x^2 - 11x + 5}{4x+5} = x^2 - 3x + 1$$

11.

$$\begin{array}{r}
x^3 + 3x^2 - 1 \\
x + 2 \overline{) \, x^4 + 5x^3 + 6x^2 - x - 2} \\
-(x^4 + 2x^3) \\
\hline
3x^3 + 6x^2 \\
-(3x^3 + 6x^2) \\
\hline
-x - 2 \\
-(-x - 2) \\
\hline
0
\end{array}$$

$$\frac{x^4 + 5x^3 + 6x^2 - x - 2}{x + 2} = x^3 + 3x^2 - 1$$

13.

$$\begin{array}{r}
7 \\
x + 2 \overline{) \, 7x + 3} \\
-(7x + 14) \\
\hline
-11
\end{array}$$

$$\frac{7x + 3}{x + 2} = 7 - \frac{11}{x + 2}$$

15.

$$\begin{array}{r}
3x + 5 \\
2x^2 + 0x + 1 \overline{) \, 6x^3 + 10x^2 + x + 8} \\
-(6x^3 + 0x^2 + 3x) \\
\hline
10x^2 - 2x + 8 \\
-(10x^2 + 0x + 5) \\
\hline
-2x + 3
\end{array}$$

$$\frac{6x^3 + 10x^2 + x + 8}{2x^2 + 1} = 3x + 5 - \frac{2x - 3}{2x^2 + 1}$$

17.

$$\begin{array}{r}
x^2 + 2x + 4 \\
x^2 - 2x + 3 \overline{) \, x^4 + 0x^3 + 3x^2 + 0x + 1} \\
-(x^4 - 2x^3 + 3x^2) \\
\hline
2x^3 + 0x^2 + 0x \\
-(2x^3 - 4x^2 + 6x) \\
\hline
4x^2 - 6x + 1 \\
-(4x^2 - 8x + 12) \\
\hline
2x - 11
\end{array}
\Rightarrow$$

$$\frac{x^4 + 3x^2 + 1}{x^2 - 2x + 3} = x^2 + 2x + 4 + \frac{2x - 11}{x^2 - 2x + 3}$$

19.

$$\begin{array}{r}
x + 3 \\
x^3 - 3x^2 + 3x - 1 \overline{) \, x^4 + 0x^3 + 0x^2 + 0x + 0} \\
-(x^4 - 3x^3 + 3x^2 - x) \\
\hline
3x^3 - 3x^2 + x + 0 \\
-(3x^3 - 9x^2 + 9x - 3) \\
\hline
6x^2 - 8x + 3
\end{array}$$

$$\frac{x^4}{(x - 1)^3} = x + 3 + \frac{6x^2 - 8x + 3}{(x - 1)^3}$$

21.
$$
\begin{array}{r}
x^{2n} + 6x^n + 9 \\
x^n + 3 \overline{\big) \; x^{3n} + 9x^{2n} + 27x^n + 27} \\
-(x^{3n} + 3x^{2n}) \qquad\qquad\quad \\
\hline
6x^{2n} + 27x^n \qquad\; \\
-(6x^{2n} + 18x^n) \qquad\; \\
\hline
9x^n + 27 \\
-(9x^n + 27) \\
\hline
0
\end{array}
$$

$$\frac{x^{3n} + 9x^{2n} + 27x^n + 27}{x^n + 3} = x^{2n} + 6x^n + 9$$

23.
$$
\begin{array}{r|rrrr}
5 & 3 & -17 & 15 & -25 \\
 & & 15 & -10 & 25 \\
\hline
 & 3 & -2 & 5 & 0
\end{array}
$$

$$\frac{3x^3 - 17x^2 + 15x - 25}{x - 5} = 3x^2 - 2x + 5$$

25.
$$
\begin{array}{r|rrrr}
-2 & 4 & 8 & -9 & -18 \\
 & & -8 & 0 & 18 \\
\hline
 & 4 & 0 & -9 & 0
\end{array}
$$

$$\frac{4x^3 + 8x^2 - 9x - 18}{x + 2} = 4x^2 - 9$$

27.
$$
\begin{array}{r|rrrr}
-10 & -1 & 0 & 75 & -250 \\
 & & 10 & -100 & 250 \\
\hline
 & -1 & 10 & -25 & 0
\end{array}
$$

$$\frac{-x^3 + 75x - 250}{x + 10} = -x^2 + 10x - 25$$

29.
$$
\begin{array}{r|rrrr}
4 & 5 & -6 & 0 & 8 \\
 & & 20 & 56 & 224 \\
\hline
 & 5 & 14 & 56 & 232
\end{array}
$$

$$\frac{5x^3 - 6x^2 + 8}{x - 4} = 5x^2 + 14x + 56 + \frac{232}{x - 4}$$

31.
$$
\begin{array}{r|rrrrr}
6 & 10 & -50 & 0 & 0 & -800 \\
 & & 60 & 60 & 360 & 2160 \\
\hline
 & 10 & 10 & 60 & 360 & 1360
\end{array}
$$

$$\frac{10^4 - 50x^3 - 800}{x - 6} = 10x^3 + 10x^2 + 60x + 360 + \frac{1360}{x - 6}$$

33.
$$
\begin{array}{r|rrrr}
-8 & 1 & 0 & 0 & 512 \\
 & & -8 & 64 & -512 \\
\hline
 & 1 & -8 & 64 & 0
\end{array}
$$

$$\frac{x^3 + 512}{x + 8} = x^2 - 8x + 64$$

35.
$$
\begin{array}{r|rrrrr}
2 & -3 & 0 & 0 & 0 & 0 \\
 & & -6 & -12 & -24 & -48 \\
\hline
 & -3 & -6 & -12 & -24 & -48
\end{array}
$$

$$\frac{-3x^4}{x - 2} = -3x^3 - 6x^2 - 12x - 24 - \frac{48}{x - 2}$$

37. $$\begin{array}{r|rrrrr} 6 & -1 & 0 & 0 & 180 & 0 \\ & & -6 & -36 & -216 & -216 \\ \hline & -1 & -6 & -36 & -36 & -216 \end{array}$$

$$\frac{180x - x^4}{x - 6} = -x^3 - 6x^2 - 36x - 36 - \frac{216}{x - 6}$$

39. $$\begin{array}{r|rrrr} -\frac{1}{2} & 4 & 16 & -23 & -15 \\ & & -2 & -7 & 15 \\ \hline & 4 & 14 & -30 & 0 \end{array}$$

$$\frac{4x^3 + 16x^2 - 23x - 15}{x + \frac{1}{2}} = 4x^2 + 14x - 30$$

41. A divisor divided evenly into the dividend if the remainder is zero.

43. $$\begin{array}{r|rrrr} 5 & 1 & 4 & -3 & c \\ & & 5 & 45 & 210 \\ \hline & 1 & 9 & 42 & c + 210 \end{array}$$

For $c + 210$ to equal zero, c must equal -210.

45. $f(x) = x^3 - x^2 - 14x + 11,\ k = 4$

$$\begin{array}{r|rrrr} 4 & 1 & -1 & -14 & 11 \\ & & 4 & 12 & -8 \\ \hline & 1 & 3 & -2 & 3 \end{array}$$

$f(x) = (x - 4)(x^2 + 3x - 2) + 3$

$f(4) = (0)(26) + 3 = 3$

47. $f(x) = x^3 + 3x^2 - 2x - 14,\ k = \sqrt{2}$

$$\begin{array}{r|rrrr} \sqrt{2} & 1 & 3 & -2 & -14 \\ & & \sqrt{2} & 2 + 3\sqrt{2} & 6 \\ \hline & 1 & 3 + \sqrt{2} & 3\sqrt{2} & -8 \end{array}$$

$f(x) = (x - \sqrt{2})\left[x^2 + (3 + \sqrt{2})x + 3\sqrt{2}\right] - 8$

$f(\sqrt{2}) = (0)(4 + 6\sqrt{2}) - 8 = -8$

49. $f(x) = 4x^3 - 13x + 10$

(a) $$\begin{array}{r|rrrr} 1 & 4 & 0 & -13 & 10 \\ & & 4 & 4 & -9 \\ \hline & 4 & 4 & -9 & \underline{1} = f(1) \end{array}$$

(b) $$\begin{array}{r|rrrr} -2 & 4 & 0 & -13 & 10 \\ & & -8 & 16 & -6 \\ \hline & 4 & -8 & 3 & 4 \quad \underline{4} = f(-2) \end{array}$$

(c) $$\begin{array}{r|rrrr} \frac{1}{2} & 4 & 0 & -13 & 10 \\ & & 2 & 1 & -6 \\ \hline & 4 & 2 & -12 & \underline{4} = f\left(\frac{1}{2}\right) \end{array}$$

(d) $$\begin{array}{r|rrrr} 8 & 4 & 0 & -13 & 10 \\ & & 32 & 256 & 1944 \\ \hline & 4 & 32 & 243 & \underline{1954} = f(8) \end{array}$$

51. $h(x) = 3x^3 + 5x^2 - 10x + 1$

(a) $$\begin{array}{r|rrrr} 3 & 3 & 5 & -10 & 1 \\ & & 9 & 42 & 96 \\ \hline & 3 & 14 & 32 & \underline{97} = h(3) \end{array}$$

(b) $$\begin{array}{r|rrrr} \frac{1}{3} & 3 & 5 & -10 & 1 \\ & & 1 & 2 & -\frac{8}{3} \\ \hline & 3 & 6 & -8 & \underline{-\frac{5}{3}} = h\left(\frac{1}{3}\right) \end{array}$$

(c) $$\begin{array}{r|rrrr} -2 & 3 & 5 & -10 & 1 \\ & & -6 & 2 & 16 \\ \hline & 3 & -1 & -8 & \underline{17} = h(-2) \end{array}$$

(d) $$\begin{array}{r|rrrr} -5 & 3 & 5 & -10 & 1 \\ & & -15 & 50 & -200 \\ \hline & 3 & -10 & 40 & \underline{-199} = h(-5) \end{array}$$

53. $f(x) = (x+3)^2(x-3)(x+1)^3$

The remainder when $k = -3$ is zero since $(x+3)$ is a factor of $f(x)$.

55.
$$\begin{array}{r|rrrr} 2 & 1 & 0 & -7 & 6 \\ & & 2 & 4 & -6 \\ \hline & 1 & 2 & -3 & 0 \end{array}$$

$$\begin{aligned} x^3 - 7x + 6 &= (x-2)(x^2+2x-3) \\ &= (x-2)(x+3)(x-1) \end{aligned}$$

Zeros: $2, -3, 1$

57.
$$\begin{array}{r|rrrr} \frac{1}{2} & 2 & -15 & 27 & -10 \\ & & 1 & -7 & 10 \\ \hline & 2 & -14 & 20 & 0 \end{array}$$

$$\begin{aligned} 2x^3 - 15x^2 + 27x - 10 & \\ &= \left(x - \tfrac{1}{2}\right)(2x^2 - 14x + 20) \\ &= (2x-1)(x-2)(x-5) \end{aligned}$$

Zeros: $\frac{1}{2}, 2, 5$

59.
$$\begin{array}{r|rrrr} \sqrt{3} & 1 & 2 & -3 & -6 \\ & & \sqrt{3} & 3+2\sqrt{3} & 6 \\ \hline & 1 & 2+\sqrt{3} & 2\sqrt{3} & 0 \end{array}$$

$$\begin{array}{r|rrr} -\sqrt{3} & 1 & 2+\sqrt{3} & 2\sqrt{3} \\ & & -\sqrt{3} & -2\sqrt{3} \\ \hline & 1 & 2 & 0 \end{array}$$

$$x^3 + 2x^2 - 3x - 6 = \left(x - \sqrt{3}\right)\left(x + \sqrt{3}\right)(x+2)$$

Zeros: $\pm\sqrt{3}, -2$

61.
$$\begin{array}{r|rrrr} 1+\sqrt{3} & 1 & -3 & 0 & 2 \\ & & 1+\sqrt{3} & 1-\sqrt{3} & -2 \\ \hline & 1 & -2+\sqrt{3} & 1-\sqrt{3} & 0 \end{array}$$

$$\begin{array}{r|rrr} 1-\sqrt{3} & 1 & -2+\sqrt{3} & 1-\sqrt{3} \\ & & 1-\sqrt{3} & -1+\sqrt{3} \\ \hline & 1 & -1 & 0 \end{array}$$

$$\begin{aligned} x^3 - 3x^2 + 2 &= \left[x - \left(1+\sqrt{3}\right)\right]\left[x - \left(1-\sqrt{3}\right)\right](x-1) \\ &= (x-1)\left(x - 1 - \sqrt{3}\right)\left(x - 1 + \sqrt{3}\right) \end{aligned}$$

Zeros: $1, 1 \pm \sqrt{3}$

63. $f(x) = x^3 - 2x^2 - 5x + 10$

(a) The zeros of f are 2 and $\approx \pm 2.236$.

(b)
$$\begin{array}{r|rrrr} 2 & 1 & -2 & -5 & 10 \\ & & 2 & 0 & -10 \\ \hline & 1 & 0 & -5 & 0 \end{array}$$

$$\begin{aligned} f(x) &= (x-2)(x^2-5) \\ &= (x-2)(x-\sqrt{5})(x+\sqrt{5}) \end{aligned}$$

65. $h(t) = t^3 - 2t^2 - 7t + 2$

(a) The zeros of h are -2, ≈ 3.732, ≈ 0.268.

(b)
$$\begin{array}{r|rrrr} -2 & 1 & -2 & -7 & 2 \\ & & -2 & 8 & -2 \\ \hline & 1 & -4 & 1 & 0 \end{array}$$

$h(t) = (t+2)(t^2 - 4t + 1)$

By the Quadratic Formula, the zeros of

$t^2 - 4t + 1$ are $2 \pm \sqrt{3}$.

Thus, $h(t) = (t+2)\left[t - \left(2+\sqrt{3}\right)\right]\left[t - \left(2-\sqrt{3}\right)\right]$.

67.
$$\begin{array}{r|rrrr} \frac{3}{2} & 4 & -8 & 1 & 3 \\ & & 6 & -3 & -3 \\ \hline & 4 & -2 & -2 & 0 \end{array}$$

$$\begin{aligned} 4x^3 - 8x^2 + x + 3 &= \left(x - \tfrac{3}{2}\right)(4x^2 - 2x - 2) \\ &= \left(x - \tfrac{3}{2}\right)(2)(2x^2 - x - 1) \\ &= (2x-3)(2x^2 - x - 1) \end{aligned}$$

Thus,

$$\frac{4x^3 - 8x^2 + x + 3}{2x - 3} = 2x^2 - x - 1 = (2x+1)(x-1).$$

69.
$$\begin{array}{r|rrrr} -1 & 1 & 3 & -1 & -3 \\ & & -1 & -2 & 3 \\ \hline & 1 & 2 & -3 & 0 \end{array}$$

$$\begin{aligned} \frac{x^3 + 3x^2 - x - 3}{x+1} &= x^2 + 2x - 3 \\ &= (x+3)(x-1) \end{aligned}$$

71.
$$\begin{array}{r|rrrrr} -1 & 1 & 6 & 11 & 6 & 0 \\ & & -1 & -5 & -6 & 0 \\ \hline & 1 & 5 & 6 & 0 & 0 \end{array}$$

$$\begin{array}{r|rrrr} -2 & 1 & 5 & 6 & 0 \\ & & -2 & -6 & 0 \\ \hline & 1 & 3 & 0 & 0 \end{array}$$

$$\begin{aligned} \frac{x^4 + 6x^3 + 11x^2 + 6x}{(x+1)(x+2)} &= x^2 + 3x \\ &= x(x+3) \end{aligned}$$

73. Since $y = 110$ when $x = 5$, we have $x = 5$ as a zero to the equation $f(x) = y - 110$.

$$f(x) = y - 110, \ 1 \le x \le 5$$
$$= -1.42x^3 + 5.04x^2 + 32.45x - 110.75$$

$$\begin{array}{r|rrrr} 5 & -1.42 & 5.04 & 32.45 & -110.75 \\ & & -7.10 & -10.30 & 110.75 \\ \hline & -1.42 & -2.06 & 22.15 & 0 \end{array}$$

Thus, $f(x) = (x - 5)(-1.42x^2 - 2.06x + 22.15)$. Using the Quadratic Formula to find the other two zeros yields

$$x = \frac{-(-2.06) \pm \sqrt{(-2.06)^2 - 4(-1.42)(22.15)}}{2(-1.42)}$$

$$= \frac{2.06 \pm \sqrt{130.0556}}{-2.84}$$

$x \approx -4.74$ or $x \approx 3.29$.

Since $1 \le x \le 5$, we choose $x = 3.29$ which corresponds to 3290 rpm.

75. $f(x) = (x - k)q(x) + r$

(a) $k = 2, \ r = 5, \ q(x) =$ any quadratic $ax^2 + bx + c$ where $a > 0$.

One example: $f(x) = (x - 2)x^2 + 5 = x^3 - 2x^2 + 5$

(b) $k = -3, \ r = 1, \ q(x) =$ any quadratic $ax^2 + bx + c$ where $a < 0$.

One example: $f(x) = (x + 3)(-x^2) + 1 = -x^3 - 3x^2 + 1$

77. $f(x) = \sqrt[3]{x + 2}$

$y = \sqrt[3]{x + 2}$

$x = \sqrt[3]{y + 2}$

$x^3 = y + 2$

$x^3 - 2 = y$

$f^{-1}(x) = x^3 - 2$

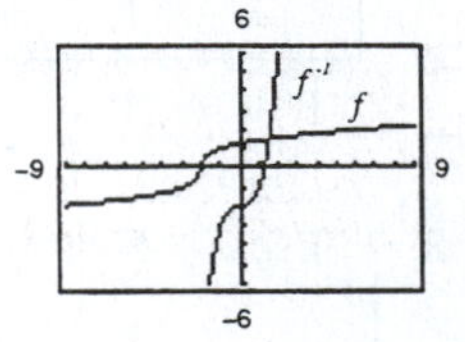

79. $f(x) = \sqrt{x}, \ g(x) = x^2 - 25$

$(f \circ g)(x) = f(g(x)) = f(x^2 - 25) = \sqrt{x^2 - 25}$

Domain: $x^2 - 25 \ge 0$

$(x + 5)(x - 5) \ge 0$

Critical numbers: $x = \pm 5$

Test intervals: $(-\infty, \ -5)(-5, \ 5)(5, \ \infty)$

Test: Is $x^2 - 25 \ge 0$?

Solution set: $[-\infty, -5] \cup [5, \infty)$ or $x \le -5, x \ge 5$

2.4 Real Zeros of Polynomial Functions

- You should know Descartes's Rule of Signs.
  - (a) The number of positive real zeros of f is either equal to the number of variations of sign of f or is less than that number by an even integer.
  - (b) The number of negative real zeros of f is either equal to the number of variations in sign of $f(-x)$ or is less than that number by an even integer.
  - (c) When there is only one variation in sign, there is exactly one positive (or negative) real zero.
- You should know the Rational Zero Test.
- You should know shortcuts for the Rational Zero Test.
  - (a) Use a graphing or programmable calculator.
  - (b) Sketch a graph.
  - (c) After finding a root, use synthetic division to reduce the degree of the polynomial.
- You should be able to observe the last row obtained from synthetic division in order to determine upper or lower bounds.
  - (a) If the test value is positive and all of the entries in the last row are positive or zero, then the test value is an upper bound.
  - (b) If the test value is negative and the entries in the last row alternate from positive to negative, then the test value is a lower bound. (Zero entries count as positive or negative.)

Solutions to Odd-Numbered Exercises

1. $f(x) = x^3 + 3$

Sign variations: 0, positive zeros: 0

$f(-x) = -x^3 + 3$

Sign variations: 1, negative zeros: 1

3. $g(x) = 5x^5 + 10x = 5x(x^4 + 2)$

Let $f(x) = x^4 + 2$.

Sign variations: 0, positive zeros: 0

$f(-x) = x^4 + 2$

Sign variations: 0, negative zeros: 0

5. $h(x) = 3x^4 + 2x^2 + 1$

Sign variations: 0, positive zeros: 0

$h(-x) = 3x^4 + 2x^2 + 1$

Sign variations: 0, negative zeros: 0

7. $g(x) = 2x^3 - 3x^2 - 3$

Sign variations: 1, positive zeros: 1

$g(-x) = -2x^3 - 3x^2 - 3$

Sign variations: 0, negative zeros: 0

9. $f(x) = -5x^3 + x^2 - x + 5$

Sign variations: 3, positive zeros: 3 or 1

$f(-x) = 5x^3 + x^2 + x + 5$

Sign variations: 0, negative zeros: 0

11. $f(x) = x^3 + 3x^2 - x - 3$

Possible rational zeros: $\pm 1, \pm 3$

Zeros shown on graph: $-3, -1, 1$

13. $f(x) = 2x^4 - 17x^3 + 35x^2 + 9x - 45$

Possible rational zeros: $\pm 1, \pm 3, \pm 5, \pm 9, \pm 15, \pm 45, \pm\frac{1}{2}, \pm\frac{3}{2}, \pm\frac{5}{2}, \pm\frac{9}{2}, \pm\frac{15}{2}, \pm\frac{45}{2}$

Zeros shown on graph: $-1, \frac{3}{2}, 3, 5$

15. $f(x) = 20x^4 + 144x^3 - 253x^2 - 900x + 800$

Possible rational zeros: $\pm 1, \pm 2, \pm 4, \pm 5, \pm 8, \pm 10, \pm 16, \pm 20, \pm 25,$
$\pm 32, \pm 40, \pm 50, \pm 80, \pm 100, \pm 160, \pm 200, \pm 400,$
$\pm 800, \pm\frac{1}{2}, \pm\frac{5}{2}, \pm\frac{25}{2}, \pm\frac{1}{4}, \pm\frac{5}{4}, \pm\frac{25}{4}, \pm\frac{1}{5}, \pm\frac{2}{5}, \pm\frac{4}{5},$
$\pm\frac{8}{5}, \pm\frac{16}{5}, \pm\frac{32}{5}, \pm\frac{1}{10}, \pm\frac{1}{20}$

Zeros shown on graph: $-8, -\frac{5}{2}, \frac{4}{5}, \frac{5}{2}$

17. $f(x) = x^3 - 6x^2 + 11x - 6$

Possible rational zeros: $\pm 1, \pm 2, \pm 3$

1	1	−6	11	−6
		1	−5	6
	1	−5	6	0

$x^3 - 6x^2 + 11x - 6 = (x - 1)(x^2 - 5x + 6) = (x - 1)(x - 2)(x - 3).$

Thus, the real zeros are 1, 2, and 3.

19. $g(x) = x^3 - 4x^2 - x + 4 = x^2(x - 4) - 1(x - 4) = (x - 4)(x^2 - 1)$
$= (x - 4)(x - 1)(x + 1)$

Thus, the zeros of $g(x)$ are 4 and ± 1.

21. $h(t) = t^3 + 12t^2 + 21t + 10$

Possible rational zeros: $\pm 1, \pm 2, \pm 5, \pm 10$

−1	1	12	21	10
		−1	−11	−10
	1	11	10	0

$t^3 + 12t^2 + 21t + 10 = (t + 1)(t^2 + 11t + 10)$
$= (t + 1)(t + 1)(t + 10)$
$= (t + 1)^2(t + 10)$

Thus, the zeros are −1 and −10.

23. $f(x) = x^3 - 4x^2 + 5x - 2$

Possible rational zeros: $\pm 1, \pm 2$

1	1	−4	5	−2
		1	−3	2
	1	−3	2	0

$x^3 - 4x^2 + 5x - 2 = (x - 1)(x^2 - 3x + 2)$
$= (x - 1)(x - 1)(x - 2)$
$= (x - 1)^2(x - 2)$

Thus, the zeros are 1 and 2.

25. $C(x) = 2x^3 + 3x^2 - 1$

Possible rational zeros: $\pm 1, \pm\frac{1}{2}$

−1	2	3	0	−1
		−2	−1	1
	2	1	−1	0

$2x^3 + 3x^2 - 1 = (x + 1)(2x^2 + x - 1)$
$= (x + 1)(x + 1)(2x - 1)$
$= (x + 1)^2(2x - 1)$

Thus, the zeros are −1 and $\frac{1}{2}$.

27. $f(x) = 9x^4 - 9x^3 - 58x^2 + 4x + 24$

Possible rational zeros: $\pm 1, \pm 2, \pm 3, \pm 4, \pm 6, \pm 8, \pm 12, \pm 24, \pm\frac{1}{3}, \pm\frac{2}{3},$
$\pm\frac{4}{3}, \pm\frac{8}{3}, \pm\frac{1}{9}, \pm\frac{2}{9}, \pm\frac{4}{9}$

$$\begin{array}{r|rrrrr} -2 & 9 & -9 & -58 & 4 & 24 \\ & & -18 & 54 & 8 & -24 \\ \hline & 9 & -27 & -4 & 12 & 0 \end{array}$$

$$\begin{array}{r|rrrr} 3 & 9 & -27 & -4 & 12 \\ & & 27 & 0 & -12 \\ \hline & 9 & 0 & -4 & 0 \end{array}$$

$$\begin{aligned} 9x^4 - 9x^3 - 58x^2 + 4x - 24 &= (x + 2)(x - 3)(9x^2 - 4) \\ &= (x + 2)(x - 3)(3x - 2)(3x + 2) \end{aligned}$$

Thus, the zeros are -2, 3, and $\pm\frac{2}{3}$.

29. $z^4 - z^3 - 2z - 4 = 0$

Possible rational zeros: $\pm 1, \pm 2, \pm 4$

$$\begin{array}{r|rrrrr} -1 & 1 & -1 & 0 & -2 & -4 \\ & & -1 & 2 & -2 & 4 \\ \hline & 1 & -2 & 2 & -4 & 0 \end{array}$$

$$\begin{array}{r|rrrr} 2 & 1 & -2 & 2 & -4 \\ & & 2 & 0 & 4 \\ \hline & 1 & 0 & 2 & 0 \end{array}$$

$z^4 - z^3 - 2z - 4 = (x + 1)(x - 2)(x^2 + 2)$

The only real zeros are -1 and 2.

31. $2y^4 + 7y^3 - 26y^2 + 23y - 6 = 0$

Possible rational zeros: $\pm 1, \pm 2, \pm 3, \pm 6, \pm\frac{1}{2}, \pm\frac{3}{2}$

$$\begin{array}{r|rrrrr} 1 & 2 & 7 & -26 & 23 & -6 \\ & & 2 & 9 & -17 & 6 \\ \hline & 2 & 9 & -17 & 6 & 0 \end{array}$$

$$\begin{array}{r|rrrr} -6 & 2 & 9 & -17 & 6 \\ & & -12 & 18 & -6 \\ \hline & 2 & -3 & 1 & 0 \end{array}$$

$$\begin{aligned} 2y^4 + 7y^3 - 26y^2 + 23y - 6 &= (y - 1)(y + 6)(2y^2 - 3y + 1) \\ &= (y - 1)(y + 6)(2y - 1)(y - 1) \\ &= (y - 1)^2(y + 6)(2y - 1) \end{aligned}$$

The only real zeros are $1, -6$, and $\frac{1}{2}$.

33. $f(x) = x^3 + x^2 - 4x - 4$

(a) Possible rational zeros: $\pm 1, \pm 2, \pm 4$

(b)

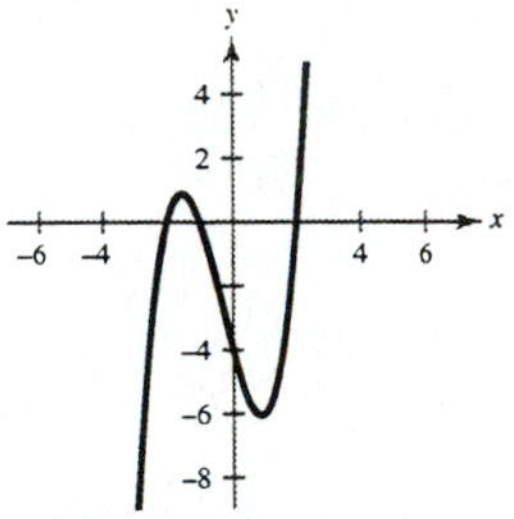

(c) $-2, -1, 2$ on graph

35. $f(x) = -4x^3 + 15x^2 - 8x - 3$

(a) Possible rational zeros: $\pm\frac{1}{4}, \pm\frac{1}{2}, \pm\frac{3}{4}, \pm 1, \pm\frac{3}{2}, \pm 3$

(b)

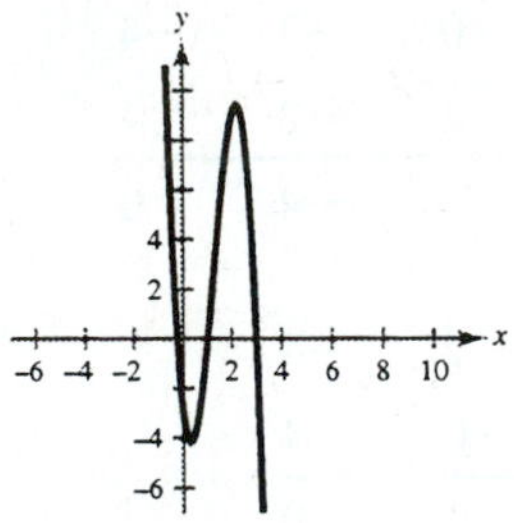

(c) $-\frac{1}{4}$, 1, 3 on graph

37. $f(x) = -2x^4 + 13x^3 - 21x^2 + 2x + 8$

(a) Possible rational zeros: $\pm\frac{1}{2}, \pm 1, \pm 2, \pm 4, \pm 8$

(b)

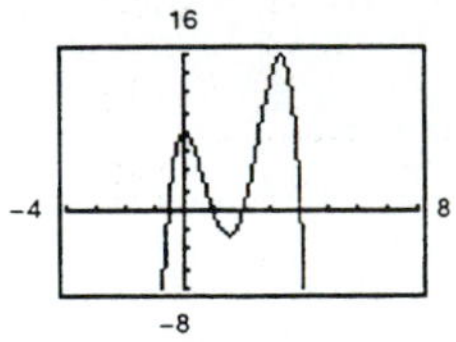

(c) $-\frac{1}{2}$, 1, 2, 4 on graph

39. $f(x) = 32x^3 - 52x^2 + 17x + 3$

(a) Possible rational zeros: $\pm 1, \pm 3, \pm\frac{1}{2}, \pm\frac{3}{2}, \pm\frac{1}{4}, \pm\frac{3}{4}, \pm\frac{1}{8}, \pm\frac{3}{8}, \pm\frac{1}{16}, \pm\frac{3}{16}, \pm\frac{1}{32}, \pm\frac{3}{32}$

(b)

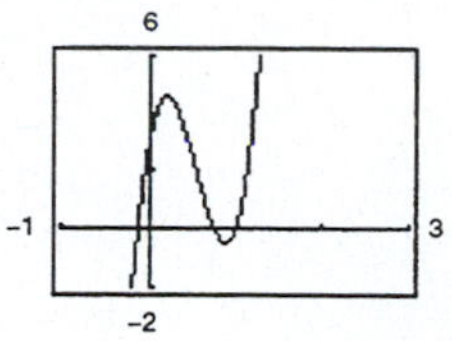

(c) $-\frac{1}{8}, \frac{3}{4}$, 1 on graph

41. $f(x) = x^4 - 3x^2 + 2$

(a) From the calculator we have

$x = \pm 1$ and $x \approx \pm 1.414$.

(b)

1	1	0	−3	0	2
		1	1	−2	−2
	1	1	−2	−2	0

−1	1	1	−2	−2
		−1	0	2
	1	0	−2	0

$$f(x) = (x - 1)(x + 1)(x^2 - 2)$$
$$= (x - 1)(x + 1)(x - \sqrt{2})(x + \sqrt{2})$$

The exact roots are $x = \pm 1, \pm\sqrt{2}$.

43. $h(x) = x^5 - 7x^4 + 10x^3 + 14x^2 - 24x$

(a) $h(x) = x(x^4 - 7x^3 + 10x^2 + 14x - 24)$

From the calculator we have

$x = 0, 3, 4$ and $x \approx \pm 1.414$.

(b)

3	1	−7	10	14	−24
		3	−12	−6	24
	1	−4	−2	8	0

4	1	−4	−2	8
		4	0	−8
	1	0	−2	0

$$f(x) = x(x - 3)(x - 4)(x^2 - 2)$$
$$= x(x - 3)(x - 4)(x - \sqrt{2})(x + \sqrt{2})$$

The exact roots are $x = 0, 3, 4, \pm\sqrt{2}$.

45. $f(x) = x^4 - 4x^3 + 15$

(a)

4	1	−4	0	0	15
		4	0	0	0
	1	0	0	0	15

4 is an upper bound.

(b)

−1	1	−4	0	0	15
		−1	5	−5	5
	1	−5	5	−5	20

−1 is a lower bound.

47. $f(x) = x^4 - 4x^3 + 16x - 16$

(a)

5	1	−4	0	16	−16
		5	5	25	205
	1	1	5	41	189

5 is an upper bound.

(b)

−3	1	−4	0	16	−16
		−3	21	−63	141
	1	−7	21	−47	125

−3 is a lower bound.

49. $f(x) = 4x^3 - 3x - 1$

Possible rational zeros: $\pm 1, \pm\frac{1}{2}, \pm\frac{1}{4}$

$$\begin{array}{r|rrrr} 1 & 4 & 0 & -3 & -1 \\ & & 4 & 4 & 1 \\ \hline & 4 & 4 & 1 & 0 \end{array}$$

$4x^3 - 3x - 1 = (x - 1)(4x^2 + 4x + 1) = (x - 1)(2x + 1)^2$

Thus, the zeros are 1 and $-\frac{1}{2}$.

51. $f(y) = 4y^3 + 3y^2 + 8y + 6$

Possible rational zeros: $\pm 1, \pm 2, \pm 3, \pm 6, \pm\frac{1}{2}, \pm\frac{3}{2}, \pm\frac{1}{4}, \pm\frac{3}{4}$

$$\begin{array}{r|rrrr} -\frac{3}{4} & 4 & 3 & 8 & 6 \\ & & -3 & 0 & -6 \\ \hline & 4 & 0 & 8 & 0 \end{array}$$

$4y^3 + 3y^2 + 8y + 6 = \left(y + \frac{3}{4}\right)(4y^2 + 8) = \left(y + \frac{3}{4}\right)4(y^2 + 2) = (4y + 3)(y^2 + 2)$

Thus, the only real zero is $-\frac{3}{4}$.

53. $P(x) = x^4 - \frac{25}{4}x^2 + 9$

$= \frac{1}{4}(4x^4 - 25x^2 + 36)$

$= \frac{1}{4}(4x^2 - 9)(x^2 - 4)$

$= \frac{1}{4}(2x + 3)(2x - 3)(x + 2)(x - 2)$

The zeros are $\pm\frac{3}{2}$ and ± 2.

55. $f(x) = x^3 - \frac{1}{4}x^2 - x + \frac{1}{4}$

$= \frac{1}{4}(4x^3 - x^2 - 4x + 1)$

$= \frac{1}{4}[x^2(4x - 1) - 1(4x - 1)]$

$= \frac{1}{4}(4x - 1)(x^2 - 1)$

$= \frac{1}{4}(4x - 1)(x + 1)(x - 1)$

The zeros are $\frac{1}{4}$ and ± 1.

57. $f(x) = x^3 - 1 = (x - 1)(x^2 + x + 1)$

Rational zeros: 1 $(x = 1)$

Irrational zeros: 0

Matches (d).

59. $f(x) = x^3 - x = x(x + 1)(x - 1)$

Rational zeros: 3 $(x = 0, \pm 1)$

Irrational zeros: 0

Matches (b).

61. (a)

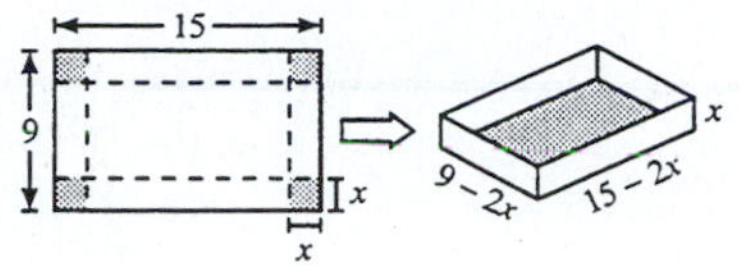

(b) $V = l \cdot w \cdot h = (15 - 2x)(9 - 2x)x$
$= x(9 - 2x)(15 - 2x)$

Since length, width, and height cannot be negative, we have $0 < x < \frac{9}{2}$ for the domain.

(c)

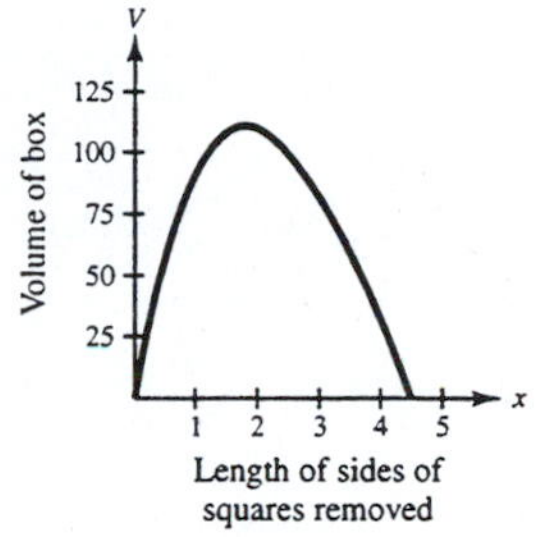

The volume is maximum when $x \approx 1.82$.

The dimensions are: length $= 15 - 2(1.82) = 11.36$

width $= 9 - 2(1.82) = 5.36$

height $= x = 1.82$

1.82 cm × 5.36 cm × 11.36 cm

(d) $56 = x(9 - 2x)(15 - 2x)$

$56 = 135x - 48x^2 + 4x^3$

$0 = 4x^3 - 48x^2 + 135x - 56$

The zeros of this polynomial are $\frac{1}{2}, \frac{7}{2}$, and 8. x cannot equal 8 since it is not in the domain of V. [The length cannot equal -1 and the width cannot equal -7. The product of $(8)(-1)(-7) = 56$ so it showed up as an extraneous solution.]

63. $g(x) = -f(x)$. This function would have the same zeros as $f(x)$ so r_1, r_2, and r_3 are also zeros of $g(x)$.

65. $g(x) = f(x - 5)$. The graph of $g(x)$ is a horizontal shift of the graph of $f(x)$ five units to the right so the zeros of $g(x)$ are $5 + r_1$, $5 + r_2$, and $5 + r_3$.

67. $g(x) = 3 + f(x)$. Since $g(x)$ is a vertical shift of the graph of $f(x)$, the zeros of $g(x)$ cannot be determined.

69.
$$P = -76x^3 + 4830x^2 - 320{,}000,\ 0 \le x \le 60$$
$$2{,}500{,}000 = -76x^3 + 4830x^2 - 320{,}000$$
$$76x^3 - 4830x^2 + 2{,}820{,}000 = 0$$

The zeros of this equation are $x \approx 46.1$, $x \approx 38.4$, and $x \approx -21.0$. Since $0 \le x \le 60$, we disregard $x \approx -21.0$. The smaller remaining solution is $x \approx 38.4$.

71. $C = 100\left(\dfrac{200}{x^2} + \dfrac{x}{x + 30}\right),\ 1 \le x$

C is minimum when $3x^3 - 40x^2 - 2400x - 36000 = 0$. The only real zero is $x \approx 40$.

73. $g(x) = f(x - 2)$

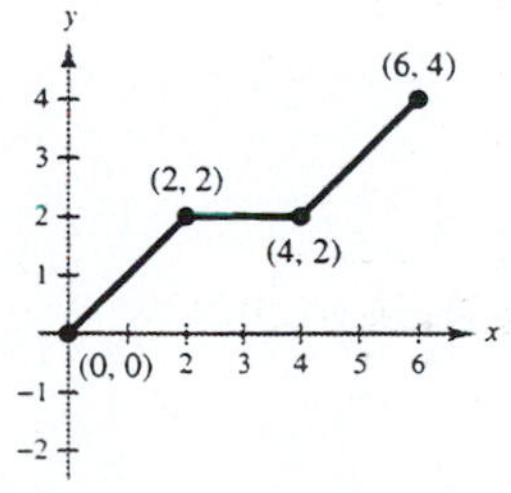

Horizontal shift two units to the right

75. $g(x) = 2f(x)$

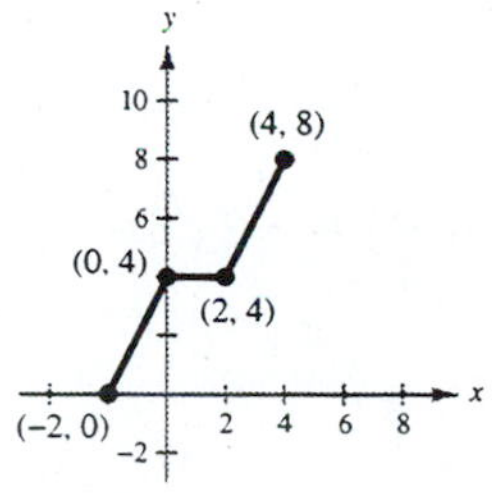

Vertical stretch

77. $g(x) = f(2x)$

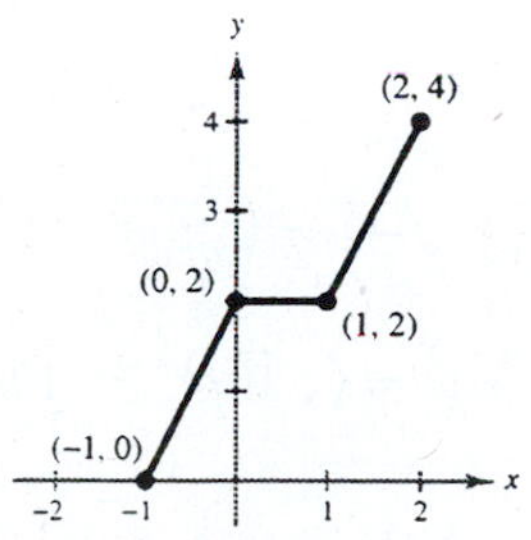

Horizontal shrink

Section 2.5 Complex Numbers

- $a + bi = c + di$ if and only if $a = c$ and $b = d$
- Operations on complex numbers
  - (a) Addition: $(a + bi) + (c + di) = (a + c) + (b + d)i$
  - (b) Subtraction: $(a + bi) - (c + di) = (a - c) + (b - d)i$
  - (c) Multiplication: $(a + bi)(c + di) = (ac - bd) + (ad + bc)i$
  - (d) Division: $\dfrac{a + bi}{c + di} = \dfrac{a + bi}{c + di} \cdot \dfrac{c - di}{c - di} = \dfrac{ac + bd}{c^2 + d^2} + \dfrac{bc - ad}{c^2 + d^2}i$
- The complex conjugate of $a + bi$ is $a - bi$:
  $(a + bi)(a - bi) = a^2 + b^2$
- $\sqrt{-a} = \sqrt{a}\, i$ for $a > 0$.

Solutions to Odd-Numbered Exercises

1. $a + bi = -10 + 6i$

$a = -10$

$b = 6$

3. $(a - 1) + (b + 3)i = 5 + 8i$

$a - 1 = 5 \implies a = 6$

$b + 3 = 8 \implies b = 5$

5. $4 + \sqrt{-9} = 4 + 3i$

7. $2 - \sqrt{-27} = 2 - \sqrt{27}i = 2 - 3\sqrt{3}i$

9. $\sqrt{-75} = \sqrt{75}\, i = 5\sqrt{3}\, i$

11. $-6i + i^2 = -6i - 1 = -1 - 6i$

13. $8 = 8 + 0i = 8$

15. $\sqrt{-0.09} = \sqrt{0.09}\, i = 0.3i$

17. $(5 + i) + (6 - 2i) = 11 - i$

19. $(8 - i) - (4 - i) = 8 - i - 4 + i = 4$

21. $\left(-2 + \sqrt{-8}\right)\left(5 - \sqrt{50}\right) = -2 + 2\sqrt{2}i + 5 - 5\sqrt{2}i$

$= 3 - 3\sqrt{2}\, i$

23. $13i - (14 - 7i) = 13i - 14 + 7i = -14 + 20i$

25. $-\left(\frac{3}{2} + \frac{5}{2}i\right) + \left(\frac{5}{3} + \frac{11}{3}i\right) = -\frac{3}{2} - \frac{5}{2}i + \frac{5}{3} + \frac{11}{3}i$

$= -\frac{9}{6} - \frac{15}{6}i + \frac{10}{6} + \frac{22}{6}i$

$= \frac{1}{6} + \frac{7}{6}i$

27. $\sqrt{-6} \cdot \sqrt{-2} = \left(\sqrt{6}i\right)\left(\sqrt{2}i\right) = \sqrt{12}\, i^2 = \left(2\sqrt{3}\right)(-1) = -2\sqrt{3}\, i$

29. $\left(\sqrt{-10}\right)^2 = \left(\sqrt{10}\, i\right)^2 = 10i^2 = -10$

31. $(1 + i)(3 - 2i) = 3 - 2i + 3i - 2i^2$

$= 3 + i + 2$

$= 5 + i$

33. $6i(5 - 2i) = 30i - 12i^2 = 30i + 12 = 12 + 30i$

35. $(\sqrt{14} + \sqrt{10}\,i)(\sqrt{14} - \sqrt{10}\,i) = 14 - 10i^2 = 14 + 10 = 24$

37. $(4 + 5i)^2 = 16 + 40i + 25i^2 = 16 + 40i - 25$

$= -9 + 40i$

39. $(2 + 3i)^2 + (2 - 3i)^2 = 4 + 12i + 9i^2 + 4 - 12i + 9i^2$

$= 4 + 12i - 9 + 4 - 12i - 9$

$= -10$

41. $\sqrt{-6}\sqrt{-6} = \sqrt{6}i\sqrt{6}i = 6i^2 = -6$

43. The complex conjugate of $5 + 3i$ is $5 - 3i$.

$(5 + 3i)(5 - 3i) = 25 - 9i^2 = 25 + 9 = 34$

45. The complex conjugate of $-2 - \sqrt{5}\,i$ is $-2 + \sqrt{5}\,i$.

$(-2 - \sqrt{5}\,i)(-2 + \sqrt{5}\,i) = 4 - 5i^2 = 4 + 5 = 9$

47. The complex conjugate of $20i$ is $-20i$.

$(20i)(-20i) = -400i^2 = 400$

49. The complex conjugate of $\sqrt{8}$ is $\sqrt{8}$.

$(\sqrt{8})(\sqrt{8}) = 8$

51. $\dfrac{6}{i} = \dfrac{6}{i} \cdot \dfrac{-i}{-i} = \dfrac{-6i}{-i^2} = \dfrac{-6i}{1} = -6i$

53. $\dfrac{4}{4 - 5i} = \dfrac{4}{4 - 5i} \cdot \dfrac{4 + 5i}{4 + 5i} = \dfrac{4(4 + 5i)}{16 + 25} = \dfrac{16 + 20i}{41} = \dfrac{16}{41} + \dfrac{20}{41}i$

55. $\dfrac{2 + i}{2 - i} = \dfrac{2 + i}{2 - i} \cdot \dfrac{2 + i}{2 + i} = \dfrac{4 + 4i + i^2}{4 + 1} = \dfrac{3 + 4i}{5} = \dfrac{3}{5} + \dfrac{4}{5}i$

57. $\dfrac{6 - 7i}{i} = \dfrac{6 - 7i}{i} \cdot \dfrac{-i}{-i} = \dfrac{-6i + 7i^2}{1} = -7 - 6i$

59. $\dfrac{1}{(4 - 5i)^2} = \dfrac{1}{16 - 40i + 25i^2} = \dfrac{1}{-9 - 40i} \cdot \dfrac{-9 + 40i}{-9 + 40i}$

$= \dfrac{-9 + 40i}{81 + 1600} = \dfrac{-9 + 40i}{1681} = -\dfrac{9}{1681} + \dfrac{40}{1681}i$

61. $\dfrac{2}{1 + i} - \dfrac{3}{1 - i} = \dfrac{2(1 - i) - 3(1 + i)}{(1 + i)(1 - i)}$

$= \dfrac{2 - 2i - 3 - 3i}{1 + 1}$

$= \dfrac{-1 - 5i}{2}$

$= -\dfrac{1}{2} - \dfrac{5}{2}i$

63. $\dfrac{i}{3-2i}+\dfrac{2i}{3+8i}=\dfrac{i(3+8i)+2i(3-2i)}{(3-2i)(3+8i)}$

$$=\frac{3i+8i^2+6i-4i^2}{9+24i-6i-16i^2}$$

$$=\frac{4i^2+9i}{9+18i+16}$$

$$=\frac{-4+9i}{25+18i}\cdot\frac{25-18i}{25-18i}$$

$$=\frac{-100+72i+225i-162i^2}{625+324}$$

$$=\frac{-100+297i+162}{949}$$

$$=\frac{62+297i}{949}$$

$$=\frac{62}{949}+\frac{297}{949}i$$

65. $x^2-2x+2=0;\ a=1,\ \ b=-2,\ \ c=2$

$$x=\frac{-(-2)\pm\sqrt{(-2)^2-4(1)(2)}}{2(1)}=\frac{2\pm\sqrt{-4}}{2}=\frac{2\pm 2i}{2}=1\pm i$$

67. $4x^2+16x+17=0;\ a=4,\ \ b=16,\ \ c=17$

$$x=\frac{-16\pm\sqrt{(16)^2-4(4)(17)}}{2(4)}$$

$$=\frac{-16\pm\sqrt{-16}}{8}=\frac{-16+4i}{8}$$

$$=-2\pm\frac{1}{2}i$$

69. $4x^2+16x+15=0;\ a=4,\ \ b=16,\ \ c=15$

$$x=\frac{-16\pm\sqrt{(16)^2-4(4)(15)}}{2(4)}=\frac{-16\pm\sqrt{16}}{8}=\frac{-16\pm 4}{8}$$

$$x=-\frac{12}{8}=-\frac{3}{2}\quad\text{or}\quad x=\frac{-20}{8}=-\frac{5}{2}$$

71. $16t^2-4t+3=0;\ a=16,\ \ b=-4,\ \ c=3$

$$t=\frac{-(-4)\pm\sqrt{(-4)^2-4(16)(3)}}{2(16)}$$

$$=\frac{4\pm\sqrt{-176}}{32}=\frac{4\pm 4\sqrt{11}i}{32}$$

$$=\frac{1}{8}\pm\frac{\sqrt{11}}{8}i$$

73. $y = \frac{1}{4}(4x^2 - 20x + 25)$

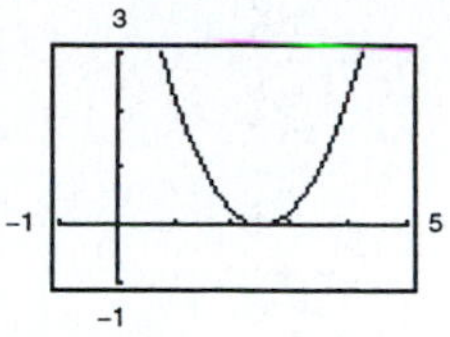

$$\begin{aligned} 0 &= \tfrac{1}{4}(4x^2 - 20x + 25) \\ 0 &= 4x^2 - 20x + 25 \\ 0 &= (2x - 5)^2 \\ 0 &= 2x - 5 \\ -2x &= -5 \\ x &= \tfrac{5}{2} \end{aligned}$$

x-intercept: $\left(\frac{5}{2}, 0\right)$

75. $y = -(x^2 - 4x + 5)$

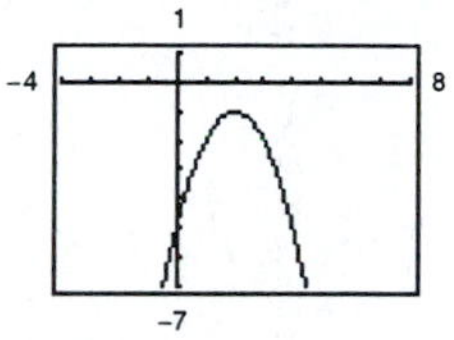

$$\begin{aligned} 0 &= -(x^2 - 4x + 5) \\ 0 &= x^2 - 4x + 5 \\ x &= \frac{-(-4) \pm \sqrt{(-4)^2 - 4(1)(5)}}{2(1)} \\ &= \frac{4 \pm 2i}{2} = 2 \pm i \end{aligned}$$

No x-intercepts

No real solutions

77. The number of x-intercepts of the graph of $y = ax^2 + bx + c$ corresponds to the number of real solutions of the equation $0 = ax^2 + bx + c$. If there are no x-intercepts, the quadratic equation has two complex solutions.

79. $$\begin{aligned} -6i^3 + i^2 &= -6i^2 i + i^2 \\ &= -6(-1)i + (-1) \\ &= 6i - 1 \\ &= -1 + 6i \end{aligned}$$

81. $$\begin{aligned} -5i^5 &= -5i^2 i^2 i \\ &= -5(-1)(-1)i \\ &= -5i \end{aligned}$$

83. $$\begin{aligned} &\left(\sqrt{-75}\right)^3\left(5\sqrt{3}\,i\right)^3 = 5^3\left(\sqrt{3}\right)^3 i^3 \\ &125\left(3\sqrt{3}\right)(-1) \\ &= -375\sqrt{3}\,i \end{aligned}$$

85. $\dfrac{1}{i^3} = \dfrac{1}{-i} = \dfrac{1}{-i} \cdot \dfrac{i}{i} = \dfrac{i}{-i^2} = \dfrac{i}{1} = i$

87. $(2)^3 = 8$

$$\begin{aligned} (1 + \sqrt{3}i)^3 &= (-1)^3 + 3(-1)^2(\sqrt{3}i) + 3(-1)(\sqrt{3}i)^2 + (\sqrt{3}i)^3 \\ &= -1 + 3\sqrt{3}i - 9i^2 + 3\sqrt{3}i^3 \\ &= -1 + 3\sqrt{3}\,i + 9 - 3\sqrt{3}i \\ &= 8 \end{aligned}$$

$$\begin{aligned} \left(-1 - \sqrt{3}\,i\right)^3 &= (-1)^3 + 3(-1)^2\left(-\sqrt{3}\,i\right) + 3(-1)\left(-\sqrt{3}\,i\right)^2 + \left(-\sqrt{3}\,i\right)^3 \\ &= -1 - 3\sqrt{3}i - 9i^2 - 3\sqrt{3}\,i^3 \\ &= -1 - 3\sqrt{3}\,i + 9 + 3\sqrt{3}\,i \\ &= 8 \end{aligned}$$

89. $(a + bi) + (a - bi) = 2a$ which is a real number.

91. $(a + bi)(a - bi) = a^2 - (bi)^2 = a^2 - b^2 i^2$

$= a^2 + b^2$ which is a real number.

93. $(a_1 + b_1 i) + (a_2 + b_2 i) = (a_1 + a_2) + (b_1 + b_2)i$

The complex conjugate of the sum is $(a_1 + a_2) - (b_1 + b_2)i$, and the sum of the conjugates is

$$\begin{aligned}(a_1 - b_1 i) + (a_2 - b_2 i) &= (a_1 + a_2) + (-b_1 - b_2)i \\ &= (a_1 + a_2) - (b_1 + b_2)i.\end{aligned}$$

Thus, the conjugate of the sum is the sum of the conjugates.

95. $(4 + 3x) + (8 - 6x - x^2) = -x^2 - 3x + 12$

97. $\begin{aligned}(2x - 5)^2 &= (2x)^2 - 2(2x)(5) + (5)^2 \\ &= 4x^2 - 20x + 25\end{aligned}$

99.
$$\begin{aligned}V &= \frac{4}{3}\pi a^2 b \\ 3V &= 4\pi a^2 b \\ \frac{3V}{4\pi b} &= a^2 \\ \sqrt{\frac{3V}{4\pi b}} &= a \\ a &= \frac{1}{2}\sqrt{\frac{3V}{\pi b}}\end{aligned}$$

Section 2.6 The Fundamental Theorem of Algebra

- You should know that if f is a polynomial of degree $n > 0$, then f has at least one zero in the complex number system.
- You should know that if $a + bi$ is a complex zero of a polynomial f, with real coefficients, then $a - bi$ is also a complex zero of f.
- You should know the difference between a factor that is irreducible over the rationals (such as $x^2 - 7$) and a factor that is irreducible over the reals (such as $x^2 + 9$).

Solutions to Odd-Numbered Exercises

1. $f(x) = x(x - 6)^2 = x(x - 6)(x - 6)$

The three zeros are: $x = 0$, $x = 6$, and $x = 6$.

3. $h(t) = (t - 3)(t - 2)(t - 3i)(t + 3i)$

The four zeros are:
$t = 3$, $t = 2$, $t = 3i$, and $t = -3i$.

5. $\begin{aligned}f(x) &= x^3 - 4x^2 + x - 4 = x^2(x - 4) + 1(x - 4) \\ &= (x - 4)(x^2 + 1)\end{aligned}$

The only real zero of $f(x)$ is $x = 4$. This corresponds to the x-intercept of $(4, 0)$ on the graph.

7. $f(x) = x^4 + 4x^2 + 4 = (x^2 + 2)^2$

$f(x)$ has no real zeros and the graph of $f(x)$ has no x-intercepts.

9. $\begin{aligned}f(x) &= x^2 + 25 \\ &= (x + 5i)(x - 5i)\end{aligned}$

The zeros of $f(x)$ are $x = \pm 5i$.

11. $h(x) = x^2 - 4x + 1$

h has no rational zeros.

By the Quadratic Formula, the zeros are $x = \dfrac{4 \pm \sqrt{16 - 4}}{2} = 2 \pm \sqrt{3}$.

$h(x) = \left[x - \left(2 + \sqrt{3}\right)\right]\left[x - \left(2 - \sqrt{3}\right)\right] = \left(x - 2 - \sqrt{3}\right)\left(x - 2 + \sqrt{3}\right)$

13. $f(x) = x^4 - 81$

$= (x^2 - 9)(x^2 + 9)$

$= (x + 3)(x - 3)(x + 3i)(x - 3i)$

The zeros of $f(x)$ are $x = \pm 3$ and $x = \pm 3i$.

15. $f(z) = z^2 - 2z + 2$

f has no rational zeros.

By the Quadratic Formula, the zeros are $z = \dfrac{2 \pm \sqrt{4 - 8}}{2} = 1 \pm i$.

$f(z) = [z - (1 + i)][z - (1 - i)] = (z - 1 - i)(z - 1 + i)$

17. $g(x) = x^3 - 6x^2 + 13x - 10$

Possible rational zeros: $\pm 1, \pm 2, \pm 5, \pm 10$

2	1	−6	13	−10
		2	−8	10
	1	−4	5	0

By the Quadratic Formula, the zeros of $x^2 - 4x + 5$ are $x = \dfrac{4 \pm \sqrt{16 - 20}}{2} = 2 \pm i$.

The zeros of $g(x)$ are $x = 2$ and $x = 2 \pm i$.

$g(x) = (x - 2)[x - (2 + i)][x - (2 - i)]$

$= (x - 2)(x - 2 - i)(x - 2 + i)$

19. $f(t) = t^3 - 3t^2 - 15t + 125$

Possible rational zeros: $\pm 1, \pm 5, \pm 25, \pm 125$

−5	1	−3	−15	125
		−5	40	−125
	1	−8	25	0

By the Quadratic Formula, the zeros of $t^2 - 8t + 25$ are $t = \dfrac{8 \pm \sqrt{64 - 100}}{2} = 4 \pm 3i$.

The zeros of $f(t)$ are $t = -5$ and $t = 4 \pm 3i$.

$f(t) = [t - (-5)][t - (4 + 3i)][t - (4 - 3i)]$

$= (t + 5)(t - 4 - 3i)(t - 4 + 3i)$

21. $h(x) = x^3 - x + 6$

Possible rational zeros: $\pm 1, \pm 2, \pm 3, \pm 6$

−2	1	0	−1	6
		−2	4	−6
	1	−2	3	0

By the Quadratic Formula, the zeros of $x^2 - 2x + 3$ are

$x = \dfrac{2 \pm \sqrt{4 - 12}}{2} = 1 \pm \sqrt{2}\,i$.

The zeros of $h(x)$ are $x = -2$ and $x = 1 \pm \sqrt{2}\,i$.

$h(x) = [x - (-2)]\left[x - \left(1 + \sqrt{2}\,i\right)\right]\left[x - \left(1 - \sqrt{2}\,i\right)\right]$

$= (x + 2)\left(x - 1 - \sqrt{2}\,i\right)\left(x - 1 + \sqrt{2}\,i\right)$

23. $f(x) = 5x^3 - 9x^2 + 28x + 6$

Possible rational zeros: $\pm 1, \pm 2, \pm 3, \pm 6, \pm \frac{1}{5}, \pm \frac{2}{5}, \pm \frac{3}{5}, \pm \frac{6}{5}$

$-\frac{1}{5}$	5	-9	28	6
		-1	2	-6
	5	-10	30	0

By the Quadratic Formula, the zeros of $5x^2 - 10x + 30 = 5(x^2 - 2x + 6)$ are

$$x = \frac{2 \pm \sqrt{4 - 24}}{2} = 1 \pm \sqrt{5}\, i.$$

The zeros of $f(x)$ are $x = -\frac{1}{5}$ and $x = 1 \pm \sqrt{5}\, i$.

$$f(x) = \left[x - \left(-\tfrac{1}{5}\right)\right](5)\left[x - \left(1 + \sqrt{5}\, i\right)\right]\left[x - \left(1 - \sqrt{5}\, i\right)\right]$$
$$= (5x + 1)\left(x - 1 - \sqrt{5}\, i\right)\left(x - 1 + \sqrt{5}\, i\right)$$

25. $g(x) = x^4 - 4x^3 + 8x^2 - 16x + 16$

Possible rational zeros: $\pm 1, \pm 2, \pm 4, \pm 8, \pm 16$

2	1	-4	8	-16	16
		2	-4	8	-16
2	1	-2	4	-8	
		2	0	8	
	1	0	4	0	

$g(x) = (x - 2)(x - 2)(x^2 + 4) = (x - 2)^2(x + 2i)(x - 2i)$

The zeros of $g(x)$ are 2 and $\pm 2i$.

27. $f(x) = x^4 + 10x^2 + 9$

$$= (x^2 + 1)(x^2 + 9)$$
$$= (x + i)(x - i)(x + 3i)(x - 3i)$$

The zeros of $f(x)$ are $x = \pm i$ and $x = \pm 3i$.

29. $f(x) = x^3 + 24x^2 + 214x + 740$

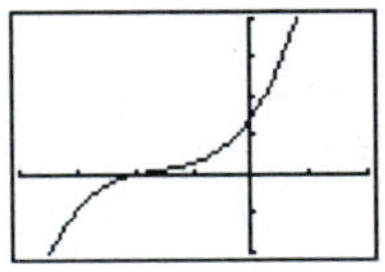

The graph reveals one zero at $x = -10$.

-10	1	24	214	740
		-10	-140	-740
	1	14	74	0

By the Quadratic Formula, the zeros of $x^2 + 14x + 74$ are

$$x = \frac{-14 \pm \sqrt{196 - 296}}{2} = -7 \pm 5i.$$

The zeros of $f(x)$ are $x = -10$ and $x = -7 \pm 5i$.

31. $f(x) = 16x^3 - 20x^2 - 4x + 15$

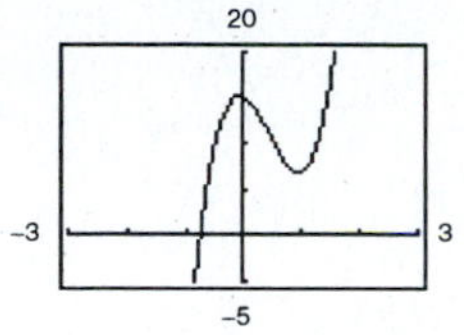

The graph reveals one zero at $x = -\frac{3}{4}$.

$-\frac{3}{4}$	16	-20	-4	15
		-12	24	-15
	16	-32	20	0

By the Quadratic Formula, the zeros of $16x^2 - 32x + 20 = 4(4x^2 - 8x + 5)$ are

$$x = \frac{8 \pm \sqrt{64 - 80}}{8} = 1 \pm \frac{1}{2}i.$$

The zeros of $f(x)$ are $x = -\frac{3}{4}$ and $x = 1 \pm \frac{1}{2}i$.

33. $f(x) = 2x^4 + 5x^3 + 4x^2 + 5x + 2$

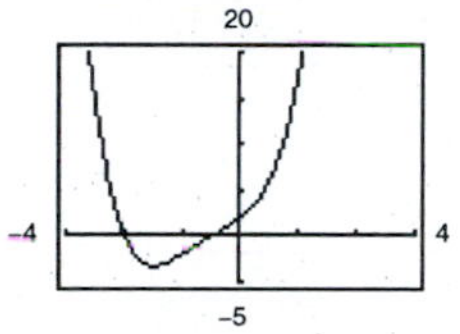

The graph reveals zeros at $x = -2$ and $x = -\frac{1}{2}$.

-2	2	5	4	5	2
		-4	-2	-4	-2
	2	1	2	1	0

$-\frac{1}{2}$	2	1	2	1
		-1	0	-1
	2	0	2	0

The zeros of $2x^2 + 2 = 2(x^2 + 1)$ are $x = \pm i$.

The zeros of $f(x)$ are $x = -2$, $x = -\frac{1}{2}$, and $x = \pm i$.

35. $f(x) = (x - 1)(x - 5i)(x + 5i)$

$= (x - 1)(x^2 + 25)$

$= x^3 - x^2 + 25x - 25$

Note: $f(x) = a(x^3 - x^2 + 25x - 25)$,

where a is any nonzero real number, has the zeros 1 and $\pm 5i$.

37. $f(x) = (x - 6)[x - (-5 + 2i)][x - (-5 - 2i)]$

$= (x - 6)[(x + 5) - 2i][(x + 5) + 2i]$

$= (x - 6)[(x + 5)^2 - (2i)^2]$

$= (x - 6)(x^2 + 10x + 25 + 4)$

$= (x - 6)(x^2 + 10x + 29)$

$= x^3 + 4x^2 - 31x - 174$

Note: $f(x) = a(x^3 + 4x^2 - 31x - 174)$, where a is any nonzero real number, has the zeros 6, and $-5 \pm 2i$.

39. $f(x) = (x - i)(x + i)(x - 6i)(x + 6i)$
$= (x^2 + 1)(x^2 + 36)$
$= x^4 + 37x^2 + 36$

Note: $f(x) = a(x^4 + 37x^2 + 36)$, where a is any nonzero real number, has the zeros $\pm i$ and $\pm 6i$.

41. If $3 + \sqrt{2}\,i$ is a zero, so is its conjugate, $3 - \sqrt{2}i$.

$$\begin{aligned} f(x) &= (3x - 2)(x + 1)\left[x - \left(3 + \sqrt{2}\,i\right)\right]\left[x - \left(3 - \sqrt{2}\,i\right)\right] \\ &= (3x - 2)(x + 1)\left[(x - 3) - \sqrt{2}\,i\right]\left[(x - 3) + \sqrt{2}\,i\right] \\ &= (3x^2 + x - 2)\left[(x - 3)^2 - \left(\sqrt{2}\,i\right)^2\right] \\ &= (3x^2 + x - 2)(x^2 - 6x + 9 + 2) \\ &= (3x^2 + x - 2)(x^2 - 6x + 11) \\ &= 3x^4 - 17x^3 + 25x^2 + 23x - 22 \end{aligned}$$

Note: $f(x) = a(3x^4 - 17x^3 + 25x^2 + 23x - 22)$, where a is any nonzero real number, has the zeros $\frac{2}{3}$, -1, and $3 \pm \sqrt{2}\,i$.

43. If $-\frac{1}{2} + i$ is a zero, so is its conjugate, $-\frac{1}{2} - i$.

$$\begin{aligned} f(x) &= 4\left(x - \tfrac{3}{4}\right)(x + 2)\, 2\left[x - \left(-\tfrac{1}{2} + i\right)\right] 2\left[x - \left(-\tfrac{1}{2} - i\right)\right] \\ &= (4x - 3)(x + 2)[(2x + 1) - 2i][(2x + 1) + 2i] \\ &= (4x^2 + 5x - 6)[(2x + 1)^2 - (2i)^2] \\ &= (4x^2 + 5x - 6)(4x^2 + 4x + 1 + 4) \\ &= (4x^2 + 5x - 6)(4x^2 + 4x + 5) \\ &= 16x^4 + 36x^3 + 16x^2 + x - 30 \end{aligned}$$

Note: $f(x) = a(16x^4 + 36x^3 + 16x^2 + x - 30)$, where a is any nonzero real number, has the zeros $x = \frac{3}{4}$, $x = -2$, and $x = -\frac{1}{2} \pm i$. In fact, we used constant multiples of the linear factors in this equation to simplify the fractions. Without these constants, you would have

$$\begin{aligned} f(x) &= \left(x - \tfrac{3}{4}\right)(x + 2)\left(x + \tfrac{1}{2} - i\right)\left(x + \tfrac{1}{2} + i\right) \\ &= x^4 + \tfrac{9}{4}x^3 + x^2 + \tfrac{1}{16}x - \tfrac{15}{8}. \end{aligned}$$

45. $f(x) = x^4 + 6x^2 - 27$

(a) $f(x) = (x^2 + 9)(x^2 - 3)$

(b) $f(x) = (x^2 + 9)\left(x + \sqrt{3}\right)\left(x - \sqrt{3}\right)$

(c) $f(x) = (x + 3i)(x - 3i)\left(x + \sqrt{3}\right)\left(x - \sqrt{3}\right)$

47.

$$\begin{array}{r}
x^2 - 2x + 3 \\
x^2 - 2x - 2\overline{)x^4 - 4x^3 + 5x^2 - 2x - 6} \\
\underline{x^4 - 2x^3 - 2x^2} \\
-2x^3 + 7x^2 - 2x \\
\underline{-2x^3 + 4x^2 + 4x} \\
3x^2 - 6x - 6 \\
\underline{3x^2 - 6x - 6} \\
0
\end{array}$$

$f(x) = (x^2 - 2x - 2)(x^2 - 2x + 3)$

(a) $f(x) = (x^2 - 2x - 2)(x^2 - 2x + 3)$

(b) $f(x) = \left(x - 1 + \sqrt{3}\right)\left(x - 1 - \sqrt{3}\right)(x^2 - 2x + 3)$

(c) $f(x) = \left(x - 1 + \sqrt{3}\right)\left(x - 1 - \sqrt{3}\right)\left(x - 1 + \sqrt{2}\,i\right)\left(x - 1 - \sqrt{2}\,i\right)$

Note: Use the Quadratic Formula for (b) and (c).

49. $f(x) = 2x^3 + 3x^2 + 50x + 75$

Since $5i$ is a zero, so is $-5i$.

$$\begin{array}{r|rrrr}
5i & 2 & 3 & 50 & 75 \\
 & & 10i & -50 + 15i & -75 \\
\hline
 & 2 & 3 + 10i & 15i & 0
\end{array}$$

$$\begin{array}{r|rrr}
-5i & 2 & 3 + 10i & 15i \\
 & & -10i & -15i \\
\hline
 & 2 & 3 & 0
\end{array}$$

The zero of $2x + 3$ is $x = -\frac{3}{2}$.

The zeros of $f(x)$ are $x = -\frac{3}{2}$ and $x = \pm 5i$.

Alternate Solution

Since $x = \pm 5i$ are zeros of $f(x)$, $(x + 5i)(x - 5i) = x^2 + 25$ is a factor of $f(x)$. By long division we have:

$$\begin{array}{r}
2x + 3 \\
x^2 + 0x + 25\overline{)2x^3 + 3x^2 + 50x + 75} \\
\underline{2x^3 + 0x^2 + 50x} \\
3x^2 + 0x + 75 \\
\underline{3x^2 + 0x + 75} \\
0
\end{array}$$

Thus, $f(x) = (x^2 + 25)(2x + 3)$ and the zeros of f are $x = \pm 5i$ and $x = -\frac{3}{2}$.

51. $f(x) = 2x^4 - x^3 + 7x^2 - 4x - 4$

Since $2i$ is a zero, so is $-2i$.

$$\begin{array}{r|rrrrr} 2i & 2 & -1 & 7 & -4 & -4 \\ & & 4i & -8-2i & 4-2i & 4 \\ \hline & 2 & -1+4i & -1-2i & -2 & 0 \end{array}$$

$$\begin{array}{r|rrrr} -2i & 2 & -1+4i & -1-2i & -2i \\ & & -4i & 2i & 2i \\ \hline & 2 & -1 & -1 & 0 \end{array}$$

The zeros of $2x^2 - x - 1 = (2x + 1)(x - 1)$ are $x = -\frac{1}{2}$ and $x = 1$.
The zeros of $f(x)$ are $x = \pm 2i,\ x = -\frac{1}{2},$ and $x = 1$.

Alternate Solution

Since $x = \pm 2i$ are zeros of $f(x)$, $(x + 2i)(x - 2i) = x^2 + 4$ is a factor of $f(x)$.
By long division we have:

$$\begin{array}{r} 2x^2 - \phantom{x^3+{}} x - 1 \\ x^2 + 0x + 4 \overline{)\, 2x^4 - x^3 + 7x^2 - 4x - 4} \\ \underline{2x^4 + 0x^3 + 8x^2} \phantom{{} - 4x - 4} \\ -x^3 - x^2 - 4x \phantom{{} - 4} \\ \underline{-x^3 + 0x^2 - 4x} \phantom{{} - 4} \\ -x^2 + 0x - 4 \\ \underline{-x^2 + 0x - 4} \\ 0 \end{array}$$

Thus, $f(x) = (x^2 + 4)(2x^2 - x - 1)$

$= (x + 2i)(x - 2i)(2x + 1)(x - 1)$

and the zeros of $f(x)$ are $x = \pm 2i,\ x = -\frac{1}{2},$ and $x = 1$.

53. $g(x) = 4x^3 + 23x^2 + 34x - 10$

Since $-3 + i$ is a zero, so is $-3 - i$.

$$\begin{array}{r|rrrr} -3+i & 4 & 23 & 34 & -10 \\ & & -12+4i & -37-i & 10 \\ \hline & 4 & 11+4i & -3-i & 0 \end{array}$$

$$\begin{array}{r|rrr} -3-i & 4 & 11+4i & -3-i \\ & & -12-4i & 3+i \\ \hline & 4 & -1 & 0 \end{array}$$

The zero of $4x - 1$ is $x = \frac{1}{4}$. The zeros of $g(x)$ are $x = -3 \pm i$ and $x = \frac{1}{4}$.

Alternate Solution

Since $-3 \pm i$ are zeros of $g(x)$,

$$\begin{aligned}[x - (-3 + i)][x - (-3 - i)] &= [(x + 3) - i][(x + 3) + i] \\ &= (x + 3)^2 - i^2 \\ &= x^2 + 6x + 10\end{aligned}$$

is a factor of $g(x)$. By long division we have:

$$\begin{array}{r} 4x - 1 \\ x^2 + 6x + 10\overline{)4x^3 + 23x^2 + 34x - 10} \\ \underline{4x^3 + 24x^2 + 40x} \\ -x^2 - 6x - 10 \\ \underline{-x^2 - 6x - 10} \\ 0 \end{array}$$

Thus, $g(x) = (x^2 + 6x + 10)(4x - 1)$ and the zeros of $g(x)$ are $x = -3 \pm i$ and $x = \frac{1}{4}$.

55. Since $-3 + \sqrt{2}\,i$ is a zero, so is $-3 - \sqrt{2}\,i$, and

$$\begin{aligned}&\left[x - \left(-3 + \sqrt{2}\,i\right)\right]\left[x - \left(-3 - \sqrt{2}\,i\right)\right] \\ &\quad = \left[(x + 3) - \sqrt{2}\,i\right]\left[(x + 3) + \sqrt{2}\,i\right] \\ &\quad = (x + 3)^2 - \left(\sqrt{2}\,i\right)^2 \\ &\quad = x^2 + 6x + 11\end{aligned}$$

is a factor of $f(x)$. By long division, we have:

$$\begin{array}{r} x^2 - 3x + 2 \\ x^2 + 6x + 11\overline{)x^4 + 3x^3 - 5x^2 - 21x + 22} \\ \underline{x^4 + 6x^3 + 11x^2} \\ -3x^3 - 16x^2 - 21x \\ \underline{-3x^3 - 18x^2 - 33x} \\ 2x^2 + 12x + 22 \\ \underline{2x^2 + 12x + 22} \\ 0 \end{array}$$

Thus, $f(x) = (x^2 + 6x + 11)(x^2 - 3x + 2)$
$= (x^2 + 6x + 11)(x - 1)(x - 2)$

and the zeros of f are $x = -3 \pm \sqrt{2}\,i$, $x = 1$, and $x = 2$.

57. Since $\frac{1}{2}(1 - \sqrt{5}\,i)$ is a zero, so is $\frac{1}{2}(1 + \sqrt{5}\,i)$, and

$$\begin{aligned}&\left[x - \tfrac{1}{2}(1 - \sqrt{5}\,i)\right]\left[x - \tfrac{1}{2}(1 + \sqrt{5}\,i)\right] \\ &\quad = \left[\left(x - \tfrac{1}{2}\right) + \tfrac{1}{2}\sqrt{5}\,i\right]\left[\left(x - \tfrac{1}{2}\right) - \tfrac{1}{2}\sqrt{5}\,i\right] \\ &\quad = \left(x - \tfrac{1}{2}\right)^2 - \left(\tfrac{1}{2}\sqrt{5}\,i\right)^2 \\ &\quad = x^2 - x + \tfrac{1}{4} + \tfrac{5}{4} \\ &\quad = x^2 - x + \tfrac{3}{2}\end{aligned}$$

is a factor of $h(x)$. By long division, we have:

$$\begin{array}{r} 8x - 6 \\ x^2 - x + \tfrac{3}{2}\overline{)8x^3 - 14x^2 + 18x - 9} \\ \underline{8x^3 - 8x^2 + 12x} \\ -6x^2 + 6x - 9 \\ \underline{-6x^2 + 6x - 9} \\ 0 \end{array}$$

Thus, $h(x) = \left(x^2 - x + \frac{3}{2}\right)(8x - 6)$ and the zeros of $h(x)$ are

$x = \dfrac{3}{4}$ and $x = \dfrac{1}{2} \pm \dfrac{\sqrt{5}}{2}i$.

59. $f(x) = x^3 + ix^2 + ix - 1$

(a)
$$\begin{array}{c|cccc} i & 1 & i & i & -1 \\ & & i & -2 & -1 - 2i \\ \hline & 1 & 2i & -2 + i & -2 - 2i \end{array}$$

Since the remainder is not zero, $x = i$ is not a zero of f.

(b) The theorem that states that complex zeros occur in conjugate pairs has the condition that the coefficients of $f(x)$ must be real numbers. This polynomial has complex coefficients for x^2 and x.

61. $f(x) = x^4 - 4x^2 + k$

$$x^2 = \frac{-(-4) \pm \sqrt{(-4)^2 - 4(1)(k)}}{2(1)} = \frac{4 \pm 2\sqrt{4 - k}}{2} = 2 \pm \sqrt{4 - k}$$

$$x = \pm\sqrt{2 \pm \sqrt{4 - k}}$$

(a) For there to be four distinct real roots, both $4 - k$ and $2 \pm \sqrt{4 - k}$ must be positive. This occurs when $0 < k < 4$. Thus, some possible k-values are $k = 1$, $k = 2$, $k = 3$, $k = \frac{1}{2}$, $k = \sqrt{2}$, etc.

(b) For there to be two real roots, each of multiplicity 2, $4 - k$ must equal zero. Thus, $k = 4$.

(c) For there to be two real zeros and two complex zeros, $2 + \sqrt{4 - k}$ must be positive and $2 - \sqrt{4 - k}$ must be negative. This occurs when $k < 0$. Thus, some possible k-values are $k = -1$, $k = -2$, $k = -\frac{1}{2}$, etc.

(d) For there to be four complex zeros, $2 \pm \sqrt{4 - k}$ must be complex. This occurs when $k > 4$. Some possible k-values are $k = 5$, $k = 6$, $k = 7.4$, etc.

63. $h = -16t^2 + 48t, \ 0 \le t \le 3$

$$\begin{aligned} &= -16(t^2 - 3t) \\ &= -16\left(t^2 - 3t + \tfrac{9}{4} - \tfrac{9}{4}\right) \\ &= -16\left[\left(t - \tfrac{3}{2}\right)^2 - \tfrac{9}{4}\right] \\ &= -16\left(t - \tfrac{3}{2}\right)^2 + 36 \end{aligned}$$

The maximum height that the baseball reaches is 36 feet when $t = 1.5$ seconds. No, it is not possible for the ball to reach a height of 64 feet.

Alternate Solution

Let $h = 64$ and solve for t.

$$64 = -16t^2 + 48t$$

$$16t^2 - 48t + 64 = 0$$

$$16(t^2 - 3t + 4) = 0$$

$$t^2 - 3t + 4 = 0$$

$$t = \frac{3 \pm \sqrt{9 - 16}}{2} = \frac{3 \pm \sqrt{7}\,i}{2}$$

No, it is not possible since solving this equation yields imaginary roots.

65. (a) $f(x) = (x - \sqrt{b}i)(x + \sqrt{b}i) = x^2 + b$

(b) $f(x) = [x - (a + bi)][x - (a - bi)]$

$= [(x - a) - bi][(x - a) + bi]$

$= (x - a)^2 - (bi)^2$

$= x^2 - 2ax + a^2 + b^2$

67.

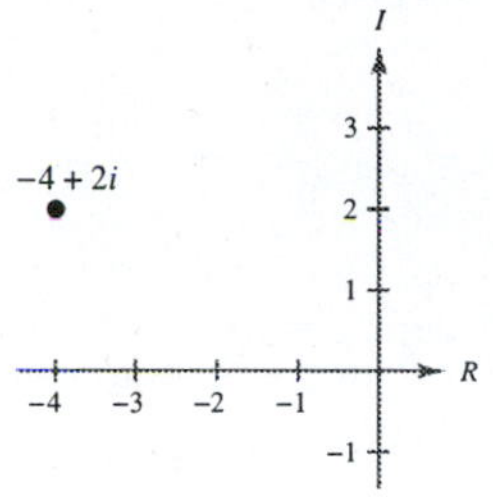

69. $(1.50)(420) = 630$

71. $\dfrac{x}{10} = \dfrac{3}{1.25}$

$1.25x = 30$

$x = 24$ feet

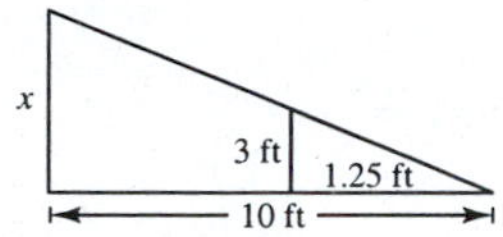

Section 2.7 Rational Functions

■ You should know the following basic facts about rational functions.

(a) A function of the form $f(x) = P(x)/Q(x)$, $Q(x) \neq 0$, where $P(x)$ and $Q(x)$ are polynomials, is called a rational function.

(b) The domain of a rational function is the set of all real numbers except those which make the denominator zero.

(c) If $f(x) = P(x)/Q(x)$ is in reduced form, and a is a value such that $Q(a) = 0$, then the line $x = a$ is a vertical asymptote of the graph of f. $f(x) \to \infty$ or $f(x) \to -\infty$ as $x \to a$.

(d) The line $y = b$ is a horizontal asymptote of the graph of f if $f(x) \longrightarrow b$ as $x \longrightarrow \infty$ or $x \longrightarrow -\infty$.

(e) Let $f(x) = \dfrac{P(x)}{Q(x)} = \dfrac{a_nx^n + a_{n-1}x^{n-1} + \cdots + a_1x + a_0}{b_mx^m + b_{m-1}x^{m-1} + \cdots + b_1x + b_0}$ where $P(x)$ and $Q(x)$ have no common factors.

1. If $n < m$, then the x-axis ($y = 0$) is a horizontal asymptote.
2. If $n = m$, then $y = \dfrac{a_n}{b_m}$ is a horizontal asymptote.
3. If $n > m$, then there are no horizontal asymptotes.

■ You should be able to graph $f(x) = \dfrac{p(x)}{q(x)}$.

(a) Find the x-and y-intercepts.

(b) Find any vertical or horizontal asymptotes.

(c) Plot additional points.

(d) If the degree of the numerator is one more than the degree of the denominator, use long division to find the slant asymptote.

Solutions to Odd-Numbered Exercises

1. $f(x) = \frac{1}{x^2}$

Domain: all real numbers except $x = 0$

Vertical asymptote: $x = 0$

Horizontal asymptote: $y = 0$

[Degree of $p(x) <$ degree of $q(x)$]

3. $f(x) = \frac{2 + x}{2 - x} = \frac{x + 2}{-x + 2}$

Domain: all real numbers except $x = 2$

Vertical asymptote: $x = 2$

Horizontal asymptote: $y = -1$

[Degree of $p(x) =$ degree of $q(x)$]

5. $f(x) = \frac{x^3}{x^2 - 1}$

Domain: all real numbers except $x = \pm 1$

Vertical asymptote: $x = \pm 1$

Horizontal asymptotes: None

[Degree of $p(x) >$ degree of $q(x)$]

7. $f(x) = \frac{2}{x + 2}$

Vertical asymptote: $x = -2$

Horizontal asymptote: $y = 0$

Matches graph (d).

9. $f(x) = \frac{4x + 1}{x}$

Vertical asymptote: $x = 0$

Horizontal asymptote: $y = 4$

Matches graph (f).

11. $f(x) = \frac{x - 2}{x - 4}$

Vertical asymptote: $x = 4$

Horizontal asymptote: $y = 1$

Matches graph (e).

13. $f(x) = \frac{x^2 - 4}{x + 2}, \; g(x) = x - 2$

(a) Domain of f: all real numbers except $x = -2$
Domain of g: all real numbers

(b) Since $x + 2$ is a common factor of both the numerator and the denominator of $f(x)$, $x = -2$ is not a vertical asymptote of f. f has no vertical asymptotes.

(c)

x	-4	-3	-2.5	-2	-1.5	-1	0
$f(x)$	-6	-5	-4.5	undef.	-3.5	-3	-2
$g(x)$	-6	-5	-4.5	-4	-3.5	-3	-2

(d) f and g differ only where f is undefined.

15. $f(x) = \frac{1}{(x + 2)(x - 1)} = \frac{1}{x^2 + x - 2}$

17. $f(x) = \frac{2x^2}{x^2 + 1}$

19. $f(x) = 4 - \frac{1}{x}$

(a) As $x \to \pm\infty, f(x) \to 4$

(b) As $x \to \infty, f(x) \to 4$ but is less than 4.

(c) As $x \to -\infty, f(x) \to 4$ but is greater than 4.

21. $f(x) = \frac{2x - 1}{x - 3}$

(a) As $x \to \pm\infty, f(x) \to 2$

(b) As $x \to \infty, f(x) \to 2$ but is greater than 2.

(c) As $x \to -\infty, f(x) \to 2$ but is less than 2.

23. $f(x) = \dfrac{x^2 - 4}{x + 4} = \dfrac{(x + 2)(x - 2)}{x + 1}$

The zeros of f correspond to the zeros of the numerator and are $x = \pm 2$.

25. $f(x) = 1 - \dfrac{2}{x - 3} = \dfrac{x - 5}{x - 3}$

The zero of f corresponds to the zero of the numerator and is $x = 5$.

27. $t = \dfrac{38M + 16{,}965}{10(M + 5000)}$

M	200	400	600	800	1000
t	0.472	0.596	0.710	0.817	0.916

M	1200	1400	1600	1800	2000
t	1.009	1.096	1.178	1.255	1.328

The greater the mass, the more time required per oscillation. Also, the model is a "good fit" to the actual data.

29. $g(x) = \dfrac{2}{x} + 1$

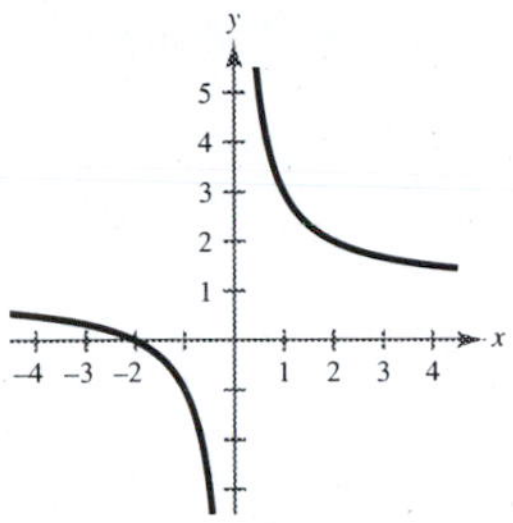

Vertical shift one unit upward

31. $g(x) = -\dfrac{2}{x}$

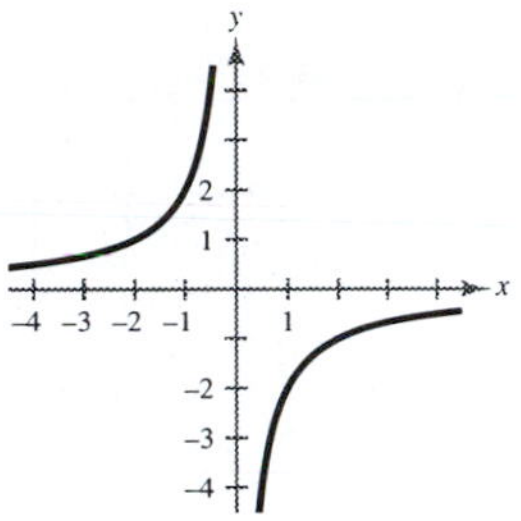

Reflection in the x-axis

33. $f(x) = \dfrac{1}{x + 2}$

y-intercept: $\left(0, \dfrac{1}{2}\right)$

Vertical asymptote: $x = -2$

Horizontal asymptote: $y = 0$

x	-4	-3	-1	0	1
y	$-\frac{1}{2}$	-1	1	$\frac{1}{2}$	$\frac{1}{3}$

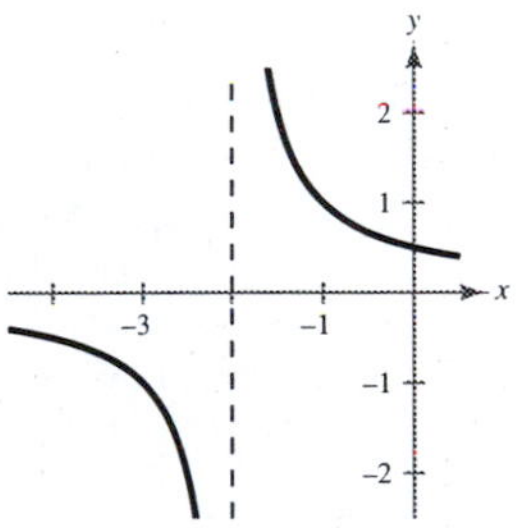

35. $h(x) = -\dfrac{1}{x + 2}$

y-intercept: $\left(0, -\dfrac{1}{2}\right)$

Vertical asymptote: $x = -2$

Horizontal asymptote: $y = 0$

x	-4	-3	-1	0
y	$\frac{1}{2}$	1	-1	$-\frac{1}{2}$

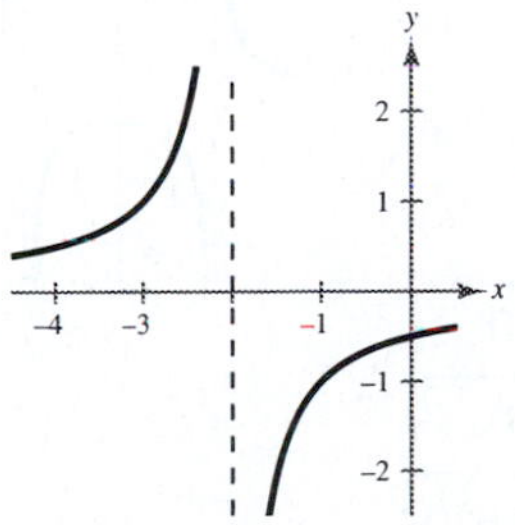

Note: This is the graph of $f(x) = \dfrac{1}{x + 2}$ (Exercise 33) reflected about the x-axis.

37. $C(x) = \dfrac{5 + 2x}{1 + x} = \dfrac{2x + 5}{x + 1}$

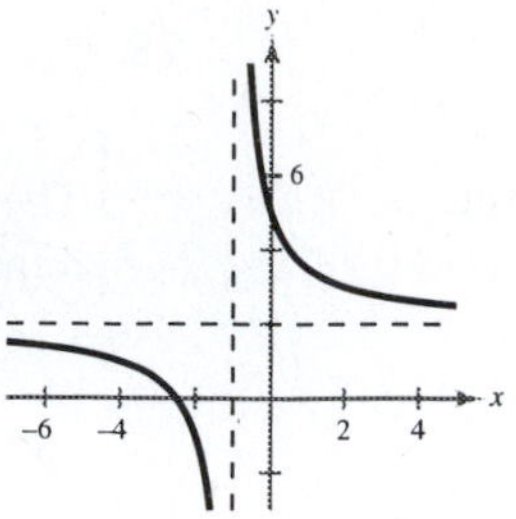

y-intercept: $\left(-\dfrac{5}{2}, 0\right)$

x-intercept: (0, 5)

Vertical asymptote: $x = -1$

Horizontal asymptote: $y = 2$

x	-4	-3	-1	0	1	2
$C(x)$	1	$\frac{1}{2}$	-1	5	$\frac{7}{2}$	3

39. $g(x) = \dfrac{1}{x + 2} + 2 = \dfrac{2x + 5}{x + 2}$

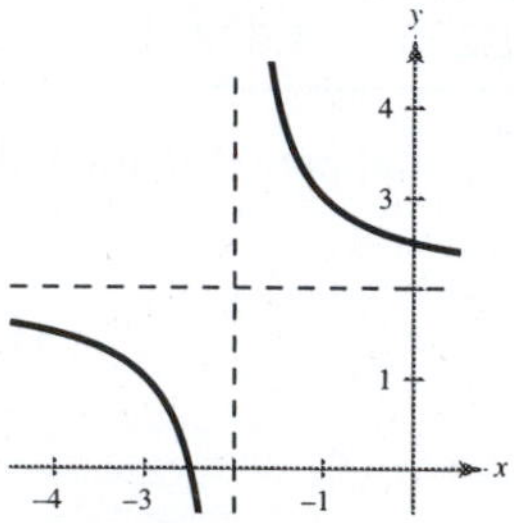

Intercepts: $\left(-\dfrac{5}{2}, 0\right), \left(0, \dfrac{5}{2}\right)$

Vertical asymptote: $x = -2$

Horizontal asymptote: $y = 2$

x	-4	-3	-1	0	1
y	$\frac{3}{2}$	1	3	$\frac{5}{2}$	$\frac{7}{3}$

Note: This is the graph of $f(x) = \dfrac{1}{x + 2}$ (Exercise 33) shifted upward two units.

41. $f(x) = \dfrac{x^2}{x^2 + 9}$

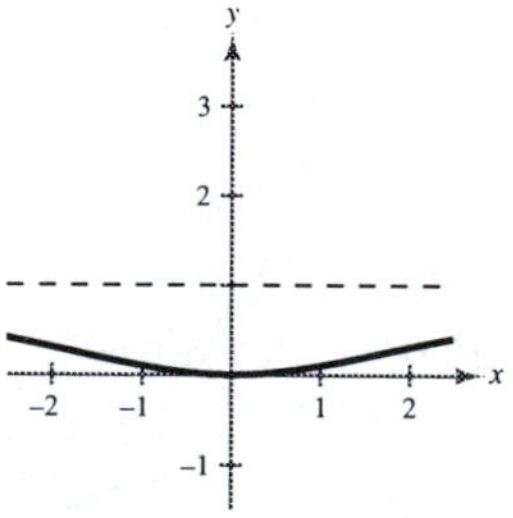

Intercept: (0, 0)

Horizontal asymptote: $y = 1$

y-axis symmetry

x	± 1	± 2	± 3
y	$\frac{1}{10}$	$\frac{4}{13}$	$\frac{1}{2}$

43. $h(x) = \dfrac{x^2}{x^2 - 9}$

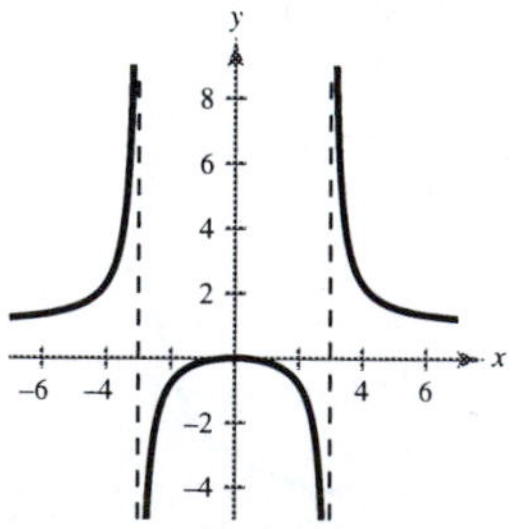

Intercept: (0, 0)

Vertical asymptotes: $x = \pm 3$

Horizontal asymptote: $y = 1$

y-axis symmetry

x	± 5	± 4	± 2	± 1	0
y	$\frac{25}{16}$	$\frac{16}{7}$	$-\frac{4}{5}$	$-\frac{1}{8}$	0

45. $g(s) = \dfrac{s}{s^2 + 1}$

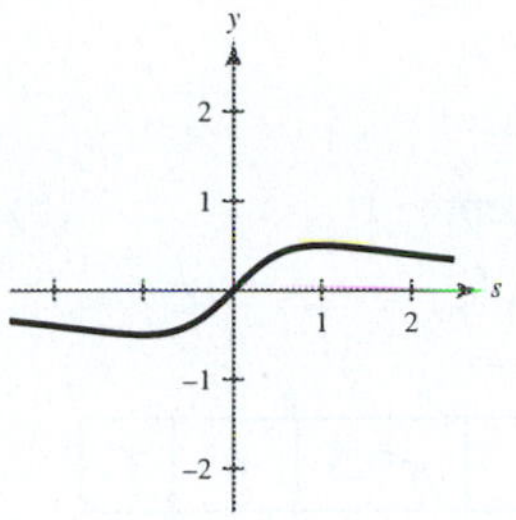

Intercept: $(0, 0)$

Horizontal asymptote: $y = 0$

Origin symmetry

x	-2	-1	0	1	2
$g(s)$	$-\frac{2}{5}$	$-\frac{1}{2}$	0	$\frac{1}{2}$	$\frac{2}{5}$

47. $g(x) = \dfrac{4(x + 1)}{x(x - 4)}$

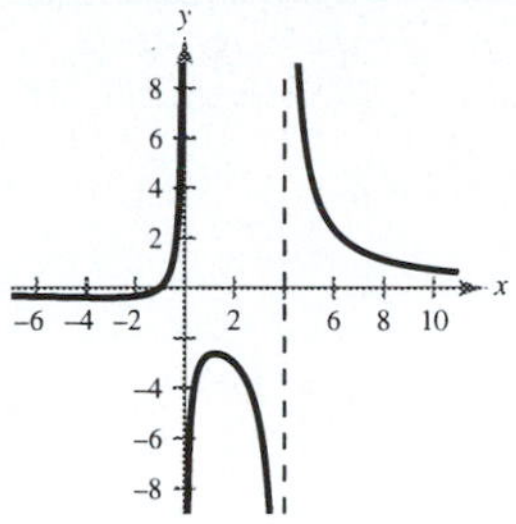

Intercept: $(-1, 0)$

Vertical asymptotes: $x = 0$ and $x = 4$

Horizontal asymptote: $y = 0$

x	-2	-1	1	2	3	5	6
y	$-\frac{1}{3}$	0	$-\frac{8}{3}$	-3	$-\frac{16}{3}$	$\frac{24}{5}$	$\frac{7}{3}$

49. $f(x) = \dfrac{3x}{x^2 - x - 2} = \dfrac{3x}{(x + 1)(x - 2)}$

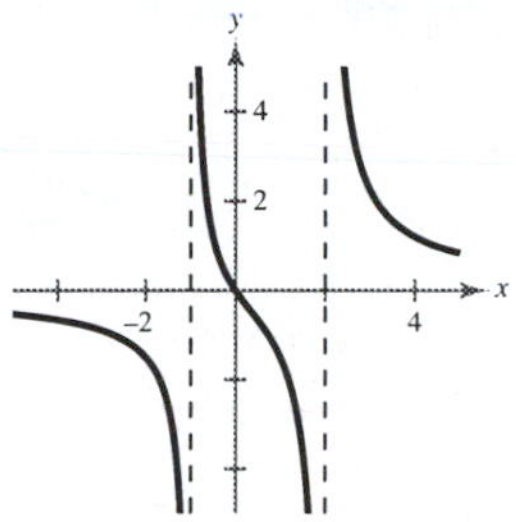

Intercept: $(0, 0)$

Vertical asymptotes: $x = -1, 2$

Horizontal asymptote: $y = 0$

x	-3	0	1	3	4
y	$-\frac{9}{10}$	0	$-\frac{3}{2}$	$\frac{9}{4}$	$\frac{6}{5}$

51. $f(x) = \dfrac{2 + x}{1 - x} = -\dfrac{x + 2}{x - 1}$

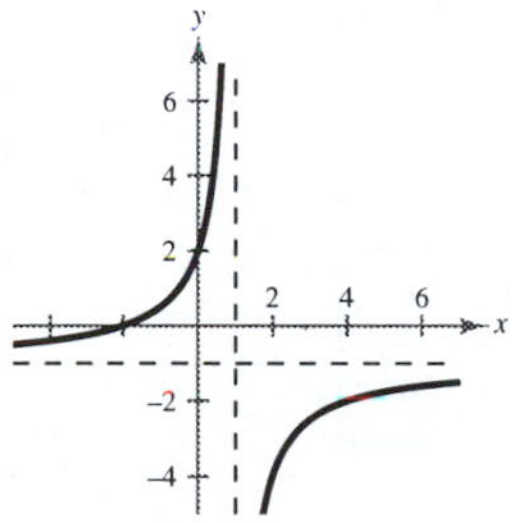

x-intercept: $(-2, 0)$

y-intercept: $(0, 2)$

Vertical asymptotes: $x = 1$

Horizontal asymptote: $y = -1$

x	-2	-1	0	2	3
y	0	$\frac{1}{2}$	2	-4	$-\frac{5}{2}$

53. $f(t) = \dfrac{3t + 1}{t}$

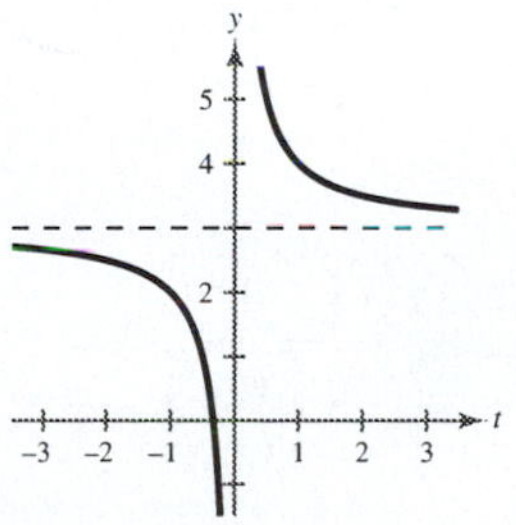

t-intercept: $\left(-\dfrac{1}{3}, 0\right)$

Vertical asymptotes: $t = 0$

Horizontal asymptote: $y = 3$

t	-2	-1	1	2
$f(t)$	$\frac{5}{2}$	2	4	$\frac{7}{2}$

55. $f(x) = \dfrac{x^2 - 1}{x + 1}$, $g(x) = x - 1$

(a) Domain of f: all real numbers except -1
Domain of g: all real numbers

(b) Since $(x + 1)$ is a factor of both the numerator and the denominator of f, $x = -1$ is not a vertical asymptote. f has no vertical asymptotes.

(c)

x	-3	-2	-1.5	-1	-0.5	0	1
$f(x)$	-4	-3	-2.5	undef.	-1.5	-1	0
$g(x)$	-4	-3	-2.5	-2	-1.5	-1	0

(d)

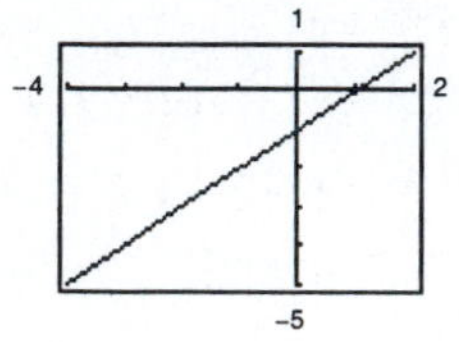

(e) Since there are only a finite number of pixels, the utility may not attempt to evaluate the function where it does not exist.

57. $h(t) = \dfrac{4}{t^2 + 1}$

Domain: all real numbers OR $(-\infty, \infty)$
Horizontal asymptote: $y = 0$

t	± 2	± 1	0
$h(t)$	$\frac{4}{5}$	2	4

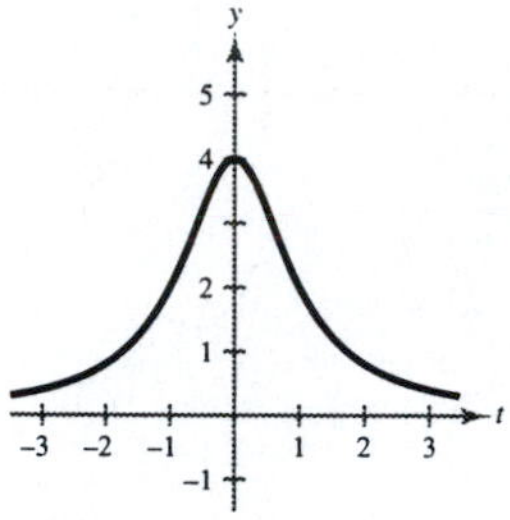

59. $f(t) = \dfrac{2t^2}{t^2 - 4}$

Domain: all real numbers except ± 2,
OR $(-\infty, -2) \cup (-2, 2) \cup (2, \infty)$
Vertical asymptotes: $x = \pm 2$
Horizontal asymptote: $y = 2$

t	± 4	± 3	± 1	0
$f(t)$	$\frac{8}{3}$	$\frac{18}{5}$	$-\frac{2}{3}$	0

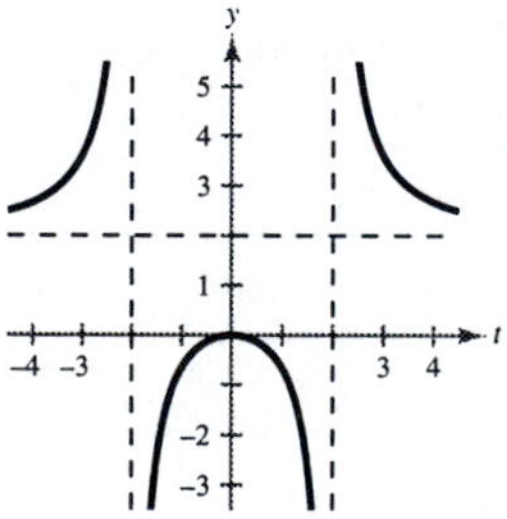

61. $h(x) = \dfrac{6x}{\sqrt{x^2 + 1}}$

Horizontal asymptotes: $y = \pm 6$

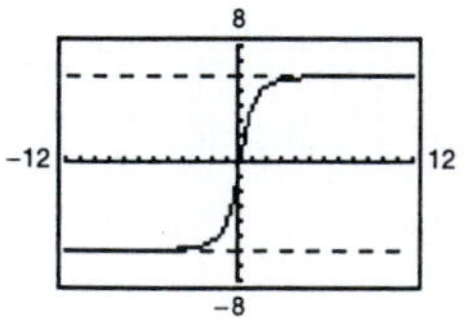

63. $f(x) = \dfrac{4(x - 1)^2}{x^2 - 4x + 5}$

Horizontal asymptote: $y = 4$

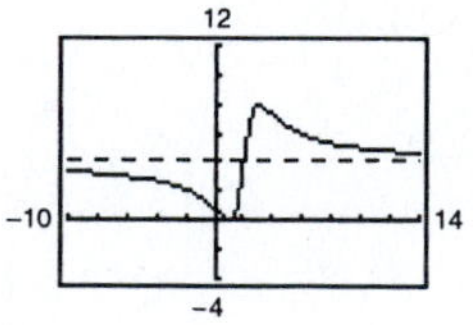

65. $h(x) = \dfrac{6 - 2x}{3 - x} = \dfrac{2(3 - x)}{3 - x}$

Since $h(x)$ is not reduced and $(3 - x)$ is a factor of both the numerator and the denominator, $x = 3$ is not a horizontal asymptote.

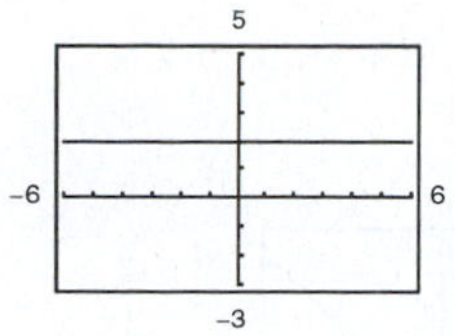

67. False. The graph would have two distinct branches that are separated by the vertical asymptote.

69. $f(x) = \dfrac{2x^2 + 1}{x} = 2x + \dfrac{1}{x}$

Vertical asymptote: $x = 0$

Slant asymptote: $y = 2x$

Origin symmetry

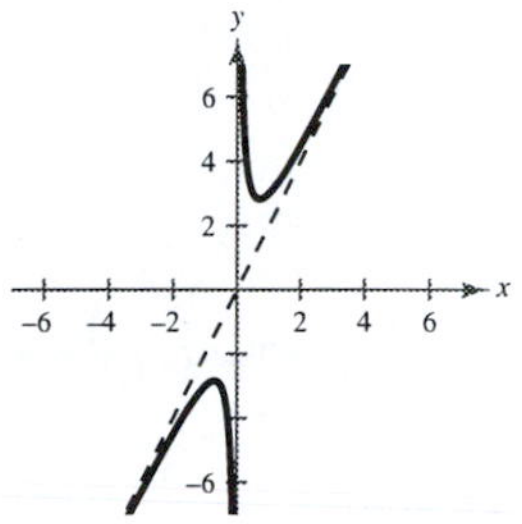

71. $g(x) = \dfrac{x^2 + 1}{x} = x + \dfrac{1}{x}$

Vertical asymptote: $x = 0$

Slant asymptote: $y = x$

Origin symmetry

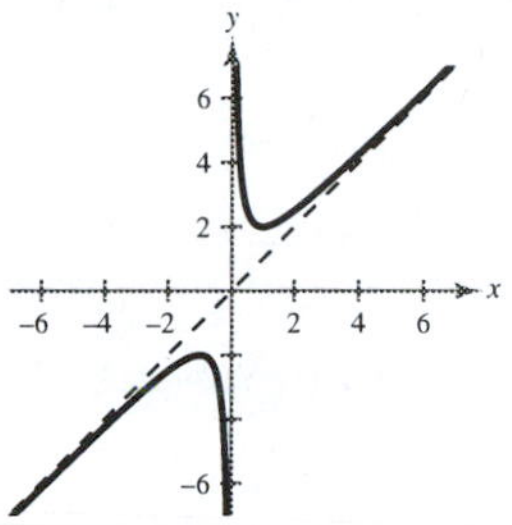

73. $f(x) = \dfrac{x^3}{x^2 - 1} = x + \dfrac{x}{x^2 - 1}$

Intercept: $(0, 0)$

Vertical asymptotes: $x = \pm 1$

Slant asymptote: $y = x$

Origin symmetry

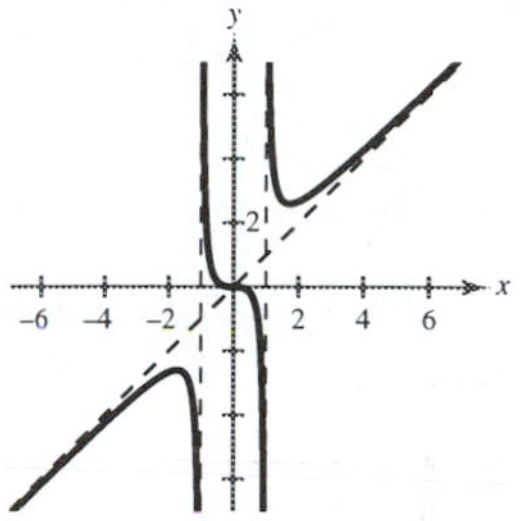

75. $f(x) = \dfrac{x^2 + 5x + 8}{x + 3} = x + 2 + \dfrac{2}{x + 3}$

Domain: all real numbers except -3
OR $(-\infty, -3) \cup (-3, \infty)$

y-intercept: $\left(0, \frac{8}{3}\right)$

Vertical asymptote: $x = -3$

Slant asymptote: $y = x + 2$

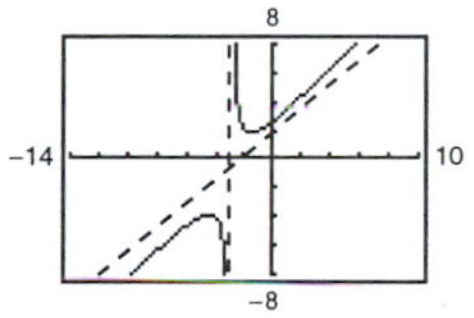

77. $y = \dfrac{1}{x + 5} + \dfrac{4}{x}$

(a)

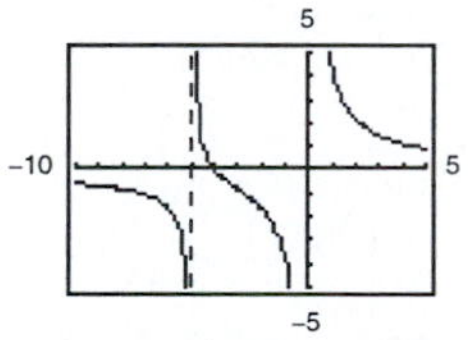

x-intercept: $(-4, 0)$

(b)

$$0 = \frac{1}{x + 5} + \frac{4}{x}$$

$$-\frac{4}{x} = \frac{1}{x + 5}$$

$$-4(x + 5) = x$$

$$-4x - 20 = x$$

$$-5x = 20$$

$$x = -4$$

79. $y = x - \dfrac{6}{x-1}$

(a)

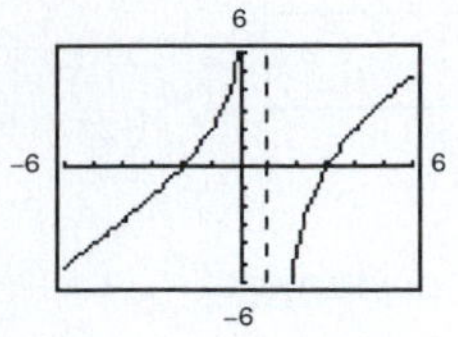

x-intercept: $(-2, 0), (3, 0)$

(b)

$$0 = x - \frac{6}{x-1}$$
$$\frac{6}{x-1} = x$$
$$6 = x(x-1)$$
$$0 = x^2 - x - 6$$
$$0 = (x+2)(x-3)$$
$$x = -2, \quad x = 3$$

81. (a) $0.25(50) + 0.75(x) = C(50 + x)$

$$C = \frac{12.50 + 0.75x}{50 + x} \cdot \frac{4}{4}$$

$$C = \frac{50 + 3x}{4(50 + x)} = \frac{3x + 50}{4(x + 50)}$$

(b) Domain: $x > 0$ and $x \le 1000 - 50$
Thus, $0 \le x \le 950$ OR $[0, 950]$.

(c)

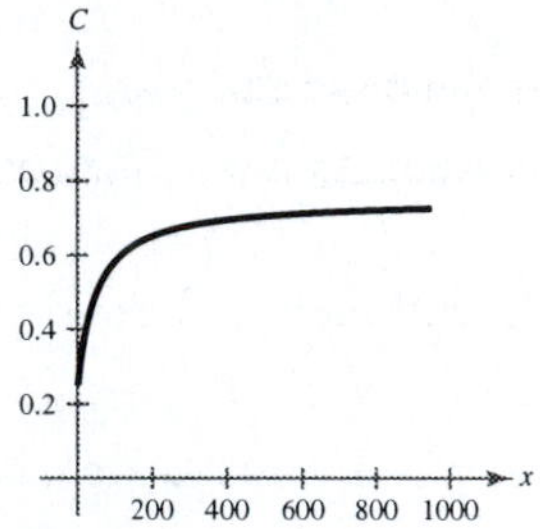

As the tank is filled, the rate at which the concentration is increasing slows down. It approaches the horizontal asymptote of $C = \frac{3}{4} = 0.75$.

Section 2.8 Partial Fractions

- You should know how to decompose a rational function $\dfrac{N(x)}{D(x)}$ into partial fractions.
  (a) If the fraction is improper, divide to obtain
  $$\frac{N(x)}{D(x)} = p(x) + \frac{N_1(x)}{D(x)}$$
  where $p(x)$ is a polynomial.
  (b) Factor the denominator completely into linear and irreducible (over the reals) quadratic factors.
  (c) For each factor of the form $(px + q)^m$, the partial fraction decomposition includes the terms
  $$\frac{A_1}{(px+q)} + \frac{A_2}{(px+q)^2} + \cdots + \frac{A_m}{(px+q)^m}.$$
  (d) For each factor of the form $(ax^2 + bx + c)^n$, the partial fraction decomposition includes the terms
  $$\frac{B_1x + C_1}{ax^2+bx+c} + \frac{B_2x + C_2}{(ax^2+bx+c)^2} + \cdots + \frac{B_nx + C_n}{(ax^2+bx+c)^n}.$$
- You should know how to determine the values of the constants in the numerators.
  (a) Set $\dfrac{N_1(x)}{D(x)} =$ partial fraction decomposition.
  (b) Multiply both sides by $D(x)$. This is called the basic equation.
  (c) For distinct linear factors, substitute the roots of the distinct linear factors into the basic equation.
  (d) For repeated linear factors, use the coefficients found in part (c) to rewrite the basic equation. Then use other values of x to solve for the remaining coefficients.
  (e) For quadratic factors, expand the basic equation, collect like terms, and then equate the coefficients of like terms.

Solutions to Odd-Numbered Exercises

1. $\dfrac{7}{x^2 - 14x} = \dfrac{7}{x(x-14)} = \dfrac{A}{x} + \dfrac{B}{x-14}$

3. $\dfrac{12}{x^3 - 10x^2} = \dfrac{12}{x^2(x-10)} = \dfrac{A}{x} + \dfrac{B}{x^2} + \dfrac{C}{x-10}$

5. $\dfrac{2x-3}{x^3 + 10x} = \dfrac{2x-3}{x(x^2+10)} = \dfrac{A}{x} + \dfrac{Bx + C}{x^2+10}$

7. $\dfrac{1}{x^2 - 1} = \dfrac{A}{x+1} + \dfrac{B}{x-1}$

$$1 = A(x-1) + B(x+1)$$

Let $x = -1$: $1 = -2A \implies A = -\dfrac{1}{2}$

Let $x = 1$: $1 = 2B \implies B = \dfrac{1}{2}$

$$\frac{1}{x^2-1} = \frac{1/2}{x-1} - \frac{1/2}{x+1} = \frac{1}{2}\left(\frac{1}{x-1} - \frac{1}{x+1}\right)$$

9. $\dfrac{1}{x^2 + x} = \dfrac{A}{x} + \dfrac{B}{x+1}$

$$1 = A(x+1) + Bx$$

Let $x = 0$: $1 = A$

Let $x = -1$: $1 = -B \implies B = -1$

$$\frac{1}{x^2+x} = \frac{1}{x} - \frac{1}{x+1}$$

11. $\dfrac{1}{2x^2 + x} = \dfrac{A}{2x + 1} + \dfrac{B}{x}$

$1 = Ax + B(2x + 1)$

Let $x = -\frac{1}{2}$: $1 = -\frac{1}{2}A \implies A = -2$

Let $x = 0$: $1 = B$

$\dfrac{1}{2x^2 + x} = \dfrac{1}{x} - \dfrac{2}{2x + 1}$

13. $\dfrac{3}{x^2 + x - 2} = \dfrac{A}{x - 1} + \dfrac{B}{x + 2}$

$3 = A(x + 2) + B(x - 1)$

Let $x = 1$: $3 = 3A \implies A = 1$

Let $x = -2$: $3 = -3B \implies B = -1$

$\dfrac{3}{x^2 + x - 2} = \dfrac{1}{x - 1} - \dfrac{1}{x + 2}$

15. $\dfrac{x^2 + 12x + 12}{x^3 - 4x} = \dfrac{A}{x} + \dfrac{B}{x + 2} + \dfrac{C}{x - 2}$

$x^2 + 12x + 12 = A(x + 2)(x - 2) + Bx(x - 2) + Cx(x + 2)$

Let $x = 0$: $12 = -4A \implies A = -3$

Let $x = -2$: $-8 = 8B \implies B = -1$

Let $x = 2$: $40 = 8C \implies C = 5$

$\dfrac{x^2 + 12x + 12}{x^3 - 4x} = -\dfrac{3}{x} - \dfrac{1}{x + 2} + \dfrac{5}{x - 2}$

17. $\dfrac{4x^2 + 2x - 1}{x^2(x + 1)} = \dfrac{A}{x} + \dfrac{B}{x^2} + \dfrac{C}{x + 1}$

$4x^2 + 2x - 1 = Ax(x + 1) + B(x + 1) + Cx^2$

Let $x = 0$: $-1 = B$

Let $x = -1$: $1 = C$

Let $x = 1$: $5 = 2A + 2B + C$

$5 = 2A - 2 + 1$

$6 = 2A$

$3 = A$

$\dfrac{4x^2 + 2x - 1}{x^2(x + 1)} = \dfrac{3}{x} - \dfrac{1}{x^2} + \dfrac{1}{x + 1}$

19. $\dfrac{3x}{(x - 3)^2} = \dfrac{A}{x - 3} + \dfrac{B}{(x - 3)^2}$

$3x = A(x - 3) + B$

Let $x = 3$: $9 = B$

Let $x = 0$: $0 = -3A + B$

$0 = -3A + 9$

$3 = A$

$\dfrac{3x}{(x - 3)^2} = \dfrac{3}{x - 3} + \dfrac{9}{(x - 3)^2}$

21. $\dfrac{x^2 - 1}{x(x^2 + 1)} = \dfrac{A}{x} + \dfrac{Bx + C}{x^2 + 1}$

$x^2 - 1 = A(x^2 + 1) + (Bx + C)x$

Let $x = 0$: $-1 = A$

$x^2 - 1 = Ax^2 + A + Bx^2 + Cx$

$= -x^2 - 1 + Bx^2 + Cx$

$= x^2(B - 1) + Cx - 1$

Equating coefficients of like powers:

$1 = B - 1$

$2 = B$ and $0 = C$

$\dfrac{x^2 - 1}{x(x^2 + 1)} = -\dfrac{1}{x} + \dfrac{2x}{x^2 + 1}$

23. $\dfrac{x^2}{x^4 - 2x^2 - 8} = \dfrac{x^2}{(x^2 - 4)(x^2 + 2)} = \dfrac{A}{x + 2} + \dfrac{B}{x - 2} + \dfrac{Cx + D}{x^2 + 2}$

$x^2 = A(x - 2)(x^2 + 2) + B(x + 2)(x^2 + 2) + (Cx + D)(x^2 - 4)$

Let $x = -2$: $4 = -24A \implies A = -\dfrac{1}{6}$

Let $x = 2$: $4 = 24B \implies B = \dfrac{1}{6}$

$$x^2 = -\frac{1}{6}(x - 2)(x^2 + 2) + \frac{1}{6}(x + 2)(x^2 + 2) + (Cx + D)(x^2 - 4)$$

$$x^2 = -\frac{1}{6}x^3 + \frac{1}{3}x^2 - \frac{1}{3}x + \frac{2}{3} + \frac{1}{6}x^3 + \frac{1}{3}x^2 + \frac{1}{3}x + \frac{2}{3} + Cx^3 + Dx^2 - 4Cx - 4D$$

$$x^2 = Cx^3 + \left(\frac{2}{3} + D\right)x^2 - 4Cx + \left(\frac{4}{3} - 4D\right)$$

Equating coefficients of like powers:

$C = 0$

$1 = \dfrac{2}{3} + D \implies D = \dfrac{1}{3}$

$$\frac{x^2}{x^4 - 2x^2 - 8} = -\frac{1}{6(x + 2)} + \frac{1}{6(x - 2)} + \frac{1}{3(x^2 + 2)}$$

25. $\dfrac{x}{16x^4 - 1} = \dfrac{A}{2x + 1} + \dfrac{B}{2x - 1} + \dfrac{Cx + D}{4x^2 + 1}$

$x = A(2x - 1)(4x^2 + 1) + B(2x + 1)(4x^2 + 1) + (Cx + D)(2x + 1)(2x - 1)$

Let $x = -\dfrac{1}{2}$: $-\dfrac{1}{2} = -4A \implies A = \dfrac{1}{8}$

Let $x = \dfrac{1}{2}$: $\dfrac{1}{2} = 4B \implies B = \dfrac{1}{8}$

Let $x = 0$: $0 = -A + B - D$

$0 = -\dfrac{1}{8} + \dfrac{1}{8} - D$

$0 = D$

Let $x = 1$: $1 = 5A + 15B + 3C + 3D$

$1 = \dfrac{5}{8} + \dfrac{15}{8} + 3C + 0$

$-\dfrac{1}{2} = C$

$$\frac{x}{16x^4 - 1} = \frac{1/8}{2x + 1} + \frac{1/8}{2x - 1} - \frac{x/2}{4x^2 + 1} = \frac{1}{8(2x + 1)} + \frac{1}{8(2x - 1)} - \frac{x}{2(4x^2 + 1)}$$

27. $$\frac{x^2+5}{(x+1)(x^2-2x+3)} = \frac{A}{x+1} + \frac{Bx+C}{x^2-2x+3}$$

$$x^2+5 = A(x^2-2x+3) + (Bx+C)(x+1)$$

Let $x=-1$: $6 = 6A \implies A = 1$

$$x^2+5 = x^2-2x+3+Bx^2+Bx+Cx+C$$
$$= x^2(1+B) + x(-2+B+C) + (3+C)$$

Equating coefficients of like powers:

$1 = 1 + B$, $\quad 0 = -2 + B + C$, and $\quad 5 = 3 + C$

$0 = B \qquad 0 = -2 + 0 + C \qquad 2 = C$

$2 = C$

$$\frac{x^2+5}{(x+1)(x^2-2x+3)} = \frac{1}{x+1} + \frac{2}{x^2-2x+3}$$

29. $$\frac{x^4}{(x-1)^3} = \frac{x^4}{x^3-3x^2+3x-1} = x+3+\frac{6x^2-8x+3}{(x-1)^3}$$

$$\frac{6x^2-8x+3}{(x-1)^3} = \frac{A}{x-1} + \frac{B}{(x-1)^2} + \frac{C}{(x-1)^3}$$

$$6x^2-8x+3 = A(x-1)^2 + B(x-1) + C$$

Let $x = 1$: $1 = C$

$$6x^2-8x+3 = Ax^2-2Ax+A+Bx-B+1$$

$$6x^2-8x+3 = Ax^2+(-2A+B)x+(A-B+1)$$

Equating coefficients of like powers:

$6 = A$, $\quad -8 = -2A + B$ and $\quad 3 = A - B + 1$

$-8 = -12 + B \qquad 3 = 6 - B + 1$

$4 = B \qquad 4 = B$

$$\frac{x^4}{(x-1)^3} = x+3+\frac{6}{x-1}+\frac{4}{(x-1)^2}+\frac{1}{(x-1)^3}$$

31. $$\frac{5-x}{2x^2+x-1} = \frac{A}{2x-1} + \frac{B}{x+1}$$

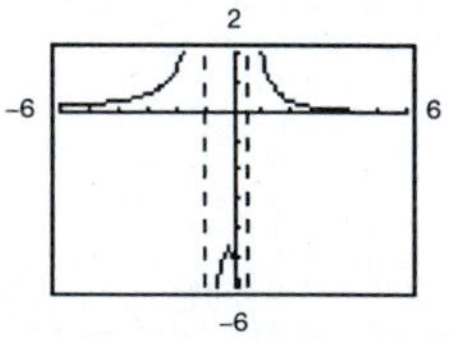

$$-x+5 = A(x+1) + B(2x-1)$$

Let $x = \frac{1}{2}$: $\frac{9}{2} = \frac{3}{2}A \implies A = 3$

Let $x = -1$: $6 = -3B \implies B = -2$

$$\frac{5-x}{2x^2+x-1} = \frac{3}{2x-1} - \frac{2}{x+1}$$

33. $\dfrac{x-1}{x^3+x^2} = \dfrac{A}{x} + \dfrac{B}{x^2} + \dfrac{C}{x+1}$

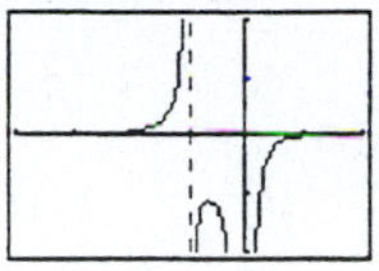

$x - 1 = Ax(x+1) + B(x+1) + Cx^2$

Let $x = -1$: $-2 = C$

Let $x = 0$: $-1 = B$

Let $x = 1$: $0 = 2A + 2B + C$

$0 = 2A - 2 - 2$

$2 = A$

$$\frac{x-1}{x^3+x^2} = \frac{2}{x} - \frac{1}{x^2} - \frac{2}{x+1}$$

35. $\dfrac{x^2+x+2}{(x^2+2)^2} = \dfrac{Ax+B}{x^2+2} + \dfrac{Cx+D}{(x^2+2)^2}$

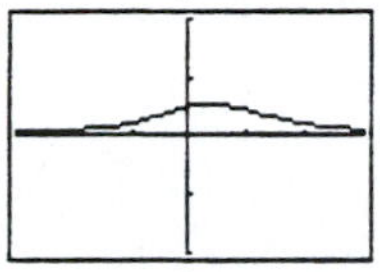

$x^2 + x + 2 = (Ax+B)(x^2+2) + Cx + D$

$x^2 + x + 2 = Ax^3 + Bx^2 + (2A + C)x + (2B + D)$

Equating coefficients of like powers:

$0 = A$

$1 = B$

$1 = 2A + C \implies C = 1$

$2 = 2B + D \implies D = 0$

$$\frac{x^2+x+2}{(x^2+2)^2} = \frac{1}{x^2+2} + \frac{x}{(x^2+2)^2}$$

37. $\dfrac{2x^3-4x^2-15x+5}{x^2-2x-8} = 2x + \dfrac{x+5}{(x+2)(x-4)}$

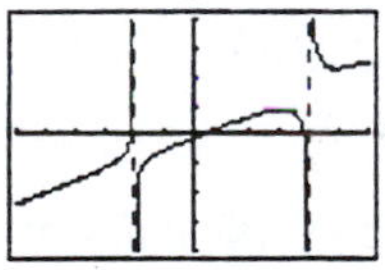

$\dfrac{x+5}{(x+2)(x-4)} = \dfrac{A}{x+2} + \dfrac{B}{x-4}$

$x + 5 = A(x-4) + B(x+2)$

Let $x = -2$: $3 = -6A \implies A = -\dfrac{1}{2}$

Let $x = 4$: $9 = 6B \implies B = \dfrac{3}{2}$

$$\frac{2x^3-4x^2-15x+5}{x^2-2x-8} = 2x + \frac{1}{2}\left(\frac{3}{x-4} - \frac{1}{x+2}\right)$$

39. $\dfrac{1}{a^2-x^2} = \dfrac{A}{a+x} + \dfrac{B}{a-x}$, a is a constant.

$1 = A(a-x) + B(a+x)$

Let $x = -a$: $1 = 2aA \implies A = \dfrac{1}{2a}$

Let $x = a$: $1 = 2aB \implies B = \dfrac{1}{2a}$

$$\frac{1}{a^2-x^2} = \frac{1}{2a}\left(\frac{1}{a+x} + \frac{1}{a-x}\right)$$

41. $\dfrac{1}{y(a-y)} = \dfrac{A}{y} + \dfrac{B}{a-y}$

$$1 = A(a-y) + By$$

Let $y = 0$: $1 = aA \implies A = \dfrac{1}{a}$

Let $y = a$: $1 = aB \implies B = \dfrac{1}{a}$

$\dfrac{1}{y(a-y)} = \dfrac{1}{a}\left(\dfrac{1}{y} + \dfrac{1}{a-y}\right)$

43. $\dfrac{x-12}{x(x-4)} = \dfrac{A}{x} + \dfrac{B}{x-4}$

$x - 12 = A(x-4) + Bx$

Let $x = 0$: $-12 = -4A \implies A = 3$

Let $x = 4$: $-8 = 4B \implies B = -2$

$\dfrac{x-12}{x(x-4)} = \dfrac{3}{x} - \dfrac{2}{x-4}$

$y = \dfrac{x-12}{x(x-4)}$

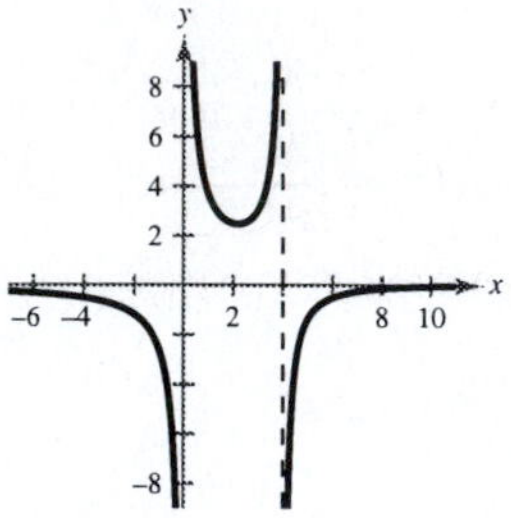

Vertical asymptotes: $x = 0$ and $x = 4$

$y = \dfrac{3}{x}$

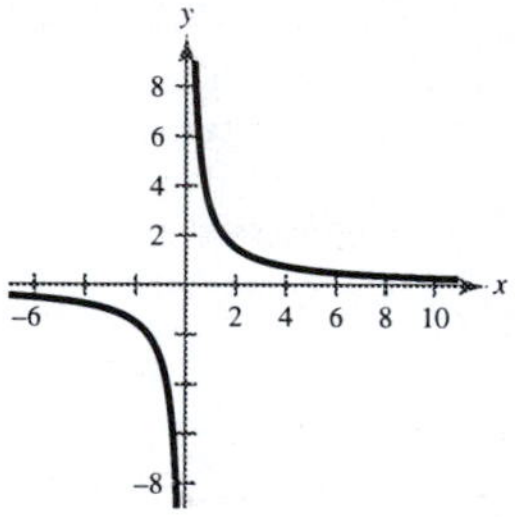

Vertical asymptote: $x = 0$

$y = -\dfrac{2}{x-4}$

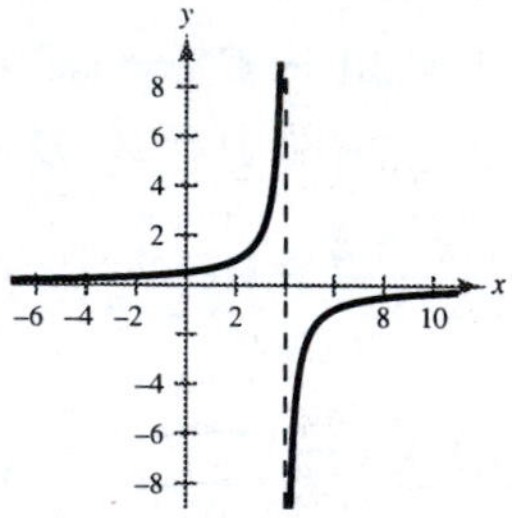

Vertical asymptote: $x = 4$

The combination of the vertical asymptotes of the terms of the decompositions are the same as the vertical asymptotes of the rational function.

45. $\dfrac{2(4x-3)}{x^2-9} = \dfrac{A}{x-3} + \dfrac{B}{x+3}$

$2(4x-3) = A(x+3) + B(x-3)$

Let $x = 3$: $18 = 6A \implies A = 3$

Let $x = -3$: $-30 = -6B \implies B = 5$

$\dfrac{2(4x-3)}{x^2-9} = \dfrac{3}{x-3} + \dfrac{5}{x+3}$

$y = \dfrac{2(4x-3)}{x^2-9}$

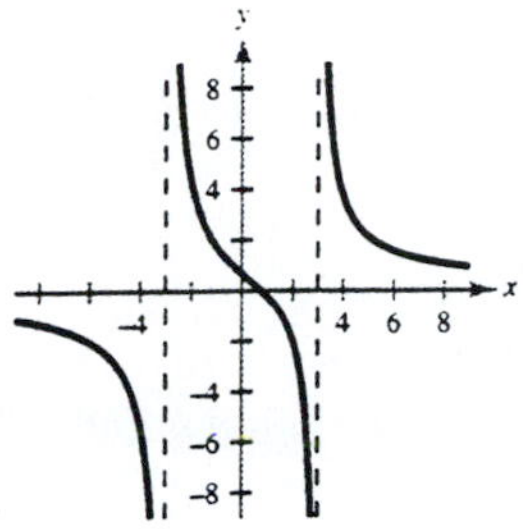

Vertical asymptotes: $x = \pm 3$

$y = \dfrac{3}{x-3}$

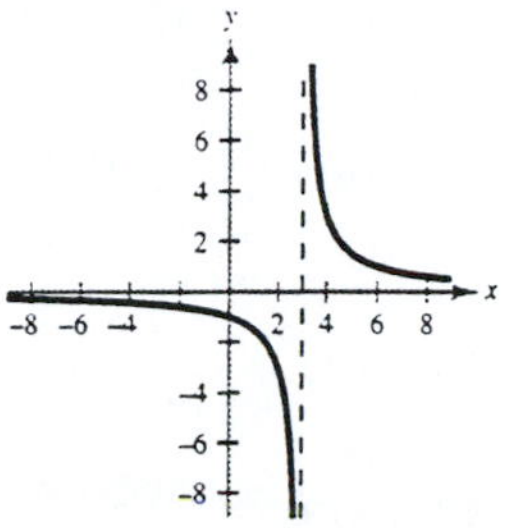

Vertical asymptote: $x = 3$

$y = \dfrac{5}{x+3}$

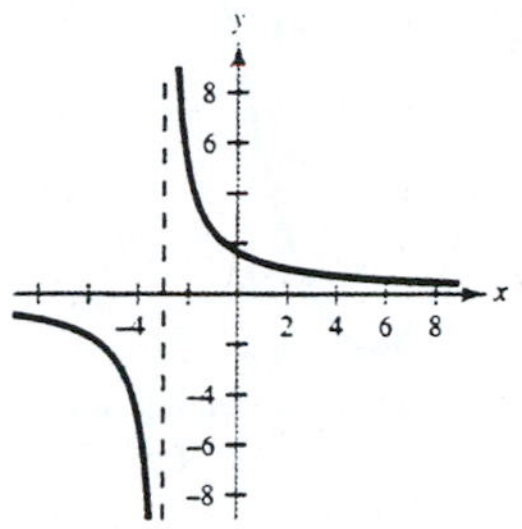

Vertical asymptote: $x = -3$

The combination of the vertical asymptotes of the terms of the decompositions are the same as the vertical asymptotes of the rational function.

47. (a) $\dfrac{2000(4-3x)}{(11-7x)(7-4x)} = \dfrac{A}{11-7x} + \dfrac{B}{7-4x},\ 0 \le x \le 1$

$2000(4-3x) = A(7-4x) + B(11-7x)$

Let $x = \dfrac{11}{7}$: $-\dfrac{10{,}000}{7} = \dfrac{5}{7}A \implies A = -2000$

Let $x = \dfrac{7}{4}$: $-2500 = -\dfrac{5}{4}B \implies B = 2000$

$\dfrac{2000(4-3x)}{(11-7x)(7-4x)} = \dfrac{-2000}{11-7x} + \dfrac{2000}{7-4x} = \dfrac{2000}{7-4x} - \dfrac{2000}{11-7x},\ 0 \le x \le 1$

(b) $y_1 = \dfrac{2000}{7-4x}$

$y_2 = \dfrac{2000}{11-7x}$

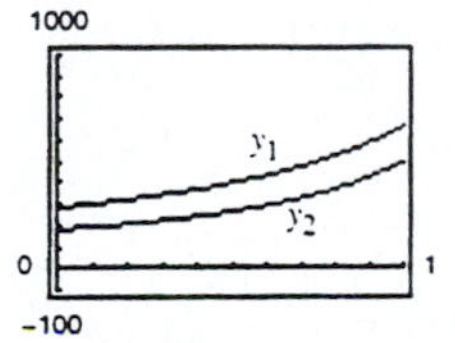

❑ Review Exercises for Chapter 2

Solutions to Odd-Numbered Exercises

1. $f(x) = \left(x + \frac{3}{2}\right)^2 + 1$

Vertex: $\left(-\frac{3}{2}, 1\right)$

y-intercept: $\left(0, \frac{13}{4}\right)$

No x-intercepts

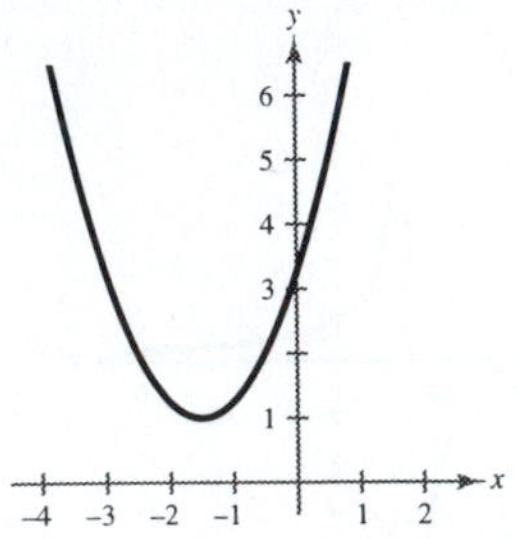

3. $f(x) = \frac{1}{3}(x^2 + 5x - 4)$

$$= \frac{1}{3}\left(x^2 + 5x + \frac{25}{4} - \frac{25}{4} - 4\right)$$

$$= \frac{1}{3}\left[\left(x - \frac{5}{2}\right)^2 - \frac{41}{4}\right]$$

$$= \frac{1}{3}\left(x - \frac{5}{2}\right)^2 - \frac{41}{12}$$

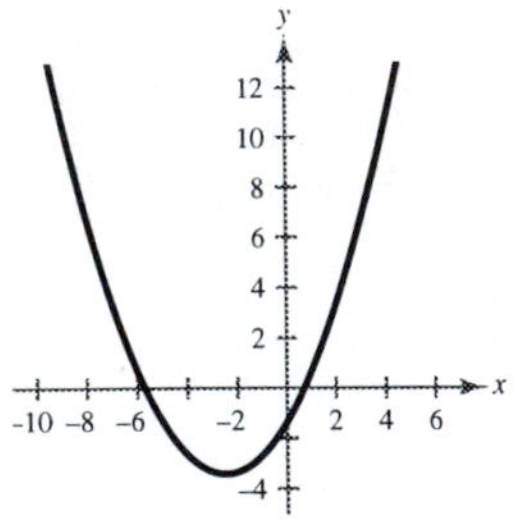

Vertex: $\left(\frac{5}{2}, -\frac{41}{12}\right)$

y-intercept: $\left(0, -\frac{4}{3}\right)$

x-intercept: $0 = \frac{1}{3}(x^2 + 5x - 4)$

$$0 = x^2 + 5x - 4$$

$$x = \frac{-5 \pm \sqrt{41}}{2}$$

Use the Quadratic Formula.

$$\left(\frac{-5 \pm \sqrt{41}}{2}, 0\right)$$

5. Vertex: $(1, -4)$ $\Rightarrow$ $f(x) = a(x - 1)^2 - 4$

Point: $(2, -3)$ $\Rightarrow$ $-3 = a(2 - 1)^2 - 4$

$$1 = a$$

Thus, $f(x) = (x - 1)^2 - 4$.

7. $g(x) = x^2 - 2x$

$$= x^2 - 2x + 1 - 1$$

$$= (x - 1)^2 - 1$$

The minimum occurs at the vertex $(1, -1)$.

9. $f(x) = 6x - x^2$

$$= -(x^2 - 6x + 9 - 9)$$

$$= -(x - 3)^2 + 9$$

The maximum occurs at the vertex $(3, 9)$.

11. $f(t) = -2t^2 + 4t + 1$

$$= -2(t^2 - 2t + 1 - 1) + 1$$

$$= -2[(t - 1)^2 - 1] + 1$$

$$= -2(t - 1)^2 + 3$$

The maximum occurs at the vertex $(1, 3)$.

13. (a)

x	y	Area
1	$4 - \frac{1}{2}(1)$	$(1)[4 - \frac{1}{2}(1)] = \frac{7}{2}$
2	$4 - \frac{1}{2}(2)$	$(2)[4 - \frac{1}{2}(2)] = 6$
3	$4 - \frac{1}{2}(3)$	$(3)[4 - \frac{1}{2}(3)] = \frac{15}{2}$
4	$4 - \frac{1}{2}(4)$	$(4)[4 - \frac{1}{2}(4)] = 8$
5	$4 - \frac{1}{2}(5)$	$(5)[4 - \frac{1}{2}(5)] = \frac{15}{2}$
6	$4 - \frac{1}{2}(6)$	$(6)[4 - \frac{1}{2}(6)] = 6$

(b) The dimensions that will produce a maximum area are $x = 4$ and $y = 2$.

(c) $A = xy = x\left(\frac{8 - x}{2}\right)$ since $x + 2y - 8 = 0 \implies y = \frac{8 - x}{2}$.

Since the figure is in the first quadrant and x and y must be positive,

the domain of $A = x\left(\frac{8 - x}{2}\right)$ is $0 < x < 8$.

(d)

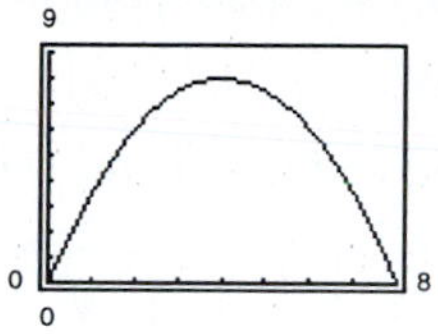

The maximum area of 8 occurs at the vertex when $x = 4$ and $y = \frac{8 - 4}{2} = 2$.

(e) $A = x\left(\frac{8 - x}{2}\right)$

$= \frac{1}{2}(8x - x^2)$

$= -\frac{1}{2}(x^2 - 8x)$

$= -\frac{1}{2}(x^2 - 8x + 16 - 16)$

$= -\frac{1}{2}[(x - 4)^2 - 16]$

$= -\frac{1}{2}(x - 4)^2 + 8$

The maximum area of 8 occurs when $x = 4$ and $y = \frac{8 - 4}{2} = 2$.

15. $f(x) = -x^2 + 6x + 9$

The degree is even and the leading coefficient is negative. The graph falls to the left and right.

17. $f(x) = \frac{3}{4}(x^4 + 3x^2 + 2)$

The degree is even and the leading coefficient is positive. The graph rises to the left and right.

19. $f(x) = \frac{1}{2}x^3 - 2x + 1;\ g(x) = \frac{1}{2}x^3$

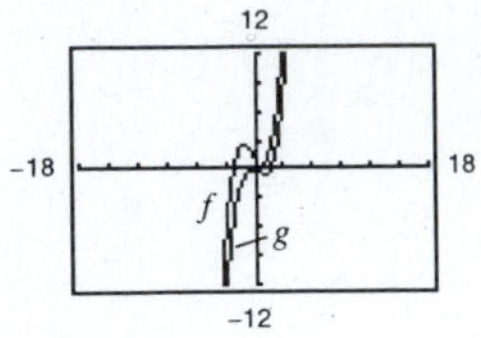

21. $g(x) = x^4 - x^3 - 2x^2$

$= x^2(x + 1)(x - 2)$

Intercepts: $(0, 0)$, $(-1, 0)$, $(2, 0)$

The graph rises to the left and to the right.

x	-2	-1	$-\frac{1}{2}$	0	1	2	3
y	16	0	-0.31	0	-2	0	36

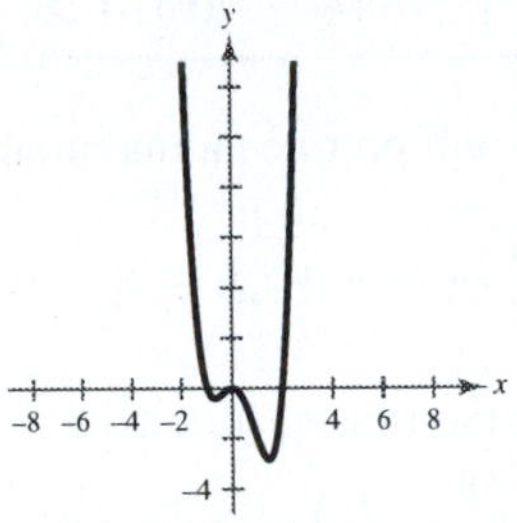

23. $f(t) = t^3 - 3t = t(t^2 - 3)$

Intercepts: $(0, 0)$, $(\pm\sqrt{3}, 0)$

The graph rises to the right and falls to the left.

x	-2	-1	0	1	2
y	-2	2	0	-2	2

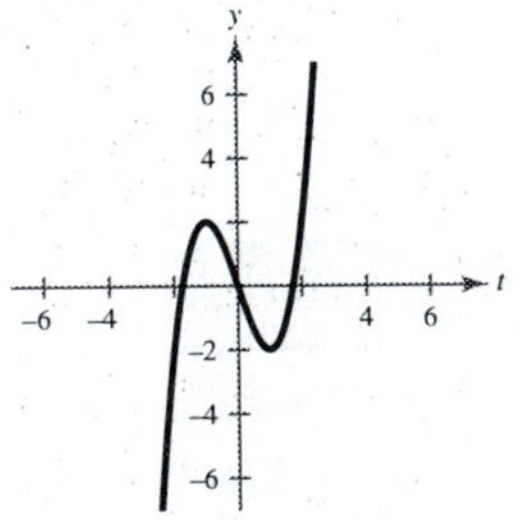

25. $f(x) = x(x + 3)^2$

Intercepts: $(0, 0)$, $(-3, 0)$

The graph falls to the left and rises to the right.

x	-4	-3	-2	-1	0	1
y	-4	0	-2	-4	0	16

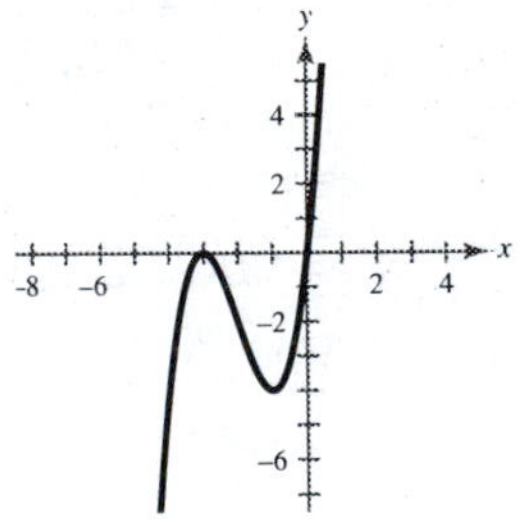

27. (a) The combined length and girth is

$$y + 4x = 216$$

$$y = 216 - 4x.$$

(b) The volume is $V = x^2y = x^2(216 - 4x)$.

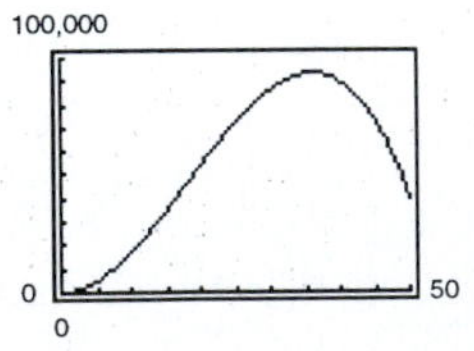

3

The volume is maximum when $x = 36$ centimeters and $y = 216 - 4(36) = 72$ centimeters.

29. $y_1 = \dfrac{x^2}{x-2}$

$$\begin{aligned} y_2 &= x + 2 + \frac{4}{x-2} \\ &= \frac{(x+2)(x-2)}{x-2} + \frac{4}{x-2} \\ &= \frac{x^2-4}{x-2} + \frac{4}{x-2} \\ &= \frac{x^2}{x-2} \\ &= y_1 \end{aligned}$$

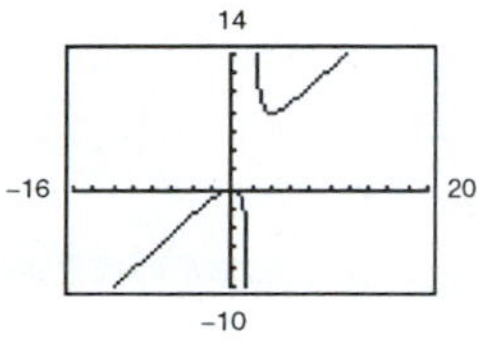

31.

$$\begin{array}{r|l} & 8x + 5 \\ \hline 3x-2 & 24x^2 - x - 8 \\ & \underline{24x^2 - 16x} \\ & \qquad 15x - 8 \\ & \qquad \underline{15x - 10} \\ & \qquad\qquad 2 \end{array}$$

Thus, $\dfrac{24x^2 - x - 8}{3x-2} = 8x + 5 + \dfrac{2}{3x-2}$.

33.

$$\begin{array}{r|l} & x^2 \qquad\qquad - 2 \\ \hline x^2 + 0x - 1 & x^4 + 0x^3 - 3x^2 + 0x + 2 \\ & \underline{x^4 + 0x^3 - x^2} \\ & \qquad -2x^2 + 0x + 2 \\ & \qquad \underline{-2x^2 + 0x + 2} \\ & \qquad\qquad 0 \end{array}$$

Thus, $\dfrac{x^4 - 3x^2 + 2}{x^2 - 1} = x^2 - 2$.

35.

$$\begin{array}{r|l} & x^2 - x + 1 \\ \hline x^2 + 2x & x^4 + x^3 - x^2 + 2x + 0 \\ & \underline{x^4 + 2x^3} \\ & \quad -x^3 - x^2 \\ & \quad \underline{-x^3 - 2x^2} \\ & \qquad x^2 + 2x \\ & \qquad \underline{x^2 + 2x} \\ & \qquad\qquad 0 \end{array}$$

Thus, $\dfrac{x^4 + x^3 - x^2 + 2x}{x^2 + 2x} = x^2 - x + 1$.

37.

$$\begin{array}{r|rrrrr} 2 & 0.25 & -4 & 0 & 0 & 0 \\ & & 0.50 & -7 & -14 & -28 \\ \hline & 0.25 & -3.50 & -7 & -14 & -28 \end{array}$$

Thus, $\dfrac{0.25x^4 - 4x^3}{x-2} = 0.25x^3 - 3.50x^2 - 7x - 14 - \dfrac{28}{x-2}$.

39.

$$\begin{array}{r|rrrrr} \frac{2}{3} & 6 & -4 & -27 & 18 & 0 \\ & & 4 & 0 & -18 & 0 \\ \hline & 6 & 0 & -27 & 0 & 0 \end{array}$$

Thus, $\dfrac{6x^4 - 4x^3 - 27x^2 + 18x}{x - (2/3)} = 6x^3 - 27x$.

41. $f(x) = 2x^3 + 3x^2 - 20x - 21$

(a) 4

4	2	3	−20	−21
		8	44	96
	2	11	24	75

No, $x = 4$ is not a zero of f.

(b) −1

−1	2	3	−20	−21
		−2	−1	21
	2	1	−21	0

Yes, $x = -1$ is a zero of f.

(c) $-\frac{7}{2}$

$-\frac{7}{2}$	2	3	−20	−21
		−7	14	21
	2	−4	−6	0

Yes, $x = -\frac{7}{2}$ is a zero of f.

(d) 0

0	2	3	−20	−21
		0	0	0
	2	3	−20	−21

No, $x = 0$ is not a zero of f.

43. $(7 + 5i) + (-4 + 2i) = (7 - 4) + (5i + 2i) = 3 + 7i$

45. $5i(13 - 8i) = 65i - 40i^2 = 40 + 65i$

47. $\frac{6 + i}{i} = \frac{6 + i}{i} \cdot \frac{-i}{-i} = \frac{-6i - i^2}{-i^2} = \frac{-6i + 1}{1} = 1 - 6i$

49. $f(x) = 6(x + 1)^2\left(x - \frac{1}{3}\right)\left(x + \frac{1}{2}\right)$ Multiply by 6 to clear the fractions.

$= (x + 1)^2\, 3\left(x - \frac{1}{3}\right) 2\left(x + \frac{1}{2}\right)$

$= (x^2 + 2x + 1)(3x - 1)(2x + 1)$

$= (x^2 + 2x + 1)(6x^2 + x - 1)$

$= 6x^4 + 13x^3 + 7x^2 - x - 1$

Note: $f(x) = a(6x^4 + 13x^3 + 7x^2 - x - 1)$, where a is any real nonzero number, has zeros $-1, -1, \frac{1}{3}$, and $-\frac{1}{2}$.

51. $f(x) = 4x^3 - 11x^2 + 10x - 3$

Possible rational zeros: $\pm 1, \pm 3, \pm\frac{1}{2}, \pm\frac{3}{2}, \pm\frac{1}{4}, \pm\frac{3}{4}$

1	4	−11	10	−3
		4	−7	3
	4	−7	3	0

$4x^3 - 11x^2 + 10x - 3 = (x - 1)(4x^2 - 7x + 3) = (x - 1)^2(4x - 3)$

Thus, the zeros of $f(x)$ are $x = 1$ and $x = \frac{3}{4}$.

53. $f(x) = 6x^3 - 5x^2 + 24x - 20$

$= x^2(6x - 5) + 4(6x - 5)$

$= (6x - 5)(x^2 + 4)$

$= (6x - 5)(x + 2i)(x - 2i)$

Thus, the zeros of f are $x = \frac{5}{6}$ and $x = \pm 2i$.

55. $f(x) = 6x^4 - 25x^3 + 14x^2 + 27x - 18$

Possible rational zeros: $\pm 1, \pm 2, \pm 3, \pm 6, \pm 9, \pm 18, \pm\frac{1}{2}, \pm\frac{3}{2}, \pm\frac{9}{2}, \pm\frac{1}{3}, \pm\frac{2}{3}, \pm\frac{1}{6}$

$$\begin{array}{r|rrrrr} -1 & 6 & -25 & 14 & 27 & -18 \\ & & -6 & 31 & -45 & 18 \\ \hline & 6 & -31 & 45 & -18 & 0 \end{array}$$

$$\begin{array}{r|rrrr} 3 & 6 & -31 & 45 & -18 \\ & & 18 & -39 & 18 \\ \hline & 6 & -13 & 6 & 0 \end{array}$$

$$\begin{aligned} 6x^4 - 25x^3 + 14x^2 + 27x - 18 &= (x + 1)(x - 3)(6x^2 - 13x + 6) \\ &= (x + 1)(x - 3)(3x - 2)(2x - 3) \end{aligned}$$

Thus, the zeros of $f(x)$ are $x = -1$, $x = 3$, $x = \frac{2}{3}$, and $x = \frac{3}{2}$.

57. $f(x) = x^4 + 2x + 1$

(a)

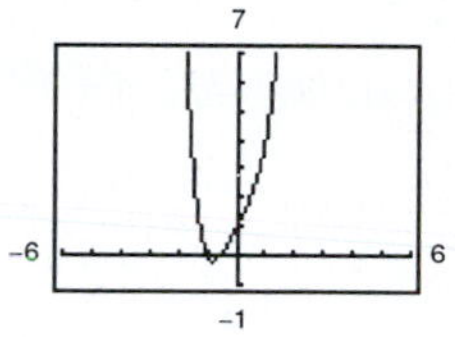

(b) The graph has two x-intercepts, so there are two real zeros.

(c) The zeros are $x = -1$ and $x \approx -0.54$.

59. $h(x) = x^3 - 6x^2 + 12x - 10$

(a)

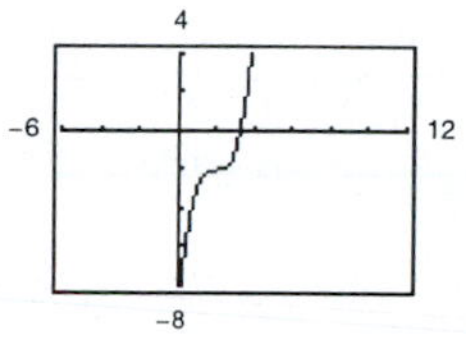

(b) The graph has one x-intercept, so there is one real zero.

(c) $x \approx 3.26$

61. (a) $S = 1.209 + 0.290t + 0.176t^2 - 0.031t^3 + 0.0013t^4$

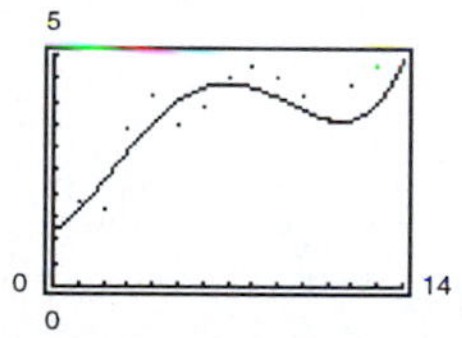

The model is a fairly "good fit."

(b) One explanation may be a recession. The model also shows a downturn in sales.

(c) $S(9) - S(11) \approx 0.54$
The actual decrease of 0.90 was more than this.

(d) $S(15) \approx 6.35$

63. $f(x) = \dfrac{-5}{x^2}$

y-axis symmetry

Vertical asymptote: $x = 0$

Horizontal asymptote: $y = 0$

x	± 3	± 2	± 1
y	$-\frac{5}{9}$	$-\frac{5}{4}$	-5

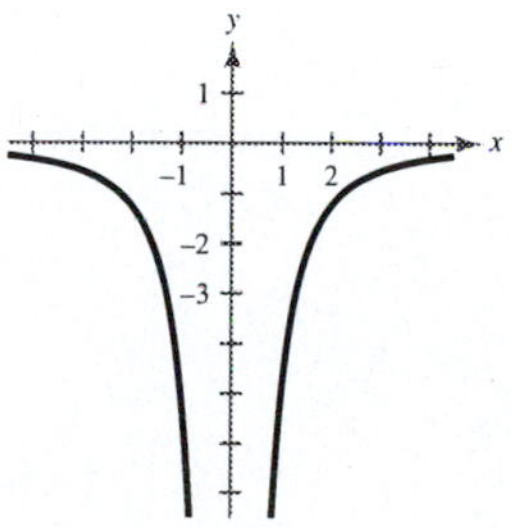

65. $p(x) = \dfrac{x^2}{x^2 + 1}$

Intercept: $(0, 0)$

y-axis symmetry

Horizontal asymptote: $y = 1$

x	± 3	± 2	± 1	0
y	$\frac{9}{10}$	$\frac{4}{5}$	$\frac{1}{2}$	0

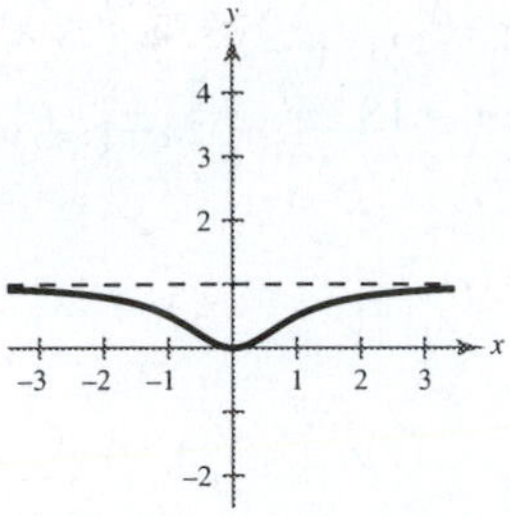

67. $f(x) = \dfrac{x}{x^2 + 1}$

Intercept: $(0, 0)$

Origin symmetry

Horizontal asymptote: $y = 0$

x	-2	-1	0	1	2
y	$-\frac{2}{5}$	$-\frac{1}{2}$	0	$\frac{1}{2}$	$\frac{2}{5}$

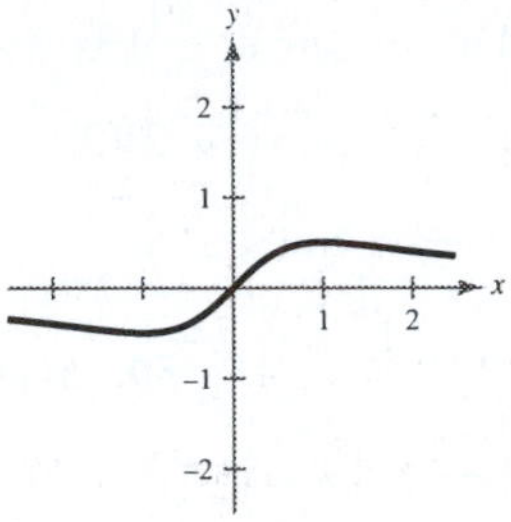

69. $f(x) = \dfrac{2x^3}{x^2 + 1} = 2x - \dfrac{2x}{x^2 + 1}$

Intercept: $(0, 0)$

Origin symmetry

Slant asymptote: $y = 2x$

x	-2	-1	0	1	2
y	$-\frac{16}{5}$	-1	0	1	$\frac{16}{5}$

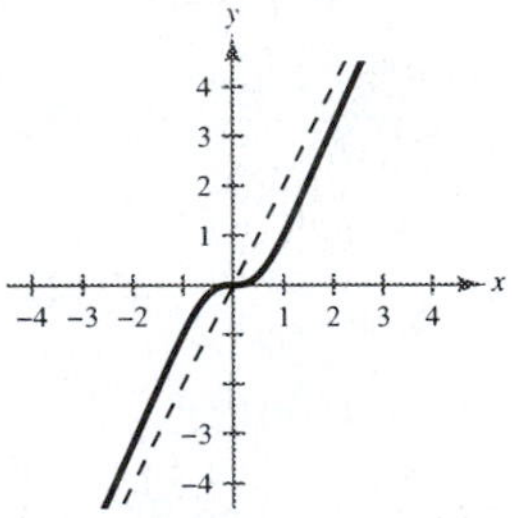

71. $s(x) = \dfrac{8x^2}{x^2 + 4}$

Intercept: $(0, 0)$

Horizontal asymptote: $y = 8$

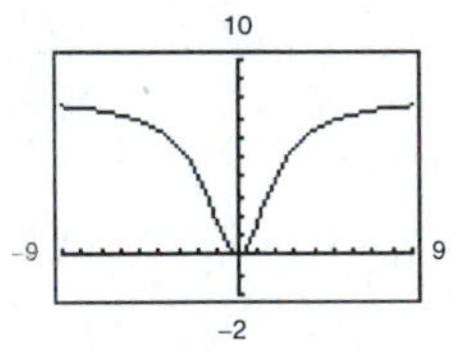

73. $g(x) = \dfrac{x^2 + 1}{x + 1} = x - 1 + \dfrac{2}{x + 1}$

Intercept: $(0, 1)$

Slant asymptote: $y = x - 1$

Vertical asymptote: $x = -1$

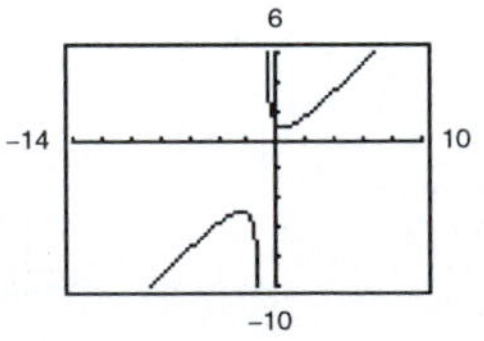

75. $\dfrac{4-x}{x^2+6x+8} = \dfrac{A}{x+2} + \dfrac{B}{x+4}$

$4 - x = A(x+4) + B(x+2)$

Let $x = -2$: $6 = 2A \Rightarrow A = 3$

Let $x = -4$: $8 = -2B \Rightarrow B = -4$

$\dfrac{4-x}{x^2+6x+8} = \dfrac{3}{x+2} - \dfrac{4}{x+4}$

77. $\dfrac{x^2}{x^2+2x-15} = 1 - \dfrac{2x-15}{x^2+2x-15}$

$\dfrac{-2x+15}{(x+5)(x-3)} = \dfrac{A}{x+5} + \dfrac{B}{x-3}$

$-2x + 15 = A(x-3) + B(x+5)$

Let $x = -5$: $25 = -8A \Rightarrow A = -\dfrac{25}{8}$

Let $x = 3$: $9 = 8B \Rightarrow B = \dfrac{9}{8}$

$\dfrac{x^2}{x^2+2x-15} = 1 + \dfrac{9}{8(x-3)} - \dfrac{25}{8(x+5)}$

79. $\dfrac{x^2+2x}{x^3-x^2+x-1} = \dfrac{A}{x-1} + \dfrac{Bx+C}{x^2+1}$

$x^2 + 2x = A\,(x^2+1) + (Bx+C)(x-1)$

Let $x = 1$: $3 = 2A \Rightarrow A = \dfrac{3}{2}$

Let $x = 0$: $0 = A - C \Rightarrow C = \dfrac{3}{2}$

Let $x = 2$: $8 = 5A + 2B + C$

$8 = \left(\dfrac{15}{2}\right) + 2B + \left(\dfrac{3}{2}\right) \Rightarrow B = -\dfrac{1}{2}$

$$\dfrac{x^2+2x}{x^3-x^2+x-1} = \dfrac{3/2}{x-1} + \dfrac{-(1/2)x + 3/2}{x^2+1}$$

$$= \dfrac{1}{2}\left(\dfrac{3}{x-1} - \dfrac{x-3}{x^2+1}\right)$$

❑ Practice Test for Chapter 2

1. Sketch the graph of $f(x) = x^2 - 6x + 5$ and identify the vertex and the intercepts.
2. Find the number of units x that produce a minimum cost C if $C = 0.01x^2 - 90x + 15{,}000$.
3. Find the quadratic function that has a maximum at $(1, 7)$ and passes through the point $(2, 5)$.
4. Find two quadratic functions that have x-intercepts $(2, 0)$ and $\left(\frac{4}{3}, 0\right)$.
5. Use the leading coefficient test to determine the right and left end behavior of the graph of the polynomial function $f(x) = -3x^5 + 2x^3 - 17$.
6. Find all the real zeros of $f(x) = x^5 - 5x^3 + 4x$.
7. Find a polynomial function with 0, 3, and -2 as zeros.
8. Sketch $f(x) = x^3 - 12x$.
9. Divide $3x^4 - 7x^2 + 2x - 10$ by $x - 3$ using long division.
10. Divide $x^3 - 11$ by $x^2 + 2x - 1$.
11. Use synthetic division to divide $3x^5 + 13x^4 + 12x - 1$ by $x + 5$.
12. Use synthetic division to find $f(-6)$ given $f(x) = 7x^3 + 40x^2 - 12x + 15$.
13. Find the real zeros of $f(x) = x^3 - 19x - 30$.
14. Find the real zeros of $f(x) = x^4 + x^3 - 8x^2 - 9x - 9$.
15. List all possible rational zeros of the function $f(x) = 6x^3 - 5x^2 + 4x - 15$.
16. Find the rational zeros of the polynomial $f(x) = x^3 - \frac{20}{3}x^2 + 9x - \frac{10}{3}$.
17. Write $f(x) = x^4 + x^3 + 5x - 10$ as a product of linear factors.
18. Find a polynomial with real coefficients that has 2, $3 + i$, and $3 - 2i$ as zeros.
19. Use synthetic division to show that $3i$ is a zero of $f(x) = x^3 + 4x^2 + 9x + 36$.
20. Sketch the graph of $f(x) = \dfrac{x - 1}{2x}$ and label all intercepts and asymptotes.
21. Find all the asymptotes of $f(x) = \dfrac{8x^2 - 9}{x^2 + 1}$.
22. Find all the asymptotes of $f(x) = \dfrac{4x^2 - 2x + 7}{x - 1}$.
23. Given $z_1 = 4 - 3i$ and $z_2 = -2 + i$, find the following:

    (a) $z_1 - z_2$

    (b) z_1z_2

    (c) z_1/z_2

For Exercises 24–25, write the partial fraction decomposition for the rational expression.

24. $\dfrac{1 - 2x}{x^2 + x}$

25. $\dfrac{6x - 17}{(x - 3)^2}$

CHAPTER 3
Exponential and Logarithmic Functions

CHAPTER 3
Exponential and Logarithmic Functions

Section 3.1 Exponential Functions and Their Graphs

- You should know that a function of the form $y = a^x$, where $a > 0$, $a \neq 1$, is called an exponential function with base a.
- You should be able to graph exponential functions.
- You should know formulas for compound interest.

  (a) For n compoundings per year: $A = P\left(1 + \frac{r}{n}\right)^{nt}$.

  (b) For continuous compoundings: $A = Pe^{rt}$.

Solutions to Odd-Numbered Exercises

1. $(3.4)^{5.6} \approx 946.852$

3. $(1.005)^{400} \approx 7.352$

5. $5^{-\pi} \approx 0.006$

7. $100^{\sqrt{2}} \approx 673.639$

9. $e^{-3/4} \approx 0.472$

11. $f(x) = 3^{x-2}$

$= 3^x 3^{-2}$

$= 3^x\left(\frac{1}{3^2}\right)$

$= \frac{1}{9}(3^x)$

$= h(x)$

Thus, $f(x) \neq g(x)$, but $f(x) = h(x)$.

13. $f(x) = 16(4^{-x})$

$= 4^2(4^{-x})$

$= 4^{2-x}$

$= \left(\frac{1}{4}\right)^{-(2-x)}$

$= \left(\frac{1}{4}\right)^{x-2}$

$= g(x)$

and $f(x) = 16(4^{-x})$

$= 16(2^2)^{-x}$

$= 16(2^{-2x})$

$= h(x)$

Thus, $f(x) = g(x) = h(x)$.

15. $f(x) = 2^x$ rises to the right.

Asymptote: $y = 0$

Intercept: $(0, 1)$

Matches graph (d).

17. $f(x) = 2^{-x}$ falls to the right.

Asymptote: $y = 0$

Intercept: $(0, 1)$

Matches graph (a).

19. $g(x) = 5^x$

x	-2	-1	0	1	2
$g(x)$	$\frac{1}{25}$	$\frac{1}{5}$	1	5	25

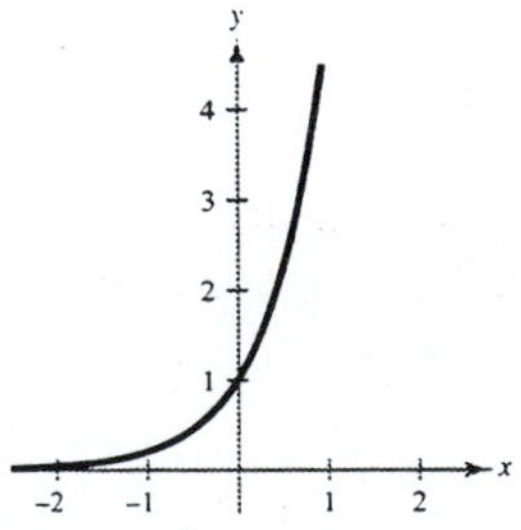

21. $f(x) = \left(\frac{1}{5}\right)^x = 5^{-x}$

x	-2	-1	0	1	2
y	25	5	1	$\frac{1}{5}$	$\frac{1}{25}$

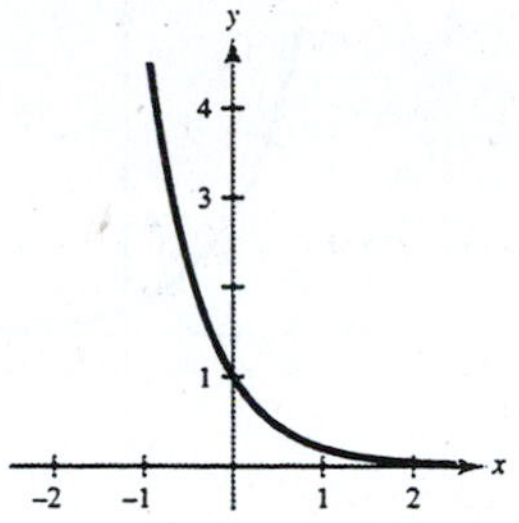

23. $h(x) = 5^{x-2}$

x	-1	0	1	2	3
y	$\frac{1}{125}$	$\frac{1}{25}$	$\frac{1}{5}$	1	5

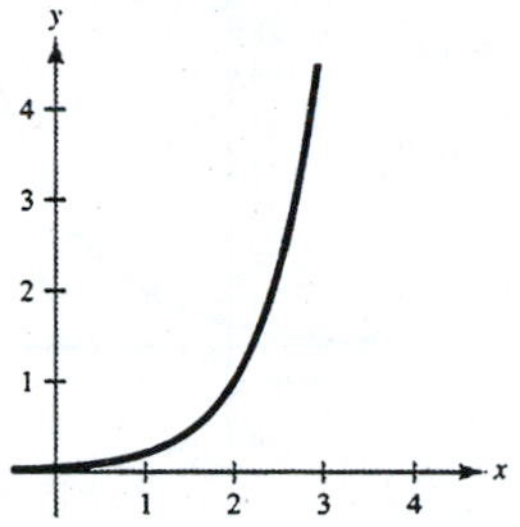

25. $g(x) = 5^{-x} - 3$

x	-1	0	1	2
y	2	-2	$-2\frac{4}{5}$	$-2\frac{24}{25}$

Asymptote: $y = -3$

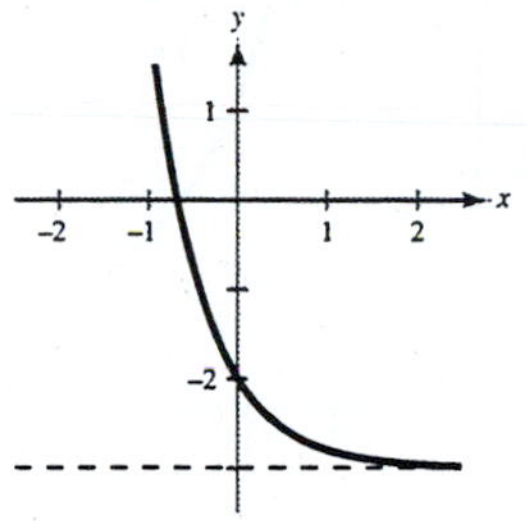

27. $y = 2^{-x^2}$

x	± 2	± 1	0
y	$\frac{1}{16}$	$\frac{1}{2}$	1

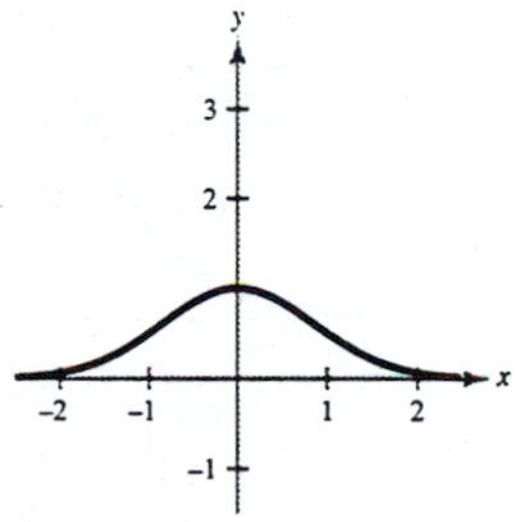

29. $f(x) = 3^{x-2} + 1$

x	-1	0	1	2	3	4
y	$1\frac{1}{27}$	$1\frac{1}{9}$	$1\frac{1}{3}$	2	4	10

Asymptote: $y = 1$

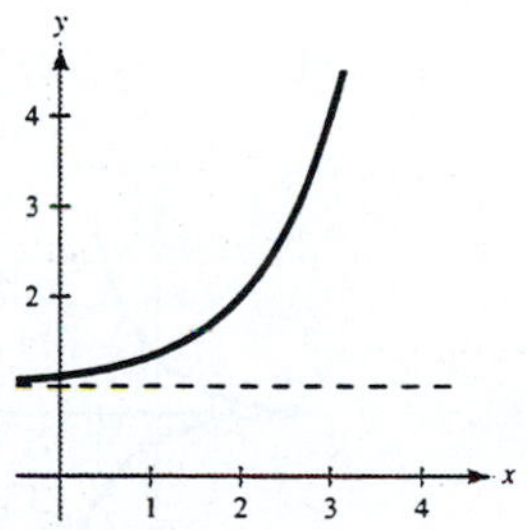

31. $y = 1.08^{-5x}$

x	-1	0	1	2
y	1.47	1	0.68	0.46

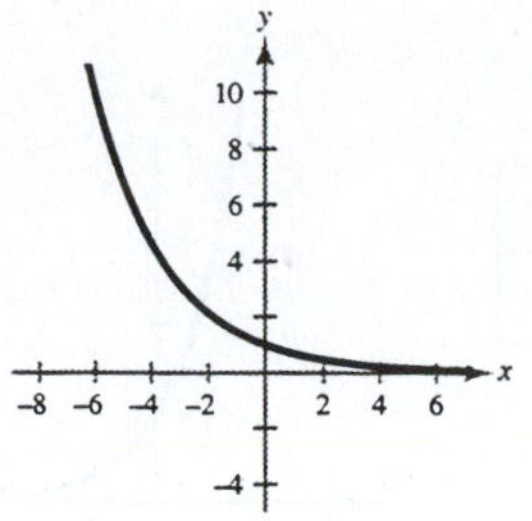

33. $s(t) = 2e^{0.12t}$

t	-4	0	4	8
$s(t)$	1.24	2	3.23	5.22

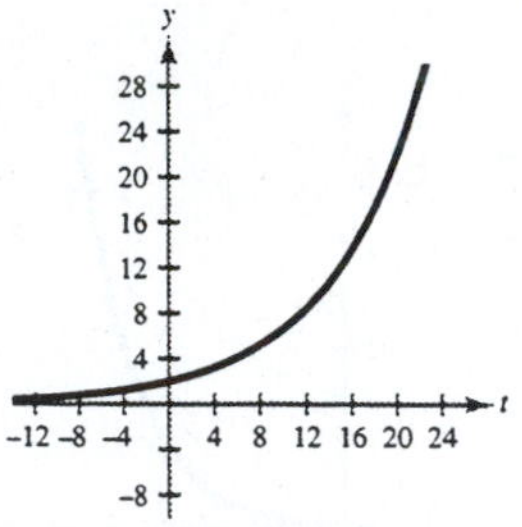

35. $g(x) = 1 + e^{-x}$

x	-2	-1	0	1	2
y	8.39	3.72	2	1.37	1.14

Asymptote: $y = 1$

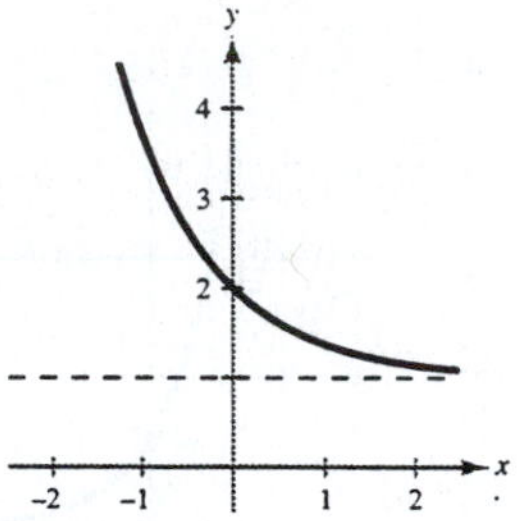

37. $y = 3^x$ and $y = 4^x$

x	-2	-1	0	1	2
3^x	$\frac{1}{9}$	$\frac{1}{3}$	1	3	9
4^x	$\frac{1}{16}$	$\frac{1}{4}$	1	4	16

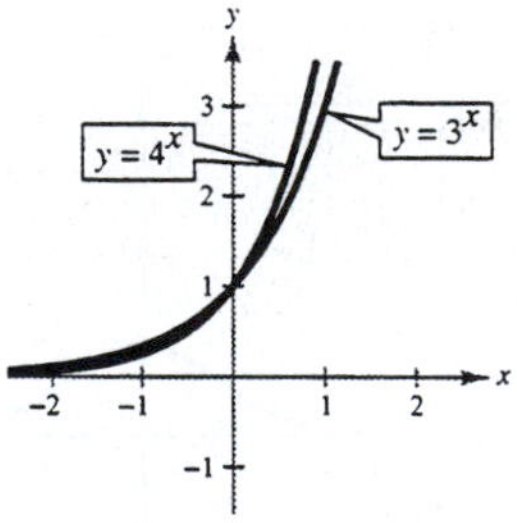

(a) $4^x < 3^x$ when $x < 0$.

(b) $4^x > 3^x$ when $x > 0$.

39. $f(x) = 3^x$

(a) $g(x) = f(x - 2) = 3^{x-2}$

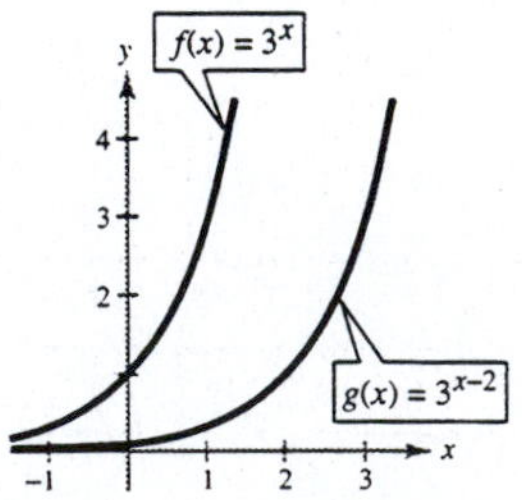

Horizontal shift 2 units to the right

(b) $h(x) = -\frac{1}{2}f(x) = -\frac{1}{2}(3^x)$

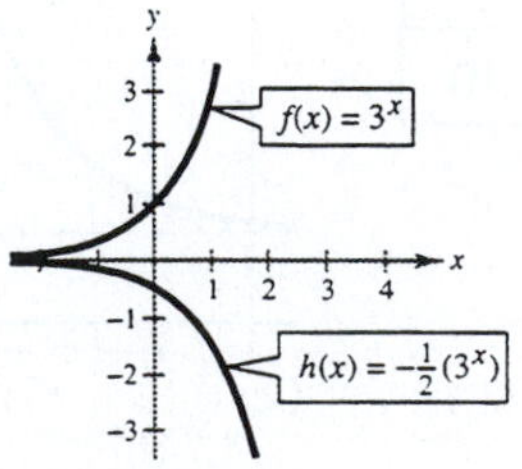

Vertical shrink and a reflection about the x-axis

(c) $q(x) = f(-x) + 3 = 3^{-x} + 3$

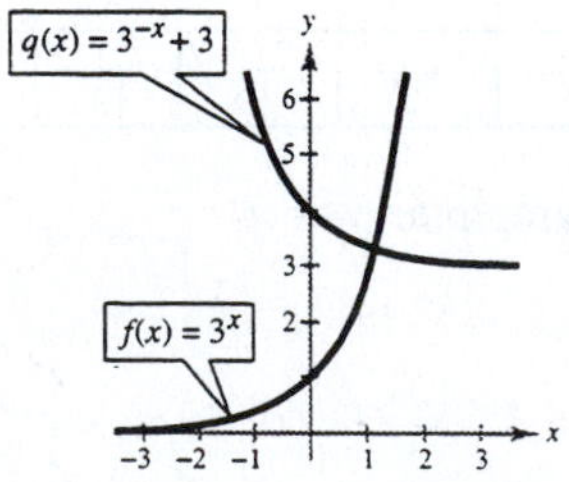

Reflection about the y-axis and a vertical translation 3 units upward

41. (a) $f(x) = x^2e^{-x}$

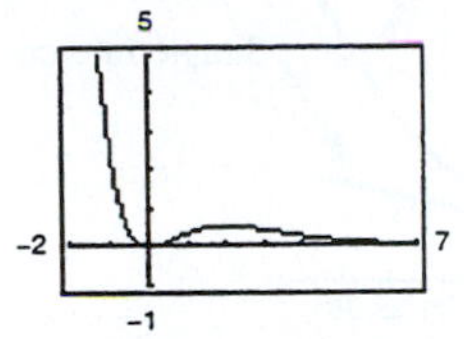

Decreasing: $(-\infty, 0)$, $(2, \infty)$
Increasing: $(0, 2)$
Relative maximum: $(2, 4e^{-2})$
Relative minimum: $(0, 0)$

(b) $g(x) = x2^{3-x}$

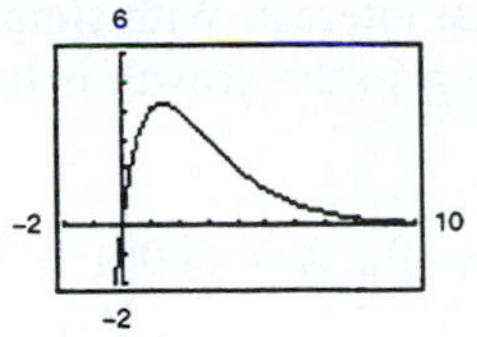

Decreasing: $(1.44, \infty)$
Increasing: $(-\infty, 1.44)$
Relative maximum: $(1.44, 4.25)$

43. The exponential function, $y = e^x$, increases at a faster rate than the polynomial function, $y = x^n$.

45. $f(x) = \left(1 + \dfrac{0.5}{x}\right)^x$ and $g(x) = e^{0.5}$ (Horizontal line)

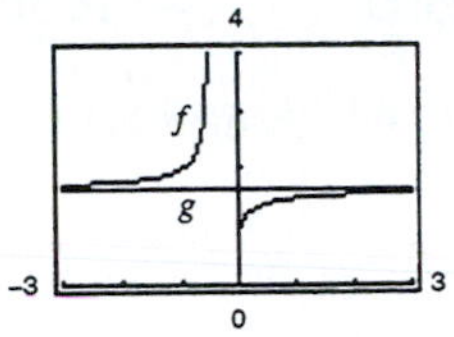

As $x \to \infty, f(x) \to g(x)$.

47. $P = \$2500$, $r = 12\%$, $t = 10$ years

Compounded n times per year: $A = 2500\left(1 + \dfrac{0.12}{n}\right)^{10n}$

Compounded continuously: $A = 2500e^{0.12(10)}$

n	1	2	4	12	365	Continuous Compounding
A	\$7,764.62	\$8,017.84	\$8,155.09	\$8250.97	\$8298.66	\$8,300.29

49. $P = \$2500$, $r = 12\%$, $t = 20$ years

Compounded n times per year: $A = 2500\left(1 + \dfrac{0.12}{n}\right)^{20n}$

Compounded continuously: $A = 2500e^{0.12(20)}$

n	1	2	4	12	365	Continuous Compounding
A	\$24,115.73	\$25,714.29	\$26,602.23	\$27,231.38	\$27,547.07	\$27,557.94

51. $A = Pe^{rt}$

$100{,}000 = Pe^{0.09t}$

$\dfrac{100{,}000}{e^{0.09t}} = P$

$P = 100{,}000e^{-0.09t}$

t	1	10	20	30	40	50
P	\$91,393.12	\$40,656.97	\$16,529.89	\$6,720.55	\$2,732.37	\$1,110.90

53. $A = 25{,}000e^{(0.0875)(25)} \approx \$222{,}822.57$

55. (a) The graph that is increasing faster represents 7% compounded annually. When interest is compounded, you earn interest on that interest. With simple interest there is no compounding so the growth is linear.

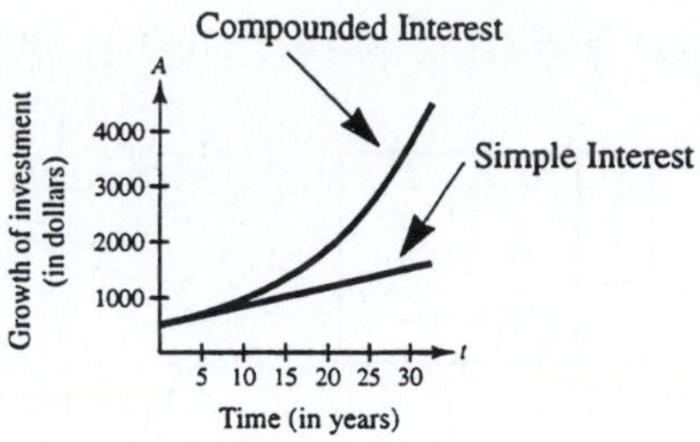

(b) Compound interest formula: $A = 500\left(1 + \dfrac{0.07}{1}\right)^{(1)t}$

$= 500(1.07)^t$

Simple interest formula: $A = Prt + P$

$= 500(0.07)t + 500$

57. $C(10) = 23.95(1.04)^{10} \approx \$35.45$

59. $P(t) = 100e^{0.2197t}$

(a) $P(0) = 100$

(b) $P(5) \approx 300$

(c) $P(10) \approx 900$

61. $Q = 25\left(\frac{1}{2}\right)^{t/1620}$

(a) When $t = 0$, $Q = 25\left(\frac{1}{2}\right)^{0/1620} = 25(1) = 25$ units.

(b) When $t = 1000$, $Q = 25\left(\frac{1}{2}\right)^{1000/1620} \approx 16.30$ units.

(c)

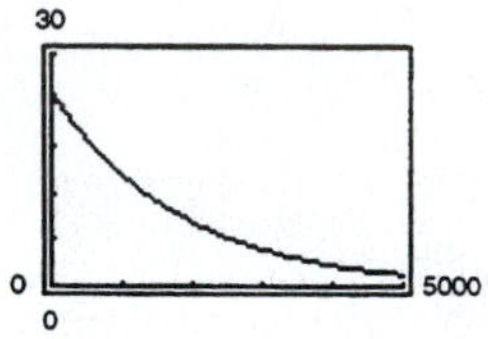

63. $P = 10{,}958e^{-0.15h}$

(a)

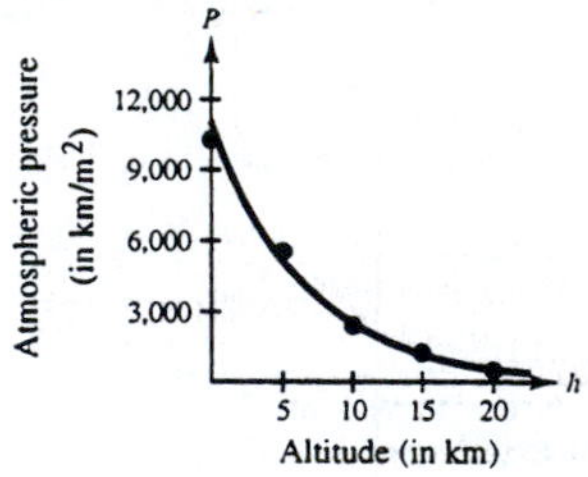

(b)

h	0	5	10	15	20
P	10,958	5176	2445	1155	546

The model is a "good fit."

(c) $P(8) \approx 3300 \text{ kg/m}^2$

(d) $2000 = 10{,}958e^{-0.15h}$ when $x \approx 11.3$ km.

65. False, $e \neq \dfrac{271{,}801}{99{,}990}$.

Since e is an irrational number it cannot equal a rational number.

67. Since $\sqrt{2} \approx 1.414$ we know that $1 < \sqrt{2} < 2$.

Thus, $2^1 < 2^{\sqrt{2}} < 2^2$

$2 < 2^{\sqrt{2}} < 4$.

69. $y_4 = 1 + \dfrac{x}{1!} + \dfrac{x^2}{2!} + \dfrac{x^3}{3!} + \dfrac{x^4}{4!}$

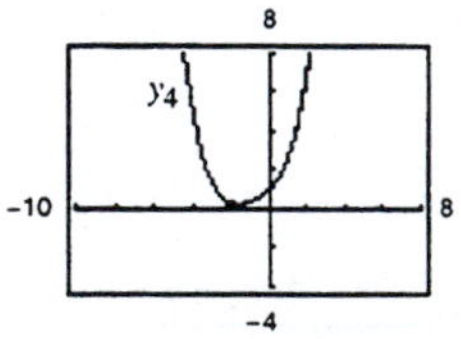

$e^x = 1 + \dfrac{x}{1!} + \dfrac{x^2}{2!} + \dfrac{x^3}{3!} + \dfrac{x^4}{4!} + \dfrac{x^5}{5!} + \cdots$

71. $2x - 7y + 14 = 0$

$2x + 14 = 7y$

$\frac{1}{7}(2x + 14) = y$

73. $x^2 + y^2 = 25$

$y^2 = 25 - x^2$

$y = \pm\sqrt{25 - x^2}$

Section 3.2 Logarithmic Functions and Their Graphs

- You should know that a function of the form $y = \log_a x$, where $a > 0$, $a \neq 1$, and $x > 0$, is called a logarithm of x to base a.
- You should be able to convert from logarithmic form to exponential form and vice versa.
  $y = \log_a x \iff a^y = x$
- You should know the following properties of logarithms.
  (a) $\log_a 1 = 0$ since $a^0 = 1$.
  (b) $\log_a a = 1$ since $a^1 = a$.
  (c) $\log_a a^x = x$ since $a^x = a^x$.
  (d) If $\log_a x = \log_a y$, then $x = y$.
- You should know the definition of the natural logarithmic function.
  $\log_e x = \ln x, x > 0$
- You should know the properties of the natural logarithmic function.
  (a) $\ln 1 = 0$ since $e^0 = 1$.
  (b) $\ln e = 1$ since $e^1 = e$.
  (c) $\ln e^x = x$ since $e^x = e^x$.
  (d) If $\ln x = \ln y$, then $x = y$.
- You should be able to graph logarithmic functions.

Solutions to Odd-Numbered Exercises

1. $\log_4 64 = 3 \implies 4^3 = 64$

3. $\log_7 \frac{1}{49} = -2 \implies 7^{-2} = \frac{1}{49}$

5. $\log_{32} 4 = \frac{2}{5} \implies 32^{2/5} = 4$

7. $\ln 1 = 0 \implies e^0 = 1$

9. $5^3 = 125 \implies \log_5 125 = 3$

11. $81^{1/4} = 3 \implies \log_{81} 3 = \frac{1}{4}$

13. $6^{-2} = \frac{1}{36} \implies \log_6 \frac{1}{36} = -2$

15. $e^3 = 20.0855\ldots \implies \ln 20.0855\ldots = 3$

17. $e^x = 4 \implies \ln 4 = x$

19. $\log_2 16 = \log_2 2^4 = 4$

21. $\log_{16} 4 = \log_{16} 16^{1/2} = \frac{1}{2}$

23. $\log_7 1 = \log_7 7^0 = 0$

25. $\log_{10} 0.01 = \log_{10} 10^{-2} = -2$

27. $\log_8 32 = \log_8 8^{5/3} = \frac{5}{3}$

29. $\ln e^3 = 3$

31. $\log_{10} 345 \approx 2.538$

33. $\log_{10} 145 \approx 2.161$

35. $\ln 18.42 \approx 2.913$

37. $\ln(1 + \sqrt{3}) \approx 1.005$

39. $\ln 0.32 \approx -1.139$

41. $f(x) = 3^x,\ g(x) = \log_3 x$

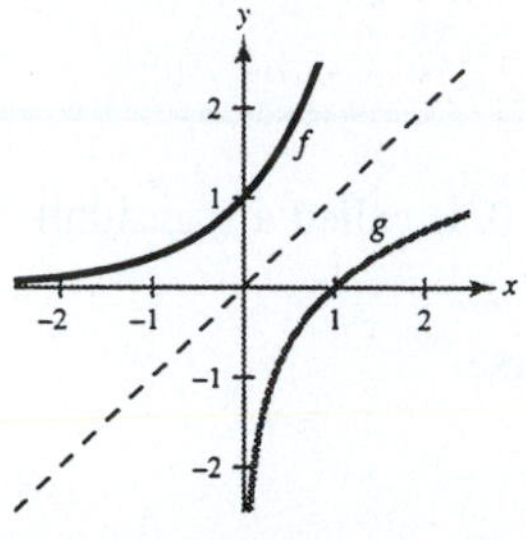

f and g are inverses. Their graphs are reflected about the line $y = x$.

43. $f(x) = e^x,\ g(x) = \ln x$

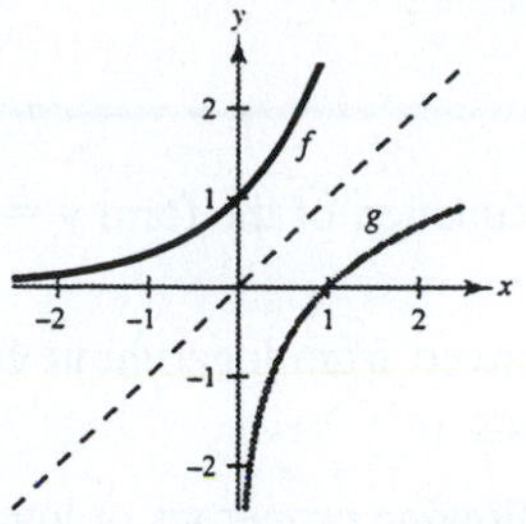

f and g are inverses. Their graphs are reflected about the line $y = x$.

45. $f(x) = \log_3 x + 2$
Asymptote: $x = 0$
Point on graph: $(1, 2)$
Matches graph (c).

47. $f(x) = -\log_3(x + 2)$
Asymptote: $x = -2$
Point on graph: $(-1, 0)$
Matches graph (d).

49. $f(x) = \log_3(1 - x)$
Asymptote: $x = 1$
Point on graph: $(0, 0)$
Matches graph (b).

51. $f(x) = \log_4 x$
Domain: $x > 0 \implies$ The domain is $(0, \infty)$.
Vertical asymptote: $x = 0$
x-intercept: $(1, 0)$
$y = \log_4 x \implies 4^y = x$

x	$\frac{1}{4}$	1	4	2
y	-1	0	1	$\frac{1}{2}$

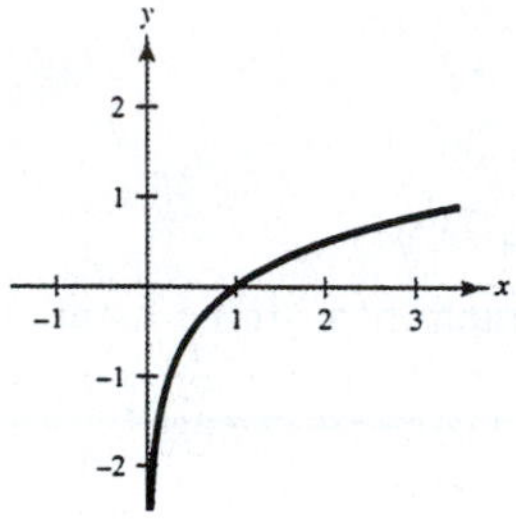

53. $y = -\log_3 x + 2$
Domain: $(0, \infty)$
Vertical asymptote: $x = 0$
x-intercept: $-\log_3 x + 2 = 0$

$$2 = \log_3 x$$
$$3^2 = x$$
$$9 = x$$

The x-intercept is $(9, 0)$.

$y = -\log_2 x + 2$

$\log_3 x = 2 - y \implies 3^{2-y} = x$

x	27	9	3	1	$\frac{1}{3}$
y	-1	0	1	2	3

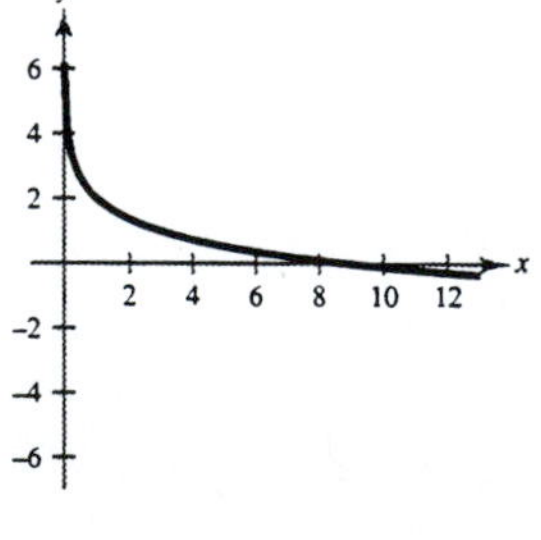

55. $f(x) = -\log_6(x + 2)$

Domain: $x + 2 > 0 \Rightarrow x > -2$

The domain is $(-2, \infty)$.

Vertical asymptote: $x + 2 = 0 \Rightarrow x = -2$

x-intercept:
$$\begin{aligned} 0 &= -\log_6(x + 2) \\ 0 &= \log_6(x + 2) \\ 6^0 &= x + 2 \\ 1 &= x + 2 \\ -1 &= x \end{aligned}$$

The x-intercept is $(-1, 0)$.

$$\begin{aligned} y &= -\log_6(x + 2) \\ -y &= \log_6(x + 2) \\ 6^{-y} - 2 &= x \end{aligned}$$

x	4	-1	$-1\frac{5}{6}$	$-1\frac{35}{36}$
y	-1	0	1	2

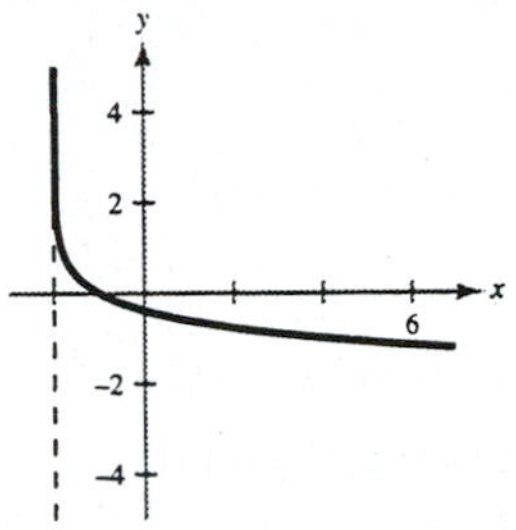

57. $y = \log_{10}\left(\dfrac{x}{5}\right)$

Domain: $\dfrac{x}{5} > 0 \Rightarrow x > 0$

The domain is $(0, \infty)$.

Vertical asymptote: $\dfrac{x}{5} = 0 \Rightarrow x = 0$

The vertical asymptote is the y-axis.

x-intercept: $\log_{10}\left(\dfrac{x}{5}\right) = 0$

$$\frac{x}{5} = 10^0$$

$$\frac{x}{5} = 1 \Rightarrow x = 5$$

The x-intercept is $(5, 0)$.

x	1	2	3	4	5	6	7
y	-0.70	-0.40	-0.22	-0.10	0	0.08	0.15

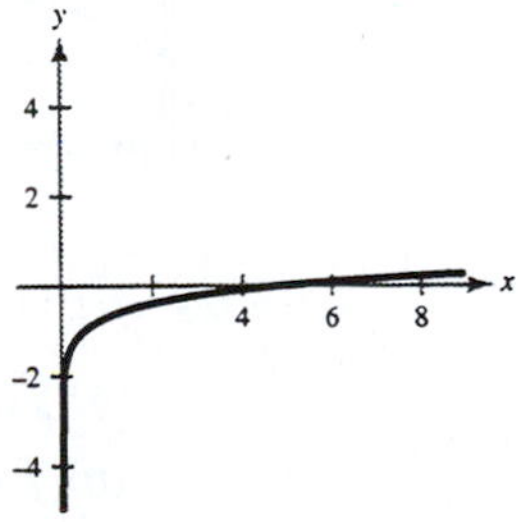

59. $f(x) = \ln(x - 2)$

Domain: $x - 2 > 0 \Rightarrow x > 2$

The domain is $(2, \infty)$.

Vertical asymptote: $x - 2 = 0 \Rightarrow x = 2$

x-intercept:
$$\begin{aligned} 0 &= \ln(x - 2) \\ e^0 &= x - 2 \\ 3 &= x \end{aligned}$$

The x-intercept is $(3, 0)$.

x	2.5	3	4	5
y	-0.69	0	0.69	1.10

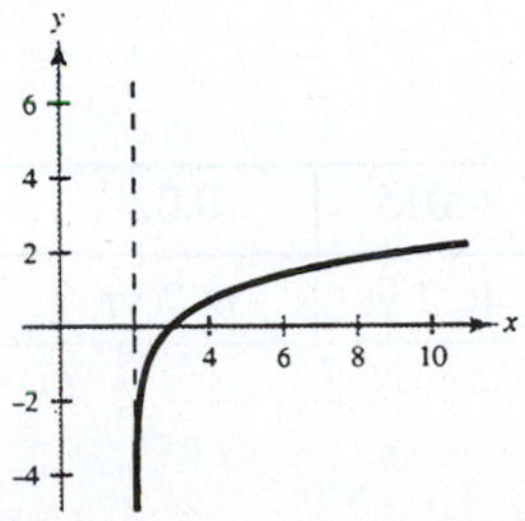

61. $g(x) = \ln(-x)$

Domain: $-x > 0 \implies x < 0$

The domain is $(-\infty, 0)$.

Vertical asymptote: $-x = 0 \implies x = 0$

x-intercept: $0 = \ln(-x)$

$e^0 = -x$

$-1 = x$

The x-intercept is $(-1, 0)$.

x	-0.5	-1	-2	-3
y	-0.69	0	0.69	1.10

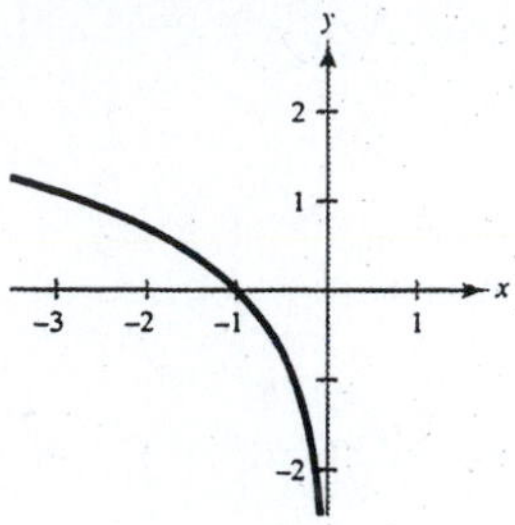

63. $f(x) = |\ln x|$

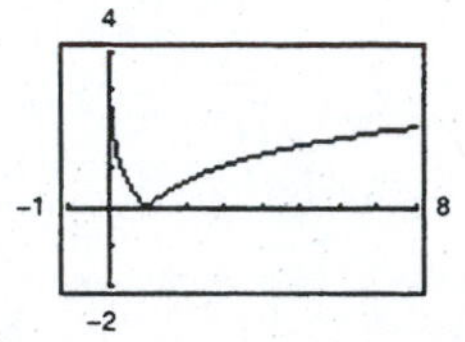

Increasing on $(1, \infty)$

Decreasing on $(0, 1)$

Relative minimum: $(1, 0)$

65. $f(x) = \dfrac{x}{2} - \ln \dfrac{x}{4}$

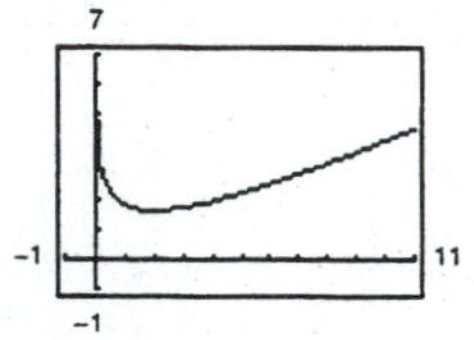

Increasing on $(2, \infty)$

Decreasing on $(0, 2)$

Relative minimum: $\left(2, 1 - \ln \frac{1}{2}\right)$

67. (a) $f(x) = \ln x$

$g(x) = \sqrt{x}$

The natural log function grows at a slower rate than the square root function.

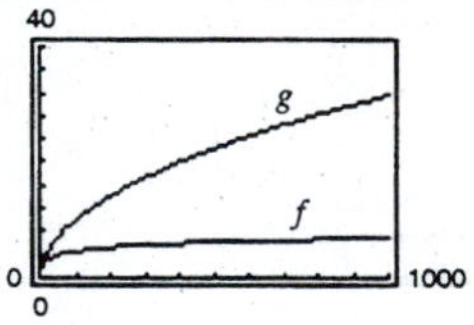

(b) $f(x) = \ln x$

$g(x) = \sqrt[4]{x}$

The natural log function grows at a slower rate than the fourth root function.

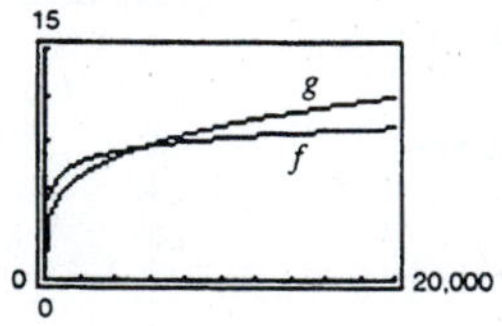

69. $y_1 = \ln x$

$y_2 = x - 1$

$y_3 = (x - 1) - \frac{1}{2}(x - 1)^2$

$y_4 = (x - 1) - \frac{1}{2}(x - 1)^2 + \frac{1}{3}(x - 1)^3$

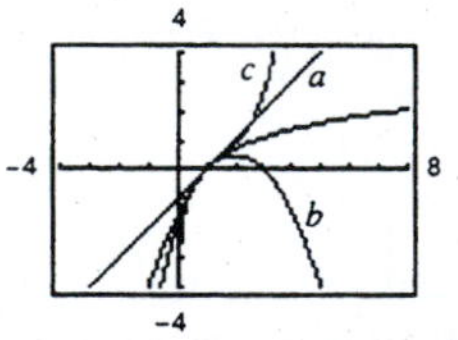

71. $f(t) = 80 - 17 \log_{10}(t + 1), 0 \le t \le 12$

(a) $f(0) = 80 - 17 \log_{10} 1 = 80.0$

(b) $f(4) = 80 - 17 \log_{10} 5 \approx 68.1$

(c) $f(10) = 80 - 17 \log_{10} 11 \approx 62.3$

73. $t = \dfrac{\ln 2}{r}$

r	0.005	0.01	0.015	0.02	0.025	0.03
t	138.6 yr	69.3 yr	46.2 yr	34.7 yr	27.7 yr	23.1 yr

75. $y = 80.4 - 11 \ln x$

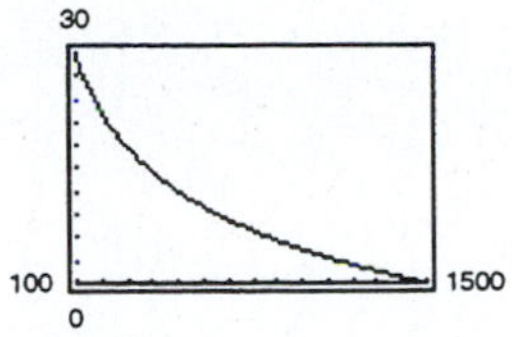

$y(300) = 80.4 - 11 \ln 300 \approx 17.66 \text{ ft}^3/\text{min}$

77. $W = 19{,}440(\ln 9 - \ln 3) \approx 21{,}357$ ft-lb

79. $t = 10.042 \ln\left(\dfrac{1316.35}{1316.35 - 1250}\right) \approx 30$ years

81. Total amount $= (1316.35)(30)(12) = \$473{,}886$

Interest $= 473{,}886 - 150{,}000 = \$323{,}886$

83. $f(x) = \dfrac{\ln x}{x}$

(a)

x	1	5	10	10^2	10^4	10^6
$f(x)$	0	0.322	0.230	0.046	0.00092	0.0000138

(b) As $x \to \infty$, $f(x) \to 0$.

(c)

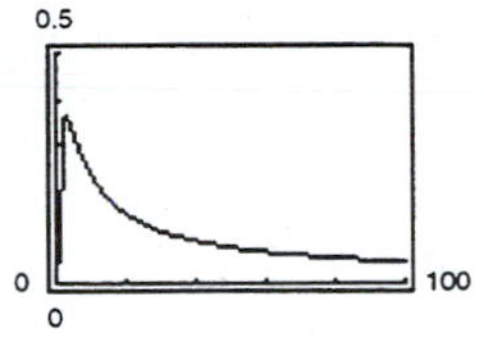

85. $8n - 3$

87. $83.95 + 37.50t$ Parts and labor

Section 3.3 Properties of Logarithms

- You should know the following properties of logarithms.

  (a) $\log_a x = \dfrac{\log_b x}{\log_b a}$

  (b) $\log_a(uv) = \log_a u + \log_a v$ $\ln(uv) = \ln u + \ln v$

  (c) $\log_a(u/v) = \log_a u - \log_a v$ $\ln(u/v) = \ln u - \ln v$

  (d) $\log_a u^n = n \log_a u$ $\ln u^n = n \ln u$
- You should be able to rewrite logarithmic expressions using these properties.

Solutions to Odd-Numbered Exercises

1. $f(x) = \log_{10} x$

$g(x) = \dfrac{\ln x}{\ln 10}$

$f(x) = g(x)$

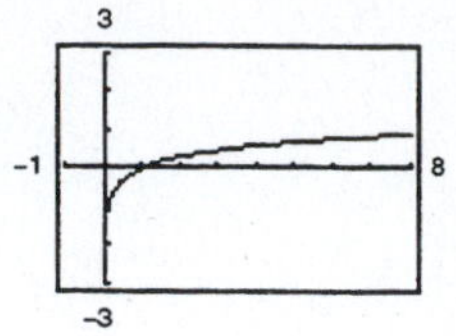

3. $\log_3 5 = \dfrac{\log_{10} 5}{\log_{10} 3}$

5. $\log_2 x = \dfrac{\log_{10} x}{\log_{10} 2}$

7. $\log_3 5 = \dfrac{\ln 5}{\ln 3}$

9. $\log_2 x = \dfrac{\ln x}{\ln 2}$

11. $\log_3 7 = \dfrac{\log_{10} 7}{\log_{10} 3} = \dfrac{\ln 7}{\ln 3} \approx 1.771$

13. $\log_{1/2} 4 = \dfrac{\log_{10} 4}{\log_{10} (1/2)} = \dfrac{\ln 4}{\ln (1/2)} = -2.000$

15. $\log_9(0.4) = \dfrac{\log_{10} 0.4}{\log_{10} 9} = \dfrac{\ln 0.4}{\ln 9} \approx -0.417$

17. $\log_{15} 1250 = \dfrac{\log_{10} 1250}{\log_{10} 15} = \dfrac{\ln 1250}{\ln 15} \approx 2.633$

19. $\log_{10} 5x = \log_{10} 5 + \log_{10} x$

21. $\log_{10} \dfrac{5}{x} = \log_{10} 5 - \log_{10} x$

23. $\log_8 x^4 = 4 \log_8 x$

25. $\ln\sqrt{z} = \ln z^{1/2} = \frac{1}{2} \ln z$

27. $\ln xyz = \ln x + \ln y + \ln z$

29. $\ln\sqrt{a-1} = \frac{1}{2} \ln(a-1)$

31. $\ln z(z-1)^2 = \ln z + \ln(z-1)^2$

$= \ln z + 2\ln(z-1)$

33. $\ln \sqrt[3]{\dfrac{x}{y}} = \dfrac{1}{3} \ln \dfrac{x}{y}$

$= \dfrac{1}{3}[\ln x - \ln y]$

$= \dfrac{1}{3} \ln x - \dfrac{1}{3} \ln y$

35. $\ln\left(\dfrac{x^4\sqrt{y}}{z^5}\right) = \ln x^4\sqrt{y} - \ln z^5$

$= \ln x^4 + \ln\sqrt{y} - \ln z^5$

$= 4\ln x + \dfrac{1}{2}\ln y - 5\ln z$

37. $\log_b\left(\dfrac{x^2}{y^2z^3}\right) = \log_b x^2 - \log_b y^2z^3$

$= \log_b x^2 - [\log_b y^2 + \log_b z^3]$

$= 2\log_b x - 2\log_b y - 3\log_b z$

39. $y_1 = \ln[x^3(x+4)]$

$y_2 = 3\ln x + \ln(x+4)$

$y_1 = y_2$

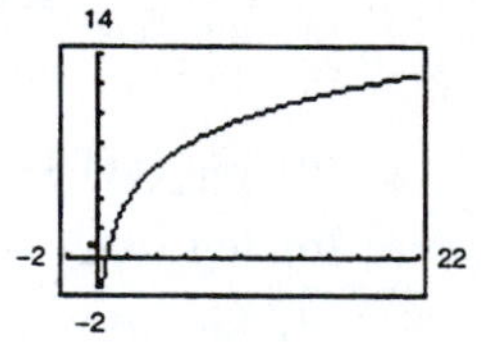

41. $\ln x + \ln 2 = \ln 2x$

43. $\log_4 z - \log_4 y = \log_4 \dfrac{z}{y}$

45. $2\log_2(x+4) = \log_2(x+4)^2$

47. $\frac{1}{3}\log_3 5x = \log_3(5x)^{1/3} = \log_3\sqrt[3]{5x}$

49. $\ln x - 3\ln(x+1) = \ln x - \ln(x+1)^3$

$= \ln \dfrac{x}{(x+1)^3}$

51. $\ln(x-2) - \ln(x+2) = \ln\left(\dfrac{x-2}{x+2}\right)$

53. $\ln x - 2[\ln(x+2) + \ln(x-2)] = \ln x - 2\ln(x+2)(x-2)$

$$= \ln x - 2\ln(x^2-4)$$

$$= \ln x - \ln(x^2-4)^2$$

$$= \ln \frac{x}{(x^2-4)^2}$$

55. $\frac{1}{3}[2\ln(x+3) + \ln x - \ln(x^2-1)] = \frac{1}{3}[\ln(x+3)^2 + \ln x - \ln(x^2-1)]$

$$= \frac{1}{3}[\ln x(x+3)^2 - \ln(x^2-1)]$$

$$= \frac{1}{3}\ln\frac{x(x+3)^2}{x^2-1}$$

$$= \ln\sqrt[3]{\frac{x(x+3)^2}{x^2-1}}$$

57. $\frac{1}{3}[\ln y + 2\ln(y+4)] - \ln(y-1) = \frac{1}{3}[\ln y + \ln(y+4)^2] - \ln(y-1)$

$$= \frac{1}{3}\ln y(y+4)^2 - \ln(y-1)$$

$$= \ln\sqrt[3]{y(y+4)^2} - \ln(y-1)$$

$$= \ln\frac{\sqrt[3]{y(y+4)^2}}{y-1}$$

59. $2\ln 3 - \frac{1}{2}\ln(x^2+1) = \ln 3^2 - \ln\sqrt{x^2+1}$

$$= \ln\frac{9}{\sqrt{x^2+1}}$$

61. $y_1 = 2[\ln 8 - \ln(x^2+1)]$

$y_2 = \ln\left[\frac{64}{(x^2+1)^2}\right]$

$y_1 = y_2$

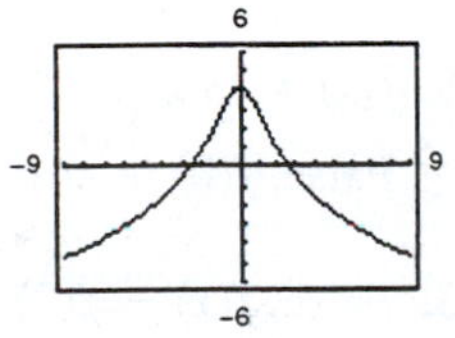

63. $y_1 = \ln x^2$

$y_2 = 2\ln x$

$y_1 = y_2$ for $x > 0$

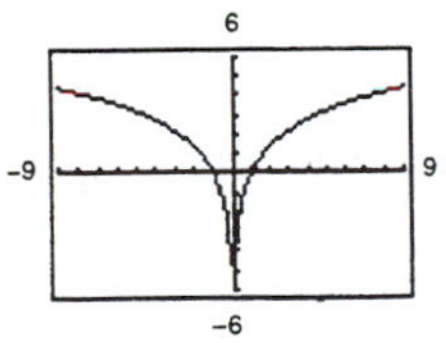

They are not equivalent. The domain of $f(x)$ is all real numbers except 0. The domain of $g(x)$ is $x > 0$.

65. $\log_2(\frac{32}{4}) = \log_2 32 - \log_2 4$ by Property 2.

67. $f(x) = \ln\frac{x}{2}$, $g(x) = \frac{\ln x}{\ln 2}$, $h(x) = \ln x - \ln 2$

$f(x) = h(x)$ by Property 2.

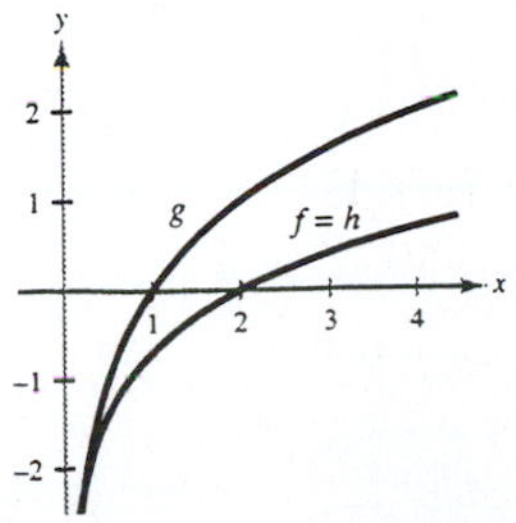

69. $\log_3 9 = 2\log_3 3 = 2$

71. $\log_4 16^{1.2} = 1.2(\log_4 16) = 1.2(2) = 2.4$

73. $\log_3(-9)$ is undefined. -9 is not in the domain of $\log_3 x$.

75. $\log_5 75 - \log_5 3 = \log_5 \frac{75}{3} = \log_5 25 = \log_5 5^2 = 2\log_5 5 = 2$

77. $\ln e^2 - \ln e^5 = 2 - 5 = -3$

79. $\log_{10} 0$ is undefined. 0 is not in the domain of $\log_{10} x$.

81. $\ln e^{4.5} = 4.5$

83. $\log_4 8 = \log_4 2^3 = 3\log_4 2 = 3\log_4 \sqrt{4} = 3\log_4 4^{1/2} = 3\left(\frac{1}{2}\right)\log_4 4 = \frac{3}{2}$

85.
$$\begin{aligned}\log_5 \tfrac{1}{250} &= \log_5 1 - \log_5 250 = 0 - \log_5(125 \cdot 2)\\ &= -\log_5(5^3 \cdot 2) = -[\log_5 5^3 + \log_5 2]\\ &= -[3\log_5 5 + \log_5 2] = -3 - \log_5 2\end{aligned}$$

87. $\ln(5e^6) = \ln 5 + \ln e^6 = \ln 5 + 6 = 6 + \ln 5$

89. $f(t) = 90 - 15\log_{10}(t+1),\ 0 \le t \le 12$

(a) $f(0) = 90$

(b) $f(6) \approx 77$

(c) $f(12) \approx 73$

(d)
$$\begin{aligned}75 &= 90 - 15\log_{10}(t+1)\\ -15 &= -15\log_{10}(t+1)\\ 1 &= \log_{10}(t+1)\\ 10^1 &= t+1\\ t &= 9 \text{ months}\end{aligned}$$

(e) $f(t) = 90 - \log_{10}(t+1)^{15}$

(f)

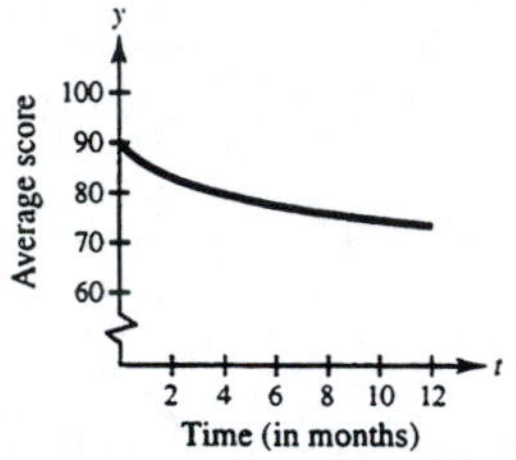

91. $f(x) = \ln x$

False, $f(0) \neq 0$ since 0 is not in the domain of $f(x)$. $f(1) = \ln 1 = 0$

93. False. $f(x) - f(2) = \ln x - \ln 2 = \ln \dfrac{x}{2} \neq \ln(x-2)$

95. False. $f(u) = 2f(v) \implies \ln u = 2\ln v \implies \ln u = \ln v^2 \implies u = v^2$

97. Let $x = \log_b u$ and $y = \log_b v$, then $b^x = u$ and $b^y = v$.

$$\frac{u}{v} = \frac{b^x}{b^y} = b^{x-y}$$

$$\log_b\left(\frac{u}{v}\right) = \log_b(b^{x-y}) = x - y = \log_b u - \log_b v$$

99. $\dfrac{24xy^{-2}}{16x^{-3}y} = \dfrac{24xx^3}{16yy^2} = \dfrac{3x^4}{2y^3},\ x \neq 0$

101. $(18x^3y^4)^{-3}(18x^3y^4)^3 = \dfrac{(18x^3y^4)^3}{(18x^3y^4)^3} = 1$ if $x \neq 0,\ y \neq 0$.

Section 3.4 Exponential and Logarithmic Equations

- To solve an exponential equation, isolate the exponential expression, then take the logarithm of both sides. Then solve for the variable.
  1. $\log_a a^x = x$
  2. $\ln e^x = x$
- To solve a logarithmic equation, rewrite it in exponential form. Then solve for the variable.
  1. $a^{\log_a x} = x$
  2. $e^{\ln x} = x$
- If $a > 0$ and $a \neq 1$ we have the following:
  1. $\log_a x = \log_a y \implies x = y$
  2. $a^x = a^y \implies x = y$

Solutions to Odd-Numbered Exercises

1. $4^{2x-7} = 64$

(a) $x = 5$

$4^{2(5)-7} = 4^3 = 64$

Yes, $x = 5$ is a solution.

(b) $x = 2$

$4^{2(2)-7} = 4^{-3} = \frac{1}{64} \neq 64$

No, $x = 2$ is not a solution.

3. $3e^{x+2} = 75$

(a) $x = -2 + e^{25}$

$3e^{(-2+e^{25})+2} = 3e^{e^{25}} \neq 75$

No, $x = -2 + e^{25}$ is not a solution.

(b) $x = -2 + \ln 25$

$3e^{(-2+\ln 25)+2} = 3e^{\ln 25} = 3(25) = 75$

Yes, $x = -2 + \ln 25$ is a solution.

(c) $x \approx 1.2189$

$3e^{1.2189+2} = 3e^{3.2189} \approx 75$

Yes, $x \approx 1.2189$ is a solution.

5. $\log_4(3x) = 3 \implies 3x = 4^3 \implies 3x = 64$

(a) $x \approx 20.3560$

$3(20.3560) = 61.0680 \neq 64$

No, $x \approx 20.3560$ is not a solution.

(b) $x = -4$

$3(-4) = -12 \neq 64$

No, $x = -4$ is not a solution.

(c) $x = \frac{64}{3}$

$3\left(\frac{64}{3}\right) = 64$

Yes, $x = \frac{64}{3}$ is a solution.

7. $f(x) = g(x)$

$2^x = 8$

$2^x = 2^3$

$x = 3$

Point of intersection: $(3, 8)$

9. $f(x) = g(x)$

$\log_3 x = 2$

$x = 3^2$

$x = 9$

Point of intersection: $(9, 2)$

11. $4^x = 16$

$4^x = 4^2$

$x = 2$

13. $7^x = \frac{1}{49}$

$7^x = 7^{-2}$

$x = -2$

15. $\left(\frac{3}{4}\right)^x = \frac{27}{64}$

$\left(\frac{3}{4}\right)^x = \left(\frac{3}{4}\right)^3$

$x = 3$

17. $\log_4 x = 3$

$x = 4^3$

$x = 64$

19. $\log_{10} x = -1$

$x = 10^{-1}$

$x = \frac{1}{10}$

21. $\log_{10} 10^{x^2} = x^2$

23. $e^{\ln(5x+2)} = 5x + 2$

25. $e^{\ln x^2} = x^2$

27. $e^x = 10$

$x = \ln 10 \approx 2.303$

29. $7 - 2e^x = 5$

$-2e^x = -2$

$e^x = 1$

$x = \ln 1 = 0$

31. $e^{3x} = 12$

$3x = \ln 12$

$x = \frac{\ln 12}{3} \approx 0.828$

33. $500e^{-x} = 300$

$e^{-x} = \frac{3}{5}$

$-x = \ln \frac{3}{5}$

$x = -\ln \frac{3}{5} = \ln \frac{5}{3} \approx 0.511$

35. $e^{2x} - 4e^x - 5 = 0$

$(e^x - 5)(e^x + 1) = 0$

$e^x = 5$ or $e^x = -1$ (No solution)

$x = \ln 5 \approx 1.609$

37. $20(100 - e^{x/2}) = 500$

$100 - e^{x/2} = 25$

$-e^{x/2} = -75$

$e^{x/2} = 75$

$\frac{x}{2} = \ln 75$

$x = 2 \ln 75 \approx 8.635$

39. $10^x = 42$

$x = \log_{10} 42 \approx 1.623$

41. $3^{2x} = 80$

$\ln 3^{2x} = \ln 80$

$2x \ln 3 = \ln 80$

$x = \frac{\ln 80}{2 \ln 3} \approx 1.994$

43. $5^{-t/2} = 0.20$

$5^{-t/2} = \frac{1}{5}$

$5^{-t/2} = 5^{-1}$

$-\frac{t}{2} = -1$

$t = 2$

45. $2^{3-x} = 565$

$\ln 2^{3-x} = \ln 565$

$(3 - x) \ln 2 = \ln 565$

$3 \ln 2 - x \ln 2 = \ln 565$

$-x \ln 2 = \ln 565 - \ln 2^3$

$x \ln 2 = \ln 8 - \ln 565$

$x = \frac{\ln 8 - \ln 565}{\ln 2} \approx -6.142$

47. $g(x) = 6e^{1-x} - 25$

The zero is $x \approx -0.427$.

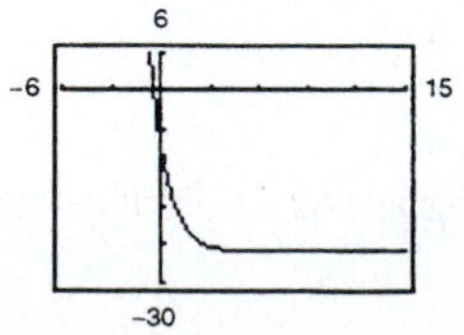

49. $g(t) = e^{0.09t} - 3$

The zero is $x \approx 12.207$.

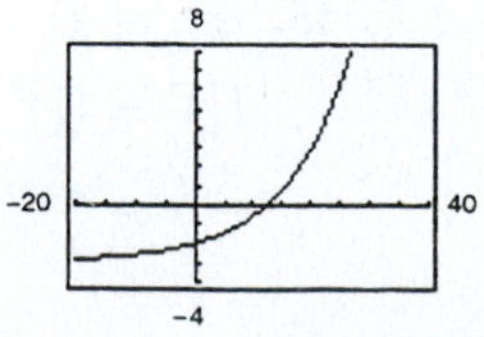

51. $8(10^{3x}) = 12$

$$10^{3x} = \frac{12}{8}$$

$$3x = \log_{10}\left(\frac{3}{2}\right)$$

$$x = \tfrac{1}{3}\log_{10}\left(\frac{3}{2}\right) \approx 0.059$$

53. $\left(1 + \frac{0.065}{365}\right)^{365t} = 4$

$$\ln\left(1 + \frac{0.065}{365}\right)^{365t} = \ln 4$$

$$365t \ln\left(1 + \frac{0.065}{365}\right) = \ln 4$$

$$t = \frac{\ln 4}{365 \ln\left(1 + \frac{0.065}{365}\right)} \approx 21.330$$

55. $\ln x = -3$

$$x = e^{-3} \approx 0.050$$

57. $\ln 2x = 2.4$

$$2x = e^{2.4}$$

$$x = \frac{e^{2.4}}{2} \approx 5.512$$

59. $\ln \sqrt{x + 2} = 1$

$$\sqrt{x + 2} = e^1$$

$$x + 2 = e^2$$

$$x = e^2 - 2 \approx 5.389$$

61. $\log_{10}(z - 3) = 2$

$$z - 3 = 10^2$$

$$z = 10^2 + 3 = 103$$

63. $\ln x + \ln(x - 2) = 1$

$$\ln[x(x - 2)] = 1$$

$$x(x - 2) = e^1$$

$$x^2 - 2x - e = 0$$

$$x = \frac{2 \pm \sqrt{4 + 4e}}{2}$$

$$= \frac{2 \pm 2\sqrt{1 + e}}{2}$$

$$= 1 \pm \sqrt{1 + e}$$

Using the positive value for x, we have $x = 1 + \sqrt{1 + e} \approx 2.928$.

65. $\log_{10}(x + 4) - \log_{10} x = \log_{10}(x + 2)$

$$\log_{10}\left(\frac{x + 4}{x}\right) = \log_{10}(x + 2)$$

$$\frac{x + 4}{x} = x + 2$$

$$x + 4 = x^2 + 2x$$

$$0 = x^2 + x - 4 \qquad \text{Quadratic Formula}$$

$$x = \frac{-1 \pm \sqrt{17}}{2}$$

Choosing the positive value of x (the negative value is extraneous), we have

$$x = \frac{-1 + \sqrt{17}}{2} \approx 1.562.$$

67. $\log_3 x + \log_3(x^2 - 8) = \log_3 8x$

$$\log_3 x(x^2 - 8) = \log_3 8x$$

$$x(x^2 - 8) = 8x$$

$$x^3 - 8x = 8x$$

$$x^3 - 16x = 0$$

$$x(x + 4)(x - 4) = 0$$

$$x = 0,\ x = -4, \text{ or } x = 4$$

The only solution that is in the domain is $x = 4$. Both $x = 0$ and $x = -4$ are extraneous.

69. $\ln(x + 5) = \ln(x - 1) - \ln(x + 1)$

$$\ln(x + 5) = \ln\left(\frac{x - 1}{x + 1}\right)$$

$$x + 5 = \frac{x - 1}{x + 1}$$

$$(x + 5)(x + 1) = x - 1$$

$$x^2 + 6x + 5 = x - 1$$

$$x^2 + 5x + 6 = 0$$

$$(x + 2)(x + 3) = 0$$

$$x = -2 \text{ or } x = -3$$

Both of these solutions are extraneous, so the equation has no solution.

71. $6\log_3(0.5x) = 11$

$\log_3(0.5x) = \frac{11}{6}$

$0.5x = 3^{11/6}$

$x = 2(3^{11/6}) \approx 14.988$

73. $2\ln x = 7$

$\ln x = \frac{7}{2}$

$x = e^{7/2} \approx 33.115$

75. $\ln x + \ln(x^2 + 1) = 8$

$\ln x(x^2 + 1) = 8$

$x(x^2 + 1) = e^8$

$x^3 + x - e^8 = 0$

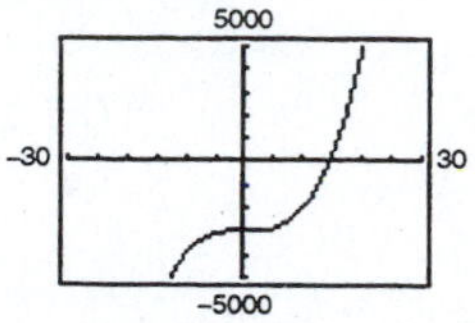

From the graph we have $x \approx 14.369$.

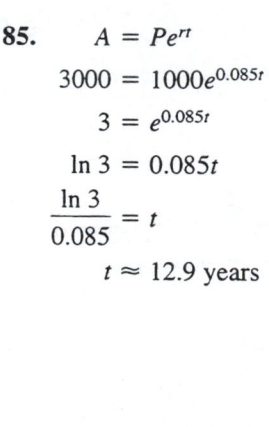

77. $y_1 = 7$

$y_2 = 2^x$

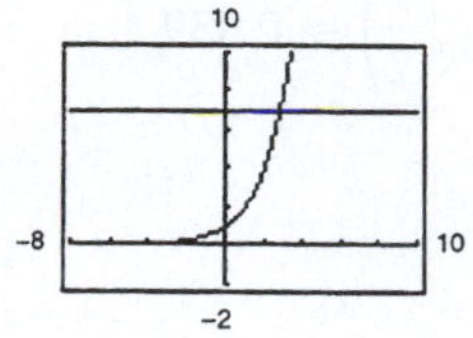

From the graph we have $x \approx 2.807$ when $y = 7$.

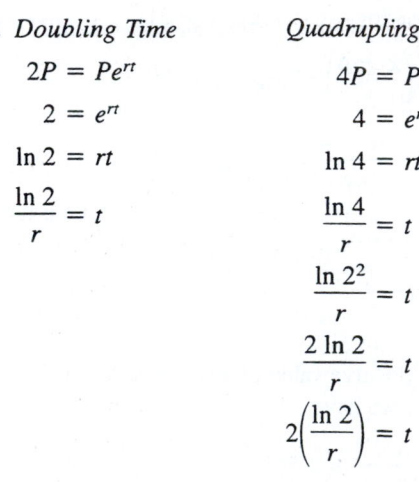

79. $y_1 = 3$

$y_2 = \ln x$

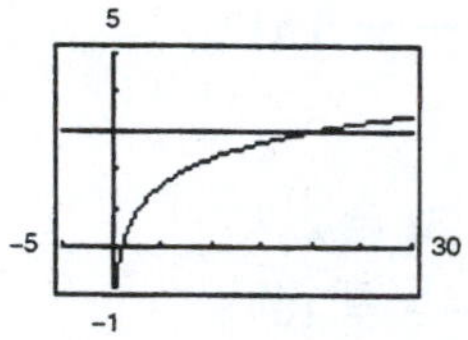

From the graph we have $x \approx 20.806$ when $y = 3$.

81. $A = Pe^{rt}$

$2000 = 1000e^{0.085t}$

$2 = e^{0.085t}$

$\ln 2 = 0.085t$

$\frac{\ln 2}{0.085} = t$

$t \approx 8.2$ years

83. *Doubling Time*

$2P = Pe^{rt}$

$2 = e^{rt}$

$\ln 2 = rt$

$\frac{\ln 2}{r} = t$

Quadrupling Time

$4P = Pe^{rt}$

$4 = e^{rt}$

$\ln 4 = rt$

$\frac{\ln 4}{r} = t$

$\frac{\ln 2^2}{r} = t$

$\frac{2\ln 2}{r} = t$

$2\left(\frac{\ln 2}{r}\right) = t$

Yes, it takes twice as long to quadruple.

85. $A = Pe^{rt}$

$3000 = 1000e^{0.085t}$

$3 = e^{0.085t}$

$\ln 3 = 0.085t$

$\frac{\ln 3}{0.085} = t$

$t \approx 12.9$ years

87. $p = 500 - 0.5(e^{0.004x})$

(a) $p = 350$

$350 = 500 - 0.5(e^{0.004x})$

$300 = e^{0.004x}$

$0.004x = \ln 300$

$x \approx 1426$ units

(b) $p = 300$

$300 = 500 - 0.5(e^{0.004x})$

$400 = e^{0.004x}$

$0.004x = \ln 400$

$x \approx 1498$ units

11. Since $A = 500e^{rt}$ and $A = 1292.85$ when $t = 10$, we have the following.

$$1292.85 = 500e^{10r}$$

$$r = \frac{\ln(1292.85/500)}{10} \approx 0.9095 = 9.5\%$$

The time to double is given by

$$1000 = 500e^{0.095t}$$

$$t = \frac{\ln 2}{0.095} \approx 7.30 \text{ years.}$$

13. Since $A = Pe^{0.045t}$ and $A = 10{,}000.00$ when $t = 10$, we have the following.

$$10{,}000.00 = Pe^{0.045(10)}$$

$$\frac{10{,}000.00}{e^{0.045(10)}} = P \approx \$6376.28$$

The time to double is given by

$$t = \frac{\ln 2}{0.045} \approx 15.40 \text{ years.}$$

15. $500{,}000 = P\left(1 + \dfrac{0.075}{12}\right)^{12(20)}$

$$P = \frac{500{,}000}{\left(1 + \dfrac{0.075}{12}\right)^{12(20)}} = \$112{,}087.09$$

17. $P = 1000$, $r = 11\%$

(a) $n = 1$

$$t = \frac{\ln 2}{\ln(1 + 0.11)} \approx 6.642 \text{ years}$$

(b) $n = 12$

$$t = \frac{\ln 2}{12 \ln\left(1 + \frac{0.11}{12}\right)} \approx 6.330 \text{ years}$$

(c) $n = 365$

$$t = \frac{\ln 2}{365 \ln\left(1 + \frac{0.11}{365}\right)} \approx 6.302 \text{ years}$$

(d) Continuously

$$t = \frac{\ln 2}{0.11} \approx 6.301 \text{ years}$$

19.

$$3P = Pe^{rt}$$
$$3 = e^{rt}$$
$$\ln 3 = rt$$
$$\frac{\ln 3}{r} = t$$

r	2%	4%	6%	8%	10%	12%
$t = \dfrac{\ln 3}{r}$	54.93	27.47	18.31	13.73	10.99	9.16

21.

$$3P = P(1 + r)^t$$
$$3 = (1 + r)^t$$
$$\ln 3 = \ln(1 + r)^t$$
$$\ln 3 = t \ln(1 + r)$$
$$\frac{\ln 3}{\ln(1 + r)} = t$$

r	2%	4%	6%	8%	10%	12%
$t = \dfrac{\ln 3}{\ln(1 + r)}$	55.48	28.01	18.85	14.27	11.53	9.69

23. Continuous compounding results in faster growth.

$A = 1 + 0.075[\![t]\!]$ and $A = e^{0.07t}$

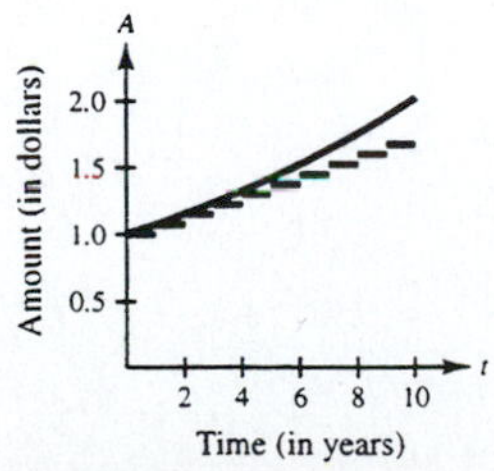

25. $\frac{1}{2}C = Ce^{k(1620)}$

$k = \frac{\ln 0.5}{1620}$

Given $C = 10$ grams, after 1000 years we have

$y = 10e^{[(\ln 0.5)/1620](1000)}$

≈ 6.52 grams.

27. $\frac{1}{2}C = Ce^{k(5730)}$

$k = \frac{\ln 0.5}{5730}$

Given $y = 2$ grams after 1000 years, we have

$2 = Ce^{[(\ln 0.5)/5730](1000)}$

$C \approx 2.26$ grams.

29. $\frac{1}{2}C = Ce^{k(24,360)}$

$k = \frac{\ln 0.5}{24,360}$

Given $y = 2.1$ grams after 1000 years, we have

$2.1 = Ce^{[(\ln 0.5)/24,360](1000)}$

$C \approx 2.16$ grams.

31. $y = ae^{bx}$

$1 = ae^{b(0)} \Rightarrow 1 = a$

$10 = e^{b(3)}$

$\ln 10 = 3b$

$\frac{\ln 10}{3} = b \quad \Rightarrow b \approx 0.7675$

Thus, $y = e^{0.7675x}$.

33. $y = ae^{bx}$

$1 = ae^{b(0)} \Rightarrow 1 = a$

$\frac{1}{4} = e^{b(3)}$

$\ln\left(\frac{1}{4}\right) = 3b$

$\frac{\ln(\frac{1}{4})}{3} = b \quad \Rightarrow b \approx -0.4621$

Thus, $y = e^{-0.4621x}$.

35. $P = 105,300e^{0.015t}$

$150,000 = 105,300e^{0.015t}$

$\ln \frac{1500}{1053} = 0.015t$

$t \approx 23.59$

The population will reach 150,000 during 2013. [Note: 1990 + 23.59]

37. For 1945, use $t = -45$.

$1350 = 2500d^{k(-45)}$

$\ln(\frac{1350}{2500}) = -45k \Rightarrow k \approx 0.0137$

For 2010, use $t = 20$.

$P = 2500e^{0.0137(20)} \approx 3288$ people

39. $y = ae^{bt}$

$4.22 = ae^{b(0)} \Rightarrow a = 4.22$

$6.49 = 4.22e^{b(10)}$

$\frac{6.49}{4.22} = e^{10b}$

$\ln\left(\frac{6.49}{4.22}\right) = 10b \quad \Rightarrow b \approx 0.0430$

$y = 4.22e^{0.0430t}$

When $t = 20$,
$y = 4.22e^{0.0430(20)} \approx 9.97$ million.

41. $y = ae^{bt}$

$3.00 = ae^{b(0)} \Rightarrow a = 3$

$2.74 = 3e^{b(10)}$

$\frac{2.74}{3} = e^{10b}$

$\ln\left(\frac{2.74}{3}\right) = 10b \quad \Rightarrow b \approx -0.0091$

$y = 3e^{-0.0091t}$

When $t = 20$,
$y = 3e^{-0.0091(20)} \approx 2.50$ million.

43. b is determined by the growth rate. The greater the rate of growth, the greater the value of b.

45. $N = 100e^{kt}$

$300 = 100e^{5k}$

$k = \frac{\ln 3}{5} \approx 0.2197$

$N = 100e^{0.2197t}$

$200 = 100e^{0.2197t}$

$t = \frac{\ln 2}{0.2197} \approx 3.15$ hours

47. $y = Ce^{kt}$

$$\frac{1}{2}C = Ce^{(1620)k}$$

$$\ln\frac{1}{2} = 1620k$$

$$k = \frac{\ln(1/2)}{1620}$$

When $t = 100$, we have

$y = Ce^{[\ln(1/2)/1620](100)} \approx 0.958C = 95.8\%C.$

After 100 years, approximately 95.8% of the radioactive radium will remain.

49. (0, 22,000), (2, 13,000)

(a) $m = \dfrac{13{,}000 - 22{,}000}{2 - 0} = -4500$

$b = 22{,}000$

Thus, $V = -4500t + 22{,}000$.

(b) $a = 22{,}000$

$$13{,}000 = 22{,}000e^{k(2)}$$

$$\frac{13}{22} = e^{2k}$$

$$\ln\left(\frac{13}{22}\right) = 2k \Rightarrow k \approx -0.263$$

Thus, $V = 22{,}000e^{-0.263t}$.

(c) The exponential model depreciates faster in the first two years.

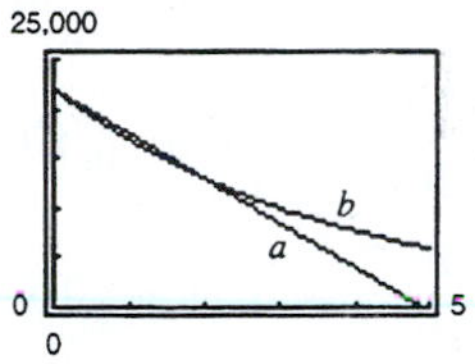

(d)

t	1	3
$V = -4500t + 22{,}000$	\$17,500	\$8500
$V = 22{,}000e^{-0.263t}$	\$16,912	\$9995

(e) The slope of the linear model means that the car depreciates \$4500 per year.

51. $S(t) = 100(1 - e^{kt})$

(a) $15 = 100(1 - e^{k(1)})$

$$-85 = -100e^k$$

$$k = \ln 0.85$$

$$k \approx -0.1625$$

$$S(t) = 100(1 - e^{-0.1625t})$$

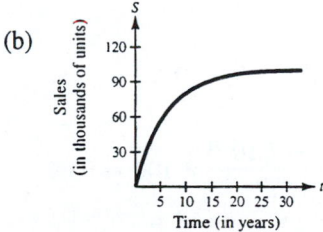

(b)

(c) $S(5) = 100(1 - e^{-0.1625(5)})$

$\approx 55.625 = 55{,}625$ units

53. $S = 10(1 - e^{kx})$

$x = 5$ (in hundreds), $S = 2.5$ (in thousands)

(a) $2.5 = 10(1 - e^{k(5)})$

$0.25 = 1 - e^{5k}$

$e^{5k} = 0.75$

$5k = \ln 0.75$

$k \approx -0.0575$

$S = 10(1 - e^{-0.0575x})$

(b) When $x = 7$,

$S = 10(1 - e^{-0.0575(7)}) \approx 3.314$

which corresponds to 3314 units.

55. $N = 30(1 - e^{kt})$

(a) $N = 19,\ t = 20$

$19 = 30(1 - e^{20k})$

$20k = \ln \frac{11}{30}$

$k \approx -0.050$

$N = 30(1 - e^{-0.050t})$

(b) $N = 25$

$25 = 30(1 - e^{-0.05t})$

$\frac{5}{30} = e^{-0.05t}$

$t = -\frac{1}{0.05} \ln \frac{5}{30} \approx 36$ days

(c) No, this is not a linear function.

57. $R = \log_{10} \dfrac{I}{I_0} = \log_{10} I$ since $I_0 = 1$.

(a) $R = \log_{10} 80{,}500{,}000 \approx 7.91$

(b) $R = \log_{10} 48{,}275{,}000 \approx 7.68$

59. $\beta(I) = 10 \log_{10} \dfrac{I}{I_0}$ where $I_0 = 10^{-16}$ watt/cm^2.

(a) $\beta(10^{-14}) = 10 \log_{10} \dfrac{10^{-14}}{10^{-16}} = 10 \log_{10} 10^2 = 20$ decibels

(b) $\beta(10^{-9}) = 10 \log_{10} \dfrac{10^{-9}}{10^{-16}} = 10 \log_{10} 10^7 = 70$ decibels

(c) $\beta(10^{-6.5}) = 10 \log_{10} \dfrac{10^{-6.5}}{10^{-16}} = 10 \log_{10} 10^{9.5} = 95$ decibels

(d) $\beta(10^{-4}) = 10 \log_{10} \dfrac{10^{-4}}{10^{-16}} = 10 \log_{10} 10^{12} = 120$ decibels

61. $\beta = 10 \log_{10} \dfrac{I}{I_0}$

$10^{\beta/10} = \dfrac{I}{I_0}$

$I = I_0 10^{\beta/10}$

$\%\text{ decrease} = \dfrac{I_0 10^{9.3} - I_0 10^{8.0}}{I_0 10^{9.3}} \times 100 \approx 95\%$

63. $\text{pH} = -\log_{10}[H^+] = -\log_{10}[2.3 \times 10^{-5}] \approx 4.64$

65. $5.8 = -\log_{10}[H^+]$

$10^{-5.8} = H^+$

$H^+ \approx 1.58 \times 10^{-6}$ moles per liter

67. $2.5 = -\log_{10}[H^+]$

$10^{-2.5} = H^+$ for the fruit.

$9.5 = -\log_{10}[H^+]$

$10^{-9.5} = H^+$ for the antacid tablet.

$\dfrac{10^{-2.5}}{10^{-9.5}} = 10^7$

69. Interest: $u = M - \left(M - \frac{Pr}{12}\right)\left(1 + \frac{r}{12}\right)^{12t}$

Principle: $v = \left(M - \frac{Pr}{12}\right)\left(1 + \frac{r}{12}\right)^{12t}$

(a) $P = 120{,}000,\ t = 35,\ r = 0.095,\ M = 985.93$

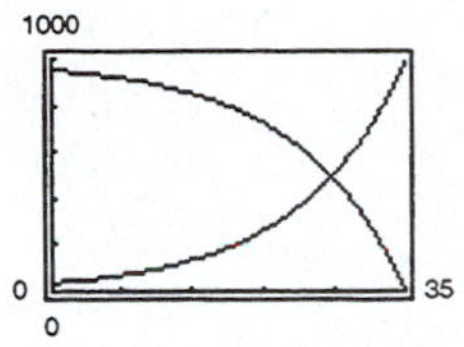

(b) In the early years of the mortgage, the majority of the monthly payment goes toward interest. The principle and interest are nearly equal when $t \approx 27.676 \approx 28$ years.

(c) $P = 120{,}000,\ t = 20,\ r = 0.095,\ M = 1118.56$

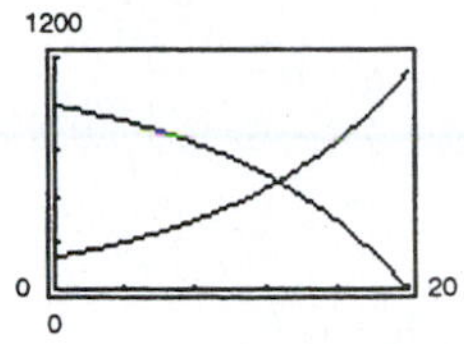

The interest is still the majority of the monthly payment in the early years. Now the principle and interest are nearly equal when $t \approx 12.675 \approx 12.7$ years.

71. $t_1 = 40.757 + 0.556s - 15.817 \ln s$

$t_2 = 1.2259 + 0.0023s^2$

(a) Linear model: $t_3 \approx 0.2729s - 6.0143$

Exponential model: $t_4 \approx 1.5385e^{1.0291s}$

(b)

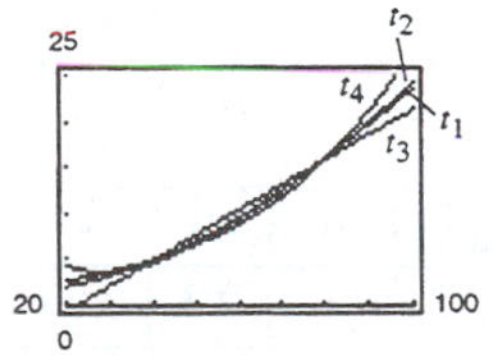

(c)

s	30	40	50	60	70	80	90
t_1	3.6	4.6	6.7	9.4	12.5	15.9	19.6
t_2	3.3	4.9	7.0	9.5	12.5	15.9	19.9
t_3	2.2	4.9	7.6	10.4	13.1	15.8	18.5
t_4	3.7	4.9	6.6	8.8	11.8	15.8	21.2

(d) Model t_1: $S_1 = |3.4 - 3.6| + |5 - 4.6| + |7 - 6.7| + |9.3 - 9.4| + |12 - 12.5| + |15.8 - 15.9| + |20 - 19.6| = 2.0$

Model t_2: $S_2 = |3.4 - 3.3| + |5 - 4.9| + |7 - 7| + |9.3 - 9.5| + |12 - 12.5| + |15.8 - 15.9| + |20 - 19.9| = 1.1$

Model t_3: $S_3 = |3.4 - 2.2| + |5 - 4.9| + |7 - 7.6| + |9.3 - 10.4| + |12 - 13.1| + |15.8 - 15.8| + |20 - 18.5| = 5.6$

Model t_4: $S_4 = |3.4 - 3.7| + |5 - 4.9| + |7 - 6.6| + |9.3 - 8.8| + |12 - 11.8| + |15.8 - 15.8| + |20 - 21.2| = 2.7$

t_2, the Quadratic model, is the best fit with the data.

73. Answers will vary.

75.

−4	4	4	−39	36
		−16	48	−36
	4	−12	9	0

Thus, $\dfrac{4x^3 + 4x^2 - 39x + 36}{x + 4} = 4x^2 - 12x + 9.$

77.

4	2	−8	3	−9
		8	0	12
	2	0	3	3

Thus, $\dfrac{2x^3 - 8x^2 + 3x - 9}{x - 4} = 2x^2 + 3 + \dfrac{3}{x - 4}.$

❑ Review Exercises for Chapter 3

Solutions to Odd-Numbered Exercises

1. $f(x) = 4^x$
Intercept: $(0, 1)$
Horizontal asymptote: x-axis
Increasing on: $(-\infty, \infty)$
Matches graph (e)

3. $f(x) = -4^x$
Intercept: $(0, -1)$
Horizontal asymptote: x-axis
Decreasing on: $(-\infty, \infty)$
Matches graph (a)

5. $f(x) = \log_4 x$
Intercept: $(1, 0)$
Vertical asymptote: y-axis
Increasing on: $(0, \infty)$
Matches graph (d)

7. $f(x) = 0.3^x$

x	−2	−1	0	1	2
y	11.11	3.33	1	0.3	0.09

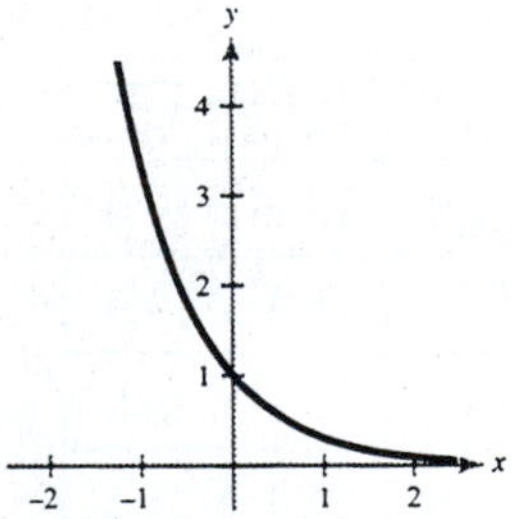

9. $h(x) = e^{-x/2}$

x	−2	−1	0	1	2
y	2.72	1.65	1	0.61	0.37

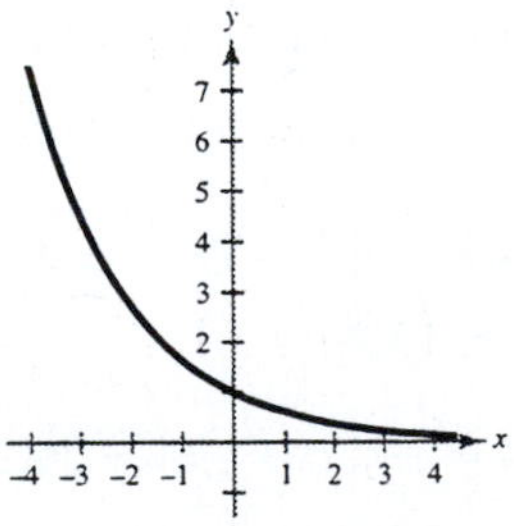

11. $f(x) = e^{x+2}$

x	-3	-2	-1	0	1
y	0.37	1	2.72	7.39	20.09

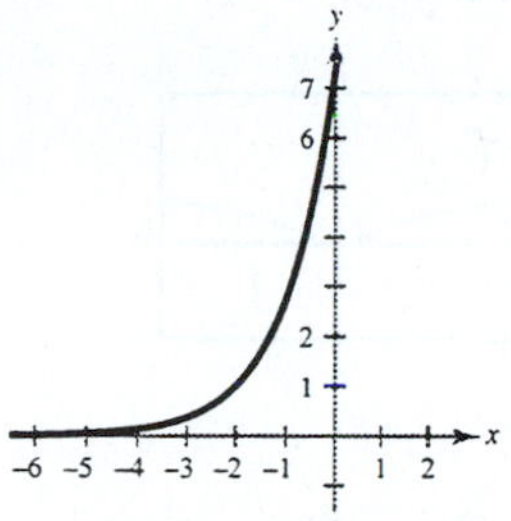

13. $g(x) = 200e^{4/x}$

As $x \to \infty$, $g(x) \to 200$ so we have a horizontal asymptote at $y = 200$.

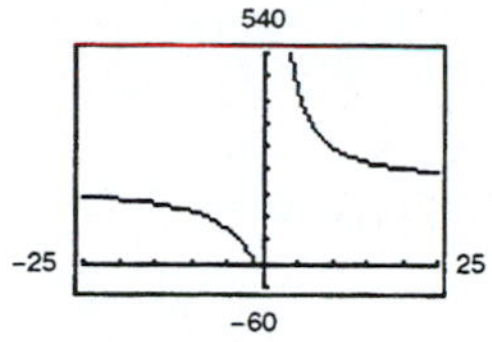

15. $A = 3500\left(1 + \frac{0.105}{n}\right)^{10n}$ or $A = 3500e^{(0.105)(10)}$

n	1	2	4	12	365	Continuous Compounding
A	\$9,499.28	\$9,738.91	\$9,867.22	\$9,956.20	\$10,000.27	\$10,001.78

17. $200{,}000 = Pe^{0.08t}$

$$P = \frac{200{,}000}{e^{0.08t}}$$

t	1	10	20	30	40	50
P	\$184,623.27	\$89,865.79	\$40,379.30	\$18,143.59	\$8,152.44	\$3,663.13

19. $F(t) = 1 - e^{-t/3}$

(a) $F\left(\frac{1}{2}\right) \approx 0.154$

(b) $F(2) \approx 0.487$

(c) $F(5) \approx 0.811$

21. (a) $A = 50{,}000e^{(0.0875)(35)} \approx \$1{,}069{,}047.14$

(b) The doubling time is

$$\frac{\ln 2}{0.0875} \approx 7.9 \text{ years.}$$

23. $g(x) = \log_2 x \implies 2^y = x$

Domain: $(0, \infty)$

Vertical asymptote: $x = 0$

x	$\frac{1}{4}$	$\frac{1}{2}$	1	2	4
y	-2	-1	0	1	2

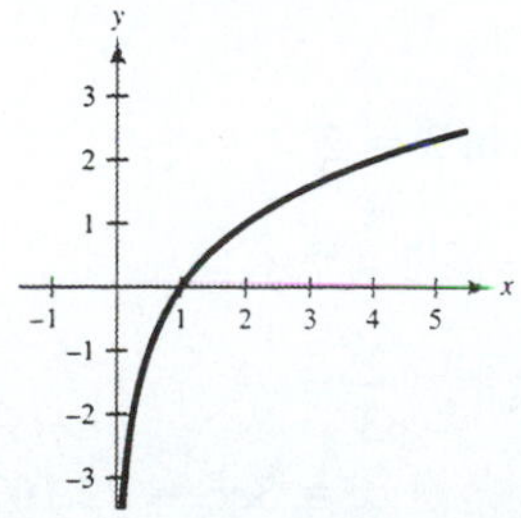

25. $f(x) = \ln x + 3$

Domain: $(0, \infty)$.

Vertical asymptote: $x = 0$

x	1	2	3	$\frac{1}{2}$	$\frac{1}{4}$
$f(x)$	3	3.69	4.10	2.31	1.61

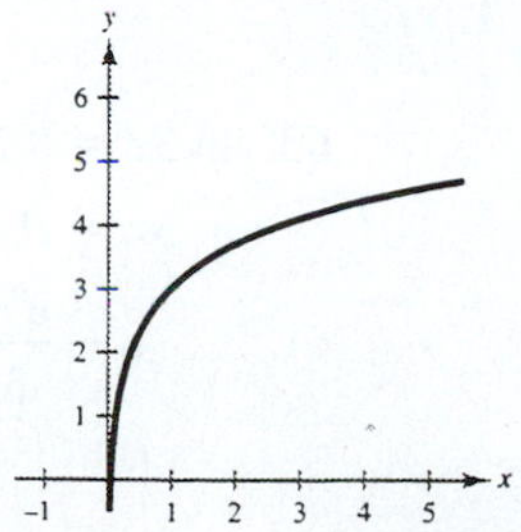

27. $h(x) = \ln(e^{x-1})$
$= (x - 1) \ln e$
$= x - 1$

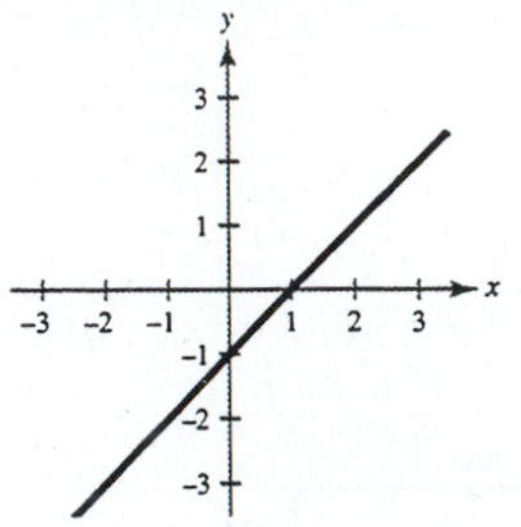

29. $y = \log_{10}(x^2 + 1)$

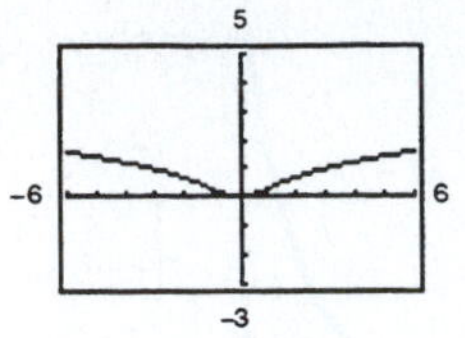

31. $4^3 = 64$
$\log_4 64 = 3$

33. $\log_{10} 1000 = \log_{10} 10^3 = 3$

35. $\ln e^7 = 7$

37. $\log_4 9 = \dfrac{\log_{10} 9}{\log_{10} 4} \approx 1.585$

$\log_4 9 = \dfrac{\ln 9}{\ln 4} \approx 1.585$

39. $\log_{12} 200 = \dfrac{\log_{10} 200}{\log_{10} 12} \approx 2.132$

$\log_{12} 200 = \dfrac{\ln 200}{\ln 12} \approx 2.132$

41. $\log_5 5x^2 = \log_5 5 + \log_5 x^2$
$= 1 + 2 \log_5|x|$

43. $\log_{10} \dfrac{5\sqrt{y}}{x^2} = \log_{10} 5\sqrt{y} - \log_{10} x^2$
$= \log_{10} 5 + \log_{10} \sqrt{y} - \log_{10} x^2$
$= \log_{10} 5 + \dfrac{1}{2}\log_{10} y - 2 \log_{10}|x|$

45. $\log_2 5 + \log_2 x = \log_2 5x$

47. $\dfrac{1}{2}\ln|2x - 1| - 2 \ln|x + 1| = \ln \sqrt{|2x - 1|} - \ln|x + 1|^2$
$= \ln \dfrac{\sqrt{|2x - 1|}}{(x + 1)^2}$

49. True; by the inverse properties, $\log_b b^{2x} = 2x$.

51. False; $\ln x + \ln y = \ln(xy) \neq \ln(x + y)$

53. True, $\log\left(\dfrac{10}{x}\right) = \log 10 - \log x = 1 - \log x$.

55. $S = 25 - \dfrac{13 \ln(10/12)}{\ln 3} \approx 27.16$ miles

57. $e^x = 12$
$x = \ln 12 \approx 2.485$

59. $3e^{-5x} = 132$
$e^{-5x} = 44$
$-5x = \ln 44$
$x = \dfrac{\ln 44}{-5} \approx -0.757$

61. $e^{2x} - 7e^x + 10 = 0$
$(e^x - 2)(e^x - 5) = 0$
$e^x = 2$ or $e^x = 5$
$x = \ln 2$ $x = \ln 5$
$x \approx 0.693$ $x \approx 1.609$

63. $\ln 3x = 8.2$
$3x = e^{8.2}$
$x = \dfrac{e^{8.2}}{3} \approx 1213.650$

65. $\ln x - \ln 3 = 2$
$\ln \dfrac{x}{3} = 2$
$\dfrac{x}{3} = e^2$
$x = 3e^2 \approx 22.167$

67. $$\log(x-1) = \log(x-2) - \log(x+2)$$
$$\log(x-1) = \log\left(\frac{x-2}{x+2}\right)$$
$$x - 1 = \frac{x-2}{x+2}$$
$$(x-1)(x+2) = x - 2$$
$$x^2 + x - 2 = x - 2$$
$$x^2 = 0$$
$$x = 0$$

Since $x = 0$ is not in the domain of $\ln(x-1)$ or of $\ln(x-2)$, it is an extraneous solution. The equation has no solution.

69. $2^{0.6x} - 3x = 0$

Graph $y_1 = 2^{0.6x} - 3x$

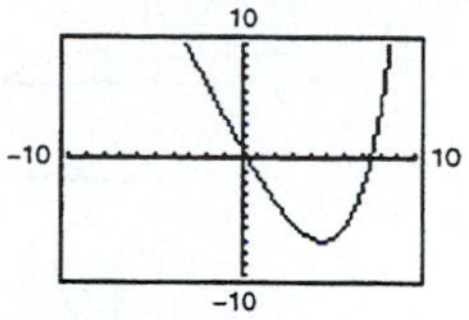

The x-intercepts are at $x \approx 0.39$ and at $x \approx 7.48$.

71. $2\ln(x+3) + 3x = 8$

Graph $y_1 = 2\ln(x+3) + 3x - 8$

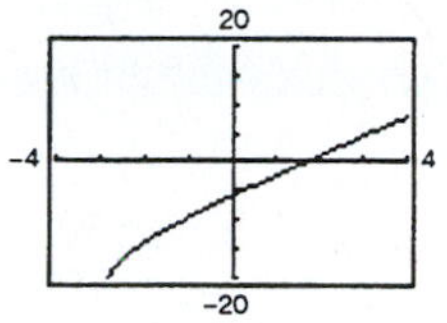

The x-intercept is at $x \approx 1.64$.

73. $$y = ae^{bx}$$
$$2 = ae^{b(0)} \implies a = 2$$
$$3 = 2e^{b(4)}$$
$$1.5 = e^{4b}$$
$$\ln 1.5 = 4b \implies b \approx 0.1014$$

Thus, $y \approx 2e^{0.1014x}$

75. $p = 500 - 0.5e^{0.004x}$

(a) $p = 450$
$$450 = 500 - 0.5e^{0.004x}$$
$$0.5e^{0.004x} = 50$$
$$e^{0.004x} = 100$$
$$0.004x = \ln 100$$
$$x \approx 1151 \text{ units}$$

(b) $p = 400$
$$400 = 500 - 0.5e^{0.004x}$$
$$0.5e^{0.004x} = 100$$
$$e^{0.004x} = 200$$
$$0.004x = \ln 200$$
$$x \approx 1325 \text{ units}$$

77. (a) $$\frac{\ln 2}{r} = 5$$
$$\ln 2 = 5r$$
$$r = \frac{\ln 2}{5} \approx 0.1386 = 13.86\%$$

(b) $A = 10{,}000e^{0.1386(1)} \approx \$11{,}486.65$

79. $R = \log_{10} I$ since $I_0 = 1$.

(a) $\log_{10} I = 8.4$
$$I = 10^{8.4}$$

(b) $\log_{10} I = 6.85$
$$I = 10^{6.85}$$

(c) $\log_{10} I = 9.1$
$$I = 10^{9.1}$$

Practice Test for Chapter 3

1. Solve for x: $x^{3/5} = 8$.

2. Solve for x: $3^{x-1} = \frac{1}{81}$.

3. Graph $f(x) = 2^{-x}$.

4. Graph $g(x) = e^x + 1$.

5. If \$5000 is invested at 9% interest, find the amount after three years if the interest is compounded

(a) monthly (b) quarterly (c) continuously.

6. Write the equation in logarithmic form: $7^{-2} = \frac{1}{49}$.

7. Solve for x: $x - 4 = \log_2 \frac{1}{64}$.

8. Given $\log_b 2 = 0.3562$ and $\log_b 5 = 0.8271$, evaluate $\log_b \sqrt[4]{8/25}$.

9. Write $5 \ln x - \frac{1}{2} \ln y + 6 \ln z$ as a single logarithm.

10. Using your calculator and the change of base formula, evaluate $\log_9 28$.

11. Use your calculator to solve for N: $\log_{10} N = 0.6646$

12. Graph $y = \log_4 x$.

13. Determine the domain of $f(x) = \log_3(x^2 - 9)$.

14. Graph $y = \ln(x - 2)$.

15. True or false: $\dfrac{\ln x}{\ln y} = \ln(x - y)$

16. Solve for x: $5^x = 41$

17. Solve for x: $x - x^2 = \log_5 \frac{1}{25}$

18. Solve for x: $\log_2 x + \log_2(x - 3) = 2$

19. Solve for x: $\dfrac{e^x + e^{-x}}{3} = 4$

20. Six thousand dollars is deposited into a fund at an annual interest rate of 13%. Find the time required for the investment to double if the interest is compounded continuously.

CHAPTER 4
Trigonometry

CHAPTER 4
Trigonometry

Section 4.1 Radian and Degree Measure

You should know the following basic facts about angles, their measurement, and their applications.

- Types of Angles:
  - (a) Acute: Measure between 0° and 90°.
  - (b) Right: Measure 90°.
  - (c) Obtuse: Measure between 90° and 180°.
  - (d) Straight: Measure 180°.
- α and β are complementary if $\alpha + \beta = 90°$. They are supplementary if $\alpha + \beta = 180°$.
- Two angles in standard position that have the same terminal side are called coterminal angles.
- To convert degrees to radians, use $1° = \pi/180$ radians.
- To convert radians to degrees, use 1 radian $= (180/\pi)°$.
- $1' =$ one minute $= 1/60$ of $1°$.
- $1'' =$ one second $= 1/60$ of $1' = 1/3600$ of $1°$.
- The length of a circular arc is $s = r\theta$ where θ is measured in radians.
- Speed = distance/time
- Angular speed $= \theta/t = s/rt$

Solutions to Odd-Numbered Exercises

1.

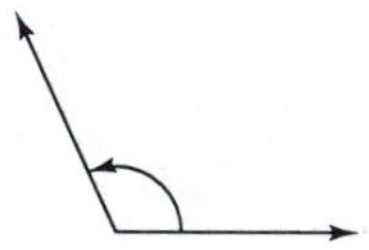

The angle shown is approximately 2 radians.

3.

The angle shown is approximately -3 radians.

5. (a) Since $0 < \frac{\pi}{5} < \frac{\pi}{2}$; $\frac{\pi}{5}$ lies in Quadrant I.

(b) Since $\pi < \frac{7\pi}{5} < \frac{3\pi}{2}$; $\frac{7\pi}{5}$ lies in Quadrant III.

7. (a) Since $-\frac{\pi}{2} < -\frac{\pi}{12} < 0$; $-\frac{\pi}{12}$ lies in Quadrant IV.

(b) Since $-\frac{3\pi}{2} < -\frac{11\pi}{9} < -\pi$; $-\frac{11\pi}{9}$ lies in Quadrant II.

9. (a) Since $\pi < 3.5 < \frac{3\pi}{2}$; 3.5 lies in Quadrant III.

(b) Since $\frac{\pi}{2} < 2.25 < \pi$; 2.25 lies in Quadrant II.

11. (a)

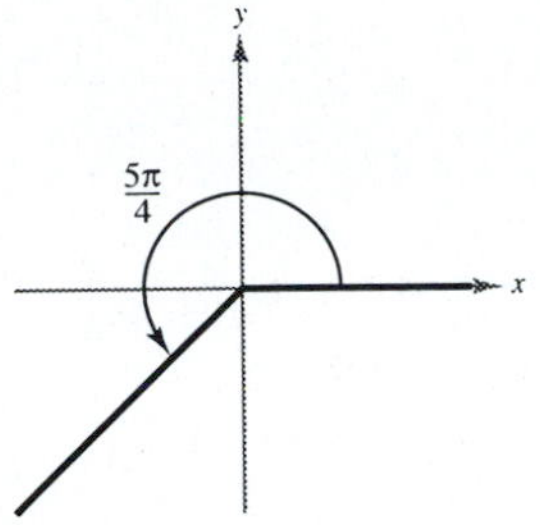

(b)

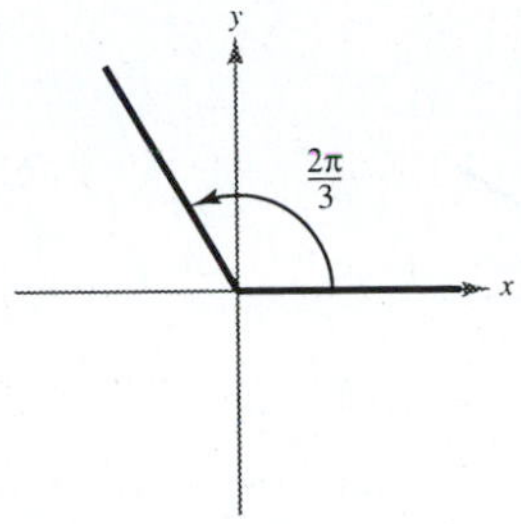

13. (a)

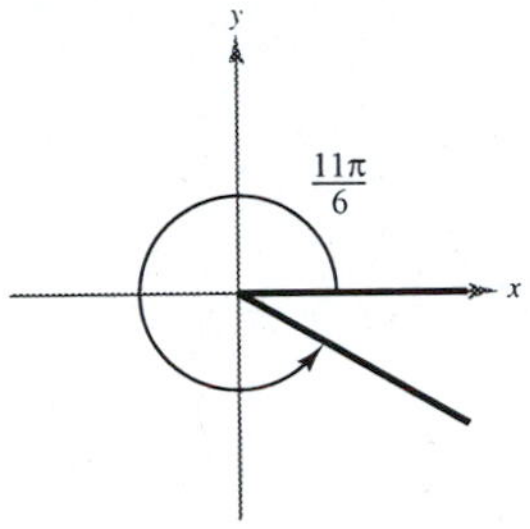

(b)

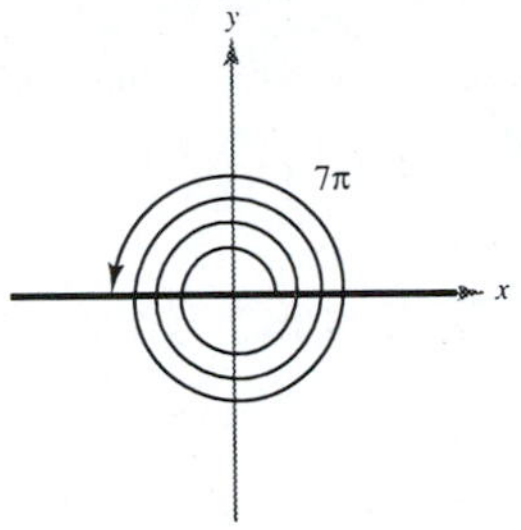

15. (a) Coterminal angles for $\dfrac{\pi}{12}$

$$\frac{\pi}{12} + 2\pi = \frac{25\pi}{12}$$

$$\frac{\pi}{12} - 2\pi = -\frac{23\pi}{12}$$

(b) Coterminal angles for $\dfrac{2\pi}{3}$

$$\frac{2\pi}{3} + 2\pi = \frac{8\pi}{3}$$

$$\frac{2\pi}{3} - 2\pi = -\frac{4\pi}{3}$$

17. (a) Coterminal angles for $-\dfrac{9\pi}{4}$

$$-\frac{9\pi}{4} + 4\pi = \frac{7\pi}{4}$$

$$\frac{7\pi}{4} - 2\pi = -\frac{\pi}{4}$$

(b) Coterminal angles for $-\dfrac{2\pi}{15}$

$$-\frac{2\pi}{15} + 2\pi = \frac{28\pi}{15}$$

$$-\frac{2\pi}{15} - 2\pi = -\frac{32\pi}{15}$$

19. (a) Complement: $\dfrac{\pi}{2} - \dfrac{\pi}{3} = \dfrac{\pi}{6}$

Supplement: $\pi - \dfrac{\pi}{3} = \dfrac{2\pi}{3}$

(b) Complement: Not possible; $\dfrac{3\pi}{4}$ is greater than $\dfrac{\pi}{2}$.

Supplement: $\pi - \dfrac{3\pi}{4} = \dfrac{\pi}{4}$

21.

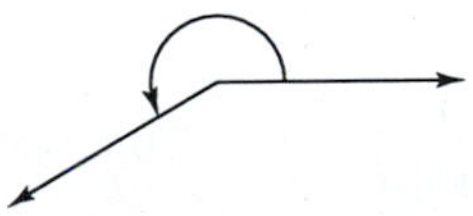

The angle shown is approximately 210°.

23.

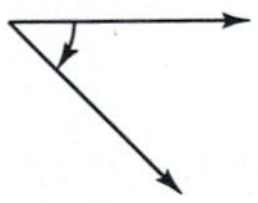

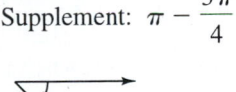

The angle shown is approximately $-45°$.

25. (a) Since $90° < 130° < 180°$; $130°$ lies in Quadrant II.

(b) Since $270° < 285° < 360°$; $285°$ lies in Quadrant IV.

27. (a) Since $-180° < -132°50' < -90°$; $-132°\,50'$ lies in Quadrant III.

(b) Since $-360° < -336° < -270°$; $-336°$ lies in Quadrant I.

29. (a)

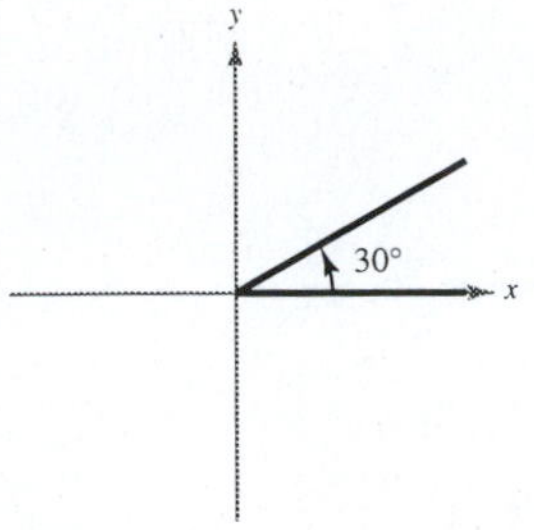

(b)

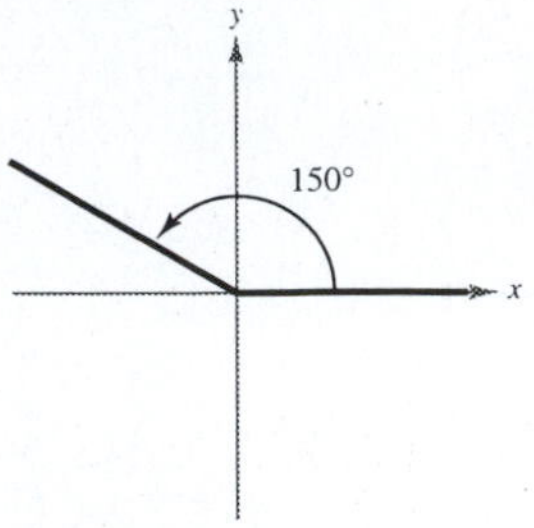

31. (a)

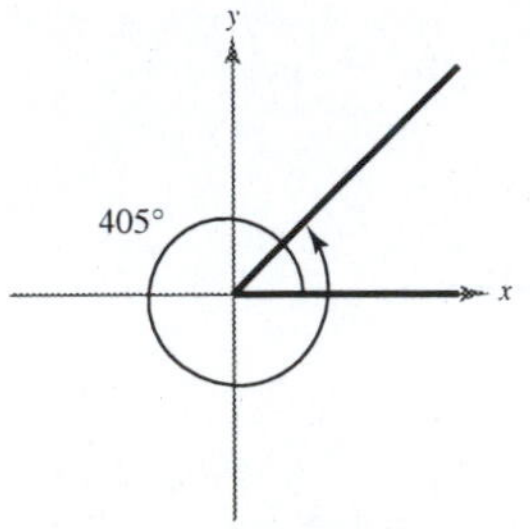

(b)

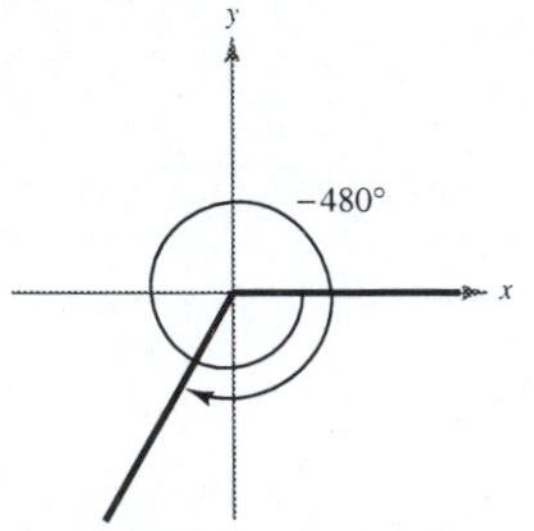

33. (a) Coterminal angles for 45°

$45° + 360° = 405°$

$45° - 360° = -315°$

(b) Coterminal angles for −36°

$-36° + 360° = 324°$

$-36° - 360° = -396°$

35. (a) Coterminal angles for 300°

$300° + 360° = 660°$

$300° - 360° = -60°$

(b) Coterminal angles for 740°

$740° - 2(360°) = 20°$

$20° - 360° = -340°$

37. (a) Complement: $90° - 18° = 72°$

Supplement: $180° - 18° = 162°$

(b) Complement: Not possible; 115° is greater than 90°.

Supplement: $180° - 115° = 65°$

39. (a) $30° = 30\left(\frac{\pi}{180}\right) = \frac{\pi}{6}$

(b) $150° = 150\left(\frac{\pi}{180}\right) = \frac{5\pi}{6}$

41. (a) $-20° = -20\left(\frac{\pi}{180}\right) = -\frac{\pi}{9}$

(b) $-240° = -240\left(\frac{\pi}{180}\right) = -\frac{4\pi}{3}$

43. (a) $\frac{3\pi}{2} = \frac{3\pi}{2}\left(\frac{180}{\pi}\right)^{\circ} = 270°$

(b) $\frac{7\pi}{6} = \frac{7\pi}{6}\left(\frac{180}{\pi}\right)^{\circ} = 210°$

45. (a) $\frac{7\pi}{3} = \frac{7\pi}{3}\left(\frac{180}{\pi}\right)^{\circ} = 420°$

(b) $-\frac{11\pi}{30} = -\frac{11\pi}{30}\left(\frac{180}{\pi}\right)^{\circ} = -66°$

47. $115° = 115\left(\frac{\pi}{180}\right) \approx 2.007$ radians

49. $-216.35° = -216.35\left(\frac{\pi}{180}\right) \approx -3.776$ radians

51. $532° = 532\left(\frac{\pi}{180}\right) \approx 9.285$ radians

53. $-0.83° = -0.83\left(\frac{\pi}{180}\right) \approx -0.014$ radian

55. $\frac{\pi}{7} = \frac{\pi}{7}\left(\frac{180}{\pi}\right) \approx 25.714°$

57. $\frac{15\pi}{8} = \frac{15\pi}{8}\left(\frac{180}{\pi}\right) = 337.5°$

59. $-4.2\pi = -4.2\pi\left(\frac{180}{\pi}\right) = -756°$

61. $-2 = -2\left(\frac{180}{\pi}\right) \approx -114.592°$

63. (a) $54° 45' = 54° + \left(\frac{45}{60}\right)° = 54.75°$

(b) $-128° 30' = -128° - \left(\frac{30}{60}\right)° = -128.5°$

65. (a) $85° 18' 30'' = \left(85 + \frac{18}{60} + \frac{30}{3600}\right)° \approx 85.308°$

(b) $330° 25'' = \left(330 + \frac{25}{3600}\right)° \approx 330.007°$

67. (a) $240.6° = 240° + 0.6(60)' = 240° 36'$

(b) $-145.8° = -[145° + 0.8(60')] = -145° 48'$

69. (a) $2.5 = 2.5\left(\frac{180}{\pi}\right)° \approx 143.23945° \approx 143° 14' 22''$

(b) $-3.58 = -3.58\left(\frac{180}{\pi}\right)° \approx -205.11889° \approx -205° 7' 8''$

71. $s = r\theta$

$6 = 5\theta$

$\theta = \frac{6}{5}$ radians

73. $s = r\theta$

$32 = 7\theta$

$\theta = \frac{32}{7} = 4\frac{4}{7}$ radians

75. $s = r\theta$

$4 = 15\theta$

$\theta = \frac{4}{15}$ radian

77. $s = r\theta$

$25 = 14.5\theta$

$\theta = \frac{25}{14.5} \approx 1.724$ radians

79. $s = r\theta$, θ in radians

$s = 15(180)\left(\frac{\pi}{180}\right) = 15\pi$ inches

≈ 47.12 inches

81. $s = r\theta$, θ in radians

$s = 6(2) = 12$ meters

83. $\theta = 41° 15' 42'' - 32° 47' 9'' = 8° 28' 33'' \approx 8.47583° \approx 0.14793$ radian

$s = r\theta = 4000(0.14793) \approx 591.72$ miles

85. $\theta = 42° 7' 15'' - 25° 46' 37'' = 16° 20' 38'' \approx 0.285255$ radian

$s = r\theta = 4000(0.285255) \approx 1141.02$ miles

87. $\theta = \frac{s}{r} = \frac{600}{6378} \approx 0.094 \text{ radian} \approx 5.39°$

89. $\theta = \frac{s}{r} = \frac{2.5}{6} = \frac{25}{60} = \frac{5}{12}$ radian

91. (a) 50 miles per hour $= 50(5280)/60 = 4400$ feet per minute

The circumference of the tire is $C = 2.5\pi$ feet.

The number of revolutions per minute is $r = 4400/2.5\pi \approx 560.2$ rev/min.

(b) The angular speed is θ/t.

$$\theta = \frac{4400}{2.5\pi}(2\pi) = 3520 \text{ radians}$$

$$\text{Angular speed} = \frac{3520 \text{ radians}}{1 \text{ minute}} = 3520 \text{ rad/min}$$

93. $1 \text{ Radian} = \left(\frac{180}{\pi}\right)° \approx 57.3°$, so one radian is much larger than one degree.

95. Circumference: $C = 2\pi(1.68) = 3.36\pi$ inches

360 rev/min = 6 rev/sec

Linear speed: $(3.36\pi)(6) = 20.16$ inches/sec

97. The area of a circle is $A = \pi r^2 \Rightarrow \pi = \frac{A}{r^2}$.

The circumference of a circle is $C = 2\pi r$.

$$C = 2\left(\frac{A}{r^2}\right)r$$

$$C = \frac{2A}{r}$$

$$\frac{Cr}{2} = A$$

For a sector, $C = s = r\theta$

Thus, $A = \frac{(r\theta)r}{2} = \frac{1}{2}\theta r^2$ for a sector.

99. $s = r\theta = (4000 \text{ miles})(0.031)\left(\frac{\pi}{180}\right) \approx 2.164$ miles

Section 4.2 Trigonometric Functions: The Unit Circle

- You should know the definition of the trigonometric functions in terms of the unit circle. Let t be a real number and (x, y) the point on the unit circle corresponding to t.

  $\sin t = y$ $\qquad$ $\csc t = \frac{1}{y}, \quad y \neq 0$

  $\cos t = x$ $\qquad$ $\sec t = \frac{1}{x}, \quad x \neq 0$

  $\tan t = \frac{y}{x}, \quad x \neq 0$ $\qquad$ $\cot t = \frac{x}{y}, \quad y \neq 0$

- The cosine and secant functions are even.

  $\cos(-t) = \cos t$ $\qquad$ $\sec(-t) = \sec t$

- The other four trigonometric functions are odd.

  $\sin(-t) = -\sin t$ $\qquad$ $\csc(-t) = -\csc t$

  $\tan(-t) = -\tan t$ $\qquad$ $\cot(-t) = -\cot t$

- Be able to evaluate the trigonometric functions with a calculator.

Solutions to Odd-Numbered Exercises

1. $x = -\frac{3}{5}, \quad y = \frac{4}{5}$

$\sin t = y = \frac{4}{5}$ $\qquad$ $\csc t = \frac{1}{y} = \frac{5}{4}$

$\cos t = x = -\frac{3}{5}$ $\qquad$ $\sec t = \frac{1}{x} = -\frac{5}{3}$

$\tan t = \frac{y}{x} = -\frac{4}{3}$ $\qquad$ $\cot t = \frac{x}{y} = -\frac{3}{4}$

3. $x = \frac{8}{17}, \quad y = -\frac{15}{17}$

$\sin t = y = -\frac{15}{17}$ $\qquad$ $\csc t = \frac{1}{y} = -\frac{17}{15}$

$\cos t = x = \frac{8}{17}$ $\qquad$ $\sec t = \frac{1}{x} = \frac{17}{8}$

$\tan t = \frac{y}{x} = -\frac{15}{8}$ $\qquad$ $\cot t = \frac{x}{y} = -\frac{8}{15}$

5. $t = \frac{\pi}{4}$ corresponds to $\left(\frac{\sqrt{2}}{2}, \frac{\sqrt{2}}{2}\right)$.

7. $t = \frac{5\pi}{6}$ corresponds to $\left(-\frac{\sqrt{3}}{2}, \frac{1}{2}\right)$.

9. $t = \frac{4\pi}{3}$ corresponds to $\left(-\frac{1}{2}, -\frac{\sqrt{3}}{2}\right)$.

11. $t = \frac{3\pi}{2}$ corresponds to $(0, -1)$.

13. $t = \frac{\pi}{4}$ corresponds to $\left(\frac{\sqrt{2}}{2}, \frac{\sqrt{2}}{2}\right)$.

$\sin t = y = \frac{\sqrt{2}}{2}$

$\cos t = x = \frac{\sqrt{2}}{2}$

$\tan t = \frac{y}{x} = 1$

15. $t = -\frac{\pi}{6}$ corresponds to $\left(\frac{\sqrt{3}}{2}, -\frac{1}{2}\right)$.

$\sin t = y = -\frac{1}{2}$

$\cos t = x = \frac{\sqrt{3}}{2}$

$\tan t = \frac{y}{x} = -\frac{1}{\sqrt{3}}$

17. $t = -\frac{5\pi}{4}$ corresponds to $\left(-\frac{\sqrt{2}}{2}, \frac{\sqrt{2}}{2}\right)$.

$\sin t = y = \frac{\sqrt{2}}{2}$

$\cos t = x = -\frac{\sqrt{2}}{2}$

$\tan t = \frac{y}{x} = -1$

19. $t = \frac{11\pi}{6}$ corresponds to $\left(\frac{\sqrt{3}}{2}, -\frac{1}{2}\right)$.

$\sin t = y = -\frac{1}{2}$

$\cos t = x = \frac{\sqrt{3}}{2}$

$\tan t = \frac{y}{x} = -\frac{1}{\sqrt{3}}$

21. $t = \frac{4\pi}{3}$ corresponds to $\left(-\frac{1}{2}, -\frac{\sqrt{3}}{2}\right)$.

$\sin t = y = -\frac{\sqrt{3}}{2}$

$\cos t = x = -\frac{1}{2}$

$\tan t = \frac{y}{x} = \sqrt{3}$

23. $t = -\frac{3\pi}{2}$ corresponds to $(0, 1)$.

$\sin t = y = 1$

$\cos t = x = 0$

$\tan t = \frac{y}{x}$ is undefined.

25. $t = \frac{3\pi}{4}$ corresponds to $\left(-\frac{\sqrt{2}}{2}, \frac{\sqrt{2}}{2}\right)$.

$\sin t = y = \frac{\sqrt{2}}{2}$ $\qquad \csc t = \frac{1}{y} = \sqrt{2}$

$\cos t = x = -\frac{\sqrt{2}}{2}$ $\qquad \sec t = \frac{1}{x} = -\sqrt{2}$

$\tan t = \frac{y}{x} = -1$ $\qquad \cot t = \frac{x}{y} = -1$

27. $t = \frac{\pi}{2}$ corresponds to $(0, 1)$.

$\sin t = y = 1$ $\qquad \csc t = \frac{1}{y} = 1$

$\cos t = x = 0$ $\qquad \sec t = \frac{1}{x}$ is undefined.

$\tan t = \frac{y}{x}$ is undefined. $\qquad \cot t = \frac{x}{y} = 0$

29. $t = -\frac{4\pi}{3}$ corresponds to $\left(-\frac{1}{2}, \frac{\sqrt{3}}{2}\right)$.

$\sin t = y = \frac{\sqrt{3}}{2}$ $\qquad \csc t = \frac{1}{y} = \frac{2\sqrt{3}}{3}$

$\cos t = x = -\frac{1}{2}$ $\qquad \sec t = \frac{1}{x} = -2$

$\tan t = \frac{y}{x} = -\sqrt{3}$ $\qquad \cot t = \frac{x}{y} = -\frac{\sqrt{3}}{3}$

31. $\sin 3\pi = \sin \pi = 0$

33. $\cos\frac{8\pi}{3} = \cos\frac{2\pi}{3} = -\frac{1}{2}$

35. $\cos\frac{19\pi}{6} = \cos\frac{7\pi}{6} = -\frac{\sqrt{3}}{2}$

37. $\sin\left(-\frac{9\pi}{4}\right) = \sin\left(-\frac{\pi}{4}\right) = -\frac{\sqrt{2}}{2}$

39. $\sin t = \frac{1}{3}$

(a) $\sin(-t) = -\sin t = -\frac{1}{3}$

(b) $\csc(-t) = -\csc t = -3$

41. $\cos(-t) = -\frac{7}{8}$

(a) $\cos t = \cos(-t) = -\frac{7}{8}$

(b) $\sec(-t) = \frac{1}{\cos(-t)} = -\frac{8}{7}$

43. $\sin y = \frac{4}{5}$

(a) $\sin(\pi - t) = \sin t = \frac{4}{5}$

(b) $\sin(t + \pi) = -\sin t = -\frac{4}{5}$

45. $\sin\frac{\pi}{4} \approx 0.7071$

47. $\cos(-3) \approx -0.9900$

49. $\cos(-1.7) \approx -0.1288$

51. $\csc 0.8 = \frac{1}{\sin 0.8} \approx 1.3940$

53. $\sec 22.8 = \frac{1}{\cos 22.8} \approx -1.4486$

55. (a) $\sin 5 \approx -1$

(b) $\cos 2 \approx -0.4$

57. (a) $\sin t = 0.25$

$t \approx 0.25$ or 2.89

(b) $\cos t = -0.25$

$t \approx 1.82$ or 4.46

59. $\cos 1.5 \approx 0.0707$

$2\cos 0.75 \approx 1.4634$

$\cos 2t \neq 2\cos t$

61. (a) The points have y-axis symmetry.

(b) $\sin t_1 = \sin(\pi - t_1)$ since they have the same y-value.

(c) $-\cos t_1 = \cos(\pi - t_1)$ since the x-values have the opposite signs.

63. $y(t) = \frac{1}{4}\cos 6t$

(a) $y(0) = \frac{1}{4}\cos 0 = 0.2500$ feet

(b) $y\left(\frac{1}{4}\right) = \frac{1}{4}\cos\frac{3}{2} \approx 0.0177$ feet

(c) $y\left(\frac{1}{2}\right) = \frac{1}{4}\cos 3 \approx -0.2475$ feet

65. $I = 5e^{-2(0.7)}\sin(0.7) \approx 0.794$

67. Let $h(t) = f(t)g(t)$

$= \sin t \cos t.$

Then, $h(-t) = \sin(-t)\cos(-t)$

$= -\sin t \cos t$

$= -h(t).$

Thus, $h(t)$ is odd.

69.

$f(x) = \frac{1}{2}(3x - 2)$

$y = \frac{1}{2}(3x - 2)$

$x = \frac{1}{2}(3y - 2)$

$2x = 3y - 2$

$2x + 2 = 3y$

$\frac{2}{3}(x + 1) = y$

$f^{-1}(x) = \frac{2}{3}(x + 1)$

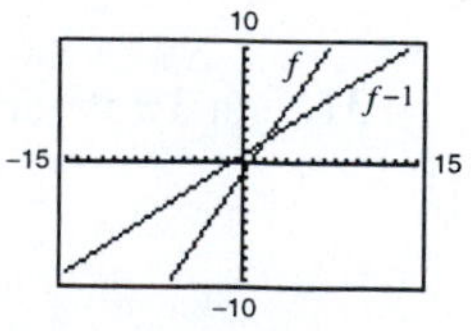

71.

$$f(x) = \sqrt{x^2 - 4}, \quad x \geq 2, \quad y \geq 0$$
$$y = \sqrt{x^2 - 4}$$
$$x = \sqrt{y^2 - 4}$$
$$x^2 = y^2 - 4$$
$$x^2 + 4 = y^2$$
$$\sqrt{x^2 + 4} = y, \quad x \geq 0$$
$$f^{-1}(x) = \sqrt{x^2 + 4}, \quad x \geq 0$$

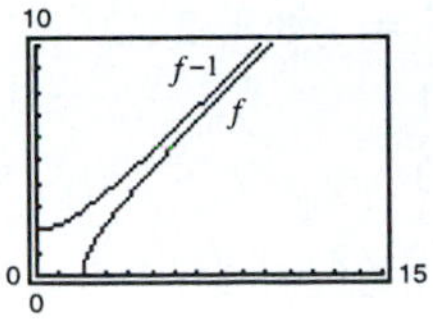

Note: The domain of $f^{-1}(x)$ equals the range of $f(x)$.

Section 4.3 Right Triangle Trigonometry

- You should know the right triangle definition of trigonometric functions.

  (a) $\sin\theta = \dfrac{\text{opp}}{\text{hyp}}$ (b) $\cos\theta = \dfrac{\text{adj}}{\text{hyp}}$ (c) $\tan\theta = \dfrac{\text{opp}}{\text{adj}}$

  (d) $\csc\theta = \dfrac{\text{hyp}}{\text{opp}}$ (e) $\sec\theta = \dfrac{\text{hyp}}{\text{adj}}$ (f) $\cot\theta = \dfrac{\text{adj}}{\text{opp}}$

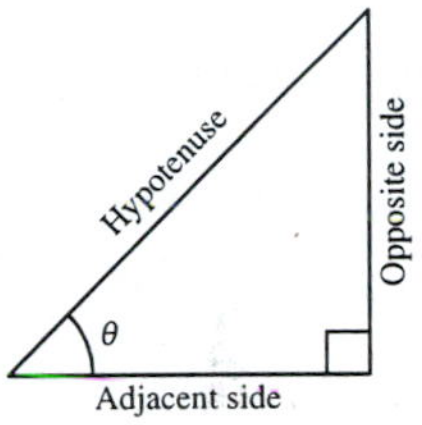

- You should know the following identities.

  (a) $\sin\theta = \dfrac{1}{\csc\theta}$ (b) $\csc\theta = \dfrac{1}{\sin\theta}$ (c) $\cos\theta = \dfrac{1}{\sec\theta}$

  (d) $\sec\theta = \dfrac{1}{\cos\theta}$ (e) $\tan\theta = \dfrac{1}{\cot\theta}$ (f) $\cot\theta = \dfrac{1}{\tan\theta}$

  (g) $\tan\theta = \dfrac{\sin\theta}{\cos\theta}$ (h) $\cot\theta = \dfrac{\cos\theta}{\sin\theta}$ (i) $\sin^2\theta + \cos^2\theta = 1$

  (j) $1 + \tan^2\theta = \sec^2\theta$ (k) $1 + \cot^2\theta = \csc^2\theta$

- You should know that two acute angles α and β are complementary if $\alpha + \beta = 90°$, and that cofunctions of complementary angles are equal.
- You should know the trigonometric function values of 30°, 45°, and 60°, or be able to construct triangles from which you can determine them.

Solutions to Odd-Numbered Exercises

1.

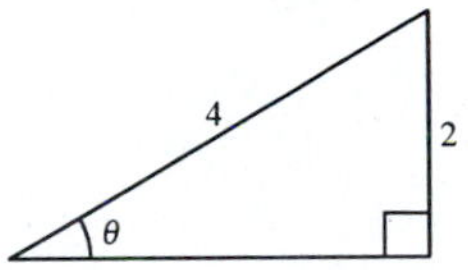

$$\text{adj} = \sqrt{4^2 - 2^2} = \sqrt{12} = 2\sqrt{3}$$

$$\sin\theta = \frac{\text{opp}}{\text{hyp}} = \frac{2}{4} = \frac{1}{2} \qquad \csc\theta = \frac{\text{hyp}}{\text{opp}} = \frac{4}{2} = 2$$

$$\cos\theta = \frac{\text{adj}}{\text{hyp}} = \frac{2\sqrt{3}}{4} = \frac{\sqrt{3}}{2} \qquad \sec\theta = \frac{\text{hyp}}{\text{adj}} = \frac{4}{2\sqrt{3}} = \frac{2\sqrt{3}}{3}$$

$$\tan\theta = \frac{\text{opp}}{\text{adj}} = \frac{2}{2\sqrt{3}} = \frac{\sqrt{3}}{3} \qquad \cot\theta = \frac{\text{adj}}{\text{opp}} = \frac{2\sqrt{3}}{2} = \sqrt{3}$$

3.

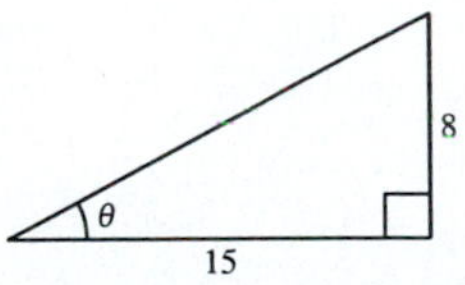

$$\text{hyp} = \sqrt{8^2 + 15^2} = 17$$

$$\sin\theta = \frac{\text{opp}}{\text{hyp}} = \frac{8}{17} \qquad \csc\theta = \frac{\text{hyp}}{\text{opp}} = \frac{17}{8}$$

$$\cos\theta = \frac{\text{adj}}{\text{hyp}} = \frac{15}{17} \qquad \sec\theta = \frac{\text{hyp}}{\text{adj}} = \frac{17}{15}$$

$$\tan\theta = \frac{\text{opp}}{\text{adj}} = \frac{8}{15} \qquad \cot\theta = \frac{\text{adj}}{\text{opp}} = \frac{15}{8}$$

5. 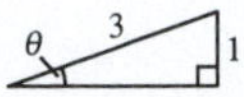

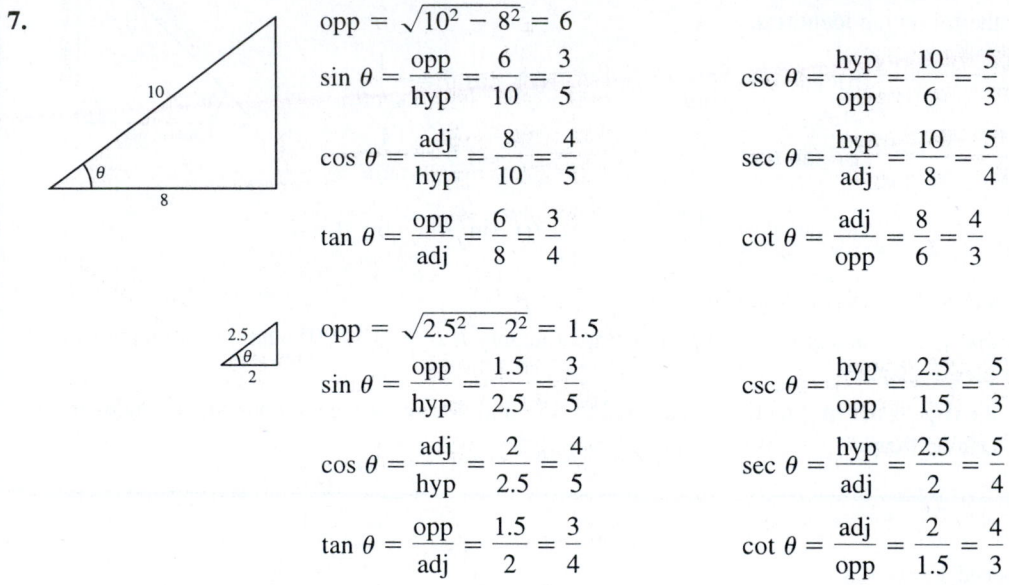

$$\text{adj} = \sqrt{3^2 - 1^2} = \sqrt{8} = 2\sqrt{2}$$

$$\sin\theta = \frac{\text{opp}}{\text{hyp}} = \frac{1}{3} \qquad \csc\theta = \frac{\text{hyp}}{\text{opp}} = 3$$

$$\cos\theta = \frac{\text{adj}}{\text{hyp}} = \frac{2\sqrt{2}}{3} \qquad \sec\theta = \frac{\text{hyp}}{\text{adj}} = \frac{3}{2\sqrt{2}} = \frac{3\sqrt{2}}{4}$$

$$\tan\theta = \frac{\text{opp}}{\text{adj}} = \frac{1}{2\sqrt{2}} = \frac{\sqrt{2}}{4} \qquad \cot\theta = \frac{\text{adj}}{\text{opp}} = 2\sqrt{2}$$

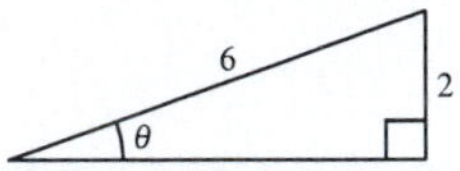

$$\text{adj} = \sqrt{6^2 - 2^2} = \sqrt{32} = 4\sqrt{2}$$

$$\sin\theta = \frac{\text{opp}}{\text{hyp}} = \frac{2}{6} = \frac{1}{3} \qquad \csc\theta = \frac{\text{hyp}}{\text{opp}} = \frac{6}{2} = 3$$

$$\cos\theta = \frac{\text{adj}}{\text{hyp}} = \frac{4\sqrt{2}}{6} = \frac{2\sqrt{2}}{3} \qquad \sec\theta = \frac{\text{hyp}}{\text{adj}} = \frac{6}{4\sqrt{2}} = \frac{3}{2\sqrt{2}} = \frac{3\sqrt{2}}{4}$$

$$\tan\theta = \frac{\text{opp}}{\text{adj}} = \frac{2}{4\sqrt{2}} = \frac{1}{2\sqrt{2}} = \frac{\sqrt{2}}{4} \qquad \cot\theta = \frac{\text{adj}}{\text{opp}} = \frac{4\sqrt{2}}{2} = 2\sqrt{2}$$

The function values are the same since the triangles are similar and the corresponding sides are proportional.

7. 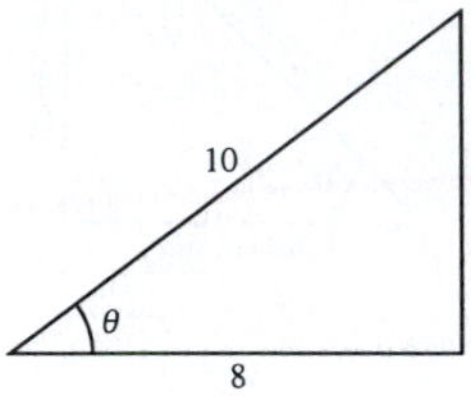

$$\text{opp} = \sqrt{10^2 - 8^2} = 6$$

$$\sin\theta = \frac{\text{opp}}{\text{hyp}} = \frac{6}{10} = \frac{3}{5} \qquad \csc\theta = \frac{\text{hyp}}{\text{opp}} = \frac{10}{6} = \frac{5}{3}$$

$$\cos\theta = \frac{\text{adj}}{\text{hyp}} = \frac{8}{10} = \frac{4}{5} \qquad \sec\theta = \frac{\text{hyp}}{\text{adj}} = \frac{10}{8} = \frac{5}{4}$$

$$\tan\theta = \frac{\text{opp}}{\text{adj}} = \frac{6}{8} = \frac{3}{4} \qquad \cot\theta = \frac{\text{adj}}{\text{opp}} = \frac{8}{6} = \frac{4}{3}$$

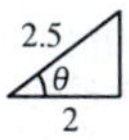

$$\text{opp} = \sqrt{2.5^2 - 2^2} = 1.5$$

$$\sin\theta = \frac{\text{opp}}{\text{hyp}} = \frac{1.5}{2.5} = \frac{3}{5} \qquad \csc\theta = \frac{\text{hyp}}{\text{opp}} = \frac{2.5}{1.5} = \frac{5}{3}$$

$$\cos\theta = \frac{\text{adj}}{\text{hyp}} = \frac{2}{2.5} = \frac{4}{5} \qquad \sec\theta = \frac{\text{hyp}}{\text{adj}} = \frac{2.5}{2} = \frac{5}{4}$$

$$\tan\theta = \frac{\text{opp}}{\text{adj}} = \frac{1.5}{2} = \frac{3}{4} \qquad \cot\theta = \frac{\text{adj}}{\text{opp}} = \frac{2}{1.5} = \frac{4}{3}$$

The function values are the same since the triangles are similar and the corresponding sides are proportional.

9. Given: $\sin\theta = \frac{2}{3} = \frac{\text{opp}}{\text{hyp}}$

$$2^2 + (\text{adj})^2 = 3^2$$
$$\text{adj} = \sqrt{5}$$

$\cos\theta = \frac{\sqrt{5}}{3}$

$\tan\theta = \frac{2\sqrt{5}}{5}$

$\cot\theta = \frac{\sqrt{5}}{2}$

$\sec\theta = \frac{3\sqrt{5}}{5}$

$\csc\theta = \frac{3}{2}$

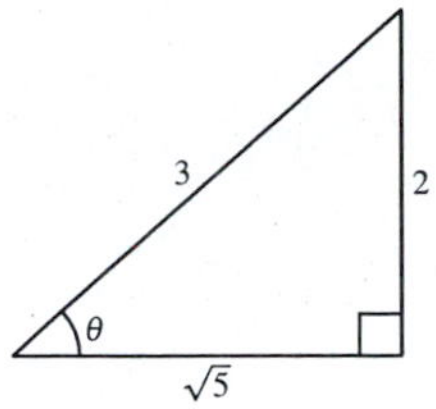

11. Given: $\sec\theta = 2 = \frac{2}{1} = \frac{\text{hyp}}{\text{adj}}$

$$(\text{opp})^2 + 1^2 = 2^2$$
$$\text{opp} = \sqrt{3}$$

$\sin\theta = \frac{\sqrt{3}}{2}$

$\cos\theta = \frac{1}{2}$

$\tan\theta = \sqrt{3}$

$\cot\theta = \frac{\sqrt{3}}{3}$

$\csc\theta = \frac{2\sqrt{3}}{3}$

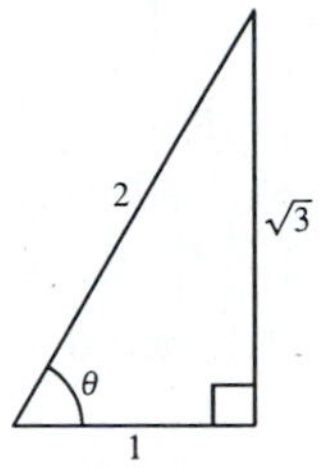

13. Given: $\tan\theta = 3 = \frac{3}{1} = \frac{\text{opp}}{\text{adj}}$

$$3^2 + 1^2 = (\text{hyp})^2$$
$$\text{hyp} = \sqrt{10}$$

$\sin\theta = \frac{3\sqrt{10}}{10}$

$\cos\theta = \frac{\sqrt{10}}{10}$

$\cot\theta = \frac{1}{3}$

$\sec\theta = \sqrt{10}$

$\csc\theta = \frac{\sqrt{10}}{3}$

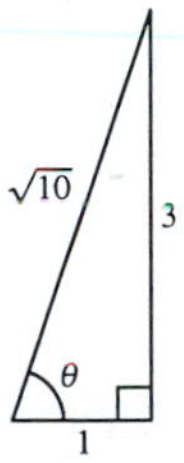

15. Given: $\cot\theta = \frac{3}{2} = \frac{\text{adj}}{\text{opp}}$

$$2^2 + 3^2 = (\text{hyp})^2$$
$$\text{hyp} = \sqrt{13}$$

$\sin\theta = \frac{2}{\sqrt{13}} = \frac{2\sqrt{13}}{13}$

$\cos\theta = \frac{3}{\sqrt{13}} = \frac{3\sqrt{13}}{13}$

$\tan\theta = \frac{2}{3}$

$\csc\theta = \frac{\sqrt{13}}{2}$

$\sec\theta = \frac{\sqrt{13}}{3}$

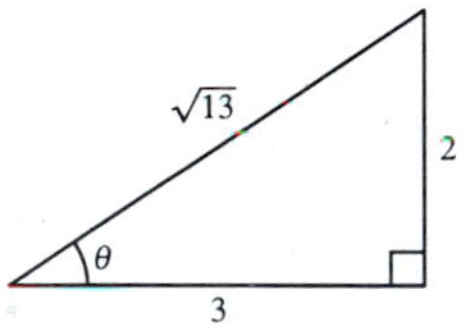

17. $\sin 60^\circ = \frac{\sqrt{3}}{2}$, $\cos 60^\circ = \frac{1}{2}$

(a) $\tan 60^\circ = \frac{\sin 60^\circ}{\cos 60^\circ} = \sqrt{3}$

(b) $\sin 30^\circ = \cos 60^\circ = \frac{1}{2}$

(c) $\cos 30^\circ = \sin 60^\circ = \frac{\sqrt{3}}{2}$

(d) $\cot 60^\circ = \frac{\cos 60^\circ}{\sin 60^\circ} = \frac{1}{\sqrt{3}} = \frac{\sqrt{3}}{3}$

19. $\csc\theta = 3$, $\sec\theta = \frac{3\sqrt{2}}{4}$

(a) $\sin\theta = \frac{1}{\csc\theta} = \frac{1}{3}$

(b) $\cos\theta = \frac{1}{\sec\theta} = \frac{2\sqrt{2}}{3}$

(c) $\tan\theta = \frac{\sin\theta}{\cos\theta} = \frac{1/3}{(2\sqrt{2})/3} = \frac{\sqrt{2}}{4}$

(d) $\sec(90^\circ - \theta) = \csc\theta = 3$

21. $\cos\alpha = \dfrac{1}{4}$

(a) $\sec\alpha = \dfrac{1}{\cos\alpha} = 4$

(b) $\sin^2\alpha + \cos^2\alpha = 1$

$$\sin^2\alpha + \left(\frac{1}{4}\right)^2 = 1$$

$$\sin^2\alpha = \frac{15}{16}$$

$$\sin\alpha = \frac{\sqrt{15}}{4}$$

(c) $\cot\alpha = \dfrac{\cos\alpha}{\sin\alpha} = \dfrac{1/4}{\sqrt{15}/4} = \dfrac{1}{\sqrt{15}} = \dfrac{\sqrt{15}}{15}$

(d) $\sin(90^\circ - \alpha) = \cos\alpha = \dfrac{1}{4}$

23. $\tan\theta\cot\theta = \tan\theta\left(\dfrac{1}{\tan\theta}\right) = 1$

25. $\tan\alpha\cos\alpha = \left(\dfrac{\sin\alpha}{\cos\alpha}\right)\cos\alpha = \sin\alpha$

27.
$$\begin{aligned}(1+\cos\theta)(1-\cos\theta) &= 1-\cos^2\theta\\ &= (\sin^2\theta+\cos^2\theta)-\cos^2\theta\\ &= \sin^2\theta\end{aligned}$$

29.
$$\begin{aligned}(\sec\theta+\tan\theta)(\sec\theta-\tan\theta) &= \sec^2\theta-\tan^2\theta\\ &= (1+\tan^2\theta)-\tan^2\theta\\ &= 1\end{aligned}$$

31.
$$\begin{aligned}\frac{\sin\theta}{\cos\theta}+\frac{\cos\theta}{\sin\theta} &= \frac{\sin^2\theta+\cos^2\theta}{\sin\theta\cos\theta}\\ &= \frac{1}{\sin\theta\cos\theta}\\ &= \frac{1}{\sin\theta}\cdot\frac{1}{\cos\theta}\\ &= \csc\theta\sec\theta\end{aligned}$$

33. (a) $\cos 60^\circ = \dfrac{1}{2}$

(b) $\tan\dfrac{\pi}{6} = \dfrac{1}{\sqrt{3}} = \dfrac{\sqrt{3}}{3}$

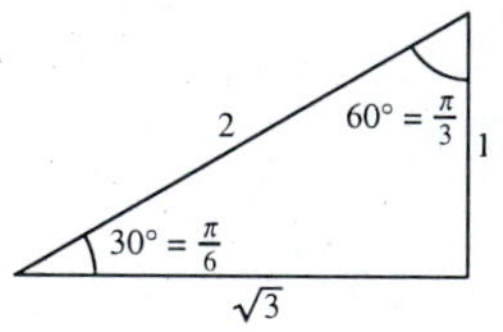

35. (a) $\cot 45^\circ = 1$

(b) $\cos 45^\circ = \dfrac{1}{\sqrt{2}} = \dfrac{\sqrt{2}}{2}$

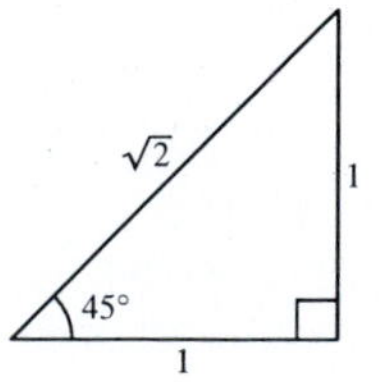

37. (a) $\sin 10^\circ \approx 0.1736$

(b) $\cos 80^\circ \approx 0.1736$

Note: $\cos 80^\circ = \sin(90^\circ - 80^\circ) = \sin 10^\circ$

39. (a) $\sin 16.35° \approx 0.2815$

(b) $\csc 16.35° = \dfrac{1}{\sin 16.35°} \approx 3.5523$

41. (a) $\sec 42° 12' = \sec 42.2° = \dfrac{1}{\cos 42.2°} \approx 1.3499$

(b) $\csc 48° 7' = \dfrac{1}{\sin\left(48 + \frac{7}{60}\right)°} \approx 1.3432$

43. Make sure that your calculator is in radian mode.

(a) $\cot \dfrac{\pi}{16} = \dfrac{1}{\tan(\pi/16)} \approx 5.0273$ (b) $\tan \dfrac{\pi}{16} \approx 0.1989$

45. Make sure that your calculator is in radian mode.

(a) $\csc 1 = \dfrac{1}{\sin 1} \approx 1.1884$ (b) $\tan \dfrac{1}{2} \approx 0.5463$

47. (a) $\sin \theta = \dfrac{1}{2} \implies \theta = 30° = \dfrac{\pi}{6}$

(b) $\csc \theta = 2 \implies \theta = 30° = \dfrac{\pi}{6}$

49. (a) $\sec \theta = 2 \implies \theta = 60° = \dfrac{\pi}{3}$

(b) $\cot \theta = 1 \implies \theta = 45° = \dfrac{\pi}{4}$

51. (a) $\csc \theta = \dfrac{2\sqrt{3}}{3} \implies \theta = 60° = \dfrac{\pi}{3}$

(b) $\sin \theta = \dfrac{\sqrt{2}}{2} \implies \theta = 45° = \dfrac{\pi}{4}$

53. (a) $\sin \theta = 0.8191 \implies \theta \approx 55° \approx 0.960$ radian

(b) $\cos \theta = 0.0175 \implies \theta \approx 89° \approx 1.553$ radians

55. (a) $\tan \theta = 1.1920 \implies \theta \approx 50° \approx 0.873$ radian

(b) $\tan \theta = 0.4663 \implies \theta \approx 25° \approx 0.436$ radian

57.
$$\tan 30° = \frac{y}{75}$$
$$\frac{\sqrt{3}}{3} = \frac{y}{75}$$
$$75\left(\frac{\sqrt{3}}{3}\right) = y$$
$$25\sqrt{3} = y$$

59.
$$\cot 60° = \frac{x}{32}$$
$$\frac{\sqrt{3}}{3} = \frac{x}{32}$$
$$\frac{32\sqrt{3}}{3} = x$$

61.
$$\sin 40° = \frac{15}{r}$$
$$r = \frac{15}{\sin 40°} \approx 23.3$$

63.
$$\sin 50° = \frac{y}{8}$$
$$y = 8 \sin 50° \approx 6.1$$

65.
$$\frac{h}{23} = \frac{6}{8}$$
$$h = \frac{138}{8} = \frac{69}{4} = 17\tfrac{1}{4} \text{ ft}$$

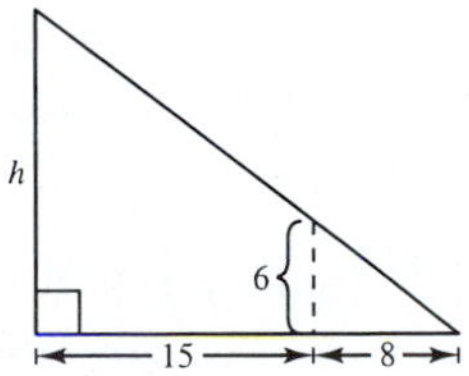

67. (a)

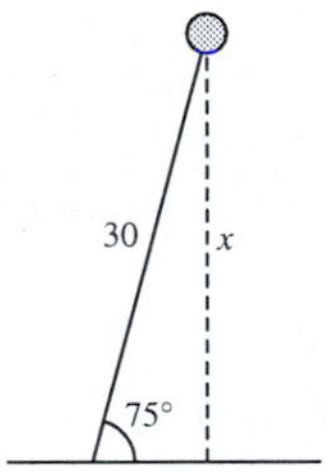

(b) $\sin 75° = \dfrac{x}{30}$

(c) $x = 30 \sin 75° \approx 29$ meters

69. Let $x =$ distance from the boat to the shoreline

$$\tan 3° = \frac{60}{x}$$
$$x = \frac{60}{\tan 3°} \approx 1144.9 \text{ feet}$$

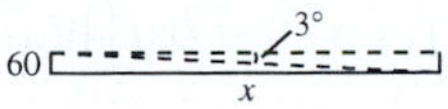

71.

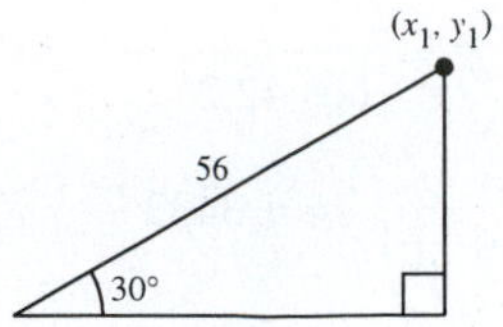

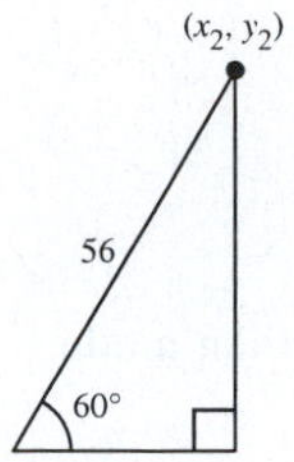

$\sin 30° = \dfrac{y_1}{56}$

$y_1 = (\sin 30°)(56) = \left(\dfrac{1}{2}\right)(56) = 28$

$\cos 30° = \dfrac{x_1}{56}$

$x_1 = \cos 30°(56) = \dfrac{\sqrt{3}}{2}(56) = 28\sqrt{3}$

$(x_1, y_1) = (28\sqrt{3}, 28)$

$\sin 60° = \dfrac{y_2}{56}$

$y_2 = \sin 60°(56) = \left(\dfrac{\sqrt{3}}{2}\right)(56) = 28\sqrt{3}$

$\cos 60° = \dfrac{x_2}{56}$

$x_2 = (\cos 60°)(56) = \left(\dfrac{1}{2}\right)(56) = 28$

$(x_2, y_2) = (28, 28\sqrt{3})$

73. $x \approx 9.397, \; y \approx 3.420$

$\sin\theta = \dfrac{y}{10} \approx 0.34$

$\cos\theta = \dfrac{x}{10} \approx 0.94$

$\tan\theta = \dfrac{y}{x} \approx 0.36$

$\cot\theta = \dfrac{x}{y} \approx 2.75$

$\sec\theta = \dfrac{10}{x} \approx 1.06$

$\csc\theta = \dfrac{10}{y} \approx 2.92$

75. (a)

θ	0	0.1	0.2	0.3	0.4	0.5
$\sin\theta$	0	0.0998	0.1987	0.2955	0.3894	0.4794

(b) In the interval $(0, 0.5]$, $\theta > \sin\theta$

(c) As $\theta \to 0$, $\sin\theta \to 0$

77. True, $\csc x = \dfrac{1}{\sin x} \implies \sin 60° \csc 60° = \sin 60°\left(\dfrac{1}{\sin 60°}\right) = 1$

79. False, $\dfrac{\sqrt{2}}{2} + \dfrac{\sqrt{2}}{2} = \sqrt{2} \neq 1$

81. False, $\dfrac{\sin 60°}{\sin 30°} = \dfrac{\cos 30°}{\sin 30°} = \cot 30° \approx 1.7321$; $\sin 2° \approx 0.0349$

83. $\dfrac{x^2 - 6x}{x^2 + 4x - 12} \cdot \dfrac{x^2 + 12x + 36}{x^2 - 36} = \dfrac{x(x - 6)}{(x + 6)(x - 2)} \cdot \dfrac{(x + 6)(x + 6)}{(x + 6)(x - 6)}$

$= \dfrac{x}{x - 2}$

85. $\dfrac{3}{x + 2} - \dfrac{2}{x - 2} + \dfrac{x}{x^2 + 4x + 4} = \dfrac{3(x + 2)(x - 2) - 2(x + 2)^2 + x(x - 2)}{(x - 2)(x + 2)^2}$

$= \dfrac{3(x^2 - 4) - 2(x^2 + 4x + 4) + x^2 - 2x}{(x - 2)(x + 2)^2}$

$= \dfrac{2x^2 - 10x - 20}{(x - 2)(x + 2)^2} = \dfrac{2(x^2 - 5x - 10)}{(x - 2)(x + 2)^2}$

Section 4.4 Trigonometric Functions of Any Angle

- Know the Definitions of Trigonometric Functions of Any Angle.
  If θ is in standard position, (x, y) a point on the terminal side and $r = \sqrt{x^2 + y^2} \neq 0$, then

  $\sin \theta = \dfrac{y}{r}$ $\csc \theta = \dfrac{r}{y},\ y \neq 0$

  $\cos \theta = \dfrac{x}{r}$ $\sec \theta = \dfrac{r}{x},\ x \neq 0$

  $\tan \theta = \dfrac{y}{x},\ x \neq 0$ $\cot \theta = \dfrac{x}{y},\ y \neq 0$
- You should know the signs of the trigonometric functions in each quadrant.
- You should know the trigonometric function values of the quadrant angles 0, $\dfrac{\pi}{2}$, π, and $\dfrac{3\pi}{2}$.
- You should be able to find reference angles.
- You should be able to evaluate trigonometric functions of any angle. (Use reference angles.)
- You should know that the period of sine and cosine is 2π.

Solutions to Odd-Numbered Exercises

1. (a) $(x, y) = (4, 3)$

$r = \sqrt{16 + 9} = 5$

$\sin \theta = \dfrac{y}{r} = \dfrac{3}{5}$ $\csc \theta = \dfrac{r}{y} = \dfrac{5}{3}$

$\cos \theta = \dfrac{x}{r} = \dfrac{4}{5}$ $\sec \theta = \dfrac{r}{x} = \dfrac{5}{4}$

$\tan \theta = \dfrac{y}{x} = \dfrac{3}{4}$ $\cot \theta = \dfrac{x}{y} = \dfrac{4}{3}$

(b) $(x, y) = (-8, -15)$

$r = \sqrt{64 + 225} = 17$

$\sin \theta = \dfrac{y}{r} = -\dfrac{15}{17}$ $\csc \theta = \dfrac{r}{y} = -\dfrac{17}{15}$

$\cos \theta = \dfrac{x}{r} = -\dfrac{8}{17}$ $\sec \theta = \dfrac{r}{x} = -\dfrac{17}{8}$

$\tan \theta = \dfrac{y}{x} = \dfrac{15}{8}$ $\cot \theta = \dfrac{x}{y} = \dfrac{8}{15}$

3. (a) $(x, y) = \left(-\sqrt{3}, -1\right)$

$r = \sqrt{3 + 1} = 2$

$\sin \theta = \dfrac{y}{r} = -\dfrac{1}{2}$ $\quad \csc \theta = \dfrac{r}{y} = -2$

$\cos \theta = \dfrac{x}{r} = -\dfrac{\sqrt{3}}{2}$ $\quad \sec \theta = \dfrac{r}{x} = -\dfrac{2\sqrt{3}}{3}$

$\tan \theta = \dfrac{y}{x} = \dfrac{\sqrt{3}}{3}$ $\quad \cot \theta = \dfrac{x}{y} = \sqrt{3}$

(b) $(x, y) = (-2, 2)$

$r = \sqrt{4 + 4} = 2\sqrt{2}$

$\sin \theta = \dfrac{y}{r} = \dfrac{\sqrt{2}}{2}$ $\quad \csc \theta = \dfrac{r}{y} = \sqrt{2}$

$\cos \theta = \dfrac{x}{r} = -\dfrac{\sqrt{2}}{2}$ $\quad \sec \theta = \dfrac{r}{x} = -\sqrt{2}$

$\tan \theta = \dfrac{y}{x} = -1$ $\quad \cot \theta = \dfrac{x}{y} = -1$

5. (a) $(x, y) = (7, 24)$

$r = \sqrt{49 + 576} = 25$

$\sin \theta = \dfrac{y}{r} = \dfrac{24}{25}$ $\quad \csc \theta = \dfrac{r}{y} = \dfrac{25}{24}$

$\cos \theta = \dfrac{x}{r} = \dfrac{7}{25}$ $\quad \sec \theta = \dfrac{r}{x} = \dfrac{25}{7}$

$\tan \theta = \dfrac{y}{x} = \dfrac{24}{7}$ $\quad \cot \theta = \dfrac{x}{y} = \dfrac{7}{24}$

(b) $(x, y) = (7, -24)$

$r = \sqrt{49 + 576} = 25$

$\sin \theta = \dfrac{y}{r} = -\dfrac{24}{25}$ $\quad \csc \theta = \dfrac{r}{y} = -\dfrac{25}{24}$

$\cos \theta = \dfrac{x}{r} = \dfrac{7}{25}$ $\quad \sec \theta = \dfrac{r}{x} = \dfrac{25}{7}$

$\tan \theta = \dfrac{y}{x} = -\dfrac{24}{7}$ $\quad \cot \theta = \dfrac{x}{y} = -\dfrac{7}{24}$

7. (a) $(x, y) = (-4, 10)$

$r = \sqrt{16 + 100} = 2\sqrt{29}$

$\sin \theta = \dfrac{y}{r} = \dfrac{5\sqrt{29}}{29}$ $\quad \csc \theta = \dfrac{r}{y} = \dfrac{\sqrt{29}}{5}$

$\cos \theta = \dfrac{x}{r} = -\dfrac{2\sqrt{29}}{29}$ $\quad \sec \theta = \dfrac{r}{x} = -\dfrac{\sqrt{29}}{2}$

$\tan \theta = \dfrac{y}{x} = -\dfrac{5}{2}$ $\quad \cot \theta = \dfrac{x}{y} = -\dfrac{2}{5}$

(b) $(x, y) = (3, -5)$

$r = \sqrt{9 + 25} = \sqrt{34}$

$\sin \theta = \dfrac{y}{r} = -\dfrac{5\sqrt{34}}{34}$ $\quad \csc \theta = \dfrac{r}{y} = -\dfrac{\sqrt{34}}{5}$

$\cos \theta = \dfrac{x}{r} = \dfrac{3\sqrt{34}}{34}$ $\quad \sec \theta = \dfrac{r}{x} = \dfrac{\sqrt{34}}{3}$

$\tan \theta = \dfrac{y}{x} = -\dfrac{5}{3}$ $\quad \cot \theta = \dfrac{x}{y} = -\dfrac{3}{5}$

9. (a) $\sin \theta < 0 \implies \theta$ lies in Quadrant III or in Quadrant IV.

$\cos \theta < 0 \implies \theta$ lies in Quadrant II or in Quadrant III.

$\sin \theta < 0$ *and* $\cos \theta < 0 \implies \theta$ lies in Quadrant III.

(b) $\sin \theta > 0 \implies \theta$ lies in Quadrant I or in Quadrant II.

$\cos \theta < \theta \implies \theta$ lies in Quadrant II or in Quadrant III.

$\sin \theta > 0$ *and* $\cos \theta < 0 \implies \theta$ lies in Quadrant II.

11. (a) $\sin \theta > 0 \implies \theta$ lies in Quadrant I or in Quadrant II.

$\tan \theta < 0 \implies \theta$ lies in Quadrant II or in Quadrant IV.

$\sin \theta > 0$ *and* $\tan \theta < 0 \implies \theta$ lies in Quadrant II.

(b) $\cos \theta > 0 \implies \theta$ lies in Quadrant I or in Quadrant IV.

$\tan \theta < 0 \implies \theta$ lies in Quadrant II or in Quadrant IV.

$\cos \theta > 0$ *and* $\tan \theta < 0 \implies \theta$ lies in Quadrant IV.

13. $\sin\theta = \frac{y}{r} = \frac{3}{5} \implies x^2 = 25 - 9 = 16$

θ in Quadrant II $\implies x = -4$

$\sin\theta = \frac{y}{r} = \frac{3}{5}$ $\quad\csc\theta = \frac{r}{y} = \frac{5}{3}$

$\cos\theta = \frac{x}{r} = -\frac{4}{5}$ $\quad\sec\theta = \frac{r}{x} = -\frac{5}{4}$

$\tan\theta = \frac{y}{x} = -\frac{3}{4}$ $\quad\cot\theta = \frac{x}{y} = -\frac{4}{3}$

15. $\sin\theta < 0 \implies y < 0$

$\tan\theta = \frac{y}{x} = \frac{-15}{8} \implies r = 17$

$\sin\theta = \frac{y}{r} = -\frac{15}{17}$ $\quad\csc\theta = \frac{r}{y} = -\frac{17}{15}$

$\cos\theta = \frac{x}{r} = \frac{8}{17}$ $\quad\sec\theta = \frac{r}{x} = \frac{17}{8}$

$\tan\theta = \frac{y}{x} = -\frac{15}{8}$ $\quad\cot\theta = \frac{x}{y} = -\frac{8}{15}$

17. $\cot\theta = \frac{x}{y} = -\frac{3}{1} = \frac{3}{-1}$

$\cos\theta > 0 \implies x$ is positive; $x = 3, y = -1, r = \sqrt{10}$

$\sin\theta = \frac{y}{r} = -\frac{\sqrt{10}}{10}$ $\quad\csc\theta = \frac{r}{y} = -\sqrt{10}$

$\cos\theta = \frac{x}{r} = \frac{3\sqrt{10}}{10}$ $\quad\sec\theta = \frac{r}{x} = \frac{\sqrt{10}}{3}$

$\tan\theta = \frac{y}{x} = -\frac{1}{3}$ $\quad\cot\theta = \frac{x}{y} = -3$

19. $\sec\theta = \frac{r}{x} = \frac{2}{-1} \implies y^2 = 4 - 1 = 3$

$\sin\theta > 0 \implies y = \sqrt{3}$

$\sin\theta = \frac{y}{r} = \frac{\sqrt{3}}{2}$ $\quad\csc\theta = \frac{r}{y} = \frac{2\sqrt{3}}{3}$

$\cos\theta = \frac{x}{r} = -\frac{1}{2}$ $\quad\sec\theta = \frac{r}{x} = -2$

$\tan\theta = \frac{y}{x} = -\sqrt{3}$ $\quad\cot\theta = \frac{x}{y} = -\frac{\sqrt{3}}{3}$

21. $\sin\theta = 0 \implies \theta = n\pi$

$\sec\theta = -1 \implies \theta = \pi$

$\sin\theta = \frac{y}{r} = \frac{0}{r} = 0$ $\quad\csc\theta = \frac{r}{y}$ is undefined.

$\cos\theta = \frac{x}{r} = \frac{-r}{r} = -1$ $\quad\sec\theta = \frac{r}{x} = -1$

$\tan\theta = \frac{y}{x} = \frac{0}{x} = 0$ $\quad\cot\theta = \frac{x}{y}$ is undefined.

23. To find a point on the terminal side of θ, use any point on the line $y = -x$ that lies in Quadrant II. $(-1, 1)$ is one such point.

$x = -1, y = 1, r = \sqrt{2}$

$\sin\theta = \frac{1}{\sqrt{2}} = \frac{\sqrt{2}}{2}$ $\quad\csc\theta = \sqrt{2}$

$\cos\theta = -\frac{1}{\sqrt{2}} = -\frac{\sqrt{2}}{2}$ $\quad\sec\theta = -\sqrt{2}$

$\tan\theta = -1$ $\quad\cot\theta = -1$

25. To find a point on the terminal side of θ, use any point on the line $y = 2x$ that lies in Quadrant III. $(-1, -2)$ is one such point.

$x = -1, y = -2, r = \sqrt{5}$

$\sin\theta = -\frac{2}{\sqrt{5}} = -\frac{2\sqrt{5}}{5}$ $\quad\csc\theta = \frac{\sqrt{5}}{-2} = -\frac{\sqrt{5}}{2}$

$\cos\theta = -\frac{1}{\sqrt{5}} = -\frac{\sqrt{5}}{5}$ $\quad\sec\theta = \frac{\sqrt{5}}{-1} = -\sqrt{5}$

$\tan\theta = \frac{-2}{-1} = 2$ $\quad\cot\theta = \frac{-1}{-2} = \frac{1}{2}$

27. $(x, y) = (-1, 0)$

$\cos\pi = \frac{x}{r} = \frac{-1}{1} = -1$

29. $(x, y) = (-1, 0)$

$\sec\pi = \frac{r}{x} = \frac{1}{-1} = -1$

31. $(x, y) = (0, 1)$

$\tan\frac{\pi}{2} = \frac{y}{x} = \frac{1}{0}$ undefined

33. $(x, y) = (0, 1)$

$\cot\frac{\pi}{2} = \frac{x}{y} = \frac{0}{1} = 0$

35. (a) $\theta = 203°$

$\theta' = 203° - 180° = 23°$

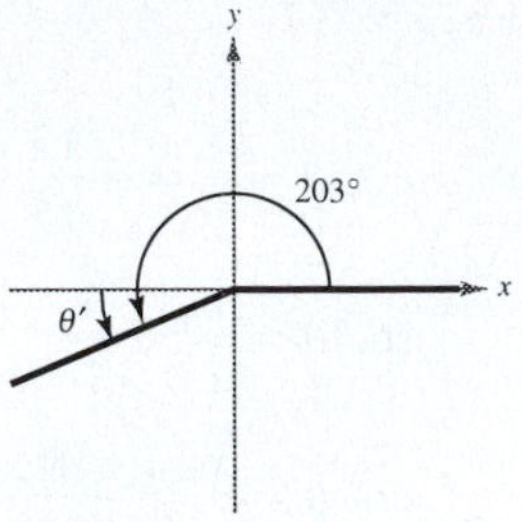

(b) $\theta = 127°$

$\theta' = 180° - 127° = 53°$

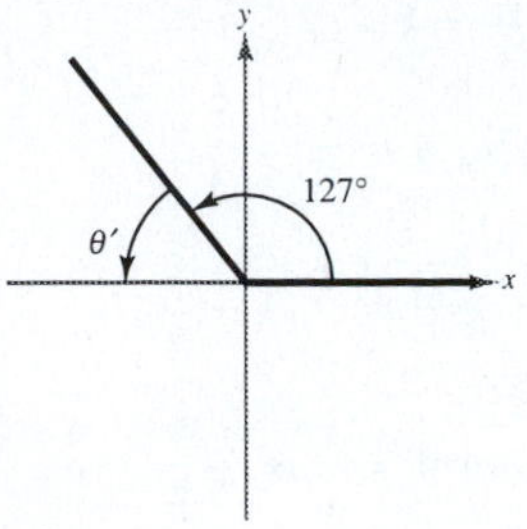

37. (a) $\theta = -245°$

$360° - 245° = 115°$ (coterminal angle)

$\theta' = 180° - 115° = 65°$

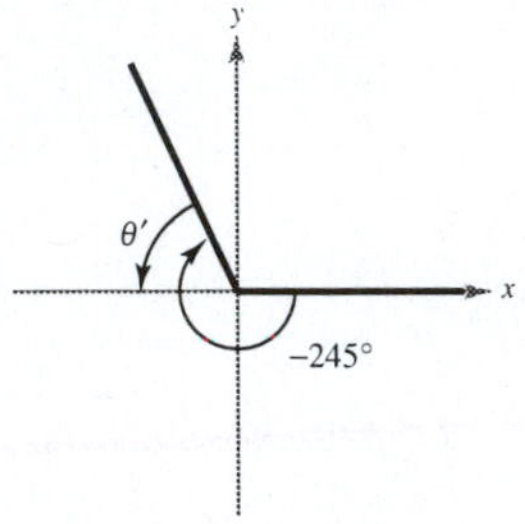

(b) $\theta = -72°$

$\theta' = 72°$

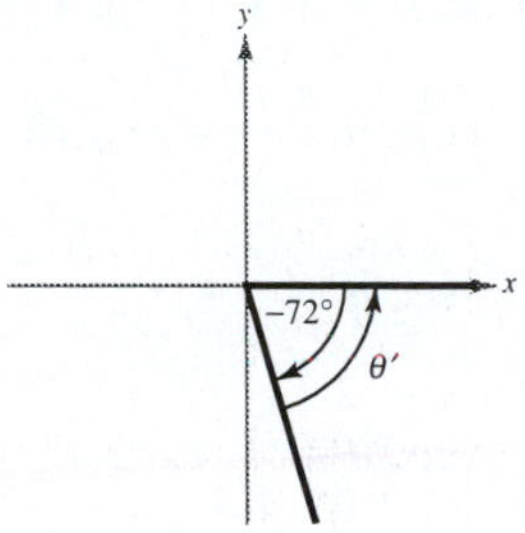

39. (a) $\theta = \frac{2\pi}{3}$

$\theta' = \pi - \frac{2\pi}{3} = \frac{\pi}{3}$

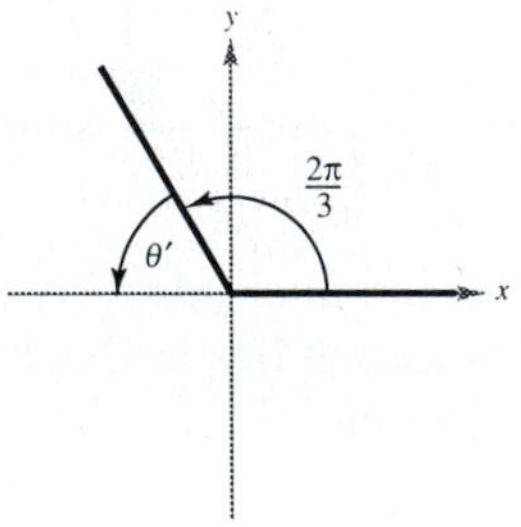

(b) $\theta = \frac{7\pi}{6}$

$\theta' = \frac{7\pi}{6} - \pi = \frac{\pi}{6}$

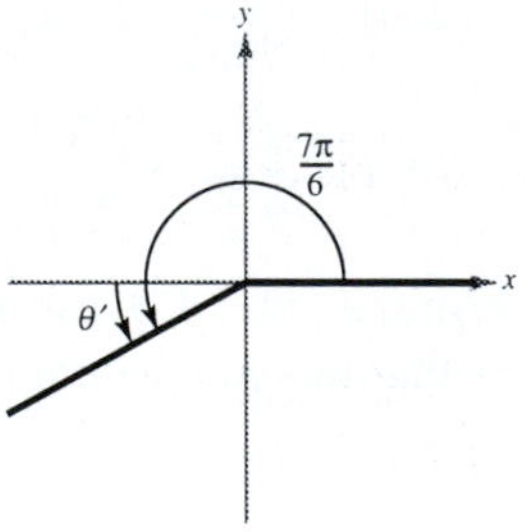

41. (a) $\theta = 3.5$

$\theta' = 3.5 - \pi$

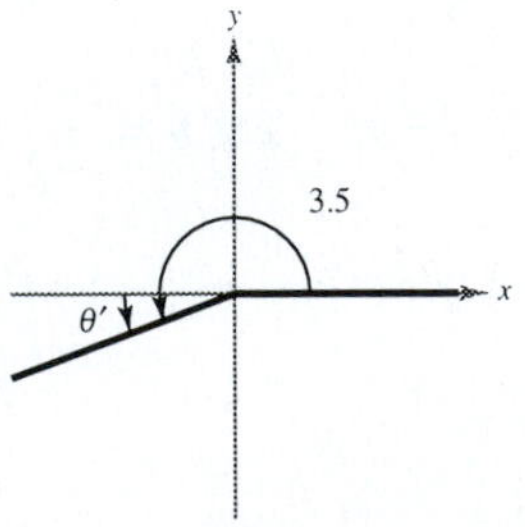

(b) $\theta = 5.8$

$\theta' = 2\pi - 5.8$

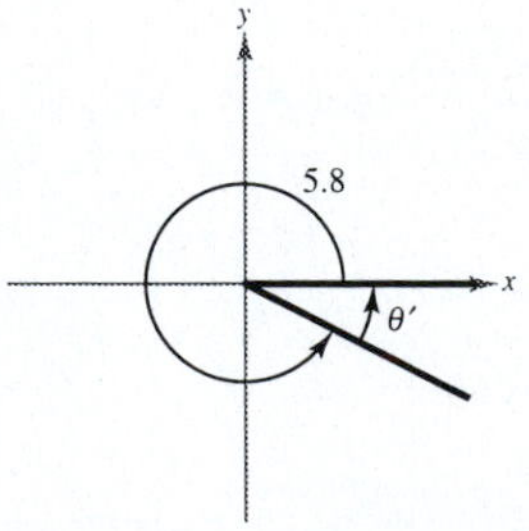

43. (a) $\theta' = 45°$, Quadrant III

$\sin 225° = -\sin 45° = -\dfrac{\sqrt{2}}{2}$

$\cos 225° = -\cos 45° = -\dfrac{\sqrt{2}}{2}$

$\tan 225° = \tan 45° = 1$

(b) $\theta' = 45°$, Quadrant II

$\sin(-225°) = \sin 45° = \dfrac{\sqrt{2}}{2}$

$\cos(-225°) = -\cos 45° = -\dfrac{\sqrt{2}}{2}$

$\tan(-225°) = -\tan 45° = -1$

45. (a) $\theta' = 30°$, Quadrant I

$\sin 750° = \sin 30° = \dfrac{1}{2}$

$\cos 750° = \cos 30° = \dfrac{\sqrt{3}}{2}$

$\tan 750° = \tan 30° = \dfrac{\sqrt{3}}{3}$

(b) $\theta' = 30°$, Quadrant II

$\sin 510° = \sin 30° = \dfrac{1}{2}$

$\cos 510° = -\cos 30° = -\dfrac{\sqrt{3}}{2}$

$\tan 510° = -\tan 30° = -\dfrac{\sqrt{3}}{3}$

47. (a) $\theta' = \dfrac{\pi}{3}$, Quadrant III

$\sin \dfrac{4\pi}{3} = -\sin \dfrac{\pi}{3} = -\dfrac{\sqrt{3}}{2}$

$\cos \dfrac{4\pi}{3} = -\cos \dfrac{\pi}{3} = -\dfrac{1}{2}$

$\tan \dfrac{4\pi}{3} = \tan \dfrac{\pi}{3} = \sqrt{3}$

(b) $\theta' = \dfrac{\pi}{3}$, Quadrant II

$\sin \dfrac{2\pi}{3} = \sin \dfrac{\pi}{3} = \dfrac{\sqrt{3}}{2}$

$\cos \dfrac{2\pi}{3} = -\cos \dfrac{\pi}{3} = -\dfrac{1}{2}$

$\tan \dfrac{2\pi}{3} = -\tan \dfrac{\pi}{3} = -\sqrt{3}$

49. (a) $\theta' = \dfrac{\pi}{6}$, Quadrant IV

$\sin\left(-\dfrac{\pi}{6}\right) = -\sin \dfrac{\pi}{6} = -\dfrac{1}{2}$

$\cos\left(-\dfrac{\pi}{6}\right) = \cos \dfrac{\pi}{6} = \dfrac{\sqrt{3}}{2}$

$\tan\left(-\dfrac{\pi}{6}\right) = -\tan \dfrac{\pi}{6} = -\dfrac{\sqrt{3}}{3}$

(b) $\theta' = \dfrac{\pi}{6}$, Quadrant II

$\sin \dfrac{5\pi}{6} = \sin \dfrac{\pi}{6} = \dfrac{1}{2}$

$\cos \dfrac{5\pi}{6} = -\cos \dfrac{\pi}{6} = -\dfrac{\sqrt{3}}{2}$

$\tan \dfrac{5\pi}{6} = -\tan \dfrac{\pi}{6} = -\dfrac{\sqrt{3}}{3}$

51. (a) $\theta' = \dfrac{\pi}{4}$, Quadrant II

$\sin \dfrac{11\pi}{4} = \sin \dfrac{\pi}{4} = \dfrac{\sqrt{2}}{2}$

$\cos \dfrac{11\pi}{4} = -\cos \dfrac{\pi}{4} = -\dfrac{\sqrt{2}}{2}$

$\tan \dfrac{11\pi}{4} = -\tan \dfrac{\pi}{4} = -1$

(b) $\theta' = \dfrac{\pi}{6}$, Quadrant IV

$\sin\left(-\dfrac{13\pi}{6}\right) = -\sin \dfrac{\pi}{6} = -\dfrac{1}{2}$

$\cos\left(-\dfrac{13\pi}{6}\right) = \cos \dfrac{\pi}{6} = \dfrac{\sqrt{3}}{2}$

$\tan\left(-\dfrac{13\pi}{6}\right) = -\tan \dfrac{\pi}{6} = -\dfrac{\sqrt{3}}{3}$

53. (a) $\sin 10° \approx 0.1736$

(b) $\csc 10° = \dfrac{1}{\sin 10°} \approx 5.7588$

55. (a) $\cos(-110°) \approx -0.3420$

(b) $\cos 250° \approx -0.3420$

57. (a) $\tan 240° \approx 1.7321$

(b) $\cot 210° = \dfrac{1}{\tan 210°} \approx 1.7321$

59. (a) $\tan \dfrac{\pi}{9} \approx 0.3640$

(b) $\tan \dfrac{10\pi}{9} \approx 0.3640$

61. (a) $\sin 0.65 \approx 0.6052$

(b) $\sin(-5.63) \approx 0.6077$

63. (a) $\sin \theta = \dfrac{1}{2} \implies$ reference angle is 30° or $\dfrac{\pi}{6}$ and θ is in Quadrant I or Quadrant II.

Values in degrees: 30°, 150°

Values in radian: $\dfrac{\pi}{6}, \dfrac{5\pi}{6}$

(b) $\sin \theta = -\dfrac{1}{2} \implies$ reference angle is 30° or $\dfrac{\pi}{6}$ and θ is in Quadrant III or Quadrant IV.

Values in degrees: 210°, 330°

Values in radians: $\dfrac{7\pi}{6}, \dfrac{11\pi}{6}$

65. (a) $\csc \theta = \dfrac{2\sqrt{3}}{3} \implies$ reference angle is 60° or $\dfrac{\pi}{3}$ and θ is in Quadrant I or Quadrant II.

Values in degrees: 60°, 120°

Values in radians: $\dfrac{\pi}{3}, \dfrac{2\pi}{3}$

(b) $\cot \theta = -1 \implies$ reference angle is 45° or $\dfrac{\pi}{4}$ and θ is in Quadrant II or Quadrant IV.

Values in degrees: 135°, 315°

Values in radians: $\dfrac{3\pi}{4}, \dfrac{7\pi}{4}$

67. (a) $\tan \theta = 1 \implies$ reference angle is 45° or $\dfrac{\pi}{4}$ and θ is in Quadrant I or Quadrant III.

Values in degrees: 45°, 225°

Values in radians: $\dfrac{\pi}{4}, \dfrac{5\pi}{4}$

(b) $\cot \theta = -\sqrt{3} \implies$ reference angle is 30° or $\dfrac{\pi}{6}$ and θ is in Quadrant II or Quadrant IV.

Values in degrees: 150°, 330°

Values in radians: $\dfrac{5\pi}{6}, \dfrac{11\pi}{6}$

69. (a) $\sin \theta = 0.8191 \implies \theta' \approx 54.99°$

Quadrant I: $\theta = \sin^{-1} 0.8191 \approx 54.99°$

Quadrant II: $\theta = 180° - \sin^{-1} 0.8191 \approx 125.01°$

(b) $\theta' = \sin^{-1} 0.2589 \approx 15.00°$

Quadrant III: $\theta = 180° + 15° = 195°$

Quadrant IV: $\theta = 360° - 15° = 345°$

71. (a) $\cos\theta = 0.9848 \implies \theta' \approx 0.175$

Quadrant I: $\theta = \cos^{-1}(0.9848) \approx 0.175$

Quadrant IV: $\theta = 2\pi - \theta' \approx 6.109$

(b) $\theta' = \cos^{-1} 0.5890 \approx 0.941$

Quadrant II: $\theta = \pi - 0.941 \approx 2.201$

Quadrant III: $\theta = \pi + 0.941 \approx 4.083$

73. (a) $\tan\theta = 1.192 \implies \theta' \approx 0.873$

Quadrant I: $\theta = \tan^{-1} 1.192 \approx 0.873$

Quadrant III: $\theta = \pi + \theta' \approx 4.014$

(b) $\theta' = \tan^{-1} 8.144 \approx 1.4486$

Quadrant II: $\theta = \pi - 1.4486 \approx 1.693$

Quadrant IV: $\theta = 2\pi - 1.4486 \approx 4.835$

75. $\sin\theta = -\frac{3}{5}$

$$\sin^2\theta + \cos^2\theta = 1$$

$$\cos^2\theta = 1 - \sin^2\theta$$

$$\cos^2\theta = 1 - \left(-\tfrac{3}{5}\right)^2$$

$$\cos^2\theta = 1 - \tfrac{9}{25}$$

$$\cos^2\theta = \tfrac{16}{25}$$

$\cos\theta > 0$ in Quadrant IV.

$\cos\theta = \frac{4}{5}$

77. $\tan\theta = \dfrac{3}{2}$

$$\sec^2\theta = 1 + \tan^2\theta$$

$$\sec^2\theta = 1 + \left(\frac{3}{2}\right)^2$$

$$\sec^2\theta = 1 + \frac{9}{4}$$

$$\sec^2\theta = \frac{13}{4}$$

$\sec\theta < 0$ in Quadrant III.

$\sec\theta = -\dfrac{\sqrt{13}}{2}$

79. $\cos\theta = \dfrac{5}{8}$

$$\cos\theta = \frac{1}{\sec\theta} \implies \sec\theta = \frac{1}{\cos\theta}$$

$$\sec\theta = \frac{1}{5/8} = \frac{8}{5}$$

81. (a) $t = 1$

$$T = 45 - 23\cos\left[\frac{2\pi}{365}(1 - 32)\right] \approx 25.2°\text{ F}$$

(b) $t = 185$

$$T = 45 - 23\cos\left[\frac{2\pi}{365}(185 - 32)\right] \approx 65.1°\text{ F}$$

(c) $t = 291$

$$T = 45 - 23\cos\left[\frac{2\pi}{365}(291 - 32)\right] \approx 50.8°\text{ F}$$

83. $\sin\theta = \dfrac{6}{d} \implies d = \dfrac{6}{\sin\theta}$

(a) $d = \dfrac{6}{\sin 30°} = 12$ miles

(b) $d = \dfrac{6}{\sin 90°} = 6$ miles

(c) $d = \dfrac{6}{\sin 120°} \approx 6.9$ miles

85. $y = 2^{x-1}$

x	-1	0	1	2	3
y	$\frac{1}{4}$	$\frac{1}{2}$	1	2	4

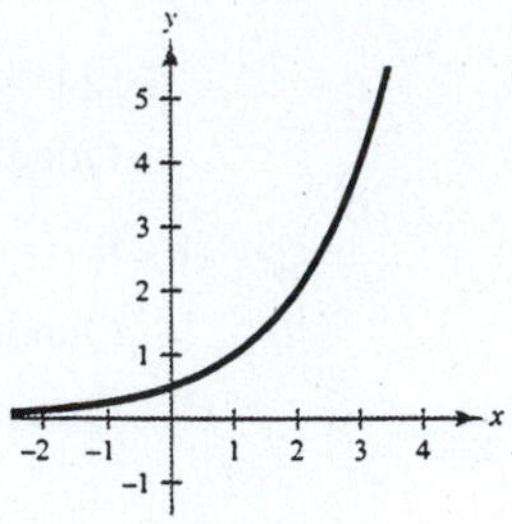

87. $y = \ln(x - 1)$

Domain: $x - 1 > 0 \implies x > 1$

x	1.1	1.5	2	3	4
y	-2.30	-0.69	0	0.69	1.10

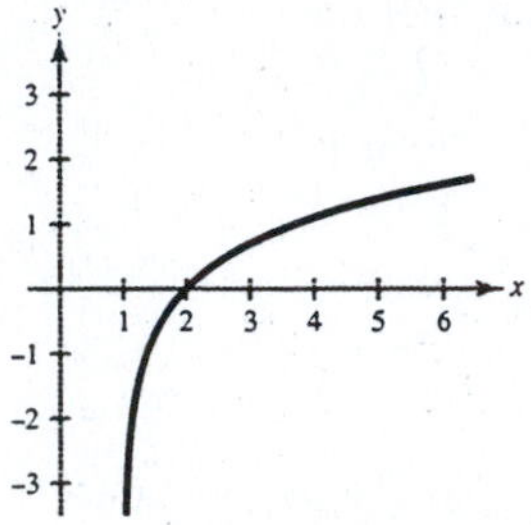

Section 4.5 Graphs of Sine and Cosine Functions

- You should be able to graph $y = a\sin(bx - c)$ and $y = a\cos(bx - c)$.
- Amplitude: $|a|$
- Period: $\dfrac{2\pi}{|b|}$
- Shift: Solve $bx - c = 0$ and $bx - c = 2\pi$.
- Key Increments: $\dfrac{1}{4}$ (period)

Solutions to Odd-Numbered Exercises

1. $y = 3\sin 2x$

Period: $\dfrac{2\pi}{2} = \pi$

Amplitude: $|3| = 3$

3. $y = \dfrac{5}{2}\cos\dfrac{x}{2}$

Period: $\dfrac{2\pi}{1/2} = 4\pi$

Amplitude: $\left|\dfrac{5}{2}\right| = \dfrac{5}{2}$

5. $y = \dfrac{2}{3}\sin \pi x$

Period: $\dfrac{2\pi}{\pi} = 2$

Amplitude: $\left|\dfrac{2}{3}\right| = \dfrac{2}{3}$

7. $y = -2\sin x$

Period: $\dfrac{2\pi}{1} = 2\pi$

Amplitude: $|-2| = 2$

9. $y = 3\sin 10x$

Period: $\dfrac{2\pi}{10} = \dfrac{\pi}{5}$

Amplitude: $|3| = 3$

11. $y = \dfrac{1}{2}\cos\dfrac{2\pi}{3}$

Period: $\dfrac{2\pi}{2/3} = 3\pi$

Amplitude: $\left|\dfrac{1}{2}\right| = \dfrac{1}{2}$

13. $y = 3\sin 4\pi x$

Period: $\dfrac{2\pi}{4\pi} = \dfrac{1}{2}$

Amplitude: $|3| = 3$

15. $f(x) = \sin x$

$g(x) = \sin(x - \pi)$

The graph of g is a horizontal shift to the right π units of the graph of f (a phase shift).

17. $f(x) = \cos 2x$

$g(x) = -\cos 2x$

The graph of g is a reflection in the x-axis of the graph of f.

19. $f(x) = \cos x$

$g(x) = \cos 2x$

The period of f is twice that of g.

21. $f(x) = \sin x$

$f(x) = 2 + \sin x$

The graph of g is a vertical shift 2 units upward of the graph of f.

23. The graph of g has twice the amplitude as the graph of f. The period is the same.

25. The graph of g is a horizontal shift π units to the right of the graph of f.

27. $y_1 = \frac{1}{2}\sin x;\ y_2 = \frac{3}{2}\sin x;\ y_3 = -3\sin x$

Changing the value of a changes the amplitude.

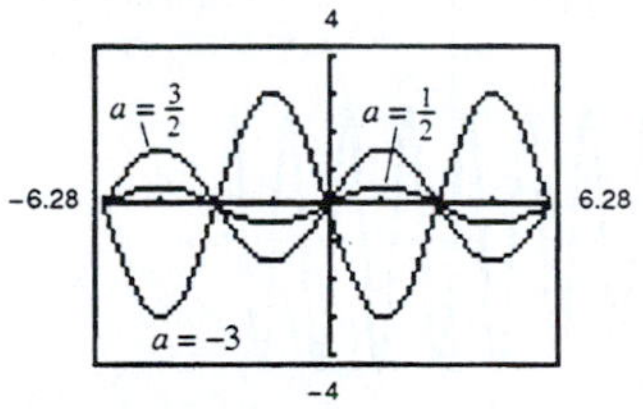

29. $y_1 = \sin\left(\frac{1}{2}x\right);\ y_2 = \left(\frac{3}{2}x\right);\ y_3 = \sin(4x)$

Changing the value of b changes the period.

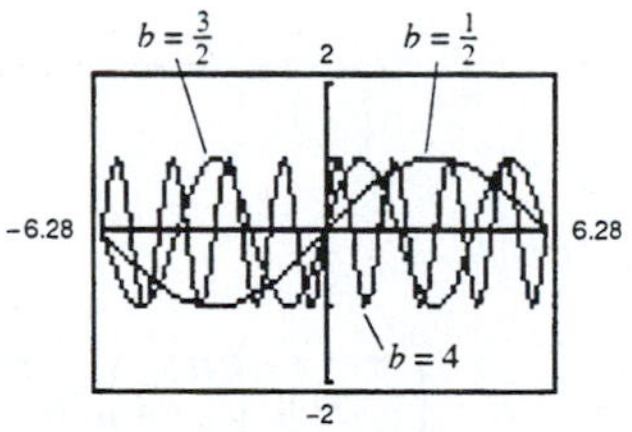

31. $f(x) = -2\sin x$

Period: 2π

Amplitude: 2

$g(x) = 4\sin x$

Period: 2π

Amplitude: 4

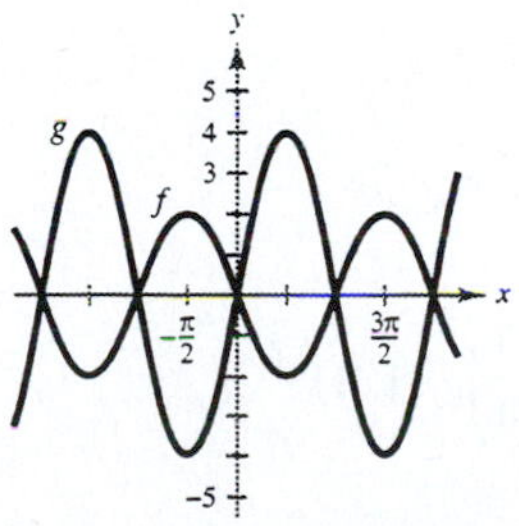

33. $f(x) = \cos x$

Period: 2π

Amplitude: 1

$g(x) = 1 + \cos x$

is a vertical shift of the graph of $f(x)$ one unit upward.

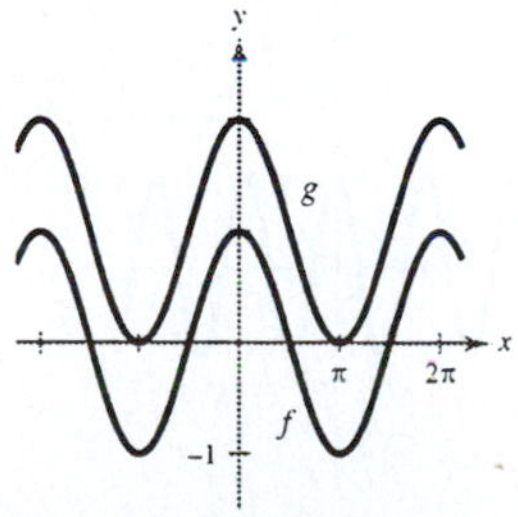

35. $f(x) = -\frac{1}{2}\sin\frac{x}{2}$

Period: 4π

Amplitude: $\frac{1}{2}$

$g(x) = 3 - \frac{1}{2}\sin\frac{x}{2}$ is the graph of $f(x)$ shifted vertically three units upward.

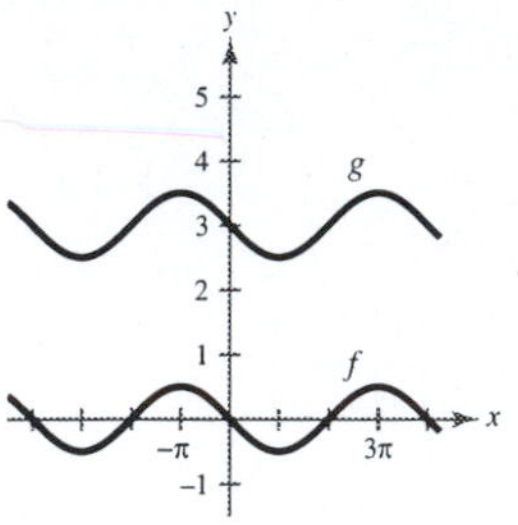

37. $f(x) = 2\cos x$

Period: 2π

Amplitude: 2

$g(x) = 2\cos(x + \pi)$ is the graph of $f(x)$ shifted π units to the left.

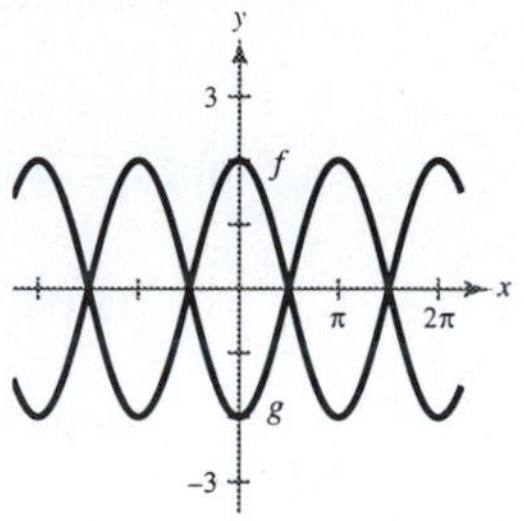

39. Since sine and cosine are cofunctions and x and $x - (\pi/2)$ are complementary, we have

$\sin x = \cos\left(x - \frac{\pi}{2}\right)$.

Period: 2π

Amplitude: 1

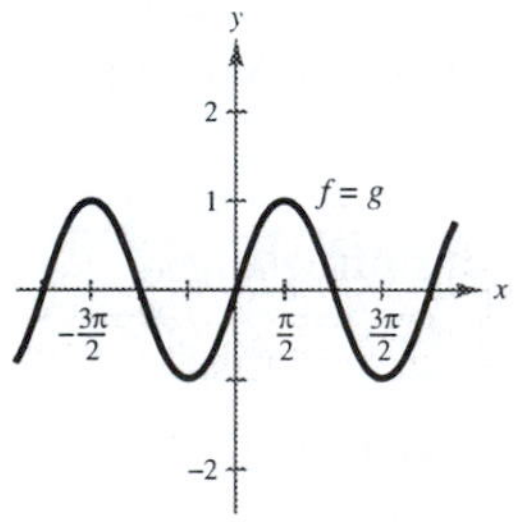

41. $f(x) = \cos x$

$g(x) = -\sin\left(x - \frac{\pi}{2}\right) = \sin\left(\frac{\pi}{2} - x\right) = \cos x$

Thus, $f(x) = g(x)$.

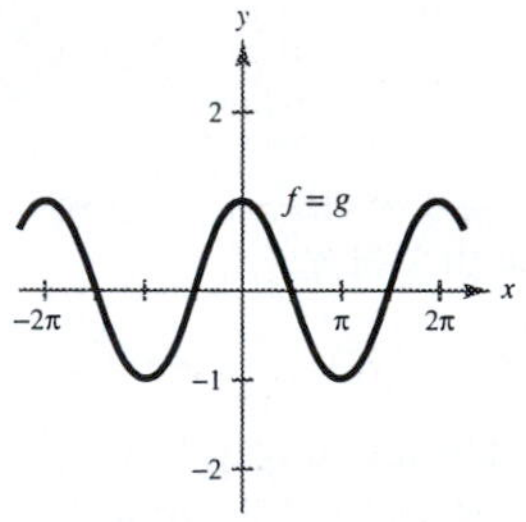

43. $y = -2\sin 6x;\ a = -2,\ b = 6,\ c = 0$

Period: $\frac{2\pi}{6} = \frac{\pi}{3}$

Amplitude: $|-2| = 2$

Key points: $(0, 0), \left(\frac{\pi}{12}, -2\right), \left(\frac{\pi}{6}, 0\right), \left(\frac{\pi}{4}, 2\right), \left(\frac{\pi}{3}, 0\right)$

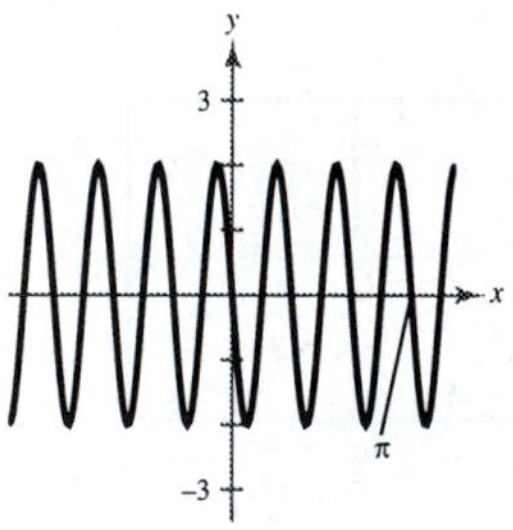

45. $y = \cos 2\pi x$

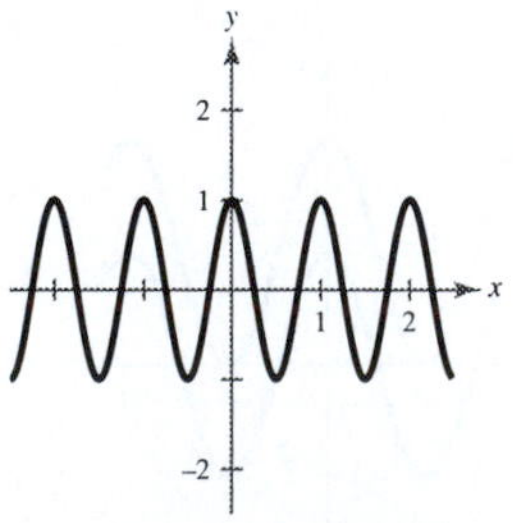

Period: $\frac{2\pi}{2\pi} = 1$

Amplitude: 1

Key points: $(1, 0), \left(0, \frac{1}{4}\right), \left(-1, \frac{1}{2}\right), \left(0, \frac{3}{4}\right), (1, 1)$

47. $y = -\sin \dfrac{2\pi x}{3}$; $a = -1, b = \dfrac{2\pi}{3}, c = 0$

Period: $\dfrac{2\pi}{2\pi/3} = 3$

Amplitude: 1

Key points: $(0, 0), \left(\dfrac{3}{4}, -1\right), \left(\dfrac{3}{2}, 0\right), \left(\dfrac{9}{4}, 1\right), (3, 0)$

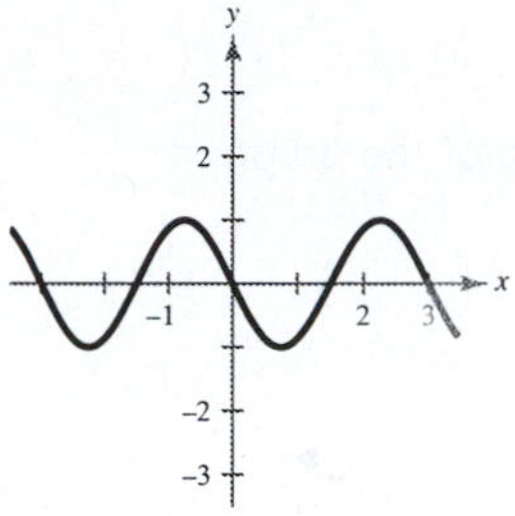

49. $y = \sin\left(x - \dfrac{\pi}{4}\right)$; $a = 1,\ b = 1,\ c = \dfrac{\pi}{4}$

Period: 2π

Amplitude: 1

Shift: Set $x - \dfrac{\pi}{4} = 0$ and $x - \dfrac{\pi}{4} = 2\pi$

$x = \dfrac{\pi}{4}$ $\qquad$ $x = \dfrac{9\pi}{4}$

Key points: $\left(\dfrac{\pi}{4}, 0\right), \left(\dfrac{3\pi}{4}, 1\right), \left(\dfrac{5\pi}{4}, 0\right), \left(\dfrac{7\pi}{4}, -1\right), \left(\dfrac{9\pi}{4}, 0\right)$

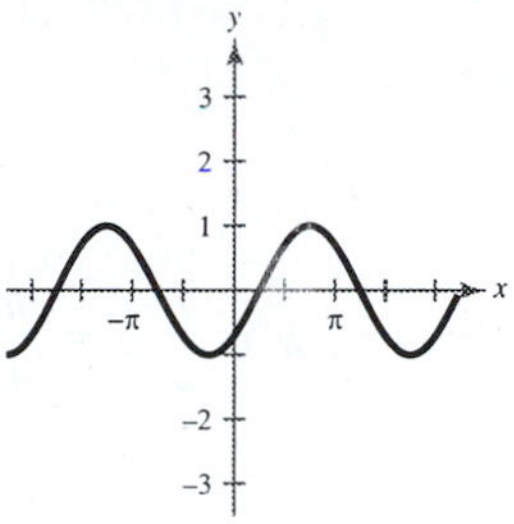

51. $y = 3\cos(x + \pi)$

Period: 2π

Amplitude: 3

Shift: Set $x + \pi = 0$ and $x + \pi = 2\pi$

$x = -\pi$ $\qquad$ $x = \pi$

Key points: $(-\pi, 3), \left(-\dfrac{\pi}{2}, 0\right), (0, -3), \left(\dfrac{\pi}{2}, 0\right), (\pi, 3)$

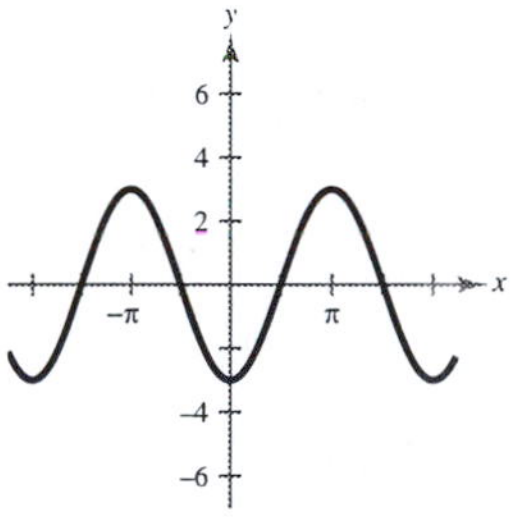

53. $y = \dfrac{1}{10}\cos(60\pi x)$; $a = \dfrac{1}{10},\ b = 60\pi,\ c = 0$

Period: $\dfrac{2\pi}{60\pi} = \dfrac{1}{30}$

Amplitude: $\dfrac{1}{10}$

Key points: $\left(0, \dfrac{1}{10}\right), \left(\dfrac{1}{120}, 0\right), \left(\dfrac{1}{60}, -\dfrac{1}{10}\right), \left(\dfrac{1}{40}, 0\right), \left(\dfrac{1}{30}, \dfrac{1}{10}\right)$

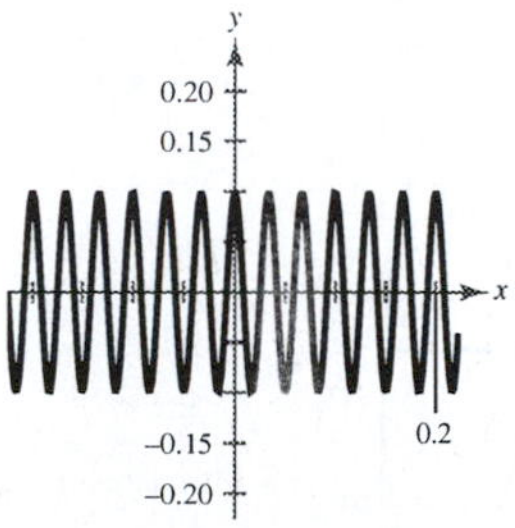

55. $y = 2 - \sin \dfrac{2\pi x}{3}$

Vertical shift 2 units upward of the graph in Exercise 47.

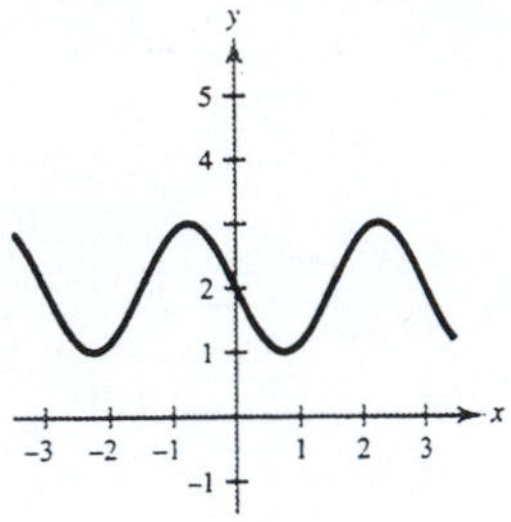

57. $y = 3\cos(x + \pi) - 3$

Vertical shift 3 units downward of the graph in Exercise 51.

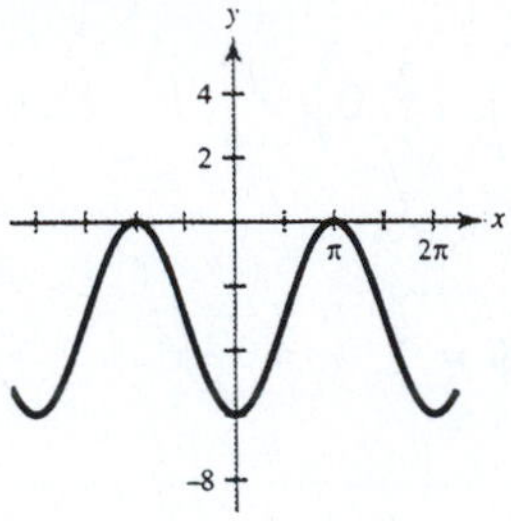

59. $y = \dfrac{2}{3}\cos\left(\dfrac{x}{2} - \dfrac{\pi}{4}\right)$; $a = \dfrac{2}{3}$, $b = \dfrac{1}{2}$, $c = \dfrac{\pi}{4}$

Period: 4π

Amplitude: $\dfrac{2}{3}$

$$\frac{x}{2} - \frac{\pi}{4} = 0 \quad \text{and} \quad \frac{x}{2} - \frac{\pi}{4} = 2\pi$$

$$x = \frac{\pi}{2} \qquad\qquad x = \frac{9\pi}{2}$$

Key points: $\left(\dfrac{\pi}{2}, \dfrac{2}{3}\right)$, $\left(\dfrac{3\pi}{2}, 0\right)$, $\left(\dfrac{5\pi}{2}, \dfrac{-2}{3}\right)$, $\left(\dfrac{7\pi}{2}, 0\right)$, $\left(\dfrac{9\pi}{2}, \dfrac{2}{3}\right)$

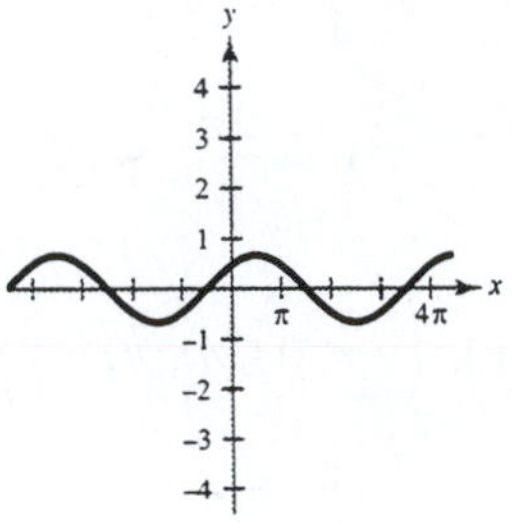

61. $y = -2\sin(4x + \pi)$

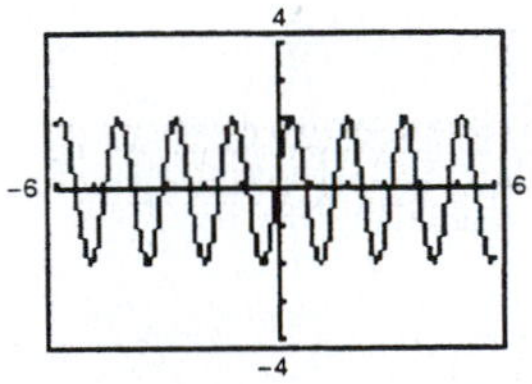

63. $y = \cos\left(2\pi x - \dfrac{\pi}{2}\right) + 1$

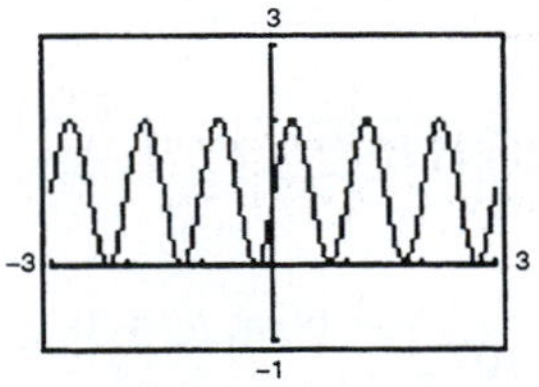

65. $y = -0.1\sin\left(\dfrac{\pi x}{10} + \pi\right)$

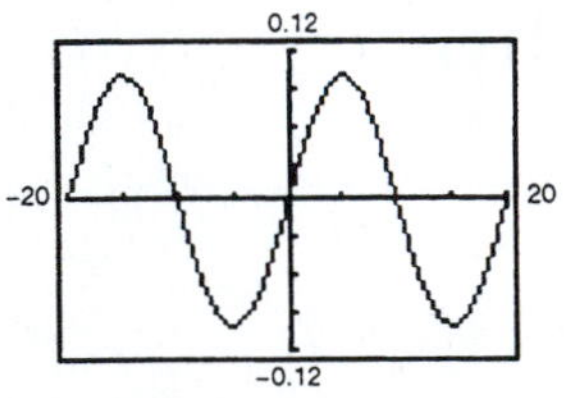

67. $y = 5\cos(\pi - 2x) + 2$

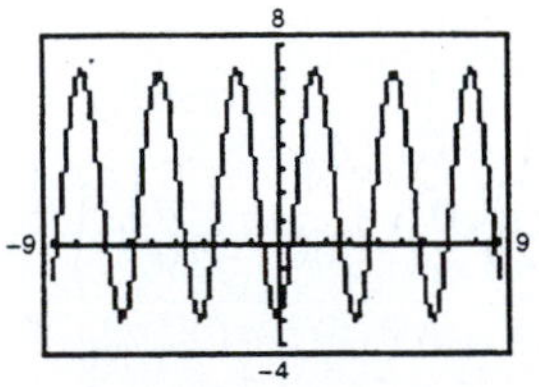

69. $f(x) = a\cos x + d$

Amplitude: $\frac{1}{2}[5 - (-1)] = 3 \Rightarrow a = 3$

Vertical shift 2 units upward of $g(x) = 3\cos x \Rightarrow d = 2$.
Thus, $f(x) = 3\cos x + 2 = 2 + 3\cos x$.

71. $f(x) = a\cos x + d$

Amplitude: $\frac{1}{2}[8 - 0] = 4$

Since $f(x)$ is the graph of $g(x) = 4\cos x$ reflected about the x-axis and shifted vertically 4 units upward, we have $a = -4$ and $d = 4$.

Thus, $f(x) = -4\cos x + 4$.

73. $y = a\sin(bx - c)$

Amplitude: $|a| = |3|$ Since the graph is reflected about the x-axis, we have $a = -3$.

Period: $\frac{2\pi}{b} = \pi \Rightarrow b = 2$

Phase shift: $c = 0$

Thus, $y = -3\sin 2x$.

75. $y = a\sin(bx + c)$

Amplitude: $a = 1$

Period: $2\pi \Rightarrow b = 1$

Phase shift: $bx + c = 0$ when $x = \frac{\pi}{4}$

$$(1)\left(\frac{\pi}{4}\right) + c = 0 \Rightarrow c = -\frac{\pi}{4}$$

Thus, $y = \sin\left(x - \frac{\pi}{4}\right)$.

77. $y_1 = \sin x$

$y_2 = -\frac{1}{2}$

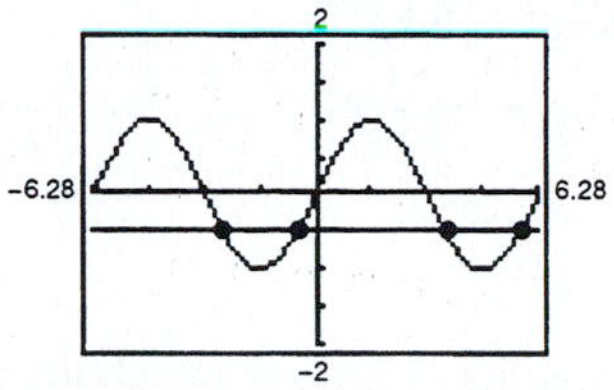

In the interval $[-2\pi, 2\pi]$, $\sin x = -\frac{1}{2}$ when $x = -\frac{5\pi}{6}, -\frac{\pi}{6}, \frac{7\pi}{6}, \frac{11\pi}{6}$.

79. $y_1 = \cos x$

$y_2 = \frac{\sqrt{2}}{2}$

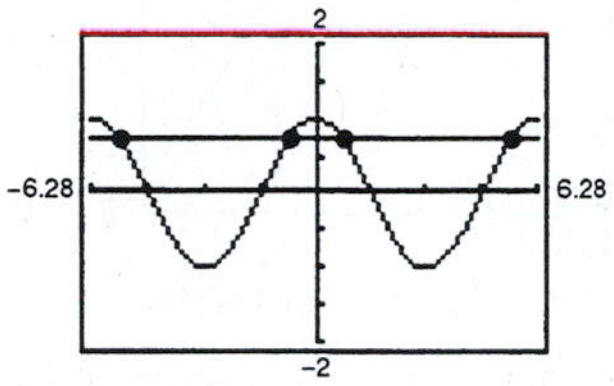

In the interval $[-2\pi, 2\pi]$, $\cos x = \frac{\sqrt{2}}{2}$ when $x = \pm\frac{\pi}{4}, \pm\frac{7\pi}{4}$.

81. (a) $h(x) = \cos^2 x$ is even.

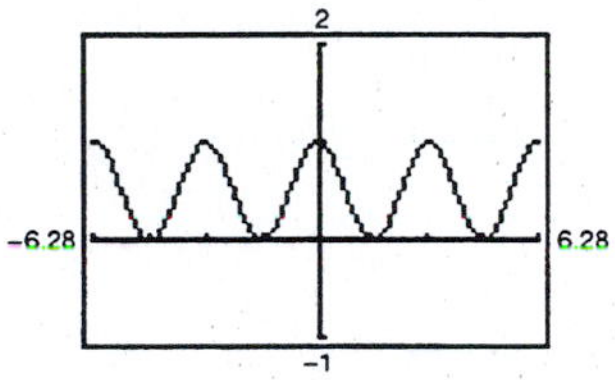

(b) $g(x) = \sin^2 x$ is even.

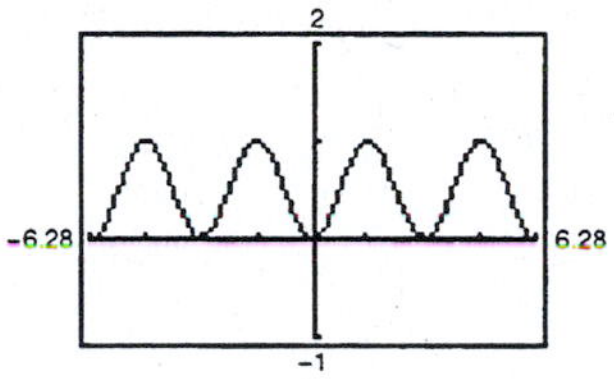

83. $y = 0.85\sin\frac{\pi t}{3}$

(a) Time for one cycle = one period $= \frac{2\pi}{\pi/3} = 6$ sec

(b) Cycles per min $= \frac{60}{6} = 10$ cycles per min

(c) Amplitude: 0.85

Period: 6

Key points: $(0, 0)$, $\left(\frac{3}{2}, 0.85\right)$, $(3, 0)$, $\left(\frac{9}{2}, -0.85\right)$, $(6, 0)$

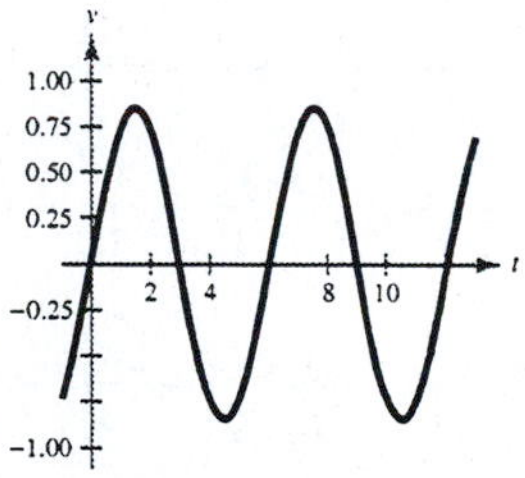

85. $y = 0.001\sin 880\pi t$

(a) Period: $\frac{2\pi}{880\pi} = \frac{1}{440}$ seconds

(b) $f = \frac{1}{p} = 440$ cycles per second

87. $S = 22.3 - 3.4\cos\frac{\pi t}{6}$, $1 \le t \le 12$

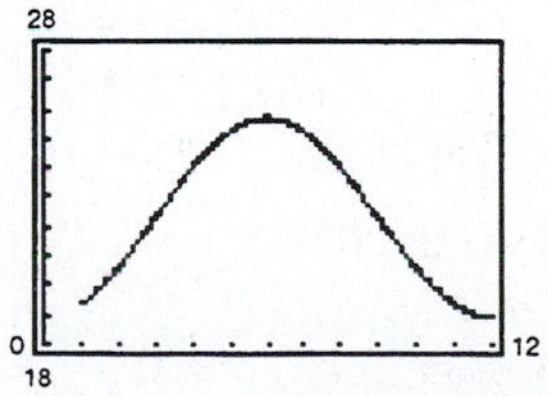

89. (a) $C(t) = 56.35 + 27.35 \sin\left(\frac{\pi t}{6} + 4.19\right)$

(b)

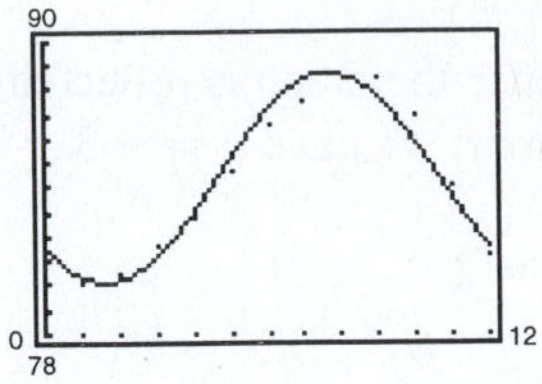

The model is a good fit for most months.

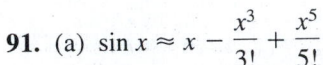

(c)

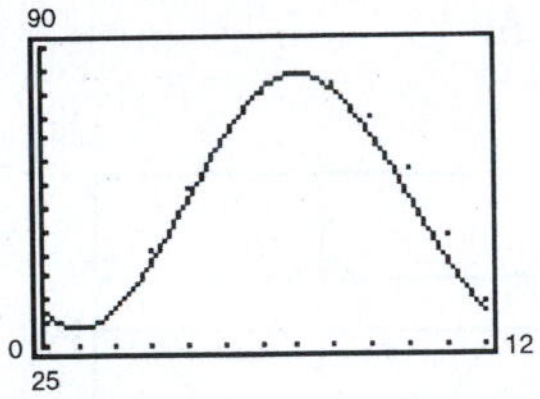

The model is a good fit.

(d) Use the constant term of each model to estimate the average annual temperature.

Honolulu: 84.40°

Chicago: 58.50°

(e) Each model has a period of 12. This corresponds to the 12 months in a year.

(f) Chicago has a greater variability in temperatures during the year. The amplitude of each model indicates this variability.

91. (a) $\sin x \approx x - \frac{x^3}{3!} + \frac{x^5}{5!}$

Near $x = 0$ the graphs are approximately the same. They appear to coincide from $-\pi/2$ to $\pi/2$.

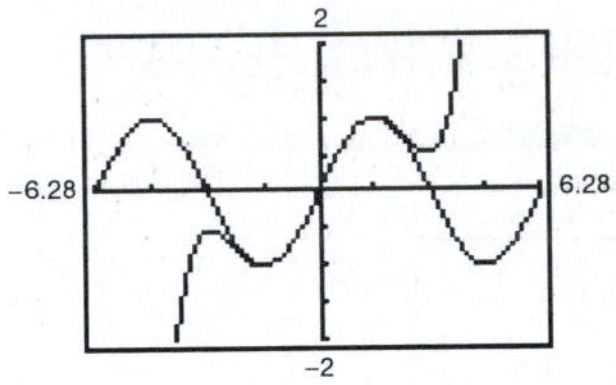

(b) $\cos x \approx 1 - \frac{x^2}{2!} + \frac{x^4}{4!}$

Near $x = 0$ the graphs are approximately the same. They appear to coincide from $-\pi/2$ to $\pi/2$.

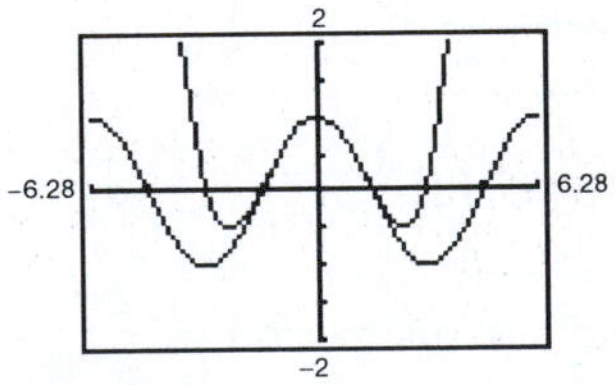

(c) $\sin x \approx x - \frac{x^3}{3!} + \frac{x^5}{5!} - \frac{x^7}{7!}$

$\cos x \approx 1 - \frac{x^2}{2!} + \frac{x^4}{4!} - \frac{x^6}{6!}$

The accuracy is increased.

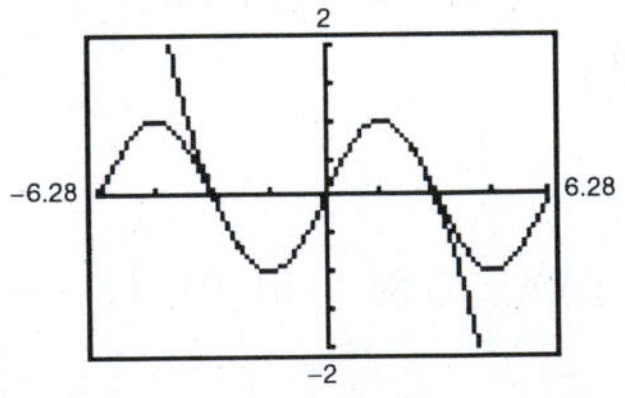

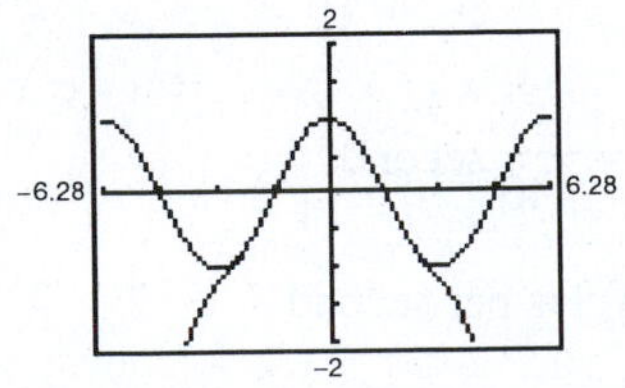

93. (a)

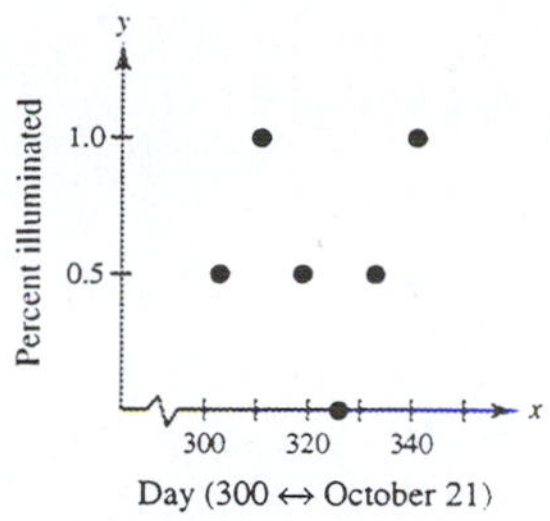

(b) $y = \frac{1}{2} + \frac{1}{2}\sin\left[\frac{\pi}{15}(t - 303)\right]$

(c)

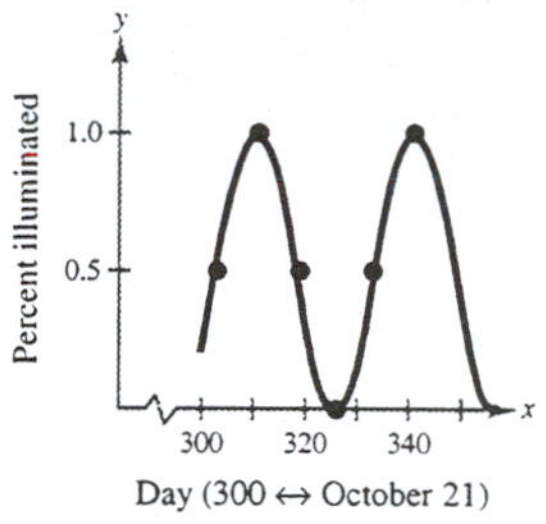

The model is a good fit.

(d) $y(356) \approx 0.003 \approx 0$

95. $\log_2 [x^2(x - 3)] = \log_2 x^2 + \log_2(x - 3)$

$= 2 \log_2 x + \log_2(x - 3)$

97. $\ln\sqrt{\frac{z}{z^2 + 1}} = \frac{1}{2}\ln\left(\frac{z}{z^2 + 1}\right) = \frac{1}{2}[\ln z - \ln(z^2 + 1)]$

$= \frac{1}{2}\ln z - \frac{1}{2}\ln(z^2 + 1)$

Section 4.6 Graphs of Other Trigonometric Functions

- You should be able to graph

  $y = a \tan (bx - c)$ $\quad$ $y = a \cot (bx - c)$

  $y = a \sec (bx - c)$ $\quad$ $y = a \csc (bx - c)$

- When graphing $y = a \sec (bx - c)$ or $y = a \csc (bx - c)$ you should first graph $y = a \cos (bx - c)$ or $y = a \sin (bx - c)$ because
  (a) The x-intercepts of sine and cosine are the vertical asymptotes of cosecant and secant.
  (b) The maximums of sine and cosine are the local minimums of cosecant and secant.
  (c) The minimums of sine and cosine are the local maximums of cosecant and secant.
- You should be able to graph using a damping factor.

Solutions to Odd-Numbered Exercises

1. $y = \sec \dfrac{x}{2}$

Period: $\dfrac{2\pi}{1/2} = 4\pi$

Matches graph (g).

3. $y = \tan 2x$

Period: $\dfrac{\pi}{2}$

Matches graph (f).

5. $y = \cot \dfrac{\pi x}{2}$

Period: $\dfrac{\pi}{\pi/2} = 2$

Matches graph (b).

7. $y = -\csc x$

Period: 2π

Matches graph (e).

9. $y = \dfrac{1}{3} \tan x$

Period: π

Two consecutive asymptotes:

$$x = -\frac{\pi}{2} \text{ and } x = \frac{\pi}{2}$$

x	$-\frac{\pi}{4}$	0	$\frac{\pi}{4}$
y	$-\frac{1}{3}$	0	$\frac{1}{3}$

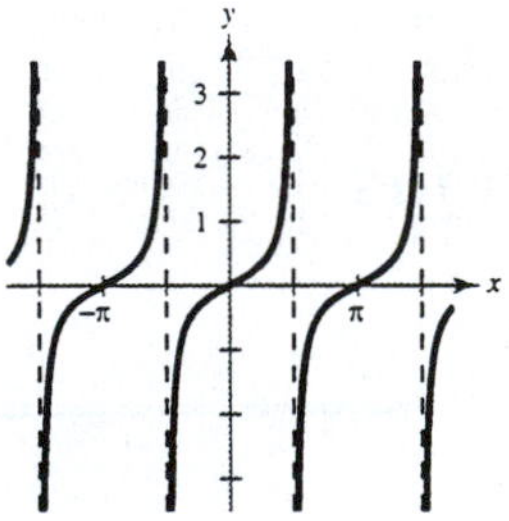

11. $y = \tan 2x$

Period: $\dfrac{\pi}{2}$

Two consecutive asymptotes: $2x = -\dfrac{\pi}{2} \Rightarrow x = -\dfrac{\pi}{4}$

$2x = \dfrac{\pi}{2} \Rightarrow x = \dfrac{\pi}{4}$

x	$-\frac{\pi}{8}$	0	$\frac{\pi}{8}$
y	-1	0	1

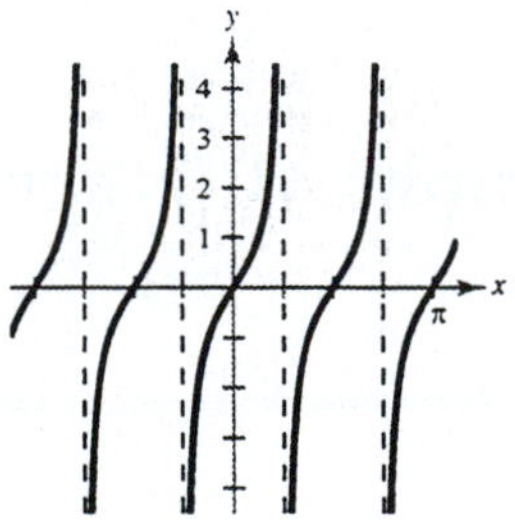

13. $y = -\dfrac{1}{2} \sec x$

Graph $y = -\dfrac{1}{2} \cos x$ first.

Period: 2π

One cycle: 0 to 2π

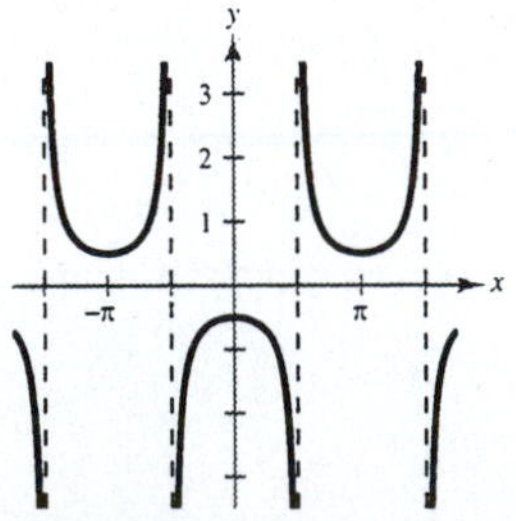

15. $y = \sec \pi x$

Graph $y = \cos \pi x$ first.

Period: $\dfrac{2\pi}{\pi} = 2$

One cycle: 0 to 2

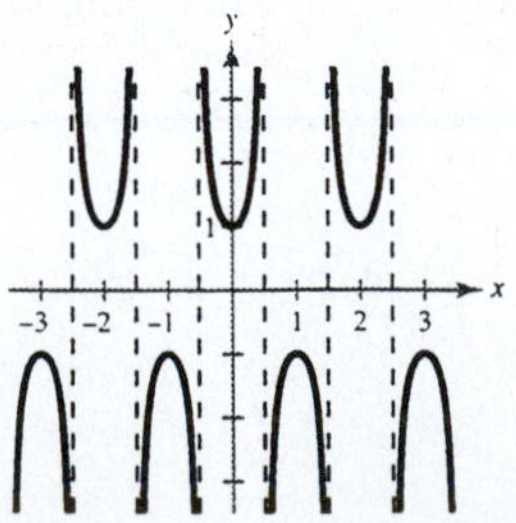

17. $y = \sec \pi x - 1$

Reflect the graph in Exercise 15 about the x-axis and then shift it vertically down one unit.

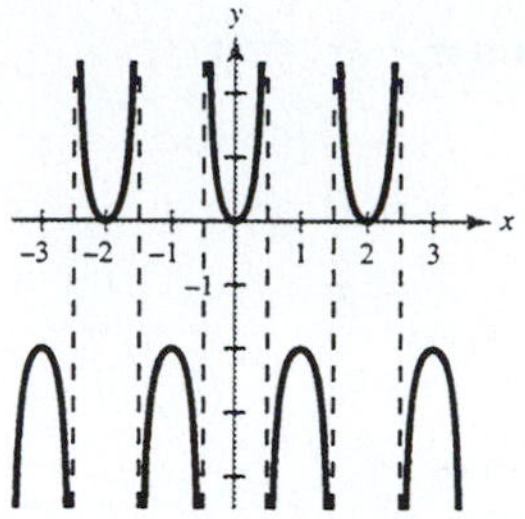

19. $y = \csc \frac{x}{2}$

Graph $y = \sin \frac{x}{2}$ first.

Period: $\frac{2\pi}{1/2} = 4\pi$

One cycle: 0 to 4π

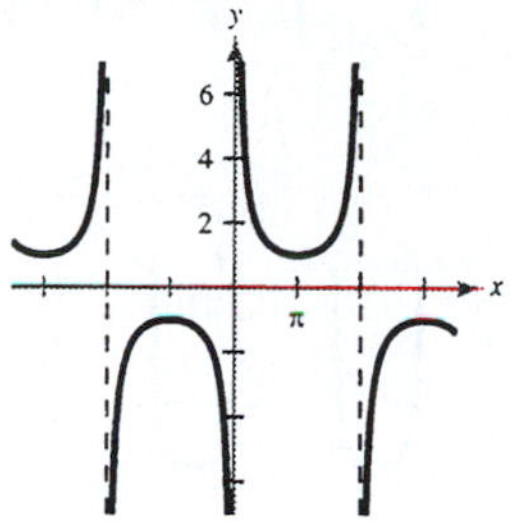

21. $y = \cot \frac{x}{2}$

Period: $\frac{\pi}{1/2} = 2\pi$

Two consecutive asymptotes: $\frac{x}{2} = 0 \Rightarrow x = 0$

$\frac{x}{2} = \pi \Rightarrow x = 2\pi$

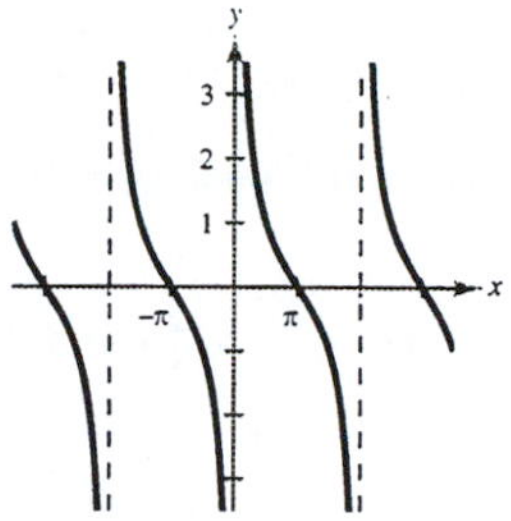

x	$\frac{\pi}{2}$	π	$\frac{3\pi}{2}$
y	1	0	-1

23. $y = \frac{1}{2} \sec 2x$

Graph $y = \frac{1}{2} \cos 2x$ first.

Period: $\frac{2\pi}{2} = \pi$

One cycle: 0 to π

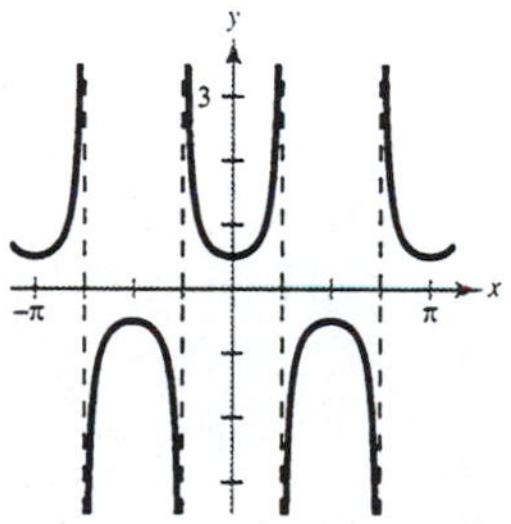

25. $y = \tan \frac{\pi x}{4}$

Period: $\frac{\pi}{\pi/4} = 4$

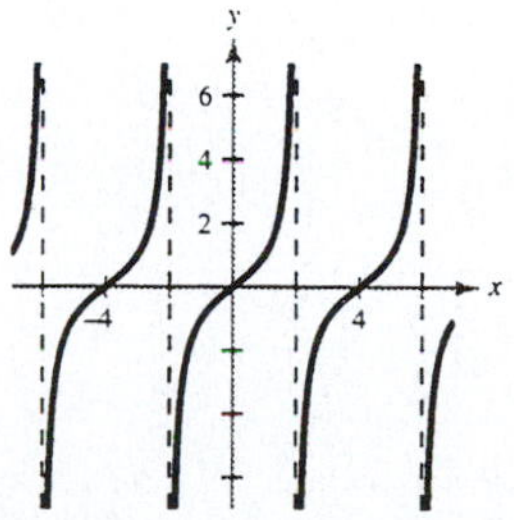

Two consecutive asymptotes: $\frac{\pi x}{4} = -\frac{\pi}{2} \Rightarrow x = -2$

$\frac{\pi x}{4} = \frac{\pi}{2} \Rightarrow x = 2$

x	-1	0	1
y	-1	0	1

27. $y = \csc(\pi - x)$

Graph $y = \sin(\pi - x)$ first.

Period: 2π

Shift: Set $\pi - x = 0$ and $\pi - x = 2\pi$

$x = \pi$ $\qquad x = -\pi$

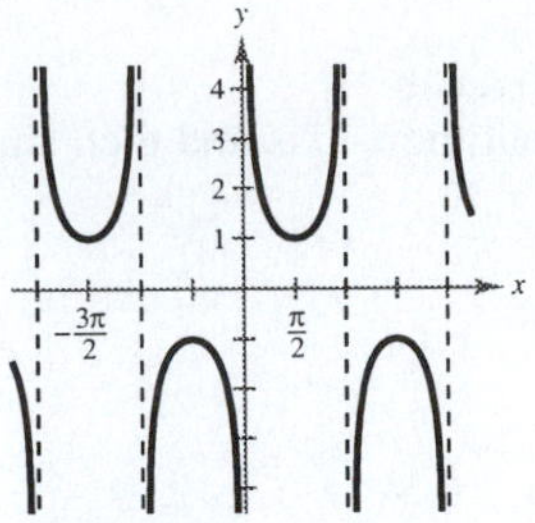

29. $y = \frac{1}{4}\csc\left(x + \frac{\pi}{4}\right)$

Graph $y = \frac{1}{4}\sin\left(x + \frac{\pi}{4}\right)$ first.

Period: 2π

Shift: Set $x + \frac{\pi}{4} = 0$ and $x + \frac{\pi}{4} = 2\pi$

$x = -\frac{\pi}{4}$ to $x = \frac{7\pi}{4}$

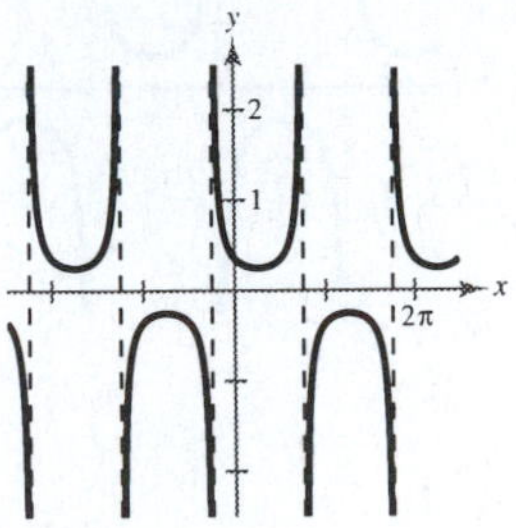

31. $y = \tan\frac{x}{3}$

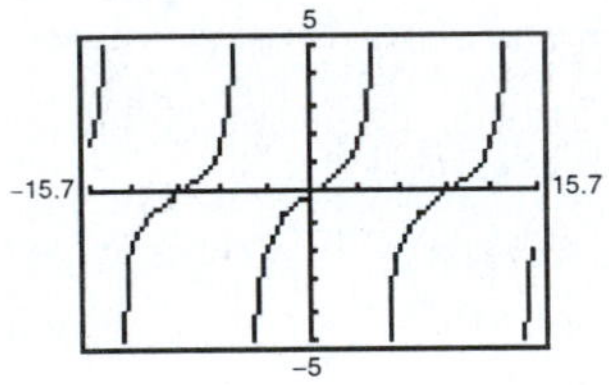

33. $y = -2\sec 4x$

$= \frac{-2}{\cos 4x}$

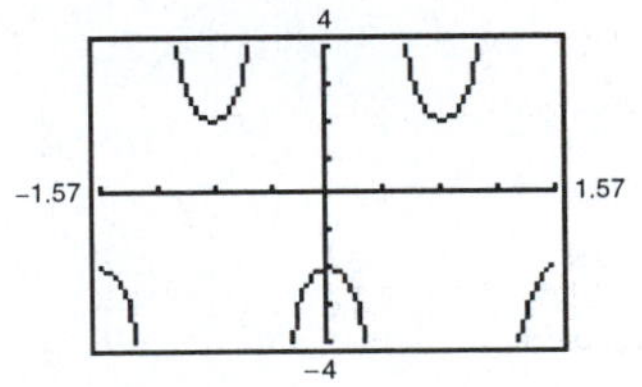

35. $y = \tan\left(x - \frac{\pi}{4}\right)$

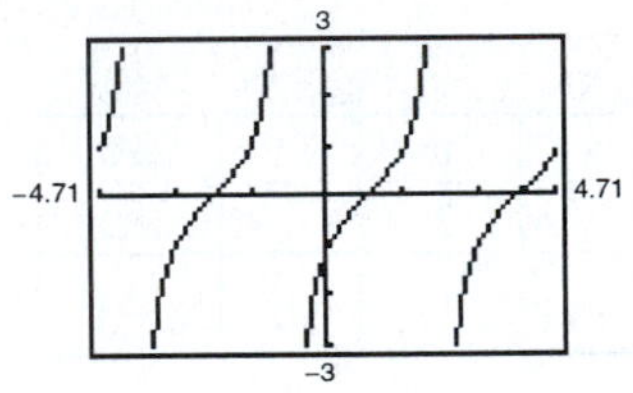

37. $y = \frac{1}{4}\cot\left(x - \frac{\pi}{2}\right)$

$= \frac{1}{4\tan(x - \pi/2)}$

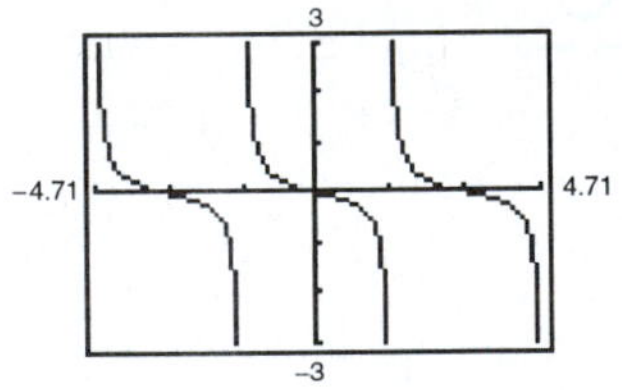

39. $y = 2\sec(2x - \pi)$

$y = \frac{2}{\cos(2x - \pi)}$

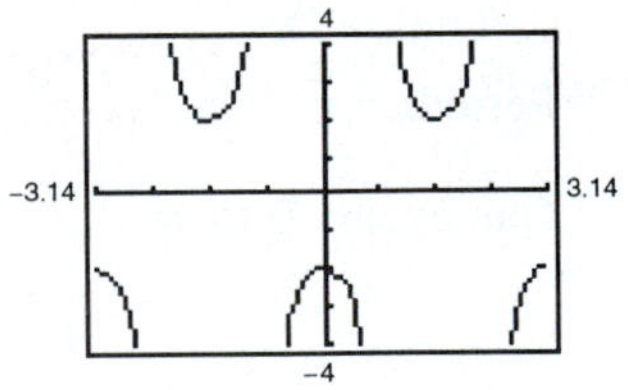

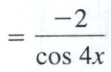

41. $\tan x = 1$

$$x = -\frac{7\pi}{4}, -\frac{3\pi}{4}, \frac{\pi}{4}, \frac{5\pi}{4}$$

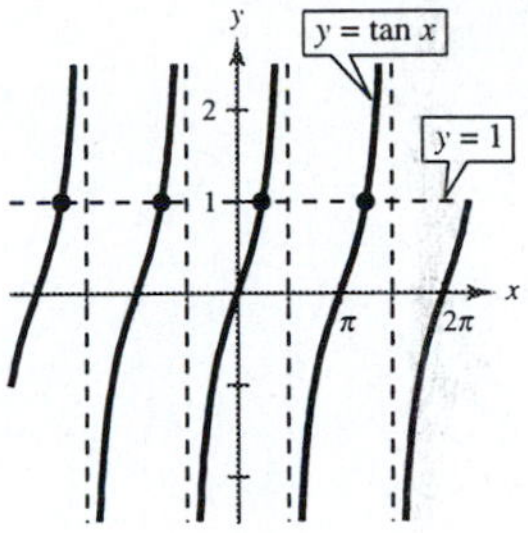

43. $\sec x = -2$

$$x = \pm\frac{2\pi}{3}, \pm\frac{4\pi}{3}$$

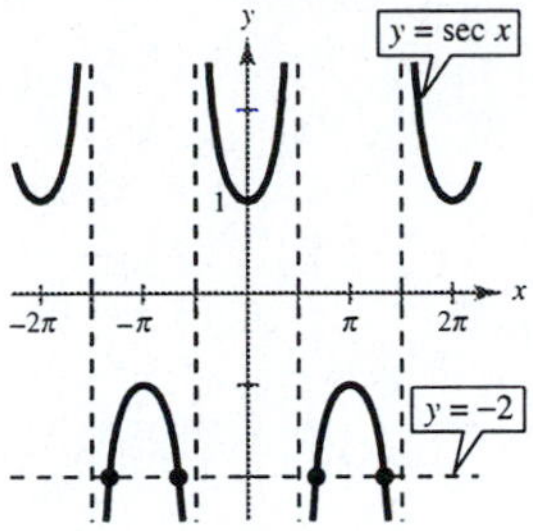

45. Thus graph of $f(x) = \sec x$ has y-axis symmetry. Thus, the function is even.

47. As $x \longrightarrow \frac{\pi}{2}$ from the left, $f(x) = \tan x \longrightarrow \infty$.

As $x \longrightarrow \frac{\pi}{2}$ from the right, $f(x) = \tan x \longrightarrow -\infty$.

49. $f(x) = 2 \sin x$

$g(x) = \frac{1}{2} \csc x$

(a)

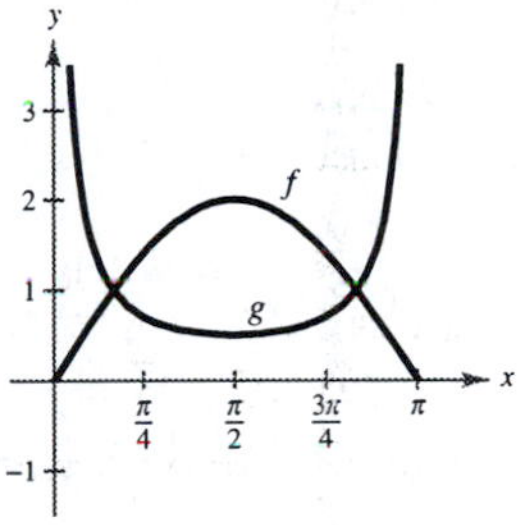

(b) $f > g$ on the interval, $\frac{\pi}{6} < x < \frac{5\pi}{6}$

(c) As $x \longrightarrow \pi, f(x) = 2 \sin x \longrightarrow 0$ and $g(x) = \frac{1}{2} \csc x \longrightarrow \pm\infty$ since $g(x)$ is the reciprocal of $f(x)$.

51. $y_1 = \sin x \csc x$ and $y_2 = 1$

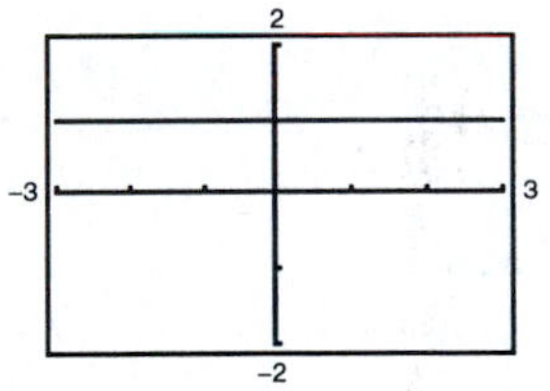

$$\sin x \csc x = \sin x \left(\frac{1}{\sin x}\right) = 1, \sin x \neq 0$$

53. $y_1 = \frac{\cos x}{\sin x}$ and $y_2 = \cot x = \frac{1}{\tan x}$

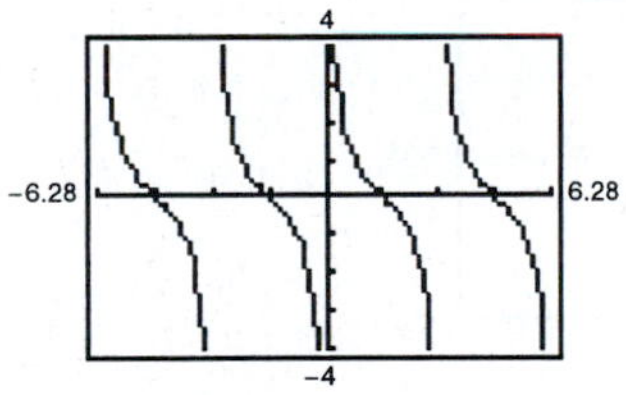

$$\cot x = \frac{\cos x}{\sin x}$$

55. $f(x) = x \cos x$

As $x \longrightarrow 0, f(x) \longrightarrow 0$.

Matches graph (d).

57. $g(x) = |x| \sin x$

As $x \longrightarrow 0$, $g(x) \longrightarrow 0$.

Matches graph (b).

59. $f(x) = \sin x + \cos\left(x + \dfrac{\pi}{2}\right),\ g(x) = 0$

$f(x) = g(x)$ The graph is the line $y = 0$.

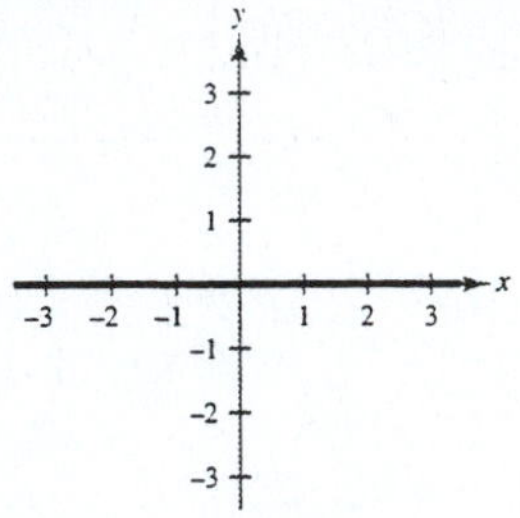

61. $f(x) = \sin^2 x,\ g(x) = \dfrac{1}{2}(1 - \cos 2x)$

$f(x) = g(x)$

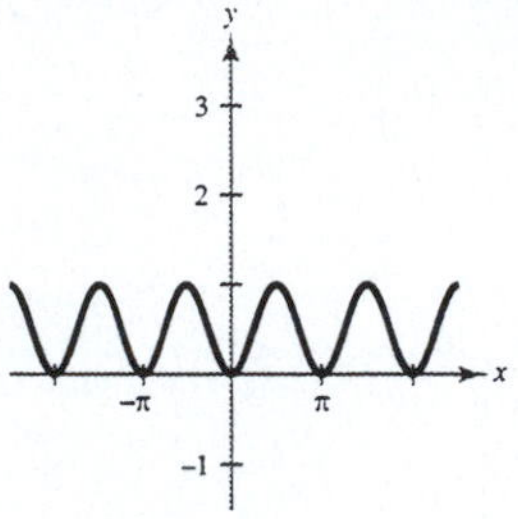

63. $f(x) = 2^{-x/4} \cos \pi x$

$-2^{-x/4} \le f(x) \le 2^{-2x/4}$

The damping factor is $y = 2^{-x/4}$.

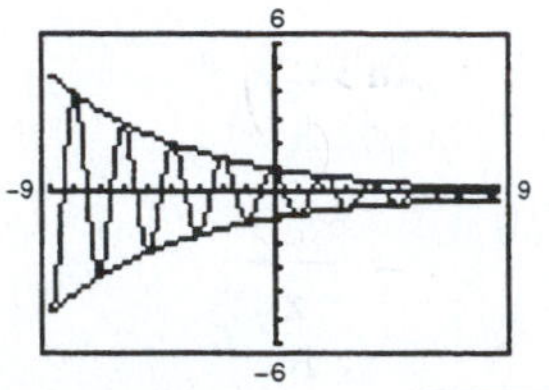

65. $g(x) = e^{-x^2/2} \sin x$

$-e^{-x^2/2} \le g(x) \le e^{-x^2/2}$

The damping factor is $y = e^{-x^2/2}$.

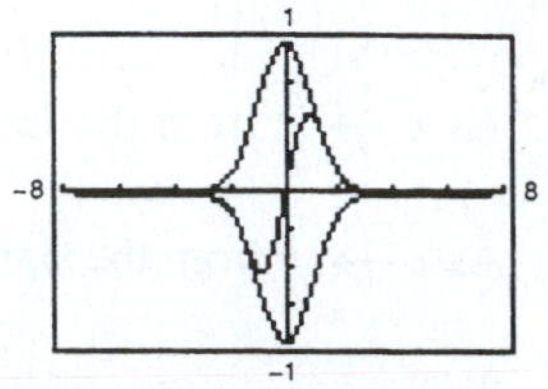

67. $\tan x = \dfrac{5}{d}$

$d = \dfrac{5}{\tan x} = 5 \cot x$

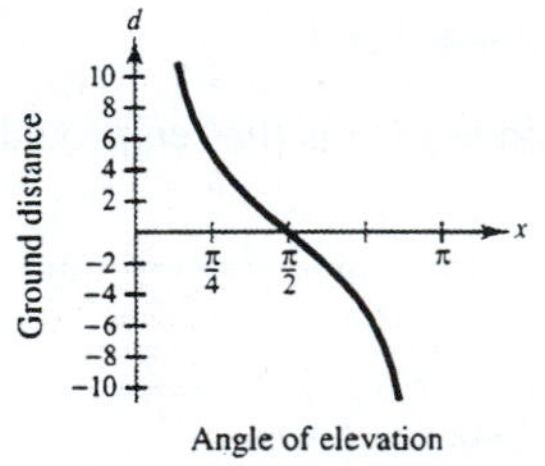

69. As the predator population increases, the number of prey decrease. When the number of prey is small, the number of predators decreases.

71. (a)

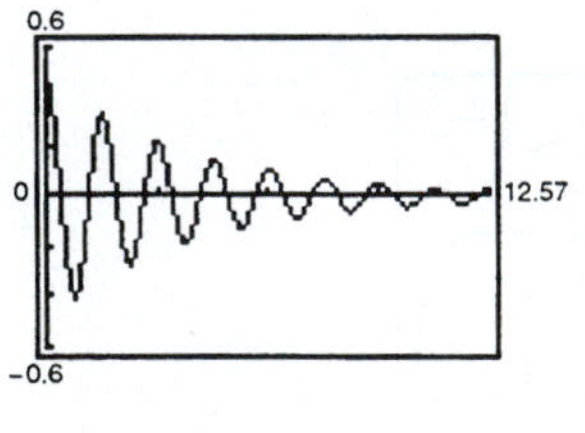

(b) The displacement function is a damped sine wave. It approaches 0 as t increases.

73. $\tan x \approx x + \dfrac{2x^3}{3!} + \dfrac{16x^5}{5!}$

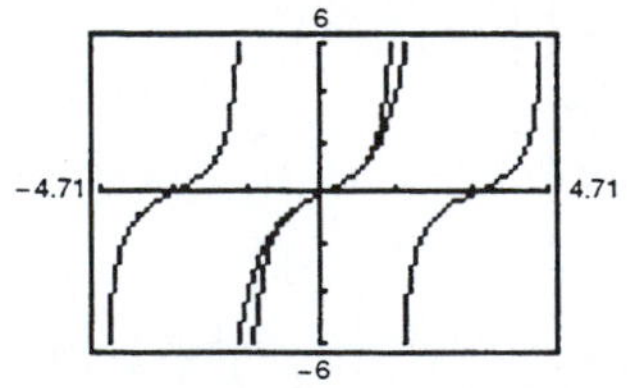

The graphs are approximately the same when x is near zero. As x gets larger, the graphs are further apart.

75. (a) $y_1 = \dfrac{4}{\pi}\left(\sin \pi x + \dfrac{1}{3}\sin 3\pi x\right)$

$y_2 = \dfrac{4}{\pi}\left(\sin \pi x + \dfrac{1}{3}\sin 3\pi x + \dfrac{1}{5}\sin 5\pi x\right)$

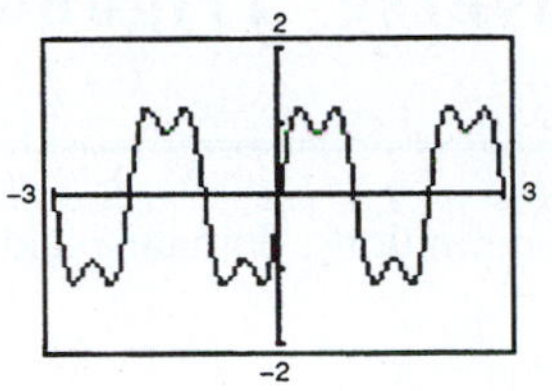

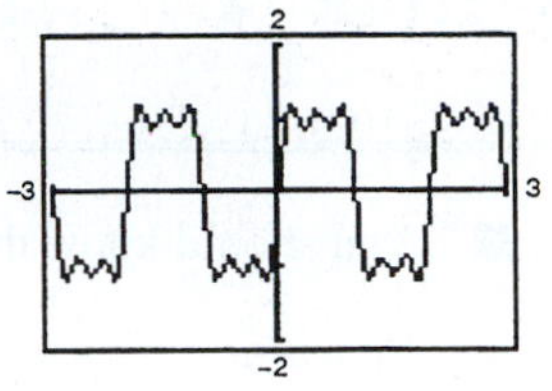

(b) $y_3 = \dfrac{4}{\pi}\left(\sin \pi x + \dfrac{1}{3}\sin 3\pi x + \dfrac{1}{5}\sin 5\pi x + \dfrac{1}{7}\sin 7\pi x\right)$

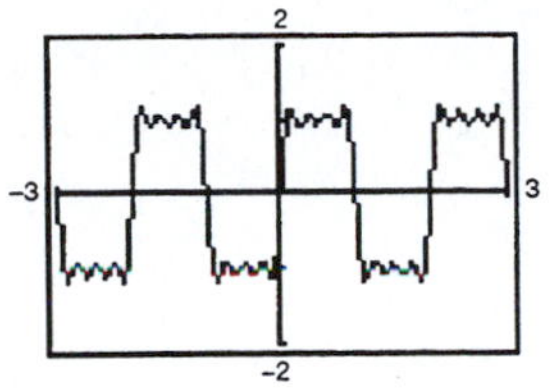

(c) $y_4 = \dfrac{4}{\pi}\left(\sin \pi x + \dfrac{1}{3}\sin 3\pi x + \dfrac{1}{5}\sin 5\pi x + \dfrac{1}{7}\sin 7\pi x + \dfrac{1}{9}\sin 9\pi x\right)$

77. $y = \dfrac{6}{x} + \cos x,\ x > 0$

As $x \longrightarrow 0,\ y \longrightarrow \infty$.

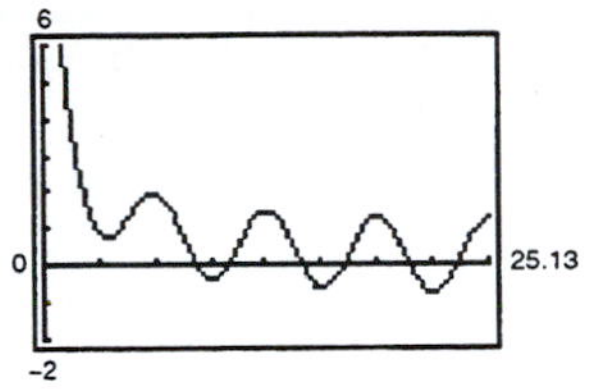

79. $g(x) = \dfrac{\sin x}{x}$

As $x \longrightarrow 0,\ g(x) \longrightarrow 1$.

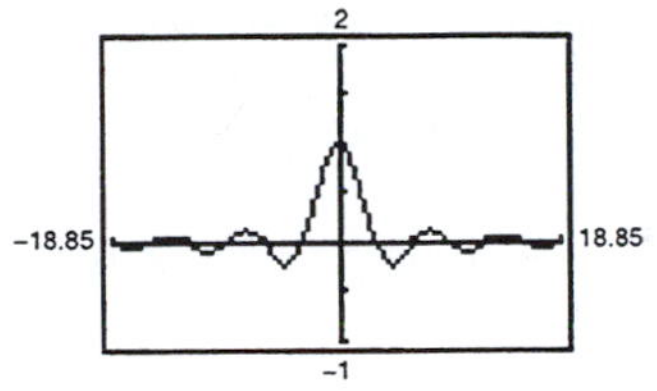

81. $f(x) = \sin \dfrac{1}{x}$

As $x \longrightarrow 0, f(x)$ oscillates between -1 and 1.

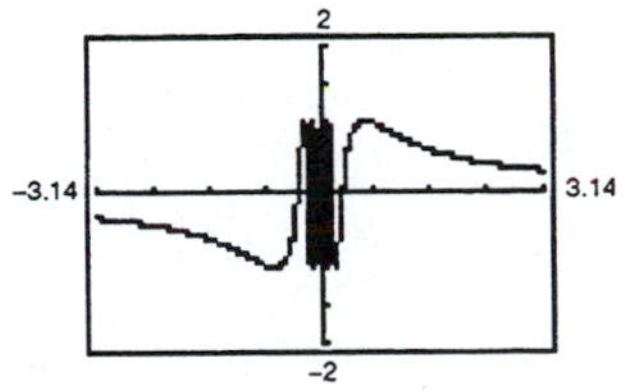

83. $e^{2x} = 54$

$2x = \ln 54$

$x = \dfrac{\ln 54}{2} \approx 1.994$

85. $\ln(x^2 + 1) = 3.2$

$x^2 + 1 = e^{3.2}$

$x^2 = e^{3.2} - 1$

$x = \pm\sqrt{e^{3.2} - 1} \approx \pm 4.851$

Section 4.7 Inverse Trigonometric Functions

- You should know the definitions, domains, and ranges of $y = \arcsin x$, $y = \arccos x$, and $y = \arctan x$.

Function	Domain	Range
$y = \arcsin x \Rightarrow x = \sin y$	$-1 \le x \le 1$	$-\frac{\pi}{2} \le y \le \frac{\pi}{2}$
$y = \arccos x \Rightarrow x = \cos y$	$-1 \le x \le 1$	$0 \le y \le \pi$
$y = \arctan x \Rightarrow x = \tan y$	$-\infty < x < \infty$	$-\frac{\pi}{2} < x < \frac{\pi}{2}$

- You should know the inverse properties of the inverse trigonometric functions.

  $\sin(\arcsin x) = x$ and $\arcsin(\sin y) = y, -\frac{\pi}{2} \le y \le \frac{\pi}{2}$

  $\cos(\arccos x) = x$ and $\arccos(\cos y) = y, 0 \le y \le \pi$

  $\tan(\arctan x) = x$ and $\arctan(\tan y) = y, -\frac{\pi}{2} < y < \frac{\pi}{2}$

- You should be able to use the triangle technique to convert trigonometric functions of inverse trigonometric functions into algebraic expressions.

Solutions to Odd-Numbered Exercises

1. False, $\frac{5\pi}{6}$ is not in the range of the arcsine function.

3. $y = \arcsin \frac{1}{2} \Rightarrow \sin y = \frac{1}{2}$ for $-\frac{\pi}{2} \le y \le \frac{\pi}{2} \Rightarrow y = \frac{\pi}{6}$

5. $y = \arccos \frac{1}{2} \Rightarrow \cos y = \frac{1}{2}$ for $0 \le y \le \pi \Rightarrow y = \frac{\pi}{3}$

7. $y = \arctan \frac{\sqrt{3}}{3} \Rightarrow \tan y = \frac{\sqrt{3}}{3}$ for $-\frac{\pi}{2} < y < \frac{\pi}{2} \Rightarrow y = \frac{\pi}{6}$

9. $y = \arccos\left(-\frac{\sqrt{3}}{2}\right) \Rightarrow \cos y = -\frac{\sqrt{3}}{2}$ for $0 \le y \le \pi \Rightarrow y = \frac{5\pi}{6}$

11. $y = \arctan(-\sqrt{3}) \Rightarrow \tan y = -\sqrt{3}$ for $-\frac{\pi}{2} < y < \frac{\pi}{2} \Rightarrow y = -\frac{\pi}{3}$

13. $y = \arccos\left(-\frac{1}{2}\right) \Rightarrow \cos y = -\frac{1}{2}$ for $0 \le y \le \pi \Rightarrow y = \frac{2\pi}{3}$

15. $y = \arcsin \frac{\sqrt{3}}{2} \Rightarrow \sin y = \frac{\sqrt{3}}{2}$ for $-\frac{\pi}{2} \le y \le \frac{\pi}{2} \Rightarrow y = \frac{\pi}{3}$

17. $y = \arctan 0 \Rightarrow \tan y = 0$ for $-\frac{\pi}{2} < y < \frac{\pi}{2} \Rightarrow y = 0$

19. $\arccos 0.28 = \cos^{-1} 0.28 \approx 1.29$

21. $\arcsin(-0.75) = \sin^{-1}(-0.75) \approx -0.85$

23. $\arctan(-3) = \tan^{-1}(-3) \approx -1.25$

25. $\arcsin 0.31 = \sin^{-1} 0.31 \approx 0.32$

27. $\arccos(-0.41) = \cos^{-1}(-0.41) \approx 1.99$

29. $\arctan 0.92 = \tan^{-1} 0.92 \approx 0.74$

31. This is the graph of $y = \arctan x$. The coordinates are $\left(-\sqrt{3}, -\frac{\pi}{3}\right)$, $\left(-\frac{1}{\sqrt{3}}, -\frac{\pi}{6}\right)$, and $\left(1, \frac{\pi}{4}\right)$.

33. $f(x) = \tan x$ and $g(x) = \arctan x$

Graph $y_1 = \tan x$

$y_2 = \tan^{-1} x$

$y_3 = x$

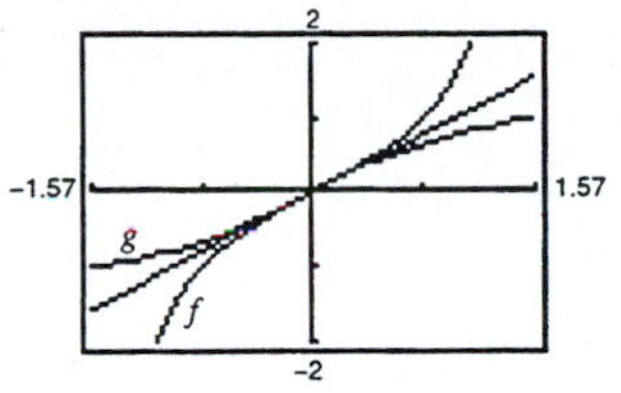

35. $\tan\theta = \frac{x}{4}$

$\theta = \arctan \frac{x}{4}$

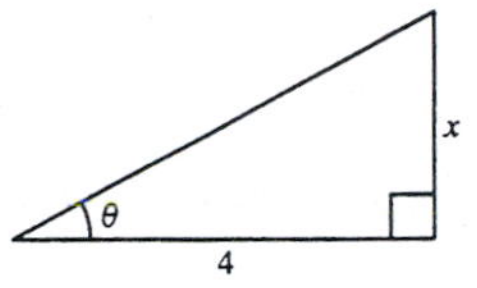

37. $\sin\theta = \frac{x+2}{5}$

$\theta = \arcsin\left(\frac{x+2}{5}\right)$

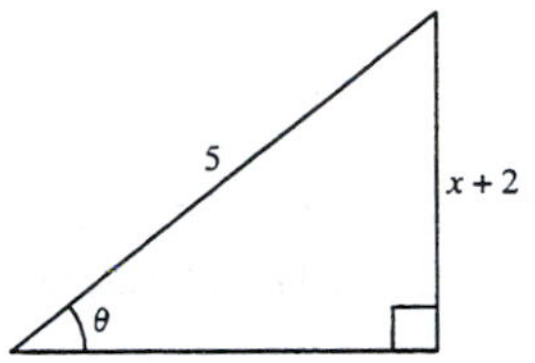

39. $\sin(\arcsin 0.3) = 0.3$

41. $\cos[\arccos(-0.1)] = -0.1$

43. $\arcsin(\sin 3\pi) = \arcsin(0) = 0$

Note: 3π is not in the range of the arcsine function.

45. Let $y = \arctan \frac{3}{4}$. Then,

$\tan y = \frac{3}{4}, \; 0 < y < \frac{\pi}{2}$

and $\sin y = \frac{3}{5}$.

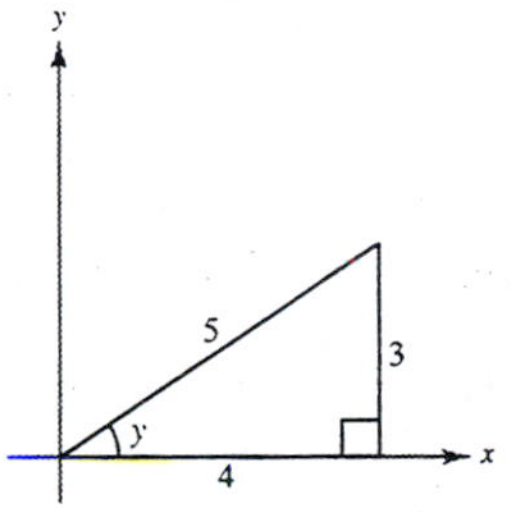

47. Let $y = \arctan 2$. Then,

$\tan y = 2 = \frac{2}{1}, \; 0 < y < \frac{\pi}{2}$

and $\cos y = \frac{1}{\sqrt{5}} = \frac{\sqrt{5}}{5}$.

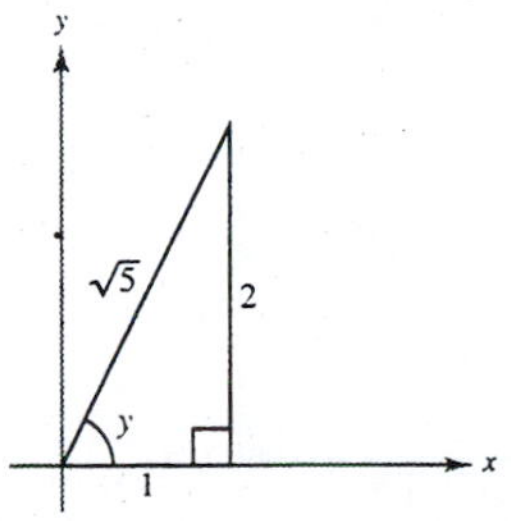

49. Let $y = \arcsin \frac{5}{13}$. Then,

$$\sin y = \frac{5}{13},\ 0 < y < \frac{\pi}{2}$$

and $\cos y = \frac{12}{13}$.

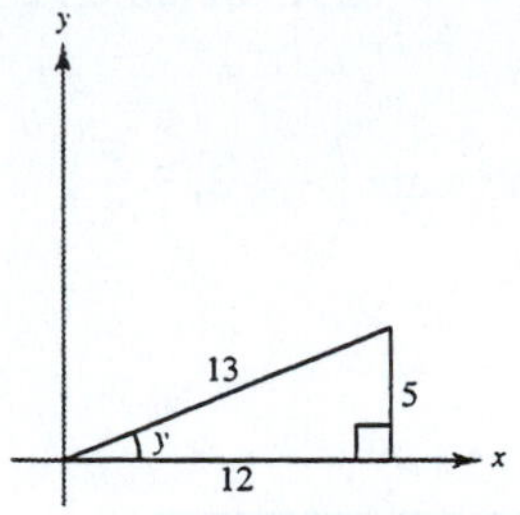

51. Let $y = \arctan\left(-\frac{3}{5}\right)$. Then,

$$\tan y = -\frac{3}{5},\ -\frac{\pi}{2} < y < 0$$

and $\sec y = \frac{\sqrt{34}}{5}$.

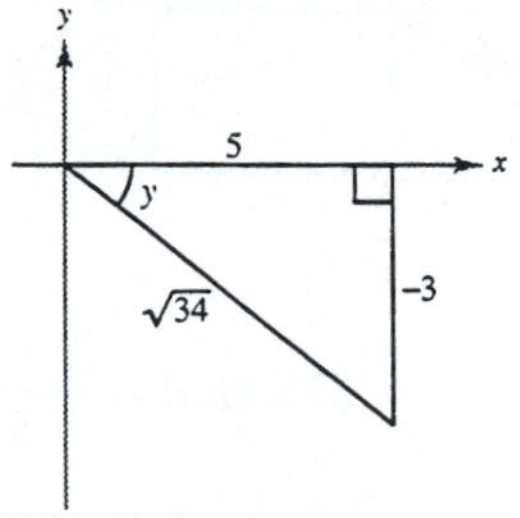

53. Let $y = \arccos\left(-\frac{2}{3}\right)$. Then,

$$\cos y = -\frac{2}{3},\ \frac{\pi}{2} < y < \pi$$

and $\sin y = \frac{\sqrt{5}}{3}$.

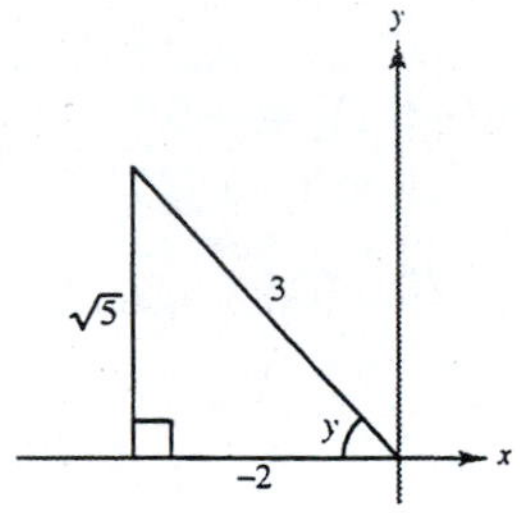

55. Let $y = \arctan x$. Then,

$$\tan y = x = \frac{x}{1}$$

and $\cot y = \frac{1}{x}$.

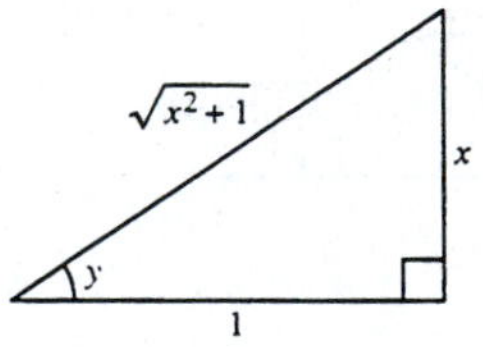

57. Let $y = \arcsin(2x)$. Then,

$$\sin y = 2x = \frac{2x}{1}$$

and $\cos y = \sqrt{1 - 4x^2}$.

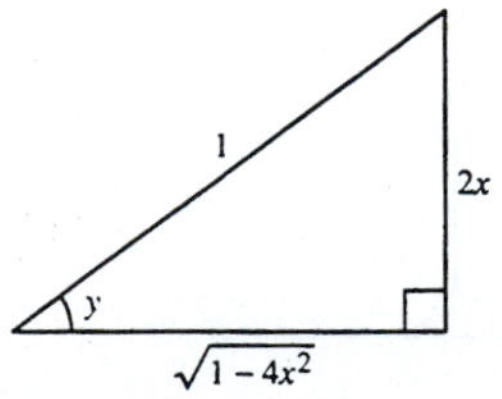

59. Let $y = \arccos x$. Then,

$$\cos y = x = \frac{x}{1}$$

and $\sin y = \sqrt{1 - x^2}$.

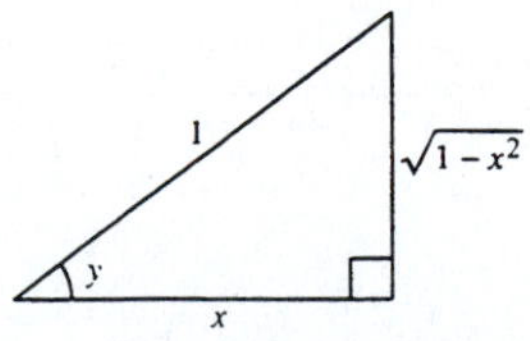

61. Let $y = \arccos\left(\frac{x}{3}\right)$. Then,

$$\cos y = \frac{x}{3}$$

and $\tan y = \dfrac{\sqrt{9 - x^2}}{x}$.

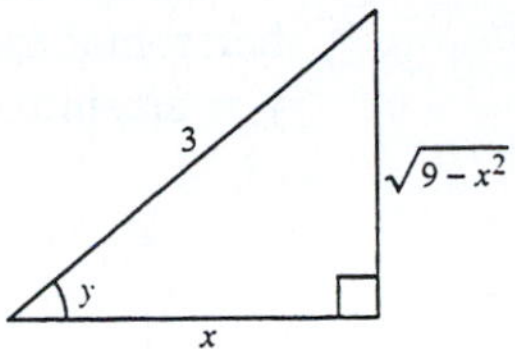

63. Let $y = \arctan \dfrac{x}{\sqrt{2}}$. Then,

$$\tan y = \frac{x}{\sqrt{2}}$$

and $\csc y = \dfrac{\sqrt{x^2 + 2}}{x}$.

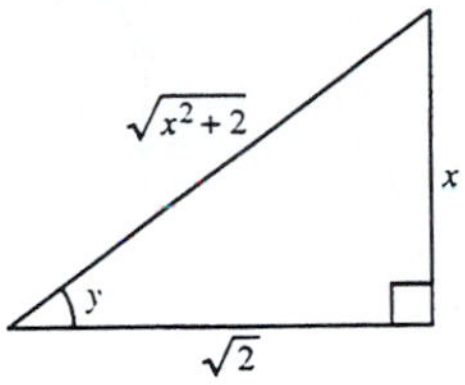

65. $f(x) = \sin(\arctan 2x)$, $g(x) = \dfrac{2x}{\sqrt{1 - 4x^2}}$

Let $y = \arctan 2x$. Then,

$$\tan y = 2x = \frac{2x}{1}$$

and $\sin y = \dfrac{2x}{\sqrt{1 + 4x^2}}$.

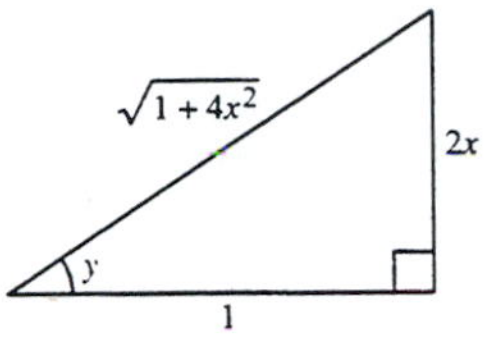

$$g(x) = \frac{2x}{\sqrt{1 + 4x^2}} = f(x)$$

The graph has horizontal asymptotes at $y = \pm 1$.

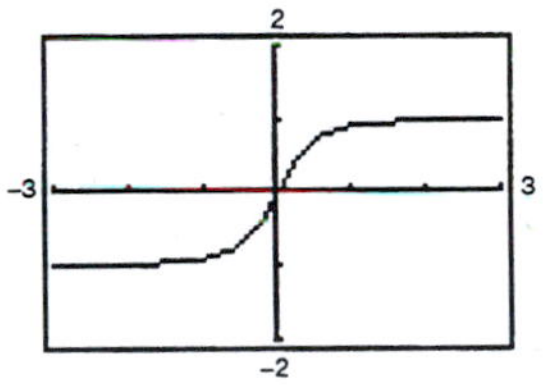

67. Let $y = \arctan \dfrac{9}{x}$. Then,

$$\tan y = \frac{9}{x} \text{ and } \sin y = \frac{9}{\sqrt{x^2 + 81}}.$$

Thus, $\arcsin y = \dfrac{9}{\sqrt{x^2 + 81}}$.

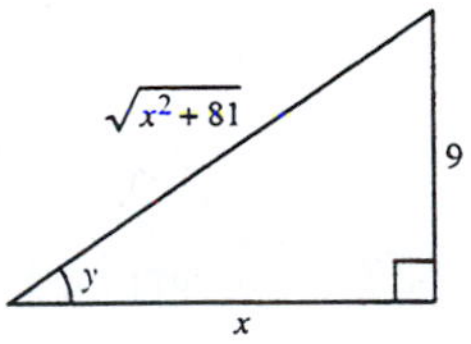

69. Let $y = \arccos \dfrac{3}{\sqrt{x^2 - 2x + 10}}$. Then,

$$\cos y = \frac{3}{\sqrt{x^2 - 2x + 10}} = \frac{3}{\sqrt{(x - 1)^2 + 9}}$$

and $\sin y = \dfrac{|x - 1|}{\sqrt{(x - 1)^2 + 9}}$.

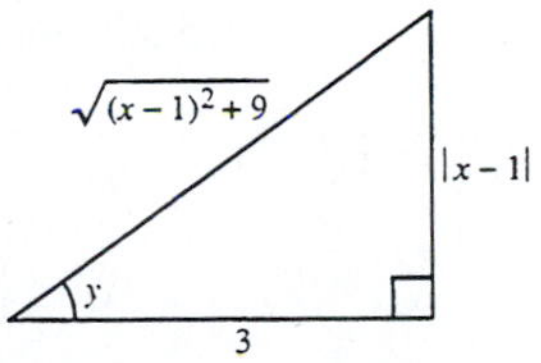

Thus, $\arcsin y = \dfrac{|x - 1|}{\sqrt{(x - 1)^2 + 9}} = \arcsin \dfrac{|x - 1|}{\sqrt{x^2 - 2x + 10}}$.

71. $y = 2 \arccos x$

Domain: $-1 \le x \le 1$

Range: $0 \le y \le 2\pi$

Vertical stretch of $f(x) = \arccos x$

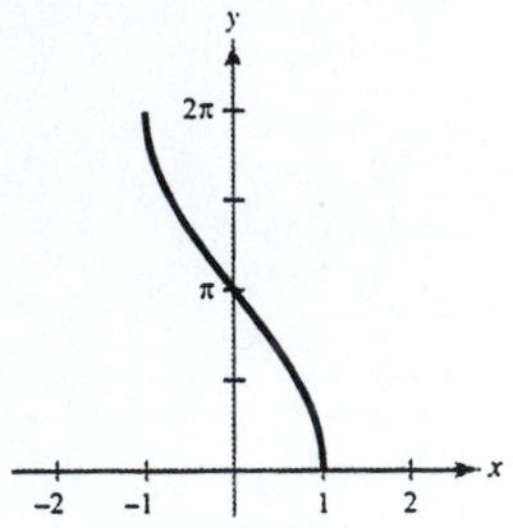

73. The graph of $f(x) = \arcsin(x - 1)$ is a horizontal translation of the graph of $y = \arcsin x$ by one unit.

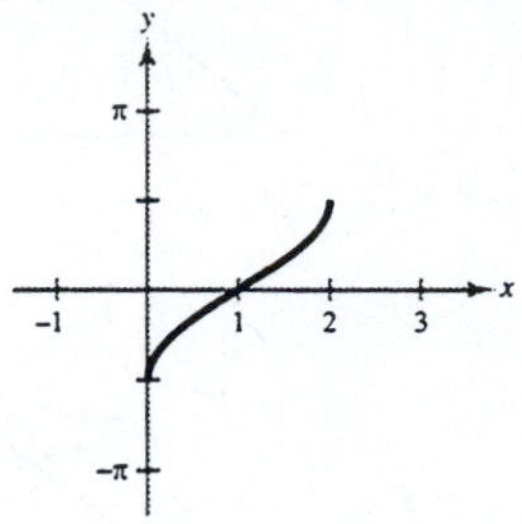

75. $f(x) = \arctan 2x$

Domain: all real numbers

Range: $-\frac{\pi}{2} < y < \frac{\pi}{2}$

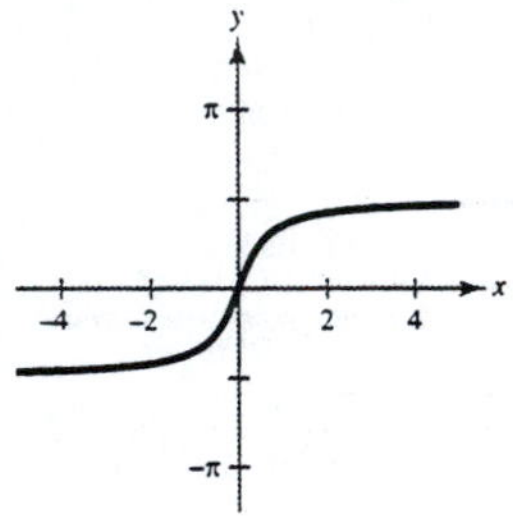

77. $h(v) = \tan(\arccos v) = \dfrac{\sqrt{1 - v^2}}{v}$

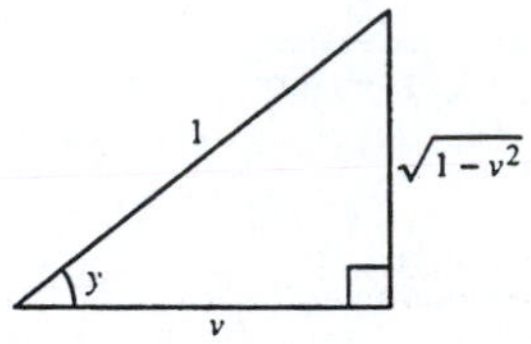

Domain: $-1 \le v \le 1, v \ne 0$

Range: all real numbers

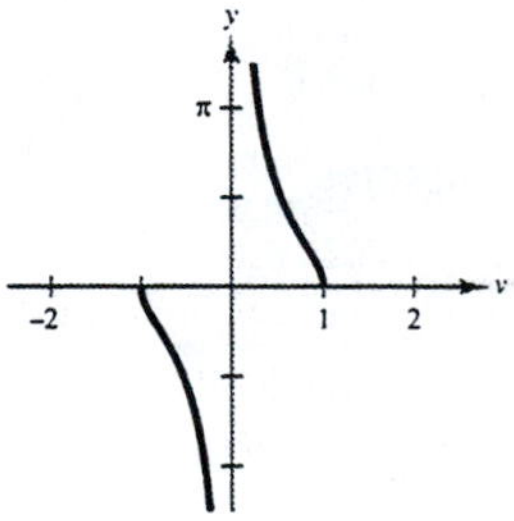

79. $f(t) = 3 \cos 2t + 3 \sin 2t = \sqrt{3^2 + 3^2} \sin\left(3t + \arctan \frac{3}{3}\right)$

$= 3\sqrt{2} \sin(3t + \arctan 1)$

$= 3\sqrt{2} \sin\left(3t + \frac{\pi}{4}\right)$

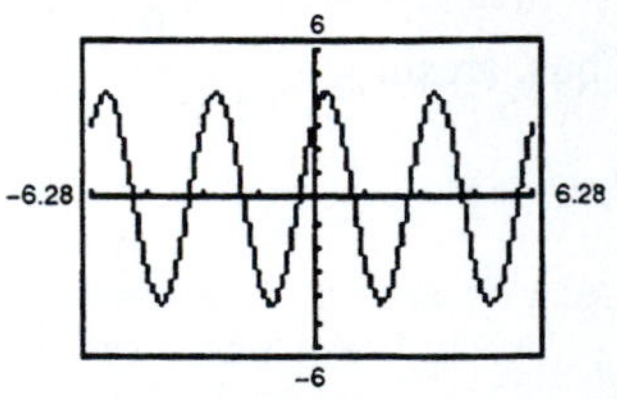

The graphs are the same.

81. $f(x) = \sin x,\ f^{-1}(x) = \arcsin x$

(a) $f \circ f^{-1} = f(f^{-1}(x)) = f(\arcsin x) = \sin(\arcsin x)$

$f^{-1} \circ f = f^{-1}(f(x)) = f^{-1}(\sin x) = \arcsin(\sin x)$

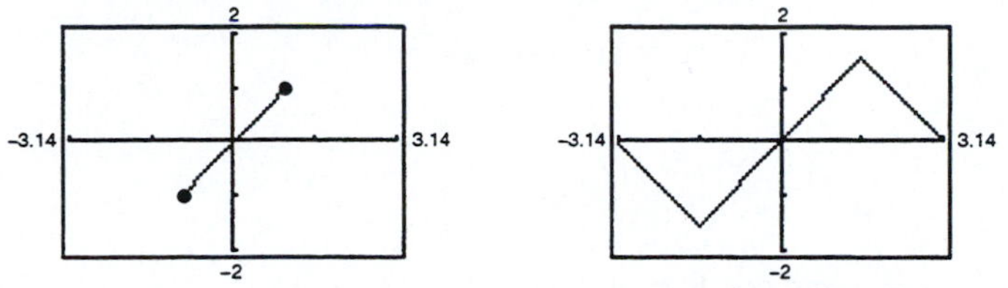

(b) Both the domain and range of $f \circ f^{-1} = \sin(\arcsin x)$ are the intervals of $[-1, 1]$.

The domain of $f^{-1} \circ f$ is all real numbers. The range is the interval $\left[-\frac{\pi}{2}, \frac{\pi}{2}\right]$.

Neither graph is the line $y = x$ because of these domain/range restrictions.

83. (a) $\sin\theta = \frac{10}{s}$

$\theta = \arcsin\frac{10}{s}$

(b) $s = 48$: $\theta = \arcsin\frac{10}{48} \approx 0.21$

$s = 24$: $\theta = \arcsin\frac{10}{24} \approx 0.43$

85. $\beta = \arctan\frac{3x}{x^2 + 4}$

(a)

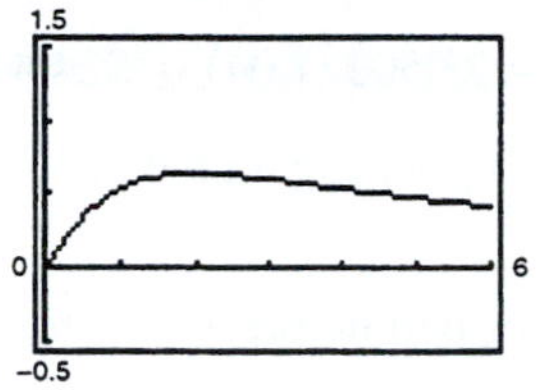

(b) β is maximum when $x = 2$.

(c) The graph has a horizontal asymptote at $\beta = 0$. As x increases, β decreases.

87. (a) $\tan\theta = \frac{5}{x}$

$\theta = \arctan\frac{5}{x}$

(b) $x = 10$: $\theta = \arctan\frac{5}{10} \approx 26.6°$

$x = 3$: $\theta = \arctan\frac{5}{3} \approx 59.0°$

89. $y = \operatorname{arccot} x$ if and only if $\cot y = x$.

Domain: $-\infty < x < \infty$

Range: $0 < x < \pi$

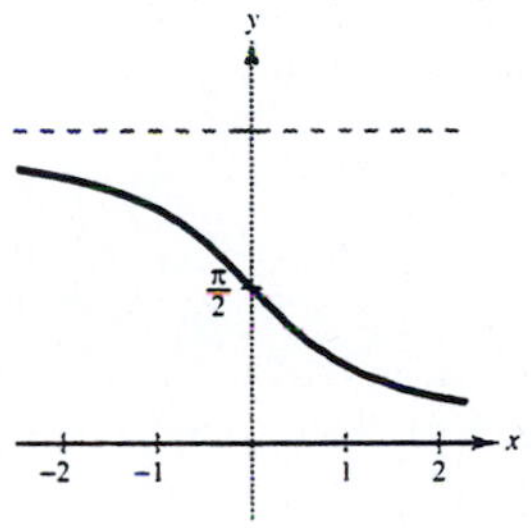

91. $y = \operatorname{arccsc} x$ if and only if $\csc y = x$.

Domain: $(-\infty, -1] \cup [1, \infty)$

Range: $\left[-\frac{\pi}{2}, 0\right) \cup \left(0, \frac{\pi}{2}\right]$

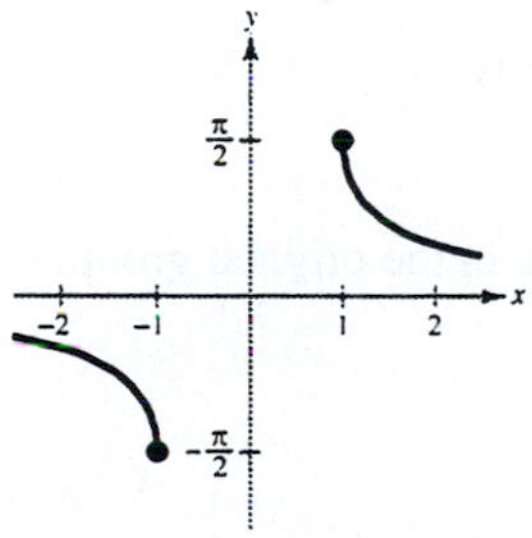

93. Let $y = \arcsin(-x)$. Then,

$$\sin y = -x$$
$$-\sin y = x$$
$$\sin(-y) = x$$
$$-y = \arcsin x$$
$$y = -\arcsin x.$$

Therefore, $\arcsin(-x) = -\arcsin x$.

95.
$$y = \pi - \arccos x$$
$$\cos y = \cos(\pi - \arccos x)$$
$$\cos y = \cos \pi \cos(\arccos x) + \sin \pi \sin(\arccos x)$$
$$\cos y = -x$$
$$y = \arccos(-x)$$

97. Let $\alpha = \arcsin x$ and $\beta = \arccos x$, then $\sin \alpha = x$ and $\cos \beta = x$. Thus, $\sin \alpha = \cos \beta$ which implies that α and β are complementary angles and we have

$$\alpha + \beta = \frac{\pi}{2}$$
$$\arcsin x + \arccos x = \frac{\pi}{2}.$$

99. Now: Cost = 23,500 + 725 = \$24,225

Wait a month: Cost = 23,500 (1.04) = \$24,440

The customer should buy now and save \$215.

101. Let x = the number of people presently in the group. Each person's share is now $\dfrac{250{,}000}{x}$.

If two more join the group, each person's share would then be $\dfrac{250{,}000}{x+2}$.

$$\text{Share per person with two more people} = \text{Original share per person} - 6250$$

$$\frac{250{,}000}{x+2} = \frac{250{,}000}{x} - 6250$$
$$250{,}000x = 250{,}000(x+2) - 6250x(x+2)$$
$$250{,}000x = 250{,}000x + 500{,}000 - 6250x^2 - 12500x$$

$$6250x^2 + 12500x - 500{,}000 = 0$$
$$6250(x^2 + 2x - 80) = 0$$
$$6250(x+10)(x-8) = 0$$
$$x = -10 \quad \text{or} \quad x = 8$$

Not possible

There were 8 people in the original group.

Section 4.8 Applications and Models

Solutions to Odd-Numbered Exercises

- You should be able to solve right triangles.
- You should be able to solve right triangle applications.
- You should be able to solve applications of simple harmonic motion.

1. Given: $A = 20°$, $b = 10$

$$\tan A = \frac{a}{b} \implies a = b \tan A = 10 \tan 20° \approx 3.64$$

$$\cos A = \frac{a}{c} \implies c = \frac{a}{\cos A} = \frac{10}{\cos 20°} \approx 10.64$$

$$B = 90° - 20° = 70°$$

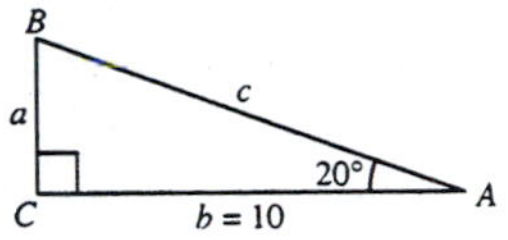

3. Given: $B = 71°$, $b = 24$

$$\tan B = \frac{b}{a} \implies a = \frac{b}{\tan B} = \frac{24}{\tan 71°} \approx 8.26$$

$$\sin B = \frac{b}{c} \implies c = \frac{b}{\sin B} = \frac{24}{\sin 71°} \approx 25.38$$

$$A = 90° - 71° = 19°$$

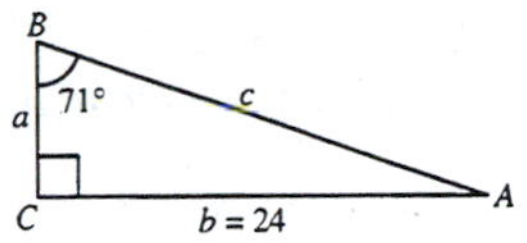

5. Given: $a = 6$, $b = 10$

$$c^2 = a^2 + b^2 \implies c = \sqrt{36 + 100}$$
$$= 2\sqrt{34} \approx 11.66$$

$$\tan A = \frac{a}{b} = \frac{6}{10} \implies A = \arctan\frac{3}{5} \approx 30.96°$$

$$B = 90° - 30.96° = 59.04°$$

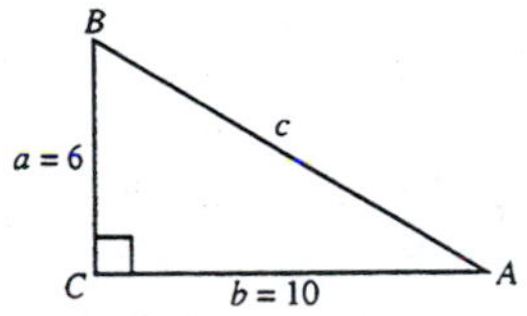

7. $b = 16$, $c = 52$

$$a = \sqrt{52^2 - 16^2}$$
$$= \sqrt{2448} = 12\sqrt{17} \approx 49.48$$

$$\cos A = \frac{16}{52}$$

$$A = \arccos\frac{16}{52} \approx 72.08°$$

$$B = 90° - 72.08° \approx 17.92°$$

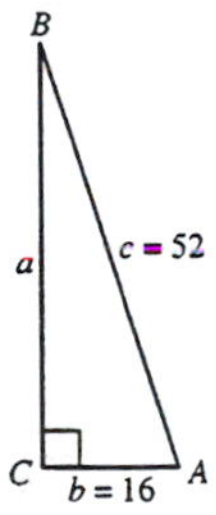

9. $A = 12°15'$, $c = 430.5$

$$B = 90° - 12°15' = 77°45'$$

$$\sin 12°15' = \frac{a}{430.5}$$

$$a = 430.5 \sin 12°15' \approx 91.34$$

$$\cos 12°15' = \frac{b}{430.5}$$

$$b = 430.5 \cos 12°15' \approx 420.70$$

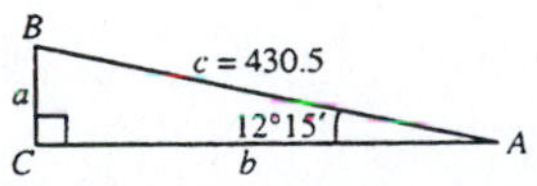

11. $\tan\theta = \dfrac{h}{1/2b} \implies h = \dfrac{1}{2}b\tan\theta$

$$h = \frac{1}{2}(4)\tan 52° \approx 2.56 \text{ in.}$$

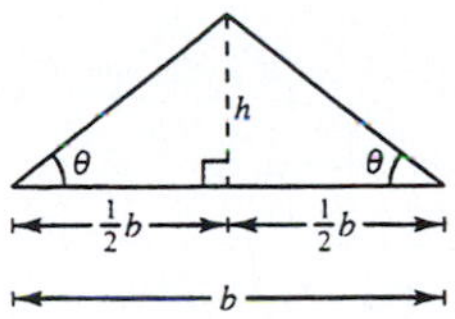

13. $\tan 30° = \dfrac{60}{x}$

$\dfrac{1}{\sqrt{3}} = \dfrac{60}{x}$

$x = 60\sqrt{3}$

≈ 103.9 feet

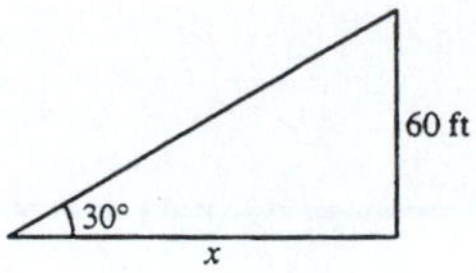

15. $\sin 74° = \dfrac{h}{16}$

$16 \sin 74° = h$

$h \approx 15.4$ feet

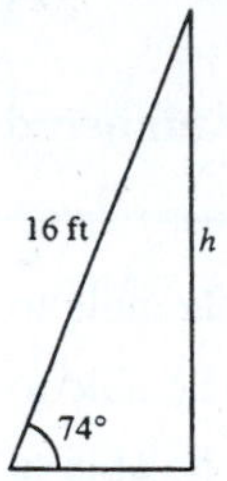

17. (a)

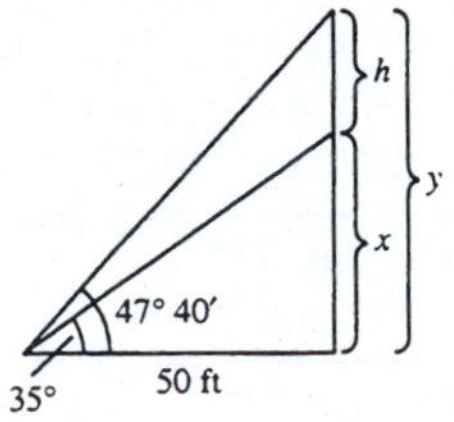

(b) Let the height of the church = x and the height of the church and steeple = y. Then,

$\tan 35° = \dfrac{x}{50}$ and $\tan 47°40' = \dfrac{y}{50}$

$x = 50 \tan 35°$ and $y = 50 \tan 47°40'$

$h = y - x = 50\,(\tan 47°40' - \tan 35°)$.

(c) $h \approx 19.9$ feet

19. $\sin 34° = \dfrac{x}{4000}$

$x = 4000 \sin 34°$

≈ 2236.8 feet

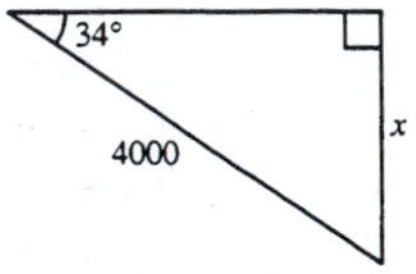

21. $\tan \theta = \dfrac{75}{50}$

$\theta = \arctan \dfrac{3}{2} \approx 56.3°$

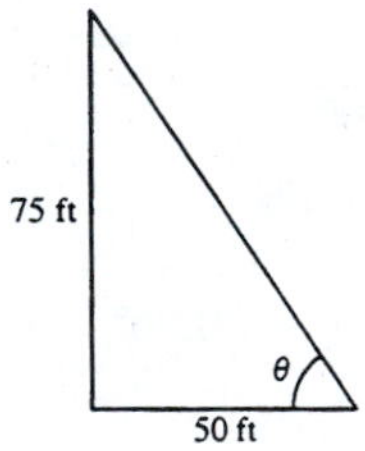

23. $\sin \theta = \dfrac{4000}{4150}$

$\theta = \arcsin\left(\dfrac{4000}{4150}\right)$

$\theta \approx 74.5°$

$\alpha = 90° - 74.5° = 15.5°$

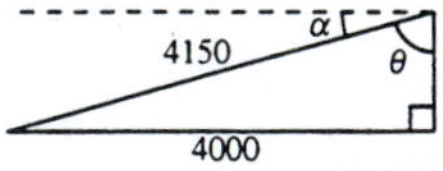

25. Since the airplane speed is

$\left(275\dfrac{\text{ft}}{\text{sec}}\right)\left(60\dfrac{\text{sec}}{\text{min}}\right) = 16{,}500\dfrac{\text{ft}}{\text{min}},$

after one minute its distance travelled in 16,500 feet.

$\sin 18° = \dfrac{a}{16{,}500}$

$a = 16{,}500 \sin 18°$

≈ 5099 ft

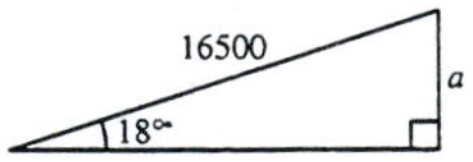

27. $\sin 10.5° = \dfrac{x}{4}$

$x = 4 \sin 10.5°$

≈ 0.73 mile

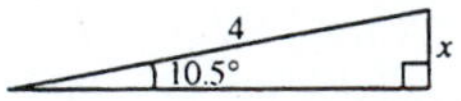

29. The plane has traveled 1.5 (550)= 825 miles.

$$\sin 38° = \frac{a}{825} \implies a \approx 508 \text{ miles north}$$

$$\cos 38° = \frac{b}{825} \implies b \approx 650 \text{ miles east}$$

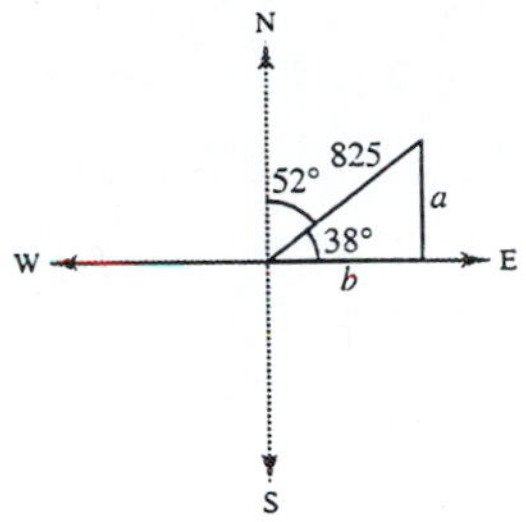

31. $\theta = 32°,\ \phi = 68°$

(a) $\alpha = 90° - 32° = 58°$

Bearing from A to C: N 58° E

(b) $\beta = \theta = 32°$

$\gamma = 90° - \phi = 22°$

$C = \beta + \gamma = 54°$

$$\tan C = \frac{d}{50} \implies \tan 54° = \frac{d}{50} \implies d \approx 68.82 \text{ meters}$$

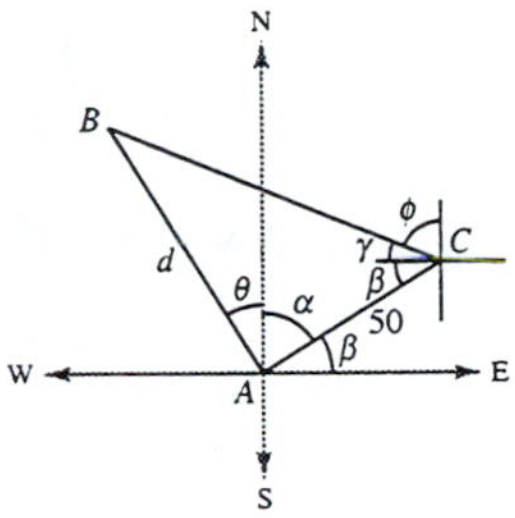

33. $\tan \theta = \frac{45}{30} \implies \theta \approx 56.3°$

Bearing: N 56.3° W

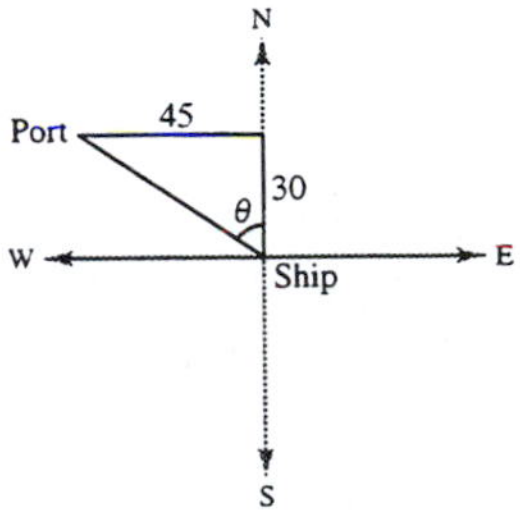

35. $\tan 6.5° = \dfrac{350}{d} \implies d \approx 3071.91$ ft

$\tan 4° = \dfrac{350}{D} \implies D \approx 5005.23$ ft

Distance between ships: $D - d \approx 1933.32$ ft

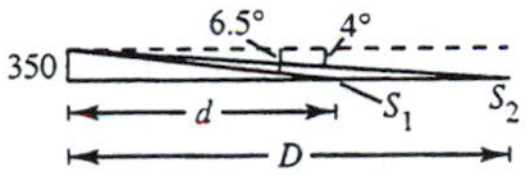

37.

$$\tan 57° = \frac{a}{x} \implies x = a \cot 57°$$

$$\tan 16° = \frac{a}{x + (55/6)}$$

$$\tan 16° = \frac{a}{a \cot 57° + (55/6)}$$

$$\cot 16° = \frac{a \cot 57° + (55/6)}{a}$$

$$a \cot 16° - a \cot 57° = \frac{55}{6} \implies a \approx 3.23 \text{ miles}$$

$$\approx 17,054 \text{ ft}$$

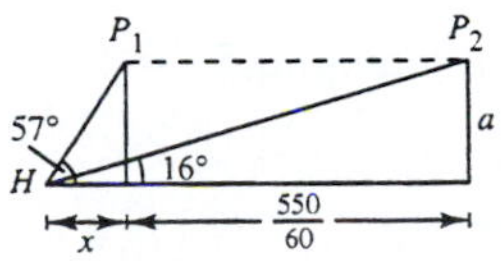

39. L_1: $3x - 2y = 5 \Rightarrow y = \frac{3}{2}x - \frac{5}{2} \Rightarrow m_1 = \frac{3}{2}$

L_2: $x - y = 1 \Rightarrow y = -x + 1 \Rightarrow m_2 = -1$

$$\tan\alpha = \left|\frac{-1 - \frac{3}{2}}{1 + (-1)\left(\frac{3}{2}\right)}\right| = \left|\frac{-\frac{5}{2}}{-\frac{1}{2}}\right| = 5$$

$\alpha = \arctan 5 \approx 78.7°$

41. The diagonal of the base has a length of $\sqrt{a^2 + a^2} = \sqrt{2}a$.

Now, we have $\tan\theta = \dfrac{a}{\sqrt{2}a} = \dfrac{1}{\sqrt{2}}$

$\theta = \arctan\dfrac{1}{\sqrt{2}}$

$\theta \approx 35.3°$.

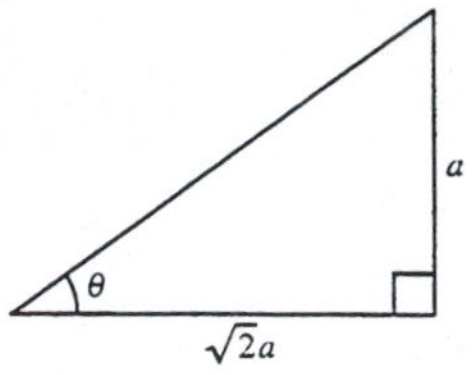

43. $\cos 30° = \dfrac{b}{r}$

$b = \cos 30° r$

$b = \dfrac{\sqrt{3}r}{2}$

$y = 2b = 2\left(\dfrac{\sqrt{3}r}{2}\right) = \sqrt{3}r$

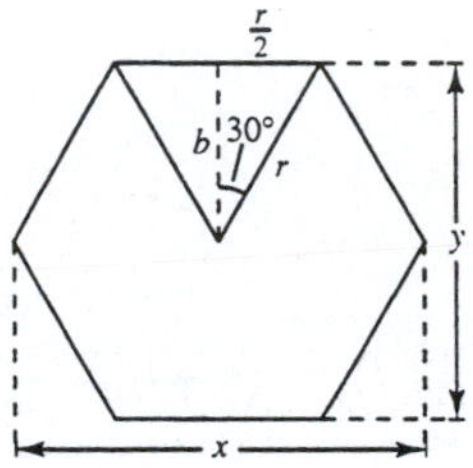

45. $\sin 36° = \dfrac{d}{25} \Rightarrow d \approx 14.69$

Length of side: $2d \approx 29.38$ inches

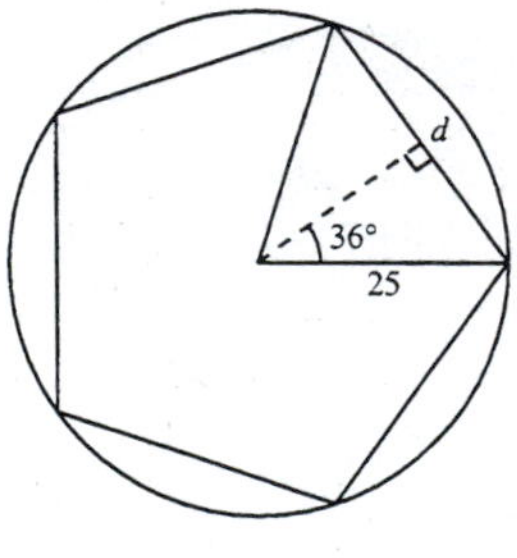

47. $\tan 35° = \dfrac{a}{10}$

$a = 10 \tan 35° \approx 7$

$\cos 33° = \dfrac{10}{c}$

$c = \dfrac{10}{\cos 35°} \approx 12.2$

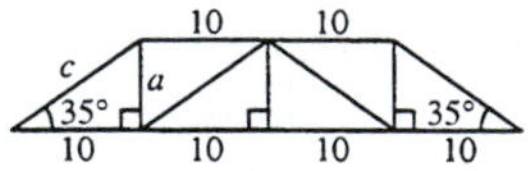

49. $d = 4 \cos 8\pi t$

(a) Maximum displacement = amplitude = 4

(b) Frequency $= \dfrac{\omega}{2\pi} = \dfrac{8\pi}{2\pi}$

$= 4$ cycles per unit of time

(c) $8\pi t = \dfrac{\pi}{2} \Rightarrow t = \dfrac{1}{16}$

51. $d = \dfrac{1}{16} \sin 120\pi t$

(a) Maximum displacement = amplitude $= \dfrac{1}{16}$

(b) Frequency $= \dfrac{\omega}{2\pi} = \dfrac{120\pi}{2\pi}$

$= 60$ cycles per unit of time

(c) $120\pi t = \pi \Rightarrow t = \dfrac{1}{120}$

53. $d = 0$ when $t = 0$, $a = 4$, Period $= 2$

Use $d = a \sin \omega t$ since $d = 0$ when $t = 0$.

$$\frac{2\pi}{\omega} = 2 \implies \omega = \pi$$

Thus, $d = 4 \sin \pi t$.

55. $d = 3$ when $t = 0$, $a = 3$, Period $= 1.5$

Use $d = a \cos \omega t$ since $d = 3$ when $t = 0$.

$$\frac{2\pi}{\omega} = 1.5 \implies \omega = \frac{4\pi}{3}$$

Thus, $d = 3 \cos\left(\frac{4\pi}{3}t\right) = 3 \cos\left(\frac{4\pi t}{3}\right)$.

57. $d = a \sin \omega t$

$$\text{Period} = \frac{2\pi}{\omega} = \frac{1}{\text{frequency}}$$

$$\frac{2\pi}{\omega} = \frac{1}{264}$$

$$\omega = 2\pi(264) = 528\pi$$

59. $y = \frac{1}{4} \cos 16t,\ t > 0$

(a)

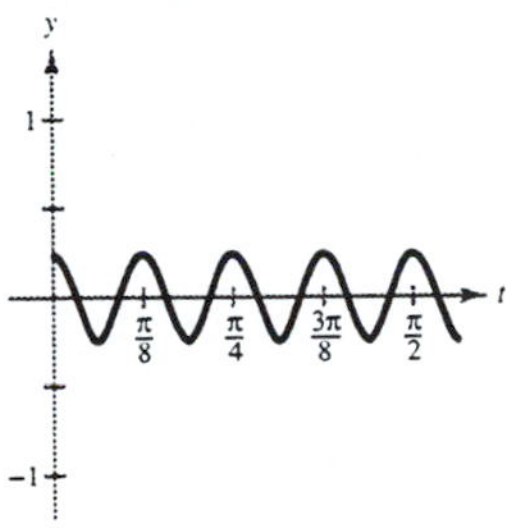

(b) Period: $\frac{2\pi}{16} = \frac{\pi}{8}$

(c) $\frac{1}{4} \cos 16t = 0$ when $16t = \frac{\pi}{2} \implies t = \frac{\pi}{32}$

61. (a) & (b)

Base 1	*Base 2*	*Altitude*	*Area*
8	$8 + 16 \cos 10°$	$8 \sin 10°$	22.1
8	$8 + 16 \cos 20°$	$8 \sin 20°$	42.5
8	$8 + 16 \cos 30°$	$8 \sin 30°$	59.7
8	$8 + 16 \cos 40°$	$8 \sin 40°$	72.7
8	$8 + 16 \cos 50°$	$8 \sin 50°$	80.5
8	$8 + 16 \cos 60°$	$8 \sin 60°$	83.1
8	$8 + 16 \cos 70°$	$8 \sin 70°$	80.7

The maximum occurs when $\theta = 60°$ and is approximately 83.1 square feet.

(c) $A(\theta) = [8 + (8 + 16 \cos \theta)]\left[\frac{8 \sin \theta}{2}\right]$

$$= (16 + 16 \cos \theta)(4 \sin \theta)$$

$$= 64\,(1 + \cos \theta)(\sin \theta)$$

(d)

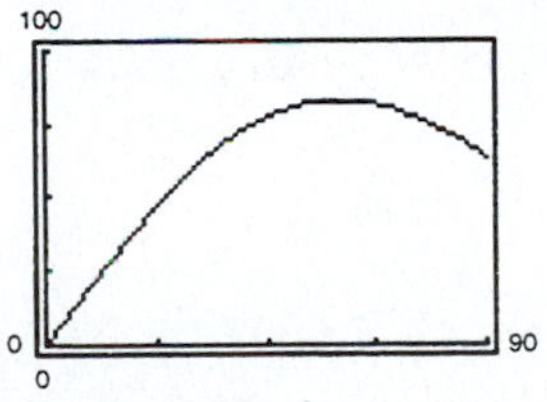

The maximum occurs when $\theta = \frac{\pi}{3} = 60°$.

63. (a)

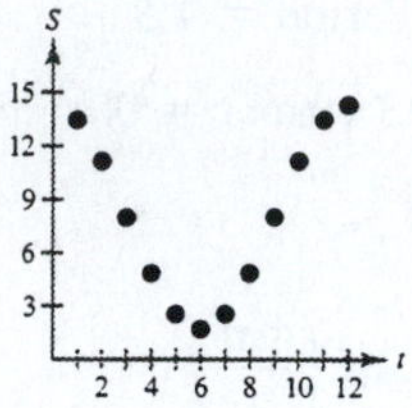

(b) $a = \frac{1}{2}(14.3 - 1.7) = 6.3$

$\frac{2\pi}{b} = 12 \Rightarrow b = \frac{\pi}{6}$

Shift: $d = 14.3 - 6.3 = 8$

$S = d + a \cos bt$

$S = 8 + 6.3 \cos\left(\frac{\pi t}{6}\right)$

The model is a good fit.

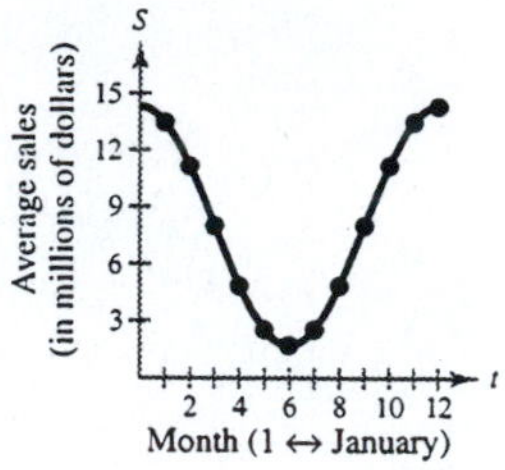

(c) Period: $\frac{2\pi}{\pi/6} = 12$

This corresponds to the 12 months in a year. Since the sales of outerwear is seasonal this is reasonable.

(d) The amplitude represents the maximum displacement from average sales of 8 million dollars. Sales are greatest in December (cold weather + Christmas) and least in June.

65. $(y - 2)^2 = 8(x + 2)$

Parabola

Vertex: $(-2, 2)$

$p = 2 > 0$, opens to the right

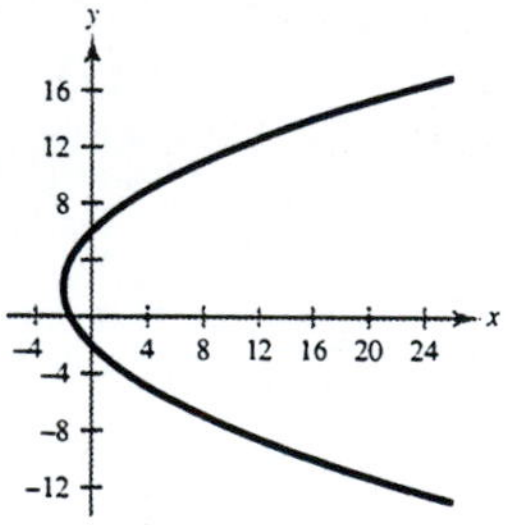

67. $\frac{x^2}{4} - y^2 = 1$

Hyperbola

Horizontal major axis

Center: $(0, 0)$

Vertices: $(\pm 2, 0)$

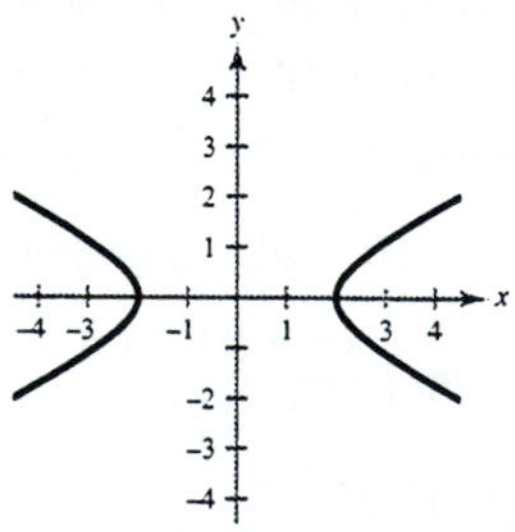

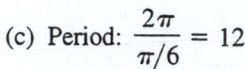

Review Exercises for Chapter 4

Solutions to Odd-Numbered Exercises

1. $\theta = \dfrac{11\pi}{4}$

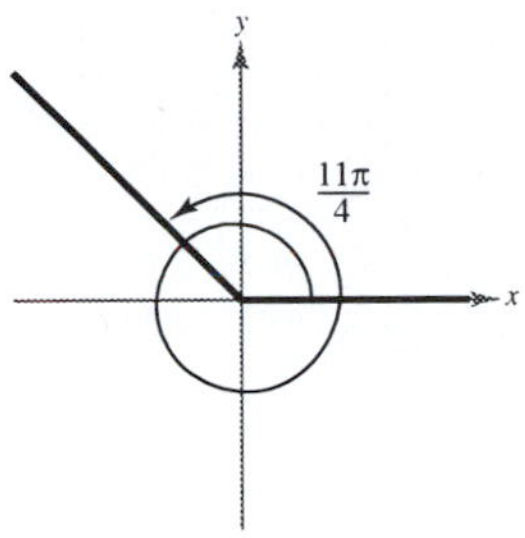

Coterminal angles: $\dfrac{11\pi}{4} - 2\pi = \dfrac{3\pi}{4}$

$\dfrac{3\pi}{4} - 2\pi = -\dfrac{5\pi}{4}$

3. $\theta = -110°$

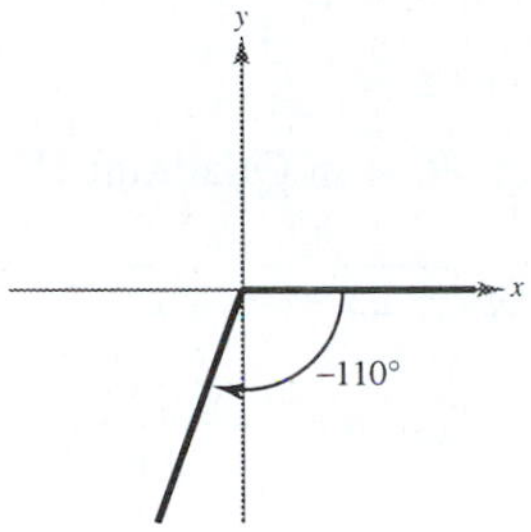

Coterminal angles: $-110° + 360° = 250°$

$-110° - 360° = -470°$

5. $135°\ 16'45'' = \left(135 + \frac{16}{60} + \frac{45}{3600}\right)° \approx 135.28°$

7. $5°\ 22'\ 53'' = \left(5 + \frac{22}{60} + \frac{53}{3600}\right)° \approx 5.38°$

9. $135.27° = 135° + (0.27)(60)'$

$= 135° + 16' + 0.2(60)''$

$= 135°\ 16'\ 12''$

11. $-85.15° = -[85° + (0.15)(60)'] = -85°\ 9'$

13. $\dfrac{5\pi \text{ rad}}{7} = \dfrac{5\pi \text{ rad}}{7} \cdot \dfrac{180°}{\pi \text{ rad}} \approx 128.57°$

15. $-3.5 \text{ rad} = -3.5 \text{ rad} \cdot \dfrac{180°}{\pi \text{ rad}} \approx -200.54°$

17. $480° = 480° \cdot \dfrac{\pi \text{ rad}}{180°} = \dfrac{8\pi}{3} \text{ rad} \approx 8.3776 \text{ rad}$

19. $-33°\ 45' = -33.75° = -33.75° \cdot \dfrac{\pi \text{ rad}}{180°} = -\dfrac{3\pi}{16} \text{ rad} \approx -0.5890 \text{ rad}$

21. $252°$ is in Quadrant III.

Reference angle $= 252° - 180° = 72°$

23. $-\dfrac{6\pi}{5}$ is in Quadrant II and is coterminal to $\dfrac{4\pi}{5}$.

Reference angle $= \pi - \dfrac{4\pi}{5} = \dfrac{\pi}{5}$

25. $t = \dfrac{7\pi}{6}$ corresponds to the point $\left(-\dfrac{\sqrt{3}}{2}, -\dfrac{1}{2}\right)$.

$\sin \dfrac{7\pi}{6} = -\dfrac{1}{2}$

$\cos \dfrac{7\pi}{6} = -\dfrac{\sqrt{3}}{2}$

$\tan \dfrac{7\pi}{6} = \dfrac{1}{\sqrt{3}}$

27. $t = -\dfrac{\pi}{3}$ corresponds to the point $\left(\dfrac{1}{2}, -\dfrac{\sqrt{3}}{2}\right)$.

$\sin\left(-\dfrac{\pi}{3}\right) = -\dfrac{\sqrt{3}}{2}$

$\cos\left(-\dfrac{\pi}{3}\right) = \dfrac{1}{2}$

$\tan\left(-\dfrac{\pi}{3}\right) = -\sqrt{3}$

29. $x = 12, y = 16, r = \sqrt{144 + 256} = \sqrt{400} = 20$

$\sin \theta = \frac{y}{r} = \frac{4}{5}$ $\quad \csc \theta = \frac{r}{y} = \frac{5}{4}$

$\cos \theta = \frac{x}{r} = \frac{3}{5}$ $\quad \sec \theta = \frac{r}{x} = \frac{5}{3}$

$\tan \theta = \frac{y}{x} = \frac{4}{3}$ $\quad \cot \theta = \frac{x}{y} = \frac{3}{4}$

31. $\sec \theta = \frac{6}{5}, \tan \theta < 0 \implies \theta$ is in Quadrant IV.

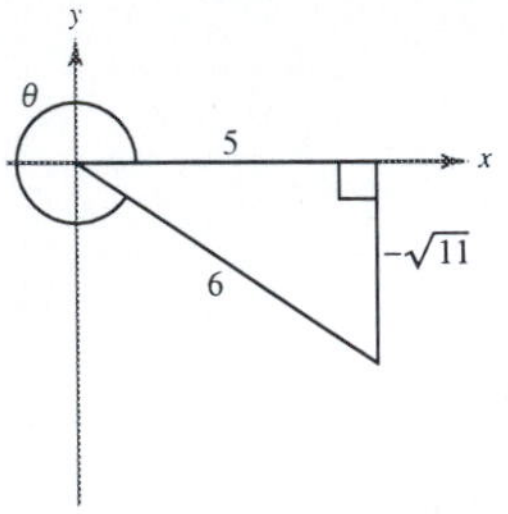

$r = 6, x = 5, y = -\sqrt{36 - 25} = -\sqrt{11}$

$\sin \theta = \frac{y}{r} = -\frac{\sqrt{11}}{6}$ $\quad \csc \theta = -\frac{6\sqrt{11}}{11}$

$\cos \theta = \frac{x}{r} = \frac{5}{6}$ $\quad \sec \theta = \frac{6}{5}$

$\tan \theta = \frac{y}{x} = -\frac{\sqrt{11}}{5}$ $\quad \cot \theta = -\frac{5\sqrt{11}}{11}$

33. $\tan \frac{\pi}{3} = \sqrt{3}$

35. $\cos 495° = -\cos 45° = -\frac{\sqrt{2}}{2}$

37. $\tan 33° \approx 0.65$

39. $\sec \frac{12\pi}{5} = \frac{1}{\cos\left(\frac{12\pi}{5}\right)} \approx 3.24$

41. $\cos \theta = -\frac{\sqrt{2}}{2} \implies \theta$ is in Quadrant II or III.

Reference angle: $\frac{\pi}{4}$

$\theta = \frac{3\pi}{4}, \frac{5\pi}{4}$ or $\theta = 135°, 225°$

43. $\sin \theta = 0.8387 \implies \theta$ is in Quadrant I or II.

Reference angle: $\arcsin 0.8387 \approx 0.9949$

$\theta \approx 0.9949$ rad or $0.9949 \cdot \frac{180}{\pi} \approx 57.0°$

$\theta \approx \pi - 0.9949 \approx 2.1467$ rad or

$2.1467 \cdot \frac{180}{\pi} \approx 123.0°$

45. $\theta = 120°$

$m = \tan 120° = -\sqrt{3}$

47. $x + y - 10 = 0$

$y = -x + 1 \implies m = -1$

$\tan \theta = -1$

$\theta = 135°$

49. $y = 3 \cos 2\pi x$

Amplitude: 3

Period: $\frac{2\pi}{2\pi} = 1$

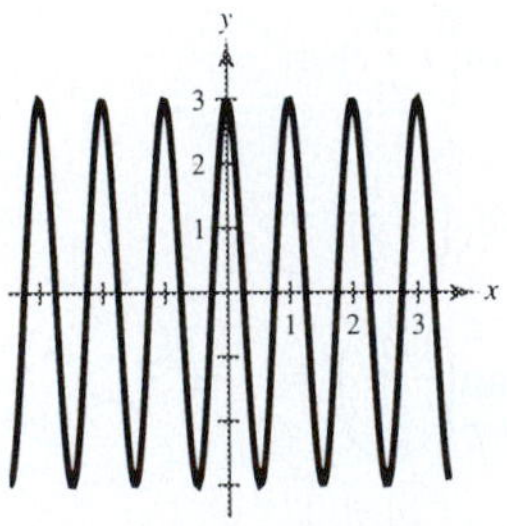

51. $f(x) = 5 \sin \frac{2x}{5}.$

Amplitude: 5

Period: $\frac{2\pi}{2/5} = 5\pi$

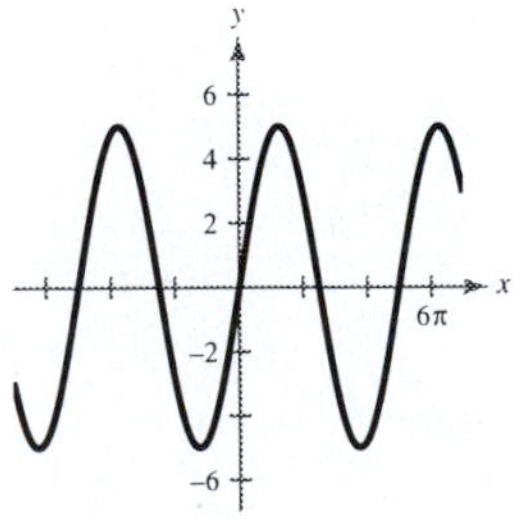

53. $f(x) = -\frac{1}{4}\cos\frac{\pi x}{4}$

Amplitude: $\left|-\frac{1}{4}\right| = \frac{1}{4}$

Period: $\frac{2\pi}{\pi/4} = 8$

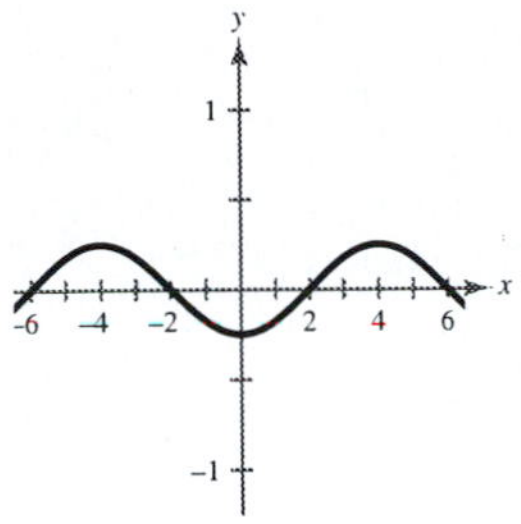

55. $g(t) = \frac{5}{2}\sin(t - \pi)$

Amplitude: $\frac{5}{2}$

Period: 2π

Shift: $t - \pi = 0$ to $t - \pi = 2\pi$

$t = \pi \qquad t = 3\pi$

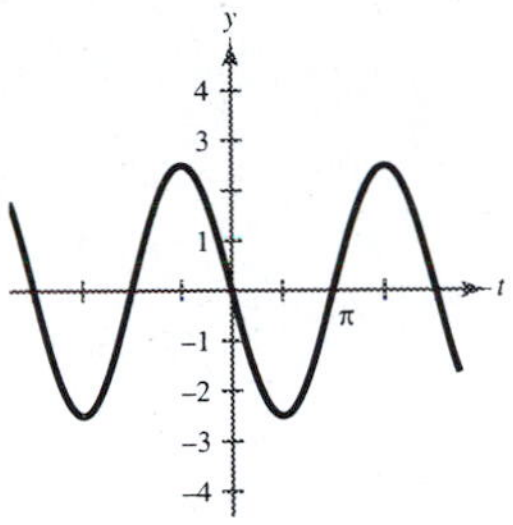

57. $h(t) = \tan\left(t - \frac{\pi}{4}\right)$

Period: π

Two consecutive asymptotes: $t - \frac{\pi}{4} = -\frac{\pi}{2}$ and $t - \frac{\pi}{4} = \frac{\pi}{2}$

$t = -\frac{\pi}{4} \qquad t = \frac{3\pi}{4}$

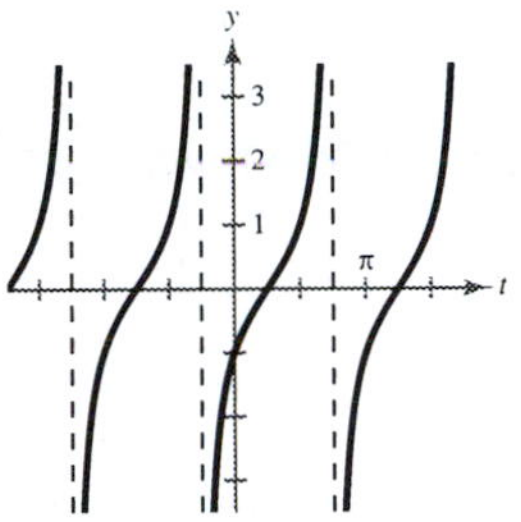

t	0	$\frac{\pi}{4}$	$\frac{\pi}{2}$
$h(t)$	-1	0	1

59. $y = \arcsin\frac{x}{2}$

Domain: $-2 \le x \le 2$

Range: $-\frac{\pi}{2} \le y \le \frac{\pi}{2}$

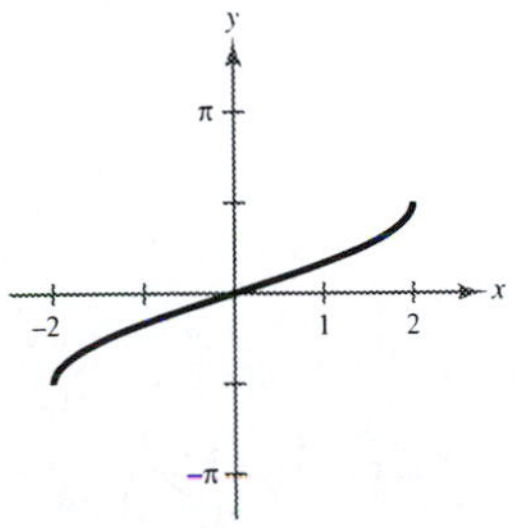

61. $f(x) = \frac{x}{4} - \sin x$

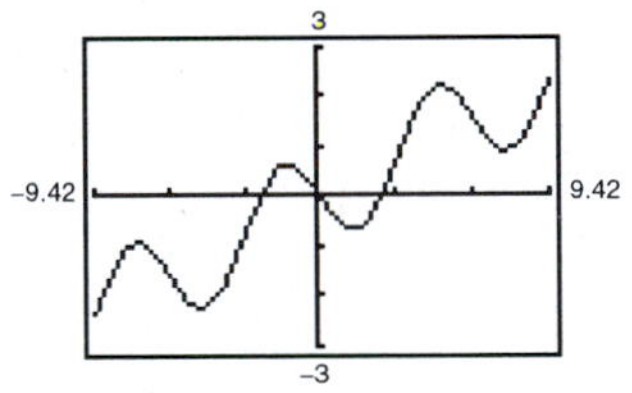

Not periodic.

63. $f(x) = \dfrac{\pi}{2} + \arctan x$

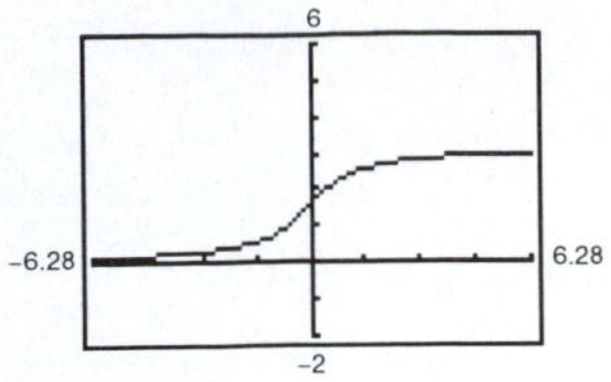

Not periodic.

65. $h(\theta) = \theta \sin \pi\theta$

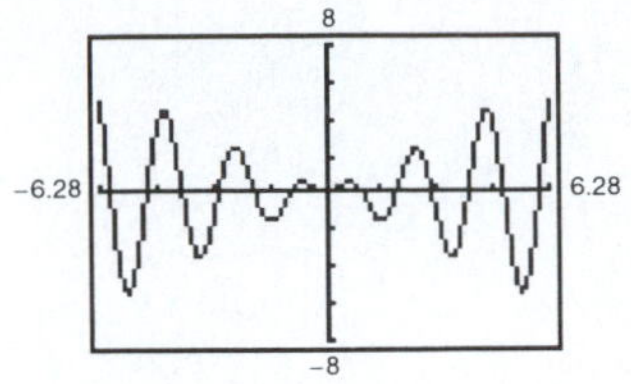

Not periodic.

67. $f(t) = 2.5e^{-t/4} \sin 2\pi t$

$-2.5e^{-t/4} \le f(t) \le 2.5e^{-t/4}$

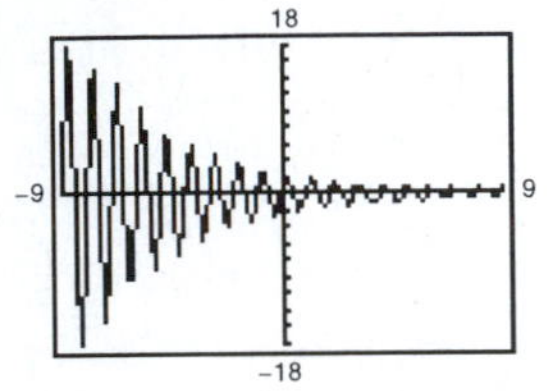

Not periodic.

69. $f(x) = e^{\sin x}$

The graph is periodic. The period is 2π.

Maximum: $\left(\dfrac{\pi}{2}, e\right)$

Minimum: $\left(\dfrac{3\pi}{2}, e^{-1}\right)$

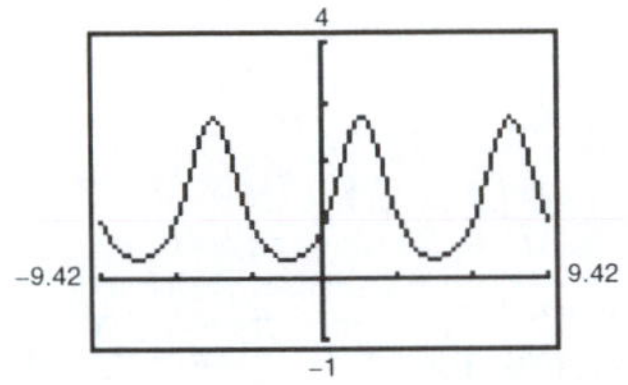

71. $g(x) = 2 \sin x \cos^2 x$

The graph is periodic. The period is 2π.

Relative Minimum: $\left(\dfrac{\pi}{2}, 0\right)$, $(3.76, -0.77)$, $(5.67, -0.77)$

Relative Maximum: $(0.61, 0.77)$, $(2.53, 0.77)$, $\left(\dfrac{3\pi}{2}, 0\right)$

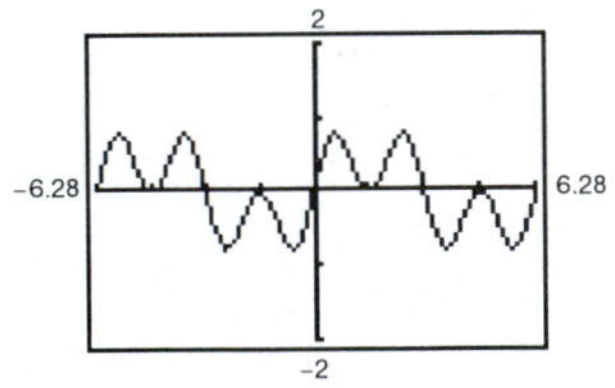

73. Let $y = \arcsin(x - 1)$. Then, $\sin y = (x - 1) = \dfrac{x - 1}{1}$, and

$$\sec y = \frac{1}{\sqrt{1 - (x - 1)^2}} = \frac{1}{\sqrt{-x^2 + 2x}} = \frac{\sqrt{-x^2 + 2x}}{-x^2 + 2x}.$$

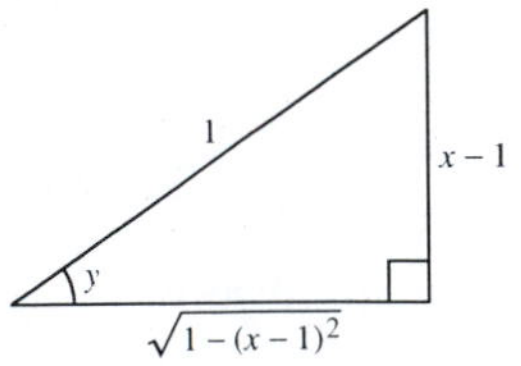

75. Let $y = \arccos \dfrac{x^2}{4 - x^2}$. Then $\cos y = \dfrac{x}{4 - x^2}$, and

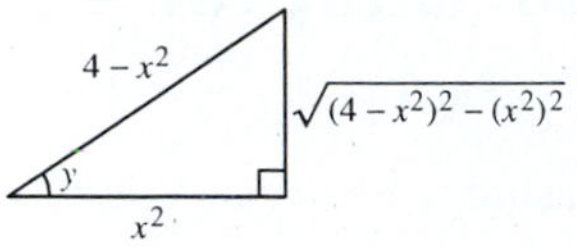

$$\sin y = \frac{\sqrt{(4 - x^2)^2 - (x^2)^2}}{4 - x^2}$$

$$= \frac{\sqrt{16 - 8x^2}}{4 - x^2}$$

$$= \frac{2\sqrt{4 - 2x^2}}{4 - x^2}.$$

77. $\sin 50° = \dfrac{h}{12}$

$h = 12 \sin 50°$

≈ 9.2 m

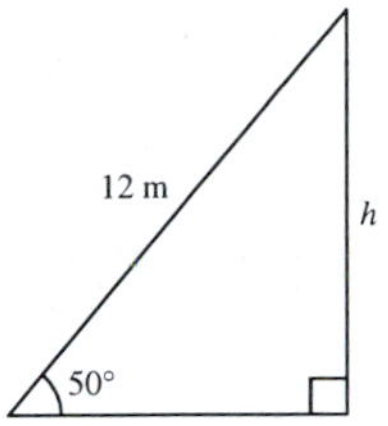

79. $\tan 25° = \dfrac{h}{2.5}$

$h = 2.5 \tan 25°$

≈ 1.2 miles

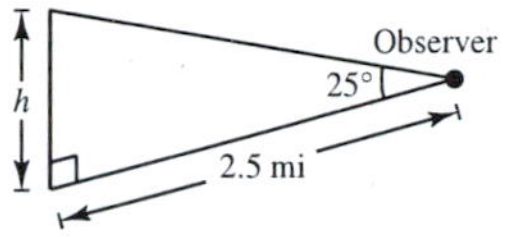

81. $\tan 1°\ 10' = \dfrac{a}{3.5}$

$a = 3.5 \tan 1°\ 10'$

≈ 0.07 km

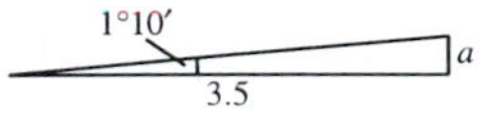

83. (a) $\tan\theta = \dfrac{x}{12}$

$x = 12 \tan\theta$

Area = Area of triangle − Area of sector

$= \left(\frac{1}{2}bh\right) - \left(\frac{1}{2}r^2\theta\right)$

$= \frac{1}{2}(12)(12 \tan \theta) - \frac{1}{2}(12^2)(\theta)$

$= 72 \tan\theta - 72\theta$

$= 72(\tan\theta - \theta)$

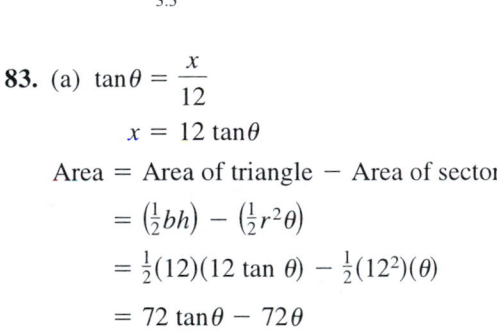

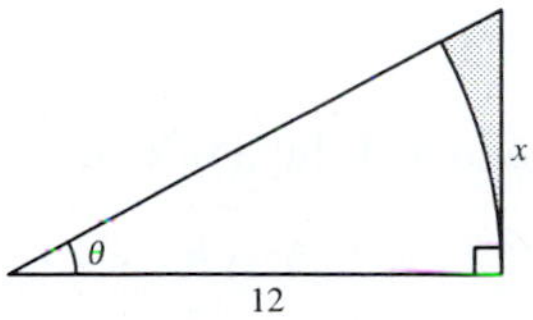

(b)

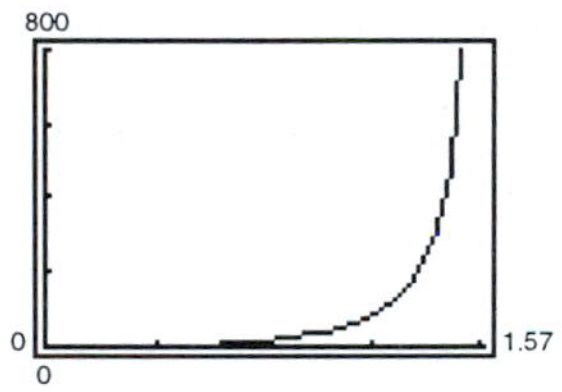

As $\theta \longrightarrow \dfrac{\pi}{2}$, $A \longrightarrow \infty$. The area increases without bound as θ approaches $\dfrac{\pi}{2}$.

❑ Practice Test for Chapter 4

1. Express 350° in radian measure.

2. Express $(5\pi)/9$ in degree measure.

3. Convert $135°\ 14'\ 12''$ to decimal form.

4. Convert $-22.569°$ to D° M′ S″ form.

5. If $\cos\theta = \frac{2}{3}$, use the trigonometric identities to find $\tan\theta$.

6. Find θ given $\sin\theta = 0.9063$, $0 \le \theta < 2\pi$.

7. Solve for x in the figure below.

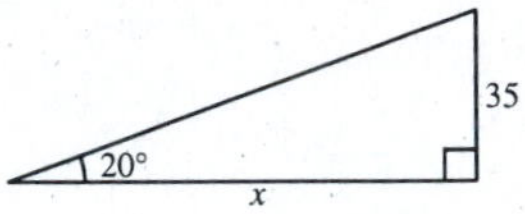

8. Find the magnitude of the reference angle for $\theta = (6\pi)/5$.

9. Evaluate csc 3.92.

10. Find $\sec\theta$ given that θ lies in Quadrant III and $\tan\theta = 6$.

11. Graph $y = 3\sin\dfrac{x}{2}$.

12. Graph $y = -2\cos(x - \pi)$.

13. Graph $y = \tan 2x$.

14. Graph $y = -\csc\left(x + \dfrac{\pi}{4}\right)$.

15. Graph $y = 2x + \sin x$, using a graphing calculator.

16. Graph $y = 3x\cos x$, using a graphing calculator.

17. Evaluate arcsin 1.

18. Evaluate $\arctan(-3)$.

19. Evaluate $\sin\left(\arccos\dfrac{4}{\sqrt{35}}\right)$.

20. Write an algebraic expression for $\cos\left(\arcsin\dfrac{x}{4}\right)$.

For Exercises 21–23, solve the right triangle.

21. $A = 40°,\ c = 12$

22. $B = 6.84°,\ a = 21.3$

23. $a = 5,\ b = 9$

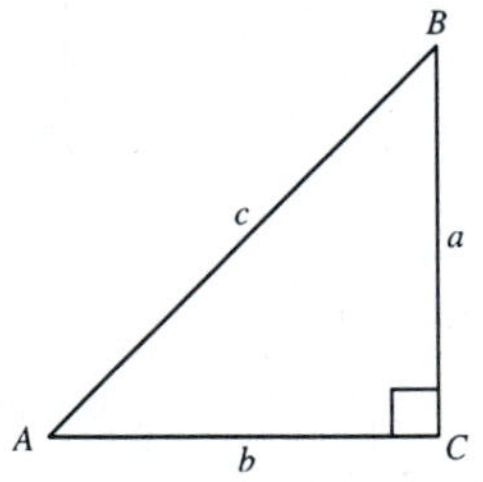

24. A 20-foot ladder leans against the side of a barn. Find the height of the top of the ladder if the angle of elevation of the ladder is 67°.

25. An observer in a lighthouse 250 feet above sea level spots a ship off the shore. If the angle of depression to the ship is 5°, how far out is the ship?

CHAPTER 5
Analytic Trigonometry

CHAPTER 5
Analytic Trigonometry

Section 5.1 Using Fundamental Identities

- You should know the fundamental trigonometric identities.

(a) Reciprocal Identities

$$\sin u = \frac{1}{\csc u} \qquad \csc u = \frac{1}{\sin u}$$

$$\cos u = \frac{1}{\sec u} \qquad \sec u = \frac{1}{\cos u}$$

$$\tan u = \frac{1}{\cot u} = \frac{\sin u}{\cos u} \qquad \cot u = \frac{1}{\tan u} = \frac{\cos u}{\sin u}$$

(b) Pythagorean Identities

$$\sin^2 u + \cos^2 u = 1$$

$$1 + \tan^2 u = \sec^2 u$$

$$1 + \cot^2 u = \csc^2 u$$

(c) Cofunction Identities

$$\sin\left(\frac{\pi}{2} - u\right) = \cos u \qquad \cos\left(\frac{\pi}{2} - u\right) = \sin u$$

$$\tan\left(\frac{\pi}{2} - u\right) = \cot u \qquad \cot\left(\frac{\pi}{2} - u\right) = \tan u$$

$$\sec\left(\frac{\pi}{2} - u\right) = \csc u \qquad \csc\left(\frac{\pi}{2} - u\right) = \sec u$$

(d) Negative Angle Identities

$$\sin(-x) = -\sin x \qquad \csc(-x) = -\csc x$$

$$\cos(-x) = \cos x \qquad \sec(-x) = \sec x$$

$$\tan(-x) = -\tan x \qquad \cot(-x) = -\cot x$$

- You should be able to use these fundamental identities to find function values.
- You should be able to convert trigonometric expressions to equivalent forms by using the fundamental identities.

Solutions to Odd-Numbered Exercises

1. $\sin x = \frac{1}{2}$, $\cos x = \frac{\sqrt{3}}{2} \implies x$ is in Quadrant I.

$$\tan x = \frac{\sin x}{\cos x} = \frac{1/2}{\sqrt{3}/2} = \frac{1}{\sqrt{3}} = \frac{\sqrt{3}}{3}$$

$$\cot x = \frac{1}{\tan x} = \sqrt{3}$$

$$\sec x = \frac{1}{\cos x} = \frac{2}{\sqrt{3}} = \frac{2\sqrt{3}}{3}$$

$$\csc x = \frac{1}{\sin x} = 2$$

3. $\sec\theta = \sqrt{2}$, $\sin\theta = -\frac{\sqrt{2}}{2} \implies \theta$ is in Quadrant IV.

$$\cos\theta = \frac{1}{\sec\theta} = \frac{1}{\sqrt{2}} = \frac{\sqrt{2}}{2}$$

$$\tan\theta = \frac{\sin\theta}{\cos\theta} = \frac{-\sqrt{2}/2}{\sqrt{2}/2} = -1$$

$$\cot\theta = \frac{1}{\tan\theta} = -1$$

$$\csc\theta = \frac{1}{\sin\theta} = -\sqrt{2}$$

5. $\tan x = \frac{5}{12}$, $\sec x = -\frac{13}{12} \implies x$ is in Quadrant III.

$$\cos x = \frac{1}{\sec x} = -\frac{12}{13}$$

$$\sin x = -\sqrt{1-\cos^2 x} = -\sqrt{1-\frac{144}{169}} = -\frac{5}{13}$$

$$\cot x = \frac{1}{\tan x} = \frac{12}{5}$$

$$\csc x = \frac{1}{\sin x} = -\frac{13}{5}$$

7. $\sec\phi = -1$, $\sin\phi = 0 \implies \phi = \pi$

$\cos\phi = -1$

$\tan\phi = 0$

$\cot\phi$ is undefined.

$\csc\phi$ is undefined.

9. $\sin(-x) = -\sin x = -\frac{2}{3} \implies \sin x = \frac{2}{3}$

$\sin x = \frac{2}{3}$, $\tan x = -\frac{2\sqrt{5}}{5} \implies x$ is in Quadrant II.

$$\cos x = -\sqrt{1-\sin^2 x} = -\sqrt{1-\frac{4}{9}} = -\frac{\sqrt{5}}{3}$$

$$\cot x = \frac{1}{\tan x} = -\frac{\sqrt{5}}{2}$$

$$\sec x = \frac{1}{\cos x} = -\frac{3\sqrt{5}}{5}$$

$$\csc x = \frac{1}{\sin x} = \frac{3}{2}$$

11. $\tan\theta = 2$, $\sin\theta < 0 \implies \theta$ is in Quadrant III.

$$\sec\theta = -\sqrt{\tan^2\theta + 1} = -\sqrt{4+1} = -\sqrt{5}$$

$$\cos\theta = \frac{1}{\sec\theta} = -\frac{1}{\sqrt{5}} = -\frac{\sqrt{5}}{5}$$

$$\sin\theta = -\sqrt{1-\cos^2\theta}$$

$$= -\sqrt{1-\frac{1}{5}} = -\frac{2}{\sqrt{5}} = -\frac{2\sqrt{5}}{5}$$

$$\csc\theta = \frac{1}{\sin\theta} = -\frac{\sqrt{5}}{2}$$

$$\cot\theta = \frac{1}{\tan\theta} = \frac{1}{2}$$

13. $\sin\theta = -1$, $\cot\theta = 0 \implies \theta = \frac{3\pi}{2}$

$\cos\theta = \sqrt{1-\sin^2\theta} = 0$

$\sec\theta$ is undefined.

$\tan\theta$ is undefined.

$\csc\theta = -1$

15. By looking at the basic graphs of $\sin x$ and $\csc x$, we see that as $x \longrightarrow \frac{\pi}{2}^-$, $\sin x \longrightarrow 1$ and $\csc x \longrightarrow 1$.

17. By looking at the basic graphs of tan x and cot x, we see that as $x \longrightarrow \frac{\pi^-}{2}$, $\tan x \longrightarrow \infty$ and $\cot x \longrightarrow 0$.

19. $\sec x \cos x = \sec x \cdot \frac{1}{\sec x} = 1$

The expression is matched with (d).

21. $\tan^2 x - \sec^2 x = \tan^2 x - (\tan^2 x + 1) = -1$

The expression is matched with (a).

23. $\frac{\sin(-x)}{\cos(-x)} = \frac{-\sin x}{\cos x} = -\tan x$

The expression is matched with (e).

25. $\sin x \sec x = \sin x \cdot \frac{1}{\cos x} = \tan x$

The expression is matched with (b).

27. $\sec^4 x - \tan^4 x = (\sec^2 x + \tan^2 x)(\sec^2 x - \tan^2 x)$

$= (\sec^2 x + \tan^2 x)(1) = \sec^2 x + \tan^2 x$

The expression is matched with (f).

29. $\frac{\sec^2 x - 1}{\sin^2 x} = \frac{\tan^2 x}{\sin^2 x} = \frac{\sin^2 x}{\cos^2 x} \cdot \frac{1}{\sin^2 x} = \sec^2 x$

The expression is matched with (e).

31. $\tan \phi \csc \phi = \frac{\sin \phi}{\cos \phi} \cdot \frac{1}{\sin \phi} = \frac{1}{\cos \phi} = \sec \phi$

33. $\cos \beta \tan \beta = \cos \beta\left(\frac{\sin \beta}{\cos \beta}\right)$

$= \sin\beta$

35. $\frac{\cot x}{\csc x} = \frac{\cos x/\sin x}{1/\sin x}$

$= \frac{\cos x}{\sin x} \cdot \frac{\sin x}{1} = \cos x$

37. $\sec \alpha \frac{\sin \alpha}{\tan \alpha} = \frac{1}{\cos \alpha}(\sin \alpha) \cot \alpha$

$= \frac{1}{\cos \alpha}(\sin \alpha)\left(\frac{\cos \alpha}{\sin \alpha}\right) = 1$

39. $\frac{\sin(-x)}{\cos x} = -\frac{\sin x}{\cos x} = -\tan x$

41. $\cos\left(\frac{\pi}{2} - x\right)\sec x = (\sin x)(\sec x)$

$= (\sin x)\left(\frac{1}{\cos x}\right)$

$= \frac{\sin x}{\cos x}$

$= \tan x$

43. $\frac{\cos^2 y}{1 - \sin y} = \frac{1 - \sin^2 y}{1 - \sin y}$

$= \frac{(1 + \sin y)(1 - \sin y)}{1 - \sin y}$

$= 1 + \sin y$

45. $\tan^2 x - \tan^2 x \sin^2 x = \tan^2 x(1 - \sin^2 x)$

$= \tan^2 x \cos^2 x$

$= \frac{\sin^2 x}{\cos^2 x} \cdot \cos^2 x$

$= \sin^2 x$

47. $\sin^2 x \sec^2 x - \sin^2 x = \sin^2 x(\sec^2 x - 1)$

$= \sin^2 x \tan^2 x$

49. $\tan^4 x + 2\tan^2 x + 1 = (\tan^2 x + 1)^2$

$= (\sec^2 x)^2$

$= \sec^4 x$

51. $\sin^4 x - \cos^4 x = (\sin^2 x + \cos^2 x)(\sin^2 x - \cos^2 x)$

$= (1)(\sin^2 x - \cos^2 x)$

$= \sin^2 x - \cos^2 x$

53. $(\sin x + \cos x)^2 = \sin^2 x + 2\sin x \cos x + \cos^2 x$

$= (\sin^2 x + \cos^2 x) + 2\sin x \cos x$

$= 1 + 2\sin x \cos x$

55. $(\sec x + 1)(\sec x - 1) = \sec^2 x - 1 = \tan^2 x$

57. $\dfrac{1}{1 + \cos x} + \dfrac{1}{1 - \cos x} = \dfrac{1 - \cos x + 1 + \cos x}{(1 + \cos x)(1 - \cos x)}$

$= \dfrac{2}{1 - \cos^2 x}$

$= \dfrac{2}{\sin^2 x}$

$= 2\csc^2 x$

59. $\dfrac{\cos x}{1 + \sin x} + \dfrac{1 + \sin x}{\cos x} = \dfrac{\cos^2 x + (1 + \sin x)^2}{\cos x(1 + \sin x)} = \dfrac{\cos^2 x + 1 + 2\sin x + \sin^2 x}{\cos x(1 + \sin x)}$

$= \dfrac{2 + 2\sin x}{\cos x(1 + \sin x)}$

$= \dfrac{2(1 + \sin x)}{\cos x(1 + \sin x)}$

$= \dfrac{2}{\cos x}$

$= 2\sec x$

61. $\dfrac{\sin^2 y}{1 - \cos y} = \dfrac{1 - \cos^2 y}{1 - \cos y}$

$= \dfrac{(1 + \cos y)(1 - \cos y)}{1 - \cos y}$

$= 1 + \cos y$

63. $\dfrac{3}{\sec x - \tan x} \cdot \dfrac{\sec x + \tan x}{\sec x + \tan x} = \dfrac{3(\sec x + \tan x)}{\sec^2 x - \tan^2 x}$

$= \dfrac{3(\sec x + \tan x)}{1}$

$= 3(\sec x + \tan x)$

65. $y_1 = \cos\left(\frac{\pi}{2} - x\right),\ y_2 = \sin x$

x	0.2	0.4	0.6	0.8	1.0	1.2	1.4
y_1	0.1987	0.3894	0.5646	0.7174	0.8415	0.9320	0.9854
y_2	0.1987	0.3894	0.5646	0.7174	0.8415	0.9320	0.9854

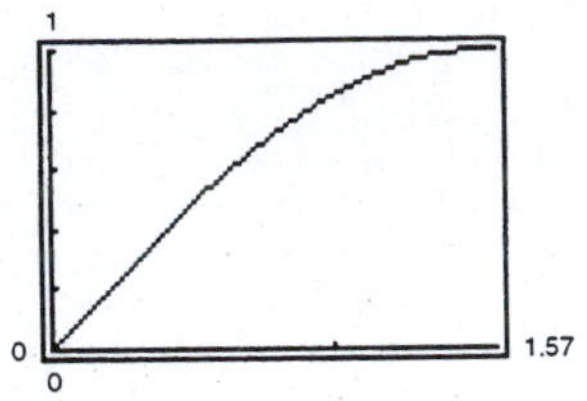

Conclusion: $y_1 = y_2$

67. $y_1 = \dfrac{\cos x}{1 - \sin x},\ y_2 = \dfrac{1 + \sin x}{\cos x}$

x	0.2	0.4	0.6	0.8	1.0	1.2	1.4
y_1	1.2230	1.5085	1.8958	2.4650	3.4082	5.3319	11.6814
y_2	1.2230	1.5085	1.8958	2.4650	3.4082	5.3319	11.6814

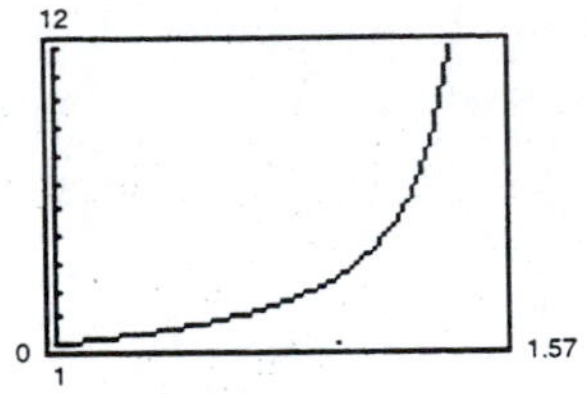

Conclusion: $y_1 = y_2$

69. $y_1 = \cos x \cot x + \sin x = \csc x$

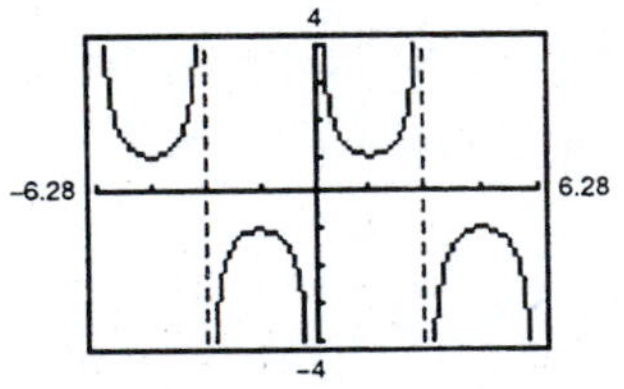

71. $\sqrt{25 - x^2} = \sqrt{25 - (5\sin\theta)^2},\ x = 5\sin\theta$

$$= \sqrt{25 - 25\sin^2\theta}$$
$$= \sqrt{25(1 - \sin^2\theta)}$$
$$= \sqrt{25\cos^2\theta}$$
$$= 5\cos\theta$$

73. $\sqrt{x^2 - 9} = \sqrt{(3\sec\theta)^2 - 9},\ x = 3\sec\theta$

$$= \sqrt{9\sec^2\theta - 9}$$
$$= \sqrt{9(\sec^2\theta - 1)}$$
$$= \sqrt{9\tan^2\theta}$$
$$= 3\tan\theta$$

75. $\sqrt{x^2 + 25} = \sqrt{(5\tan\theta)^2 + 25},\ x = 5\tan\theta$

$$= \sqrt{25\tan^2\theta + 25}$$
$$= \sqrt{25(\tan^2\theta + 1)}$$
$$= \sqrt{25\sec^2\theta}$$
$$= 5\sec\theta$$

77. $\sin\theta = \sqrt{1 - \cos^2\theta}$

Let $y_1 = \sin x$ and $y_2 = \sqrt{1 - \cos^2 x},\ 0 \le x \le 2\pi$.

$y_1 = y_2$ for $0 \le x \le \pi$, so we have

$\sin\theta = \sqrt{1 - \cos^2\theta}$ for $0 \le \theta \le \pi$.

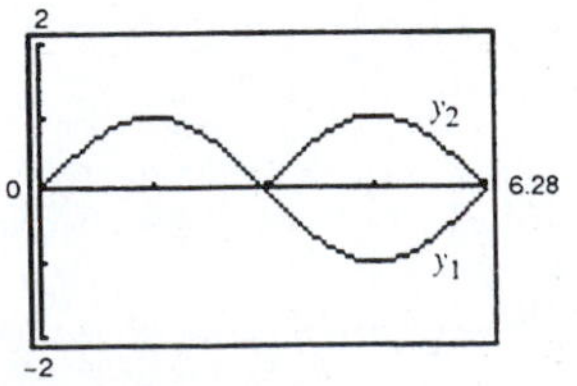

79. $\sec\theta = \sqrt{1+\tan^2\theta}$

Let $y_1 = \dfrac{1}{\cos x}$ and $y_2 = \sqrt{1+\tan^2 x}$, $0 \le x \le 2\pi$.

$y_1 = y_2$ for $0 \le x < \dfrac{\pi}{2}$ and $\dfrac{3\pi}{2} < x \le 2\pi$, so we have

$\sec\theta = \sqrt{1+\tan^2\theta}$ for $0 \le \theta < \dfrac{\pi}{2}$ and $\dfrac{3\pi}{2} < \theta \le 2\pi$.

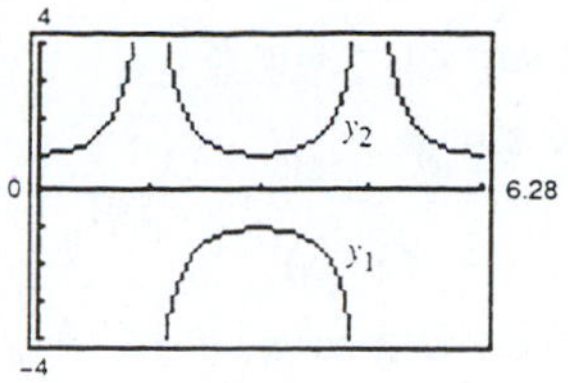

81. $\ln|\cos\theta| - \ln|\sin\theta| = \ln\dfrac{|\cos\theta|}{|\sin\theta|} = \ln|\cot\theta|$

83. False; $\dfrac{\sin k\theta}{\cos k\theta} = \tan k\theta$

85. True; $\sin\theta\csc\theta = \sin\theta\left(\dfrac{1}{\sin\theta}\right) = 1$, provided $\sin\theta \neq 0$.

87. (a) $\csc^2 132° - \cot^2 132° \approx 1.8107 - 0.8107 = 1$

(b) $\csc^2\dfrac{2\pi}{7} - \cot^2\dfrac{2\pi}{7} \approx 1.6360 - 0.6360 = 1$

89. $\cos\left(\dfrac{\pi}{2} - \theta\right) = \sin\theta$

(a) $\theta = 80°$

$\cos(90° - 80°) = \sin 80°$

$0.9848 = 0.9848$

(b) $\theta = 0.8$

$\cos\left(\dfrac{\pi}{2} - 0.8\right) = \sin 0.8$

$0.7174 = 0.7174$

91. Since $\sin^2\theta + \cos^2\theta = 1$ and $\cos^2\theta = 1 - \sin^2\theta$:

$\cos\theta = \pm\sqrt{1-\sin^2\theta}$

$\tan\theta = \dfrac{\sin\theta}{\cos\theta} = \pm\dfrac{\sin\theta}{\sqrt{1-\sin^2\theta}}$

$\cot\theta = \dfrac{1}{\tan\theta} = \pm\dfrac{\sqrt{1-\sin^2\theta}}{\sin\theta}$

$\sec\theta = \dfrac{1}{\cos\theta} = \pm\dfrac{1}{\sqrt{1-\sin^2\theta}}$

$\csc\theta = \dfrac{1}{\sin\theta}$

93. $(\sqrt{x}+5)(\sqrt{x}-5) = (\sqrt{x})^2 - (5)^2 = x - 25$

95. $(2\sqrt{z}+3)^2 = (2\sqrt{z})^2 + 2(2\sqrt{z})(3) + (3)^2$

$= 4z + 12\sqrt{z} + 9$

Section 5.2 Verifying Trigonometric Identities

- You should know the difference between an expression, a conditional equation, and an identity.
- You should be able to solve trigonometric identities, using the following techniques.
  - (a) Work with *one* side at a time. Do not "cross" the equal sign.
  - (b) Use algebraic techniques such as combining fractions, factoring expressions, rationalizing denominators, and squaring binomials.
  - (c) Use the fundamental identities.
  - (d) Convert all the terms into sines and cosines.

Solutions to Odd-Numbered Exercises

1. $\sin t\csc t = \sin t\left(\dfrac{1}{\sin t}\right) = 1$

3. $(1+\sin\alpha)(1-\sin\alpha) = (1-\sin^2\alpha) = \cos^2\alpha$

5. $\cos^2\beta - \sin^2\beta = (1 - \sin^2\beta) - \sin^2\beta$
$= 1 - 2\sin^2\beta$

7. $\tan^2\theta + 4 = (\sec^2\theta - 1) + 4$
$= \sec^2\theta + 3$

9. $\sin^2\alpha - \sin^4\alpha = \sin^2\alpha(1 - \sin^2\alpha)$
$= (1 - \cos^2\alpha)(\cos^2\alpha)$
$= \cos^2\alpha - \cos^4\alpha$

11. $\dfrac{\sec^2 x}{\tan x} = \sec^2 x \cdot \dfrac{1}{\tan x} = \dfrac{1}{\cos^2 x} \cdot \dfrac{\cos x}{\sin x}$
$= \dfrac{1}{\cos x} \cdot \dfrac{1}{\sin x}$
$= \sec x \csc x$

13. $\dfrac{\cot^2 t}{\csc t} = \dfrac{\cos^2 t}{\sin^2 t} \cdot \sin t$
$= \dfrac{\cos^2 t}{\sin t}$
$= \dfrac{1 - \sin^2 t}{\sin t} = \dfrac{1}{\sin t} - \dfrac{\sin^2 t}{\sin t}$
$= \csc t - \sin t$

15. $\sin^{1/2} x \cos x - \sin^{5/2} x \cos x = \sin^{1/2} x \cos x(1 - \sin^2 x) = \sin^{1/2} x \cos x \cdot \cos^2 x = \cos^3 x\sqrt{\sin x}$

17. $\dfrac{1}{\sec x \tan x} = \cos x \cot x = \cos x \cdot \dfrac{\cos x}{\sin x}$
$= \dfrac{\cos^2 x}{\sin x}$
$= \dfrac{1 - \sin^2 x}{\sin x}$
$= \dfrac{1}{\sin x} - \sin x$
$= \csc x - \sin x$

19. $\csc x - \sin x = \dfrac{1}{\sin x} - \sin x$
$= \dfrac{1 - \sin^2 x}{\sin x}$
$= \dfrac{\cos^2 x}{\sin x}$
$= \cos x \cdot \dfrac{\cos x}{\sin x}$
$= \cos x \cot x$

21. $\cos x + \sin x \tan x = \cos x + \sin x \cdot \dfrac{\sin x}{\cos x}$
$= \dfrac{\cos^2 x + \sin^2 x}{\cos x}$
$= \dfrac{1}{\cos x}$
$= \sec x$

23. $\dfrac{1}{\tan x} + \dfrac{1}{\cot x} = \dfrac{\cot x + \tan x}{\tan x \cot x}$
$= \dfrac{\cot x + \tan x}{1}$
$= \tan x + \cot x$

25. $\dfrac{\cos\theta\cot\theta}{1 - \sin\theta} - 1 = \dfrac{\cos\theta\cot\theta - (1 - \sin\theta)}{1 - \sin\theta}$
$= \dfrac{\cos\theta(\cos\theta/\sin\theta) - 1 + \sin\theta}{1 - \sin\theta} \cdot \dfrac{\sin\theta}{\sin\theta}$
$= \dfrac{\cos^2\theta - \sin\theta + \sin^2\theta}{\sin\theta(1 - \sin\theta)}$
$= \dfrac{1 - \sin\theta}{\sin\theta(1 - \sin\theta)}$
$= \dfrac{1}{\sin\theta}$
$= \csc\theta$

27. $\dfrac{1}{\cot x + 1} + \dfrac{1}{\tan x + 1} = \dfrac{\tan x + 1 + \cot x + 1}{(\cot x + 1)(\tan x + 1)}$

$= \dfrac{\tan x + \cot x + 2}{\cot x \tan x + \cot x + \tan x + 1}$

$= \dfrac{\tan x + \cot x + 2}{\tan x + \cot x + 2}$

$= 1$

29. $\cos\left(\dfrac{\pi}{2} - x\right)\csc x = \sin x\left(\dfrac{1}{\sin x}\right) = 1$

31. $\dfrac{\csc(-x)}{\sec(-x)} = \dfrac{1/\sin(-x)}{1/\cos(-x)}$

$= \dfrac{\cos(-x)}{\sin(-x)}$

$= \dfrac{\cos x}{-\sin x}$

$= -\cot x$

33. $\dfrac{\cos(-\theta)}{1 + \sin(-\theta)} = \dfrac{\cos\theta}{1 - \sin\theta} \cdot \dfrac{1 + \sin\theta}{1 + \sin\theta}$

$= \dfrac{\cos\theta(1 + \sin\theta)}{1 - \sin^2\theta}$

$= \dfrac{\cos\theta(1 + \sin\theta)}{\cos^2\theta}$

$= \dfrac{1 + \sin\theta}{\cos\theta}$

$= \dfrac{1}{\cos\theta} + \dfrac{\sin\theta}{\cos\theta}$

$= \sec\theta + \tan\theta$

35. $\dfrac{\sin x \cos y + \cos x \sin y}{\cos x \cos y - \sin x \sin y} = \dfrac{\dfrac{\sin x \cos y}{\cos x \cos y} + \dfrac{\cos x \sin y}{\cos x \cos y}}{\dfrac{\cos x \cos y}{\cos x \cos y} - \dfrac{\sin x \sin y}{\cos x \cos y}} = \dfrac{\tan x + \tan y}{1 - \tan x \tan y}$

37. $\dfrac{\tan x + \cot y}{\tan x \cot y} = \dfrac{\dfrac{1}{\cot x} + \dfrac{1}{\tan y}}{\dfrac{1}{\cot x} \cdot \dfrac{1}{\tan y}} \cdot \dfrac{\cot x \tan y}{\cot x \tan y} = \tan y + \cot x$

39. $\sqrt{\dfrac{1 + \sin\theta}{1 - \sin\theta}} = \sqrt{\dfrac{1 + \sin\theta}{1 - \sin\theta} \cdot \dfrac{1 + \sin\theta}{1 + \sin\theta}}$

$= \sqrt{\dfrac{(1 + \sin\theta)^2}{1 - \sin^2\theta}}$

$= \sqrt{\dfrac{(1 + \sin\theta)^2}{\cos^2\theta}}$

$= \dfrac{1 + \sin\theta}{|\cos\theta|}$

41. $\sin^2 x + \sin^2\left(\dfrac{\pi}{2} - x\right) = \sin^2 x + \cos^2 x = 1$

43. $\csc x \cos\left(\dfrac{\pi}{2} - x\right) = \dfrac{1}{\sin x} \cdot \sin x = 1$

45. $2\sec^2 x - 2\sec^2 x \sin^2 x - \sin^2 x - \cos^2 x = 2\sec^2 x(1 - \sin^2 x) - (\sin^2 x + \cos^2 x)$

$$= 2\sec^2 x(\cos^2 x) - 1$$

$$= 2 \cdot \frac{1}{\cos^2 x} \cdot \cos^2 x - 1$$

$$= 2 - 1$$

$$= 1$$

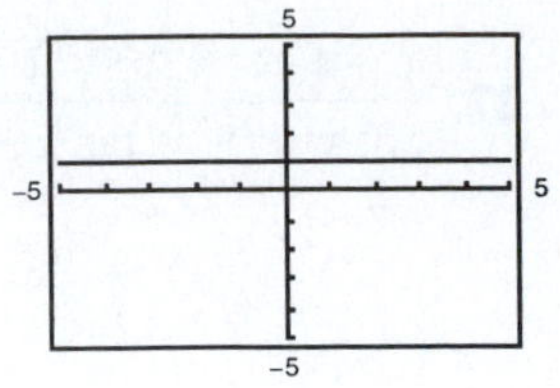

47. $2 + \cos^2 x - 3\cos^4 x = (1 - \cos^2 x)(2 + 3\cos^2 x)$

$$= \sin^2 x(2 + 3\cos^2 x)$$

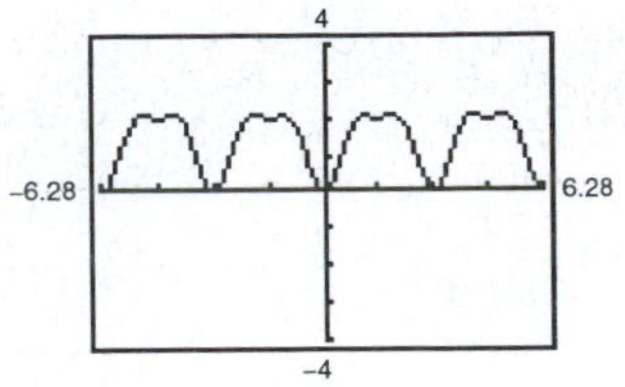

49. $\csc^4 x - 2\csc^2 x + 1 = (\csc^2 x - 1)^2$

$$= (\cot^2 x)^2 = \cot^4 x$$

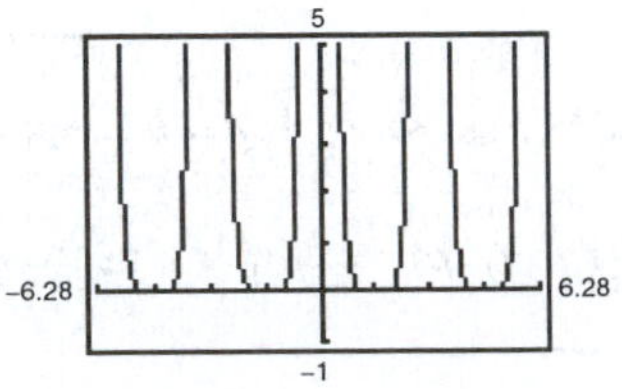

51. $\sec^4 \theta - \tan^4 \theta = (\sec^2 \theta + \tan^2 \theta)(\sec^2 \theta - \tan^2 \theta)$

$$= (1 + \tan^2 \theta + \tan^2 \theta)(1)$$

$$= 1 + 2\tan^2 \theta$$

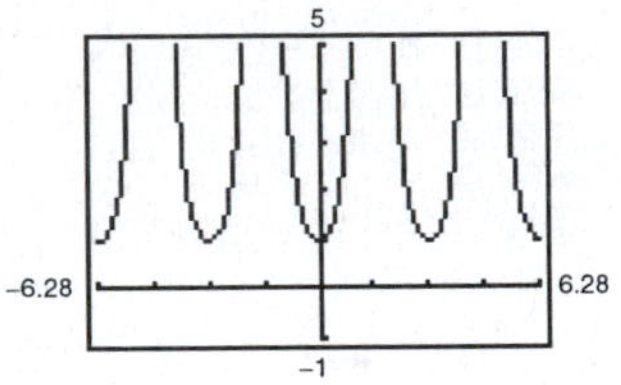

53. $\dfrac{\sin \beta}{1 - \cos \beta} \cdot \dfrac{1 + \cos \beta}{1 + \cos \beta} = \dfrac{\sin \beta(1 + \cos \beta)}{1 - \cos^2 \beta}$

$$= \frac{1 + \cos \beta}{\sin \beta}$$

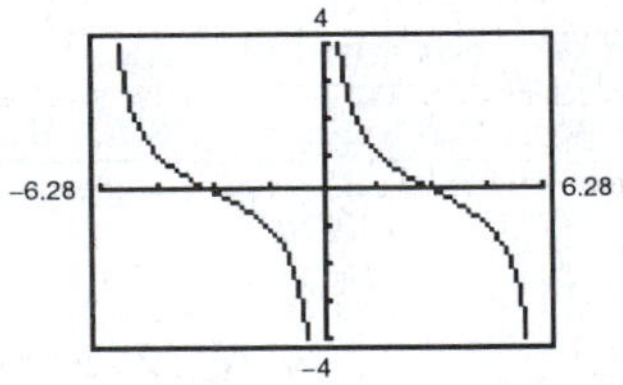

55. $\dfrac{\tan^3 \alpha - 1}{\tan \alpha - 1} = \dfrac{(\tan \alpha - 1)(\tan^2 \alpha + \tan \alpha + 1)}{\tan \alpha - 1} = \tan^2 \alpha + \tan \alpha + 1$

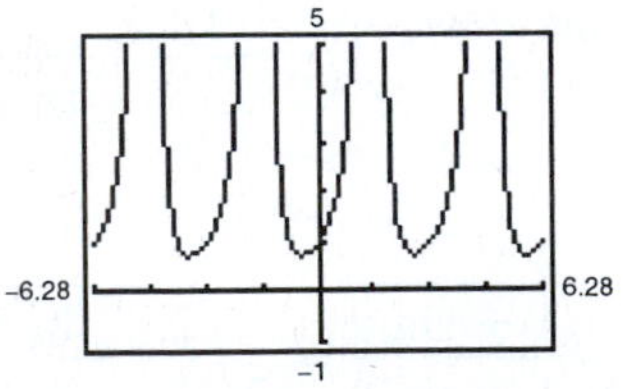

57. $\ln|\tan \theta| = \ln\left|\dfrac{\sin \theta}{\cos \theta}\right|$

$$= \ln\frac{|\sin \theta|}{|\cos \theta|}$$

$$= \ln|\sin \theta| - \ln|\cos \theta|$$

59. $-\ln(1 + \cos \theta) = \ln(1 + \cos \theta)^{-1}$

$$= \ln \frac{1}{1 + \cos \theta} \cdot \frac{1 - \cos \theta}{1 - \cos \theta}$$

$$= \ln \frac{1 - \cos \theta}{1 - \cos^2 \theta}$$

$$= \ln \frac{1 - \cos \theta}{\sin^2 \theta}$$

$$= \ln(1 - \cos \theta) - \ln \sin^2 \theta$$

$$= \ln(1 - \cos \theta) - 2\ln|\sin \theta|$$

61. Since $\sin^2\theta = 1 - \cos^2\theta$, then
$\sin\theta = \pm\sqrt{1-\cos^2\theta}$; $\sin\theta \neq \sqrt{1-\cos\theta}$
if θ lies in Quadrant III or IV.
One such angle is $\theta = \dfrac{7\pi}{4}$.

63. $\sqrt{\tan^2\theta} = |\tan\theta|$
$\sqrt{\tan^2\theta} \neq \tan\theta$ if θ lies in Quadrant II or IV.
One such angle is $\theta = \dfrac{3\pi}{4}$.

65. $\sin^2 25° + \cos^2 25° = 1$

67.
$$\begin{aligned}\cos^2 20° + \cos^2 52° + \cos^2 38° + \cos^2 70° &= \cos^2 20° + \cos^2 52^2 + \sin^2(90° - 38°) + \sin^2(90° - 70°)\\ &= \cos^2 20° + \cos^2 52^2 + \sin^2 52° + \sin^2 20°\\ &= (\cos^2 20° + \sin^2 20°) + (\cos^2 52° + \sin^2 52°)\\ &= 1 + 1\\ &= 2\end{aligned}$$

69. When n is even,
$$\cos\left[\frac{(2n+1)\pi}{2}\right] = \cos\frac{\pi}{2} = 0.$$
When n is odd,
$$\cos\left[\frac{(2n+1)\pi}{2}\right] = \cos\frac{3\pi}{2} = 0.$$
Thus, $\cos\left[\dfrac{(2n+1)\pi}{2}\right] = 0$ for all integers n.

71.
$$\begin{aligned}\cos x - \csc x\cot x &= \cos x - \frac{1}{\sin x}\frac{\cos x}{\sin x}\\ &= \cos x\left(1 - \frac{1}{\sin^2 x}\right)\\ &= \cos x(1 - \csc^2 x)\\ &= -\cos x(\csc^2 x - 1)\\ &= -\cos x\cot^2 x\end{aligned}$$

73.

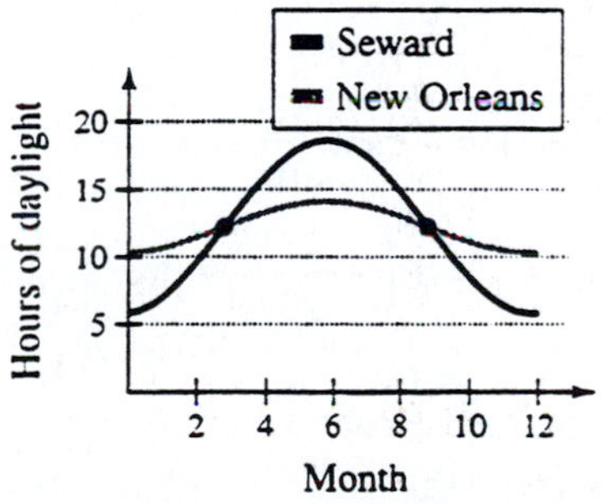

From the graph, you can see that Seward has the greater variation in the number of daylight hours.

Constant in Seward model: 6.4

Constant in New Orleans model: 1.9

75. $(2 + 3i) - \sqrt{-26} = 2 + 3i - \sqrt{26}i = 2 + (3 - \sqrt{26})i$

77. $\sqrt{-16}(1 + \sqrt{-4}) = 4i(1 + 2i) = 4i + 8i^2 = 4i - 8$

Section 5.3 Solving Trigonometric Equations

- You should be able to identify and solve trigonometric equations.
- A trigonometric equation is a conditional equation. It is true for a specific set of values.
- To solve trigonometric equations, use algebraic techniques such as collecting like terms, taking square roots, factoring, squaring, converting to quadratic form, using formulas, and using inverse functions. Study the examples in this section.

Solutions to Odd-Numbered Exercises

1. $y = \sin\dfrac{\pi x}{2} + 1$

From the graph in the textbook we see that the curve has x-intercepts at $x = -1$ and at $x = 3$.

3. $y = \tan^2\left(\dfrac{\pi x}{6}\right) - 3$

From the graph in the textbook we see that the curve has x-intercepts at $x = \pm 2$.

5. $2\cos x - 1 = 0$

(a) $2\cos\dfrac{\pi}{3} - 1 = 2\left(\dfrac{1}{2}\right) - 1 = 0$

(b) $2\cos\dfrac{5\pi}{3} - 1 = 2\left(\dfrac{1}{2}\right) - 1 = 0$

7. $3\tan^2 2x - 1 = 0$

(a) $$\begin{aligned}3\left[\tan 2\left(\frac{\pi}{12}\right)\right]^2 - 1 &= 3\tan^2\frac{\pi}{6} - 1\\ &= 3\left(\frac{1}{\sqrt{3}}\right)^2 - 1\\ &= 0\end{aligned}$$

(b) $$\begin{aligned}3\left[\tan 2\left(\frac{5\pi}{12}\right)\right]^2 - 1 &= 3\tan^2\frac{5\pi}{6} - 1\\ &= 3\left(-\frac{1}{\sqrt{3}}\right)^2 - 1\\ &= 0\end{aligned}$$

9. $2\sin^2 x - \sin x - 1 = 0$

(a) $$\begin{aligned}2\sin^2\frac{\pi}{2} - \sin\frac{\pi}{2} - 1 &= 2(1)^2 - 1 - 1\\ &= 0\end{aligned}$$

(b) $$\begin{aligned}2\sin^2\frac{7\pi}{6} - \sin\frac{7\pi}{6} - 1 &= 2\left(-\frac{1}{2}\right)^2 - \left(-\frac{1}{2}\right) - 1\\ &= \frac{1}{2} + \frac{1}{2} - 1\\ &= 0\end{aligned}$$

11. $2\cos x + 1 = 0$

$$2\cos x = -1$$
$$\cos x = -\frac{1}{2}$$
$$x = \frac{2\pi}{3} + 2n\pi$$
$$\text{or } x = \frac{4\pi}{3} + 2n\pi$$

13. $\sqrt{3}\csc x - 2 = 0$

$$\sqrt{3}\csc x = 2$$
$$\csc x = \frac{2}{\sqrt{3}}$$
$$x = \frac{\pi}{3} + 2n\pi$$
$$\text{or } x = \frac{2\pi}{3} + 2n\pi$$

15. $3\sec^2 x - 4 = 0$

$$\sec x = \pm\frac{2}{\sqrt{3}}$$
$$x = \frac{\pi}{6} + n\pi$$
$$\text{or } x = \frac{5\pi}{6} + n\pi$$

17. $2\sin^2 2x = 1$

$$\sin 2x = \pm\frac{1}{\sqrt{2}} = \pm\frac{\sqrt{2}}{2}$$
$$2x = \frac{\pi}{4} + 2n\pi,\ 2x = \frac{3\pi}{4} + 2n\pi,\ 2x = \frac{5\pi}{4} + 2n\pi,\ 2x = \frac{7\pi}{4} + 2n\pi,$$
$$2x = \frac{9\pi}{4} + 2n\pi,\ 2x = \frac{11\pi}{4} + 2n\pi,\ 2x = \frac{13\pi}{4} + 2n\pi,\ 2x = \frac{15\pi}{4} + 2n\pi,$$

Thus, $x = \frac{\pi}{8} + n\pi, \frac{3\pi}{8} + n\pi, \frac{5\pi}{8} + n\pi, \frac{7\pi}{8} + n\pi, \frac{9\pi}{8} + n\pi, \frac{11\pi}{8} + n\pi,$
$\frac{13\pi}{8} + n\pi, \frac{15\pi}{8} + n\pi.$

19. $4\sin^2 x - 3 = 0$

$$\sin x = \pm\frac{\sqrt{3}}{2}$$
$$x = \frac{\pi}{3} + n\pi$$
$$\text{or } x = \frac{2\pi}{3} + n\pi$$

21.

$$\sin^2 x = 3\cos^2 x$$
$$\sin^2 x - 3(1 - \sin^2 x) = 0$$
$$4\sin^2 x = 3$$
$$\sin x = \pm\frac{\sqrt{3}}{2}$$
$$x = \frac{\pi}{3} + n\pi$$
$$\text{or } x = \frac{2\pi}{3} + n\pi$$

23. $(3\tan^2 x - 1)(\tan^2 x - 3) = 0$

$3\tan^2 x - 1 = 0$ or $\tan^2 x - 3 = 0$

$$\tan x = \pm\frac{1}{\sqrt{3}} \qquad \tan x = \pm\sqrt{3}$$
$$x = \frac{\pi}{6} + n\pi \qquad x = \frac{\pi}{3} + n\pi$$
$$\text{or } x = \frac{5\pi}{6} + n\pi \qquad \text{or } x = \frac{2\pi}{3} + n\pi$$

25. $\cos^3 x = \cos x$

$\cos^3 x - \cos x = 0$

$\cos x(\cos^2 x - 1) = 0$

$\cos x = 0$ or $\cos^2 x - 1 = 0$

$x = \frac{\pi}{2}, \frac{3\pi}{2}$ $\quad\quad$ $\cos x = \pm 1$

$x = 0, \pi$

27. $3\tan^3 x - \tan x = 0$

$\tan x(3\tan^2 x - 1) = 0$

$\tan x = 0$ or $3\tan^2 x - 1 = 0$

$x = 0, \pi$ $\quad\quad$ $\tan x = \pm\frac{\sqrt{3}}{3}$

$x = \frac{\pi}{6}, \frac{5\pi}{6}, \frac{7\pi}{6}, \frac{11\pi}{6}$

29. $\sec^2 x - \sec x - 2 = 0$

$(\sec x - 2)(\sec x + 1) = 0$

$\sec x - 2 = 0$ or $\sec x + 1 = 0$

$\sec x = 2$ $\quad\quad$ $\sec x = -1$

$x = \frac{\pi}{3}, \frac{5\pi}{3}$ $\quad\quad$ $x = \pi$

31. $2\sin x + \csc x = 0$

$2\sin x + \frac{1}{\sin x} = 0$

$2\sin^2 x + 1 = 0$

$\sin^2 x = -\frac{1}{2} \Longrightarrow$ No solution

33. $\csc x + \cot x = 1$

$\frac{1}{\sin x} + \frac{\cos x}{\sin x} = 1$

$1 + \cos x = \sin x$

$(1 + \cos x)^2 = \sin^2 x$

$1 + 2\cos x + \cos^2 x = 1 - \cos^2 x$

$2\cos^2 x + 2\cos x = 0$

$2\cos x(\cos x + 1) = 0$

$\cos x = 0$ or $\cos x = -1$

$x = \frac{\pi}{2}, \frac{3\pi}{2}$ $\quad\quad$ $x = \pi$

($3\pi/2$ is extraneous.) $\quad$ (π is extraneous.)

$x = \pi/2$ is the only solution.

35. $\cos\left(\frac{x}{2}\right) = \frac{\sqrt{2}}{2}$

$\frac{x}{2} = \frac{\pi}{4}$ or $\frac{x}{2} = \frac{7\pi}{4}$

$x = \frac{\pi}{2}$ $\quad\quad$ $x = \frac{7\pi}{2}$

(Disregard since it is outside the interval $[0, 2\pi)$.)

37. $\frac{1 + \cos x}{1 - \cos x} = 0$

$1 + \cos x = 0$

$\cos x = -1$

$x = \pi$

39. $2\sec^2 x + \tan^2 x - 3 = 0$

$2(\tan^2 x + 1) + \tan^2 x - 3 = 0$

$3\tan^2 x - 1 = 0$

$\tan x = \pm\frac{\sqrt{3}}{3}$

$x = \frac{\pi}{6}, \frac{5\pi}{6}, \frac{7\pi}{6}, \frac{11\pi}{6}$

41. $6y^2 - 13y + 6 = 0$

$(3y - 2)(2y - 3) = 0$

$3y - 2 = 0$ or $2y - 3 = 0$

$y = \frac{2}{3}$ $\qquad$ $y = \frac{3}{2}$

$6\cos^2 x - 13\cos x + 6 = 0$

$(3\cos x - 2)(2\cos x - 3) = 0$

$3\cos x - 2 = 0$ or $2\cos x - 3 = 0$

$\cos x = \frac{2}{3}$ $\qquad$ $\cos x = \frac{3}{2}$

$x \approx 0.8411, 5.4421$ $\qquad$ (No solution)

43. $2\cos x - \sin x = 0$

$$2 = \frac{\sin x}{\cos x}$$

$$2 = \tan x$$

$$x = \arctan 2 \approx 1.1071$$

$$\text{or } x = \arctan 2 + \pi \approx 4.2487$$

Graph $y_1 = 2\cos x - \sin x$.

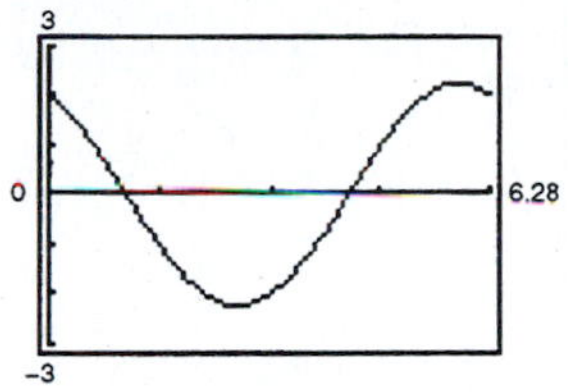

The x-intercepts occur at $x \approx 1.1071$ and $x \approx 4.2387$.

45.
$$\frac{1 + \sin x}{\cos x} + \frac{\cos x}{1 + \sin x} = 4$$

$$\frac{(1 + \sin x)^2 + \cos^2 x}{\cos x(1 + \sin x)} = 4$$

$$\frac{1 + 2\sin x + \sin^2 x + \cos^2 x}{\cos x(1 + \sin x)} = 4$$

$$\frac{2 + 2\sin x}{\cos x(1 + \sin x)} = 4$$

$$\frac{2}{\cos x} = 4$$

$$\cos x = \frac{1}{2}$$

$$x = \frac{\pi}{3}, \frac{5\pi}{3}$$

Graph $y_1 = \dfrac{1 + \sin x}{\cos x} + \dfrac{\cos x}{1 + \sin x} - 4$.

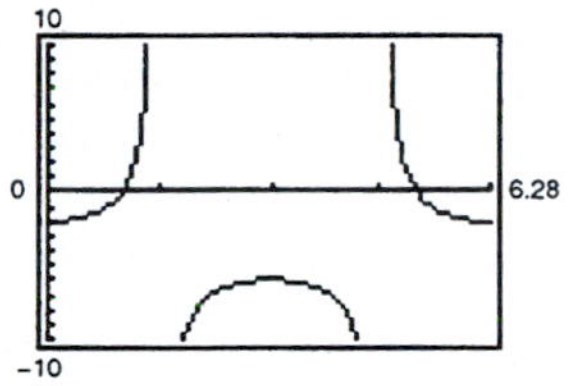

The x-intercepts occur at $x = \dfrac{\pi}{3} \approx 1.0472$ and $x = \dfrac{5\pi}{3} \approx 5.2360$.

47. $2\sin x - x = 0$

Graph $y_1 = 2\sin x - x$.

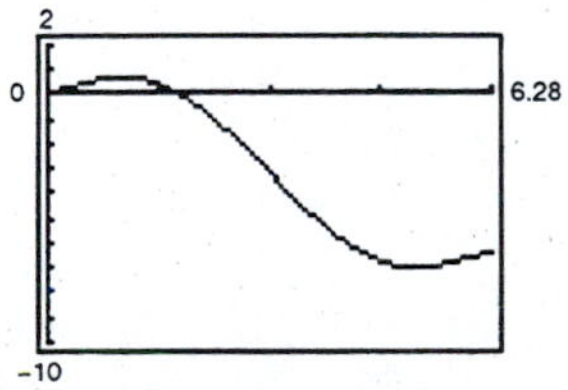

The x-intercepts occur at $x = 0$ and $x \approx 1.8955$.

49. $\sec^2 x + 0.5\tan x - 1 = 0$

Graph $y_1 = \dfrac{1}{(\cos x)^2} + 0.5\tan x - 1$.

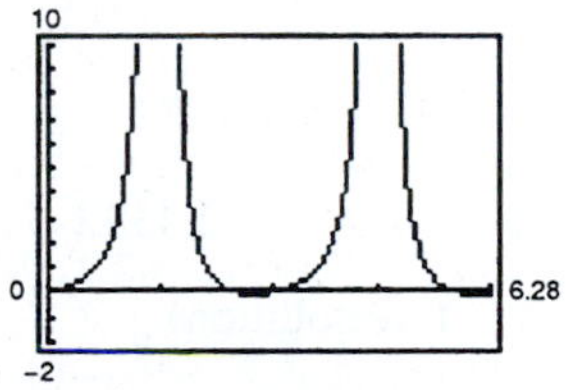

The x-intercepts occur at $x = 0$, $x \approx 2.6779$, $x = \pi \approx 3.1416$, and $x \approx 5.8195$.

51. $2\tan^2 x + 7\tan x - 15 = 0$

$(2\tan x - 3)(\tan x + 5) = 0$

$2\tan x - 3 = 0$ or $\tan x + 5 = 0$

$\tan x = 1.5$ $\qquad$ $\tan x = -5$

$x \approx 0.9828, 4.1244$ $\qquad$ $x \approx 1.7682, 4.9098$

Graph $y_1 = 2\tan^2 x + 7\tan x - 15$.

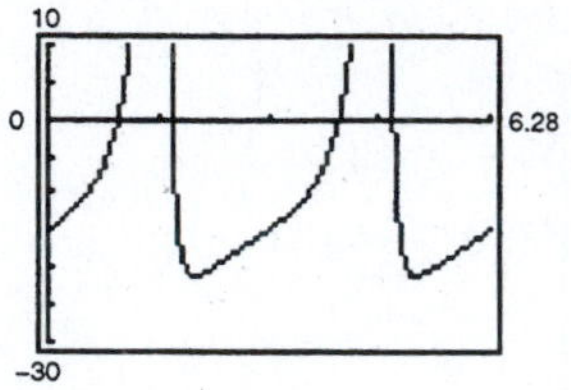

The x-intercepts occur at $x \approx 0.9828$, $x \approx 1.7682$, $x \approx 4.1244$, and $x \approx 4.9098$.

53. $12\sin^2 x - 13\sin x + 3 = 0$

$(3\sin x - 1)(4\sin x - 3) = 0$

$3\sin x - 1 = 0$ or $4\sin x - 3 = 0$

$\sin x = \frac{1}{3}$ $\qquad$ $\sin x = \frac{3}{4}$

$x \approx 0.3398, 2.8018$ $\qquad$ $x \approx 0.8481, 2.2935$

Graph $y_1 = 12\sin^2 x - 13\sin x + 3$.

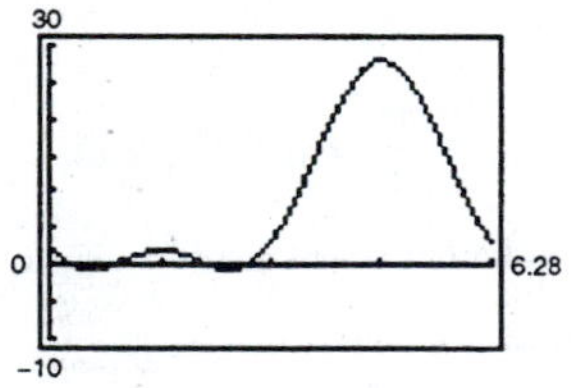

The x-intercepts occur at $x \approx 0.3398$, $x \approx 0.8481$, $x \approx 2.2935$, and $x \approx 2.8018$.

55. $\sin^2 x + 2\sin x - 1 = 0$

$$\sin x = \frac{-2 \pm \sqrt{4 + 4}}{2} \approx 0.4142, \ -2.4142$$

$\sin x \approx 0.4142$ or $\sin x \approx -2.4142$

$x \approx 0.4271, 2.7145$ $\qquad$ (No solution)

Graph $y_1 = \sin^2 x + 2\sin x - 1$.

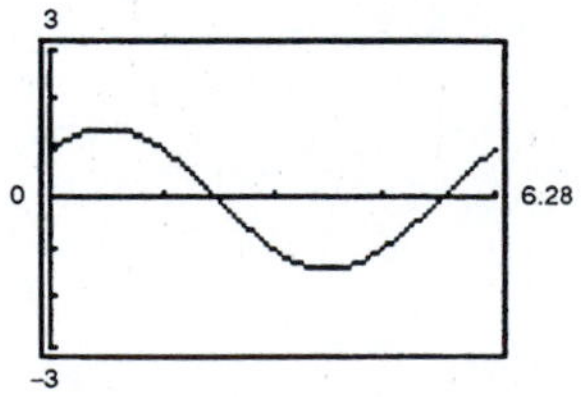

The x-intercepts occur at $x \approx 0.4271$ and $x \approx 2.7145$.

57. (a) $f(x) = \sin x + \cos x$

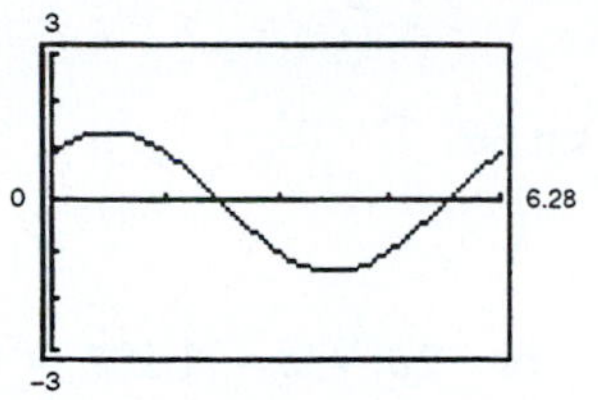

Maximum: $\left(\frac{\pi}{4}, \sqrt{2}\right)$

Minimum: $\left(\frac{5\pi}{4}, -\sqrt{2}\right)$

(b) $\cos x - \sin x = 0$

$$\cos x = \sin x$$

$$1 = \frac{\sin x}{\cos x}$$

$$\tan x = 1$$

$$x = \frac{\pi}{4}, \frac{5\pi}{4}$$

$$f\left(\frac{\pi}{4}\right) = \sin\frac{\pi}{4} + \cos\frac{\pi}{4} = \frac{\sqrt{2}}{2} + \frac{\sqrt{2}}{2} = \sqrt{2}$$

$$f\left(\frac{5\pi}{4}\right) = \sin\frac{5\pi}{4} + \cos\frac{5\pi}{4} = -\sin\frac{\pi}{4} + \left(-\cos\frac{\pi}{4}\right) = -\frac{\sqrt{2}}{2} - \frac{\sqrt{2}}{2} = -\sqrt{2}$$

Therefore, the maximum point in the interval $[0, 2\pi)$ is $(\pi/4, \sqrt{2})$ and the minimum point is $(5\pi/4, -\sqrt{2})$.

59. $f(x) = \tan\frac{\pi x}{4}$

Since $\tan 0 = 0$, $x = 1$ is the smallest nonnegative fixed point.

61. $f(x) = \cos\frac{1}{x}$

(a) The domain of $f(x)$ is all real numbers except 0.

(b) The graph has y-axis symmetry and a horizontal asymptote at $y = 1$.

(c) As $x \longrightarrow 0$, $f(x)$ oscillates between -1 and 1.

(d) There are infinitely many solutions in the interval $[-1, 1]$.

(e) The greatest solution appears to occur at $x \approx 0.6366$.

63. $y = \frac{1}{12}(\cos 8t - 3 \sin 8t)$

$$\frac{1}{12}(\cos 8t - 3 \sin 8t) = 0$$

$$\cos 8t = 3 \sin 8t$$

$$\frac{1}{3} = \tan 8t$$

$$8t \approx 0.32175 + n\pi$$

$$t \approx 0.04 + \frac{n\pi}{8}$$

In the interval $0 \le t \le 1$, $t \approx 0.04, 0.43$, and 0.83.

65. $r = \frac{1}{32}v_0^2 \sin 2\theta$, $r = 300$, $v_0 = 100$

$$300 = \frac{1}{32}(100)^2 \sin 2\theta$$

$$\sin 2\theta = 0.96$$

$2\theta \approx 1.2870$ or $2\theta \approx \pi - 1.287 \approx 1.855$

$\theta \approx 0.6435 \approx 37°$ $\quad$ $\theta \approx 0.9275 \approx 53°$

67. $A = 2x \cos x,\ 0 < x < \frac{\pi}{2}$

(a) 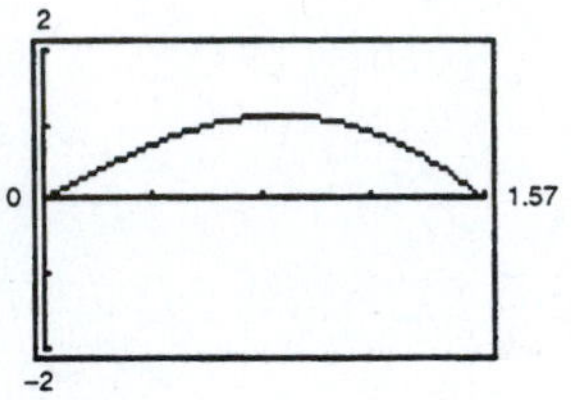

The maximum area of $A \approx 1.12$ occurs when $x \approx 0.86$.

(b) $A \ge 1$ for $0.6 < x < 1.1$

69. (a) 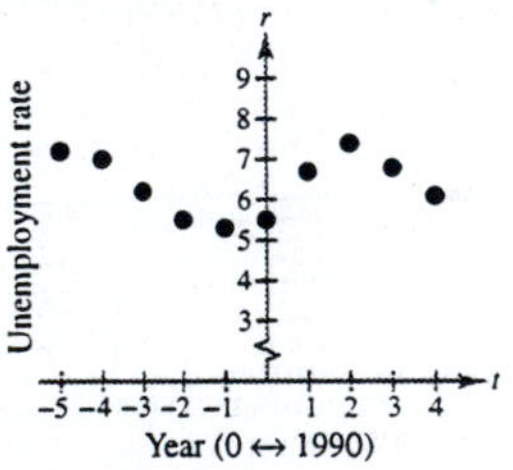

(b) By checking the graphs we see that $(iii)\ r = 1.05 \sin (0.95)(t + 6.32) + 6.20$ best fits the data.

(c) The constant term gives the rate of 6.20%.

(d) Period: $\frac{2\pi}{0.95} \approx 7$ years.

(e) $r \approx 6.00$ when $t \approx 10$ which corresponds to 2000.

Section 5.4 Sum and Difference Formulas

- You should memorize the sum and difference formulas.

$$\sin(u \pm v) = \sin u \cos v \pm \cos u \sin v$$

$$\cos(u \pm v) = \cos u \cos v \mp \sin u \sin v$$

$$\tan(u \pm v) = \frac{\tan u \pm \tan v}{1 \mp \tan u \tan v}$$

- You should be able to use these formulas to find the values of the trigonometric functions of angles whose sums or differences are special angles.
- You should be able to use these formulas to solve trigonometric equations.

Solutions to Odd-Numbered Exercises

1. (a) $\cos\left(\frac{\pi}{4} + \frac{\pi}{3}\right) = \cos\frac{\pi}{4}\cos\frac{\pi}{3} - \sin\frac{\pi}{4}\sin\frac{\pi}{3}$

$$= \frac{\sqrt{2}}{2} \cdot \frac{1}{2} - \frac{\sqrt{2}}{2} \cdot \frac{\sqrt{3}}{2}$$

$$= \frac{\sqrt{2} - \sqrt{6}}{4}$$

(b) $\cos\frac{\pi}{4} + \cos\frac{\pi}{3} = \frac{\sqrt{2}}{2} + \frac{1}{2} = \frac{\sqrt{2} + 1}{2}$

3. (a) $\sin\left(\frac{7\pi}{6} - \frac{\pi}{3}\right) = \sin\frac{5\pi}{6} = \sin\frac{\pi}{6} = \frac{1}{2}$

(b) $\sin\frac{7\pi}{6} - \sin\frac{\pi}{3} = -\frac{1}{2} - \frac{\sqrt{3}}{2} = \frac{-1 - \sqrt{3}}{2}$

5. Both statements are false. Parts (a) and (b) are unequal in Exercises 1–4.

7. $\sin 75° = \sin(30° + 45°)$

$= \sin 30° \cos 45° + \sin 45° \cos 30°$

$= \frac{1}{2} \cdot \frac{\sqrt{2}}{2} + \frac{\sqrt{2}}{2} \cdot \frac{\sqrt{3}}{2}$

$= \frac{\sqrt{2}}{4}(1 + \sqrt{3})$

$\cos 75° = \cos(30° + 45°)$

$= \cos 30° \cos 45° - \sin 30° \sin 45°$

$= \frac{\sqrt{3}}{2} \cdot \frac{\sqrt{2}}{2} - \frac{1}{2} \cdot \frac{\sqrt{2}}{2}$

$= \frac{\sqrt{2}}{4}(\sqrt{3} - 1)$

$\tan 75° = \tan(30° + 45°)$

$= \frac{\tan 30° + \tan 45°}{1 - \tan 30° \tan 45°}$

$= \frac{(\sqrt{3}/3) + 1}{1 - (\sqrt{3}/3)} = \frac{\sqrt{3} + 3}{3 - \sqrt{3}} \cdot \frac{3 + \sqrt{3}}{3 + \sqrt{3}}$

$= \frac{6\sqrt{3} + 12}{6} = \sqrt{3} + 2$

9. $\sin 105° = \sin(60° + 45°)$

$= \sin 60° \cos 45° + \sin 45° \cos 60°$

$= \frac{\sqrt{3}}{2} \cdot \frac{\sqrt{2}}{2} + \frac{\sqrt{2}}{2} \cdot \frac{1}{2}$

$= \frac{\sqrt{2}}{4}(\sqrt{3} + 1)$

$\cos 105° = \cos(60° + 45°)$

$= \cos 60° \cos 45° - \sin 60° \sin 45°$

$= \frac{1}{2} \cdot \frac{\sqrt{2}}{2} - \frac{\sqrt{3}}{2} \cdot \frac{\sqrt{2}}{2}$

$= \frac{\sqrt{2}}{4}(1 - \sqrt{3})$

$\tan 105° = \tan(60° + 45°)$

$= \frac{\tan 60° + \tan 45°}{1 - \tan 60° \tan 45°}$

$= \frac{\sqrt{3} + 1}{1 - \sqrt{3}} = \frac{\sqrt{3} + 1}{1 - \sqrt{3}} \cdot \frac{1 + \sqrt{3}}{1 + \sqrt{3}}$

$= \frac{4 + 2\sqrt{3}}{-2} = -2 - \sqrt{3}$

11. $\sin 195° = \sin(225° - 30°)$

$= \sin 225° \cos 30° - \sin 30° \cos 225°$

$= -\sin 45° \cos 30° + \sin 30° \cos 45°$

$= -\frac{\sqrt{2}}{2} \cdot \frac{\sqrt{3}}{2} + \frac{1}{2} \cdot \frac{\sqrt{2}}{2}$

$= \frac{\sqrt{2}}{4}(\sqrt{3} - 1)$

$\cos 195° = \cos(225° - 30°)$

$= \cos 225° \cos 30° + \sin 225° \sin 30°$

$= -\cos 45° \cos 30° - \sin 45° \sin 30°$

$= -\frac{\sqrt{2}}{2} \cdot \frac{\sqrt{3}}{2} - \frac{\sqrt{2}}{2} \cdot \frac{1}{2}$

$= -\frac{\sqrt{2}}{4}(\sqrt{3} + 1)$

$\tan 195° = \tan(225° - 30°)$

$= \frac{\tan 225° - \tan 30°}{1 + \tan 225° \tan 30°}$

$= \frac{\tan 45° - \tan 30°}{1 + \tan 45° \tan 30°}$

$= \frac{1 - (\sqrt{3}/3)}{1 + (\sqrt{3}/3)} = \frac{3 - \sqrt{3}}{3 + \sqrt{3}} \cdot \frac{3 - \sqrt{3}}{3 - \sqrt{3}}$

$= \frac{12 - 6\sqrt{3}}{6} = 2 - \sqrt{3}$

13. $$\sin\frac{11\pi}{12} = \sin\left(\frac{3\pi}{4} + \frac{\pi}{6}\right)$$
$$= \sin\frac{3\pi}{4}\cos\frac{\pi}{6} + \sin\frac{\pi}{6}\cos\frac{3\pi}{4}$$
$$= \frac{\sqrt{2}}{2} \cdot \frac{\sqrt{3}}{2} + \frac{1}{2}\left(-\frac{\sqrt{2}}{2}\right)$$
$$= \frac{\sqrt{2}}{4}(\sqrt{3} - 1)$$

$$\cos\frac{11\pi}{12} = \cos\left(\frac{3\pi}{4} + \frac{\pi}{6}\right)$$
$$= \cos\frac{3\pi}{4}\cos\frac{\pi}{6} - \sin\frac{3\pi}{4}\sin\frac{\pi}{6}$$
$$= -\frac{\sqrt{2}}{2} \cdot \frac{\sqrt{3}}{2} - \frac{\sqrt{2}}{2} \cdot \frac{1}{2}$$
$$= -\frac{\sqrt{2}}{4}(\sqrt{3} + 1)$$

$$\tan\frac{11\pi}{4} = \tan\left(\frac{3\pi}{4} + \frac{\pi}{6}\right)$$
$$= \frac{\tan(3\pi/4) + \tan(\pi/6)}{1 - \tan(3\pi/4)\tan(\pi/6)}$$
$$= \frac{-1 + (\sqrt{3}/3)}{1 - (-1)(\sqrt{3}/3)}$$
$$= \frac{-3 + \sqrt{3}}{3 + \sqrt{3}} \cdot \frac{3 - \sqrt{3}}{3 - \sqrt{3}}$$
$$= \frac{-12 + 6\sqrt{3}}{6} = -2 + \sqrt{3}$$

15. $$\sin\frac{17\pi}{12} = \sin\left(\frac{9\pi}{4} - \frac{5\pi}{6}\right)$$
$$= \sin\frac{9\pi}{4}\cos\frac{5\pi}{6} - \sin\frac{5\pi}{6}\cos\frac{9\pi}{4}$$
$$= \frac{\sqrt{2}}{2}\left(-\frac{\sqrt{3}}{2}\right) - \left(\frac{1}{2}\right)\left(\frac{\sqrt{2}}{2}\right)$$
$$= -\frac{\sqrt{2}}{4}(\sqrt{3} + 1)$$

$$\cos\frac{17\pi}{12} = \cos\left(\frac{9\pi}{4} - \frac{5\pi}{6}\right)$$
$$= \cos\frac{9\pi}{4}\cos\frac{5\pi}{6} + \sin\frac{9\pi}{4}\sin\frac{5\pi}{6}$$
$$= \frac{\sqrt{2}}{2}\left(-\frac{\sqrt{3}}{2}\right) + \frac{\sqrt{2}}{2}\left(\frac{1}{2}\right)$$
$$= \frac{\sqrt{2}}{4}(1 - \sqrt{3})$$

$$\tan\frac{17\pi}{12} = \tan\left(\frac{9\pi}{4} - \frac{5\pi}{6}\right)$$
$$= \frac{\tan(9\pi/4) - \tan(5\pi/6)}{1 + \tan(9\pi/4)\tan(5\pi/6)}$$
$$= \frac{1 - (-\sqrt{3}/3)}{1 + (-\sqrt{3}/3)}$$
$$= \frac{3 + \sqrt{3}}{3 - \sqrt{3}} \cdot \frac{3 + \sqrt{3}}{3 + \sqrt{3}}$$
$$= \frac{12 + 6\sqrt{3}}{6} = 2 + \sqrt{3}$$

17. $$285^\circ = 225^\circ + 60^\circ$$
$$\sin 285^\circ = \sin(225^\circ + 60^\circ)$$
$$= \sin 225^\circ \cos 60^\circ + \cos 225^\circ \sin 60^\circ$$
$$= -\frac{\sqrt{2}}{2}\left(\frac{1}{2}\right) - \frac{\sqrt{2}}{2}\left(\frac{\sqrt{3}}{2}\right) = -\frac{\sqrt{2}}{4}(1 + \sqrt{3})$$

$$\cos 285^\circ = \cos(225^\circ + 60^\circ)$$
$$= \cos 225^\circ \cos 60^\circ - \sin 225^\circ \sin 60^\circ$$
$$= -\frac{\sqrt{2}}{2}\left(\frac{1}{2}\right) - \left(-\frac{\sqrt{2}}{2}\right)\left(\frac{\sqrt{3}}{2}\right) = \frac{\sqrt{2}}{4}(-1 + \sqrt{3})$$
$$= \frac{\sqrt{2}}{4}(\sqrt{3} - 1)$$

$$\tan 285^\circ = \tan(225^\circ + 60^\circ)$$
$$= \frac{\tan 225^\circ + \tan 60^\circ}{1 - \tan 225^\circ \tan 60^\circ} = \frac{1 + \sqrt{3}}{1 - \sqrt{3}} \cdot \frac{1 + \sqrt{3}}{1 + \sqrt{3}}$$
$$= \frac{4 + 2\sqrt{3}}{-2} = -2 - \sqrt{3} = -(2 + \sqrt{3})$$

19. $$-\frac{13\pi}{12} = -\left(\frac{3\pi}{4} + \frac{\pi}{3}\right)$$

$$\sin\left[-\left(\frac{3\pi}{4} + \frac{\pi}{3}\right)\right] = -\sin\left(\frac{3\pi}{4} + \frac{\pi}{3}\right)$$

$$= -\left[\sin\frac{3\pi}{4}\cos\frac{\pi}{3} + \cos\frac{3\pi}{4}\sin\frac{\pi}{3}\right]$$

$$= -\left[\frac{\sqrt{2}}{2}\left(\frac{1}{2}\right) + \left(-\frac{\sqrt{2}}{2}\right)\left(\frac{\sqrt{3}}{2}\right)\right]$$

$$= -\frac{\sqrt{2}}{4}(1 - \sqrt{3}) = \frac{\sqrt{2}}{4}(\sqrt{3} - 1)$$

$$\cos\left[-\left(\frac{3\pi}{4} + \frac{\pi}{3}\right)\right] = \cos\left(\frac{3\pi}{4} + \frac{\pi}{3}\right)$$

$$= \cos\frac{3\pi}{4}\cos\frac{\pi}{3} - \sin\frac{3\pi}{4}\sin\frac{\pi}{3}$$

$$= -\frac{\sqrt{2}}{2}\left(\frac{1}{2}\right) - \frac{\sqrt{2}}{2}\left(\frac{\sqrt{3}}{2}\right) = -\frac{\sqrt{2}}{4}(1 + \sqrt{3})$$

$$\tan\left[-\left(\frac{3\pi}{4} + \frac{\pi}{3}\right)\right] = -\tan\left(\frac{3\pi}{4} + \frac{\pi}{3}\right)$$

$$= -\frac{\tan(3\pi/4) + \tan(\pi/3)}{1 - \tan(3\pi/4)\tan(\pi/3)} = -\frac{-1 + \sqrt{3}}{1 - (-\sqrt{3})}$$

$$= \frac{1 - \sqrt{3}}{1 + \sqrt{3}} \cdot \frac{1 - \sqrt{3}}{1 - \sqrt{3}} = \frac{4 - 2\sqrt{3}}{-2} = -2 + \sqrt{3}$$

21. $\cos 25° \cos 15° - \sin 25° \sin 15° = \cos(25° + 15°) = \cos 40°$

23. $\sin 230° \cos 30° - \cos 230° \sin 30° = \sin(230° - 30°) = \sin 200°$

25. $\dfrac{\tan 325° - \tan 86°}{1 + \tan 325° \tan 86°} = \tan(325° - 86°) = \tan 239°$

27. $\sin 3 \cos 1.2 - \cos 3 \sin 1.2 = \sin(3 - 1.2) = \sin 1.8$

29. $\dfrac{\tan 2x + \tan x}{1 - \tan 2x \tan x} = \tan(2x + x) = \tan 3x$

For Exercises 31 – 37, we have:

$\sin u = \frac{5}{13}$, u in Quadrant II $\Rightarrow$ $\cos u = -\frac{12}{13}$

$\cos v = -\frac{3}{5}$, v in Quadrant II $\Rightarrow$ $\sin v = \frac{4}{5}$

31. $\sin(u + v) = \sin u \cos v + \cos u \sin v$

$= \left(\frac{5}{13}\right)\left(-\frac{3}{5}\right) + \left(-\frac{12}{13}\right)\left(\frac{4}{5}\right)$

$= -\frac{63}{65}$

33. $\cos(u + v) = \cos u \cos v - \sin u \sin v$

$= \left(-\frac{12}{13}\right)\left(-\frac{3}{5}\right) - \left(\frac{5}{13}\right)\left(\frac{4}{5}\right)$

$= \frac{16}{65}$

35. $\sec(u+v) = \dfrac{1}{\cos(u+v)} = \dfrac{1}{16/65} = \dfrac{65}{16}$

Use Exercise 33 for $\cos(u + v)$.

37. $\tan(u-v) = \dfrac{\tan u - \tan v}{1 + \tan u \tan v} = \dfrac{-5/12 - (-4/3)}{1 + (-5/12)(-4/3)}$

$= \dfrac{11/12}{14/9} = \dfrac{33}{56}$

For Exercises 39–43, we have:

$\sin u = -\frac{7}{25}$, u in Quadrant III $\Longrightarrow \cos u = -\frac{24}{25}$

$\cos v = -\frac{4}{5}$, v in Quadrant III $\Longrightarrow \sin v = -\frac{3}{5}$

39. $\cos(u+v) = \cos u \cos v - \sin u \sin v$

$= \left(-\frac{24}{25}\right)\left(-\frac{4}{5}\right) - \left(-\frac{7}{25}\right)\left(-\frac{3}{5}\right)$

$= \frac{3}{5}$

41. $\sin(v-u) = \sin v \cos u - \cos v \sin u$

$= \left(-\frac{3}{5}\right)\left(-\frac{24}{25}\right) - \left(-\frac{4}{5}\right)\left(-\frac{7}{25}\right)$

$= \frac{44}{125}$

43. $\csc(u+v) = \dfrac{1}{\sin(u+v)} = \dfrac{1}{\sin u \cos v + \cos u \sin v}$

$= \dfrac{1}{\left(-\frac{7}{25}\right)\left(-\frac{4}{5}\right) + \left(-\frac{24}{25}\right)\left(-\frac{3}{5}\right)} = \dfrac{1}{\frac{4}{5}} = \dfrac{5}{4}$

45. $\sin(3\pi - x) = \sin 3\pi \cos x - \sin x \cos 3\pi = (0)(\cos x) - (\sin x)(-1) = \sin x$

47. $\sin\left(\dfrac{\pi}{6} + x\right) = \sin\dfrac{\pi}{6}\cos x + \sin x \cos\dfrac{\pi}{6} = \dfrac{1}{2}(\cos x + \sqrt{3}\sin x)$

49. $\cos(\pi - \theta) + \sin\left(\dfrac{\pi}{2} + \theta\right) = \cos\pi\cos\theta + \sin\pi\sin\theta + \sin\dfrac{\pi}{2}\cos\theta + \sin\theta\cos\dfrac{\pi}{2}$

$= (-1)(\cos\theta) + (0)(\sin\theta) + (1)(\cos\theta) + (\sin\theta)(0)$

$= -\cos\theta + \cos\theta$

$= 0$

51. $\cos(x+y)\cos(x-y) = (\cos x \cos y - \sin x \sin y)(\cos x \cos y + \sin x \sin y)$

$= \cos^2 x \cos^2 y - \sin^2 x \sin^2 y$

$= \cos^2 x(1 - \sin^2 y) - \sin^2 x \sin^2 y$

$= \cos^2 x - \cos^2 x \sin^2 y - \sin^2 x \sin^2 y$

$= \cos^2 x - \sin^2 y(\cos^2 x + \sin^2 x)$

$= \cos^2 x - \sin^2 y$

53. $\sin(x + y) + \sin(x - y) = \sin x \cos y + \sin y \cos x + \sin x \cos y - \sin y \cos x$

$= 2 \sin x \cos y$

55. $\cos(n\pi + \theta) = \cos n\pi \cos \theta - \sin n\pi \sin \theta$

$= (-1)^n (\cos \theta) - (0)(\sin \theta)$

$= (-1)^n (\cos \theta)$, where n is an integer.

57. $C = \arctan \dfrac{b}{a} \implies \sin C = \dfrac{b}{\sqrt{a^2 + b^2}}, \cos C = \dfrac{a}{\sqrt{a^2 + b^2}}$

$$\sqrt{a^2 + b^2} \sin(B\theta + C) = \sqrt{a^2 + b^2}\left(\sin B\theta \cdot \frac{a}{\sqrt{a^2 + b^2}} + \frac{b}{\sqrt{a^2 + b^2}} \cdot \cos \beta\theta\right) = a \sin B\theta + b \cos B\theta$$

59. $\cos\left(\dfrac{3\pi}{2} - x\right) = \cos \dfrac{3\pi}{2} \cos x + \sin \dfrac{3\pi}{2} \sin x$

$= (0)(\cos x) + (-1)(\sin x)$

$= -\sin x$

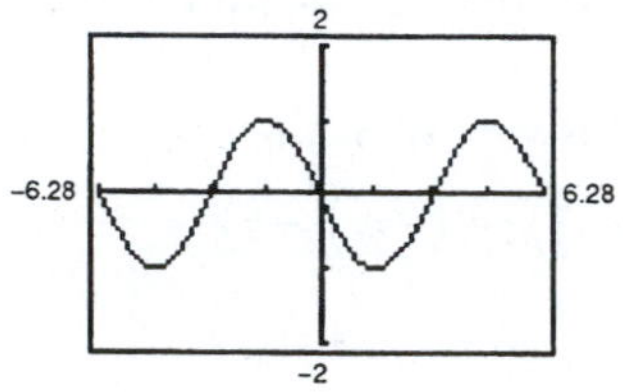

61. $\sin\left(\dfrac{3\pi}{2} + \theta\right) + \sin(\pi - \theta)$

$= \sin \dfrac{3\pi}{2} \cos \theta + \cos \dfrac{3\pi}{2} \sin \theta + \sin \pi \cos \theta - \cos \pi \sin \theta$

$= (-1)(\cos \theta) + (0)(\sin \theta) + (0)(\cos \theta) - (-1)(\sin \theta)$

$= -\cos \theta + \sin \theta$

$= \sin \theta - \cos \theta$

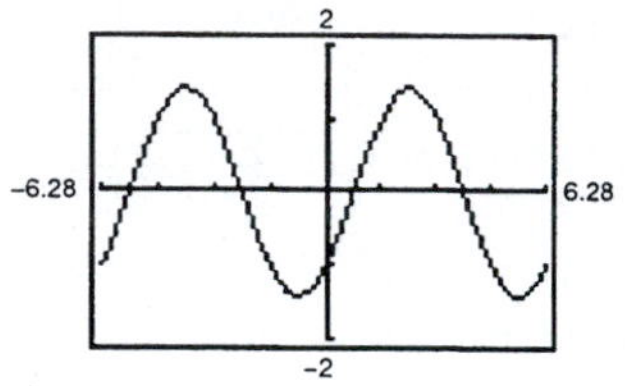

63. $\sin \theta + \cos \theta$

$a = 1,\ b = 1,\ B = 1$

(a) $C = \arctan \dfrac{b}{a} = \arctan 1 = \dfrac{\pi}{4}$

$\sin \theta + \cos \theta = \sqrt{a^2 + b^2} \sin(B\theta + C)$

$= \sqrt{2} \sin\left(\theta + \dfrac{\pi}{4}\right)$

(b) $C = \arctan \dfrac{a}{b} = \arctan 1 = \dfrac{\pi}{4}$

$\sin \theta + \cos \theta = \sqrt{a^2 + b^2} \cos(B\theta - C)$

$= \sqrt{2} \cos\left(\theta - \dfrac{\pi}{4}\right)$

65. $12 \sin 3\theta + 5 \cos 3\theta$

$a = 12,\ b = 5,\ B = 3$

(a) $C = \arctan \dfrac{b}{a} = \arctan \dfrac{5}{12} \approx 0.3948$

$$12 \sin 3\theta + 5 \cos 3\theta = \sqrt{a^2 + b^2} \sin(B\theta + C)$$
$$\approx 13 \sin(3\theta + 0.3948)$$

(b) $C = \arctan \dfrac{a}{b} = \arctan \dfrac{12}{5} \approx 1.1760$

$$12 \sin 3\theta + 5 \cos 3\theta = \sqrt{a^2 + b^2} \cos(B\theta - C)$$
$$\approx 13 \cos(3\theta - 1.1760)$$

67. $C = \arctan \dfrac{b}{a} = \dfrac{\pi}{2} \implies a = 0$

$\sqrt{a^2 + b^2} = 2 \implies b = 2$

$B = 1$

$$2 \sin\left(\theta + \frac{\pi}{2}\right) = (0)(\sin\theta) + (2)(\cos\theta) = 2 \cos \theta$$

69.
$$\begin{aligned}\sin(\arcsin x + \arccos x) &= \sin(\arcsin x) \cos(\arccos x) + \sin(\arccos x) \cos(\arcsin x)\\ &= x \cdot x + \sqrt{1 - x^2} \cdot \sqrt{1 - x^2}\\ &= x^2 + 1 - x^2\\ &= 1\end{aligned}$$

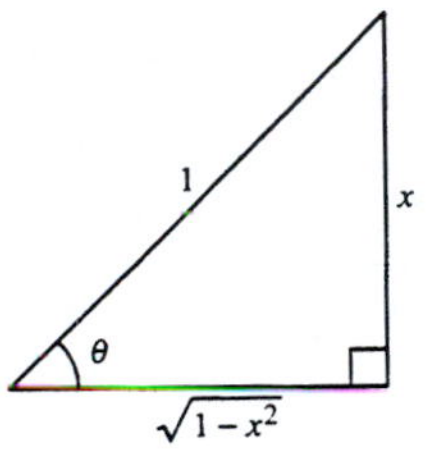

$\theta = \arcsin x$

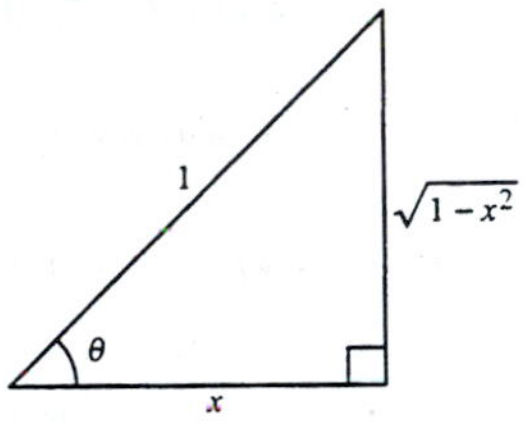

$\theta = \arccos x$

71.
$$\sin\left(x + \frac{\pi}{3}\right) + \sin\left(x - \frac{\pi}{3}\right) = 1$$

$$\sin x \cos \frac{\pi}{3} + \cos x \sin \frac{\pi}{3} + \sin x \cos \frac{\pi}{3} - \cos x \sin \frac{\pi}{3} = 1$$
$$2 \sin x(0.5) = 1$$
$$\sin x = 1$$
$$x = \frac{\pi}{2}$$

73. $\cos\left(x + \frac{\pi}{4}\right) - \cos\left(x - \frac{\pi}{4}\right) = 1$

$$\cos x \cos \frac{\pi}{4} - \sin x \sin \frac{\pi}{4} - \left(\cos x \cos \frac{\pi}{4} + \sin x \sin \frac{\pi}{4}\right) = 1$$

$$-2 \sin x\left(\frac{\sqrt{2}}{2}\right) = 1$$

$$-\sqrt{2} \sin x = 1$$

$$\sin x = -\frac{1}{\sqrt{2}}$$

$$\sin x = -\frac{\sqrt{2}}{2}$$

$$x = \frac{5\pi}{4}, \frac{7\pi}{4}$$

75. Analytically: $\cos\left(x + \frac{\pi}{4}\right) + \cos\left(x - \frac{\pi}{4}\right) = 1$

$$\cos x \cos \frac{\pi}{4} - \sin x \sin \frac{\pi}{4} + \cos x \cos \frac{\pi}{4} + \sin x \sin \frac{\pi}{4} = 1$$

$$2 \cos x\left(\frac{\sqrt{2}}{2}\right) = 1$$

$$\sqrt{2} \cos x = 1$$

$$\cos x = \frac{1}{\sqrt{2}}$$

$$\cos x = \frac{\sqrt{2}}{2}$$

$$x = \frac{\pi}{4}, \frac{7\pi}{4}$$

Graphically: Graph $y_1 = \cos\left(x + \frac{\pi}{4}\right) + \cos\left(x - \frac{\pi}{4}\right)$ and $y_2 = 1$.

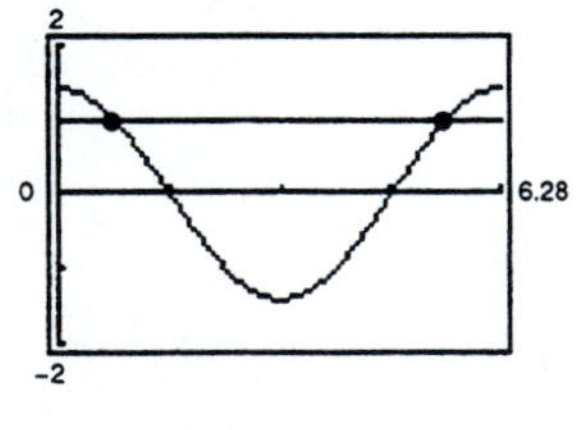

The points of intersection occur at $x = \frac{\pi}{4}$ and $x = \frac{7\pi}{4}$.

77. $\sin^2\left(\theta + \frac{\pi}{4}\right) + \sin^2\left(\theta - \frac{\pi}{4}\right) = \left[\sin\theta\cos\frac{\pi}{4} + \cos\theta\sin\frac{\pi}{4}\right]^2 + \left[\sin\theta\cos\frac{\pi}{4} - \cos\theta\sin\frac{\pi}{4}\right]^2$

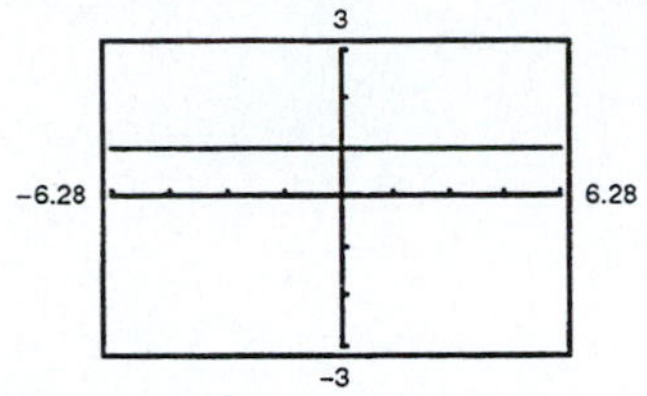

$$= \left[\frac{\sin\theta}{\sqrt{2}} + \frac{\cos\theta}{\sqrt{2}}\right]^2 + \left[\frac{\sin\theta}{\sqrt{2}} - \frac{\cos\theta}{\sqrt{2}}\right]^2$$

$$= \frac{\sin^2\theta}{2} + \sin\theta\cos\theta + \frac{\cos^2\theta}{2} + \frac{\sin^2\theta}{2} - \sin\theta\cos\theta + \frac{\cos^2\theta}{2}$$

$$= \sin^2\theta + \cos^2\theta$$

$$= 1$$

79. $y = \frac{1}{3}\sin 2t + \frac{1}{4}\cos 2t$

(a) $a = \frac{1}{3}, \ b = \frac{1}{4}, \ B = 2$

$$C = \arctan\frac{b}{a} = \arctan\frac{3}{4} \approx 0.6435$$

$$y \approx \sqrt{\left(\tfrac{1}{3}\right)^2 + \left(\tfrac{1}{4}\right)^2}\,\sin(2t + 0.6435)$$

$$= \frac{5}{12}\sin(2t + 0.6435)$$

(b) Amplitude: $\frac{5}{12}$ feet

(c) Frequency: $\frac{1}{\text{period}} = \frac{b}{2\pi} = \frac{2}{2\pi} = \frac{1}{\pi}$ cycles per second

Section 5.5 Multiple-Angle and Product-to-Sum Formulas

- You should know the following double-angle formulas.
  (a) $\sin 2u = 2 \sin u \cos u$
  (b) $\cos 2u = \cos^2 u - \sin^2 u$
  $= 2\cos^2 u - 1$
  $= 1 - 2\sin^2 u$
  (c) $\tan 2u = \dfrac{2 \tan u}{1 - \tan^2 u}$
- You should be able to reduce the power of a trigonometric function.
  (a) $\sin^2 u = \dfrac{1 - \cos 2u}{2}$
  (b) $\cos^2 u = \dfrac{1 + \cos 2u}{2}$
  (c) $\tan^2 u = \dfrac{1 - \cos 2u}{1 + \cos 2u}$
- You should be able to use the half-angle formulas.
  (a) $\sin \dfrac{u}{2} = \pm\sqrt{\dfrac{1 - \cos u}{2}}$
  (b) $\cos \dfrac{u}{2} = \pm\sqrt{\dfrac{1 + \cos u}{2}}$
  (c) $\tan \dfrac{u}{2} = \dfrac{1 - \cos u}{\sin u} = \dfrac{\sin u}{1 + \cos u}$
- You should be able to use the product-sum formulas.
  (a) $\sin u \sin v = \dfrac{1}{2}[\cos(u - v) - \cos(u + v)]$
  (b) $\cos u \cos v = \dfrac{1}{2}[\cos(u - v) + \cos(u + v)]$
  (c) $\sin u \cos v = \dfrac{1}{2}[\sin(u + v) + \sin(u - v)]$
  (d) $\cos u \sin v = \dfrac{1}{2}[\sin(u + v) - \sin(u - v)]$
- You should be able to use the sum-product formulas.
  (a) $\sin x + \sin y = 2 \sin\left(\dfrac{x + y}{2}\right)\cos\left(\dfrac{x - y}{2}\right)$
  (b) $\sin x - \sin y = 2 \cos\left(\dfrac{x + y}{2}\right)\sin\left(\dfrac{x - y}{2}\right)$
  (c) $\cos x + \cos y = 2 \cos\left(\dfrac{x + y}{2}\right)\cos\left(\dfrac{x - y}{2}\right)$
  (d) $\cos x - \cos y = -2 \sin\left(\dfrac{x + y}{2}\right)\sin\left(\dfrac{x - y}{2}\right)$

Solutions to Odd-Numbered Exercises

Figure for Exercises 1–7

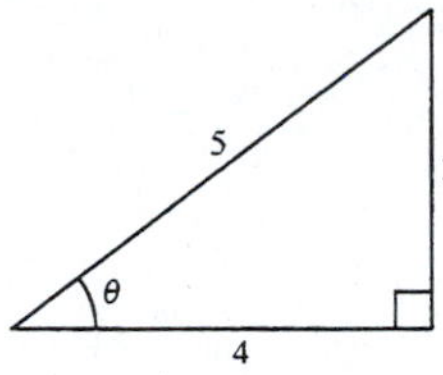

$\sin\theta = \frac{3}{5}$

$\cos\theta = \frac{4}{5}$

$\tan\theta = \frac{3}{4}$

1. $\sin\theta = \frac{3}{5}$

3. $\cos 2\theta = 2\cos^2\theta - 1$
$= 2\left(\frac{4}{5}\right)^2 - 1$
$= \frac{32}{25} - \frac{25}{25}$
$= \frac{7}{25}$

5. $\tan 2\theta = \dfrac{2\tan\theta}{1 - \tan^2\theta}$
$= \dfrac{2(3/4)}{1 - (3/4)^2}$
$= \dfrac{3/2}{1 - 9/16}$
$= \dfrac{3}{2} \cdot \dfrac{16}{7}$
$= \dfrac{24}{7}$

7. $\csc 2\theta = \dfrac{1}{\sin 2\theta}$
$= \dfrac{1}{2\sin\theta\cos\theta}$
$= \dfrac{1}{2(3/5)(4/5)}$
$= \dfrac{25}{24}$

9. $\sin 2x - \sin x = 0$
$2\sin x\cos x - \sin x = 0$
$\sin x(2\cos x - 1) = 0$
$\sin x = 0$ or $2\cos x - 1 = 0$
$x = 0, \pi$ $\qquad \cos x = \dfrac{1}{2}$
$\qquad x = \dfrac{\pi}{3}, \dfrac{5\pi}{3}$
$x = 0, \dfrac{\pi}{3}, \pi, \dfrac{5\pi}{3}$

11. $4\sin x\cos x = 1$
$2\sin 2x = 1$
$\sin 2x = \dfrac{1}{2}$
$2x = \dfrac{\pi}{6} + 2n\pi$ or $2x = \dfrac{5\pi}{6} + 2n\pi$
$x = \dfrac{\pi}{12} + n\pi \qquad x = \dfrac{5\pi}{12} + n\pi$
$x = \dfrac{\pi}{12}, \dfrac{13\pi}{12} \qquad x = \dfrac{5\pi}{12}, \dfrac{17\pi}{12}$

13.

$$\cos 2x = \cos x$$
$$\cos^2 x - \sin^2 x = \cos x$$
$$\cos^2 x - (1 - \cos^2 x) - \cos x = 0$$
$$2\cos^2 x - \cos x - 1 = 0$$
$$(2\cos x + 1)(\cos x - 1) = 0$$

$2\cos x + 1 = 0$ or $\cos x - 1 = 0$

$\cos x = -\dfrac{1}{2}$ $\qquad$ $\cos x = 1$

$x = \dfrac{2\pi}{3}, \dfrac{4\pi}{3}$ $\qquad$ $x = 0$

15.

$$\tan 2x - \cot x = 0$$
$$\frac{2\tan x}{1 - \tan^2 x} = \cot x$$
$$2\tan x = \cot x(1 - \tan^2 x)$$
$$2\tan x = \cot x - \cot x \tan^2 x$$
$$2\tan x = \cot x - \tan x$$
$$3\tan x = \cot x$$
$$3\tan x - \cot x = 0$$
$$3\tan x - \frac{1}{\tan x} = 0$$
$$\frac{3\tan^2 x - 1}{\tan x} = 0$$
$$\frac{1}{\tan x}(3\tan^2 x - 1) = 0$$
$$\cot x(3\tan^2 x - 1) = 0$$

$\cot x = 0$ or $3\tan^2 x - 1 = 0$

$x = \dfrac{\pi}{2}, \dfrac{3\pi}{2}$ $\qquad$ $\tan^2 x = \dfrac{1}{3}$

$\tan x = \pm\dfrac{\sqrt{3}}{3}$

$x = \dfrac{\pi}{6}, \dfrac{5\pi}{6}, \dfrac{7\pi}{6}, \dfrac{11\pi}{6}$

$$x = \frac{\pi}{6}, \frac{\pi}{2}, \frac{5\pi}{6}, \frac{7\pi}{6}, \frac{3\pi}{2}, \frac{11\pi}{6}$$

17.

$$\sin 4x = -2\sin 2x$$
$$\sin 4x + 2\sin 2x = 0$$
$$2\sin 2x\cos 2x + 2\sin 2x = 0$$
$$2\sin 2x(\cos 2x + 1) = 0$$

$2\sin 2x = 0$ or $\cos 2x + 1 = 0$

$\sin 2x = 0$ $\qquad$ $\cos 2x = -1$

$2x = n\pi$ $\qquad$ $2x = \pi + 2n\pi$

$x = \dfrac{n}{2}\pi$ $\qquad$ $x = \dfrac{\pi}{2} + n\pi$

$x = 0, \dfrac{\pi}{2}, \pi, \dfrac{3\pi}{2}$ $\qquad$ $x = \dfrac{\pi}{2}, \dfrac{3\pi}{2}$

19.

$$f(x) = 6\sin x\cos x$$
$$= 3(2\sin x\cos x)$$
$$= 3\sin 2x$$

21.

$$g(x) = 4 - 8\sin^2 x$$
$$= 4(1 - 2\sin^2 x)$$
$$= 4\cos 2x$$

23. $\sin u = \frac{3}{5},\ 0 < u < \frac{\pi}{2} \implies \cos u = \frac{4}{5}$

$\sin 2u = 2 \sin u \cos u = 2 \cdot \frac{3}{5} \cdot \frac{4}{5} = \frac{24}{25}$

$\cos 2u = \cos^2 u - \sin^2 u = \frac{16}{25} - \frac{9}{25} = \frac{7}{25}$

$\tan 2u = \frac{2 \tan u}{1 - \tan^2 u} = \frac{2(3/4)}{1 - (9/16)} = \frac{24}{7}$

25. $\tan u = \frac{1}{2},\ \pi < u < \frac{3\pi}{2} \implies \sin u = -\frac{1}{\sqrt{5}}$ and

$\cos u = -\frac{2}{\sqrt{5}}$

$\sin 2u = 2 \sin u \cos u = 2\left(-\frac{1}{\sqrt{5}}\right)\left(-\frac{2}{\sqrt{5}}\right) = \frac{4}{5}$

$\cos 2u = \cos^2 u - \sin^2 u = \left(-\frac{2}{\sqrt{5}}\right)^2 - \left(-\frac{1}{\sqrt{5}}\right)^2 = \frac{3}{5}$

$\tan 2u = \frac{2 \tan u}{1 - \tan^2 u} = \frac{2(1/2)}{1 - (1/4)} = \frac{4}{3}$

27. $\sec u = -\frac{5}{2},\ \frac{\pi}{2} < u < \pi \implies \sin u = \frac{\sqrt{21}}{5}$ and $\cos u = -\frac{2}{5}$

$\sin 2u = 2 \sin u \cos u = 2\left(\frac{\sqrt{21}}{5}\right)\left(-\frac{2}{5}\right) = -\frac{4\sqrt{21}}{25}$

$\cos 2u = \cos^2 u - \sin^2 u = \left(-\frac{2}{5}\right)^2 - \left(\frac{\sqrt{21}}{5}\right)^2 = -\frac{17}{25}$

$\tan 2u = \frac{2 \tan u}{1 - \tan^2 u} = \frac{2(-\sqrt{21}/2)}{1 - (-\sqrt{21}/2)^2}$

$= \frac{-\sqrt{21}}{1 - 21/4} = \frac{4\sqrt{21}}{17}$

29. $\cos^4 x = (\cos^2 x)(\cos^2 x) = \left(\frac{1 + \cos 2x}{2}\right)\left(\frac{1 + \cos 2x}{2}\right) = \frac{1 + 2\cos 2x + \cos^2 2x}{4}$

$= \frac{1 + 2\cos 2x + (1 + \cos 4x)/2}{4}$

$= \frac{2 + 4\cos 2x + 1 + \cos 4x}{8}$

$= \frac{3 + 4\cos 2x + \cos 4x}{8}$

$= \frac{1}{8}(3 + 4\cos 2x + \cos 4x)$

31. $(\sin^2 x)(\cos^2 x) = \left(\frac{1 - \cos 2x}{2}\right)\left(\frac{1 + \cos 2x}{2}\right)$

$= \frac{1 - \cos^2 2x}{4}$

$= \frac{1}{4}\left(1 - \frac{1 + \cos 4x}{2}\right)$

$= \frac{1}{8}(2 - 1 - \cos 4x)$

$= \frac{1}{8}(1 - \cos 4x)$

33. $\sin^2 x \cos^4 x = \sin^2 x \cos^2 x \cos^2 x = \left(\frac{1 - \cos 2x}{2}\right)\left(\frac{1 + \cos 2x}{2}\right)\left(\frac{1 + \cos 2x}{2}\right)$

$$= \frac{1}{8}(1 - \cos 2x)(1 + \cos 2x)(1 + \cos 2x)$$

$$= \frac{1}{8}(1 - \cos^2 2x)(1 + \cos 2x)$$

$$= \frac{1}{8}(1 + \cos 2x - \cos^2 2x - \cos^3 2x)$$

$$= \frac{1}{8}\left[1 + \cos 2x - \left(\frac{1 + \cos 4x}{2}\right) - \cos 2x\left(\frac{1 + \cos 4x}{2}\right)\right]$$

$$= \frac{1}{16}[2 + 2\cos 2x - 1 - \cos 4x - \cos 2x - \cos 2x \cos 4x]$$

$$= \frac{1}{16}\left[1 + \cos 2x - \cos 4x - \left(\frac{1}{2}\cos 2x + \frac{1}{2}\cos 6x\right)\right]$$

$$= \frac{1}{32}(2 + 2\cos 2x - 2\cos 4x - \cos 2x - \cos 6x)$$

$$= \frac{1}{32}(2 + \cos 2x - 2\cos 4x - \cos 6x)$$

Figure for Exercises 35 – 39

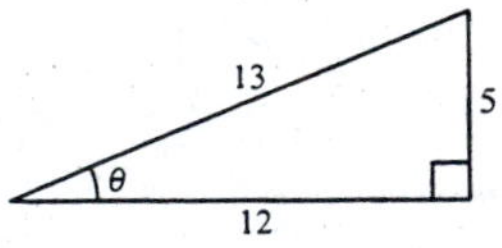

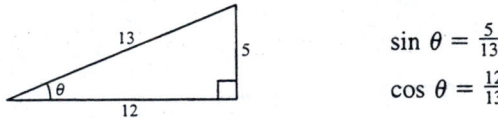

$\sin\theta = \frac{5}{13}$

$\cos\theta = \frac{12}{13}$

35. $\cos\frac{\theta}{2} = \sqrt{\frac{1 + \cos\theta}{2}} = \sqrt{\frac{1 + 12/13}{2}} = \sqrt{\frac{25}{26}} = \frac{5}{\sqrt{26}}$

37. $\tan\frac{\theta}{2} = \frac{\sin\theta}{1 + \cos\theta} = \frac{5/13}{1 + 12/13} = \frac{5}{25} = \frac{1}{5}$

39. $\csc\frac{\theta}{2} = \frac{1}{\sin(\theta/2)} = \frac{1}{\sqrt{(1 - \cos\theta)/2}} = \frac{1}{\sqrt{\left(1 - \frac{12}{13}\right)/2}} = \frac{1}{\sqrt{1/26}} = \sqrt{26}$

41. $\sin 105° = \sin\left(\frac{1}{2}\cdot 210°\right) = \sqrt{\frac{1 - \cos 210°}{2}} = \sqrt{\frac{1 + (\sqrt{3}/2)}{2}} = \frac{1}{2}\sqrt{2 + \sqrt{3}}$

$\cos 105° = \cos\left(\frac{1}{2}\cdot 210°\right) = -\sqrt{\frac{1 + \cos 210°}{2}} = -\sqrt{\frac{1 - (\sqrt{3}/2)}{2}} = -\frac{1}{2}\sqrt{2 - \sqrt{3}}$

$\tan 105° = \tan\left(\frac{1}{2}\cdot 210°\right) = \frac{\sin 210°}{1 + \cos 210°} = \frac{-1/2}{1 - (\sqrt{3}/2)} = -2 - \sqrt{3}$

43. $\sin 112° 30' = \sin\left(\frac{1}{2} \cdot 225°\right) = \sqrt{\frac{1 - \cos 225°}{2}} = \sqrt{\frac{1 + (\sqrt{2}/2)}{2}} = \frac{1}{2}\sqrt{2 + \sqrt{2}}$

$\cos 112° 30' = \cos\left(\frac{1}{2} \cdot 225°\right) = -\sqrt{\frac{1 + \cos 225°}{2}} = -\sqrt{\frac{1 - (\sqrt{2}/2)}{2}} = -\frac{1}{2}\sqrt{2 - \sqrt{2}}$

$\tan 112° 30' = \tan\left(\frac{1}{2} \cdot 225°\right) = \frac{\sin 225°}{1 + \cos 225°} = \frac{-\sqrt{2}/2}{1 - (\sqrt{2}/2)} = -1 - \sqrt{2}$

45. $\sin \frac{\pi}{8} = \sin\left[\frac{1}{2}\left(\frac{\pi}{4}\right)\right] = \sqrt{\frac{1 - \cos(\pi/4)}{2}} = \frac{1}{2}\sqrt{2 - \sqrt{2}}$

$\cos \frac{\pi}{8} = \cos\left[\frac{1}{2}\left(\frac{\pi}{4}\right)\right] = \sqrt{\frac{1 + \cos(\pi/4)}{2}} = \frac{1}{2}\sqrt{2 + \sqrt{2}}$

$\tan \frac{\pi}{8} = \tan\left[\frac{1}{2}\left(\frac{\pi}{4}\right)\right] = \frac{\sin(\pi/4)}{1 + \cos(\pi/4)} = \frac{\sqrt{2}/2}{1 + (\sqrt{2}/2)} = \sqrt{2} - 1$

47. $\sin u = \frac{5}{13}, \ \frac{\pi}{2} < u < \pi \implies \cos u = -\frac{12}{13}$

$\sin\left(\frac{u}{2}\right) = \sqrt{\frac{1 - \cos u}{2}} = \sqrt{\frac{1 + (12/13)}{.2}} = \frac{5\sqrt{26}}{26}$

$\cos\left(\frac{u}{2}\right) = \sqrt{\frac{1 + \cos u}{2}} = \sqrt{\frac{1 - (12/13)}{2}} = \frac{\sqrt{26}}{26}$

$\tan\left(\frac{u}{2}\right) = \frac{\sin u}{1 + \cos u} = \frac{5/13}{1 - (12/13)} = \frac{5}{1} = 5$

49. $\tan u = -\frac{5}{8}, \ \frac{3\pi}{2} < u < 2\pi \implies \sin u = -\frac{5}{\sqrt{89}}$ and $\cos u = \frac{8}{\sqrt{89}}$

$\sin\left(\frac{u}{2}\right) = \sqrt{\frac{1 - \cos u}{2}} = \sqrt{\frac{1 - (8/\sqrt{89})}{2}} \sqrt{\frac{\sqrt{89} - 8}{2\sqrt{89}}} = \sqrt{\frac{89 - 8\sqrt{89}}{178}}$

$\cos\left(\frac{u}{2}\right) = -\sqrt{\frac{1 + \cos u}{2}} = -\sqrt{\frac{1 + (8/\sqrt{89})}{2}} = -\sqrt{\frac{\sqrt{89} + 8}{2\sqrt{89}}} = -\sqrt{\frac{89 + 8\sqrt{89}}{178}}$

$\tan\left(\frac{u}{2}\right) = \frac{1 - \cos u}{\sin u} = \frac{1 - (8/\sqrt{89})}{-5/\sqrt{89}} = \frac{8 - \sqrt{89}}{5}$

51. $\csc u = -\frac{5}{3}, \ \pi < u < \frac{3\pi}{2} \implies \sin u = -\frac{3}{5}$ and $\cos u = -\frac{4}{5}$

$\sin\left(\frac{u}{2}\right) = \sqrt{\frac{1 - \cos u}{2}} = \sqrt{\frac{1 + (4/5)}{2}} = \frac{3\sqrt{10}}{10}$

$\cos\left(\frac{u}{2}\right) = -\sqrt{\frac{1 + \cos u}{2}} = -\sqrt{\frac{1 - (4/5)}{2}} = -\frac{\sqrt{10}}{10}$

$\tan\left(\frac{u}{2}\right) = \frac{1 - \cos u}{\sin u} = \frac{1 + (4/5)}{-3/5} = -3$

53. $\sqrt{\dfrac{1-\cos 6x}{2}} = |\sin 3x|$

55. $-\sqrt{\dfrac{1-\cos 8x}{1+\cos 8x}} = -\dfrac{\sqrt{(1-\cos 8x)/2}}{\sqrt{(1+\cos 8x)/2}}$

$$= -\left|\frac{\sin 4x}{\cos 4x}\right|$$

$$= -|\tan 4x|$$

57. $\sin\dfrac{x}{2} + \cos x = 0$

$$\pm\sqrt{\frac{1-\cos x}{2}} = -\cos x$$

$$\frac{1-\cos x}{2} = \cos^2 x$$

$$0 = 2\cos^2 x + \cos x - 1$$

$$= (2\cos x - 1)(\cos x + 1)$$

$\cos x = \dfrac{1}{2}$ or $\cos x = -1$

$x = \dfrac{\pi}{3}, \dfrac{5\pi}{3}$ $\quad x = \pi$

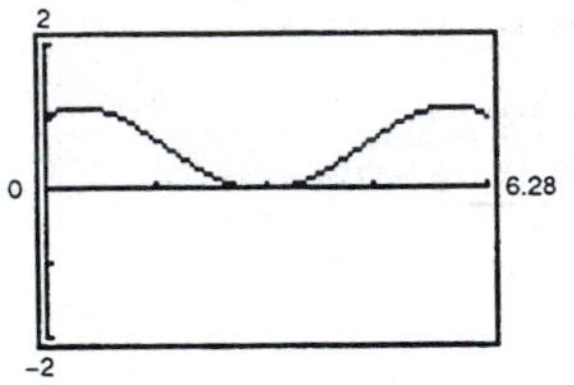

By checking these values in the original equation, we see that $x = \pi/3$ and $x = 5\pi/3$ are extraneous, and $x = \pi$ is the only solution.

59. $\cos\dfrac{x}{2} - \sin x = 0$

$$\pm\sqrt{\frac{1+\cos x}{2}} = \sin x$$

$$\frac{1+\cos x}{2} = \sin^2 x$$

$$1 + \cos x = 2\sin^2 x$$

$$1 + \cos x = 2 - 2\cos^2 x$$

$$2\cos^2 x + \cos x - 1 = 0$$

$$(2\cos x - 1)(\cos x + 1) = 0$$

$2\cos x - 1 = 0$ or $\cos x + 1 = 0$

$\cos x = \dfrac{1}{2}$ $\quad \cos x = -1$

$x = \dfrac{\pi}{3}, \dfrac{5\pi}{3}$ $\quad x = \pi$

$$x = \frac{\pi}{3}, \pi, \frac{5\pi}{3}$$

$\pi/3$, π, and $5\pi/3$ are all solutions to the equation.

61. $6\sin\dfrac{\pi}{4}\cos\dfrac{\pi}{4} = 6 \cdot \dfrac{1}{2}\left[\sin\left(\dfrac{\pi}{4}+\dfrac{\pi}{4}\right) + \sin\left(\dfrac{\pi}{4}-\dfrac{\pi}{4}\right)\right] = 3\left(\sin\dfrac{\pi}{2} + \sin 0\right)$

63. $\sin 5\theta \cos 3\theta = \frac{1}{2}[\sin(5\theta + 3\theta) + \sin(5\theta - 3\theta)] = \frac{1}{2}(\sin 8\theta + \sin 2\theta)$

65. $5\cos(-5\beta)\cos 3\beta = 5 \cdot \frac{1}{2}[\cos(-5\beta - 3\beta) + \cos(-5\beta + 3\beta)] = \frac{5}{2}[\cos(-8\beta) + \cos(-2\beta)]$

$$= \tfrac{5}{2}(\cos 8\beta + \cos 2\beta)$$

67. $\sin(x+y)\sin(x-y) = \frac{1}{2}(\cos 2y - \cos 2x)$

69. $\sin(\theta + \pi)\cos(\theta - \pi) = \frac{1}{2}(\sin 2\theta + \sin 2\pi)$

71. $\sin 60^\circ + \sin 30^\circ = 2\sin\left(\dfrac{60^\circ + 30^\circ}{2}\right)\cos\left(\dfrac{60^\circ - 30^\circ}{2}\right) = 2\sin 45^\circ \cos 15^\circ$

73. $\cos\dfrac{3\pi}{4} - \cos\dfrac{\pi}{4} = -2\sin\left(\dfrac{(3\pi/4) + (\pi/4)}{2}\right)\sin\left(\dfrac{(3\pi/4) - (\pi/4)}{2}\right) = -2\sin\dfrac{\pi}{2}\sin\dfrac{\pi}{4}$

75. $\cos 6x + \cos 2x = 2\cos\left(\dfrac{6x + 2x}{2}\right)\cos\left(\dfrac{6x - 2x}{2}\right) = 2\cos 4x\cos 2x$

77. $\sin(\alpha + \beta) - \sin(\alpha - \beta) = 2\cos\left(\frac{\alpha + \beta + \alpha - \beta}{2}\right)\sin\left(\frac{\alpha + \beta - \alpha + \beta}{2}\right) = 2\cos\alpha\sin\beta$

79. $\cos(\phi + 2\pi) + \cos\phi = 2\cos\left(\frac{\phi + 2\pi + \phi}{2}\right)\cos\left(\frac{\phi + 2\pi - \phi}{2}\right) = 2\cos(\phi + \pi)\cos\pi$

81.

$$\sin 6x + \sin 2x = 0$$
$$2\sin\left(\frac{6x + 2x}{2}\right)\cos\left(\frac{6x - 2x}{2}\right) = 0$$
$$2(\sin 4x)\cos 2x = 0$$

$\sin 4x = 0$ or $\cos 2x = 0$

$4x = n\pi$ $\qquad 2x = \frac{\pi}{2} + n\pi$

$x = \frac{n\pi}{4}$ $\qquad x = \frac{\pi}{4} + \frac{n\pi}{2}$

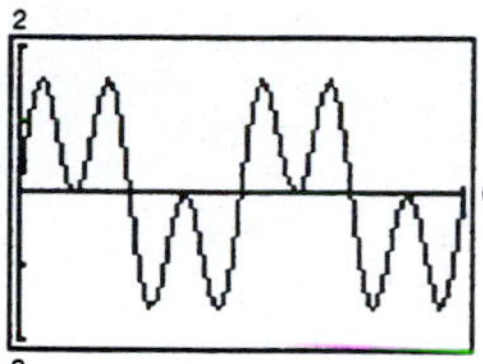

In the interval we have

$x = 0, \frac{\pi}{4}, \frac{\pi}{2}, \frac{3\pi}{4}, \pi, \frac{5\pi}{4}, \frac{3\pi}{2}, \frac{7\pi}{4}.$

83.

$$\frac{\cos 2x}{\sin 3x - \sin x} - 1 = 0$$
$$\frac{\cos 2x}{\sin 3x - \sin x} = 1$$
$$\frac{\cos 2x}{2\cos 2x\sin x} = 1$$
$$2\sin x = 1$$
$$\sin x = \frac{1}{2}$$
$$x = \frac{\pi}{6}, \frac{5\pi}{6}$$

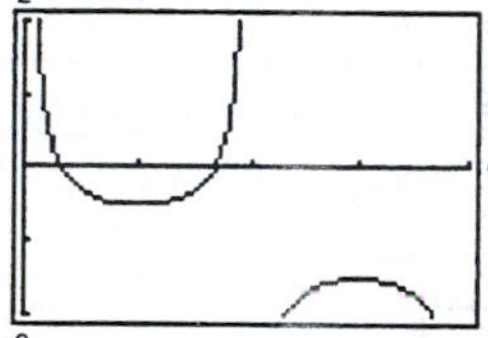

85. $\sin^2\alpha = \left(\frac{5}{13}\right)^2 = \frac{25}{169}$

$\sin^2\alpha = 1 - \cos^2\alpha = 1 - \left(\frac{12}{13}\right)^2 = 1 - \frac{144}{169} = \frac{25}{169}$

87. $\sin\alpha\cos\beta = \left(\frac{5}{13}\right)\left(\frac{4}{5}\right) = \frac{4}{13}$

$\sin\alpha\cos\beta = \cos\left(\frac{\pi}{2} - \alpha\right)\sin\left(\frac{\pi}{2} - \beta\right) = \left(\frac{5}{13}\right)\left(\frac{4}{5}\right) = \frac{4}{13}$

89.

$$\csc 2\theta = \frac{1}{\sin 2\theta}$$
$$= \frac{1}{2\sin\theta\cos\theta}$$
$$= \frac{1}{\sin\theta}\cdot\frac{1}{2\cos\theta}$$
$$= \frac{\csc\theta}{2\cos\theta}$$

91.

$$\cos^2 2\alpha - \sin^2 2\alpha = \cos[2(2\alpha)]$$
$$= \cos 4\alpha$$

93. $(\sin x + \cos x)^2 = \sin^2 x + 2\sin x \cos x + \cos^2 x$
$= (\sin^2 x + \cos^2 x) + 2\sin x \cos x$
$= 1 + \sin 2x$

95. $1 + \cos 10y = 1 + \cos^2 5y - \sin^2 5y$
$= 1 + \cos^2 5y - (1 - \cos^2 5y)$
$= 2\cos^2 5y$

97. $\sec \frac{u}{2} = \frac{1}{\cos(u/2)}$

$= \pm\sqrt{\frac{2}{1 + \cos u}}$

$= \pm\sqrt{\frac{2\sin u}{\sin u(1 + \cos u)}}$

$= \pm\sqrt{\frac{2\sin u}{\sin u + \sin u \cos u}}$

$= \pm\sqrt{\frac{(2\sin u)/(\cos u)}{(\sin u)/(\cos u) + (\sin u \cos u)/(\cos u)}}$

$= \pm\sqrt{\frac{2\tan u}{\tan u + \sin u}}$

99. $\frac{\cos 4x + \cos 2x}{\sin 4x + \sin 2x} = \frac{2\cos\left(\frac{4x + 2x}{2}\right)\cos\left(\frac{4x - 2x}{2}\right)}{2\sin\left(\frac{4x + 2x}{2}\right)\cos\left(\frac{4x - 2x}{2}\right)}$

$= \frac{2\cos 3x \cos x}{2\sin 3x \cos x}$

$= \cot 3x$

101. $\frac{\cos t + \cos 3t}{\sin 3t - \sin t} = \frac{2\cos\left(\frac{4t}{2}\right)\cos\left(-\frac{2t}{2}\right)}{2\cos\left(\frac{4t}{2}\right)\sin\left(\frac{2t}{2}\right)}$

$= \frac{\cos(-t)}{\sin(t)}$

$= \frac{\cos(t)}{\sin(t)}$

$= \cot t$

103. $\cos 3\beta = \cos(2\beta + \beta)$
$= \cos 2\beta \cos\beta - \sin 2\beta \sin \beta$
$= (\cos^2 \beta - \sin^2 \beta)\cos \beta - 2\sin \beta \cos \beta \sin \beta$
$= \cos^3 \beta - \sin^2 \beta \cos \beta - 2\sin^2 \beta \cos \beta$
$= \cos^3 \beta - 3\sin^2 \beta \cos \beta$

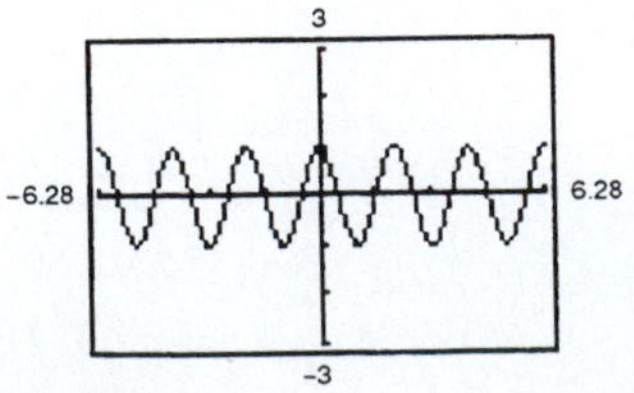

105. $$\frac{\cos 4x - \cos 2x}{2\sin 3x} = \frac{-2\sin\left(\frac{4x+2x}{2}\right)\sin\left(\frac{4x-2x}{2}\right)}{2\sin 3x} = \frac{-2\sin 3x \sin x}{2\sin 3x} = -\sin x$$

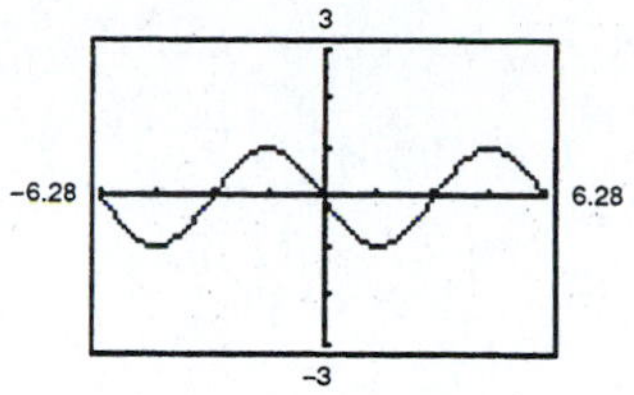

107. $\sin^2 x = \dfrac{1-\cos 2x}{2} = \dfrac{1}{2} - \dfrac{\cos 2x}{2}$

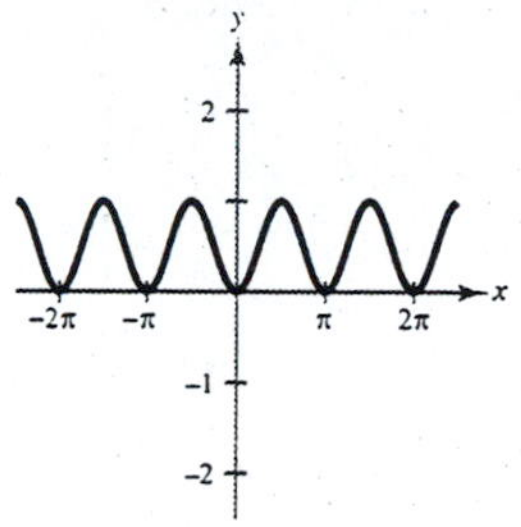

109. (a) $y = 4\sin\dfrac{x}{2} + \cos x$

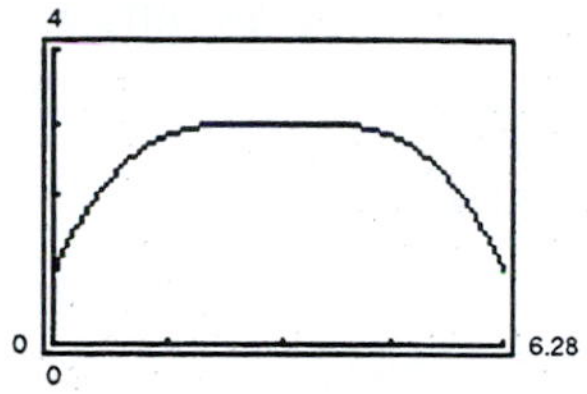

Maximum: $(\pi, 3)$

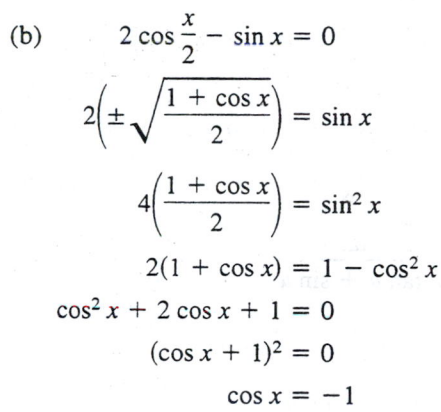

(b)
$$\begin{aligned} 2\cos\frac{x}{2} - \sin x &= 0 \\ 2\left(\pm\sqrt{\frac{1+\cos x}{2}}\right) &= \sin x \\ 4\left(\frac{1+\cos x}{2}\right) &= \sin^2 x \\ 2(1+\cos x) &= 1 - \cos^2 x \\ \cos^2 x + 2\cos x + 1 &= 0 \\ (\cos x + 1)^2 &= 0 \\ \cos x &= -1 \\ x &= \pi \end{aligned}$$

111. $f(x) = 2\sin x\,[2\cos^2(x/2) - 1]$

(a)

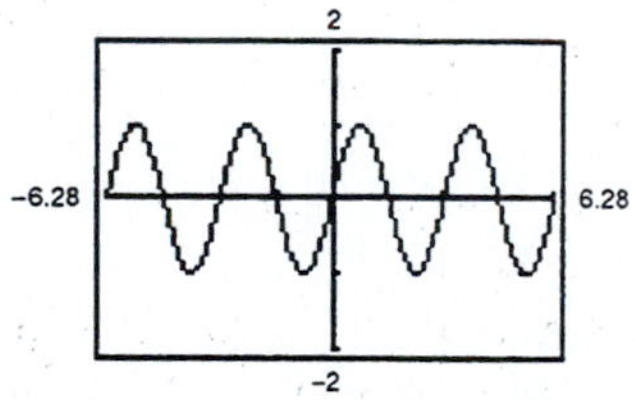

(b) and (c)

$$\begin{aligned} &2\sin x[2\cos^2(x/2) - 1] \\ &= 2\sin x[(1+\cos x) - 1] \\ &= 2\sin x(\cos x) \\ &= 2\sin x\cos x \end{aligned}$$

113. $\sin(2\arcsin x) = 2\sin(\arcsin x)\cos(\arcsin x) = 2x\sqrt{1-x^2}$

115. (a) $A = \frac{1}{2}bh$

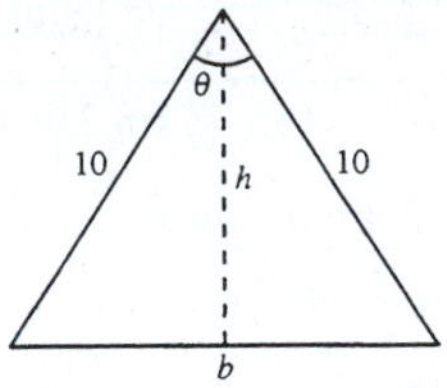

$\cos\frac{\theta}{2} = \frac{h}{10} \implies h = 10\cos\frac{\theta}{2}$

$\sin\frac{\theta}{2} = \frac{(1/2)b}{10} \implies \frac{1}{2}b = 10\sin\frac{\theta}{2}$

$A = 10\sin\frac{\theta}{2}10\cos\frac{\theta}{2} \implies A = 100\sin\frac{\theta}{2}\cos\frac{\theta}{2}$

(b) $A = 100\sin\frac{\theta}{2}\cos\frac{\theta}{2}$

$A = 50\left(2\sin\frac{\theta}{2}\cos\frac{\theta}{2}\right)$

$A = 50\sin\theta$

When $\theta = \pi/2$, $\sin\theta = 1$ $\implies$ the area is a maximum.

$A = 50\sin\frac{\pi}{2} = 50(1) = 50$ square feet

117. Let x = profit for September, then $x + 0.16x$ = profit for October.

$x + (x + 0.16x) = 507{,}600$

$2.16x = 507{,}600$

$x = 235{,}000$

$x + 0.16x = 272{,}600$

Profit for September: \$235,000

Profit for October: \$276,600

119. Let x = number of gallons of 100% concentrate.

$0.30(55 - x) + 1.00x = 0.50(55)$

$16.50 - 0.30x + x = 27.50$

$0.70x = 11$

$x \approx 15.7$ gallons

Review Exercises for Chapter 5

Solutions to Odd-Numbered Exercises

1. $\dfrac{1}{\cot^2 x + 1} = \dfrac{1}{\csc^2 x} = \sin^2 x$

3. $\dfrac{\sin^2 \alpha - \cos^2 \alpha}{\sin^2 \alpha - \sin \alpha \cos \alpha} = \dfrac{(\sin \alpha + \cos \alpha)(\sin \alpha - \cos \alpha)}{\sin \alpha(\sin \alpha - \cos \alpha)}$

$= \dfrac{\sin \alpha + \cos \alpha}{\sin \alpha}$

$= 1 + \cot \alpha$

5. $\tan^2 \theta(\csc^2 \theta - 1) = \tan^2 \theta(\cot^2 \theta)$

$= \tan^2 \theta\left(\dfrac{1}{\tan^2 \theta}\right)$

$= 1$

7. $\dfrac{2 \tan(x + 1)}{1 - \tan^2(x + 1)} = \tan[2(x + 1)]$

$= \tan(2x + 2)$

9. $\tan x(1 - \sin^2 x) = \tan x \cos^2 x$

$= \dfrac{\sin x}{\cos x} \cdot \cos^2 x$

$= \sin x \cos x$

$= \dfrac{1}{2}(2 \sin x \cos x)$

$= \dfrac{1}{2} \sin 2x$

11. $\sec^2 x \cot x - \cot x = \cot x(\sec^2 x - 1)$

$= \cot x \tan^2 x$

$= \dfrac{1}{\tan x} \tan^2 x$

$= \tan x$

13. $\sin^5 x \cos^2 x = \sin^4 x \cos^2 x \sin x$

$= (1 - \cos^2 x)^2 \cos^2 x \sin x$

$= (1 - 2 \cos^2 x + \cos^4 x) \cos^2 x \sin x$

$= (\cos^2 x - 2 \cos^4 x + \cos^6 x)\sin x$

15. $\sin 3\theta \sin \theta = \frac{1}{2}[\cos(3\theta - \theta) - \cos(3\theta + \theta)]$

$= \frac{1}{2}(\cos 2\theta - \cos 4\theta)$

17. $\sqrt{\dfrac{1 - \sin \theta}{1 + \sin \theta}} = \sqrt{\dfrac{1 - \sin \theta}{1 + \sin \theta} \cdot \dfrac{1 - \sin \theta}{1 - \sin \theta}}$

$= \sqrt{\dfrac{(1 - \sin \theta)^2}{1 - \sin^2 \theta}} = \sqrt{\dfrac{(1 - \sin \theta)^2}{\cos^2 \theta}} = \dfrac{|1 - \sin \theta|}{|\cos \theta|} = \dfrac{1 - \sin \theta}{|\cos \theta|}$

Note: We can drop the absolute value on $1 - \sin \theta$ since it is always nonnegative.

19. $\cos 3x = \cos(2x + x)$
$= \cos 2x \cos x - \sin 2x \sin x$
$= (\cos^2 x - \sin^2 x) \cos x - 2 \sin x \cos x \sin x$
$= \cos^3 x - 3 \sin^2 x \cos x$
$= \cos^3 x - 3 \cos x(1 - \cos^2 x)$
$= \cos^3 x - 3 \cos x + 3 \cos^3 x$
$= 4 \cos^3 x - 3 \cos x$

21. $\cot\left(\frac{\pi}{2} - x\right) = \frac{\cos[(\pi/2) - x]}{\sin[(\pi/2) - x]}$
$= \frac{\cos(\pi/2) \cos x + \sin(\pi/2)\sin x}{\sin(\pi/2) \cos x - \sin x \cos(\pi/2)}$
$= \frac{\sin x}{\cos x}$
$= \tan x$

23. $\frac{\sec x - 1}{\tan x} = \frac{(1/\cos x) - 1}{\sin x/\cos x} = \frac{1 - \cos x}{\sin x} = \tan\frac{x}{2}$

25. $2 \sin y \cos y \sec 2y = (\sin 2y)(\sec 2y)$
$= \frac{\sin 2y}{\cos 2y}$
$= \tan 2y$

27. $\sin\left(x - \frac{3\pi}{2}\right) = \sin x \cos\frac{3\pi}{2} - \sin\frac{3\pi}{2} \cos x$
$= (\sin x)(0) - (-1)(\cos x)$
$= \cos x$

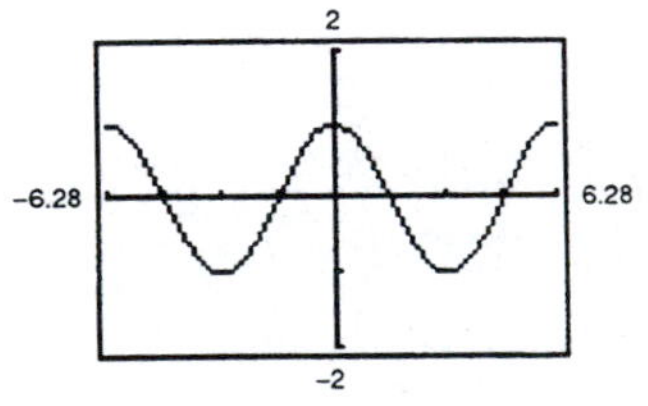

29. $\frac{1 - \cos 2x}{1 + \cos 2x} = \frac{1 - (1 - 2 \sin^2 x)}{1 + (2 \cos x^2 - 1)}$
$= \frac{2 \sin^2 x}{2 \cos^2 x}$
$= \tan^2 x$

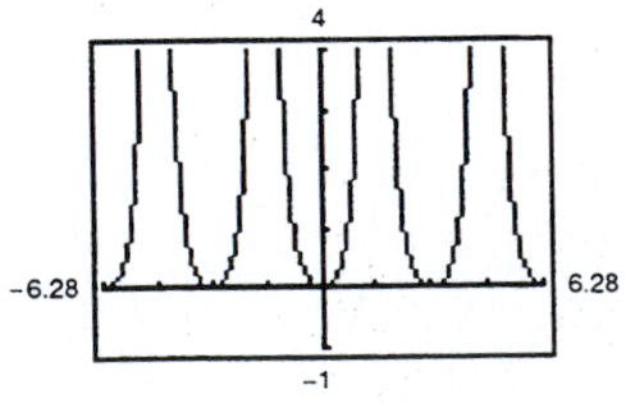

31. $\sin\frac{5\pi}{12} = \sin\left(\frac{2\pi}{3} - \frac{\pi}{4}\right)$
$= \sin\left(\frac{2\pi}{3}\right)\cos\left(\frac{\pi}{4}\right) - \cos\frac{2\pi}{3}\sin\left(\frac{\pi}{4}\right)$
$= \left(\frac{\sqrt{3}}{2}\right)\left(\frac{\sqrt{2}}{2}\right) - \left(-\frac{1}{2}\right)\left(\frac{\sqrt{2}}{2}\right)$
$= \frac{\sqrt{2}}{4}(\sqrt{3} + 1)$

33. $\cos(157° 30') = \cos\frac{315°}{2} = -\sqrt{\frac{1 + \cos 315°}{2}} = -\sqrt{\frac{1 + \cos 45°}{2}}$
$= -\sqrt{\frac{1 + \sqrt{2}/2}{2}} = -\sqrt{\frac{2 + \sqrt{2}}{4}} = -\frac{\sqrt{2 + \sqrt{2}}}{2}$

For Exercises 35–39

$$\sin u = \frac{3}{4},\ u \text{ in Quadrant II} \implies \cos u = -\frac{\sqrt{7}}{4}$$

$$\cos v = -\frac{5}{13},\ v \text{ in Quadrant II} \implies \sin v = \frac{12}{13}$$

35.
$$\begin{aligned}\sin(u+v) &= \sin u \cos v + \cos u \sin v\\ &= \left(\frac{3}{4}\right)\left(-\frac{5}{13}\right) + \left(-\frac{\sqrt{7}}{4}\right)\left(\frac{12}{13}\right)\\ &= -\frac{15}{52} - \frac{12\sqrt{7}}{52}\\ &= \frac{-3(5+4\sqrt{7})}{52}\end{aligned}$$

37.
$$\begin{aligned}\cos(u-v) &= \cos u \cos v + \sin u \sin v\\ &= \left(-\frac{\sqrt{7}}{4}\right)\left(-\frac{5}{13}\right) + \left(\frac{3}{4}\right)\left(\frac{12}{13}\right)\\ &= \frac{5\sqrt{7}+36}{52}\end{aligned}$$

39.
$$\begin{aligned}\cos\frac{u}{2} &= \sqrt{\frac{1+\cos u}{2}}\\ &= \sqrt{\frac{1+(-\sqrt{7}/4)}{2}}\\ &= \sqrt{\frac{4-\sqrt{7}}{8}}\\ &= \frac{1}{4}\sqrt{2(4-\sqrt{7})}\end{aligned}$$

41. If $\frac{\pi}{2} < \theta < \pi$, then $\cos\frac{\theta}{2} < 0$. False, if

$$\frac{\pi}{2} < \theta < \pi \implies \frac{\pi}{4} < \frac{\theta}{2} < \frac{\pi}{2},$$

which is in Quadrant I $\implies \cos(\theta/2) > 0$.

43. $4\sin(-x)\cos(-x) = -2\sin 2x$. True.

$$4\sin(-x)\cos(-x) = 4(-\sin x)(\cos x) = -4\sin x\cos x = -2(2\sin x\cos x) = -2\sin 2x$$

45.
$$\begin{aligned}\sin x - \tan x &= 0\\ \sin x - \frac{\sin x}{\cos x} &= 0\\ \sin x\cos x - \sin x &= 0\\ \sin x(\cos x - 1) &= 0\end{aligned}$$

$\sin x = 0$ or $\cos x - 1 = 0$

$x = 0, \pi$ $\qquad$ $\cos x = 1$

$\qquad\qquad\qquad x = 0$

47.
$$\begin{aligned}\sin 2x + \sqrt{2}\sin x &= 0\\ 2\sin x\cos x + \sqrt{2}\sin x &= 0\\ \sin x\left(2\cos x + \sqrt{2}\right) &= 0\end{aligned}$$

$\sin x = 0$ or $2\cos x + \sqrt{2} = 0$

$x = 0, \pi$ $\qquad$ $\cos x = -\frac{\sqrt{2}}{2}$

$\qquad\qquad\qquad x = \frac{3\pi}{4}, \frac{5\pi}{4}$

49.
$$\begin{aligned}\cos^2 x + \sin x &= 1\\ 1 - \sin^2 x + \sin x &= 1\\ \sin x(\sin x - 1) &= 0\end{aligned}$$

$\sin x = 0$ or $\sin x = 1$

$x = 0, \pi$ $\qquad$ $x = \frac{\pi}{2}$

51. $\dfrac{1+\sin x}{\cos x}+\dfrac{\cos x}{1+\sin x}=4$

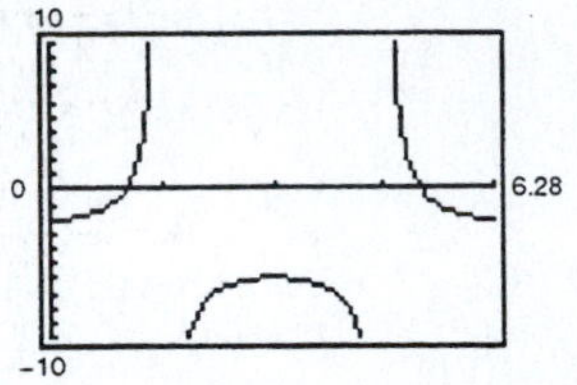

$$(1+\sin x)^2+\cos^2 x=4\cos x(1+\sin x)$$
$$1+2\sin x+\sin^2 x+\cos^2 x=4\cos x(1+\sin x)$$
$$2+2\sin x-4\cos x(1+\sin x)=0$$
$$2(1+\sin x)(1-2\cos x)=0$$
$$1+\sin x=0 \quad \text{or} \quad 1-2\cos x=0$$
$$\sin x=-1 \qquad \cos x=\frac{1}{2}$$
$$x=\frac{3\pi}{2} \qquad x=\frac{\pi}{3},\frac{5\pi}{3}$$

(extraneous solution)

53. $\tan^3 x-\tan^2 x+3\tan x-3=0$

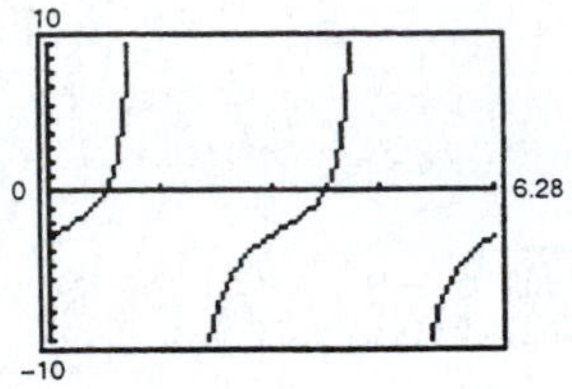

$$\tan^2 x(\tan x-1)+3(\tan x-1)=0$$
$$(\tan^2 x+3)(\tan x-1)=0$$
$$\tan^2 x+3=0 \quad \text{or} \quad \tan x-1=1$$
$$\text{(No solution)} \quad \text{or} \quad \tan x=1$$
$$x=\frac{\pi}{4},\frac{5\pi}{4}$$

55. False, $\sin\theta=\frac{1}{2}$ has an infinite number of solutions but is not an identity.

57. $\cos 3\theta+\cos 2\theta=2\cos\left(\dfrac{3\theta+2\theta}{2}\right)\cos\left(\dfrac{3\theta-2\theta}{2}\right)$

$$=2\cos\frac{5\theta}{2}\cos\frac{\theta}{2}$$

59. $\sin 3\alpha\sin 2\alpha=\frac{1}{2}[\cos(3\alpha-2\alpha)-\cos(3\alpha+2\alpha)]$

$$=\frac{1}{2}(\cos\alpha-\cos 5\alpha)$$

61. $\cos(2\arccos 2x)=\cos 2\theta$

$$=\cos^2\theta-\sin^2\theta$$
$$=(2x)^2-\left(\sqrt{1-4x^2}\right)^2$$
$$=4x^2-(1-4x^2)$$
$$=8x^2-1$$

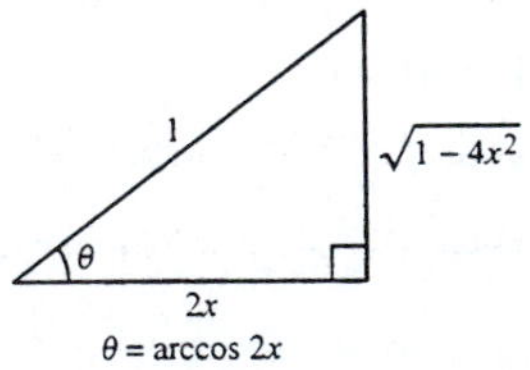

63. $\sin^{-1/2} \cos x = \dfrac{\cos x}{\sin^{1/2} x}$

$$= \frac{\cos x}{\sqrt{\sin x}} \cdot \frac{\sqrt{\sin x}}{\sqrt{\sin x}}$$

$$= \frac{\cos x}{\sin x}\sqrt{\sin x}$$

$$= \cot x\sqrt{\sin x}$$

65. $y = 1.5 \sin 8t - 0.5 \cos 8t$

(a) $a = \dfrac{3}{2},\ b = -\dfrac{1}{2},\ B = 8,\ c = \arctan\left(-\dfrac{1/2}{3/2}\right)$

$$y = \sqrt{(3/2)^2 + (-1/2)^2} \sin\left(8t + \arctan -\frac{1}{3}\right)$$

$$y = \frac{1}{2}\sqrt{10} \sin\left(8t - \arctan\frac{1}{3}\right)$$

(b) Amplitude: $\dfrac{1}{2}\sqrt{10}$ sin feet

(c) Frequency: $\dfrac{8}{2\pi} = \dfrac{4}{\pi}$ cycles per second

Practice Test for Chapter 5

1 Find the value of the other five trigonometric functions, given $\tan x = \frac{4}{11}$, $\sec x < 0$.

2. Simplify $\dfrac{\sec^2 x + \csc^2 x}{\csc^2 x(1 + \tan^2 x)}$.

3. Rewrite as a single logarithm and simplify $\ln|\tan \theta| - \ln|\cot \theta|$.

4. True or false:

$$\cos\left(\frac{\pi}{2} - x\right) = \frac{1}{\csc x}$$

5. Factor and simplify: $\sin^4 x + (\sin^2 x)\cos^2 x$

6. Multiply and simplify: $(\csc x + 1)(\csc x - 1)$

7. Rationalize the denominator and simplify:

$$\frac{\cos^2 x}{1 - \sin x}$$

8. Verify:

$$\frac{1 + \cos \theta}{\sin \theta} + \frac{\sin \theta}{1 + \cos \theta} = 2 \csc \theta$$

9. Verify:

$$\tan^4 x + 2\tan^2 x + 1 = \sec^4 x$$

10. Use the sum or difference formulas to determine:

(a) $\sin 105°$ (b) $\tan 15°$

11. Simplify: $(\sin 42°)\cos 38° - (\cos 42°)\sin 38°$

12. Verify $\tan\left(\theta + \dfrac{\pi}{4}\right) = \dfrac{1 + \tan \theta}{1 - \tan \theta}$.

13. Write $\sin(\arcsin x - \arccos x)$ as an algebraic expression in x.

14. Use the double-angle formulas to determine:

(a) $\cos 120°$ (b) $\tan 300°$

15. Use the half-angle formulas to determine:

(a) $\sin 22.5°$ (d) $\tan \dfrac{\pi}{12}$

16. Given $\sin = 4/5$, θ lies in Quadrant II, find $\cos(\theta/2)$.

17. Use the power-reducing identities to write $(\sin^2 x)\cos^2 x$ in terms of the first power of cosine.

18. Rewrite as a sum: $6(\sin 5\theta)\cos 2\theta$.

19. Rewrite as a product:

$\sin(x + \pi) + \sin(x - \pi)$.

20. Verify $\dfrac{\sin 9x + \sin 5x}{\cos 9x - \cos 5x} = -\cot 2x$.

21. Verify:

$(\cos u)\sin v = \frac{1}{2}[\sin(u + v) - \sin(u - v)]$.

22. Find all solutions in the interval $[0, 2\pi)$:

$4\sin^2 x = 1$

23. Find all solutions in the interval $[0, 2\pi)$:

$\tan^2 \theta + (\sqrt{3} - 1)\tan\theta - \sqrt{3} = 0$

24. Find all solutions in the interval $[0, 2\pi)$:

$\sin 2x = \cos x$

25. Use the quadratic formula to find all solutions in the interval $[0, 2\pi)$:

$\tan^2 x - 6\tan x + 4 = 0$

CHAPTER 6
Additional Topics in Trigonometry

CHAPTER 6
Additional Topics in Trigonometry

Section 6.1 Law of Sines

- If ABC is any oblique triangle with sides a, b, and c, then

$$\frac{a}{\sin A} = \frac{b}{\sin B} = \frac{c}{\sin C}.$$

- You should be able to use the Law of Sines to solve an oblique triangle for the remaining three parts, given:
  (a) Two angles and any side (AAS or ASA)
  (b) Two sides and an angle opposite one of them (SSA)
    1. If A is acute and $h = b \sin A$:
       (a) $a < h$, no triangle is possible.
       (b) $a = h$ or $a > b$, one triangle is possible.
       (c) $h < a < b$, two triangles are possible.
    2. If A is obtuse and $h = b \sin A$:
       (a) $a \le b$, no triangle is possible.
       (b) $a > b$, one triangle is possible.
- The area of any triangle equals one-half the product of the lengths of two sides times the sine of their included angle.

$$A = \tfrac{1}{2}ab \sin C = \tfrac{1}{2}ac \sin B = \tfrac{1}{2}bc \sin A$$

Solutions to Odd-Numbered Exercises

1. Given: $A = 30°$, $B = 45°$, $a = 20$

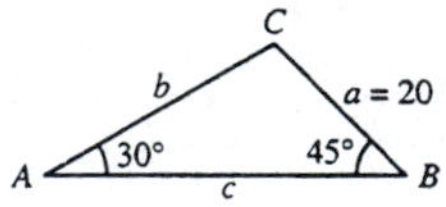

$C = 180° - A - B = 105°$

$$b = \frac{a}{\sin A}(\sin B) = \frac{20 \sin 45°}{\sin 30°} = 20\sqrt{2} \approx 28.28$$

$$c = \frac{a}{\sin A}(\sin C) = \frac{20 \sin 105°}{\sin 30°} \approx 38.64$$

3. Given: $A = 10°$, $B = 60°$, $a = 7.5$

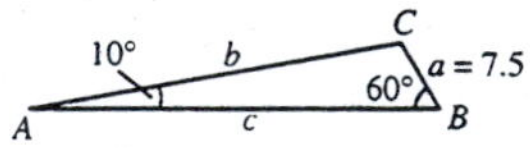

$C = 180° - A - B = 110°$

$$b = \frac{a}{\sin A}(\sin B) = \frac{7.5}{\sin 10°}(\sin 60°) \approx 37.40$$

$$c = \frac{a}{\sin A}(\sin C) = \frac{7.5}{\sin 10°}(\sin 110°) \approx 40.59$$

5. Given: $A = 36°$, $a = 8$, $b = 5$

$$\sin B = \frac{b \sin A}{a} = \frac{5 \sin 36°}{8} \approx 0.36737 \implies B \approx 21.55°$$

$$C = 180° - A - B \approx 180° - 36° - 21.55 = 122.45°$$

$$c = \frac{a}{\sin A}(\sin C) = \frac{8}{\sin 36°}(\sin 122.45°) \approx 11.49$$

7. Given: $A = 150°$, $C = 20°$, $a = 200$

$$B = 180° - A - C = 180° - 150° - 20° = 10°$$

$$b = \frac{a}{\sin A}(\sin B) = \frac{200}{\sin 150°}(\sin 10°) \approx 69.46$$

$$c = \frac{a}{\sin A}(\sin C) = \frac{200}{\sin 150°}(\sin 20°) \approx 136.81$$

9. Given: $A = 83°20'$, $C = 54.6°$, $c = 18.1$

$$B = 180° - A - C = 180° - 80°20' - 54°36' = 42°4'$$

$$a = \frac{c}{\sin C}(\sin A) = \frac{18.1}{\sin 54.6°}(\sin 83°20') \approx 22.05$$

$$b = \frac{c}{\sin C}(\sin B) = \frac{18.1}{\sin 54.6°}(\sin 42°4') \approx 14.88$$

11. Given: $B = 15°30'$, $a = 4.5$, $b = 6.8$

$$\sin A = \frac{a \sin B}{b} = \frac{4.5 \sin 15°30'}{6.8} \approx 0.17685 \implies A \approx 10°11'$$

$$C = 180° - A - B \approx 180° - 10°11' - 15°30' = 154°19'$$

$$c = \frac{b}{\sin B}(\sin C) = \frac{6.8}{\sin 15°30'}(\sin 154°19') \approx 11.03$$

13. Given: $C = 145°$, $b = 4$, $c = 14$

$$\sin B = \frac{b \sin C}{c} = \frac{4 \sin 145°}{14} \approx 0.1639 \implies B \approx 9.43°$$

$$A = 180° - B - C \approx 180° - 9.43° - 145° = 25.57°$$

$$a = \frac{c}{\sin C}(\sin A) \approx \frac{14}{\sin 145°}(\sin 25.57°) \approx 10.5$$

15. Given: $B = 110°15'$, $a = 48$, $b = 16$

$$\sin B = \frac{b \sin A}{a} = \frac{16 \sin 110°15'}{48} \approx 0.31273 \implies B \approx 18°13'$$

$$C = 180° - A - B \approx 180° - 110°15' - 18°13' = 51°32'$$

$$c = \frac{a}{\sin A}(\sin C) = \frac{48}{\sin 110°15'}(\sin 51°32') \approx 40.06$$

17. Given: $a = 4.5,\ b = 12.8,\ A = 58°$

$h = 12.8 \sin 58° \approx 10.86$

Since $a < h$, no triangle is formed.

19. Given: $a = 4.5, b = 5,\ A = 58°$

$$\sin B = \frac{b \sin A}{a} = \frac{5 \sin 58°}{4.5} \approx 0.9423 \implies B = 70.4° \text{ or } B = 109.6°$$

Case 1

$B \approx 70.4°$

$C \approx 180° - 70.4° - 58° = 51.6°$

$c \approx \dfrac{4.5}{\sin 58°}(\sin 51.6°) \approx 4.16$

Case 2

$B \approx 109.6°$

$C \approx 180° - 109.6° - 58° = 12.4°$

$c \approx \dfrac{4.5}{\sin 58°}(\sin 12.4°) \approx 1.14$

21. Given: $a = 125,\ b = 200,\ A = 110°$

No triangle is formed because A is obtuse and $a < b$.

23. Given: $A = 36°,\ a = 5$

(a) One solution if $b \le 5$ or $b = \dfrac{5}{\sin 36°}$

(b) Two solutions if $5 < b < \dfrac{5}{\sin 36°}$

(c) No solution is $b > \dfrac{5}{\sin 36°}$

25. (a)

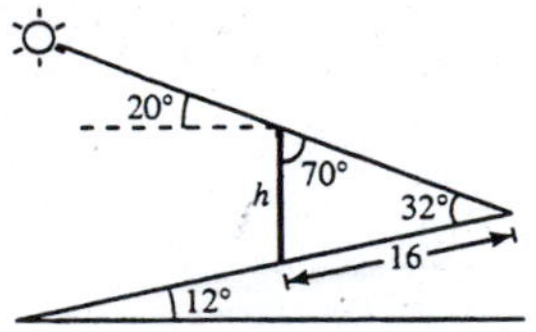

(b) $\dfrac{16}{\sin 70°} = \dfrac{h}{\sin 32°}$

(c) $h = \dfrac{16 \sin 32°}{\sin 70°} \approx 9$ meters

27. $\dfrac{\sin(42° - \theta)}{10} = \dfrac{\sin 48°}{17}$

$\sin(42° - \theta) \approx 0.43714$

$\theta \approx 16.1°$

29. Given: $A = 74° - 28° = 46°,$

$B = 180° - 41° - 74° = 65°,\ c = 100$

$C = 180° - 46° - 65° = 69°$

$a = \dfrac{c}{\sin C}(\sin A) = \dfrac{100}{\sin 69°}(\sin 46°) \approx 77$ meters

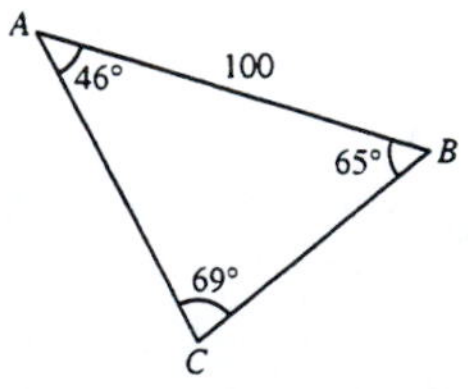

31. (a)

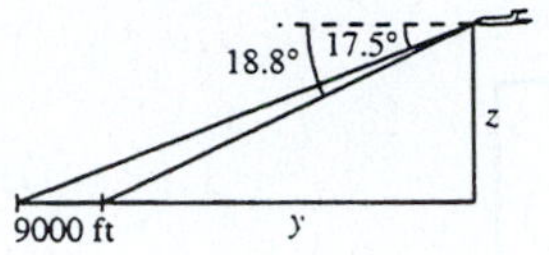

(b) $\dfrac{x}{\sin 17.5°} = \dfrac{9000}{\sin 1.3°}$

$x \approx 119{,}289.1261 \text{ feet} \approx 22.6 \text{ miles}$

(c) $\dfrac{y}{\sin 71.2°} = \dfrac{x}{\sin 90°}$

$y = x \sin 71.2° \approx 119{,}289.1261 \sin 71.2°$

$\approx 112{,}924.963 \text{ feet} \approx 21.4 \text{ miles}$

(d) $z = x \sin 18.8° \approx 119{,}289.1261 \sin 18.8° \approx 37{,}443 \text{ feet}$

33. $A = 65° - 28° = 37°$

$c = 30$

$B = 180° - 16.5° - 65° = 98.5°$

$C = 180° - 37° - 98.5° = 44.5°$

$a = \dfrac{c}{\sin C}(\sin A) = \dfrac{30}{\sin 44.5°}(\sin 37°) \approx 25.8 \text{ km to } B$

$b = \dfrac{c}{\sin C}(\sin B) = \dfrac{30}{\sin 44.5°}(\sin 98.5°) \approx 42.3 \text{ km to } A$

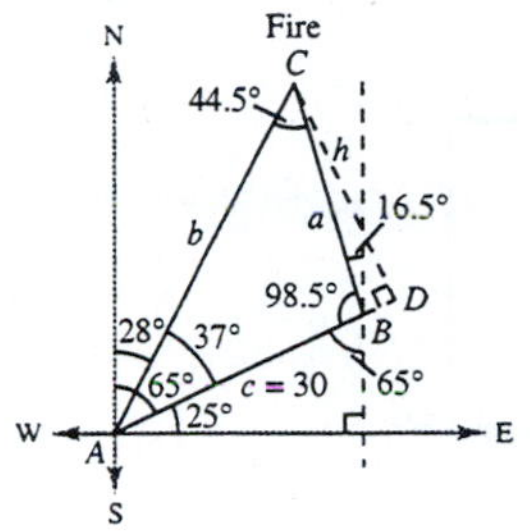

35. $A = 90° - 62° = 28°$

$B = 90° + 38° = 128°, \; c = 5$

$C = 180° - 128° - 28° = 24°$

$a = \dfrac{c}{\sin C}(\sin A) = \dfrac{5}{\sin 24°}(\sin 28°) \approx 5.77$

$d = a \sin(90° - 38°) \approx 5.77 \sin 52° \approx 4.55 \text{ miles}$

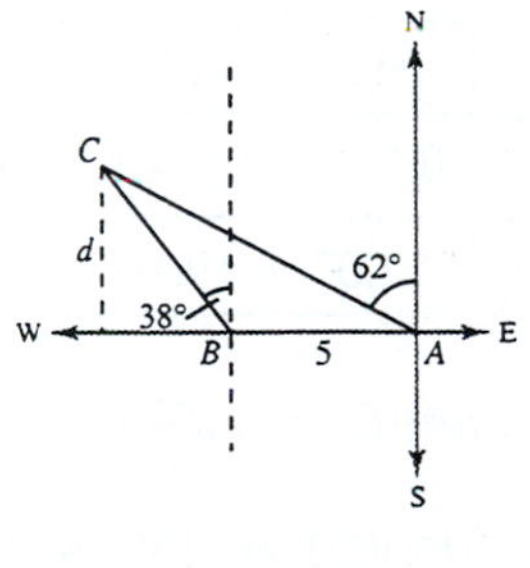

37. (a) $\dfrac{\sin\alpha}{9} = \dfrac{\sin\beta}{18}$

$\sin\alpha = 0.5\sin\beta$

$\alpha = \arcsin(0.5\sin\beta)$

(b) 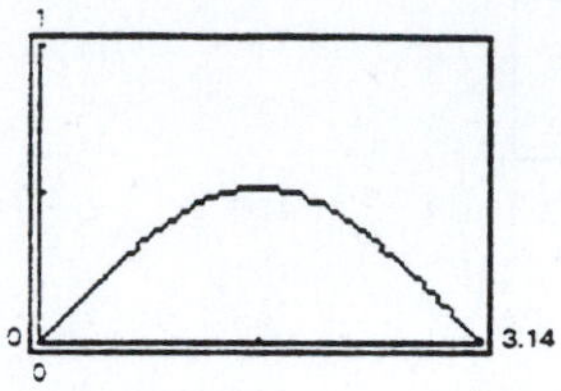

Domain: $0 < \beta < \pi$

Range: $0 < \alpha \le \dfrac{\pi}{6}$

(c) $\gamma = \pi - \alpha - \beta = \pi - \beta - \arcsin(0.5\sin\beta)$

$\dfrac{c}{\sin\gamma} = \dfrac{18}{\sin\beta}$

$c = \dfrac{18\sin\gamma}{\sin\beta} = \dfrac{18\sin[\pi - \beta - \arcsin(0.5\sin\beta)]}{\sin\beta}$

(d) 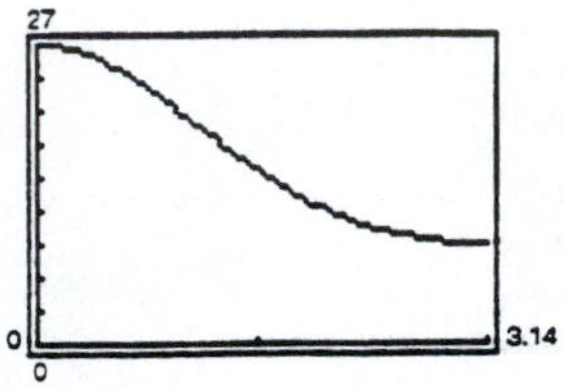

Domain: $0 < \beta < \pi$

Range: $9 < \alpha < 27$

(e)

β	0	0.4	0.8	1.2	1.6	2.0	2.4	2.8
α	0	0.1960	0.3669	0.4848	0.5234	0.4720	0.3445	0.1683
c	27	25.92	27.07	19.19	15.33	12.29	10.31	9.27

As $\beta \to 0,\ c \to 27$

As $\beta \to \pi,\ c \to 9$

39. Area $= \frac{1}{2}ab\sin C = \frac{1}{2}(4)(6)\sin 120° \approx 10.4$

41. Area $= \frac{1}{2}bc\sin A = \frac{1}{2}(57)(85)\sin 43°45' \approx 1675.2$

43. Area $= \frac{1}{2}ac\sin B = \frac{1}{2}(62)(20)\sin 130° \approx 474.9$

45. (a) $A = \dfrac{1}{2}(30)(20)\sin\left(\theta + \dfrac{\theta}{2}\right) - \dfrac{1}{2}(8)(20)\sin\dfrac{\theta}{2} - \dfrac{1}{2}(8)(30)\sin\theta$

$= 300\sin\dfrac{3\theta}{2} - 80\sin\dfrac{\theta}{2} - 120\sin\theta$

$= 20\left[15\sin\dfrac{3\theta}{2} - 4\sin\dfrac{\theta}{2} - 6\sin\theta\right]$

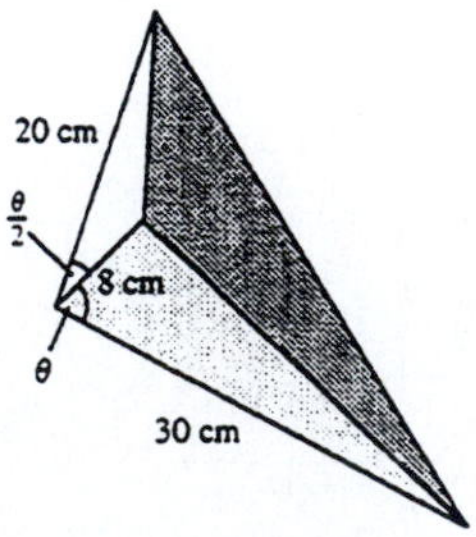

(b) 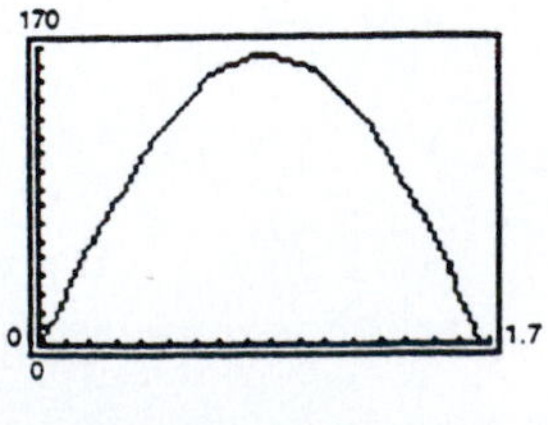

(c) Domain: $0 \le \theta \le 1.6690$

The domain would increase in length and the area would increase if the 8-centimeter line segment were decreased.

Section 6.2 Law of Cosines

- If ABC is any oblique triangle with sides a, b, and c, the following equations are valid.

  (a) $a^2 = b^2 + c^2 - 2bc \cos A$ or $\cos A = \dfrac{b^2 + c^2 - a^2}{2bc}$

  (b) $b^2 = a^2 + c^2 - 2ac \cos B$ or $\cos B = \dfrac{a^2 + c^2 - b^2}{2ac}$

  (c) $c^2 = a^2 + b^2 - 2ab \cos C$ or $\cos C = \dfrac{a^2 + b^2 - c^2}{2ab}$

- You should be able to use the Law of Cosines to solve an oblique triangle for the remaining three parts, given:

  (a) Three sides (SSS)

  (b) Two sides and their included angle (SAS)

- Given any triangle with sides of length a, b, and c, then the area of the triangle is

  $$\text{Area} = \sqrt{s(s-a)(s-b)(s-c)}, \text{ where } s = \frac{a+b+c}{2}. \qquad \text{(Heron's Formula)}$$

Solutions to Odd-Numbered Exercises

1. Given: $a = 6,\ b = 8,\ c = 12$

$$\cos A = \frac{b^2 + c^2 - a^2}{2bc} = \frac{64 + 144 - 36}{2(8)(12)} \approx 0.8958 \implies A \approx 26.4°$$

$$\sin B = \frac{b \sin A}{a} \approx \frac{8 \sin 26.4°}{6} \approx 0.5928 \implies B \approx 36.3°$$

$$C \approx 180° - 26.4° - 36.3° = 117.3°$$

3. Given: $A = 30°,\ b = 15,\ c = 30$

$$a^2 = b^2 + c^2 - 2bc \cos A$$

$$= 225 + 900 - 2(15)(30) \cos 30° \approx 18.5897$$

$$a \approx 18.6$$

$$\cos B = \frac{a^2 + c^2 - b^2}{2ac} \approx \frac{(18.6)^2 + 900 - 225}{2(18.6)(30)} \approx 0.9148$$

$$B \approx 23.8°$$

$$C \approx 180° - 30° - 23.8° = 126.2°$$

5. Given: $a = 9,\ b = 12,\ c = 15$

$$\cos C = \frac{a^2 + b^2 - c^2}{2ab} = \frac{81 + 144 - 225}{2(9)(12)} = 0 \implies C = 90°$$

$$\sin A = \frac{9}{15} = \frac{3}{5} \implies A \approx 36.9°$$

$$B \approx 180° - 90° - 36.9° = 53.1°$$

7. Given: $a = 75.4,\ b = 52,\ c = 52$

$$\cos A = \frac{b^2 + c^2 - a^2}{2bc} = \frac{52^2 + 52^2 - 75.4^2}{2(52)(52)} = -0.05125 \implies A \approx 92.94°$$

$$\sin B = \frac{b \sin A}{a} \approx \frac{52(0.9987)}{75.4} \approx 0.68875 \implies B \approx 43.53°$$

$$C = B \approx 43.53°$$

9. Given: $A = 120°,\ b = 3,\ c = 10$

$$a^2 = b^2 + c^2 - 2bc \cos A = 9 + 100 - 60 \cos 120° = 139 \implies a \approx 11.79$$

$$\sin B = \frac{b \sin A}{a} \approx \frac{3 \sin 120°}{11.79} \approx 0.2204 \implies B \approx 12.7°$$

$$C = 180° - 120° - 12.7° = 47.3°$$

11. Given: $B = 8°45',\ a = 25,\ c = 15$

$$b^2 = a^2 + c^2 - 2ac \cos B \approx 625 + 225 - 2(25)(15)(0.9884) \approx 108.7 \implies b \approx 10.4$$

$$\sin C = \frac{c \sin B}{b} \approx \frac{15(0.1521)}{10.43} \approx 0.2188 \implies C \approx 12.64° \approx 12°38'$$

$$A \approx 180° - 8°45' - 12°38' = 158°37'$$

13. Given: $C = 125°40',\ a = 32,\ b = 32$

$$c^2 = a^2 + b^2 - 2ab \cos C \approx 32^2 + 32^2 - 2(32)(32)(-0.5831) \approx 3242.1 \implies c \approx 56.9$$

$$A = B \implies 2A = 180° - 125°40' = 54°20' \implies A = B = 27°10'$$

15.

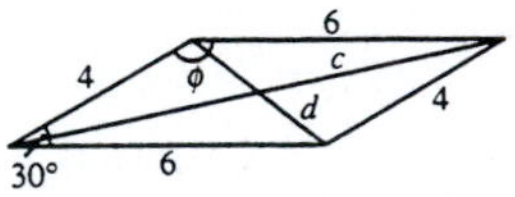

$$d^2 = 4^2 + 6^2 - 2(4)(6) \cos 30°$$

$$d \approx 3.23$$

$$2\phi = 360° - 2(30°)$$

$$\phi = 150°$$

$$c^2 = 4^2 + 6^2 - 2(4)(6) \cos 150°$$

$$c \approx 9.67$$

17.

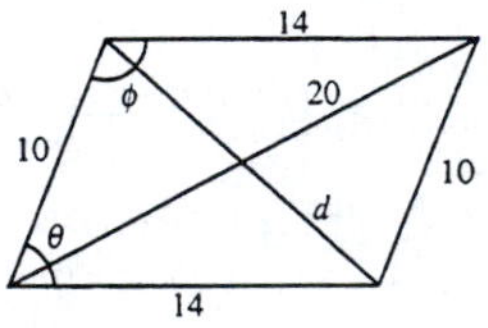

$$\cos \phi = \frac{10^2 + 14^2 - 20^2}{2(10)(14)}$$

$$\phi \approx 111.8°$$

$$2\theta \approx 360° - 2(111.8°)$$

$$\theta = 68.2°$$

$$d^2 = 10^2 + 14^2 - 2(10)(14) \cos 68.2°$$

$$d \approx 13.86$$

19. $\cos\alpha = \dfrac{(9)^2 + (10)^2 - (6)^2}{2(9)(10)}$

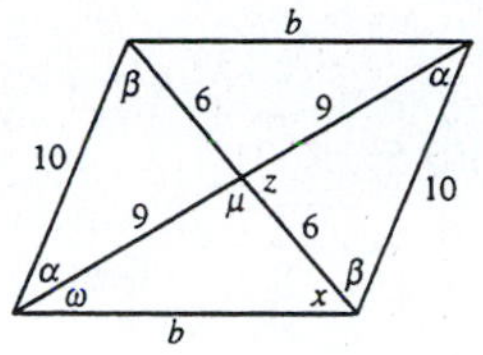

$\alpha \approx 36.3°$

$\cos\beta = \dfrac{6^2 + 10^2 - 9^2}{2(6)(10)}$

$\beta \approx 62.7°$

$z = 180° - \alpha - \beta \approx 81.0°$

$\mu = 180° - z \approx 99.0°$

$b^2 = 9^2 + 6^2 - 2(9)(6)(\cos 99.0°)$

$b \approx 11.58$

$\cos\omega \approx \dfrac{9^2 + 11.58^2 - 6^2}{2(9)(11.58)}$

$\omega \approx 30.8°$

$\theta = \alpha + \omega \approx 67.1°$

$\cos x \approx \dfrac{6^2 + 11.58^2 - 9^2}{2(6)(11.58)}$

$x \approx 50.2°$

$\phi = \beta + x \approx 112.9°$

21. $\cos B = \dfrac{1100^2 + 2500^2 - 2000^2}{2(1100)(2500)}$

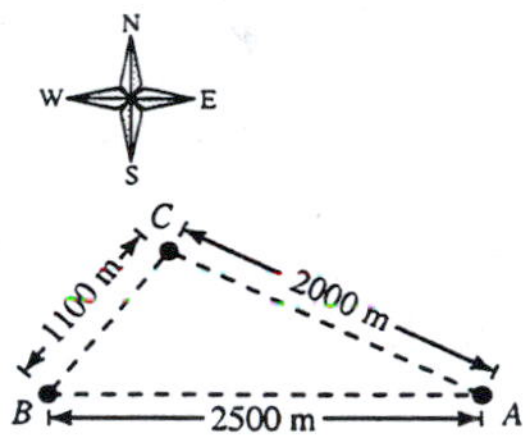

$B \approx 51° \implies$ Bearing of N 39° E

$\cos C = \dfrac{1100^2 + 2000^2 - 2500^2}{2(1100)(2000)}$

$C \approx 103.7°$

$\alpha = 180° - 51° - 103.7° = 25.3° \implies$ Bearing of S 64.7° E

23.

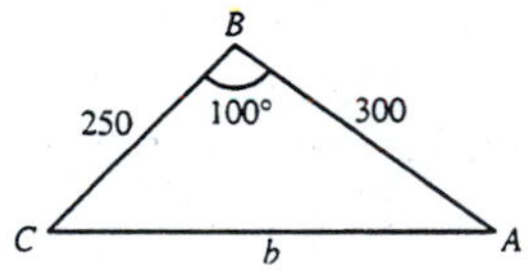

$b^2 = 250^2 + 300^2 - 2(250)(300)\cos 100°$

$b \approx 422.5$ meters

25.

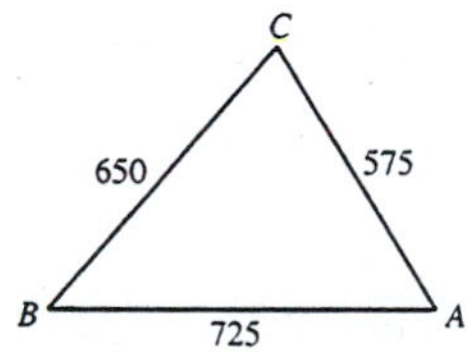

The largest angle is across from the largest side.

$\cos C = \dfrac{650^2 + 575^2 - 725^2}{2(650)(575)}$

$C \approx 72.3°$

27. $C = 180° - 53° - 67° = 60°$

$c^2 = a^2 + b^2 - 2ab\cos C$

$= 36^2 + 48^2 - 2(36)(48)(0.5)$

$= 1872$

$c \approx 43.3$ mi

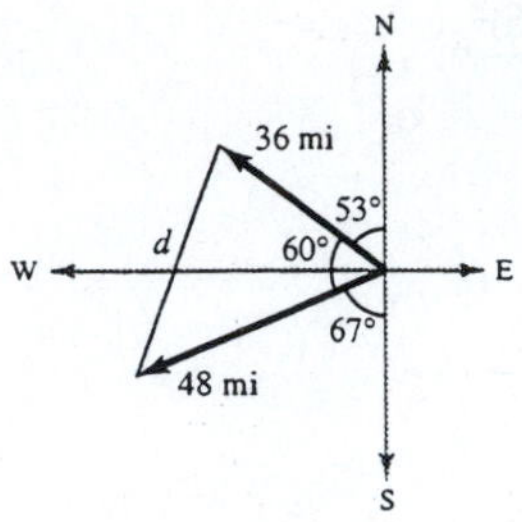

29. (a) $\cos\theta = \dfrac{273^2 + 178^2 - 235^2}{2(273)(178)}$

$\theta \approx 58.4°$

Bearing: N 58.4° W

(b) $\cos\phi = \dfrac{235^2 + 178^2 - 273^2}{2(235)(178)}$

$\phi \approx 81.5°$

Bearing: S 81.5° W

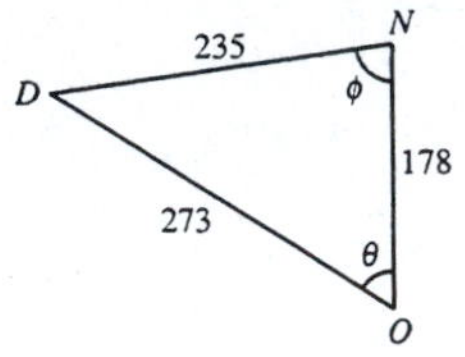

31. $d^2 = 60.5^2 + 90^2 - 2(60.5)(90)\cos 45° \approx 4059.9 \implies d \approx 63.7$ ft

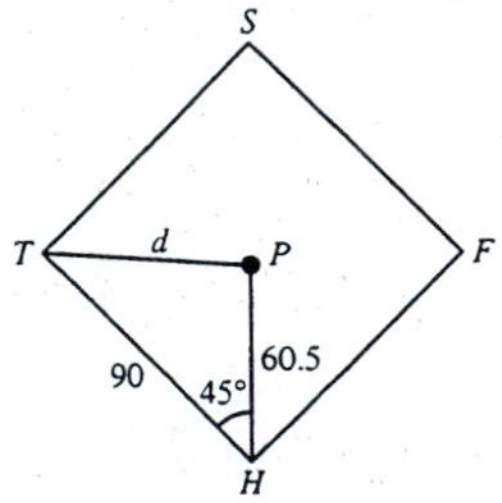

33. $\overline{RS} = \sqrt{8^2 + 10^2} = \sqrt{164} = 2\sqrt{41} \approx 12.8$ ft

$\overline{PQ} = \frac{1}{2}\sqrt{16^2 + 10^2} = \frac{1}{2}\sqrt{356} = \sqrt{89} \approx 9.4$ ft

$\tan P = \frac{10}{16}$

$P = \arctan\frac{5}{8} \approx 32.0°$

$\overline{QS} \approx \sqrt{8^2 + 9.4^2 - 2(8)(9.4)\cos 32°} \approx \sqrt{24.81} \approx 5.0$ ft

35. (a) $7^2 = 1.5^2 + x^2 - 2(1.5)(x)\cos\theta$

$49 = 2.25 + x^2 - 3x\cos\theta$

(b) $$x^2 - 3x\cos\theta = 46.75$$

$$x^2 - 3x\cos\theta + \left(\frac{3\cos\theta}{2}\right)^2 = 46.75 + \left(\frac{3\cos\theta}{2}\right)^2$$

$$\left[x - \frac{3\cos\theta}{2}\right]^2 = \frac{187}{4} - \frac{9\cos^2\theta}{4}$$

$$x - \frac{3\cos\theta}{2} = \pm\sqrt{\frac{187 + 9\cos^2\theta}{4}}$$

Choosing the positive values of x, we have $x = \frac{1}{2}\left[3\cos\theta + \sqrt{9\cos^2\theta + 187}\right]$.

(c)

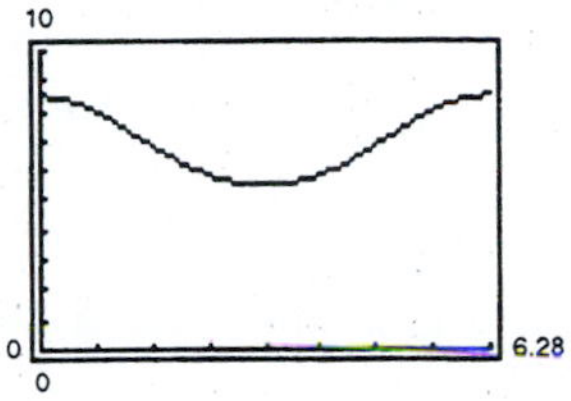

(d) When $\theta = 2\pi$, the piston has moved $2(2)(1.5) = 6$ inches.

37. $A = 180° - 40° - 20° = 120°$

$$\frac{x}{\sin 20°} = \frac{7}{\sin 120°}$$

$$x = \frac{7\sin 20°}{\sin 120°}$$

$x = 2.76$ feet

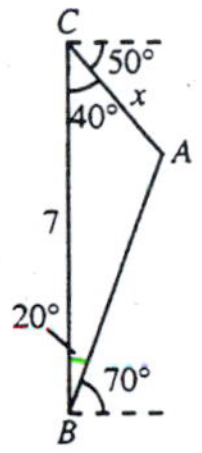

39. $a = 25,\ b = 55,\ c = 72,\ s = \dfrac{25 + 55 + 72}{2} = 76$

(a) $A = \sqrt{(76)(76-25)(76-55)(76-72)} \approx 570.60$

(b) $\cos C = \dfrac{a^2 + b^2 - c^2}{2ab} = \dfrac{25^2 + 55^2 - 72^2}{2(25)(55)} \implies C \approx 123.905°$

$2R = \dfrac{c}{\sin C} \approx \dfrac{72}{\sin 123.905°} \implies R \approx 43.3754$

$A = \pi R^2 \approx 5910.68$

(c) $r = \sqrt{\dfrac{(s-a)(s-b)(s-c)}{s}} = \sqrt{\dfrac{(51)(21)(4)}{76}} \approx 7.5079$

$A = \pi r^2 \approx 177.09$

41. $a = 5,\ b = 7,\ c = 10 \implies s = \dfrac{a+b+c}{2} = 11$

Area $= \sqrt{s(s-a)(s-b)(s-c)} = \sqrt{11(6)(4)(1)} \approx 16.25$

43. $a = 12,\ b = 15,\ c = 9 \implies s = \dfrac{12 + 15 + 9}{2} = 18$

Area $= \sqrt{18(6)(3)(9)} = 54$

45. $a = 20,\ b = 20,\ c = 10 \implies s = \dfrac{20 + 20 + 10}{2} = 25$

Area $= \sqrt{25(5)(5)(15)} \approx 96.82$

47. $a = 200,\ b = 500,\ c = 600 \implies s = \dfrac{200 + 500 + 600}{2} = 650$

Area $= \sqrt{650(450)(150)(50)} \approx 46{,}837.5$ sq ft

49.
$$\begin{aligned}\frac{1}{2}bc(1 + \cos A) &= \frac{1}{2}bc\left[1 + \frac{b^2 + c^2 - a^2}{2bc}\right]\\ &= \frac{1}{2}bc\left[\frac{2bc + b^2 + c^2 - a^2}{2bc}\right]\\ &= \frac{1}{4}[(b + c)^2 - a^2]\\ &= \frac{1}{4}[(b + c) + a][(b + c) - a]\\ &= \frac{b + c + a}{2} \cdot \frac{b + c - a}{2}\\ &= \frac{a + b + c}{2} \cdot \frac{-a + b + c}{2}\end{aligned}$$

51. Let $\theta = \arcsin 2x$, then

$\sin \theta = 2x = \dfrac{2x}{1}$ and $\sec \theta = \dfrac{1}{\sqrt{1 - 4x^2}}$.

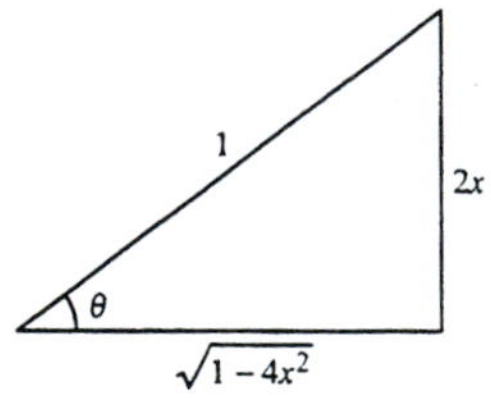

53. Let $\theta = \arctan(x - 2)$, then

$\tan \theta = x - 2 = \dfrac{x - 2}{1}$ and $\cot \theta = \dfrac{1}{x - 2}$.

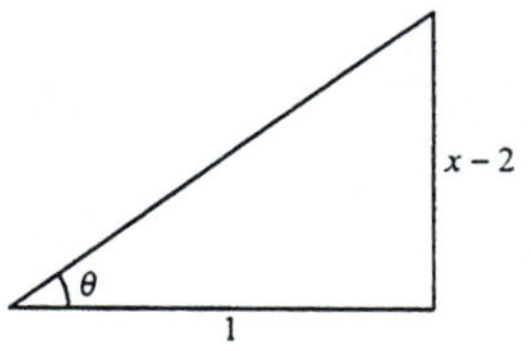

Section 6.3 Vectors in the Plane

- A vector $\mathbf{v}$ is the collection of all directed line segments that are equivalent to a given directed line segment $\overrightarrow{PQ}$.
- You should be able to *geometrically* perform the operations of vector addition and scalar multiplication.
- The component form of the vector with initial point $P = (p_1, p_2)$ and terminal point $Q = (q_1, q_2)$ is

  $$\overrightarrow{PQ} = \langle q_1 - p_1, q_2 - p_2 \rangle = \langle v_1, v_2 \rangle = \mathbf{v}.$$
- The magnitude of $\mathbf{v} = \langle v_1, v_2 \rangle$ is given by $\|\mathbf{v}\| = \sqrt{v_1^2 + v_2^2}$.
- You should be able to perform the operations of scalar multiplication and vector addition in component form.
- You should know the following properties of vector addition and scalar multiplication.
  - (a) $\mathbf{u} + \mathbf{v} = \mathbf{v} + \mathbf{u}$
  - (b) $(\mathbf{u} + \mathbf{v}) + \mathbf{w} = \mathbf{u} + (\mathbf{v} + \mathbf{w})$
  - (c) $\mathbf{u} + \phi = \mathbf{u}$
  - (d) $\mathbf{u} + (-\mathbf{u}) = \phi$
  - (e) $c(d\mathbf{u}) = (cd)\mathbf{u}$
  - (f) $(c + d)\mathbf{u} = c\mathbf{u} + d\mathbf{u}$
  - (g) $c(\mathbf{u} + \mathbf{v}) = c\mathbf{u} + c\mathbf{v}$
  - (h) $1(\mathbf{u}) = \mathbf{u}, 0\mathbf{u} = \phi$
  - (i) $\|c\mathbf{v}\| = |c|\,\|\mathbf{v}\|$
- A unit vector in the direction of $\mathbf{v}$ is given $\mathbf{u} = \dfrac{\mathbf{v}}{\|\mathbf{v}\|}$.
- The standard unit vectors are $\mathbf{i} = \langle 1, 0 \rangle$ and $\mathbf{j} = \langle 0, 1 \rangle$. $\mathbf{v} = \langle v_1, v_2 \rangle$ can be written as $\mathbf{v} = v_1\mathbf{i} + v_2\mathbf{j}$.
- A vector $\mathbf{v}$ with magnitude $\|\mathbf{v}\|$ and direction θ can be written as $\mathbf{v} = a\mathbf{i} + b\mathbf{j} = \|\mathbf{v}\|(\cos\theta)\mathbf{i} + \|\mathbf{v}\|(\sin\theta)\mathbf{j}$ where $\tan\theta = b/a$.

Solutions to Odd-Numbered Exercises

1. Initial point: $(0, 0)$

Terminal point: $(4, 3)$

$\mathbf{v} = \langle 4 - 0, 3 - 0 \rangle = \langle 4, 3 \rangle$

$\|\mathbf{v}\| = \sqrt{4^2 + 3^2} = 5$

3. Initial point: $(2, 2)$

Terminal point: $(-1, 4)$

$\mathbf{v} = \langle -1 - 2, 4 - 2 \rangle = \langle -3, 2 \rangle$

$\|\mathbf{v}\| = \sqrt{(-3)^2 + 2^2} = \sqrt{13}$

5. Initial point: $(3, -2)$

Terminal point: $(3, 3)$

$\mathbf{v} = \langle 3 - 3, 3 - (-2) \rangle = \langle 0, 5 \rangle$

$\|\mathbf{v}\| = \sqrt{0^2 + 5^2} = \sqrt{25} = 5$

7. Initial point: $(-1, 5)$

Terminal point: $(15, 12)$

$\mathbf{v} = \langle 15 - (-1), 12 - 5 \rangle = \langle 16, 7 \rangle$

$\|\mathbf{v}\| = \sqrt{16^2 + 7^2} = \sqrt{305}$

9. Initial point: $(-3, -5)$

Terminal point: $(5, 1)$

$\mathbf{v} = \langle 5 - (-3), 1 - (-5) \rangle = \langle 8, 6 \rangle$

$\|\mathbf{v}\| = \sqrt{8^2 + 6^2} = \sqrt{100} = 10$

11.

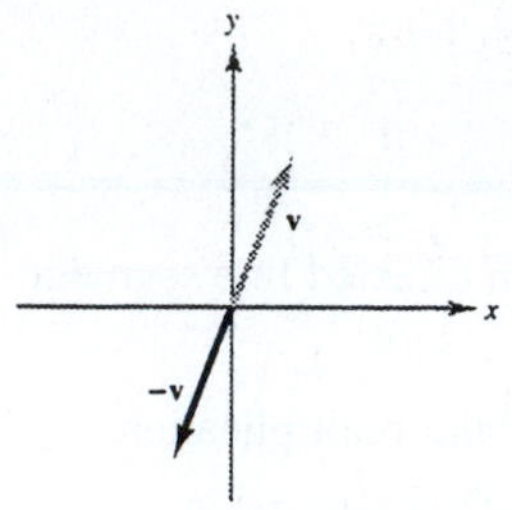

13.

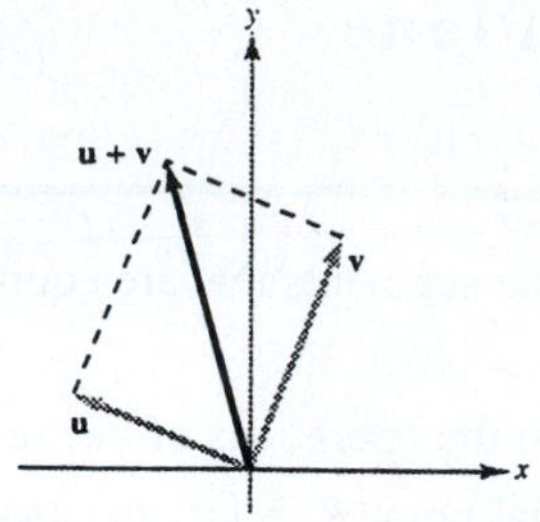

15.

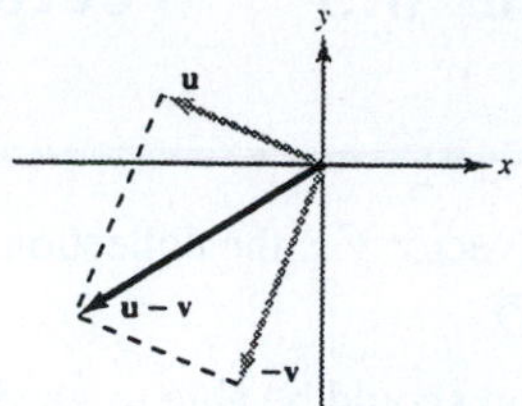

17. $\mathbf{u} = \langle 1, 2\rangle, \mathbf{v} = \langle 3, 1\rangle$

(a) $\mathbf{u} + \mathbf{v} = \langle 4, 3\rangle$

(b) $\mathbf{u} - \mathbf{v} = \langle -2, 1\rangle$

(c) $2\mathbf{u} - 3\mathbf{v} = \langle 2, 4\rangle - \langle 9, 3\rangle = \langle -7, 1\rangle$

19. $\mathbf{u} = \langle 4, -2\rangle, \mathbf{v} = \langle 0, 0\rangle$

(a) $\mathbf{u} + \mathbf{v} = \langle 4, -2\rangle$

(b) $\mathbf{u} - \mathbf{v} = \langle 4, -2\rangle$

(c) $2\mathbf{u} - 3\mathbf{v} = \langle 8, -4\rangle - \langle 0, 0\rangle = \langle 8, -4\rangle$

21. $\mathbf{u} = \mathbf{i} + \mathbf{j}, \mathbf{v} = 2\mathbf{i} - 3\mathbf{j}$

(a) $\mathbf{u} + \mathbf{v} = 3\mathbf{i} - 2\mathbf{j}$

(b) $\mathbf{u} - \mathbf{v} = -\mathbf{i} + 4\mathbf{j}$

(c) $2\mathbf{u} - 3\mathbf{v} = (2\mathbf{i} + 2\mathbf{j}) - (6\mathbf{i} - 9\mathbf{j})$
$= -4\mathbf{i} + 11\mathbf{j}$

23. $\mathbf{u} = 2\mathbf{i}, \mathbf{v} = \mathbf{j}$

(a) $\mathbf{u} + \mathbf{v} = 2\mathbf{i} + \mathbf{j}$

(b) $\mathbf{u} - \mathbf{v} = 2\mathbf{i} - \mathbf{j}$

(c) $2\mathbf{u} - 3\mathbf{v} = 4\mathbf{i} - 3\mathbf{j}$

25.
$$\mathbf{u} = \frac{1}{\|\mathbf{v}\|}\mathbf{v} = \frac{1}{\sqrt{5^2 + 0^2}}\langle 5, 0\rangle = \frac{1}{5}\langle 5, 0\rangle = \langle 1, 0\rangle$$

27.
$$\mathbf{u} = \frac{1}{\|\mathbf{v}\|}\mathbf{v} = \frac{1}{\sqrt{(-2)^2 + 2^2}}\langle -2, 2\rangle = \frac{1}{2\sqrt{2}}\langle -2, 2\rangle = \left\langle -\frac{1}{\sqrt{2}}, \frac{1}{\sqrt{2}}\right\rangle$$

29.
$$\mathbf{u} = \frac{1}{\|\mathbf{v}\|}\mathbf{v} = \frac{1}{\sqrt{16 + 9}}(4\mathbf{i} - 3\mathbf{j}) = \frac{1}{5}(4\mathbf{i} - 3\mathbf{j}) = \frac{4}{5}\mathbf{i} - \frac{3}{5}\mathbf{j}$$

31.
$$\mathbf{u} = \frac{1}{\|\mathbf{v}\|}\mathbf{v} = \frac{1}{2}(2\mathbf{j}) = \mathbf{j}$$

33.
$$5\left(\frac{1}{\|\mathbf{v}\|}\mathbf{v}\right) = 5\left(\frac{1}{\sqrt{3^2 + 3^2}}\langle 3, 3\rangle\right) = 5\left(\frac{1}{3\sqrt{2}}\langle 3, 3\rangle\right) = \left\langle \frac{5}{\sqrt{2}}, \frac{5}{\sqrt{2}}\right\rangle$$

35.
$$7\left(\frac{1}{\|\mathbf{v}\|}\mathbf{v}\right) = 7\left(\frac{1}{\sqrt{(-3)^2 + 4^2}}\langle -3, 4\rangle\right) = \frac{7}{5}\langle -3, 4\rangle = \left\langle -\frac{21}{5}, \frac{28}{5}\right\rangle$$

37. $\mathbf{v} = \frac{3}{2}\mathbf{u}$

$= \frac{3}{2}(2\mathbf{i} - \mathbf{j})$

$= 3\mathbf{i} - \frac{3}{2}\mathbf{j} = \langle 3, -\frac{3}{2}\rangle$

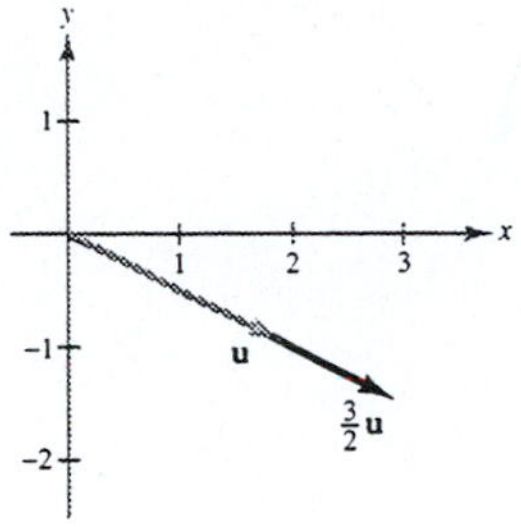

39. $\mathbf{v} = \mathbf{u} + 2\mathbf{w}$

$= (2\mathbf{i} - \mathbf{j}) + 2(\mathbf{i} + 2\mathbf{j})$

$= 4\mathbf{i} + 3\mathbf{j} = \langle 4, 3\rangle$

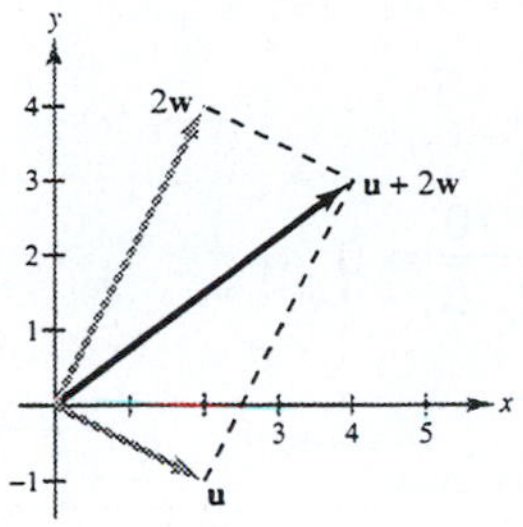

41. $\mathbf{v} = \frac{1}{2}(3\mathbf{u} + \mathbf{w})$

$= \frac{1}{2}(6\mathbf{i} - 3\mathbf{j} + \mathbf{i} + 2\mathbf{j})$

$= \frac{7}{2}\mathbf{i} - \frac{1}{2}\mathbf{j} = \langle \frac{7}{2}, -\frac{1}{2}\rangle$

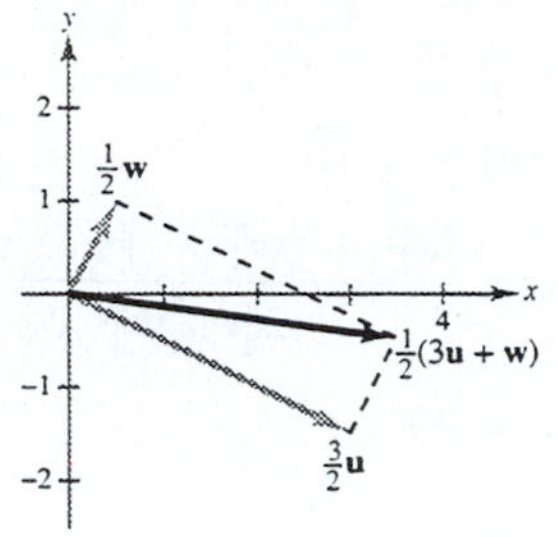

43. $\mathbf{v} = 5(\cos 30°\mathbf{i} + \sin 30°\mathbf{j})$

$\|\mathbf{v}\| = 5, \ \theta = 30°$

45. $\mathbf{v} = 6\mathbf{i} - 6\mathbf{j}$

$\|\mathbf{v}\| = \sqrt{6^2 + (-6)^2} = \sqrt{72} = 6\sqrt{2}$

$\tan\theta = \dfrac{-6}{6} = -1$

Since $\mathbf{v}$ lies in Quadrant IV, $\theta = 315°$.

47. $\mathbf{v} = \langle 3\cos 0°, 3\sin 0°\rangle$

$= \langle 3, 0\rangle$

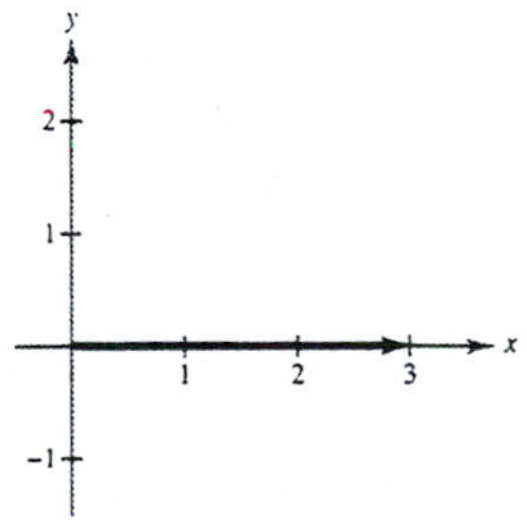

49. $\mathbf{v} = \langle \cos 150°, \sin 150°\rangle$

$= \left\langle -\dfrac{\sqrt{3}}{2}, \dfrac{1}{2}\right\rangle$

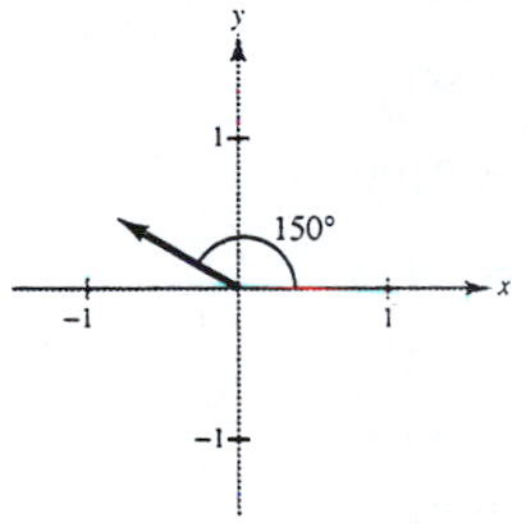

51. $\mathbf{v} = \langle 3\sqrt{2}\cos 150°, \ 3\sqrt{2}\sin 150°\rangle$

$= \left\langle -\dfrac{3\sqrt{6}}{2}, \dfrac{3\sqrt{2}}{2}\right\rangle$

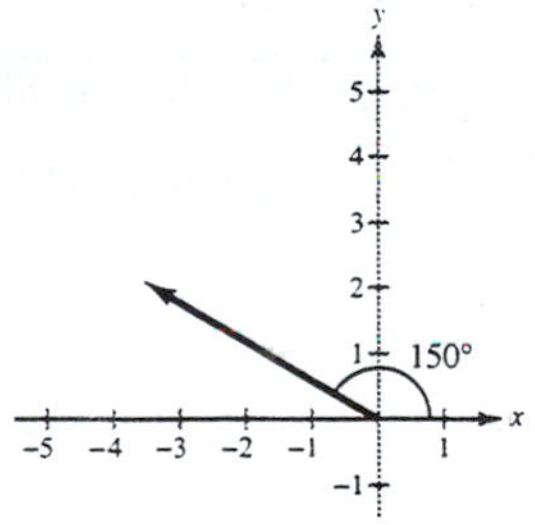

53. $\mathbf{v} = 2\left(\dfrac{1}{\sqrt{3^2 + 1^2}}\right)(\mathbf{i} + 3\mathbf{j})$

$= \dfrac{2}{\sqrt{10}}(\mathbf{i} + 3\mathbf{j})$

$= \dfrac{\sqrt{10}}{5}\mathbf{i} + \dfrac{3\sqrt{10}}{5}\mathbf{j} = \left\langle \dfrac{\sqrt{10}}{5}, \dfrac{3\sqrt{10}}{5}\right\rangle$

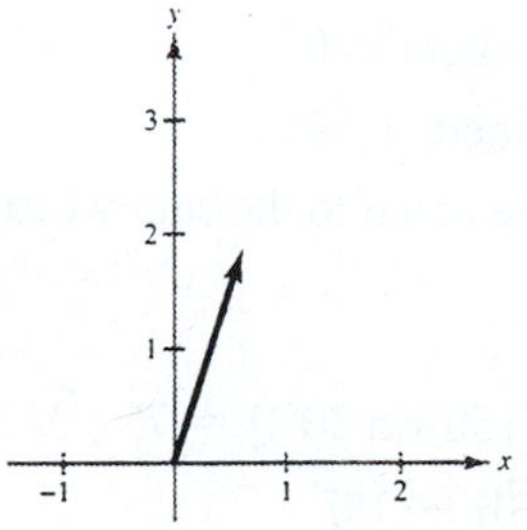

55. $\mathbf{u} = \langle 5\cos 0°, 5\sin 0°\rangle = \langle 5, 0\rangle$

$\mathbf{v} = \langle 5\cos 90°, 5\sin 90°\rangle = \langle 0, 5\rangle$

$\mathbf{u} + \mathbf{v} = \langle 5, 5\rangle$

57. $\mathbf{u} = \langle 20\cos 45°, 20\sin 45°\rangle = \langle 10\sqrt{2}, 10\sqrt{2}\rangle$

$\mathbf{v} = \langle 50\cos 180°, 50\sin 180°\rangle = \langle -50, 0\rangle$

$\mathbf{u} + \mathbf{v} = \langle 10\sqrt{2} - 50, 10\sqrt{2}\rangle$

59. $\mathbf{v} = \mathbf{i} + \mathbf{j}$

$\mathbf{w} = 2(\mathbf{i} - \mathbf{j})$

$\mathbf{u} = \mathbf{v} - \mathbf{w} = -\mathbf{i} + 3\mathbf{j}$

$\|\mathbf{v}\| = \sqrt{2}$

$\|\mathbf{w}\| = 2\sqrt{2}$

$\|\mathbf{v} - \mathbf{w}\| = \sqrt{10}$

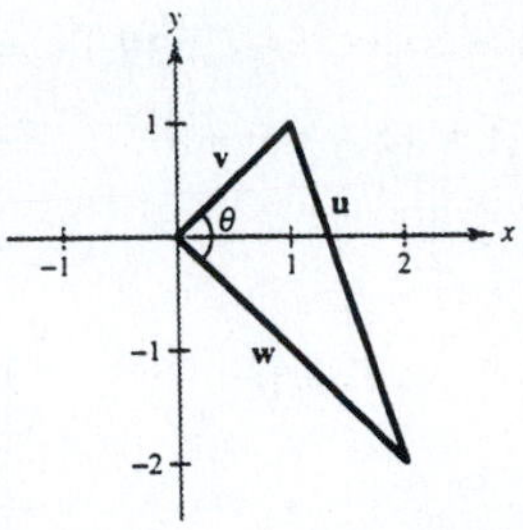

$$\cos\alpha = \frac{\|\mathbf{v}\|^2 + \|\mathbf{w}\|^2 - \|\mathbf{v} - \mathbf{w}\|^2}{2\|\mathbf{v}\|\,\|\mathbf{w}\|} = \frac{2 + 8 - 10}{2\sqrt{2} \cdot 2\sqrt{2}} = 0$$

$\alpha = 90°$

61. $\mathbf{v} = \mathbf{i} + \mathbf{j}$

$\mathbf{w} = 3\mathbf{i} - \mathbf{j}$

$\mathbf{u} = \mathbf{v} - \mathbf{w} = -2\mathbf{i} + 2\mathbf{j}$

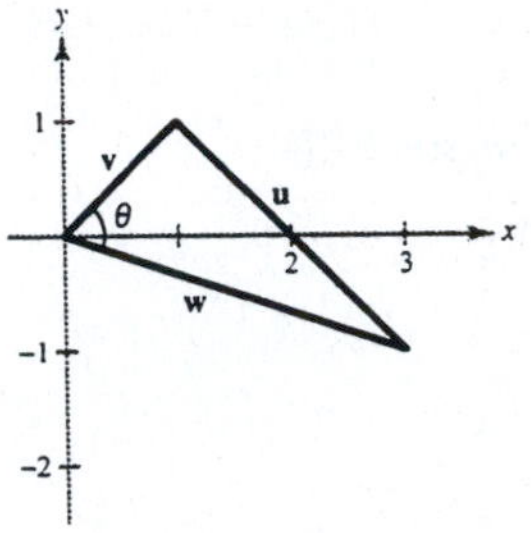

$$\cos\alpha = \frac{\|\mathbf{v}\|^2 + \|\mathbf{w}\|^2 - \|\mathbf{v} - \mathbf{w}\|^2}{2\|\mathbf{v}\|\,\|\mathbf{w}\|} = \frac{2 + 10 - 8}{2\sqrt{2}\,\sqrt{10}} \approx 0.4472$$

$\alpha = 63.4°$

63. Force One: $\mathbf{u} = 45\mathbf{i}$

Force Two: $\mathbf{v} = 60\cos\theta\mathbf{i} + 60\sin\theta\mathbf{j}$

Resultant Force: $\mathbf{u} + \mathbf{v} = (45 + 60\cos\theta)\mathbf{i} + 60\sin\theta\mathbf{j}$

$$\|\mathbf{u} + \mathbf{v}\| = \sqrt{(45 + 60\cos\theta)^2 + (60\sin\theta)^2} = 90$$

$$2025 + 5400\cos\theta + 3600 = 8100$$

$$5400\cos\theta = 2475$$

$$\cos\theta = \frac{2475}{5400} \approx 0.4583$$

$$\theta \approx 62.7°$$

65. (a) The angle between them is 0°.

(b) The angle between them is 180°.

(c) No. At most it can be equal to the sum when the angle between them is 0°.

67. $\mathbf{u} = 220\mathbf{i}$

$\mathbf{v} = (150\cos 30°)\mathbf{i} + (150\sin 30°)\mathbf{j} = 75\sqrt{3}\mathbf{i} + 75\mathbf{j}$

$\mathbf{u} + \mathbf{v} = (220 + 75\sqrt{3})\mathbf{i} + 75\mathbf{j}$

$\|\mathbf{u} + \mathbf{v}\| = \sqrt{\left(220 + 75\sqrt{3}\right)^2 + 75^2} \approx 357.85$ newtons

$$\tan\theta = \frac{75}{220 + 75\sqrt{3}} \implies \theta \approx 12.1°$$

69. $\mathbf{u} = (75 \cos 30°)\mathbf{i} + (75 \sin 30°)\mathbf{j} \approx 64.95\mathbf{i} + 37.5\mathbf{j}$

$\mathbf{v} = (100 \cos 45°)\mathbf{i} + (100 \sin 45°)\mathbf{j} \approx 70.71\mathbf{i} + 70.71\mathbf{j}$

$\mathbf{w} = (125 \cos 120°)\mathbf{i} + (125 \sin 120°)\mathbf{j} \approx -62.5\mathbf{i} + 108.3\mathbf{j}$

$\mathbf{u} + \mathbf{v} + \mathbf{w} \approx 73.16\mathbf{i} + 216.5\mathbf{j}$

$\|\mathbf{u} + \mathbf{v} + \mathbf{w}\| \approx 228.5$ pounds

$$\tan \theta \approx \frac{216.5}{73.16} \approx 2.9592$$

$$\theta \approx 71.3°$$

71. Horizontal component of velocity: $80 \cos 40° \approx 61.28$ ft/sec

Vertical component of velocity: $80 \sin 40° \approx 51.42$ ft/sec

73. Cable $\overrightarrow{AC}$: $\mathbf{u} = \|\mathbf{u}\|(\cos 50°\mathbf{i} - \sin 50°\mathbf{j})$

Cable $\overrightarrow{BC}$: $\mathbf{u} = \|\mathbf{u}\|(\cos 30°\mathbf{i} - \sin 30°\mathbf{j})$

Resultant: $\mathbf{u} + \mathbf{v} = -1000\mathbf{j}$

$\|\mathbf{u}\| \cos 50° - \|\mathbf{v}\| \cos 30° = 0$

$-\|\mathbf{u}\| \sin 50° - \|\mathbf{v}\| \sin 30° = -2000$

Solving this system of equations yields:

$T_{AC} = \|\mathbf{u}\| \approx 1758.8$ pounds

$T_{BC} = \|\mathbf{v}\| \approx 1305.4$ pounds

75. Towline 1: $\mathbf{u} = \|\mathbf{u}\|(\cos 18°\mathbf{i} + \sin 18°\mathbf{j})$

Towline 2: $\mathbf{v} = \|\mathbf{u}\|(\cos 18°\mathbf{i} - \sin 18°\mathbf{j})$

Resultant: $\mathbf{u} + \mathbf{v} = 6000\mathbf{i}$

$\|\mathbf{u}\| \cos 18° + \|\mathbf{u}\| \cos 18° = 6000$

$\|\mathbf{u}\| \approx 3154.4$

Therefore, the tension on each towline is $\|\mathbf{u}\| \approx 3154.4$ pounds.

77. Airspeed: $\mathbf{u} = (875 \cos 32°)\mathbf{i} - (875 \sin 32°)\mathbf{j}$

Groundspeed: $\mathbf{v} = (800 \cos 40°)\mathbf{i} - (800 \sin 40°)\mathbf{j}$

Wind: $\mathbf{w} = \mathbf{v} - \mathbf{u} = (800 \cos 40° - 875 \cos 32°)\mathbf{i} + (-800 \sin 40° + 875 \sin 32°)\mathbf{j}$

$\approx -129.2065\mathbf{i} - 50.5507\mathbf{j}$

Wind speed: $\|\mathbf{w}\| \approx \sqrt{(-129.2065)^2 + (-50.5507)^2}$

≈ 138.7 kilometers per hour

Wind direction: $\tan \theta \approx \dfrac{-50.5507}{-129.2065}$

$\theta \approx 21.4°$

N 21.4° E

79. $W = FD = (85 \cos 60°)(20) = 850$ ft/lb

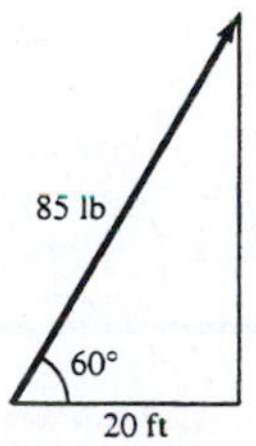

81. True

83. False, $a = b = 0$.

85. Let $\mathbf{v} = (\cos\theta)\mathbf{i} + (\sin\theta)\mathbf{j}$.

$\|\mathbf{v}\| = \sqrt{\cos^2\theta + \sin^2\theta} = \sqrt{1} = 1$

Therefore, $\mathbf{v}$ is a unit vector for any value of θ.

87. $\mathbf{u} = \langle 5 - 1, 2 - 6\rangle = \langle 4, -4\rangle$

$\mathbf{v} = \langle 9 - 4, 4 - 5\rangle = \langle 5, -1\rangle$

$\mathbf{u} - \mathbf{v} = \langle -1, -3\rangle$ or $\mathbf{v} - \mathbf{u} = \langle 1, 3\rangle$

89. Let d be the distance from the bridge deck to the water level.

$$\frac{d}{\sin 55} = \frac{17.779}{\sin 90}$$

$$d \approx 14.6 \text{ meters}$$

91.
$$\begin{aligned}\sqrt{x^2 - 64} &= \sqrt{(8\sec\theta)^2 - 64}\\ &= \sqrt{64(\sec^2\theta - 1)}\\ &= 8\sqrt{\tan^2\theta}\\ &= 8\tan\theta \quad \text{for } 0 < \theta < \frac{\pi}{2}\end{aligned}$$

93.
$$\begin{aligned}\sqrt{x^2 + 36} &= \sqrt{(6\tan\theta)^2 + 36}\\ &= \sqrt{36(\tan^2\theta + 1)}\\ &= 6\sqrt{\sec^2\theta}\\ &= 6\sec\theta \quad \text{for } 0 < \theta < \frac{\pi}{2}\end{aligned}$$

Section 6.4 Vectors and Dot Products

- Know the definition of the dot product of $\mathbf{u} = \langle u_1, u_2\rangle$ and $\mathbf{v} = \langle v_1, v_2\rangle$.

  $\mathbf{u}\cdot\mathbf{v} = u_1v_1 + u_2v_2$
- Know the following properties of the dot product:
  1. $\mathbf{u}\cdot\mathbf{v} = \mathbf{v}\cdot\mathbf{u}$
  2. $\mathbf{0}\cdot\mathbf{v} = 0$
  3. $\mathbf{u}\cdot(\mathbf{v} + \mathbf{w}) = \mathbf{u}\cdot\mathbf{v} + \mathbf{u}\cdot\mathbf{w}$
  4. $\mathbf{v}\cdot\mathbf{v} = \|\mathbf{v}\|^2$
  5. $c(\mathbf{u}\cdot\mathbf{v}) = c\mathbf{u}\cdot\mathbf{v} = \mathbf{u}\cdot c\mathbf{v}$
- If θ is the angle between two nonzero vectors $\mathbf{u}$ and $\mathbf{v}$, then

  $$\cos\theta = \frac{\mathbf{u}\cdot\mathbf{v}}{\|\mathbf{u}\|\,\|\mathbf{v}\|}.$$
- The vectors $\mathbf{u}$ and $\mathbf{v}$ are orthogonal if $\mathbf{u}\cdot\mathbf{v} = 0$.
- Know the definition of vector components.

  $\mathbf{u} = \mathbf{w}_1 + \mathbf{w}_2$ where $\mathbf{w}_1$ and $\mathbf{w}_2$ are orthogonal, and $\mathbf{w}_1$ is parallel to $\mathbf{v}$. $\mathbf{w}_1$ is called the projection of $\mathbf{u}$ onto $\mathbf{v}$ and is denoted by $\mathbf{w}_1 = \text{proj}_{\mathbf{v}}\mathbf{u} = \left(\frac{\mathbf{u}\cdot\mathbf{v}}{\|\mathbf{v}\|^2}\right)\mathbf{v}$. Then we have $\mathbf{w}_2 = \mathbf{u} - \mathbf{w}_1$.
- Know the definition of work.
  1. Projection form: $w = \|\text{proj}_{\overrightarrow{PQ}}\mathbf{F}\|\,\|\overrightarrow{PQ}\|$
  2. Dot product form: $w = \mathbf{F}\cdot\overrightarrow{PQ}$

Solutions to Odd-Numbered Exercises

1. $\mathbf{u} = \langle 3, 4\rangle$, $\mathbf{v} = \langle 2, -3\rangle$

$\mathbf{u}\cdot\mathbf{v} = 3(2) + 4(-3) = -6$

3. $\mathbf{u} = 4\mathbf{i} - 2\mathbf{j}$, $\mathbf{v} = \mathbf{i} - \mathbf{j}$

$\mathbf{u}\cdot\mathbf{v} = 4(1) + (-2)(-1) = 6$

5. $\mathbf{u} = \langle 2, 2\rangle$

$\mathbf{u}\cdot\mathbf{u} = 2(2) + 2(2) = 8$

The result is a scalar.

7. $\mathbf{u} = \langle 2, 2\rangle,\ \mathbf{v} = \langle -3, 4\rangle$

$$(\mathbf{u} \cdot \mathbf{v})\mathbf{v} = [(2)(-3) + 2(4)]\langle -3, 4\rangle$$
$$= 2\langle -3, 4\rangle = \langle -6, 8\rangle$$

The result is a vector.

9. $\mathbf{u} = \langle -5, 12\rangle$

$$\|\mathbf{u}\| = \sqrt{\mathbf{u} \cdot \mathbf{u}} = \sqrt{(-5)^2 + 12^2} = 13$$

11. $\mathbf{u} = 20\mathbf{i} + 25\mathbf{j}$

$$\|\mathbf{u}\| = \sqrt{(20)^2 + (25)^2} = \sqrt{1025} = 5\sqrt{41}$$

13. $\mathbf{u} = \langle 1245, 2600\rangle,\ \mathbf{v} = \langle 12.20, 8.50\rangle$

$$\mathbf{u} \cdot \mathbf{v} = 1245(12.20) + 2600(8.50) = \$37{,}289$$

This gives the total revenue that can be earned by selling all of the units.

15. $\mathbf{u} = \langle 1, 0\rangle,\ \mathbf{v} = (0, -2\rangle$

$$\cos\theta = \frac{\mathbf{u} \cdot \mathbf{v}}{\|\mathbf{u}\|\,\|\mathbf{v}\|} = \frac{0}{(1)(2)} = 0$$

$$\theta = 90^\circ$$

17. $\mathbf{u} = 3\mathbf{i} + 4\mathbf{j},\ \mathbf{v} = -2\mathbf{j}$

$$\cos\theta = \frac{\mathbf{u} \cdot \mathbf{v}}{\|\mathbf{u}\|\,\|\mathbf{v}\|} = -\frac{8}{(5)(2)}$$

$$\theta = \arccos\left(-\frac{4}{5}\right)$$

$$\theta \approx 143.13^\circ$$

19. $\mathbf{u} = \left(\cos\frac{\pi}{3}\right)\mathbf{i} + \left(\sin\frac{\pi}{3}\right)\mathbf{j} = \frac{1}{2}\mathbf{i} + \frac{\sqrt{3}}{2}\mathbf{j}$

$$\mathbf{v} = \left(\cos\frac{3\pi}{4}\right)\mathbf{i} + \left(\sin\frac{3\pi}{4}\right)\mathbf{j} = -\frac{\sqrt{2}}{2}\mathbf{i} + \frac{\sqrt{2}}{2}\mathbf{j}$$

$$\|\mathbf{u}\| = \|\mathbf{v}\| = 1$$

$$\cos\theta = \frac{\mathbf{u} \cdot \mathbf{v}}{\|\mathbf{u}\|\,\|\mathbf{v}\|} = \mathbf{u} \cdot \mathbf{v} = \left(\frac{1}{2}\right)\left(-\frac{\sqrt{2}}{2}\right) + \left(\frac{\sqrt{3}}{2}\right)\left(\frac{\sqrt{2}}{2}\right) = \frac{-\sqrt{2} + \sqrt{6}}{4}$$

$$\theta = \arccos\left(\frac{-\sqrt{2} + \sqrt{6}}{4}\right) = 75^\circ = \frac{5\pi}{12}$$

21. $\mathbf{u} = 3\mathbf{i} + 4\mathbf{j},\ \mathbf{v} = -7\mathbf{i} + 5\mathbf{j}$

$$\cos\theta = \frac{\mathbf{u} \cdot \mathbf{v}}{\|\mathbf{u}\|\,\|\mathbf{v}\|} = -\frac{1}{(5)(\sqrt{74})} \implies \theta \approx 91.33^\circ$$

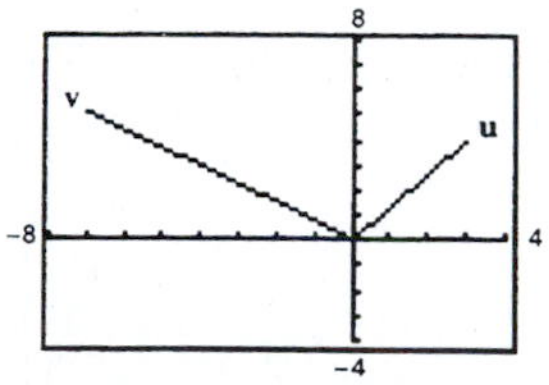

23. $\mathbf{u} = 5\mathbf{i} + 5\mathbf{j},\ \mathbf{v} = -6\mathbf{i} + 6\mathbf{j}$

$$\cos\theta = \frac{\mathbf{u} \cdot \mathbf{v}}{\|\mathbf{u}\|\,\|\mathbf{v}\|} = 0 \implies \theta = 90^\circ$$

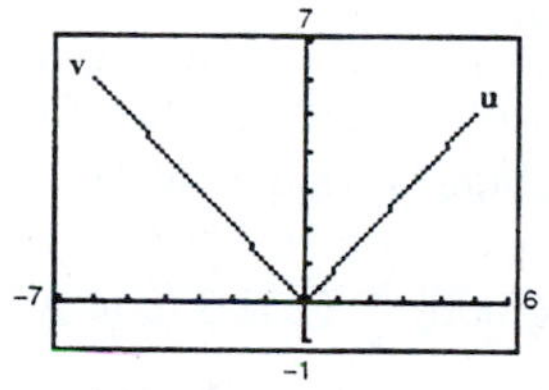

25. $P = (1, 2),\ Q = (3, 4),\ R = (2, 5)$

$$\overrightarrow{PQ} = \langle 2, 2\rangle,\ \overrightarrow{PR} = \langle 1, 3\rangle,\ \overrightarrow{QR} = \langle -1, -1\rangle$$

$$\cos\alpha = \frac{\overrightarrow{PQ} \cdot \overrightarrow{PR}}{\|\overrightarrow{PQ}\|\,\|\overrightarrow{PR}\|} = \frac{8}{(2\sqrt{2})(\sqrt{10})} \implies \alpha = \arccos\frac{2}{\sqrt{5}} \approx 26.6^\circ$$

$$\cos\beta = \frac{\overrightarrow{PQ} \cdot \overrightarrow{QR}}{\|\overrightarrow{PQ}\|\,\|\overrightarrow{QR}\|} = 0 \implies \beta = 90^\circ.\text{ Thus, } \gamma = 180^\circ - 26.6^\circ - 90^\circ = 63.4^\circ.$$

27. $\mathbf{u} \cdot \mathbf{v} = \|\mathbf{u}\|\,\|\mathbf{v}\| \cos\theta$

$$= (4)(10)\cos\frac{2\pi}{3}$$

$$= 40\left(-\frac{1}{2}\right)$$

$$= -20$$

29. $\mathbf{u} = \langle -12, 30\rangle,\ \mathbf{v} = \left\langle \frac{1}{2}, -\frac{5}{4}\right\rangle$

$\mathbf{u} = -24\mathbf{v} \implies$ **u** and **v** are parallel.

31. $\mathbf{u} = \frac{1}{4}(3\mathbf{i} - \mathbf{j}),\ \mathbf{v} = 5\mathbf{i} + 6\mathbf{j}$

$\mathbf{u} \neq k\mathbf{v} \implies$ Not parallel

$\mathbf{u} \cdot \mathbf{v} \neq 0 \implies$ Not orthogonal

Neither

33. $\mathbf{u} = 2\mathbf{i} - 2\mathbf{j},\ \mathbf{v} = -\mathbf{i} - \mathbf{j}$

$\mathbf{u} \cdot \mathbf{v} = 0 \implies$ **u** and **v** are orthogonal.

35. $\mathbf{u} = \langle 3, 4\rangle,\ \mathbf{v} = \langle 8, 2\rangle$

$$\mathbf{w}_1 = \text{proj}_{\mathbf{v}}\mathbf{u} = \left(\frac{\mathbf{u} \cdot \mathbf{v}}{\|\mathbf{v}\|^2}\right)\mathbf{v} = \left(\frac{32}{68}\right)\mathbf{v} = \frac{8}{17}\langle 8, 2\rangle = \frac{16}{17}\langle 4, 1\rangle$$

$$\mathbf{w}_2 = \mathbf{u} - \mathbf{w}_1 = \langle 3, 4\rangle - \frac{16}{17}\langle 4, 1\rangle = \frac{13}{17}\langle -1, 4\rangle$$

37. $\mathbf{u} = \langle 0, 3\rangle,\ \mathbf{v} = \langle 2, 15\rangle$

$$\mathbf{w}_1 = \text{proj}_{\mathbf{v}}\mathbf{u} = \left(\frac{\mathbf{u} \cdot \mathbf{v}}{\|\mathbf{v}\|^2}\right)\mathbf{v} = \frac{45}{229}\langle 2, 15\rangle$$

$$\mathbf{w}_2 = \mathbf{u} - \mathbf{w}_1 = \langle 0, 3\rangle - \frac{45}{229}\langle 2, 15\rangle = \left\langle -\frac{90}{229}, \frac{12}{229}\right\rangle = \frac{6}{229}\langle -15, 2\rangle$$

39. $\mathbf{u} = \langle 3, 5\rangle$

For **v** to be orthogonal to **u**, $\mathbf{u} \cdot \mathbf{v}$ must equal 0.

Two possibilities: $\langle -5, 3\rangle$ and $\langle 5, -3\rangle$

41. $\mathbf{u} = \frac{1}{2}\mathbf{i} - \frac{2}{3}\mathbf{j}$

For **u** and **v** to be orthogonal, $\mathbf{u} \cdot \mathbf{v}$ must equal 0.

Two possibilities: $\frac{2}{3}\mathbf{i} + \frac{1}{2}\mathbf{j}$ and $-\frac{2}{3}\mathbf{i} - \frac{1}{2}\mathbf{j}$

43. (a) $\mathbf{F} = -36{,}000\mathbf{j}$ Gravitational force

$\mathbf{v} = (\cos 10°)\mathbf{i} + (\sin 10°)\mathbf{j}$

$$\mathbf{w}_1 = \text{proj}_{\mathbf{v}}\mathbf{F} = \left(\frac{\mathbf{F} \cdot \mathbf{v}}{\|\mathbf{v}\|^2}\right)\mathbf{v} = (\mathbf{F} \cdot \mathbf{v})\,\mathbf{v} \approx -6251.3\mathbf{v}$$

The magnitude of this force is 6251.3, therefore a force of 6251.3 pounds is needed to keep the truck from rolling down the hill.

(b) $\mathbf{w}_2 = \mathbf{F} - \mathbf{w}_1 = -36{,}000\mathbf{j} + 6251.3(\cos 10°\mathbf{i} + \sin 10°\mathbf{j})$

$$= [(6251.3\cos 10°)\mathbf{i} + (6251.3\sin 10° - 36{,}000)\mathbf{j}]$$

$\|\mathbf{w}_2\| \approx 35{,}453.1$ pounds

45. (a) $\mathbf{u} \cdot \mathbf{v} = 0 \implies$ **u** and **v** are orthogonal and $\theta = \frac{\pi}{2}$.

(b) $\mathbf{u} \cdot \mathbf{v} > 0 \implies \cos\theta > 0 \implies 0 \le \theta < \frac{\pi}{2}$

(c) $\mathbf{u} \cdot \mathbf{v} < 0 \implies \cos\theta < 0 \implies \frac{\pi}{2} < \theta \le \pi$

47. $w = (245)(3) = 735$ Newton-meters

49. $w = (\cos 30°)(45)(20) \approx 779.4$ foot-pounds

51. $\mathbf{w} = \|\text{proj}_{\overrightarrow{PQ}}\mathbf{v}\|\|\overrightarrow{PQ}\|$ where $\overrightarrow{PQ} = \langle 4, 7\rangle$ and $\mathbf{v} = \langle 1, 4\rangle$.

$$\text{proj}_{\overrightarrow{PQ}}\mathbf{v} = \left(\frac{\mathbf{v} \cdot \overrightarrow{PQ}}{\|\overrightarrow{PQ}\|^2}\right)\overrightarrow{PQ} = \left(\frac{32}{65}\right)\langle 4, 7\rangle$$

$$\mathbf{w} = \|\text{proj}_{\overrightarrow{PQ}}\mathbf{v}\|\|\overrightarrow{PQ}\| = \left(\frac{32\sqrt{65}}{65}\right)(\sqrt{65}) = 32$$

53. In a rhombus, $\|\mathbf{u}\| = \|\mathbf{v}\|$. The diagonals are $\mathbf{u} + \mathbf{v}$ and $\mathbf{u} - \mathbf{v}$.

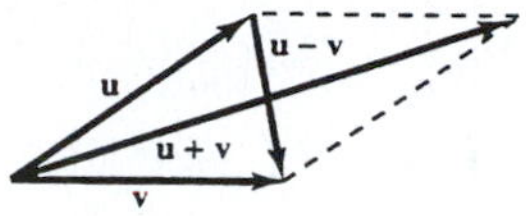

$$\begin{aligned}(\mathbf{u} + \mathbf{v}) \cdot (\mathbf{u} - \mathbf{v}) &= (\mathbf{u} + \mathbf{v}) \cdot \mathbf{u} - (\mathbf{u} + \mathbf{v}) \cdot \mathbf{v} \\ &= \mathbf{u} \cdot \mathbf{u} + \mathbf{v} \cdot \mathbf{u} - \mathbf{u} \cdot \mathbf{v} - \mathbf{v} \cdot \mathbf{v} \\ &= \|\mathbf{u}\|^2 - \|\mathbf{v}\|^2 = 0\end{aligned}$$

Therefore, the diagonals are orthogonal.

55. (a) Let $\mathbf{v} = \langle v_1, v_2\rangle$.

$$\mathbf{0} \cdot \mathbf{v} = 0(v_1) + 0(v_2) = 0$$

(b) Let $\mathbf{u} = \langle u_1, u_2\rangle$, $\mathbf{v} = \langle v_1, v_2\rangle$ and $\mathbf{w} = \langle w_1, w_2\rangle$.

$$\begin{aligned}\mathbf{u} \cdot (\mathbf{v} + \mathbf{w}) &= \langle u_1, u_2\rangle \cdot \langle v_1 + w_1, v_2 + w_2\rangle \\ &= u_1(v_1 + w_1) + u_2(v_2 + w_2) \\ &= u_1v_1 + u_1w_1 + u_2v_2 + u_2w_2 \\ &= (u_1v_1 + u_2v_2) + (u_1w_1 + u_2w_2) \\ &= \mathbf{u} \cdot \mathbf{v} + \mathbf{u} \cdot \mathbf{w}\end{aligned}$$

(c) Let $\mathbf{u} = \langle u_1, u_2\rangle$ and $\mathbf{v} = \langle v_1, v_2\rangle$.

$$\begin{aligned}c(\mathbf{u} \cdot \mathbf{v}) &= c(u_1v_1 + u_2v_2) \\ &= c(u_1v_1) + c(u_2v_2) \\ &= u_1(cv_1) + u_2(cv_2) \\ &= \mathbf{u} \cdot (c\mathbf{v})\end{aligned}$$

Section 6.5 DeMoivre's Theorem

- You should be able to graphically represent complex numbers and know the following facts about them.
- The absolute value of the complex numbers $z = a + bi$ is $|z| = \sqrt{a^2 + b^2}$.
- The trigonometric form of the complex number $z = a + bi$ is $z = r(\cos\theta + i\sin\theta)$ where
  (a) $a = r\cos\theta$
  (b) $b = r\sin\theta$
  (c) $r = \sqrt{a^2 + b^2}$; r is called the modulus of z.
  (d) $\tan\theta = \frac{b}{a}$; θ is called the argument of z.
- Given $z_1 = r_1(\cos\theta_1 + i\sin\theta_1)$ and $z_2 = r_2(\cos\theta_2 + i\sin\theta_2)$:
  (a) $z_1z_2 = r_1r_2[\cos(\theta_1 + \theta_2) + i\sin(\theta_1 + \theta_2)]$
  (b) $\dfrac{z_1}{z_2} = \dfrac{r_1}{r_2}[\cos(\theta_1 - \theta_2) + i\sin(\theta_1 - \theta_2)]$, $z_2 \neq 0$
- You should know DeMoivre's Theorem: If $z = r(\cos\theta + i\sin\theta)$, then for any positive integer n,
  $z^n = r^n(\cos n\theta + i\sin n\theta)$.
- You should know that for any positive integer n, $z = r(\cos\theta + i\sin\theta)$ has n distinct nth roots given by
  $$\sqrt[n]{r}\left[\cos\left(\frac{\theta + 2\pi k}{n}\right) + i\sin\left(\frac{\theta + 2\pi k}{n}\right)\right]$$
  where $k = 0, 1, 2, \ldots, n - 1$.

Solutions to Odd-Numbered Exercises

1. $|-5i| = \sqrt{0^2 + (-5)^2}$
$= \sqrt{25} = 5$

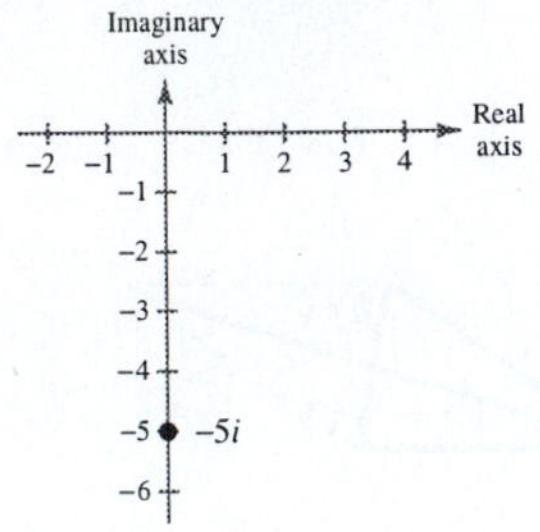

3. $|-4 + 4i| = \sqrt{(-4)^2 + (4)^2}$
$= \sqrt{32} = 4\sqrt{2}$

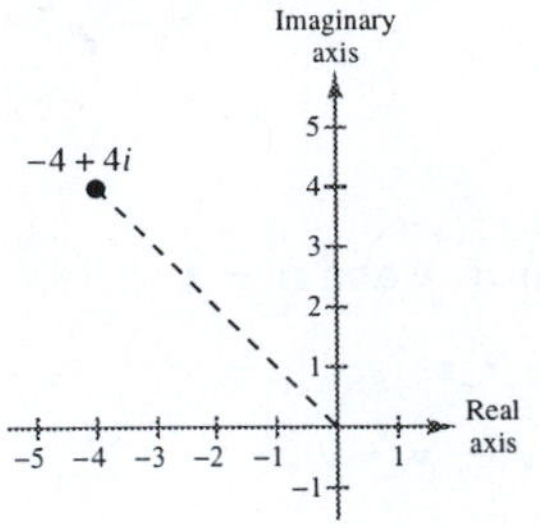

5. $|6 - 7i| = \sqrt{6^2 + (-7)^2}$
$= \sqrt{85}$

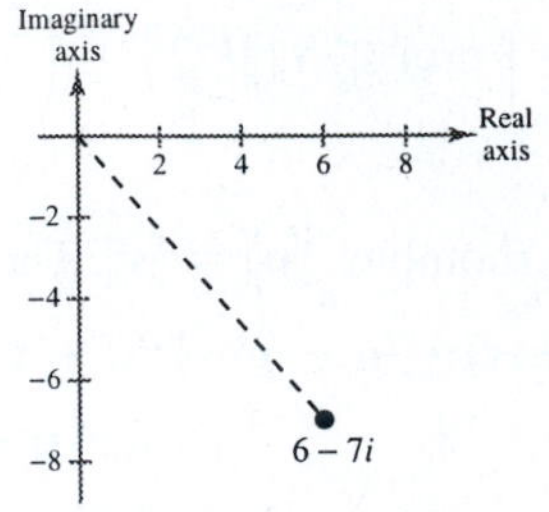

7. $z = 3i$

$r = \sqrt{0^2 + 3^2} = \sqrt{9} = 3$

$\tan\theta = \dfrac{3}{0}$, undefined $\Rightarrow \theta = \dfrac{\pi}{2}$

$z = 3\left(\cos\dfrac{\pi}{2} + i\sin\dfrac{\pi}{2}\right)$

9. $z = -2 - 2i$

$r = \sqrt{(-2)^2 + (-2)^2} = \sqrt{8} = 2\sqrt{2}$

$\tan\theta = \dfrac{-2}{-2} = 1$, θ is in Quadrant III.

$\theta = 225°$ or $\dfrac{5\pi}{4}$

$z = 2\sqrt{2}\left(\cos\dfrac{5\pi}{4} + i\sin\dfrac{5\pi}{4}\right)$

11. $z = 3 - 3i$

$r = \sqrt{3^2 + (-3)^2} = \sqrt{18} = 3\sqrt{2}$

$\tan\theta = \dfrac{-3}{3} = -1$, θ is in Quadrant IV $\Rightarrow \theta = \dfrac{7\pi}{4}$.

$z = 3\sqrt{2}\left(\cos\dfrac{7\pi}{4} + i\sin\dfrac{7\pi}{4}\right)$

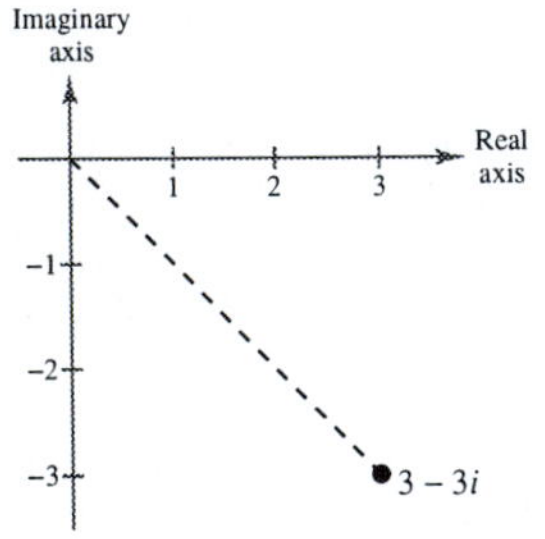

13. $z = \sqrt{3} + i$

$r = \sqrt{\left(\sqrt{3}\right)^2 + 1^2} = \sqrt{4} = 2$

$\tan\theta = \dfrac{1}{\sqrt{3}} = \dfrac{\sqrt{3}}{3} \Rightarrow \theta = \dfrac{\pi}{6}$

$z = 2\left(\cos\dfrac{\pi}{6} + i\sin\dfrac{\pi}{6}\right)$

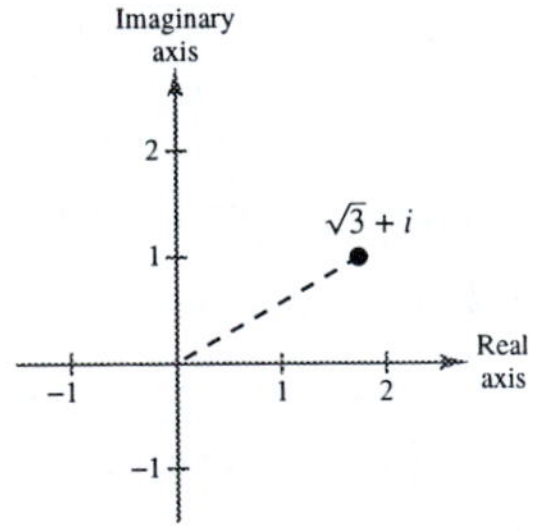

15. $z = -2(1 + \sqrt{3}i)$

$r = \sqrt{(-2)^2 + (-2\sqrt{3})^2} = \sqrt{16} = 4$

$\tan\theta = \dfrac{\sqrt{3}}{1} = \sqrt{3}$, θ is in Quadrant III $\Rightarrow$ $\theta = \dfrac{4\pi}{3}$.

$z = 4\left(\cos\dfrac{4\pi}{3} + i\sin\dfrac{4\pi}{3}\right)$

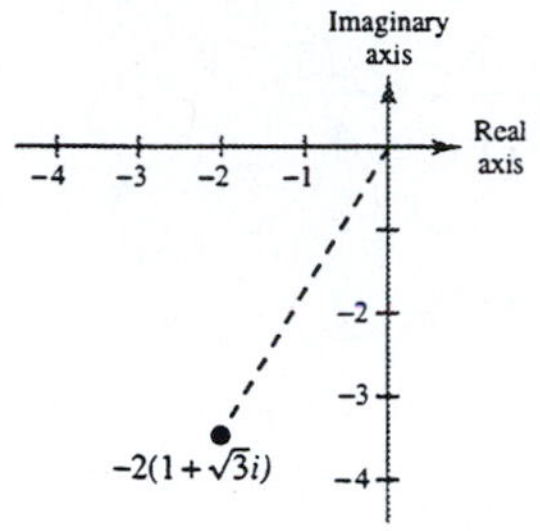

17. $z = 0 + 6i$

$r = \sqrt{0^2 + (6)^2} = \sqrt{36} = 6$

$\tan\theta = \dfrac{6}{0}$, undefined $\Rightarrow$ $\theta = \dfrac{\pi}{2}$

$z = 6\left(\cos\dfrac{\pi}{2} + i\sin\dfrac{\pi}{2}\right)$

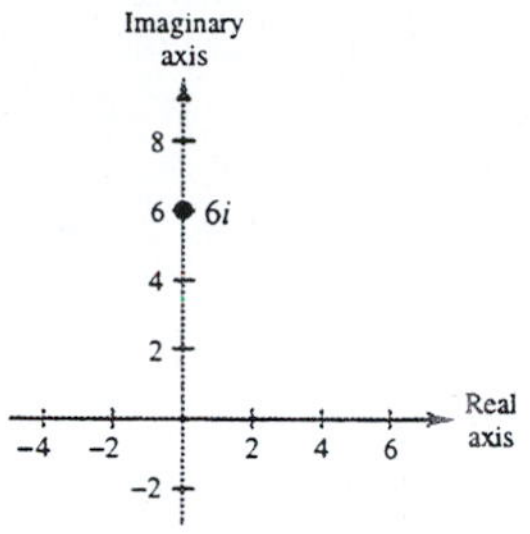

19. $z = -7 + 4i$

$r = \sqrt{(-7)^2 + (4)^2} = \sqrt{65}$

$\tan\theta = \dfrac{4}{-7}$, θ is in Quadrant II $\Rightarrow$ $\theta \approx 2.62$.

$z \approx \sqrt{65}\,(\cos 2.62 + i\sin 2.62)$

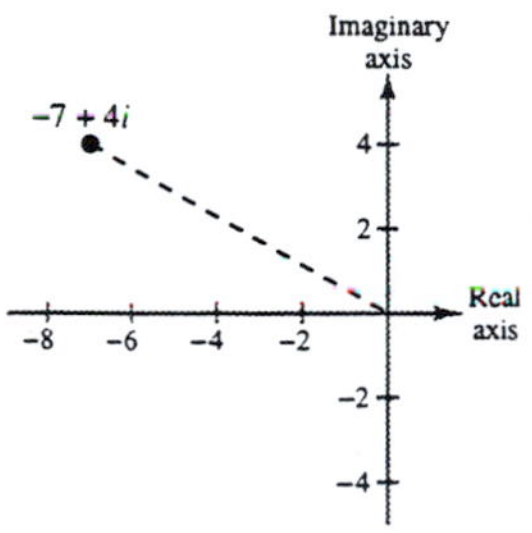

21. $z = 7 + 0i$

$r = \sqrt{(7)^2 + (0)^2} = \sqrt{49} = 7$

$\tan\theta = \dfrac{0}{7} = 0 \Rightarrow \theta = 0$

$z = 7\,(\cos 0 + i\sin 0)$

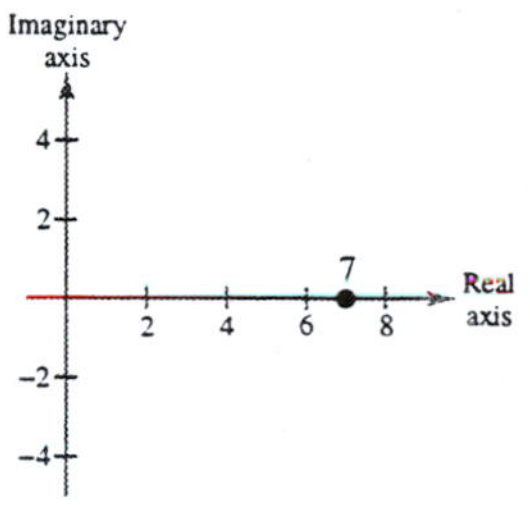

23. $z = 1 + 6i$

$r = \sqrt{1^2 + (6)^2} = \sqrt{37}$

$\tan\theta = \dfrac{6}{1} = 6 \Rightarrow \theta \approx 1.41$

$z \approx \sqrt{37}\,(\cos 1.41 + i\sin 1.41)$

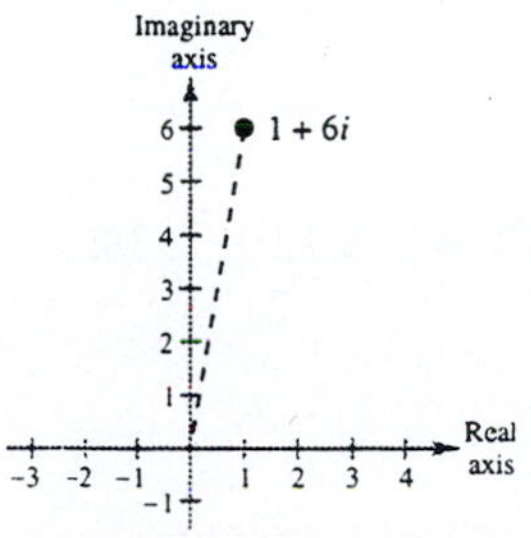

25. $z = -3 - i$

$r = \sqrt{(-3)^2 + (-1)^2} = \sqrt{10}$

$\tan\theta = \dfrac{-1}{-3} = \dfrac{1}{3}$, θ is in Quadrant III $\Rightarrow$ $\theta \approx 3.46$.

$z \approx \sqrt{10}\,(\cos 3.46 + i\sin 3.46)$

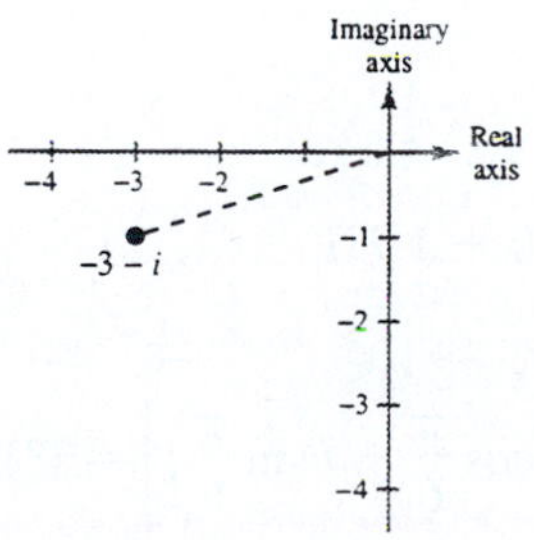

27. $z = 5 + 2i$

$r \approx 5.39$

$\theta \approx 0.38$

$z \approx 5.39(\cos 0.38 + i \sin 0.38)$

29. $z = 3\sqrt{2} - 7i$

$r \approx 8.19$

$\theta \approx -1.03 + 2\pi \approx 5.25$

$z \approx 8.19(\cos 5.25 + i \sin 5.25)$

31. $2(\cos 150° + i \sin 150°) = 2\left[-\frac{\sqrt{3}}{2} + i\left(\frac{1}{2}\right)\right]$

$= -\sqrt{3} + i$

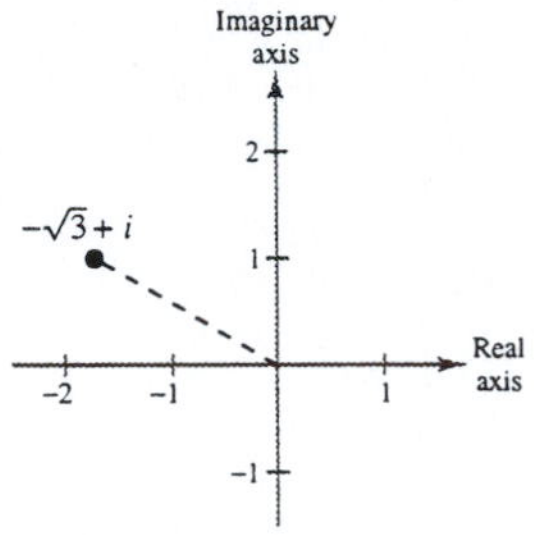

33. $\frac{3}{2}(\cos 300° + i \sin 300°) = \frac{3}{2}\left[\frac{1}{2} + i\left(-\frac{\sqrt{3}}{2}\right)\right]$

$= \frac{3}{4} - \frac{3\sqrt{3}}{4}i$

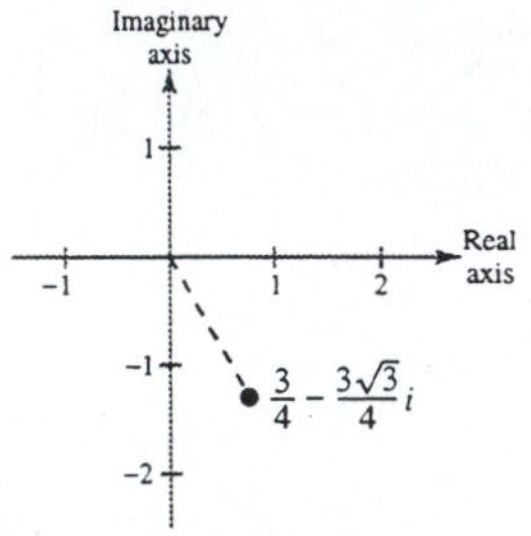

35. $3.75\left(\cos\frac{3\pi}{4} + i \sin\frac{3\pi}{4}\right) = -\frac{15\sqrt{2}}{8} + \frac{15\sqrt{2}}{8}i$

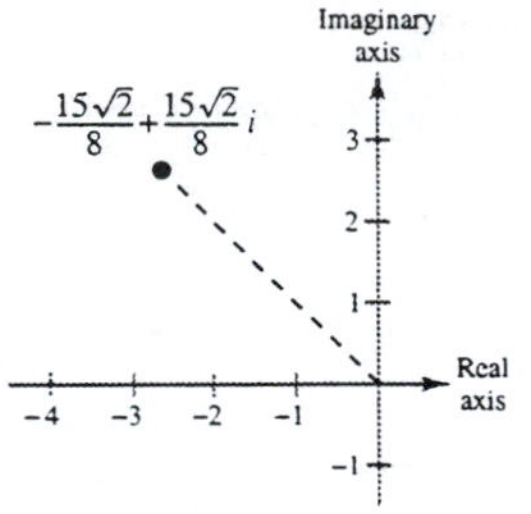

37. $4\left(\cos\frac{3\pi}{2} + i \sin\frac{3\pi}{2}\right) = 4(0 - i) = -4i$

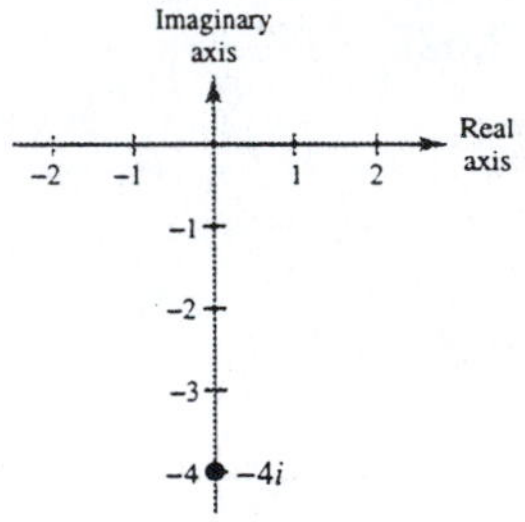

39. $3[\cos(18°45') + i \sin(18°45')] \approx 2.8408 + 0.9643i$

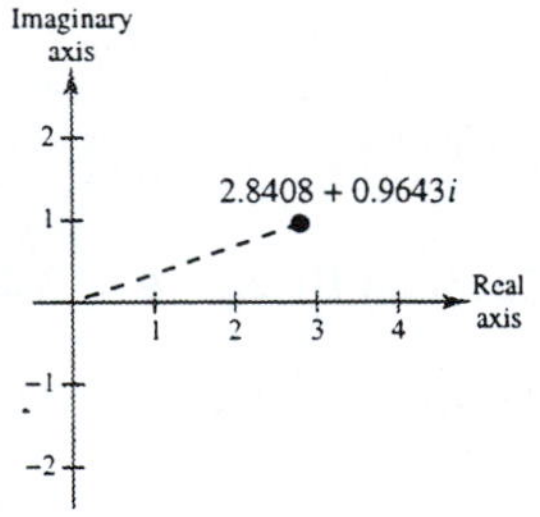

41. $5\left(\cos\frac{\pi}{9} + i \sin\frac{\pi}{9}\right) \approx 4.70 + 1.71i$

43. $4(\cos 216.5° + i \sin 216.5°) \approx -3.22 - 2.38i$

45. $\left[3\left(\cos\frac{\pi}{3} + i \sin\frac{\pi}{3}\right)\right]\left[4\left(\cos\frac{\pi}{6} + i \sin\frac{\pi}{6}\right)\right] = (3)(4)\left[\cos\left(\frac{\pi}{3} + \frac{\pi}{6}\right) + i \sin\left(\frac{\pi}{6} + \frac{\pi}{3}\right)\right]$

$= 12\left(\cos\frac{\pi}{2} + i \sin\frac{\pi}{2}\right)$

47. $\left[\frac{5}{3}(\cos 140° + i \sin 140°)\right]\left[\frac{2}{3}(\cos 60° + i \sin 60°)\right] = \left(\frac{5}{3}\right)\left(\frac{2}{3}\right)[\cos(140° + 60°) + i \sin(140° + 60°)]$
$= \frac{10}{9}(\cos 200° + i \sin 200°)$

49. $[0.45(\cos 310° + i \sin 310°)][0.60(\cos 200° + i \sin 200°)] = (0.45)(0.60)[\cos(310° + 200°) + i \sin(310° + 200°)]$
$= 0.27(\cos 510° + i \sin 510°)$
$= 0.27(\cos 150° + i \sin 150°)$

51. $\dfrac{\cos 40° + i \sin 40°}{\cos 10° + i \sin 10°} = \cos(40° - 10°) + i \sin(40° - 10°) = \cos 30° + i \sin 30°$

53. $\dfrac{\cos(5\pi/3) + i \sin(5\pi/3)}{\cos \pi + i \sin \pi} = \cos\left(\frac{5\pi}{3} - \pi\right) + i \sin\left(\frac{5\pi}{3} - \pi\right) = \cos\left(\frac{2\pi}{3}\right) + i \sin\left(\frac{2\pi}{3}\right)$

55. $\dfrac{12(\cos 52° + i \sin 52°)}{3(\cos 110° + i \sin 110°)} = 4[\cos(52° - 110°) + i \sin(52° - 110°)]$
$= 4[\cos(-58°) + i \sin(-58°)]$

57. (a) $2 + 2i = 2\sqrt{2}(\cos 45° + i \sin 45°)$
$1 - i = \sqrt{2}[\cos(-45°) + i \sin(-45°)]$

(b) $(2 + 2i)(1 - i) = [2\sqrt{2}(\cos 45° + i \sin 45°)][\sqrt{2}(\cos(-45°) + i \sin(-45°))] = 4(\cos 0° + i \sin 0°) = 4$

(c) $(2 + 2i)(1 - i) = 2 - 2i + 2i - 2i^2 = 2 + 2 = 4$

59. (a) $-2i = 2[\cos(-90°) + i \sin(-90°)]$
$1 + i = \sqrt{2}(\cos 45° + i \sin 45°)$

(b) $-2i(1 + i) = 2[\cos(-90°) + i \sin(-90°)][\sqrt{2}(\cos 45° + i \sin 45°)]$
$= 2\sqrt{2}[\cos(-45°) + i \sin(-45°)]$
$= 2\sqrt{2}\left[\frac{1}{\sqrt{2}} - \frac{1}{\sqrt{2}}i\right] = 2 - 2i$

(c) $-2i(1 + i) = -2i - 2i^2 = -2i + 2 = 2 - 2i$

61. (a) $5 = 5(\cos 0° + i \sin 0°)$
$2 + 3i \approx \sqrt{13}(\cos 56.31° + i \sin 56.31°)$

(b) $\dfrac{5}{2 + 3i} \approx \dfrac{5(\cos 0° + i \sin 0°)}{\sqrt{13}(\cos 56.3° + i \sin 56.3°)} = \dfrac{5\sqrt{13}}{13}[\cos(-56.3°) + i \sin(-56.3°)] \approx 0.7694 - 1.154i$

(c) $\dfrac{5}{2 + 3i} = \dfrac{5}{2 + 3i} \cdot \dfrac{2 - 3i}{2 - 3i} = \dfrac{10 - 15i}{13} = \dfrac{10}{13} - \dfrac{15}{13}i \approx 0.7694 - 1.154i$

63. $\dfrac{z_1}{z_2} = \dfrac{r_1(\cos \theta_1 + i \sin \theta_1)}{r_2(\cos \theta_2 + i \sin \theta_2)} \cdot \dfrac{\cos \theta_2 + i \sin \theta_2}{\cos \theta_2 + i \sin \theta_2}$

$= \dfrac{r_1}{r_2(\cos^2 \theta_2 + \sin^2 \theta_2)}[\cos \theta_1 \cos \theta_2 + \sin \theta_1 \sin \theta_2 + i(\sin \theta_1 \cos \theta_2 - \sin \theta_2 \cos \theta_1)]$

$= \dfrac{r_1}{r_2}[\cos(\theta_1 - \theta_2) + i \sin(\theta_1 - \theta_2)]$

65. (a) $z\bar{z} = [r(\cos\theta + i\sin\theta)][r(\cos(-\theta) + i\sin(-\theta))]$
$= r^2[\cos(\theta - \theta) + i\sin(\theta - \theta)]$
$= r^2[\cos 0 + i\sin 0]$
$= r^2$

(b) $\dfrac{z}{\bar{z}} = \dfrac{r(\cos\theta + i\sin\theta)}{r[\cos(-\theta) + i\sin(-\theta)]}$
$= \dfrac{r}{r}[\cos(\theta - (-\theta)) + i\sin(\theta - (-\theta))]$
$= \cos 2\theta + i\sin 2\theta$

67. Let $z = x + iy$ such that:

$|z| = 2 \Rightarrow 2 = \sqrt{x^2 + y^2}$
$\Rightarrow 4 = x^2 + y^2$: circle with radius of 2

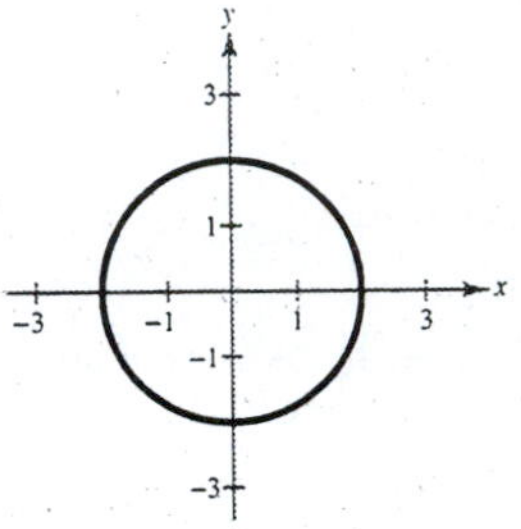

69. $(1 + i)^5 = \left[\sqrt{2}\left(\cos\dfrac{\pi}{4} + i\sin\dfrac{\pi}{4}\right)\right]^5$
$= (\sqrt{2})^5\left(\cos\dfrac{5\pi}{4} + i\sin\dfrac{5\pi}{4}\right)$
$= 4\sqrt{2}\left(-\dfrac{\sqrt{2}}{2} - \dfrac{\sqrt{2}}{2}i\right)$
$= -4 - 4i$

71. $(-1 + i)^{10} = \left[\sqrt{2}\left(\cos\dfrac{3\pi}{4} + i\sin\dfrac{3\pi}{4}\right)\right]^{10}$
$= (\sqrt{2})^{10}\left(\cos\dfrac{30\pi}{4} + i\sin\dfrac{30\pi}{4}\right)$
$= 32\left[\cos\left(\dfrac{3\pi}{2} + 6\pi\right) + i\sin\left(\dfrac{3\pi}{2} + 6\pi\right)\right]$
$= 32\left(\cos\dfrac{3\pi}{2} + i\sin\dfrac{3\pi}{2}\right)$
$= 32[0 + i(-1)]$
$= -32i$

73. $2(\sqrt{3} + i)^7 = 2\left[2\left(\cos\dfrac{\pi}{6} + i\sin\dfrac{\pi}{6}\right)\right]^7$
$= 2\left[2^7\left(\cos\dfrac{7\pi}{6} + i\sin\dfrac{7\pi}{6}\right)\right]$
$= 256\left(-\dfrac{\sqrt{3}}{2} - \dfrac{1}{2}i\right)$
$= -128\sqrt{3} - 128i$

75. $[5(\cos 20° + i\sin 20°)]^3 = 5^3(\cos 60° + i\sin 60°) = \dfrac{125}{2} + \dfrac{125\sqrt{3}}{2}i$

77. $\left(\cos\frac{5\pi}{4} + i\sin\frac{5\pi}{4}\right)^{10} = \cos\frac{25\pi}{2} + i\sin\frac{25\pi}{2}$

$= \cos\left(12\pi + \frac{\pi}{2}\right) + i\sin\left(12\pi + \frac{\pi}{2}\right) = \cos\frac{\pi}{2} + i\sin\frac{\pi}{2} = i$

79. $[5(\cos 3.2 + i\sin 3.2)]^4 = 5^4(\cos 12.8 + i\sin 12.8)$

$\approx 608.02 + 144.69i$

81. $(3 - 2i)^5 = -597 - 122i$

83. $[3(\cos 15° + i\sin 15°)]^4 = 81(\cos 60° + i\sin 60°)$

$= \frac{81}{2} + \frac{8\sqrt{3}}{2}i$

85. $\left[-\frac{1}{2}(1 + \sqrt{3}i)\right]^6 = \left[\cos\frac{4\pi}{3} + i\sin\frac{4\pi}{3}\right]^6$

$= \cos 8\pi + i\sin 8\pi$

$= 1$

87. (a) In trigonometric form we have:

$2(\cos 30° + i\sin 30°)$

$2(\cos 150° + i\sin 150°)$

$2(\cos 270° + i\sin 270°)$

(b) There are three roots evenly spaced around a circle of radius 2. Therefore, they represent the cube roots of some number of modulus 8. Cubing them shows that they are all cube roots of $8i$.

(c) $[2(\cos 30° + i\sin 30°)]^3 = 8i$

$[2(\cos 150° + i\sin 150°)]^3 = 8i$

$[2(\cos 270° + i\sin 270°)]^3 = 8i$

89. (a) Square roots of $5(\cos 120° + i\sin 120°)$:

$\sqrt{5}\left[\cos\left(\frac{120° + 360°k}{2}\right) + i\sin\left(\frac{120° + 360°k}{2}\right)\right],\ k = 0, 1$

$k = 0$: $\sqrt{5}(\cos 60° + i\sin 60°)$

$k = 1$: $\sqrt{5}(\cos 240° + i\sin 240°)$

(b)

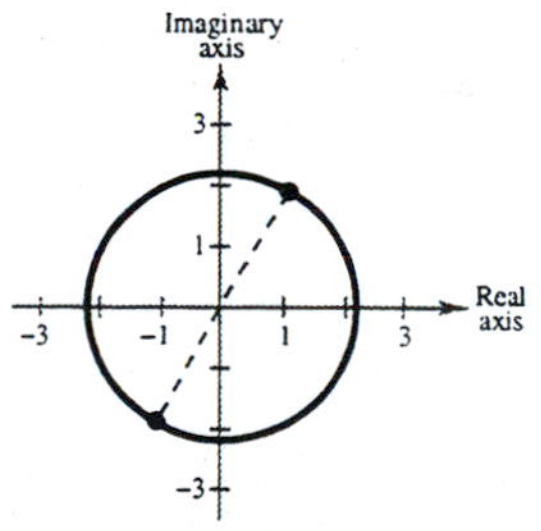

(c) $\frac{\sqrt{5}}{2} + \frac{\sqrt{15}}{2}i,\ -\frac{\sqrt{5}}{2} - \frac{\sqrt{15}}{2}i$

91. (a) Fourth roots of $16\left(\cos\dfrac{4\pi}{3} + i\sin\dfrac{4\pi}{3}\right)$:

$$\sqrt[4]{16}\left[\cos\left(\frac{(4\pi/3) + 2k\pi}{4}\right) + i\sin\left(\frac{(4\pi/3) + 2k\pi}{4}\right)\right],\ k = 0, 1, 2, 3$$

$$k = 0\text{: } 2\left(\cos\frac{\pi}{3} + i\sin\frac{\pi}{3}\right)$$

$$k = 1\text{: } 2\left(\cos\frac{5\pi}{6} + i\sin\frac{5\pi}{6}\right)$$

$$k = 2\text{: } 2\left(\cos\frac{4\pi}{3} + i\sin\frac{4\pi}{3}\right)$$

$$k = 3\text{: } 2\left(\cos\frac{11\pi}{6} + i\sin\frac{11\pi}{6}\right)$$

(b)

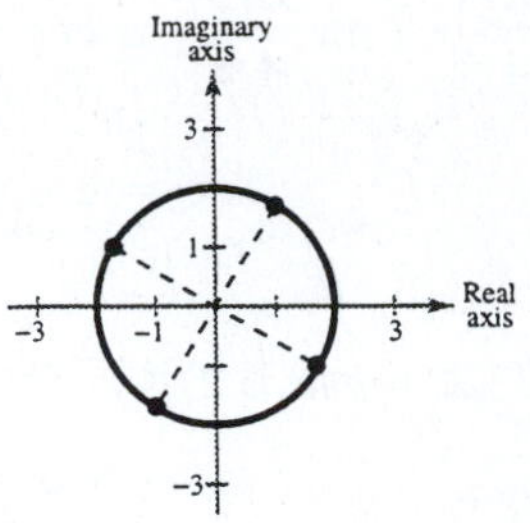

(c) $1 + \sqrt{3}i,\ -\sqrt{3} + i,\ -1 - \sqrt{3}i,\ \sqrt{3} - i$

93. (a) Square roots of $-25i = 25\left(\cos\dfrac{3\pi}{2} + i\sin\dfrac{3\pi}{2}\right)$:

$$\sqrt{25}\left[\cos\left(\frac{(3\pi/2) + 2k\pi}{2}\right) + i\sin\left(\frac{(3\pi/2) + 2k\pi}{2}\right)\right],\ k = 0, 1$$

$$k = 0\text{: } 5\left(\cos\frac{3\pi}{4} + i\sin\frac{3\pi}{4}\right)$$

$$k = 1\text{: } 5\left(\cos\frac{7\pi}{4} + i\sin\frac{7\pi}{4}\right)$$

(b)

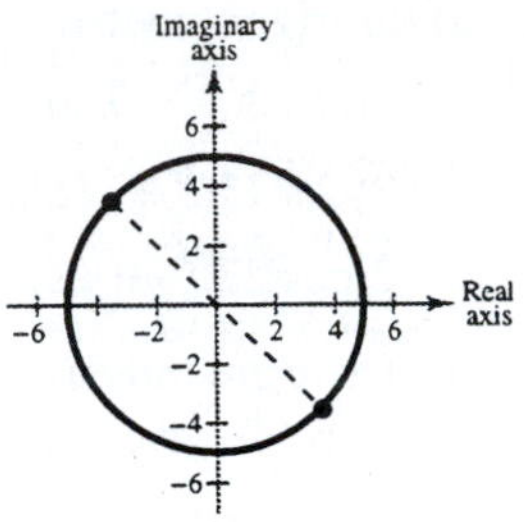

(c) $-\dfrac{5\sqrt{2}}{2} + \dfrac{5\sqrt{2}}{2}i,\ \dfrac{5\sqrt{2}}{2} - \dfrac{5\sqrt{2}}{2}i$

95. (a) Cube roots of $-\dfrac{125}{3}(1 + \sqrt{3}i) = 125\left(\cos\dfrac{4\pi}{3} + i\sin\dfrac{4\pi}{3}\right)$:

$$\sqrt[3]{125}\left[\cos\left(\frac{(4\pi/3) + 2k\pi}{3}\right) + i\sin\left(\frac{(4\pi/3) + 2k\pi}{3}\right)\right],\ k = 0, 1, 2$$

$$k = 0\text{: } 5\left(\cos\frac{4\pi}{9} + i\sin\frac{4\pi}{9}\right)$$

$$k = 1\text{: } 5\left(\cos\frac{10\pi}{9} + i\sin\frac{10\pi}{9}\right)$$

$$k = 2\text{: } 5\left(\cos\frac{16\pi}{9} + i\sin\frac{16\pi}{9}\right)$$

(b)

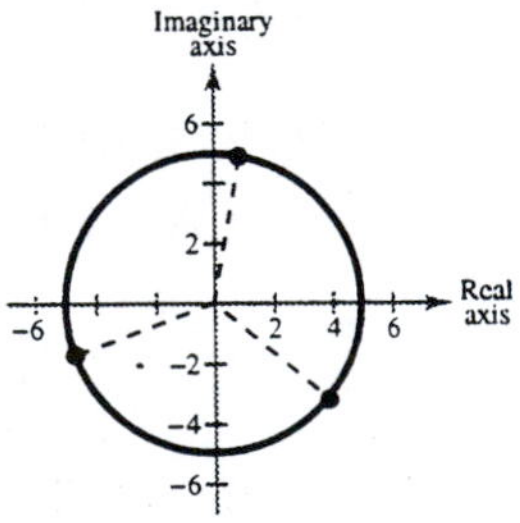

(c) $0.8682 + 4.924i,\ -4.698 - 1.710i,\ 3.830 - 3.214i$

97. (a) Cube roots of $8 = 8(\cos 0 + i \sin 0)$:

$$\sqrt[3]{8}\left[\cos\left(\frac{2k\pi}{3}\right) + i \sin\left(\frac{2k\pi}{3}\right)\right],\ k = 0, 1, 2$$

$k = 0$: $2(\cos 0 + i \sin 0)$

$k = 1$: $2\left(\cos \frac{2\pi}{3} + i \sin \frac{2\pi}{3}\right)$

$k = 2$: $2\left(\cos \frac{4\pi}{3} + i \sin \frac{4\pi}{3}\right)$

(b)

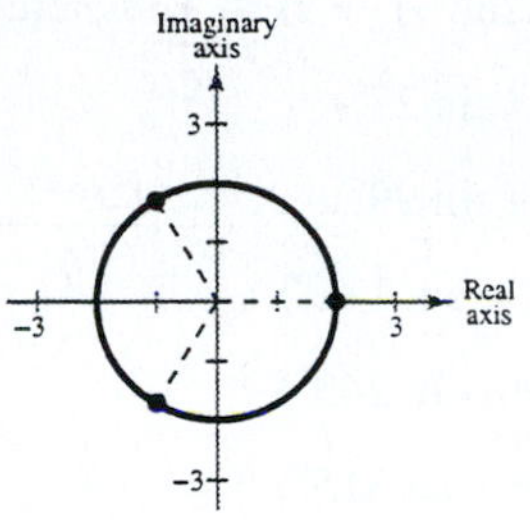

(c) $2, -1 + \sqrt{3}i, -1 - \sqrt{3}i$

99. (a) Fifth roots of $1 = \cos 0 + i \sin 0$:

$$\cos\left(\frac{2k\pi}{5}\right) + i \sin\left(\frac{2k\pi}{5}\right),\ k = 0, 1, 2, 3, 4$$

$k = 0$: $\cos 0 + i \sin 0$

$k = 1$: $\cos \frac{2\pi}{5} + i \sin \frac{2\pi}{5}$

$k = 2$: $\cos \frac{4\pi}{5} + i \sin \frac{4\pi}{5}$

$k = 3$: $\cos \frac{6\pi}{5} + i \sin \frac{6\pi}{5}$

$k = 4$: $\cos \frac{8\pi}{5} + i \sin \frac{8\pi}{5}$

(b)

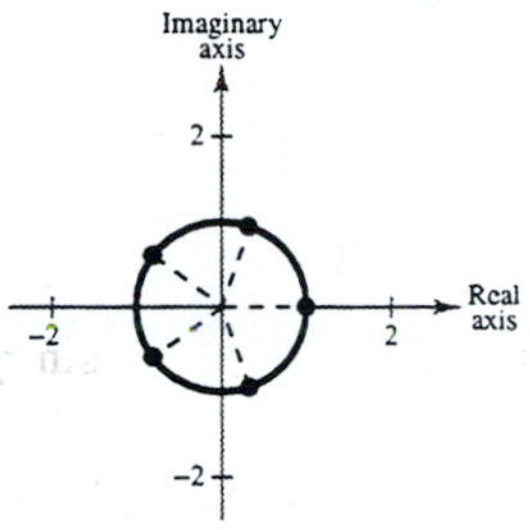

(c) $1, 0.3090 + 0.9511i, -0.8090 + 0.5878i, -0.8090 - 0.5878i, 0.3090 - 0.9511i$

101. (a) The cube roots of $-125 = 125(\cos 180° + i \sin 180°)$ are:

$5(\cos 60° + i \sin 60°)$

$5(\cos 180° + i \sin 180°)$

$5(\cos 300° + i \sin 300°)$

(b)

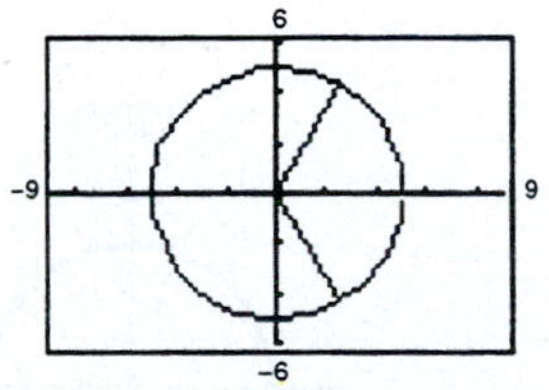

(c) $\frac{5}{2} + \frac{5\sqrt{3}}{2}i, -5, \frac{5}{2} - \frac{5\sqrt{3}}{2}i$

103. (a) The fifth roots of $128(-1 + i) = 128\sqrt{2}(\cos 135° + i \sin 135°)$ are:

$2\sqrt[5]{4\sqrt{2}}(\cos 27° + i \sin 27°)$

$2\sqrt[5]{4\sqrt{2}}(\cos 99° + i \sin 99°)$

$2\sqrt[5]{4\sqrt{2}}(\cos 171° + i \sin 171°)$

$2\sqrt[5]{4\sqrt{2}}(\cos 243° + i \sin 243°)$

$2\sqrt[5]{4\sqrt{2}}(\cos 315° + i \sin 315°)$

(b)

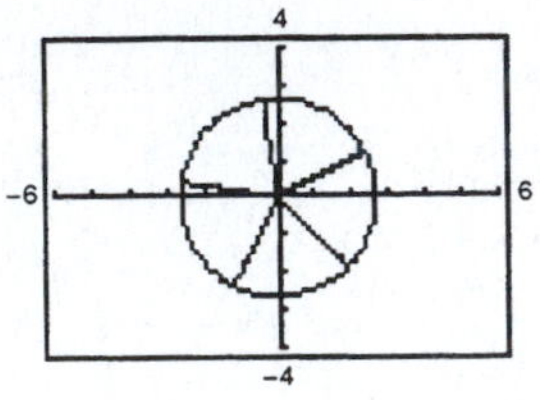

(c) $2.52 + 1.28i, -0.44 + 2.79i, -2.79 + 0.44i, -1.28 - 2.52i, 2 - 2i$

105. $x^4 - i = 0$

$x^4 = i$

The solutions are the fourth roots of $i = \cos \frac{\pi}{2} + i \sin \frac{\pi}{2}$:

$$\sqrt[4]{1}\left[\cos\left(\frac{(\pi/2) + 2k\pi}{4}\right) + i \sin\left(\frac{(\pi/2) + 2k\pi}{4}\right)\right], \quad k = 0, 1, 2, 3$$

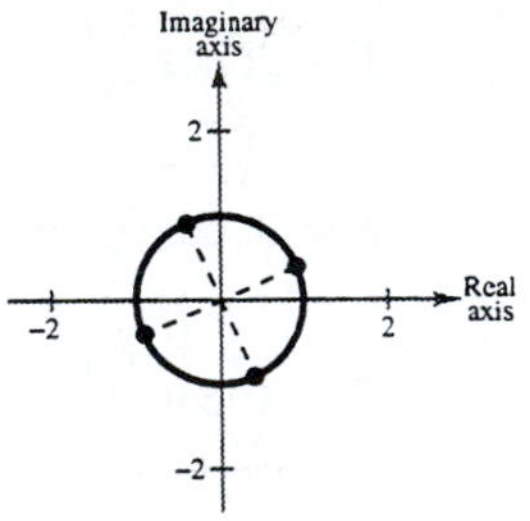

$k = 0: \cos \frac{\pi}{8} + i \sin \frac{\pi}{8}$

$k = 1: \cos \frac{5\pi}{8} + i \sin \frac{5\pi}{8}$

$k = 2: \cos \frac{9\pi}{8} + i \sin \frac{9\pi}{8}$

$k = 3: \cos \frac{13\pi}{8} + i \sin \frac{13\pi}{8}$

107. $x^5 + 243 = 0$

$x^5 = -243$

The solutions are the fifth roots of $-243 = 243(\cos \pi + i \sin \pi)$:

$$\sqrt[5]{243}\left[\cos\left(\frac{\pi + 2k\pi}{5}\right) + i \sin\left(\frac{\pi + 2k\pi}{5}\right)\right], \quad k = 0, 1, 2, 3, 4$$

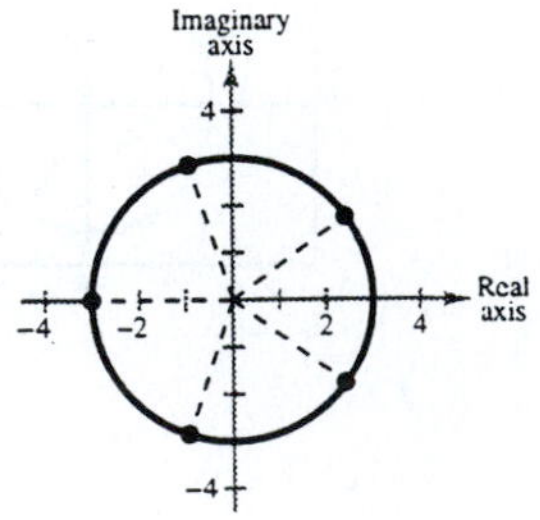

$k = 0: 3\left(\cos \frac{\pi}{5} + i \sin \frac{\pi}{5}\right)$

$k = 1: 3\left(\cos \frac{3\pi}{5} + i \sin \frac{3\pi}{5}\right)$

$k = 2: 3(\cos \pi + i \sin \pi) = -3$

$k = 3: 3\left(\cos \frac{7\pi}{5} + i \sin \frac{7\pi}{5}\right)$

$k = 4: 3\left(\cos \frac{9\pi}{5} + i \sin \frac{9\pi}{5}\right)$

109. $x^3 + 64i = 0$

$$x^3 = -64i$$

The solutions are the cube roots of $-64i = 64\left(\cos\frac{3\pi}{2} + i\sin\frac{3\pi}{2}\right)$:

$$\sqrt[3]{64}\left[\cos\left(\frac{(3\pi/2) + 2k\pi}{3}\right) + i\sin\left(\frac{(3\pi/2) + 2k\pi}{3}\right)\right],\ k = 0, 1, 2$$

$k = 0$: $4\left(\cos\frac{\pi}{2} + i\sin\frac{\pi}{2}\right) = 4i$

$k = 1$: $4\left(\cos\frac{7\pi}{6} + i\sin\frac{7\pi}{6}\right) = -2\sqrt{3} - 2i$

$k = 2$: $4\left(\cos\frac{11\pi}{6} + i\sin\frac{11\pi}{6}\right) = 2\sqrt{3} - 2i$

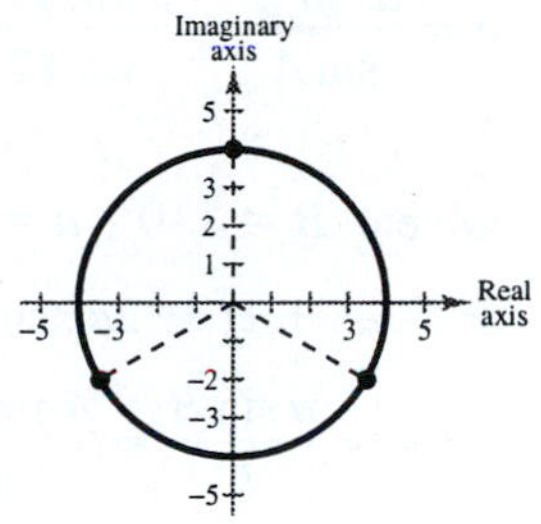

111. $x^3 - (1 - i) = 0$

$$x^3 = 1 - i = \sqrt{2}(\cos 315° + i\sin 315°)$$

The solutions are the cube roots of $1 - i$:

$$\sqrt[3]{\sqrt{2}}\left[\cos\left(\frac{315° + 360°k}{3}\right) + i\sin\left(\frac{315° + 360°k}{3}\right)\right],\ k = 0, 1, 2$$

$k = 0$: $\sqrt[6]{2}(\cos 105° + i\sin 105°)$

$k = 1$: $\sqrt[6]{2}(\cos 225° + i\sin 225°)$

$k = 2$: $\sqrt[6]{2}(\cos 345° + i\sin 345°)$

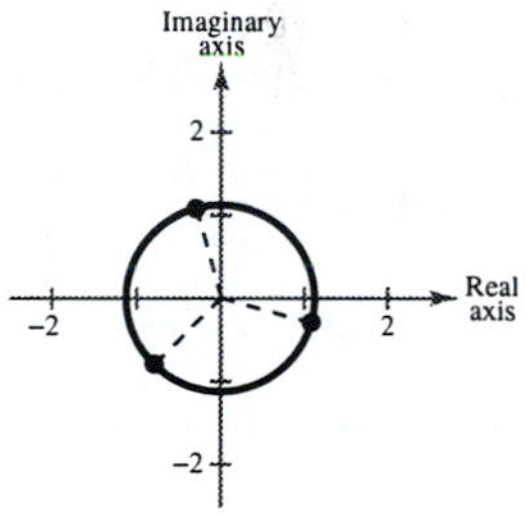

❑ Review Exercises for Chapter 6

Solutions to Odd-Numbered Exercises

1. Given: $a = 5,\ b = 8,\ c = 10$

$$\cos C = \frac{a^2 + b^2 - c^2}{2ab} = \frac{25 + 64 - 100}{80} \approx -0.1375 \implies C \approx 97.9°$$

$$\sin A = \frac{a\sin C}{c} \approx \frac{5(0.9905)}{10} \approx 0.4953 \implies A \approx 29.7°$$

$$B = 180° - A - C = 180° - 29.7° - 97.9° = 52.4°$$

3. Given: $A = 12°, B = 58°, a = 5$

$C = 180° - A - B = 180° - 12° - 58° = 110°$

$$b = \frac{a \sin B}{\sin A} = \frac{5 \sin 58°}{\sin 12°} \approx \frac{5(0.8480)}{0.2079} \approx 20.4$$

$$c = \frac{a \sin C}{\sin A} = \frac{5 \sin 110°}{\sin 12°} \approx \frac{5(0.9397)}{0.2079} \approx 22.6$$

5. Given: $B = 110°, a = 4, c = 4$

$b^2 = a^2 + c^2 - 2ac \cos B \approx 16 + 16 - 2(4)(4)(-0.3420) \approx 42.94 \implies b \approx 6.6$

$$\sin A = \frac{a \sin B}{b} \approx \frac{4 \sin 110°}{6.6} \approx \frac{4(0.9397)}{6.6} \approx 0.5736 \implies A \approx 35°$$

$c = a \implies C = A \approx 35°$

7. Given: $A = 75°, a = 2.5, b = 16.5$

$$\sin B = \frac{b \sin A}{a} = \frac{16.5 \sin 75°}{2.5} \approx \frac{16.5(0.9659)}{2.5} \approx 5.375 \implies \text{no triangle formed}$$

No solution

9. Given: $B = 115°, a = 7, b = 14.5$

$$\sin A = \frac{a \sin B}{b} = \frac{7 \sin 115°}{14.5} \approx \frac{7(0.9063)}{14.5} \approx 0.4375 \implies A \approx 25.9°$$

$C \approx 180° - 115° - 25.9° = 39.1°$

$c^2 = a^2 + b^2 - 2ab \cos C \approx 7^2 + 14.5^2 - 2(7)(14.5)(0.7760) \approx 101.7 \implies c \approx 10.1$

11. Given: $A = 15°, a = 5, b = 10$

$$\sin B = \frac{b \sin A}{a} = \frac{10 \sin 15°}{5} \approx \frac{10(0.2588)}{5} \approx 0.5176 \implies B \approx 31.2° \text{ or } 148.8°$$

Case 1: $B \approx 31.2°$

$C \approx 180° - 15° - 31.2° = 133.8°$

$$c = \frac{a \sin C}{\sin A} \approx 13.9$$

Case 2: $B \approx 148.8°$

$C \approx 180° - 15° - 148.8° = 16.2°$

$$c = \frac{a \sin C}{\sin A} \approx 5.39$$

13. Given: $B = 150°, a = 10, c = 20$

$b^2 = a^2 + c^2 - 2ac \cos B \approx 100 + 400 - 400(-0.8660) \approx 846.4 \implies b \approx 29.1$

$$\sin C = \frac{c \sin B}{b} \approx \frac{20(0.5)}{29.09} \approx 0.3437 \implies C \approx 20.1°$$

$A = 180° - B - C \approx 180° - 150° - 20.1° = 9.9°$

15. Given: $B = 25°,\ a = 6.2,\ b = 4$

$$\sin A = \frac{a \sin B}{b} \approx 0.6551 \implies A \approx 40.9° \text{ or } 139.1°$$

Case 1: $A \approx 40.9°$

$C \approx 180° - 25° - 40.9° = 114.1°$

$c \approx 8.6$

Case 2: $A \approx 139.1°$

$C \approx 180° - 25° - 139.1° = 15.9°$

$c \approx 2.6$

17. $a = 4,\ b = 5,\ c = 7$

$$s = \frac{a+b+c}{2} = \frac{4+5+7}{2} = 8$$

$$\text{Area} = \sqrt{s(s-a)(s-b)(s-c)}$$

$$= \sqrt{8(4)(3)(1)} \approx 9.798$$

19. $A = 27°,\ b = 5,\ c = 8$

$$\text{Area} = \frac{1}{2}bc \sin A \approx \frac{1}{2}(5)(8)(0.4540) \approx 9.08$$

21. $\alpha = 180° - 31° = 149°$

$\phi = 180° - 149° - 17° = 14°$

$$x = \frac{50 \sin 17°}{\sin \phi} = \frac{50 \sin 17°}{\sin 14°} \approx 60.43$$

$h = x \sin 31°$

$\approx 60.43(0.5150) \approx 31.1$ meters

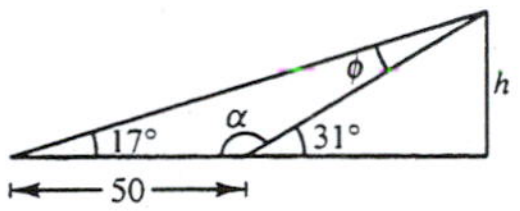

23. $\sin 28° = \dfrac{h}{75}$

$h = 75 \sin 28° \approx 35.21$ feet

$\cos 28° = \dfrac{x}{75}$

$x = 75 \cos 28° \approx 66.22$ feet

$\tan 45° = \dfrac{H}{x}$

$H = x \tan 45° \approx 66.22$ feet

Height of tree: $H - h \approx 31$ feet

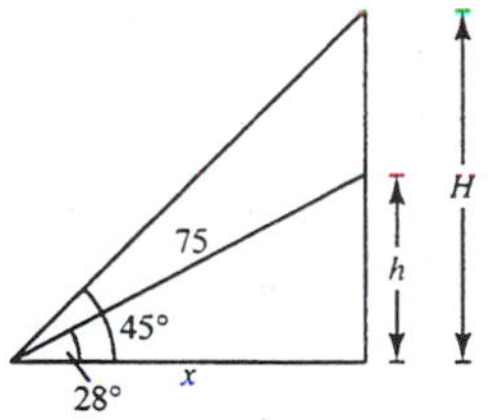

25. $d^2 = 850^2 + 1060^2 - 2(850)(1060) \cos 72°$

$\approx 1{,}289{,}251$

$d \approx 1135$ miles

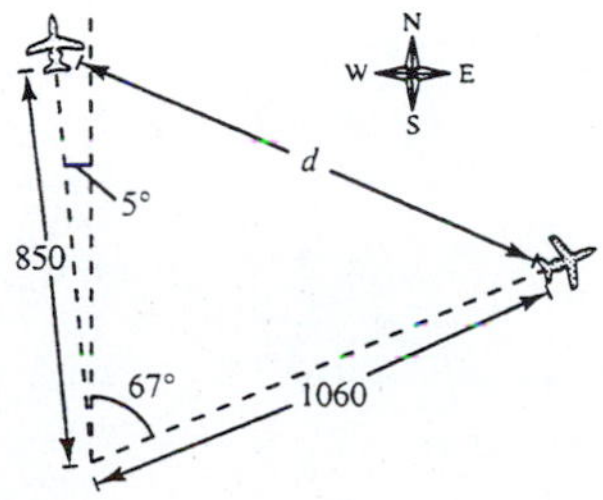

27. Initial point: $(-5, 4)$

Terminal point: $(2, -1)$

$\mathbf{v} = \langle 2 - (-5),\ -1 - 4 \rangle = \langle 7, -5 \rangle$

29. Initial point: $(0, 10)$

Terminal point: $(7, 3)$

$\mathbf{v} = \langle 7 - 0,\ 3 - 10 \rangle = \langle 7, -7 \rangle$

31. $\langle 8 \cos 120°,\ 8 \sin 120° \rangle = \langle -4, 4\sqrt{3} \rangle$

33. $\mathbf{v} = -10\mathbf{i} + 10\mathbf{j}$

$\|\mathbf{v}\| = \sqrt{(-10)^2 + (10)^2} = \sqrt{200} = 10\sqrt{2}$

$\tan\theta = \dfrac{10}{-10} = -1 \implies \theta = 135°$ since

$\mathbf{v}$ is in Quadrant II.

$\mathbf{v} = 10\sqrt{2}(\mathbf{i}\sin 135° + \mathbf{j}\cos 135°)$

35. $\mathbf{u} = 6\mathbf{i} - 5\mathbf{j}$

$$\frac{1}{\|\mathbf{u}\|}\mathbf{u} = \frac{1}{\sqrt{6^2 + 5^2}}(6\mathbf{i} - 5\mathbf{j}) = \frac{6}{\sqrt{61}}\mathbf{i} - \frac{5}{\sqrt{61}}\mathbf{j}$$

$$= \left\langle \frac{6}{\sqrt{61}}, -\frac{5}{\sqrt{61}} \right\rangle$$

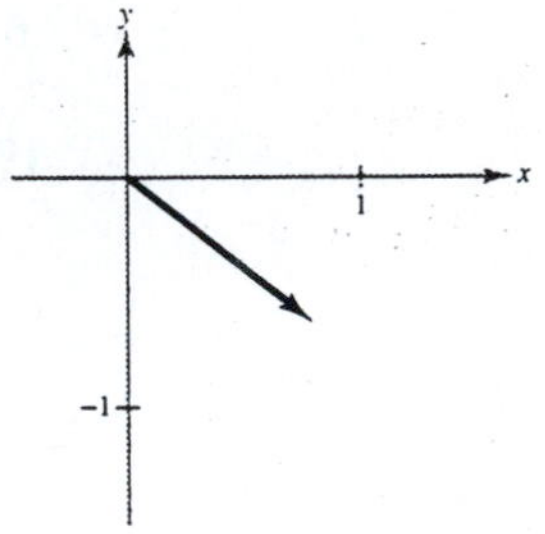

37. $\mathbf{u} = 6\mathbf{i} - 5\mathbf{j}$, $\mathbf{v} = 10\mathbf{i} + 3\mathbf{j}$

$4\mathbf{u} - 5\mathbf{v} = (24\mathbf{i} - 20\mathbf{j}) - (50\mathbf{i} + 15\mathbf{j}) = -26\mathbf{i} - 35\mathbf{j}$

$= \langle -26, -35 \rangle$

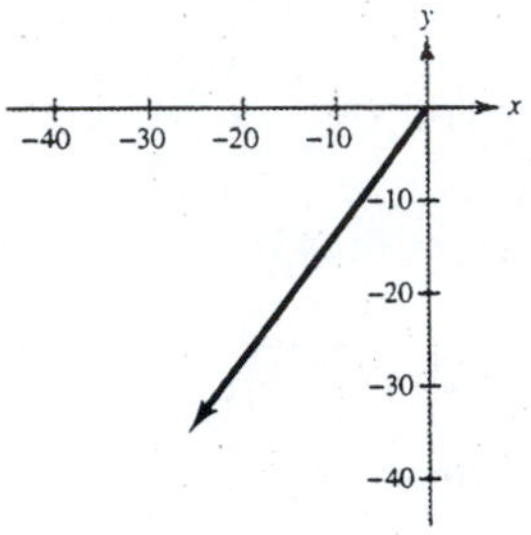

39. $\mathbf{u} = 15[(\cos 20°)\mathbf{i} + (\sin 20°)\mathbf{j}]$

$\mathbf{v} = 20[(\cos 63°)\mathbf{i} + (\sin 63°)\mathbf{j}]$

$\mathbf{u} + \mathbf{v} \approx 23.1752\mathbf{i} + 22.9504\mathbf{j}$

$\|\mathbf{u} + \mathbf{v}\| \approx 32.6161$

$\tan\theta \approx \dfrac{22.9504}{23.1752} \implies \theta \approx 44.72°$

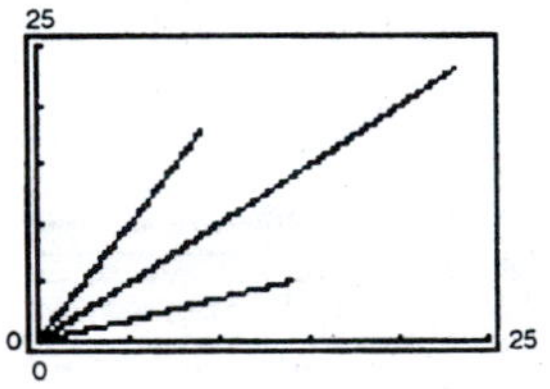

41. $\tan\alpha = \frac{12}{5} \implies \sin\alpha = \frac{12}{13}$ and $\cos\alpha = \frac{5}{13}$

$\tan\beta = \frac{3}{4} \implies \sin(180° - \beta) = \frac{3}{5}$ and $\cos(180° - \beta) = -\frac{4}{5}$

$\mathbf{u} = 250\left(\frac{5}{13}\mathbf{i} + \frac{12}{13}\mathbf{j}\right)$

$\mathbf{v} = 100\left(-\frac{4}{5}\mathbf{i} + \frac{3}{5}\mathbf{j}\right)$

$\mathbf{w} = 200(0\mathbf{i} - \mathbf{j})$

$\mathbf{r} = \mathbf{u} + \mathbf{v} + \mathbf{w} = \left(\frac{1250}{13} - 80 + 0\right)\mathbf{i} + \left(\frac{3000}{13} + 60 - 200\right)\mathbf{j} = \frac{210}{13}\mathbf{i} + \frac{1180}{13}\mathbf{j}$

$\|\mathbf{r}\| = \sqrt{\left(\frac{210}{13}\right)^2 + \left(\frac{1180}{13}\right)^2} \approx 92.2$ lb

$\tan\theta = \frac{1180}{210} \implies \theta \approx 79.9°$

43. Rope One: $\mathbf{u} = \|\mathbf{u}\|(\cos 30°\mathbf{i} - \sin 30°\mathbf{j}) = \|\mathbf{u}\|\left(\frac{\sqrt{3}}{2}\mathbf{i} - \frac{1}{2}\mathbf{j}\right)$

Rope Two: $\mathbf{v} = \|\mathbf{u}\|(-\cos 30°\mathbf{i} - \sin 30°\mathbf{j}) = \|\mathbf{u}\|\left(-\frac{\sqrt{3}}{2}\mathbf{i} - \frac{1}{2}\mathbf{j}\right)$

Resultant: $\mathbf{u} + \mathbf{v} = -\|\mathbf{u}\|\mathbf{j} = -180\mathbf{j}$

$\|\mathbf{u}\| = 180$

Therefore, the tension on each rope is $\|\mathbf{u}\| = 180$ lb.

45. Airspeed: $\mathbf{u} = 724(\cos 60°\mathbf{i} + \sin 60°\mathbf{j})$

$= 362(\mathbf{i} + \sqrt{3}\mathbf{j})$

Wind: $\mathbf{w} = 32\mathbf{i}$

Groundspeed $= \mathbf{u} + \mathbf{w} = (394\mathbf{i} + 362\sqrt{3}\mathbf{j})$

$\|\mathbf{u} + \mathbf{w}\| = \sqrt{(394)^2 + (326\sqrt{3})^2} \approx 740.5$ km/hr

$\tan\theta = \frac{362\sqrt{3}}{394} \implies \theta \approx 57.9°$

Bearing: N 32.1° E

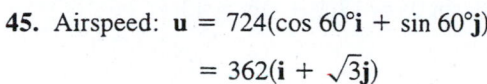

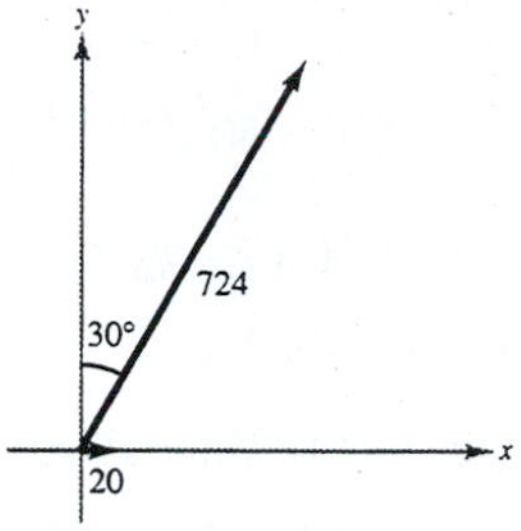

47. $P(7, -4), Q(-3, 2)$

$\overrightarrow{PQ} = \langle -3 - 7, 2 - (-4)\rangle = \langle -10, 6\rangle$

$\|\overrightarrow{PQ}\| = \sqrt{(-10)^2 + (6)^2} = \sqrt{136} = 2\sqrt{34}$

$$\frac{\overrightarrow{PQ}}{\|\overrightarrow{PQ}\|} = \frac{1}{2\sqrt{34}}\langle -10, 6\rangle = \frac{1}{\sqrt{34}}\langle -5, 3\rangle$$

49. $\mathbf{u} = \langle 39, -12\rangle, \mathbf{v} = \langle -26, 8\rangle$

$\mathbf{u} \cdot \mathbf{v} = 39(-26) + (-12)(8) = -1110 \neq 0 \implies$ **u** and **v** are not orthogonal.

$\mathbf{v} = -\frac{2}{3}\mathbf{u} \implies$ **u** and **v** are parallel.

51. $\mathbf{u} = \cos\frac{7\pi}{4}\mathbf{i} + \sin\frac{7\pi}{4}\mathbf{j} = \left\langle \frac{1}{\sqrt{2}}, -\frac{1}{\sqrt{2}}\right\rangle$

$\mathbf{v} = \cos\frac{5\pi}{6}\mathbf{i} + \sin\frac{5\pi}{6}\mathbf{j} = \left\langle -\frac{\sqrt{3}}{2}, \frac{1}{2}\right\rangle$

$$\cos\theta = \frac{\mathbf{u} \cdot \mathbf{v}}{\|\mathbf{u}\|\,\|\mathbf{v}\|} = \frac{-\sqrt{3} - 1}{2\sqrt{2}} \Rightarrow \theta = \frac{11\pi}{12}$$

53. $\mathbf{u} = \langle 2\sqrt{2}, -4\rangle, \mathbf{v} = \langle -\sqrt{2}, 1\rangle$

$$\cos\theta = \frac{\mathbf{u} \cdot \mathbf{v}}{\|\mathbf{u}\|\,\|\mathbf{v}\|} = \frac{-8}{(\sqrt{24})(\sqrt{3})} \implies \theta \approx 160.5°$$

55. $\mathbf{u} = \langle -4, 3\rangle, \mathbf{v} = \langle -8, -2\rangle$

$$\text{proj}_{\mathbf{v}}\mathbf{u} = \left(\frac{\mathbf{u} \cdot \mathbf{v}}{\|\mathbf{v}\|^2}\right)\mathbf{v} = \left(\frac{26}{68}\right)\langle -8, -2\rangle$$

$$= -\frac{13}{17}\langle 4, 1\rangle$$

57. $\mathbf{u} = \langle 2, 7\rangle, \mathbf{v} = \langle 1, -1\rangle$

$$\text{proj}_{\mathbf{v}}\mathbf{u} = \left(\frac{\mathbf{u} \cdot \mathbf{v}}{\|\mathbf{v}\|^2}\right)\mathbf{v} = -\frac{5}{2}\langle 1, -1\rangle$$

59. $5 - 5i$

$r = \sqrt{5^2 + (-5)^2} = \sqrt{50} = 5\sqrt{2}$

$\tan\theta = \frac{-5}{5} = -1 \implies \theta \approx 315°$ since the complex number is in Quadrant IV.

$5 - 5i = 5\sqrt{2}(\cos 315° + i \sin 315°)$

61. $5 + 12i$

$r = \sqrt{5^2 + 12^2} = \sqrt{169} = 13$

$\tan\theta = \frac{12}{5} \implies \theta \approx 67.38°$ since the complex number is in Quadrant I.

$5 + 12i \approx 13(\cos 67.38° + i \sin 67.38°)$

63. $100(\cos 240° + i \sin 240°) = 100\left(-\frac{1}{2} - \frac{\sqrt{3}}{2}i\right)$

$= -50 - 50\sqrt{3}i$

65. $13(\cos 0 + i \sin 0) = 13(1 + 0i) = 13$

67. (a) $z_1 = 2\sqrt{3} - 2i = 4(\cos 330° + i \sin 330°)$

$z_2 = -10i = 10(\cos 270° + i \sin 270°)$

(b) $z_1z_2 = [4(\cos 330° + i \sin 330°)][10(\cos 270° + i \sin 270°)]$

$= 40(\cos 600° + i \sin 600°)$

$= 40(\cos 240° + i \sin 240°)$

$\approx -20.00 - 34.64i$

$\frac{z_1}{z_2} = \frac{4(\cos 330° + i \sin 330°)}{10(\cos 270° + i \sin 270°)}$

$= \frac{2}{5}(\cos 60° + i \sin 60°)$

69. $\left[5\left(\cos\frac{\pi}{12} + i \sin\frac{\pi}{12}\right)\right]^4 = 5^4\left(\cos\frac{4\pi}{12} + i \sin\frac{4\pi}{12}\right)$

$= 625\left(\cos\frac{\pi}{3} + i \sin\frac{\pi}{3}\right)$

$= 625\left(\frac{1}{2} + \frac{\sqrt{3}}{2}i\right)$

$= \frac{625}{2} + \frac{625\sqrt{3}}{2}i$

71. $(2 + 3i)^6 \approx [\sqrt{13}(\cos 56.3° + i \sin 56.3°)]^6$

$= 13^3(\cos 337.9° + i \sin 337.9°)$

$\approx 13^3(0.9263 - 0.3769i)$

$\approx 2035 - 828i$

73. (a) The trigonometric form of the three roots shown is:

$4(\cos 60° + i \sin 60°)$

$4(\cos 180° + i \sin 180°)$

$4(\cos 300° + i \sin 300°)$

(b) Since there are three evenly spaced roots on the circle of radius 4, they are cube roots of a complex number of modulus $4^3 = 64$. Cubing them yields -64.

(c) $[4(\cos 60° + i \sin 60°)]^3 = -64$

$[4(\cos 180° + i \sin 180°)]^3 = -64$

$[4(\cos 300° + i \sin 300°)]^3 = -64$

75. Sixth roots of $-729i = 729\left(\cos\frac{3\pi}{2} + i\sin\frac{3\pi}{2}\right)$:

$$\sqrt[6]{729}\left(\cos\frac{(3\pi/2) + 2k\pi}{6} + i\sin\frac{(3\pi/2) + 2k\pi}{6}\right), k = 0, 1, 2, 3, 4, 5$$

$$k = 0: 3\left(\cos\frac{\pi}{4} + i\sin\frac{\pi}{4}\right)$$

$$k = 1: 3\left(\cos\frac{7\pi}{12} + i\sin\frac{7\pi}{12}\right)$$

$$k = 2: 3\left(\cos\frac{11\pi}{12} + i\sin\frac{11\pi}{12}\right)$$

$$k = 3: 3\left(\cos\frac{5\pi}{4} + i\sin\frac{5\pi}{4}\right)$$

$$k = 4: 3\left(\cos\frac{19\pi}{12} + i\sin\frac{19\pi}{12}\right)$$

$$k = 5: 3\left(\cos\frac{23\pi}{12} + i\sin\frac{23\pi}{12}\right)$$

77. $x^4 + 81 = 0$

$x^4 = -81$ Solve by finding the fourth roots of -81.

$-81 = 81(\cos\pi + i\sin\pi)$

$$\sqrt[4]{-81} = \sqrt[4]{81}\left[\cos\left(\frac{\pi + 2\pi k}{4}\right) + i\sin\left(\frac{\pi + 2\pi k}{4}\right)\right], k = 0, 1, 2, 3$$

$$k = 0: 3\left(\cos\frac{\pi}{4} + i\sin\frac{\pi}{4}\right) = \frac{3\sqrt{2}}{2} + \frac{3\sqrt{2}}{2}i$$

$$k = 1: 3\left(\cos\frac{3\pi}{4} + i\sin\frac{3\pi}{4}\right) = -\frac{3\sqrt{2}}{2} + \frac{3\sqrt{2}}{2}i$$

$$k = 2: 3\left(\cos\frac{5\pi}{4} + i\sin\frac{5\pi}{4}\right) = -\frac{3\sqrt{2}}{2} - \frac{3\sqrt{2}}{2}i$$

$$k = 3: 3\left(\cos\frac{7\pi}{4} + i\sin\frac{7\pi}{4}\right) = \frac{3\sqrt{2}}{2} - \frac{3\sqrt{2}}{2}i$$

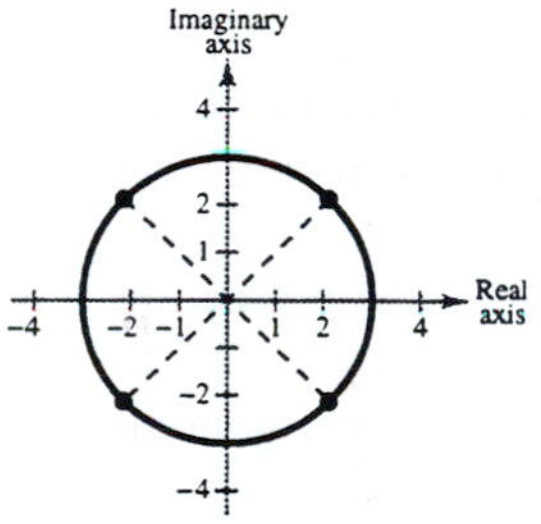

79. $x^3 + 8i = 0$

$x^3 = -8i$ Solve by finding the cube roots of $-8i$.

$$-8i = 8\left(\cos\frac{3\pi}{2} + i\sin\frac{3\pi}{2}\right)$$

$$\sqrt[3]{-8i} = \sqrt[3]{8}\left[\cos\frac{(3\pi/2) + 2\pi k}{3} + i\sin\frac{(3\pi/2) + 2\pi k}{3}\right], k = 0, 1, 2$$

$$k = 0: 2\left(\cos\frac{\pi}{2} + i\sin\frac{\pi}{2}\right) = 2i$$

$$k = 1: 2\left(\cos\frac{7\pi}{6} + i\sin\frac{7\pi}{6}\right) = -\sqrt{3} - i$$

$$k = 2: 2\left(\cos\frac{11\pi}{6} + i\sin\frac{11\pi}{6}\right) = \sqrt{3} - i$$

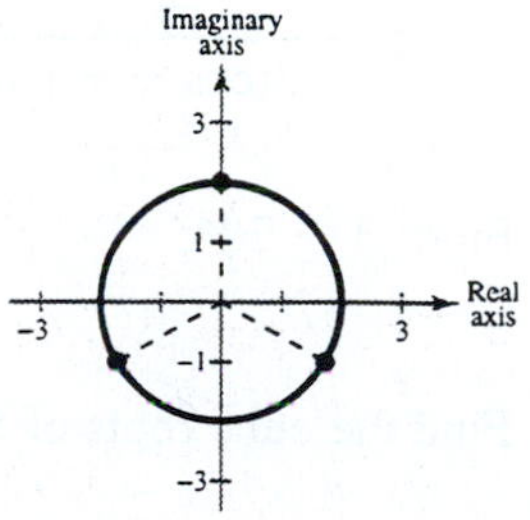

❑ Practice Test for Chapter 6

For Exercises 1 and 2, use the Law of Sines to find the remaining sides and angles of the triangle.

1. $A = 40°,\ B = 12°,\ b = 100$

2. $C = 150°,\ a = 5,\ c = 20$

3. Find the area of the triangle: $a = 3,\ b = 6,\ C = 130°$.

4. Determine the number of solutions to the triangle: $a = 10,\ b = 35,\ A = 22.5°$.

For Exercises 5 and 6, use the Law of Cosines to find the remaining sides and angles of the triangle.

5. $a = 49,\ b = 53,\ c = 38$

6. $C = 29°,\ a = 100,\ b = 300$

7. Use Heron's Formula to find the area of the triangle: $a = 4.1,\ b = 6.8,\ c = 5.5$.

8. A ship travels 40 miles due east, then adjusts its course 12° southward. After traveling 70 miles in that direction, how far is the ship from its point of departure?

9. $\mathbf{w} = 4\mathbf{u} - 7\mathbf{v}$ where $\mathbf{u} = 3\mathbf{i} + \mathbf{j}$ and $\mathbf{v} = -\mathbf{i} + 2\mathbf{j}$. Find $\mathbf{w}$.

10. Find a unit vector in the direction of $\mathbf{v} = 5\mathbf{i} - 3\mathbf{j}$.

11. Find the dot product and the angle between $\mathbf{u} = 6\mathbf{i} + 5\mathbf{j}$ and $\mathbf{v} = 2\mathbf{i} - 3\mathbf{j}$.

12. $\mathbf{v}$ is a vector of magnitude 4 making an angle of 30° with the positive x-axis. Find $\mathbf{v}$ in component form.

13. Find the projection of $\mathbf{u}$ onto $\mathbf{v}$ given $\mathbf{u} = \langle 3, -1\rangle$ and $\mathbf{v} = \langle -2, 4\rangle$.

14. Give the trigonometric form of $z = 5 - 5i$.

15. Give the standard form of $z = 6(\cos 225° + i \sin 225°)$

16. Multiply $[7(\cos 23° + i \sin 23°)][4(\cos 7° + i \sin 7°)]$.

17. Divide $\dfrac{9\left(\cos \dfrac{5\pi}{4} + i \sin \dfrac{5\pi}{4}\right)}{3(\cos \pi + i \sin \pi)}$.

18. Find $(2 + 2i)^8$.

19. Find the cube roots of $8\left(\cos \dfrac{\pi}{3} + i \sin \dfrac{\pi}{3}\right)$.

20. Find all the solutions to $x^4 + i = 0$.

CHAPTER 7
Systems of Equations and Inequalities

CHAPTER 7
Systems of Equations and Inequalities

Section 7.1 Solving Systems of Equations

- You should be able to solve systems of equations by the method of substitution.
  1. Solve one of the equations for one of the variables.
  2. Substitute this expression into the other equation and solve.
  3. Back-substitute into the first equation to find the value of the other variable.
  4. Check your answer in each of the original equations.
- You should be able to find solutions graphically. (See Example 5 in textbook.)

Solutions to Odd-Numbered Exercises

1. $2x + y = 6$ Equation 1

$-x + y = 0$ Equation 2

Solve for y in Equation 1: $y = 6 - 2x$

Substitute for y in Equation 2: $-x + (6 - 2x) = 0$

Solve for x: $-3x + 6 = 0 \implies x = 2$

Back-substitute $x = 2$: $y = 6 - 2(2) = 2$

Answer: $(2, 2)$

3. $x - y = -4$ Equation 1

$x^2 - y = -2$ Equation 2

Solve for y in Equation 1: $y = x + 4$

Substitute for y in Equation 2: $x^2 - (x + 4) = -2$

Solve for x: $x^2 - x - 2 = 0 \implies (x + 1)(x - 2) = 0 \implies x = -1, 2$

Back-substitute $x = -1$: $y = -1 + 4 = 3$

Back-substitute $x = 2$: $y = 2 + 4 = 6$

Answers: $(-1, 3), (2, 6)$

5. $x - 3y = 15$ Equation 1

$x^2 + y^2 = 25$ Equation 2

Solve for x in Equation 1: $x = 3y + 15$

Substitute for x in Equation 2: $(3y + 15)^2 + y^2 = 25$

Solve for y: $10y^2 + 90y + 200 = 0 \implies y^2 + 9y + 20 = 0 \implies (y + 5)(y + 4) = 0 \implies y = -5, -4$

Back-substitute $y = -5$: $x = 3(-5) + 15 = 0$

Back-substitute $y = -4$: $x = 3(-4) + 15 = 3$

Answers: $(0, -5), (3, -4)$

7. $x^2 + y = 0$ Equation 1

$x^2 - 4x - y = 0$ Equation 2

Solve for y in Equation 1: $y = -x^2$

Substitute for y in Equation 2: $x^2 - 4x - (-x^2) = 0$

Solve for x: $2x^2 - 4x = 0 \implies 2x(x - 2) = 0 \implies x = 0, 2$

Back-substitute $x = 0$: $y = -0^2 = 0$

Back-substitute $x = 2$: $y = -2^2 = -4$

Answers: $(0, 0), (2, -4)$

9. $x - 6y = -8$ Equation 1

$x^2 - 4y^3 = 0$ Equation 2

Solve for x in Equation 1: $x = 6y - 8$

Substitute for x in Equation 2: $(6y - 8)^2 - 4y^3 = 0$

Solve for y: $-4y^3 + 36y^2 - 96y + 64 = 0$

$$y^3 - 9y^2 + 24y - 16 = 0$$

$$(y - 1)(y - 4)^2 = 0 \implies y = 1, 4$$

Back-substitute $y = 1$: $x = 6(1) - 8 = -2$

Back-substitute $y = 4$: $x = 6(4) - 8 = 16$

Answers: $(-2, 1), (16, 4)$

11. $x - y = 0$ Equation 1

$5x - 3y = 10$ Equation 2

Solve for y in Equation 1: $y = x$

Substitute for y in Equation 2: $5x - 3x = 10$

Solve for x: $2x = 10 \implies x = 5$

Back-substitute in Equation 1: $y = x = 5$

Answer: $(5, 5)$

13. $2x - y + 2 = 0$ Equation 1

$4x + y - 5 = 0$ Equation 2

Solve for y in Equation 1: $y = 2x + 2$

Substitute for y in Equation 2: $4x + (2x + 2) - 5 = 0$

Solve for x: $4x + (2x + 2) - 5 = 0 \Rightarrow 6x - 3 = 0 \Rightarrow x = \frac{1}{2}$

Back-substitute $x = \frac{1}{2}$: $y = 2x + 2 = 2\left(\frac{1}{2}\right) + 2 = 3$

Answer: $\left(\frac{1}{2}, 3\right)$

15. $30x - 40y - 33 = 0$ Equation 1

$10x + 20y - 21 = 0$ Equation 2

Solve for x in Equation 2: $x = -2y + \frac{21}{10}$

Substitute for x in Equation 1: $30\left(-2y + \frac{21}{10}\right) - 40y - 33 = 0$

Solve for y: $-100y + 30 = 0 \Rightarrow y = \frac{3}{10}$

Back-substitute $y = \frac{3}{10}$: $x = -2y + \frac{21}{10} = -2\left(\frac{3}{10}\right) + \frac{21}{10} = \frac{3}{2}$

Answer: $\left(\frac{3}{2}, \frac{3}{10}\right)$

17. $\frac{1}{5}x + \frac{1}{2}y = 8$ Equation 1

$x + y = 20$ Equation 2

Solve for x in Equation 2: $x = 20 - y$

Substitute for x in Equation 1: $\frac{1}{5}(20 - y) + \frac{1}{2}y = 8$

Solve for y: $4 + \frac{3}{10}y = 8 \Rightarrow y = \frac{40}{3}$

Back-substitute $y = \frac{40}{3}$: $x = 20 - y = 20 - \frac{40}{3} = \frac{20}{3}$

Answer: $\left(\frac{20}{3}, \frac{40}{3}\right)$

19. $2x + y = 4$ Equation 1

$-4x + 2y = -12$ Equation 2

Solve for y in Equation 1: $y = 2x - 4$

Substitute for y in Equation 2: $-4x + 2(2x - 4) = -12$

Solve for x: $-8 \neq -12$ Inconsistent

No Solution.

21. $x - y = 0$ Equation 1

$2x + y = 0$ Equation 2

Solve for y in Equation 1: $y = x$

Substitute for y in Equation 2: $2x + x = 0$

Solve for x: $3x = 0 \Rightarrow x = 0$

Back-substitute $x = 0$: $y = x = 0$

Answer: $(0, 0)$

23. $-x + 2y = 2$

$3x + y = 15$

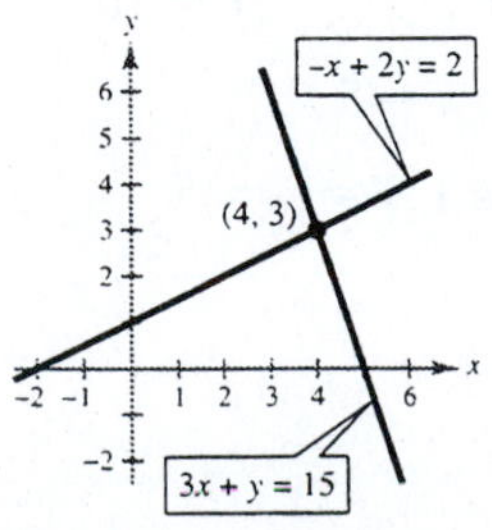

Point of intersection: $(4, 3)$

25. $x - 3y = -2$

$5x + 3y = 17$

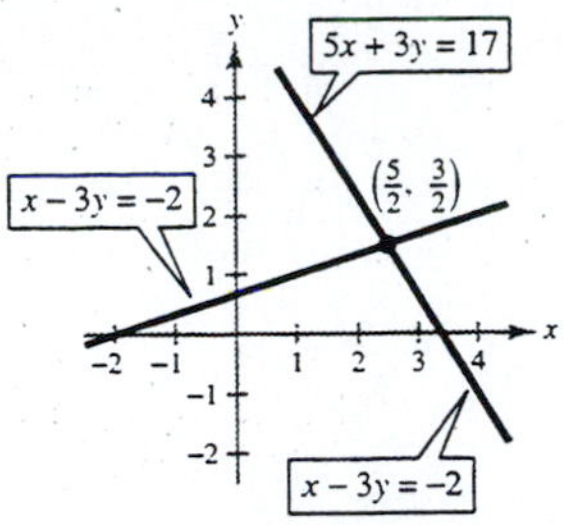

Point of intersection: $\left(\frac{5}{2}, \frac{3}{2}\right)$

27. $x + y = 4$

$x^2 + y^2 - 4x = 0$

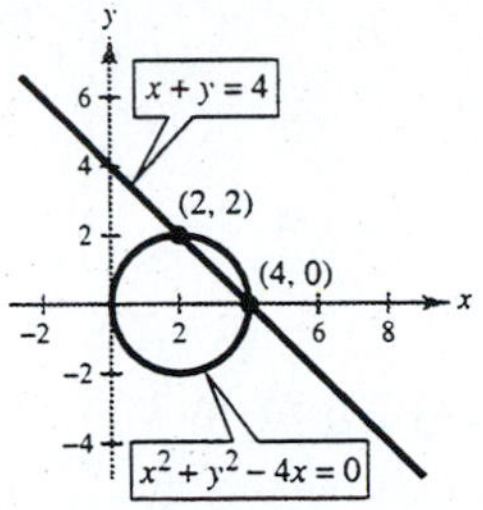

Points of intersection: (2, 2), (4, 0)

29. $7x + 8y = 24 \Rightarrow y_1 = -\frac{7}{8}x + 3$

$x - 8y = 8 \Rightarrow y_2 = \frac{1}{8}x - 1$

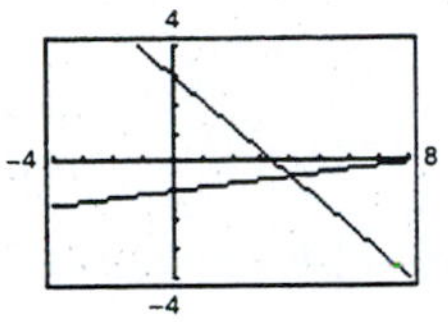

Point of intersection: $\left(4, -\frac{1}{2}\right)$

31. $3x - 2y = 0 \Rightarrow y_1 = \frac{3}{2}x$

$x^2 - y^2 = 4 \Rightarrow y_2 = \sqrt{x^2 - 4}$

$y_3 = -\sqrt{x^2 - 4}$

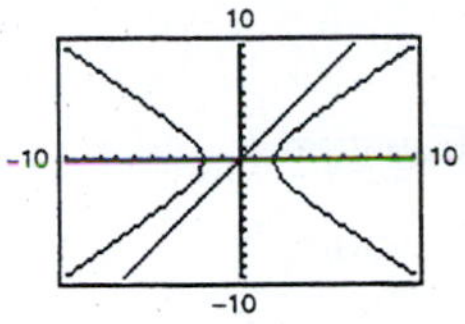

No points of intersection
Inconsistent
No solution

33. $x^2 + y^2 = 8 \Rightarrow y_1 = \sqrt{8 - x^2}$ and $y_2 = -\sqrt{8 - x^2}$

$y = x^2 \Rightarrow y_3 = x^2$

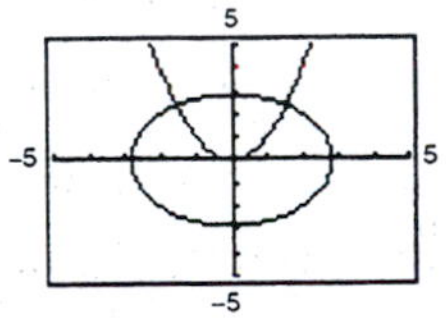

Points of intersection: $\left(\pm\sqrt{\frac{-1 + \sqrt{33}}{2}}, \frac{-1 + \sqrt{33}}{2}\right)$

$\approx (\pm 1.54, 2.37)$

Algebraically we have:

$$x^2 + (x^2)^2 = 8$$

$$x^4 + x^2 - 8 = 0$$

$$x^2 = \frac{-1 \pm \sqrt{1^2 - 4(1)(-8)}}{2(1)}$$

Use the positive value $\pm\sqrt{\frac{1 + \sqrt{33}}{2}}$

$$y = x^2 = \frac{-1 + \sqrt{33}}{2}$$

35. $y = e^x$

$x - y + 1 = 0 \Rightarrow y = x + 1$

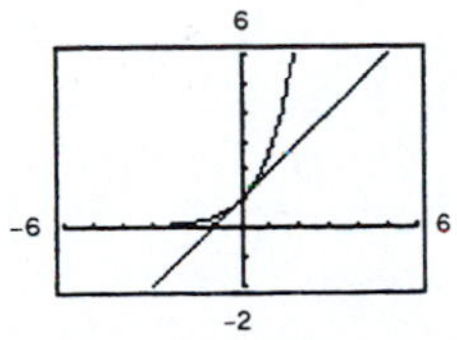

Point of intersection: (0, 1)

37. $y = \sqrt{x}$

$y = x$

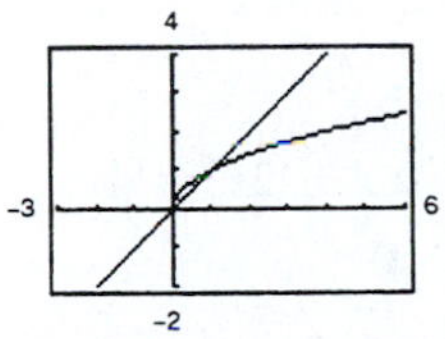

Points of intersection: (0, 0), (1, 1)

39. $x^2 + y^2 = 169 \Rightarrow y_1 = \sqrt{169 - x^2}$ and $y_2 = -\sqrt{169 - x^2}$

$x^2 - 8y = 104 \Rightarrow y_3 = \frac{1}{8}x^2 - 13$

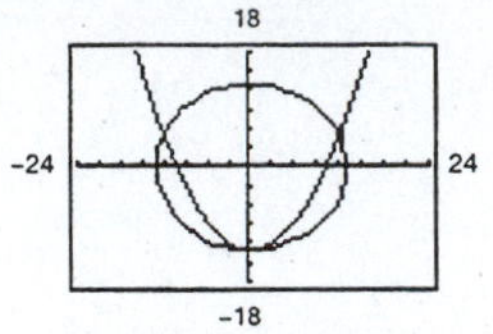

Points of intersection: $(0, -13), (\pm 12, 5)$

41. $y = 2x$ Equation 1

$y = x^2 + 1$ Equation 2

Substitute for y in Equation 2: $2x = x^2 + 1$

Solve for x: $x^2 - 2x + 1 = (x - 1)^2 = 0 \Rightarrow x = 1$

Back-substitute $x = 1$ in Equation 1: $y = 2x = 2$

Answer: $(1, 2)$

43. $3x - 7y + 6 = 0$ Equation 1

$x^2 - y^2 = 4$ Equation 2

Solve for y in Equation 1: $y = \dfrac{3x + 6}{7}$

Solve for y in Equation 2: $x^2 - \left(\dfrac{3x + 6}{7}\right)^2 = 4$

Solve for x: $x^2 - \left(\dfrac{9x^2 + 36x + 36}{49}\right) = 4$

$$49x^2 - (9x^2 + 36x + 36) = 196$$

$$40x^2 - 36x - 232 = 0$$

$4(10x - 29)(x + 2) = 0 \Rightarrow x = \dfrac{29}{10}, -2$

Back-substitute $x = \dfrac{29}{10}$: $y = \dfrac{3x + 6}{7} = \dfrac{3(29/10) + 6}{7} = \dfrac{21}{10}$

Back-substitute $x = -2$: $y = \dfrac{3x + 6}{7} = 0$

Answers: $\left(\dfrac{29}{10}, \dfrac{21}{10}\right), (-2, 0)$

45. $x - 2y = 4$ Equation 1

$x^2 - y = 0$ Equation 2

Solve for y in Equation 2: $y = x^2$

Substitute for y in Equation 1: $x - 2x^2 = 4$

Solve for x: $0 = 2x^2 - x + 4$

No real solutions, the discriminant in the Quadratic Formula is negative.

Inconsistent

No real solution

47. $y - e^{-x} = 1 \Rightarrow y = e^{-x} + 1$

$y - \ln x = 3 \Rightarrow y = \ln x + 3$

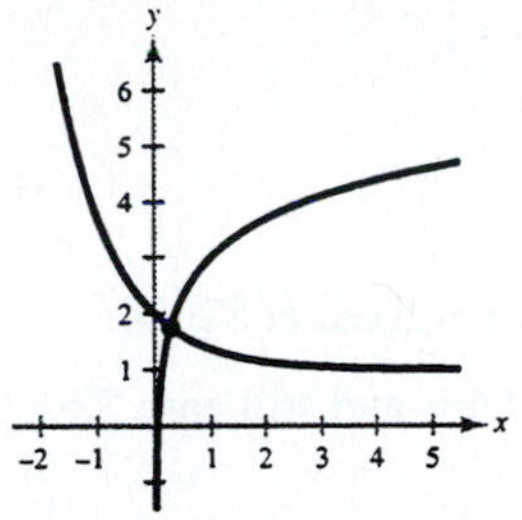

Point of intersection: Approximately (0.287, 1.75)

49. $y = x^4 - 2x^2 + 1$ Equation 1

$y = 1 - x^2$ Equation 2

Substitute for y in Equation 1: $1 - x^2 = x^4 - 2x^2 + 1$

Solve for x: $x^4 - x^2 = 0 \Rightarrow x^2(x^2 - 1) = 0$

$\Rightarrow x = 0, \pm 1$

Back-substitute $x = 0$: $1 - x^2 = 1$

Back-substitute $x = 1$: $1 - x^2 = 1 - 1^2 = 0$

Back-substitute $x = -1$: $1 - x^2 = 1 - (-1)^2 = 0$

Answers: $(0, 1), (\pm 1, 0)$

51. $xy - 1 = 0$ Equation 1

$2x - 4y + 7 = 0$ Equation 2

Solve for y in Equation 1: $y = \dfrac{1}{x}$

Substitute for y in Equation 2: $2x - 4\left(\dfrac{1}{x}\right) + 7 = 0$

Solve for x: $2x^2 - 4 + 7x = 0 \Rightarrow (2x - 1)(x + 4) = 0 \Rightarrow x = \dfrac{1}{2}, -4$

Back-substitute $x = \dfrac{1}{2}$: $y = \dfrac{1}{1/2} = 2$

Back-substitute $x = -4$: $y = \dfrac{1}{-4} = -\dfrac{1}{4}$

Answers: $\left(\dfrac{1}{2}, 2\right), \left(-4, -\dfrac{1}{4}\right)$

53. The system has no solution if you arrive at a false statement, ie. $4 = 8$, or you have a quadratic equation with a negative discriminant, which would yield imaginary solutions.

55. $C = 8650x + 250{,}000,\ R = 9950x$

$R = C$

$9950x = 8650x + 250{,}000$

$1300x = 250{,}000$

$x \approx 192$ units

57. $C = 2.65x + 350{,}000,\ R = 4.15x$

$R = C$

$4.15x = 2.65x + 350{,}000$

$1.50x = 350{,}000$

$x \approx 233{,}333$ units

59. $C = 3.45x + 16{,}000,\ R = 5.95x$

(a) $R = C$

$5.95x = 3.45x + 16{,}000$

$2.50x = 16{,}000$

$x \approx 6400$ units

(b) $P = R - C$

$6000 = 5.95x - (3.45x + 16{,}000)$

$6000 = 2.5x - 16{,}000$

$22000 = 2.5x$

$x \approx 8800$ units

61. (a) $x + y = 25{,}000$

$0.06x + 0.085y = 2000$

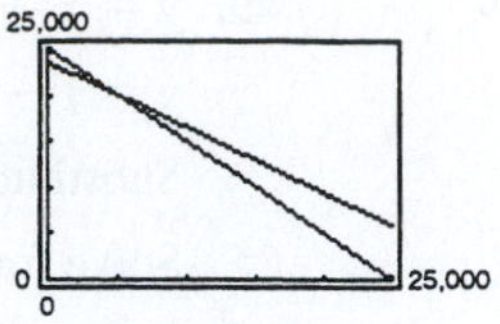

(b) $y_1 = 25{,}000 - x$

$$y_2 = \frac{2000 - 0.06x}{0.085}$$

As the amount at 6% increases, the amount at 8.5% decreases. The amount of interest is fixed at \$2000.

(c) The point of intersection occurs when $x = 5000$, so the most that can be invested at 6% and still earn \$2000 per year in interest is \$5000.

63. $0.06x = 0.03x + 250$

$0.03x = 250$

$x \approx \$8333.33$

To make the straight commission offer the better offer, you would have to sell more than \$8333.33 per week.

65. $V = (D - 4)^2,\ 5 \le D \le 40$ Doyle Log Rule

$V = 0.79D^2 - 2D - 4,\ 5 \le D \le 40$ Scribner Log Rule

(a)

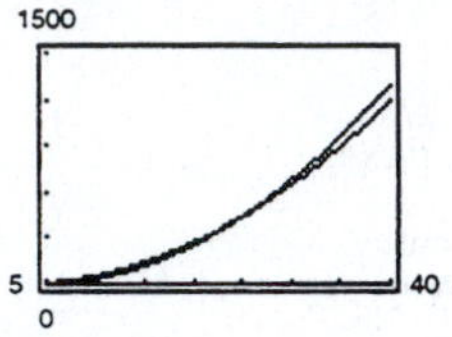

(b) The graphs intersect when $D \approx 24.7$ inches.

(c) For large logs, Doyle's Log Rule gives a greater volume for a given diameter.

67. $2l + 2w = 30 \implies l + w = 15$

$l = w + 3 \implies (w + 3) + 2 = 15$

$2w = 12$

$w = 6$

$l = w + 3 = 9$

Dimensions: 6 meters × 9 meters

69. $2l + 2w = 42 \implies l + w = 21$

$w = \frac{3}{4}l \implies l + \frac{3}{4}l = 21$

$\frac{7}{4}l = 21$

$l = 12$

$w = \frac{3}{4}l = 9$

Dimensions: 9 inches × 12 inches

71. $2l + 2w = 40 \implies l + w = 20 \implies w = 20 - l$

$lw = 96 \implies l(20 - l) = 96$

$20l - l^2 = 96$

$0 = l^2 - 20l + 96$

$0 = (l - 8)(l - 12)$

$l = 8 \text{ or } l = 12$

$w = 12,\ w = 8$

Since the length is supposed to be greater than the width, we have $l = 12$ kilometers and $w = 8$ kilometers.

73. The point lies on the line through (0, 10) and $(\sqrt{61}, 0)$. This line's equation is $y = -\dfrac{10}{\sqrt{61}}x + 10$.

Substitute for y in $\dfrac{x^2}{25} - \dfrac{y^2}{36} = 1$.

$$\frac{x^2}{25} - \frac{(-10/\sqrt{61}x + 10)^2}{36} = 1$$

$$36x^2 - 25\left(-\frac{10}{\sqrt{61}}x + 10\right)^2 = 900$$

$$36x^2 - 25\left(\frac{100}{61}x^2 - \frac{200}{\sqrt{61}}x + 100\right) = 900$$

$$-\frac{304}{61}x^2 + \frac{5000}{\sqrt{61}}x - 3400 = 0$$

By the Quadratic Formula we have $x \approx 122.91$ or $x \approx 5.55$. Only $x \approx 5.55$ agrees with the graph. The point of intersection is approximately (5.55, 2.89).

75. (0, 6.6), (1, 6.8), (2, 7.1), (3, 7.1)

(a) Linear model: $f(t) = 0.18t + 6.63$

Quadratic model: $g(t) = -0.05t^2 + 0.33t + 6.58$

(b)

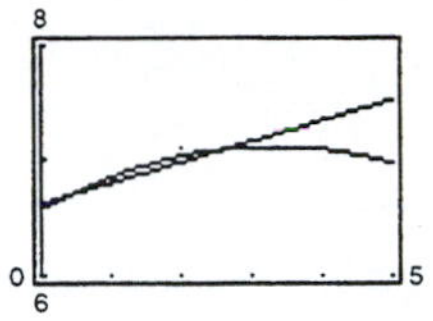

(c) Points of intersection: (0.382, 6.70), (2.618, 7.10)

(d) Linear model when $t = 4$: $f(4) = 7.35$ million short tons.

Quadratic model when $t = 4$: $g(4) = 7.1$ million short tons.

Since the sale of newsprint has been slowly decreasing (more papers are on-line now and many people would rather listen to the news than read it) the Quadratic model is probably more accurate.

77. (−2, 7), (5, 5)

$$m = \frac{5 - 7}{5 - (-2)} = -\frac{2}{7}$$

$$y - 7 = -\frac{2}{7}(x - (-2))$$

$$7y - 49 = -2x - 4$$

$$2x + 7y - 45 = 0$$

79. (6, 3), (10, 3)

$$m = \frac{3 - 3}{10 - 6} = 0 \implies \text{The line is horizontal.}$$

$$y = 3$$

81. $\left(\frac{3}{5}, 0\right)$, (4, 6)

$$m = \frac{6 - 0}{4 - \frac{3}{5}} = \frac{6}{\frac{17}{5}} = \frac{30}{17}$$

$$y - 6 = \frac{30}{17}(x - 4)$$

$$17y - 102 = 30x - 120$$

$$0 = 30x - 17y - 18$$

Section 7.2 Two-Variable Linear Systems

- You should be able to solve a linear system by the method of elimination.
  1. Obtain coefficients for either x or y that differ only in sign. This is done by multiplying all the terms of one or both equations by appropriate constants.
  2. Add the equations to eliminate one of the variables and then solve for the remaining variable.
  3. Use back-substitution into either original equation and solve for the other variable.
  4. Check your answer.
- You should know that for a system of two linear equations, one of the following is true.
  1. There are infinitely many solutions; the lines are identical. The system is consistent.
  2. There is no solution; the lines are parallel. The system is inconsistent.
  3. There is one solution; the lines intersect at one point. The system is consistent.

Solutions to Odd-Numbered Exercises

1. $2x + y = 5$ Equation 1

$x - y = 5$ Equation 2

Add to eliminate y: $3x = 6 \Rightarrow x = 2$

Substitute $x = 2$ in Equation 2: $2 - y = 1 \Rightarrow y = 1$

Answer: $(2, 1)$

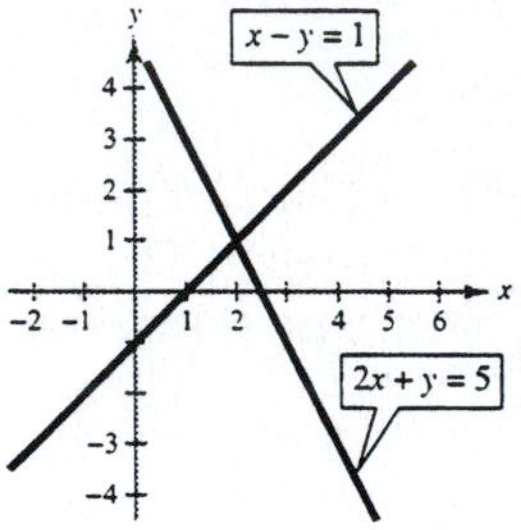

3. $x + y = 0$ Equation 1

$3x + 2y = 1$ Equation 2

Multiply Equation 1 by -2: $-2x - 2y = 0$

Add this to Equation 2 to eliminate y: $x = 1$

Substitute $x = 1$ in Equation 1: $1 + y = 0 \Rightarrow y = -1$

Answer: $(1, -1)$

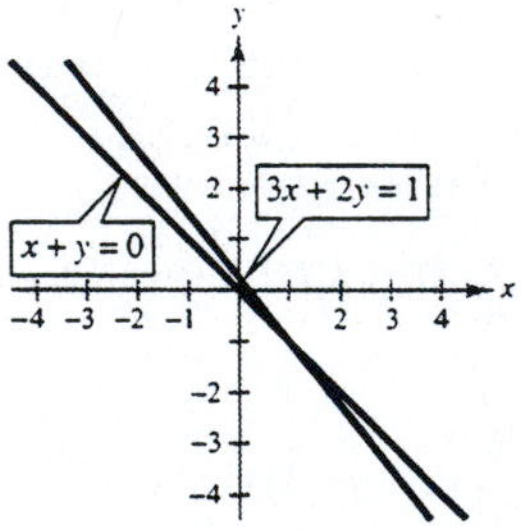

5. $x - y = 2$ Equation 1

$-2x + 2y = 5$ Equation 2

Multiply Equation 1 by 2: $2x - 2y = 4$

Add this to Equation 2: $0 = 9$

There are no solutions.

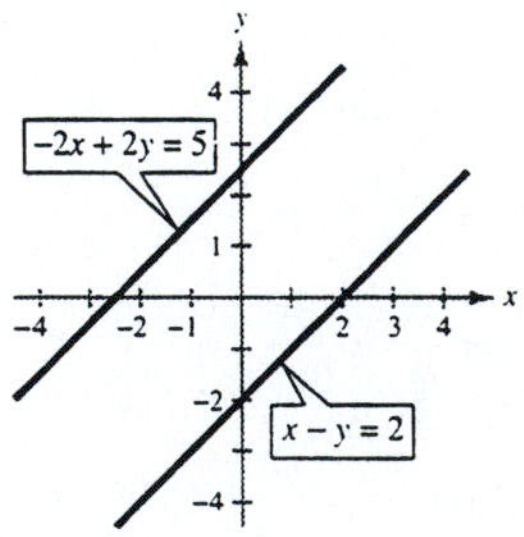

7. $3x - 2y = 5$ Equation 1

$-6x + 4y = -10$ Equation 2

Multiply Equation 1 by 2 and add to Equation 2: $0 = 0$

The equations are dependent. There are infinitely many solutions.

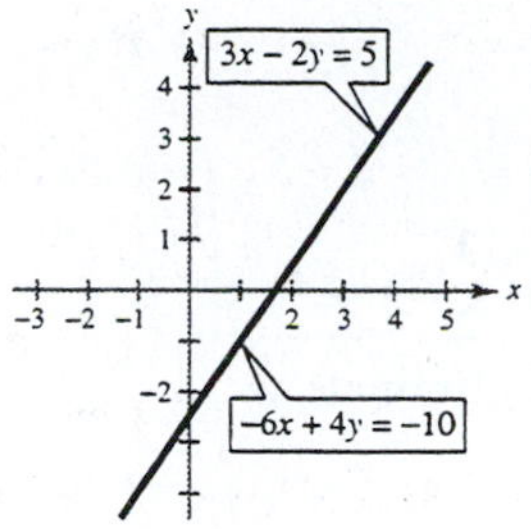

9. $9x + 3y = 1$ Equation 1

$3x - 6y = 5$ Equation 2

Multiply Equation 2 by (-3): $9x + 3y = 1$

$-9x + 18y = -15$

Add to eliminate x: $21y = -14 \implies y = -\frac{2}{3}$

Substitute $y = -\frac{2}{3}$ in Equation 1: $9x + 3\left(-\frac{2}{3}\right) = 1$

$x = \frac{1}{3}$

Answer: $\left(\frac{1}{3}, -\frac{2}{3}\right)$

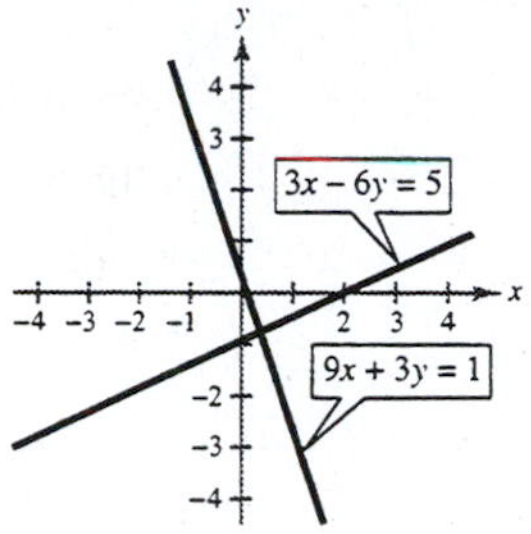

11. $x + 2y = 4$ Equation 1

$x - 2y = 1$ Equation 2

Add to eliminate y:

$2x = 5$

$x = \frac{5}{2}$

Substitute $x = \frac{5}{2}$ in Equation 1:

$\frac{5}{2} + 2y = 4 \implies y = \frac{3}{4}$

Answer: $\left(\frac{5}{2}, \frac{3}{4}\right)$

13. $2x + 3y = 18$ Equation 1

$5x - y = 11$ Equation 2

Multiply Equation 2 by 3: $15x - 3y = 33$

Add this to Equation 1 to eliminate y:

$17x = 51 \implies x = 3$

Substitute $x = 3$ in Equation 1:

$6 + 3y = 18 \implies y = 4$

Answer: $(3, 4)$

15. $3x + 2y = 10$ Equation 1

$2x + 5y = 3$ Equation 2

Multiply Equation 1 by 2 and Equation 2 by (-3):

$6x + 4y = 20$

$-6x - 15y = -9$

Add to eliminate x: $-11y = 11 \implies y = -1$

Substitute $y = -1$ in Equation 1:

$3x - 2 = 10 \implies x = 4$

Answer: $(4, -1)$

17. $2u + v = 120$ Equation 1

$u + 2v = 120$ Equation 2

Multiply Equation 2 by (-2):
$-2u - 4v = -240$

Add this to Equation 1 to eliminate u:

$-3v = -120$

$v = 40$

Substitute $v = 40$ in Equation 2:

$u + 80 = 120 \implies u = 40$

Answer: $(40, 40)$

19. $6r - 5s = 3$ Equation 1

$-12r + 10s = 5$ Equation 2

Multiply Equation 1 by 2: $12r - 10s = 6$

Add this to Equation 2 to eliminate r: $0 = 11$

Inconsistent

No solution

21. $\frac{x}{4} + \frac{y}{6} = 1$ Equation 1

$x - y = 3$ Equation 2

Multiply Equation 1 by 6: $\frac{3}{2}x + y = 6$

Add this to Equation 2 to eliminate y:

$\frac{5}{2}x = 9 \Rightarrow x = \frac{18}{5}$

Substitute $x = \frac{18}{5}$ in Equation 2:

$\frac{18}{5} - y = 3$

$y = \frac{3}{5}$

Answer: $\left(\frac{18}{5}, \frac{3}{5}\right)$

23. $\frac{x+3}{4} + \frac{y-1}{3} = 1$ Equation 1

$2x - y = 12$ Equation 2

Multiply Equation 1 by 12 and Equation 2 by 4:

$3x + 4y = 7$

$8x - 4y = 48$

Add to eliminate y: $11x = 55 \Rightarrow x = 5$

Substitute $x = 5$ into Equation 2:

$2(5) - y = 12 \Rightarrow y = -2$

Answer: $(5, -2)$

25. $2.5x - 3y = 1.5$ Equation 1

$10x - 12y = 6$ Equation 2

Multiply Equation 1 by (-4):

$-10x + 12y = -6$

Add this to Equation 2 to eliminate x:

$0 = 0$ (Dependent)

The solution set consists of all points lying on the line $10x - 12y = 6$.

Let $x = a$, then $y = \frac{5}{6}a - \frac{1}{2}$.

Answer: $\left(a, \frac{5}{6}a - \frac{1}{2}\right)$, where a is any real number.

27. $0.05x - 0.03y = 0.21$ Equation 1

$0.07x + 0.02y = 0.16$ Equation 2

Multiply Equation 1 by 200 and Equation 2 by 300: $10x - 6y = 42$

$21x + 6y = 48$

Add to eliminate y: $31x = 90$

$x = \frac{90}{31}$

Substitute $x = \frac{90}{31}$ in Equation 2:

$0.07\left(\frac{90}{31}\right) + 0.02y = 0.16$

$y = -\frac{67}{31}$

Answer: $\left(\frac{90}{31}, -\frac{67}{31}\right)$

29. $4b + 3m = 3$ Equation 1

$3b + 11m = 13$ Equation 2

Multiply Equation 1 by 3 and Equation 2 by (-4):

$12b + 9m = 9$

$-12b - 44m = -52$

Add to eliminate b: $-35m = -43$

$m = \frac{43}{35}$

Substitute $m = \frac{43}{35}$ in Equation 1: $5b + 3\left(\frac{43}{35}\right) = 3 \Rightarrow b = -\frac{6}{35}$

Answer: $\left(-\frac{6}{35}, \frac{43}{35}\right)$

31. $\frac{1}{5}x - \frac{1}{3}y = 1$

$-3x + 5y = 9$

The lines are parallel. The system is inconsistent.

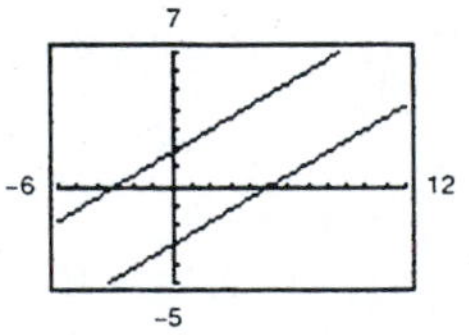

33. $2x - 5y = 0$

$x - y = 3$

The system is consistent. There is one solution.

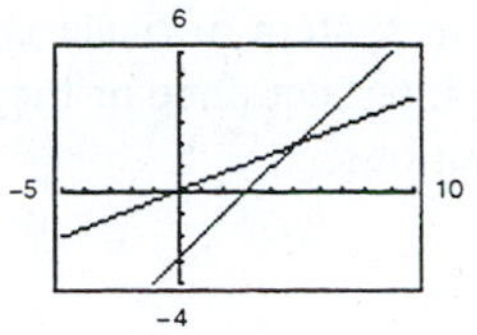

35. $8x + 9y = 42$

$6x - y = 16$

Answer: $(3, 2)$

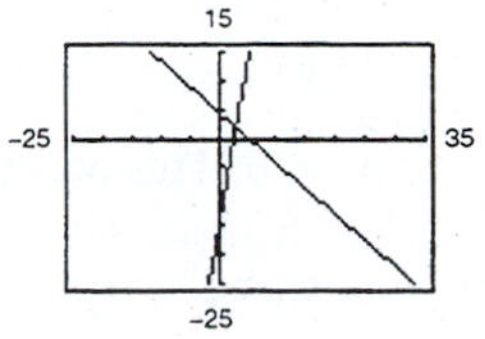

37. $4y = -8$

$7x - 2y = 25$

Answer: $(3, -2)$

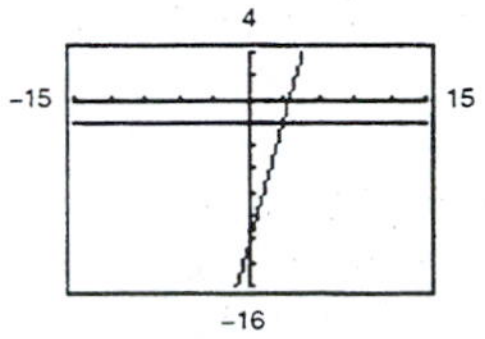

39. $3x - 5y = 7$ Equation 1

$2x + y = 9$ Equation 2

Multiply Equation 2 by 5:

$10x + 5y = 45$

Add this to Equation 1:

$13x = 52 \implies x = 4$

Back-substitute $x = 4$ into Equation 2:

$2(4) + y = 9 \implies y = 1$

Answer: $(4, 1)$

41. $y = 2x - 5$ Equation 1

$y = 5x - 11$ Equation 2

Since both equations are solved for y, set them equal to one another and solve for x.

$$2x - 5 = 5x - 11$$
$$6 = 3x$$
$$2 = x$$

Back-substitute $x = 2$ into Equation 1:

$y = 2(2) - 5 = -1$

Answer: $(2, -1)$

43. There are infinitely many systems that have the solution $\left(3, \frac{5}{2}\right)$. One possible system is:

$2(3) + 2\left(\frac{5}{2}\right) = 11 \implies 2x + 2y = 11$

$3 - 4\left(\frac{5}{2}\right) = -7 \implies x - 4y = -7$

45. $100y - x = 200$ Equation 1

$99y - x = -198$ Equation 2

Subtract Equation 2 from Equation 1 to eliminate x: $y = 398$

Substitute $y = 398$ into Equation 1: $100(398) - x = 200 \implies x = 39{,}600$

Answer: $(39{,}600, 398)$

The lines are not parallel. The scale on the axes must be changed to see the point of intersection.

47. No, it is not possible for a consistent system of linear equations to have exactly two solutions. Either the lines will intersect once or they will coincide and then the system would have infinite solutions.

49. $4x - 8y = -3$ Equation 1

$2x + ky = 16$ Equation 2

Multiply Equation 2 by -2: $-4x - 2ky = -32$

Add this to Equation 1: $-8y - 2ky = -35$

The system in inconsistent if $-8y - 2ky = 0$. This occurs when $k = -4$.

51. Let $x =$ the air speed and $y =$ the wind speed.

$3.6(x - y) = 1800$ Equation 1

$6(x + y) = 1800$ Equation 2

$$\begin{aligned} x - y &= 500 \\ x + y &= 600 \\ 2x &= 1100 \\ x &= 550 \\ 550 + y &= 600 \\ y &= 50 \end{aligned}$$

Answer: $x = 550$ mph, $y = 50$ mph

53. Let $x =$ the number of liters at 20%

$y =$ the number of liters at 50%.

(a) $x + y = 10$

$0.2x + 0.5y = 0.3(10)$

(b) 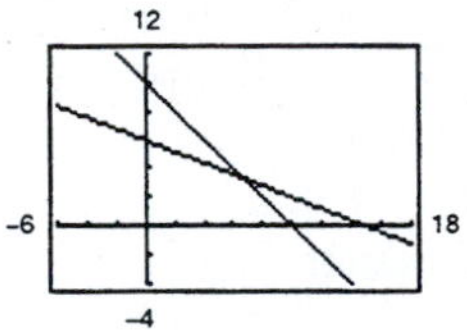

As x increases, y decreases.

(c) -2 Equation 1 $\quad -2x - 2y = -20$

10 Equation 2 $\quad 2x + 5y = 30$

$$\begin{aligned} 3y &= 10 \\ y &= \tfrac{10}{3} \\ x + \tfrac{10}{3} &= 10 \\ x &= \tfrac{20}{3} \end{aligned}$$

Answer: $x = 6\frac{2}{3}$ liters at 20%

$y = 3\frac{1}{3}$ liters at 50%

55. Let $x =$ amount invested at 10.5%

$y =$ amount invested at 12%.

$x + y = 12{,}000$ Equation 1

$0.105x + 0.12y = 1350$ Equation 2

$$\begin{aligned} -12x - 12y &= -144{,}000 \\ 10.5x + 12y &= 135{,}000 \\ -1.5x \quad y &= -9000 \\ x \quad y &= 6000 \\ 6000 + \quad y &= 12{,}000 \\ y &= 6000 \end{aligned}$$

Answer: $y = \$6000$ at 12%

$x = \$6000$ at 10.5%

57. Let x = number of adult tickets sold

y = number of child tickets sold.

$$x + y = 500 \quad \text{Equation 1}$$
$$7.5x + 4y = \$3312.50 \quad \text{Equation 2}$$

$$\begin{aligned} -4x - 4y &= -2000.00 \\ 7.5x + 4y &= 3312.50 \\ \hline 3.5x &= 1312.50 \\ x &= 375 \\ 375 + y &= 500 \\ y &= 125 \end{aligned}$$

Answer: $x = 375$ adult tickets

$y = 125$ child tickets

59. Let x = distance one person drives

y = distance other person drives.

$$\begin{aligned} x + y &= 300 && \text{Equation 1} \\ y &= 3x && \text{Equation 2} \\ x + 3x &= 300 && \text{Use substitution} \\ 4x &= 300 \\ x &= 75 \\ y &= 3x = 225 \end{aligned}$$

Answer: 75 km and 225 km

61. Demand = Supply

$$\begin{aligned} 50 - 0.5x &= 0.125x \\ 50 &= 0.625x \\ x &= 80 \text{ units} \\ p &= \$10 \\ 4x &= 300 \end{aligned}$$

Answer: (80, 10)

63. Demand = Supply

$$\begin{aligned} 140 - 0.00002x &= 80 + 0.00001x \\ 60 &= 0.00003x \\ x &= 2{,}000{,}000 \text{ units} \\ p &= \$100.00 \end{aligned}$$

Answer: (2,000,000, 100)

65.

$$\begin{aligned} 5b + 10a = 20.2 &\implies -10b - 20a = -40.4 \\ 10b + 30a = 50.1 &\implies 10b + 30a = 50.1 \\ & \hline 10a = 9.7 \\ & a = 0.97 \\ & b = 2.10 \end{aligned}$$

Least squares regression line:

$y = 0.97x + 2.10$

67.

$$\begin{aligned} 7b + 21a = 35.1 &\implies -21b - 63a = -105.3 \\ 21b + 91a = 114.2 &\implies 21b + 91a = 114.2 \\ & \hline 28a = 8.9 \\ & a = \tfrac{89}{280} \\ & b = \tfrac{1137}{280} \end{aligned}$$

Least squares regression line:

$y = \frac{1}{280}(89x + 1137)$

$y \approx 0.318x + 4.061$

69. $(-2, 0), (0, 1), (2, 3)$

$3b = 4 \implies b = \frac{4}{3}$

$8a = 6 \implies a = \frac{3}{4}$

Least squares regression line:

$y = \frac{3}{4}x + \frac{4}{3}$

71. $(0, 4), (1, 3), (1, 1), (2, 0)$

$$\begin{aligned} 4b + 4a = 8 &\implies 4b + 4a = 8 \\ 4b + 6a = 4 &\implies \underline{-4b - 6a = -4} \\ & \qquad -2a = 4 \\ & \qquad a = -2 \\ & \qquad b = 4 \end{aligned}$$

Least squares regression line: $y = -2x + 4$

73. $(1.00, 450), (1.25, 375), (1.50, 330)$

$3b + 3.75a = 1155$

$3.75b + 4.8125a = 1413.75$

By elimination we have $a = -240$ and $b = 685$ and the least squares regression line is $y = -240x + 685$. When $x = 1.40$, we have $y = -240(1.40) + 685 = 349$ units.

Section 7.3 Multivariable Linear Systems

- You should know the operations that lead to equivalent systems of equations:
  (a) Interchange any two equations.
  (b) Multiply all terms of an equation by a nonzero constant.
  (c) Replace an equation by the sum of itself and a constant multiple of any other equation in the system.
- You should be able to use the method of elimination.

Solutions to Odd-Numbered Exercises

1. $2x - y + 5z = 24$ Equation 1

$y + 2z = 6$ Equation 2

$z = 4$ Equation 3

Back-substitute $z = 4$ into Equation 2.

$y + 2(4) = 6$

$z = -2$

Back-substitute $y = -2$ and $z = 4$ into Equation 1.

$2x - (-2) + 5(4) = 24$

$2x + 22 = 24$

$x = 1$

Answer: $(1, -2, 4)$

3. $2x + y - 3z = 10$ Equation 1

$y = 2$ Equation 2

$y - z = 4$ Equation 3

Back-substitute $y = 2$ into Equation 3.

$2 - z = 4$

$z = -2$

Back-substitute $y = 2$ and $z = -2$ into Equation 1.

$2x + 2 - 3(-2) = 10$

$2x + 8 = 10$

$x = 1$

Answer: $(1, 2, -2)$

5. $4x - 2y + z = 8$ Equation 1

$2z = 4$ Equation 2

$-y + z = 4$ Equation 3

From Equation 2 we have $z = 2$. Back-substitute $z = 2$ into Equation 3.

$$-y + 2 = 4$$
$$y = -2$$

Back-substitute $y = -2$ and $z = 2$ into Equation 1.

$$4x - 2(-2) + 2 = 8$$
$$4x + 6 = 8$$
$$x = \tfrac{1}{2}$$

Answer: $\left(\frac{1}{2}, -2, 2\right)$

7. $x - 2y + 3z = 5$ Equation 1

$-x + 3y - 5z = 4$ Equation 2

$2x \quad - 3z = 0$ Equation 3

Add Equation 1 to Equation 2.

$$y - 2z = 9$$

This is the first step in putting the system in row-echelon form.

9. $x + y + z = 6$ Equation 1

$2x - y + z = 3$ Equation 2

$3x \quad - z = 0$ Equation 3

$x + y + z = 6$

$-3y - z = -9$ -2Eq.1 + Eq.2

$-3y - 4z = -18$ -3Eq.1 + Eq.3

$x + y + z = 6$

$-3y - z = -9$

$-3z = -9$ $-$Eq.2 + Eq.3

$-3z = -9 \Rightarrow z = 3$

$-3y - 3 = -9 \Rightarrow y = 2$

$x + 2 + 3 = 6 \Rightarrow x = 1$

Answer: $(1, 2, 3)$

11. $2x \quad + 2z = 2$ Equation 1

$5x + 3y \quad = 4$ Equation 2

$3x \quad - z = 0$ Equation 3

$x + y + z = 6$

$x + 3y - 4z = 0$ -2Eq.1 + Eq.2

$2x \quad + 2z = 2$ Interchange equations.

$3y - 4z = 4$

$x + 3y - 4z = 0$

$-6y + 10z = 2$ -2Eq.1 + Eq.2

$3y - 4z = 4$

$-6y + 10z = 2$

$z = 5$ $\frac{1}{2}$Eq.2 + Eq.3

$z = 5$

$-6y + 10(5) = 2 \Rightarrow y = 8$

$x + 3(8) - 4(5) = 0 \Rightarrow x = -4$

Answer: $(-4, 8, 5)$

13.
$$\begin{aligned} 3x + 3y &= 9 \quad \text{Interchange equations.} \\ 2x - 3z &= 10 \\ 6y + 4z &= -12 \end{aligned}$$

$$\begin{aligned} x + y &= 3 \quad \tfrac{1}{3}\text{Eq.1} \\ 2x - 3z &= 10 \\ 6y + 4z &= -12 \end{aligned}$$

$$\begin{aligned} x + y &= 3 \\ -2y - 3z &= 4 \quad -2\text{Eq.1} + \text{Eq.2} \\ 6y + 4z &= -12 \end{aligned}$$

$$\begin{aligned} x + y &= 3 \\ -2y - 3z &= 4 \\ -5z &= 0 \quad 3\text{Eq.2} + \text{Eq.3} \end{aligned}$$

$$\begin{aligned} -5x = 0 \Rightarrow z &= 0 \\ -2y - 3(0) = 4 \Rightarrow y &= -2 \\ x - 2 = 3 \Rightarrow x &= 5 \end{aligned}$$

Answer: $(5, -2, 0)$

15.
$$\begin{aligned} x + y - 2z &= 3 \quad \text{Interchange equations.} \\ 3x - 2y + 4z &= 1 \\ 2x - 3y + 6z &= 8 \end{aligned}$$

$$\begin{aligned} x + y - 2z &= 3 \\ -5y + 10z &= -8 \quad -3\text{Eq.1} + \text{Eq.2} \\ -5y + 10z &= 2 \quad -2\text{Eq.1} + \text{Eq.3} \end{aligned}$$

$$\begin{aligned} x + y - 2z &= 3 \\ -5y + 10z &= -8 \\ 0 &= 10 \rightarrow \leftarrow -\text{Eq.2} + \text{Eq.3} \end{aligned}$$

No solution, inconsistent

17.
$$\begin{aligned} 3x + 3y + 5z &= 1 \\ 3x + 5y + 9z &= 0 \\ 5x + 9y + 17z &= 0 \end{aligned}$$

$$\begin{aligned} 6x + 6y + 10z &= 2 \quad 2\text{ Eq.1} \\ 3x + 5y + 9z &= 0 \\ 5x + 9y + 17z &= 0 \end{aligned}$$

$$\begin{aligned} x - 3y - 7z &= 2 \quad -\text{Eq.3} + \text{Eq.1} \\ 3x + 5y + 9z &= 0 \\ 5x + 9y + 17z &= 0 \end{aligned}$$

$$\begin{aligned} x - 3y - 7z &= 2 \\ 14y + 30z &= -6 \quad -3\text{Eq.1} + \text{Eq.2} \\ 24y + 52z &= -10 \quad -5\text{Eq.1} + \text{Eq.3} \end{aligned}$$

$$\begin{aligned} x - 3y - 7z &= 2 \\ 84y + 180z &= -36 \quad 6\text{Eq.2} \\ 84y + 182z &= -35 \quad 3.5\text{Eq.3} \end{aligned}$$

$$\begin{aligned} x - 3y - 7z &= 2 \\ 84y + 180z &= -36 \\ 2z &= 1 \quad -\text{Eq.2} + \text{Eq.3} \end{aligned}$$

$$\begin{aligned} 2z &= 1 \Rightarrow x = \tfrac{1}{2} \\ 84y + 180\left(\tfrac{1}{2}\right) &= -36 \Rightarrow y = \tfrac{1}{2} \\ x - 3\left(-\tfrac{3}{2}\right) - 7\left(\tfrac{1}{2}\right) &= 2 \Rightarrow x = 1 \end{aligned}$$

Answer: $\left(1, -\tfrac{3}{2}, \tfrac{1}{2}\right)$

19.
$$\begin{aligned} x + 2y - 7z &= -4 \\ 2x + y + z &= 13 \\ 3x + 9y - 36z &= -33 \end{aligned}$$

$$\begin{aligned} x + 2y - 7z &= -4 & & -2\text{Eq.}2 + \text{Eq.}2 \\ -3y + 15z &= 21 & & -3\text{Eq.}1 + \text{Eq.}3 \\ 3y - 15z &= -21 \end{aligned}$$

$$\begin{aligned} x + 2y - 7z &= -4 \\ -3y + 15z &= 21 \\ 0 &= 0 & & \text{Eq.}2 + \text{Eq.}3 \end{aligned}$$

$$\begin{aligned} x + 2y - 7z &= -4 \\ y - 5z &= -7 & & \tfrac{1}{3}\text{Eq.}2 \end{aligned}$$

$$\begin{aligned} x + 3z &= 10 & & -2\text{Eq.}2 + \text{Eq.}1 \\ y - 5z &= -7 \end{aligned}$$

Let $z = a$, then:

$y = 5a - 7$

$x = -3a + 10$

Answer: $(-3a + 10, 5a - 7, a)$

21.
$$\begin{aligned} 3x - 3y + 6z &= 6 \\ x + 2y - z &= 5 \\ 5x - 8y + 13z &= 7 \end{aligned}$$

$$\begin{aligned} x - y + 2z &= 2 & & \tfrac{1}{3}\text{Eq.}1 \\ 3y - 3z &= 3 & & -\text{Eq.}1 + \text{Eq.}2 \\ -3y + 3z &= -3 & & -5\text{Eq.}1 + \text{Eq.}3 \end{aligned}$$

$$\begin{aligned} x - y + 2z &= 2 \\ y - z &= 1 & & \tfrac{1}{3}\text{Eq.}2 \\ 0 &= 0 & & \text{Eq.}2 + \text{Eq.}3 \end{aligned}$$

$$\begin{aligned} x + z &= 3 & & \text{Eq.}2 + \text{Eq.}1 \\ y - z &- 1 \end{aligned}$$

Let $z = a$, then:

$y = a + 1$

$x = -a + 3$

Answer: $(-a + 3, a + 1, a)$

23.
$$\begin{aligned} x - 2y + 5z &= 2 \\ 4x - z &= 0 \end{aligned}$$

Let $z = a$, then $x = \frac{1}{4}a$.

$\frac{1}{4}a - 2y + 5a = 2$

$a - 8y + 20a = 8$

$-8y = -21a + 8$

$y = \frac{21}{8}a - 1$

Answer: $\left(\frac{1}{4}a, \frac{21}{8}a - 1, a\right)$

To avoid fractions, we could go back and let $z = 8a$, then $4x - 8a = 0 \Rightarrow x = 2a$.

$2a - 2y + 5(8a) = 2$

$-2y + 42a = 2$

$y = 21a - 1$

Answer: $(2a, 21a - 1, 8a)$

25.
$$\begin{aligned} 2x - 3y + z &= -2 \\ -4x + 9y &= 7 \end{aligned}$$

$$\begin{aligned} 2x - 3y + z &= -2 \\ 3y + 2z &= 3 & & 2\text{Eq.}1 + \text{Eq.}2 \end{aligned}$$

$$\begin{aligned} 2x + 3z &= 1 & & \text{Eq.}2 + \text{Eq.}1 \\ 3y + 2z &= 3 \end{aligned}$$

Let $x = a$, then:

$y = -\frac{2}{3}a + 1$

$x = -\frac{3}{2}a + \frac{1}{2}$

Answer: $\left(-\frac{3}{2}a + \frac{1}{2}, -\frac{2}{3}a + 1, a\right)$

27. $\begin{aligned} x \quad\quad\quad + 3w &= 4 \\ 2y - z - w &= 0 \\ 3y \quad - 2w &= 1 \\ 2x - y + 4z \quad &= 5 \end{aligned}$

$$\begin{aligned} x \quad\quad\quad + 3w &= 4 \\ 2y - z - w &= 0 \\ 3y \quad - 2w &= 1 \\ -y + 4z - 6w &= -3 \quad -2\text{Eq.}1 + \text{Eq.}4 \end{aligned}$$

$$\begin{aligned} x \quad\quad\quad + 3w &= 4 \\ y - 4z + 6w &= 3 \quad -\text{Eq.}4 \text{ and} \\ 2y - z - w &= 0 \quad \text{interchange} \\ 3y \quad - 2w &= 1 \quad \text{the equations.} \end{aligned}$$

$$\begin{aligned} x \quad\quad\quad + 3w &= 4 \\ y - 4z + 6w &= 3 \\ 7z - 13w &= -6 \quad -\text{Eq.}2 + \text{Eq.}3 \\ 12z - 20w &= -8 \quad -3\text{Eq.}2 + \text{Eq.}4 \end{aligned}$$

$$\begin{aligned} x \quad\quad\quad + 3w &= 4 \\ y - 4z + 6w &= 3 \\ z - 3w &= -2 \quad -\tfrac{1}{2}\text{Eq.}4 + \text{Eq.}3 \\ 12z - 20w &= -8 \end{aligned}$$

$$\begin{aligned} x \quad\quad\quad + 3w &= 4 \\ y - 4z + 6w &= 3 \\ z - 3w &= -2 \\ 16w &= 16 \quad -12\text{Eq.}3 + \text{Eq.}4 \end{aligned}$$

$$\begin{aligned} 16w &= 16 \Rightarrow w = 1 \\ z - 3(1) &= -2 \Rightarrow z = 1 \\ y - 4(1) + 6(1) &= 3 \Rightarrow y = 1 \\ x + 3(1) &= 4 \Rightarrow x = 1 \end{aligned}$$

Answer: (1, 1, 1, 1)

29. $\begin{aligned} x \quad + 4z &= 1 \\ x + y + 10z &= 10 \\ 2x - y + 2z &= -5 \end{aligned}$

$$\begin{aligned} x \quad + 4z &= 1 \\ y + 6z &= 9 \quad -\text{Eq.}1 + \text{Eq.}2 \\ -y - 6z &= -7 \quad -2\text{Eq.}1 + \text{Eq.}3 \end{aligned}$$

$$\begin{aligned} x \quad + 4z &= 1 \\ y + 6z &= 9 \\ 0 &= 2 \quad \rightarrow\leftarrow \text{Eq.}2 + \text{Eq.}3 \end{aligned}$$

No solution, inconsistent

31. $\begin{aligned} 2x + 3y \quad &= 0 \\ 4x + 3y - z &= 0 \\ 8x + 3y + 3z &= 0 \end{aligned}$

$$\begin{aligned} 2x + 3y \quad &= 0 \\ -3y - z &= 0 \quad -2\text{Eq.}1 + \text{Eq.}2 \\ -9y + 3z &= 0 \quad -4\text{Eq.}1 + \text{Eq.}3 \end{aligned}$$

$$\begin{aligned} 2x + 3y \quad &= 0 \\ -3y - z &= 0 \\ 6z &= 0 \quad -3\text{Eq.}2 + \text{Eq.}3 \end{aligned}$$

$$\begin{aligned} 6z = 0 &\Rightarrow z = 0 \\ -3y - 0 = 0 &\Rightarrow y = 0 \\ 2x + 3(0) = 0 &\Rightarrow x = 0 \end{aligned}$$

Answer: (0, 0, 0)

33. $\begin{aligned} 23x + 4y - z &= 0 \quad \text{Interchange equations.} \\ 12x + 5y + z &= 0 \end{aligned}$

$$\begin{aligned} x + 6y + 3z &= 0 \quad 2\text{Eq.}2 - \text{Eq.}1 \\ -67y - 35z &= 0 \quad -12\text{Eq.}1 + \text{Eq.}2 \end{aligned}$$

To avoid fractions, let $z = 67a$, then:

$$\begin{aligned} -67y - 35(67a) &= 0 \\ y &= -35a \\ x + 6(-35a) + 3(67a) &= 0 \\ x &= 9a \end{aligned}$$

Answer: $(9a, -35a, 67a)$

35. No, they are not equivalent. The constant in the second equation should be -11 and the coefficient of z in the third equation should be 2.

37. There are an infinite number of linear systems that have $(4, -1, 2)$ as their solution. One such system is as follows:

$$3(4) + (-1) - (2) = 9 \Longrightarrow 3x + y - z = 9$$
$$(4) + 2(-1) - (2) = 0 \Longrightarrow x + 2y - z = 0$$
$$-(4) + (-1) + 3(2) = 1 \Longrightarrow -x + y + 3z = 1$$

39. $y = ax^2 + bx + c$ passing through $(0, 0), (2, -2), (4, 0)$

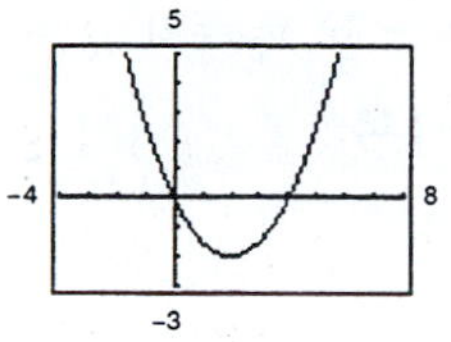

$(0, 0)$: $0 = c$

$(2, -2)$: $-2 = 4a + 2b + c \Rightarrow -1 = 2a + b$

$(4, 0)$: $0 = 16a + 4b + c \Rightarrow 0 = 4a + b$

Answer: $a = \frac{1}{2}, b = -2, c = 0$

The equation of the parabola is $y = \frac{1}{2}x^2 - 2x$.

41. $y = ax^2 + bx + c$ passing through $(2, 0), (3, -1), (4, 0)$

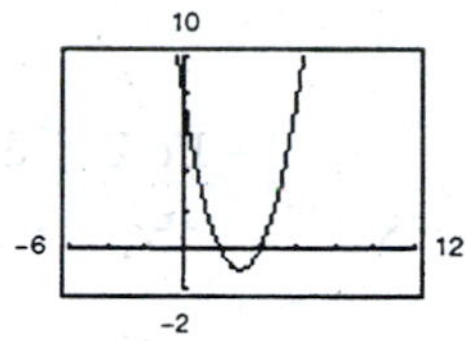

$(2, 0)$: $0 = 4a + 2b + c \Rightarrow 0 = -4a - 2b - c$

$(3, -1)$: $-1 = 9a + 3b + c \Rightarrow -1 = 5a + b$

$(4, 0)$: $0 = 16a + 4b + c \Rightarrow 0 = 12a + 2b$

Answer: $a = 1, b = -6, c = 8$

The equation of the parabola is $y = x^2 - 6x + 8$.

43. $x^2 + y^2 + Dx + Ey + F = 0$ passing through $(0, 0), (2, 2), (4, 0)$

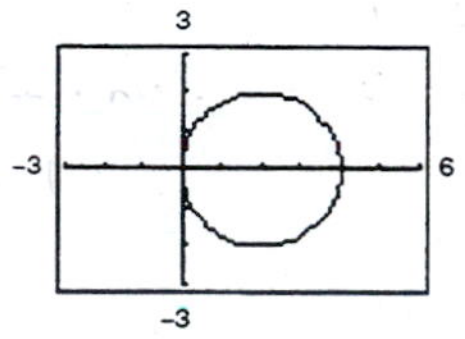

$(0, 0)$: $F = 0$

$(2, 2)$: $8 + 2D + 2E + F = 0 \Rightarrow D + E = -4$

$(4, 0)$: $16 + 4D + F = 0 \Rightarrow D = -4$ and $E = 0$

The equation of the circle is $x^2 + y^2 - 4x = 0$.

To graph, let $y_1 = \sqrt{4x - x^2}$ and $y_2 = -\sqrt{4x - x^2}$.

45. $x^2 + y^2 + Dx + Ey + F = 0$ passing through $(-3, -1), (2, 4), (-6, 8)$

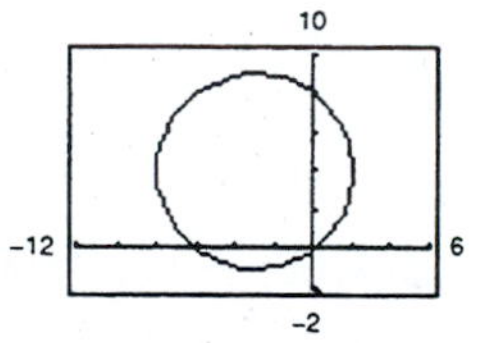

$(-3, -1)$: $10 - 3D - E + F = 0 \Rightarrow 10 = 3D + E - F$

$(2, 4)$: $20 + 2D + 4E + F = 0 \Rightarrow 20 = -2D - 4E - F$

$(-6, 8)$: $100 - 6D + 8E + F = 0 \Rightarrow 100 = 6D - 8E - F$

Answer: $D = 6, E = -8, F = 0$

The equation of the circle is $x^2 + y^2 + 6x - 8y = 0$.
To graph, complete the squares first, then solve for y.

$$(x^2 + 6x + 9) + (y^2 - 8y + 16) = 0 + 9 + 16$$
$$(x + 3)^2 + (y - 4)^2 = 25$$
$$(y - 4)^2 = 25 - (x + 3)^2$$
$$y - 4 = \pm\sqrt{25 - (x + 3)^2}$$
$$y = 4 \pm \sqrt{25 - (x + 3)^2}$$

Let $y_1 = 4 + \sqrt{25 - (x + 3)^2}$ and $y_2 = 4 - \sqrt{25 - (x + 3)^2}$.

47. $s = \frac{1}{2}at^2 + v_0t + s_0$

(1, 128), (2, 80), (3, 0)

$$128 = \tfrac{1}{2}a + v_0 + s_0 \Rightarrow a + 2v_0 + 2s_0 = 256$$
$$80 = 2a + 2v_0 + s_0 \Rightarrow 2a + 2v_0 + s_0 = 80$$
$$0 = \tfrac{9}{2}a + 3v_0 + s_0 \Rightarrow 9a + 6v_0 + 2s_0 = 0$$

Solving this system yields $a = -32$, $v_0 = 0$, $s_0 = 144$.

Thus, $s = \frac{1}{2}(-32)t^2 + (0)t + 144$

$= -16t^2 + 144.$

49. $s = \frac{1}{2}at^2 + v_0t + s_0$

(1, 452), (2, 372), (3, 260)

$$452 = \tfrac{1}{2}a + v_0 + s_0 \Rightarrow a + 2v_0 + 2s_0 = 904$$
$$372 = 2a + 2v_0 + s_0 \Rightarrow 2a + 2v_0 + s_0 = 372$$
$$260 = \tfrac{9}{2}a + 3v_0 + s_0 \Rightarrow 9a + 6v_0 + 2s_0 = 520$$

Solving this system yields $a = -32$, $v_0 = -32$, $s_0 = 500$.

Thus, $s = \frac{1}{2}(-32)t^2 + (-32)t + 500$

$= -16t^2 - 32t + 500.$

51. Let x = amount at 5%

Let y = amount at 6%

Let z = amount at 7%

$$x + y + z = 16{,}000$$
$$0.05x + 0.06y + 0.07z = 990$$
$$x + 3000 = z$$
$$y + 2000 = z$$

$$(z - 3000) + (z - 2000) + z = 16{,}000$$
$$3z = 21{,}000$$
$$z = 7000$$

$x = 4000$, $y = 5000$

Check: $0.05(4000) + 0.06(5000) + 0.07(7000) = 990$

Answer: $x = \$4000$ at 5%

$y = \$5000$ at 6%

$z = \$7000$ at 7%

53. Let x = amount at 8%
Let y = amount at 9%
Let z = amount at 10%

$$\begin{aligned} x + y + z &= 775{,}000 \\ 0.08x + 0.09y + 0.10z &= 67{,}500 \\ x &= 4z \end{aligned}$$

$$\begin{aligned} y + 5z &= 775{,}000 \\ 0.09y + 0.42z &= 67{,}500 \end{aligned}$$

$z = 75{,}000$

$y = 775{,}000 - 5z = 400{,}000$

$x = 4z = 300{,}000$

Answer: x = \$300,000 at 8%
y = \$400,000 at 9%
z = \$75,000 at 10%

55. Let C = amount in certificates of deposit
Let M = amount in municipal bonds
Let B = amount in blue-chip stocks
Let G = amount in growth or speculative stocks

$$\begin{aligned} C + M + B + G &= 500{,}000 \\ 0.10C + 0.08M + 0.12B + 0.13G &= 0.10(500{,}000) \\ B + G &= \tfrac{1}{4}(500{,}000) \end{aligned}$$

This system has infinitely many solutions.

Let $G = s$, then $B = 125{,}000 - s$

$M = 125{,}000 + \frac{1}{2}s$

$C = 250{,}000 - \frac{1}{2}s$

Answer:
$(250{,}000 - \frac{1}{2}s, 125{,}000 + \frac{1}{2}s, 125{,}000 - s, s)$,
where $0 \le s \le 125{,}000$

One possible solution is to let $s = 50{,}000$.
Certificates of deposit: \$225,000
Municipal bonds: \$150,000
Blue-chip stocks: \$75,000
Growth or speculative stocks: \$50,000

57. Let x = gallons of spray X
Let y = gallons of spray Y
Let z = gallons of spray Z

$$\left.\begin{aligned} \text{Chemical A: } \tfrac{1}{5}x + \tfrac{1}{2}z &= 12 \\ \text{Chemical B: } \tfrac{2}{5}x + \tfrac{1}{2}z &= 16 \end{aligned}\right\} \Rightarrow x = 20, z = 16$$

$$\text{Chemical C: } \tfrac{2}{5}x + y = 26 \quad \Rightarrow \quad y = 18$$

Answer: 20 liters of spray X
18 liters of spray Y
16 liters of spray Z

59.

	Product	
Truck	A	B
Large	6	3
Medium	4	4
Small	0	3

Possible solutions:
(1) 4 medium trucks
(2) 2 large trucks, 1 medium truck, 2 small trucks
(3) 3 large trucks, 1 medium truck, 1 small truck
(4) 3 large trucks, 3 small trucks

61. $t_1 - 2t_2 = 0$

$t_1 - 2a = 128 \Rightarrow 2t_2 - 2a = 128$

$t_2 + a = 32 \Rightarrow -2t_2 - 2a = -64$

$$\begin{aligned} -4a &= 64 \\ a &= -16 \\ t_2 &= 48 \\ t_1 &= 96 \end{aligned}$$

Answer: $t_1 = 96$ lb
$t_2 = 48$ lb
$a = -16$ ft/sec^2

63. $\dfrac{1}{x^3 - x} = \dfrac{A}{x} + \dfrac{B}{x - 1} + \dfrac{C}{x + 1}$

$$1 = A(x + 1)(x - 1) + Bx(x + 1) + Cx(x - 1)$$

$$1 = Ax^2 - A + B^2 + Bx + Cx^2 - Cx$$

$$1 = (A + B + C)x^2 + (B - C)x - A$$

By equating coefficients, we have:

$$0 = A + B + C$$

$$0 = B - C$$

$$1 = -A \quad \Rightarrow \quad A = -1$$

$$B + C = 1$$

$$B - C = 0$$

$$2B = 1 \Rightarrow B = \frac{1}{2}$$

$$C = \frac{1}{2}$$

$$\frac{A}{x} + \frac{B}{x - 1} + \frac{C}{x + 1} = \frac{-1}{x} + \frac{1/2}{x - 1} + \frac{1/2}{x + 1} = \frac{1}{2}\left(-\frac{2}{x} + \frac{1}{x - 1} + \frac{1}{x + 1}\right)$$

65. $\dfrac{x^2 - 3x - 3}{x(x - 2)(x + 3)} = \dfrac{A}{x} + \dfrac{B}{x - 2} + \dfrac{C}{x + B}$

$$x^2 - 3x - 3 = A(x - 2)(x + 3) + Bx(x + 3) + Cx(x - 2)$$

$$x^2 - 3x - 3 = Ax^2 + Ax - 6A + Bx^2 + 3Bx + Cx^2 - 2Cx$$

$$x^2 - 3x - 3 = (A + B + C)x^2 + (A + 3B - 2C)x - 6A$$

By equating coefficients, we have:

$$1 = A + B + C$$

$$-3 = A + 3B - 2C$$

$$-3 = -6A \quad \Rightarrow \quad A = \tfrac{1}{2}$$

$$\tfrac{1}{2} = B + C \Rightarrow 1 = 2B + 2C$$

$$-\tfrac{7}{2} = 3B - 2C \Rightarrow -\tfrac{7}{2} = 3B - 2C$$

$$-\tfrac{5}{2} = 5B$$

$$B = -\tfrac{1}{2}$$

$$C = 1$$

$$\frac{x^2 - 3x - 3}{x(x - 2)(x + 3)} = \frac{1/2}{x} - \frac{1/2}{x - 2} + \frac{1}{x + 3} = \frac{1}{2}\left(\frac{1}{x} - \frac{1}{x - 2} + \frac{2}{x + 3}\right)$$

67. Least squares regression parabola through $(-4, 5), (-2, 6), (2, 6), (4, 2)$

$n = 4$

$\sum x_i = 0$ $\quad\sum y_i = 19$

$\sum x_i^2 = 40$ $\quad\sum x_i^3 = 0$

$\sum x_i^4 = 544$ $\quad\sum x_i y_i = -12$

$\sum x_i^2 y_i = 160$

$$\begin{aligned} 4c \quad + 40a &= 19 \\ 40b \quad &= -12 \\ 40c \quad + 544a &= 160 \end{aligned}$$

Solving this system yields $a = -\frac{5}{24}$, $b = -\frac{3}{10}$, and $c = \frac{41}{6}$. Thus, $y = -\frac{5}{24}x^2 - \frac{3}{10}x + \frac{41}{6}$.

69. Least squares regression parabola through $(0, 0), (2, 2), (3, 6), (4, 12)$

$n = 4$

$\sum x_i = 9$ $\quad\sum y_i = 20$

$\sum x_i^2 = 29$ $\quad\sum x_i^3 = 99$

$\sum x_i^4 = 353$ $\quad\sum x_i y_i = 70$

$\sum x_i^2 y_i = 254$

$$\begin{aligned} 4c + 9b + 29a &= 20 \\ 9c + 29b + 99a &= 70 \\ 29c + 99b + 353a &= 254 \end{aligned}$$

Solving this system yields $a = 1$, $b = -1$, and $c = 0$. Thus, $y = x^2 - x$.

71. (a) Least squares regression parabola through $(20, 25), (30, 55), (40, 105), (50, 188), (60, 300)$

$n = 5$

$\sum x_i = 200$ $\quad\sum y_i = 673$

$\sum x_i^2 = 9000$ $\quad\sum x_i^3 = 440{,}000$

$\sum x_i^4 = 22{,}740{,}000$ $\quad\sum x_i y_i = 33{,}750$

$\sum x_i^2 y_i = 1{,}777{,}500$

$$\begin{aligned} 5c + 200b + 9000a &= 673 \\ 200c + 9000b + 440{,}000a &= 33{,}750 \\ 9000c + 440{,}000b + 22{,}740{,}000a &= 1{,}777{,}500 \end{aligned}$$

Solving this system yields $a \approx 0.14$, $b \approx -4.43$, and $c \approx 58.40$.
Thus, $y = 0.14x^2 - 4.43x + 58.40$.

(b)

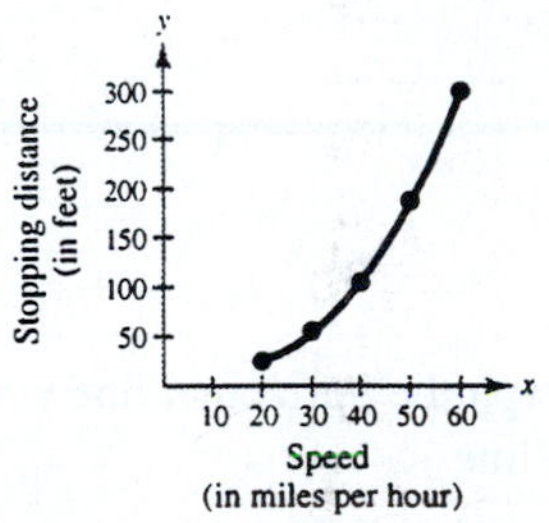

(c) When $x = 70$, $y \approx 434.3$ feet.

73. $\left.\begin{aligned} y + \quad \lambda &= 0 \\ x + \lambda &= 0 \end{aligned}\right\} \Rightarrow x = y = -\lambda$

$x + y - 10 = 0 \quad \Rightarrow \quad 2x - 10 = 0$

$x = 5$

$y = 5$

$\lambda = -5$

75. $2x - 2x\lambda = 0 \Rightarrow 2x(1 - \lambda) = 0 \Rightarrow \lambda = 1 \text{ or } x = 0$

$-2y + \lambda = 0 \Rightarrow \quad 2y = \lambda \Rightarrow y = \dfrac{1}{2}$

$y - x^2 = 0 \Rightarrow \quad x^2 = y \Rightarrow x = \pm\sqrt{\dfrac{1}{2}} = \pm\dfrac{\sqrt{2}}{2}$

Answer: $x = \pm\dfrac{\sqrt{2}}{2}$ or $x = 0$

$y = \dfrac{1}{2}$ or $y = 0$

$\lambda = 1$ or $\lambda = 0$

77. The slope represents the average increase in sales per year.

79. $(0.075)(85) = 6.375$

81. $(0.005)(n) = 400$

$n = 80{,}000$

Section 7.4 Systems of Inequalities

- You should be able to sketch the graph of an inequality in two variables.
  - (a) Replace the inequality with an equal sign and graph the equation. Use a dashed line for < or >, a solid line for ≤ or ≥.
  - (b) Test a point in each region formed by the graph. If the point satisfies the inequality, shade the whole region.

Solutions to Odd-Numbered Exercises

1. $x \geq 2$

Using a solid line, graph the vertical line $x = 2$ and shade to the right of this line.

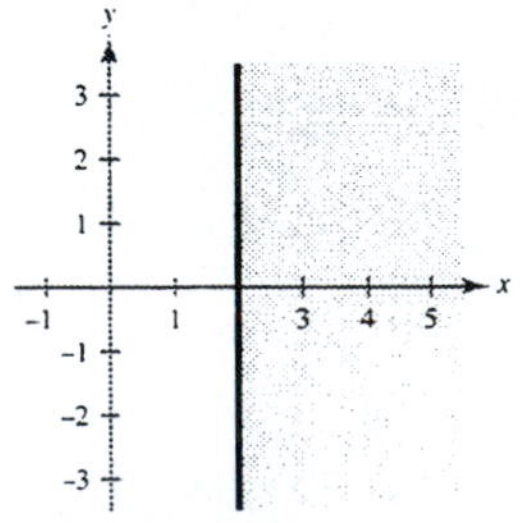

3. $y \geq -1$

Using a solid line, graph the horizontal line $y = -1$ and shade above this line.

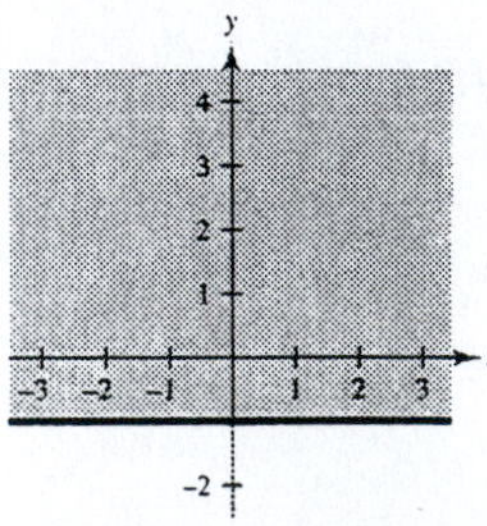

5. $y < 2 - x$

Using a dashed line, graph $y = 2 - x$, and then shade below the line. (Use (0, 0) as a test point.)

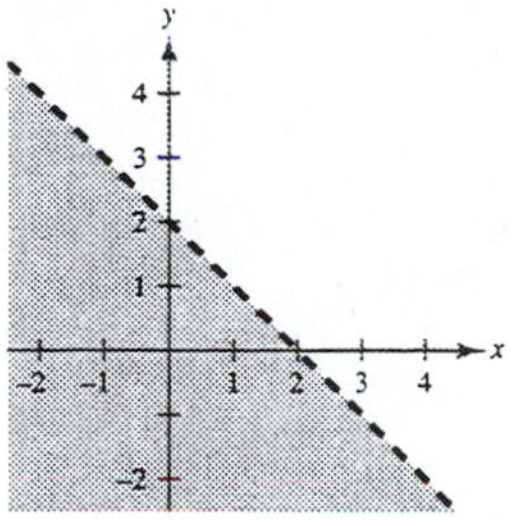

7. $2y - x \geq 4$

Using a solid line, graph $2y - x = 4$, and then shade above the line. (Use (0, 0) as a test point.)

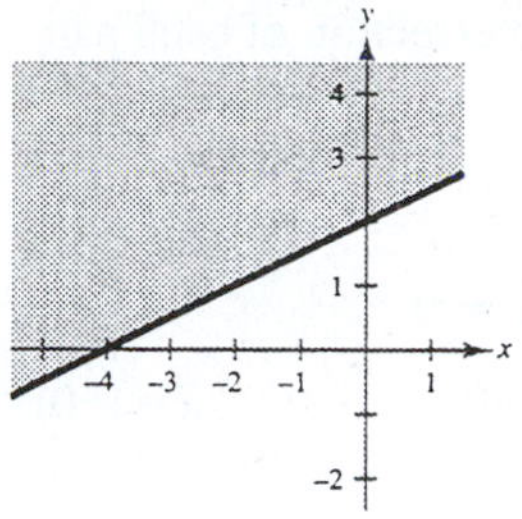

9. $(x + 1)^2 + (y - 2)^2 < 9$

Using a dashed line, sketch the circle $(x + 1)^2 + (y - 2)^2 = 9$.

Center: $(-1, 2)$

Radius: 3

Test point: (0, 0). Shade the inside of the circle.

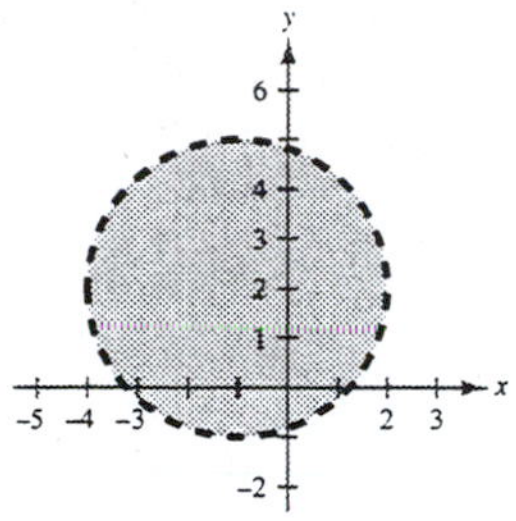

11. $y \leq \dfrac{1}{1 + x^2}$

Using a solid line, graph $y = \dfrac{1}{1 + x^2}$, and then shade below the curve. (Use (0, 0) as a test point.)

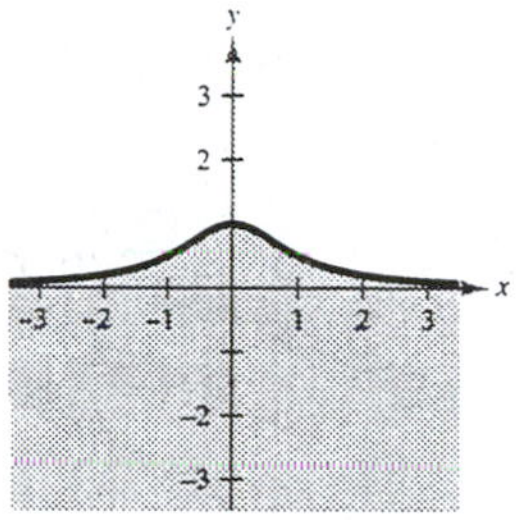

13. $y \geq \dfrac{2}{3}x - 1$

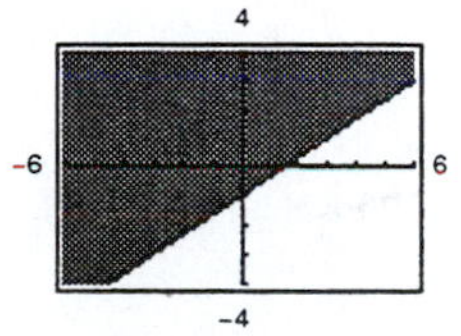

15. $x^2 + 5y - 10 \leq 0$

$$y \leq 2 - \frac{x^2}{5}$$

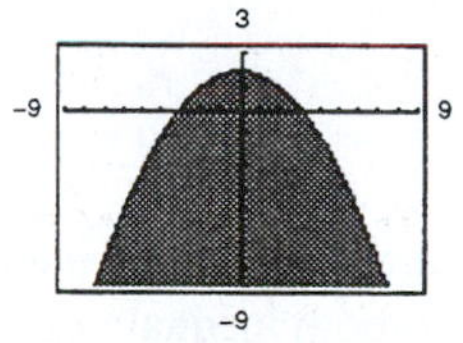

17. The line through $(-4, 0)$ and $(0, 2)$ is $y = \frac{1}{2}x + 2$. For the shaded region below the line, we have $y \leq \frac{1}{2}x + 2$.

19. The line through (0, 2) and (3, 0) is $y = -\frac{2}{3}x + 2$. For the shaded region above the line, we have

$$y \geq -\frac{2}{3}x + 2$$

$$3y \geq -2x + 6$$

$$2x + 3y \geq 6$$

$$\frac{x}{3} + \frac{y}{2} \geq 1$$

21. $x + y \le 1$

$-x + y \le 1$

$y \ge 0$

First, find the points of intersection of each pair of equations.

Vertex A	Vertex B	Vertex C
$x + y = 1$	$x + y = 1$	$-x + y = 1$
$-x + y = 1$	$y = 0$	$y = 0$
$(0, 1)$	$(1, 0)$	$(-1, 0)$

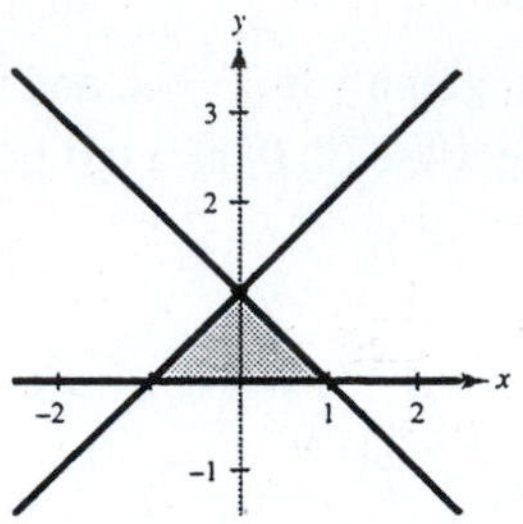

23. $x + y \le 5$

$x \ge 2$

$y \ge 0$

First, find the points of intersection of each pair of equations.

Vertex A	Vertex B	Vertex C
$x + y = 5$	$x + y = 5$	$x = 2$
$x = 2$	$y = 0$	$y = 0$
$(2, 3)$	$(5, 0)$	$(2, 0)$

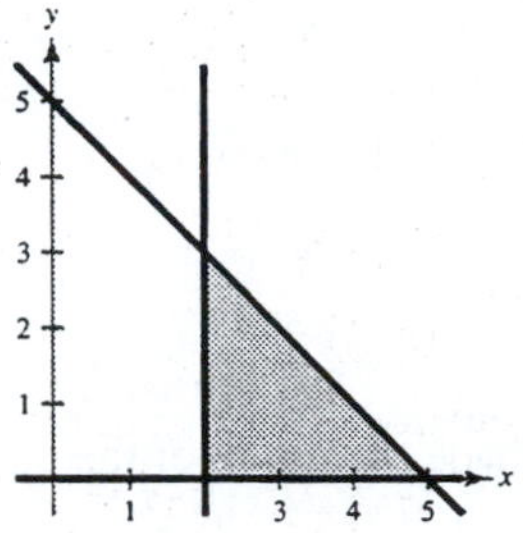

25. $-3x + 2y < 6$

$x - 4y > -2$

$2x + y < 3$

First, find the points of intersection of each pair of equations.

Vertex *A*	Vertex *B*	Vertex *C*
$-3x + 2y = 6$	$-3x + 2y = 6$	$x - 4y = -2$
$x - 4y = -2$	$2x + y = 3$	$2x + y = 3$
$(-2, 0)$	$(0, 3)$	$\left(\frac{10}{9}, \frac{7}{9}\right)$

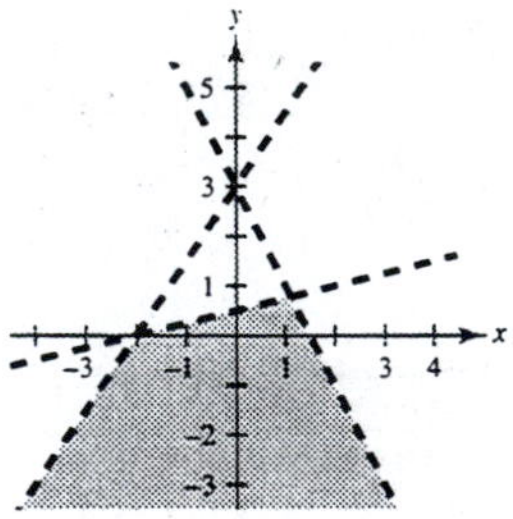

27. $2x + y > 2$

$6x + 3y < 2$

The lines are parallel. There are no points of intersection. There is no region common to both inequalities.

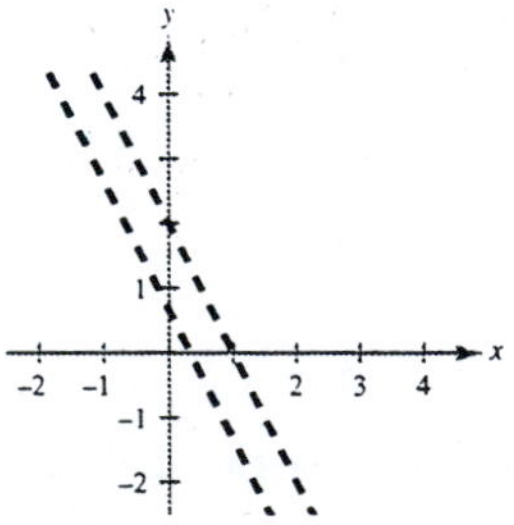

29. $x \geq 1$

$x - 2y \leq 3$

$3x + 2y \geq 9$

$x + y \leq 6$

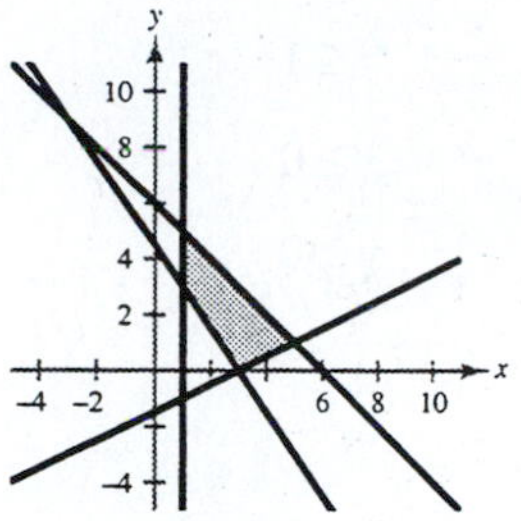

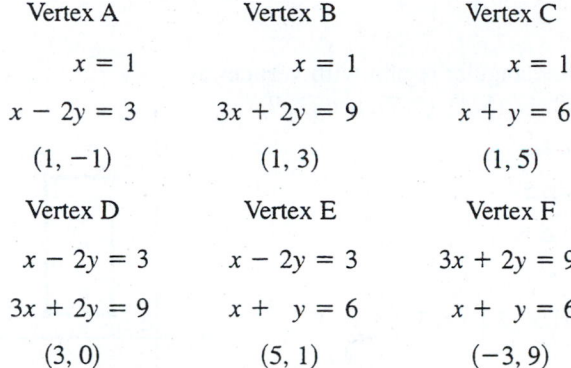

First, find the points of intersection of each pair of equations.

Vertex A	Vertex B	Vertex C
$x = 1$	$x = 1$	$x = 1$
$x - 2y = 3$	$3x + 2y = 9$	$x + y = 6$
$(1, -1)$	$(1, 3)$	$(1, 5)$

Vertex D	Vertex E	Vertex F
$x - 2y = 3$	$x - 2y = 3$	$3x + 2y = 9$
$3x + 2y = 9$	$x + y = 6$	$x + y = 6$
$(3, 0)$	$(5, 1)$	$(-3, 9)$

By shading each inequality, we find that the vertices of the region are $(1, 5)$, $(1, 3)$, $(3, 0)$, and $(5, 1)$.

31. $x^2 + y^2 \leq 9$

$x^2 + y^2 \geq 1$

There are no points of intersection. The region common to both inequalities is the region between the circles.

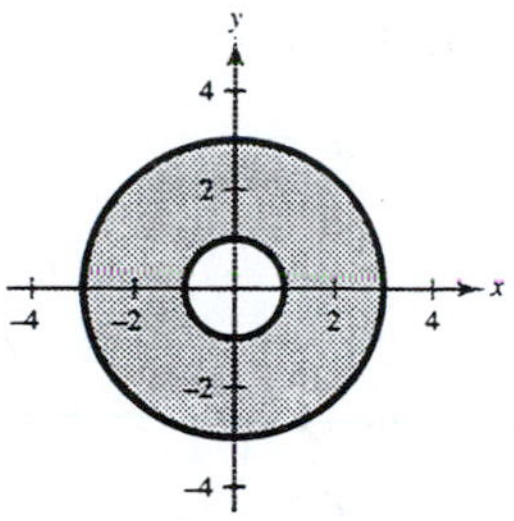

33. $x > y^2$

$x < y + 2$

Points of intersection:

$$y^2 = y + 2$$

$$y^2 - y - 2 = 0$$

$$(y + 1)(y - 2) = 0$$

$$y = -1, 2$$

$(1, -1)$, $(4, 2)$

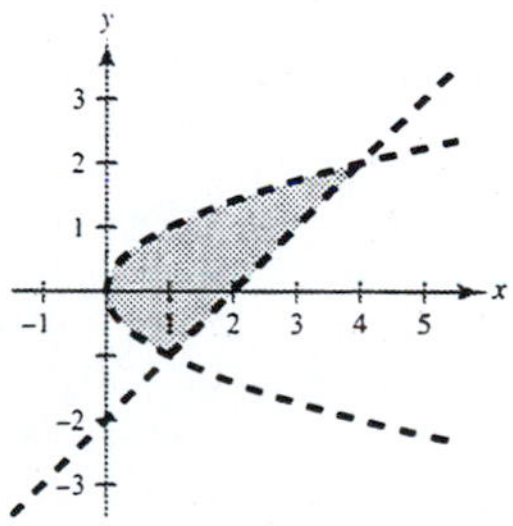

35. $y \leq \sqrt{3x} + 1$

$y \geq x^2 + 1$

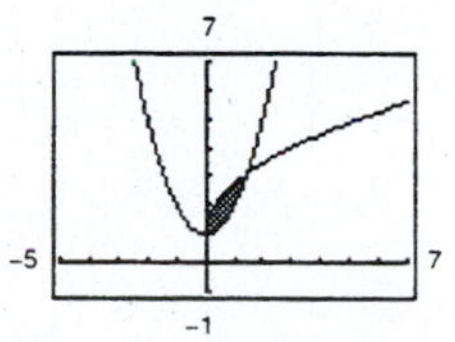

37. $y < x^3 - 2x + 1$

$y > -2x$

$x \leq 1$

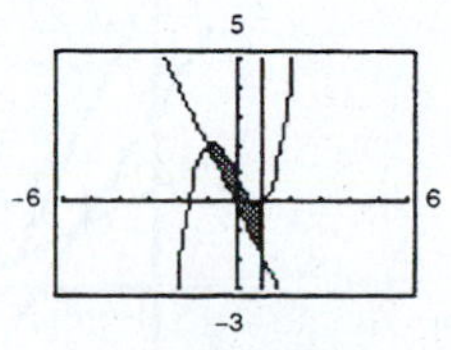

39. $x^2y \geq 1$

$0 < x \leq 4$

$y \leq 4$

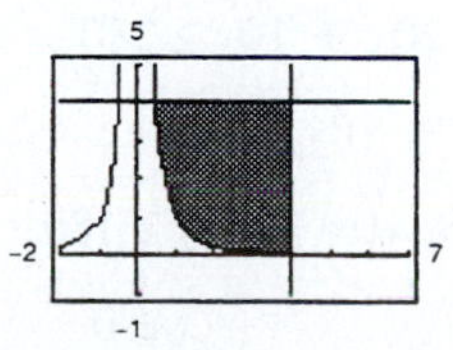

41. $y \le -x + 4 \Rightarrow \frac{x}{4} + \frac{y}{4} \le 1$

$x \ge 0 \qquad x \ge 0$

$y \ge 0 \qquad y \ge 0$

43. Line through points (0, 4) and (4, 0): $y = 4 - x$

Line through points (0, 2) and (8, 0): $y = 2 - \frac{1}{4}x$

$y \ge 4 - x$

$y \ge 2 - \frac{1}{4}x$

$x \ge 0$

$y \ge 0$

45. $x^2 + y^2 \le 16$

$x \ge 0$

$y \ge 0$

47. Rectangular region with vertices at (2, 1), (5, 1), (5, 7), and (2, 7)

$x \ge 2$

$x \le 5$

$y \ge 1$

$y \le 7$

Thus, $2 \le x \le 5$, $1 \le y \le 7$.

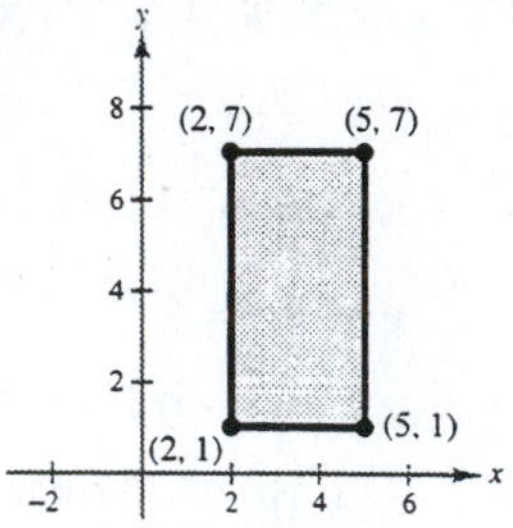

49. Triangle with vertices at (0, 0), (5, 0), (2, 3)

(0, 0), (5, 0) Line: $y \ge 0$

(0, 0), (2, 3) Line: $y \le \frac{3}{2}x$

(2, 3), (5, 0) Line: $y \le -x + 5$

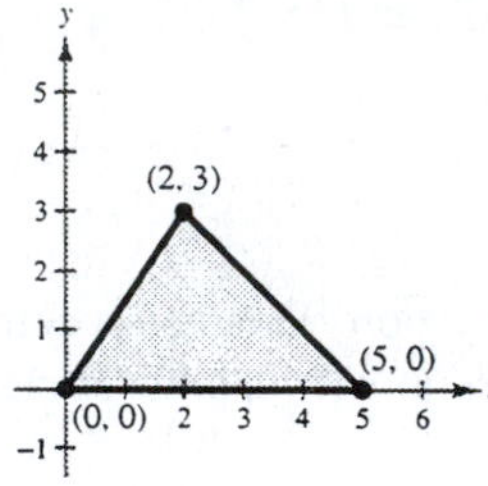

51. Assembly center constraint: $x + \frac{3}{2}y \le 12$

Finishing center constraint: $\frac{4}{3}x + \frac{3}{2}y \le 15$

Point of intersection: (9, 2)

Physical constraints: $x \ge 0$ and $y \ge 0$

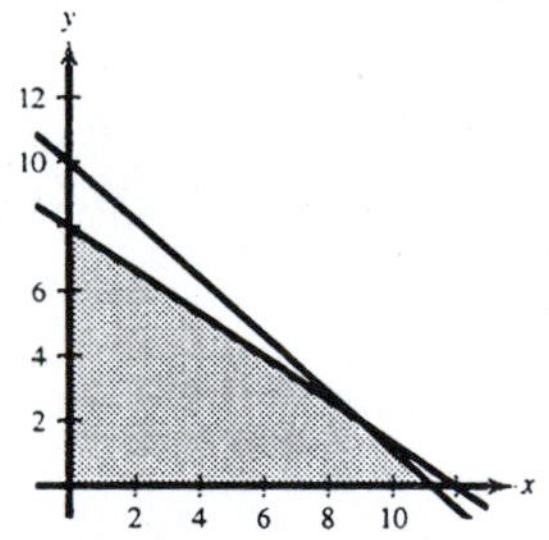

53. Account constraints:

$x \ge 5000$

$y \ge 5000$

$2x \le y$

$x + y \le 20{,}000$

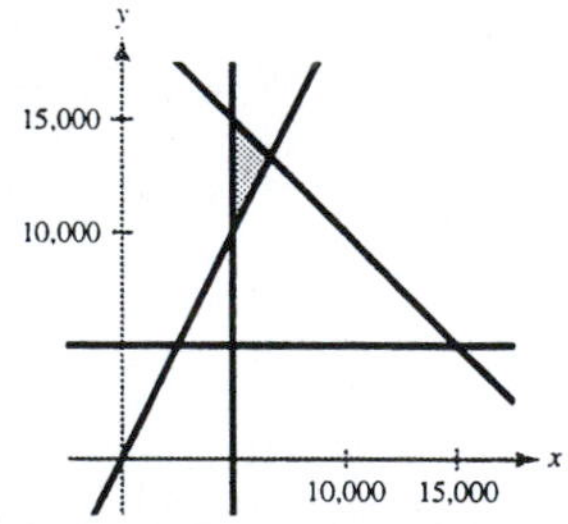

55. x = number of ounces of food X

y = number of ounces of food Y

Calcium: $20x + 10y \ge 280$

Iron: $15x + 10y \ge 160$

Vitamin B: $10x + 20y \ge 180$

$x \ge 0$

$y \ge 0$

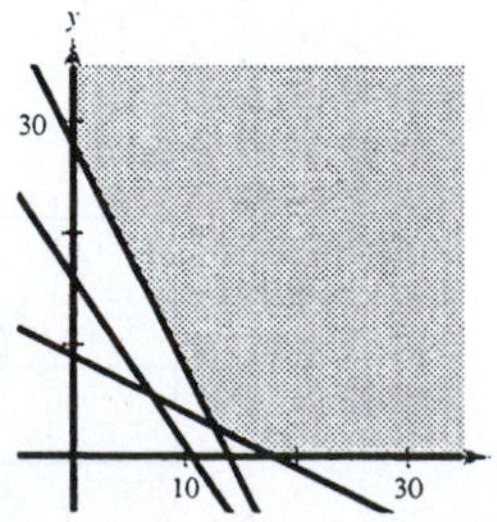

57. $xy \geq 500$ Body-building space

$2x + \pi y \geq 125$ Track (Two semi-circles and two lengths)

$x \geq 0$ Physical constraint

$y \geq 0$ Physical constraint

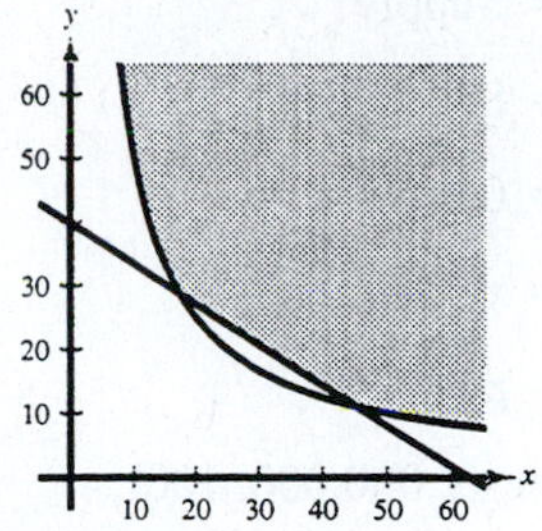

59. Demand = Supply

$50 - 0.5x = 0.125x$

$50 = 0.625x$

$80 = x$

$10 = p$

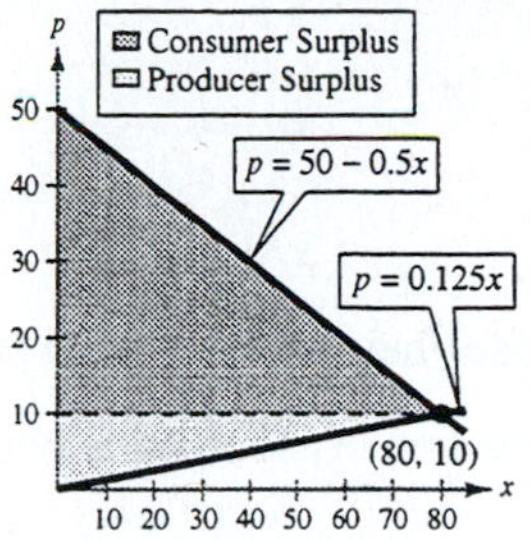

Point of equilibrium: (80, 10)

The consumer surplus is the area of the triangle bounded by

$p \leq 50 - 0.5x$

$p \geq 10$

$x \geq 0.$

Consumer surplus $= \frac{1}{2}(\text{base})(\text{height}) = \frac{1}{2}(80)(40) = \$1600$

The producer surplus is the area of the triangle bounded by

$p \geq 0.125x$

$p \leq 10$

$x \geq 0.$

Producer surplus $= \frac{1}{2}(\text{base})(\text{height}) = \frac{1}{2}(80)(10) = \$400$

61. $$\text{Demand} = \text{Supply}$$

$$140 - 0.00002x = 80 + 0.00001x$$

$$60 = 0.00003x$$

$$2{,}000{,}000 = x$$

$$100 = p$$

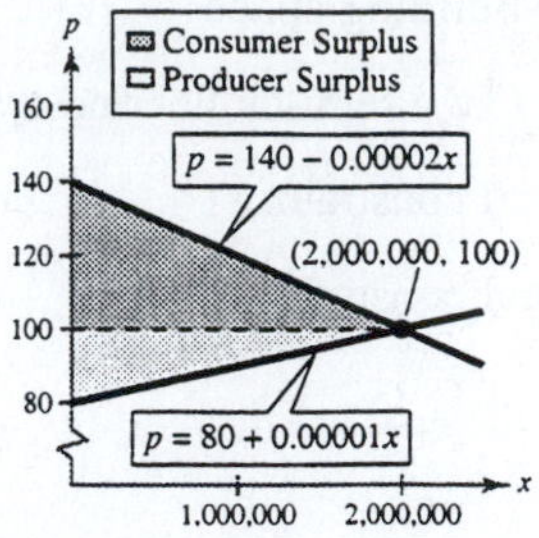

Point of equilibrium: (2,000,000, 100)

The consumer surplus is the area of the triangle bounded by

$p \le 140 - 0.00002x$

$p \ge 100$

$x \ge 0.$

Consumer surplus $= \frac{1}{2}(\text{base})(\text{height}) = \frac{1}{2}(2{,}000{,}000)(40) =$ \$40,000,000 or \$40 million

The producer surplus is the area of the triangle bounded by

$p \ge 80 + 0.00001x$

$p \le 100$

$x \ge 0.$

Producer surplus $= \frac{1}{2}(\text{base})(\text{height}) = \frac{1}{2}(2{,}000{,}000)(20) =$ \$20,000,000 or \$20 million

63. Test a point on either side of the boundary.

Section 7.5 Linear Programming

- To solve a linear programming problem:
  1. Sketch the solution set for the system of constraints.
  2. Find the vertices of the region.
  3. Test the objective function at each of the vertices.

Solutions to Odd-Numbered Exercises

1. $z = 4x + 5y$

At (0, 6): $z = 4(0) + 5(6) = 30$

At (0, 0): $z = 4(0) + 5(0) = 0$

At (6, 0): $z = 4(6) + 5(0) = 24$

The minimum value is 0 at (0, 0).
The maximum value is 30 at (0, 6).

3. $z = 10x + 6y$

At (0, 6): $z = 10(0) + 6(6) = 36$

At (0, 0): $z = 10(0) + 6(0) = 0$

At (6, 0): $z = 10(6) + 6(0) = 60$

The minimum value is 0 at (0, 0).
The maximum value is 60 at (6, 0).

5. $z = 3x + 2y$

At (0, 5): $z = 3(0) + 2(5) = 10$

At (4, 0): $z = 3(4) + 2(0) = 12$

At (3, 4): $z = 3(3) + 2(4) = 17$

At (0, 0): $z = 3(0) + 2(0) = 0$

The minimum value is 0 at (0, 0).
The maximum value is 17 at (3, 4).

7. $z = 5x + 0.5y$

At (0, 5): $z = 5(0) + \frac{5}{2} = \frac{5}{2}$

At (4, 0): $z = 5(4) + \frac{0}{2} = 20$

At (3, 4): $z = 5(3) + \frac{4}{2} = 17$

At (0, 0): $z = 5(0) + \frac{0}{2} = 0$

The minimum value is 0 at (0, 0).
The maximum value is 20 at (4, 0).

9. $z = 10x + 7y$

At (0, 45): $z = 10(0) + 7(45) = 315$

At (30, 45): $z = 10(30) + 7(45) = 615$

At (60, 20): $z = 10(60) + 7(20) = 740$

At (60, 0): $z = 10(60) + 7(0) = 600$

At (0, 0): $z = 10(0) + 7(0) = 0$

The minimum value is 0 at (0, 0).
The maximum value is 740 at (60, 20).

11. $z = 25x + 30y$

At (0, 45): $z = 25(0) + 30(45) = 1350$

At (30, 45): $z = 25(30) + 30(45) = 2100$

At (60, 20): $z = 25(60) + 30(20) = 2100$

At (60, 0): $z = 25(60) + 30(0) = 1500$

At (0, 0): $z = 25(0) + 30(0) = 0$

The minimum value is 0 at (0, 0).
The maximum value is 2100 at any point along the line segment connecting (30, 45) and (60, 20).

13. $z = 6x + 10y$

At (0, 2): $z = 6(0) + 10(2) = 20$

At (5, 0): $z = 6(5) + 10(0) = 30$

At (0, 0): $z = 6(0) + 10(0) = 0$

The minimum value is 0 at (0, 0).
The maximum value is 30 at (5, 0).

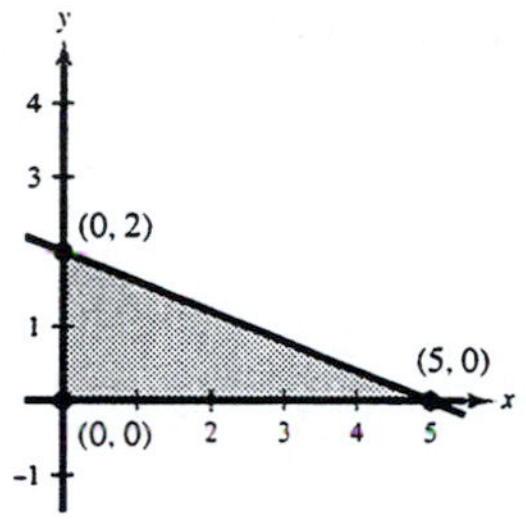

15. $z = 9z + 24y$

At (0, 2): $z = 9(0) + 24(2) = 48$

At (5, 0): $z = 9(5) + 24(0) = 45$

At (0, 0): $z = 9(0) + 24(0) = 0$

The minimum value is 0 at (0, 0).
The maximum value is 48 at (0, 2).

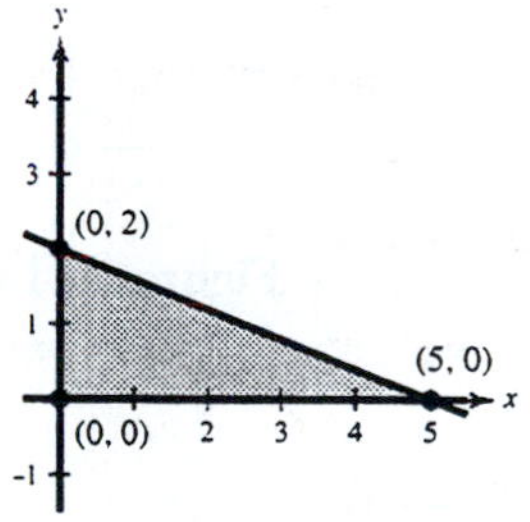

17. $z = 4x + 5y$

At (10, 0): $z = 4(10) + 5(0) = 40$

At (5, 3): $z = 4(5) + 5(3) = 35$

At (0, 8): $z = 4(0) + 5(8) = 40$

The minimum value is 35 at (5, 3).
C is unbounded. Therefore, there is no maximum.

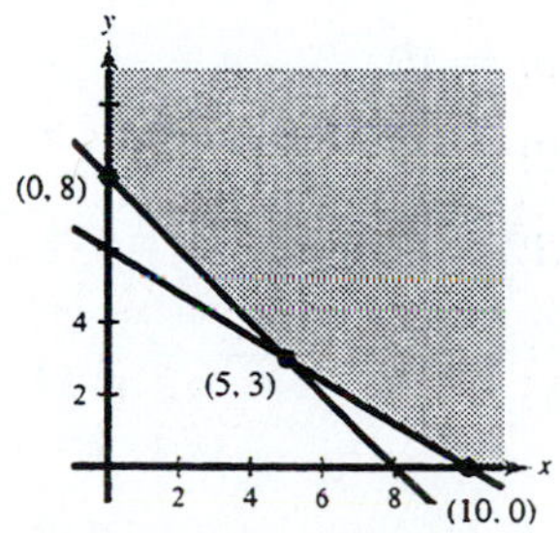

19. $z = 2x + 7y$

At $(10, 0)$: $z = 2(10) + 7(0) = 20$

At $(5, 3)$: $z = 2(5) + 7(3) = 31$

At $(0, 8)$: $z = 2(0) + 7(8) = 56$

The minimum value is 20 at $(10, 0)$.
C is unbounded. Therefore, there is no maximum.

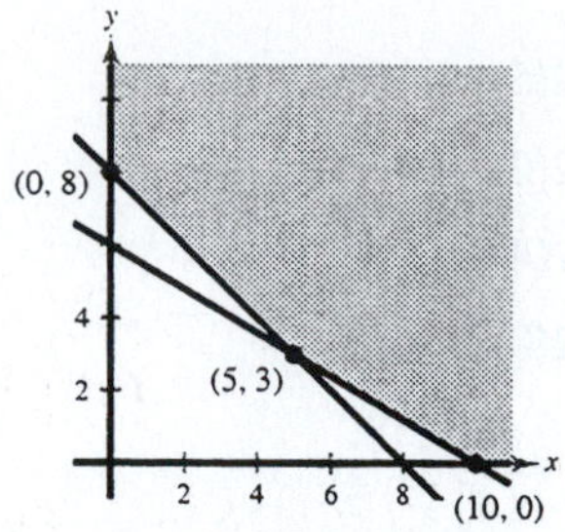

21. $z = 4x + y$

At $(36, 0)$: $z = 4(36) + 0 = 144$

At $(40, 0)$: $z = 4(40) + 0 = 160$

At $(24, 8)$: $z = 4(24) + 8 = 104$

The minimum value is 104 at $(24, 8)$.
The maximum value is 160 at $(40, 0)$.

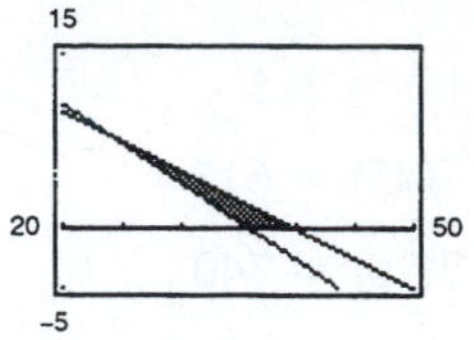

23. $z = x + 4y$

At $(36, 0)$: $z = 36 + 4(0) = 36$

At $(40, 0)$: $z = 40 + 4(0) = 40$

At $(24, 8)$: $z = 24 + 4(8) = 56$

The minimum value is 36 at $(36, 0)$.
The maximum value is 56 at $(24, 8)$.

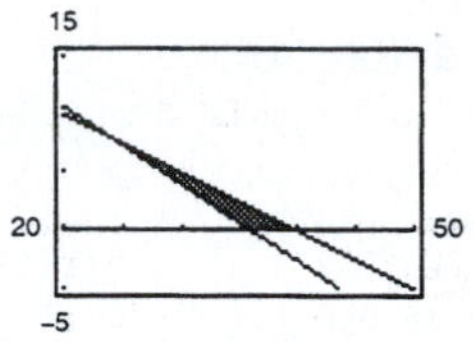

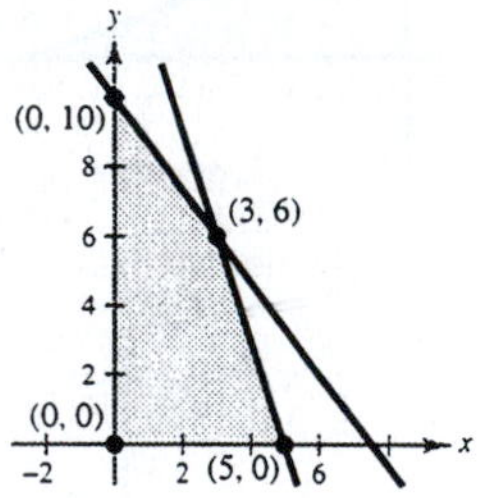

Figure for Exercises 25 and 27

25. $z = 2x + y$

At $(0, 10)$: $z = 2(0) + (10) = 10$

At $(3, 6)$: $z = 2(3) + (6) = 12$

At $(5, 0)$: $z = 2(5) + (0) = 10$

At $(0, 0)$: $z = 2(0) + (0) = 0$

The maximum value is 12 at $(3, 6)$.

27. $z = x + y$

At $(0, 10)$: $z = (0) + (10) = 10$

At $(3, 6)$: $z = (3) + (6) = 9$

At $(5, 0)$: $z = (5) + (0) = 5$

At $(0, 0)$: $z = (0) + (0) = 0$

The maximum value is 10 at $(0, 10)$.

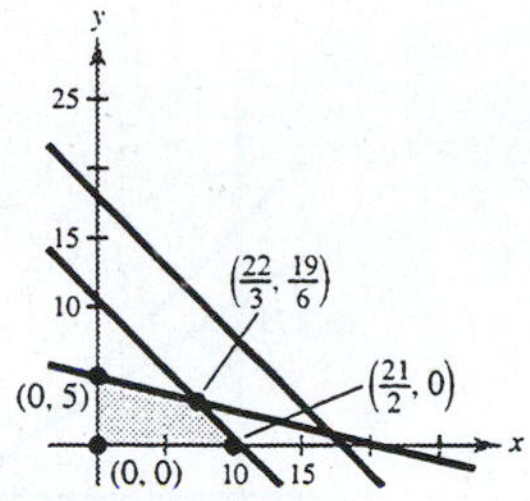

Figure for Exercises 29 and 31

29. $z = x + 5y$

At $(0, 5)$: $z = 0 + 5(5) = 25$

At $\left(\frac{22}{3}, \frac{19}{6}\right)$: $z = \frac{22}{3} + 5\left(\frac{19}{6}\right) = \frac{139}{6}$

At $\left(\frac{21}{2}, 0\right)$: $z = \frac{21}{2} + 5(0) = \frac{21}{2}$

At $(0, 0)$: $z = 0 + 5(0) = 0$

The maximum value is 25 at $(0, 5)$.

31. $z = 4x + 5y$

At $(0, 5)$: $z = 4(0) + 5(5) = 25$

At $\left(\frac{22}{3}, \frac{19}{6}\right)$: $z = 4\left(\frac{22}{3}\right) + 5\left(\frac{19}{6}\right) = \frac{271}{6}$

At $\left(\frac{21}{2}, 0\right)$: $z = 4\left(\frac{21}{2}\right) + 5(0) = 42$

At $(0, 0)$: $z = 4(0) + 5(0) = 0$

The maximum value is $\frac{271}{6}$ at $\left(\frac{22}{3}, \frac{19}{6}\right)$.

33. There are an infinite number of objective functions that would have a maximum at $(0, 4)$. One such objective function is $z = x + 5y$.

35. There are an infinite number of objective functions that would have a maximum at $(5, 0)$. One such objective function is $z = 4x + y$.

37. x = number of Model A

y = number of Model B

Constraints: $2x + 2.5y \leq 4000$

$4x + y \leq 4800$

$x + 0.75y \leq 1500$

$x \geq 0$

$y \geq 0$

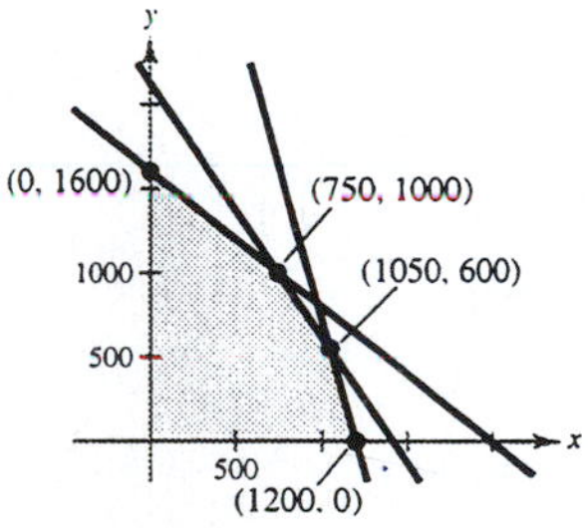

Objective function: $P = 45x + 50y$

Vertices: $(0, 0)$, $(0, 1600)$, $(750, 1000)$, $(1050, 600)$, $(1200, 0)$

At $(0, 0)$: $P = 45(0) + 50(0) = 0$

At $(0, 1600)$: $P = 45(0) + 50(1600) = 80{,}000$

At $(750, 1000)$: $P = 45(750) + 50(1000) = 83{,}750$

At $(1050, 600)$: $P = 45(1050) + 50(600) = 77{,}250$

At $(1200, 0)$: $P = 45(1200) + 50(0) = 54{,}000$

The maximum profit of $83,750 occurs when 750 units of Model A and 1000 units of Model B are produced.

39. $x =$ number of \$250 models

$y =$ number of \$400 models

Constraints: $250x + 400y \le 70{,}000$

$x + y \le 250$

$x \ge 0$

$y \ge 0$

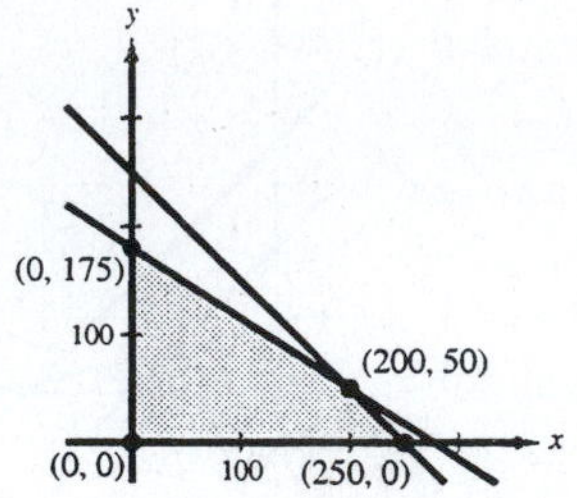

Objective function: $P = 45x + 50y$

Vertices: (0, 175), (200, 50), (250, 0), (0, 0)

At (0, 175): $P = 45(0) + 50(175) = 8750$

At (200, 50): $P = 45(200) + 50(50) = 11{,}500$

At (250, 0): $P = 45(250) + 50(0) = 11{,}250$

At (0, 0): $P = 45(0) + 50(0) = 0$

To maximize the profit, the merchant should stock 200 units of the model costing \$250 and 50 units of the model costing \$400. Then the maximum profit would be \$11,500.

41. $x =$ number of bags of Brand X

$y =$ number of bags of Brand Y

Constraints: $2x + y \ge 12$

$2x + 9y \ge 36$

$2x + 3y \ge 24$

$x \ge 0$

$y \ge 0$

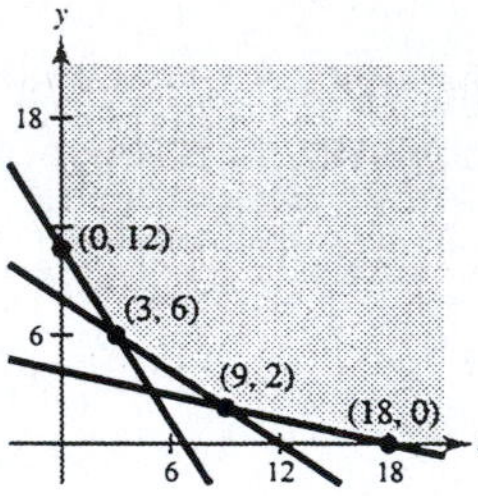

Objective function: $C = 25x + 20y$

Vertices: (0, 12), (3, 6), (9, 2), (18, 0)

At (0, 12): $C = 25(0) + 20(12) = 240$

At (3, 6): $C = 25(3) + 20(6) = 195$

At (9, 2): $C = 25(9) + 20(2) = 265$

At (18, 0): $C = 25(18) + 20(0) = 450$

To minimize cost, use three bags of Brand X and six bags of Brand Y for a total cost of \$195.

43. x = the number of audits, and y = the number of tax returns.

Objective function: Maximize $R = 2000x + 300y$.

Constraints: $100x + 12.5y \le 900$

$$10x + 2.5y \le 100$$

$$x \ge 0, y \ge 0$$

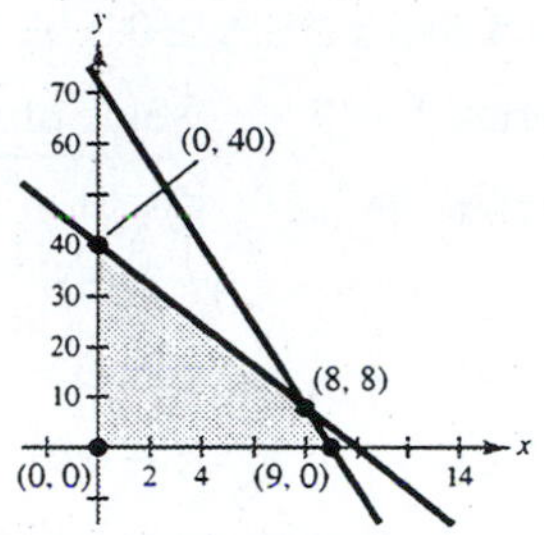

Vertex	Value of $R = 2000x + 300y$
(0, 0)	$R = 2000(0) + 300(0) = 0$
(0, 40)	$R = 2000(0) + 300(40) = 12{,}000$
(8, 8)	$R = 2000(8) + 300(8) = 18{,}400$, maximum value
(9, 0)	$R = 2000(9) + 300(0) = 18{,}000$

The revenue will be maximum if the firm does 8 audits and 8 tax returns each week. Maximum revenue is \$18,400.

45. Objective function: $z = 2.5x + y$

Constraints: $x \ge 0, y \ge 0, 3x + 5y \le 15, 5x + 2y \le 10$

At (0, 0): $z = 0$

At (2, 0): $z = 5$

At $\left(\frac{20}{19}, \frac{45}{19}\right)$: $z = \frac{95}{19} = 5$

At (0, 3): $z = 3$

z is the maximum at any point on the line $5x + 2y = 10$ between the points (2, 0) and $\left(\frac{20}{19}, \frac{45}{19}\right)$.

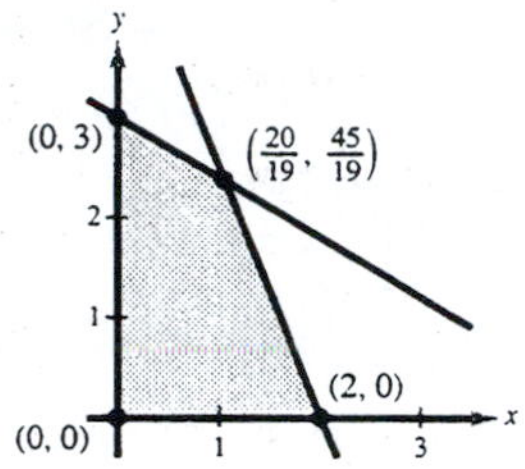

47. Objective function: $z = -x + 2y$

Constraints: $x \ge 0, y \ge 0, x \le 10, x + y \le 7$

At (0, 0): $z = -0 + 2(0) = 0$

At (0, 7): $z = -0 + 2(7) = 14$

At (7, 0): $z = -7 + 2(0) = -7$

The constraint $x \le 10$ is extraneous.

The maximum value of 14 occurs at (0, 7).

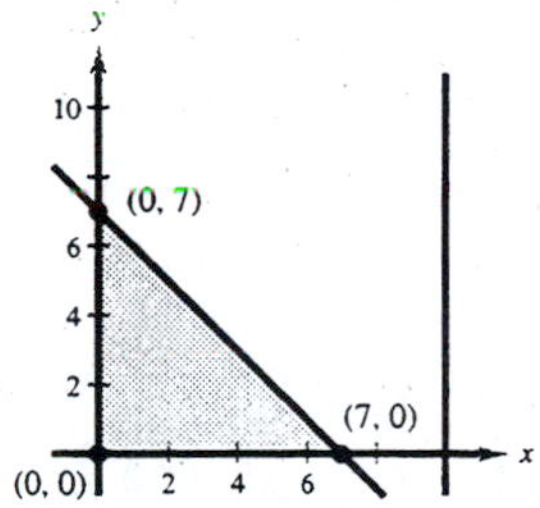

49. Objective function: $z = 3x + 4y$

Constraints: $x \ge 0, y \ge 0, x + y \le 1, 2x + y \le 4$

The constraint $2x + y \le 4$ is extraneous.
The maximum value of $z = 4$ occurs at (0, 1).

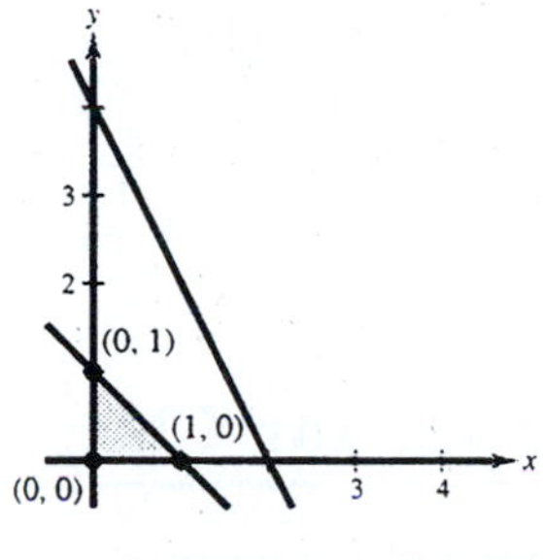

51. Constraints: $x \geq 0$, $y \geq 0$, $x + 3y \leq 15$, $4x + y \leq 16$

Vertex	Value of $z = 3x + ty$
(0, 0)	$z = 0$
(0, 5)	$z = 5t$
(3, 4)	$z = 9 + 4t$
(4, 0)	$z = 12$

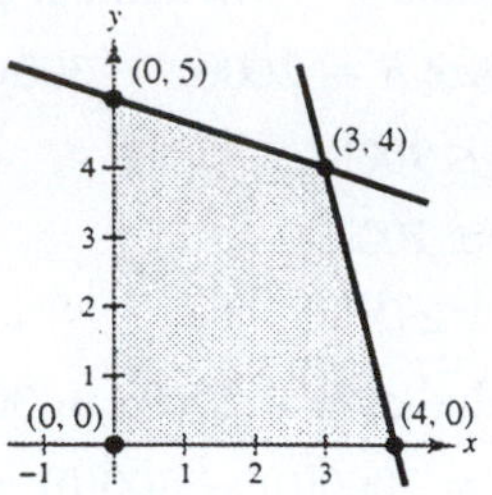

(a) For the maximum value to be at (0, 5), $z = 5t$ must be greater than or equal to $z = 9 + 4t$ and $z = 12$.

$5t \geq 9 + 4t$ and $5t \geq 12$

$t \geq 9$ $\qquad t \geq \frac{12}{5}$

Thus, $t \geq 9$.

(b) For the maximum value to be at (3, 4), $z = 9 + 4t$ must be greater than or equal to $z = 5t$ and $z = 12$.

$9 + 4t \geq 5t$ and $9 + 4t \geq 12$

$9 \geq t$ $\qquad 4t \geq 3$

$t \geq \frac{3}{4}$

Thus, $\frac{3}{4} \leq t \leq 9$.

53. $$\frac{\left(\frac{9}{x}\right)}{\left(\frac{6}{x} + 2\right)} = \frac{\frac{9}{x}}{\frac{6 + 2x}{x}} = \frac{9}{x} \cdot \frac{x}{2(3 + x)} = \frac{9}{2(3 + x)} = \frac{9}{2(x + 3)}, x \neq 0$$

55. $$\frac{\left(\frac{4}{x^2 - 9} + \frac{2}{x - 2}\right)}{\left(\frac{1}{x + 3} + \frac{1}{x - 3}\right)} = \frac{\frac{4(x - 2) + 2(x^2 - 9)}{(x - 2)(x^2 - 9)}}{\frac{(x - 3) + (x + 3)}{x^2 - 9}}$$

$$= \frac{2x^2 + 4x - 26}{(x - 2)(x^2 - 9)} \cdot \frac{x^2 - 9}{2x}$$

$$= \frac{2(x^2 + 2x - 13)}{(x - 2)(2x)}$$

$$= \frac{x^2 + 2x - 13}{x(x - 2)}, x \neq \pm 3$$

❑ Review Exercises for Chapter 7

Solutions to Odd-Numbered Exercises

1. $x + y = 2 \Rightarrow y = 2 - x$

$x - y = 0 \Rightarrow x - (2 - x) = 0$

$2x - 2 = 0$

$x = 1$

$y = 2 - 1 = 1$

Solution: (1, 1)

3. $x^2 - y^2 = 9$

$x - y = 1 \Rightarrow x = y + 1$

$(y + 1)^2 - y^2 = 9$

$2y + 1 = 9$

$y = 4$

$x = 5$

Solution: (5, 4)

5. $y = 2x^2$

$y = x^4 - 2x^2 \Rightarrow 2x^2 = x^4 - 2x^2$

$0 = x^4 - 4x^2$

$0 = x^2(x^2 - 4)$

$0 = x^2(x + 2)(x - 2)$

$x = 0, x = -2, x = 2$

$y = 0, y = 8, y = 8$

Solutions: $(0, 0), (-2, 8), (2, 8)$

7. $y^2 - 2y + x = 0 \Rightarrow (y - 1)^2 = 1 - x \Rightarrow y = 1 \pm \sqrt{1 - x}$

$x + y = 0 \Rightarrow y = -x$

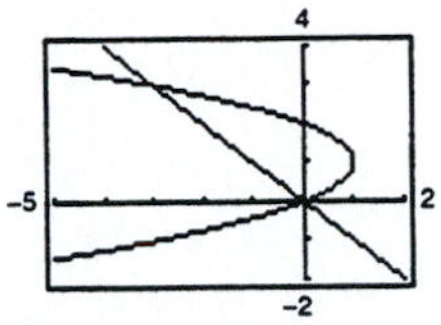

Points of intersection: $(0, 0)$ and $(-3, 3)$

9. $y = 2(6 - x)$

$y = 2^{x-2}$

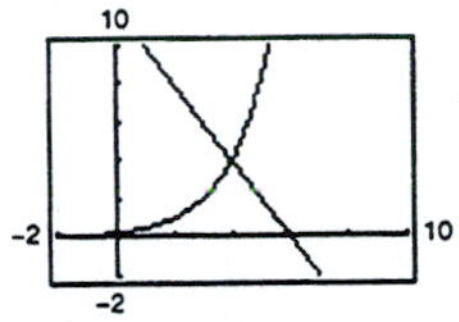

Point of intersection: $(4, 4)$

11.
$$\begin{aligned} 2x - y &= 2 \Rightarrow 16x - 8y = 16 \\ 6x + 8y &= 39 \Rightarrow 6x + 8y = 39 \\ & \quad\;\; 22x = 55 \\ & \quad\;\; x = \tfrac{55}{22} = \tfrac{5}{2} \\ & \quad\;\; y = 3 \end{aligned}$$

Solution: $\left(\frac{5}{2}, 3\right)$

13.
$$\begin{aligned} 0.2x + 0.3y = 0.14 &\Rightarrow 20x + 30y = 14 \Rightarrow 20x + 30y = 14 \\ 0.4x + 0.5y = 0.20 &\Rightarrow 4x + 5y = 2 \Rightarrow -20x - 25y = -10 \\ & \qquad 5y = 4 \\ & \qquad y = \tfrac{4}{5} \\ & \qquad x = -\tfrac{1}{2} \end{aligned}$$

Solution: $\left(-\frac{1}{2}, \frac{4}{5}\right) = (-0.5, 0.8)$

15.
$$\begin{aligned} 3x - 2y = 0 &\Rightarrow 3x - 2y = 0 \\ 3x + 2(y + 5) = 10 &\Rightarrow 3x + 2y = 0 \\ & \qquad 6x = 0 \\ & \qquad x = 0 \\ & \qquad y = 0 \end{aligned}$$

Solution: $(0, 0)$

17.
$$\begin{aligned} 1.25x - 2y = 3.5 &\Rightarrow 5x - 8y = 14 \\ 5x - 8y = 14 &\Rightarrow -5x + 8y = -14 \\ & \qquad 0 = 0 \end{aligned}$$

Infinite solutions

Let $y = a$, then $5x - 8a = 14 \Rightarrow x = \frac{14}{5} + \frac{8}{5}a.$

Solution: $\left(\frac{14}{5} + \frac{8}{5}a, a\right)$

19. There are infinite linear systems with the solution $\left(\frac{4}{3}, 3\right)$.One possible solution is:

$$3\left(\tfrac{4}{3}\right) + 3 = 7 \Rightarrow 3x + y = 7$$

$$-6\left(\tfrac{4}{3}\right) + 3(3) = 1 \Rightarrow -6x + 3y = 1$$

21. Revenue $= 4.95x$

Cost $= 2.85x + 10{,}000$

Break even when Revenue = Cost

$$4.95x = 2.85x + 10{,}000$$

$$2.10x = 10{,}000$$

$$x \approx 4762 \text{ units}$$

23. Let $x =$ the amount of 75% solution, and $y =$ amount of 50% solution.

$$x + y = 100 \Rightarrow y = 100 - x$$

$$0.75x + 0.50y = 0.60(100)$$

$$0.75x + 0.50(100 - x) = 60$$

$$0.75x + 50 - 0.50x = 60$$

$$0.25x = 10$$

$$x = 40$$

$$y = 100 - x = 60$$

Answer: 40 liters of 75% solution, 60 liters of 60% solution.

25. Let $x =$ speed of the slower plane

Let $y =$ speed of the faster plane

Then, distance of first plane + distance of second plane = 275 miles

(rate of first plane)(time) + (rate of second plane)(time) = 275 miles

$$x\left(\tfrac{40}{60}\right) + y\left(\tfrac{40}{60}\right) = 275$$

$$y = x + 25$$

$$\tfrac{2}{3}x + \tfrac{2}{3}(x + 25) = 275$$

$$4x + 50 = 825$$

$$4x = 775$$

$$x = 193.75 \text{ mph}$$

$$y = x + 25 = 218.75 \text{ mph}$$

27. Demand = Supply

$$37 - 0.0002x = 22 + 0.00001x$$
$$15 = 0.00021x$$
$$x = \frac{500{,}000}{7}, p = \frac{159}{7}$$

Point of equilibrium: $\left(\frac{500{,}000}{7}, \frac{159}{7}\right)$

29.
$$\begin{aligned} x + 2y + 6z &= 4 \\ -3x + 2y - z &= -4 \\ 4x + 2z &= 16 \end{aligned}$$

$$\begin{aligned} x + 2y + 6z &= 4 \\ 8y + 17z &= 8 \quad 3\text{Eq.1} + \text{Eq.2} \\ -8y - 22z &= 0 \quad -4\text{Eq.1} + \text{Eq3} \end{aligned}$$

$$\begin{aligned} x + 2y + 6z &= 4 \\ 8y + 17z &= 8 \\ -5z &= 8 \quad \text{Eq.2} + \text{Eq.3} \end{aligned}$$

$$z = -\tfrac{8}{5} = -1.6$$
$$8y + 17(-1.6) = 8 \implies y = 4.4$$
$$x + 2(4.4) + 6(-1.6) = 4 \implies x = 4.8$$

Solution: $(4.8, 4.4, -1.6)$

31.
$$\begin{aligned} x - 2y + z &= -6 \\ 2x - 3y &= -7 \\ -x + 3y - 3z &= 11 \end{aligned}$$

$$\begin{aligned} x - 2y + z &= -6 \\ y - 2z &= 5 \quad -2\text{Eq.1} + \text{Eq.2} \\ y - 2z &= 5 \quad \text{Eq.1} + \text{Eq.3} \end{aligned}$$

$$\begin{aligned} x - 2y + z &= -6 \\ y - 2z &= 5 \\ 0 &= 0 \quad -\text{Eq.2} + \text{Eq.3} \end{aligned}$$

Let $z = a$, then:

$$y = 2a + 5$$
$$x - 2(2a + 5) + a = -6$$
$$x - 3a - 10 = -6$$
$$x = 3a + 4$$

Solution: $(3a + 4, 2a + 5, a)$ where a is any real number.

33.
$$\begin{aligned} 2x + 5y - 19z = 34 &\implies 6x + 15y - 57z = 102 \\ 3x + 8y - 31z = 54 &\implies -6x - 16y + 62z = -108 \\ & \qquad\quad -y + 5z = -6 \end{aligned}$$

Let $z = a$. Then:

$$y = 5a + 6$$
$$x = \tfrac{1}{2}[34 - 5(5a + 6) + 19a] = -3a + 2$$

Solution: $(-3a + 2, 5a + 6, a)$ where a is any real number.

35. There are an infinite number of linear systems with the solution $(4, -1, 3)$. One possible system is as follows:

$$\begin{aligned} 2(4) + (-1) - 2(3) = 1 &\Rightarrow 2x + y - 2z = 1 \\ (4) + (-1) - (3) = 0 &\Rightarrow x + y - z = 0 \\ 2(4) - 3(-1) - 2(3) = 5 &\Rightarrow 2x - 3y - 2z = 5 \end{aligned}$$

37. $y = ax^2 + bx + c$ through $(0, -5)$, $(1, -2)$, and $(2, 5)$.

$$\begin{aligned} (0,-5)&: -5 = c \Rightarrow c = -5 \\ (1,-2)&: -2 = a + b + c \Rightarrow a + b = 3 \\ (2,\ 5)&: 5 = 4a + 2b + c \Rightarrow 2a + b = 5 \end{aligned}$$

$$\begin{aligned} 2a + b &= 5 \\ -a - b &= -3 \\ a &= 2 \\ b &= 1 \end{aligned}$$

The equation of the parabola is $y = 2x^2 + x - 5$.

39. $x^2 + y^2 + Dx + Ey + F = 0$ through $(-1, -2)$, $(5, -2)$ and $(2, 1)$.

$$\begin{aligned} (-1,-2)&: 5 - D - 2E + F = 0 \Rightarrow D + 2E - F = 5 \\ (5,-2)&: 29 + 5D - 2E + F = 0 \Rightarrow 5D - 2E + F = -29 \\ (2,\ 1)&: 5 + 2D + 2E + F = 0 \Rightarrow 2D + E + F = -5 \end{aligned}$$

From the first two equations we have

$$\begin{aligned} 6D &= -24 \\ D &= -4. \end{aligned}$$

Substituting $D = -4$ into the second and third equations yields:

$$\begin{aligned} -20 - 2E + F = -29 &\Rightarrow -2E + F = -9 \\ -8 + E + F = -5 &\Rightarrow -E - F = -3 \\ &\quad -3E = -12 \\ &\quad E = 4 \\ &\quad F = -1 \end{aligned}$$

The equation of the circle is $x^2 + y^2 - 4x + 4y - 1 = 0$.

41. From the following chart we obtain our system of equations.

	A	B	C
Mixture X	$\frac{1}{5}$	$\frac{2}{5}$	$\frac{2}{5}$
Mixture Y	0	0	1
Mixture Z	$\frac{1}{3}$	$\frac{1}{3}$	$\frac{1}{3}$
Desired Mixture	$\frac{6}{27}$	$\frac{8}{27}$	$\frac{13}{27}$

$$\left.\begin{aligned} \tfrac{1}{5}x + \tfrac{1}{3}z &= \tfrac{6}{27} \\ \tfrac{2}{5}x + \tfrac{1}{3}z &= \tfrac{8}{27} \end{aligned}\right\} x = \tfrac{10}{27},\ z = \tfrac{12}{27}$$

$$\tfrac{2}{5}x + y + \tfrac{1}{3}z = \tfrac{13}{27} \Rightarrow y = \tfrac{5}{27}$$

To obtain the desired mixture, use 10 gallons of X, 5 gallons of Y, and 12 gallons of Z.

43. $5b + 10a = 17.8 \Rightarrow -10b - 20a = -35.6$

$10b + 30a = 45.7 \Rightarrow 10b + 30a = 45.7$

$$10a = 10.1$$
$$a = 1.01$$
$$b = 1.54$$

Least squares regression line: $y = 1.01x + 1.54$

45. (a) $7b + 156.8a = 169.5$

$156.8b + 3522.78a = 3806.8$

By elimination we have $a \approx 0.956$, $b \approx 2.799$.

The least squares regression line is $y = 0.956x + 2.779$.

(b)

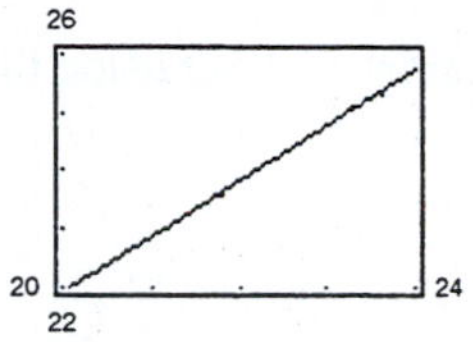

(c) The line is a good model for the data.

(d) A one year change in x (the women's ages) results in y (men's ages) changing 0.956 year.

47. $x + 2y \le 160$

$3x + y \le 180$

$x \ge 0$

$y \ge 0$

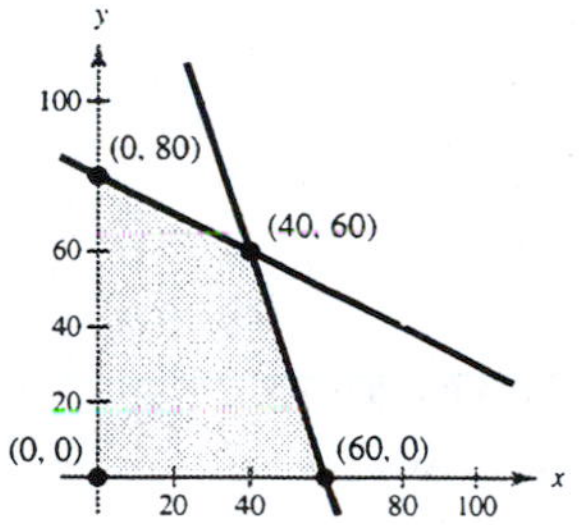

Vertex A	Vertex B	Vertex C	Vertex D	Vertex E	Vertex F
$x + 2y = 160$	$x + 2y = 160$	$3x + y = 180$	$x = 0$	$x + 2y = 160$	$3x + y = 180$
$3x + y = 180$	$x = 0$	$y = 0$	$y = 0$	$y = 0$	$x = 0$
(40, 60)	(0, 80)	(60, 0)	(0, 0)	(160, 0)	(0, 180)
				Outside the region	Outside the region

49. $3x + 2y \geq 24$

$x + 2y \geq 12$

$2 \leq x \leq 15$

$y \leq 15$

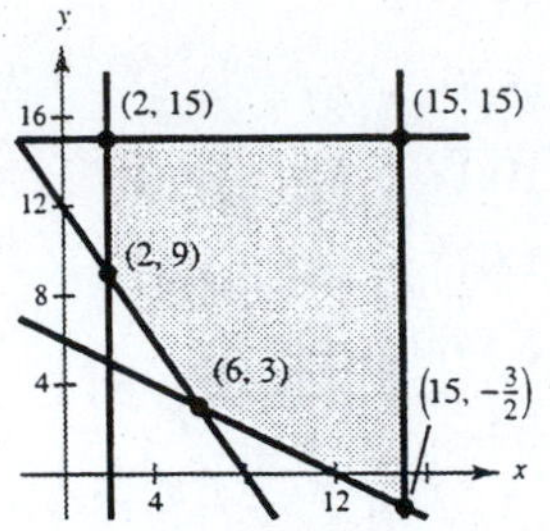

Vertex A	Vertex B	Vertex C	Vertex D	Vertex E
$3x + 2y = 24$	$3x + 2y = 24$	$3x + 2y = 24$	$3x + 2y = 24$	$x + 2y = 12$
$x + 2y = 12$	$x = 2$	$x = 15$	$y = 15$	$x = 2$
$(6, 3)$	$(2, 9)$	$(15, -\frac{21}{2})$	$(-2, 15)$	$(2, 5)$
		Outside the region	Outside the region	Outside the region

Vertex F	Vertex G	Vertex H	Vertex I
$x + 2y = 12$	$x + 2y = 12$	$x = 2$	$x = 15$
$x = 15$	$y = 15$	$y = 15$	$y = 15$
$(15, -\frac{3}{2})$	$(-18, 15)$	$(2, 15)$	$(15, 15)$
	Outside the region		

51. $y < x + 1$

$y > x^2 - 1$

Vertices:

$x + 1 = x^2 - 1$

$0 = x^2 - x - 2 = (x + 1)(x - 2)$

$x = -1$ or $x = 2$

$y = 0 \qquad y = 3$

$(-1, 0) \qquad (2, 3)$

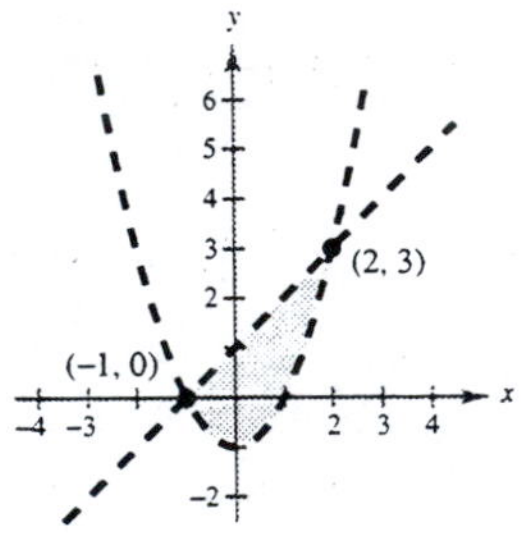

53. $2x - 3y \geq 0$

$2x - y \leq 8$

$y \geq 0$

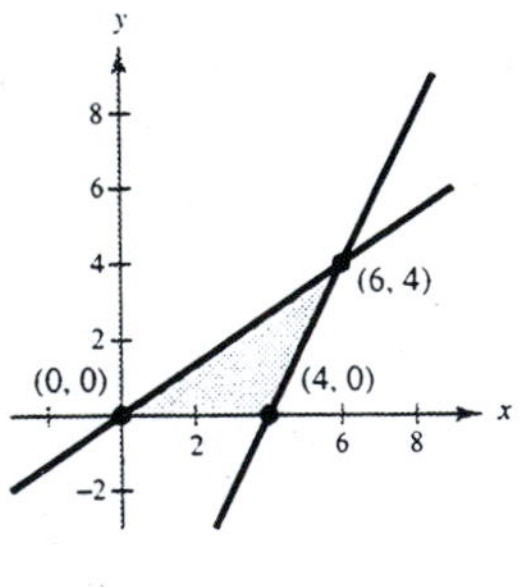

Vertex A	Vertex B	Vertex C
$2x - 3y = 0$	$2x - 3y = 0$	$2x - y = 8$
$2x - y = 8$	$y = 0$	$y = 0$
$(6, 4)$	$(0, 0)$	$(4, 0)$

55.

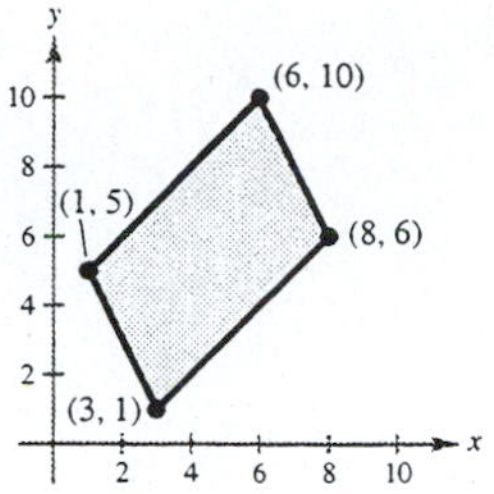

Line through (1, 5), (3, 1): $2x + y = 7$

Line through (1, 5), (6, 10): $-x + y = 4$

Line through (6, 10), (8, 6): $2x + y = 22$

Line through (8, 6), (3, 1): $-x + y = -2$

System of inequalities:

$$-x + y \leq 4$$
$$2x + y \leq 22$$
$$-x + y \geq -2$$
$$2x + y \geq 7$$

57. Let x = the number of bushels for Harrisburg, and y = the number of bushels for Philadelphia.

$$x \geq 400$$
$$y \geq 600$$
$$x + y \leq 1500$$

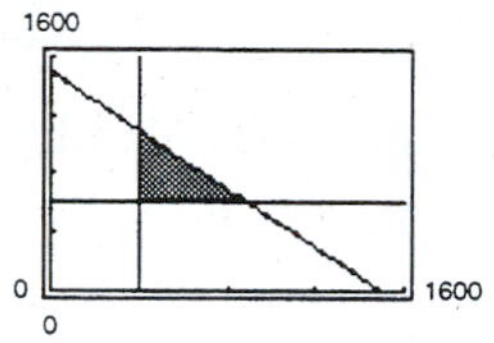

59. Demand = Supply

$$160 - 0.0001x = 70 + 0.0002x$$
$$90 = 0.0003x$$
$$x = 300{,}000 \text{ units}$$
$$p = \$130$$

Point of equilibrium: (300,000, 130)

Consumer surplus: $\frac{1}{2}(300{,}000)(30) = \$4{,}500{,}000$

Producer surplus: $\frac{1}{2}(300{,}000)(60) = \$9{,}000{,}000$

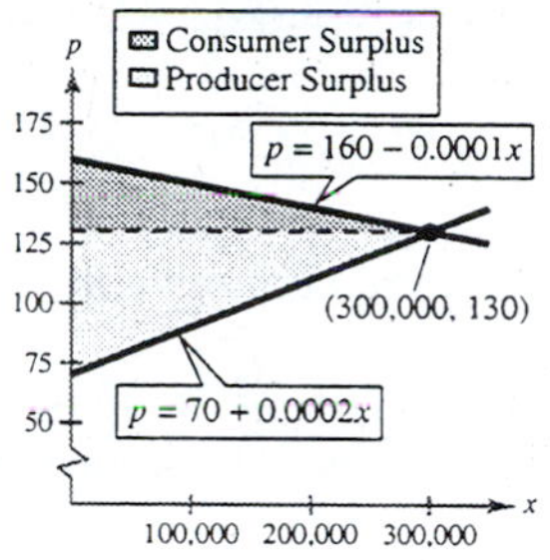

61. Maximize $z = 3x + 4y$ subject to the following constraints.

$$x \geq 0$$
$$y \geq 0$$
$$2x + 5y \leq 50$$
$$4x + y \leq 28$$

Vertex	Value of $z = 3x + 4y$
(0, 0)	$z = 0$
(0, 10)	$z = 40$
(5, 8)	$z = 47$, maximum value
(7, 0)	$z = 21$

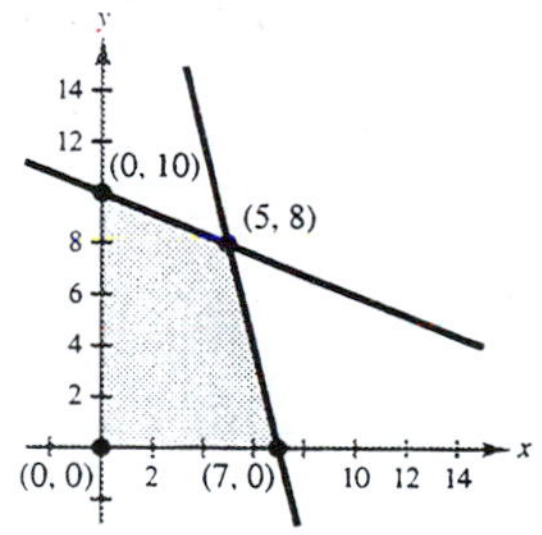

63. Minimize $z = 1.75x + 2.25y$ subject to the following constraints.

$$2x + y \geq 25$$
$$3x + 2y \geq 45$$
$$x \geq 0$$
$$y \geq 0$$

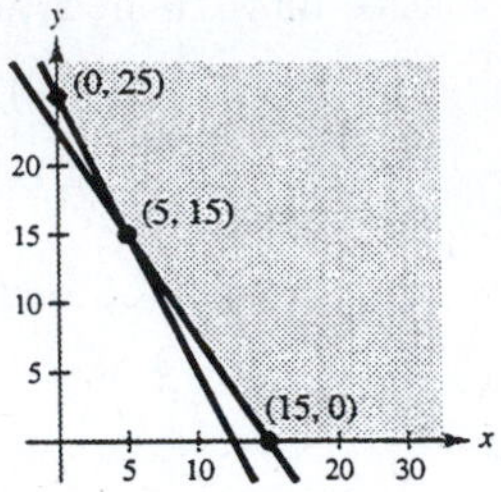

Vertex	Value of $z = 1.75x + 2.25y$
(0, 25)	$z = 56.25$
(5, 15)	$z = 42.5$
(15, 0)	$z = 26.25$, minimum value

65. Let x = number of haircuts
Let y = number of perms
Maximize $R = 17x + 60y$ subject to the following constraints.

$$x \geq 0$$
$$y \geq 0$$
$$\left(\frac{20}{60}\right)x + \left(\frac{70}{60}\right)y \leq 24 \implies 2x + 7y \leq 144$$

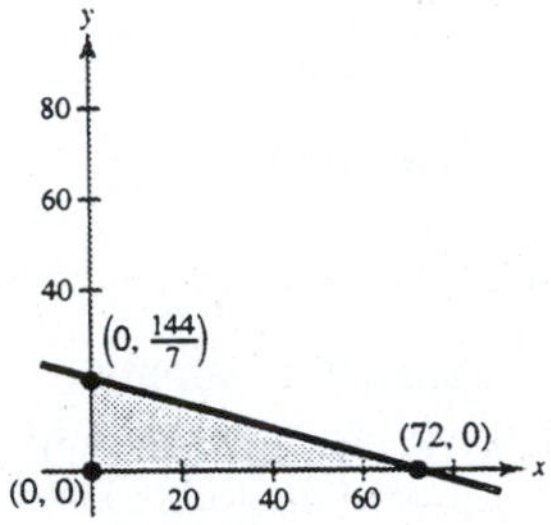

Vertex	Value of $R = 17x + 60y$
(0, 0)	$R = 0$
(72, 0)	$R = 1224$
$\left(0, \frac{144}{7}\right)$	$R \approx 1234.29$, maximum value

The revenue is maximum when $y = \frac{144}{7} \approx 20$ perms. (Round down since the student cannot work more than 24 hours. Note: Since we rounded down, the student would have enough time left to do 2 haircuts.)

67. Let x = the number of bags of Brand X, and y = the number of bags of Brand Y.
Objective function: Minimize $C = 15x + 30y$
Constraints: $8x + 2y \geq 16$

$$x + y \geq 5$$
$$2x + 7y \geq 20$$
$$x \geq 0, y \geq 0$$

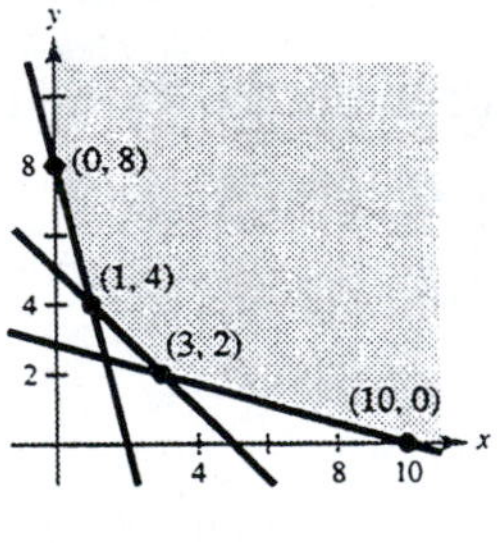

Vertex	Value of $C = 15x + 30y$
(0, 8)	$C = 15(0) + 30(8) = 240$
(1, 4)	$C = 15(1) + 30(4) = 135$
(3, 2)	$C = 15(3) + 30(2) = 105$, minimum value
(10, 0)	$C = 15(10) + 30(0) = 150$

To minimize cost, use three bags of Brand X and two bags of Brand Y.

The cost is \$105.

❑ Practice Test for Chapter 7

For Exercises 1–3, solve the given system by the method of substitution.

1. $x + y = 1$

$3x - y = 15$

2. $x - 3y = -3$

$x^2 + 6y = 5$

3. $x + y + z = 6$

$2x - y + 3z = 0$

$5x + 2y - z = -3$

4. Find the two numbers whose sum is 110 and product is 2800.

5. Find the dimensions of a rectangle if its perimeter is 170 feet and its area is 2800 square feet.

For Exercises 6–8, solve the linear system by elimination.

6. $2x + 15y = 4$

$x - 3y = 23$

7. $x + y = 2$

$38x - 19y = 7$

8. $0.4x + 0.5y = 0.112$

$0.3x - 0.7y = -0.131$

9. Herbert invests \$17,000 in two funds that pay 11% and 13% simple interest, respectively. If he receives \$2080 in yearly interest, how much is invested in each fund?

10. Find the least squares regression line for the points (4, 3), (1, 1), (−1, −2), and (−2, −1).

For Exercises 11–13, solve the system of equations.

11. $$\begin{aligned} x + y \quad\quad &= -2 \\ 2x - y + z &= 11 \\ 4y - 3z &= -20 \end{aligned}$$

12. $$\begin{aligned} 4x - y + 5z &= 4 \\ 2x + y - z &= 0 \\ 2x + 4y + 8z &= 0 \end{aligned}$$

13. $$\begin{aligned} 3x + 2y - z &= 5 \\ 6x - y + 5z &= 2 \end{aligned}$$

14. Find the equation of the parabola $y = ax^2 + bx + c$ passing through the points $(0, -1)$, $(1, 4)$ and $(2, 13)$.

15. Find the position equation $s = \frac{1}{2}at^2 + v_0t + s_0$ given that $s = 12$ feet after 1 second, $s = 5$ feet after 2 seconds, and $s = 4$ after 3 seconds.

16. Graph $x^2 + y^2 \geq 9$.

17. Graph the solution of the system.

$$\begin{aligned} x + y &\leq 6 \\ x &\geq 2 \\ y &\geq 0 \end{aligned}$$

18. Derive a set of inequalities to describe the triangle with vertices $(0, 0)$, $(0, 7)$, and $(2, 3)$.

19. Find the maximum value of the objective function, $z = 30z + 26y$, subject to the following constraints.

$$\begin{aligned} x &\geq 0 \\ y &\geq 0 \\ 2x + 3y &\leq 21 \\ 5x + 3y &\leq 30 \end{aligned}$$

20. Graph the system of inequalities.

$$\begin{aligned} x^2 + y^2 &\leq 4 \\ (x - 2)^2 + y^2 &\geq 4 \end{aligned}$$

CHAPTER 8
Matrices and Determinants

CHAPTER 8
Matrices and Determinants

Section 8.1 Matrices and Systems of Equations

- You should be able to use elementary row operations to produce a row-echelon form (or reduced row-echelon form) of a matrix.
  1. Interchange two rows.
  2. Multiply a row by a nonzero constant.
  3. Add a multiple of one row to another row.
- You should be able to use either Gaussian elimination with back-substitution or Gauss-Jordan elimination to solve a system of linear equations.

Solutions to Odd-Numbered Exercises

1. Since the matrix has three rows and two columns, its order is 3×2.

3. Since the matrix has three rows and one column, its order is 3×1.

5. Since the matrix has two rows and two columns, its order is 2×2.

7.
$$\begin{aligned} 4x - 3y &= -5 \\ -x + 3y &= 12 \end{aligned}$$
$$\left[\begin{array}{rr:r} 4 & -3 & -5 \\ -1 & 3 & 12 \end{array}\right]$$

9.
$$\begin{aligned} x + 10y - 2z &= 2 \\ 5x - 3y + 4z &= 0 \\ 2x + y &= 6 \end{aligned}$$
$$\left[\begin{array}{rrr:r} 1 & 10 & -2 & 2 \\ 5 & -3 & 4 & 0 \\ 2 & 1 & 0 & 6 \end{array}\right]$$

11.
$$\left[\begin{array}{rr:r} 1 & 2 & 7 \\ 2 & -3 & 4 \end{array}\right]$$
$$\begin{aligned} x + 2y &= 7 \\ 2x - 3y &= 4 \end{aligned}$$

13.
$$\left[\begin{array}{rrr:r} 2 & 0 & 5 & -12 \\ 0 & 1 & -2 & 7 \\ 6 & 3 & 0 & 2 \end{array}\right]$$
$$\begin{aligned} 2x + 5z &= -12 \\ y - 2z &= 7 \\ 6x + 3y &= 2 \end{aligned}$$

15.
$$\begin{bmatrix} 1 & 0 & 0 & 0 \\ 0 & 1 & 1 & 5 \\ 0 & 0 & 0 & 0 \end{bmatrix}$$
This matrix is in reduced row-echelon form.

17.
$$\begin{bmatrix} 2 & 0 & 4 & 0 \\ 0 & -1 & 3 & 6 \\ 0 & 0 & 1 & 5 \end{bmatrix}$$
The first nonzero entries in rows one and two are not one. The matrix is not in row-echelon form.

19.
$$\begin{bmatrix} 1 & 4 & 3 \\ 2 & 10 & 5 \end{bmatrix}$$
$$-2R_1 + R_2 \rightarrow \begin{bmatrix} 1 & 4 & 3 \\ 0 & \boxed{2} & -1 \end{bmatrix}$$

21.
$$\begin{bmatrix} 1 & 1 & 4 & -1 \\ 3 & 8 & 10 & 3 \\ -2 & 1 & 12 & 6 \end{bmatrix}$$
$$\begin{matrix} \\ -3R_1 + R_2 \rightarrow \\ 2R_1 + R_3 \rightarrow \end{matrix} \begin{bmatrix} 1 & 1 & 4 & -1 \\ 0 & 5 & \boxed{-2} & \boxed{6} \\ 0 & 3 & \boxed{20} & \boxed{4} \end{bmatrix}$$
$$\tfrac{1}{5}R_2 \rightarrow \begin{bmatrix} 1 & 1 & 4 & -1 \\ 0 & 1 & -\frac{2}{5} & \frac{6}{5} \\ 0 & 3 & 20 & 4 \end{bmatrix}$$

23. $\begin{bmatrix} 1 & 2 & 3 \\ 2 & -1 & -4 \\ 3 & 1 & -1 \end{bmatrix}$

(a) $\begin{bmatrix} 1 & 2 & 3 \\ 0 & -5 & -10 \\ 3 & 1 & -1 \end{bmatrix}$ (b) $\begin{bmatrix} 1 & 2 & 3 \\ 0 & -5 & -10 \\ 0 & -5 & -10 \end{bmatrix}$ (c) $\begin{bmatrix} 1 & 2 & 3 \\ 0 & -5 & -10 \\ 0 & 0 & 0 \end{bmatrix}$

(d) $\begin{bmatrix} 1 & 2 & 3 \\ 0 & 1 & 2 \\ 0 & 0 & 0 \end{bmatrix}$ (e) $\begin{bmatrix} 1 & 0 & -1 \\ 0 & 1 & 2 \\ 0 & 0 & 0 \end{bmatrix}$ This matrix is in reduced row-echelon form.

25. $\begin{bmatrix} 1 & 1 & 0 & 5 \\ -2 & -1 & 2 & -10 \\ 3 & 6 & 7 & 14 \end{bmatrix}$

$$\begin{array}{r} \\ 2R_1 + R_2 \rightarrow \\ -3R_1 + R_3 \rightarrow \end{array} \begin{bmatrix} 1 & 1 & 0 & 5 \\ 0 & 1 & 2 & 0 \\ 0 & 3 & 7 & -1 \end{bmatrix}$$

$$\begin{array}{r} \\ \\ -3R_2 + R_3 \rightarrow \end{array} \begin{bmatrix} 1 & 1 & 0 & 5 \\ 0 & 1 & 2 & 0 \\ 0 & 0 & 1 & -1 \end{bmatrix}$$

27. $\begin{bmatrix} 1 & -1 & -1 & 1 \\ 5 & -4 & 1 & 8 \\ -6 & 8 & 18 & 0 \end{bmatrix}$

$$\begin{array}{r} \\ -5R_1 + R_2 \rightarrow \\ 6R_1 + R_3 \rightarrow \end{array} \begin{bmatrix} 1 & -1 & -1 & 1 \\ 0 & 1 & 6 & 3 \\ 0 & 2 & 12 & 6 \end{bmatrix}$$

$$\begin{array}{r} \\ \\ -2R_2 + R_3 \rightarrow \end{array} \begin{bmatrix} 1 & -1 & -1 & 1 \\ 0 & 1 & 6 & 3 \\ 0 & 0 & 0 & 0 \end{bmatrix}$$

29. $\begin{bmatrix} 3 & 3 & 3 \\ -1 & 0 & -4 \\ 2 & 4 & -2 \end{bmatrix}$

$$\begin{array}{r} \frac{1}{3}R_1 \rightarrow \\ \\ \\ \end{array} \begin{bmatrix} 1 & 1 & 1 \\ -1 & 0 & -4 \\ 2 & 4 & -2 \end{bmatrix}$$

$$\begin{array}{r} \\ R_1 + R_2 \rightarrow \\ -2R_1 + R_3 \rightarrow \end{array} \begin{bmatrix} 1 & 1 & 1 \\ 0 & 1 & -3 \\ 0 & 2 & -4 \end{bmatrix}$$

$$\begin{array}{r} -R_2 + R_1 \rightarrow \\ \\ -2R_2 + R_3 \rightarrow \end{array} \begin{bmatrix} 1 & 0 & 4 \\ 0 & 1 & -3 \\ 0 & 0 & 2 \end{bmatrix}$$

$$\begin{array}{r} \\ \\ \frac{1}{2}R_3 \rightarrow \end{array} \begin{bmatrix} 1 & 0 & 4 \\ 0 & 1 & -3 \\ 0 & 0 & 1 \end{bmatrix}$$

$$\begin{array}{r} -4R_3 + R_1 \rightarrow \\ 3R_3 + R_2 \rightarrow \\ \\ \end{array} \begin{bmatrix} 1 & 0 & 0 \\ 0 & 1 & 0 \\ 0 & 0 & 1 \end{bmatrix}$$

31. $\begin{bmatrix} 1 & 2 & 3 & -5 \\ 1 & 2 & 4 & -9 \\ -2 & -4 & -4 & 3 \\ 4 & 8 & 11 & -14 \end{bmatrix}$

$$\begin{array}{r} \\ -R_1 + R_2 \rightarrow \\ 2R_1 + R_3 \rightarrow \\ -4R_1 + R_4 \rightarrow \end{array} \begin{bmatrix} 1 & 2 & 3 & -5 \\ 0 & 0 & 1 & -4 \\ 0 & 0 & 2 & -7 \\ 0 & 0 & -1 & 6 \end{bmatrix}$$

$$\begin{array}{r} -3R_2 + R_1 \rightarrow \\ \\ -R_2 + R_3 \rightarrow \\ R_2 + R_4 \rightarrow \end{array} \begin{bmatrix} 1 & 2 & 0 & 7 \\ 0 & 0 & 1 & -4 \\ 0 & 0 & 0 & 1 \\ 0 & 0 & 0 & 2 \end{bmatrix}$$

$$\begin{array}{r} -7R_3 + R_1 \rightarrow \\ 4R_3 + R_2 \rightarrow \\ \\ -2R_3 + R_4 \rightarrow \end{array} \begin{bmatrix} 1 & 2 & 0 & 0 \\ 0 & 0 & 1 & 0 \\ 0 & 0 & 0 & 1 \\ 0 & 0 & 0 & 0 \end{bmatrix}$$

33.
$$\begin{aligned} x - 2y &= 4 \\ y &= -3 \\ x - 2(-3) &= 4 \\ x &= -2 \end{aligned}$$

Answer: $(-2, -3)$

35.
$$\begin{aligned} x - y + 2z &= 4 \\ y - z &= 2 \\ z &= -2 \\ y - (-2) &= 2 \\ y &= 0 \\ x - 0 + 2(-2) &= 4 \\ x &= 8 \end{aligned}$$

Answer: $(8, 0, -2)$

37. $\begin{bmatrix} 1 & 0 & \vdots & 7 \\ 0 & 1 & \vdots & -5 \end{bmatrix}$

$x = 7$

$y = -5$

Answer: $(7, -5)$

39. $\begin{bmatrix} 1 & 0 & 0 & \vdots & -4 \\ 0 & 1 & 0 & \vdots & -8 \\ 0 & 0 & 1 & \vdots & 2 \end{bmatrix}$

$x = -4$

$y = -8$

$z = 2$

Answer: $(-4, -8, 2)$

41. $x + 2y = 7$

$2x + y = 8$

$$\begin{bmatrix} 1 & 2 & \vdots & 7 \\ 2 & 1 & \vdots & 8 \end{bmatrix}$$

$$-2R_1 + R_2 \rightarrow \begin{bmatrix} 1 & 2 & \vdots & 7 \\ 0 & -3 & \vdots & -6 \end{bmatrix}$$

$$-\tfrac{1}{3}R_2 \rightarrow \begin{bmatrix} 1 & 2 & \vdots & 7 \\ 0 & 1 & \vdots & 2 \end{bmatrix}$$

$y = 2$

$x + 2(2) = 7 \Rightarrow x = 3$

Answer: $(3, 2)$

43. $-3x + 5y = -22$

$3x + 4y = 4$

$4x - 8y = 32$

$$\begin{bmatrix} -3 & 5 & \vdots & -22 \\ 3 & 4 & \vdots & 4 \\ 4 & -8 & \vdots & 32 \end{bmatrix}$$

$$R_3 + R_1 \rightarrow \begin{bmatrix} 1 & -3 & \vdots & 10 \\ 3 & 4 & \vdots & 4 \\ 4 & -8 & \vdots & 32 \end{bmatrix}$$

$$\begin{matrix} \\ -3R_1 + R_2 \rightarrow \\ -4R_1 + R_3 \rightarrow \end{matrix} \begin{bmatrix} 1 & -3 & \vdots & 10 \\ 0 & 13 & \vdots & -26 \\ 0 & 4 & \vdots & -8 \end{bmatrix}$$

$$\begin{matrix} \\ \tfrac{1}{13}R_2 \rightarrow \\ -4R_2 + R_3 \rightarrow \end{matrix} \begin{bmatrix} 1 & -3 & \vdots & 10 \\ 0 & 1 & \vdots & -2 \\ 0 & 0 & \vdots & 0 \end{bmatrix}$$

$y = -2$

$x - 3(-2) = 10 \Rightarrow x = 4$

Answer: $(4, -2)$

45. $8x - 4y = 7$

$5x + 2y = 1$

$$\begin{bmatrix} 8 & -4 & \vdots & 7 \\ 5 & 2 & \vdots & 1 \end{bmatrix}$$

$$\begin{matrix} 3R_1 \rightarrow \\ 5R_2 \rightarrow \end{matrix} \begin{bmatrix} 24 & -12 & \vdots & 21 \\ 25 & 10 & \vdots & 5 \end{bmatrix}$$

$$-R_2 + R_1 \rightarrow \begin{bmatrix} -1 & -22 & \vdots & 16 \\ 25 & 10 & \vdots & 5 \end{bmatrix}$$

$$25R_1 + R_2 \rightarrow \begin{bmatrix} -1 & -22 & \vdots & 16 \\ 0 & -540 & \vdots & 405 \end{bmatrix}$$

$$\begin{matrix} -R_1 \rightarrow \\ -\tfrac{1}{540}R_2 \rightarrow \end{matrix} \begin{bmatrix} 1 & 22 & \vdots & -16 \\ 0 & 1 & \vdots & -\tfrac{3}{4} \end{bmatrix}$$

$y = -\tfrac{3}{4}$

$x + 22\left(-\tfrac{3}{4}\right) = -16 \Rightarrow x = \tfrac{1}{2}$

Answer: $\left(\tfrac{1}{2}, -\tfrac{3}{4}\right)$

47. $-x + 2y = 1.5$

$2x - 4y = 3.0$

$$\begin{bmatrix} -1 & 2 & \vdots & 1.5 \\ 2 & -4 & \vdots & 3.0 \end{bmatrix}$$

$$2R_1 + R_2 \rightarrow \begin{bmatrix} -1 & 2 & \vdots & 1.5 \\ 0 & 0 & \vdots & 6.0 \end{bmatrix}$$

The system is inconsistent and there is no solution.

49. $x \quad - 3z = -2$

$3x + y - 2z = 5$

$2x + 2y + z = 4$

$$\begin{bmatrix} 1 & 0 & -3 & \vdots & -2 \\ 3 & 1 & -2 & \vdots & 5 \\ 2 & 2 & 1 & \vdots & 4 \end{bmatrix}$$

$$\begin{matrix} \\ -3R_1 + R_2 \rightarrow \\ -2R_1 + R_3 \rightarrow \end{matrix} \begin{bmatrix} 1 & 0 & -3 & \vdots & -2 \\ 0 & 1 & 7 & \vdots & 11 \\ 0 & 2 & 7 & \vdots & 8 \end{bmatrix}$$

$$\begin{matrix} \\ \\ -2R_2 + R_3 \rightarrow \end{matrix} \begin{bmatrix} 1 & 0 & -3 & \vdots & -2 \\ 0 & 1 & 7 & \vdots & 11 \\ 0 & 0 & -7 & \vdots & -14 \end{bmatrix}$$

$$\begin{matrix} \\ \\ -\frac{1}{7}R_3 \rightarrow \end{matrix} \begin{bmatrix} 1 & 0 & -3 & \vdots & -2 \\ 0 & 1 & 7 & \vdots & 11 \\ 0 & 0 & 1 & \vdots & 2 \end{bmatrix}$$

$z = 2$

$y + 7(2) = 11 \implies y = -3$

$x - 3(2) = -2 \implies x = 4$

Answer: $(4, -3, 2)$

51. $x + y - 5z = 3$

$x \quad - 2z = 1$

$2x - y - z = 0$

$$\begin{bmatrix} 1 & 1 & -5 & \vdots & 3 \\ 1 & 0 & -2 & \vdots & 1 \\ 2 & -1 & -1 & \vdots & 0 \end{bmatrix}$$

$$\begin{matrix} \\ -R_1 + R_2 \rightarrow \\ -2R_1 + R_3 \rightarrow \end{matrix} \begin{bmatrix} 1 & 1 & -5 & \vdots & 3 \\ 0 & -1 & 3 & \vdots & -2 \\ 0 & -3 & 9 & \vdots & -6 \end{bmatrix}$$

$$\begin{matrix} \\ \\ -3R_2 + R_3 \rightarrow \end{matrix} \begin{bmatrix} 1 & 1 & -5 & \vdots & 3 \\ 0 & -1 & 3 & \vdots & -2 \\ 0 & 0 & 0 & \vdots & 0 \end{bmatrix}$$

$$\begin{matrix} R_2 + R_1 \rightarrow \\ -R_2 \rightarrow \\ \end{matrix} \begin{bmatrix} 1 & 0 & -2 & \vdots & 1 \\ 0 & 1 & -3 & \vdots & 2 \\ 0 & 0 & 0 & \vdots & 0 \end{bmatrix}$$

$z = a$

$y - 3a = 2 \implies y = 3a + 2$

$x - 2a = 1 \implies x = 2a + 1$

Answer: $(2a + 1, 3a + 2, a)$

53. $x + 2y + z = 8$

$3x + 7y + 6z = 26$

$$\begin{bmatrix} 1 & 2 & 1 & \vdots & 8 \\ 3 & 7 & 6 & \vdots & 26 \end{bmatrix}$$

$$\begin{matrix} \\ -3R_1 + R_2 \rightarrow \end{matrix} \begin{bmatrix} 1 & 2 & 1 & \vdots & 8 \\ 0 & 1 & 3 & \vdots & 2 \end{bmatrix}$$

$$\begin{matrix} -2R_2 + R_1 \rightarrow \\ \end{matrix} \begin{bmatrix} 1 & 0 & -5 & \vdots & 4 \\ 0 & 1 & 3 & \vdots & 2 \end{bmatrix}$$

$z = a$

$y + 3a = 2 \implies y = -3a + 2$

$x - 5a = 4 \implies x = 5a + 4$

Answer: $(5a + 4, -3a + 2, a)$

55. $x + 2y = 0$

$-x - y = 0$

$$\begin{bmatrix} 1 & 2 & \vdots & 0 \\ -1 & -1 & \vdots & 0 \end{bmatrix}$$

$$\begin{matrix} \\ R_1 + R_2 \rightarrow \end{matrix} \begin{bmatrix} 1 & 2 & \vdots & 0 \\ 0 & 1 & \vdots & 0 \end{bmatrix}$$

$y = 0,\ x + 2(0) = 0 \implies x = 0$

Answer: $(0, 0)$

57. $3x + 3y + 12z = 6$

$x + y + 4z = 2$

$2x + 5y + 20z = 10$

$-x + 2y + 8z = 4$

$$\begin{bmatrix} 3 & 3 & 12 & \vdots & 6 \\ 1 & 1 & 4 & \vdots & 2 \\ 2 & 5 & 20 & \vdots & 10 \\ -1 & 2 & 8 & \vdots & 4 \end{bmatrix} \Rightarrow \begin{bmatrix} 1 & 0 & 0 & \vdots & 0 \\ 0 & 0 & 0 & \vdots & 0 \\ 0 & 1 & 4 & \vdots & 2 \\ 0 & 0 & 0 & \vdots & 0 \end{bmatrix}$$

$z = a$

$y = -4a + 2$

$x = 0$

Answer: $(0, -4a + 2, a)$

59. $\begin{aligned} 2x + y - z + 2w &= -6 \\ 3x + 4y + w &= 1 \\ x + 5y + 2z + 6w &= -3 \\ 5x + 2y - z - w &= 3 \end{aligned}$ $\quad \left[\begin{array}{cccc:c} 2 & 1 & -1 & 2 & -6 \\ 3 & 4 & 0 & 1 & 1 \\ 1 & 5 & 2 & 6 & -3 \\ 5 & 2 & -1 & -1 & 3 \end{array}\right] \Rightarrow \left[\begin{array}{cccc:c} 1 & 5 & 2 & 6 & -3 \\ 0 & 1 & -1 & -3 & 2 \\ 0 & 0 & 238 & 629 & -306 \\ 0 & 0 & 0 & -71 & 142 \end{array}\right]$

$x = 1$
$y = 0$
$z = 4$
$w = -2$

Answer: $(1, 0, 4, -2)$

61. $\begin{aligned} x + y + z &= 0 \\ 2x + 3y + z &= 0 \\ 3x + 5y + z &= 0 \end{aligned}$ $\quad \left[\begin{array}{ccc:c} 1 & 1 & 1 & 0 \\ 2 & 3 & 1 & 0 \\ 3 & 5 & 1 & 0 \end{array}\right] \Rightarrow \left[\begin{array}{ccc:c} 1 & 0 & 2 & 0 \\ 0 & 1 & -1 & 0 \\ 0 & 0 & 0 & 0 \end{array}\right]$

$z = a$
$y = a$
$x = -2a$

Answer: $(-2a, a, a)$

63. $z = a$
$y = -4a + 1$
$x = -3a - 2$

One possible system is:

$$\begin{aligned} x + y + 7z &= (-3a - 2) + (-4a + 1) + 7a = -1 \\ x + 2y + 11z &= (-3a - 2) + 2(-4a + 1) + 11a = -0 \\ 2x + y + 10z &= 2(-3a - 2) + (-4a + 1) + 10a = -3 \end{aligned}$$

65. $x =$ amount at 8%, $y =$ amount at 9%, $z =$ amount at 12%

$$\begin{aligned} x + y + z &= 1{,}500{,}000 \\ 0.08x + 0.09y + 0.12z &= 133{,}000 \\ x - 4z &= 0 \end{aligned}$$

$$\left[\begin{array}{ccc:c} 1 & 1 & 1 & 1{,}500{,}000 \\ 0.08 & 0.09 & 0.12 & 133{,}000 \\ 1 & 0 & -4 & 0 \end{array}\right]$$

$$\begin{array}{r} \\ -0.08R_1 + R_2 \rightarrow \\ -R_1 + R_3 \rightarrow \end{array} \left[\begin{array}{ccc:c} 1 & 1 & 1 & 1{,}500{,}000 \\ 0 & 0.01 & 0.04 & 13{,}000 \\ 0 & -1 & -5 & -1{,}500{,}000 \end{array}\right]$$

$$\begin{array}{r} \\ 100R_2 \rightarrow \\ R_2 + R_3 \rightarrow \end{array} \left[\begin{array}{ccc:c} 1 & 1 & 1 & 1{,}500{,}000 \\ 0 & 1 & 4 & 1{,}300{,}000 \\ 0 & 0 & -1 & -200{,}000 \end{array}\right]$$

$-z = -200{,}000 \Rightarrow z = 200{,}000$
$y + 4(200{,}000) = 1{,}300{,}000 \Rightarrow y = 500{,}000$
$x + (500{,}000) + (200{,}000) = 1{,}500{,}000 \Rightarrow x = 800{,}000$

Answer: \$800,000 at 8%, \$500,000 at 9%, \$200,000 at 12%

67. $\dfrac{4x^2}{(x+1)^2(x-1)} = \dfrac{A}{x-1} + \dfrac{B}{x+1} + \dfrac{C}{(x+1)^2}$

$4x^2 = A(x+1)^2 + B(x+1)(x-1) + C(x-1)$

Let $x = 1$: $4 = 4A \implies A = 1$

Let $x = -1$: $4 = -2C \implies C = -2$

Let $x = 0$: $0 = A - B - C \implies 0 = 1 - B - (-2) \implies B = 3$

Thus, $\dfrac{4x^2}{(x+1)^2(x-1)} = \dfrac{1}{x-1} + \dfrac{3}{x+1} - \dfrac{2}{(x+1)^2}$.

69. $f(x) = ax^2 + bx + c$

$f(1) = a + b + c = 8$

$f(2) = 4a + 2b + c = 13$

$f(3) = 9a + 3b + c = 20$

$$\begin{bmatrix} 1 & 1 & 1 & \vdots & 8 \\ 4 & 2 & 1 & \vdots & 13 \\ 9 & 3 & 1 & \vdots & 20 \end{bmatrix}$$

$$\begin{matrix} \\ -4R_1 + R_2 \rightarrow \\ -9R_1 + R_3 \rightarrow \end{matrix} \begin{bmatrix} 1 & 1 & 1 & \vdots & 8 \\ 0 & -2 & -3 & \vdots & -19 \\ 0 & -6 & -8 & \vdots & -52 \end{bmatrix}$$

$$\begin{matrix} \\ -\frac{1}{2}R_2 \rightarrow \\ -3R_2 + R_3 \rightarrow \end{matrix} \begin{bmatrix} 1 & 1 & 1 & \vdots & 8 \\ 0 & 1 & \frac{3}{2} & \vdots & \frac{19}{2} \\ 0 & 0 & 1 & \vdots & 5 \end{bmatrix}$$

$c = 5$

$b + \frac{3}{2}(5) = \frac{19}{2} \implies b = 2$

$a + 2 + 5 = 8 \implies a = 1$

Answer: $y = x^2 + 2x + 5$

71. (a) (0, 5.0), (15, 9.6), (30, 12.4)

$f(x) = ax^2 + bx + c$

$f(0) = c = 5$

$f(15) = 225a + 15b + c = 9.6 \implies 225a + 15b = 4.6$

$f(30) = 900a + 30b + c = 12.4 \implies 900a + 30b = 7.4$

$$\begin{bmatrix} 225 & 15 & \vdots & 4.6 \\ 900 & 30 & \vdots & 7.4 \end{bmatrix}$$

$$\begin{matrix} \frac{1}{225}R_1 \rightarrow \\ -900R_1 + R_2 \rightarrow \end{matrix} \begin{bmatrix} 1 & \frac{1}{15} & \vdots & \frac{23}{1125} \\ 0 & -30 & \vdots & -11 \end{bmatrix}$$

$-30b = -11 \implies b = \dfrac{11}{30} \approx 0.367$

$1 + \dfrac{1}{15}\left(\dfrac{11}{30}\right) = \dfrac{23}{1125} \implies a = -\dfrac{1}{250} = -0.004$

Thus, $y = -0.004x^2 + 0.367x + 5$.

— CONTINUED —

71. — CONTINUED —

(b) 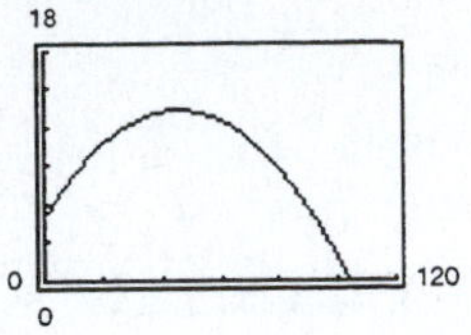

The maximum height is approximately 13 feet and the ball strikes the ground at approximately 104 feet.

(c) The maximum occurs at the vertex.

$$-\frac{b}{2a} = \frac{-0.367}{2(-0.004)} = 45.875$$

$$f(45.875) = -0.004(45.875)^2 + 0.367(45.875) + 5 = 13.418 \text{ feet}$$

The ball strikes the ground when $y = 0$.

$$-0.004x^2 + 0.367x + 5 = 0$$

By the Quadratic Formula and using the positive value for x we have $x \approx 103.793$ feet.

73. (a) $x_1 + x_3 = 600$

$x_1 = x_2 + x_4 \Rightarrow x_1 - x_2 - x_4 = 0$

$x_2 + x_5 = 500$

$x_3 + x_6 = 600$

$x_4 + x_7 = x_6 \Rightarrow x_4 - x_6 + x_7 = 0$

$x_5 + x_7 = 500$

$$\left[\begin{array}{ccccccc:c} 1 & 0 & 1 & 0 & 0 & 0 & 0 & 600 \\ 1 & -1 & 0 & -1 & 0 & 0 & 0 & 0 \\ 0 & 1 & 0 & 0 & 1 & 0 & 0 & 500 \\ 0 & 0 & 1 & 0 & 0 & 1 & 0 & 600 \\ 0 & 0 & 0 & 1 & 0 & -1 & 1 & 0 \\ 0 & 0 & 0 & 0 & 1 & 0 & 1 & 500 \end{array}\right]$$

$$\begin{array}{r} \\ -R_1 + R_2 \rightarrow \\ R_2 + R_3 \rightarrow \\ R_3 + R_4 \rightarrow \\ R_4 + R_5 \rightarrow \\ -R_5 + R_6 \rightarrow \end{array} \left[\begin{array}{ccccccc:c} 1 & 0 & 1 & 0 & 0 & 0 & 0 & 600 \\ 0 & -1 & -1 & -1 & 0 & 0 & 0 & -600 \\ 0 & 0 & -1 & -1 & 1 & 0 & 0 & -100 \\ 0 & 0 & 0 & -1 & 1 & 1 & 0 & 500 \\ 0 & 0 & 0 & 0 & 1 & 0 & 1 & 500 \\ 0 & 0 & 0 & 0 & 0 & 0 & 0 & 0 \end{array}\right]$$

$$\begin{array}{r} \\ -R_3 + R_2 \rightarrow \\ -R_4 + R_3 \rightarrow \\ -R_4 \rightarrow \\ \\ \\ \end{array} \left[\begin{array}{ccccccc:c} 1 & 0 & 1 & 0 & 0 & 0 & 0 & 600 \\ 0 & -1 & 0 & 0 & -1 & 0 & 0 & -500 \\ 0 & 0 & -1 & 0 & 0 & -1 & 0 & -600 \\ 0 & 0 & 0 & 1 & -1 & -1 & 0 & -500 \\ 0 & 0 & 0 & 0 & 1 & 0 & 1 & 500 \\ 0 & 0 & 0 & 0 & 0 & 0 & 0 & 0 \end{array}\right]$$

Let $x_7 = t$ and $x_6 = s$, then $x_5 = 500 - t$,
$x_4 = -500 + s + (500 - t) = s - t$,
$x_3 = 600 - s$, $x_2 = 500 - (500 - t) = t$, $x_1 = 600 - (600 - s) = s$.

Answer: $(s, t, 600 - s, s - t, 500 - t, s, t)$

(b) $s = 0, t = 0$: $x_1 = 0, x_2 = 0, x_3 = 600, x_4 = 0, x_5 = 500, x_6 = 0, x_7 = 0$

(c) $s = 0, t = -500$: $x_1 = 0, x_2 = -500, x_3 = 600, x_4 = 500, x_5 = 1000, x_6 = 0, x_7 = -500$

75. $f(x) = 2^{x-1}$

x	-1	0	1	2	3
y	$\frac{1}{4}$	$\frac{1}{2}$	1	2	4

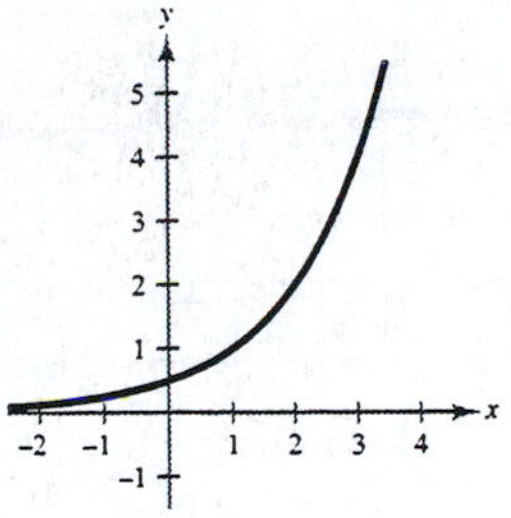

77. $h(x) = \log_2(x - 1) \implies 2^y = x - 1 \implies 2^y + 1 = x$

x	$\frac{3}{2}$	2	3	5	9
y	-1	0	1	2	3

Vertical asymptote: $x = 1$

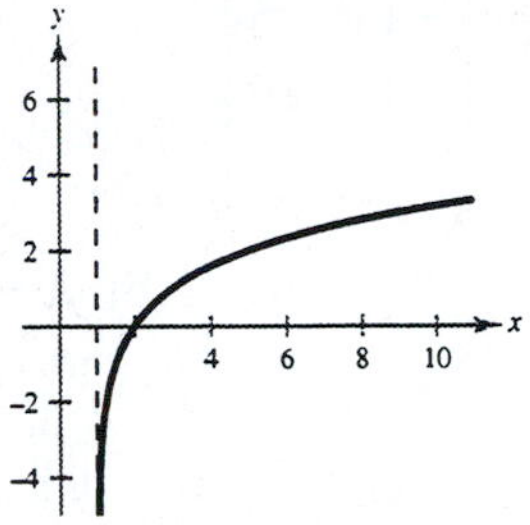

Section 8.2 Operations with Matrices

- $A = B$ if and only if they have the same order and $a_{ij} = b_{ij}$.
- You should be able to perform the operations of matrix addition, scalar multiplication, and matrix multiplication.
- Some properties of matrix addition and scalar multiplication are:
  (a) $A + B = B + A$
  (b) $A + (B + C) = (A + B) + C$
  (c) $(cd)A = c(dA)$
  (d) $1A = A$
  (e) $c(A + B) = cA + cB$
  (f) $(c + d)A = cA + dA$
- You should remember that $AB \neq BA$ in general.

Solutions to Odd-Numbered Exercises

1. $x = -4, \ y = 22$

3. $2x + 1 = 5, \ 3y - 5 = 4$

$x = 2, \ y = 3$

5. (a) $A + B = \begin{bmatrix} 1 & -1 \\ 2 & -1 \end{bmatrix} + \begin{bmatrix} 2 & -1 \\ -1 & 8 \end{bmatrix} = \begin{bmatrix} 1+2 & -1-1 \\ 2-1 & -1+8 \end{bmatrix} = \begin{bmatrix} 3 & -2 \\ 1 & 7 \end{bmatrix}$

(b) $A - B = \begin{bmatrix} 1 & -1 \\ 2 & -1 \end{bmatrix} - \begin{bmatrix} 2 & -1 \\ -1 & 8 \end{bmatrix} = \begin{bmatrix} 1-2 & -1+1 \\ 2+1 & -1-8 \end{bmatrix} = \begin{bmatrix} -1 & 0 \\ 3 & -9 \end{bmatrix}$

— CONTINUED —

5. — CONTINUED —

(c) $3A = 3\begin{bmatrix} 1 & -1 \\ 2 & -1 \end{bmatrix} = \begin{bmatrix} 3(1) & 3(-1) \\ 3(2) & 3(-1) \end{bmatrix} = \begin{bmatrix} 3 & -3 \\ 6 & -3 \end{bmatrix}$

(d) $3A - 2B = \begin{bmatrix} 3 & -3 \\ 6 & -3 \end{bmatrix} - 2\begin{bmatrix} 2 & -1 \\ -1 & 8 \end{bmatrix} = \begin{bmatrix} 3 & -3 \\ 6 & -3 \end{bmatrix} + \begin{bmatrix} -4 & 2 \\ 2 & -16 \end{bmatrix} = \begin{bmatrix} -1 & -1 \\ 8 & -19 \end{bmatrix}$

7. $A = \begin{bmatrix} 6 & -1 \\ 2 & 4 \\ -3 & 5 \end{bmatrix}, B = \begin{bmatrix} 1 & 4 \\ -1 & 5 \\ 1 & 10 \end{bmatrix}$

(a) $A + B = \begin{bmatrix} 7 & 3 \\ 1 & 9 \\ -2 & 15 \end{bmatrix}$ (b) $A - B = \begin{bmatrix} 5 & -5 \\ 3 & -1 \\ -4 & -5 \end{bmatrix}$ (c) $3A = \begin{bmatrix} 18 & -3 \\ 6 & 12 \\ -9 & 15 \end{bmatrix}$

(d) $3A - 2B = \begin{bmatrix} 18 & -3 \\ 6 & 12 \\ -9 & 15 \end{bmatrix} - \begin{bmatrix} 2 & 8 \\ -2 & 10 \\ 2 & 20 \end{bmatrix} = \begin{bmatrix} 16 & -11 \\ 8 & 2 \\ -11 & -5 \end{bmatrix}$

9. $A = \begin{bmatrix} 2 & 2 & -1 & 0 & 1 \\ 1 & 1 & -2 & 0 & -1 \end{bmatrix}, B = \begin{bmatrix} 1 & 1 & -1 & 1 & 0 \\ -3 & 4 & 9 & -6 & -7 \end{bmatrix}$

(a) $A + B = \begin{bmatrix} 3 & 3 & -2 & 1 & 1 \\ -2 & 5 & 7 & -6 & -8 \end{bmatrix}$

(b) $A - B = \begin{bmatrix} 1 & 1 & 0 & -1 & 1 \\ 4 & -3 & -11 & 6 & 6 \end{bmatrix}$

(c) $3A = \begin{bmatrix} 6 & 6 & -3 & 0 & 3 \\ 3 & 3 & -6 & 0 & -3 \end{bmatrix}$

(d) $3A - 2B = \begin{bmatrix} 6 & 6 & -3 & 0 & 3 \\ 3 & 3 & -6 & 0 & -3 \end{bmatrix} - \begin{bmatrix} 2 & 2 & -2 & 2 & 0 \\ -6 & 8 & 18 & -12 & -14 \end{bmatrix} = \begin{bmatrix} 4 & 4 & -1 & -2 & 3 \\ 9 & -5 & -24 & 12 & 11 \end{bmatrix}$

11. $X = 3\begin{bmatrix} -2 & -1 \\ 1 & 0 \\ 3 & 4 \end{bmatrix} - 2\begin{bmatrix} 0 & 3 \\ 2 & 0 \\ -4 & -1 \end{bmatrix} = \begin{bmatrix} -6 & -3 \\ 3 & 0 \\ 9 & -12 \end{bmatrix} - \begin{bmatrix} 0 & 6 \\ 4 & 0 \\ -8 & -2 \end{bmatrix} = \begin{bmatrix} -6 & -9 \\ -1 & 0 \\ 17 & -10 \end{bmatrix}$

13. $X = -\frac{3}{2}A + \frac{1}{2}B = -\frac{3}{2}\begin{bmatrix} -2 & -1 \\ 1 & 0 \\ 3 & -4 \end{bmatrix} + \frac{1}{2}\begin{bmatrix} 0 & 3 \\ 2 & 0 \\ -4 & -1 \end{bmatrix} = \begin{bmatrix} 3 & 3 \\ -\frac{1}{2} & 0 \\ -\frac{13}{2} & \frac{11}{2} \end{bmatrix}$

15. (a) $AB = \begin{bmatrix} 1 & 2 \\ 4 & 2 \end{bmatrix}\begin{bmatrix} 2 & -1 \\ -1 & 8 \end{bmatrix} = \begin{bmatrix} 2-2 & -1+16 \\ 8-2 & -4+16 \end{bmatrix} = \begin{bmatrix} 0 & 15 \\ 6 & 12 \end{bmatrix}$

(b) $BA = \begin{bmatrix} 2 & -1 \\ -1 & 8 \end{bmatrix}\begin{bmatrix} 1 & 2 \\ 4 & 2 \end{bmatrix} = \begin{bmatrix} 2-4 & 4-2 \\ -1+32 & -2+16 \end{bmatrix} = \begin{bmatrix} -2 & 2 \\ 31 & 14 \end{bmatrix}$

(c) $A^2 = \begin{bmatrix} 1 & 2 \\ 4 & 2 \end{bmatrix}\begin{bmatrix} 1 & 2 \\ 4 & 2 \end{bmatrix} = \begin{bmatrix} 1+8 & 2+4 \\ 4+8 & 8+4 \end{bmatrix} = \begin{bmatrix} 9 & 6 \\ 12 & 12 \end{bmatrix}$

17. (a) $AB = \begin{bmatrix} 3 & -1 \\ 1 & 3 \end{bmatrix}\begin{bmatrix} 1 & -3 \\ 3 & 1 \end{bmatrix} = \begin{bmatrix} 3-3 & -9-1 \\ 1+9 & -3+3 \end{bmatrix} = \begin{bmatrix} 0 & -10 \\ 10 & 0 \end{bmatrix}$

(b) $BA = \begin{bmatrix} 1 & -3 \\ 3 & 1 \end{bmatrix}\begin{bmatrix} 3 & -1 \\ 1 & 3 \end{bmatrix} = \begin{bmatrix} 3-3 & -1-9 \\ 9+1 & -3+3 \end{bmatrix} = \begin{bmatrix} 0 & -10 \\ 10 & 0 \end{bmatrix}$

(c) $A^2 = \begin{bmatrix} 3 & -1 \\ 1 & 3 \end{bmatrix}\begin{bmatrix} 3 & -1 \\ 1 & 3 \end{bmatrix} = \begin{bmatrix} 9-1 & -3-3 \\ 3+3 & -1+9 \end{bmatrix} = \begin{bmatrix} 8 & -6 \\ 6 & 8 \end{bmatrix}$

19. (a) $AB = \begin{bmatrix} 1 & -1 & 7 \\ 2 & -1 & 8 \\ 3 & 1 & -1 \end{bmatrix}\begin{bmatrix} 1 & 1 & 2 \\ 2 & 1 & 1 \\ 1 & -3 & 2 \end{bmatrix} = \begin{bmatrix} 1-2+7 & 1-1-21 & 2-1+14 \\ 2-2+8 & 2-1-24 & 4-1+16 \\ 3+2-1 & 3+1+3 & 6+1-2 \end{bmatrix} = \begin{bmatrix} 6 & -21 & 15 \\ 8 & -23 & 19 \\ 4 & 7 & 5 \end{bmatrix}$

(b) $BA = \begin{bmatrix} 1 & 1 & 2 \\ 2 & 1 & 1 \\ 1 & -3 & 2 \end{bmatrix}\begin{bmatrix} 1 & -1 & 7 \\ 2 & -1 & 8 \\ 3 & 1 & -1 \end{bmatrix} = \begin{bmatrix} 1+2+6 & -1-1+2 & 7+8-2 \\ 2+2+3 & -2-1+1 & 14+8-1 \\ 1-6+6 & -1+3+2 & 7-24-2 \end{bmatrix} = \begin{bmatrix} 9 & 0 & 13 \\ 7 & -2 & 21 \\ 1 & 4 & -19 \end{bmatrix}$

(c) $A^2 = \begin{bmatrix} 1 & -1 & 7 \\ 2 & -1 & 8 \\ 3 & 1 & -1 \end{bmatrix}\begin{bmatrix} 1 & -1 & 7 \\ 2 & -1 & 8 \\ 3 & 1 & -1 \end{bmatrix} = \begin{bmatrix} 1-2+21 & -1+1+7 & 7-8-7 \\ 2-2+24 & -2+1+8 & 14-8-8 \\ 3+2-3 & -3-1-1 & 21+8+1 \end{bmatrix} = \begin{bmatrix} 20 & 7 & -8 \\ 24 & 7 & -2 \\ 2 & -5 & 30 \end{bmatrix}$

21. A is 3×2 and B is $3 \times 3 \implies AB$ is not defined.

23. A is 3×2, B is $2 \times 2 \implies AB$ is 3×2.

$$AB = \begin{bmatrix} -1 & 3 \\ 4 & -5 \\ 0 & 2 \end{bmatrix}\begin{bmatrix} 1 & 2 \\ 0 & 7 \end{bmatrix} = \begin{bmatrix} -1 & 19 \\ 4 & -27 \\ 0 & 14 \end{bmatrix}$$

25. A is 3×3, B is $3 \times 3 \implies AB$ is 3×3.

$$AB = \begin{bmatrix} 5 & 0 & 0 \\ 0 & -8 & 0 \\ 0 & 0 & 7 \end{bmatrix}\begin{bmatrix} \frac{1}{5} & 0 & 0 \\ 0 & -\frac{1}{8} & 0 \\ 0 & 0 & \frac{1}{2} \end{bmatrix} = \begin{bmatrix} 1 & 0 & 0 \\ 0 & 1 & 0 \\ 0 & 0 & \frac{7}{2} \end{bmatrix}$$

27. A is 2×1, B is $1 \times 4 \implies AB$ is 2×4.

$$\begin{bmatrix} 10 \\ 12 \end{bmatrix}\begin{bmatrix} 6 & -2 & 1 & 6 \end{bmatrix} = \begin{bmatrix} 60 & -20 & 10 & 60 \\ 72 & -24 & 12 & 72 \end{bmatrix}$$

29. $\begin{bmatrix} 5 & 6 & -3 \\ -2 & 5 & 1 \\ 10 & -5 & 5 \end{bmatrix}\begin{bmatrix} 1 & -1 & 2 \\ 8 & 1 & 4 \\ 4 & -2 & 9 \end{bmatrix} = \begin{bmatrix} 41 & 7 & 7 \\ 42 & 5 & 25 \\ -10 & -25 & 45 \end{bmatrix}$

31. $\begin{bmatrix} -3 & 8 & -6 & 8 \\ -12 & 15 & 9 & 6 \\ 5 & -1 & 1 & 5 \end{bmatrix}\begin{bmatrix} 3 & 1 & 6 \\ 24 & 15 & 14 \\ 16 & 10 & 21 \\ 8 & -4 & 10 \end{bmatrix} = \begin{bmatrix} 151 & 25 & 48 \\ 516 & 279 & 387 \\ 47 & -20 & 87 \end{bmatrix}$

33. A is 2×4 and B is $2 \times 4 \implies AB$ is not defined.

35. $A = \begin{bmatrix} -1 & 1 \\ -2 & 1 \end{bmatrix}, X = \begin{bmatrix} x \\ y \end{bmatrix}, B = \begin{bmatrix} 4 \\ 0 \end{bmatrix}$

By Gauss-Jordan elimination on

$$\left[\begin{array}{cc:c} -1 & 1 & 4 \\ -2 & 1 & 0 \end{array}\right]$$

$$\begin{array}{r} -R_1 \rightarrow \\ 2R_1 + R_2 \rightarrow \end{array} \left[\begin{array}{cc:c} 1 & -1 & -4 \\ 0 & -1 & -8 \end{array}\right]$$

$$\begin{array}{r} R_2 + R_1 \rightarrow \\ -R_2 \rightarrow \end{array} \left[\begin{array}{cc:c} 1 & 0 & 4 \\ 0 & 1 & 8 \end{array}\right]$$

we have $x = 4$ and $y = 8$.

37. $A = \begin{bmatrix} 2 & 3 \\ 1 & 4 \end{bmatrix}, X = \begin{bmatrix} x \\ y \end{bmatrix}, B = \begin{bmatrix} 5 \\ 10 \end{bmatrix}$

By Gauss-Jordan elimination on

$$\left[\begin{array}{cc:c} 1 & 4 & 10 \\ 2 & 3 & 5 \end{array}\right]$$

$$\begin{array}{r} \\ -2R_1 + R_2 \rightarrow \end{array} \left[\begin{array}{cc:c} 1 & 4 & 10 \\ 0 & -5 & -15 \end{array}\right]$$

$$\begin{array}{r} -4R_2 + R_1 \rightarrow \\ -\frac{1}{5}R_2 \rightarrow \end{array} \left[\begin{array}{cc:c} 1 & 0 & -2 \\ 0 & 1 & 3 \end{array}\right]$$

we have $x = -2$ and $y = 3$.

39. $A = \begin{bmatrix} 2 & 0 \\ 4 & 5 \end{bmatrix}$

$$f(A) = A^2 - 5A + 2 = \begin{bmatrix} 2 & 0 \\ 4 & 5 \end{bmatrix}\begin{bmatrix} 2 & 0 \\ 4 & 5 \end{bmatrix} - 5\begin{bmatrix} 2 & 4 \\ 0 & 5 \end{bmatrix} + 2\begin{bmatrix} 1 & 0 \\ 0 & 1 \end{bmatrix} = \begin{bmatrix} -4 & 0 \\ 8 & 2 \end{bmatrix}$$

41. $A = \begin{bmatrix} 3 & 1 & 4 \\ 0 & 2 & 6 \\ 0 & 0 & 5 \end{bmatrix}$

$$f(A) = \begin{bmatrix} 3 & 1 & 4 \\ 0 & 2 & 6 \\ 0 & 0 & 5 \end{bmatrix}^3 - 10\begin{bmatrix} 3 & 1 & 4 \\ 0 & 2 & 6 \\ 0 & 0 & 5 \end{bmatrix}^2 + 31\begin{bmatrix} 3 & 1 & 4 \\ 0 & 2 & 6 \\ 0 & 0 & 5 \end{bmatrix} - 30\begin{bmatrix} 1 & 0 & 0 \\ 0 & 1 & 0 \\ 0 & 0 & 1 \end{bmatrix} = \begin{bmatrix} 0 & 0 & 0 \\ 0 & 0 & 0 \\ 0 & 0 & 0 \end{bmatrix}$$

43. $AC = \begin{bmatrix} 0 & 1 \\ 0 & 1 \end{bmatrix}\begin{bmatrix} 2 & 3 \\ 2 & 3 \end{bmatrix} = \begin{bmatrix} 2 & 3 \\ 2 & 3 \end{bmatrix}$

$BC = \begin{bmatrix} 1 & 0 \\ 1 & 0 \end{bmatrix}\begin{bmatrix} 2 & 3 \\ 2 & 3 \end{bmatrix} = \begin{bmatrix} 2 & 3 \\ 2 & 3 \end{bmatrix}$

Thus, $AC = BC$ even though $A \neq B$.

For 45–53, A is of order 2×3, B is of order 2×3, C is of order 3×2 and D is of order 2×2.

45. $A + 2C$ is not possible. A and C are not of the same order.

47. AB is not possible. The number of columns of A does not equal the number of rows of B.

49. $BC - D$ is possible. The resulting order is 2×2.

51. (CA) is 3×3 so $(CA)D$ is not possible.

53. $D(A - 3B)$ is possible. The resulting order is 2×3.

55. $1.20\begin{bmatrix} 60 & 40 & 20 \\ 30 & 90 & 60 \end{bmatrix} = \begin{bmatrix} 72 & 48 & 24 \\ 36 & 108 & 72 \end{bmatrix}$

57. $BA = [3.75 \quad 7.00]\begin{bmatrix} 100 & 75 & 75 \\ 125 & 150 & 100 \end{bmatrix} = [\$1250.00 \quad \$1331.25 \quad \$981.25]$

The entries in the last matrix represent the profits for both crops at the three outlets.

59. $ST = \begin{bmatrix} 3 & 2 & 2 & 3 & 0 \\ 0 & 2 & 3 & 4 & 3 \\ 4 & 2 & 1 & 3 & 2 \end{bmatrix} \begin{bmatrix} 840 & 1100 \\ 1200 & 1350 \\ 1450 & 1650 \\ 2650 & 3000 \\ 3050 & 3200 \end{bmatrix} = \begin{bmatrix} \$15{,}770 & \$18{,}300 \\ \$26{,}500 & \$29{,}250 \\ \$21{,}260 & \$24{,}150 \end{bmatrix}$

The entries represent the wholesale and retail inventory values of the inventories at the three outlets.

61. $ST = \begin{bmatrix} 1 & 0.5 & 0.2 \\ 1.6 & 1.0 & 0.2 \\ 2.5 & 2.0 & 0.4 \end{bmatrix} \begin{bmatrix} 12 & 10 \\ 9 & 8 \\ 6 & 5 \end{bmatrix} = \begin{bmatrix} \$17.70 & \$15.00 \\ \$29.40 & \$25.00 \\ \$50.40 & \$43.00 \end{bmatrix}$

This represents the labor cost for each boat size at each plant.

63. $A^2 = \begin{bmatrix} i & 0 \\ 0 & i \end{bmatrix} \begin{bmatrix} i & 0 \\ 0 & i \end{bmatrix} = \begin{bmatrix} -1 & 0 \\ 0 & -1 \end{bmatrix}$ and $i^2 = -1$

$A^3 = A^2A = \begin{bmatrix} -1 & 0 \\ 0 & -1 \end{bmatrix} \begin{bmatrix} i & 0 \\ 0 & i \end{bmatrix} = \begin{bmatrix} -i & 0 \\ 0 & -i \end{bmatrix}$ and $i^3 = -i$

$A^4 = A^3A = \begin{bmatrix} -i & 0 \\ 0 & -i \end{bmatrix} \begin{bmatrix} i & 0 \\ 0 & i \end{bmatrix} = \begin{bmatrix} 1 & 0 \\ 0 & 1 \end{bmatrix}$ and $i^4 = 1$

65. The product of two diagonal matrices of the same order is a diagonal matrix whose entries are the products of the corresponding diagonal entries of A and B.

Section 8.3 The Inverse of a Square Matrix

- You should be able to find the inverse, if it exists, of a square matrix.
  - (a) Write the $n \times 2n$ matrix that consists of the given matrix A on the left and the $n \times n$ identity matrix I on the right to obtain $[A \;\vdots\; I]$. Note that we separate the matrices A and I by a dotted line. We call this process **adjoining** the matrices A and I.
  - (b) If possible, row reduce A to I using elementary row operations of the *entire* matrix $[A \;\vdots\; I]$. The result will be the matrix $[I \;\vdots\; A^{-1}]$. If this is not possible, then A is not invertible.
  - (c) Check your work by multiplying to see that $AA^{-1} = I = A^{-1}A$.
- You should be able to use inverse matrices to solve systems of equation.

Solutions to Odd-Numbered Exercises

1. $AB = \begin{bmatrix} 2 & 1 \\ 5 & 3 \end{bmatrix} \begin{bmatrix} 3 & -1 \\ -5 & 2 \end{bmatrix} = \begin{bmatrix} 2(3) + 1(-5) & 2(-1) + 1(2) \\ 5(3) + 3(-5) & 5(-1) + 3(2) \end{bmatrix} = \begin{bmatrix} 1 & 0 \\ 0 & 1 \end{bmatrix}$

$BA = \begin{bmatrix} 3 & -1 \\ -5 & 2 \end{bmatrix} \begin{bmatrix} 2 & 1 \\ 5 & 3 \end{bmatrix} = \begin{bmatrix} 3(2) + (-1)(5) & 3(1) + (-1)(3) \\ -5(2) + 2(5) & -5(1) + 2(3) \end{bmatrix} = \begin{bmatrix} 1 & 0 \\ 0 & 1 \end{bmatrix}$

3. $$AB = \begin{bmatrix} 1 & 2 \\ 3 & 4 \end{bmatrix}\begin{bmatrix} -2 & 1 \\ \frac{3}{2} & -\frac{1}{2} \end{bmatrix} = \begin{bmatrix} -2+3 & 1-1 \\ -6+6 & 3-2 \end{bmatrix} = \begin{bmatrix} 1 & 0 \\ 0 & 1 \end{bmatrix}$$

$$BA = \begin{bmatrix} -2 & 1 \\ \frac{3}{2} & -\frac{1}{2} \end{bmatrix}\begin{bmatrix} 1 & 2 \\ 3 & 4 \end{bmatrix} = \begin{bmatrix} -2+3 & -4+4 \\ \frac{3}{2}-\frac{3}{2} & 3-2 \end{bmatrix} = \begin{bmatrix} 1 & 0 \\ 0 & 1 \end{bmatrix}$$

5. $$AB = \frac{1}{3}\begin{bmatrix} -2 & 2 & 3 \\ 1 & -1 & 0 \\ 0 & 1 & 4 \end{bmatrix}\begin{bmatrix} -4 & -5 & 3 \\ -4 & -8 & 3 \\ 1 & 2 & 0 \end{bmatrix} = \frac{1}{3}\begin{bmatrix} -8+8+3 & 10-16+6 & -6+6 \\ -4+4 & -5+8 & 3-3 \\ -4+4 & -8+8 & 3 \end{bmatrix}$$

$$= \frac{1}{3}\begin{bmatrix} 3 & 0 & 0 \\ 0 & 3 & 0 \\ 0 & 0 & 3 \end{bmatrix} = \begin{bmatrix} 1 & 0 & 0 \\ 0 & 1 & 0 \\ 0 & 0 & 1 \end{bmatrix}$$

$$BA = \frac{1}{3}\begin{bmatrix} -4 & -5 & 3 \\ -4 & -8 & 3 \\ 1 & 2 & 0 \end{bmatrix}\begin{bmatrix} -2 & 2 & 3 \\ 1 & -1 & 0 \\ 0 & 1 & 4 \end{bmatrix} = \frac{1}{3}\begin{bmatrix} 8-5 & -8+5+3 & -12+12 \\ 8-8 & -8+8+3 & -12+12 \\ -2+2 & 2-2 & 3 \end{bmatrix} = \begin{bmatrix} 1 & 0 & 0 \\ 0 & 1 & 0 \\ 0 & 0 & 1 \end{bmatrix}$$

7. $$AB = \begin{bmatrix} 2 & 0 & 1 & 1 \\ 3 & 0 & 0 & 1 \\ -1 & 1 & -2 & 1 \\ 4 & -1 & 1 & 0 \end{bmatrix}\begin{bmatrix} -1 & 2 & -1 & -1 \\ -4 & 9 & -5 & -6 \\ 0 & 1 & -1 & -1 \\ 3 & -5 & 3 & 3 \end{bmatrix} = \begin{bmatrix} 1 & 0 & 0 & 0 \\ 0 & 1 & 0 & 0 \\ 0 & 0 & 1 & 0 \\ 0 & 0 & 0 & 1 \end{bmatrix}$$

$$BA = \begin{bmatrix} -1 & 2 & -1 & -1 \\ -4 & 9 & -5 & -6 \\ 0 & 1 & -1 & -1 \\ 3 & -5 & 3 & 3 \end{bmatrix}\begin{bmatrix} 2 & 0 & 1 & 1 \\ 3 & 0 & 0 & 1 \\ -1 & 1 & -2 & 1 \\ 4 & -1 & 1 & 0 \end{bmatrix} = \begin{bmatrix} 1 & 0 & 0 & 0 \\ 0 & 1 & 0 & 0 \\ 0 & 0 & 1 & 0 \\ 0 & 0 & 0 & 1 \end{bmatrix}$$

9. $$[A \;\vdots\; I] = \left[\begin{array}{cc:cc} 2 & 0 & 1 & 0 \\ 0 & 3 & 0 & 1 \end{array}\right]$$

$$\begin{array}{l} \frac{1}{2}R_1 \rightarrow \\ \frac{1}{3}R_2 \rightarrow \end{array} \left[\begin{array}{cc:cc} 1 & 0 & \frac{1}{2} & 0 \\ 0 & 1 & 0 & \frac{1}{3} \end{array}\right] = [I \;\vdots\; A^{-1}]$$

$$A^{-1} = \begin{bmatrix} \frac{1}{2} & 0 \\ 0 & \frac{1}{3} \end{bmatrix} = \frac{1}{6}\begin{bmatrix} 3 & 0 \\ 0 & 2 \end{bmatrix}$$

11. $$[A \;\vdots\; I] = \left[\begin{array}{cc:cc} 1 & -2 & 1 & 0 \\ 2 & -3 & 0 & 1 \end{array}\right]$$

$$-2R_1 + R_2 \rightarrow \left[\begin{array}{cc:cc} 1 & -2 & 1 & 0 \\ 0 & 1 & -2 & 1 \end{array}\right]$$

$$2R_2 + R_1 \rightarrow \left[\begin{array}{cc:cc} 1 & 0 & -3 & 2 \\ 0 & 1 & -2 & 1 \end{array}\right] = [I \;\vdots\; A^{-1}]$$

$$A^{-1} = \begin{bmatrix} -3 & 2 \\ -2 & 1 \end{bmatrix}$$

13. $[A \;\vdots\; I] = \begin{bmatrix} -1 & 1 & \vdots & 1 & 0 \\ -2 & 1 & \vdots & 0 & 1 \end{bmatrix}$

$-2R_1 + R_2 \rightarrow \begin{bmatrix} -1 & 1 & \vdots & 1 & 0 \\ 0 & -1 & \vdots & -2 & 1 \end{bmatrix}$

$R_2 + R_1 \rightarrow \begin{bmatrix} -1 & 0 & \vdots & -1 & 1 \\ 0 & -1 & \vdots & -2 & 1 \end{bmatrix}$

$\begin{matrix} -R_1 \rightarrow \\ -R_2 \rightarrow \end{matrix} \begin{bmatrix} 1 & 0 & \vdots & 1 & -1 \\ 0 & 1 & \vdots & 2 & -1 \end{bmatrix} = [I \;\vdots\; A^{-1}]$

$A^{-1} = \begin{bmatrix} 1 & -1 \\ 2 & -1 \end{bmatrix}$

15. $[A \;\vdots\; I] = \begin{bmatrix} 2 & 4 & \vdots & 1 & 0 \\ 4 & 8 & \vdots & 0 & 1 \end{bmatrix}$

$-2R_1 + R_2 \rightarrow \begin{bmatrix} 2 & 4 & \vdots & 1 & 0 \\ 0 & 0 & \vdots & -2 & 1 \end{bmatrix}$

The two zeros in the second row imply that the inverse does not exist.

17. $A = \begin{bmatrix} 2 & 7 & 1 \\ -3 & -9 & 2 \end{bmatrix}$

A has no inverse because it is not square.

19. $\begin{bmatrix} 1 & 1 & 1 & \vdots & 1 & 0 & 0 \\ 3 & 5 & 4 & \vdots & 0 & 1 & 0 \\ 3 & 6 & 5 & \vdots & 0 & 0 & 1 \end{bmatrix}$

$\begin{matrix} \\ -3R_1 + R_2 \rightarrow \\ -3R_1 + R_3 \rightarrow \end{matrix} \begin{bmatrix} 1 & 1 & 1 & \vdots & 1 & 0 & 0 \\ 0 & 2 & 1 & \vdots & -3 & 1 & 0 \\ 0 & 3 & 2 & \vdots & -3 & 0 & 1 \end{bmatrix}$

$\begin{matrix} -R_2 + R_1 \\ \frac{1}{2}R_2 \rightarrow \\ -3R_2 + R_3 \rightarrow \end{matrix} \begin{bmatrix} 1 & 0 & \frac{1}{2} & \vdots & \frac{5}{2} & -\frac{1}{2} & 0 \\ 0 & 1 & \frac{1}{2} & \vdots & -\frac{3}{2} & \frac{1}{2} & 0 \\ 0 & 0 & \frac{1}{2} & \vdots & \frac{3}{2} & -\frac{3}{2} & 1 \end{bmatrix}$

$\begin{matrix} -R_3 + R_1 \rightarrow \\ -R_3 + R_2 \rightarrow \\ 2R_3 \rightarrow \end{matrix} \begin{bmatrix} 1 & 0 & 0 & \vdots & 1 & 1 & -1 \\ 0 & 1 & 0 & \vdots & -3 & 2 & -1 \\ 0 & 0 & 1 & \vdots & 3 & -3 & 2 \end{bmatrix}$

$A^{-1} = \begin{bmatrix} 1 & 1 & -1 \\ -3 & 2 & -1 \\ 3 & -3 & 2 \end{bmatrix}$

21. $[A \;\vdots\; I] = \left[\begin{array}{ccc:ccc} 1 & 0 & 0 & 1 & 0 & 0 \\ 3 & 4 & 0 & 0 & 1 & 0 \\ 2 & 5 & 5 & 0 & 0 & 1 \end{array}\right]$

$$\begin{array}{r} \\ -3R_1 + R_2 \rightarrow \\ -2R_1 + R_3 \rightarrow \end{array} \left[\begin{array}{ccc:ccc} 1 & 0 & 0 & 1 & 0 & 0 \\ 0 & 4 & 0 & -3 & 1 & 0 \\ 0 & 5 & 5 & -2 & 0 & 1 \end{array}\right]$$

$$\begin{array}{r} \\ \\ -\frac{5}{4}R_2 + R_3 \rightarrow \end{array} \left[\begin{array}{ccc:ccc} 1 & 0 & 0 & 1 & 0 & 0 \\ 0 & 4 & 0 & -3 & 1 & 0 \\ 0 & 0 & 5 & \frac{7}{4} & -\frac{5}{4} & 1 \end{array}\right]$$

$$\begin{array}{r} \\ \frac{1}{4}R_2 \rightarrow \\ \frac{1}{5}R_3 \rightarrow \end{array} \left[\begin{array}{ccc:ccc} 1 & 0 & 0 & 1 & 0 & 0 \\ 0 & 1 & 0 & -\frac{3}{4} & \frac{1}{4} & 0 \\ 0 & 0 & 1 & \frac{7}{20} & -\frac{1}{4} & \frac{1}{5} \end{array}\right] = [I \;\vdots\; A^{-1}]$$

$$A^{-1} = \frac{1}{20}\begin{bmatrix} 20 & 0 & 0 \\ -15 & 5 & 0 \\ 7 & -5 & 4 \end{bmatrix} = \begin{bmatrix} 1 & 0 & 0 \\ -0.75 & 0.25 & 0 \\ 0.35 & -0.25 & 0.2 \end{bmatrix}$$

23. $[A \;\vdots\; I] = \left[\begin{array}{cccc:cccc} -8 & 0 & 0 & 0 & 1 & 0 & 0 & 0 \\ 0 & 1 & 0 & 0 & 0 & 1 & 0 & 0 \\ 0 & 0 & 4 & 0 & 0 & 0 & 1 & 0 \\ 0 & 0 & 0 & -5 & 0 & 0 & 0 & 1 \end{array}\right]$

$$\begin{array}{r} -\frac{1}{8}R_1 \rightarrow \\ \\ \frac{1}{4}R_3 \rightarrow \\ -\frac{1}{5}R_4 \rightarrow \end{array} \left[\begin{array}{cccc:cccc} 1 & 0 & 0 & 0 & -\frac{1}{8} & 0 & 0 & 0 \\ 0 & 1 & 0 & 0 & 0 & 1 & 0 & 0 \\ 0 & 0 & 1 & 0 & 0 & 0 & \frac{1}{4} & 0 \\ 0 & 0 & 0 & 1 & 0 & 0 & 0 & -\frac{1}{5} \end{array}\right] = [I \;\vdots\; A^{-1}]$$

$$A^{-1} = \begin{bmatrix} -\frac{1}{8} & 0 & 0 & 0 \\ 0 & 1 & 0 & 0 \\ 0 & 0 & \frac{1}{4} & 0 \\ 0 & 0 & 0 & -\frac{1}{5} \end{bmatrix}$$

25. $A = \begin{bmatrix} 1 & 2 & -1 \\ 3 & 7 & -10 \\ -5 & -7 & -15 \end{bmatrix}$

$$A^{-1} = \begin{bmatrix} -175 & 37 & -13 \\ 95 & -20 & 7 \\ 14 & -3 & 1 \end{bmatrix}$$

27. $A = \begin{bmatrix} 1 & 1 & 2 \\ 3 & 1 & 0 \\ -2 & 0 & 3 \end{bmatrix}$

$$A^{-1} = \frac{1}{2}\begin{bmatrix} -3 & 3 & 2 \\ 9 & -7 & -6 \\ -2 & 2 & 2 \end{bmatrix}$$

29. $A = \begin{bmatrix} 0.1 & 0.2 & 0.3 \\ -0.3 & 0.2 & 0.2 \\ 0.5 & 0.4 & 0.4 \end{bmatrix}$

$$A^{-1} = \frac{5}{11}\begin{bmatrix} 0 & -4 & 2 \\ -22 & 11 & 11 \\ 22 & -6 & -8 \end{bmatrix}$$

31. $A = \begin{bmatrix} 1 & 0 & 3 & 0 \\ 0 & 2 & 0 & 4 \\ 1 & 0 & 3 & 0 \\ 0 & 2 & 0 & 4 \end{bmatrix}$

A^{-1} does not exist.

33. $A = \begin{bmatrix} 1 & -2 & -1 & -2 \\ 3 & -5 & -2 & -3 \\ 2 & -5 & -2 & -5 \\ -1 & 4 & 4 & 11 \end{bmatrix}$

$A^{-1} = \begin{bmatrix} -24 & 7 & 1 & -2 \\ -10 & 3 & 0 & -1 \\ -29 & 7 & 3 & -2 \\ 12 & -3 & -1 & 1 \end{bmatrix}$

35. $AA^{-1} = \begin{bmatrix} a & b \\ c & d \end{bmatrix}\left(\dfrac{1}{ad - bc}\right)\begin{bmatrix} d & -b \\ -c & a \end{bmatrix} = \dfrac{1}{ad - bc}\begin{bmatrix} a & b \\ c & d \end{bmatrix}\begin{bmatrix} d & -b \\ -c & a \end{bmatrix}$

$= \dfrac{1}{ad - bc}\begin{bmatrix} ad - bc & 0 \\ 0 & ad - bc \end{bmatrix} = \begin{bmatrix} 1 & 0 \\ 0 & 1 \end{bmatrix}$

$A^{-1}A = \dfrac{1}{ad - bc}\begin{bmatrix} d & -b \\ -c & a \end{bmatrix}\begin{bmatrix} a & b \\ c & d \end{bmatrix} = \dfrac{1}{ad - bc}\begin{bmatrix} ad - bc & 0 \\ 0 & ad - bc \end{bmatrix} = \begin{bmatrix} 1 & 0 \\ 0 & 1 \end{bmatrix}$

37. $\begin{bmatrix} x \\ y \end{bmatrix} = \begin{bmatrix} -3 & 2 \\ -2 & 1 \end{bmatrix}\begin{bmatrix} 5 \\ 10 \end{bmatrix} = \begin{bmatrix} 5 \\ 0 \end{bmatrix}$

Answer: (5, 0)

39. $\begin{bmatrix} x \\ y \end{bmatrix} = \begin{bmatrix} -3 & 2 \\ -2 & 1 \end{bmatrix} = \begin{bmatrix} 4 \\ 2 \end{bmatrix} = \begin{bmatrix} -8 \\ -6 \end{bmatrix}$

Answer: $(-8, -6)$

41. $\begin{bmatrix} x \\ y \\ z \end{bmatrix} = \begin{bmatrix} 1 & 1 & -1 \\ -3 & 2 & -1 \\ 3 & -3 & 2 \end{bmatrix}\begin{bmatrix} 0 \\ 5 \\ 2 \end{bmatrix} = \begin{bmatrix} 3 \\ 8 \\ -11 \end{bmatrix}$

Answer: $(3, 8, -11)$

43. $\begin{bmatrix} x_1 \\ x_2 \\ x_3 \\ x_4 \end{bmatrix} = \begin{bmatrix} -24 & 7 & 1 & -2 \\ -10 & 3 & 0 & -1 \\ -29 & 7 & 3 & -2 \\ 12 & -3 & -1 & 1 \end{bmatrix}\begin{bmatrix} 0 \\ 1 \\ -1 \\ 2 \end{bmatrix} = \begin{bmatrix} 2 \\ 1 \\ 0 \\ 0 \end{bmatrix}$

Answer: (2, 1, 0, 0)

45. $A = \begin{bmatrix} 3 & 4 \\ 5 & 3 \end{bmatrix}$

$A^{-1} = \dfrac{1}{9 - 20}\begin{bmatrix} 3 & -4 \\ -5 & 3 \end{bmatrix}$

$\begin{bmatrix} x \\ y \end{bmatrix} = -\dfrac{1}{11}\begin{bmatrix} 3 & -4 \\ -5 & 3 \end{bmatrix}\begin{bmatrix} -2 \\ 4 \end{bmatrix} = -\dfrac{1}{11}\begin{bmatrix} -22 \\ 22 \end{bmatrix} = \begin{bmatrix} 2 \\ -2 \end{bmatrix}$

Answer: $(2, -2)$

47. $A = \begin{bmatrix} -0.4 & 0.8 \\ 2 & -4 \end{bmatrix}$

$A^{-1} = \dfrac{1}{1.6 - 1.6}\begin{bmatrix} -4 & -0.8 \\ -2 & -0.4 \end{bmatrix}$

A^{-1} does not exist.

No solution

49. $A = \begin{bmatrix} 3 & 6 \\ 6 & 14 \end{bmatrix}$

$A^{-1} = \dfrac{1}{42 - 36}\begin{bmatrix} 14 & -6 \\ -6 & 3 \end{bmatrix}$

$\begin{bmatrix} x \\ y \end{bmatrix} = \dfrac{1}{6}\begin{bmatrix} 14 & -6 \\ -6 & 3 \end{bmatrix}\begin{bmatrix} 6 \\ 11 \end{bmatrix} = \dfrac{1}{6}\begin{bmatrix} 18 \\ -3 \end{bmatrix} = \begin{bmatrix} 3 \\ -\frac{1}{2} \end{bmatrix}$

Answer: $(3, -\frac{1}{2})$

51. $A = \begin{bmatrix} 4 & -1 & 1 \\ 2 & 2 & 3 \\ 5 & -2 & 6 \end{bmatrix}$

$A^{-1} = \frac{1}{55}\begin{bmatrix} 18 & 4 & -5 \\ 3 & 19 & -10 \\ -14 & 3 & 10 \end{bmatrix}$

$$\begin{bmatrix} x \\ y \\ z \end{bmatrix} = \frac{1}{55}\begin{bmatrix} 18 & 4 & -5 \\ 3 & 19 & -10 \\ -14 & 3 & 10 \end{bmatrix}\begin{bmatrix} -5 \\ 10 \\ 1 \end{bmatrix} = \frac{1}{55}\begin{bmatrix} -55 \\ 165 \\ 110 \end{bmatrix} = \begin{bmatrix} -1 \\ 3 \\ 2 \end{bmatrix}$$

Answer: $(-1, 3, 2)$

53. $A = \begin{bmatrix} 5 & -3 & 2 \\ 2 & 2 & -3 \\ 1 & -7 & 8 \end{bmatrix}$ A^{-1} does not exist.

No solution

55. $A = \begin{bmatrix} 7 & -3 & 0 & 2 \\ -2 & 1 & 0 & -1 \\ 4 & 0 & 1 & -2 \\ -1 & 1 & 0 & -1 \end{bmatrix}$

$A^{-1} = \begin{bmatrix} 0 & -1 & 0 & 1 \\ -1 & -5 & 0 & 3 \\ -2 & -4 & 1 & -2 \\ -1 & -4 & 0 & 1 \end{bmatrix}$

$$\begin{bmatrix} x \\ y \\ z \\ w \end{bmatrix} = \begin{bmatrix} 0 & -1 & 0 & 1 \\ -1 & -5 & 0 & 3 \\ -2 & -4 & 1 & -2 \\ -1 & -4 & 0 & 1 \end{bmatrix}\begin{bmatrix} 41 \\ -13 \\ 12 \\ -8 \end{bmatrix} = \begin{bmatrix} 5 \\ 0 \\ -2 \\ 3 \end{bmatrix}$$

Answer: $(5, 0, -2, 3)$

For 57–59 use $A = \begin{bmatrix} 1 & 1 & 1 \\ 0.065 & 0.07 & 0.09 \\ 0 & 2 & -1 \end{bmatrix}$. Using the methods of this section, we have $A^{-1} = \frac{1}{11}\begin{bmatrix} 50 & -600 & -4 \\ -13 & 200 & 5 \\ -26 & 400 & -1 \end{bmatrix}$.

57. $X = A^{-1}B = \frac{1}{11}\begin{bmatrix} 50 & -600 & -4 \\ -13 & 200 & 5 \\ -26 & 400 & -1 \end{bmatrix}\begin{bmatrix} 25{,}000 \\ 1900 \\ 0 \end{bmatrix} = \begin{bmatrix} 10{,}000 \\ 5000 \\ 10{,}000 \end{bmatrix}$

Answer: \$10,000 in AAA bonds, \$5000 in A bonds, \$10,000 in B bonds

59. $X = A^{-1}B = \frac{1}{11}\begin{bmatrix} 50 & -600 & -4 \\ -13 & 200 & 5 \\ -26 & 400 & -1 \end{bmatrix}\begin{bmatrix} 12{,}000 \\ 835 \\ 0 \end{bmatrix} = \begin{bmatrix} 9000 \\ 1000 \\ 2000 \end{bmatrix}$

Answer: \$9000 in AAA bonds, \$1000 in A bonds, \$2000 in B bonds

61. The inverse matrix remained the same for each system.

63. $A = \begin{bmatrix} 2 & 0 & 4 \\ 0 & 1 & 4 \\ 1 & 1 & -1 \end{bmatrix}$

$A^{-1} = \frac{1}{14}\begin{bmatrix} 5 & -4 & 4 \\ -4 & 6 & 8 \\ 1 & 2 & -2 \end{bmatrix}$

$\begin{bmatrix} I_1 \\ I_2 \\ I_3 \end{bmatrix} = \frac{1}{14}\begin{bmatrix} 5 & -4 & 4 \\ -4 & 6 & 8 \\ 1 & 2 & -2 \end{bmatrix}\begin{bmatrix} 14 \\ 10 \\ 0 \end{bmatrix} = \begin{bmatrix} -3 \\ 8 \\ 5 \end{bmatrix}$

Answer: $I_1 = -3$ amps, $I_2 = 8$ amps, $I_3 = 5$ amps

65. (a) Given $A = \begin{bmatrix} a_{11} & 0 \\ 0 & a_{22} \end{bmatrix}$, $A^{-1} = \begin{bmatrix} \frac{1}{a_{11}} & 0 \\ 0 & \frac{1}{a_{22}} \end{bmatrix}$.

Given $A = \begin{bmatrix} a_{11} & 0 & 0 \\ 0 & a_{22} & 0 \\ 0 & 0 & a_{33} \end{bmatrix}$, $A^{-1} = \begin{bmatrix} \frac{1}{a_{11}} & 0 & 0 \\ 0 & \frac{1}{a_{22}} & 0 \\ 0 & 0 & \frac{1}{a_{33}} \end{bmatrix}$.

(b) In general, the inverse of the diagonal matrix A is

$$\begin{bmatrix} \frac{1}{a_{11}} & 0 & 0 & \cdots & 0 \\ 0 & \frac{1}{a_{22}} & 0 & \cdots & 0 \\ 0 & 0 & \frac{1}{a_{33}} & \cdots & 0 \\ \vdots & \vdots & \vdots & \cdots & \vdots \\ 0 & 0 & 0 & \cdots & \frac{1}{a_{33}} \end{bmatrix}.$$

67. Men: $s = 1.279 - 0.0049(22) \approx 1.171$ minutes

Women: $s = 1.411 - 0.0078(22) \approx 1.239$ minutes

69. $3^{x/2} = 315$

$\ln 3^{x/2} = \ln 315$

$\frac{x}{2} \ln 3 = \ln 315$

$x = \frac{2 \ln 315}{\ln 3} \approx 10.47$

71. $\log_2 x - 2 = 4.5$

$\log_2 x = 6.5$

$x = 2^{6.5} \approx 90.51$

Section 8.4 The Determinant of a Square Matrix

- You should be able to determine the determinant of a matrix of order 2×2 by using the products of the diagonals.
- You should be able to use expansion by cofactors to find the determinant of a matrix of order 3 or greater.
- The determinant of a triangular matrix equals the product of the entries on the main diagonal.

Solutions to Odd-Numbered Exercises

1. 5

3. $\begin{vmatrix} 2 & 1 \\ 3 & 4 \end{vmatrix} = 2(4) - 1(3) = 8 - 3 = 5$

5. $\begin{vmatrix} 5 & 2 \\ -6 & 3 \end{vmatrix} = 5(3) - 2(-6) = 15 + 12 = 27$

7. $\begin{vmatrix} -7 & 6 \\ \frac{1}{2} & 3 \end{vmatrix} = -7(3) - 6(\frac{1}{2}) = -21 - 3 = -24$

9. $\begin{vmatrix} 2 & 6 \\ 0 & 3 \end{vmatrix} = 2(3) - 6(0) = 6$

11. $\begin{vmatrix} 2 & -1 & 0 \\ 4 & 2 & 1 \\ 4 & 2 & 1 \end{vmatrix} = 2\begin{vmatrix} 2 & 1 \\ 2 & 1 \end{vmatrix} - 4\begin{vmatrix} -1 & 0 \\ 2 & 1 \end{vmatrix} + 4\begin{vmatrix} -1 & 0 \\ 2 & 1 \end{vmatrix} = 2(0) - 4(-1) + 4(-1) = 0$

13. $\begin{vmatrix} 6 & 3 & -7 \\ 0 & 0 & 0 \\ 4 & -6 & 3 \end{vmatrix} = 0\begin{vmatrix} 3 & -7 \\ -6 & 3 \end{vmatrix} - 0\begin{vmatrix} 6 & -7 \\ 4 & 3 \end{vmatrix} + 0\begin{vmatrix} 6 & 3 \\ 4 & -6 \end{vmatrix} = 0$

15. $\begin{vmatrix} -1 & 2 & 5 \\ 0 & 3 & 4 \\ 0 & 0 & 3 \end{vmatrix} = (-1)(3)(3) = -9$ (Upper Triangular)

17. $\begin{vmatrix} 0.3 & 0.2 & 0.2 \\ 0.2 & 0.2 & 0.2 \\ -0.4 & 0.4 & 0.3 \end{vmatrix} = -0.002$

19. $\begin{vmatrix} 1 & 4 & -2 \\ 3 & 6 & -6 \\ -2 & 1 & 4 \end{vmatrix} = 0$

21. $\begin{bmatrix} 3 & 4 \\ 2 & -5 \end{bmatrix}$

(a) $M_{11} = -5$
$M_{12} = 2$
$M_{21} = 4$
$M_{22} = 3$

(b) $C_{11} = M_{11} = -5$
$C_{12} = -M_{12} = -2$
$C_{21} = -M_{21} = -4$
$C_{22} = M_{22} = 3$

23. $\begin{bmatrix} 3 & -2 & 8 \\ 3 & 2 & -6 \\ -1 & 3 & 6 \end{bmatrix}$

(a) $M_{11} = \begin{vmatrix} 2 & -6 \\ 3 & 6 \end{vmatrix} = 12 + 18 = 30$

$M_{12} = \begin{vmatrix} 3 & -6 \\ -1 & 6 \end{vmatrix} = 18 - 6 = 12$

$M_{13} = \begin{vmatrix} 3 & 2 \\ -1 & 3 \end{vmatrix} = 9 + 2 = 11$

$M_{21} = \begin{vmatrix} -2 & 8 \\ 3 & 6 \end{vmatrix} = -12 - 24 = -36$

$M_{22} = \begin{vmatrix} 3 & 8 \\ -1 & 6 \end{vmatrix} = 18 + 8 = 26$

$M_{23} = \begin{vmatrix} 3 & -2 \\ -1 & 3 \end{vmatrix} = 9 - 2 = 7$

$M_{31} = \begin{vmatrix} -2 & 8 \\ 2 & -6 \end{vmatrix} = 12 - 16 = -4$

$M_{32} = \begin{vmatrix} 3 & 8 \\ 3 & -6 \end{vmatrix} = -18 - 24 = -42$

$M_{33} = \begin{vmatrix} 3 & -2 \\ 3 & 2 \end{vmatrix} = 6 + 6 = 12$

(b) $C_{11} = (-1)^2 M_{11} = 30$

$C_{12} = (-1)^3 M_{12} = -12$

$C_{13} = (-1)^4 M_{13} = 11$

$C_{21} = (-1)^3 M_{21} = 36$

$C_{22} = (-1)^4 M_{22} = 26$

$C_{23} = (-1)^5 M_{23} = -7$

$C_{31} = (-1)^4 M_{31} = -4$

$C_{32} = (-1)^5 M_{32} = 42$

$C_{33} = (-1)^6 M_{33} = 12$

25. (a) $\begin{vmatrix} -3 & 2 & 1 \\ 4 & 5 & 6 \\ 2 & -3 & 1 \end{vmatrix} = -3\begin{vmatrix} 5 & 6 \\ -3 & 1 \end{vmatrix} - 2\begin{vmatrix} 4 & 6 \\ 2 & 1 \end{vmatrix} + \begin{vmatrix} 4 & 5 \\ 2 & -3 \end{vmatrix} = -3(23) - 2(-8) - 22 = -75$

(b) $\begin{vmatrix} -3 & 2 & 1 \\ 4 & 5 & 6 \\ 2 & -3 & 1 \end{vmatrix} = -2\begin{vmatrix} 4 & 6 \\ 2 & 1 \end{vmatrix} + 5\begin{vmatrix} -3 & 1 \\ 2 & 1 \end{vmatrix} + 3\begin{vmatrix} -3 & 1 \\ 4 & 6 \end{vmatrix} = -2(-8) + 5(-5) + 3(-22) = -75$

27. (a) $\begin{vmatrix} 5 & 0 & -3 \\ 0 & 12 & 4 \\ 1 & 6 & 3 \end{vmatrix} = 0\begin{vmatrix} 0 & -3 \\ 6 & 3 \end{vmatrix} + 12\begin{vmatrix} 5 & -3 \\ 1 & 3 \end{vmatrix} - 4\begin{vmatrix} 5 & 0 \\ 1 & 6 \end{vmatrix} = 0(18) + 12(18) - 4(30) = 96$

(b) $\begin{vmatrix} 5 & 0 & -3 \\ 0 & 12 & 4 \\ 1 & 6 & 3 \end{vmatrix} = 0\begin{vmatrix} 0 & 4 \\ 1 & 3 \end{vmatrix} + 12\begin{vmatrix} 5 & -3 \\ 1 & 3 \end{vmatrix} - 6\begin{vmatrix} 5 & -3 \\ 0 & 4 \end{vmatrix} = 0(-4) + 12(18) - 6(20) = 96$

29. (a) $\begin{vmatrix} 6 & 0 & -3 & 5 \\ 4 & 13 & 6 & -8 \\ -1 & 0 & 7 & 4 \\ 8 & 6 & 0 & 2 \end{vmatrix} = -4\begin{vmatrix} 0 & -3 & 5 \\ 0 & 7 & 4 \\ 6 & 0 & 2 \end{vmatrix} + 13\begin{vmatrix} 6 & -3 & 5 \\ -1 & 7 & 4 \\ 8 & 0 & 2 \end{vmatrix} - 6\begin{vmatrix} 6 & 0 & 5 \\ -1 & 0 & 4 \\ 8 & 6 & 2 \end{vmatrix} - 8\begin{vmatrix} 6 & 0 & -3 \\ -1 & 0 & 7 \\ 8 & 6 & 0 \end{vmatrix}$

$= -4(-282) + 13(-298) - 6(-174) - 8(-234) = 170$

(b) $\begin{vmatrix} 6 & 0 & -3 & 5 \\ 4 & 13 & 6 & -8 \\ -1 & 0 & 7 & 4 \\ 8 & 6 & 0 & 2 \end{vmatrix} = 0\begin{vmatrix} 4 & 6 & -8 \\ -1 & 7 & 4 \\ 8 & 0 & 2 \end{vmatrix} + 13\begin{vmatrix} 6 & -3 & 5 \\ -1 & 7 & 4 \\ 8 & 0 & 2 \end{vmatrix} + 0\begin{vmatrix} 6 & -3 & 5 \\ 4 & 6 & -8 \\ 8 & 0 & 2 \end{vmatrix} + 6\begin{vmatrix} 6 & -3 & 5 \\ 4 & 6 & -8 \\ -1 & 7 & 4 \end{vmatrix}$

$= 0 + 13(-298) + 0 + 6(674) = 170$

31. Expand by Column 3.

$\begin{vmatrix} 1 & 4 & -2 \\ 3 & 2 & 0 \\ -1 & 4 & 3 \end{vmatrix} = -2\begin{vmatrix} 3 & 2 \\ -1 & 4 \end{vmatrix} + 3\begin{vmatrix} 1 & 4 \\ 3 & 2 \end{vmatrix} = -2(14) + 3(-10) = -58$

33. $\begin{vmatrix} 2 & 4 & 6 \\ 0 & 3 & 1 \\ 0 & 0 & -5 \end{vmatrix} = (2)(3)(-5) = -30$ (Upper Triangular)

35. Expand by Column 3.

$\begin{vmatrix} 2 & 6 & 6 & 2 \\ 2 & 7 & 3 & 6 \\ 1 & 5 & 0 & 1 \\ 3 & 7 & 0 & 7 \end{vmatrix} = 6\begin{vmatrix} 2 & 7 & 6 \\ 1 & 5 & 1 \\ 3 & 7 & 7 \end{vmatrix} - 3\begin{vmatrix} 2 & 6 & 2 \\ 1 & 5 & 1 \\ 3 & 7 & 7 \end{vmatrix} = 6(-20) - 3(16) = -168$

37. Expand by Column 1.

$\begin{vmatrix} 5 & 3 & 0 & 6 \\ 4 & 6 & 4 & 12 \\ 0 & 2 & -3 & 4 \\ 0 & 1 & -2 & 2 \end{vmatrix} = 5\begin{vmatrix} 6 & 4 & 12 \\ 2 & -3 & 4 \\ 1 & -2 & 2 \end{vmatrix} - 4\begin{vmatrix} 3 & 0 & 6 \\ 2 & -3 & 4 \\ 1 & -2 & 2 \end{vmatrix} = 5(0) - 4(0) = 0$

39. Expand by Column 2, then by Column 4.

$\begin{vmatrix} 3 & 2 & 4 & -1 & 5 \\ -2 & 0 & 1 & 3 & 2 \\ 1 & 0 & 0 & 4 & 0 \\ 6 & 0 & 2 & -1 & 0 \\ 3 & 0 & 5 & 1 & 0 \end{vmatrix} = -2\begin{vmatrix} -2 & 1 & 3 & 2 \\ 1 & 0 & 4 & 0 \\ 6 & 2 & -1 & 0 \\ 3 & 5 & 1 & 0 \end{vmatrix} = (-2)(-2)\begin{vmatrix} 1 & 0 & 4 \\ 6 & 2 & -1 \\ 3 & 5 & 1 \end{vmatrix} = 4(103) = 412$

41. $\begin{vmatrix} 3 & 8 & -7 \\ 0 & -5 & 4 \\ 8 & 1 & 6 \end{vmatrix} = -126$

43. $\begin{vmatrix} 7 & 0 & -14 \\ -2 & 5 & 4 \\ -6 & 2 & 12 \end{vmatrix} = 0$

45. $\begin{vmatrix} 1 & -1 & 8 & 4 \\ 2 & 6 & 0 & -4 \\ 2 & 0 & 2 & 6 \\ 0 & 2 & 8 & 0 \end{vmatrix} = -336$

47. $\begin{vmatrix} 3 & -2 & 4 & 3 & 1 \\ -1 & 0 & 2 & 1 & 0 \\ 5 & -1 & 0 & 3 & 2 \\ 4 & 7 & -8 & 0 & 0 \\ 1 & 2 & 3 & 0 & 2 \end{vmatrix} = 410$

49. $\begin{vmatrix} w & x \\ y & z \end{vmatrix} = wz - xy$

$-\begin{vmatrix} y & z \\ w & x \end{vmatrix} = -(xy - wz) = wz - xy$

Thus, $\begin{vmatrix} w & x \\ y & z \end{vmatrix} = -\begin{vmatrix} y & z \\ w & x \end{vmatrix}$.

51. $\begin{vmatrix} w & x \\ y & z \end{vmatrix} = wz - xy$

$\begin{vmatrix} w & x + cw \\ y & z + cy \end{vmatrix} = w(z + cy) - y(x + cw) = wz - xy$

Thus, $\begin{vmatrix} w & x \\ y & z \end{vmatrix} = \begin{vmatrix} w & x + cw \\ y & z + cy \end{vmatrix}$.

53. $\begin{vmatrix} 1 & x & x^2 \\ 1 & y & y^2 \\ 1 & z & z^2 \end{vmatrix} = \begin{vmatrix} y & y^2 \\ z & z^2 \end{vmatrix} - \begin{vmatrix} x & x^2 \\ z & z^2 \end{vmatrix} + \begin{vmatrix} x & x^2 \\ y & y^2 \end{vmatrix}$

$= (yz^2 - y^2z) - (xz^2 - x^2z) + (xy^2 - x^2y)$
$= yz^2 - xz^2 - y^2z + x^2z + xy(y - x)$
$= z^2(y - x) - z(y^2 - x^2) + xy(y - x)$
$= z^2(y - x) - z(y - x)(y + x) + xy(y - x)$
$= (y - x)[z^2 - z(y + x) + xy]$
$= (y - x)[z^2 - zy - zx + xy]$
$= (y - x)[z^2 - zx - zy + xy]$
$= (y - x)[z(z - x) - y(z - x)]$
$= (y - x)(z - x)(z - y)$

55. $= \begin{vmatrix} x - 1 & 2 \\ 3 & x - 2 \end{vmatrix} = 0$

$(x - 1)(x - 2) - 6 = 0$
$x^2 - 3x - 4 = 0$
$(x + 1)(x - 4) = 0$
$x = -1$ or $x = 4$

57. $\begin{vmatrix} 4u & -1 \\ -1 & 2v \end{vmatrix} = 8uv - 1$

59. $\begin{vmatrix} e^{2x} & e^{3x} \\ 2e^{2x} & 3e^{3x} \end{vmatrix} = 3e^{5x} - 2e^{5x} = e^{5x}$

61. $\begin{vmatrix} x & \ln x \\ 1 & \frac{1}{x} \end{vmatrix} = 1 - \ln x$

63. (a) $\begin{vmatrix} -1 & 0 \\ 0 & 3 \end{vmatrix} = -3$ (b) $\begin{vmatrix} 2 & 0 \\ 0 & -1 \end{vmatrix} = -2$

(c) $\begin{bmatrix} -1 & 0 \\ 0 & 3 \end{bmatrix}\begin{bmatrix} 2 & 0 \\ 0 & -1 \end{bmatrix} = \begin{bmatrix} -2 & 0 \\ 0 & -3 \end{bmatrix}$ (d) $\begin{vmatrix} -2 & 0 \\ 0 & -3 \end{vmatrix} = 6$

65. (a) $\begin{vmatrix} -1 & 2 & 1 \\ 1 & 0 & 1 \\ 0 & 1 & 0 \end{vmatrix} = 2$ (b) $\begin{vmatrix} -1 & 0 & 0 \\ 0 & 2 & 0 \\ 0 & 0 & 3 \end{vmatrix} = -6$

(c) $\begin{bmatrix} -1 & 2 & 1 \\ 1 & 0 & 1 \\ 0 & 1 & 0 \end{bmatrix}\begin{bmatrix} -1 & 0 & 0 \\ 0 & 2 & 0 \\ 0 & 0 & 3 \end{bmatrix} = \begin{bmatrix} 1 & 4 & 3 \\ -1 & 0 & 3 \\ 0 & 2 & 0 \end{bmatrix}$

(d) $\begin{vmatrix} 1 & 4 & 3 \\ -1 & 0 & 3 \\ 0 & 2 & 0 \end{vmatrix} = -12$

67. Let $A = \begin{bmatrix} 1 & 3 \\ -2 & 4 \end{bmatrix}$ and $B = \begin{bmatrix} -4 & 0 \\ 3 & 5 \end{bmatrix}$

$|A| = \begin{vmatrix} 1 & 3 \\ -2 & 4 \end{vmatrix} = 10,\ |B| = \begin{vmatrix} -4 & 0 \\ 3 & 5 \end{vmatrix} = -20,\ |A| + |B| = -10$

$A + B = \begin{bmatrix} -3 & 3 \\ 1 & 9 \end{bmatrix},\ |A + B| = \begin{vmatrix} -3 & 3 \\ 1 & 9 \end{vmatrix} = -30$

Thus, $|A + B| \neq |A| + |B|$. Your answer may differ, depending on how you choose A and B.

69. A square matrix is a square array of numbers. A determinant of a square matrix is a real number.

71. Parabola

Vertex: $(0, 3)$

Focus: $(2, 3)$

Horizontal axis: $(y - k)^2 = 4p(x - h)$

$p = 2$

$(y - 3)^2 = 4(2)(x - 0)$

$(y - 3)^2 = 8x$

73. Ellipse

Vertices: $(\pm 8, 0)$

Foci: $(\pm 6, 0)$

Horizontal major axis

Center: $(0, 0)$

$a = 8, c = 6, b = \sqrt{64 - 36} = \sqrt{28}$

$\frac{x^2}{a^2} + \frac{y^2}{b^2} = 1$

$\frac{x^2}{64} + \frac{y^2}{28} = 1$

Section 8.5 Applications of Matrices and Determinants

- You should be able to use Cramer's Rule to solve a system of linear equations.
- Now you should be able to solve a system of linear equations by substitution, elimination, elementary row operations on an augmented matrix, using the inverse matrix, or Cramer's Rule.
- You should be able to find the area of a triangle with vertices (x_1, y_1), (x_2, y_2), and (x_3, y_3).

$$\text{Area} = \pm\frac{1}{2}\begin{vmatrix} x_1 & y_1 & 1 \\ x_2 & y_2 & 1 \\ x_3 & y_3 & 1 \end{vmatrix}$$

The $\pm$ symbol indicates that the appropriate sign should be chosen so that the area is positive.

- You should be able to test to see if three points, (x_1, y_1), (x_2, y_2), and (x_3, y_3), are collinear.

$$\begin{vmatrix} x_1 & y_1 & 1 \\ x_2 & y_2 & 1 \\ x_3 & y_3 & 1 \end{vmatrix} = 0, \text{ if and only if they are collinear.}$$

- You should be able to find the equation of the line through (x_1, y_1) and (x_2, y_2) by evaluating.

$$\begin{vmatrix} x & y & 1 \\ x_1 & y_1 & 1 \\ x_2 & y_2 & 1 \end{vmatrix} = 0$$

- You should be able to encode and decode messages by using an invertible $n \times n$ matrix.

Solutions to Odd-Numbered Exercises

1. $3x + 4y = -2$

$5x + 3y = 4$

$$x = \frac{\begin{vmatrix} -2 & 4 \\ 4 & 3 \end{vmatrix}}{\begin{vmatrix} 3 & 4 \\ 5 & 3 \end{vmatrix}} = \frac{-22}{-11} = 2$$

$$y = \frac{\begin{vmatrix} 3 & -2 \\ 5 & 4 \end{vmatrix}}{\begin{vmatrix} 3 & 4 \\ 5 & 3 \end{vmatrix}} = -\frac{22}{11} = -2$$

Answer: $(2, -2)$

3. $4x - y + z = -5$

$2x + 2y + 3z = 10$

$5x - 2y + 6z = 1$

$$D = \begin{vmatrix} 4 & -1 & 1 \\ 2 & 2 & 3 \\ 5 & -2 & 6 \end{vmatrix} = 55$$

$$x = \frac{\begin{vmatrix} -5 & -1 & 1 \\ 10 & 2 & 3 \\ 1 & -2 & 6 \end{vmatrix}}{55} = \frac{-55}{55} = -1$$

$$y = \frac{\begin{vmatrix} 4 & -5 & 1 \\ 2 & 10 & 3 \\ 5 & 1 & 6 \end{vmatrix}}{55} = \frac{165}{55} = 3$$

$$z = \frac{\begin{vmatrix} 4 & -1 & -5 \\ 2 & 2 & 10 \\ 5 & -2 & 1 \end{vmatrix}}{55} = \frac{110}{55} = 2$$

Answer: $(-1, 3, 2)$

5. $3x + 3y + 5z = 1$
$3x + 5y + 9z = 2$
$5x + 9y + 17z = 4$

$$D = \begin{vmatrix} 3 & 3 & 5 \\ 3 & 5 & 9 \\ 5 & 9 & 17 \end{vmatrix} = 4$$

$$x = \frac{\begin{vmatrix} 1 & 3 & 5 \\ 2 & 5 & 9 \\ 4 & 9 & 17 \end{vmatrix}}{4} = 0,\ y = \frac{\begin{vmatrix} 3 & 1 & 5 \\ 3 & 2 & 9 \\ 5 & 4 & 17 \end{vmatrix}}{4} = -\frac{1}{2},\ z = \frac{\begin{vmatrix} 3 & 3 & 1 \\ 3 & 5 & 2 \\ 5 & 9 & 4 \end{vmatrix}}{4} = \frac{1}{2}$$

Answer: $\left(0, -\frac{1}{2}, \frac{1}{2}\right)$

7. Vertices: $(0, 0)$, $(3, 1)$, $(1, 5)$

$$\text{Area} = \frac{1}{2}\begin{vmatrix} 0 & 0 & 1 \\ 3 & 1 & 1 \\ 1 & 5 & 1 \end{vmatrix} = \frac{1}{2}\begin{vmatrix} 3 & 1 \\ 1 & 5 \end{vmatrix} = 7 \text{ square units}$$

9. Vertices: $(-2, -3)$, $(2, -3)$, $(0, 4)$

$$\text{Area} = \frac{1}{2}\begin{vmatrix} -2 & -3 & 1 \\ 2 & -3 & 1 \\ 0 & 4 & 1 \end{vmatrix} = \frac{1}{2}\left(-2\begin{vmatrix} -3 & 1 \\ 4 & 1 \end{vmatrix} - 2\begin{vmatrix} -3 & 1 \\ 4 & 1 \end{vmatrix}\right) = \frac{1}{2}(14 + 14) = 14 \text{ square units}$$

11. Vertices: $\left(0, \frac{1}{2}\right)$, $\left(\frac{5}{2}, 0\right)$, $(4, 3)$

$$\text{Area} = \frac{1}{2}\begin{vmatrix} 0 & \frac{1}{2} & 1 \\ \frac{5}{2} & 0 & 1 \\ 4 & 3 & 1 \end{vmatrix} = \frac{1}{2}\left(2 + \frac{15}{2} - \frac{5}{4}\right) = \frac{33}{8} \text{ square units}$$

13. Vertices: $(-2, 4)$, $(2, 3)$, $(-1, 5)$

$$\text{Area} = \frac{1}{2}\begin{vmatrix} -2 & 4 & 1 \\ 2 & 3 & 1 \\ -1 & 5 & 1 \end{vmatrix} = \frac{1}{2}\left[\begin{vmatrix} 2 & 3 \\ -1 & 5 \end{vmatrix} - \begin{vmatrix} -2 & 4 \\ -1 & 5 \end{vmatrix} + \begin{vmatrix} -2 & 4 \\ 2 & 3 \end{vmatrix}\right] = \frac{5}{2} \text{ square units}$$

15. Vertices: $(-3, 5)$, $(2, 6)$, $(3, -5)$

$$\text{Area} = -\frac{1}{2}\begin{vmatrix} -3 & 5 & 1 \\ 2 & 6 & 1 \\ 3 & -5 & 1 \end{vmatrix} = -\frac{1}{2}\left[\begin{vmatrix} 2 & 6 \\ 3 & -5 \end{vmatrix} - \begin{vmatrix} -3 & 5 \\ 3 & -5 \end{vmatrix} + \begin{vmatrix} -3 & 5 \\ 2 & 6 \end{vmatrix}\right] = 28 \text{ square units}$$

17. $4 = \pm\frac{1}{2}\begin{vmatrix} -5 & 1 & 1 \\ 0 & 2 & 1 \\ -2 & x & 1 \end{vmatrix}$

$\pm 8 = -5\begin{vmatrix} 2 & 1 \\ x & 1 \end{vmatrix} - 2\begin{vmatrix} 1 & 1 \\ 2 & 1 \end{vmatrix}$

$\pm 8 = -5(2 - x) - 2(-1)$

$\pm 8 = 5x - 8$

$x = \dfrac{8 \pm 8}{5}$

$x = \dfrac{16}{5}$ OR $x = 0$

19. Vertices: $(0, 25)$, $(10, 0)$, $(28, 5)$

$$\text{Area} = \frac{1}{2}\begin{vmatrix} 0 & 25 & 1 \\ 10 & 0 & 1 \\ 28 & 5 & 1 \end{vmatrix} = 250 \text{ square miles}$$

21. Points: $(3, -1)$, $(0, -3)$, $(12, 5)$

$$\begin{vmatrix} 3 & -1 & 1 \\ 0 & -3 & 1 \\ 12 & 5 & 1 \end{vmatrix} = 3\begin{vmatrix} -3 & 1 \\ 5 & 1 \end{vmatrix} + 12\begin{vmatrix} -1 & 1 \\ -3 & 1 \end{vmatrix} = 0$$

The points are collinear.

23. Points: $\left(2, -\frac{1}{2}\right)$, $(-4, 4)$, $(6, -3)$

$$\begin{vmatrix} 2 & -\frac{1}{2} & 1 \\ -4 & 4 & 1 \\ 6 & -3 & 1 \end{vmatrix} = \begin{vmatrix} -4 & 4 \\ 6 & -3 \end{vmatrix} - \begin{vmatrix} 2 & -\frac{1}{2} \\ 6 & -3 \end{vmatrix} + \begin{vmatrix} 2 & -\frac{1}{2} \\ -4 & 4 \end{vmatrix} = -3 \neq 0$$

The points are not collinear.

25. Points: $(0, 2)$, $(1, 2.4)$, $(-1, 1.6)$

$$\begin{vmatrix} 0 & 2 & 1 \\ 1 & 2.4 & 1 \\ -1 & 1.6 & 1 \end{vmatrix} = -2\begin{vmatrix} 1 & 1 \\ -1 & 1 \end{vmatrix} + \begin{vmatrix} 1 & 2.4 \\ -1 & 1.6 \end{vmatrix} = 0$$

The points are collinear.

27. Points: $(0, 0)$, $(5, 3)$

$$\text{Equation: } \begin{vmatrix} x & y & 1 \\ 0 & 0 & 1 \\ 5 & 3 & 1 \end{vmatrix} = 5y - 3x = 0 \Longrightarrow 3x - 5y = 0$$

29. Points: $(-4, 3)$, $(2, 1)$

$$\text{Equation: } \begin{vmatrix} x & y & 1 \\ -4 & 3 & 1 \\ 2 & 1 & 1 \end{vmatrix} = 2x - 6y - 10 = 0 \Longrightarrow x + 3y - 5 = 0$$

31. Points: $\left(-\frac{1}{2}, 3\right), \left(\frac{5}{2}, 1\right)$

$$\text{Equation: } \begin{vmatrix} x & y & 1 \\ -\frac{1}{2} & 3 & 1 \\ \frac{5}{2} & 1 & 1 \end{vmatrix} = 2x + 3y - 8 = 0$$

33. $\begin{vmatrix} 2 & -5 & 1 \\ 4 & x & 1 \\ 5 & -2 & 1 \end{vmatrix} = 0$

$$2\begin{vmatrix} x & 1 \\ -2 & 1 \end{vmatrix} + 5\begin{vmatrix} 4 & 1 \\ 5 & 1 \end{vmatrix} + \begin{vmatrix} 4 & x \\ 5 & -2 \end{vmatrix} = 0$$

$$2(x + 2) + 5(-1) + (-8 - 5x) = 0$$

$$-3x - 9 = 0$$

$$x = -3$$

35. The uncoded row matrices are the rows of the 7×3 matrix on the left.

$$\begin{matrix} \text{T} & \text{R} & \text{O} \\ \text{U} & \text{B} & \text{L} \\ \text{E} & & \text{I} \\ \text{N} & & \text{R} \\ \text{I} & \text{V} & \text{E} \\ \text{R} & & \text{C} \\ \text{I} & \text{T} & \text{Y} \end{matrix} \quad \begin{bmatrix} 20 & 18 & 15 \\ 21 & 2 & 12 \\ 5 & 0 & 9 \\ 14 & 0 & 18 \\ 9 & 22 & 5 \\ 18 & 0 & 3 \\ 9 & 20 & 25 \end{bmatrix} \begin{bmatrix} 1 & -1 & 0 \\ 1 & 0 & -1 \\ -6 & 2 & 3 \end{bmatrix} = \begin{bmatrix} -52 & 10 & 27 \\ -49 & 3 & 34 \\ -49 & 13 & 27 \\ -94 & 22 & 54 \\ 1 & 1 & -7 \\ 0 & -12 & 9 \\ -121 & 41 & 55 \end{bmatrix}$$

Answer: $[-52, 10, 27], [-49, 3, 34], [-49, 13, 27], [-94, 22, 54], [1, 1, -7], [0, -12, 9], [-121, 41, 55]$

In Exercises 37–39, use the matrix $A = \begin{bmatrix} 1 & 2 & 2 \\ 3 & 7 & 9 \\ -1 & -4 & -7 \end{bmatrix}$

37. L A N   D I N   G _ S   U C C   E S S   F U L

[12 1 14]   [4 9 14]   [7 0 19]   [21 3 3]   [5 19 19]   [6 21 12]

$$[12 \quad 1 \quad 14]A = [1 \quad -25 \quad -65]$$
$$[4 \quad 9 \quad 14]A = [17 \quad 15 \quad -9]$$
$$[7 \quad 0 \quad 19]A = [-12 \quad -62 \quad -119]$$
$$[21 \quad 3 \quad 3]A = [27 \quad 51 \quad 48]$$
$$[5 \quad 19 \quad 19]A = [43 \quad 67 \quad 48]$$
$$[6 \quad 21 \quad 12]A = [57 \quad 111 \quad 117]$$

Cryptogram: 1 −25 −65 17 15 −9 −12 −62 −119 27 51 48 43 67 48 57 111 117

39. H A P P Y _ B I R T H D A Y _

[8 1 16] [16 25 0] [2 9 18] [20 8 4] [1 25 0]

$[8 \quad 1 \quad 16]\,A = [5 \quad -41 \quad -87]$

$[16 \quad 25 \quad 0]\,A = [91 \quad 207 \quad 257]$

$[2 \quad 9 \quad 18]\,A = [11 \quad -5 \quad -41]$

$[20 \quad 8 \quad 4]\,A = [40 \quad 80 \quad 84]$

$[1 \quad 25 \quad 0]\,A = [76 \quad 177 \quad 227]$

Cryptogram: −5 −41 −87 91 207 257 11 −5 −41 40 80 84 76 177 227

41. $A^{-1} = \begin{bmatrix} 1 & 2 \\ 3 & 5 \end{bmatrix}^{-1} = \begin{bmatrix} -5 & 2 \\ 3 & -1 \end{bmatrix}$

$$\begin{bmatrix} 11 & 21 \\ 64 & 112 \\ 25 & 50 \\ 29 & 53 \\ 23 & 46 \\ 40 & 75 \\ 55 & 92 \end{bmatrix} \begin{bmatrix} -5 & 2 \\ 3 & -1 \end{bmatrix} = \begin{bmatrix} 8 & 1 \\ 16 & 16 \\ 25 & 0 \\ 14 & 5 \\ 23 & 0 \\ 25 & 5 \\ 1 & 18 \end{bmatrix} \begin{matrix} \text{H} & \text{A} \\ \text{P} & \text{P} \\ \text{Y} & \\ \text{N} & \text{E} \\ \text{W} & \\ \text{Y} & \text{E} \\ \text{A} & \text{R} \end{matrix}$$

Message: HAPPY NEW YEAR

43. $A^{-1} = \begin{bmatrix} 1 & 2 & 2 \\ 3 & 7 & 9 \\ -1 & -4 & -7 \end{bmatrix}^{-1} = \begin{bmatrix} -13 & 6 & 4 \\ 12 & -5 & -3 \\ -5 & 2 & 1 \end{bmatrix}$

$$\begin{bmatrix} 20 & 17 & -15 \\ -12 & -56 & -104 \\ 1 & -25 & -65 \\ 62 & 143 & 181 \end{bmatrix} \begin{bmatrix} -13 & 6 & 4 \\ 12 & -5 & -3 \\ -5 & 2 & 1 \end{bmatrix} = \begin{bmatrix} 19 & 5 & 14 \\ 4 & 0 & 16 \\ 12 & 1 & 14 \\ 5 & 19 & 0 \end{bmatrix} \begin{matrix} \text{S} & \text{E} & \text{N} \\ \text{D} & & \text{P} \\ \text{L} & \text{A} & \text{N} \\ \text{E} & \text{S} & \end{matrix}$$

Message: SEND PLANES

45. Let A be the 2×2 matrix needed to decode the message.

$$\begin{bmatrix} -18 & -18 \\ 1 & 16 \end{bmatrix} A = \begin{bmatrix} 0 & 18 \\ 15 & 14 \end{bmatrix} \begin{matrix} & \text{R} \\ \text{O} & \text{N} \end{matrix}$$

$$A = \begin{bmatrix} -18 & -18 \\ 1 & 16 \end{bmatrix}^{-1} \begin{bmatrix} 0 & 18 \\ 15 & 14 \end{bmatrix} = \begin{bmatrix} -\frac{8}{135} & -\frac{1}{15} \\ \frac{1}{270} & \frac{1}{15} \end{bmatrix} \begin{bmatrix} 0 & 18 \\ 15 & 14 \end{bmatrix} = \begin{bmatrix} -1 & -2 \\ 1 & 1 \end{bmatrix}$$

$$\begin{bmatrix} 8 & 21 \\ -15 & -10 \\ -13 & -13 \\ 5 & 10 \\ 5 & 25 \\ 5 & 19 \\ -1 & 6 \\ 20 & 40 \\ -18 & -18 \\ 1 & 16 \end{bmatrix} \begin{bmatrix} -1 & -2 \\ 1 & 1 \end{bmatrix} = \begin{bmatrix} 13 & 5 \\ 5 & 20 \\ 0 & 13 \\ 5 & 0 \\ 20 & 15 \\ 14 & 9 \\ 7 & 8 \\ 20 & 0 \\ 0 & 18 \\ 15 & 14 \end{bmatrix} \begin{matrix} \text{M} & \text{E} \\ \text{E} & \text{T} \\ & \text{M} \\ \text{E} & \\ \text{T} & \text{O} \\ \text{N} & \text{I} \\ \text{G} & \text{H} \\ \text{T} & \\ & \text{R} \\ \text{O} & \text{N} \end{matrix}$$

Message: MEET ME TONIGHT RON

Review Exercises for Chapter 8

Solutions to Odd-Numbered Exercises

1. $\left[\begin{array}{rr|r} 3 & -10 & 15 \\ 5 & 4 & 22 \end{array}\right]$

3. $\left[\begin{array}{rrr|r} 5 & 1 & 7 & -9 \\ 4 & 2 & 0 & 10 \\ 9 & 4 & 2 & 3 \end{array}\right]$ $\quad \begin{aligned} 5x + y + 7z &= -9 \\ 4x + 2y \quad &= 10 \\ 9x + 4y + 2z &= 3 \end{aligned}$

5. $\begin{bmatrix} 0 & 1 & 1 \\ 1 & 2 & 3 \\ 2 & 2 & 2 \end{bmatrix}$

$\begin{array}{r} R_1 + R_2 \rightarrow \\ -R_1 + R_2 \rightarrow \\ -2R_1 + R_3 \rightarrow \end{array} \begin{bmatrix} 1 & 3 & 4 \\ 0 & -1 & -1 \\ 0 & -4 & -6 \end{bmatrix}$

$\begin{array}{r} 3R_2 + R_1 \rightarrow \\ -R_2 \rightarrow \\ -4R_2 + R_3 \rightarrow \end{array} \begin{bmatrix} 1 & 0 & 1 \\ 0 & 1 & 1 \\ 0 & 0 & -2 \end{bmatrix}$

$\begin{array}{r} -R_3 + R_1 \rightarrow \\ -R_3 + R_2 \rightarrow \\ -\frac{1}{2}R_3 \rightarrow \end{array} \begin{bmatrix} 1 & 0 & 0 \\ 0 & 1 & 0 \\ 0 & 0 & 1 \end{bmatrix}$

7. $\left[\begin{array}{rr|r} 5 & 4 & 2 \\ -1 & 1 & -22 \end{array}\right]$

$\begin{array}{r} 4R_2 + R_1 \rightarrow \\ R_1 + R_2 \rightarrow \end{array} \left[\begin{array}{rr|r} 1 & 8 & -86 \\ 0 & 9 & -108 \end{array}\right]$

$\begin{array}{r} -8R_2 + R_1 \rightarrow \\ \frac{1}{9}R_2 \rightarrow \end{array} \left[\begin{array}{rr|r} 1 & 0 & 10 \\ 0 & 1 & -12 \end{array}\right]$

$x = 10,\ y = -12$

Answer: $(10, -12)$

9. $\left[\begin{array}{rr|r} 2 & 1 & 0.3 \\ 3 & -1 & -1.3 \end{array}\right]$

$\begin{array}{r} R_2 - R_1 \rightarrow \\ -3R_1 + R_2 \rightarrow \end{array} \left[\begin{array}{rr|r} 1 & -2 & -1.6 \\ 0 & 5 & 3.5 \end{array}\right]$

$\begin{array}{r} 2R_2 + R_1 \rightarrow \\ \frac{1}{5}R_2 \rightarrow \end{array} \left[\begin{array}{rr|r} 1 & 0 & -0.2 \\ 0 & 1 & 0.7 \end{array}\right]$

$x = -0.2,\ y = 0.7$

Answer: $(-0.2, 0.7)$

11. $\left[\begin{array}{rrr|r} -1 & 1 & 2 & 1 \\ 2 & 3 & 1 & -2 \\ 5 & 4 & 2 & 4 \end{array}\right]$

$\begin{array}{r} -R_1 \rightarrow \\ 2R_1 + R_2 \rightarrow \\ 5R_1 + R_3 \rightarrow \end{array} \left[\begin{array}{rrr|r} 1 & -1 & -2 & -1 \\ 0 & 5 & 5 & 0 \\ 0 & 9 & 12 & 9 \end{array}\right]$

$\begin{array}{r} R_2 + R_1 \rightarrow \\ \frac{1}{5}R_2 \rightarrow \\ -9R_2 + R_3 \rightarrow \end{array} \left[\begin{array}{rrr|r} 1 & 0 & -1 & -1 \\ 0 & 1 & 1 & 0 \\ 0 & 0 & 3 & 9 \end{array}\right]$

$\begin{array}{r} R_3 + R_1 \rightarrow \\ -R_3 + R_2 \rightarrow \\ \frac{1}{3}R_3 \rightarrow \end{array} \left[\begin{array}{rrr|r} 1 & 0 & 0 & 2 \\ 0 & 1 & 0 & -3 \\ 0 & 0 & 1 & 3 \end{array}\right]$

$x = 2, y = -3, z = 3$

Answer: $(2, -3, 3)$

13. $\left[\begin{array}{rrr|r} 4 & 4 & 4 & 5 \\ 4 & -2 & -8 & 1 \\ 5 & 3 & 8 & 6 \end{array}\right]$

$\begin{array}{r} R_3 - R_1 \rightarrow \\ -4R_1 + R_2 \rightarrow \\ -5R_1 + R_3 \rightarrow \end{array} \left[\begin{array}{rrr|r} 1 & -1 & 4 & 1 \\ 0 & 2 & -24 & -3 \\ 0 & 8 & -12 & 1 \end{array}\right]$

$\begin{array}{r} R_2 + R_1 \rightarrow \\ \frac{1}{2}R_2 \rightarrow \\ -8R_2 + R_3 \rightarrow \end{array} \left[\begin{array}{rrr|r} 1 & 0 & -8 & -\frac{1}{2} \\ 0 & 1 & -12 & -\frac{3}{2} \\ 0 & 0 & 84 & 13 \end{array}\right]$

$\begin{array}{r} 8R_3 + R_1 \rightarrow \\ 12R_3 + R_2 \rightarrow \\ \frac{1}{84}R_3 \rightarrow \end{array} \left[\begin{array}{rrr|r} 1 & 0 & 0 & \frac{31}{42} \\ 0 & 1 & 0 & \frac{5}{14} \\ 0 & 0 & 1 & \frac{13}{84} \end{array}\right]$

$x = \frac{31}{42}, y = \frac{5}{14}, z = \frac{13}{84}$

Answer: $\left(\frac{31}{42}, \frac{5}{14}, \frac{13}{84}\right)$

15. $\begin{bmatrix} 2 & 1 & 2 & \vdots & 4 \\ 2 & 2 & 0 & \vdots & 5 \\ 2 & -1 & 6 & \vdots & 2 \end{bmatrix}$

$$\begin{matrix} \\ -R_1 + R_2 \rightarrow \\ -R_1 + R_3 \rightarrow \end{matrix} \begin{bmatrix} 2 & 1 & 2 & \vdots & 4 \\ 0 & 1 & -2 & \vdots & 1 \\ 0 & -2 & 4 & \vdots & -2 \end{bmatrix}$$

$$\begin{matrix} -R_2 + R_1 \rightarrow \\ \\ 2R_2 + R_3 \rightarrow \end{matrix} \begin{bmatrix} 2 & 0 & 4 & \vdots & 3 \\ 0 & 1 & -2 & \vdots & 1 \\ 0 & 0 & 0 & \vdots & 0 \end{bmatrix}$$

Let $z = a$, then:

$y - 2a = 1 \Rightarrow y = 2a + 1$

$2x + 4a = 3 \Rightarrow x = -2a + \frac{3}{2}$

Answer: $\left(-2a + \frac{3}{2}, 2a + 1, a\right)$

17. $\begin{bmatrix} 1 & 2 & 6 & \vdots & 1 \\ 2 & 5 & 15 & \vdots & 4 \\ 3 & 1 & 3 & \vdots & -6 \end{bmatrix}$

$$\begin{matrix} \\ -2R_1 + R_2 \rightarrow \\ -3R_1 + R_3 \rightarrow \end{matrix} \begin{bmatrix} 1 & 2 & 6 & \vdots & 1 \\ 0 & 1 & 3 & \vdots & 2 \\ 0 & -5 & -15 & \vdots & -9 \end{bmatrix}$$

$$\begin{matrix} -2R_2 + R_1 \rightarrow \\ \\ 5R_2 + R_3 \rightarrow \end{matrix} \begin{bmatrix} 1 & 0 & 0 & \vdots & -3 \\ 0 & 1 & 3 & \vdots & 2 \\ 0 & 0 & 0 & \vdots & 1 \end{bmatrix}$$

$x = -3$

$y + 3z = 2$

$0 = 1$

Inconsistent, no solution

19. If a system of linear equations has a unique solution, the augmented matrix reduces to a form in which the number of rows with nonzero entries on the coefficient side of the matrix equals the number of variables.

21. $\begin{bmatrix} 2 & 1 & 0 \\ 0 & 5 & -4 \end{bmatrix} - 3\begin{bmatrix} 5 & 3 & -6 \\ 0 & -2 & 5 \end{bmatrix} = \begin{bmatrix} 2 & 1 & 0 \\ 0 & 5 & -4 \end{bmatrix} - \begin{bmatrix} 15 & 9 & -18 \\ 0 & -6 & 15 \end{bmatrix}$

$$= \begin{bmatrix} -13 & -8 & 18 \\ 0 & 11 & -19 \end{bmatrix}$$

23. $\begin{bmatrix} 1 & 2 \\ 5 & -4 \\ 6 & 0 \end{bmatrix}\begin{bmatrix} 6 & -2 & 8 \\ 4 & 0 & 0 \end{bmatrix} = \begin{bmatrix} 1(6) + 2(4) & 1(-2) + 2(0) & 1(8) + 2(0) \\ 5(6) + (-4)(4) & 5(-2) + (-4)(0) & 5(8) + (-4)(0) \\ 6(6) + (0)(4) & 6(-2) + (0)(0) & 6(8) + (0)(0) \end{bmatrix}$

$$= \begin{bmatrix} 14 & -2 & 8 \\ 14 & -10 & 40 \\ 36 & -12 & 48 \end{bmatrix}$$

25. $\begin{bmatrix} 1 & 5 & 6 \\ 2 & -4 & 0 \end{bmatrix}\begin{bmatrix} 6 & 4 \\ -2 & 0 \\ 8 & 0 \end{bmatrix} = \begin{bmatrix} 1(6) + 5(-2) + 6(8) & 1(4) + 5(0) + 6(0) \\ 2(6) - 4(-2) + 0(8) & 2(4) - 4(0) + 0(0) \end{bmatrix}$

$$= \begin{bmatrix} 44 & 4 \\ 20 & 8 \end{bmatrix}$$

27. $\begin{bmatrix} 1 & 3 & 2 \\ 0 & 2 & -4 \\ 0 & 0 & 3 \end{bmatrix}\begin{bmatrix} 4 & -3 & 2 \\ 0 & 3 & -1 \\ 0 & 0 & 2 \end{bmatrix} = \begin{bmatrix} 1(4) & 1(-3) + 3(3) & 1(2) + 3(-1) + 2(2) \\ 0 & 2(3) & 2(-1) + (-4)(2) \\ 0 & 0 & 3(2) \end{bmatrix}$

$$= \begin{bmatrix} 4 & 6 & 3 \\ 0 & 6 & -10 \\ 0 & 0 & 6 \end{bmatrix}$$

29. $3\begin{bmatrix} 8 & -2 & 5 \\ 1 & 3 & -1 \end{bmatrix} + 6\begin{bmatrix} 4 & -2 & -3 \\ 2 & 7 & 6 \end{bmatrix} = \begin{bmatrix} 48 & -18 & -3 \\ 15 & 51 & 33 \end{bmatrix}$

31. $\begin{bmatrix} 4 & 1 \\ 11 & -7 \\ 12 & 3 \end{bmatrix}\begin{bmatrix} 3 & -5 & 6 \\ 2 & -2 & -2 \end{bmatrix} = \begin{bmatrix} 14 & -22 & 22 \\ 19 & -41 & 80 \\ 42 & -66 & 66 \end{bmatrix}$

33. $X = 3A - 2B = 3\begin{bmatrix} -4 & 0 \\ 1 & -5 \\ -3 & 2 \end{bmatrix} - 2\begin{bmatrix} 1 & 2 \\ -2 & 1 \\ 4 & 4 \end{bmatrix} = \begin{bmatrix} -14 & -4 \\ 7 & -17 \\ -17 & -2 \end{bmatrix}$

35. $X = \frac{1}{3}[B - 2A] = \frac{1}{3}\left(\begin{bmatrix} 1 & 2 \\ -2 & 1 \\ 4 & 4 \end{bmatrix} - 2\begin{bmatrix} -4 & 0 \\ 1 & -5 \\ -3 & 2 \end{bmatrix}\right) = \frac{1}{3}\begin{bmatrix} 9 & 2 \\ -4 & 11 \\ 10 & 0 \end{bmatrix}$

37. $\begin{bmatrix} 5 & 4 \\ -1 & 1 \end{bmatrix}\begin{bmatrix} x \\ y \end{bmatrix} = \begin{bmatrix} 2 \\ -22 \end{bmatrix}$

$\begin{bmatrix} 5x + 4y \\ -x + y \end{bmatrix} = \begin{bmatrix} 2 \\ -22 \end{bmatrix}$

$5x + 4y = 2$

$-x + 4y = -22$

39. $\begin{bmatrix} 2 & 6 \\ 3 & -6 \end{bmatrix}^{-1} = \begin{bmatrix} \frac{1}{5} & \frac{1}{5} \\ \frac{1}{10} & -\frac{1}{15} \end{bmatrix}$

41. $\begin{bmatrix} 2 & 0 & 3 \\ -1 & 1 & 1 \\ 2 & -2 & 1 \end{bmatrix}^{-1} = \begin{bmatrix} \frac{1}{2} & -1 & -\frac{1}{2} \\ \frac{1}{2} & -\frac{2}{3} & -\frac{5}{6} \\ 0 & \frac{2}{3} & \frac{1}{3} \end{bmatrix}$

43. $\begin{vmatrix} 50 & -30 \\ 10 & 5 \end{vmatrix} = 50(5) - (-30)(10) = 550$

45. $\begin{vmatrix} 3 & 0 & -4 & 0 \\ 0 & 8 & 1 & 2 \\ 6 & 1 & 8 & 2 \\ 0 & 3 & -4 & 1 \end{vmatrix} = 3\begin{vmatrix} 8 & 1 & 2 \\ 1 & 8 & 2 \\ 3 & -4 & 1 \end{vmatrix} + (-4)\begin{vmatrix} 0 & 8 & 2 \\ 6 & 1 & 2 \\ 0 & 3 & 1 \end{vmatrix}$ (Expansion along Row 1)

$= 3[8(8 - (-8)) - 1(1 - 6) + 2(-4 - 24)] - 4[0 - 6(8 - 6) + 0]$

$= 3[128 + 5 - 56] - 4[-12]$

$= 279$

47. $x + 2y = -1$

$3x + 4y = -5$

$\begin{bmatrix} 1 & 2 \\ 3 & 4 \end{bmatrix}^{-1} = \begin{bmatrix} -2 & 1 \\ \frac{3}{2} & -\frac{1}{2} \end{bmatrix}$

$\begin{bmatrix} x \\ y \end{bmatrix} = \begin{bmatrix} -2 & 1 \\ \frac{3}{2} & -\frac{1}{2} \end{bmatrix}\begin{bmatrix} -1 \\ -5 \end{bmatrix} = \begin{bmatrix} -3 \\ 1 \end{bmatrix}$

$x = -3, y = 1$

Answer: $(-3, 1)$

49. $-3x - 3y - 4z = 2$

$y + z = -1$

$4x + 3y + 4z = -1$

$\begin{bmatrix} -3 & -3 & -4 \\ 0 & 1 & 1 \\ 4 & 3 & 4 \end{bmatrix}^{-1} = \begin{bmatrix} 1 & 0 & 1 \\ 4 & 4 & 3 \\ -4 & -3 & -3 \end{bmatrix}$

$\begin{bmatrix} x \\ y \\ z \end{bmatrix} = \begin{bmatrix} 1 & 0 & 1 \\ 4 & 4 & 3 \\ -4 & -3 & -3 \end{bmatrix}\begin{bmatrix} 2 \\ -1 \\ -1 \end{bmatrix} = \begin{bmatrix} 1 \\ 1 \\ -2 \end{bmatrix}$

$x = 1, y = 1, z = -2$

Answer: $(1, 1, -2)$

51.
$$\begin{aligned} x + 3y + 2z &= 2 \\ -2x - 5y - z &= 10 \\ 2x + 4y \quad &= -12 \end{aligned}$$

$$\begin{bmatrix} 1 & 3 & 2 \\ -2 & -5 & -1 \\ 2 & 4 & 0 \end{bmatrix}^{-1} = \begin{bmatrix} 2 & 4 & \frac{7}{2} \\ -1 & -2 & -\frac{3}{2} \\ 1 & 1 & \frac{1}{2} \end{bmatrix}$$

$$\begin{bmatrix} x \\ y \\ z \end{bmatrix} = \begin{bmatrix} 1 & 0 & 1 \\ 4 & 4 & 3 \\ -4 & -3 & -3 \end{bmatrix} \begin{bmatrix} 2 \\ -1 \\ -1 \end{bmatrix} = \begin{bmatrix} 1 \\ 1 \\ -2 \end{bmatrix}$$

$x = 2, y = -4, z = 6$

Answer: $(2, -4, 6)$

53.
$$\begin{aligned} -x + y + z &= 6 \\ 4x - 3y + z &= 20 \\ 2x - y + 3z &= 8 \end{aligned}$$

$$\begin{bmatrix} -1 & 1 & 1 \\ 4 & -3 & 1 \\ 2 & -1 & 3 \end{bmatrix}^{-1} \text{ does not exist}$$

The system is inconsistent and has no solution.

55. x = number of carnations, y = number of roses

$x + y = 12$

$0.75x + 1.50y = 12.00$

$$x = \frac{\begin{vmatrix} 12 & 1 \\ 12 & 1.50 \end{vmatrix}}{\begin{vmatrix} 1 & 1 \\ 0.75 & 1.50 \end{vmatrix}} = \frac{6}{0.75} = 8$$

$$y = \frac{\begin{vmatrix} 1 & 12 \\ 0.75 & 12 \end{vmatrix}}{\begin{vmatrix} 1 & 1 \\ 0.75 & 1.50 \end{vmatrix}} = \frac{3}{0.75} = 4$$

Answer: 8 carnations, 4 roses

57. $(-1, 2),\ (0, 3),\ (1, 6)$

$f(x) = ax^2 + bx + c$

$f(-1) = a - b + c = 2$

$f(0) = c = 3$

$f(1) = a + b + c = 6$

$$D = \begin{vmatrix} 1 & -1 & 1 \\ 0 & 0 & 1 \\ 1 & 1 & 1 \end{vmatrix} = -2$$

$$a = \frac{\begin{vmatrix} 2 & -1 & 1 \\ 3 & 0 & 1 \\ 6 & 1 & 1 \end{vmatrix}}{-2} = \frac{-2}{-2} = 1$$

$$b = \frac{\begin{vmatrix} 1 & 2 & 1 \\ 0 & 3 & 1 \\ 1 & 6 & 1 \end{vmatrix}}{-2} = \frac{-4}{-2} = 2;\ c = 3$$

Thus, $y = x^2 + 2x + 3$.

59. $13a + 91b = 1107$

$91a + 819b = 8404.7$

(a) $a \approx 59.9,\ b \approx 3.6$

$y = 59.9 + 3.6t$

(b)

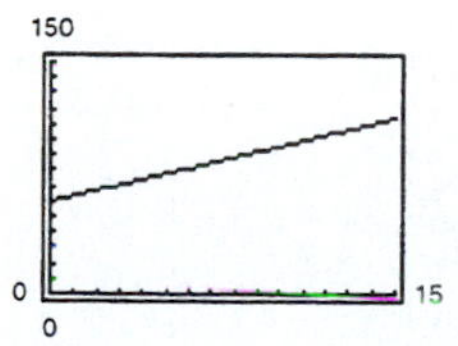

(c) The median price of one-family homes sold in the United States has been increasing by an average of 3.6 thousand dollars ($3600) each year.

(d) $y(15) = 59.9 + 3.6(15) = 113.9$ which corresponds to a price of $113,900.

61. $(1, 0),\ (5, 0),\ (5, 8)$

$$\text{Area} = \frac{1}{2}\begin{vmatrix} 1 & 0 & 1 \\ 5 & 0 & 1 \\ 5 & 8 & 1 \end{vmatrix} = \frac{1}{2}(32)$$

$= 16$ square units

63. $(1, 2),\ (4, -5),\ (3, 2)$

$$\text{Area} = \frac{1}{2}\begin{vmatrix} 1 & 2 & 1 \\ 4 & -5 & 1 \\ 3 & 1 & 1 \end{vmatrix} = \frac{1}{2}(14)$$

$= 7$ square units

65. $(-4, 0),\ (4, 4)$

$$\begin{vmatrix} x & y & 1 \\ -4 & 0 & 1 \\ 4 & 4 & 1 \end{vmatrix} = 0$$

$$-4x + 8y - 16 = 0$$

$$x - 2y + 4 = 0$$

67. $\left(-\frac{5}{2}, 3\right),\ \left(\frac{7}{2}, 1\right)$

$$\begin{vmatrix} x & y & 1 \\ -\frac{5}{2} & 3 & 1 \\ \frac{7}{2} & 1 & 1 \end{vmatrix} = 0$$

$$2x + 6y - 13 = 0$$

69. Expansion by Row 3

$$\begin{vmatrix} a_{11} & a_{12} & a_{13} \\ a_{21} & a_{22} & a_{23} \\ a_{31} + c_1 & a_{32} + c_2 & a_{33} + c_3 \end{vmatrix} = (a_{31} + c_1)\begin{vmatrix} a_{12} & a_{13} \\ a_{22} & a_{23} \end{vmatrix} - (a_{32} + c_2)\begin{vmatrix} a_{11} & a_{13} \\ a_{21} & a_{23} \end{vmatrix} + (a_{33} + c_3)\begin{vmatrix} a_{11} & a_{12} \\ a_{21} & a_{22} \end{vmatrix}$$

$$= a_{31}\begin{vmatrix} a_{12} & a_{13} \\ a_{22} & a_{23} \end{vmatrix} - a_{32}\begin{vmatrix} a_{11} & a_{13} \\ a_{21} & a_{23} \end{vmatrix} + a_{33}\begin{vmatrix} a_{11} & a_{12} \\ a_{21} & a_{22} \end{vmatrix} + c_1\begin{vmatrix} a_{12} & a_{13} \\ a_{22} & a_{23} \end{vmatrix}$$

$$- c_2\begin{vmatrix} a_{11} & a_{13} \\ a_{21} & a_{23} \end{vmatrix} + c_3\begin{vmatrix} a_{11} & a_{12} \\ a_{21} & a_{22} \end{vmatrix}$$

$$= \begin{vmatrix} a_{11} & a_{12} & a_{13} \\ a_{21} & a_{22} & a_{23} \\ a_{31} & a_{32} & a_{33} \end{vmatrix} + \begin{vmatrix} a_{11} & a_{12} & a_{13} \\ a_{21} & a_{22} & a_{23} \\ c_1 & c_2 & c_3 \end{vmatrix}$$

Note: Expand each of these matrices by Row 3 to see the previous step.

Practice Test for Chapter 8

1. Put the matrix in reduced echelon form.

$$\begin{bmatrix} 1 & -2 & 4 \\ 3 & -5 & 9 \end{bmatrix}$$

For Exercises 2–4, use matrices to solve the system of equations.

2. $\begin{aligned} 3x + 5y &= 3 \\ 2x - y &= -11 \end{aligned}$

3. $\begin{aligned} 2x + 3y &= -3 \\ 3x + 2y &= 8 \\ x + y &= 1 \end{aligned}$

4. $\begin{aligned} x \quad\quad + 3z &= -5 \\ 2x + y \quad\quad &= 0 \\ 3x + y - z &= 3 \end{aligned}$

5. Multiply $\begin{bmatrix} 1 & 4 & 5 \\ 2 & 0 & -3 \end{bmatrix}\begin{bmatrix} 1 & 6 \\ 0 & -7 \\ -1 & 2 \end{bmatrix}$.

6. Given $A = \begin{bmatrix} 9 & 1 \\ -4 & 8 \end{bmatrix}$ and $B = \begin{bmatrix} 6 & -2 \\ 3 & 5 \end{bmatrix}$, find $3A - 5B$.

7. Find $f(A)$:

$f(x) = x^2 - 7x + 8,\ A = \begin{bmatrix} 3 & 0 \\ 7 & 1 \end{bmatrix}$.

8. True or false:

$(A + B)(A + 3B) = A^2 + 4AB + 3B^2$ where A and B are matrices.

(Assume that A^2, AB, and B^2 exist.)

For Exercises 9–10, find the inverse of the matrix, if it exists.

9. $\begin{bmatrix} 1 & 2 \\ 3 & 5 \end{bmatrix}$

10. $\begin{bmatrix} 1 & 1 & 1 \\ 3 & 6 & 5 \\ 6 & 10 & 8 \end{bmatrix}$

11. Use an inverse matrix to solve the systems.

(a) $\begin{aligned} x + 2y &= 4 \\ 3x + 5y &= 1 \end{aligned}$

(b) $\begin{aligned} x + 2y &= 3 \\ 3x + 5y &= -2 \end{aligned}$

For Exercises 12–14, find the determinant of the matrix.

12. $\begin{bmatrix} 6 & -1 \\ 3 & 4 \end{bmatrix}$

13. $\begin{bmatrix} 1 & 3 & -1 \\ 5 & 9 & 0 \\ 6 & 2 & -5 \end{bmatrix}$

14. $\begin{bmatrix} 1 & 4 & 2 & 3 \\ 0 & 1 & -2 & 0 \\ 3 & 5 & -1 & 1 \\ 2 & 0 & 6 & 1 \end{bmatrix}$

15. Evaluate $\begin{vmatrix} 6 & 4 & 3 & 0 & 6 \\ 0 & 5 & 1 & 4 & 8 \\ 0 & 0 & 2 & 7 & 3 \\ 0 & 0 & 0 & 9 & 2 \\ 0 & 0 & 0 & 0 & 1 \end{vmatrix}$.

16. Use a determinant to find the area of the triangle with vertices (0, 7), (5, 0), and (3, 9).

17. Find the equation of the line through (2, 7) and (−1, 4).

For Exercises 18–20, use Cramer's Rule to find the indicated value.

18. Find x.

$$\begin{aligned} 6x - 7y &= 4 \\ 2x + 5y &= 11 \end{aligned}$$

19. Find z.

$$\begin{aligned} 3x \qquad + z &= 1 \\ y + 4z &= 3 \\ x - y \qquad &= 2 \end{aligned}$$

20. Find y.

$$\begin{aligned} 721.4x - 29.1y &= 33.77 \\ 45.9x + 105.6y &= 19.85 \end{aligned}$$

CHAPTER 9
Sequences and Probability

CHAPTER 9
Sequences and Probability

Section 9.1 Sequences and Summation Notation

- Given the general nth term in a sequence, you should be able to find, or list, some of the terms.
- You should be able to find an expression for the nth term of a sequence.
- You should be able to use and evaluate factorials.
- You should be able to use sigma notation for a sum.

Solutions to Odd-Numbered Exercises

1. $a_n = 2n + 1$

$a_1 = 2(1) + 1 = 3$

$a_2 = 2(2) + 1 = 5$

$a_3 = 2(3) + 1 = 7$

$a_4 = 2(4) + 1 = 9$

$a_5 = 2(5) + 1 = 11$

3. $a_n = 2^n$

$a_1 = 2^1 = 2$

$a_2 = 2^2 = 4$

$a_3 = 2^3 = 8$

$a_4 = 2^4 = 16$

$a_5 = 2^5 = 32$

5. $a_n = (-2)^n$

$a_1 = (-2)^1 = -2$

$a_2 = (-2)^2 = 4$

$a_3 = (-2)^3 = -8$

$a_4 = (-2)^4 = 16$

$a_5 = (-2)^5 = -32$

7. $a_n = \dfrac{n+1}{n}$

$a_1 = \dfrac{1+1}{1} = 2$

$a_2 = \dfrac{3}{2}$

$a_3 = \dfrac{4}{3}$

$a_4 = \dfrac{5}{4}$

$a_5 = \dfrac{6}{5}$

9. $a_n = \dfrac{6n}{3n^2 - 1}$

$a_1 = \dfrac{6(1)}{3(1)^2 - 1} = 3$

$a_2 = \dfrac{6(2)}{3(2)^2 - 1} = \dfrac{12}{11}$

$a_3 = \dfrac{6(3)}{3(3)^2 - 1} = \dfrac{9}{13}$

$a_4 = \dfrac{6(4)}{3(4)^2 - 1} = \dfrac{24}{47}$

$a_5 = \dfrac{6(5)}{3(5)^2 - 1} = \dfrac{15}{37}$

11. $a_n = \dfrac{1 + (-1)^n}{n}$

$a_1 = 0$

$a_2 = \dfrac{2}{2} = 1$

$a_3 = 0$

$a_4 = \dfrac{2}{4} = \dfrac{1}{2}$

$a_5 = 0$

13. $a_n = 3 - \dfrac{1}{2^n}$

$a_1 = 3 - \dfrac{1}{2} = \dfrac{5}{2}$

$a_2 = 3 - \dfrac{1}{4} = \dfrac{11}{4}$

$a_3 = 3 - \dfrac{1}{8} = \dfrac{23}{8}$

$a_4 = 3 - \dfrac{1}{16} = \dfrac{47}{16}$

$a_5 = 3 - \dfrac{1}{32} = \dfrac{95}{32}$

15. $a_n = \dfrac{1}{n^{3/2}}$

$a_1 = \dfrac{1}{1} = 1$

$a_2 = \dfrac{1}{2^{3/2}}$

$a_3 = \dfrac{1}{3^{3/2}}$

$a_4 = \dfrac{1}{4^{3/2}} = \dfrac{1}{8}$

$a_5 = \dfrac{1}{5^{3/2}}$

17. $a_n = \dfrac{3^n}{n!}$

$a_1 = \dfrac{3^1}{1!} = \dfrac{3}{1} = 3$

$a_2 = \dfrac{3^2}{2!} = \dfrac{9}{2}$

$a_3 = \dfrac{27}{6} = \dfrac{9}{2}$

$a_4 = \dfrac{81}{24} = \dfrac{27}{8}$

$a_5 = \dfrac{243}{120} = \dfrac{81}{40}$

19. $a_n = \dfrac{(-1)^n}{n^2}$

$a_1 = -\dfrac{1}{1} = -1$

$a_2 = \dfrac{1}{4}$

$a_3 = -\dfrac{1}{9}$

$a_4 = \dfrac{1}{16}$

$a_5 = -\dfrac{1}{25}$

21. $a_n = \dfrac{2}{3}$

$a_1 = \dfrac{2}{3}$

$a_2 = \dfrac{2}{3}$

$a_3 = \dfrac{2}{3}$

$a_4 = \dfrac{2}{3}$

$a_5 = \dfrac{2}{3}$

23. $a_{25} = (-1)^{25}(3(25) - 2) = -73$

25. $a_1 = 28$ and $a_{k+1} = a_k - 4$

$a_1 = 28$

$a_2 = a_1 - 4 = 28 - 4 = 24$

$a_3 = a_2 - 4 = 24 - 4 = 20$

$a_4 = a_3 - 4 = 20 - 4 = 16$

$a_5 = a_4 - 4 = 16 - 4 = 12$

27. $a_1 = 3$ and $a_{k+1} = 2(a_k - 1)$

$a_1 = 3$

$a_2 = 2(a_1 - 1) = 2(3 - 1) = 4$

$a_3 = 2(a_2 - 1) = 2(4 - 1) = 6$

$a_4 = 2(a_3 - 1) = 2(6 - 1) = 10$

$a_5 = 2(a_4 - 1) = 2(10 - 1) = 18$

29. $a_n = \frac{2}{3}n$

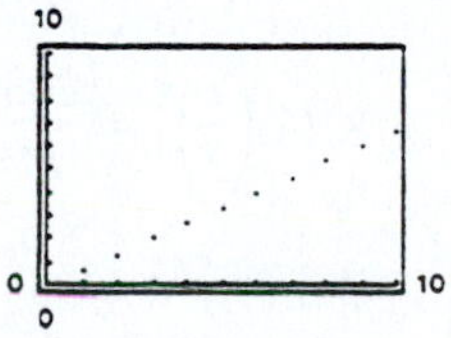

31. $a_n = 16(-0.5)^{n-1}$

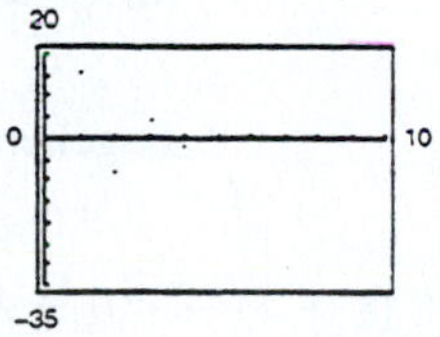

33. $a_n = \dfrac{2n}{n+1}$

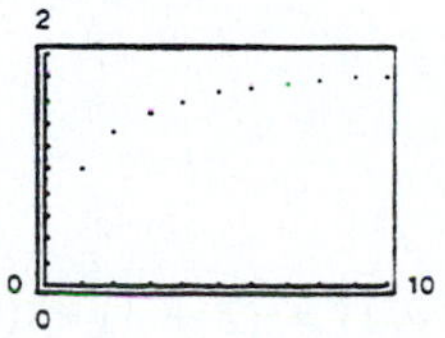

35. $a_n = \dfrac{8}{n+1}$

$a_n \to 0$ as $n \to \infty$

$a_1 = 4,\ a_{10} = \dfrac{8}{11}$

Matches graph (c).

37. $a_n = 4(0.5)^{n-1}$

$a_n \to 0$ as $n \to \infty$

$a_1 = 4,\ a_{10} \approx 0.008$

Matches graph (d).

39. $\dfrac{4!}{6!} = \dfrac{4!}{6 \cdot 5 \cdot 4!} = \dfrac{1}{30}$

41. $\dfrac{10!}{8!} = \dfrac{10 \cdot 9 \cdot 8!}{8!} = 90$

43. $\dfrac{(n+1)!}{n!} = \dfrac{(n+1)n!}{n!} = n + 1$

45. $\dfrac{(2n-1)!}{(2n+1)!} = \dfrac{(2n-1)!}{(2n+1)(2n)(2n-1)!}$

$= \dfrac{1}{2n(2n+1)}$

47. 1, 4, 7, 10, 13, . . .

$a_n = 1 + (n-1)3 = 3n - 2$

49. 0, 3, 8, 15, 24, . . .

$a_n = n^2 - 1$

51. $\dfrac{2}{3}, \dfrac{3}{4}, \dfrac{4}{5}, \dfrac{5}{6}, \dfrac{6}{7}, \ldots$

$a_n = \dfrac{n+1}{n+2}$

53. $\dfrac{1}{2}, \dfrac{-1}{4}, \dfrac{1}{8}, \dfrac{-1}{16}, \ldots$

$a_n = \dfrac{(-1)^{n+1}}{2^n}$

55. $1 + \dfrac{1}{1}, 1 + \dfrac{1}{2}, 1 + \dfrac{1}{3}, 1 + \dfrac{1}{4}, 1 + \dfrac{1}{5}, \ldots$

$a_n = 1 + \dfrac{1}{n}$

57. $1, \dfrac{1}{2}, \dfrac{1}{6}, \dfrac{1}{24}, \dfrac{1}{120}, \ldots$

$a_n = \dfrac{1}{n!}$

59. $1, -1, 1, -1, 1, \ldots$

$a_n = (-1)^{n+1}$

61. $a_1 = 6$ and $a_{k+1} = a_k + 2$

$a_1 = 6$

$a_2 = a_1 + 2 = 6 + 2 = 8$

$a_3 = a_2 + 2 = 8 + 2 = 10$

$a_4 = a_3 + 2 = 10 + 2 = 12$

$a_5 = a_4 + 2 = 12 + 2 = 14$

In general, $a_n = 2n + 4$.

63. $a_1 = 81$ and $a_{k+1} = \dfrac{1}{3}a_k$

$a_1 = 81$

$a_2 = \dfrac{1}{3}a_1 = \dfrac{1}{3}(81) = 27$

$a_3 = \dfrac{1}{3}a_2 = \dfrac{1}{3}(27) = 9$

$a_4 = \dfrac{1}{3}a_3 = \dfrac{1}{3}(9) = 3$

$a_5 = \dfrac{1}{3}a_4 = \dfrac{1}{3}(3) = 1$

In general, $a_n = 81\left(\dfrac{1}{3}\right)^{n-1} = 81(3)\left(\dfrac{1}{3}\right)^n = \dfrac{243}{3^n}$.

65. $\displaystyle\sum_{i=1}^{5}(2i+1) = (2+1) + (4+1) + (6+1) + (8+1) + (10+1) = 35$

67. $\displaystyle\sum_{k=1}^{4} 10 = 10 + 10 + 10 + 10 = 40$

69. $\displaystyle\sum_{i=0}^{4} i^2 = 0^2 + 1^2 + 2^2 + 3^2 + 4^2 = 30$

71. $\displaystyle\sum_{k=0}^{3} \frac{1}{k^2+1} = \frac{1}{1} + \frac{1}{1+1} + \frac{1}{4+1} + \frac{1}{9+1} = \frac{9}{5}$

73. $\displaystyle\sum_{i=1}^{4}[(i-1)^2 + (i+1)^3] = [(0)^2 + (2)^3] + [(1)^2 + (3)^3] + [(2)^2 + (4)^3] + [(3)^2 + (5)^3] = 238$

75. $\displaystyle\sum_{i=1}^{4} 2^i = 2^1 + 2^2 + 2^3 + 2^4 = 30$

77. $\displaystyle\sum_{j=1}^{6}(24 - 3j) = 81$

79. $\displaystyle\sum_{k=0}^{4} \frac{(-1)^k}{k+1} = 1 - \frac{1}{2} + \frac{1}{3} - \frac{1}{4} + \frac{1}{5} = \frac{47}{60}$

81. $\displaystyle\frac{1}{3(1)} + \frac{1}{3(2)} + \frac{1}{3(3)} + \cdots + \frac{1}{3(9)} = \sum_{i=1}^{9} \frac{1}{3i}$

83. $\displaystyle\left[2\left(\frac{1}{8}\right) + 3\right] + \left[2\left(\frac{2}{8}\right) + 3\right] + \left[2\left(\frac{3}{8}\right) + 3\right] + \cdots + \left[2\left(\frac{8}{8}\right) + 3\right] = \sum_{i=1}^{8}\left[2\left(\frac{i}{8}\right) + 3\right]$

85. $\displaystyle 3 - 9 + 27 - 81 + 243 - 729 = \sum_{i=1}^{6}(-1)^{i+1}3^i$

87. $\displaystyle\frac{1}{1^2} - \frac{1}{2^2} + \frac{1}{3^2} - \frac{1}{4^2} + \cdots - \frac{1}{20^2} = \sum_{i=1}^{20} \frac{(-1)^{i+1}}{i^2}$

89. $\displaystyle\frac{1}{4} + \frac{3}{8} + \frac{7}{16} + \frac{15}{32} + \frac{31}{64} = \sum_{i=1}^{5} \frac{2^i - 1}{2^{i+1}}$

91. $\displaystyle A_n = 5000\left(1 + \frac{0.08}{4}\right)^n, n = 1, 2, 3, \ldots$

(a) $A_1 = \$5100.00$
$A_2 = \$5202.00$
$A_3 = \$5306.04$
$A_4 = \$5412.16$
$A_5 = \$5520.40$
$A_6 = \$5630.81$
$A_7 = \$5743.43$
$A_8 = \$5858.30$

(b) $A_{40} = \$11{,}040.20$

93. $a_n = 510.13 + 16.37n + 3.23n^2, n = 1, \ldots, 11$

$a_1 = 529.73$
$a_2 = 555.79$
$a_3 = 588.31$
$a_4 = 627.29$
$a_5 = 672.73$
$a_6 = 724.63$
$a_7 = 782.99$
$a_8 = 847.81$
$a_9 = 919.09$
$a_{10} = 996.83$
$a_{11} = 1081.03$

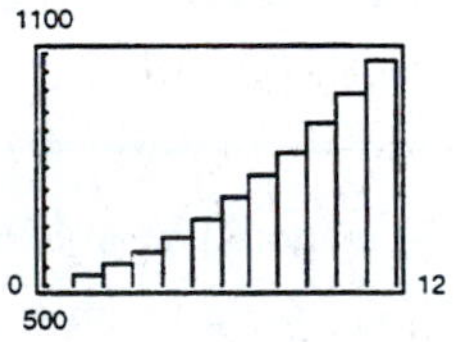

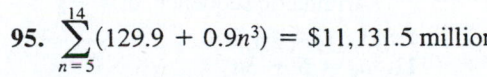

95. $\displaystyle\sum_{n=5}^{14}(129.9 + 0.9n^3) = \$11{,}131.5$ million

97. $a_1 = 1, a_2 = 1, a_{k+2} = a_{k+1} + a_k$

$a_1 = 1$	$b_1 = \frac{1}{1} = 1$
$a_2 = 1$	$b_2 = \frac{2}{1} = 2$
$a_3 = 1 + 1 = 2$	$b_3 = \frac{3}{2}$
$a_4 = 2 + 1 = 3$	$b_4 = \frac{5}{3}$
$a_5 = 3 + 2 = 5$	$b_5 = \frac{8}{5}$
$a_6 = 5 + 3 = 8$	$b_6 = \frac{13}{8}$
$a_7 = 8 + 5 = 13$	$b_7 = \frac{21}{13}$
$a_8 = 13 + 8 = 21$	$b_8 = \frac{34}{21}$
$a_9 = 21 + 13 = 34$	$b_9 = \frac{55}{34}$
$a_{10} = 34 + 21 = 55$	$b_{10} = \frac{89}{55}$
$a_{11} = 55 + 34 = 89$	
$a_{12} = 89 + 55 = 144$	

99. $\dfrac{327.15 + 785.69 + 433.04 + 265.38 + 604.12 + 590.30}{6} \approx \$500.95$

101.
$$\sum_{i=1}^{n}(x_i - \bar{x}) = \sum_{i=1}^{n} x_i - \sum_{i=1}^{n} \bar{x}$$
$$= \sum_{i=1}^{n} x_i - n\bar{x}$$
$$= \sum_{i=1}^{n} x_i - n\left(\frac{1}{n}\sum_{i=1}^{n} x_i\right)$$
$$= 0$$

103. True, $\sum_{i=1}^{4}(i^2 + 2i) = \sum_{i=1}^{4} i^2 + 2\sum_{i=1}^{4} i$

by the properties of sums.

Section 9.2 Arithmetic Sequences

- You should be able to recognize an arithmetic sequence, find its common difference, and find its nth term.
- You should be able to find the nth partial sum of an arithmetic sequence by using the formula

$$S_n = \frac{n}{2}(a_1 + a_n).$$

Solutions to Odd-Numbered Exercises

1. 10, 8, 6, 4, 2, . . .

Arithmetic sequence, $d = -2$

3. 1, 2, 4, 8, 16, 32, . . .

Not an arithmetic sequence

5. $\frac{9}{4}, 2, \frac{7}{4}, \frac{3}{2}, \frac{5}{4}, 1, \ldots$

Arithmetic sequence, $d = -\frac{1}{4}$

7. $-12, -8, -4, 0, 4, \ldots$

Arithmetic sequence, $d = 4$

9. 5.3, 5.7, 6.1, 6.5, 6.9, . . .

Arithmetic sequence, $d = 0.4$

11. $a_n = 5 + 3n$

8, 11, 14, 17, 20

Arithmetic sequence, $d = 3$

13. $a_n = \dfrac{1}{n+1}$

$\dfrac{1}{2}, \dfrac{1}{3}, \dfrac{1}{4}, \dfrac{1}{5}, \dfrac{1}{6}$

Not an arithmetic sequence

15. $a_n = 100 - 3n$

97, 94, 91, 88, 85

Arithmetic sequence, $d = -3$

17. $a_n = 3 + \dfrac{(-1)^n 2}{n}$

$1, 4, \dfrac{7}{3}, \dfrac{7}{2}, \dfrac{13}{5}$

Not an arithmetic sequence

19. $a_1 = 15,\ a_{k+1} = a_k + 4$

$a_2 = 15 + 4 = 19$

$a_3 = 19 + 4 = 23$

$a_4 = 23 + 4 = 27$

$a_5 = 27 + 4 = 31$

$a_n = 11 + 4n$

21. $a_1 = 200,\ a_{k+1} = a_k - 10$

$a_2 = 200 - 10 = 190$

$a_3 = 190 - 10 = 180$

$a_4 = 180 - 10 = 170$

$a_5 = 170 - 10 = 160$

$a_n = 210 - 10n$

23. $a_1 = \frac{3}{2},\ a_{k+1} = a_k - \frac{1}{4}$

$a_2 = \frac{3}{2} - \frac{1}{4} = \frac{5}{4}$

$a_3 = \frac{5}{4} - \frac{1}{4} = 1$

$a_4 = 1 - \frac{1}{4} = \frac{3}{4}$

$a_5 = \frac{3}{4} - \frac{1}{4} = \frac{1}{2}$

$a_n = \frac{7}{4} - \frac{1}{4}n$

25. $a_1 = 5,\ d = 6$

$a_1 = 5$

$a_2 = 5 + 6 = 11$

$a_3 = 11 + 6 = 17$

$a_4 = 17 + 6 = 23$

$a_5 = 23 + 6 = 29$

27. $a_1 = -2.6,\ d = -0.4$

$a_1 = -2.6$

$a_2 = -2.6 + (-0.4) = -3.0$

$a_3 = -3.0 + (-0.4) = -3.4$

$a_4 = -3.4 + (-0.4) = -3.8$

$a_5 = -3.8 + (-0.4) = -4.2$

29. $a_1 = 2,\ a_{12} = 46$

$46 = 2 + (12 - 1)d$

$44 = 11d$

$4 = d$

$a_1 = 2$

$a_2 = 2 + 4 = 6$

$a_3 = 6 + 4 = 10$

$a_4 = 10 + 4 = 14$

$a_5 = 14 + 4 = 18$

31. $a_8 = 26,\ a_{12} = 42$

$26 = a_8 = a_1 + (n - 1)d = a_1 + 7d$

$42 = a_{12} = a_1 + (n - 1)d = a_1 + 11d$

Answer: $d = 4,\ a_1 = -2$

$a_1 = -2$

$a_2 = -2 + 4 = 2$

$a_3 = 2 + 4 = 6$

$a_4 = 6 + 4 = 10$

$a_5 = 10 + 4 = 14$

33. $a_1 = 1,\ d = 3$

$a_n = a_1 + (n - 1)d = 1 + (n - 1)(3)$

35. $a_1 = 100,\ d = -8$

$a_n = a_1 + (n - 1)d = 100 + (n - 1)(-8)$

37. $a_1 = x,\ d = 2x$

$a_n = a_1 + (n - 1)d = x + (n - 1)(2x)$

39. $4, \frac{3}{2}, -1, -\frac{7}{2}, \ldots$

$d = -\frac{5}{2}$

$a_n = a_1 + (n - 1)d = 4 + (n - 1)\left(-\frac{5}{2}\right)$

41. $a_1 = 5,\ a_4 = 15$

$a_4 = a_1 + 3d \implies 15 = 5 + 3d \implies d = \frac{10}{3}$

$a_n = a_1 + (n - 1)d = 5 + (n - 1)\left(\frac{10}{3}\right)$

43. $a_3 = 94,\ a_6 = 85$

$a_6 = a_3 + 3d \implies 85 = 94 + 3d \implies d = -3$

$a_1 = a_3 - 2d \implies a_1 = 94 - 2(-3) = 100$

$a_n = a_1 + (n - 1)d = 100 + (n - 1)(-3)$

45. $a_n = -\frac{2}{3}n + 6$

$d = -\frac{2}{3}$ so the sequence is decreasing, and $a_1 = 5\frac{1}{3}$. Matches (b).

47. $a_n = 2 + \frac{3}{4}n$

$d = \frac{3}{4}$ so the sequence is increasing, and $a_1 = 2\frac{3}{4}$. Matches (c).

49. $a_n = 15 - \frac{3}{2}n$

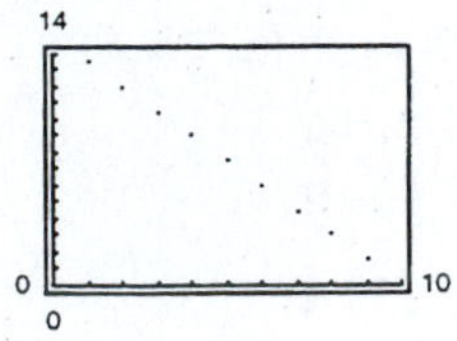

51. $a_n = 0.2n + 3$

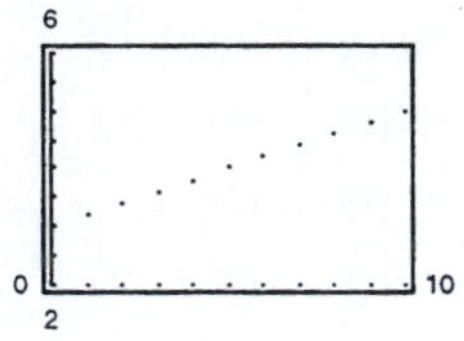

53. Since $a_n = dn + c$, its geometric pattern is linear.

55. 8, 20, 32, 44, . . .

$a_1 = 8,\ d = 12,\ n = 10$

$a_{10} = 8 + 9(12) = 116$

$S_{10} = \frac{10}{2}(8 + 116) = 620$

57. −6, −2, 2, 6, . . .

$a_1 = -6,\ d = 4,\ n = 50$

$a_{50} = -6 + 49(4) = 190$

$S_{50} = \frac{50}{2}(-6 + 190) = 4600$

59. 40, 37, 34, 31, . . .

$a_1 = 40, d = -3, n = 10$

$a_{10} = 40 + 9(-3) = 13$

$S_{10} = \frac{10}{2}(40 + 13) = 265$

61. $a_1 = 100,\ a_{25} = 220,\ n = 25$

$$S_n = \frac{n}{2}[a_1 + a_n]$$

$$S_{25} = \frac{25}{2}(100 + 220) = 4000$$

63. $a_1 = 1,\ a_{50} = 50,\ n = 50$

$$\sum_{n=1}^{50} n = \frac{50}{2}(1 + 50) = 1275$$

65. $a_1 = 5,\ a_{100} = 500, n = 100$

$$\sum_{n=1}^{100} 5n = \frac{100}{2}(5 + 500) = 25{,}250$$

67. $\sum_{n=11}^{30} n - \sum_{n=1}^{10} n = \frac{20}{2}(11 + 30) - \frac{10}{2}(1 + 10) = 355$

69. $a_1 = 4,\ a_{500} = 503, n = 500$

$$\sum_{n=1}^{500} (n + 3) = \frac{500}{2}(4 + 503) = 126{,}750$$

71. $a_1 = 7, a_{20} = 45,\ n = 20$

$$\sum_{n=1}^{20} (2n + 5) = \frac{20}{2}(7 + 45) = 520$$

73. $a_0 = 1000,\ a_{50} = 750,\ n = 51$

$$\sum_{n=0}^{50} (100 - 5n) = \frac{51}{2}(1000 + 750) = 44{,}625$$

75. $a_1 = \frac{742}{3},\ a_{60} = 90,\ n = 60$

$$\sum_{i=1}^{60} \left(250 - \frac{8}{3}i\right) = \frac{60}{2}\left(\frac{742}{3} + 90\right) = 10{,}120$$

77. $a_1 = 1,\ a_{100} = 199,\ n = 100$

$$\sum_{n=1}^{100} (2n - 1) = \frac{100}{2}(1 + 199) = 10{,}000$$

79. (a) $a_1 = 32{,}500,\ d = 1500$

$a_6 = a_1 + 5d = 32{,}500 + 5(1500) = \$40{,}000$

(b) $S_6 = \frac{6}{2}[32{,}500 + 40{,}000] = \$217{,}500$

81. $a_1 = 20,\ d = 4,\ n = 30$

$a_{30} = 20 + 29(4) = 136$

$S_{30} = \frac{30}{2}(20 + 136) = 2340$ seats

83. $a_1 = 14, a_{18} = 31$

$S_{18} = \frac{18}{2}(14 + 31) = 405$ bricks

85. (a) $1 + 3 = 4$

$1 + 3 + 5 = 9$

$1 + 3 + 5 + 7 = 16$

$1 + 3 + 5 + 7 + 9 = 25$

$1 + 3 + 5 + 7 + 9 + 11 = 36$

(b) $S_n = n^2$

$S_7 = 1 + 3 + 5 + 7 + 9 + 11 + 13 = 49 = 7^2$

(c) $S_n = \frac{n}{2}[1 + (2n - 1)] = \frac{n}{2}(2n) = n^2$

87. $S_{20} = \frac{20}{2}\{a_1 + [a_1 + (20 - 1)(3)]\} = 650$

$10(2a_1 + 57) = 650$

$2a_1 + 57 = 65$

$2a_1 = 8$

$a_1 = 4$

Section 9.3 Geometric Sequences

- You should be able to identify a geometric sequence, find its common ratio, and find the nth term.
- You should be able to find the nth partial sum of a geometric sequence with common ratio r using the formula.

  $$S_n = a_1\left(\frac{1 - r^n}{1 - r}\right)$$

- You should know that if $|r| < 1$, then

  $$\sum_{n=1}^{\infty} a_1 r^{n-1} = \frac{a_1}{1 - r}.$$

Solutions to Odd-Numbered Exercises

1. $5, 15, 45, 135, \ldots$

Geometric sequence, $r = 3$

3. $3, 12, 21, 30, \ldots$

Not a geometric sequence

Note: It is an arithmetic sequence with $d = 9$.

5. $1, -\frac{1}{2}, \frac{1}{4}, -\frac{1}{8}, \ldots$

Geometric sequence, $r = -\frac{1}{2}$

7. $\frac{1}{2}, \frac{2}{3}, \frac{3}{4}, \frac{4}{5}, \ldots$

Not a geometric sequence

9. $1, \frac{1}{2}, \frac{1}{3}, \frac{1}{4}, \ldots$

Not a geometric sequence

11. $a_1 = 2, \ r = 3$

$a_1 = 2$

$a_2 = 2(3) = 6$

$a_3 = 6(3) = 18$

$a_4 = 18(3) = 54$

$a_5 = 54(3) = 162$

13. $a_1 = 1, \ r = \frac{1}{2}$

$a_1 = 1$

$a_2 = 1\left(\frac{1}{2}\right) = \frac{1}{2}$

$a_3 = \frac{1}{2}\left(\frac{1}{2}\right) = \frac{1}{4}$

$a_4 = \frac{1}{4}\left(\frac{1}{2}\right) = \frac{1}{8}$

$a_5 = \frac{1}{8}\left(\frac{1}{2}\right) = \frac{1}{16}$

15. $a_1 = 5, \ r = -\frac{1}{10}$

$a_1 = 5$

$a_2 = 5\left(-\frac{1}{10}\right) = -\frac{1}{2}$

$a_3 = \left(-\frac{1}{2}\right)\left(-\frac{1}{10}\right) = \frac{1}{20}$

$a_4 = \frac{1}{20}\left(-\frac{1}{10}\right) = -\frac{1}{200}$

$a_5 = \left(-\frac{1}{200}\right)\left(-\frac{1}{10}\right) = \frac{1}{2000}$

17. $a_1 = 1,\ r = e$

$a_1 = 1$

$a_2 = 1(e) = e$

$a_3 = (e)(e) = e^2$

$a_4 = (e^2)(e) = e^3$

$a_5 = (e^3)(e) = e^4$

19. $a_1 = 3,\ r = \dfrac{x}{2}$

$a_1 = 3$

$a_2 = 3\left(\dfrac{x}{2}\right) = \dfrac{3x}{2}$

$a_3 = \left(\dfrac{3x}{2}\right)\left(\dfrac{x}{2}\right) = \dfrac{3x^2}{4}$

$a_4 = \left(\dfrac{3x^2}{4}\right)\left(\dfrac{x}{2}\right) = \dfrac{3x^3}{8}$

$a_5 = \left(\dfrac{3x^3}{8}\right)\left(\dfrac{x}{2}\right) = \dfrac{3x^4}{16}$

21. $a_1 = 64,\ a_{k+1} = \dfrac{1}{2}a_k$

$a_1 = 64$

$a_2 = \dfrac{1}{2}(64) = 32$

$a_3 = \dfrac{1}{2}(32) = 16$

$a_4 = \dfrac{1}{2}(16) = 8$

$a_5 = \dfrac{1}{2}(8) = 4$

$a_n = 64\left(\dfrac{1}{2}\right)^{n-1} = 128\left(\dfrac{1}{2}\right)^n$

23. $a_1 = 4,\ a_{k+1} = 3a_k$

$a_1 = 4$

$a_2 = 3(4) = 12$

$a_3 = 3(12) = 36$

$a_4 = 3(36) = 108$

$a_5 = 3(108) = 324$

$a_n = 4(3)^{n-1} = \frac{4}{3}(3)^n$

25. $a_k = 6,\ a_{k+1} = -\frac{3}{2}a_k$

$a_1 = 6$

$a_2 = -\frac{3}{2}(6) = -9$

$a_3 = -\frac{3}{2}(-9) = \frac{27}{2}$

$a_4 = -\frac{3}{2}\left(\frac{27}{2}\right) = -\frac{81}{4}$

$a_5 = -\frac{3}{2}\left(-\frac{81}{4}\right) = \frac{243}{8}$

$a_n = 6\left(-\frac{3}{2}\right)^{n-1}$ or $a_n = -4\left(-\frac{3}{2}\right)^n$

27. $a_1 = 4,\ r = \frac{1}{2},\ n = 10$

$a_n = a_1r^{n-1}$

$a_{10} = 4\left(\frac{1}{2}\right)^9 = \left(\frac{1}{2}\right)^7 = \frac{1}{128}$

29. $a_1 = 6,\ r = -\dfrac{1}{3},\ n = 12$

$a_n = a_1r^{n-1}$

$a_{12} = 6\left(-\dfrac{1}{3}\right)^{11} = \dfrac{-2}{3^{10}}$

31. $a_1 = 100,\ r = e^x,\ n = 9$

$a_n = a_1r^{n-1}$

$a_9 = 100(e^x)^8 = 100e^{8x}$

33. $a_1 = 500,\ r = 1.02,\ n = 40$

$a_n = a_1r^{n-1}$

$a_{40} = 500(1.02)^{39} \approx 1082.37$

35. $a_1 = 16,\ a_4 = \dfrac{27}{4},\ n = 3$

$\dfrac{27}{4} = 16r^3 \Rightarrow r = \dfrac{3}{4}$

$a_n = a_1r^{n-1}$

$a_3 = 16\left(\dfrac{3}{4}\right)^2 = 9$

37. $a_2 = a_1r = -18 \Rightarrow a_1 = \dfrac{-18}{r}$

$a_5 = a_1r^4 = (a_1r)r^3 = -18r^3 = \dfrac{2}{3} \Rightarrow r = -\dfrac{1}{3}$

$a_1 = \dfrac{-18}{r} = \dfrac{-18}{-1/3} = 54$

$a_6 = a_1r^5 = 54\left(\dfrac{-1}{3}\right)^5 = -\dfrac{54}{243} = -\dfrac{2}{9}$

39. $a_n = 18\left(\frac{2}{3}\right)^{n-1}$

$r = \frac{2}{3} < 1$, so the sequence is decreasing.

Matches (a).

41. $a_n = 18\left(\frac{3}{2}\right)^{n-1}$

$r = \frac{3}{2} > 1$, so the sequence is increasing.

Matches (b).

43. $a_n = 12(-0.75)^{n-1}$

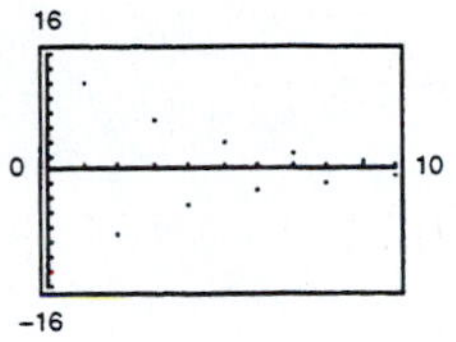

45. $a_n = 2(1.3)^{n-1}$

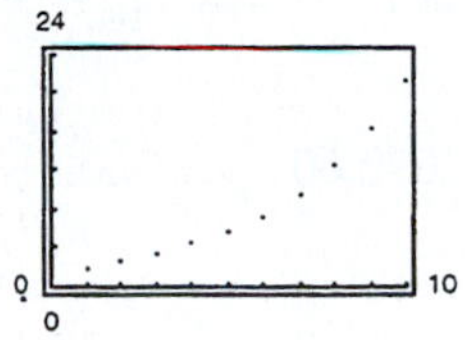

47. Given real numbers r between -1 and 1, as the exponent increases, r^n approaches zero.

49. $A = P\left(1 + \dfrac{r}{n}\right)^{nt} = 1000\left(1 + \dfrac{0.10}{n}\right)^{n(10)}$

(a) $n = 1,\ A = 1000(1 + 0.10)^{10} \approx \$2593.74$

(b) $n = 2,\ A = 1000\left(1 + \dfrac{0.10}{2}\right)^{2(10)} \approx \$2653.30$

(c) $n = 4,\ A = 1000\left(1 + \dfrac{0.10}{4}\right)^{4(10)} \approx \$2685.06$

(d) $n = 12,\ A = 1000\left(1 + \dfrac{0.10}{12}\right)^{12(10)} \approx \$2707.04$

(e) $n = 365,\ A = 1000\left(1 + \dfrac{0.10}{365}\right)^{365(10)} \approx \$2717.91$

51. $V_5 = 135{,}000(0.70)^5 = \$22{,}689.45$

53. $8,\ -4,\ 2,\ -1,\ \frac{1}{2}, \ldots$

$S_1 = 8$

$S_2 = 8 + (-4) = 4$

$S_3 = 8 + (-4) + 2 = 6$

$S_4 = 8 + (-4) + 2 + (-1) = 5$

55. $\displaystyle\sum_{n=1}^{9} 2^{n-1} \Rightarrow a_1 = 1,\ r = 2$

$S_9 = \dfrac{1(1 - 2^9)}{1 - 2} = 511$

57. $\displaystyle\sum_{i=1}^{7} 64\left(-\frac{1}{2}\right)^{i-1} \Rightarrow a_1 = 64,\ r = -\frac{1}{2}$

$S_7 = 64\left[\dfrac{1 - \left(-\frac{1}{2}\right)^7}{1 - \left(-\frac{1}{2}\right)}\right] = \dfrac{128}{3}\left[1 - \left(-\dfrac{1}{2}\right)^7\right] = 43$

59. $\displaystyle\sum_{n=0}^{20} 3\left(\frac{3}{2}\right)^n = \sum_{n=1}^{21} 3\left(\frac{3}{2}\right)^{n-1} \Rightarrow a_1 = 3,\ r = \frac{3}{2}$

$S_{21} = 3\left[\dfrac{1 - \left(\frac{3}{2}\right)^{21}}{1 - \frac{3}{2}}\right] = -6\left[1 - \left(\dfrac{3}{2}\right)^{21}\right] \approx 29{,}921.31$

61. $\sum_{i=1}^{10} 8\left(-\frac{1}{4}\right)^{i-1} \Rightarrow a_1 = 8,\ r = -\frac{1}{4}$

$$S_{10} = 8\left[\frac{1-\left(-\frac{1}{4}\right)^{10}}{1-\left(-\frac{1}{4}\right)}\right] = \frac{32}{5}\left[1-\left(-\frac{1}{4}\right)^{10}\right] \approx 6.4$$

63. $\sum_{n=0}^{5} 300(1.06)^n = \sum_{n=1}^{6} 300(1.06)^{n-1} \Rightarrow a_1 = 300,\ r = 1.06$

$$S_6 = 300\left[\frac{1-(1.06)^6}{1-1.06}\right] \approx 2092.60$$

65. $5 + 15 + 45 + \cdots + 3645$

$r = 3$ and $3645 = 5(3)^{n-1} \Rightarrow n = 7$

Thus, the sum can be written as $\sum_{n=1}^{7} 5(3)^{n-1}$

67. $$A = \sum_{n=1}^{60} 100\left(1+\frac{0.10}{12}\right)^n = 100\left(1+\frac{0.10}{12}\right)\cdot\frac{\left[1-\left(1+\frac{0.10}{12}\right)^{60}\right]}{\left[1-\left(1+\frac{0.10}{12}\right)\right]} \approx \$7808.24$$

69. Let $N = 12t$ be the total number of deposits.

$$\begin{aligned}
A &= P\left(1+\frac{r}{12}\right) + P\left(1+\frac{r}{12}\right)^2 + \cdots + P\left(1+\frac{r}{12}\right)^N \\
&= \left(1+\frac{r}{12}\right)\left[P + P\left(r+\frac{r}{12}\right) + \cdots + P\left(1+\frac{r}{12}\right)^{N-1}\right] \\
&= P\left(1+\frac{r}{12}\right)\sum_{n=1}^{N}\left(1+\frac{r}{12}\right)^{n-1} \\
&= P\left(1+\frac{r}{12}\right)\frac{1-\left(1+\frac{r}{12}\right)^N}{1-\left(1+\frac{r}{12}\right)} \\
&= P\left(1+\frac{r}{12}\right)\left(-\frac{12}{r}\right)\left[1-\left(1+\frac{r}{12}\right)^N\right] \\
&= P\left(\frac{12}{r}+1\right)\left[-1+\left(1+\frac{r}{12}\right)^N\right] \\
&= P\left[\left(1+\frac{r}{12}\right)^N - 1\right]\left(1+\frac{12}{r}\right) \\
&= P\left[\left(1+\frac{r}{12}\right)^{12t} - 1\right]\left(1+\frac{12}{r}\right)
\end{aligned}$$

71. $P = \$50,\ r = 7\%,\ t = 20$ years

(a) Compounded monthly: $A = 50\left[\left(1+\frac{0.07}{12}\right)^{12(20)} - 1\right]\left(1+\frac{12}{0.07}\right) \approx \$26{,}198.27$

(b) Compounded continuously: $A = \dfrac{50e^{0.07/12}(e^{0.07(20)}-1)}{e^{0.07/12}-1} \approx \$26{,}263.88$

73. $P = \$100,\ r = 10\%,\ t = 40$ years

(a) Compounded monthly: $A = 100\left[\left(1 + \frac{0.10}{12}\right)^{12(40)} - 1\right]\left(1 + \frac{12}{0.10}\right) \approx \$637{,}678.02$

(b) Compounded continuously: $A = \dfrac{100e^{0.10/12}(e^{(0.10)(40)} - 1)}{e^{0.10/12} - 1} \approx \$645{,}861.43$

75. $P = W\sum_{n=1}^{12t}\left[\left(1 + \frac{r}{12}\right)^{-1}\right]^n$

$$= W\left(1 + \frac{r}{12}\right)^{-1}\left[\frac{1 - \left(1 + \frac{r}{12}\right)^{-12t}}{1 - \left(1 - \frac{r}{12}\right)^{-1}}\right]$$

$$= W\left(\frac{1}{1 + \frac{r}{12}}\right)\frac{\left[1 - \left(1 + \frac{r}{12}\right)^{-12t}\right]}{1 - \frac{1}{\left(1 + \frac{r}{12}\right)}}$$

$$= W\frac{\left[1 - \left(1 + \frac{r}{12}\right)^{-12t}\right]}{\left(1 + \frac{r}{12}\right) - 1}$$

$$= W\left(\frac{12}{r}\right)\left[1 - \left(1 - \frac{r}{12}\right)^{-12t}\right]$$

77. $64 + 32 + 16 + 8 + 4 + 2 = 126$
Total area of shaded region is approximately 126 square inches.

79. $S_n = \sum_{i=1}^{n} 0.01(2)^{i-1}$

$S_{29} = \$5{,}368{,}709.11$

$S_{30} = \$10{,}737{,}418.23$

$S_{31} = \$21{,}474{,}836.47$

81. $a_1 = 1,\ r = \frac{1}{2}$

$$\sum_{n=0}^{\infty}\left(\frac{1}{2}\right)^n = \frac{a_1}{1 - r} = \frac{1}{1 - (1/2)} = 2$$

83. $a_1 = 1,\ r = -\frac{1}{2}$

$$\sum_{n=1}^{\infty}\left(-\frac{1}{2}\right)^{n-1} = \frac{a_1}{1 - r} = \frac{1}{1 - (-1/2)} = \frac{2}{3}$$

85. $a_1 = 4,\ r = \frac{1}{4}$

$$\sum_{n=0}^{\infty}4\left(\frac{1}{4}\right)^n = \frac{a_1}{1 - r} = \frac{4}{1 - (1/4)} = \frac{16}{3}$$

87. $8 + 6 + \frac{9}{2} + \frac{27}{8} + \cdots = \sum_{n=0}^{\infty}8\left(\frac{3}{4}\right)^n = \frac{8}{1 - 3/4} = 32$

89. $0.\overline{36} = \sum_{n=0}^{\infty}0.36(0.01)^n = \frac{0.36}{1 - 0.01} = \frac{0.36}{0.99} = \frac{36}{99} = \frac{4}{11}$

91. $0.3\overline{18} = 0.3 + \sum_{n=0}^{\infty} 0.018(0.01)^n = \frac{3}{10} + \frac{0.018}{1 - 0.01}$

$= \frac{3}{10} + \frac{0.018}{0.99} = \frac{3}{10} + \frac{18}{990} = \frac{3}{10} + \frac{2}{110}$

$= \frac{35}{110} = \frac{7}{22}$

93. $f(x) = 6\left[\frac{1 - (0.5)^x}{1 - (0.5)}\right], \sum_{n=0}^{\infty} 6\left(\frac{1}{2}\right)^n = \frac{6}{1 - 1/2} = 12$

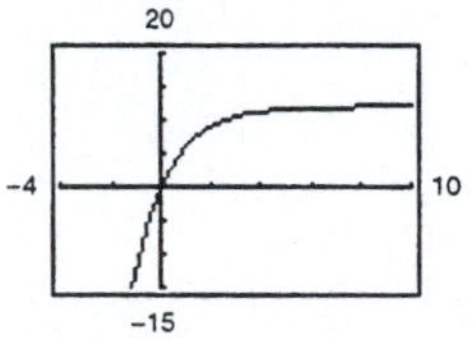

The horizontal asymptote of $f(x)$ is $y = 12$. This corresponds to the sum of the series.

95. (a) Total distance $= \left[\sum_{n=0}^{\infty} 32(0.81)^n\right] - 16 = \frac{32}{1 - 0.81} - 16 \approx 152.42$ feet

(b) $t = 1 + 2\sum_{n=1}^{\infty} (0.9)^n = 1 + 2\left[\frac{0.9}{1 - 0.9}\right] = 19$ seconds

Section 9.4 Mathematical Induction

- You should be sure that you understand the principle of mathematical induction. If P_n is a statement involving the positive integer n, where P_1 is true and the truth of P_k implies the truth of P_{k+1}, then P_n is true for all positive integers n.
- You should be able to verify (by induction) the formulas for the sums of powers of integers and be able to use these formulas.

Solutions to Odd-Numbered Exercises

1. $P_k = \frac{5}{k(k + 1)}$

$P_{k+1} = \frac{5}{(k + 1)((k + 1) + 1)} = \frac{5}{(k + 1)(k + 2)}$

3. $P_k = \frac{k^2(k + 1)^2}{4}$

$P_{k+1} = \frac{(k + 1)^2((k + 1) + 1)^2}{4} = \frac{(k + 1)^2(k + 2)^2}{4}$

5. 1. When $n = 1$, $S_1 = 2 = 1(1 + 1)$.

2. Assume that

$S_k = 2 + 4 + 6 + 8 + \cdots + 2k = k(k + 1).$

Then,

$S_{k+1} = 2 + 4 + 6 + 8 + \cdots + 2k + 2(k + 1)$

$= S_k + 2(k + 1) = k(k + 1) + 2(k + 1) = (k + 1)(k + 2).$

We conclude by mathematical induction that the formula is valid for all positive integer values of n.

7. 1. When $n = 1$, $S_1 = 2 = \frac{1}{2}(5(1) - 1)$.

2. Assume that

$$S_k = 2 + 7 + 12 + 17 + \cdots + (5k - 3) = \frac{k}{2}(5k - 1).$$

Then,

$$\begin{aligned} S_{k+1} &= 2 + 7 + 12 + 17 + \cdots + (5k - 3) + [5(k + 1) - 3] \\ &= S_k + (5k + 5 - 3) = \frac{k}{2}(5k - 1) + 5k + 2 \\ &= \frac{5k^2 - k + 10k + 4}{2} = \frac{5k^2 + 9k + 4}{2} \\ &= \frac{(k + 1)(5k + 4)}{2} = \frac{(k + 1)}{2}[5(k + 1) - 1]. \end{aligned}$$

We conclude by mathematical induction that the formula is valid for all positive integer values of n.

9. 1. When $n = 1$, $S_1 = 1 = 2^1 - 1$.

2. Assume that

$$S_k = 1 + 2 + 2^2 + 2^3 + \cdots + 2^{k-1} = 2^k - 1.$$

Then,

$$\begin{aligned} S_{k+1} &= 1 + 2 + 2^2 + 2^3 + \cdots + 2^{k-1} + 2^k \\ &= S_k + 2^k = 2^k - 1 + 2^k = 2(2^k) - 1 = 2^{k+1} - 1. \end{aligned}$$

Therefore, by mathematical induction, the formula is valid for all positive integer values of n.

11. 1. When $n = 1$, $S_1 = 1 = \frac{1(1 + 1)}{2}$.

2. Assume that

$$S_k = 1 + 2 + 3 + 4 + \cdots + k = \frac{k(k + 1)}{2}.$$

Then,

$$\begin{aligned} S_{k+1} &= 1 + 2 + 3 + 4 + \cdots + k + (k + 1) \\ &= S_k + (k + 1) = \frac{k(k + 1)}{2} + \frac{2(k + 1)}{2} = \frac{(k + 1)(k + 2)}{2}. \end{aligned}$$

Therefore, we conclude that this formula holds for all positive integer values of n.

13. 1. When $n = 1$, $S_1 = 1^3 = 1 = \frac{1(1 + 1)^2}{4}$.

2. Assume that

$$S_k = 1^3 + 2^3 + 3^3 + 4^3 + \cdots + k^3 = \frac{k^2(k + 1)^2}{4}.$$

Then,

$$\begin{aligned} S_{k+1} &= 1^3 + 2^3 + 3^3 + 4^3 + \cdots + k^3 + (k + 1)^3 \\ &= S_k + (k + 1)^3 = \frac{k^2(k + 1)^2}{4} + (k + 1)^3 = \frac{k^2(k + 1)^2 + 4(k + 1)^3}{4} \\ &= \frac{(k + 1)^2[k^2 + 4(k + 1)]}{4} = \frac{(k + 1)^2(k^2 + 4k + 4)}{4} = \frac{(k + 1)^2(k + 2)^2}{4}. \end{aligned}$$

Therefore, we conclude that this formula holds for all positive integer values of n.

15. 1. When $n = 1$, $S_1 = 1 = \dfrac{(1)^2(1 + 1)^2(2(1)^2 + 2(1) - 1)}{12}$.

2. Assume that

$$S_k = \sum_{i=1}^{k} i^5 = \frac{k^2(k + 1)^2(2k^2 + 2k - 1)}{12}.$$

Then,

$$\begin{aligned}
S_{k+1} &= \sum_{i=1}^{k+1} i^5 = \sum_{i=1}^{k} i^5 + (k + 1)^5 \\
&= \frac{k^2(k + 1)^2(2k^2 + 2k - 1)}{12} + \frac{12(k + 1)^5}{12} \\
&= \frac{(k + 1)^2[k^2(2k^2 + 2k - 1) + 12(k + 1)^3]}{12} \\
&= \frac{(k + 1)^2[2k^4 + 2k^3 - k^2 + 12(k^3 + 3k^2 + 3k + 1)]}{12} \\
&= \frac{(k + 1)^2[2k^4 + 14k^3 + 35k^2 + 36k + 12]}{12} \\
&= \frac{(k + 1)^2(k^2 + 4k + 4)(2k^2 + 6k + 3)}{12} \\
&= \frac{(k + 1)^2(k + 2)^2[2(k + 1)^2 + 2(k + 1) - 1]}{12}.
\end{aligned}$$

Therefore, we conclude that this formula holds for all positive integer values of n.

17. 1. When $n = 1$, $S_1 = 2 = \dfrac{1(2)(3)}{3}$.

2. Assume that,

$$S_k = 1(2) + 2(3) + 3(4) + \cdots + k(k + 1) = \frac{k(k + 1)(k + 2)}{3}.$$

Then,

$$\begin{aligned}
S_{k+1} &= 1(2) + 2(3) + 3(4) + \cdots + k(k + 1) + (k + 1)(k + 2) \\
&= S_k + (k + 1)(k + 2) = \frac{k(k + 1)(k + 2)}{3} + \frac{3(k + 1)(k + 2)}{3} \\
&= \frac{(k + 1)(k + 2)(k + 3)}{3}.
\end{aligned}$$

Thus, this formula is valid for all positive integer values of n.

19. $\displaystyle\sum_{n=1}^{20} n = \frac{20(20 + 1)}{2} = 210$

21. $\displaystyle\sum_{n=1}^{6} n^2 = \frac{6(6 + 1)(2(6) + 1)}{6} = 91$

23. $\displaystyle\sum_{n=1}^{5} n^4 = \frac{5(5 + 1)(2(5) + 1)(3(5)^2 + 3(5) - 1)}{30} = 979$

25. $\displaystyle\sum_{n=1}^{6} (n^2 - n) = \sum_{n=1}^{6} n^2 - \sum_{n=1}^{6} n = \frac{6(6 + 1)(2(6) + 1)}{6} - \frac{6(6 + 1)}{2} = 91 - 21 = 70$

27. $\sum_{i=1}^{6}(6i - 8i^3) = 6\sum_{i=1}^{6} i - 8\sum_{i=1}^{6} i^3 = 6\left[\frac{6(6+1)}{2}\right] - 8\left[\frac{(6)^2(6+1)^2}{4}\right] = 6(21) - 8(441) = -3402$

29. $1 + 5 + 9 + 13 + \cdots = n(2n - 1)$

31. $1 + \frac{9}{10} + \frac{81}{100} + \frac{729}{1000} + \cdots = 10 - 10\left(\frac{9}{10}\right)^n$

33. $\frac{1}{4} + \frac{1}{12} + \frac{1}{24} + \frac{1}{40} + \cdots + \frac{1}{2n(n-1)} + \cdots = \frac{n}{2(n+1)}$

35. 1. When $n = 4$, $4! = 24$ and $2^4 = 16$, thus $4! > 2^4$.

2. Assume

$k! > 2^k,\ k > 4.$

Then,

$(k + 1)! = k!(k + 1) > 2^k(2)$ since $k + 1 > 2$.

Thus, $(k + 1)! > 2^{k+1}$.

Therefore, by mathematical induction, the formula is valid for all integers n such that $n \geq 4$.

37. 1. When $n = 2$, $\frac{1}{\sqrt{1}} + \frac{1}{\sqrt{2}} \approx 1.707$ and $\sqrt{2} \approx 1.414$, thus $\frac{1}{\sqrt{1}} + \frac{1}{\sqrt{2}} > \sqrt{2}$.

2. Assume

$$\frac{1}{\sqrt{1}} + \frac{1}{\sqrt{2}} + \frac{1}{\sqrt{3}} + \cdots + \frac{1}{\sqrt{k}} > \sqrt{k},\ k > 2.$$

Then,

$$\frac{1}{\sqrt{1}} + \frac{1}{\sqrt{2}} + \frac{1}{\sqrt{3}} + \cdots + \frac{1}{\sqrt{k}} + \frac{1}{\sqrt{k+1}} > \sqrt{k} + \frac{1}{\sqrt{k+1}}.$$

Now we need to show that

$$\sqrt{k} + \frac{1}{\sqrt{k+1}} > \sqrt{k+1},\ k > 2.$$

This is true since

$$\sqrt{k(k+1)} > k$$

$$\sqrt{k(k+1)} + 1 > k + 1$$

$$\frac{\sqrt{k(k+1)} + 1}{\sqrt{k+1}} > \frac{k+1}{\sqrt{k+1}}$$

$$\sqrt{k} + \frac{1}{\sqrt{k+1}} > \sqrt{k+1}.$$

Therefore,

$$\frac{1}{\sqrt{1}} + \frac{1}{\sqrt{2}} + \frac{1}{\sqrt{3}} + \ldots + \frac{1}{\sqrt{k}} + \frac{1}{\sqrt{k+1}} > \sqrt{k+1}.$$

Therefore, by mathematical induction, the formula is valid for all integers n such that $n \geq 2$.

39. 1. When $n = 1$, $(ab)^1 = a^1b^1 = ab$.

2. Assume that $(ab)^k = a^kb^k$.

Then, $(ab)^{k+1} = (ab)^k(ab)$

$= a^kb^kab$

$= a^{k+1}b^{k+1}$.

Thus, $(ab)^n = a^nb^n$.

41. 1. When $n = 1$, $(x_1)^{-1} = x_1^{-1}$.

2. Assume that

$(x_1x_2x_3 \cdots x_k)^{-1} = x_1^{-1}x_2^{-1}x_3^{-1} \cdots x_k^{-1}$.

Then,

$$\begin{aligned}(x_1x_2x_3 \cdots x_kx_{k+1})^{-1} &= [(x_1x_2x_3 \cdots x_k)x_{k+1}]^{-1}\\ &= (x_1x_2x_3 \ldots x_k)^{-1}x_{k+1}^{-1}\\ &= x_1^{-1}x_2^{-1}x_3^{-1} \cdots x_k^{-1}x_{k+1}^{-1}.\end{aligned}$$

Thus, the formula is valid.

43. 1. When $n = 1$, $x(y_1) = xy_1$.

2. Assume that

$x(y_1 + y_2 + \cdots + y_k) = xy_1 + xy_2 + \cdots + xy_k$.

Then,

$$\begin{aligned}xy_1 + xy_2 + \cdots + xy_k + xy_{k+1} &= x(y_1 + y_2 + \cdots + y_k) + xy_{k+1}\\ &= x[(y_1 + y_2 + \cdots + y_k) + y_{k+1}]\\ &= x(y_1 + y_2 + \cdots + y_k + y_{k+1}).\end{aligned}$$

Hence, the formula holds.

45. 1. When $n = 1$, $\sin(x + n) = (-1)^1 \sin x$

2. Assume that $\sin(x + k\pi) = (-1)^k \sin x$

Then, $\sin[x + (k + 1)\pi] = \sin[(x + k\pi) + \pi] = (-1)^1 \sin(x + k\pi)$

$= (-1)^1(-1)^k \sin x = (-1)^{k+1} \sin x$.

Thus, the formula is valid for all positive interger values of n.

47. 1. When $n = 1$, $(1^3 + 3(1)^2 + 2(1)) = 6$ and 3 is a factor.

2. Assume that 3 is a factor of $(k^3 + 3k^2 + 2k)$.

Then,

$$\begin{aligned}&[(k + 1)^3 + 3(k + 1)^2 + 2(k + 1)]\\ &= k^3 + 3k^2 + 3k + 1 + 3k^2 + 6k + 3 + 2k + 2\\ &= (k^3 + 3k^2 + 2k) + (3k^2 + 9k + 6)\\ &= (k^3 + 3k^2 + 2k) + 3(k^2 + 3k + 2).\end{aligned}$$

Since 3 is a factor of $(k^3 + 3k^2 + 2k)$, our assumption, and 3 is a factor of $3(k^2 + 3k + 2)$, then 3 is a factor of the whole sum.

Thus, 3 is a factor of $(n^3 + 3n^2 + 2n)$ for every positive integer n.

49. See page 622 in the textbook.

51. $a_0 = 1, a_n = a_{n-1} + 2$

$a_0 = 1$

$a_1 = a_0 + 2 = 1 + 2 = 3$

$a_2 = a_1 + 2 = 3 + 2 = 5$

$a_3 = a_2 + 2 = 5 + 2 = 7$

$a_4 = a_3 + 2 = 7 + 2 = 9$

53. $a_0 = 4, a_1 = 2, a_n = a_{n-1} - a_{n-2}$

$a_0 = 4$

$a_1 = 2$

$a_2 = a_1 - a_0 = 2 - 4 = -2$

$a_3 = a_2 - a_1 = -2 - 2 = -4$

$a_4 = a_3 - a_2 = -4-(-2) = -2$

55. $f(1) = 0, a_n = a_{n-1} + 3$

$a_1 = f(1) = 0$

$a_2 = a_1 + 3 = 0 + 3 = 3$

$a_3 = a_2 + 3 = 3 + 3 = 6$

$a_4 = a_3 + 3 = 6 + 3 = 9$

$a_5 = a_4 + 3 = 9 + 3 = 12$

a_n : 0 3 6 9 12

First differences : 3 3 3 3

Second differences : 0 0 0

Since the first differences are equal, the sequence has a linear model.

57. $f(1) = 3, a_n = a_{n-1} - n$

$a_1 = f(1) = 3$

$a_2 = a_1 - 2 = 3 - 2 = 1$

$a_3 = a_2 - 3 = 1 - 3 = -2$

$a_4 = a_3 - 4 = -2 - 4 = -6$

$a_5 = a_4 - 5 = -6 - 5 = -11$

a_n : 3 1 −2 −6 −11

First differences : −2 −3 −4 −5

Second differences : −1 −1 −1

Since the second differences are all the same, the sequence has a quadratic model.

59. $a_0 = 0, a_n = a_{n-1} + n$

$a_0 = 0$

$a_1 = a_0 + 1 = 0 + 1 = 1$

$a_2 = a_1 + 2 = 1 + 2 = 3$

$a_3 = a_2 + 3 = 3 + 3 = 6$

$a_4 = a_3 + 4 = 6 + 4 = 10$

a_n : 0 1 3 6 10

First differences : 1 2 3 4

Second differences : 1 1 1

Since the second differences are equal, the sequence has a quadratic model.

61. $f(1) = 2, a_n = a_{n-1} + 2$

$a_1 = f(1) = 2$

$a_2 = a_1 + 2 = 2 + 2 = 4$

$a_3 = a_2 + 2 = 4 + 2 = 6$

$a_4 = a_3 + 2 = 6 + 2 = 8$

$a_5 = a_4 + 2 = 8 + 2 = 10$

a_n : 2 4 6 8 10

First differences : 2 2 2 2

Second differences : 0 0 0

Since the first differences are equal, the sequence has a linear model.

63. $a_0 = 1, a_n = a_{n-1} + n^2$

$a_0 = 1$

$a_1 = 1 + 1^2 = 2$

$a_2 = 2 + 2^2 = 6$

$a_3 = 6 + 3^2 = 15$

$a_4 = 15 + 4^2 = 31$

a_n : 1 2 6 15 31

First differences : 1 4 9 16

Second differences : 3 5 7

Since neither the first differences nor the second differences are equal, the sequence does not have a linear or a quadratic model.

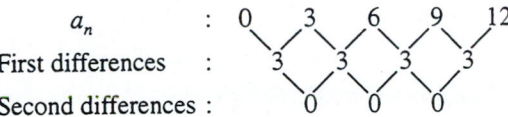

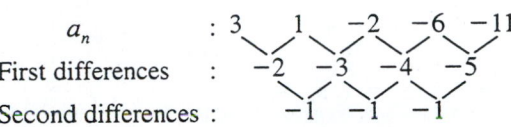

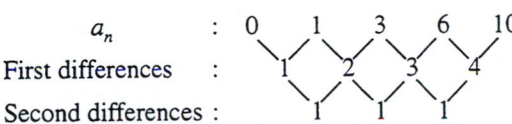

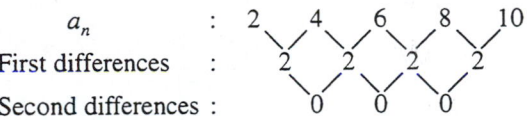

65. $a_0 = 3, a_1 = 3, a_4 = 15$

Let $a_n = an^2 + bn + c$.

Thus: $a_0 = a(0)^2 + b(0) + c = 3 \Rightarrow c = 3$

$a_1 = a(1)^2 + b(1) + c = 3 \Rightarrow a + b + c = 3$

$a + b = 0$

$a_4 = a(4)^2 + b(4) + c = 15 \Rightarrow 16a + 4b + c = 15$

$16a + 4b = 12$

$4a + b = 3$

By elimination: $-a - b = 0$

$4a + b = 3$

$3a \quad = 3$

$a = 1 \Rightarrow b = -1$

Thus, $a_n = n^2 - n + 3$.

67. $a_0 = -3, a_2 = 1, a_4 = 9$

Let $a_n = an^2 + bn + c$.

Then: $a_0 = a(0)^2 + b(0) + c = -3 \Rightarrow c = -3$

$a_2 = a(2)^2 + b(2) + c = 1 \Rightarrow 4a + 2b + c = 1$

$4a + 2b = 4$

$2a + b = 2$

$a_4 = a(4)^2 + b(4) + c = 9 \Rightarrow 16a + 4b + c = 9$

$16a + 4b = 12$

$4a + b = 3$

By elimination: $-2a - b = -2$

$4a + b = 3$

$2a \quad = -1$

$a = \frac{1}{2} \Rightarrow b = 1$

Thus, $a_n = \frac{1}{2}n^2 + n - 3$.

69. $y = x^2$

$-3x + 2y = 2 \Rightarrow -3x + 2x^2 = 2$

$2x^2 - 3x - 2 = 0$

$(2x + 1)(x - 2) = 0$

$x = -\frac{1}{2}$ or $x = 2$

$y = \frac{1}{4} \qquad y = 4$

Points of intersection: $\left(-\frac{1}{2}, \frac{1}{4}\right), (2, 4)$

71.
$$\begin{aligned} x - y &= -1 \\ x + 2y - 2z &= 3 \\ 3x - y + 2z &= 3 \end{aligned}$$

Using an augmented matrix, we have:

$$\begin{bmatrix} 1 & -1 & 0 & \vdots & -1 \\ 1 & 2 & -2 & \vdots & 3 \\ 3 & -1 & 2 & \vdots & 3 \end{bmatrix}$$

$$\begin{matrix} \\ -R_1 + R_2 \rightarrow \\ -3R_1 + R_3 \rightarrow \end{matrix} \begin{bmatrix} 1 & -1 & 0 & \vdots & -1 \\ 0 & 3 & -2 & \vdots & 4 \\ 0 & 2 & 2 & \vdots & 6 \end{bmatrix}$$

$$\begin{matrix} \\ -R_3 + R_2 \rightarrow \\ \frac{1}{2}R_3 \rightarrow \end{matrix} \begin{bmatrix} 1 & -1 & 0 & \vdots & -1 \\ 0 & 1 & -4 & \vdots & -2 \\ 0 & 1 & 1 & \vdots & 3 \end{bmatrix}$$

$$\begin{matrix} R_2 + R_1 \rightarrow \\ \\ -R_2 + R_3 \rightarrow \end{matrix} \begin{bmatrix} 1 & 0 & -4 & \vdots & -3 \\ 0 & 1 & -4 & \vdots & -2 \\ 0 & 0 & 5 & \vdots & 5 \end{bmatrix}$$

$$\begin{matrix} 4R_3 + R_1 \rightarrow \\ 4R_3 + R_2 \rightarrow \\ \frac{1}{5}R_3 \rightarrow \end{matrix} \begin{bmatrix} 1 & 0 & 0 & \vdots & 1 \\ 0 & 1 & 0 & \vdots & 2 \\ 0 & 0 & 1 & \vdots & 1 \end{bmatrix}$$

Thus, $x = 1, y = 2, z = 1$.

Answer: $(1, 2, 1)$

Section 9.5 The Binomial Theorem

- You should be able to use the formula

$$(x + y)^n = x^n + nx^{n-1}y + \frac{n(n-1)}{2!}x^{n-2}y^2 + \cdots + {}_nC_r x^{n-r}y^r + \cdots + y^n$$

where ${}_nC_r = \dfrac{n!}{(n-r)!r!}$, to expand$(x + y)^n$.

- You should be able to use Pascal's Triangle in binomial expansion.

Solutions to Odd-Numbered Exercises

1. ${}_5C_3 = \dfrac{5!}{3!2!} = \dfrac{5 \cdot 4}{2 \cdot 1} = 10$

3. ${}_{12}C_0 = \dfrac{12!}{0!12!} = 1$

5. ${}_{20}C_{15} = \dfrac{20!}{15!5!} = \dfrac{20 \cdot 19 \cdot 18 \cdot 17 \cdot 16}{5 \cdot 4 \cdot 3 \cdot 2 \cdot 1} = 15{,}504$

7. ${}_{100}C_{98} = \dfrac{100!}{98!2!} = \dfrac{100 \cdot 99}{2 \cdot 1} = 4950$

9. ${}_{100}C_2 = \dfrac{100!}{2!98!} = \dfrac{100 \cdot 99}{2 \cdot 1} = 4950$

11. The first and last number in each row is 1. Every other number is found by adding the two numbers immediately above it.

13.

1
1 1
1 2 1
1 3 3 1
1 4 6 4 1
1 5 10 10 5 1
1 6 15 20 15 6 1
1 7 21 35 (35) 21 7 1

${}_7C_4 = 35$, the 5th entry in the 8th row.

15.

1
1 1
1 2 1
1 3 3 1
1 4 6 4 1
1 5 10 10 5 1
1 6 15 20 15 6 1
1 7 21 35 35 21 7 1
1 8 28 56 70 (56) 28 8 1

${}_8C_5 = 56$, the 6th entry in the 9th row.

17. $$\begin{aligned}(x+1)^4 &= {}_4C_0x^4 + {}_4C_1x^3(1) + {}_4C_2x^2(1)^2 + {}_4C_3x(1)^3 + {}_4C_4(1)^4\\ &= x^4 + 4x^3 + 6x^2 + 4x + 1\end{aligned}$$

19. $$\begin{aligned}(a+2)^3 &= {}_3C_0a^3 + {}_3C_1a^2(2) + {}_3C_2a(2)^2 + {}_3C_3(2)^3\\ &= a^3 + 3a^2(2) + 3a(2)^2 + (2)^3\\ &= a^3 + 6a^2 + 12a + 8\end{aligned}$$

21. $$\begin{aligned}(y-2)^4 &= {}_4C_0y^4 - {}_4C_1y^3(2) + {}_4C_2y^2(2)^2 - {}_4C_3y(2)^3 + {}_4C_4(2)^4\\ &= y^4 - 4y^3(2) + 6y^2(4) - 4y(8) + 16\\ &= y^4 - 8y^3 + 24y^2 - 32y + 16\end{aligned}$$

23. $$\begin{aligned}(x+y)^5 &= {}_5C_0x^5 + {}_5C_1x^4y + {}_5C_2x^3y^2 + {}_5C_3x^2y^3 + {}_5C_4xy^4 + {}_5C_5y^5\\ &= x^5 + 5x^4y + 10x^3y^2 + 10x^2y^3 + 5xy^4 + y^5\end{aligned}$$

25. $$\begin{aligned}(r+3s)^6 &= {}_6C_0r^6 + {}_6C_1r^5(3s) + {}_6C_2r^4(3s)^2 + {}_6C_3r^3(3s)^3 + {}_6C_4r^2(3s)^4\\ &\quad + {}_6C_5r(3s)^5 + {}_6C_6(3s)^6\\ &= r^6 + 18r^5s + 135r^4s^2 + 540r^3s^3 + 1215r^2s^4 + 1458rs^5 + 729s^6\end{aligned}$$

27. $$\begin{aligned}(x-y)^5 &= {}_5C_0x^5 - {}_5C_1x^4y + {}_5C_2x^3y^2 - {}_5C_3x^2y^3 + {}_5C_4xy^4 - {}_5C_5y^5\\ &= x^5 - 5x^4y + 10x^3y^2 - 10x^2y^3 + 5xy^4 - y^5\end{aligned}$$

29. $$\begin{aligned}(1-2x)^3 &= {}_3C_0 1^3 - {}_3C_1 1^2(2x) + {}_3C_2 1(2x)^2 - {}_3C_3(2x)^3\\ &= 1 - 3(2x) + 3(2x)^2 - (2x)^3\\ &= 1 - 6x + 12x^2 - 8x^3\end{aligned}$$

31. $$\begin{aligned}(x^2+5)^4 &= {}_4C_0(x^2)^4 + {}_4C_1(x^2)^3(5) + {}_4C_2(x^2)^2(5)^2 + {}_4C_3(x^2)(5)^3 + {}_4C_4(5)^4\\ &= x^8 + 4x^6(5) + 6x^4(25) + 4x^2(125) + 625\\ &= x^8 + 20x^6 + 150x^4 + 500x^2 + 625\end{aligned}$$

33. $$\begin{aligned}\left(\frac{1}{x}+y\right)^5 &= {}_5C_0\left(\frac{1}{x}\right)^5 + {}_5C_1\left(\frac{1}{x}\right)^4y + {}_5C_2\left(\frac{1}{x}\right)^3y^2 + {}_5C_3\left(\frac{1}{x}\right)^2y^3 + {}_5C_4\left(\frac{1}{x}\right)y^4 + {}_5C_5y^5\\ &= \frac{1}{x^5} + \frac{5y}{x^4} + \frac{10y^2}{x^3} + \frac{10y^3}{x^2} + \frac{5y^4}{x} + y^5\end{aligned}$$

35. $2(x-3)^4 + 5(x-3)^2 = 2[x^4 - 4(x^3)(3) + 6(x^2)(3^2) - 4(x)(3^3) + 3^4] + 5[x^2 - 2(x)(3) + 3^2]$
$= 2(x^4 - 12x^3 + 54x^2 - 108x + 81) + 5(x^2 - 6x + 9)$
$= 2x^4 - 24x^3 + 113x^2 - 246x + 207$

37. 5th Row of Pascal's Triangle: 1 5 10 10 5 1

$(2t-s)^5 = 1(2t)^5 + 5(2t)^4(-s) + 10(2t)^3(-s)^2 + 10(2t)^2(-s)^3 + 5(2t)(-s)^4 + 1(-s)^5$
$= 32t^5 - 80t^4s + 80t^3s^2 - 40t^2s^3 + 10ts^4 - s^5$

39. 4th Row of Pascal's Triangle: 1 4 6 4 1

$(3-2z)^4 = 3^4 - 4(3)^3(2z) + 6(3)^2(2z)^2 - 4(3)(2z)^3 + (2z)^4$
$= 81 - 216z + 216z^2 - 96z^3 + 16z^4$

41. The term involving x^5 in the expansion of $(x+3)^{12}$ is

${}_{12}C_7x^5(3)^7 = \dfrac{12!}{7!5!} \cdot 3^7x^5 = 1{,}732{,}104x^5$. The coefficient is 1,732,104.

43. The term involving x^8y^2 in the expansion of $(x-2y)^{10}$ is

${}_{10}C_2x^8(-2y)^2 = \dfrac{10!}{2!8!} \cdot 4x^8y^2 = 180x^8y^2$. The coefficient is 180.

45. The coefficient of x^4y^5 in the expansion of $(3x-2y)^9$ is

${}_9C_5(3)^4(-2)^5 = \dfrac{9!}{5!4!}(81)(-32) = -326{,}592.$

47. The coefficient of $x^8y^6 = (x^2)^4y^6$ in the expansion of $(x^2+y)^{10}$ is ${}_{10}C_6 = 210$.

49. There are $n+1$ terms in the expansion of $(x+y)^n$.

51. $(\sqrt{x}+3)^4 = (\sqrt{x})^4 + 4(\sqrt{x})^3(3) + 6(\sqrt{x})^2(3)^2 + 4(\sqrt{x})(3)^3 + (3)^4$
$= x^2 + 12x\sqrt{x} + 54x + 108\sqrt{x} + 81$
$= x^2 + 12x^{3/2} + 54x + 108x^{1/2} + 81$

53. $(x^{2/3} - y^{1/3})^3 = (x^{2/3})^3 - 3(x^{2/3})^2(y^{1/3}) + 3(x^{2/3})(y^{1/3})^2 - (y^{1/3})^3$
$= x^2 - 3x^{4/3}y^{1/3} + 3x^{2/3}y^{2/3} - y$

55.
$$\frac{f(x+h)-f(x)}{h} = \frac{(x+h)^3 - x^3}{h} = \frac{x^3 + 3x^2h + 3xh^2 + h^3 - x^3}{h} = \frac{h(3x^2 + 3xh + h^2)}{h} = 3x^2 + 3xh + h^2$$

57.
$$\frac{f(x+h)-f(x)}{h} = \frac{\sqrt{x+h}-\sqrt{x}}{h} = \frac{\sqrt{x+h}-\sqrt{x}}{h} \cdot \frac{\sqrt{x+h}+\sqrt{x}}{\sqrt{x+h}+\sqrt{x}} = \frac{(x+h)-x}{h(\sqrt{x+h}+\sqrt{x})} = \frac{1}{\sqrt{x+h}+\sqrt{x}}$$

59. $(1+i)^4 = {}_4C_0 1^4 + {}_4C_1(1)^3i + {}_4C_2(1)^2i^2 + {}_4C_3 1 \cdot i^3 + {}_4C_4 i^4$
$= 1 + 4i - 6 - 4i + 1$
$= -4$

61. $(2 - 3i)^6 = {}_6C_0 2^6 - {}_6C_1 2^5(3i) + {}_6C_2 2^4(3i)^2 - {}_6C_3 2^3(3i)^3 + {}_6C_4 2^2(3i)^4 - {}_6C_5 2(3i)^5 + {}_6C_6(3i)^6$

$= 64 - 576i - 2160 + 4320i + 4860 - 2916i - 729$

$= 2035 + 828i$

63. $\left(-\frac{1}{2} + \frac{\sqrt{3}}{2}i\right)^3 = \frac{1}{8}(-1 + \sqrt{3}i)^3$

$= \frac{1}{8}\left[(-1)^3 + 3(-1)^2(\sqrt{3}i) + 3(-1)(\sqrt{3}i)^2 + (\sqrt{3}i)^3\right]$

$= \frac{1}{8}\left[-1 + 3\sqrt{3}i + 9 - 3\sqrt{3}i\right]$

$= 1$

65. ${}_7C_4\left(\frac{1}{2}\right)^4\left(\frac{1}{2}\right)^3 = 35\left(\frac{1}{16}\right)\left(\frac{1}{8}\right) \approx 0.273$

67. ${}_8C_4\left(\frac{1}{3}\right)^4\left(\frac{2}{3}\right)^4 = 70\left(\frac{1}{81}\right)\left(\frac{16}{81}\right) \approx 0.171$

69. $(1.02)^8 = (1 + 0.02)^8 = 1 + 8(0.02) + 28(0.02)^2 + 56(0.02)^3 + 70(0.02)^4 + 56(0.02)^5$

$+ 28(0.02)^6 + 8(0.02)^7 + (0.02)^8$

$= 1 + 0.16 + 0.0112 + 0.000448 + \cdots \approx 1.172$

71. $(2.99)^{12} = (3 - 0.01)^{12}$

$= 3^{12} - 12(3)^{11}(0.01) + 66(3)^{10}(0.01)^2 - 220(3)^9(0.01)^3 + 495(3)^8(0.01)^4$

$- 792(3)^7(0.01)^5 + 924(3)^6(0.01)^6 - 792(3)^5(0.01)^7 + 495(3)^4(0.01)^8$

$- 220(3)^3(0.01)^9 + 66(3)^2(0.01)^{10} - 12(3)(0.01)^{11} + (0.01)^{12} \approx 510{,}568.785$

73. $f(x) = x^3 - 4x$

$g(x) = f(x + 4)$

$= (x + 4)^3 - 4(x + 4)$

$= x^3 + 3x^2(4) + 3x(4)^2 + (4)^3 - 4x - 16$

$= x^3 + 12x^2 + 48x + 64 - 4x - 16$

$= x^3 + 12x^2 + 44x + 48$

The graph of g is the same as the graph of f shifted 4 units to the left.

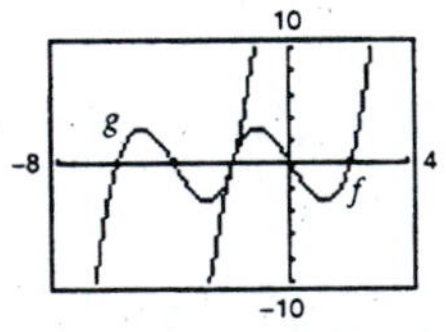

75. ${}_nC_{n-r} = \frac{n!}{(n - (n - r))!(n - r)!}$

$= \frac{n!}{r!(n - r)!}$

$= \frac{n!}{(n - r)!r!}$

$= {}_nC_r$

77. $_nC_r + {_nC_{r-1}} = \dfrac{n!}{(n-r)!r!} + \dfrac{n!}{(n-r+1)!(r-1)!}$

$= \dfrac{n!(n-r+1)!(r-1)! + n!(n-r)!r!}{(n-r)!r!(n-r+1)!(r-1)!}$

$= \dfrac{n![(n-r+1)!(r-1)! + r!(n-r)!]}{(n-r)!r!(n-r+1)!(r-1)!}$

$= \dfrac{n!(r-1)![(n-r+1)! + r(n-r)!]}{(n-r)!r!(n-r+1)!(r-1)!}$

$= \dfrac{n!(n-r)![(n-r+1) + r]}{(n-r)!r!(n-r+1)!}$

$= \dfrac{n![n+1]}{r!(n-r+1)!}$

$= \dfrac{(n+1)!}{[(n+1)-r]!r!}$

$= {_{n+1}C_r}$

79. $f(t) = 0.1506t^2 + 0.7361t + 21.1374, 0 \le t \le 22$

(a) $g(t) = f(t+10)$

$= 0.1506(t+10)^2 + 0.7361(t+10) + 21.1374$

$= 0.1506(t^2 + 20t + 100) + 0.7361t + 7.361 + 21.1374$

$= 0.1506t^2 + 3.7481t + 43.5584$

(b)

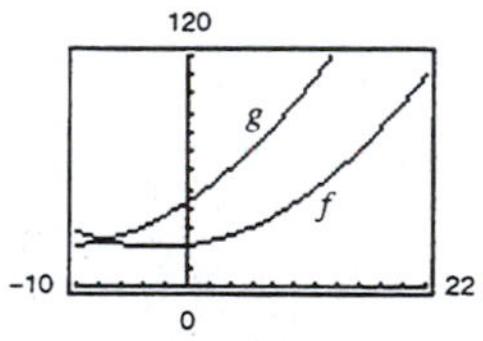

81. $f(x) = (1-x)^3$

$g(x) = 1 - 3x$

$h(x) = 1 - 3x + 3x^2$

$p(x) = 1 - 3x + 3x^2 - x^3$

Since $p(x)$ is the expansion of $f(x)$, they have the same graph.

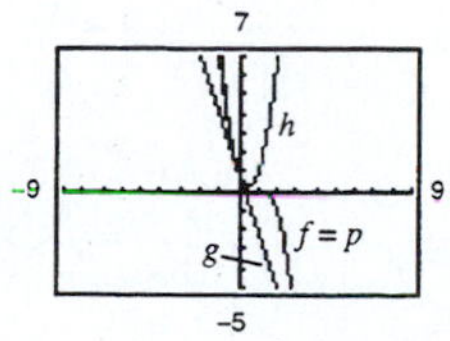

83. $g(x) = f(x-3)$

$g(x)$ is shifted 3 units to the right of $f(x)$.

85. $g(x) = -f(x)$

$g(x)$ is the reflection of $f(x)$ in the x-axis.

Section 9.6 Counting Principles

- You should know The Fundamental Principle of Counting.
- $_nP_r = \dfrac{n!}{(n-r)!}$ is the number of permutations of n elements taken r at a time.
- Given a set of n objects that has n_1 of one kind, n_2 of a second kind, and so on, the number of distinguishable permutations is

$$\frac{n!}{n_1!n_2! \ldots n_k!}.$$

- $_nC_r = \dfrac{n!}{(n-r)!r!}$ is the number of combinations of n elements taken r at a time.

Solutions to Odd-Numbered Exercises

1. Odd integers: 1, 3, 5, 7, 9, 11

6 ways

3. Prime integers: 2, 3, 5, 7, 11

5 ways

5. Divisible by 4: 4, 8, 12

3 ways

7. Sum is 8: $1 + 7,\ 2 + 6,\ 3 + 5,\ 4 + 4$

4 ways

9. Amplifiers: 2 choices

Compact disc players: 4 choices

Speakers: 6 choices

Total: $2 \cdot 4 \cdot 6 = 48$ ways

11. Chemist: 3 choices

Statistician: 4 choices

Total: $3 \cdot 4 = 12$ ways

13. $2^6 = 64$

15. 1st Position: 2 choices

2nd Position: 3 choices

3rd Position: 2 choices

4th Position: 1 choices

Total: $2 \cdot 3 \cdot 2 \cdot 1 = 12$ ways

Label the four people A, B, C, and D and suppose that A and B are willing to take the first position. The twelve combinations are as follows

ABCD	BACD
ABDC	BADC
ACBD	BCAD
ACDB	BCDA
ADBC	BDAC
ADCB	BDCA

17. $26 \cdot 26 \cdot 10 \cdot 10 \cdot 10 \cdot 10 = 6{,}760{,}000$

19. (a) $9 \cdot 10 \cdot 10 = 900$

(b) $9 \cdot 9 \cdot 8 = 648$

(c) $9 \cdot 10 \cdot 2 = 180$

(d) $10 \cdot 10 \cdot 10 - 400 = 600$

21. $40^3 = 64{,}000$

23. (a) $6 \cdot 5 \cdot 4 \cdot 3 \cdot 2 \cdot 1 = 720$

(b) $6 \cdot 1 \cdot 4 \cdot 1 \cdot 2 \cdot 1 = 48$

25. ${}_nP_r = \dfrac{n!}{(n-r)!}$

So, ${}_4P_4 = \dfrac{4!}{0!} = 4! = 24.$

27. ${}_8P_3 = \dfrac{8!}{5!} = 8 \cdot 7 \cdot 6 = 336$

29. ${}_5P_4 = \dfrac{5!}{1!} = 120$

31. $14 \cdot {}_nP_3 = {}_{n+2}P_4$ Note: $n \geq 3$ for this to be defined.

$$14\left(\frac{n!}{(n-3)!}\right) = \frac{(n+2)!}{(n-2)!}$$

$$14n(n-1)(n-2) = (n+2)(n+1)n(n-1)$$ (We can divide here by $n(n-1)$ since $n \neq 0, n \neq 1$.)

$$14n(n-2) = (n+2)(n+1)$$

$$14n - 28 = n^2 + 3n + 2$$

$$0 = n^2 - 11n + 30$$

$$0 = (n-5)(n-6)$$

$$n = 5 \quad \text{or} \quad n = 6$$

33. ${}_{20}P_5 = 1{,}860{,}480$

35. ${}_{100}P_3 = 970{,}200$

37. ${}_{20}C_5 = 15{,}504$

39. ${}_{100}P_{80} \approx 3.836 \times 10^{139}$

This number is too large for some calculators to evaluate.

41.

ABCD	BACD	CABD	DABC
ABDC	BADC	CADB	DACB
ACBD	BCAD	CBAD	DBAC
ACDB	BCDA	CBDA	DBCA
ADBC	BDAC	CDAB	DCAB
ADCB	BDCA	CDBA	DCBA

43. $5! = 120$ ways

45. ${}_{12}P_4 = \dfrac{12!}{8!} = 12 \cdot 11 \cdot 10 \cdot 9 = 11{,}880$ ways

47. $\dfrac{7!}{2!1!3!1!} = \dfrac{7!}{2!3!} = 420$

49. $\dfrac{7!}{2!1!1!1!1!1!} = \dfrac{7!}{2!} = 7 \cdot 6 \cdot 5 \cdot 4 \cdot 3 = 2520$

51. ${}_6C_2 = 15$

The 15 ways are listed below.

AB, AC, AD, AE, AF,

BC, BD, BE, BF, CD,

CE, CF, DE, DF, EF

53. ${}_{20}C_4 = 4845$ groups

55. ${}_{40}C_6 = 3{,}838{,}380$ ways

57. ${}_{100}C_4 = 3{,}921{,}225$ subsets

59. ${}_7C_2 = 21$ lines

61. (a) ${}_8C_4 = \dfrac{8!}{(8-4)!4!} = \dfrac{8!}{4!4!} = \dfrac{8 \cdot 7 \cdot 6 \cdot 5}{4 \cdot 3 \cdot 2} = 70$ ways

(b) ${}_3C_2 \cdot {}_5C_2 = \dfrac{3!}{(3-2)!2!} \cdot \dfrac{5!}{(5-2)!2!} = 3 \cdot 10 = 30$ ways

63. (a) ${}_8C_4 = \dfrac{8!}{4!4!} = 70$ ways

(b) There are 10 ways that a group of four can be formed without any couples in the group. Therefore, if at least one couple is to be in the group, there are $70 - 10 = 60$ ways that could occur.

(c) $2 \cdot 2 \cdot 2 \cdot 2 = 16$ ways

65. ${}_5C_2 - 5 = 10 - 5 = 5$ diagonals

67. ${}_8C_2 - 8 = 28 - 8 = 20$ diagonals

69. ${}_nP_{n-1} = \dfrac{n!}{(n-(n-1))!} = \dfrac{n!}{1!} = \dfrac{n!}{0!} = {}_nP_n$

71. ${}_nC_{n-1} = \dfrac{n!}{(n-(n-1))!(n-1)!} = \dfrac{n!}{(1)!(n-1)!} = \dfrac{n!}{(n-1)!1!} = {}_nC_1$

Section 9.7 Probability

You should know the following basic principles of probability.

- If an event E has $n(E)$ equally likely outcomes and its sample space has $n(S)$ equally likely outcomes, then the probability of event E is

$$P(E) = \frac{n(E)}{n(S)}, \text{ where } 0 \le P(E) \le 1.$$

- If A and B are mutually exclusive events, then $P(A \cup B) = P(A) + P(B)$.

  If A and B are not mutually exclusive events, then $P(A \cup B) = P(A) + P(B) - P(A \cap B)$.
- If A and B are independent events, then the probability that both A and B will occur is $P(A)P(B)$.
- The complement of an event E is $P(E') = 1 - P(E)$.

Solutions to Odd-Numbered Exercises

1. $\{(h, 1), (h, 2), (h, 3), (h, 4), (h, 5), (h, 6), (t, 1), (t, 2), (t, 3), (t, 4), (t, 5), (t, 6)\}$

3. $\{ABC, ACB, BAC, BCA, CAB, CBA\}$

5. $\{(A, B), (A, C), (A, D), (A, E), (B, C), (B, D), (B, E), (C, D), (C, E), (D, E)\}$

7. $E = \{HHT, HTH, THH\}$

$$P(E) = \frac{n(E)}{n(S)} = \frac{3}{8}$$

9. $E = \{HHH, HHT, HTH, HTT, THH, THT, TTH\}$

$$P(E) = \frac{n(E)}{n(S)} = \frac{7}{8}$$

11. $E = \{K, K, K, K, Q, Q, Q, Q, J, J, J, J\}$

$$P(E) = \frac{n(E)}{n(S)} = \frac{12}{52} = \frac{3}{13}$$

13. $E = \{K, K, Q, Q, J, J\}$

$$P(E) = \frac{n(E)}{n(S)} = \frac{6}{52} = \frac{3}{26}$$

15. $E = \{(1, 3), (2, 2), (3, 1)\}$

$$P(E) = \frac{n(E)}{n(S)} = \frac{3}{36} = \frac{1}{12}$$

17. not $E = \{(5, 6), (6, 5), (6, 6)\}$

$n(E) = n(S) - n(\text{not } E) = 36 - 3 = 33$

$$P(E) = \frac{n(E)}{n(S)} = \frac{33}{36} = \frac{11}{12}$$

19. $E_3 = \{(1, 2), (2, 1)\},\ n(E_3) = 2$

$E_5 = \{(1, 4), (2, 3), (3, 2), (4, 1)\},\ n(E_5) = 4$

$E_7 = \{(1, 6), (2, 5), (3, 4), (4, 3), (5, 2), (6, 1)\},\ n(E_7) = 6$

$E = E_3 \cup E_5 \cup E_7$

$n(E) = 2 + 4 + 6 = 12$

$$P(E) = \frac{n(E)}{n(S)} = \frac{12}{36} = \frac{1}{3}$$

21. $P(E) = \dfrac{{}_3C_2}{{}_6C_2} = \dfrac{3}{15} = \dfrac{1}{5}$

23. $P(E) = \dfrac{{}_4C_2}{{}_6C_2} = \dfrac{6}{15} = \dfrac{2}{5}$

25. $1 - p = 1 - 0.7 = 0.3$

27. $1 - p = 1 - 0.15 = 0.85$

29. (a) $0.37(2.5) = 0.925$ million $= 925{,}000$

(b) 18%

(c) $11\% + 6\% = 17\%$

31. (a) $\frac{290}{500} = 0.58 = 58\%$

(b) $\frac{478}{500} = 0.956 = 95.6\%$

(c) $\frac{2}{500} = 0.004 = 0.4\%$

33. (a) $\dfrac{672}{1254}$

(b) $\dfrac{582}{1254}$

(c) $\dfrac{672 - 124}{1254} = \dfrac{548}{1254}$

35. $p + p + 2p = 1$

$p = 0.25$

Taylor: $0.50 = \frac{1}{2}$

Moore: $0.25 = \frac{1}{4}$

Jenkins: $0.25 = \frac{1}{4}$

37. (a) $\dfrac{{}_{15}C_{10}}{{}_{20}C_{10}} = \dfrac{3003}{184{,}756} = \dfrac{21}{1292} \approx 0.016$

(b) $\dfrac{{}_{15}C_8 \cdot {}_5C_2}{{}_{20}C_{10}} = \dfrac{64{,}350}{184{,}756} = \dfrac{225}{646} \approx 0.348$

(c) $\dfrac{{}_{15}C_9 \cdot {}_5C_1}{{}_{20}C_{10}} + \dfrac{{}_{15}C_{10}}{{}_{20}C_{10}} = \dfrac{25{,}025 + 3003}{184{,}756} = \dfrac{28{,}028}{184{,}756} = \dfrac{49}{323} \approx 0.152$

39. Total ways to insert letters: $4! = 24$ ways

4 correct: 1 way

3 correct: not possible

2 correct: 6 ways

1 correct: 8 ways

0 correct: 9 ways

(a) $\dfrac{8}{24} = \dfrac{1}{3}$

(b) $\dfrac{8 + 6 + 1}{24} = \dfrac{15}{24} = \dfrac{5}{8}$

41. (a) $\dfrac{1}{{}_5P_5} = \dfrac{1}{120}$

(b) $\dfrac{1}{{}_4P_4} = \dfrac{1}{24}$

43. (a) $\dfrac{4}{52} \cdot \dfrac{4}{52} = \dfrac{1}{169}$

(b) $\dfrac{4}{52} \cdot \dfrac{3}{51} = \dfrac{1}{221}$

45. (a) $\dfrac{{}_9C_4}{{}_{12}C_4} = \dfrac{126}{495} = \dfrac{14}{55}$ (4 good units)

(b) $\dfrac{({}_9C_2)\,({}_3C_2)}{{}_{12}C_4} = \dfrac{108}{495} = \dfrac{12}{55}$ (2 good units)

(c) $\dfrac{({}_9C_3)({}_3C_1)}{{}_{12}C_4} = \dfrac{252}{495} = \dfrac{28}{55}$ (3 good units)

At least 2 good units: $\dfrac{12}{55} + \dfrac{28}{55} + \dfrac{14}{55} = \dfrac{54}{55}$

47. (a) $P(EE) = \frac{15}{30} \cdot \frac{15}{30} = \frac{1}{4}$

(b) $P(EO \text{ or } OE) = 2\left(\frac{15}{30}\right)\left(\frac{15}{30}\right) = \frac{1}{2}$

(c) $P(N_1 < 10, N_2 < 10) = \frac{9}{30} \cdot \frac{9}{30} = \frac{9}{100}$

(d) $P(N_1N_1) = \frac{30}{30} \cdot \frac{1}{30} = \frac{1}{30}$

49. (a) $P(SS) = (0.985)^2 \approx 0.9702$

(b) $P(S) = 1 - P(FF) = 1 - (0.015)^2 \approx 0.9998$

(c) $P(FF) = (0.015)^2 \approx 0.0002$

51. (a) $\left(\frac{1}{4}\right)^5 = \frac{1}{1024}$

(b) $\left(\frac{3}{4}\right)^5 = \frac{243}{1024}$

(c) $1 - \frac{243}{1024} = \frac{781}{1024}$

53. $(0.32)^2 = 0.1024$

55. $1 - \dfrac{(45)^2}{(60)^2} = 1 - \left(\dfrac{45}{60}\right)^2 = 1 - \left(\dfrac{3}{4}\right)^2 = 1 - \dfrac{9}{16} = \dfrac{7}{16}$

57. (a) As you consider successive people with distinct birthdays, the probabilities must decrease to take into account the birth dates already used. Because the birth dates of people are independent events, multiply the respective probabilities of distinct birthdays.

(b) $\frac{365}{365} \cdot \frac{364}{365} \cdot \frac{363}{365} \cdot \frac{362}{365}$

(c) $P_1 = \frac{365}{365} = 1$

$$P_2 = \frac{365}{365} \cdot \frac{364}{365} = \frac{364}{365} P_1 = \frac{365 - (2 - 1)}{365} P_1$$

$$P_3 = \frac{365}{365} \cdot \frac{364}{365} \cdot \frac{363}{365} = \frac{363}{365} P_2 = \frac{365 - (3 - 1)}{365} P_2$$

$$P_n = \frac{365}{365} \cdot \frac{364}{365} \cdot \frac{363}{365} \cdot \cdots \frac{365 - (n - 1)}{365} = \frac{365 - (n - 1)}{365} P_{n-1}$$

(d) Q_n is the probability that the birthdays are *not* distinct which is equivalent to at least 2 people having the same birthday.

(e)

n	10	15	20	23	30	40	50
P_n	0.88	0.75	0.59	0.49	0.29	0.11	0.03
Q_n	0.12	0.25	0.41	0.51	0.71	0.89	0.97

(f) 23, See the chart above.

59. $1 - 0.546 = 0.454$

Review Exercises for Chapter 9

Solutions to Odd-Numbered Exercises

1. $a_n = 2 + \frac{6}{n}$

$a_1 = 2 + \frac{6}{1} = 8$

$a_2 = 2 + \frac{6}{2} = 5$

$a_3 = 2 + \frac{6}{3} = 4$

$a_4 = 2 + \frac{6}{4} = \frac{7}{2}$

$a_5 = 2 + \frac{6}{5} = \frac{16}{5}$

3. $a_n = \frac{72}{n!}$

$a_1 = \frac{72}{1!} = 72$

$a_2 = \frac{72}{2!} = 36$

$a_3 = \frac{72}{3!} = 12$

$a_4 = \frac{72}{4!} = 3$

$a_5 = \frac{72}{5!} = \frac{3}{5}$

5. $a_n = \frac{3}{2}n$

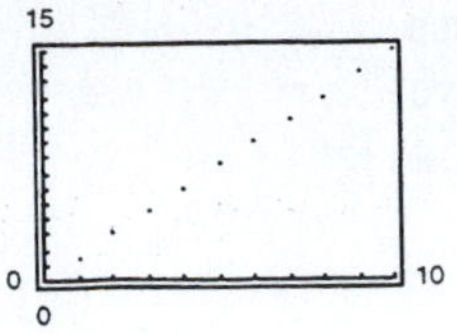

7. $a_n = \frac{3n}{n+2}$

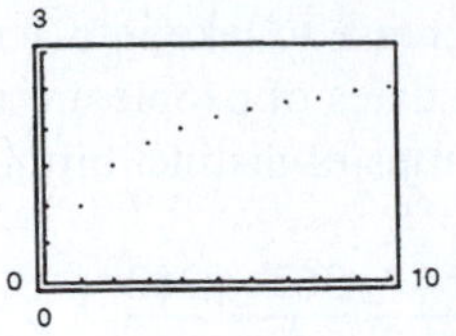

9. $\frac{1}{2(1)} + \frac{1}{2(2)} + \frac{1}{2(3)} + \cdots + \frac{1}{2(20)} = \sum_{k=1}^{20} \frac{1}{2k}$

11. $\frac{1}{2} + \frac{2}{3} + \frac{3}{4} + \cdots + \frac{9}{10} = \sum_{k=1}^{9} \frac{k}{k+1}$

13. $\sum_{i=1}^{6} 5 = 6(5) = 30$

15. $\sum_{j=1}^{4} \frac{6}{j^2} = \frac{6}{1^2} + \frac{6}{2^2} + \frac{6}{3^2} + \frac{6}{4^2} = 6 + \frac{3}{2} + \frac{2}{3} + \frac{3}{8} = \frac{205}{24}$

17. $\sum_{k=1}^{10} 2k^3 = 2(1)^3 + 2(2)^3 + 2(3)^3 + \cdots + 2(10)^3 = 6050$

19. $\sum_{n=0}^{10} (n^2 + 3) = \sum_{n=0}^{10} n^2 + \sum_{n=0}^{10} 3 = \frac{10(11)(21)}{6} + 11(3) = 418$

21. $a_1 = 3, d = 4$

$a_1 = 3$

$a_2 = 3 + 4 = 7$

$a_3 = 7 + 4 = 11$

$a_4 = 11 + 4 = 15$

$a_5 = 15 + 4 = 19$

23. $a_4 = 10 \quad a_{10} = 28$

$a_{10} = a_4 + 6d$

$28 = 10 + 6d$

$18 = 6d$

$3 = d$

$a_1 = a_4 - 3d$

$a_1 = 10 - 3(3)$

$a_1 = 1$

$a_2 = 1 + 3 = 4$

$a_3 = 4 + 3 = 7$

$a_4 = 7 + 3 = 10$

$a_5 = 10 + 3 = 13$

25. $a_1 = 35, a_{k+1} = a_k - 3$

$a_1 = 35$

$a_2 = a_1 - 3 = 35 - 3 = 32$

$a_3 = a_2 - 3 = 32 - 3 = 29$

$a_4 = a_3 - 3 = 29 - 3 = 26$

$a_5 = a_4 - 3 = 26 - 3 = 23$

$a_n = 35 + (n - 1)(-3) = 38 - 3n$

27. $a_1 = 9, a_{k+1} = a_k + 7$

$a_1 = 9$

$a_2 = a_1 + 7 = 9 + 7 = 16$

$a_3 = a_2 + 7 = 16 + 7 = 23$

$a_4 = a_3 + 7 = 23 + 7 = 30$

$a_5 = a_4 + 7 = 30 + 7 = 37$

$a_n = 9 + (n - 1)(7) = 2 + 7n$

29. $a_n = 100 + (n - 1)(-3) = 103 - 3n,\ a_1 = 100,\ a_{20} = 43,\ S_{20} = \frac{20}{2}(100 + 43) = 1430$

31. $\sum_{j=1}^{10}(2j - 3)$ is arithmetic. Therefore, $a_1 = -1,\ a_{10} = 17,\ S_{10} = \frac{10}{2}[-1 + 17] = 80$.

33. $\sum_{k=1}^{11}\left(\frac{2}{3}k + 4\right)$ is arithmetic. Therefore, $a_1 = \frac{14}{3},\ a_{11} = \frac{34}{3},\ s_{11} = \frac{11}{2}\left[\frac{14}{3} + \frac{34}{3}\right] = 88$.

35. $\sum_{k=1}^{100} 5k$ is arithmetic. Therefore, $a_1 = 5,\ a_{100} = 500,\ s_{500} = \frac{100}{2}(5 + 500) = 25{,}250$.

37. (a) $34{,}000 + 4(2250) = \$43{,}000$

(b) $\sum_{k=1}^{5}[34{,}000 + (k - 1)(2250)] = \sum_{k=1}^{5}(31{,}750 + 2250k) = \$192{,}500$

39. $a_1 = 4,\ r = -\frac{1}{4}$

$a_1 = 4$

$a_2 = 4\left(-\frac{1}{4}\right) = -1$

$a_3 = -1\left(-\frac{1}{4}\right) = \frac{1}{4}$

$a_4 = \frac{1}{4}\left(-\frac{1}{4}\right) = -\frac{1}{16}$

$a_5 = -\frac{1}{16}\left(-\frac{1}{4}\right) = \frac{1}{64}$

41. $a_1 = 9,\ a_3 = 4$

$a_3 = a_1r^2$

$4 = 9r^2$

$\frac{4}{9} = r^2 \Rightarrow r = \pm\frac{2}{3}$

$a_1 = 9$		$a_1 = 9$
$a_2 = 9\left(\frac{2}{3}\right) = 6$		$a_2 = 9\left(-\frac{2}{3}\right) = -6$
$a_3 = 6\left(\frac{2}{3}\right) = 4$	OR	$a_3 = -6\left(-\frac{2}{3}\right) = 4$
$a_4 = 4\left(\frac{2}{3}\right) = \frac{8}{3}$		$a_4 = 4\left(-\frac{2}{3}\right) = -\frac{8}{3}$
$a_5 = \frac{8}{3}\left(\frac{2}{3}\right) = \frac{16}{9}$		$a_5 = -\frac{8}{3}\left(-\frac{2}{3}\right) = \frac{16}{9}$

43. $a_1 = 120,\ a_{k+1} = \frac{1}{3}a_k$

$a_1 = 120$

$a_2 = \frac{1}{3}(120) = 40$

$a_3 = \frac{1}{3}(40) = \frac{40}{3}$

$a_4 = \frac{1}{3}\left(\frac{40}{3}\right) = \frac{40}{9}$

$a_5 = \frac{1}{3}\left(\frac{40}{9}\right) = \frac{40}{27}$

$a_n = 120\left(\frac{1}{3}\right)^{n-1}$

45. $a_1 = 25,\ a_{k+1} = -\frac{3}{5}a_k$

$a_1 = 25$

$a_2 = -\frac{3}{5}(25) = -15$

$a_3 = -\frac{3}{5}(-15) = 9$

$a_4 = -\frac{3}{5}(9) = -\frac{27}{5}$

$a_5 = -\frac{3}{5}\left(-\frac{27}{5}\right) = \frac{81}{25}$

$a_n = 25\left(-\frac{3}{5}\right)^{n-1}$

47. $a_2 = a_1 r$

$-8 = 16r$

$-\frac{1}{2} = r$

$a_n = 16\left(-\frac{1}{2}\right)^{n-1}$

$\sum_{n=1}^{20} 16\left(-\frac{1}{2}\right)^{n-1} = 16\left[\frac{1-(-\frac{1}{2})^{20}}{1-(-\frac{1}{2})}\right] \approx 10.67$

49. $\sum_{i=1}^{7} 2^{i-1} = \frac{1-2^7}{1-2} = 127$

51. $\sum_{i=1}^{\infty} \left(\frac{7}{8}\right)^{i-1} = \frac{1}{1-\frac{7}{8}} = 8$

53. $\sum_{k=1}^{\infty} 4\left(\frac{2}{3}\right)^{k-1} = \frac{4}{1-\frac{2}{3}} = 12$

55. $\sum_{i=1}^{10} 10\left(\frac{3}{5}\right)^{i-1} \approx 24.849$

57. (a) $a_t = 120{,}000(0.7)^t$

(b) $a_5 = 120{,}000(0.7)^5 = \$20{,}168.40$

59. $A = \sum_{i=1}^{24} 200\left(1 + \frac{0.06}{12}\right)^t \approx \$5111.82$

61. 1. When $n = 1$, $1 = \frac{1}{2}(3(1) - 1)$.

2. Assume that

$S_k = 1 + 4 + \cdots + (3k-2) = \frac{k}{2}(3k-1).$

Then,

$$S_{k+1} = 1 + 4 + \cdots + (3k-2) + (3(k+1)-2) = S_k + (3k+1)$$

$$= \frac{k}{2}(3k-1) + (3k+1) = \frac{k(3k-1) + 2(3k+1)}{2}$$

$$= \frac{3k^2+5k+2}{2} = \frac{(k+1)(3k+2)}{2} = \frac{(k+1)}{2}(3(k+1)-1).$$

Therefore, by mathematical induction, the formula is valid for all positive integer values of n.

37. Vertex: $(0, 0) \Rightarrow h = 0, k = 0$

Directrix: $y = 2 \Rightarrow p = -2$

$(x - h)^2 = 4p(y - k)$

$(x - 0)^2 = 4(-2)(y - 0)$

$x^2 = -8y$ or $y = -\frac{1}{8}x^2$

39. Vertex: $(0, 0) \Rightarrow h = 0, k = 0$

Horizontal axis and passes through the point (4, 6)

$(y - k)^2 = 4p(x - h)$

$(y - 0)^2 = 4p(x - 0)$

$y^2 = 4px$

$6^2 = 4p(4)$

$36 = 16p \Rightarrow p = \frac{9}{4}$

$y^2 = 4\left(\frac{9}{4}\right)x$

$y^2 = 9x$

41. Vertex: (3, 1) and opens downward. Passes through (2, 0) and (4, 0).

$y = -(x - 2)(x - 4)$

$= -x^2 + 6x - 8$

$= -(x - 3)^2 + 1$

$(x - 3)^2 = -(y - 1)$

43. Vertex: $(-2, 0)$ and opens to the right. Passes through (0, 2).

$(y - 0)^2 = 4p(x + 2)$

$2^2 = 4p(0 + 2)$

$\frac{1}{2} = p$

$y^2 = 4\left(\frac{1}{2}\right)(x + 2)$

$y^2 = 2(x + 2)$

45. Vertex: (3, 2)

Focus: (1, 2)

Horizontal axis

$p = 1 - 3 = -2$

$(y - 2)^2 = 4(-2)(x - 3)$

$(y - 2)^2 = -8(x - 3)$

47. Vertex: (0, 4)

Directrix: $y = 2$

Vertical axis

$p = 4 - 2 = 2$

$(x - 0)^2 = 4(2)(y - 4)$

$x^2 = 8(y - 4)$

49. Focus: (2, 2)

Directrix: $x = -2$

Horizontal axis

Vertex: (0, 2)

$p = 2 - 0 = 2$

$(y - 2)^2 = 4(2)(x - 0)$

$(y - 2)^2 = 8x$

51. $(y - 3)^2 = 6(x + 1)$

For the upper half of the parabola:

$y - 3 = +\sqrt{6(x + 1)}$

$y = \sqrt{6(x + 1)} + 3$

53. Vertex: $(0, 0) \Rightarrow h = 0, k = 0$

Focus: $(0, 3.5) \Rightarrow p = 3.5$

$(x - h)^2 = 4p(y - k)$

$(x - 0)^2 = 4(3.5)(y - 0)$

$x^2 = 14y$ or $y = \frac{1}{14}x^2$

55. (a) Converting 16 meters to 1600 centimeters, and superimposing the coordinate plane over the parabola so that its vertex is (0, 0), shows us that the points $(\pm 800, 3)$ are on the parabola.

$(x - 0)^2 = 4p(y - 0)$

$x^2 = 4py$

At $(\pm 800, 3)$ we have:

$640{,}000 = 12p$

$p = \dfrac{640{,}000}{12}$

$x^2 = 4\left(\dfrac{640{,}000}{12}\right)y$

$y = \dfrac{3x^2}{640{,}000}$

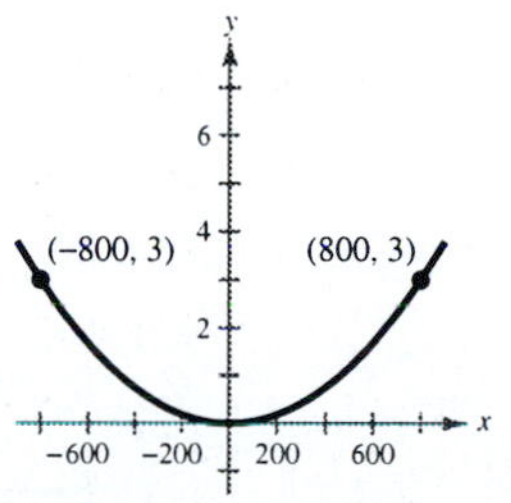

—CONTINUED—

55. —CONTINUED—

(b) Let $y = 1$, then:

$$1 = \frac{3x^2}{640{,}000}$$

$$x^2 = \frac{640{,}000}{3}$$

$$x = \frac{800\sqrt[3]{3}}{3} \approx 462 \text{ centimeters}$$

57. $R = 375x - \frac{3}{2}x^2$

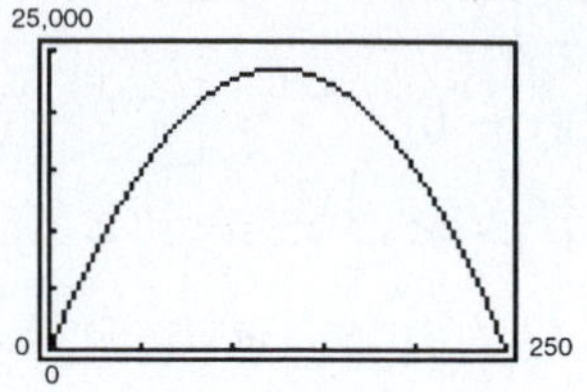

The revenue is maximum when $x = 125$ units.

59. $y = -0.08x^2 + x + 4$

(a)

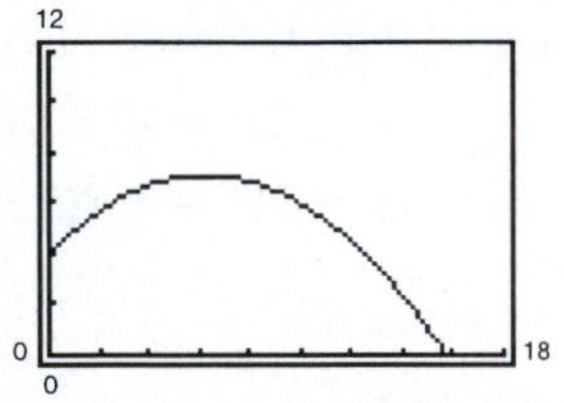

(b) The maximum occurs at the point (6.25, 7.125) and the graph crosses the x-axis when $x \approx 15.69$ feet.

61. The slope of the line $y - y_1 = \frac{x_1}{2p}(x - x_1)$ is $m = \frac{x_1}{2p}$.

63. $x^2 = 2y \Rightarrow p = \frac{1}{2}$

Point: $(x_1, y_1) = (4, 8)$

Use: $y - y_1 = \frac{x_1}{2p}(x - x_1)$

$$y - 8 = \frac{4}{2(1/2)}(x - 4)$$

$$y - 8 = 4x - 16$$

$$y = 4x - 8 \Rightarrow 0 = 4x - y - 8$$

x-intercept: $(2, 0)$

65. $y = -2x^2 \Rightarrow x^2 = -\frac{1}{2}y \Rightarrow p = -\frac{1}{8}$

Point: $(x_1, y_1) = (-1, -2)$

Use: $y - y_1 = \frac{x_1}{2p}(x - x_1)$

$$y + 2 = \frac{-1}{2\,(-1/8)}(x + 1)$$

$$y + 2 = 4(x + 1)$$

$$y = 4x + 2 \Rightarrow 0 = 4x - y + 2$$

x-intercept: $\left(-\frac{1}{2}, 0\right)$

67. $f(x) = (x - 3)[x - (2 + i)][x - (2 - i)]$

$= (x - 3)(x - 2 - i)(x - 2 + i)$

$= (x - 3)(x^2 - 4x + 5)$

$= x^3 - 7x^2 + 17x - 15$

69. $g(x) = 6x^4 + 7x^3 - 29x^2 - 28x + 20$

Possible rational roots: $\pm 1, \pm 2, \pm 4, \pm 5, \pm 10, \pm 20, \pm\frac{1}{2}, \pm\frac{5}{2}, \pm\frac{1}{3}, \pm\frac{2}{3}, \pm\frac{4}{3}, \pm\frac{5}{3}, \pm\frac{10}{3}, \pm\frac{20}{3}, \pm\frac{1}{6}, \pm\frac{5}{6}$

$x = \pm 2$ are both solutions.

2	6	7	−29	−28	20
		12	38	18	−20
−2	6	19	9	−10	0
		−12	−14	10	
	6	7	−5	0	

$g(x) = (x - 2)(x + 2)(6x^2 + 7x - 5)$

$= (x - 2)(x + 2)(2x - 1)(3x + 5)$

The zeros of $g(x)$ are $x = \pm 2, x = \frac{1}{2}, x = -\frac{5}{3}$.

Section 10.3 Ellipses

- An **ellipse** is the set of all points (x, y) the sum of whose distances from two distinct fixed points **(foci)** is constant.
- The standard equation of an ellipse with center (h, k) and major and minor axes of lengths $2a$ and $2b$ is:
  (a) $\frac{(x-h)^2}{a^2} + \frac{(y-k)^2}{b^2} = 1$ if the major axis is horizontal.
  (b) $\frac{(x-h)^2}{b^2} + \frac{(y-k)^2}{a^2} = 1$ if the major axis is vertical.
- $c^2 = a^2 - b^2$ where c is the distance from the center to a focus.
- The eccentricity of an ellipse is $e = \frac{c}{a}$

Solutions to Odd-Numbered Exercises

1. $\frac{x^2}{4} + \frac{y^2}{9} = 1$

Center: $(0, 0)$

$a = 3, b = 2$

Vertical major axis

Matches graph (b).

3. $\frac{x^2}{4} + \frac{y^2}{25} = 1$

Center: $(0, 0)$

$a = 5, b = 2$

Vertical major axis

Matches graph (d).

5. $\frac{(x-2)^2}{16} + (y+1)^2 = 1$

Center: $(2, -1)$

$a = 4, b = 1$

Horizontal major axis

Matches graph (a).

7. $\frac{x^2}{25} + \frac{y^2}{16} = 1$

Center: $(0, 0)$

$a = 5, b = 4, c = 3$

Foci: $(\pm 3, 0)$

Vertices: $(\pm 5, 0)$

$e = \frac{3}{5}$

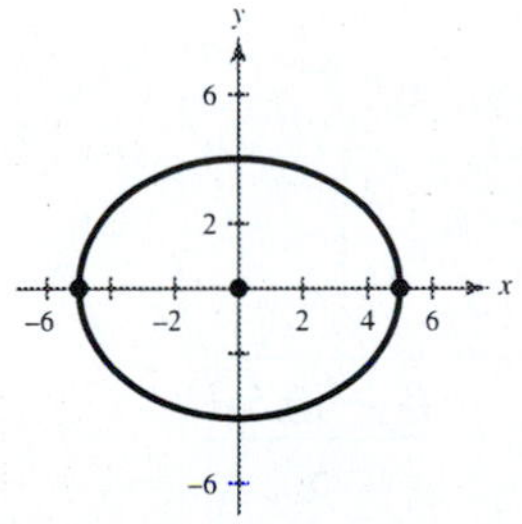

9. $\frac{x^2}{16} + \frac{y^2}{25} = 1$

$a = 5, b = 4, c = 3$

Center: $(0, 0)$

Foci: $(0, \pm 3)$

Vertices: $(0, \pm 5)$

$e = \frac{3}{5}$

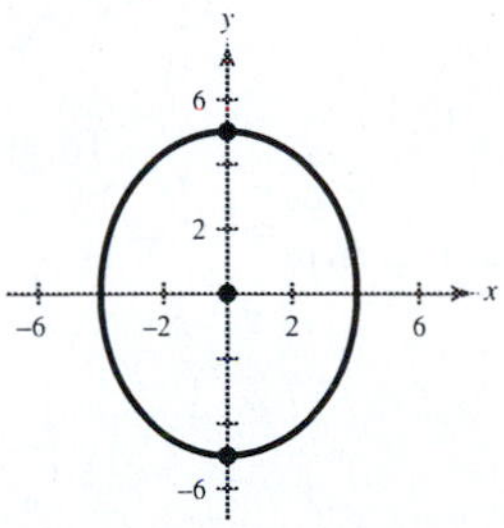

11. $\frac{x^2}{9} + \frac{y^2}{5} = 1$

Center: $(0, 0)$

$a = 3, b = \sqrt{5}, c = 2$

Foci: $(\pm 2, 0)$

Vertices: $(\pm 3, 0)$

$e = \frac{2}{3}$

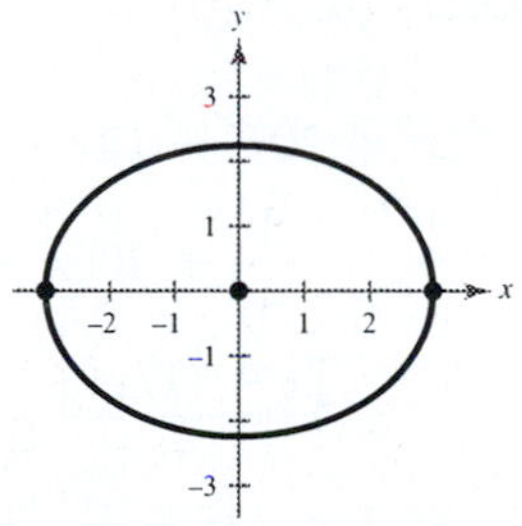

13. $\frac{(x-1)^2}{9} + \frac{(y-5)^2}{25} = 1$

$a = 5, b = 3, c = 4$

Center: $(1, 5)$

Foci: $(1, 9), (1, 1)$

Vertices: $(1, 10), (1, 0)$

$e = \frac{4}{5}$

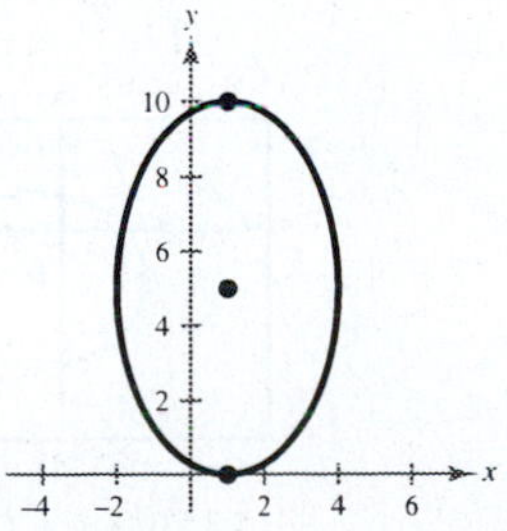

15. $9x^2 + 4y^2 + 36x - 24y + 36 = 0$

$$9(x^2 + 4x + 4) + 4(y^2 - 6y + 9) = -36 + 36 + 36$$

$$\frac{(x+2)^2}{4} + \frac{(y-3)^2}{9} = 1$$

$a = 3, b = 2, c = \sqrt{5}$

Center: $(-2, 3)$

Foci: $\left(-2, 3 \pm \sqrt{5}\right)$

Vertices: $(-2, 6), (-2, 0)$

$e = \dfrac{\sqrt{5}}{3}$

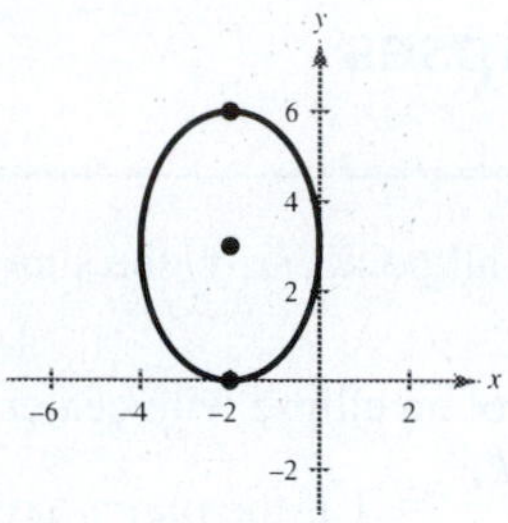

17. $16x^2 + 25y^2 - 32x + 50y + 16 = 0$

$$16(x^2 - 2x + 1) + 25(y^2 + 2y + 1) = -16 + 16 + 25$$

$$\frac{(x-1)^2}{25/16} + (y+1)^2 = 1$$

$a = \dfrac{5}{4}, b = 1, c = \dfrac{3}{4}$

Center: $(1, -1)$

Foci: $\left(\dfrac{7}{4}, -1\right), \left(\dfrac{1}{4}, -1\right)$

Vertices: $\left(\dfrac{9}{4}, -1\right), \left(-\dfrac{1}{4}, -1\right)$

$e = \dfrac{3}{5}$

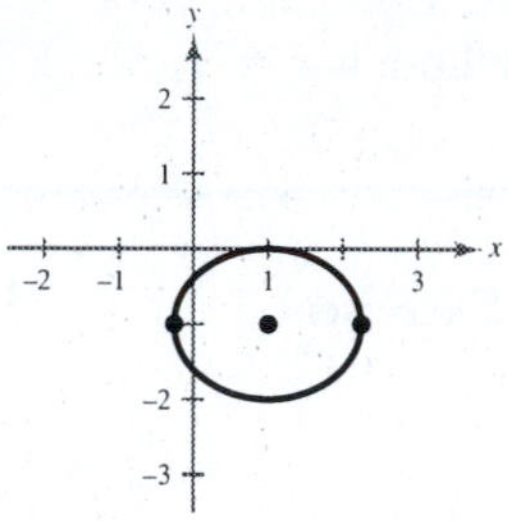

19. $5x^2 + 3y^2 = 15$

$$\frac{x^2}{3} + \frac{y^2}{5} = 1$$

Center: $(0, 0)$

$a = \sqrt{5}, b = \sqrt{3}, c = \sqrt{2}$

Foci: $\left(0, \pm\sqrt{2}\right)$

Vertices: $\left(0, \pm\sqrt{5}\right)$

To graph, solve for y.

$$y^2 = \frac{15 - 5x^2}{3}$$

$$y_1 = \sqrt{\frac{15 - 5x^2}{3}}$$

$$y_2 = -\sqrt{\frac{15 - 5x^2}{3}}$$

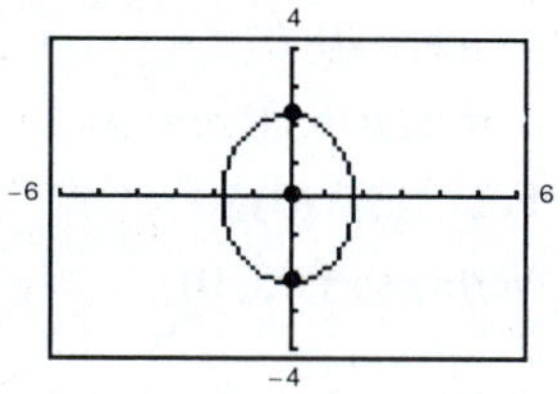

21. $12x^2 + 20y^2 - 12x + 40y - 37 = 0$

$$12\left(x^2 - x + \frac{1}{4}\right) + 20(y^2 + 2y + 1) = 37 + 3 + 20$$

$$\frac{[x - (1/2)]^2}{5} + \frac{(y+1)^2}{3} = 1$$

$a = \sqrt{5}, b = \sqrt{3}, c = \sqrt{2}$

Center: $\left(\dfrac{1}{2}, -1\right)$

Foci: $\left(\dfrac{1}{2} \pm \sqrt{2}, -1\right)$

Vertices: $\left(\dfrac{1}{2} \pm \sqrt{5}, -1\right)$

$e = \dfrac{\sqrt{10}}{5}$

To graph, solve for y.

$$(y+1)^2 = 3\left[1 - \frac{(x-0.5)^2}{5}\right]$$

$$y_1 = -1 + \sqrt{1 - \frac{(x-0.5)^2}{5}}$$

$$y_1 = -1 - \sqrt{1 - \frac{(x-0.5)^2}{5}}$$

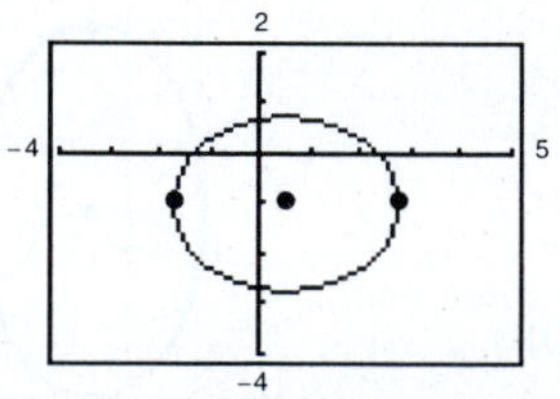

23. For the right half of the ellipse, solve for x and use the positive square root.

$$\frac{(x-3)^2}{9} + \frac{y^2}{4} = 1$$

$$4(x-3)^2 + 9y^2 = 36$$

$$4(x-3)^2 = 36 - 9y^2$$

$$(x-3)^2 = \frac{9(4-y^2)}{4}$$

$$x - 3 = \frac{2}{3}\sqrt{4-y^2}$$

$$x = 3 + \frac{2}{3}\sqrt{4-y^2}$$

$$= \frac{3}{2}\left(2 + \sqrt{4-y^2}\right)$$

25. Center: $(0, 0)$

$a = 2, b = 1$

Vertical major axis

$$\frac{(x-h)^2}{b^2} + \frac{(y-k)^2}{a^2} = 1$$

$$\frac{x^2}{1} + \frac{y^2}{4} = 1$$

27. Vertices: $(\pm 5, 0)$

$a = 5, c = 2 \Rightarrow b = \sqrt{21}$

Foci: $(\pm 2, 0)$

Horizontal major axis

Center: $(0, 0)$

$$\frac{(x-h)^2}{a^2} + \frac{(y-k)^2}{b^2} = 1$$

$$\frac{x^2}{25} + \frac{y^2}{21} = 1$$

29. Foci: $(\pm 5, 0) \Rightarrow c = 5$

Center: $(0, 0)$

Horizontal major axis

Major axis of length 12 $\Rightarrow 2a = 12$

$a = 6$

$6^2 - b^2 = 5^2 \Rightarrow b^2 = 11$

$$\frac{(x-h)^2}{a^2} + \frac{(y-k)^2}{b^2} = 1$$

$$\frac{x^2}{36} + \frac{y^2}{11} = 1$$

31. Vertices: $(0, \pm 5) \Rightarrow a = 5$

Center: $(0, 0)$

Vertical major axis

$$\frac{(x-h)^2}{b^2} + \frac{(y-k)^2}{a^2} = 1$$

$$\frac{x^2}{b^2} + \frac{y^2}{25} = 1$$

Point: $(4, 2)$

$$\frac{4^2}{b^2} + \frac{2^2}{25} = 1$$

$$\frac{16}{b^2} = 1 - \frac{4}{25} = \frac{21}{25}$$

$$400 = 21b^2$$

$$\frac{400}{21} = b^2$$

$$\frac{x^2}{400/21} + \frac{y^2}{25} = 1$$

$$\frac{21x^2}{400} + \frac{y^2}{25} = 1$$

33. Center: $(2, 3)$

$a = 3, \quad b = 1$

Vertical major axis

$$\frac{(x-h)^2}{b^2} + \frac{(y-k)^2}{a^2} = 1$$

$$\frac{(x-2)^2}{1} + \frac{(y-3)^2}{9} = 1$$

35. Center: $(2, 2)$

$a = 3, \quad b = 2$

Horizontal major axis

$$\frac{(x-h)^2}{a^2} + \frac{(y-k)^2}{b^2} = 1$$

$$\frac{(x-2)^2}{9} + \frac{(y-2)^2}{4} = 1$$

37. Vertices: $(0, 2), (4, 2) \Rightarrow a = 2$

Minor axis of length 2 $\Rightarrow b = 1$

Center: $(2, 2) = (h, k)$

$$\frac{(x-h)^2}{a^2} + \frac{(y-k)^2}{b^2} = 1$$

$$\frac{(x-2)^2}{4} + \frac{(y-2)^2}{1} = 1$$

39. Foci: $(0, 0), (0, 8) \Rightarrow c = 4$

Major axis of length 16 $\Rightarrow a = 8$

$b^2 = a^2 - c^2 = 64 - 16 = 48$

Center: $(0, 4) = (h, k)$

$$\frac{(x-h)^2}{b^2} + \frac{(y-k)^2}{a^2} = 1$$

$$\frac{x^2}{48} + \frac{(y-4)^2}{64} = 1$$

41. Vertices: $(3, 1), (3, 9) \Rightarrow a = 4$

Center: $(3, 5)$

Minor axis of length 6 $\Rightarrow b = 3$

Vertical major axis

$$\frac{(x-h)^2}{b^2} + \frac{(y-k)^2}{a^2} = 1$$

$$\frac{(x-3)^2}{9} + \frac{(y-5)^2}{16} = 1$$

43. Center: $(0, 4)$

Vertices: $(-4, 4), (4, 4) \Rightarrow a = 4$

$a = 2c \Rightarrow 4 = 2c \Rightarrow c = 2$

$2^2 = 4^2 - b^2 \Rightarrow b^2 = 12$

Horizontal major axis

$$\frac{(x-h)^2}{a^2} + \frac{(y-k)^2}{b^2} = 1$$

$$\frac{x^2}{16} + \frac{(y-4)^2}{12} = 1$$

45. (a) The length of the string is $2a$.

(b) The path is an ellipse because the sum of the distances from the two thumbtacks is always the length of the string, that is, it is constant.

47.

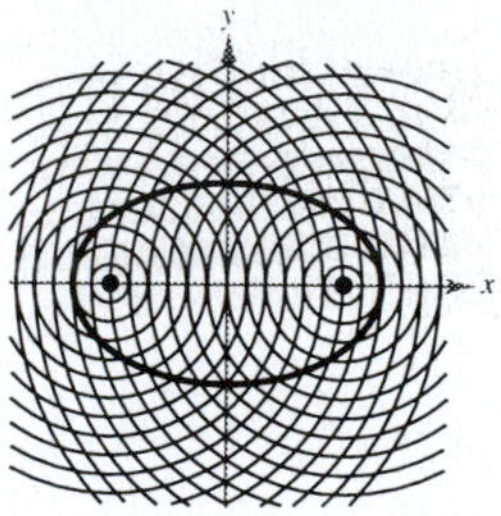

49. $\dfrac{x^2}{a^2} + \dfrac{y^2}{b^2} = 1$

(a) $a + b = 20 \Rightarrow b = 20 - a$

$A = \pi ab = \pi a(20 - a)$

(b) $264 = \pi a(20 - a)$

$0 = -\pi a^2 + 20\pi a - 264$

$0 = \pi a^2 - 20\pi a + 264$

$a = 14$ or $a = 6$. The equation of an ellipse with an area of 264 is $\dfrac{x^2}{196} + \dfrac{y^2}{36} = 1$.

(c)

a	8	9	10	11	12	13
A	301.6	311.0	314.2	311.0	301.6	285.9

The area is maximum when $a = 10$ and the ellipse is a circle.

(d)

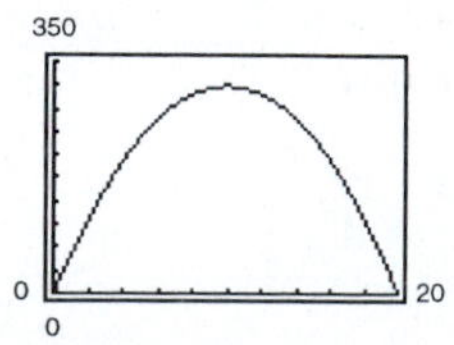

The area is maximum (314.16) when $a = b = 10$ and the ellipse is a circle.

51. Vertices: $(\pm 5, 0) \Rightarrow a = 5$

Eccentricity: $\frac{3}{5} \Rightarrow c = \frac{3}{5}a = 3$

$b^2 = a^2 - c^2 = 25 - 9 = 16$

Center: $(0, 0) = (h, k)$

$$\frac{(x-h)^2}{a^2} + \frac{(y-k)^2}{b^2} = 1$$

$$\frac{x^2}{25} + \frac{y^2}{16} = 1$$

53. $2a = 36.18 \Rightarrow a = 18.09$

$e = \frac{c}{a} = 0.97 \Rightarrow c = (18.09)(0.97) = 17.5473$

$b^2 = a^2 - c^2 = (18.09)^2 - (17.5473)^2 \approx 19.34$

The equation of the ellipse is:

$$\frac{x^2}{(18.09)^2} + \frac{y^2}{19.34} = 1$$

$$\frac{x^2}{327.25} + \frac{y^2}{19.34} = 1$$

55. apogee $= 938 + 6378 = 7316$

perigee $= 212 + 6378 = 6590$

$$e = \frac{7316 - 6590}{7316 + 6590} \approx 0.052$$

57. False: The graph of $\frac{x^2}{4} + y^4 = 1$ is not an ellipse. The degree on y is 4, not 2.

59. $\frac{x^2}{4} + \frac{y^2}{1} = 1$

$a = 2, b = 1, c = \sqrt{3}$

Points on the ellipse: $(\pm 2, 0), (0, \pm 1)$

Length of latus recta: $\frac{2b^2}{a} = \frac{2(1)^2}{2} = 1$

Additional points: $\left(-\sqrt{3}, \pm\frac{1}{2}\right), \left(\sqrt{3}, \pm\frac{1}{2}\right)$

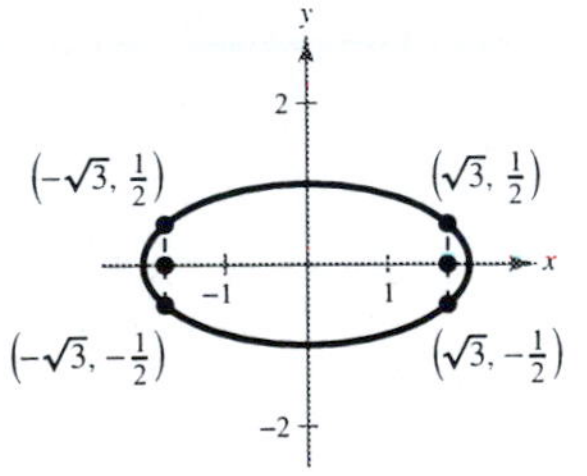

61. $9x^2 + 4y^2 = 36$

$\frac{x^2}{4} + \frac{y^2}{9} = 1$

$a = 3, b = 2, c = \sqrt{5}$

Points on the ellipse: $(\pm 2, 0), (0, \pm 3)$

Length of latus recta: $\frac{2b^2}{a} = \frac{2 \cdot 2^2}{3} = \frac{8}{3}$

Additional points: $\left(\pm\frac{4}{3}, -\sqrt{5}\right), \left(\pm\frac{4}{3}, \sqrt{5}\right)$

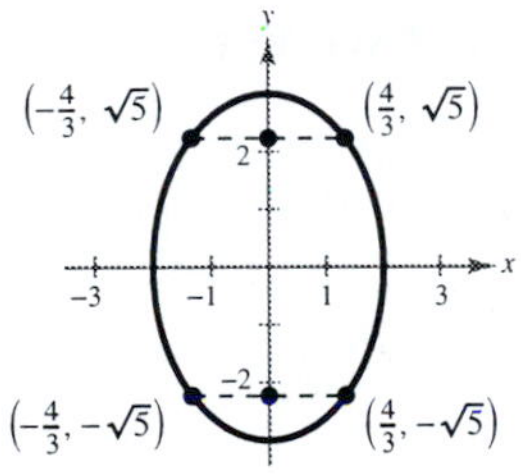

Section 10.4 Hyperbolas

- A **hyperbola** is the set of all points (x, y) the difference of whose distances from two distinct fixed points **(foci)** is constant.
- The standard equation of a hyperbola with center (h, k) and transverse and conjugate axes of lengths $2a$ and $2b$ is:
  (a) $\dfrac{(x-h)^2}{a^2} - \dfrac{(y-k)^2}{b^2} = 1$ if the traverse axis is horizontal.
  (b) $\dfrac{(y-k)^2}{a^2} - \dfrac{(x-h)^2}{b^2} = 1$ if the traverse axis is vertical.
- $c^2 = a^2 + b^2$ where c is the distance fromthe center to a focus.
- The asymptotes of a hyperbola are:
  (a) $y = k \pm \dfrac{b}{a}(x - h)$ if the transverse axis is horizontal.
  (b) $y = k \pm \dfrac{a}{b}(x - h)$ the transverse axis is vertical.
- The eccentricity of a hyperbola is $e = \dfrac{c}{a}$.
- To classify a nondegenerate conic from its general equation $Ax^2 + Cy^2 + Dx + Ey + F = 0$:
  (a) If $A = C$ $(A \neq 0, C \neq 0)$, then it is a circle.
  (b) If $AC = 0$ $(A = 0$ or $C = 0$, but not both), then it is a parabola.
  (c) If $AC > 0$, then it is an ellipse.
  (d) If $AC < 0$, then it is a hyperbola.

Solutions to Odd Numbered Exercises

1. $\dfrac{x^2}{16} - \dfrac{y^2}{4} = 1$

Center: $(0, 0)$

$a = 4, b = 2$

Horizontal transverse axis

Matches graph (b).

3. $\dfrac{y^2}{9} - \dfrac{x^2}{16} = 1$

Center: $(0, 0)$

$a = 3, b = 4$

Vertical transverse axis

Matches graph (e).

5. $\dfrac{(x-1)^2}{16} - \dfrac{y^2}{4} = 1$

Center: $(1, 0)$

$a = 4, b = 2$

Horizontal transverse axis

Matches graph (a).

7. $x^2 - y^2 = 1$

$a = 1, b = 1, c = \sqrt{2}$

Center: $(0, 0)$

Vertices: $(\pm 1, 0)$

Foci: $(\pm\sqrt{2}, 0)$

Asymptotes: $y = \pm x$

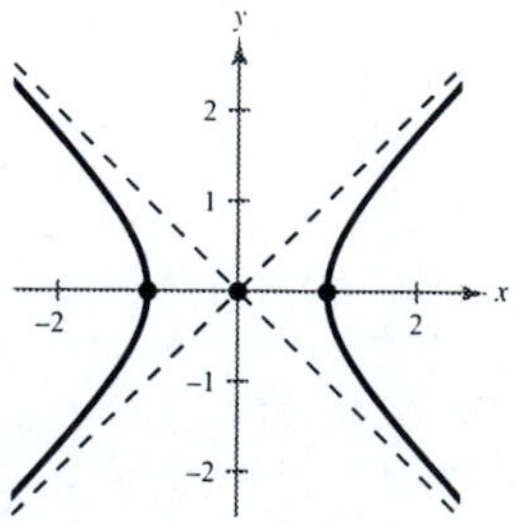

9. $\dfrac{y^2}{1} - \dfrac{x^2}{4} = 1$

$a = 1, b = 2, c = \sqrt{5}$

Center: $(0, 0)$

Vertices: $(0, \pm 1)$

Foci: $\left(0, \pm\sqrt{5}\right)$

Asymptotes: $y = \pm\dfrac{1}{2}x$

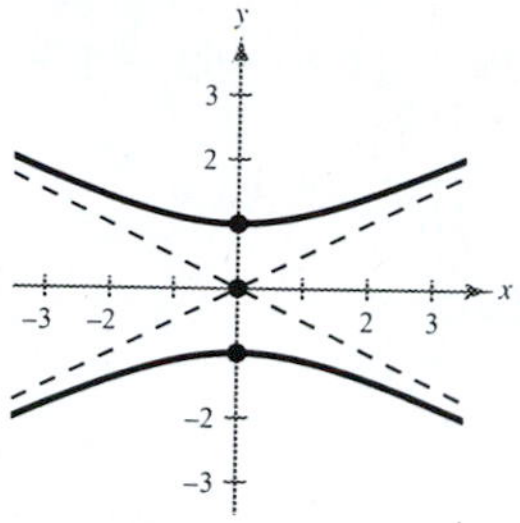

11. $\dfrac{y^2}{25} - \dfrac{x^2}{144} = 1$

$a = 5, b = 12, c = 13$

Center: $(0, 0)$

Vertices: $(0, \pm 5)$

Foci: $(0, \pm 13)$

Asymptotes: $y = \pm\dfrac{5}{12}x$

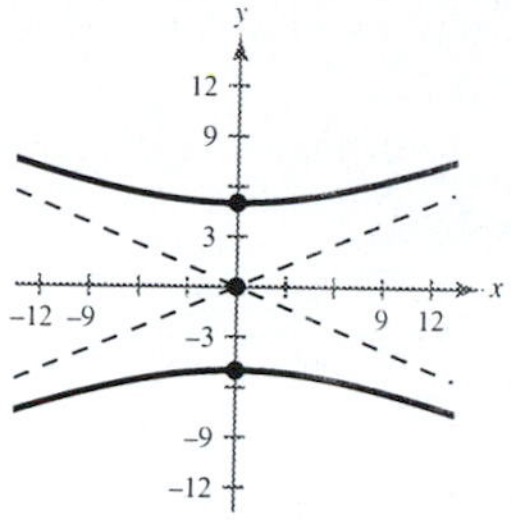

13. $\dfrac{(x-1)^2}{4} - \dfrac{(y+2)^2}{1} = 1$

$a = 2, b = 1, c = \sqrt{5}$

Center: $(1, -2)$

Vertices: $(-1, -2), (3, -2)$

Foci: $\left(1 \pm \sqrt{5}, -2\right)$

Asymptotes: $y = -2 \pm \dfrac{1}{2}(x - 1)$

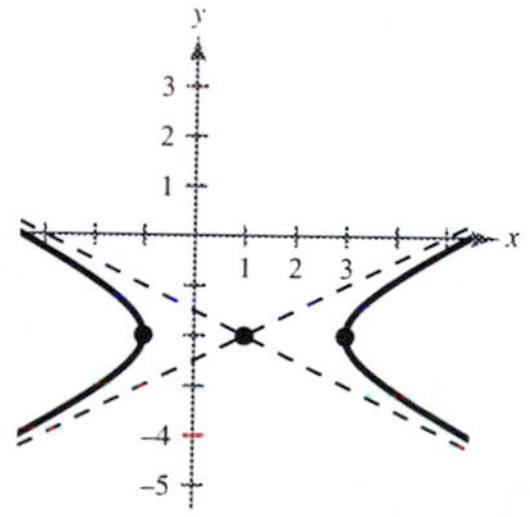

15. $(y + 6)^2 - (x - 2)^2 = 1$

$a = 1, b = 1, c = \sqrt{2}$

Center: $(2, -6)$

Vertices: $(2, -5), (2, -7)$

Foci: $\left(2, -6 \pm \sqrt{2}\right)$

Asymptotes: $y = -6 \pm (x - 2)$

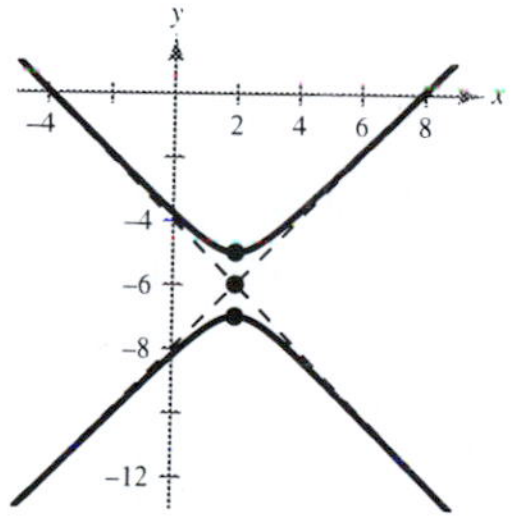

17.

$$9x^2 - y^2 - 36x - 6y + 18 = 0$$

$$9\left(x^2 - 4x + 4\right) - \left(y^2 + 6y + 9\right) = -18 + 36 - 9$$

$$\frac{(x-2)^2}{1} - \frac{(y+3)^2}{9} = 1$$

$a = 1, b = 3, c = \sqrt{10}$

Center: $(2, -3)$

Vertices: $(1, -3), (3, -3)$

Foci: $\left(2 \pm \sqrt{10}, -3\right)$

Asymptotes: $y = -3 \pm 3(x - 2)$

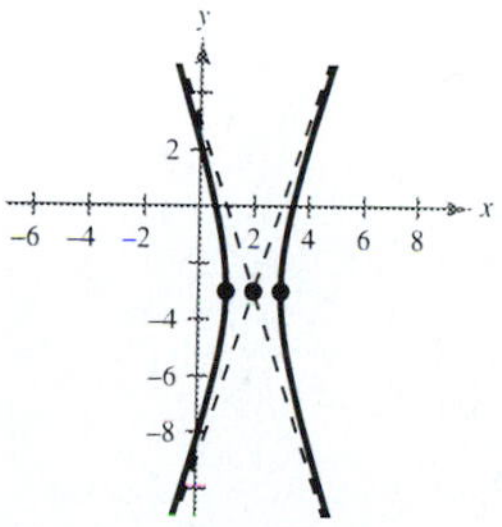

19.
$$x^2 - 9y^2 + 2x - 54y - 80 = 0$$
$$(x^2 + 2x + 1) - 9(y^2 + 6y + 9) = 80 + 1 - 81$$
$$(x + 1)^2 - 9(y + 3)^2 = 0$$
$$y + 3 = \pm\frac{1}{3}(x + 1)$$

Degenerate hyperbola is two lines intersecting at $(-1, -3)$.

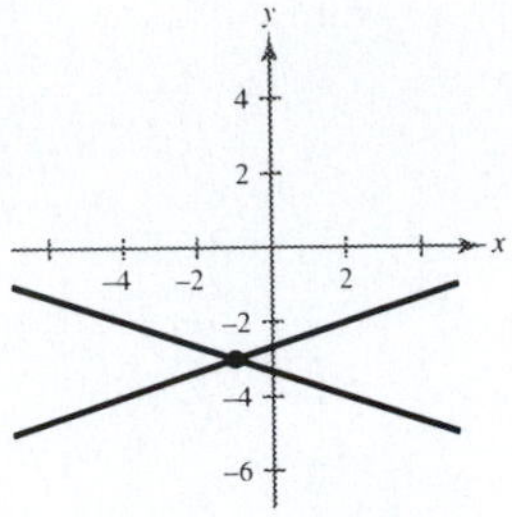

21. $2x^2 - 3y^2 = 6$

$$\frac{x^2}{3} - \frac{y^2}{2} = 1$$

$a = \sqrt{3}, b = \sqrt{2}, c = \sqrt{5}$

Center: $(0, 0)$

Vertices: $(\pm\sqrt{3}, 0)$

Foci: $(\pm\sqrt{5}, 0)$

Asymptotes: $y = \pm\sqrt{\frac{2}{3}}x$

To use a graphing calculator, solve first for y.

$$y^2 = \frac{2x^2 - 6}{3}$$

$$\left.\begin{aligned} y_1 &= \sqrt{\frac{2x^2 - 6}{3}} \\ y_2 &= -\sqrt{\frac{2x^2 - 6}{3}} \end{aligned}\right\} \text{Hyperbola}$$

$$\left.\begin{aligned} y_3 &= \sqrt{\frac{2}{3}}x \\ y_4 &= -\sqrt{\frac{2}{3}}x \end{aligned}\right\} \text{Asymptotes}$$

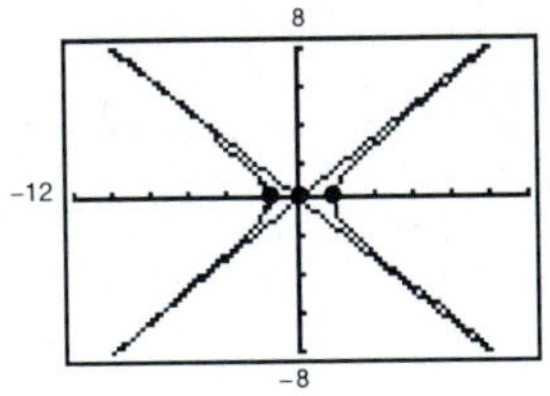

23.
$$9y^2 - x^2 + 2x + 54y + 62 = 0$$
$$9(y^2 + 6y + 9) - (x^2 - 2x + 1) = -62 - 1 + 81$$
$$\frac{(y + 3)^2}{2} - \frac{(x - 1)^2}{18} = 1$$

$a = \sqrt{2}, b = 3\sqrt{2}, c = 2\sqrt{5}$

Center: $(1, -3)$

Vertices: $(1, -3 \pm \sqrt{2})$

Foci: $(1, -3 \pm 2\sqrt{5})$

Asymptotes: $y = -3 \pm \frac{1}{3}(x - 1)$

To use a graphing calculator, solve for y first.

$$9(y + 3)^2 = 18 + (x - 1)^2$$

$$y = -3 \pm \sqrt{\frac{18 + (x - 1)^2}{9}}$$

$$\left.\begin{aligned} y_1 &= -3 + \frac{1}{3}\sqrt{18 + (x - 1)^2} \\ y_2 &= -3 - \frac{1}{3}\sqrt{18 + (x - 1)^2} \end{aligned}\right\} \text{Hyperbola}$$

$$\left.\begin{aligned} y_3 &= -3 + \frac{1}{3}(x - 1) \\ y_4 &= -3 - \frac{1}{3}(x - 1) \end{aligned}\right\} \text{Asymptotes}$$

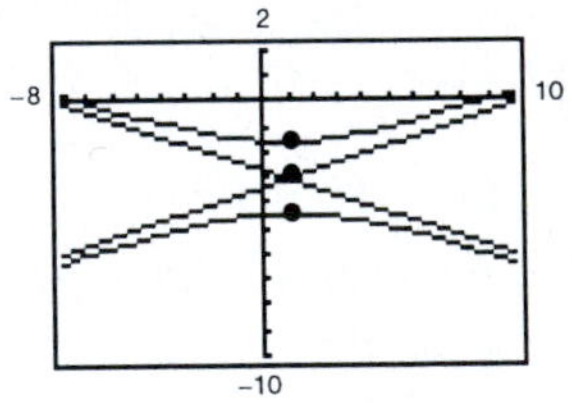

25. Vertices: $(0, \pm3) \Rightarrow a = 3$

Solution point: $(-2, 5)$

Center: $(0, 0) = (h, k)$

$$\frac{(y - k)^2}{a^2} - \frac{(x - h)^2}{b^2} = 1$$

$$\frac{y^2}{9} - \frac{x^2}{b^2} = 1 \Rightarrow b^2 = \frac{9x^2}{y^2 - 9} = \frac{9(-2)^2}{5^2 - 9} = \frac{36}{16} = \frac{9}{4}$$

$$\frac{y^2}{9} - \frac{x^2}{9/4} = 1$$

27. Vertices: $(0, \pm 2) \Rightarrow a = 2$

Foci: $(0, \pm 4) \Rightarrow c = 4$

$b^2 = c^2 - a^2 = 16 - 4 = 12$

Center: $(0, 0) = (h, k)$

$$\frac{(y-k)^2}{a^2} - \frac{(x-h)^2}{b^2} = 1$$

$$\frac{y^2}{4} - \frac{x^2}{12} = 1$$

29. Vertices: $(\pm 1, 0) \Rightarrow a = 1$

Asymptotes: $y = \pm 3x \Rightarrow \frac{b}{a} = 3, b = 3$

Center: $(0, 0) = (h, k)$

$$\frac{(x-h)^2}{a^2} - \frac{(y-k)^2}{b^2} = 1$$

$$\frac{x^2}{1} - \frac{y^2}{9} = 1$$

31. Foci: $(0, \pm 8) \Rightarrow c = 8$

Asymptotes: $y = \pm 4x \Rightarrow \frac{a}{b} = 4 \Rightarrow a = 4b$

Center: $(0, 0) = (h, k)$

$c^2 = a^2 + b^2 \Rightarrow 64 = 16b^2 + b^2$

$$\frac{64}{17} = b^2 \Rightarrow a^2 = \frac{1024}{17}$$

$$\frac{(y-k)^2}{a^2} - \frac{(x-h)^2}{b^2} = 1$$

$$\frac{y^2}{1024/17} - \frac{x^2}{64/17} = 1$$

$$\frac{17y^2}{1024} - \frac{17x^2}{65} = 1$$

33. Vertices: $(0, 0), (0, 2) \Rightarrow a = 1$

Solution point: $(\sqrt{3}, 3)$

Center: $(0, 1) = (h, k)$

$$\frac{(y-k)^2}{a^2} - \frac{(x-h)^2}{b^2} = 1$$

$$(y-1)^2 - \frac{x^2}{b^2} = 1 \Rightarrow 4 - \frac{3}{b^2} = 1 \Rightarrow b^2 = 1$$

$$(y-1)^2 - x^2 = 1$$

35. Center: $(3, 2) = (h, k)$

Vertices: $(1, 2), (5, 2) \Rightarrow a = 2$

Solution point: $(0, 0)$

$$\frac{(x-h)^2}{a^2} - \frac{(y-k)^2}{b^2} = 1$$

$$\frac{(x-3)^2}{4} - \frac{(y-2)^2}{b^2} = 1 \Rightarrow \frac{9}{4} - \frac{4}{b^2} = 1 \Rightarrow b^2 = \frac{16}{5}$$

$$\frac{(x-3)^2}{4} - \frac{(y-2)^2}{16/5} = 1$$

37. Vertices: $(2, 0), (6, 0) \Rightarrow a = 2$

Foci: $(0, 0), (8, 0) \Rightarrow c = 4$

$b^2 = c^2 - a^2 = 16 - 4 = 12$

Center: $(4, 0) = (h, k)$

$$\frac{(x-h)^2}{a^2} - \frac{(y-k)^2}{b^2} = 1$$

$$\frac{(x-4)^2}{4} - \frac{y^2}{12} = 1$$

39. Vertices: $(4, 1), (4, 9) \Rightarrow a = 4$

Foci: $(4, 0), (4, 10) \Rightarrow c = 5$

$b^2 = c^2 - a^2 = 25 - 16 = 9$

Center: $(4, 5) = (h, k)$

$$\frac{(y-k)^2}{a^2} - \frac{(x-h)^2}{b^2} = 1$$

$$\frac{(y-4)^2}{16} - \frac{(x-4)^2}{9} = 1$$

41. Vertices: $(2, 3), (2, -3) \Rightarrow a = 3$

Solution point: $(0, 5)$

Center: $(2, 0) = (h, k)$

$$\frac{(y-k)^2}{a^2} - \frac{(x-h)^2}{b^2} = 1$$

$$\frac{y^2}{9} - \frac{(x-2)^2}{b^2} = 1 \Rightarrow b^2 = \frac{9(x-2)^2}{y^2-9} = \frac{9(-2)^2}{25-9} = \frac{36}{16} = \frac{9}{4}$$

$$\frac{y^2}{9} - \frac{(x-2)^2}{9/4} = 1$$

43. Vertices: $(0, 2), (6, 2) \Rightarrow a = 3$

Asymptotes: $y = \frac{2}{3}x,\ y = 4 - \frac{2}{3}x$

$$\frac{b}{a} = \frac{2}{3} \Rightarrow b = 2$$

Center: $(3, 2) = (h, k)$

$$\frac{(x-h)^2}{a^2} - \frac{(y-k)^2}{b^2} = 1$$

$$\frac{(x-3)^2}{9} - \frac{(y-2)^2}{4} = 1$$

45. $x = 3 - \frac{2}{3}\sqrt{1 + (y-1)^2}$ represents the left branch of the hyperbola.

47. Since $\overline{AB} = 1100$ feet and the sound takes one second longer to reach B than A, the explosion must occur on the vertical line through A and B below A.

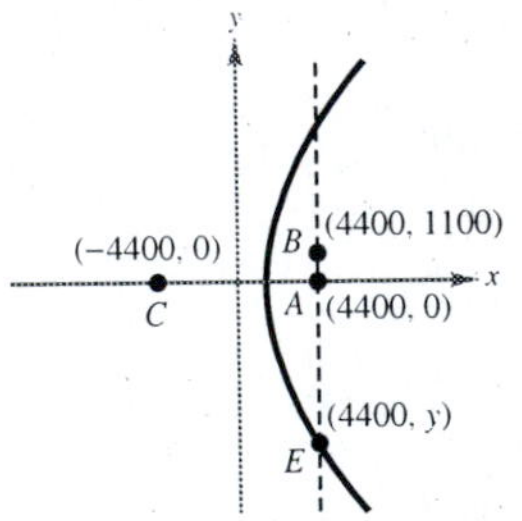

Foci: $(\pm 4400, 0) \Rightarrow c = 4400$

Center: $(0, 0) = (h, k)$

$$\frac{\overline{CE}}{1100} - \frac{\overline{AE}}{1100} = 5 \Rightarrow 2a = 5500,\ a = \frac{5500}{2} = 2750$$

$b^2 = c^2 - a^2 = (4400)^2 - (2750)^2 = 11{,}797{,}500$

$$\frac{x^2}{(2750)^2} - \frac{y^2}{11{,}797{,}500} = 1$$

$$y^2 = 11{,}797{,}500\left(\frac{x^2}{(2750)^2} - 1\right)$$

$$y^2 = 11{,}797{,}500\left(\frac{(4400)^2}{(2750)^2} - 1\right) = 18{,}404{,}100$$

$$y = -4290$$

The explosion occurs at $(4400, -4290)$.

49. Center: $(0, 0) = (h, k)$

Focus: $(24, 0) \Rightarrow c = 24$

Solution point: $(24, 24)$

$24^2 = a^2 + b^2 \Rightarrow b^2 = 24^2 - a^2$

$$\frac{(x-h)^2}{a^2} - \frac{(y-k)^2}{b^2} = 1$$

$$\frac{x^2}{a^2} - \frac{y^2}{24^2 - a^2} = 1 \Rightarrow \frac{24^2}{a^2} - \frac{24^2}{24^2 - a^2} = 1$$

Solving yields $a^2 = \dfrac{(3 - \sqrt{5})24^2}{2} \approx 220.0124$ and $b^2 \approx 355.9876$.

Thus, we have $\dfrac{x^2}{220.0124} - \dfrac{y^2}{355.9876} = 1$.

The right vertex is at $(a, 0) \approx (14.83, 0)$.

51. $x^2 + y^2 - 6x + 4y + 9 = 0$

$A = 1, C = 1$

$A = C \Rightarrow$ Circle

53. $4x^2 - y^2 - 4x - 3 = 0$

$A = 4, C = -1$

$AC = 4(-1)$

$= -4 < 0 \Rightarrow$ Hyperbola

55. $4x^2 + 3y^2 + 8x - 24y + 51 = 0$

$A = 4, C = 3$

$AC = 4(3) = 12 > 0 \Rightarrow$ Ellipse

57. $25x^2 - 10x - 200y - 119 = 0$

$A = 25, C = 0$

$AC = 25(0) = 0 \Rightarrow$ Parabola

59. $(x^3 - 3x^2) - (6 - 2x - 4x^2) = x^3 - 3x^2 - 6 + 2x + 4x^2$

$= x^3 + x^2 + 2x - 6$

61.
$$\begin{array}{r l}
 & x^2 - 2x + 1 + \dfrac{2}{x+2} \\
x + 2 \,\big)\!\!\! & \overline{x^3 + 0x^2 - 3x + 4} \\
 & \underline{x^3 + 2x^2} \\
 & \quad -2x^2 - 3x \\
 & \quad \underline{-2x^2 - 4x} \\
 & \qquad\qquad x + 4 \\
 & \qquad\qquad \underline{x + 2} \\
 & \qquad\qquad\quad 2
\end{array}$$

Thus, $\dfrac{x^3 - 3x + 4}{x + 2} = x^2 - 2x + 1 + \dfrac{2}{x + 2}$.

Section 10.5 Rotation of Conics

- The general second-degree equation $Ax^2 + Bxy + Cy^2 + Dx + Ey + F = 0$ can be rewritten as $A'(x')^2 + C'(y')^2 + D'x' + E'y' + F' = 0$ by rotating the coordinate axes through the angle θ, where $\cot 2\theta = (A - C)/B$.
- $x = x'\cos\theta - y'\sin\theta$
  $y = x'\sin\theta + y'\cos\theta$
- The graph of the nondegenerate equation $Ax^2 + Bxy + Cy^2 + Dx + Ey + F = 0$ is:
  (a) An ellipse or circle if $B^2 - 4AC < 0$.
  (b) A parabola if $B^2 - 4AC = 0$.
  (c) A hyperbola if $B^2 - 4AC > 0$.

Solutions to Odd-Numbered Exercises

1. $\theta = 90°$; Point: $(0, 3)$

$x' = x\cos\theta - y\sin\theta = 0(\cos 90°) - 3(\sin 90°) = -3$

$y' = x\sin\theta + y\cos\theta = 0(\sin 90°) + 3(\cos 90°) = 0$

Thus, $(x', y') = (-3, 0)$.

3. $\theta = 30°$; Point: $(1, 4)$

$$x' = x\cos\theta - y\sin\theta = 1(\cos 30°) - 4(\sin 30°) = \frac{\sqrt{3}}{2} - \frac{4}{2} = \frac{1}{2}(\sqrt{3} - 4)$$

$$y' = x\sin\theta + y\cos\theta = 1(\sin 30°) + 4(\cos 30°) = \frac{1}{2} + \frac{4\sqrt{3}}{2} = \frac{1}{2}(1 + 4\sqrt{3})$$

Thus, $(x', y') = \left(\frac{1}{2}(\sqrt{3} - 4), \frac{1}{2}(1 + 4\sqrt{3})\right)$.

5. $xy + 1 = 0$

$A = 0, B = 1, C = 0$

$$\cot 2\theta = \frac{A - C}{B} = 0 \Rightarrow 2\theta = \frac{\pi}{2} \Rightarrow \theta = \frac{\pi}{4}$$

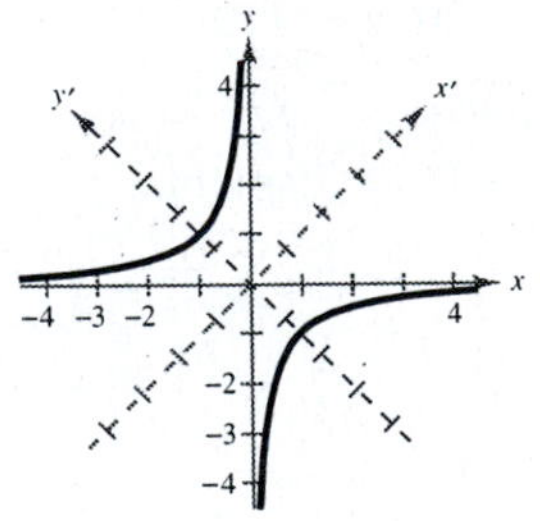

$$x = x'\cos\frac{\pi}{4} - y'\sin\frac{\pi}{4} = x'\left(\frac{\sqrt{2}}{2}\right) - y'\left(\frac{\sqrt{2}}{2}\right) = \frac{x' - y'}{\sqrt{2}}$$

$$y = x'\sin\frac{\pi}{4} + y'\cos\frac{\pi}{4} = x'\left(\frac{\sqrt{2}}{2}\right) + y'\left(\frac{\sqrt{2}}{2}\right) = \frac{x' + y'}{\sqrt{2}}$$

$$xy + 1 = 0$$

$$\left(\frac{x' - y'}{\sqrt{2}}\right)\left(\frac{x' + y'}{\sqrt{2}}\right) + 1 = 0$$

$$\frac{(y')^2}{2} - \frac{(x')^2}{2} = 1$$

7. $x^2 - 10xy + y^2 + 1 = 0$

$A = 1, B = -10, C = 1$

$$\cot 2\theta = \frac{A - C}{B} = 0 \Rightarrow 2\theta = \frac{\pi}{2} \Rightarrow \theta = \frac{\pi}{4}$$

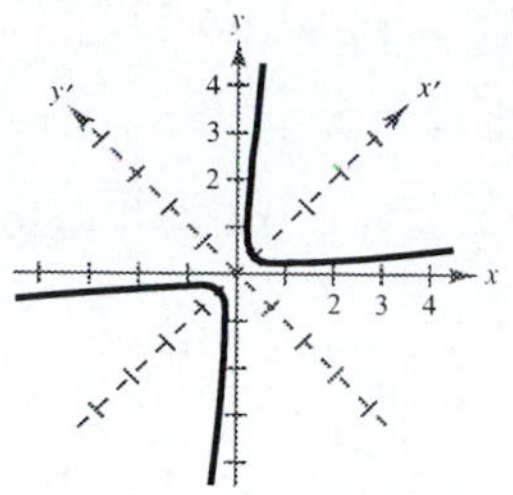

$$x = x'\cos\frac{\pi}{4} - y'\sin\frac{\pi}{4} = x'\left(\frac{\sqrt{2}}{2}\right) - y'\left(\frac{\sqrt{2}}{2}\right) = \frac{x' - y'}{\sqrt{2}}$$

$$y = x'\sin\frac{\pi}{4} + y'\cos\frac{\pi}{4} = x'\left(\frac{\sqrt{2}}{2}\right) + y'\left(\frac{\sqrt{2}}{2}\right) = \frac{x' + y'}{\sqrt{2}}$$

$x^2 - 10xy + y^2 + 1 = 0$

$$\left(\frac{x' - y'}{\sqrt{2}}\right)^2 - 10\left(\frac{x' - y'}{\sqrt{2}}\right)\left(\frac{x' + y'}{\sqrt{2}}\right) + \left(\frac{x' + y'}{\sqrt{2}}\right)^2 + 1 = 0$$

$$\frac{(x')^2}{2} - x'y' + \frac{(y')^2}{2} - 5(x')^2 + 5(y')^2 + \frac{(x')^2}{2} + x'y' + \frac{(y')^2}{2} + 1 = 0$$

$$-4(x')^2 + 6(y')^2 = -1$$

$$\frac{(x')^2}{1/4} - \frac{(y')^2}{1/6} = 1$$

9. $xy - 2y - 4x = 0$

$A = 0, B = 1, C = 0$

$$\cot 2\theta = \frac{A - C}{B} = 0 \Rightarrow 2\theta = \frac{\pi}{2} \Rightarrow \theta = \frac{\pi}{4}$$

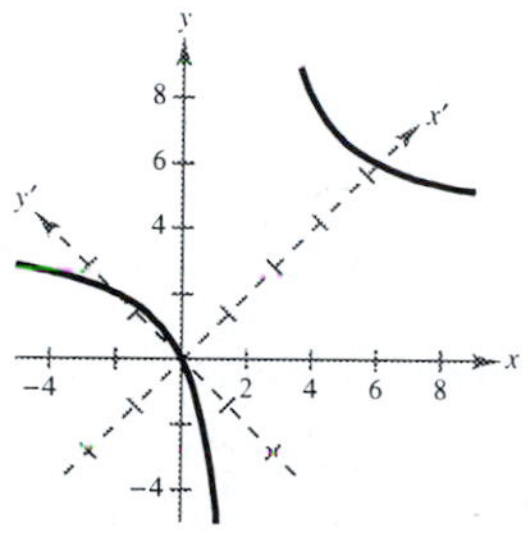

$$x = x'\cos\frac{\pi}{4} - y'\sin\frac{\pi}{4} = x'\left(\frac{\sqrt{2}}{2}\right) - y'\left(\frac{\sqrt{2}}{2}\right) = \frac{x' - y'}{\sqrt{2}}$$

$$y = x'\sin\frac{\pi}{4} + y'\cos\frac{\pi}{4} = x'\left(\frac{\sqrt{2}}{2}\right) + y'\left(\frac{\sqrt{2}}{2}\right) = \frac{x' + y'}{\sqrt{2}}$$

$xy - 2y - 4x = 0$

$$\left(\frac{x' - y'}{\sqrt{2}}\right)\left(\frac{x' + y'}{\sqrt{2}}\right) - 2\left(\frac{x' + y'}{\sqrt{2}}\right) - 4\left(\frac{x' - y'}{\sqrt{2}}\right) = 0$$

$$\frac{(x')^2}{2} - \frac{(y')^2}{2} - \sqrt{2}x' - \sqrt{2}y' - 2\sqrt{2}x' + 2\sqrt{2}y' = 0$$

$$\left[(x')^2 - 6\sqrt{2}x' + (3\sqrt{2})^2\right] - \left[(y')^2 - 2\sqrt{2}y' + (\sqrt{2})^2\right] = 0 + (3\sqrt{2})^2 - (\sqrt{2})^2$$

$$(x' - 3\sqrt{2})^2 - (y' - \sqrt{2})^2 = 16$$

$$\frac{(x' - 3\sqrt{2})^2}{16} - \frac{(y' - \sqrt{2})^2}{16} = 1$$

11. $5x^2 - 2xy + 5y^2 - 12 = 0$

$A = 5, B = -2, C = 5$

$\cot 2\theta = \dfrac{A - C}{B} = 0 \Rightarrow 2\theta = \dfrac{\pi}{2} \Rightarrow \theta = \dfrac{\pi}{4}$

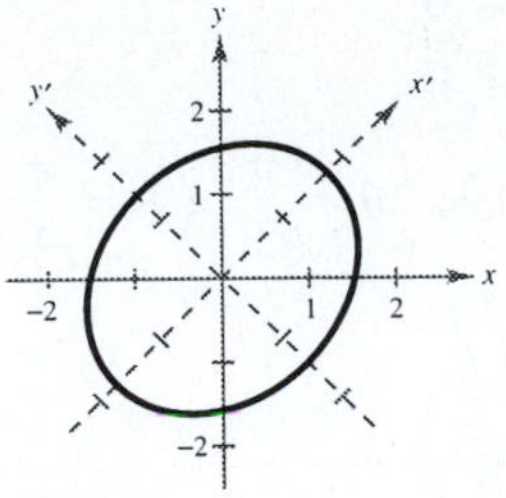

$$x = x'\cos\frac{\pi}{4} - y'\sin\frac{\pi}{4} = x'\left(\frac{\sqrt{2}}{2}\right) - y'\left(\frac{\sqrt{2}}{2}\right) = \frac{x' - y'}{\sqrt{2}}$$

$$y = x'\sin\frac{\pi}{4} + y'\cos\frac{\pi}{4} = x'\left(\frac{\sqrt{2}}{2}\right) + y'\left(\frac{\sqrt{2}}{2}\right) = \frac{x' + y'}{\sqrt{2}}$$

$$5x^2 - 2xy + 5y^2 - 12 = 0$$

$$5\left(\frac{x' - y'}{\sqrt{2}}\right)^2 - 2\left(\frac{x' - y'}{\sqrt{2}}\right)\left(\frac{x' + y'}{\sqrt{2}}\right) + 5\left(\frac{x' + y}{\sqrt{2}}\right)^2 - 12 = 0$$

$$\frac{5(x')^2}{2} - 5x'y' + \frac{5(y')^2}{2} - (x')^2 + (y')^2 + \frac{5(x')^2}{2} + 5x'y' + \frac{5(y')^2}{2} - 12 = 0$$

$$4(x')^2 + 6(y')^2 = 12$$

$$\frac{(x')^2}{3} + \frac{(y')^2}{2} = 1$$

13. $3x^2 - 2\sqrt{3}xy + y^2 + 2x + 2\sqrt{3}y = 0$

$A = 3, B = -2\sqrt{3}, C = 1$

$\cot 2\theta = \dfrac{A - C}{B} = -\dfrac{1}{\sqrt{3}} \Rightarrow \theta = 60°$

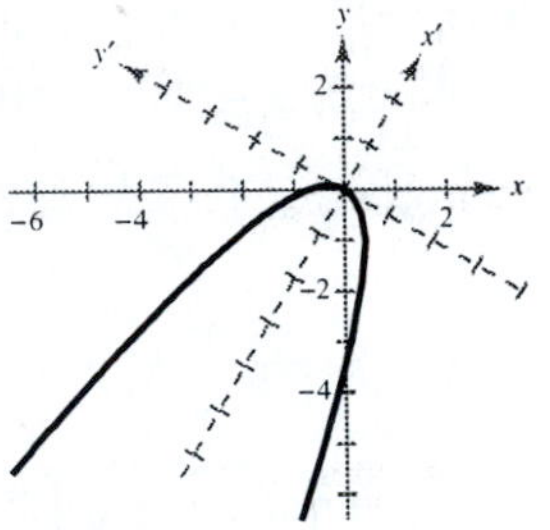

$$x = x'\cos 60° - y'\sin 60° = x'\left(\frac{1}{2}\right) - y'\left(\frac{\sqrt{3}}{2}\right) = \frac{x' - \sqrt{3}y'}{2}$$

$$y = x'\sin 60° + y'\cos 60° = x'\left(\frac{\sqrt{3}}{2}\right) + y'\left(\frac{1}{2}\right) = \frac{\sqrt{3}x' - y'}{2}$$

$$3x^2 - 2\sqrt{3}xy + y^2 + 2x + 2\sqrt{3}y = 0$$

$$3\left(\frac{x' - \sqrt{3}y'}{2}\right)^2 - 2\sqrt{3}\left(\frac{x' - \sqrt{3}y'}{2}\right)\left(\frac{\sqrt{3}x' + y'}{2}\right) + \left(\frac{\sqrt{3}x' + y'}{2}\right)^2 + 2\left(\frac{x' - \sqrt{3}y'}{2}\right) + 2\sqrt{3}\left(\frac{\sqrt{3}x' + y'}{2}\right) = 0$$

$$\frac{3(x')^2}{4} - \frac{6\sqrt{3}x'y'}{4} + \frac{9(y')^2}{4} - \frac{6(x')^2}{4} + \frac{4\sqrt{3}x'y'}{4} + \frac{6(y')^2}{4} + \frac{3(x')^2}{4} + \frac{2\sqrt{3}x'y'}{4} + \frac{(y')^2}{4}$$

$$+ x' - \sqrt{3}y' + 3x' + \sqrt{3}y' = 0$$

$$4(y')^2 + 4x' = 0$$

$$x' = -(y')^2$$

15. $9x^2 + 24xy + 16y^2 + 90x - 130y = 0$

$A = 9, B = 24, C = 16$

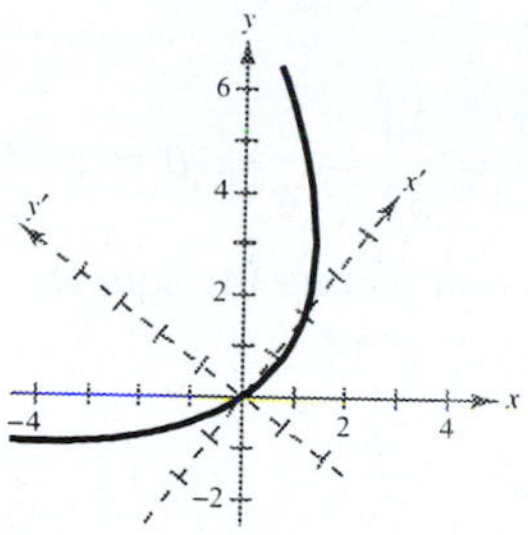

$$\cot 2\theta = \frac{A - C}{B} = -\frac{7}{24} \Rightarrow \theta \approx 53.13°$$

$$\cos 2\theta = -\frac{7}{25}$$

$$\sin \theta = \sqrt{\frac{1 - \cos \theta}{2}} = \sqrt{\frac{1 - (-7/25)}{2}} = \frac{4}{5}$$

$$\cos \theta = \sqrt{\frac{1 + \cos 2\theta}{2}} = \sqrt{\frac{1 + (-7/25)}{2}} = \frac{3}{5}$$

$$x = x' \cos \theta - y' \sin \theta = x'\left(\frac{3}{5}\right) - y'\left(\frac{4}{5}\right) = \frac{3x' - 4y'}{5}$$

$$y = x' \sin \theta + y' \cos \theta = x'\left(\frac{4}{5}\right) + y'\left(\frac{3}{5}\right) = \frac{4x' + 3y'}{5}$$

$$9x^2 + 24xy + 16y^2 + 90x - 130y = 0$$

$$9\left(\frac{3x' - 4y'}{5}\right)^2 + 24\left(\frac{3x' - 4y'}{5}\right)\left(\frac{4x' + 3y'}{5}\right) + 16\left(\frac{3x' - 4y'}{5}\right)^2 + 90\left(\frac{3x' - 4y'}{5}\right) - 130\left(\frac{4x' + 3y'}{5}\right) = 0$$

$$\frac{81(x')^2}{25} - \frac{216x'y'}{25} + \frac{144(y')^2}{25} + \frac{288(x')^2}{25} - \frac{168x'y'}{25} - \frac{288(y')^2}{25} + \frac{256(x')^2}{25} + \frac{384x'y'}{25}$$

$$+ \frac{144(y')^2}{25} + 54x' - 72y' - 104x' - 78y' = 0$$

$$25(x')^2 - 50x' - 150y' = 0$$

$$(x')^2 - 2x' + 1 = 6y' + 1$$

$$y' = \frac{(x')^2}{6} - \frac{x'}{3}$$

17. $x^2 + xy + y^2 = 10$

$$\cot 2\theta = \frac{A - C}{B} = \frac{1 - 1}{0} = 0 \Rightarrow \theta = \frac{\pi}{4} \text{ or } 45°$$

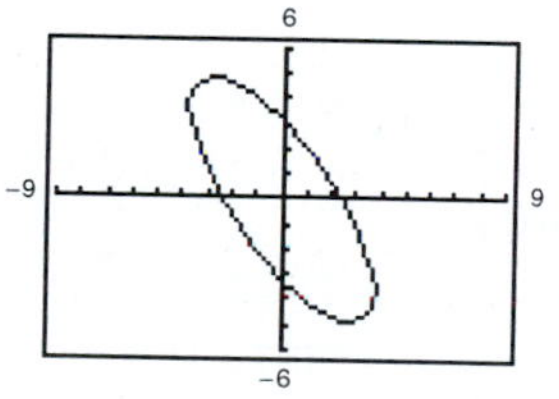

To graph the conic using a graphing calculator, we need to solve for y in terms of x.

$$y^2 + xy = 10 - x^2$$

$$y^2 + xy + \frac{x^2}{4} = 10 - x^2 + \frac{x^2}{4}$$

$$\left(y + \frac{x}{2}\right)^2 = \frac{40 - 3x^2}{4}$$

$$y = -\frac{x}{2} \pm \frac{\sqrt{40 - 3x^2}}{2}$$

Enter $y_1 = \dfrac{-x + \sqrt{40 - 3x^2}}{2}$

and $y_2 = \dfrac{-x - \sqrt{40 - 3x^2}}{2}$.

19. $17x^2 + 32xy - 7y^2 = 75$

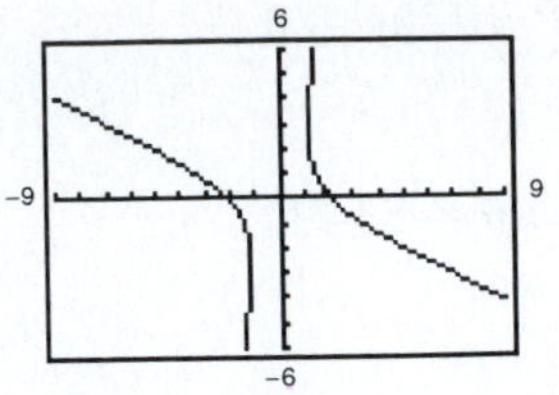

$$\cot 2\theta = \frac{A - C}{B} = \frac{17 + 7}{32} = \frac{24}{32} = \frac{3}{4} \Rightarrow \theta \approx 26.57°$$

Solve for y in terms of x by completing the square.

$$-7y^2 + 32xy = -17x^2 + 75$$

$$y^2 - \frac{32}{7}xy = \frac{17}{7}x^2 - \frac{75}{7}$$

$$y^2 - \frac{32}{7}xy + \frac{256}{49}x^2 = \frac{119}{49}x^2 - \frac{525}{49} + \frac{256}{49}x^2$$

$$\left(y - \frac{16}{7}x\right)^2 = \frac{375x^2 - 525}{49}$$

$$y = \frac{16}{7}x \pm \sqrt{\frac{375x^2 - 525}{49}}$$

$$y = \frac{16x \pm 5\sqrt{15x^2 - 21}}{7}$$

Use $y_1 = \dfrac{16x + 5\sqrt{15x^2 - 21}}{7}$

and $y_2 = \dfrac{16x - 5\sqrt{15x^2 - 21}}{7}$.

21. $32x^2 + 50xy + 7y^2 = 52$

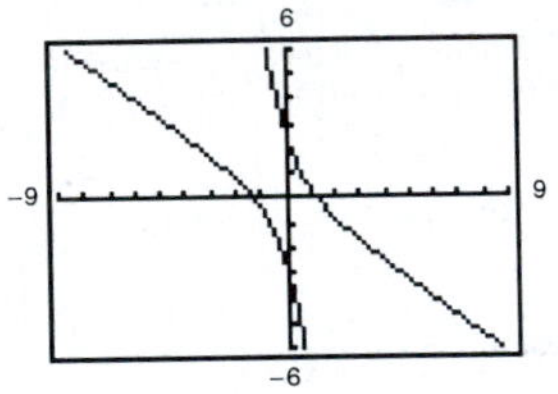

$$\cot 2\theta = \frac{A - C}{B} = \frac{32 - 7}{50} = \frac{1}{2} \Rightarrow \theta \approx 31.72°$$

Solve for y in terms of x by completing the square.

$$7y^2 + 50xy = 52 - 32x^2$$

$$y^2 + \frac{50}{7}xy = \frac{52 - 32x^2}{7}$$

$$y^2 + \frac{50}{7}xy + \frac{625}{49}x^2 = \frac{52 - 32x^2}{7} + \frac{625x^2}{49}$$

$$\left(y + \frac{25}{7}x\right)^2 = \frac{364 + 401x^2}{49}$$

$$y = -\frac{25x}{7} \pm \frac{\sqrt{364 + 401x^2}}{7}$$

Enter $y_1 = \dfrac{-25x + \sqrt{364 + 401x^2}}{7}$

and $y_2 = \dfrac{-25x - \sqrt{364 + 401x^2}}{7}$.

23. $xy + 3 = 0$

$B^2 - 4AC = 1 \Rightarrow$ The graph is a hyperbola.

$$\cot 2\theta = \frac{A - C}{B} = 0 \Rightarrow \theta = 45°$$

Matches graph (e).

25. $-2x^2 + 3xy + 2y^2 + 3 = 0$

$B^2 - 4AC = (3)^2 - 4(-2)(2) = 25 \Rightarrow$ The graph is a hyperbola.

$\cot 2\theta = \dfrac{A - C}{B} = -\dfrac{4}{3} \Rightarrow \theta \approx -18.43°$

Matches graph (b).

27. $3x^2 + 2xy + y^2 - 10 = 0$

$B^2 - 4AC = (2)^2 - 4(3)(1) = -8 \Rightarrow$ The graph is an ellipse or circle.

$\cot 2\theta = \dfrac{A - C}{B} = 1 \Rightarrow \theta = 22.5°$

Matches graph (d).

29. $16x^2 - 24xy + 9y^2 - 30x - 40y = 0$

$B^2 - 4AC = (-24)^2 - 4(16)(9) = 0$

Parabola

31. $13x^2 - 8xy + 7y^2 - 45 = 0$

$B^2 - 4AC = (-8)^2 - 4(13)(7) = -300$

Ellipse or circle

33. $x^2 - 6xy - 5y^2 + 4x - 22 = 0$

$B^2 - 4AC = (-6)^2 - 4(1)(-5) = 56$

Hyperbola

35. $x^2 + 4xy + 4y^2 - 5x - y - 3 = 0$

$B^2 - 4AC = (4)^2 - 4(1)(4) = 0$

Parabola

37. $y^2 - 4x^2 = 0$

$y^2 = 4x^2$

$y = \pm 2x$

Two intersecting lines

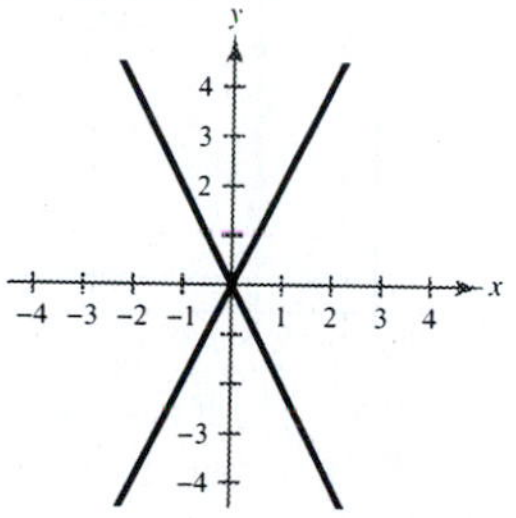

39. $x^2 + 2xy + y^2 - 1 = 0$

$(x + y)^2 - 1 = 0$

$(x + y)^2 = 1$

$x + y = \pm 1$

$y = -x \pm 1$

Two parallel lines

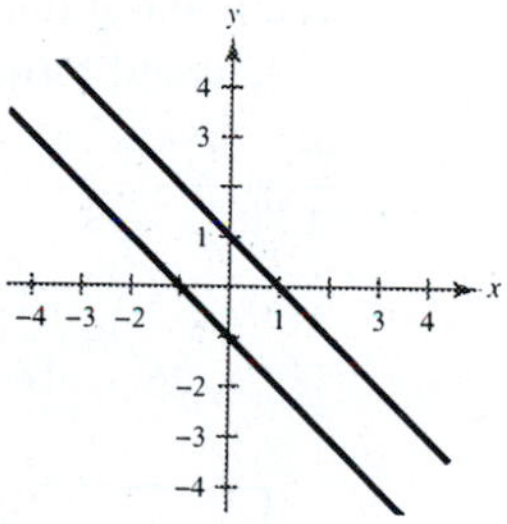

41.

$-x^2 + y^2 + 4x - 6y + 4 = 0 \Rightarrow (y - 3)^2 - (x - 2)^2 = 1$

$x^2 + y^2 - 4x - 6y + 12 = 0 \Rightarrow (x - 2)^2 + (y - 3)^2 = 1$

$2y^2 - 12y + 16 = 0$

$2(y - 2)(y - 4) = 0$

$y = 2$ or $y = 4$

For $y = 2$: $x^2 + 2^2 - 4x - 6(2) + 12 = 0$

$x^2 - 4x + 4 = 0$

$(x - 2)^2 = 0$

$x = 2$

For $y = 4$: $x^2 + 4^2 - 4x - 6(4) + 12 = 0$

$x^2 - 4x + 4 = 0$

$(x - 2)^2 = 0$

$x = 2$

The points of intersection are (2, 2) and (2, 4).

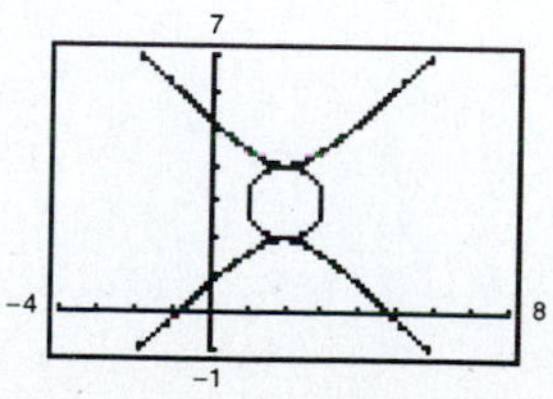

43.
$$\begin{array}{r} -4x^2 - y^2 - 32x + 24y - \ \ 64 = 0 \\ 4x^2 + y^2 + 56x - 24y + 304 = 0 \\ \hline 24x \qquad\qquad + 240 = 0 \\ x = -10 \end{array}$$

When $x = -10$: $4(-10)^2 + y^2 + 56(-10) - 24y + 304 = 0$

$$y^2 - 24y + 144 = 0$$
$$(y - 12)^2 = 0$$
$$y = 12$$

The point of intersection is $(-10, 12)$.
In standard form the equations are:

$$\frac{(x + 4)^2}{36} + \frac{(y - 12)^2}{144} = 1$$
$$\frac{(x + y)^2}{9} + \frac{(y - 12)^2}{36} = 1$$

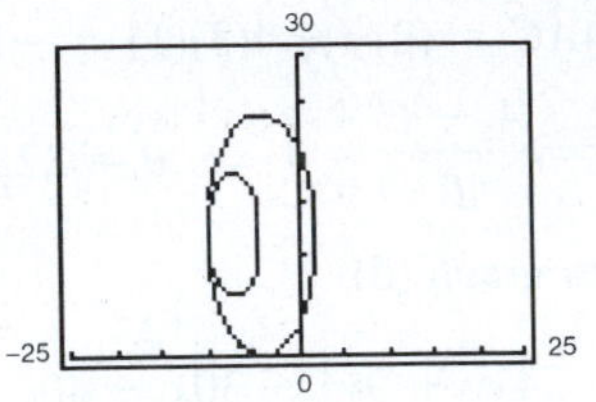

45.
$$\begin{array}{r} x^2 - y^2 - 12x + 12y - 36 = 0 \\ x^2 + y^2 - 12x - 12y + 36 = 0 \\ \hline 2x^2 - 24x = 0 \\ 2x(x - 12) = 0 \\ x = 0 \text{ or } x = 12 \end{array}$$

When $x = 0$: $y^2 - 12y + 36 = 0$

$$(y - 6)^2 = 0$$
$$y = 6$$

When $x = 12$: $12^2 + y^2 - 12(12) - 12y + 36 = 0$

$$y^2 - 12y + 36 = 0$$
$$(y - 6)^2 = 0$$
$$y = 6$$

The points of intersection are $(0, 6)$ and $(12, 6)$.
In standard form the equations are:

$$\frac{(x - 6)^2}{36} + \frac{(y - 6)^2}{36} = 1$$
$$(x - 6)^2 + (y - 6)^2 = 36$$

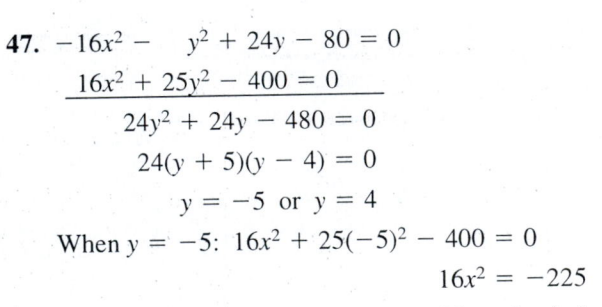

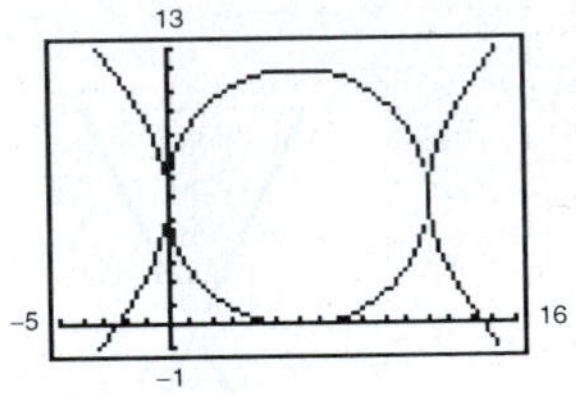

47.
$$\begin{array}{r} -16x^2 - \quad y^2 + 24y - 80 = 0 \\ 16x^2 + 25y^2 - 400 = 0 \\ \hline 24y^2 + 24y - 480 = 0 \\ 24(y + 5)(y - 4) = 0 \\ y = -5 \text{ or } y = 4 \end{array}$$

When $y = -5$: $16x^2 + 25(-5)^2 - 400 = 0$

$$16x^2 = -225$$

No real solution

When $y = 4$: $16x^2 + 25(4)^2 - 400 = 0$

$$16x^2 = 0$$
$$x = 0$$

The point of intersection is $(0, 4)$.
In standard form the equations are:

$$\frac{x^2}{4} + \frac{(y - 12)^2}{64} = 1$$
$$\frac{x^2}{25} + \frac{y^2}{16} = 1$$

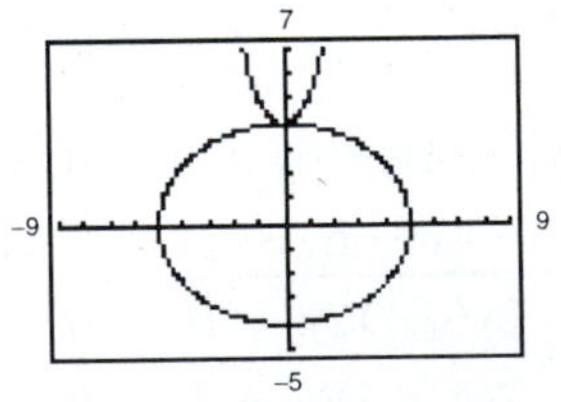

49. $x^2 + y^2 - 25 = 0 \Rightarrow y^2 = 25 - x^2$

$$9x - 4y^2 = 0 \Rightarrow 9x - 4(25 - x^2) = 0$$
$$4x^2 + 9x - 100 = 0$$
$$(4x + 25)(x - 4) = 0$$
$$x = -\tfrac{25}{4} \text{ or } x = 4$$

When $x = -\frac{25}{4}$: $y^2 = 25 - \left(-\frac{25}{4}\right)^2$

$$y^2 = -\tfrac{225}{16}$$

No real solution

When $x = 4$: $y^2 = 25 - 4^2$

$$y^2 = 9$$
$$y = \pm 3$$

The points of intersection are $(4, 3)$, $(4, -3)$.
In standard form the equations are:

$$x^2 + y^2 = 25$$
$$y^2 = \tfrac{9}{4}x$$

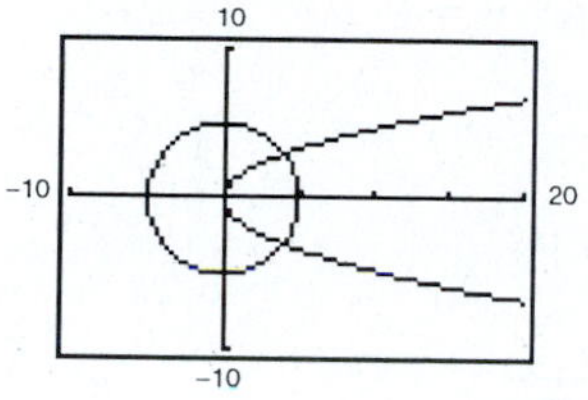

51.

$$x^2 + 2y^2 - 4x + 6y - 5 = 0$$
$$x + y + 5 = 0 \Rightarrow y = -x - 5$$
$$x^2 + 2(-x - 5)^2 - 4x + 6(-x - 5) - 5 = 0$$
$$x^2 + 2x^2 + 20x + 50 - 4x - 6x - 30 - 5 = 0$$
$$3x^2 + 10x + 15 = 0$$

No real solution

No points of intersection
In standard form we have:

$$\frac{(x-2)^2}{27/2} + \frac{(y+3/2)^2}{27/4} = 1$$
$$x + y = -5$$

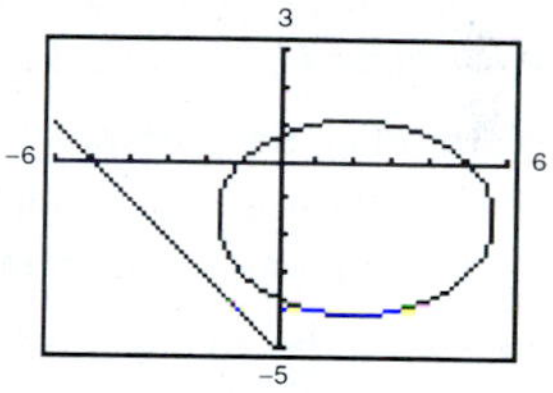

53.

$$xy + x - 2y + 3 = 0 \Rightarrow y = \frac{-x-3}{x-2}$$
$$x^2 + 4y^2 - 9 = 0$$
$$x^2 + 4\left(\frac{-x-3}{x-2}\right)^2 = 9$$
$$x^2(x-2)^2 + 4(-x-3)^2 = 9(x-2)^2$$
$$x^2(x^2 - 4x + 4) + 4(x^2 + 6x + 9) = 9(x^2 - 4x + 4)$$
$$x^4 - 4x^3 + 4x^2 + 4x^2 + 24x + 36 = 9x^2 - 36x + 36$$
$$x^4 - 4x^3 - x^2 + 60x = 0$$
$$x(x+3)(x^2 - 7x + 20) = 0$$
$$x = 0 \text{ or } x = -3$$

Note: $x^2 - 7x + 20 = 0$ has no real solution.

When $x = 0$: $y = \dfrac{-0-3}{0-2} = \dfrac{3}{2}$

When $x = -3$: $y = \dfrac{-(-3)-3}{-3-2} = 0$

The points of intersection are $\left(0, \frac{3}{2}\right)$, $(-3, 0)$.

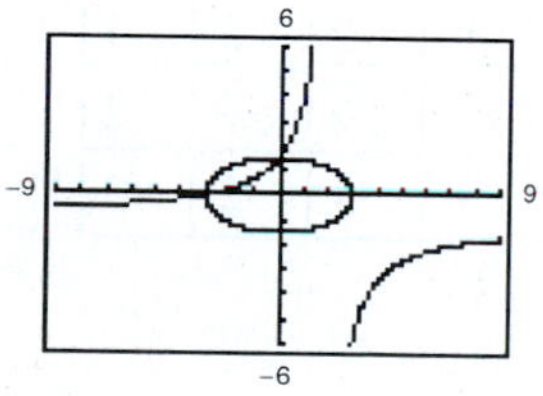

55. $(x') + (y')^2 = (x\cos\theta + y\sin\theta)^2 + (y\cos\theta - x\sin\theta)^2$

$$= x^2\cos^2\theta + 2xy\cos\theta\sin\theta + y^2\sin^2\theta + y^2\cos^2\theta - 2xy\cos\theta\sin\theta + x^2\sin^2\theta$$
$$= x^2(\cos^2\theta + \sin^2\theta) + y^2(\sin^2\theta + \cos^2\theta) = x^2 + y^2 = r^2$$

57. $g(x) = \dfrac{2}{2 - x}$

y-intercept: $(0, 1)$

Vertical asymptote: $x = 2$

Horizontal asymptote: $y = 0$

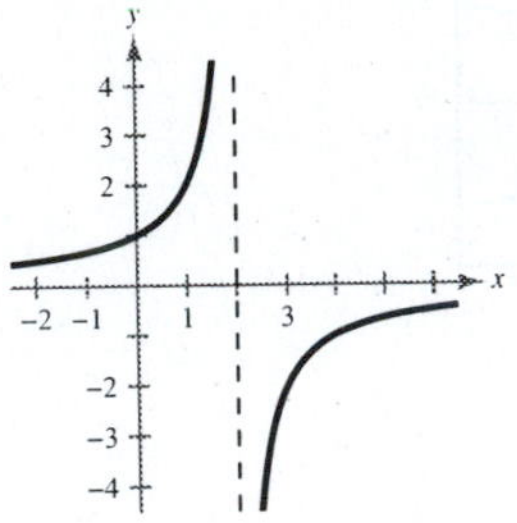

59. $h(t) = \dfrac{t^2}{2 - t} = -t - 2 + \dfrac{4}{2 - t}$

Intercept: $(0, 0)$

Vertical asymptote: $t = 2$

Slant asymptote: $y = -t - 2$

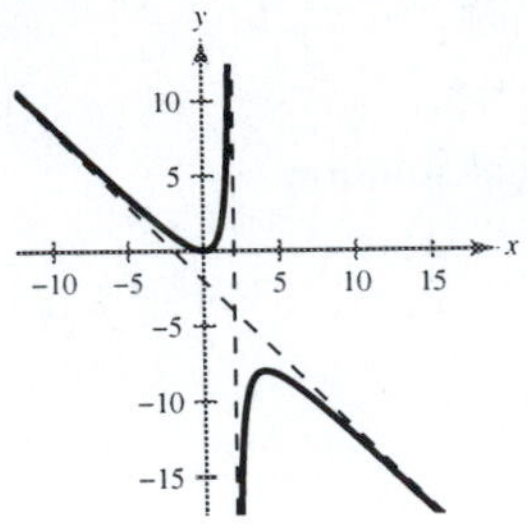

Section 10.6 Parametric Equations

- If f and g are continuous functions of t on an interval I, then the set of ordered pairs $(f(t), g(t))$ is a *plane curve* C. The equations $x = f(t)$ and $y = g(t)$ are *parametric equations* for C and t is the *parameter.*
- To eliminate the parameter:
  (a) Solve for t in one equation and substitute into the second equation.
  (b) Use trigonometric identities.
- You should be able to find the parametric equations for a graph.

Solutions to Odd-Numbered Exercises

1. $x = \sqrt{t}, y = 1 - t$

(a)

t	0	1	2	3	4
x	0	1	$\sqrt{2}$	$\sqrt{3}$	2
y	1	0	-1	-2	-3

(b)

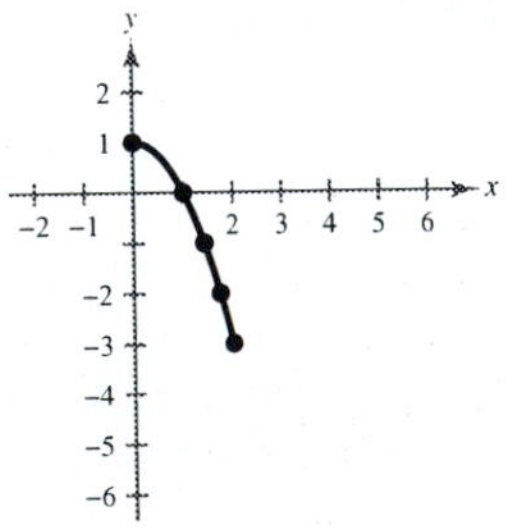

(c) $x = \sqrt{t} \Rightarrow x^2 = t$
$y = 1 - t \Rightarrow y = 1 - x^2$

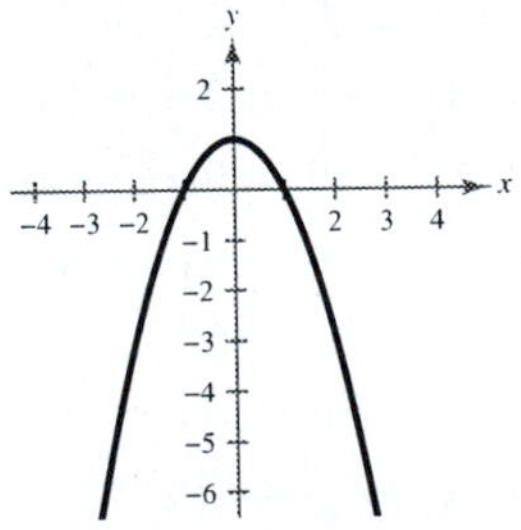

The graph of the parametric equations only shows the right half of the parabola, whereas the rectangular equation yields the entire parabola.

3. $x = t$

$y = -2t \Rightarrow y = -2x$

$2x + y = 0$

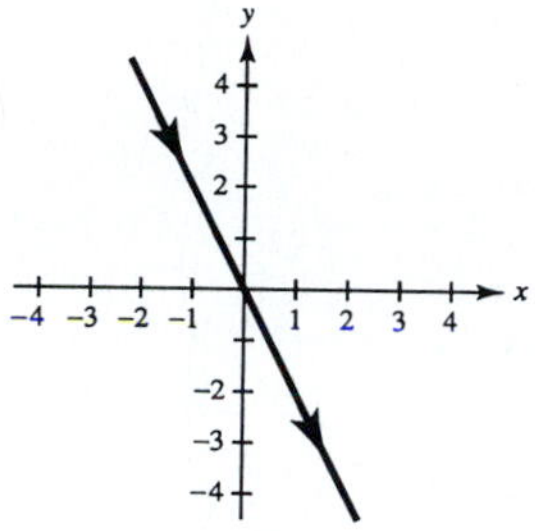

5. $x = 3t - 1 \Rightarrow t = \dfrac{x+1}{3}$

$y = 2t + 1 \Rightarrow y = 2\left(\dfrac{x+1}{3}\right) + 1$

$2x - 3y + 5 = 0$

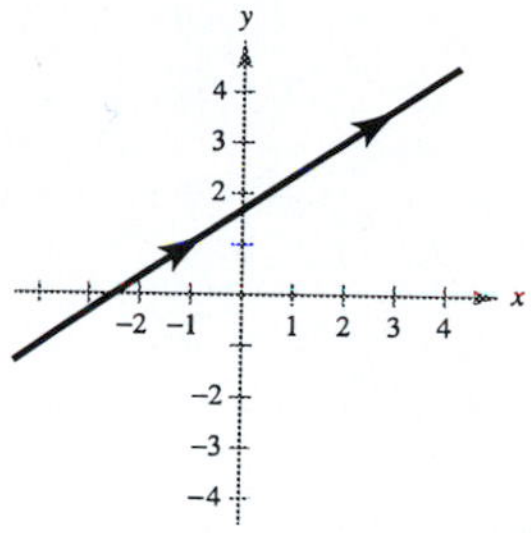

7. $x = \frac{1}{4}t \Rightarrow t = 4x$

$y = t^2 \Rightarrow y = 16x^2$

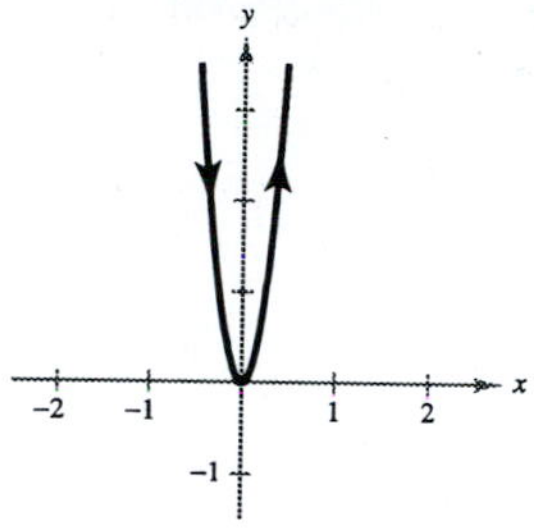

9. $x = t + 1 \Rightarrow t = x - 1$

$y = t^2 \Rightarrow y = (x-1)^2$

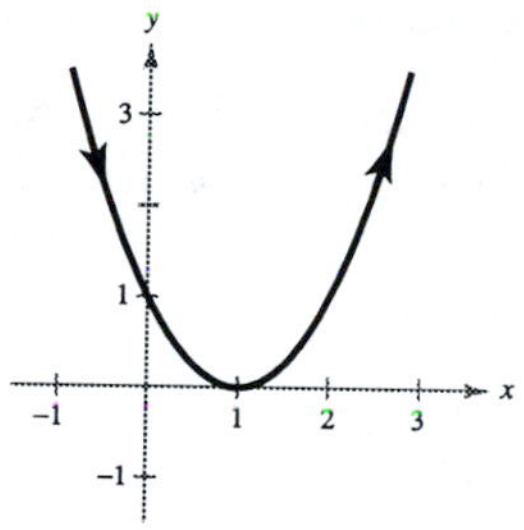

11. $x = t^3 \Rightarrow t = x^{1/3}$

$y = \frac{1}{2}t^2 \Rightarrow y = \frac{1}{2}x^{2/3}$

$y = \frac{1}{2}\sqrt[3]{x^2}$

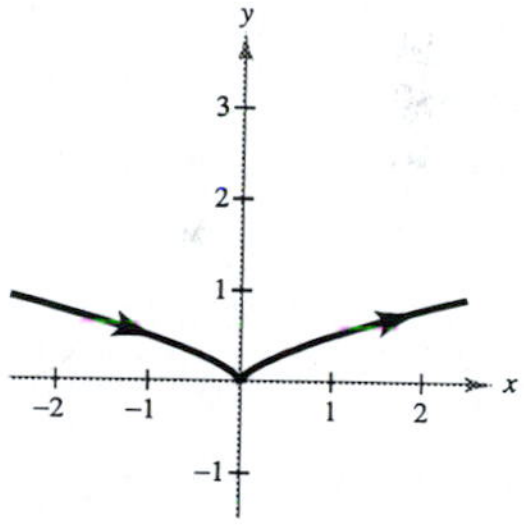

13. $x = 2t \Rightarrow t = \dfrac{x}{2}$

$y = |t - 2| \Rightarrow y = \left|\dfrac{x}{2} - 2\right|$

$y = \dfrac{1}{2}|x - 4|$

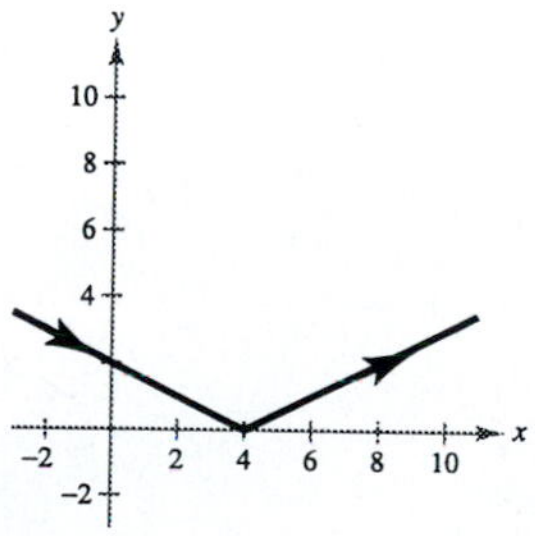

15. $x = 3\cos\theta \Rightarrow \left(\dfrac{x}{3}\right)^2 = \cos^2\theta$

$y = 3\sin\theta \Rightarrow \left(\dfrac{y}{3}\right)^2 = \sin^2\theta$

$\left(\dfrac{x}{3}\right)^2 + \left(\dfrac{y}{3}\right)^2 = 1$

$x^2 + y^2 = 9$

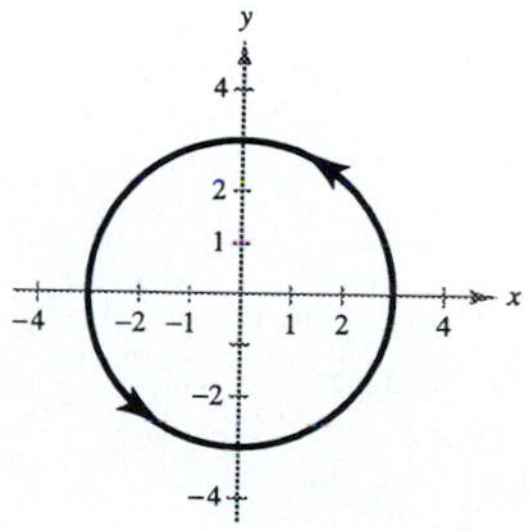

17. $x = 4\sin 2\theta \Rightarrow \left(\frac{x}{4}\right)^2 = \sin^2 2\theta$

$y = 2\cos 2\theta \Rightarrow \left(\frac{y}{2}\right)^2 = \cos^2 2\theta$

$\left(\frac{x}{4}\right)^2 + \left(\frac{y}{3}\right)^2 = 1$

$\frac{x^2}{16} + \frac{y^2}{4} = 1$

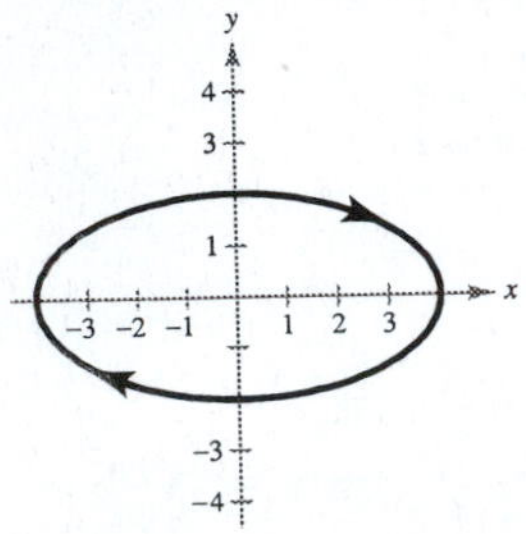

19. $x = 4 + 2\cos\theta \Rightarrow \left(\frac{x-4}{2}\right)^2 = \cos^2\theta$

$y = -1 + \sin\theta \Rightarrow (y+1)^2 = \sin^2\theta$

$\frac{(x-4)^2}{4} + \frac{(y+1)^2}{1} = 1$

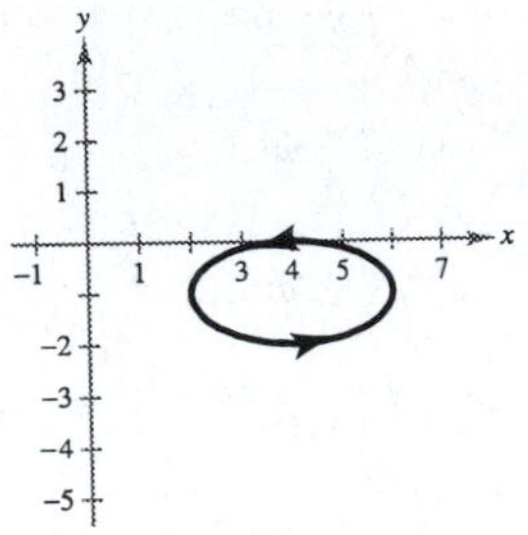

21. $x = 4 + 2\cos\theta \Rightarrow \left(\frac{x-4}{2}\right)^2 = \cos^2\theta$

$y = -1 + 4\sin\theta \Rightarrow \left(\frac{y+1}{4}\right)^2 = \sin^2\theta$

$\frac{(x-4)^2}{4} + \frac{(y+1)^2}{16} = 1$

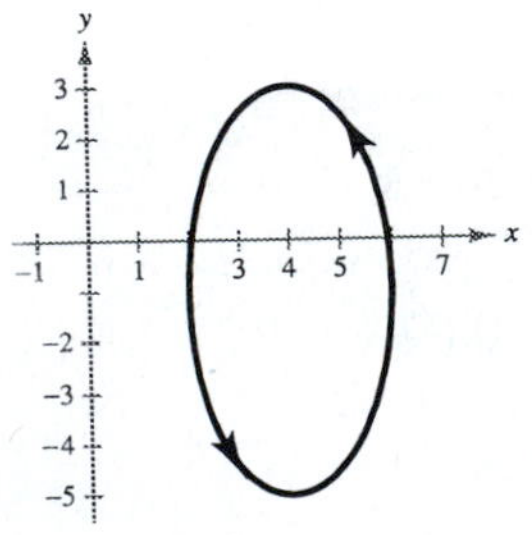

23. $x = 4\sec\theta \Rightarrow \left(\frac{x}{4}\right)^2 = \sec^2\theta$

$y = 3\tan\theta \Rightarrow \left(\frac{y}{3}\right)^2 = \tan^2\theta$

$1 + \left(\frac{y}{3}\right)^2 = \left(\frac{x}{4}\right)^2$

$\frac{x^2}{16} - \frac{y^2}{9} = 1$

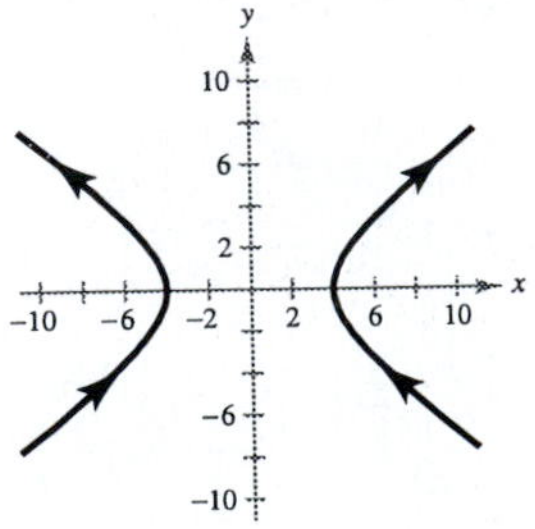

25. $x = e^{-1} \Rightarrow \frac{1}{x} = e^t$

$y = e^{3t} \Rightarrow y = (e^t)^3$

$y = \left(\frac{1}{x}\right)^3$

$y = \frac{1}{x^3},\ x > 0, y > 0$

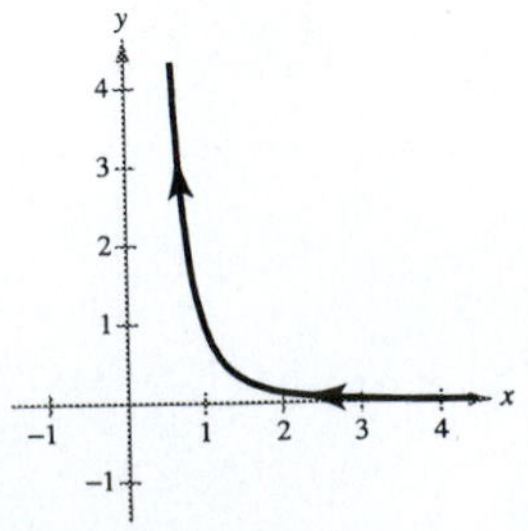

27. $x = t^3 \Rightarrow x^{1/3} = t$

$y = 3\ln t \Rightarrow y = \ln t^3$

$y = \ln(x^{1/3})^3$

$y = \ln x$

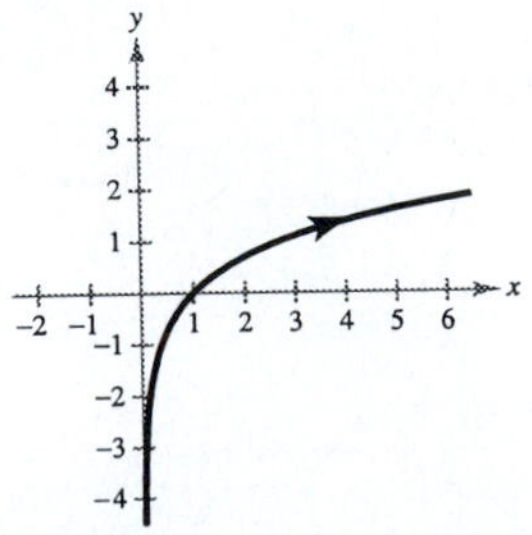

29. By eliminating the parameter, each curve becomes $y = 2x + 1$.

(a) $x = t$
$y = 2t + 1$
There are no restrictions on x and y.
Domain: $(-\infty, \infty)$
Orientation: Left to right

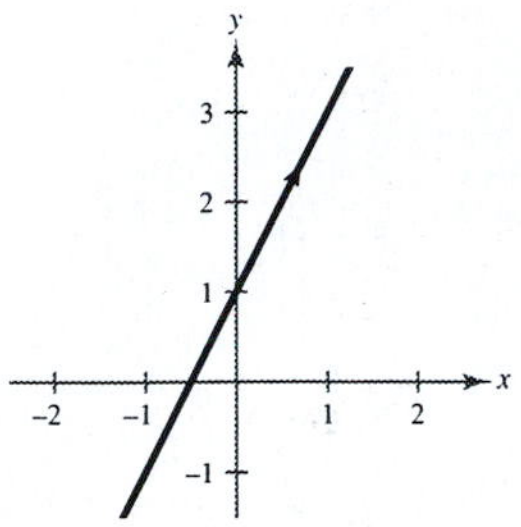

(b) $x = \cos\theta \Rightarrow -1 \le x \le 1$
$y = 2\cos\theta + 1 \Rightarrow -1 \le y \le 3$
The graph oscillates.
Domain: $[-1, 1]$
Orientation: Depends on θ

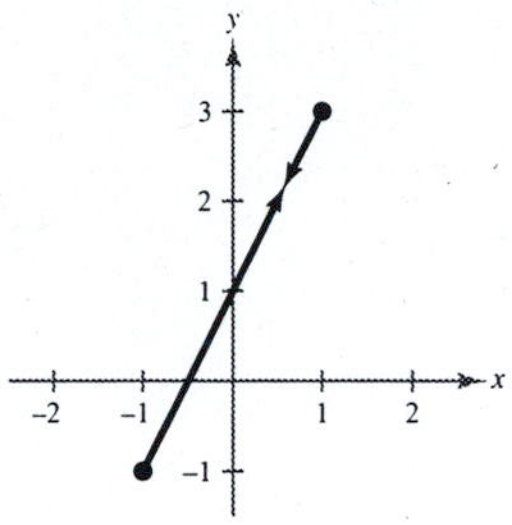

(c) $x = e^{-t} \Rightarrow x > 0$
$y = 2e^{-t} + 1 \Rightarrow y > 1$
Domain: $(0, \infty)$
Orientation: Downward or right to left

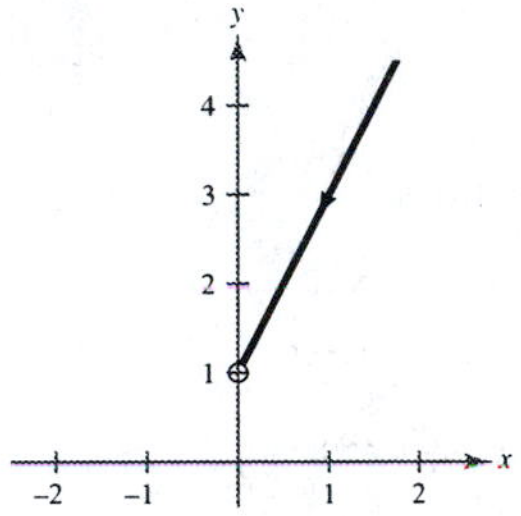

(d) $x = e^{t} \Rightarrow x > 0$
$y = 2e^{t} + 1 \Rightarrow y > 1$
Domain: $(0, \infty)$
Orientation: Upward or left to right

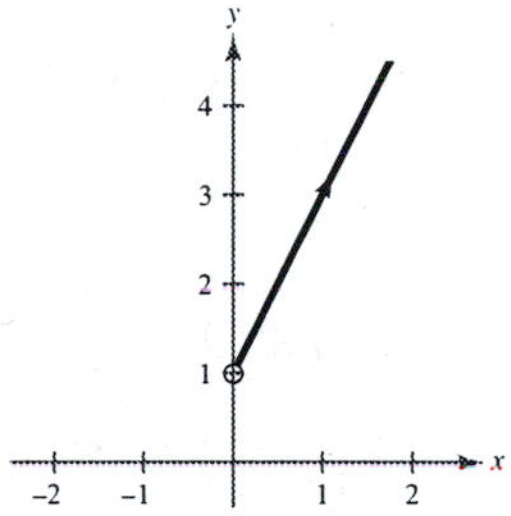

31. (a) $x = \cos\theta$
$y = \sin\theta$
$x^2 + y^2 = 1$

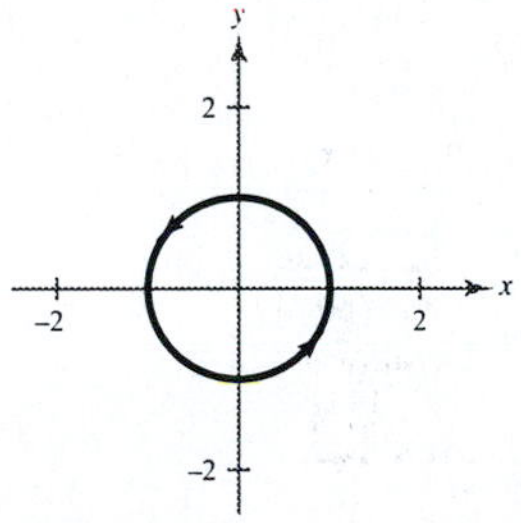

Oriented counterclockwise

(b) $x = \sin\theta$
$y = \cos\theta$
$x^2 + y^2 = 1$

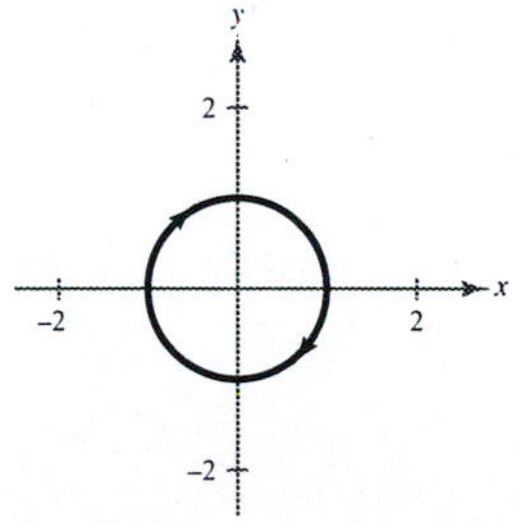

Oriented clockwise

—CONTINUED—

31. —CONTINUED—

(c) $x = \sin^2 \theta$

$y = \cos^2 \theta$

$x + y = 1, 0 \le x \le 1, 0 \le y \le 1$

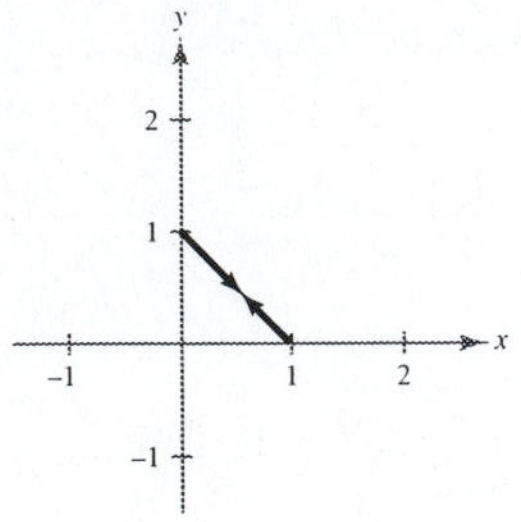

Oscillates

(d) $x = -\cos \theta$

$y = \sin \theta$

$x^2 + y^2 = 1$

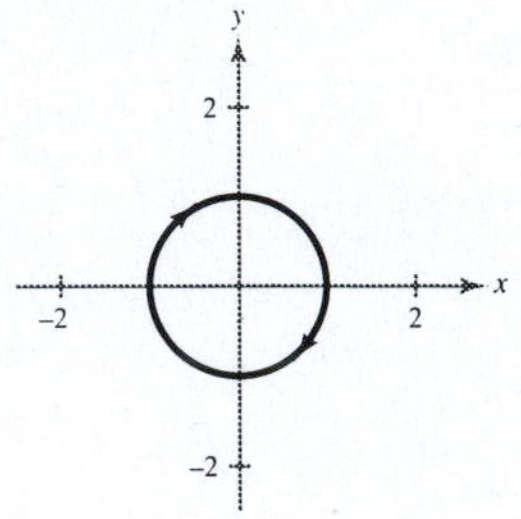

Oriented clockwise

33. $x = x_1 + t(x_2 - x_1), y = y_1 + t(y_2 - y_1)$

$$\frac{x - x_1}{x_2 - x_1} = t$$

$$y = y_1 + \left(\frac{x - x_1}{x_2 - x_1}\right)(y_2 - y_1)$$

$$y - y_1 = \frac{y_2 - y_1}{x_2 - x_1}(x - x_1) = m(x - x_1)$$

35. $x = h + a \cos \theta, y = k + b \sin \theta$

$$\frac{x - h}{a} = \cos \theta, \frac{y - k}{b} = \sin \theta$$

$$\frac{(x - h)^2}{a^2} + \frac{(y - k)^2}{b^2} = 1$$

37. From Exercise 33 we have:

$x = 0 + t(5 - 0) = 5t$

$y = 0 + t(-2 - 0) = -2t$

39. From Exercise 34 we have:

$x = 2 + 4 \cos \theta$

$y = 1 + 4 \sin \theta$

41. Vertices: $(\pm 5, 0) \Rightarrow (h, k) = (0, 0)$ and $a = 5$

Foci: $(\pm 4, 0) \Rightarrow c = 4$

$c^2 = a^2 - b^2 \Rightarrow 16 = 25 - b^2 \Rightarrow b = 3$

From Exercise 35 we have:

$x = 5 \cos \theta$

$y = 3 \sin \theta$

43. Vertices: $(\pm 4, 0) \Rightarrow (h, k) = (0, 0)$ and $a = 4$

Foci: $(\pm 5, 0) \Rightarrow c = 5$

$c^2 = a^2 + b^2 \Rightarrow 25 = 16 + b^2 \Rightarrow b = 3$

From Exercise 36 we have:

$x = 4 \sec \theta$

$y = 3 \tan \theta$

45. $y = 3x - 2$

Examples

$x = t, \quad y = 3t - 2$

$x = \frac{t}{3}, \quad y = t - 2$

$x = \frac{t + 2}{3}, \quad y = t$

$x = 2t, \quad y = 6t - 2$

47. $y = x^3$

Examples

$x = t, \quad y = t^3$

$x = \sqrt[3]{t}, \quad y = t$

$x = \tan t, \quad y = \tan^3 t$

49. $x = 2(\theta - \sin \theta)$

$y = 2(1 - \cos \theta)$

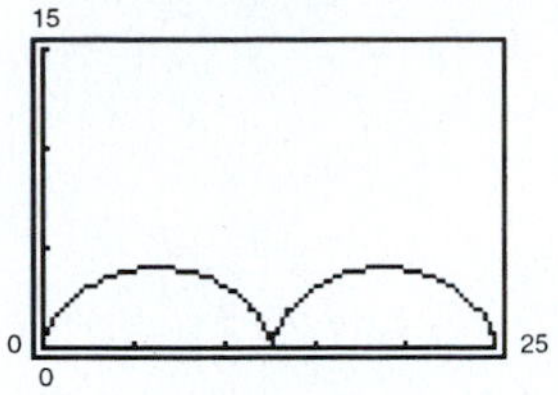

51. $x = \theta - \dfrac{3}{2}\sin \theta$

$y = 1 - \dfrac{3}{2}\cos \theta$

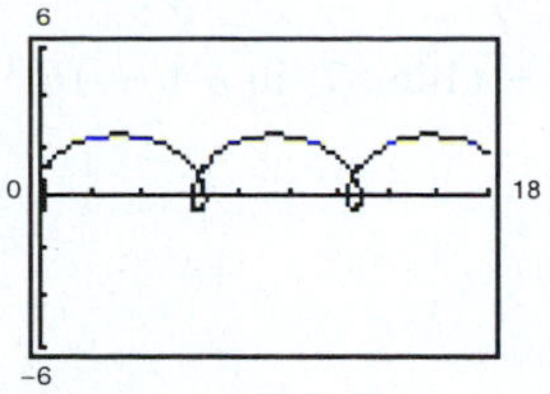

53. $x = 3\cos^3 \theta$

$y = 3\sin^3 \theta$

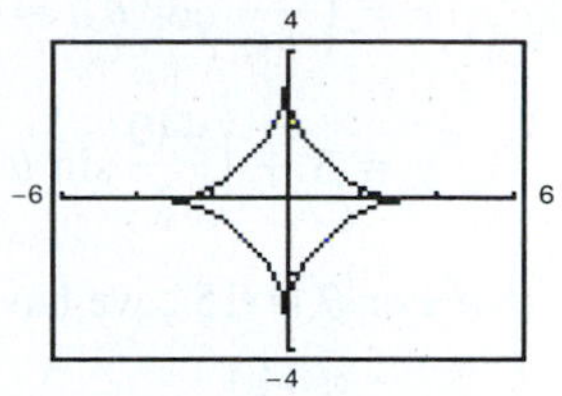

55. $x = 2\cot \theta$

$y = 2\sin^2 \theta$

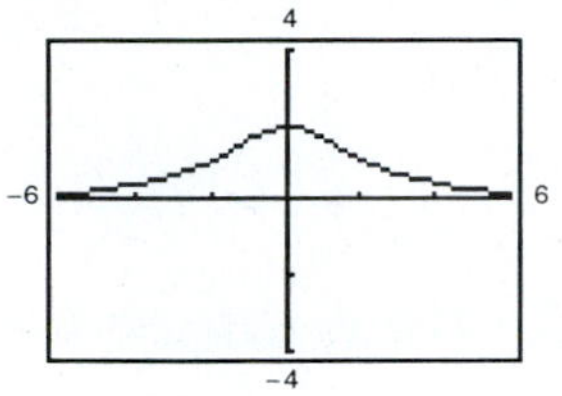

57. $x = 2\cos\theta \Rightarrow -2 \le x \le 2$

$y = \sin 2\theta \Rightarrow -1 \le y \le 1$

Matches graph (b).

Domain: $[-2, 2]$

Range: $[-1, 1]$

59. $x = \dfrac{1}{2}(\cos \theta + \theta \sin \theta)$

$y = \dfrac{1}{2}(\sin \theta - \theta \cos \theta)$

Matches graph (d).

Domain: $(-\infty, \infty)$

Range: $(-\infty, \infty)$

61. $x = (v_0 \cos \theta)t$

$y = h + (v_0 \sin \theta)t - 16t^2$

(a) $\theta = 20°$, $v_0 = 88$ ft/sec
Maximum height: 14.2 ft
Range: 155.6 ft

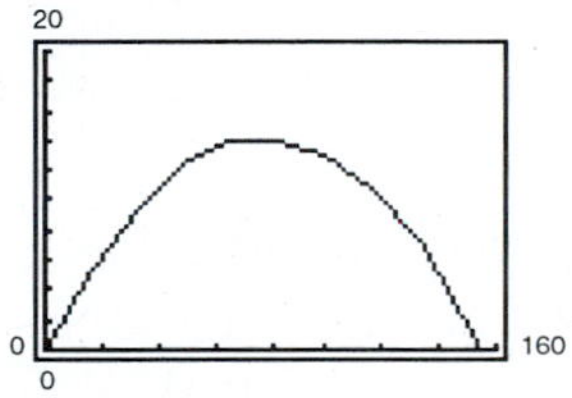

(b) $\theta = 20°$, $v_0 = 132$ ft/sec
Maximum height: 31.8 ft
Range: 350.0 ft

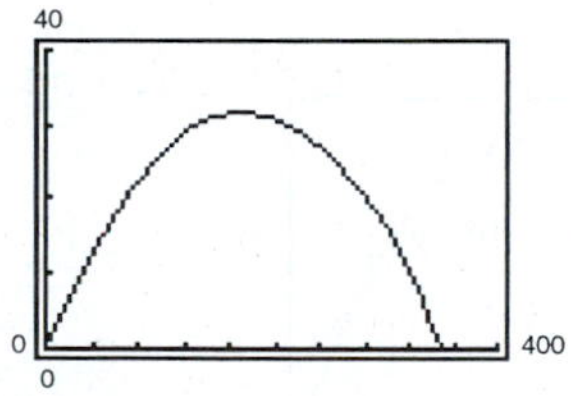

(c) $\theta = 245°$, $v_0 = 88$ ft/sec
Maximum height: 60.5 ft
Range: 242.0 ft

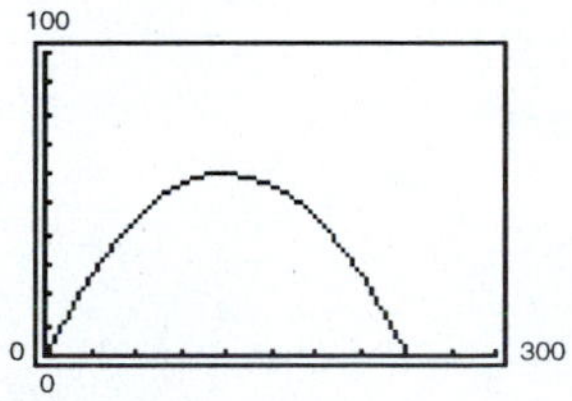

(d) $\theta = 45°$, $v_0 = 132$ ft/sec
Maximum height: 136.1 ft
Range: 544.5 ft

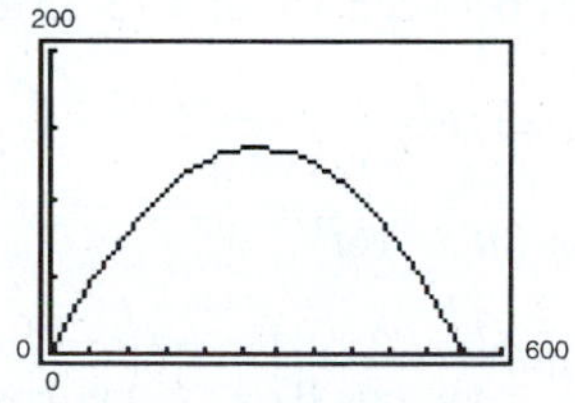

63. (a) 100 miles per hour $= 100\left(\frac{5280}{3600}\right)$ ft/sec $= \frac{440}{3}$ ft/sec

$$x = \left(\frac{440}{3}\cos\theta\right)t \approx (146.67\cos\theta)t$$

$$y = 3 + \left(\frac{440}{3}\sin\theta\right)t - 16t^2 \approx 3 + (146.67\sin\theta)t - 16t^2$$

(b) For $\theta = 15°$, we have:

$$x = \left(\frac{440}{3}\cos 15°\right)t \approx 141.7t$$

$$y = 3 + \left(\frac{440}{3}\sin 15°\right)t - 16t^2 \approx 3 + 38.0t - 16t^2$$

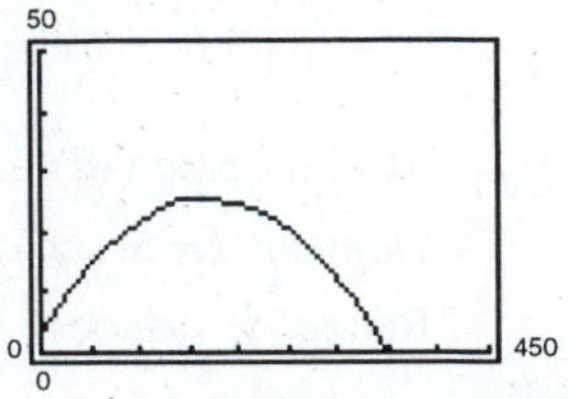

The ball hits the ground inside the ballpark, so it is not a home run.

(c) For $\theta = 23°$, we have:

$$x = \left(\frac{440}{3}\cos 23°\right)t \approx 135.0t$$

$$y = 3 + \left(\frac{440}{3}\sin 23°\right)t - 16t^2 \approx 3 + 57.3t - 16t^2$$

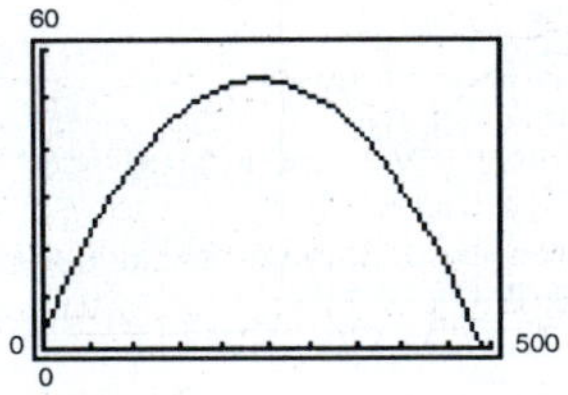

The ball easily clears the 10-foot fence at 400 feet so it is a home run.

(d) Find θ so that $y = 10$ when $x = 400$ by graphing the parametric equations for θ values between 15° and 23°. This occurs when $\theta \approx 19.4°$.

65. $x = (v_0\cos\theta)t \Rightarrow t = \dfrac{x}{v_0\cos\theta}$

$$\begin{aligned} y &= h + (v_0\sin\theta)t - 16t^2 \\ &= h + (v_0\sin\theta)\left(\frac{x}{v_0\cos\theta}\right) - 16\left(\frac{x}{v_0\cos\theta}\right)^2 \\ &= h + (\tan\theta)x - \frac{16x^2}{v_0^2\cos^2\theta} \\ &= -\frac{16\sec^2\theta}{v_0^2}x^2 + (\tan\theta)x + h \end{aligned}$$

67. When the circle has rolled θ radians, the center is at $(a\theta, a)$.

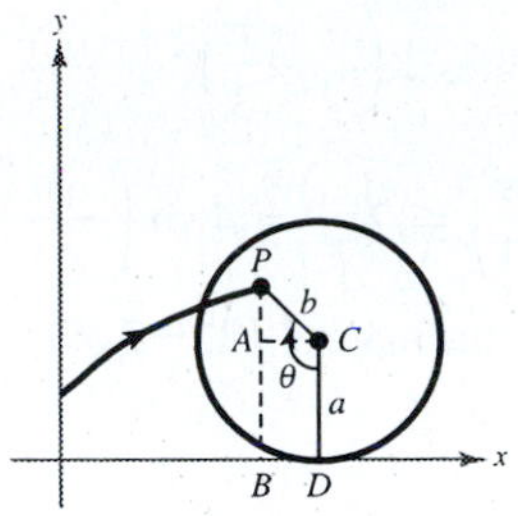

$$\sin\theta = \sin(180° - \theta)$$
$$= \frac{|AC|}{b} = \frac{|BD|}{b} \Rightarrow |BD| = b\sin\theta$$
$$\cos\theta = -\cos(180° - \theta)$$
$$= \frac{|AP|}{-b} \Rightarrow |AP| = -b\cos\theta$$

Therefore, $x = a\theta - b\sin\theta$ and $y = a - b\cos\theta$.

69. $x = t$

$y = t^2 + 1 \Rightarrow y = x^2 + 1$

$x = 3t$

$y = 9t^2 + 1 \Rightarrow y = x^2 + 1$

True

71.
$$\begin{aligned} 5x - 7y &= 11 \Rightarrow & 5x - 7y &= 11 \\ -3x + y &= -13 \Rightarrow & -21x + 7y &= -91 \\ & & -16x &= -80 \\ & & x &= 5 \end{aligned}$$
$$5(5) - 7y = 11 \Rightarrow y = 2$$

Solution: $(5, 2)$

73.
$$\begin{aligned} 3a - 2b + c &= 8 \Rightarrow & 9a - 6b + 3c &= 24 \\ 2a + b - 3c &= -3 \Rightarrow & 2a + b - 3c &= -3 \\ & & 11a - 5b &= 21 \end{aligned}$$
$$\begin{aligned} 2a + b - 3c &= -3 \Rightarrow & 6a + 3b - 9c &= -9 \\ a - 3b + 9c &= 16 \Rightarrow & a - 3b + 9c &= 16 \\ & & 7a &= 7 \\ & & a &= 1 \end{aligned}$$

$a = 11(1) - 5b = 21 \Rightarrow b = -2$

$3(1) - 2(-2) + c = 8 \Rightarrow c = 1$

Solution: $(1, -2, 1)$

Section 10.7 Polar Coordinates

- In polar coordinates you do not have unique representation of points. The point (r, θ) can be repsresented by $(r, \theta \pm 2n\pi)$ or by $(-r, \theta \pm (2n + 1)\pi)$ where n is any integer. The pole is represented by $(0, \theta)$ where θ is any angle.
- To convert from polar coordinates to rectangular coordinates, use the following relationships.
  $x = r\cos\theta$
  $y = r\sin\theta$
- To convert from rectangular coordinates to polar coordinates, use the following relationships.
  $r = \pm\sqrt{x^2 + y^2}$
  $\tan\theta = y/x$
  If θ is in the same quadrant as the point (x, y), then r is positive. If θ is in the opposite quadrant as the point (x, y), then r is negative.
- You should be able to convert rectangular equations to polar form and vice versa.

Solutions to Odd-Numbered Exercises

1. Polar coordinates: $\left(4, \frac{3\pi}{6}\right)$

$x = 4\cos\left(\frac{3\pi}{6}\right) = 0, y = 4\sin\left(\frac{3\pi}{6}\right) = 4$

Rectangular coordinates: $(0, 4)$

3. Polar coordinates: $\left(-1, \frac{5\pi}{4}\right)$

$x = -1\cos\left(\frac{5\pi}{4}\right) = \frac{\sqrt{2}}{2}, y = -1\sin\left(\frac{5\pi}{4}\right) = \frac{\sqrt{2}}{2}$

Rectangular coordinates: $\left(\frac{\sqrt{2}}{2}, \frac{\sqrt{2}}{2}\right)$

5. Polar coordinates: $\left(4, -\frac{\pi}{3}\right)$

$x = 4\cos\left(-\frac{\pi}{3}\right) = 2,\ y = 4\sin\left(-\frac{\pi}{3}\right) = -2\sqrt{3}$

Rectangular coordinates: $(2, -2\sqrt{3})$

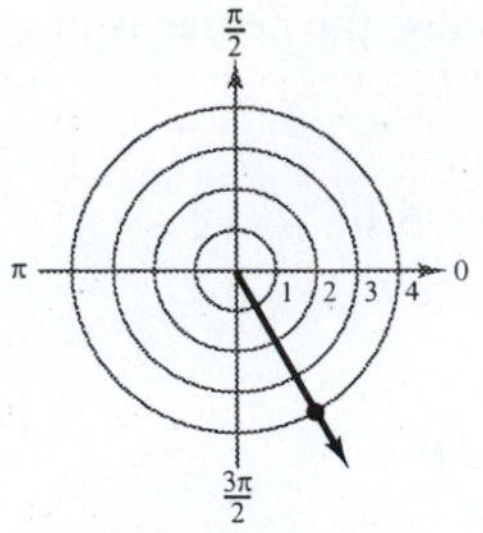

7. Polar coordinates: $\left(0, -\frac{7\pi}{6}\right)$

$x = 0\cos\left(-\frac{7\pi}{6}\right) = 0,\ y = 0\sin\left(-\frac{7\pi}{6}\right) = 0$

Rectangular coordinates: $(0, 0)$

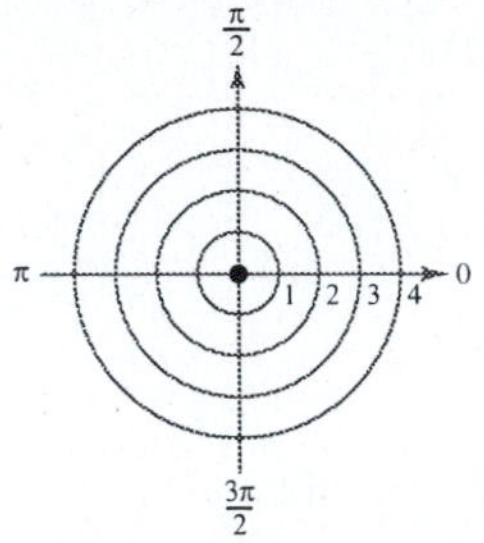

9. Polar coordinates: $(\sqrt{2}, 2.36)$

$x = \sqrt{2}\cos(2.36) \approx -1.004$

$y = \sqrt{2}\sin(2.36) \approx 0.996$

Rectangular coordinates: $(-1.004, 0.996)$

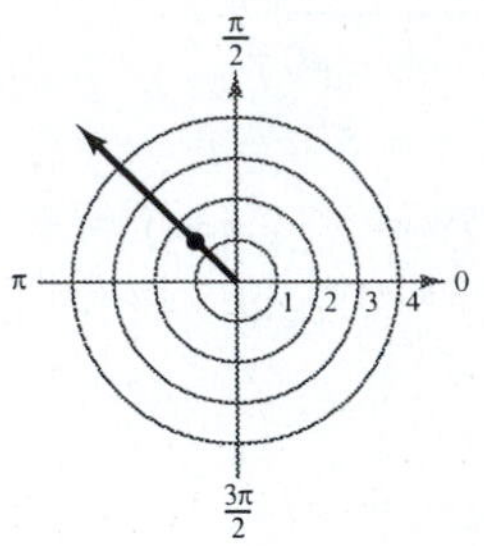

11. Polar coordinates: $\left(2, \frac{3\pi}{4}\right)$

$\left(2 < \frac{3\pi}{4}\right)$ ▶ Rec

$\approx (-1.4142, 1.4142)$

$= (-\sqrt{2}, \sqrt{2})$

13. Polar coordinates: $(-4.5, 1.3)$

$(-4.5 < 1.3)$ ▶ Rec

$\approx (-1.204, -4.336)$

15. Rectangular coordinates: $(1, 1)$

$r = \pm\sqrt{2},\ \tan\theta = 1,\ \theta = \frac{\pi}{4}$ or $\frac{5\pi}{4}$

Polar coordinates: $\left(\sqrt{2}, \frac{\pi}{4}\right), \left(-\sqrt{2}, \frac{5\pi}{4}\right)$

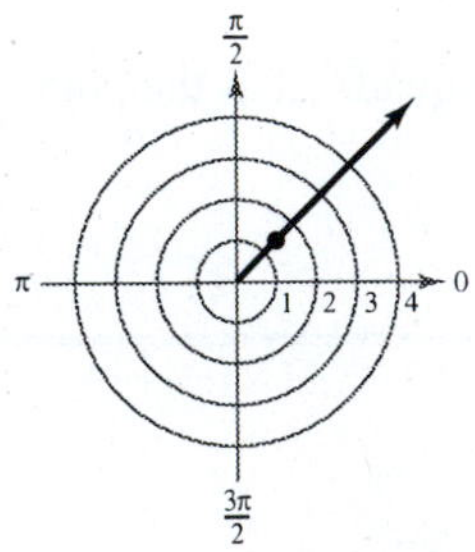

17. Rectangular coordinates: $(-6, 0)$

$r = \pm 6,\ \tan\theta = 0,\ \theta = 0$ or π

Polar coordinates: $(6, \pi), (-6, 0)$

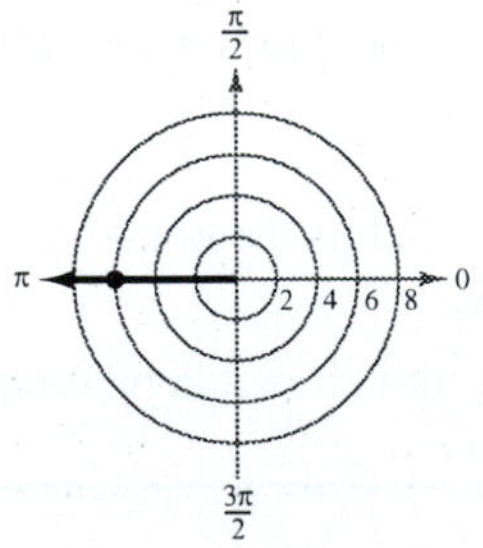

19. Rectangular coordinates: $(-3, 4)$

$r = \pm\sqrt{9 + 16} = \pm 5$, $\tan\theta = -\frac{4}{3}$, $\theta \approx 2.214, 5.356$

Polar coordinates: $(5, 2.214)$, $(-5, 5.356)$

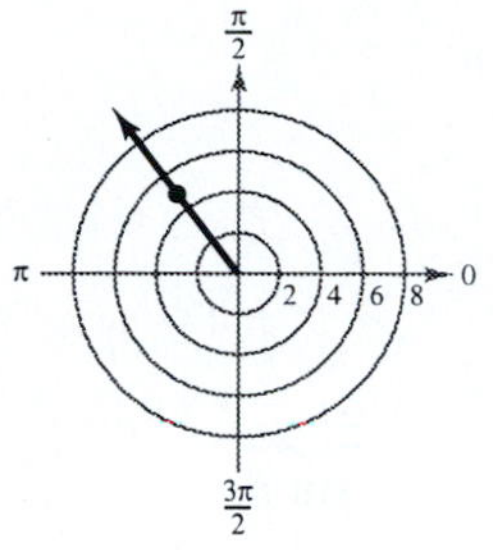

21. Rectangular coordinates: $\left(-\sqrt{3}, -\sqrt{3}\right)$

$r = \pm\sqrt{3 + 3} = \pm\sqrt{6}$, $\tan\theta = 1$, $\theta = \frac{\pi}{4}$ or $\frac{5\pi}{4}$

Polar coordinates: $\left(\sqrt{6}, \frac{5\pi}{4}\right)$, $\left(-\sqrt{6}, \frac{\pi}{4}\right)$

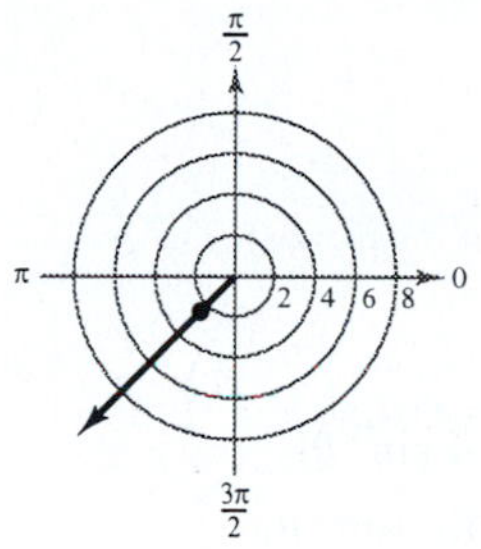

23. Rectangular coordinates: $(4, 6)$

$r = \pm\sqrt{4^2 + 6^2} = \pm\sqrt{52} = \pm 2\sqrt{13}$

$\tan\theta = \frac{6}{4} \Rightarrow \theta \approx 0.983, 4.124$

Polar coordinates: $\left(2\sqrt{13}, 0.983\right)$, $\left(-2\sqrt{13}, 4.124\right)$

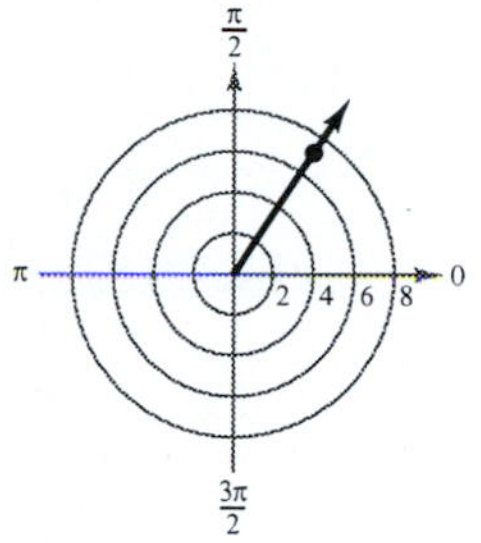

25. Rectangular: $(3, -2)$

$(3, -2)$ ▶ Pol

$\approx (3.606, -0.588)$

27. Rectangular: $\left(\sqrt{3}, 2\right)$

$\left(\sqrt{3}, 2\right)$ ▶ Pol

$\approx (2.646, 0.857)$

29. Rectangular: $\left(\frac{5}{2}, \frac{4}{3}\right)$

$\left(\frac{5}{2}, \frac{4}{3}\right)$ ▶ Pol

$\approx (2.833, 0.490)$

31. True, $|r_1| = |r_2|$

33. $x^2 + y^2 = 9$

$r = 3$

35.
$$x^2 + y^2 - 2ax = 0$$
$$r^2 - 2ar\cos\theta = 0$$
$$r(r - 2a\cos\theta) = 0$$
$$r = 2a\cos\theta$$

37.
$$y = 4$$
$$r\sin\theta = 4$$
$$r = 4\csc\theta$$

39.
$$x = 10$$
$$r\cos\theta = 10$$
$$r = 10\sec\theta$$

41.
$$3x - y + 2 = 0$$
$$3r\cos\theta - r\sin\theta + 2 = 0$$
$$r(3\cos\theta - \sin\theta) = -2$$
$$r = \frac{-2}{3\cos\theta - \sin\theta}$$

43.
$$xy = 4$$
$$(r\cos\theta)(r\sin\theta) = 4$$
$$r^2 = 4\sec\theta\csc\theta = 8\csc 2\theta$$

45.
$$(x^2 + y^2)^2 - 9(x^2 - y^2) = 0$$
$$(r^2) - 9(r^2\cos^2\theta - r^2\sin^2\theta) = 0$$
$$r^2\left[r^2 - 9(\cos 2\theta)\right] = 0$$
$$r^2 = 9\cos 2\theta$$

47.
$$r = 4\sin\theta$$
$$r^2 = 4r\sin\theta$$
$$x^2 + y^2 = 4y$$
$$x^2 + y^2 - 4y = 0$$

49.
$$\theta = \frac{\pi}{6}$$
$$\tan\theta = \frac{\sqrt{3}}{3}$$
$$\frac{y}{x} = \frac{\sqrt{3}}{3}$$
$$y = \frac{\sqrt{3}}{3}x$$
$$\sqrt{3}\,x - 3y = 0$$

51.
$$r = 2\csc\theta$$
$$r\sin\theta = 2$$
$$y = 2$$

53.
$$r = 2\sin 3\theta$$
$$r = 2(3\sin\theta - 4\sin^3\theta)$$
$$r^4 = 6r^3\sin\theta - 8r^3\sin^3\theta$$
$$(x^2 + y^2)^2 = 6(x^2 + y^2)y - 8y^3$$
$$(x^2 + y^2)^2 = 6x^2y - 2y^3$$

55.
$$r = \frac{6}{2 - \sin\theta}$$
$$r(2 - \sin\theta) = 6$$
$$2r = 6 + r\sin\theta$$
$$2\left(\pm\sqrt{x^2 + y^2}\right) = 6 + 3y$$
$$4(x^2 + y^2) = (6 + 3y)^2$$
$$4x^2 + 4y^2 = 36 + 36y + 9y^2$$
$$4x^2 - 5y^2 - 36y - 36 = 0$$

57.
$$r = 3$$
$$r^2 = 9$$
$$x^2 + y^2 = 9$$

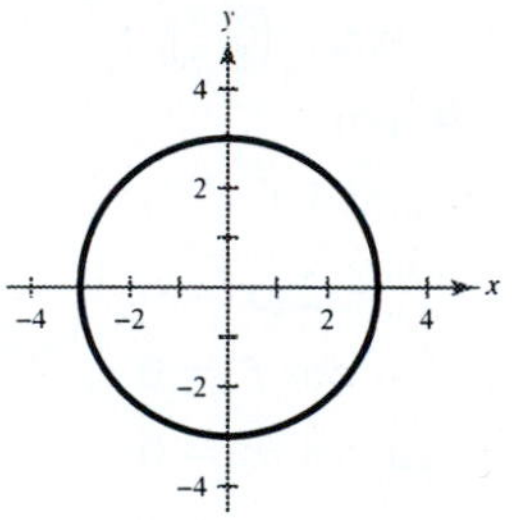

59.
$$\theta = \frac{\pi}{4}$$
$$\tan\theta = \tan\frac{\pi}{4}$$
$$\frac{y}{x} = 1$$
$$y = x$$
$$x - y = 0$$

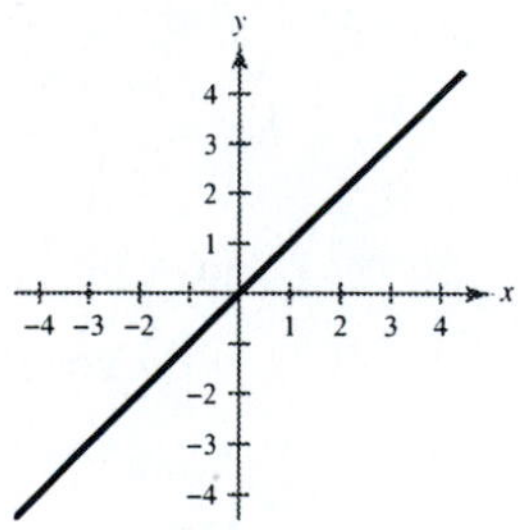

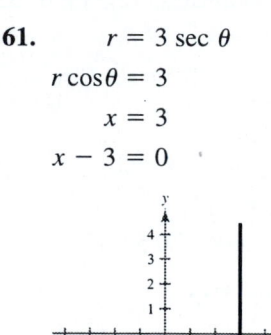

61.
$$r = 3\sec\theta$$
$$r\cos\theta = 3$$
$$x = 3$$
$$x - 3 = 0$$

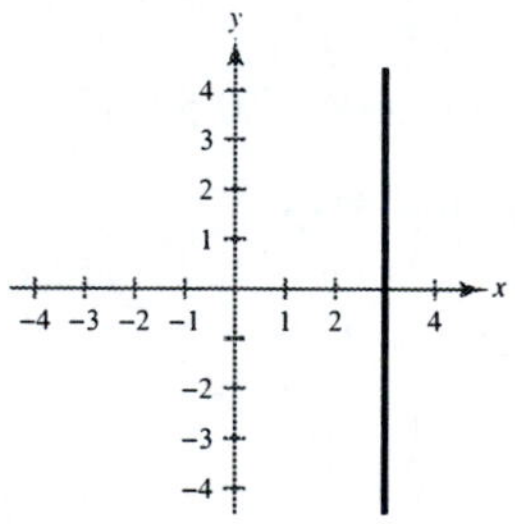

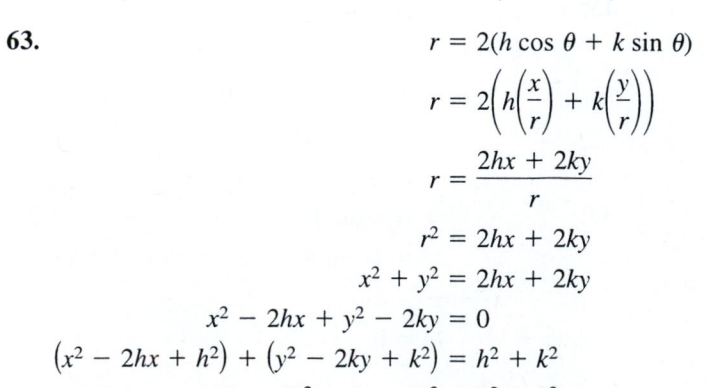

63.
$$r = 2(h\cos\theta + k\sin\theta)$$
$$r = 2\left(h\left(\frac{x}{r}\right) + k\left(\frac{y}{r}\right)\right)$$
$$r = \frac{2hx + 2ky}{r}$$
$$r^2 = 2hx + 2ky$$
$$x^2 + y^2 = 2hx + 2ky$$
$$x^2 - 2hx + y^2 - 2ky = 0$$
$$(x^2 - 2hx + h^2) + (y^2 - 2ky + k^2) = h^2 + k^2$$
$$(x - h)^2 + (y - k)^2 = h^2 + k^2$$

Center: (h, k)

Radius: $\sqrt{h^2 + k^2}$

65. (a) $(r_1, \theta_1) = (x_1, y_1)$ where $x_1 = r_1 \cos \theta_1$ and $y_1 = r_1 \sin \theta_1$.

$(r_2, \theta_2) = (x_2, y_2)$ where $x_2 = r_2 \cos \theta_2$ and $y_2 = r_2 \sin \theta_2$.

$$\begin{aligned} d &= \sqrt{(x_1 - x_2)^2 + (y_1 - y_2)^2} \\ &= \sqrt{x_1^2 - 2x_1x_2 + x_2^2 + y_1^2 - 2y_1y_2 + y_2^2} \\ &= \sqrt{(x_1^2 + y_1^2) + (x_2^2 + y_2 + y_2^2) - 2(x_1x_2 + y_1y_2)} \\ &= \sqrt{r_1^2 + r_2^2 - 2\,(r_1r_2 \cos \theta_1 \cos \theta_2 + r_1r_2 \sin \theta_1 \sin \theta_2)} \\ &= \sqrt{r_1^2 + r_2^2 - 2r_1r_2 \cos(\theta_1 - \theta_2)} \end{aligned}$$

(b) If $\theta_1 = \theta_2$, then

$$\begin{aligned} d &= \sqrt{r_1^2 + r_2^2 - 2r_1r_2} \\ &= \sqrt{(r_1 - r_2)^2} \\ &= |r_1 - r_2|. \end{aligned}$$

This represents the distance between two points on the line $\theta = \theta_1 = \theta_2$.

(c) If $\theta_1 - \theta_2 = 90°$, then

$d = \sqrt{r_1^2 + r_2^2}$.

This is the result of the Pythagorean Theorem.

(d) The results should be the same. For example, use the points

$$\left(3, \frac{\pi}{6}\right) \text{ and } \left(4, \frac{\pi}{3}\right).$$

The distance is $d \approx 2.053$.

Now use the representations

$$\left(-3, \frac{7\pi}{6}\right) \text{ and } \left(-4, \frac{4\pi}{3}\right).$$

The distance is still $d \approx 2.053$.

67. $5x - 7y = -11$

$-3x + y = -3$

By Cramer's Rule we have:

$$x = \frac{\begin{vmatrix} -11 & -7 \\ -3 & 1 \end{vmatrix}}{\begin{vmatrix} 5 & -7 \\ -3 & 1 \end{vmatrix}} = \frac{-32}{-16} = 2$$

$$y = \frac{\begin{vmatrix} 5 & -11 \\ -3 & -3 \end{vmatrix}}{\begin{vmatrix} 5 & -7 \\ -3 & 1 \end{vmatrix}} = \frac{-48}{-16} = 3$$

Solution: $(2, 3)$

69. $3a - 2b + c = 0$

$2a + b - 3c = 0$

$a - 3b + 9c = 8$

$$\begin{vmatrix} 3 & -2 & 1 \\ 2 & 1 & -3 \\ 1 & -3 & 9 \end{vmatrix} = 35$$

By Cramer's Rule we have:

$$x = \frac{\begin{vmatrix} 0 & -2 & 1 \\ 0 & 1 & -3 \\ 8 & -3 & 9 \end{vmatrix}}{35} = \frac{40}{35} = \frac{8}{7}$$

$$y = \frac{\begin{vmatrix} 3 & 0 & 1 \\ 2 & 0 & -3 \\ 1 & 8 & 9 \end{vmatrix}}{35} = \frac{88}{35}$$

$$z = \frac{\begin{vmatrix} 3 & -2 & 0 \\ 2 & 1 & 0 \\ 1 & -3 & 8 \end{vmatrix}}{35} = \frac{56}{35} = \frac{8}{5}$$

Solution: $\left(\frac{8}{7}, \frac{88}{35}, \frac{8}{5}\right)$

71.
$$\begin{aligned} x + y + z - 3w &= -8 \\ 3x - y - 2z + w &= 7 \\ -x + y - z + 2w &= -2 \\ 2y + w &= -6 \end{aligned}$$

$$\begin{vmatrix} 1 & 1 & 1 & -3 \\ 3 & -1 & -2 & 1 \\ -1 & 1 & -1 & 2 \\ 0 & 2 & 0 & 1 \end{vmatrix} = 20$$

By Cramer's Rule we have:

$$x = \frac{\begin{vmatrix} -8 & 1 & 1 & -3 \\ 7 & -1 & -2 & 1 \\ -2 & 1 & -1 & 2 \\ -6 & 2 & 0 & 1 \end{vmatrix}}{20} = \frac{20}{20} = 1$$

$$y = \frac{\begin{vmatrix} 1 & -8 & 1 & -3 \\ 3 & 7 & -2 & 1 \\ -1 & -2 & -1 & 2 \\ 0 & -6 & 0 & 1 \end{vmatrix}}{20} = \frac{-80}{20} = -4$$

$$z = \frac{\begin{vmatrix} 1 & 1 & -8 & -3 \\ 3 & -1 & 7 & 1 \\ -1 & 1 & -2 & 2 \\ 0 & 2 & -6 & 1 \end{vmatrix}}{20} = \frac{20}{20} = 1$$

$$w = \frac{\begin{vmatrix} 1 & 1 & 1 & -8 \\ 3 & -1 & -2 & 7 \\ -1 & 1 & -1 & -2 \\ 0 & 2 & 0 & -6 \end{vmatrix}}{20} = \frac{40}{20} = 2$$

Solution: $(1, -4, 1, 2)$

Section 10.8 Graphs of Polar Equations

- When graphing polar equations:
  1. Test for symmetry
     (a) $\theta = \pi/2$: Replace (r, θ) by $(r, \pi - \theta)$ or $(-r, -\theta)$.
     (b) Polar axis: Replace (r, θ) by $(r, -\theta)$ or $(-r, \pi - \theta)$.
     (c) Pole: Replace (r, θ) by $(r, \pi + \theta)$ or $(-r, \theta)$.
     (d) $r = f(\sin \theta)$ is symmetric with repsect to the line $\theta = \pi/2$.
     (e) $r = f(\cos \theta)$ is symmetric with respect to the polar axis.
  2. Find the θ values for which $|r|$ is maximum.
  3. Find the θ values for which $r = 0$.
  4. Know the different types of polar graphs.
     (a) Limacons
     $r = a \pm b \cos \theta$
     $r = a \pm b \sin \theta$
     (b) Rose Curves, $n \geq 2$
     $r = a \cos n\theta$
     $r = a \sin n\theta$
     (c) Circles
     $r = a \cos \theta$
     $r = a \sin \theta$
     $r = a$
     (d) Lemniscates
     $r^2 = a^2 \cos 2\theta$
     $r^2 = a^2 \sin 2\theta$
  5. Plot additional points.

Solutions to Odd-Numbered Exercises

1. $r = 3\cos 2\theta$

Rose curve with 4 petals

3. $r = 2 - \cos\theta$

Convex limaçon

5. $r = 6\sin 2\theta$

Rose curve with 4 petals

7. $r = 10 + 6\cos\theta$

$\theta = \frac{\pi}{2}$: $-r = 10 + 6\cos(-\theta)$

$-r = 10 + 6\cos\theta$

Not an equivalent equation

Polar axis: $r = 10 + 6\cos(-\theta)$

$r = 10 + 6\cos\theta$

Equivalent equation

Pole: $-r = 10 + 6\cos\theta$

Not an equivalent equation

Answer: Symmetric with respect to polar axis

9. $r = \dfrac{2}{1 + \sin\theta}$

$\theta = \frac{\pi}{2}$: $r = \dfrac{2}{1 + \sin(\pi - \theta)}$

$r = \dfrac{2}{1 + \sin\pi\cos\theta - \cos\pi\sin\theta}$

$r = \dfrac{2}{1 + \sin\theta}$

Equivalent equation

Polar axis: $r = \dfrac{2}{1 + \sin(-\theta)}$

$r = \dfrac{2}{1 - \sin\theta}$

Not an equivalent equation

Pole: $-r = \dfrac{2}{1 + \sin\theta}$

Answer: Symmetric with respect to $\theta = \pi/2$

11. $r = 4\sec\theta\csc\theta$

$\theta = \frac{\pi}{2}$: $-r = 4\sec(-\theta)\csc(-\theta)$

$-r = -4\sec\theta\csc\theta$

$r = 4\sec\theta\csc\theta$

Equivalent equation

Polar axis: $-r = 4\sec(\pi - \theta)\csc(\pi - \theta)$

$-r = 4(-\sec\theta)\csc\theta$

$r = 4\sec\theta\csc\theta$

Equivalent equation

Pole: $r = 4\sec(\pi + \theta)\csc(\pi + \theta)$

$r = 4(-\sec\theta)(-\csc\theta)$

$r = 4\sec\theta\csc\theta$

Equivalent equation

Answer: Symmetric with respect to $\theta = \pi/2$, polar axis, and pole

13. $|r| = |10(1 - \sin\theta)| = 10|1 - \sin\theta| \le 10(2) = 20$

$|1 - \sin\theta| = 2$

$1 - \sin\theta = 2$ or $1 - \sin\theta = -2$

$\sin\theta = -1$ | $\sin\theta = 3$

$\theta = \frac{3\pi}{2}$ | Not possible

Maximum: $|r| = 20$ when $\theta = \frac{3\pi}{2}$.

$0 = 10(1 - \sin\theta)$

$\sin\theta = 1$

$\theta = \frac{\pi}{2}$

Zero: $r = 0$ when $\theta = \frac{\pi}{2}$.

15. $|r| = |4\cos 3\theta| = 4|\cos 3\theta| \le 4$

$|\cos 3\theta| = 1$

$\cos 3\theta = \pm 1$

$\theta = 0, \frac{\pi}{3}, \frac{2\pi}{3}$

Maximum: $|r| = 4$ when $\theta = 0, \frac{\pi}{3}, \frac{2\pi}{3}$.

$0 = 4\cos 3\theta$

$\cos 3\theta = 0$

$\theta = \frac{\pi}{6}, \frac{\pi}{2}, \frac{5\pi}{6}$

Zero: $r = 0$ when $\theta = \frac{\pi}{6}, \frac{\pi}{2}, \frac{5\pi}{6}$.

17. Circle: $r = 5$

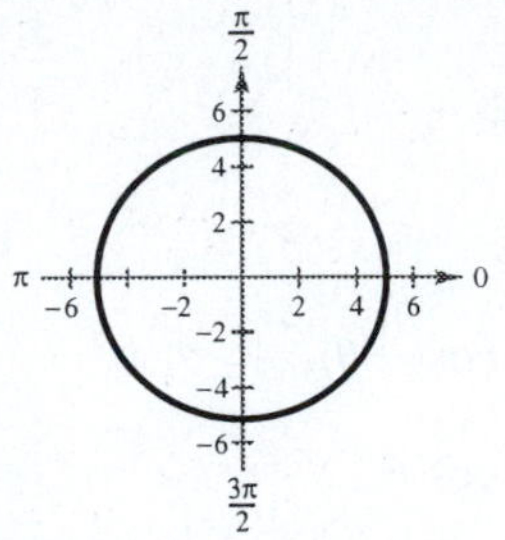

19. Circle: $r = \frac{\pi}{6}$

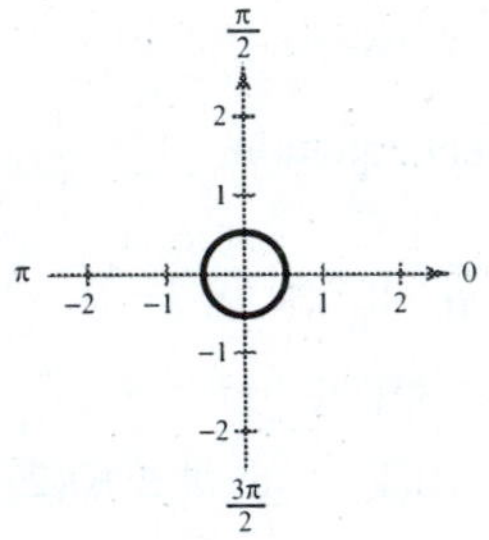

21. $r = 3\sin\theta$

Symmetric with respect to $\theta = \frac{\pi}{2}$

Circle with a radius of $\frac{3}{2}$

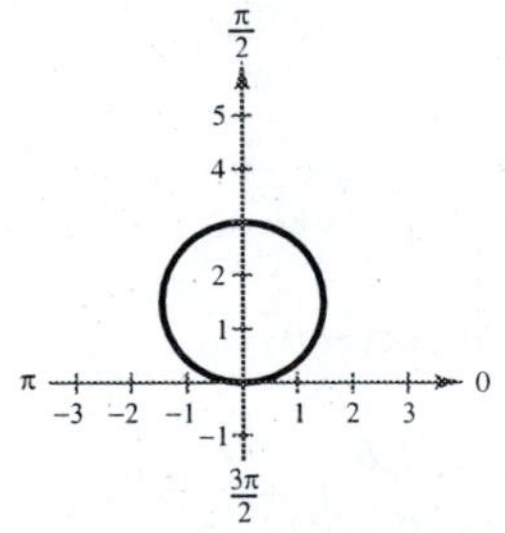

23. $r = 3 - 3\cos\theta$

Symmetric with respect to polar axis

$\frac{a}{b} = \frac{3}{3} = 1 \Rightarrow$ Cardioid

$|r| = 6$ when $\theta = \pi$.

$r = 0$ when $\pi = 0$.

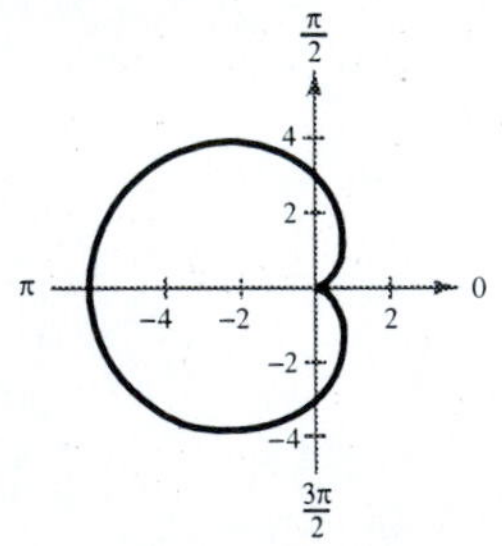

25. $r = 4 + 4\sin\theta$

Symmetric with respect to

$\theta = \frac{\pi}{2}$

$\frac{a}{b} = \frac{4}{4} = 1 \Rightarrow$ Cardioid

$|r| = 8$ when $\theta = \frac{\pi}{2}$.

$r = 0$ when $\theta = \frac{3\pi}{2}$.

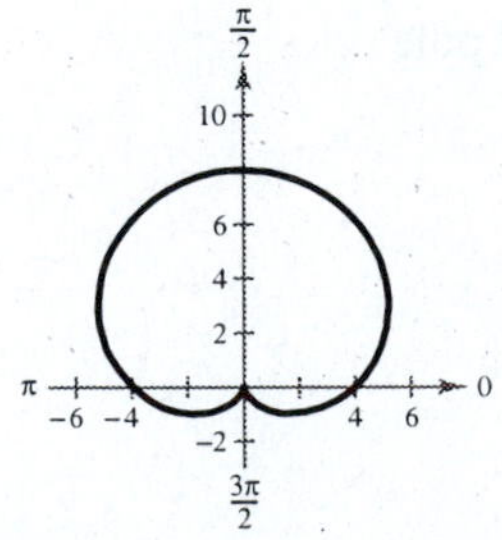

27. $r = 3 - 2 \cos \theta$

Symmetric with respect to polar axis

$\frac{a}{b} = \frac{3}{2} > 1 \Rightarrow$ Dimpled limaçon

$|r| = 5$ when $\theta = \pi$.

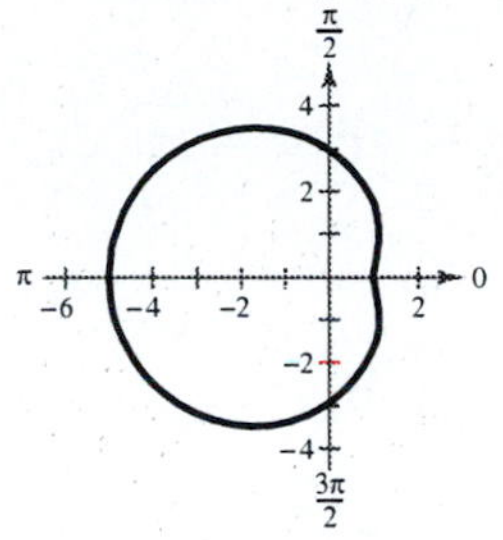

29. $r = 2 + \sin \theta$

Symmetric with respect to $\theta = \frac{\pi}{2}$

$\frac{a}{b} = \frac{2}{1} \geq 2 \Rightarrow$ Convex limaçon

$|r| = 3$ when $\theta = \frac{\pi}{2}$.

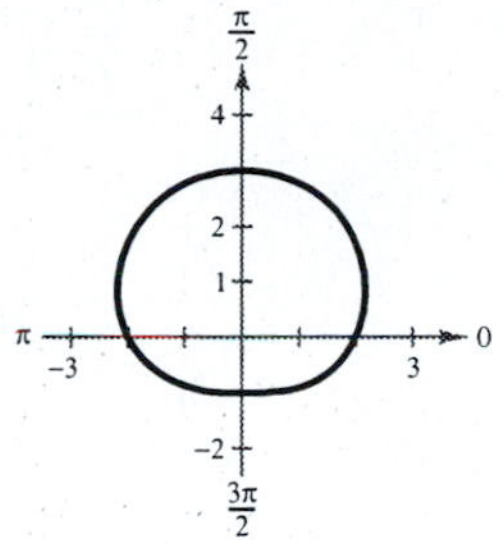

31. $r = 2 + 4 \sin \theta$

Symmetric with respect to $\theta = \frac{\pi}{2}$

$\frac{a}{b} = \frac{2}{4} < 1 \Rightarrow$ Limaçon with inner loop

$|r| = 6$ when $\theta = \frac{\pi}{2}$.

$r = 0$ when $\theta = \frac{7\pi}{6}, \frac{11\pi}{6}$.

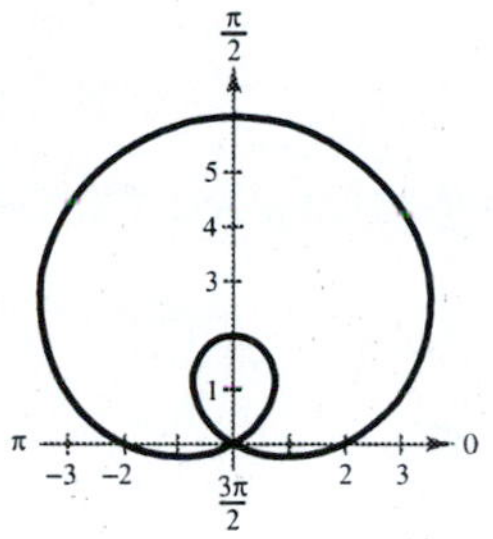

33. $r = 3 - 4 \cos \theta$

Symmetric with respect to polar axis

$\frac{a}{b} = \frac{2}{4} < 1 \Rightarrow$ Limaçon with inner loop

$|r| = 7$ when $\theta = \pi$.

$r = 0$ when $\cos \theta = \frac{3}{4}$ or

$\theta \approx 0.723, 5.560$

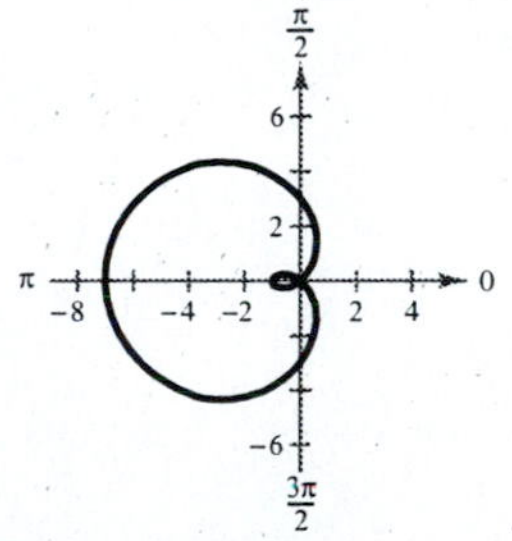

35. $r = 2 \cos 3\theta$

Symmetric with respect to polar axis

Rose curve ($n = 3$) with 3 petals

$|r| = 2$ when $\theta = 0, \frac{2\pi}{3}, \frac{4\pi}{3}$.

$r = 0$ when $\theta = \frac{\pi}{6}, \frac{\pi}{2}, \frac{5\pi}{6}$.

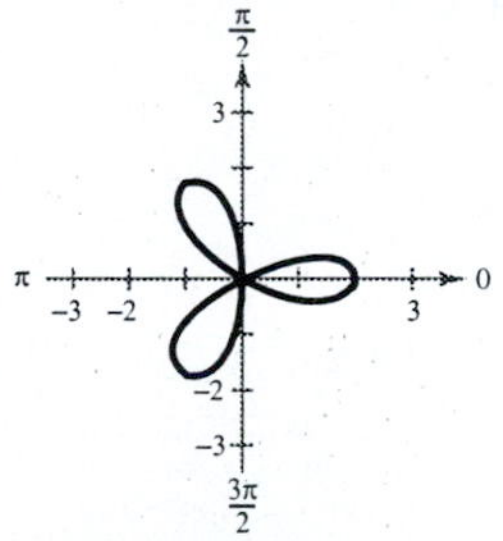

37. $r = 3 \sin 2\theta$

Symmetric with respect to $\theta = \dfrac{\pi}{2}$

Rose curve ($n = 2$) with 4 petals

$|r| = 3$ when $\theta = \dfrac{\pi}{4}, \dfrac{3\pi}{4}, \dfrac{5\pi}{4}, \dfrac{7\pi}{4}$.

$r = 0$ when $\theta = 0, \dfrac{\pi}{2}, \pi$.

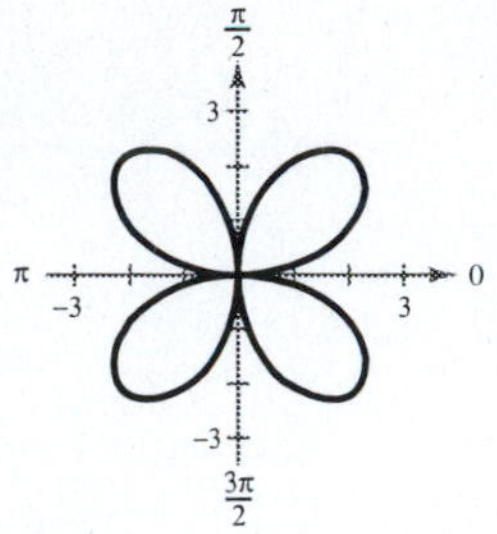

39. $r = 2 \sec \theta$

$r = \dfrac{2}{\cos \theta}$

$r \cos \theta = 2$

$x = 2 \Rightarrow$ Line

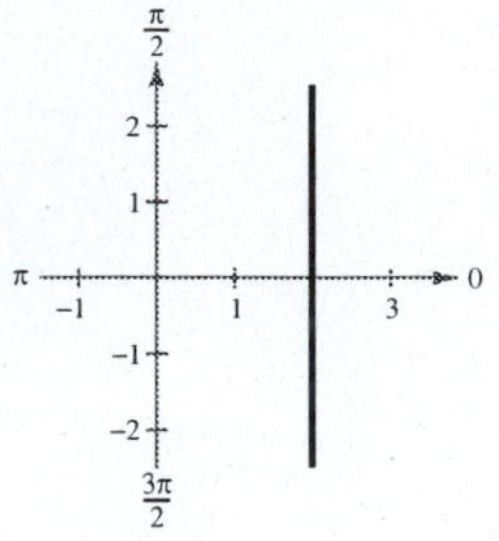

41. $r = \dfrac{3}{\sin \theta - 2 \cos \theta}$

$r(\sin \theta - 2 \cos \theta) = 3$

$y - 2x = 3$

$y = 2x + 3 \Rightarrow$ Line

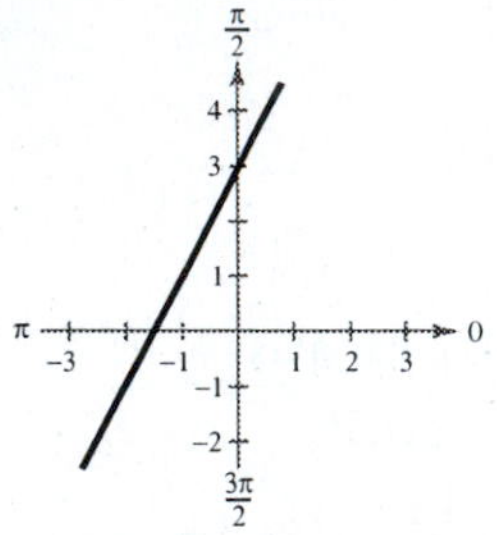

43. $r^2 = 4 \cos 2\theta$

Symmetric with respect to polar axis

Lemniscate

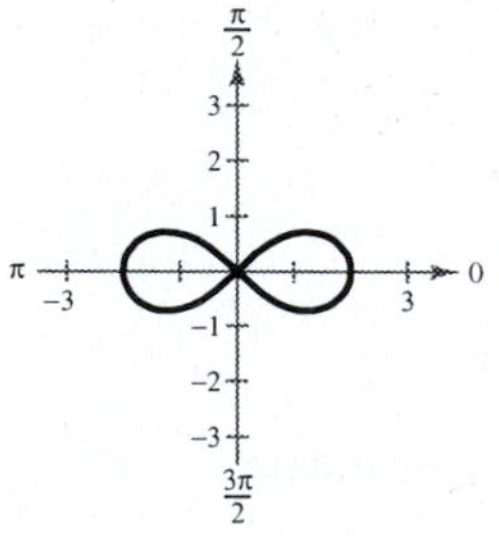

45. $r = \dfrac{\theta}{2}$

Symmetric with respect to $\theta = \dfrac{\pi}{2}$

Spiral

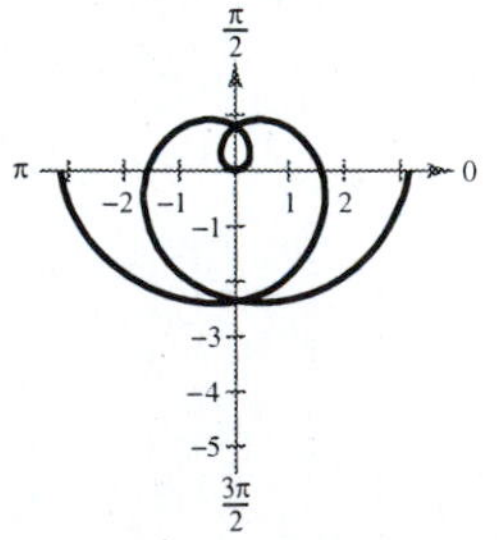

47. $r = 6 \cos \theta$

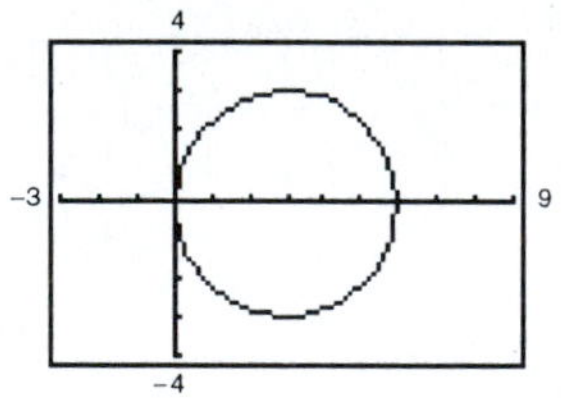

49. $r = 3(2 - \sin\theta)$

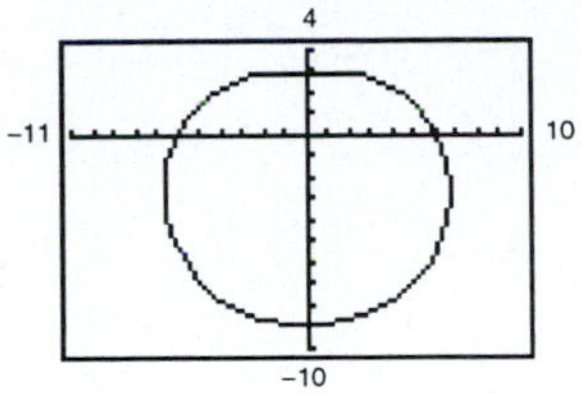

51. $r = 4\sin\theta\cos^2\theta$

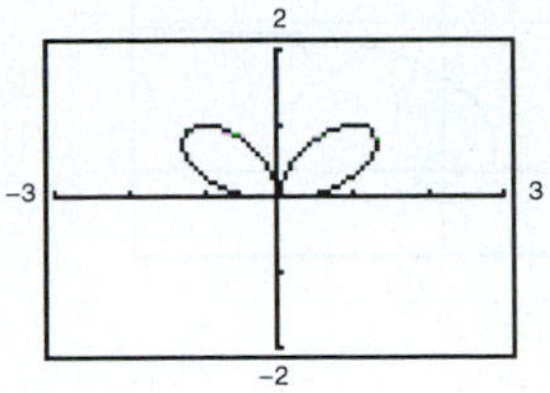

53. $r = 2\csc\theta + 5 = \dfrac{2}{\sin\theta} + 5$

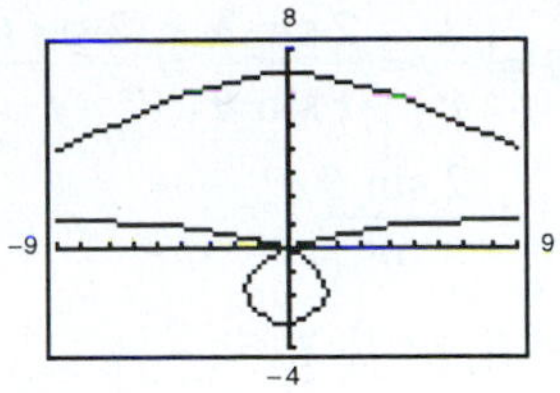

55. $r = 3 - 4\cos\theta$

$0 \le \theta < 2\pi$

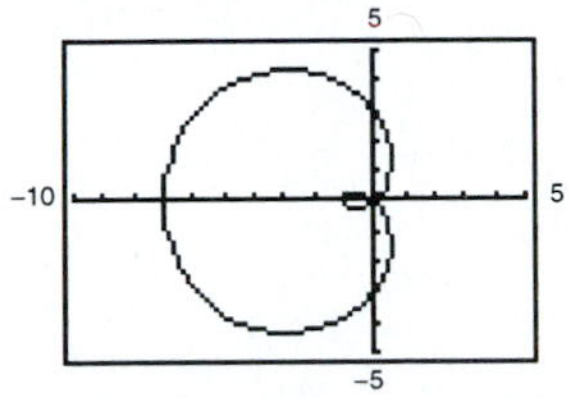

57. $r = 2 + \sin\theta$

$0 \le \theta < 2\pi$

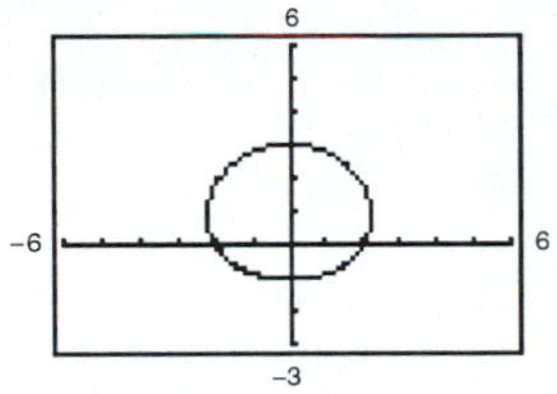

59. $r = 2\cos\left(\dfrac{3\theta}{2}\right)$

$0 \le \theta < 4\pi$

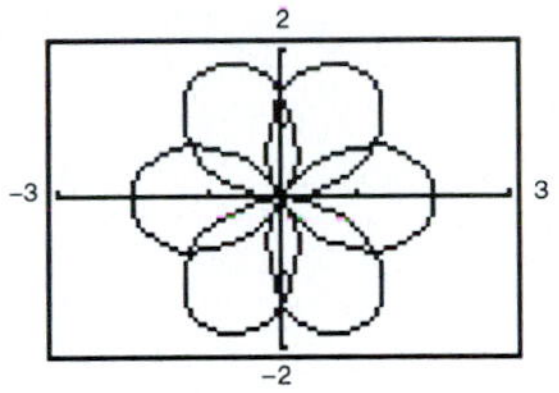

61. $r^2 = 4\sin 2\theta$

$0 \le \theta < \pi$

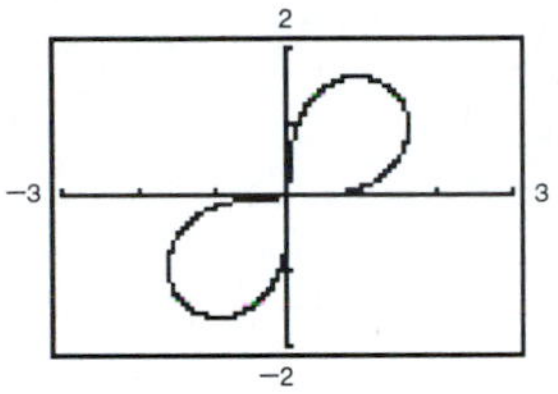

63.

$$\begin{aligned}
r &= 2 - \sec\theta = 2 - \frac{1}{\cos\theta} \\
r\cos\theta &= 2\cos\theta - 1 \\
r(r\cos\theta) &= 2r\cos\theta - r \\
\left(\pm\sqrt{x^2 + y^2}\right)x &= 2x - \left(\pm\sqrt{x^2 + y^2}\right) \\
\left(\pm\sqrt{x^2 + y^2}\right)(x + 1) &= 2x \\
\left(\pm\sqrt{x^2 + y^2}\right) &= \frac{2x}{x + 1} \\
x^2 + y^2 &= \frac{4x^2}{(x + 1)^2} \\
y^2 &= \frac{4x^2}{(x + 1)^2} - x^2 \\
&= \frac{4x^2 - x^2(x + 1)^2}{(x + 1)^2} = \frac{4x^2 - x^2(x^2 + 2x + 1)}{(x + 1)^2} \\
&= \frac{-x^4 - 2x^3 + 3x^2}{(x + 1)^2} = \frac{-x^2(x^2 + 2x - 3)}{(x + 1)^2} \\
y &= \pm\sqrt{\frac{x^2(3 - 2x - x^2)}{(x + 1)^2}} = \pm\left|\frac{x}{x + 1}\right|\sqrt{3 - 2x - x^2}
\end{aligned}$$

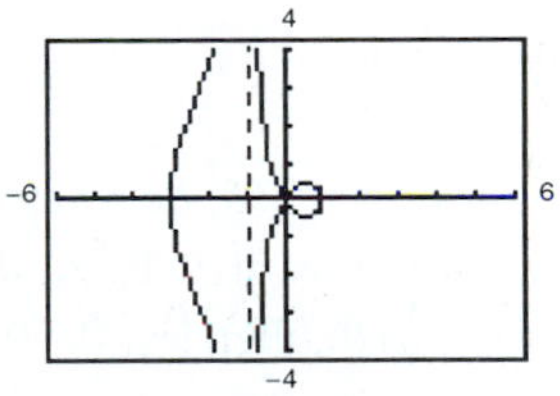

The graph has an asymptote at $x = -1$.

65. $r = \dfrac{2}{\theta}$

$$\theta = \frac{2}{r} = \frac{2\sin\theta}{r\sin\theta} = \frac{2\sin\theta}{y}$$

$$y = \frac{2\sin\theta}{\theta}$$

As $\theta \Rightarrow 0$, $y \Rightarrow 2$.

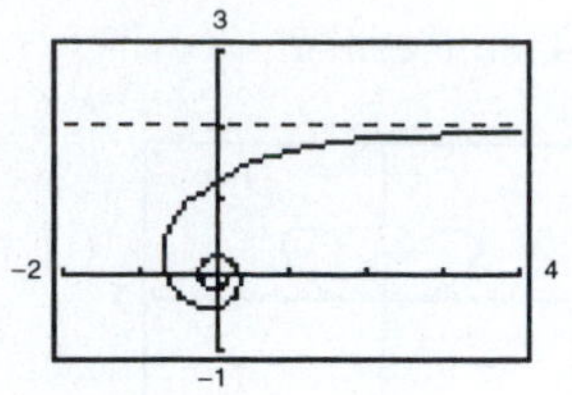

67. $r = 4\sin\theta$

(a) $0 \le \theta \le \dfrac{\pi}{2}$

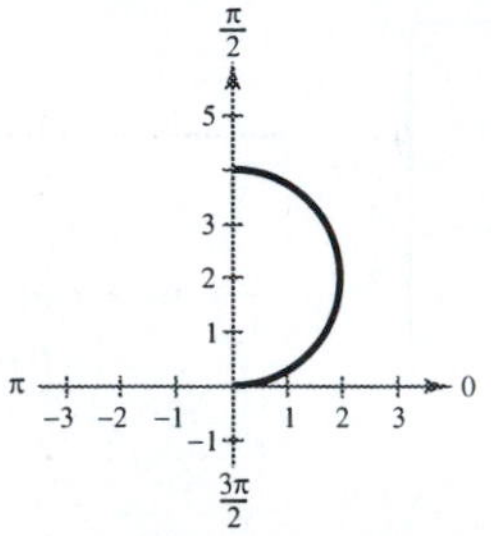

Right half of the circle

(b) $\dfrac{\pi}{2} \le \theta \le \pi$

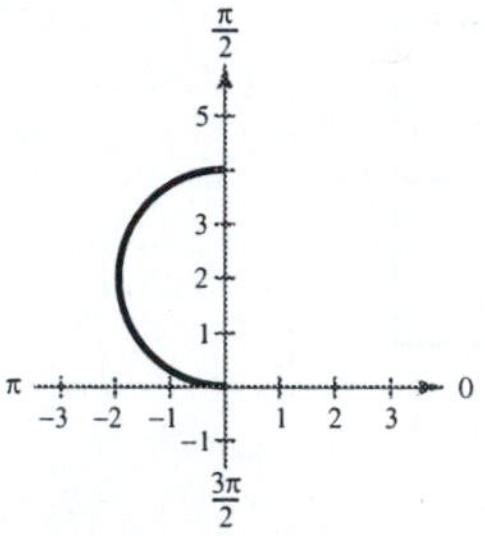

Left half of the circle

(c) $-\dfrac{\pi}{2} \le \theta \le \dfrac{\pi}{2}$

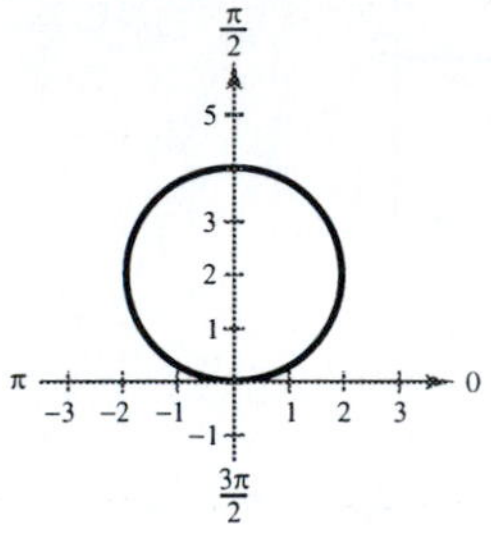

Entire circle

(d) $\dfrac{\pi}{4} \le \theta \le \dfrac{3\pi}{4}$

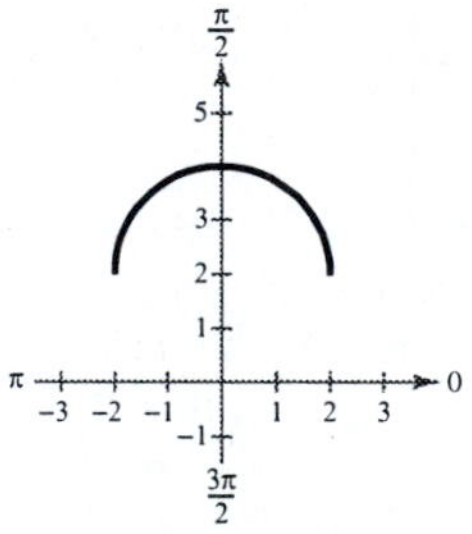

Top half of the circle

69. Let the curve $r = f(\theta)$ be rotated by ϕ to form the curve $r = g(\theta)$. If (r_1, θ_1) is a point on $r = f(\theta)$, then $(r_1, \theta_1 + \phi)$ is on $r = g(\theta)$. That is, $g(\theta_1 + \phi) = r_1 = f(\theta_1)$. Letting $\theta = \theta_1 + \phi$, or $\theta_1 = \theta - \phi$, we see that $g(\theta) = g(\theta_1 + \phi) = f(\theta_1) = f(\theta - \phi)$.

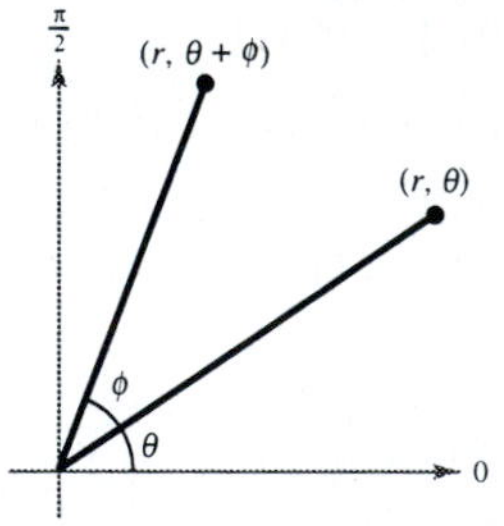

71. (a) $r = 2 - \sin\left(\theta - \dfrac{\pi}{4}\right)$

$$= 2 - \frac{\sqrt{2}}{2}(\sin\theta - \cos\theta)$$

(b) $r = 2 - \sin\left(\theta - \dfrac{\pi}{2}\right)$

$$= 2 + \cos\theta$$

(c) $r = 2 - \sin(\theta - \pi)$

$$= 2 + \sin\theta$$

(d) $r = 2 - \sin\left(\theta - \dfrac{3\pi}{2}\right)$

$$= 2 - \cos\theta$$

73. (a) $r = 1 - \sin\theta$

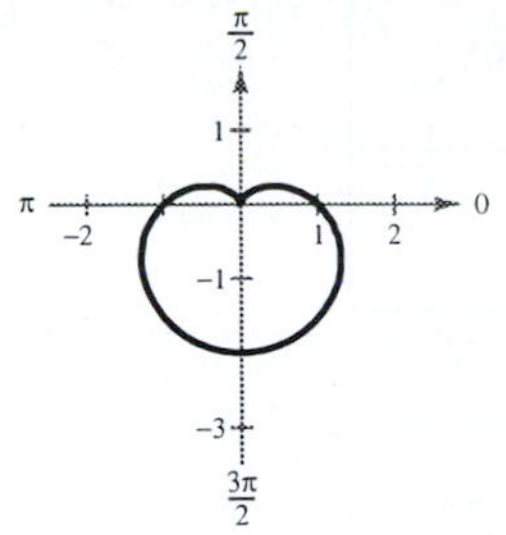

(b) $r = 1 - \sin\left(\theta - \dfrac{\pi}{4}\right)$

Rotate the graph in part (a) through the angle $\dfrac{\pi}{4}$.

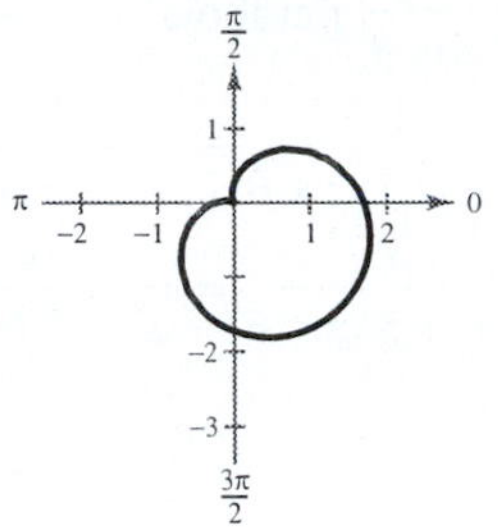

75. $r = 2 + k\cos\theta$

$k = 0$: $r = 2$

Circle

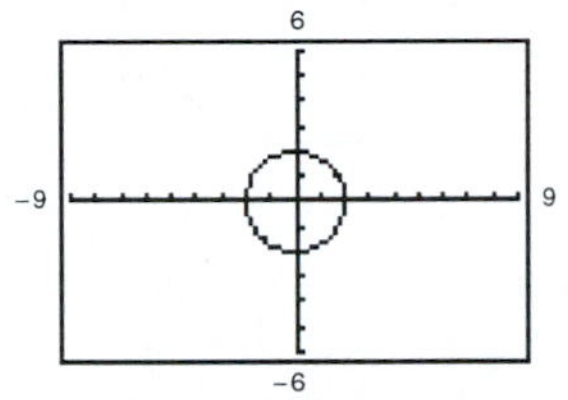

$k = 1$: $r = 2 + \cos\theta$

Convex limaçon

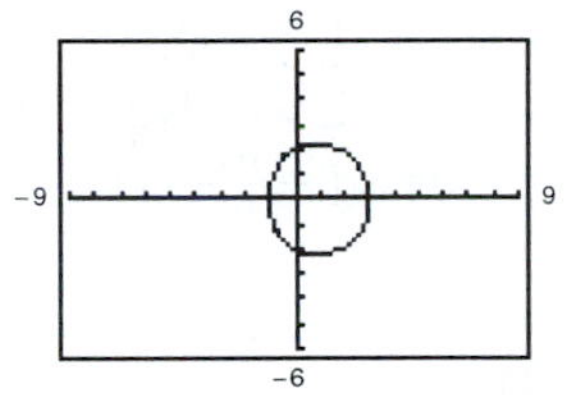

$k = 2$: $r = 2 + 2\cos\theta$

Cardioid

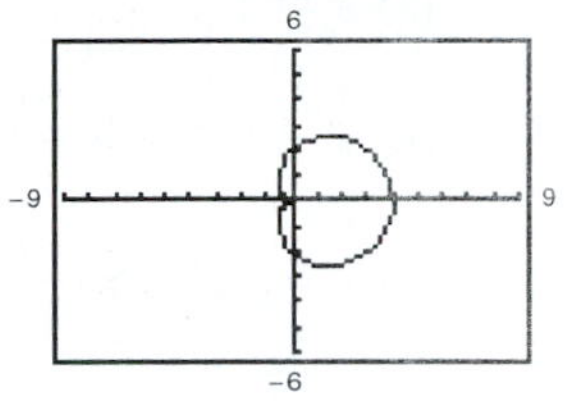

$k = 3$: $r = 2 + 3\cos\theta$

Limaçon with inner loop

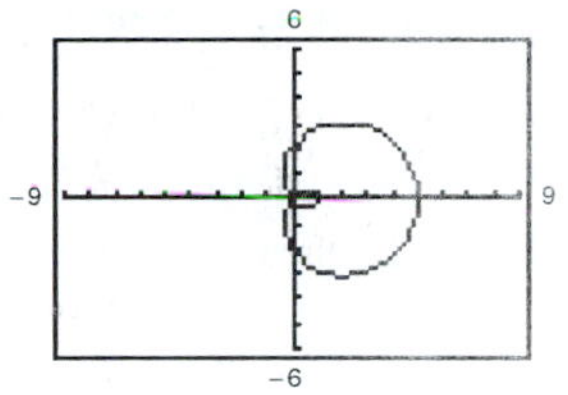

Section 10.9 Polar Equations of Conics

- The graph of a polar equation of the form

  $$r = \frac{ep}{1 \pm e\cos\theta} \quad \text{or} \quad r = \frac{ep}{1 \pm e\sin\theta}$$

  is a conic, *where $e > 0$ is the eccentricity and $|p|$ is the distance between the focus (pole) and the directrix.*

  (a) If $e < 1$, the graph is an ellipse.
  (b) If $e = 1$, the graph is a parabola.
  (c) If $e > 1$, the graph is a hyperbola.

- Guidelines for finding polar equations of conics:

  (a) Horizontal directrix above the pole: $r = \dfrac{ep}{1 + e\sin\theta}$

  (b) Horizontal directrix below the pole: $r = \dfrac{ep}{1 - e\sin\theta}$

  (c) Vertical directrix to the right of the pole: $r = \dfrac{ep}{1 + e\cos\theta}$

  (d) Vertical directrix to the left of the pole: $r = \dfrac{ep}{1 - e\cos\theta}$

Solutions to Odd-Numbered Exercises

1. $r = \dfrac{2e}{1 + e\cos\theta}$

(a) $e = 1,\ r = \dfrac{2}{1 + \cos\theta}$, parabola

(b) $e = 0.5,\ r = \dfrac{1}{1 + 0.5\cos\theta} = \dfrac{2}{2 + \cos\theta}$, ellipse

(c) $e = 1.5,\ r = \dfrac{3}{1 + 1.5\cos\theta} = \dfrac{6}{2 + 3\cos\theta}$, hyperbola

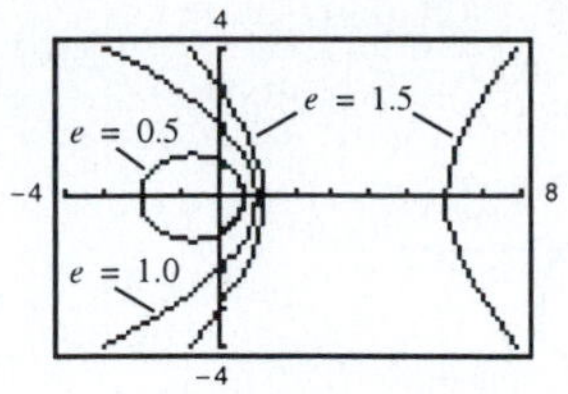

3. $r = \dfrac{2e}{1 - e\sin\theta}$

(a) $e = 1,\ r = \dfrac{2}{1 - \sin\theta}$, parabola

(b) $e = 0.5,\ r = \dfrac{1}{1 - 0.5\sin\theta} = \dfrac{2}{2 - \sin\theta}$, ellipse

(c) $e = 1.5,\ r = \dfrac{3}{1 - 1.5\sin\theta} = \dfrac{6}{2 - 3\sin\theta}$, hyperbola

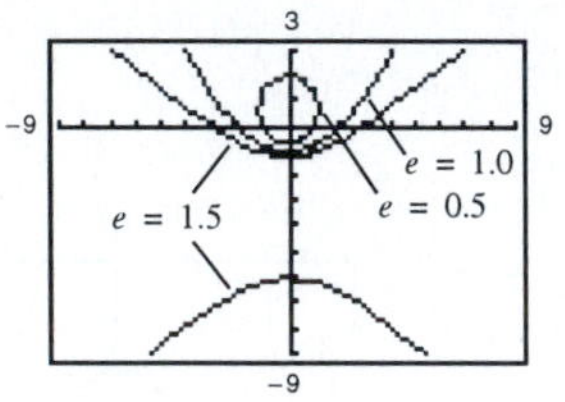

5. $r = \dfrac{4}{1 - \cos\theta}$

$e = 1 \Rightarrow$ Parabola

Vertical directrix to the left of the pole

Matches graph (b).

7. $r = \dfrac{3}{1 + 2\sin\theta}$

$e = 2 \Rightarrow$ Hyperbola

Matches graph (d).

9. $r = \dfrac{2}{1 - \cos\theta}$

$e = 1$, the graph is a parabola.

Vertex: $(1, \pi)$

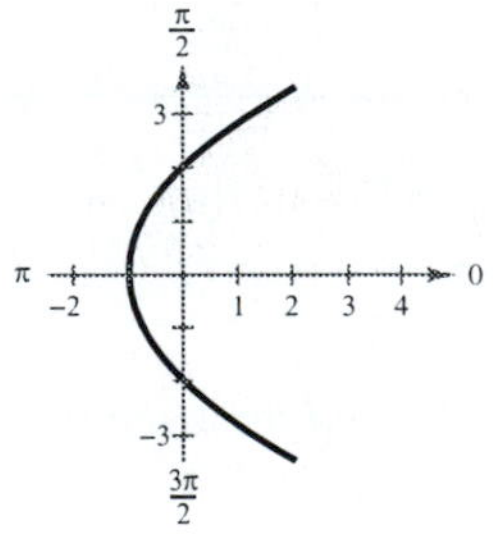

11. $r = \dfrac{5}{1 + \sin\theta}$

$e = 1$, the graph is a parabola.

Vertex: $\left(\dfrac{5}{2}, \dfrac{\pi}{2}\right)$

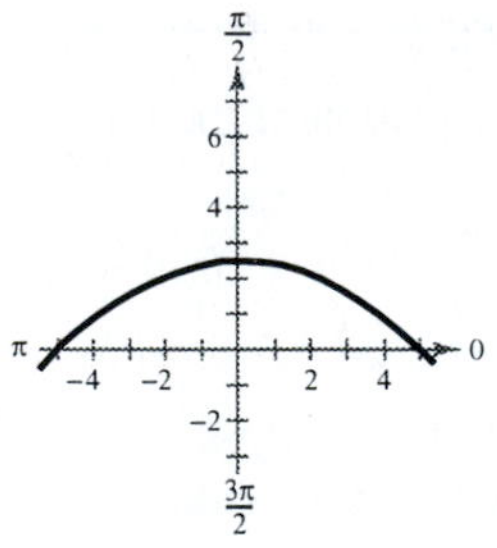

13. $r = \dfrac{2}{2 - \cos\theta} = \dfrac{1}{1 - \frac{1}{2}\cos\theta}$

$e = \dfrac{1}{2} < 1$, the graph is an ellipse.

Vertices: $(2, 0), \left(\dfrac{2}{3}, \pi\right)$

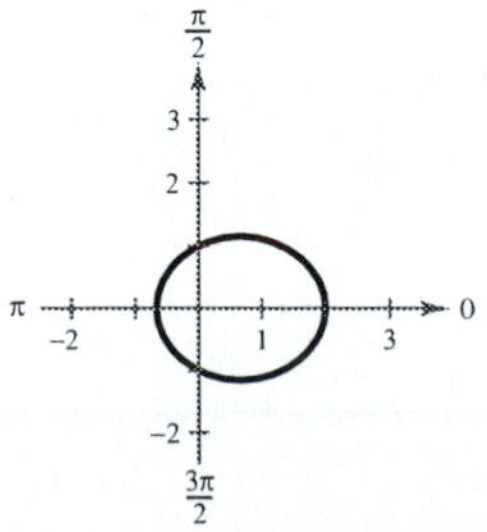

15. $r = \dfrac{4}{2 + \sin\theta} = \dfrac{2}{1 + \frac{1}{2}\sin\theta}$

$e = \dfrac{1}{2} < 1$, the graph is an ellipse.

Vertices: $\left(\dfrac{4}{3}, \dfrac{\pi}{2}\right), \left(4, \dfrac{3\pi}{2}\right)$

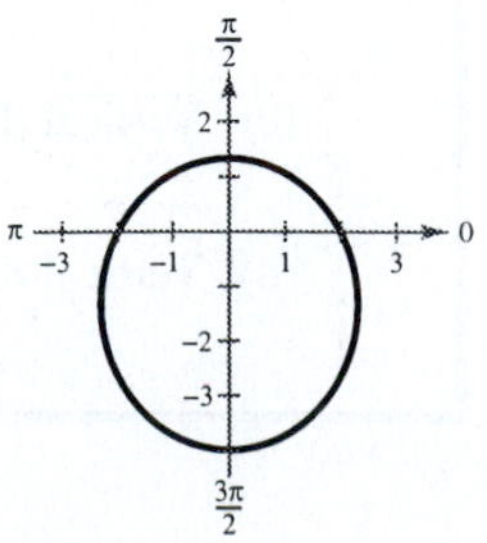

17. $r = \dfrac{3}{2 + 4\sin\theta} = \dfrac{3/2}{1 + 2\sin\theta}$

$e = 2 > 1$, the graph is a hyperbola.

Vertices: $\left(\dfrac{1}{2}, \dfrac{\pi}{2}\right), \left(-\dfrac{3}{2}, \dfrac{3\pi}{2}\right)$

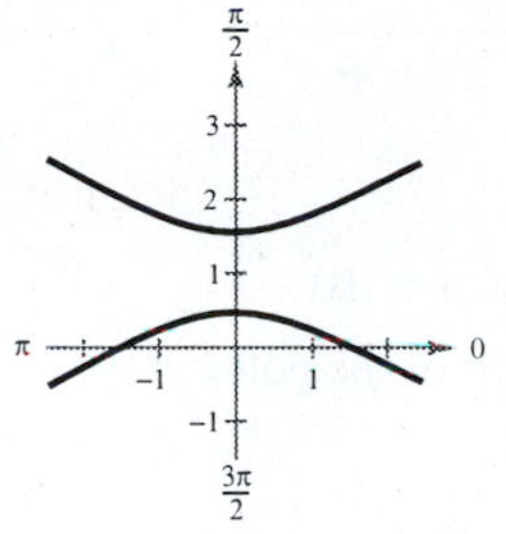

19. $r = \dfrac{3}{2 - 6\cos\theta} = \dfrac{3/2}{1 - 3\cos\theta}$

$e = 3 > 1$, the graph is a hyperbola.

Vertices: $\left(-\dfrac{3}{4}, 0\right), \left(\dfrac{3}{8}, \pi\right)$

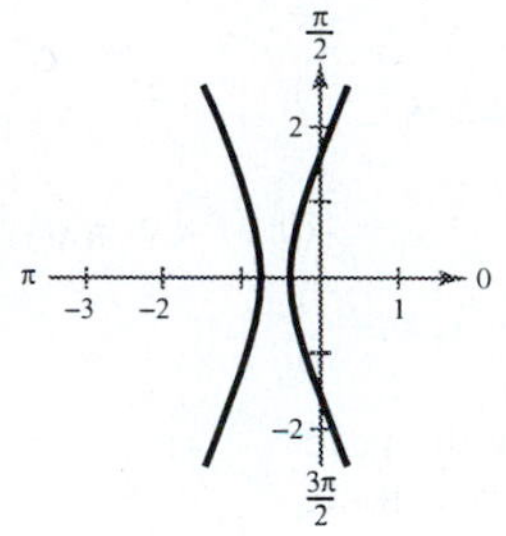

21. $r = \dfrac{6}{2 - \cos\theta} = \dfrac{3}{1 - \frac{1}{2}\cos\theta}$

$e = \dfrac{1}{2} < 1$, the graph is an ellipse.

Vertices: $(6, 0), (2, \pi)$

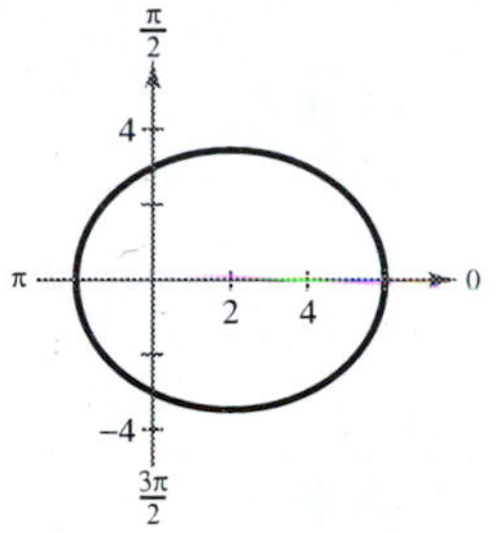

23. $r = \dfrac{-1}{1 - \sin\theta}$

$e = 1 \Rightarrow$ Parabola

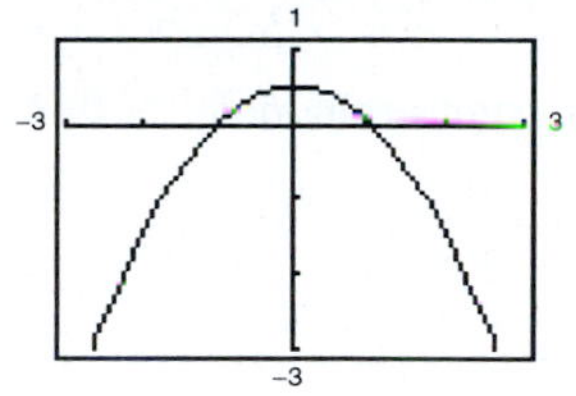

25. $r = \dfrac{3}{-4 + 2\cos\theta}$

$e = \dfrac{1}{2} \Rightarrow$ Ellipse

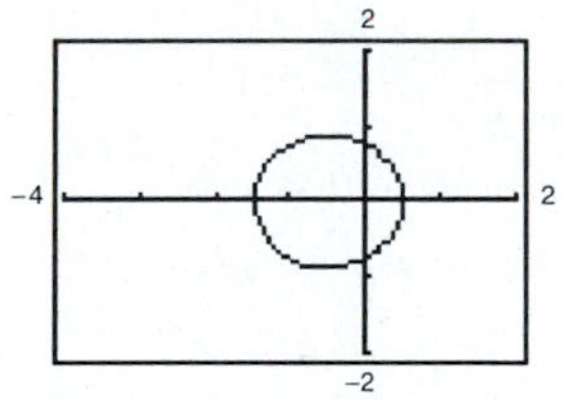

27. $r = \dfrac{2}{1 - \cos(\theta - \pi/4)}$

Rotate the graph in Exercise 9 through the angle $\dfrac{\pi}{4}$.

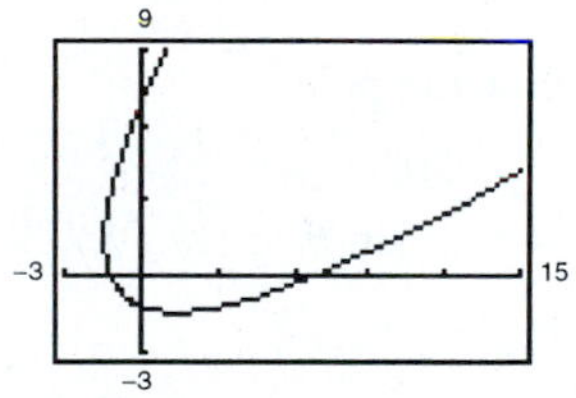

29. $r = \dfrac{4}{2 + \sin(\theta + \pi/6)}$

Rotate the graph in Exercise 15 through the angle $-\pi/6$.

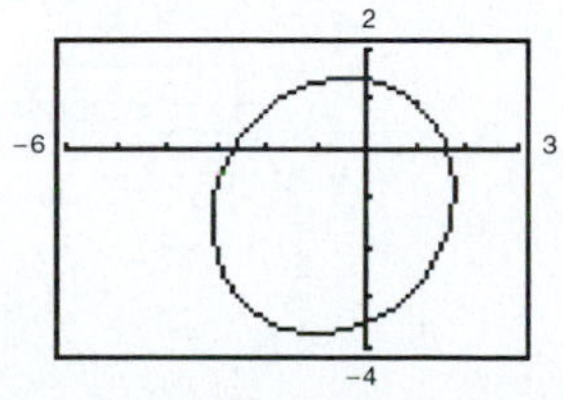

31. Parabola: $e = 1$

Directrix: $x = -1$

Vertical directrix to the left of the pole

$r = \dfrac{1(1)}{1 - 1\cos\theta} = \dfrac{1}{1 - \cos\theta}$

33. Ellipse: $e = \dfrac{1}{2}$

Directrix: $y = 1$

$p = 1$

Horizontal directrix above the pole

$$r = \frac{\frac{1}{2}(1)}{1 + \frac{1}{2}\sin\theta} = \frac{1}{2 + \sin\theta}$$

35. Hyperbola: $e = 2$

Directrix: $x = 1$

$p = 1$

Vertical directrix to the right of the pole

$$r = \frac{2(1)}{1 + 2\cos\theta} = \frac{2}{1 + 2\cos\theta}$$

37. Parabola

Vertex: $\left(1, -\dfrac{\pi}{2}\right) \Rightarrow e = 1, p = 2$

Horizontal directrix below the pole

$$r = \frac{1(2)}{1 - 1\sin\theta} = \frac{2}{1 - \sin\theta}$$

39. Parabola

Vertex: $(5, \pi) \Rightarrow e = 1, p = 10$

Vertical directrix to the left of the pole

$$r = \frac{1(10)}{1 - 1\cos\theta} = \frac{10}{1 - \cos\theta}$$

41. Ellipse: Vertices $(2, 0), (8, \pi)$

Center: $(3, \pi)$; $c = 3, a = 5, e = \dfrac{3}{5}$

Vertical directrix to the right of the pole

$$r = \frac{\frac{3}{5}p}{1 + \frac{3}{5}\cos\theta} = \frac{3p}{5 + 3\cos\theta}$$

$$2 = \frac{3p}{5 + 3\cos 0}$$

$$p = \frac{16}{3}$$

$$r = \frac{3\left(\frac{16}{3}\right)}{5 + 3\cos\theta} = \frac{16}{5 + 3\cos\theta}$$

43. Ellipse: Vertices $(20, 0), (4, \pi)$

Center: $(8, 0)$; $c = 8, a = 12, e = \dfrac{2}{3}$

Vertical directrix to the left of the pole

$$r = \frac{\frac{2}{3}p}{1 - \frac{2}{3}\cos\theta} = \frac{2p}{3 - 2\cos\theta}$$

$$20 = \frac{2p}{3 - 2\cos 0}$$

$$p = 10$$

$$r = \frac{2(10)}{3 - 2\cos\theta} = \frac{20}{3 - 2\cos\theta}$$

45. Hyperbola: Vertices $\left(1, \dfrac{3\pi}{2}\right), \left(9, \dfrac{3\pi}{2}\right)$

Center: $\left(5, \dfrac{3\pi}{2}\right)$; $c = 5, a = 4, e = \dfrac{5}{4}$

Horizontal directrix below the pole

$$r = \frac{5/4p}{1 - 5/4\sin\theta} = \frac{5p}{4 - 5\sin\theta}$$

$$1 = \frac{5p}{4 - 5\sin(3\pi/2)}$$

$$p = \frac{9}{5}$$

$$r = \frac{5(9/5)}{4 - 5\sin\theta} = \frac{9}{4 - 5\sin\theta}$$

47.

$$\frac{x^2}{a^2} + \frac{y^2}{b^2} = 1$$

$$\frac{r^2\cos^2\theta}{a^2} + \frac{r^2\sin^2\theta}{b^2} = 1$$

$$\frac{r^2\cos^2\theta}{a^2} + \frac{r^2(1 - \cos^2\theta)}{b^2} = 1$$

$$r^2b^2\cos^2\theta + r^2a^2 - r^2a^2\cos^2\theta = a^2b^2$$

$$r^2(b^2 - a^2)\cos^2\theta + r^2a^2 = a^2b^2$$

$$b^2 - a^2 = -c^2$$

$$-r^2c^2\cos^2\theta + r^2a^2 = a^2b^2$$

$$-r^2\left(\frac{c}{a}\right)^2\cos^2\theta + r^2 = b^2, \quad e = \frac{c}{a}$$

$$-r^2e^2\cos^2\theta + r^2 = b^2$$

$$r^2(1 - e^2\cos^2\theta) = b^2$$

$$r^2 = \frac{b^2}{1 - e^2\cos^2\theta}$$

49. $\dfrac{x^2}{169} + \dfrac{y^2}{144} = 1$

$a = 13, b = 12, c = 5, e = \dfrac{5}{13}$

$$r^2 = \frac{144}{1 - \left(\frac{25}{169}\right)\cos^2\theta} = \frac{24{,}336}{169 - 25\cos^2\theta}$$

51. $\dfrac{x^2}{9} - \dfrac{y^2}{16} = 1$

$a = 3, b = 4, c = 5, e = \dfrac{5}{3}$

$$r^2 = \frac{-16}{1 - \left(\frac{25}{9}\right)\cos^2\theta} = \frac{144}{25\cos^2\theta - 9}$$

53. One focus: $\left(5, \dfrac{\pi}{2}\right)$

Vertices: $\left(4, \dfrac{\pi}{2}\right), \left(4, -\dfrac{\pi}{2}\right)$

$a = 4, c = 5 \Rightarrow b = 3$ and $e = \dfrac{5}{4}$

$$\frac{y^2}{16} - \frac{x^2}{9} = 1$$

$$\frac{r^2\sin^2\theta}{16} - \frac{r^2\cos^2\theta}{9} = 1$$

$$9r^2\sin^2\theta - 16r^2(1 - \sin^2\theta) = 144$$

$$25r^2\sin^2\theta - 16r^2 = 144$$

$$r^2 = \frac{144}{25\sin^2\theta - 16} = \frac{-144}{16 - 25\sin^2\theta}$$

55. When $\theta = 0$, $r = c + a = ea + a = a(1 + e)$. Therefore,

$$a(1 + e) = \frac{ep}{1 - e\cos\theta}$$

$$a(1 + e)(1 - e) = ep$$

$$a(1 - e^2) = ep.$$

Thus, $r = \dfrac{ep}{1 - e\cos\theta} = \dfrac{(1 - e^2)a}{1 - e\cos\theta}$.

57. $r = \dfrac{[1 - (0.0167)^2](92.957 \times 10^6)}{1 - 0.0167\cos\theta}$

$\approx \dfrac{9.2931 \times 10^7}{1 - 0.0167\cos\theta}$

Perihelion distance:
$r = 92.957 \times 10^6(1 - 0.0167) \approx 9.1405 \times 10^7$

Aphelion distance:
$r = 92.957 \times 10^6(1 + 0.0167) \approx 9.4509 \times 10^7$

59. $r = \dfrac{[1 - (0.2481)^2](5.9 \times 10^9)}{1 - 0.2481\cos\theta}$

$\approx \dfrac{5.5368 \times 10^9}{1 - 0.2481\cos\theta}$

Perihelion distance:
$r = 5.9 \times 10^9(1 - 0.2481) \approx 4.4362 \times 10^9$

Aphelion distance:
$r = 5.9 \times 10^9(1 + 0.2481) \approx 7.3638 \times 10^9$

61. Vertex: $\left(4100, \dfrac{\pi}{2}\right)$

Focus: $(0, 0)$

$e = 1, p = 8200$

$$r = \frac{ep}{1 + e\sin\theta} = \frac{8200}{1 + \sin\theta}$$

When $\theta = 30^\circ$, $r = 8200/1.5 \approx 5466.67$.
Distance between the surface of the earth and the satellite is $r - 4000 \approx 1467$ miles.

63. $f(x) = 4^{x-2}$

Exponential function

x	0	1	2	3
y	$\frac{1}{16}$	$\frac{1}{4}$	1	4

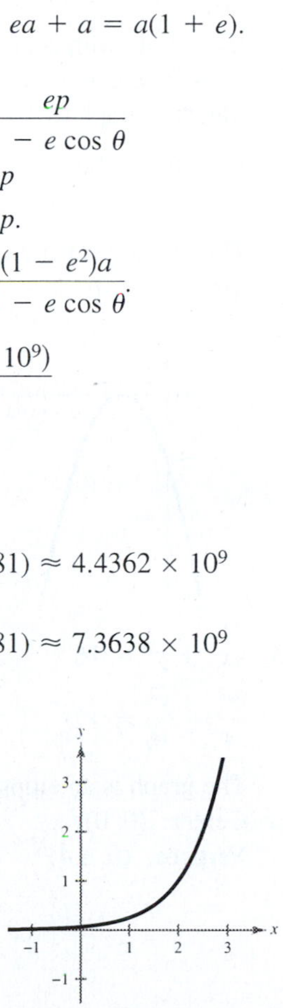

65. $h(t) = \log_4(t - 2)$

Logarithmic function

$4^y = t - 2$

$4^y + 2 = t$

t	$2\frac{1}{4}$	3	6
y	-1	0	1

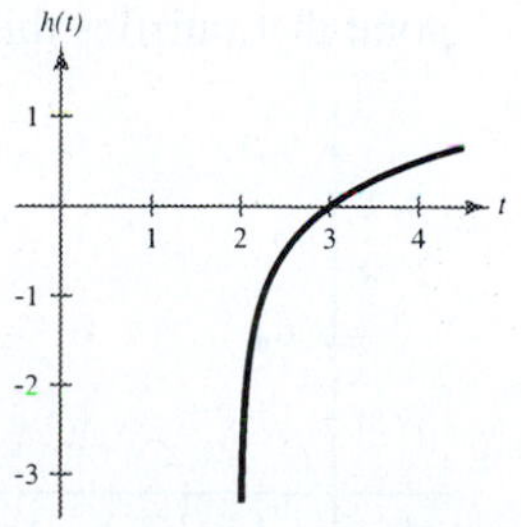

Vertical asymptote: $t = 2$

Review Exercises for Chapter 10

Solutions to Odd-Numbered Exercises

1. $m = \tan 120°$
$= -\sqrt{3}$

3. $y = -x + 10$
$m = -1 = \tan \theta$
$\theta = \arctan(-1) = 135°$

5. $(1, 2) \Rightarrow x_1 = 1, y_1 = 2$
$x - y - 3 = 0 \Rightarrow A = 1, B = -1, C = -3$
$$d = \frac{|1(1) + (-1)(2) + (-3)|}{\sqrt{1^2 + (-1)^2}} = \frac{4}{\sqrt{2}} = 2\sqrt{2}$$

7. $4x^2 + y^2 = 4$
$$\frac{x^2}{1} + \frac{y^2}{4} = 1$$
Ellipse with center (0, 0) and a vertical major axis
Matches graph (g).

9. $4x^2 - y^2 = 4$
$$\frac{x^2}{1} - \frac{y^2}{4} = 1$$
Hyperbola with center (0, 0) and a horizontal transverse axis
Matches graph (f).

11. $x^2 + 4y^2 = 4$
$$\frac{x^2}{4} + \frac{y^2}{1} = 1$$
Ellipse with center (0, 0) and a horizontal major axis.
Matches graph (d)

13. $x^2 = -6y$
$$y = -\frac{1}{6}x^2$$
Parabola with vertex (0, 0) and opening downward
Matches graph (b).

15. $x^2 - 5y^2 = -5$
$$\frac{y^2}{1} - \frac{x^2}{5} = 1$$
Hyperbola with center (0, 0) and a vertical transverse axis
Matches graph (h).

17. $4x - y^2 = 0$
$y^2 = 4x$
The graph is a parabola.
Vertex: (0, 0)

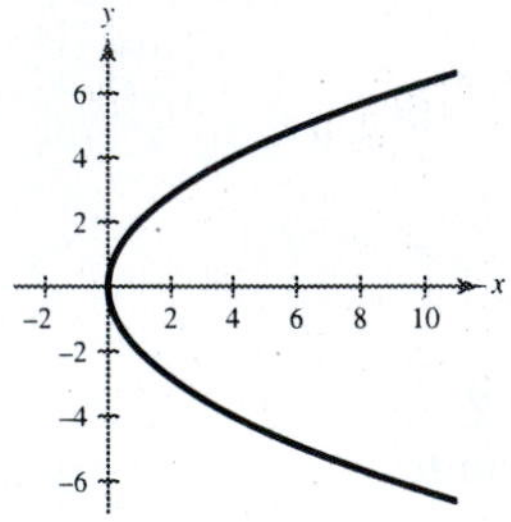

19. $x^2 - 6x + 2y + 9 = 0$
$(x - 3)^2 = -2y$
The graph is a parabola.
Vertex: (3, 0)

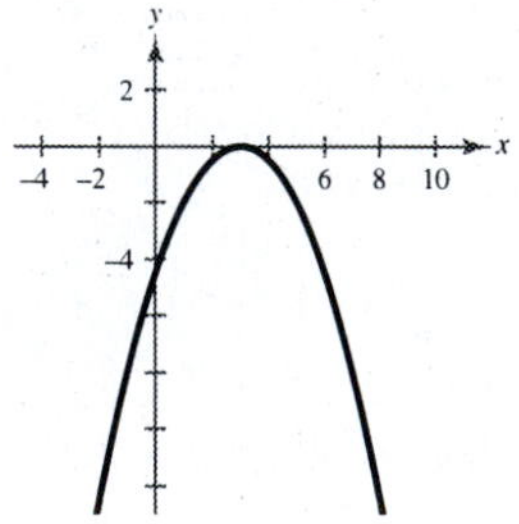

21. $x^2 + y^2 - 2x - 4y + 5 = 0$
$(x - 1)^2 + (y - 2)^2 = 0$
The graph is a degenerate circle. (1, 2) is the only point that satisfies this equation.

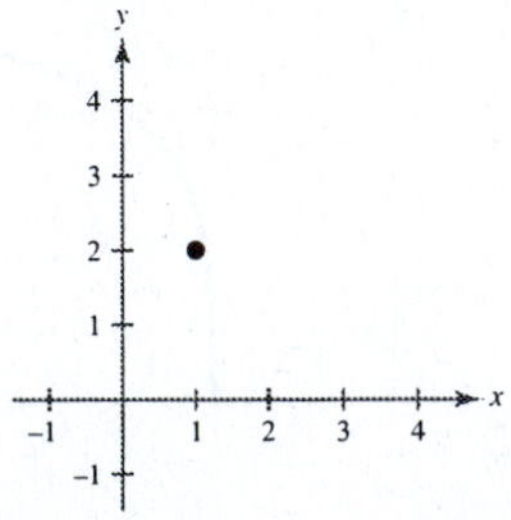

23. $4x^2 + y^2 = 16$
$$\frac{x^2}{4} + \frac{y^2}{16} = 1$$
The graph is an ellipse.
Center: (0, 0)
Vertices: $(0, \pm 4)$

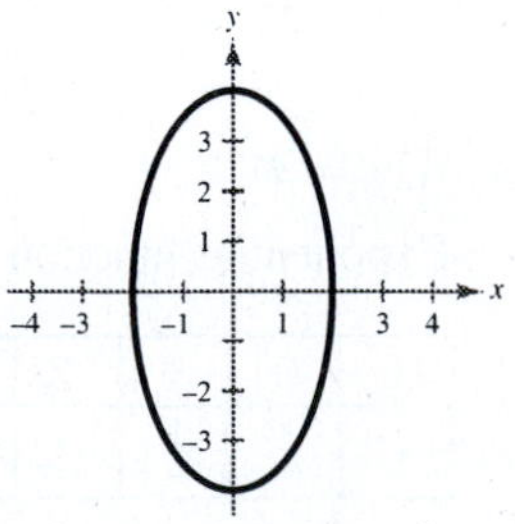

25. $x^2 + 9y^2 + 10x - 18y + 25 = 0$

$$(x + 5)^2 + 9(y - 1)^2 = 9$$

$$\frac{(x + 5)^2}{9} + \frac{(y - 1)^2}{1} = 1$$

The graph is an ellipse.

Center: $(-5, 1)$

Vertices: $(-8, 1)$ $(-2, 1)$

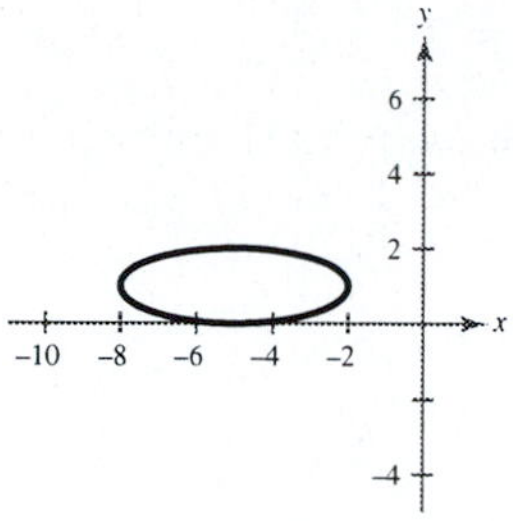

27. $5y^2 - 4x^2 = 20$

$$\frac{y^2}{4} - \frac{x^2}{5} = 1$$

The graph is a hyperbola.

Center: $(0, 0)$

Vertices: $(\pm 2, 0)$

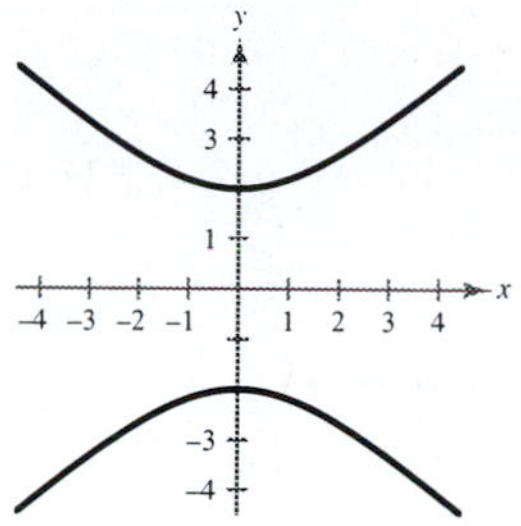

29. $B^2 - 4AC = 2^2 - 4(1)(1) = 0$

The graph is a parabola.

$$\cot 2\theta = \frac{A - C}{B} = 0 \Rightarrow 2\theta = \frac{\pi}{2} \Rightarrow \theta = \frac{\pi}{4}$$

$$x = x'\cos\frac{\pi}{4} - y'\sin\frac{\pi}{4} = \frac{x' - y'}{\sqrt{2}}$$

$$y = x'\sin\frac{\pi}{4} + y'\cos\frac{\pi}{4} = \frac{x' + y'}{\sqrt{2}}$$

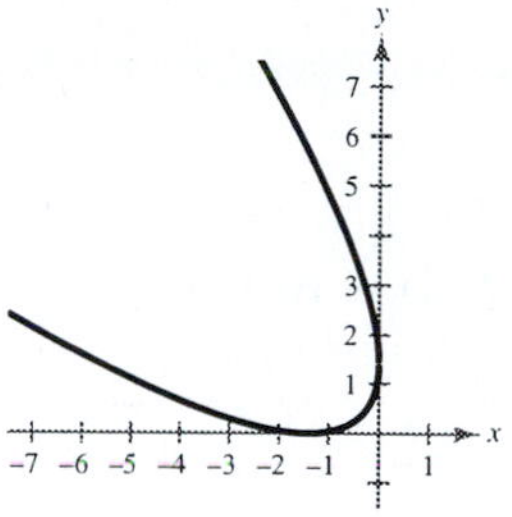

$$\left(\frac{x' - y'}{\sqrt{2}}\right)^2 + \left(\frac{x' + y'}{\sqrt{2}}\right)^2 + 2\left(\frac{x' - y'}{\sqrt{2}}\right)\left(\frac{x' + y'}{\sqrt{2}}\right) + 2\sqrt{2}\left(\frac{x' - y'}{\sqrt{2}}\right) - 2\sqrt{2}\left(\frac{x' + y'}{\sqrt{2}}\right) + 2 = 0$$

$$2(x')^2 - 4y' + 2 = 0$$

$$(x')^2 = 2y' - 1$$

Vertex: $(x', y') = \left(0, \frac{1}{2}\right)$, $\theta = 45°$

31. $AC = 3(2) = 6 > 0$

The graph is an ellipse.

$$3x^2 + 2y^2 - 12x + 12y + 29 = 0$$

$$\frac{(x - 2)^2}{1/3} + \frac{(y + 3)^2}{1/2} = 1$$

Center: $(2, -3)$

Vertices: $\left(2, -3 \pm \frac{\sqrt{2}}{2}\right)$

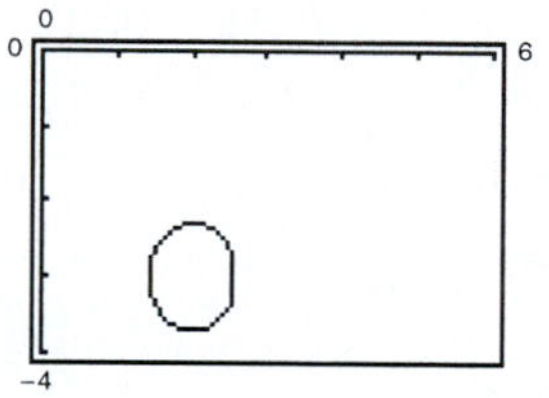

To use a graphing calculator, we need to solve for y in terms of x.

$$2(y + 3)^2 = 1 - 3(x - 2)^2|$$

$$y = -3 \pm \sqrt{\frac{1 - (x - 2)^2}{2}}$$

33. $x^2 - 10xy + y^2 + 1 = 0$

Since $B^2 - 4AC = (-10)^2 - 4(1)(1) > 0$, the graph is a hyperbola.

To use a graphing calculator, we need to solve for y in terms of x.

$$(y^2 - 10xy + 25x^2) = -x^2 - 1 + 25x^2$$
$$(y - 5x)^2 = 24x^2 - 1$$
$$y = 5x \pm \sqrt{24x^2 - 1}$$

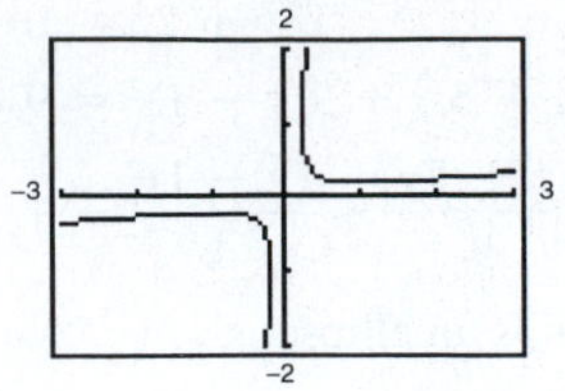

35. Vertex: $(4, 2) = (h, k)$

Focus: $(4, 0) \Rightarrow p = -2$

$(x - h)^2 = 4p(y - k)$

$(x - 4)^2 = -8(y - 2)$

37. Vertex: $(0, 2) = (h, k)$

Directrix: $x = -3 \Rightarrow p = 3$

$(y - k^{2)} = 4p(x - h)$

$(y - 2)^2 = 12x$

39. Vertices: $(-3, 0), (7, 0) \Rightarrow a = 5$

$(h, k) = (2, 0)$

Foci: $(0, 0), (4, 0) \Rightarrow c = 2$

$b^2 = a^2 - c^2 = 25 - 4 = 21$

$$\frac{(x - h)^2}{a^2} + \frac{(y - k)^2}{b^2} = 1$$
$$\frac{(x - 2)^2}{25} + \frac{y^2}{21} = 1$$

41. Vertices: $(0, \pm 6) \Rightarrow a = 6, (h, k) = (0, 0)$

Passes through $(2, 2)$

$$\frac{(x - h)^2}{b^2} + \frac{(y - k)^2}{a^2} = 1$$
$$\frac{x^2}{b^2} + \frac{y^2}{36} = 1 \Rightarrow b^2 = \frac{36(4)}{36 - 4} = \frac{36x^2}{36 - y^2} = \frac{9}{2}$$
$$\frac{x^2}{9/2} + \frac{y^2}{36} = 1$$
$$\frac{2x^2}{9} + \frac{y^2}{36} = 1$$

43. Vertices: $(0, \pm 1) \Rightarrow a = 1, (h, k) = (0, 0)$

Foci: $(0, \pm 3) \Rightarrow c = 3$

$b^2 = c^2 - a^2 = 9 - 1 = 8$

$$\frac{(y - k)^2}{a^2} - \frac{(x - h)^2}{b^2} = 1$$
$$y^2 - \frac{x^2}{8} = 1$$

45. Foci: $(0, 0), (8, 0) \Rightarrow c = 4, (h, k) = (4, 0)$

Asymptotes: $y = \pm 2(x - 4) \Rightarrow \dfrac{b}{a} = 2, b = 2a$

47. Parabola

Opens downward

Vertex: (0, 12)

$(x - h)^2 = 4p(y - k)$

$x^2 = 4p(y - 12)$

Solution points: $(\pm 4, 10)$

$16 = 4p(10 - 12)$

$16 = -8p$

$-2 = p$

$x^2 = -8(y - 12)$

To find the x-intercepts, let $y = 0$.

$x^2 = 96$

$x = \pm\sqrt{96} = \pm 4\sqrt{6}$

At the base,z the archway is $2(4\sqrt{6}) = 8\sqrt{6}$ meters wide.

49. $BD = AD + 6\left(\frac{1100}{5280}\right)$

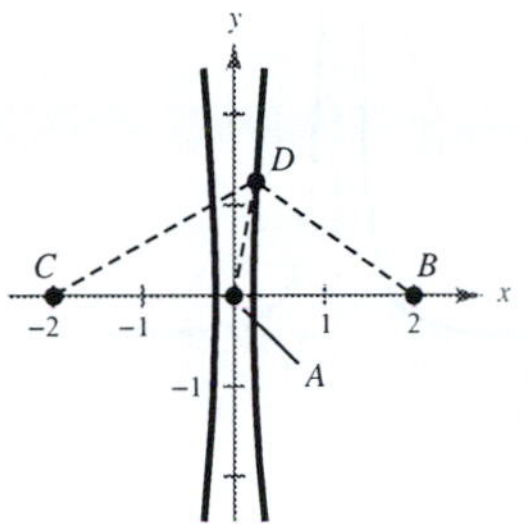

$CD = AD + 8\left(\frac{1100}{5280}\right)$

$2a = CD - BD = 2\left(\frac{1100}{5280}\right)$

$a = \frac{5}{24}, c = 2 \Rightarrow b^2 = \frac{2279}{576}$

Thus, we have $\frac{576x^2}{25} - \frac{576y^2}{2279} = 1$ OR:

$CD = AD + 8\left(\frac{1100}{5280}\right)$

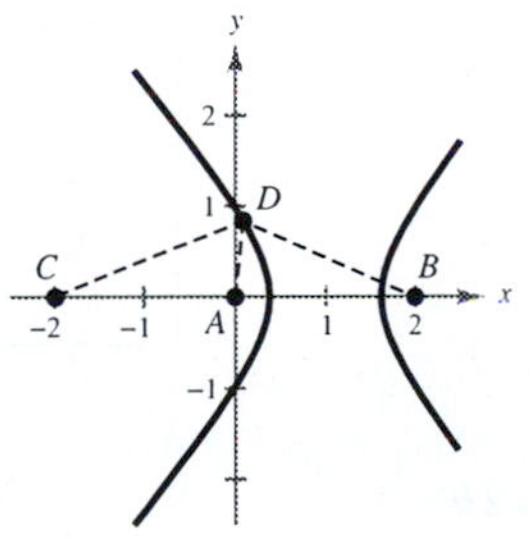

$BD = AD + 6\left(\frac{1100}{5280}\right)$

$2a = BD - AD = 6\left(\frac{1100}{5280}\right)$

$a = 3\left(\frac{5}{24}\right) = \frac{5}{8}, c = 1 \Rightarrow b^2 = \frac{39}{64}$

Center: (1, 0)

$\frac{64(x-1)^2}{25} - \frac{64y^2}{39} = 1$

$CD - 8\left(\frac{1100}{5280}\right) = AD$

$$BD = CD - 8\left(\frac{1100}{5280}\right) + 6\left(\frac{1100}{5280}\right) = CD - 2\left(\frac{1100}{5280}\right)$$

Thus, the friend to the west hears the explosion two seconds after the friend to the east.

51. $x = 2t \Rightarrow \frac{x}{2} = t$

$y = 4t \Rightarrow y = 4\left(\frac{x}{2}\right) = 2x$

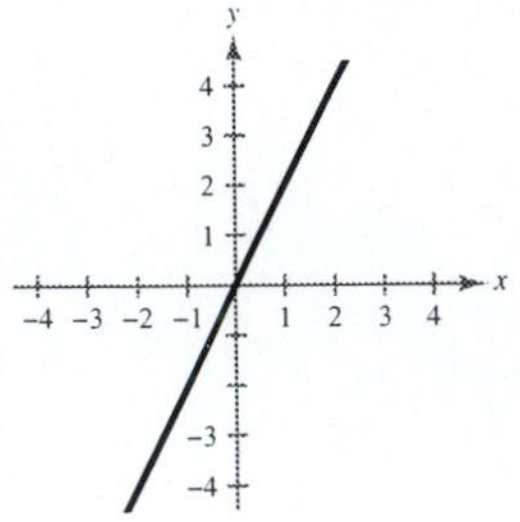

53. $x = 1 + 4t, y = 2 - 3t$

$t = \frac{x-1}{4}$

$y = 2 - 3\left(\frac{x-1}{4}\right)$

$3x + 4y = 11$

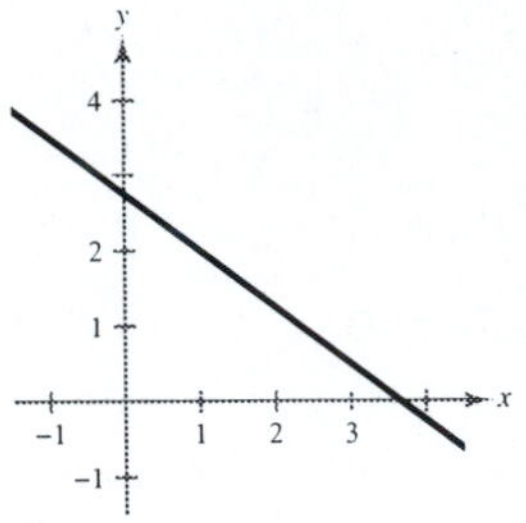

55. $x = \frac{1}{t}, y = t^2$

$t = \frac{1}{x}$

$y = \frac{1}{x^2}$

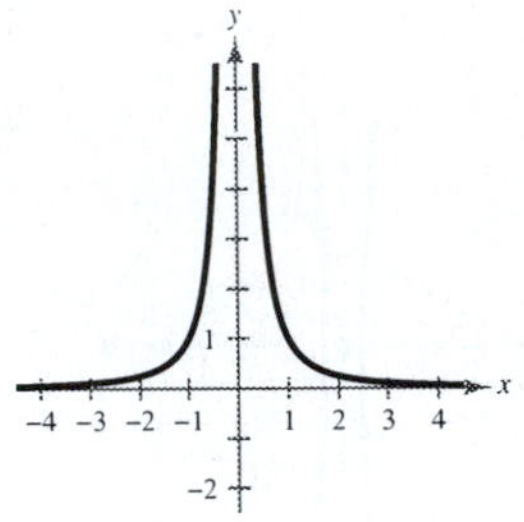

57. $x = 6\cos\theta, y = 6\sin\theta$

$\cos\theta = \frac{x}{6}, \sin\theta = \frac{y}{6}$

$\frac{x^2}{36} + \frac{y^2}{36} = 1$

$x^2 + y^2 = 36$

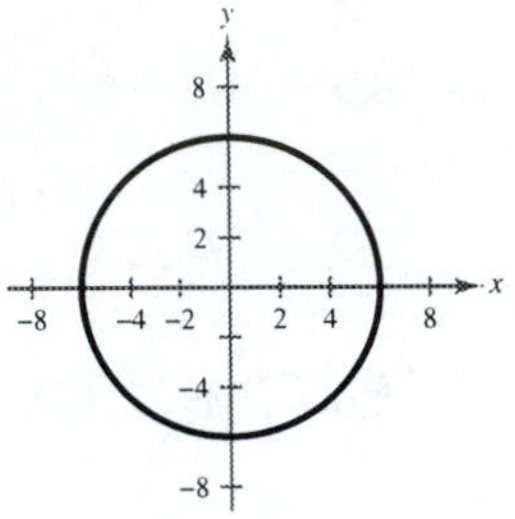

59. $(h, k) = (-3, 4)$

$2a = 8 \Rightarrow a = 4$

$2b = 6 \Rightarrow b = 3$

$\frac{(x+3)^2}{16} + \frac{(y-4)^2}{9} = 1$

$x = -3 + 4\cos\theta$

$y = 4 + 3\sin\theta$

This solution is not unique.

61. $x = \cos 3\theta + 5\cos\theta$

$y = \sin 3\theta + 5\sin\theta$

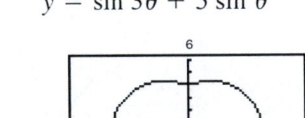

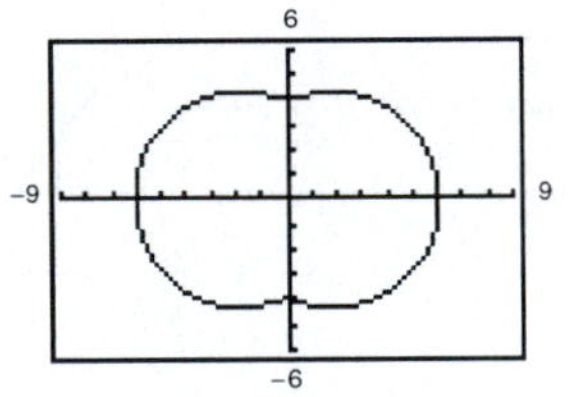

63.

$r = 3\cos\theta$

$r^2 = 3r\cos\theta$

$x^2 + y^2 = 3x$

65.

$r^2 = \cos 2\theta$

$r^2 = 1 - 2\sin^2\theta$

$r^4 = r^2 - 2r^2\sin^2\theta$

$(x^2 + y^2)^2 = x^2 + y^2 - 2y^2$

$(x^2 + y^2)^2 - x^2 + y^2 = 0$

67. $(x^2 + y^2)^2 = ax^2y$

$(r^2)^2 = ar^2\cos^2\theta\, r\sin\theta$

$r = a\cos^2\theta\sin\theta$

69. $r = 4$

Circle of radius 4 centered at the pole

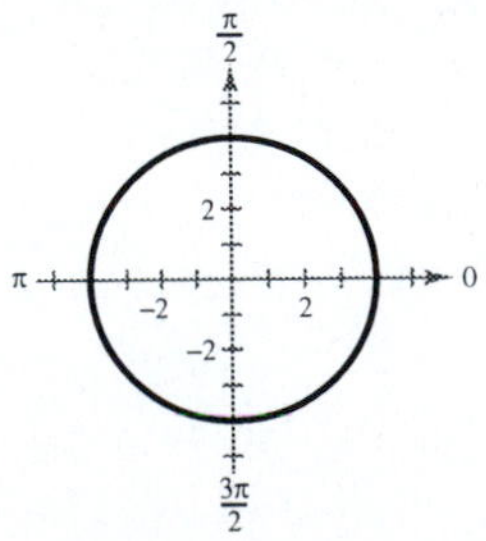

71. $r = 4\sin 2\theta$

Symmetric with respect to $\theta = \pi/2$

Rose curve ($n = 2$) with 4 petals

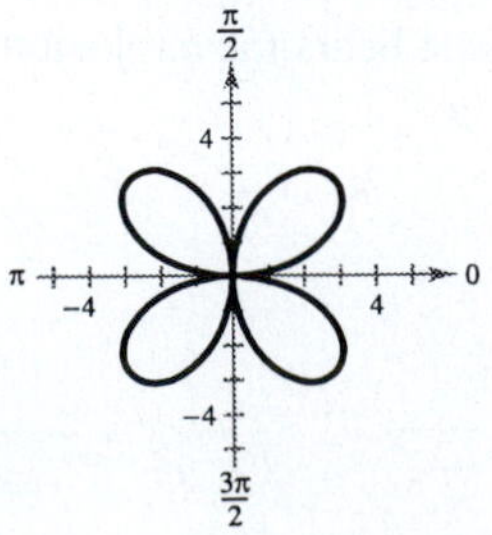

73. $r = -2 - 2\cos\theta$

Symmetric with respect to polar axis

$\frac{a}{b} = \frac{2}{2} = 1 \Rightarrow$ Cardioid

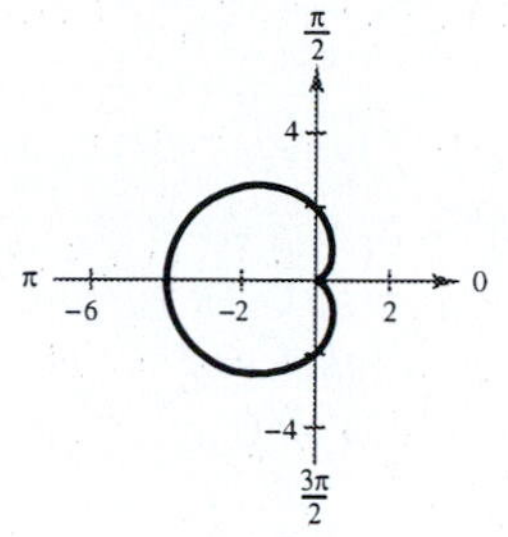

75. $r = 4 - 3\cos\theta$

Symmetric with respect to polar axis

$\frac{a}{b} = \frac{4}{3} > 0 \Rightarrow$ Dimpled limaçon

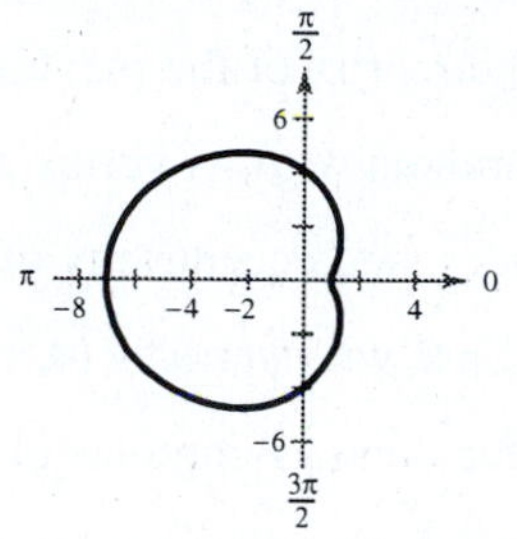

77. $r = -3\cos 2\theta$

Symmetric with respect to the polar axis, $\theta = \pi/2$, and the pole

Rose curve with 4 petals

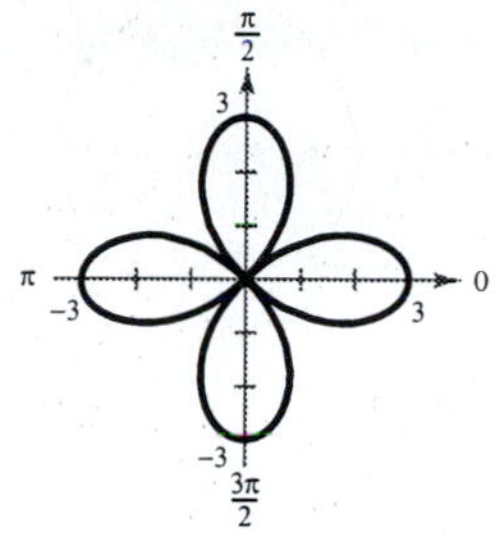

79. $r = \dfrac{2}{1 - \sin\theta}$, $e = 1$

Parabola symmetric with $\theta = \pi/2$ and the vertex at $(1, 3\pi/2)$

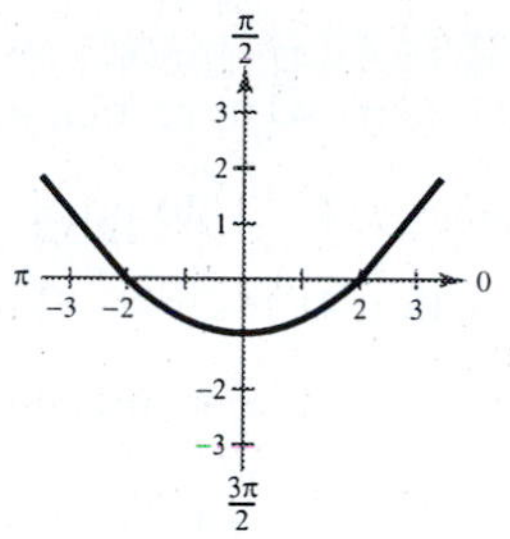

81. $r^2 = 4\sin^2 2\theta \Rightarrow r = \pm 2\sin 2\theta$

Symmetric with respect to $\theta = \pi/2$, polar axis, and pole

Rose curve ($n = 2$) with 4 petals

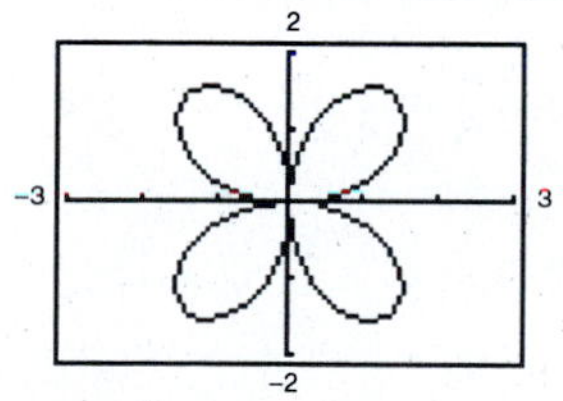

83. $r = \dfrac{3}{\cos(\theta - \pi/4)}$

The graph is a line.

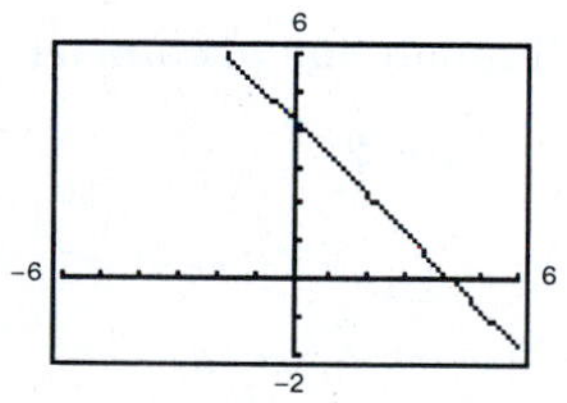

85. Center: $(5, \pi/2)$

Solution point: $(0, 0) \Rightarrow$ Radius $= 5 \Rightarrow a = 10$

$r = a\sin\theta$

$r = 10\sin\theta$

87. Parabola: $r = \dfrac{ep}{1 - e\cos\theta}$, $e = 1$

Vertex: $(2, \pi)$

Focus: $(0, 0) \Rightarrow p = 4$

$$r = \frac{4}{1 - \cos\theta}$$

89. Ellipse: $r = \dfrac{ep}{1 - e\cos\theta}$; Vertices: $(5, 0), (1, \pi) \Rightarrow a = 3$; One focus: $(0, 0) \Rightarrow c = 2$

$$e = \frac{c}{a} = \frac{2}{3}, p = \frac{5}{2}$$

$$r = \frac{(2/3)(5/2)}{1 - (2/3)\cos\theta} = \frac{5/3}{1 - (2/3)\cos\theta} = \frac{5}{3 - 2\cos\theta}$$

❑ Practice Test for Chapter 10

1. Find the angle, θ, between the lines $3x + 4y = 12$ and $4x - 3y = 12$.

2. Find the distance between the point $(5, -9)$ and the line $3x - 7y = 21$.

3. Find the vertex, focus and directrix of the parabola $x^2 - 6x - 4y + 1 = 0$.

4. Find an equation of the parabola with its vertex at $(2, -5)$ and focus at $(2, -6)$.

5. Find the center, foci, vertices, and eccentricity of the ellipse $x^2 + 4y^2 - 2x + 32y + 61 = 0$.

6. Find an equation of the ellipse with vertices $(0, \pm 6)$ and eccentricity $e = \frac{1}{2}$.

7. Find the center, vertices, foci, and asymptotes of the hyperbola $16y^2 - x^2 - 6x - 128y + 231 = 0$.

8. Find an equation of the hyperbola with vertices at $(\pm 3, 2)$ and foci at $(\pm 5, 2)$.

9. Rotate the axes to eliminate the xy-term. Sketch the graph of the resulting equation, showing both sets of axes.

   $5x^2 + 2xy + 5y^2 - 10 = 0$

10. Use the discriminant to determine whether the graph of the equation is a parabola, ellipse, or hyperbola.

    (a) $6x^2 - 2xy + y^2 = 0$ (b) $x^2 + 4xy + 4y^2 - x - y + 17 = 0$

11. Convert the polar point $(\sqrt{2}, (3\pi)/4)$ to rectangular coordinates.

12. Convert the rectangular point $(\sqrt{3}, -1)$ to polar coordinates.

13. Convert the rectangular equation $4x - 3y = 12$ to polar form.

14. Convert the polar equation $r = 5 \cos \theta$ to rectangular form.

15. Sketch the graph of $r = 1 - \cos \theta$.

16. Sketch the graph of $r = 5 \sin 2\theta$.

17. Sketch the graph of $r = \dfrac{3}{6 - \cos \theta}$.

18. Find a polar equation of the parabola with its vertex at $(6, \pi/2)$ and focus at $(0, 0)$.

For Exercises 19 and 20, eliminate the parameter and write the corresponding rectangular equation.

19. $x = 3 - 2 \sin \theta,\ y = 1 + 5 \cos \theta$

20. $x = e^{2t},\ y = e^{4t}$

❑ Chapter P Practice Test Solutions

1. $\frac{|-42| - 20}{15 - |-4|} = \frac{42 - 20}{15 - 4} = \frac{22}{11} = 2$

2. $\frac{x}{z} - \frac{z}{y} = \frac{x}{z} \cdot \frac{y}{y} - \frac{z}{y} \cdot \frac{z}{z} = \frac{xy - z^2}{yz}$

3. $|x - 7| \leq 4$

4. $10(-5)^3 = 10(-125) = -1250$

5. $(-4x^3)(-2x^{-5})\left(\frac{1}{16}x\right) = (-4)(-2)\left(\frac{1}{16}\right)x^{3+(-5)+1} = \frac{8}{16}x^{-1} = \frac{1}{2x}$

6. $0.0000412 = 4.12 \times 10^{-5}$

7. $125^{2/3} = (\sqrt[3]{125})^2 = (5)^2 = 25$

8. $\sqrt[4]{64x^7y^9} = \sqrt[4]{16 \cdot 4x^4x^3y^8y} = 2xy^2\sqrt[4]{4x^3y}$

9. $\frac{6}{\sqrt{12}} = \frac{6}{2\sqrt{3}} \cdot \frac{\sqrt{3}}{\sqrt{3}} = \frac{6\sqrt{3}}{6} = \sqrt{3}$

10. $3\sqrt{80} - 7\sqrt{500} = 3(4\sqrt{5}) - 7(10\sqrt{5}) = 12\sqrt{5} - 70\sqrt{5} = -58\sqrt{5}$

11. $(8x^4 - 9x^2 + 2x - 1) - (3x^3 + 5x + 4) = 8x^4 - 3x^3 - 9x^2 - 3x - 5$

12. $(x - 3)(x^2 + x - 7) = x^3 + x^2 - 7x - 3x^2 - 3x + 21 = x^3 - 2x^2 - 10x + 21$

13. $[(x - 2) - y]^2 = (x - 2)^2 - 2y(x - 2) + y^2$

$= x^2 - 4x + 4 - 2xy + 4y + y^2 = x^2 + y^2 - 2xy - 4x + 4y + 4$

14. $16x^4 - 1 = (4x^2 + 1)(4x^2 - 1) = (4x^2 + 1)(2x + 1)(2x - 1)$

15. $6x^2 + 5x - 4 = (2x - 1)(3x + 4)$

16. $x^3 - 64 = x^3 - 4^3 = (x - 4)(x^2 + 4x + 16)$

17. $-\frac{3}{x} + \frac{x}{x^2 + 2} = \frac{-3(x^2 + 2) + x^2}{x(x^2 + 2)} = \frac{-2x^2 - 6}{x(x^2 + 2)} = \frac{-2(x^2 + 3)}{x(x^2 + 2)}$

18. $\frac{x - 3}{4x} \div \frac{x^2 - 9}{x^2} = \frac{x - 3}{4x} \cdot \frac{x^2}{(x + 3)(x - 3)} = \frac{x}{4(x + 3)}$

19. $\frac{1 - \frac{1}{x}}{1 - \frac{1}{1 - (1/x)}} = \frac{\frac{x - 1}{x}}{1 - \frac{1}{(x - 1)/x}} = \frac{\frac{x - 1}{x}}{1 - \frac{x}{x - 1}} = \frac{\frac{x - 1}{x}}{\frac{-1}{x - 1}} = \frac{x - 1}{x} \cdot \frac{x - 1}{-1} = \frac{-(x - 1)^2}{x}$

20. (a)

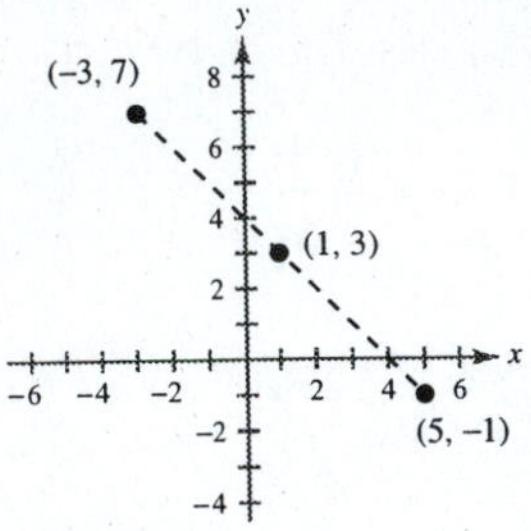

(b) $d = \sqrt{[5-(-3)]^2 + (-1-7)^2}$

$= \sqrt{(8)^2 + (-8)^2}$

$= \sqrt{64 + 64}$

$= \sqrt{128}$

$= 8\sqrt{2}$

(c) $\left(\frac{-3+5}{2}, \frac{7+(-1)}{2}\right)$

$= (1, 3)$

21. (a) $x^2 - 2x - 35 = 0$

$(x-7)(x+5) = 0$

$x - 7 = 0$ or $x + 5 = 0$

$x = 7$ or $x = -5$

(b) $x^2 - 2x - 35 = 0$

$x^2 - 2x = 35$

$x^2 - 2x + 1 = 35 + 1$

$(x-1)^2 = 36$

$x - 1 = \pm\sqrt{36}$

$x = 1 \pm 6$

$x = 7$ or $x = -5$

(c) $x^2 - 2x - 35 = 0$

$x = \frac{-(-2) \pm \sqrt{(-2)^2 - 4(1)(-35)}}{2(1)}$

$= \frac{2 \pm \sqrt{4 + 140}}{2}$

$= \frac{2 \pm 12}{2}$

$= 1 \pm 6$

Thus, $x = 7$ or $x = -5$.

22. $x^5 - 5x^3 + 4x = 0$

$x(x^4 - 5x^2 + 4) = 0$

$x(x^2 - 1)(x^2 - 4) = 0$

$x(x+1)(x-1)(x+2)(x-2) = 0$

$x = 0, x = \pm 1, x = \pm 2$

23. $x = 2\sqrt{x+3}$

$x^2 = 4(x+3)$

$x^2 = 4x + 12$

$x^2 - 4x - 12 = 0$

$(x-6)(x+2) = 0$

$x - 6 = 0 \Rightarrow x = 6$

$x + 2 = 0 \Rightarrow x = -2$, extraneous

24. $x^2 - 16 \le 0$

$(x+4)(x-4) \le 0$

Critical numbers: $x = \pm 4$

Test intervals: $(-\infty, -4), (-4, 4), (4, \infty)$

Test: Is $(x+4)(x-4) \le 0$?

Solution set: $[-4, 4]$

25. $\left|\frac{4-x}{3}\right| > 2$

$\frac{4-x}{3} < -2$ or $\frac{4-x}{3} > 2$

$4 - x < -6$ $\quad$ $4 - x > 6$

$-x < -10$ $\quad$ $-x > 2$

$x > 10$ $\quad$ $x < -2$

❑ Chapter 1 Practice Test Solutions

1. $y = \sqrt{7 - x}$

x	7	6	3	-2
y	0	1	2	3

2. $y = \sqrt{25 - x^2}$

Domain: $25 - x^2 \geq 0$

$(5 - x)(5 + x) \geq 0$

Critical numbers: $x = \pm 5$

Test intervals: $(-\infty, -5), (-5, 5), (5, \infty)$

Solution set: $[-5, 5]$

3. $[x - (-3)]^2 + (y - 5)^2 = 6^2$

$(x + 3)^2 + (y - 5)^2 = 36$

4. $m = \dfrac{-1 - 4}{3 - 2} = -5$

$y - 4 = -5(x - 2)$

$y - 4 = -5x + 10$

$y = -5x + 14$

5. $y = \dfrac{4}{3}x - 3$

6. $2x + 3y = 0$

$y = -\dfrac{2}{3}x$

$m_1 = -\dfrac{2}{3}$

$\perp m_2 = \dfrac{3}{2}$ through $(4, 1)$

$y - 1 = \dfrac{3}{2}(x - 4)$

$y - 1 = \dfrac{3}{2}x - 6$

$y = \dfrac{3}{2}x - 5$

7. $(5, 32)$ and $(9, 44)$

$m = \dfrac{44 - 32}{9 - 5} = \dfrac{12}{4} = 3$

$y - 32 = 3(x - 5)$

$y - 32 = 3x - 15$

$y = 3x + 17$

When $x = 20$, $y = 3(20) + 17$

$y = \$77.$

8. $f(x - 3) = (x - 3)^2 - 2(x - 3) + 1$

$= x^2 - 6x + 9 - 2x + 6 + 1$

$= x^2 - 8x + 16$

9. $f(3) = 12 - 11 = 1$

$\dfrac{f(x) - f(3)}{x - 3} = \dfrac{(4x - 11) - 1}{x - 3}$

$= \dfrac{4x - 12}{x - 3}$

$= \dfrac{4(x - 3)}{x - 3} = 4, x \neq 3$

10. $f(x) = \sqrt{36 - x^2} = \sqrt{(6 + x)(6 - x)}$

Domain: $[-6, 6]$

Range: $[0, 6]$, because

$(6 + x)(6 - x) \geq 0$ on this interval

11. (a) $6x - 5y + 4 = 0$

$y = \dfrac{6x + 4}{5}$ is a function of x.

(b) $x^2 + y^2 = 9$

$y = \pm\sqrt{9 - x^2}$ is not a function of x.

(c) $y^3 = x^2 + 6$

$y = \sqrt[3]{x^2 + 6}$ is a function of x.

12. Parabola

Vertex: $(0, -5)$

Intercepts: $(0, -5)$, $(\pm\sqrt{5}, 0)$

y-axis symmetry

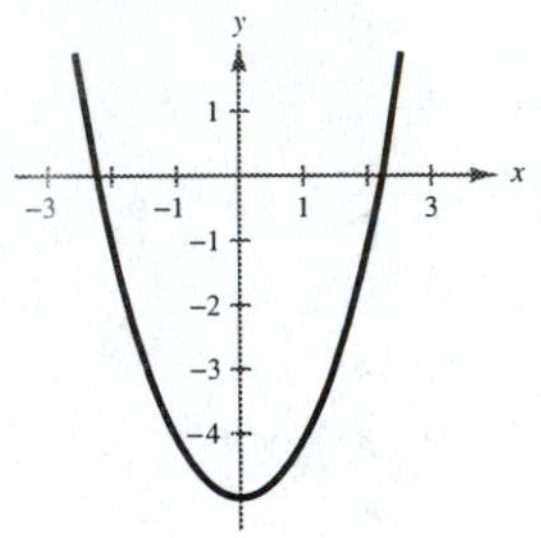

13. Intercepts: $(0, 3)$, $(-3, 0)$

x	0	1	-1	2	-2	-3	-4
y	3	4	2	5	1	0	1

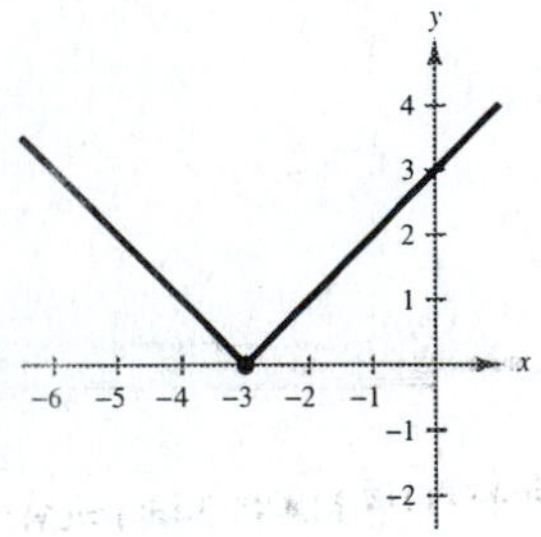

14.

x	0	1	2	3	-1	-2	-3
y	1	3	5	7	2	6	12

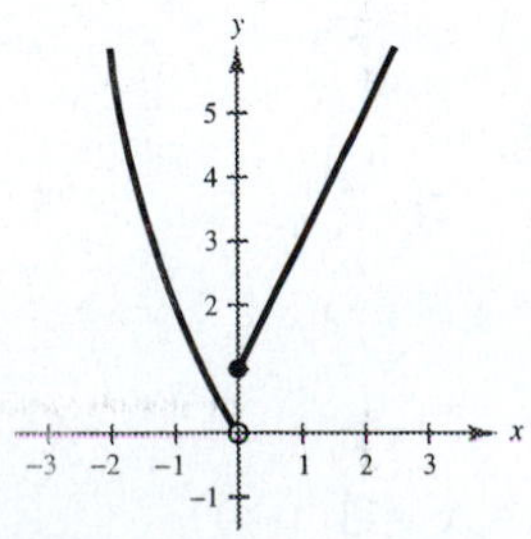

15. (a) $f(x + 2)$

Horizontal shift two units to the left

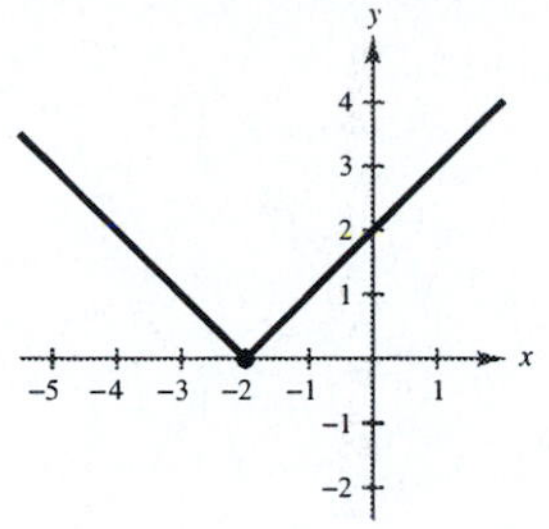

(b) $-f(x) + 2$

Reflection in the x-axis and a vertical shift two units upward

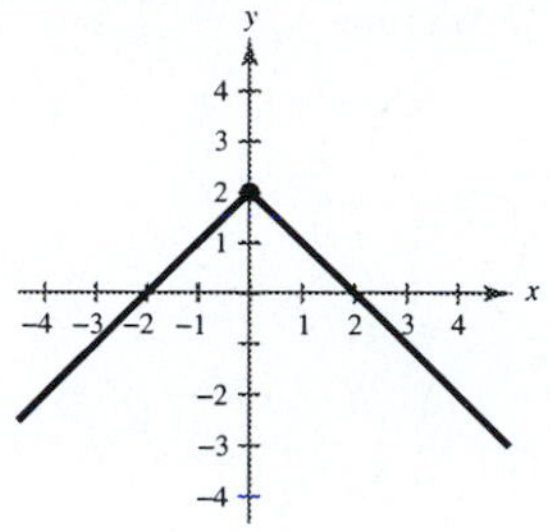

16. (a) $(g - f)(x) = g(x) - f(x)$

$$= (2x^2 - 5) - (3x + 7)$$

$$= 2x^2 - 3x - 12$$

(b) $(fg)(x) = f(x)g(x)$

$$= (3x + 7)(2x^2 - 5)$$

$$= 6x^3 + 14x^2 - 15x - 35$$

17. $f(g(x)) = f(2x + 3)$

$$= (2x + 3)^2 - 2(2x + 3) + 16$$

$$= 4x^2 + 12x + 9 - 4x - 6 + 16$$

$$= 4x^2 + 8x + 19$$

18.

$$f(x) = x^3 + 7$$

$$y = x^3 + 7$$

$$x = y^3 + 7$$

$$x - 7 = y^3$$

$$\sqrt[3]{x - 7} = y$$

$$f^{-1}(x) = \sqrt[3]{x - 7}$$

19. (a) $f(x) = |x - 6|$ does not have an inverse.
Its graph does not pass the horizontal line test.

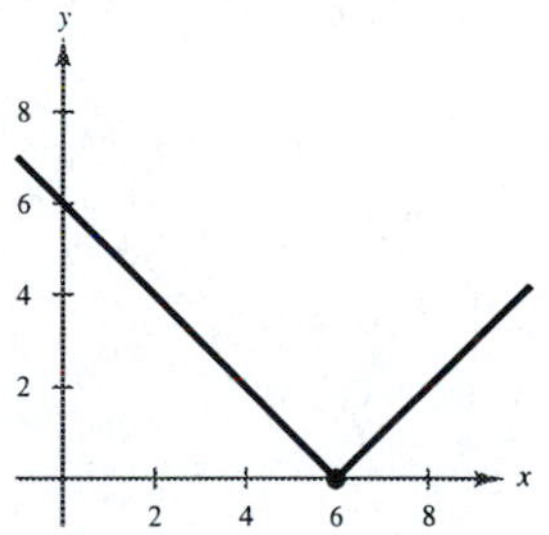

(b) $f(x) = ax + b, a \neq 0$ does have an inverse.

$$y = ax + b$$

$$x = ay + b$$

$$\frac{x - b}{a} = y$$

$$f^{-1}(x) = \frac{x - b}{a}$$

(c) $f(x) = x^3 - 19$ does have an inverse.

$$y = x^3 - 19$$

$$x = y^3 - 19$$

$$x + 19 = y^3$$

$$\sqrt[3]{x + 19} = y$$

$$f^{-1}(x) = \sqrt[3]{x + 19}$$

20. $f(x) = \sqrt{\dfrac{3-x}{x}}, \quad 0 < x \le 3, y \ge 0$

$$y = \sqrt{\frac{3-x}{x}}$$

$$x = \sqrt{\frac{3-y}{y}}$$

$$x^2 = \frac{3-y}{y}$$

$$x^2y = 3 - y$$

$$x^2y + y = 3$$

$$y(x^2 + 1) = 3$$

$$y = \frac{3}{x^2+1}$$

$$f^{-1}(x) = \frac{3}{x^2+1}, \quad x \ge 0$$

21. False. The slopes of 3 and $\frac{1}{3}$ are not **negative** reciprocals.

22. True. Let $y = (f \circ g)(x)$. Then $x = (f \circ g)^{-1}(y)$.

Also,

$$(f \circ g)(x) = y$$

$$f(g(x)) = y$$

$$g(x) = f^{-1}(y)$$

$$x = g^{-1}(f^{-1}(y))$$

$$x = (g^{-1} \circ f^{-1})(y)$$

Since $x = x$, we have $(f \circ g)^{-1}(y) = (g^{-1} \circ f^{-1})(y)$.

23. True. It must pass the vertical line test to be a function and it must pass the horizontal line test to have an inverse.test to have an inverse.

24. $z = \dfrac{cx^3}{\sqrt{y}}$

$$-1 = \frac{c(-1)^3}{\sqrt{25}}$$

$$-1 = \frac{-c}{5}$$

$$5 = c$$

$$z = \frac{5x^3}{\sqrt{y}}$$

25. $y \approx 0.669x + 2.669$

❑ Chapter 2 Practice Test Solutions

1. x-intercepts: (1, 0), (5, 0)
y-intercepts: (0, 5)
Vertex: (3, −4)

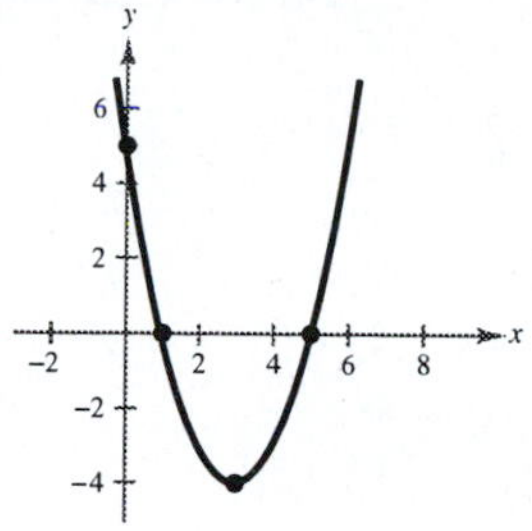

2. $a = 0.01, b = -90$

$$\frac{-b}{2a} = \frac{90}{2(.01)} = 4500 \text{ units}$$

3. Vertex (1, 7) opening downward through (2, 5)

$y = a(x - 1)^2 + 7$ Standard form
$5 = a(2 - 1)^2 + 7$
$5 = a + 7$
$a = -2$
$y = -2(x - 1)^2 + 7$
$= -2(x^2 - 2x + 1) + 7$
$= -2x^2 + 4x + 5$

4. $y = \pm a(x - 2)(3x - 4)$ where a is any real number

$y = \pm(3x^2 - 10x + 8)$

5. Leading coefficient: −3
Degree: 5
Moves down to the right and up to the left

6. $0 = x^5 - 5x^3 + 4x$
$= x(x^4 - 5x^2 + 4)$
$= x(x^2 - 1)(x^2 - 4)$
$= x(x + 1)(x - 1)(x + 2)(x - 2)$
$x = 0, x = \pm 1, x = \pm 2$

7. $f(x) = x(x - 3)(x + 2)$
$= x(x^2 - x - 6)$
$= x^3 - x^2 - 6x$

8. Intercepts: $(0, 0), (\pm 2\sqrt{3}, 0)$
Moves up to the right.
Moves down to the left.

x	−2	−1	0	1	2
y	16	11	0	−11	−16

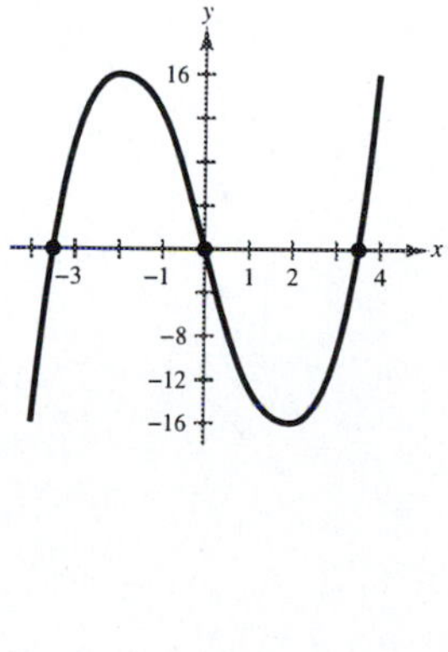

9.

$$3x^3 + 9x^2 + 20x + 62 + \frac{176}{x-3}$$

$$\begin{array}{r}
x - 3\overline{)3x^4 + 0x^3 - 7x^2 + 2x - 10} \\
\underline{3x^4 - 9x^3 } \\
9x^3 - 7x^2 \\
\underline{9x^3 - 27x^2 } \\
20x^2 + 2x \\
\underline{20x^2 - 60x } \\
62x - 10 \\
\underline{62x - 186} \\
176
\end{array}$$

10.

$$x - 2 + \frac{5x - 13}{x^2 + 2x - 1}$$

$$\begin{array}{r}
x^2 + 2x - 1\overline{)x^3 + 0x^2 + 0x - 11} \\
\underline{x^3 + 2x^2 - x } \\
-2x^2 + x - 11 \\
\underline{-2x^2 - 4x + 2} \\
5x - 13
\end{array}$$

11.

$$\begin{array}{r|rrrrrr}
-5 & 3 & 13 & 0 & 0 & 12 & -1 \\
 & & -15 & 10 & -50 & 250 & -1310 \\
\hline
 & 3 & -2 & 10 & -50 & 262 & -1311
\end{array}$$

$$\frac{3x^5 + 13x^4 + 12x - 1}{x + 5} = 3x^4 - 2x^3 + 10x^2 - 50x + 262 - \frac{1311}{x + 5}$$

12.

$$\begin{array}{r|rrrr}
-6 & 7 & 40 & -12 & 15 \\
 & & -42 & 12 & 0 \\
\hline
 & 7 & -2 & 0 & 15
\end{array}$$

$f(-6) = 15$

13. $0 = x^3 - 19x - 30$

Possible rational roots: $\pm1, \pm2, \pm3, \pm5, \pm6, \pm10, \pm15, \pm30$

$$\begin{array}{r|rrrr}
-2 & 1 & 0 & -19 & -30 \\
 & & -2 & 4 & 30 \\
\hline
 & 1 & -2 & -15 & 0
\end{array} \quad x = -2 \text{ is a zero.}$$

$0 = (x + 2)(x^2 - 2x - 15)$

$0 = (x + 2)(x + 3)(x - 5)$

Zeros: $x = -2, x = -3, x = 5$

14. $0 = x^4 + x^3 - 8x^2 - 9x - 9$

Possible rational roots: $\pm 1, \pm 3, \pm 9$

3	1	1	−8	−9	−9
		3	12	12	9
	1	4	4	3	0

$x = 3$ is a zero.

$0 = (x - 3)(x^3 + 4x^2 + 4x + 3)$

Possible Rational Roots of $x^3 + 4x^2 + 4x + 3$: $\pm 1, \pm 3$

−3	1	4	4	3
		−3	−3	−3
	1	1	1	0

$x = -3$ is a zero.

$0 = (x - 3)(x + 3)(x^2 + x + 1)$

The zeros of $x^2 + x + 1$ are $x = \dfrac{-1 \pm \sqrt{3}i}{2}$.

Zeros: $x = 3, x = -3, x = -\dfrac{1}{2} + \dfrac{\sqrt{3}}{2}i, x = -\dfrac{1}{2} - \dfrac{\sqrt{3}}{2}i$

15. $0 = 6x^3 - 5x^2 + 4x - 15$

Possible rational roots: $\pm 1, \pm 3, \pm 5, \pm 15, \pm\frac{1}{2}, \pm\frac{3}{2}, \pm\frac{5}{2}, \pm\frac{15}{2}, \pm\frac{1}{3}, \pm\frac{5}{3}, \pm\frac{1}{6}, \pm\frac{5}{6}$

16. $0 = x^3 - \frac{20}{3}x^2 + 9x - \frac{10}{3}$

$0 = 3x^3 - 20x^2 + 27x - 10$

Possible Rational Roots: $\pm 1, \pm 2, \pm 5, \pm 10, \pm\frac{1}{3}, \pm\frac{2}{3}, \pm\frac{5}{3}, \pm\frac{10}{3}$

1	3	−20	27	−10
		3	−17	10
	3	−17	10	0

$0 = (x - 1)(3x^2 - 17x + 10)$
$0 = (x - 1)(3x - 2)(x - 5)$
Zeros: $x = 1, x = \frac{2}{3}, x = 5$

17. Possible rational roots: $\pm 1, \pm 2, \pm 5, \pm 10$

1	1	1	3	5	−10
		1	2	5	10
	1	2	5	10	0

$x = 1$ is a zero.

−2	1	2	5	10
		−2	0	−10
	1	0	5	0

$x = -2$ is a zero.

$$f(x) = (x - 1)(x + 2)(x^2 + 5)$$
$$= (x - 1)(x + 2)(x + \sqrt{5}i)(x - \sqrt{5}i)$$

18. $f(x) = (x - 2)[x - (3 + i)][x - (3 - i)]$

$= (x - 2)[(x - 3) - i][(x - 3) + i]$

$= (x - 2)[(x - 3)^2 - i^2]$

$= (x - 2)[x^2 - 6x + 10]$

$= x^3 - 8x^2 + 22x - 20$

19.

3i	1	4	9	36
		$3i$	$12i - 9$	-36
	1	$4 + 3i$	$12i$	0

20. Vertical asymptote: $x = 0$

Horizontal asymptote: $y = \frac{1}{2}$

x-intercept: $(1, 0)$

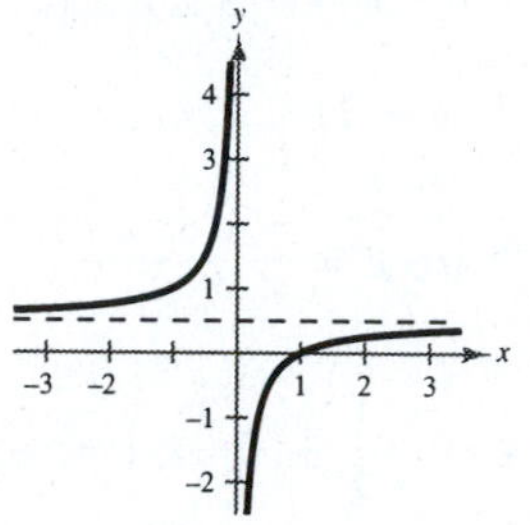

21. $y = 8$ is a horizontal asymptote since the degree on the numerator equals the degree of the denominator. There are no vertical asymptotes.

22. $x = 1$ is a vertical asymptote.

$$\frac{4x^2 - 2x + 7}{x - 1} = 4x + 2 + \frac{9}{x - 1}$$

so $y = 4x + 2$ is a slant asymptote.

23. (a) $(4 - 3i) - (-2 + i) = 4 - 3i + 2 - i = 6 - 4i$

(b) $(4 - 3i)(-2 + i) = -8 + 4i + 6i - 3i^2 = -8 + 10i + 3 = -5 + 10i$

(c) $\frac{4 - 3i}{-2 + i} = \frac{4 - 3i}{-2 + i} \cdot \frac{-2 - i}{-2 - i} = \frac{-8 - 4i + 6i + 3i^2}{4 + 1}$

$= \frac{-11 + 2i}{5} = -\frac{11}{5} + \frac{2}{5}i$

24. $\frac{1 - 2x}{x^2 + x} = \frac{1 - 2x}{x(x + 1)} = \frac{A}{x} + \frac{B}{x + 1}$

$1 - 2x = A(x + 1) + Bx$

When $x = 0$, $1 = A$.

When $x = -1$, $3 = -B \Rightarrow B = -3$.

$\frac{1 - 2x}{x^2 + x} = \frac{1}{x} - \frac{3}{x + 1}$

25. $\frac{6x - 17}{(x - 3)^2} = \frac{A}{x - 3} + \frac{B}{(x - 3)^2}$

$6x - 17 = A(x - 3) + B$

When $x = 3$, $1 = B$.

When $x = 0$, $-17 = -3A + B \Rightarrow A = 6$.

$\frac{6x - 17}{(x - 3)^2} = \frac{6}{x - 3} + \frac{1}{(x - 3)^2}$

❑ Chapter 3 Practice Test Solutions

1. $x^{3/5} = 8$

$x = 8^{5/3} = (\sqrt[3]{8})^5 = 2^5 = 32$

2. $3^{x-1} = \frac{1}{81}$

$3^{x-1} = 3^{-4}$

$x - 1 = -4$

$x = -3$

3. $f(x) = 2^{-x} = \left(\frac{1}{2}\right)^x$

x	-2	-1	0	1	2
$f(x)$	4	2	1	$\frac{1}{2}$	$\frac{1}{4}$

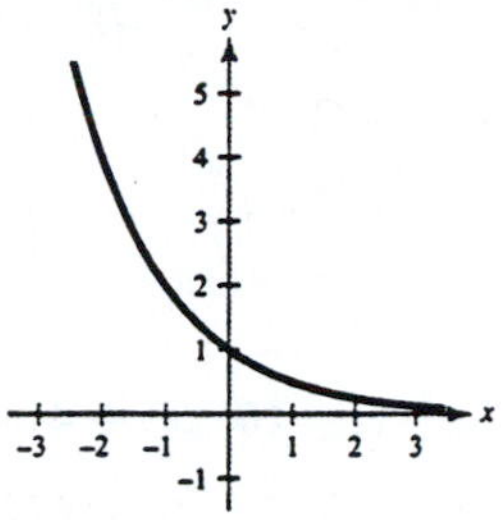

4. $g(x) = e^x + 1$

x	-2	-1	0	1	2
$g(x)$	1.14	1.37	2	3.72	8.39

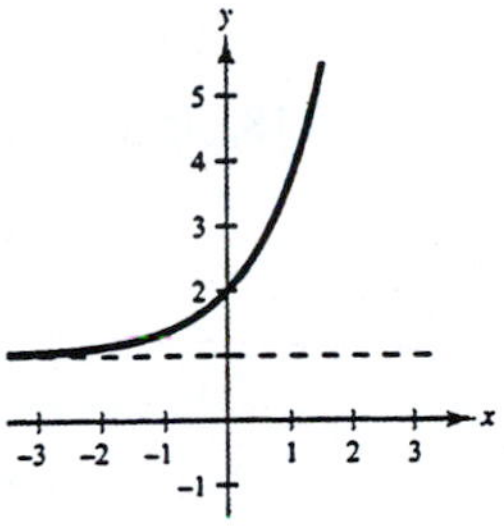

5. $A = P\left(1 + \frac{r}{n}\right)^{nt}$ OR $A = Pe^{rt}$

(a) $A = 5000\left(1 + \frac{0.09}{12}\right)^{12(3)} \approx \$6543.23$

(b) $A = 5000\left(1 + \frac{0.09}{4}\right)^{4(3)} \approx \$6530.25$

(c) $A = 5000e^{(0.09)(3)} \approx \$6549.82$

6. $7^{-2} = \frac{1}{49}$

$\log_7 \frac{1}{49} = -2$

7. $x - 4 = \log_2 \frac{1}{64}$

$2^{x-4} = \frac{1}{64}$

$2^{x-4} = 2^{-6}$

$x - 4 = -6$

$x = -2$

8. $\log_b \sqrt[4]{\frac{8}{25}} = \frac{1}{4} \log_b \frac{8}{25}$

$= \frac{1}{4}[\log_b 8 - \log_b 25]$

$= \frac{1}{4}[\log_b 2^3 - \log_b 5^2]$

$= \frac{1}{4}[3 \log_b 2 - 2 \log_b 5]$

$= \frac{1}{4}[3(0.3562) - 2(0.8271)]$

$= -0.1464$

9. $5 \ln x - \frac{1}{2} \ln y + 6 \ln z = \ln x^5 - \ln \sqrt{y} + \ln z^6 = \ln\left(\frac{x^5 z^6}{\sqrt{y}}\right)$

10. $\log_9 28 = \dfrac{\log 28}{\log 9} \approx 1.5166$

11. $\log N = 0.6646$

$N = 10^{0.6646} \approx 4.62$

12.

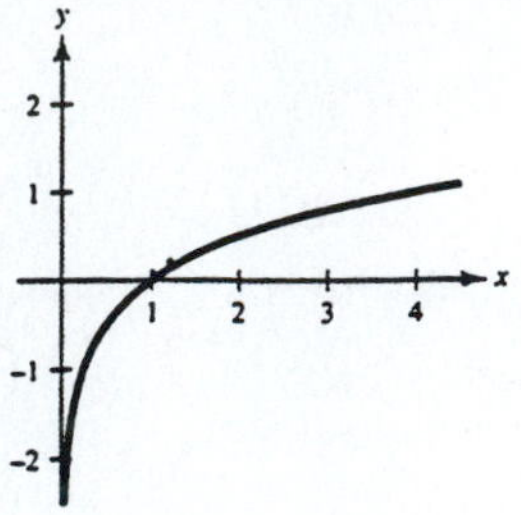

13. Domain:

$x^2 - 9 > 0$

$(x + 3)(x - 3) > 0$

$x < -3$ or $x > 3$

14.

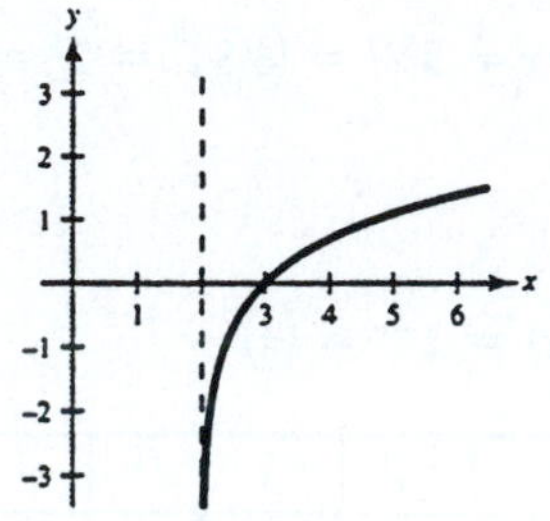

15. $\dfrac{\ln x}{\ln y} \neq \ln(x - y)$ since $\dfrac{\ln x}{\ln y} = \log_y x.$

16. $5^3 = 41$

$x = \log_5 41 = \dfrac{\ln 41}{\ln 5} \approx 2.3074$

17. $x - x^2 = \log_5 \frac{1}{25}$

$5^{x-x^2} = \frac{1}{25}$

$5^{x-x^2} = 5^{-2}$

$x - x^2 = -2$

$0 = x^2 - x - 2$

$0 = (x + 1)(x - 2)$

$x = -1$ or $x = 2$

18. $\log_2 x + \log_2(x - 3) = 2$

$\log_2[x(x - 3)] = 2$

$x(x - 3) = 2^2$

$x^2 - 3x = 4$

$x^2 - 3x - 4 = 0$

$(x + 1)(x - 4) = 0$

$x = 4$

$x = -1$ (extraneous)

$x = 4$ is the only solution.

19. $\dfrac{e^x + e^{-x}}{3} = 4$

$e^x(e^x + e^{-x}) = 12e^x$

$e^{2x} + 1 = 12e^x$

$e^{2x} - 12e^x + 1 = 0$

$e^x = \dfrac{12 \pm \sqrt{144 - 4}}{2}$

$e^x \approx 11.9161$ or $e^x \approx 0.0839$

$x = \ln 11.9161$ $\quad x = \ln 0.0839$

$x \approx 2.478$ $\quad x \approx -2.478$

20. $A = Pe^{rt}$

$12{,}000 = 6000e^{0.13t}$

$2 = e^{0.13t}$

$0.13t = \ln 2$

$t = \dfrac{\ln 2}{0.13}$

$t \approx 5.3319$ years or 5 years 4 months

❑ Chapter 4 Practice Test Solutions

1. $350° = 350\left(\frac{\pi}{180}\right) = \frac{35\pi}{18}$

2. $\frac{5\pi}{9} = \frac{5\pi}{9} \cdot \frac{180}{\pi} = 100°$

3. $135°\, 14'\, 12'' = \left(135 + \frac{14}{60} + \frac{12}{3600}\right)°$

$\approx 135.2367°$

4. $-22.569° = -(22° + 0.569(60)')$

$= -22°\, 34.14'$

$= -(22°\, 34' + 0.14(60)'')$

$\approx -22°\, 34'\, 8''$

5. $\cos\theta = \frac{2}{3}$

$x = 2, r = 3, y = \pm\sqrt{9-4} = \pm\sqrt{5}$

$\tan\theta = \frac{y}{x} = \pm\frac{\sqrt{5}}{2}$

6. $\sin\theta = 0.9063$

$\theta = \arcsin(0.9063)$

$\theta = 65° = \frac{13\pi}{36}$

or

$\theta = 180° - 65° = 115° = \frac{23\pi}{36}$

7. $\tan 20° = \frac{35}{x}$

$x = \frac{35}{\tan 20°} \approx 96.1617$

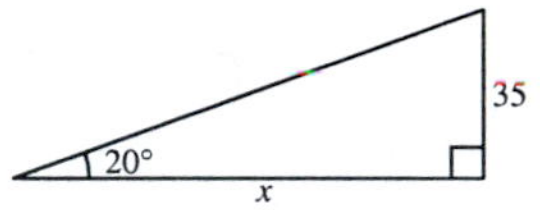

8. $\theta = \frac{6\pi}{5}$, θ is in Quadrant III.

Reference angle: $\frac{6\pi}{5} - \pi = \frac{\pi}{5}$ or 36°

9. $\csc 3.92 = \frac{1}{\sin 3.92} \approx -1.4242$

10. $\tan\theta = 6 = \frac{6}{1}$, θ lies in Quadrant III.

$y = -6, x = -1, r = \sqrt{36+1} = \sqrt{37}$,

so $\sec\theta = \frac{\sqrt{37}}{-1} \approx -6.0828$.

11. Period: 4π

Amplitude: 3

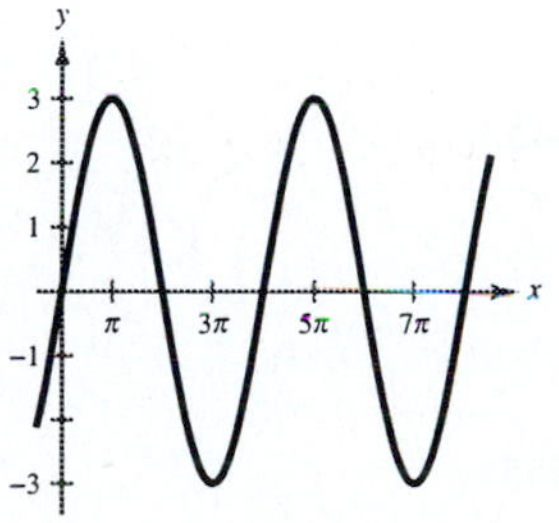

12. Period: 2π

Amplitude: 2

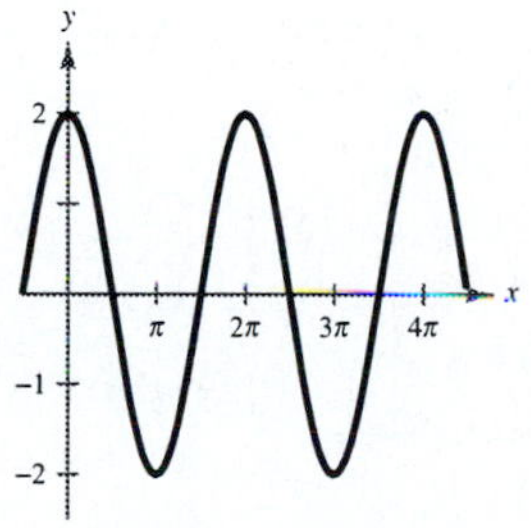

13. Period: $\frac{\pi}{2}$

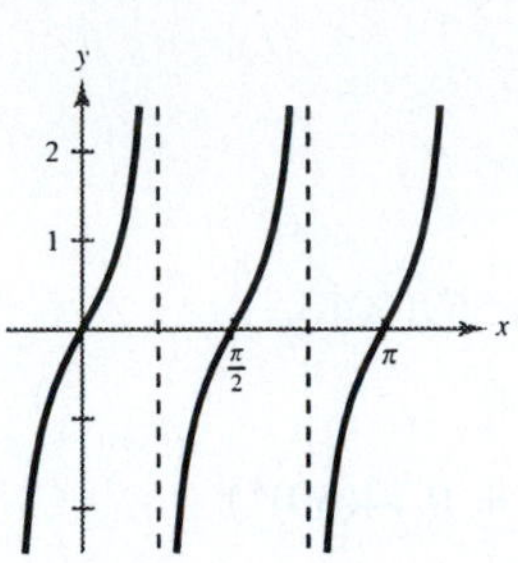

14. Period: 2π

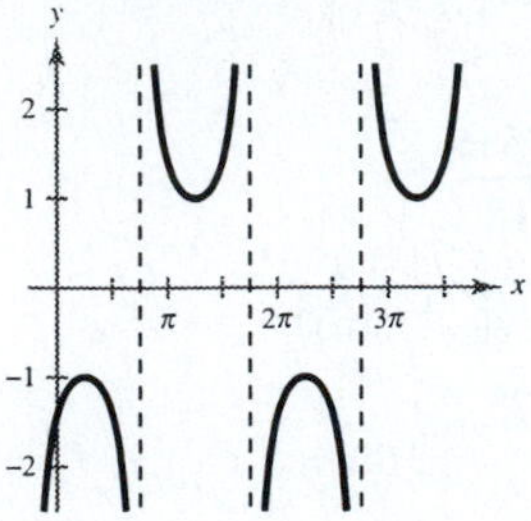

15.

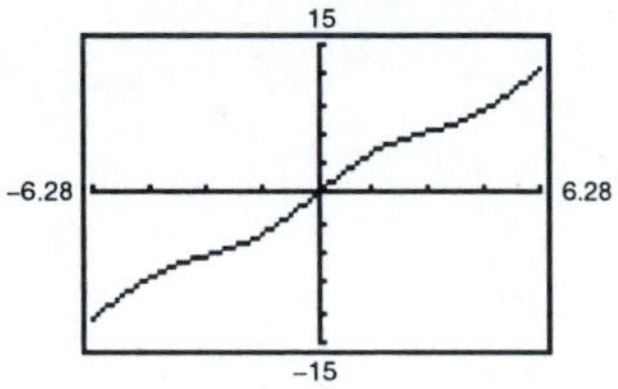

16.

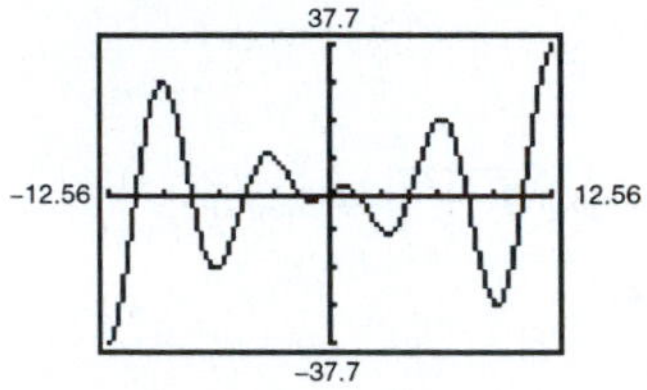

17. $\theta = \arcsin 1$

$\sin\theta = 1$

$\theta = \frac{\pi}{2}$

18. $\theta = \arctan(-3)$

$\tan\theta = -3$

$\theta \approx -1.249 \approx -71.565°$

19. $\sin\left(\arccos\frac{4}{\sqrt{35}}\right)$

$\sin\theta = \frac{\sqrt{19}}{\sqrt{35}} \approx 0.7368$

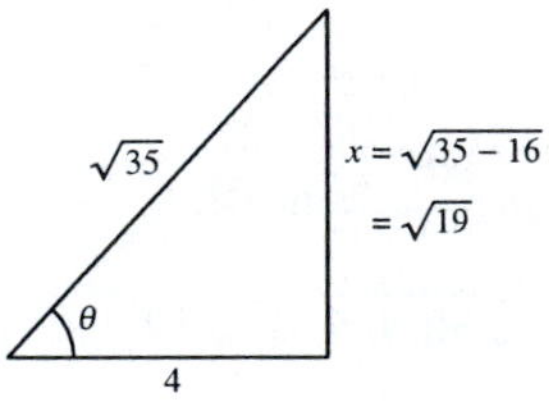

20. $\cos\left(\arcsin\frac{x}{4}\right)$

$\cos\theta = \frac{\sqrt{16 - x^2}}{4}$

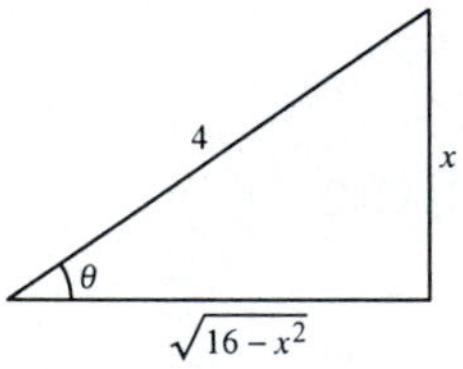

21. Given $A = 40°$, $c = 12$

$B = 90° - 40° = 50°$

$\sin 40° = \frac{a}{12}$

$a = 12 \sin 40° \approx 7.713$

$\cos 40° = \frac{b}{12}$

$b = 12 \cos 40° \approx 9.193$

22. Given $B = 6.84°$, $a = 21.3$

$A = 90° - 6.84° = 83.16°$

$\sin 83.16° = \frac{21.3}{c}$

$c = \frac{21.3}{\sin 83.16°} \approx 21.453$

$\tan 83.16° = \frac{21.3}{b}$

$b = \frac{21.3}{\tan 83.16°} \approx 2.555$

23. Given $a = 5,\ b = 9$

$c = \sqrt{25 + 81} = \sqrt{106} \approx 10.296$

$\tan A = \frac{5}{9}$

$A = \arctan \frac{5}{9} \approx 29.055°$

$B = 90° - 29.055° = 60.945°$

24. $\sin 67° = \dfrac{x}{20}$

$x = 20 \sin 67° \approx 18.41$ feet

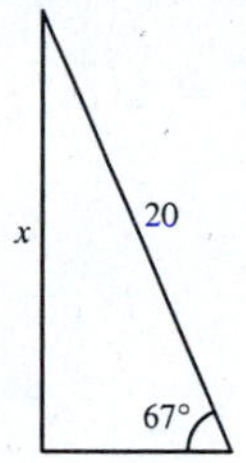

25. $\tan 5° = \dfrac{250}{x}$

$x = \dfrac{250}{\tan 5°}$

≈ 2857.513 feet

≈ 0.541 mi

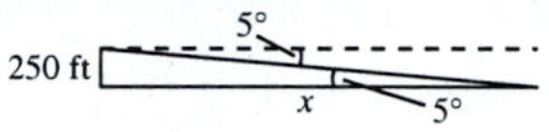

❑ Chapter 5 Practice Test Solutions

1. $\tan x = \frac{4}{11}$, $\sec x < 0 \implies x$ is in Quadrant III.

$y = -4,\ x = -11,\ r = \sqrt{16 + 121} = \sqrt{137}$

$\sin x = -\frac{4}{\sqrt{137}} = -\frac{4\sqrt{137}}{137}$ $\qquad \csc x = -\frac{\sqrt{137}}{4}$

$\cos x = -\frac{11}{\sqrt{137}} = -\frac{11\sqrt{137}}{137}$ $\qquad \sec x = -\frac{\sqrt{137}}{11}$

$\tan x = \frac{4}{11}$ $\qquad \cot x = \frac{11}{4}$

2. $$\frac{\sec^2 x + \csc^2 x}{\csc^2 x(1 + \tan^2 x)} = \frac{\sec^2 x + \csc^2 x}{\csc^2 x + (\csc^2 x)\tan^2 x} = \frac{\sec^2 x + \csc^2 x}{\csc^2 x + \frac{1}{\sin^2 x} \cdot \frac{\sin^2 x}{\cos^2 x}}$$

$$= \frac{\sec^2 x + \csc^2 x}{\csc^2 x + \frac{1}{\cos^2 x}} = \frac{\sec^2 x + \csc^2 x}{\csc^2 x + \sec^2 x} = 1$$

3. $\ln|\tan\theta| - \ln|\cot\theta| = \ln\frac{|\tan\theta|}{|\cot\theta|} = \ln\left|\frac{\sin\theta/\cos\theta}{\cos\theta/\sin\theta}\right| = \ln\left|\frac{\sin^2\theta}{\cos^2\theta}\right| = \ln|\tan^2\theta| = 2\ln|\tan\theta|$

4. $\cos\left(\frac{\pi}{2} - x\right) = \frac{1}{\csc x}$ is true since $\cos\left(\frac{\pi}{2} - x\right) = \sin x = \frac{1}{\csc x}$.

5. $\sin^4 x + (\sin^2 x)\cos^2 x = \sin^2 x(\sin^2 x + \cos^2 x) = \sin^2 x(1) = \sin^2 x$

6. $(\csc x + 1)(\csc x - 1) = \csc^2 x - 1 = \cot^2 x$

7. $\frac{\cos^2 x}{1 - \sin x} \cdot \frac{1 + \sin x}{1 + \sin x} = \frac{\cos^2 x(1 + \sin x)}{1 - \sin^2 x} = \frac{\cos^2 x(1 + \sin x)}{\cos^2 x} = 1 + \sin x$

8. $$\frac{1 + \cos\theta}{\sin\theta} + \frac{\sin\theta}{1 + \cos\theta} = \frac{(1 + \cos\theta)^2 + \sin^2\theta}{\sin\theta(1 + \cos\theta)}$$

$$= \frac{1 + 2\cos\theta + \cos^2\theta + \sin^2\theta}{\sin\theta(1 + \cos\theta)} = \frac{2 + 2\cos\theta}{\sin\theta(1 + \cos\theta)} = \frac{2}{\sin\theta} = 2\csc\theta$$

9. $\tan^4 x + 2\tan^2 x + 1 = (\tan^2 x + 1)^2 = (\sec^2 x)^2 = \sec^4 x$

10. (a) $\sin 105° = \sin(60° + 45°) = \sin 60° \cos 45° + \cos 60° \sin 45°$

$$= \frac{\sqrt{3}}{2} \cdot \frac{\sqrt{2}}{2} + \frac{1}{2} \cdot \frac{\sqrt{2}}{2} = \frac{\sqrt{2}}{4}(\sqrt{3} + 1)$$

(b) $\tan 15° = \tan(60° - 45°) = \dfrac{\tan 60° - \tan 45°}{1 + \tan 60° \tan 45°}$

$$= \frac{\sqrt{3} - 1}{1 + \sqrt{3}} \cdot \frac{1 - \sqrt{3}}{1 - \sqrt{3}} = \frac{2\sqrt{3} - 1 - 3}{1 - 3} = \frac{2\sqrt{3} - 4}{-2} = 2 - \sqrt{3}$$

11. $(\sin 42°) \cos 38° - (\cos 42°) \sin 38° = \sin(42° - 38°) = \sin 4°$

12. $\tan\left(\theta + \dfrac{\pi}{4}\right) = \dfrac{\tan\theta + \tan(\pi/4)}{1 - (\tan\theta)\tan(\pi/4)} = \dfrac{\tan\theta + 1}{1 - \tan\theta(1)} = \dfrac{1 + \tan\theta}{1 - \tan\theta}$

13. $\sin(\arcsin x - \arccos x) = \sin(\arcsin x)\cos(\arccos x) - \cos(\arcsin x)\sin(\arccos x)$

$$= (x)(x) - \left(\sqrt{1 - x^2}\right)\left(\sqrt{1 - x^2}\right) = x^2 - (1 - x^2) = 2x^2 - 1$$

14. (a) $\cos(120°) = \cos[2(60°)] = 2\cos^2 60° - 1 = 2\left(\dfrac{1}{2}\right)^2 - 1 = -\dfrac{1}{2}$

(b) $\tan(300°) = \tan[2(150°)] = \dfrac{2\tan 150°}{1 - \tan^2 150°} = \dfrac{-2\sqrt{3}/3}{1 - (1/3)} = -\sqrt{3}$

15. (a) $\sin 22.5° = \sin\dfrac{45°}{2} = \sqrt{\dfrac{1 - \cos 45°}{2}} = \sqrt{\dfrac{1 - \sqrt{2}/2}{2}} = \dfrac{\sqrt{2 - \sqrt{2}}}{2}$

(b) $\tan\dfrac{\pi}{12} = \tan\dfrac{\pi/6}{2} = \dfrac{\sin(\pi/6)}{1 + \cos(\pi/6)} = \dfrac{1/2}{1 + \sqrt{3}/2} = \dfrac{1}{2 + \sqrt{3}} = 2 - \sqrt{3}$

16. $\sin\theta = \dfrac{4}{5}$, θ lies in Quadrant II $\Rightarrow$ $\cos\theta = -\dfrac{3}{5}$.

$$\cos\frac{\theta}{2} = \sqrt{\frac{1 + \cos\theta}{2}} = \sqrt{\frac{1 - 3/5}{2}} = \sqrt{\frac{2}{10}} = \frac{1}{\sqrt{5}} = \frac{\sqrt{5}}{5}$$

17. $(\sin^2 x)\cos^2 x = \dfrac{1 - \cos 2x}{2} \cdot \dfrac{1 + \cos 2x}{2} = \dfrac{1}{4}[1 - \cos^2 2x] = \dfrac{1}{4}\left[1 - \dfrac{1 + \cos 4x}{2}\right]$

$$= \frac{1}{8}[2 - (1 + \cos 4x)] = \frac{1}{8}[1 - \cos 4x]$$

18. $6(\sin 5\theta)\cos 2\theta = 6\{\tfrac{1}{2}[\sin(5\theta + 2\theta) + \sin(5\theta - 2\theta)]\} = 3[\sin 7\theta + \sin 3\theta]$

19. $\sin(x + \pi) + \sin(x - \pi) = 2\left(\sin\dfrac{[(x + \pi) + (x - \pi)]}{2}\right)\cos\dfrac{[(x + \pi) - (x - \pi)]}{2} = 2\sin x \cos\pi = -2\sin x$

20. $\dfrac{\sin 9x + \sin 5x}{\cos 9x - \cos 5x} = \dfrac{2\sin 7x \cos 2x}{-2\sin 7x \sin 2x} = -\dfrac{\cos 2x}{\sin 2x} = -\cot 2x$

21. $\frac{1}{2}[\sin(u+v) - \sin(u-v)] = \frac{1}{2}\{(\sin u)\cos v + (\cos u)\sin v - [(\sin u)\cos v - (\cos u)\sin v]\}$

$= \frac{1}{2}[2(\cos u)\sin v] = (\cos u)\sin v$

22. $4\sin^2 x = 1$

$\sin^2 x = \frac{1}{4}$

$\sin x = \pm\frac{1}{2}$

$\sin x = \frac{1}{2}$ or $\sin x = -\frac{1}{2}$

$x = \frac{\pi}{6}$ or $\frac{5\pi}{6}$ $\quad x = \frac{7\pi}{6}$ or $\frac{11\pi}{6}$

23. $\tan^2\theta + (\sqrt{3} - 1)\tan\theta - \sqrt{3} = 0$

$(\tan\theta - 1)(\tan\theta + \sqrt{3}) = 0$

$\tan\theta = 1$ or $\tan\theta = -\sqrt{3}$

$\theta = \frac{\pi}{4}$ or $\frac{5\pi}{4}$ $\quad \theta = \frac{2\pi}{3}$ or $\frac{5\pi}{3}$

24. $\sin 2x = \cos x$

$2(\sin x)\cos x - \cos x = 0$

$\cos x(2\sin x - 1) = 0$

$\cos x = 0$ or $\sin x = \frac{1}{2}$

$x = \frac{\pi}{2}$ or $\frac{3\pi}{2}$ $\quad x = \frac{\pi}{6}$ or $\frac{5\pi}{6}$

25. $\tan^2 x - 6\tan x + 4 = 0$

$\tan x = \frac{-(-6) \pm \sqrt{(-6)^2 - 4(1)(4)}}{2(1)}$

$\tan x = \frac{6 \pm \sqrt{20}}{2} = 3 \pm \sqrt{5}$

$\tan x = 3 + \sqrt{5}$ or $\tan x = 3 - \sqrt{5}$

$x \approx 1.3821$ or 4.5237 $\quad x = 0.6524$ or 3.7940

❑ Chapter 6 Practice Test Solutions

1. $C = 180° - (40° + 12°) = 128°$

$a = \sin 40°\left(\frac{100}{\sin 12°}\right) \approx 309.164$

$c = \sin 128°\left(\frac{100}{\sin 12°}\right) \approx 379.012$

2. $\sin A = 5\left(\frac{\sin 150°}{20}\right) = 0.125$

$A \approx 7.181°$

$B \approx 180° - (150° + 7.181°) = 22.819°$

$b = \sin 22.819°\left(\frac{20}{\sin 150°}\right) \approx 15.513$

3. Area $= \frac{1}{2}ab \sin C$

$= \frac{1}{2}(3)(5)\sin 130°$

≈ 5.745 square units

4. $h = b \sin A$

$= 35 \sin 22.5°$

≈ 13.394

$a = 10$

Since $a < h$ and A is acute, the triangle has no solution.

5. $\cos A = \dfrac{(53)^2 + (38)^2 - (49)^2}{2(53)(38)} \approx 0.4598$

$A \approx 62.627°$

$\cos B = \dfrac{(49)^2 + (38)^2 - (53)^2}{2(49)(38)} \approx 0.2782$

$B \approx 73.847°$

$C \approx 180° - (62.627° + 73.847°)$

$= 43.526°$

6. $c^2 = (100)^2 + (300)^2 - 2(100)(300)\cos 29°$

≈ 47522.8176

$c \approx 218$

$\cos A = \dfrac{(300)^2 + (218)^2 - (100)^2}{2(300)(218)} \approx 0.97495$

$A \approx 12.85°$

$B \approx 180° - (12.85° + 29°) = 138.15°$

7. $s = \dfrac{a + b + c}{2} = \dfrac{4.1 + 6.8 + 5.5}{2} = 8.2$

Area $= \sqrt{s(s - a)(s - b)(s - c)}$

$= \sqrt{8.2(8.2 - 4.1)(8.2 - 6.8)(8.2 - 5.5)}$

$= 11.273$ square units

8. $x^2 = (40)^2 + (70)^2 - 2(40)(70)\cos 168°$

≈ 11977.6266

$x \approx 190.442$ miles

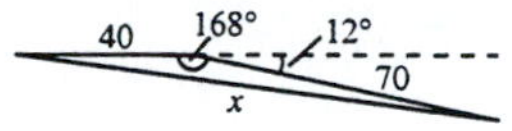

9. $\mathbf{w} = 4(3\mathbf{i} + \mathbf{j}) - 7(-\mathbf{i} + 2\mathbf{j})$

$= 19\mathbf{i} - 10\mathbf{j}$

10. $\dfrac{\mathbf{v}}{\|\mathbf{v}\|} = \dfrac{5\mathbf{i} + 3\mathbf{j}}{\sqrt{25 + 9}} = \dfrac{5}{\sqrt{34}}\mathbf{i} - \dfrac{3}{\sqrt{34}}\mathbf{j}$

$= \dfrac{5\sqrt{34}}{34}\mathbf{i} - \dfrac{3\sqrt{34}}{34}\mathbf{j}$

11. $\mathbf{u} = 6\mathbf{i} + 5\mathbf{j} \qquad \mathbf{v} = 2\mathbf{i} - 3\mathbf{j}$

$\mathbf{u} \cdot \mathbf{v} = 6(2) + 5(-3) = -3$

$\|\mathbf{u}\| = \sqrt{61} \qquad \|\mathbf{v}\| = \sqrt{13}$

$\cos\theta = \dfrac{-3}{\sqrt{61}\sqrt{13}}$

$\theta \approx 96.116°$

12. $4(\mathbf{i} \cos 30° + \mathbf{j} \sin 30°)$

$= 4\left(\dfrac{\sqrt{3}}{2}\mathbf{i} + \dfrac{1}{2}\mathbf{j}\right)$

$= \langle 2\sqrt{3}, 2\rangle$

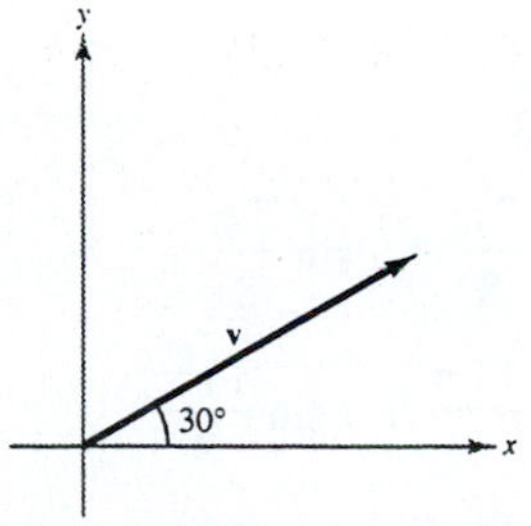

13. $\text{proj}_{\mathbf{v}}\mathbf{u} = \left(\dfrac{\mathbf{u} \cdot \mathbf{v}}{\|\mathbf{v}\|^2}\right)\mathbf{v} = \dfrac{-10}{20}\langle -2, 4\rangle = \langle 1, -2\rangle$

14. $r = \sqrt{25 + 25} = \sqrt{50} = 5\sqrt{2}$

$\tan\theta = \dfrac{-5}{5} = -1$

Since z is in Quadrant IV,

$\theta = 315°$

$z = 5\sqrt{2}(\cos 315° + i \sin 315°).$

15. $\cos 225° = -\dfrac{\sqrt{2}}{2}$ $\sin 225° = -\dfrac{\sqrt{2}}{2}$

$$z = 6\left(-\frac{\sqrt{2}}{2} - i\frac{\sqrt{2}}{2}\right)$$

$$= -3\sqrt{2} - 3\sqrt{2}i$$

16. $[7(\cos 23° + i \sin 23°)][4(\cos 7° + i \sin 7°)] = 7(4)[\cos(23° + 7°) + i \sin(23° + 7°)]$

$$= 28(\cos 30° + i \sin 30°)$$

17. $\dfrac{9\left(\cos\dfrac{5\pi}{4} + i \sin\dfrac{5\pi}{4}\right)}{3(\cos\pi + i \sin\pi)} = \dfrac{9}{3}\left[\cos\left(\dfrac{5\pi}{4} - \pi\right) + i \sin\left(\dfrac{5\pi}{4} - \pi\right)\right] = 3\left(\cos\dfrac{\pi}{4} + i \sin\dfrac{\pi}{4}\right)$

18. $(2 + 2i)^8 = [2\sqrt{2}(\cos 45° + i \sin 45°)]^8 = \left(2\sqrt{2}\right)^8[\cos(8)(45°) + i \sin(8)(45°)]$

$$= 4096[\cos 360° + i \sin 360°] = 4096$$

19. $z = 8\left(\cos\dfrac{\pi}{3} + i \sin\dfrac{\pi}{3}\right),\ n = 3$

The cube roots of z are: $\sqrt[3]{8}\left[\cos\dfrac{(\pi/3) + 2\pi k}{3} + i \sin\dfrac{(\pi/3) + 2\pi k}{3}\right], k = 0, 1, 2$

For $k = 0$, $\sqrt[3]{8}\left[\cos\dfrac{\pi/3}{3} + i \sin\dfrac{\pi/3}{3}\right] = 2\left(\cos\dfrac{\pi}{9} + i \sin\dfrac{\pi}{9}\right)$

For $k = 1$, $\sqrt[3]{8}\left[\cos\dfrac{\pi/3 + 2\pi}{3} + i \sin\dfrac{\pi/3 + 2\pi}{3}\right] = 2\left(\cos\dfrac{7\pi}{9} + i \sin\dfrac{7\pi}{9}\right)$

For $k = 2$, $\sqrt[3]{8}\left[\cos\dfrac{\pi/3 + 4\pi}{3} + i \sin\dfrac{\pi/3 + 4\pi}{3}\right] = 2\left(\cos\dfrac{13\pi}{9} + i \sin\dfrac{13\pi}{9}\right)$

20. $x^4 = -i = 1\left(\cos\dfrac{3\pi}{2} + i \sin\dfrac{3\pi}{2}\right)$

The fourth roots are: $\sqrt[4]{1}\left[\cos\dfrac{(3\pi/2) + 2\pi k}{4} + i \sin\dfrac{(3\pi/2) + 2\pi k}{4}\right], k = 0, 1, 2, 3$

For $k = 0$, $\cos\dfrac{3\pi/2}{4} + i \sin\dfrac{3\pi/2}{4} = \cos\dfrac{3\pi}{8} + i \sin\dfrac{3\pi}{8}$

For $k = 1$, $\cos\dfrac{3\pi/2 + 2\pi}{4} + i \sin\dfrac{3\pi/2 + 2\pi}{4} = \cos\dfrac{7\pi}{8} + i \sin\dfrac{7\pi}{8}$

For $k = 2$, $\cos\dfrac{3\pi/2 + 4\pi}{4} + i \sin\dfrac{3\pi/2 + 4\pi}{4} = \cos\dfrac{11\pi}{8} + i \sin\dfrac{11\pi}{8}$

For $k = 3$, $\cos\dfrac{3\pi/2 + 6\pi}{4} + i \sin\dfrac{3\pi/2 + 6\pi}{4} = \cos\dfrac{15\pi}{8} + i \sin\dfrac{15\pi}{8}$

❑ Chapter 7 Practice Test Solutions

1. $x + y = 1$

$3x - y = 15 \Rightarrow y = 3x - 15$

$x + (3x - 15) = 1$

$4x = 16$

$x = 4$

$y = -3$

Answer: $(4, -3)$

2. $x - 3y = -3 \Rightarrow x = 3y - 3$

$x^2 + 5y = 5$

$(3y - 3)^2 + 6y = 5$

$9y^2 - 18y + 9 + 6y = 5$

$9y^2 - 12y + 4 = 0$

$(3y - 2)^2 = 0$

$y = \frac{2}{3}$

$x = -1$

Answer: $\left(-1, \frac{2}{3}\right)$

3. $x + y + z = 6 \Rightarrow z = 6 - x - y$

$2x - y + 3z = 0 \Rightarrow 2x - y + 3(6 - x - y) = 0 \Rightarrow -x - 4y = -18 \Rightarrow x = 18 - 4y$

$5x + 2y - z = -3 \Rightarrow 5x + 2y - (6 - x - y) = -3 \Rightarrow 6x + 3y = 3$

$6(18 - 4y) + 3y = 3$

$-21y = -105$

$y = 5$

$x = 18 - 4y = -2$

$z = 6 - x - y = 3$

Answer: $(-2, 5, 3)$

4. $x + y = 110 \Rightarrow y = 110 - x$

$xy = 2800$

$x(110 - x) = 2800$

$0 = x^2 - 110x + 2800$

$0 = (x - 40)(x - 70)$

$x = 40$ or $x = 70$

$y = 70$ $\quad y = 40$

Answer: The two numbers are 40 and 70.

5. $2x + 2y = 170 \Rightarrow y = \dfrac{170 - 2x}{2} = 85 - x$

$xy = 2800$

$x(85 - x) = 2800$

$0 = x^2 - 85x + 2800$

$0 = (x - 25)(x - 60)$

$x = 25$ or $x = 60$

$y = 60$ $\quad y = 25$

Dimensions: $60' \times 25'$

6. $2x + 15y = 4 \Rightarrow 2x + 15y = 4$

$x - 3y = 23 \Rightarrow 5x - 15y = 115$

$7x = 119$

$x = 17$

$y = \dfrac{x - 23}{3}$

$= -2$

Answer: $(17, -2)$

7. $x + y = 2 \Rightarrow 19x + 19y = 38$

$38x - 19y = 7 \Rightarrow 38x - 19y = 7$

$57x = 45$

$x = \dfrac{45}{57} = \dfrac{15}{19}$

$y = 2 - x = \dfrac{38}{19} - \dfrac{15}{19} = \dfrac{23}{19}$

Answer: $\left(\dfrac{15}{19}, \dfrac{23}{19}\right)$

8. $$\begin{aligned} 0.4x + 0.5y &= 0.112 &\Rightarrow\ 0.28x + 0.35y &= 0.0784 \\ 0.3x - 0.7y &= -0.131 &\Rightarrow\ 0.15x - 0.35y &= -0.0655 \\ & & 0.43x \qquad &= 0.0129 \end{aligned}$$

$$x = \frac{0.0129}{0.43} = 0.03$$

$$y = \frac{0.112 - 0.4x}{0.5} = 0.20$$

Answer: (0.03, 0.20)

9. Let x = amount in 11% fund and y = amount in 13% fund.

$$x + y = 17000 \Rightarrow y = 17000 - x$$

$$0.11x + 0.13y = 2080$$

$$0.11x + 0.13(17000 - x) = 2080$$

$$-0.02x = -130$$

$$x = \$6500 \quad \text{at } 11\%$$

$$y = \$10{,}500 \text{ at } 13\%$$

10. $(4, 3), (1, 1), (-1, -2), (-2, -1)$

$$n = 4, \sum_{i=1}^{4} x_i = 2, \sum_{i=1}^{4} y_i = 1, \sum_{i=1}^{4} x_i^2 = 22, \sum_{i=1}^{4} x_i y_i = 17$$

$$\begin{aligned} 4b + 2a &= 1 &\Rightarrow\ 4b + 2a &= 1 \\ 2b + 22a &= 17 &\Rightarrow\ -4b - 44a &= -34 \\ & & -42a &= -33 \end{aligned}$$

$$a = \tfrac{33}{42} = \tfrac{11}{14}$$

$$b = \tfrac{1}{4}\left(1 - 2\left(\tfrac{33}{42}\right)\right) = -\tfrac{1}{7}$$

$$y = ax + b = \tfrac{11}{14}x - \tfrac{1}{7}$$

11. $$\begin{aligned} x + y \qquad &= -2 &\Rightarrow\ -2x - 2y \qquad &= 4 & -9y + 3z &= 45 \\ 2x - y + z &= 11 & 2x - y + z &= 11 & 4y - 3z &= -20 \\ 4y - 3z &= -20 & -3y + z &= 15 & -5y \qquad &= 25 \\ & & & & y &= -5 \\ & & & & x &= 3 \\ & & & & z &= 0 \end{aligned}$$

Answer: $(3, -5, 0)$

12. $4x - y + 5z = 4 \Rightarrow 4x - y + 5z = 4$

$2x + y - z = 0 \Rightarrow -4x - 2y + 2z = 0$

$2x + 4y + 8z = 0$

$$\begin{aligned} -3y + 7z &= 4 \\ 2x + 4y + 8z &= 0 \\ -2x - y + z &= 0 \\ \hline 3y + 9z &= 0 \\ -3y + 7z &= 4 \\ \hline 16z &= 4 \\ z &= \tfrac{1}{4} \\ y &= -\tfrac{3}{4} \\ x &= \tfrac{1}{2} \end{aligned}$$

Answer: $\left(\frac{1}{2}, -\frac{3}{4}, \frac{1}{4}\right)$

13. $3x + 2y - z = 5 \Rightarrow 6x + 4y - 2z = 10$

$6x - y + 5z = 2 \Rightarrow -6x + y - 5z = -2$

$$5y - 7z = 8$$

$$y = \frac{8 + 7z}{5}$$

$$\begin{aligned} 3x + 2y - z &= 5 \\ 12x - 2y + 10z &= 4 \\ \hline 15x + 9z &= 9 \end{aligned}$$

$$x = \frac{9 - 9z}{15} = \frac{3 - 3z}{5}$$

Let $z = a$, then $x = \dfrac{3 - 3a}{5}$ and $y = \dfrac{8 + 7a}{5}$.

Answer: $\left(\dfrac{3 - 3a}{5}, \dfrac{8 + 7a}{5}, a\right)$ where a is any real number.

14. $y = ax^2 + bx + c$ passes through $(0, -1)$, $(1, 4)$, and $(2, 13)$.

At $(0, -1)$: $-1 = a(0)^2 + b(0) + c \Rightarrow c = -1$

At $(1, 4)$: $4 = a(1)^2 + b(1) - 1 \Rightarrow 5 = a + b \Rightarrow 5 = a + b$

At $(2, 13)$: $13 = a(2)^2 + b(2) - 1 \Rightarrow 14 = 4a + 2b \Rightarrow -7 = -2a - b$

$$\begin{aligned} -2 &= -a \\ a &= 2 \\ b &= 3 \end{aligned}$$

Thus, $y = 2x^2 + 3x - 1$.

15. $s = \frac{1}{2}at^2 + v_0t + s_0$ passes through (1, 12), (2, 5), and (3, 4).

At (1, 12): $12 = \frac{1}{2}a + v_0 + s_0 \Rightarrow 24 = a + 2v_0 + 2s_0$

At (2, 5): $5 = 2a + 2v_0 + s_0 \Rightarrow -5 = -2a - 2v_0 - s_0$

At (3, 4): $4 = \frac{9}{2}a + 3v_0 + s_0 \Rightarrow$

$$19 = -a + s_0$$

$$15 = 6a + 6v_0 + 3s_0$$

$$-8 = -9a - 6v_0 - 2s_0$$

$$7 = -3a + s_0$$

$$-19 = a - s_0$$

$$-12 = -2a$$

$$a = 6$$

$$s_0 = 25$$

$$v_0 = -16$$

Thus, $s = \frac{1}{2}(6)t^2 - 16t + 25 = 3t^2 - 16t + 25$.

16. $x^2 + y^2 \geq 9$

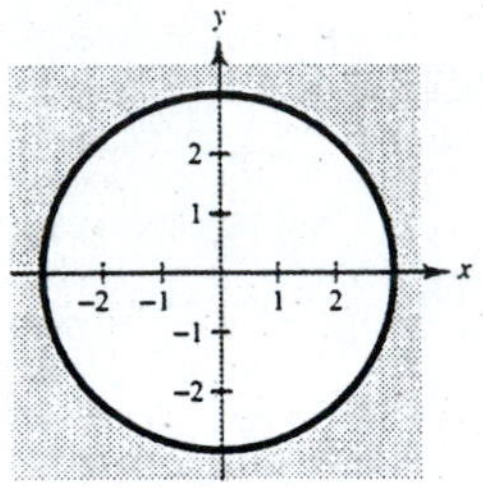

17. $x + y \leq 6$

$x \geq 2$

$y \geq 0$

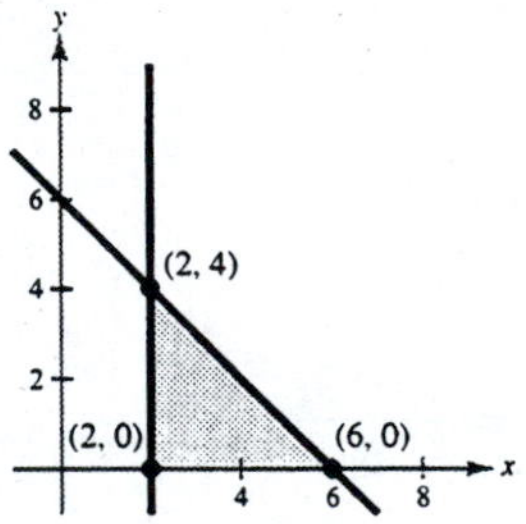

18. Line through (0, 0) and (0, 7):

$x = 0$

Line through (0, 0) and (2, 3):

$y = \frac{3}{2}x$ or $3x - 2y = 0$

Line through (0, 7) and (2, 3):

$y = -2x + 7$ or $2x + y = 7$

Inequalities: $x \geq 0$

$3x - 2y \leq 0$

$2x + y \leq 7$

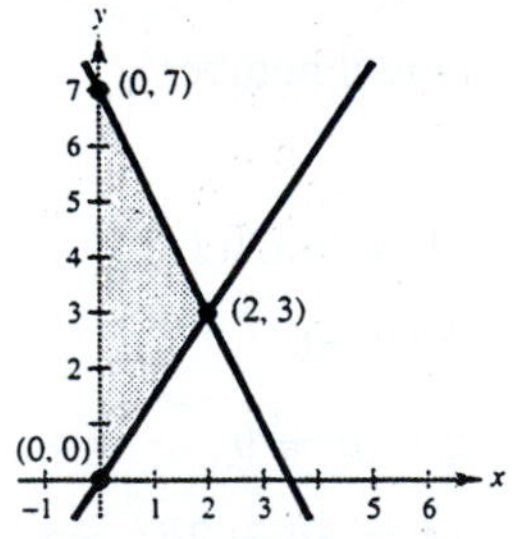

19. Vertices: (0, 0), (0, 7), (6, 0), (3, 5)

$z = 30x + 26y$

At (0, 0): $z = 0$

At (0, 7): $z = 182$

At (6, 0): $z = 180$

At (3, 5): $z = 220$

The maximum value of z is 220.

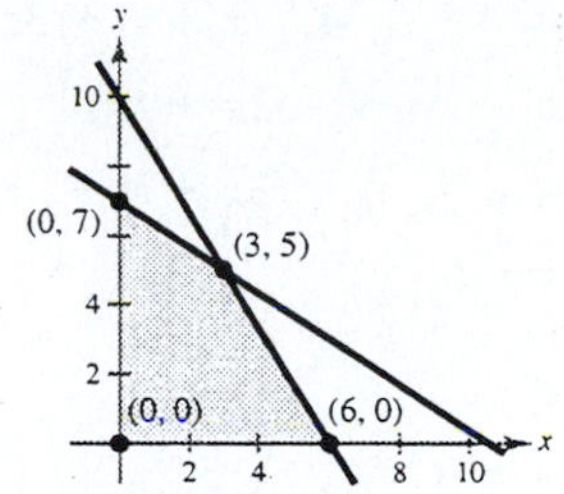

20. $x^2 + y^2 \le 4$

$(x - 2)^2 + y^2 \ge 4$

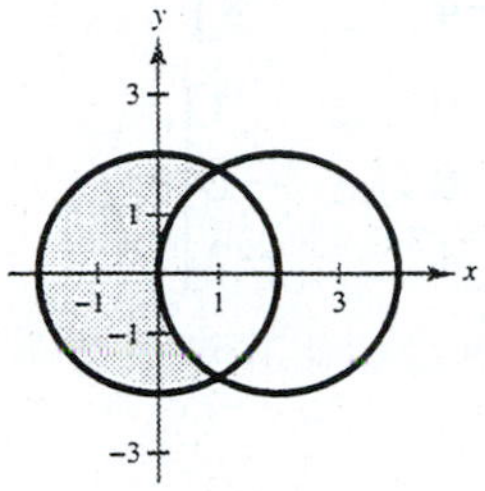

❑ Chapter 8 Practice Test Solutions

1.
$$\begin{bmatrix} 1 & -2 & 4 \\ 3 & -5 & 9 \end{bmatrix}$$

$$-3R_1 + R_2 \rightarrow \begin{bmatrix} 1 & -2 & 4 \\ 0 & 1 & -3 \end{bmatrix}$$

$$2R_2 + R_1 \rightarrow \begin{bmatrix} 1 & 0 & -2 \\ 0 & 1 & -3 \end{bmatrix}$$

2. $3x + 5y = 3$

$2x - y = -11$

$$\begin{bmatrix} 3 & 5 & \vdots & 3 \\ 2 & -1 & \vdots & -11 \end{bmatrix}$$

$$-R_2 + R_1 \rightarrow \begin{bmatrix} 1 & 6 & \vdots & 14 \\ 2 & -1 & \vdots & -11 \end{bmatrix}$$

$$-2R_1 + R_2 \rightarrow \begin{bmatrix} 1 & 6 & \vdots & 14 \\ 0 & -13 & \vdots & -39 \end{bmatrix}$$

$$-\tfrac{1}{13}R_2 \rightarrow \begin{bmatrix} 1 & 6 & \vdots & 14 \\ 0 & 1 & \vdots & 3 \end{bmatrix}$$

$$-6R_2 + R_1 \rightarrow \begin{bmatrix} 1 & 0 & \vdots & -4 \\ 0 & 1 & \vdots & 3 \end{bmatrix}$$

$x = -4, y = 3$

Answer: $(-4, 3)$

3. $2x + 3y = -3$

$3x - 2y = 8$

$x + y = 1$

$$\left[\begin{array}{cc|c} 2 & 3 & -3 \\ 3 & 2 & 8 \\ 1 & 1 & 1 \end{array}\right]$$

$$\begin{array}{r} R_3 \rightarrow \\ \\ R_1 \rightarrow \end{array}\left[\begin{array}{cc|c} 1 & 1 & 1 \\ 3 & 2 & 8 \\ 2 & 3 & -3 \end{array}\right]$$

$$\begin{array}{r} \\ -3R_1 + R_2 \rightarrow \\ -2R_1 + R_3 \rightarrow \end{array}\left[\begin{array}{cc|c} 1 & 1 & 1 \\ 0 & -1 & 5 \\ 0 & 1 & -5 \end{array}\right]$$

$$\begin{array}{r} R_2 + R_1 \rightarrow \\ -R_2 \rightarrow \\ -R_2 + R_3 \rightarrow \end{array}\left[\begin{array}{cc|c} 1 & 0 & 6 \\ 0 & 1 & -5 \\ 0 & 0 & 0 \end{array}\right]$$

$x = 6, y = -5$

Answer: $(6, -5)$

4. $x + 3z = -5$

$2x + y = 0$

$3x + y - z = -3$

$$\left[\begin{array}{ccc|c} 1 & 0 & 3 & -5 \\ 2 & 1 & 0 & 0 \\ 3 & 1 & -1 & 3 \end{array}\right]$$

$$\begin{array}{r} \\ -2R_1 + R_2 \rightarrow \\ -3R_1 + R_3 \rightarrow \end{array}\left[\begin{array}{ccc|c} 1 & 0 & 3 & -5 \\ 0 & 1 & -6 & 10 \\ 0 & 1 & -10 & 18 \end{array}\right]$$

$$\begin{array}{r} \\ \\ -R_2 + R_3 \rightarrow \end{array}\left[\begin{array}{ccc|c} 1 & 0 & 3 & -5 \\ 0 & 1 & -6 & 10 \\ 0 & 0 & -4 & 8 \end{array}\right]$$

$$\begin{array}{r} -3R_3 + R_1 \rightarrow \\ 6R_3 + R_2 \rightarrow \\ -\frac{1}{4}R_4 \rightarrow \end{array}\left[\begin{array}{ccc|c} 1 & 0 & 0 & 1 \\ 0 & 1 & 0 & -2 \\ 0 & 0 & 1 & -2 \end{array}\right]$$

$x = 1, y = -2, z = -2$

Answer: $(1, -2, -2)$

5. $$\begin{bmatrix} 1 & 4 & 5 \\ 2 & 0 & -3 \end{bmatrix}\begin{bmatrix} 1 & 6 \\ 0 & -7 \\ -1 & 2 \end{bmatrix} = \begin{bmatrix} -4 & -12 \\ 5 & 6 \end{bmatrix}$$

$$= \begin{bmatrix} 27 & 3 \\ -12 & 24 \end{bmatrix} - \begin{bmatrix} 30 & -10 \\ 15 & 25 \end{bmatrix}$$

6. $$3A - 5B = 3\begin{bmatrix} 9 & 1 \\ -4 & 8 \end{bmatrix} - 5\begin{bmatrix} 6 & -2 \\ 3 & 5 \end{bmatrix}$$

$$= \begin{bmatrix} -3 & 13 \\ -27 & -1 \end{bmatrix}$$

7. $$f(A) = \begin{bmatrix} 3 & 0 \\ 7 & 1 \end{bmatrix}^2 - 7\begin{bmatrix} 3 & 0 \\ 7 & 1 \end{bmatrix} + 8\begin{bmatrix} 1 & 0 \\ 0 & 1 \end{bmatrix}$$

$$= \begin{bmatrix} 3 & 0 \\ 7 & 1 \end{bmatrix}\begin{bmatrix} 3 & 0 \\ 7 & 1 \end{bmatrix} - \begin{bmatrix} 21 & 0 \\ 49 & 7 \end{bmatrix} + \begin{bmatrix} 8 & 0 \\ 0 & 8 \end{bmatrix}$$

$$= \begin{bmatrix} 9 & 0 \\ 28 & 1 \end{bmatrix} - \begin{bmatrix} 21 & 0 \\ 49 & 7 \end{bmatrix} + \begin{bmatrix} 8 & 0 \\ 0 & 8 \end{bmatrix}$$

$$= \begin{bmatrix} -4 & 0 \\ -21 & 2 \end{bmatrix}$$

8. False since

$(A + B)(A + 3B) = A(A + 3B) + B(A + 3B)$

$= A^2 + 3AB + BA + 3B^2$ and, in general, $AB \neq BA$.

9.
$$\left[\begin{array}{cc:cc} 1 & 2 & 1 & 0 \\ 3 & 5 & 0 & 1 \end{array}\right]$$

$$\begin{array}{r} \\ -3R_1 + R_2 \rightarrow \end{array}\left[\begin{array}{cc:cc} 1 & 2 & 1 & 0 \\ 0 & -1 & -3 & 1 \end{array}\right]$$

$$\begin{array}{r} 2R_2 + R_1 \rightarrow \\ -R_2 \rightarrow \end{array}\left[\begin{array}{cc:cc} 1 & 0 & -5 & 2 \\ 0 & 1 & 3 & -1 \end{array}\right]$$

$$A^{-1} = \begin{bmatrix} -5 & 2 \\ 3 & -1 \end{bmatrix}$$

10.
$$\left[\begin{array}{ccc:ccc} 1 & 1 & 1 & 1 & 0 & 0 \\ 3 & 6 & 5 & 0 & 1 & 0 \\ 6 & 10 & 8 & 0 & 0 & 1 \end{array}\right]$$

$$\begin{array}{r} \\ -3R_1 + R_2 \rightarrow \\ -6R_1 + R_3 \rightarrow \end{array}\left[\begin{array}{ccc:ccc} 1 & 1 & 1 & 1 & 0 & 0 \\ 0 & 3 & 2 & -3 & 1 & 0 \\ 0 & 4 & 2 & -6 & 0 & 1 \end{array}\right]$$

$$\begin{array}{r} -R_2 + R_1 \rightarrow \\ \frac{1}{3}R_2 \rightarrow \\ -4R_2 + R_3 \rightarrow \end{array}\left[\begin{array}{ccc:ccc} 1 & 0 & \frac{1}{3} & 2 & -\frac{1}{3} & 0 \\ 0 & 1 & \frac{2}{3} & -1 & \frac{1}{3} & 0 \\ 0 & 0 & -\frac{2}{3} & -2 & -\frac{4}{3} & 1 \end{array}\right]$$

$$\begin{array}{r} \frac{1}{2}R_3 + R_1 \rightarrow \\ R_3 + R_2 \rightarrow \\ -\frac{3}{2}R_3 \rightarrow \end{array}\left[\begin{array}{ccc:ccc} 1 & 0 & 0 & 1 & -1 & \frac{1}{2} \\ 0 & 1 & 0 & -3 & -1 & 1 \\ 0 & 0 & 1 & 3 & 2 & -\frac{3}{2} \end{array}\right]$$

$$A^{-1} = \begin{bmatrix} 1 & -1 & \frac{1}{2} \\ -3 & -1 & 1 \\ 3 & 2 & -\frac{3}{2} \end{bmatrix}$$

11. (a) $x + 2y = 4$

$3x + 5y = 1$

$$\left[\begin{array}{cc:cc} 1 & 2 & 1 & 0 \\ 3 & 5 & 0 & 1 \end{array}\right]$$

$$\begin{array}{r} \\ -3R_1 + R_2 \rightarrow \end{array}\left[\begin{array}{cc:cc} 1 & 2 & 1 & 0 \\ 0 & -1 & -3 & 1 \end{array}\right]$$

$$\begin{array}{r} -2R_2 + R_1 \rightarrow \\ -R_2 \rightarrow \end{array}\left[\begin{array}{cc:cc} 1 & 0 & -5 & 2 \\ 0 & 1 & 3 & -1 \end{array}\right]$$

$$X = A^{-1}B = \begin{bmatrix} -5 & 2 \\ 3 & -1 \end{bmatrix}\begin{bmatrix} 4 \\ 1 \end{bmatrix} = \begin{bmatrix} -18 \\ 11 \end{bmatrix}$$

$x = -18, y = 11$

Answer: $(-18, 11)$

(b) $x + 2y = 3$

$3x + 5y = -2$

$$X = A^{-1}B = \begin{bmatrix} -5 & 2 \\ 3 & -1 \end{bmatrix}\begin{bmatrix} 3 \\ -2 \end{bmatrix} = \begin{bmatrix} -19 \\ 11 \end{bmatrix}$$

$x = -19, y = 11$

Answer: $(-19, 11)$

12. $\begin{vmatrix} 6 & -1 \\ 3 & 4 \end{vmatrix} = 24 - (-3) = 27$

13. $\begin{vmatrix} 1 & 3 & -1 \\ 5 & 9 & 0 \\ 6 & 2 & -5 \end{vmatrix} - 1\begin{vmatrix} 5 & 9 \\ 6 & 2 \end{vmatrix} - 5\begin{vmatrix} 1 & 3 \\ 5 & 9 \end{vmatrix} = 74$

14. $\begin{vmatrix} 1 & 4 & 2 & 3 \\ 0 & 1 & -2 & 0 \\ 3 & 5 & -2 & 1 \\ 2 & 0 & 6 & 1 \end{vmatrix} = \begin{vmatrix} 1 & 2 & 3 \\ 3 & -1 & 1 \\ 2 & 6 & 1 \end{vmatrix} + 2\begin{vmatrix} 1 & 4 & 3 \\ 3 & 5 & 1 \\ 2 & 0 & 1 \end{vmatrix}$

$= 51 + 2(-29) = -7$ (Expansion along Row 2.)

15. $\begin{vmatrix} 6 & 4 & 3 & 0 & 6 \\ 0 & 5 & 1 & 4 & 8 \\ 0 & 0 & 2 & 7 & 3 \\ 0 & 0 & 0 & 9 & 2 \\ 0 & 0 & 0 & 0 & 1 \end{vmatrix} = 6(5)(2)(9)(1) = 540$ (Upper triangular)

16. $\text{Area} = \frac{1}{2}\begin{vmatrix} 0 & 7 & 1 \\ 5 & 0 & 1 \\ 3 & 9 & 1 \end{vmatrix} = \frac{1}{2}(31)$

17. $\begin{vmatrix} x & y & 1 \\ 2 & 7 & 1 \\ -1 & 4 & 1 \end{vmatrix} = 3x - 3y + 15 = 0 \text{ OR } = x - y + 5 = 0$

18. $x = \dfrac{\begin{vmatrix} 4 & -7 \\ 11 & 5 \end{vmatrix}}{\begin{vmatrix} 6 & -7 \\ 2 & 5 \end{vmatrix}} = \dfrac{97}{44}$

19. $z = \dfrac{\begin{vmatrix} 3 & 0 & 1 \\ 0 & 1 & 3 \\ 1 & -1 & 2 \end{vmatrix}}{\begin{vmatrix} 3 & 0 & 1 \\ 0 & 1 & 4 \\ 1 & -1 & 0 \end{vmatrix}} = \dfrac{14}{11}$

20. $y = \dfrac{\begin{vmatrix} 721.4 & 33.77 \\ 45.9 & 19.85 \end{vmatrix}}{\begin{vmatrix} 721.4 & -29.1 \\ 45.9 & 105.6 \end{vmatrix}} = \dfrac{12{,}769.747}{77{,}515.530} \approx 0.1647$

❑ Chapter 9 Practice Test Solutions

1. $a_n = \dfrac{2n}{(n+2)!}$

$a_1 = \dfrac{2(1)}{3!} = \dfrac{2}{6} = \dfrac{1}{3}$

$a_2 = \dfrac{2(2)}{4!} = \dfrac{4}{24} = \dfrac{1}{6}$

$a_3 = \dfrac{2(3)}{5!} = \dfrac{6}{120} = \dfrac{1}{20}$

$a_4 = \dfrac{2(4)}{6!} = \dfrac{8}{720} = \dfrac{1}{90}$

$a_5 = \dfrac{2(5)}{7!} = \dfrac{10}{5040} = \dfrac{1}{504}$

Terms: $\dfrac{1}{3}, \dfrac{1}{6}, \dfrac{1}{20}, \dfrac{1}{90}, \dfrac{1}{504}$

2. $a_n = \dfrac{n+3}{3^n}$

3. $\sum_{i=1}^{6}(2i - 1) = 1 + 3 + 5 + 7 + 9 + 11 = 36$

4. $a_1 = 23,\ d = -2$

$a_2 = 23 + (-2) = 21$

$a_3 = 21 + (-2) = 19$

$a_4 = 19 + (-2) = 17$

$a_5 = 17 + (-2) = 15$

Terms: 23, 21, 19, 17, 15

5. $a_1 = 12, d = 3, n = 50$

$a_n = a_1 + (n - 1)d$

$a_{50} = 12 + (50 - 1)3 = 159$

6. $a_1 = 1$

$a_{200} = 200$

$S_n = \frac{n}{2}(a_1 + a_n)$

$S_{200} = \frac{200}{2}(1 + 200) = 20{,}100$

7. $a_1 = 7,\ r = 2$

$a_2 = 7(2) = 14$

$a_3 = 7(2)^2 = 28$

$a_4 = 7(2)^3 = 56$

$a_5 = 7(2)^4 = 112$

Terms: 7, 14, 28, 56, 112

8. $\sum_{n=0}^{9} 6\left(\frac{2}{3}\right)^n,\ a_1 = 6,\ r = \frac{2}{3},\ n = 10$

$$S_n = \frac{a_1(1 - r^n)}{1 - r} = \frac{6\left(1 - \left(\frac{2}{3}\right)^{10}\right)}{1 - \frac{2}{3}} \approx 17.6879$$

9. $\sum_{n=0}^{\infty} (0.03)^n,\ a_1 = 1,\ r = 0.03$

$$S = \frac{a_1}{1 - r} = \frac{1}{1 - 0.03} = \frac{1}{0.97} = \frac{100}{97} \approx 1.0309$$

10. For $n = 1$, $1 = \frac{1(1 + 1)}{2}$.

Assume that $1 + 2 + 3 + 4 + \cdots + k = \frac{k(k + 1)}{2}$.

Now for $n = k + 1$,

$$\begin{aligned} 1 + 2 + 3 + 4 + \cdots + k + (k + 1) &= \frac{k(k + 1)}{2} + k + 1 \\ &= \frac{k(k + 1)}{2} + \frac{2(k + 1)}{2} \\ &= \frac{(k + 1)(k + 2)}{2}. \end{aligned}$$

Thus, $1 + 2 + 3 + 4 + \cdots + n = \frac{n(n + 1)}{2}$ for all integers $n \geq 1$.

11. For $n = 4$, $4! > 2^4$. Assume that $k! > 2^k$.

Then $(k + 1)! = (k + 1)(k!) > (k + 1)2^k > 2 \cdot 2^k = 2^{k+1}$.

Thus, $n! > 2^n$ for all integers $n \geq 4$.

12. ${}_{13}C_4 = \frac{13!}{(13 - 4)!4!} = 715$

13. $$\begin{aligned} (x + 3)^5 &= x^5 + 5x^4(3) + 10x^3(3)^2 + 10x^2(3)^3 + 5x(3)^4 + (3)^5 \\ &= x^5 + 15x^4 + 90x^3 + 270x^2 + 405x + 243 \end{aligned}$$

14. ${}_{12}C_5 x^7(-2)^5 = -25{,}344x^7$

15. ${}_{30}P_4 = \frac{30!}{(30 - 4)!} = 657{,}720$

16. $6! = 720$ ways

17. ${}_{12}P_3 = 1320$

18. $P(2) + P(3) + P(4) = \frac{1}{36} + \frac{2}{36} + \frac{3}{36}$

$$= \frac{6}{36} = \frac{1}{6}$$

19. $P(K, B10) = \frac{4}{52} \cdot \frac{2}{51} = \frac{2}{663}$

20. Let A = probability of no faulty units.

$$P(A) = \left(\frac{997}{1000}\right)^{50} \approx 0.8605$$

$$P(A') = 1 - P(A) \approx 0.1395$$

❑ Chapter 10 Practice Test Solutions

1. $3x + 4y = 12 \Rightarrow y = -\frac{3}{4}x + 3 \Rightarrow m_1 = -\frac{3}{4}$

$4x - 3y = 12 \Rightarrow y = \frac{4}{3}x - 4 \Rightarrow m_2 = \frac{4}{3}$

$$\tan\theta = \left|\frac{\frac{4}{3} - \left(-\frac{3}{4}\right)}{1 + \left(\frac{4}{3}\right)\left(-\frac{3}{4}\right)}\right| = \left|\frac{\frac{25}{12}}{0}\right|$$

Since $\tan\theta$ is undefined, the lines are perpendicular (note that $m_2 = -1/m_1$) and $\theta = 90°$.

2. $x_1 = 5, x_2 = -9, A = 3, B = -7, C = -21$

$$d = \frac{|3(5) + (-7)(-9) + (-21)|}{\sqrt{3^2 + (-7)^2}} = \frac{57}{\sqrt{58}} \approx 7.484$$

3.
$$\begin{aligned} x^2 - 6x - 4y + 1 &= 0 \\ x^2 - 6x + 9 &= 4y - 1 + 9 \\ (x - 3)^2 &= 4y + 8 \\ (x - 3)^2 &= 4(1)(y + 2) \Rightarrow p = 1 \end{aligned}$$

Vertex: $(3, -2)$

Focus: $(3, -1)$

Directrix: $y = -3$

4. Vertex: $(2, -5)$

Focus: $(2, -6)$

Vertical axis; opens downward with $p = -1$

$$\begin{aligned} (x - h)^2 &= 4p(y - k) \\ (x - 2)^2 &= 4(-1)(y + 5) \\ x^2 - 4x + 4 &= -4y - 20 \end{aligned}$$

$x^2 - 4x + 4y + 24 = 0$

5.
$$\begin{aligned} x^2 + 4y^2 - 2x + 32y + 61 &= 0 \\ (x^2 - 2x + 1) + 4(y^2 + 8y + 16) &= -61 + 1 + 64 \\ (x - 1)^2 + 4(y + 4)^2 &= 4 \\ \frac{(x - 1)^2}{4} + \frac{(y + 4)^2}{1} &= 1 \end{aligned}$$

$a = 2, b = 1, c = \sqrt{3}$

Horizontal major axis

Center: $(1, -4)$

Foci: $\left(1 \pm \sqrt{3}, -4\right)$

Vertices: $(3, -4), (-1, -4)$

Eccentricity: $e = \frac{\sqrt{3}}{2}$

6. Vertices: $(0, \pm 6)$

Eccentricity: $e = \frac{1}{2}$

Center: $(0, 0)$

Vertical major axis

$a = 6, e = \frac{c}{a} = \frac{c}{6} = \frac{1}{2} \Rightarrow c = 3$

$b^2 = (6)^2 - (3)^2 = 27$

$\frac{x^2}{27} + \frac{y^2}{36} = 1$

7.
$$\begin{aligned} 16y^2 - x^2 - 6x - 128y + 231 &= 0 \\ 16(y^2 - 8y + 16) - (x^2 + 6x + 9) &= -231 + 256 - 9 \\ 16(y - 4)^2 - (x + 3)^2 &= 16 \\ \frac{(y - 4)^2}{1} - \frac{(x + 3)^2}{16} &= 1 \end{aligned}$$

$a = 1, b = 4, c = \sqrt{17}$

Center: $(-3, 4)$

Vertical transverse axis

Vertices: $(-3, 5), (-3, 3)$

Foci: $\left(-3, 4 \pm \sqrt{17}\right)$

Asymptotes: $y = 4 \pm \frac{1}{4}(x + 3)$

8. Vertices: $(\pm 3, 2)$

Foci: $(\pm 5, 2)$

Center: $(0, 2)$

Horizontal transverse axis

$a = 3, c = 5, b = 4$

$$\frac{(x-0)^2}{9} - \frac{(y-2)^2}{16} = 1$$

$$\frac{x^2}{9} - \frac{(y-2)^2}{16} = 1$$

9. $5x^2 + 2xy + 5y^2 - 10 = 0$

$A = 5, B = 2, C = 5$

$$\cot 2\theta = \frac{5-5}{2} = 0$$

$$2\theta = \frac{\pi}{2} \Rightarrow \theta = \frac{\pi}{4}$$

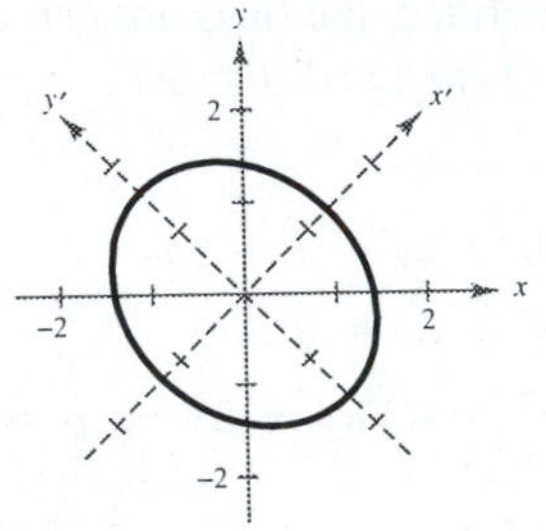

$$x = x' \cos\frac{\pi}{4} - y' \sin\frac{\pi}{4} = \frac{x'-y'}{\sqrt{2}}$$

$$x = x' \cos\frac{\pi}{4} + y' \sin\frac{\pi}{4} = \frac{x'+y'}{\sqrt{2}}$$

$$5\left(\frac{x'-y'}{\sqrt{2}}\right)^2 + 2\left(\frac{x'-y'}{\sqrt{2}}\right)\left(\frac{x'+y'}{\sqrt{2}}\right) + 5\left(\frac{x'+y'}{\sqrt{2}}\right)^2 - 10 = 0$$

$$\frac{5(x')^2}{2} - \frac{10x'y'}{2} + \frac{5(y')^2}{2} + (x')^2 - (y')^2 + \frac{5(x')^2}{2} + \frac{10x'y'}{2} + \frac{5(y')^2}{2} - 10 = 0$$

$$6(x')^2 + 4(y')^2 - 10 = 0$$

$$\frac{3(x')^2}{5} + \frac{2(y')^2}{5} = 1$$

$$\frac{(x')^2}{5/3} + \frac{(y')^2}{5/2} = 1$$

Ellipse centered at the origin

10. (a) $6x^2 - 2xy + y^2 = 0$

$A = 6, B = -2, C = 1$

$B^2 - 4AC = (-2)^2 - 4(6)(1) = -20 < 0$

Ellipse

(b) $x^2 + 4xy + 4y^2 - x - y + 17 = 0$

$A = 1, B = 4, C = 4$

$B^2 - 4AC = (4)^2 - 4(1)(4) = 0$

Parabola

11. Polar: $\left(\sqrt{2}, \frac{3\pi}{4}\right)$

$$x = \sqrt{2}\cos\frac{3\pi}{4} = \sqrt{2}\left(-\frac{1}{\sqrt{2}}\right) = -1$$

$$y = \sqrt{2}\sin\frac{3\pi}{4} = \sqrt{2}\left(\frac{1}{\sqrt{2}}\right) = 1$$

Rectangular: $(-1, 1)$

12. Rectangular: $(\sqrt{3}, -1)$

$$r = \pm\sqrt{(\sqrt{3})^2 + (-1)^2} = \pm 2$$

$$\tan\theta = \frac{\sqrt{3}}{-1} = -\sqrt{3}$$

$$\theta = \frac{2\pi}{3} \text{ or } \theta = \frac{5\pi}{3}$$

Polar: $\left(-2, \frac{2\pi}{3}\right)$ or $\left(2, \frac{5\pi}{3}\right)$

13. Rectangular: $4x - 3y = 12$

Polar: $4r\cos\theta - 3r\sin\theta = 12$

$$r(4\cos\theta - 3\sin\theta) = 12$$

$$r = \frac{12}{4\cos\theta - 3\sin\theta}$$

14. Polar: $r = 5\cos\theta$

$$r^2 = 5r\cos\theta$$

Rectangular: $x^2 + y^2 = 5x$

$$x^2 + y^2 - 5x = 0$$

15. $r = 1 - \cos\theta$

Cardioid

Symmetry: Polar axis

Maximum value of $|r|$: $r = 2$ when $\theta = \pi$.

Zero of r: $r = 0$ when $\theta = 0$

θ	0	$\frac{\pi}{2}$	π	$\frac{3\pi}{2}$
r	0	1	2	1

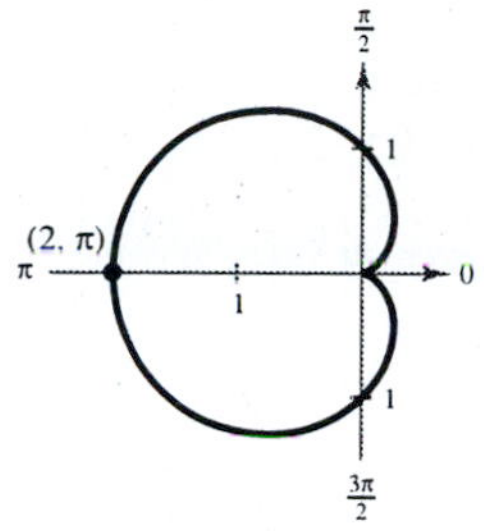

16. $r = 5\sin 2\theta$

Rose curve with four petals

Symmetry: Polar axis, $\theta = \frac{\pi}{2}$, and pole

Maximum value of $|r|$: $|r| = 5$ when $\theta = \frac{\pi}{4}, \frac{3\pi}{4}, \frac{5\pi}{4}, \frac{7\pi}{4}$

Zeros of r: $r = 0$ when $\theta = 0, \frac{\pi}{2}, \pi, \frac{3\pi}{2}$

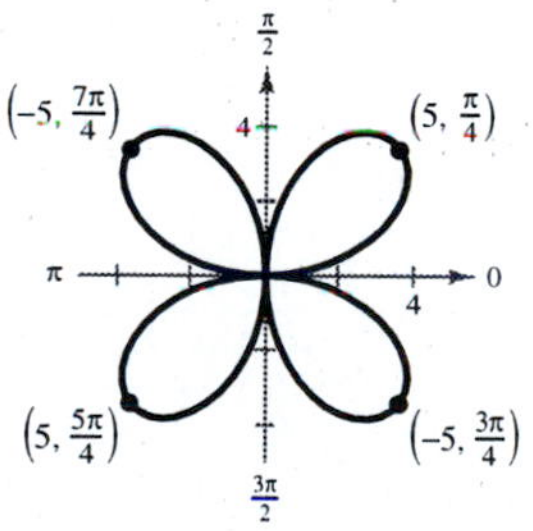

17. $r = \frac{3}{6 - \cos\theta}$

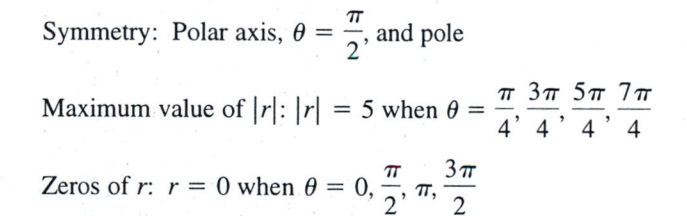

$$r = \frac{\frac{1}{2}}{1 - \frac{1}{6}\cos\theta}$$

$e = \frac{1}{6} < 1$, so the graph is an ellipse.

θ	0	$\frac{\pi}{2}$	π	$\frac{3\pi}{2}$
r	$\frac{3}{5}$	$\frac{1}{2}$	$\frac{3}{7}$	$\frac{1}{2}$

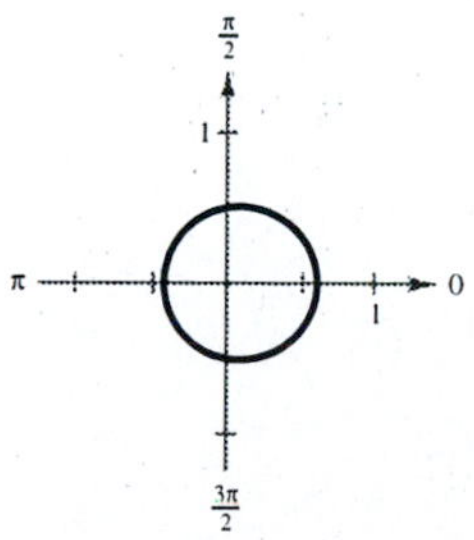

18. Parabola

Vertex: $\left(6, \frac{\pi}{2}\right)$

Focus: (0, 0)

$e = 1$

$$r = \frac{ep}{1 + e \sin \theta}$$

$$r = \frac{p}{1 + \sin \theta}$$

$$6 = \frac{p}{1 + \sin(\pi/2)}$$

$$6 = \frac{p}{2}$$

$$12 = p$$

$$r = \frac{12}{1 + \sin \theta}$$

19. $x = 3 - 2 \sin \theta, y = 1 + 5 \cos \theta$

$$\frac{x - 3}{-2} = \sin \theta, \frac{y - 1}{5} = \cos \theta$$

$$\left(\frac{x - 3}{-2}\right)^2 + \left(\frac{y - 1}{5}\right)^2 = 1$$

$$\frac{(x - 3)^2}{4} + \frac{(y - 1)^2}{25} = 1$$

20. $x = e^{2t}, y = e^{4t}$

$x > 0, y > 0$

$y = (e^{2t})^2 = (x)^2 = x^2, x < 0, y > 0$

PART II Solutions to Chapter Tests and Cumulative Tests

❑ Chapter Test Solutions for Chapter P

1. $\frac{5}{18} \div \frac{15}{8} = \frac{\overset{1}{\cancel{5}}}{\underset{9}{\cancel{18}}} \cdot \frac{\overset{4}{\cancel{8}}}{\underset{3}{\cancel{15}}} = \frac{4}{27}$

2. $\sqrt{5} \cdot \sqrt{125} = \sqrt{625} = 25$

3. $3z^2(2z^3)^2 = 3z^2(4z^6) = 12z^8$

4. $9z\sqrt{8z} - 3\sqrt{2z^3} = 18z\sqrt{2z} - 3z\sqrt{2z} = 15z\sqrt{2z}$

5. $(x^2 + 3) - [3x + (8 - x^2)] = x^2 + 3 - 3x - 8 + x^2$
$= 2x^2 - 3x - 5$

6. $(3x - 2)^2 = (3x)^2 - 2(3x)(2) + (2)^2 = 9x^2 - 12x + 14$

7. $\frac{8x}{x-3} + \frac{24}{3-x} = \frac{8x}{x-3} - \frac{24}{x-3} = \frac{8x - 24}{x-3} = \frac{8(x-3)}{x-3} = 8, x \neq 3$

8. $\left(\frac{2}{x} - \frac{2}{x+1}\right) \div \left(\frac{4}{x^2 - 1}\right) = \left(\frac{2}{x(x+1)}\right)\left(\frac{(x+1)(x-1)}{4}\right) = \frac{x-1}{2x}, x \neq \pm 1$

9. $x^3 + 2x^2 - 4x - 8 = x^2(x + 2) - 4(x + 2)$
$= (x + 2)(x^2 - 4)$
$= (x + 2)^2(x - 2)$

10. $A = \frac{1}{2}(3x)(\sqrt{3}x) - \frac{1}{2}(2x)\left(\frac{2}{3}\sqrt{3}x\right)$
$= \frac{3\sqrt{3}x^2}{2} - \frac{2\sqrt{3}x^2}{3} = \frac{9\sqrt{3}x^2 - 4\sqrt{3}x^2}{6} = \frac{5\sqrt{3}x^2}{6} = \frac{5}{6}\sqrt{3}x^2$

11. $\frac{2}{3}(x - 1) + \frac{1}{4}x = 10$
$12\left[\frac{2}{3}(x - 1) + \frac{1}{4}x\right] = 12(10)$
$8(x - 1) + 3x = 120$
$11x - 8 = 120$
$11x = 128$
$x = \frac{128}{11}$

12. $\frac{x-2}{x+2} + \frac{4}{x+2} + 4 = 0$
$\frac{x+2}{x+2} + 4 = 0$
$5 \neq 0$
No solution

13. $3x^2 + 6x + 2 = 0$
$x = \frac{-6 \pm \sqrt{6^2 - 4(3)(2)}}{2(3)}$
$= \frac{-6 \pm \sqrt{12}}{6}$
$= \frac{-6 \pm 2\sqrt{3}}{6}$
$= \frac{-3 \pm \sqrt{3}}{3}$

14. $x^4 + x^2 - 6 = 0$
$(x^2 - 2)(x^2 + 3) = 0$
$x^2 - 2 = 0 \Rightarrow x = \pm\sqrt{2}$
$x^2 + 3 = 0 \Rightarrow x = \pm\sqrt{3}i$

15. $2\sqrt{x} - \sqrt{2x+1} = 1$

$$-\sqrt{2x+1} = 1 - 2\sqrt{x}$$

$$2x + 1 = 1 - 4\sqrt{x} + 4x$$

$$4\sqrt{x} = 2x$$

$$16x = 4x^2$$

$$16x - 4x^2 = 0$$

$$4x(4 - x) = 0$$

$4x = 0$ or $4 - x = 0$

$x = 0$ $\quad x = 4$

$x = 4$ is the only solution which checks in the orginal equation. $x = 0$ is extraneous.

16. $|3x - 1| = 7$

$3x - 1 = -7$ or $3x - 1 = 7$

$3x = -6 \qquad 3x = 8$

$x = -2 \qquad x = \frac{8}{3}$

17. $-3 \le 2(x + 4) < 14$

$$-3 \le 2x + 8 < 14$$

$$-11 \le 2x < 6$$

$$-\frac{11}{2} \le x < 3$$

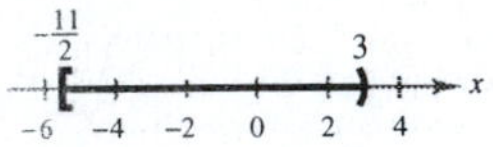

18. $\dfrac{2}{x} > \dfrac{5}{x+6}$

$$\frac{2}{x} - \frac{5}{x+6} > 0$$

$$\frac{12 - 3x}{x(x+6)} > 0$$

Critical numbers: $x = -6, x = 0, x = 4$

Test intervals: $(-\infty, -6), (-6, 0), (0, 4), (4, \infty)$

Test: Is $\dfrac{3(4 - x)}{x(x+6)} > 0$?

Solution set: $(-\infty, -6) \cup (0, 4)$

Inequality notation: $x < -6, 0 < x < 4$

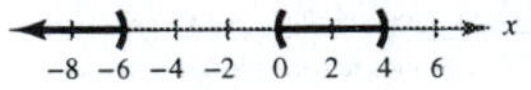

19. $(100 \text{ km/hr})(2\frac{1}{4} \text{ hr}) + (x \text{ km/hr})(1\frac{1}{3} \text{ hr}) = 350 \text{ km}$

$$225 + \frac{4}{3}x = 250$$

$$\frac{4}{3}x = 125$$

$$x = \frac{375}{4} = 93\frac{3}{4} \text{ km/hr}$$

20.

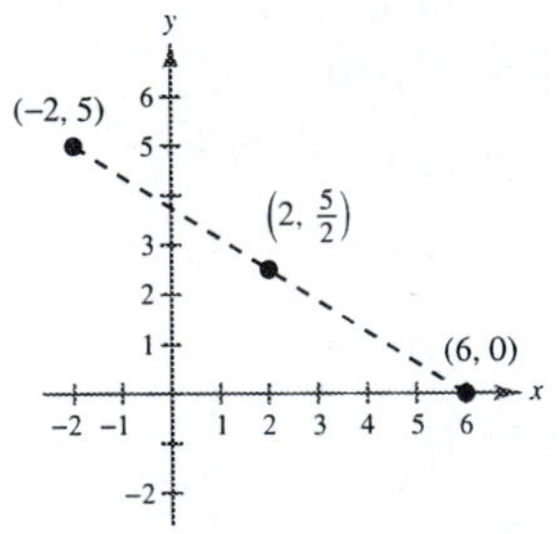

Midpoint: $\left(\dfrac{-2+6}{2}, \dfrac{5+0}{2}\right) = \left(2, \dfrac{5}{2}\right)$

Distance:

$d = \sqrt{(-2-6)^2 + (5-0)^2} = \sqrt{64+25} = \sqrt{89}$

❑ Chapter Test Solutions for Chapter 1

1. $y = 4 - \frac{3}{4}x$

No symmetry

x- intercept: $\left(\frac{16}{3}, 0\right)$

y- intercept: $(0, 4)$

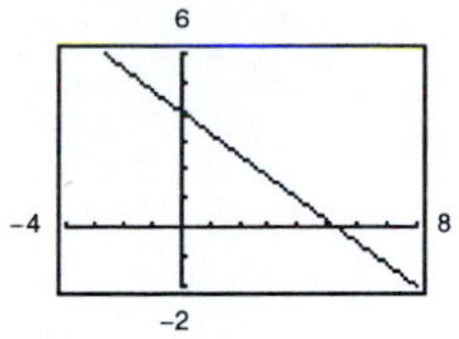

2. $y = 4 - \frac{3}{4}|x|$

y- axis symmetry

x- intercepts: $\left(\pm\frac{16}{3}, 0\right)$

y- intercept: $(0, 4)$

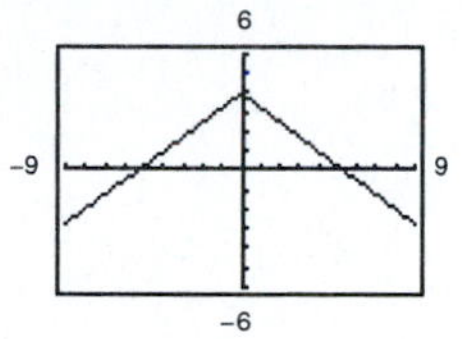

3. $y = 4 - (x - 2)^2$

Parabola; vertex: $(2, 4)$

No symmetry

x- intercepts: $(0, 0)$ and $(4, 0)$

y- intercept: $(0, 0)$

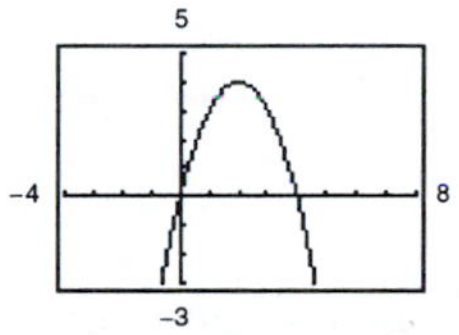

4. $y = \sqrt{3 - x}$

Domain: $x \le 3$

No symmetry

x- intercept: $(3, 0)$

y- intercept: $\left(0, \sqrt{3}\right)$

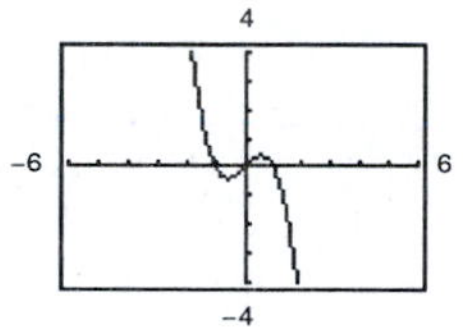

5. $m = \dfrac{2 - 6}{3 - (-3)} = \dfrac{-4}{6} = -\dfrac{2}{3}$

6. $m = \dfrac{3}{2}$

$(3 - 2, -1 - 3) = (1, -4)$

$(3 + 2, -1 + 3) = (5, 2)$

$(5 + 2, 2 + 3) = (7, 5)$

7. $5x + 2y = 3 \implies y = -\frac{5}{2}x + \frac{3}{2} \implies m_1 = -\frac{5}{2}$

Perpendicular slope: $m_2 = \frac{2}{5}$

$y - 4 = \frac{2}{5}(x - 0)$

$5y - 20 = 2x$

$0 = 2x - 5y + 20$

8. No. For some x- values there are two y- values. The graph fails the vertical line test.

9. $f(-6) = 10 - \sqrt{3 - (-6)} = 10 - 3 = 7$

10. $f(t - 3) = 10 - \sqrt{3 - (t - 3)} = 10 - \sqrt{6 - t}$

11. $3 - x \geq 0$

$-x \geq -3$

$x \leq 3$

Domain: $(-\infty, 3]$

12. $\dfrac{f(x) - f(2)}{x - 2} = \dfrac{\left(10 - \sqrt{3 - x}\right) - \left(10 - \sqrt{3 - 2}\right)}{x - 2}$

$= \dfrac{1 - \sqrt{3 - x}}{x - 2} = \dfrac{\sqrt{3 - x} - 1}{2 - x}$

13. $h(x) = \frac{1}{4}x^4 - 2x^2$

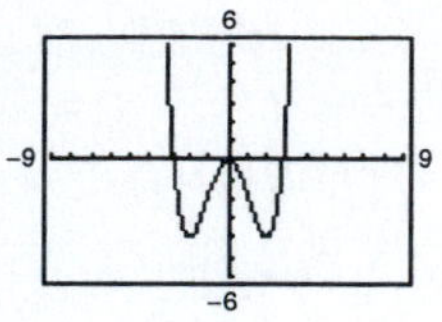

Even

Increasing on $(-2, 0), (2, \infty)$

Decreasing on $(-\infty, -2), (0, 2)$

14. $g(t) = |t + 2| - |t - 2|$

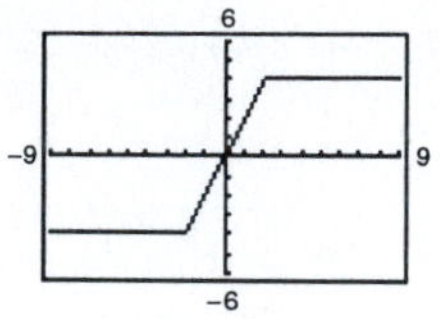

Odd

Increasing on $(-2, 2)$

Constant on $(-\infty, -2), (2, \infty)$

15. $(f - g)(x) = f(x) - g(x) = x^2 - \sqrt{2 - x}$

Domain: $(-\infty, 2]$

16. $\left(\dfrac{f}{g}\right)(x) = \dfrac{f(x)}{g(x)} = \dfrac{x^2}{\sqrt{2 - x}}$

Domain: $(-\infty, 2)$

17. $(f \circ g)(x) = f(g(x)) = f\left(\sqrt{2 - x}\right)$

$= \left(\sqrt{2 - x}\right)^2 = 2 - x$

Domain: $(-\infty, \infty)$

18. $g(x) = \sqrt{2 - x}$

Domain: $(-\infty, 2]$

Range: $[0. \infty)$

The domain of $g^{-1}(x)$ is the range of $g(x)$ and the range of $g^{-1}(x)$ is the domain of $g(x)$.

$y = \sqrt{2 - x}$

$x = \sqrt{2 - y}$

$x^2 = 2 - y$

$x^2 + 2 = y$

$g^{-1}(x) = x^2 + 2, \ x \geq 0$

Domain: $[0, \infty)$

19. (a) $\frac{1}{2}g(x - 2)$

Horizontal shift two units to the right Vertrical shrink by $\frac{1}{2}$

x	$g(x)$
-2	4
0	0
2	2
4	1

x	$\frac{1}{2}g(x - 2)$
0	2
2	0
4	1
6	$\frac{1}{2}$

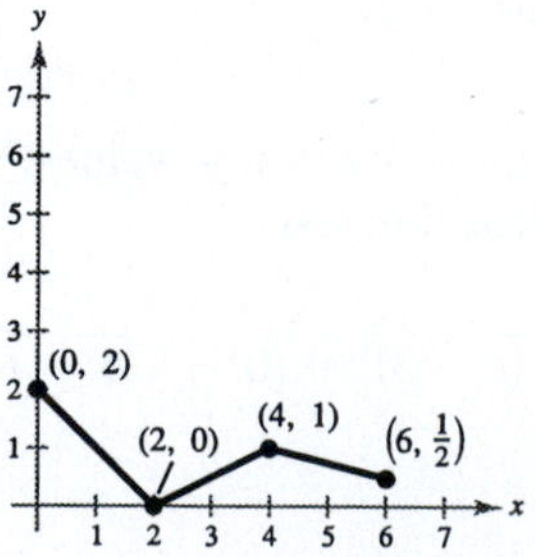

(b) $g\left(\frac{1}{2}x\right) - 1$

Change in x by a factor of $\frac{1}{2}$ Vertical shift one unit down

x	$g\left(\frac{1}{2}x\right) - 1$
-4	$4 - 1 = 3$
0	$0 - 1 = -1$
4	$2 - 1 = 1$
8	$1 - 1 = 0$

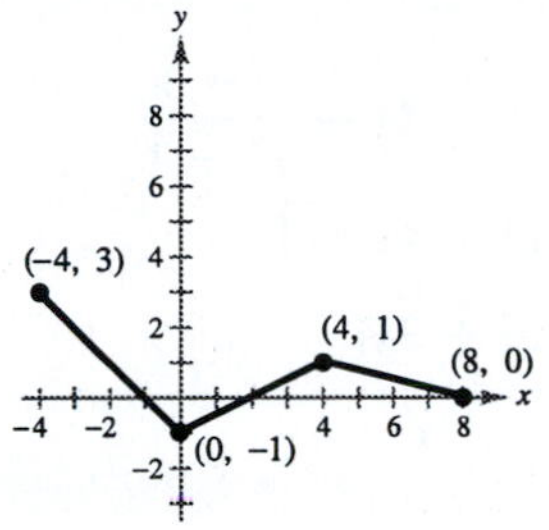

20. $d = Ks^2$

$35 = K(70)^2$

$K = \frac{1}{140}$

$d = \frac{s^2}{140}$

$d = \frac{(100)^2}{140} = \frac{500}{7} \approx 71.4$ meters

Chapter Test Solutions for Chapter 2

1. $f(x) = x^2$

(a) $g(x) = 2 - x^2$

Reflection in the x-axis followed by a vertical shift two units upward

(b) $h(x) = \left(x - \frac{3}{2}\right)^2$

Horizontal shift $\frac{3}{2}$ units to the right

2. Vertex: $(3, -6)$

$y = a(x - 3)^2 - 6$

Point on graph: $(0, 3)$

$3 = a(0 - 3)^2 - 6$

$9 = 9a \implies a = 1$

Thus, $y = (x - 3)^2 - 6$

3. $h(t) = -\frac{3}{4}t^5 + 2t^2$

The degree is odd and the leading coefficient is negative. The graph rises to the left and falls to the right.

4.

$$\begin{array}{r} 3x + \dfrac{x-1}{x^2+1} \\ x^2 + 0x + 1 \overline{)\,3x^3 + 0x^2 + 4x - 1} \\ -(3x^3 + 0x^2 + 3x) \\ \hline x - 1 \end{array}$$

Thus, $\dfrac{3x^3 + 4x - 1}{x^2 + 1} = 3x + \dfrac{x - 1}{x^2 + 1}$.

5. $f(x) = x^3 + 5x^2 - 4x - 20$

Possible rational zeros: $\pm 1, \pm 2, \pm 4, \pm 5, \pm 10, \pm 20$

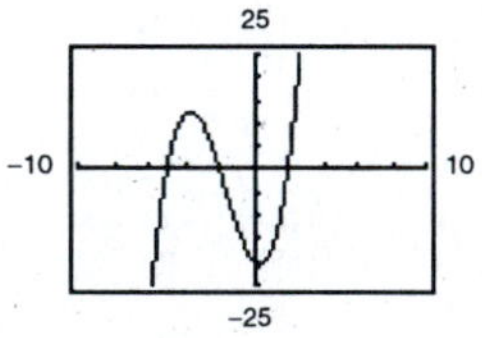

From the graph, we have $x = -5$ and $x = \pm 2$.

6. $g(t) = 2t^4 - 3t^3 + 16t - 24$

Possible rational zeros: $\pm 1, \pm 2, \pm 3, \pm 4, \pm 6, \pm 8, \pm 12, \pm 24, \pm \frac{1}{2}, \pm \frac{3}{2}$

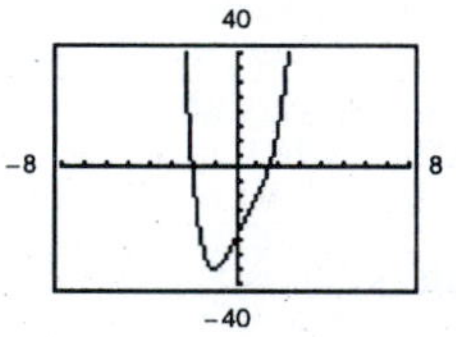

From the graph, we have $x = -2$ and $x = \frac{3}{2}$.

7. $f(x) = 3x^5 + 2x^4 - 12x - 8$

$x \approx \pm 1.414$ and $x \approx -0.667$

8. $g(v) = 2v^3 - 11v^2 + 22v - 15$

Possible rational zeros: $\pm 1, \pm 3, \pm 5, \pm 15, \pm \frac{1}{2}, \pm \frac{3}{2}, \pm \frac{5}{2}, \pm \frac{15}{2}$

By checking, we find that $x = \frac{3}{2}$ is a zero.

$$\begin{array}{r|rrrr} \frac{3}{2} & 2 & -11 & 22 & -15 \\ & & 3 & -12 & 15 \\ \hline & 2 & -8 & 10 & 0 \end{array}$$

$g(v) = \left(v - \frac{3}{2}\right)(2v^2 - 8v + 10) = (2v - 3)(x^2 - 4v + 5)$

Using the Quadratic Formula on $v^2 - 4v + 5 = 0$ yields $v = 2 \pm i$. The zeros of $g(v)$ are $v = \frac{3}{2}$ and $v = 2 \pm i$.

9. $$f(x) = \left[x - (1 + \sqrt{3}i)\right]\left[x - (1 - \sqrt{3}i)\right](x - 2)(x - 2)$$
$$= \left[(x - 1) - \sqrt{3}i\right]\left[(x - 1) + \sqrt{3}i\right](x^2 - 4x + 4)$$
$$= \left[(x - 1)^2 - 3i^2\right](x^2 - 4x + 4)$$
$$= (x^2 - 2x + 4)(x^2 - 4x + 4)$$
$$= x^4 - 6x^3 + 16x^2 - 24x + 16$$

10. $x = \pm 3$ must make the demoninator zero. The numerator must have the same degree as the denominator and a leading coefficient of 4 (assuming that the leading coefficient of the denominator is one.) One possible function is

$$f(x) = \frac{4x^2}{x^2 - 9}$$

11. $$h(x) = \frac{4}{x^2} - 1 = \frac{4 - x^2}{x^2}$$

Domain: all real numbers except $x = 0$

Interval notation: $(-\infty, 0) \cup (0, \infty)$

Vertical asymptote: $x = 0$

Horizontal asymptote: $y = -1$

x-intercepts: $(\pm 2, 0)$

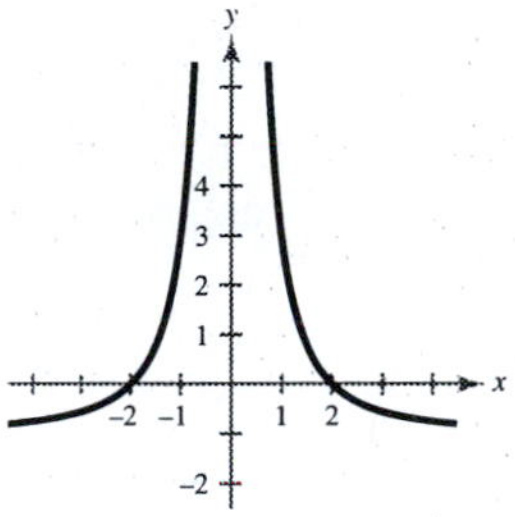

12. $$g(x) = \frac{x^2 + 2}{x - 1} = x + 1 + \frac{3}{x - 1}$$

Domain: all real numbers except $x = 1$

Interval notation: $(-\infty, 1) \cup (1, \infty)$

Vertical asymptote: $x = 1$

Horizontal asymptote: None

Slant asymptote: $y = x + 1$

y-intercept: $(0, -2)$

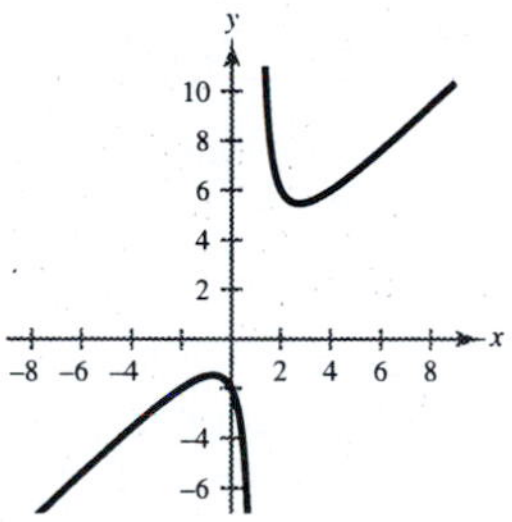

13. (a) Equate the slopes.

$$\frac{y - 1}{0 - 2} = \frac{1 - 0}{2 - x}$$

$$\frac{y - 1}{-2} = \frac{1}{2 - x}$$

$$y - 1 = -2\left(\frac{1}{2 - x}\right)$$

$$y = 1 + \frac{2}{x - 2}$$

(b) $$A = \frac{1}{2}xy = \frac{1}{2}x\left(1 + \frac{2}{x - 2}\right)$$

$$= \frac{x}{2} + \frac{x}{x - 2} = \frac{x^2}{2(x - 2)}$$

In context, we have $x > 2$ for the domain.

(c)

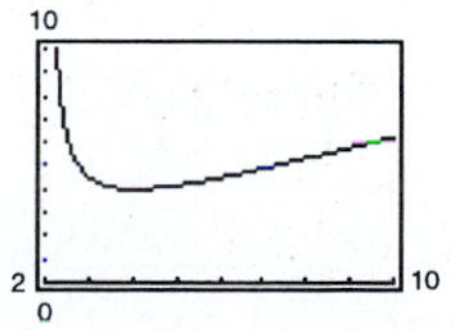

The minimum area of $A = 4$ occurs at $x = 4$.

14. (a) $\dfrac{2x+5}{(x-2)(x+1)} = \dfrac{A}{x-2} + \dfrac{B}{x+1}$

$$2x + 5 = A(x+1) + B(x-2)$$

Let $x = 2 : 9 = 3A \implies A = 3$

Let $x = -1 : 3 = -3B \implies B = -1$

Thus, $\dfrac{2x+5}{x^2-x-2} = \dfrac{3}{x-2} - \dfrac{1}{x+1}$.

(b) $\dfrac{3x^2-2x+4}{x^2(2-x)} = \dfrac{A}{x} + \dfrac{B}{x^2} + \dfrac{C}{2-x}$

$$3x^2 - 2x + 4 = Ax(2-x) + B(2-x) + Cx^2$$

Let $x = 0 : 4 = 2B \implies B = 2$

Let $x = 2 : 12 = 4C \implies C = 3$

Let $x = 1 : 5 = A + B + C \implies A = 0$

Thus, $\dfrac{3x^2-2x+4}{x^2(2-x^2)} = \dfrac{2}{x^2} + \dfrac{3}{2-x} = \dfrac{2}{x^2} - \dfrac{3}{x-2}$.

(c) $\dfrac{x^2-1}{x(x^2+1)} = \dfrac{A}{x} + \dfrac{Bx+C}{x^2+1}$

$$x^2 - 1 = A(x^2+1) + (Bx+C)x$$
$$= (A+B)x^2 + Cx + A$$

By equating coefficients we have:

$1 = A + B$

$0 = C$

$-1 = A \implies B = 2$

Thus, $\dfrac{x^2-1}{x^3+x} = -\dfrac{1}{x} + \dfrac{2x}{x^2+1}$.

❑ Cumulative Test Solutions for Chapters 1–3

1. $x - 3y + 12 = 0$

Line

x-intercept: $(-12, 0)$

y-intercept: $(0, 4)$

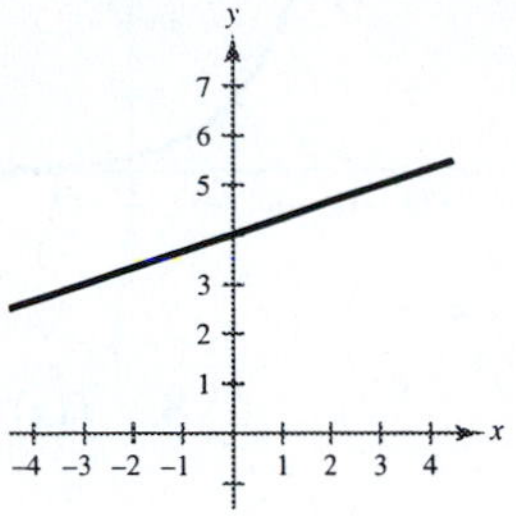

2. $h(x) = -(x^2 + 4x)$

Parabola: Vertex $(-2, 4)$

x-intercepts: $(-4, 0), (0, 0)$

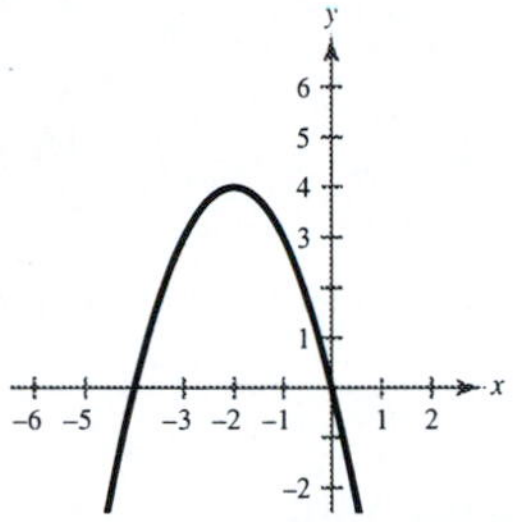

3. $y = \sqrt{4 - x}$

Domain: $(-\infty, 4]$

x-intercept: $(4, 0)$

y-intercept: $(0, 2)$

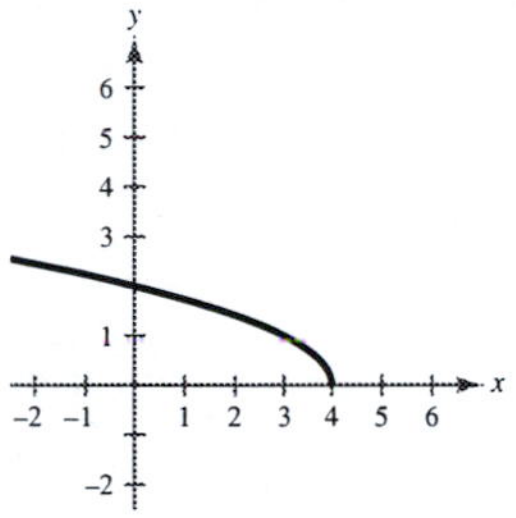

4. $y = \dfrac{2x}{x - 3}$

Vertical asymptote: $x = 3$

Horizontal asymptote: $y = 2$

Intercept: $(0, 0)$

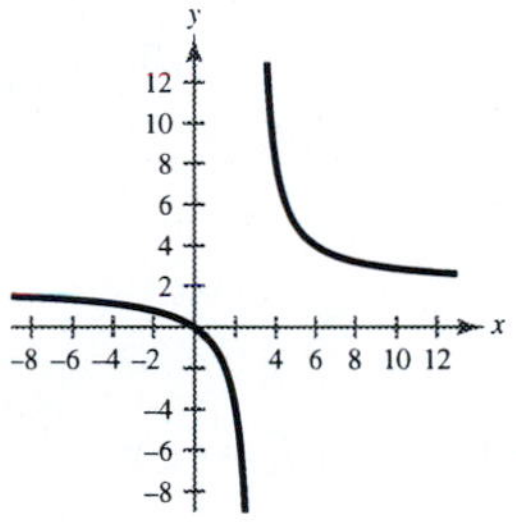

5. $g(x) = \log_3(x - 2) \Rightarrow 3^y + 2 = x$

Domain: $(2, \infty)$

Vertical asymptote: $x = 2$

Intercept: $(3, 0)$

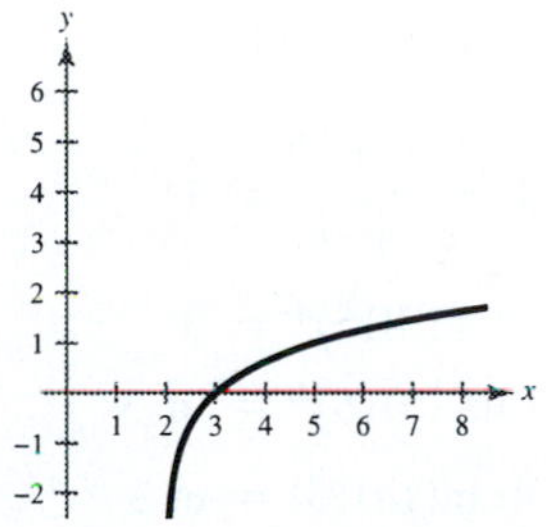

6. $f(x) = 2^{-x/3}$

Exponential function

Horizontal asymptote: $x = 0$

Intercept: $(0, 1)$

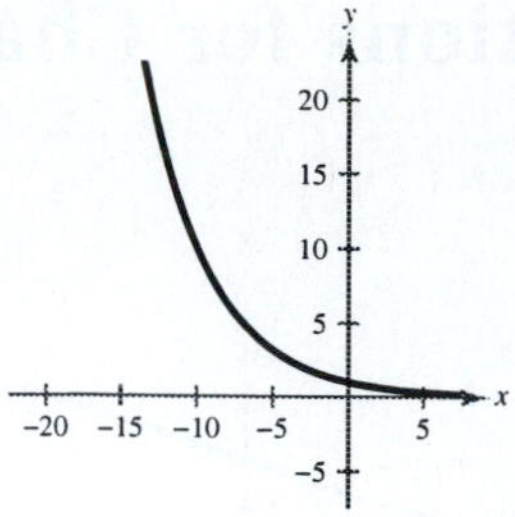

7. $m = \dfrac{8 - 1}{3 - (-1/2)} = \dfrac{7}{7/2} = 2$

$$y - 8 = 2(x - 3)$$
$$y - 8 = 2x - 6$$
$$0 = 2x - y + 2$$

8. $f(x) = \dfrac{x}{x - 2}$

(a) $f(6) = \dfrac{6}{6 - 2} = \dfrac{6}{4} = \dfrac{3}{2}$

(b) $f(2) = \dfrac{2}{2 - 2} = \dfrac{2}{0}$ which is undefined.

(c) $f(s + 2) = \dfrac{s + 2}{(s + 2) - 2} = \dfrac{s + 2}{s}$

9. $f(x) = x^3$

$g(x) = x^3 - 2$ is a vertical shift two units downward.

$h(x) = (x - 2)^3$ is a horizontal shift two units to the right.

10.
$$y = 5x - 2$$
$$x = 5y - 2$$
$$5y = x + 2$$
$$y = \tfrac{1}{5}(x + 2)$$
$$h^{-1}(x) = \tfrac{1}{5}(x + 2)$$

11. $f(x) = x^3 + 2x^2 + 4x + 8$

$= x^2(x + 2) + 4(x + 2)$

$= (x + 2)(x^2 + 4)$

$x + 2 = 0 \Rightarrow x = -2$

$x^2 + 4 = 0 \Rightarrow x = \pm 2i$

The zeros of $f(x)$ are -2 and $\pm 2i$.

12. $200{,}000 = P\left(1 + \dfrac{0.08}{365}\right)^{(365)(20)}$

$$P = \frac{200{,}000}{\left(1 + \dfrac{0.08}{365}\right)^{(365)(20)}} \approx \$40{,}386.38$$

13. $\log_4 64 = 3$

$4^3 = 64$

14. $3 \ln z - [\ln(z + 1) + \ln(z - 1)] = \ln z^3 - \ln(z + 1)(z - 1)$

$$= \ln\left[\frac{z^3}{(z + 1)(z - 1)}\right]$$

15. (a) $e^{x/2} = 450$

$$\frac{x}{2} = \ln 450$$
$$x = 2 \ln 450 \approx 12.218$$

(b) $\left(1 + \dfrac{0.06}{4}\right)^{4t} = 3$

$$(1.015)^{4t} = 3$$
$$\ln(1.015)^{4t} = \ln 3$$
$$4t \ln(1.015) = \ln 3$$
$$t = \frac{\ln 3}{4 \ln 1.015}$$
$$\approx 18.447$$

(c) $5 \ln(x + 4) = 22$

$$\ln(x + 4) = 4.4$$
$$x + 4 = e^{4.4}$$
$$x = -4 + e^{4.4}$$
$$\approx 77.451$$

16. $C = 28{,}000$

$$20{,}000 = 28{,}000e^{K(1)}$$

$$\frac{5}{7} = e^K$$

$$\ln\left(\frac{5}{7}\right) = K$$

$$y = 28{,}000e^{[\ln(5/7)](3)} \approx \$10{,}204.08$$

17. $p(t) = \dfrac{1200}{1 + 3e^{-t/5}}$

(a) $p(0) = \dfrac{1200}{1 + 3(1)} = 300$

(b)
$$800 = \frac{1200}{1 + 3e^{-t/5}}$$

$$1 + 3e^{-t/5} = \frac{1200}{800}$$

$$3e^{-t/5} = 0.5$$

$$-\frac{t}{5} = \ln\left(\frac{0.5}{3}\right)$$

$$t = -5\ln\left(\frac{0.5}{3}\right) \approx 8.96 \text{ years}$$

(c) As $t \to \infty$, $p(t) \to \dfrac{1200}{1 + 3(0)} = 1200.$

Horizontal asymptote: $p(t) = 1200$

18. The graph has a horizontal asymptote at $y = 6$, (this eliminates (a)), and has y-axis symmetry (his eliminates (b) and (c)). The graph matches (d).

❑ Chapter Test Solutions for Chapter 4

1. $\theta = \dfrac{5\pi}{4}$

(a)

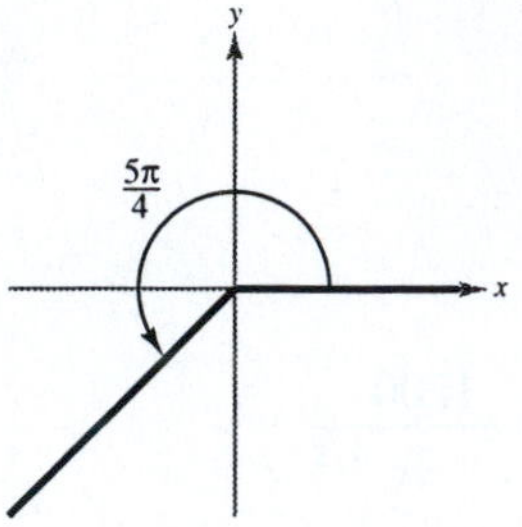

(b) $\dfrac{5\pi}{4} + 2\pi = \dfrac{13\pi}{4}$

$\dfrac{5\pi}{4} - 2\pi = -\dfrac{3\pi}{4}$

(c) $\dfrac{5\pi}{4} \cdot \dfrac{180}{\pi} = 225°$

2. 90 kilometers per hour $= \dfrac{90(1000)}{60} = 1500$ meters per minute

Circumference of wheel $= \pi$ meters

Number of revolutions per minute $= \dfrac{1500}{\pi}$

$\theta = \left(\dfrac{1500}{\pi}\right)(2\pi) = 3000$ radians

Angular speed $= \dfrac{\theta}{t} = \dfrac{3000 \text{ radians}}{1 \text{ minute}} = 3000$ radians per minute

3. $x = -2, y = 6,$

$r = \sqrt{(-2)^2 + 6^2} = \sqrt{40} = 2\sqrt{10}$

$\sin\theta = \dfrac{y}{r} = \dfrac{6}{2\sqrt{10}} = \dfrac{3}{\sqrt{10}}$

$\cos\theta = \dfrac{x}{r} = \dfrac{-2}{2\sqrt{10}} = -\dfrac{1}{\sqrt{10}}$

$\tan\theta = \dfrac{y}{x} = \dfrac{6}{-2} = -3$

$\cot\theta = \dfrac{x}{y} = \dfrac{-2}{6} = -\dfrac{1}{3}$

$\sec\theta = \dfrac{r}{x} = \dfrac{2\sqrt{10}}{-2} = -\sqrt{10}$

$\csc\theta = \dfrac{r}{y} = \dfrac{2\sqrt{10}}{6} = \dfrac{\sqrt{10}}{3}$

4. $\tan\theta = \dfrac{3}{2}$

Assuming that θ is in Quadrant I, we have $y = 3$, $x = 2$, and $r = \sqrt{3^3 + 2^2} = \sqrt{13}$.

$\sin\theta = \dfrac{y}{r} = \dfrac{3}{\sqrt{13}}$

$\cos\theta = \dfrac{x}{r} = \dfrac{2}{\sqrt{13}}$

$\tan\theta = \dfrac{y}{x} = \dfrac{3}{2}$

$\cot\theta = \dfrac{x}{y} = \dfrac{2}{3}$

$\sec\theta = \dfrac{r}{x} = \dfrac{\sqrt{13}}{2}$

$\csc\theta = \dfrac{r}{y} = \dfrac{\sqrt{13}}{3}$

5. $\theta' = 360° - 290° = 70°$

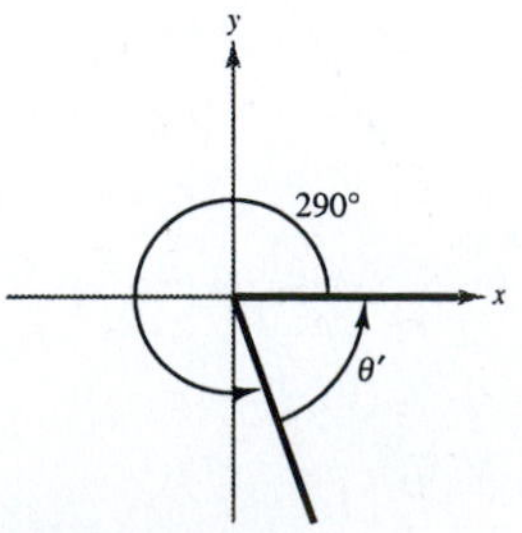

6. $\sec\theta < 0 \Longrightarrow \theta$ is in Quadrant II or III.

$\tan\theta > 0 \Longrightarrow \theta$ is in Quadrant I or III.

Both are true in Quandrant III.

7. $\cos\theta = -\dfrac{\sqrt{3}}{2}$

Reference angle is 30° and θ is in Quadrant II or III.

$\theta = 150°$ or $210°$

8. $\csc\theta = 1.030$

$\dfrac{1}{\sin\theta} = 1.030$

$\sin\theta = \dfrac{1}{1.030}$

$\theta = \arcsin\dfrac{1}{1.030}$

$\theta \approx 1.33$ and $\pi - 1.33 \approx 1.81$

9. $g(x) = -2\sin\left(x - \dfrac{\pi}{4}\right)$

Period: 2π

Amplitude: $|-2| = 2$

Shifted to the right by $(\pi/4)$ units and reflected in the x-axis.

x	0	$\frac{\pi}{4}$	$\frac{\pi}{2}$	$\frac{3\pi}{4}$	π
y	$\sqrt{2}$	0	$-\sqrt{2}$	-2	$-\sqrt{2}$

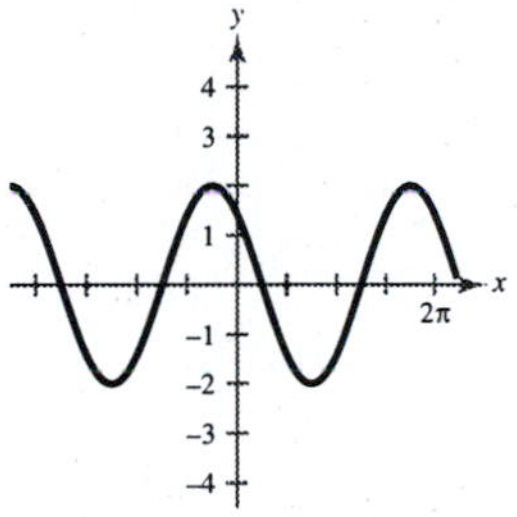

10. $f(\alpha) = \dfrac{1}{2}\tan 2\alpha$

Period: $\dfrac{\pi}{2}$

Asymptotes: $x = -\dfrac{\pi}{4}, x = \dfrac{\pi}{4}$

α	$-\frac{\pi}{8}$	0	$\frac{\pi}{8}$
$f(\alpha)$	$-\frac{1}{2}$	0	$\frac{1}{2}$

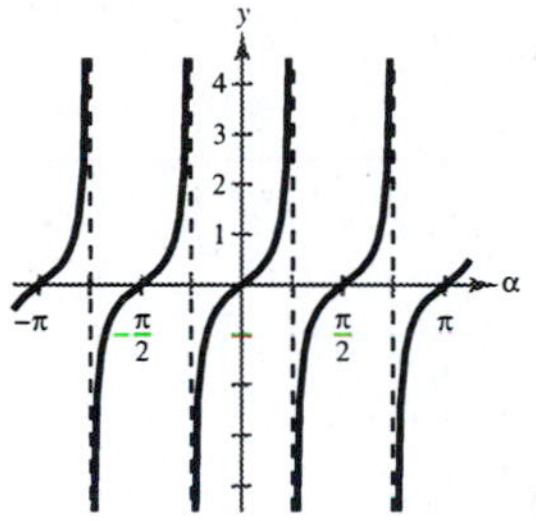

11. $y = \sin 2\pi x + 2\cos \pi x$

Periodic: period $= 2$

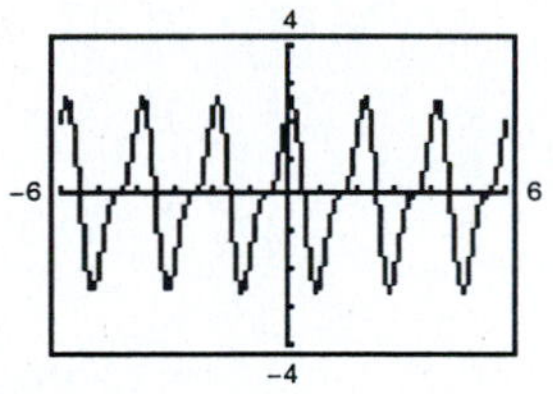

12. $y = 6e^{-0.12t}\cos(0.25t),\ 0 \le t \le 32$

Not periodic

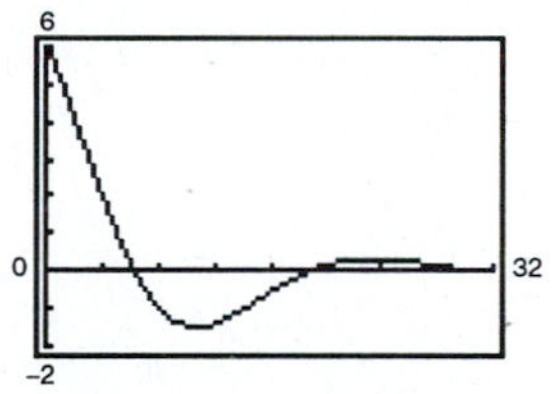

13. $f(x) = a\sin(bx + c)$

Amplitude: $2 \Longrightarrow |a| = 2$

Reflected in the x-axis: $a = -2$

Period: $4\pi = \dfrac{2\pi}{b} \Longrightarrow b = \dfrac{1}{2}$

Phase shift: $\dfrac{c}{b} = -\dfrac{\pi}{2} \Longrightarrow c = -\dfrac{\pi}{4}$

$f(x) = -2\sin\left(\dfrac{x}{2} - \dfrac{\pi}{4}\right)$

14. Let $u = \arccos\dfrac{2}{3}$,

$\cos u = \dfrac{2}{3}$.

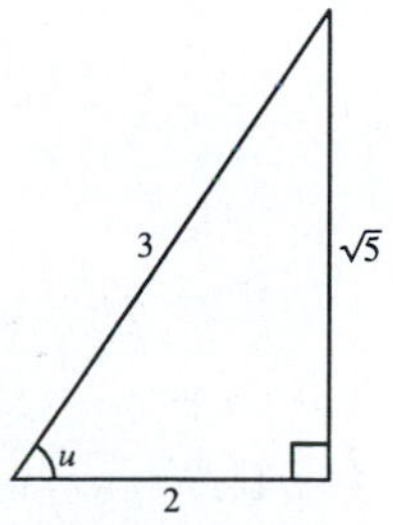

$\tan\left(\arccos\dfrac{2}{3}\right) = \tan u = \dfrac{\sqrt{5}}{2}$

15. $f(x) = 2\arcsin\left(\frac{1}{2}x\right)$

Domain: $-2 \le x \le 2$

Range: $-\pi \le y \le \pi$

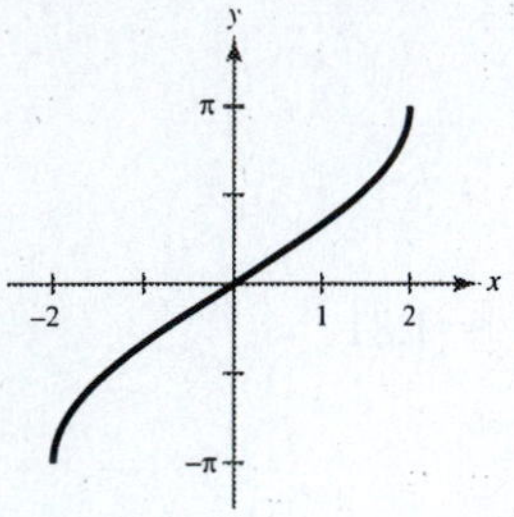

16. $x = (18)(3)\cos(90° + 16°) \approx -14.88$

$y = (18)(3)\sin(90° + 16°) \approx 51.91$

Coordinates: $(-14.88, 51.91)$

❑ Chapter Test Solutions for Chapter 5

1. $\tan\theta = \dfrac{3}{2}$ and $\cos\theta < 0$

θ is in Quadrant III.

$$\sec\theta = -\sqrt{1+\tan^2\theta} = -\sqrt{1+\left(\frac{3}{2}\right)^2} = -\frac{\sqrt{13}}{2}$$

$$\cos\theta = \frac{1}{\sec\theta} = -\frac{2}{\sqrt{13}}$$

$$\sin\theta = \tan\theta\cos\theta = \left(\frac{3}{2}\right)\left(-\frac{2}{\sqrt{13}}\right) = -\frac{3}{\sqrt{13}}$$

$$\csc\theta = \frac{1}{\sin\theta} = -\frac{\sqrt{13}}{3}$$

$$\cot\theta = \frac{1}{\tan\theta} = \frac{2}{3}$$

2. $\csc^2\beta(1-\cos^2\beta) = \dfrac{1}{\sin^2\beta}(\sin^2\beta) = 1$

3.
$$\frac{\sec^4 x - \tan^4 x}{\sec^2 x + \tan^2 x} = \frac{(\sec^2 x + \tan^2 x)(\sec^2 x - \tan^2 x)}{\sec^2 x + \tan^2 x}$$
$$= \sec^2 x - \tan^2 x = 1$$

4.
$$\frac{\cos\theta}{\sin\theta} + \frac{\sin\theta}{\cos\theta} = \frac{\cos^2\theta + \sin^2\theta}{\sin\theta\cos\theta} = \frac{1}{\sin\theta\cos\theta}$$

5. $y = \tan\theta,\ y = -\sqrt{\sec^2\theta - 1}$

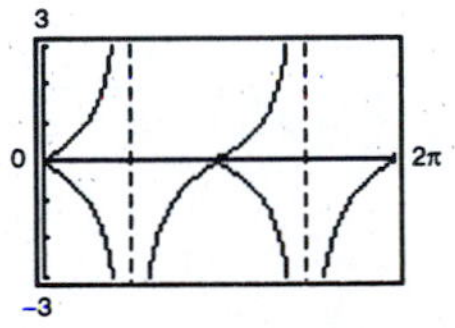

$\tan\theta = -\sqrt{\sec^2\theta - 1}$ on

$\dfrac{\pi}{2} < \theta \le \pi,\ \dfrac{3\pi}{2} < \theta < 2\pi.$

6. $y_1 = \cos x + \sin x \tan x,\ y_2 = \sec x$

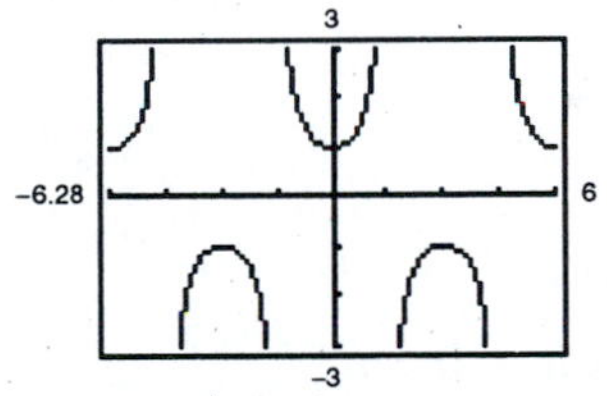

It appears that $y_1 = y_2$.

$$\cos x + \sin x\tan x = \cos + \sin x\frac{\sin x}{\cos x}$$
$$= \cos + \frac{\sin^2 x}{\cos x}$$
$$= \frac{\cos^2 x + \sin^2 x}{\cos x}$$
$$= \frac{1}{\cos x} = \sec x$$

7. $\sin\theta\sec\theta = \sin\theta\dfrac{1}{\cos\theta} = \dfrac{\sin\theta}{\cos\theta} = \tan\theta$

8. $\sec^2 x\tan^2 x + \sec^2 x = \sec^2 x(\sec^2 x - 1) + \sec^2 x$
$= \sec^4 x - \sec^2 x + \sec^2 x$
$= \sec^4 x$

9. $\dfrac{\csc\alpha + \sec\alpha}{\sin\alpha + \cos\alpha} = \dfrac{\dfrac{1}{\sin\alpha} + \dfrac{1}{\cos\alpha}}{\sin\alpha + \cos\alpha} = \dfrac{\dfrac{\cos\alpha + \sin\alpha}{\sin\alpha\cos\alpha}}{\sin\alpha + \cos\alpha} = \dfrac{1}{\sin\alpha\cos\alpha}$

$= \dfrac{\cos^2\alpha + \sin^2\alpha}{\sin\alpha\cos\alpha} = \dfrac{\cos^2\alpha}{\sin\alpha\cos\alpha} + \dfrac{\sin^2\alpha}{\sin\alpha\cos\alpha}$

$= \dfrac{\cos\alpha}{\sin\alpha} + \dfrac{\sin\alpha}{\cos\alpha} = \cot\alpha + \tan\alpha$

10. $\cos\left(x + \dfrac{\pi}{2}\right) = \cos\left(\dfrac{\pi}{2} - (-x)\right) = \sin(-x) = -\sin x$

11. $\sin(n\pi + \theta) = (-1)^n\sin\theta$, n is an integer.

For n odd: $\sin(n\pi + \theta) = \sin n\pi\cos\theta + \cos n\pi\sin\theta$
$= (0)\cos\theta + (-1)\sin\theta = -\sin\theta$

For n even: $\sin(n\pi + \theta) = \sin n\pi\cos\theta + \cos n\pi\sin\theta$
$= (0)\cos\theta + (1)\sin\theta = \sin\theta$

When n is odd, $(-1)^n = -1$. When n is even $(-1)^n = 1$. Thus, $\sin(n\pi + \theta) = (-1)^n\sin\theta$ for n is and integer.

12. $(\sin x + \cos x)^2 = \sin^2 x + 2\sin x\cos x + \cos^2 x$
$= 1 + 2\sin x\cos x$
$= 1 + \sin^2 2x$

13. $\tan^2 x + \tan x = 0$
$\tan x(1 + \tan x) = 0$
$\tan x = 0$ or $\tan x + 1 = 0$
$x = 0, \pi$ $\quad$ $\tan x = -1$
$x = \dfrac{3\pi}{4}, \dfrac{7\pi}{4}$

14. $\sin 2\alpha - \cos\alpha = 0$
$2\sin\alpha\cos\alpha - \cos\alpha = 0$
$\cos\alpha(2\sin\alpha - 1) = 0$
$\cos\alpha = 0$ or $2\sin\alpha - 1 = 0$
$\alpha = \dfrac{\pi}{2}, \dfrac{3\pi}{2}$ $\quad$ $\sin\alpha = \dfrac{1}{2}$
$\alpha = \dfrac{2\pi}{6}, \dfrac{5\pi}{6}$

15. $4\cos^2 x - 3 = 0$
$\cos^2 x = \dfrac{3}{4}$
$\cos x = \pm\sqrt{\dfrac{3}{4}} \pm \dfrac{\sqrt{3}}{2}$
$x = \dfrac{\pi}{6}, \dfrac{5\pi}{6}, \dfrac{7\pi}{6}, \dfrac{11\pi}{6}$

16. $\csc^2 x - \csc x - 2 = 0$

$(\csc x - 2)(\csc x + 1) = 0$

$\csc x - 2 = 0$ or $\csc x + 1 = 0$

$\csc x = 2$ $\csc x = -1$

$\dfrac{1}{\sin x} = 2$ $\dfrac{1}{\sin x} = -1$

$\sin x = \dfrac{1}{2}$ $\sin x = -1$

$x = \dfrac{\pi}{6}, \dfrac{5\pi}{6}$ $x = \dfrac{3\pi}{2}$

17. $3\cos x - x = 0$

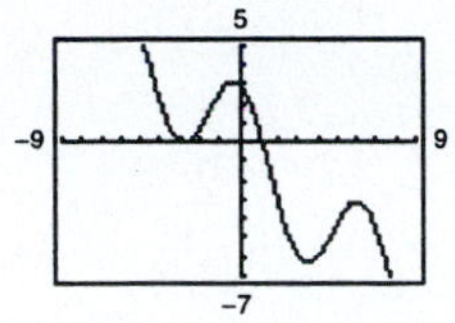

$x \approx -2.9381, -2.6632, 1.1701$

18. $\cos^2 x + \cos x - 6 = 0$

$\cos^2 x + \cos x = 6$

But the maximum value of $\cos^2 x$ is 1 and the maximum value of $\cos x$ is 1. Thus, $|\cos^2 x + \cos x| \le 2$ for all x and $\cos^2 x + \cos x$ can never equal 6.

19. $105° = 135° - 30°$

$$\begin{aligned}\cos 105° &= \cos(135° - 30°)\\ &= \cos 135° \cos 30° + \sin 45° \sin 30°\\ &= -\cos 45° \cos 30° + \sin 45° \sin 30°\\ &= \left(-\frac{\sqrt{2}}{2}\right)\left(\frac{\sqrt{3}}{2}\right) + \left(\frac{\sqrt{2}}{2}\right)\left(\frac{1}{2}\right)\\ &= \frac{-\sqrt{6} + \sqrt{2}}{4} = \frac{\sqrt{2} - \sqrt{6}}{4}\end{aligned}$$

20. $\sin 2u = 2 \sin u \cos u$

$$= 2\left(\frac{2}{\sqrt{5}}\right)\left(\frac{1}{\sqrt{5}}\right) = \frac{4}{5}$$

$$\tan 2u = \frac{2 \tan u}{1 - \tan^2 u} = \frac{2(2)}{1 - (2)^2} = \frac{4}{-3} = -\frac{4}{3}$$

❑ Cumulative Test Solutions for Chapters 4-6

1. (a)

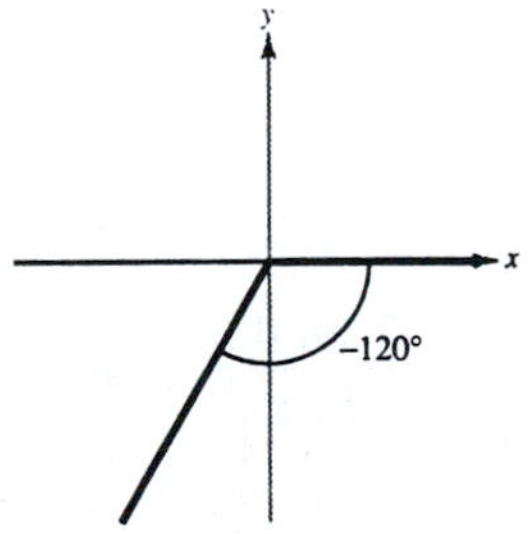

(b) $-120° + 360° = 240°$

(c) $-120°\left(\dfrac{\pi}{180°}\right) = -\dfrac{2\pi}{3}$

(d) $-120°$ is located in Quadrant III.
$240° - 180° = 60°$

(e) $\sin(-120°) = -\sin 60° = -\dfrac{\sqrt{3}}{2}$

$\cos(-120°) = -\cos 60° = -\dfrac{1}{2}$

$\tan(-120°) = \tan 60° = \sqrt{3}$

$\csc(-120°) = \dfrac{1}{-\sin 60°} = -\dfrac{2\sqrt{3}}{3}$

$\sec(-120°) = \dfrac{1}{-\cos 60°} = -2$

$\cot(-120°) = \dfrac{1}{\tan 60°} = \dfrac{\sqrt{3}}{3}$

2. $2.35\left(\frac{180°}{\pi}\right) \approx 134.6°$

3. $\tan\theta = \frac{y}{x} = -\frac{4}{3} \Rightarrow r = 5$

θ is in Quadrant IV $\Rightarrow x = 3$.

$\cos\theta = \frac{x}{r} = \frac{3}{5}$

4. (a) $f(x) = 3 - 2\sin\pi x$

Period: $\frac{2\pi}{\pi} = 2$

Amplitude: $|a| = |-2| = 2$

Upward shift of 3 units (reflected in x-axis prior to shift)

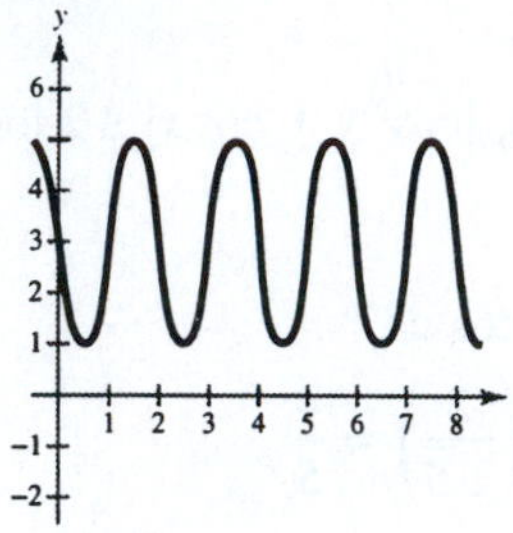

(b) $g(x) = \frac{1}{2}\tan\left(x - \frac{\pi}{2}\right)$

Period: π

Asymptotes: $x = 0, x = \pi$

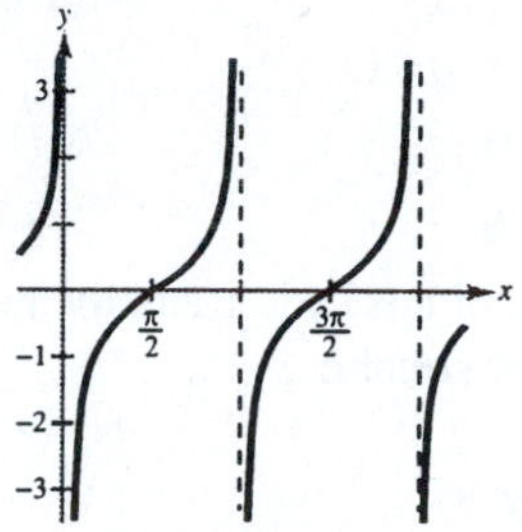

5. $h(x) = a\cos(bx + c)$

Graph is reflected in x-axis.

Amplitude: $a = -3$

Period: $2 = \frac{2\pi}{\pi} \Rightarrow b = \pi$

No phase shift: $c = 0$

$h(x) = -3\cos(\pi x)$

6. $y = \arccos(2x)$

$\sin y = \sin(\arccos(2x)) = \sqrt{1 - 4x^2}$

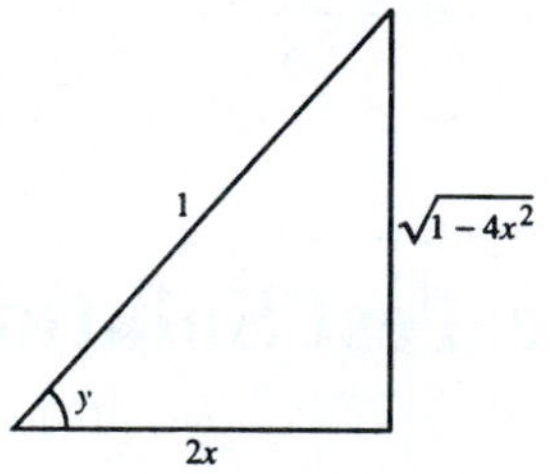

7. $$\frac{\sin\theta - 1}{\cos\theta} - \frac{\cos\theta}{\sin\theta - 1} = \frac{\sin\theta - 1}{\cos\theta} - \frac{\cos\theta(\sin\theta + 1)}{\sin^2 - 1}$$

$$= \frac{\sin\theta - 1}{\cos\theta} + \frac{\cos\theta(\sin\theta + 1)}{\cos\theta} = \frac{\sin\theta - 1}{\cos\theta} - \frac{\sin\theta + 1}{\cos\theta} = \frac{2\sin\theta}{\cos\theta} = 2\tan\theta$$

8. (a) $\cot^2 \alpha(\sec^2 \alpha - 1) = \cot^2 \alpha \tan^2 \alpha = 1$

(b) $\sin(x + y)\sin(x - y) = \frac{1}{2}[\cos(x + y - (x - y)) - \cos(x + y + x - y)]$

$= \frac{1}{2}[\cos 2y - \cos 2x] = \frac{1}{2}[1 - 2\sin^2 y - (1 - 2\sin^2 x)] = \sin^2 x - \sin^2 y$

(c) $\sin^2 x \cos^2 x = \left(\frac{1 - \cos 2x}{2}\right)\left(\frac{1 + \cos 2x}{2}\right)$

$= \frac{1}{4}(1 - \cos 2x)(1 + \cos 2x)$

$= \frac{1}{4}(1 - \cos^2 2x)$

$= \frac{1}{4}\left(1 - \frac{1 + \cos 4x}{2}\right)$

$= \frac{1}{8}(2 - (1 + \cos 4x))$

$= \frac{1}{8}(1 - \cos 4x)$

9. (a) $2\cos^2 \beta - \cos \beta = 0$

$\cos \beta(2\cos \beta - 1) = 0$

$\cos \beta = 0$ $2\cos \beta - 1 = 0$

$\beta = \frac{\pi}{2}, \frac{3\pi}{2}$ $\cos \beta = \frac{1}{2}$

$\beta = \frac{\pi}{3}, \frac{5\pi}{3}$

Answer: $\frac{\pi}{3}, \frac{\pi}{2}, \frac{3\pi}{2}, \frac{5\pi}{3}$

(b) $3\tan \theta - \cot \theta = 0$

$3\tan \theta - \frac{1}{\tan \theta} = 0$

$\frac{3\tan^2 \theta - 1}{\tan \theta} = 0$

$3\tan^2 \theta - 1 = 0$

$\tan^2 \theta = \frac{1}{3}$

$\tan \theta = \pm\frac{\sqrt{3}}{3}$

$0 = \frac{\pi}{6}, \frac{5\pi}{6}, \frac{7\pi}{6}, \frac{11\pi}{6}$

10. (a) $\frac{\sin B}{8} = \frac{\sin 30}{9}$

$\sin B = \frac{8}{9}\left(\frac{1}{2}\right)$

$B = \arcsin\left(\frac{4}{9}\right)$

$B \approx 26.4°$

$C = 180° - A - B \approx 123.6°$

$\frac{c}{\sin 123.6} = \frac{9}{\sin 30}$

$c \approx 15.0$

(b) $a^2 = 8^2 + 10^2 - 2(8)(10)\cos 30$

$a^2 \approx 25.4$

$a \approx 5.0$

$\cos B = \frac{5.0^2 + 10^2 - 8^2}{2(5.0)(10)}$

$\cos B = 0.61$

$B \approx 52.4°$

$C = 180° - A - B \approx 97.6°$

11. $r = |-2 + 2i| = \sqrt{(-2)^2 + (2)^2} = 2\sqrt{2}$

$\tan\theta = \frac{2}{-2} = -1$

Since $\tan 135° = -1$ and $-2 + 2i$ lies in Quadrant II, $\theta = 135°$. Thus, $-2 + 2i = 2\sqrt{2}(\cos 135° + i \sin 135°)$.

12. $[4(\cos 30° + i \sin 30°)][6(\cos 120° + i \sin 120°)] = (4)(6)[\cos(30° + 120°) + i \sin(30° + 120°)]$
$= 24(\cos 150° + i \sin 150°)$

13. $1 = 1(\cos 0 + i \sin 0)$

$n = 3, r = 1$

$$\sqrt[3]{1}\left(\cos\left(\frac{0 + 2\pi(0)}{3}\right) + i \sin\left(\frac{0 + 2\pi(0)}{3}\right)\right) = 1$$

$$\sqrt[3]{1}\left(\cos\left(\frac{0 + 2\pi(1)}{3}\right) + i \sin\left(\frac{0 + 2\pi(1)}{3}\right)\right) = \cos\frac{2\pi}{3} + i \sin\frac{2\pi}{3} = -\frac{1}{2} + \frac{\sqrt{3}}{2}i$$

$$\sqrt[3]{1}\left(\cos\left(\frac{0 + 2\pi(2)}{3}\right) + i \sin\left(\frac{0 + 2\pi(2)}{3}\right)\right) = \cos\frac{4\pi}{3} + i \sin\frac{4\pi}{3} = -\frac{1}{2} - \frac{\sqrt{3}}{2}i$$

14. Height of smaller triangle:

$\tan 16° 45' = \frac{h_1}{200}$

$h_1 = 200 \tan 16.75° \approx 60.2$ feet

Height of larger triangle:

$\tan 18° = \frac{h_2}{200}$

$h_2 = 200 \tan 18° \approx 65.0$

Height of flag:

$h_2 - h_1 = 65.0 - 60.2 \approx 5$ feet

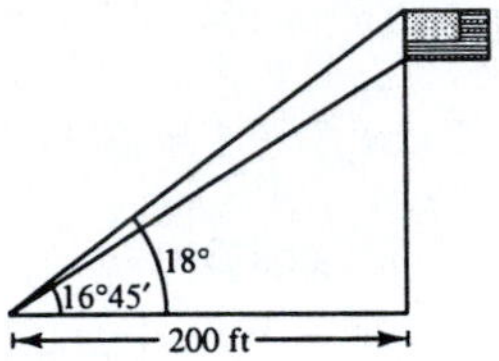

Not to scale.

15. $\mathbf{v}_1 = 500\langle\cos 30, \sin 30\rangle = \langle 250\sqrt{3}, 250\rangle$

$\mathbf{v}_2 = 50\langle\cos 60, \sin 60\rangle = \langle 25, 25\sqrt{3}\rangle$

$\mathbf{v} = \mathbf{v}_1 + \mathbf{v}_2 = \langle 250\sqrt{3} + 25, 250 + 25\sqrt{3}\rangle \approx \langle 458.0, 293.3\rangle$

$\|\mathbf{v}\| = \sqrt{(458.0)^2 + (293.3)^2} \approx 543.9$

$\tan\theta = \frac{293.3}{458.0} \approx 0.6404 \Rightarrow \theta \approx 32.6°$

The plane is traveling N 32.6° E at 543.9 kilometers per hour.

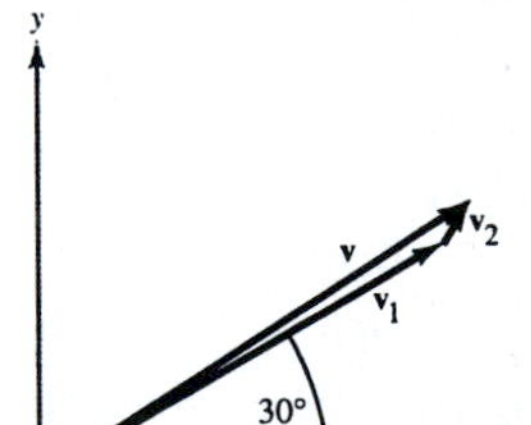

❑ Chapter Test Solutions for Chapter 7

1. $x - y = 4 \Longrightarrow y = x - 4$

$3x + 2y = 2 \Longrightarrow 3x + 2(x - 4) = 2$

$5x - 8 = 2$

$5x = 10$

$x = 2 \Longrightarrow y = -2$

Solution: $(2, -2)$

2. $y = x - 1$

$y = (x - 1)^3$

$x - 1 = (x - 1)^3$

$x - 1 = x^3 - 3x^2 + 3x - 1$

$0 = x^3 - 3x^2 + 2x$

$0 = x(x - 1)(x - 2)$

$x = 0, \quad x = 1, \quad x = 2$

$y = -1, \quad y = 0, \quad y = 1$

Solutions: $(0, -1), (1, 0), (2, 1)$

3. $x - y = 4 \Longrightarrow x = y + 4$

$2x - y^2 = 0 \Longrightarrow 2(y + 4) - y^2 = 0$

$0 = y^2 - 2y - 8$

$0 = (y + 2)(y - 4)$

$y = -2$ or $y = 4$

$x = 2 \qquad x = 8$

Solutions: $(2, -2), (8, 4)$

4. $2x - 3y = 0$

$2x + 3y = 12$

Solution: $(3, 2)$

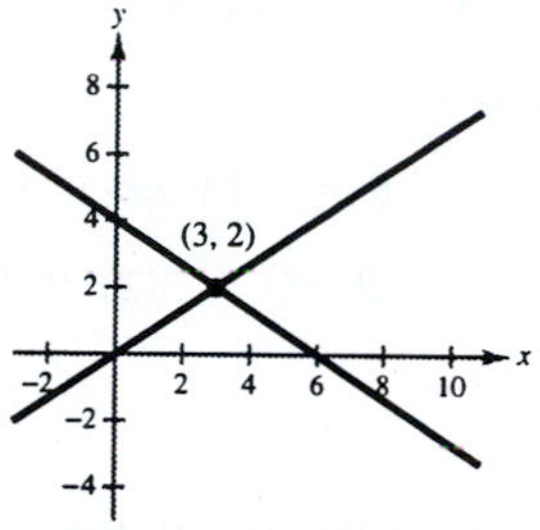

5. $y = 9 - x^2$

$y = x + 3$

Solutions: $(-3, 0), (2, 5)$

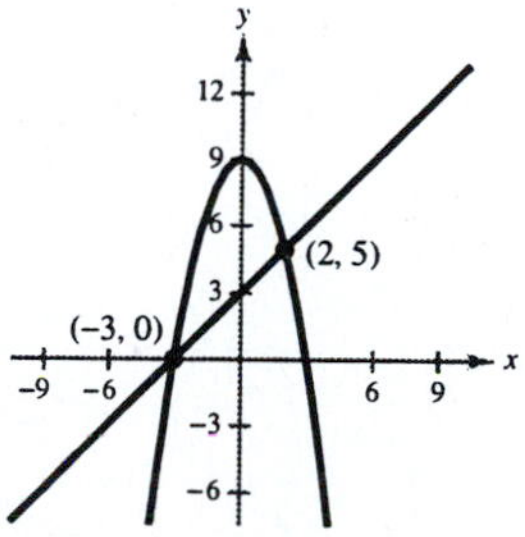

6. $y = \log_3 x \Longrightarrow 3^y = x$

$y = -\frac{1}{3}x + 2$

Solutions: $(3, 1)$

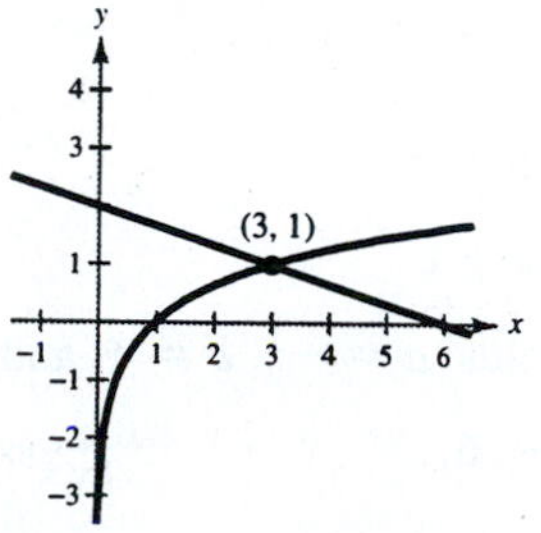

7. $2x + 3y = 17 \implies 8x + 12y = 68$

$5x - 4y = -15 \implies 15x - 12y = -45$

$23x = 23$

$x = 1 \implies y = 5$

Solution: $(1, 5)$

8. $x - 2y + 3z = 11$

$2x - z = 3$

$3y + z = -8$

$x - 2y + 3z = 11$

$4y - 7z = -19 \quad -2\text{ Eq.1} + \text{Eq. 2}$

$3y + z = -8$

$x - 2y + 3z = 11$

$y - 8z = -11 \quad -\text{Eq. 3} + \text{Eq. 2}$

$25z = 25 \implies z = 1$

$y - 8 = -11 \implies y = -3$

$x - 2(-3) + 3(1) = 11 \implies x = 2$

Solution: $(2, -3, 1)$

9. There are infinitely many systems with the solution $\left(\frac{4}{3}, -5\right)$. One possibility is:

$3\left(\frac{4}{3}\right) - (-5) = 9 \implies 3x - y = 9$

$6\left(\frac{4}{3}\right) + (-5) = 3 \implies 6x + y = 3$

10. $y = ax^2 + bx + c$

$(0, 6)$: $6 = c$

$(-2, 2)$: $2 = 4a - 2b + c$

$\left(3, \frac{9}{2}\right)$: $\frac{9}{2} = 9a + 3b + c$

Solving this system yields: $a = -\frac{1}{2}$, $b = 1$, and $c = 6$.

Thus, $y = -\frac{1}{2}x^2 + x + 6$.

11. $2x + y \leq 4$

$2x - y \geq 0$

$x \geq 0$

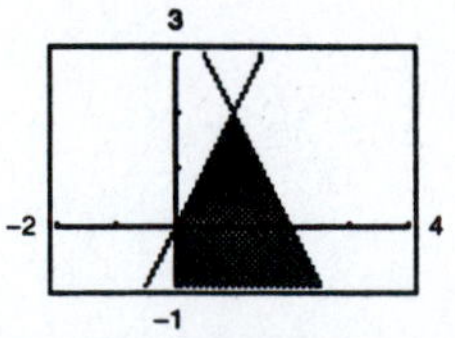

12. $y < -x^4 + x^2 + 4$

$y > 4x$

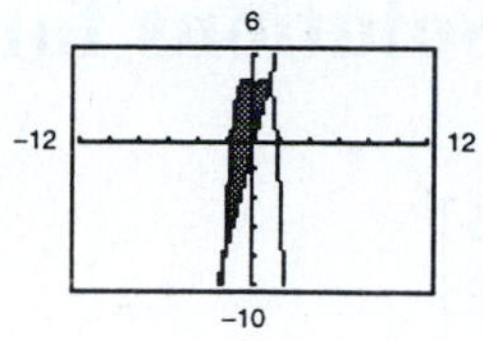

13. Line through (0, 15) and (9, 12): $x + 3y = 45$

Line through (9, 12) and (12, 5): $7x + 3y = 99$

Line through (12, 5) and (12, 0): $x = 12$

Inequalities: $x + 3y \leq 45$

$7x + 3y \leq 99$

$x \leq 12$

$x \geq 0$

$y \geq 0$

14. Maximize $z = 20x + 12y$ subject to:

$x \geq 0, \quad y \geq 0$

$x + 4y \leq 32$

$3x + 2y \leq 36$

At (0, 0) we have $z = 0$.

At (0, 8) we have $z = 96$.

At (8, 6) we have $z = 232$.

At (12, 0) we have $z = 240$.

The maximum value, $z = 240$, occurs at (12, 0).

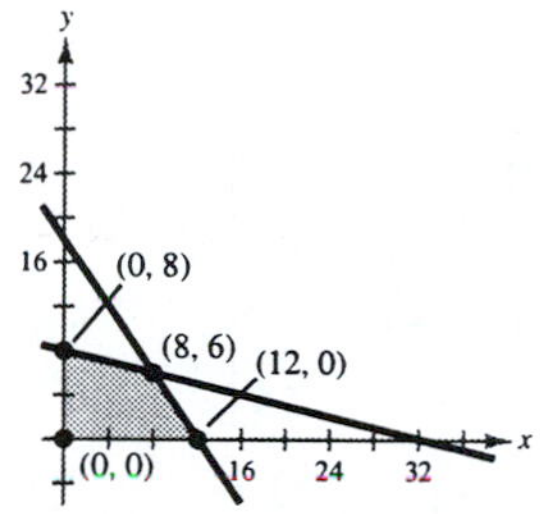

15. Maximize $P = 55x + 75y$ subject to:

$x \geq 0, \quad y \geq 0$

$x + y \leq 300$

$275x + 400y \leq 100{,}000$

At (0, 0) we have $z = 0$.

At (0, 250) we have $z = 18{,}750$.

At (160, 140) we have $z = 19{,}300$.

At (300, 0) we have $z = 16{,}500$.

The merchant should stock 160 units of the \$275 model and 140 units of the \$400 model to maximize profit.

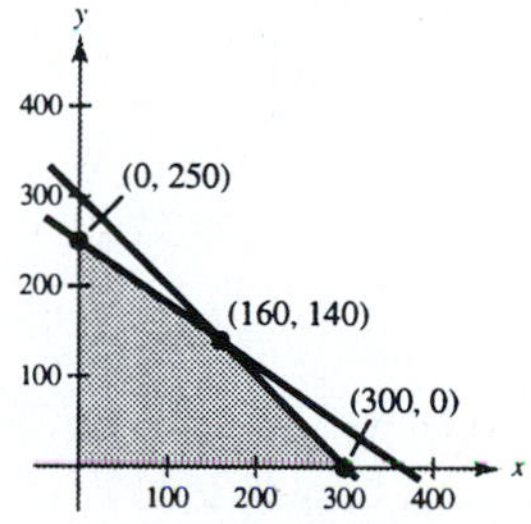

❑ Chapter Test Solutions for Chapter 8

1. $$\begin{bmatrix} 1 & -1 & 5 \\ 6 & 2 & 3 \\ 5 & 3 & -3 \end{bmatrix}$$

$$\begin{matrix} \\ -6R_1 + R_2 \rightarrow \\ -5R_1 + R_3 \rightarrow \end{matrix} \begin{bmatrix} 1 & -1 & 5 \\ 0 & 8 & -27 \\ 0 & 8 & -28 \end{bmatrix}$$

$$\begin{matrix} \\ \\ -R_2 + R_3 \rightarrow \end{matrix} \begin{bmatrix} 1 & -1 & 5 \\ 0 & 8 & -27 \\ 0 & 0 & -1 \end{bmatrix}$$

$$\begin{matrix} R_2 + R_1 \rightarrow \\ \frac{1}{8}R_2 \rightarrow \\ -R_3 \rightarrow \end{matrix} \begin{bmatrix} 1 & 0 & \frac{13}{8} \\ 0 & 1 & -\frac{27}{8} \\ 0 & 0 & 1 \end{bmatrix}$$

$$\begin{matrix} -\frac{13}{8}R_3 + R_1 \rightarrow \\ \frac{27}{8}R_3 + R_2 \rightarrow \\ \\ \end{matrix} \begin{bmatrix} 1 & 0 & 0 \\ 0 & 1 & 0 \\ 0 & 0 & 1 \end{bmatrix}$$

2. $$\begin{bmatrix} 1 & 0 & -1 & 2 \\ -1 & 1 & 1 & -3 \\ 1 & 1 & -1 & 1 \\ 3 & 2 & -3 & 4 \end{bmatrix}$$

$$\begin{matrix} \\ R_1 + R_2 \rightarrow \\ -R_1 + R_3 \rightarrow \\ -3R_1 + R_4 \rightarrow \end{matrix} \begin{bmatrix} 1 & 0 & -1 & 2 \\ 0 & 1 & 0 & -1 \\ 0 & 1 & 0 & -1 \\ 0 & 2 & 0 & -2 \end{bmatrix}$$

$$\begin{matrix} \\ \\ -R_2 + R_3 \rightarrow \\ -2R_2 + R_4 \rightarrow \end{matrix} \begin{bmatrix} 1 & 0 & -1 & 2 \\ 0 & 1 & 0 & -1 \\ 0 & 0 & 0 & 0 \\ 0 & 0 & 0 & 0 \end{bmatrix}$$

3. $\begin{bmatrix} 4 & 3 & -2 & \vdots & 14 \\ -1 & -1 & 2 & \vdots & -5 \\ 3 & 1 & -4 & \vdots & 8 \end{bmatrix} \rightarrow \begin{bmatrix} 1 & 0 & 0 & \vdots & 1 \\ 0 & 1 & 0 & \vdots & 3 \\ 0 & 0 & 1 & \vdots & -\frac{1}{2} \end{bmatrix}$

Solution: $\left(1, 3, -\frac{1}{2}\right)$

4. $y = ax^2 + bx + c$

$(-2, -2)$: $-2 = 4a - 2b + c$

$(2, 2)$: $2 = 4a + 2b + c$

$(4, -2)$: $-2 = 16a + 4b + c$

Solving this system yields $a = -\frac{1}{2}$, $b = 1$, and $c = 2$.

Thus, $y = -\frac{1}{2}x^2 + x + 2$.

5. (a) $A - B = \begin{bmatrix} 5 & 4 & 4 \\ -4 & -4 & 0 \end{bmatrix} - \begin{bmatrix} 4 & -1 & 6 \\ -4 & 0 & -3 \end{bmatrix}$

$= \begin{bmatrix} 1 & 5 & -2 \\ 0 & -4 & 3 \end{bmatrix}$

(b) $3A = 3\begin{bmatrix} 5 & 4 & 4 \\ -4 & -4 & 0 \end{bmatrix} = \begin{bmatrix} 15 & 12 & 12 \\ -12 & -12 & 0 \end{bmatrix}$

(c) $3A - 2B = 3\begin{bmatrix} 5 & 4 & 4 \\ -4 & -4 & 0 \end{bmatrix} - 2\begin{bmatrix} 4 & -1 & 6 \\ -4 & 0 & -3 \end{bmatrix}$

$= \begin{bmatrix} 15 & 12 & 12 \\ -12 & -12 & 0 \end{bmatrix} - \begin{bmatrix} 8 & -2 & 12 \\ -8 & 0 & -6 \end{bmatrix}$

$= \begin{bmatrix} 7 & 14 & 0 \\ -4 & -12 & 6 \end{bmatrix}$

6. $\begin{bmatrix} 2 & -2 & 6 \\ 3 & -1 & 7 \\ 2 & 0 & -2 \end{bmatrix}\begin{bmatrix} 4 & 4 \\ 3 & 2 \\ 1 & -2 \end{bmatrix} = \begin{bmatrix} 8 & -8 \\ 16 & -4 \\ 6 & 12 \end{bmatrix}$

7. $\begin{bmatrix} -6 & 4 \\ 10 & -5 \end{bmatrix}^{-1} = \begin{bmatrix} \frac{1}{2} & \frac{2}{5} \\ 1 & \frac{3}{5} \end{bmatrix}$

8. $\begin{bmatrix} \frac{1}{2} & \frac{2}{5} \\ 1 & \frac{3}{5} \end{bmatrix}\begin{bmatrix} 10 \\ 20 \end{bmatrix} = \begin{bmatrix} 13 \\ 22 \end{bmatrix}$

Solution: $(13, 22)$

9. $\begin{vmatrix} 4 & 0 & 3 \\ 1 & -8 & 2 \\ 3 & 2 & 2 \end{vmatrix} = -2$

10. $A = -\frac{1}{2}\begin{vmatrix} -5 & 0 & 1 \\ 4 & 4 & 1 \\ 3 & 2 & 1 \end{vmatrix} = -\frac{1}{2}(-14) = 7$

❑ Cumulative Test Solutions for Chapters 7-9

1. $y = 3 - x^2$

$$2(y - 2) = x - 1 \Longrightarrow 2(3 - x^2 - 2) = x - 1$$
$$2(1 - x^2) = x - 1$$
$$2 - 2x^2 = x - 1$$
$$0 = 2x^2 + x - 3$$
$$0 = (2x + 3)(x - 1)$$
$$x = -\tfrac{3}{2} \quad \text{or} \quad x = 1$$
$$y = \tfrac{3}{4} \qquad\qquad y = 2$$

Solutions: $\left(-\frac{3}{2}, \frac{3}{4}\right)$, $(1, 2)$

2. $x + 3y = -1 \Longrightarrow 4x + 12y = -4$

$2x + 4y = 0 \Longrightarrow -6x - 12y = 0$

$-2x = -4$

$x = 2 \Longrightarrow y = -1$

Solution: $(2, -1)$

3.
$$-2x + 4y - z = 3$$
$$x - 2y + 2z = -6$$
$$x - 3y - z = 1$$

$$-2x + 4y - z = 3$$
$$x - 2y + 2z = -6$$
$$x - 3y - z = 1$$
$$x - 2y + 2z = -6$$
$$3z = -9 \qquad 2\text{ Eq. }1 + \text{Eq. }2$$
$$-y - 3z = 7 \qquad -\text{Eq. }1 + \text{Eq. }3$$

From Equation 2 we have $z = -3$. Substituting this into Equation 3 yields $y = 2$. Using these in Equation 1 yields $x = 4$.

Solution: $(4, 2, -3)$

4. $x + 3y - 2z = -7$

$-2x + y - z = -5$

$4x + y + z = 3$

$$\begin{aligned} x + 3y - 2z &= -7 \\ 7y - 5z &= -19 \quad && 2\text{ Eq. }1 + \text{Eq. }2 \\ -11y + 9 &= 31 && -4\text{ Eq. }1 + \text{Eq. }3 \end{aligned}$$

$$\begin{aligned} x \quad + \tfrac{1}{7}z &= \tfrac{8}{7} \quad && -3\text{ Eq. }2 + \text{Eq. }1 \\ y - \tfrac{5}{7}z &= -\tfrac{19}{7} && \tfrac{1}{7}\text{ Eq. }2 \\ \tfrac{8}{7}z &= \tfrac{8}{7} && 11\text{ Eq. }2 + \text{Eq. }3 \end{aligned}$$

$$\begin{aligned} x &= 1 \quad && -\tfrac{1}{7}\text{ Eq. }3 + \text{Eq. }1 \\ y &= -2 && \tfrac{5}{7}\text{ Eq. }3 + \text{Eq. }2 \\ z &= 1 && \tfrac{7}{8}\text{ Eq. }3 \end{aligned}$$

Solution: $(1, -2, 1)$

5. Maximize $z = 3x + 2y$.

Subject to: $x + 4y \le 20$

$2x + y \le 12$

$x \ge 0, y \ge 0$

At $(0, 0)$: $z = 0$

At $(0, 5)$: $z = 10$

At $(4, 4)$: $z = 20$

At $(6, 0)$: $z = 18$

Maximum of $z = 20$ at $(4, 4)$

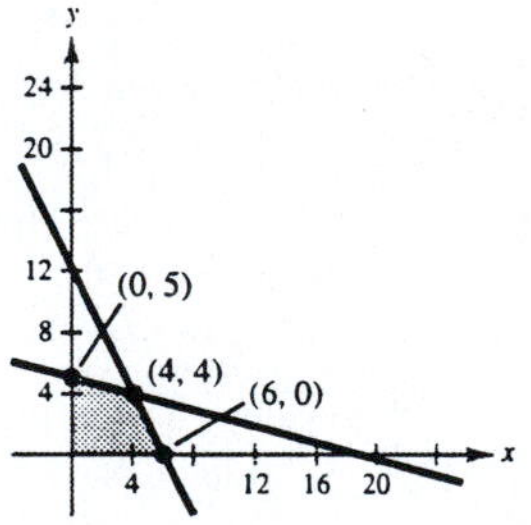

6. $$\begin{bmatrix} 4 & -3 \\ 2 & 1 \\ 5 & 0 \end{bmatrix}\begin{bmatrix} 3 & -2 \\ 1 & -3 \end{bmatrix} = \begin{bmatrix} 9 & 1 \\ 7 & -7 \\ 15 & -10 \end{bmatrix}$$

7. $$\begin{bmatrix} 1 & 2 & -1 \\ 3 & 7 & -10 \\ -5 & -7 & -15 \end{bmatrix}^{-1} = \begin{bmatrix} -175 & 37 & -13 \\ 95 & -20 & 7 \\ 14 & -3 & 1 \end{bmatrix}$$

8. $a_n = 4n + 4$

$a_1 = 8, \quad a_{20} = 84$

$S_{20} = \frac{20}{2}(8 + 84) = 920$

9. $$\sum_{i=0}^{\infty} 3\left(\frac{1}{2}\right)^i = \frac{3}{1 - (1/2)} = 6$$

10. $S_1 = 3 = 1(2(1) + 1)$

Assume that $S_k = 3 + 7 + 11 + 15 + \cdots + (4k - 1) = k(2k + 1)$.

Then, $S_{k+1} = 3 + 7 + 11 + 15 + \cdots + (4k - 1) + [4(k + 1) - 1]$

$= S_k + (4k + 3)$

$= k(2k + 1) + (4k + 3)$

$= 2k^2 + 5k + 3$

$= (k + 1)(2k + 3)$

$= (k + 1)[2(k + 1) + 1]$.

Therefore, the formula is valid for all $n \geq 1$.

11. $(z - 3)^4 = z^4 - 4z^3(3) + 6z^2(3)^2 - 4z(3)^3 + (3)^4$

$= z^4 - 12z^3 + 54z^2 - 108z + 81$

12. ${}_{10}P_3 = 720$

13. The first digit is 4 or 5, so the probability of picking it correctly is $\frac{1}{2}$. Then there are two numbers left for the second digit so its probability is also $\frac{1}{2}$. If these two are correct, then the third digit must be the remaining number. The probability of winning is:

$\left(\frac{1}{2}\right)\left(\frac{1}{2}\right)(1) = \frac{1}{4}$

Chapter Test Solutions for Chapter 10

1. $3x - 9y + 4 = 0$

$$y = \frac{1}{3}x + \frac{4}{9}$$

$$m = \frac{1}{3}$$

$$\tan \theta = \frac{1}{3}$$

$$\theta = \arctan \frac{1}{3} \approx 18.4°$$

2. $(7, 5)$, $y = 4 - 3x$

$$3x + y - 4 = 0$$

$$d = \frac{|3(7) + (1)(5) + -4|}{\sqrt{3^2 + 1^2}}$$

$$= \frac{22}{\sqrt{10}} = \frac{11\sqrt{10}}{5}$$

3. $y^2 - 4x + 4 = 0$

$$y^2 = 4(x - 1)$$

Vertex: $(1, 0)$

Focus: $(2, 0)$

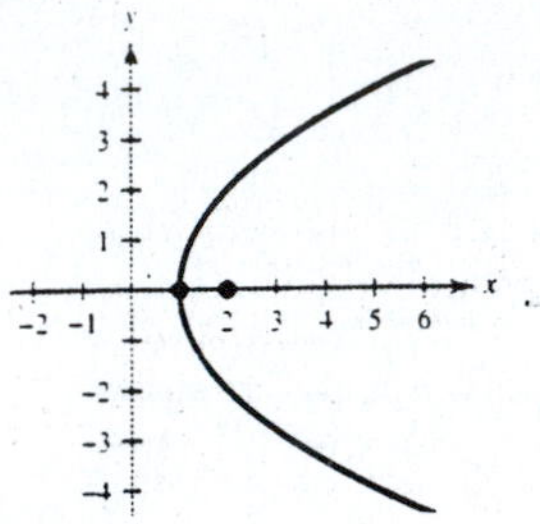

4. $\dfrac{(x - 2)^2}{4} + \dfrac{(y + 1)^2}{9} = 1$

Center: $(2, -1)$

$a = 3, b = 2,$

$c = 9 - 4 = 5 \implies c = \sqrt{5}$

Vertices: $(2, 2), (2, -4)$

Foci: $\left(2, -1 \pm \sqrt{5}\right)$

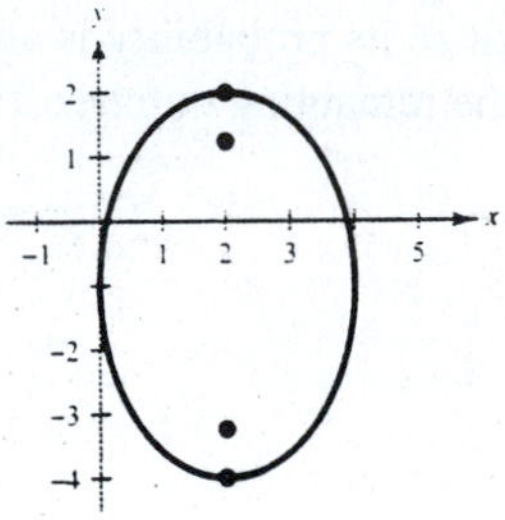

5. $x^2 - \dfrac{y^2}{4} = 1$

Center: $(0, 0)$

Horizontal transverse axis

$a = 1, b = 2.$

$c = 1 + 4 = 5 \implies c = \sqrt{5}$

Vertices: $(\pm 1, 0)$

Foci: $\left(\pm\sqrt{5}, 0\right)$

Asymptotes: $y = \pm 2x$

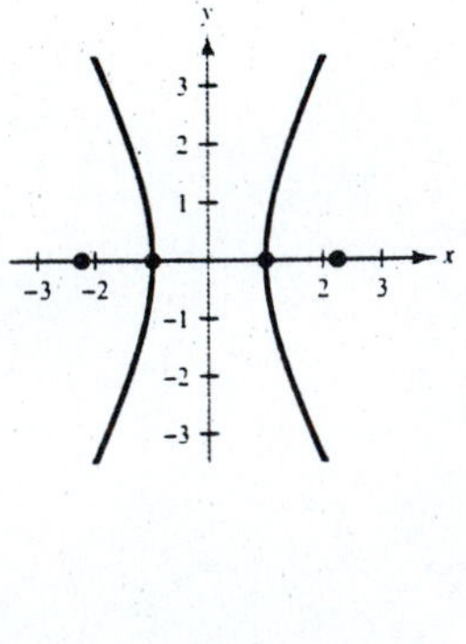

6. $x^2 - 4y^2 - 4x = 0$

$$(x - 2)^2 - 4y^2 = 4$$

$$\frac{(x - 2)^2}{4} - \frac{y^2}{1} = 1$$

Center: $(2, 0)$

Horizontal transverse axis

$a = 2, b = 1, c^2 = 1 + 4 = 5 \implies c = \sqrt{5}$

Vertices: $(0, 0), (4, 0)$

Foci: $\left(2 \pm \sqrt{5}, 0\right)$

Asymptotes: $y = \pm\frac{1}{2}(x - 2)$

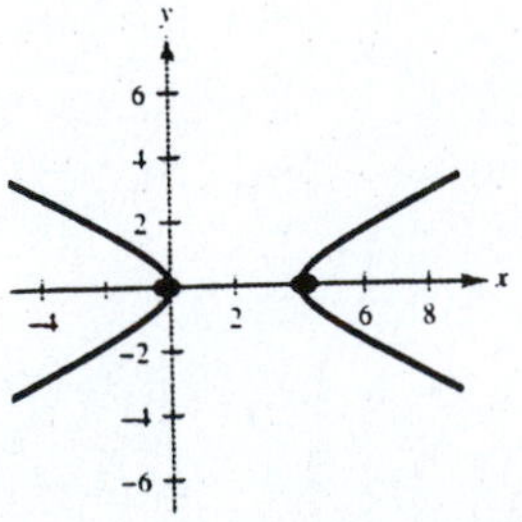

7. Vertex: $(3, -2)$,
Vertical axis
Point: $(0, 4)$

$$(x - h)^2 = 4p(y - k)$$
$$(x - 3)^2 = 4p(y + 2)$$
$$(0 - 3)^2 = 4p(4 + 2)$$
$$p = 0.375$$

Equation: $(x - 3)^2 = 1.5(y + 2)$

$$2x^2 - 12x - 3y + 12 = 0$$

8. Foci: $(0, 0), (0, 4) \Rightarrow c = 2$
Vertical transverse axis
Center: $(0, 2) \Rightarrow k = 2, h = 0$

Asymptotes: $y = \pm\frac{1}{2}x + 2 \Rightarrow \frac{a}{b} = \frac{1m}{2m}$

$$c^2 = a^2 + b^2 \Rightarrow 2^2 = (1m)^2 + (2m)^2$$
$$4 = m^2 + 4m^2$$
$$4 = 5m^2$$
$$m = \sqrt{\frac{4}{5}} = \frac{2\sqrt{5}}{5}$$
$$a = 1\left(\frac{2\sqrt{5}}{5}\right) = \frac{2\sqrt{5}}{5} \Rightarrow a^2 = \frac{4}{5}$$
$$b = 2\left(\frac{2\sqrt{5}}{5}\right) = \frac{4\sqrt{5}}{5} \Rightarrow b^2 = \frac{16}{5}$$
$$\frac{(y - 2)^2}{4/5} - \frac{x^2}{16/5} = 1$$
$$4(y - 2)^2 - x^2 = \frac{16}{5}$$
$$4y^2 - 16y + 16 - x^2 = \frac{16}{5}$$
$$20y^2 - 80y + 80 - 5x^2 = 16$$
$$5x^2 - 20y^2 + 80y - 64 = 0$$

9. $\frac{(x + 2)^2}{4} + \frac{y^2}{9} = 1$

$$(x + 2)^2 = \frac{4}{9}(9 - y^2)$$
$$x + 2 = \frac{2}{3}\sqrt{9 - y^2}$$
$$x = -2 \pm \frac{2}{3}\sqrt{9 - y^2}$$

The right half of the ellipse is $x = -2 + \frac{2}{3}\sqrt{9 - y^2}$.

10. (a) $x^2 + 6xy + y^2 - 6 = 0$

$A = 1, B = 6, C = 1$

$\cot 2\theta = \dfrac{1 - 1}{6} = 0$

$\dfrac{1}{\tan 2\theta} = 0$

$\theta = 45°$

(b) $y^2 + 6xy + 9x^2 = 6 - x^2 + 9x^2$

$(y + 3x)^2 = 6 + 8x^2$

$y + 3x = \pm\sqrt{6 + 8x^2}$

$y = -3x \pm \sqrt{6 + 8x}$

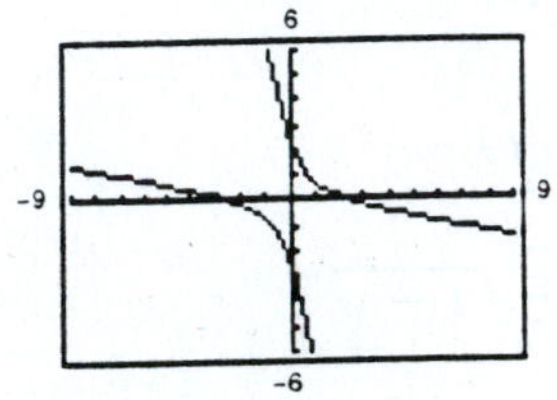

11. $x = 2 + 3\cos\theta$

$y = 2\sin\theta$

θ	0	$\pi/2$	π	$3\pi/2$
x	5	2	-1	2
y	0	2	0	-2

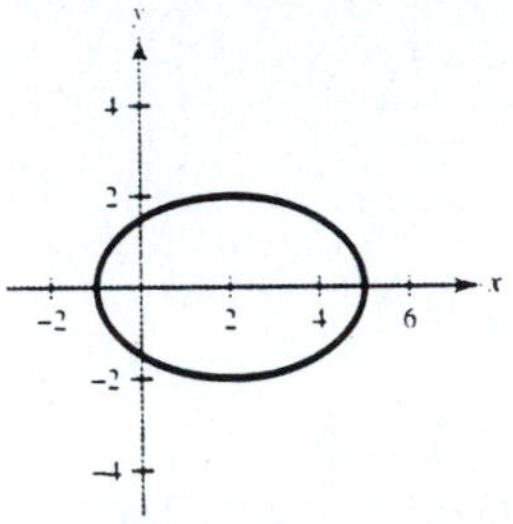

$x = 2 + 3\cos\theta \Rightarrow \dfrac{x - 2}{3} = \cos\theta$

$y = 2\sin\theta \Rightarrow \dfrac{y}{2} = \sin\theta$

$\cos^2\theta + \sin^2\theta = 1$

$\dfrac{(x - 2)^2}{9} + \dfrac{y^2}{4} = 1$

12. $(6, 4), (2, -3)$

$x = x_1 + t(x_2 - x_1) = 6 + t(2 - 6) = 6 - 4t$

$y = y_1 + t(y_2 - y_1) = 4 + t(-3 - 4) = 4 - 7t$

Answers are not unique. Another possible set:

$x = 6 + 4t$

$y = 4 + 7t$

13. $x^2 + y^2 - 6y = 0$

$r^2 - 6r\sin\theta = 0$

$r^2 = 6r\sin\theta$

$r = 6\sin\theta$

14. (a) $r = \dfrac{4}{1 + \cos\theta}$

$e = 1 \Rightarrow$ Parabola

Vertex: $(2, 0)$

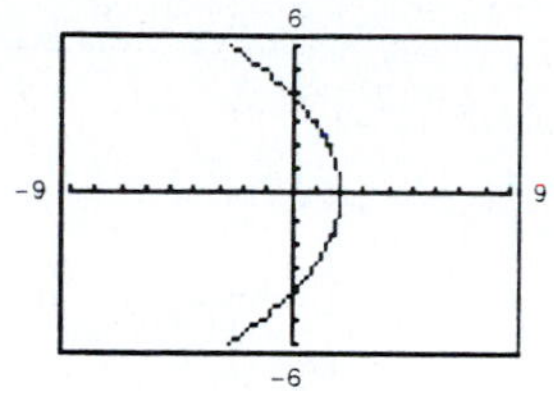

(b) $r = \dfrac{4}{2 + \cos\theta}$

$e = \dfrac{1}{2} \Rightarrow$ Ellipse

Vertex: $\left(\dfrac{4}{3}, 0\right), (4, \pi)$

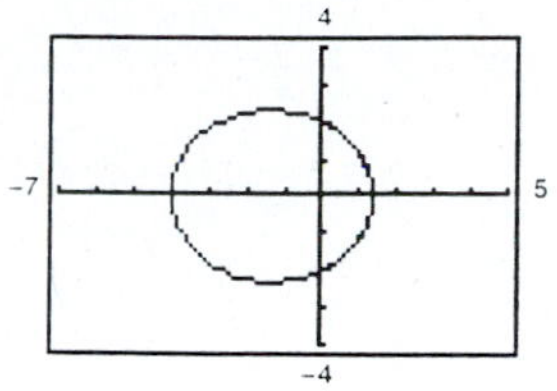

15. (a) $r = 2 + 3\sin\theta$

$\dfrac{a}{b} = \dfrac{2}{3} < 1$ Limaçon with inner loop

Matches graph (iii).

(b) $r = 3\sin\theta$

Circle

Matches graph (i).

(c) $r = 3\sin 2\theta$

Rose curve $(n = 2)$ with four petals

Matches graph (ii).

PART III Solutions to Even-Numbered Exercises and Focus on Concepts

CHAPTER P
Prerequisites

CHAPTER P
Prerequisites

Section P.1 Real Numbers

Solutions to Even-Numbered Exercises

2. $\sqrt{5}, -7, -\frac{7}{3}, 0, 3.12, \frac{5}{4}$

(a) Natural numbers: none

(b) Integers: $-7, 0$

(c) Rational numbers: $-7, -\frac{7}{3}, 0, 3.12, \frac{5}{4}$

(d) Irrational numbers: $\sqrt{5}$

4. $2.30300030003\ldots, 0.7575, -4.63, \sqrt{10}$

(a) Natural numbers: none

(b) Integers: none

(c) Rational numbers: $0.7575, -4.63, \sqrt{10}$

(d) Irrational numbers: $2.30300030003\ldots, \sqrt{10}$

6. $25, -17, -\frac{12}{5}, \sqrt{9}, 3.12, \frac{1}{2}\pi$

(a) Natural numbers: $25, \sqrt{9}$

(b) Integers: $25, -17, \sqrt{9}$

(c) Rational numbers: $25, -17, -\frac{12}{5}, \sqrt{9}, 3.12$

(d) Irrational numbers: $\frac{1}{2}\pi$

8. $\frac{1}{3} = 0.\overline{3}$

10. $\frac{6}{11} = 0.\overline{54}$

12. $-6 < -2.5$

14. $-3.5 < 1$

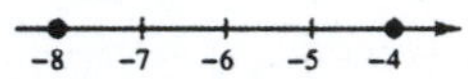

16. $1 < \frac{16}{3}$

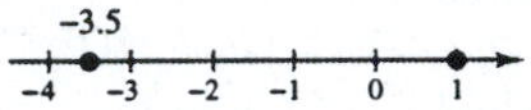

18. $-\frac{8}{7} < -\frac{3}{7}$

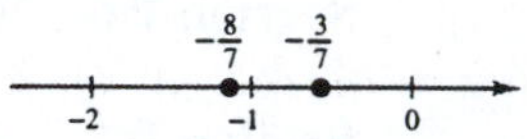

20. The inequality $x \geq -2$ is the set of all real numbers greater than or equal to -2. The interval is unbounded.

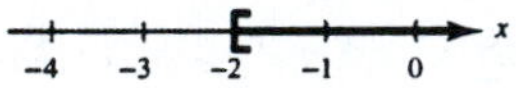

22. The inequality $x > 3$ is the set of all real numbers greater than 3. The interval is unbounded.

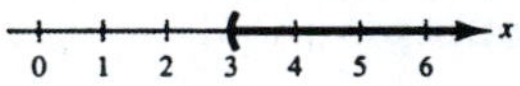

24. The inequality $x < 2$ is the set of all real numbers less than 2. The interval is unbounded.

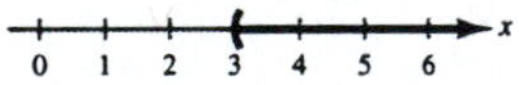

26. The inequality $0 \leq x \leq 5$ is the set of all real numbers greater than or equal to zero and less than or equal to 5. The interval is bounded.

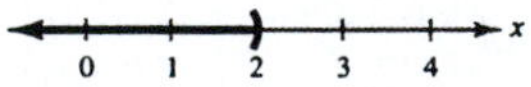

28. The inequality $0 < x \leq 6$ is the set of all real numbers greater than zero and less than or equal to 6. The interval is bounded.

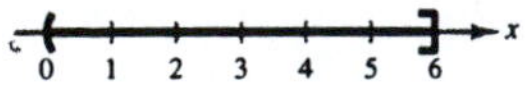

30. $\frac{381}{220} = 1.73\overline{18}$, 1.7320, $\sqrt{3} \approx 1.73205$, $\frac{26}{15} = 1.7\overline{3}$, $\sqrt{10} - \sqrt{2} \approx 1.74806$

32. $z \geq 10$

34. $y \leq 25$

36. $2.5\% \leq r \leq 5\%$

38. $|0| = 0$

40. $|4 - \pi| = 4 - \pi \approx 0.8584$

42. $-3 - |-3| = -3 - (3) = -6$

44. $|-1| - |-2| = (1) - (2) = -1$

46. $2|33| = 2(33) = 66$

48. $|-4| = |4|$ since $|-4| = 4$ and $|4| = 4$.

50. $-|-6| < |-6|$ since $|-6| = 6$ and $-|-6| = -(6) = -6$.

52. $-(-2) > -2$ since $-(-2) = 2$.

54. $d(-4, -\frac{3}{2}) = |-4 - (-\frac{3}{2})|$
$= |-4 + \frac{3}{2}|$
$= |-\frac{5}{2}| = \frac{5}{2}$

56. $d(\frac{1}{4}, \frac{11}{4}) = |\frac{1}{4} - \frac{11}{4}|$
$= |-\frac{10}{4}| = \frac{10}{4} = \frac{5}{2}$

58. $d(-126, -75) = |-126 - (-75)|$
$= |-126 + 75|$
$= |-51| = 51$

60. $d(9.34, -5.65) = |9.34 - (-5.65)|$
$= |9.34 + 5.65|$
$= |14.99| = 14.99$

62. $d(x, -10) = |x - (-10)| = |x + 1|$, and $d(x, -10) \geq 6$. Thus, $|x + 10| \geq 6$.

64. $d(103, 86) = |103 - 86| = 17$

66. $d(y, a) = |y - a|$, and $d(y, a) \leq 2$. Thus, $|y - a| \leq 2$.

68.

Budgeted Expense, b	*Actual Expense, a*	$\lvert a - b\rvert$	*0.05b*
\$9400	\$9772	\$372	0.05(9400) = \$470

Because the difference between the actual expenses and the budget is less than \$500 and less than 5% of the budgeted amount, there is compliance with the "budget variance test."

70.

Budgeted Expense, b	*Actual Expense, a*	$\lvert a - b\rvert$	*0.05b*
\$2575	\$2613	\$38	0.05(2575) = \$128.75

Because the difference between the actual expenses and the budget is less than \$500 and less than 5% of the budgeted amount, there is compliance with the "budget variance test."

72. $|520 - 590.9| = 70.9$ billion deficit for 1980

74. $|1153.5 - 1408.2| = \$254.7$ billion deficit for 1993

76. Yes, y is nonnegative if $y \geq 0$; y is positive if $y > 0$.

78. $3x^2 - 8x - 11$
Terms: $4x^2, -8x, -11$

80. $3x^4 + 3x^3$
Terms: $3x^4, 3x^3$

82. $9 - 7x$
(a) $9 - 7(-3) = 9 + 21 = 30$
(b) $9 - 7(3) = 9 - 21 = -12$

84. $-x^2 + 5x - 4$
(a) $-(-1)^2 + 5(-1) - 4 = -1 - 5 - 4 = -10$
(b) $-(1)^2 + 5(1) - 4 = -1 + 5 - 4 = 0$

86. $\dfrac{x}{x + 2}$
(a) $\dfrac{2}{2 + 2} = \dfrac{2}{4} = \dfrac{1}{2}$
(b) $\dfrac{-2}{-2 + 2} = \dfrac{2}{0}$
Division by 0 is undefined.

88. $2(\frac{1}{2}) = 1$
Multiplicative Inverse Property

90. $(x + 3) - (x + 3) = 0$
Additive Inverse Property

92. $(z - 2) + 0 = z - 2$
Additive Identity Property

94. $x + (y + 10) = (x + y) + 10$
Associative Property of Addition

96. $\frac{1}{7}(7 \cdot 12) = (\frac{1}{7} \cdot 7)12$ Associative Property of Multiplication
$= 1 \cdot 12$ Multiplicative Inverse Property
$= 12$ Multiplicative Identity Property

98. $10(23 - 30 + 7) + 10(-7 + 7) = 10(0) = 0$

100. $15 - \frac{3 - 3}{5} = 15 - \frac{0}{5} = 15 - 0 = 15$

102. $\frac{27 - 35}{4} = \frac{-8}{4} = -2$

104. $\frac{6}{7} - \frac{4}{7} = \frac{6 - 4}{7} = \frac{2}{7}$

106. $\frac{10}{11} + \frac{6}{33} - \frac{13}{66} = \frac{6 \cdot 10}{6 \cdot 11} + \frac{2 \cdot 6}{2 \cdot 33} - \frac{13}{66}$

$= \frac{60}{66} + \frac{12}{66} - \frac{13}{66}$

$= \frac{59}{66}$

108. $\frac{11}{16} \div \frac{3}{4} = \frac{11}{16} \cdot \frac{4}{3} = \frac{44}{48} = \frac{11}{12}$

110. $(\frac{3}{5} \div 3) - (6 \cdot \frac{4}{8}) = (\frac{3}{5} \cdot \frac{1}{3}) - (3)$

$= \frac{1}{5} - 3$

$= \frac{1}{5} - \frac{15}{5}$

112. $3(-\frac{5}{12} + \frac{3}{8}) = -0.125 \approx -0.13$

114. $\frac{\frac{1}{5}(-8 - 9)}{-\frac{1}{3}} = 10.20$

116. Because $5/n$ becomes larger and larger as n becomes smaller and smaller, you can conjecture that $5/n$ approaches infinity as n approaches 0.

118. Because $5/n$ becomes smaller and smaller as n becomes larger and larger, you can conjecture that $5/n$ approaches 0 as n approaches infinity.

Section P.2 Exponents and Radicals

Solutions to Even-Numbered Exercises

2. $(-2)^7 = (-2) \times (-2) \times (-2) \times (-2) \times (-2) \times (-2) \times (-2)$

4. $-\left(\frac{3}{2} \times \frac{3}{2} \times \frac{3}{2} \times \frac{3}{2}\right) = -\left(\frac{3}{2}\right)^4$

6. (a) $\frac{5^5}{5^2} = 5^{5-2} = 5^3 = 125$

(b) $\frac{3^2}{3^4} = 3^{2-4} = 3^{-2} = \frac{1}{3^2} = \frac{1}{9}$

8. (a) $(2^3 \cdot 3^2)^2 = 2^{3 \cdot 2} \cdot 3^{2 \cdot 2}$

$= 2^6 \cdot 3^4$

$= 64 \cdot 81 = 5184$

(b) $\left(-\frac{3}{5}\right)^3\left(\frac{5}{3}\right)^2 = (-1)^3 \frac{3^3}{5^3} \cdot \frac{5^2}{3^2}$

$= -1 \cdot 3^{3-2} \cdot 5^{2-3}$

$= -3 \cdot 5^{-1} = -\frac{3}{5}$

10. (a) $\frac{4 \cdot 3^{-2}}{2^{-2} \cdot 3^{-1}} = 4 \cdot 2^2 \cdot 3^{-2-(-1)}$

$= 4 \cdot 4 \cdot 3^{-1} = \frac{16}{3}$

(b) $(-2)^0 = 1$

12. $(8^{-4})(10^3) \approx 0.244$

14. $\frac{4^3}{3^{-4}} = 5184$

16. When $x = 4$,

$7x^{-2} = 7(4)^{-2} = \frac{7}{4^2} = \frac{7}{16}.$

18. When $x = 3$,

$5(-x)^3 = 5(-3)^3 = 5(-27) = -135.$

20. (a) $(3x)^2 = 3^2x^2 = 9x^2$

(b) $(4x^3)^2 = 4^2x^{3 \cdot 2} = 16x^6$

22. (a) $(-z)^3(3z^4) = (-1)^3(z^3)3z^4$
$= -1 \cdot 3 \cdot z^{3+4} = -3z^7$

(b) $\frac{25y^8}{10y^4} = \frac{5}{2}y^{8-4} = \frac{5}{2}y^4$

24. (a) $\frac{r^4}{r^6} = r^{4-6} = r^{-2} = \frac{1}{r^2}$

(b) $\left(\frac{4}{y}\right)^3\left(\frac{3}{y}\right)^4 = \frac{4^3}{y^3} \cdot \frac{3^4}{y^4} = \frac{64 \cdot 81}{y^{3+4}} = \frac{5184}{y^7}$

26. (a) $(2x^5)^0 = 1,\ x \neq 0$

(b) $(z+2)^{-3}(z+2)^{-1} = (z+2)^{-3+(-1)}$
$= (z+2)^{-4} = \frac{1}{(z+2)^4}$

28. (a) $(4y^{-2})(8y^4) = 32y^{4-2} = 32y^2$

(b) $\left(\frac{x^{-3}y^4}{5}\right)^{-3} = \frac{x^{-3(-3)}y^{4(-3)}}{5^{-3}}$
$= \frac{x^9y^{-12}}{5^{-3}} = \frac{125x^9}{y^{12}}$

30. (a) $[(x^2y^{-2})^{-1}]^{-1} = [x^{2(-1)}y^{-2(-1)}]^{-1}$
$= x^{2(-1)(-1)}y^{-2(-1)(-1)}$
$= x^2y^{-2} = \frac{x^2}{y^2}$

(b) $(5x^2z^6)^3(5x^2z^6)^{-3} = (5x^2z^6)^{3-3}$
$= (5x^2z^6)^0 = 1$

32. (a) $\frac{x^2 \cdot x^n}{x^3 \cdot x^n} = \frac{x^{2+n}}{x^{3+n}} = x^{2+n-(2+n)} = x^{-1} = \frac{1}{x}$

(b) $\left(\frac{a^{-3}}{b^{-3}}\right)\left(\frac{a}{b}\right)^3 = \left(\frac{a}{b}\right)^{-3}\left(\frac{a}{b}\right)^3 = \left(\frac{a}{b}\right)^{-3+3} = \left(\frac{a}{b}\right)^0 = 1$

	Radical Form	*Rational Exponent Form*
34.	$\sqrt[3]{64} = 4$, Given	$64^{1/3} = 4$, Answer
36.	$-\sqrt{144} = -12$, Answer	$-(144^{1/2}) = -12$, Given
38.	$\sqrt[3]{614.125} = 8.5$, Given	$(614.125)^{1/3} = 8.5$, Answer
40.	$\sqrt[5]{-243} = -3$, Answer	$(-243)^{1/5} = -3$, Given
42.	$(\sqrt[4]{81})^3 = 27$, Given	$81^{3/4} = 27$, Answer
44.	$\sqrt[4]{16^5} = (\sqrt[4]{16})^5 = 32$, Answer	$16^{5/4} = 32$, Given

46. (a) $\sqrt{49} = 7$ because $7^2 = 49$.

(b) $\sqrt[3]{\frac{27}{8}} = \frac{3}{2}$ because $\left(\frac{3}{2}\right)^3 = \frac{3^3}{2^3} = \frac{27}{8}$.

48. (a) $\sqrt[3]{0} = 0$

(b) $\frac{\sqrt[4]{81}}{3} = \frac{3}{3} = 1$ because $3^4 = 81$.

50. (a) $\sqrt[4]{562^4} = 562$ because $562^{4/4} = 562$.

(b) $36^{3/2} = 216$ because $36^{3/2} = (\sqrt{36})^3 = 6^3 = 216$.

52. (a) $100^{-3/2} = (\sqrt{100})^{-3} = 10^{-3} = \frac{1}{1000}$

(b) $\left(\frac{9}{4}\right)^{-1/2} = \left(\frac{4}{9}\right)^{1/2} = \frac{4^{1/2}}{9^{1/2}} = \frac{2}{3}$

54. (a) $\left(-\frac{125}{27}\right)^{-1/3} = \left(-\frac{27}{125}\right)^{1/3} = \frac{(-27)^{1/3}}{(125)^{1/3}} = \frac{-3}{5} = -\frac{3}{5}$

(b) $-\left(\frac{1}{125}\right)^{-4/3} = -(125)^{4/3} = -(125^{1/3})^4 = -(5)^4 = -625$

56. (a) $\sqrt[3]{45^2} \approx 12.651$

(b) $\sqrt[6]{125} \approx 2.236$

58. (a) $(15.25)^{-1.4} \approx 0.022$

(b) $(3.4)^{2.5} \approx 21.316$

60. (a) $\sqrt[3]{\frac{16}{27}} = \frac{\sqrt[3]{2^3 \cdot 2}}{\sqrt[3]{3^3}} = \frac{2\sqrt[3]{2}}{3}$

(b) $\sqrt{\frac{75}{4}} = \frac{\sqrt{5^2 \cdot 3}}{\sqrt{2^2}} = \frac{5\sqrt{3}}{2}$

62. (a) $\sqrt{54xy^4} = \sqrt{6 \cdot 3^2 \cdot x \cdot (y^2)^2} = 3y^2\sqrt{6x}$

(b) $\sqrt{\frac{32a^4}{b^2}} = \frac{\sqrt{(2^2)^2 \cdot 2 \cdot (a^2)^2}}{\sqrt{b^2}} = \frac{4a^2\sqrt{2}}{|b|}$

64. (a) $\sqrt[4]{(3x^2)^4} = 3x^2$

(b) $\sqrt[5]{96x^5} = \sqrt[5]{3 \cdot 2^5 \cdot x^5} = 2x\sqrt[5]{3}$

66. $\dfrac{8^{12/5}}{8^{2/5}} = 8^{(12/5)-(2/5)} = 8^{10/5} = 8^2 = 64$

68. $\dfrac{x^{4/3}y^{2/3}}{(xy)^{1/3}} = \dfrac{x^{4/3}y^{2/3}}{x^{1/3}y^{1/3}} = x^{(4/3)-(1/3)}y^{(2/3)-(1/3)} = xy^{1/3}$

70. $\dfrac{5^{-1/2} \cdot 5x^{5/2}}{(5x)^{3/2}} = \dfrac{5^{-1/2} \cdot 5x^{5/2}}{5^{3/2}x^{3/2}} = 5^{-1}x = \dfrac{x}{5},\ x > 0$

72. (a) $\dfrac{5}{\sqrt{10}} = \dfrac{5}{\sqrt{10}} \cdot \dfrac{\sqrt{10}}{\sqrt{10}} = \dfrac{5\sqrt{10}}{10} = \dfrac{\sqrt{10}}{2}$

(b) $\dfrac{5}{\sqrt[3]{(5x)^2}} = \dfrac{5}{\sqrt[3]{(5x)^2}} \cdot \dfrac{\sqrt[3]{5x}}{\sqrt[3]{5x}} = \dfrac{5\sqrt[3]{5x}}{5x} = \dfrac{\sqrt[3]{5x}}{x}$

74. (a) $\dfrac{5}{\sqrt{14}-2} = \dfrac{5}{\sqrt{14}-2} \cdot \dfrac{\sqrt{14}+2}{\sqrt{14}+2} = \dfrac{5(\sqrt{14}+2)}{14-4} = \dfrac{5(\sqrt{14}+2)}{10} = \dfrac{\sqrt{14}+2}{2}$

(b) $\dfrac{5}{2\sqrt{10}-5} = \dfrac{5}{2\sqrt{10}-5} \cdot \dfrac{2\sqrt{10}+5}{2\sqrt{10}+5} = \dfrac{5(2\sqrt{10}+5)}{40-25} = \dfrac{5(2\sqrt{10}+5)}{15} = \dfrac{2\sqrt{10}+5}{3}$

76. (a) $\dfrac{\sqrt{2}}{3} = \dfrac{\sqrt{2}}{3} \cdot \dfrac{\sqrt{2}}{\sqrt{2}} = \dfrac{2}{3\sqrt{2}}$

(b) $\sqrt[4]{\dfrac{5}{4}} = \dfrac{\sqrt[4]{5}}{\sqrt[4]{4}} \cdot \dfrac{\sqrt[4]{5^3}}{\sqrt[4]{5^3}} = \dfrac{5}{\sqrt[4]{4 \cdot 125}} = \dfrac{5}{\sqrt[4]{500}}$

78. (a) $\dfrac{\sqrt{3}-\sqrt{2}}{2} = \dfrac{\sqrt{3}-\sqrt{2}}{2} \cdot \dfrac{\sqrt{3}+\sqrt{2}}{\sqrt{3}+\sqrt{2}} = \dfrac{3-2}{2(\sqrt{3}+\sqrt{2})} = \dfrac{1}{2(\sqrt{3}+\sqrt{2})}$

(b) $\dfrac{2\sqrt{3}+\sqrt{3}}{3} = \dfrac{3\sqrt{3}}{3} \cdot \dfrac{\sqrt{3}}{\sqrt{3}} = \dfrac{9}{3\sqrt{3}} = \dfrac{3}{\sqrt{3}}$

80. (a) $\sqrt[6]{x^3} = x^{3/6} = x^{1/2} = \sqrt{x}$

(b) $\sqrt[4]{(3x^2)^4} = 3x^2$

82. (a) $\sqrt{\sqrt{243(x+1)}} = [(243(x+1)]^{1/2})^{1/2}$
$= (243(x+1))^{1/4}$
$= \sqrt[4]{243(x+1)}$
$= \sqrt[4]{3 \cdot 3^4(x+1)}$
$= 3\sqrt[4]{3(x+1)}$

(b) $\sqrt{\sqrt[3]{10a^7b}} = ((10a^7b)^{1/3})^{1/2}$
$= (10a^7b)^{1/6}$
$= \sqrt[6]{10a \cdot a^6 \cdot b}$
$= a\sqrt[6]{10ab}$

84. (a) $4\sqrt{27} - \sqrt{75} = 4\sqrt{3^2 \cdot 3} - \sqrt{5^2 \cdot 3}$
$= 4 \cdot 3\sqrt{3} - 5\sqrt{3}$
$= 12\sqrt{3} - 5\sqrt{3}$
$= 7\sqrt{3}$

(b) $\sqrt[3]{16} + 3\sqrt[3]{54} = \sqrt[3]{2 \cdot 2^3} + 3\sqrt[3]{2 \cdot 3^3}$
$= 2\sqrt[3]{2} + 3 \cdot 3\sqrt[3]{2}$
$= 2\sqrt[3]{2} + 9\sqrt[3]{2}$
$= 11\sqrt[3]{2}$

86. (a) $3\sqrt{x+1} + 10\sqrt{x+1} = 13\sqrt{x+1}$

(b) $7\sqrt{80x} - 2\sqrt{125x} = 7\sqrt{4^2 \cdot 5x} - 2\sqrt{5^2 \cdot 5x}$
$= 7 \cdot 4\sqrt{5x} - 2 \cdot 5\sqrt{5x}$
$= 28\sqrt{5x} - 10\sqrt{5x}$
$= 18\sqrt{5x}$

88. $\sqrt{\dfrac{3}{11}} = \dfrac{\sqrt{3}}{\sqrt{11}}$

90. $\sqrt{3^2 + 4^2} = \sqrt{9 + 16} = \sqrt{25} = 5$
Thus, $5 = \sqrt{3^2 + 4^2}$.

92. $9{,}461{,}000{,}000{,}000{,}000 = 9.461 \times 10^{15}$ kilometers

94. $0.00003937 = 3.937 \times 10^{-5}$ inch

96. $1.3 \times 10^7 = 13{,}000{,}000$ degrees Celsius

98. $9.0 \times 10^{-4} = 0.0009$ meter

100. (a) $(9.3 \times 10^6)^3(6.1 \times 10^{-4}) \approx 4.907 \times 10^{17}$

(b) $\dfrac{(2.414 \times 10^4)^6}{(1.68 \times 10^5)^5} \approx 1.479$

102. (a) $(2.65 \times 10^{-4})^{1/3} \approx 0.064$

(b) $\sqrt{9 \times 10^{-4}} = 0.030$

104. $\left(\dfrac{2}{\sqrt{5}}\right)^2 = \dfrac{2^2}{(\sqrt{5})^2} = \dfrac{4}{5}$

This is not equivalent to rationalizing the denominator because rationalizing the denominator produces a number equivalent to the original fraction but squaring does not.

106. $t = 0.03[12^{5/2} - (12 - h)^{5/2}],\ 0 \le h \le 12$

For $h = 7$:

$t = 0.03[12^{5/2} - (12 - 7)^{5/2}]$

$= 0.03[12^{5/2} - 5^{5/2}] \approx 13.29$ seconds

108. Size $= 0.03\sqrt{v}$; For $v = \dfrac{3}{4}$:

Size $= 0.03\sqrt{\dfrac{3}{4}}$

$= 0.03 \cdot \dfrac{\sqrt{3}}{\sqrt{4}}$

$= 0.03\dfrac{\sqrt{3}}{2}$

≈ 0.026 inch

110. Paper: $(0.375)(1.957 \times 10^8 \text{ tons}) = 7.34 \times 10^7$ tons
Metals: $(0.083)(1.957 \times 10^8 \text{ tons}) = 1.62 \times 10^7$ tons
Glass: $(0.067)(1.957 \times 10^8 \text{ tons}) = 1.31 \times 10^7$ tons
Plastics: $(0.083)(1.957 \times 10^8 \text{ tons}) = 1.62 \times 10^7$ tons
Yard waste: $(0.179)(1.957 \times 10^8 \text{ tons}) = 3.50 \times 10^7$ tons
Other: $(0.213)(1.957 \times 10^8 \text{ tons}) = 4.17 \times 10^7$ tons

Section P.3 Polynomials and Factoring

Solutions to Even-Numbered Exercises

2. $(2x^2+1)-(x^2-2x+1)=2x^2+1-x^2+2x-1=x^2+2x$

4. $-(5x^2-1)-(-3x^2+5)=-5x^2+1+3x^2-5=-2x^2-4$

6. $$\begin{aligned}(15x^4-18x-19)-(13x^4-5x+15)&=15x^4-18x-19-13x^4+5x-15\\&=2x^4-13x-34\end{aligned}$$

8. $$\begin{aligned}(y^3+1)-[(y^2+1)+(3y-7)]&=y^3+1-(y^2+1)-(3y-7)\\&=y^3+1-y^2-1-3y+7\\&=y^3-y^2-3y+8\end{aligned}$$

10. $$\begin{aligned}y^2(4y^2+2y-3)&=4y^2(y^2)+2y(y^2)-3(y^2)\\&=4y^4+2y^3-3y^2\end{aligned}$$

12. $$\begin{aligned}-4x(3-x^3)&=3(-4x)-x^3(-4x)\\&=-12x+4x^4\\&=4x^4-12x\end{aligned}$$

14. $$\begin{aligned}(-2x)(-3x)(5x+2)&=6x^2(5x+2)\\&=6x^2(5x)+6x^2(2)\\&=30x^3+12x^2\end{aligned}$$

16. Add: $$\begin{array}{r}2x^5-3x^3+2x+3\\4x^3+x-6\\\hline 2x^5+x^3+3x-3\end{array}$$

18. Subtract: $$\begin{array}{r}0.6t^4-2\ t^2\qquad\\-t^4+0.5t^2-5.6\\\hline 1.6t^4-2.5t^2+5.6\end{array}$$

20. Multiply: $$\begin{array}{r}4x^4+x^3-6x^2+9\\x^2+2x+3\\\hline 4x^6+x^5-6x^4+9x^2\\8x^5+2x^4-12x^3+18x\\12x^4+3x^3-18x^2+27\\\hline 4x^6+9x^5+8x^4-9x^3-9x^2+18x+27\end{array}$$

22. Multiply: $$\begin{array}{r}x^2+2x+4\\x-4\\\hline x^3+2x^2+4x\\-2x^2-4x-8\\\hline x^3\qquad\qquad-8\end{array}$$

24. Multiply: $$\begin{array}{r}x^2+3x-2\\x^2-3x-2\\\hline x^4+3x^3-2x^2\\-3x^3-9x^2+6x\\-2x^2-6x+4\\\hline x^4\qquad-13x^2\qquad+4\end{array}$$

26. $$\begin{aligned}(x-5)(x+10)&=x^2+10x-5x-50\\&=x^2+5x-50\end{aligned}$$

28. $$\begin{aligned}(7x-2)(4x-3)&=28x^2-21x-8x+6\\&=28x^2-29x+6\end{aligned}$$

30. $$\begin{aligned}(4x+5)^2&=(4x+5)(4x+5)\\&=16x^2+20x+20x+25\\&=16x^2+40x+25\end{aligned}$$

32. $$\begin{aligned}(5-8x)^2&=(5)^2+2(5)(-8x)+(-8x)^2\\&=25-80x+64x^2\end{aligned}$$

34. $$\begin{aligned}[(x+1)-y]^2&=(x+1)^2+2(x+1)(-y)+(-y)^2\\&=x^2+2x+1-2xy-2y+y^2\\&=x^2-2xy+y^2+2x-2y+1\end{aligned}$$

36. $$\begin{aligned}(2x+3)(2x-3)&=(2x)^2-(3)^2\\&=4x^2-9\end{aligned}$$

38. $$\begin{aligned}(2x+3y)(2x-3y)&=(2x)^2-(3y)^2\\&=4x^2-9y^2\end{aligned}$$

40. $$\begin{aligned}[(x+y)+1][(x+y)-1]&=(x+y)^2-1^2\\&=x^2+2xy+y^2-1\end{aligned}$$

42. $(3a^3 - 4b^2)(3a^3 + 4b^2) = (3a^3)^2 - (4b^2)^2$
$= 9a^6 - 16b^4$

44. $(x - 2)^3 = x^3 - 3x^2(2) + 3x(2)^2 - (2)^3$
$= x^3 - 6x^2 + 12x - 8$

46. $(3x + 2y)^3 = (3x)^3 + 3(3x)^2(2y) + 3(3x)(2y)^2 + (2y)^3$
$= 27x^3 + 54x^2y + 36xy^2 + 8y^3$

48. $(8x + 3)^2 = (8x)^2 + 2(8x)(3) + 3^2$
$= 64x^2 + 48x + 9$

50. $(2x - 1)(x + 3) + 3(x + 3) = 2x^2 + 5x - 3 + 3x + 9$
$= 2x^2 + 8x + 6$

52. $(x + y)(x - y)(x^2 + y^2) = (x^2 - y^2)(x^2 + y^2)$
$= (x^2)^2 - (y^2)^2$
$= x^4 - y^4$

54. No, the product of two binomials isn't always a binomial. For example, $(x + 1)(x + 1)$ $= x^2 + 2x + 1$, which is a trinomial.

56. To find the unknown polynomial, add:
$-x^3 + 3x^2 + 2x - 1 + (5x^2 + 8) = -x^3 + 8x^2 + 2x + 7$

58. (a) $1200(1 + r)^3 = 1200(1^3 + 3(1)^2r + 3(1)(r)^2 + r^3)$
$= 1200r^3 + 3600r^2 + 3600r + 1200$

(b)

r	6%	7%	$7\frac{1}{2}$%	8%	$8\frac{1}{2}$%
$1200(1 + r)^3$	\$1429.22	\$1470.05	\$1490.76	\$1511.65	\$1532.75

60. Volume $= x(26 - 2x)(18 - 2x)$
$= (x)(-2)(x - 13)(-2)(x - 9)$
$= 4x(x - 13)(x - 9)$

x (cm)	1	2	3
V (cubic cm)	384	616	720

62. (a) From the figure $S_8 \approx 570$ and $S_6 \approx 355$. The difference of these estimates is 355 pounds.

(b) As the span increases, the difference in the safe loads decreases in magnitude.

64. $(x + a)(x + a) = x(x + a) + a(x + a)$
Distributive Property

66. $5y - 30 = 5(y - 6)$

68. $4x^3 - 6x^2 + 12x = 2x(2x^2 - 3x + 6)$

70. $x^2 - \frac{1}{4} = \left(x + \frac{1}{2}\right)\left(x - \frac{1}{2}\right)$

72. $49 - 9y^2 = (7 + 3y)(7 - 3y)$

74. $25 - (z + 5)^2 = [5 + (z + 5)][5 - (z + 5)]$
$= (z + 10)(-z)$

76. $x^2 + 10x + 25 = (x + 5)^2$

78. $9x^2 - 12x + 4 = (3x - 2)^2$

80. $x^2 + 5x + 6 = (x + 3)(x + 2)$

82. $(t^2 - t - 6) = (t - 3)(t + 2)$

84. $24 + 5z - z^2 = (-1)(z^2 - 5z - 24)$
$= (-1)(z - 8)(z + 3)$
$= (8 - z)(3 + z)$

86. $2x^2 - x - 1 = (2x + 1)(x - 1)$

88. $-5u^2 - 13u + 6 = (-1)(5u^2 + 13u - 6)$
$= (-1)(5u - 2)(u + 3)$
$= (-5u + 2)(u + 3)$

90. $x^3 - 27 = x^3 - 3^3$
$= (x - 3)(x^2 + 3x + 9)$

92. $z^3 + 125 = z^3 + 5^3$
$= (x + 5)(x^2 - 5x + 25)$

94. $x^3 + 5x^2 - 5x - 25 = x^2(x + 5) - 5(x + 5)$
$= (x^2 - 5)(x + 5)$

96. $5x^3 - 10x^2 + 3x - 6 = 5x^2(x - 2) + 3(x - 2)$
$= (x - 2)(5x^3 + 3)$

98. $12x^2 - 48 = 12(x^2 - 4)$
$= 12(x + 2)(x - 2)$

100. $6x^2 - 54 = 6(x^2 - 9)$
$= 6(x + 3)(x - 3)$

102. $16 + 6x - x^2 = (8 - x)(2 + x)$

104. $-9x^2 + 6x - 1 = (-3x + 1)(3x - 1)$

106. $2y^3 - 7y^2 - 15y = y(2y^2 - 7y - 15)$
$= y(2y + 3)(y - 5)$

108. $13x + 6 + 5x^2 = 5x^2 + 13x + 6$
$= (5x + 3)(x + 2)$

110. $5 - x + 5x^2 - x^3 = (5 - x) + x^2(5 - x)$
$= (5 - x)(1 + x^2)$

112. $3u - 2u^2 + 6 - u^3 = -u^3 - 2u^2 + 3u + 6$
$= -u^2(u + 2) + 3(u + 2)$
$= (u + 2)(3 - u^2)$

114. $(t - 1)^2 - 49 = [(t - 1) + 7][(t - 1) - 7]$
$= (t + 6)(t - 8)$

116. $(x^2 + 8)^2 - 36x^2 = [(x^2 + 8) - 6x][(x^2 + 8) + 6x]$
$= (x^2 - 6x + 8)(x^2 + 6x + 8)$
$= (x - 4)(x - 2)(x + 4)(x + 2)$

118. $5x^3 + 40 = 5(x^3 + 8)$
$= 5(x^3 + 2^3)$
$= 5(x + 2)(x^2 - 2x + 4)$

120. $5(3 - 4x)^2 - 8(3 - 4x)(5x - 1) = (3 - 4x)[5(3 - 4x) - 8(5x - 1)]$
$= (3 - 4x)(23 - 60x)$

122. $7(3x + 2)^2(1 - x)^2 + (3x + 2)(1 - x)^3$
$(3x + 2)(1 - x)^2[7(3x + 2) + (1 - x)] = (3x + 2)(1 - x)^2(20x + 15)$
$= (3x + 2)(1 - x)^2(5)(4x + 3)$
$= 5(1 - x)^2(3x + 2)(4x + 3)$

124. $3(x - 2)^2(x + 1)^4 + (x - 2)^3(4)(x + 1)^3 = (x - 2)^2(x + 1)^3[3(x + 1) + 4(x - 2)]$
$= (x - 2)^2(x + 1)^3(7x - 5)$

126. $5(x^6 + 1)^4(6x^5)(3x + 2)^3 + 3(3x + 2)^2(3)(x^6 + 1)^5 = (x^6 + 1)^4(3x + 2)^2[5(6x^5)(3x + 2) + 3(3)(x^6 + 1)]$
$= [(x^2)^3 + 1^3]^4(3x + 2)^2[90x^6 + 60x^5 + 9x^6 + 9]$
$= [(x^2 + 1)(x^4 - x^2 + 1)]^4(3x + 2)^2(3)[33x^6 + 20x^5 + 3]$
$= 3(x^2 + 1)^4(x^4 - x^2 + 1)^4(3x + 2)^2(33x^6 + 20x^5 + 3)$

128. $5w^3(9w + 1)^4(9) + (2w + 1)^5(3w^2) = 3w^2[15w(9w + 1)^4 + (2w + 1)^5$

130. $a^2 + 2ab + b^2 = (a + b)^2$
Matches graph (c).

132. $ab + a + b + 1 = (a + 1)(b + 1)$
Matches graph (d).

134. $x^2 + 4x + 3 = (x + 3)(x + 1)$

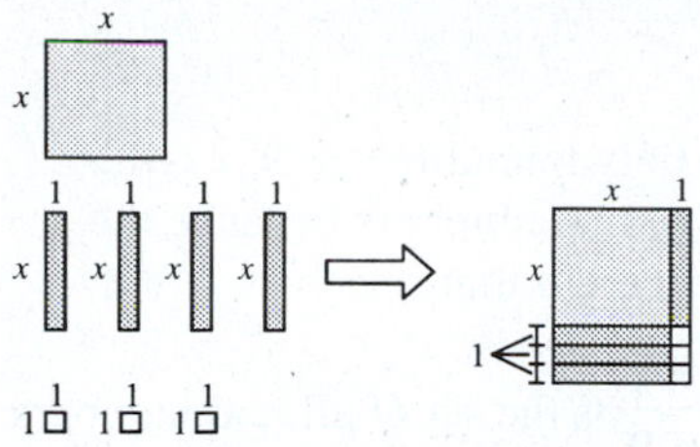

136. $x^2 + 3x + 2 = (x + 2)(x + 1)$

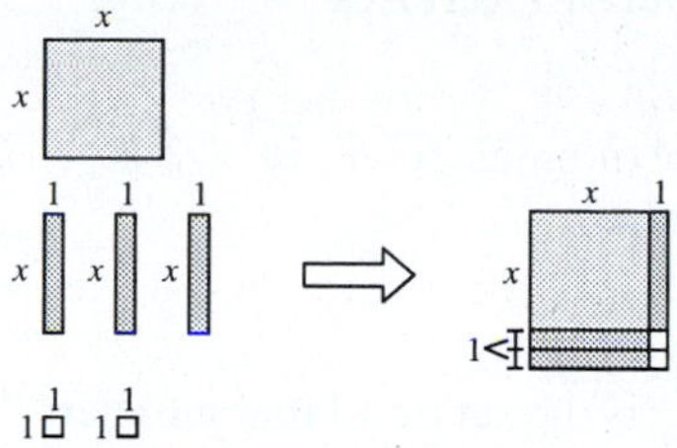

138. Area of square − Area of circle: $(2r)^2 - \pi r^2 = 4r^2 - \pi r^2 = r^2(4 - \pi)$

140. (Area of larger triangle) − (Area of smaller triangle) $= \frac{1}{2}\left(\frac{5}{4}\right)(x + 3)(x + 3) - \frac{1}{2}(5)(4)$

$$= \tfrac{5}{8}(x^2 + 6x + 9) - 10$$
$$= \tfrac{5}{8}x^2 + \tfrac{30}{8}x + \tfrac{45}{8} - 10$$
$$= \tfrac{5}{8}x^2 + \tfrac{30}{8}x - \tfrac{35}{8}$$
$$= \tfrac{5}{8}(x^2 + 6x - 7)$$
$$= \tfrac{5}{8}(x + 7)(x - 1)$$

142. For $x^2 + bx + 50$ to be factorable, b must equal $m + n$ where $mn = 50$.

Factors of 50	Sum of Factors
(1)(50)	$1 + 50 = 51$
(−1)(−50)	$-1 + (-50) = -51$
(2)(25)	$2 + 25 = 27$
(−2)(−25)	$-2 + (-25) = -27$
(5)(10)	$5 + 10 = 15$
(−5)(−10)	$-5 + (-10) = 15$

The possible values of b are 51, −51, 27, −27, 15, and −15.

144. For $3x^2 - 10x + c$ to be factorable, the factors of $3c$ must add up to −10.

Possible c values	$3c$	Factors of c that add up to −10
3	9	$(-1)(-9) = 9$ and $-1 + (-9) = -10$
−8	−24	$(-12)(2) = -24$ and $-12 + 2 = -10$

These are a few possible c values. There are many correct answers.
If $c = 3$, $3x^2 - 10x + 3 = (3x - 1)(x - 3)$.
If $c = -8$, $3x^2 - 10x - 8 = (3x + 2)(x - 4)$.

146. No, $(3x - 6)(x + 1)$ is not completely factored.
A factor of 3 remains to be factored out.
Completely factored form is: $3(x - 2)(x + 1)$.

148. $kQx - kx^2 = kx(Q - x)$

Section P.4 Fractional Expressions

Solutions to Even-Numbered Exercises

2. The domain of the polynomial $2x^2 + 5x - 2$ is the set of all real numbers.

4. The domain of the polynomial $6x^2 - 9, x > 0$ is the set of all positive real numbers because the polynomial is restricted to that set.

6. The domain of $\dfrac{x+1}{2x+1}$ is the set of all real numbers such that $x \neq -\dfrac{1}{2}$.

8. The domain of $\dfrac{2x+1}{x^2-9}$ is the set of all real numbers such that $x \neq 3$ and $x \neq -3$.

10. The domain of $\dfrac{1}{\sqrt{x+1}}$ is the set of all real numbers such that

12. $\dfrac{3}{4} = \dfrac{3(x+1)}{4(x+1)}$

The missing factor is $(x + 1)$, where $x \neq -1$.

14. $\dfrac{3y-4}{y+1} = \dfrac{(3y-4)(y-1)}{(y+1)(y-1)} = \dfrac{(3y-4)(y-1)}{y^2-1}$

The missing factor is $(y - 1)$, where $y \neq 1$.

16. $\dfrac{1-z}{z^2} = \dfrac{(1-z)(z+1)}{z^2(z+1)} = \dfrac{(1-z)(z+1)}{z^3+z^2}$

The missing factor is $(z + 1)$, where $z \neq -1$.

18. $\dfrac{18y^2}{60y^5} = \dfrac{6y^2(3)}{6y^2(10y^3)} = \dfrac{3}{10y^3}$

20. $\dfrac{9x^2+9x}{2x+2} = \dfrac{9x(x+1)}{2(x+1)} = \dfrac{9x}{2}, \quad x \neq -1$

22. $\dfrac{x^2-25}{5-x} = \dfrac{(x+5)(x-5)}{-1(x-5)} = -(x+5), \quad x \neq 5$

24. $\dfrac{x^2+8x-20}{x^2+11x+10} = \dfrac{(x+10)(x-2)}{(x+10)(x+1)} = \dfrac{x-2}{x-1}, \quad x \neq -10$

26. $\dfrac{3-x}{x^2+11x+10} = \dfrac{3-x}{(x+10)(x+1)}$

28. $\dfrac{x^2-9}{x^3+x^2-9x-9} = \dfrac{x^2-9}{(x^2-9)(x+1)} = \dfrac{1}{x+1}, \quad x \neq \pm 3$

30. $\dfrac{y^3-2y^2-3y}{y^3+1} = \dfrac{y(y-3)(y+1)}{(y+1)(y^2-y+1)} = \dfrac{y(y-3)}{y^2-y+1}, \quad y \neq -1$

32.

x	0	1	2	3	4	5	6
$\dfrac{x-3}{x^2-x-6}$	$\frac{1}{2}$	$\frac{1}{3}$	$\frac{1}{4}$	Undef.	$\frac{1}{6}$	$\frac{1}{7}$	$\frac{1}{8}$
$\dfrac{1}{x+2}$	$\frac{1}{2}$	$\frac{1}{3}$	$\frac{1}{4}$	$\frac{1}{5}$	$\frac{1}{6}$	$\frac{1}{7}$	$\frac{1}{8}$

The expressions are equivalent except at $x = 3$.

34. $\dfrac{(ax-b)}{(b-ax)} = \dfrac{(ax-b)}{-1(ax-b)} = -1$

Yes, the statement is true for all nonzero numbers a and b.

36. Area of shaded portion: $\left(\dfrac{x+5}{2}\right)^2 = \dfrac{(x+5)^2}{4}$

Area of total figure: $(2x + 3)(x + 5)$

Ratio: $\dfrac{\dfrac{(x+5)^2}{4}}{(2x+3)(x+5)} = \dfrac{\dfrac{(x+5)}{4}}{(2x+3)} = \dfrac{x+5}{4(2x+3)}$

38. $\dfrac{x+13}{x^3(3-x)} \cdot \dfrac{x(x-3)}{5} = \dfrac{x+13}{x^3(x-3)(-1)} \cdot \dfrac{x(x-3)}{5} = \dfrac{x+13}{-5x^2} = -\dfrac{x+13}{5x^2}, x \neq 3$

40. $\dfrac{(x-9)(x+7)}{x+1}\cdot\dfrac{x}{9-x}=\dfrac{(x-9)(x+7)}{x+1}\cdot\dfrac{x}{(-1)(x-9)}$

$=\dfrac{x(x+7)}{(-1)(x+1)}=-\dfrac{x(x+7)}{x+1},\ x\neq 9$

42. $\dfrac{4y-16}{5y+15}\cdot\dfrac{2y+6}{4-y}=\dfrac{4(y-4)}{5(y+3)}\cdot\dfrac{2(y+3)}{(-1)(y-4)}$

$=\dfrac{8}{-5}=-\dfrac{8}{5},\ y\neq -3, 4$

44. $\dfrac{y^3-8}{2y^3}\cdot\dfrac{4y}{y^2-5y+6}=\dfrac{(y-2)(y^2+2y+4)}{2y^3}\cdot\dfrac{4y}{(y-2)(y-3)}$

$=\dfrac{2(y^2+2y+4)}{y^2(y-3)},\ y\neq 2$

46. $\dfrac{x^3-1}{x+1}\cdot\dfrac{x^2+1}{x^2-1}=\dfrac{(x-1)(x^2+x+1)}{x+1}\cdot\dfrac{x^2+1}{(x+1)(x-1)}$

$=\dfrac{(x^2+x+1)(x^2+1)}{(x+1)^2},\ x\neq 1$

48. $\dfrac{x+2}{5(x-3)}\div\dfrac{x-2}{5(x-3)}=\dfrac{x+2}{5(x-3)}\cdot\dfrac{5(x-3)}{x-2}=\dfrac{x+2}{x-2},\ x\neq 3$

50. $\dfrac{\left(\dfrac{x^2-1}{x}\right)}{\left[\dfrac{(x-1)^2}{x}\right]}=\dfrac{x^2-1}{x}\cdot\dfrac{x}{(x-1)^2}=\dfrac{(x+1)(x-1)}{x}\cdot\dfrac{x}{(x-1)(x-1)}=\dfrac{(x+1)}{(x-1)},\ x\neq 0, 1$

52. $\dfrac{2x-1}{x+3}+\dfrac{1-x}{x+3}=\dfrac{2x-1+1-x}{x+3}=\dfrac{x}{x+3}$

54. $\dfrac{3}{x-1}-5=\dfrac{3}{x-1}-\dfrac{5(x-1)}{x-1}=\dfrac{3-5(x-1)}{x-1}=\dfrac{3-5x+5}{x-1}=\dfrac{8-5x}{x-1}$

56. $\dfrac{2x}{x-5}-\dfrac{5}{5-x}=\dfrac{2x}{x-5}-\dfrac{5(-1)}{(-1)(5-x)}=\dfrac{2x}{x-5}-\dfrac{-5}{x-5}=\dfrac{2x+5}{x-5}$

58. $\dfrac{x}{x^2+x-2}-\dfrac{1}{x+2}=\dfrac{x}{(x+2)(x-1)}-\dfrac{1(x-1)}{(x+2)(x-1)}$

$=\dfrac{x-(x-1)}{(x+2)(x-1)}=\dfrac{x-x+1}{(x+2)(x-1)}=\dfrac{1}{(x+2)(x-1)}$

60. $\dfrac{2}{x^2-x-2}+\dfrac{10}{x^2+2x-8}=\dfrac{2}{(x-2)(x+1)}+\dfrac{10}{(x+4)(x-2)}$

$=\dfrac{2(x+4)}{(x-2)(x+1)(x+4)}+\dfrac{10(x+1)}{(x-2)(x+1)(x+4)}$

$=\dfrac{2x+8+10x+10}{(x-2)(x+1)(x+4)}=\dfrac{12x+19}{(x-2)(x+1)(x+4)}=\dfrac{6(2x+3)}{(x-2)(x+1)(x+4)}$

62. $\dfrac{2}{x+1}+\dfrac{2}{x-1}+\dfrac{1}{x^2-1}=\dfrac{2}{x+1}+\dfrac{2}{x-1}+\dfrac{1}{(x+1)(x-1)}$

$=\dfrac{2(x-1)}{(x+1)(x-1)}+\dfrac{2(x+1)}{(x+1)(x-1)}+\dfrac{1}{(x+1)(x-1)}$

$=\dfrac{2x-2+2x+2+1}{(x+1)(x-1)}=\dfrac{4x+1}{(x+1)(x-1)}$

64. $2x(x-5)^{-3} - 4x^2(x-5)^{-4} = \dfrac{2x}{(x-5)^3} - \dfrac{4x^2}{(x-5)^4} = \dfrac{2x(x-5)}{(x-5)^4} - \dfrac{4x^2}{(x-5)^4}$

$$= \frac{2x^2 - 10x - 4x^2}{(x-5)^4} = \frac{-2x^2 - 10x}{(x-5)^4} = \frac{-2x(x+5)}{(x-5)^4}$$

66. $\dfrac{6-x}{x(x+2)} + \dfrac{x+2}{x^2} + \dfrac{8}{x^2(x+2)} = \dfrac{x(6-x)}{x^2(x+2)} + \dfrac{(x+2)^2}{x^2(x+2)} + \dfrac{8}{x^2(x+2)}$

$$= \frac{6x - x^2 + x^2 + 4x + 4 + 8}{x^2(x+2)} = \frac{10x + 12}{x^2(x+2)} = \frac{2(5x+6)}{x^2(x+2)}$$

The error was an incorrect expansion of $(x+2)^2$ in the numerator.

68. $\dfrac{(x-4)}{\left(\dfrac{x}{4} - \dfrac{4}{x}\right)} = \dfrac{\left(\dfrac{x-4}{1}\right)}{\left(\dfrac{x^2}{4x} - \dfrac{16}{4x}\right)} = \dfrac{\left(\dfrac{x-4}{1}\right)}{\left(\dfrac{x^2-16}{4x}\right)} = \dfrac{x-4}{1} \cdot \dfrac{4x}{x^2-16} = \dfrac{x-4}{1} \cdot \dfrac{4x}{(x+4)(x-4)} = \dfrac{4x}{x+4},\ x \neq 0, 4$

70. $\dfrac{\left(\dfrac{5}{y} - \dfrac{6}{2y+1}\right)}{\left(\dfrac{5}{y} + 4\right)} = \dfrac{\left(\dfrac{5(2y+1)}{y(2y+1)} - \dfrac{6y}{y(2y+1)}\right)}{\left(\dfrac{5}{y} + \dfrac{4y}{y}\right)} = \dfrac{\left(\dfrac{10y + 5 - 6y}{y(2y+1)}\right)}{\left(\dfrac{5+4y}{y}\right)} = \dfrac{4y+5}{y(2y+1)} \cdot \dfrac{y}{5+4y} = \dfrac{1}{2y+1},\ y \neq 0, -\dfrac{5}{4}$

72. $\dfrac{\left(\dfrac{x+4}{x+5} - \dfrac{x}{x+1}\right)}{4} = \dfrac{\left(\dfrac{(x+4)(x+1)}{(x+5)(x+1)} - \dfrac{x(x+5)}{(x+5)(x+1)}\right)}{\dfrac{4}{1}} = \left(\dfrac{(x+4)(x+1)}{(x+5)(x+1)} - \dfrac{x(x+5)}{(x+5)(x+1)}\right) \cdot \dfrac{1}{4}$

$$= \left(\frac{x^2 + 5x + 4 - x^2 - 5x}{(x+5)(x+1)}\right) \cdot \frac{1}{4} = \frac{4}{(x+5)(x+1)} \cdot \frac{1}{4} = \frac{1}{(x+5)(x+1)}$$

74. $\dfrac{\left(\dfrac{x+h}{x+h+1} - \dfrac{x}{x+1}\right)}{h} = \dfrac{\left(\dfrac{(x+h)(x+1)}{(x+h+1)(x+1)} - \dfrac{x(x+h+1)}{(x+h+1)(x+1)}\right)}{\dfrac{h}{1}}$

$$= \left(\frac{(x+h)(x+1)}{(x+h+1)(x+1)} - \frac{x(x+h+1)}{(x+h+1)(x+1)}\right) \cdot \frac{1}{h}$$

$$= \left(\frac{x^2 + x + hx + h - x^2 - xh - x}{(x+h+1)(x+1)}\right) \cdot \frac{1}{h}$$

$$= \frac{h}{(x+h+1)(x+1)} \cdot \frac{1}{h} = \frac{1}{(x+h+1)(x+1)},\quad h \neq 0$$

76. $\dfrac{3x^{1/3} - x^{-2/3}}{3x^{-2/3}} = \dfrac{3x^{1/3} - \dfrac{1}{x^{2/3}}}{\dfrac{3}{x^{2/3}}} = \left(3x^{1/3} - \dfrac{1}{x^{2/3}}\right) \cdot \dfrac{x^{2/3}}{3}$

$$= \left(\frac{3x^{1/3} \cdot x^{2/3}}{x^{2/3}} - \frac{1}{x^{2/3}}\right) \cdot \frac{x^{2/3}}{3} = \left(\frac{3x - 1}{x^{2/3}}\right) \cdot \frac{x^{2/3}}{3} = \frac{3x-1}{3}$$

78. $$\frac{-x^3(1-x^2)^{-1/2} - 2x(1-x^2)^{1/2}}{x^4} = \frac{\dfrac{-x^3}{(1-x^2)^{1/2}} - 2x(1-x^2)^{1/2}}{x^4}$$

$$= \frac{\dfrac{-x^3}{(1-x^2)^{1/2}} - \dfrac{2x(1-x^2)^{1/2}(1-x^2)^{1/2}}{(1-x^2)^{1/2}}}{x^4} = \frac{\dfrac{-x^3 - 2x(1-x^2)}{(1-x^2)^{1/2}}}{x^4}$$

$$= \frac{-x^3 - 2x + 2x^3}{(1-x^2)^{1/2}} \cdot \frac{1}{x^4} = \frac{x^3 - 2x}{(1-x^2)^{1/2}} \cdot \frac{1}{x^4}$$

$$= \frac{x(x^2-2)}{x^4(1-x^2)^{1/2}} = \frac{x^2-2}{x^3(1-x^2)^{1/2}}$$

80. $$\frac{(2x+1)^{1/3} - \dfrac{4x}{3(2x+1)^{2/3}}}{(2x+1)^{2/3}} = \frac{\dfrac{3(2x+1)^{1/3}(2x-1)^{2/3}}{3(2x+1)^{2/3}} - \dfrac{4x}{3(2x+1)^{2/3}}}{(2x+1)^{2/3}} = \frac{\dfrac{3(2x+1)-4x}{3(2x+1)^{2/3}}}{(2x+1)^{2/3}}$$

$$= \frac{3(2x+1)-4x}{3(2x+1)^{2/3}} \cdot \frac{1}{(2x+1)^{2/3}} = \frac{6x+3-4x}{3(2x+1)^{2/3}} \cdot \frac{1}{(2x+1)^{2/3}} = \frac{2x+3}{3(2x+1)^{4/3}}$$

82. $$\frac{\sqrt{z-3}-\sqrt{z}}{3} = \frac{\sqrt{z-3}-\sqrt{z}}{3} \cdot \frac{\sqrt{z-3}+\sqrt{z}}{\sqrt{z-3}+\sqrt{z}}$$

$$= \frac{(z-3)-z}{3(\sqrt{z-3}+\sqrt{z})} = \frac{-3}{3(\sqrt{z-3}+\sqrt{z})} = \frac{-1}{\sqrt{z-3}+\sqrt{z}}$$

84. $$\frac{t}{3} + \frac{t}{5} = \frac{5(t)}{5(3)} + \frac{3(t)}{3(5)} = \frac{5t}{15} + \frac{3t}{15} = \frac{8t}{15} = \frac{8}{15}t$$

86. Space in each part: $$\frac{\dfrac{3x}{4} - \dfrac{x}{3}}{4} = \frac{\dfrac{3(3x)}{3(4)} - \dfrac{4(x)}{4(3)}}{4} = \frac{\dfrac{9x}{12} - \dfrac{4x}{12}}{4} = \frac{\dfrac{5x}{12}}{4} = \frac{5x}{12} \cdot \frac{1}{4} = \frac{5x}{48}$$

First number: $$\frac{x}{3} + \frac{5x}{48} = \frac{16(x)}{16(3)} + \frac{5x}{48} = \frac{16x}{48} + \frac{5x}{48} = \frac{21x}{48} = \frac{7x}{16}$$

Second number: $$\frac{7x}{16} + \frac{5x}{48} = \frac{21x}{48} + \frac{5x}{48} = \frac{26x}{48} = \frac{13x}{24}$$

Third number: $$\frac{13x}{24} + \frac{5x}{48} = \frac{26x}{48} + \frac{5x}{48} = \frac{31x}{48}$$

The three numbers are $\dfrac{7x}{16}, \dfrac{13x}{24}$, and $\dfrac{31x}{48}$.

88. (a) $N = 5 \times 12 = 60$ payments, $M = \$400, P = \$18{,}000$

$$r = \frac{\left[\dfrac{24(NM-P)}{N}\right]}{\left(P + \dfrac{NM}{12}\right)} = \frac{\left[\dfrac{24(60 \cdot 400 - 18{,}000)}{60}\right]}{\left(18{,}000 + \dfrac{60 \cdot 400}{12}\right)} = 0.12 \text{ or } 12\%$$

(b) $$r = \frac{\left[\dfrac{24(NM-P)}{N}\right]}{\left(P + \dfrac{NM}{12}\right)} = \frac{\left[\dfrac{24(NM-P)}{N}\right]}{\left(\dfrac{12P+NM}{12}\right)} = \frac{24(NM-P)}{N} \cdot \frac{12}{12P+NM} = \frac{288(NM-P)}{N(12P+NM)}$$

$$r = \frac{288(60 \cdot 400 - 18{,}000)}{60(12 \cdot 18{,}000 + 60 \cdot 400)} = 0.12 \text{ or } 12\%$$

90. (a)

Year	1988	1989	1990	1991	1992
Gold–estimated	$434	$397	$375	$360	$349
Gold–actual	$438	$383	$385	$363	$345
Silver–estimated	$6.57	$5.46	$4.73	$4.23	$3.85
Silver–actual	$6.54	$5.50	$4.82	$4.04	$3.94

The estimates are quite close to the actual prices.

(b) Ratio of the price of gold to the price of silver:

$$\frac{\dfrac{5301t + 37{,}498}{19t + 100}}{\dfrac{237t + 4734}{176t + 1000}} = \frac{5301t + 37{,}498}{19t + 100} \cdot \frac{176t + 1000}{237t + 4734}$$

$$= \frac{932{,}976t^2 + 11{,}900{,}648t + 37{,}498{,}000}{3(19t + 100)(79t + 1578)} = \frac{8(116{,}622t^2 + 1{,}487{,}581t + 4{,}687{,}250)}{3(19t + 100)(79t + 1578)}$$

Year	1988	1989	1990	1991	1992
Ratio	65.99	72.83	79.21	85.08	90.48

Because the ratio over this period of time is increasing, the price of gold became more expensive relative to the price of silver.

92. $$\text{Probability} = \frac{\text{Shaded area}}{\text{(area of triangle)}} = \frac{\frac{1}{2} \cdot \frac{4}{x}(x + 2)(x + x + 4)}{\frac{1}{2}(x + 4)\left[(x + 2) + \frac{4}{x}(x + 2)\right]}$$

$$= \frac{\dfrac{4(x + 2)(2x + 4)}{x}}{(x + 4)(x + 2)\left(1 + \dfrac{4}{x}\right)} = \frac{\dfrac{4 \cdot 2(x + 2)^2}{x}}{(x + 4)(x + 2)\left(1 + \dfrac{4}{x}\right)}$$

$$= \frac{8(x + 2)^2}{x} \cdot \frac{1}{(x + 4)(x + 2)\left(1 + \dfrac{4}{x}\right)}$$

$$= \frac{8(x + 2)^2}{(x + 4)(x + 2)(x + 4)} = \frac{8(x + 2)}{(x + 4)^2}$$

Section P.5 Solving Equations

Solutions to Even-Numbered Exercises

2. $7 - 3x = 5x - 17$

(a) $x = -3$

$$7 - 3(-3) \stackrel{?}{=} 5(-3) - 17$$
$$7 + 9 \stackrel{?}{=} -15 - 17$$
$$16 \neq -32$$

No, $x = -3$ is *not* a solution.

(b) $x = 0$

$$7 - 3(0) \stackrel{?}{=} 5(0) - 17$$
$$7 \neq -17$$

No, $x = 0$ is *not* a solution.

(c) $x = 8$

$$7 - 3(8) \stackrel{?}{=} 5(8) - 17$$
$$7 - 24 \stackrel{?}{=} 40 - 17$$
$$-17 \neq 23$$

No, $x = 8$ is *not* a solution.

(d) $x = 3$

$$7 - 3(3) \stackrel{?}{=} 5(3) - 17$$
$$7 - 9 \stackrel{?}{=} 15 - 17$$
$$-2 = -2$$

Yes, $x = 3$ *is* a solution.

4. $5x^3 + 2x - 3 = 4x^3 + 2x - 11$

(a) $x = 2$

$5(2)^3 + 2(2) - 3 \stackrel{?}{=} 4(2)^3 + 2(2) - 11$

$40 + 4 - 3 \stackrel{?}{=} 32 + 4 - 11$

$41 \neq 25$

No, $x = 2$ is *not* a solution.

(b) $x = -2$

$5(-2)^3 + 2(-2) - 3 \stackrel{?}{=} 4(-2)^3 + 2(-2) - 11$

$-40 - 4 - 3 \stackrel{?}{=} -32 - 4 - 11$

$-47 = -47$

Yes, $x = -2$ *is* a solution.

(c) $x = 0$

$5(0)^3 + 2(0) - 3 \stackrel{?}{=} 4(0)^3 + 2(0) - 11$

$-3 \neq -11$

No, $x = 0$ is *not* a solution.

(d) $x = 10$

$5(10)^3 + 2(10) - 3 \stackrel{?}{=} 4(10)^3 + 2(10) - 11$

$5017 \neq 4009$

No, $x = 10$ is *not* a solution.

6. $\sqrt[3]{x - 8} = 3$

(a) $x = 2$

$\sqrt[3]{2 - 8} \stackrel{?}{=} 3$

$\sqrt[3]{-6} \neq 3$

No, $x = 2$ is *not* a solution.

(b) $x = -5$

$\sqrt[3]{-5 - 8} \stackrel{?}{=} 3$

$\sqrt[3]{-13} \neq 3$

No, $x = -5$ is *not* a solution.

(c) $x = 35$

$\sqrt[3]{35 - 8} \stackrel{?}{=} 3$

$\sqrt[3]{27} \stackrel{?}{=} 3$

$3 = 3$

Yes, $x = 35$ is a solution.

(d) $x = 8$

$\sqrt[3]{8 - 8} \stackrel{?}{=} 3$

$0 \neq 3$

No, $x = 8$ is *not* a solution.

8. $3(x + 2) = 5x + 4$ is a conditional equation because for $x = 0$, $3(0 + 2) = 3$ but $5(0) + 4 = 4$.

10. $3(x + 2) - 5 = 3x + 1$ is an identity because

$$\begin{aligned} 3(x + 2) - 5 &= 3x + 6 - 5 \\ &= 3x + 1 \end{aligned}$$

and is true for all x.

12. $3 + \dfrac{1}{x + 1} = \dfrac{4x}{x + 1}$ is a conditional equation because when $x = 0$, $3 + \dfrac{1}{0 + 1} = 4$ but $\dfrac{4(0)}{0 + 1} = 0$.

14.

$3(x - 4) + 10 = 7$	Given equation
$3x - 12 + 10 = 7$	Distributive Property
$3x - 2 = 7$	Associative Property of Addition
$3x - 2 + 2 = 7 + 2$	Addition Property of Equality
$3x = 9$	Additive Inverse Property
$\dfrac{3x}{3} = \dfrac{9}{3}$	Multiplicative Property of Equality
$x = 3$	Multiplicative Inverse Property

16. $2(13t - 15) + 3(t - 19) = 0$

$26t - 30 + 3t - 57 = 0$

$29t = 87$

$t = 3$

18. $\dfrac{x}{5} - \dfrac{x}{2} = 3$

$2x - 5x = 30$

$-3x = 30$

$x = -10$

20. $0.60x + 0.40(100 - x) = 50$

$6x + 4(100 - x) = 500$

$2x = 100$

$x = 50$

22. $3(x + 3) = 5(1 - x) - 1$

$3x + 9 = 4 - 5x$

$8x = -5$

$x = -\frac{5}{8}$

24. $\dfrac{17+y}{y}+\dfrac{32+y}{y}=100$

$17+y+32+y=100y$

$49=98y$

$y=\dfrac{1}{2}$

26. $\dfrac{10x+3}{5x+6}=\dfrac{1}{2}$

$2(10x+3)=5x+6$

$15x=0$

$x=0$

28. $\dfrac{15}{x}-4=\dfrac{6}{x}+3$

$\dfrac{9}{x}=7$

$x=\dfrac{9}{7}$

30. $\dfrac{1}{x-2}+\dfrac{3}{x+3}=\dfrac{4}{x^2+x-6}$

$x+3+3(x-2)=4$

$4x-3=4$

$4x=7$

$x=\dfrac{7}{4}$

32. $\dfrac{2}{(x-4)(x-2)}=\dfrac{1}{x-4}+\dfrac{2}{x-2}$

$2=x-2+2(x-4)$

$2=3x-10$

$x=4$, extraneous solution

No solution

34. $\dfrac{4}{u-1}+\dfrac{6}{3u+1}=\dfrac{15}{3u+1}$

$4(3u+1)+6(u-1)=15(u-1)$

$18u-2=15u-15$

$3u=-13$

$u=-\dfrac{13}{3}$

36. $(x+1)^2+2(x-2)=(x+1)(x-2)$

$x^2+2x+1+2x-4=x^2-x-2$

$5x=1$

$x=\frac{1}{5}$

38. $(2x+1)^2=4(x^2+x+1)$

$4x^2+4x+1=4x^2+4x+4$

$1=4$, not a true statement

No solution

40. $5+ax=12-bx$

$ax+bx=7$

$x=\dfrac{7}{a+b},\quad a+b\neq 0\Rightarrow a\neq -b$

42. $y=0.43t+30.86$

$30=0.43t+30.86$

$-0.86=0.43t$

$t=-2$

The number of married women in the work force reached 30 million in $1990-2=1988$.

44. $9x^2-1=0$

$(3x+1)(3x-1)=0$

$3x+1=0\Rightarrow x=-\frac{1}{3}$

$3x-1=0\Rightarrow x=\frac{1}{3}$

46. $x^2+10x+25=0$

$(x+5)(x+5)=0$

$x+5=0\Rightarrow x=-5$

48. $16x^2+56x+49=0$

$(4x+7)(4x+7)=0$

$4x+7=0\Rightarrow x=-\frac{7}{4}$

50. $(x+a)^2-b^2=0$

$[(x+a)+b][(x+a)-b]=0$

$(x+a)+b=0\Rightarrow x=-a-b$

$(x+a)-b=0\Rightarrow x=b-a$

52. $20x^3-125x=0$

$5x(4x^2-25)=0$

$5x(2x-5)(2x+5)=0$

$5x=0\Rightarrow x=0$

$2x-5=0\Rightarrow x=\frac{5}{2}$

$2x+5=0\Rightarrow x=-\frac{5}{2}$

54. $x^3-3x^2-x+3=0$

$x^2(x-3)-(x-3)=0$

$(x-3)(x+1)(x-1)=0$

$x-3=0\Rightarrow x=3$

$x+1=0\Rightarrow x=-1$

$x-1=0\Rightarrow x=1$

56. $x^3 + 2x^2 - 3x - 6 = 0$

$$x^2(x + 2) - 3(x + 2) = 0$$
$$(x^2 - 3)(x + 2) = 0$$
$$(x + \sqrt{3})(x - \sqrt{3})(x + 2) = 0$$
$$x + \sqrt{3} = 0 \Rightarrow x = -\sqrt{3}$$
$$x - \sqrt{3} = 0 \Rightarrow x = \sqrt{3}$$
$$x + 2 = 0 \Rightarrow x = -2$$

58. $x^2 = 144$

$$x = \pm\sqrt{144} = \pm 12, \pm 12.00$$

60. $9x^2 = 25$

$$x^2 = \frac{25}{9}$$
$$x = \pm\sqrt{\frac{25}{9}} = \pm\frac{5}{3} \approx \pm 1.67$$

62. $(x + 13)^2 = 21$

$$x + 13 = \pm\sqrt{21}$$
$$x = -13 \pm \sqrt{21} \approx -8.42, -17.58$$

64. $(x - 5)^2 = 20$

$$x - 5 = \pm\sqrt{20}$$
$$x = 5 \pm 2\sqrt{5} \approx 0.53, 9.47$$

66. $x^2 + 4x = 0$

$$x^2 + 4x + 4 = 4$$
$$(x + 2)^2 = 4$$
$$x + 2 = \pm 2$$
$$x = -2 \pm 2 = 0, -4$$

68. $x^2 + 8x + 14 = 0$

$$x^2 + 8x + 16 = 2$$
$$(x + 4)^2 = 2$$
$$x + 4 = \pm\sqrt{2}$$
$$x = -4 \pm \sqrt{2}$$

70. $4x^2 - 4x - 99 = 0$

$$x^2 - x - \frac{99}{4} = 0$$
$$x^2 - x + \frac{1}{4} = \frac{100}{4}$$
$$\left(x - \frac{1}{2}\right)^2 = \frac{100}{4}$$
$$x - \frac{1}{2} = \pm 5$$
$$x = \frac{1}{2} \pm 5 = \frac{11}{2}, -\frac{9}{2}$$

72. $2x^2 - x - 1 = 0$

$$a = 2, b = -1, c = -1$$
$$x = \frac{-(-1) \pm \sqrt{(-1)^2 - 4(2)(-1)}}{2(2)}$$
$$x = 1, -\frac{1}{2}$$

74. $4x^2 - 4x - 4 = 0$

$$x^2 - x - 1 = 0$$
$$a = 1, b = -1, c = -1$$
$$x = \frac{-(-1) \pm \sqrt{(-1)^2 - 4(1)(-1)}}{2(1)}$$
$$x = \frac{1}{2} \pm \frac{\sqrt{5}}{2}$$

76. $16x^2 + 22 = 40x$

$$8x^2 - 20x + 11 = 0$$
$$a = 8, b = -20, c = 11$$
$$x = \frac{-(-20) \pm \sqrt{(-20)^2 - 4(8)(11)}}{2(8)}$$
$$x = \frac{5}{4} \pm \frac{\sqrt{3}}{4}$$

78. $36x^2 + 24x - 7 = 0$

$$a = 36, b = 24, c = -7$$
$$x = \frac{-24 \pm \sqrt{24^2 - 4(36)(-7)}}{2(36)}$$
$$x = -\frac{1}{3} \pm \frac{\sqrt{11}}{6}$$

80. $9x^2 + 24x + 16 = 0$

$$a = 9, b = 24, c = 16$$
$$x = \frac{-24 \pm \sqrt{24^2 - 4(9)(16)}}{2(9)}$$
$$x = -\frac{4}{3}$$

82. $25h^2 + 80h + 61 = 0$

$$a = 25, b = 80, c = 61$$
$$x = \frac{-80 \pm \sqrt{80^2 - 4(25)(61)}}{2(25)}$$
$$x = -\frac{8}{5} \pm \frac{\sqrt{3}}{5}$$

84. $3(x + 4)^2 + (x + 4) - 2 = 0$

(a) Let $u = x + 4$.

$$3u^2 + u - 2 = 0$$
$$(3u - 2)(u + 1) = 0$$
$$u = \tfrac{2}{3}, -1$$
$$x + 4 = \tfrac{2}{3} \Rightarrow x = -\tfrac{10}{3}$$
$$x + 4 = -1 \Rightarrow x = -5$$

(b) $3(x^2 + 8x + 16) + x + 4 - 2 = 0$

$$3x^2 + 24x + 48 + x + 4 - 2 = 0$$
$$3x^2 + 25x + 50 = 0$$
$$(3x + 10)(x + 5) = 0$$
$$3x + 10 = 0 \Rightarrow x = -\tfrac{10}{3}$$
$$x + 5 = 0 \Rightarrow x = -5$$

(c) The method of part (a) reduces the number of algebraic steps.

86. (a)

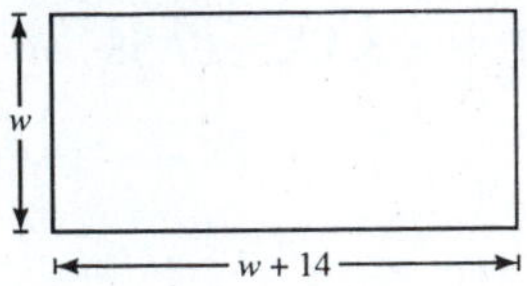

(b) $w(w + 14) = 1632$

$$w^2 + 14w - 1632 = 0$$

(c) $(w - 34)(w + 48) = 0$

$$w = 34, -48$$

$w = 34$ feet, $l = 34 + 14 = 48$ feet

88. $4x^4 - 65x^2 + 16 = 0$

$$(4x^2 - 1)(x^2 - 16) = 0$$
$$4x^2 - 1 = 0 \Rightarrow x = \pm\tfrac{1}{2}$$
$$x^2 - 16 = 0 \Rightarrow x = \pm 4$$

90. $6\left(\dfrac{s}{s + 1}\right)^2 + 5\left(\dfrac{s}{s + 1}\right) - 6 = 0$

Let $u = \dfrac{s}{s + 1}$.

$$6u^2 + 5u - 6 = 0$$
$$(3u - 2)(2u + 3) = 0$$
$$u = \frac{2}{3}, -\frac{3}{2}$$
$$\frac{s}{s + 1} = \frac{2}{3} \Rightarrow s = 2$$

92. $6x - 7\sqrt{x} - 3 = 0$

$$(3\sqrt{x} + 1)(2\sqrt{x} - 3) = 0$$

$3\sqrt{x} + 1 = 0$, no solution

$$2\sqrt{x} - 3 = 0 \Rightarrow x = \frac{9}{4}$$

94. $9t^{2/3} + 24t^{1/3} + 16 = 0$

$$(3t^{1/3} + 4)^2 = 0$$
$$t^{1/3} = -\tfrac{4}{3}$$
$$t = -\tfrac{64}{27}$$

96. $\sqrt{5 - x} - 3 = 0$

$$\sqrt{5 - x} = 3$$
$$5 - x = 9$$
$$-4 = x$$

98. $\sqrt[3]{3x + 1} - 5 = 0$

$$\sqrt[3]{3x + 1} = 5$$
$$3x + 1 = 125$$
$$3x = 124$$
$$x = \tfrac{124}{3}$$

100. $2x - \sqrt{15 - 4x} = 0$

$$2x = \sqrt{15 - 4x}$$
$$4x^2 = 15 - 4x$$
$$4x^2 + 4x - 15 = 0$$
$$(2x + 5)(2x - 3) = 0$$
$$2x + 5 = 0 \Rightarrow x = -\tfrac{5}{2}$$
$$2x - 3 = 0 \Rightarrow x = \tfrac{3}{2}$$

102. $\sqrt{x+5} = \sqrt{x-5}$

$x + 5 = x - 5$

$10 = 0$

No solution

104. $\sqrt{x} + \sqrt{x-20} = 10$

$\sqrt{x} - 10 = \sqrt{x-20}$

$x - 20\sqrt{x} + 100 = x - 20$

$20\sqrt{x} = 120$

$\sqrt{x} = 6$

$x = 36$

106. $3\sqrt{x} - \dfrac{4}{\sqrt{x}} = 4$

$3x - 4 - 4\sqrt{x} = 0$

$(3\sqrt{x} + 2)(\sqrt{x} - 2) = 0$

$3\sqrt{x} + 2 = 0 \Rightarrow$ extraneous

$\sqrt{x} - 2 = 0 \Rightarrow x = 4$

108. $(x+3)^{3/4} = 27$

$x + 3 = 27^{4/3}$

$x + 3 = 81$

$x = 78$

110. $34.70 = 40 - \sqrt{0.0001x + 1}$

$\sqrt{0.0001x + 1} = 5.3$

$0.0001x + 1 = 28.09$

$0.0001x = 27.09$

$x = 270{,}900$

The demand is 270,900 units at a price of $34.70.

112. $i = \pm\sqrt{\dfrac{1}{LC}}\sqrt{Q^2 - 9}$

$i^2 = \dfrac{Q^2 - q}{LC}$

$i^2LC = Q^2 - q$

$q + i^2LC = Q^2$

$\pm\sqrt{q + i^2LC} = Q$

114. To solve an equation involving radicals, first isolate the radical by subtracting x from both sides of the equation. Square both sides and solve the resulting equation.

116. $|x - 2| = 3$

$x - 2 = 3$ or $x - 2 = -3$

$x = 5$ $\quad$ $x = -1$

118. $|3x + 2| = 7$

$3x + 2 = 7$ or $3x + 2 = -7$

$3x = 5$ $\quad$ $3x = -9$

$x = \frac{5}{3}$ $\quad$ $x = -3$

120. $|x - 10| = x^2 - 10x$

$x - 10 = x^2 - 10x$

$0 = x^2 - 11x + 10$

$0 = (x-1)(x-10)$

$0 = x - 1 \Rightarrow x = 1,$ extraneous

$0 = x - 10 \Rightarrow x = 10$

or $-(x - 10) = x^2 - 10x$

$0 = x^2 - 9x - 10$

$0 = (x - 10)(x + 1)$

$x = 10, -1$

Solutions: $-1, 10$

122. Solutions: $0, 2, \frac{5}{2}$

$x(x-2)\left(x - \frac{5}{2}\right) = x\left(x^2 - \frac{9}{2}x + 5\right)$

$= x^3 - \frac{9}{2}x^2 + 5x$

124. $2x^2 = 200$

$x^2 = 100$

$x = 10$

Length: $x + 4 = 14$ centimeters

Width: $x + 4 = 14$ centimeters

Section P.6 Solving Inequalities

Solutions to Even-Numbered Exercises

2. $(4, 10]$

Inequality: $4 < x \le 10$

Bounded

4. $[-6, \infty)$

Inequality: $-6 \le x < \infty$

Unbounded

6. $x \ge 5$

Matches graph (h).

8. $0 \le x \le \frac{9}{2}$

Matches graph (e).

10. $|x| > 4$

$x < -4$ or $x > 4$

Matches graph (a).

12. $|x + 5| < 3$

$-3 < x + 5 < 3$

$-8 < x < -2$

Matches graph (d).

14. $x + 1 < \dfrac{2x}{3}$

(a) When $x = 0$:

$0 + 1 \overset{?}{<} \dfrac{2(0)}{3}$

$1 \not< 0$

No, $x = 0$ is *not* a solution.

(b) When $x = 4$:

$4 + 1 \overset{?}{<} \dfrac{2(4)}{3}$

$5 \not< \dfrac{8}{3}$

No, $x = 4$ is *not* a solution.

(c) When $x = -4$:

$-4 + 1 \overset{?}{<} \dfrac{2(-4)}{3}$

$-3 < -\dfrac{8}{3}$

Yes, $x = -4$ *is* a solution.

(d) When $x = -3$:

$-3 + 1 \overset{?}{<} \dfrac{2(-3)}{3}$

$-2 \not< -2$

No, $x = -3$ is *not* a solution.

16. $|2x - 3| < 15$

(a) When $x = -6$:

$|2(-6) - 3| \overset{?}{<} 15$

$15 \not< 15$

No, $x = -6$ is *not* a solution.

(b) When $x = 0$:

$|2(0) - 3| \overset{?}{<} 15$

$3 < 15$

Yes, $x = 0$ *is* a solution.

(c) When $x = 12$:

$|2(12) - 3| \overset{?}{<} 15$

$21 \not< 15$

No, $x = 12$ is *not* a solution.

(d) When $x = 7$:

$|2(7) - 3| \overset{?}{<} 15$

$11 < 15$

Yes, $x = 7$ *is* a solution.

18. $x^2 - x - 12 \ge 0$

(a) When $x = 5$:

$5^2 - 5 - 12 \overset{?}{\ge} 0$

$8 \ge 0$

Yes, $x = 5$ *is* a solution.

(b) When $x = 0$:

$0^2 - 0 - 12 \overset{?}{\ge} 0$

$-12 \not\ge 0$

No, $x = 0$ is *not* a solution.

(c) When $x = -4$:

$(-4)^2 - (-4) - 12 \overset{?}{\ge} 0$

$8 \ge 0$

Yes, $x = -4$ *is* a solution.

(d) When $x = -3$:

$(-3)^2 - (-3) - 12 \overset{?}{\ge} 0$

$0 \ge 0$

Yes, $x = -3$ *is* a solution.

20. $\dfrac{3x^2}{x^2+4} < 1$

(a) When $x = -2$:

$\dfrac{3(-2)}{(-2)^2+4} \overset{?}{<} 1$

$\dfrac{3}{2} \not< 1$

No, $x = -2$ is *not* a solution.

(b) When $x = -1$:

$\dfrac{3(-1)}{(-1)^2+4} \overset{?}{<} 1$

$\dfrac{3}{5} < 1$

Yes, $x = -1$ *is* a solution.

(c) When $x = 0$:

$\dfrac{3(0)}{(0)^2+4} \overset{?}{<} 1$

$0 < 1$

Yes, $x = 0$ *is* a solution.

(d) When $x = 3$:

$\dfrac{3(3)}{3^2+4} \overset{?}{<} 1$

$\dfrac{27}{13} \not< 1$

No, $x = 3$ *is not* a solution.

22. $2x > 3$

$x > \frac{3}{2}$

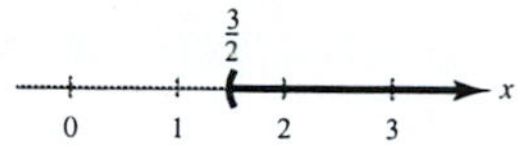

24. $-6x > 15$

$x < -\frac{15}{6}$

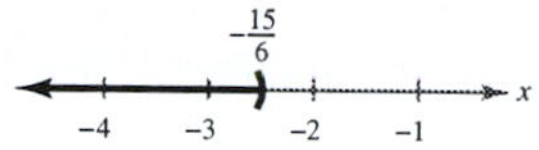

26. $x + 7 \le 12$

$x \le 5$

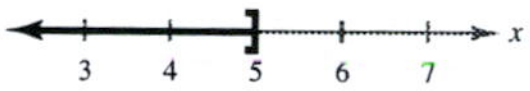

28. $2x + 7 < 3$

$2x < -4$

$x < -2$

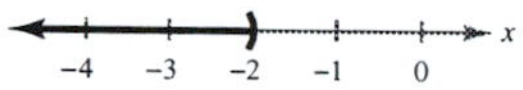

30. $6x - 4 \le 2$

$6x \le 6$

$x \le 1$

32. $-8 \le 1 - 3(x-2) < 13$

$-9 \le -3(x-2) < 12$

$3 \ge x - 2 > -4$

$5 \ge x > -2$

$-2 < x \le 5$

34. $0 \le \dfrac{x+3}{2} < 5$

$0 \le x + 3 < 10$

$-3 \le x < 7$

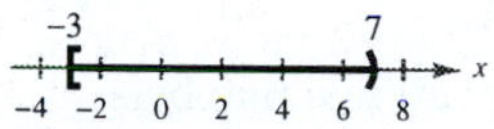

36. $-1 < -\dfrac{x}{3} < 1$

$3 > x > -3$

$-3 < x < 3$

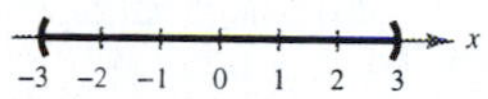

38. $|2x| < 6$

$-6 < 2x < 6$

$-3 < x < 3$

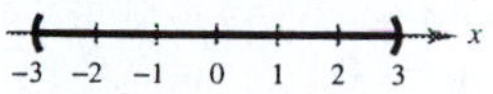

40. $|5x| > 10$

$5x > 10$ or $5x < -10$

$x > 2$ $\quad$ $x < -2$

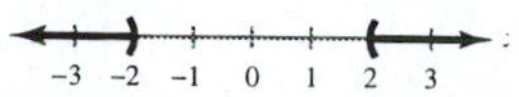

42. $|x - 7| < 6$

$-6 < x - 7 < 6$

$1 < x < 13$

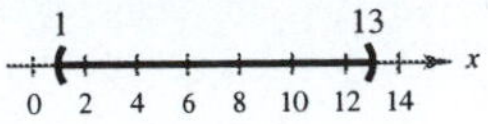

44. $|x + 14| + 3 > 17$

$|x + 14| > 14$

$x + 14 > 14$ or $x + 14 < -14$

$x > 0$ $\quad$ $x < -28$

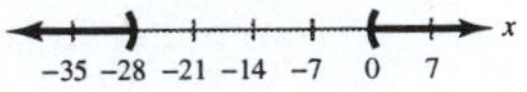

46. $|1 - 2x| < 5$

$-5 < 1 - 2x < 5$

$-6 < -2x < 4$

$3 > x > -2$

$-2 < x < 3$

48. $\left|1 - \frac{2x}{3}\right| < 1$

$-1 < 1 - \frac{2x}{3} < 1$

$-2 < -\frac{2x}{3} < 0$

$-6 < -2x < 0$

$3 > x > 0$

$0 < x < 3$

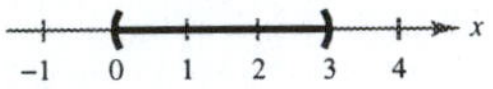

50. $3|4 - 5x| \le 9$

$|4 - 5x| \le 3$

$-3 \le 4 - 5x \le 3$

$-7 \le -5x \le -1$

$\frac{7}{5} \ge x \ge \frac{1}{5}$

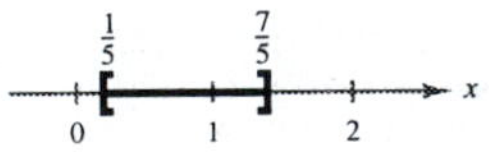

52. $|x - 5| \ge 0$

$x - 5 \ge 0$ or $x - 5 \le 0$

All real numbers x

54. 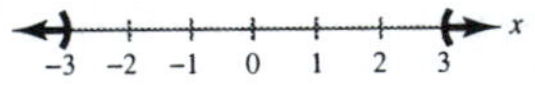

$x < -3$ or $x > 3$

$|x| > 3$

56. 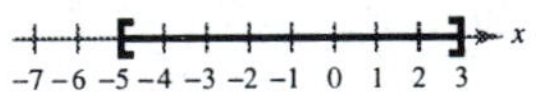

$-5 \le x \le 3$

$-5 + 1 \le x + 1 \le 3 + 1$

$-4 \le x + 1 \le 4$

$|x + 1| \le 4$

58. $|x - 8| \ge 5$

60. $|x + 6| \le 7$

62. $|x - 8| > 4$

Description: all real numbers that are more than four units from 8.

64. $x^2 < 5$

$x^2 - 5 < 0$

$(x + \sqrt{5})(x - \sqrt{5}) = 0$

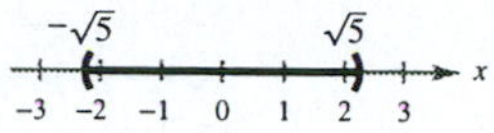

Critical Numbers: $x = \pm\sqrt{5}$

Test Intervals: $(-\infty, -\sqrt{5}) \Rightarrow (x + \sqrt{5})(x - \sqrt{5}) > 0$

$(-\sqrt{5}, \sqrt{5}) \Rightarrow (x + \sqrt{5})(x - \sqrt{5}) < 0$

$(\sqrt{5}, \infty) \Rightarrow (x + \sqrt{5})(x - \sqrt{5}) > 0$

Solution Interval: $(-\sqrt{5}, \sqrt{5})$

66. $(x - 3)^2 \geq 1$

$x^2 - 6x + 8 \geq 0$

$(x - 2)(x - 4) \geq 0$

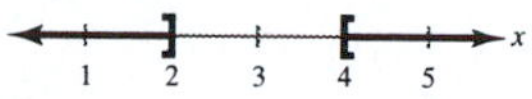

Critical Numbers: $x = 2, x = 4$

Test Intervals: $(-\infty, 2) \Rightarrow (x - 2)(x - 4) > 0$

$(2, 4) \Rightarrow (x - 2)(x - 4) < 0$

$(4, \infty) \Rightarrow (x - 2)(x - 4) > 0$

Solution Intervals: $(-\infty, 2] \cup [4, \infty)$

68. $(x + 6)^2 \leq 8$

$x^2 + 12x + 28 \leq 0$

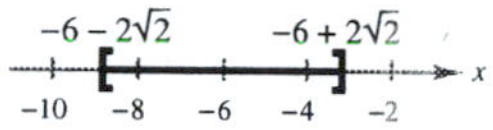

Zeros: $x = \dfrac{-12 \pm \sqrt{12^2 - 4(1)(28)}}{2(1)} = -6 \pm 2\sqrt{2}$

Critical Numbers: $-6 - 2\sqrt{2}, -6 + 2\sqrt{2}$

Test Intervals: $(-\infty, -6 - 2\sqrt{2}) \Rightarrow x^2 + 12x + 28 > 0$

$(-6 - 2\sqrt{2}, -6 + 2\sqrt{2}) \Rightarrow x^2 + 12x + 28 < 0$

$(-6 + 2\sqrt{2}, \infty) \Rightarrow x^2 + 12x + 28 > 0$

Solution Interval: $[-6 - 2\sqrt{2}, -6 + 2\sqrt{2}]$

70. $x^2 - 6x + 9 < 16$

$x^2 - 6x - 7 < 0$

$(x + 1)(x - 7) < 0$

Critical Numbers: $x = -1, x = 7$

Test Intervals: $(-\infty, -1) \Rightarrow (x + 1)(x - 7) > 0$

$(-1, 7) \Rightarrow (x + 1)(x - 7) < 0$

$(7, \infty) \Rightarrow (x + 1)(x - 7) > 0$

Solution Interval: $(-1, 7)$

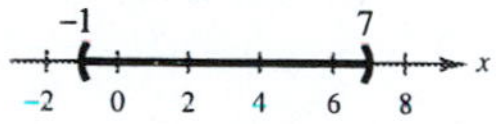

72. $6(x + 2)(x - 1) < 0$

Critical Numbers: $x = -2, x = 1$

Test Intervals: $(-\infty, -2) \Rightarrow 6(x + 2)(x - 1) > 0$

$(-2, 1) \Rightarrow 6(x + 2)(x - 1) < 0$

$(1, \infty) \Rightarrow 6(x + 2)(x - 1) > 0$

Solution Interval: $(-2, 1)$

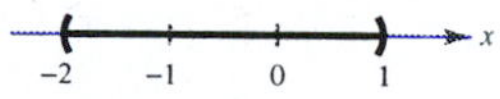

74. $x^2 - 4x - 1 > 0$

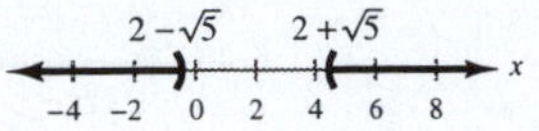

$x = \dfrac{4 \pm \sqrt{16 + 4}}{2} = 2 \pm \sqrt{5}$

Critical Numbers: $x = 2 - \sqrt{5}, x = 2 + \sqrt{5}$

Test Intervals: $(-\infty, 2 - \sqrt{5}) \Rightarrow x^2 - 4x - 1 > 0$

$(2 - \sqrt{5}, 2 + \sqrt{5}) \Rightarrow x^2 - 4x - 1 < 0$

$(2 + \sqrt{5}, \infty) \Rightarrow x^2 - 4x - 1 > 0$

Solution Intervals: $(-\infty, 2 - \sqrt{5}) \cup (2 + \sqrt{5}, \infty)$

76. $4x^3 - 12x^2 > 0$

$4x^2(x - 3) > 0$

Critical Numbers: $x = 0, x = 3$

Test Intervals: $(-\infty, 0) \Rightarrow 4x^2(x - 3) < 0$

$(0, 3) \Rightarrow 4x^2(x - 3) < 0$

$(3, \infty) \Rightarrow 4x^2(x - 3) > 0$

Solution Interval: $(3, \infty)$

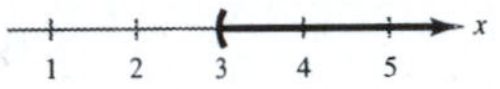

78. $x^4(x - 3) \le 0$

Critical Numbers: $x = 0, x = 3$

Test Intervals: $(-\infty, 0) \Rightarrow x^4(x - 3) < 0$

$(0, 3) \Rightarrow x^4(x - 3) < 0$

$(3, \infty) \Rightarrow x^4(x - 3) > 0$

Solution Intervals: $(-\infty, 0] \cup [0, 3]$ or $(-\infty, 3]$

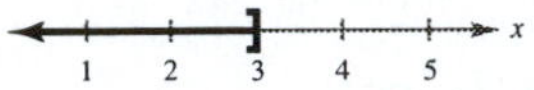

80. $\dfrac{1}{x} - 4 < 0$

$\dfrac{1 - 4x}{x} < 0$

Critical Numbers: $x = 0, x = \dfrac{1}{4}$

Test Intervals: $(-\infty, 0) \Rightarrow \dfrac{1 - 4x}{x} < 0$

$\left(0, \dfrac{1}{4}\right) \Rightarrow \dfrac{1 - 4x}{x} > 0$

$\left(\dfrac{1}{4}, \infty\right) \Rightarrow \dfrac{1 - 4x}{x} < 0$

Solution Interval: $(-\infty, 0) \cup \left(\dfrac{1}{4}, \infty\right)$

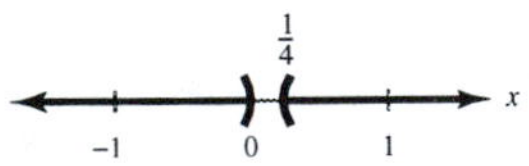

82. $\dfrac{x + 12}{x + 2} - 3 \ge 0$

$\dfrac{x + 12 - 3(x + 2)}{x + 2} \ge 0$

$\dfrac{6 - 2x}{x + 2} \ge 0$

Critical Numbers: $x = -2, x = 3$

Test Intervals: $(-\infty, -2) \Rightarrow \dfrac{6 - 2x}{x + 2} < 0$

$(-2, 3) \Rightarrow \dfrac{6 - 2x}{x + 2} > 0$

$(3, \infty) \Rightarrow \dfrac{6 - 2x}{x + 2} < 0$

Solution Intervals: $(-2, 3]$

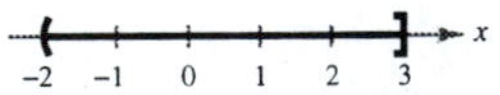

84. $$\frac{5}{x-6} > \frac{3}{x+2}$$

$$\frac{5(x+2)-3(x-6)}{(x-6)(x+2)} > 0$$

$$\frac{2x+28}{(x-6)(x+2)} > 0$$

Critical Numbers: $x = -14, x = -2, x = 6$

Test Intervals: $(-\infty, -14) \Rightarrow \frac{2x+28}{(x-6)(x+2)} < 0$

$(-14, -2) \Rightarrow \frac{2x+28}{(x-6)(x+2)} > 0$

$(-2, 6) \Rightarrow \frac{2x+28}{(x-6)(x+2)} < 0$

$(6, \infty) \Rightarrow \frac{2x+28}{(x-6)(x+2)} > 0$

Solution Intervals: $(-14, -2) \cup (6, \infty)$

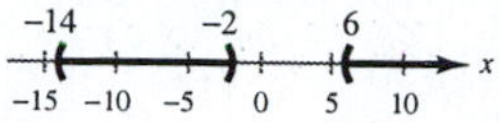

86. $6x + 15 \geq 0$

$6x \geq -15$

$x \geq -\frac{5}{2}$

Domain: $\left[-\frac{5}{2}, \infty\right)$

88. $x^2 - 4 \geq 0$

$(x+2)(x-2) \geq 0$

Critical Numbers: $x = -2, x = 2$

Test Intervals: $(-\infty, -2) \Rightarrow (x+2)(x-2) > 0$

$(-2, 2) \Rightarrow (x+2)(x-2) < 0$

$(2, \infty) \Rightarrow (x+2)(x-2) > 0$

Domain: $(-\infty, -2] \cup [2, \infty)$

90. $144 - 9x^2 \geq 0$

$9(4-x)(4+x) \geq 0$

Critical Numbers: $x = -4, x = 4$

Test Intervals: $(-\infty, -4) \Rightarrow 9(4-x)(4+x) < 0$

$(-4, 4) \Rightarrow 9(4-x)(4+x) > 0$

$(4, \infty) \Rightarrow 9(4-x)(4+x) < 0$

Domain: $[-4, 4]$

92. Company A: \$250 per week

Company B: $150 + 0.25x$

$150 + 0.25x > 250$

$0.25x > 100$

$x > 400$

You must drive more than 400 miles for the fee of Company B to be greater than the fee of Company A.

94. $0.32m + 2300 < 10{,}000$

$0.32m < 7700$

$m < 24{,}062.5$

The number of miles that yields an operating cost less than \$10,000 is 24,062.5 miles.

96. $S = 15.812 + 1.472t$

$15.812 + 1.472t > 40$

$1.472t > 24.188$

$t > 16.432$

The average salary will exceed \$40,000 when $t \geq 17$.

98. $|x - 10.4| \leq \frac{1}{16}$

$-\frac{1}{16} \leq x - 10.4 \leq \frac{1}{16}$

$10.3375 \leq x \leq 10.4625$

$(10.3375)^2 \leq x^2 \leq (10.4625)^2$

$106.864 \leq x^2 \leq 109.464$

$106.864 \leq \text{area} \leq 109.464$

100.

$$2L + 2W = 100 \Rightarrow W = 50 - L$$
$$LW \geq 500$$
$$L(50 - L) \geq 500$$
$$-L^2 + 50L - 500 \geq 0$$
$$\left[L - \left(25 - 5\sqrt{5}\right)\right]\left[L + \left(25 + 5\sqrt{5}\right)\right] \leq 0$$

13.8 meters $\approx 25 - 5\sqrt{5} \leq L \leq 25 + 5\sqrt{5} \approx$ 36.2 meters

(Use the Quadratic Formula to find the critical numbers.)

102.

$$\frac{1}{R} = \frac{1}{R_1} + \frac{1}{2}$$
$$2R_1 = 2R + RR_1$$
$$2R_1 = R(2 + R_1)$$
$$\frac{2R_1}{2 + R_1} = R$$

Since $R \geq 1$, we have

$$\frac{2R_1}{2 + R_1} \geq 1$$
$$\frac{2R_1}{2 + R_1} - 1 \geq 0$$
$$\frac{R_1 - 2}{2 + R_1} \geq 0.$$

Since $R_1 > 0$, the only critical number is $R_1 = 2$. The inequality is satisfied when $R_1 \geq 2$ ohms.

Section P.7 Errors and the Algebra of Calculus

Solutions to Even-Numbered Exercises

2. $\dfrac{4}{16x - (2x + 1)} = \dfrac{4}{14x - 1}$

Distribute the minus sign.

4. $\dfrac{x - 1}{(5 - x)(-x)} = \dfrac{x - 1}{-(x)(5 - x)} = \dfrac{-(x - 1)}{x(5 - x)} = \dfrac{1 - x}{x(5 - x)}$

Therefore, there is no error.

6. $x(yz) = (xy)z$

Use the Associative Property of Multiplication.

8. $(5z)(6z) = (5 \cdot 6)(z \cdot z) = 30z^2$

Add exponents when multiplying powers with like bases.

10. $\left(\dfrac{x}{y}\right)^3 = \dfrac{x}{y} \cdot \dfrac{x}{y} \cdot \dfrac{x}{y} = \dfrac{x^3}{y^3}$

Cube the numerator and denominator.

12. $\sqrt{25 - x^2}$ does not simplify. Do not apply the radical to the terms.

14. $\dfrac{2x^2+1}{5x}$ does not simplify. Cancel common factors, not common terms.

16. $\dfrac{1}{a^{-1}+b^{-1}} = \dfrac{1}{(1/a)+(1/b)} \cdot \dfrac{ab}{ab} = \dfrac{ab}{b+a}$

The exponents apply to each term in the denominator, not the entire fraction.

18. $(x^2+5x)^{1/2} = [x(x+5)]^{1/2} = x^{1/2}(x+5)^{1/2}$

Apply exponent to each factor.

20. $(3x^2-6x)^3 = [3x(x-2)]^3 = 27x^3(x-2)^3$

Apply exponent to each factor.

22. $\dfrac{7+5(x+3)}{x+3} = \dfrac{7+5x+15}{x+3} = \dfrac{5x+22}{x+3}$

Cancel common factors, not common terms.

24. $\dfrac{2x+3x^2}{4x} = \dfrac{x(2+3x)}{4x} = \dfrac{2+3x}{4}$

Factor before canceling.

26. $\dfrac{7x^2}{10} = \dfrac{7}{10}(x^2)$

The required factor is x^2.

28. $\dfrac{3}{4}x + \dfrac{1}{2} = \dfrac{3}{4}x + \dfrac{2}{4} = \dfrac{1}{4}(3x+2)$

The required factor is $3x+2$.

30. $\frac{5}{2}z^2 - \frac{1}{4}z + 2 = \left(\frac{1}{4}\right)\left(4 \cdot \frac{5}{2}z^2 - 4 \cdot \frac{1}{4}z + 4 \cdot 2\right)$

$= \left(\frac{1}{4}\right)(10z^2 - z + 8)$

The required factor is $\frac{1}{4}$.

32. $x^2(x^3-1)^4 = \frac{1}{3}(x^3-1)^4(3x^2)$

The required factor is $\frac{1}{3}$.

34. $5x\sqrt[3]{1+x^2} = \left(\frac{5}{2}\right)\sqrt[3]{1+x^2}(2x)$

The required factor is $\frac{5}{2}$.

36. $\dfrac{4x+6}{(x^2+3x+7)^3} = \dfrac{2(2x+3)}{(x^2+3x+7)^3} = \dfrac{2}{1} \cdot \dfrac{(2x+3)}{1} \cdot \dfrac{1}{(x^2+3x+7)^3} = (2)\dfrac{1}{(x^2+3x+7)^3}(2x+3)$

The required factor is 2.

38. $\dfrac{1}{(x-1)\sqrt{(x-1)^4-4}} = \dfrac{(x-1)}{(x-1)(x-1)\sqrt{(x-1)^4-4}} = \dfrac{(x-1)}{(x-1)^2\sqrt{(x-1)^4-4}}$

The required factor is $(x-1)$.

40. $\dfrac{(x-1)^2}{169} + (y+5)^2 = \dfrac{(x-1)(x-1)^2}{(x-1)(169)} + (y+5)^2 = \dfrac{(x-1)^3}{169(x-1)} + (y+5)^2$

The required factor is $(x-1)$.

42. $\dfrac{3x^2}{4} - \dfrac{9y^2}{16} = \dfrac{\left(\frac{1}{3}\right)3x^2}{\left(\frac{1}{3}\right)4} - \dfrac{\left(\frac{1}{9}\right)9y^2}{\left(\frac{1}{9}\right)16} = \dfrac{x^2}{\frac{4}{3}} - \dfrac{y^2}{\frac{16}{9}}$

The required factors are $\frac{4}{3}$ and $\frac{16}{9}$.

44. $\dfrac{x^2}{4/9} + \dfrac{y^2}{7/8} = x^2\left(\dfrac{9}{4}\right) + y^2\left(\dfrac{8}{7}\right) = \dfrac{9x^2}{4} + \dfrac{8y^2}{7}$

The required factors are 4 and 7.

46. $x^{1/3} - 5x^{4/3} = x^{1/3}(1-5x)$

The required factor is $(1-5x)$.

48. $(1-3x)^{4/3} - 4x(1-3x)^{1/3} = (1-3x)^{1/3}[(1-3x)^{3/3} - 4x]$

$= (1-3x)^{1/3}(1-3x-4x)$

$= (1-3x)^{1/3}(1-7x)$

The required factor is $(1-7x)$.

50. $\dfrac{1}{2\sqrt{x}} + 5x^{3/2} - 10x^{5/2} = \dfrac{1}{2\sqrt{x}} + \dfrac{5x^{3/2}(2\sqrt{x})}{2\sqrt{x}} - \dfrac{10x^{5/2}(2\sqrt{x})}{2\sqrt{x}}$

$= \dfrac{1}{2\sqrt{x}}\left(1 + 5x^{3/2}\sqrt{x} - 10x^{5/2}\sqrt{x}\right)$

$= \dfrac{1}{2\sqrt{x}}(1 + 10x^2 - 20x^3)$

The required factor is $(1 + 10x^2 - 20x^3)$.

52. $\frac{3}{7}(t+1)^{7/3} - \frac{3}{4}(t+1)^{4/3} = \frac{12}{28}(t+1)^{4/3}(t+1)^{3/3} - \frac{21}{28}(t+1)^{4/3}$

$$= \frac{3(t+1)^{4/3}}{28}[4(t+1) - 7]$$

$$= \frac{3(t+1)^{4/3}}{28}(4t - 3)$$

The required factor is $(4t - 3)$.

54. $\frac{x^3 - 5x^2 + 4}{x^2} = \frac{x^3}{x^2} - \frac{5x^2}{x^2} + \frac{4}{x^2} = x - 5 + 4x^{-2}$

56. $\frac{2x^5 - 3x^3 + 5x - 1}{x^{3/2}} = \frac{2x^5}{x^{3/2}} - \frac{3x^3}{x^{3/2}} + \frac{5x}{x^{3/2}} - \frac{1}{x^{3/2}}$

$$= 2x^{5-3/2} - 3x^{3-3/2} + 5x^{1-3/2} - x^{-3/2}$$

$$= 2x^{7/2} - 3x^{3/2} + 5x^{-1/2} - x^{-3/2}$$

58. $\frac{x^3 - 5x^4}{3x^2} = \frac{x^3}{3x^2} - \frac{5x^4}{3x^2} = \frac{x}{3} = \frac{5x^2}{3}$

60. $\frac{x^5(-3)(x^2+1)^{-4}(2x) - (x^2+1)^{-3}(5)x^4}{(x^5)^2} = \frac{x^4(x^2+1)^{-3}[-6x^2(x^2+1)^{-1} - 5]}{x^{10}}$

$$= \frac{-6x^2(x^2+1)^{-1} - 5}{(x^2+1)^3x^6} = \frac{\frac{-6x^2}{x^2+1} - 5}{x^6(x^2+1)^3} = \frac{\frac{-6x^2}{x^2+1} - \frac{5(x^2+1)}{x^2+1}}{x^6(x^2+1)^3}$$

$$= \frac{(x^2+1)^{-1}(-6x^2 - 5x^2 - 5)}{x^6(x^2+1)^3} = \frac{-11x^2 - 5}{x^6(x^2+1)^4}$$

62. $\frac{(4x^2+9)^{1/2} - (2x+3)(\frac{1}{2})(4x^2+9)^{-1/2}(8x)}{[(4x^2+9)^{1/2}]^2} = \frac{2(4x^2+9)^{-1/2}[(4x^2+9) - 2x(2x+3)]}{(4x^2+9)}$

$$= \frac{2(4x^2 + 9 - 4x^2 - 6x)}{(4x^2+9)^{3/2}} = \frac{2(9 - 6x)}{(4x^2+9)^{3/2}} = \frac{-6(2x - 3)}{(4x^2+9)^{3/2}}$$

64. $\frac{\sqrt{2x-1} - \frac{x+2}{\sqrt{2x-1}}}{2x-1} = \frac{\sqrt{2x-1} - \frac{x+2}{\sqrt{2x-1}}}{2x-1} \cdot \frac{\sqrt{2x-1}}{\sqrt{2x-1}}$

$$= \frac{(2x-1) - (x+2)}{(2x-1)\sqrt{2x-1}} = \frac{x-3}{(2x-1)^{3/2}}$$

66. $\frac{(x+1)(\frac{1}{2})(2x-3x^2)^{-1/2}(2-6x) - (2x-3x^2)^{1/2}}{(x+1)^2} = \frac{(x+1)(2x-3x^2)^{-1/2}(1-3x) - (2x-3x^2)^{1/2}}{(x+1)^2}$

$$= \frac{(2x-3x^2)^{-1/2}[(x+1)(1-3x) - (2x-3x^2)]}{(x+1)^2}$$

$$= \frac{x - 3x^2 + 1 - 3x - 2x + 3x^2}{(2x-3x^2)^{1/2}(x+1)^2}$$

$$= \frac{-6(2x-3)}{(4x^2+9)^{3/2}}$$

68. (a) $y_1 = \dfrac{-\sqrt{9-x^2}}{x^2} - \dfrac{1}{\sqrt{9-x^2}} = \dfrac{-\sqrt{9-x^2}\left(\sqrt{9-x^2}\right)}{x^2\sqrt{9-x^2}} - \dfrac{x^2}{x^2\sqrt{9-x^2}}$

$= \dfrac{-(9-x^2)-x^2}{x^2\sqrt{9-x^2}} = \dfrac{-9+x^2-x^2}{x^2\sqrt{9-x^2}} = \dfrac{-9}{x^2\sqrt{9-x^2}} = y_2$

(b)

x	-2	-1	$-\frac{1}{2}$	$\frac{1}{4}$	1	2	$\frac{5}{2}$
y_1	-1.01	-3.18	-12.17	-48.17	-3.18	-1.01	-0.87
y_2	-1.01	-3.18	-12.17	-48.17	-3.18	-1.01	-0.87

Section P.8 Graphical Representation of Data

Solutions to Even-Numbered Exercises

2.

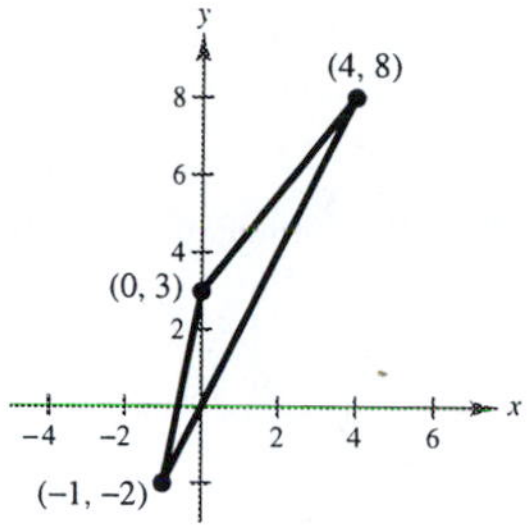

4.

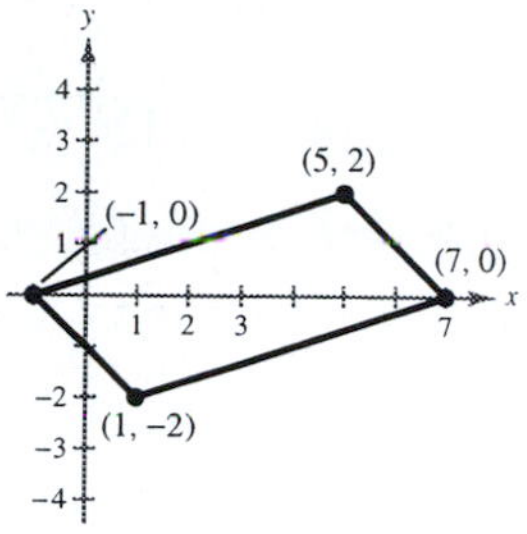

6. A: $\left(\frac{3}{2}, -4\right)$; B: $(0, -2)$;
C: $\left(-3, \frac{5}{2}\right)$, D: $(-6, 0)$

8. $(4, -8)$

10. $(-12, 0)$

12. No, it is not true that the scales on the x- and y-axis must be the same. The scales depend on the magnitude of the coordinates.

14. $x < 0$ and $y < 0$ in Quadrant III.

16. $x > 2$ and $y = 3$ in Quadrant I.

18. $x > 4$ in Quadrants I and IV.

20. If $(-x, y)$ is in Quadrant IV, then (x, y) must be in Quadrant III.

22. If $xy < 0$, then x and y have opposite signs. This happens in Quadrants II and IV.

24. $(-3 + 6, 6 - 3) = (3, 3)$

$(-5 + 6, 3 - 3) = (1, 0)$

$(-3 + 6, 0 - 3) = (3, -3)$

$(-1 + 6, 3 - 3) = (5, 0)$

26.

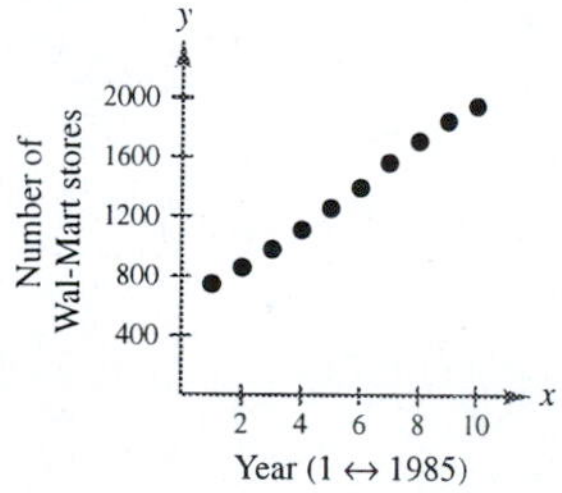

28. $y = 2 - \frac{1}{2}x^2$

x	-2	-1	$-\frac{1}{2}$	0	$\frac{1}{2}$	1	2
y	0	$\frac{3}{2}$	$\frac{15}{8}$	2	$\frac{15}{8}$	$\frac{3}{2}$	0

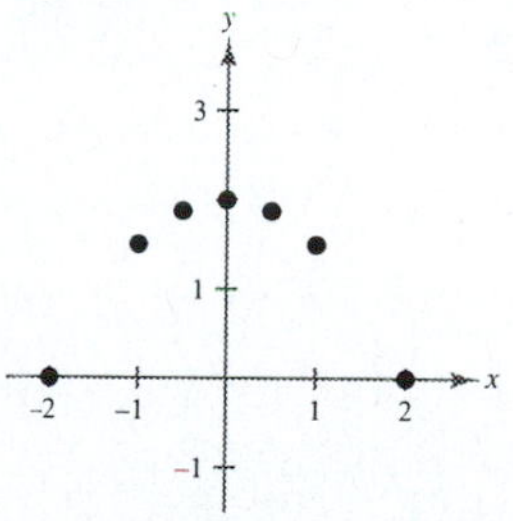

30. Approximate highest price = \$13.70
Approximate price paid in 1993 = \$12.80

$$\text{Percent drop} = \frac{13.70 - 12.80}{13.70} \approx 0.07 \text{ or } 7\%$$

32. (a) Cost during Super Bowl XV (1981) = \$275,000
Cost during Super Bowl V (1971) = \$75,000
Increase = \$275,000 − \$75,000 = \$200,000

(b) Cost during Super Bowl XXV (1991) = \$800,000
Increase = \$800,000 − \$275,000 = \$525,000

34. Minimum wage in 1990 = \$3.80
Minimum wage in 1994 = \$4.25

$$\text{Percent increase} = \frac{\$4.25 - 3.80}{3.80} \approx 0.118 \text{ or } 11.8\%$$

36. No, there are many variables that will affect the final exam score.

38. $d = |1 - 8| = |-7| = 7$

40. $d = |-4 - 6| = |-10| = 10$

42. (a) (1, 0), (13, 5)
$\text{Distance} = \sqrt{(13-1)^2 + (5-0)^2} = \sqrt{12^2 + 5^2} = \sqrt{169} = 13$
(13, 5), (13, 0)
$\text{Distance} = |5 - 0| = |5| = 5$
(1, 0), (13, 0)
$\text{Distance} = |1 - 13| = |-12| = 12$

(b) $5^2 + 12^2 = 25 + 144 = 169 = 13^2$

44. (a) (1, 5), (5, −2)
$\text{Distance} = \sqrt{(1-5)^2 + (5-(-2))^2} = \sqrt{(-4)^2 + (7)^2} = \sqrt{16 + 49} = \sqrt{65}$
(1, 5), (1, −2)
$\text{Distance} = |5 - (-2)| = |5 + 2| = |7| = 7$
(1, −2), (5, −2)
$\text{Distance} = |1 - 5| = |-4| = 4$

(b) $4^2 + 7^2 = 16 + 49 = 65 = (\sqrt{65})^2$

46. (a)

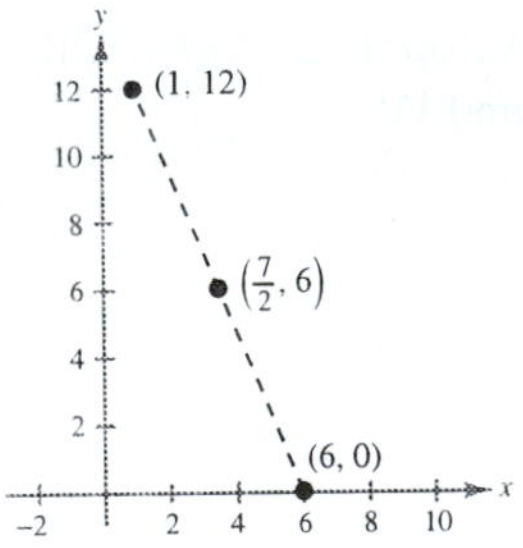

(b) $d = \sqrt{(1-6)^2 + (12-0)^2}$
$= \sqrt{25 + 144} = 13$

(c) $\left(\frac{1+6}{2}, \frac{12+0}{2}\right) = \left(\frac{7}{2}, 6\right)$

48. (a)

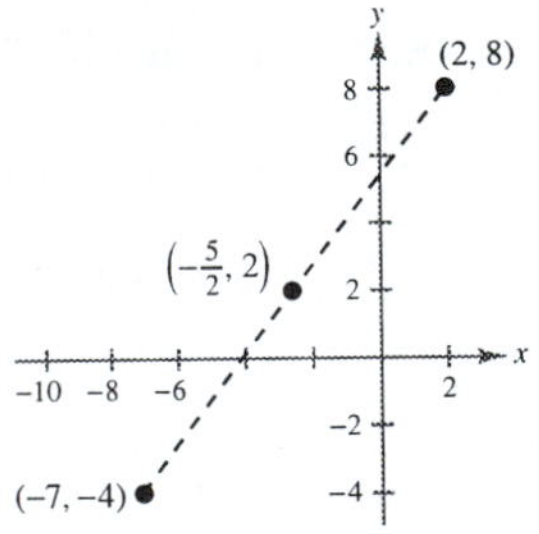

(b) $d = \sqrt{(-7-2)^2 + (-4-8)^2}$
$= \sqrt{81 + 144} = 15$

(c) $\left(\frac{-7+2}{2}, \frac{-4+8}{2}\right) = \left(-\frac{5}{2}, 2\right)$

50. (a)

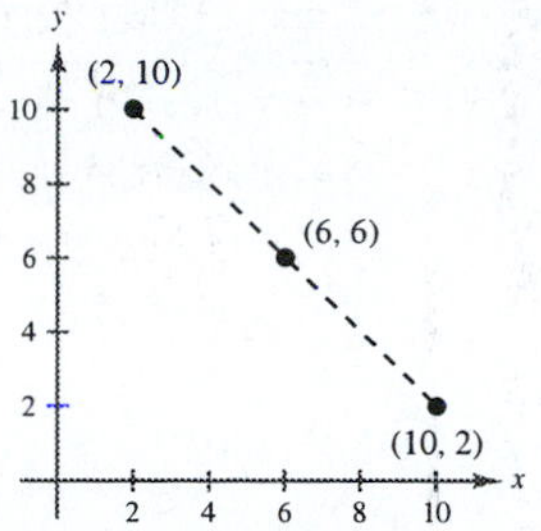

(b) $d = \sqrt{(2-10)^2 + (10-2)^2}$
$= \sqrt{64 + 64} = 8\sqrt{2}$

(c) $\left(\dfrac{2+10}{2}, \dfrac{10+2}{2}\right) = (6, 6)$

52. (a)

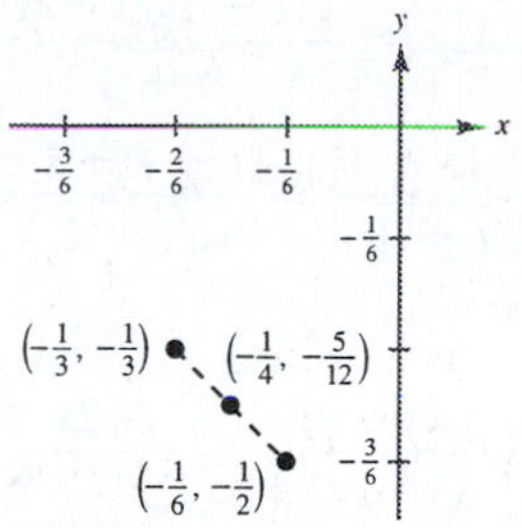

(b) $d = \sqrt{\left(-\dfrac{1}{3} + \dfrac{1}{6}\right)^2 + \left(-\dfrac{1}{3} + \dfrac{1}{2}\right)^2}$
$= \sqrt{\dfrac{1}{36} + \dfrac{1}{36}} = \dfrac{\sqrt{2}}{6}$

54. (a)

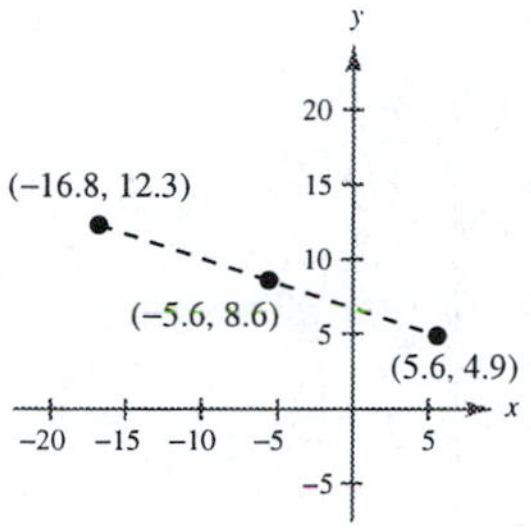

(b) $d = \sqrt{(-16.8 - 5.6)^2 + (12.3 - 4.9)^2}$
$= \sqrt{501.76 + 54.76} = \sqrt{556.52}$

(c) $\left(\dfrac{-16.8 + 5.6}{2}, \dfrac{12.3 + 4.9}{2}\right) = (-5.6, 8.6)$

56. (a)

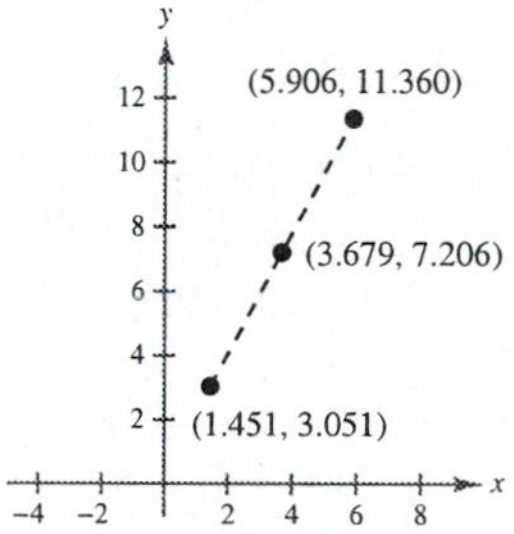

(b) $d = \sqrt{(1.451 - 5.906)^2 + (3.051 - 11.360)^2}$
$\approx \sqrt{88.887}$

(c) $\left(\dfrac{1.451 + 5.906}{2}, \dfrac{3.051 + 11.360}{2}\right) \approx (3.679, 7.206)$

58. $\left(\dfrac{1991 + 1995}{2}, \dfrac{\$4{,}200{,}000 + \$5{,}650{,}000}{2}\right) = (1993, \$4{,}925{,}000)$

60. $d_1 = \sqrt{(1-3)^2 + (-3-2)^2} = \sqrt{4 + 25} = \sqrt{29}$
$d_2 = \sqrt{(3+2)^2 + (2-4)^2} = \sqrt{25 + 4} = \sqrt{29}$
$d_3 = \sqrt{(1+2)^2 + (-3-4)^2} = \sqrt{9 + 49} = \sqrt{58}$
$d_1 = d_2$

62. $d_1 = \sqrt{(4-0)^2 + (0-6)^2} = \sqrt{16 + 36} = \sqrt{52} = 2\sqrt{13}$
$d_2 = \sqrt{(0+4)^2 + (6-0)^2} = \sqrt{16 + 36} = \sqrt{52} = 2\sqrt{13}$
$d_3 = \sqrt{(-4-0)^2 + (0+6)^2} = \sqrt{16 + 36} = \sqrt{52} = 2\sqrt{13}$
$d_4 = \sqrt{(4-0)^2 + (0+6)^2} = \sqrt{16 + 36} = \sqrt{52} = 2\sqrt{13}$
$d_1 = d_2 = d_3 = d_4$

64. $d_1 = \sqrt{(0-3)^2 + (1-7)^2} = \sqrt{9 + 36} = \sqrt{45} = 3\sqrt{5}$
$d_2 = \sqrt{(3-4)^2 + (7-4)^2} = \sqrt{1 + 9} = \sqrt{10}$
$d_3 = \sqrt{(4-1)^2 + (4+2)^2} = \sqrt{9 + 36} = \sqrt{45} = 3\sqrt{5}$
$d_4 = \sqrt{(0-1)^2 + (1+2)^2} = \sqrt{1 + 9} = \sqrt{10}$
Opposite sides have equal lengths of $3\sqrt{5}$ and $\sqrt{10}$.

62. $\dfrac{t}{\sqrt{t+1}} - \sqrt{t+1} = \dfrac{t}{\sqrt{t+1}} - \dfrac{\sqrt{t+1}\sqrt{t+1}}{\sqrt{t+1}}$

$= \dfrac{t - (t+1)}{\sqrt{t+1}} = \dfrac{t - t - 1}{\sqrt{t+1}} = \dfrac{-1}{\sqrt{t+1}}$

The missing factor is -1.

64. $\dfrac{4x-6}{(x-1)^2} \div \dfrac{2x^2-3x}{x^2+2x-3} = \dfrac{4x-6}{(x-1)^2} \cdot \dfrac{x^2+2x-3}{2x^2-3x}$

$= \dfrac{2(2x-3)}{(x-1)^2} \cdot \dfrac{(x+3)(x-1)}{x(2x-3)} = \dfrac{2(x+3)}{x(x-1)}$

66. $\dfrac{1}{x} - \dfrac{x-1}{x^2+1} = \dfrac{x^2+1}{x(x^2+1)} - \dfrac{(x-1)x}{x(x^2+1)} = \dfrac{x^2+1-x^2+x}{x(x^2+1)} = \dfrac{x+1}{x(x^2+1)}$

68. $\dfrac{1}{L}\left(\dfrac{1}{y} - \dfrac{1}{L-y}\right) = \dfrac{1}{L}\left(\dfrac{L-y}{y(L-y)} - \dfrac{y}{y(L-y)}\right)$

$= \dfrac{1}{L}\left(\dfrac{L-y-y}{y(L-y)}\right) = \dfrac{1}{L}\left(\dfrac{L-2y}{y(L-y)}\right) = \dfrac{L-2y}{Ly(L-y)}$

70. $\dfrac{\left(\dfrac{1}{2x-3} - \dfrac{1}{2x+3}\right)}{\left(\dfrac{1}{2x} - \dfrac{1}{2x+3}\right)} = \dfrac{\left(\dfrac{2x+3}{(2x-3)(2x+3)} - \dfrac{2x-3}{(2x-3)(2x+3)}\right)}{\left(\dfrac{2x+3}{2x(2x+3)} - \dfrac{2x}{2x(2x+3)}\right)}$

$= \dfrac{\dfrac{2x+3-2x+3}{(2x-3)(2x+3)}}{\dfrac{2x+3-2x}{2x(2x+3)}} = \dfrac{6}{(2x-3)(2x+3)} \cdot \dfrac{2x(2x+3)}{3} = \dfrac{4x}{2x-3}$

72. $4x + 2(7-x) = 5$

$4x + 14 - 2x = 5$

$2x = -9$

$x = -\frac{9}{2}$

74. $\frac{1}{2}(x-3) - 2(x+1) = 5$

$\frac{1}{2}x - \frac{3}{2} - 2x - 2 = 5$

$-\frac{3}{2}x = \frac{17}{2}$

$x = -\frac{17}{3}$

76. $\dfrac{1}{x-2} = 3$

$1 = 3(x-2)$

$1 = 3x - 6$

$7 = 3x$

$\dfrac{7}{3} = x$

78. $15 + x - 2x^2 = 0$

$(5+2x)(3-x) = 0$

$5 + 2x = 0 \Rightarrow x = -\dfrac{5}{2}$

$3 - x = 0 \Rightarrow x = 3$

80. $16x^2 = 25$

$x^2 = \dfrac{25}{16}$

$x = \pm\sqrt{\dfrac{25}{16}} = \pm\dfrac{5}{4}$

82. $x^2 + 6x - 3 = 0$

$a = 1,\ b = 6,\ c = -3$

$x = \dfrac{-6 \pm \sqrt{6^2 - 4(1)(-3)}}{2(1)}$

$= \dfrac{-6 \pm \sqrt{48}}{2} = -3 \pm 2\sqrt{3}$

66. (a) $(x_2, y_2) = (2x_m - x_1, 2y_m - y_1)$
$= (2 \cdot 4 - 1, 2(-1) - (-2)) = (7, 0)$

(b) $(x_2, y_2) = (2x_m - x_1, 2y_m - y_1)$
$= (2 \cdot 2 - (-5), 2 \cdot 4 - 11) = (9, -3)$

68. (a) $\left(\frac{3x_1 + x_2}{4}, \frac{3y_1 + y_2}{4}\right) = \left(\frac{3 \cdot 1 + 4}{4}, \frac{3(-2) - 1}{4}\right) = \left(\frac{7}{4}, -\frac{7}{4}\right)$

$\left(\frac{x_1 + x_2}{2}, \frac{y_1 + y_2}{2}\right) = \left(\frac{1 + 4}{2}, \frac{-2 - 1}{2}\right) = \left(\frac{5}{2}, -\frac{3}{2}\right)$

$\left(\frac{x_1 + 3x_2}{4}, \frac{y_1 + 3y_2}{4}\right) = \left(\frac{1 + 3 \cdot 4}{4}, \frac{-2 + 3(-1)}{4}\right) = \left(\frac{13}{4}, -\frac{5}{4}\right)$

(b) $\left(\frac{3x_1 + x_2}{4}, \frac{3y_1 + y_2}{4}\right) = \left(\frac{3(-2) + 0}{4}, \frac{3(-3) + 0}{4}\right) = \left(-\frac{3}{2}, -\frac{9}{4}\right)$

$\left(\frac{x_1 + x_2}{2}, \frac{y_1 + y_2}{2}\right) = \left(\frac{-2 + 0}{2}, \frac{-3 + 0}{2}\right) = \left(-1, -\frac{3}{2}\right)$

$\left(\frac{x_1 + 3x_2}{4}, \frac{y_1 + 3y_2}{4}\right) = \left(\frac{-2 + 0}{4}, \frac{-3 + 0}{4}\right) = \left(-\frac{1}{2}, -\frac{3}{4}\right)$

70. Let (0, 0) represent the coordinates of the point of departure and let (100, 150) represent the coordinates of the destination.

Distance $= \sqrt{(0 - 100)^2 + (0 - 150)^2}$
$= \sqrt{10{,}000 + 22{,}500}$
$= \sqrt{325{,}000}$
$= 50\sqrt{13} \approx 180.28$ miles

72. The midpoint of the diagonal connecting $(0, 0)$ and $(a + b, c)$ is $\big((a + b)/2, c/2\big)$. The midpoint of the diagonal connecting $(a, 0)$ and (b, c) is $\big((a + b)/2, c/2\big)$.
Thus, the diagonals bisect each other.

❑ Focus on Concepts

1.

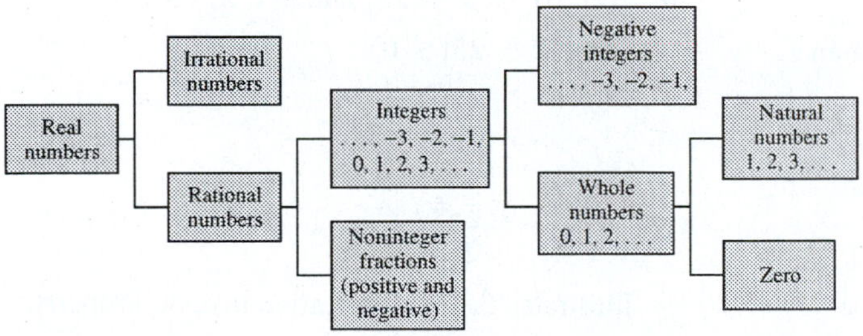

2. (a) Negative (b) Positive
(c) Negative (d) Positive

3. If $a < 0$, then $|a| = -a$. For example, $a = -7$ in $|-7|$. Therefore, $|-7| = -(-7) = 7$.

4. (a) The base for the exponent -1 is $3x$. Therefore,

$$(3x)^{-1} = \frac{1}{3x}.$$

(b) When multiplying, add exponents. $y^3 \cdot y^2 = y^5$.

(c) Multiply the exponents to obtain $(a^2b^3)^4 = a^8b^{12}$.

(d) The square of a binomial contains a cross-product term. $(a + b)^2 = a^2 + 2ab + b^2$

(e) If $x < 0$, then $\sqrt{4x^2} > 0$ but $2x < 0$. $\sqrt{4x^2} = 2|x|$.

(f) Radicals cannot be combined unless the index and the radicand are the same.

5. No. A number written in scientific notation has the form $c \times 10^n$, where $1 \le c < 10$ and n is an integer. In true scientific notation, the number 52.7×10^5 is 5.27×10^6.

6. (a) Yes. $(x^4 + 2x - 2) + (x^3 + 2) = x^4 + 3x$

(b) No. When third- and fourth-degree polynomials are added, the fourth-degree term of the fourth-degree polynomial will be in the sum.

(c) No. the sum will be of fourth degree. The *product* of the the two polynomials would be of seventh degree.

7. The polynomial is written as a product.

8. Factor the numerator and the denominator and cancel all common factors.

9. b

10. c

11. d

12. a

❑ Review Exercises for Chapter P

Solutions to Even-Numbered Exercises

2. $\{\sqrt{15}, -22, -\frac{10}{3}, 0, 5.2, \frac{3}{7}\}$

(a) Natural numbers: none

(b) Integers: $-22, 0$

(c) Rational numbers: $-22, -\frac{10}{3}, 0, 5.2, \frac{3}{7}$

(d) Irrational numbers: $\sqrt{15}$

4. (a) $\frac{9}{25} = 0.36$ (b) $\frac{5}{7} = 0.\overline{714285}$

6. $x > 1$

The set consists of all real numbers greater than 1.

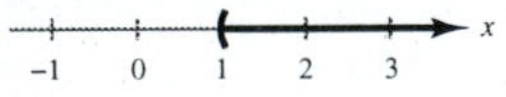

8. $d(x, 25) = |x - 25|$ and $d(x, 25) \le 10$

Thus $|x - 25| \le 10$.

10. $\left|y - \frac{1}{2}\right| > 2$

12. $\frac{2}{y + 4} \cdot \frac{y + 4}{2} = 1, \quad y \ne -4$

Illustrates the Multiplicative Inverse Property.

14. $(t + 4)(2t) = (2t)(t + 4)$ illustrates the Commutative Property of Multiplication.

16. (a) $\frac{(8y)^0}{y^2} = \frac{1}{y^2}$

(b) $\frac{40(b - 3)^5}{75(b - 3)^2} = \frac{40}{75}(b - 3)^{5-2}$

$$= \frac{8}{15}(b - 3)^3$$

18. (a) $(x + y^{-1})^{-1} = \left(x + \dfrac{1}{y}\right)^{-1}$

$= \left(\dfrac{xy + 1}{y}\right)^{-1}$

$= \dfrac{y}{xy + 1}$

(b) $\left(\dfrac{x^{-3}}{y}\right)\left(\dfrac{x}{y}\right)^{-1} = \left(\dfrac{1}{x^3 y}\right)\left(\dfrac{y}{x}\right)$

$= \dfrac{1}{x^4}$

$= x^{-4}$

20. $0.3048 = 3.048 \times 10^{-1}$

22. $2.74 \times 10^{-3} = 0.00274$

24. (a) $50{,}000\left(1 + \dfrac{0.075}{12}\right)^{48} = 67{,}429.958$

(b) $\dfrac{28{,}000{,}000 + 34{,}000{,}000}{87{,}000{,}000} = 0.713$

26. Radical form: $\sqrt[4]{16} = 2$, Answer

Rational exponent form: $16^{1/4} = 2$

28. (a) $\sqrt[3]{\dfrac{2x^3}{27}} = \sqrt[3]{2\left(\dfrac{x}{3}\right)^3} = \dfrac{x}{3}\sqrt[3]{2}$

(b) $\sqrt[5]{64x^6} = \sqrt{2 \cdot 2^5 \cdot x \cdot x^5} = 2x\sqrt{2x}$

30. $\dfrac{1}{\sqrt{x} - 1} = \dfrac{1}{\sqrt{x} - 1} \cdot \dfrac{\sqrt{x} + 1}{\sqrt{x} + 1} = \dfrac{\sqrt{x} + 1}{x - 1}$

32. $\sqrt{8x^3} + \sqrt{2x} = \sqrt{(2x)^2 2x} + \sqrt{2x} = 2x\sqrt{2x} + \sqrt{2x} = (2x + 1)\sqrt{2x}$

34.

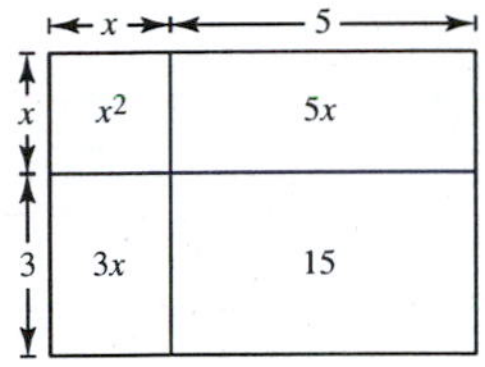

Using $A = l \cdot w$ we have $(x + 5)(x + 3)$.
We could also add up the areas of the four inner rectangles, $x^2 + 3x + 5x + 15$. Notice that $(x + 5)(x + 3) = x^2 + 3x + 5x + 15$ by the Distributive Property (FOIL).

36. $\left(\frac{1}{3}x\right)\left(\frac{1}{3}y\right) = \frac{1}{9}xy$

The denominators must be multiplied.

38. $\frac{2}{9} \times \frac{4}{9} = \frac{8}{81}$

The denominators must be multiplied.

40. $(2x)^4 = 2^4x^4 = 16x^4$

The exponent must be applied to the entire quantity inside parentheses.

42. $(3^4)^4 = 3^{4 \cdot 4} = 3^{16}$

To raise a power to a power, the exponents must be multiplied.

44. $(5 + 8)^2 = 5^2 + 2(5)(8) + 8^2 = 25 + 80 + 64 = 169$

The middle term was omitted.

46. $\sqrt{7x}\,\sqrt[3]{2} = (7x)^{1/2}(2)^{1/3} = (7x)^{3/6}(2)^{2/6}$

$= \sqrt[6]{(7x)^3}\;\sqrt[6]{(2)^2} = \sqrt[6]{343x^3}\,\sqrt[6]{4}$

$= \sqrt[6]{1372x^3}$

The indices must be the same to multiply the radicands.

48. $8y - [2y^2 - (3y - 8)] = 8y - 2y^2 + (3y - 8)$

$= -2y^2 + 11y - 8$

50. $(3\sqrt{5} + 2)(3\sqrt{5} - 2) = (3\sqrt{5})^2 - 2^2$

$= 45 - 4 = 41$

52. $\left(x - \dfrac{1}{x}\right)(x + 2) = x^2 + 2x - 1 - \dfrac{2}{x}$

54. $x(x - 3) + 4(x - 3) = (x - 3)(x + 4)$

56. $3x^2 + 14x + 8 = (3x + 2)(x + 4)$

58. $x^3 - 1 = (x - 1)(x^2 + x + 1)$

60. $R = 1600x - 0.50x^2$

$= x(1600 - 0.50x)$

$p = 1600 - 0.50x$

62. $$\frac{t}{\sqrt{t+1}} - \sqrt{t+1} = \frac{t}{\sqrt{t+1}} - \frac{\sqrt{t+1}\sqrt{t+1}}{\sqrt{t+1}}$$
$$= \frac{t-(t+1)}{\sqrt{t+1}} = \frac{t-t-1}{\sqrt{t+1}} = \frac{-1}{\sqrt{t+1}}$$

The missing factor is -1.

64. $$\frac{4x-6}{(x-1)^2} \div \frac{2x^2-3x}{x^2+2x-3} = \frac{4x-6}{(x-1)^2} \cdot \frac{x^2+2x-3}{2x^2-3x}$$
$$= \frac{2(2x-3)}{(x-1)^2} \cdot \frac{(x+3)(x-1)}{x(2x-3)} = \frac{2(x+3)}{x(x-1)}$$

66. $$\frac{1}{x} - \frac{x-1}{x^2+1} = \frac{x^2+1}{x(x^2+1)} - \frac{(x-1)x}{x(x^2+1)} = \frac{x^2+1-x^2+x}{x(x^2+1)} = \frac{x+1}{x(x^2+1)}$$

68. $$\frac{1}{L}\left(\frac{1}{y} - \frac{1}{L-y}\right) = \frac{1}{L}\left(\frac{L-y}{y(L-y)} - \frac{y}{y(L-y)}\right)$$
$$= \frac{1}{L}\left(\frac{L-y-y}{y(L-y)}\right) = \frac{1}{L}\left(\frac{L-2y}{y(L-y)}\right) = \frac{L-2y}{Ly(L-y)}$$

70. $$\frac{\left(\frac{1}{2x-3} - \frac{1}{2x+3}\right)}{\left(\frac{1}{2x} - \frac{1}{2x+3}\right)} = \frac{\left(\frac{2x+3}{(2x-3)(2x+3)} - \frac{2x-3}{(2x-3)(2x+3)}\right)}{\left(\frac{2x+3}{2x(2x+3)} - \frac{2x}{2x(2x+3)}\right)}$$
$$= \frac{\frac{2x+3-2x+3}{(2x-3)(2x+3)}}{\frac{2x+3-2x}{2x(2x+3)}} = \frac{6}{(2x-3)(2x+3)} \cdot \frac{2x(2x+3)}{3} = \frac{4x}{2x-3}$$

72. $4x + 2(7-x) = 5$

$4x + 14 - 2x = 5$

$2x = -9$

$x = -\frac{9}{2}$

74. $\frac{1}{2}(x-3) - 2(x+1) = 5$

$\frac{1}{2}x - \frac{3}{2} - 2x - 2 = 5$

$-\frac{3}{2}x = \frac{17}{2}$

$x = -\frac{17}{3}$

76. $\frac{1}{x-2} = 3$

$1 = 3(x-2)$

$1 = 3x - 6$

$7 = 3x$

$\frac{7}{3} = x$

78. $15 + x - 2x^2 = 0$

$(5+2x)(3-x) = 0$

$5 + 2x = 0 \Rightarrow x = -\frac{5}{2}$

$3 - x = 0 \Rightarrow x = 3$

80. $16x^2 = 25$

$x^2 = \frac{25}{16}$

$x = \pm\sqrt{\frac{25}{16}} = \pm\frac{5}{4}$

82. $x^2 + 6x - 3 = 0$

$a = 1,\ b = 6,\ c = -3$

$$x = \frac{-6 \pm \sqrt{6^2 - 4(1)(-3)}}{2(1)}$$
$$= \frac{-6 \pm \sqrt{48}}{2} = -3 \pm 2\sqrt{3}$$

84. $4x^3 - 6x^2 = 0$

$x^2(4x - 6) = 0$

$x^2 = 0 \Rightarrow x = 0$

$4x - 6 = 0 \Rightarrow x = \frac{3}{2}$

86. $\frac{1}{(t+1)^2} = 1$

$1 = (t + 1)^2$

$0 = t^2 + 2t$

$0 = t(t + 2)$

$0 = t \Rightarrow t = 0$

$0 = t + 2 \Rightarrow t = -2$

88. $\sqrt{x - 2} - 8 = 0$

$\sqrt{x - 2} = 8$

$x - 2 = 64$

$x = 66$

90. $\sqrt{3x - 2} = 4 - x$

$3x - 2 = (4 - x)^2$

$3x - 2 = 16 - 8x + x^2$

$0 = 18 - 11x + x^2$

$0 = (x - 9)(x - 2)$

$0 = x - 9 \Rightarrow x = 9$, extraneous

$0 = x - 2 \Rightarrow x = 2$

92. $5\sqrt{x} - \sqrt{x - 1} = 6$

$5\sqrt{x} = 6 + \sqrt{x - 1}$

$25x = 36 + 12\sqrt{x - 1} + x - 1$

$24x - 35 = 12\sqrt{x - 1}$

$576x^2 - 1680x + 1225 = 144(x - 1)$

$576x^2 - 1824x + 1369 = 0$

$$x = \frac{-(-1824) \pm \sqrt{(-1864)^2 - 4(576)(1369)}}{2(576)}$$

$$= \frac{1824 \pm \sqrt{172{,}800}}{1152} = \frac{1824 \pm 240\sqrt{3}}{1152}$$

$$x = \frac{38 + 5\sqrt{3}}{24}$$

$$x = \frac{38 - 5\sqrt{3}}{25}, \text{ extraneous}$$

94. $(x + 2)^{3/4} = 27$

$x + 2 = 27^{3/4}$

$x + 2 = 81$

$x = 79$

96. $8x^2(x^2 - 4)^{1/3} + (x^2 - 4)^{4/3} = 0$

$(x^2 - 4)^{1/3}[8x^2 + x^2 - 4] = 0$

$(x^2 - 4)^{1/3}(9x^2 - 4) = 0$

$(x - 2)^{1/3}(x + 2)^{1/3}(3x - 2)(3x + 2) = 0$

$x - 2 = 0 \Rightarrow x = 2$

$x + 2 = 0 \Rightarrow x = -2$

$3x - 2 = 0 \Rightarrow x = \frac{2}{3}$

$3x + 2 = 0 \Rightarrow x = -\frac{2}{3}$

98. $|2x + 3| = 7$

$2x + 3 = 7$ or $2x + 3 = -7$

$2x = 4$ or $2x = -10$

$x = 2$ $\quad x = -5$

100. $|x^2 - 6| = x$

$x^2 - 6 = x$ or $-(x^2 - 6) = x$

$x^2 - x - 6 = 0$ $\quad x^2 + x - 6 = 0$

$(x - 3)(x + 2) = 0$ $\quad (x + 3)(x + 2) = 0$

$x - 3 = 0 \Longrightarrow x = 3$ $\quad x - 2 = 0 \Longrightarrow x = 2$

$x + 2 = 0 \Longrightarrow x = -2$, extraneous $\quad x + 3 = 0 \Longrightarrow x = -3$, extraneous

102. $Z = \sqrt{R^2 - X^2}$

$Z^2 = R^2 - X^2$

$X^2 = R^2 - Z^2$

$X^2 = \pm\sqrt{R^2 - Z^2}$

104.

$$E = 2kw\left(\frac{v}{2}\right)^2$$

$$\frac{E}{2kw} = \left(\frac{v}{2}\right)^2$$

$$\pm\sqrt{\frac{E}{2kw}} = \frac{v}{2}$$

$$\pm 2\sqrt{\frac{E}{2kw}} = v$$

$$\pm\sqrt{\frac{4E}{2kw}} = v$$

$$\pm\sqrt{\frac{2E}{kw}} = v$$

106. $\frac{1}{2}(3 - x) > \frac{1}{3}(2 - 3x)$

$3(3 - x) > 2(2 - 3x)$

$9 - 3x > 4 - 6x$

$3x > -5$

$x > -\frac{5}{3}, \quad \left(-\frac{5}{3}, \infty\right)$

108.

$$\frac{2}{x + 1} \leq \frac{3}{x - 1}$$

$$\frac{2(x - 1) - 3(x + 1)}{(x + 1)(x - 1)} \leq 0$$

$$\frac{2x - 2 - 3x - 3}{(x + 1)(x - 1)} \leq 0$$

$$\frac{-(x + 5)}{(x + 1)(x - 1)} \leq 0$$

Test intervals: $(-\infty, -5) \Longrightarrow \frac{-(x + 5)}{(x + 1)(x - 1)} > 0$

$(-5, -1) \Longrightarrow \frac{-(x + 5)}{(x + 1)(x - 1)} < 0$

$(-1, 1) \Longrightarrow \frac{-(x + 5)}{(x + 1)(x - 1)} > 0$

$(1, -\infty) \Longrightarrow \frac{-(x + 5)}{(x + 1)(x - 1)} < 0$

Solution intervals: $[-5, -1) \cup (1, \infty)$

110. $|x| \leq 4$

$-4 \leq x \leq 4$

$[-4, 4]$

112. $|x - 3| > 4$

$x - 3 > 4$ or $x - 3 < -4$

$x > 7$ $\quad x < -1$

$(-\infty, -1) \cup (7, \infty)$

114. $2x^2 + x \geq 15$

$2x^2 + x - 15 \geq 0$

$(2x - 5)(x + 3) \geq 0$

Test intervals: $(-\infty, -3) \Longrightarrow (2x - 5)(x + 3) > 0$

$\left(-3, \frac{5}{2}\right) \Longrightarrow (2x - 5)(x + 3) < 0$

$\left(\frac{5}{2}, \infty\right) \Longrightarrow (2x - 5)(x + 3) > 0$

Solution interval: $(-\infty, -3] \cup \left[\frac{5}{2}, \infty\right)$

116. $|x(x-6)| < 5$

$x(x-6) < 5$ or $x(x-6) > -5$

$x^2 - 6x - 5 < 0$ $\quad$ $x^2 - 6x + 5 > 0$

Critical numbers: $-0.74, 6.74, 1, 5$

Test intervals: $(-\infty, -0.74) \Rightarrow |x(x-6)| > 5$

$(-0.74, 1) \Rightarrow |x(x-6)| < 5$

$(1, 5) \Rightarrow |x(x-6)| > 5$

$(5, 6.74) \Rightarrow |x(x-6)| < 5$

$(6.74, \infty) \Rightarrow |x(x-6)| > 5$

Solution interval: $(-0.74, 1) \cup (5, 6.74)$

118. $\sqrt{x(x-4)}$

$x(x-4) \ge 0$

Critical numbers: $0, 4$

Test intervals: $(-\infty, 0) \Rightarrow x(x-4) > 0$

$(0, 4) \Rightarrow x(x-4) < 0$

$(4, \infty) \Rightarrow x(x-4) > 0$

Domain: $(-\infty, 0] \cup [4, \infty)$

120.

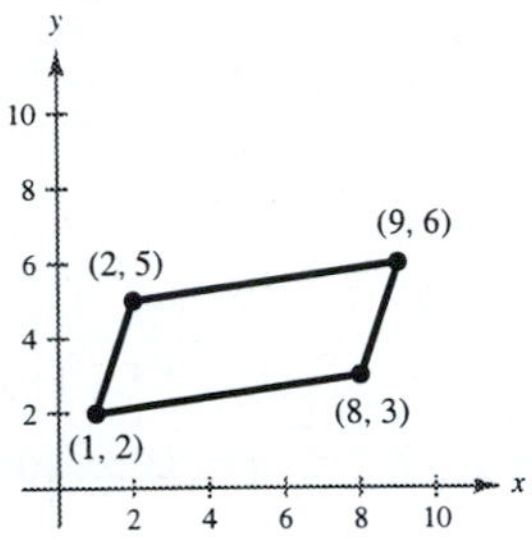

$$d_1 = \sqrt{(1-8)^2 + (2-3)^2} = \sqrt{49+1} = \sqrt{50} = 5\sqrt{2}$$
$$d_2 = \sqrt{(8-9)^2 + (3-6)^2} = \sqrt{1+9} = \sqrt{10}$$
$$d_3 = \sqrt{(9-2)^2 + (6-5)^2} = \sqrt{49+1} = \sqrt{50} = 5\sqrt{2}$$
$$d_4 = \sqrt{(1-2)^2 + (2-5)^2} = \sqrt{1+9} = \sqrt{10}$$

Opposite sides have equal lengths of $\sqrt{10}$ and $5\sqrt{2}$.

122. $y > 0$ in Quadrants I and II.

124. If $xy = 4$ then the coordinates have the same sign. This happens in Quadrants I and III.

126. (a)

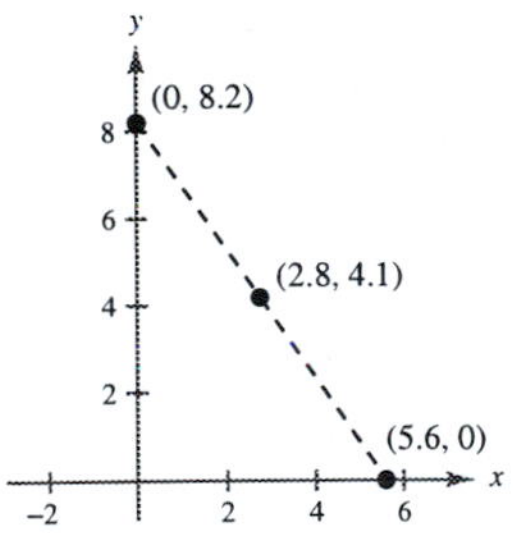

(b) $d = \sqrt{(5.6-0)^2 + (0-8.2)^2}$

$= \sqrt{31.36 + 67.24} = \sqrt{98.6} \approx 9.9$

(c) $\left(\dfrac{5.6+0}{2}, \dfrac{0+8.2}{2}\right) = (2.8, 4.1)$

128.

n	1	10	10^2	10^4	10^6	10^{10}
$5/\sqrt{n}$	5	1.5811	0.5	0.05	0.0005	0.00005

$\dfrac{5}{\sqrt{n}}$ approaches 0 as n increases without bound.

130. September's profit + October's profit = 689,000

Let x = September's profit.

Then $x + 0.12x$ = October's profit.

$x + (x + 0.12x) = 689{,}000$

$2.12x = 689{,}000$

$x = 325{,}000$

$x + 0.12x = 364{,}000$

September: \$325,000

October: \$364,000

132. Let $x =$ the number of liters of pure antifreeze.

$$30\% \text{ of } (10 - x) + 100\% \text{ of } x = 50\% \text{ of } 10$$
$$0.30(10 - x) + 1.00x = 0.50(10)$$
$$3 - 0.30x + 1.00x = 5$$
$$0.70x = 2$$
$$x = \frac{2}{0.70} = \frac{20}{7} = 2\frac{6}{7} \text{ liters}$$

134. Let $x =$ the number of farmers in the group.

$$\text{Cost per farmer} = \frac{48{,}000}{x}$$

If two more farmers join the group, the cost per farmer will be $\frac{48{,}000}{x + 2}$.

Since this new cost is \$4000 less than the original cost,

$$\frac{48{,}000}{x} - 4000 = \frac{48{,}000}{x + 2}$$
$$48{,}000(x + 2) - 4000x(x + 2) = 48{,}000x$$
$$12(x + 2) - x(x + 2) = 12x \quad \text{Divide both sides by 4000.}$$
$$12x + 24 - x^2 - 2x = 12x$$
$$0 = x^2 + 2x - 24$$
$$0 = (x + 6)(x - 4)$$
$$x = -6, \text{ extraneous} \quad \text{or} \quad x = 4$$
$$x = 4 \text{ farmers.}$$

136.

	Rate	Time	Distance
To work	r	$\frac{56}{r}$	56
From work	$r + 8$	$\frac{56}{r + 8}$	56

$$\text{time} = \frac{\text{distance}}{\text{rate}}$$

time to work = time from work + 10 minutes

$$\frac{56}{r} = \frac{56}{r + 8} + \frac{1}{6} \quad \text{Convert minutes to portion of an hour.}$$
$$6(r + 8)(56) = 6r(56) + r(r + 8)$$
$$336r + 2688 = 336r + r^2 + 8r$$
$$0 = r^2 + 8r - 2688$$
$$0 = (r - 48)(r + 56)$$

Using the positive value for r, we have $r = 48$ miles per hour. The average speed on the trip home was $r + 8 = 56$ miles per hour.

CHAPTER 1
Functions and Their Graphs

CHAPTER 1
Equations and Inequalities

Section 1.1 Graphs and Graphing Utilities

Solutions to Even-Numbered Exercises

2. $y = x^2 - 3x + 2$

(a) $(2, 0)$: $(2)^2 - 3(2) + 2 \stackrel{?}{=} 0$

$4 - 6 + 2 \stackrel{?}{=} 0$

$0 = 0$

Yes, the point *is* on the graph.

(b) $(-2, 8)$: $(-2)^2 - 3(-2) + 2 \stackrel{?}{=} 8$

$4 + 6 + 2 \stackrel{?}{=} 8$

$12 \neq 8$

No, the point *is not* on the graph.

4. $y = \frac{1}{3}x^3 - 2x^2$

(a) $\left(2, -\frac{16}{3}\right)$: $\frac{1}{3}(2)^3 - 2(2)^2 \stackrel{?}{=} -\frac{16}{3}$

$\frac{1}{3} \cdot 8 - 2 \cdot 4 \stackrel{?}{=} -\frac{16}{3}$

$\frac{8}{3} - 8 \stackrel{?}{=} -\frac{16}{3}$

$\frac{8}{3} - \frac{24}{3} \stackrel{?}{=} -\frac{16}{3}$

$-\frac{16}{3} = -\frac{16}{3}$

Yes, the point *is* on the graph.

(b) $(-3, 9)$: $\frac{1}{3}(-3)^3 - 2(-3)^2 \stackrel{?}{=} 9$

$\frac{1}{3}(-27) - 2(9) \stackrel{?}{=} 9$

$-9 - 18 \stackrel{?}{=} 9$

$-27 \neq 9$

No, the point *is not* on the graph.

6. $x^2 + y^2 = 20$

(a) $(3, -2)$: $3^2 + (-2)^2 \stackrel{?}{=} 20$

$9 + 4 \stackrel{?}{=} 20$

$13 \neq 20$

No, the point *is not* on the graph.

(b) $(-4, 2)$: $(-4)^2 + 2^2 \stackrel{?}{=} 20$

$16 + 4 \stackrel{?}{=} 20$

$20 = 20$

Yes, the point *is* on the graph.

8. $y = \dfrac{1}{x^2 + 1}$

(a) $(0, 0)$: $\dfrac{1}{0^2 + 1} \stackrel{?}{=} 0$

$\dfrac{1}{1} \stackrel{?}{=} 0$

$1 \neq 0$

No, the point *is not* on the graph.

(b) $(3, 0.1)$: $\dfrac{1}{3^2 + 1} \stackrel{?}{=} 0.1$

$\dfrac{1}{9 + 1} \stackrel{?}{=} 0.1$

$\dfrac{1}{10} \stackrel{?}{=} 0.1$

$0.1 = 0.1$

Yes, the point *is* on the graph.

10. $y = \frac{3}{2}x - 1$

x	−2	0	$\frac{2}{3}$	1	2
y	−4	−1	0	$\frac{1}{2}$	2

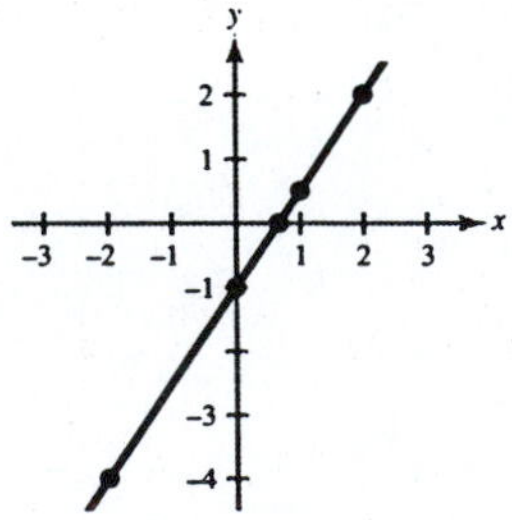

12. $y = 4 - x^2$

x	-2	-1	0	1	2
y	0	3	4	3	0

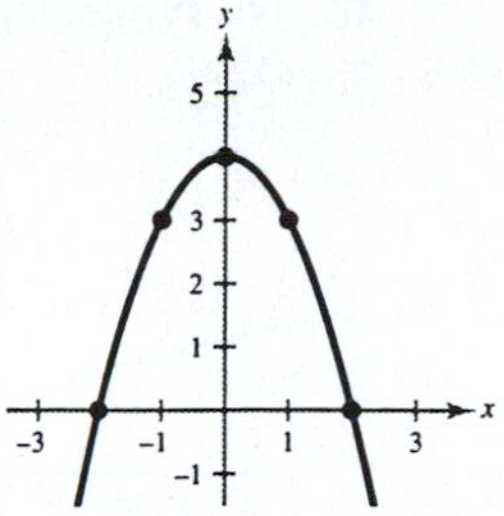

14.

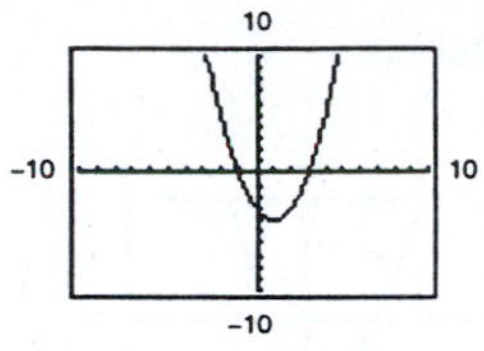

Intercepts: $(-1, 0), (0, -3), (3, 0)$

16.

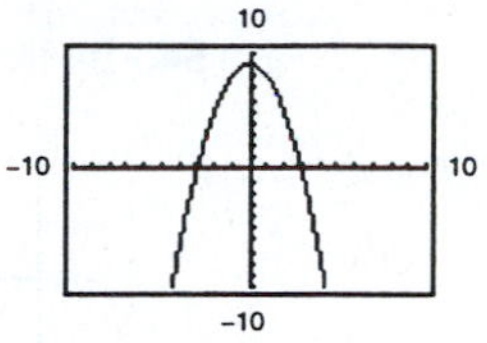

Intercepts: $(-3, 0), (0, 9), (3, 0)$

18.

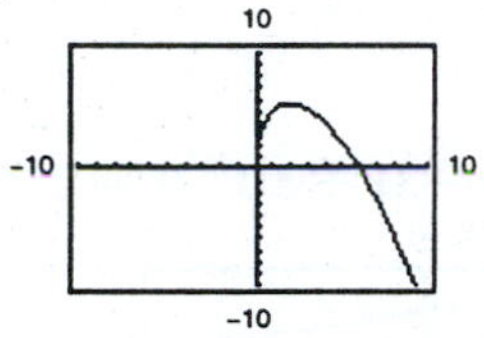

Intercepts: $(0, 0), (6, 0)$

20.

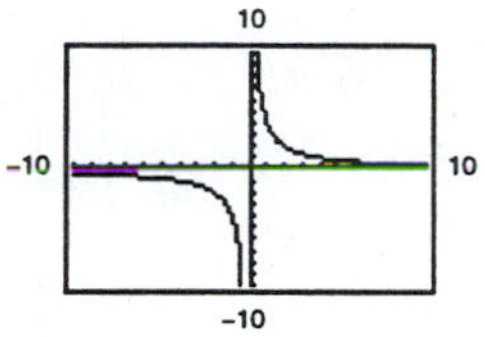

No intercepts

22. $xy^2 + 10 = 0$
$x(-y)^2 + 10 = 0$
$xy^2 + 10 = 0$
x-axis symmetry

24. $y = \sqrt{9 - x^2}$
$y = \sqrt{9 - (-x)^2}$
$y = \sqrt{9 - x^2}$
y-axis symmetry

26. $xy = 4$
$(-x)(-y) = 4$
$xy = 4$
Origin symmetry

28. $y = x^4 - x^2 + 3$
$y = (-x)^4 - (-x)^2 + 3$
$y = x^4 - x^2 + 3$
y-axis symmetry

30.

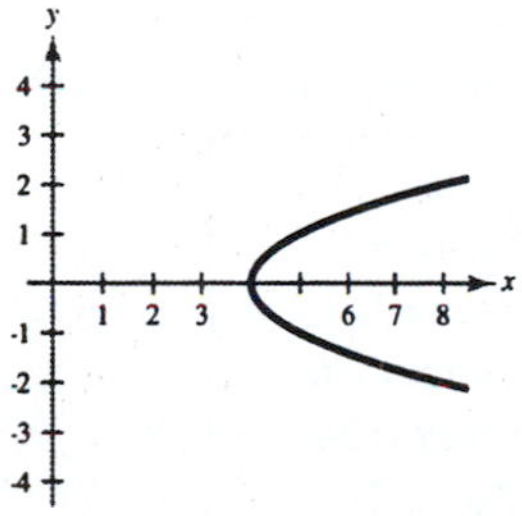

32.

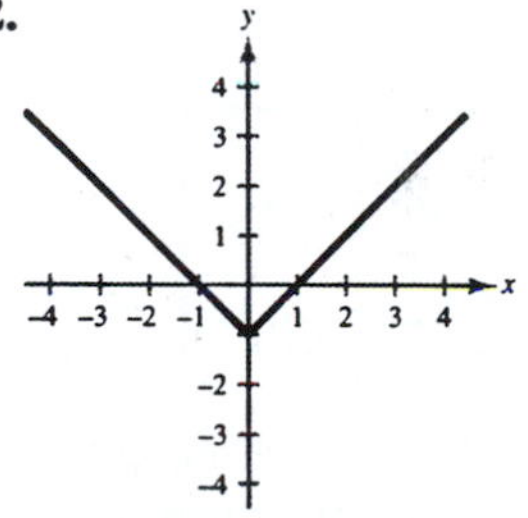

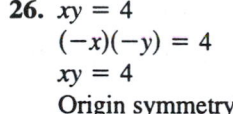

34. $y = x^2 - 2x$ is a parabola. Matches (a).

36. $y = 2\sqrt{x}$ does through the origin. Matches (e).

38. $y = |x| - 3$ involves an absolute value. Matches (d).

40. No symmetry

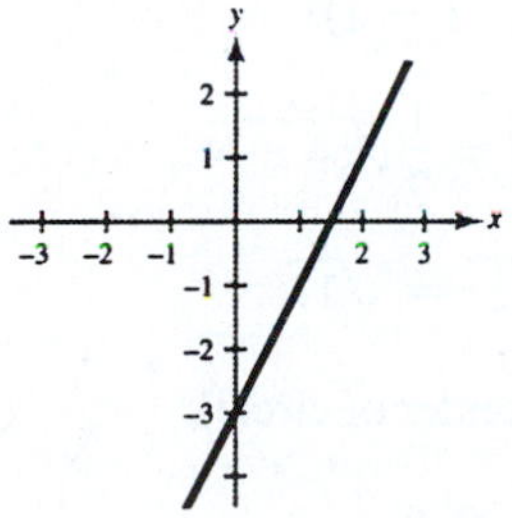

42. y-axis symmetry

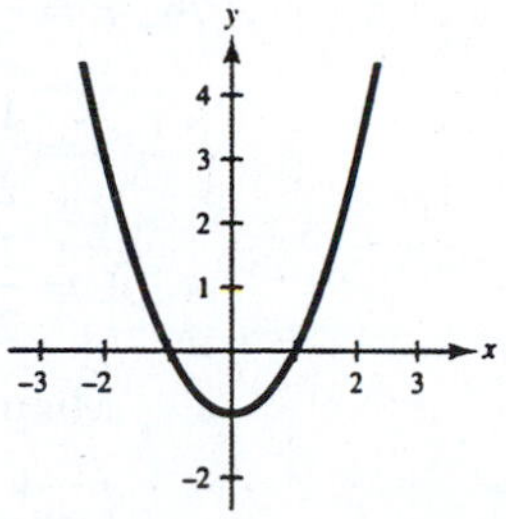

44. No symmetry

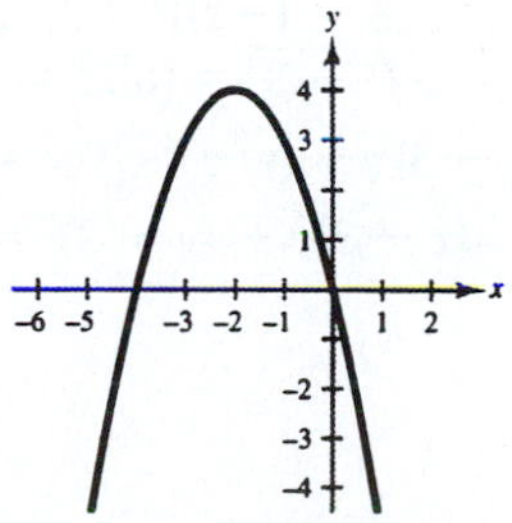

46. No symmetry

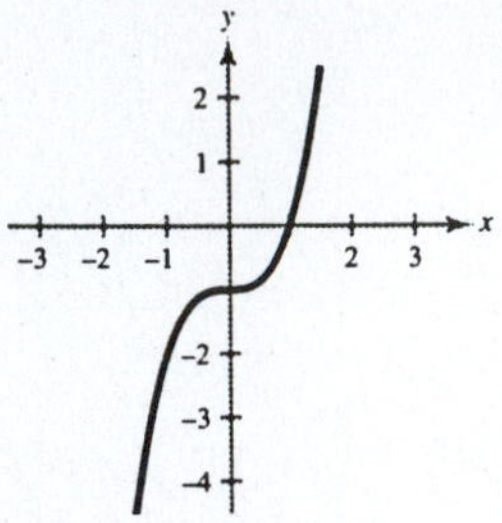

48. No symmetry

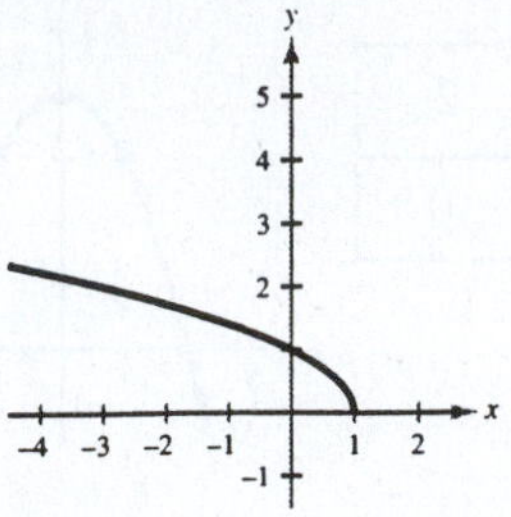

50. y-axis symmetry

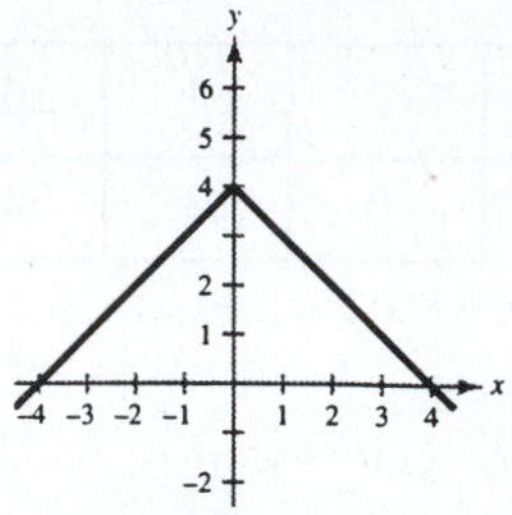

52. x-axis symmetry

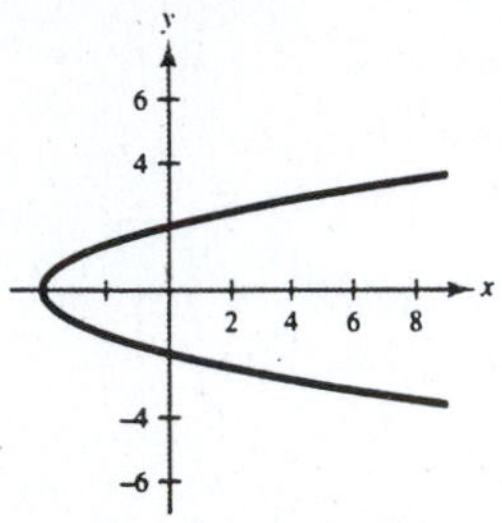

54. Intercepts: $(0, -1), \left(\frac{3}{2}, 0\right)$

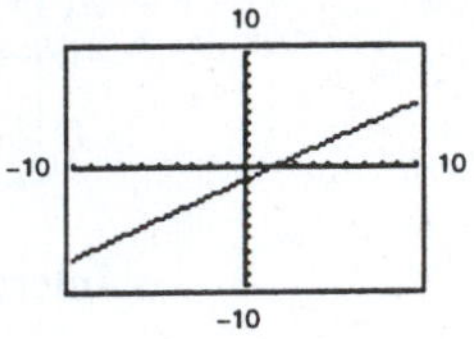

56. Intercepts: $(-4, 0), (2, 0), (0, -4)$

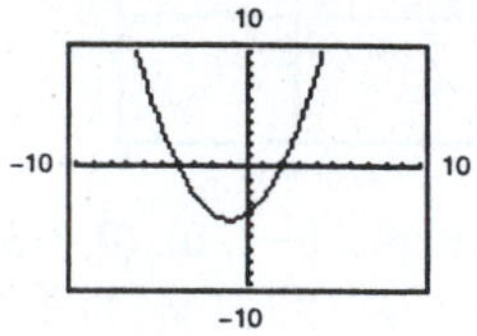

58. Intercept: $(0, 4)$

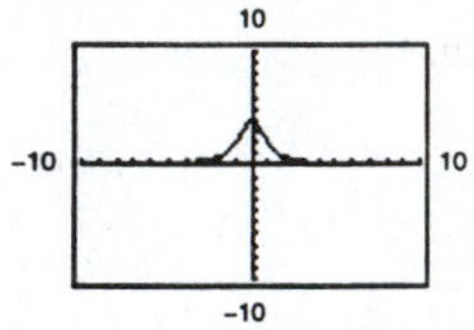

60. Intercepts: $(-1, 0), (0, 1)$

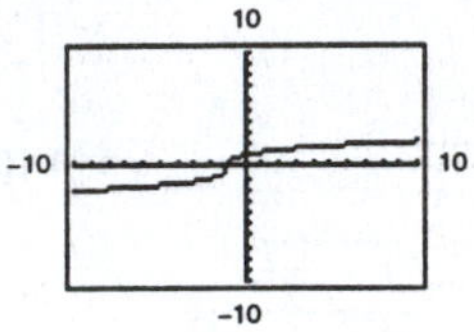

62. $y = -3x + 50$

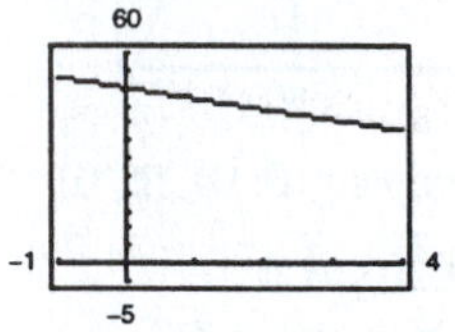

The specified setting gives a more complete graph. (The y-intercept is visible.)

64. $y = 4(x + 5)\sqrt{4 - x}$

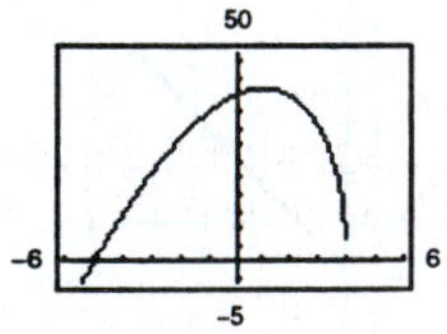

The specified setting gives a more complete graph.

66. $y = x^3 - 3x^2 + 4$

Range/Window

Xmin = -3
Xmax = 4
Xscl = 1
Ymin = -20
Ymax = 20
Yscl = 5

68. $y = 8\sqrt[3]{x - 6}$

Range/Window

Xmin = -5
Xmax = 15
Xscl = 2
Ymin = -20
Ymax = 20
Yscl = 4

70. $(x - 0)^2 + (y - 0)^2 = 5^2$

$$x^2 + y^2 = 25$$

72. $(x - 0)^2 + \left(y - \frac{1}{3}\right)^2 = \left(\frac{1}{3}\right)^2$

$$x^2 + \left(y - \frac{1}{3}\right)^2 = \frac{1}{9}$$

74. $r = \sqrt{(3 - (-1))^2 + (-2 - 1)^2}$

$$= \sqrt{4^2 + 3^2} = \sqrt{25} = 5$$

$$(x - 3)^2 + (y - (-2))^2 = 5^2$$

$$(x - 3)^2 + (y + 2)^2 = 25$$

76. $r = \frac{1}{2}\sqrt{(-4 - 4)^2 + (-1 - 1)^2}$

$$= \frac{1}{2}\sqrt{(-8)^2 + (-2)^2} = \frac{1}{2}\sqrt{64 + 4}$$

$$= \frac{1}{2}\sqrt{68} = \left(\frac{1}{2}\right)(2)\sqrt{17} = \sqrt{17}$$

Midpoint of diameter (center of circle):

$$\left(\frac{-4 + 4}{2}, \frac{-1 + 1}{2}\right) = (0, 0)$$

$$(x - 0)^2 + (y - 0)^2 = (\sqrt{17})^2$$

$$x^2 + y^2 = 17$$

78. Center: (1, 0), radius: 6

$(x - 1)^2 + (y - 0)^2 = 6^2$

$(x - 1)^2 + y^2 = 36$

80. $x^2 + y^2 = 16$

Center: (0, 0), radius: 4

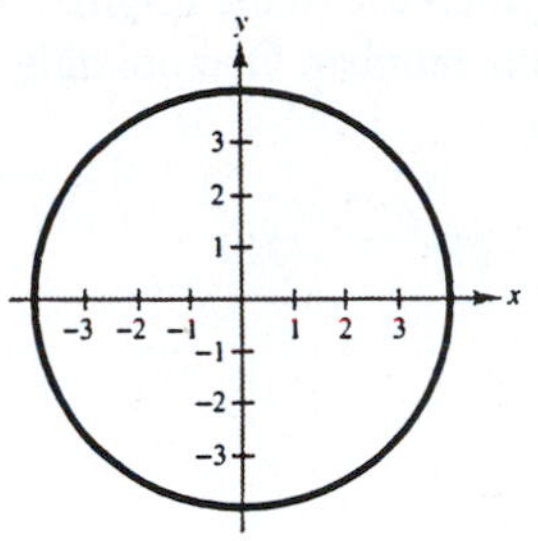

82. $x^2 + (y - 1)^2 = 1$

Center: (0, 1), radius: 1

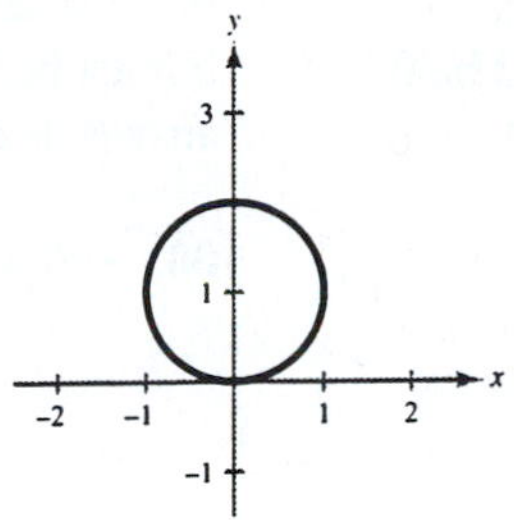

84. $(x - 2)^2 + (y + 1)^2 = 2$

Center: $(2, -1)$, radius: $\sqrt{2}$

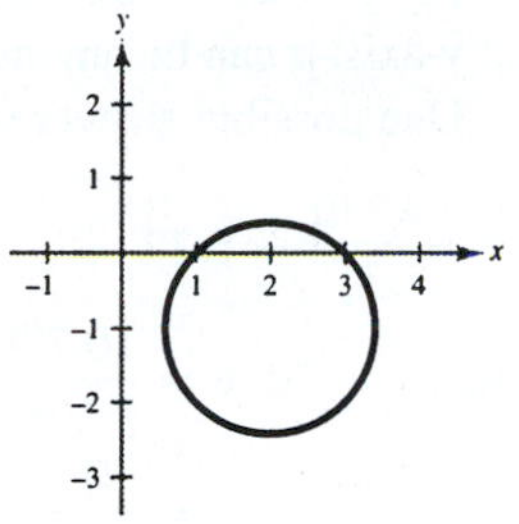

86. $y_1 = 2 + \sqrt{16 - (x - 1)^2}$

$y_2 = 2 - \sqrt{16 - (x - 1)^2}$

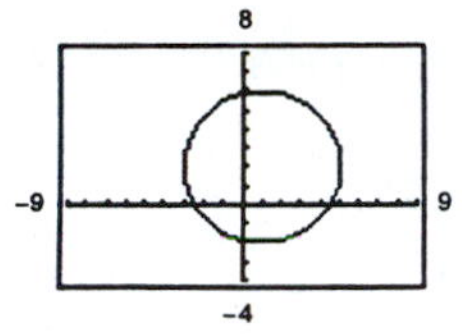

A circle is bounded by their graphs.

88. $y_1 = \frac{1}{2}x + (x + 1)$

$y_2 = \frac{3}{2}x + 1$

Graphing these with a graphing utility shows that their graphs are identical. The Associative Property is illustrated.

90. $y_1 = (x - 3) \cdot \dfrac{1}{x - 3}$

$y_2 = 1$

Graphing these with a graphing utility shows that their graphs

92. (a)

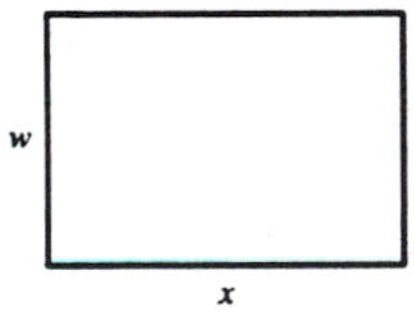

(b) Perimeter: $12 = 2x + 2w$

$12 = 2(x + w)$

$6 = x + w$

Thus, $w = 6 - x$.

Area $= xw = x(6 - x)$.

(c)

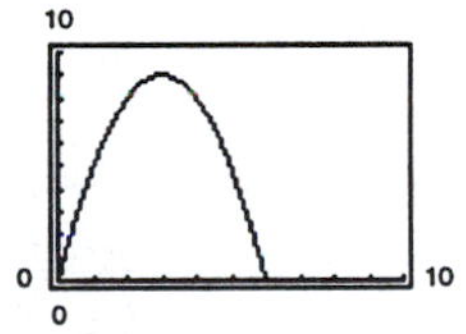

(d) The maximum area corresponds to the highest point on the graph, which appears to be (3, 9). Thus, $x = 3$ and $w = 6 - x = 6 - 3 = 3$.

94. (a)

Year	1950	1960	1970	1980	1990	1994
Per Capita Debt	\$1688	\$1572	\$1807	\$3981	\$12,848	\$15,750
Model	\$1570	\$1416	\$1972	\$4769	\$11,337	\$15,362

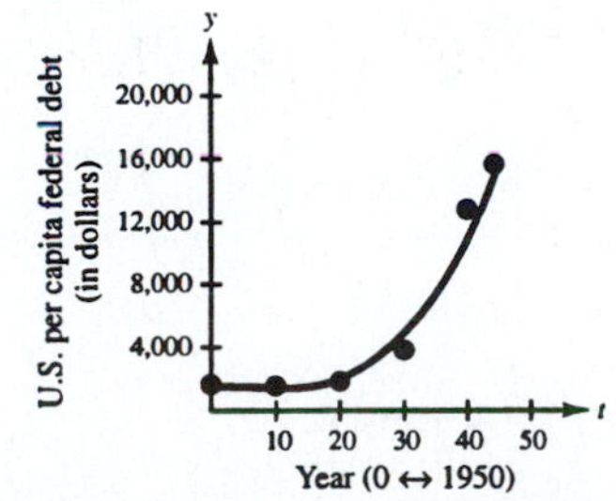

(b) When $t = 48$, $y = 0.255(48)^3 - 4.096(48)^2 + 1570.417 \approx \$20{,}334$.

(c) When $t = 50$, $y = 0.255(50)^3 - 4.096(50)^2 + 1570.417 \approx \$23{,}205$.

96. (a)

$$y = ax^2 + bx^3$$
$$ax^2 + bx^3 = a(-x)^2 + b(-x)^3$$
$$ax^2 + bx^3 = ax^2 - bx^3$$
$$2bx^3 = 0$$
$$b = 0$$

For $y = ax^2 + bx^3$ to be symmetric to the y-axis, a can be any number and b should be 0. One possible answer is $a = 1$ and $b = 0$.

(b)

$$y = ax^2 + bx^3$$
$$-y = a(-x)^2 + b(-x)^3$$
$$-y = ax^2 - bx^3$$
$$y = -ax^2 + bx^3$$

For $y = ax^2 + bx^3$ to be symmetric to the origin, a must be 0 and b can be any number. One possible answer is $a = 0$ and $b = 1$.

98. $y = \dfrac{10{,}770}{x^2} - 0.37$

When $x = 50$, $y = \dfrac{10{,}770}{50^2} - 0.37$

$= \dfrac{10{,}770}{2500} - 0.37 \approx 3.9$ ohms.

100. $-(7 \times 7 \times 7 \times 7) = -(7)^4 = -7^4$

102. $(3 + 4)^2 = (7)^2 = 49$

$3^2 + 4^2 = 9 + 16 = 25$

$49 \neq 25$

False

104. $\sqrt[4]{x^5} = \sqrt[4]{x \cdot x^4} = |x|\sqrt[4]{x}$

106. $\dfrac{55}{\sqrt{20} - 3} = \dfrac{55}{\sqrt{20} - 3} \cdot \dfrac{\sqrt{20} + 3}{\sqrt{20} + 3}$

$= \dfrac{55(\sqrt{20} + 3)}{20 - 9} = \dfrac{55(\sqrt{20} + 3)}{11}$

$= 5(\sqrt{20} + 3) = 5(2\sqrt{5} + 3)$

108. $\sqrt[3]{\sqrt{y}} = (y^{1/2})^{1/3} = y^{1/6} = \sqrt[6]{y}$

Section 1.2 Lines in the Plane and Slope

Solutions to Even-Numbered Exercises

2. (a) $m = 0$. The line is horizontal. Matches L_2.

(b) $m = -\frac{3}{4}$. Because the slope is negative, the line falls. Matches L_1.

(c) $m = 1$. Because the slope is positive, the line rises. Matches L_3.

4.

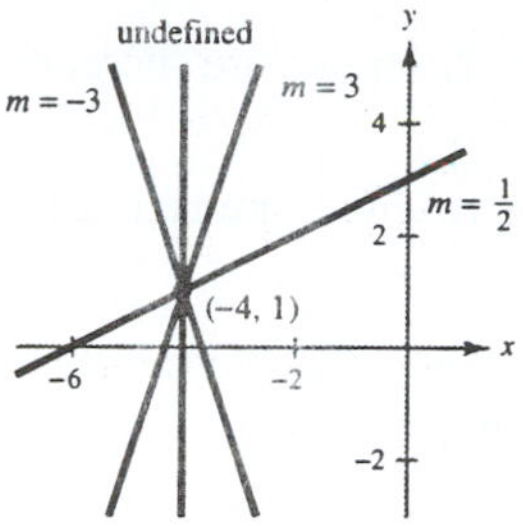

6. The line appears to go through (1, 0) and (4, 8).

$$\text{Slope} = \frac{y_2 - y_1}{x_2 - x_1} = \frac{8 - 0}{4 - 1} = \frac{8}{3}$$

8. The line appears to go through (0, 7) and (7, 0).

$$\text{Slope} = \frac{y_2 - y_1}{x_2 - x_1} = \frac{0 - 7}{7 - 0} = -1$$

10. The line appears to go through (0, 1) and (8, 7).

$$\text{Slope} = \frac{y_2 - y_1}{x_2 - x_1} = \frac{7 - 1}{8 - 0} = \frac{3}{4}$$

12. $\text{Slope} = \dfrac{-4 - 4}{4 - 2} = -4$

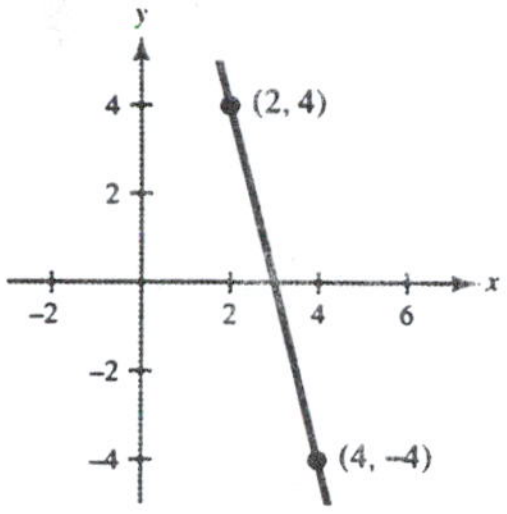

14. $\text{Slope} = \dfrac{0 - (-10)}{-4 - 0} = -\dfrac{5}{2}$

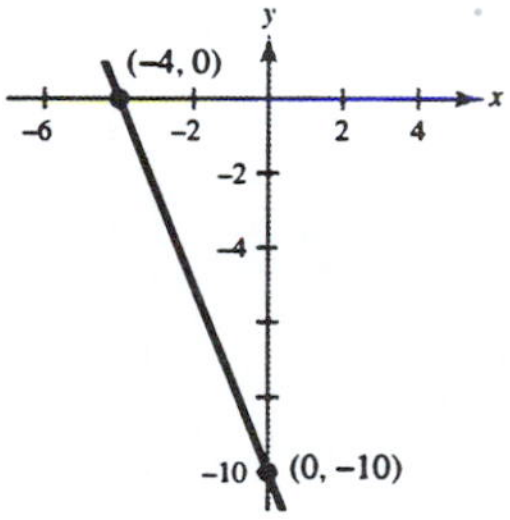

16. $\text{Slope} = \dfrac{-\frac{1}{4} - \frac{3}{4}}{\frac{5}{4} - \frac{7}{8}} = \dfrac{-1}{\frac{3}{8}} = -\dfrac{8}{3}$

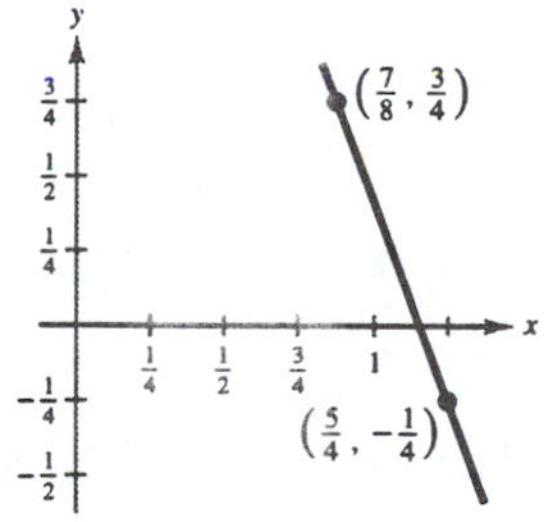

18. Because m is undefined, x does not change. Three other points are: $(-4, 0)$, $(-4, 3)$, $(-4, 5)$.

20. Because $m = -1$, y decreases by 1 for every one unit increase in x. Three other points are: $(0, 4)$, $(9, -5)$, $(11, -7)$.

22. Because $m = 0$, y does not change. Three other points are: $(-4, -1)$, $(-2, -1)$, $(0, -1)$.

24. L_1: $(-2, -1)$, $(1, 5)$

$$m_1 = \frac{5 - (-1)}{1 - (-2)} = \frac{6}{3} = 2$$

L_2: $(1, 3)$, $(5, -5)$

$$m_2 = \frac{-5 - 3}{5 - 1} = \frac{-8}{4} = -2$$

The lines are neither parallel nor perpendicular.

26. L_1: $(4, 8), (-4, 2)$

$$m_1 = \frac{2-8}{-4-4} = \frac{-6}{-8} = \frac{3}{4}$$

L_2: $(3, -5), \left(-1, \frac{1}{3}\right)$

$$m_2 = \frac{\frac{1}{3}-(-5)}{-1-3} = \frac{\frac{16}{3}}{-4} = -\frac{4}{3}$$

The lines are perpendicular.

28. No, the slopes of two perpendicular lines have opposite signs (assume that neither line is vertical or horizontal).

30. (a) $m = 400$. The revenues are increasing \$400 per day.

(b) $m = 100$. The revenues are increasing \$100 per day.

(c) $m = 0$. There is no change in revenue. (Revenue remains constant.)

32. The steepest portion of the graph is between 1993 and 1994.

34. $\frac{\text{rise}}{\text{run}} = \frac{3}{4} = \frac{x}{\frac{1}{2}(32)}$

$$\frac{3}{4} = \frac{x}{16}$$

$$4x = 48$$

$$x = 12$$

The maximum height in the attic is 12 feet.

36. $2x + 3y - 9 = 0$

$$3y = -2x + 9$$

$$y = -\frac{2}{3}x + 3$$

Slope: $m = -\frac{2}{3}$

y-intercept: $(0, 3)$

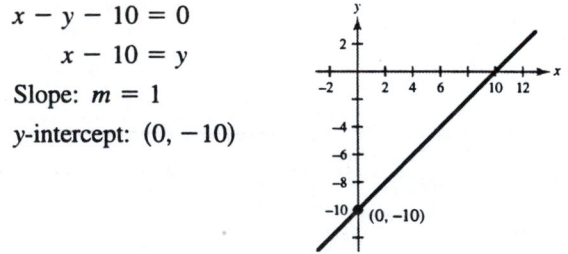

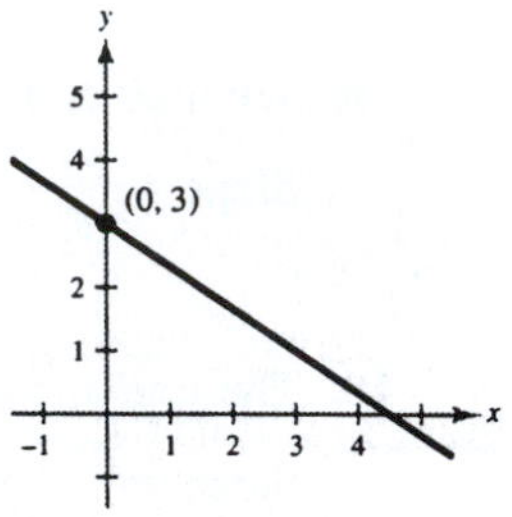

38. $3y +.5 = 0$

$$3y = -5$$

$$y = -\tfrac{5}{3}$$

Slope: $m = 1$

y-intercept: $(0, -10)$

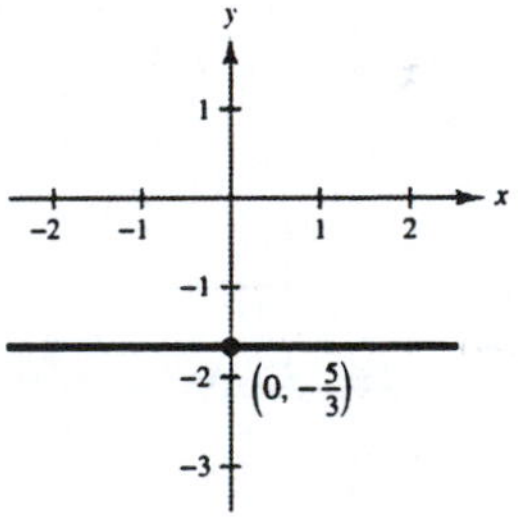

40. $x - y - 10 = 0$

$$x - 10 = y$$

Slope: $m = 1$

y-intercept: $(0, -10)$

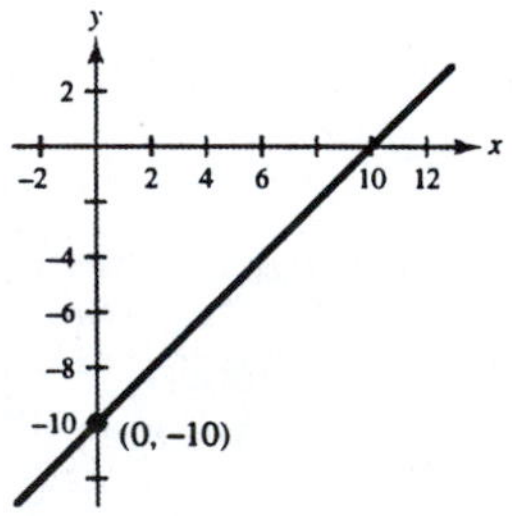

42. $(4, 3), (-4, -4)$

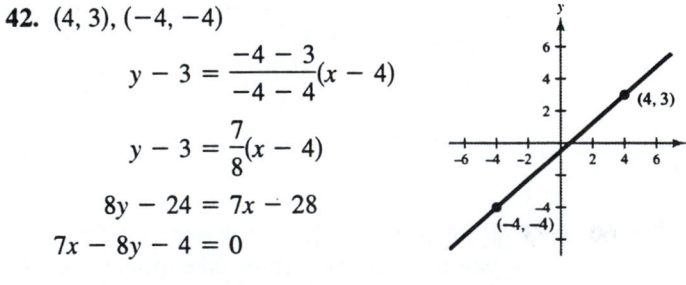

$$y - 3 = \frac{-4-3}{-4-4}(x-4)$$

$$y - 3 = \frac{7}{8}(x-4)$$

$$8y - 24 = 7x - 28$$

$$7x - 8y - 4 = 0$$

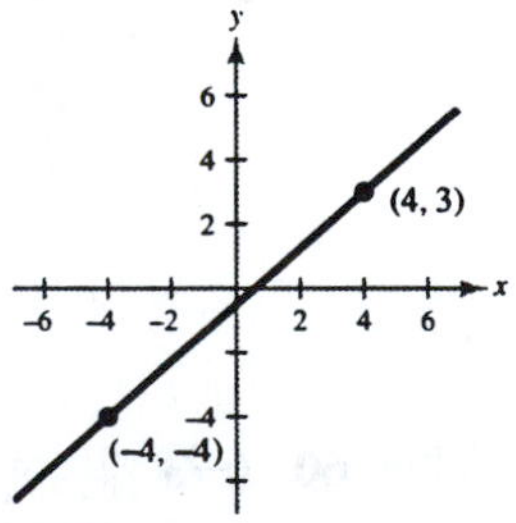

44. $(-1, 4), (6, 4)$

$$y - 4 = \frac{4-4}{6-(-1)}(x+1)$$

$$y - 4 = 0(x+1)$$

$$y - 4 = 0$$

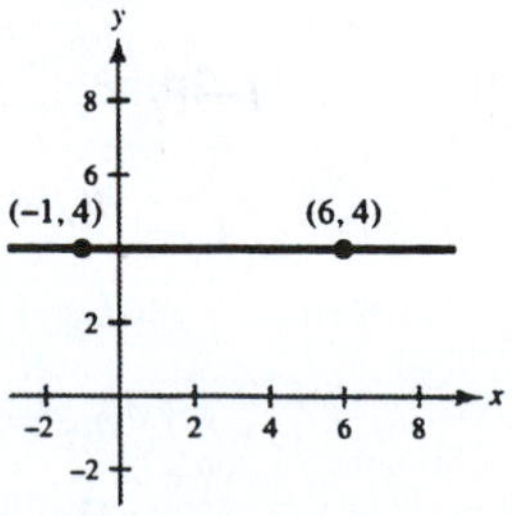

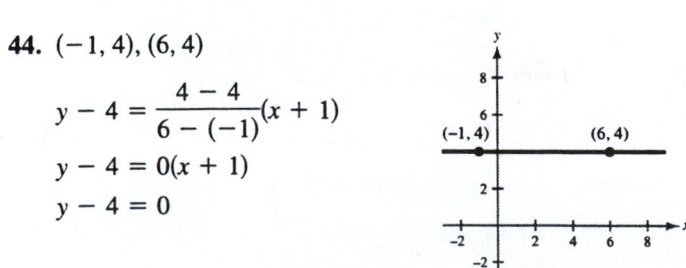

46. $(1, 1), \left(6, -\frac{2}{3}\right)$

$$y - 1 = \frac{-\frac{2}{3} - 1}{6 - 1}(x - 1)$$

$$y - 1 = -\frac{1}{3}(x - 1)$$

$$y - 1 = -\frac{1}{3}x + \frac{1}{3}$$

$$3y - 3 = -x + 1$$

$$x + 3y - 4 = 0$$

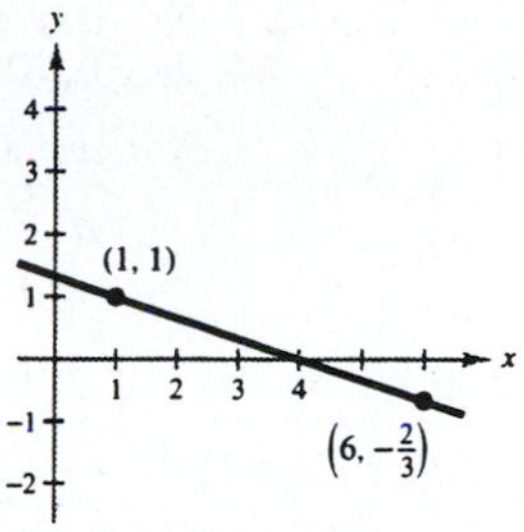

48. $(-8, 0.6), (2, -2.4)$

$$y - 0.6 = \frac{-2.4 - 0.6}{2 - (-8)}(x + 8)$$

$$y - 0.6 = -\frac{3}{10}(x + 8)$$

$$10y - 6 = -3(x + 8)$$

$$10y - 6 = -3x - 24$$

$$3x + 10y + 18 = 0$$

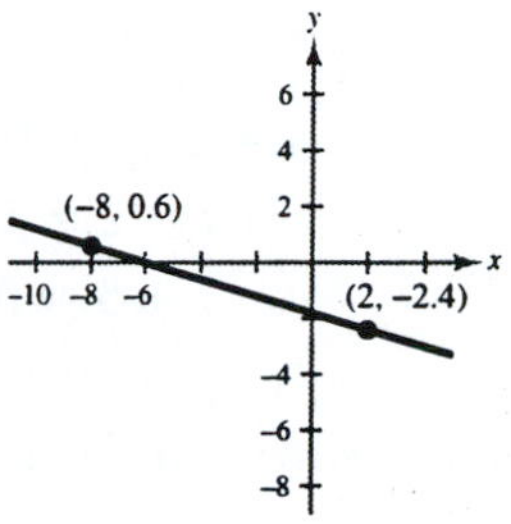

50. $m = -1, (0, 10)$

$$y - 10 = -1(x - 0)$$

$$y - 10 = -x$$

$$x + y - 10 = 0$$

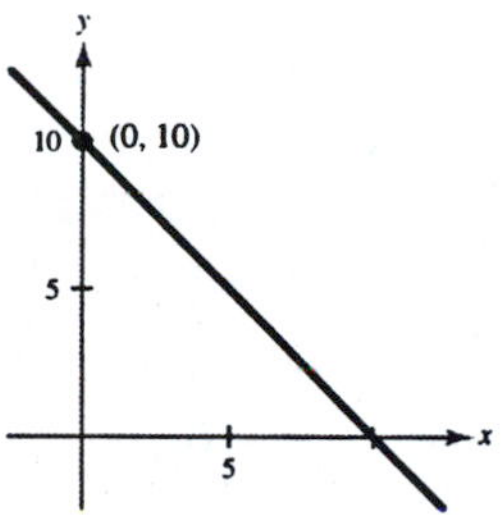

52. $m = 4, (0, 0)$

$$y - 0 = 4(x - 0)$$

$$y = 4x$$

$$4x - y = 0$$

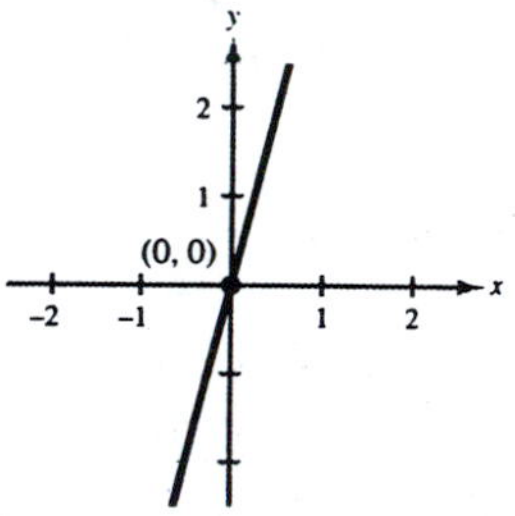

54. $m = \frac{3}{4}, (-2, -5)$

$$y + 5 = \frac{3}{4}(x + 2)$$

$$4y + 20 = 3x + 6$$

$$0 = 3x - 4y - 14$$

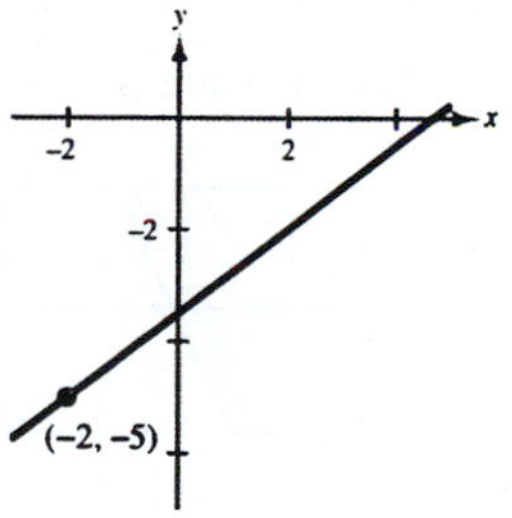

56. $m = 0, (-10, 4)$

$$y - 4 = 0(x + 10)$$

$$y - 4 = 0$$

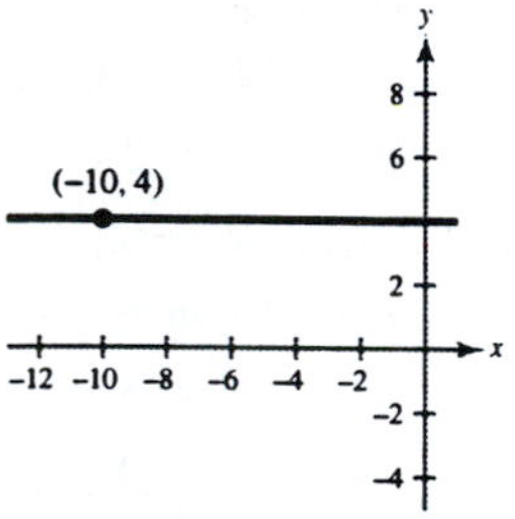

58. $m = -3, \left(-\frac{1}{2}, \frac{3}{2}\right)$

$$y - \frac{3}{2} = -3\left(x + \frac{1}{2}\right)$$

$$y - \frac{3}{2} = -3x - \frac{3}{2}$$

$$3x + y = 0$$

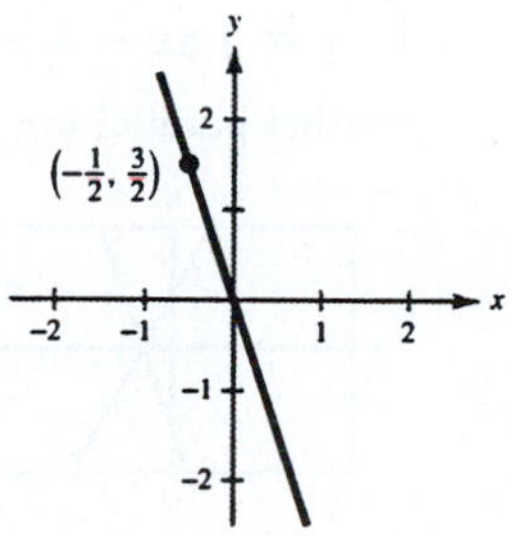

60. $(-3, 0), (0, 4)$

$$\frac{x}{-3} + \frac{y}{4} = 1$$

$$(-12)\frac{x}{-3} + (-12)\frac{y}{4} = (-12) \cdot 1$$

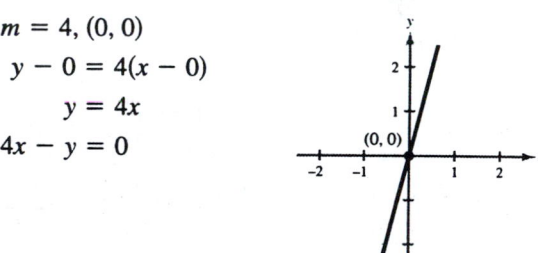

$$4x - 3y + 12 = 0$$

62. $\left(\frac{2}{3}, 0\right), (0, -2)$

$$\frac{x}{2/3} + \frac{y}{-2} = 1$$

$$\frac{3x}{2} - \frac{y}{2} = 1$$

$$3x - y - 2 = 0$$

64. $(a, 0), (0, a)$

$$\frac{x}{a} + \frac{y}{a} = 1$$

$$x + y = a$$

$$-3 + 4 = a$$

$$1 = a$$

$$x + y = 1$$

$$x + y - 1 = 0$$

66. $x + y = 7$

$$y = -x + 7$$

Slope: $m = -1$

(a) $m = -1, (-3, 2)$

$$y - 2 = -1(x + 3)$$

$$y - 2 = -x - 3$$

$$x + y + 1 = 0$$

(b) $m = 1, (-3, 2)$

$$y - 2 = 1(x + 3)$$

$$x - y + 5 = 0$$

68. $5x + 3y = 0$

$$3y = -5x$$

$$y = -\tfrac{5}{3}x$$

Slope: $m = -\frac{5}{3}$

(a) $m = -\frac{5}{3}, \left(\frac{7}{8}, \frac{3}{4}\right)$

$$y - \tfrac{3}{4} = -\tfrac{5}{3}\left(x - \tfrac{7}{8}\right)$$

$$24y - 18 = -40\left(x - \tfrac{7}{8}\right)$$

$$24y - 18 = -40x + 35$$

$$40x + 24y - 53 = 0$$

(b) $m = \frac{3}{5}, \left(\frac{7}{8}, \frac{3}{4}\right)$

$$y - \tfrac{3}{4} = \tfrac{3}{5}\left(x - \tfrac{7}{8}\right)$$

$$40y - 30 = 24\left(x - \tfrac{7}{8}\right)$$

$$40y - 30 = 24x - 21$$

$$24x - 40y + 9 = 0$$

70. $x = 4$

m is undefined.

(a) $(2, 5)$, m is undefined.

$$x = 2$$

$$x - 2 = 0$$

(b) $(2, 5), m = 0$

$$y = 5$$

$$y - 5 = 0$$

72. $L_1: y = 2x - 1$

$L_2: y = 2x + 1$

The lines are parallel.

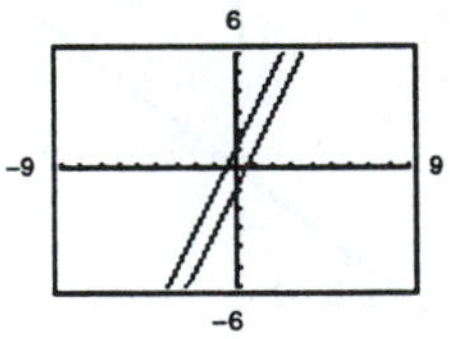

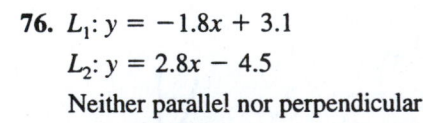

74. $L_1: y = -\frac{4}{5}x - 5$

$L_2: y = \frac{5}{4}x + 1$

The lines are perpendicular.

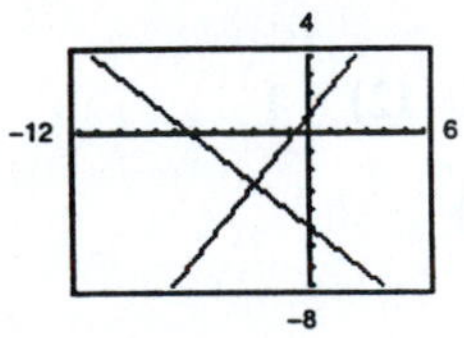

76. $L_1: y = -1.8x + 3.1$

$L_2: y = 2.8x - 4.5$

Neither parallel nor perpendicular

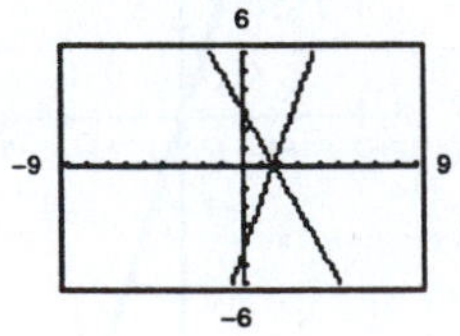

78.

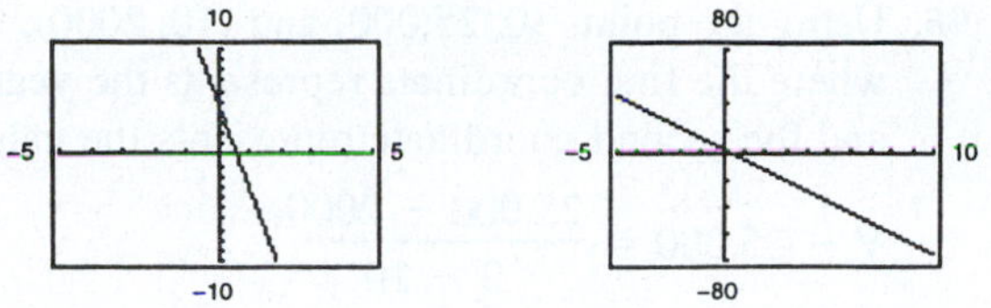

The first setting shows the x- and y-intercepts more clearly.

80. (a) $y = \frac{2}{3}x$ (b) $y = -\frac{3}{2}x$ (c) $y = \frac{2}{3}x + 2$

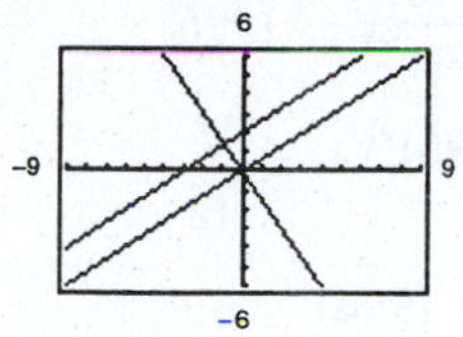

(a) is parallel to (c). (b) is perpendicular to (a) and (c).

82. (a) $y = x - 8$ (b) $y = x + 1$ (c) $y = -x + 3$

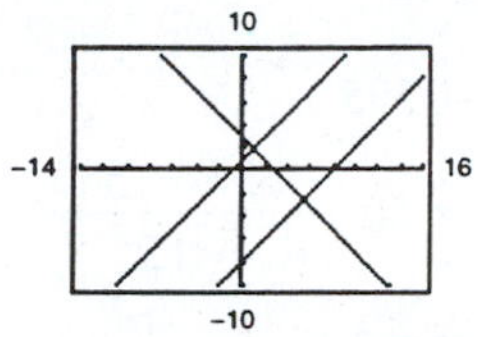

(a) is parallel to (b). (c) is perpendicular to (a) and (b).

84. One point on the line is (6, 156), and the slope is $m = 4.5$.

$$V - 156 = 4.5(t - 6)$$
$$V - 156 = 4.5t - 27$$
$$V = 4.5t + 129$$

86. The y-intercept is 8.5 and the slope is 2, which represents the increase in hourly wage per unit produced. Matches graph (c).

88. The y-intercepts is 750 and the slope is -100, which represents the decrease in the value of the word processor each year. Matches graph (d).

90. Set the distance between $\left(3, \frac{5}{2}\right)$ and (x, y) equal to the distance between $(-7, 1)$ and (x, y).

$$\sqrt{(x - 3)^2 + \left(y - \frac{5}{2}\right)^2} = \sqrt{[x - (-7)]^2 + (y - 1)^2}$$
$$(x - 3)^2 + \left(y - \frac{5}{2}\right)^2 = (x + 7)^2 + (y - 1)^2$$
$$x^2 - 6x + 9 + y^2 - 5y + \frac{25}{4} = x^2 + 14x + 49 + y^2 - 2y + 1$$
$$-6x - 5y + \frac{61}{4} = 14x - 2y + 50$$
$$-24x - 20y + 61 = 56x - 8y + 200$$
$$80x + 12y + 139 = 0$$

This line is the perpendicular bisector of the line segment connecting $\left(3, \frac{5}{2}\right)$ and $(-7, 1)$.

92. $F = \frac{9}{5}C + 32$

$F = 0°$; $\quad 0 = \frac{9}{5}C + 32$
$$-32 = \frac{9}{5}C$$
$$-17.9 \approx C$$

$C = -10°$; $\quad F = \frac{9}{5}(-10) + 32$
$$F = -18 + 32$$
$$F = 14$$

$C = 10°$; $\quad F = \frac{9}{5}(10) + 32$
$$F = 18 + 32$$
$$F = 50$$

$F = 68°$; $\quad 68 = \frac{9}{5}C + 32$
$$36 = \frac{9}{5}C$$
$$20 = C$$

$F = 90°$; $\quad 90 = \frac{9}{5}C + 32$
$$58 = \frac{9}{5}C$$
$$32.2 \approx C$$

$C = 177°$; $\quad F = \frac{9}{5}(177) + 32$
$$F = 318.6 + 32$$
$$F = 350.6$$

C	$-17.8°$	$-10°$	$10°$	$20°$	$32.2°$	$177°$
F	$0°$	$14°$	$50°$	$68°$	$90°$	$350.6°$

94. Using the points (1994, 2546) and (1996, 2702),

$$y - 2702 = \frac{2702 - 2546}{1996 - 1994}(x - 1996)$$

$$y - 2702 = 78(x - 1996)$$

When $x = 2000$:

$$y - 2702 = 78(2000 - 1996)$$

$$y = 312 + 2702 = 3014$$

The college will have 3014 students in 2000.

96. Using the points (0, 25,000) and (10, 2000), where the first coordinate represents the year t and the second coordinate represents the value V,

$$V - 25{,}000 = \frac{25{,}000 - 2000}{0 - 10}t$$

$$V - 25{,}000 = -2300t$$

$$V = 25{,}000 - 2300t$$

98. (Hourly wage) = (Base pay) + (Piecework pay)

$W = 11.5 + 0.75x$

100. (a) (580, 50) and (625, 47)

$$x - 50 = \frac{47 - 50}{625 - 580}(p - 580)$$

$$x - 50 = -\frac{1}{15}(p - 580)$$

$$x = 50 - \frac{p - 580}{15}$$

(b) $p = 655$

$$x = 50 - \frac{655 - 580}{15}$$

$$= 45$$

If the rent is \$655, 45 units will be rented.

(c) $p = 595$

$$x = 50 - \frac{595 - 580}{15}$$

$$= 49$$

If the rent is \$595, 49 units will be rented.

102. (Monthly wage) = (Salary) + (Commission)

$W = 2500 + 0.07S$

104. (a) $12{,}000 - x$ is invested in the fund paying 8% interest.

(b) Model: $\begin{pmatrix}\text{Interest from}\\ 5\frac{1}{2}\%\text{ fund}\end{pmatrix} + \begin{pmatrix}\text{Interest from}\\ 8\%\text{ fund}\end{pmatrix} = \begin{pmatrix}\text{Annual}\\ \text{interest}\end{pmatrix}$

Labels: Interest from $5\frac{1}{2}\%$ fund $= 0.055x$

Interest from 8% fund $= 0.08(12{,}000 - x)$

Annual interest $= y$

Equation: $y = 0.055x + 0.08(12{,}000 - x)$

$$y = 0.055x + 960 - 0.08x$$

$$y = -0.025x + 960$$

(c)

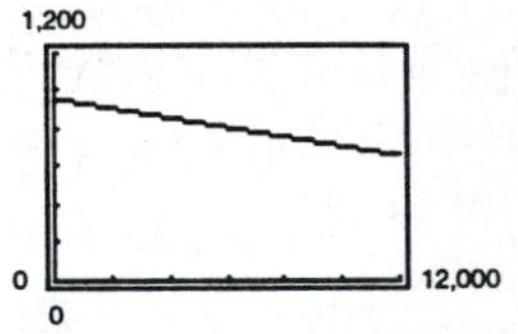

(d) As the amount invested at the lower interest increases, the annual interest decreases.

106. (a) and (b)

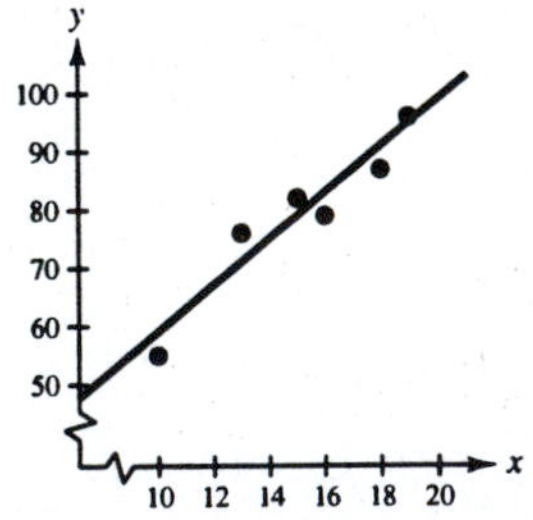

(c) Answers will vary somewhat depending on how students drew their best-fitting lines. An approximate equation using least squares regression is $y = 4x + 19$.

(d) If $x = 17$, $y = 4(17) + 19 = 87$.

(e) If the instructor added 4 points to the average test score of each student, this is the graphical equivalent of a vertical shift of four units upward.

108. $y = 8 - \sqrt{x}$ is a root equation. Matches graph (c).

110. $y = |x + 2| - 1$ is an absolute value equation. Matches graph (b).

Section 1.3 Functions

Solutions to Even-Numbered Exercises

2. No, it is not a function. The domain value of -1 is matched with two output values.

4. Yes, it is a function. Each domain value is matched with only one range value.

6. No, the table does not represent a function. The input values of 0 and 1 are each matched with two different output values.

8. Yes, the table does represent a function. Each input value is matched with only one output value.

10. (a) The element c in A is matched with two elements, 2 and 3 of B, so it is not a function.

(b) Each element of A is matched with exactly one element of B, so it does represent a function.

(c) This is not a function from A to B (it represents a function from B to A instead).

(d) Each element of A is matched with exactly one element of B, so it does represent a function.

12. Reading from the graph, $f(1988)$ is approximately 22 million.

14. $x = y^2 \Rightarrow y = \pm\sqrt{x}$

Thus, y *is not* a function of x.

16. $x + y^2 = 4 \Rightarrow y = \pm\sqrt{4 - x}$

Thus, y *is not* a function of x.

18. $(x - 2)^2 + y^2 \Rightarrow y = \pm\sqrt{4 - (x - 2)^2}$

Thus, y *is not* a function of x.

20. $y = \sqrt{x + 5}$

This is a function of x.

22. $|y| = 4 - x \Rightarrow y = 4 - x$ or $y = -(4 - x)$

Thus, y *is not* a function of x.

24. $g(x) = x^2 - 2x$

(a) $g(2) = (2)^2 - 2(2)$

(b) $g(-3) = (-3)^2 - 2(-3)$

(c) $g(t + 1) = (t + 1)^2 - 2(t + 1)$

(d) $g(x + c) = (x + c)^2 - 2(x + c)$

26. $g(y) = 7 - 3y$

(a) $g(0) = 7 - 3(0) = 7$

(b) $g(\frac{7}{3}) = 7 - 3(\frac{7}{3}) = 0$

(c) $g(s + 2) = 7 - 3(s + 2)$

$= 7 - 3s - 6 = 1 - 35$

28. $V(r) = \dfrac{4}{3}\pi r^3$

(a) $V(3) = \dfrac{4}{3}\pi(3)^3 = 36\pi$

(b) $V\left(\dfrac{3}{2}\right) = \dfrac{4}{3}\pi\left(\dfrac{3}{2}\right)^3 = \dfrac{4}{3} \cdot \dfrac{27}{8}\pi = \dfrac{9\pi}{2}$

(c) $V(2r) = \dfrac{4}{3}\pi(2r)^3 = \dfrac{32\pi r^3}{3}$

30. $f(x) = \sqrt{x + 8} + 2$

(a) $f(-8) = \sqrt{(-8) + 8} + 2 = 2$

(b) $f(1) = \sqrt{(1) + 8} + 2 = 5$

(c) $f(x - 8) = \sqrt{(x - 8) + 8} + 2 = \sqrt{x} + 2$

32. $q(t) = \dfrac{2t^2 + 3}{t^2}$

(a) $q(2) = \dfrac{2(2)^2 + 3}{(2)^2} = \dfrac{8 + 3}{4} = \dfrac{11}{4}$

(b) $q(0) = \dfrac{2(0)^2 + 3}{(0)^2}$

Division by zero is undefined.

(c) $q(-x) = \dfrac{2(-x)^2 + 3}{(-x)^2} = \dfrac{2x^2 + 3}{x^2}$

34. $f(x) = |x| + 4$

(a) $f(2) = |2| + 4 = 6$

(b) $f(-2) = |-2| + 4 = 6$

(c) $f(x^2) = |x^2| + 4 = x^2 + 4$

36. $f(x) = \begin{cases} x^2 + 2, & x \le 1 \\ 2x^2 + 2, & x > 1 \end{cases}$

(a) $f(-2) = (-2)^2 + 2 = 6$

(b) $f(1) = (1)^2 + 2 = 3$

(c) $f(2) = 2(2)^2 + 2 = 10$

38. $g(x) = \sqrt{x-3}$

$g(3) = \sqrt{3-3} = 0$

$g(4) = \sqrt{4-3} = 1$

$g(5) = \sqrt{5-3} = \sqrt{2}$

$g(6) = \sqrt{6-3} = \sqrt{3}$

$g(7) = \sqrt{7-3} = 2$

x	3	4	5	6	7
$g(x)$	0	1	$\sqrt{2}$	$\sqrt{3}$	2

40. $f(s) = \dfrac{|s-2|}{s-2}$

$f(0) = \dfrac{|0-2|}{0-2} = \dfrac{2}{-2} = -1$

$f(1) = \dfrac{|1-2|}{1-2} = \dfrac{1}{-1} = -1$

$f\left(\dfrac{3}{2}\right) = \dfrac{\left|\frac{3}{2}-2\right|}{\frac{3}{2}-2} = \dfrac{\frac{1}{2}}{-\frac{1}{2}} = -1$

$f\left(\dfrac{5}{2}\right) = \dfrac{\left|\frac{5}{2}-2\right|}{\frac{5}{2}-2} = \dfrac{\frac{1}{2}}{\frac{1}{2}} = 1$

$f(4) = \dfrac{|4-2|}{4-2} = \dfrac{2}{2} = 1$

s	0	1	$\frac{3}{2}$	$\frac{5}{2}$	4
$f(s)$	-1	-1	-1	1	1

42. $h(x) = \begin{cases} 9 - x^2, & x < 3 \\ x - 3, & x \ge 3 \end{cases}$

$h(1) = 9 - (1)^2 = 8$

$h(2) = 9 - (2)^2 = 5$

$h(3) = (3) - 3 = 0$

$h(4) = (4) - 3 = 1$

$h(5) = (5) - 3 = 2$

s	1	2	3	4	5
$h(x)$	8	5	0	1	2

44.
$$f(x) = 0$$
$$\frac{3x-4}{5} = 0$$
$$3x - 4 = 0$$
$$3x = 4$$
$$x = \frac{4}{3}$$

46.
$$f(x) = 0$$
$$x^3 - x = 0$$
$$x(x^2 = 1) = 0$$
$$x(x+1)(x-1) = 0$$
$$x = 0$$
$$x + 1 = 0 \Rightarrow x = -1$$
$$x - 1 = 0 \Rightarrow x = 1$$

48.
$$f(x) = g(x)$$
$$x^2 + 2x + 1 = 3x + 3$$
$$x^2 - x - 2 = 0$$
$$(x-2)(x+1) = 0$$
$$x - 2 = 0 \Rightarrow x = 2$$
$$x + 1 = 0 \Rightarrow x = -1$$

50.
$$f(x) = g(x)$$
$$x^4 - 2x^2 = 2x^2$$
$$x^4 - 4x^2 = 0$$
$$x^2(x^2 - 4) = 0$$
$$x^2(x+2)(x-2) = 0$$
$$x^2 = 0 \Rightarrow x = 0$$
$$x + 2 = 0 \Rightarrow x = -2$$
$$x - 2 = 0 \Rightarrow x = 2$$

52. $f(x) = 1 - 2x^2$

Because $f(x)$ is a polynomial, the domain is all real numbers x.

54. $s(y) = \dfrac{3y}{y+5}$

$y + 5 \neq 0$

$y \neq -5$

The domain is all real numbers $y \neq 5$.

56. $f(t) = \sqrt[3]{t+4}$

Because $f(t)$ is a cube root, the domain is all real numbers t.

58. $h(x) = \dfrac{10}{x^2 - 2x}$

$x^2 - 2x \neq 0$

$x(x - 2) \neq 0$

The domain is all real numbers $x \neq 0$ and $x \neq 2$.

60. $f(s) = \dfrac{\sqrt{s-1}}{s-4}$

$s - 1 \geq 0$ and $s - 4 \neq 0$

$s \geq 1$ and $s \neq 4$

The domain is $s \geq 1$, $s \neq 4$.

62. $f(x) = \dfrac{2x}{x^2 + 1}$

$\left\{\left(-2, -\frac{4}{5}\right), (-1, -1), (0, 0), (1, 1), \left(2, \frac{4}{5}\right)\right\}$

64. $f(x) = |x + 1|$

$\{(-2, 1), (-1, 0), (0, 1), (1, 2), (2, 3)\}$

66. An advantage to function notation is that it gives a name to the relationship so it can be easily referenced. When evaluating a function, you see both the input and output values.

68. By plotting the data, you can see that it represents a line, or $f(x) = cx$. Because $(0, 0)$ and $\left(1, \frac{1}{4}\right)$ are on the line, the slope is $\frac{1}{4}$. Thus, $f(x) = \frac{1}{4}x$.

70. By plotting the data, you can see that it represents $h(x) = c\sqrt{|x|}$. Because $\sqrt{|-4|} = 2$ and $\sqrt{|-1|} = 1$ but the corresponding y values are 6 and 3. Thus, $c = 3$ and $h(x) = 3\sqrt{|x|}$.

72.

$$\begin{aligned}
f(x) &= 5x - x^2 \\
f(5th) &= 5(5th) - (5th)^2 \\
&= 25 + 5h - (25 + 10h + h^2) \\
&= 25 + 5h - 25 - 10h - h^2 \\
&= -h^2 - 5h \\
f(5) &= 5(5) - (5)^2 \\
&= 25 - 25 = 0 \\
\frac{f(5th) - f(5)}{h} &= \frac{-h^2 - 5h}{h} \\
&= \frac{-h(h + 5)}{h} = -(h + 5)
\end{aligned}$$

74.

$$\begin{aligned}
f(x) &= 2x \\
f(x + c) &= 2(x + c) = 2x + 2c \\
f(x) &= 2x \\
\frac{f(x + c) - f(x)}{c} &= \frac{2x + 2c - 2x}{c} = 2
\end{aligned}$$

76.

$$\begin{aligned}
f(t) &= \frac{1}{t} \\
f(1) &= \frac{1}{1} = 1 \\
\frac{f(t) - f(1)}{t - 1} &= \frac{\frac{1}{t} - 1}{t - 1} = \frac{\frac{1}{t} - \frac{t}{t}}{t - 1} = \frac{\frac{1 - t}{t}}{t - 1} = \frac{\left(-\frac{1}{t}\right)(t - 1)}{t - 1} = -\frac{1}{t}
\end{aligned}$$

78. $A = \frac{1}{2}bh$, in an equilateral triangle $b = s$ and

$$s^2 = h^2 + \left(\frac{s}{2}\right)^2$$

$$h = \sqrt{s^2 - \left(\frac{s}{2}\right)^2}$$

$$h = \sqrt{\frac{4s^2}{4} - \frac{s^2}{4}} = \frac{\sqrt{3}s}{2}$$

$$A = \frac{1}{2}s \cdot \frac{\sqrt{3}s}{2} = \frac{\sqrt{3}s^2}{4}$$

80. (a)

Units x	*Price*	*Profit P*
102	$90 - 2(0.15)$	$102[90 - 2(0.15)] - 102(60) = 3029.40$
104	$90 - 4(0.15)$	$104[90 - 4(0.15)] - 104(60) = 3057.60$
106	$90 - 6(0.15)$	$106[90 - 6(0.15)] - 106(60) = 3084.60$
108	$90 - 8(0.15)$	$108[90 - 8(0.15)] - 108(60) = 3110.40$
110	$90 - 10(0.15)$	$110[90 - 10(0.15)] - 110(60) = 3135.00$
112	$90 - 12(0.15)$	$112[90 - 112(0.15)] - 12(60) = 3158.40$

(b)

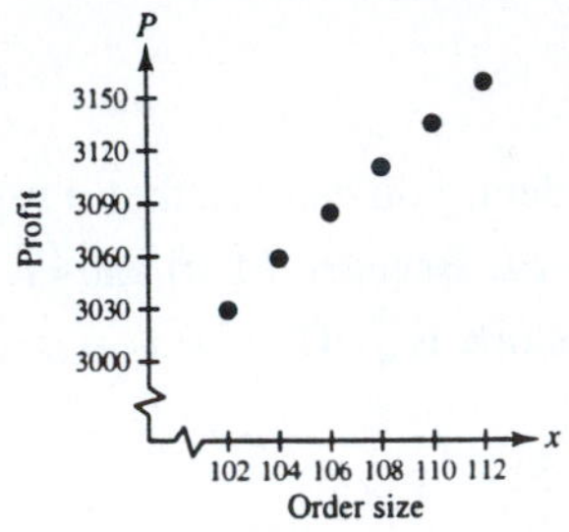

Yes

(c) Profit = Revenue − Cost

$= \text{(price per unit)(number of units)} - \text{(cost)(number of units)}$

$= [90 - (x - 100)(0.15)]x - 60x, \ x > 100$

$= (90 - 0.15x + 15)x - 60x$

$= (105 - 0.15x)x - 60x$

$= 105x - 0.15x^2 - 60x$

$= 45x - 0.15x^2, \ x > 100$

82. $A = l \cdot w = (2x)y = 2xy$

But $y = \sqrt{36 - x^2}$, so $A = 2x\sqrt{36 - x^2}, \ 0 < x < 6$.

84. In 1978: $p(-2) = 19.247 + 1.694(-2) = 15.859$

The average mobile home price in 1978 was $15,859.

In 1988: $p(8) = 19.305 + 0.427(8) + 0.033(8)^2 = 24.833$

The average mobile home price in 1988 was $24,833.

In 1993: $p(13) = 19.305 + 0.427(13) + 0.033(13)^2 = 30.433$

The average mobile home price in 1993 was $30,433.

86. (a) Model: (Total cost) = (Fixed costs) + (Variable costs)

Labels: Total cost = C

Fixed cost = 6000

Variable costs = $0.95x$

Equation: $C = 6000 + 0.95x$

(b) $\overline{C} = \dfrac{C}{x} = \dfrac{6000 + 0.95x}{x}$

88. $F(y) = 149.76\sqrt{10}y^{5/2}$

y	5	10	20	30	40
$F(y)$	26,474.08	149,760.00	847,170.49	2,334,527.36	4,792,320

(a) The force in tons of the water against the dam increases with the depth of the water.

(b) It appears that approximately 21 feet of water would produce 1,000,000 tons of force. You can find a better estimate by creating a new table with y at smaller intervals between 20 and 25.

90. (a) $f(1992) = 28$

(b) $\dfrac{f(1994) - f(1991)}{1994 - 1991} = \dfrac{9 - 60}{3} = -17$

Over the years 1991 through 1994 there was an average decrease in the lynx population of 17 lynx per year.

(c)

t	1988	1989	1990	1991	1992	1993	1994	1995
N	9.0	19.8	43.9	53.6	30.9	16.7	10.2	6.9
Actual data	10	16	50	60	28	15	9	8

The model yields values that are similar to the actual data.

92. $\dfrac{3}{t} + \dfrac{5}{t} = 1$

$\dfrac{8}{t} = 1$

$8 = t$

94. $\dfrac{12}{x} - 3 = \dfrac{4}{x} + 9$

$\dfrac{12}{x} - \dfrac{4}{x} = 9 + 3$

$\dfrac{8}{x} = 12$

$\dfrac{8}{12} = x$

$x = \dfrac{2}{3}$

Section 1.4 Analyzing Graphs of Functions

Solutions to Even-Numbered Exercises

2. $f(x) = \sqrt{x - 1}$

Domain: $x - 1 \geq 0$

$x \geq 1$

$[1, \infty)$

Range: $[0, \infty)$

4. $f(x) = \frac{1}{2}|x - 2|$

Domain: $(-\infty, \infty)$

Range: $[0, \infty)$

6. $g(x) = \dfrac{|x - 1|}{x - 1}$

Domain: $x - 1 \neq 0$

$x \neq 1$

$(-\infty, 1), (1, \infty)$

Range: $-1, 1$

8. $y = \frac{1}{4}x^3$

A vertical line intersects the graph no more than once, so y is a function of x.

10. $x^2 + y^2 = 25$

A vertical line intersects the graph more than once, so y is not a function of x.

12. $x = |y + 2|$

A vertical line intersects the graph more than once, so y is not a function of x.

14. No, the graph in Exercise 10 does not represent x as a function of y because for some values of y there correspond more than one value of x.

16. $f(x) = 6[x - (0.1x)^5]$

The third setting shows the most complete graph.

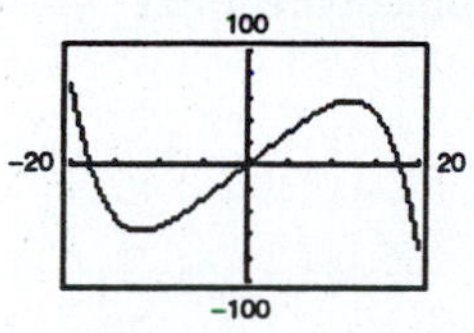

18. $f(x) = 10x\sqrt{400 - x^2}$

The third setting shows the most complete graph.

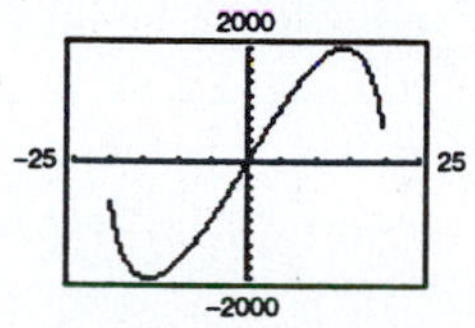

20. $f(x) = x^2 - 4x$

(a) The graph is decreasing on $(-\infty, 2)$ and increasing on $(2, \infty)$.

(b) $f(-x) = (-x)^2 - 4(-x) = x^2 + 4x$

$x^2 + 4x \neq f(x)$

$x^2 + 4x \neq -f(x)$

The function is neither odd nor even.

22. $f(x) = \sqrt{x^2 - 1}$

(a) The graph is decreasing on $(-\infty, -1)$ and increasing on $(1, \infty)$.

(b) $f(-x) = \sqrt{(-x)^2 - 1} = \sqrt{x^2 - 1} = f(x)$

The function is even.

24. $f(x) = x^{2/3}$

(a)

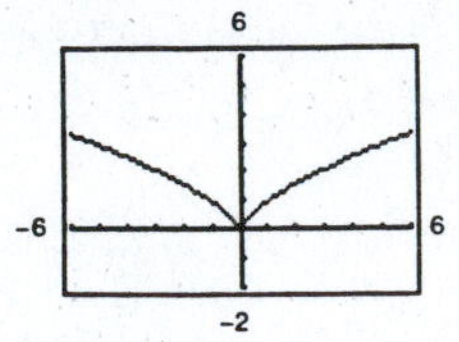

(b) The graph is decreasing on $(-\infty, 0)$ and increasing on $(0, \infty)$.

(c) $f(-x) = (-x)^{2/3} = x^{2/3} = f(x)$

The function is even.

26. $f(x) = |x + 1| + |x - 1|$

(a)

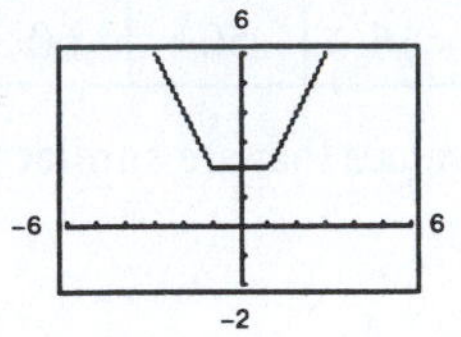

(b) The graph is decreasing on $(-\infty, -1)$, constant on $(-1, 1)$, and increasing on $(1, \infty)$.

(c) $f(-x) = |-x + 1| + |x - 1|$

$= |(-1)(x - 1)| + |(-1)(x + 1)|$

$= |x - 1| + |x + 1|$

$= f(x)$

The function is even.

28. $h(x) = x^3 - 5$

$h(-x) = (-x)^3 - 5$

$= -x^3 - 5$

$\neq h(x)$

$\neq -h(x)$

The function is neither odd nor even.

30. $f(x) = x\sqrt{1 - x^2}$

$f(-x) = -x\sqrt{1 - (-x)^2}$

$= -x\sqrt{1 - x^2}$

$= -f(x)$

The function is odd.

32. $g(s) = 4s^{2/3}$

$g(-s) = 4(-s)^{2/3}$

$= 4s^{2/3}$

$= g(s)$

The function is even.

34. $(4, 9)$

(a) If f is even, another point is $(-4, 9)$.

(b) If f is odd, another point is $(-4, -9)$.

36. $g(x) = x$

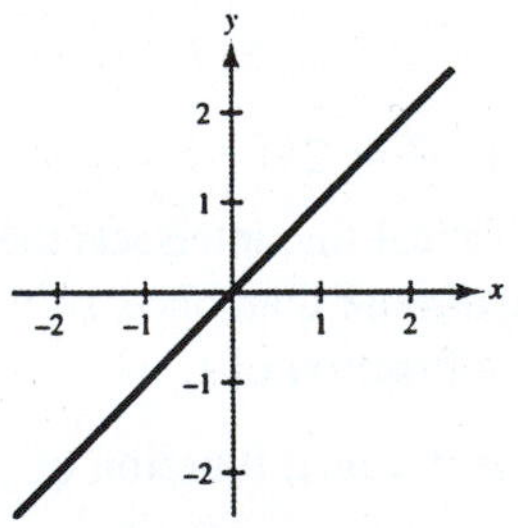

$g(-x) = -x = -g(x)$

The function is odd.

38. $h(x) = x^2 - 4$

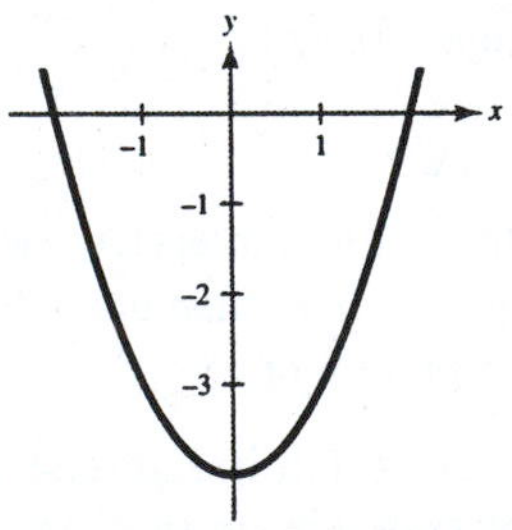

$h(-x) = (-x)^2 - 4$

$= x^2 - 4 = h(x)$

The function is even.

40. $f(t) = -t^4$

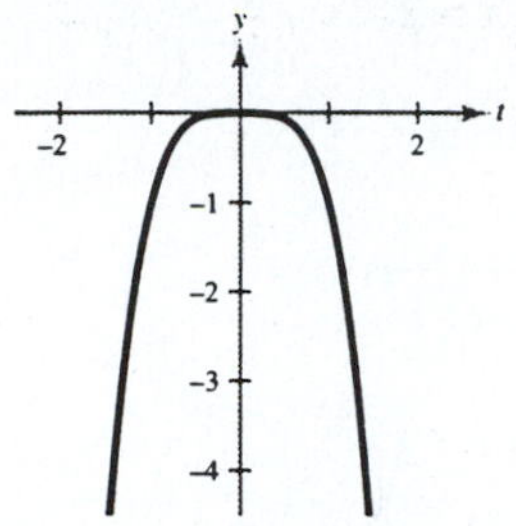

$f(-t) = -(-t)^4$

$= -t^4 = f(t)$

The function is even.

42. $f(x) = x^{3/2}$

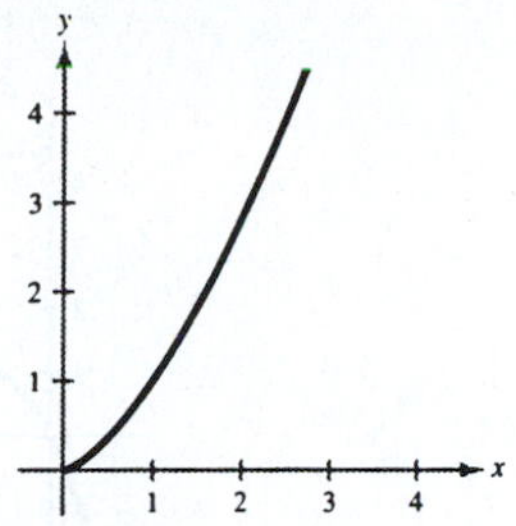

The graph is neither odd nor even.

44. $f(x) = |x + 2|$

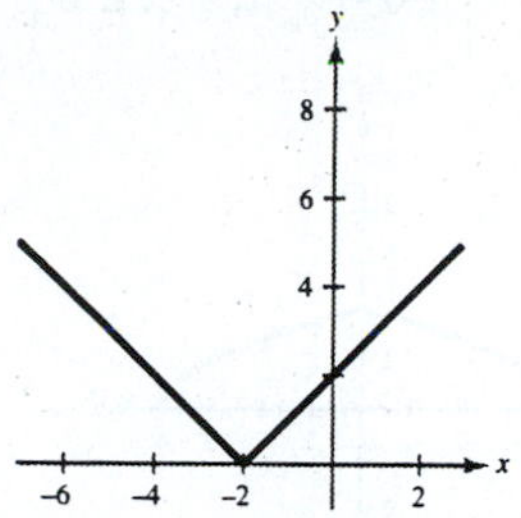

The graph is neither odd nor even.

46. $f(x) = \begin{cases} 2x + 1, & x \le -1 \\ x^2 - 2, & x > -1 \end{cases}$

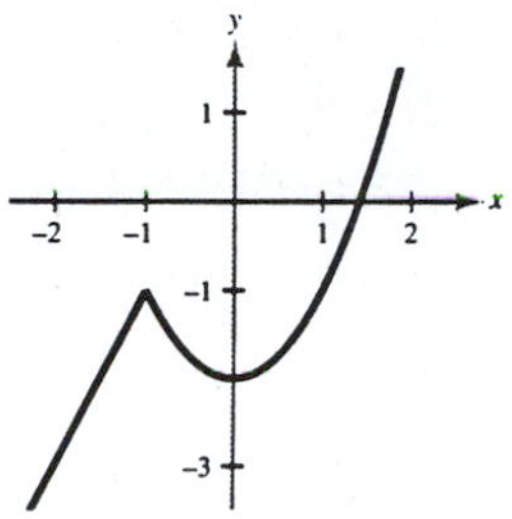

The graph is neither odd nor even.

48. $f(x) = 4x + 2$

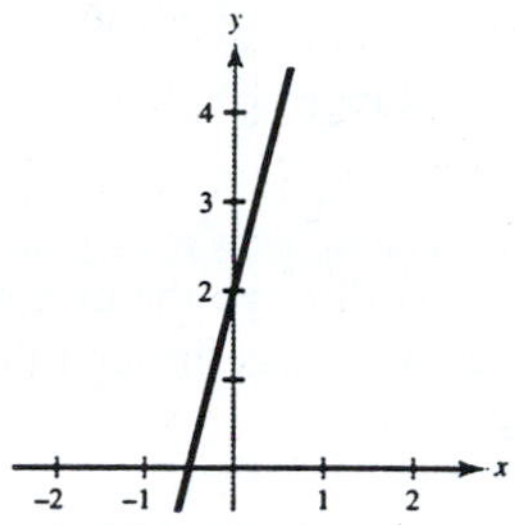

$f(x) \ge 0$

$4x + 2 \ge 0$

$4x \ge -2$

$x \ge -\frac{1}{2}$

$\left[-\frac{1}{2}, \infty\right)$

50. $f(x) = x^2 - 4x$

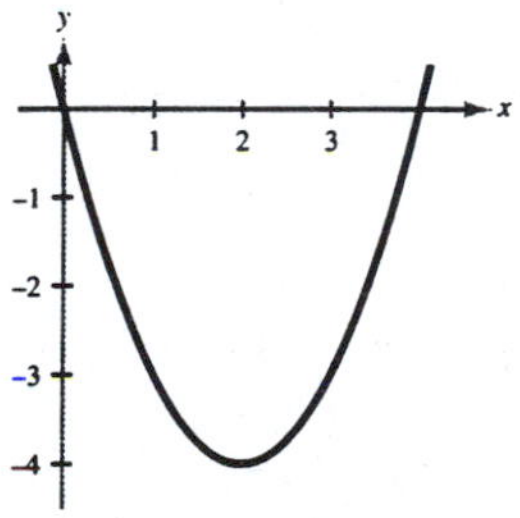

$f(x) \ge 0$

$x^2 - 4x \ge 0$

$x(x - 4) \ge 0$

$(-\infty, 0), (4, \infty)$

52. $f(x) = \sqrt{x + 2}$

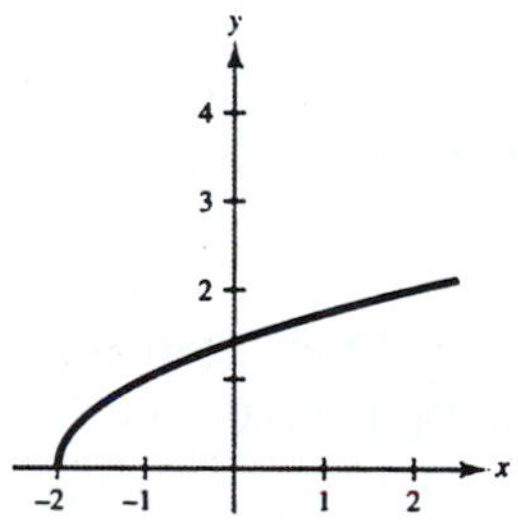

$f(x) \ge 0$

$\sqrt{x + 2} \ge 0$

$x + 2 \ge 0$

$x \ge -2$

$[-2, \infty)$

54. $f(x) = -(1 + |x|)$

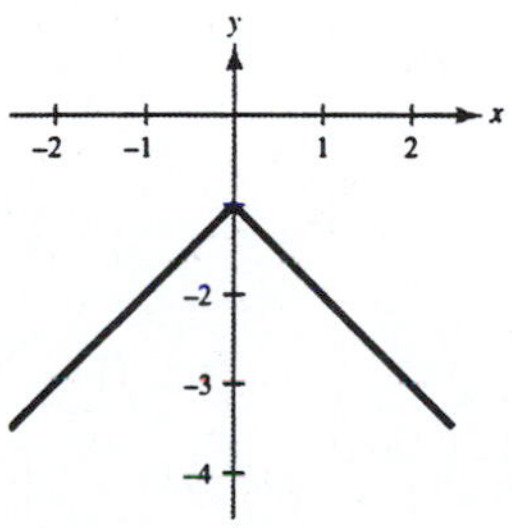

$f(x)$ is never greater than 0. ($f(x) < 0$ for all x.)

56. $f(x) = \frac{1}{2}(2 + |x|)$

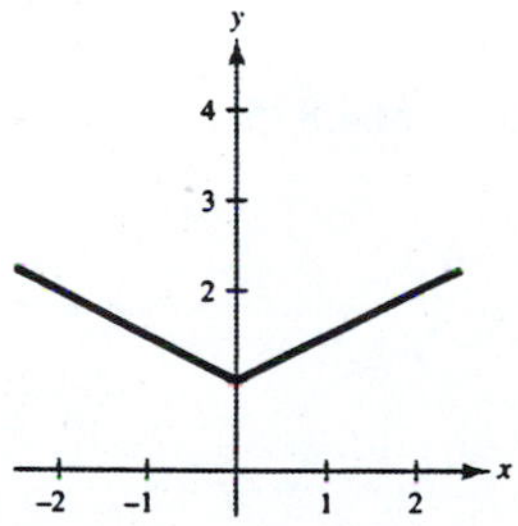

$f(x)$ is always greater than 0.

$(-\infty, \infty)$

58. $f(x) = \begin{cases} \sqrt{4+x}, & x < 0 \\ \sqrt{4-x}, & x \geq 0 \end{cases}$

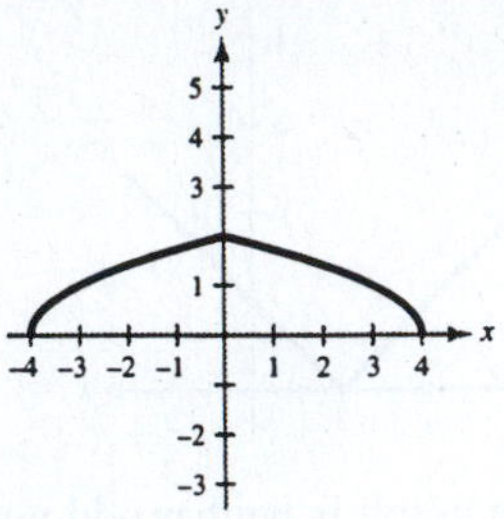

60. $f(x) = \begin{cases} 1-(x-1)^2, & x \leq 2 \\ \sqrt{x-2}, & x > 2 \end{cases}$

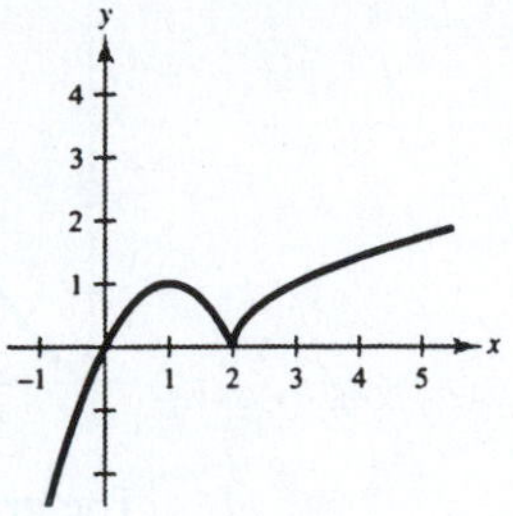

62. $h(t) = \sqrt{4-t^2}$

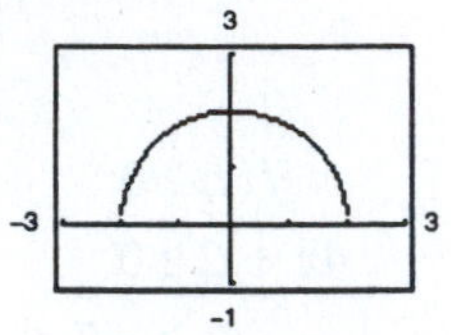

Domain: $4 - t^2 \geq 0$

$4 \geq t^2$

$[-2, 2]$

Range: $[0, 2]$

64. $g(x) = 2\left(\frac{1}{4}x - [\![\frac{1}{4}x]\!]\right)^2$

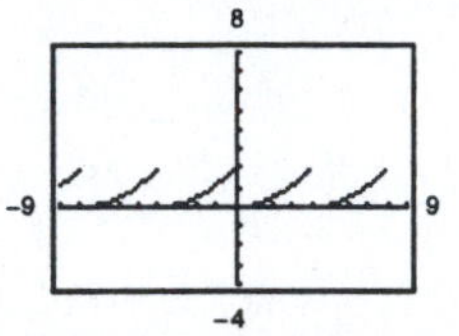

Domain: $(-\infty, \infty)$

Range: $(0, 2)$

Pattern: Sawtooth

66. The graph of $y = x^7$ will pass through the origin and will be symmetric with the origin. The graph of $y = x^8$ will pass through the origin and will be symmetric with respect to the y-axis.

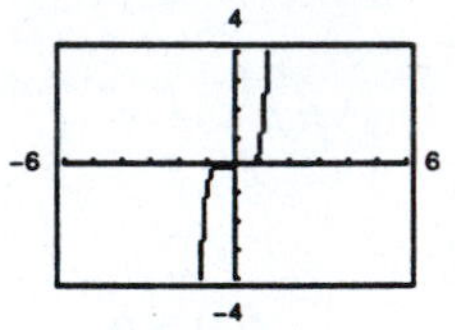

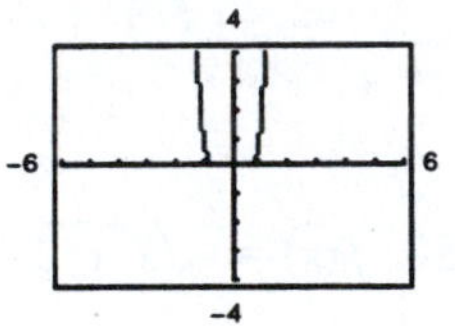

68. Model: (Total cost) = (Flat rate) + (Rate per pound)

Labels: Total cost = C

Flat rate = 9.80

Rate per pound = $2.50[\![x]\!],\ x > 0$

Equation: $C = 9.80 + 2.50[\![x]\!],\ x > 0$

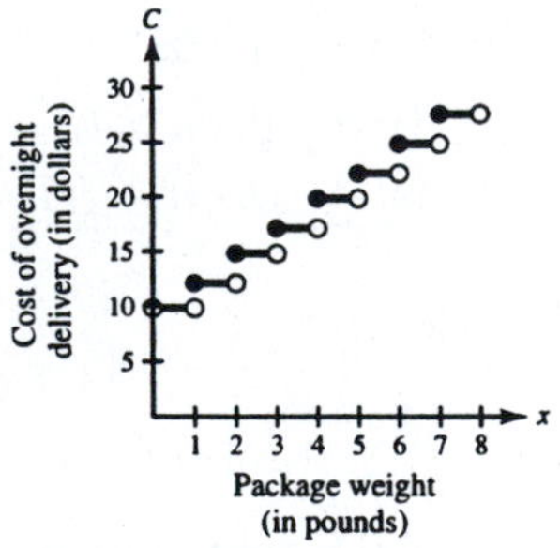

70. $L = -0.294x^2 + 97.744x - 664.875$

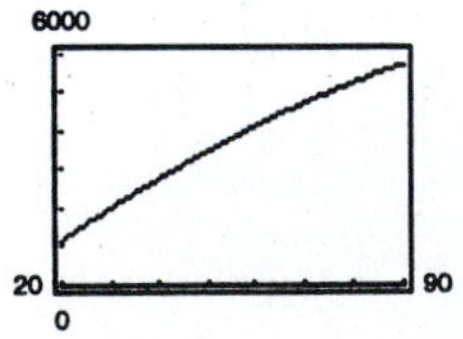

A bulb of approximately 30 watts is necessary to obtain 2000 lumens.

72. $h = \text{top} - \text{bottom}$

$= 3 - (4x - x^2)$

$= 3 - 4x + x^2$

74. $h = \text{top} - \text{bottom}$

$= 2 - \sqrt[3]{x}$

76. $L = \text{right} - \text{left}$

$= 2 - \sqrt[3]{2y}$

78. $L = \text{right} - \text{left}$

$= \dfrac{2}{y} - 0 = \dfrac{2}{y}$

80. (a) $A = \text{(area of square)} - \text{(area of corners)}$
$= 64 - 2x^2, \; 0 \le x \le 4$

(b)

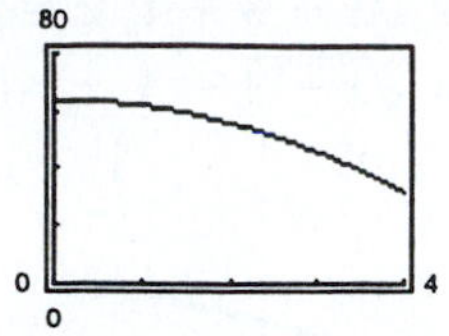

Over the domain of $0 \le x \le 4$, the range is $32 \le A \le 64$.

(c) If x is the maximum value of 4, the resulting figure would be a square (with the hypotenuses of the corner triangles as its sides). The length of the sides of this figure would be $\sqrt{4^2 + 4^2} = 4\sqrt{2}$.

82.

Interval	*Intake Pipe*	*Drainpipe* 1	*Drainpipe* 2
[0, 5]	Open	Closed	Closed
[5, 10]	Open	Open	Closed
[10, 20]	Closed	Closed	Closed
[20, 30]	Closed	Closed	Open
[30, 40]	Open	Open	Open
[40, 45]	Open	Closed	Open
[45, 50]	Open	Open	Open
[50, 60]	Open	Open	Closed

84. $f(x) = a_{2n}x^{2n} + a_{2n-2}x^{2n-2} + \cdots + a_2x^2 + a_0$

$f(-x) = a_{2n}(-x)^{2n} + a_{2n-2}(-x)^{2n-2} + \cdots + a_2(-x)^2 + a_0$

$= a_{2n}x^{2n} + a_{2n} - 2x^{2n-2} + \cdots + a_2x^2 + a_0$

$= f(x)$

$f(-x) = f(x)$; thus, $f(x)$ is even.

86.
$$100 - (x - 5)^2 = 0$$
$$100 - (x^2 - 10x + 25) = 0$$
$$100 - x^2 + 10x - 25 = 0$$
$$x^2 - 10x - 75 = 0$$
$$(x - 15)(x + 5) = 0$$
$$x - 15 = 0 \Rightarrow x = 15$$
$$x + 5 = 0 \Rightarrow x = -5$$

88.
$$16x^2 - 40x + 25 = 0$$
$$(4x - 5)(4x - 5) = 0$$
$$4x - 5 = 0 \Rightarrow x = \tfrac{5}{4}$$

Section 1.5 Translations and Combinations

Solutions to Even-Numbered Exercises

2. (a) $f(x) = x^2 - 2, \; c = -2$
$f(x) = x^2, \; c = 0$
$f(x) = x^2 + 2, \; c = 2$

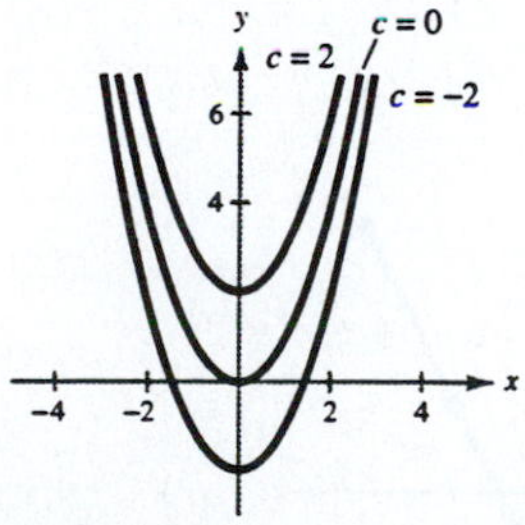

(b) $f(x) = (x^2 + 2), \; c = -2$
$f(x) = x^2, \; c = 0$
$f(x) = (x^2 - 2), \; c = 2$

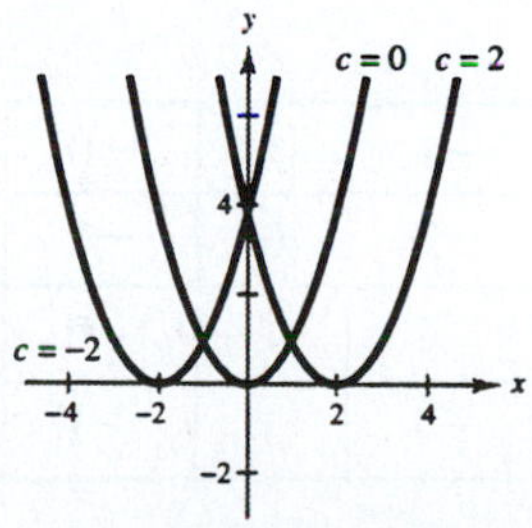

4. (a) $f(x) = \sqrt{x} - 3,\ c = -3$
$f(x) = \sqrt{x} - 1,\ c = -1$
$f(x) = \sqrt{x} + 1,\ c = 1$
$f(x) = \sqrt{x} + 3,\ c = 3$

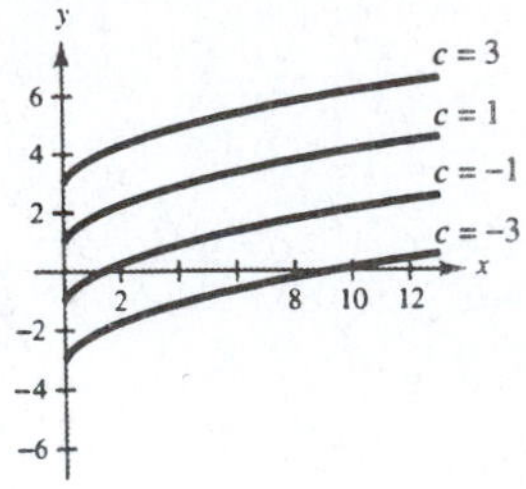

(b) $f(x) = \sqrt{x + 3},\ c = -3$
$f(x) = \sqrt{x + 1},\ c = -1$
$f(x) = \sqrt{x - 1},\ c = 1$
$f(x) = \sqrt{x - 3},\ c = 3$

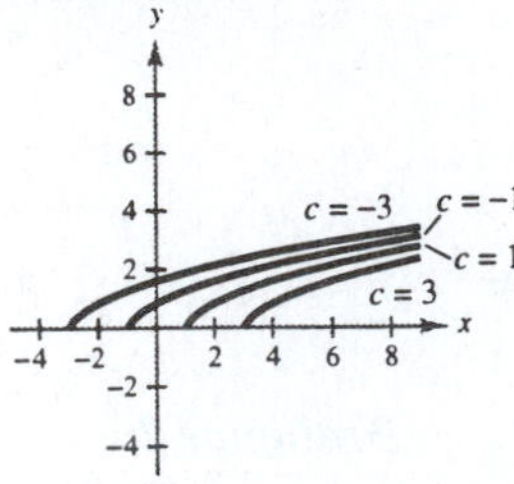

(c) $f(x) = \sqrt{x - 3} - 3,\ c = -3$
$f(x) = \sqrt{x - 3} - 1,\ c = -1$
$f(x) = \sqrt{x - 3} + 1,\ c = 1$
$f(x) = \sqrt{x - 3} + 3,\ c = 3$

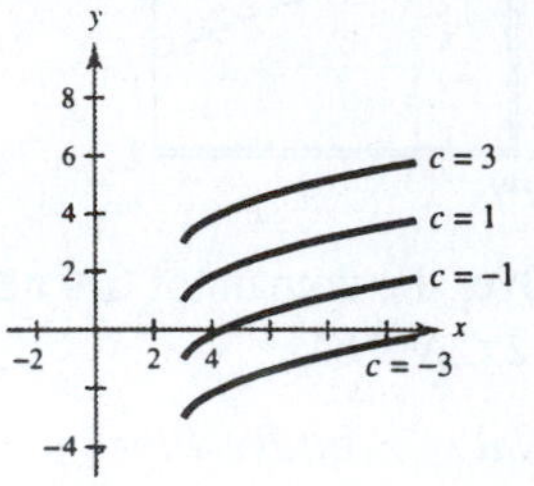

6. (a) $y = f(x) - 1$

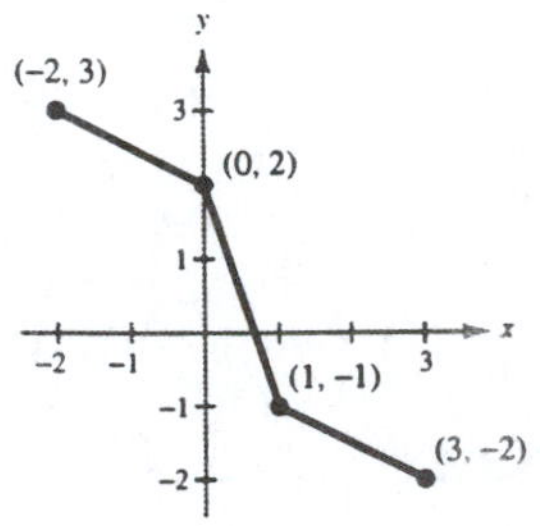

(b) $y = f(x + 1)$

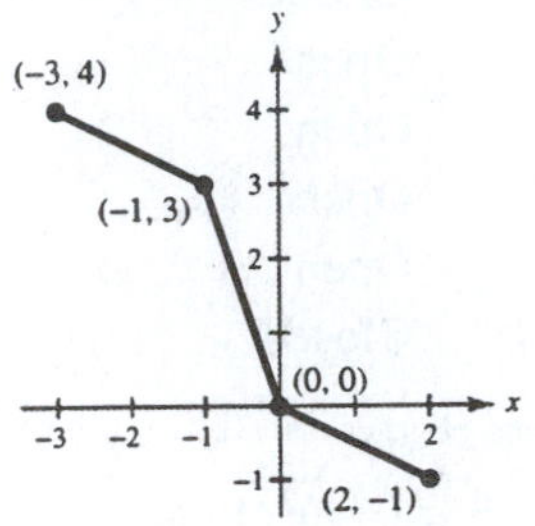

(c) $y = f(x - 1)$

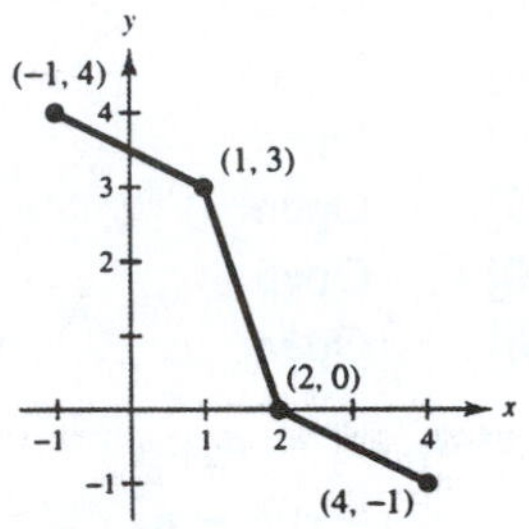

(d) $y = -f(x - 2)$

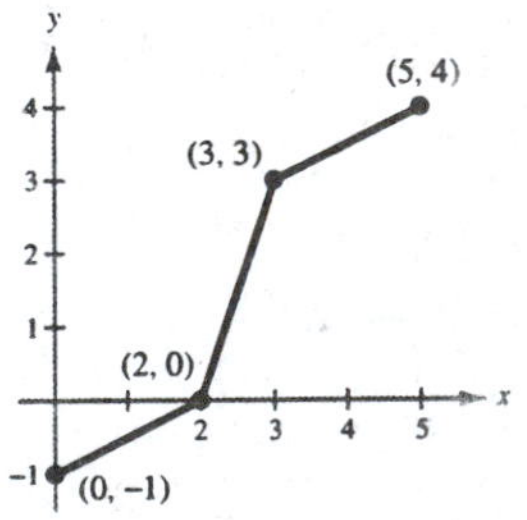

(e) $y = f(-x)$

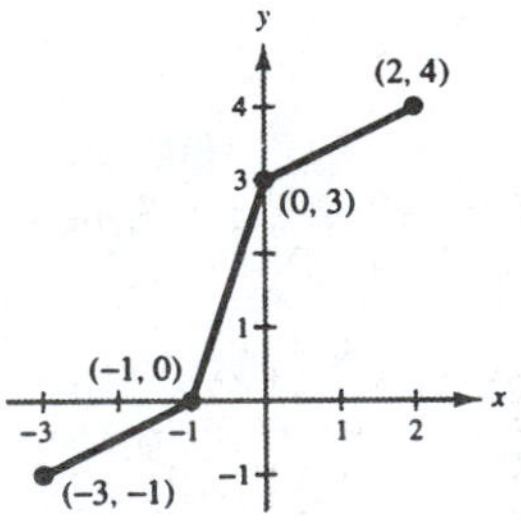

(f) $y = \frac{1}{2}f(x)$

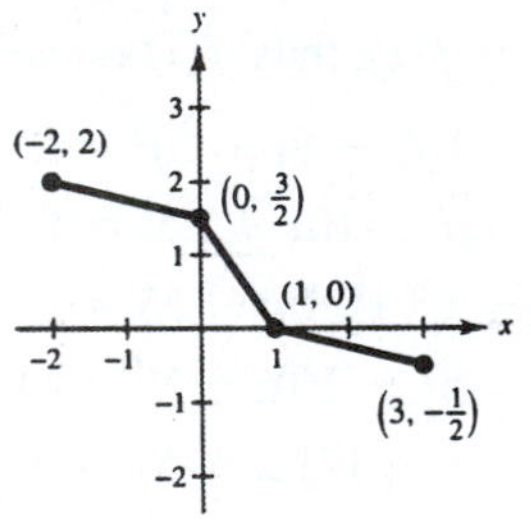

8. (a) The graph of $f(x) = x^3$ was reflected in the y-axis and shifted upward one unit.

$y = -x^3 + 1 = 1 - x^3$

(b) The graph of $f(x) = x^3$ was shifted upward one unit and to the right one unit.

$y = (x - 1)^3 + 1$

10. Common function $y = x$

Transformation: vertical shrink

Formula: $y = \frac{1}{2}x$

12. Common function: constant function

Formula: $y = 7$

14. Common function: $y = |x|$

Transformation: horizontal shift

Formula: $y = |x + 2|$

16.

x	-2	-1	0	1	2
$f(x)$	-2	0	-1	-1	1
$g(x)$	1	1	0	2	2
$h(x) = (f + g)(x)$	-1	1	-1	1	3

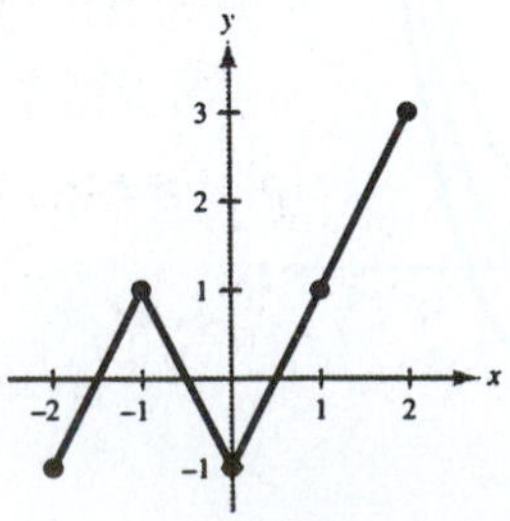

18.

x	0	1	2	3	4
$f(x)$	2	2	2	2	2
$g(x)$	0	1	$\sqrt{2}$	$\sqrt{3}$	2
$h(x) = (f + g)(x)$	2	3	$2 + \sqrt{2}$	$2 + \sqrt{3}$	4

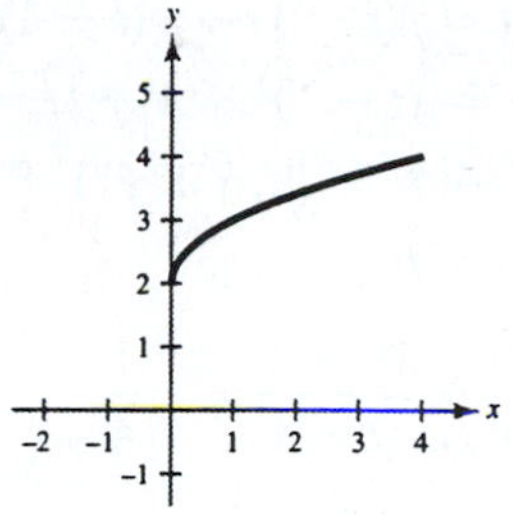

20. $f(x) = 2x - 5,\ g(x) = 1 - x$

(a) $(f + g)(x) = 2x - 5 + 1 - x$
$= x - 4$

(b) $(f - g)(x) = 2x - 5 - (1 - x)$
$= 2x - 5 - 1 + x$
$= 3x - 6$

(c) $(fg)(x) = (2x - 5)(1 - x)$
$= 2x - 2x^2 - 5 + 5x$
$= -2x^2 + 7x - 5$

(d) $\left(\dfrac{f}{g}\right)(x) = \dfrac{2x - 5}{1 - x}$

Domain: $1 - x \neq 0$
$x \neq 1$

22. $f(x) = 2x - 5,\ g(x) = 5$

(a) $(f + g)(x) = 2x - 5 + 5 = 2x$

(b) $(f - g)(x) = 2x - 5 - 6 = 2x - 10$

(c) $(fg)(x) = (2x - 5)(5) = 10x - 25$

(d) $\left(\dfrac{f}{g}\right)(x) = \dfrac{2x - 5}{5} = \dfrac{2}{5}x - 1$

Domain: $-\infty < x < \infty$

24. $f(x) = \sqrt{x^2 - 4},\ g(x) = \dfrac{x^2}{x^2 + 1}$

(a) $(f + g)(x) = \sqrt{x^2 - 4} + \dfrac{x^2}{x^2 + 1}$

(b) $(f - g)(x) = \sqrt{x^2 - 4} - \dfrac{x^2}{x^2 + 1}$

(c) $(fg)(x) = \sqrt{x^2 - 4} - \dfrac{x^2}{x^2 + 1}$

(d) $\left(\dfrac{f}{g}\right)(x) = \sqrt{x^2 - 4} \div \dfrac{x^2}{x^2 + 1}$
$= \dfrac{(x^2 + 1)\sqrt{x^2 - 4}}{x^2}$

Domain: $x^2 - 4 \geq 0$
$x^2 \geq 4 \Longrightarrow x \geq 2$ or $x \leq -2$

Domain: $|x| \geq 2$

26. $f(x) = \dfrac{x}{x + 1},\ g(x) = x^3$

(a) $(f + g)(x) = \dfrac{x}{x + 1} + x^3 = \dfrac{x + x^4 + x^3}{x + 1}$

(b) $(f - g)(x) = \dfrac{x}{x + 1} - x^3 = \dfrac{x - x^4 + x^3}{x + 1}$

(c) $(fg)(x) = \dfrac{x}{x + 1} \cdot x^3 = \dfrac{x^4}{x + 1}$

(d) $\left(\dfrac{f}{g}\right)(x) = \dfrac{x}{x + 1} \div x^3$
$= \dfrac{x}{x + 1} \cdot \dfrac{1}{x^3} = \dfrac{1}{x^2(x + 1)}$

Domain: $x \neq 0, x \neq -1$

28. $(f - g)(-2) = f(-2) - g(-2)$
$= (-2)^2 + 1 - (-2 - 4)$
$= 4 + 1 - (-6)$
$= 11$

30. $(f + g)(1) = f(1) + g(1)$
$= (1)^2 + 1 + (1) - 4$
$= -1$

32. $(f + g)(t - 1) = f(t - 1) + g(t - 1)$
$= (t - 1)^2 + 1 + (t - 1) - 4$
$= t^2 - 2t + 1 + 1 + t - 1 - 4$
$= t^2 - t - 3$

34. $(fg)(-6) = f(-6) \cdot g(-6)$
$= [(-6)^2 + 1][(-6) - 4]$
$= (37)(-10)$
$= -370$

36. $\left(\frac{f}{g}\right)(0) = \frac{f(0)}{g(0)} = \frac{0^2 + 1}{0 - 4} = -\frac{1}{4}$

38. $(2f)(5) = 2 \cdot f(5) = 2(5^2 + 1) = 52$

40. $f(x) = \frac{1}{3}x,\ g(x) = -x + 4$
$(f + g)(x) = \frac{1}{3}x - x + 4 = -\frac{2}{3}x + 4$

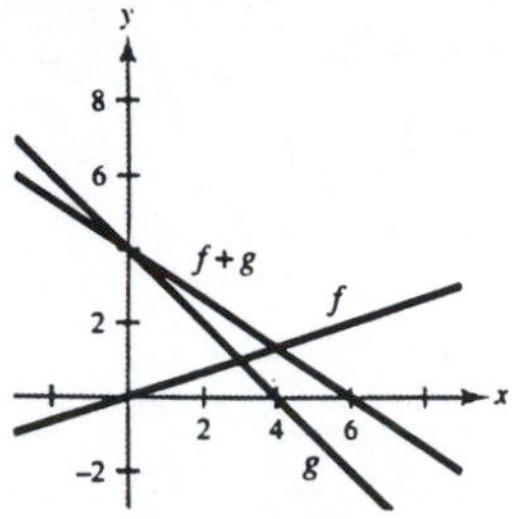

42. $f(x) = 4 - x^2,\ g(x) = x$
$(f + g)(x) = 4 - x^2 + x = 4 + x - x^2$

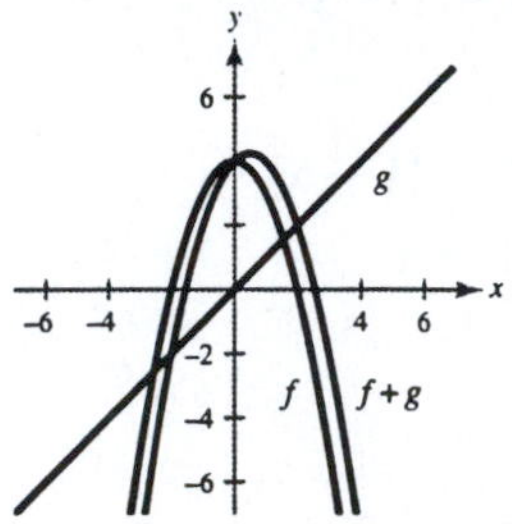

44. $f(x) = \frac{x}{2},\ g(x) = \sqrt{x}$
$(f + g)(x) = \frac{x}{2} + \sqrt{x}$

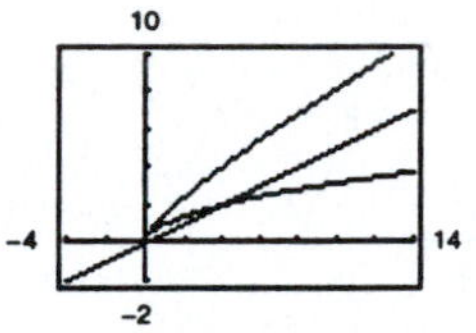

$f(x)$ contributes most to the magnitude of the sum for $0 \le x \le 2$. $f(x)$ also contributes most to the magnitude of the sum for $x > 5$.

46. (a) Total sales $= R_1 + R_2$
$= 480 - 8t - 0.8t^2 + 254 + 0.78t$
$= 734 - 7.22t - 0.8t^2$

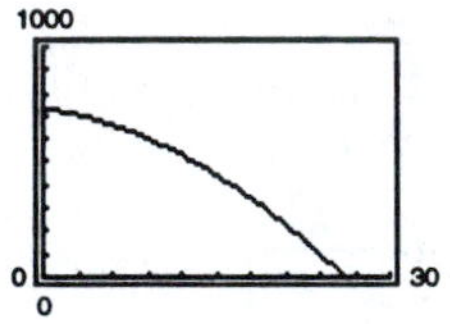

(b) Total sales have been increasing.

48. $y_1 = 0.16t^2 - 2.43t + 13.96$
$y_2 = 0.17t + 0.38$
$y_3 = 0.04t + 0.44$
$y_1 + y_2 + y_3 = 0.16t^2 - 2.43t + 13.96 + 0.17t$
$+ 0.38 + 0.04t + 0.44$
$= 0.16t^2 - 2.22t + 14.78$

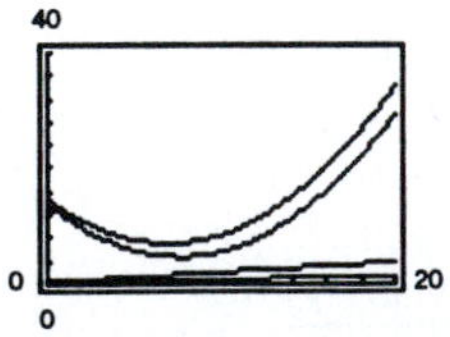

$(y_1 + y_2 + y_3)(15) = 0.16(15)^2 - 2.22(15) + 14.78$
$= 17.48$

The total variable costs per mile in 1995 was 17.49 cents.

50. From 0 to 6, $T(t) = 60$. From 6 to 6.5, $T(t)$ goes from 60 to 72.

$T(t) - 60 = \frac{72 - 60}{6.5 - 6}(t - 6)$
$T(t) - 60 = 24t - 144$
$T(t) = 24t - 84$

From 6.5 to 20.5, $T(t) = 72$. From 20.5 to 21, $T(t)$ goes from 72 to 60.

$T(t) - 60 = \frac{72 - 60}{20.5 - 21}(t - 21)$
$T(t) - 60 = -24(t - 21)$
$T(t) = -24t + 564$

From 21 to 24, $T(t) = 60$.

$$T(t) = \begin{cases} 60, & 0 \le t \le 6 \\ 24t - 84, & 6 < t < 6.5 \\ 72, & 6.5 \le t \le 20.5 \\ -24t + 564, & 20.5 < t < 21 \\ 60, & 21 \le t \le 24 \end{cases}$$

52. $f(x) = \sqrt[3]{x-1},\ g(x) = x^3 + 1$

(a) $(f \circ g)(x) = f(g(x))$
$= f(x^3 + 1)$
$= \sqrt[3]{(x^3 + 1) - 1}$
$= \sqrt[3]{x^3} = x$

(b) $(g \circ f)(x) = g(f(x))$
$= g(\sqrt[3]{x-1})$
$= (\sqrt[3]{x-1})^3 + 1$
$= (x - 1) + 1 = x$

(c) $(f \circ f)(x) = f(f(x))$
$= f(\sqrt[3]{x-1})$
$= \sqrt[3]{\sqrt[3]{x-1} - 1}$

54. $f(x) = x^3,\ g(x) = \dfrac{1}{x}$

(a) $(f \circ g)(x) = f(g(x))$
$= f\left(\dfrac{1}{x}\right)$

(b) $(g \circ f)(x) = g(f(x))$
$= g(x^3)$
$= \dfrac{1}{x^3}$

(c) $(f \circ f)(x) = f(f(x))$
$= f(x^3)$
$= (x^3)^3 = x^9$

56. $f(x) = \sqrt[3]{x-1},\ g(x) = x^3 + 1$

(a) $(f \circ g)(x) = f(g(x))$
$= f(x^3 + 1)$
$= \sqrt[3]{x^3 + 1 - 1}$
$= x$

(b) $(g \circ f)(x) = g(f(x))$
$= g(\sqrt[3]{x-1})$
$= (\sqrt[3]{x-1})^3 + 1$
$= x - 1 + 1 = x$

58. $f(x) = x^4,\ g(x) = x^4$

(a) $(f \circ g)(x) = f(g(x))$
$= f(x^4)$
$= (x^4)^4 = x^{16}$

(b) $(g \circ f)(x) = g(f(x))$
$= g(x^4)$
$= (x^4)^4 = x^{16}$

60. $f(x) = 2x - 3,\ g(x) = 2x - 3$

(a) $(f \circ g)(x) = f(g(x))$
$= f(2x - 3)$
$= 2(2x - 3) - 3$
$= 4x - 9$

(b) $(g \circ f)(x) = g(f(x))$
$= g(2x - 3)$
$= 2(2x - 3) - 3$
$= 4x - 9$

62. $f(x) = x^{2/3},\ g(x) = x^6$

(a) $(f \circ g)(x) = f(g(x))$
$= f(x^6)$
$= (x^6)^{2/3} = x^4$

(b) $(g \circ f)(x) = g(f(x))$
$= g(x^{2/3})$
$= (x^{2/3})^6 = x^4$

64. (a) $(f - g)(1) = f(1) - g(1)$
$= 2 - 3 = -1$

(b) $(fg)(4) = f(4) \cdot g(4)$

66. (a) $(f \circ g)(1) = f(g(1))$
$= f(3) = 2$

(b) $(g \circ f)(3) = g(f(3))$
$= g(2) = 2$

68. $g(x) = f(x) - 1$

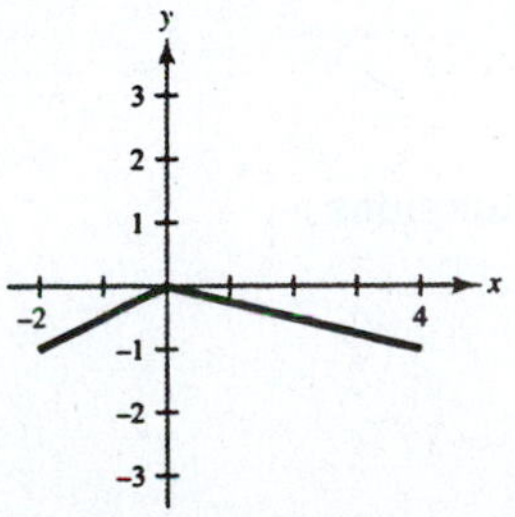

70. $g(x) = -2f(x)$

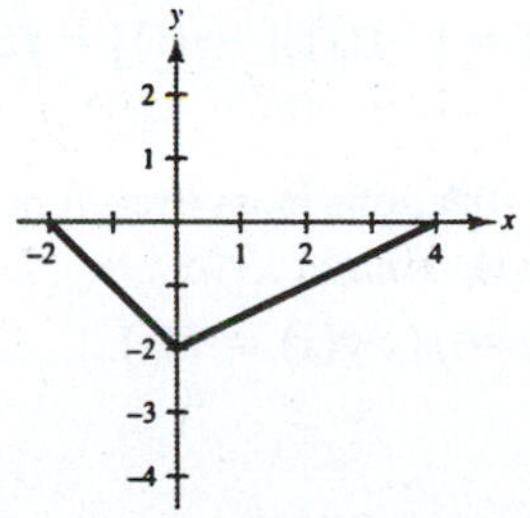

72. $h(x) = (1 - x)^3$

One possibility: Let $g(x) = 1 - x$ and $f(x) = x^3$.

$(f \circ g)(x) = f(1 - x) = (1 - x)^3 = h(x)$

74. $h(x) = \sqrt{9 - x}$

One possibility: Let $g(x) = 9 - x$ and $f(x) = \sqrt{x}$.

$(f \circ g)(x) = f(9 - x) = \sqrt{9 - x} = h(x)$

76. $h(x) = \dfrac{4}{(5x + 2)^2}$

One possibility: Let $g(x) = 5x + 2$ and $f(x) = \dfrac{4}{x^2}$.

$(f \circ g)(x) = f(5x + 2)$

78. $f(x) = \dfrac{1}{x}$, $g(x) = x + 3$

(a) Domain of f: all real numbers $x \neq 0$

(b) Domain of g: all real numbers

(c) $(f \circ g)(x) = f(x + 3) = \dfrac{1}{x + 3}$

Domain: $x + 3 \neq 0$

$x \neq -3$

All real numbers $x \neq -3$

80. $f(x) = 2x + 3$, $g(x) = \dfrac{x}{2}$

(a) Domain of f: all real numbers

(b) Domain of g: all real numbers

(c) $(f \circ g)(x) = f\left(\dfrac{x}{2}\right) = 2\left(\dfrac{x}{2}\right) + 3 = x + 3$

Domain: all real numbers

82.

$$\begin{aligned} f(x) &= 1 - x^2 \\ f(x + h) &= 1 - (x + h)^2 \\ &= 1 - (x^2 + 2hx + h^2) \\ &= 1 - x^2 - 2hx - h^2 \end{aligned}$$

$$\frac{f(x + h) - f(x)}{h} = \frac{1 - x^2 - 2hx - h^2 - (1 - x^2)}{h} = \frac{-2hx - h^2}{h} = -2x - h$$

84.

$$\begin{aligned} f(x) &= \sqrt{2x + 1} \\ f(x + h) &= \sqrt{2(x + h) + 1} \\ \frac{f(x + h) - f(x)}{h} &= \frac{\sqrt{2(x + h) + 1} - \sqrt{2x + 1}}{h} \\ &= \frac{\sqrt{2(x + h) + 1} - \sqrt{2x + 1}}{h} \cdot \frac{\sqrt{2(x + h) + 1} + \sqrt{2x + 1}}{\sqrt{2(x + h) + 1} + \sqrt{2x + 1}} \\ &= \frac{[2(x + h) + 1] - (2x + 1)}{h\left(\sqrt{2(x + h) + 1} + \sqrt{2x + 1}\right)} \\ &= \frac{2x + 2h + 1 - 2x - 1}{h\left(\sqrt{2(x + h) + 1} + \sqrt{2x + 1}\right)} \end{aligned}$$

86. $(A \circ r)(t) = A(r(t)) = A(0.6t) = \pi(0.6t)^2 = 0.36\pi t^2$

$A \circ r$ represents the area of the circle at time t.

88. (a) $f(g(x)) = f(0.03x) = 0.03x - 500{,}000$

(b) $g(f(x)) = g(x - 500{,}000) = 0.03(x - 500{,}000)$

$g(f(x))$ represents 3% of an amount over \$500,000.

90. The product of two odd functions is an even function. Let $f(x)$ and $g(x)$ be two odd functions and define $h(x) = f(x)g(x)$. Then

$$h(-x) = f(-x)g(-x) = [-f(x)][-g(x)] = f(x)g(x) = h(x).$$

Thus, h is even.

The product of two even functions is an even function. Let $f(x)$ and $g(x)$ be two even functions and define $h(x) = f(x)g(x)$. Then

$$h(-x) = f(-x)g(-x) = f(x)g(x) = h(x).$$

Thus, h is even.

Section 1.6 Inverse Functions

Solutions to Even-Numbered Exercises

2. The inverse is a line through (0, 6) and (6, 0). Matches graph (b).

4. The inverse is a third-degree equation through (0, 0). Matches graph (d).

6. $f(x) = \frac{1}{5}x$

$f^{-1}(x) = 5x$

$f(f^{-1}(x)) = f(5x) = \frac{1}{5}(5x) = x$

8. $f(x) = x - 5$

$f^{-1}(x) = x + 5$

$f(f^{-1}(x)) = f(x + 5) = (x + 5) - 5 = x$

$f^{-1}(f(x)) = f^{-1}(x - 5) = (x - 5) + 5 = x$

10. $f(x) = x^5$

$f^{-1}(x) = \sqrt[5]{x}$

$f(f^{-1}(x)) = f(\sqrt[5]{x}) = (\sqrt[5]{x})^5 = x$

$f^{-1}(f(x)) = f^{-1}(x^5) = \sqrt[5]{x^5} = x$

12. (a) $f(x) = x - 5,\ g(x) = x + 5$

$f(g(x)) = f(x + 5) = (x + 5) - 5 = x$

$g(f(x)) = g(x - 5) = (x - 5) + 5 = x$

(b)

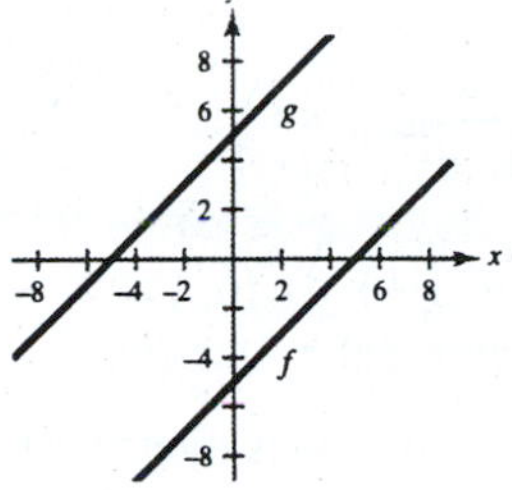

14. (a) $f(x) = 3 - 4x,\ g(x) = \dfrac{3 - x}{4}$

$$f(g(x)) = f\left(\frac{3 - x}{4}\right) = 3 - 4\left(\frac{3 - x}{4}\right) = 3 - (3 - x) = x$$

$$g(f(x)) = g(3 - 4x) = \frac{3 - (3 - 4x)}{4} = \frac{4x}{4} = x$$

(b)

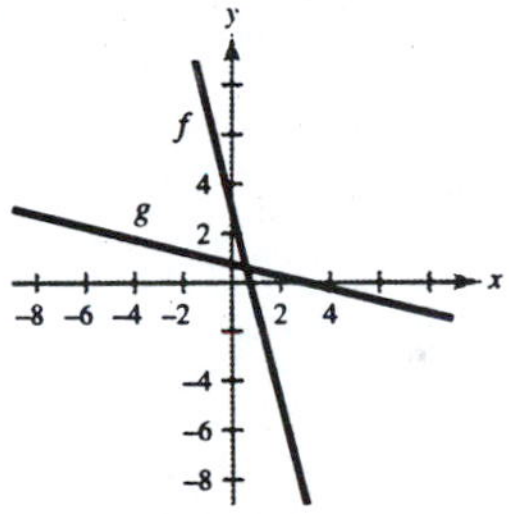

16. (a) $f(x) = \dfrac{1}{x},\ g(x) = \dfrac{1}{x}$

$$f(g(x)) = f\left(\frac{1}{x}\right) = \frac{1}{1/x} = 1 \div \frac{1}{x} = 1 \cdot \frac{x}{1} = x$$

$$g(f(x)) = g\left(\frac{1}{x}\right) = \frac{1}{1/x} = 1 \div \frac{1}{x} = 1 \cdot \frac{x}{1} = x$$

(b)

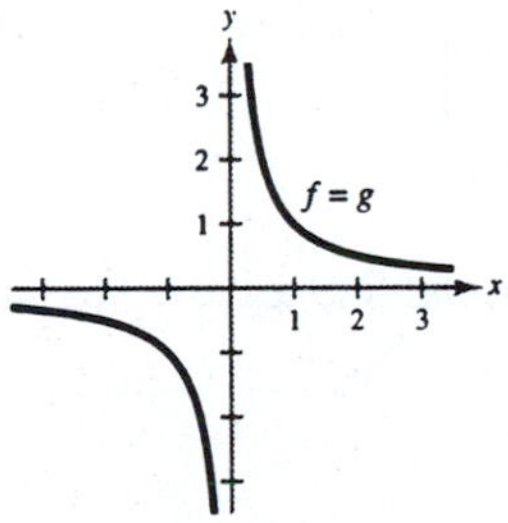

18. (a) $f(x) = 1 - x^3,\ g(x) = \sqrt[3]{1 - x}$

$f(g(x)) = f(\sqrt[3]{1 - x}) = 1 - (\sqrt[3]{1 - x})^3 = 1 - (1 - x) = x$

$g(f(x)) = g(1 - x^3) = \sqrt[3]{1 - (1 - x^3)} = \sqrt[3]{x^3} = x$

(b)

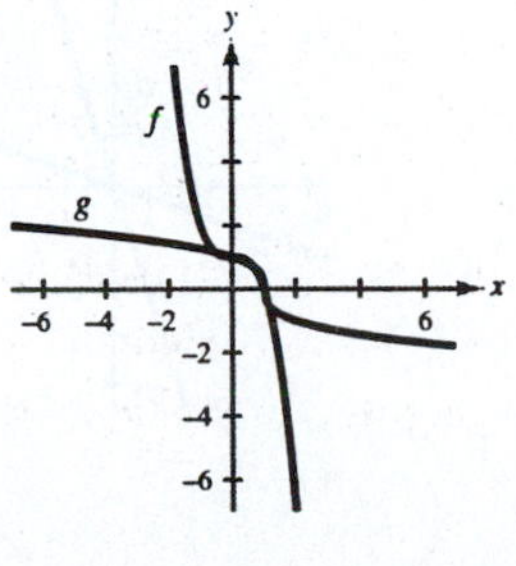

20. (a) $f(x) = \dfrac{1}{1+x},\ x \geq 0;\ g(x) = \dfrac{1-x}{x},\ 0 < x \leq 1$

$$f(g(x)) = f\left(\frac{1-x}{x}\right) = \frac{1}{1+\left(\frac{1-x}{x}\right)} = \frac{1}{\frac{x}{x}+\frac{1-x}{x}} = \frac{1}{\frac{1}{x}} = x$$

$$g(f(x)) = g\left(\frac{1}{1+x}\right) = \frac{1-\left(\frac{1}{1+x}\right)}{\left(\frac{1}{1+x}\right)} = \frac{\frac{1+x}{1+x}-\frac{1}{1+x}}{\frac{1}{1+x}} = \frac{\frac{x}{1+x}}{\frac{1}{1+x}} = \frac{x}{1+x}\cdot\frac{x+1}{1} = x$$

(b)

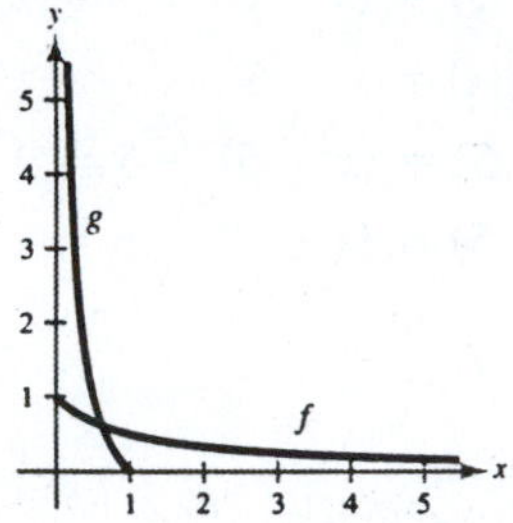

22. No, $\{(-3, 10), (-2, 6), (-1, 4), (0, 1), (2, -3), (2, -10)\}$ does not represent a function.

24. No, because some horizontal lines intersect the graph twice, f does not have an inverse.

26. Yes, because no horizontal lines intersect the graph at more than one point, f has an inverse.

28. $f(x) = 10$

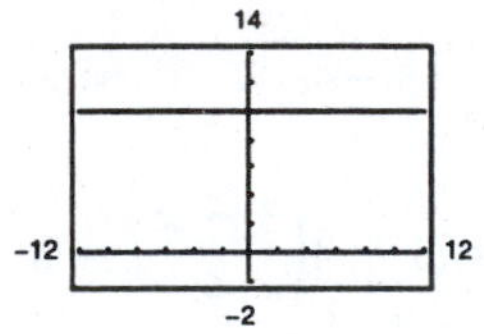

f does not pass the horizontal line test, so f has no inverse.

30. $g(x) = (x + 5)^3$

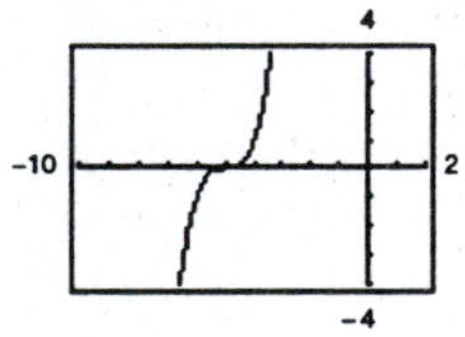

g passes the horizontal line test, so g has an inverse.

32. $f(x) = \frac{1}{8}(x + 2)^2 - 1$

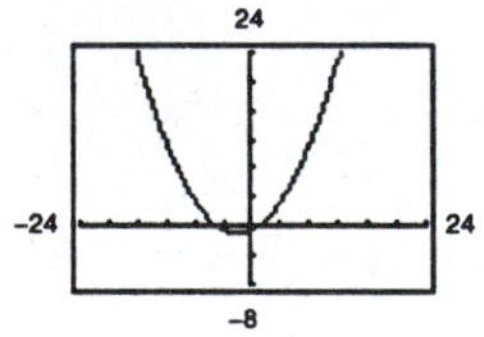

f does not pass the horizontal line test, so f has no inverse.

34.

$$f(x) = 3x$$
$$y = 3x$$
$$x = 3y$$
$$\frac{x}{3} = y$$
$$f^{-1}(x) = \frac{x}{3}$$

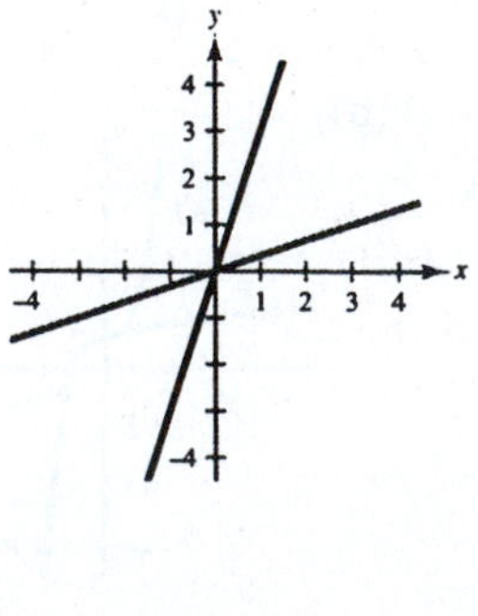

36.

$$f(x) = x^3 + 1$$
$$y = x^3 + 1$$
$$x = y^3 + 1$$
$$x - 1 = y^3$$
$$\sqrt[3]{x-1} = y$$
$$f^{-1}(x) = \sqrt[3]{x-1}$$

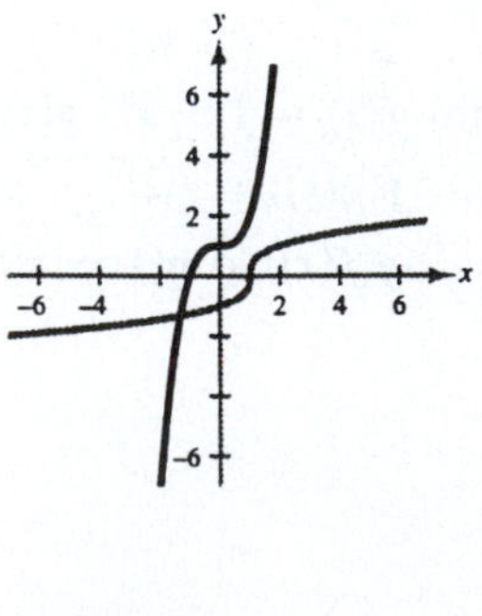

38. $f(x) = x^2, \; x \geq 0$

$y = x^2$

$x = y^2$

$\sqrt{x} = y$

$f^{-1}(x) = \sqrt{x}$

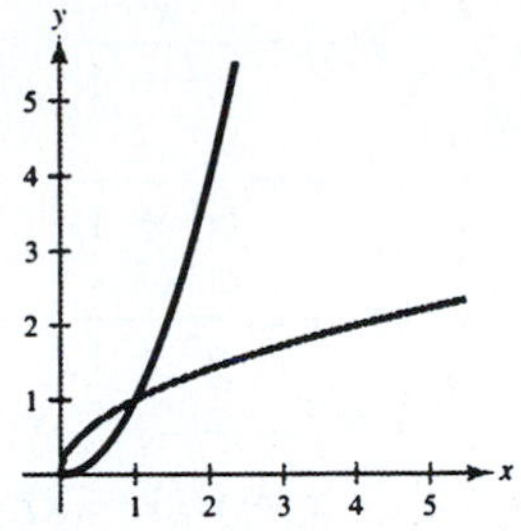

40. $f(x) = \dfrac{4}{x}$

$y = \dfrac{4}{x}$

$x = \dfrac{4}{y}$

$y = \dfrac{4}{x}$

$f^{-1}(x) = \dfrac{4}{x}$

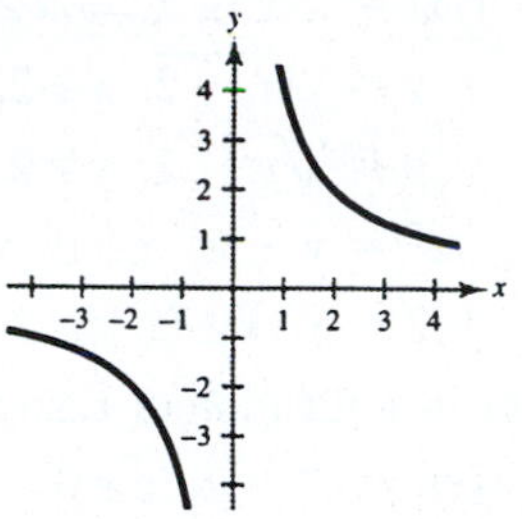

42. $f(x) = x^{3/5}$

$y = x^{3/5}$

$x = y^{3/5}$

$x^{3/5} = (y^{3/5})^{5/3}$

$x^{5/3} = y$

$f^{-1}(x) = x^{5/3}$

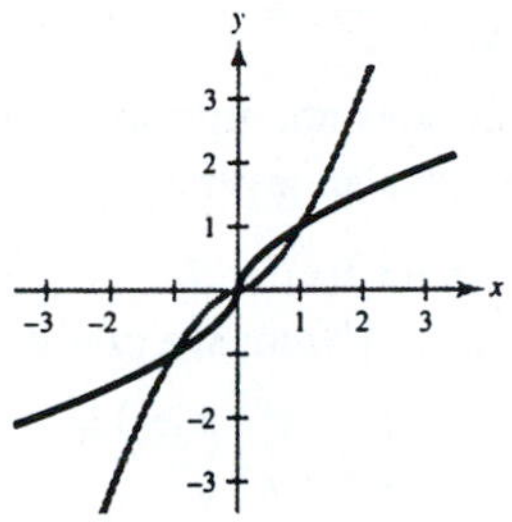

44. $f(x) = \dfrac{1}{x^2}$

$y = \dfrac{1}{x^2}$

$x = \dfrac{1}{y^2}$

$y^2 = \dfrac{1}{x}$

$y = \pm\sqrt{\dfrac{1}{x}}$

This does not represent y as a function of x. f does not have an inverse.

46. $f(x) = 3x + 5$

$y = 3x + 5$

$x = 3y + 5$

$x - 5 = 3y$

$\dfrac{x - 5}{3} = y$

This is a function of x, so f has an inverse.

$f^{-1}(x) = \dfrac{x - 5}{3}$

48. $f(x) = \dfrac{3x + 4}{5}$

$y = \dfrac{3x + 4}{5}$

$x = \dfrac{3y + 4}{5}$

$5x = 3y + 4$

$5x - 4 = 3y$

$\dfrac{5x - 4}{3} = y$

This is a function of x, so f has an inverse.

$f^{-1}(x) = \dfrac{5x - 4}{3}$

50. $q(x) = (x - 5)^2$

$y = (x - 5)^2$

$x = (y - 5)^2$

$\pm\sqrt{x} = y - 5$

$5 \pm \sqrt{x} = y$

This does not represent y as a function of x, so q does not have an inverse.

52. $f(x) = |x - 2|, \; x \leq 2 \Rightarrow y \geq 0$

$y = |x - 2|, \; x \leq 2, \; y \geq 0$

$x = |y - 2|, \; y \leq 2, \; x \geq 0$

$x = y - 2$ or $-x = y - 2$

$2 + x = y$ or $2 - x = y$

The portion that satisfies the conditions $y \leq 2$ and $x \geq 0$ is $2 - x = y$. This is a function of x, so f has an inverse.

$f^{-1}(x) = 2 - x, \; x \geq 0$

54. $f(x) = \sqrt{x - 2} \Rightarrow x \geq 2,\ y \geq 0$

$y = \sqrt{x - 2},\ x \geq 2,\ y \geq 0$

$x = \sqrt{y - 2},\ y \geq 2,\ x \geq 0$

$x^2 = y - 2,\ x \geq 0,\ y \geq 2$

$x^2 + 2 = y,\ x \geq 0,\ y \geq 2$

This is a function of x, so f has an inverse.

$f^{-1}(x) = x^2 + 2,\ x \geq 0$

56. $f(x) = \dfrac{x^2}{x^2 + 1}$

$y = \dfrac{x^2}{x^2 + 1}$

$x = \dfrac{y^2}{y^2 + 1}$

$xy^2 + x = y^2$

$x = y^2(1 - x)$

$\dfrac{x}{1 - x} = y^2$

$\pm\sqrt{\dfrac{x}{1 - x}} = y$

This does not represent y as a function of x, so f does not have an inverse.

58. $f(x) = ax + b,\ a \neq 0$

$y = ax + b$

$x = ay + b$

$x - b = ay$

$\dfrac{x - b}{a} = y$

This is a function of x, so f has an inverse.

$f^{-1}(x) = \dfrac{x - b}{a}$

60. If we let $f(x) = 1 - x^4,\ x \geq 0$, then f has an inverse. [Note: we could also let $x \leq 0$.]

$f(x) = 1 - x^4,\ x \geq 0 \Rightarrow y \leq 1$

$y = 1 - x^4,\ x \geq 0,\ y \leq 1$

$x = 1 - y^4,\ y \geq 0,\ x \leq 1$

$y^4 = 1 - x,\ y \geq 0,\ x \leq 1$

$y = \sqrt[4]{1 - x},\ x \leq 1,\ y \geq 0$

Thus, $f^{-1}(x) = \sqrt[4]{1 - x},\ x \leq 1$.

62. If we let $f(x) = |x - 2|,\ x \geq 2$, then f has an inverse. [Note: we could also let $x \leq 2$.]

$f(x) = |x - 2|,\ x \geq 2$

$f(x) = x - 2$ when $x \geq 2$.

$y = x - 2,\ x \geq 2,\ y \geq 0$

$x = y - 2,\ x \geq 0,\ y \geq 2$

$x + 2 = y,\ x \geq 0,\ y \geq 2$

Thus, $f^{-1}(x) = x + 2,\ x \geq 0$.

64.

x	$f(x)$
4	−3
3	−2
−1	0
−2	6

x	$f^{-1}(x)$
−3	4
−2	3
0	−1
6	−2

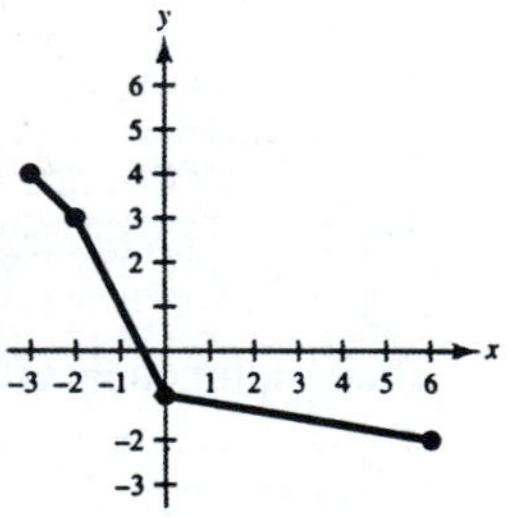

66. True

68. False, $f(x) = \dfrac{1}{x}$ has an inverse $f^{-1}(x) = \dfrac{1}{x}$.

In Exercises 70, 72 , and 74, $f(x) = \frac{1}{8}x - 3,\ g(x) - x^3,\ f^{-1}(x) = 8(x + 3),\ g^{-1}(x) = \sqrt[3]{x}$.

70. $(g^{-1} \circ f^{-1})(-3) = g^{-1}(f^{-1}(-3))$

$= g^{-1}(8(-3 + 3))$

$= g^{-1}(0) = \sqrt[3]{0} = 0$

72. $(g^{-1} \circ g^{-1})(-4) = g^{-1}(g^{-1}(-4))$

$= g^{-1}(\sqrt[3]{-4})$

74. $g^{-1} \circ f^{-1} = g^{-1}(f^{-1}(x))$

$= g^{-1}(8(x + 3))$

$= \sqrt[3]{8(x + 3)}$

$= 2^3\sqrt{x + 3}$

In Exercises 76 and 78, $f(x) = x + 4,\ g(x) = 2x - 5,\ f^{-1}(x) = x - 4,\ g^{-1}(x) = \dfrac{x+5}{2}$.

76.
$$\begin{aligned} f^{-1} \circ g^{-1}(x) &= f^{-1}(g^{-1}(x)) \\ &= f^{-1}\left(\frac{x+5}{2}\right) \\ &= \frac{x+5}{2} - 4 \\ &= \frac{x+5-8}{2} \\ &= \frac{x-3}{2} \end{aligned}$$

78.
$$\begin{aligned} (g \circ f)(x) &= g(f(x)) \\ &= g(x+4) \\ &= 2(x+4) - 5 \\ &= 2x + 8 - 5 \\ &= 2x + 3 \\ y &= 2x + 3 \\ x &= 2y + 3 \\ x - 3 &= 2y \\ \frac{x-3}{2} &= y \\ (g \circ f)^{-1}(x) &= \frac{x-3}{2} \end{aligned}$$

80. (a) $(\text{Total cost}) = \left(\begin{array}{c}\text{Cost of}\\ \text{first commodity}\end{array}\right) + \left(\begin{array}{c}\text{Cost of}\\ \text{second commodity}\end{array}\right)$

Labels: Total cost $= y$
Amount of first commodity $= x$
Amount of second commodity $= 50 - x$
Cost of first commodity $= 1.25x$
Cost of second commodity $= 1.60(50 - x)$

Equation: $y = 1.25x + 1.60(50 - x)$

(b)
$$\begin{aligned} x &= 1.25y + 1.60(50 - y) \\ x &= 1.25y + 80 - 1.60y \\ x - 80 &= -0.35y \\ \frac{x-80}{-0.35} &= y \\ y &= \frac{80-x}{0.35} \end{aligned}$$

$x =$ total cost

$y =$ number of pounds of less expensive commodity

(c) To keep the number of pounds of less expensive commodity nonnegative, $x \le 80$.

(d) $\dfrac{80-73}{0.35} = y = 20$ pounds

82. If $f(x) = k(2 - x - x^3)$ has an inverse and $f^{-1}(3) = -2$, then $f(-2) = 3$. Thus,
$$\begin{aligned} f(-2) &= k(2 - (-2) - (-2)^3) = 3 \\ k(2 + 2 + 8) &= 3 \\ 12k &= 3 \\ k &= \tfrac{3}{12} = \tfrac{1}{4} \end{aligned}$$
Thus, $k = \frac{1}{4}$.

84. No, if (12, 21.02) were added to the table, the graph of f would not pass the horizontal line test.

86.
$$\begin{aligned} (x-5)^2 &= 8 \\ x - 5 &= \pm\sqrt{8} \\ x &= 5 \pm 2\sqrt{2} \end{aligned}$$

88.
$$\begin{aligned} 9x^2 + 12x + 3 &= 0 \\ (9x + 3)(x + 1) &= 0 \\ 9x + 3 = 0 &\Rightarrow x = -\tfrac{1}{3} \\ x + 1 = 0 &\Rightarrow x = -1 \end{aligned}$$

90.
$$\begin{aligned} 2x^2 - 4x - 6 &= 0 \\ 2(x^2 - 2x - 3) &= 0 \\ 2(x + 1)(x - 3) &= 0 \\ x + 1 = 0 &\Rightarrow x = -1 \\ x - 3 = 0 &\Rightarrow x = 3 \end{aligned}$$

92.
$$\begin{aligned} 2x^2 + 4x - 9 &= 2(x-1)^2 \\ 2x^2 + 4x - 9 &= 2(x^2 - 2x + 1) \\ 2x^2 + 4x - 9 &= 2x^2 - 4x + 2 \\ 8x - 11 &= 0 \\ 8x &= 11 \\ x &= \tfrac{11}{8} \end{aligned}$$

94. $(200 - 2x)(100 - 2x) = \frac{1}{4}(100)(200)$

$20{,}000 - 600x + 4x^2 = 5000$

$4x^2 - 600x + 15{,}000 = 0$

$4(x^2 - 150x + 3750) = 0$

Thus, $a = 1$, $b = -150$, and $c = 3750$.

$$x = \frac{-(-150) \pm \sqrt{(-150)^2 - 4(1)(3750)}}{2(1)}$$

$$x \approx \frac{150 + 86.6025}{2} \approx 118.301 \text{ ft. (not possible since lot is only 100 ft. wide)}$$

$$x \approx \frac{150 - 86.6025}{2} \approx 31.669 \text{ ft.}$$

The first person must mow an approximately 31.7-foot wide strip along each side. The person must go around the lot approximately

$$\frac{31.7}{24 \text{ inches}} = \frac{31.7}{2 \text{ feet}} \approx 16 \text{ times.}$$

96. Given $h = 2b$ and $A = 10$

$A = \frac{1}{2}bh$

$10 = \frac{1}{2}b(2b)$

$10 = b^2$

$\sqrt{10} = b$ and $h = 2b = 2\sqrt{10}$

The base is $\sqrt{10}$ feet and the height is $2\sqrt{10}$ feet.

Section 1.7 Mathematical Modeling

Solutions to Even-Numbered Exercises

2. $y = 5.4 - 0.03t,\ 8 \le t \le 52$

$t = 0$ represents 1940.

Year	1948	1952	1956	1960	1964	1968	1972	1976	1980	1984	1988	1992
Actual Number	5.30	5.20	4.91	4.84	4.72	4.53	4.32	4.16	4.15	4.12	4.06	4.12
Model	5.16	5.04	4.92	4.80	4.68	4.56	4.44	4.32	4.20	4.08	3.96	3.84

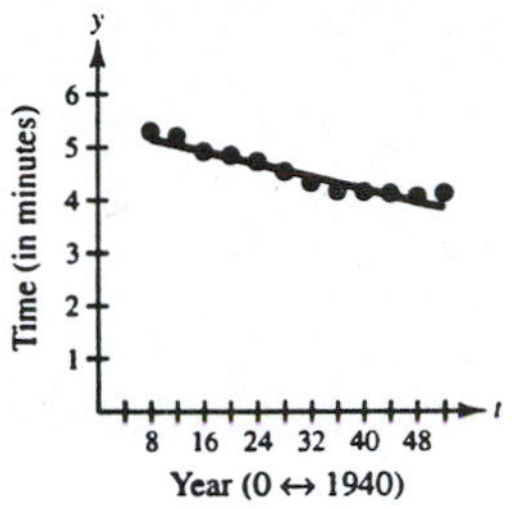

The model is a "good fit" for the actual data.

4. The graph appears to represent $y = \frac{3}{2}x$ which is a direct variation.

6.

x	2	4	6	8	10
$y = 2x^2$	8	32	72	128	200

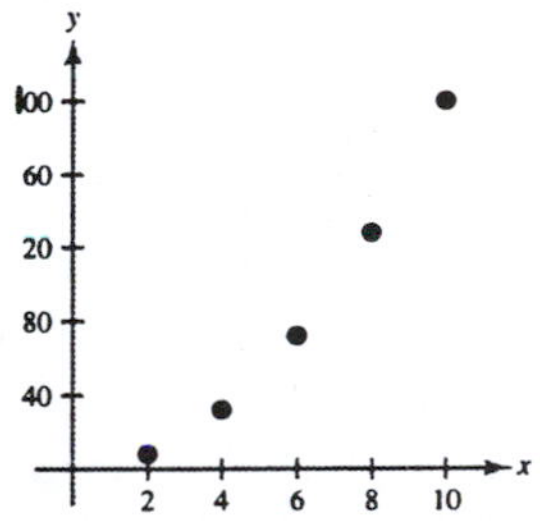

8.

x	2	4	6	8	10
$y = \frac{1}{4}x^2$	1	4	9	16	25

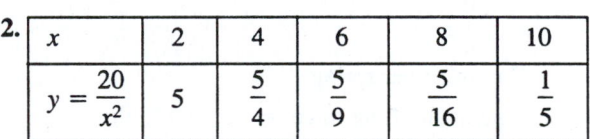

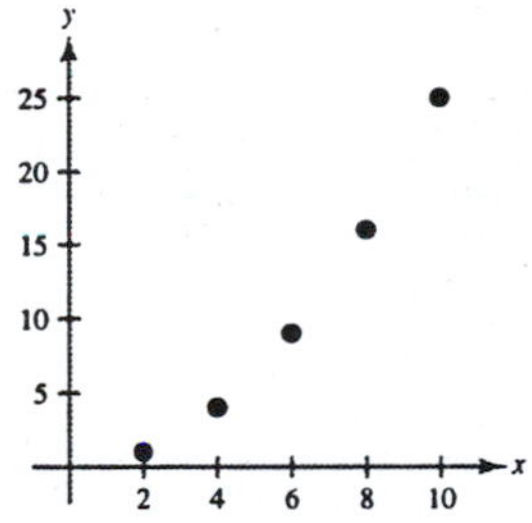

10.

x	2	4	6	8	10
$y = \frac{5}{x^2}$	$\frac{5}{4}$	$\frac{5}{16}$	$\frac{5}{36}$	$\frac{5}{64}$	$\frac{1}{20}$

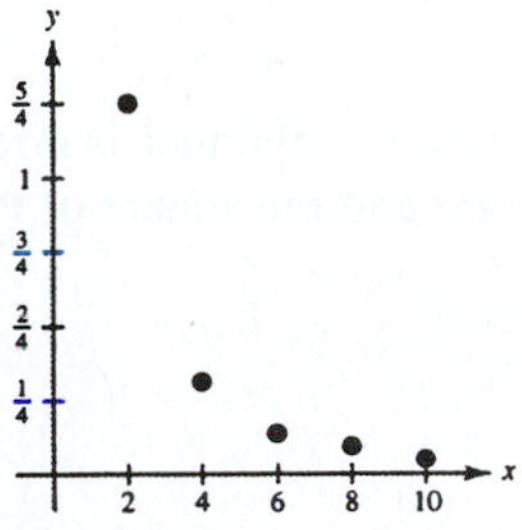

12.

x	2	4	6	8	10
$y = \frac{20}{x^2}$	5	$\frac{5}{4}$	$\frac{5}{9}$	$\frac{5}{16}$	$\frac{1}{5}$

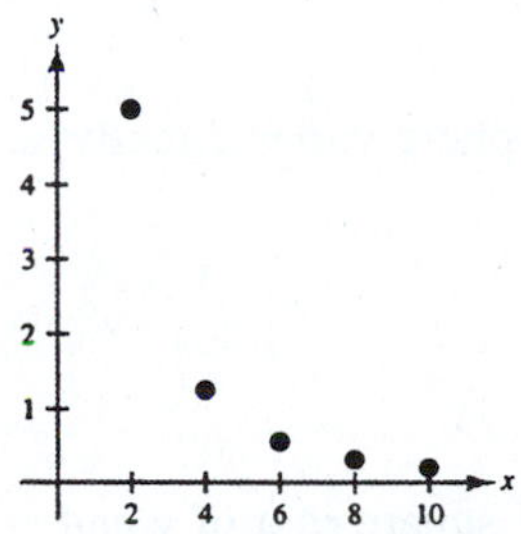

14. The chart represents the equation $y = \frac{2}{5}x$.

16. $y = \frac{k}{x}$

$24 = \frac{k}{5}$

$120 = k$

Thus, $y = \frac{120}{x}$. This equation checks with the other points given in the chart.

18. $y = mx$

$14 = m(2)$

$7 = m$

$y = 7x$

20. $y = mx$

$580 = m(6)$

$\frac{290}{3} = m$

$y = \frac{290}{3}x$

22. $I = kP$

$337.50 = k(5000)$

$0.0675 = k$

$I = 0.0675P$

24. $y = kx$

$10.22 = k(145.99)$

$0.07 \approx k$

$y = 0.07x$

$y = 0.07(540.50)$

$y \approx 37.84$

The sales tax is \$37.84.

26. $y = kx$

$53 = k(14)$

$\frac{53}{14} = k$

$y = \frac{53}{14}x$

Gallons	−2	−1	0	1	2
Liters	18.9	37.9	75.7	94.6	113.6

28. $d = kF$

$0.12 = k(220)$

$\frac{3}{5500} = k$

$d = \frac{3}{5500}F$

$0.16 = \frac{3}{5500}F$

$\frac{880}{3} = F$

The required force is $293\frac{1}{3}$ newtons.

30. $d = kF$

$1 = k(15)$

$k = \frac{1}{15}$

$d = \frac{1}{15}F$

$\frac{8}{2} = \frac{1}{15}F$

$F = 60$ lb per spring

Combined lifting force $= 2F = 120$ lbs.

32. $V = ke^3$

34. $h = \frac{k}{\sqrt{5}}$

36. $x = \frac{k}{t + 1}$

38. $V = klwh$

40. $z = kx^2y^3$

42. $R = k(T - T_e)$

44. $R = kS(S - L)$

46. $S = 4\pi r^2$

The surface area of a sphere varies directly as the square of the radius r.

48. $V = \pi r^2 h$

The volume of a right circular cylinder is directly proportional to the height and the square of the radius.

50. $\omega = \sqrt{\frac{kg}{W}}$

ω varies directly as the square root of g and inversely as the square root of W.

52. $s = kt^2$

$64 = k2^2$

$64 = 4k$

$16 = k$

$s = 16t^2$

54. $y = \frac{k}{x}$

$7 = \frac{k}{4}$

$28 = k$

$y = \frac{28}{x}$

56. $z = kxy$

$32 = k(10)(16)$

$32 = 160k$

$\frac{1}{5} = k$

$z = \frac{1}{5}xy$

58. $R = \frac{k}{s^2}$

$80 = \frac{k}{\left(\frac{1}{5}\right)^2}$

$80 = 25k$

$\frac{16}{5} = k$

$R = \frac{16}{5s^2}$

60. $P = \frac{kx}{y^2}$

$\frac{28}{3} = \frac{k(42)}{9^2}$

$\frac{28}{3} \cdot \frac{81}{42} = k$

$\frac{2 \cdot 27}{3} = k$

$18 = k$

$P = \frac{18x}{y^2}$

62. $v = \frac{kpq}{s^2}$

$1.5 = \frac{k(4.1)(6.3)}{(1.2)^2}$

$\frac{(1.5)(1.44)}{(4.1)(6.3)} = k$

$\frac{2.16}{25.83} = k$

$k = \frac{24}{287}$

$v = \frac{24pq}{287s^2}$

64. $P = kS(L - S)$

$10 = k(4)(6 - 4)$

$10 = k(8)$

$\frac{5}{4} = k$

$P = \frac{5}{4}S(L - S)$

66. $d = kv^2$

If the velocity is doubled:

$d = k(2v)^2$

$d = k \cdot 4v^2$

$\frac{4kv^2}{kv^2} = 4$

68. From Exercise 67, $k \approx 5.73 \times 10^{-8}$.

$r = \frac{4(5.73 \times 10^{-8})l}{\pi d^2}$

$d = \sqrt{\frac{4(5.73 \times 10^{-8})l}{\pi r}}$

$d = \sqrt{\frac{4(5.73 \times 10^{-8})(14)}{\pi(0.05)}}$

$d \approx 0.0045$ feet $= 0.054$ inches

70. $d = ks^2$

$75 = k(30)^2$

$\frac{1}{12} = k$

$d = \frac{1}{12}s^2$

$d = \frac{1}{12}(50)^2$

$d = 208\frac{1}{3}$ feet

72. $d = \frac{k}{p}$

$500 = \frac{k}{3.75}$

$1875 = k$

$d = \frac{1875}{p}$

$d = \frac{1875}{4.25} \approx 441$ units

74. Load $= \frac{kwd^2}{l}$

(a) load $= \frac{k(2w)d^2}{2l} = \frac{kwd^2}{l}$

The safe load is unchanged.

(b) load $= \frac{k(2w)(2d)^2}{l} = \frac{8kwd^2}{l}$

The safe load is 8 times as great.

(c) load $= \frac{k(2w)(2d)^2}{2l} = \frac{4kwd^2}{l}$

The safe load is 4 times as great.

(d) load $= \frac{kw\left(\frac{d}{2}\right)^2}{l} = \frac{\frac{1}{4}kwd^2}{l}$

The safe load is one-fourth as great.

76. (a)

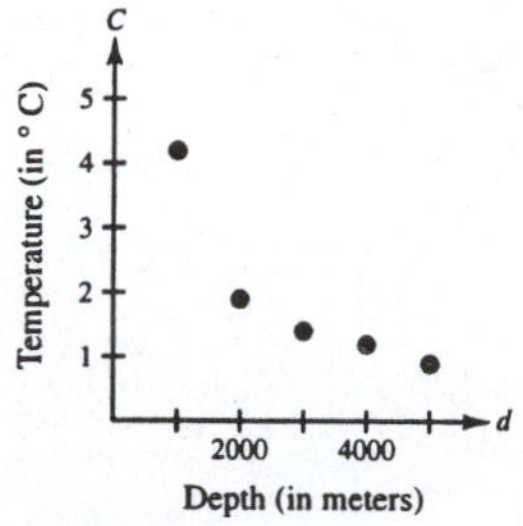

(b) Yes, the data appears to be modeled by the inverse proportion model.

$4.2 = \frac{k}{1000}$

$k = 4200$

(c)

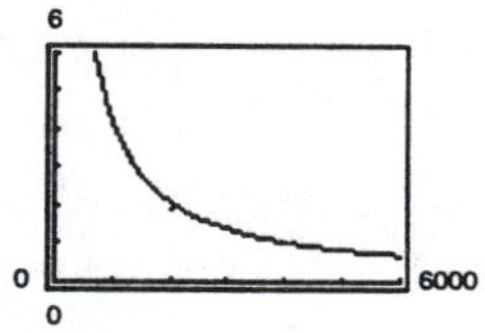

(d) $3 = \frac{4200}{d}$

$d = 1400$ meters

78. $I = \frac{k}{d^2}$

When the distance is doubled:

$I = \frac{k}{(2d)^2} = \frac{k}{4d^2}$

The illumination is one-fourth as great.

The model given in Exercise 78 is very close to $I = \frac{k}{d^2}$ The difference is probably due to measurement error.

80. The points do not follow a linear pattern. A linear model would not be a good approximation.

82. The points follow a linear pattern. A linear model would be a good approximation.

84.

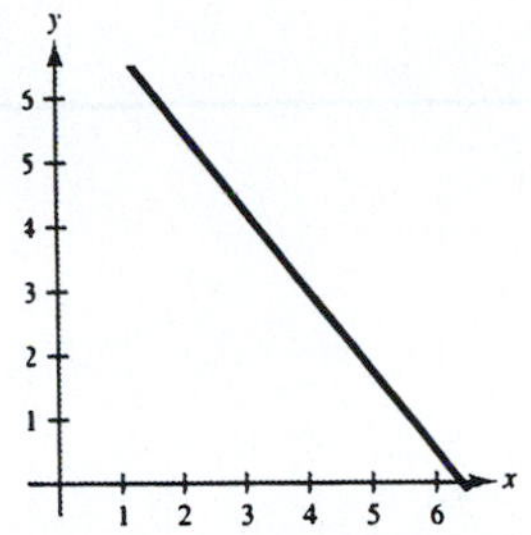

Using the points (4, 3) and (2, 5.5) we have

$$y - 3 = \frac{3 - 5.5}{4 - 2}(x - 4)$$

$$y - 3 = -\frac{5}{4}(x - 4)$$

$$y = -\frac{5}{4}x + 8$$

86.

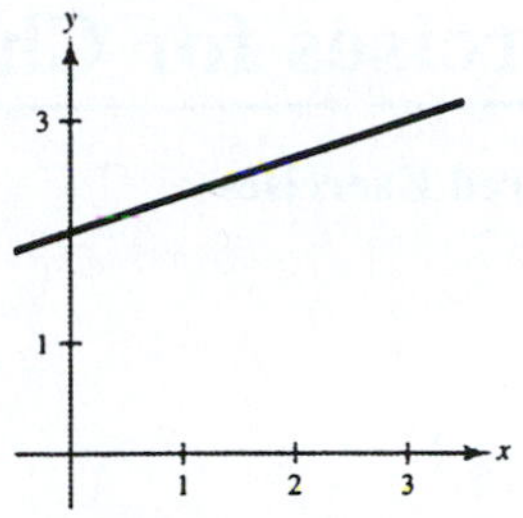

Using the points (0, 2)and (3, 3), we have

$$y - 2 = \frac{3 - 2}{3 - 0}(x)$$

$$y = \frac{1}{3}x + 2$$

88. The accuracy of the model in predicting prize winnings is questionable because the model is based on limited data.

90. (a) Using the linear regression capabilities of a calculator yields $y \approx 2284.9t - 4872.9$

(b)

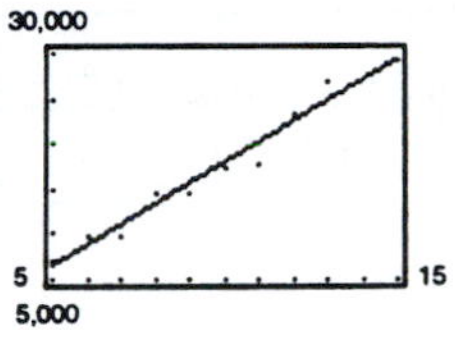

(c) Using the model with $t = 16$ yields $31,685.5 million.

92. (a) Using the linear regression capabilities of a calculator yields $y \approx 0.80x + 1.87$

(b)

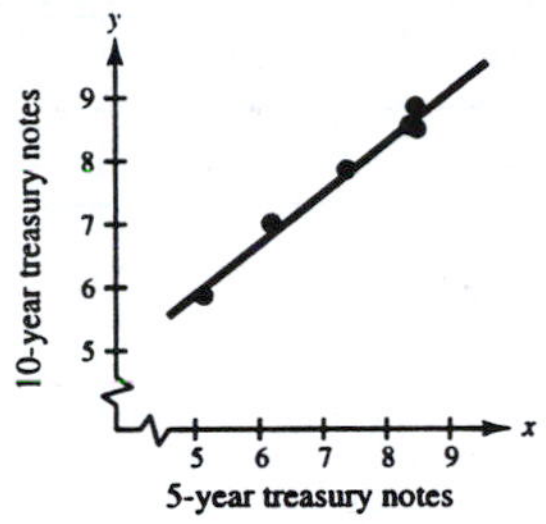

(c) No, a 1 percent increase in the yield of a 5-year note corresponds to an increase of 0.8 percent in the yield of a 10-year note, on average.

❑ Focus on Concepts

1. No. The slope cannot be determined without knowing the scale on the y-axis. The slopes could be the same.
2. -4. The slope with the greatest magnitude corresponds to the steepest line.
3. V-intercept: Initial cost; Slope: Annual depreciation
4. No. The element 3 in the domain corresponds to two elements in the range.
5. (a)

Xmin = -15
Xmax = 6
Xscl = 3
Ymin = -18
Ymax = 6
Yscl = 3

(b)

Xmin = -24
Xmax = 36
Xscl = 6
Ymin = -54
Ymax = 12
Yscl = 6

6. (a) Even. The graph is a reflection in the x-axis.
(b) Even. The graph is a reflection in the y-axis.
(c) Even. The graph is a vertical translation of f.
(d) Neither. The graph is a horizontal translation of f.

7. (a) $g(t) = \frac{3}{4}f(t)$
(b) $g(t) = f(t) + 10{,}000$
(c) $g(t) = f(t - 2)$

❑ Review Exercises for Chapter 1

Solutions to Even-Numbered Exercises

2. $3x + 2y + 6 = 0$

$$2x = -3x - 6$$

$$y = -\frac{3}{2}x - 3$$

Line with x-intercept $(-2, 0)$ and y-intercept $(0, -3)$

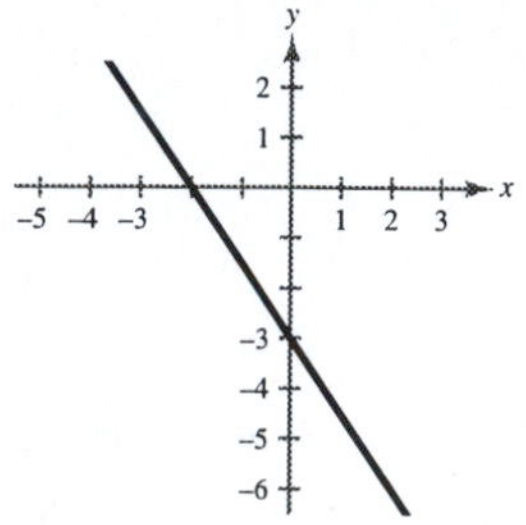

4. $y = 8 - |x|$

For $x > 0$, graph $y = 8 - x$;

For $x < 0$, graph $y = 8 + x$.

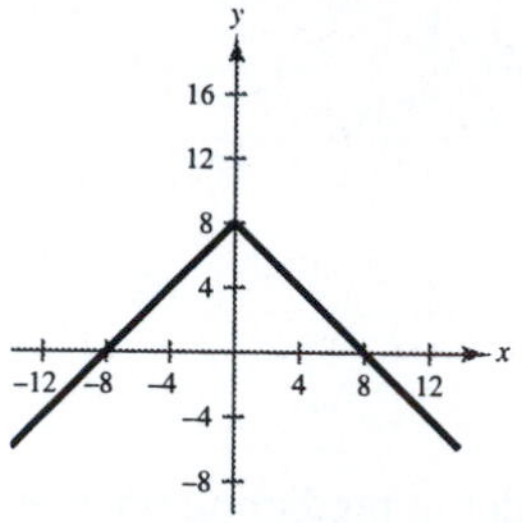

6. $y = \sqrt{x + 2}$, domain: $[-2, \infty)$

x	-2	0	2	7
y	0	$\sqrt{2}$	2	3

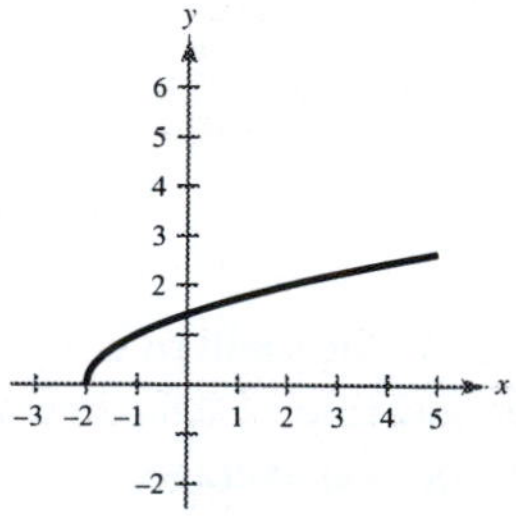

8. $y = x^2 - 4x$ is a parabola.

x	-1	0	1	2	3
y	5	0	-3	-4	-3

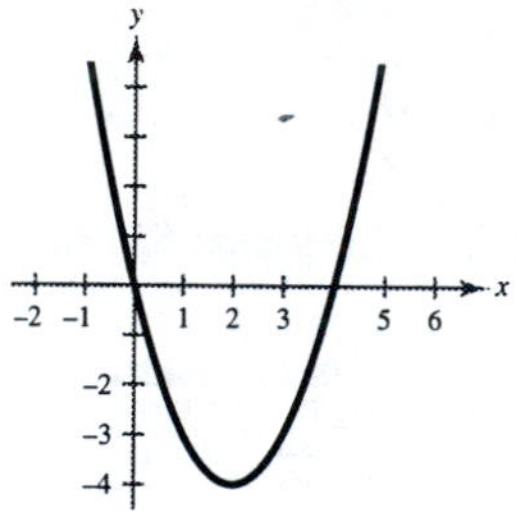

10. $x^2 + y^2 = 10 \Rightarrow y = \pm\sqrt{10 - x^2}$

x	0	1	-1	3	-3
y	$\pm\sqrt{10}$	± 3	± 3	± 1	± 1

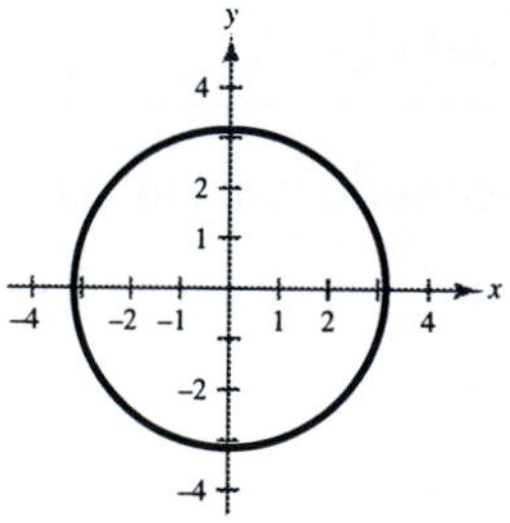

12. $y = 4 - (x - 4)^2$

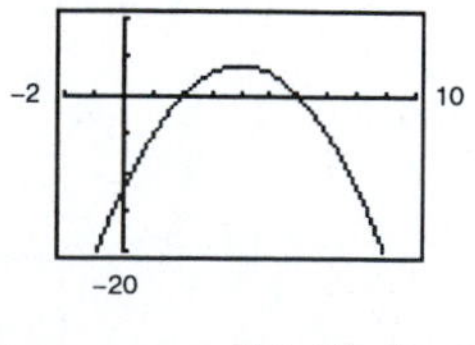

Intercepts: $(6, 0)$, $(2, 0)$, $(0, -12)$

14. $y = \frac{1}{4}x^3 - 3x$

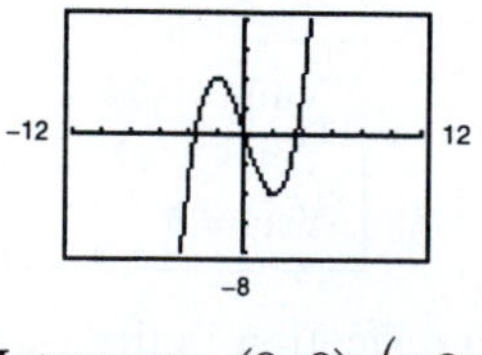

Intercepts: $(0, 0)$, $(\pm 2\sqrt{3}, 0)$

16. $y = x\sqrt{x+3}$

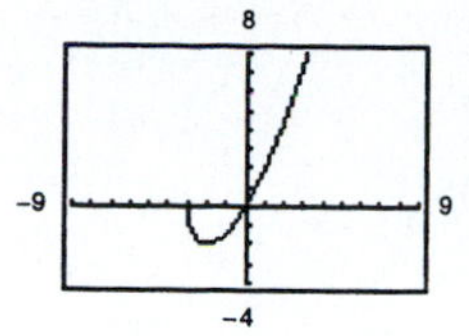

Intercepts: $(0, 0)$, $(-3, 0)$

18. $y = |x+2| + |3-x|$

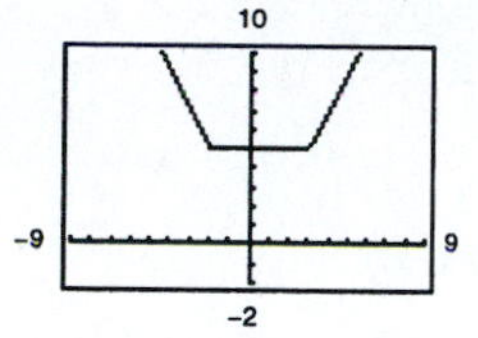

Intercept: $(0, 5)$

20. $(x+2)^2 + y^2 = 16$

$$y^2 = 16 - (x+2)^2$$

$$y = \pm\sqrt{16-(x+2)^2}$$

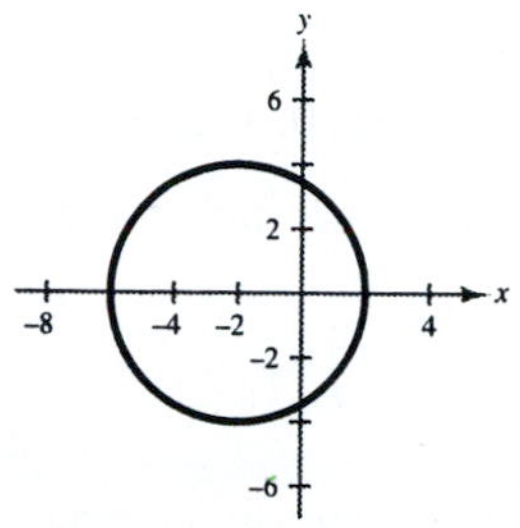

22. $y = 0.002x^2 - 0.06x - 1$

Xmin = -20
Xmax = 50
Xscl = 10
Ymin = -2
Ymax = 1
Yscl = 0.5

24. Endpoints of diameter: $(0, 0)$ and $(4, -6)$

Midpoint is center: $\left(\dfrac{0+4}{2}, \dfrac{0-6}{2}\right) = (2, -3)$

Distance is diameter:

$$\sqrt{(4-0)^2 + (-6-0)^2} = \sqrt{52} = 2\sqrt{13}$$

Radius: $\dfrac{2\sqrt{13}}{2} = \sqrt{13}$

Equation: $(x-2)^2 + (y+3)^2 = 13$

26. (a) m is undefined $\Longrightarrow$ The line is vertical. Matches L_2.

(b) $m = -1 \Longrightarrow$ The line falls. Matches L_3.

(c) $m = \frac{5}{2} \Longrightarrow$ The line rises. Matches L_1.

28. $(-3, 2)$, $(8, 2)$

$$m = \frac{2-2}{-3-8} = \frac{0}{-11} = 0$$

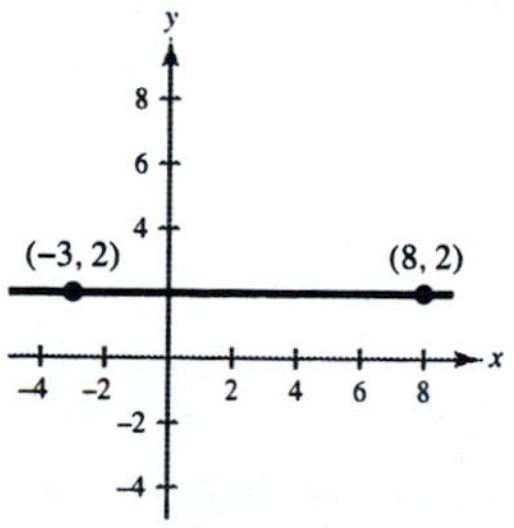

30. $(-6, 1)$, $(1, t)$, $(10, 5)$ are collinear.

$$\frac{t-1}{1-(-6)} = \frac{5-1}{10-(-6)}$$

$$\frac{t-1}{7} = \frac{4}{16}$$

$$t - 1 = \frac{7}{4}$$

$$t = \frac{11}{4}$$

32. $(-1, 4)$, $(2, 0)$

$$y - 0 = \frac{4-0}{-1-2}(x-2)$$

$$y = -\frac{4}{3}(x-2)$$

34. $y - 6 = 0(x - (-2))$

$$y - 6 = 0$$

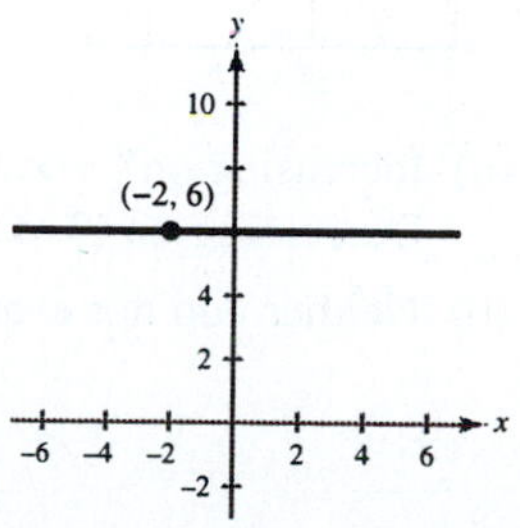

36. $(-8, 3),\ 2x + 3y = 5$

$3y = 5 - 2x$

$y = \frac{5}{3} - \frac{2}{3}x$

(a) Parallel slope: $m = -\frac{2}{3}$

$y - 3 = -\frac{2}{3}(x + 8)$

$3y - 9 = -2x - 16$

$2x + 3y + 7 = 0$

(b) Perpendicular slope: $m = \frac{3}{2}$

$y - 3 = \frac{3}{2}(x - (-8))$

$2y - 6 = 3x + 24$

$0 = 3x - 2y + 30$

38. $(6, 72.95),\ m = 5.15$

$y - 72.95 = 5.15(x - 6)$

$y - 72.95 = 5.15x - 30.9$

$y = 5.15x + 42.05$

40. (a) $(5, 85),\ m = 3.75$

$V - 85 = 3.75(x - 5)$

$V - 85 = 3.75x - 18.75$

$V = 3.75x + 66.25$

(b)

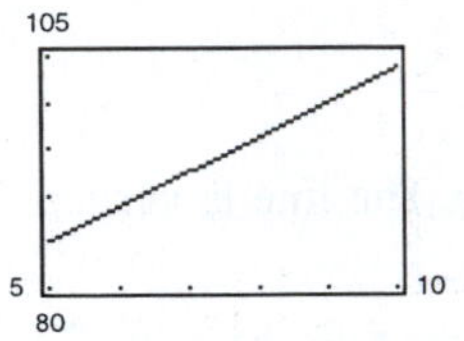

(c) $V = \$103.75$

42. $A = \{u, v, w\}$ and $B = \{-2, -1, 0, 1, 2\}$

(a) u is matched with two elements in the range so it is not a function.

(b) Function

(c) Function

(d) w is matched with two elements in the range so it is not a function.

44. $2x - y - 3 = 0$

$2x - 3 = y$

Yes, the equation represents y as a function of x.

46. $|y| = x + 2$ corresponds to $y = x + 2$ or $-y = x + 2$. y is not a function of x. Each x-value corresponds to two y-values.

48. $g(x) = x^{4/3}$

(a) $g(8) = 8^{4/3} = 2^4 = 16$

(b) $g(t + 1) = (t + 1)^{4/3}$

(c) $\dfrac{g(8) - g(1)}{8 - 1} = \dfrac{16 - 1}{7} = \dfrac{15}{7}$

(d) $g(-x) = (-x)^{4/3} = x^{4/3}$

50. $h(x) = \dfrac{x}{x^2 - x - 6}$

$= \dfrac{x}{(x - 3)(x + 2)}$

Domain: All real numbers except $x = 3, -2$

52. $h(t) = |t + 1|$

Domain: All real numbers

54. $h(x) = 4x^3 - x^4$

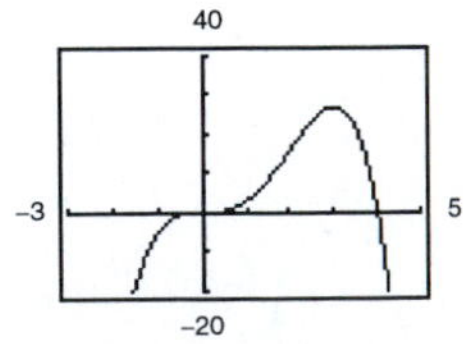

(a) Increasing on $(-\infty, 3)$
Decreasing on $(3, \infty)$

(b) Neither odd nor even

56. (a) Increasing: $(-\infty, -3), (-1, \infty)$
Decreasing: $(-3, -1)$

(b) $f(-x) = \sqrt[3]{-x(-x + 3)^2}$
The function is neither odd nor even.

58.

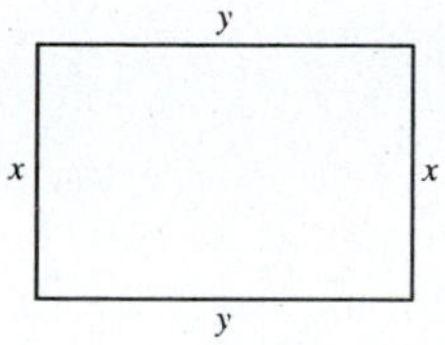

(a) $2x + 2y = 24$

$$y = 12 - x$$

$$A = xy = x(12 - x)$$

(b) Since x and y cannot be negative, we have $0 < x < 12$. The domain is $(0, 12)$.

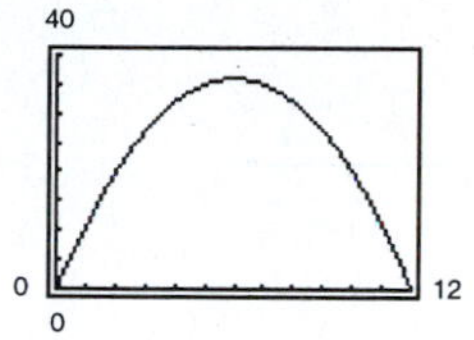

(c) The maximum area of 36 occurs when $x = 6$ and the rectangle is a 6×6 square

60. $(fh)(1) = f(1) \cdot h(1)$
$= (3 - 2(1))(3(1)^2 + 2)$
$= 1 \cdot 5 = 5$

62. $(g \circ f)(-2) = g(f(-2))$
$= g(3 - 2(-2))$
$= g(7)$
$= \sqrt{7}$

64. $f(x) = 5x - 7$

(a) $y = 5x - 7$
$x = 5y - 7$
$x + 7 = 5y$
$\dfrac{x+7}{5} = y$
$f^{-1}(x) = \dfrac{x+7}{5}$

(b)

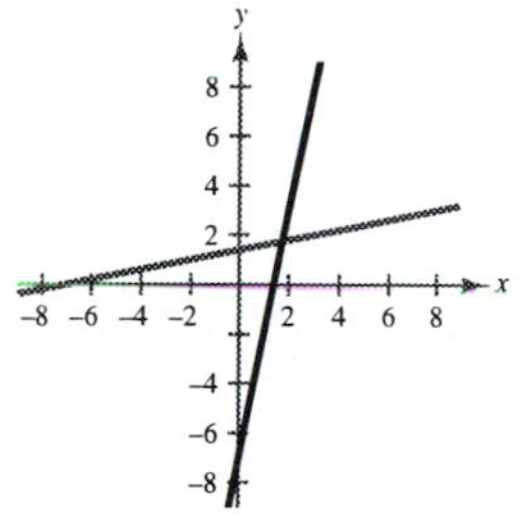

(c) $f^{-1}(f(x)) = \dfrac{(5x - 7) + 7}{5}$
$= \dfrac{5x}{5} = x$
$f(f^{-1}(x)) = 5\left(\dfrac{x+7}{5}\right) - 7$
$= x + 7 - 7 = x$

66. $f(x) = x^3 + 2$

(a) $y = x^3 + 2$
$x = y^3 + 2$
$x - 2 = y^3$
$\sqrt[3]{x - 2} = y$
$f^{-1}(x) = \sqrt[3]{x - 2}$

(b)

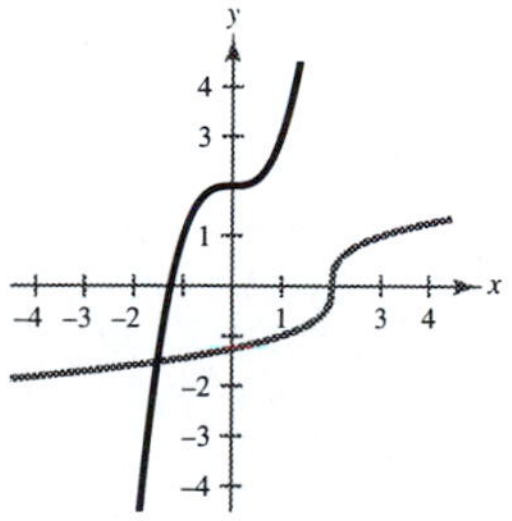

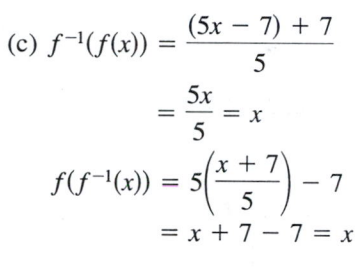

(c) $f^{-1}(f(x)) = \sqrt[3]{(x^3 + 2) - 2}$
$= \sqrt[3]{x^3} = x$
$f(f^{-1})(x)) = \left(\sqrt[3]{x - 2}\right)^3 + 2$
$= x - 2 + 2$

68. $R = \dfrac{k}{x^3}$
$128 = \dfrac{k}{2^3}$
$1024 = k$
$R = \dfrac{1024}{x^3}$

70. $w = \dfrac{kxy}{z^3}$
$\dfrac{44}{9} = \dfrac{k(12)(11)}{6^3}$
$\dfrac{44(216)}{9(132)} = k$
$k = 8$
$w = \dfrac{8xy}{z^3}$

72. (a) Using the least squares regression capabilities of a calculator:

$y_1 = 13.78 + 0.31t$

$y_2 = 13.78 + 0.19t$

(b)

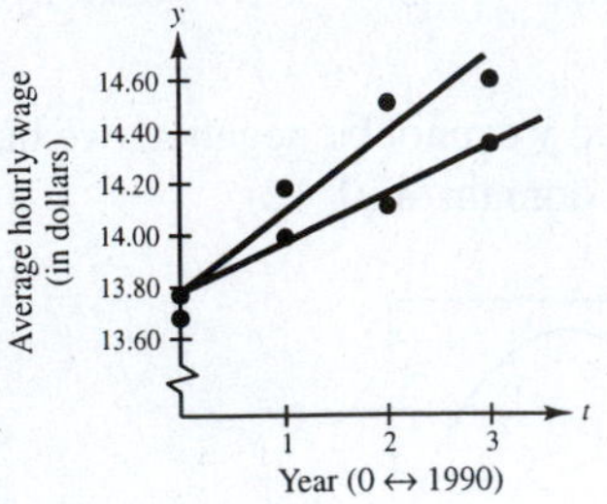

(c) The slope in y_1 means that the hourly wages in the mining industry are increasing by an average of \$0.31 per year. The slope in y_2 means that the hourly wages in the construction industry are increasing by an average of \$0.19 per year.

(d) For $t = 10$ in 2000

$y_1 = 13.78 + 0.31(10) = \$16.88$ in mining

$y_2 = 13.78 + 0.19(10) = \$15.68$ in construction

CHAPTER 2
Polynomial and Rational Functions

CHAPTER 2
Polynomial and Rational Functions

Section 2.1 Quadratic Functions

Solutions to Even-Numbered Exercises

2. $f(x) = (x + 4)^2$ opens upward and has vertex $(-4, 0)$. Matches graph (c).

4. $f(x) = 3 - x^2$ opens downward and has vertex $(0, 3)$. Matches graph (h).

6. $f(x) = (x + 1)^2 - 2$ opens upward and has vertex $(-1, -2)$. Matches graph (a).

8. $f(x) = -(x - 4)^2$ opens downward and has vertex $(4, 0)$. Matches graph (d).

10. (a) $y = x^2 + 1$

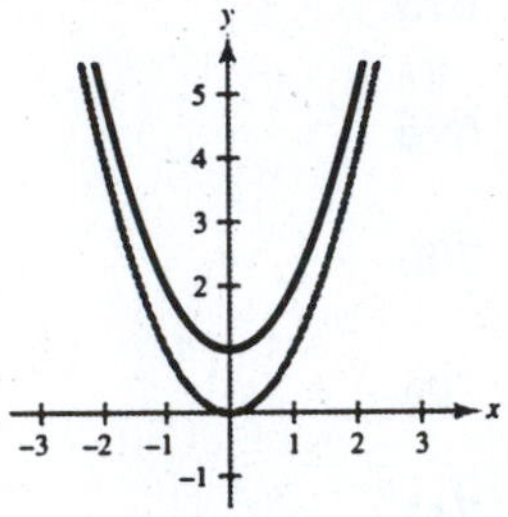

Vertical translation 1 unit upward

(b) $y = x^2 - 1$

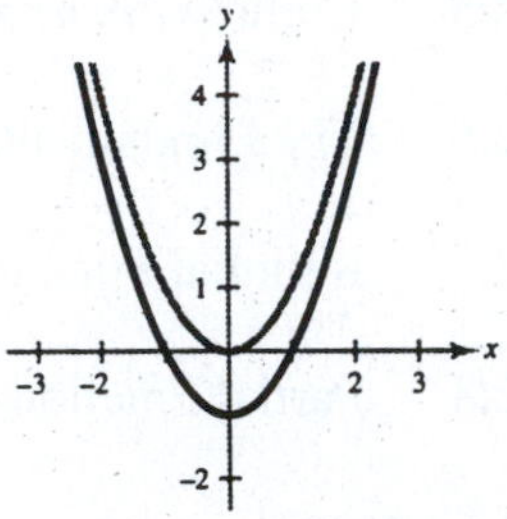

Vertical translation 1 unit downward

(c) $y = x^2 + 3$

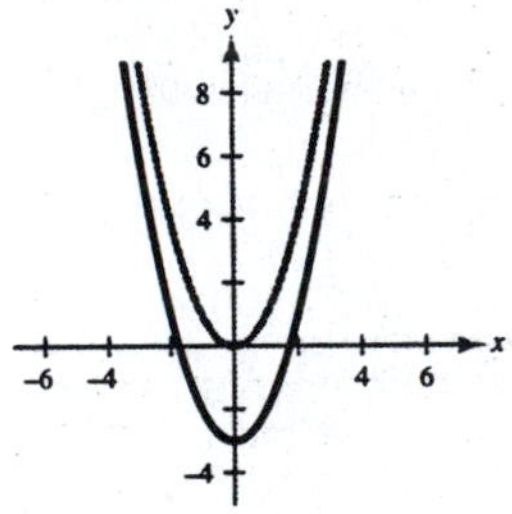

Vertical translation 3 units upward

(d) $y = x^2 - 3$

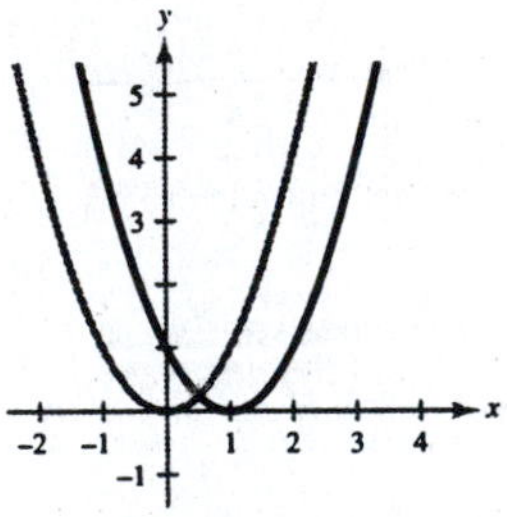

Vertical translation 3 units downward

12. (a) $y = -\frac{1}{2}(x - 2)^2 + 1$

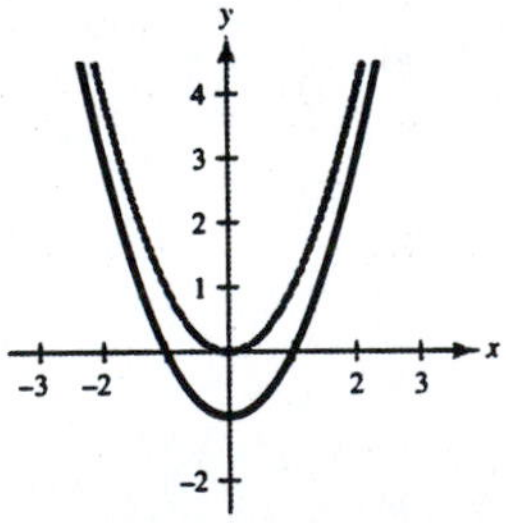

Horizontal translation 2 units to right, vertical shrink by $\frac{1}{2}$, reflection in the x-axis, and vertical translation 1 unit upward

(b)

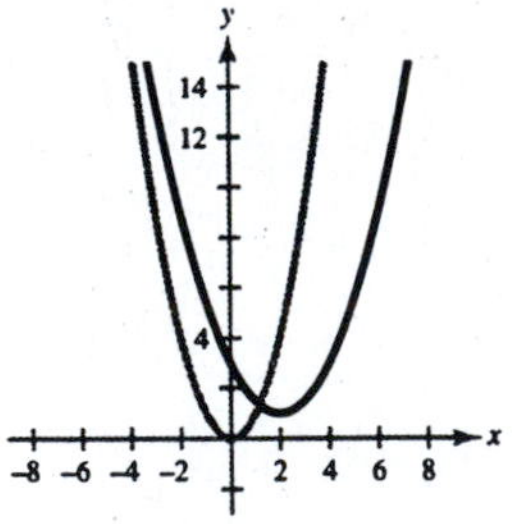

14. $f(x) = \frac{1}{2}x^2 - 4$
Vertex: $(0, -4)$
Intercepts: $(\pm 2\sqrt{2}, 0), (0, -4)$

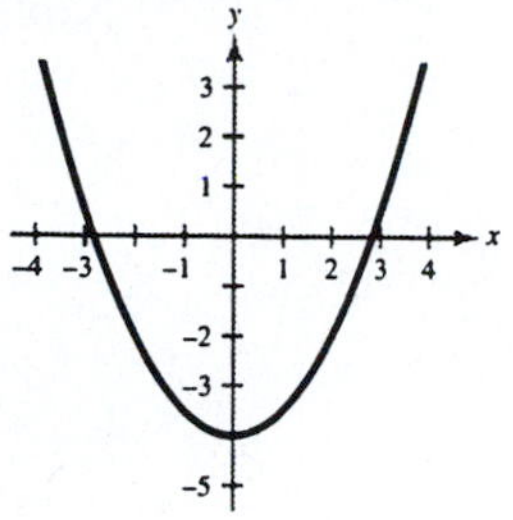

16. $h(x) = 25 - x^2$
Vertex: $(0, 25)$
Intercepts: $(\pm 5, 0), (0, 25)$

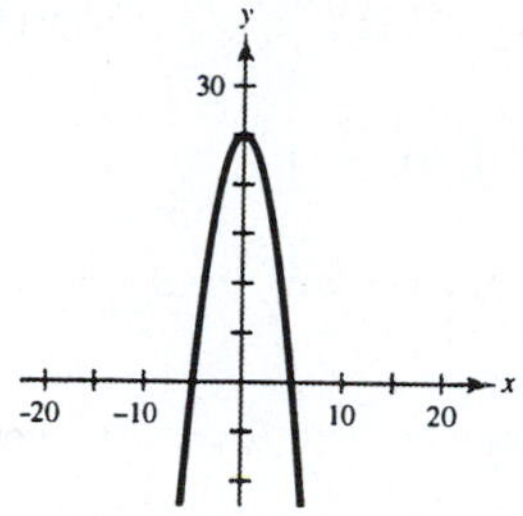

18. $f(x) = (x - 6)^2 + 3$
Vertex: $(6, 3)$
Intercept: $(0, 39)$

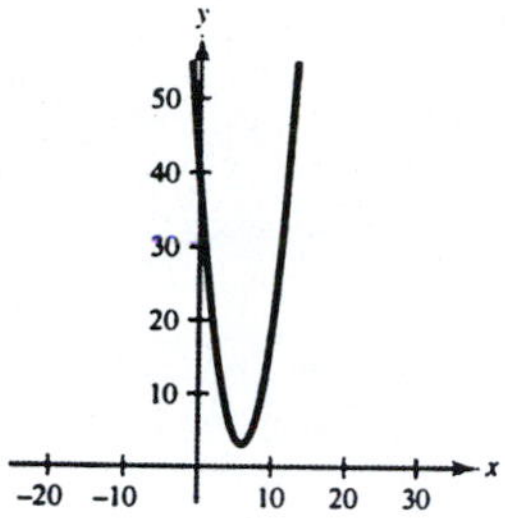

20. $g(x) = x^2 + 2x + 1 = (x + 1)^2$
Vertex: $(-1, 0)$
Intercepts: $(-1, 0), (0, 1)$

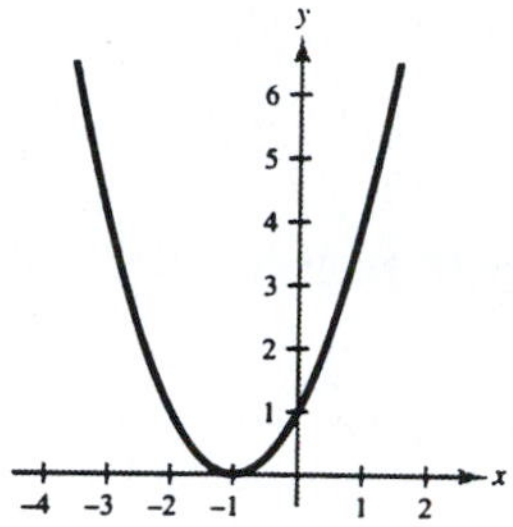

22. $f(x) = x^2 + 3x + \frac{1}{4} = \left(x + \frac{3}{2}\right)^2 - 2$
Vertex: $\left(-\frac{3}{2}, -2\right)$
Intercepts: $\left(-\frac{3}{2} \pm \sqrt{2}, 0\right), \left(0, \frac{1}{4}\right)$

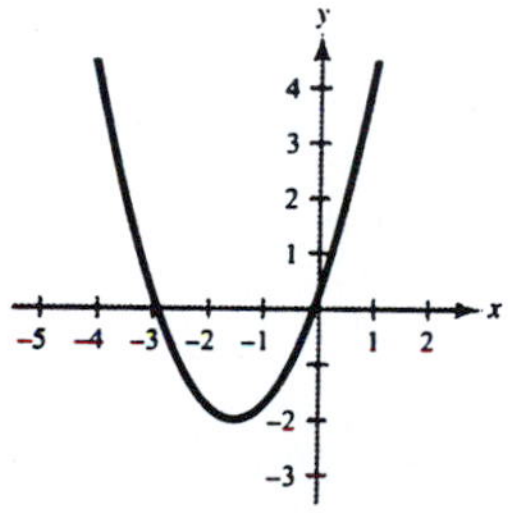

24. $f(x) = -x^2 - 4x + 1 = -1(x^2 + 4x - 1)$
$= -1[(x + 2)^2 - 5]$
Vertex: $(-2, 5)$
Intercepts: $\left(-2 \pm \sqrt{5}, 0\right), (0, 1)$

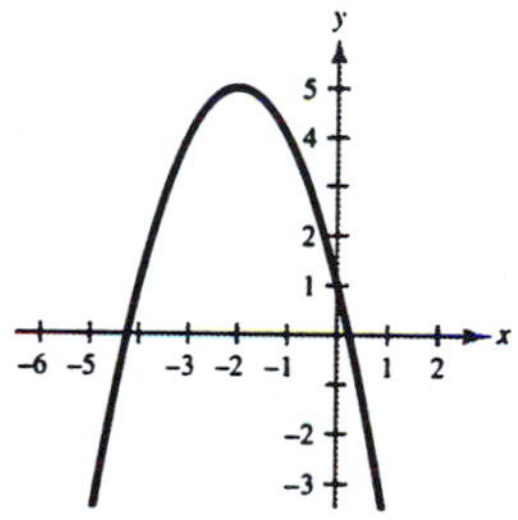

26. $f(x) = 2x^2 - x + 1$
$= 2\left(x^2 - \frac{1}{2}x\right) + 1$
$= 2\left(x - \frac{1}{4}\right)^2 - \frac{1}{8} + 1$
$= 2\left(x - \frac{1}{4}\right)^2 + \frac{7}{8}$
Vertex: $\left(\frac{1}{4}, \frac{7}{8}\right)$
Intercept: $(0, 1)$

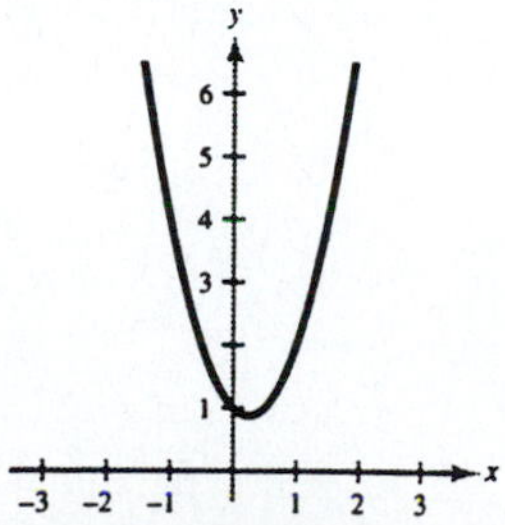

28. $g(x) = x^2 + 8x + 11$
$= (x + 4)^2 - 5$
Vertex: $(-4, -5)$
Intercepts: $\left(-4 \pm \sqrt{5}, 0\right), (0, 11)$

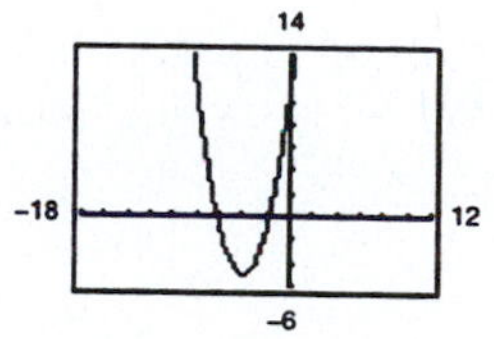

30. $g(x) = \frac{1}{2}(x^2 + 4x - 2)$
$= \frac{1}{2}(x^2 + 4x) - 1$
$= \frac{1}{2}(x + 2)^2 - 3$

Vertex: $(-2, -3)$

Intercepts: $\left(-2 \pm \sqrt{6}, 0\right), (0, -1)$

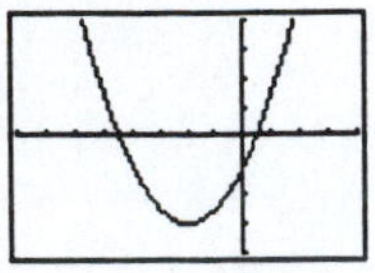

32. (0, 1) is the vertex.

$f(x) = a(x - 0)^2 + 1 = ax^2 + 1$

Since the graph passes through (1, 0),

$0 = a(1)^2 + 1$

$-1 = a.$

Thus, $y = -x^2 + 1.$

34. $(-2, -1)$ is the vertex.

$f(x) = a(x + 2)^2 - 1$

Since the graph passes through (0, 3),

$3 = a(0 + 2)^2 - 1$

$3 = 4a - 1$

$4 = 4a$

$1 = a.$

Thus, $y = (x + 2)^2 - 1.$

36. (2, 0) is the vertex.

$f(x) = a(x - 2)^2 + 0 = a(x - 2)^2$

Since the graph passes through (3, 2),

$2 = a(3 - 2)^2$

$2 = a.$

Thus, $y = 2(x - 2)^2.$

38. $(4, -1)$ is the vertex.

$f(x) = a(x - 4)^2 - 1$

Since the graph passes through (2, 3),

$3 = a(2 - 4)^2 - 1$

$3 = 4a - 1$

$4 = 4a$

$1 = a.$

Thus, $f(x) = (x - 4)^2 - 1.$

40. (2, 3) is the vertex.

$f(x) = a(x - 2)^2 + 3$

Since the graph passes through (0, 2),

$2 = a(0 - 2)^2 + 3$

$2 = 4a + 3$

$-1 = 4a$

$-\frac{1}{4} = a.$

Thus, $f(x) = -\frac{1}{4}(x - 2)^2 + 3.$

42. $(-2, -2)$ is the vertex.

$f(x) = a(x + 2)^2 - 2$

Since the graph passes through $(-1, 0)$,

$0 = a(-1 + 2)^2 - 2$

$0 = a - 2$

$2 = a.$

Thus, $f(x) = 2(x + 2)^2 - 2.$

44. $y = x^2 - 6x + 9$

x-intercept: (3, 0)

$0 = x^2 - 6x + 9$

$0 = (x - 3)^2$

$x = 3$

46. $y = 2x^2 + 5x - 3$

x-intercepts: $\left(\frac{1}{2}, 0\right), (-3, 0)$

$0 = 2x^2 + 5x - 3$

$0 = (2x - 1)(x + 3)$

$x = \frac{1}{2}, -3$

48. $y = x^2 - 9x + 18$

x-intercepts: (3, 0), (6, 0)

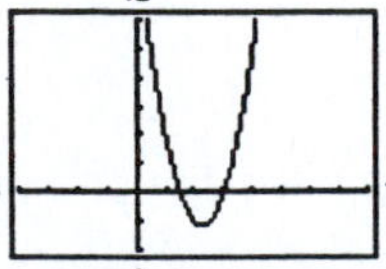

$0 = x^2 - 9x + 18$

$0 = (x - 6)(x - 3)$

$x = 3, 6$

50. $y = -\frac{1}{2}(x^2 - 6x - 7)$

x-intercepts: $(7, 0), (-1, 0)$

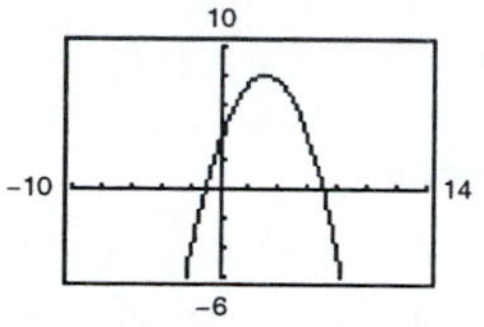

$0 = -\frac{1}{2}(x^2 - 6x - 7)$

$0 = x^2 - 6x - 7$

$0 = (x - 7)(x + 1)$

$x = 7, -1$

52. $f(x) = 2\left[x - \left(-\frac{5}{2}\right)\right](x - 2)$

$= 2\left(x + \frac{5}{2}\right)(x - 2)$

$= 2\left(x^2 + \frac{1}{2}x - 5\right)$

$= 2x^2 + x - 10$, opens upward

$g(x) = -f(x)$, opens downward

$g(x) = -2x^2 - x + 10$

54. $f(x) = (x - 4)(x - 8)$

$= x^2 - 12x + 32$, opens upward

$g(x) = -f(x)$, opens downward

$g(x) = -x^2 + 12x - 32$

56. $f(x) = [x - (-5)](x - 5)$

$= (x + 5)(x - 5)$

$= x^2 - 25$, opens upward

$g(x) = -f(x)$, opens downward

$g(x) = -x^2 + 25$

58. Let x = first number and y = second number. Then, $x + y = S$, $y = S - x$. The product is $P(x) = xy = x(S - x)$.

$P(x) = Sx - x^2$

$= -x^2 + Sx$

$= -\left(x^2 - Sx + \frac{S^2}{4} - \frac{S^2}{4}\right)$

$= -\left(x - \frac{S}{2}\right)^2 + \frac{S^2}{4}$

The maximum value of the product occurs at the vertex of $P(x)$ and is $S^2/4$. This happens when $x = y = S/2$.

60. Let x = first number and y = second number. Then, $x + y = 50$, $y = 50 - x$. The product is $P(x) = xy = x(50 - x)$.

$P(x) = 50x - x^2$

$= -x^2 + 50x$

$= -(x^2 - 50x + 625 - 625)$

$= -(x - 25)^2 + 625$

The maximum value of the product occurs at the vertex of $P(x)$ and is 625. This happens when $x = y = 25$.

62. Let x = length of rectangle and y = width of rectangle.

$2x + 2y = 36$

$y = 18 - x$

(a) $A(x) = xy = x(18 - x)$

Domain: $0 < x < 18$

(b)

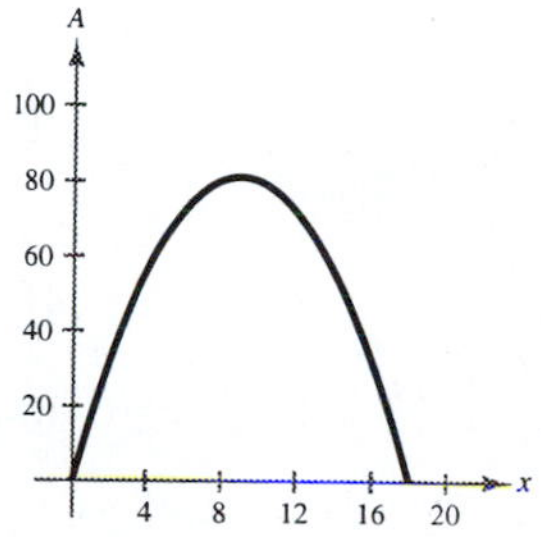

(c) The area is maximum (81 square meters) when $x = y = 9$ meters. The rectangle has dimensions 9 meters × 9 meters.

64. (a)

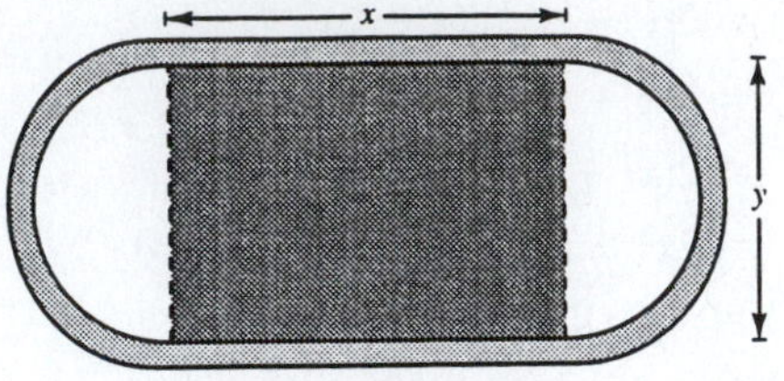

(c) Distance traveled around track in one lap:

$$d = \pi y + 2x = 200$$
$$\pi y = 200 - 2x$$
$$y = \frac{200 - 2x}{\pi}$$

(b) Radius of semicircular ends of track: $r = \frac{1}{2}y$

distance around two semicircular parts of track:

$$d = 2\pi r = 2\pi\left(\frac{1}{2}y\right) = \pi y$$

(d) Area of rectangular region:

$$A = xy = x\left(\frac{200 - 2x}{\pi}\right)$$
$$= \frac{1}{\pi}(200x - 2x^2)$$
$$= -\frac{2}{\pi}(x^2 - 100x)$$
$$= -\frac{2}{\pi}(x^2 - 100x + 2500 - 2500)$$
$$= -\frac{2}{\pi}(x - 50)^2 + \frac{5000}{\pi}$$

The area is maximum when $x = 50$ and

$$y = \frac{200 - 2(50)}{\pi} = \frac{100}{\pi}.$$

66. $R = 100x - 0.0002x^2 = -0.0002x^2 + 100x$

The vertex occurs at

$$x = -\frac{b}{2a} = -\frac{100}{2(-0.0002)} = 250{,}000.$$

The revenue is maximum when $x = 250{,}000$ units.

68. $C = 10{,}000 - 110x + 0.045x^2$

The vertex occurs at

$$x = -\frac{-110}{2(0.045)} \approx 1222.$$

The cost is minimum when $x \approx 1222$ units.

70. $P = 230 + 20x - 0.5x^2$

The vertex occurs at

$$x = -\frac{b}{2a} = -\frac{20}{2(-0.5)} = 20$$

Because x is in hundreds of dollars, $20 \times 100 =$ 2000 dollars is the amount spent on advertising that gives maximum profit.

72. $y = -\frac{4}{9}x^2 + \frac{24}{9}x + 12$

The maximum height of the dive occurs at the vertex,

$$x = -\frac{b}{2a} = -\frac{\frac{24}{9}}{2\left(-\frac{4}{9}\right)} = 3.$$

The height at $x = 3$ is

$$-\frac{4}{9}(3)^2 + \frac{24}{9}(3) + 12 = 16.$$

The maximum height of the dive is 16 feet.

74. (a)

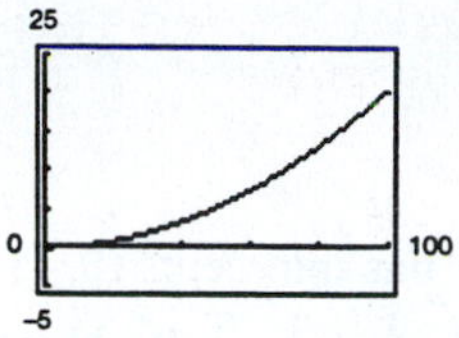

(b) $0.002s^2 + 0.005s - 0.029 = 10$

$$2s^2 + 5s - 29 = 10{,}000$$

$$2s^2 + 5s - 10{,}029 = 0$$

$$a = 2, b = 5, c = -10{,}029$$

$$s = \frac{-5 \pm \sqrt{5^2 - 4(2)(-10{,}029)}}{2(2)}$$

$$s = \frac{-5 \pm \sqrt{80{,}257}}{4}$$

$$s \approx -72.1, 69.6$$

The maximum speed if power is not to exceed 10 horsepower is 69.6 miles per hour.

76. (a) and (b)

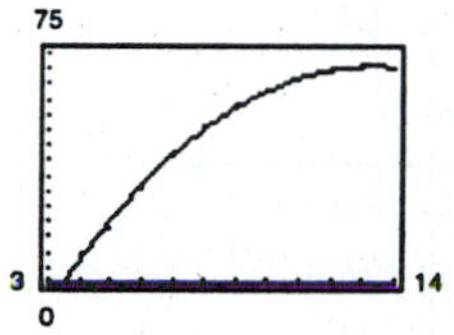

(c) No, the model would probably not give accurate predictions in 2000 because the model will begin to decrease and will eventually become negative.

78. (a)

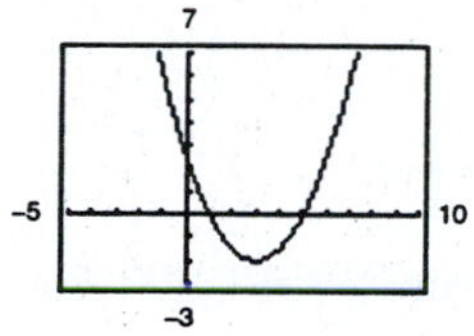

(b)

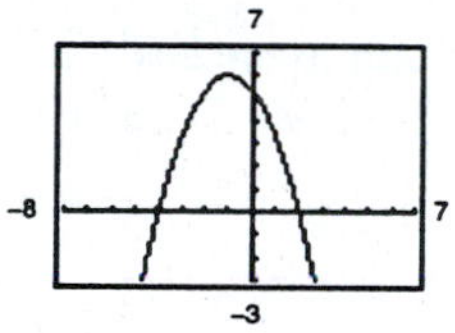

80. $\left(\frac{7}{2}, 2\right), m = \frac{3}{2}$

$$y - 2 = \frac{3}{2}\left(x - \frac{7}{2}\right)$$

$$2y - 4 = 3\left(x - \frac{7}{2}\right)$$

$$2y - 4 = 3x - \frac{21}{2}$$

$$3x - 2y - \frac{13}{2} = 0$$

82. $2x - 3y = 4$

$$2x - 4 = 3y$$

$$\frac{2}{3}x - \frac{4}{3} = y$$

$$-4x + 6y = 15$$

$$6y = 4x + 15$$

$$y = \frac{4}{6}x + \frac{15}{6}$$

$$y = \frac{2}{3}x + \frac{5}{2}$$

These lines have the same slope $m = \frac{2}{3}$. Thus, the lines are parallel.

Section 2.2 Polynomial Functions of Higher Degree

Solutions to Even-Numbered Exercises

2. $f(x) = x^2 - 4x$ is a parabola with intercepts $(0, 0)$ and $(4, 0)$ and opens upward. Matches graph (g).

4. $f(x) = 2x^3 - 3x + 1$ has intercepts $(0, 1)$, $(1, 0)$, $\left(-\frac{1}{2} - \frac{1}{2}\sqrt{3}, 0\right)$ and $\left(-\frac{1}{2} + \frac{1}{2}\sqrt{3}, 0\right)$. Matches graph (f).

6. $f(x) = -\frac{1}{3}x^3 + x^2 - \frac{4}{3}$ has y-intercept $\left(0, -\frac{4}{3}\right)$. Matches graph (e).

8. $f(x) = \frac{1}{5}x^5 - 2x^3 + \frac{9}{5}x$ has intercepts $(0, 0)$, $(1, 0)$, $(-1, 0)$, $(3, 0)$, $(-3, 0)$. Matches (b).

10. $y = x^5$

(a) $f(x) = (x + 1)^5$

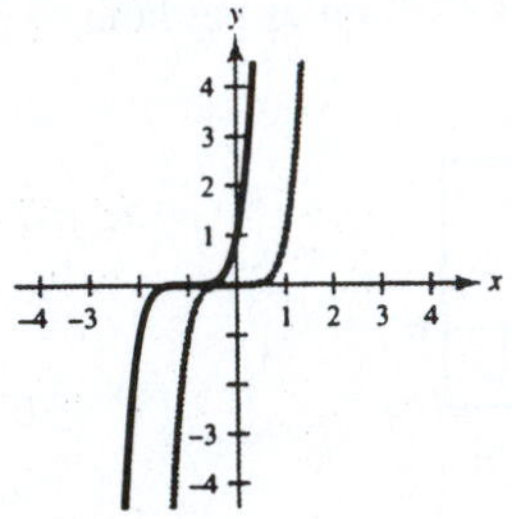

Horizontal shift one unit to the left

(b) $f(x) = x^5 + 1$

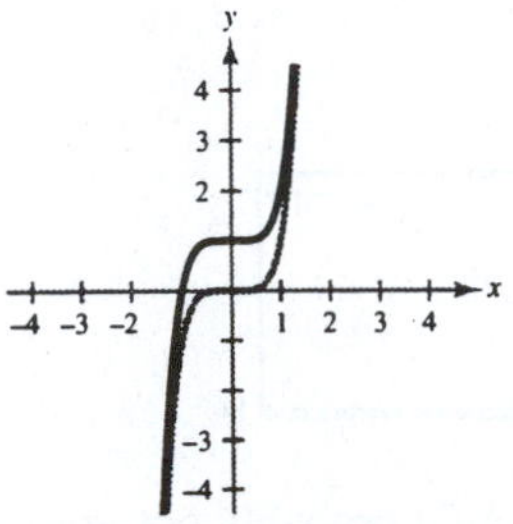

Vertical shift one unit upward

(c) $f(x) = 1 - \frac{1}{2}x^5$

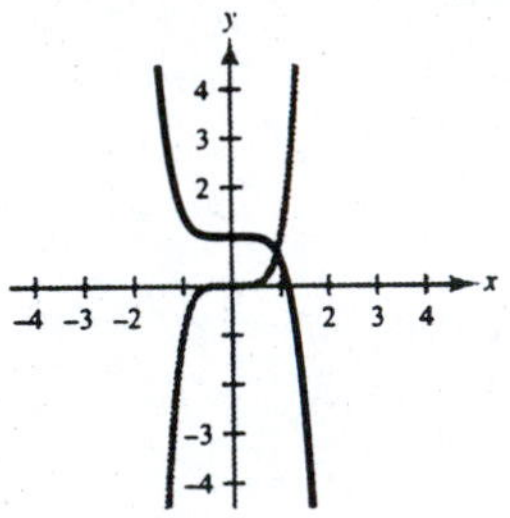

Reflection in the x-axis, vertical shrink and vertical shift one unit upward

(d) $f(x) = -\frac{1}{2}(x + 1)^5$

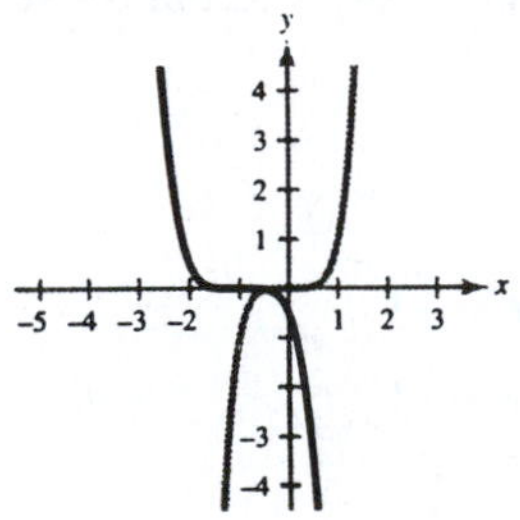

Reflection in the x-axis, vertical shrink and horizontal shift one unit to the left

12. $y = x^6$

(a) $f(x) = -\frac{1}{8}x^6$

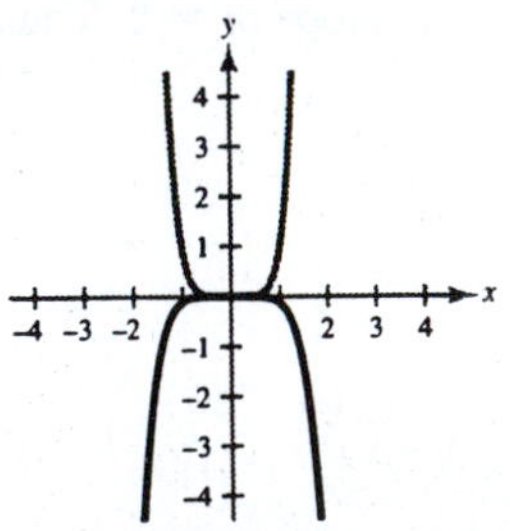

Vertical shrink and reflection in the x-axis

(b) $f(x) = (x + 2)^6 - 4$

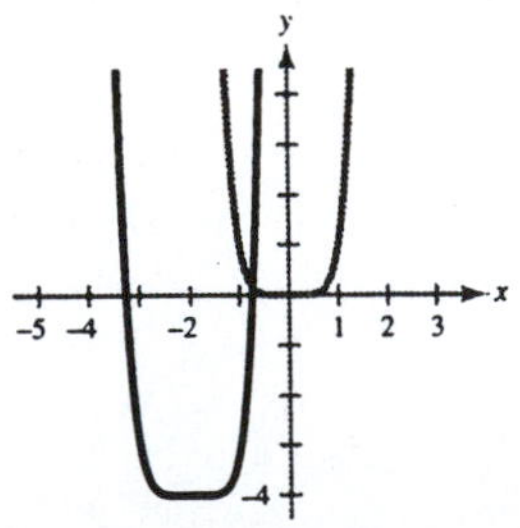

Horizontal shift two units to the left and vertical shift 4 units downward

— **CONTINUED** —

12. — CONTINUED —

(c) $f(x) = x^6 - 4$

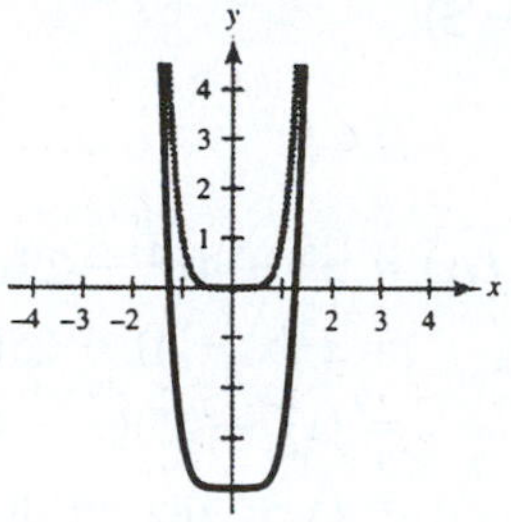

Vertical shift 4 units downward

(d) $f(x) = -\frac{1}{4}x^6 + 1$

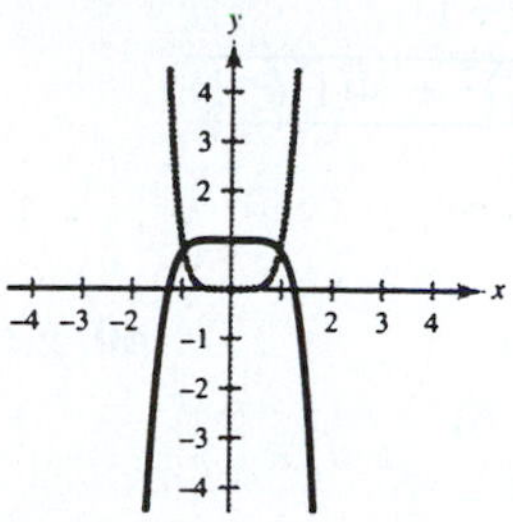

Vertical shrink, vertical shift upward one unit, and reflection in the x-axis

14. $f(x) = 2x^2 - 3x + 1$

Degree: 2

Leading coefficient: 2

The degree is even and the leading coefficient is positive. The graph rises to the left and right.

16. $h(x) = 1 - x^6$

Degree: 6

Leading coefficient: -1

The degree is even and the leading coefficient is negative. The graph falls to the left and right.

18. $f(x) = 2x^5 - 5x + 7.5$

Degree: 5

Leading coefficient: 2

The degree is odd and the leading coefficient is positive. The graph falls to the left and rises to the right.

20. $f(x) = \dfrac{3x^4 - 2x + 5}{4}$

Degree: 4

Leading coefficient: $\frac{3}{4}$

The degree is even and the leading coefficient is positive. The graph rises to the left and right.

22. $f(s) = -\frac{7}{8}(s^3 + 5s^2 - 7s + 1)$

Degree: 3

Leading coefficient: $-\frac{7}{8}$

The degree is odd and the leading coefficient is negative. The graph rises to the left and falls to the right.

24. $f(x) = -\frac{1}{3}(x^3 - 3x + 2)$, $g(x) = -\frac{1}{3}x^3$

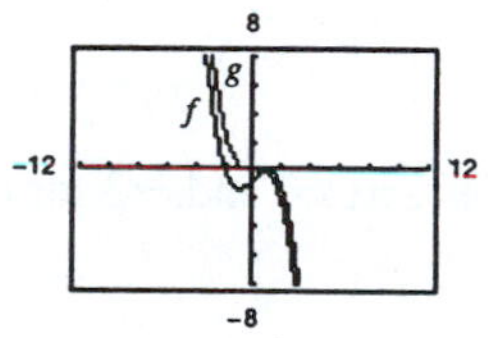

26. $f(x) = 3x^4 - 6x^2$, $g(x) = 3x^4$

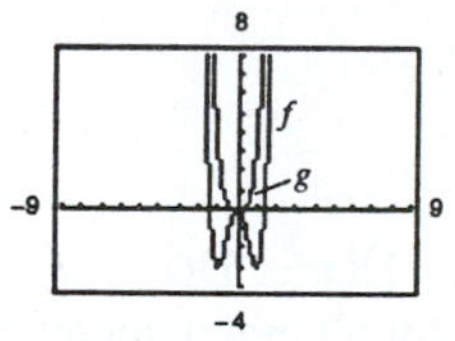

28. $f(x) = 49 - x^2$

$= (7 - x)(7 + x)$

$x = \pm 7$

30. $f(x) = x^2 + 10x + 25$

$= (x + 5)^2$

$x = -5$

32. $f(x) = \frac{1}{2}x^2 + \frac{5}{2}x - \frac{3}{2}$

$a = \frac{1}{2}, b = \frac{5}{2}, c = -\frac{3}{2}$

$$x = \frac{-\frac{5}{2} \pm \sqrt{\left(\frac{5}{2}\right)^2 - 4\left(\frac{1}{2}\right)\left(-\frac{3}{2}\right)}}{1} = -\frac{5}{2} \pm \sqrt{\frac{37}{4}}$$

$$x = -\frac{5}{2} + \frac{1}{2}\sqrt{37} = \frac{-5 + \sqrt{37}}{2}$$

$$x = -\frac{5}{2} - \frac{1}{2}\sqrt{37} = \frac{-5 - \sqrt{37}}{2}$$

34. $g(x) = 5(x^2 - 2x - 1)$

$a = 1, b = -2, c = -1$

$x = \dfrac{-(-2) \pm \sqrt{(-2)^2 - 4(1)(-1)}}{2}$

$x = 1 \pm \sqrt{2}$

36. $f(x) = x^4 - x^3 - 20x^2$

$= x^2(x^2 - x - 20)$

$= x^2(x + 4)(x - 5)$

$x = 0, -4, 5$

38. $f(x) = x^5 + x^3 - 6x$

$= x(x^4 + x^2 - 6)$

$= x(x^2 + 3)(x^2 - 2)$

$x = 0, \pm\sqrt{2}$

40. $g(t) = t^5 - 6t^3 + 9t$

$= t(t^4 - 6t^2 + 9)$

$= t(t^2 - 3)^2$

$t = 0, \pm\sqrt{3}$

42. $f(x) = x^3 - 4x^2 - 25x + 100$

$= x^2(x - 4) - 25(x - 4)$

$= (x^2 - 25)(x - 4)$

$= (x + 5)(x - 5)(x - 4)$

$x = \pm 5, 4$

44. $y = 4x^3 + 4x^2 - 7x + 2$

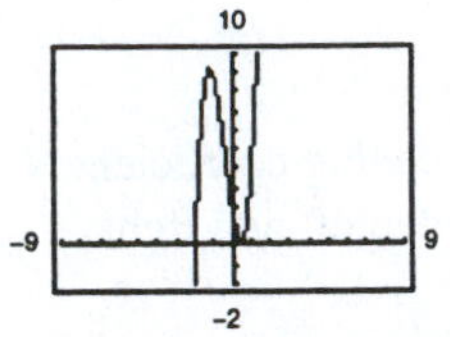

$0 = 4x^3 + 4x^2 - 7x + 2$

$= (2x - 1)(2x^2 + 3x - 2)$

$= (2x - 1)(2x - 1)(x + 2)$

$x = -2, \frac{1}{2}$

x-intercepts: $(-2, 0), \left(\frac{1}{2}, 0\right)$

46. $y = \frac{1}{4}x^3(x^2 - 9)$

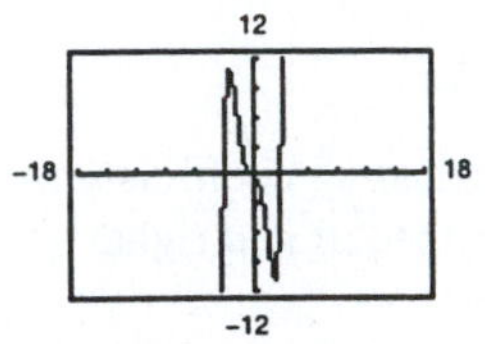

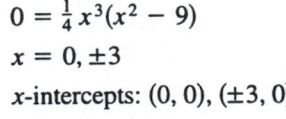

$0 = \frac{1}{4}x^3(x^2 - 9)$

$x = 0, \pm 3$

x-intercepts: $(0, 0), (\pm 3, 0)$

48. $f(x) = (x - 0)(x - (-3))$

$= x(x + 3)$

$= x^2 + 3x$

Note: $f(x) = ax(x + 3)$ has zeros 0 and -3 for all real numbers a.

50. $f(x) = (x - (-4))(x - 5)$

$= (x + 4)(x - 5)$

$= x^2 - x - 20$

Note: $f(x) = a(x + 4)(x - 5)$ has zeros -4 and 5 for all real numbers a.

52. $f(x) = (x - 0)(x - 2(x - 5)$

$= x(x - 2)(x - 5)$

$= x(x^2 - 7x + 10)$

$= x^3 - 7x^2 + 10x$

Note: $f(x) = ax(x - 2)(x - 5)$ has zeros 0, 2, 5 for all real numbers a.

54. $f(x) = (x - (-2))(x - (-1))(x - 0)(x - 1)(x - 2)$

$= x(x + 2)(x + 1)(x - 1)(x - 2)$

$= x(x^2 - 4)(x^2 - 1)$

$= x(x^4 - 5x^2 + 4)$

$= x^5 - 5x^3 + 4x$

Note: $f(x) = ax(x + 2)(x + 1)(x - 1)(x - 2)$ has zeros $-2, -1, 0, 1, 2$ for all real numbers a.

56. $f(x) = (x - 2)\left[x - (4 + \sqrt{5})\right]\left[x - (4 - \sqrt{5})\right]$

$= (x - 2)\left[x - (4 + \sqrt{5})\right]\left[(x - 4) - \sqrt{5}\right]$

$= (x - 2)\left[(x - 4)^2 - 5\right]$

$= x(x - 4)^2 - 5x - 2(x - 4)^2 + 10$

$= x^3 - 8x^2 + 16x - 5x - 2x^2 + 16x - 32 + 10$

$= x^3 - 10x^2 + 27x - 22$

Note: $f(x) = a(x^3 - 10x^2 + 27x - 22)$ has these zeros for all real numbers x.

58. $f(x) = 0.11x^3 - 2.07x^2 + 9.81x - 6.88$

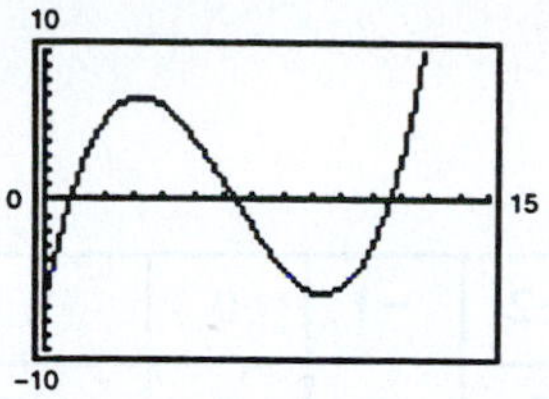

The function has three zeros. They are in the intervals $(0, 1)$, $(6, 7)$, and $(11, 12)$.

60. $h(x) = x^4 - 10x^2 + 3$

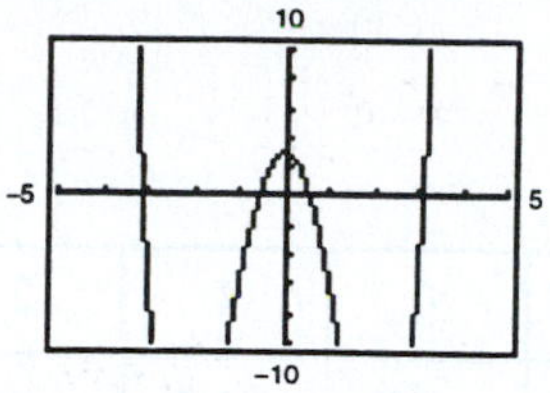

The function has four zeros. They are in the intervals $(-4, -3)$, $(-1, 0)$, $(0, 1)$, and $(3, 4)$.

62. $h(x) = \frac{1}{3}x - 3$

Line with y-intercept $(0, 3)$ and x-intercept $(9, 0)$

Right: moves up

Left: moves down

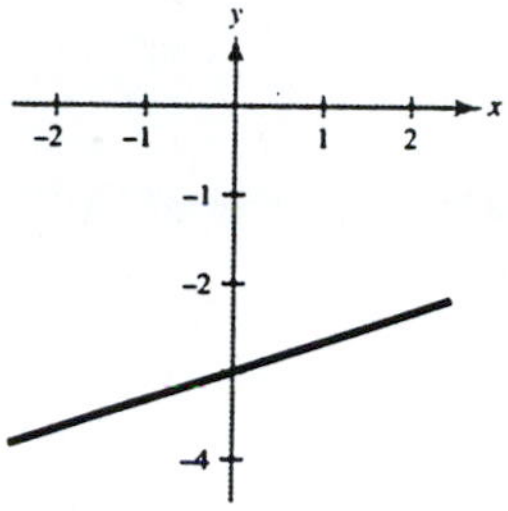

64. $g(x) = -x^2 + 10x - 16$

$= -(x - 5)^2 + 9$

Parabola; opens downward

Vertex: $(5, 9)$

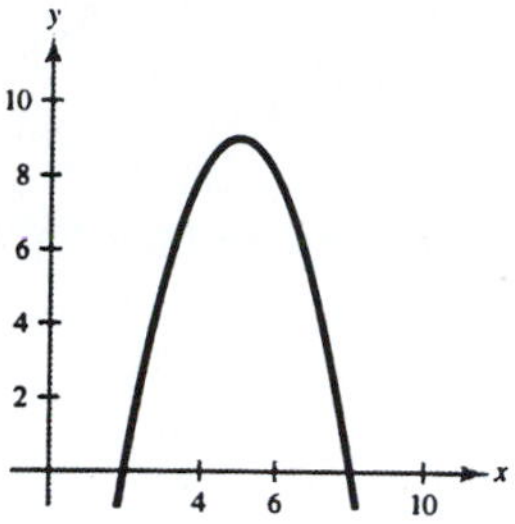

66. $f(x) = 1 - x^3$

Zero: 1

Right: moves down

Left: moves up

x	-2	-1	0	1	2
$f(x)$	9	2	1	0	-7

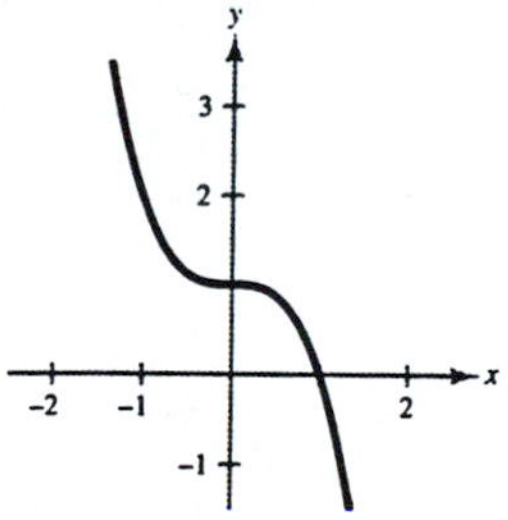

68. $f(x) = x^2(x - 4)$

Zeros: 0, 4

Right: moves up

Left: moves down

x	-1	0	1	2	3	5
$f(x)$	-5	0	-3	-8	-9	25

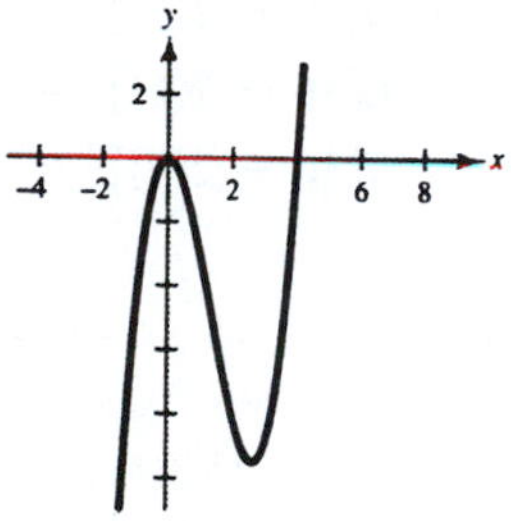

70. $g(x) = \frac{1}{10}(x + 1)^2(x - 3)^3$

Zeros: $-1, 3$

Right: moves up

Left: moves down

x	-2	-1	0	1	2
$f(x)$	-12.5	0	-2.7	-3.2	-0.9

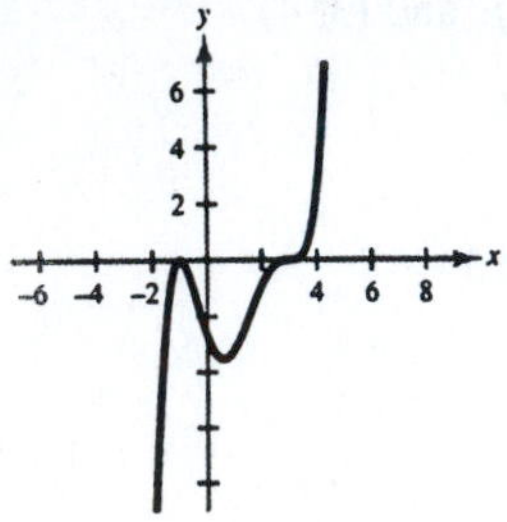

72. $g(x) = 1 - (x + 1)^6$

Zeros: $-2, 0$

Right: moves down

Left: moves down

x	-3	-2	-1	0	1
$f(x)$	-63	0	1	0	-63

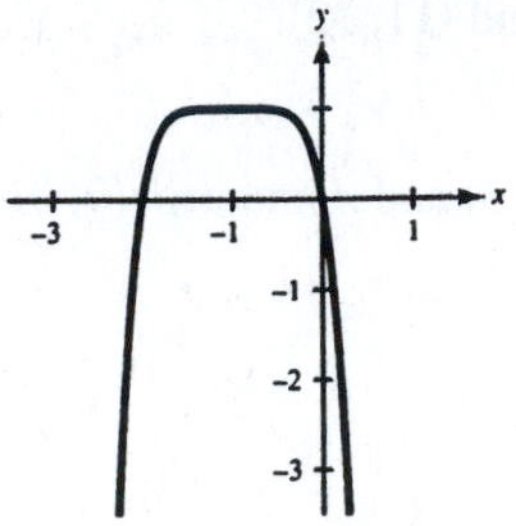

74. $f(x) = \frac{1}{4}x^4 - 2x^2$

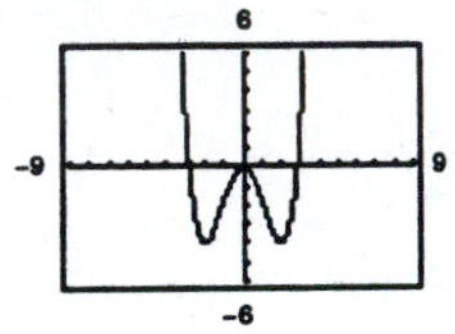

76. $h(x) = \frac{1}{5}(x + 2)^2(3x - 5)^2$

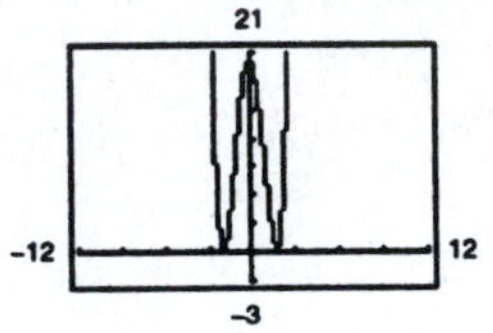

78. (a) and (b)

Height, x	Length and Width	Volume, V
1	$36 - 2(1)$	$1[36 - 2(1)]^2 = 1156$
2	$36 - 2(2)$	$2[36 - 2(2)]^2 = 2048$
3	$36 - 2(3)$	$3[36 - 2(3)]^2 = 2700$
4	$36 - 2(4)$	$4[36 - 2(4)]^2 = 3136$
5	$36 - 2(5)$	$5[36 - 2(5)]^2 = 3380$
6	$36 - 2(6)$	$6[36 - 2(6)]^2 = 3456$
7	$36 - 2(7)$	$7[36 - 2(7)]^2 = 3388$

(c) Volume = length × width × height

Because box is made from a square, length = width. Thus:

Volume = $(\text{length})^2 \times$ height

$= (36 - 2x)^2x$

Domain: $0 < 36 - 2x < 36$

$-36 < -2x < 0$

$18 > x > 0$

(d)

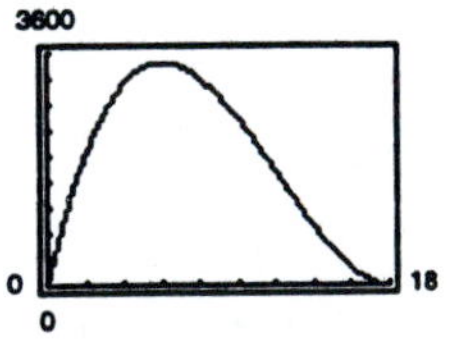

$x = 6$ when $V(x)$ is maximum.

80. (a) Volume $= l \cdot w \cdot h = (24 - 2x)(24 - 4x)x$

$= 2(12 - x) \cdot 4(6 - x)x$

$= 8x(12 - x)(6 - x)$

(b) $x > 0, \quad 12 - x > 0, \quad 6 - x > 0$

$x < 12 \qquad x < 6$

(c)

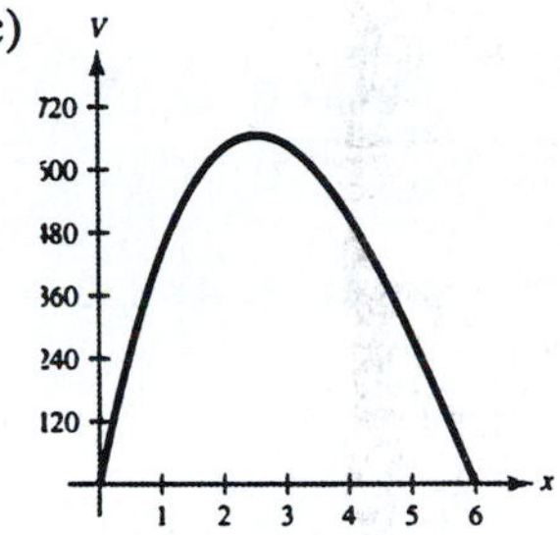

$x \approx 2.55$ when $V(x)$ is maximum.

82. $G = -0.003t^3 + 0.137t^2 + 0.458t - 0.839$

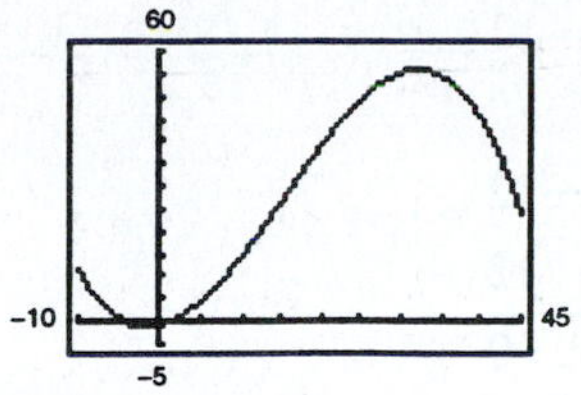

The tree is growing most rapidly at $t \approx 15$.

84. $3x - y^2 = 4$

$3x - 4 = y^2$

$\pm\sqrt{3x - 4} = y$

No, $3x - y^2 = 4$ does not determine y as a function of x. For some x there correspond two values of y.

86. $f(x) = \dfrac{x}{\sqrt{4 - x}}$

$4 - x > 0$

$4 > x$

Domain: $(-\infty, 4)$

Section 2.3 Polynomial and Synthetic Division

Solutions to Even-Numbered Exercises

2. $y_2 = 3 + \dfrac{4}{x - 3}$

$= \dfrac{3(x - 3) + 4}{x - 3}$

$= \dfrac{3x - 9 + 4}{x - 3}$

$= \dfrac{3x - 5}{x - 3}$

$= y_1$

4. $y_2 = x - 2 + \dfrac{4}{x + 2}$

$= \dfrac{x(x + 2) - 2(x + 2) + 4}{x + 2}$

$= \dfrac{x^2 + 2x - 2x - 4 + 4}{x + 2}$

$= \dfrac{x^2}{x + 2}$

$= y_1$

6.

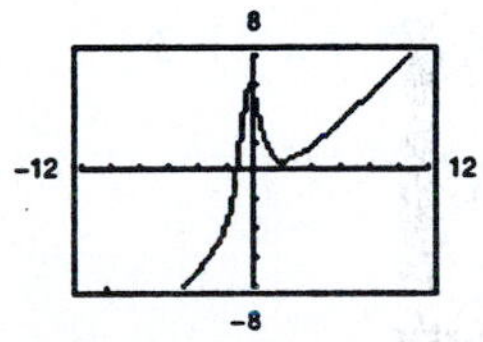

$y_2 = x - 3 + \dfrac{2(x + 4)}{x^2 + x + 1}$

$= \dfrac{x(x^2 + x + 1) - 3(x^2 + x + 1) + 2(x + 4)}{x^2 + x + 1}$

$= \dfrac{x^3 + x^2 + x - 3x^2 - 3x - 3 + 2x + 8}{x^2 + x + 1}$

$= \dfrac{x^3 - 2x^2 + 5}{x^2 + x + 1}$

$= y_1$

8.
$$\begin{array}{r} 5x+3 \\ x-4\,\overline{)\,5x^2-17x-12} \\ \underline{5x^2-20x} \\ 3x-12 \\ \underline{3x-12} \\ 0 \end{array}$$
$$\frac{5x^2-17x-12}{x-4} = 5x+3$$

10.
$$\begin{array}{r} 2x^2-4x+3 \\ 3x-2\,\overline{)\,6x^3-16x^2+17x-6} \\ \underline{6x^3-4x^2} \\ -12x^2+17x \\ \underline{-12x^2+8x} \\ 9x-6 \\ \underline{9x-6} \\ 0 \end{array}$$
$$\frac{6x^3-16x^2+17x-6}{3x-2} = 2x^2-4x+3$$

12.
$$\begin{array}{r} x+4 \\ x^2-3\,\overline{)\,x^3+4x^2-3x-12} \\ \underline{x^3-3x} \\ 4x^2-12 \\ \underline{4x^2-12} \\ 0 \end{array}$$
$$\frac{x^3+4x^2-3x-12}{x^2-3} = x+4$$

14.
$$\begin{array}{r} 4 \\ 2x+1\,\overline{)\,8x-5} \\ \underline{8x+4} \\ -9 \end{array}$$
$$\frac{8x-5}{2x+1} = 4-\frac{9}{2x+1}$$

16.
$$\begin{array}{r} x \\ x^2+1\,\overline{)\,x^3+0x^2+0x-9} \\ \underline{x^3+x} \\ -x-9 \end{array}$$
$$\frac{x^3-9}{x^2+1} = x-\frac{x+9}{x^2+1}$$

18.
$$\begin{array}{r} x^2 \\ x^3-1\,\overline{)\,x^5+0x^4+0x^3+0x^2+0x+7} \\ \underline{x^5-x^2} \\ x^2+7 \end{array}$$
$$\frac{x^5+7}{x^3-1} = x^2+\frac{x^2+7}{x^3-1}$$

20.
$$\begin{array}{r} 2x \\ x^2-2x+1\,\overline{)\,2x^3-4x^2-15x+5} \\ \underline{2x^3-4x^2+2x} \\ -17x+5 \end{array}$$
$$\frac{2x^3-4x^2-15x+5}{(x-1)^2} = 2x-\frac{17x-5}{x^2-2x+1}$$

22.
$$\begin{array}{r} x^{2n}-x^n+3 \\ x^n-2\,\overline{)\,x^{3n}-3x^{2n}+5x^n-6} \\ \underline{x^{3n}-2x^{2n}} \\ -x^{2n}+5x^n \\ \underline{-x^{2n}+2x^n} \\ 3x^n-6 \\ \underline{3x^n-6} \\ 0 \end{array}$$
$$\frac{x^{3n}-3x^{2n}+5x^n-6}{x^n-2} = x^{2n}-x^n+3$$

24.
$$\begin{array}{r|rrrr} -3 & 5 & 18 & 7 & -6 \\ & & -15 & -9 & 6 \\ \hline & 5 & 3 & -2 & 0 \end{array}$$
$$\frac{5x^3+18x^2+7x-6}{x+3} = 5x^2+3x-2$$

26.
$$\begin{array}{r|rrrr} 2 & 9 & -18 & -16 & 32 \\ & & 18 & 0 & -32 \\ \hline & 9 & 0 & -16 & 0 \end{array}$$
$$\frac{9x^3-18x^2-16x+32}{x-2} = 9x^2-16$$

28.

6	3	−16	0	−72
		18	12	72
	3	2	12	0

$$\frac{3x^3 - 16x^2 - 72}{x - 6} = 3x^2 + 2x + 12$$

30.

−2	5	0	6	8
		−10	20	−52
	5	−10	26	−44

$$\frac{5x^3 + 6x + 8}{x + 2} = 5x^2 - 10x + 26 - \frac{44}{x + 2}$$

32.

−3	1	−13	0	0	−120	80
		−3	48	−144	432	−936
	1	−16	48	−144	312	−856

$$\frac{x^5 - 13x^4 - 120x + 80}{x + 3} = x^4 - 16x^3 + 48x^2 - 144x + 312 - \frac{856}{x + 3}$$

34.

−3	5	0	0	0
		−15	45	−135
	5	−15	45	−135

$$\frac{5x^3}{x + 3} = 5x^2 - 15x + 45 - \frac{135}{x + 3}$$

36.

−2	−3	0	0	0	0
		6	−12	24	−48
	−3	6	−12	24	−48

$$\frac{-3x^4}{x + 2} = -3x^3 + 6x^2 - 12x + 24 - \frac{48}{x + 2}$$

38.

−1	−1	2	−3	5
		1	−3	6
	−1	3	−6	11

$$\frac{5 - 3x + 2x^2 - x^3}{x + 1} = -x^2 + 3x - 6 + \frac{11}{x + 1}$$

40.

$\frac{3}{2}$	3	−4	0	5
		$\frac{9}{2}$	$\frac{3}{4}$	$\frac{9}{8}$
	3	$\frac{1}{2}$	$\frac{3}{4}$	$\frac{49}{8}$

$$\frac{3x^3 - 4x^2 + 5}{x - \frac{3}{2}} = 3x^2 + \frac{1}{2}x + \frac{3}{4} + \frac{49}{8x - 12}$$

42. You can check polynomial division by multiplying the quotient by the divisor. This should yield the original dividend if the multiplication was performed correctly.

44.

−2	1	0	0	−2	1	c
		−2	4	−8	20	−42
	1	−2	4	−10	21	c−42

To divide evenly, $c-42$ must equal zero. Thus, c must equal 42.

46. $f(x) = 15x^4 + 10x^3 - 6x^2 + 14, \quad k = -\frac{2}{3}$

$-\frac{2}{3}$	15	10	−6	0	14
		−10	0	4	$-\frac{8}{3}$
	15	0	−6	4	$\frac{34}{3}$

$f(x) = \left(x + \frac{2}{3}\right)(15x^3 - 6x + 4) + \frac{34}{3}$

$f\left(-\frac{2}{3}\right) = \frac{34}{3}$

48. $f(x) = 4x^3 - 6x^2 - 12x - 4, \quad k = 1 - \sqrt{3}$

$1 - \sqrt{3}$	4	−6	−12	−4
		$4 - 4\sqrt{3}$	$10 - 2\sqrt{3}$	4
	4	$-2 - 4\sqrt{3}$	$-2 - 2\sqrt{3}$	0

$f(x) = \left(x - 1 + \sqrt{3}\right)\left[4x^2 - \left(2 + 4\sqrt{3}\right)x - \left(2 + 2\sqrt{3}\right)\right]$

$f\left(1 - \sqrt{3}\right) = 0$

50. $g(x) = x^6 - 4x^4 + 3x^2 + 2$

(a)
$$\begin{array}{r|rrrrrrr} 2 & 1 & 0 & -4 & 0 & 3 & 0 & 2 \\ & & 2 & 4 & 0 & 0 & 6 & 12 \\ \hline & 1 & 2 & 0 & 0 & 3 & 6 & 14 = g(2) \end{array}$$

(b)
$$\begin{array}{r|rrrrrrr} -4 & 1 & 0 & -4 & 0 & 3 & 0 & 2 \\ & & -4 & 16 & -48 & 192 & -780 & 3120 \\ \hline & 1 & -4 & 12 & -48 & 195 & -780 & 3122 = g(-4) \end{array}$$

(c)
$$\begin{array}{r|rrrrrrr} 3 & 1 & 0 & -4 & 0 & 3 & 0 & 2 \\ & & 3 & 9 & 15 & 45 & 144 & 432 \\ \hline & 1 & 3 & 5 & 15 & 48 & 144 & 434 = g(3) \end{array}$$

(d)
$$\begin{array}{r|rrrrrrr} -1 & 1 & 0 & -4 & 0 & 3 & 0 & 2 \\ & & -1 & 1 & 3 & -3 & 0 & 0 \\ \hline & 1 & -1 & -3 & 3 & 0 & 0 & 2 = g(-1) \end{array}$$

52. $f(x) = 0.4x^4 - 1.6x^3 + 0.7x^2 - 2$

(a)
$$\begin{array}{r|rrrrr} 1 & 0.4 & -1.6 & 0.7 & 0 & -2 \\ & & 0.4 & -1.2 & -0.5 & -0.5 \\ \hline & 0.4 & -1.2 & -0.5 & -0.5 & -2.5 = f(1) \end{array}$$

(b)
$$\begin{array}{r|rrrrr} -2 & 0.4 & -1.6 & 0.7 & 0 & -2 \\ & & -0.8 & 4.8 & -11 & 22 \\ \hline & 0.4 & -2.4 & 5.5 & -11 & 20 = f(-2) \end{array}$$

(c)
$$\begin{array}{r|rrrrr} 5 & 0.4 & -1.6 & 0.7 & 0 & -2 \\ & & 2.0 & 2.0 & 13.5 & 67.5 \\ \hline & 0.4 & 0.4 & 2.7 & 13.5 & 67.5 = f(5) \end{array}$$

(d)
$$\begin{array}{r|rrrrr} -10 & 0.4 & -1.6 & 0.7 & 0 & -2 \\ & & -4.0 & 56.0 & -567 & 5670 \\ \hline & 0.4 & -5.6 & 56.7 & -567 & 5668 = f(-10) \end{array}$$

54. In this case it is easier to evaluate $f(2)$ directly because $f(x)$ is in factored form. To evaluate using synthetic division you would have to expand each factor and then multiply it all out.

56.
$$\begin{array}{r|rrrr} -4 & 1 & 0 & -28 & -48 \\ & & -4 & 16 & 48 \\ \hline & 1 & -4 & -12 & 0 \end{array}$$

$$\begin{aligned} x^3 - 28x - 48 &= (x + 4)(x^2 - 4x - 12) \\ &= (x + 4)(x - 6)(x + 2) \end{aligned}$$

Zeros: $-4, -2, 6$

58.
$$\begin{array}{r|rrrr} \frac{2}{3} & 48 & -80 & 41 & -6 \\ & & 32 & -32 & 6 \\ \hline & 48 & -48 & 9 & 0 \end{array}$$

$$\begin{aligned} 48x^3 - 80x^2 + 41x - 6 &= \left(x - \tfrac{2}{3}\right)(48x^2 - 48x + 9) \\ &= \left(x - \tfrac{2}{3}\right)(4x - 3)(12x - 3) \\ &= (3x - 2)(4x - 3)(4x - 1) \end{aligned}$$

Zeros: $\frac{2}{3}, \frac{3}{4}, \frac{1}{4}$

60.
$$\begin{array}{r|rrrr} \sqrt{2} & 1 & 2 & -2 & -4 \\ & & \sqrt{2} & 2\sqrt{2} + 2 & 4 \\ \hline & 1 & 2 + \sqrt{2} & 2\sqrt{2} & 0 \end{array}$$

$$\begin{aligned} x^3 + 2x^2 - 2x - 4 &= \left(x - \sqrt{2}\right)\left[x^2 + \left(2 + \sqrt{2}\right)x + 2\sqrt{2}\right] \\ &= \left(x - \sqrt{2}\right)(x + 2)\left(x + \sqrt{2}\right) \end{aligned}$$

Zeros: $-2, -\sqrt{2}, \sqrt{2}$

62. $$\begin{array}{r|rrrr} 2-\sqrt{5} & 1 & -1 & -13 & -3 \\ & & 2-\sqrt{5} & 7-3\sqrt{5} & 3 \\ \hline & 1 & 1-\sqrt{5} & -6-3\sqrt{5} & 0 \end{array}$$

$$x^3 - x^2 - 13x - 3 = \left[x - (2 - \sqrt{5})\right]\left[x^2 + (1 - \sqrt{5})x - (6 + 3\sqrt{5})\right]$$
$$= (x - 2 + \sqrt{5})(x - 2 - \sqrt{5})(x + 3)$$

Zeros: $2 - \sqrt{5}, 2 + \sqrt{5}, -3$

64. $g(x) = x^3 - 4x^2 - 2x + 8$

(a) The zeros of g are $x = 4, x \approx -1.414$, $x \approx 1.414$.

(b) $$\begin{array}{r|rrrr} 4 & 1 & -4 & -2 & 8 \\ & & 4 & 0 & -8 \\ \hline & 1 & 0 & -2 & 0 \end{array}$$

$f(x) = (x - 4)(x^2 - 2)$
$= (x - 4)(x - \sqrt{2})(x + \sqrt{2})$

66. $f(s) = s^3 - 12s^2 + 40s - 24$

(a) The zeros of f are $s = 6, s \approx 0.764, s \approx 5.236$

(b) $$\begin{array}{r|rrrr} 6 & 1 & -12 & 40 & -24 \\ & & 6 & -36 & 24 \\ \hline & 1 & -6 & 4 & 0 \end{array}$$

$f(s) = (s - 6)(s^2 - 6s + 4)$
$= (s - 6)\left[s - (3 + \sqrt{5})\right]\left[s - (3 - \sqrt{5})\right]$

68. $$\begin{array}{r|rrrr} -8 & 1 & 1 & -64 & -64 \\ & & -8 & 56 & 64 \\ \hline & 1 & -7 & -8 & 0 \end{array}$$

$$\frac{x^3 + x^2 - 64x - 64}{x + 8} = x^2 - 7x - 8$$

70. $$\begin{array}{r|rrrr} 1 & 2 & 3 & -3 & -2 \\ & & 2 & 5 & 2 \\ \hline & 2 & 5 & 2 & 0 \end{array}$$

$$\frac{2x^3 + 3x^2 - 3x - 2}{x - 1} = 2x^2 + 5x + 2$$

72. $$\frac{x^4 + 9x^3 - 5x^2 - 36x + 4}{(x + 2)(x - 2)}$$

$$\begin{array}{r|rrrrr} 2 & 1 & 9 & -5 & -36 & 4 \\ & & 2 & 22 & 34 & -4 \\ \hline & 1 & 11 & 17 & -2 & 0 \end{array}$$

$$\begin{array}{r|rrrr} -2 & 1 & 11 & 17 & -2 \\ & & -2 & -18 & 2 \\ \hline & 1 & 9 & -1 & 0 \end{array}$$

$$\frac{x^4 + 9x^3 - 5x^2 - 36x + 4}{x^2 - 4} = x^2 + 9x - 1$$

74.

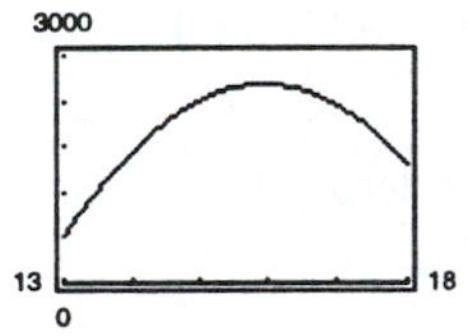

$y \approx 16.89$

76. (a) and (b)

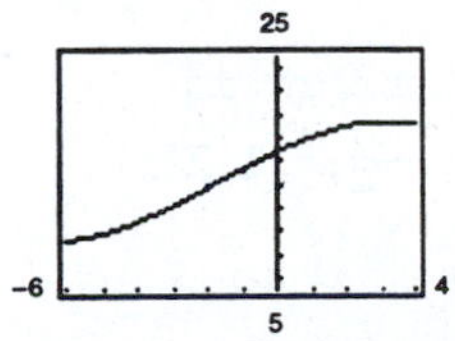

(c)

t	-5	-4	-3	-2	-1
R	9.75	10.80	12.16	13.72	15.31

t	0	1	2	3
R	16.82	18.10	19.01	19.41

The estimated values are quite close to the actual data.

(d) $$\begin{array}{r|rrrr} 6 & -0.023 & -0.115 & 1.415 & 16.823 \\ & & -0.138 & -1.518 & -0.618 \\ \hline & -0.023 & -0.253 & -0.103 & 16.205 \end{array}$$

$R(6) = 16.205$. No, the model will not be accurate in predicting future cable rates because the model will fall to the right and eventually become negative (which doesn't make sense in this context).

78. $f(g(x)) = f\left(\frac{1}{x-2}\right) = 2 + \left(\frac{1}{\frac{1}{x-2}}\right)$

$= 2 + (x - 2)$

$= x$

$g(f(x)) = g\left(2 + \frac{1}{x}\right) = \frac{1}{\left(2 + \frac{1}{x}\right) - 2}$

$= \frac{1}{\frac{1}{x}}$

$= x$

80. No, $f(x) = x^3 + 3x^2$ does not have an inverse. Its graph does not pass the horizontal line test.

Section 2.4 Real Zeros of Polynomial Functions

Solutions to Even-Numbered Exercises

2. $g(x) = x^3 + 3x^2$

Sign variations: 0, positive zeros: 0

$g(-x) = -x^3 + 3x^2$

Sign variations: 1, negative zeros: 1

4. $h(x) = 4x^2 - 8x + 3$

Sign variations: 2, positive zeros: 2 or 0

$h(-x) = 4x^2 + 8x + 3$

Sign variations: 0, negative zeros: 0

6. $h(x) = 2x^4 - 3x + 2$

Sign variations: 2, positive zeros: 2 or 0

$h(-x) = 2x^4 + 3x + 2$

Sign variations: 0, negative zeros: 0

8. $f(x) = 4x^3 - 3x^2 + 2x - 1$

Sign variations: 3, positive zeros: 3 or 1

$f(-x) = -4x^3 - 3x^2 - 2x - 1$

Sign variations: 0, negative zeros: 0

10. $f(x) = 3x^3 + 2x^2 + x + 3$

Sign variations: 0, positive zeros: 0

$f(-x) = -3x^3 + 2x^2 - x + 3$

Sign variations: 3, negative zeros: 3 or 1

12. $f(x) = x^3 - 4x^2 - 4x + 16$

$p =$ factor of 16

$q =$ factor of 1

Possible rational zeros: $\pm 1, \pm 2, \pm 4, \pm 8, \pm 16$

Zeros shown on graph: $-2, 2, 4$

14. $f(x) = 6x^3 - 71x^2 - 13x + 12$

$p =$ factor of 12

$q =$ factor of 6

Possible rational zeros: $\pm 1, \pm 2, \pm 3, \pm 4, \pm 6, \pm 12,$
$\pm\frac{1}{2}, \pm\frac{1}{3}, \pm\frac{1}{6}, \pm\frac{2}{3}, \pm\frac{3}{2}, \pm\frac{4}{3}$

Zeros shown on graph: $-\frac{1}{2}, \frac{1}{3}, 12$

16. $f(x) = 4x^5 - 8x^4 - 5x^3 + 10x^2 + x - 2$

$p =$ factor of -2

$q =$ factor of 4

Possible rational zeros: $\pm 1, \pm 2, \pm\frac{1}{2}, \pm\frac{1}{4}$

Zeros shown on graph: $-1, -\frac{1}{2}, \frac{1}{2}, 1, 2$

18. $f(x) = x^3 - 7x - 6$

Possible rational zeros: $\pm 1, \pm 2, \pm 3, \pm 6$

$$\begin{array}{r|rrrr} 3 & 1 & 0 & -7 & -6 \\ & & 3 & 9 & 6 \\ \hline & 1 & 3 & 2 & \end{array}$$

$f(x) = (x - 3)(x^2 + 3x + 2) = (x - 3)(x + 2)(x + 1)$

Thus, the real zeros are $-2, -1, 3$.

20. $h(x) = x^3 - 9x^2 + 20x - 12$

Possible rational zeros: $\pm 1, \pm 2, \pm 3, \pm 4, \pm 6, \pm 12$

$$\begin{array}{r|rrrr} 1 & 1 & -9 & 20 & -12 \\ & & 1 & -8 & 12 \\ \hline & 1 & -8 & 12 & 0 \end{array}$$

$h(x) = (x - 1)(x^2 - 8x + 12)$

$= (x - 1)(x - 2)(x - 6)$

Thus, the real zeros are 1, 2, 6.

22. $f(x) = x^3 + 6x^2 + 12x + 8$

Possible rational zeros: $\pm 1, \pm 2, \pm 4, \pm 8$

$$\begin{array}{r|rrrr} -2 & 1 & 6 & 12 & 8 \\ & & -2 & -8 & -8 \\ \hline & 1 & 4 & 4 & 0 \end{array}$$

$f(x) = (x + 2)(x^2 + 4x + 4)$

Thus, the real zero is -2.

24. $p(x) = x^3 - 9x^2 + 27x - 27$

Possible rational zeros: $\pm 1, \pm 3, \pm 9, \pm 27$

$$\begin{array}{r|rrrr} 3 & 1 & -9 & 27 & -27 \\ & & 3 & -18 & 27 \\ \hline & 1 & -6 & 9 & 0 \end{array}$$

$f(x) = (x - 3)(x^2 - 6x + 9)$

$= (x - 3)(x - 3)(x - 3)$

Thus, the real zero is 3.

26. $f(x) = 3x^2 - 19x^2 + 33x - 9$

Possible rational zeros: $\pm 1, \pm 3, \pm 9, \pm \frac{1}{3}$

$$\begin{array}{r|rrrr} 3 & 3 & -19 & 33 & -9 \\ & & 9 & -30 & 9 \\ \hline & 3 & -10 & 3 & 0 \end{array}$$

$f(x) = (x - 3)(3x^2 - 10x + 3)$

$= (x - 3)(3x - 1)(x - 3)$

Thus, the real zeros are $3, \frac{1}{3}$.

28. $f(x) = 2x^4 - 15x^3 + 23x^2 + 15x - 25$

Possible rational zeros: $\pm 1, \pm 5, \pm 25, \pm \frac{1}{2}, \pm \frac{5}{2}, \pm \frac{25}{2}$

$$\begin{array}{r|rrrrr} 5 & 2 & -15 & 23 & 15 & -25 \\ & & 10 & -25 & -10 & 25 \\ \hline & 2 & -5 & -2 & 5 & 0 \end{array}$$

$$\begin{array}{r|rrrr} 1 & 2 & -5 & -2 & 5 \\ & & 2 & -3 & -5 \\ \hline & 2 & -3 & -5 & 0 \end{array}$$

$$\begin{array}{r|rrr} -1 & 2 & -3 & -5 \\ & & -2 & 5 \\ \hline & 2 & -5 & 0 \end{array}$$

$f(x) = (x - 5)(x - 1)(x + 1)(2x - 5)$

Thus, the real zeros are $5, 1, -1, \frac{5}{2}$.

30. $x^4 - 13x^2 - 12x = 0$

$x(x^3 - 13x - 12) = 0$

$$\begin{array}{r|rrrr} -1 & 1 & 0 & -13 & -12 \\ & & -1 & 1 & 12 \\ \hline & 1 & -1 & -12 & 0 \end{array}$$

$x(x + 1)(x^2 - x - 12) = 0$

$x(x + 1)(x - 4)(x + 3) = 0$

The real zeros are $0, -1, 4, -3$.

32. $x^5 - x^4 - 3x^3 + 5x^2 - 2x = 0$

$x(x^4 - x^3 - 3x^2 + 5x - 2) = 0$

$$\begin{array}{r|rrrrr} 1 & 1 & -1 & -3 & 5 & -2 \\ & & 1 & 0 & -3 & 2 \\ \hline & 1 & 0 & -3 & 2 & 0 \end{array}$$

$$\begin{array}{r|rrrr} -2 & 1 & 0 & -3 & 2 \\ & & -2 & 4 & -2 \\ \hline & 1 & -2 & 1 & 0 \end{array}$$

$x(x - 1)(x + 2)(x^2 - 2x + 1) = 0$

$x(x - 1)(x + 2)(x - 1)(x - 1) = 0$

The real zeros are $-2, 0, 1$.

34. $f(x) = -3x^3 + 20x^2 - 36x + 16$

(a) Possible real zeros: $\pm 1, \pm 2, \pm 4, \pm 8, \pm 16, \pm \frac{1}{3}, \pm \frac{2}{3}, \pm \frac{4}{3}, \pm \frac{8}{3}, \pm \frac{16}{3}$

(b)

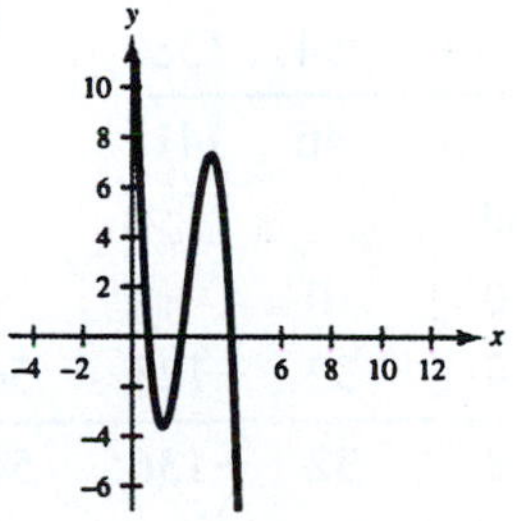

(c) Real zeros: $\frac{2}{3}, 2, 4$

36. $f(x) = 4x^3 - 12x^2 - x + 15$

(a) Possible real zeros: $\pm 1, \pm 3, \pm 5, \pm 15, \pm\frac{1}{2}, \pm\frac{3}{2}, \pm\frac{5}{2}, \pm\frac{15}{2}, \pm\frac{1}{4}, \pm\frac{3}{4}, \pm\frac{5}{4}, \pm\frac{15}{4}$

(b)

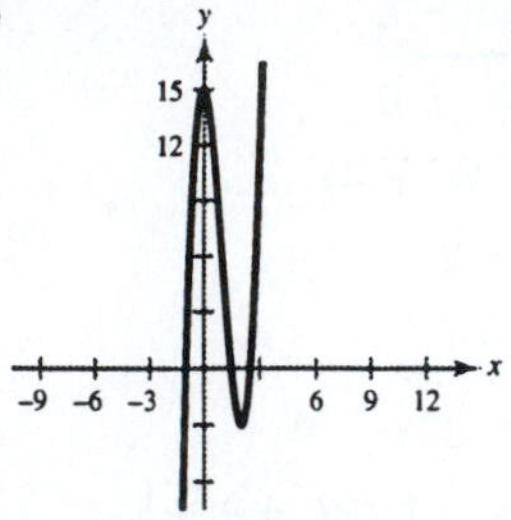

(c) Real zeros: $-1, \frac{3}{2}, \frac{5}{2}$

38. $f(x) = 4x^4 - 17x^2 + 4$

(a) Possible real zeros: $\pm 1, \pm 2, \pm 4, \pm\frac{1}{2}, \pm\frac{1}{4}$

(b)

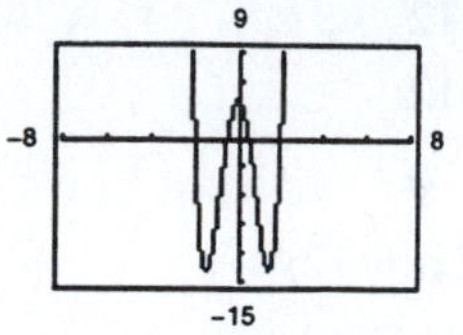

(c) Real zeros: $\pm 2, \pm\frac{1}{2}$

40. $f(x) = 4x^3 + 7x^2 - 11x - 18$

(a) Possible real zeros: $\pm 1, \pm 2, \pm 3, \pm 6, \pm 9, \pm 18, \pm\frac{1}{2}, \pm\frac{3}{2}, \pm\frac{9}{2}, \pm\frac{1}{4}, \pm\frac{3}{4}, \pm\frac{9}{4}$

(b)

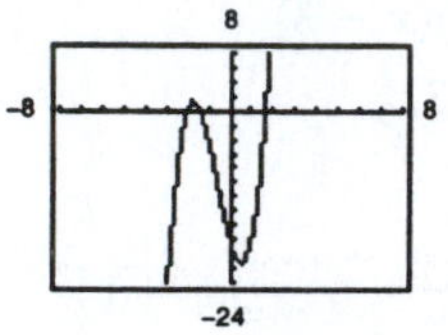

(c) Real zeros: $-2, \dfrac{1}{8} \pm \dfrac{\sqrt{145}}{8}$

42. $P(t) = t^4 - 7t^2 + 12$

(a) $t = \pm 2, \pm 1.732$

(b) $P(t) = t^4 - 7t^2 + 12$

$= (t^2 - 3)(t^2 - 4)$

$= (t - \sqrt{3})(t + \sqrt{3})(t - 2)(t + 2)$

44. $g(x) = 6x^4 - 11x^3 - 51x^2 + 99x - 27$

(a) $x = \pm 3, \frac{3}{2}, \frac{1}{3}$

(b)

3	6	−11	−51	99	−27
		18	21	−90	27
	6	7	−30	9	0

−3	6	7	−30	9
		−18	33	−9
	6	−11	3	0

$g(x) = (x - 3)(x + 3)(6x^2 - 11x + 3)$

$= (x - 3)(x + 3)(3x - 1)(2x - 3)$

46. $f(x) = 2x^3 - 3x^2 - 12x + 8$

(a)

4	2	−3	−12	8
		8	20	32
	2	5	8	40

4 is an upper bound.

(b)

−3	2	−3	−12	8
		−6	27	−45
	2	−9	15	−37

−3 is a lower bound.

48. $f(x) = 2x^4 - 8x + 3$

(a)

3	2	0	0	−8	3
		6	18	54	138
	2	6	18	46	141

3 is an upper bound.

(b)

−4	2	0	0	−8	3
		−8	32	−128	544
	2	−8	32	−136	547

−4 is a lower bound.

50. $f(z) = 12z^3 - 4x^2 - 27z + 9$

Possible rational zeros: $\pm 1, \pm 3, \pm 9, \pm\frac{1}{2}, \pm\frac{3}{2}, \pm\frac{9}{2}, \pm\frac{1}{3}, \pm\frac{1}{4}, \pm\frac{3}{4}, \pm\frac{9}{4}, \pm\frac{1}{6}, \pm\frac{1}{12}$

$\frac{3}{2}$	12	−4	−27	9
		18	21	−9
	12	14	−6	0

$f(z) = \left(z - \frac{3}{2}\right)(6z^2 + 7z - 3)$

$= \left(z - \frac{3}{2}\right)(3z - 1)(2z + 3)$

Real zeros: $-\frac{3}{2}, \frac{1}{3}, \frac{3}{2}$

52. $g(x) = 3x^2 - 2x^2 + 15x - 10$

Possible rational zeros: $\pm 1, \pm 2, \pm 5, \pm 10, \pm \frac{1}{3}, \pm \frac{2}{3}, \pm \frac{5}{3}, \pm \frac{10}{3}$

$$\begin{array}{r|rrrr} \frac{2}{3} & 3 & -2 & 15 & -10 \\ & & 2 & 0 & 10 \\ \hline & 3 & 0 & 15 & 0 \end{array}$$

$g(x) = \left(x - \frac{2}{3}\right)(3x^2 + 15)$

Real zeros: $\frac{2}{3}$

54. $f(x) = \frac{1}{2}(2x^3 - 3x^2 - 23x + 12)$

Possible rational zeros: $\pm 1, \pm 2, \pm 3, \pm 4, \pm 6, \pm 12, \pm \frac{1}{2}, \pm \frac{3}{2}$

$$\begin{array}{r|rrrr} 4 & 2 & -3 & -23 & 12 \\ & & 8 & 20 & -12 \\ \hline & 2 & 5 & -3 & 0 \end{array}$$

$f(x) = \frac{1}{2}(x - 4)(2x^2 + 5x - 3)$

$= \frac{1}{2}(x - 4)(2x - 1)(x + 3)$

Rational zeros: $-3, \frac{1}{2}, 4$

56. $f(x) = \frac{1}{6}(6z^3 + 11z^2 - 3z - 2)$

Possible rational zeros: $\pm 1, \pm 2, \pm \frac{1}{2}, \pm \frac{1}{3}, \pm \frac{2}{3}, \pm \frac{1}{6}$

$$\begin{array}{r|rrrr} -2 & 6 & 11 & -3 & -2 \\ & & -12 & 2 & 2 \\ \hline & 6 & -1 & -1 & 0 \end{array}$$

$f(x) = \frac{1}{6}(x + 2)(6x^2 - x - 1)$

$= \frac{1}{6}(x + 2)(3x + 1)(2x - 1)$

Rational zeros: $-2, -\frac{1}{3}, \frac{1}{2}$

58. $f(x) = x^3 - 2$

$= \left(x - \sqrt[3]{2}\right)\left(x^2 + \sqrt[3]{2}x + \sqrt[3]{4}\right)$

Rational zeros: 0

Irrational zeros: 1 $\left(x = \sqrt[3]{2}\right)$

Matches (a).

60. $f(x) = x^3 - 2x$

$= x(x^2 - 2)$

$= x\left(x + \sqrt{2}\right)\left(x - \sqrt{2}\right)$

Rational zeros: 1 $(x = 0)$

Irrational zeros: 2 $\left(x = \pm\sqrt{2}\right)$

Matches (c).

62. (a) Combined length and width:

$4x + y = 120 \Rightarrow y = 120 - 4x$

Volume $= l \cdot w \cdot h = x^2y$

$= x^2(120 - 4x)$

$= 4x^2(30 - x)$

(b)

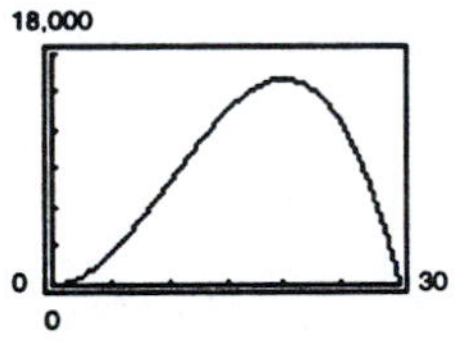

Dimensions with maximum volume:
$20 \times 20 \times 40$

(c) $13{,}500 = 4x^2(30 - x)$

$4x^3 - 120x^2 + 13{,}500 = 0$

$x^3 - 30x^2 + 3375 = 0$

$$\begin{array}{r|rrrr} 15 & 1 & -30 & 0 & 3375 \\ & & 15 & -225 & -3375 \\ \hline & 1 & -15 & -225 & 0 \end{array}$$

$(x - 15)(x^2 - 15x - 225) = 0$

Using the Quadratic equation,

$x = 15, \dfrac{15 \pm 15\sqrt{5}}{2}$.

The value of $\dfrac{15 - 15\sqrt{5}}{2}$ is not possible because it is negative.

64. $g(x) = 3f(x)$. This function would also have the same zeros as $f(x)$ because $g(x) = 3f(x) = 3(x - r_1)(x - r_2)(x - r_3)$.

The roots are r_1, r_2, and r_3.

66. $g(x) = f(2x)$

$= (2x - r_1)(2x - r_2)(2x - r_3)$

The roots are $\frac{r_1}{2}, \frac{r_2}{2}$, and $\frac{r_3}{2}$.

68. $g(x) = f(-x)$

$= (-x - r_1)(-x - r_2)(-x - r_3)$

The roots are $-r_1, -r_2, -r_3$.

70.

$$P = -45x^3 + 2500x^2 - 275{,}000$$
$$800{,}000 = -45x^3 + 2500x^2 - 275{,}000$$
$$0 = 45x^3 - 2500x^2 + 1{,}075{,}000$$
$$0 = 9x^3 - 500x^2 + 215{,}000$$

The zeros of this equation are $x \approx -18.0$, $x \approx 31.5$, and $x \approx 42.0$. Because $0 \le x \le 50$, disregard $x \approx -18.02$. The smaller remaining solution is $x \approx 31.5$, or \$315,000.

72. (a)

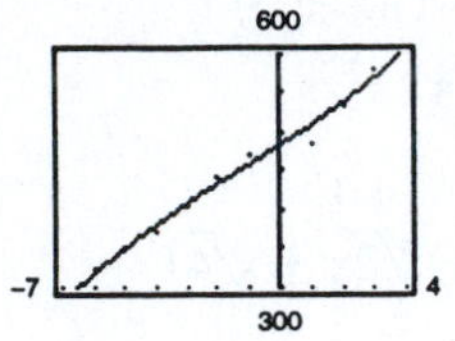

(b) $525 = 0.161t^3 + 0.626t^2 + 25.646t + 484.137$

$0 = 0.161t^3 + 0.626t^2 + 25.646t - 40.863$

The zero of this equation is $t \approx 1.515$. Thus, according to the model, the annual value of imports reached \$525 billion in 1991.

(c) Because the degree of the model is odd and the leading coefficient is positive, the right-hand behavior of this model is increasing. Yes, the value of imports will continue

74. $g(x) = f(x) - 2$

Vertical shift down two units

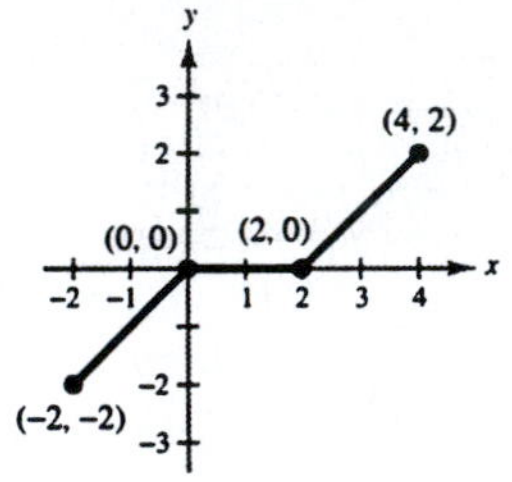

76. $g(x) = f(-x)$

Reflection in the y-axis

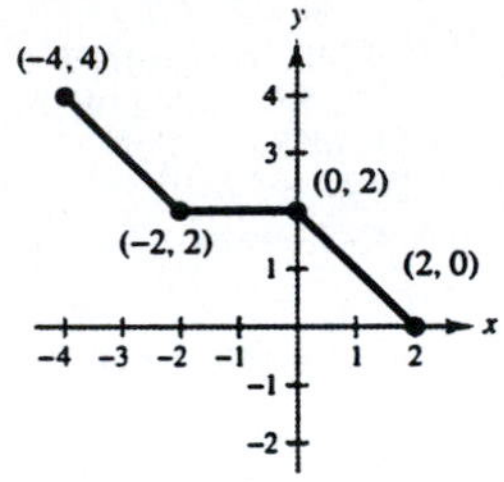

78. $g(x) = f\left(\frac{1}{2}x\right)$

Horizontal stretch

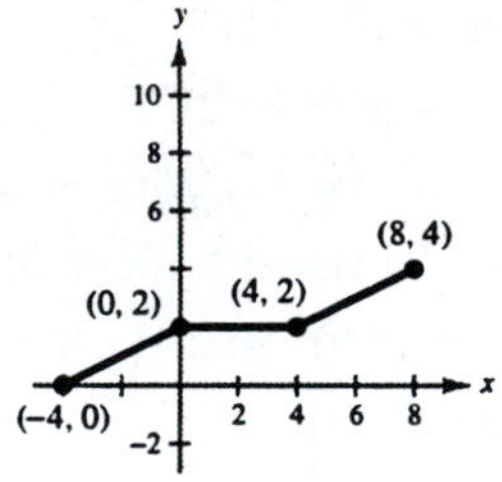

Section 2.5 Complex Numbers

Solutions to Even-Numbered Exercises

2. $a + bi = 13 + 4i$

$a = 13$

$b = 4$

4. $(a + 6) + 2bi = 6 - 5i$

$2b = -5$

$b = -\frac{5}{2}$

$a + 6 = 6$

$a = 0$

6. $3 + \sqrt{-16} = 3 + 4i$

22. $(8 + \sqrt{-18}) - (4 + 3\sqrt{2}i) = 8 + 3\sqrt{2}i - 4 - 3\sqrt{2}i = 4$

24. $22 + (-5 + 8i) + 10i = 17 + 18i$

26. $(1.6 + 3.2i) + (-5.8 + 4.3i) = -4.2 + 7.5i$

28. $\sqrt{-5} \cdot \sqrt{-10} = (\sqrt{5}i)(\sqrt{10}i)$

$= \sqrt{50}i^2 = 5\sqrt{2}(-1) = -5\sqrt{2}$

30. $(\sqrt{-75})^2 = (\sqrt{75}i)^2 = 75i^2 = -75$

32. $(6 - 2i)(2 - 3i) = 12 - 18i - 4i + 6i^2$

$= 12 - 22i - 6$

$= 6 - 22i$

34. $-8i(9 + 4i) = -72i - 32i^2$

$= 32 - 72i$

36. $(3 + \sqrt{-5})(7 - \sqrt{-10}) = (3 + \sqrt{5}i)(7 - \sqrt{10}i)$

$= 21 - 3\sqrt{10}i + 7\sqrt{5}i - \sqrt{50}i^2$

$= 21 + \sqrt{50} + 7\sqrt{5}i - 3\sqrt{10}i$

$= (21 + 5\sqrt{2}) + (7\sqrt{5} - 3\sqrt{10})i$

38. $(2 - 3i)^2 = 4 - 12i + 9i^2$

$= 4 - 9 - 12i$

$= -5 - 12i$

40. $(1 - 2i)^2 - (1 + 2i)^2 = 1 - 4i + 4i^2 - (1 + 4i + 4i^2)$

$= 1 - 4i + 4i^2 - 1 - 4i - 4i^2$

$= -8i$

42. False. If the complex number is real, the number equals its conjugate.

44. The complex conjugate of $9 - 12i$ is $9 + 12i$.

$(9 - 12i)(9 + 12i) = 81 - 144i^2$

$= 81 - (-144)$

$= 225$

46. The complex conjugate of $-4 + \sqrt{2}i$ is $-4 - \sqrt{2}i$.

$(-4 + \sqrt{2}i)(-4 - \sqrt{2}i) = 16 - 2i^2$

$= 16 - (-2)$

$= 18$

48. The complex conjugate of $\sqrt{-15} = \sqrt{15}i$ is $-\sqrt{15}i$.

$(\sqrt{15}i)(-\sqrt{15}i) = -15i^2 = -(-15) = 15$

50. The complex conjugate of $1 + \sqrt{8}$ is $1 + \sqrt{8}$.

$(1 + \sqrt{8})(1 + \sqrt{8}) = 1 + 2\sqrt{8} + 8$

$= 9 + 4\sqrt{2}$

52. $-\frac{10}{2i} \cdot \frac{-2i}{-2i} = \frac{20i}{-4i} = \frac{20i}{4} = 5i$

54. $\frac{3}{1 - i} \cdot \frac{1 + i}{1 + i} = \frac{3 + 3i}{1 - i^2} = \frac{3 + 3i}{2} = \frac{3}{2} + \frac{3}{2}i$

56. $\frac{8 - 7i}{1 - 2i} \cdot \frac{1 + 2i}{1 + 2i} = \frac{8 + 16i - 7i - 14i^2}{1 - 4i^2}$

$= \frac{22 + 9i}{5} = \frac{22}{5} + \frac{9}{5}i$

58. $\dfrac{8+20i}{2i} \cdot \dfrac{-2i}{-2i} = \dfrac{-16i-40i^2}{-4i^2} = 10-4i$

60. $\dfrac{(2-3i)(5i)}{2+3i} \cdot \dfrac{2-3i}{2-3i} = \dfrac{(10i-15i^2)(2-3i)}{4-9i^2}$

$= \dfrac{20i-30i^2-30i^2+45i^3}{13}$

$= \dfrac{60+20i-45i}{13}$

$= \dfrac{60-25i}{13} = \dfrac{60}{13} - \dfrac{25}{13}i$

62. $\dfrac{2i}{2+i} + \dfrac{5}{2-i} = \dfrac{2i(2-i)}{(2+i)(2-i)} + \dfrac{5(2+i)}{(2+i)(2-i)}$

$= \dfrac{4i-2i^2+10+5i}{4-i^2}$

$= \dfrac{12+9i}{5}$

$= \dfrac{12}{5} + \dfrac{9}{5}i$

64. $\dfrac{1+i}{i} - \dfrac{3}{4-i} = \dfrac{(1+i)(4-i)-3i}{i(4-i)}$

$= \dfrac{4-i+4i-i^2-3i}{4i-i^2}$

$= \dfrac{5}{1+4i} \cdot \dfrac{1-4i}{1-4i}$

$= \dfrac{5-20i}{1-16i^2}$

$= \dfrac{5}{17} - \dfrac{20}{17}i$

66. $x^2+6x+10=0; \quad a=1, b=6, c=10$

$x = \dfrac{-6 \pm \sqrt{6^2-4(1)(10)}}{2(1)}$

$= \dfrac{-6 \pm \sqrt{-4}}{2}$

$= -3 \pm i$

68. $9x^2-6x+37=0$

$x = \dfrac{-(-6) \pm \sqrt{(-6)^2-4(9)(37)}}{2(9)}$

$= \dfrac{6 \pm \sqrt{-1296}}{18}$

$= \dfrac{1}{3} \pm \dfrac{36i}{18}$

$= \dfrac{1}{3} \pm 2i$

70. $9x^2-6x-35=0$

$x = \dfrac{-(-6) \pm \sqrt{(-6)^2-4(9)(-35)}}{2(9)}$

$= \dfrac{6 \pm \sqrt{1296}}{18}$

$= \dfrac{6 \pm 36}{18}$

$= -\dfrac{5}{3}, \dfrac{7}{3}$

72. $5s^2+6s+3=0$

$s = \dfrac{-6 \pm \sqrt{36-4(5)(3)}}{2(5)}$

$= \dfrac{-6 \pm \sqrt{-24}}{10}$

$= -\dfrac{3}{5} \pm \dfrac{\sqrt{6}i}{5}$

74. $y = -(x^2-4x+3)$

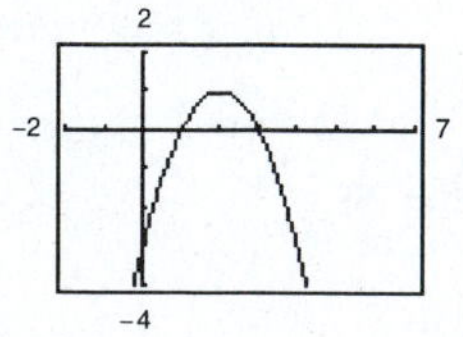

$0 = -(x^2-4x+3)$

$0 = x^2-4x+3$

$0 = (x-3)(x-1)$

$0 = x-3 \Rightarrow x=3$

$0 = x-1 \Rightarrow x=1$

x-intercepts: $(1, 0)$ and $(3, 0)$

76. $y = \frac{1}{4}(x^2 - 2x + 9)$

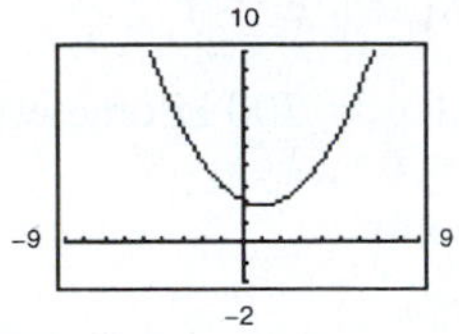

$0 = \frac{1}{4}(x^2 - 2x + 9)$

$0 = x^2 - 2x + 9$

$x = \frac{-(-2) \pm \sqrt{(-2)^2 - 4(1)(9)}}{2(1)}$

$= \frac{2 \pm \sqrt{-32}}{2}$

$= 1 \pm 2\sqrt{2}i$

No real solutions, no x-intercepts

78. (a) $i^{40} = i^4 \cdot i^4 \cdot i^4 \cdot i^4 \cdot i^4 \cdot i^4 \cdot i^4 \cdot i^4 \cdot i^4 \cdot i^4$
$= 1 \cdot 1 \cdot 1 \cdot 1 \cdot 1 \cdot 1 \cdot 1 \cdot 1 \cdot 1 \cdot 1$
$= 1$

(b) $i^{25} = i^4 \cdot i^4 \cdot i^4 \cdot i^4 \cdot i^4 \cdot i^4 \cdot i$
$= 1 \cdot 1 \cdot 1 \cdot 1 \cdot 1 \cdot 1 \cdot i$
$= i$

(c) $i^{50} = i^{25} \cdot i^{25} = i \cdot i = i^2 = -1$

(d) $i^{67} = i^{50} \cdot i^{17} = -1 \cdot i^4 \cdot i^4 \cdot i^4 \cdot i^4 \cdot i = -i$

80. $4i^2 - 2i^3 = -4 + 2i$

82. $(-i)^3 = (-1)(i^3) = (-1)(-i) = i$

84. $(\sqrt{-2})^6 = (\sqrt{2}i)^6 = 8i^6 - 8i^4i^2 = -8$

86. $\frac{1}{(2i)^3} = \frac{1}{8i^3} = \frac{1}{-8i} \cdot \frac{8i}{8i} = \frac{8i}{-64i^2} = \frac{1}{8}i$

88. $2^4 = 16,\ (-2)^4 = 16$

$(2i)^4 = 2^4i^4 = 16(1) = 16$

$(-2i)^4 = (-2)^4i^4 = 16(1) = 16$

90. $(a + bi) - (a - bi) = a + bi - a + bi = 2bi$, which is an imaginary number.

$(a - bi) - (a + bi) = 1 - bi - a - bi = -2bi$, which is an imaginary number.

92. $(a_1 + b_1i)(a_2 + b_2i) = a_1a_2 + a_1b_2i + a_2b_1i + b_1b_2i^2$
$= (a_1a_2 - b_1b_2) + (a_1b_2 + a_2b_1)i$

The conjugate of this product is $(a_1a_2 - b_1b_2) - (a_1b_2 + a_2b_1)i$.

The product of the conjugates is:

$(a_1 - b_1i)(a_2 - b_2i) = a_1a_2 - a_1b_2i - a_2b_1i + b_1b_2i$
$= (a_1a_2 - b_1b_2) - (a_1b_2 + a_2b_1)i$

Thus, the conjugate of the product of two complex numbers is the product of their conjugates.

94. $(x^3 - 3x^2) - (6 - 2x - 4x^2) = x^3 - 3x^2 - 6 + 2x + 4x^2$
$= x^3 + x^2 + 2x - 6$

96. $(3x - \frac{1}{2})(x + 4) = 3x^2 + 12x - \frac{1}{2}x - 2$
$= 3x^2 + 11\frac{1}{2}x - 2$

98. $[(x + y) + 3]^2 = (x + y)^2 + 6(x + y) + 9$
$= x^2 + 2xy + y^2 + 6x + 6y + 9$

100. $F = \alpha\frac{m_1m_2}{r^2}$

$r^2 = \alpha\frac{m_1m_2}{F}$

$r^2 = \sqrt{\frac{\alpha m_1m_2}{F}}$

102. $r \times t = d$

Model: $(\text{Average speed}) = \dfrac{(\text{Distance on first leg}) + (\text{Distance on second leg})}{(\text{Time for first leg}) + (\text{Time for second leg})}$

Labels: Average speed $= S$, distance on first leg $= 200$ kilometers, distance on second leg $= 200$ kilometers,

time for first leg $= \dfrac{d}{r} = \dfrac{200}{100} = 2$ hours, time for second leg $= \dfrac{d}{r} = \dfrac{200}{80} = \dfrac{5}{2}$ hours

Expression: $S = \dfrac{200 + 200}{2 + 2.5} \approx 88.89$ kilometers per hour

Section 2.6 The Fundamental Theorem of Algebra

Solutions to Even-Numbered Exercises

2. $g(x) = (x - 2)(x + 4)^3$
$= (x - 2)(x + 4)(x + 4)(x + 4)$

The four zeros are: $2, -4, -4, -4$

4. $h(m) = (m - 4)^2(m - 2 + 4i)(m - 2 - 4i)$
$= (m - 4)(m - 4)(m - 2 + 4i)(m - 2 - 4i)$

The four zeros are: $4, 4, 2 - 4i, 2 + 4i$

6. $f(x) = x^3 - 4x^2 - 4x + 16$
$= x^2(x - 4) - 4(x - 4)$
$= (x^2 - 4)(x - 4)$
$= (x + 2)(x - 2)(x - 4)$

The zeros are: $x = 2, -2$, and 4. This corresponds to the x-intercepts of $(-2, 0)$, $(2, 0)$, and $(4, 0)$ on the graph.

8. $f(x) = x^4 - 3x^2 - 4$
$= (x^2 - 4)(x^2 + 1)$
$= (x + 2)(x - 2)(x^2 + 1)$

The only real zeros are $x = -2, 2$. This corresponds to the x-intercepts of $(-2, 0)$ and $(2, 0)$ on the graph.

10. $f(x) = x^2 - x + 56$

Zeros: $x = \dfrac{1 \pm \sqrt{223}i}{2}$

$f(x) = \left(x - \dfrac{1 - \sqrt{223}i}{2}\right)\left(x - \dfrac{1 + \sqrt{223}i}{2}\right)$

12. $g(x) = x^2 + 10x + 23$

Zeros: $x = \dfrac{-10 \pm \sqrt{8}}{2} = -5 \pm \sqrt{2}$

$g(x) = \left(x + 5 + \sqrt{2}\right)\left(x + 5 - \sqrt{2}\right)$

14. $f(y) = y^4 - 625$

Zeros: $x = \pm 5, \pm 5i$

$f(y) = (y + 5)(y - 5)(y + 5i)(y - 5i)$

16. $h(x) = x^3 - 3x^2 + 4x - 2$

$$\begin{array}{r|rrrr} 1 & 1 & -3 & 4 & -2 \\ & & 1 & -2 & 2 \\ \hline & 1 & -2 & 2 & 0 \end{array}$$

Zeros: $x = 1, \dfrac{2 \pm \sqrt{4}i}{2} = 1 \pm i$

$h(x) = (x - 1)(x - 1 - i)(x - 1 + i)$

18. $f(x) = x^3 - 2x^2 - 11x + 52$

$$\begin{array}{r|rrrr} -4 & 1 & -2 & -11 & 52 \\ & & -4 & 24 & -52 \\ \hline & 1 & -6 & 13 & 0 \end{array}$$

Zeros: $x = -4, \dfrac{6 \pm \sqrt{16}i}{2} = 3 \pm 2i$

$f(x) = (x + 4)(x - 3 - 2i)(x - 3 + 2i)$

20. $f(x) = x^3 + 11x^2 + 39x + 29$

$$\begin{array}{r|rrrr} -1 & 1 & 11 & 39 & 29 \\ & & -1 & -10 & -29 \\ \hline & 1 & 10 & 29 & 0 \end{array}$$

Zeros: $x = -1, \dfrac{-10 \pm \sqrt{16}i}{2} = -5 \pm 2i$

$f(x) = (x + 1)(x + 5 + 2i)(x + 5 - 2i)$

22. $h(x) = x^3 + 9x^2 + 27x + 35$

−5	1	9	27	35
		−5	−20	−35
	1	4	7	0

Zeros: $x = -5, \dfrac{-4 \pm \sqrt{12}i}{2} = -2 \pm \sqrt{3}i$

$h(x) = (x + 5)(x + 2 + \sqrt{3}i)(x + 2 - \sqrt{3}i)$

24. $g(x) = 3x^3 - 4x^2 + 8x + 8$

$-\frac{2}{3}$	3	−4	8	8
		−2	4	−8
	3	−6	12	0

Zeros: $x = -\dfrac{2}{3}, \dfrac{6 \pm \sqrt{108}i}{6} = 1 \pm \sqrt{3}i$

$g(x) = (3x + 2)(x - 1 + \sqrt{3}i)(x - 1 - \sqrt{3}i)$

26. $h(x) = x^4 + 6x^3 + 10x^2 + 6x + 9$

−3	1	6	10	6	9
		−3	−9	−3	−9
−3	1	3	1	3	0
		−3	0	−3	
	1	0	1	0	

Zeros: $x = -3, \pm i$

$h(x) = (x + 3)^2(x + i)(x - i)$

28. $f(x) = x^4 + 29x^2 + 100$

$= (x^2 + 25)(x^2 + 4)$

Zeros: $x = \pm 2i, \pm 5i$

$f(x) = (x + 2i)(x - 2i)(x + 5i)(x - 5i)$

30. $f(s) = 2s^3 - 5s^2 + 12s - 5$

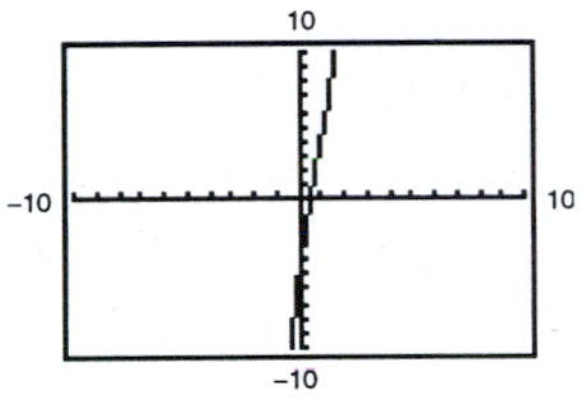

The graph reveals one zero at $x = \frac{1}{2}$.

$\frac{1}{2}$	2	−5	12	−5
		1	−2	5
	2	−4	10	0

Zeros: $s = \dfrac{1}{2}, \dfrac{4 \pm \sqrt{64}i}{4} = 1 \pm 2i$

$f(s) = (2s - 1)(s - 1 + 2i)(s - 1 - 2i)$

32. $f(x) = 9x^3 - 15x^2 + 11x - 5$

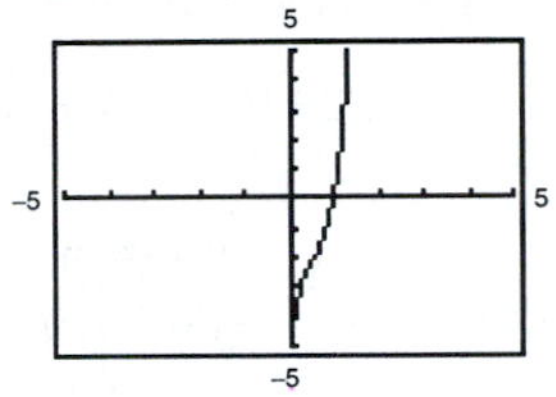

The graph reveals one zero at $x = 1$.

1	9	−15	11	−5
		9	−6	5
	9	−6	5	0

Zeros: $x = 1, \dfrac{6 \pm \sqrt{144}i}{18} = \dfrac{1}{3} \pm \dfrac{2}{3}i$

$f(x) = (x - 1)(3x - 1 + 2i)(3x - 1 - 2i)$

34. $g(x) = x^5 - 8x^4 + 28x^3 - 56x^2 + 64x - 32$

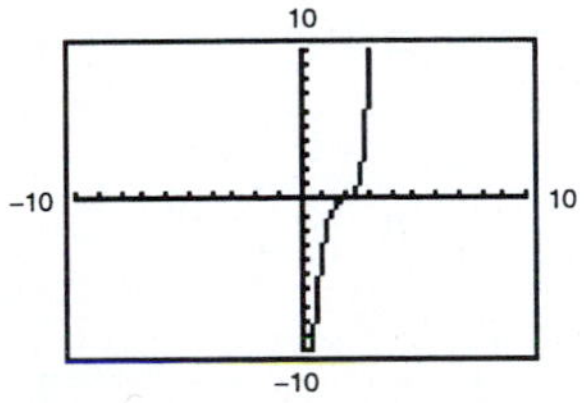

The graph reveals one zero at $x = 2$.

2	1	−8	28	−56	64	−32
		2	−12	32	−48	32
2	1	−6	16	−24	16	0
		2	−8	16	−16	
2	1	−4	8	−8	0	
		2	−4	8		
	1	−2	4	0		

Zeros: $x = 2, \dfrac{2 \pm \sqrt{12}i}{2} = 1 \pm \sqrt{3}i$

$g(x) = (x - 2)^3(x - 1 + \sqrt{3}i)(x - 1 - \sqrt{3}i)$

36. $f(x) = (x - 4)(x - 3i)(x + 3i)$

$= (x - 4)(x^2 + 9)$

$= x^3 - 4x^2 + 9x - 36$

Note: $f(x) = a(x^3 - 4x^2 + 9x - 36)$, where a is any real number, has the zeros 4, $3i$ and $-3i$.

38. $f(x) = (x - 2)(x - 4 - i)(x - 4 + i)$

$= (x - 2)(x^2 - 8x + 17)$

$= x^3 - 10x^2 + 33x - 34$

Note: $f(x) = a(x^3 - 10x^2 + 33x - 34)$ where a is any real number, has the zeros 2, $4 \pm i$.

40. $f(x) = (x - 2)^3(x - 4i)(x + 4i)$

$= (x^3 - 6x^2 + 12x - 8)(x^2 + 16)$

$= x^5 - 6x^4 + 28x^3 - 104x^2 + 192x - 128$

Note: $f(x) = a(x^5 - 6x^4 + 28x^3 - 104x^2 + 192x - 128)$, where a is any real number, has the zeros $2, 2, 2, 4i, -4i$.

42. $f(x) = (x + 5)^2(x - 1 - \sqrt{3}i)(x - 1 + \sqrt{3}i)$

$= (x^2 + 10x + 25)(x^2 - 2x + 4)$

$= x^4 + 8x^3 + 9x^2 - 10x + 100$

Note: $f(x) = a(x^4 + 8x^3 + 9x^2 - 10x + 100)$, where a is any real number, has the zeros $-5, -5, 1 \pm \sqrt{3}i$.

44. $f(x) = x^2(x - 4)(x - 1 - i)(x - 1 + i)$

$= (x^3 - 4x^2)(x^2 - 2x + 2)$

$= x^5 - 6x^4 + 10x^3 - 8x^2$

Note: $f(x) = a(x^5 - 6x^4 + 10x^3 - 8x^2)$, where a is any real number, has the zeros $0, 0, 4, 1 + i$.

46. $f(x) = x^4 - 2x^3 - 3x^2 + 12x - 18$

(a) $f(x) = (x^2 - 6)(x^2 - 2x + 3)$

(b) $f(x) = (x + \sqrt{6})(x - \sqrt{6})(x^2 - 2x + 3)$

(c) $f(x) = (x + \sqrt{6})(x - \sqrt{6})(x - 1 - \sqrt{2}i)(x - 1 + \sqrt{2}i)$

48. $f(x) = x^4 - 3x^3 - x^2 - 12x - 20$

(a) $f(x) = (x^2 + 4)(x^2 - 3x - 5)$

(b) $f(x) = (x^2 + 4)\left(x - \frac{3 + \sqrt{29}}{2}\right)\left(x - \frac{3 - \sqrt{29}}{2}\right)$

(c) $f(x) = (x + 2i)(x - 2i)\left(x - \frac{3 + \sqrt{29}}{2}\right)\left(x - \frac{3 - \sqrt{29}}{2}\right)$

50. $f(x) = x^3 + x^2 + 9x + 9$

Since $3i$ is a zero, so is $-3i$.

$3i$	1	1	9	9
		$3i$	$-9 + 3i$	-9
$-3i$	1	$1 + 3i$	$3i$	0
		$-3i$	$-3i$	
	1	1	0	

The zero of $x + 1$ is $x = -1$.

The zeros of f are $x = -1, \pm 3i$.

52. $g(x) = x^3 - 7x^2 - x + 87$

Since $5 + 2i$ is a zero, so is $5 - 2i$.

$5 + 2i$	1	-7	-1	87
		$5 + 2i$	$-14 + 6i$	-87
$5 - 2i$	1	$-2 + 2i$	$-15 + 6i$	0
		$5 - 2i$	$15 - 6i$	
	1	3	0	

The zero of $x + 3$ is $x = -3$.

The zeros of f are $x = -3, 5 \pm 2i$.

54. $h(x) = 3x^2 - 4x^2 + 8x + 8$

Since $1 - \sqrt{3}i$ is a zero, so is $1 + \sqrt{3}i$.

$1 - \sqrt{3}i$	3	-4	8	8
		$3 - 3\sqrt{3}i$	$-10 - 2\sqrt{3}i$	-8
$1 + \sqrt{3}i$	3	$-1 - 3\sqrt{3}i$	$-2 - 2\sqrt{3}i$	0
		$3 + 3\sqrt{3}i$	$2 + 2\sqrt{3}i$	
	3	2	0	

The zero of $x + 1$ is $x = -1$.

The zeros of f are $x = -1, \pm 3i$.

56. $f(x) = x^3 + 4x^2 + 14x + 20$

Since $-1 - 3i$ is a zero, so is $-1 + 3i$.

$$\begin{array}{r|rrrr} -1-3i & 1 & 4 & 14 & 20 \\ & & -1-3i & -12-6i & -20 \\ \hline & 1 & 3-3i & 2-6i & 0 \end{array}$$

$$\begin{array}{r|rrr} -1+3i & 1 & 3-3i & 2-6i \\ & & -1+3i & -2+6i \\ \hline & 1 & 2 & 0 \end{array}$$

The zero of $x + 2$ is $x = -2$.

The zeros of f are $x = -2, -1 \pm 3i$.

58. $f(x) = 25x^3 - 55x^2 - 54x - 18$

Since $\frac{1}{5}(-2 + \sqrt{2}i) = \frac{-2 + \sqrt{2}i}{5}$ is a zero, so is $\frac{-2 - \sqrt{2}i}{5}$.

$$\begin{array}{r|rrrr} \frac{-2+\sqrt{2}i}{5} & 25 & -55 & -54 & -18 \\ & & -10+5\sqrt{2}i & 24-15\sqrt{2}i & 18 \\ \hline \frac{-2-\sqrt{2}i}{5} & 25 & -65+5\sqrt{2}i & -30-15\sqrt{2}i & 0 \\ & & -10-5\sqrt{2}i & 30+15\sqrt{2}i & \\ \hline & 25 & -75 & 0 & \end{array}$$

The zero of $25x - 75$ is $x = 3$.

The zeros of f are $x = 3, \frac{-2 \pm \sqrt{2}i}{5}$.

60. No, there does not exist a third-degree polynomial function with integer coefficients that has no real zeros because the most complex factors it can have is two and the Linear Factorization Theorem guarantees that there are 3 linear factors, so one factor *must* be real.

62. (a) $f(x - 2) = (x - 2)^4 - 4(x - 2)^2 + k$

$$(x-2)^2 = 2 \pm \sqrt{4-k}$$
$$x - 2 = \pm\sqrt{2 \pm \sqrt{4-k}}$$
$$x = 2 \pm \sqrt{2 \pm \sqrt{4-k}}$$

Because the number and type of roots are decided by the radicands of the solution (which are the same radicands as in the solution for $f(x)$), the answers to Exercise 61 do not change for the function $g(x) = f(x - 2)$.

(b) $g(x) = f(2x) = (2x)^4 - 4(2x)^2 + k$

$$(2x)^2 = 2 \pm \sqrt{4-k}$$
$$2x = \pm\sqrt{2 \pm \sqrt{4-k}}$$
$$x = \pm\tfrac{1}{2}\sqrt{2 \pm \sqrt{4-k}}$$

Because the number and type of roots are decided by the radicands in the solution, the answers to Exercise 61 do not change for the function $g(x) = f(2x)$.

64. $P = R - C = xp - C$

$$= x(140 - 0.0001x) - (80x + 150{,}000)$$
$$= -0.0001x^2 + 60x - 150{,}000$$
$$= 9{,}000{,}000$$

Thus, $0 = 0.0001x^2 - 60x + 9{,}150{,}000$.

$$x = \frac{60 \pm \sqrt{-60}}{0.0002} = 300{,}000 \pm 10{,}000\sqrt{15}i$$

Since the zeros are both complex, it is not possible to determine a price p that would yield a profit of 9 million dollars.

66. (a) The graph is not of $f(x) = x^2(x + 2)(x - 3.5)$ because this function has a zero at $x = 0$ and the given graph does not go through (0, 0).

(b) The graph is not of $f(x) = (x + 2)(x - 3.5)$ because this function's graph is a parabola and the graph given is not.

(c) This is the graph of $h(x) = (x + 2)(x - 3.5)(x^2 + 1)$

(d) The graph is not of $k(x) = (x + 1)(x + 2)(x - 3.5)$ because this function has $(-1, 0)$ as an intercept and the given graph does not.

68. $\frac{3}{2} + \frac{11}{2}i$

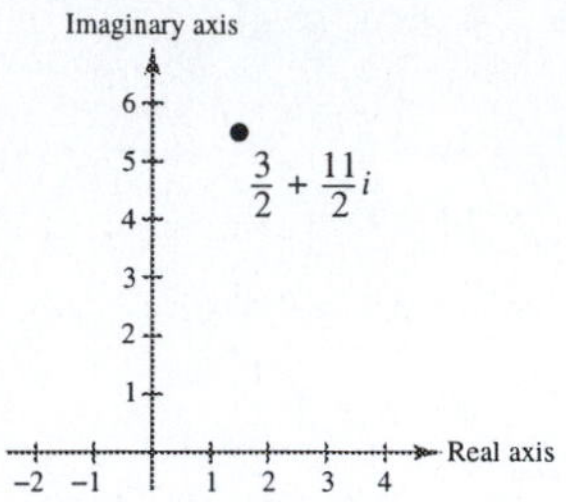

70. *Model:* (Distance traveled by 2nd car) − (Distance traveled by 1st car) = 12

Labels: Time traveled = t

Distance traveled by 2nd car = $100t$

Distance traveled by 1st car = $84t$

Equation: $100t - 84t = 12$

$$16t = 12$$

$$t = \frac{12}{16} = \frac{3}{4}$$

The cars are 12 kilometers apart after $\frac{3}{4}$ hour or 45 minutes.

72. (a) $d^2 = x^2 + 90^2$

(b) $x = 20$

$$d^2 = 20^2 + 90^2$$

$$d^2 = 8500$$

$$d = \sqrt{8500} \approx 92.2 \text{ feet}$$

Section 2.7 Rational Functions

Solutions to Even-Numbered Exercises

2. $f(x) = \dfrac{4}{(x-2)^3}$

Domain: all real numbers $x \neq 2$
Vertical asymptote: $x = 2$
Horizontal asymptote: $y = 0$
[Degree of $p(x)$ < degree of $q(x)$]

4. $f(x) = \dfrac{1-5x}{1+2x}$

Domain: all real numbers $x \neq -\dfrac{1}{2}$

Vertical asymptote: $x = -\dfrac{1}{2}$

Horizontal asymptote: $y = -\dfrac{5}{2}$

[Degree of $p(x)$ = degree of $q(x)$]

6. $f(x) = \dfrac{2x^2}{x+1}$

Domain: all real numbers $x \neq -1$
Vertical asymptote: $x = -1$
Horizontal asymptote: none
[Degree of $p(x)$ > degree of $q(x)$]

8. $f(x) = \dfrac{1}{x-3}$

Vertical asymptote: $x = 3$
Horizontal asymptote: $y = 0$
Matches graph (a).

10. $f(x) = \dfrac{1-x}{x}$

Vertical asymptote: $x = 0$
Horizontal asymptote: $y = -1$
Matches graph (c).

12. $f(x) = \dfrac{x+2}{x+4}$

Vertical asymptote: $x = -4$
Horizontal asymptote: $y = 1$
Matches graph (b).

14. $f(x) = \dfrac{x^2(x-3)}{x^2-3x}$, $g(x) = x$

(a) Domain of f: all real numbers $x \neq 0$ and $x \neq 3$
Domain of g: all real numbers

(b) Vertical asymptotes of f: none
Vertical asymptotes of g: none

(c)

x	-1	0	1	2	3	3.5	4
$f(x)$	-1	Undef.	1	2	Undef.	3.5	4
$g(x)$	-1	0	1	2	3	3.5	4

(d) The two functions differ only where f is undefined.

16. $f(x) = \dfrac{1}{1+x^2}$

18. $f(x) = \dfrac{-3x^2}{x(2x-5)}$

20. $f(x) = 2 + \dfrac{1}{x - 3}$

(a) As $x \longrightarrow \pm\infty, f(x) \longrightarrow 2$.

(b) As $x \longrightarrow \infty, f(x) \longrightarrow 2$ but is greater than 2.

(c) As $x \longrightarrow -\infty, f(x) \longrightarrow 2$ but is less than 2.

22. $f(x) = \dfrac{2x - 1}{x^2 + 1}$

(a) As $x \longrightarrow \pm\infty, f(x) \longrightarrow 0$.

(b) As $x \longrightarrow \infty, f(x) \longrightarrow 0$ but is greater than 0.

(c) As $x \longrightarrow -\infty, f(x) \longrightarrow 0$ but is less than 0.

24. $h(x) = 4 + \dfrac{5}{x^2 + 2} = \dfrac{4x^2 + 13}{x^2 + 2}$

The zeros of h correspond to the zeros of the numerator and there are no real zeros.

26. $g(x) = \dfrac{x^3 - 8}{x^2 + 1}$

The zeros of g correspond to the zeros of the numerator and are $x = 2$.

28. When $t = 1.056$:

$$\begin{aligned} 1.056 &= \frac{38M + 16{,}965}{10(M + 5000)} \\ 10.56M + 52{,}800 &= 38M + 16{,}965 \\ 35{,}835 &= 27.44M \\ 1306 \text{ grams} &\approx M \end{aligned}$$

30. $g(x) = \dfrac{2}{x - 1}$

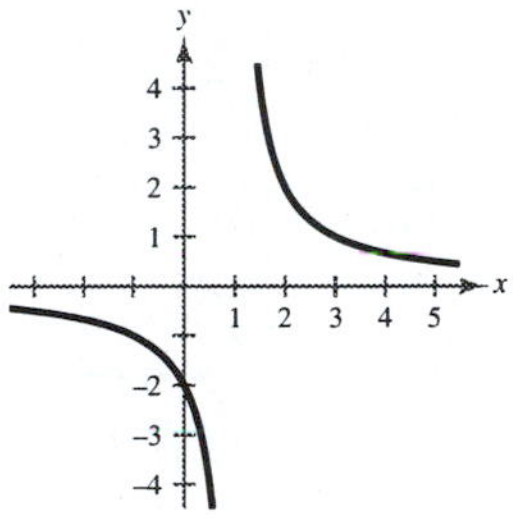

Horizontal shift one unit to the right

32. $g(x) = \dfrac{1}{x + 2}$

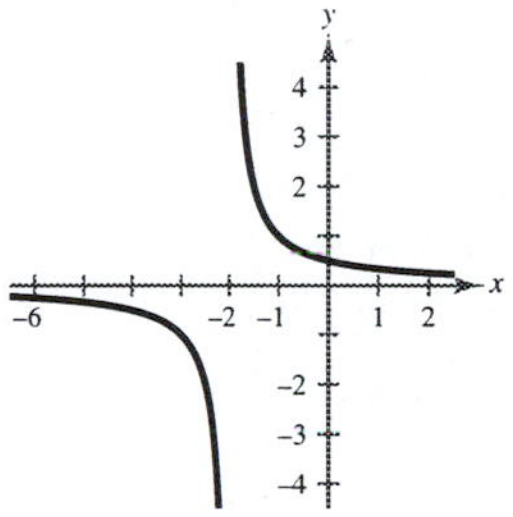

Horizontal shift two units to the left

34. $f(x) = \dfrac{1}{x - 3}$

y-intercept: $\left(0, -\dfrac{1}{3}\right)$

Vertical asymptote: $x = 3$

Horizontal asymptote: $y = 0$

x	0	1	2	4	5
y	$-\frac{1}{3}$	$-\frac{1}{2}$	-1	1	$\frac{1}{2}$

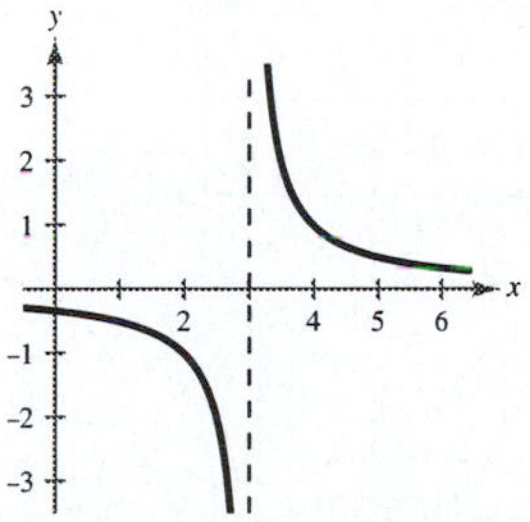

36. $g(x) = \dfrac{1}{3 - x}$

y-intercept: $\left(0, \dfrac{1}{3}\right)$

Vertical asymptote: $x = 3$

Horizontal asymptote: $y = 0$

x	0	1	2	4	5
y	$\frac{1}{3}$	$\frac{1}{2}$	1	-1	$-\frac{1}{2}$

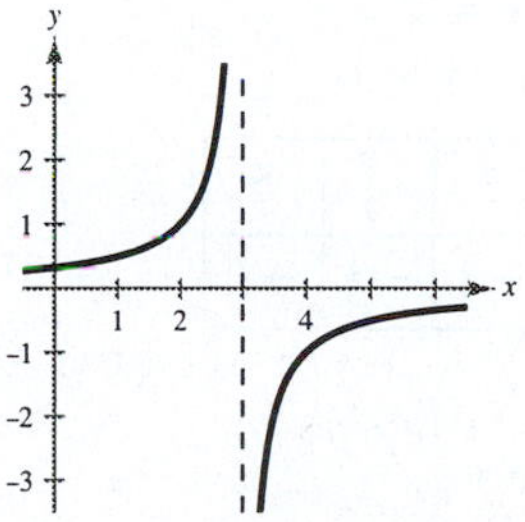

38. $P(x) = \dfrac{1 - 3x}{1 - x}$

y-intercept: $(0, 1)$

x-intercept: $\left(\dfrac{1}{3}, 0\right)$

Vertical asymptote: $x = 1$

Horizontal asymptote: $y = 3$

x	-1	0	2	3
y	2	1	5	4

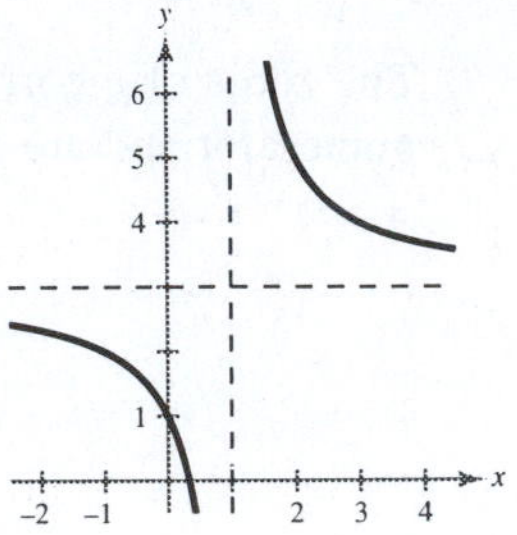

40. $f(t) = \dfrac{1 - 2t}{t}$

x-intercept: $\left(\dfrac{1}{2}, 0\right)$

Vertical asymptote: $x = 0$

Horizontal asymptote: $y = -2$

x	-2	-1	1	2	3
y	$-\frac{5}{2}$	-3	-1	$-\frac{3}{2}$	$-\frac{5}{3}$

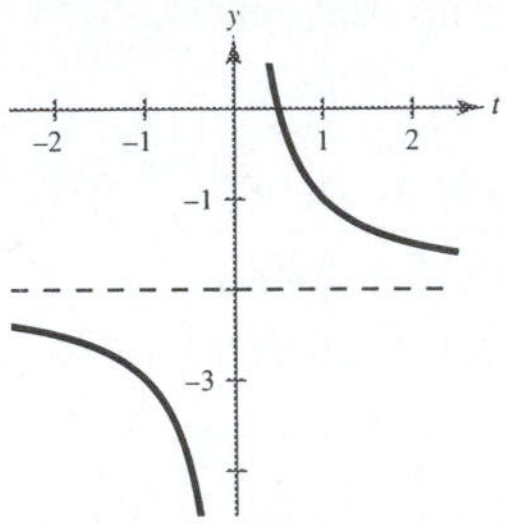

42. $f(x) = 2 - \dfrac{3}{x^2}$

x-intercepts: $(-1.225, 0), (1.225, 0)$

Vertical asymptote: $x = 0$

Horizontal asymptote: $y = 2$

y-axis symmetry

x	-2	-1	1	2	3
y	$\frac{5}{4}$	$-\frac{1}{2}$	-1	$\frac{5}{4}$	$1\frac{2}{3}$

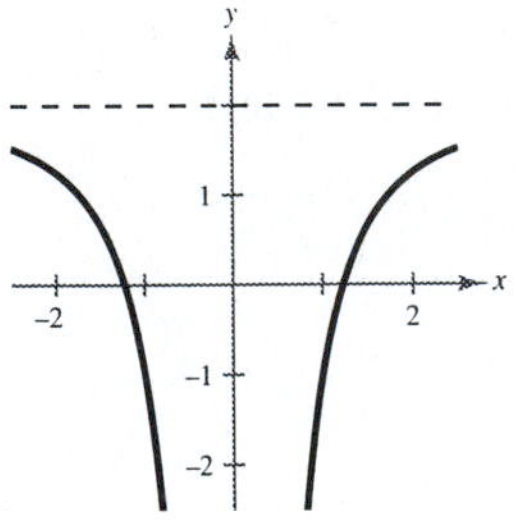

44. $g(x) = \dfrac{x}{x^2 - 9}$

Intercept: $(0, 0)$

Vertical asymptotes: $x = 3, x = -3$

Horizontal asymptote: $y = 0$

x	-6	-5	-4	-2	-1	0	1	2	4
y	$-\frac{2}{9}$	$-\frac{5}{16}$	$-\frac{4}{7}$	$\frac{2}{5}$	$\frac{1}{8}$	0	$-\frac{1}{8}$	$-\frac{2}{5}$	$\frac{4}{7}$

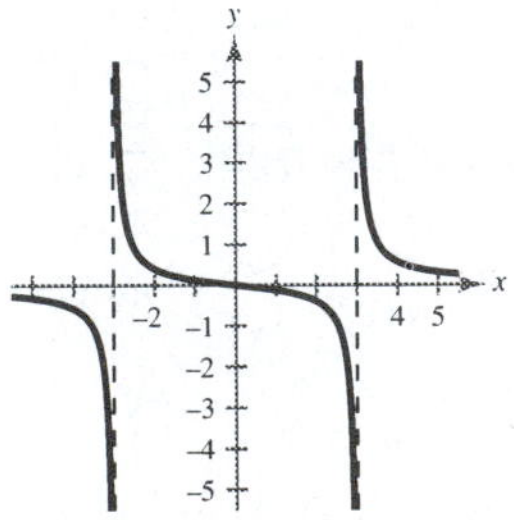

46. $f(x) = -\dfrac{1}{(x - 2)^2}$

y-intercepts: $\left(0, -\dfrac{1}{4}\right)$

Vertical asymptote: $x = 2$

Horizontal asymptote: $y = 0$

x	-1	0	1	3	4	5
y	$-\frac{1}{9}$	$-\frac{1}{4}$	-1	-1	$-\frac{1}{4}$	$-\frac{1}{9}$

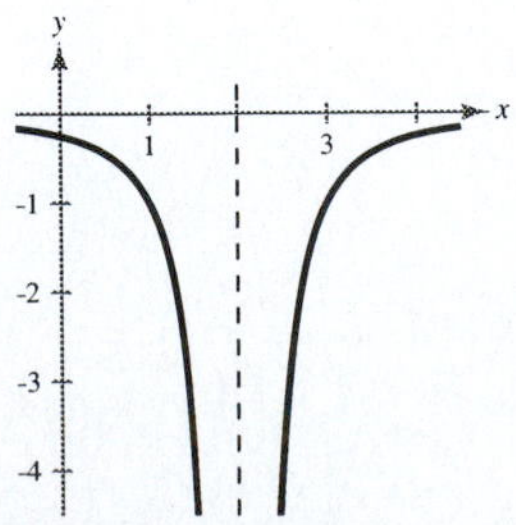

48. $h(x) = \dfrac{2}{x^2(x - 2)}$

Vertical asymptotes: $x = 0, x = 2$

Horizontal asymptote: $y = 0$

x	-2	-1	$\frac{1}{2}$	1	$\frac{3}{2}$	3
y	$-\frac{1}{8}$	$-\frac{2}{3}$	$-\frac{16}{3}$	-2	$-\frac{16}{9}$	$\frac{2}{9}$

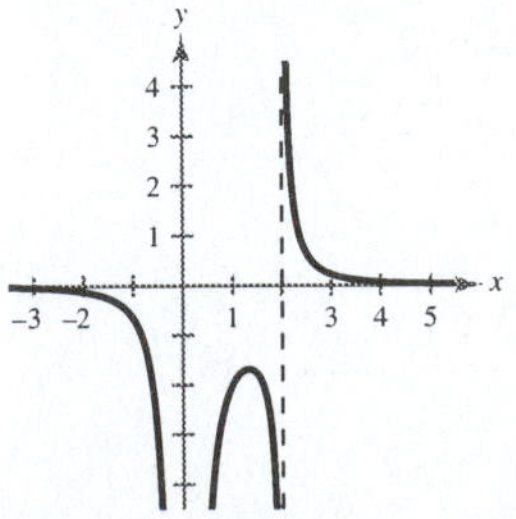

50. $f(x) = \dfrac{2x}{x^2 + x - 2} = \dfrac{2x}{(x + 2)(x - 1)}$

Intercepts: $(0, 0)$
Vertical asymptotes: $x = -2, x = 1$
Horizontal asymptote: $y = 0$

x	-4	-3	-1	0	2	3	4
y	$-\frac{4}{5}$	$-\frac{3}{2}$	1	0	1	$\frac{3}{5}$	$\frac{4}{9}$

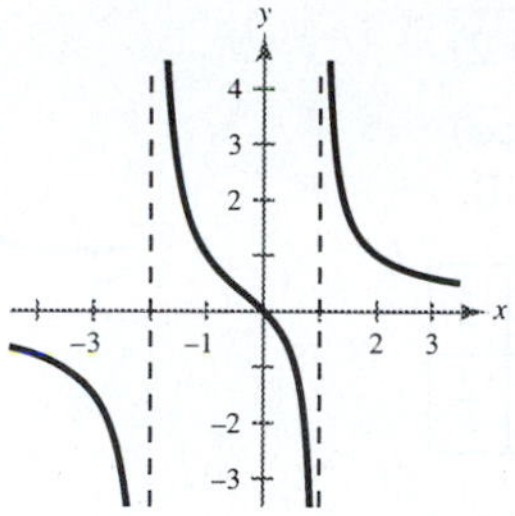

52. $f(x) = \dfrac{3 - x}{2 - x}$

y-intercept: $\left(0, \dfrac{3}{2}\right)$
x-intercept: $(3, 0)$

Vertical asymptote: $x = 2$
Horizontal asymptote: $y = 1$

x	-1	0	1	3	4	5
y	$\frac{4}{3}$	$\frac{3}{2}$	2	0	$\frac{1}{2}$	$\frac{2}{3}$

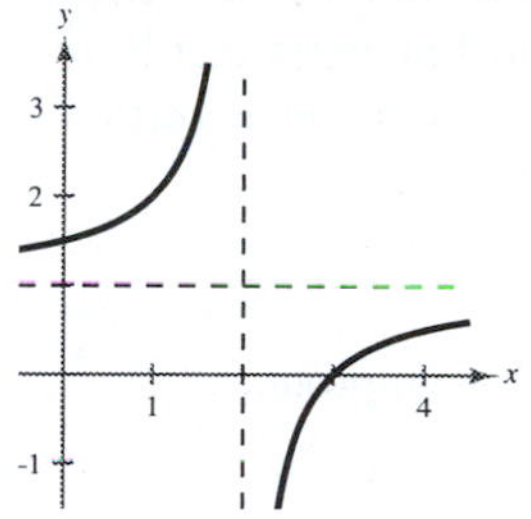

54. $h(x) = \dfrac{1}{x - 3} + 1 = \dfrac{x - 2}{x - 3}$

y-intercept: $\left(0, \dfrac{2}{3}\right)$
x-intercept: $(2, 0)$

Vertical asymptote: $x = 3$
Horizontal asymptote: $y = 1$

x	-1	0	1	2	4	5	6
y	$\frac{3}{4}$	$\frac{2}{3}$	$\frac{1}{2}$	0	2	$\frac{3}{2}$	$\frac{4}{3}$

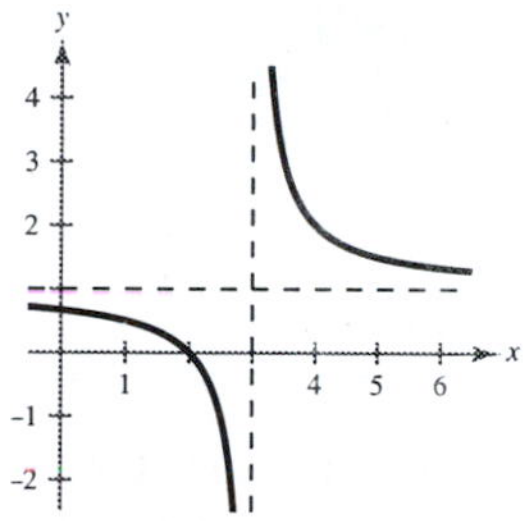

56. $f(x) = \dfrac{x^2(x - 2)}{x^2 - 2x}$

$g(x) = x$

(a) Domain of f: all $x \neq 0, x \neq 2$
Domain of g: all real numbers

(b) The function f has no vertical asymptotes.

(c)

x	-1	0	1	1.5	2	2.5	3
$f(x)$	-1	Undef.	1	1.5	Undef.	2.5	3
$g(x)$	-1	0	1	1.5	2	2.5	3

(d)

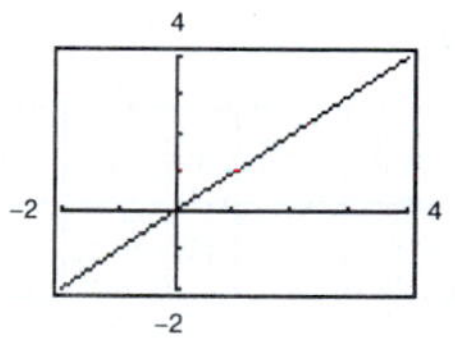

(e) Because there are only a finite number of pixels, the graphing utility may not attempt to evaluate the function where it does not exist.

58. $g(x) = -\dfrac{x}{(x - 2)^2}$

Domain: $(-\infty, 2), (2, \infty)$
Asymptotes: $x = 2, y = 0$

x	-2	-1	0	1	3	4
y	$\frac{1}{8}$	$\frac{1}{9}$	0	-1	-3	-1

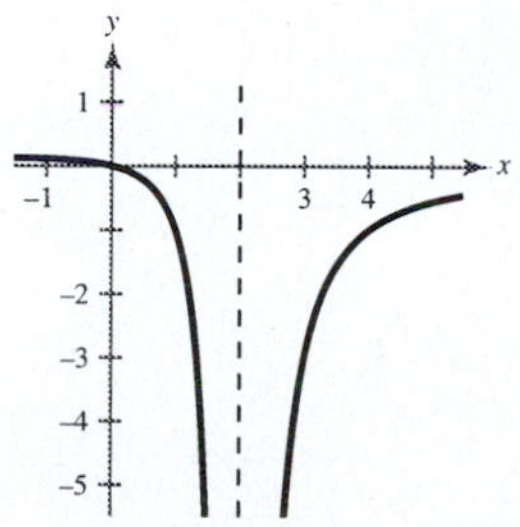

60. $f(x) = \dfrac{x+4}{x^2+x-6} = \dfrac{x+4}{(x+3)(x-2)}$

Domain: $(-\infty, -3), (-3, 2), (2, \infty)$

Asymptotes: $x = -3, x = 2, y = 0$

x	-4	-2	-1	0	1	3
y	0	$-\frac{1}{2}$	$-\frac{1}{2}$	$-\frac{2}{3}$	$-\frac{5}{4}$	$\frac{7}{6}$

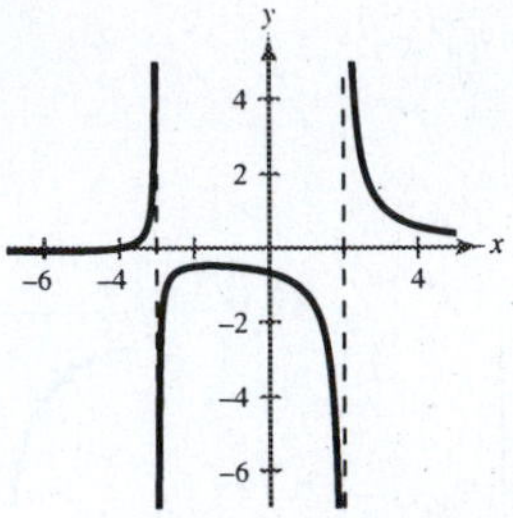

62. $g(x) = \dfrac{4|x-2|}{x+1}$

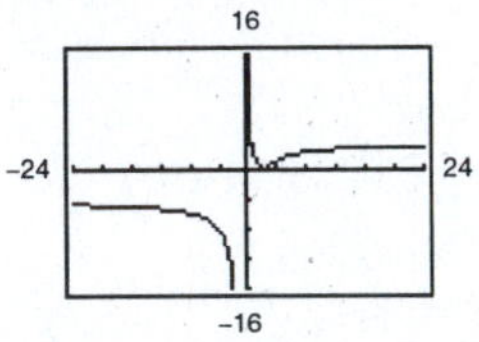

There appears to be two horizontal asymptotes, one at $y = -4$ and one at $y = 4$.

64. $g(x) = \dfrac{3x^4 - 5x + 3}{x^4 + 1}$

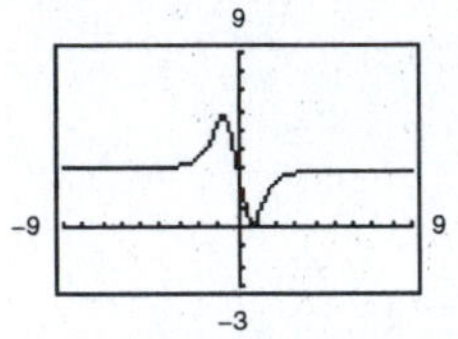

The graph appears to cross its horizontal asymptote of $y = 3$.

66. $g(x) = \dfrac{x^2 + x - 2}{x - 1} = \dfrac{(x-1)(x+2)}{x-1}$

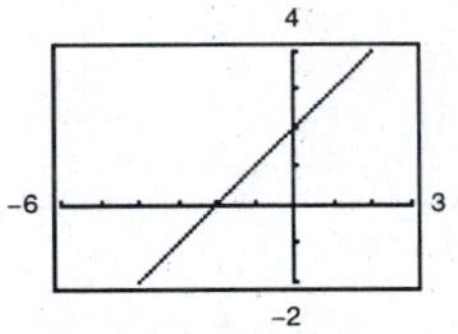

There is no vertical asymptote because the fraction in the original function was not reduced.

68. No, not every rational function has a vertical asymptote. Only rational functions that have denominators having real roots have vertical asymptotes. For instance

$$f(x) = \frac{1}{x^2 + 1}$$

does not have a vertical asymptote.

70. $f(x) = \dfrac{1 - x^2}{x} = -x + \dfrac{1}{x}$

Intercepts: $(-1, 0), (1, 0)$

Slant asymptote: $y = -x$

x	-4	-3	-2	-1	1	2	3	4
y	$\frac{15}{4}$	$\frac{8}{3}$	$\frac{3}{2}$	0	0	$-\frac{3}{2}$	$-\frac{8}{3}$	$-\frac{15}{4}$

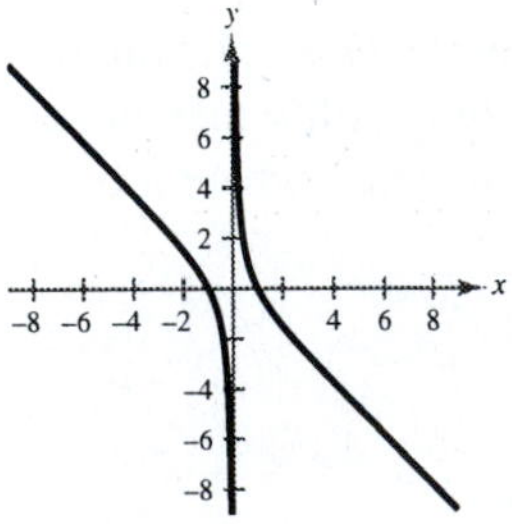

72. $h(x) = \dfrac{x^2}{x-1} = x + 1 + \dfrac{1}{x-1}$

Intercept: $(0, 0)$

Vertical asymptote: $x = 1$

Slant asymptote: $y = x + 1$

x	-2	-1	0	2	3
y	$-\frac{4}{3}$	$-\frac{1}{2}$	0	4	$\frac{9}{2}$

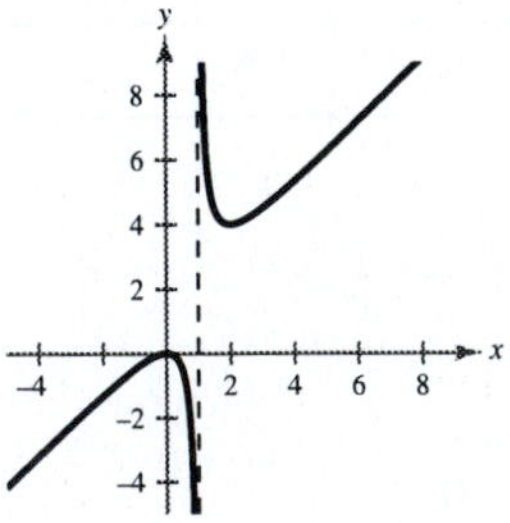

74. $g(x) = \dfrac{x^3}{2x^2 - 8} = \dfrac{1}{2}x + \dfrac{2x}{x^2 - 4}$

Intercept: $(0, 0)$

Vertical asymptotes: $x = 2, x = -2$

Slant asymptote: $y = \dfrac{1}{2}$

x	-3	-1	0	1	3
y	-2.7	$\frac{1}{6}$	0	$\frac{1}{6}$	2.7

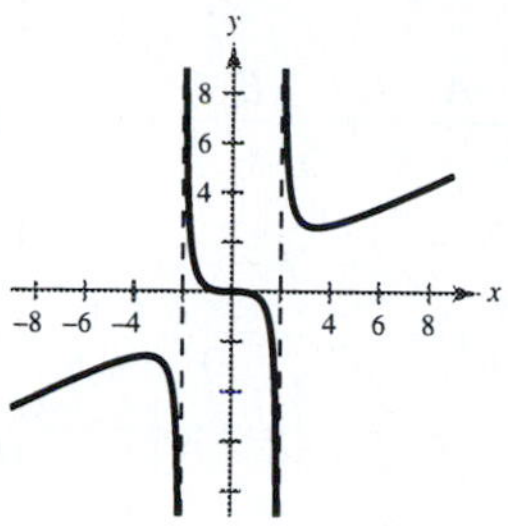

76. $f(x) = \dfrac{2x^2 + x}{x + 1}$

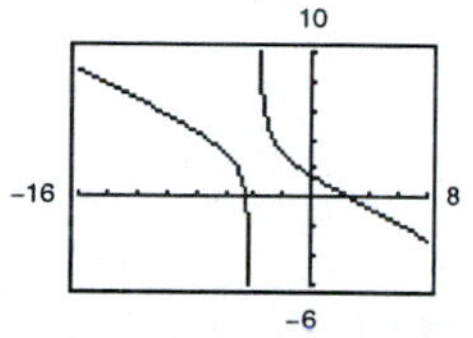

Domain: $(-\infty, -1), (-1, \infty)$

Asymptotes: $x = -1, y = 2x - 1$

The line is the slant asymptote $y = 2x - 1$.

78. $y = 20\left(\dfrac{2}{x + 1} - \dfrac{3}{x}\right)$

(a)

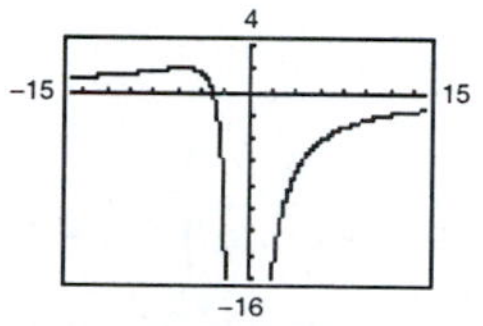

x-intercept: $(-3, 0)$

(b) $0 = 20\left(\dfrac{2}{x + 1} - \dfrac{3}{x}\right)$

$$\frac{2}{x + 1} = \frac{3}{x}$$

$$2x = 3x + 3$$

$$-3 = x$$

80. $y = x - \dfrac{9}{x}$

(a)

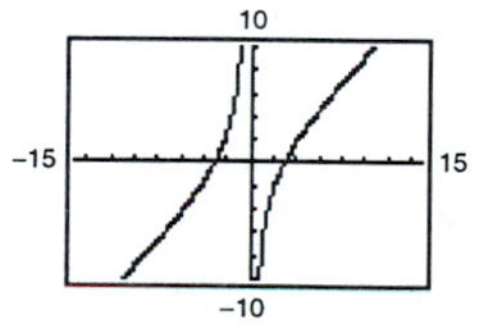

x-intercepts: $(-3, 0), (3, 0)$

(b) $0 = x - \dfrac{9}{x}$

$$x = \frac{9}{x}$$

$$x^2 = 9$$

$$x = \pm 3$$

82. (a) $A = xy$ and

$(x - 4)(y - 2) = 30$

$$y - 2 = \frac{30}{x - 4}$$

$$y = 2 + \frac{30}{x - 4} = \frac{2x + 22}{x - 4}$$

Thus, $A = xy = x\left(\dfrac{2x + 2}{x - 4}\right) = \dfrac{2x(x + 11)}{x - 4}$.

(b) Domain: Since the margins on the left and right are each 2 inches, $x > 4$, OR $(4, \infty)$.

(c)

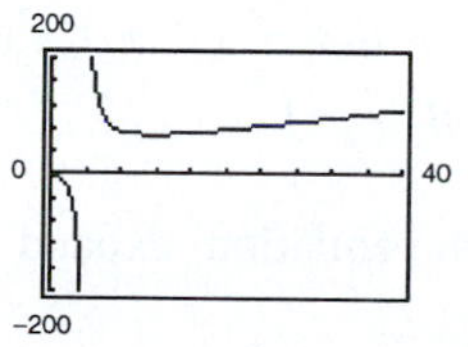

The area is minimum when $x \approx 11.75$ in. and $y \approx 5.9$ in.

Section 2.8 Partial Fractions

Solutions to Even-Numbered Exercises

2. $\dfrac{x-2}{x^2+4x+3} = \dfrac{x-2}{(x+3)(x+1)} = \dfrac{A}{x+3} + \dfrac{B}{x+1}$

4. $\dfrac{4x^2+3}{(x-5)^3} = \dfrac{A}{x-5} + \dfrac{B}{(x-5)^2} + \dfrac{C}{(x-5)^3}$

6. $\dfrac{x-1}{x(x^2+1)^2} = \dfrac{A}{x} + \dfrac{Bx+C}{x^2+1} + \dfrac{Dx+E}{(x^2+1)^2}$

8. $\dfrac{1}{4x^2-9} = \dfrac{A}{2x+3} + \dfrac{B}{2x-3}$

$1 = A(2x-3) + B(2x+3)$

Let $x = -\dfrac{3}{2}$: $1 = -6A \Rightarrow A = -\dfrac{1}{6}$

Let $x = \dfrac{3}{2}$: $1 = 6B \Rightarrow B = \dfrac{1}{6}$

$\dfrac{1}{4x^2-9} = \dfrac{1}{6}\left[\dfrac{1}{2x-3} - \dfrac{1}{2x+3}\right]$

10. $\dfrac{3}{x^2-3x} = \dfrac{A}{x-3} + \dfrac{B}{x}$

$3 = Ax + B(x-3)$

Let $x = 3$: $3 = 3A \Rightarrow A = 1$

Let $x = 0$: $3 = -3B \Rightarrow B = -1$

$\dfrac{3}{x^2-3x} = \dfrac{1}{x-3} - \dfrac{1}{x}$

12. $\dfrac{5}{x^2+x-6} = \dfrac{A}{x+3} + \dfrac{B}{x-2}$

$5 = A(x-2) + B(x+3)$

Let $x = -3$: $5 = -5A \Rightarrow A = -1$

Let $x = 2$: $5 = 5B \Rightarrow B = 1$

$\dfrac{5}{x^2+x-6} = \dfrac{1}{x-2} - \dfrac{1}{x+3}$

14. $\dfrac{x+1}{x^2+4x+3} = \dfrac{x+1}{(x+3)(x+1)}$

$= \dfrac{1}{x+3},\ x \neq -1$

16. $\dfrac{x+2}{x(x-4)} = \dfrac{A}{x} + \dfrac{B}{x-4}$

$x + 2 = A(x-4) + Bx$

Let $x = 0$: $2 = -4A \Rightarrow A = -\dfrac{1}{2}$

Let $x = 4$: $6 = 4B \Rightarrow B = \dfrac{3}{2}$

$\dfrac{x+2}{x(x-4)} = \dfrac{1}{2}\left[\dfrac{3}{x-4} - \dfrac{1}{x}\right]$

18. $\dfrac{2x-3}{(x-1)^2} = \dfrac{A}{x-1} + \dfrac{B}{(x-1)^2}$

$2x - 3 = A(x-1) + B$

Let $x = 1$: $-1 = B$

Let $x = 0$: $-3 = -A + B$

$-3 = -A - 1$

$2 = A$

$\dfrac{2x-3}{(x-1)^2} = \dfrac{2}{x-1} - \dfrac{1}{(x-1)^2}$

20. $\dfrac{6x^2+1}{x^2(x-1)^3} = \dfrac{A}{x} + \dfrac{B}{x^2} + \dfrac{C}{x-1} + \dfrac{D}{(x-1)^2} + \dfrac{E}{(x-1)^3}$

$6x^2 + 1 = Ax(x-1)^3 + B(x-1)^3 + Cx^2(x-1)^2 + Dx^2(x-1) + Ex^2$

Let $x = 0$: $1 = -B \Rightarrow B = -1$

Let $x = 1$: $7 = E$

Substitute B and E into the equation, expand the binomials, collect like terms, and equate the coefficients of like terms.

$x^3 - 4x^2 + 3x = (A + C)x^4 - (3A + 2C - D)x^3 + (3A + C - D)x^2 - Ax$

$-A = 3 \Rightarrow A = -3$

$A + C = 0 \Rightarrow C = 3$

$-3A - 2C + D = 1$

$9 - 6 + D = 1 \Rightarrow D = -2$

$\dfrac{6x^2+1}{x^2(x-1)^3} = -\dfrac{3}{x} - \dfrac{1}{x^2} + \dfrac{3}{x-1} - \dfrac{2}{(x-1)^2} + \dfrac{7}{(x-1)^3}$

22. $$\frac{x}{(x-1)(x^2+x+1)} = \frac{A}{x-1} + \frac{Bx+C}{x^2+x+1}$$

$$x = A(x^2+x+1) + (Bx+C)(x-1)$$

Let $x = 1$: $1 = 3A \implies A = \frac{1}{3}$

$$x = \frac{1}{3}(x^2+x+1) + (Bx+C)(x-1)$$

$$= \frac{1}{3}x^2 + \frac{1}{3}x + \frac{1}{3} + Bx^2 - Bx + Cx - C$$

$$= \left(\frac{1}{3} + B\right)x^2 + \left(\frac{1}{3} - B + C\right)x + \frac{1}{3} - C$$

Equating coefficients of like powers:

$0 = \frac{1}{3} + B \implies B = -\frac{1}{3}$

$0 = \frac{1}{3} - C \implies C = \frac{1}{3}$

$$\frac{x}{(x-1)(x^2+x+1)} = \frac{1}{3}\left[\frac{1}{x-1} - \frac{x-1}{x^2+x+1}\right]$$

24. $$\frac{2x^2+x+8}{(x^2+4)^2} = \frac{Ax+B}{x^2+4} + \frac{Cx+D}{(x^2+4)^2}$$

$2x^2 + x + 8 = (Ax+B)(x^2+4) + Cx + D$

$2x^2 + x + 8 = Ax^3 + Bx^2 + (4A+C)x + (4B+D)$

Equating coefficients of like powers:

$0 = A$

$2 = B$

$1 = 4A + C \implies C = 1$

$8 = 4B + D \implies D = 0$

$$\frac{2x^2+x+8}{(x^2+4)^2} = \frac{2}{x^2+4} + \frac{x}{(x^2+4)^2}$$

26. $$\frac{x^2-4x+7}{(x+1)(x^2-2x+3)} = \frac{A}{x+1} + \frac{Bx+C}{x^2-2x+3}$$

$$x^2 - 4x + 7 = A(x^2-2x+3) + Bx(x+1) + C(x+1)$$

Let $x = -1$: $12 = 6A \implies A = 2$

Let $x = 0$: $7 = 3A + C \implies C = 1$

Let $x = 1$: $4 = 2A + 2B + 2C$

$$4 = 4 + 2B + 2 \implies B = -1$$

$$\frac{x^2-4x+7}{(x+1)(x^2-2x+3)} = \frac{2}{x+1} - \frac{x-1}{x^2-2x+3}$$

28. $\dfrac{x+1}{x^3+x} = \dfrac{A}{x} + \dfrac{Bx+C}{x^2+1}$

$x + 1 = A(x^2 + 1) + Bx^2 + Cx$

Let $x = 0$: $1 = A$

$x + 1 = x^2 + 1 + Bx^2 + Cx$

$x + 1 = (1 + B)x^2 + Cx + 1$

Equating coefficients of like powers:

$0 = 1 + B \Rightarrow B = -1$

$1 = C$

$\dfrac{x+1}{x^3+x} = \dfrac{1}{x} - \dfrac{x-1}{x^2+1}$

30. $\dfrac{x^2-x}{x^2+x+1} = 1 - \dfrac{2x+1}{x^2+x+1}$

32. $\dfrac{3x^2-7x-2}{x^3-x} = \dfrac{A}{x} + \dfrac{B}{x+1} + \dfrac{C}{x-1}$

$3x^2 - 7x - 2 = A(x^2 - 1) + Bx(x - 1) + Cx(x + 1)$

Let $x = 0$: $-2 = -A \Rightarrow A = 2$

Let $x = -1$: $8 = 2B \Rightarrow B = 4$

Let $x = 1$: $-6 = 2C \Rightarrow C = -3$

$\dfrac{3x^2-7x-2}{x^3-x} = \dfrac{2}{x} + \dfrac{4}{x+1} - \dfrac{3}{x-1}$

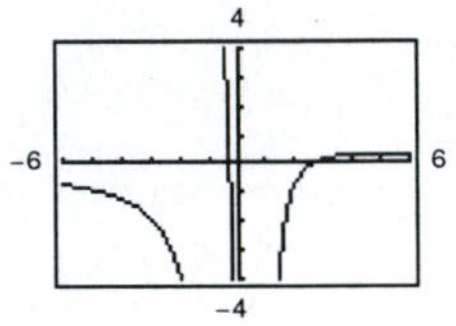

34. $\dfrac{4x^2-1}{2x(x+1)^2} = \dfrac{A}{2x} + \dfrac{B}{x+1} + \dfrac{C}{(x+1)^2}$

$4x^2 - 1 = A(x + 1)^2 + 2Bx(x + 1) + 2Cx$

Let $x = 0$: $-1 = A$

Let $x = -1$: $3 = -2C \Rightarrow C = -\dfrac{3}{2}$

Let $x = 1$: $3 = 4A + 4B + 2C$

$3 = -4 + 4B - 3$

$\dfrac{5}{2} = B$

$\dfrac{4x^2-1}{2x(x+1)^2} = \dfrac{1}{2}\left[-\dfrac{1}{x} + \dfrac{5}{x+1} - \dfrac{3}{(x+1)^2}\right]$

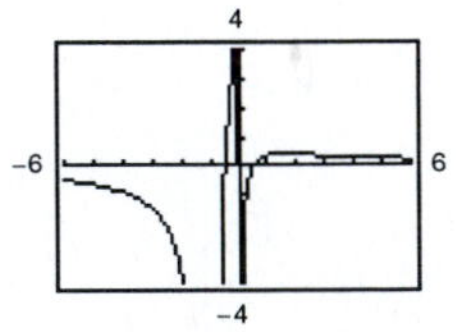

36. $\dfrac{x^3}{(x+2)^2(x-2)^2} = \dfrac{A}{x+2} + \dfrac{B}{(x+2)^2} + \dfrac{C}{x-2} + \dfrac{D}{(x-2)^2}$

$$x^3 = A(x+2)(x-2)^2 + B(x-2)^2 + C(x+2)^2(x-2) + D(x+2)^2$$

Let $x = -2$: $-8 = 16B \Rightarrow B = -\dfrac{1}{2}$

Let $x = 2$: $8 = 16D \Rightarrow D = \dfrac{1}{2}$

$x^3 = A(x+2)(x-2)^2 - \dfrac{1}{2}(x-2)^2 + C(x+2)^2(x-2) + \dfrac{1}{2}(x+2)^2$

$x^3 - 4 = (A + C)x^3 + (-2A + 2C)x^2 + (-4A - 4C)x + (8A - 8C)$

Equating coefficients of like powers:

$0 = -2A + 2C \Rightarrow A = C$

$1 = A + C$

$1 = 2A \Rightarrow A = \dfrac{1}{2} \Rightarrow C = \dfrac{1}{2}$

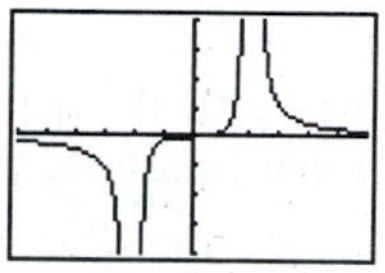

$\dfrac{x^3}{(x+2)^2(x-2)^2} = \dfrac{1}{2}\left[\dfrac{1}{x+2} - \dfrac{1}{(x+2)^2} + \dfrac{1}{x-2} + \dfrac{1}{(x-2)^2}\right]$

38. $\dfrac{x^3 - x + 3}{x^2 + x - 2} = x - 1 + \dfrac{2x+1}{(x+2)(x-1)}$

$\dfrac{2x+1}{(x+2)(x-1)} = \dfrac{A}{x+2} + \dfrac{B}{x-1}$

$$2x + 1 = A(x-1) + B(x+2)$$

Let $x = -2$: $-3 = -3A \Rightarrow A = 1$

Let $x = 1$: $3 = 3B \Rightarrow B = 1$

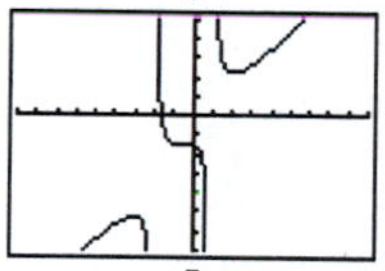

$\dfrac{x^3 - x + 3}{x^2 + x - 2} = x - 1 + \dfrac{1}{x+2} + \dfrac{1}{x-1}$

40. $\dfrac{1}{x(x+a)} = \dfrac{A}{x} + \dfrac{B}{x+a}$, a is a constant.

$$1 = A(x+a) + Bx$$

Let $x = 0$: $1 = aA \Rightarrow A = \dfrac{1}{a}$

Let $x = -a$: $1 = -aB \Rightarrow B = -\dfrac{1}{a}$

$\dfrac{1}{x(x+a)} = \dfrac{1}{a}\left[\dfrac{1}{x} - \dfrac{1}{x+a}\right]$

42. $\dfrac{1}{(x+1)(a-x)} = \dfrac{A}{x+1} + \dfrac{B}{a-x}$, a is a positive integer.

$$1 = A(a-x) + B(x+1)$$

Let $x = -1$: $1 = A(a+1) \Rightarrow A = \dfrac{1}{a+1}$

Let $x = a$: $1 = B(a+1) \Rightarrow B = \dfrac{1}{a+1}$

$\dfrac{1}{(x+1)(a-x)} = \dfrac{1}{a+1}\left[\dfrac{1}{x+1} + \dfrac{1}{a-x}\right]$

44. $y = \dfrac{2(x+1)^2}{x(x^2+1)} = \dfrac{A}{x} + \dfrac{Bx + C}{x^2+1}$

$2(x+1)^2 = A(x^2+1) + Bx^2 + Cx$

Let $x = 0$: $2 = A$

$2x^2 + 4x + 2 = 2x^2 + 2 + Bx^2 + Cx$

$2x^2 + 4x + 2 = (2 + B)x^2 + Cx + 2$

Equating coefficients of like powers:

$2 = 2 + B \Rightarrow B = 0$

$4 = C$

$\dfrac{2(x+1)^2}{x(x^2+1)} = \dfrac{2}{x} + \dfrac{4}{x^2+1}$

$\dfrac{2(x+1)^2}{x(x^2+1)}$

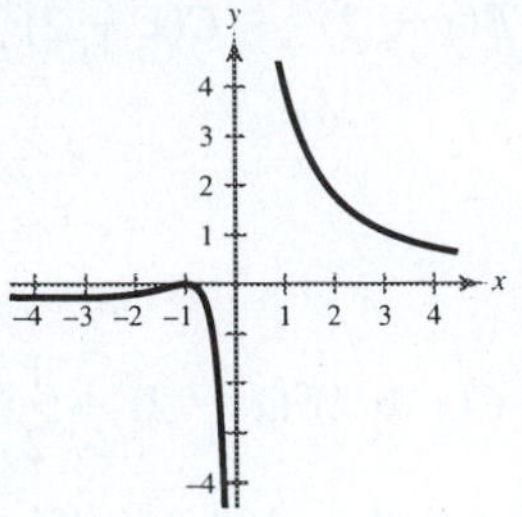

Vertical asymptote at $x = 0$

$y = \dfrac{2}{x}$ and $y = \dfrac{4}{x^2+1}$

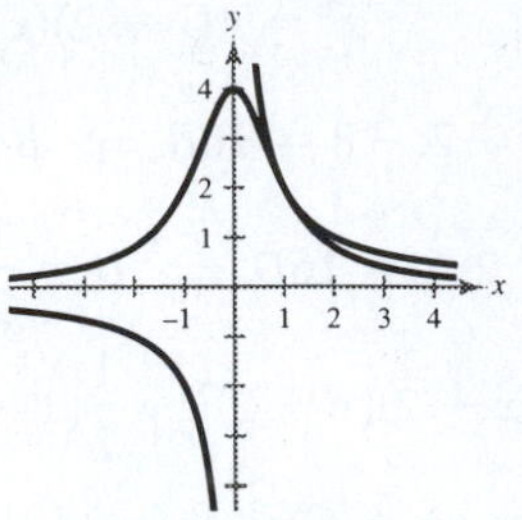

$y = \dfrac{2}{x}$ has vertical asymptote $x = 0$.

The combination of the vertical asymptotes of the terms of the decomposition are the same as the vertical asymptotes of the rational function.

46. $y = \dfrac{2(4x^2 - 15x + 39)}{x^2(x^2 - 10x + 26)} = \dfrac{A}{x} + \dfrac{Bx}{x^2} + \dfrac{Cx + D}{x^2 - 10x + 26}$

$2(4x^2 - 15x + 39) = Ax(x^2 - 10x + 26) + B(x^2 - 10x + 26) + Cx^3 + Dx^2$

Let $x = 0$: $78 = 26B$

$3 = B$

$8x^2 - 30x + 78 = Ax^3 - 10Ax^2 + 26Ax + 3x^2 - 30x + 78 + Cx^3 + Dx^2$

$8x^2 - 30x + 78 = (A + C)x^3 + (-10A + 3 + D)x^2 + (26A - 30)x + 78$

Equating coefficients of like powers:

$-30 = 26A - 30 \Rightarrow A = 0$

$0 = A + C \Rightarrow 0 = 0 + C \Rightarrow C = 0$

$8 = -10A + 3 + D \Rightarrow 8 = 0 + 3 + D \Rightarrow D = 5$

$\dfrac{2(4x^2 - 15x + 39)}{x^2(x^2 - 10x + 26)} = \dfrac{3}{x^2} + \dfrac{5}{x^2 - 10x + 26}$

$\dfrac{2(4x^2 - 15x + 39)}{x^2(x^2 - 10x + 26)}$

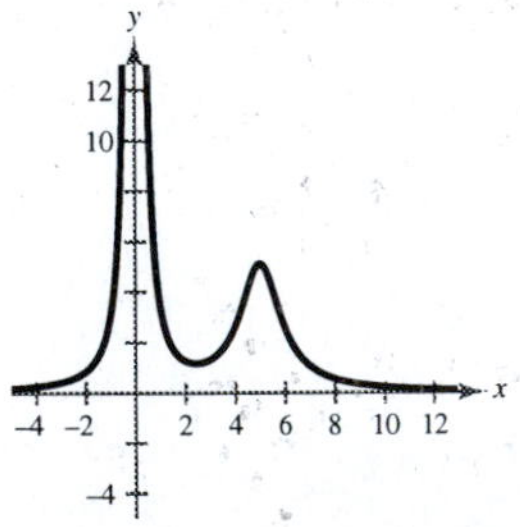

Vertical asymptote is $x = 0$.

$\dfrac{3}{x^2}$ and $\dfrac{5}{x^2 - 10x + 26}$

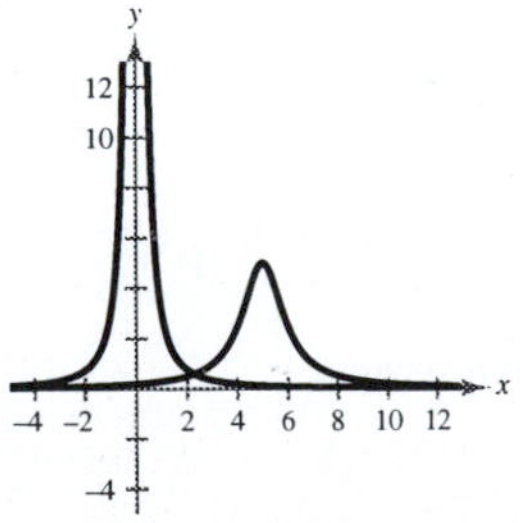

$y = \dfrac{3}{x^2}$ has vertical asymptote $x = 0$.

The combination of the vertical asymptotes of the terms of the decomposition are the same as the vertical asymptotes of the rational function.

Focus on Concepts

1. Prefer the conditions (a) and (b) because profits would be increasing.

2. (a) Degree: 3; leading coefficient: positive
(b) Degree: 2; leading coefficient: positive
(c) Degree: 4; leading coefficient: positive
(d) Degree: 5; leading coefficient: positive

3. (a) No

(b) Yes

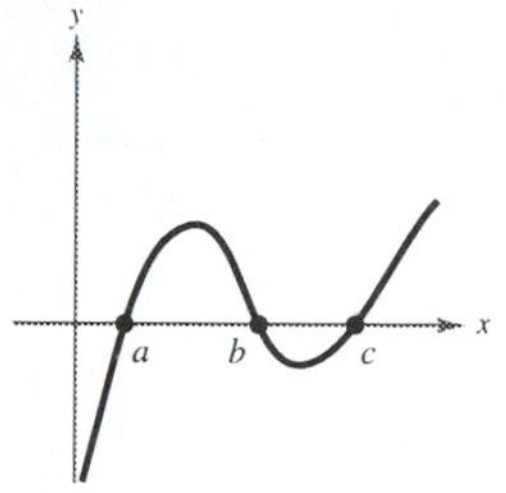

(c) Yes

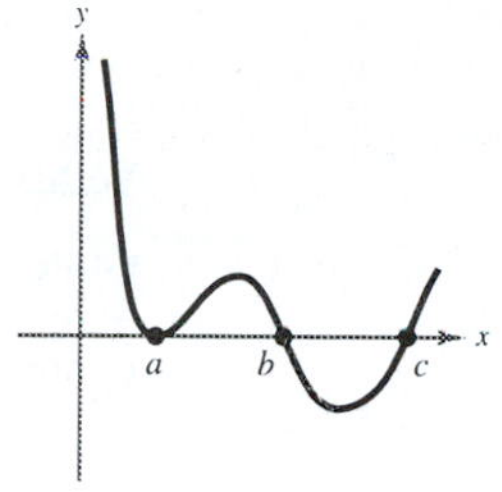

(d) Yes

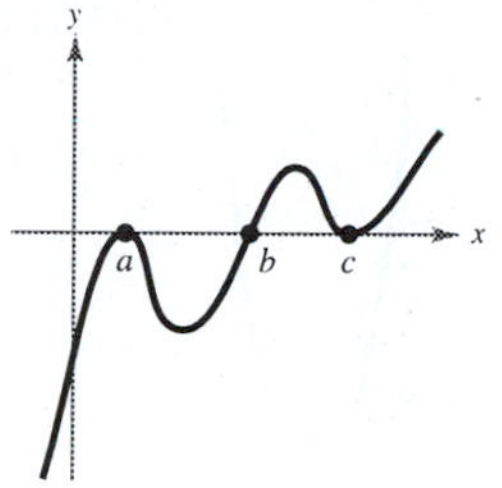

4. $y = \dfrac{2(x-3)}{x-2}$

5. $y = \dfrac{2}{x^2-2}$

6. $y = \dfrac{2}{x-2}$

7. $y = \dfrac{2x}{x-2}$

❑ Review Exercises for Chapter 2

Solutions to Even-Numbered Exercises

2. $f(x) = (x - 4)^2 - 4$

Vertex: $(4, -4)$

y-intercept: $(0, 12)$

x-intercepts: $(2, 0), (6, 0)$

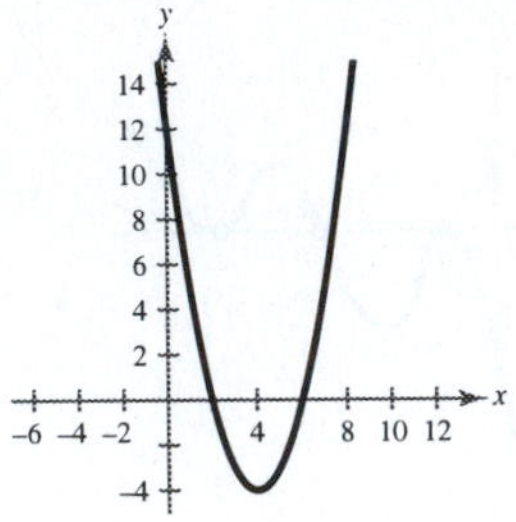

4. $f(x) = 3x^2 - 12x + 11$

$$= 3\left(x^2 - 4x + 4 - 4 + \frac{11}{3}\right)$$

$$= 3\left[(x - 2)^2 - \frac{1}{3}\right]$$

$$= 3(x - 2)^2 - 1$$

Vertex: $(2, -1)$

y-intercept: $(0, 11)$

x-intercepts: $x = \dfrac{12 \pm \sqrt{12}}{6}$

$$= 2 \pm \frac{1}{3}\sqrt{3}$$

$$\left(2 + \frac{1}{3}\sqrt{3}, 0\right), \left(2 - \frac{1}{3}\sqrt{3}, 0\right)$$

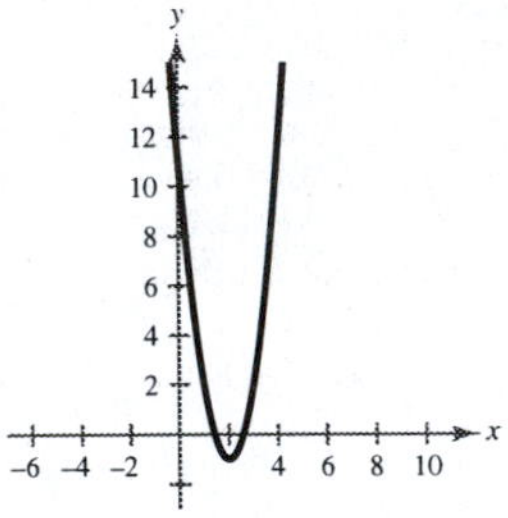

6. Vertex: $(2, 3) \Longrightarrow f(x) = a(x - 2)^2 + 3$

Point: $(-1, 6) \Longrightarrow 6 = a(-1 - 2)^2 + 3$

$$6 = 9a + 3$$

$$3 = 9a$$

$$\frac{1}{3} = a$$

$f(x) = \frac{1}{3}(x - 2)^2 + 3$

8. $f(x) = x^2 + 8x + 10$

$$= x^2 + 8x + 16 - 16 + 10$$

$$= (x + 4)^2 - 6$$

The minimum occurs at the vertex $(-4, -6)$.

10. $h(x) = 3 + 4x - x^2$

$$= -(x^2 - 4x - 3)$$

$$= -(x^2 - 4x + 4 - 4 - 3)$$

$$= -[(x - 2)^2 - 7]$$

$$= -(x - 2)^2 + 7$$

The maximum occurs at the vertex $(2, 7)$.

12. $h(x) = 4x^2 + 4x + 13$

$$= 4\left(x^2 + x + \tfrac{1}{4} - \tfrac{1}{4} + \tfrac{13}{4}\right)$$

$$= 4\left[\left(x + \tfrac{1}{2}\right)^2 + 3\right]$$

$$= 4\left(x + \tfrac{1}{2}\right)^2 + 12$$

The minimum occurs at the vertex $\left(-\frac{1}{2}, 12\right)$.

14. $P = -\frac{1}{2}x^2 + 20x + 230$

$$= -\tfrac{1}{2}(x^2 - 40x) + 230$$

$$= -\tfrac{1}{2}(x^2 - 40x + 400) + 200 + 230$$

$$= -\tfrac{1}{2}(x - 20)^2 + 430$$

The vertex of the parabola is at (20, 30).

The maximum profit will be achieved with \$2000 dollars spent on advertising.

16. $f(x) = \frac{1}{2}x^3 + 2x$

The degree is odd and the leading coefficient is positive. The graph falls to the left and rises to the right.

18. $h(x) = -x^5 - 7x^2 + 10x$

The degree is odd and the leading coefficient is negative. The graph rises to the left and falls to the right.

20. $f(x) = -x^4 + 2x^3$; $g(x) = -x^4$

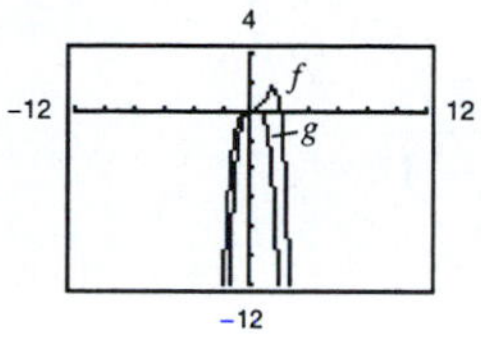

22. $h(x) = -2x^2 - x^2 + x$

$= x(-2x^2 - x + 1)$

$= x(-2x + 1)(x + 1)$

Intercepts: $(0, 0), \left(\frac{1}{2}, 0\right), (-1, 0)$

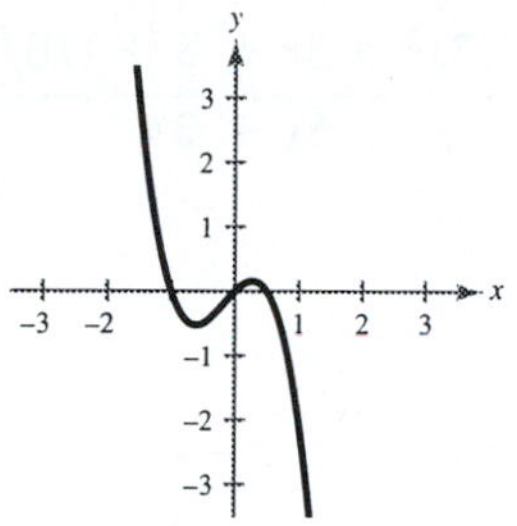

24. $f(x) = -x^3 + 3x - 2$

$= -(x + 2)(x^2 - 2x + 1)$

$= -(x + 2)(x - 1)^2$

Intercepts: $(-2, 0), (1, 0)$

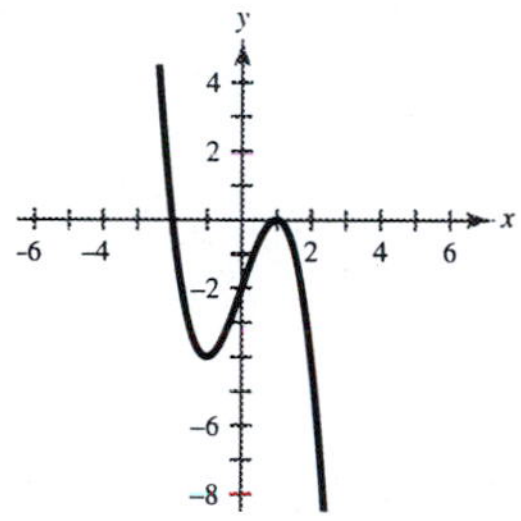

26. $f(t) = t^4 - 4t^2$

$= t^2(t^2 - 4)$

$= t^2(t - 2)(t + 2)$

Intercepts: $(0, 0), (2, 0), (-2, 0)$

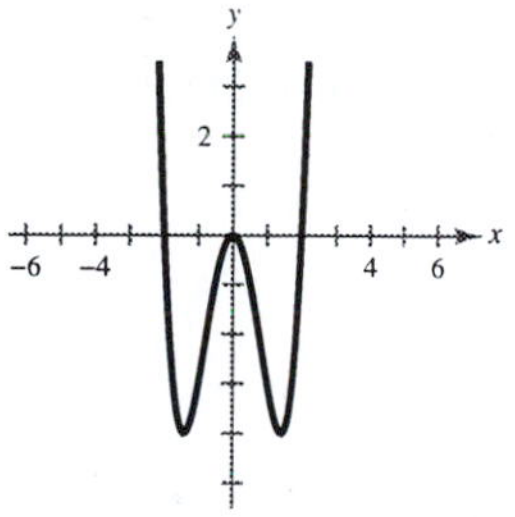

28. (a) Combined length and girth:

$2\pi r + h = 216 \Longrightarrow h = 216 - 2\pi r$

Volume $= \pi r^2 h$

$= \pi r^2(216 - 2\pi r)$

(b)

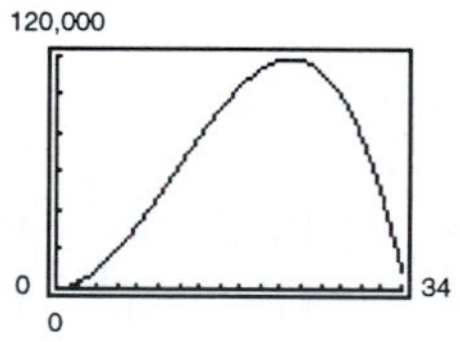

The volume is maximum when $r \approx 22.9$,

$h \approx 216 - 2\pi(22.9) = 72.1$

30. $y_1 = \dfrac{x^4 + 1}{x^2 + 2}$

$y_2 = x^2 - 2 + \dfrac{5}{x^2 + 2}$

$= \dfrac{x^2(x^2 + 2)}{x^2 + 2} - \dfrac{2(x^2 + 2)}{x^2 + 2} + \dfrac{5}{x^2 + 2}$

$= \dfrac{x^4 + 2x^2 - 2x^2 - 4 + 5}{x^2 + 2}$

$= \dfrac{x^4 + 1}{x^2 + 2} = y_1$

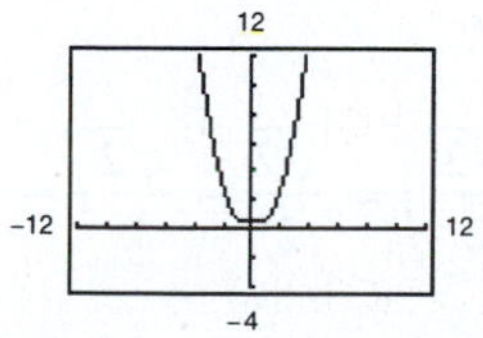

32.
$$\begin{array}{r}
\frac{4}{3} \\
3x-2\,\overline{)\,4x+7} \\
\underline{4x-\frac{8}{3}} \\
\frac{29}{3}
\end{array}$$

$$\frac{4x+7}{3x-2}=\frac{4}{3}+\frac{29}{3(3x-2)}$$

34.
$$\begin{array}{r}
3x^2 \qquad\quad +3 \\
x^2-1\,\overline{)\,3x^4+0x^3+0x^2+0x+0} \\
\underline{3x^4 \qquad\;\; -3x^2 \qquad\qquad} \\
3x^2 \qquad +0 \\
\underline{3x^2 \qquad -3} \\
3
\end{array}$$

$$\frac{3x^4}{x^2-1}=3x^2+3+\frac{3}{x^2-1}$$

36.
$$\begin{array}{r}
3x^2+5x+8+(10/2x^2-1) \\
2x^2-1\,\overline{)\,6x^4+10x^3+13x^2-5x+2} \\
\underline{6x^4 \qquad\qquad -3x^2 \qquad\qquad\quad} \\
10x^3+16x^2 \qquad\quad \\
\underline{10x^3 \qquad\qquad -5x \quad} \\
16x^2 \qquad -8 \\
\overline{10}
\end{array}$$

38.
$$\begin{array}{r|rrrr}
\frac{1}{2} & 2 & 2 & -1 & 2 \\
 & & 1 & \frac{3}{2} & \frac{1}{4} \\
\hline
 & 2 & 3 & \frac{1}{2} & \frac{9}{4}
\end{array}$$

$$(2x^3+2x^2-x+2)\div\left(x-\frac{1}{2}\right)=2x^2+3x+\frac{1}{2}+\frac{9}{2(2x-1)}$$

40.
$$\begin{array}{r|rrrr}
5 & 0.1 & 0.3 & 0 & -0.5 \\
 & & 0.5 & 4 & 20 \\
\hline
 & 0.1 & 0.8 & 4 & 19.5
\end{array}$$

$$(0.1x^3+0.3x^2-0.5)\div(x-5)=0.1x^2+0.8x+4+\frac{19.5}{x-5}$$

42. $f(x)=20x^4+9x^3-14x^2-3x$

(a)
$$\begin{array}{r|rrrrr}
-1 & 20 & 9 & -14 & -3 & 0 \\
 & & -20 & 11 & 3 & 0 \\
\hline
 & 20 & -11 & -3 & 0 & 0
\end{array}$$

Yes, $x=-1$ is a zero of f.

(b)
$$\begin{array}{r|rrrrr}
\frac{3}{4} & 20 & 9 & -14 & -3 & 0 \\
 & & 15 & 18 & 3 & 0 \\
\hline
 & 20 & 24 & 4 & 0 & 0
\end{array}$$

Yes, $x=\frac{3}{4}$ is a zero of f.

(c)
$$\begin{array}{r|rrrrr}
0 & 20 & 9 & -14 & -3 & 0 \\
 & & 0 & 0 & 0 & 0 \\
\hline
 & 20 & 9 & -14 & -3 & 0
\end{array}$$

Yes, $x=0$ is a zero of f.

(d)
$$\begin{array}{r|rrrrr}
1 & 20 & 9 & -14 & -3 & 0 \\
 & & 20 & 29 & 15 & 12 \\
\hline
 & 20 & 29 & 15 & 12 & 12
\end{array}$$

No, $x=1$ is not a zero of f.

44. $\left(\frac{\sqrt{2}}{2}-\frac{\sqrt{2}}{2}i\right)-\left(\frac{\sqrt{2}}{2}+\frac{\sqrt{2}}{2}i\right)=\frac{\sqrt{2}}{2}-\frac{\sqrt{2}}{2}i-\frac{\sqrt{2}}{2}-\frac{\sqrt{2}}{2}i=-2\frac{\sqrt{2}}{2}i=-\sqrt{2}i$

46. $i(6 + i)(3 - 2i) = i(18 - 12i + 3i - 2i^2)$
$= i(20 - 9i)$
$= 20i - 9i^2$
$= 9 + 20i$

48. $\dfrac{3 + 2i}{5 + i} = \dfrac{3 + 2i}{5 + i} \cdot \dfrac{5 - i}{5 - i}$
$= \dfrac{15 + 7i - 2i^2}{25 - i^2}$
$= \dfrac{17 + 7i}{26}$
$= \dfrac{17}{26} + \dfrac{7i}{26}$

50. $(x - 2)(x + 3)[x - (1 - 2i)][x - (1 + 2i)]$
$(x^2 + x - 6)[x^2 - x - 2ix - x + 2ix + (1 - 4i^2)]$
$(x^2 + x - 6)(x^2 - 2x + 5)$
$x^4 - x^3 - 3x^2 + 17x - 30$

52. $f(x) = 10x^3 + 21x^2 - x - 6$
$= (x + 2)(10x^2 + x - 3)$
$= (x + 2)(5x + 3)(2x - 1)$
$x + 2 = 0 \Rightarrow x = -2$
$5x + 3 = 0 \Rightarrow x = -\frac{3}{5}$
$2x - 1 = 0 \Rightarrow x = \frac{1}{2}$

54. $f(x) = x^3 - 1.3x^2 - 1.7x + 0.6$

2	1	−1.3	−1.7	0.6
		2	1.4	−0.6
	1	0.7	−0.3	0

−1	1	0.7	−0.3
		−1	0.3
	1	−0.3	0

Thus, $f(x) = (x - 2)(x + 1)(x - 0.3)$ and the zeros of f are $x = 2, -1, 0, 3$.

56. $f(x) = 5x^4 + 126x^2 + 25$
$f(x) = (5x^2 + 1)(x^2 + 25)$
$5x^2 + 1 = 0$
$x^2 = -\dfrac{1}{5}$
$x = \pm\dfrac{\sqrt{5}}{5}i$
$x^2 + 25 = 0$
$x^2 = -25$
$x = \pm 5i$

58. $g(x) = x^3 - 3x^2 + 3x + 2$

(a)

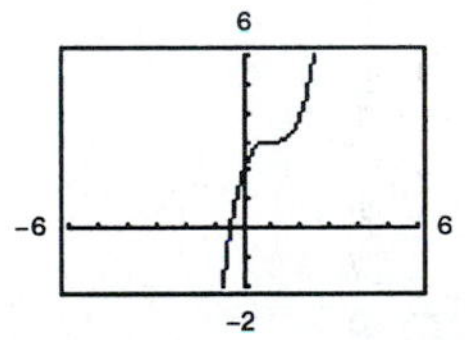

(b) One real zero because the graph has only one x-intercept.

(c) The zero is $x \approx -0.44$.

60. $f(x) = x^5 + 2x^3 - 3x - 20$

(a)

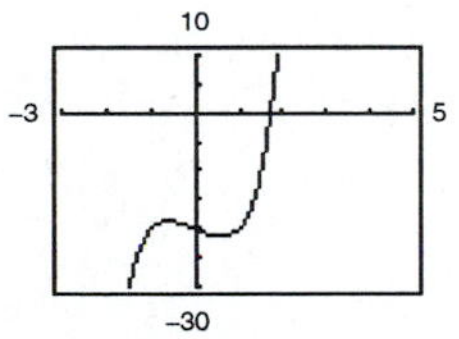

(b) One real zero because the graph has only one x-intercept.

(c) The zero is $x \approx 1.72$.

62. $30 = -0.00428x^2 + 1.442x - 3.136$
$0 = -0.00428x^2 + 1.442x - 33.136$

$$x = \frac{-1.442 \pm \sqrt{(1.442)^2 - 4(-0.00428)(-33.136)}}{2(-0.00428)}$$

$x \approx 24.8, 312.11$

The age of the bride is approximately 24.8 years when the age of the groom is 30 years.

64. $h(x) = \dfrac{x-3}{x-2}$

x-intercept: $(3, 0)$

y-intercept: $\left(0, \dfrac{3}{2}\right)$

Vertical asymptote: $x = 2$

Horizontal asymptote: $y = 1$

x	-1	0	1	3	4	5
y	$\frac{4}{3}$	$\frac{3}{2}$	2	0	$\frac{1}{2}$	$\frac{2}{3}$

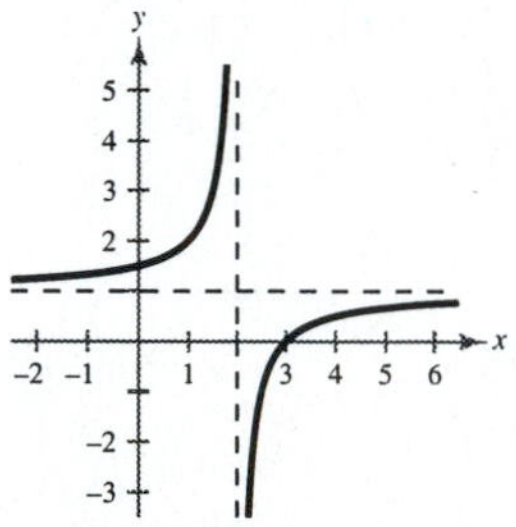

66. $f(x) = \dfrac{2x}{x^2+4}$

Intercept: $(0, 0)$

Origin symmetry

Horizontal asymptote: $y = 0$

x	-2	-1	0	1	2
y	$-\frac{1}{2}$	$-\frac{2}{5}$	0	$\frac{2}{5}$	$\frac{1}{2}$

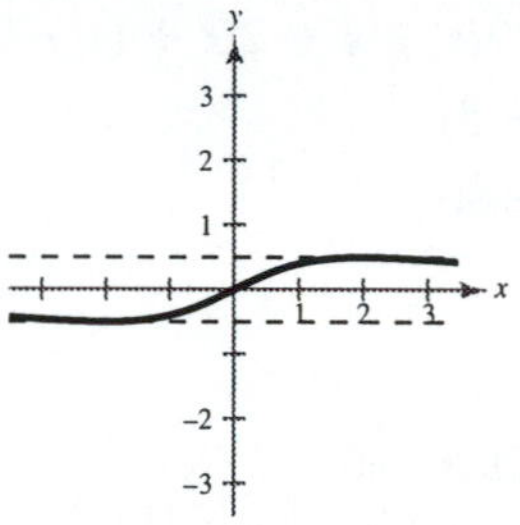

68. $h(x) = \dfrac{4}{(x-1)^2}$

y-intercept: $(0, 4)$

Vertical asymptote: $x = 1$

Horizontal asymptote: $y = 0$

x	-2	-1	0	2	3	4
y	$\frac{4}{9}$	1	4	4	1	$\frac{4}{9}$

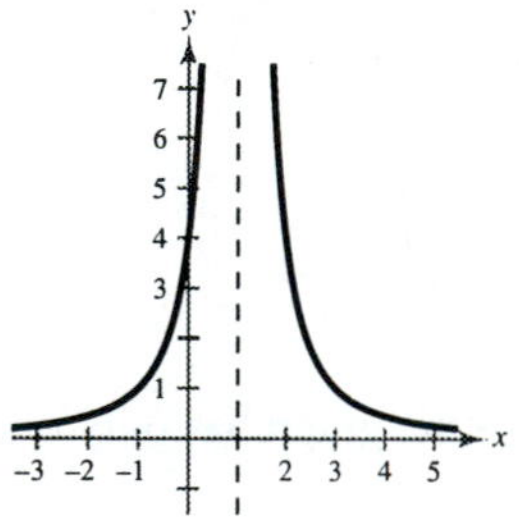

70. $y = \dfrac{2x^2}{x^2-4}$

Intercept: $(0, 0)$

y-axis symmetry

Vertical asymptotes: $x = 2, x = -2$

Horizontal asymptote: $y = 2$

x	± 5	± 4	± 3	± 1	0
y	$\frac{50}{21}$	$\frac{8}{3}$	$\frac{18}{5}$	$-\frac{2}{3}$	0

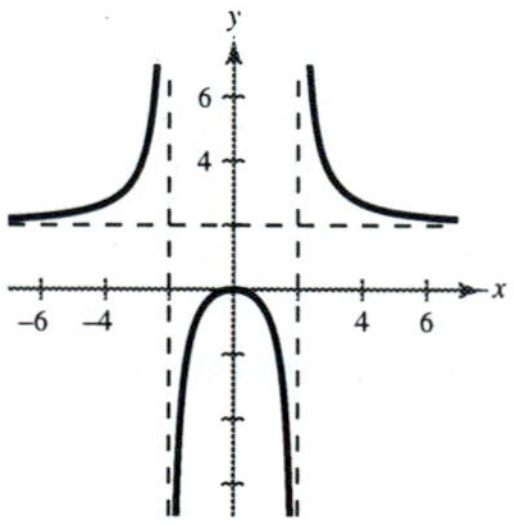

72. $y = \dfrac{5x}{x^2-4}$

Intercept: $(0, 0)$

Vertical asymptotes: $x = 2, x = -2$

Horizontal asymptote: $y = 0$

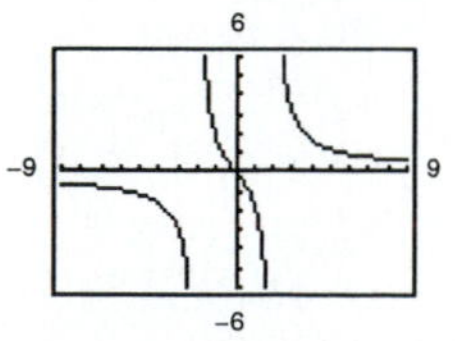

74. $y = \dfrac{1}{x+3} + 2 = \dfrac{2x+7}{x+3}$

Intercepts: $(-3.5, 0), \left(0, 2\frac{1}{3}\right)$

Vertical asymptote: $x = -3$

Horizontal asymptote: $y = 2$

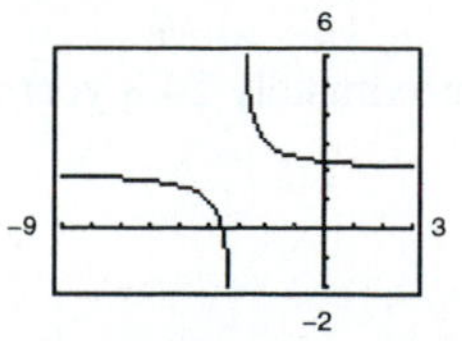

76. $\dfrac{-x}{x^2 + 3x + 2} = \dfrac{A}{x+1} + \dfrac{B}{x+2}$

$-x = A(x + 2) + B(x + 1)$

Let $x = -1$: $1 = A$

Let $x = -2$: $2 = -B \Rightarrow B = -2$

$\dfrac{-x}{x^2 + 3x + 2} = \dfrac{1}{x+1} - \dfrac{2}{x+2}$

78. $\dfrac{9}{x^2 - 9} = \dfrac{A}{x-3} + \dfrac{B}{x+3}$

$9 = A(x + 3) + B(x - 3)$

Let $x = 3$: $9 = 6A \Rightarrow A = \dfrac{3}{2}$

Let $x = -3$: $9 = -6B \Rightarrow B = -\dfrac{3}{2}$

$\dfrac{9}{x^2 - 9} = \dfrac{1}{2}\left(\dfrac{3}{x-3} - \dfrac{3}{x+3}\right)$

80. $\dfrac{4x - 2}{3(x-1)^2} = \dfrac{A}{x-1} + \dfrac{B}{(x-1)^2}$

$\dfrac{4}{3}x - \dfrac{2}{3} = A(x - 1) + B$

Let $x = 1$: $\dfrac{2}{3} = B$

Let $x = 2$: $2 = A + \dfrac{2}{3} \Rightarrow A = \dfrac{4}{3}$

$\dfrac{4x - 2}{3(x-1)^2} = \dfrac{4}{3(x-1)} + \dfrac{2}{3(x-1)^2}$

CHAPTER 3
Exponential and Logarithmic Functions

CHAPTER 3
Exponential and Logarithmic Functions

Section 3.1 Exponential Functions and Their Graphs

Solutions to Even-Numbered Exercises

2. $5000(2^{-1.5}) \approx 1767.767$

4. $8^{2\pi} \approx 472{,}369.379$

6. $\sqrt[3]{4395} \approx 16.380$

8. $e^{1/2} \approx 1.649$

10. $e^{3.2} \approx 24.533$

12. $g(x) = 2^{2x+6}$
$= 2^{2x} \cdot 2^6$
$= 64(2^{2x})$
$= 64(2^2)^x$
$= 64(4^x)$
$= h(x)$
Thus, $g(x) = h(x)$ but $g(x) \neq f(x)$.

14. $f(x) = 5^{-x} + 3$
$g(x) = 5^{3-x} = 5^3 \cdot 5^{-x}$
$h(x) = -5^{x-3} = -(5^x \cdot 5^{-3})$
Thus, $f(x)$, $g(x)$, and $h(x)$ are all distinct.

16. $f(x) = 2^x + 1$ rises to the right.
Asymptote: $y = 1$
Intercept: $(0, 2)$
Matches graph (c).

18. $f(x) = 2^{x-2}$ rises to the right.
Asymptote: $y = 0$
Intercept: $\left(0, \frac{1}{4}\right)$
Matches graph (b).

20. $f(x) = \left(\frac{3}{2}\right)^x$

x	-2	-1	0	1	2
y	$\frac{4}{9}$	$\frac{2}{3}$	1	$\frac{3}{2}$	$\frac{9}{4}$

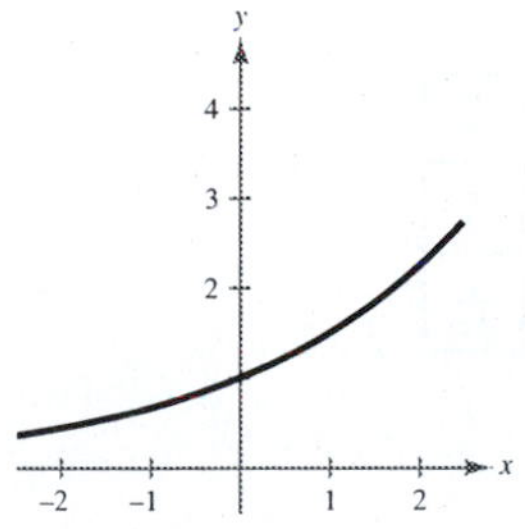

Asymptote: $y = 0$
Intercept: $(0, 1)$
Increasing

22. $h(x) = \left(\frac{3}{2}\right)^{-x}$

x	-2	-1	0	1	2
y	$\frac{9}{4}$	$\frac{3}{2}$	1	$\frac{2}{3}$	$\frac{4}{9}$

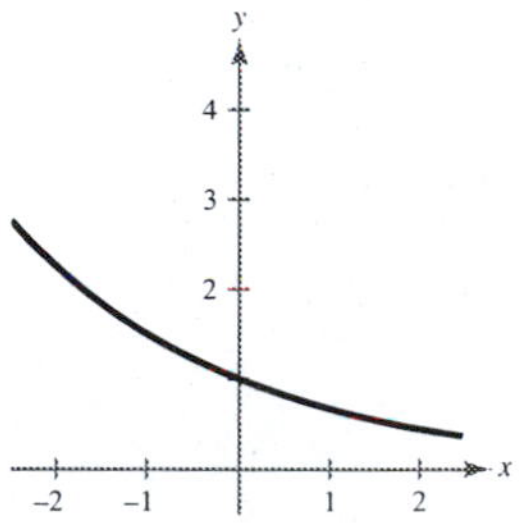

Asymptote: $y = 0$
Intercept: $(0, 1)$
Decreasing

24. $g(x) = \left(\frac{3}{2}\right)^{x+2}$

x	-4	-3	-2	-1	0
y	$\frac{4}{9}$	$\frac{2}{3}$	1	$\frac{3}{2}$	$\frac{9}{4}$

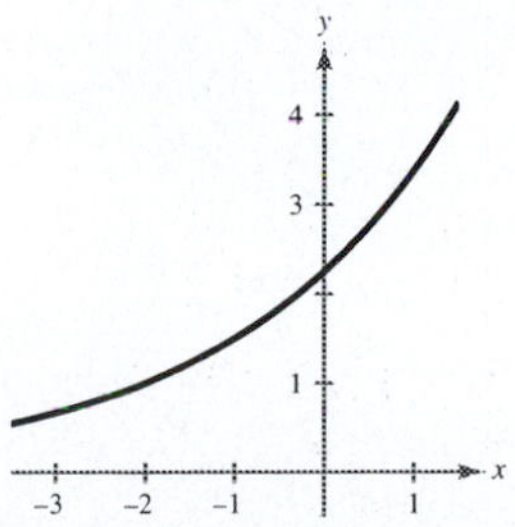

Asymptote: $y = 0$
Intercept: $\left(0, \frac{9}{4}\right)$
Increasing

26. $f(x) = \left(\frac{3}{2}\right)^{-x} + 2$

x	-2	-1	0	1	2
y	$\frac{17}{4}$	$\frac{7}{2}$	3	$\frac{8}{3}$	$\frac{22}{9}$

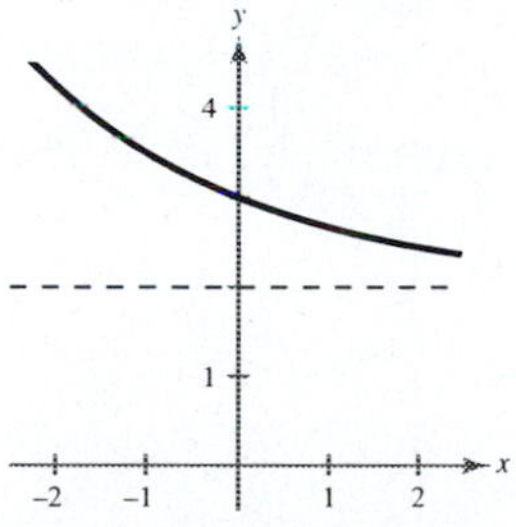

Asymptote: $y = 2$
Intercept: $(0, 3)$
Decreasing

28. $y = 3^{-|x|}$

x	-2	-1	0	1	2
y	$\frac{1}{9}$	$\frac{1}{3}$	1	$\frac{1}{3}$	$\frac{1}{9}$

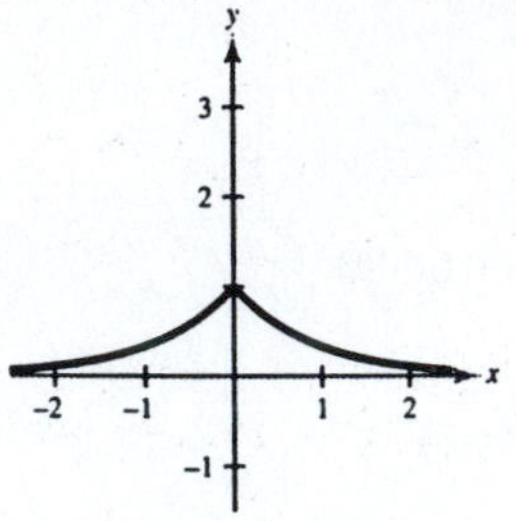

30. $y = 4^{x+1} - 2$

x	-2	-1	0	1
y	$-\frac{7}{4}$	-1	2	14

Asymptote: $y = -2$

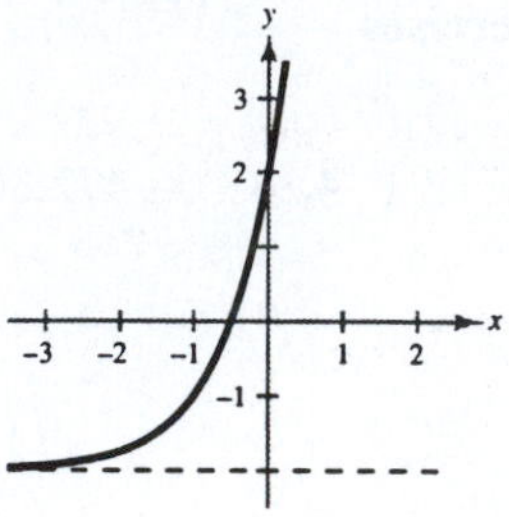

32. $y = 1.08^{5x}$

x	-1	0	1	2
y	0.68	1	1.47	2.16

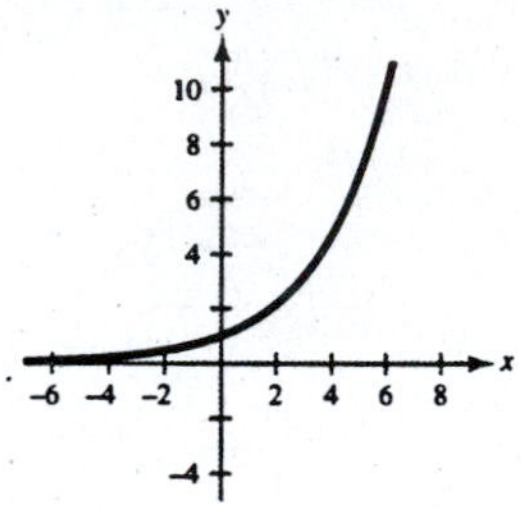

34. $s(t) = 3e^{-0.2t}$

x	-8	-4	0	4	8
y	14.86	6.68	3	1.35	0.61

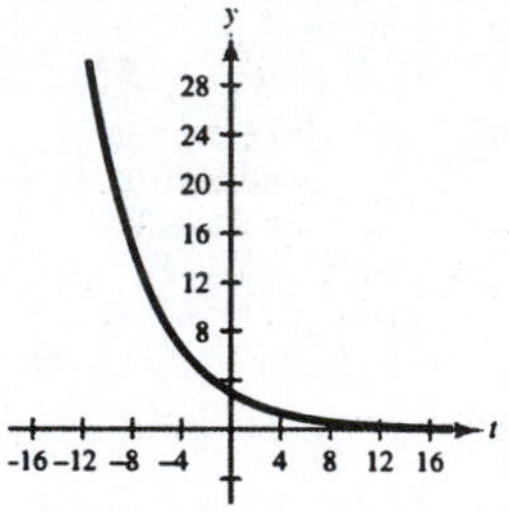

36. $h(x) = e^{x-2}$

x	-1	0	1	2	3
y	0.05	0.14	0.37	1	2.72

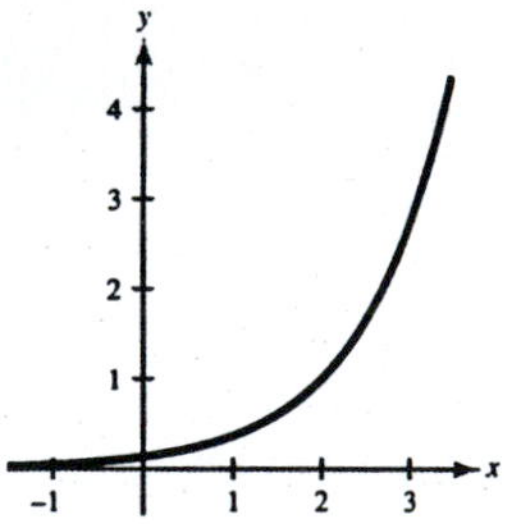

38. $y = \left(\frac{1}{2}\right)^x$ and $y = \left(\frac{1}{4}\right)^x$

x	-2	-1	0	1	2
$\left(\frac{1}{2}\right)^x$	4	2	1	$\frac{1}{2}$	$\frac{1}{4}$
$\left(\frac{1}{4}\right)^x$	16	4	1	$\frac{1}{4}$	$\frac{1}{16}$

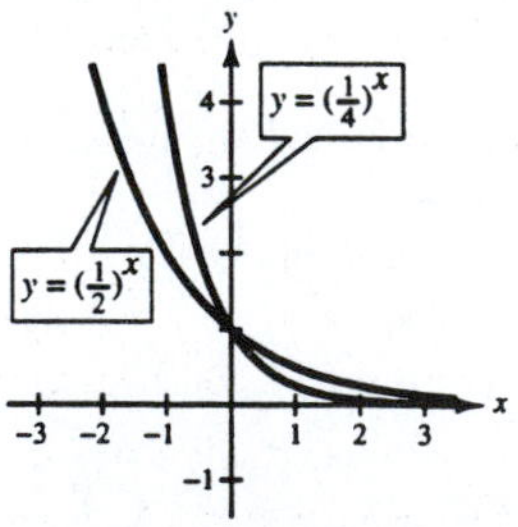

(a) $\left(\frac{1}{4}\right)^x < \left(\frac{1}{2}\right)^x$ when $x > 0$.

(b) $\left(\frac{1}{4}\right)^x > \left(\frac{1}{2}\right)^x$ when $x < 0$.

40. (a) $f(x) = \dfrac{8}{1 + e^{-0.5x}}$

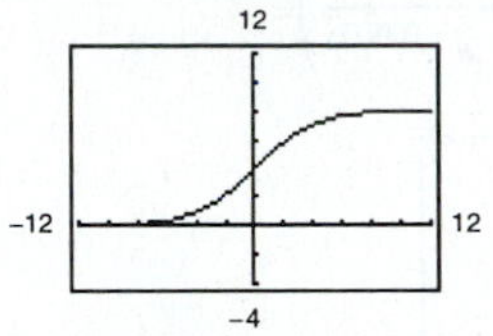

Horizontal asymptotes at $y = 0$ and $y = 8$

(b) $g(x) = \dfrac{8}{1 + e^{-0.5/x}}$

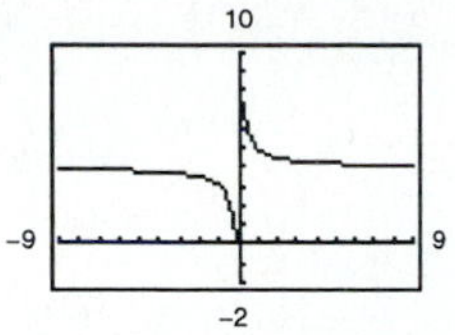

Horizontal asymptote at $y = 4$
Vertical asymptote at $x = 0$

42.

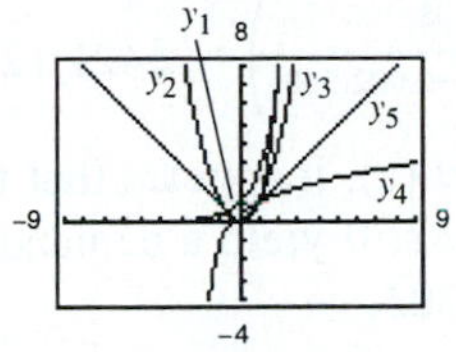

The function that increases at the fastest rate for "large" values of x is $y_1 = e^x$.

44. For a quantity that is described as growing exponentially, it usually implies rapid growth.

46. In Exercise 45 $f(x) = [1 + (0.5/x)^x]$ appears to approach $g(x) = e^{0.5}$ as x increases without bound. Therefore, the value of $[1 + (r/x)]^x$ approaches e^r as x increases without bound. For example, if $r = 1$:

x	1	10	100	200	500	1100	10,000
$\left[1 + \left(\frac{1}{x}\right)\right]^x$	2	2.5937	2.7048	2.7115	2.7156	2.7170	2.718

$e^1 \approx 2.718281828 \ldots$

48. $P > \$1000$, $r = 10\%$, $t = 10$ years

Compounded n times per year: $A = 1000\left(1 + \dfrac{0.10}{n}\right)^{10n}$

Compounded continuously: $A = 1000e^{0.10(10)}$

n	1	2	4	12	365	Continuous
A	\$2,593.74	\$2,653.30	\$2,685.06	\$2,707.04	\$2,717.90	\$2,718.28

50. $P = \$1000$, $r = 10\%$, $t = 40$ years

Compounded n times per year: $A = 1000\left(1 + \dfrac{0.10}{n}\right)^{40n}$

Compounded continuously: $A = 1000e^{0.10(40)}$

n	1	2	4	12	365	Continuous
A	\$45,259.26	\$49,561.44	\$51,977.87	\$53,700.66	\$54,568.25	\$54.598.15

52.
$$A = Pe^{rt}$$
$$100{,}000 = Pe^{0.12t}$$
$$\frac{100{,}000}{e^{0.12t}} = P$$
$$P = 100{,}000e^{-0.12t}$$

t	1	10	20	30	40	50
P	\$88,692.04	\$30,119.42	\$9071.80	\$2732.37	\$822.97	\$247.88

54. $A = 5000e^{(0.075)(50)} \approx \$212{,}605.41$

56. $V(t) = 20{,}000\left(\frac{3}{4}\right)^t$

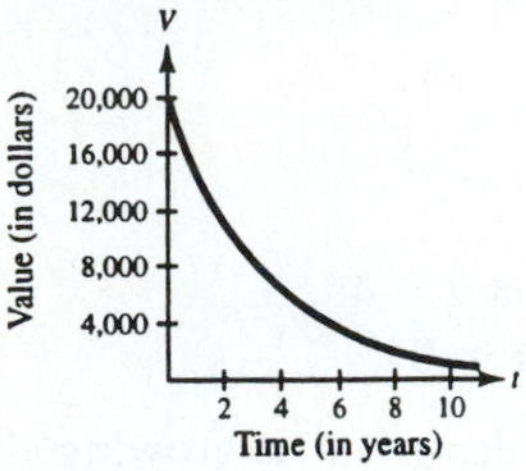

$V(2) = 20{,}000\left(\frac{3}{4}\right)^2 = \$11{,}250$

58. $p = 5000\left(1 - \frac{4}{4 + e^{-0.002x}}\right)$

(a)

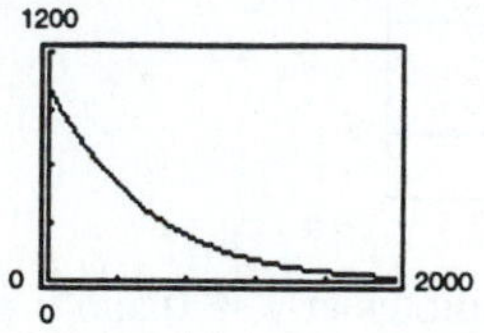

(b) When $x = 500$:

$p = 5000\left(1 - \frac{4}{4 + e^{-0.002(500)}}\right) \approx \$421.12$

(c) From the graph in part (a), it appears that the greatest price that will still yield a demand of at least 600 units is \$303.

60. $P(t) = 2500e^{0.0293t}$

(a) $P(10) \approx 3351$

(b) $P(20) \approx 4492$

62. $Q = 10\left(\frac{1}{2}\right)^{t/5730}$

(a) When $t = 0$: $Q = 10\left(\frac{1}{2}\right)^{0/5730}$
$= 10(1) = 10$ units

(b) When $t = 2000$: $Q = 10\left(\frac{1}{2}\right)^{2000/5730}$
≈ 7.85 units

(c)

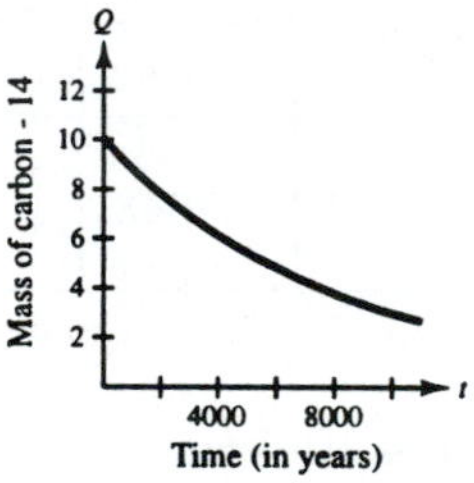

64. $y = \frac{300}{3 + 17e^{-0.065x}}$

(a)

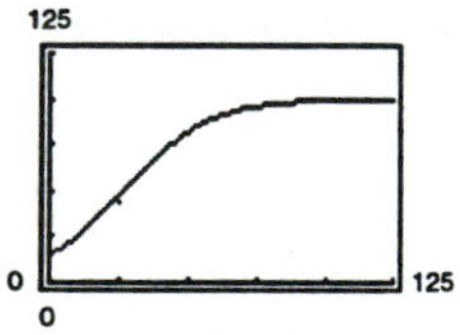

(b)

x	0	25	50	75	100
y	15	47	82	96	99

The model is a "good fit".

(c) When $x = 36$: $y = \frac{300}{3 + 17e^{-0.065(36)}} \approx 64.7\%$

(d) $\frac{2}{3} = \frac{300}{3 + 17e^{-0.065x}}$ when $x \approx 36.9$.

66. The functions (c) 3^x and (d) 2^{-x} are exponential.

68.

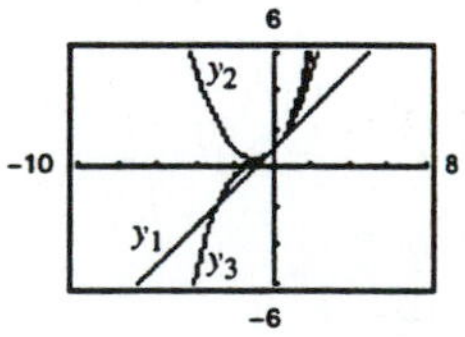

70. (a) $f(u + v) = a^{u+v}$
$= a^u \cdot a^v$
$= f(u) \cdot f(v)$

(b) $f(2x) = a^{2x}$
$= (a^x)^2$
$= [f(x)]^2$

72. $x^2 + 3y = 4$
$3y = 4 - x^2$
$y = \frac{1}{3}(4 - x^2)$

74. $x - |y| = 2$
$x - 2 = |y|$
$$y = \begin{cases} x - 2, & x \geq 2 \\ -(x - 2), & x < 2 \end{cases}$$

Section 3.2 Logarithmic Functions and Their Graphs

Solutions to Even-Numbered Exercises

2. $\log_3 81 = 4 \Rightarrow 3^4 = 81$

4. $\log_{10} \frac{1}{1000} = -3 \Rightarrow 10^{-3} = \frac{1}{1000}$

6. $\log_{16} 8 = \frac{3}{4} \Rightarrow 16^{3/4} = 8$

8. $\ln 4 = 1.386\ldots \Rightarrow e^{1.386\ldots} = 4$

10. $8^2 = 64 \Rightarrow \log_8 64 = 2$

12. $9^{3/2} = 27 \Rightarrow \log_9 27 = \frac{3}{2}$

14. $10^{-3} = 0.001 \Rightarrow \log_{10} 0.001 = -3$

16. $e^0 = 1 \Rightarrow \ln 1 = 0$

18. $u^v = w \Rightarrow \log_u w = v$

20. $\log_2 \frac{1}{8} = \log_2 2^{-3} = -3$

22. $\log_{27} 9 = \log_{27} 27^{2/3} = \frac{2}{3}$

24. $\log_{10} 1000 = \log_{10} 10^3 = 3$

26. $\log_{10} 10 = \log_{10} 10^1 = 1$

28. $\log_9 243 = \log_9 9^{5/2} = \frac{5}{2}$

30. $\log_a a^2 = 2$

32. $\log_{10} \frac{4}{5} \approx -0.097$

34. $\log_{10} 12.5 \approx 1.097$

36. $\ln \sqrt{42} \approx 1.869$

38. $\ln(\sqrt{5} - 2) \approx -1.444$

40. $\ln 0.75 \approx -0.288$

42. $f(x) = 5^x, g(x) = \log_5 x$

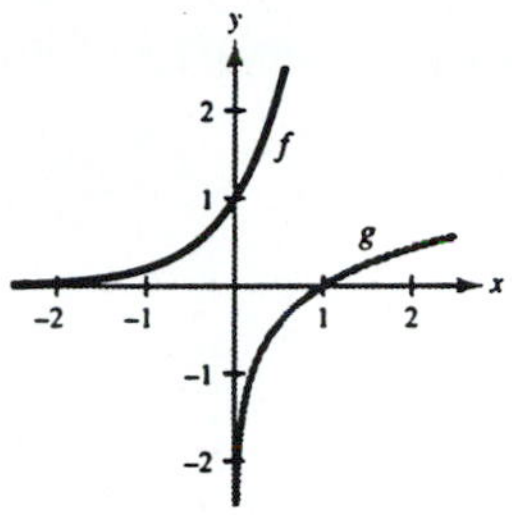

f and g are inverses. Their graphs are reflected about the line $y = x$.

44. $f(x) = 10^x, g(x) = \log_{10} x$

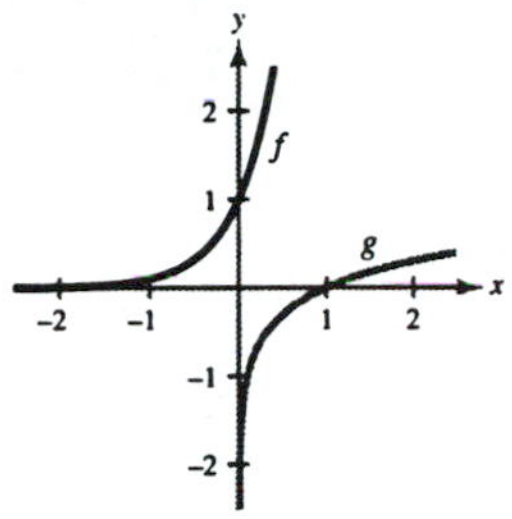

f and g are inverses. Their graphs are reflected about the line $y = x$.

46. $f(x) = -\log_3 x$
Asymptote: $x = 0$
Point on graph: $(1, 0)$
Matches graph (f).

48. $f(x) = \log_3(x - 1)$
Asymptote: $x = 1$
Point on graph: $(2, 0)$
Matches graph (e).

50. $f(x) = -\log_3(-x)$
Asymptote: $x = 0$
Point on graph: $(-1, 0)$
Matches graph (a).

52. $g(x) = \log_6 x$
Domain: $(0, \infty)$
Vertical asymptote: $x = 0$
x-intercept: $(1, 0)$
$y = \log_6 x \Rightarrow 6^y = x$

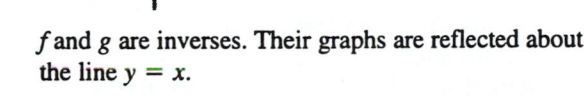

x	$\frac{1}{6}$	1	$\sqrt{6}$	36
y	-1	0	$\frac{1}{2}$	2

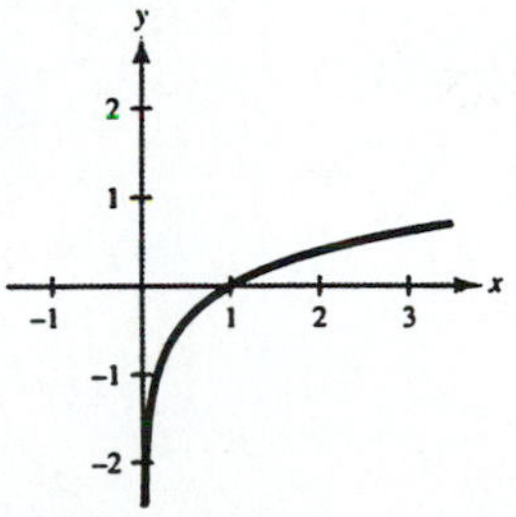

54. $h(x) = \log_4(x - 3)$
Domain: $x - 3 > 0 \Rightarrow x > 3$
The domain is $(3, \infty)$.
Vertical asymptote: $x - 3 = 0 \Rightarrow x = 3$
x-intercept: $\log_4(x - 3) = 0$

$$4^0 = x - 3$$
$$1 = x - 3$$
$$4 = x$$

The x-intercept is $(4, 0)$.
$y = \log_4(x - 3) \Rightarrow 4^y + 3 = x$

x	$3\frac{1}{4}$	4	7	19
y	-1	0	1	2

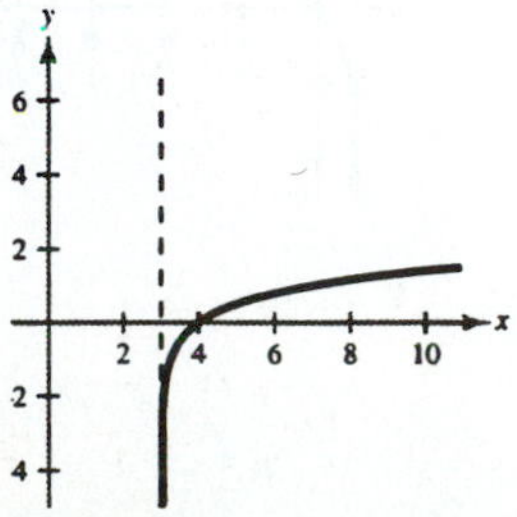

56. $y = \log_5(x - 1) + 4$

Domain: $x - 1 > 0 \Rightarrow x > 1$

The domain is $(1, \infty)$.

Vertical asymptote: $x - 1 = 0 \Rightarrow x = 1$

x-intercept: $\log_5(x - 1) + 4 = 0$

$$\log_5(x - 1) = -4$$
$$5^{-4} = x - 1$$
$$\frac{1}{625} = x - 1$$
$$\frac{626}{625} = x$$

The x-intercept is $\left(\frac{626}{625}, 0\right)$.

$y = \log_5(x - 1) + 4 \Rightarrow 5^{y-4} + 1 = x$

x	1.00032	1.0016	1.008	1.04	1.2
y	-1	0	1	2	3

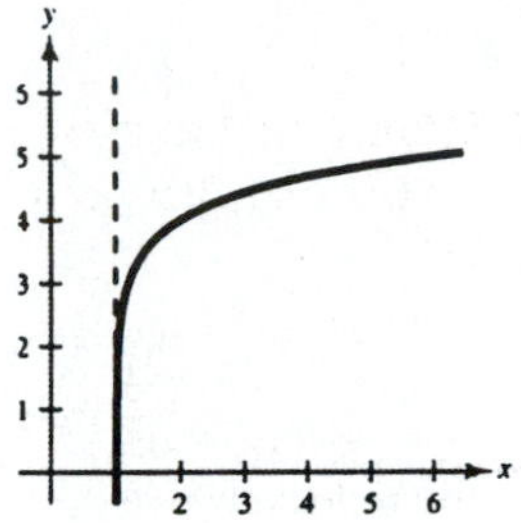

58. $y = \log_{10}(-x)$

Domain: $-x > 0 \Rightarrow x < 0$

The domain is $(-\infty, 0)$.

Vertical asymptote: $x = 0$

x-intercept: $\log_{10}(-x) = 0$

$$10^0 = -x$$
$$-1 = x$$

The x-intercept is $(-1, 0)$.

$y = \log_{10}(-x) \Rightarrow -10^y = x$

x	$-\frac{1}{100}$	$-\frac{1}{10}$	-1	-10
y	-2	-1	0	1

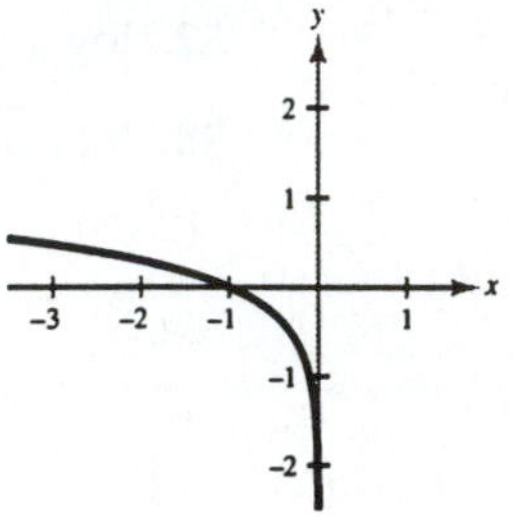

60. $h(x) = \ln(x + 1)$

Domain: $x + 1 > 0 \Rightarrow x > -1$

The domain is $(-1, \infty)$.

Vertical asymptote: $x + 1 = 0 \Rightarrow x = -1$

x-intercept: $\ln(x + 1) = 0$

$$e^0 = x + 1$$
$$1 = x + 1$$
$$0 = x$$

The x-intercept is $(0, 0)$.

$y = \ln(x + 1) \Rightarrow e^y - 1 = x$

x	-0.39	0	1.72	6.39	19.09
y	$-\frac{1}{2}$	0	1	2	3

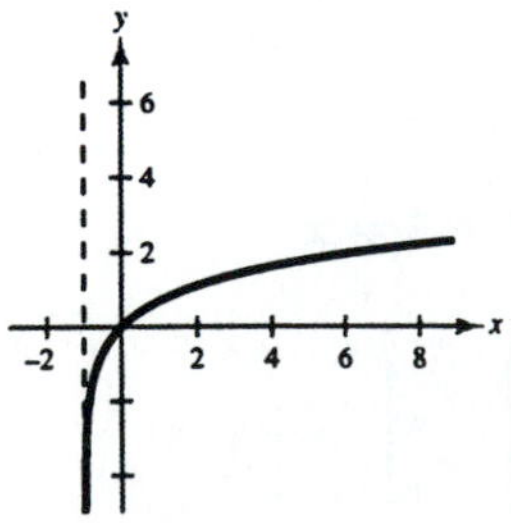

62. $f(x) = \ln(3 - x)$

Domain: $3 - x > 0 \Rightarrow x < 3$

The domain is $(-\infty, 3)$.

Vertical asymptote: $3 - x = 0 \Rightarrow x = 3$

x-intercept: $\ln(3 - x) = 0$

$$e^0 = 3 - x$$
$$1 = 3 - x$$
$$2 = x$$

The x-intercept is $(2, 0)$.

$y = \ln(3 - x) \Rightarrow x = 3 - e^y$

x	2.95	2.86	2.63	2	0.28
y	-3	-2	-1	0	1

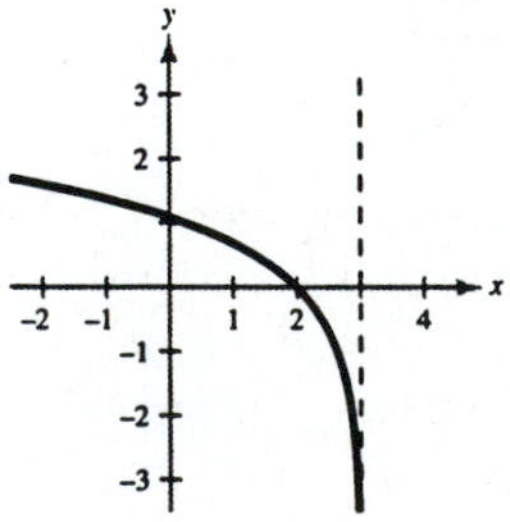

64. $h(x) = \ln(x^2 + 1)$

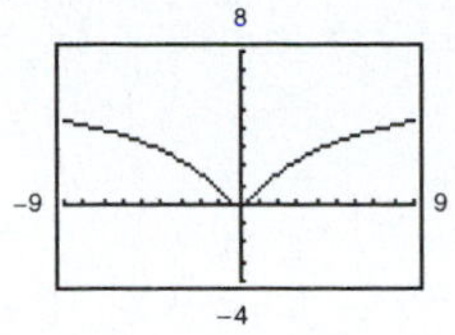

Increasing on $(0, \infty)$.
Decreasing on $(-\infty, 0)$.
Relative minimum: $(0, 0)$

66. $g(x) = \dfrac{12 \ln x}{x}$

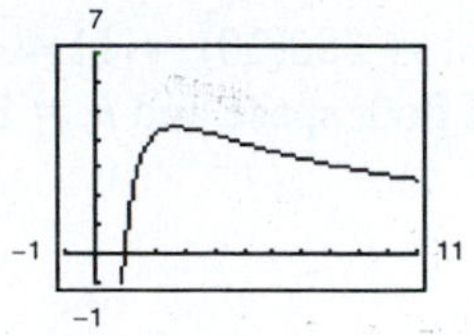

Domain: $(0, \infty)$
Increasing on $(0, 2.72)$.
Decreasing on $(2.72, \infty)$.
Relative maximum: $(2.72, 4.41)$

68. (a) False. If y were an exponential function of x, then $y = a^x$, but $a^1 = a$, not 0. Because one point is $(1, 0)$, y is not an exponential function of x.

(b) True. $y = \log_a x$

For $a = 2$, $y = \log_2 x$.

$x = 1$, $\log_2 1 = 0$

$x = 2$, $\log_2 2 = 1$

$x = 8$, $\log_2 8 = 3$

(c) True. $x = a^y$

For $a = 2$, $x = 2^y$.

$y = 0$, $2^0 = 1$

$y = 1$, $2^1 = 2$

$y = 3$, $2^2 = 8$

(d) False. If y were a linear function of x, the slope between $(1, 0)$ and $(2, 1)$ and the slope between $(2, 1)$ and $(8, 3)$ would be the same.

However,

$$m_1 = \frac{1-0}{2-1} = 1 \text{ and } m_2 = \frac{3-1}{8-2} = \frac{2}{6} = \frac{1}{3}.$$

Therefore, y is not a linear function of x.

70. $y_5 = (x - 1) - \frac{1}{2}(x - 1)^2 + \frac{1}{3}(x - 1)^3 - \frac{1}{4}(x - 1)^4$

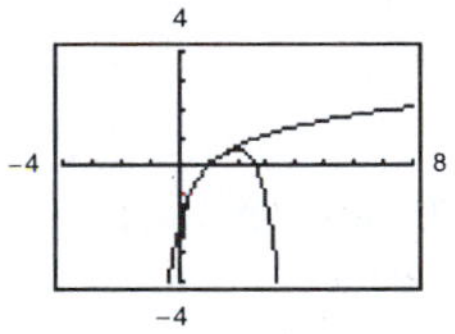

The pattern implies that $\ln x = (x - 1) - \frac{1}{2}(x - 1)^2 + \frac{1}{3}(x - 1)^3 - \frac{1}{4}(x - 1)^4 + \cdots$.

72. $t = \dfrac{10 \ln 2}{\ln 67 - \ln 50} \approx 23.68$ years

74. $t = \dfrac{\ln k}{0.095}$

(a)

K	1	2	4	6	8	10	12
t	0	7.3	14.6	18.9	21.9	24.2	26.2

The number of years required to multiply the original investment by K increases with K. However, the larger the value of K, the fewer the years required to increase the value of the investment by an additional multiple of the original investment.

(b)

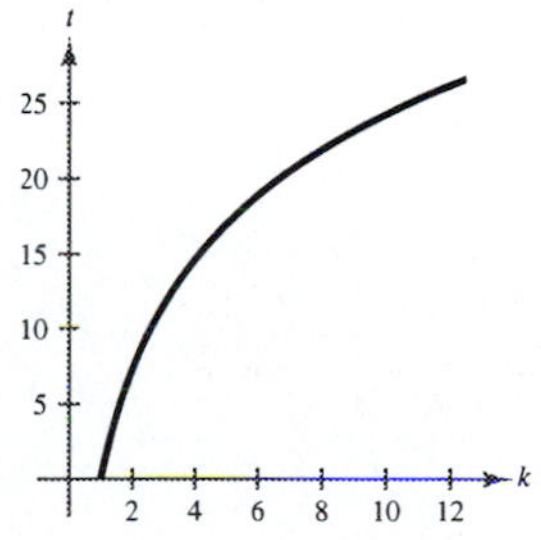

76. (a) $\frac{450}{30} = 15$ cubic feet per minute

(b) 380 cubic feet

(c) Total air space required $= 380(30) = 11{,}400$ cubic feet

Let $x =$ square feet of floor space and $h = 30$ feet.

$$V = xh$$
$$11{,}400 = x(30)$$
$$380 = x$$

The minimum number of square feet of floor space required when the ceiling height of 30 feet is 380 square feet.

78. $\beta = 10 \log_{10}\left(\frac{I}{10^{-16}}\right)$

(a) $\beta = 10 \log_{10}\left(\frac{10^{-4}}{10^{-16}}\right) = 10 \log_{10}(10^{12}) = 10(12)$

$\beta = 120$ decibels

(b) $\beta = 10 \log_{10}\left(\frac{10^{-6}}{10^{-16}}\right) = 10 \log_{10}(10^{10}) = 10(10)$

$= 100$ decibels

(c) No, 120 decibels is not 100 times 100 decibels. The difference is due to the logarithmic relationship between intensity and number of decibels.

80. $t = 10.042 \ln\left(\frac{1982.26}{1982.26 - 1250}\right) \approx 10$ years

82. Total amount $= (1982.26)(10)(12) \approx \$237{,}871$

Interest $= \$237{,}871 - 150{,}000 = \$87{,}871$

84. $f(x) = \log_{10} x$

(a) Domain: $(0, \infty)$

(b)
$$y = \log_{10} x$$
$$x = \log_{10} y$$
$$10^x = y$$
$$f^{-1}(x) = 10^x$$

(c) Since $\log_{10} 1000 = 3$ and $\log_{10} 10{,}000 = 4$, the interval in which $f(x)$ will be found is $(3, 4)$.

(d) When $f(x)$ is negative, x is in the interval $(0, 1)$.

(e) $0 = \log_{10} 1$

$1 = \log_{10} 10$

$2 = \log_{10} 100$

$3 = \log_{10} 1000$

When $f(x)$ is increased by one unit, x is increased by a factor of 10.

(f) $f(x_1) = 3n$ $\quad$ $f(x_2) = n$

$\log_{10} x_1 = 3n$ $\quad$ $\log_{10} x_2 = n$

$x_1 = 10^{3n}$ $\quad$ $x_2 = 10^n$

$x_1 : x_2 = 10^{3n} : 10^n = 10^{2n} : 1$

86. $9.25 + 0.75q$

88. $A = l \cdot w$

$= (10 + w)w$

$= 10w + w^2$

Section 3.3 Properties of Logarithms

Solutions to Even-Numbered Exercises

2. $f(x) = \ln x$

$g(x) = \dfrac{\log_{10} x}{\log_{10} e}$

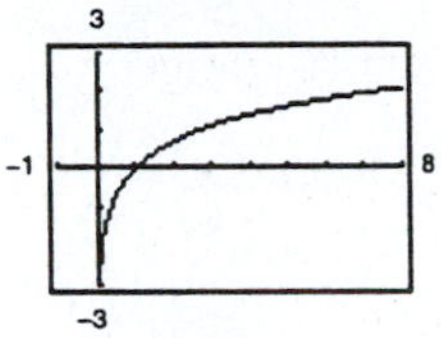

$f(x) = g(x)$

4. $\log_4 10 = \dfrac{\log_{10} 10}{\log_{10} 4} = \dfrac{1}{\log_{10} 4}$

6. $\ln 5 = \dfrac{\log_{10} 5}{\log_{10} e}$

8. $\log_4 10 = \dfrac{\ln 10}{\ln 4}$

10. $\log_{10} 5 = \dfrac{\ln 5}{\ln 10}$

12. $\log_7 4 = \dfrac{\log_{10} 4}{\log_{10} 7} = \dfrac{\ln 4}{\ln 7} \approx 0.712$

14. $\log_4 0.55 = \dfrac{\log_{10} 0.55}{\log_{10} 4} = \dfrac{\ln 0.55}{\ln 4} \approx -0.431$

16. $\log_{20} 125 = \dfrac{\log_{10} 125}{\log_{10} 20} = \dfrac{\ln 125}{\ln 20} \approx 1.612$

18. $\log_{1/3} 0.015 = \dfrac{\log_{10} 0.015}{\log_{10} (1/3)} = \dfrac{\ln 0.015}{\ln (1/3)} \approx 3.823$

20. $\log_{10} 10z = \log_{10} 10 + \log_{10} z$

22. $\log_{10} \dfrac{y}{2} = \log_{10} y - \log_{10} 2$

24. $\log_6 z^{-3} = -3 \log_6 z$

26. $\ln \sqrt[3]{t} = \ln t^{1/3} = \dfrac{1}{3} \ln t$

28. $\ln \dfrac{xy}{z} = \ln x + \ln y - \ln z$

30.
$$\begin{aligned}\ln\left(\frac{x^2-1}{x^3}\right) &= \ln(x^2-1) - \ln x^3\\ &= \ln[(x+1)(x-1)] - \ln x^3\\ &= \ln(x+1) + \ln(x-1) - 3\ln x\end{aligned}$$

32.
$$\begin{aligned}\ln\sqrt{\frac{x^2}{y^3}} = \ln\left(\frac{x^2}{y^3}\right)^{1/2} &= \frac{1}{2}\ln\left(\frac{x^2}{y^3}\right)\\ &= \frac{1}{2}(\ln x^2 - \ln y^3)\\ &= \frac{1}{2}(2\ln x - 3\ln y)\end{aligned}$$

34.
$$\begin{aligned}\ln\left(\frac{x}{\sqrt{x^2+1}}\right) &= \ln x - \ln\sqrt{x^2+1}\\ &= \ln x - \ln(x^2+1)^{1/2}\\ &= \ln x - \frac{1}{2}\ln(x^2+1)\end{aligned}$$

36.
$$\begin{aligned}\ln\sqrt{x^2(x+2)} &= \ln[x^2(x+2)]^{1/2}\\ &= \ln[x(x+2)^{1/2}]\\ &= \ln x + \ln(x+2)^{1/2}\\ &= \ln x + \frac{1}{2}\ln(x+2)\end{aligned}$$

38.
$$\begin{aligned}\log_b \frac{\sqrt{x}\,y^4}{z^4} &= \log_b \sqrt{x}\,y^4 - \log_b z^4\\ &= \log_b x^{1/2} + \log_b y^4 - \log_b z^4\\ &= \frac{1}{2}\log_b x + 4\log_b y - 4\log_b z\end{aligned}$$

40. $y_1 = \ln\left(\dfrac{\sqrt{x}}{x-2}\right)$

$y_2 = \dfrac{1}{2}\ln x - \ln(x-2)$

$y_1 = y_2$

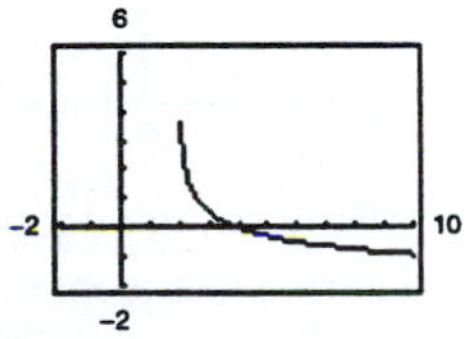

42. $\ln y + \ln z = \ln yz$

44. $\log_5 8 - \log_5 t = \log_5 \dfrac{8}{t}$

46. $-4\log_6 2x = \log_6 (2x)^{-4} = \log_6 \dfrac{1}{16x^4}$

48. $\dfrac{3}{2}\log_7(z-2) = \log_7(z-2)^{3/2}$

50. $2 \ln 8 + 5 \ln z = \ln 8^2 + \ln z^5$
$= \ln 64z^5$

52. $3 \ln x + 2 \ln y - 4 \ln z = \ln x^3 + \ln y^2 - \ln z^4$
$= \ln x^3y^2 - \ln z^4$
$= \ln \dfrac{x^3y^2}{z^4}$

54. $4[\ln z + \ln(z + 5)] - 2 \ln(z - 5) = 4[\ln z(x + 5)] - \ln(z - 5)^2$
$= \ln[z(z + 5)]^4 - \ln(z - 5)^2$
$= \ln \dfrac{z^4(z + 5)^4}{(z - 5)^2}$

56. $2[\ln x - \ln(x + 1) - \ln(x - 1)] = 2\left[\ln \dfrac{x}{x + 1} - \ln(x - 1)\right]$
$= 2\left[\ln \dfrac{x}{(x + 1)(x - 1)}\right]$
$= 2\left[\ln \dfrac{x}{x^2 - 1}\right]$
$= \ln\left(\dfrac{x}{x^2 - 1}\right)^2$

58. $\frac{1}{2}[\ln(x + 1) + 2 \ln(x - 1)] + 3 \ln x = \frac{1}{2}[\ln(x + 1) + \ln(x - 1)^2] + \ln x^3$
$= \frac{1}{2}[\ln(x + 1)(x - 1)^2] + \ln x^3$
$= \ln[(x + 1)(x - 1)^2]^{1/2} + \ln x^3$
$= \ln[(x + 1)^{1/2}(x - 1)] + \ln x^3$
$= \ln\left[x^3(x - 1)\sqrt{x + 1}\right]$

60. $\dfrac{3}{2} \ln 5t^6 - \dfrac{3}{4} \ln t^4 = \ln(5t^6)^{3/2} - \ln(t^4)^{3/4}$
$= \ln 5^{3/2}\, t^9 - \ln t^3$
$= \ln \dfrac{5\sqrt{5}t^9}{t^3}$
$= \ln 5\sqrt{5}\, t^6$

62. $y_1 = \ln x + \dfrac{1}{3} \ln(x + 1)$
$y_2 = \ln\left(x\sqrt[3]{x + 1}\right)$
$y_1 = y_2$

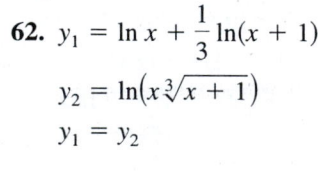

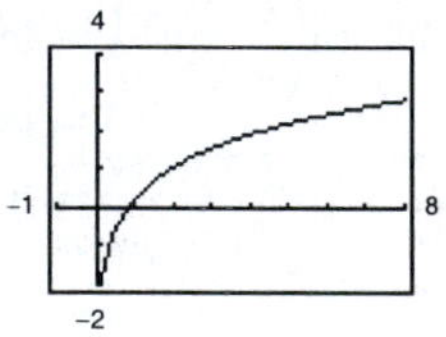

64. $y_1 = \frac{1}{4} \ln[x^4(x^2 + 1)]$
$y_2 = \ln x + \frac{1}{4} \ln(x^2 + 1)$

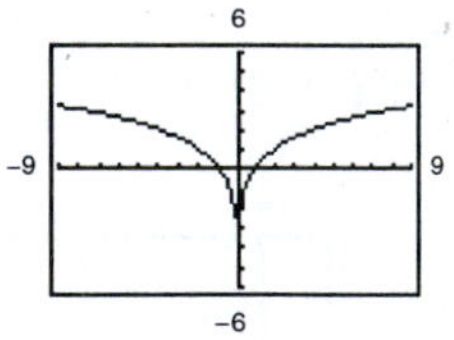

They are not equivalent. The domain of y_1 is all real numbers except 0. The domain of y_2 is $x > 0$.

66. $\log_7\sqrt{70} = \frac{1}{2} \log_7 70$
$= \frac{1}{2}[\log_7 7 + \log_7 10]$
$= \frac{1}{2}[1 + \log_7 10]$
$= \frac{1}{2} + \frac{1}{2} \log_7 10$
$= \frac{1}{2} + \log_7\sqrt{10}$
by Property 1 and Property 3.

68. $\ln 2 \approx 0.6931$, $\ln 3 \approx 1.0986$, $\ln 5 \approx 1.6094$

$\ln 2 \approx 0.6931$

$\ln 3 \approx 1.0986$

$\ln 4 = \ln 2 + \ln 2 \approx 0.6931 + 0.6931 = 1.3862$

$\ln 5 \approx 1.6094$

$\ln 6 = \ln 2 + \ln 3 \approx 0.6931 + 1.0986 = 1.7917$

$\ln 8 = \ln 2^3 = 3 \ln 2 \approx 3(0.6931) = 2.0793$

$\ln 9 = \ln 3^2 = 2 \ln 3 \approx 2(1.0986) = 2.1972$

$\ln 10 = \ln 5 + \ln 2 \approx 1.6094 + 0.6931 = 2.3025$

$\ln 12 = \ln 2^2 + \ln 3 = 2 \ln 2 + \ln 3 \approx 2(0.6931) + 1.0986 = 2.4848$

$\ln 15 = \ln 5 + \ln 3 \approx 1.6094 + 1.0986 = 2.7080$

$\ln 16 = \ln 2^4 = 4 \ln 2 \approx 4(0.6931) = 2.7724$

$\ln 18 = \ln 3^2 + \ln 2 = 2 \ln 3 + \ln 2 \approx 2(1.0986) + 0.6931 = 2.8903$

$\ln 20 = \ln 5 + \ln 2^2 = \ln 5 + 2 \ln 2 \approx 1.6094 + 2(0.6931) = 2.9956$

70. $\log_6 \sqrt[3]{6} = \log_6 6^{1/3} = \frac{1}{3} \log_6 6 = \frac{1}{3}(1) = \frac{1}{3}$

72. $\log_5 \frac{1}{125} = \log_5 5^{-3} = -3 \log_5 5 = -3(1) = -3$

74. $\log_2(-16)$ is undefined because -16 is not in the domain of $\log_2 x$.

76. $\log_4 2 + \log_4 32 = \log_4 4^{1/2} + \log_4 4^{5/2}$

$= \frac{1}{2} \log_4 4 + \frac{5}{2} \log_4 4$

$= \frac{1}{2}(1) + \frac{5}{2}(1)$

$= 3$

78. $3 \ln e^4 = (3)(4) \ln e$

$= 12(1) = 12$

80. $\ln 1 = 0$

82. $\ln \sqrt[4]{e^3} = \ln e^{3/4}$

$= \frac{3}{4} \ln e$

$= \frac{3}{4}(1) = \frac{3}{4}$

84. $\log_2(4^2 \cdot 3^4) = \log_2 4^2 + \log_2 3^4$

$= 2 \log_2 4 + 4 \log_2 3$

$= 2 \log_2 2^2 + 4 \log_2 3$

$= 4 \log_2 2 + 4 \log_2 3$

$= 4 + 4 \log_2 3$

86. $\log_{10} \frac{9}{300} = \log_{10} \frac{3}{100}$

$= \log_{10} 3 - \log_{10} 100$

$= \log_{10} 3 - \log_{10} 10^2$

$= \log_{10} 3 - 2 \log_{10} 10$

$= \log_{10} 3 - 2$

88. $\ln \dfrac{6}{e^2} = \ln 6 - \ln e^2$

$= \ln 6 - 2 \ln e$

$= \ln 6 - 2$

90. $\beta = 10 \log_{10}\left(\dfrac{I}{10^{-16}}\right)$

$= 10(\log_{10} I - \log_{10} 10^{-16})$

$= 10(\log_{10} I - (-16)\log_{10} 10)$

$= 10(\log_{10} I + 16)$

When $I = 10^{-10}$:

$\beta = 10(\log_{10} 10^{-10} + 16)$

$= 10(-10 \log_{10} 10 + 16)$

$= 10(-10 + 16)$

$= 10(6) = 60$ decibels

92. $f(ax) = f(a) + f(x)$, $a > 0$, $x > 0$

True, because

$f(ax) = \ln ax = \ln a + \ln x$

$= f(a) + f(x)$.

94. $\sqrt{f(x)} = \frac{1}{2}f(x)$; False.

$\sqrt{f(x)} = \sqrt{\ln x}$ can't be simplified further.

$f(\sqrt{x}) = \ln\sqrt{x} = \ln x^{1/2} = \frac{1}{2}\ln x = \frac{1}{2}f(x)$

96. If $f(x) < 0$, then $0 < x < 1$.

True.

98. Let $x = \log_b u$, then $u = b^x$ and $u^n = b^{nx}$.

$\log_b u^n = \log_b b^{nx} = nx = n\log_b u$

100. $\left(\dfrac{2x^2}{3y}\right)^{-3} = \left(\dfrac{3y}{2x^2}\right)^3$

$= \dfrac{(3y)^3}{(2x^2)^3}$

$= \dfrac{27y^3}{8x^6}$

102. $xy(x^{-1} + y^{-1})^{-1} = \dfrac{xy}{x^{-1} + y^{-1}}$

$= \dfrac{xy}{\frac{1}{x} + \frac{1}{y}}$

$= \dfrac{xy}{\frac{y + x}{xy}}$

$= \dfrac{(xy)^2}{x + y}$

Section 3.4 Exponential and Logarithmic Equations

Solutions to Even-Numbered Exercises

2. $2^{3x+1} = 32$

(a) $x = -1$

$2^{3(-1)+1} = 2^{-2} = \frac{1}{4}$

No, $x = -1$ is not a solution.

(b) $x = 2$

$2^{3(2)+1} = 2^7 = 128$

No, $x = 2$ is not a solution.

4. $5^{2x+3} = 812$

(a) $x = -1.5 + \log_5\sqrt{812}$

$5^{2(-1.5+\log_5\sqrt{812})+3} = 5^{-3+2\log_5\sqrt{812}+3}$

$= 5^{2\log_5 812^{1/2}}$

$= 5^{\log_5(812^{1/2})^2}$

$= 5^{\log_5 812} = 812$

Yes, $x = -1.5 + \log_5\sqrt{812}$ is a solution.

(b) $x \approx 0.5813$

$5^{2(0.5813)+3} = 5^{4.1626} \approx 812$

Yes, $x \approx 0.5813$ is a solution.

(c) $x = \dfrac{1}{2}\left(-3 + \dfrac{\ln 812}{\ln 5}\right)$

$5^{2[1/2(-3+(\ln 812/\ln 5))]+3} = 5^{-3+(\ln 812/\ln 5)+3}$

$= 5^{\ln 812/\ln 5}$

$= 5^{\log_5 812}$

$= 812$

Yes, $x = \dfrac{1}{2}\left(-3 + \dfrac{\ln 812}{\ln 5}\right)$ is a solution.

6. $\ln(x - 1) = 3.8$

(a) $x = 1 + e^{3.8}$

$\ln(1 + e^{3.8} - 1) = \ln e^{3.8} = 3.8$

Yes, $x = 1 + e^{3.8}$ is a solution.

(b) $x \approx 45.7012$

$\ln(45.7012 - 1) = \ln(44.7012) \approx 3.8$

Yes, $x \approx 45.7012$ is a solution.

(c) $x = 1 + \ln 3.8$

$\ln(1 + \ln 3.8 - 1) = \ln(\ln 3.8) \approx 0.289$

No, $x = 1 + \ln 3.8$ is not a solution.

8. $f(x) = g(x)$
$$27^x = 9$$
$$27^x = 27^{2/3}$$
$$x = \tfrac{2}{3}$$
Point of intersection: $\left(\tfrac{2}{3}, 9\right)$

10. $f(x) = g(x)$
$$\ln(x - 4) = 0$$
$$x - 4 = e^0$$
$$x - 4 = 1$$
$$x = 5$$
Point of intersection: $(5, 0)$

12. $3^x = 243$
$$3^x = 3^5$$
$$x = 5$$

14. $8^x = 4$
$$8^x = 8^{2/3}$$
$$x = \tfrac{2}{3}$$

16. $3^{x-1} = 27$
$$3^{x-1} = 3^3$$
$$x - 1 = 3$$
$$x = 4$$

18. $\log_x 625 = 4$
$$625 = x^4$$
$$\sqrt[4]{625} = x$$
$$x = 5$$

20. $\ln(2x - 1) = 0$
$$e^0 = 2x - 1$$
$$1 = 2x - 1$$
$$2 = 2x$$
$$1 = x$$

22. $\log_6 6^{2x-1} = 2x - 1$

24. $-1 + \ln e^{2x} = -1 + 2x = 2x - 1$

26. $-8 + e^{\ln x^3} = -8 + x^3 = x^3 - 8$

28. $4e^x = 91$
$$e^x = \frac{91}{4}$$
$$\ln e^x = \ln\frac{91}{4}$$
$$x = \ln\frac{91}{4} \approx 3.125$$

30. $-14 + 3e^x = 11$
$$3e^x = 25$$
$$e^x = \frac{25}{3}$$
$$\ln e^x = \ln\frac{25}{3}$$
$$x = \ln\frac{25}{3} \approx 2.120$$

32. $e^{2x} = 50$
$$\ln e^{2x} = \ln 50$$
$$2x = \ln 50$$
$$x = \frac{\ln 50}{2} \approx 1.956$$

34. $1000e^{-4x} = 75$
$$e^{-4x} = \tfrac{3}{40}$$
$$\ln e^{-4x} = \ln\tfrac{3}{40}$$
$$-4x = \ln\tfrac{3}{40}$$
$$x = -\tfrac{1}{4}\ln\tfrac{3}{40} \approx 0.648$$

36. $e^{2x} - 5e^x + 6 = 0$
$$(e^x - 2)(e^x - 3) = 0$$
$$e^x = 2 \text{ or } e^x = 3$$
$$x = \ln 2 \approx 0.693 \text{ or } x = \ln 3 \approx 1.099$$

38. $\dfrac{400}{1 + e^{-x}} = 350$
$$400 = 350(1 + e^{-x})$$
$$\frac{8}{7} = 1 + e^{-x}$$
$$\frac{8}{7} - 1 = e^{-x}$$
$$\frac{1}{7} = e^{-x}$$
$$\ln\frac{1}{7} = \ln e^{-x}$$
$$-x = \ln\frac{1}{7}$$
$$-x = \ln 7^{-1}$$
$$-x = -\ln 7$$
$$x = \ln 7 \approx 1.946$$

40. $10^x = 570$
$$\log_{10} 10^x = \log_{10} 570$$
$$x = \log_{10} 570 \approx 2.756$$

42.
$$6^{5x} = 3000$$
$$\ln 6^{5x} = \ln 3000$$
$$(5x)\ln 6 = \ln 3000$$
$$5x = \frac{\ln 3000}{\ln 6}$$
$$x = \frac{\ln 3000}{5 \ln 6} \approx 0.894$$

44.
$$4^{-3t} = 0.10$$
$$\ln 4^{-3t} = \ln 0.10$$
$$(-3t)\ln 4 = \ln 0.10$$
$$-3t = \frac{\ln 0.10}{\ln 4}$$
$$t = -\frac{\ln 0.10}{3 \ln 4} \approx 0.554$$

46.
$$\left(1 + \frac{0.10}{12}\right)^{12t} = 2$$
$$\ln\left(1 + \frac{0.10}{12}\right)^{12t} = \ln 2$$
$$(12t)\ln\left(1 + \frac{0.10}{12}\right) = \ln 2$$
$$t = \frac{\ln 2}{12\ln\left(1 + \frac{0.10}{12}\right)} \approx 6.960$$

48. $f(x) = 3e^{3x/2} - 962$

The zero is $x \approx 3.847$.

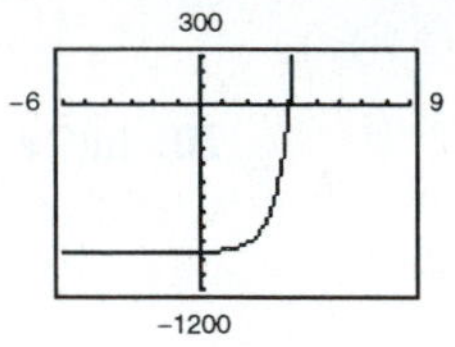

50. $h(t) = e^{0.125t} - 8$

The zero is $t \approx 16.636$.

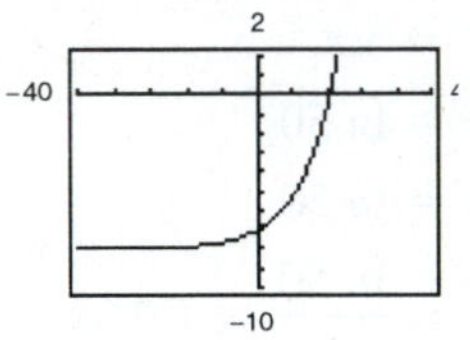

52.
$$3(5^{x-1}) = 21$$
$$5^{x-1} = 7$$
$$\ln 5^{x-1} = \ln 7$$
$$(x-1)\ln 5 = \ln 7$$
$$x - 1 = \frac{\ln 7}{\ln 5}$$
$$x = \frac{\ln 7}{\ln 5} + 1 \approx 2.209$$

54.
$$\frac{3000}{2 + e^{2x}} = 2$$
$$1500 = 2 + e^{2x}$$
$$1498 = e^{2x}$$
$$\ln 1498 = \ln e^{2x}$$
$$\ln 1498 = 2x$$
$$\frac{\ln 1498}{2} = x \approx 3.656$$

56. $\ln x = 2$
$$e^{\ln x} = e^2$$
$$x = e^2 \approx 7.389$$

58. $3 \ln 5x = 10$
$$\ln 5x = \frac{10}{3}$$
$$e^{\ln 5x} = e^{10/3}$$
$$5x = e^{10/3}$$
$$x = \frac{e^{10/3}}{5} \approx 5.606$$

60. $\ln(x + 1)^2 = 2$
$$e^{\ln(x+1)^2} = e^2$$
$$(x + 1)^2 = e^2$$
$$x + 1 = e \text{ or } x + 1 = -e$$
$$x = e - 1 \approx 1.718$$
$$\text{or } x = -e - 1 \approx -3.718$$

62. $\log_{10} x^2 = 6$
$$10^{\log_{10} x^2} = 10^6$$
$$x^2 = 10^6$$
$$x = \pm\sqrt{10^6} = \pm 1000$$

64. $\ln x + \ln(x + 3) = 1$
$$\ln[x(x + 3)] = 1$$
$$e^{\ln[x(x+3)]} = e^1$$
$$x(x + 3) = e^1$$
$$x^2 + 3x - e = 0$$
$$x = \frac{-3 \pm \sqrt{9 + 4e}}{2}$$

Using the positive value for x, we have $x = \dfrac{-3 + \sqrt{9 + 4e}}{2} \approx 0.729$.

66. $\log_4 x - \log_4(x-1) = \dfrac{1}{2}$

$$\log_4\left(\frac{x}{x-1}\right) = \frac{1}{2}$$
$$4^{\log_4(x/x-1)} = 4^{1/2}$$
$$\frac{x}{x-1} = 2$$
$$x = 2(x-1)$$
$$x = 2x - 2$$
$$2 = x$$

68. $\log_2 x + \log_2(x+2) = \log_2(x+6)$

$$\log_2[x(x+2)] = \log_2(x+6)$$
$$x^2 + 2x = x + 6$$
$$x^2 + x - 6 = 0$$
$$x = \frac{-1 \pm \sqrt{1^2 - 4(1)(-6)}}{2}$$
$$= \frac{-1 \pm 5}{2} = -3, 2$$

Choosing the positive value of x, we have $x = 2$.

70. $\ln(x+1) - \ln(x-2) = \ln x^2$

$$\ln\left(\frac{x+1}{x-2}\right) = \ln x^2$$
$$\frac{x+1}{x-2} = x^2$$
$$x + 1 = x^3 - 2x^2$$
$$0 = x^3 - 2x^2 - x - 1$$

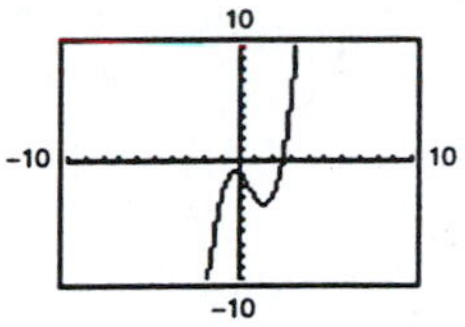

From the graph, we have $x \approx 2.547$.

72. $5\log_{10}(x-2) = 11$

$$\log_{10}(x-2) = \frac{11}{5}$$
$$x - 2 = 10^{11/5}$$
$$= 10^{11/5} + 2 \approx 160.489$$

74. $\ln 4x = 1$

$$e^{\ln 4x} = e^1$$
$$4x = e$$
$$x = \frac{e}{4} \approx 0.680$$

76. $\log_{10} 8x - \log_{10}(1 + \sqrt{x}) = 2$

$$\log_{10}\frac{8x}{1+\sqrt{x}} = 2$$
$$\frac{8x}{1+\sqrt{x}} = 10^2$$
$$8x = 100 + 100\sqrt{x}$$
$$8x - 10\sqrt{x} - 100 = 0$$
$$2x - 25\sqrt{x} - 25 = 0$$
$$\sqrt{x} = \frac{25 \pm \sqrt{25^2 - 4(2)(-25)}}{4}$$
$$= \frac{25 \pm 5\sqrt{33}}{4}$$

Choosing the positive value, we have

$\sqrt{x} \approx 13.431$

$x \approx 180.384.$

78. $y_1 = 500$

$y_2 = 1500e^{-x/2}$

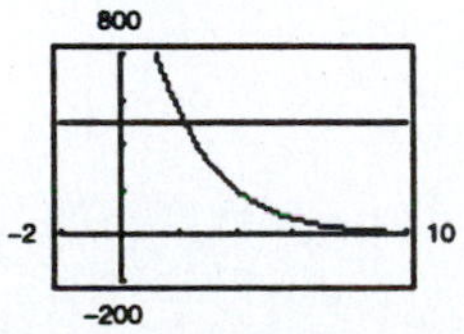

From the graph, we have $x \approx 2.197$.

80. $y_1 = 10$

$y_2 = 4\ln(x-2)$

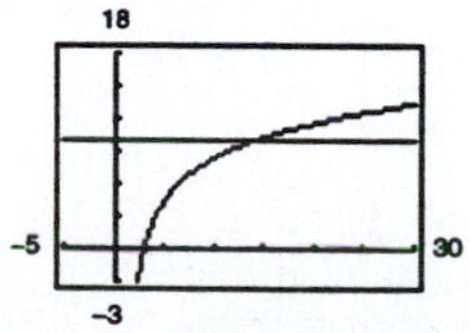

From the graph, we have $x \approx 14.182$.

82.
$$r = 0.12$$
$$A = Pe^{rt}$$
$$2000 = 1000e^{0.12t}$$
$$2 = e^{0.12t}$$
$$\ln 2 = \ln e^{0.12t}$$
$$\ln 2 = 10.12t$$
$$\frac{\ln 2}{0.12} = t$$
$$t \approx 5.8 \text{ years}$$

84. To find the length of time it takes for an investment P to double to $2P$, solve
$$2P = Pe^{rt}$$
$$2 = e^{rt}$$
$$\ln 2 = rt$$
$$\frac{\ln 2}{r} = t.$$
Thus, you can see that the time is not dependent on the size of the investment, but rather the interest rate.

86.
$$r = 0.12$$
$$A = Pe^{rt}$$
$$3000 = 1000e^{0.12t}$$
$$3 = e^{0.12t}$$
$$\ln 3 = \ln e^{0.12t}$$
$$\ln 3 = 0.12t$$
$$\frac{\ln 3}{0.12} = t$$
$$t = 9.2 \text{ years}$$

88. $p = 5000\left(1 - \dfrac{4}{4 + e^{-0.002x}}\right)$

(a) When $p = \$600$:
$$600 = 5000\left(1 - \frac{4}{4 + e^{-0.002x}}\right)$$
$$0.12 = 1 - \frac{4}{4 + e^{-0.002x}}$$
$$\frac{4}{4 + e^{-0.002x}} = 0.88$$
$$4 = 3.52 + 0.88e^{-0.002x}$$
$$0.48 = 0.88e^{-0.002x}$$
$$\frac{6}{11} = e^{-0.002x}$$
$$\ln\frac{6}{11} = \ln e^{-0.002x}$$
$$\ln\frac{6}{11} = -0.002x$$
$$x = \frac{\ln(6/11)}{0.002} \approx 303 \text{ units}$$

(b) When $p = \$400$:
$$400 = 5000\left(1 - \frac{4}{4 + e^{-0.002x}}\right)$$
$$0.08 = 1 - \frac{4}{4 + e^{-0.002x}}$$
$$\frac{4}{4 + e^{-0.002x}} = 0.92$$
$$4 = 3.68 + 0.92e^{-0.002x}$$
$$0.32 = 0.92e^{-0.002x}$$
$$\frac{8}{23} = e^{-0.002x}$$
$$\ln\frac{8}{23} = \ln e^{-0.002x}$$
$$x = \frac{\ln(8/23)}{0.002} \approx 528 \text{ units}$$

90. $N = 68(10^{-0.04x})$

When $N = 21$:

$$21 = 68(10^{-0.04x})$$

$$\frac{21}{68} = 10^{-0.04x}$$

$$\log_{10}\frac{21}{68} = -0.04x$$

$$x = -\frac{\log(21/68)}{0.04} \approx 12.76 \text{ inches}$$

92. $P = \dfrac{0.83}{1 + e^{-0.2n}}$

(a)

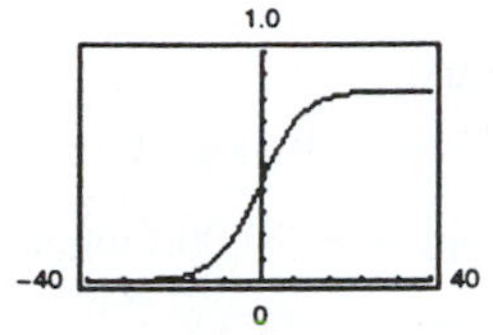

(b) Horizontal asymptotes: $y = 0$, $y = 0.83$
The upper asymptote, $y = 0.83$, indicates that the proportion of correct responses will approach 0.83 as the number of trials increases.

(c) When $P = 60\%$ or $P = 0.60$:

$$0.60 = \frac{0.83}{1 + e^{-0.2n}}$$

$$1 + e^{-0.2n} = \frac{0.83}{0.60}$$

$$e^{-0.2n} = \frac{0.83}{0.60} - 1$$

$$\ln e^{-0.2n} = \ln\left(\frac{0.83}{0.60} - 1\right)$$

$$-0.2n = \ln\left(\frac{0.83}{0.60} - 1\right)$$

$$n = -\frac{\ln\left(\frac{0.83}{0.60} - 1\right)}{0.2} \approx 5 \text{ trials}$$

94. $y = -3.00 + 11.88 \ln x + \dfrac{36.94}{x}$

(a)

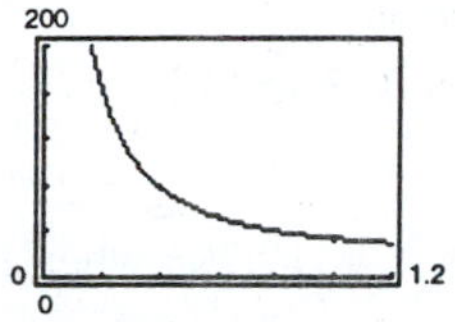

The model seems to fit the data well.

(b) When $y = 30$:

$$30 = -3.00 + 11.88 \ln x + \frac{36.94}{x}$$

Add the graph of $y = 30$ to the graph in part (a) and estimate the point of intersection of the two graphs. We find that $x \approx 1.20$ meters.

(c) No, it is probably not practical to lower the number of *g*s experienced during impact to less than 23 because the required distance traveled at $y = 23$ is $x \approx 2.27$ meters. It is probably not practical to design a car allowing a passenger to move forward 2.27 meters (or 7.45 feet) during an impact.

96. $\sqrt{32} - 2\sqrt{25} = 4\sqrt{2} - 10$

98. $$\frac{3}{\sqrt{10} - 2} = \frac{3}{\sqrt{10} - 2} \cdot \frac{\sqrt{10} + 2}{\sqrt{10} + 2} = \frac{3\sqrt{10} + 6}{6} = \frac{\sqrt{10} + 2}{2}$$

Section 3.5 Exponential and Logarithmic Models

Solutions to Even-Numbered Exercises

2. $y = 6e^{-x/4}$
This is an exponential decay model. Matches graph (e).

4. $y = \dfrac{12}{x+4}$
This is a rational function. Matches graph (b).

6. $y = \sqrt{x}$
This is a square root. Matches graph (f).

8. Since $A = 20{,}000e^{0.105t}$, the time to double is given by $40{,}000 = 20{,}000e^{0.105t}$, and we have

$$t = \frac{\ln 2}{0.105} \approx 6.60 \text{ years.}$$

Amount after 10 years:

$$A = 20{,}000e^{0.105(10)} \approx \$57{,}153.02$$

10. Since $A = 10{,}000e^{rt}$ and $A = 20{,}000$ when $t = 5$, we have

$$20{,}000 = 10{,}000e^{5r}$$
$$r = \frac{\ln 2}{5} \approx 0.1386 \text{ or } 13.86\%.$$

Amount after 10 years:

$$A = 10{,}000e^{0.1386(10)} \approx \$39{,}988.23$$

12. Since $A = 600e^{rt}$ and $A = 19{,}205$ when $t = 10$, we have

$$19{,}205 = 600e^{10r}$$
$$r = \frac{\ln(19{,}205/600)}{10} \approx 0.3466 \text{ or } 34.66\%.$$

The time to double is given by

$$1200 = 600e^{0.3466t}$$
$$t = \frac{\ln 2}{0.3466} \approx 2 \text{ years.}$$

14. Since $A = Pe^{0.08t}$ and $A = 20{,}000$ when $t = 10$, we have

$$20{,}000 = Pe^{0.08(10)}$$
$$P = \frac{20{,}000}{e^{0.08(10)}} \approx \$8986.58.$$

The time to double is given by

$$t = \frac{\ln 2}{0.08} \approx 8.66 \text{ years.}$$

16.
$$A = P\left(1 + \frac{r}{n}\right)^{nt}$$
$$500{,}000 = P\left(1 + \frac{0.12}{12}\right)^{12(40)}$$
$$P = \$4214.16$$

18. $P = 1000$, $r = 10.5\% = 0.105$

(a) $n = 1$

$$t = \frac{\ln 2}{\ln(1 + 0.105)} \approx 6.94 \text{ years}$$

(b) $n = 12$

$$t = \frac{\ln 2}{12 \ln\left(1 + \frac{0.105}{12}\right)} \approx 6.63 \text{ years}$$

(c) $n = 365$

$$t = \frac{\ln 2}{365 \ln\left(1 + \frac{0.105}{365}\right)} \approx 6.602 \text{ years}$$

(d) Compounded continuously

$$t = \frac{\ln 2}{0.105} \approx 6.601 \text{ years}$$

20.

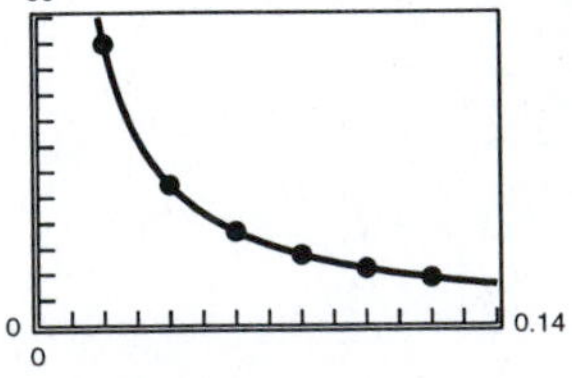

Using the power regression feature of a graphing utility, $t = 1.099r^{-1}$.

22.

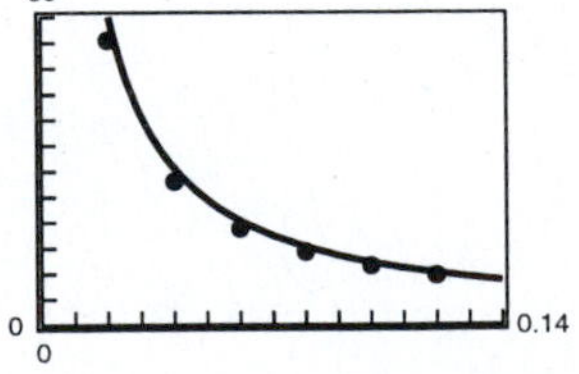

Using the power regression feature of a graphing utility, $t = 1.222r^{-1}$.

24.

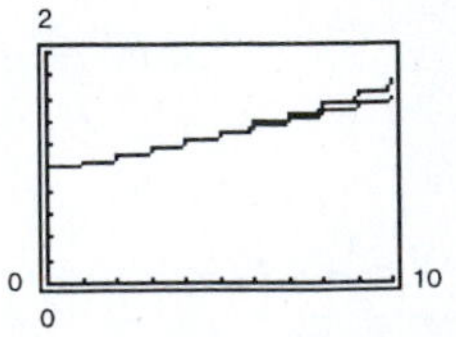

From the graph, $5\frac{1}{2}\%$ compounded daily grows faster than 6% simple interest.

26. $\frac{1}{2}C = Ce^{k(1620)}$

$k = \frac{\ln 0.5}{1620}$

Given $y = 1.5$ grams after 1000 years, we have

$1.5 = Ce^{[(\ln 0.5)/1620](1000)}$

$C \approx 2.30$ grams.

28. $\frac{1}{2}C = Ce^{k(5730)}$

$k = \frac{\ln 0.5}{5730}$

Given $C = 3$ grams, after 1000 years we have

$y = 3e^{[(\ln 0.5)/5730](1000)}$

$y \approx 2.66$ grams.

30. $\frac{1}{2}C = Ce^{k(24,360)}$

$k = \frac{\ln 0.5}{24,360}$

Given $y = 0.4$ grams after 1000 years, we have

$0.4 = Ce^{[(\ln 0.5)/24,360](1000)}$

$C \approx 0.41$ grams.

32. $y = ae^{bx}$

$\frac{1}{2} = ae^{b(0)} \Rightarrow a = \frac{1}{2}$

$5 = \frac{1}{2}e^{b(4)}$

$10 = e^{4b}$

$\ln 10 = 4b$

$\frac{\ln 10}{4} = b \Rightarrow b \approx 0.5756$

Thus, $y = \frac{1}{2}e^{0.5756x}$.

34. $y = ae^{bx}$

$5 = ae^{b(0)} \Rightarrow a = 5$

$1 = 5e^{b(4)}$

$\frac{1}{5} = e^{4b}$

$\ln \frac{1}{5} = 4b$

$b = \frac{\ln (1/5)}{r} \Rightarrow b \approx -0.4024$

Thus, $y = 5e^{-0.4024x}$.

36. $P = 240,360e^{0.012t}$

$275,000 = 240,360e^{0.012t}$

$\ln \frac{27,500}{24,036} = 0.012t$

$t = \frac{\ln (27,500/24,036)}{0.012} \approx 11$

The population will reach 275,000 in 2001.

38. For 1960, we use $t = -30$.

$100,250 = 140,500e^{k(-30)}$

$k = -\frac{\ln(100,250/140,500)}{30} \approx 0.0113$

For 2000, we use $t = 10$.

$P = 140,500e^{-[(\ln(100,250/140,500))/30](10)} \approx 157,232$

40. $y = ae^{bt}$

$2.30 = ae^{b(0)} \Rightarrow a \approx 2.30$

$2.65 = 2.30e^{b(10)}$

$\frac{2.65}{2.30} = e^{10b}$

$\ln\left(\frac{2.65}{2.30}\right) = 10b \Rightarrow b \approx 0.0142$

Thus, $y = 2.30e^{0.0142t}$.

When $t = 20$

$y = 230e^{(0.0142)(20)} \approx 3.05$ million.

42. $y = ae^{bt}$

$9.17 = ae^{b(0)} \Rightarrow a = 9.17$

$8.57 = 9.17e^{b(10)}$

$\frac{8.57}{9.17} = e^{10b}$

$\ln\left(\frac{8.57}{9.17}\right) = 10b \Rightarrow b \approx -0.068$

Thus, $y = 9.17e^{-0.0068t}$.

When $t = 20$, $y = 9.17e^{-0.0068(20)} \approx 8.01$ million.

44. The constant b in the equation $y = ae^{bt}$ determines whether the population is increasing ($b > 0$) or is decreasing ($b < 0$).

46. $N = 250e^{kt}$

$280 = 250e^{k(10)}$

$k = \frac{\ln 1.12}{10}$

$N = 250e^{[(\ln 1.12)/10]t}$

$500 = 250e^{[(\ln 1.12)/10]t}$

$t = \frac{\ln 2}{(\ln 1.12)/10} \approx 61.16$ hours

48. $y = Ce^{kt}$

$\frac{1}{2}C = Ce^{5730k}$

$\ln\frac{1}{2} = 5730k$

$k = \frac{\ln (1/2)}{5730}$

The ancient charcoal has only 15% as much radioactive carbon.

$0.15C = Ce^{[(\ln 0.5)/5730]t}$

$\ln 0.15 = \frac{\ln 0.5}{5730}t$

$t = \frac{5730 \ln 0.15}{\ln 0.5} \approx 15{,}683$ years

50. $y = ae^{bt}$

$3000 = 4600e^{b(2)}$

$\frac{15}{23} = e^{2b}$

$\ln\frac{15}{23} = 2b$

$b = \frac{\ln (15/23)}{2}$

$y = 4600e^{[(\ln(15/23))/2]t}$

For $t = 3$:

$y = 4600e^{[(\ln(15/23))/2](3)} \approx \$2423$

52. $S = \frac{500{,}000}{1 + 0.6e^{kt}}$

(a) $300{,}000 = \frac{500{,}000}{1 + 0.6e^{2k}}$

$1 + 0.6e^{2k} = \frac{5}{3}$

$0.6e^{2k} = \frac{2}{3}$

$e^{2k} = \frac{10}{9}$

$2k = \ln\left(\frac{10}{9}\right)$

$k = \frac{1}{2}\ln\left(\frac{10}{9}\right) \approx 0.053$

$S = \frac{500{,}000}{1 + 0.6e^{0.053t}}$

(b) When $t = 5$:

$S = \frac{500{,}000}{1 + 0.6e^{[0.5 \ln(10/9)](5)}} \approx \$280{,}771$

54.
$$y = ae^{bt}$$
$$632{,}000 = 742{,}000e^{b(2)}$$
$$\frac{632}{742} = e^{2b}$$
$$b = \frac{1}{2}\ln\left(\frac{632}{742}\right)$$
$$y = 742{,}000e^{0.5(3)\ln(632/742)} \approx \$583{,}275$$

56. $p(t) = \dfrac{1000}{1 + 9e^{-0.1656t}}$

(a)

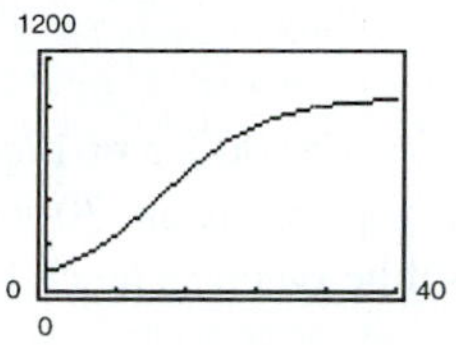

The horizontal asymptotes are $y = 0$ and $y = 1000$. The asymptote with the larger p-value, $y = 1000$, indicates that the population size will approach 1000 as time increases.

(b) $p(5) = \dfrac{1000}{1 + 9e^{-0.1656(5)}} \approx 203$ animals

(c)
$$500 = \frac{1000}{1 + 9e^{-0.1656t}}$$
$$1 + 9e^{-0.1656t} = 2$$
$$9e^{-0.1656t} = 1$$
$$e^{-0.1656t} = \frac{1}{9}$$
$$t = -\frac{\ln(1/9)}{0.1656} \approx 13 \text{ months}$$

58. $R = \log_{10}\dfrac{I}{I_0} = \log_{10} I$ since $I_0 = 1$.

(a) $8.6 = \log_{10} I$

$10^{8.6} = I \approx 398{,}107{,}171$

(b) $6.7 = \log_{10} I$

$10^{6.7} = I \approx 5{,}011{,}872$

60. $\beta(I) = 10\log_{10}\dfrac{I}{I_0}$ where $I_0 = 10^{-16}$ watt/cm^2.

(a) $\beta(10^{-13}) = 10\log_{10}\dfrac{10^{-13}}{10^{-16}} = 10\log_{10}10^3 = 30$ decibels

(b) $\beta(10^{-7.5}) = 10\log_{10}\dfrac{10^{-7.5}}{10^{-16}} = 10\log_{10}10^{8.5} = 85$ decibels

(c) $\beta(10^{-7}) = 10\log_{10}\dfrac{10^{-7}}{10^{-16}} = 10\log_{10}10^{9} = 90$ decibels

(d) $\beta(10^{-4.5}) = 10\log_{10}\dfrac{10^{-4.5}}{10^{-16}} = 10\log_{10}10^{11.5} = 115$ decibels

62.
$$\beta = 10\log_{10}\frac{I}{I_0}$$
$$10^{\beta/10} = \frac{I}{I_0}$$
$$I = I_0\,10^{\beta/10}$$
$$\%\text{ decrease} = \frac{I_0\,10^{8.8} - I_0\,10^{7.2}}{I_0\,10^{8.8}} \times 100 \approx 97\%$$

64. $\text{pH} = -\log_{10}[\text{H}^+] = -\log_{10}[11.3 \times 10^{-6}] \approx 4.95$

66. $3.2 = -\log_{10}[H^+]$

$19^{-3.2} = H^+$

$H^+ \approx 6.3 \times 10^{-4}$ moles per liter

68. $pH - 1 = \log_{10}[H^+]$

$-(pH - 1) = \log_{10}[H^+]$

$10^{-(pH-1)} = [H^+]$

$10^{-pH+1} = [H^+]$

$10^{-pH} \cdot 10 = [H^+]$

The hydrogen ion concentration is increased by a factor of 10.

70. $u = 120{,}000\left[\dfrac{0.095t}{1 - \left(1 + \dfrac{0.095}{12}\right)^{12t}} - 1\right]$

(a) 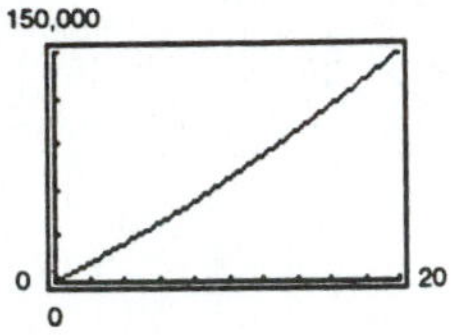

(b) From the graph, when $u = 120{,}000$, $t \approx 17$ years. Yes, if a person takes an approximately 30-year long mortgage the interest will be twice as much as the mortgage.

72. Answers will vary.

74. $t = -2.5 \ln \dfrac{T - 70}{98.6 - 70}$

At 9:00 A.M. we have:

$t = -2.5 \ln \dfrac{85.7 - 70}{98.6 - 70} \approx 1.5$ hours

From this you can conclude that the person died at 7:30 A.M.

76.

$\frac{3}{2}$	8	−36	54	−27
		12	−36	27
	8	−24	18	0

Thus, $\dfrac{8x^3 - 36x^2 + 54x - 27}{x - (3/2)} = 8x^2 - 24x + 18.$

78.

−5	1	0	0	−3	1
		−5	25	−125	640
	1	−5	25	−128	641

Thus,

$\dfrac{x^4 - 3x + 1}{x + 5} = x^3 - 5x^2 + 25x - 128 + \dfrac{641}{x + 5}.$

❑ Focus on Concepts

1. $b < d < a < c$

b and d are negative.

2. (a) True. $\log_b uv = \log_b u + \log_b v$

(b) False. $2.04 \approx \log_{10}(10 + 100) \neq (\log_{10} 10)(\log_{10} 100) = 2$

(c) False. $1.95 \approx \log_{10}(100 - 10) \neq \log_{10} 100 - \log_{10} 10 = 1$

(d) True. $\log_b \dfrac{u}{v} = \log_b u - \log_b v$

3. Double the interest rate or time because it doubles the exponent in the exponential function.

4. (a) Logarithmic (b) Logistic (c) Exponential

(d) Linear (e) None of the above (f) Exponential

Review Exercises for Chapter 3

Solutions to Even-Numbered Exercises

2. $f(x) = 4^{-x}$

Intercept: (0,1)
Horizontal asymptote: x-axis
Decreasing on: $(-\infty, \infty)$
Matches graph (f).

4. $f(x) = 4^x + 1$

Intercept: (0, 2)
Horizontal asymptote: $y = 1$
Increasing on: $(-\infty, \infty)$
Matches graph (b).

6. $f(x) = \log_4(x - 1)$

Intercept: (2, 0)
Vertical asymptote: $x = 1$
Increasing on: $(1, \infty)$
Matches graph (c).

8. $g(x) = 0.3^{-x}$

x	-2	-1	0	1	2
y	0.09	0.3	1	$3\frac{1}{3}$	$11\frac{1}{9}$

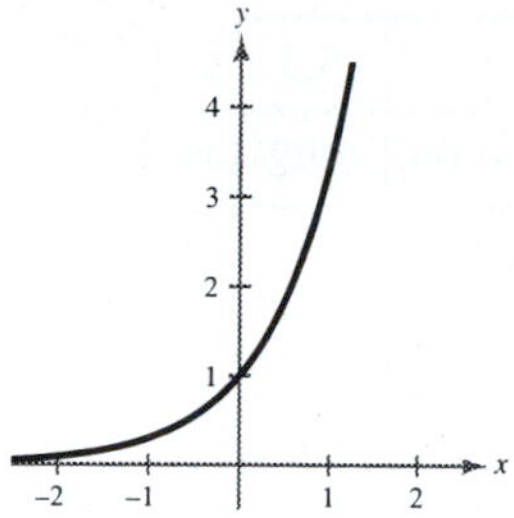

10. $h(x) = 2 - e^{-x/2}$

x	-2	-1	0	1	2
y	-0.72	0.35	1	1.39	1.63

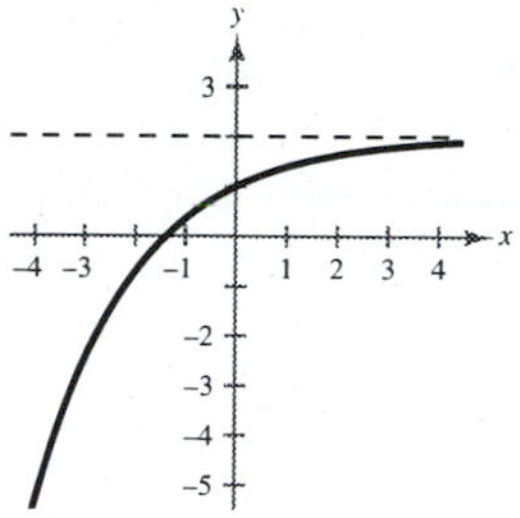

12. $s(t) = 4e^{-2/t}, t > 0$

t	$\frac{1}{2}$	1	2	3	4
y	0.07	0.54	1.47	2.05	2.43

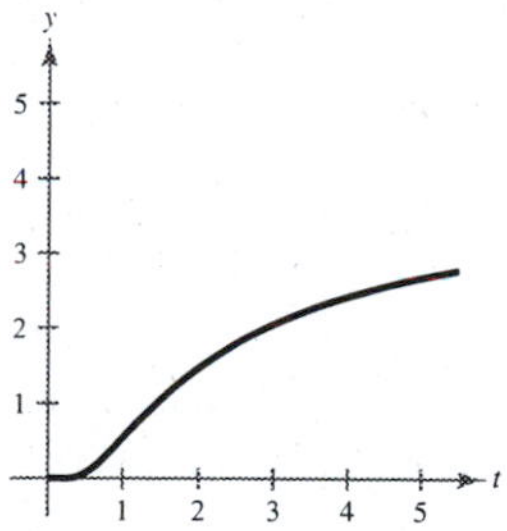

14. $f(x) = \dfrac{10}{1 + 2^{-0.05x}}$

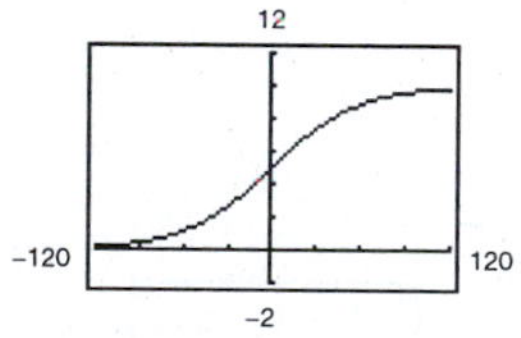

As $x \to +\infty, f(x) \to 10$, so $y = 10$ is an asymptote.

As $x \to -\infty, f(x) \to 0$, so $y = 0$ is an asymptote.

16. $A = 2000\left(1 + \dfrac{0.12}{n}\right)^{30n}$ or $A = 2000e^{(0.12)(30)}$

n	1	2	4	12	365	Continuous
A	\$59,919.84	\$65,975.38	\$60,421.97	\$71,899.28	\$73,153.17	\$73,196.47

18. $200{,}000 = P\left(1 + \dfrac{0.10}{12}\right)^{12t}$

$$P = \frac{200{,}000}{\left(1 + \dfrac{0.10}{12}\right)^{12t}}$$

t	1	10	20	30	40	50
P	\$181.042.49	\$73,881.39	\$27,292.30	\$10,081.97	\$3724.35	\$1375.80

20. $V(t) = 14{,}000\left(\frac{3}{4}\right)^t$

(a) 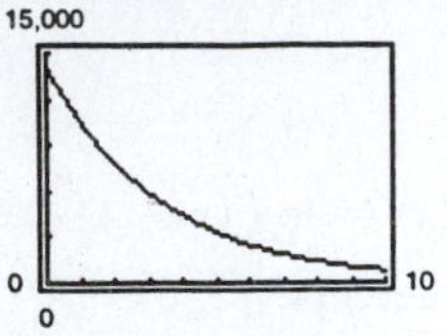

(b) $V(2) = 14{,}000\left(\frac{3}{4}\right)^2 = \$7875$

(c) According to the model, the car depreciates most rapidly at the beginning. Yes, this is realistic.

22. $y = 28e^{0.6-0.012s},\ s \geq 50$

Speed	50	55	60	65	70
Miles per gallon	28	26.4	24.8	23.4	22.0

24. $g(x) = \log_5 x \Rightarrow 5^y = x$

Domain: $(0, \infty)$

Vertical asymptote: $x = 0$

x	$\frac{1}{25}$	$\frac{1}{5}$	1	5	25
y	-2	-1	0	1	2

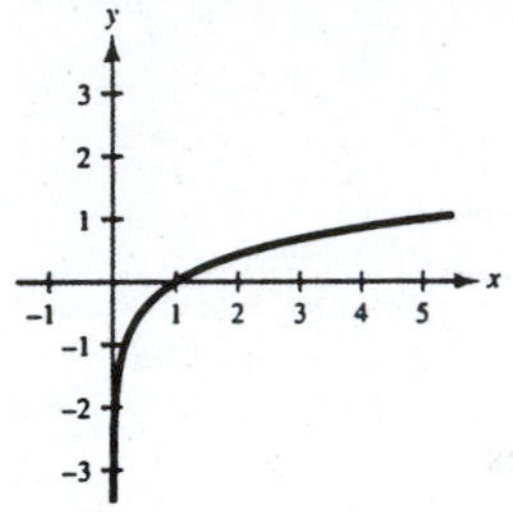

26. $f(x) = \ln(x - 3)$

Domain: $(3, \infty)$

Vertical asymptote: $x = 3$

x	3.5	4	4.5	5	5.5
y	-0.69	0	0.41	0.69	0.92

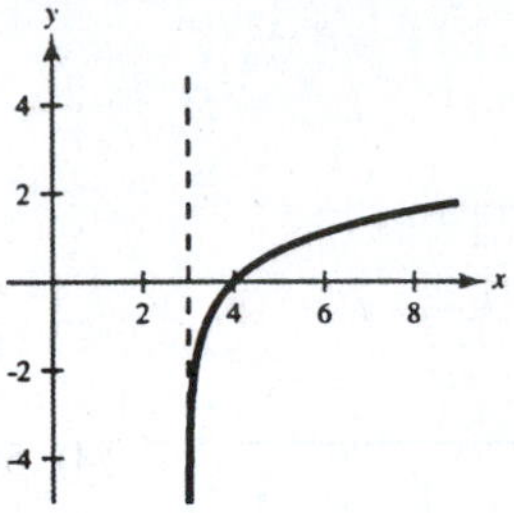

28. $f(x) = \frac{1}{4}\ln x$

Domain: $(0, \infty)$

Vertical asymptote: $x = 0$

x	$\frac{1}{2}$	1	$\frac{3}{2}$	2	$\frac{5}{2}$	3
y	-0.17	0	0.10	0.17	0.23	0.27

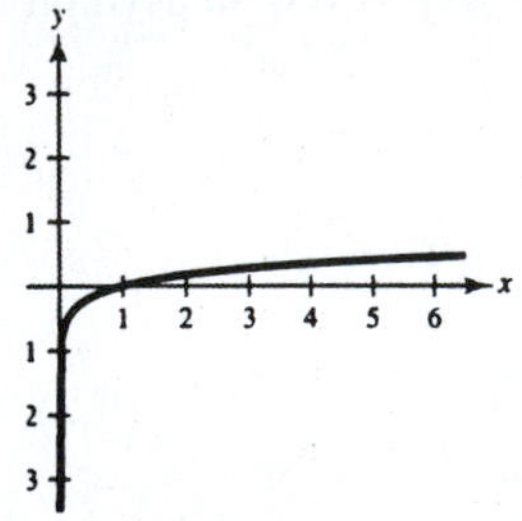

30. $y\sqrt{x}\ln(x + 1)$

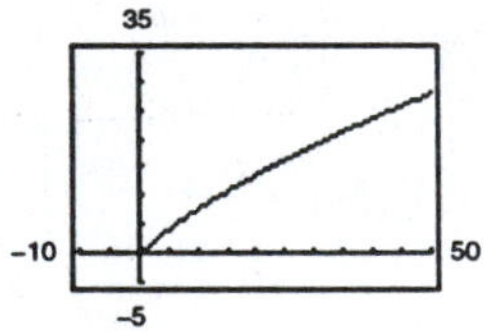

32. $25^{3/2} = 125$

$\log_{25} 125 = \dfrac{3}{2}$

34. $\log_9 3 = \log_9 9^{1/2} = \dfrac{1}{2}$

36. $\log_a \dfrac{1}{a} = \log_a a^{-1} = -1$

38. $\log_{1/2} 5 = \dfrac{\log_{10} 5}{\log_{10}(1/2)} \approx -2.322$

$\log_{1/2} 5 = \dfrac{\ln 5}{\ln(1/2)} \approx -2.322$

40. $\log_3 0.28 = \dfrac{\log_{10} 0.28}{\log_{10} 3} \approx -1.159$

$\log_3 0.28 = \dfrac{\ln 0.28}{\ln 3} \approx -1.159$

42. $\log_7 \dfrac{\sqrt{x}}{4} = \log_7 \sqrt{x} - \log_7 4$

$= \log_7 x^{1/2} - \log^7 4$

$= \dfrac{1}{2}\log_7 x - \log_7 4$

44. $\ln\left|\dfrac{x-1}{x+1}\right| = \ln|x-1| - \ln|x+1|$

46. $\log_6 y - 2\log_6 z = \log_6 y - \log_6 z^2$

$= \log_6 \dfrac{y}{z^2}$

48. $5\ln|x-2| - \ln|x+2| - 3\ln|x| = \ln|x-2|^5 - \ln|x+2| - \ln|x|^3$

$= \ln\left|\dfrac{(x-2)^5}{(x+2)x^3}\right|$

50. $e^{x-1} = e^x \cdot e^{-1} = \dfrac{e^x}{e}$

True (by properties of exponents).

52. $\ln(x+y) = \ln(x \cdot y)$ False

$\ln(x \cdot y) = \ln x + \ln y \neq \ln(x+y)$

54. The domain of the function $f(x) = \ln x$ is the set of all real numbers.

False; the domain of $f(x) = \ln x$ is $(0, \infty)$.

56. $t = 50\log_{10} \dfrac{18{,}000}{18{,}000 - h}$

(a) Domain: $0 \le x < 18{,}000$

(b)

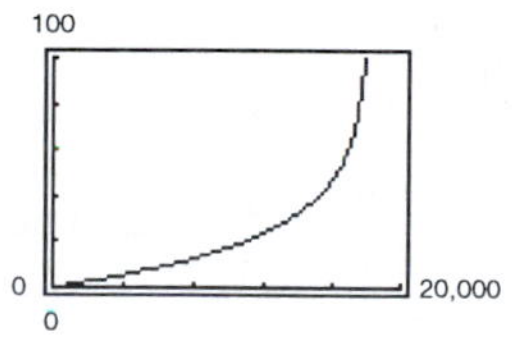

Vertical asymptote: $x = 18{,}000$

(c) As the plane approaches its absolute ceiling, it climbs at a slower rate.

(d) $50\log_{10} \dfrac{18{,}000}{18{,}000 - 4000} \approx 5.46$ minutes

58.
$$\begin{aligned} e^{3x} &= 25 \\ \ln e^{3x} &= \ln 25 \\ 3x &= \ln 25 \\ x &= \frac{\ln 25}{3} \approx 1.073 \end{aligned}$$

60.
$$\begin{aligned} 14e^{3x+2} &= 560 \\ e^{3x+2} &= 40 \\ \ln e^{3x+2} &= \ln 40 \\ 3x + 2 &= \ln 40 \\ x &= \frac{(\ln 40) - 2}{3} \approx 0.563 \end{aligned}$$

62.
$$e^{2x} - 6e^x + 8 = 0$$
$$(e^x - 4)(e^x - 2) = 0$$
$e^x = 4$ or $e^x = 2$

$x = \ln 4$ $\quad x = \ln 2$

$x \approx 1.386$ $\quad x \approx 0.693$

64.
$$\begin{aligned} 2\ln 4x &= 15 \\ \ln 4x &= \tfrac{15}{2} \\ 4x &= e^{7.5} \\ x &= \tfrac{1}{4}e^{7.5} \approx 452.011 \end{aligned}$$

66.
$$\begin{aligned} \ln\sqrt{x+1} &= 2 \\ \tfrac{1}{2}\ln(x+1) &= 2 \\ \ln(x+1) &= 4 \\ x + 1 &= e^4 \\ x &= e^4 - 1 \approx 53.598 \end{aligned}$$

68.
$$\begin{aligned} \log(1 - x) &= -1 \\ 1 - x &= 10^{-1} \\ 1 - \tfrac{1}{10} &= x \\ x &= \tfrac{9}{10} \end{aligned}$$

70. $25e^{-0.3x} = 12$

$25e^{-0.3x} - 12 = 0$

Graph $y_1 = 25e^{-0.3x} - 12$.

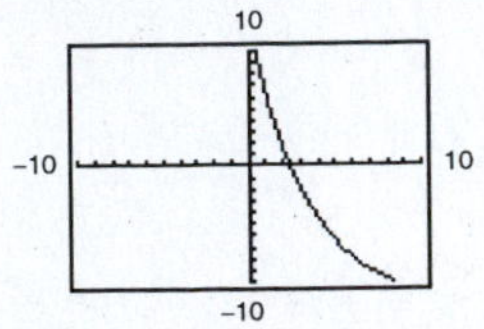

The x-intercept is at $x \approx 2.45$.

72. $6 \log_{10}(x^2 + 1) - x = 0$

Graph

$y_1 = 6 \log_{10}(x^2 + 1) - x$.

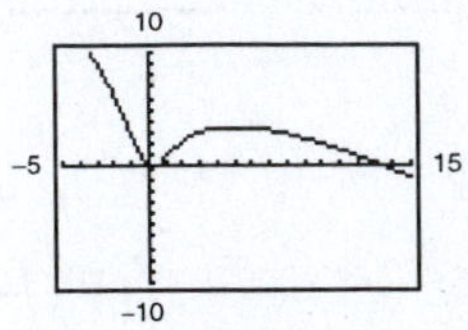

The x-intercepts are at $x = 0$, $x = 0.416$, and $x \approx 13.63$.

74.
$$y = ae^{bx}$$
$$\tfrac{1}{2} = ae^{b(0)} \Rightarrow a = \tfrac{1}{2}$$
$$5 = \tfrac{1}{2}e^{b(5)}$$
$$10 = e^{5b}$$
$$\ln 10 = 5b \Rightarrow b \approx 0.4605$$
Thus, $y = \tfrac{1}{2}e^{0.4605x}$.

76. $N = \dfrac{157}{1 + 5.4e^{-0.12t}}$

(a) When $N = 50$:
$$50 = \frac{157}{1 + 5.4e^{-0.12t}}$$
$$1 + 5.4e^{-0.12t} = \frac{157}{50}$$
$$5.4e^{-0.12t} = \frac{107}{50}$$
$$e^{-0.12t} = \frac{107}{270}$$
$$-0.12t = \ln \frac{107}{270}$$
$$t = \frac{\ln(107/270)}{-0.12} \approx 7.7 \text{ weeks}$$

(b) When $N = 75$:
$$75 = \frac{157}{1 + 5.4e^{-0.12t}}$$
$$1 + 5.4e^{-0.12t} = \frac{157}{75}$$
$$5.4e^{-0.12t} = \frac{82}{75}$$
$$e^{-0.12t} = \frac{82}{405}$$
$$-0.12^t = \ln \frac{82}{405}$$
$$t = \frac{\ln(82/405)}{-0.12} \approx 13.3 \text{ weeks}$$

78.
$$\beta = 10 \log_{10}\left(\frac{I}{10^{-16}}\right)$$
$$125 = 10 \log_{10}\left(\frac{I}{10^{-16}}\right)$$
$$12.5 = \log_{10}\left(\frac{I}{10^{-16}}\right)$$
$$10^{12.5} = \frac{I}{10^{-16}}$$
$$I = 10^{-3.5} \text{ watts/cm}^2$$

CHAPTER 4
Trigonometry

CHAPTER 4
Trigonometry

Section 4.1 Radian and Degree Measure

Solutions to Even-Numbered Exercises

2. 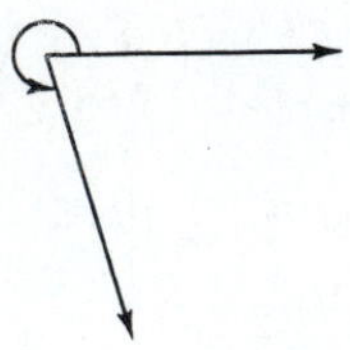

The angle shown is approximately 5 radians.

4. 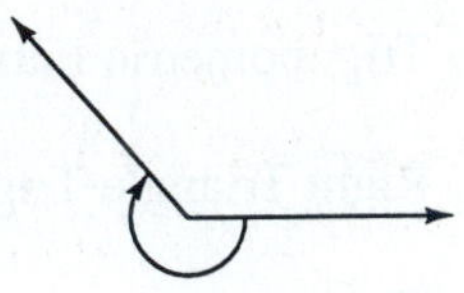

The angle shown is approximately -4 radians.

6. (a) Since $\pi < \frac{5\pi}{4} < \frac{3\pi}{2}$; $\frac{5\pi}{4}$ lies in Quadrant III.

(b) Since $\frac{3\pi}{2} < \frac{7\pi}{4} < 2\pi$, $\frac{7\pi}{4}$ lies in Quadrant IV.

8. (a) Since $-\frac{\pi}{2} < -1 < 0$; -1 lies in Quadrant IV.

(b) Since $-\pi < -2 < -\frac{\pi}{2}$; -2 lies in Quadrant III.

10. (a) Since $\frac{3\pi}{2} < 5.63 < 2\pi$; 5.63 lies in Quadrant IV.

(b) Since $-\pi < -2.25 < -\frac{\pi}{2}$; -2.25 lies in Quadrant III.

12. (a) $-\frac{7\pi}{4}$

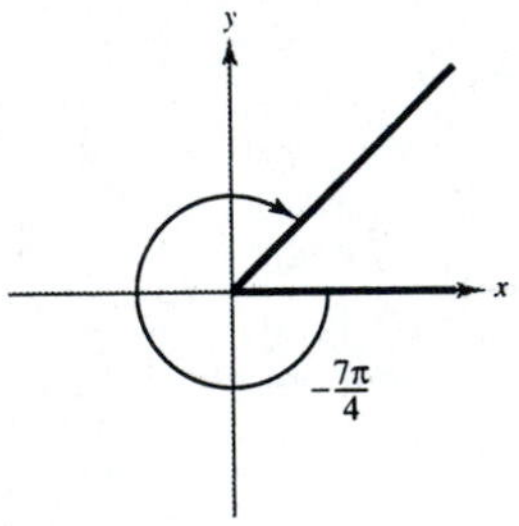

(b) $-\frac{5\pi}{2}$

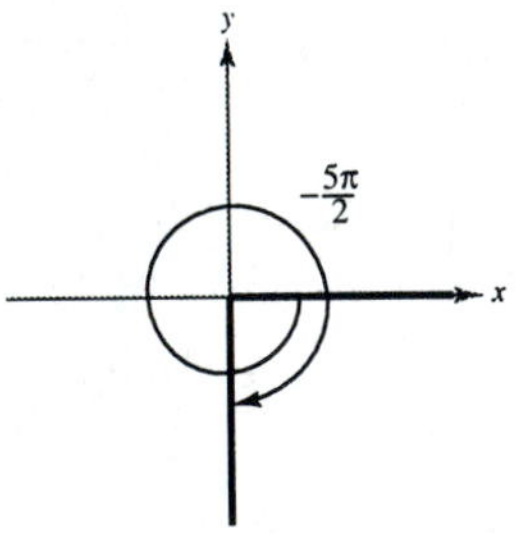

14. (a) 4

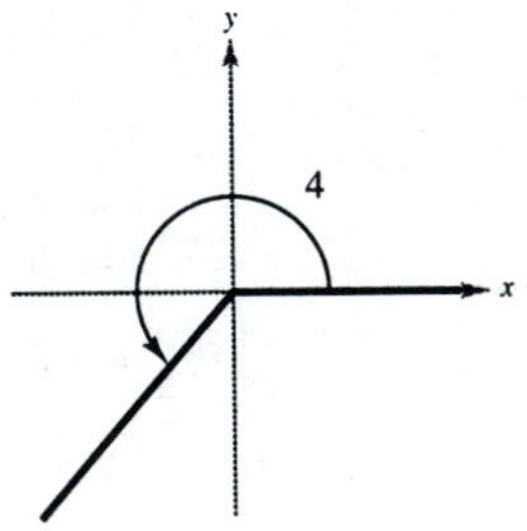

(b) -3

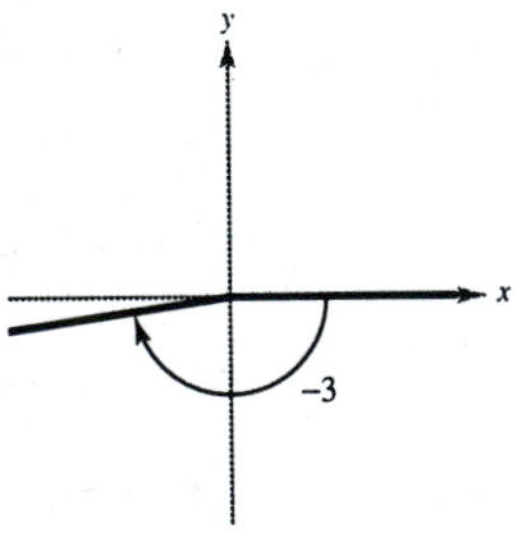

16. (a) $\frac{7\pi}{6} + 2\pi = \frac{19\pi}{6}$

$\frac{7\pi}{6} - 2\pi = -\frac{5\pi}{6}$

(b) $-\frac{11\pi}{6} + 2\pi = \frac{\pi}{6}$

$-\frac{11\pi}{6} - 2\pi = -\frac{23\pi}{6}$

18. (a) $\frac{8\pi}{9} + 2\pi = \frac{26\pi}{9}$

$\frac{8\pi}{9} - 2\pi = -\frac{10\pi}{9}$

(b) $\frac{8\pi}{45} + 2\pi = \frac{98\pi}{45}$

$\frac{8\pi}{45} - 2\pi = -\frac{82\pi}{45}$

20. (a) Complement: $\frac{\pi}{2} - 1 \approx 0.57$

Supplement: $\pi - 1 \approx 2.14$

(b) Complement: none $\left(2 > \frac{\pi}{2}\right)$

Supplement: $\pi - 2 \approx 1.14$

22.

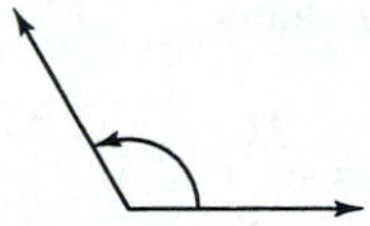

The angle shown is approximately 120°.

24.

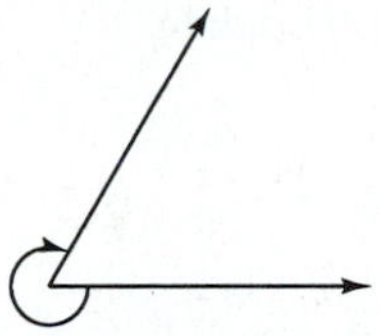

The angle shown is approximately −300°.

26. (a) Since $0° < 8.3° < 90°$; 8.3° lies in Quadrant I.

(b) Since $180° < 257°\ 30' < 270°$; 257° 30′ lies in Quadrant III.

28. (a) Since $-270° < -260° < -180°$; −260° lies in Quadrant II.

(b) Since $-90° < -3.4° < 0°$; −3.4° lies in Quadrant IV.

30. (a) −270°

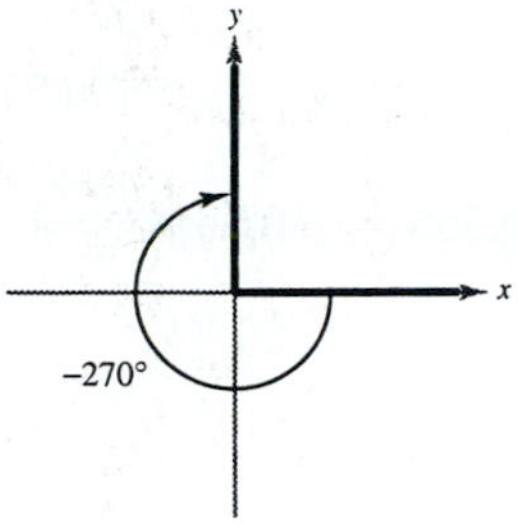

(b) −120°

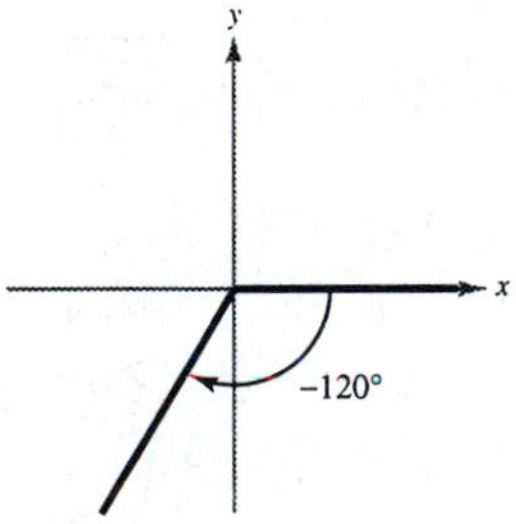

32. (a) 750°

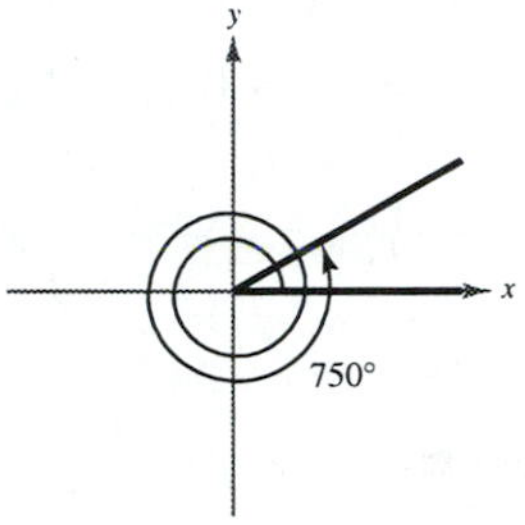

(b) −600°

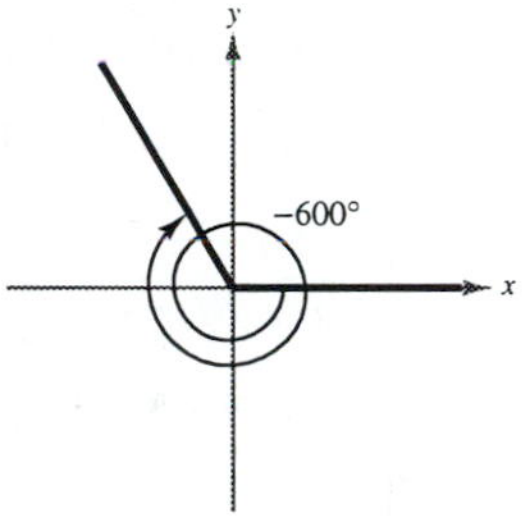

34. (a) $120° + 360° = 480°$

$120° - 360° = -240°$

(b) $-390° + 720° = 330°$

$-390° + 360° = -30°$

36. (a) $-420° + 720° = 300°$

$-420° + 360° = -60°$

(b) $230° - 360° = -130°$

$230° + 360° = 590°$

38. (a) Complement: $90° - 79° = 11°$

Supplement: $180° - 79° = 101°$

(b) Complement: none $(150° > 90°)$

Supplement: $180° - 150° = 30°$

40. (a) $315° = 315°\left(\frac{\pi}{180°}\right) = \frac{7\pi}{4}$

(b) $120° = 120°\left(\frac{\pi}{180°}\right) = \frac{2\pi}{3}$

42. (a) $-270° = -270°\left(\frac{\pi}{180°}\right) = -\frac{3\pi}{2}$

(b) $144° = 144°\left(\frac{\pi}{180°}\right) = \frac{4\pi}{5}$

44. (a) $-\frac{7\pi}{12} = -\frac{7\pi}{12}\left(\frac{180°}{\pi}\right) = -105°$

(b) $\frac{\pi}{9} = \frac{\pi}{9}\left(\frac{180°}{\pi}\right) = 20°$

46. (a) $\frac{11\pi}{6} = \frac{11\pi}{6}\left(\frac{180°}{\pi}\right) = 330°$

(b) $\frac{34\pi}{15} = \frac{34\pi}{15}\left(\frac{180°}{\pi}\right) = 408°$

48. $87.4° = 87.4°\left(\frac{\pi}{180°}\right) \approx 1.525$ radians

50. $-48.27° = -48.27°\left(\frac{\pi}{180°}\right) \approx -0.842$ radian

52. $0.54° = 0.54°\left(\frac{\pi}{180°}\right) \approx 0.009$ radian

54. $345° = 345°\left(\frac{\pi}{180°}\right) \approx 6.021$ radians

56. $\frac{5\pi}{11} = \frac{5\pi}{11}\left(\frac{180°}{\pi}\right) \approx 81.818°$

58. $6.5\pi = 6.5\pi\left(\frac{180°}{\pi}\right) = 1170°$

60. $4.8 = 4.8\left(\frac{180°}{\pi}\right) \approx 275.020°$

62. $-0.57 = -0.57\left(\frac{180°}{\pi}\right) \approx -32.659°$

64. (a) $245°\,10' = 245° + \left(\frac{10}{60}\right)°$
$\approx 245° + 0.167°$
$= 245.167°$

(b) $2°\,12' = 2° + \left(\frac{12}{60}\right)°$
$= 2° + 0.2° = 2.2°$

66. (a) $-135°\,36'' = -135° - \left(\frac{36}{3600}\right)°$
$= -135° - 0.01° = -135.01°$

(b) $-408°\,16'20'' = -\left(408° + \left(\frac{16}{60}\right)° + \left(\frac{20}{3600}\right)°\right)$
$\approx -(408° + 0.267° + 0.006°)$
$= -408.272°$

68. (a) $-345.12° = -(345° + (0.12)(60'))$
$= -(345° + 7' + 0.2(60''))$
$= -345°\,7'\,12''$

(b) $0.45 = 0.45\left(\frac{180}{\pi}\right)°$
$\approx 25.7831°$
$= 25° + (0.7831)(60')$
$\approx 25° + 46' + (0.986)(60'')$
$\approx 25°\,46'\,59''$

70. (a) $-0.355 = -0.355\left(\frac{180}{\pi}\right)°$
≈ -20.34
$= -(20° + (0.34)(60'))$
$= -(20° + 20' + 0.4(60''))$
$= -20°\,20'\,24''$

(b) $0.7865 = 0.7865\left(\frac{180}{\pi}\right)°$
≈ 45.0631
$= 45° + (0.0631)(60')$
$= 45° + 3' + 0.786(60'')$
$\approx 45°\,3'\,47''$

72. $S = r\theta$
$31 = 12\theta$
$\theta = \frac{31}{12} = 2\frac{7}{12}$ radians

74. $S = r\theta$
$60 = 75\theta$
$\theta = \frac{60}{75} = \frac{4}{5}$ radian

Because the angle represented is clockwise, this angle is $-\frac{4}{5}$ radian.

76. $r = 16$ feet, $s = 10$ feet

$\theta = \frac{s}{r} = \frac{10}{16} = \frac{5}{8}$ radian

78. $r = 80$ kilometers, $s = 160$ kilometers

$\theta = \frac{s}{r} = \frac{160}{80} = 2$ radians

80. $r = 9$ feet, $\theta = 60° = \frac{\pi}{3}$

$s = r\theta = 9\left(\frac{\pi}{3}\right) = 3\pi$ feet

82. $r = 40$ centimeters, $\theta = \frac{3\pi}{4}$

$s = r\theta = 40\left(\frac{3\pi}{4}\right) = 30\pi$ centimeters

84. $r = 4000$ miles

$\theta = 47°\,36'\,32'' - 37°\,46'\,39'' = 9°\,49'\,53''$

≈ 0.1716 radian

$s = r\theta \approx (4000)(0.1716) \approx 686.4$ miles

86. $r = 4000$ miles

$\theta = 31°\,47' + 26°\,10' = 57°\,57'$

≈ 1.0114 radians

$s = r\theta \approx (4000)(1.0114) \approx 4045.7$ miles

88. $r = 6378$ kilometers, $s = 800$ kilometers

$$\theta = \frac{s}{r} = \frac{800}{6378} \approx 0.125 \text{ radian} = 0.125\left(\frac{180°}{\pi}\right) \approx 7.19°$$

90. $\theta = \frac{s}{r} = \frac{12}{5} = 2.4 \text{ radians} = 2.4\left(\frac{180°}{\pi}\right) \approx 137.5°$

92. Linear velocity for either pulley: $1700(2\pi) = 3400\pi$ in/min

(a) Angular speed of motor pulley: $\omega = \frac{v}{r} = \frac{3400\pi}{1} = 3400\pi$ rad/min

Angular speed of the saw arbor: $\omega = \frac{v}{r} = \frac{3400\pi}{2} = 1700\pi$ rad/min

(b) Revolutions per minute of the saw arbor: $\frac{1700\pi}{2\pi} = 850$ rev/min

94. If θ is constant, the length of the arc is proportional to the radius ($s = r\theta$).

96. (a) Arc length of larger sprocket in feet:

$s = r\theta$

$s = \frac{1}{3}(2\pi) = \frac{2\pi}{3}$ feet

Therefore, the chain moves $2\pi/3$ feet as does the smaller rear sprocket. Thus, the angle θ of the smaller sprocket is ($r = 2$ inches $= 2/12$ feet).

$$\theta = \frac{s}{r} = \frac{(2\pi/3)\text{ ft}}{(2/12)\text{ ft}} = 4\pi$$

and the arc length of the tire in feet is:

$s = \theta r$

$s = (4\pi)\left(\frac{14}{12}\right) = \frac{14\pi}{3}$ feet

$$\text{Speed} = \frac{s}{t} = \frac{14\pi/3}{1 \text{ sec}} = \frac{14\pi}{3} \text{ feet per second}$$

(b) $\frac{14\pi \text{ feet}}{3 \text{ seconds}} \times \frac{3600 \text{ seconds}}{1 \text{ hour}} \times \frac{1 \text{ mile}}{5280 \text{ feet}} \approx 10$ miles per hour

98. $s = 1.2$ miles, $r = 4000$ miles

$$\theta = \frac{s}{r} = \frac{1.2}{4000} = 0.0003 \text{ radians} = 0.0003\left(\frac{180°}{\pi}\right) \approx 0.0172°$$

Section 4.2 Trigonometric Functions: The Unit Circle

Solutions to Even-Numbered Exercises

2. $(x, y) = \left(\frac{12}{13}, \frac{5}{13}\right)$

$\sin t = y = \frac{5}{13}$

$\cos t = x = \frac{12}{13}$

$\tan t = \frac{y}{x} = \frac{5/13}{12/13} = \frac{5}{12}$

$\csc t = \frac{1}{y} = \frac{1}{5/13} = \frac{13}{5}$

$\sec t = \frac{1}{x} = \frac{1}{12/13} = \frac{13}{12}$

$\cot t = \frac{x}{y} = \frac{12/13}{5/13} = \frac{12}{5}$

4. $(x, y) = \left(-\frac{4}{5}, -\frac{3}{5}\right)$

$\sin t = y = -\frac{3}{5}$

$\cos t = x = -\frac{4}{5}$

$\tan t = \frac{y}{x} = \frac{-3/5}{-4/5} = \frac{3}{4}$

$\csc t = \frac{1}{y} = \frac{1}{-3/5} = -\frac{5}{3}$

$\sec t = \frac{1}{x} = \frac{1}{-4/5} = -\frac{5}{4}$

$\cot t = \frac{x}{y} = \frac{-4/5}{-3/5} = \frac{4}{3}$

6. $t = \frac{\pi}{3} \Rightarrow \left(\frac{1}{2}, \frac{\sqrt{3}}{2}\right)$

8. $t = \frac{5\pi}{4} \Rightarrow \left(-\frac{\sqrt{2}}{2}, -\frac{\sqrt{2}}{2}\right)$

10. $t = \frac{11\pi}{6} \Rightarrow \left(\frac{\sqrt{3}}{2}, -\frac{1}{2}\right)$

12. $t = \pi \Rightarrow (-1, 0)$

14. $t = -\frac{\pi}{4}$ corresponds to the point:

$(x, y) = \left(\frac{\sqrt{2}}{2}, -\frac{\sqrt{2}}{2}\right)$

$\sin\left(-\frac{\pi}{4}\right) = y = -\frac{\sqrt{2}}{2}$

$\cos\left(-\frac{\pi}{4}\right) = x = \frac{\sqrt{2}}{2}$

$\tan\left(-\frac{\pi}{4}\right) = \frac{y}{x} = \frac{-\sqrt{2}/2}{\sqrt{2}/2} = -1$

16. $t = \frac{\pi}{3}$ corresponds to the point:

$(x, y) = \left(\frac{1}{2}, \frac{\sqrt{3}}{2}\right)$

$\sin \frac{\pi}{3} = y = \frac{\sqrt{3}}{2}$

$\cos \frac{\pi}{3} = x = \frac{1}{2}$

$\tan \frac{\pi}{3} = \frac{y}{x} = \frac{\sqrt{3}/2}{1/2} = \sqrt{3}$

18. $t = -\frac{5\pi}{6}$ corresponds to the point:

$(x, y) = \left(-\frac{\sqrt{3}}{2}, -\frac{1}{2}\right)$

$\sin\left(-\frac{5\pi}{6}\right) = y = -\frac{1}{2}$

$\cos\left(-\frac{5\pi}{6}\right) = x = \frac{\sqrt{3}}{2}$

$\tan\left(-\frac{5\pi}{6}\right) = \frac{y}{x} = \frac{-1/2}{-\sqrt{3}/2} = \frac{1}{\sqrt{3}}$

20. $t = \frac{2\pi}{3}$ corresponds to the point:

$(x, y) = \left(-\frac{1}{2}, \frac{\sqrt{3}}{2}\right)$

$\sin \frac{2\pi}{3} = y = \frac{\sqrt{3}}{2}$

$\cos \frac{2\pi}{3} = x = -\frac{1}{2}$

$\tan \frac{2\pi}{3} = \frac{y}{x} = \frac{\sqrt{3}/2}{-1/2} = -\sqrt{3}$

22. $t = \frac{7\pi}{4}$ corresponds to the point:

$$(x, y) = \left(\frac{\sqrt{2}}{2}, -\frac{\sqrt{2}}{2}\right)$$

$$\sin\frac{7\pi}{4} = y = -\frac{\sqrt{2}}{2}$$

$$\cos\frac{7\pi}{4} = x = \frac{\sqrt{2}}{2}$$

$$\tan\frac{7\pi}{4} = \frac{y}{x} = \frac{-\sqrt{2}/2}{\sqrt{2}/2} = -1$$

24. $t = -2\pi$ corresponds to the point: $(x, y) = (1, 0)$

$$\sin(-2\pi) = y = 0$$

$$\cos(-2\pi) = x = 1$$

$$\tan(-2\pi) = \frac{y}{x} = \frac{0}{1} = 0$$

26. $t = -\frac{2\pi}{3}$ corresponds to the point:

$$(x, y) = \left(-\frac{1}{2}, -\frac{\sqrt{3}}{2}\right)$$

$$\sin\left(-\frac{2\pi}{3}\right) = y = -\frac{\sqrt{3}}{2}$$

$$\cos\left(-\frac{2\pi}{3}\right) = x = -\frac{1}{2}$$

$$\tan\left(-\frac{2\pi}{3}\right) = \frac{y}{x} = \frac{-\sqrt{3}/2}{-1/2} = \sqrt{3}$$

$$\csc\left(-\frac{2\pi}{3}\right) = \frac{1}{y} = \frac{1}{-\sqrt{3}/2} = -\frac{2\sqrt{3}}{3}$$

$$\sec\left(-\frac{2\pi}{3}\right) = \frac{1}{x} = \frac{1}{-1/2} = -2$$

$$\cot\left(-\frac{2\pi}{3}\right) = \frac{x}{y} = \frac{-1/2}{-\sqrt{3}/2} = \frac{\sqrt{3}}{3}$$

28. $t = \frac{3\pi}{2}$ corresponds to the point: $(x, y) = (0, -1)$

$$\sin\frac{3\pi}{2} = y = -1$$

$$\cos\frac{3\pi}{2} = x = 0$$

$$\tan\frac{3\pi}{2} = \frac{y}{x} = \frac{-1}{0} \Rightarrow \text{undefined}$$

$$\csc\frac{3\pi}{2} = \frac{1}{y} = \frac{1}{-1} = -1$$

$$\sec\frac{3\pi}{2} = \frac{1}{x} = \frac{1}{0} \Rightarrow \text{undefined}$$

$$\cot\frac{3\pi}{2} = \frac{x}{y} = \frac{0}{-1} = 0$$

30. $t = -\frac{11\pi}{6}$ corresponds to the point:

$$(x, y) = \left(\frac{\sqrt{3}}{2}, \frac{1}{2}\right)$$

$$\sin\left(-\frac{11\pi}{6}\right) = y = \frac{1}{2}$$

$$\cos\left(-\frac{11\pi}{6}\right) = x = \frac{\sqrt{3}}{2}$$

$$\tan\left(-\frac{11\pi}{6}\right) = \frac{y}{x} = \frac{1/2}{\sqrt{3}/2} = \frac{\sqrt{3}}{3}$$

$$\csc\left(-\frac{11\pi}{6}\right) = \frac{1}{y} = \frac{1}{1/2} = 2$$

$$\sec\left(-\frac{11\pi}{6}\right) = \frac{1}{x} = \frac{1}{\sqrt{3}/2} = \frac{2\sqrt{3}}{3}$$

$$\cot\left(-\frac{11\pi}{6}\right) = \frac{x}{y} = \frac{\sqrt{3}/2}{1/2} = \sqrt{3}$$

32. Because $3\pi = 2\pi + \pi$:

$$\cos 3\pi = \cos(2\pi + \pi) = \cos \pi = -1$$

34. Because $\frac{9\pi}{4} = 2\pi + \frac{\pi}{4}$:

$$\sin\frac{9\pi}{4} = \sin\left(2\pi + \frac{\pi}{4}\right) = \sin\frac{\pi}{4} = \frac{\sqrt{2}}{2}$$

36. Because $-\frac{13\pi}{6} = -2\pi - \frac{\pi}{6}$:

$$\sin\left(-\frac{13\pi}{6}\right) = \sin\left(-2\pi - \frac{\pi}{6}\right) = \sin\left(-\frac{\pi}{6}\right) = -\sin\left(\frac{\pi}{6}\right) = -\frac{1}{2}$$

38. Because $-\frac{8\pi}{3} = -4\pi + \frac{4\pi}{3}$:

$$\cos\left(-\frac{8\pi}{3}\right) = \cos\left(-4\pi + \frac{4\pi}{3}\right) = \cos\frac{4\pi}{3} = -\frac{1}{2}$$

40. $\sin(-t) = \frac{2}{5}$

(a) $\sin t = -\sin(-t) = -\frac{2}{5}$

(b) $\csc t = \frac{1}{\sin(t)} = \frac{1}{-\sin(-t)} = -\frac{5}{2}$

42. $\cos t = -\frac{3}{4}$

(a) $\cos(-t) = \cos t = -\frac{3}{4}$

(b) $\sec(-t) = \frac{1}{\cos(-t)} = \frac{1}{\cos t} = -\frac{4}{3}$

44. $\cos t = \frac{4}{5}$

(a) $\cos(\pi - t) = -\cos t = -\frac{4}{5}$

(b) $\cos(t + \pi) = -\cos t = -\frac{4}{5}$

46. $\tan \pi = 0$

48. $\cos 1 = \frac{1}{\tan 1} \approx 0.6421$

50. $\csc 2.3 = \frac{1}{\sin 2.3} \approx 1.3410$

52. $\sec 1.8 = \frac{1}{\cos 1.8} \approx -4.4014$

54. $\sin(-0.9) \approx -0.7833$

56. (a) $\sin 0.75 = y \approx 0.7$

(b) $\cos 2.5 = x \approx -0.8$

58. (a) $\sin t = -0.75$

$t \approx 4.0$ or $t \approx 5.4$

(b) $\cos t = 0.75$

$t \approx 0.72$ or $t \approx 5.56$

60. $\sin 0.25 \approx 0.2474$

$\sin 0.75 \approx 0.6816$

$\sin 1 \approx 0.8415$

$\sin(0.25 + 0.75) = \sin 1 \neq \sin 0.25 + \sin 0.75$

because, $0.8415 \neq 0.2474 + 0.6816 = 0.9290$.

62. (a) The points (x_1, y_1) and (x_2, y_2) are symmetric about the origin.

(b) Because of the symmetry of the points, you can make the conjecture that $\sin(t_1 - \pi) = -\sin t_1$.

(c) Because of the symmetry of the points, you can make the conjecture that $\cos(t_1 - \pi) = -\cos t_1$.

64. $y(t) = \frac{1}{4}e^{-t} \cos 6t$

(a) When $t = 0$: $y(0) = \frac{1}{4}e^{-0} \cos 0 = 0.2500$ foot

(b) When $t = \frac{1}{4}$: $y\left(\frac{1}{4}\right) = \frac{1}{4}e^{-1/4} \cos\left(6 \cdot \frac{1}{4}\right) \approx 0.0138$ foot

(c) When $t = \frac{1}{2}$: $y\left(\frac{1}{2}\right) = \frac{1}{2}e^{-1/2} \cos\left(6 \cdot \frac{1}{2}\right) \approx -0.1501$ foot

66. $\cos\theta = x = \cos(-\theta)$

$\sin\theta = \dfrac{1}{x} = \sec(-\theta)$

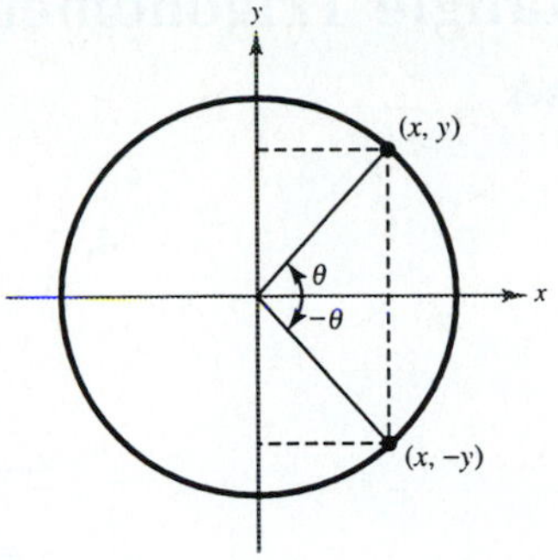

$$\sin\theta = y$$
$$\sin(-\theta) = -y = -\sin\theta$$
$$\sec\theta = \frac{1}{y}$$
$$\sec(-\theta) = -\frac{1}{y} = -\sec\theta$$
$$\tan\theta = \frac{y}{x}$$
$$\tan(-\theta) = \frac{-y}{x} = -\tan\theta$$
$$\cot\theta = \frac{x}{y}$$
$$\cot(-\theta) = \frac{x}{-y} = -\cot\theta$$

68. $f(t) = \sin t$ and $g(t) = \tan t$

Both f and g are odd functions.

$h(t) = f(t)g(t) = \sin t \tan t$

$$h(-t) = \sin(-t)\tan(-t)$$
$$= (-\sin t)(-\tan t)$$
$$= \sin t \tan t = h(t)$$

The function $h(t) = f(t)g(t)$ is even.

70. $f(x) = \frac{1}{4}x^3 + 1$

$$y = \tfrac{1}{4}x^3 + 1$$
$$x = \tfrac{1}{4}y^3 + 1$$
$$x - 1 = \tfrac{1}{4}y^3$$
$$4(x-1) = y^3$$
$$y = \sqrt[3]{4(x-1)}$$
$$f^{-1}(x) = \sqrt[3]{4(x-1)}$$

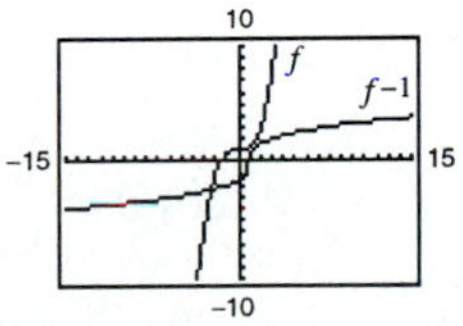

72. $f(x) = \dfrac{2x}{x+1},\ x > -1$

$$y = \frac{2x}{x+1},\ x > -1$$
$$x = \frac{2y}{y+1}$$
$$xy + x = 2y$$
$$x = 2y - xy$$
$$x = y(2 - x)$$
$$\frac{x}{2-x} = y,\ x < 2$$
$$f^{-1}(x) = \frac{x}{2-x},\ x < 2$$

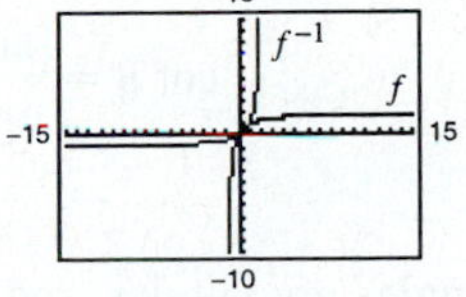

Section 4.3 Right Triangle Trigonometry

Solutions to Even-Numbered Exercises

2.

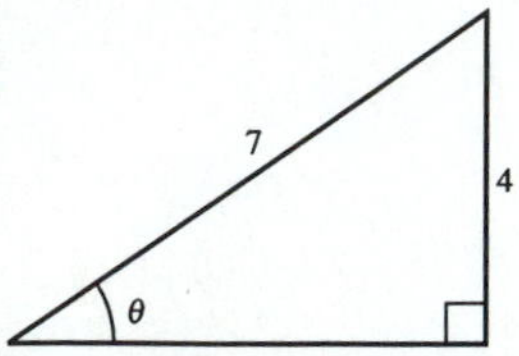

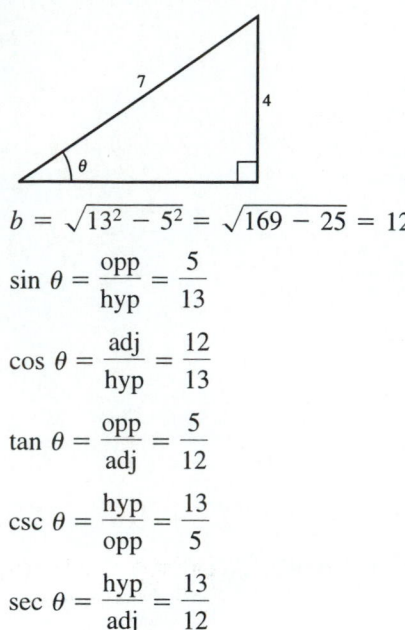

$b = \sqrt{13^2 - 5^2} = \sqrt{169 - 25} = 12$

$\sin\theta = \dfrac{\text{opp}}{\text{hyp}} = \dfrac{5}{13}$

$\cos\theta = \dfrac{\text{adj}}{\text{hyp}} = \dfrac{12}{13}$

$\tan\theta = \dfrac{\text{opp}}{\text{adj}} = \dfrac{5}{12}$

$\csc\theta = \dfrac{\text{hyp}}{\text{opp}} = \dfrac{13}{5}$

$\sec\theta = \dfrac{\text{hyp}}{\text{adj}} = \dfrac{13}{12}$

$\cot\theta = \dfrac{\text{adj}}{\text{opp}} = \dfrac{12}{5}$

4.

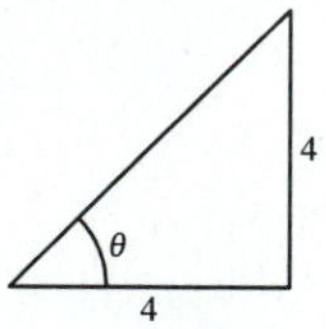

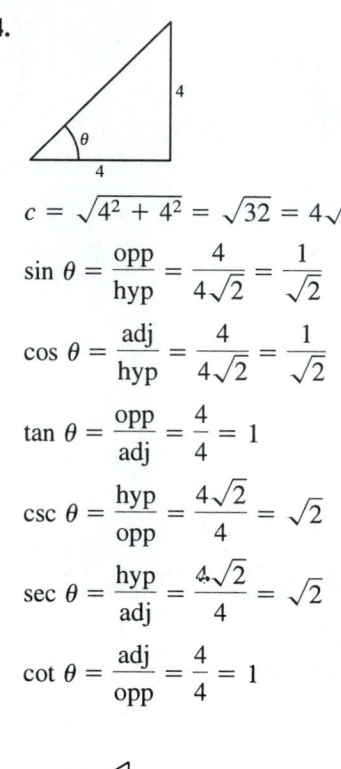

$c = \sqrt{4^2 + 4^2} = \sqrt{32} = 4\sqrt{2}$

$\sin\theta = \dfrac{\text{opp}}{\text{hyp}} = \dfrac{4}{4\sqrt{2}} = \dfrac{1}{\sqrt{2}}$

$\cos\theta = \dfrac{\text{adj}}{\text{hyp}} = \dfrac{4}{4\sqrt{2}} = \dfrac{1}{\sqrt{2}}$

$\tan\theta = \dfrac{\text{opp}}{\text{adj}} = \dfrac{4}{4} = 1$

$\csc\theta = \dfrac{\text{hyp}}{\text{opp}} = \dfrac{4\sqrt{2}}{4} = \sqrt{2}$

$\sec\theta = \dfrac{\text{hyp}}{\text{adj}} = \dfrac{4\sqrt{2}}{4} = \sqrt{2}$

$\cot\theta = \dfrac{\text{adj}}{\text{opp}} = \dfrac{4}{4} = 1$

6.

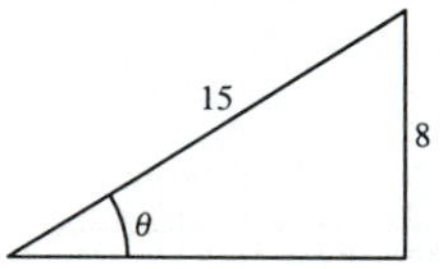

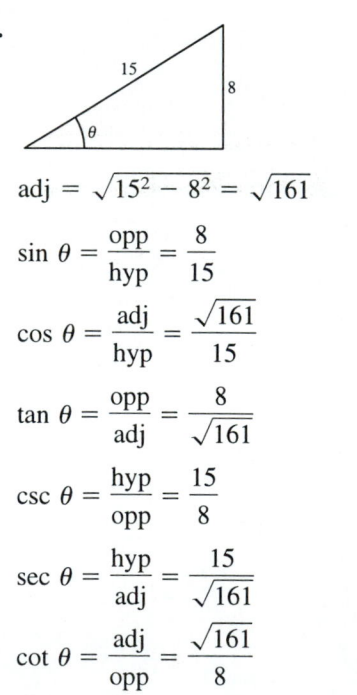

$\text{adj} = \sqrt{15^2 - 8^2} = \sqrt{161}$

$\sin\theta = \dfrac{\text{opp}}{\text{hyp}} = \dfrac{8}{15}$

$\cos\theta = \dfrac{\text{adj}}{\text{hyp}} = \dfrac{\sqrt{161}}{15}$

$\tan\theta = \dfrac{\text{opp}}{\text{adj}} = \dfrac{8}{\sqrt{161}}$

$\csc\theta = \dfrac{\text{hyp}}{\text{opp}} = \dfrac{15}{8}$

$\sec\theta = \dfrac{\text{hyp}}{\text{adj}} = \dfrac{15}{\sqrt{161}}$

$\cot\theta = \dfrac{\text{adj}}{\text{opp}} = \dfrac{\sqrt{161}}{8}$

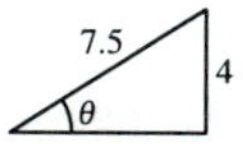

$\text{adj} = \sqrt{7.5^2 - 4^2} = \dfrac{\sqrt{161}}{2}$

$\sin\theta = \dfrac{\text{opp}}{\text{hyp}} = \dfrac{4}{7.5} = \dfrac{8}{15}$

$\cos\theta = \dfrac{\text{adj}}{\text{hyp}} = \dfrac{\sqrt{161}}{2 \cdot 7.5} = \dfrac{\sqrt{161}}{15}$

$\tan\theta = \dfrac{\text{opp}}{\text{adj}} = \dfrac{4}{(\sqrt{161}/2)} = \dfrac{8}{\sqrt{161}}$

$\csc\theta = \dfrac{\text{hyp}}{\text{opp}} = \dfrac{7.5}{4} = \dfrac{15}{8}$

$\sec\theta = \dfrac{\text{hyp}}{\text{adj}} = \dfrac{7.5}{(\sqrt{161}/2)} = \dfrac{15}{\sqrt{161}}$

$\cot\theta = \dfrac{\text{adj}}{\text{opp}} = \dfrac{\sqrt{161}}{2 \cdot 4} = \dfrac{\sqrt{161}}{8}$

The function values are the same because the triangles are similar, and corresponding sides are proportional.

8. 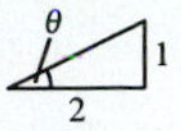

$\text{hyp} = \sqrt{1^2 + 2^2} = \sqrt{5}$

$\sin\theta = \dfrac{\text{opp}}{\text{hyp}} = \dfrac{1}{\sqrt{5}}$

$\cos\theta = \dfrac{\text{adj}}{\text{hyp}} = \dfrac{2}{\sqrt{5}}$

$\tan\theta = \dfrac{\text{opp}}{\text{adj}} = \dfrac{1}{2}$

$\csc\theta = \dfrac{\text{hyp}}{\text{opp}} = \dfrac{\sqrt{5}}{1} = \sqrt{5}$

$\sec\theta = \dfrac{\text{hyp}}{\text{adj}} = \dfrac{\sqrt{5}}{2}$

$\cot\theta = \dfrac{\text{adj}}{\text{opp}} = \dfrac{2}{1} = 2$

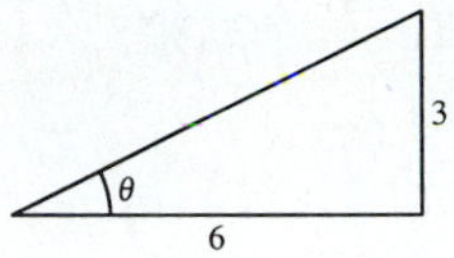

$\text{hyp} = \sqrt{3^2 + 6^2} = 3\sqrt{5}$

$\sin\theta = \dfrac{3}{3\sqrt{5}} = \dfrac{1}{\sqrt{5}}$

$\cos\theta = \dfrac{6}{3\sqrt{5}} = \dfrac{2}{\sqrt{5}}$

$\tan\theta = \dfrac{3}{6} = \dfrac{1}{2}$

$\csc\theta = \dfrac{3\sqrt{5}}{3} = \sqrt{5}$

$\sec\theta = \dfrac{3\sqrt{5}}{6} = \dfrac{\sqrt{5}}{2}$

$\cot\theta = \dfrac{6}{3} = 2$

The function values are the same because the triangles are similar, and corresponding sides are proportional.

10. 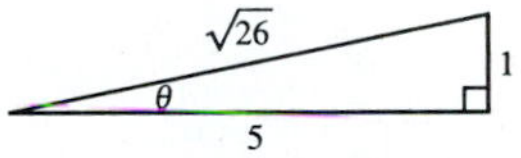

$\text{opp} = \sqrt{5^2 + 1^2} = \sqrt{26}$

$\sin\theta = \dfrac{\text{opp}}{\text{hyp}} = \dfrac{1}{\sqrt{26}} = \dfrac{\sqrt{26}}{26}$

$\cos\theta = \dfrac{\text{adj}}{\text{hyp}} = \dfrac{5}{\sqrt{26}} = \dfrac{5\sqrt{26}}{26}$

$\tan\theta = \dfrac{\text{opp}}{\text{adj}} = \dfrac{1}{5}$

$\csc\theta = \dfrac{\text{hyp}}{\text{opp}} = \dfrac{\sqrt{26}}{1} = \sqrt{26}$

$\sec\theta = \dfrac{\text{hyp}}{\text{adj}} = \dfrac{\sqrt{26}}{5}$

12. 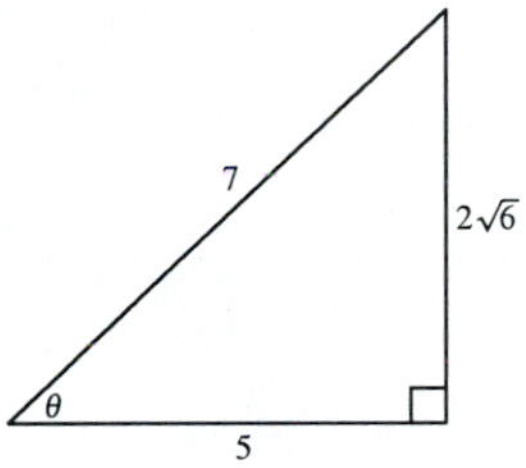

$\text{opp} = \sqrt{7^2 - 5^2} = \sqrt{24} = 2\sqrt{6}$

$\sin\theta = \dfrac{\text{opp}}{\text{hyp}} = \dfrac{2\sqrt{6}}{7}$

$\tan\theta = \dfrac{\text{opp}}{\text{adj}} = \dfrac{2\sqrt{6}}{5}$

$\csc\theta = \dfrac{1}{\sin\theta} = \dfrac{7}{2\sqrt{6}} = \dfrac{7\sqrt{6}}{12}$

$\sec\theta = \dfrac{1}{\cos\theta} = \dfrac{7}{5}$

$\cot\theta = \dfrac{1}{\tan\theta} = \dfrac{5}{2\sqrt{6}} = \dfrac{5\sqrt{6}}{12}$

14.

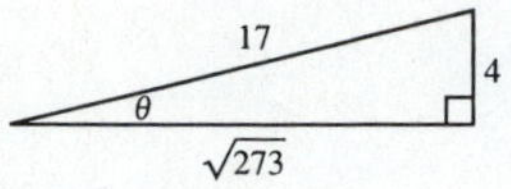

$\text{adj} = \sqrt{17^2 - 4^2} = \sqrt{237}$

$\sin \theta = \dfrac{\text{opp}}{\text{hyp}} = \dfrac{4}{17}$

$\cos \theta = \dfrac{\text{adj}}{\text{hyp}} = \dfrac{\sqrt{273}}{17}$

$\tan \theta = \dfrac{\text{opp}}{\text{adj}} = \dfrac{4}{\sqrt{273}} = \dfrac{4\sqrt{273}}{273}$

$\sec \theta = \dfrac{1}{\cos \theta} = \dfrac{17}{\sqrt{273}} = \dfrac{17\sqrt{273}}{273}$

$\cot \theta = \dfrac{1}{\tan \theta} = \dfrac{\sqrt{273}}{4}$

16.

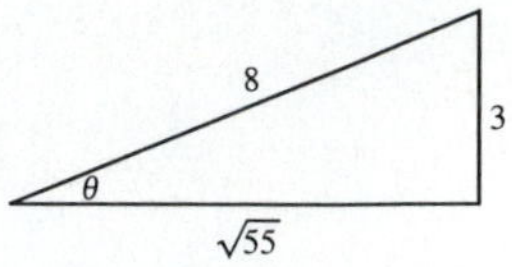

$\text{adj} = \sqrt{8^2 - 3^2} = \sqrt{55}$

$\cos \theta = \dfrac{\text{adj}}{\text{hyp}} = \dfrac{\sqrt{55}}{8}$

$\tan \theta = \dfrac{\text{opp}}{\text{adj}} = \dfrac{3}{\sqrt{55}} = \dfrac{3\sqrt{55}}{55}$

$\csc \theta = \dfrac{1}{\sin \theta} = \dfrac{8}{3}$

$\sec \theta = \dfrac{1}{\cos \theta} = \dfrac{8}{\sqrt{55}} = \dfrac{8\sqrt{55}}{55}$

$\cot \theta = \dfrac{1}{\tan \theta} = \dfrac{\sqrt{55}}{3}$

18. $\sin 30° = \dfrac{1}{2}$, $\tan 30° = \dfrac{\sqrt{3}}{3}$

(a) $\csc 30° = \dfrac{1}{\sin 30°} = 2$

(b) $\cot 60° = \tan(90° - 60°) = \tan 30° = \dfrac{\sqrt{3}}{3}$

(c) $\cos 30° = \dfrac{\sin 30°}{\tan 30°} = \dfrac{(1/2)}{(\sqrt{3}/3)} = \dfrac{3}{2\sqrt{3}} = \dfrac{\sqrt{3}}{2}$

(d) $\cot 30° = \dfrac{1}{\tan 30°} = \dfrac{3}{\sqrt{3}} = \dfrac{3\sqrt{3}}{3} = \sqrt{3}$

20. $\sec \theta = 5$, $\tan \theta = 2\sqrt{6}$

(a) $\cos \theta = \dfrac{1}{\sec \theta} = \dfrac{1}{5}$

(b) $\cot \theta = \dfrac{1}{\tan \theta} = \dfrac{1}{2\sqrt{6}} = \dfrac{\sqrt{6}}{12}$

(c) $\cot(90° - \theta) = \tan \theta = 2\sqrt{6}$

(d) $\sin \theta = \tan \theta \cos \theta = (2\sqrt{6})\left(\dfrac{1}{5}\right) = \dfrac{2\sqrt{6}}{5}$

22. $\tan \beta = 5$

(a) $\cot \beta = \dfrac{1}{\tan \beta} = \dfrac{1}{5}$

(b) $\cos \beta = \dfrac{1}{\sqrt{1 + \tan^2 \beta}} = \dfrac{1}{\sqrt{1 + 5^2}}$

$= \dfrac{1}{\sqrt{26}}$

(c) $\tan(90° - \beta) = \cot \beta = \dfrac{1}{\tan \beta} = \dfrac{1}{5}$

(d) $\csc \beta = \sqrt{1 + \cot^2 \beta}$

$= \sqrt{1 + \left(\dfrac{1}{5}\right)^2}$

$= \sqrt{1 + \dfrac{1}{25}} = \sqrt{\dfrac{26}{25}} = \dfrac{\sqrt{26}}{5}$

24. $\cos \theta \sec \theta = \cos \theta \dfrac{1}{\cos \theta} = 1$

26. $\cot \alpha \sin \alpha = \dfrac{\cos \alpha}{\sin \alpha} \sin \alpha = \cos \alpha$

28. $(1 + \sin \theta)(1 - \sin \theta) = 1 - \sin^2 \theta = \cos^2 \theta$

30. $\sin^2 \theta - \cos^2 \theta = \sin^2 \theta - (1 - \sin^2 \theta)$

$= \sin^2 \theta - 1 + \sin^2 \theta$

$= 2 \sin^2 \theta - 1$

32. $\dfrac{\tan \beta + \cot \beta}{\tan \beta} = \dfrac{\tan \beta}{\tan \beta} + \dfrac{\cot \beta}{\tan \beta}$

$= 1 + \dfrac{\cot \beta}{(1/\cot \beta)}$

$= 1 + \cot^2 \beta = \csc^2 \beta$

34. (a) $\csc 30° = \dfrac{1}{\sin 30°} = \dfrac{1}{(1/2)} = 2$

(b) $\sin \dfrac{\pi}{4} = \dfrac{\sqrt{2}}{2}$

36. (a) $\sin \dfrac{\pi}{3} = \dfrac{\sqrt{3}}{2}$

(b) $\csc 45° = \dfrac{1}{\sin 45°} = \dfrac{1}{(\sqrt{2}/2)} = \sqrt{2}$

38. (a) $\tan 23.5° \approx 0.4348$

(b) $\cot 66.5° = \dfrac{1}{\tan 66.5°} \approx 0.4348$

40. (a) $\cos 16°\, 18' = \cos\left(16 + \dfrac{18}{60}\right)^{\circ} \approx 0.9598$

(b) $\sin 73°\, 56' = \sin\left(73 + \dfrac{56}{60}\right)^{\circ} \approx 0.9609$

42. (a) $\cos 4°\, 50'\, 15'' = \cos\left(4 + \dfrac{50}{60} + \dfrac{15}{3600}\right)^{\circ}$

≈ 0.9964

(b) $\sec 4°\, 50'\, 15'' = \dfrac{1}{\cos 4°\, 50'\, 15''}$

≈ 1.0036

44. (a) $\sec 0.75 = \dfrac{1}{\cos 0.75} \approx 1.3667$

(Note: 0.75 is in radians.)

(b) $\cos 0.75 \approx 0.7317$

46. (a) $\sec\left(\dfrac{\pi}{2} - 1\right) = \dfrac{1}{\cos\left(\dfrac{\pi}{2} - 1\right)} \approx 1.1884$

(b) $\cot\left(\dfrac{\pi}{2} - \dfrac{1}{2}\right) = \dfrac{1}{\tan\left(\dfrac{\pi}{2} - \dfrac{1}{2}\right)} \approx 0.5463$

48. (a) $\cos \theta = \dfrac{\sqrt{2}}{2} \Rightarrow \theta = 45° = \dfrac{\pi}{4}$

(b) $\tan \theta = 1 \Rightarrow \theta = 45° = \dfrac{\pi}{4}$

50. (a) $\tan \theta = \sqrt{3} \Rightarrow \theta = 60° = \dfrac{\pi}{3}$

(b) $\cos \theta = \dfrac{1}{2} \Rightarrow \theta = 60° = \dfrac{\pi}{3}$

52. (a) $\cot \theta = \dfrac{\sqrt{3}}{3}$

$\tan \theta = \dfrac{3}{\sqrt{3}} = \sqrt{3} \Rightarrow \theta = 60° = \dfrac{\pi}{3}$

(b) $\sec \theta = \sqrt{2}$

$\cos \theta = \dfrac{1}{\sqrt{2}} = \dfrac{\sqrt{2}}{2} \Rightarrow \theta = 45° = \dfrac{\pi}{4}$

54. (a) $\cos \theta = 0.9848 \Rightarrow \theta \approx 10° \approx 0.175$

(b) $\cos \theta = 0.8746 \Rightarrow \theta \approx 29° \approx 0.506$

56. (a) $\sin \theta = 0.3746 \Rightarrow \theta \approx 22° \approx 0.384$

(b) $\cos \theta = 0.3746 \Rightarrow \theta \approx 68° \approx 1.187$

58. $\cos 60° = \dfrac{x}{12}$

$x = 12 \cos 60° = 6$

60. $\sin 45° = \dfrac{20}{r}$

$r = \dfrac{20}{\sin 45°} = \dfrac{20}{\sqrt{2}/2} = 20\sqrt{2}$

62. $\tan 20° = \dfrac{25}{x}$

$x = \dfrac{25}{\tan 20°} \approx 68.7$

64. $\cos 75° = \dfrac{25}{r}$

$r = \dfrac{25}{\cos 75°} \approx 96.6$

66. (a)

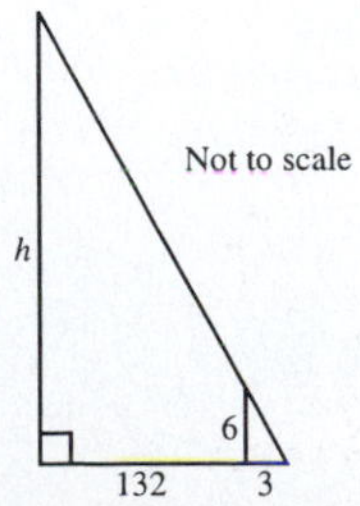

(b) $\tan \theta = \dfrac{6}{3}$ and $\tan \theta = \dfrac{h}{135}$

Thus, $\dfrac{6}{3} = \dfrac{h}{135}$.

(c) $\dfrac{135 \cdot 6}{3} = h = 270$ feet

68. $\tan\theta = \dfrac{\text{opp}}{\text{adj}}$

$\tan 54° = \dfrac{w}{100}$

$w = 100\tan 54° \approx 137.6$ feet

70. (a)

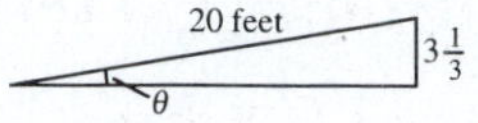

(b) $\sin\theta = \dfrac{\text{opp}}{\text{hyp}}$

$\sin\theta = \dfrac{3\frac{1}{3}}{20}$

(c) $\sin\theta = \dfrac{1}{6} \Rightarrow \theta = 9.59°$

72. $\tan 3° = \dfrac{x}{15}$

$x = 15\tan 3°$

$d = 5 + 2x$

$= 5 + 2(15\tan 3°) \approx 6.57$ centimeters

74. $x \approx 2.588,\ y \approx 9.659$

$\sin\theta = \dfrac{y}{10} \approx 0.97$

$\cos\theta = \dfrac{x}{10} \approx 0.26$

$\tan\theta = \dfrac{y}{x} \approx 3.73$

$\csc\theta = \dfrac{10}{y} \approx 1.04$

$\sec\theta = \dfrac{10}{x} \approx 3.86$

$\cot\theta = \dfrac{x}{y} \approx 0.27$

76. (a)

θ	0	0.3	0.6	0.9	1.2	1.5
$\sin\theta$	0	0.2955	0.5646	0.7833	0.9320	0.9975
$\cos\theta$	1	0.9553	0.8253	0.6216	0.3624	0.0707

(b) On [0, 1.5], $\sin\theta$ is an increasing function.

(c) On [0, 1.5], $\cos\theta$ is a decreasing function.

(d) As the angle increases the length of the side opposite the angle increases relative to the length of the hypotenuse and the length of the side adjacent to the angle decreases relative to the length of the hypotenuse. Thus the sine increases and the cosine decreases.

78. $\sec 30° = \csc 60°$

True, because

$\sec(90° - \theta) = \csc\theta.$

80. $\cot^2 10° - \csc^2 10° = -1$

True, because

$1 + \cot^2\theta = \csc^2\theta$

$\cot^2\theta = \csc^2\theta - 1$

$\cot^2\theta - \csc^2\theta = -1.$

82. $\tan[(0.8)^2] = \tan^2(0.8)$

False.

$\tan[(0.8)^2] = \tan 0.64 \approx 0.745$

$\tan^2(0.8) = (\tan 0.8)^2 \approx 1.060$

84. $$\frac{2t^2+5t-12}{9-4t^2} \div \frac{t^2-16}{4t^2+12t+9} = \frac{2t^2+5t-12}{9-4t^2}\cdot\frac{4t^2+12t+9}{t^2-16}$$

$$= \frac{(2t-3)(t+4)}{(3+2t)(3-2t)}\cdot\frac{(2t+3)(2t+3)}{(t+4)(t-4)} = -\frac{(2t+3)}{(t-4)}$$

$$= \frac{2t+3}{4-t}$$

86. $$\frac{\left(\frac{3}{x}-\frac{1}{4}\right)}{\left(\frac{12}{x}-1\right)} = \frac{\frac{12-x}{4x}}{\frac{12-x}{x}} = \frac{12-x}{4x}\cdot\frac{x}{12-x} = \frac{1}{4}$$

Section 4.4 Trigonometric Functions of Any Angle

Solutions to Even-Numbered Exercises

2. (a) $x = 12, y = -5$

$r = \sqrt{12^2 + (-5)^2} = 13$

$\sin \theta = \dfrac{y}{r} = \dfrac{-5}{13} = -\dfrac{5}{13}$

$\cos \theta = \dfrac{x}{r} = \dfrac{12}{13}$

$\tan \theta = \dfrac{y}{x} = \dfrac{-5}{12} = -\dfrac{5}{12}$

$\csc \theta = \dfrac{r}{y} = \dfrac{13}{-5} = -\dfrac{13}{5}$

$\sec \theta = \dfrac{r}{x} = \dfrac{13}{12}$

$\cot \theta = \dfrac{x}{y} = \dfrac{12}{-5} = -\dfrac{12}{5}$

(b) $x = -1, y = 1$

$r = \sqrt{(-1)^2 + 1^2} = \sqrt{2}$

$\sin \theta = \dfrac{y}{r} = \dfrac{1}{\sqrt{2}} = \dfrac{\sqrt{2}}{2}$

$\cos \theta = \dfrac{x}{r} = \dfrac{-1}{\sqrt{2}} = -\dfrac{\sqrt{2}}{2}$

$\tan \theta = \dfrac{y}{x} = \dfrac{1}{-1} = -1$

$\csc \theta = \dfrac{r}{y} = \dfrac{\sqrt{2}}{1} = \sqrt{2}$

$\sec \theta = \dfrac{r}{x} = \dfrac{\sqrt{2}}{-1} = -\sqrt{2}$

$\cot \theta = \dfrac{x}{y} = \dfrac{-1}{1} = -1$

4. (a) $x = 3, y = 1$

$r = \sqrt{3^2 + 1^2} = \sqrt{10}$

$\sin \theta = \dfrac{y}{r} = \dfrac{1}{\sqrt{10}} = \dfrac{\sqrt{10}}{10}$

$\cos \theta = \dfrac{x}{r} = \dfrac{3}{\sqrt{10}} = \dfrac{3\sqrt{10}}{10}$

$\tan \theta = \dfrac{y}{x} = \dfrac{1}{3}$

$\csc \theta = \dfrac{r}{y} = \dfrac{\sqrt{10}}{1} = \sqrt{10}$

$\sec \theta = \dfrac{r}{x} = \dfrac{\sqrt{10}}{3}$

$\cot \theta = \dfrac{x}{y} = \dfrac{3}{1} = 3$

(b) $x = 2, y = -4$

$r = \sqrt{2^2 + (-4)^2} = 2\sqrt{5}$

$\sin \theta = \dfrac{y}{r} = \dfrac{-4}{2\sqrt{5}} = -\dfrac{2\sqrt{5}}{5}$

$\cos \theta = \dfrac{x}{r} = \dfrac{2}{2\sqrt{5}} = \dfrac{\sqrt{5}}{5}$

$\tan \theta = \dfrac{y}{x} = \dfrac{-4}{2} = -2$

$\csc \theta = \dfrac{r}{y} = \dfrac{2\sqrt{5}}{-4} = -\dfrac{\sqrt{5}}{2}$

$\sec \theta = \dfrac{r}{x} = \dfrac{2\sqrt{5}}{2} = \sqrt{5}$

$\cot \theta = \dfrac{x}{y} = \dfrac{2}{-4} = -\dfrac{1}{2}$

6. (a) $x = 8, y = 15$

$r = \sqrt{8^2 + 15^2} = 17$

$\sin\theta = \frac{y}{r} = \frac{15}{17}$

$\cos\theta = \frac{x}{r} = \frac{8}{17}$

$\tan\theta = \frac{y}{x} = \frac{15}{8}$

$\csc\theta = \frac{r}{y} = \frac{17}{15}$

$\sec\theta = \frac{r}{x} = \frac{17}{8}$

$\cot\theta = \frac{x}{y} = \frac{8}{15}$

(b) $x = -9, y = -40$

$r = \sqrt{(-9)^2 + (-40)^2} = \sqrt{1681} = 41$

$\sin\theta = \frac{y}{r} = \frac{-40}{41} = -\frac{40}{41}$

$\cos\theta = \frac{x}{r} = \frac{-9}{41} = -\frac{9}{41}$

$\tan\theta = \frac{y}{x} = \frac{-40}{-9} = \frac{40}{9}$

$\csc\theta = \frac{r}{y} = \frac{41}{-40} = -\frac{41}{40}$

$\sec\theta = \frac{r}{x} = \frac{41}{-9} = -\frac{41}{9}$

$\cot\theta = \frac{x}{y} = \frac{-9}{-40} = \frac{9}{40}$

8. (a) $x = -5, y = -2$

$r = \sqrt{(-5)^2 + (-2)^2} = \sqrt{29}$

$\sin\theta = \frac{y}{r} = \frac{-2}{\sqrt{29}} = -\frac{2\sqrt{29}}{29}$

$\cos\theta = \frac{x}{r} = \frac{-5}{\sqrt{29}} = -\frac{5\sqrt{29}}{29}$

$\tan\theta = \frac{y}{x} = \frac{-2}{-5} = \frac{2}{5}$

$\csc\theta = \frac{r}{y} = \frac{\sqrt{29}}{-2} = -\frac{\sqrt{29}}{2}$

$\sec\theta = \frac{r}{x} = \frac{\sqrt{29}}{-5} = -\frac{\sqrt{29}}{5}$

$\cot\theta = \frac{x}{y} = \frac{-5}{-2} = \frac{5}{2}$

(b) $x = -\frac{3}{2}, y = 3$

$r = \sqrt{\left(-\frac{3}{2}\right)^2 + 3^2} = \frac{3\sqrt{5}}{2}$

$\sin\theta = \frac{y}{r} = \frac{3}{3\sqrt{5}/2} = \frac{2\sqrt{5}}{5}$

$\cos\theta = \frac{x}{r} = \frac{-3/2}{3\sqrt{5}/2} = -\frac{\sqrt{5}}{5}$

$\tan\theta = \frac{y}{x} = \frac{3}{-3/2} = -2$

$\csc\theta = \frac{r}{y} = \frac{3\sqrt{5}/2}{3} = \frac{\sqrt{5}}{2}$

$\sec\theta = \frac{r}{x} = \frac{3\sqrt{5}/2}{-3/2} = -\sqrt{5}$

$\cot\theta = \frac{x}{y} = \frac{-3/2}{3} = -\frac{1}{2}$

10. (a) $\sin\theta > 0$ and $\cos\theta > 0$

$\frac{y}{r} > 0$ and $\frac{x}{r} > 0$

Quadrant I

(b) $\sin\theta < 0$ and $\cos\theta > 0$

$\frac{y}{r} < 0$ and $\frac{x}{r} > 0$

Quadrant IV

12. (a) $\sec\theta > 0$ and $\cot\theta < 0$

$\frac{r}{x} > 0$ and $\frac{x}{y} < 0$

Quadrant IV

(b) $\csc\theta < 0$ and $\tan\theta > 0$

$\frac{r}{y} < 0$ and $\frac{y}{x} > 0$

Quadrant III

14. $\cos\theta = \frac{x}{r} = \frac{-4}{5} \Longrightarrow y = |3|$

θ in Quadrant III $\Longrightarrow y = -3$

$\sin\theta = \frac{y}{r} = -\frac{3}{5}$ $\qquad \csc\theta = -\frac{5}{3}$

$\cos\theta = \frac{x}{r} = -\frac{4}{5}$ $\qquad \sec\theta = -\frac{5}{4}$

$\tan\theta = \frac{y}{x} = \frac{3}{4}$ $\qquad \cot\theta = \frac{4}{3}$

16. $\cos\theta = \frac{x}{r} = \frac{8}{17} \Longrightarrow y = |15|$

$\tan\theta < 0 \Longrightarrow y = -15$

$\sin\theta = \frac{y}{r} = \frac{-15}{17} = -\frac{15}{17}$ $\qquad \csc\theta = -\frac{17}{15}$

$\cos\theta = \frac{x}{r} = \frac{8}{17}$ $\qquad \sec\theta = \frac{17}{8}$

$\tan\theta = \frac{y}{x} = \frac{-15}{8} = -\frac{15}{8}$ $\qquad \tan\theta = -\frac{8}{15}$

18. $\csc\theta = \frac{r}{y} = \frac{4}{1} \Longrightarrow x = \left|\sqrt{15}\right|$

$\cot\theta < 0 \Longrightarrow x = -\sqrt{15}$

$\sin\theta = \frac{y}{r} = \frac{1}{4}$ $\qquad \csc\theta = 4$

$\cos\theta = \frac{x}{r} = -\frac{\sqrt{15}}{4}$ $\qquad \sec\theta = -\frac{4\sqrt{15}}{15}$

$\tan\theta = \frac{y}{x} = -\frac{\sqrt{15}}{15}$ $\qquad \cot\theta = -\sqrt{15}$

20. $\cot\theta$ is undefined $\Longrightarrow \theta = n\pi$

$\frac{\pi}{2} \le \theta \le \frac{3\pi}{2} \Longrightarrow \theta = \pi, y = 0, x = -r$

$\sin\theta = \frac{y}{r} = \frac{0}{r} = 0$ $\qquad \csc\theta = \frac{r}{y}$ is undefined.

$\cos\theta = \frac{x}{r} = \frac{-r}{r} = -1$ $\qquad \sec\theta = \frac{r}{x} = -1$

$\tan\theta = \frac{y}{x} = \frac{0}{x} = 0$ $\qquad \cot\theta = \frac{x}{y}$ is undefined.

22. $\tan\theta$ is undefined $\Longrightarrow \theta = n\pi + \frac{\pi}{2}$

$\pi \le \theta \le 2\pi \Longrightarrow \theta = \frac{3\pi}{2}, x = 0, y = -r$

$\sin\theta = \frac{y}{r} = \frac{-r}{r} = -1$ $\qquad \csc\theta = \frac{r}{y} = -1$

$\cos\theta = \frac{x}{r} = \frac{0}{r} = 0$ $\qquad \sec\theta = \frac{r}{x}$ is undefined.

$\tan\theta = \frac{y}{x}$ is undefined. $\qquad \cot\theta = \frac{x}{y} = \frac{0}{y} = 0$

24. $\left(-x, -\frac{1}{3}x\right)$, Quadrant III

$r = \sqrt{x^2 + \frac{1}{9}x^2} = \frac{\sqrt{10}x}{3}$

$\sin\theta = \frac{y}{r} = \frac{(1/3)-x}{(\sqrt{10}x)/3} = -\frac{\sqrt{10}}{10}$

$\cos\theta = \frac{x}{r} = \frac{-x}{(\sqrt{10}x)/3} = -\frac{3\sqrt{10}}{10}$

$\tan\theta = \frac{y}{x} = \frac{(-1/3)x}{-x} = \frac{1}{3}$

$\csc\theta = \frac{r}{y} = \frac{(\sqrt{10}x)/3}{(-1/3)x} = -\sqrt{10}$

$\sec\theta = \frac{r}{x} = \frac{(\sqrt{10}x)/3}{-x} = -\frac{\sqrt{10}}{3}$

$\cot\theta = \frac{x}{y} = \frac{-x}{(-1/3)x} = 3$

26. $4x + 3y = 0 \Longrightarrow y = -\frac{4}{3}x$

$\left(x, -\frac{4}{3}x\right)$, Quadrant IV

$r = \sqrt{x^2 + \frac{16}{5}x^2} = \frac{5}{3}x$

$\sin\theta = \frac{y}{r} = \frac{(-4/3)x}{(5/3)x} = -\frac{4}{5}$

$\cos\theta = \frac{x}{r} = \frac{x}{(5/3)x} = \frac{3}{5}$

$\tan\theta = \frac{y}{x} = \frac{(-4/3)x}{x} = -\frac{4}{3}$

$\csc\theta = -\frac{5}{4}$

$\sec\theta = \frac{5}{3}$

$\tan\theta = -\frac{3}{4}$

28. $\cos \frac{3\pi}{2} = \frac{x}{r} = \frac{0}{1} = 0$

since $(3\pi/2)$ corresponds to $(0, -1)$.

30. $\sec \frac{3\pi}{2} = \frac{r}{x} = \frac{1}{0} \Longrightarrow$ undefined

since $(3\pi/2)$ corresponds to $(0, -1)$.

32. $\tan \pi = \frac{y}{x} = \frac{0}{-1} = 0$

since π corresponds to $(-1, 0)$.

34. $\csc \pi = \frac{r}{y} = \frac{1}{0} \Longrightarrow$ undefined

since π corresponds to $(-1, 0)$.

36. (a) $\theta = 309°$

$\theta' = 360° - 309° = 51°$

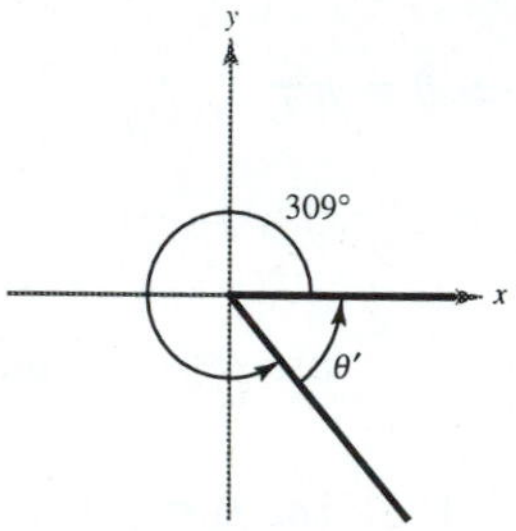

(b) $\theta = 226°$

$\theta' = 226° - 180° = 46°$

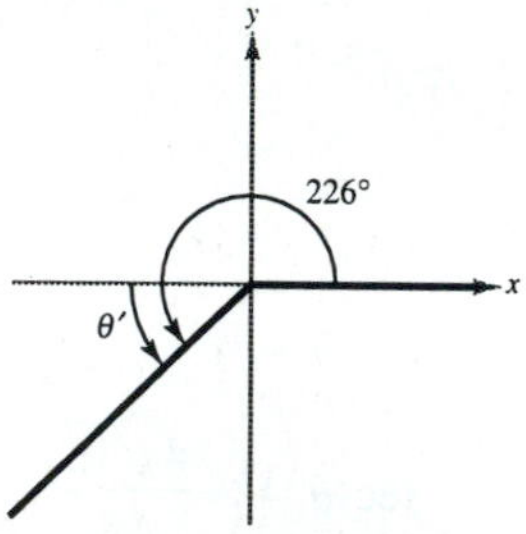

38. (a) $\theta = -145°$ is coterminal with $215°$.

$\theta' = 215° - 180° = 35°$

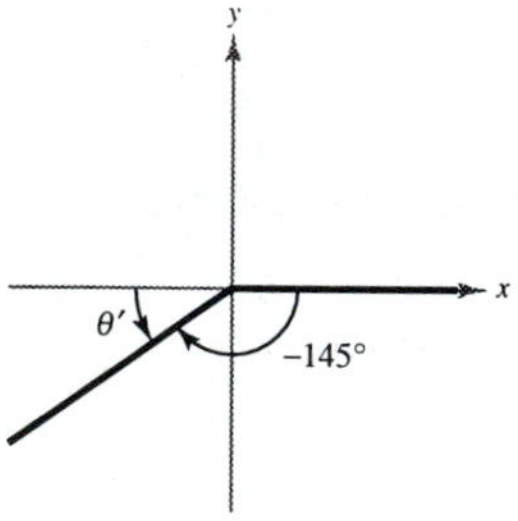

(b) $\theta = -239°$ is coterminal with $121°$.

$\theta' = 180° - 121° = 59°$

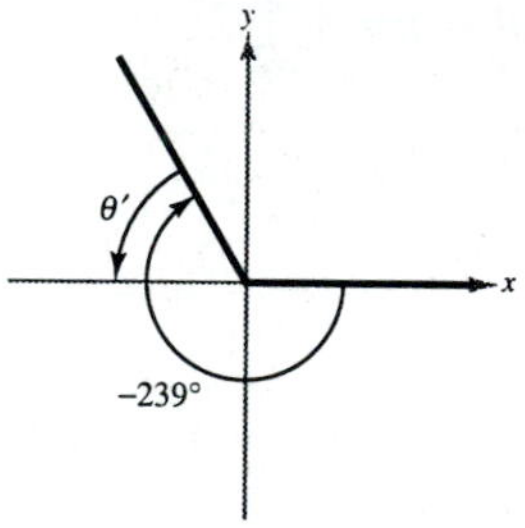

40. (a) $\theta = \frac{7\pi}{4}$

$\theta' = 2\pi - \frac{7\pi}{4} = \frac{\pi}{4}$

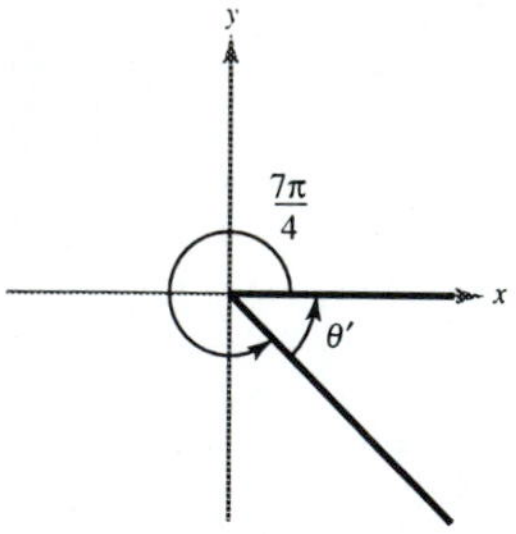

(b) $\theta = \frac{8\pi}{9}$

$\theta' = \pi - \frac{8\pi}{9} = \frac{\pi}{9}$

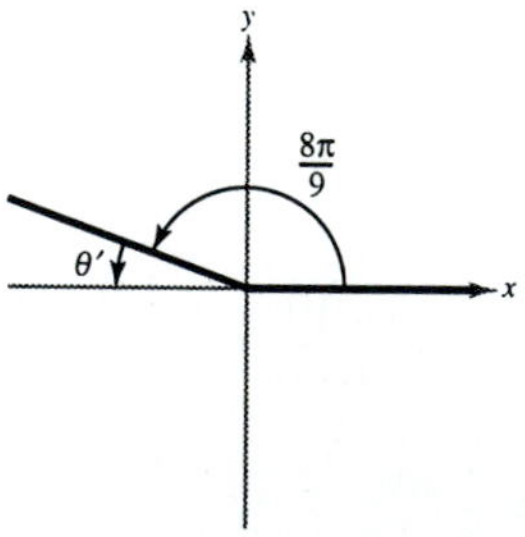

42. (a) $\theta = \frac{11\pi}{3}$ is coterminal with $\frac{5\pi}{3}$.

$\theta' = 2\pi - \frac{5\pi}{3} = \frac{\pi}{3}$

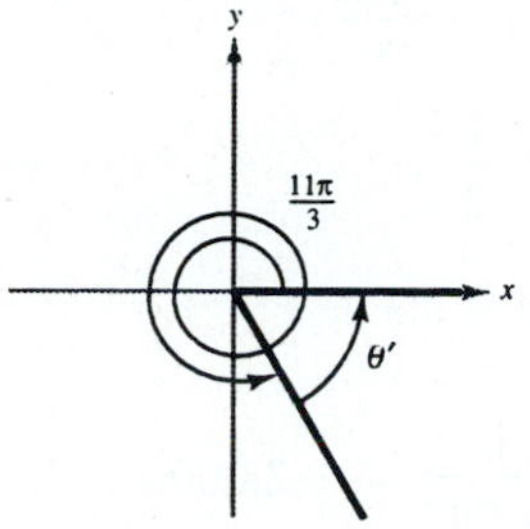

(b) $\theta = -\frac{7\pi}{10}$ is coterminal with $\frac{13\pi}{10}$.

$\theta' = \frac{13\pi}{10} - \pi = \frac{3\pi}{10}$

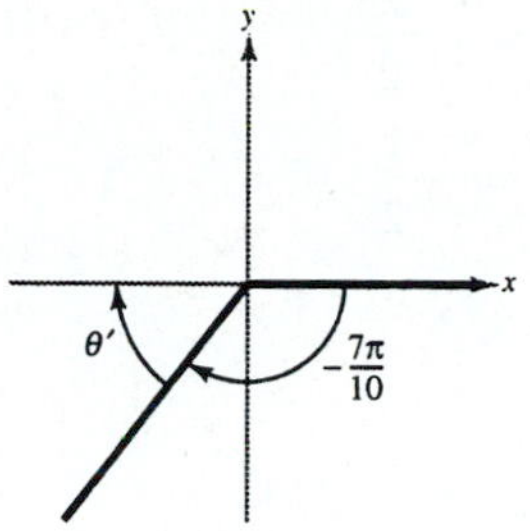

44. (a) $\theta = 300°$, $\theta' = 360° - 300° = 60°$ in Quadrant IV.

$\sin 300° = -\sin 60° = -\frac{\sqrt{3}}{2}$

$\cos 300° = \cos 60° = \frac{1}{2}$

$\tan 300° = -\tan 60° = -\sqrt{3}$

(b) $\theta = 330°$, $\theta' = 360° - 330° = 30°$ in Quadrant IV.

$\sin 330° = -\sin 30° = -\frac{1}{2}$

$\cos 330° = \cos 30° = \frac{\sqrt{3}}{2}$

$\tan 330° = -\tan 30° = -\frac{\sqrt{3}}{3}$

46. (a) $\theta = -405°$, $\theta' = 405° - 360° = 45°$ in Quadrant IV.

$\sin(-405°) = -\sin 45° = -\frac{\sqrt{2}}{2}$

$\cos(-405°) = \cos 45° = \frac{\sqrt{2}}{2}$

$\tan(-405°) = -\tan 45° = -1$

(b) $\theta = -120°$ is coterminal with $240°$.

$\theta' = 240° - 180° = 60°$ in Quadrant III.

$\sin(-120°) = -\sin 60° = -\frac{\sqrt{3}}{2}$

$\cos(-120°) = -\cos 60° = -\frac{1}{2}$

$\tan(-120°) = \tan 60° = \sqrt{3}$

48. (a) $\theta = \frac{\pi}{4}$, $\theta' = \frac{\pi}{4}$ in Quadrant I.

$\sin\frac{\pi}{4} = \frac{\sqrt{2}}{2}$

$\cos\frac{\pi}{4} = \frac{\sqrt{2}}{2}$

$\tan\frac{\pi}{4} = 1$

(b) $\theta = \frac{5\pi}{4}$, $\theta' = \frac{5\pi}{4} - \pi = \frac{\pi}{4}$ in Quadrant III.

$\sin\frac{5\pi}{4} = -\sin\frac{\pi}{4} = -\frac{\sqrt{2}}{2}$

$\cos\frac{5\pi}{4} = -\cos\frac{\pi}{4} = -\frac{\sqrt{2}}{2}$

$\tan\frac{5\pi}{4} = \tan\frac{\pi}{4} = 1$

50. (a) $\theta = -\frac{\pi}{2}$ is coterminal with $\frac{3\pi}{2}$.

$\sin\left(-\frac{\pi}{2}\right) = \sin\frac{3\pi}{2} = -1$

$\cos\left(-\frac{\pi}{2}\right) = \cos\frac{3\pi}{2} = 0$

$\tan\left(-\frac{\pi}{2}\right) = \tan\frac{3\pi}{2}$ is undefined.

(b) $\theta = \frac{\pi}{2}$

$\sin\frac{\pi}{2} = 1$

$\cos\frac{\pi}{2} = 0$

$\tan\frac{\pi}{2}$ is undefined.

52. (a) $\theta = \dfrac{10\pi}{3}$ is coterminal with $\dfrac{4\pi}{3}$.

$\theta' = \dfrac{4\pi}{3} - \pi = \dfrac{\pi}{3}$ in Quadrant III.

$\sin\dfrac{10\pi}{3} = -\sin\dfrac{\pi}{3} = -\dfrac{\sqrt{3}}{2}$

$\cos\dfrac{10\pi}{3} = -\cos\dfrac{\pi}{3} = -\dfrac{1}{2}$

$\tan\dfrac{10\pi}{3} = \tan\dfrac{\pi}{3} = \sqrt{3}$

(b) $\theta = \dfrac{17\pi}{3}$ is coterminal with $\dfrac{5\pi}{3}$.

$\theta' = 2\pi - \dfrac{5\pi}{3} = \dfrac{\pi}{3}$ in Quadrant IV.

$\sin\dfrac{17\pi}{3} = -\sin\dfrac{\pi}{3} = -\dfrac{\sqrt{3}}{2}$

$\cos\dfrac{17\pi}{3} = \cos\dfrac{\pi}{3} = \dfrac{1}{2}$

$\tan\dfrac{17\pi}{3} = \tan\dfrac{\pi}{3} = -\sqrt{3}$

54. (a) $\sec 225° = \dfrac{1}{\cos 225°} \approx -1.4142$

(b) $\sec 135° = \dfrac{1}{\cos 135°} \approx -1.4142$

56. (a) $\csc 330° = \dfrac{1}{\sin 330°} = -2.0000$

(b) $\csc 150° = \dfrac{1}{\sin 150°} = 2.0000$

58. (a) $\cot 1.35 = \dfrac{1}{\tan 1.35} \approx 0.2245$

(b) $\tan 1.35 \approx 4.4552$

60. (a) $\tan\left(-\dfrac{\pi}{9}\right) \approx -0.3640$

(b) $\tan\left(-\dfrac{10\pi}{9}\right) \approx -0.3640$

62. (a) $\sin(-0.65) \approx -0.6052$

(b) $\sin(5.63) \approx -0.6077$

64. (a) $\cos\theta = \dfrac{\sqrt{2}}{2} \Rightarrow$ reference angle is 45° or $\dfrac{\pi}{4}$ and θ is in Quadrant I or IV.

Values in degrees: 45°, 315°

Values in radians: $\dfrac{\pi}{4}, \dfrac{7\pi}{4}$

(b) $\cos\theta = -\dfrac{\sqrt{2}}{2} \Rightarrow$ reference angle is 45° or $\dfrac{\pi}{4}$ and θ is in Quadrant II or III.

Values in degrees: 135°, 225°

Values in radians: $\dfrac{3\pi}{4}, \dfrac{5\pi}{4}$

66. (a) $\sec\theta = 2 \Rightarrow$ reference angle is 60° or $\dfrac{\pi}{3}$ and θ is in Quadrant I or IV.

Values in degrees: 60°, 300°

Values in radians: $\dfrac{\pi}{3}, \dfrac{5\pi}{3}$

(b) $\sec\theta = -2 \Rightarrow$ reference angle is 60° or $\dfrac{\pi}{3}$ and θ is in Quadrant II or III.

Values in degrees: 120°, 240°

Values in radians: $\dfrac{2\pi}{3}, \dfrac{4\pi}{3}$

68. (a) $\sin\theta = \frac{\sqrt{3}}{2} \Rightarrow$ reference angle is $60°$ or $\frac{\pi}{3}$ and θ is in Quadrant I or II.

Values in degrees: $60°, 120°$

Values in radians: $\frac{\pi}{3}, \frac{2\pi}{3}$

(b) $\sin\theta = -\frac{\sqrt{3}}{2} \Rightarrow$ reference angle is $60°$ or $\frac{\pi}{3}$ and θ is in Quadrant III or IV.

Values in degrees: $240°, 300°$

Values in radians: $\frac{4\pi}{3}, \frac{5\pi}{3}$

70. (a) $\cos\theta = 0.8746$

Quadrant I: $\theta = \cos^{-1} 0.8746 \approx 29.00°$
Quadrant IV: $\theta = 360° - 29.00° = 331.00°$

(b) $\cos\theta = -0.2419$

Quadrant II:
$\theta = 180° - \cos^{-1} 0.2419 \approx 104.00°$
Quadrant III:
$\theta = 180° + \cos^{-1} 0.2419 \approx 256.00°$

72. (a) $\sin\theta = 0.0175$

Quadrant I: $\theta = \sin^{-1} 0.0175 \approx 0.018$
Quadrant II: $\theta = \pi - 0.018 = 3.124$

(b) $\sin\theta = -0.6691$

Quadrant III: $\theta = \pi + \sin^{-1} 0.6691 \approx 3.875$

Quadrant IV: $\theta = 2\pi - \sin^{-1} 0.6691 \approx 5.550$

74. (a) $\cot\theta = 5.671 \Rightarrow \frac{1}{\tan\theta} = 5.671$

Quadrant I: $\theta = \tan^{-1}\left(\frac{1}{5.671}\right) \approx 0.175$

Quadrant III: $\theta = \pi + 0.175 = 3.316$

(b) $\cot\theta = -1.280 \Rightarrow \frac{1}{\tan\theta} = -1.280$

Quadrant II: $\theta = \pi - \tan^{-1}\left(\frac{1}{1.280}\right) \approx 2.478$

Quadrant IV: $\theta = 2\pi - \tan^{-1}\left(\frac{1}{1.280}\right) \approx 5.620$

76. $\cot\theta = -3$

$1 + \cot^2\theta = \csc^2\theta$

$1 + (-3)^2 = \csc^2\theta$

$10 = \csc^2\theta$

$\csc\theta > 0$ in Quadrant II.

$\sqrt{10} = \csc\theta$

$\csc\theta = \frac{1}{\sin\theta}$

$\sin\theta = \frac{1}{\csc\theta} = \frac{1}{\sqrt{10}} = \frac{\sqrt{10}}{10}$

78. $\csc\theta = -2$

$1 + \cot^2\theta = \csc^2\theta$

$\cot^2\theta = \csc^2\theta - 1$

$\cot^2\theta = (-2)^2 - 1$

$\cot^2\theta = 3$

$\cot\theta < 0$ in Quadrant IV.

$\cot\theta = -\sqrt{3}$

80. $\sec\theta = -\frac{9}{4}$

$1 + \tan^2\theta = \sec^2\theta$

$\tan^2\theta = \sec^2\theta - 1$

$\tan^2\theta = \left(-\frac{9}{2}\right)^2 - 1$

$\tan^2\theta = \frac{65}{16}$

$\tan\theta > 0$ in Quadrant III.

$\tan\theta = \frac{\sqrt{65}}{4}$

82. $S = 23.1 + 0.442t + 4.3 \sin \frac{\pi t}{6}$

(a) February 1996 $\Rightarrow t = 2$

$$S = 23.1 + 0.442(2) + 4.3 \sin \frac{2\pi}{6}$$

≈ 27.7 thousand or 27,700 units

(b) February 1997 $\Rightarrow t = 14$

$$S = 23.1 + 0.442(14) + 4.3 \sin \frac{14\pi}{6}$$

≈ 33.0 thousand or 33,000 units

(c) September 1996 $\Rightarrow t = 9$

$$S = 23.1 + 0.442(9) + 4.3 \sin \frac{9\pi}{6}$$

≈ 22.8 thousand or 22,800 units

(d) September 1997 $\Rightarrow t = 21$

$$S = 23.1 + 0.442(21) + 4.3 \sin \frac{21\pi}{6}$$

≈ 28.1 thousand or 28,100 units

84. As θ increases from 0° to 90°, x decreases from 12 cm to 0 cm and y increases from 0 cm to 12 cm. Therefore, $\sin \theta = y/12$ increases from 0 to 1 and $\cos \theta = x/12$ decreases from 1 to 0. Thus, $\tan \theta = y/x$ increases without bound, and when $\theta = 90°$ the tangent is undefined.

86. $y = 3^{-x/2}$

x	-4	-2	0	2	4
y	9	3	1	$\frac{1}{3}$	$\frac{1}{9}$

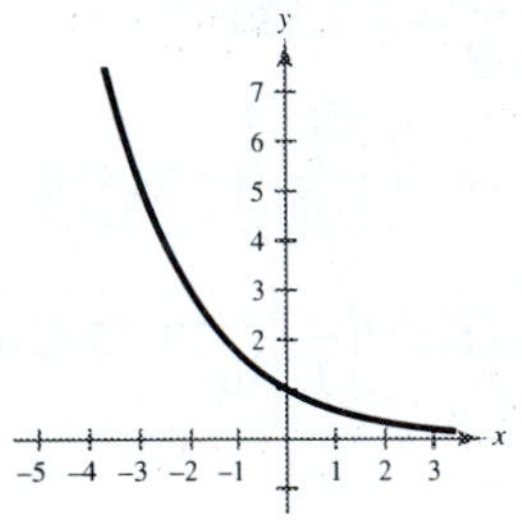

88. $y = \ln x^4$

x	± 6	± 4	± 2	± 1	0
y	7.17	5.55	2.77	0	$-\infty$

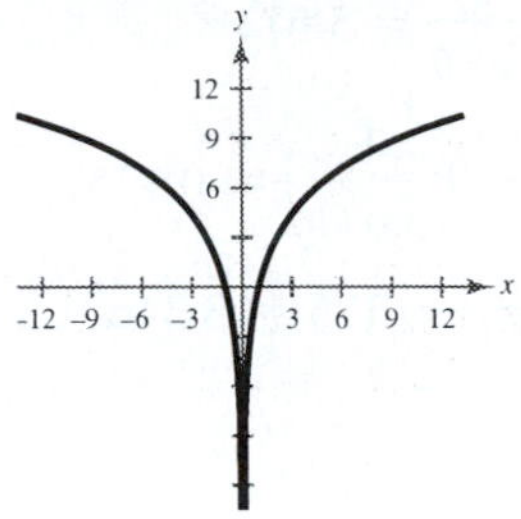

56. $y = 2\cos x - 3$

Period $= 2\pi$

Amplitude $= 2$

x	0	$\frac{\pi}{2}$	π	$\frac{3\pi}{2}$	2π
y	-1	-3	-5	-3	-1

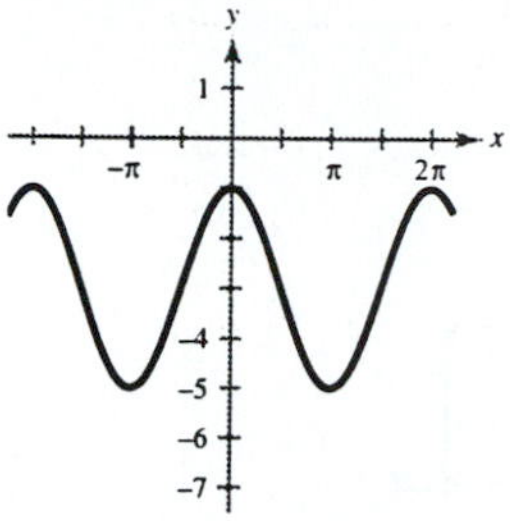

58. $y = 4\cos\left(x + \frac{\pi}{4}\right) + 4$

Period $= 2\pi$

Amplitude $= 4$

x	$-\frac{\pi}{4}$	$\frac{\pi}{4}$	$\frac{3\pi}{4}$	$\frac{5\pi}{4}$	$\frac{7\pi}{4}$
y	8	4	0	4	8

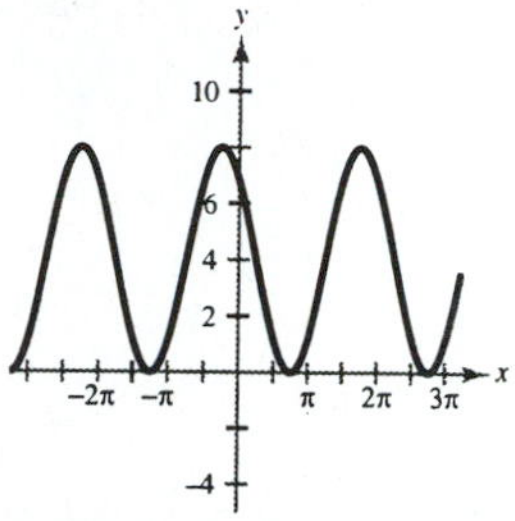

60. $y = -3\cos(6x + \pi)$

Period $= \frac{2\pi}{6} = \frac{\pi}{3}$

Amplitude $= 3$

x	0	$\frac{\pi}{12}$	$\frac{\pi}{6}$	$\frac{3\pi}{12}$	$\frac{\pi}{3}$
y	3	0	-3	0	3

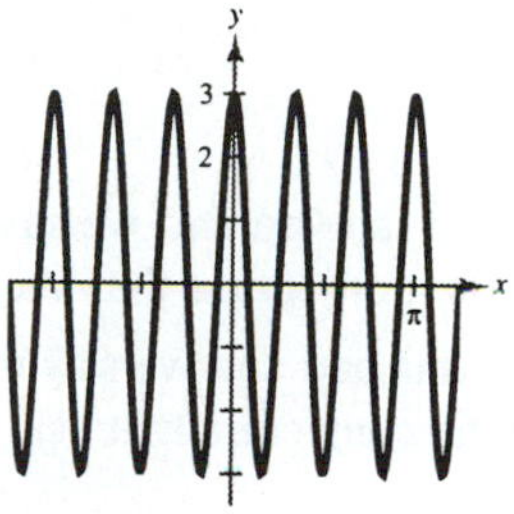

62. $y = -4\sin\left(\frac{2}{3}x - \frac{\pi}{3}\right)$

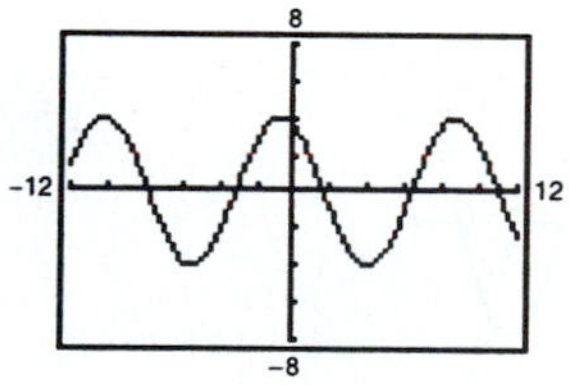

64. $y = 3\cos\left(\frac{\pi x}{2} + \frac{\pi}{2}\right) - 2$

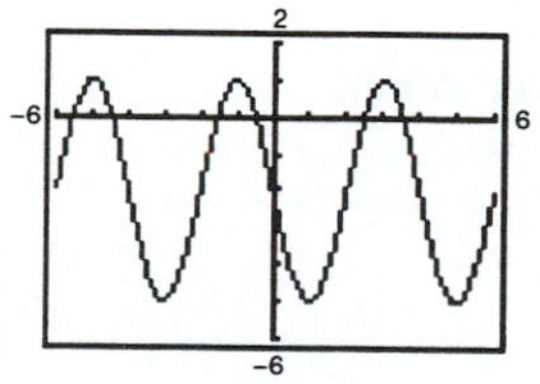

66. $y = 5\sin(\pi - 2x) + 10$

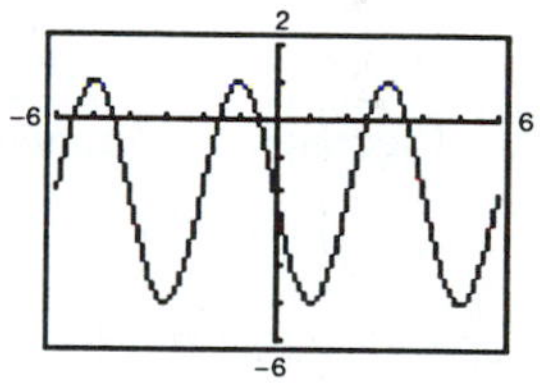

68. $y = \frac{1}{100}\sin 120\,\pi t$

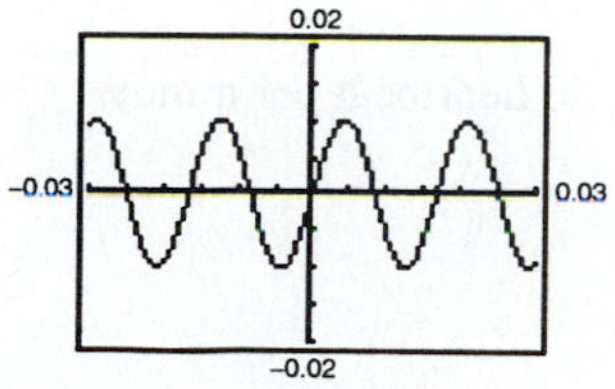

70. $f(x) = a\cos x + d$

Amplitude $= \frac{1 - (-3)}{2} = 2$

$1 = 2\cos 0 + d$

$d = 1 - 2 = -1$

$y = -1 + 2\cos x$

72. $f(x) = a\cos x + d$

Amplitude $= \frac{-2 - (-4)}{2} = 1$

Reflected in the x-axis: $a = -1$

$-4 = -1\cos 0 + d$

$d = -3$

$y = -3 - \cos x$

74. $y = a\sin(bx - c)$

Amplitude $= 2 \Longrightarrow a = 2$

Period $= 4\pi$

$\frac{2\pi}{b} = 4\pi \Longrightarrow b = \frac{1}{2}$

Phase shift: $c = 0$

$y = 2\sin\left(\frac{x}{2}\right)$

76. $y = a\sin(bx - c)$

Amplitude $= 2 \Longrightarrow a = 2$

Period $= 4$

$\frac{2\pi}{b} = 4 \Longrightarrow b = \frac{\pi}{2}$

Phase shift: $\frac{c}{b} = -1 \Longrightarrow c = -\frac{\pi}{2}$

$y = 2\sin\left(\frac{\pi x}{2} + \frac{\pi}{2}\right)$

78. $y_1 = \cos x$

$y_2 = -1$

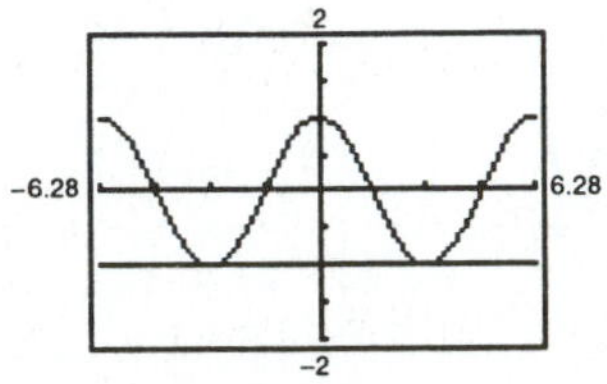

$y_1 = y_2$ when $x = \pi, -\pi$

80. $y_1 = \sin x$

$y_2 = \frac{\sqrt{3}}{2}$

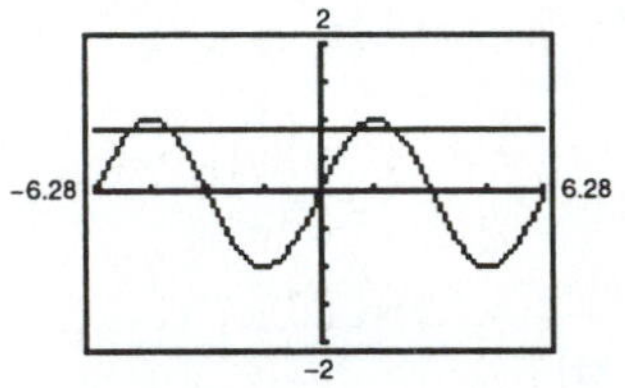

$y_1 = y_2$ when $x = \frac{\pi}{3}, \frac{2\pi}{3}, -\frac{4\pi}{3}, -\frac{5\pi}{3}$.

82. (a) In Exercise 81, $f(x) = \cos x$ is even and we saw that $h(x) = \cos^2 x$ is even. Therefore, for $f(x)$ even and $h(x) = [f(x)]^2$, we make the conjecture that $h(x)$ is even.

(b) In Exercise 81, $g(x) = \sin x$ is odd and we saw that $h(x) = \sin^2 x$ is even. Therefore, for $g(x)$ odd and $h(x) = [g(x)]^2$, we make the conjecture that $h(x)$ is even.

84. $v = 1.75\sin\frac{\pi t}{2}$

(a) Period $= \frac{2\pi}{(\pi/2)} = 4$ seconds

(b) $\frac{1 \text{ cycle}}{4 \text{ seconds}} \cdot \frac{60 \text{ seconds}}{1 \text{ minute}} = 15$ cycles per minute

(c)

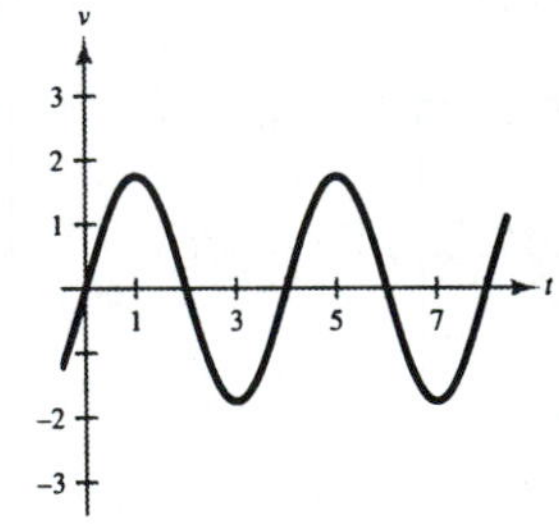

86. $P = 100 - 20\cos\frac{5\pi t}{3}$

(a) Period $= \frac{2\pi}{(5\pi/3)} = \frac{6}{5}$ seconds

(b) $\frac{1 \text{ heartbeat}}{(6/5) \text{ seconds}} \cdot \frac{60 \text{ seconds}}{1 \text{ minute}} = 50$ heartbeat per minute

88. $S = 74.50 + 43.75\sin\frac{\pi t}{6}$

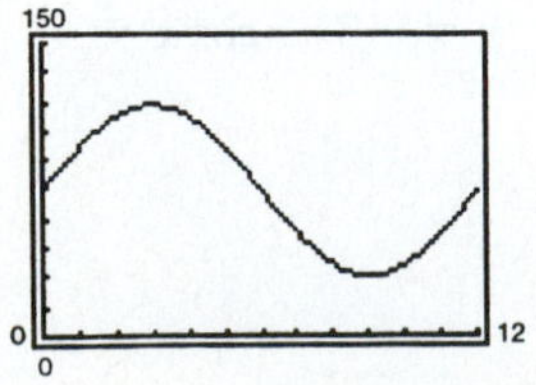

90. $c = 30.3 + 21.6 \sin\left(\frac{2\pi t}{365} + 10.9\right)$

(a) Period $= \frac{2\pi}{(2\pi/365)} = 365$

Yes, this is what is expected because there are 365 days in a year.

(b) The average daily fuel consumption is given by the amount of the vertical shift (from 0) which is given by the constant 30.3.

(c)

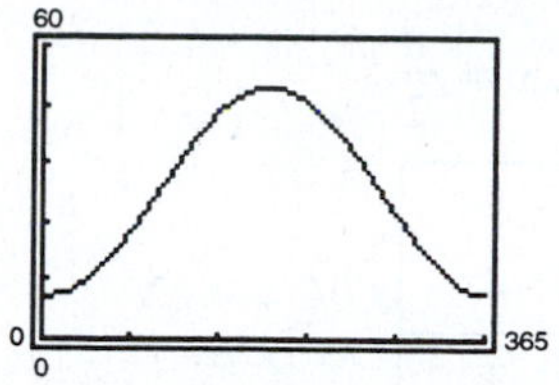

The consumption exceeds 40 gallons per day when $124 \le x \le 252$.

92. (a) $\sin\frac{1}{2} \approx \frac{1}{2} - \frac{(1/2)^3}{3!} + \frac{(1/2)^5}{5!} \approx 0.4794$

$\sin\frac{1}{2} \approx 0.4794$ (by calculator)

(b) $\sin 1 \approx 1 - \frac{1}{3!} + \frac{1}{5!} \approx 0.8417$

$\sin 1 \approx 0.8415$ (by calculator)

(c) $\sin\frac{\pi}{6} \approx 1 - \frac{(\pi/6)^3}{3!} + \frac{(\pi/6)^5}{5!} \approx 0.5000$

$\sin\frac{\pi}{6} = 0.5$ (by calculator)

(d) $\cos(-0.5) \approx 1 - \frac{(-0.5)^2}{2!} + \frac{(-0.5)^4}{4!} \approx 0.8776$

$\cos(-0.5) \approx 0.8776$ (by calculator)

(e) $\cos 1 \approx 1 - \frac{1}{2!} + \frac{1}{4!} \approx 0.5417$

$\cos 1 \approx 0.5403$ (by calculator)

(f) $\cos\frac{\pi}{4} \approx 1 - \frac{(\pi/4)^2}{2!} + \frac{(\pi/4)^4}{4!} = 0.7074$

$\cos\frac{\pi}{4} \approx 0.7071$

The error in the approximation is not the same in each case. The error appears to increase as x moves farther away from 0.

94. $\log_{10}\sqrt{x-2} = \log_{10}(x-2)^{1/2} = \frac{1}{2}\log_{10}(x-2) = \frac{1}{2}\log(x-2)$

96. $\ln\frac{t^3}{t-1} = \ln t^3 - \ln(t-1) = 3\ln t - \ln(t-1)$

Section 4.6 Graphs of Other Trigonometric Functions

Solutions to Even-Numbered Exercises

2. $y = \tan\frac{x}{2}$

Period $= \frac{\pi}{b} = \frac{\pi}{(1/2)} = 2\pi$

Asymptotes: $x = -\pi, x = \pi$

Matches graph (d).

4. $y = 2\csc x$

Period $= \frac{2\pi}{b} = \frac{2\pi}{1} = 2\pi$

Asymptotes: $x = 0, x = \pi$

Matches graph (a).

6. $y = \frac{1}{2}\sec\frac{\pi x}{2}$

Period $= \frac{2\pi}{b} = \frac{2\pi}{(\pi/2)} = 4$

Asymptotes: $x = -1, x = 1$

Matches graph (h).

8. $y = -2\sec 2\pi x$

Period $= \frac{2\pi}{2\pi} = 1$

Asymptotes: $x = -\frac{1}{4}, x = \frac{1}{4}$

Reflected in x-axis

Matches graph (c).

10. $y = \frac{1}{4}\tan x$

Period $= \pi$

Asymptotes: $x = -\frac{\pi}{2}, x = \frac{\pi}{2}$

x	$-\frac{\pi}{4}$	0	$\frac{\pi}{4}$
y	$-\frac{1}{4}$	0	$\frac{1}{4}$

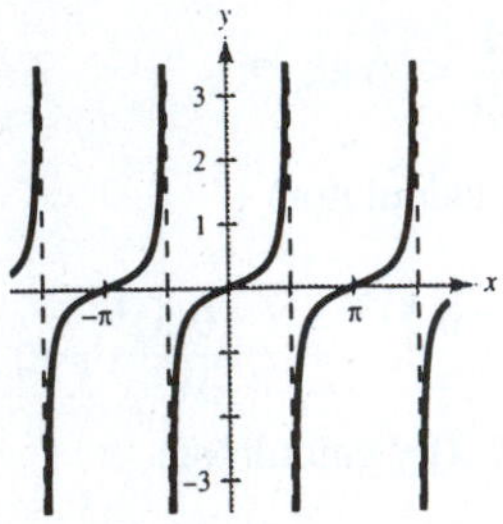

12. $y = -3\tan \pi x$

Period $= \frac{\pi}{\pi} = 1$

Asymptotes: $x = -\frac{1}{2}, x = \frac{1}{2}$

x	$-\frac{1}{4}$	0	$\frac{1}{4}$
y	3	0	-3

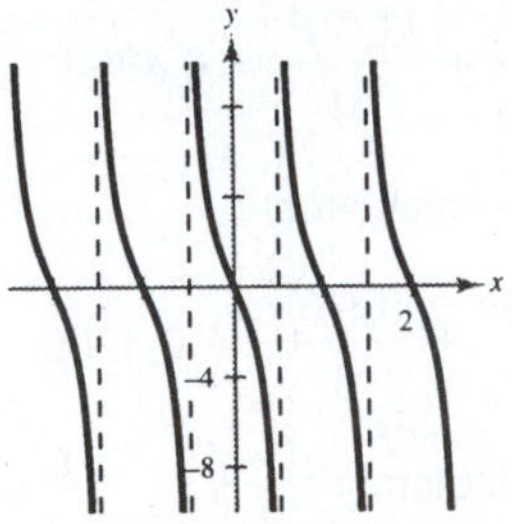

14. $y = \frac{1}{4}\sec x$

Period $= 2\pi$

Asymptotes: $x = -\frac{\pi}{2}, x = \frac{\pi}{2}$

x	$-\frac{\pi}{4}$	0	$\frac{\pi}{4}$
y	0.354	$\frac{1}{4}$	0.354

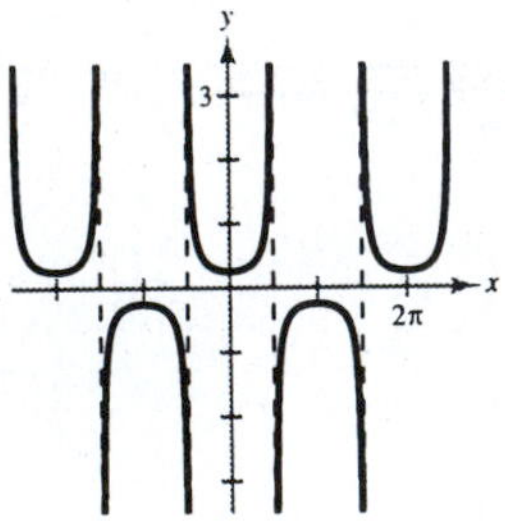

16. $y = 2\sec 4x$

Period $= \frac{2\pi}{4} = \frac{\pi}{2}$

Asymptotes: $x = -\frac{\pi}{8}, x = \frac{\pi}{8}$

x	$-\frac{\pi}{16}$	0	$\frac{\pi}{16}$
y	2.828	2	2.828

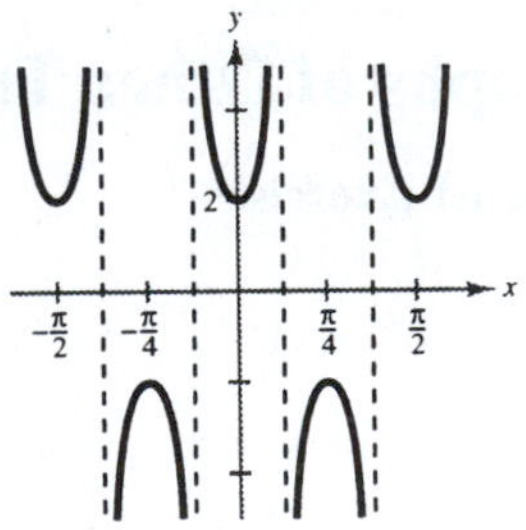

18. $y = -2\sec 4x + 2$

Period $= \frac{2\pi}{4} = \frac{\pi}{2}$

Asymptotes: $x = -\frac{\pi}{8}, x = \frac{\pi}{8}$

x	$-\frac{\pi}{16}$	0	$\frac{\pi}{16}$
y	-0.828	0	-0.828

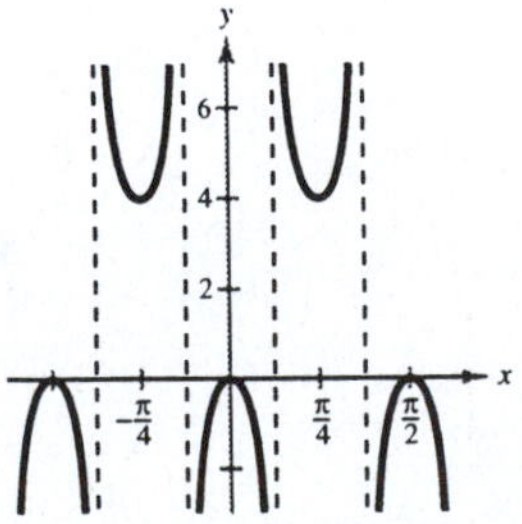

20. $y = \csc \dfrac{x}{3}$

Period $= \dfrac{2\pi}{(1/3)} = 6\pi$

Asymptotes: $x = 0$, $x = 3\pi$

x	π	2π	4π
y	1.155	1.155	-1.155

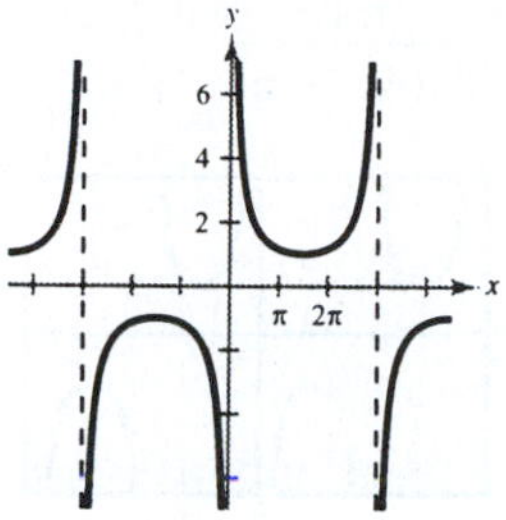

22. $y = 3 \cot \dfrac{\pi x}{2}$

Period $= \dfrac{\pi}{(\pi/2)} = 2$

Asymptotes: $x = 0$, $x = 2$

x	$\frac{1}{4}$	1	$\frac{3}{2}$
y	7.243	0	-3

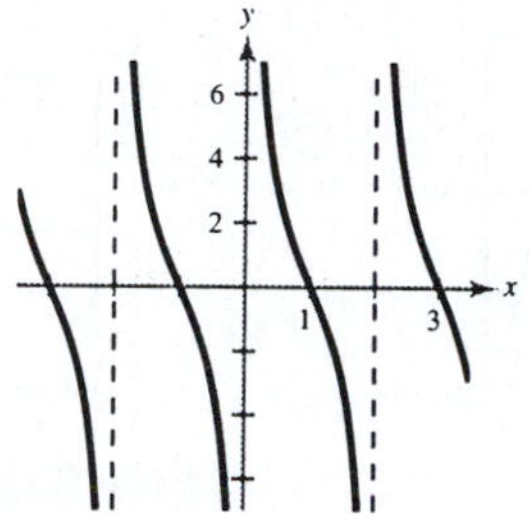

24. $y = -\dfrac{1}{2} \tan x$

Period $= \pi$

Asymptotes: $x = -\dfrac{\pi}{2}$, $x = \dfrac{\pi}{2}$

x	$-\frac{\pi}{4}$	0	$\frac{\pi}{4}$
y	$\frac{1}{2}$	0	$-\frac{1}{2}$

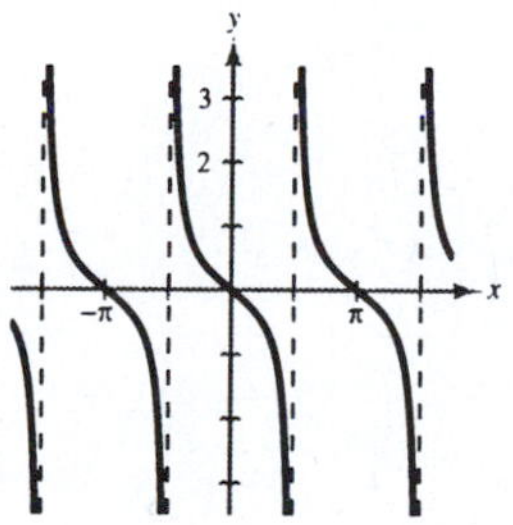

26. $y = \sec(x + \pi)$

Period $= 2\pi$

Asymptotes: $x = -\dfrac{\pi}{2}$, $x = \dfrac{\pi}{2}$

x	$-\frac{\pi}{4}$	0	$\frac{\pi}{4}$
y	-1.414	-1	-1.414

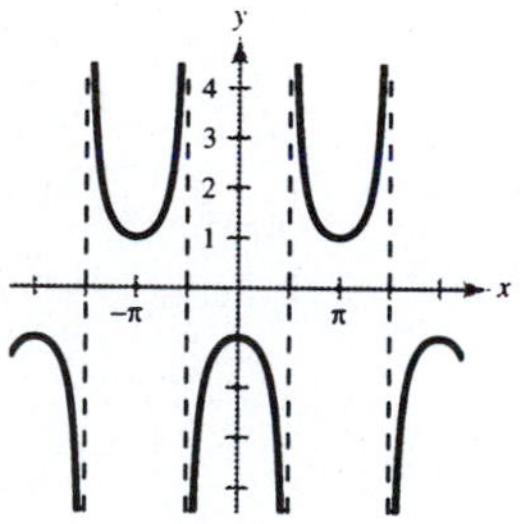

28. $y = \sec(\pi - x)$

Period $= 2\pi$

Asymptotes: $x = -\dfrac{\pi}{2}$, $x = \dfrac{\pi}{2}$

x	$-\frac{\pi}{4}$	0	$\frac{\pi}{4}$
y	-1.414	-1	-1.414

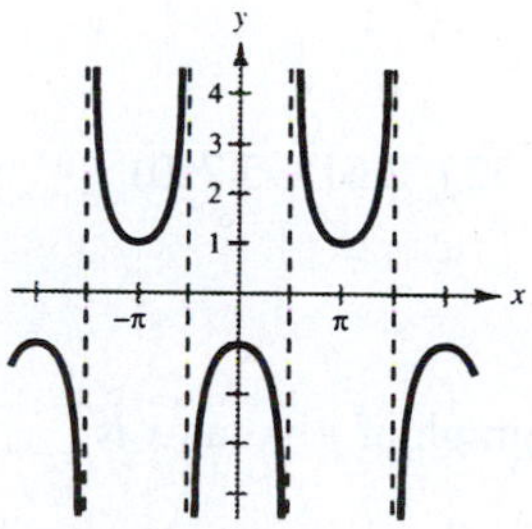

30. $y = 2\cot\left(x + \frac{\pi}{2}\right)$

Period $= \pi$

Asymptotes: $x = -\frac{\pi}{2}, x = \frac{\pi}{2}$

x	$-\frac{\pi}{4}$	0	$\frac{\pi}{4}$
y	2	0	-2

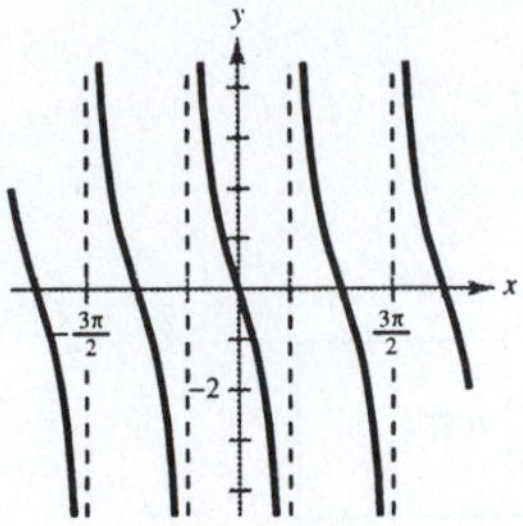

32. $y = -\tan 2x$

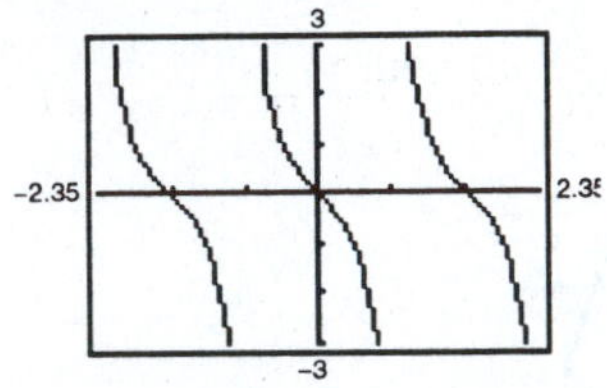

34. $y = \sec \pi x \Longrightarrow y = \frac{1}{\cos(\pi x)}$

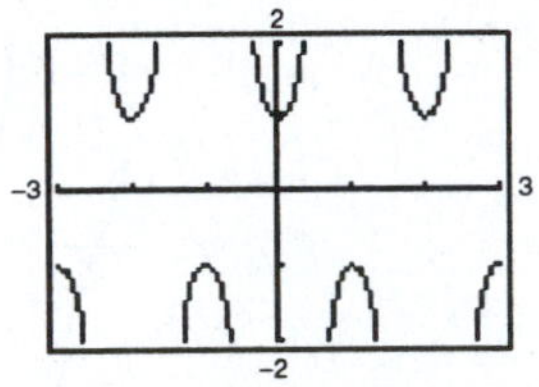

36. $y = -\csc(4x - \pi)$

$y = \frac{-1}{\sin(4x - \pi)}$

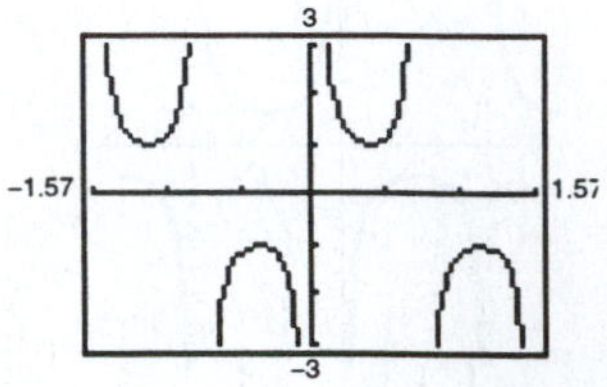

38. $y = 0.1\tan\left(\frac{\pi x}{4} + \frac{\pi}{4}\right)$

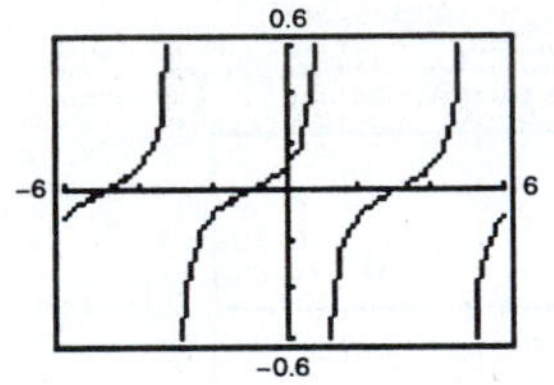

40. $y = \frac{1}{3}\sec\left(\frac{\pi x}{2} + \frac{\pi}{2}\right) \Longrightarrow y = \frac{1}{3\cos\left(\frac{\pi x}{2} + \frac{\pi}{2}\right)}$

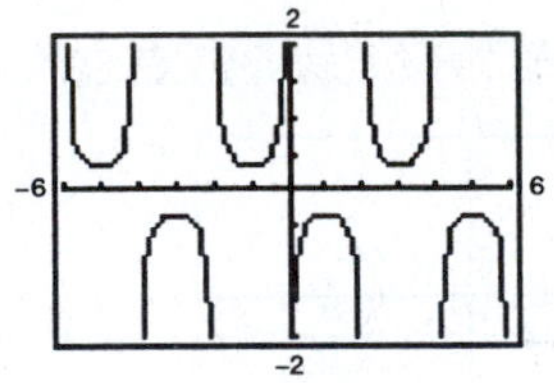

42.

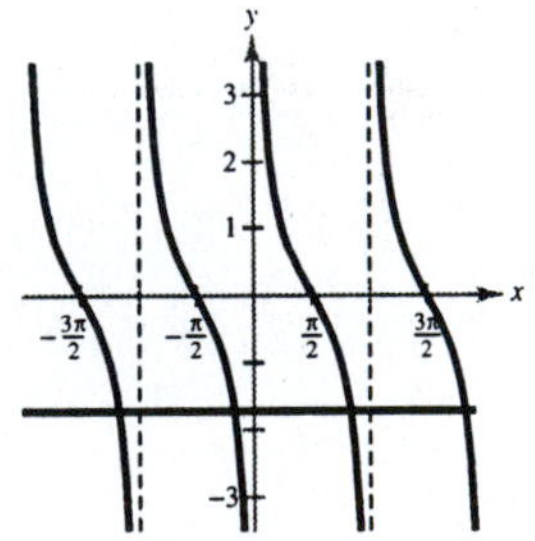

The solutions appear to be:

$x = -\frac{7\pi}{6}, -\frac{\pi}{6}, \frac{5\pi}{6}, \frac{11\pi}{6}$

(or in decimal form: $-3.665, -0.524, 2.618, 5.760$)

44.

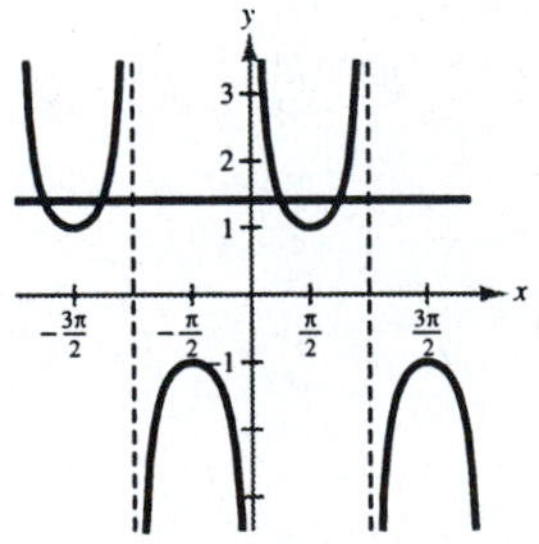

The solutions appear to be:

$-\frac{7\pi}{4}, -\frac{5\pi}{4}, \frac{\pi}{4}, \frac{3\pi}{4}$

(or in decimal form: $-5.498, -3.927, 0.785, 2.356$)

46. $f(x) = \tan x$

$\tan(-x) = -\tan x$

Thus, the function is odd and the graph of $y = \tan x$ is symmetric with the origin.

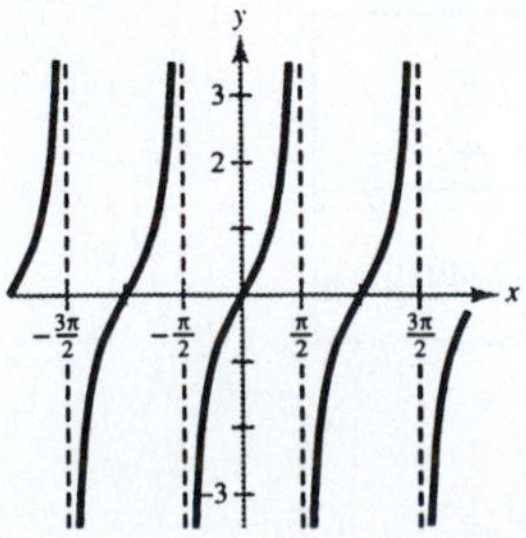

48. For $f(x) = \csc x$, as x approaches π from the left, f approaches ∞. As x approaches π from the right, f approaches $-\infty$.

50. $f(x) = \tan \dfrac{\pi x}{2},\ g(x) = \dfrac{1}{2} \sec \dfrac{\pi x}{2}$

(a)

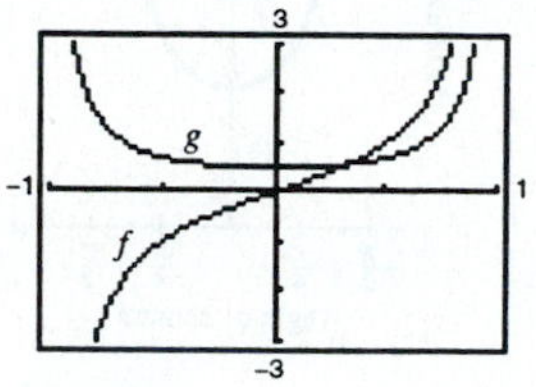

(b) The interval in which $f < g$ is $(-1, 1/3)$.

(c) The interval in which $2f < 2g$ is $(-1, 1/3)$, which is the same interval as part (b).

52. $y_1 = \sin x \sec x,\ y_2 = \tan x$

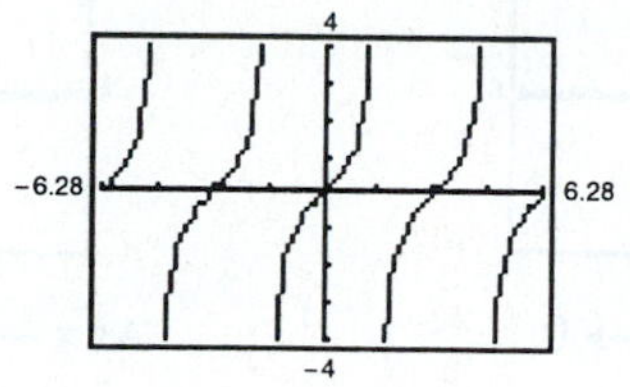

It appears that $y_1 = y_2$.

$$\sin x \sec x = \sin x \frac{1}{\cos x} = \frac{\sin x}{\cos x} = \tan x$$

54. $y_1 = \sec^2 x - 1,\ y_2 = \tan^2 x$

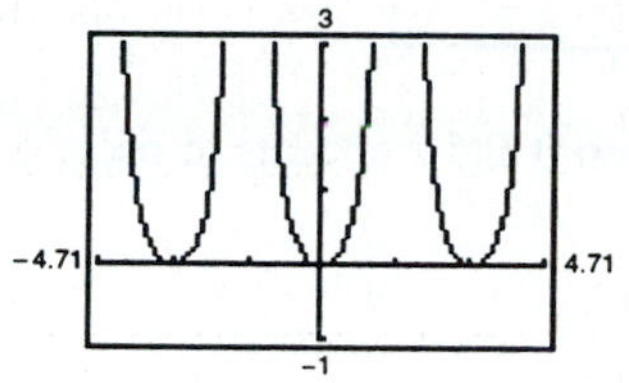

It appears that $y_1 = y_2$.

$1 + \tan^2 x = \sec^2 x$

$\tan^2 x = \sec^2 x - 1$

56. $f(x) = |x \sin x|$

Matches graph (a) as $x \longrightarrow 0, f(x) \longrightarrow 0$.

58. $g(x) = |x| \cos x$

Matches graph (c) as $x \longrightarrow 0, g(x) \longrightarrow 0$.

60. $f(x) = \sin x - \cos\left(x + \dfrac{\pi}{2}\right)$

$g(x) = 2 \sin x$

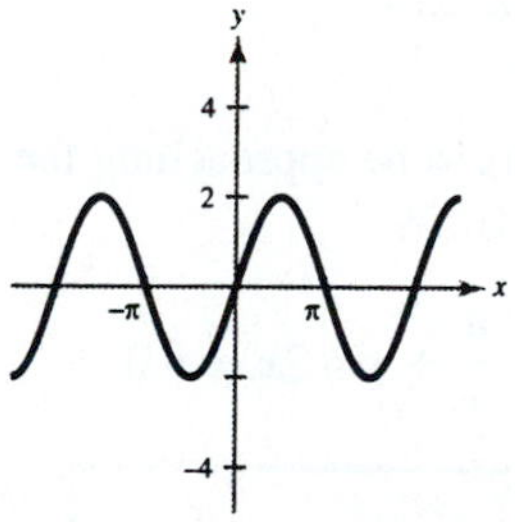

It appears that $f(x) = g(x)$. That is, that

$$\sin x - \cos\left(x + \frac{\pi}{2}\right) = 2 \sin x.$$

62. $f(x) = \cos^2 \dfrac{\pi x}{2}$

$g(x) = \dfrac{1}{2}(1 + \cos \pi x)$

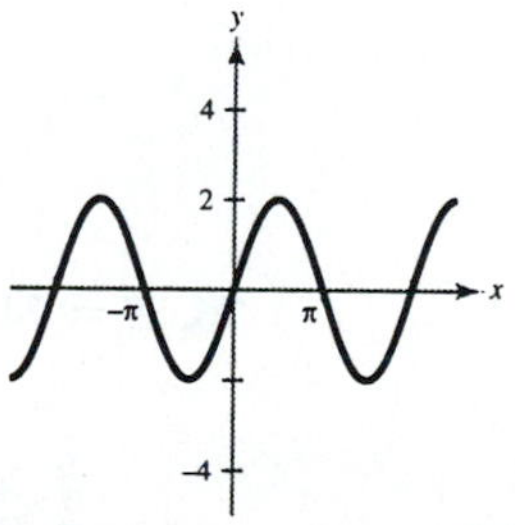

It appears that $f(x) = g(x)$. That is, that

$$\cos^2 \frac{\pi x}{2} = \frac{1}{2}(1 + \cos \pi x).$$

64. $f(x) = e^{-x}\cos x$

Damping factor: e^{-x}

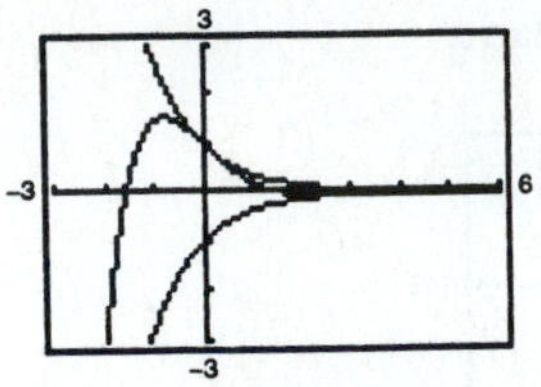

As $x \longrightarrow \infty, f(x) \longrightarrow 0$.

66. $h(x) = 2^{-x^2/4}\sin x$

Damping factor: $2^{-x^2/4}$

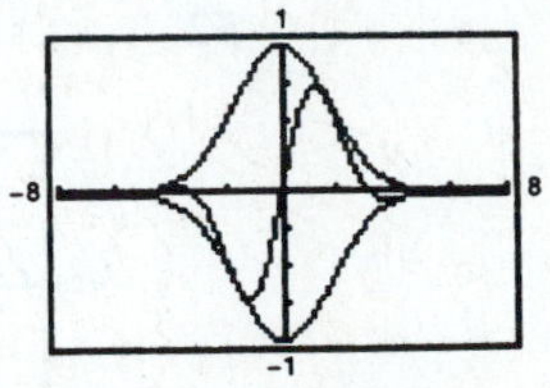

As $x \longrightarrow \infty, h(x) \longrightarrow 0$.

68. $\cos x = \dfrac{36}{d} \Longrightarrow d = \dfrac{36}{\cos x}$

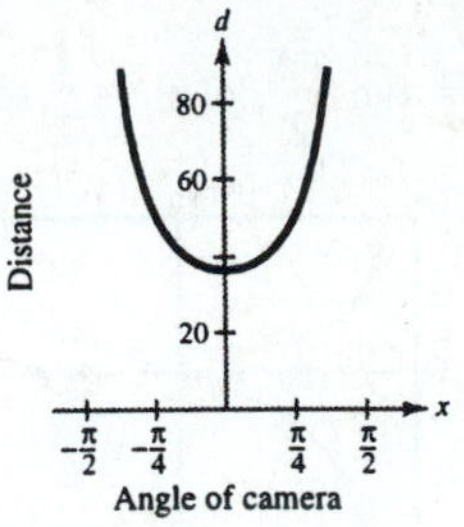

70. $H(t) = 54.33 - 20.38\cos\dfrac{\pi t}{6} - 15.69\sin\dfrac{\pi t}{6}$

$L(t) = 39.36 - 15.70\cos\dfrac{\pi t}{6} - 14.16\sin\dfrac{\pi t}{6}$

(a) Period of $\cos\dfrac{\pi t}{6}$: $\dfrac{2\pi}{(\pi/6)} = 12$

Period of $\sin\dfrac{\pi t}{6}$: $\dfrac{2\pi}{(\pi/6)} = 12$

Period of $H(t)$: 12

Period of $L(t)$: 12

(b) From the graph, it appears that the greatest difference between high and low temperatures occurs in summer. The smallest difference occurs in winter.

(c) The highest high and low temperatures appear to occur around the middle of July, roughly one month after the time when the sun is northernmost in the sky.

72. $f(x) = x - \cos x$

(a)

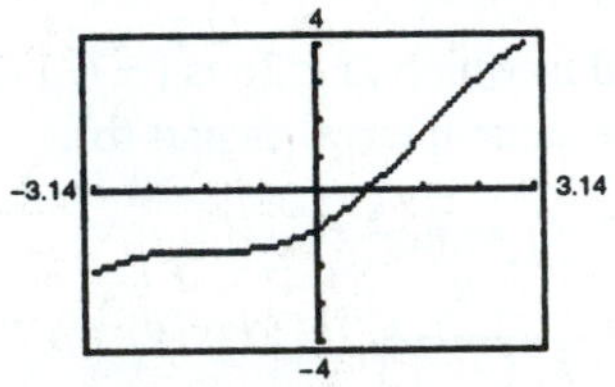

The zero between 0 and 1 appears to occur at $x \approx 0.739$.

(b) $x_n = \cos(x_{n-1})$

$x_0 = 1$

$x_1 = \cos 1 \approx 0.5403$

$x_2 = \cos 0.5403 \approx 0.8576$

$x_3 = \cos 0.8576 \approx 0.6543$

$x_4 = \cos 0.6543 \approx 0.7935$

$x_5 = \cos 0.7935 \approx 0.7014$

$x_6 = \cos 0.7014 \approx 0.7640$

$x_7 = \cos 0.7640 \approx 0.7221$

$x_8 = \cos 0.7221 \approx 0.7504$

$x_9 = \cos 0.7504 \approx 0.7314$

$\vdots$

This sequence appears to be approaching the zero of the zero $x = 0.739$.

74. $y_1 = \sec x$

$y_2 = 1 + \dfrac{x^2}{2!} + \dfrac{5x^4}{4!}$

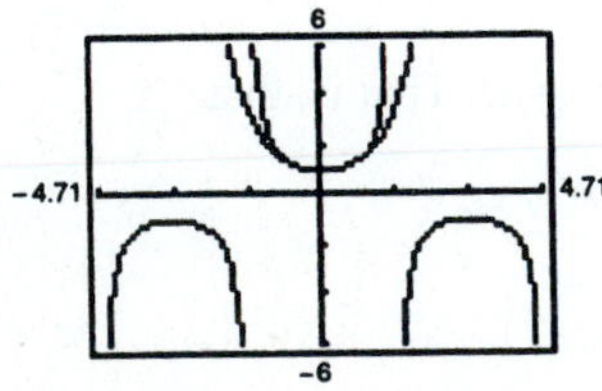

The approximation appears to be good for roughly $[-1.1, 1.1]$.

76. $S = 74 + 3x + 40\sin\dfrac{\pi t}{6}$

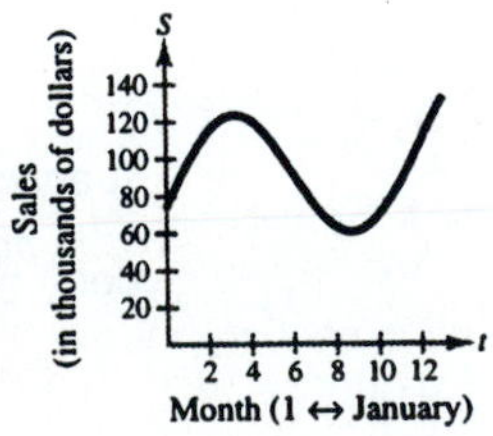

78. $y = \dfrac{4}{x} + \sin 2x, x > 0$

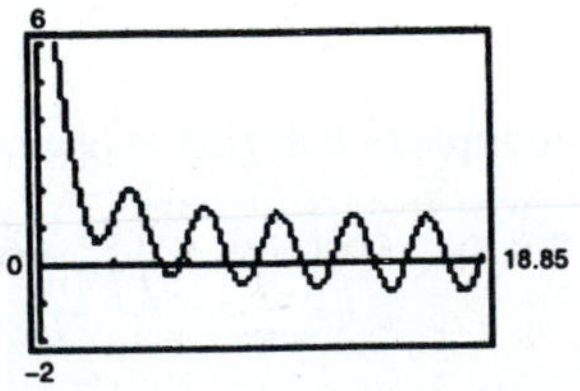

As $x \longrightarrow 0, f(x) \longrightarrow \infty$.

80. $f(x) = \dfrac{1 - \cos x}{x}$

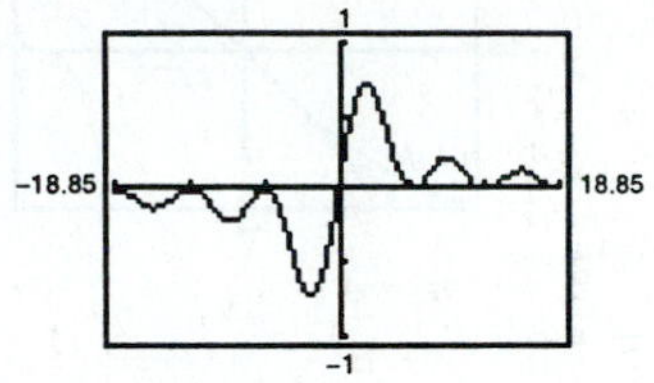

As $x \longrightarrow 0, f(x) \longrightarrow 0.$

82. $h(x) = x \sin \dfrac{1}{x}$

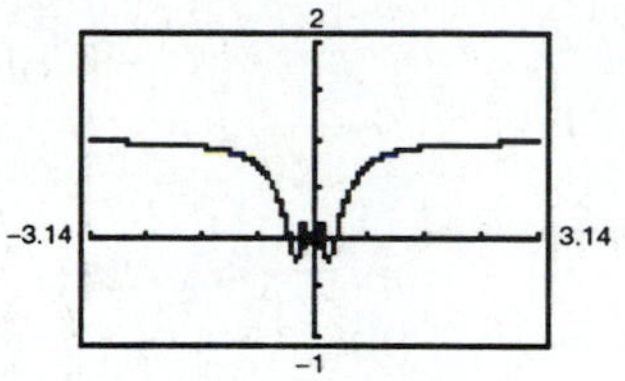

As $x \longrightarrow 0$, $h(x)$ oscillates.

84. $\dfrac{300}{1 + e^{-x}} = 100$

$$\frac{300}{100} = 1 + e^{-x}$$
$$3 = 1 + e^{-x}$$
$$2 = e^{-x}$$
$$\ln 2 = -x$$
$$x = -\ln 2 \approx 0.693$$

86. $\log_8 x + \log_8(x - 1) = \dfrac{1}{3}$

$$\log_8[x(x - 1)] = \frac{1}{3}$$
$$x(x - 1) = 8^{1/3}$$
$$x^2 - x = 2$$
$$x^2 - x - 2 = 0$$
$$(x - 2)(x + 1) = 0$$
$$x = 2, -1$$

Section 4.7 Inverse Trigonometric Functions

Solutions to Even-Numbered Exercises

2. The statement $\tan(5\pi/4) = 1 \Longrightarrow \arctan 1 = (5\pi/4)$ is false because $(5\pi/4)$ is not in the range of the arctangent function.

4. $y = \arcsin 0 \Longrightarrow \sin y = 0$ for $-\dfrac{\pi}{2} \le y \le \dfrac{\pi}{2} \Longrightarrow y = 0$

6. $y = \arccos 0 \Longrightarrow \cos y = 0$ for $0 \le y \le \pi \Longrightarrow y = \dfrac{\pi}{2}$

8. $y = \arctan(-1) \Longrightarrow \tan y = -1$ for $-\dfrac{\pi}{2} < y < \dfrac{\pi}{2} \Longrightarrow y = -\dfrac{\pi}{4}$

10. $y = \arcsin\left(-\dfrac{\sqrt{2}}{2}\right) \Longrightarrow \sin y = -\dfrac{\sqrt{2}}{2}$ for $-\dfrac{\pi}{2} \le y \le \dfrac{\pi}{2} \Longrightarrow y = -\dfrac{\pi}{4}$

12. $y = \arctan\left(\sqrt{3}\right) \Longrightarrow \tan y = \sqrt{3}$ for $-\dfrac{\pi}{2} < y < \dfrac{\pi}{2} \Longrightarrow y = \dfrac{\pi}{3}$

14. $y = \arcsin \dfrac{\sqrt{2}}{2} \Longrightarrow \sin y = \dfrac{\sqrt{2}}{2}$ for $-\dfrac{\pi}{2} \le y \le \dfrac{\pi}{2} \Longrightarrow y = \dfrac{\pi}{4}$

16. $y = \arctan\left(-\dfrac{\sqrt{3}}{3}\right) \Longrightarrow \tan y = -\dfrac{\sqrt{3}}{3}$ for $-\dfrac{\pi}{2} < y < \dfrac{\pi}{2} \Longrightarrow y = -\dfrac{\pi}{6}$

18. $y = \arccos 1 \Longrightarrow \cos y = 1$ for $0 \le y \le \pi \Longrightarrow y = 0$

20. $\arcsin 0.45 \approx 0.47$

22. $\arccos(-0.7) \approx 2.35$

24. $\arctan 15 \approx 1.50$

26. $\arccos 0.26 \approx 1.31$

28. $\arcsin(-0.125) \approx -0.13$

30. $\arctan 2.8 \approx 1.23$

32. $\arccos(-1) = \pi$

$\arccos\left(-\frac{1}{2}\right) = \frac{2\pi}{3}$

$\cos\left(\frac{\pi}{6}\right) = \frac{\sqrt{3}}{2}$

34. $f(x) = \sin x$

$g(x) = \arcsin x$

$y = x$

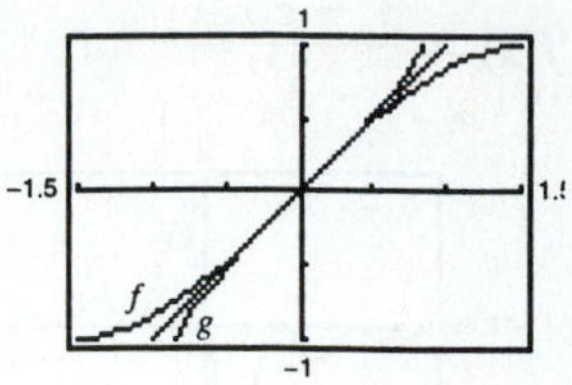

36. $\cos\theta = \frac{4}{x}$

$\theta = \arccos\frac{4}{x}$

38. $\tan\theta = \frac{x+1}{10}$

$\theta = \arctan\left(\frac{x+1}{10}\right)$

40. $\tan(\arctan 25) = 25$

42. $\sin[\arcsin(-0.2)] = -0.2$

44. $\arccos\left(\cos\frac{7\pi}{2}\right) = \arccos 0 = \frac{\pi}{2}$

Note: $(7\pi/2)$ is not in the range of the arccosine function.

46. Let $u = \arcsin\frac{4}{5}$,

$\sin u = \frac{4}{5}, 0 < u < \frac{\pi}{2}$.

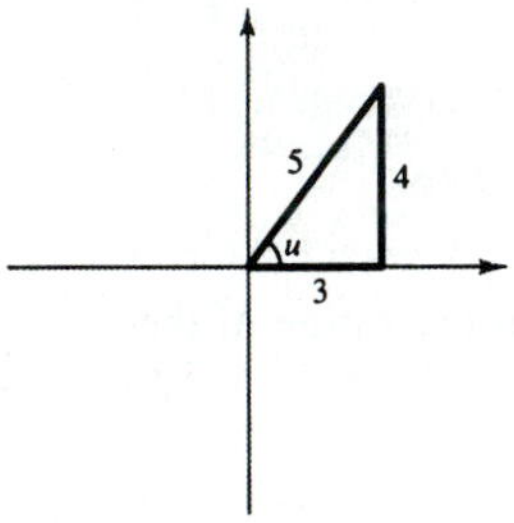

$\sec\left(\arcsin\frac{4}{5}\right) = \sec u = \frac{\text{hyp}}{\text{adj}} = \frac{5}{3}$

48. Let $u = \arccos\frac{\sqrt{5}}{5}$,

$\cos u = \frac{\sqrt{5}}{5}, 0 < u < \frac{\pi}{2}$.

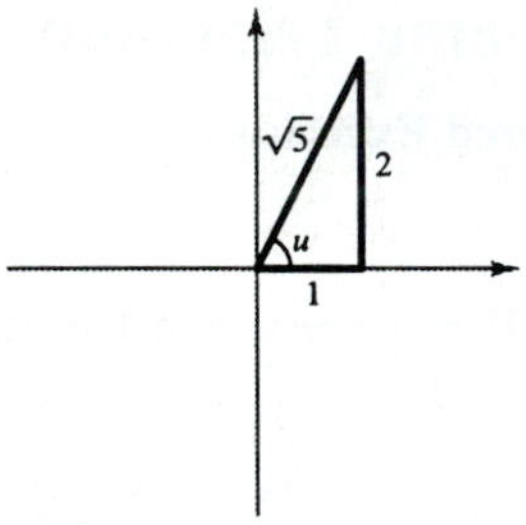

$\sin\left(\arccos\frac{\sqrt{5}}{5}\right) = \sin u = \frac{2}{\sqrt{5}} = \frac{2\sqrt{5}}{5}$

50. Let $u = \arctan\left(-\frac{5}{12}\right)$,

$\tan u = -\frac{5}{12}, -\frac{\pi}{2} < u < 0$.

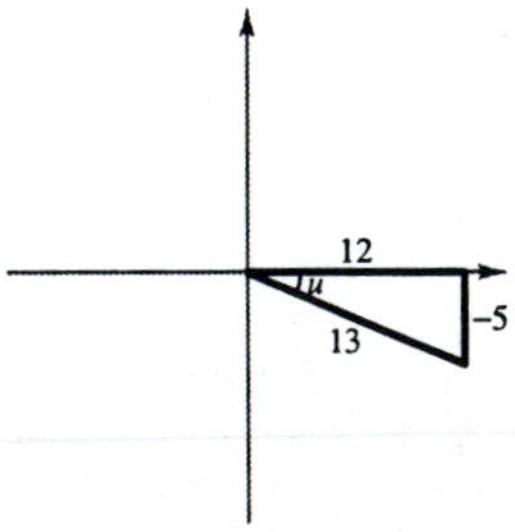

$\csc\left[\arctan\left(-\frac{5}{12}\right)\right] = \csc u = \frac{\text{hyp}}{\text{opp}} = -\frac{13}{5}$

52. Let $u = \arcsin\left(-\frac{3}{4}\right)$,

$\sin u = -\frac{3}{4}, -\frac{\pi}{2} < u < 0$.

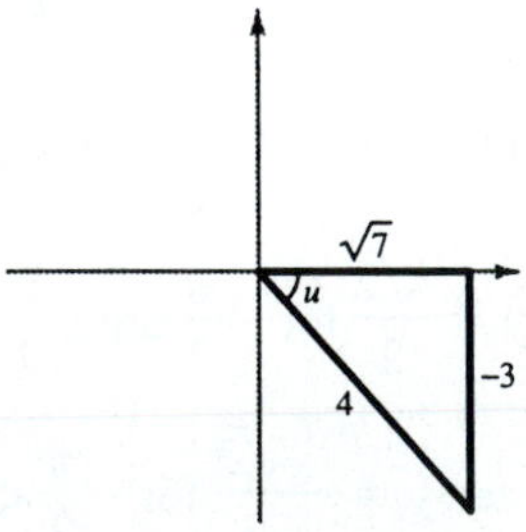

$\tan\left[\arcsin\left(-\frac{3}{4}\right)\right] = \tan u = -\frac{3}{\sqrt{7}} = -\frac{3\sqrt{7}}{7}$

54. Let $u = \arctan \frac{5}{8}$,

$\tan u = \frac{5}{8}, 0 < u < \frac{\pi}{2}$.

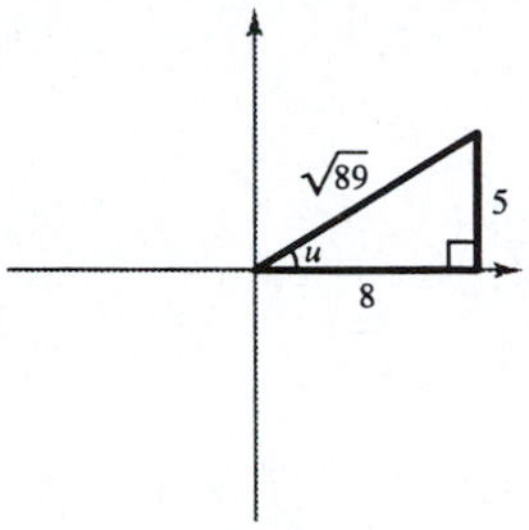

$$\cot\left(\arctan \frac{5}{8}\right) = \cot u = \frac{\text{adj}}{\text{opp}} = \frac{8}{5}$$

56. Let $u = \arctan x$,

$\tan u = x = \frac{x}{1}$.

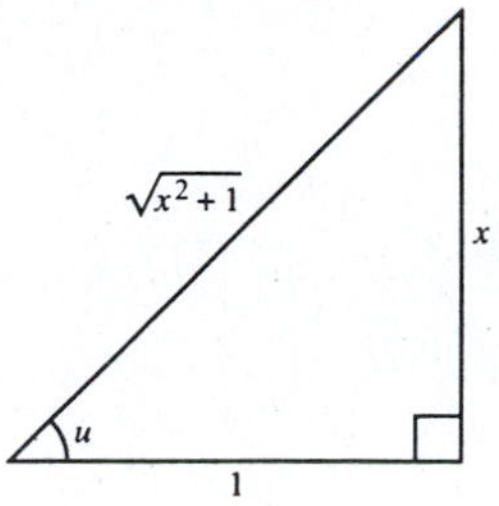

$$\sin(\arctan x) = \sin u = \frac{\text{opp}}{\text{hyp}} = \frac{x}{\sqrt{x^2 + 1}}$$

58. Let $u = \arctan 3x$,

$\tan u = 3x = \frac{3x}{1}$.

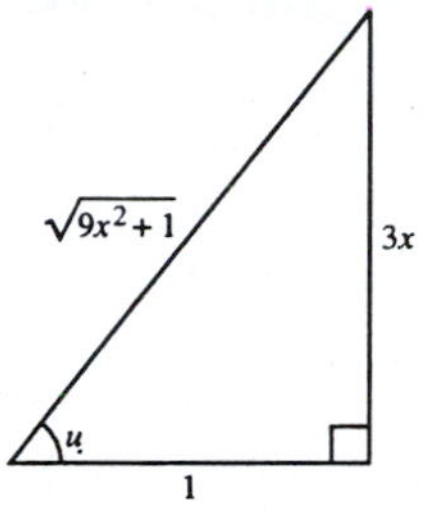

$$\sec(\arctan 3x) = \sec u = \frac{\text{hyp}}{\text{adj}} = \sqrt{9x^2 + 1}$$

60. Let $u = \arcsin(x - 1)$,

$\sin u = x - 1 = \frac{x - 1}{1}$.

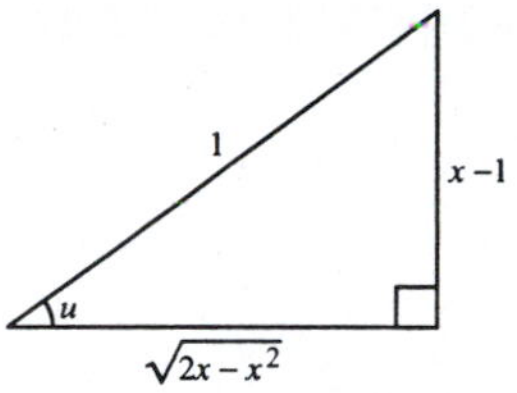

$$\sec[\arcsin(x - 1)] = \sec u = \frac{\text{hyp}}{\text{adj}} = \frac{1}{\sqrt{2x - x^2}}$$

62. Let $u = \arctan \frac{1}{x}$,

$\tan u = \frac{1}{x}$.

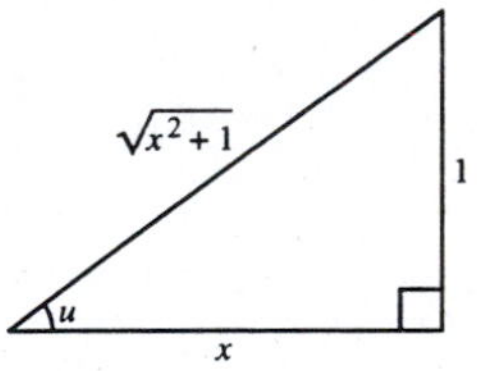

$$\cot\left(\arctan \frac{1}{x}\right) = \cot u = \frac{\text{adj}}{\text{opp}} = x$$

64. Let $u = \arcsin \frac{x - h}{r}$,

$\sin u = \frac{x - h}{r}$.

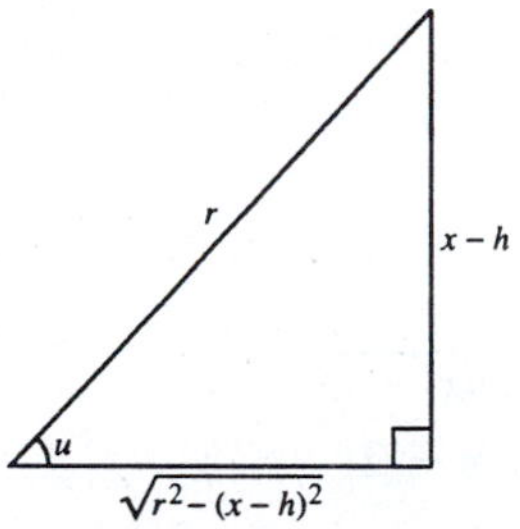

$$\cos\left(\arcsin \frac{x - h}{r}\right) = \cos u = \frac{\sqrt{r^2 - (x - h)^2}}{r}$$

66. $f(x) = \tan\left(\arccos \dfrac{x}{2}\right)$

$g(x) = \dfrac{\sqrt{4 - x^2}}{x}$

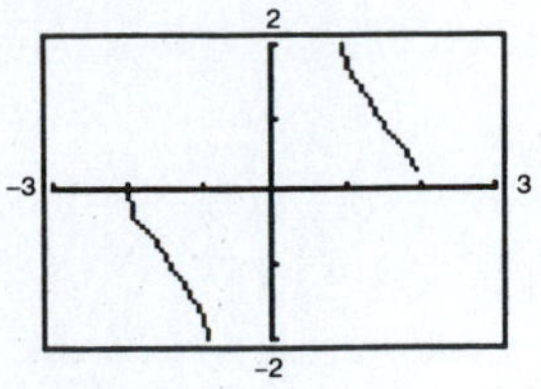

Asymptote: $x = 0$

These are equal because:

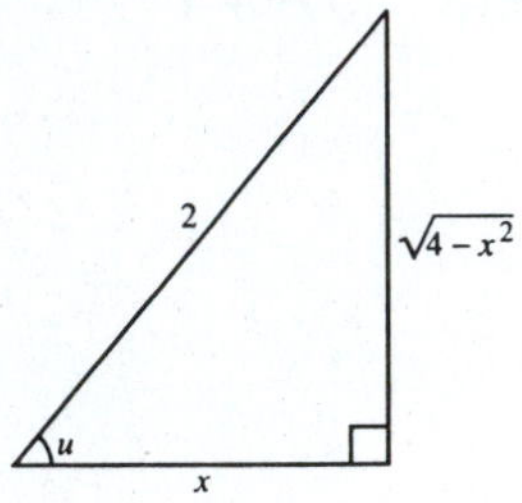

Let $u = \arccos \dfrac{x}{2}$.

$\tan\left(\arccos \dfrac{x}{2}\right) = \tan u = \dfrac{\sqrt{4 - x^2}}{x}$

68. If $\arcsin \dfrac{\sqrt{36 - x^2}}{6} = u$,

then $\sin u = \dfrac{\sqrt{36 - x^2}}{6}$.

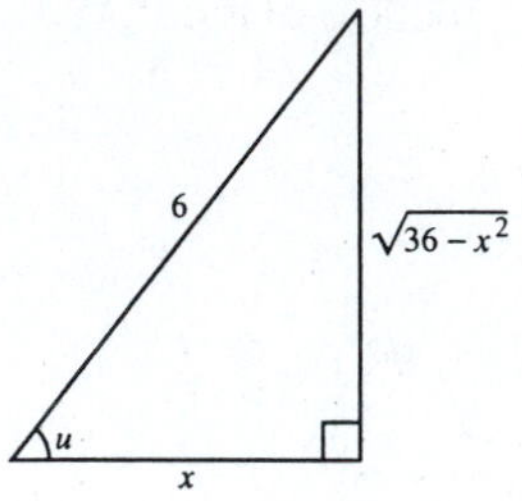

$\arcsin \dfrac{\sqrt{36 - x^2}}{6} = \arccos \dfrac{x}{6}$

70. If $\arccos \dfrac{x - 2}{2} = u$,

then $\cos u = \dfrac{x - 2}{2}$.

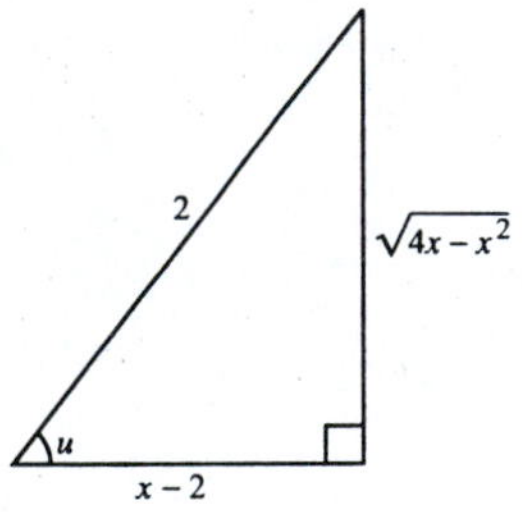

$\arccos \dfrac{x - 2}{2} = \arctan \dfrac{\sqrt{4x - x^2}}{x - 2}$

72. $y = \arcsin \dfrac{x}{2}$

Domain: $-2 \le x \le 2$

Range: $-\dfrac{\pi}{2} \le y \le \dfrac{\pi}{2}$

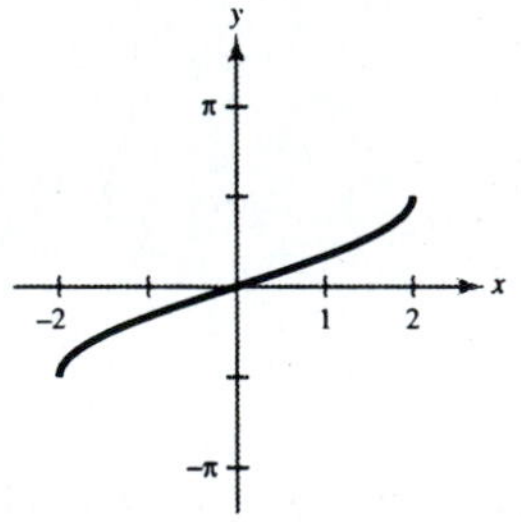

74. $g(t) = \arccos(t + 2)$

Domain: $-3 \le t \le -1$

This is the graph of $y = \arccos t$ shifted two units to the left.

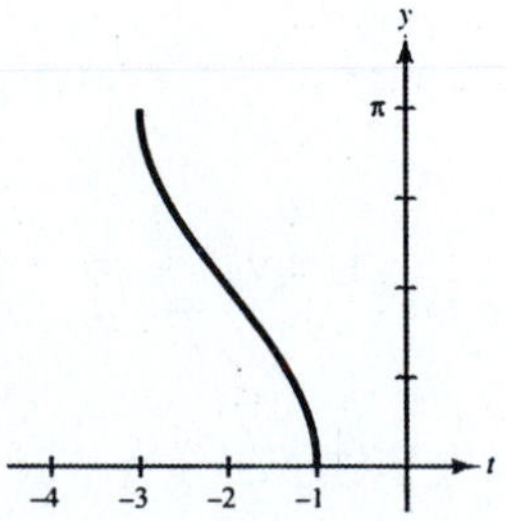

76. $f(x) = \dfrac{\pi}{2} + \arctan x$

Domain: $(-\infty, \infty)$

Range: $(0, \pi)$

This is the graph of $y = \arctan x$ shifted upward $\dfrac{\pi}{2}$ units.

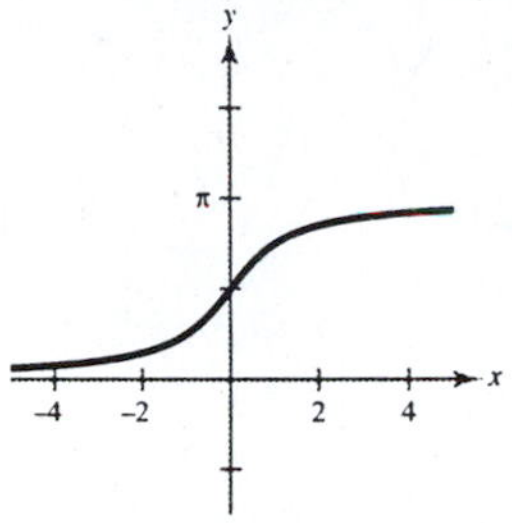

78. $f(x) = \arccos \dfrac{x}{4}$

Domain: $[-4, 4]$

Range: $[0, \pi]$

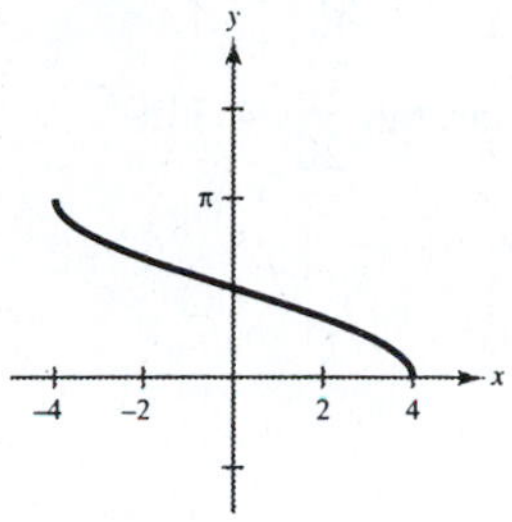

80. $f(t) = 4 \cos \pi t + 3 \sin \pi t$

$$= \sqrt{4^2 + 3^2} \sin\left(\pi t + \arctan \frac{4}{3}\right)$$

$$= 5 \sin\left(\pi t + \arctan \frac{4}{3}\right)$$

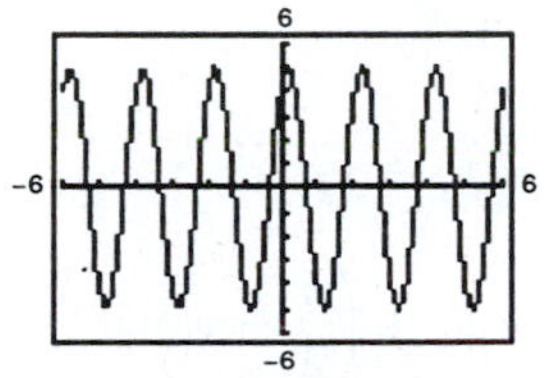

The graph implies that $A \cos \omega t + B \sin \omega t = \sqrt{A^2 + b^2} \sin\left(\omega t + \arctan \dfrac{A}{B}\right)$ is true.

82. $f(x) = \sqrt{x}$

$g(x) = 6 \arctan x$

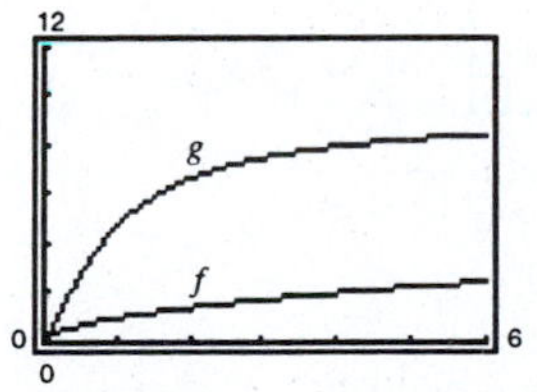

As x increases to infinity, g approaches 3π, but f has no maximum. Using the solve feature of the graphing utility, you find $a \approx 87.54$.

84. (a) $\tan \theta = \dfrac{s}{750}$

$\theta = \arctan \dfrac{s}{750}$

(b) When $s = 300$,

$\theta = \arctan \dfrac{300}{750} \approx 21.8°$.

When $s = 1200$,

$\theta = \arctan \dfrac{1200}{750} \approx 58.0°$.

86. Area $= \arctan b - \arctan a$

(a) $a = 0, b = 1$

Area $= \arctan 1 - \arctan 0 = \dfrac{\pi}{4} - 0 = \dfrac{\pi}{4}$

(b) $a = -1, b = 1$

Area $= \arctan 1 - \arctan(-1)$

$= \dfrac{\pi}{4} - \left(-\dfrac{\pi}{4}\right) = \dfrac{\pi}{2}$

(c) $a = 0, b = 3$

Area $= \arctan 3 - \arctan 0$

$\approx 1.25 - 0 = 1.25$

$= 1.25$

(d) $a = -1, b = 3$

Area $= \arctan 3 - \arctan(-1)$

$\approx 1.25 - \left(-\dfrac{\pi}{4}\right) \approx 2.03$

88. (a) $\tan \theta = \dfrac{x}{20}$

$\theta = \arctan \dfrac{x}{20}$

(b) When $x = 5$, $\theta = \arctan \dfrac{5}{20} \approx 14.0°$.

When $x = 12$, $\theta = \arctan \dfrac{12}{20} \approx 31.0°$.

90. $y = \text{arcsec}\, x$ if and only if $\sec y = x$ where $x \le -1 \cup x \ge 1$ and $0 \le y \le \pi/2$ and $\pi/2 < y \le \pi$. The domain of y arcsec x is $(-\infty, -1] \cup [1, \infty)$ and the range is $[0, \pi/2) \cup (\pi/2, \pi]$.

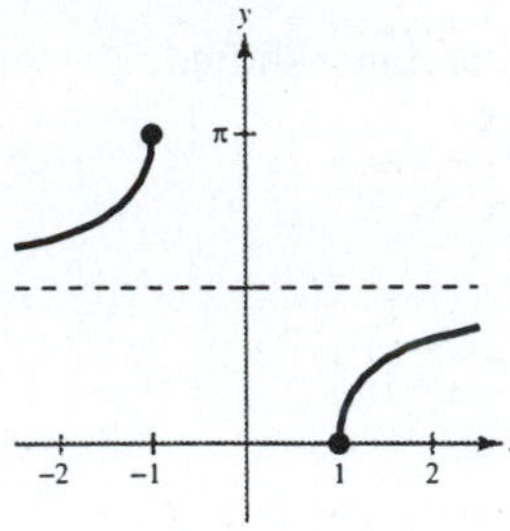

92. (a) $y = \text{arcsec}\, \sqrt{2} \Longrightarrow \sec y = \sqrt{2}$ and $0 \le y < \dfrac{\pi}{2} \cup \dfrac{\pi}{2} < y \le \pi \Longrightarrow y = \dfrac{\pi}{4}$

(b) $y = \text{arcsec}\, 1 \Longrightarrow \sec y = 1$ and $0 \le y < \dfrac{\pi}{2} \cup \dfrac{\pi}{2} < y \le \pi \Longrightarrow y = 0$

(c) $y = \text{arccot}\left(-\sqrt{3}\right) \Longrightarrow \cot y = -\sqrt{3}$ and $0 < y < \pi \Longrightarrow y = \dfrac{5\pi}{6}$

(d) $y = \text{arccsc}\, 2 \Longrightarrow \csc y = 2$ and $-\dfrac{\pi}{2} \le y < 0 \cup 0 < y \le \dfrac{\pi}{2} \Longrightarrow y = \dfrac{\pi}{6}$

94.
$$y = \arctan(-x)$$
$$\tan y = -x,\ -\frac{\pi}{2} < y < \frac{\pi}{2}$$
$$-\tan y = x$$
$$\tan(-y) = x,\ -\frac{\pi}{2} < -y < \frac{\pi}{2}$$
$$\arctan(\tan(-y)) = \arctan x$$
$$-y = \arctan x$$
$$y = -\arctan x$$

96. $y_2 = \dfrac{\pi}{2} - y_1$

$$\arctan x + \arctan \frac{1}{x} = y_1 + y_2$$
$$= y_1 + \left(\frac{\pi}{2} - y_1\right) = \frac{\pi}{2}$$

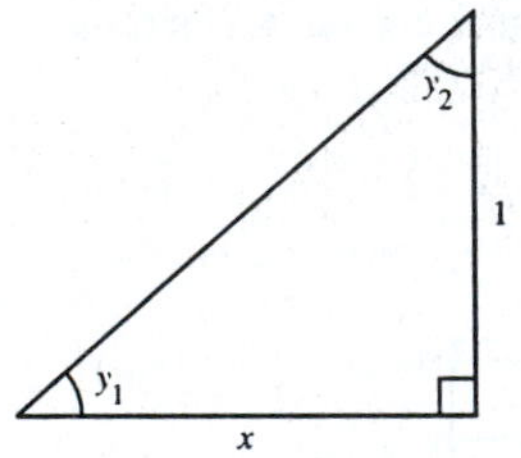

98. $\arcsin x = \text{arsin}\, \dfrac{x}{1} = \arctan \dfrac{x}{\sqrt{1 - x^2}}$

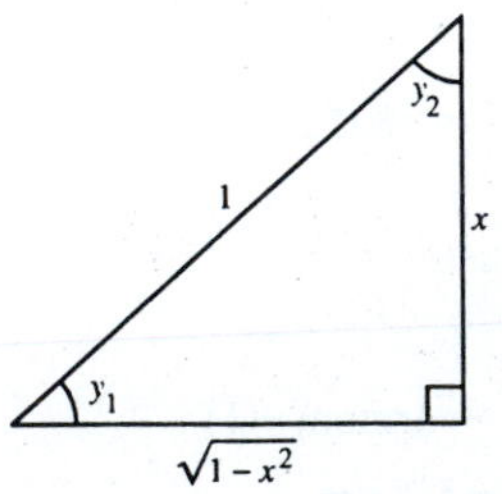

100. $739(1.3) = \$960.70$

102. Rate downstream: $18 + x$

Rate upstream: $18 - x$

$$\text{rate} \times \text{time} = \text{distance} \Longrightarrow t = \frac{d}{r}$$

(Time to go upstream) + (Time to go downstream) = 4

$$\frac{35}{18 - x} + \frac{35}{18 + x} = 4$$

$$35(18 + x) + 35(18 - x) = 4(18 - x)(18 + x)$$

$$630 + 35x + 630 - 35x = 4(324 - x^2)$$

$$1260 = 4(324 - x^2)$$

$$315 = 324 - x^2$$

$$x^2 = 9$$

$$x = \pm 3$$

The speed of the current is 3 miles per hour.

Section 4.8 Applications and Models

Solutions to Even-Numbered Exercises

2. $B = 54°, c = 15$

$$A = 90° - B = 90° - 54° = 36°$$

$$\sin B = \frac{b}{c} \Longrightarrow b = c \sin B$$

$$= 15 \sin 54° \approx 12.14$$

$$\cos B = \frac{a}{c} \Longrightarrow a = c \cos B$$

$$= 15 \cos 54° \approx 8.82$$

4. $A = 8.4°, a = 40.5$

$$B = 90° - A = 90° - 8.4° = 81.6°$$

$$\tan A = \frac{a}{b} \Longrightarrow b = \frac{a}{\tan A}$$

$$= \frac{40.5}{\tan 8.4°} \approx 274.27$$

$$\sin A = \frac{a}{c} \Longrightarrow c = \frac{a}{\sin A}$$

$$= \frac{40.5}{\sin 8.4°} \approx 277.24$$

6. $a = 25, c = 35$

$$b = \sqrt{c^2 - a^2} = \sqrt{35^2 - 25^2} = \sqrt{600} \approx 24.49$$

$$\sin A = \frac{a}{c} \Longrightarrow A = \arcsin \frac{a}{c}$$

$$= \arcsin \frac{25}{35} \approx 45.58$$

$$\cos B = \frac{a}{c} \Longrightarrow B = \arccos \frac{a}{c}$$

$$= \arccos \frac{25}{35} \approx 44.42°$$

8. $b = 1.32, c = 9.45$

$$a = \sqrt{c^2 - b^2} = \sqrt{87.5601} \approx 9.36$$

$$\cos A = \frac{b}{c} \Longrightarrow A = \arccos \frac{b}{c}$$

$$= \arccos \frac{1.32}{9.45} \approx 81.97°$$

$$\sin B = \frac{b}{c} \Longrightarrow B = \arcsin \frac{b}{c}$$

$$= \arcsin \frac{1.32}{9.45} \approx 8.03°$$

10. $B = 65° \, 12', a = 14.2$

$$A = 90° - B = 90° - 65° \, 12' = 24° \, 48'$$

$$\cos B = \frac{a}{c} \Longrightarrow c = \frac{a}{\cos B}$$

$$= \frac{14.2}{\cos 65° \, 12'} \approx 33.85$$

$$\tan B = \frac{b}{a} \Longrightarrow b = a \tan B$$

$$= 14.2 \tan 65° \, 12'$$

$$\approx 30.73$$

12. $\theta = 18°, b = 10$ meters

$$\tan \theta = \frac{\text{altitude}}{b/2}$$

$$\text{altitude} = \frac{b}{2} \tan \theta$$

$$= \frac{10}{2} \tan 18° \approx 1.62 \text{ meters}$$

14. $\tan 20° = \dfrac{600}{x}$

$$x = \frac{600}{\tan 20°} \approx 1648.5 \text{ feet}$$

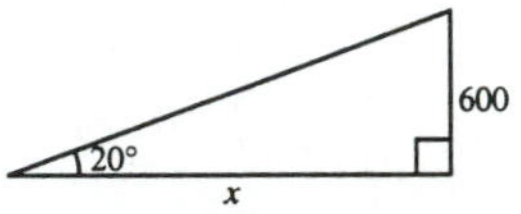

16. $\tan 33° = \dfrac{h}{125}$

$$h = 125 \tan 33° \approx 81.2 \text{ feet}$$

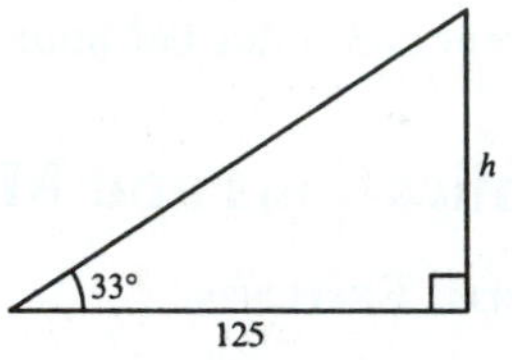

18. $\tan 28° = \dfrac{a}{100} \Longrightarrow a = 100 \tan 28°$

$$\tan 39.75° = \frac{a + s}{100}$$

$$a + s = 100 \tan 39.75°$$

$$s = 100 \tan 39.75° - a$$

$$= 100 \tan 39.75 - 100 \tan 28°$$

$$\approx 30 \text{ feet}$$

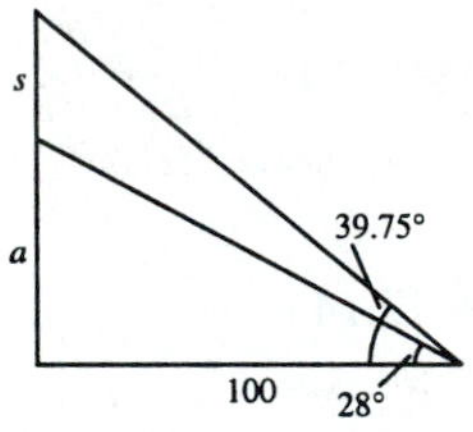

20. $\sin 50° = \dfrac{h}{100}$

$$h = 100 \sin 50° \approx 76.6 \text{ feet}$$

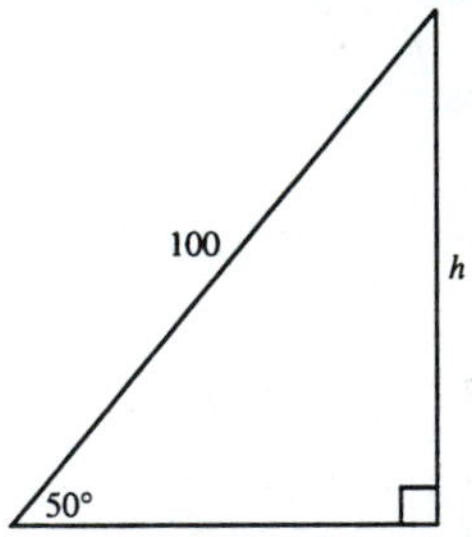

22. (a)

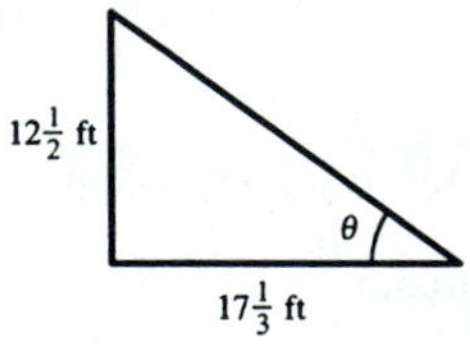

(b) $\tan \theta = \dfrac{12\frac{1}{2}}{17\frac{1}{3}}$

(c) $\theta = \arctan \dfrac{12\frac{1}{2}}{17\frac{1}{3}} \approx 35.8°$

24.

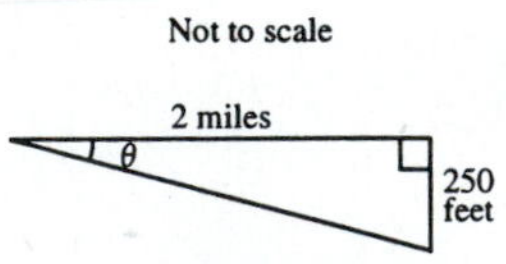

$$\tan \theta = \frac{250}{2(5280)}$$

$$\theta = \arctan \frac{250}{2(5280)} \approx 1.36°$$

26.

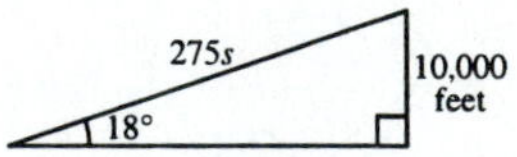

$$\sin 18° = \frac{10,000}{275 \, s}$$

$$s = \frac{10,000}{275(\sin 18°)} \approx 117.7 \text{ seconds}$$

28.

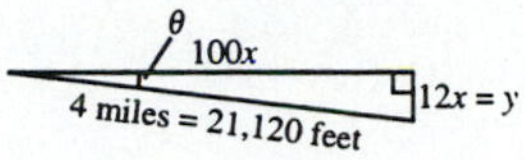

Angle of grade: $\tan \theta = \dfrac{12x}{100x}$

$\theta = \arctan 0.12 \approx 6.8°$

Change in elevation:

$\sin \theta = \dfrac{y}{21{,}120}$

$y = 21{,}120 \sin \theta$

$= 21{,}120 \sin(\arctan 0.12)$

≈ 2516.3 feet

30. $\sin 63° = \dfrac{a}{120} \Longrightarrow a \approx 107$ nautical miles south

$\cos 63° = \dfrac{b}{120} \Longrightarrow b \approx 54.5$ nautical miles west

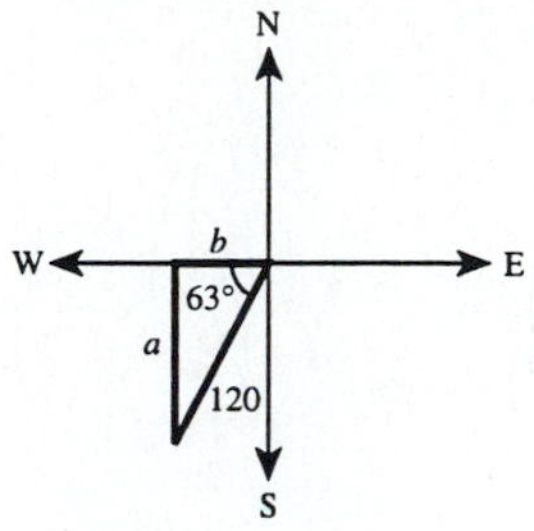

32. $\tan 14° = \dfrac{d}{x} \Longrightarrow x = d \cot 14°$

$\tan 34° = \dfrac{d}{y} \Longrightarrow \dfrac{d}{30 - x} = \dfrac{d}{30 - d \cot 14°}$

$\cot 34° = \dfrac{30 - d \cot 14°}{d}$

$d \cot 34° = 30 - d \cot 14°$

$d = \dfrac{30}{\cot 34° + \cot 14°}$

≈ 5.46 kilometers

34. $\tan \theta = \dfrac{85}{120} \Longrightarrow \theta = 35.3°$

Bearing: S 35.3° W

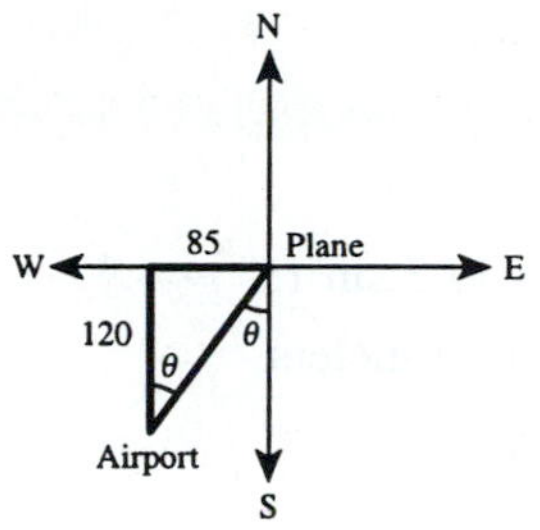

36.

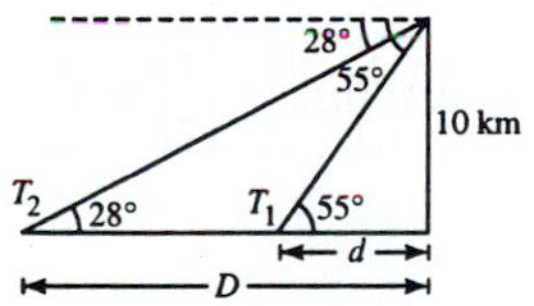

$\cot 55° = \dfrac{d}{10} \Longrightarrow d \approx 7$ kilometers

$\cot 28° = \dfrac{D}{10} \Longrightarrow D \approx 18.8$ kilometers

Distance between towns:

$D - d = 18.8 - 7 = 11.8$ kilometers

38.

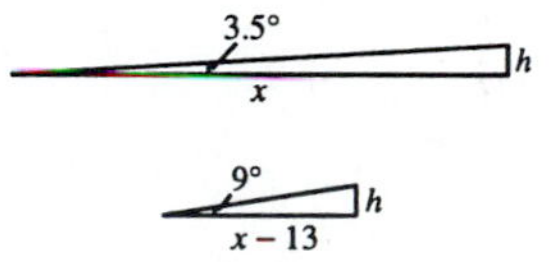

$\tan 3.5° = \dfrac{h}{x}, \tan 9° = \dfrac{h}{x - 13}$

$x = \dfrac{h}{\tan 3.5°}, x = \dfrac{h}{\tan 9°} + 13$

$\dfrac{h}{\tan 3.5°} = \dfrac{h}{\tan 9°} + 13$

$\dfrac{h}{\tan 3.5°} = \dfrac{h + 13 \tan 9°}{\tan 9°}$

$h \tan 9° = h \tan 3.5° + 13(\tan 9°)(\tan 3.5°)$

$h(\tan 9° - \tan 3.5°) = 13(\tan 9°)(\tan 3.5°)$

$h = \dfrac{13(\tan 9°)(\tan 3.5°)}{\tan 9° - \tan 3.5°} \approx 1.3 \text{ miles} \approx 6839 \text{ feet}$

40. $L_1 = 2x + y = 8 \Longrightarrow m_1 = -2$

$L_2 = x - 5y = -4 \Longrightarrow m_2 = \dfrac{1}{5}$

$\tan\alpha = \left|\dfrac{m_2 - m_1}{1 + m_2 m_1}\right|$

$\alpha = \arctan\left|\dfrac{m_2 - m_1}{1 + m_2 m_1}\right|$

$= \arctan\left|\dfrac{\frac{1}{5} - (-2)}{1 + \frac{1}{5}(-2)}\right|$

$= \arctan\left(3\tfrac{2}{3}\right) \approx 74.7°$

42.

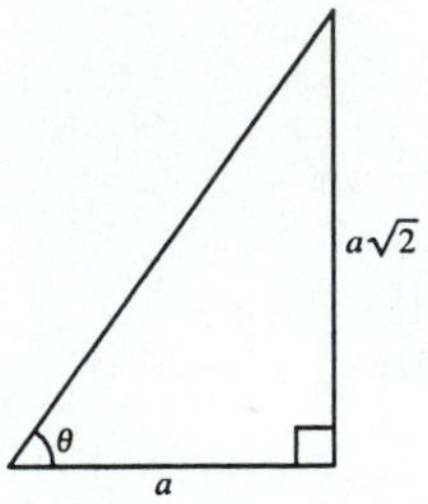

$\tan\theta = \dfrac{a\sqrt{2}}{a} = \sqrt{2}$

$\theta = \arctan\sqrt{2} \approx 54.7°$

44.

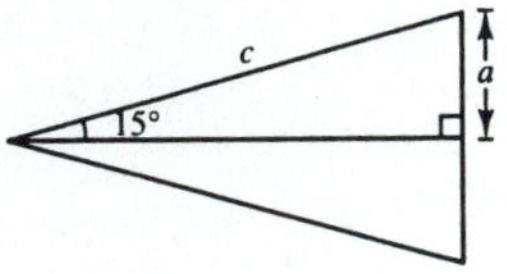

$c = \dfrac{35}{2} = 17.5$

$\sin 15° = \dfrac{a}{c}$

$a = c\sin 15° = 17.5\sin 15° \approx 4.53$

Distance $= 2a \approx 9.06$ centimeters

46.

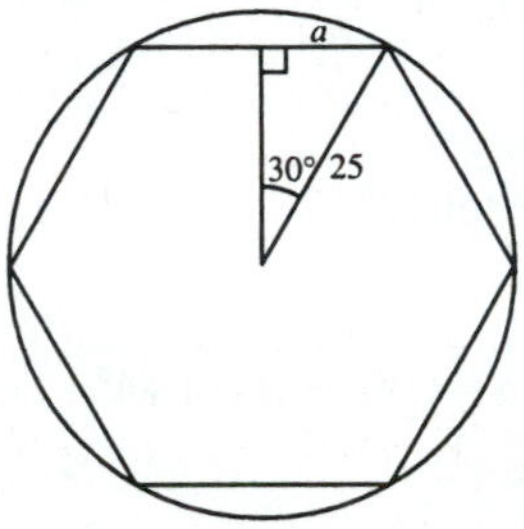

$\sin 30° = \dfrac{a}{25}$

$a = 25\sin 30° = 12.5$

Length of side $= 2a = 2(12.5) = 25$ inches

48.

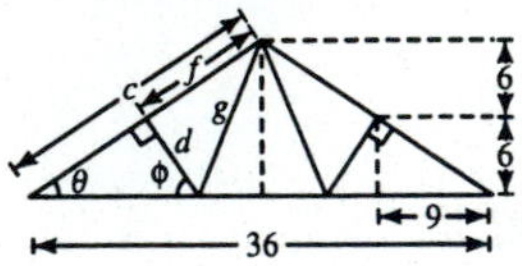

$\tan\theta = \dfrac{12}{18}$

$\theta = \arctan\dfrac{2}{3} = 0.588 \text{ rad} \approx 33.7°$

$\cos\theta = \dfrac{18}{c}$

$c = \dfrac{18}{\cos\theta} \approx 21.6$

$f \approx \dfrac{21.6}{2} = 10.8$

$\phi \approx 90 - 33.7 = 56.3°$

$\sin\phi = \dfrac{6}{d}$

$d = \dfrac{6}{\sin\phi} \approx 7.2$

$g = \sqrt{10.8^2 + 7.2^2} \approx 12.98$

50. $d = \dfrac{1}{2}\cos 20\pi t$

(a) Maximum displacement: $|a| = \left|\dfrac{1}{2}\right| = \dfrac{1}{2}$

(b) Frequency

$\dfrac{\omega}{2\pi} = \dfrac{20\pi}{2\pi} = 10$

(c) Least positive value for t for which $d = 0$

$\dfrac{1}{2}\cos 20\pi t = 0$

$\cos 20\pi t = 0$

$20\pi t = \arccos 0$

$20\pi t = \dfrac{\pi}{2}$

$t = \dfrac{\pi}{2} \cdot \dfrac{1}{20\pi} = \dfrac{1}{40}$

52. $d = \frac{1}{64} \sin 792 \pi t$

(a) Maximum displacement:

$$|a| = \left|\frac{1}{64}\right| = \frac{1}{64}$$

(b) Frequency

$$\frac{\omega}{2\pi} = \frac{792\pi}{2\pi} = 396$$

(c) Least positive value for t for which $d = 0$

$$\frac{1}{64} \sin 792 \pi t = 0$$

$$\sin 792 \pi t = 0$$

$$792 \pi t = \arcsin 0$$

$$792 \pi t = \pi$$

$$t = \frac{\pi}{792 \pi} = \frac{1}{792}$$

54. Displacement at $t = 0$ is $0 \Longrightarrow d = a \sin \omega t$

Amplitude: $|a| = 3$

Period: $\frac{2\pi}{\omega} = 6 \Longrightarrow \omega = \frac{\pi}{3}$

$d = 3 \sin\left(\frac{\pi t}{3}\right)$

56. Displacement at $t = 0$ is $2 \Longrightarrow d = a \cos \omega t$

Amplitude: $|a| = 2$

Period: $\frac{2\pi}{\omega} = 10 \Longrightarrow \omega = \frac{\pi}{5}$

$d = 2 \cos\left(\frac{\pi t}{5}\right)$

58. At $t = 0$, buoy is at its high point $\Longrightarrow d = a \cos \omega t$.

Distance from high to low $= 2|a| = 3.5$

$$|a| = \frac{7}{4}$$

Returns to high point every 10 seconds:

Period $= \frac{2\pi}{\omega} = 10 \Longrightarrow \omega = \frac{\pi}{5}$

$d = \frac{7}{4} \cos \frac{\pi t}{5}$

60. (a)

θ	L_1	L_2	$L_1 + L_2$
0.1	$\frac{2}{\sin 0.1}$	$\frac{3}{\cos 0.1}$	23.0
0.2	$\frac{2}{\sin 0.2}$	$\frac{3}{\cos 0.2}$	13.1
0.3	$\frac{2}{\sin 0.3}$	$\frac{3}{\cos 0.3}$	9.9
0.4	$\frac{2}{\sin 0.4}$	$\frac{3}{\cos 0.4}$	8.4

(b)

0.5	$\frac{2}{\sin 0.5}$	$\frac{3}{\cos 0.5}$	7.6
0.6	$\frac{2}{\sin 0.6}$	$\frac{3}{\cos 0.6}$	7.2
0.7	$\frac{2}{\sin 0.7}$	$\frac{3}{\cos 0.7}$	7.0
0.8	$\frac{2}{\sin 0.8}$	$\frac{3}{\cos 0.8}$	7.1

The minimum length of the elevator is 7.0 meters.

(c) $L = L_1 + L_2 = \frac{2}{\sin \theta} + \frac{3}{\cos \theta}$

(d)

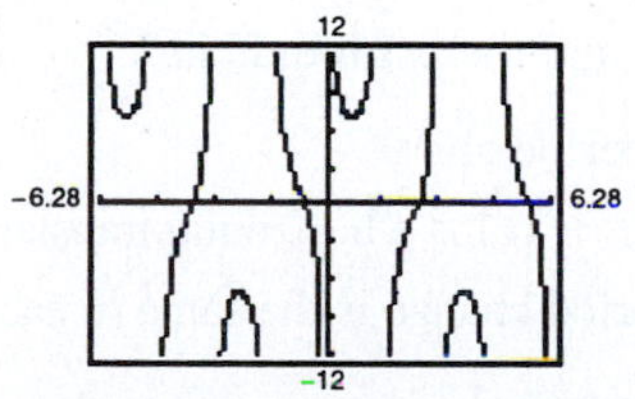

From the graph, it appears that the minimum length is 7.0 meters, which agrees with the estimate of part (b).

62. (a)

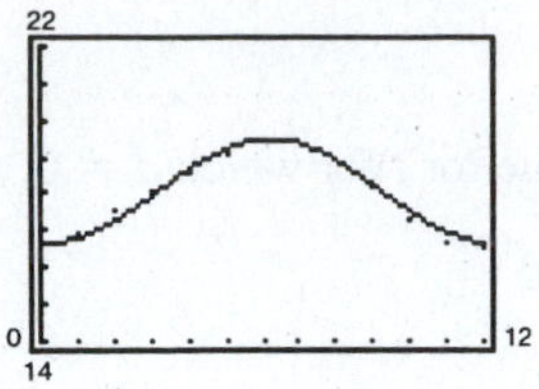

(b) Period $= \dfrac{2\pi}{b} = \dfrac{2\pi}{(\pi/6)} = 12$

Yes, this is what is expected there are 12 months in one year.

(c) Amplitude: $|a| = |1.41| = 1.41$, which represents the maximum change in time of sunset from the average time ($d = 18.09$) of sunset.

64. $3x - 2y = 4$ graph is a line.

$y = \dfrac{3}{2}x - 2$

$m = \dfrac{3}{2}$, y-intercept $= -2$

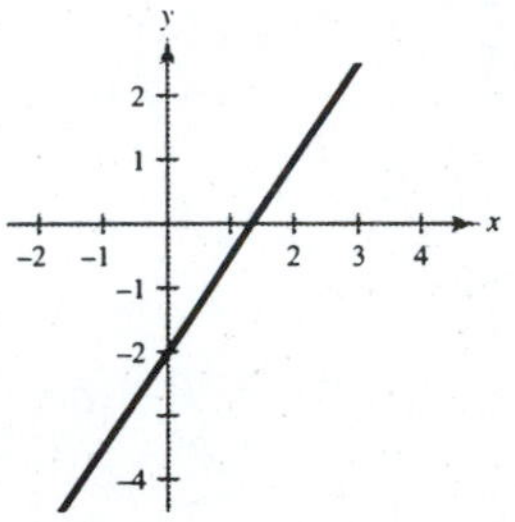

66. $\dfrac{x^2}{4} + y^2 = 1$ graph is an ellipse.

Center: (0, 0)

Major axis: x-axis, length: 4

Minor axis: y-axis, length: 2

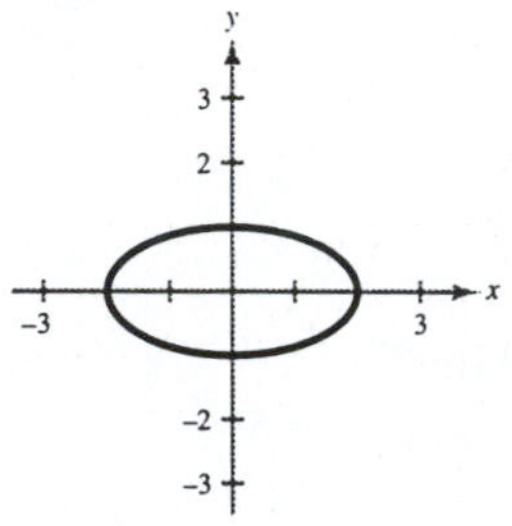

❑ Focus on Concepts

1. (a) The vertex is at the origin and the initial side is on the positive x-axis.
(b) Clockwise rotation of the terminal side.
(c) Two angles in standard position where the terminal sides coincide
(d) The magnitude of the angle is between 90° and 180°

2. Increases. The linear velocity is proportional to the radius.

3. False. For each θ there corresponds exactly one value of y.

4. Corresponding sides of similar triangles are proportional.

5. Undefined because $\sec\theta = 1/\cos\theta$.

6. Determine the trigonometric function of the reference angle and, depending on the quadrant in which the obtuse angle lies, prefix the appropriate sign.

7. d; the period is 2π and the amplitude is 3.

8. a; the period is 2π and, because $a < 0$, the graph is reflected about the x-axis.

9. b; the period is 2 and the amplitude is 2.

10. c; the period is 4π and the amplitude is 2.

11. (a) Equal; two-period shift
(b) Not equal; $f\left(t + \frac{1}{2}c\right)$ is a horizontal translation and $f\left(\frac{1}{2}t\right)$ is a period change.
(c) Equal; the period change is the same in each.

12. Their range is $(-\infty, \infty)$.

13. (a) The displacement is increased.
(b) The friction damps the oscillations more quickly.
(c) The frequency of the oscillations increases.

14. False. $3\pi/4$ is not in the range of the arctangent function.

❑ Review Exercises for Chapter 4

Solutions to Even-Numbered Exercises

2. $\frac{2\pi}{9}$

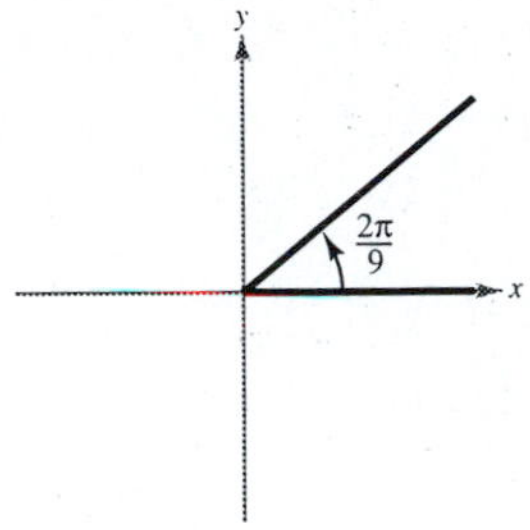

4. $-405°$

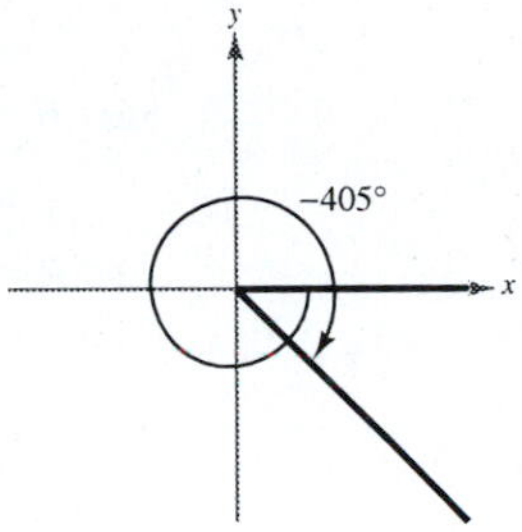

6. $-234°\ 50'' = -\left(234° + \frac{50°}{3600}\right)$
$\approx -(234° + 0.01°)$
$= -234.01°$

8. $280°\ 8'\ 50'' = 280° + \frac{8°}{60} + \frac{50°}{3600}$
$\approx 280° + 0.13° + 0.01°$
$= 280.15°$

10. $25.1° = 25° + (0.1)(60')$
$= 25°\ 6'$

12. $-327.85° = -(327° + (0.85)(60'))$
$= -327°\ 51'$

14. $-\frac{3\pi}{5} = -\frac{3\pi}{5}\left(\frac{180°}{\pi}\right) = -108°$

16. $1.75 = 1.75\left(\frac{180°}{\pi}\right) \approx 100.27°$

18. $-16.5° = -16.5°\left(\frac{\pi}{180°}\right) \approx -0.2880$

20. $84°\ 15' = 84° + \frac{15°}{60} = 84.25° = 84.25°\left(\frac{\pi}{180°}\right) \approx 1.4704$

22. $\theta = 640°$ is coterminal with $280°$.
$\theta' = 360° - 280° = 80°$

24. $\theta = \frac{17\pi}{3}$ is coterminal with $\frac{5\pi}{3}$.
$\theta' = 2\pi - \frac{5\pi}{3} = \frac{\pi}{3}$

26. $t = \frac{3\pi}{4}$ corresponds to the point:
$(x, y) = \left(-\frac{\sqrt{2}}{2}, \frac{\sqrt{2}}{2}\right)$
$\sin\frac{3\pi}{4} = y = \frac{\sqrt{2}}{2}$
$\cos\frac{3\pi}{4} = x = -\frac{\sqrt{2}}{2}$
$\tan\frac{3\pi}{4} = \frac{y}{x} = \frac{\sqrt{2}/2}{-\sqrt{2}/2} = -1$

28. $t = 2\pi$ corresponds to the point: $(x, y) = (1, 0)$
$\sin 2\pi = y = 0$
$\cos 2\pi = x = 1$
$\tan 2\pi = \frac{y}{x} = \frac{0}{1} = 0$

30. $x = 4, y = -8$

$r = \sqrt{4^2 + (-8)^2} = 4\sqrt{5}$

$\sin\theta = \dfrac{y}{r} = \dfrac{-8}{4\sqrt{5}} = -\dfrac{2\sqrt{5}}{5}$

$\cos\theta = \dfrac{x}{r} = \dfrac{4}{4\sqrt{5}} = \dfrac{\sqrt{5}}{5}$

$\tan\theta = \dfrac{y}{x} = \dfrac{-8}{4} = -2$

$\csc\theta = \dfrac{r}{y} = \dfrac{4\sqrt{5}}{-8} = -\dfrac{\sqrt{5}}{2}$

$\sec\theta = \dfrac{r}{x} = \dfrac{4\sqrt{5}}{4} = \sqrt{5}$

$\cot\theta = \dfrac{x}{y} = \dfrac{4}{-8} = -\dfrac{1}{2}$

32. $\tan\theta = \dfrac{y}{x} = -\dfrac{12}{5} \Longrightarrow r = 13$

$\sin\theta > 0 \Longrightarrow y = 12, x = -5$

$\sin\theta = \dfrac{y}{r} = \dfrac{12}{13}$

$\cos\theta = \dfrac{x}{r} = -\dfrac{5}{13}$

$\csc\theta = \dfrac{r}{y} = \dfrac{13}{12}$

$\sec\theta = \dfrac{r}{x} = \dfrac{13}{-5} = -\dfrac{13}{5}$

$\cot\theta = \dfrac{x}{y} = -\dfrac{5}{12}$

34. $\sec\dfrac{\pi}{4} = \dfrac{1}{\cos(\pi/4)} = \dfrac{1}{(\sqrt{2}/2)} = \sqrt{2}$

36. $\csc 270° = \dfrac{1}{\sin 270°} = \dfrac{1}{-1} = -1$

38. $\csc 105° = \dfrac{1}{\sin 105°} \approx 1.04$

40. $\sin\left(-\dfrac{\pi}{9}\right) \approx -0.34$

42. $\sec\theta$ is undefined $\Longrightarrow \cos\theta = 0.$

$\theta = 90° = \dfrac{\pi}{2}, \theta = 270° = \dfrac{3\pi}{2}$

44. $\cot\theta = -1.5399$

$\dfrac{1}{\tan\theta} = -1.5399$

$\theta = \tan^{-1}\left(\dfrac{1}{-1.5399}\right) \approx -33° = 327°$

or 5.7072 radians

$327° - 180° = 147°$ or 2.5656 radians

46. $\theta = 55.8°$

$\tan\theta = \tan 55.8° \approx 1.47$

48. $3x - 2y - 4 = 0$

$-2y = -3x - 4$

$y = \frac{3}{2}x + 2$

$\tan\theta = \frac{3}{2}$

$\theta \approx 56.3°$

50. $y = -2\sin\pi x$

Period $= \dfrac{2\pi}{\pi} = 2$

Amplitude: $|-2| = 2$

Reflected in x-axis

x	$-\frac{1}{2}$	0	$\frac{1}{2}$
y	2	0	-2

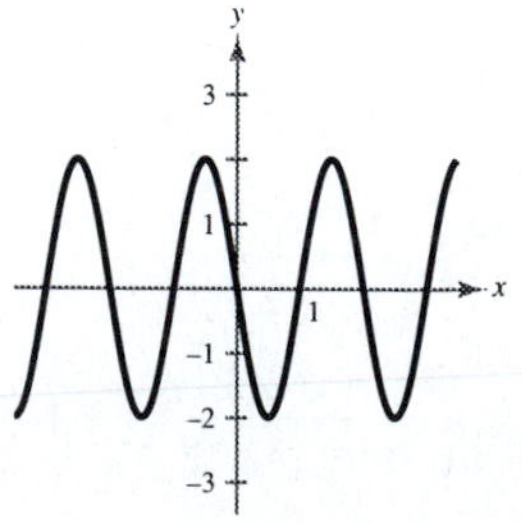

52. $f(x) = 8\cos\left(-\dfrac{x}{4}\right)$

Period $= \dfrac{2\pi}{(1/4)} = 8\pi$

Amplitude: 8

Reflected in y-axis

x	-4π	-2π	0	2π	4π
y	-8	0	8	0	-8

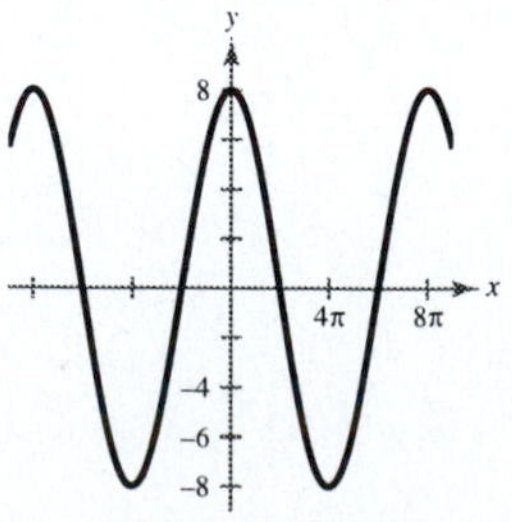

54. $f(x) = -\tan \dfrac{\pi x}{4}$

Period $= \dfrac{\pi}{(\pi/4)} = 4$

Asymptotes: $x = -2, x = 2$

Reflected in x-axis

x	-1	0	1
y	1	0	-1

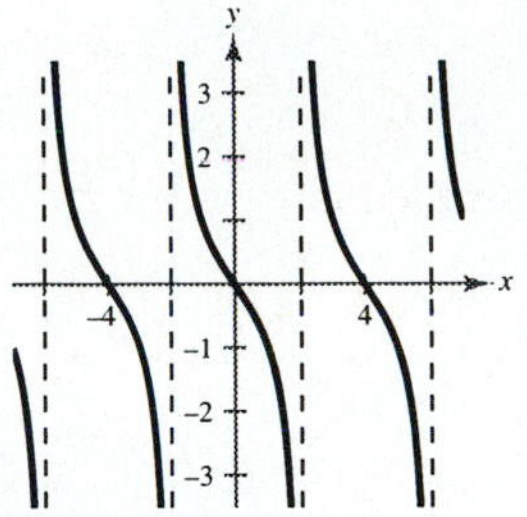

56. $g(t) = 3\cos(t + \pi)$

Period $= 2\pi$

Amplitude: 3

This is the graph of $y = 3\cos t$ shifted to the left π units.

t	$-\pi$	$-\dfrac{\pi}{2}$	0	$\dfrac{\pi}{2}$	π
$g(t)$	3	0	-3	0	3

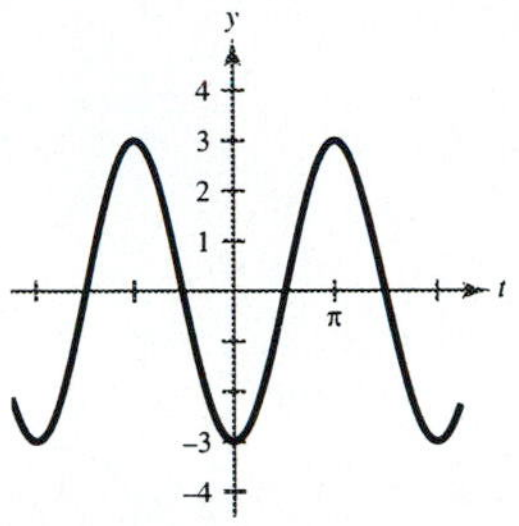

58. $h(t) = \sec\left(t - \dfrac{\pi}{4}\right)$

This is the graph of $y = \sec t$ shifted to the right $(\pi/4)$ units.

Period $= 2\pi$

Asymptotes: $x = -\dfrac{\pi}{4}, x = \dfrac{3\pi}{4}$

t	0	$\dfrac{\pi}{4}$	$\dfrac{\pi}{2}$
$h(t)$	1.414	1	1.414

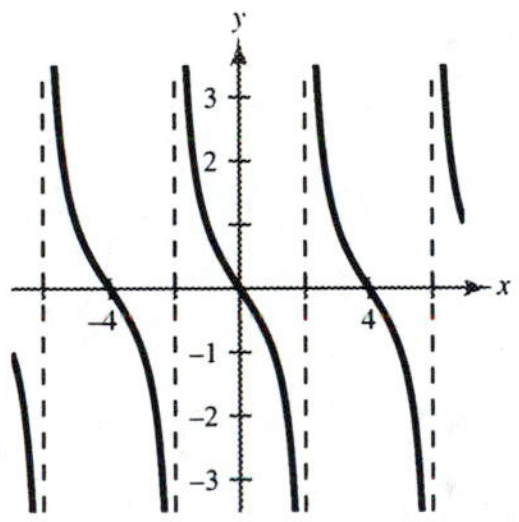

60. $y = 2 \arccos x$

Domain: $[-1, 1]$

Range: $[0, 2\pi]$

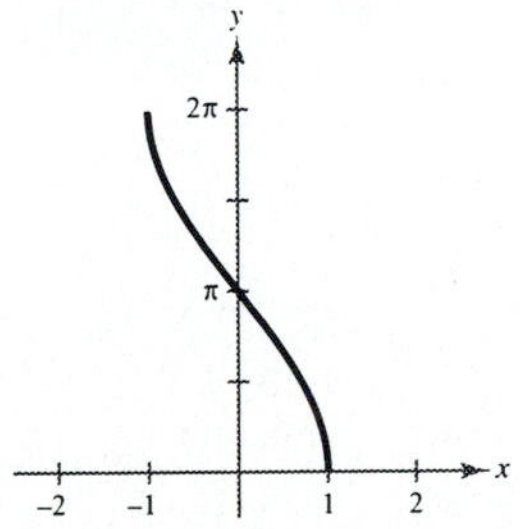

62. $y = \dfrac{x}{3} + \cos \pi x$

Not periodic

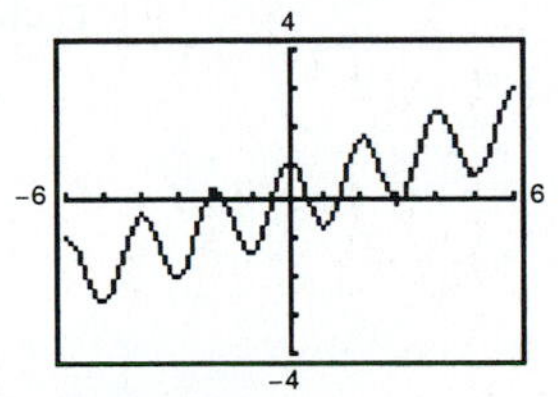

64. $y = 4 - \dfrac{x}{4} + \cos \pi x$

Not periodic

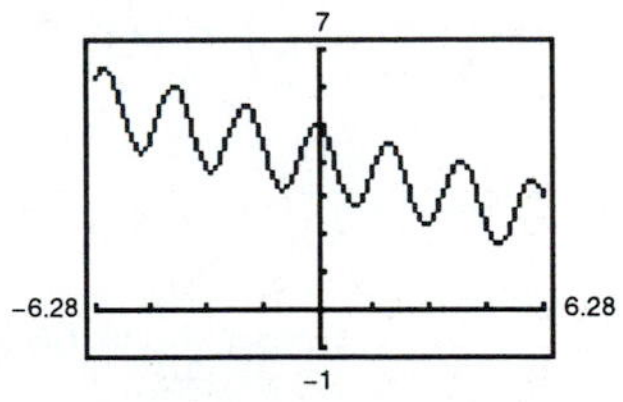

66. $f(\theta) = \cot \dfrac{\pi\theta}{8}$

Period: $\dfrac{\pi}{(\pi/8)} = 8$

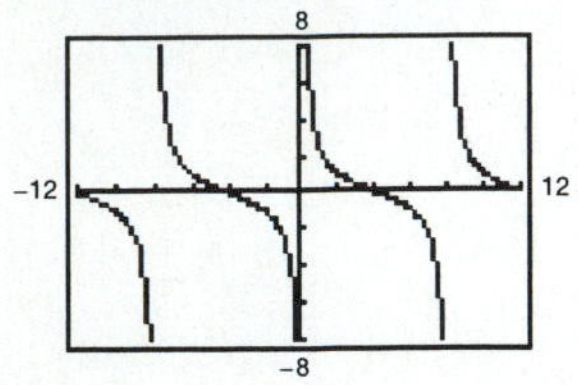

68. $f(x) = \arccos(x - \pi)$

Not periodic

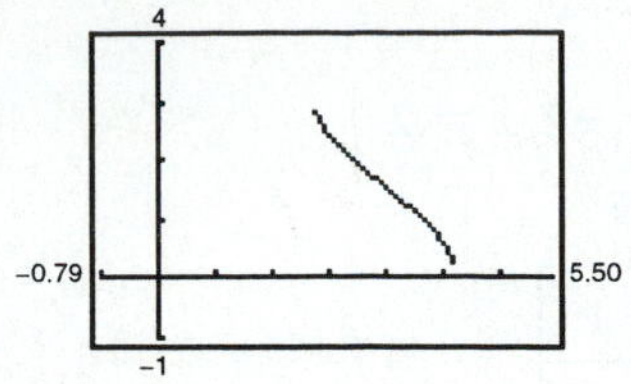

70. $g(x) = \sin e^x$

Not periodic

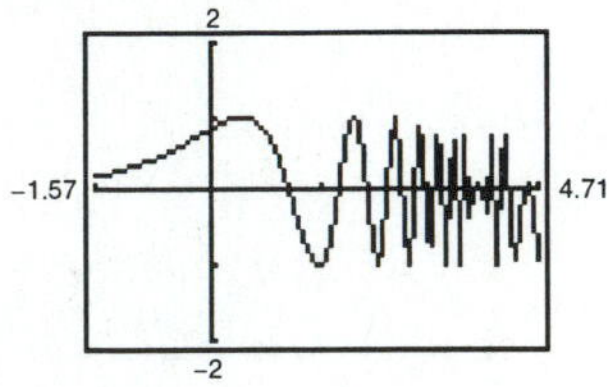

72. $h(x) = 4 \sin^2 x \cos^2 x$

Periodic

Maximum point: $\left(\dfrac{\pi}{4}, 1\right)$

Minimum point: $(0, 0)$

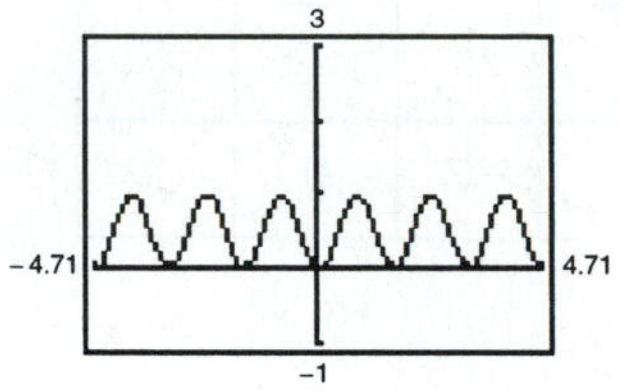

74. Let $u = \arccos \dfrac{x}{2}$,

$\cos u = \dfrac{x}{2}$.

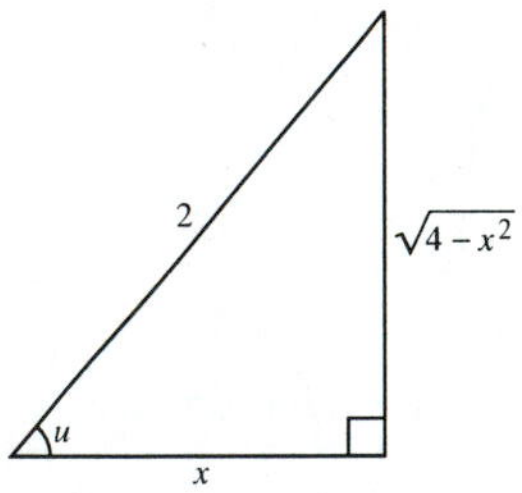

$\tan\left(\arccos \dfrac{x}{2}\right) = \tan u = \dfrac{\sqrt{4 - x^2}}{x}$

76. Let $u = \arcsin 10x$,

$\sin u = 10x$.

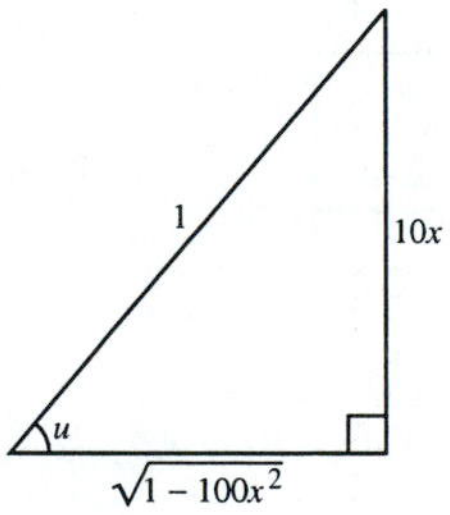

$\csc(\arcsin 10x) = \csc u = \dfrac{\text{hyp}}{\text{opp}} = \dfrac{1}{10x}$

78. $\tan \theta = \dfrac{70}{30}$

$\theta = \arctan \dfrac{70}{30} \approx 66.8°$

80. $\sin 48° = \dfrac{d_1}{650} \Longrightarrow d_1 \approx 483$
$\cos 25° = \dfrac{d_2}{810} \Longrightarrow d_2 \approx 734$ $\Big\}\; d_1 + d_2 = 1217$

$\cos 48° = \dfrac{d_3}{650} \Longrightarrow d_3 \approx 435$
$\sin 25° = \dfrac{d_4}{810} \Longrightarrow d_4 \approx 342$ $\Big\}\; d_3 - d_4 \approx 93$

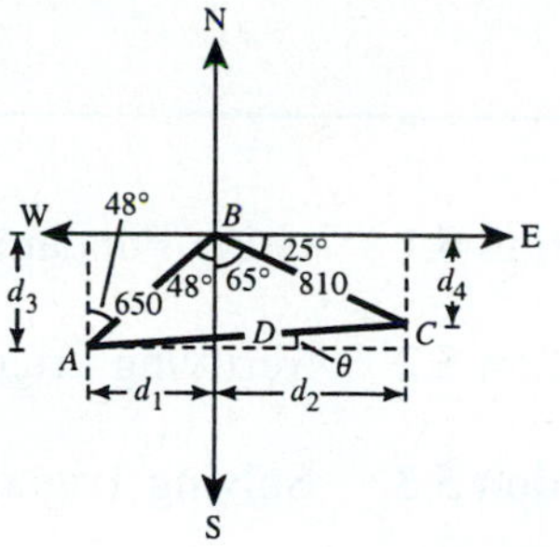

$\tan \theta \approx \dfrac{93}{1217} \Longrightarrow \theta \approx 4.4°$

$\sec 4.4° \approx \dfrac{D}{1217} \Longrightarrow D \approx 1217 \sec 4.4° \approx 1221$

The distance is 1221 miles and the bearing is N 85.6° E.

82. $\tan 14° = \dfrac{y}{37{,}000} \Longrightarrow y = 37{,}000 \tan 14° \approx 9225.1$ feet

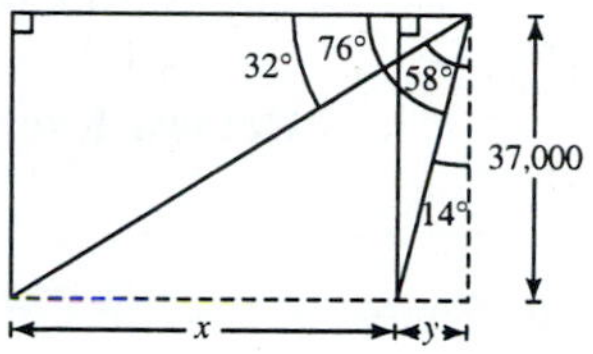

$\tan 58° = \dfrac{x + y}{37{,}000} \Longrightarrow x + y = 37{,}000 \tan 58° \approx 59{,}212.4$ feet

$x = 59{,}212.4 - 9225.1 \approx 49{,}987.2$ feet

84. (a)

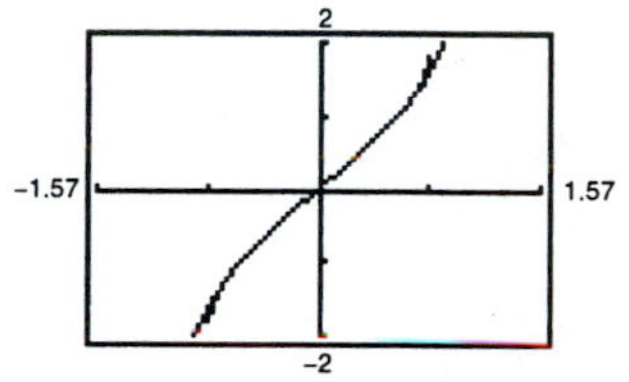

The polynomial approximation of the arcsine function is accurate over $-1 \le x \le 1$.

(b)

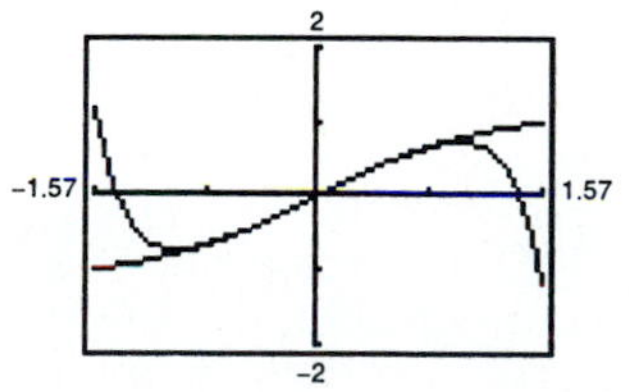

The polynomial approximation of the arctangent function accurate of $-\dfrac{1}{2} \le x \le \dfrac{1}{2}$.

(c) The next term appears to be $\dfrac{x^9}{9}$.

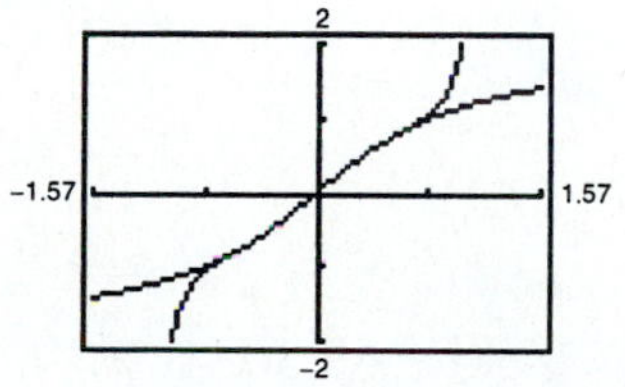

The accuracy of the approximation improved.

CHAPTER 5
Analytic Trigonometry

CHAPTER 5
Analytic Trigonometry

Section 5.1 Using Fundamental Identities

Solutions to Even-Numbered Exercises

2. $\tan x = \dfrac{\sqrt{3}}{3}, \cos x = -\dfrac{\sqrt{3}}{2}$

x is in Quadrant III.

$$\sin x = -\sqrt{1 - \left(-\frac{\sqrt{3}}{2}\right)^2} = -\sqrt{\frac{1}{4}} = -\frac{1}{2}$$

$$\csc x = \frac{1}{\sin x} = -2$$

$$\sec x = \frac{1}{\cos x} = -\frac{2}{\sqrt{3}} = -\frac{2\sqrt{3}}{3}$$

$$\cot x = \frac{1}{\tan x} = \frac{3}{\sqrt{3}} = \sqrt{3}$$

6. $\cot \phi = -3, \sin \phi = \dfrac{\sqrt{10}}{10}$

ϕ is in Quadrant II.

$$\cos \phi = \cot \phi \sin \phi = -\frac{3\sqrt{10}}{10}$$

$$\tan \phi = \frac{1}{\cot \phi} = -\frac{1}{3}$$

$$\csc \phi = \frac{1}{\sin \phi} = \sqrt{10}$$

$$\sec \phi = \frac{1}{\cos \phi} = -\frac{10}{3\sqrt{10}} = -\frac{\sqrt{10}}{3}$$

10. $\csc x = 5, \cos x > 0$

x is in Quadrant I.

$$\sin x = \frac{1}{\csc x} = \frac{1}{5}$$

$$\cos x = \sqrt{1 - \left(\frac{1}{5}\right)^2} = \frac{2\sqrt{6}}{5}$$

$$\tan x = \frac{\sin x}{\cos x} = \frac{1}{5} \cdot \frac{5}{2\sqrt{6}} = \frac{\sqrt{6}}{12}$$

$$\sec x = \frac{1}{\cos x} = \frac{5}{2\sqrt{6}} = \frac{5\sqrt{6}}{12}$$

$$\cot x = \frac{1}{\tan x} = 2\sqrt{6}$$

4. $\csc \theta = \dfrac{5}{3}, \tan \theta = \dfrac{3}{4}$

θ is in Quadrant I.

$$\sin \theta = \frac{1}{\csc \theta} = \frac{3}{5}$$

$$\cos \theta = \frac{\sin \theta}{\tan \theta} = \frac{3}{5} \cdot \frac{4}{3} = \frac{4}{5}$$

$$\sec \theta = \frac{1}{\cos \theta} = \frac{5}{4}$$

$$\cot \theta = \frac{1}{\tan \theta} = \frac{4}{3}$$

8. $\cos\left(\dfrac{\pi}{2} - x\right) = \dfrac{3}{5}, \cos x = \dfrac{4}{5}$

x is in Quadrant I.

$$\sin x = \sqrt{1 - \left(\frac{4}{5}\right)^2} = \frac{3}{5}$$

$$\tan x = \frac{\sin x}{\cos x} = \frac{3}{5} \cdot \frac{5}{4} = \frac{3}{4}$$

$$\csc x = \frac{1}{\sin x} = \frac{5}{3}$$

$$\sec x = \frac{1}{\cos x} = \frac{5}{4}$$

$$\cos x = \frac{1}{\tan x} = \frac{4}{3}$$

12. $\sec \theta = -3, \tan \theta < 0$

θ is in Quadrant II.

$$\cos \theta = \frac{1}{\sec \theta} = -\frac{1}{3}$$

$$\sin \theta = \sqrt{1 - \left(-\frac{1}{3}\right)^2} = \frac{2\sqrt{2}}{3}$$

$$\tan \theta = \frac{\sin \theta}{\cos \theta} = \frac{2\sqrt{2}}{3} \cdot -\frac{3}{1} = -2\sqrt{2}$$

$$\csc \theta = \frac{1}{\sin \theta} = \frac{3}{2\sqrt{2}} = \frac{3\sqrt{2}}{4}$$

$$\cot \theta = \frac{1}{\tan \theta} = -\frac{1}{2\sqrt{2}} = -\frac{\sqrt{2}}{4}$$

14. $\tan\theta$ is undefined, $\sin\theta > 0$.

$\theta = \dfrac{\pi}{2}$

$\tan\theta = \dfrac{\sin\theta}{\cos\theta}$ is undefined $\Longrightarrow \cos\theta = 0$

$\sin\theta = \sqrt{1 - 0^2} = 1$

$\csc\theta = \dfrac{1}{\sin\theta} = 1$

$\sec\theta = \dfrac{1}{\cos\theta}$ is undefined.

$\cot\theta = \dfrac{\cos\theta}{\sin\theta} = \dfrac{0}{1} = 0$

16. As $x \longrightarrow 0^+$,

$\cos x \longrightarrow 1$ and $\sec x = \dfrac{1}{\cos x} \longrightarrow 1.$

18. As $x \longrightarrow \pi^+$,

$\sin x \longrightarrow 0$ and $\csc x = \dfrac{1}{\sin x} \longrightarrow -\infty.$

20. $\cot x \sin x = \dfrac{\cos x}{\sin x}\sin x = \cos x$

Matches (b).

22. $(1 - \cos^2 x)(\csc x) = (\sin^2 x)\dfrac{1}{\sin x} = \sin x$

Matches (f).

24. $\dfrac{\sin\left[\left(\frac{\pi}{2}\right) - x\right]}{\cos\left[\left(\frac{\pi}{2}\right) - x\right]} = \dfrac{\cos x}{\sin x} = \cot x$

Matches (c).

26. $\cos^2 x(\sec^2 x - 1) = \cos^2 x(\tan^2 x)$

$= \cos^2 x\left(\dfrac{\sin^2 x}{\cos^2 x}\right)$

$= \sin^2 x$

Matches (c).

28. $\cot x \sec x = \dfrac{\cos x}{\sin x} \cdot \dfrac{1}{\cos x} = \dfrac{1}{\sin x} = \csc x$

Matches (a).

30. $\dfrac{\cos^2\left[\left(\frac{\pi}{2}\right) - x\right]}{\cos x} = \dfrac{\sin^2 x}{\cos x} = \dfrac{\sin x}{\cos x}\sin x$

$= \tan x \sin x$

Matches (d).

32. $\sin\phi(\csc\phi - \sin\phi) = \sin\phi\csc\phi - \sin^2\phi$

$= \sin\phi \cdot \dfrac{1}{\sin\phi} - \sin^2\phi$

$= 1 - \sin^2\phi = \cos^2\phi$

34. $\sec^2 x(1 - \sin^2 x) = \sec^2 x - \sec^2 x \sin^2 x$

$= \sec^2 x - \dfrac{1}{\cos^2 x} \cdot \sin^2 x$

$= \sec^2 x - \dfrac{\sin^2 x}{\cos^2 x}$

$= \sec^2 x - \tan^2 x$

$= 1$

36. $\dfrac{\csc\theta}{\sec\theta} = \dfrac{\frac{1}{\sin\theta}}{\frac{1}{\cos\theta}} = \dfrac{\cos\theta}{\sin\theta} = \cot\theta$

38. $\dfrac{1}{\tan^2 x + 1} = \dfrac{1}{\sec^2 x} = \dfrac{1}{\frac{1}{\cos^2 x}} = \cos^2 x$

40. $\dfrac{\tan^2\theta}{\sec^2\theta} = \dfrac{\sin^2\theta}{\cos^2\theta} \cdot \dfrac{1}{\sec^2\theta} = \dfrac{\sin^2\theta}{\cos^2\theta} \cdot \dfrac{1}{\frac{1}{\cos^2\theta}} = \dfrac{\sin^2\theta\cos^2\theta}{\cos^2\theta} = \sin^2\theta$

42. $\cot\left(\dfrac{\pi}{2} - x\right)\cos x = \tan x \cos x = \dfrac{\sin x}{\cos x} \cdot \cos x = \sin x$

44. $(\cos t)(1 + \tan^2 t) = (\cos t)(\sec^2 t) = \dfrac{\cos t}{\cos^2 t} = \dfrac{1}{\cos t} = \sec t$

46. $\sec^2 x \tan^2 x + \sec^2 x = \sec^2 x(\tan^2 x + 1) = \sec^2 x(\sec^2 x) = \sec^4 x$

48. $\dfrac{\sec^2 x - 1}{\sec x - 1} = \dfrac{(\sec x + 1)(\sec x - 1)}{\sec x - 1} = \sec x + 1$

50. $1 - 2\cos^2 x + \cos^4 x = (1 - \cos^2 x)(1 - \cos^2 x)$
$= \sin^2 x \sin^2 x = \sin^4 x$

52. $\csc^3 x - \csc^2 x - \csc x - 1 = \csc^2 x(\csc x - 1) - (\csc x - 1)$
$= (\csc^2 x - 1)(\csc x - 1)$
$= \cot^2 x(\csc x - 1)$

54. $(\cot x + \csc x)(\cot x - \csc x) = \cot^2 x - \csc^2 x$
$= -1$

56. $(3 - 3\sin x)(3 + 3\sin x) = 9 - 9\sin^2 x$
$= 9(1 - \sin^2 x)$
$= 9\cos^2 x$

58. $\dfrac{1}{\sec x + 1} - \dfrac{1}{\sec x - 1} = \dfrac{\sec x - 1 - (\sec x + 1)}{(\sec x + 1)(\sec x - 1)}$
$= \dfrac{\sec x - 1 - \sec x - 1}{\sec^2 x - 1}$
$= \dfrac{-2}{\tan^2 x}$
$= -2\left(\dfrac{1}{\tan^2 x}\right) = -2\cot^2 x$

60. $\tan x - \dfrac{\sec^2 x}{\tan x} = \dfrac{\tan^2 x - \sec^2 x}{\tan x}$
$= \dfrac{-1}{\tan x} = -\cot x$

62. $\dfrac{5}{\tan x + \sec x} \cdot \dfrac{\tan x - \sec x}{\tan x - \sec x} = \dfrac{5(\tan x - \sec x)}{\tan^2 x - \sec^2 x}$
$= \dfrac{5(\tan x - \sec x)}{-1}$
$= 5(\sec x - \tan x)$

64. $\dfrac{\tan^2 x}{\csc x + 1} \cdot \dfrac{\csc x - 1}{\csc x - 1} = \dfrac{\tan^2 x(\csc x - 1)}{\csc^2 x - 1}$
$= \dfrac{\tan^2 x(\csc x - 1)}{\cot^2 x}$
$= \tan^2 x(\csc x - 1)\tan^2 x$
$= \tan^4 x(\csc x - 1)$

66. $y_1 = \cos x + \sin x \tan x,\ y_2 = \sec x$

x	0.2	0.4	0.6	0.8	1.0	1.2	1.4
y_1	1.0203	1.0857	1.2116	1.4353	1.8508	2.7597	5.8835
y_2	1.0203	1.0857	1.2116	1.4353	1.8508	2.7597	5.8835

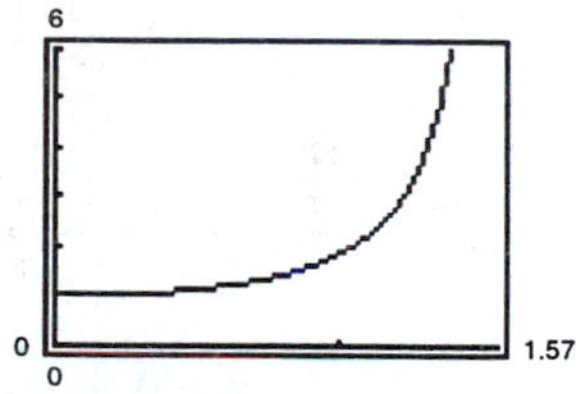

It appears that $y_1 = y_2$.

68. $y_1 = \sec^4 x - \sec^2 - x,\ y_2 = \tan^2 x + \tan^4 x$

x	0.2	0.4	0.6	0.8	1.0	1.2	1.4
y_1	0.0428	0.2107	0.6871	2.1841	8.3087	50.3869	1163.6143
y_2	0.0428	0.2107	0.6871	2.1841	8.3087	50.3869	1163.6143

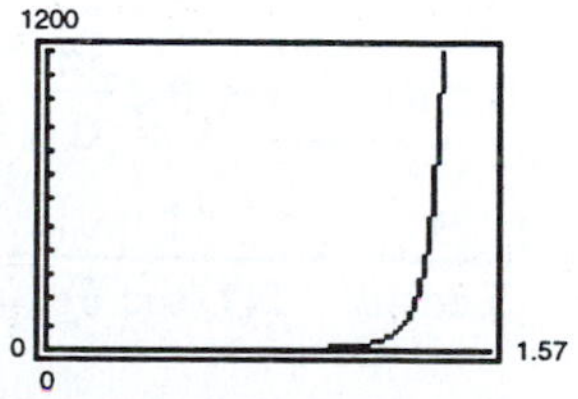

It appears that $y_1 = y_2$.

70. $y_1 = \dfrac{1}{2}\left(\dfrac{1 + \sin\theta}{\cos\theta} + \dfrac{\cos\theta}{1 + \sin\theta}\right)$

y_1 and $y_2 = \sin\theta$

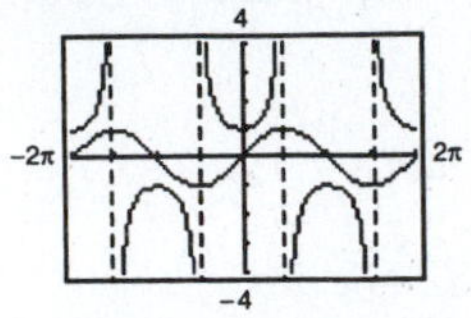

y_1 and $y_2 = \cos\theta$

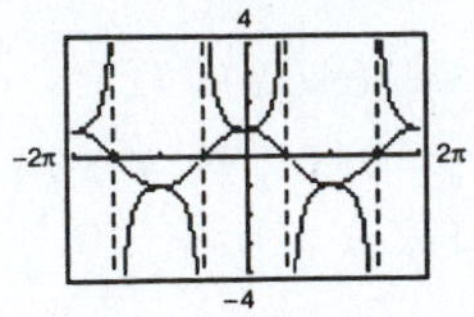

y_1 and $y_2 = \tan\theta$

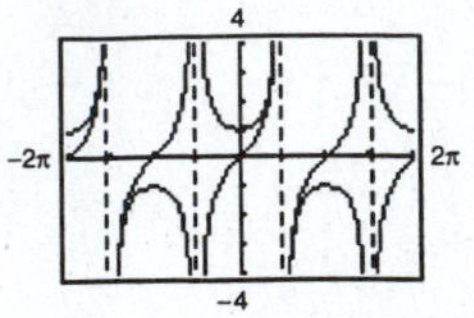

y_1 and $y_2 = \dfrac{1}{\sin\theta} = \csc\theta$

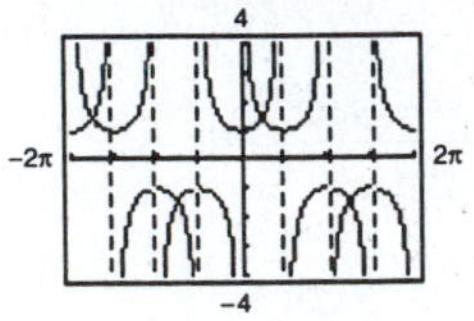

y_1 and $y_2 = \dfrac{1}{\cos\theta} = \sec\theta$

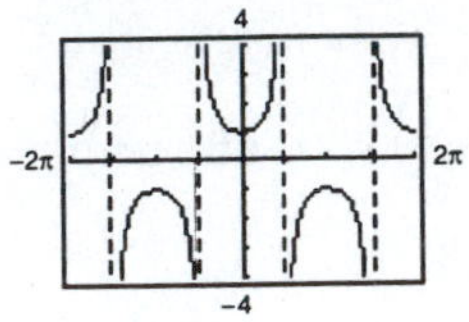

y_1 and $y_2 = \dfrac{1}{\tan\theta} = \cot\theta$

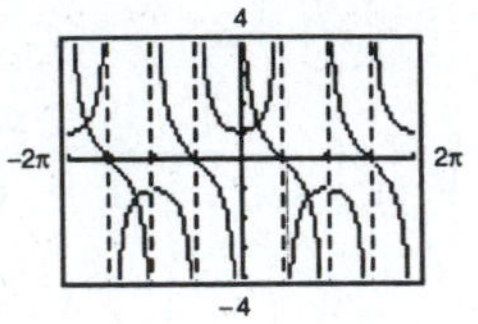

It appears that $\dfrac{1}{2}\left(\dfrac{1 + \sin\theta}{\cos\theta} + \dfrac{\cos\theta}{1 + \sin\theta}\right) = \sec\theta.$

72. Let $x = 2\sin\theta$.

$$\begin{aligned}\sqrt{16 - 4x^2} &= \sqrt{16 - 4(2\sin\theta)^2}\\ &= \sqrt{16(1 - \sin^2\theta)}\\ &= \sqrt{16\cos^2\theta}\\ &= 4\cos\theta\end{aligned}$$

74. Let $x = 2\sec\theta$.

$$\begin{aligned}\sqrt{x^2 - 4} &= \sqrt{(2\sec\theta)^2 - 4}\\ &= \sqrt{4(\sec^2\theta - 1)}\\ &= \sqrt{4\tan^2\theta}\\ &= 2\tan\theta\end{aligned}$$

76. Let $x = 10\tan\theta$.

$$\begin{aligned}\sqrt{x^2 + 100} &= \sqrt{(10\tan\theta)^2 + 100}\\ &= \sqrt{100(\tan^2\theta + 1)}\\ &= \sqrt{100\sec^2\theta}\\ &= 10\sec\theta\end{aligned}$$

78. $\cos\theta = -\sqrt{1 - \sin^2\theta}$

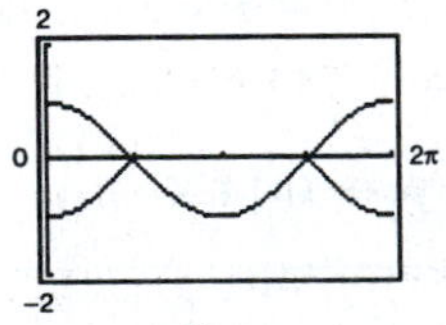

$\dfrac{\pi}{2} < \theta < \dfrac{3\pi}{2}$

80. $\tan\theta = \sqrt{\sec^2\theta - 1}$

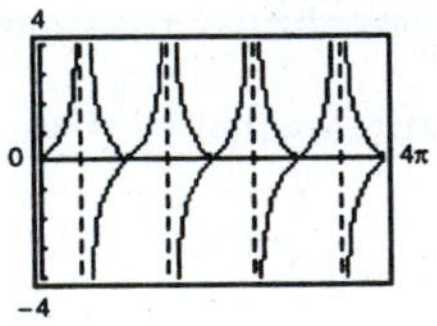

$0 \le \theta < \dfrac{\pi}{2},\ \pi \le \theta < \dfrac{3\pi}{2}$

82.
$$\begin{aligned}\ln|\cot t| + \ln(1 + \tan^2 t) &= \ln|\cot t|(1 + \tan^2 t)\\ &= \ln\frac{(1 + \tan^2 t)}{|\tan t|}\\ &= \ln\left|\frac{1}{\tan t} + \frac{\tan^2 t}{\tan t}\right|\\ &= \ln|\cot t + \tan t|\end{aligned}$$

84. $\dfrac{1}{5\cos\theta} = \dfrac{1}{5(1/\sec\theta)} = \dfrac{\sec\theta}{5} = \dfrac{1}{5}\sec\theta \neq 5\sec\theta$

Not an identity because $\dfrac{1}{5\cos\theta} \neq \dfrac{5}{\cos\theta}$.

86. $\sin\theta\csc\phi = \sin\theta\dfrac{1}{\sin\theta}$

This may be simplified only if $\theta = \phi$. Thus, $\sin\theta\csc\phi = 1$ is not an identity because θ must be equal to ϕ to be true.

88. $\tan^2\theta + 1 = \sec^2\theta$

(a) $\theta = 346°$

$(\tan 346°)^2 + 1 \approx 1.0622$

$(\sec 346°)^2 = \left(\dfrac{1}{\cos 346°}\right)^2 \approx 1.0622$

(b) $\theta = 3.1$

$(\tan 3.1)^2 + 1 \approx 1.00173$

$(\sec 3.1)^2 = \left(\dfrac{1}{\cos 3.1}\right)^2 \approx 1.00173$

90. $\sin(-\theta) = -\sin\theta$

(a) $\theta = 250°$

$\sin(-250°) \approx 0.9397$

$-(\sin 250°) \approx 0.9397$

(b) $\theta = \frac{1}{2}$

$\sin\left(-\frac{1}{2}\right) \approx -0.4794$

$-\left(\sin\frac{1}{2}\right) \approx -0.4794$

92. $\cos\theta$

$\sin\theta = \pm\sqrt{1-\cos^2\theta}$

$$\tan\theta = \frac{\sin\theta}{\cos\theta} = \pm\frac{\sqrt{1-\cos^2\theta}}{\cos\theta}$$

$$\csc\theta = \frac{1}{\sin\theta} = \pm\frac{1}{\sqrt{1-\cos^2\theta}}$$

$$\sec\theta = \frac{1}{\cos\theta}$$

$$\cot\theta = \frac{1}{\tan\theta} = \pm\frac{\cos\theta}{\sqrt{1-\cos^2\theta}}$$

94. $\sqrt{v}(\sqrt{20}-\sqrt{5}) = \sqrt{20v}-\sqrt{5v}$

$= 2\sqrt{5v}-\sqrt{5v}$

$= \sqrt{5v}$

96. $$\frac{50x}{\sqrt{30}-5} = \frac{50x}{\sqrt{30}-5}\cdot\frac{\sqrt{30}+5}{\sqrt{30}+5} = \frac{50x(\sqrt{30}+5)}{30-25} = \frac{50x(\sqrt{30}+5)}{5} = 10x(\sqrt{30}+5)$$

Section 5.2 Verifying Trigonometric Identities

Solutions to Even-Numbered Exercises

2. $\tan y\cot y = \tan y\left(\frac{1}{\tan y}\right) = 1$

4. $\cot^2 y(\sec^2 y - 1) = \cot^2 y\tan^2 y = 1$

6. $\cos^2\beta - \sin^2\beta = \cos^2\beta - (1-\cos^2\beta)$

$= 2\cos^2\beta - 1$

8. $2 - \sec^2 z = 2 - (1+\tan^2 z)$

$= 1 - \tan^2 z$

10. $\cos x + \sin x\tan x = \cos x + \sin x\left(\frac{\sin x}{\cos x}\right)$

$$= \frac{\cos^2 x + \sin^2 x}{\cos x}$$

$$= \frac{1}{\cos x}$$

$= \sec x$

12. $\cos t(\csc^2 t - 1) = \cos t\cot^2 t$

$$= \sin t\left(\frac{\cos t}{\sin t}\right)\cot^2 t$$

$$= \frac{1}{\csc t}\cot^3 t$$

$$= \frac{\cot^3 t}{\csc t}$$

14. $$\frac{1}{\sin x} - \sin x = \frac{1-\sin^2 x}{\sin x} = \frac{\cos^2 x}{\sin x}$$

16. $\sec^6 x(\sec x\tan x) - \sec^4 x(\sec x\tan x) = \sec^4 x(\sec x\tan x)(\sec^2 x - 1)$

$= \sec^4 x(\sec x\tan x)\tan^2 x$

$= \sec^5 x\tan^3 x$

18. $$\frac{\sec\theta - 1}{1-\cos\theta} = \frac{\sec\theta-1}{1-(1/\sec\theta)}\cdot\frac{\sec\theta}{\sec\theta}$$

$$= \frac{\sec\theta(\sec\theta-1)}{\sec\theta-1}$$

$= \sec\theta$

20. $$\sec x - \cos x = \frac{1}{\cos x} - \cos x = \frac{1 - \cos^2 x}{\cos x} = \frac{\sin^2 x}{\cos x} = \sin x \cdot \frac{\sin x}{\cos x} = \sin x \tan x$$

22. $$\frac{\sec x + \tan x}{\sec x - \tan x} = \frac{\sec x + \tan x}{\sec x - \tan x} \cdot \frac{\sec x + \tan x}{\sec x + \tan x} = \frac{(\sec x + \tan x)^2}{\sec^2 x - \tan^2 x} = \frac{(\sec x + \tan x)^2}{1} = (\sec x + \tan x)^2$$

24. $$\frac{1}{\sin x} - \frac{1}{\csc x} = \frac{\csc x - \sin x}{\sin x \csc x} = \frac{\csc x - \sin x}{1} = \csc x - \sin x$$

26. $$\frac{1 + \sin\theta}{\cos\theta} + \frac{\cos\theta}{1 + \sin\theta} = \frac{(1 + \sin\theta)^2 + \cos^2\theta}{\cos\theta(1 + \sin\theta)} = \frac{1 + 2\sin\theta + \sin^2\theta + \cos^2\theta}{\cos\theta(1 + \sin\theta)} = \frac{2 + 2\sin\theta}{\cos\theta(1 + \sin\theta)} = \frac{2(1 + \sin\theta)}{\cos\theta(1 + \sin\theta)} = \frac{2}{\cos\theta} = 2\sec\theta$$

28. $$\cos x - \frac{\cos x}{1 - \tan x} = \frac{\cos x(1 - \tan x) - \cos x}{1 - \tan x} = \frac{-\cos x \tan x}{1 - \tan x} = \frac{-\cos x(\sin x/\cos x)}{1 - (\sin x/\cos x)} \cdot \frac{\cos x}{\cos x} = \frac{-\sin x \cos x}{\cos x - \sin x} = \frac{\sin x \cos x}{\sin x - \cos x}$$

30. $$\frac{\cos[(\pi/2) - x]}{\sin[(\pi/2) - x]} = \frac{\sin x}{\cos x} = \tan x$$

32. $$(1 + \sin y)[1 + \sin(-y) = (1 + \sin y)(1 - \sin y) = 1 - \sin^2 y = \cos^2 y$$

34. $$\frac{1 + \sec(-\theta)}{\sin(-\theta) + \tan(-\theta)} = \frac{1 + \sec\theta}{-\sin\theta - \tan\theta} = -\frac{1 + \sec\theta}{\sin\theta + \tan\theta} = -\frac{1 + \sec\theta}{\sin\theta[1 + (1/\cos\theta)} = -\frac{1 + \sec\theta}{\sin\theta(1 + \sec\theta)} = -\frac{1}{\sin\theta} = -\csc\theta$$

36. $$\frac{\tan x + \tan y}{1 - \tan x \tan y} = \frac{\dfrac{1}{\cot x} + \dfrac{1}{\cot y}}{1 - \dfrac{1}{\cot x} \cdot \dfrac{1}{\cot y}} \cdot \frac{\cot x \cot y}{\cot x \cot y} = \frac{\cot y + \cot x}{\cot x \cot y - 1}$$

38. $$\frac{\cos x - \cos y}{\sin x + \sin y} + \frac{\sin x - \sin y}{\cos x + \cos y} = \frac{(\cos x - \cos y)(\cos x + \cos y) + (\sin x - \sin y)(\sin x + \sin y)}{(\sin x + \sin y)(\cos x + \cos y)}$$
$$= \frac{\cos^2 x - \cos^2 y + \sin^2 x - \sin^2 y}{(\sin x + \sin y)(\cos x + \cos y)}$$
$$= \frac{(\cos^2 x + \sin^2 x) - (\cos^2 y + \sin^2 y)}{(\sin x + \sin y)(\cos x + \cos y)}$$
$$= 0$$

40. $$\sqrt{\frac{1 - \cos\theta}{1 + \cos\theta}} = \sqrt{\frac{1 - \cos\theta}{1 + \cos\theta} \cdot \frac{1 - \cos\theta}{1 - \cos\theta}}$$
$$= \sqrt{\frac{(1 - \cos\theta)^2}{1 - \cos^2\theta}}$$
$$= \sqrt{\frac{(1 - \cos\theta)^2}{\sin^2\theta}}$$
$$= \frac{1 - \cos\theta}{|\sin\theta|}$$

42. $$\sec^2 y - \cot^2\left(\frac{\pi}{2} - y\right) = \sec^2 y - \tan^2 y = 1$$

44. $$\sec^2\left(\frac{\pi}{2} - x\right) - 1 = \csc^2 x - 1 = \cot^2 x$$

46. $$\csc x(\csc x - \sin x) + \frac{\sin x - \cos x}{\sin x} + \cot x = \csc^2 x - \csc x \sin x + 1 - \frac{\cos x}{\sin x} + \cot x$$
$$= \csc^2 x - 1 + 1 - \cot x + \cot x$$
$$= \csc^2 x$$

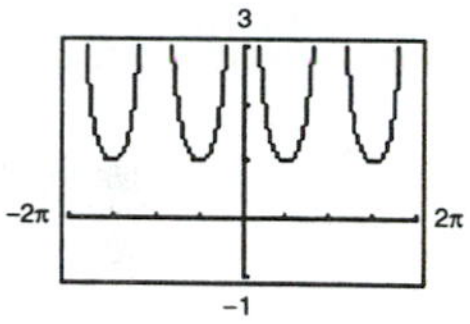

48. $$4\tan^4 x + \tan^2 x - 3 = (\tan^2 x + 1)(4\tan^2 x - 3)$$
$$= \sec^2 x(4\tan^2 x - 3)$$

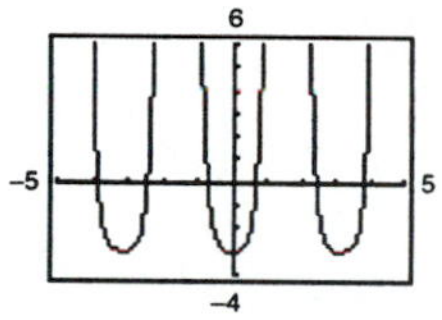

50. $$\sin x(1 - 2\cos^2 x + \cos^4 x) = \sin x(1 - \cos^2 x)^2$$
$$= \sin x(\sin^2 x)^2$$
$$= \sin^5 x$$

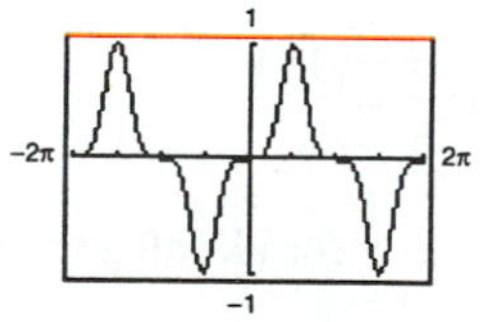

52. $$\csc^4\theta - \cot^4\theta = (\csc^2\theta - \cot^2\theta)(\csc^2\theta + \cot^2\theta)$$
$$= \csc^2\theta + \cot^2\theta$$
$$= \csc^2\theta + (\csc^2\theta - 1)$$
$$= 2\csc^2\theta - 1$$

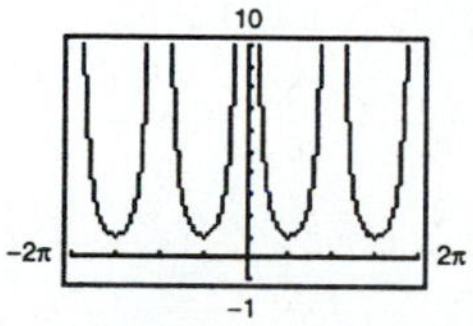

54. $$\frac{\cot\alpha}{\csc\alpha - 1} \cdot \frac{\csc\alpha + 1}{\csc\alpha + 1} = \frac{\cot\alpha(\csc\alpha + 1)}{\csc^2\alpha - 1}$$
$$= \frac{\cot\alpha(\csc\alpha + 1)}{\cot^2\alpha}$$
$$= \frac{\csc\alpha + 1}{\cot\alpha}$$

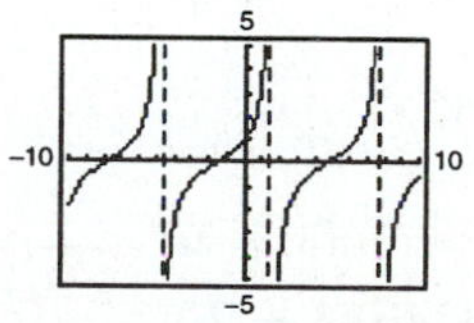

56. $$\frac{\sin^3\beta + \cos^3\beta}{\sin\beta + \cos\beta} = \frac{(\sin\beta + \cos\beta)(\sin^2\beta - \sin\beta\cos\beta + \cos^2\beta)}{\sin\beta + \cos\beta}$$
$$= \sin^2\beta + \cos^2\beta - \sin\beta\cos\beta$$
$$= 1 - \sin\beta\cos\beta$$

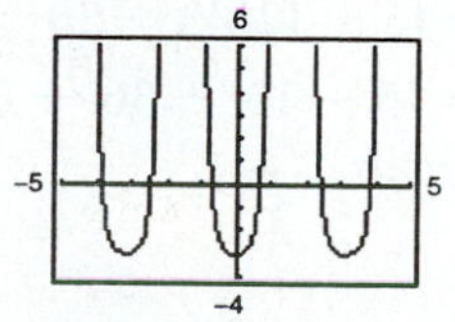

58. $$\ln|\sec\theta| = \ln\left|\frac{1}{\cos\theta}\right| = \ln|\cos\theta| = -\ln|\cos\theta|$$

60. $$\begin{aligned} -\ln\lvert\sec\theta + \tan\theta\rvert &= -\ln\left|\frac{1}{\cos\theta} + \frac{\sin\theta}{\cos\theta}\right| \\ &= \ln\left|\frac{1+\sin\theta}{\cos\theta}\right|^{-1} \\ &= \ln\left|\frac{\cos\theta}{1+\sin\theta}\cdot\frac{1-\sin\theta}{1-\sin\theta}\right| \\ &= \ln\left|\frac{\cos\theta - \cos\theta\sin\theta}{1-\sin^2\theta}\right| \\ &= \ln\left|\frac{\cos\theta - \cos\theta\sin\theta}{\cos^2\theta}\right| \\ &= \ln\lvert\sec\theta - \tan\theta\rvert \end{aligned}$$

62. $\tan\theta = \sqrt{\sec^2\theta - 1}$

True identity: $\tan\theta = \pm\sqrt{\sec^2\theta - 1}$

$\tan\theta = \pm\sqrt{\sec^2\theta - 1}$ is not true for $\pi/2 < \theta < \pi$ or $3\pi/2 < \theta < 2\pi$. Thus, the equation is not true for $\theta = 3\pi/4$.

64. $\sqrt{\sin^2 x + \cos^2 x} = \sin x + \cos x$

$\sqrt{\sin^2 x + \cos^2 x} \neq \sin x + \cos x$

The left side is 1 for any x, but the right side is not necessarily 1. The equation is not true for $x = \pi/4$.

66. $$\begin{aligned} \cos^2 18° + \cos^2 72° &= \sin^2(90° - 18°) + \cos^2 72° \\ &= \sin^2 72° + \cos^2 72° \\ &= 1 \end{aligned}$$

68. $$\begin{aligned} \sin^2 12° + \sin^2 40° + \sin^2 50° + \sin^2 78° &= \sin^2 12° + \sin^2 78° + \sin^2 40° + \sin^2 50° \\ &= \cos^2(90° - 12°) + \sin^2 78° + \cos^2(90° - 40°) + \sin^2 50° \\ &= \cos^2 78° + \sin^2 78° + \cos^2 50° + \sin^2 50° \\ &= 1 + 1 = 2 \end{aligned}$$

70. $$\begin{aligned} \sin\left[\frac{(12n+1)\pi}{6}\right] &= \sin\left[\frac{1}{6}(12n\pi + \pi)\right] \\ &= \sin\left(2n\pi + \frac{\pi}{6}\right) \\ &= \sin\frac{\pi}{6} = \frac{1}{2} \end{aligned}$$

Thus, $\sin\left[\frac{(12n+1)\pi}{6}\right] = \frac{1}{2}$ for all integers n.

72. $$\begin{aligned} \mu W\cos\theta &= W\sin\theta \\ \mu &= \frac{W\sin\theta}{W\cos\theta} \\ \mu &= \frac{\sin\theta}{\cos\theta} \\ \mu &= \tan\theta \end{aligned}$$

74. *Seward:* $D = 12.2 - 6.4\cos\left[\frac{\pi(t+0.2)}{6}\right]$

Period: $\frac{2\pi}{b} = \frac{2\pi}{\pi/6} = 12$

New Orleans: $D = 12.2 - 1.9\cos\left[\frac{\pi(t+0.2)}{6}\right]$

Period: $\frac{2\pi}{b} = \frac{2\pi}{\pi/6} = 12$

76. $$\begin{aligned} (2-5i)^2 &= (2-5i)(2-5i) \\ &= 4 - 20i + 25i^2 \\ &= 4 - 20i - 25 \\ &= -21 - 20i \end{aligned}$$

78. $$\begin{aligned} (3+2i)^3 &= (3+2i)(3+2i)(3+2i) \\ &= (9 + 12i + 4i^2)(3+2i) \\ &= (5+12i)(3+2i) \\ &= 15 + 10i + 36i + 24i^2 \\ &= -9 + 46i \end{aligned}$$

Section 5.3 Solving Trigonometric Equations

Solutions to Even-Numbered Exercises

2. $y = \sin \pi x + \cos \pi x$

$$\sin \pi x + \cos \pi x = 0$$

$$\cos \pi x = -\sin \pi x$$

$$1 = \frac{-\sin \pi x}{\cos \pi x}$$

$$1 = -\tan \pi x$$

$$-1 = \tan \pi x$$

$$\pi x = -\frac{\pi}{4}, \frac{3\pi}{4}, \frac{7\pi}{4}$$

$$x = -\frac{1}{4}, \frac{3}{4}, \frac{7}{4}$$

4. $y = \sec^4\left(\frac{\pi x}{8}\right) - 4$

$$\sec^4\left(\frac{\pi x}{8}\right) - 4 = 0$$

$$\frac{1}{\cos^4(\pi x/8)} + = 4$$

$$\cos^4\left(\frac{\pi x}{8}\right) = \frac{1}{4}$$

$$\cos\left(\frac{\pi x}{8}\right) = \sqrt[4]{\frac{1}{4}}$$

$$\cos\left(\frac{\pi x}{8}\right) = \frac{\sqrt{2}}{2}$$

$$\frac{\pi x}{8} = -\frac{\pi}{4}, \frac{\pi}{4}$$

$$x = -2, 2$$

6. $\csc x - 2 = 0$

(a) $x = \frac{\pi}{6}$

$$\csc\frac{\pi}{6} - 2 = \frac{1}{\sin(\pi/6)} - 2 = 2 - 2 = 0$$

(b) $x = \frac{5\pi}{6}$

$$\csc\frac{5\pi}{6} - 2 = \frac{1}{\sin(5\pi/6)} - 2 = 2 - 2 = 0$$

8. $2\cos^2 4x - 1 = 0$

(a) $x = \frac{\pi}{16}$

$$2\cos^2\left[4\left(\frac{\pi}{16}\right)\right] - 1 = 2\cos^2\frac{\pi}{4} - 1 = 2\left(\frac{\sqrt{2}}{2}\right)^2 - 1 = 2\left(\frac{1}{2}\right) - 1 = 1 - 1 = 0$$

(b) $x = \frac{3\pi}{16}$

$$2\cos^2\left[4\left(\frac{3\pi}{16}\right)\right] - 1 = 2\cos^2\frac{3\pi}{4} - 1 = 2\left(-\frac{\sqrt{2}}{2}\right)^2 - 1 = 2\left(\frac{1}{2}\right) - 1 = 0$$

10. $\sec^4 x - 4\sec^2 x = 0$

(a) $x = \frac{2\pi}{3}$

$$\sec^4\left(\frac{2\pi}{3}\right) - 4\sec^2\left(\frac{2\pi}{3}\right) = \frac{1}{\cos^4(2\pi/3)} - \frac{4}{\cos^2(2\pi/3)} = \frac{1}{(-1/2)^4} - \frac{4}{(-1/2)^2} = 16 - 4(4) = 0$$

—CONTINUED—

10. —CONTINUED—

(b) $x = \dfrac{5\pi}{3}$

$$\sec^4\left(\frac{5\pi}{3}\right) - 4\sec^2\left(\frac{5\pi}{3}\right) = \frac{1}{\cos^4(5\pi/3)} - \frac{4}{\cos^2(5\pi/3)}$$
$$= \frac{1}{(1/2)^4} - \frac{4}{(1/2)^2}$$
$$= 16 - 4(4) = 0$$

12. $2\sin x - 1 = 0$

$2\sin x = 1$

$\sin x = \dfrac{1}{2}$

$x = \dfrac{\pi}{6} + 2n\pi$

or $x = \dfrac{5\pi}{6} + 2n\pi$

14. $\tan x + 1 = 0$

$\tan x = -1$

$x = \dfrac{3\pi}{4} + n\pi$

or $x = \dfrac{7\pi}{4} + n\pi$

16. $\csc^2 x - 2 = 0$

$\csc x = \pm\sqrt{2}$

$x = \dfrac{\pi}{4} + \dfrac{n\pi}{2}$

18. $\tan^2 3x = 3$

$\tan 3x = \pm\sqrt{3}$

$3x = \dfrac{\pi}{3} + n\pi$

$x = \dfrac{\pi}{9} + \dfrac{n\pi}{3}$ or $3x = \dfrac{2\pi}{9} + \dfrac{n\pi}{3}$

$x = \dfrac{2\pi}{9} + \dfrac{n\pi}{3}$

20. $\sin x(\sin x + 1) = 0$

$\sin x = 0$ or $\sin x = -1$

$x = n\pi$ $\qquad x = \dfrac{3\pi}{2} + 2n\pi$

22. $\tan 3x(\tan x - 1) = 0$

$\tan 3x = 0$ or $\tan x - 1 = 0$

$3x = n\pi$ $\qquad \tan x = 1$

$x = \dfrac{n\pi}{3}$ $\qquad x = \dfrac{\pi}{4} + n\pi$

24. $\cos 2x(2\cos x + 1) = 0$

$\cos 2x = 0$ or $2\cos x + 1 = 0$

$2x = \dfrac{\pi}{2} + n\pi$ $\qquad \cos x = -\dfrac{1}{2}$

$x = \dfrac{\pi}{4} + \dfrac{n\pi}{2}$ $\qquad x = \dfrac{2\pi}{3} + 2n\pi$

or $x = \dfrac{4\pi}{3} + 2n\pi$

26. $\tan^2 x - 1 = 0$

$\tan^2 x = 1$

$\tan x = \pm 1$

$x = \dfrac{\pi}{4} + n\pi, \dfrac{3\pi}{4} + n\pi, \dfrac{5\pi}{4} + n\pi,$

$\dfrac{7\pi}{4} + n\pi$

28. $2\sin^2 x = 2 + \cos x$

$2 - 2\cos^2 x = 2 + \cos x$

$2\cos^2 x + \cos x = 0$

$\cos x(2\cos x + 1) = 0$

$\cos x = 0$ or $2\cos x + 1 = 0$

$x = \dfrac{\pi}{2}, \dfrac{3\pi}{2}$ $\qquad 2\cos x = -1$

$\cos x = -\dfrac{1}{2}$

$x = \dfrac{2\pi}{3}, \dfrac{4\pi}{3}$

30. $\sec x \csc x = 2 \csc x$

$\sec x \csc x - 2 \csc x = 0$

$\csc x(\sec x - 2) = 0$

$\csc x = 0$ or $\sec x - 2 = 0$

No solution $\quad \sec x = 2$

$x = \dfrac{\pi}{3}, \dfrac{5\pi}{3}$

32. $\sin 2x = -\dfrac{\sqrt{3}}{2}$

$2x = \dfrac{4\pi}{3} + 2n\pi$ or $2x = \dfrac{5\pi}{3} + 2n\pi$

$x = \dfrac{2\pi}{3} + n\pi \qquad x = \dfrac{5\pi}{6} + n\pi$

$x = \dfrac{2\pi}{3}, \dfrac{5\pi}{3} \qquad x = \dfrac{5\pi}{6}, \dfrac{11\pi}{6}$

34. $\tan 3x = 1$

$3x = \dfrac{\pi}{4} + 2n\pi$ or $3x = \dfrac{5\pi}{4} + 2n\pi$

$x = \dfrac{\pi}{12} + \dfrac{2n\pi}{3} \qquad x = \dfrac{5\pi}{12} + \dfrac{2n\pi}{3}$

$x = \dfrac{\pi}{12}, \dfrac{9\pi}{12}, \dfrac{17\pi}{12} \qquad x = \dfrac{5\pi}{12}, \dfrac{13\pi}{12}, \dfrac{21\pi}{12}$

36. $\sec 4x = 2$

$4x = \dfrac{\pi}{3} + 2n\pi$ or $4x = \dfrac{5\pi}{3} + 2n\pi$

$x = \dfrac{\pi}{12} + \dfrac{n\pi}{2} \qquad x = \dfrac{5\pi}{12} + \dfrac{n\pi}{2}$

$x = \dfrac{\pi}{12}, \dfrac{7\pi}{12}, \dfrac{13\pi}{12}, \dfrac{19\pi}{12} \qquad x = \dfrac{5\pi}{12}, \dfrac{11\pi}{12}, \dfrac{17\pi}{12}, \dfrac{23\pi}{12}$

38. $2 \sin^2 x + 3 \sin x + 1 = 0$

$(2 \sin x + 1)(\sin x + 1) = 0$

$2 \sin x + 1 = 0$ or $\sin x + 1 = 0$

$\sin x = -\dfrac{1}{2} \qquad \sin x = -1$

$x = \dfrac{7\pi}{6}, \dfrac{11\pi}{6} \qquad x = \dfrac{3\pi}{2}$

40. $\cos x + \sin x \tan x = 2$

$\cos x + \sin x\left(\dfrac{\sin x}{\cos x}\right) = 2$

$\dfrac{\cos^2 x + \sin^2 x}{\cos x} = 2$

$\dfrac{1}{\cos x} = 2$

$\cos x = \dfrac{1}{2}$

$x = \dfrac{\pi}{3}, \dfrac{5\pi}{3}$

42. $y^2 + y - 20 = 0$

$(y + 5)(y - 4) = 0$

$y + 5 = 0$ or $y - 4 = 0$

$y = -5 \qquad y = 4$

$\sin^2 x + \sin x - 20 = 0$

$(\sin x + 5)(\sin x - 4) = 0$

$\sin x = -5$ or $\sin x = 4$

No solution $\quad$ No solution

44. $4 \sin^3 x + 2 \sin^2 x - 2 \sin x - 1 = 0$

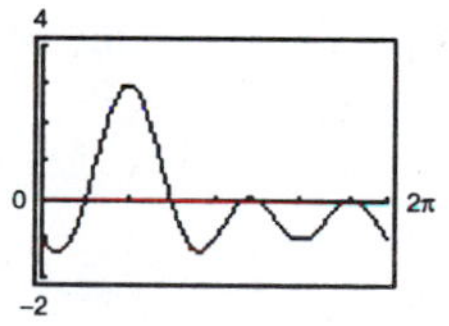

$x \approx 0.7854, 2.3562, 3.6652, 3.9270, 5.4978, 5.7596$

46. $\dfrac{\cos x \cot x}{1 - \sin x} = 3$

$y_1 = \left(\dfrac{(\cos x)/(\tan x)}{1 - \sin x}\right) - 3$

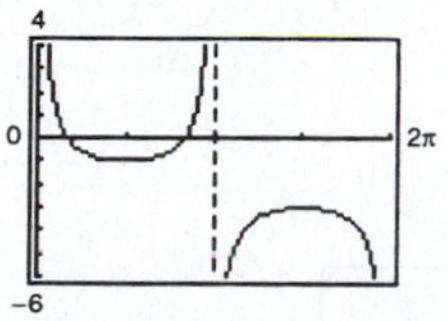

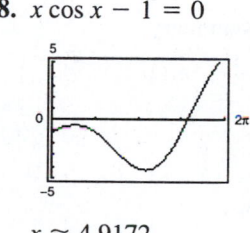

$x \approx 0.5236, 2.6180$

48. $x \cos x - 1 = 0$

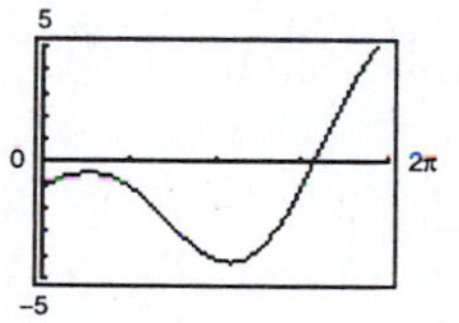

$x \approx 4.9172$

50. $\csc^2 x + 0.5 \cot x - 5 = 0$

$$y_1 = \left(\frac{1}{\sin x}\right)^2 + \frac{1}{2 \tan x} - 5$$

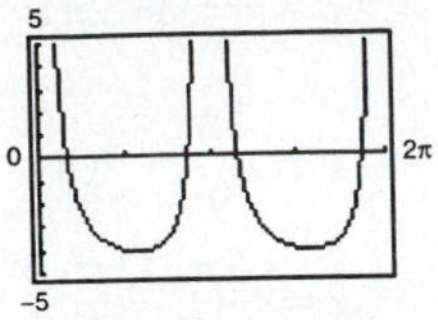

$x \approx 0.5153, 2.7259, 3.6569, 5.8675$

52. $12 \cos^2 x + 5 \cos x - 3 = 0$

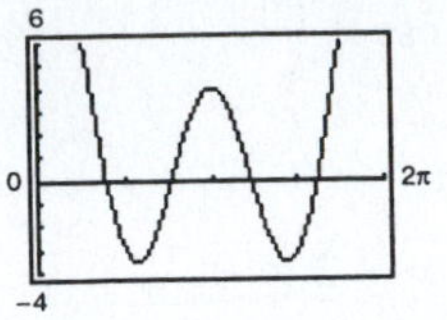

$x \approx 1.2310, 2.4189, 3.8643, 5.0522$

54. $3 \tan^2 x + 4 \tan x - 4 = 0$

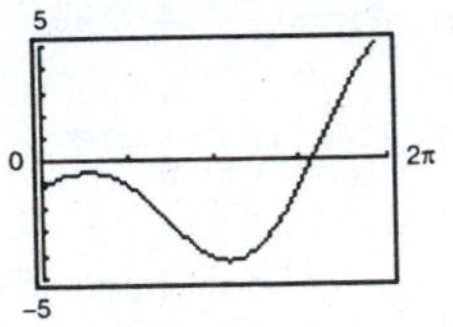

$x \approx 0.5880, 2.0344, 3.7296, 5.1760$

56. $4 \cos^2 x - 4 \cos x - 1 = 0$

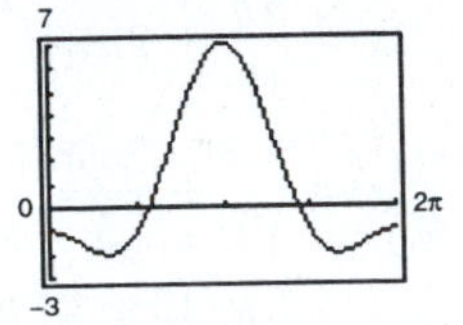

$x \approx 1.7794, 4.5038$

58. (a) $f(x) = 2 \sin x + \cos 2x$

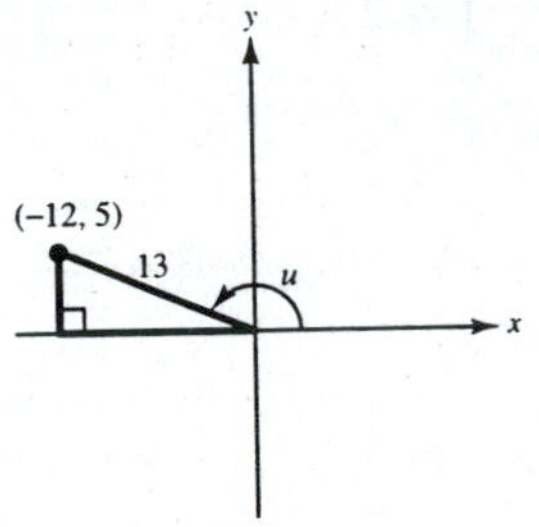

Maximum points: (0.5236, 1.5), (2.6180, 1.5)

Minimum points: (1.5708, 1), (4.7124, −3)

(b) $2 \cos - 4 \sin x \cos x = 0$

$2 \cos x(1 - 2 \sin x) = 0$

$2 \cos x = 0$ or $1 - 2 \sin x = 0$

$\cos x = 0$ $\qquad$ $\sin x = \frac{1}{2}$

$x = \frac{\pi}{2}, \frac{3\pi}{2}$ $\qquad$ $x = \frac{\pi}{6}, \frac{5\pi}{6}$

$\frac{\pi}{2} \approx 1.5708$ $\qquad$ $\frac{\pi}{6} \approx 0.5236$

$\frac{3\pi}{2} \approx 4.7124$ $\qquad$ $\frac{5\pi}{6} \approx 2.6180$

60. Graph $y = \cos x$ and $y = x$ on the same set of axes. Their point of intersection gives the value of c such that $f(c) = c \Rightarrow \cos c = c$.

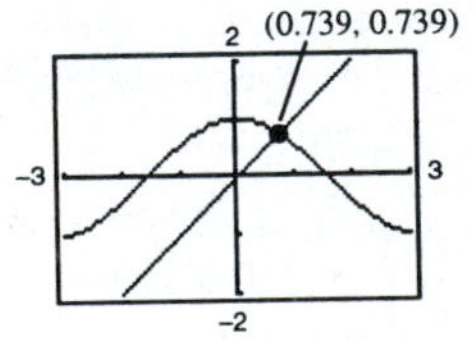

$c \approx 0.739$

62. $f(x) = \dfrac{\sin x}{x}$

(a) Domain: all real numbers except $x = 0$.

(b) The graph has y-axis symmetry.

(c) As $x \to 0, f(x) \to 1$.

(d) $\dfrac{\sin x}{x} = 0$ has four solutions in the interval $[-8, 8]$.

$$(\sin x)\left(\frac{1}{x}\right) = 0$$

$$\sin x = 0$$

$$x = -2\pi, -\pi, \pi, 2\pi$$

64. $S = 74.50 + 43.75 \sin \dfrac{\pi t}{6}$

t	1	2	3	4	5	6	7	8	9	10	11	12
S	96.4	112.4	118.3	112.4	96.4	74.5	52.6	36.6	30.8	36.6	52.6	74.5

Sales exceed 100,000 units during February, March, and April.

66. Range = 1000 yards = 3000 feet

$v_0 = 1200$ feet per second

$f = \frac{1}{32} v_0^2 \sin 2\theta$

$3000 = \frac{1}{32}(1200)^2 \sin 2\theta$

$\sin 2\theta = 0.066667$

$2\theta \approx 3.8226°$

$\theta \approx 1.9113°$

68. $y_1 = 1.56e^{-0.22t} \cos 4.9t$

$y_2 = \frac{1}{12}$ (1 inch since displacement is given in feet.)

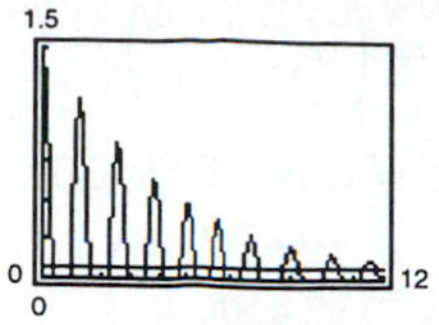

Point of intersection: $\left(10.94, \frac{1}{12}\right)$

The displacement does not exceed one inch from equilibrium after $t = 10.94$ seconds.

70. $f(x) = 3 \sin(0.6x - 2)$

(a) Zero: $\sin(0.6x - 2) = 0$

$0.6x - 2 = 0$

$0.6x = 2$

$x = \frac{2}{0.6} = \frac{10}{3}$

(b) $g(x) = -0.45x^2 + 5.52x - 13.70$

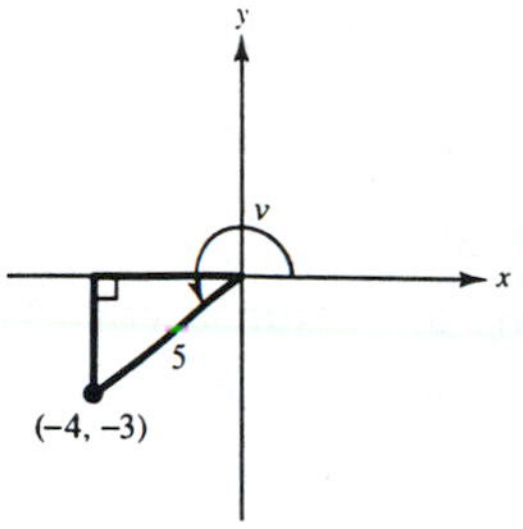

For $3.5 \le x \le 6$ the approximation appears to be good.

(c) $-0.45x^2 + 5.52x - 13.70 = 0$

$$x = \frac{-5.52 \pm \sqrt{(5.52) - 4(-0.45)(-13.70)}}{2(-0.45)}$$

$x \approx 3.46, 8.81$

The zero of g on $[0, 6]$ is 3.46. The zero is close to the zero $\frac{10}{3} \approx 3.33$ of f.

Section 5.4 Sum and Difference Formulas

Solutions to Even-Numbered Exercises

2. (a) $\sin\left(\frac{3\pi}{4} + \frac{5\pi}{6}\right) = \sin\frac{3\pi}{4}\cos\frac{5\pi}{6} + \cos\frac{3\pi}{4}\sin\frac{5\pi}{5}$

$= \left(\frac{\sqrt{2}}{2}\right)\left(-\frac{\sqrt{3}}{2}\right) + \left(-\frac{\sqrt{2}}{2}\right)\left(\frac{1}{2}\right)$

$= -\frac{\sqrt{6} + \sqrt{2}}{4}$

(b) $\sin\frac{3\pi}{4} + \sin\frac{5\pi}{6} = \frac{\sqrt{2}}{2} + \frac{1}{2} = \frac{\sqrt{2} + 1}{2}$

4. (a) $\cos\left(\frac{2\pi}{3} - \frac{\pi}{6}\right) = \cos\frac{2\pi}{3}\cos\frac{\pi}{6} + \sin\frac{2\pi}{3}\sin\frac{\pi}{6}$

$= \left(-\frac{1}{2}\right)\left(\frac{\sqrt{3}}{2}\right) + \left(\frac{\sqrt{3}}{2}\right)\left(\frac{1}{2}\right) = 0$

(b) $\cos\frac{2\pi}{3} + \cos\frac{\pi}{6} = -\frac{1}{2} + \frac{\sqrt{3}}{2} = \frac{\sqrt{3} - 1}{2}$

6. $\cos\left(x - \frac{\pi}{2}\right) = \cos x \cos\frac{\pi}{2} + \sin x \sin\frac{\pi}{2}$

$= (\cos x)(0) + (\sin x)(1)$

$= \sin x$

Thus,

$\cos\left(x - \frac{\pi}{2}\right) = -\sin x$ is false.

8. $15° = 45° - 30°$

$\sin 15° = \sin(45° - 30°) = \sin 45° \cos 30° - \cos 45° \sin 30°$

$$= \left(\frac{\sqrt{2}}{2}\right)\left(\frac{\sqrt{3}}{2}\right) - \left(\frac{\sqrt{2}}{2}\right)\left(\frac{1}{2}\right) = \frac{\sqrt{2}(\sqrt{3}-1)}{4} = \frac{\sqrt{2}}{4}(\sqrt{3}-1)$$

$\cos 15° = \cos(45° - 30°) = \cos 45° \cos 30° + \sin 45° \sin 30°$

$$= \left(\frac{\sqrt{2}}{2}\right)\left(\frac{\sqrt{3}}{2}\right) + \left(\frac{\sqrt{2}}{2}\right)\left(\frac{1}{2}\right) = \frac{\sqrt{2}(\sqrt{3}+1)}{4} = \frac{\sqrt{2}}{4}(\sqrt{3}+1)$$

$\tan 15° = \tan(45° - 30°) = \dfrac{\tan 45° - \tan 30°}{1 + \tan 45° \tan 30°}$

$$= \frac{1 - \frac{\sqrt{3}}{3}}{1 + (1)\left(\frac{\sqrt{3}}{3}\right)} = \frac{\frac{3-\sqrt{3}}{3}}{\frac{3+\sqrt{3}}{3}} = \frac{3-\sqrt{3}}{3+\sqrt{3}} \cdot \frac{3-\sqrt{3}}{3-\sqrt{3}} = \frac{12-6\sqrt{3}}{6} = 2 - \sqrt{3}$$

10. $165° = 135° + 30°$

$\sin 165° = \sin(135° + 30°)$

$= \sin 135° \cos 30° + \sin 30° \cos 135°$

$= \sin 45° \cos 30° - \sin 30° \cos 45°$

$= \dfrac{\sqrt{2}}{2} \cdot \dfrac{\sqrt{3}}{2} - \dfrac{1}{2} \cdot \dfrac{\sqrt{2}}{2}$

$= \dfrac{\sqrt{2}}{4}(\sqrt{3} - 1)$

$\cos 165° = \cos(135° + 30°)$

$= \cos 135° \cos 30° - \sin 135° \sin 30°$

$= -\cos 45° \cos 30° - \sin 45° \sin 30°$

$= -\dfrac{\sqrt{2}}{2} \cdot \dfrac{\sqrt{3}}{2} - \dfrac{\sqrt{2}}{2} \cdot \dfrac{1}{2}$

$= -\dfrac{\sqrt{2}}{4}(\sqrt{3} + 1)$

$\tan 165° = \tan(135° + 30°)$

$= \dfrac{\tan 135° + \tan 30°}{1 - \tan 135° \tan 30°}$

$= \dfrac{-\tan 45° + \tan 30°}{1 + \tan 45° \tan 30°}$

$= \dfrac{-1 + (\sqrt{3}/3)}{1 + (\sqrt{3}/3)}$

$= -2 + \sqrt{3}$

12. $255° = 300° - 45°$

$\sin 255° = \sin(300° - 45°)$

$= \sin 300° \cos 45° - \sin 45° \cos 300°$

$= -\sin 60° \cos 45° - \sin 45° \cos 60°$

$= -\dfrac{\sqrt{3}}{2} \cdot \dfrac{\sqrt{2}}{2} - \dfrac{\sqrt{2}}{2} \cdot \dfrac{1}{2}$

$= -\dfrac{\sqrt{2}}{4}(\sqrt{3} + 1)$

$\cos 255° = \cos(300° - 45°)$

$= \cos 300° \cos 45° + \sin 300° \sin 45°$

$= \cos 60° \cos 45° - \sin 60° \sin 45°$

$= \dfrac{1}{2} \cdot \dfrac{\sqrt{2}}{2} - \dfrac{\sqrt{3}}{2} \cdot \dfrac{\sqrt{2}}{2}$

$= \dfrac{\sqrt{2}}{4}(1 - \sqrt{3})$

$\tan 255° = \tan(300° - 45°)$

$= \dfrac{\tan 300° - \tan 45°}{1 + \tan 300° \tan 45°}$

$= \dfrac{-\tan 60° + \tan 45°}{1 - \tan 60° \tan 45°}$

$= \dfrac{-\sqrt{3} - 1}{1 - \sqrt{3}} = 2 + \sqrt{3}$

$= -2 + \sqrt{3}$

14. $\frac{7\pi}{12} = \frac{\pi}{3} + \frac{\pi}{4}$

$$\sin\frac{7\pi}{12} = \sin\left(\frac{\pi}{3} + \frac{\pi}{4}\right)$$
$$= \sin\frac{\pi}{3}\cos\frac{\pi}{4} + \sin\frac{\pi}{4}\cos\frac{\pi}{3}$$
$$= \frac{\sqrt{3}}{2} \cdot \frac{\sqrt{2}}{2} + \frac{\sqrt{2}}{2} \cdot \frac{1}{2}$$
$$= \frac{\sqrt{2}}{3}(\sqrt{3} + 1)$$

$$\cos\frac{7\pi}{12} = \cos\left(\frac{\pi}{3} + \frac{\pi}{4}\right)$$
$$= \cos\frac{\pi}{3}\cos\frac{\pi}{4} - \sin\frac{\pi}{3}\sin\frac{\pi}{4}$$
$$= \frac{1}{2} \cdot \frac{\sqrt{2}}{2} - \frac{\sqrt{3}}{2} \cdot \frac{\sqrt{2}}{2}$$
$$= \frac{\sqrt{2}}{4}(\sqrt{3})$$

$$\tan\frac{7\pi}{12} = \tan\left(\frac{\pi}{3} + \frac{\pi}{4}\right)$$
$$= \frac{\tan(\pi/3) + \tan(\pi/4)}{1 - \tan(\pi/3)\tan(\pi/4)}$$
$$= \frac{\sqrt{3} + 1}{1 - \sqrt{3}}$$
$$= -2 - \sqrt{3}$$

16. $-\frac{\pi}{12} = \frac{\pi}{6} - \frac{\pi}{4}$

$$\sin\left(-\frac{\pi}{12}\right) = \sin\left(\frac{\pi}{6} - \frac{\pi}{4}\right)$$
$$= \sin\frac{\pi}{6}\cos\frac{\pi}{4} - \sin\frac{\pi}{4}\cos\frac{\pi}{6}$$
$$= \frac{1}{2} \cdot \frac{\sqrt{2}}{2} - \frac{\sqrt{2}}{2} \cdot \frac{\sqrt{3}}{2}$$
$$= \frac{\sqrt{2}}{4}(1 - \sqrt{3})$$

$$\cos\left(-\frac{\pi}{12}\right) = \cos\left(\frac{\pi}{6} - \frac{\pi}{4}\right)$$
$$= \cos\frac{\pi}{6}\cos\frac{\pi}{4} + \sin\frac{\pi}{6}\sin\frac{\pi}{4}$$
$$= \frac{\sqrt{3}}{2} \cdot \frac{\sqrt{2}}{2} + \frac{1}{2} \cdot \frac{\sqrt{2}}{2}$$
$$= \frac{\sqrt{2}}{4}(\sqrt{3} + 1)$$

$$\tan\left(-\frac{\pi}{12}\right) = \tan\left(\frac{\pi}{6} - \frac{\pi}{4}\right)$$
$$= \frac{\tan(\pi/6) - \tan(\pi/4)}{1 + \tan(\pi/6)\tan(\pi/4)}$$
$$= \frac{(\sqrt{3}/3) - 1}{1 + (\sqrt{3}/3)}$$
$$= -2 + \sqrt{3}$$

18. $-105 = 30° - 135°$

$$\sin(30° - 135°) = \sin 30° \cos 135° - \cos 30° \sin 135°$$
$$= \sin 30°(-\cos 45°) - \cos 30° \sin 45°$$
$$= \left(\frac{1}{2}\right)\left(-\frac{\sqrt{2}}{2}\right) - \left(\frac{\sqrt{3}}{2}\right)\left(\frac{\sqrt{2}}{2}\right)$$
$$= -\frac{\sqrt{2}}{4}(1 + \sqrt{3})$$

$$\cos(30° - 135°) = \cos 30° \cos 135° + \sin 30° \sin 135°$$
$$= \cos 30°(-\cos 45°) + \sin 30° \sin 45°$$
$$= \left(\frac{\sqrt{3}}{2}\right)\left(-\frac{\sqrt{2}}{2}\right) + \left(\frac{1}{2}\right)\left(\frac{\sqrt{2}}{2}\right)$$
$$= \frac{\sqrt{2}}{4}(1 - \sqrt{3})$$

$$\tan(30° - 135°) = \frac{\tan 30° - \tan 135°}{1 + \tan 30° \tan 135°}$$
$$= \frac{\tan 30° - (-\tan 45°)}{1 + \tan 30°(-\tan 45°)}$$
$$= \frac{\frac{\sqrt{3}}{3} - (-1)}{1 + \left(\frac{\sqrt{3}}{3}\right)(-1)} = 2 + \sqrt{3}$$

20. $\dfrac{5\pi}{12} = \dfrac{\pi}{4} + \dfrac{\pi}{6}$

$$\sin\left(\frac{\pi}{4} + \frac{\pi}{6}\right) = \sin\frac{\pi}{4}\cos\frac{\pi}{6} + \cos\frac{\pi}{4}\sin\frac{\pi}{6} = \left(\frac{\sqrt{2}}{2}\right)\left(\frac{\sqrt{3}}{2}\right) + \left(\frac{\sqrt{2}}{2}\right)\left(\frac{1}{2}\right) = \frac{\sqrt{2}}{4}(\sqrt{3} + 1)$$

$$\cos\left(\frac{\pi}{4} + \frac{\pi}{6}\right) = \cos\frac{\pi}{4}\cos\frac{\pi}{6} - \sin\frac{\pi}{4}\sin\frac{\pi}{6} = \left(\frac{\sqrt{2}}{2}\right)\left(\frac{\sqrt{3}}{2}\right) - \left(\frac{\sqrt{2}}{2}\right)\left(\frac{1}{2}\right) = \frac{\sqrt{2}}{2}(\sqrt{3} - 1)$$

$$\tan\left(\frac{\pi}{4} + \frac{\pi}{6}\right) = \frac{\tan\frac{\pi}{4} + \tan\frac{\pi}{6}}{1 - \tan\frac{\pi}{4}\tan\frac{\pi}{6}} = \frac{1 + \frac{\sqrt{3}}{3}}{1 - (1)\left(\frac{\sqrt{3}}{3}\right)} = \sqrt{3} + 2$$

22. $\sin 140^\circ \cos 50^\circ + \cos 140^\circ \sin 50^\circ = \sin(140^\circ + 50^\circ) = \sin 190^\circ$

24. $\cos 20^\circ \cos 30^\circ + \sin 20^\circ \sin 30^\circ = \cos(30^\circ - 20^\circ) = \cos 10^\circ$

26. $\dfrac{\tan 140^\circ - \tan 60^\circ}{1 + \tan 140^\circ \tan 60^\circ} = \tan(140^\circ - 60^\circ) = \tan 80^\circ$

28. $\cos\dfrac{\pi}{7}\cos\dfrac{\pi}{5} - \sin\dfrac{\pi}{7}\sin\dfrac{\pi}{5} = \cos\left(\dfrac{\pi}{7} + \dfrac{\pi}{5}\right) = \cos\dfrac{12\pi}{35}$

30. $\cos 3x \cos 2y + \sin 3x \sin 2y = \cos(3x - 2y)$

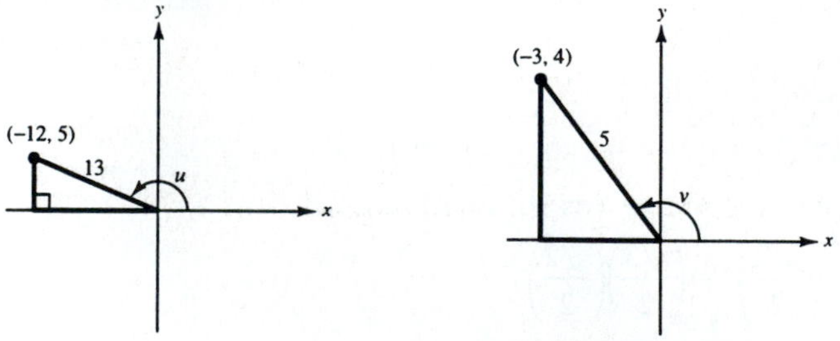

Figures for Exercises 32–38

32. $\cos(v - u) = \cos v \cos u + \sin v \sin u = \left(-\frac{3}{5}\right)\left(-\frac{12}{13}\right) + \left(\frac{4}{5}\right)\left(\frac{5}{13}\right) = \frac{36}{65} + \frac{20}{65} = \frac{56}{65}$

34. $\sin(u - v) = \sin u \cos v - \cos u \sin v = \left(\frac{5}{13}\right)\left(-\frac{3}{5}\right) - \left(-\frac{12}{13}\right)\left(\frac{4}{5}\right) = -\frac{15}{65} + \frac{48}{65} = \frac{33}{65}$

36. $\csc(u - v) = \dfrac{1}{\sin(u - v)} = \dfrac{1}{33/65} = \dfrac{65}{33}$

38. $\cot(u + v) = \dfrac{1}{\tan(u + v)} = \dfrac{1}{-63/16} = -\dfrac{16}{63}$

$$\tan(u + v) = \frac{\tan u + \tan v}{1 - \tan u \tan v} = \frac{(-5/12) - (4/3)}{1 - (-5/12)(4/3)} = \frac{-7/4}{4/9} = -\frac{63}{16}$$

$$= \frac{-7/4}{4/9} = -\frac{63}{16}$$

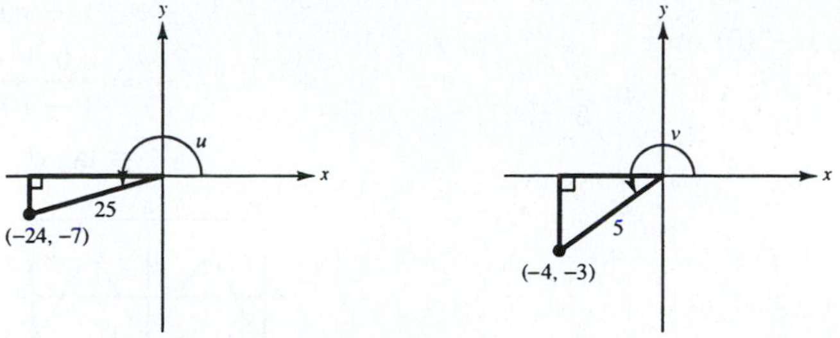

Figures for Exercises 40–44

40. $\sin(u + v) = \sin u \cos v + \cos u \sin v$
$= \left(-\tfrac{7}{25}\right)\left(-\tfrac{4}{5}\right) + \left(-\tfrac{24}{25}\right)\left(-\tfrac{3}{5}\right)$
$= \tfrac{28}{125} + \tfrac{72}{125} = \tfrac{100}{125} = \tfrac{4}{5}$

42. $\cos(u - v) = \cos u \cos v + \sin u \sin v$
$= \left(-\tfrac{24}{25}\right)\left(-\tfrac{4}{5}\right) + \left(-\tfrac{7}{25}\right)\left(-\tfrac{3}{25}\right)$
$= \tfrac{96}{125} + \tfrac{21}{125} = \tfrac{117}{125}$

44. $\sec(v - u) = \dfrac{1}{\cos(v - u)} = \dfrac{1}{\cos v \cos u + \sin v \sin u} = \dfrac{1}{\cos u \cos v + \sin u \sin v}$
$= \dfrac{1}{\cos(u - v)} = \dfrac{1}{117/125} = \dfrac{125}{117}$

46. $\sin\left(\dfrac{\pi}{2} + x\right) = \sin \dfrac{\pi}{2} \cos x + \sin x \cos \dfrac{\pi}{2}$
$= (1)(\cos x) + (\sin x)(0)$
$= \cos x$

48. $\cos\left(\dfrac{5\pi}{4} - x\right) = \cos \dfrac{5\pi}{4} \cos x + \sin \dfrac{5\pi}{4} \sin x$
$= -\dfrac{\sqrt{2}}{2}(\cos x + \sin x)$

50. $\tan\left(\dfrac{\pi}{4} - \theta\right) = \dfrac{\tan(\pi/4) - \tan \theta}{1 + \tan(\pi/4) \tan \theta} = \dfrac{1 - \tan \theta}{1 + \tan \theta}$

52. $\sin(x + y) \sin(x - y) = (\sin x \cos y + \sin y \cos x)(\sin x \cos y - \sin y \cos x)$
$= \sin^2 x \cos^2 y - \sin^2 y \cos^2 x$
$= \sin^2 x(1 - \sin^2 y) - \sin^2 y \cos^2 x$
$= \sin^2 x - \sin^2 x \sin^2 y - \sin^2 y \cos^2 x$
$= \sin^2 x - \sin^2 y(\sin^2 x + \cos^2 x)$
$= \sin^2 x - \sin^2 y$

54. $\cos(x + y) + \cos(x - y) = \cos x \cos y - \sin x \sin y + \cos x \cos y + \sin x \sin y$
$= 2 \cos x \cos y$

56. $\sin(n\pi + \theta) = \sin n\pi \cos \theta + \sin \theta \cos n\pi$
$= (0)(\cos \theta) + (\sin \theta)(-1)^n$
$= (-1)^n(\sin \theta)$, where n is an integer.

58. $C = \arctan \dfrac{a}{b} \Rightarrow \sin C = \dfrac{a}{\sqrt{a^2 + b^2}}, \cos C = \dfrac{b}{\sqrt{a^2 + b^2}}$

$\sqrt{a^2 + b^2} \cos(B\theta - C) = \sqrt{a^2 + b^2}\left(\cos B\theta \cdot \dfrac{b}{\sqrt{a^2 + b^2}} + \sin B\theta \cdot \dfrac{a}{\sqrt{a^2 + b^2}}\right)$
$= b \cos B\theta + a \sin B\theta$
$= a \sin B\theta + b \cos B\theta$

60. $\cos(\pi + x) = \cos \pi \cos x - \sin \pi \sin x$
$= (-1) \cos x - (0) \sin x$
$= -\cos x$

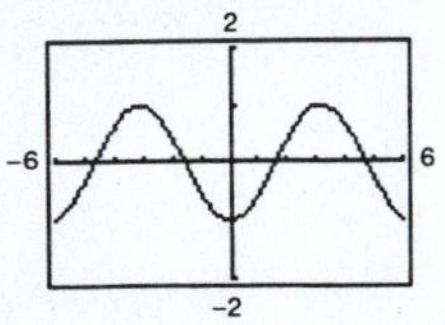

62. $\tan(\pi + \theta) = \dfrac{\tan \pi + \tan \theta}{1 - \tan \pi \tan \theta}$
$= \dfrac{0 + \tan \theta}{1 - (0) \tan \theta}$
$= \tan \theta$

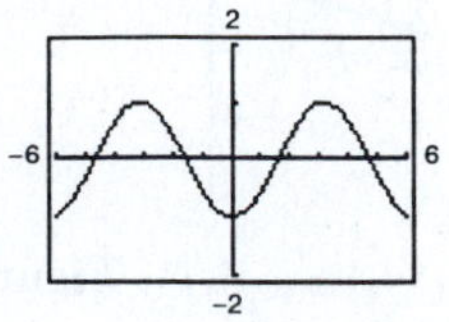

64. $3 \sin 2\theta + 4 \cos 2\theta$

$a = 3, b = 4, B = 2$

(a) $C = \arctan \dfrac{b}{a} = \arctan \dfrac{4}{3} \approx 0.9273$

$3 \sin 2\theta + 4 \cos 2\theta = \sqrt{a^2 + b^2} \sin(B\theta + C)$
$\approx 5 \sin(2\theta + 0.9273)$

(b) $C = \arctan \dfrac{a}{b} = \arctan \dfrac{3}{4} \approx 0.6435$

$3 \sin 2\theta + 4 \cos 2\theta = \sqrt{a^2 + b^2} \cos(B\theta - C)$
$\approx 5 \cos(2\theta - 0.6435)$

66. $\sin 2\theta - \cos 2\theta$

$a = 1, b = -1, B = 2$

(a) $C = \arctan \dfrac{b}{a} = \arctan(-1) = -\dfrac{\pi}{4}$

$\sin 2\theta - \cos 2\theta = \sqrt{a^2 + b^2} \sin(B\theta + C)$
$= \sqrt{2} \sin\left(2\theta - \dfrac{\pi}{4}\right)$

(b) $C = \arctan \dfrac{a}{b} = \arctan(-1) = -\dfrac{\pi}{4}$

$\sin 2\theta - \cos 2\theta = \sqrt{a^2 + b^2} \cos(B\theta - C)$
$= \sqrt{2} \cos\left(2\theta + \dfrac{\pi}{4}\right)$

68. $C = \arctan \dfrac{a}{b} = -\dfrac{3\pi}{4} \Rightarrow a = b, a < 0, b < 0$

$\sqrt{a^2 + b^2} = 5 \Rightarrow a = b = \dfrac{-5\sqrt{2}}{2}$

$B = 1$

$5 \cos\left(\theta + \dfrac{3\pi}{4}\right) = -\dfrac{5\sqrt{2}}{2} \sin \theta - \dfrac{5\sqrt{2}}{2} \cos \theta$

70. Let:

$u = \arctan 2x$ and $v = \arccos x$

$\tan u = 2x$ $\quad$ $\cos v = x$

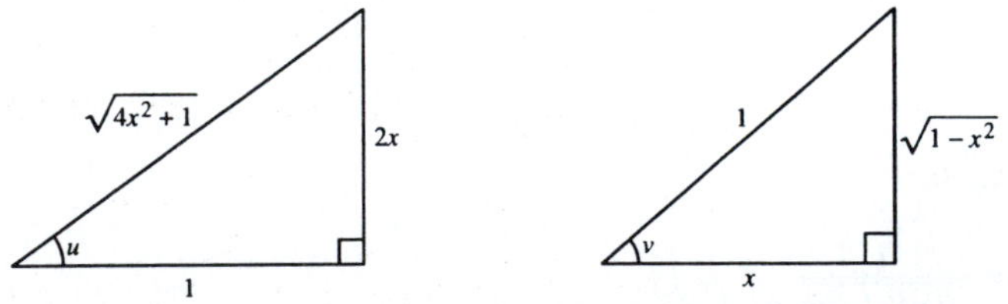

$\sin(\arctan 2x - \arccos x) = \sin(u - v)$
$= \sin u \cos v - \cos u \sin v$
$= \dfrac{2x}{\sqrt{4x^2 + 1}}(x) - \dfrac{1}{\sqrt{4x^2 + 1}}\left(\sqrt{1 - x^2}\right)$
$= \dfrac{2x^2 - \sqrt{1 - x^2}}{\sqrt{4x^2 + 1}}$

72.
$$\sin\left(x+\frac{\pi}{6}\right)-\sin\left(x-\frac{\pi}{6}\right)=\frac{1}{2}$$
$$\sin x\cos\frac{\pi}{6}+\cos x\sin\frac{\pi}{6}-\left(\sin x\cos\frac{\pi}{6}-\cos x\sin\frac{\pi}{6}\right)=\frac{1}{2}$$
$$2\cos x(0.5)=\frac{1}{2}$$
$$\cos x=\frac{1}{2}$$
$$x=\frac{\pi}{3},\frac{5\pi}{3}$$

74.
$$\tan(x+\pi)+2\sin(x+\pi)=0$$
$$\frac{\tan x+\tan\pi}{1-\tan x\tan\pi}+2(\sin x\cos\pi+\cos x\sin\pi)=0$$
$$\frac{\tan x+0}{1-\tan x(0)}+2[\sin x(-1)+\cos x(0)]=0$$
$$\frac{\tan x}{1}-2\sin x=0$$
$$\frac{\sin x}{\cos x}=2\sin x$$
$$\sin x=2\sin x\cos x$$
$$\sin x(1-2\cos x)=0$$
$$\sin x=0 \quad \text{or} \quad \cos x=\frac{1}{2}$$
$$x=0,\pi \qquad x=\frac{\pi}{3},\frac{5\pi}{3}$$

76. $\tan(x+\pi)-\cos\left(x+\frac{\pi}{2}\right)=0$

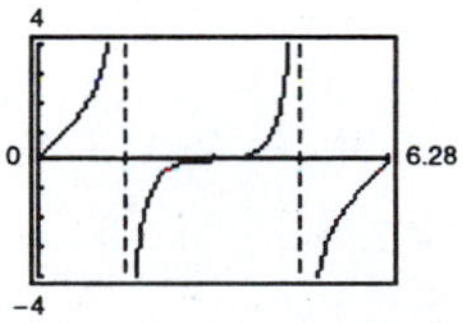

Answers: $(0, 0), (3.14, 0) \Rightarrow x = 0, \pi$

78. From the figure, it appears that $u+v=w$. Assume that u, v, and w are all in Quadrant I. From the figure:

$$\tan u=\frac{s}{3s}=\frac{1}{3}$$
$$\tan v=\frac{s}{2s}=\frac{1}{2}$$
$$\tan s=\frac{s}{s}=1$$
$$\tan(u+v)=\frac{\tan u+\tan v}{1-\tan u\tan v}=\frac{1/3+1/2}{1-(1/3)(1/2)}=\frac{5/6}{1-(1/6)}=1=\tan w.$$

Thus, $\tan(u+v)=\tan w$. Because u, v, and w are all in Quadrant I, we have
$$\arctan[\tan(u+v)]=\arctan[\tan w]$$
$$u+v=w.$$

80. $y_1 = A\cos 2\pi\left(\frac{t}{T} - \frac{x}{\lambda}\right)$

$y_2 = A\cos 2\pi\left(\frac{t}{T} + \frac{x}{\lambda}\right)$

$y_1 + y_2 = A\cos 2\pi\left(\frac{t}{T} - \frac{x}{\lambda}\right) + A\cos 2\pi\left(\frac{t}{T} + \frac{x}{\lambda}\right)$

$y_1 + y_2 = A\left[\cos 2\pi\frac{t}{T}\cos 2\pi\frac{x}{\lambda} + \sin 2\pi\frac{t}{T}\sin 2\pi\frac{x}{\lambda}\right] + A\left[\cos 2\pi\frac{t}{T}\cos 2\pi\frac{x}{\lambda} - \sin 2\pi\frac{t}{T}\sin 2\pi\frac{x}{\lambda}\right]$

$= 2A\cos 2\pi\frac{t}{T}\cos 2\pi\frac{x}{\lambda}$

Section 5.5 Multiple-Angle and Product-to-Sum Formulas

Solutions to Even-Numbered Exercises

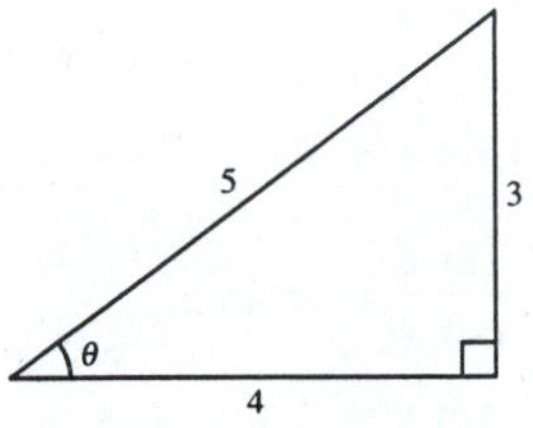

Figure for Exercises 2–8

2. $\tan\theta = \frac{3}{4}$

4. $\sin 2\theta = 2\sin\theta\cos\theta = 2\left(\frac{3}{5}\right)\left(\frac{4}{5}\right) = \frac{24}{25}$

6. $\sec 2\theta = \frac{1}{\cos 2\theta}$

$= \frac{1}{\cos^2\theta - \sin^2\theta}$

$= \frac{1}{(4/5)^2 - (3/5)^2}$

$= \frac{1}{(16/25) - (9/25)}$

$= \frac{25}{7}$

8. $\cot 2\theta = \frac{1}{\tan 2\theta}$

$= \frac{1 - \tan^2\theta}{2\tan\theta}$

$= \frac{1 - (3/4)^2}{2(3/4)}$

$= \frac{7/16}{3/2}$

$= \frac{7}{24}$

10. $\sin 2x + \cos x = 0$

$2\sin x\cos x + \cos x = 0$

$\cos x(2\sin x + 1) = 0$

$\cos x = 0$ or $2\sin x + 1 = 0$

$x = \frac{\pi}{2}, \frac{3\pi}{2}$ $\qquad \sin x = -\frac{1}{2}$

$x = \frac{7\pi}{6}, \frac{11\pi}{6}$

12. $\sin 2x\sin x = \cos x$

$2\sin x\cos x\sin x - \cos x = 0$

$\cos x(2\sin^2 x - 1) = 0$

$\cos x = 0$ or $2\sin^2 x - 1 = 0$

$x = \frac{\pi}{2}, \frac{3\pi}{2}$ $\qquad \sin^2 x = \frac{1}{2}$

$\sin x = \pm\frac{\sqrt{2}}{2}$

$x = \frac{\pi}{4}, \frac{3\pi}{4}, \frac{5\pi}{4}, \frac{7\pi}{4}$

14.
$$\cos 2x + \sin x = 0$$
$$1 - 2\sin^2 x + \sin x = 0$$
$$2\sin^2 x - \sin x - 1 = 0$$
$$(2\sin x + 1)(\sin x - 1) = 0$$
$$2\sin x + 1 = 0 \quad \text{or} \quad \sin x - 1 = 0$$
$$\sin x = -\frac{1}{2} \qquad \sin x = 1$$
$$x = \frac{7\pi}{6}, \frac{11\pi}{6} \qquad x = \frac{\pi}{2}$$

16.
$$\tan 2x - 2\cos x = 0$$
$$\frac{2\tan x}{1 - \tan^2 x} = 2\cos x$$
$$2\tan x = 2\cos x(1 - \tan^2 x)$$
$$2\tan x = 2\cos x - 2\cos x \tan^2 x$$
$$2\tan x = 2\cos x - 2\cos x\frac{\sin^2 x}{\cos^2 x}$$
$$2\tan x = 2\cos x - 2\frac{\sin^2 x}{\cos x}$$
$$\tan x = \cos x - \frac{\sin^2 x}{\cos x}$$
$$\frac{\sin x}{\cos x} = \cos x - \frac{\sin^2 x}{\cos x}$$
$$\frac{\sin x}{\cos x} + \frac{\sin^2 x}{\cos x} - \cos x = 0$$
$$\frac{\sin x + \sin^2 x - \cos^2 x}{\cos x} = 0$$
$$\frac{1}{\cos x}[\sin x + \sin^2 x - (1 - \sin^2 x)] = 0$$
$$\sec x[2\sin^2 x + \sin x - 1] = 0$$
$$\sec x(2\sin x - 1)(\sin x + 1) = 0$$
$$\sec x = 0 \quad \text{or} \quad 2\sin x - 1 = 0 \quad \text{or} \quad \sin x + 1 = 0$$
$$x = \frac{\pi}{2}, \frac{3\pi}{2} \qquad \sin x = \frac{1}{2} \qquad \sin x = -1$$
$$x = \frac{\pi}{6}, \frac{5\pi}{6} \qquad x = \frac{3\pi}{2}$$

$$x = \frac{\pi}{6}, \frac{\pi}{2}, \frac{5\pi}{6}, \frac{3\pi}{2}$$

18.
$$\begin{aligned}(\sin 2x + \cos 2x)^2 &= 1\\ \sin^2 2x + 2\sin 2x\cos 2x + \cos^2 2x &= 1\\ 2\sin 2x\cos 2x &= 0\\ \sin 4x &= 0\\ 4x &= n\pi\\ x &= \frac{n\pi}{4}\\ x &= 0, \frac{\pi}{4}, \frac{\pi}{2}, \frac{3\pi}{4}, \pi, \frac{5\pi}{4}, \frac{3\pi}{2}, \frac{7\pi}{4}\end{aligned}$$

20. $4\sin x\cos x + 2 = 2(2\sin x\cos x) + 2$
$= 2\sin 2x + 2$

22. $(\cos x + \sin x)(\cos x - \sin x) = \cos^2 x - \sin^2 x$
$= \cos 2x$

24. $\cos u = -\frac{2}{3}, \frac{\pi}{2} < u < \pi$

$\sin 2u = 2\sin u\cos u = 2 \cdot \frac{\sqrt{5}}{3}\left(-\frac{2}{3}\right) = -\frac{4\sqrt{5}}{9}$

$\cos 2u = \cos^2 u - \sin^2 u = \frac{4}{9} - \frac{5}{9} = -\frac{1}{9}$

$\tan 2u = \frac{2\tan u}{1 - \tan^2 u} = \frac{2(-\sqrt{5}/2)}{1 - (5/4)} = 4\sqrt{5}$

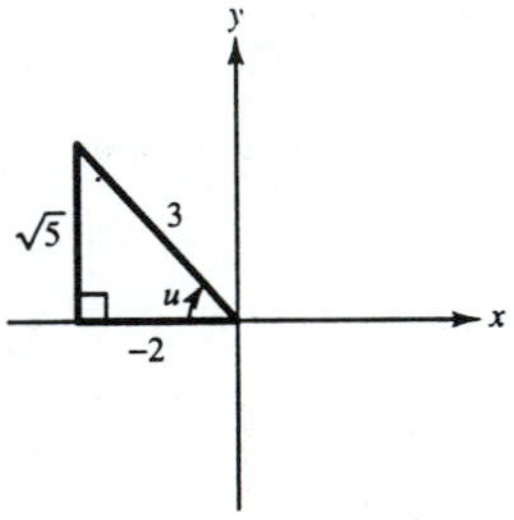

26. $\cot u = -4, \frac{3\pi}{2} < u < 2\pi$

$\sin 2u = 2\sin u\cos u = 2\left(-\frac{1}{\sqrt{17}}\right)\left(\frac{4}{\sqrt{17}}\right) = -\frac{8}{17}$

$\cos 2u = \cos^2 u - \sin^2 u$

$= \left(\frac{4}{\sqrt{17}}\right)^2 - \left(-\frac{1}{\sqrt{17}}\right)^2 = \frac{15}{17}$

$\tan 2u = \frac{2\tan u}{1 - \tan^2 u} = \frac{2(-1/4)}{1 - (-1/4)^2} = -\frac{8}{15}$

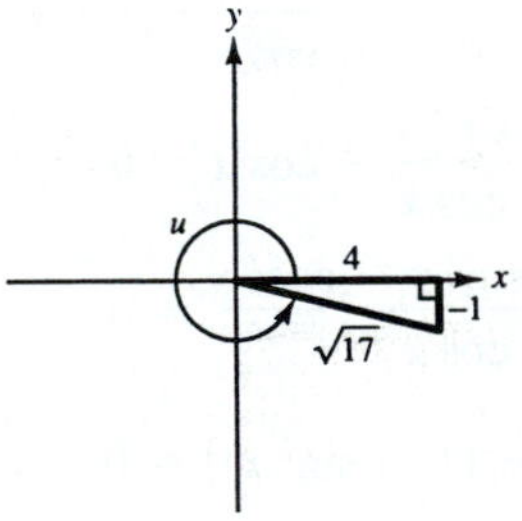

28. $\sin 2u = 2\sin u\cos u = 2 \cdot \frac{1}{3}\left(-\frac{2\sqrt{2}}{3}\right) = -\frac{4\sqrt{2}}{9}$

$\cos 2u = \cos^2 u - \sin^2 u = \left(-\frac{2\sqrt{2}}{3}\right)^2 - \left(\frac{1}{3}\right)^2 = \frac{7}{9}$

$\tan 2u = \frac{2\tan u}{1 - \tan^2 u} = \frac{2(-\sqrt{2}/4)}{1 - (-\sqrt{2}/4)^2} = -\frac{4\sqrt{2}}{7}$

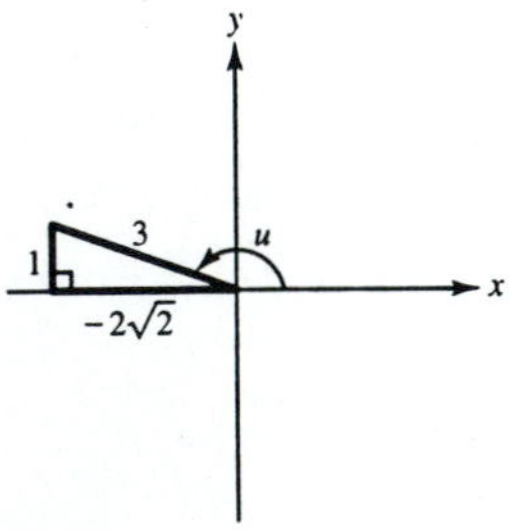

30. $\sin^4 x = (\sin^2 x)(\sin^2 x)$

$= \left(\frac{1 - \cos 2x}{2}\right)\left(\frac{1 - \cos 2x}{2}\right)$

$= \frac{1 - 2\cos 2x + \cos^2 2x}{4}$

$= \frac{1 - 2\cos 2x + \left(\frac{1 + \cos 4x}{2}\right)}{4}$

$= \frac{2 - 4\cos 2x + 1 + \cos 4x}{8}$

$= \frac{1}{8}(3 - 4\cos 2x + \cos 4x)$

32. $\cos^2 x = \dfrac{1 + \cos 2x}{2} = \dfrac{1}{2}(1 + \cos 2x)$

34. $\sin^4 x \cos^2 x = \sin^2 x \sin^2 x \cos^2 x$

$$= \left(\frac{1 - \cos 2x}{2}\right)\left(\frac{1 - \cos 2x}{2}\right)\left(\frac{1 + \cos 2x}{2}\right)$$

$$= \frac{1}{8}(1 - \cos 2x)(1 - \cos^2 2x)$$

$$= \frac{1}{8}(1 - \cos 2x - \cos^2 2x + \cos^3 2x)$$

$$= \frac{1}{8}\left[1 - \cos 2x - \left(\frac{1 + \cos 4x}{2}\right) + \cos 2x\left(\frac{1 + \cos 4x}{2}\right)\right]$$

$$= \frac{1}{16}[2 - 2\cos 2x - 1 - \cos 4x + \cos 2x + \cos 2x \cos 4x]$$

$$= \frac{1}{16}\left[1 - \cos 2x - \cos 4x + \frac{1}{2}\cos 2x + \frac{1}{2}\cos 6x\right]$$

$$= \frac{1}{32}[2 - 2\cos 2x - 2\cos 4x + \cos 2x + \cos 6x]$$

$$= \frac{1}{32}[2 - \cos 2x - 2\cos 4x + \cos 6x]$$

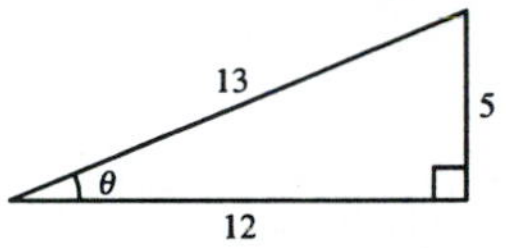

Figure for Exercises 36–40

36. $\sin\dfrac{\theta}{2} = \sqrt{\dfrac{1 - \cos\theta}{2}}$

$$= \sqrt{\frac{1 - (12/13)}{2}}$$

$$= \sqrt{\frac{1/13}{2}}$$

$$= \frac{1}{\sqrt{26}}$$

38. $2\sin\dfrac{\theta}{2}\cos\dfrac{\theta}{2} = 2\sqrt{\dfrac{1 - \cos\theta}{2}}\sqrt{\dfrac{1 + \cos\theta}{2}}$

$$= \frac{2\sqrt{1 - \cos^2\theta}}{2}$$

$$= \sqrt{1 - \left(\frac{12}{13}\right)^2}$$

$$= \sqrt{\frac{25}{169}}$$

40. $\cot\dfrac{\theta}{2} = \dfrac{1}{\tan\dfrac{\theta}{2}} = \dfrac{\sin\theta}{1 - \cos\theta} = \dfrac{\dfrac{5}{13}}{1 - \dfrac{12}{13}} = \dfrac{5}{13}\left(\dfrac{13}{1}\right) = 5$

42. $\sin 165° = \sin\left(\dfrac{1}{2}\cdot 330°\right) = \sqrt{\dfrac{1 - \cos 330°}{2}} = \sqrt{\dfrac{1 - (\sqrt{3}/2)}{2}} = \dfrac{1}{2}\sqrt{2 - \sqrt{3}}$

$\cos 165° = \cos\left(\dfrac{1}{2}\cdot 330°\right) = -\sqrt{\dfrac{1 + \cos 330°}{2}} = -\sqrt{\dfrac{1 + (\sqrt{3}/2)}{2}} = -\dfrac{1}{2}\sqrt{2 + \sqrt{3}}$

$\tan 165° = \tan\left(\dfrac{1}{2}\cdot 330°\right) = \dfrac{\sin 330°}{1 + \cos 330°} = \dfrac{-1/2}{1 + (\sqrt{3}/2)} = \dfrac{-1}{2 + \sqrt{3}} = \sqrt{3} - 2$

44. $\sin 67°\,30' = \sin\left(\frac{1}{2}\cdot 135°\right) = \sqrt{\frac{1-\cos 135°}{2}} = \sqrt{\frac{1+\left(\sqrt{2}/2\right)}{2}} = \frac{1}{2}\sqrt{2+\sqrt{2}}$

$\cos 67°\,30' = \cos\left(\frac{1}{2}\cdot 135°\right) = \sqrt{\frac{1+\cos 135°}{2}} = \sqrt{\frac{1-\left(\sqrt{2}/2\right)}{2}} = \frac{1}{2}\sqrt{2-\sqrt{2}}$

$\tan 67°\,30' = \tan\left(\frac{1}{2}\cdot 135°\right) = \frac{\sin 135°}{1+\cos 135°} = \frac{\sqrt{2}/2}{1-\left(\sqrt{2}/2\right)} = 1+\sqrt{2}$

46. $\sin\frac{\pi}{12} = \sin\left[\frac{1}{2}\left(\frac{\pi}{6}\right)\right] = \sqrt{\frac{1-\cos(\pi/6)}{2}} = \sqrt{\frac{1-\left(\sqrt{3}/2\right)}{2}} = \frac{1}{2}\sqrt{2-\sqrt{3}}$

$\cos\frac{\pi}{12} = \cos\left[\frac{1}{2}\left(\frac{\pi}{6}\right)\right] = \sqrt{\frac{1+\cos(\pi/6)}{2}} = \frac{1}{2}\sqrt{2+\sqrt{3}}$

$\tan\frac{\pi}{12} = \tan\left[\frac{1}{2}\left(\frac{\pi}{6}\right)\right] = \frac{\sin(\pi/6)}{1+\cos(\pi/6)} = \frac{1/2}{1+\left(\sqrt{3}/2\right)} = 2-\sqrt{3}$

48. $\cos u = \frac{3}{5},\ 0 < u < \frac{\pi}{2}$

$\sin\left(\frac{u}{2}\right) = \sqrt{\frac{1-\cos u}{2}} = \sqrt{\frac{1-(3/5)}{2}} = \frac{\sqrt{5}}{5}$

$\cos\left(\frac{u}{2}\right) = \sqrt{\frac{1+\cos u}{2}} = \sqrt{\frac{1+(3/5)}{2}} = \frac{2\sqrt{5}}{5}$

$\tan\left(\frac{u}{2}\right) = \frac{\sin u}{1+\cos u} = \frac{4/5}{1+(3/5)} = \frac{1}{2}$

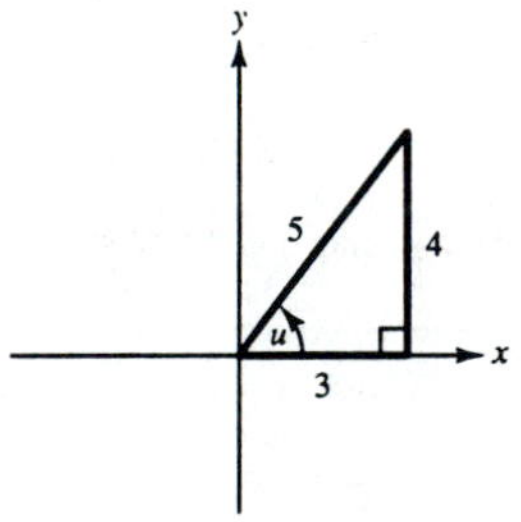

50. $\cot u = 3,\ \pi < u < \frac{3\pi}{2}$

$\sin\left(\frac{u}{2}\right) = \sqrt{\frac{1-\cos u}{2}} = \sqrt{\frac{1+\left(3/\sqrt{10}\right)}{2}} = \sqrt{\frac{10+3\sqrt{10}}{20}} = \frac{1}{2}\sqrt{\frac{10+3\sqrt{10}}{5}}$

$\cos\left(\frac{u}{2}\right) = -\sqrt{\frac{1+\cos u}{2}} = -\sqrt{\frac{1-\left(3/\sqrt{10}\right)}{2}} = -\sqrt{\frac{10-3\sqrt{10}}{20}} = -\frac{1}{2}\sqrt{\frac{10-3\sqrt{10}}{5}}$

$\tan\left(\frac{u}{2}\right) = \frac{1-\cos u}{\sin u} = \frac{1+\left(3/\sqrt{10}\right)}{-1/\sqrt{10}} = -3-\sqrt{10}$

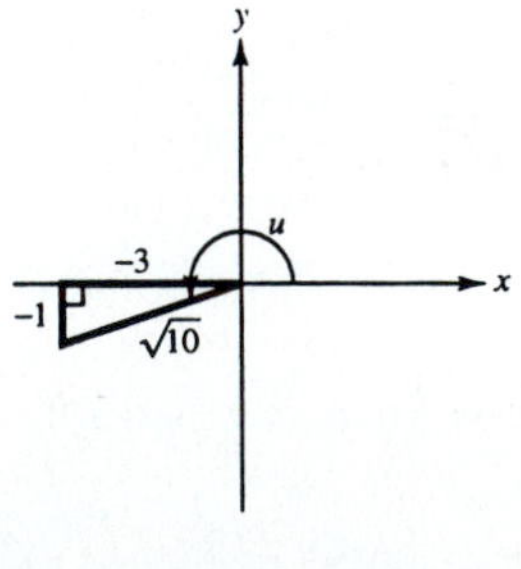

52. $\sec u = -\frac{7}{2}, \frac{\pi}{2} < u < \pi$

$$\sin\left(\frac{u}{2}\right) = \sqrt{\frac{1 - \cos u}{2}} = \sqrt{\frac{1 + (2/7)}{2}} = \frac{3\sqrt{14}}{14}$$

$$\cos\left(\frac{u}{2}\right) = \sqrt{\frac{1 + \cos u}{2}} = \sqrt{\frac{1 - (2/7)}{2}} = \frac{\sqrt{70}}{14}$$

$$\tan\left(\frac{u}{2}\right) = \frac{1 - \cos u}{\sin u} = \frac{1 + (2/7)}{3(\sqrt{5}/7)} = \frac{3\sqrt{5}}{5}$$

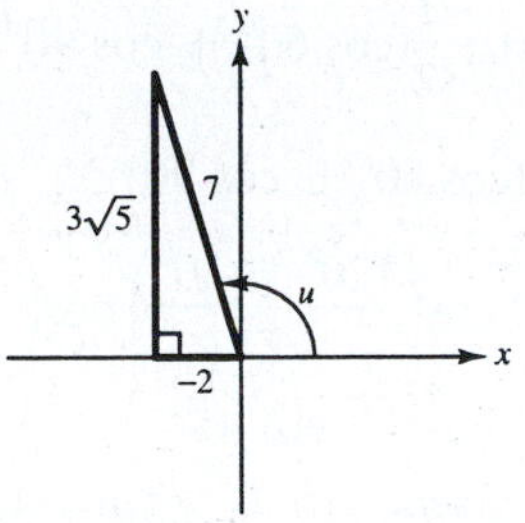

54. $\sqrt{\frac{1 + \cos 4x}{2}} = \left|\cos \frac{4x}{2}\right| = |\cos 2x|$

56. $-\sqrt{\frac{1 - \cos(x - 1)}{2}} = -\left|\sin\left(\frac{x - 1}{2}\right)\right|$

58. $h(x) = \sin\frac{x}{2} + \cos x - 1$

$$\sin\frac{x}{2} + \cos x - 1 = 0$$

$$\pm\sqrt{\frac{1 - \cos x}{2}} = 1 - \cos x$$

$$\frac{1 - \cos x}{2} = 1 - 2\cos x + \cos^2 x$$

$$1 - \cos x = 2 - 4\cos x + 2\cos^2 x$$

$$2\cos^2 x - 3\cos x + 1 = 0$$

$$(2\cos x - 1)(\cos x - 1) = 0$$

$2\cos x - 1 = 0$ or $\cos x - 1 = 0$

$\cos x = \frac{1}{2}$ $\quad$ $\cos x = 1$

$x = \frac{\pi}{3}, \frac{5\pi}{3}$ $\quad$ $x = 0$

0, $\pi/3$, and $5\pi/2$ are all solutions to the equation.

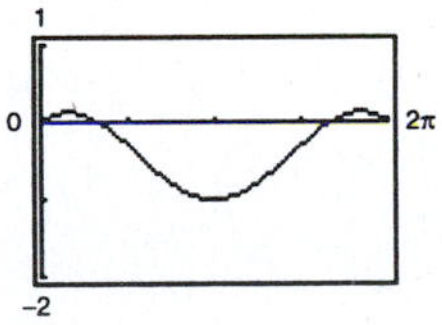

60. $g(x) = \tan\frac{x}{2} - \sin x$

$$\tan\frac{x}{2} - \sin x = 0$$

$$\frac{1 - \cos x}{\sin x} = \sin x$$

$$1 - \cos x = \sin^2 x$$

$$1 - \cos x = 1 - \cos^2 x$$

$$\cos^2 x - \cos x = 0$$

$$\cos x(\cos x - 1) = 0$$

$\cos x = 0$ or $\cos x - 1 = 0$

$x = \frac{\pi}{2}, \frac{3\pi}{2}$ $\quad$ $\cos x = 1$

$x = 0$

0, $\pi/2$, and $3\pi/2$ are all solutions to the equation.

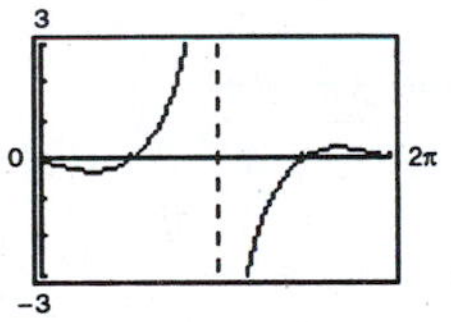

62. $4\sin\frac{\pi}{3}\cos\frac{5\pi}{6} = 4 \cdot \frac{1}{2}\left[\sin\left(\frac{\pi}{3} + \frac{5\pi}{6}\right) + \sin\left(\frac{\pi}{3} - \frac{5\pi}{6}\right)\right]$

$$= 2\left[\sin\frac{7\pi}{6} + \sin\left(-\frac{\pi}{2}\right)\right]$$

$$= 2\left(\sin\frac{7\pi}{6} - \sin\frac{\pi}{2}\right)$$

64. $3\sin 2\alpha \sin 3\alpha = 3 \cdot \frac{1}{2}[\cos(2\alpha - 3\alpha) - \cos(2\alpha + 3\alpha)]$

$$= \frac{3}{2}[\cos(-\alpha) - \cos 5\alpha]$$

$$= \frac{3}{2}(\cos\alpha - \cos 5\alpha)$$

66. $\cos 2\theta \cos 4\theta = \frac{1}{2}[\cos(2\theta - 4\theta) + \cos(2\theta + 4\theta)]$

$$= \frac{1}{2}[\cos(-2\theta) + \cos 6\theta]$$

$$= \frac{1}{2}(\cos 2\theta + \cos 6\theta)$$

68. $\sin(x + y)\cos(x - y) = \frac{1}{2}(\sin 2x + \sin 2y)$

70. $10 \cos 75° \cos 15° = 10 \cdot \frac{1}{2}(\cos 60° + \cos 90°)$

$= 5(\cos 60° + \cos 90°)$

72. $\cos 120° + \cos 30° = 2 \cos\left(\frac{120° + 30°}{2}\right) \cos\left(\frac{120° - 30°}{2}\right)$

$= 2 \cos 75° \cos 45°$

74. $\sin 5\theta - \sin 3\theta = 2 \cos\left(\frac{5\theta + 3\theta}{2}\right) \sin\left(\frac{5\theta - 3\theta}{2}\right)$

$= 2 \cos 4\theta \sin \theta$

76. $\sin x + \sin 5x = 2 \sin\left(\frac{x + 5x}{2}\right) \cos\left(\frac{x - 5x}{2}\right)$

$= 2 \sin 3x \cos(-2x)$

$= 2 \sin 3x \cos 2x$

78. $\cos\left(\theta + \frac{\pi}{2}\right) - \cos\left(\theta - \frac{\pi}{2}\right) = -2 \sin\left(\frac{\theta + (\pi/2) + \theta - (\pi/2)}{2}\right) \sin\left(\frac{\theta + (\pi/2) - \theta + (\pi/2)}{2}\right)$

$= -2 \sin 2\theta \sin \frac{\pi}{2}$

80. $\cos\left(x + \frac{\pi}{2}\right) + \sin\left(x - \frac{\pi}{2}\right) = 2 \sin\left(\frac{x + (\pi/2) + x - (\pi/2)}{2}\right) \cos\left(\frac{x + (\pi/2) - x + (\pi/2)}{2}\right)$

$= 2 \sin x \cos \frac{\pi}{2} = 0$

82. $h(x) = \cos 2x - \cos 6x$

$\cos 2x - \cos 6x = 0$

$-2 \sin 4x \sin(-2x) = 0$

$2 \sin 4x \sin 2x = 0$

$\sin 4x = 0$ or $\sin 2x = 0$

$4x = n\pi$ $\quad$ $2x = n\pi$

$x = \frac{n\pi}{4}$ $\quad$ $x = \frac{n\pi}{2}$

$x = 0, \frac{\pi}{4}, \frac{\pi}{2}, \frac{3\pi}{4}, \pi, \frac{5\pi}{4}, \frac{3\pi}{2}, \frac{7\pi}{4}$ $\quad$ $x = 0, \frac{\pi}{2}, \pi, \frac{3\pi}{2}$

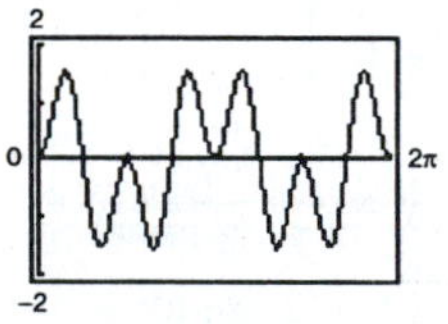

84. $f(x) = \sin^2 3x - \sin^2 x$

$\sin^2 3x - \sin^2 x = 0$

$(\sin 3x + \sin x)(\sin 3x - \sin x) = 0$

$(2 \sin 2x \cos x)(2 \cos 2x \sin x) = 0$

$\sin 2x = 0 \Rightarrow x = 0, \frac{\pi}{2}, \pi, \frac{3\pi}{2}$ or

$\cos x = 0 \Rightarrow x = \frac{\pi}{2}, \frac{3\pi}{2}$ or

$\cos 2x = 0 \Rightarrow x = \frac{\pi}{4}, \frac{3\pi}{4}, \frac{5\pi}{4}, \frac{7\pi}{4}$ or

$\sin x = 0 \Rightarrow x = 0, \pi$

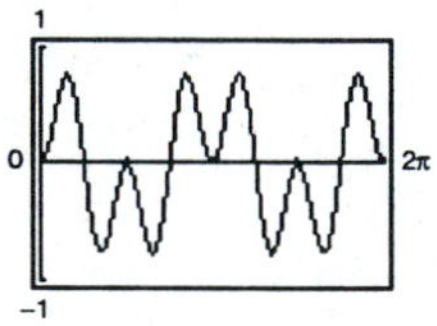

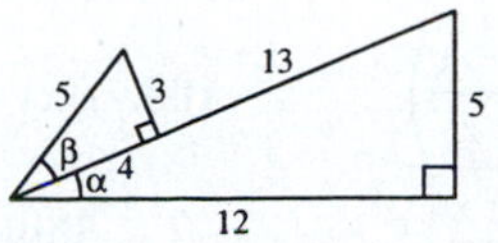

Figure for Exercises 86 and 88

86. $\cos^2 \alpha = (\cos \alpha)^2 = \left(\frac{12}{13}\right)^2 = \frac{144}{169}$

$\cos^2 \alpha = 1 - \sin^2 \alpha$

$= 1 - \left(\frac{5}{13}\right)^2$

$= 1 - \frac{25}{169} = \frac{144}{169}$

88. $\cos \alpha \sin \beta = \left(\frac{12}{13}\right)\left(\frac{3}{5}\right) = \frac{36}{65}$

$\cos \alpha \sin \beta = \sin\left(\frac{\pi}{2} - \alpha\right) \cos\left(\frac{\pi}{2} - \beta\right)$

$= \left(\frac{12}{13}\right)\left(\frac{3}{5}\right) = \frac{36}{65}$

90. $\sec 2\theta = \frac{1}{\cos 2\theta} = \frac{1}{\cos^2 \theta - \sin^2 \theta}$

$= \frac{1/\cos^2 \theta}{1 - (\sin^2 \theta/\cos^2 \theta)}$

$= \frac{\sec^2 \theta}{1 - \tan^2 \theta}$

$= \frac{\sec^2 \theta}{1 - (\sec^2 \theta - 1)}$

$= \frac{\sec^2 \theta}{2 - \sec^2 \theta}$

92. $\cos^4 x - \sin^4 x = (\cos^2 x - \sin^2 x)(\cos^2 x + \sin^2 x)$

$= (\cos 2x)(1)$

$= \cos 2x$

94. $\sin\left(\frac{\alpha}{3}\right) \cos\left(\frac{\alpha}{3}\right) = \frac{1}{2}\left[2\left(\sin\left(\frac{\alpha}{3}\right) \cos\left(\frac{\alpha}{3}\right)\right)\right]$

$= \frac{1}{2} \sin \frac{2\alpha}{3}$

96. $\frac{\cos 3\beta}{\cos \beta} = \frac{\cos^3 \beta - 3 \sin^2 \beta \cos \beta}{\cos \beta}$

$= \cos^2 \beta - 3\sin^2 \beta$

$= 1 - \sin^2 \beta - 3 \sin^2 \beta$

$= 1 - 4 \sin^2 \beta$

98. $\tan \frac{u}{2} = \frac{1 - \cos u}{\sin u}$

$= \frac{1}{\sin u} - \frac{\cos u}{\sin u}$

$= \csc u - \cot u$

100. $\frac{\sin x \pm \sin y}{\cos x + \cos y} = \frac{2 \sin\left(\frac{x \pm y}{2}\right) \cos\left(\frac{x \mp y}{2}\right)}{2 \cos\left(\frac{x + y}{2}\right) \cos\left(\frac{x - y}{2}\right)}$

$= \tan\left(\frac{x \pm y}{2}\right)$

102. $\sin\left(\frac{\pi}{6} + x\right) + \sin\left(\frac{\pi}{6} - x\right) = 2 \sin \frac{\pi}{6} \cos x$

$= 2 \cdot \frac{1}{2} \cos x$

$= \cos x$

104. $\sin 4\beta = 2 \sin 2\beta \cos 2\beta$

$= 2[2 \sin \beta \cos \beta(\cos^2 \beta - \sin^2 \beta)]$

$= 2[2 \sin \beta \cos \beta(1 - \sin^2 \beta - \sin^2 \beta)]$

$= 4 \sin \beta \cos \beta(1 - 2 \sin^2 \beta)$

Graph: $y_1 = \sin 4\beta$

$y_2 = 4 \sin \beta \cos \beta(1 - \sin^2 \beta)$

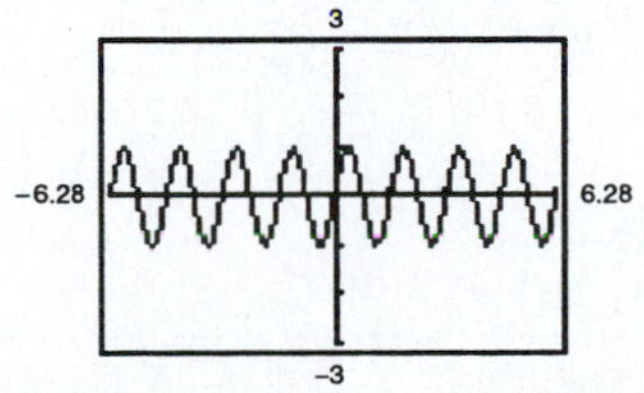

106. $$\frac{\cos 3x - \cos x}{\sin 3x - \sin x} = \frac{-2\sin\left(\frac{3x + x}{2}\right)\sin\left(\frac{3x - x}{2}\right)}{2\cos\left(\frac{3x + x}{2}\right)\sin\left(\frac{3x - x}{2}\right)}$$

$$= \frac{-2\sin 2x \sin x}{2\cos 2x \sin x}$$

$$= -\tan 2x$$

108. $f(x) = \cos^2 x = \dfrac{1 + \cos 2x}{2} = \dfrac{1}{2} + \dfrac{\cos 2x}{2}$

Shifted upward by $\frac{1}{2}$ unit.

Amplitude: $|a| = \frac{1}{2}$

Period: $\frac{2\pi}{2} = \pi$

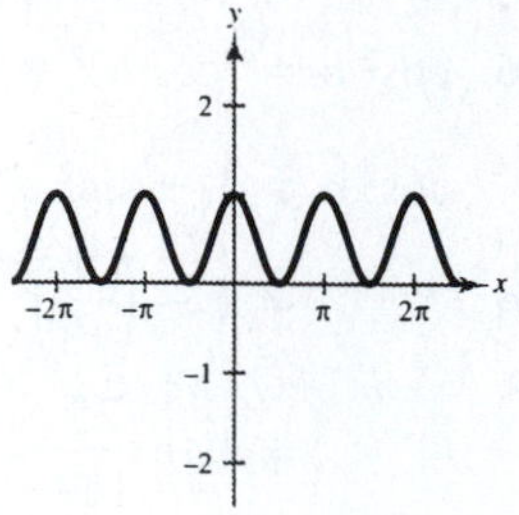

110. $f(x) = \cos 2x - 2\sin x$

(a)

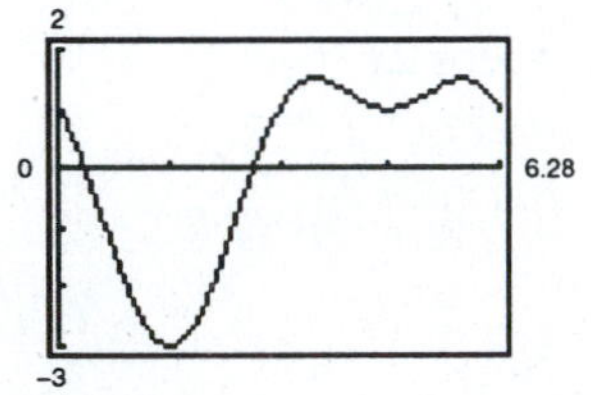

Maximum points: (3.6652, 1.5), (5.7596, 1.5)

Minimum points: (1.5708, −3), (4.7124, 1)

(b) $-2\cos x(2\sin x + 1) = 0$

$-2\cos x = 0$ or $2\sin x + 1 = 0$

$\cos x = 0$ $\qquad$ $\sin x = -\frac{1}{2}$

$x = \frac{\pi}{2}, \frac{3\pi}{2}$ $\qquad$ $x = \frac{7\pi}{6}, \frac{11\pi}{6}$

$\frac{\pi}{2} \approx 1.5708$ $\qquad$ $\frac{7\pi}{6} \approx 3.6652$

$\frac{3\pi}{2} \approx 4.7124$ $\qquad$ $\frac{11\pi}{2} \approx 5.7596$

112. $f(x) = \sin^4 x + \cos^4 x$

(a) $$\sin^4 x + \cos^4 x = (\sin^2 x)^2 + (\cos^2 x)^2$$

$$= \left(\frac{1 - \cos 2x}{2}\right)^2 + \left(\frac{1 + \cos 2x}{2}\right)^2$$

$$= \frac{1}{4}[(1 - \cos 2x)^2 + (1 + \cos 2x)^2]$$

$$= \frac{1}{4}(1 - 2\cos 2x + \cos^2 2x + 1 + 2\cos 2x + \cos^2 2x)$$

$$= \frac{1}{4}(2 + 2\cos^2 2x)$$

$$= \frac{1}{4}\left[2 + 2\left(\frac{1 + \cos 2(2x)}{2}\right)\right]$$

$$= \frac{1}{4}(3 + \cos 4x)$$

(b) $$\sin^4 x + \cos^4 x = (\sin^2 x)^2 + \cos^4 x$$

$$= (1 - \cos^2 x)^2 + \cos^4 x$$

$$= 1 - 2\cos^2 x + \cos^4 x + \cos^4 x$$

$$= 2\cos^4 x - 2\cos^2 x + 1$$

—CONTINUED—

112. —CONTINUED—

(c) $\sin^4 x + \cos^4 x = \sin^4 x + 2\sin^2 x\cos^2 x + \cos^4 x - 2\sin^2 x\cos^2 x$

$= (\sin^2 x + \cos^2 x)^2 - 2\sin^2 x\cos^2 x$

$= 1 - 2\sin^2 x\cos^2 x$

(d) $1 - 2\sin^2 x\cos^2 x = 1 - (2\sin x\cos x)(\sin x\cos x)$

$= 1 - (\sin 2x)\left(\frac{1}{2}\sin 2x\right)$

$= 1 - \frac{1}{2}\sin^2 2x$

(e) No, it does not mean that one of you is wrong. There is often more than one way to rewrite a trigonometric expression.

114. $\cos(2\arccos x) = \cos^2(\arccos x) - \sin^2(\arccos x)$

$= x^2 - (1 - x^2) = 2x^2 - 1$

116. $r = \frac{1}{32}v_0^2\sin 2\theta$

$= \frac{1}{32}v_0^2(\sin\theta\cos\theta)$

$= \frac{1}{16}v_0^2\sin\theta\cos\theta$

118. *Model:* (Distance traveled by faster car) − (Distance traveled by slower car) = 12

Label: Time $= t$

Rate of faster car $= 56$

Rate of slower car $= 48$

Distance traveled by faster car: $56t$

Distance traveled by slower car: $48t$

Equation: $56t - 48t = 12$

$8t = 12$

$t = \frac{12}{8} = \frac{3}{2} = 1.5$ hours

It takes 1.5 hours before the cars are 12 miles apart.

120.

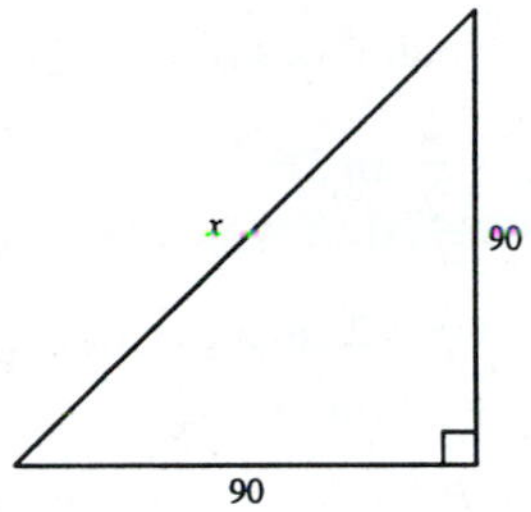

$x^2 = 90^2 + 90^2$

$x = \sqrt{2(90^2)}$

$x = 90\sqrt{2}$ feet from home plate to second base.

❑ Focus on Concepts

1. An identity is true for all values of the variable and a conditional equation is true for some values of the variable.
2. When proving an identity you use the fundamental identities and rules of algebra to transform one expression into another. To solve a trigonometric equation, use standard algebraic techniques and identities to isolate a trigonometric function involved in the equation. Find the value of the variable by using the inverse of the trigonometric function.
3. Reciprocal identities: $\csc\theta = \frac{1}{\sin\theta}$, $\sec\theta = \frac{1}{\cos\theta}$, $\cot\theta = \frac{1}{\tan\theta}$

   Quotient identities: $\tan\theta = \frac{\sin\theta}{\cos\theta}$, $\cot\theta = \frac{\cos\theta}{\sin\theta}$

   Pythagorean identities: $\sin^2\theta + \cos^2\theta = 1$, $\tan^2\theta + 1 = \sec^2\theta$, $1 + \cot^2\theta = \csc^2\theta$
4. No. $\cos\theta = \pm\sqrt{1 - \sin^2\theta}$.
5. False. The order in which algebraic operations and fundamental identities are done may vary.
6. (a) True. The period of tangent is π. (b) False. The period of cosine is 2π. (c) False. $\sec\theta\cos\theta = 1$

   (d) True. (e) True. $\sin(-\alpha) = -\sin\alpha$

7. $y_1 = y_2 + 1$

8. $y_1 = 1 - y_2$

9. 1

10. 3

11. 3

12. 4

13. Period: $\dfrac{2\pi}{d} = 12$

$d = \dfrac{2\pi}{12} = \dfrac{\pi}{6}$

❑ Review Exercises for Chapter 5

Solutions to Even-Numbered Exercises

2. $\dfrac{\sin 2\alpha}{\cos^2 \alpha - \sin^2 \alpha} = \dfrac{\sin 2\alpha}{\cos 2\alpha} = \tan 2\alpha$

4. $$\begin{aligned}\frac{\sin^3 \beta + \cos^3 \beta}{\sin \beta + \cos \beta} &= \frac{(\sin \beta + \cos \beta)(\sin^2 \beta - \sin \beta \cos \beta + \cos^2 \beta)}{\sin \beta + \cos \beta} \\ &= 1 - \sin \beta \cos \beta \\ &= 1 - \frac{1}{2} \sin 2\beta\end{aligned}$$

6. $$\begin{aligned}1 - 4 \sin^2 x \cos^2 x &= 1 - (2 \sin x \cos x)^2 \\ &= 1 - \sin^2 2x \\ &= \cos^2 2x\end{aligned}$$

8. $\sqrt{\dfrac{1 - \cos^2 x}{1 + \cos x}} = \sqrt{1 - \cos x} = \sqrt{2}\left|\sin \dfrac{x}{2}\right|$

10. $$\begin{aligned}\cos x(\tan^2 x + 1) &= \cos x \sec^2 x \\ &= \frac{1}{\sec x} \sec^2 x \\ &= \sec x\end{aligned}$$

12. $$\begin{aligned}\sin^3 \theta + \sin \theta \cos^2 \theta &= \sin \theta(\sin^2 \theta + \cos^2 \theta) \\ &= \sin \theta\end{aligned}$$

14. $$\begin{aligned}\cos^3 x \sin^2 x &= \cos x(\cos^2 x) \sin^2 x \\ &= \cos x(1 - \sin^2 x) \sin^2 x \\ &= (\sin^2 x - \sin^4 x) \cos x\end{aligned}$$

16. Using a product-sum formula, we have

$\sin 3x \cos 2x = \frac{1}{2}(\sin 5x + \sin x)$.

18. $$\begin{aligned}\sqrt{1 - \cos x} &= \sqrt{(1 - \cos x)\frac{1 + \cos x}{1 + \cos x}} \\ &= \sqrt{\frac{\sin^2 x}{1 + \cos x}} \\ &= \frac{|\sin x|}{\sqrt{1 + \cos x}}\end{aligned}$$

20. $$\begin{aligned}\cos\left(x + \frac{\pi}{2}\right) &= \cos x \cos \frac{\pi}{2} - \sin x \sin \frac{\pi}{2} \\ &= (\cos x)(0) - (\sin x)(1) \\ &= -\sin x\end{aligned}$$

22. $$\begin{aligned}\sin(\pi - x) &= \sin \pi \cos \pi - \sin x \cos \pi \\ &= (0)(\cos x) - (\sin x)(-1) \\ &= \sin x\end{aligned}$$

24. $$\begin{aligned}\frac{2 \cos 3x}{\sin 4x - \sin 2x} &= \frac{2 \cos 3x}{2 \cos 3x \sin x} \\ &= \frac{1}{\sin x} \\ &= \csc x\end{aligned}$$

26. $$\begin{aligned}\frac{\sin(\alpha + \beta)}{\cos \alpha \cos \beta} &= \frac{\sin \alpha \cos \beta + \cos \alpha \sin \beta}{\cos \alpha \cos \beta} \\ &= \frac{\sin \alpha \cos \beta}{\cos \alpha \cos \beta} + \frac{\cos \alpha \sin \beta}{\cos \alpha \cos \beta} \\ &= \tan \alpha + \tan \beta\end{aligned}$$

28. $$\begin{aligned}\sin 4x &= 2\sin 2x\cos 2x\\ &= 2[2\sin x\cos x(\cos^2 x - \sin^2 x)]\\ &= 4\sin x\cos x(2\cos^2 x - 1)\\ &= 8\cos^3 x\sin x - 4\cos x\sin x\end{aligned}$$

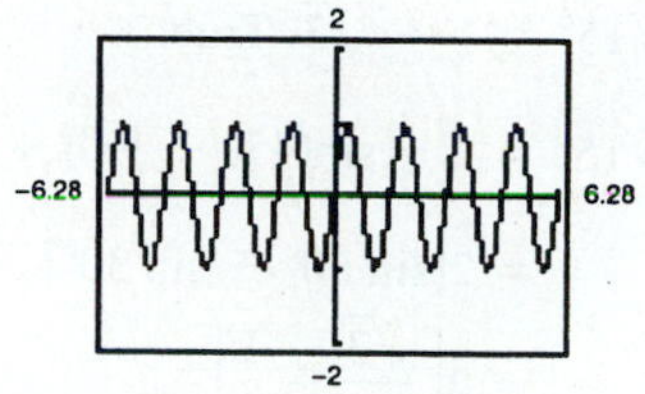

30. $$\begin{aligned}\cos^2 5x - \cos^2 x &= (\cos 5x - \cos x)(\cos 5x + \cos x)\\ &= (-2\sin 3x\sin 2x)(2\cos 3x\cos 2x)\\ &= -(2\sin 2x\cos 2x)(2\sin 3x\cos 3x)\\ &= -\sin 4x\sin 6x\end{aligned}$$

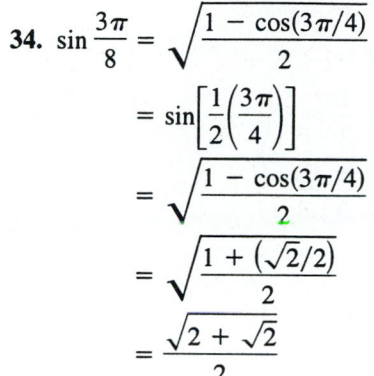

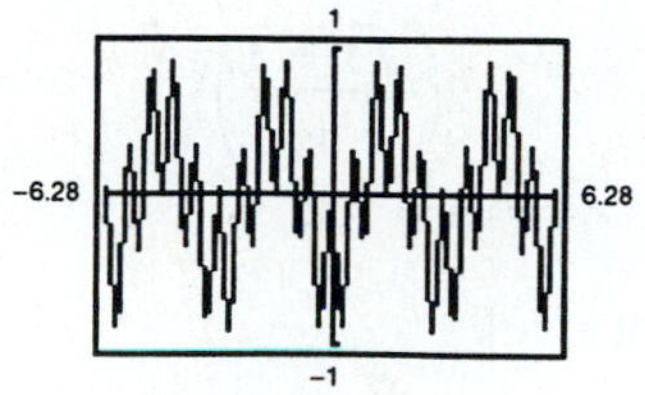

32. $$\begin{aligned}\cos(285^\circ) &= \cos(225^\circ + 60^\circ)\\ &= \cos 225^\circ\cos 60^\circ - \sin 225^\circ\sin 60^\circ\\ &= \left(-\frac{\sqrt{2}}{2}\right)\left(\frac{1}{2}\right) - \left(-\frac{\sqrt{2}}{2}\right)\left(\frac{\sqrt{3}}{2}\right)\\ &= \frac{\sqrt{2}}{4}(\sqrt{3} - 1)\end{aligned}$$

34. $$\begin{aligned}\sin\frac{3\pi}{8} &= \sqrt{\frac{1-\cos(3\pi/4)}{2}}\\ &= \sin\left[\frac{1}{2}\left(\frac{3\pi}{4}\right)\right]\\ &= \sqrt{\frac{1-\cos(3\pi/4)}{2}}\\ &= \sqrt{\frac{1+(\sqrt{2}/2)}{2}}\\ &= \frac{\sqrt{2+\sqrt{2}}}{2}\end{aligned}$$

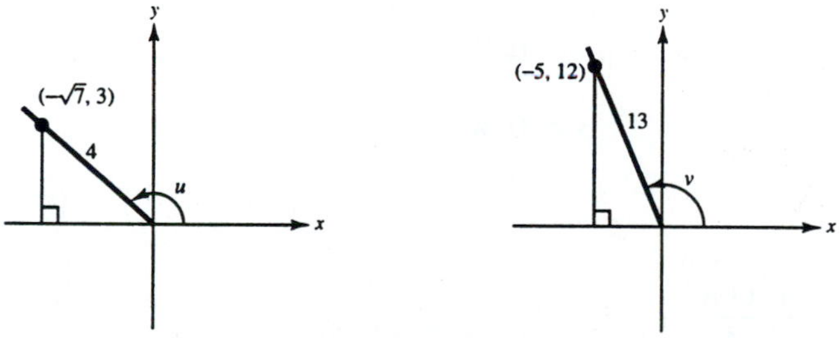

Figures for Exercises 36–40

36. $$\begin{aligned}\tan(u+v) &= \frac{\tan u + \tan v}{1 - \tan u\tan v}\\ &= \frac{-(3/\sqrt{7}) - (12/5)}{1 - (-3/\sqrt{7})(-12/5)}\\ &= \frac{15 + 12\sqrt{7}}{36 - 5\sqrt{7}}\end{aligned}$$

38. $$\begin{aligned}\sin 2v &= 2\sin v\cos v\\ &= 2\left(\frac{12}{13}\right)\left(-\frac{5}{13}\right)\\ &= -\frac{120}{169}\end{aligned}$$

40. $$\begin{aligned}\tan 2v &= \frac{\sin 2v}{\cos 2v}\\ &= \frac{2\sin v\cos v}{2\cos^2 v - 1}\\ &= \frac{2(12/13)(-5/13)}{2(-5/13)^2 - 1}\\ &= \frac{-120/169}{(50/169) - (169/169)} = \frac{120}{119}\end{aligned}$$

42. $\sin(x + y) = \sin x + \sin y$. False.
$\sin(x + y) = \sin x\cos y + \cos x\sin y$

44. $4 \sin 45° \cos 15° = 1 + \sqrt{3}$. True.

$$\begin{aligned}4 \sin 45° \cos 15° &= 4\left(\frac{1}{2}[\sin(45° + 15°) + \sin(45° - 15°)]\right)\\ &= 2[\sin 60° + \sin 30°]\\ &= 2\left[\frac{\sqrt{3}}{2} + \frac{1}{2}\right]\\ &= 2\left(\frac{\sqrt{3}+1}{2}\right)\\ &= 1 + \sqrt{3}\end{aligned}$$

46. $\csc x - 2 \cot x = 0$

$$\begin{aligned}\frac{1}{\sin x} - \frac{2 \cos x}{\sin x} &= 0\\ 1 - 2\cos x &= 0\\ \cos x &= \frac{1}{2}\\ x &= \frac{\pi}{3}, \frac{5\pi}{3}\end{aligned}$$

48.

$$\begin{aligned}\cos 4x - 7\cos 2x &= 8\\ 2\cos^2 2x - 1 - 7\cos 2x &= 8\\ 2\cos^2 2x - 7\cos 2x - 9 &= 0\\ (2\cos 2x - 9)(\cos 2x + 1) &= 0\end{aligned}$$

$2\cos 2x - 9 = 0$ or $\cos 2x + 1 = 0$

$\cos 2x = \frac{9}{2}$ $\qquad$ $\cos 2x = -1$

No solution $\qquad$ $2x = \pi + 2n\pi$

$$x = \frac{\pi}{2} + n\pi$$

$$x = \frac{\pi}{2}, \frac{3\pi}{2}$$

50. $\sin 4x - \sin 2x = 0$

$$2\cos 3x \sin x = 0$$

$\cos 3x = 0$ $\qquad$ or $\quad \sin x = 0$

$3x = \frac{\pi}{2} + n\pi$ $\qquad$ $x = 0, \pi$

$$x = \frac{\pi}{6} + \frac{n\pi}{3}$$

$$x = \frac{\pi}{6}, \frac{\pi}{2}, \frac{5\pi}{6}, \frac{7\pi}{6}, \frac{3\pi}{2}, \frac{11\pi}{6}$$

52. $y = \cos x - \cos \frac{x}{2}$

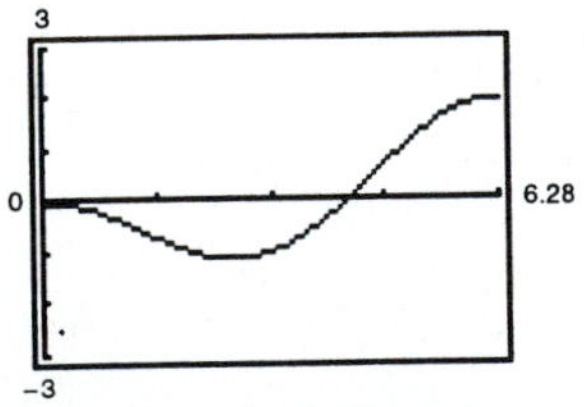

Zeros: $x \approx 0, 4.1888$

—CONTINUED—

52. —CONTINUED—

$$\cos x - \cos\frac{x}{2} = 0$$
$$\cos x = \cos\frac{x}{2}$$
$$\cos^2 x = \cos^2\frac{x}{2}$$
$$\cos^2 x - \cos^2\frac{x}{2} = 0$$
$$\left(\cos x + \cos\frac{x}{2}\right)\left(\cos x - \cos\frac{x}{2}\right) = 0$$
$$2\cos\frac{3x}{4}\cos\frac{x}{4}\left(-2\sin\frac{3x}{4}\sin\frac{x}{4}\right) = 0$$
$$-4\cos\frac{3x}{4}\cos\frac{x}{4}\sin\frac{3x}{4}\sin\frac{x}{4} = 0$$

$\cos\frac{3x}{4} = 0 \Rightarrow x = \frac{4\pi}{3}$ $\left(x = \frac{2\pi}{3} \text{ is extraneous.}\right)$ or

$\cos\frac{x}{4} = 0 \Rightarrow$ (No solution in $[0, 2\pi)$) or

$\sin\frac{3x}{4} = 0$ or $\sin\frac{x}{4} = 0$

$x = 0, \frac{4\pi}{3}$ $\qquad$ $x = 0$

54. $h(s) = \sin s + \sin 3s + \sin 5s$

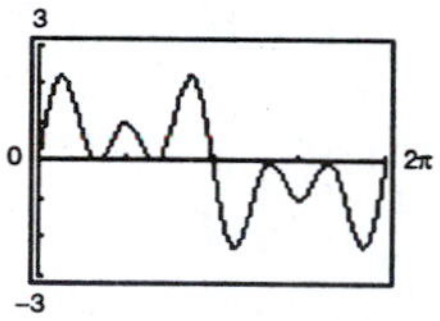

Zeros: $s \approx 0, 1.0472, 2.0944, 3.1416, 4.1888, 5.2360$

$$\sin s + \sin 3s + \sin 5s = 0$$
$$2\sin 3s\cos 2s + \sin 2s = 0$$
$$\sin 3s(2\cos 2s + 1) = 0$$

$\sin 3s = 0$ or $2\cos 2s + 1 = 0$

$3s = n\pi$ $\qquad$ $\cos 2s = -\frac{1}{2}$

$s = \frac{n\pi}{3}$ $\qquad$ $2s = \frac{2\pi}{3} + 2n\pi$ or $2s = \frac{4\pi}{3} + 2n\pi$

$s = 0, \frac{\pi}{3}, \frac{2\pi}{3}, \pi, \frac{4\pi}{3}, \frac{5\pi}{3}$ $\qquad$ $s = \frac{\pi}{3}, \frac{2\pi}{3}, \frac{4\pi}{3}, \frac{5\pi}{3}$

56. $a\sin x - b = 0$, $\sin x = \frac{b}{a}$

You know that this equation has no solution if $|a| < |b|$ because $-1 \le \sin x \le 1$ for all x. If $|a| < |b|$, then b/a is greater than 1.

58. $\sin\left(x + \frac{\pi}{4}\right) - \sin\left(x - \frac{\pi}{4}\right) = 2 \cos x \sin \frac{\pi}{4}$

$= \sqrt{2} \cos x$

60. $\cos \frac{x}{2} \cos \frac{x}{4} = \frac{1}{2}\left[\cos\left(\frac{x}{2} - \frac{x}{4}\right) + \cos\left(\frac{x}{2} + \frac{x}{4}\right)\right]$

$= \frac{1}{2}\left(\cos \frac{x}{4} + \cos \frac{3x}{4}\right)$

62. $\sin(2 \arctan x) = \sin 2\theta$

$= 2 \sin \theta \cos \theta$

$= 2\left(\frac{x}{\sqrt{x^2 + 1}}\right)\left(\frac{1}{\sqrt{x^2 + 1}}\right)$

$= \frac{2x}{x^2 + 1}$

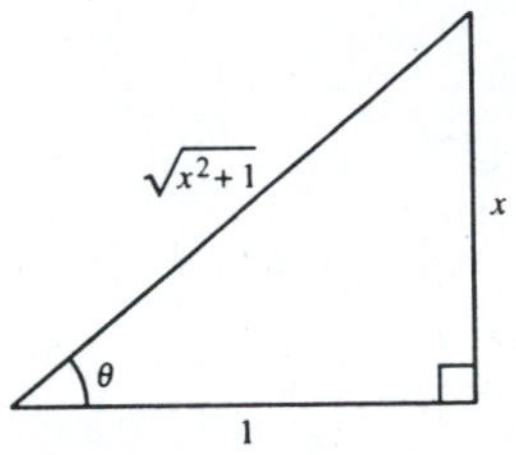

64. $r = \frac{1}{32} v_0^2 \sin 2\theta$

$100 = \frac{1}{32}(80)^2 \sin 2\theta$

$\sin 2\theta = 0.5$

$2\theta = 30°$ or $2\theta = 180° - 30° = 150°$

$\theta = 15°$ $\qquad \theta = 75°$

66. Volume V of the trough will be the area A of the isosceles triangle times the length l of the trough.

$V = A \cdot l$

(a) $A = \frac{1}{2} bh$

$\cos \frac{\theta}{2} = \frac{h}{0.5} \Rightarrow h = 0.5 \cos \frac{\theta}{2}$

$\sin \frac{\theta}{2} = \frac{b/2}{0.5} \Rightarrow \frac{b}{2} = 0.5 \sin \frac{\theta}{2}$

$A = 0.5 \sin \frac{\theta}{2} 0.5 \cos \frac{\theta}{2}$

$= (0.5)^2 \sin \frac{\theta}{2} \cos \frac{\theta}{2}$

$= 0.25 \sin \frac{\theta}{2} \cos \frac{\theta}{2}$ square meters

$V = (0.25)(4) \sin \frac{\theta}{2} \cos \frac{\theta}{2}$ cubic meters

$= \sin \frac{\theta}{2} \cos \frac{\theta}{2}$ cubic meters

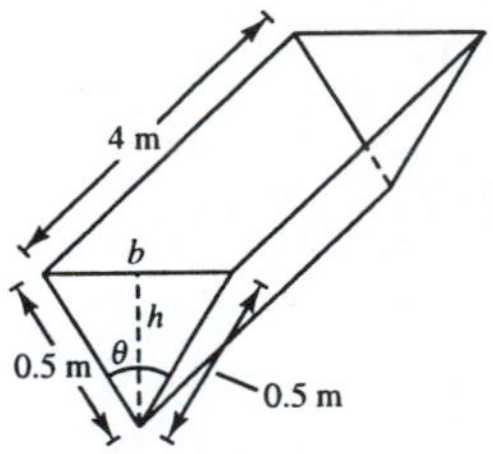

Not to scale

(b) $V = \sin \frac{\theta}{2} \cos \frac{\theta}{2}$

$= \frac{1}{2}\left(2 \sin \frac{\theta}{2} \cos \frac{\theta}{2}\right)$

$= \frac{1}{2} \sin \theta$ cubic meters

Volume is maximum when $\theta = \pi/2$.

CHAPTER 6
Additional Topics in Trigonometry

CHAPTER 6
Additional Topics in Trigonometry

Section 6.1 Law of Sines

Solutions to Even-Numbered Exercises

2.

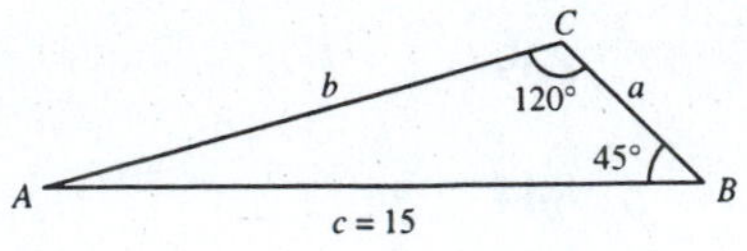

Given: $B = 45°, C = 120°, c = 15$

$A = 180° - B - C = 15°$

$$a = \frac{c}{\sin C}(\sin A) = \frac{15 \sin 15°}{\sin 120°} \approx 4.48$$

$$b = \frac{c}{\sin C}(\sin B) = \frac{15(\sin 45°)}{\sin 120°} \approx 12.25$$

4.

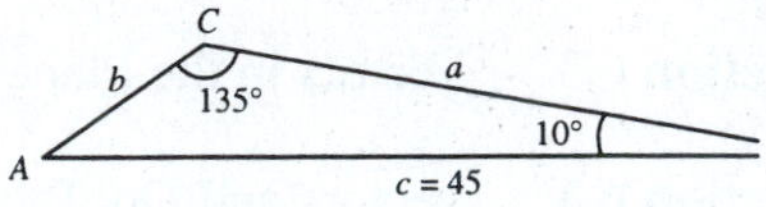

Given: $B = 10°, C = 135°, c = 45$

$A = 180° - B - C = 35°$

$$a = \frac{c}{\sin C}(\sin A) = \frac{45 \sin 35°}{\sin 135°} \approx 36.50$$

$$b = \frac{c}{\sin C}(\sin B) = \frac{45 \sin 35°}{\sin 135°} \approx 11.05$$

6. Given: $A = 60°, a = 9, c = 10$

$$\sin C = \frac{c \sin A}{a} = \frac{10 \sin 60°}{9} \approx 0.9623 \implies C \approx 74.21° \text{ or } C \approx 105.79°$$

<u>Case 1</u>

$C \approx 74.21°$

$B = 180° - A - C \approx 45.79°$

$$b = \frac{a}{\sin A}(\sin B) \approx \frac{9 \sin 45.79°}{\sin 60°} \approx 7.45$$

<u>Case 2</u>

$C \approx 105.79°$

$B = 180° - A - C \approx 14.21°$

$$b = \frac{a}{\sin A}(\sin B) \approx \frac{9 \sin 14.21°}{\sin 60°} \approx 2.55$$

8. Given: $A = 24.3°, C = 54.6°, c = 2.68$

$B = 180° - A - C = 101.1°$

$$a = \frac{c}{\sin C}(\sin A) = \frac{2.68 \sin 24.3°}{\sin 54.6°} \approx 1.35$$

$$b = \frac{c}{\sin C}(\sin B) = \frac{2.68 \sin 101.1°}{\sin 54.6°} \approx 3.23$$

10. Given: $A = 5° 40', B = 8° 15', b = 4.8$

$C = 180° - A - B = 166° 5'$

$$a = \frac{b}{\sin B}(\sin A) = \frac{4.8 \sin 5° 40'}{\sin 8° 15'} \approx 3.30$$

$$c = \frac{b}{\sin B}(\sin C) = \frac{4.8 \sin 166° 5'}{\sin 8° 15'} \approx 8.05$$

12. Given: $C = 85° 20', a = 35, c = 50$

$$\sin A = \frac{a \sin C}{c} = \frac{35 \sin 85° 20'}{50} \approx 0.6977 \implies A \approx 44.24°$$

$B = 180° - A - C \approx 50.43°$

$$b = \frac{c}{\sin C}(\sin B) \approx \frac{50 \sin 50.43°}{\sin 85° 20'} \approx 38.67$$

14. Given: $A = 100°, a = 125, c = 10$

$$\sin C = \frac{c \sin A}{a} = \frac{10 \sin 100°}{125} \approx 0.07878 \implies C \approx 4.52°$$

$B = 180° - A - C \approx 75.48°$

$$b = \frac{a}{\sin A}(\sin B) \approx \frac{125 \sin 75.48°}{\sin 100°} \approx 122.87$$

16. Given: $B = 2° 45'$, $b = 6.2$, $c = 5.8$

$$\sin C = \frac{c \sin B}{b} = \frac{5.8 \sin 2° 45'}{6.2} \approx 0.04488 \implies C \approx 2.57°$$

$$A = 180° - B - C \approx 174.68°$$

$$a = \frac{b}{\sin B}(\sin A) \approx \frac{6.2 \sin 174.68°}{\sin 2° 45'} \approx 11.99$$

18. Given: $A = 58°$, $a = 11.4$, $c = 12.8$

$$\sin B = \frac{b \sin A}{a} = \frac{12.8 \sin 58°}{11.4} \approx 0.9522 \implies B \approx 72.2° \text{ or } B \approx 107.8°$$

Case 1

$B \approx 72.2°$

$C = 180° - A - B \approx 49.8°$

$$c = \frac{a}{\sin A}(\sin C) \approx \frac{11.4 \sin 49.8°}{\sin 58°} \approx 10.27$$

Case 2

$B \approx 107.8°$

$C = 180° - A - B \approx 14.2°$

$$c = \frac{a}{\sin A}(\sin C) \approx \frac{11.4 \sin 14.2°}{\sin 58°} \approx 3.30$$

20. Given: $A = 58°$, $a = 42.4$, $b = 50$

$h = b \sin A = 50 \sin 58° \approx 42.4024$

Since $a < h$, no triangle is formed.

22. Given: $A = 110°$, $a = 125$, $b = 100$

$$\sin B = \frac{b \sin A}{a} = \frac{100 \sin 110°}{125} \approx 0.75175 \implies B \approx 48.74°$$

$$C = 180° - A - B = 21.26°$$

$$c = \frac{a}{\sin A}(\sin C) = \frac{125 \sin 21.26°}{\sin 110°} \approx 48.23$$

24. Given: $A = 60°$, $a = 10$

(a) One solution if $b \le 10$ or $b = \dfrac{10}{\sin 60^2}$.

(b) Two solutions if $10 < b < \dfrac{10}{\sin 60°}$.

(c) No solutions if $b > \dfrac{10}{\sin 60°}$.

26.

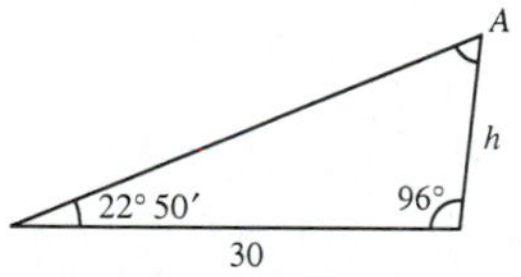

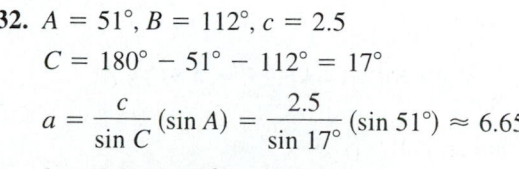

$$A = 180° - 96° - 22° 50' = 61° 10'$$

$$h = \frac{30 \sin 22° 50'}{\sin 61° 10'} \approx 13.3 \text{ meters}$$

28. $$\sin A = \frac{a \sin B}{b} = \frac{500 \sin 46°}{720} \approx 0.4995 \implies A \approx 30°$$

The bearing from C to A is S 60° W.

30. (a)

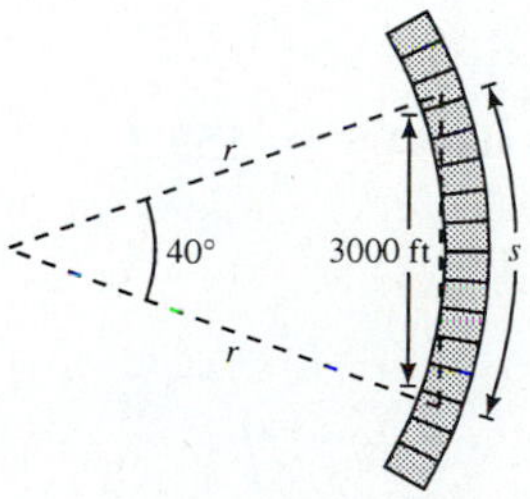

(b) $$r = \frac{3000 \sin[1/2(180° - 40°)]}{\sin 40°} \approx 4385.71 \text{ feet}$$

(c) $$s \approx 40°\left(\frac{\pi}{180°}\right)4385.71 \approx 3061.80 \text{ feet}$$

32. $A = 51°$, $B = 112°$, $c = 2.5$

$$C = 180° - 51° - 112° = 17°$$

$$a = \frac{c}{\sin C}(\sin A) = \frac{2.5}{\sin 17°}(\sin 51°) \approx 6.65$$

$$h \approx 6.65 \sin 68° \approx 6.2 \text{ mi}$$

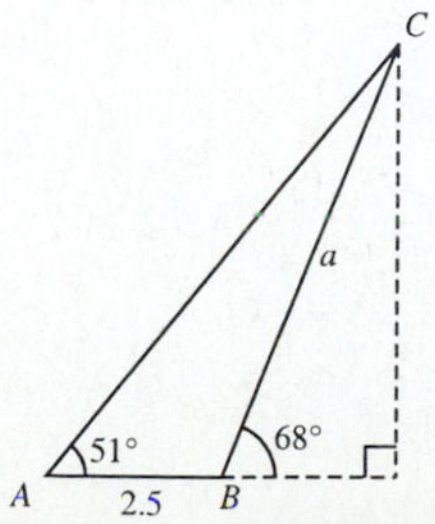

34. $A = 20, B = 90° + 63° = 153°, c = 10\left(\frac{1}{4}\right) = 2.5$

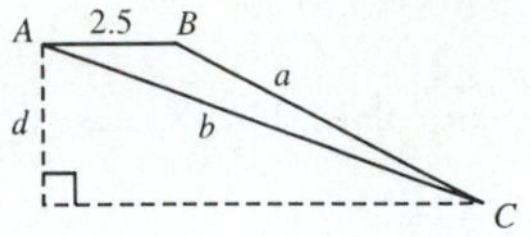

$C = 180° - 20° - 153° = 7°$

$b = \frac{c}{\sin C}(\sin B) = \frac{2.5 \sin 153°}{\sin 7°} \approx 9.31$

$d \approx b \sin A \approx 9.31 \sin 20° \approx 3.2$ miles

36. (a) $\frac{6}{\sin \theta} = \frac{1.5}{\sin C}, \sin \theta \neq 0$

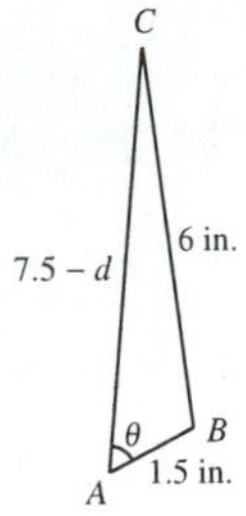

$\sin C = \frac{1.5 \sin \theta}{6} \Rightarrow C = \arcsin \frac{1.5 \sin \theta}{6}$

$B = 180° - \theta - \arcsin \frac{1.5 \sin \theta}{6}$

$\frac{7.5 - d}{\sin B} = \frac{6}{\sin \theta}$

$$d = 7.5 - \frac{6 \sin\left(180° - \theta - \arcsin \frac{1.5 \sin \theta}{6}\right)}{\sin \theta}$$

For $\theta = 0°, C = 0°, B = 180° \Rightarrow 7.5 - d = 1.5 + 6 \Rightarrow d = 0.$

θ	0°	45°	90°	135°	180°
d	0	0.5338	1.6905	2.6552	3

For $\theta = 180°, C = 0°, B = 0° \Rightarrow 7.5 - d = 6 - 1.5 \Rightarrow d = 3.$

(b) $\theta = 5°$

$$d = 7.5 - \frac{6 \sin\left(180° - 5° - \arcsin \frac{1.5 \sin 5°}{6}\right)}{\sin 5°} \approx 0.0071 \text{ inch}$$

38. $\alpha = 180 - (\phi + 180 - \theta) = \theta - \phi$

$\frac{d}{\sin \phi} = \frac{2}{\sin \alpha}$

$d = \frac{2 \sin \phi}{\sin(\phi - \theta)}$

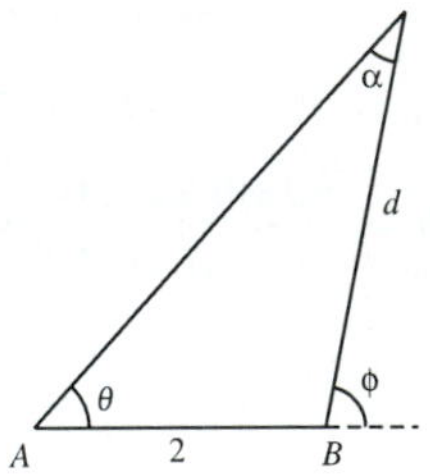

40. $B = 72° 30', a = 105, c = 64$

$\text{Area} = \frac{1}{2}ac \sin B = \left(\frac{1}{2}\right)(105)(64) \sin 72.5° \approx 3204$

42. $A = 5° 15', b = 4.5, c = 22$

$\text{Area} = \frac{1}{2}bc \sin A$

$= \left(\frac{1}{2}\right)(4.5)(22) \sin 5.25° \approx 4.529$

44. $C = 84° 30', a = 16, b = 20$

$\text{Area} = \frac{1}{2}ab \sin C$

$= \left(\frac{1}{2}\right)(16)(20) \sin 84.5° \approx 159.3$

Section 6.2 Law of Cosines

Solutions to Even-Numbered Exercises

2. Given: $a = 8, b = 3, c = 9$

$$\cos A = \frac{b^2 + c^2 - a^2}{2bc} = \frac{3^2 + 9^2 - 8^2}{2(3)(9)} \approx 0.4815 \Rightarrow A \approx 61.2°$$

$$\cos c = \frac{a^2 + b^2 - c^2}{2ab} = \frac{8^2 + 3^2 - 9^2}{2(8)(3)} \approx -0.1667 \Rightarrow c \approx 99.6°$$

$B \approx 180° - 61.2° - 99.6° \approx 19.2°$

4. Given: $C = 105°, a = 10, b = 4.5$

$c^2 = a^2 + b^2 - 2ab \cos C = 10^2 + 4.5^2 - 2(10)(4.5) \cos 105° \approx 143.5437 \Rightarrow c \approx 12.0$

$$\cos B = \frac{a^2 + c^2 - b^2}{2ac} \approx \frac{10^2 + (12.0)^2 - (4.5)^2}{2(10)(12.0)} \approx 0.93187 \Rightarrow B \approx 21.3°$$

$A = 180° - 105° - 21.3° \approx 53.7°$

6. Given: $a = 55, b = 25, c = 72$

$$\cos C = \frac{a^2 + b^2 - c^2}{2ab} = \frac{55^2 + 25^2 - 72^2}{2(55)(25)} \approx -0.5578 \Rightarrow c \approx 123.91°$$

$$\cos A = \frac{b^2 + c^2 - a^2}{2bc} = \frac{25^2 + 72^2 - 55^2}{2(25)(72)} \approx 0.7733 \Rightarrow A \approx 39.35°$$

$B = 180° - 123.91° - 39.35° \approx 16.74°$

8. Given: $a = 1.42, b = 0.75, c = 1.25$

$$\cos A = \frac{b^2 + c^2 - a^2}{2bc} = \frac{(0.75)^2 + (1.25)^2 - (1.42)^2}{2(0.75)(1.25)} = 0.05792 \Rightarrow A \approx 86.7°$$

$$\cos B = \frac{a^2 + c^2 - b^2}{2ac} = \frac{(1.42)^2 + (1.25)^2 - (0.75)^2}{2(1.42)(1.25)} \approx 0.8497 \Rightarrow B \approx 31.8°$$

$180° - 86.7° - 31.8° \approx 61.5°$

10. Given: $A = 55°, b = 3, c = 10$

$a^2 = b^2 + c^2 - 2bc \cos A = 3^2 + 10^2 - 2(3)(10) \cos 55° \approx 74.585 \Rightarrow a \approx 8.64$

$$\sin B = \frac{b \sin A}{a} \approx \frac{3 \sin 55°}{8.64} \approx 0.2844 \Rightarrow A \approx 16.5°$$

$C \approx 180° - 16.5° - 55° \approx 108.5°$

12. Given: $B = 75° 20', a = 6.2, c = 9.5$

$b^2 = a^2 + c^2 - 2ac \cos B = (6.2)^2 + (9.5)^2 - 2(6.2)(9.5) \cos 75° 20' \approx 98.8636 \Rightarrow b \approx 9.94$

$$\sin A = \frac{a \sin B}{b} \approx \frac{6.2 \sin 75° 20'}{9.94} \approx 0.6034 \Rightarrow A \approx 37.1°$$

$C \approx 180° - 75° 20' - 37.1° \approx 67.6°$

14. Given: $C = 15°, a = 6.25, b = 2.15$

$c^2 = a^2 + b^2 - 2ab \cos C \approx (6.25)^2 + (2.15)^2 - 2(6.25)(2.15) \cos 15° \approx 17.7257 \Rightarrow c \approx 4.21$

$$\cos A = \frac{b^2 + c^2 - a^2}{2bc} \approx \frac{(2.15)^2 + (4.21)^2 - (6.25)^2}{2(2.15)(4.21)} \approx -0.92338 \Rightarrow A \approx 157.4°$$

$B \approx 180° - 15° - 157.4° \approx 7.6°$

16.

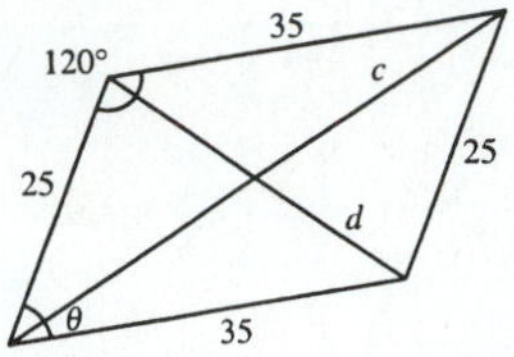

$$c^2 = 25^2 + 35^2 - 2(25)(35)\cos 120^\circ$$
$$= 2725 \Rightarrow c \approx 52.20$$
$$\theta = 360^\circ - 2(120^\circ) = 60^\circ$$
$$d^2 = 25^2 + 35^2 - 2(25)(35)\cos 60^\circ$$
$$= 975 \Rightarrow d \approx 31.22$$

18.

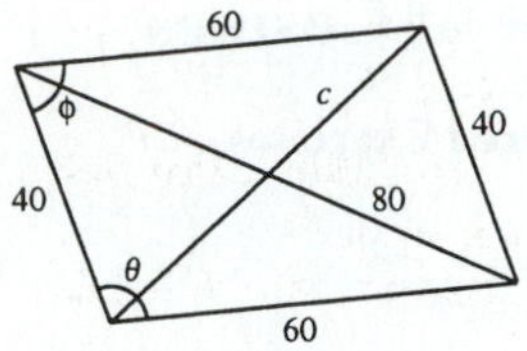

$$\cos\theta = \frac{40^2 + 60^2 - 80^2}{2(40)(60)} \approx -\frac{1}{4} \neq \theta \approx 104.5^\circ$$
$$\phi \approx 360^\circ - 2(104.5^\circ) \approx 75.5^\circ$$
$$c^2 \approx 40^2 + 60^2 - 2(40)(60)\cos 75.5^\circ = 4000$$
$$c \approx 63.25$$

20. $\cos\alpha = \dfrac{25^2 + 17.5^2 - 25^2}{2(25)(17.5)}$

$$\alpha \approx 69.512^\circ$$
$$\beta \approx 180 - \alpha \approx 110.488^\circ$$
$$a^2 = 17.5^2 + 25^2 - 2(17.5)(25)\cos 110.488^\circ$$
$$a \approx 35.18$$
$$z = 180 - 2\alpha \approx 40.976$$
$$\cos\mu = \frac{25^2 + 35.18^2 - 17.5^2}{2(25)(35.18)}$$
$$\mu \approx 27.771^\circ$$
$$\theta = \mu + z \approx 68.7^\circ$$
$$\omega = 180^\circ - \mu - \beta \approx 41.741^\circ$$
$$\phi = \omega + \alpha \approx 111.3^\circ$$

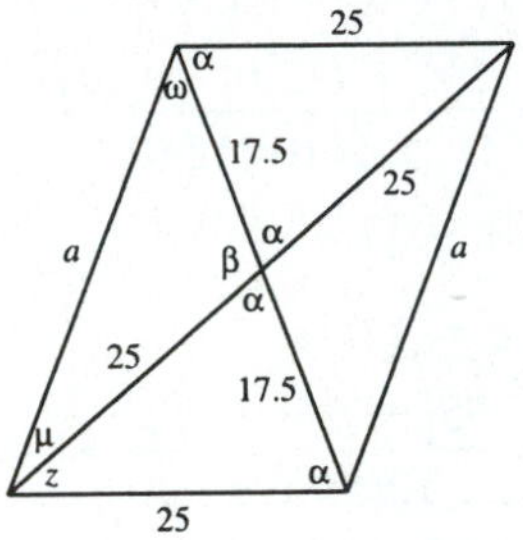

22. Given: $a = 648$, $c = 810$, $B = 180^\circ - 75^\circ + 32^\circ = 137^\circ$

The angle for the bearing from C to A is $C + 32^\circ$. We wish to find that bearing and side b.

$$b^2 = a^2 + c^2 - 2ac\cos B = 648^2 + 810^2 - 2(648)(810)\cos 137^\circ \approx 1{,}843{,}749.9$$
$$b \approx 1357.8 \text{ miles}$$
$$\sin C = \frac{c\sin B}{b} \approx \frac{810\sin 137^\circ}{1357.8} \approx 0.4068 \Rightarrow C \approx 24^\circ$$

Distance from C to A: 1357.8 miles

Bearing from C to A: S 56° W

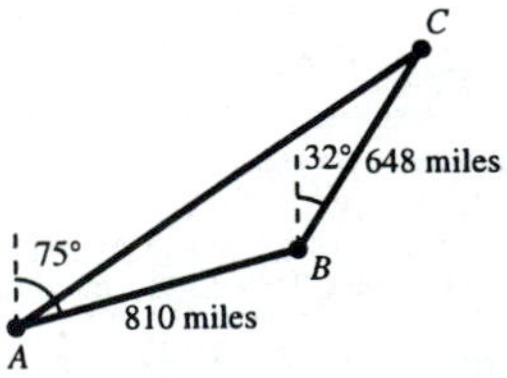

24. $\cos A = \dfrac{115^2 + 76^2 - 92^2}{2(115)(76)} \approx 0.6028 \Rightarrow A \approx 52.9^\circ$

$\cos C = \dfrac{115^2 + 92^2 - 76^2}{2(115)(92)} \approx 0.75203 \Rightarrow c \approx 41.2^\circ$

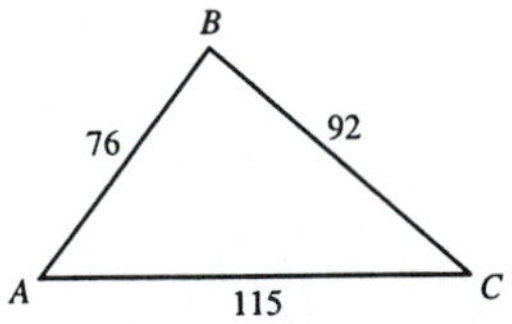

26. $\cos\theta = \dfrac{2^2 + 3^2 - (4.5)^2}{2(2)(3)} \approx -0.60417$

$$\theta \approx 127.2^\circ$$

28. The angles at the base of the tower are 96° and 84°. The longer guy wire g_1 is given by:

$$g_1{}^2 = 75^2 + 100^2 - 2(75)(100)\cos 96^\circ \approx 17{,}192.9 \Rightarrow g_1 \approx 131.1 \text{ feet}$$

The shorter guy wire g_2 is given by:

$$g_2{}^2 = 75^2 + 100^2 - 2(75)(100)\cos 84^\circ \approx 14{,}057.1 \Rightarrow g_2 \approx 118.6 \text{ feet}$$

30. Bearing of M from P: N θ E

Bearing of A from P: N ϕ E

Since M is due west of A, it follows that $\theta = M - 90°$ and $\phi = 90° - A$.

$$\cos M = \frac{165^2 + 216^2 - 368^2}{2(165)(216)} \approx -0.8634 \Rightarrow M \approx 149.7°$$

$$\cos A = \frac{165^2 + 368^2 - 216^2}{2(165)(368)} \approx 0.95515 \Rightarrow A \approx 17.2°$$

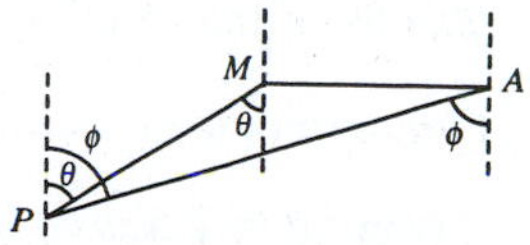

$\theta \approx 149.7° - 90° \approx 59.7° \Rightarrow$ Bearing of Minneapolis from Phoenix: N 59.7° E

$\phi \approx 90° - 17.2° \approx 72.8° \Rightarrow$ Bearing of Minneapolis from Phoenix: N 72.8° E

32. $x^2 = 330^2 + 420^2 - 2(330)(420)\cos 8°$

$\approx 10{,}797.7$

$x \approx 103.9$ feet

34. $a = 35^2 + 20^2 - 2(35)(20)\cos 42° \approx 584.6$

$a \approx 24$ miles

36. $d^2 = 10^2 + 7^2 - 2(10)(7)\cos\theta$

$$\theta = \arccos\left[\frac{10^2 + 7^2 - d^2}{2(10)(7)}\right]$$

$$s = \frac{360° - \theta}{360°}(2\pi r) = \frac{(360° - \theta)\pi}{45}$$

d (inches)	9	10	12	13	14	15	16
θ (degrees)	60.9°	69.5°	88.0°	98.2°	109.6°	122.9°	139.8°
s (inches)	20.88	20.28	18.99	18.28	17.48	16.55	15.37

38. (a) Working with $\triangle OBC$, we have $\cos a = \dfrac{a/2}{R}$.

This implies that $2R = a/\cos\alpha$. Since we know that

$$\frac{a}{\sin A} = \frac{b}{\sin B} = \frac{c}{\sin C},$$

we can complete the proof by showing that $\cos\alpha = \sin A$. The solution of the system

$$A + B + C = 180°$$
$$\alpha - C + A = \beta$$
$$\alpha + \beta = B$$

is $\alpha = 90° - A$. Therefore:

$$2R = \frac{a}{\cos\alpha} = \frac{a}{\cos(90° - A)} = \frac{a}{\sin A}.$$

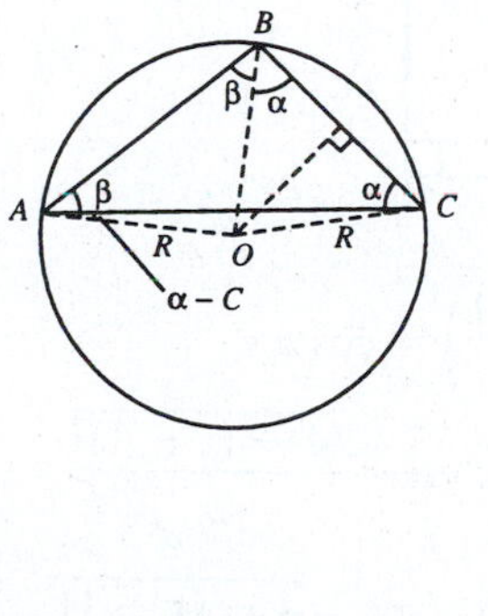

(b) By Heron's Formula, the area of the triangle is

$$\text{Area} = \sqrt{s(s-a)(s-b)(s-c)}.$$

We can also find the area by dividing the area into six triangles and using the fact that the area is 1/2 the base times the height. Using the figure as given, we have

$$\text{Area} = \frac{1}{2}xr + \frac{1}{2}xr + \frac{1}{2}yr + \frac{1}{2}yr + \frac{1}{2}zr + \frac{1}{2}zr$$
$$= r(x + y + z)$$
$$= rs.$$

Therefore: $rs = \sqrt{s(s-a)(s-b)(s-c)} \Rightarrow$

$$r = \sqrt{\frac{(s-a)(s-b)(s-c)}{s}}.$$

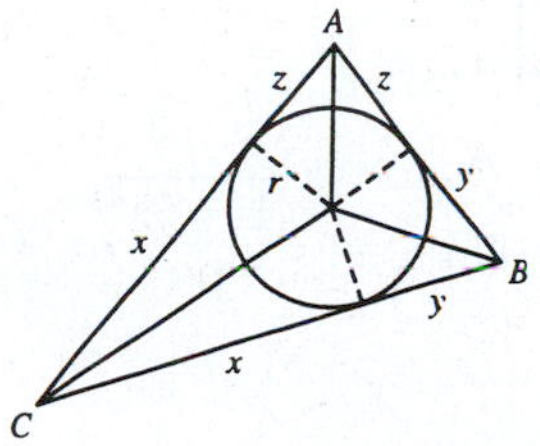

40. Given: $a = 200$ ft, $b = 250$ ft, $c = 325$ ft

$$s = \frac{200 + 250 + 325}{2} \approx 387.5$$

Radius of the inscribed circle: $r = \sqrt{\frac{(s-a)(s-b)(s-c)}{s}} = \sqrt{\frac{(187.5)(137.5)(62.5)}{387.5}} \approx 64.5$ ft

Circumference of an inscribed circle: $C = 2\pi r \approx 2\pi(64.5) \approx 405$ ft

42. Given: $a = 2.5, b = 10.2, c = 9$

$$s = \frac{2.5 + 10.2 + 9}{2} = 10.85$$

$$\text{Area} = \sqrt{s(s-a)(s-b)(s-c)}$$
$$= \sqrt{10.85(8.35)(0.65)(1.85)} \approx 10.44$$

44. Given: $a = 75.4, b = 52, c = 52$

$$s = \frac{75.4 + 52 + 52}{2} = 89.7$$

$$\text{Area} = \sqrt{s(s-a)(s-b)(s-c)}$$
$$= \sqrt{89.7(14.3)(37.7)(37.7)} \approx 1350$$

46. Given: $a = 4.25, b = 1.55, c = 3.00 \Rightarrow s = \frac{4.25 + 1.55 + 3.00}{2} = 4.4$

$$\text{Area} = \sqrt{s(s-a)(s-b)(s-c)} = \sqrt{4.4(0.15)(2.85)(1.4)} \approx 1.623$$

48.

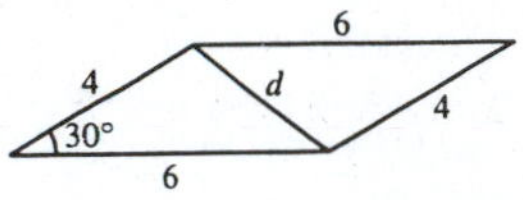

$$d^2 = 4^2 + 6^2 - 2(4)(6)\cos 30^\circ \approx 10.43$$

$$d \approx 3.23$$

$$a = 4, b = 6, c \approx 3.23$$

$$s = \frac{4 + 6 + 3.23}{2} \approx 6.615$$

$$\text{Area of triangle} = \sqrt{s(s-a)(s-b)(s-c)}$$
$$\approx \sqrt{6.615(2.615)(0.615)(3.385)}$$
$$= 6$$

Area of parallelogram

= 2 × area of triangle

= 2 × 6 = 12 square meters.

50. $\frac{1}{2}bc(1 - \cos A) = \frac{1}{2}bc\left[1 + \frac{a^2 - (b^2 + c^2)}{2bc}\right]$

$$= \frac{1}{2}bc\left[\frac{2bc + a^2 - b^2 - c^2}{2bc}\right]$$
$$= \frac{a^2 - (b^2 - 2bc + c^2)}{4}$$
$$= \frac{a^2 - (b - c)^2}{4}$$
$$= \left(\frac{a - (b - c)}{2}\right)\left(\frac{a + (b - c)}{2}\right)$$
$$= \frac{a - b + c}{2} \cdot \frac{a + b - c}{2}$$

52. Let $u = \arccos 3x$

$$\cos u = 3x = \frac{3x}{1}.$$

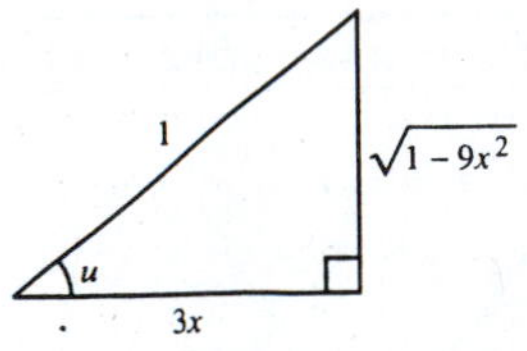

$$\tan(\arccos 3x) = \tan u = \frac{\sqrt{1 - 9x^2}}{3x}$$

54. Let $u = \arcsin \frac{x-1}{2}$

$$\sin u = \frac{x-1}{2}.$$

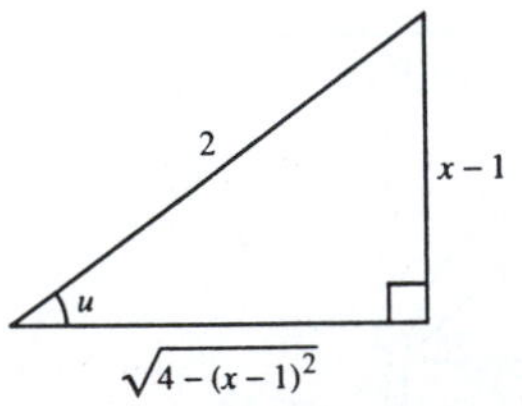

$$\cos\left(\arcsin \frac{x-1}{2}\right) = \cos u$$
$$= \frac{\sqrt{4 - (x-1)^2}}{2}$$
$$= \frac{1}{2}\sqrt{4 - (x-1)^2}$$

Section 6.3 Vectors in the Plane

Solutions to Even-Numbered Exercises

2. Initial point: $(0, 0)$

Terminal point: $(4, -2)$

$\mathbf{v} = \langle 4 - 0, -2 - 0 \rangle = \langle 4, -2 \rangle$

$\|\mathbf{v}\| = \sqrt{4^2 + (-2)^2} = \sqrt{20} = 2\sqrt{5}$

4. Initial point: $(-1, -1)$

Terminal point: $(3, 5)$

$\mathbf{v} = \langle 3 - (-1), 5 - (-1) \rangle = \langle 4, 6 \rangle$

$\|\mathbf{v}\| = \sqrt{4^2 + 6^2} = \sqrt{52} = 2\sqrt{13}$

6. Initial point: $(-4, -1)$

Terminal point: $(3, -1)$

$\mathbf{v} = \langle 3 - (-4), -1 - (-1) \rangle = \langle 7, 0 \rangle$

$\|\mathbf{v}\| = \sqrt{7^2 + 0^2} = 7$

8. Initial point: $(3.4, 0)$

Terminal point: $(0, 5.8)$

$\mathbf{v} = \langle 0 - 3.4, 5.8 - 0 \rangle = \langle -3{,}4, 5.8 \rangle$

$\|\mathbf{v}\| = \sqrt{(-3.4) + (5.8)} \approx 6.7$

10. Initial point: $(-3, 11)$

Terminal point: $(9, 40)$

$\mathbf{v} = \langle 9 - (-3), 40 - 11 \rangle = \langle 12, 29 \rangle$

$\|\mathbf{v}\| = \sqrt{12^2 + 29^2} = \sqrt{985}$

12. $3\mathbf{v}$

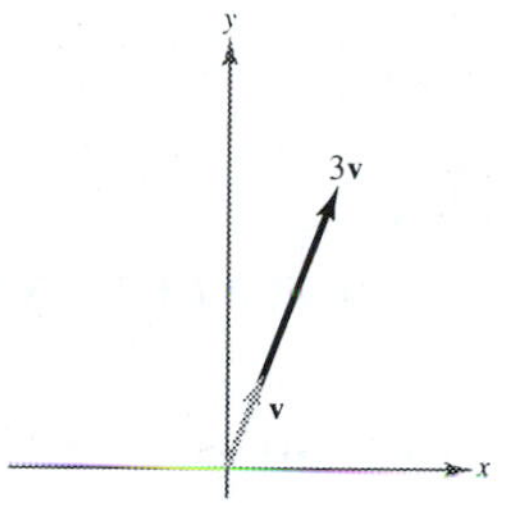

14. $\mathbf{u} + 2\mathbf{v}$

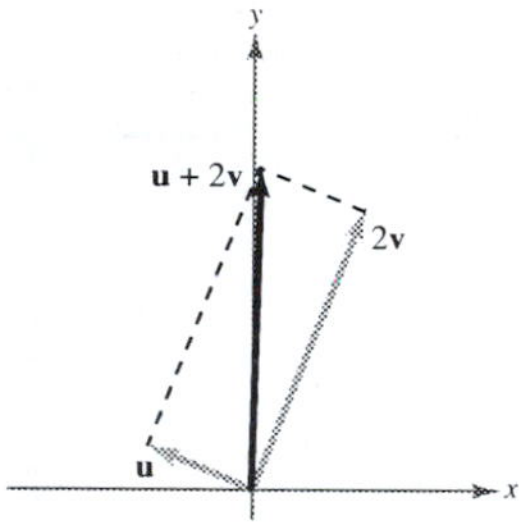

16. $\mathbf{v} - \frac{1}{2}\mathbf{u}$

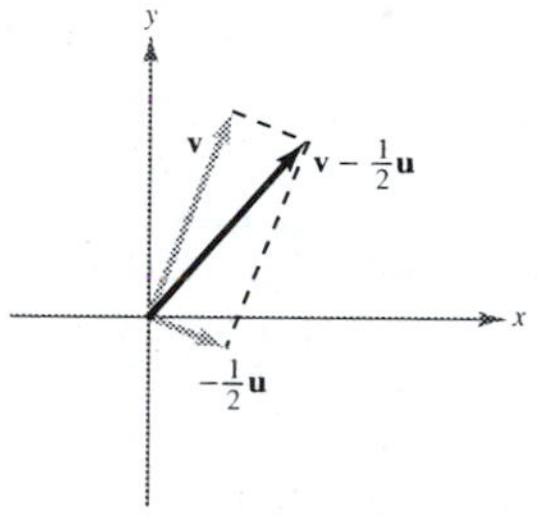

18. $\mathbf{u} = \langle 2, 3 \rangle, \mathbf{v} = \langle 4, 0 \rangle$

(a) $\mathbf{u} + \mathbf{v} = \langle 6, 3 \rangle$

(b) $\mathbf{u} - \mathbf{v} = \langle -2, 3 \rangle$

(c) $2\mathbf{u} - 3\mathbf{v} = \langle 4, 6 \rangle - \langle 12, 0 \rangle = \langle -8, 6 \rangle$

20. $\mathbf{u} = \langle 0, 0 \rangle, \mathbf{v} = \langle 2, 1 \rangle$

(a) $\mathbf{u} + \mathbf{v} = \langle 2, 1 \rangle$

(b) $\mathbf{u} - \mathbf{v} = \langle -2, -1 \rangle$

(c) $2\mathbf{u} - 3\mathbf{v} = \langle 0, 0 \rangle - \langle 6, 3 \rangle = \langle -6, -3 \rangle$

22. $\mathbf{u} = 2\mathbf{i} - \mathbf{j}, \mathbf{v} = -\mathbf{i} + \mathbf{j}$

(a) $\mathbf{u} + \mathbf{v} = \mathbf{i}$

(b) $\mathbf{u} - \mathbf{v} = 3\mathbf{i} - 2\mathbf{j}$

(c) $2\mathbf{u} - 3\mathbf{v} = (4\mathbf{i} - 2\mathbf{j}) - (-3\mathbf{i} + 3\mathbf{j}) = 7\mathbf{i} - 5\mathbf{j}$

24. $\mathbf{u} = 3\mathbf{j}, \mathbf{v} = 2\mathbf{i}$

(a) $\mathbf{u} + \mathbf{v} = 2\mathbf{i} + 3\mathbf{j}$

(b) $\mathbf{u} - \mathbf{v} = -2\mathbf{i} + 3\mathbf{j}$

(c) $2\mathbf{u} - 3\mathbf{v} = 6\mathbf{j} - 6\mathbf{i} = -6\mathbf{i} + 6\mathbf{j}$

26. $\mathbf{v} = \langle 0, -3 \rangle$

$$\mathbf{u} = \frac{1}{\|\mathbf{v}\|}\mathbf{v} = \frac{1}{\sqrt{0^2 + (-3)^2}}\langle 0, -3 \rangle$$

$$= \frac{1}{3}\langle 0, -3 \rangle$$

$$= \langle 0, -1 \rangle$$

28. $\mathbf{v} = \langle 5, -12 \rangle$

$$\mathbf{u} = \frac{1}{\|\mathbf{v}\|}\mathbf{v} = \frac{1}{\sqrt{5^2 + (-12)^2}}\langle 5, -12 \rangle$$

$$= \frac{1}{13}\langle 5, -12 \rangle$$

$$= \left\langle \frac{5}{13}, -\frac{12}{13} \right\rangle$$

30. $\mathbf{v} = \mathbf{i} + \mathbf{j}$

$$\mathbf{u} = \frac{1}{\|\mathbf{v}\|}\mathbf{v} = \frac{1}{\sqrt{1^2 + 1^2}}(\mathbf{i} + \mathbf{j}) = \frac{1}{\sqrt{2}}(\mathbf{i} + \mathbf{j}) = \frac{\sqrt{2}}{2}\mathbf{i} + \frac{\sqrt{2}}{2}\mathbf{j}$$

32. $\mathbf{w} = \mathbf{i} - 2\mathbf{j}$

$$\mathbf{u} = \frac{1}{\|\mathbf{w}\|}\mathbf{w} = \frac{1}{\sqrt{1^2 + (-2)^2}}(\mathbf{i} - 2\mathbf{j}) = \frac{1}{\sqrt{5}}(\mathbf{i} - 2\mathbf{j}) = \frac{\sqrt{5}}{5}\mathbf{i} - \frac{2\sqrt{5}}{5}\mathbf{j}$$

34. $\mathbf{v} = 3\left(\frac{1}{\|\mathbf{u}\|}\mathbf{u}\right)$

$$= 3\left(\frac{1}{\sqrt{4^2 + (-4)^2}}\langle 4, -4\rangle\right) = 3\left(\frac{1}{4\sqrt{2}}\langle 4, -4\rangle\right) = \left\langle \frac{3}{\sqrt{2}}, -\frac{3}{\sqrt{2}}\right\rangle$$

36. $\mathbf{v} = 10\left(\frac{1}{\|\mathbf{u}\|}\mathbf{u}\right)$

$$= 10\left(\frac{1}{\sqrt{0^2 + (-10)^2}}\langle -10, 0\rangle\right) = 10\left(\frac{1}{10}\langle -10, 0\rangle\right) = \langle -10, 0\rangle$$

38. $\mathbf{v} = \mathbf{u} + \mathbf{w}$

$$= (2\mathbf{i} - \mathbf{j}) + (\mathbf{i} + 2\mathbf{j}) = 3\mathbf{i} + \mathbf{j} = \langle 3, 1\rangle$$

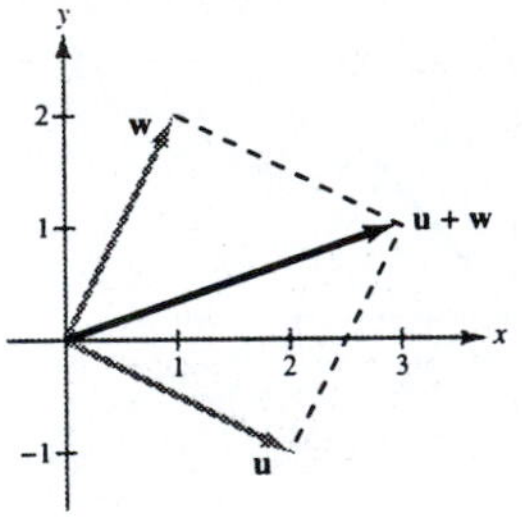

40. $\mathbf{v} = -\mathbf{u} + \mathbf{w}$

$$= -(2\mathbf{i} - \mathbf{j}) + (\mathbf{i} + 2\mathbf{j}) = -\mathbf{i} + 3\mathbf{j} = \langle -1, 3\rangle$$

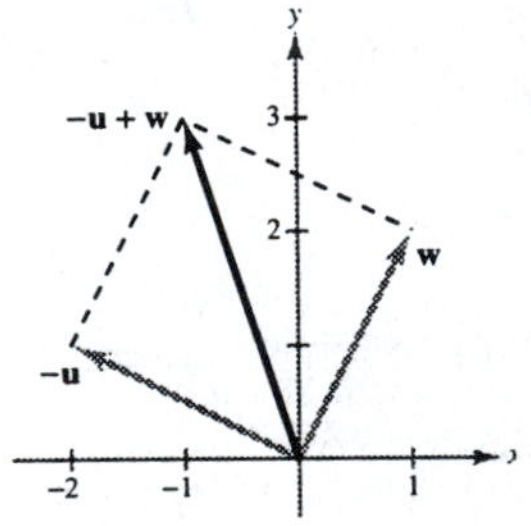

42. $\mathbf{v} = \mathbf{u} - 2\mathbf{w}$

$$= (2\mathbf{i} - \mathbf{j}) - 2(\mathbf{i} + 2\mathbf{j}) = -5\mathbf{j} = \langle 0, -5\rangle$$

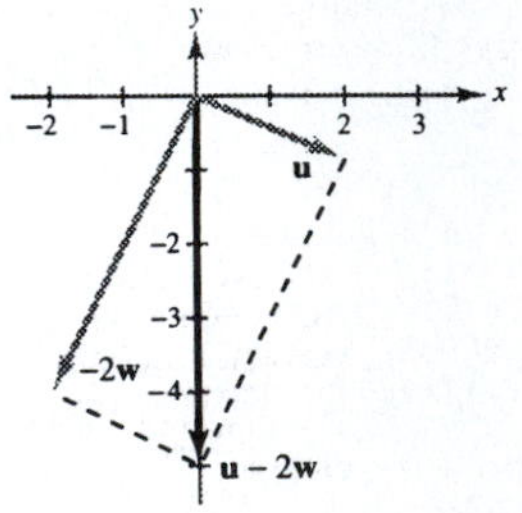

44. $\mathbf{v} = 8(\cos 135°\,\mathbf{i} + \sin 135°\,\mathbf{j})$

$\|\mathbf{v}\| = 8, \theta = 135°$

46. $\mathbf{v} = -2\mathbf{i} + 5\mathbf{j}$

$\|\mathbf{v}\| = \sqrt{(-2)^2 + 5^2} = \sqrt{29}$

$\tan\theta = -\frac{5}{2}$

Since $\mathbf{v}$ lies in Quadrant II, $\theta \approx 111.8°$.

48. $\mathbf{v} = \langle \cos 45°, \sin 45°\rangle$

$$= \left\langle \frac{\sqrt{2}}{2}, \frac{\sqrt{2}}{2}\right\rangle$$

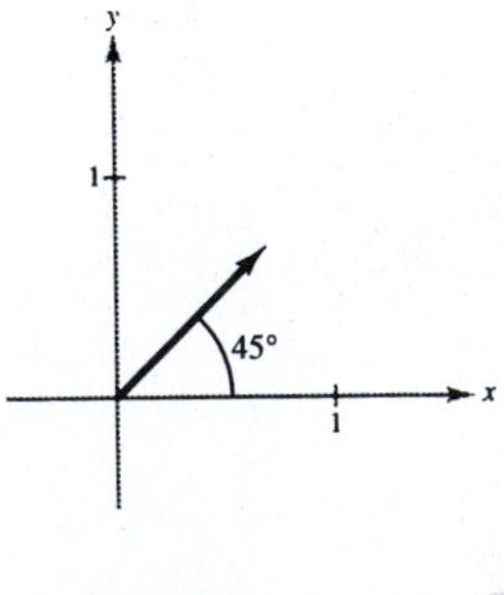

50. $\mathbf{v} = \left\langle \frac{5}{2}\cos 45°, \frac{5}{2}\sin 45°\right\rangle$

$$= \left\langle \frac{5\sqrt{2}}{4}, \frac{5\sqrt{2}}{4}\right\rangle$$

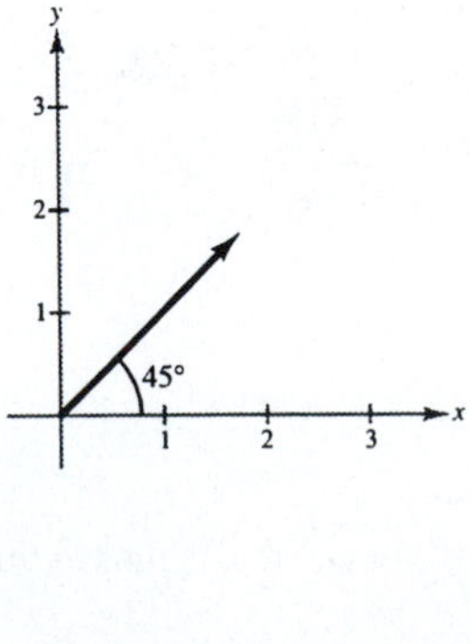

52. $\mathbf{v} = \langle 9\cos 90°, 9\sin 90°\rangle$

$$= \langle 0, 9\rangle$$

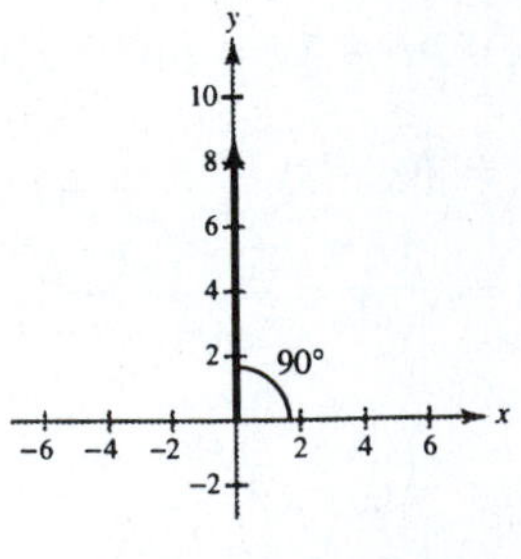

54. $\mathbf{v} = 3\left(\frac{1}{\sqrt{3^2 + 4^2}}\right)(3\mathbf{i} + 4\mathbf{j})$

$= \frac{3}{5}(3\mathbf{i} + 4\mathbf{j})$

$= \frac{9}{5}\mathbf{i} + \frac{12}{5}\mathbf{j} = \left\langle \frac{9}{5}, \frac{12}{5} \right\rangle$

56. $\mathbf{u} = \langle 2\cos 30°, 2\sin 30° \rangle = \langle \sqrt{3}, 1 \rangle$

$\mathbf{v} = \langle 2\cos 90°, 2\sin 90° \rangle = \langle 0, 2 \rangle$

$\mathbf{u} + \mathbf{v} = \langle \sqrt{3}, 3 \rangle$

58. $\mathbf{u} = \langle 35\cos 25°, 35\sin 25° \rangle = \langle 31.72, 14.79 \rangle$

$\mathbf{v} = \langle 50\cos 120°, 50\sin 120° \rangle = \langle -25, 25\sqrt{3} \rangle$

$\mathbf{u} + \mathbf{v} \approx \langle 6.72, 58.09 \rangle$

60. $\mathbf{v} = 3\mathbf{i} + \mathbf{j}$

$\mathbf{w} = 2\mathbf{i} - \mathbf{j}$

$\mathbf{u} = \mathbf{v} - \mathbf{w} = \mathbf{i} + 2\mathbf{j}$

$\cos\theta = \frac{\|\mathbf{v}\|^2 + \|\mathbf{w}\|^2 - \|\mathbf{v} - \mathbf{w}\|^2}{2\|\mathbf{v}\|\,\|\mathbf{w}\|} = \frac{10 + 5 - 5}{2\sqrt{10}\sqrt{5}} = \frac{\sqrt{2}}{2}$

$\theta = 45°$

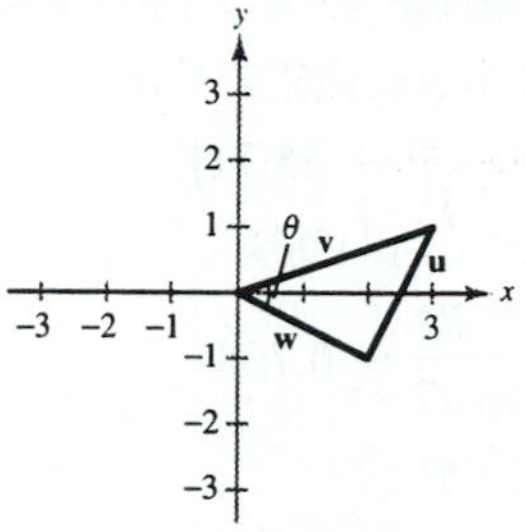

62. $\mathbf{v} = \mathbf{i} + 2\mathbf{j}$

$\mathbf{w} = 2\mathbf{i} - \mathbf{j}$

$\mathbf{u} = \mathbf{v} - \mathbf{w} = -\mathbf{i} + 3\mathbf{j}$

$\cos\theta = \frac{\|\mathbf{v}\|^2 + \|\mathbf{w}\|^2 - \|\mathbf{v} - \mathbf{w}\|^2}{2\|\mathbf{v}\|\,\|\mathbf{w}\|} = \frac{5 + 5 - 10}{2\sqrt{5}\sqrt{5}} = 0$

$\theta = 90°$

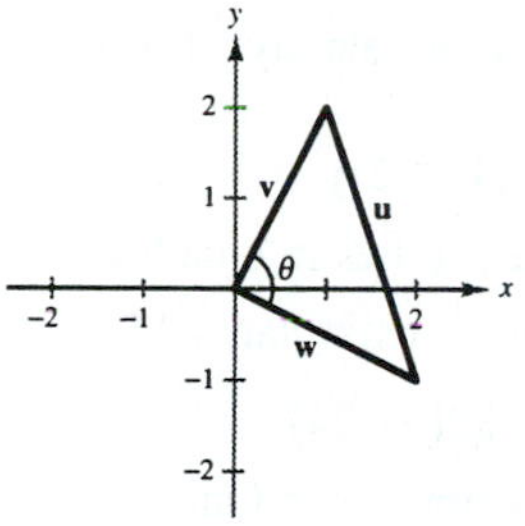

64. Force One: $\mathbf{u} = 3000\mathbf{i}$

Force Two: $\mathbf{v} = 1000\cos\theta\mathbf{i} + 1000\sin\theta\mathbf{j}$

Resultant Force: $\mathbf{u} + \mathbf{v} = (3000 + 1000\cos\theta)\mathbf{i} + 1000\sin\theta\mathbf{j}$

$\|\mathbf{u} + \mathbf{v}\| = \sqrt{(3000 + 1000\cos\theta)^2 + (1000\sin\theta)^2} = 3750$

$9{,}000{,}000 + 6{,}000{,}000\cos\theta + 1{,}000{,}000 = 14{,}062,500$

$6{,}000{,}000\cos\theta = 4{,}062,500$

$\cos\theta = \frac{4{,}062{,}500}{6{,}000{,}000} \approx 0.6771$

$\theta \approx 47.4°$

66. $\mathbf{F}_1 = \langle 10, 0 \rangle$, $\mathbf{F}_2 = 5\langle \cos\theta, \sin\theta \rangle$

(a) $\mathbf{F}_1 + \mathbf{F}_2 = \langle 10 + 5\cos\theta, 5\sin\theta \rangle$

$\|\mathbf{F}_1 + \mathbf{F}_2\| = \sqrt{(10 + 5\cos\theta)^2 + (5\sin\theta)^2}$

$= \sqrt{100 + 100\cos\theta + 25\cos^2\theta + 25\sin^2\theta}$

$= 5\sqrt{4 + 4\cos\theta + \cos^2\theta + \sin^2\theta}$

$= 5\sqrt{4 + 4\cos\theta + 1}$

$= 5\sqrt{5 + 4\cos\theta}$

(b)

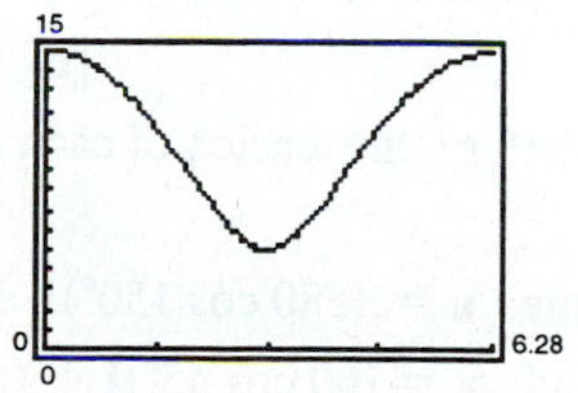

(c) Range: [5, 15]

Maximum is 15 when $\theta = 0$.

Minimum is 5 when $\theta = \pi$.

(d) The magnitude of the resultant is never 0 because the magnitudes of $\mathbf{F}_1$ and $\mathbf{F}_2$ are not the same.

68. $\mathbf{u} = (2000 \cos 30°)\,\mathbf{i} + (2000 \sin 30°\mathbf{j})$

$\approx 1732.05\mathbf{i} + 1000\mathbf{j}$

$\mathbf{v} = (900 \cos(-45°))\mathbf{i} + (900 \sin(-45°)\mathbf{j})$

$\approx 636.4\mathbf{i} + -636.4\mathbf{j}$

$\mathbf{u} + \mathbf{v} \approx 2368.4\mathbf{i} + 363.6\mathbf{j}$

$\|\mathbf{u} + \mathbf{v}\| \approx \sqrt{(2368.4)^2 + (363.6)^2} \approx 2396.19$

$\tan \theta = \dfrac{363.6}{2368.4} \approx 0.1535 \implies \theta \approx 8.7°$

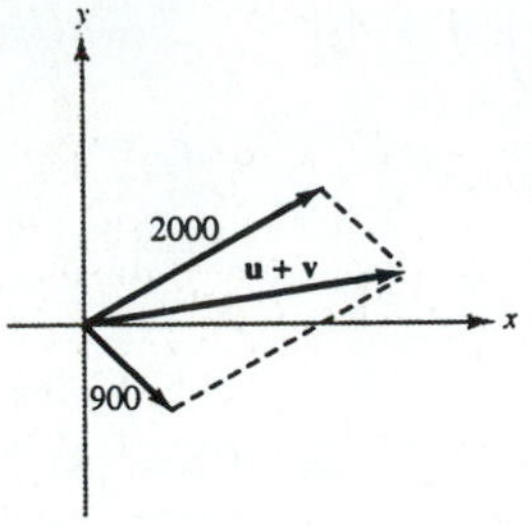

70. $\mathbf{u} = (70 \cos 30°)\mathbf{i} - (70 \sin 30°)\mathbf{j} \approx 60.62\mathbf{i} - 35\mathbf{j}$

$\mathbf{v} = (40 \cos 45°)\mathbf{i} + (40 \sin 45°)\mathbf{j} \approx 28.28\mathbf{i} + 28.28\mathbf{j}$

$\mathbf{w} = (60 \cos 135°)\mathbf{i} + (60 \sin 135°)\mathbf{j} \approx -42.43\mathbf{i} + 42.43\mathbf{j}$

$\mathbf{u} + \mathbf{v} + \mathbf{w} = 46.47\mathbf{i} + 35.71\mathbf{j}$

$\|\mathbf{u} + \mathbf{v} + \mathbf{w}\| \approx 58.61$ pounds

$\tan \theta \approx \dfrac{35.71}{46.47} \approx 0.7684$

$\theta \approx 37.5°$

72. Horizontal component of velocity: $1200 \cos 6° \approx 1193.4$ ft/sec

Vertical component of velocity: $1200 \sin 6° \approx 125.4$ ft/sec

74. Rope $\overrightarrow{AC}$: $\mathbf{u} = 10\mathbf{i} - 24\mathbf{j}$

The vector lies in Quadrant IV and its reference angle is $\arctan\left(\frac{12}{5}\right)$.

$\mathbf{u} = \|\mathbf{u}\|\left[\cos\left(\arctan \frac{12}{5}\right)\mathbf{i} - \sin\left(\arctan \frac{12}{5}\right)\mathbf{j}\right]$

Rope $\overrightarrow{BC}$: $\mathbf{v} = -20\mathbf{i} - 24\mathbf{j}$

The vector lies in Quadrant III and its reference angle is $\arctan\left(\frac{6}{5}\right)$.

$\mathbf{v} = \|\mathbf{v}\|\left[-\cos\left(\arctan \frac{6}{5}\right)\mathbf{i} - \sin\left(\arctan \frac{6}{5}\right)\mathbf{j}\right]$

Resultant: $\mathbf{u} + \mathbf{v} = -5000\mathbf{j}$

$\|\mathbf{u}\| \cos\left(\arctan \frac{12}{5}\right) - \|\mathbf{v}\| \cos\left(\arctan \frac{6}{5}\right) = 0$

$-\|\mathbf{u}\| \sin\left(\arctan \frac{12}{5}\right) - \|\mathbf{v}\| \sin\left(\arctan \frac{6}{5}\right) = -5000$

Solving this system of equations yields: $T_{AC} = \|\mathbf{u}\| \approx 3611.1$ pounds

$T_{BC} = \|\mathbf{v}\| \approx 2169.5$ pounds

76. Rope 1: $\mathbf{u} = \|\mathbf{u}\|\,(\cos 70°\mathbf{i} - \sin 70°\mathbf{j})$

Rope 2: $\mathbf{v} = \|\mathbf{u}\|\,(-\cos 70°\,\mathbf{i} - \sin 70°\mathbf{j})$

Resultant: $\mathbf{u} + \mathbf{v} = -100\mathbf{j}$

$-\|\mathbf{u}\| \sin 70° - \|\mathbf{u}\| \sin 70° = -100$

$\|\mathbf{u}\| \approx 53.2$

Therefore, the tension of each rope is $\|\mathbf{u}\| \approx 53.2$ pounds.

78. Plane: $\mathbf{u} = (580 \cos 150°)\mathbf{i} + (580 \sin 150°)\mathbf{j} \approx -502.3\mathbf{i} + 290\mathbf{j}$

Wind: $\mathbf{v} = (60 \cos 45°)\mathbf{i} + (60 \sin 45°)\mathbf{j} \approx 42.4\mathbf{i} + 42.4\mathbf{j}$

$\mathbf{u} + \mathbf{v} \approx -459.9\mathbf{i} + 332.4\mathbf{j}$

$\|\mathbf{u} + \mathbf{v}\| \approx \sqrt{(-459.9)^2 + (332.4)^2} \approx 567.4$

$\tan \theta \approx -\dfrac{332.4}{459.9} \approx -0.7229 \implies \theta \approx 144.1°$

The ground speed is 567.4 miles per hour and the heading is N 54.1° W.

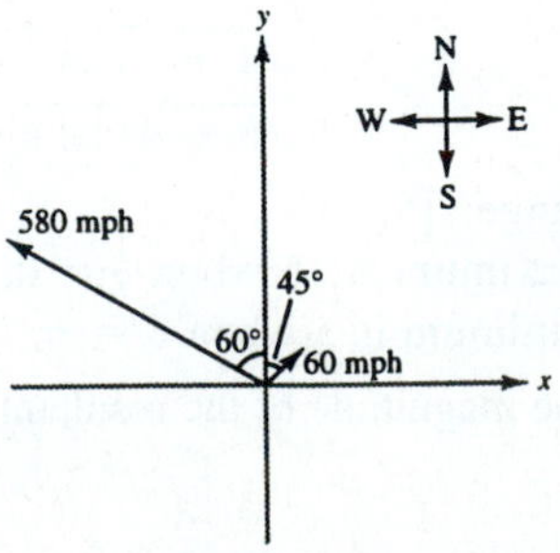

80. Horizontal force: $\mathbf{u} = \|\mathbf{u}\|\mathbf{i}$

Weight: $\mathbf{w} = -\mathbf{j}$

Rope: $\mathbf{t} = \|\mathbf{t}\|(\cos 135°\mathbf{i} + \sin 135°\mathbf{j})$

$$\mathbf{u} + \mathbf{w} + \mathbf{t} = \mathbf{0} \Rightarrow \|\mathbf{u}\| + \|\mathbf{t}\| \cos 135° = 0$$
$$-1 + \|\mathbf{t}\| \sin 135° = 0$$

$\|\mathbf{t}\| \approx \sqrt{2}$ pounds

$\|\mathbf{u}\| \approx 1$ pound

82. True

84. True

86. The following program is written for a *TI-82* or *TI-83* graphing calculator. The program sketches two vectors $\mathbf{u} = a\mathbf{i} + b\mathbf{j}$ and $\mathbf{v} = c\mathbf{i} + d\mathbf{j}$ in standard position, and then sketches the vector difference $\mathbf{u} - \mathbf{v}$ using the parallelogram law.

```
PROGRAM: SUBVECT
:Input "ENTER A", A
:Input "ENTER B", B
:Input "ENTER C", C
:Input "ENTER D", D
:Line (0, 0, A, B)
:Line (0, 0, C, D)
:Pause
:A-C→E
:B-D→F
:Line (A, B, C, D)
:Line (A, B, E, F)
:Line (0, 0, E, F)
:Pause
:ClrDraw
:Stop
```

88.
$$\mathbf{u} = \langle 80 - 10, 80 - 60\rangle = \langle 70, 20\rangle$$
$$\mathbf{v} = \langle -20 - (-100), 70 - 0\rangle = \langle 80, 70\rangle$$
$$\mathbf{u} - \mathbf{v} = \langle 70 - 80, 20 - 70\rangle = \langle -10, -50\rangle$$
$$\mathbf{v} - \mathbf{u} = \langle 80 - 70, 70 - 20\rangle = \langle 10, 50\rangle$$

90.

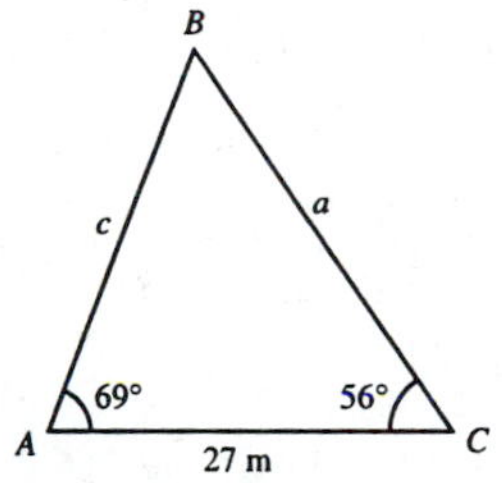

$$B = 180° - 56° - 69° = 55°$$
$$\frac{a}{\sin A} = \frac{b}{\sin B}$$
$$a = \frac{27 \sin 69°}{\sin 44°} \approx 30.8 \text{ meters}$$

92. $x = 8 \sin \theta$
$$\sqrt{64 - x^2} = \sqrt{64 - (8 \sin^2 \theta)}$$
$$= \sqrt{64 - 64 \sin^2 \theta}$$
$$= 8\sqrt{1 - \sin^2 \theta}$$
$$= 8\sqrt{\cos^2 \theta}$$
$$= 8 \cos \theta$$

94. $x = 5 \sec \theta$
$$\sqrt{(x^2 - 25)^3} = \sqrt{[(5 \sec \theta)^2 - 25]^3}$$
$$= \sqrt{(25 \sec^2 \theta - 25)^3}$$
$$= \sqrt{[25(\sec^2 \theta - 1)]^3}$$
$$= \sqrt{(25 \tan^2 \theta)^3}$$
$$= \sqrt{15{,}625 \tan^6 \theta}$$
$$= 125 \tan^3 \theta$$

Section 6.4 Vectors and Dot Products

Solutions to Even-Numbered Exercises

2. $\mathbf{u} = \langle 5, 12 \rangle, \mathbf{v} = \langle -3, 2 \rangle$

$\mathbf{u} \cdot \mathbf{v} = 5(-3) + 12(2) = 9$

4. $\mathbf{u} = 2\mathbf{i} + 5\mathbf{j}, \mathbf{v} = 9\mathbf{i} - 3\mathbf{j}$

$\mathbf{u} \cdot \mathbf{v} = 2(9) + 5(-3) = 3$

6. $\mathbf{u} = \langle 2, 2 \rangle$

$$\begin{aligned} \|\mathbf{u}\| - 2 &= \sqrt{\mathbf{u} \cdot \mathbf{u}} - 2 \\ &= \sqrt{2(2) + 2(2)} - 2 \\ &= \sqrt{8} - 2 \\ &= 2\sqrt{2} - 2, \text{ scalar} \end{aligned}$$

8. $\mathbf{u} = \langle 2, 2 \rangle, \mathbf{v} = \langle -3, 4 \rangle$

$$\begin{aligned} \mathbf{u} \cdot 2\mathbf{v} &= 2\mathbf{u} \cdot \mathbf{v} \\ &= 2[2(-3) + 2(4)] \\ &= 2(2) = 4, \text{ scalar} \end{aligned}$$

10. $\mathbf{u} = \langle 2, -4 \rangle$

$$\begin{aligned} \|\mathbf{u}\| &= \sqrt{\mathbf{u} \cdot \mathbf{u}} \\ &= \sqrt{2(2) + (-4)(-4)} \\ &= \sqrt{20} = 2\sqrt{5} \end{aligned}$$

12. $\mathbf{u} = 6\mathbf{j}$

$$\begin{aligned} \|\mathbf{u}\| &= \sqrt{\mathbf{u} \cdot \mathbf{u}} \\ &= \sqrt{6(6)} = 6 \end{aligned}$$

14. $\mathbf{u} = \langle 1245, 2600 \rangle$

$\mathbf{v} = \langle 12.20, 8.50 \rangle$

Increase prices by 5%: $1.05\mathbf{v}$

$$\begin{aligned} \mathbf{u} \cdot 1.05\mathbf{v} &= 1.05\mathbf{u} \cdot \mathbf{v} \\ &= 1.05[1245(12.20) + 2600(8.50)] \\ &= 1.05(37{,}289) \\ &= \$39{,}153.45 \end{aligned}$$

16. $\mathbf{u} = \langle 4, 4 \rangle, \mathbf{v} = \langle 2, 0 \rangle$

$$\begin{aligned} \cos\theta &= \frac{\mathbf{u} \cdot \mathbf{v}}{\|\mathbf{u}\|\,\|\mathbf{v}\|} \\ &= \frac{4(2) + 4(0)}{(4\sqrt{2})(2)} \\ &= \frac{\sqrt{2}}{2} \end{aligned}$$

$\theta = 45°$

18. $\mathbf{u} = 2\mathbf{i} - 3\mathbf{j}, \mathbf{v} = \mathbf{i} - 2\mathbf{j}$

$$\begin{aligned} \cos\theta &= \frac{\mathbf{u} \cdot \mathbf{v}}{\|\mathbf{u}\|\,\|\mathbf{v}\|} \\ &= \frac{2(1) + (-3)(-2)}{\sqrt{2^2 + 3^2}\sqrt{1^2 + 2^2}} \\ &= \frac{8}{\sqrt{65}} \approx 0.992278 \end{aligned}$$

$\theta \approx 7.13°$

20. $\mathbf{u} = \cos\left(\frac{\pi}{4}\right)\mathbf{i} + \sin\left(\frac{\pi}{4}\right)\mathbf{j} = \frac{\sqrt{2}}{2}\mathbf{i} + \frac{\sqrt{2}}{2}\mathbf{j}$

$\mathbf{v} = \cos\left(\frac{\pi}{2}\right)\mathbf{i} + \sin\left(\frac{\pi}{2}\right)\mathbf{j} = \mathbf{j}$

$$\cos\theta = \frac{\mathbf{u} \cdot \mathbf{v}}{\|\mathbf{u}\|\,\|\mathbf{v}\|} = \frac{\frac{\sqrt{2}}{2}(0) + \frac{\sqrt{2}}{2}(1)}{1 \cdot 1} = \frac{\sqrt{2}}{2}$$

$\theta = \frac{\pi}{4}$

22. $\mathbf{u} = -6\mathbf{i} - 3\mathbf{j}, \mathbf{v} = -8\mathbf{i} + 4\mathbf{j}$

$$\begin{aligned} \cos\mathbf{u} &= \frac{\mathbf{u} \cdot \mathbf{v}}{\|\mathbf{u}\|\,\|\mathbf{v}\|} = \frac{-6(-8) + (-3)(-4)}{\sqrt{45}\sqrt{80}} \\ &= \frac{36}{60} = 0.6 \end{aligned}$$

$\theta \approx 53.13°$

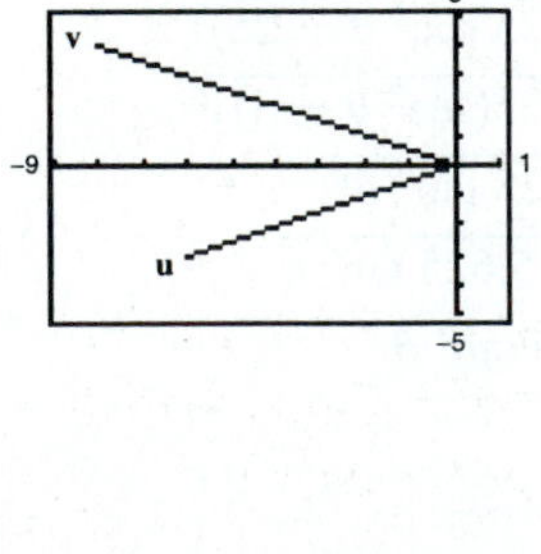

24. $\mathbf{u} = 2\mathbf{i} - 3\mathbf{j}, \mathbf{v} = 4\mathbf{i} + 3\mathbf{j}$

$$\cos\theta = \frac{\mathbf{u} \cdot \mathbf{v}}{\|\mathbf{u}\|\,\|\mathbf{v}\|} = \frac{2(4) + (-3)(3)}{\sqrt{13}\sqrt{25}} \approx -0.0555$$

$\theta \approx 93.18°$

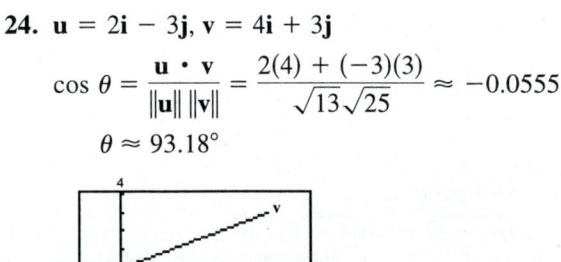

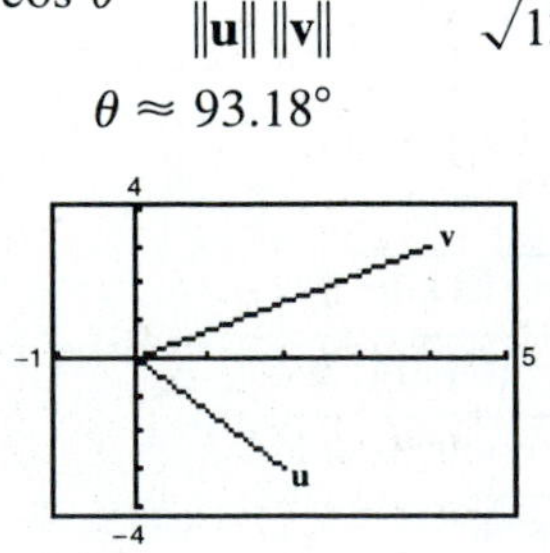

26. $P = (-3, 0), Q = (2, 2), R = (0, 6)$

$\overrightarrow{PQ} = \langle 5, 2\rangle, \overrightarrow{QR} = \langle -2, 4\rangle, \overrightarrow{PR} = \langle 3, 6\rangle,$

$\overrightarrow{QP} = \langle -5, -2\rangle$

$$\cos\alpha = \frac{\overrightarrow{PQ}\cdot\overrightarrow{PR}}{\|\overrightarrow{PR}\|\,\|\overrightarrow{PR}\|} = \frac{27}{(\sqrt{29})(\sqrt{45})} \implies \alpha \approx 41.6^\circ$$

$$\cos B = \frac{\overrightarrow{QR}\cdot\overrightarrow{QP}}{\|\overrightarrow{QR}\|\,\|\overrightarrow{QP}\|} = \frac{2}{(\sqrt{20})(\sqrt{29})} \implies \alpha \approx 85.2^\circ$$

$\phi = 180^\circ - 41.6^\circ - 85.2^\circ \approx 53.1^\circ$

28. $\|\mathbf{u}\| = 100, \|\mathbf{v}\| = 250, \theta = \frac{\pi}{6}$

$$\begin{aligned}\mathbf{u}\cdot\mathbf{v} &= \|\mathbf{u}\|\,\|\mathbf{v}\|\cos\theta\\ &= (100)(250)\cos\frac{\pi}{6}\\ &= 25{,}000\cdot\frac{\sqrt{3}}{2}\\ &= 12{,}500\sqrt{3}\end{aligned}$$

30. $\mathbf{u} = \langle 15, 45\rangle, \mathbf{v} = \langle -5, 12\rangle$

$\mathbf{u} \neq k\mathbf{v} \implies$ Not parallel

$\mathbf{u}\cdot\mathbf{v} \neq 0 \implies$ Not orthogonal

Neither

32. $\mathbf{u} = \mathbf{j}, \mathbf{v} = \mathbf{i} - 2\mathbf{j}$

$\mathbf{u} \neq k\mathbf{v} \implies$ Not parallel

$\mathbf{u}\cdot\mathbf{v} \neq 0 \implies$ Not orthogonal

Neither

34. $\mathbf{u} = \langle \cos\theta, \sin\theta\rangle$

$\mathbf{v} = \langle \sin\theta, -\cos\theta\rangle$

$\mathbf{u}\cdot\mathbf{v} = 0 \implies$ **u** and **v** are orthogonal.

36. $\mathbf{u} = \langle 4, 2\rangle, \mathbf{v} = \langle 1, -2\rangle$

$$\mathbf{w}_1 = \text{proj}_{\mathbf{v}}\mathbf{u} = \left(\frac{\mathbf{u}\cdot\mathbf{v}}{\|\mathbf{v}\|^2}\right)\mathbf{v} = 0\langle 1, -2\rangle = \langle 0, 0\rangle$$

$$\mathbf{w}_2 = \mathbf{u} - \mathbf{w}_1 = \langle 4, 2\rangle - \langle 0, 0\rangle = \langle 4, 2\rangle$$

38. $\mathbf{u} = \langle -5, -1\rangle, \mathbf{v} = \langle -1, 1\rangle$

$$\mathbf{w}_1 = \text{proj}_{\mathbf{v}}\mathbf{u} = \left(\frac{\mathbf{u}\cdot\mathbf{v}}{\|\mathbf{v}\|^2}\right)\mathbf{v} = \frac{4}{2}\langle -1, 1\rangle = 2\langle -1, 1\rangle$$

$$\begin{aligned}\mathbf{w}_2 = \mathbf{u} - \mathbf{w}_1 &= \langle -5, -1\rangle - 2\langle -1, 1\rangle\\ &= \langle -3, -3\rangle = 3\langle -1, -1\rangle\end{aligned}$$

40. $\mathbf{u} = \langle -8, 3\rangle$

For **v** to be orthogonal to **u**, $\mathbf{u}\cdot\mathbf{v}$ must be equal to 0.

Two possibilities: $\langle 3, 8\rangle, \langle -3, -8\rangle$

42. $\mathbf{u} = -\frac{5}{2}\mathbf{i} - 3\mathbf{j}$

For **v** to be orthogonal to **u**, $\mathbf{u}\cdot\mathbf{v}$ must be equal to 0.

Two possibilities: $\mathbf{v} = 3\mathbf{i} - \frac{5}{2}\mathbf{j}$

$\mathbf{v} = -3\mathbf{i} + \frac{5}{2}\mathbf{j}$

44. (a) $\mathbf{F} = -36{,}000\mathbf{j}$ Gravitational force

$\mathbf{v} = (\cos 12^\circ)\mathbf{i} + (\sin 12^\circ)\mathbf{j}$

$$\mathbf{w}_1 = \text{proj}_{\mathbf{v}}\mathbf{F} = \left(\frac{\mathbf{F}\cdot\mathbf{v}}{\|\mathbf{v}\|^2}\right)\mathbf{v} = (\mathbf{F}\cdot\mathbf{v}) \approx -7484.8\mathbf{v}$$

The magnitude of this force is 7484.8; therefore, a force of 7484.8 pounds is needed to keep the truck from rolling down the hill.

(b) $$\begin{aligned}\mathbf{w}_2 = \mathbf{F} - \mathbf{w}_1 &= -36{,}000\mathbf{j} + 7484.8[(\cos 12^\circ)\mathbf{i} + (\sin 12^\circ)\mathbf{j}]\\ &= [(7484.8\cos 12^\circ)\mathbf{i} + (7484.8\sin 12^\circ - 36{,}000)\mathbf{j}]\end{aligned}$$

$\|\mathbf{w}_2\| \approx 35{,}213.3$ pounds

46. (a) $\text{proj}_{\mathbf{v}}\mathbf{u} = \mathbf{u} \implies$ **u** and **v** are parallel.

(b) $\text{proj}_{\mathbf{v}}\mathbf{u} = 0 \implies$ **u** and **v** are orthogonal.

48. work $= (2400)(5) = 12{,}000$ foot-pounds

50. work $= (\cos 35^\circ)(15, 691)(800)$

$\approx 10{,}282{,}651$ newton $-$ meters

52. $P = (1, 3), Q = (-3, 5), \mathbf{v} = -2\mathbf{i} + 2\mathbf{j}$

work $= \|\text{proj}_{\overrightarrow{PQ}}\,\mathbf{v}\| = \|\overrightarrow{PQ}\|$ where

$\overrightarrow{PQ} = \langle -4, 2\rangle$ and $\mathbf{v} = \langle -2, 3\rangle$.

$$\text{proj}_{\overrightarrow{PQ}}\,\mathbf{v} = \left(\frac{\mathbf{v}\cdot\overrightarrow{PQ}}{\|\overrightarrow{PQ}\|}\right)\overrightarrow{PQ} = \left(\frac{14}{20}\right)\langle -4, 2\rangle$$

$$\text{work} = \|\text{proj}_{\overrightarrow{PQ}}\,\mathbf{v}\|\,\|\overrightarrow{PQ}\| = \left(\frac{4\sqrt{20}}{20}\right)(\sqrt{20}) = 14$$

54. Let $\mathbf{u} = \langle u_1, u_2\rangle$ and $\mathbf{v} = \langle v_1, v_2\rangle$.

$\mathbf{u} - \mathbf{v} = \langle u_1 - v_1, u_2 - v_2\rangle$

$$\begin{aligned}\|\mathbf{u} - \mathbf{v}\|^2 &= (u_1 - v_1)^2 + (u_2 - v_2)^2\\ &= u_1^2 - 2u_1v_1 + v_1^2 + u_2^2 - 2u_2v_2 + v_2^2\\ &= u_1^2 + u_2^2 + v_1^2 + v_2^2 - 2u_1v_1 - 2u_2v_2\\ &= \|\mathbf{u}\|^2 + \|\mathbf{v}\|^2 - 2(u_1v_1 + u_2v_2)\\ &= \|\mathbf{u}\|^2 + \|\mathbf{v}\|^2 - 2\mathbf{u}\cdot\mathbf{v}\end{aligned}$$

56. Let $\mathbf{u} \cdot \mathbf{v} = 0$ and $\mathbf{u} \cdot \mathbf{w} = 0$.

Then,
$$\begin{aligned}\mathbf{u} \cdot (c\mathbf{v} + d\mathbf{w}) &= \mathbf{u} \cdot c\mathbf{v} + \mathbf{u} \cdot d\mathbf{w}\\ &= c\mathbf{u} \cdot \mathbf{v} + d\mathbf{u} \cdot \mathbf{w}\\ &= c(0) + d(0)\\ &= 0.\end{aligned}$$

Thus for all scalars c and d, $\mathbf{u}$ is orthogonal to $c\mathbf{v} + d\mathbf{w}$.

Section 6.5 DeMoivre's Theorem

Solutions to Even-Numbered Exercises

2. $|-5| = \sqrt{5^2 + 0^2}$
$= \sqrt{25} = 5$

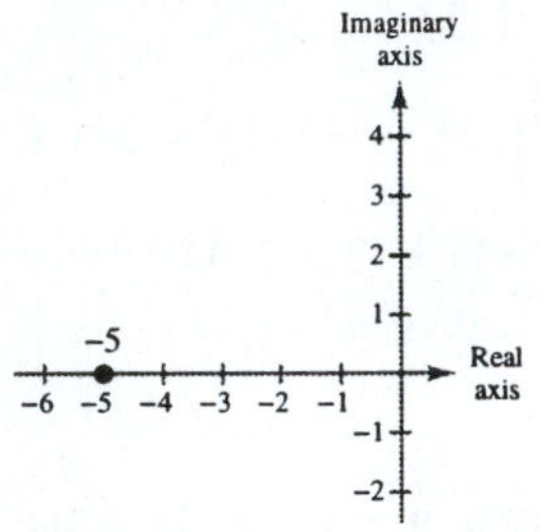

4. $|5 - 12i| = \sqrt{5^2 + (-12)^2}$
$= \sqrt{169} = 13$

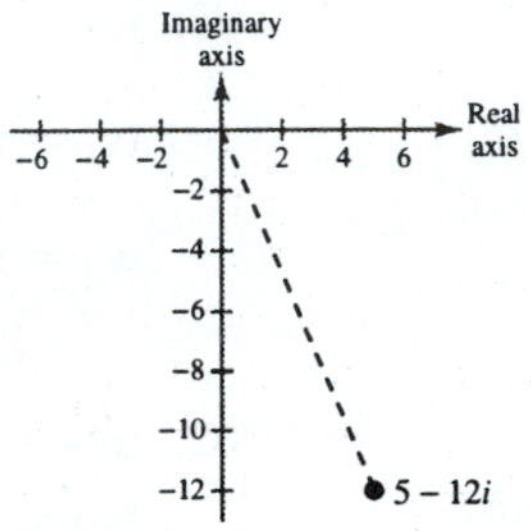

6. $|-8 + 3i| = \sqrt{(-8)^2 + (3)^2}$
$= \sqrt{73}$

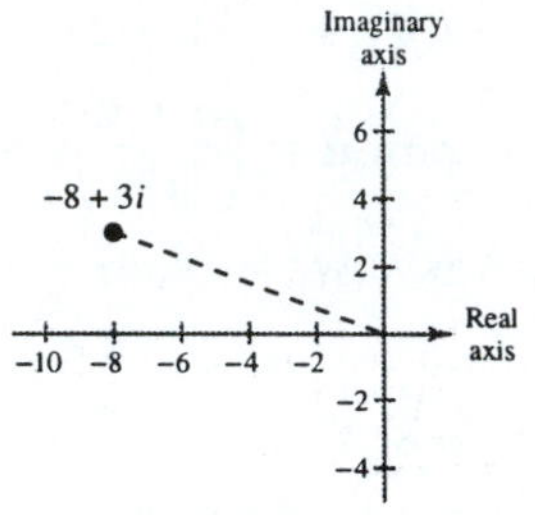

8. $z = 4$

$r = \sqrt{4^2 + 0^2} = \sqrt{16} = 4$

$\tan\theta = \dfrac{0}{4} = 0 \Rightarrow \theta = 0$

$z = 4(\cos 0 + i\sin 0)$

10. $z = -1 + \sqrt{3}i$

$r = \sqrt{(-1)^2 + (\sqrt{3})^2} = \sqrt{4} = 2$

$\tan\theta = \dfrac{\sqrt{3}}{-1} = -\sqrt{3} \Rightarrow \theta = \dfrac{2\pi}{3}$

$z = 2\left(\cos\dfrac{2\pi}{3} + i\sin\dfrac{2\pi}{3}\right)$

12. $z = 2 + 2i$

$r = \sqrt{2^2 + 2^2} = \sqrt{8} = 2\sqrt{2}$

$\tan\theta = \dfrac{2}{2} = 1 \Rightarrow \theta = \dfrac{\pi}{4}$

$z = 2\sqrt{2}\left(\cos\dfrac{\pi}{4} + i\sin\dfrac{\pi}{4}\right)$

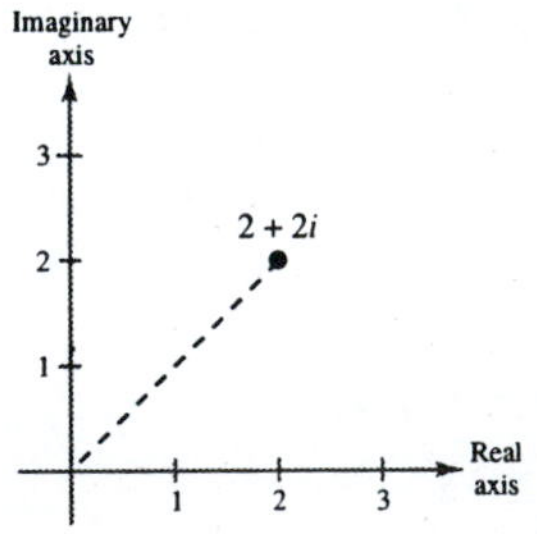

14. $z = -1 + \sqrt{3}i$

$r = \sqrt{(-1)^2 + (\sqrt{3})^2} = \sqrt{4} = 2$

$\tan\theta = \dfrac{\sqrt{3}}{-1} = -\sqrt{3} \Rightarrow 2 = \dfrac{2\pi}{3}$

$z = 2\left(\cos\dfrac{2\pi}{3} + i\sin\dfrac{2\pi}{3}\right)$

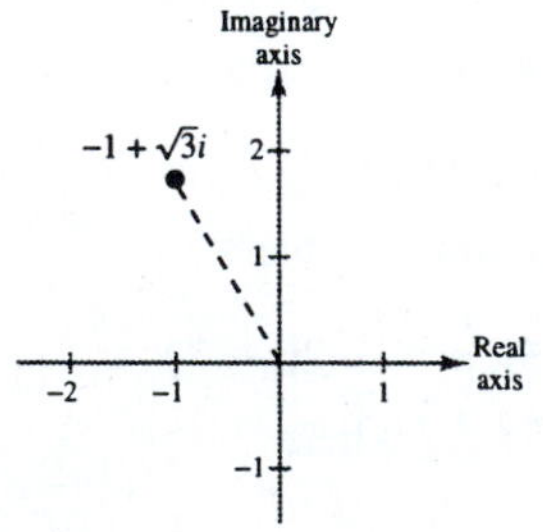

16. $z = \frac{5}{2}(\sqrt{3} - i)$

$$r = \sqrt{\left(\frac{5}{2}\sqrt{3}\right)^2 + \left(\frac{5}{2}(-1)\right)^2} = \sqrt{\frac{100}{4}} = \sqrt{25} = 5$$

$$\tan\theta = \frac{-1}{\sqrt{3}} = \frac{-\sqrt{3}}{3} \Rightarrow \theta = \frac{11\pi}{6}$$

$$z = 5\left(\cos\frac{11\pi}{6} + i\sin\frac{11\pi}{6}\right)$$

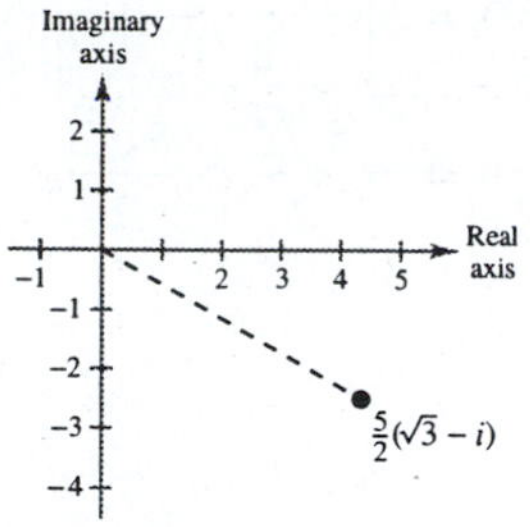

18. $z = 4 + 0i$

$$r = \sqrt{4^2 + 0^2} = \sqrt{16} = 4$$

$$\tan\theta = \frac{0}{4} = 0 \Rightarrow \theta = 0$$

$$z = 4(\cos 0 + i\sin 0)$$

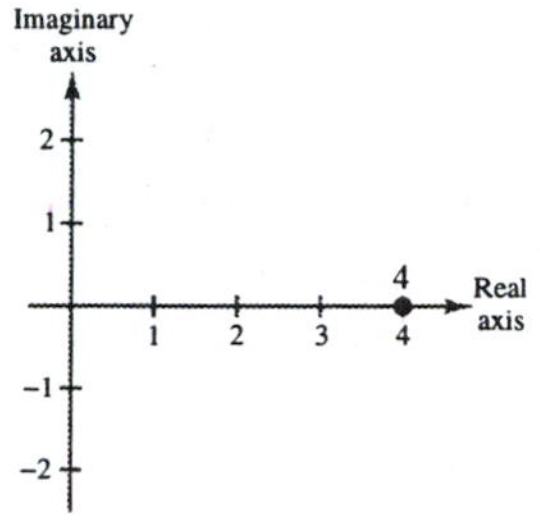

20. $z = 3 - i$

$$r = \sqrt{(3)^2 + (-1)^2} = \sqrt{10}$$

$$\tan\theta = \frac{-1}{3} = \theta \approx -18.4°$$

$$z = \sqrt{10}(\cos(-18.4°) + i\sin(-18.4°))$$

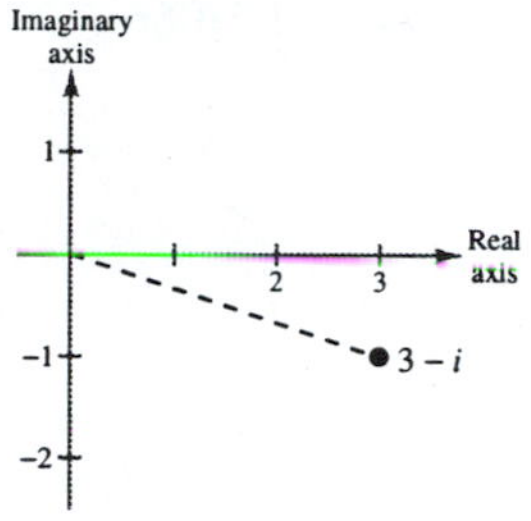

22. $z = 0 - 2i$

$$r = \sqrt{0^2 + (-2)^2} = \sqrt{4} = 2$$

$$\tan\theta = \frac{-2}{0}, \text{ undefined} \Rightarrow \theta = \frac{3\pi}{2}$$

$$z = 2\left(\cos\frac{3\pi}{2} + i\sin\frac{3\pi}{2}\right)$$

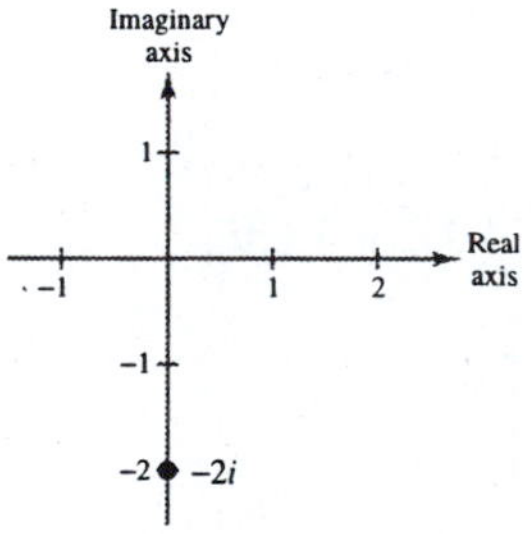

24. $z = 2\sqrt{2} - i$

$$r = \sqrt{(2\sqrt{2})^2 + (-1)^2} = \sqrt{9} = 3$$

$$\tan\theta = \frac{-1}{2\sqrt{2}} = -\frac{\sqrt{2}}{4} \Rightarrow \theta \approx (-19.5°)$$

$$z = 3(\cos(-19.5°) + i\sin(-19.5°))$$

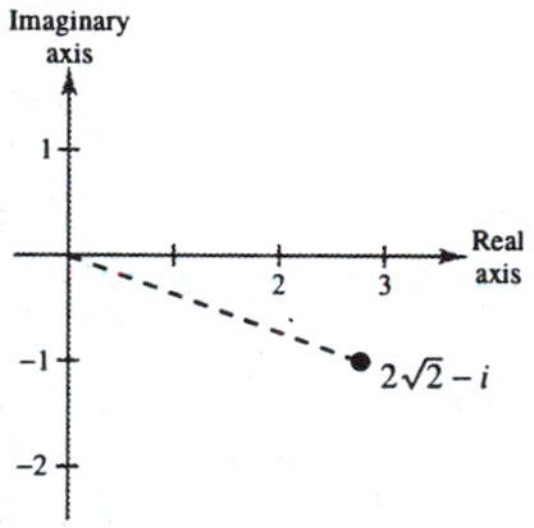

26. $z = 1 + 3i$

$$r = \sqrt{1^2 + 3^2} = \sqrt{10}$$

$$\tan\theta = \tfrac{3}{1} = 3 \Rightarrow \theta \approx 71.6°$$

$$z \approx \sqrt{10}(\cos 71.6° + i\sin 71.6°)$$

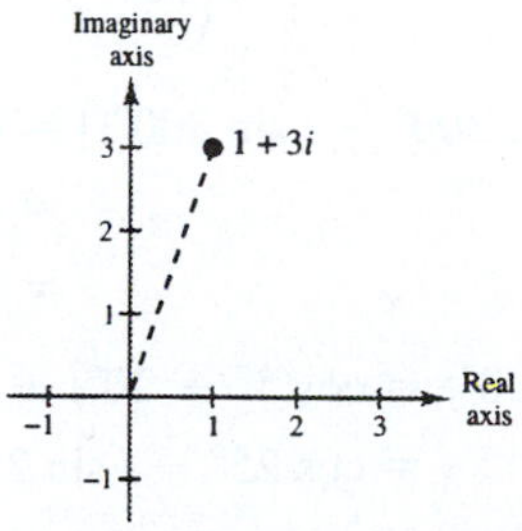

28. $-3 + i \approx 3.16 \angle 2.82$

$$= 3.16(\cos 2.82 + i\sin 2.82)$$

30. $-8 - 5\sqrt{3}i = 11.79 \angle -2.32$

or $11.79 \angle 3.97$

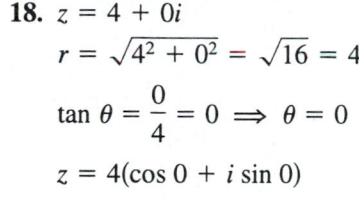

$$-8 - 5\sqrt{3}i = 11.79(\cos 3.97 + i\sin 3.97)$$

32. $5(\cos 135° + i \sin 135°) = 5\left[-\frac{\sqrt{2}}{2} + i\left(\frac{\sqrt{2}}{2}\right)\right]$

$= -\frac{5\sqrt{2}}{2} + \frac{5\sqrt{2}}{2}i$

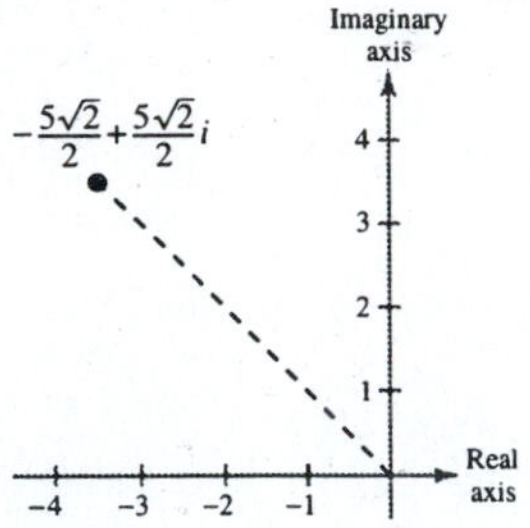

34. $\frac{3}{4}(\cos 315° + i \sin 315°) = \frac{3}{4}\left[\frac{\sqrt{2}}{2} + i\left(-\frac{\sqrt{2}}{2}\right)\right]$

$= \frac{3\sqrt{2}}{8} - \frac{3\sqrt{2}}{8}i$

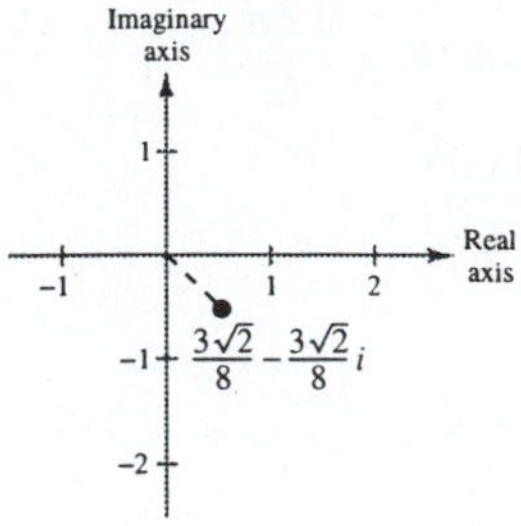

36. $8\left(\cos \frac{\pi}{12} + i \sin \frac{\pi}{12}\right) = 8(0.9659 + 0.2588i)$

$\approx 7.7274 + 2.0706i$

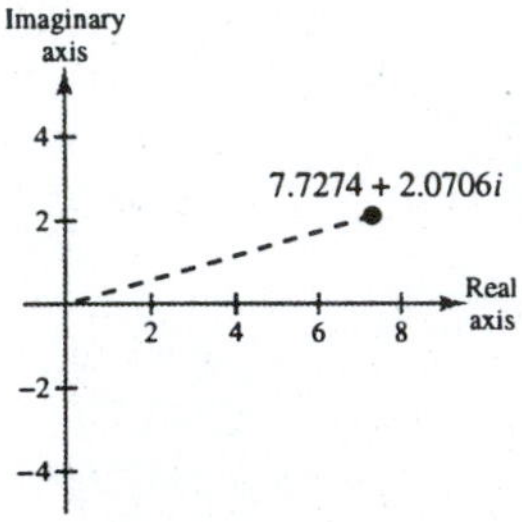

38. $7(\cos 0° + i \sin 0°) = 7$

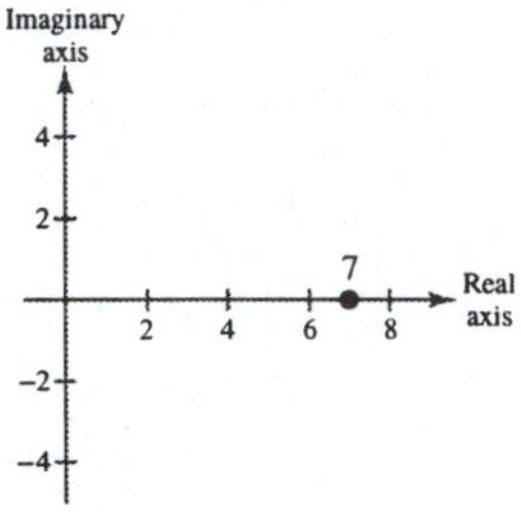

40. $6[\cos(230° 30') + i \sin(230° 30')] \approx -3.816 - 4.630i$

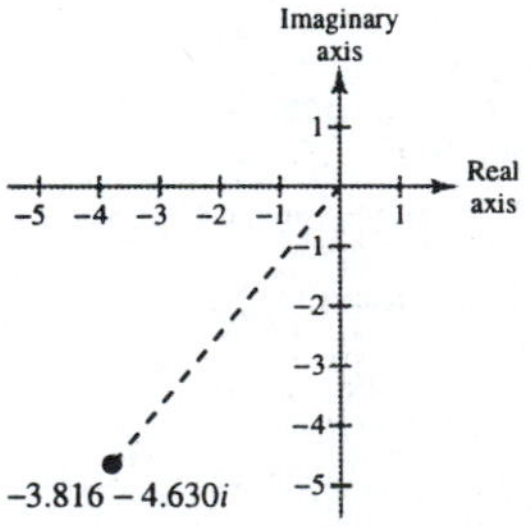

42. $12\left(\cos \frac{3\pi}{5} + i \sin \frac{3\pi}{5}\right) \approx -3.71 + 11.41i$

44. $9(\cos 58° + i \sin 58°) \approx 4.77 + 7.63i$

46. $\left[\frac{3}{2}\left(\cos \frac{\pi}{2} + i \sin \frac{\pi}{2}\right)\right]\left[6\left(\cos \frac{\pi}{4} + i \sin \frac{\pi}{4}\right)\right] = \left(\frac{3}{2}\right)(6)\left[\cos\left(\frac{\pi}{2} + \frac{\pi}{4}\right) + i \sin\left(\frac{\pi}{2} + \frac{\pi}{4}\right)\right]$

$= 9\left(\cos \frac{3\pi}{4} + i \sin \frac{3\pi}{4}\right)$

48. $[0.5(\cos 100° + i \sin 100°)][0.8(\cos 300° + i \sin 300°)] = (0.5)(0.8)[\cos(100° + 300°) + i \sin(100° + 300°)]$

$= 0.4(\cos 400° + i \sin 400°)$

$= 0.4(\cos 40° + i \sin 40°)$

50. $(\cos 5° + i \sin 5°)(\cos 20° + i \sin 20°) = \cos(5° + 20°) + i \sin(5° + 20°)$

$= \cos 25° + i \sin 25°$

52. $\frac{2(\cos 120° + i \sin 120°)}{4(\cos 40° + i \sin 40°)} = \frac{2}{4}[\cos(120° - 40°) + i \sin(120° - 40°)]$

$= \frac{1}{2}(\cos 80° + i \sin 80°)$

54. $\dfrac{5[\cos(4.3) + i\sin(4.3)]}{4[\cos(2.1) + i\sin(2.1)]} = \dfrac{5}{4}[\cos(4.3 - 2.1) + i\sin(4.3 - 2.1)]$

$= \dfrac{5}{4}[\cos(2.2) + i\sin(2.2)]$

56. $\dfrac{9(\cos 20° + i\sin 20°)}{5(\cos 75° + i\sin 75°)} = \dfrac{9}{5}[\cos(20° - 75°) + i\sin(20° - 75°)]$

$= \dfrac{9}{5}[\cos(-55°) + i\sin(55°)]$

$= \dfrac{9}{5}[\cos 305° + i\sin 305°]$

58. (a) $\sqrt{3} + i = 2(\cos 30° + i\sin 30°)$

$1 + i = \sqrt{2}(\cos 45° + i\sin 45°)$

(b) $(\sqrt{3} + i)(1 + i) = [2(\cos 30° + i\sin 30°)][\sqrt{2}(\cos 45° + i\sin 45°)]$

$= 2\sqrt{2}(\cos 75° + i\sin 75°)$

$= 2\sqrt{2}\left[\left(\dfrac{\sqrt{6} - \sqrt{2}}{4}\right) + \left(\dfrac{\sqrt{6} + \sqrt{2}}{4}\right)i\right]$

$= (\sqrt{3} - 1) + (\sqrt{3} + 1)i$

(c) $(\sqrt{3} + i)(1 + i) = \sqrt{3} + (\sqrt{3} + 1)i + i^2 = (\sqrt{3} - 1) + (\sqrt{3} + 1)i$

60. (a) $3 + 4i = 5(\cos 53.13° + i\sin 53.13°)$

$1 - \sqrt{3}i = 2(\cos 300° + i\sin 300°)$

(b) $\dfrac{3 + 4i}{1 - \sqrt{3}i} = \dfrac{5(\cos 53.13° + i\sin 53.13°)}{2(\cos 300° + i\sin 300°)}$

$= 2.5[\cos(-246.9°) + i\sin(-246.9°)]$

$= 2.5(\cos 113.13° + i\sin 113.13°)$

$= -0.9821 + 2.299i$

(c) $\dfrac{3 + 4i}{1 - \sqrt{3}i} = \dfrac{3 + 4i}{1 - \sqrt{3}i} \cdot \dfrac{1 + \sqrt{3}i}{1 + \sqrt{3}i}$

$= \dfrac{3 + (4 + 3\sqrt{3})i^2}{1 + 3}$

$= \dfrac{3 - 4\sqrt{3}}{4} + \dfrac{3 + 4\sqrt{3}}{4}i$

$\approx -0.9821 + 2.299i$

62. (a) $4i = 4(\cos 90° + i\sin 90°)$

$-4 + 2i = 2\sqrt{5}(\cos 153.4° + i\sin 153.4°)$

(b) $\dfrac{4i}{-4 + 2i} = \dfrac{4(\cos 90° + i\sin 90°)}{2\sqrt{5}(\cos 153.4° + i\sin 153.4°)}$

$= 2\sqrt{5}(\cos 296.6° + i\sin 296.6°)$

$\approx 0.400 - 0.800i$

(c) $\dfrac{4i}{-4 + 2i} = \dfrac{4i}{-4 + 2i} \cdot \dfrac{-4 - i}{-4 - 2i}$

$= \dfrac{8 - 16i}{20}$

$= \dfrac{2}{5} - \dfrac{4}{5}i$

$= 0.400 - 0.800i$

64. $z = r(\cos\theta + i\sin\theta)$

$\bar{z} = r(\cos\theta - i\sin\theta)$

$= r[\cos(-\theta) + i\sin(-\theta)]$

66. $z = r(\cos\theta + i\sin\theta)$

$-z = -r(\cos\theta + i\sin\theta)$

$= r(-\cos\theta - i\sin\theta)$

$= r[\cos(\theta + \pi) + i\sin(\theta + \pi)]$

68. Let $\theta = \dfrac{\pi}{6}$.

Let $z = x + iy$ such that:

$\tan\dfrac{\pi}{6} = \dfrac{y}{x}$

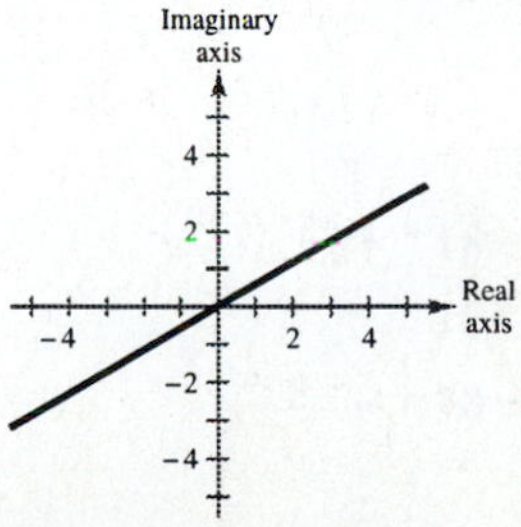

70. $(2+2i)^6 = \left[2\sqrt{2}\left(\cos\frac{\pi}{4} + i\sin\frac{\pi}{4}\right)\right]^6$

$= (2\sqrt{2})^6\left(\cos\frac{6\pi}{4} + i\sin\frac{6\pi}{4}\right)$

$= 512\left(\cos\frac{3\pi}{2} + i\sin\frac{3\pi}{2}\right)$

$= -512i$

72. $(1-i)^{12} = \left[\sqrt{2}\left(\cos\frac{7\pi}{4} + i\sin\frac{7\pi}{4}\right)\right]^{12}$

$= (\sqrt{2})^{12}(\cos 21\pi + i\sin 21\pi)$

$= 64(\cos\pi + i\sin\pi)$

$= 64(-1)$

$= -64$

74. $4(1-\sqrt{3}i)^3 = 4\left[2\left(\cos\frac{5\pi}{3} + i\sin\frac{5\pi}{3}\right)\right]^3$

$= 4[2^3(\cos 5\pi + i\sin 5\pi)]$

$= 32(-1)$

$= -32$

76. $[3(\cos 150^\circ + i\sin 150^\circ)]^4 = 3^4(\cos 600^\circ + i\sin 600^\circ)$

$= 81(\cos 240^\circ + i\sin 240^\circ)$

$= 81(-\cos 60^\circ - i\sin 60^\circ)$

$= -\frac{81}{2} - \frac{81\sqrt{3}}{2}i$

78. $\left[2\left(\cos\frac{\pi}{2} + i\sin\frac{\pi}{2}\right)\right]^8 = 2^8(\cos 4\pi + i\sin 4\pi)$

$= 256(\cos 0 + i\sin 0)$

$= 256$

80. $(\cos 0 + i\sin 0)^{20} = \cos 0 + i\sin 0$

$= 1$

82. $(\sqrt{5} - 4i)^3 = -43\sqrt{5} + 4i$

84. $\left[2\left(\cos\frac{\pi}{10} + i\sin\frac{\pi}{10}\right)\right]^5 = 32\left(\cos\frac{\pi}{2} i\sin\frac{\pi}{2}\right)$

$= 32i$

86. $2^{-1/4}(1-i)$ is a fourth root of -2 if $-2 = [2^{-1/4}(1-i)]^4$.

$[2^{-1/4}(1-i)]^4 = (2^{-1/4})^4(1-i)^4$

$= 2^{-1}(1-i)^4$

$= \frac{1}{2}(1-i)^2(1-i)^2$

$= \frac{1}{2}(-2i)(-2i)$

$= \frac{1}{2}(4i^2)$

$= \frac{1}{2}(-4) = -2$

88. (a) In trigonometric form we have:

$3(\cos 45^\circ + i\sin 45^\circ)$

$3(\cos 135^\circ + i\sin 135^\circ)$

$3(\cos 225^\circ + i\sin 225^\circ)$

$3(\cos 315^\circ + i\sin 315^\circ)$

(b) There are four roots evenly spaced around a circle of radius 3. Therefore, they represent the fourth roots of some number of modulus 81. Raising them to the fourth power shows that they are all fourth roots of -81.

(c) $[3(\cos 45^\circ + i\sin 45^\circ)]^4 = -81$

$[3(\cos 135^\circ + i\sin 135^\circ)]^4 = -81$

$[3(\cos 225^\circ + i\sin 225^\circ)]^4 = -81$

$[3(\cos 315^\circ + i\sin 315^\circ)]^4 = -81$

90. (a) Square roots of $16(\cos 60^\circ + i \sin 60^\circ)$:

$$\sqrt{16}\left[\cos\left(\frac{60^\circ + 360^\circ k}{2}\right) + i \sin\left(\frac{60^\circ + 360^\circ k}{2}\right)\right],\ k = 0, 1$$

$4(\cos 30^\circ + i \sin 30^\circ)$

$4(\cos 210^\circ + i \sin 210^\circ)$

(b)

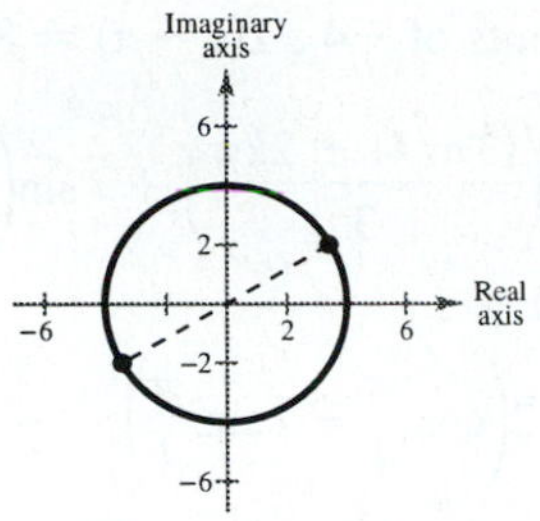

(c) $4\left(\frac{\sqrt{3}}{2} + \frac{1}{2}i\right) = 2\sqrt{3} + 2i$

$4\left(-\frac{\sqrt{3}}{2} - \frac{1}{2}i\right) = -2\sqrt{3} - 2i$

92. (a) Fifth roots of $32\left(\cos \frac{5\pi}{6} + i \sin \frac{5\pi}{6}\right)$:

$$\sqrt[5]{32}\left[\cos\left(\frac{(5\pi/6) + 2k\pi}{5}\right) + i \sin\left(\frac{(5\pi/6) + 2k\pi}{5}\right)\right]$$

$k = 0, 1, 2, 3, 4$

$k = 0$: $2\left(\cos \frac{\pi}{6} + i \sin \frac{\pi}{6}\right)$

$k = 1$: $2\left(\cos \frac{17\pi}{30} + i \sin \frac{17\pi}{30}\right)$

$k = 2$: $2\left(\cos \frac{29\pi}{30} + i \sin \frac{29\pi}{30}\right)$

$k = 3$: $2\left(\cos \frac{41\pi}{30} + i \sin \frac{41\pi}{30}\right)$

$k = 4$: $2\left(\cos \frac{53\pi}{30} + i \sin \frac{53\pi}{30}\right)$

(b)

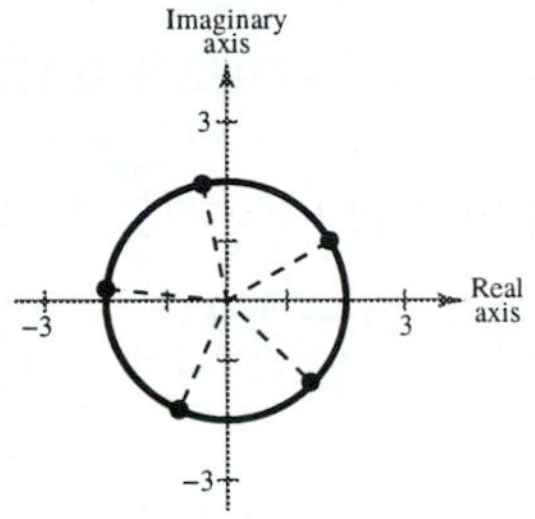

(c) $1.732 + i, -0.4158 + 1.956i, -1.989 + 0.2091i,$
$-0.8134 - 1.827i, 1.486 - 1.338i$

94. (a) Fourth roots of $625i = 625\left(\cos \frac{\pi}{2} + i \sin \frac{\pi}{2}\right)$:

$$\sqrt[4]{625}\left[\cos\left(\frac{(\pi/2) + 2k\pi}{4}\right) + i \sin\left(\frac{(\pi/2) + 2k\pi}{4}\right)\right]$$

$k = 0, 1, 2, 3$

$k = 0$: $5\left(\cos \frac{\pi}{8} + i \sin \frac{\pi}{8}\right)$

$k = 1$: $5\left(\cos \frac{5\pi}{8} + i \sin \frac{5\pi}{8}\right)$

$k = 2$: $5\left(\cos \frac{9\pi}{8} + i \sin \frac{9\pi}{8}\right)$

$k = 3$: $5\left(\cos \frac{13\pi}{8} + i \sin \frac{13\pi}{8}\right)$

(b)

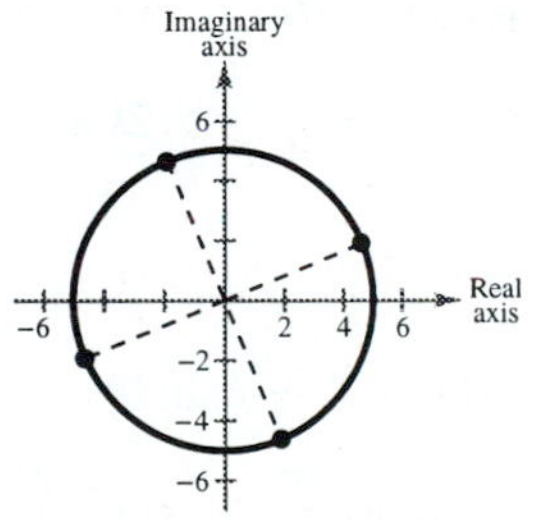

(c) $4.619 + 1.913i, -1.913 + 4.619i, -4.619 - 1.913i,$
$1.913 - 4.619i$

96. (a) Cube roots of $-4\sqrt{2}(1 - i) = 8\left(\cos\dfrac{3\pi}{4} + i\sin\dfrac{3\pi}{4}\right)$:

$$\sqrt[3]{8}\left[\cos\left(\frac{(3\pi/4) + 2k\pi}{3}\right) + i\sin\left(\frac{(3\pi/4) + 2k\pi}{3}\right)\right]$$

$k = 0, 1, 2$

$k = 0$: $2\left(\cos\dfrac{\pi}{4} + i\sin\dfrac{\pi}{4}\right)$

$k = 1$: $2\left(\cos\dfrac{11\pi}{12} + i\sin\dfrac{11\pi}{12}\right)$

$k = 2$: $2\left(\cos\dfrac{19\pi}{12} + i\sin\dfrac{19\pi}{12}\right)$

(b)

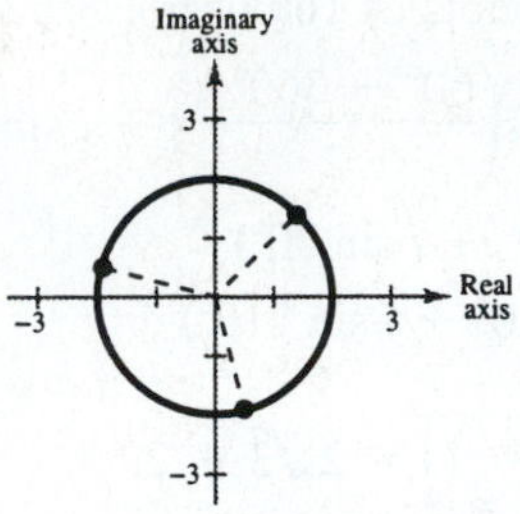

(c) $1.414 + 1.414i$, $-1.932 + 0.5176i$, $0.5176 - 1.9319i$

98. (a) Fourth roots of $i = \cos\dfrac{\pi}{2} + i\sin\dfrac{\pi}{2}$:

$$\sqrt[4]{1}\left[\cos\left(\frac{(\pi/2) + 2k\pi}{4}\right) + i\sin\left(\frac{(\pi/2) + 2k\pi}{4}\right)\right]$$

$k = 0, 1, 2, 3$

$k = 0$: $\cos\dfrac{\pi}{8} + i\sin\dfrac{\pi}{8}$

$k = 1$: $\cos\dfrac{5\pi}{8} + i\sin\dfrac{5\pi}{8}$

$k = 2$: $\cos\dfrac{9\pi}{8} + i\sin\dfrac{9\pi}{8}$

$k = 3$: $\cos\dfrac{13\pi}{8} + i\sin\dfrac{13\pi}{8}$

(b)

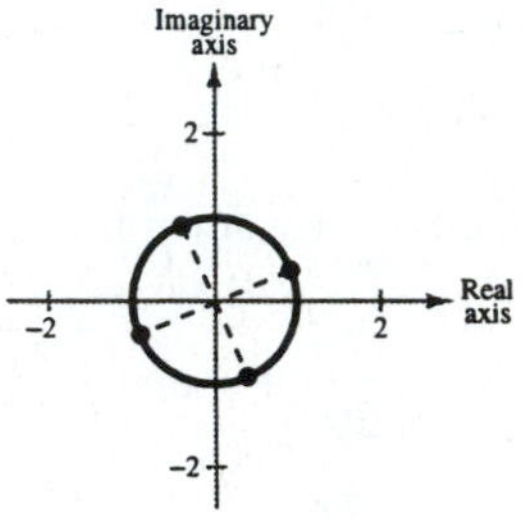

(c) $0.9239 + 0.3827i$, $-0.3827 + 0.9239i$,
$-0.9239 - 0.3827i$, $0.3827 - 0.9239i$

100. (a) Cube roots of $1000 = 1000(\cos 0 + i\sin 0)$:

$$\sqrt[3]{1000}\left(\cos\frac{2k\pi}{3} + i\sin\frac{2k\pi}{3}\right)$$

$k = 0, 1, 2$

$k = 0$: $10(\cos 0 + i\sin 0)$

$k = 1$: $10\left(\cos\dfrac{2\pi}{3} + i\sin\dfrac{2\pi}{3}\right)$

$k = 2$: $10\left(\cos\dfrac{4\pi}{3} + i\sin\dfrac{4\pi}{3}\right)$

(b)

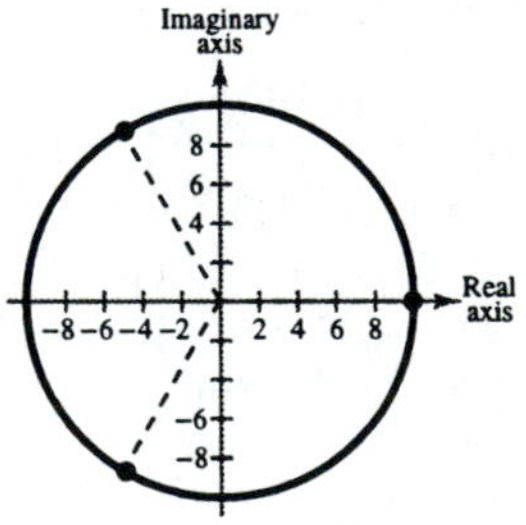

(c) 10, $-5 + 5\sqrt{3}i$, $-5 - 5\sqrt{3}i$

102. (a) The fourth roots of $-4 = 4(\cos 180° + i\sin 180°)$:

$\sqrt{2}(\cos 45° + i\sin 45°)$

$\sqrt{2}(\cos 135° + i\sin 135°)$

$\sqrt{2}(\cos 225° + i\sin 225°)$

$\sqrt{2}(\cos 315° + i\sin 315°)$

(b)

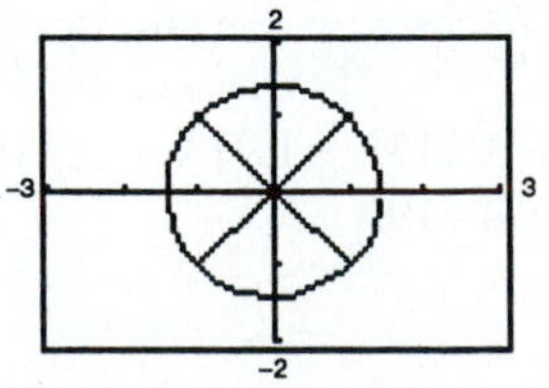

(c) $1 + i$, $-1 + i$, $-1 - i$, $1 - i$

104. (a) The sixth roots of $64i = 64(\cos 90° + i \sin 90°)$:

$2(\cos 15° + i \sin 15°)$

$2(\cos 75° + i \sin 75°)$

$2(\cos 135° + i \sin 135°)$

$2(\cos 195° + i \sin 195°)$

$2(\cos 255° + i \sin 255°)$

$2(\cos 315° + i \sin 315°)$

(b)

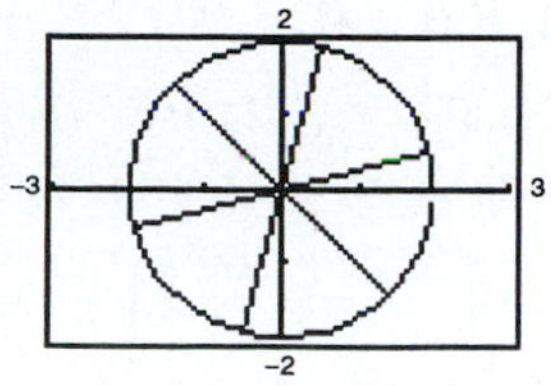

(c) $1.93 + 0.52i, 0.52 + 1.93i, -\sqrt{2} + \sqrt{2}i, -1.93 - 0.52i,$
$-0.52 - -1.93i, \sqrt{2} - \sqrt{2}i$

106. $x^3 + 1 = 0$

$x^3 = -1$

The solutions are the cube roots of $-1 = \cos \pi + i \sin \pi$:

$$\cos\left(\frac{\pi + 2k\pi}{3}\right) + i \sin\left(\frac{\pi + 2k\pi}{3}\right)$$

$k = 0, 1, 2$

$k = 0$: $\cos \frac{\pi}{3} + i \sin \frac{\pi}{3} = \frac{1}{2} + \frac{\sqrt{3}}{2}i$

$k = 1$: $\cos \pi + i \sin \pi = -1$

$k = 2$: $\cos \frac{5\pi}{3} + i \sin \frac{5\pi}{3} = \frac{1}{2} - \frac{\sqrt{3}}{2}i$

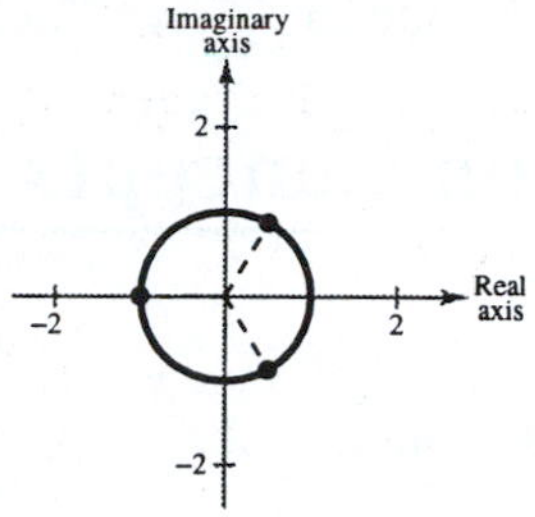

108. $x^4 - 81 = 0$

$x^4 = 81$

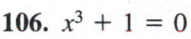

The solutions are the fourth roots of 81:

$$\sqrt[4]{81}\left(\cos \frac{0 + 2\pi k}{4} + i \sin \frac{0 + 2\pi k}{4}\right)$$

$k = 0, 1, 2, 3$

$k = 0$: $3(\cos 0 + i \sin 0) = 3$

$k = 1$: $3\left(\cos \frac{\pi}{2} + i \sin \frac{\pi}{2}\right) = 3i$

$k = 2$: $3(\cos \pi + i \sin \pi) = -3$

$k = 3$: $3\left(\cos \frac{3\pi}{2} + i \sin \frac{3\pi}{2}\right) = -3i$

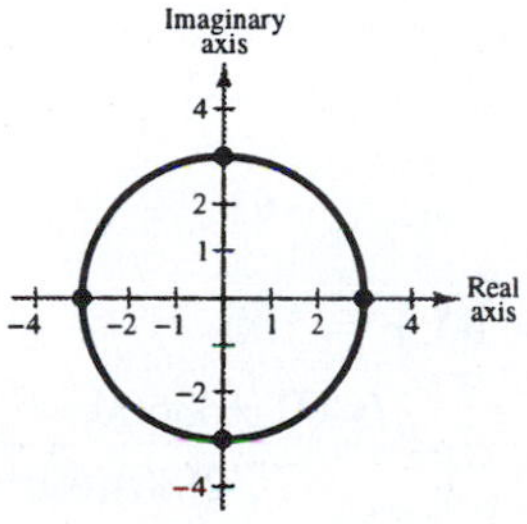

110. $x^6 - 64i = 0$

$x^6 = 64i$

The solutions are the sixth roots of $64i$:

$$\sqrt[6]{64}\left[\cos\left(\frac{(\pi/2) + 2k\pi}{6}\right) + i \sin\left(\frac{(\pi/2) + 2k\pi}{6}\right)\right]$$

$k = 0, 1, 2, 3, 4, 5$

$k = 0$: $2\left(\cos \frac{\pi}{12} + i \sin \frac{\pi}{12}\right) \approx 1.932 + 0.5176i$

$k = 1$: $2\left(\cos \frac{5\pi}{12} + i \sin \frac{5\pi}{12}\right) \approx 0.5176 + 1.932i$

$k = 2$: $2\left(\cos \frac{3\pi}{4} + i \sin \frac{3\pi}{4}\right) \approx -1.414 + 1.414i$

$k = 3$: $2\left(\cos \frac{13\pi}{12} + i \sin \frac{13\pi}{12}\right) \approx -1.932 - 0.5176i$

$k = 4$: $2\left(\cos \frac{17\pi}{12} + i \sin \frac{17\pi}{12}\right) \approx -0.5176 - 1.932i$

$k = 5$: $2\left(\cos \frac{7\pi}{4} + i \sin \frac{7\pi}{4}\right) \approx 1.414 - 1.414i$

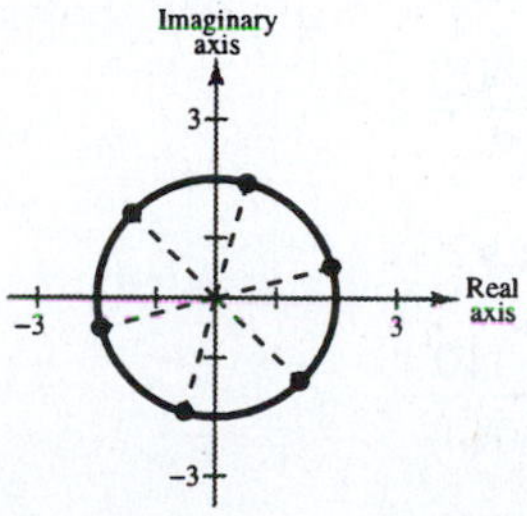

112. $x^4 + (1 + i) = 0$

$$x^4 = -1 - i = \sqrt{2}(\cos 225° + i \sin 225°)$$

The solutions are the fourth roots of $-1 - i$:

$$\sqrt[4]{\sqrt{2}}\left[\cos\left(\frac{225° + 360°k}{4}\right) + i \sin\left(\frac{225° + 360°k}{4}\right)\right]$$

$k = 0, 1, 2, 3$

$k = 0$: $\sqrt[8]{2}(\cos 56.25° + i \sin 56.25°) \approx 0.6059 + 0.9067i$

$k = 1$: $\sqrt[8]{2}(\cos 146.25° + i \sin 146.25°) \approx -0.9067 - 0.6059i$

$k = 2$: $\sqrt[8]{2}(\cos 236.25° + i \sin 236.25°) \approx -0.6059 - 0.9067i$

$k = 3$: $\sqrt[8]{2}(\cos 326.25° + i \sin 326.25°) \approx 0.9067 - 0.6059i$

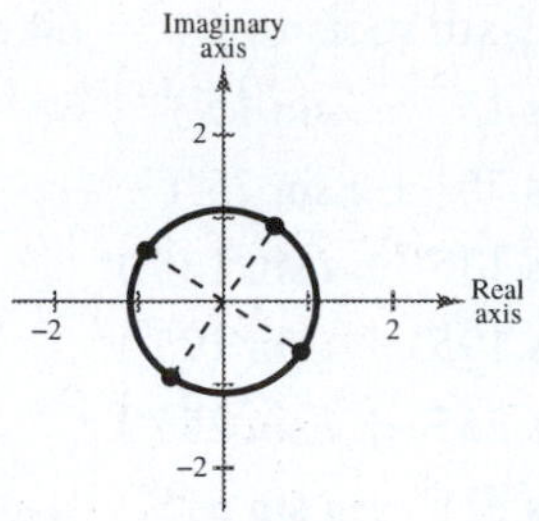

❑ Focus on Concepts

1. $\dfrac{a}{\sin A} = \dfrac{b}{\sin B} = \dfrac{c}{\sin C}$

2. $a^2 = b^2 + c^2 - 2bc \cos A$

$b^2 = a^2 + c^2 - 2ac \cos B$

$c^2 = a^2 + b^2 - 2ab \cos C$

3. True

4. Pythagorean Theorem

5. False. There may be no solution, one solution, or two solutions.

6. Direction and magnitude

7. A, C

8. a. The angle between the vectors is acute.

9. If $k > 0$, the direction is the same and the magnitude is k times as great.

If $k < 0$, the result is a vector in the opposite direction and the magnitude is k times as great.

10. The diagonal of the parallelogram with **u** and **v** as its adjacent sides.

11. b. Visualize the sum of **u** and −**v**.

12. $z_1 \cdot z_2 = -4, \dfrac{z_1}{z_2} = -i$

13. (a) 3

(b) 120°, 210°, 300°

❑ Review Exercises for Chapter 6

Solutions to Even-Numbered Exercises

2. Given: $a = 6, b = 9, c = 45°$

$c^2 = a^2 + b^2 - 2ab \cos 45° \approx 36 + 81 - 2(6)(9)(0.7071) \approx 40.63 \Longrightarrow 6.374$

$$\cos B = \frac{a^2 + c^2 - b^2}{2ac} \approx \frac{36 + 40.63 - 81}{2(6)(6.374)} \approx -0.0571 \Longrightarrow B \approx 93.3°$$

$A \approx 180° - 45° - 93.3° = 41.7°$

4. Given: $B = 110°, C = 30°, c = 10.5$

$A = 180° - 110° - 30° = 40°$

$$a = \frac{c \sin A}{\sin C} = \frac{10.5(\sin 40°)}{\sin 30°} \approx \frac{10.5(0.6428)}{0.5} \approx 13.5$$

$$b = \frac{c \sin B}{\sin C} = \frac{10.5(\sin 110°)}{\sin 30°} \approx \frac{10.5(0.9397)}{0.5} \approx 19.7$$

6. Given: $a = 80, b = 60, c = 100$

$$\cos C = \frac{a^2 + b^2 - c^2}{2ab} = \frac{6400 + 3600 - 10{,}000}{2(80)(60)} = 0 \Longrightarrow C = 90°$$

$$\sin A = \frac{80}{100} = 0.8 \Longrightarrow A \approx 53.1°$$

$$\sin B = \frac{60}{100} = 0.6 \Longrightarrow B \approx 36.9°$$

8. Given: $A = 130°, a = 50, b = 30$

$$\sin B = \frac{b \sin A}{a} = \frac{30 \sin 130°}{50} \approx \frac{30(0.7660)}{50} \approx 0.4596 \Longrightarrow B \approx 27.4°$$

$C \approx 180° - 130° - 27.4° = 22.6°$

$c^2 = a^2 + b^2 - 2ab \cos C \approx 50^2 + 30^2 - 2(50)(30)(0.9232) \approx 630.4 \Longrightarrow c \approx 25.11$

10. Given: $C = 50°, a = 25, c = 22$

$$\sin A = \frac{a \sin C}{c} = \frac{25 \sin 50°}{22} \approx \frac{25(0.7660)}{22} \approx 0.8705 \Longrightarrow A \approx 60.5° \text{ or } 119.5°$$

Case 1:

$A \approx 60.5°$

$B \approx 180° - 50° - 60.5° = 69.5°$

$$b = \frac{c \sin B}{\sin C} \approx \frac{22(0.9367)}{0.7660} \approx 26.90$$

Case 2:

$A \approx 119.5°$

$B \approx 180° - 50° - 119.5° = 10.5°$

$$b = \frac{c \sin B}{\sin C} \approx 5.234$$

12. Given: $B = 150°, a = 64, b = 10$

$$\sin A = \frac{a \sin B}{b} = \frac{64 \sin 150°}{10} \approx 3.2 \Longrightarrow \text{no triangle formed}$$

No solution

14. Given: $a = 2.5, b = 15.0, c = 4.5$

Since $a + c < b$, a triangle in not formed.

No solution

16. Given: $B = 90°, a = 5, c = 12$

$b = \sqrt{12^2 + 5^2} = \sqrt{169} = 13$

$A = \arctan \frac{5}{12} \approx 22.6°$

$C = \arctan \frac{12}{5} \approx 67.4°$

18. $a = 15, b = 8, c = 10$

$$s = \frac{15 + 8 + 10}{2} = 16.5$$

Area $= \sqrt{16.5(1.5)(8.5)(6.5)} \approx 36.98$

20. $B = 80°, a = 4, c = 8$

$$\text{Area} = \frac{1}{2}ac \sin B = \frac{1}{2}(4)(8)(0.9848) = 15.76$$

22. $a^2 = 5^2 + 8^2 - 2(5)(8) \cos 152°$

$\approx 159.6 \Longrightarrow a \approx 12.63$ ft

$b^2 = 5^2 + 8^2 - 2(5)(8) \cos 28°$

$\approx 18.36 \Longrightarrow b \approx 4.285$ ft

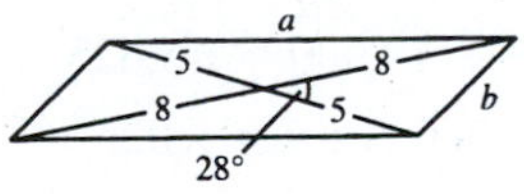

24. $b^2 = a^2 + c^2 - 2ac \cos B$

$= 300^2 + 425^2 - 2(300)(425) \cos(180° - 65°)$

≈ 378392.66

$b \approx 615.1$ meters

26. $$\frac{a}{\sin 75°} = \frac{400}{\sin 37.5°}$$

$$a = \frac{400 \sin 75°}{\sin 37.5°} \approx 634.7 \text{ ft}$$

$$\sin 67.5° = \frac{w}{a}$$

$w = 634.7 \sin 67.5° \approx 586.4$ ft

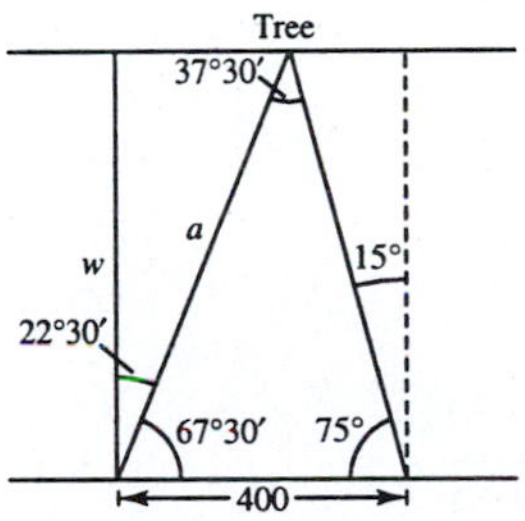

28. Initial point: $(0, 1)$

Terminal point: $\left(6, \frac{7}{2}\right)$

$\mathbf{v} = \left\langle 6 - 0, \frac{7}{2} - 1 \right\rangle = \left\langle 6, \frac{5}{2} \right\rangle$

30. Initial point: $(1, 5)$

Terminal point: $(15, 9)$

$\mathbf{v} = \langle 15 - 1, 9 - 5 \rangle = \langle 14, 4 \rangle$

32. $\left\langle \frac{1}{2}\cos 225°, \frac{1}{2}\sin 225° \right\rangle = \left\langle -\frac{\sqrt{2}}{4}, -\frac{\sqrt{2}}{4} \right\rangle$

34. $\mathbf{v} = 4\mathbf{i} - \mathbf{j}$

$\|\mathbf{v}\| = \sqrt{4^2 + (-1)^2} = \sqrt{17}$

$\tan\theta = \frac{-1}{4} = -\frac{1}{4} \Rightarrow \theta \approx 346°$, since θ is in Quadrant IV.

$\mathbf{v} = \sqrt{17}(\mathbf{i}\cos 346° + \mathbf{j}\sin 346°)$

36. $\mathbf{v} = 10\mathbf{i} + 3\mathbf{j}$

$3\mathbf{v} = 30\mathbf{i} + 9\mathbf{j}$

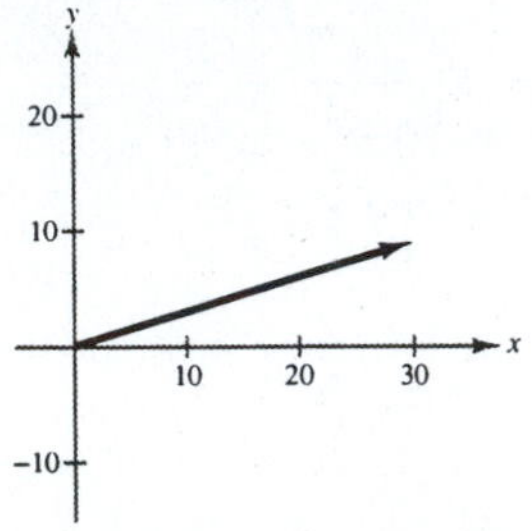

38. $\mathbf{v} = 10\mathbf{i} + 3\mathbf{j}$

$\frac{1}{2}\mathbf{v} = 5\mathbf{i} + \frac{3}{2}\mathbf{j}$

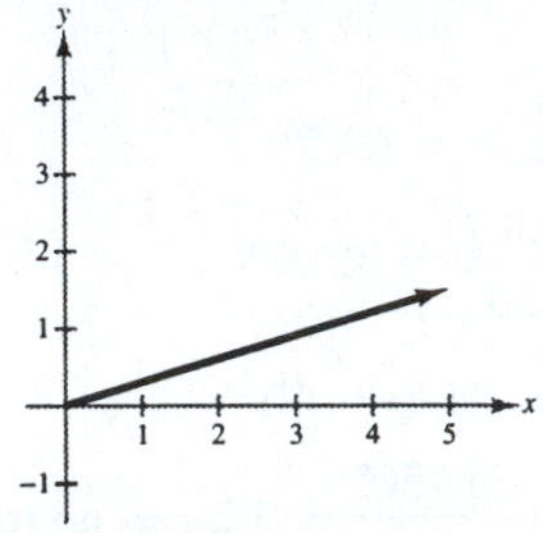

40. $\mathbf{u} = 12[(\cos 82°)\mathbf{i} + (\sin 82°)\mathbf{j}]$

$\mathbf{v} = 8[(\cos(-12°))\mathbf{i} + (\sin(-12°))\mathbf{j}]$

$\mathbf{u} + \mathbf{v} \approx 9.4953\mathbf{i} + 10.2199\mathbf{j}$

$\|\mathbf{u} + \mathbf{v}\| \approx 13.95$

$\tan\theta = \frac{10.2199}{9.4953} \Rightarrow \theta \approx 47.11°$

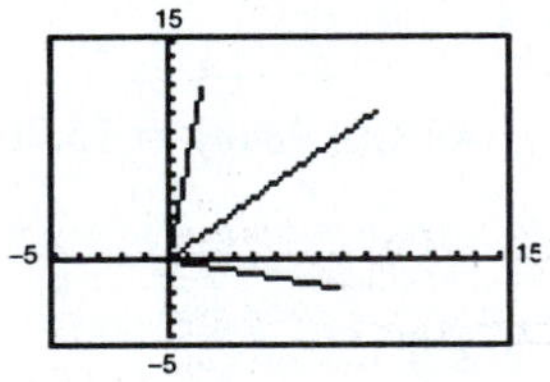

42. Force One: $\mathbf{u} = 85\mathbf{i}$

Force Two: $\mathbf{v} = 50\cos 15°\mathbf{i} + 50\sin 15°\mathbf{j}$

Resultant Force:

$\mathbf{u} + \mathbf{v} = (85 + 50\cos 15°)\mathbf{i} + (50\sin 15°)\mathbf{j}$

$\|\mathbf{u} + \mathbf{v}\| = \sqrt{(85 + 50\cos 15°)^2 + (50\sin 15°)^2}$

$= \sqrt{85^2 + 8500\cos 15° + 50^2}$

$= 133.92$ lb

$\tan\theta = \frac{50\sin 15°}{85 + 50\cos 15°} \Rightarrow \theta \approx 5.5°$ from the 85-pound force.

44. $|\overrightarrow{AC}|$ = force in direction of the slope.

$|\overrightarrow{AC}| = 500\sin 12° \approx 104.0$ lb

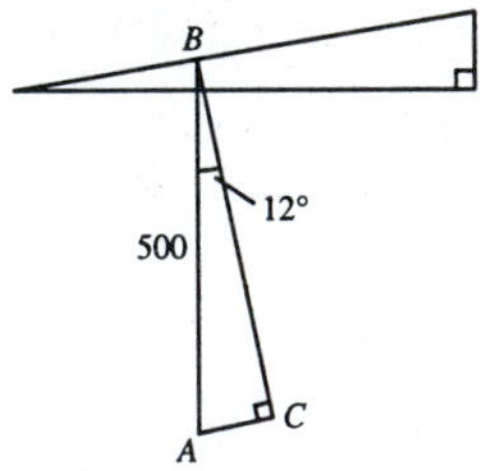

46. Force One: $\mathbf{u} = 60\mathbf{i}$

Force Two: $\mathbf{v} = 100\cos\theta\mathbf{i} + 100\sin\theta\mathbf{j}$

Resultant Force:

$\mathbf{u} + \mathbf{v} = (60 + 100\cos\theta)\mathbf{i} + 100\sin\theta\mathbf{j}$

$3{,}600 + 12{,}000\cos\theta + 10{,}000 = 15{,}625$

$12{,}000\cos\theta = 2{,}025$

$\cos\theta = \frac{2{,}025}{12{,}000}$

$= 0.16875$

$\theta \approx 80.3°$

48. $P(0, 3), Q(5, -8)$

$\overrightarrow{PQ} = \langle 5 - 0, -8 - 3\rangle = \langle 5, -11\rangle$

$\|\overrightarrow{PQ}\| = \sqrt{5^2 + (-11)^2} = \sqrt{146}$

$\frac{\overrightarrow{PQ}}{\|\overrightarrow{PQ}\|} = \frac{1}{\sqrt{146}}\langle 5, -11\rangle$

50. $\mathbf{u} = \langle 8, 5\rangle, \mathbf{v} = \langle -2, 4\rangle$

$\mathbf{u} \cdot \mathbf{v} = 8(-2) + 5(4) = 4 \neq 0$

u and **v** are not orthogonal.

$\mathbf{u} \neq k\mathbf{v} \Rightarrow$ **u** and **v** are not parallel.

Neither

52. $\mathbf{u} = \langle -6, -3\rangle, \mathbf{v} = \langle 4, 2\rangle$

$$\cos\theta = \frac{\mathbf{u}\cdot\mathbf{v}}{\|\mathbf{u}\|\,\|\mathbf{v}\|} = \frac{-30}{(\sqrt{45})(\sqrt{20})} = -1$$

$\theta = 180°$

54. $\mathbf{u} = \langle 3, 1\rangle, \mathbf{v} = \langle 4, 5\rangle$

$$\cos\theta = \frac{\mathbf{u}\cdot\mathbf{v}}{\|\mathbf{u}\|\,\|\mathbf{v}\|} = \frac{17}{(\sqrt{10})(\sqrt{41})} \Rightarrow \theta \approx 32.9°$$

56. $\mathbf{u} = \langle 5, 6\rangle, \mathbf{v} = \langle 10, 0\rangle$

$$\text{proj}_{\mathbf{v}}\mathbf{u} = \left(\frac{\mathbf{u}\cdot\mathbf{v}}{\|\mathbf{v}\|^2}\right)\mathbf{v} = \frac{50}{100}\langle 10, 0\rangle = \langle 5, 0\rangle$$

58. $\mathbf{u} = \langle -3, 5\rangle, \mathbf{v} = \langle -5, 2\rangle$

$$\text{proj}_{\mathbf{v}}\mathbf{u} = \left(\frac{\mathbf{u}\cdot\mathbf{v}}{\|\mathbf{v}\|^2}\right)\mathbf{v} = \frac{25}{29}\langle -5, 2\rangle$$

60. $r = 6, \theta = 150°$

$-3\sqrt{3} + 3i = 6(\cos 150° + i\sin 150°)$

62. $r = 7, \theta = 180°$

$-7 = 7(\cos 180° + i\sin 180°)$

64. $$24(\cos 330° + i\sin 330°) = 24\left(\frac{\sqrt{3}}{2} - \frac{1}{2}i\right) = 12\sqrt{3} - 12i$$

66. $$8\left(\cos\frac{5\pi}{6} + i\sin\frac{5\pi}{6}\right) = 8\left(-\frac{\sqrt{3}}{2} + \frac{1}{2}i\right) = -4\sqrt{3} + 4i$$

68. (a) $$z_1 = -3(1 + i) = 3\sqrt{2}\left(\cos\frac{5\pi}{4} + i\sin\frac{5\pi}{4}\right)$$

$$z_2 = 2(\sqrt{3} + i) = 4\left(\cos\frac{\pi}{6} + i\sin\frac{\pi}{6}\right)$$

(b) $$z_1z_2 = \left[3\sqrt{2}\left(\cos\frac{5\pi}{4} + i\sin\frac{5\pi}{4}\right)\right]\left[4\left(\cos\frac{\pi}{6} + i\sin\frac{\pi}{6}\right)\right] = 12\sqrt{2}\left(\cos\frac{17\pi}{12} + i\sin\frac{17\pi}{12}\right)$$

$$\frac{z_1}{z_2} = \frac{3\sqrt{2}[\cos(5\pi/4) + i\sin(5\pi/4)]}{4[\cos(\pi/6) + i\sin(\pi/6)]} = \frac{3\sqrt{2}}{4}\left(\cos\frac{13\pi}{12} + i\sin\frac{13\pi}{12}\right)$$

70. $$\left[2\left(\cos\frac{4\pi}{15} + i\sin\frac{4\pi}{15}\right)\right]^5 = 2^5\left(\cos\frac{4\pi}{3} + i\sin\frac{4\pi}{3}\right) = 32\left(-\frac{1}{2} - \frac{\sqrt{3}}{2}i\right) = -16 - 16\sqrt{3}i$$

72. $$(1 - i)^8 = \left[\sqrt{2}(\cos 315° + i\sin 315°)\right]^8 = 16(\cos 2520° + i\sin 2520°) = 16(\cos 0° + i\sin 0°) = 16$$

74. (a) The trigonometric forms of the four roots shown are:

$4(\cos 60° + i\sin 60°)$

$4(\cos 150° + i\sin 150°)$

$4(\cos 240° + i\sin 240°)$

$4(\cos 330° + i\sin 330°)$

(b) Since there are four evenly spaced roots on the circle of radius 4, they are fourth roots of a complex number of modulus 4^4. In this case, raising them to the fourth power yields $-128 - 128\sqrt{3}$.

(c) $[4(\cos 60° + i\sin 60°)]^4 = -128 - 128\sqrt{3}$

$[4(\cos 150° + i\sin 150°)]^4 = -128 - 128\sqrt{3}$

$[4(\cos 240° + i\sin 240°)]^4 = -128 - 128\sqrt{3}$

$[4(\cos 330° + i\sin 330°)]^4 = -128 - 128\sqrt{3}$

76. Fourth roots of $256 = 256(\cos 0 + i\sin 0)$:

$$\sqrt[4]{256}\left(\cos\frac{2\pi k}{4} + i\sin\frac{2\pi k}{4}\right)$$

$k = 0, 1, 2, 3$

$k = 0$: $4(\cos 0 + i\sin 0) = 4$

$k = 1$: $4(\cos\pi + i\sin\pi) = -4$

$k = 2$: $4\left(\cos\frac{\pi}{2} + i\sin\frac{\pi}{2}\right) = 4i$

$k = 3$: $4\left(\cos\frac{3\pi}{2} + i\sin\frac{3\pi}{2}\right) = -4i$

78. $x^5 - 32 = 0$

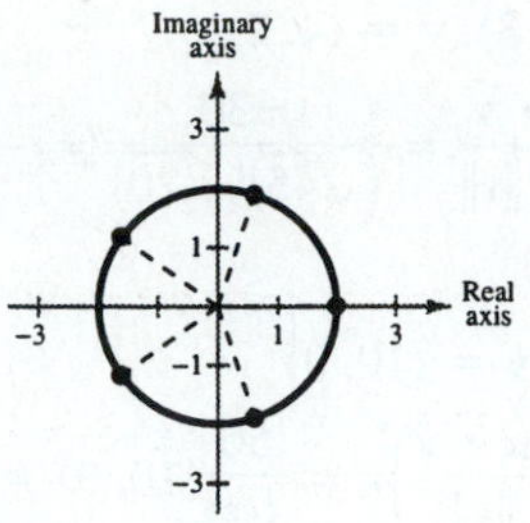

$$x^5 = 32$$

$$32 = 32(\cos 0 + i \sin 0)$$

$$\sqrt[3]{32} = \sqrt[5]{32}\left[\cos\left(0 + \frac{2\pi k}{5}\right) + i \sin\left(0 + \frac{2\pi k}{5}\right)\right)$$

$k = 0, 1, 2, 3, 4$

$k = 0$: $2(\cos 0 + i \sin 0) = 2$

$k = 1$: $2\left(\cos \frac{2\pi}{5} + i \sin \frac{2\pi}{5}\right) = 0.6180 + 1.9021i$

$k = 2$: $2\left(\cos \frac{4\pi}{5} + i \sin \frac{4\pi}{5}\right) = -1.6180 + 1.1756i$

$k = 3$: $2\left(\cos \frac{6\pi}{5} + i \sin \frac{6\pi}{5}\right) = -1.6180 - 1.1756i$

$k = 4$: $2\left(\cos \frac{8\pi}{5} + i \sin \frac{8\pi}{5}\right) = 0.6180 - 1.9021i$

80. $(x^3 - 1)(x^2 + 1) = 0$

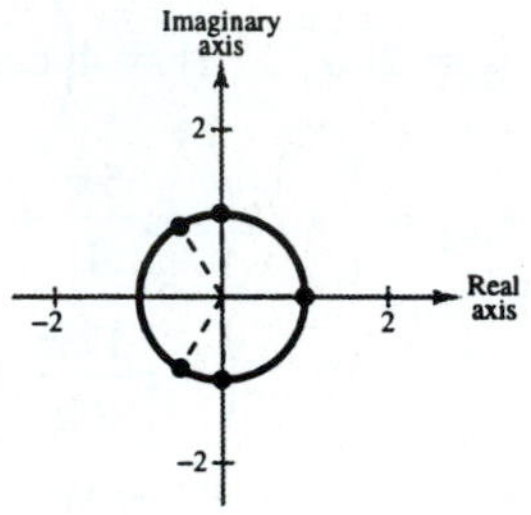

$$x^3 - 1 = 0$$

$$x^2 + 1 = 0$$

$$x^3 = 1$$

$$1 = 1(\cos 0 + i \sin 0)$$

$$\sqrt[3]{1} = \sqrt[3]{1}\left[\cos\left(\frac{0 + 2\pi k}{3}\right) + i \sin\left(\frac{0 + 2\pi k}{3}\right)\right], \quad k = 0, 1, 2$$

$$1(\cos 0 + i \sin 0) = 1$$

$$1\left(\cos \frac{2\pi}{3} + i \sin \frac{2\pi}{3}\right) = -\frac{1}{2} + \frac{\sqrt{3}}{2}i$$

$$1\left(\cos \frac{4\pi}{3} + i \sin \frac{4\pi}{3}\right) = -\frac{1}{2} - \frac{\sqrt{3}}{2}i$$

$$x^2 + 1 = 0$$

$$x^2 = -1$$

$$-1 = 1(\cos \pi + i \sin \pi)$$

$$\sqrt{-1} = \sqrt{1}\left[\cos\left(\frac{\pi + 2\pi k}{2}\right) + i \sin\left(\frac{\pi + 2\pi k}{2}\right), \quad k = 0, 1\right.$$

$k = 0, 1$

$$1\left(\cos \frac{\pi}{2} + i \sin \frac{\pi}{2}\right) = i$$

$$1\left(\cos \frac{3\pi}{2} + i \sin \frac{3\pi}{2}\right) = -i$$

CHAPTER 7
Systems of Equations and Inequalities

CHAPTER 7

Systems of Equations and Inequalities

Section 7.1 Solving Systems of Equations

Solutions to Even-Numbered Exercises

2. $x - y = -4$ Equation 1

$x + 2y = 5$ Equation 2

Solve for x in Equation 1: $x = y - 4$

Substitute for x in Equation 2: $(y - 4) + 2y = 5$

Solve for y: $3y - 4 = 5 \Rightarrow y = 3$

Back-substitute $y = 3$: $x = 3 - 4 = -1$

Answer: $(-1, 3)$

4. $3x + y = 2$ Equation 1

$x^3 + y = 0$ Equation 2

Solve for y in Equation 1: $y = 2 - 3x$

Substitute for y in Equation 2: $x^3 + (2 - 3x) = 0$

Solve for x: $x^3 - 3x + 2 = 0 \Rightarrow (x - 1)^2(x + 2) = 0 \Rightarrow x = 1, -2$

Back-substitute $x = 1$: $y = 2 - 3(1) = -1$

Back-substitute $x = -2$: $y = 2 - 3(-2) = 8$

Answers: $(1, -1), (-2, 8)$

6. $x + y = 0$ Equation 1

$x^3 - 5x - y = 0$ Equation 2

Solve for y in Equation 1: $y = -x$

Substitute for y in Equation 2: $x^3 - 5x - (-x) = 0$

Solve for x: $x^3 - 4x = 0 \Rightarrow x(x^2 - 4) = 0 \Rightarrow x = 0, \pm 2$

Back-substitute $x = 0$: $y = -0 = 0$

Back-substitute $x = 2$: $y = -2$

Back-substitute $x = -2$: $y = -(-2) = 2$

Answers: $(0, 0), (2, -2), (-2, 2)$

8. $y = -2x^2 + 2$ Equation 1

$y = 2(x^4 - 2x^2 + 1)$ Equation 2

Substitute for y in Equation 1: $2(x^4 - 2x^2 + 1) = -2x^2 + 2$

Solve for x: $x^4 - 2x^2 + 1 + x^2 - 1 = 0$

$$x^4 - x^2 = 0$$

$$x^2(x^2 - {}^1) = 0 \Rightarrow x = 0, \pm 1$$

Back-substitute $x = 0$: $y = -2(0)^2 + 2 = 2$

Back-substitute $x = 1$: $y = -2(1)^2 + 2 = 0$

Back-substitute $x = -1$: $y = -2(-1)^2 + 2 = 0$

Answers: $(0, 2), (1, 0), (-1, 0)$

10. $y = x^3 - 3x^2 + 4$ Equation 1

$y = -2x + 4$ Equation 2

Substitute for y in Equation 1: $-2x + 4 = x^3 - 3x^2 + 4$

Solve for x: $0 = x^3 - 3x^2 + 2x$

$0 = x(x^2 - 3x + 2)$

$0 = x(x - 2)(x - 1) \Rightarrow x = 0, 1, 2$

Back-substitute $x = 0$: $y = -2(0) + 4 = 4$

Back-substitute $x = 1$: $y = -2(1) + 4 = 2$

Back-substitute $x = 2$: $y = -2(2) + 4 = 0$

Answers: $(0, 4), (1, 2), (2, 0)$

12. $x + 2y = 1$ Equation 1

$5x - 4y = -23$ Equation 2

Solve for x in Equation 1: $x = 1 - 2y$

Substitute for x in Equation 2: $5(1 - 2y) - 4y = -23$

Solve for y: $-14y = -28 \Rightarrow y = 2$

Back-substitute $y = 2$: $x = 1 - 2y = 1 - 4 = -3$

Answer: $(-3, 2)$

14. $6x - 3y - 4 = 0$ Equation 1

$x + 2y - 4 = 0$ Equation 2

Solve for x in Equation 2: $x = 4 - 2y$

Substitute for x in Equation 1: $6(4 - 2y) - 3y - 4 = 0$

Solve for y: $24 - 12y - 3y - 4 = 0 \Rightarrow -15y = -20 \Rightarrow y = \frac{4}{3}$

Back-substitute $y = \frac{4}{3}$: $x = 4 - 2y = 4 - 2\left(\frac{4}{3}\right) = \frac{4}{3}$

Answer: $\left(\frac{4}{3}, \frac{4}{3}\right)$

16. $1.5x + 0.8y = 2.3$ Equation 1

$0.3x - 0.2y = 0.1$ Equation 2

Solve for y in Equation 2: $y = 1.5x - 0.5$

Substitute for y in Equation 1: $1.5x + 0.8(1.5x - 0.5) = 2.3$

Solve for x: $1.5x + 1.2x - 0.4 = 2.3 \Rightarrow 2.7x = 2.7 \Rightarrow x = 1$

Back-substitute $x = 1$: $y = 1.5x - 0.5 = 1.5 - 0.5 = 1$

Answer: $(1, 1)$

18. $\frac{1}{2}x + \frac{3}{4}y = 10$ Equation 1

$\frac{3}{4}x - y = 4$ Equation 2

Solve for y in Equation 2: $y = \frac{3}{4}x - 4$

Substitute for y in Equation 1: $\frac{1}{2}x + \frac{3}{4}\left(\frac{3}{4}x - 4\right) = 10$

Solve for x: $\frac{1}{2}x + \frac{9}{16}x - 3 = 10 \Rightarrow \frac{17}{16}x = 13 \Rightarrow x = \frac{208}{17}$

Back-substitute $x = \frac{208}{17}$: $y = \frac{3}{4}\left(\frac{208}{17}\right) - 4 = \frac{88}{17}$

Answer: $\left(\frac{208}{17}, \frac{88}{17}\right)$

20. $-\frac{2}{3}x + y = 2$ Equation 1

$2x - 3y = 6$ Equation 2

Solve for y in Equation 1: $y = \frac{2}{3}x + 2$

Substitute for y in Equation 2: $2x - 3\left(\frac{2}{3}x + 2\right) = 6$

Solve for x: $2x - 2x - 6 = 6 \Rightarrow 0 = 12$ Inconsistent

No solution

22. $x - 2y = 0$ Equation 1

$3x - y = 0$ Equation 2

Solve for x in Equation 1: $x = 2y$

Substitute for x in Equation 2: $3(2y) - y = 0$

Substitute for y: $6y - y = 0 \Rightarrow 5y = 0 \Rightarrow y = 0$

Back-substitute $y = 0$: $x = 2y = 0$

Answer: $(0, 0)$

24. $x + y = 0$

$3x - 2y = 10$

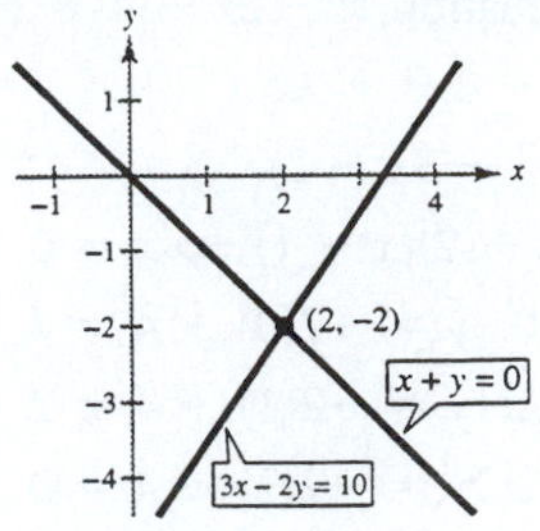

Point of intersection: $(2, -2)$

26. $-x + 2y = 1$

$x - y = 2$

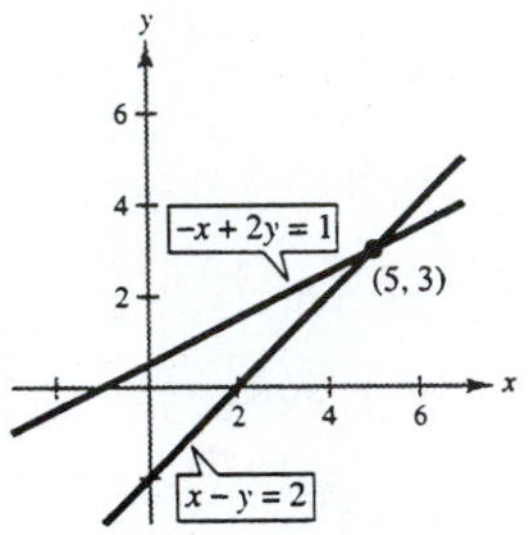

Point of intersection: $(5, 3)$

28. $x - y + 3 = 0$

$x^2 - 4x + 7 = y$

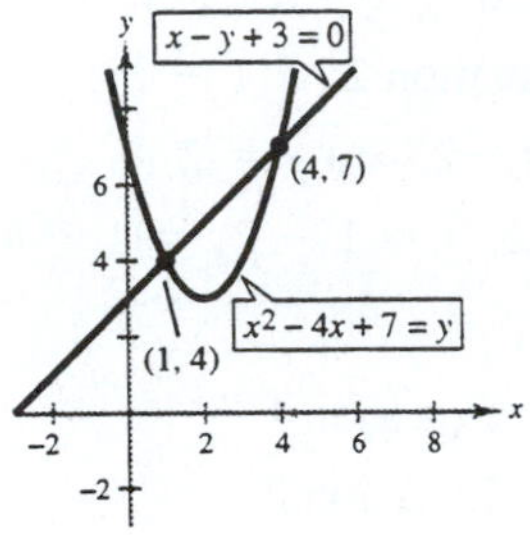

Points of intersection: $(1, 4), (4, 7)$

30. $x - y = 0 \Rightarrow y_1 = x$

$5x - 2y = 6 \Rightarrow y_2 = \frac{5}{2}x - 3$

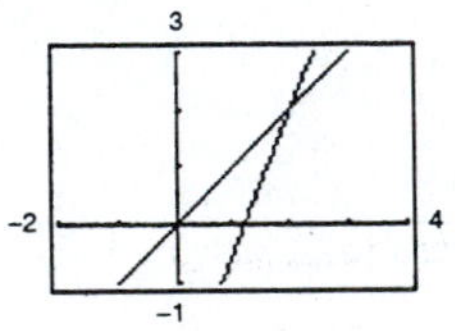

Point of intersection: $(2, 2)$

32. $2x - y + 3 = 0 \Rightarrow y_1 = 2x + 3$

$x^2 + y^2 - 4x = 0 \Rightarrow y_2 = \sqrt{4x - x^2}$

$y_3 = -\sqrt{4x - x^2}$

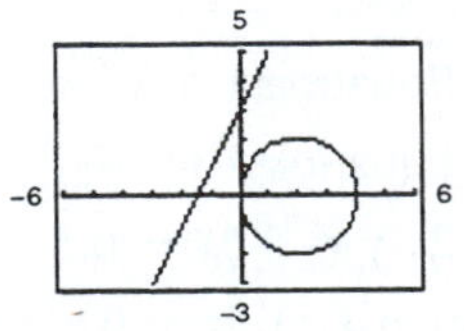

No points of intersection

34. $x^2 + y^2 = 25 \Rightarrow y_1 = \sqrt{25 - x^2}$

$y_2 = -\sqrt{25 - x^2}$

$(x - 8)^2 + y^2 = 41 \Rightarrow y_3 = \sqrt{41 - (x - 8)^2}$

$y_4 = -\sqrt{41 - (x - 8)^2}$

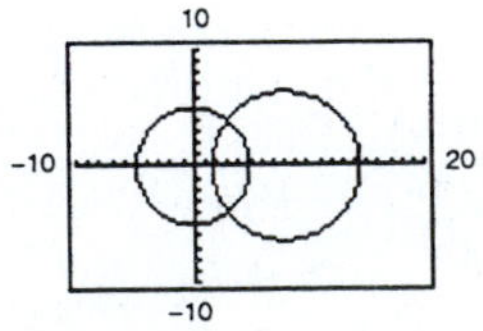

Points of intersection: $(3, 4), (3, -4)$

36. $x + 2y = 8 \Rightarrow y_1 = 4 - \frac{1}{2}x$

$y = \log_2 x \Rightarrow y_2 = \frac{\ln x}{\ln 2}$

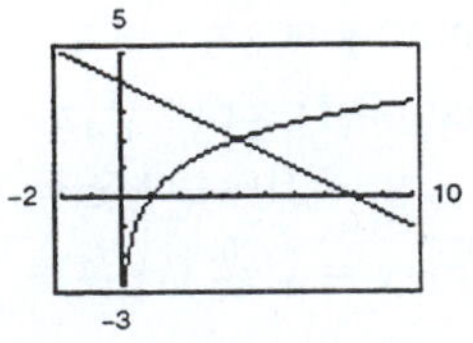

Point of intersection: $(4, 2)$

38. $x - y = 3 \Rightarrow y_1 = x - 3$

$x - y^2 = 1 \Rightarrow y_2 = \sqrt{x - 1}$

$y_3 = -\sqrt{x - 1}$

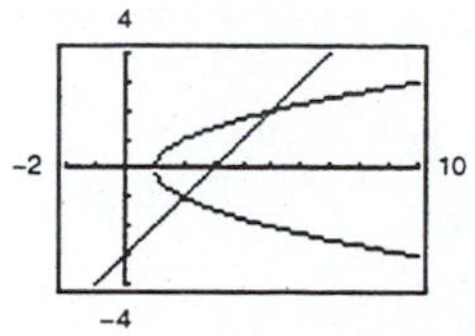

Points of intersection: $(5, 2), (2, -1)$

40. $x^2 + y^2 = 4 \Rightarrow y_1 = \sqrt{4 - x^2}$

$y_2 = -\sqrt{4 - x^2}$

$2x^2 - y = 2 \Rightarrow y_3 = 2x^2 - 2$

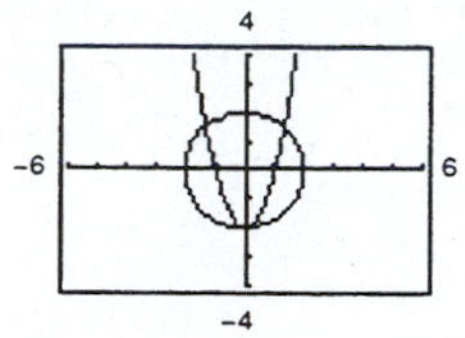

Points of intersection: $(0, -2), \left(\frac{1}{2}\sqrt{7}, \frac{3}{2}\right), \left(-\frac{1}{2}\sqrt{7}, \frac{3}{2}\right)$

42. $x + y = 4$ Equation 1

$x^2 + y = 2$ Equation 2

Solve for y in Equation 1: $y = 4 - x$

Substitute for y in Equation 2: $x^2 + (4 - x) = 2$

Solve for x: $x^2 - x + 2 = 0$

No real solutions because the discriminant in the Quadratic Formula is negative.

Inconsistent. No solution

44. $x^2 + y^2 = 25$ Equation 1

$2x + y = 10$ Equation 2

Solve for y in Equation 2: $y = 10 - 2x$

Substitute for y in Equation 1: $x^2 + (10 - 2x)^2 = 25$

Solve for x: $x^2 + 100 - 40x + 4x^2 = 25 \Rightarrow x^2 - 8x + 15 = 0$

$\Rightarrow (x - 5)(x - 3) = 0 \Rightarrow x = 3, 5$

Back-substitute $x = 3$: $y = 10 - 2(3) = 4$

Back-substitute $x = 5$: $y = 10 - 2(5) = 0$

Answers: $(3, 4), (5, 0)$

46. $y = (x + 1)^3$

$y = \sqrt{x - 1}$

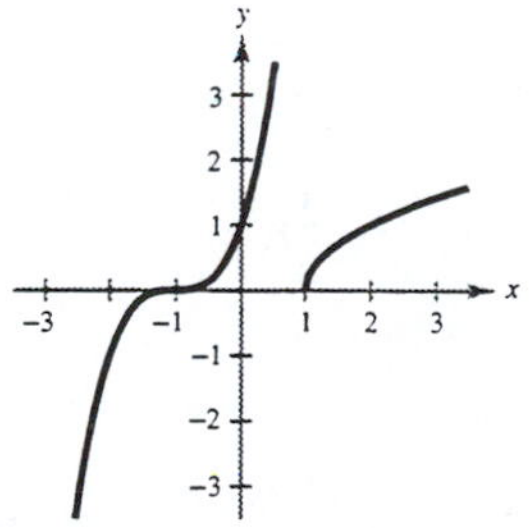

No points of intersection

48. $y = x^3 - 2x^2 + x - 1$ Equation 1

$y = -x^2 + 3x - 1$ Equation 2

Substitute for y in Equation 1:

$-x^2 + 3x - 1 = x^3 - 2x^2 + x - 1$

Solve for x: $0 = x^3 - x^2 - 2x$

$0 = x(x^2 - x - 2)$

$0 = x(x - 2)(x + 1) \Rightarrow x = 0, 2, -1$

Back-substitute $x = 0$ in Equation 2:

$y = -0^2 + 3(0) - 1 = -1$

Back-substitute $x = 2$ in Equation 2:

$y = -2^2 + 3(2) - 1 = 1$

Back-substitute $x = -1$ in Equation 2:

$y = (-1)^2 + 3(-1) - 1 = -5$

Answers: $(0, -1), (2, 1), (-1, -5)$

50. $x^2 + y = 4 \Rightarrow y = 4 - x^2$

$e^x - y = 0 \Rightarrow y = e^x$

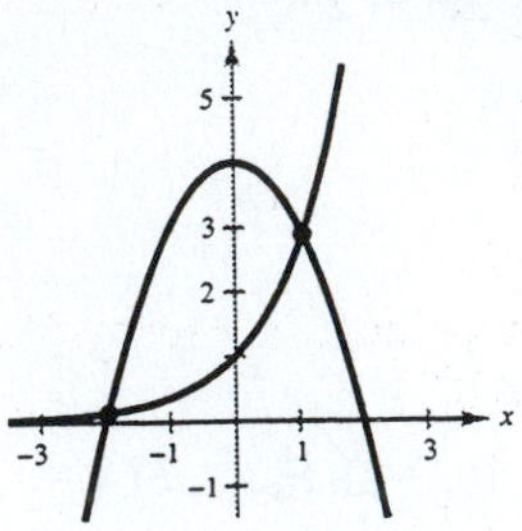

Points of intersection:

Approximately $(-1.96, 0.14)$, $(1.06, 2.88)$

52. $x - 2y = 1$ Equation 1

$y = \sqrt{x-1}$ Equation 2

Substitute for y in Equation 1: $x - 2\sqrt{x-1} = 1$

Solve for x:

$$x - 1 = 2\sqrt{x-1}$$
$$(x-1)^2 = 4(x-1)$$
$$x^2 - 2x + 1 = 4x - 4$$
$$x^2 - 6x + 5 = 0$$
$$(x-1)(x-5) = 0 \Rightarrow x = 1, 5$$

Back-substitute $x = 1$: $y = \sqrt{1-1} = 0$

Back-substitute $x = 5$: $y = \sqrt{5-1} = 2$

Answers: $(1, 0)$, $(5, 2)$

54. The advantage of the method of substitution of the graphical method is that substitution gives exact solutions but graphical solutions may only be approximate.

56. $C = 5.5\sqrt{x} + 10{,}000$, $R = 3.29x$

$$R = C$$
$$3.29x = 5.5\sqrt{x} + 10{,}000$$
$$3.29x - 10{,}000 = 5.5\sqrt{x}$$
$$10.8241x^2 - 65{,}800x + 100{,}000{,}000 = 30.25x$$
$$10.8241x^2 - 65{,}830.25x + 100{,}000{,}000 = 0$$
$$x \approx 3133 \text{ units}$$

In order for the revenue to break even with the cost, 3133 units must be sold.

58. $C = 0.08x + 50{,}000$, $R = 0.25x$

$$R = C$$
$$0.25x = 0.08x + 50{,}000$$
$$0.17x = 50{,}000$$
$$x \approx 294{,}117.6$$

In order for the revenue to break even with the cost, 294,118 units must be sold.

60. $C = 21.60x + 5000$, $R = 34.10x$

$$R = C$$
$$34.10x = 21.60x + 5000$$
$$12.5x = 5000$$
$$x = 400 \text{ units}$$

62. *Model:* $\begin{pmatrix}\text{6.5\%}\\\text{fund}\end{pmatrix} + \begin{pmatrix}\text{8.5\%}\\\text{fund}\end{pmatrix} = \begin{pmatrix}\text{Total}\\\text{investment}\end{pmatrix}$

$\begin{pmatrix}\text{6.5\%}\\\text{interest}\end{pmatrix} + \begin{pmatrix}\text{8.5\%}\\\text{interest}\end{pmatrix} = \begin{pmatrix}\text{Total}\\\text{interest}\end{pmatrix}$

Labels: Amount in 6.5% fund $= x$

Interest for 6.5% fund $= 0.065x$

Amount in 8.5% fund $= y$

Interest for 8.5% fund $= 0.085y$

Total investment $= 20{,}000$

Total interest $= 1600$

System: $x + y = 20{,}000$ Equation 1

$0.065x + 0.085y = 1600$ Equation 2

Solve for y in Equation 1: $y = 20{,}000 - x$

Substitute for y in Equation 2: $0.065x + 0.085(20{,}000 - x) = 1600$

Solve for x: $0.065x + 1700 - 0.085x = 1600$

$$100 = 0.02x$$
$$x = 5000$$

The most that can be invested at 6.5% is $5000.

64. $20{,}000 + 0.01x = 15{,}000 + 0.02x$

$$5000 = 0.01x$$
$$500{,}000 = x$$

For the second offer to be better, you would have to sell more than \$500,000 per year.

66. $p = 1.45 + 0.00014x^2$

$p = (2.388 - 0.007x)^2$

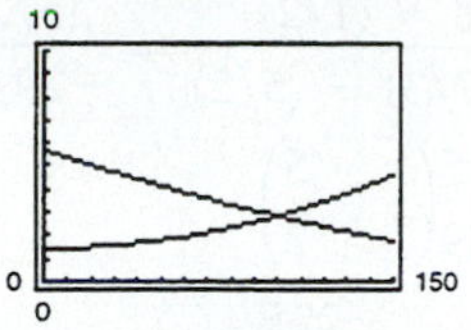

The market equilibrium (point of intersection) is approximately (100, 2.85).

68. $2l + 2w = 280 \Rightarrow l + w = 140$

$$w = l - 20 \Rightarrow l + (l - 20) = 140$$
$$2l = 160$$
$$l = 80$$

$w = l - 20 = 80 - 20 = 60$

Dimensions: 60×80 centimeters

70. $2l + 2w = 210 \Rightarrow \quad l + 2 = 105$

$$l = \tfrac{3}{2}w \Rightarrow \tfrac{3}{2}w + w = 105$$
$$\tfrac{5}{2}w = 105$$
$$w = 42$$

$l = \tfrac{3}{2}(42) = 63$

Dimensions: 42×63 feet

72. $A = \tfrac{1}{2}bh$

$$1 = \tfrac{1}{2}a^2$$
$$a^2 = 2$$
$$a = \sqrt{2}$$

The dimensions are $\sqrt{2} \times \sqrt{2} \times 2$.

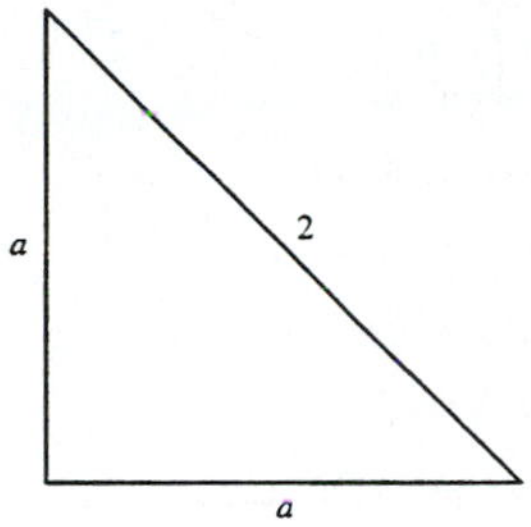

74. (a) The graph of $y = 2x$ (or the graph of $y = -5x$) intersects the graph of $y = x^2$ in two points.

(b) The graph of $y = 0$ (or the graph of $x = 1$) intersects the graph of $y = x^2$ in one point.

(c) The graph of $y = x - 2$ (or the graph of $y = -5$) intersects the graph of $y = x^2$ in no points.

76. (a)

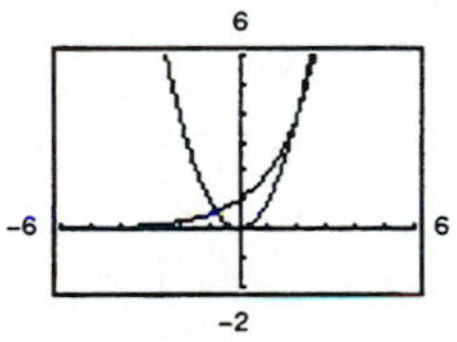

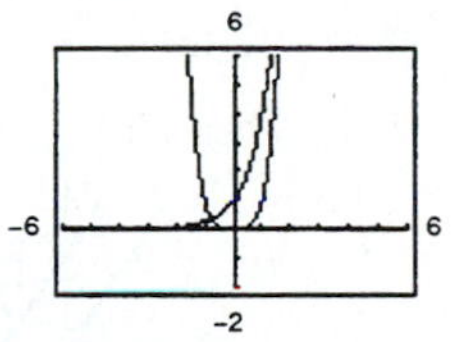

(b) Based on the graphs in part (a) it appears that for $b > 1$, there are three points of intersection for the graphs of $y = b^x$ and $y = x^b$.

78. (3.5, 4), (10, 6)

$$m = \frac{6 - 4}{10 - 3.5} = \frac{2}{6.5}$$

$$y - 6 = \frac{2}{6.5}(x - 10)$$

$$6.5y - 39 = 2x - 20$$

$$2x - 6.5y + 19 = 0$$

80. (4, −2), (4, 5)

$x = 4$

82. $\left(-\frac{7}{3}, 8\right), \left(\frac{5}{2}, \frac{1}{2}\right)$

$$m = \frac{8 - (1/2)}{-(7/3) - (5/2)} = \frac{15/2}{-29/6} = -\frac{45}{29}$$

$$y - \frac{1}{2} = -\frac{45}{29}\left(x - \frac{5}{2}\right)$$

$$29y - \frac{29}{2} = -45x + \frac{225}{2}$$

$$45x + 29y - 127 = 0$$

Section 7.2 Two-Variable Linear Systems

Solutions to Even-Numbered Exercises

2. $x + 3y = 1$ Equation 1

$-x + 2y = 4$ Equation 2

Add to eliminate x: $5y = 5 \Rightarrow y = 1$

Substitute $y = 1$ in Equation 1: $x + 3(1) = 1 \Rightarrow x = -2$

Answer: $(-2, 1)$

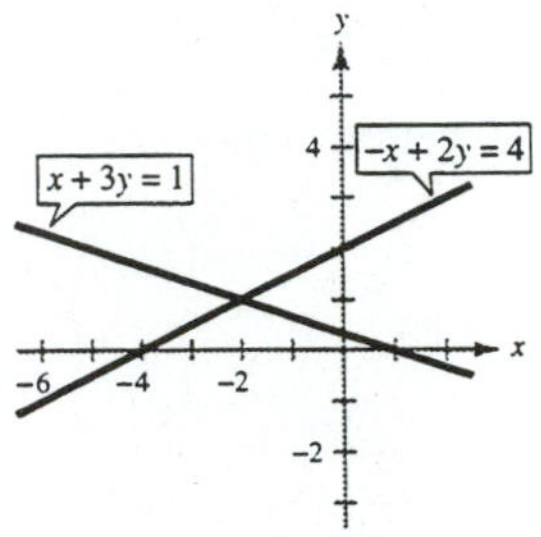

4. $2x - y = 3$ Equation 1

$4x + 3y = 21$ Equation 2

Multiply Equation 1 by 3: $6x - 3y = 9$

Add this to Equation 2 to eliminate y: $10x = 30 \Rightarrow x = 3$

Substitute $x = 3$ in Equation 1: $2(3) - y = 3 \Rightarrow y = 3$

Answer: $(3, 3)$

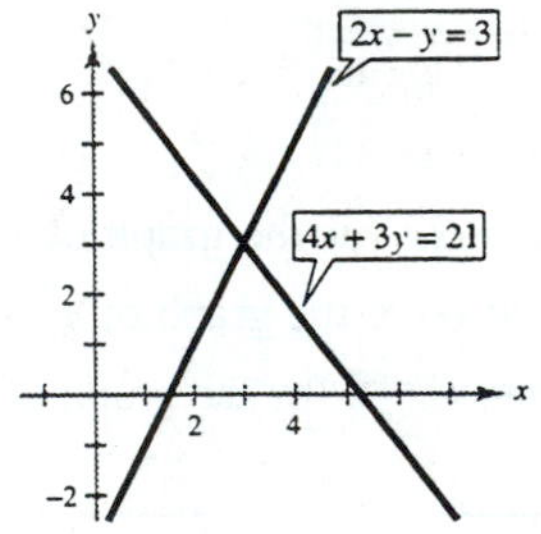

6. $3x + 2y = 3$ Equation 1

$6x + 4y = 14$ Equation 2

Multiply Equation 1 by -2: $-6x - 4y = -6$

Add this to Equation 2: $0 = 8$

There are no solutions.

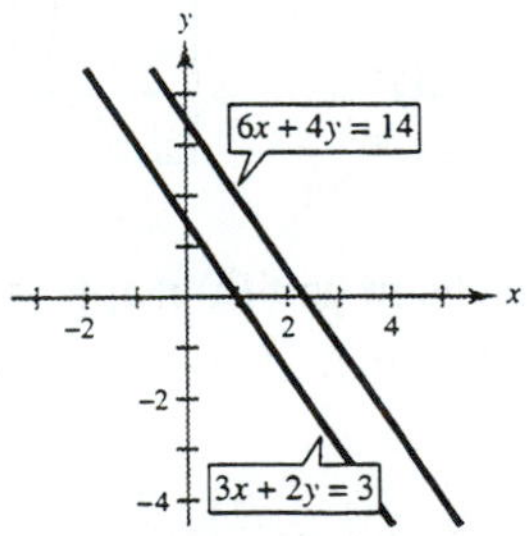

8. $x - 2y = 4$ Equation 1

$6x + 2y = 10$ Equation 2

Add to eliminate y: $7x = 14 \Rightarrow x = 2$

Substitute $x = 2$ in Equation 1: $2 - 2y = 4 \Rightarrow y = -1$

Answer: $(2, -1)$

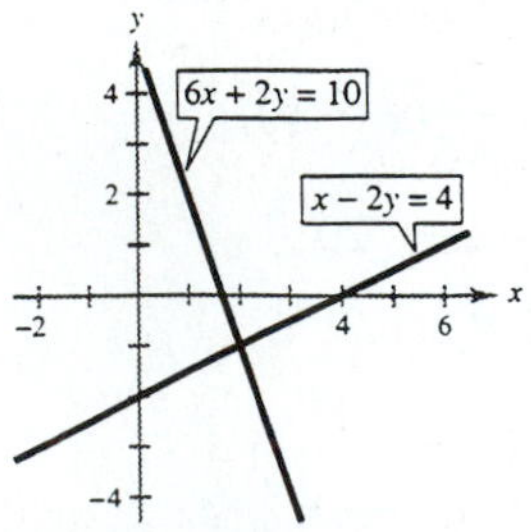

10. $5x + 3y = -18$ Equation 1

$2x - 6y = 1$ Equation 2

Multiply Equation 1 by 2: $10x + 6y = -36$

Add this to Equation 2 to eliminate y: $12x = -35 \Rightarrow x = -\frac{35}{12}$

Substitute $x = -\frac{35}{12}$ in Equation 2: $2\left(-\frac{35}{12}\right) - 6y = 1 \Rightarrow y = -\frac{41}{36}$

Answer: $\left(-\frac{35}{12}, -\frac{41}{36}\right)$

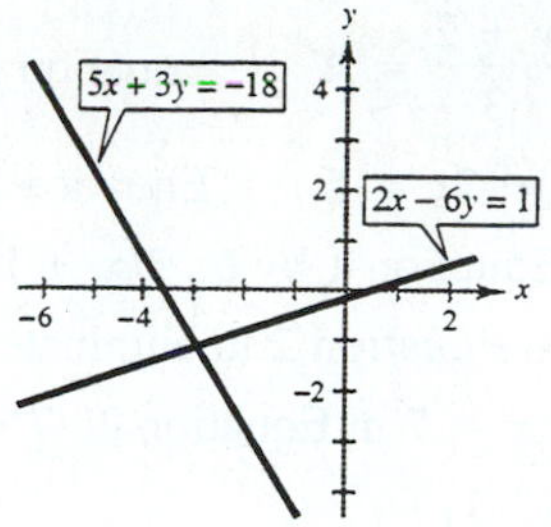

12. $3x - 5y = 2$ Equation 1

$2x + 5y = 13$ Equation 2

Add to eliminate y: $5x = 15$

$x = 3$

Substitute $x = 3$ in Equation 1: $3(3) - 5y = 2 \Rightarrow y = \frac{7}{5}$

Answer: $\left(3, \frac{7}{5}\right)$

14. $x + 7y = 12$ Equation 1

$3x - 5y = 10$ Equation 2

Multiply Equation 1 by -3: $-3x - 21y = -36$

Add this to Equation 2 to eliminate x: $-26y = -26 \Rightarrow y = 1$

Substitute $y = 1$ in Equation 1: $x + 7 = 12 \Rightarrow x = 5$

Answer: $(5, 1)$

16. $8r + 16s = 20$ Equation 1

$16r + 50s = 55$ Equation 2

Multiply Equation 1 by (-2): $-16r - 32s = -40$

Add this to Equation 2 to eliminate r: $18s = 15 \Rightarrow s = \frac{5}{6}$

Substitute $s = \frac{5}{6}$ in Equation 1: $8r + 16\left(\frac{5}{6}\right) = 20 \Rightarrow r = \frac{5}{6}$

Answer: $\left(\frac{5}{6}, \frac{5}{6}\right)$

18. $5u + 6v = 24$ Equation 1

$3u + 5v = 18$ Equation 2

Multiply Equation 1 by 3 and Equation 2 by (-5): $15u + 18v = 72$

$-15u - 25v = -90$

Add to eliminate u: $-7v = -18 \Rightarrow v = \frac{18}{7}$

Substitute $v = \frac{18}{7}$ in Equation 2: $3u + 5\left(\frac{18}{7}\right) = 18 \Rightarrow u = \frac{12}{7}$

Answer: $\left(\frac{12}{7}, \frac{18}{7}\right)$

20. $1.8x + 1.2y = 4$ Equation 1

$9x + 6y = 3$ Equation 2

Multiply Equation 1 by (-5): $-9x - 6y = -20$

Add this to Equation 2: $0 = -17$

Inconsistent; no solution

22. $\frac{2}{3}x + \frac{1}{6}y = \frac{2}{3}$ Equation 1

$4x + y = 4$ Equation 2

Multiply Equation 1 by (-6): $-4x - 6y = -4$

Add this to Equation 2: $0 = 0$ (dependent)

The solution set consists of all points lying on the line $4x + y = 4$. Let $x = a$, then $y = 4 - 4a$.

Answer: $(a, 4 - 4a)$ where a is any real number.

24. $\frac{x-1}{2} + \frac{y+2}{3} = 4$ Equation 1

$x - 2y = 5$ Equation 2

Multiply Equation 1 by 6: $3(x - 1) + 2(y + 2) = 24 \Rightarrow 3x + 2y = 23$

Add this to Equation 2 to eliminate y: $4x = 28 \Rightarrow x = 7$

Substitute $x = 7$ in Equation 2: $7 - 2y = 5 \Rightarrow y = 1$

Answer: $(7, 1)$

26. $0.02x - 0.05y = -0.19$ Equation 1

$0.03x + 0.04y = 0.52$ Equation 2

Multiply Equation 1 by 4 and Equation 2 by 5:

$0.08x - 0.2y = -0.76$

$0.15x + 0.2y = 2.6$

Add these to eliminate y: $0.23x = 1.84 \Rightarrow x = 8$

Substitute $x = 8$ in Equation 1: $0.02(8) - 0.05y = -0.19 \Rightarrow y = 7$

Answer: $(8, 7)$

28. $0.2x - 0.5y = -27.8$ Equation 1

$0.3x + 0.4y = 68.7$ Equation 2

Multiply Equation 1 by 4 and Equation 2 by 5:

$0.8x - 2y = -111.2$

$1.5x + 2y = 343.5$

Add these to eliminate y: $2.3x = 232.3 \Rightarrow x = 101$

Substitute $x = 101$ in Equation 1: $0.2(101) - 0.5y = -27.8 \Rightarrow y = 96$

Answer: $(101, 96)$

30. $3b + 3m = 7$ Equation 1

$3b + 5m = 3$ Equation 2

Subtract Equation 2 from Equation 1 to eliminate b: $-2m = 4 \Rightarrow m = -2$

Substitute $m = -2$ in Equation 1: $3b + 3(-2) = 7 \Rightarrow b = \frac{13}{3}$

Answer: $\left(\frac{13}{3}, -2\right)$

32. $2x + y = 5$

$x - 2y = -1$

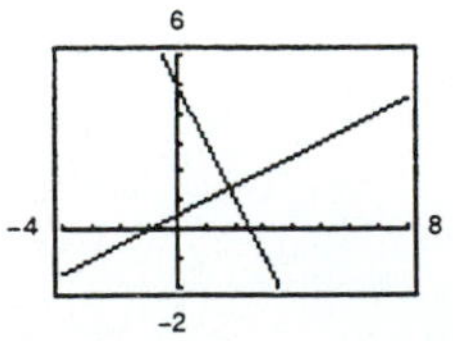

The system is consistent. There is one solution.

34. $4x - 6y = 7$

$2x - 3y = 3.5$

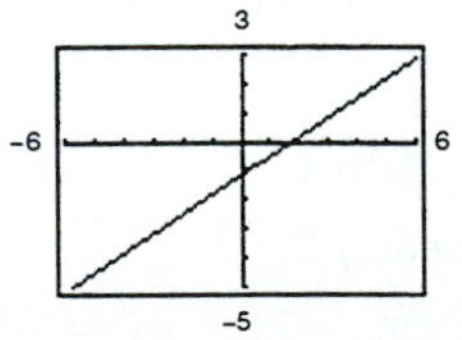

The system is consistent. There are infinitely many solutions.

36. $\frac{3}{2}x - \frac{1}{5}y = 8$

$-2x + 3y = 3$

Solution: $(6, 5)$

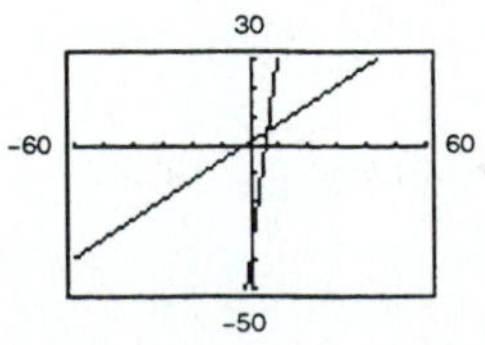

38. $0.5x + 2.2y = 9$

$6x + 0.4y = -22$

Solution: $(-4, 5)$

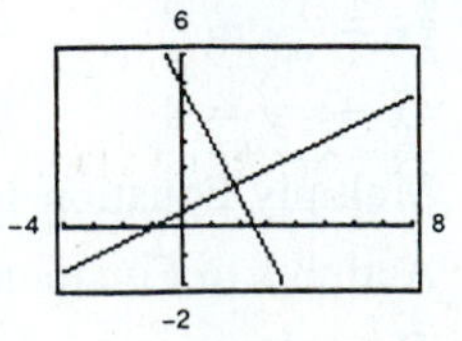

40. $-x + 3y = 17$ Equation 1

$4x + 3y = 7$ Equation 2

Subtract Equation 2 from Equation 1 to eliminate y: $-5x = 10 \Rightarrow x = -2$

Substitute $x = -2$ in Equation 1: $-(-2) + 3y = 17 \Rightarrow y = 5$

Solution: $(-2, 5)$

42. $7x + 3y = 16$ Equation 1

$y = x + 2$ Equation 2

Substitute for y in Equation 1:

$$7x + 3(x + 2) = 16$$
$$7x + 3x + 6 = 16$$
$$10x = 10 \Rightarrow x = 1$$

Substitute $x = 1$ in Equation 2: $y = 1 + 2 = 3$

Solution: $(1, 3)$

44. There are infinitely many systems that have the solution $(8, -2)$. One possible system is:

$$8 - 2 = 6 \Rightarrow x + y = 6$$
$$2(8) - (-2) = 18 \Rightarrow 2x - y = 18$$

46. $21x - 20y = 0$ Equation 1

$13x - 12y = 120$ Equation 2

Multiply Equation 2 by $\left(-\frac{5}{3}\right)$: $-\frac{65}{3}x + 20y = -200$

Add this to Equation 1 to eliminate y: $-\frac{2}{3}x = -200 \Rightarrow x = 300$

Substitute $x = 300$ in Equation 1: $21(300) - 20y = 0 \Rightarrow y = 315$

Solution: $(300, 315)$

The lines are not parallel. The scale on the axes must be changed to see the point of intersection.

48. (a) $x + y = 10$ Equation 1

$x + y = 20$ Equation 2

Subtract Equation 2 from Equation 1: $0 = -10$

System is inconsistent $\Rightarrow$ no solution

(b) $x + y = 3$ Equation 1

$2x + 2y = 6$ Equation 2

Multiply Equation 1 by (-2): $-2x - 2y = -6$

Add this to Equation 2: $0 = 0$ (dependent)

The system has an infinite number of solutions.

50. $15x + 3y = 6$ Equation 1

$-10x + ky = 9$ Equation 2

Multiply Equation 1 by $\frac{2}{3}$: $10x + 2y = 4$

Add this to Equation 2: $ky + 2y = 13$

This system is inconsistent if $ky + 2y = 0$. This occurs when $k = -2$.

52. Let $x =$ the speed of the plane that leaves first and $y =$ the speed of the plane that leaves second.

$y - x = 80$ Equation 1

$2x + \frac{3}{2}y = 3200$ Equation 2

$$-2x + 2y = 160$$
$$2x + \frac{3}{2}y = 3200$$
$$\frac{7}{2}y = 3360$$
$$y = 960$$
$$960 - x = 80$$
$$x = 880$$

Answer: First plane: 880 kilometers per hour; Second plane: 960 kilometers per hour

54. Let x = the number of 87 octane gallons; y = the number of 92 octane gallons

$$87x + 92y = 89(500) \quad \text{Equation 1}$$
$$x + y = 500 \quad \text{Equation 2}$$

$$\begin{aligned} 87x + 92y &= 44{,}500 \\ -87x - 87y &= -43{,}500 \\ \hline 5y &= 1000 \\ y &= 200 \\ x + 200 &= 500 \\ x &= 300 \end{aligned}$$

Answer: 87 octane: 300 gallons; 92 octane: 200 gallons

56. Let x = the amount invested at 5.75%; y = the amount invested at 6.25%

$$x + y = 32{,}000 \quad \text{Equation 1}$$
$$0.0575x + 0.0625y = 1900 \quad \text{Equation 2}$$

$$\begin{aligned} -5.75x - 5.75y &= -184{,}000 \\ 5.75x + 6.25y &= 190{,}000 \\ \hline 0.5y &= 6000 \\ y &= 12{,}000 \\ x + 12{,}000 &= 32{,}000 \\ x &= 20{,}000 \end{aligned}$$

The most that can be invested at 5.75% is \$20,000.

58. Let x = the number of pairs of \$66.95 shoes; y = the number of pairs of \$84.95 shoes

$$x + y = 240 \quad \text{Equation 1}$$
$$66.95x + 84.95y = 17{,}652 \quad \text{Equation 2}$$

$$\begin{aligned} -66.95x - 66.95y &= -16{,}068 \\ 66.95x + 84.95y &= 17{,}652 \\ \hline 18y &= 1584 \\ y &= 88 \\ x + 88 &= 240 \\ x &= 152 \end{aligned}$$

Answer: $x = 152$ shoes priced at \$66.95; $y = 88$ shoes priced at \$84.95

60. Let x = the amount hauled by one company; y = the amount hauled by a second company

$$x + y = 1600 \quad \text{Equation 1}$$
$$y = 4y \quad \text{Equation 2}$$

$$\begin{aligned} x + y &= 1600 \\ -x + 4y &= 0 \\ \hline 5y &= 1600 \\ y &= 320 \\ x + 320 &= 1600 \\ x &= 1280 \end{aligned}$$

Answer: One company hauled 320 tons and the other hauled 1280 tons.

62. Supply = Demand

$$25 + 0.1x = 100 = 0.05x$$
$$0.15x = 75$$
$$x = 500$$
$$p = 75$$

Equilibrium point: (500, 75)

64. Supply = Demand

$$225 + 0.0005x = 400 - 0.0002x$$
$$0.0007x = 175$$
$$x = 250{,}000$$
$$p = 350$$

Equilibrium point: (250,000, 350)

66.

$$\begin{aligned} 5b + 10a = 11.7 \Rightarrow\ & -10b - 20a = -23.4 \\ 10b + 30a = 25.6 \Rightarrow\ & 10b + 30a = 25.6 \\ & 10a = 2.2 \\ & a = 0.22 \\ & 5b + 10(0.22) = 11.7 \\ & b = 1.9 \end{aligned}$$

Least squares regression line: $y = 0.22x + 1.9$

68.

$$\begin{aligned} 6b + 15a = 23.6 \Rightarrow\ & -15b - 37.5a = -59 \\ 15b + 55a = 48.8 \Rightarrow\ & 15b + 55a = 48.8 \\ \hline & 17.5a = -10.2 \\ & a \approx -0.583 \\ & b \approx 5.390 \end{aligned}$$

Least squares regression line: $y = -0.583x + 5.390$

72. (1, 0), (2, 0), (3, 0), (3, 1), (4, 1), (4, 2), (5, 2), (6, 2)

$$\begin{aligned} 8b + 28a &= 8 \implies & -28b - 98a &= -28 \\ 28b + 116a &= 37 \implies & 28b + 116a &= 37 \\ & & 18a &= 9 \\ & & a &= \tfrac{1}{2} \\ & & 8b + 14 &= 8 \\ & & b &= -\tfrac{3}{4} \end{aligned}$$

Least squares regression line: $y = \frac{1}{2}x - \frac{3}{4}$

74. (1.0, 32), (1.5, 41), (2.0, 48), (2.5, 53)

$$\begin{aligned} 4b + 7a &= 174 \implies & -7b - 12.25a &= -304.5 \\ 7b + 13.5a &= 322 \implies & 7b + 13.5a &= 322 \\ & & 1.25a &= 17.5 \\ & & a &= 14 \\ & & 4b + 98 &= 174 \\ & & b &= 19 \end{aligned}$$

Least squares regression line: $y = 14x + 19$

When $x = 1.6$: $y = 14(1.6) + 19 = 41.4$ bushels per acre.

Section 7.3 Multivariable Linear Systems

Solutions to Even-Numbered Exercises

2. $4x - 3y - 2z = 21$ Equation 1

$6y - 5z = -8$ Equation 2

$z = -2$ Equation 3

Back-substitute $z = -2$ in Equation 2:

$6y - 5(-2) = -8$

$y = -3$

Back-substitute $z = -2$ and $y = -3$in Equation 1:

$4x - 3(-3) - 2(-2) = 21$

$4x + 13 = 21$

$x = 2$

Answer: $(2, -3, -2)$

4. $x = 8$ Equation 1

$2x + 3y = 10$ Equation 2

$x = y + 2z = 22$ Equation 3

Back-substitute $x = 8$ in Equation 2:

$2(8) + 3y = 10$

$y = -2$

Back-substitute $x = 8$ and $y = -2$ in Equation 3:

$8 - (-2) + 2z = 22$

$z = 6$

Answer: $(8, -2, 6)$

6. $5x - 8z = 22$

$3y - 5z = 10$

$z = -4$

Back-substitute $z = -4$ in Equation 2:

$3y - 5(-4) = 10 \implies y = -\frac{10}{3}$

Back-substitute $z = -4$ in Equation 1:

$5x - 8(-4) = 22 \implies x = -2$

Answer: $\left(-2, -\frac{10}{3}, -4\right)$

8. $x - 2y - 3z = 5$

$-x + 3y - 5z = 4$

$2x - 3z = 0$

Add -2 times Equation 1 to Equation 3.

$4y - 9z = -10$

This is the first step in putting the system in row-echelon form.

10.

$$\begin{array}{rcrl} x + y + z &=& 2 & \text{Equation 1} \\ -x + 3y + 2z &=& 8 & \text{Equation 2} \\ 4x + y \qquad &=& 4 & \text{Equation 3} \end{array}$$

$$\begin{array}{rcrl} x + y + z &=& 2 & \\ 4y + 3z &=& 10 & \text{Eq.1 + Eq.2} \\ -3y - 4z &=& -4 & -4\text{Eq.1 + Eq.3} \end{array}$$

$$\begin{array}{rcrl} x + y + z &=& 2 & \\ 12y + 9z &=& 30 & 3\text{Eq.2} \\ -12y - 16z &=& -16 & 4\text{Eq.3} \end{array}$$

$$\begin{array}{rcrl} x + y + z &=& 2 & \\ 12y + 9z &=& 30 & \\ -7z &=& 14 & \text{Eq.2 + Eq.3} \end{array}$$

$$\begin{array}{rcl} -7z = 14 &\Rightarrow& z = -2 \\ 12y + 9(-2) = 30 &\Rightarrow& y = 4 \\ x + 4 - 2 = 2 &\Rightarrow& x = 0 \end{array}$$

Answer: $(0, 4, -2)$

12.

$$\begin{array}{rcrl} 4x + y - 3z &=& 11 & \text{Equation 1} \\ 2x - 3y + 2z &=& 9 & \text{Equation 2} \\ x + y - z &=& -3 & \text{Equation 3} \end{array}$$

$$\begin{array}{rcrl} x + y - z &=& -3 & \text{Interchange} \\ 2x - 3y + 2z &=& 9 & \text{Equations 1 and 3} \\ 4x + y - 3z &=& 11 & \end{array}$$

$$\begin{array}{rcrl} x + y - z &=& -3 & \\ -5y + 4z &=& 15 & -2\text{Eq.1 + Eq.2} \\ -3y + z &=& 23 & -4\text{Eq.1 + Eq.3} \end{array}$$

$$\begin{array}{rcrl} x + y - z &=& -3 & \\ -15y + 12z &=& 45 & 3\text{Eq.2} \\ 15y - 5z &=& -115 & -5\text{Eq.3} \end{array}$$

$$\begin{array}{rcrl} x + y - z &=& -3 & \\ -15y + 12z &=& 45 & \\ 7z &=& -70 & \text{Eq.2 + Eq.3} \end{array}$$

$$\begin{array}{rcl} 7z = -70 &\Rightarrow& z = -10 \\ -15y + 12(-10) = 45 &\Rightarrow& y = -11 \\ x + (-11) - (-10) = -3 &\Rightarrow& x = -2 \end{array}$$

Answer: $(-2, -11, -10)$

14.

$$\begin{array}{rcrl} 2x + 4y + z &=& -4 & \text{Equation 1} \\ 2x - 4y + 6z &=& 13 & \text{Equation 2} \\ 4x - 2y + z &=& 6 & \text{Equation 3} \end{array}$$

$$\begin{array}{rcrl} 2x + 4y + z &=& -4 & \\ -8y + 5z &=& 17 & -\text{Eq.1 + Eq.2} \\ -10y - z &=& 14 & -2\text{Eq.1 + Eq.3} \end{array}$$

$$\begin{array}{rcrl} 2x + 4y + z &=& -4 & \\ -40y + 25z &=& 85 & 5\text{Eq.2} \\ -40y - 4z &=& 56 & 4\text{Eq.3} \end{array}$$

$$\begin{array}{rcrl} 2x + 4y + z &=& -4 & \\ -40y + 25z &=& 85 & \\ -29z &=& -29 & -\text{Eq.2 + Eq.3} \end{array}$$

$$\begin{array}{rcl} -29z = -29 &\Rightarrow& z = 1 \\ -40y + 25(1) = 85 &\Rightarrow& y = -\frac{3}{2} \\ 2x + 4\left(-\frac{3}{2}\right) + 1 = -4 &\Rightarrow& x = \frac{1}{2} \end{array}$$

Answer: $\left(\frac{1}{2}, -\frac{3}{2}, 1\right)$

16.

$$\begin{array}{rcrl} 5x - 3y + 2z &=& 3 & \text{Equation 1} \\ 2x + 4y - z &=& 7 & \text{Equation 2} \\ x - 11y + 4z &=& 3 & \text{Equation 3} \end{array}$$

$$\begin{array}{rcrl} x - 11y + 4z &=& 3 & \text{Interchange} \\ 5x - 3y + 2z &=& 3 & \text{Equations 1 and 3} \\ 2x + 4y - z &=& 7 & \end{array}$$

$$\begin{array}{rcrl} x - 11y + 4z &=& 3 & \\ 52y - 18z &=& -12 & -5\text{Eq.1 + Eq.2} \\ 26y - 9z &=& 1 & -2\text{Eq.1 + Eq.3} \end{array}$$

$$\begin{array}{rcrl} x - 11y + 4z &=& 3 & \\ 52y - 18z &=& -12 & \\ 0 &=& 7 & -\frac{1}{2}\text{Eq.2 + Eq.3} \end{array}$$

Inconsistent; no solution

18.
$$\begin{aligned} 2x + y + 3z &= 1 &\quad& \text{Equation 1}\\ 2x + 6y + 8z &= 3 && \text{Equation 2}\\ 6x + 8y + 18z &= 5 && \text{Equation 3}\end{aligned}$$

$$\begin{aligned} 2x + y + 3z &= 1 &\quad& \\ 5y + 5z &= 2 && -\text{Eq.1} + \text{Eq.2}\\ 5y + 9z &= 2 && -3\text{Eq.1} + \text{Eq.3}\end{aligned}$$

$$\begin{aligned} 2x + y + 3z &= 1 &\quad& \\ 5y + 5z &= 2 && \\ 4z &= 0 && -\text{Eq.2} + \text{Eq.3}\end{aligned}$$

$$4z = 0 \Rightarrow z = 0$$
$$5y + 5(0) = 2 \Rightarrow y = \tfrac{2}{5}$$
$$2x + \tfrac{2}{5} + 3(0) = 1 \Rightarrow x = \tfrac{3}{10}$$

Answer: $\left(\frac{3}{10}, \frac{2}{5}, 0\right)$

20.
$$\begin{aligned} 2x + y - 3z &= 4 &\quad& \text{Equation 1}\\ 4x + 2z &= 10 && \text{Equation 2}\\ -2x + 3y - 13z &= -8 && \text{Equation 3}\end{aligned}$$

$$\begin{aligned} 2x + y - 3z &= 4 &\quad& -2\text{Eq.1} + \text{Eq.2}\\ -2y + 8z &= 2 && \text{Eq.1} + \text{Eq.3}\\ 4y - 16z &= -4 && \end{aligned}$$

$$\begin{aligned} 2x + y - 3z &= 4 &\quad& \\ y - 4z &= -1 && -\tfrac{1}{2}\text{Eq.2}\\ 0 &= 0 && 2\text{Eq.2} + \text{Eq.3}\end{aligned}$$

$$\begin{aligned} 2x + z &= 5 &\quad& -\text{Eq.2} + \text{Eq.1}\\ y - 4z &= -1 && \end{aligned}$$

$z = a$

$y = 4a - 1$

$x = -\frac{1}{2}a + \frac{5}{2}$

Answer: $\left(-\frac{1}{2}a + \frac{5}{2}, 4a - 1, a\right)$

22.
$$\begin{aligned} x + 4z &= 13 &\quad& \text{Equation 1}\\ 4x - 2y + z &= 7 && \text{Equation 2}\\ 2x - 2y - 7z &= -19 && \text{Equation 3}\end{aligned}$$

$$\begin{aligned} x + 4z &= 13 &\quad& \\ -2y - 15z &= -45 && -4\text{Eq.1} + \text{Eq.2}\\ -2y - 15z &= -45 && -2\text{Eq.1} + \text{Eq.3}\end{aligned}$$

$$\begin{aligned} x + 4z &= 13 &\quad& \\ -2y - 15z &= -45 && \\ 0 &= 0 && -\text{Eq.2} + \text{Eq.3}\end{aligned}$$

$z = a$

$y = -\frac{15}{2}a + \frac{45}{2}$

$x = -4a + 13$

Answer: $\left(-4a + 3, -\frac{15}{2}a + \frac{45}{2}, a\right)$

24.
$$\begin{aligned} x - 3y + 2z &= 18 &\quad& \text{Equation 1}\\ 5x - 13y + 12z &= 80 && \text{Equation 2}\end{aligned}$$

$$\begin{aligned} x - 3y + 2z &= 18 &\quad& \\ 2y + 2z &= -10 && -5\text{Eq.1} + \text{Eq.2}\end{aligned}$$

$$\begin{aligned} x - 3y + 2z &= 18 &\quad& \\ y + z &= -5 && \tfrac{1}{2}\text{Eq.2}\end{aligned}$$

$$\begin{aligned} x + 5z &= 3 &\quad& 3\text{Eq.2} + \text{Eq.1}\\ y + z &= -5 && \end{aligned}$$

Let $z = a$,

then $y = -a - 5$, and $x = -5a + 3$.

Answer: $(-5a + 3, -a - 5, a)$

26.
$$\begin{aligned} 2x + 3y + 3z &= 7 &\quad& \text{Equation 1}\\ 4x + 18y + 15z &= 44 && \text{Equation 2}\end{aligned}$$

$$\begin{aligned} 2x + 3y + 3z &= 7 &\quad& \\ 12y + 9z &= 30 && -2\text{Eq.1} + \text{Eq.2}\end{aligned}$$

$$\begin{aligned} 2x + \tfrac{3}{4}z &= -\tfrac{1}{2} &\quad& -\tfrac{1}{4}\text{Eq.2}+\text{Eq.1}\\ 12y + 9z &= 30 && \end{aligned}$$

Let $z = a$, then:

$$12y + 9a = 30 \Rightarrow y = -\tfrac{3}{4}a + \tfrac{5}{2}$$
$$2x + \tfrac{3}{4}a = -\tfrac{1}{2} \Rightarrow x = -\tfrac{3}{8}a - \tfrac{1}{4}$$

Answer: $\left(-\frac{3}{8}a - \frac{1}{4}, -\frac{3}{4}a + \frac{5}{2}, a\right)$

28.
$$\begin{aligned} x + y + z + w &= 6 && \text{Equation 1}\\ 2x + 3y \quad - w &= 0 && \text{Equation 2}\\ -3x + 4y + z + 2w &= 4 && \text{Equation 3}\\ x + 2y - z + w &= 0 && \text{Equation 4}\end{aligned}$$

$$\begin{aligned} x + y + z + w &= 6 &&\\ y - 2z - 3w &= -12 && -2\text{Eq.1} + \text{Eq.2}\\ 7y + 4z + 5w &= 22 && 3\text{Eq.1} + \text{Eq.3}\\ y - 2z &= -6 && -\text{Eq.1} + \text{Eq.4}\end{aligned}$$

$$\begin{aligned} x + y + z + w &= 6 &&\\ y - 2z - 3w &= -12 &&\\ 18z + 26w &= 106 && -7\text{Eq.2} + \text{Eq.3}\\ 3w &= 6 && -\text{Eq.2} + \text{Eq.4}\end{aligned}$$

$$\begin{aligned} 3w &= 6 \Rightarrow w = 2\\ 18z + 26(2) &= 106 \Rightarrow z = 3\\ y - 2(3) - 3(2) &= -12 \Rightarrow y = 0\\ x + 0 + 3 + 2 &= 6 \Rightarrow x = 1\end{aligned}$$

Answer: $(1, 0, 3, 2)$

30.
$$\begin{aligned} 3x - 2y - 6z &= -4 && \text{Equation 1}\\ -3x + 2y + 6z &= 1 && \text{Equation 2}\\ x - y - 5z &= -3 && \text{Equation 3}\end{aligned}$$

$$\begin{aligned} x - y - 5z &= -3 && \text{Interchange the}\\ 3x - 2y - 6z &= -4 && \text{equations}\\ -3x + 2y + 6z &= 1 &&\end{aligned}$$

$$\begin{aligned} x - y - 5z &= -3 &&\\ y + 9z &= 5 && -3\text{Eq.1} + \text{Eq.2}\\ -y - 9z &= -8 && 3\text{Eq.1} + \text{Eq.3}\end{aligned}$$

$$\begin{aligned} x - y - 5z &= -3 &&\\ y + 9z &= 5 &&\\ 0 &= -3 && \text{Eq.2} + \text{Eq.3}\end{aligned}$$

Inconsistent, no solution

32.
$$\begin{aligned} 4x + 3y + 17z &= 0 && \text{Equation 1}\\ 5x + 4y + 22z &= 0 && \text{Equation 2}\\ 4x + 2y + 19z &= 0 && \text{Equation 3}\end{aligned}$$

$$\begin{aligned} 5x + 4y + 22z &= 0 && \text{Interchange the}\\ 4x + 3y + 17z &= 0 && \text{equations}\\ 4x + 2y + 19z &= 0 &&\end{aligned}$$

$$\begin{aligned} x + y + 5z &= 0 && -\text{Eq.2} + \text{Eq.1}\\ 4x + 3y + 17z &= 0 &&\\ 4x + 2y + 19z &= 0 &&\end{aligned}$$

$$\begin{aligned} x + y + 5z &= 0 &&\\ -y - 3z &= 0 && -4\text{Eq.1} + \text{Eq.2}\\ -2y - z &= 0 && -4\text{Eq.1} + \text{Eq.3}\end{aligned}$$

$$\begin{aligned} x + y + 5z &= 0 &&\\ y + 3z &= 0 && -\text{Eq.2}\\ 5z &= 0 && -2\text{Eq.2} + \text{Eq.3}\end{aligned}$$

$$\begin{aligned} 5z = 0 &\Rightarrow z = 0\\ y + 3(0) = 0 &\Rightarrow y = 0\\ x + 0 + 5(0) = 0 &\Rightarrow x = 0\end{aligned}$$

Answer: $(0, 0, 0)$

34.
$$\begin{aligned} 5x + 5y - z &= 0 && \text{Equation 1}\\ 10x + 5y + 2z &= 0 && \text{Equation 2}\\ 5x + 15y - 9z &= 0 && \text{Equation 3}\end{aligned}$$

$$\begin{aligned} 5x + 5y - z &= 0 &&\\ -5y + 4z &= 0 && -2\text{Eq.1} + \text{Eq.2}\\ 10y - 8z &= 0 && -\text{Eq.1} + \text{Eq.3}\end{aligned}$$

$$\begin{aligned} 5x \quad + 3z &= 0 && \text{Eq.2} + \text{Eq.1}\\ -5y + 4z &= 0 &&\\ 0 &= 0 && 2\text{Eq.2} + \text{Eq.3}\end{aligned}$$

$z = a$

$y = \frac{4}{5}a$

$x = -\frac{3}{5}a$

Answer: $\left(-\frac{3}{5}a, \frac{4}{5}a, a\right)$

36. When using Gaussian elimination to solve a system of linear equations, a system has no solution when there is a row representing a contradictory equation such as $0 = N$, where N is a nonzero real number.

For instance:
$$\begin{aligned} x + y &= 3 && \text{Equation 1}\\ -x - y &= 3 && \text{Equation 2}\end{aligned}$$

$$\begin{aligned} x + y &= 0 &&\\ 0 &= 6 && \text{Eq.1} + \text{Eq.2}\end{aligned}$$

No solution

38. There are an infinite number of linear systems that have $\left(-\frac{3}{2}, 4, -7\right)$ as their solution. One such system is:

$$2\left(-\tfrac{3}{2}\right) + 4 - (-7) = 8 \Rightarrow 2x + y - z = 8$$
$$4\left(-\tfrac{3}{2}\right) + 2(4) + (-7) = -5 \Rightarrow 4x + 2y + z = -5$$
$$-2\left(-\tfrac{3}{2}\right) + 5(4) - 3(-7) = 44 \Rightarrow -2x + 5y - 3z = 44$$

40. $y = ax^2 + bx + c$ passing through $(0, 3), (1, 4), (2, 3)$

$(0, 3)$: $3 = c$

$(1, 4)$: $4 = a + b + c \Rightarrow 1 = a + b$

$(2, 3)$: $3 = 4a + 2b + c \Rightarrow 0 = 2a + b$

Answer: $a = -1, b = 2, c = 3$

The equation of the parabola is $y = -x^2 + 2x + 3$.

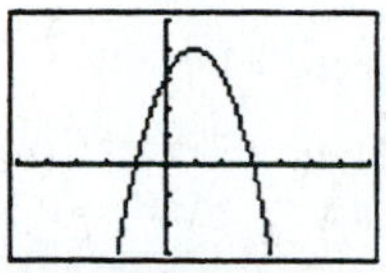

42. $y = ax^2 + bx + c$ passing through $(1, 3), (2, 2), (3, -3)$

$(1, 3)$: $3 = a + b + c$

$(2, 2)$: $2 = 4a + 2b + c \Rightarrow -1 = 3a + b$

$(3, -3)$: $-3 = 9a + 3b + c \Rightarrow -6 = 8a + 2b$

Answer: $a = -2, b = 5, c = 0$

The equation of the parabola is $y = -2x^2 + 5x$.

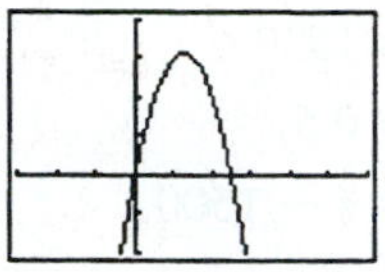

44. $x^2 + y^2 + Dx + Ey + F = 0$ passing through $(0, 0), (0, 6), (3, 3)$

$(0, 0)$: $F = 0$

$(0, 6)$: $36 + 6E + F = 0 \Rightarrow E = -6$

$(3, 3)$: $18 + 3D + 3E + F = 0 \Rightarrow D = 0$

The equation of the circle is $x^2 + y^2 - 6y = 0$.

To graph, complete the square first, then solve for y.

$$x^2 + y^2 - 6y + 9 = 9$$
$$x^2 + (y - 3)^2 = 9$$
$$(y - 3)^2 = 9 - x^2$$
$$y - 3 = \pm\sqrt{9 - x^2}$$
$$y = 3 \pm \sqrt{9 - x^2}$$

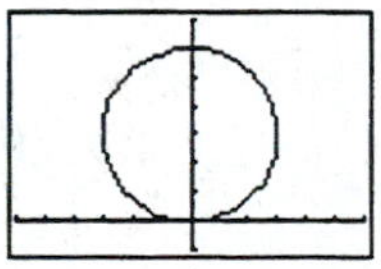

Let $y_1 = 3 + \sqrt{9 - x^2}$ and $y_2 = 3 - \sqrt{9 - x^2}$.

46. $x^2 + y^2 + Dx + Ey + F = 0$ passing through $(0, 0), (0, -2), (3, 0)$

$(0, 0)$: $F = 0$

$(0, -2)$: $4 - 2E + F = 0 \Rightarrow E = 2$

$(3, 0)$: $9 + 3D + F = 0 \Rightarrow D = -3$

The equation of the circle is $x^2 + y^2 - 3x + 2y = 0$.

To graph, complete the squares first, then solve for y.

$$x^2 - 3x + \tfrac{9}{4} + y^2 + 2y + 1 = \tfrac{9}{4} + 1$$
$$\left(x - \tfrac{3}{2}\right)^2 + (y + 1)^2 = \tfrac{13}{4}$$
$$(y + 1)^2 = \tfrac{13}{4} - \left(x - \tfrac{3}{2}\right)^2$$
$$y + 1 = \pm\sqrt{\tfrac{13}{4} - \left(x - \tfrac{3}{2}\right)^2}$$
$$y = -1 \pm \sqrt{\tfrac{13}{4} - \left(x - \tfrac{3}{2}\right)^2}$$

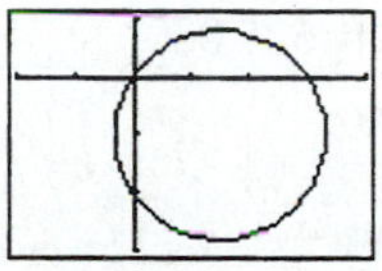

Let $y_1 = -1 + \sqrt{\frac{13}{4} - \left(x - \frac{3}{2}\right)^2}$ and $y_2 = -1 - \sqrt{\frac{13}{4} - \left(x - \frac{3}{2}\right)^2}$.

48. $s = \frac{1}{2}at^2 + v_0t + s_0$

(1, 48), (2, 64), (3, 48)

$48 = \frac{1}{2}a + v_0 + s_0 \Rightarrow a + 2v_0 + 2s_0 = 96$

$64 = 2a + 2v_0 + s_0 \Rightarrow 2a + 2v_0 + s_0 = 64$

$48 = \frac{9}{2}a + 3v_0 + s_0 \Rightarrow 9a + 6v_0 + 2s_0 = 96$

Solving this system yields $a = -32$, $v_0 = 64$, $s_0 = 0$.

Thus, $s = \frac{1}{2}(-32)t^2 + 64t + 0$

$= -16t^2 + 64t.$

50. $s = \frac{1}{2}at^2 + v_0t + s_0$

(1, 132), (2, 100), (3, 36)

$132 = \frac{1}{2}a + v_0 + s_0 \Rightarrow a + 2v_0 + 2s_0 = 264$

$100 = 2a + 2v_0 + s_0 \Rightarrow 2a + 2v_0 + s_0 = 100$

$36 = \frac{9}{2}a + 3v_0 + s_0 \Rightarrow 9a + 6v_0 + 2s_0 = 72$

Solving this system yields $a = -32$, $b = 16$, $c = 132$.

Thus, $s = \frac{1}{2}(-32)t^2 + 16t + 132$

$= -16t^2 + 16t + 132.$

52. Let x = amount at 5%
Let y = amount at 7%
Let z = amount at 8%

$0.05x + 0.07y + 0.08z = 1520$

$x = 0.5y$

$y = z - 1500$

$5(0.5y) + 7y + 8(y + 1500) = 152{,}000$

$17.5y = 140{,}000$

$y = 8000$

$x = 0.5(8000) = 4000$

$z = 8000 + 1500 = 9500$

Check: $0.05(4000) + 0.07(8000) + 0.08(9500) = 1520$

Answer: x = \$4000 at 5%
y = \$8000 at 7%
z = \$9500 at 8%

54. Let x = amount at 8%
Let y = amount at 9%
Let z = amount at 10%

$x + y + z = 800{,}000$

$0.08z + 0.09y + 0.10z = 67{,}000$

$x = 5z$

$y + 6z = 800{,}000$

$0.09y + 0.5z = 67{,}000$

$z = 125{,}000$

$y = 800{,}000 - 6(125{,}000) = 50{,}000$

$x = 5(125{,}000) = 625{,}000$

Answer: x = \$625,000 at 8%
y = \$50,000 at 9%
z = \$125,000 at 10%

56. Let C = amount in certificates of deposit
Let M = amount in municipal bonds
Let B = amount in blue-chip stocks
Let G = amount in growth or speculative stocks

$C + M + B + G = 500{,}000$

$0.09C + 0.05M + 0.12B + 0.14G = 0.10(500{,}000)$

$B + G = \frac{1}{4}(500{,}000)$

This system has infinitely many solutions.

Let $G = s$, then $B = 125{,}000 - s$

$M = \frac{1}{2}s - 31{,}250$

$C = 406{,}250 - \frac{1}{2}s.$

Answer:

$\left(406{,}250 - \frac{1}{2}s, -31{,}250 + \frac{1}{2}s, 125{,}000 - s, s\right)$

One possible solution is to let s = \$100,000.

Certificates of deposit: \$356,250
Municipal bonds: \$18,750
Blue-chip stocks: \$25,000
Growth or speculative stocks: \$100,000

58. (a) To use as little of the 50% solution as possible, the chemist should use no 10% solution.

$$\begin{aligned} x(0.20) + (10 - x)(0.50) &= 10(0.25) \\ x(0.20) + 5 - 0.50x &= 2.5 \\ -0.30x &= -2.5 \\ x &= 8\tfrac{1}{3} \text{ liters of 20\% solution} \\ 10 - x &= 1\tfrac{2}{3} \text{ liters of 50\% solution} \end{aligned}$$

(b) To use as much 50% solution as possible, the chemist should use no 20% solution.

$$\begin{aligned} x(0.10) + (10 - x)0.50 &= 10(0.25) \\ 0.10x + 5 - 0.50x &= 2.5 \\ -0.40x &= -2.5 \\ x &= 6\tfrac{1}{4} \text{ liters of 10\% solution} \\ 10 - x &= 3\tfrac{3}{4} \text{ liters of 50\% solution} \end{aligned}$$

(c) To use 2 liters of 50% solution we let $x =$ the number of liters at 10% and $y =$ the number of liters at 20%.

$$\begin{aligned} 0.10x + 0.20y + 2(0.50) &= 10(0.25) \qquad \text{Equation 1} \\ x + y &= 8 \qquad \text{Equation 2} \end{aligned}$$

Answer: $y = 7$ liters of 20% solution; $x = 1$ liter of 10% solution

60.
$$\begin{aligned} I_1 - I_2 + I_3 &= 0 \qquad \text{Equation 1} \\ 3I_1 + 2I_2 &= 7 \qquad \text{Equation 2} \\ 2I_2 + 4I_3 &= 8 \qquad \text{Equation 3} \end{aligned}$$

$$\begin{aligned} I_1 - I_2 + I_3 &= 0 \\ 5I_2 - 3I_3 &= 7 \qquad -3\text{Eq.1} + \text{Eq.2} \\ 2I_2 + 4I_3 &= 8 \end{aligned}$$

$$\begin{aligned} I_1 - I_2 + I_3 &= 0 \\ 10I_2 - 6I_3 &= 14 \qquad 2\text{Eq.2} \\ 10I_2 + 20I_3 &= 40 \qquad 5\text{Eq.3} \end{aligned}$$

$$\begin{aligned} I_1 - I_2 + I_3 &= 0 \\ 10I_2 - 6I_3 &= 14 \\ 26I_3 &= 26 \qquad -\text{Eq.2} + \text{Eq.3} \end{aligned}$$

$$\begin{aligned} 26I_3 = 26 &\Rightarrow I_3 = 1 \\ 10I_2 - 6(1) = 14 &\Rightarrow I_2 = 2 \\ I_1 - 2 + 1 = 0 &\Rightarrow I_1 = 1 \end{aligned}$$

Answer: $I_1 = 1$ ampere, $I_2 = 2$ amperes, $I_3 = 1$ ampere

62.
$$\begin{aligned} t_1 - 2t_2 &= 0 \qquad \text{Equation 1} \\ t_1 - 2a &= 128 \qquad \text{Equation 2} \\ t_2 + 2a &= 64 \qquad \text{Equation 3} \end{aligned}$$

$$\begin{aligned} t_1 - 2t_2 &= 0 \\ 2t_2 - 2a &= 128 \qquad -\text{Eq.1} + \text{Eq.2} \\ t_2 + 2a &= 64 \end{aligned}$$

$$\begin{aligned} t_1 - 2t_2 &= 0 \\ 2t_2 - 2a &= 128 \\ 3a &= 0 \qquad -\tfrac{1}{2}\text{Eq.2} + \text{Eq.3} \end{aligned}$$

$$\begin{aligned} 3a = 0 &\Rightarrow a = 0 \\ 2t_2 - 2(0) = 128 &\Rightarrow t_2 = 64 \\ t_1 - 2(64) = 0 &\Rightarrow t_1 = 128 \end{aligned}$$

Answer: $a = 0$ ft/sec^2, $t_1 = 128$ lb, $t_2 = 64$ lb

64.
$$\frac{3}{x^2 + x - 2} = \frac{A}{x - 1} + \frac{B}{x + 2}$$

$$\begin{aligned} 3 &= A(x + 2) + B(x - 1) \\ 3 &= Ax + 2A + Bx - B \\ 3 &= (A + B)x + (2A - B) \end{aligned}$$

By equating coefficients, we have: $0 = A + B \Rightarrow A = -B$

$3 = 2A - B \Rightarrow 3 = 2(-B) - B \Rightarrow B = -1$

$A = -(-1) = 1$

$$\frac{3}{x^2 + x - 2} = \frac{1}{x - 1} - \frac{1}{x + 2}$$

66. $\dfrac{12}{x(x-2)(x+3)} = \dfrac{A}{x} + \dfrac{B}{x-2} + \dfrac{C}{x+3}$

$$12 = A(x-2)(x+3) + Bx(x+3) + Cx(x-2)$$
$$12 = Ax^2 + Ax - 6A + Bx^2 + 3Bx + Cx^2 - 2Cx$$
$$12 = (A+B+C)x^2 + (A+3B-2C)x - 6A$$

By equating coefficients, we have

$0 = A + B + C$

$0 = A + 3B - 2C$

$12 = -6A \implies A = -2$

$2 = B + C \implies 4 = 2B + 2C$

$2 = 3B - 2C \qquad 2 = 3B - 2C$

$6 = 5B$

$B = \frac{6}{5}$

$C = \frac{4}{5}$

$$\frac{12}{x(x-2)(x+3)} = -\frac{2}{x} + \frac{6}{5(x-2)} + \frac{4}{5(x+3)}$$

68. Least squares regression parabola through $(-2, 0)$, $(-1, 0)$, $(0, 1)$, $(1, 2)$, $(2, 5)$

$n = 5$

$\sum x_i = 0 \qquad \sum y_i = 8 \qquad 5c + 10a = 8$

$\sum x_i^2 = 10 \qquad \sum x_i^3 = 0 \qquad 10b = 12$

$\sum x_i^4 = 34 \qquad \sum x_i y_i = 12 \qquad 10c + 34a = 22$

$\sum x_i^2 y_i = 22$

Solving this system yields $a = \frac{3}{7}$, $b = \frac{6}{5}$, $c = \frac{26}{35}$. Thus, $y = \frac{3}{7}x^2 + \frac{6}{5}x + \frac{26}{35}$.

70. Least squares regression parabola through $(0, 10)$, $(1, 9)$, $(2, 6)$, $(3, 0)$

$n = 4$

$\sum x_i = 6 \qquad \sum y_i = 25 \qquad 4c + 6b + 14a = 25$

$\sum x_i^2 = 104 \qquad \sum x_i^3 = 36 \qquad 6c + 14b + 36a = 21$

$\sum x_i^4 = 98 \qquad \sum x_i y_i = 21 \qquad 14c + 36b + 98a = 33$

$\sum x_i^2 y_i = 33$

Solving this system yields $a = -\frac{5}{4}$, $b = \frac{9}{20}$, $c = \frac{199}{20}$. Thus, $y = -\frac{5}{4}x^2 + \frac{9}{20}x + \frac{199}{20}$.

72. (a) Using the quadratic least squares regression feature, we find

$y = -0.008x^2 + 1.371x + 21.886.$

(b)

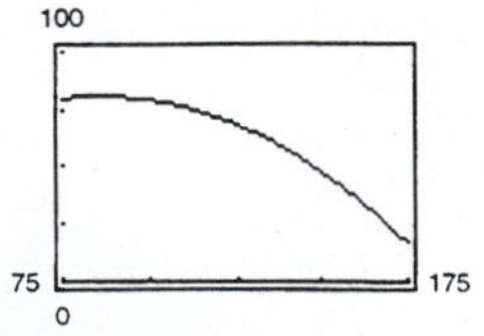

(c) For $x = 170$:

$y = -0.008(170)^2 + 1.371(170) + 21.886$

$= 23.756\%$

74. $\left.\begin{array}{l} 2x + \lambda = 0 \\ 2y + \lambda = 0 \end{array}\right\} \; x = y = -\dfrac{\lambda}{2}$

$x + y - 4 = 0 \implies 2x - 4 = 0$

$2x = 4$

$x = 2$

$y = 2$

$\lambda = -4$

76. $$\begin{aligned} 2 + 2y + 2\lambda &= 0 \\ 2x + 1 + \lambda &= 0 \Rightarrow \lambda = -2x - 1 \\ 2x + y - 100 &= 0 \end{aligned}$$

$$2 + 2y + 2(-2x - 1) = 0 \Rightarrow -4x + 2y = 0 \Rightarrow -4x + 2y = 0$$
$$2x + y - 100 = 0 \Rightarrow 2x + y = 100 \Rightarrow 4x + 2y = 200$$
$$4y = 200$$
$$y = 50$$
$$x = 25$$
$$\lambda = -2(25) - 1 = -51$$

78. $P = 47t^2 - 195t + 1109$

$S = 160t + 705$

$P = S$

$47t^2 - 195t + 1109 = 160t + 705$

$47t^2 - 355t + 404 = 0$

$$t = \frac{355 \pm \sqrt{(-355)^2 - 4(47)(404)}}{94} = \frac{355 \pm \sqrt{50{,}073}}{94} \approx 1.4, 6.2$$

Thus, compact pickup sales will again exceed compact utility sales in 1996.

80. $225 = x(150)$

$x = 1.5$ or 150%

82. $0.48x = 132$

$x = 275$

Section 7.4 Systems of Inequalities

Solutions to Even-Numbered Exercises

2. $x \le 4$

Using a solid line, graph the vertical line $x = 4$, and shade to the left of this line.

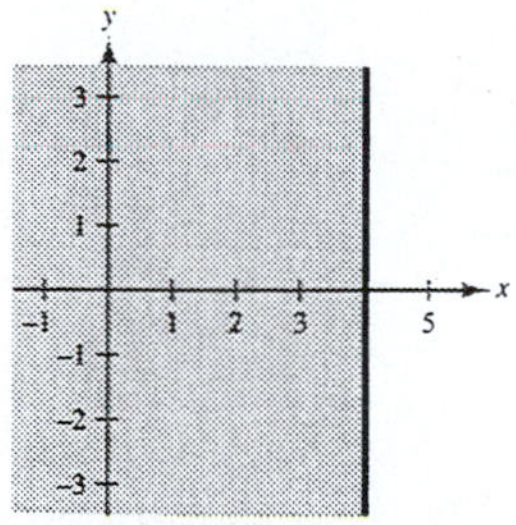

4. $y \le 3$

Using a solid line, graph the horizontal line $y = 3$, and shade below this line.

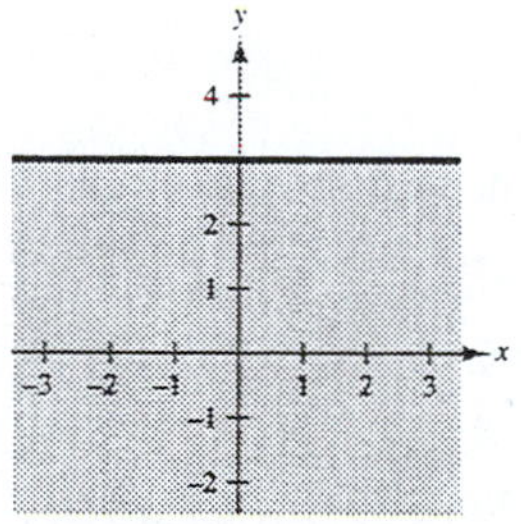

6. $y > 2x - 4$

Using a dashed line, graph $y = 2x - 4$, and shade above the line. (Use (0, 0) as a test point.)

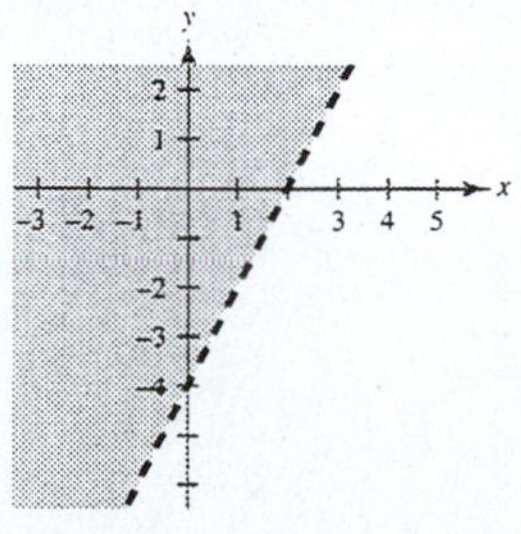

8. $5x + 3y \ge -15$

Using a solid line, graph $5x + 3y = -15$, and shade above the line. (Use (0, 0) as a test point.)

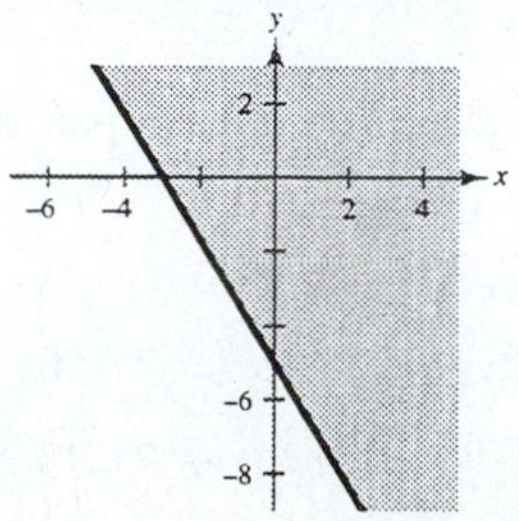

10. $y^2 - x < 0$

Using a dashed line, graph the parabola $y^2 - x = 0$, and shade to the right of this parabola. (Use $(1, 0)$ as a test point.)

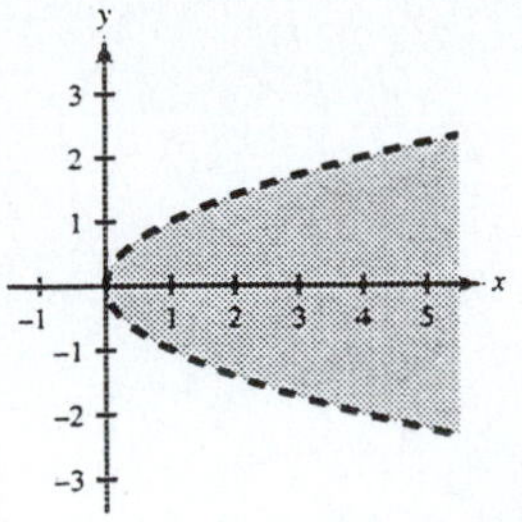

12. $y < \ln x$

Using a dashed line, graph $y = \ln x$, and shade to the right of the curve. (Use $(2, 0)$ as a test point.)

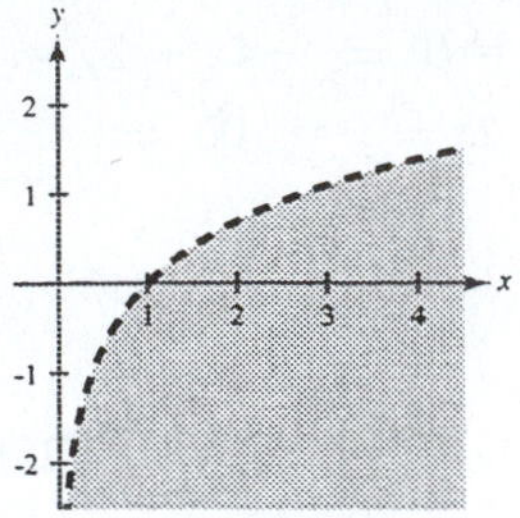

14. $y \le 6 - \frac{3}{2}x$

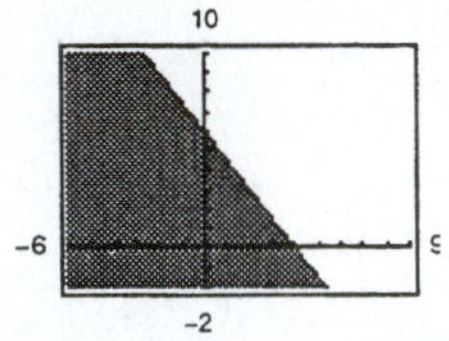

16. $2x^2 - y - 3 > 0$

$y < 2x^2 - 3$

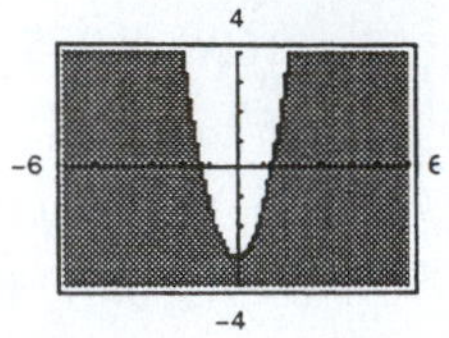

18. The parabola through $(-2, 0)$, $(0, -4)$, $(2, 0)$ is $y = x^2 - 4$. For the shaded region inside the parabola, we have $y \ge x^2 - 4$.

20. The circle shown is $x^2 + y^2 = 9$. For the shaded region inside the circle, we have $x^2 + y^2 \le 9$.

22. $3x + 2y < 6$

$x \quad > 0$

$y > 0$

First, find the points of intersection of each pair of equations.

Vertex A	Vertex B	Vertex C
$3x + 2y = 6$	$x = 0$	$3x + 2y = 6$
$x = 0$	$y = 0$	$y = 0$
$(0, 3)$	$(0, 0)$	$(2, 0)$

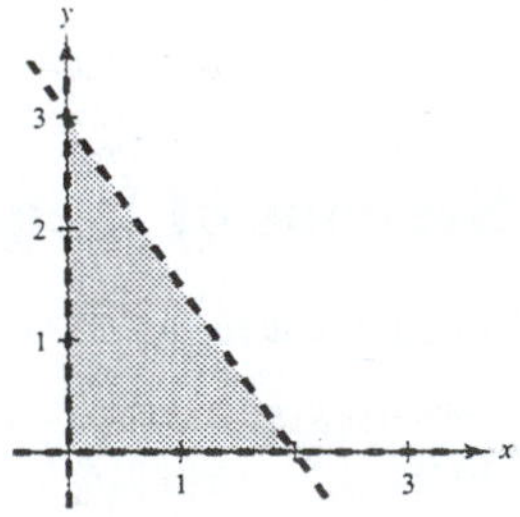

24. $2x^2 + y \ge 2$

$x \quad \le 2$

$y \le 1$

First, find the points of intersection of each pair of equations.

Vertex A	Vertex B	Vertex C
$2x + y = 2$	$x = 2$	$2x^2 + y = 2$
$x = 2$	$y = 1$	$y = 1$
$(2, -6)$	$(2, 1)$	$\left(\frac{\sqrt{2}}{2}, 1\right)$

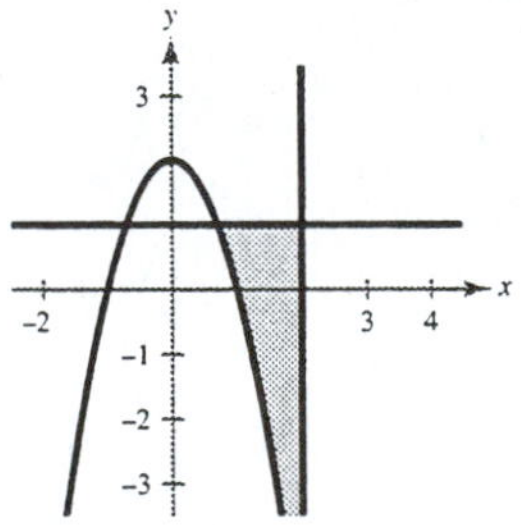

26. $x - 7y > -36$

$5x + 2y > \quad 5$

$6x - 5y > \quad 6$

First, find the points of intersection of each pair of equations.

Vertex A	Vertex B	Vertex C
$x - 7y = -36$	$5x + 2y = 5$	$x - 7y = -36$
$5x + 2y = \quad 5$	$6x - 5y = 6$	$6x - 5y = \quad 6$
$(-1, 5)$	$(1, 0)$	$(6, 6)$

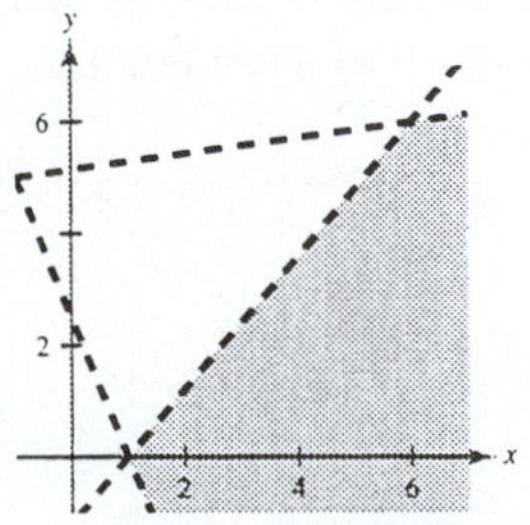

28. $x - 2y < -6$

$5x - 3y > -9$

Point of intersection: (0, 3)

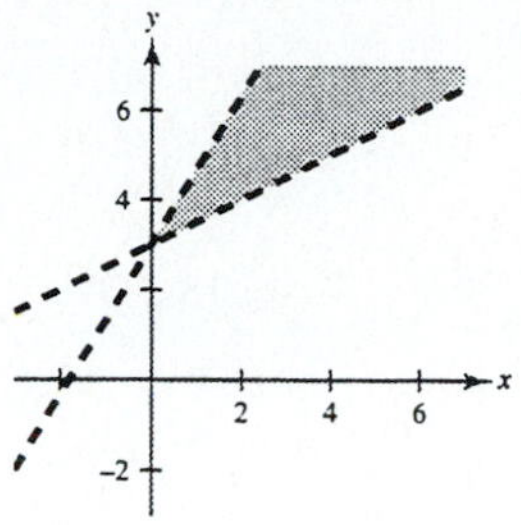

30. $x - y^2 > 0$

$x - y < 2$

Points of intersection:

$$y^2 = y + 2$$
$$y^2 - y - 2 = 0$$
$$(y + 1)(y - 2) = 0$$
$$y = -1, 2$$

(1, −1), (4, 2)

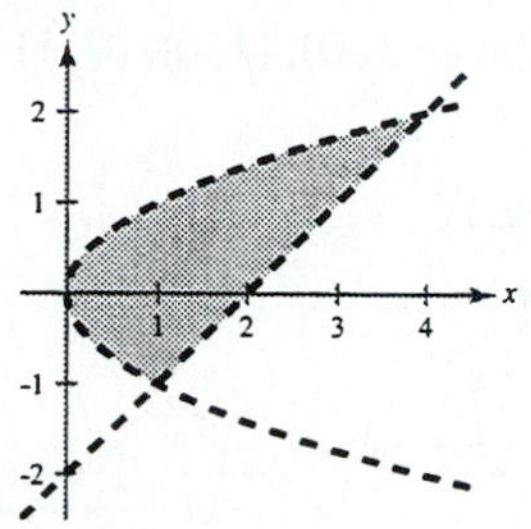

32. $x^2 + y^2 \le 25$

$4x - 3y \le 0$

Points of intersection:

$$x^2 + \left(\tfrac{4}{3}x\right)^2 = 25$$
$$\tfrac{25}{9}x^2 = 25$$
$$x = \pm 3$$

(−3, −4), (3, 4)

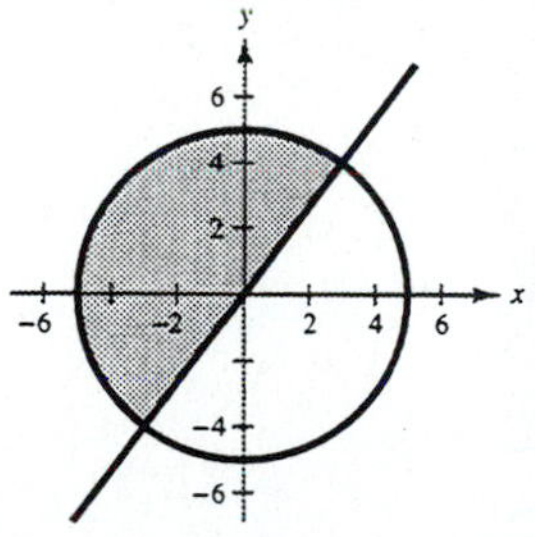

34. $x < 2y - y^2$

$0 < x + y$

Points of intersection:

$$-y = 2y - y^2$$
$$y^2 - 3y = 0$$
$$y = 0, 3$$

(0, 0), (−3, 3)

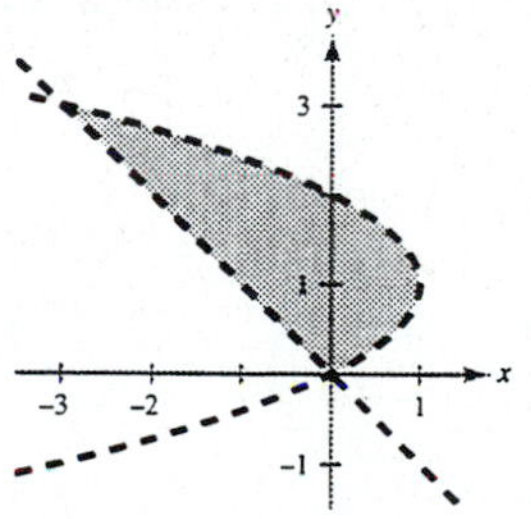

36. $y < -x^2 + 2x + 3$

$y > x^2 - 4x + 3$

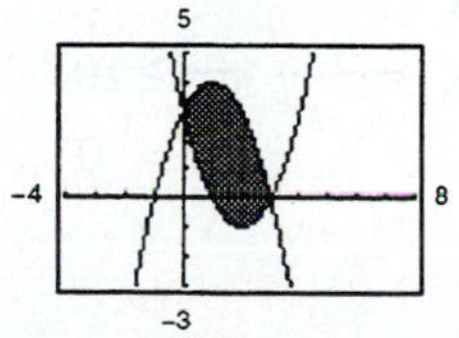

38. $y \ge x^4 - 2x^2 + 1$

$y \le 1 - x^2$

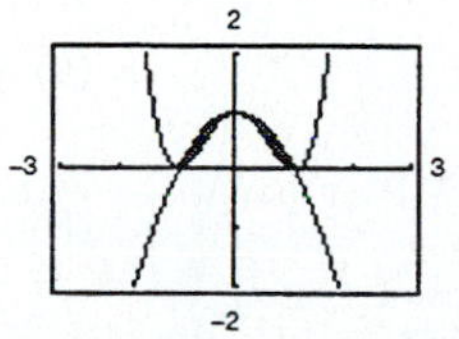

40. $y \le e^{-x^2/2}$

$y \ge 0$

$-2 \le x \le 2$

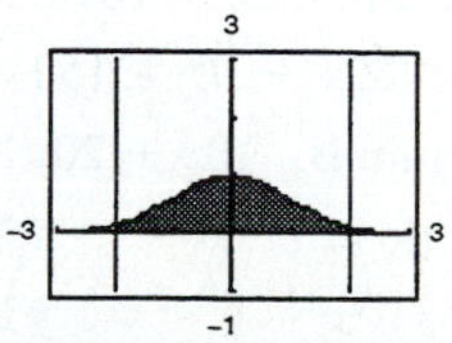

42. (0, 6), (3, 0)

Line: $y < 6 - 2x$

(0, −3), (3, 0)

Line: $y \ge x - 3$

$x \ge 1$

44. Circle: $x^2 + y^2 > 4$

46. Circle: $x^2 + y^2 \le 16$

$x \ge 0$

$y > x$

48. Parallelogram with vertices at $(0, 0)$, $(4, 0)$, $(1, 4)$, $(5, 4)$

$(0, 0), (4, 0)$: $y \geq 0$

$(4, 0), (5, 4)$: $4x - y \leq 16$

$(1, 4), (5, 4)$: $y \leq 4$

$(0, 0), (1, 4)$: $4x - y \geq 0$

$4x - y \geq 0$

$4x - y \leq 16$

$0 \leq y \leq 4$

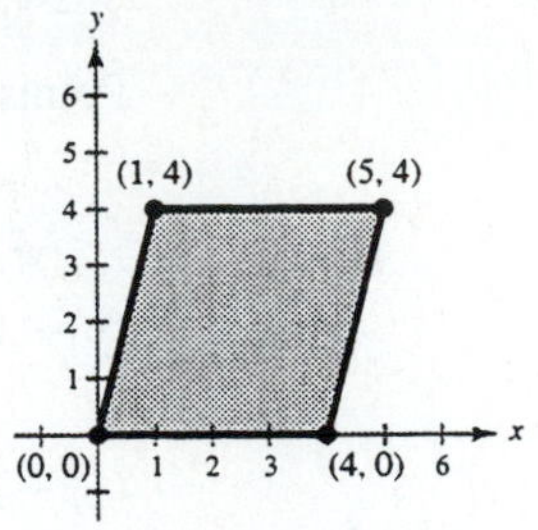

50. Triangle with vertices at $(-1, 0)$, $(1, 0)$, $(0, 1)$

$(-1, 0), (1, 0)$: $y \geq 0$

$(-1, 0), (0, 1)$: $y \leq x + 1$

$(0, 1), (1, 0)$: $y \leq -x + 1$

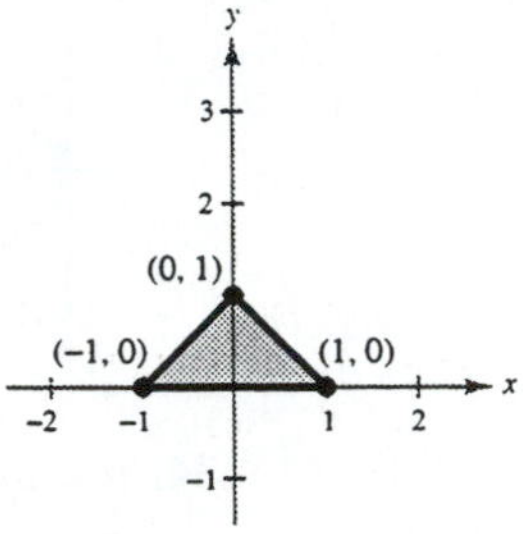

52. x = number of model A

y = number of model B

Demand: $x \geq 2y$

Cost: $8x + 12y \leq 200$

Inventory: $x \geq 4$

$y \geq 2$

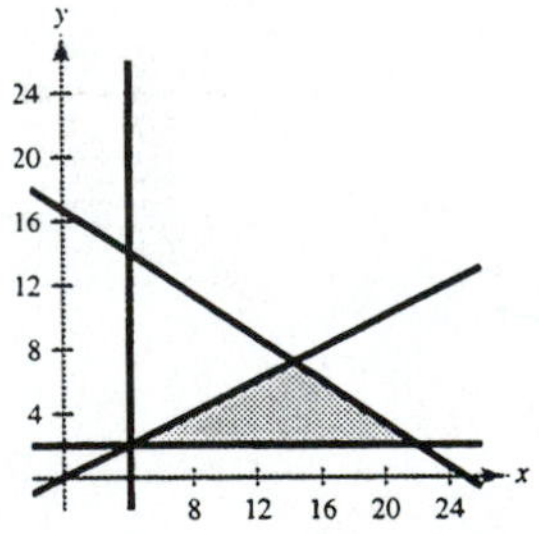

54. x = number of \$15 tickets

y = number of \$25 tickets

$x + y \geq 15{,}000$

$x \geq 8000$

$y \geq 4000$

$15x + 25y \geq 275{,}000$

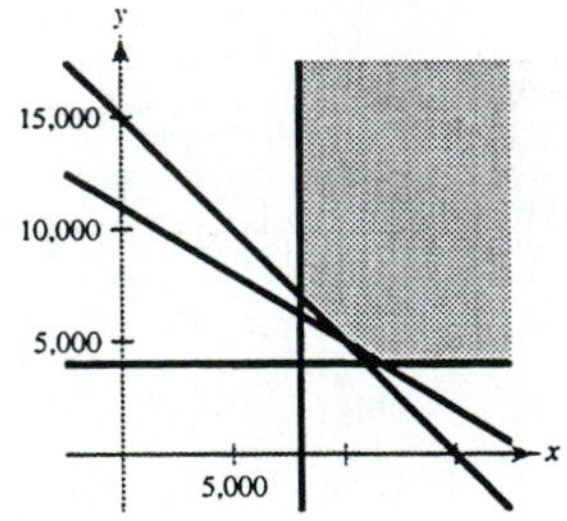

56. x = number of ounces of food X

y = number of ounces of food Y

Calcium: $20x + 10y \geq 300$

Iron: $15x + 10y \geq 150$

Vitamin B: $10x + 20y \geq 200$

$x \geq 0$

$y \geq 0$

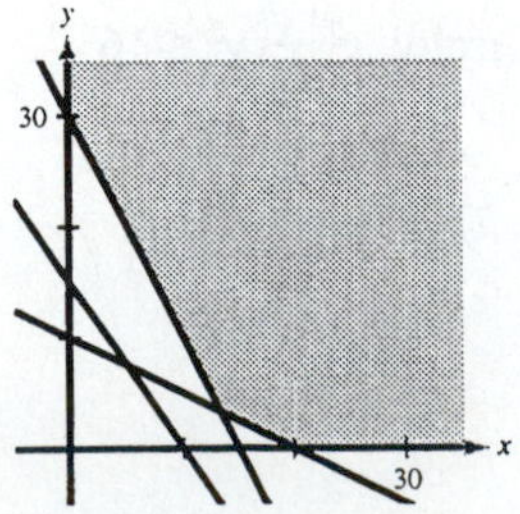

58. x = radius of smaller circle

y = radius of larger circle

(a) Constraints on circles: $\pi y^2 - \pi x^2 \geq 10$

$x > 0$

(b)

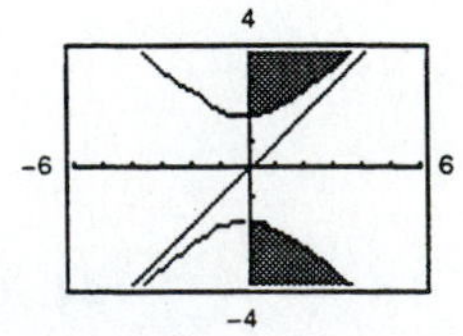

(c) The line is an asymptote to the boundary. The larger the circles, the closer the radii can be and the constraint still be satisfied.

60. Demand = Supply

$100 - 0.05x = 25 + 0.1x$

$75 = 0.15x$

$500 = x$

$75 = p$

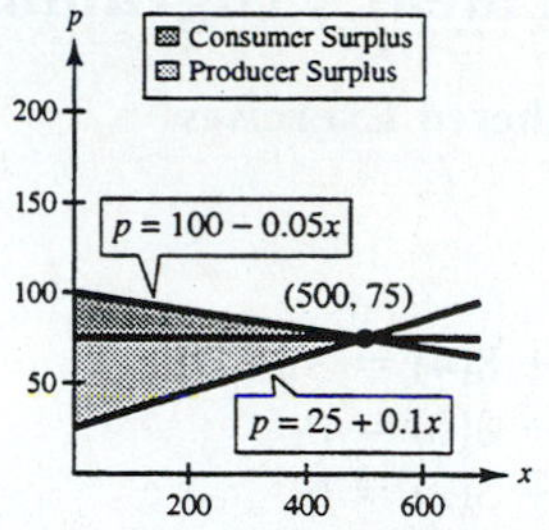

Point of equilibrium: (500, 75)

The consumer surplus is the area of the triangle bounded by

$p \le 100 - 0.05x$

$p \ge 75$

$x \ge 0.$

Consumer surplus $= \frac{1}{2}$(base)(height) $= \frac{1}{2}(500)(25) = 6250$

The producer surplus is the area of the triangle bounded by

$p \ge 25 + 0.1x$

$y \le 75$

$x \ge 0.$

Produce surplus $= \frac{1}{2}$(base)(height) $= \frac{1}{2}(500)(50) = 12{,}500$

62. Demand = Supply

$400 - 0.0002x = 225 + 0.0005x$

$175 = 0.0007x$

$250{,}000 = x$

$350 = p$

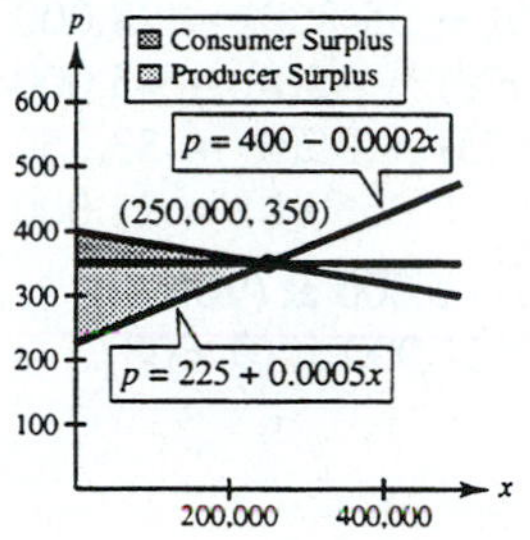

Point of equilibrium: (250,000, 350)

The consumer surplus is the area of the triangle bounded by

$p \le 400 - 0.0002x$

$p \ge 350$

$x \ge 0.$

Consumer surplus $= \frac{1}{2}$(base)(height) $= \frac{1}{2}(250{,}000)(50) = 6{,}250{,}000$

The producer surplus is the area of the triangle bounded by

$p \ge 225 + 0.0005x$

$p \le 350$

$x \ge 0.$

Produce surplus $= \frac{1}{2}$(base)(height) $= \frac{1}{2}(250{,}000)(125) = 15{,}625{,}000$

64. The graph of $x \le 4$ on the real number line is a half-line. The graph of $x \le 4$ on the rectangular coordinate system is a half-plane.

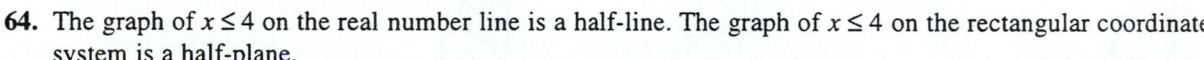

Section 7.5 Linear Programming

Solutions to Even-Numbered Exercises

2. $z = 2x + 8y$

At (0, 4): $z = 2(0) + 8(4) = 32$
At (0, 0): $z = 2(0) + 8(0) = 0$
At (2, 0): $z = 4(0) + 3(2) = 6$

The maximum value is 32 at (0, 4).
The minimum value is 0 at (0, 0).

4. $z = 7x + 3y$

At (0, 4): $z = 7(0) + 3(4) = 12$
At (0, 0): $z = 7(0) + 3(0) = 0$
At (2, 0): $z = 7(2) + 3(0) = 14$

The maximum value is 14 at (2, 0).
The minimum value is 0 at (0, 0).

6. $z = 4x + 3y$

At (0, 4): $z = 4(0) + 3(4) = 12$
At (3, 0): $z = 4(3) + 3(0) = 12$
At (5, 3): $z = 4(5) + 3(3) = 29$
At (2, 0): $z = 2(2) + 8(0) = 4$

The maximum value is 29 at (5, 3).
The minimum value is 6 at (0, 2).

8. $z = x + 6y$

At (0, 4): $z = 0 + 6(4) = 24$
At (3, 0): $z = 3 + 6(0) = 3$
At (5, 3): $z = 5 + 6(3) = 23$
At (0, 2): $z = 0 + 6(2) = 12$

The maximum value is 24 at (0, 4).
The minimum value is 3 at (3, 0).

10. $z = 50x + 35y$

At (0, 800): $z = 50(0) + 35(800) = 28{,}000$
At (900, 0): $z = 50(900) + 35(0) = 45{,}000$
At (675, 0): $z = 50(675) + 35(0) = 33{,}750$
At (0, 600): $z = 50(0) + 35(600) = 21{,}000$

The maximum value is 45,000 at (900, 0).
The minimum value is 21,000 at (0, 600).

12. $z = 16x + 18y$

At (0, 800): $z = 16(0) + 18(800) = 14{,}400$
At (900, 0): $z = 16(900) + 18(0) = 14{,}400$
At (675, 0): $z = 16(675) + 18(0) = 10{,}800$
At (0, 600): $z = 16(0) + 18(600) = 10{,}800$

The maximum value is 14,400 at any point along the line segment connecting (0,800) and (900, 0).
The minimum value is 10,800 at any point along the line segment connecting (645, 0) and (0, 600)

14. $z = 7x + 8y$

At (0, 8): $z = 7(0) + 8(8) = 64$
At (4, 0): $z = 7(4) + 8(0) = 28$
At (0, 0): $z = 7(0) + 8(0) = 0$

The maximum value is 64 at (0, 8).
The minimum value is 0 at (0, 0).

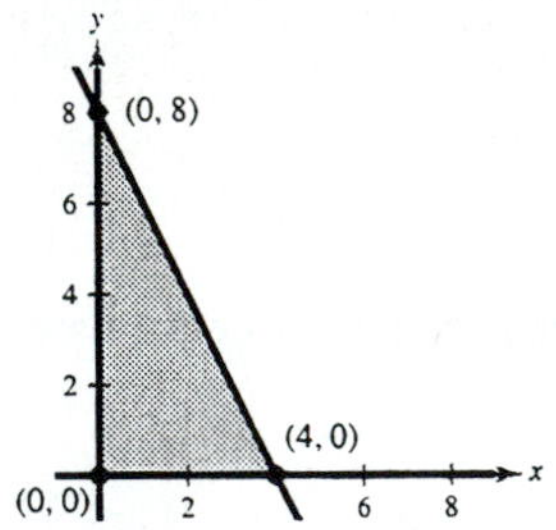

16. $z = 7x + 2y$

At (0, 8): $z = 7(0) + 2(8) = 16$
At (4, 0): $z = 7(4) + 2(0) = 28$
At (0, 0): $z = 7(0) + 2(0) = 0$

The maximum value is 28 at (4, 0).
The minimum value is 0 at (0, 0).

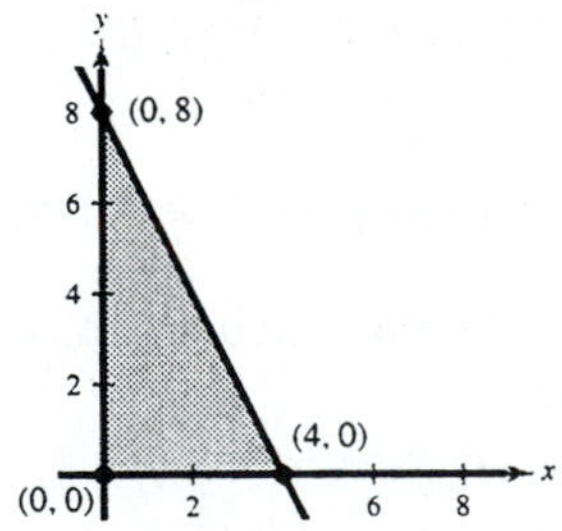

18. $z = 4x + 5y$

At $(0, 0)$: $z = 4(0) + 5(0) = 0$
At $(5, 0)$: $z = 4(5) + 5(0) = 20$
At $(4, 1)$: $z = 4(4) + 5(1) = 21$
At $(0, 3)$: $z = 4(0) + 5(3) = 15$

The maximum value is 21 at $(4, 1)$.
The minimum value is 0 at $(0, 0)$.

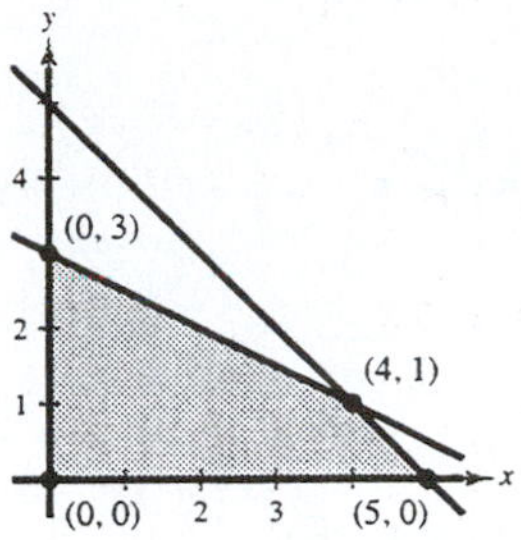

20. $z = 2x - y$

At $(0, 0)$: $z = 2(0) - 0 = 0$
At $(5, 0)$: $z = 2(5) - 0 = 10$
At $(4, 1)$: $z = 2(4) - 0 = 7$
At $(0, 3)$: $z = 2(0) - 3 = -3$

The maximum value is 10 at $(5, 0)$.
The minimum value is -3 at $(0, 3)$.

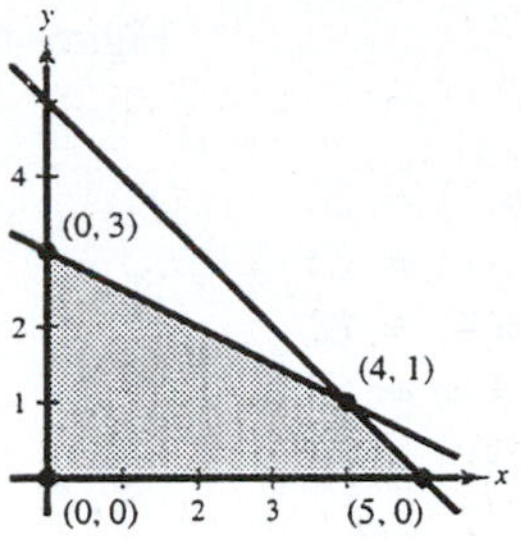

22. $z = x$

At $(0, 0)$: $z = 0$
At $(12, 0)$: $z = 12$
At $(10, 8)$: $z = 10$
At $(6, 16)$: $z = 6$
At $(0, 20)$: $z = 0$

The maximum value is 12 at $(12, 0)$.
The minimum value is 0 at any point along the line segment connecting $(0, 0)$ and $(0, 20)$.

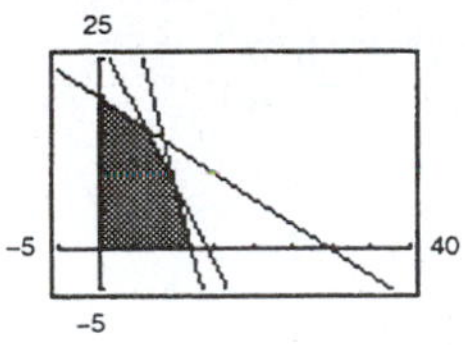

24. $z = y$

At $(0, 0)$: $z = 0$
At $(12, 0)$: $z = 0$
At $(10, 8)$: $z = 8$
At $(6, 16)$: $z = 16$
At $(0, 20)$: $z = 20$

The maximum value is 20 at $(0, 20)$.
The minimum value is 0 at any point along the line segment connecting $(0, 0)$ and $(12, 0)$.

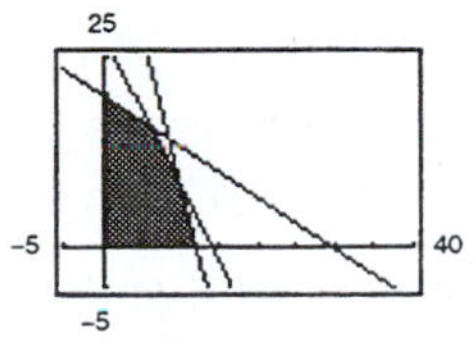

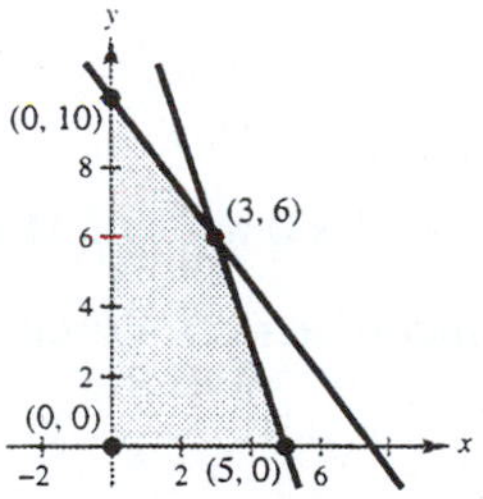

Figure for Exercises 26 and 28

26. $z = 5x + y$

At $(0, 10)$: $z = 5(0) + (10) = 10$
At $(3, 6)$: $z = 5(3) + (6) = 21$
At $(5, 0)$: $z = 5(5) + (0) = 25$
At $(0, 0)$: $z = 5(0) + (0) = 0$

The maximum value is 25 at $(5, 0)$.

28. $z = 3x + y$

At $(0, 10)$: $z = 3(0) + (10) = 10$
At $(3, 6)$: $z = 3(3) + (6) = 15$
At $(5, 0)$: $z = 3(5) + (0) = 15$
At $(0, 0)$: $z = 3(0) + (0) = 0$

The maximum value is 15 at any point along the line segment connecting $(3, 6)$ or $(5, 0)$.

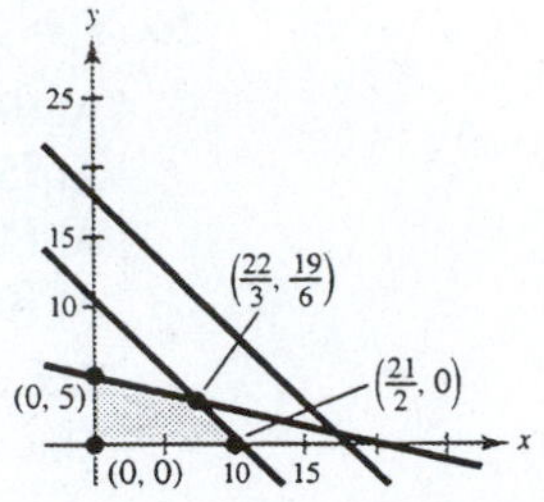

Figure for Exercises 30 and 32

30. $z = 2x + 4y$

At (0, 5): $z = 2(0) + 4(5) = 20$
At (4, 4): $z = 2(4) + 4(4) = 24$
At (5, 3): $z = 2(5) + 4(3) = 22$
At (7, 0): $z = 2(7) + 4(0) = 14$
At (0, 0): $z = 2(0) + 4(0) = 0$

The maximum value is 24 at (4, 4).

32. $z = 4x + y$

At (0, 5): $z = 4(0) + (5) = 5$
At (4, 4): $z = 4(4) + (4) = 20$
At (5, 3): $z = 4(5) + (3) = 23$
At (7, 0): $z = 4(7) + (0) = 28$
At (0, 0): $z = 4(0) + (0) = 0$

The maximum value is 28 at (7, 0).

34. There are an infinite number of objective functions that would have a maximum at (5, 0). One such objective function is $z = x + y$.

36. There are an infinite number of objective functions that would have a minimum at (5, 0). One such objective function is $z = -10x + y$.

38. x = number of Model A; y = number of Model B

Constraints:
$$\begin{aligned} 2.5x + 3y &\le 4000 \\ 2x + y &\le 2500 \\ 0.75x + 1.25y &\le 1500 \\ x &\ge 0 \\ y &\ge 0 \end{aligned}$$

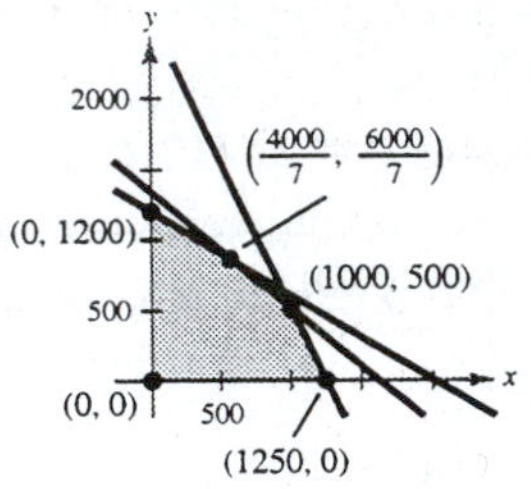

Objective function: $P = 50x + 52y$

Vertices: (0, 0), (0, 1200), $\left(\frac{4000}{7}, \frac{6000}{7}\right)$, (1000, 500), (1250, 0)

At (0, 0): $P = (50)(0) + 52(0) = 0$
At (0, 1200): $P = 50(0) + 52(1200) = 62{,}400$
At $\left(\frac{4000}{7}, \frac{6000}{7}\right)$: $P = 50\left(\frac{4000}{7}\right) + 52\left(\frac{6000}{7}\right) \approx 73{,}142.86$
At (1000, 500): $P = 50(1000) + 52(00) = 76{,}000$
At (1250, 0): $P = 50(1250) + 52(0) = 62{,}500$

The maximum profit of $76,000 occurs when 1000 units of Model A and 500 units of Model B are produced.

40. x = number of acres for crop A; y = number of acres for crop B

Constraints:
$$\begin{aligned} x + y &\le 150 \\ x + 2y &\le 240 \\ 0.3x + 0.1y &\le 30 \\ x &\ge 0 \\ y &\ge 0 \end{aligned}$$

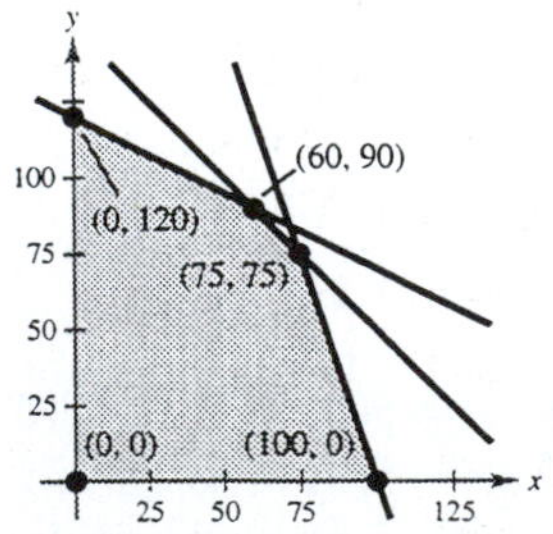

Objective function: $P = 140x + 235y$

Vertices: (0, 0), (100, 0), (0, 120), (60, 90), (75, 75)

At (0, 0): $P = 140(0) + 235(0) = 0$
At (100, 0): $P = 140(100) + 235(0) = 14{,}000$
At (0, 120): $P = 140(0) + 235(120) = 28{,}200$
At (60, 90): $P = 140(60) + 235(90) = 29{,}550$
At (75, 75): $P = 140(75) + 235(75) = 28{,}125$

To maximize the profit, the fruit grower should plant 60 acres of crop A and 90 acres of crop B. The maximum profit would be $29,550.

42. x = fraction of gallon of 80 octane

y = fraction of gallon of 92 octane

Constraints: $x + y = 1$

$$80x + 92y \geq 90$$
$$x \geq 0$$
$$y \geq 0$$

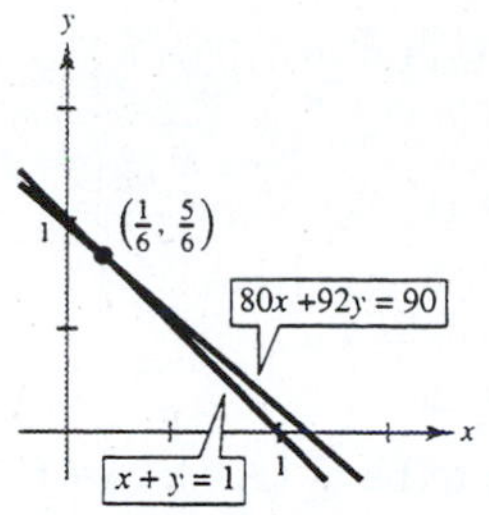

Objective function: $C = 1.13x + 1.28y$

At $\left(\frac{1}{6}, \frac{5}{6}\right)$: $C = 1.13\left(\frac{1}{6}\right) + 1.28\left(\frac{5}{6}\right) = 1.255$

To minimize cost, use $\frac{1}{6}$ gallon of 80 octane and $\frac{5}{6}$ gallon of 92 octane. The cost per gallon is \$1.255.

44. New objective function: $R = 1000x + 300y$

At (0, 0): $R = 1000(0) + 300(0) = 0$
At (0, 40): $R = 1000(0) + 300(40) = 12{,}000$
At (8, 8): $R = 1000(8) + 300(8) = 10{,}400$
At (9, 0): $R = 1000(9) + 300(0) = 900$

The revenue will be maximum (\$12,000) if the firm does no audits and 40 tax returns.

46. Objective function: $z = x + y$

Constraints: $x \geq 0, y \geq 0, -x + y \leq 1, -x + 2y \leq 4$

At (0, 0): $z = 0 + 0 = 0$
At (0, 1): $z = 0 + 1 = 1$
At (2, 3): $z = 2 + 3 = 5$

The constraints do not form a closed set of points. Therefore, $z = x + y$ is unbounded.

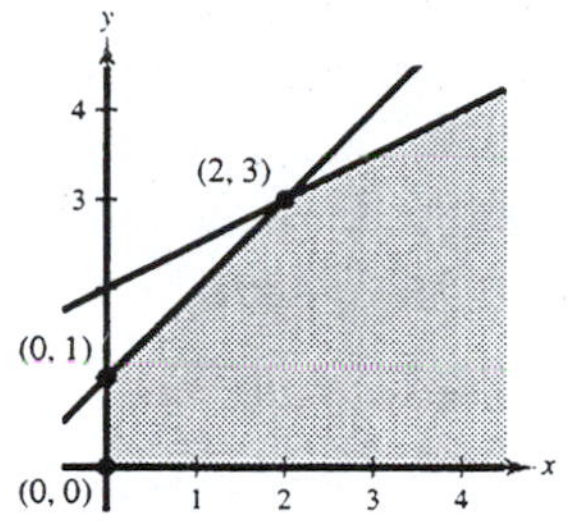

48. Objective function: $z = x + y$

Constraints: $x \geq 0, y \geq 0, -x + y \leq 0, -3x + y \leq 3$

The feasible set is empty.

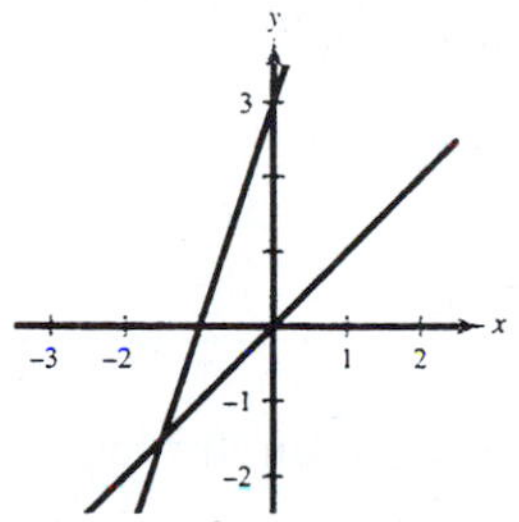

50. Objective function: $z = x + 2y$

Constraints: $x \geq 0, y \geq 0, x + 2y \leq 4, 2x + y \leq 4$

At (0, 0): $z = 0 + 2(0) = 0$
At (0, 2): $z = 0 + 2(2) = 4$
At $\left(\frac{4}{3}, \frac{4}{3}\right)$: $z = \frac{4}{3} + 2\left(\frac{4}{3}\right) = 4$
At (2, 0): $z = 2 + 2(0) = 2$

The maximum value is 4 at any point along the line segment connecting (0, 2) and $\left(\frac{4}{3}, \frac{4}{3}\right)$.

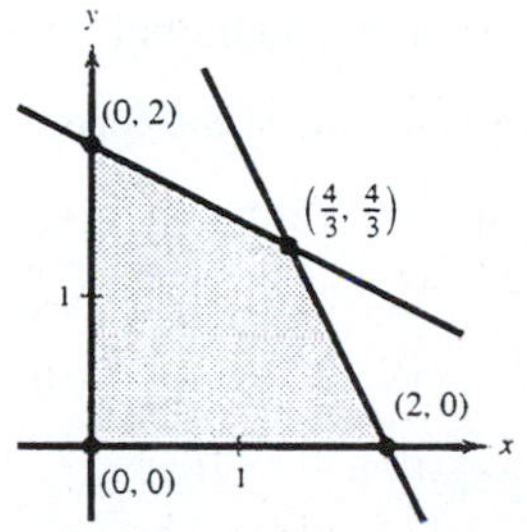

52. Constraints: $x \geq 0, y \geq 0, x + 2y \geq 4, x - y \leq 1$

$z = 3x + ty$

At (0, 0): $z = 3(0) + t(0) = 0$

At (1, 0): $z = 3(1) + t(0) = 3$

At (2, 1): $z = 3(2) + t(1) = 6 + t$

At (0, 2): $z = 3(0) + t(2) = 2t$

(a) For the maximum value to be at (2, 1), $z = 6 + t$ must be greater than $z = 2t$ and $z = 3$.

$6 + t > 2t$ and $6 + t > 3$

$6 > t$ $\quad t > -3$

Thus, $-3 < t < 6$.

(b) For maximum value to be at (0, 2), $z = 2t$ must be greater than $z = 6 + t$ and $z = 3$.

$2t > 6 + t$ and $2t > 3$

$6 > t$ $\quad t > \frac{3}{2}$

Thus, $t > 6$.

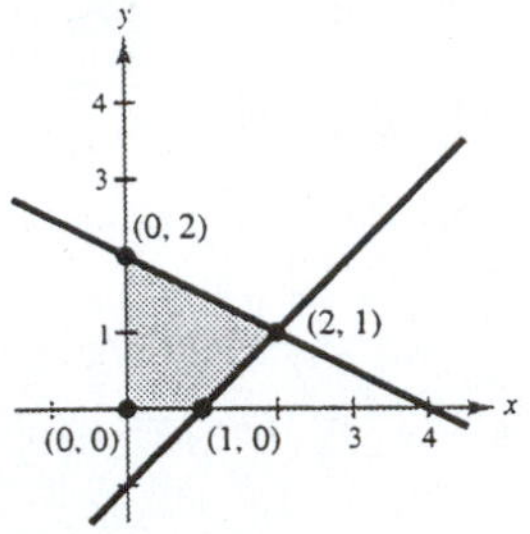

54. $$\frac{\left(1 + \frac{2}{x}\right)}{x - \frac{4}{x}} = \frac{\frac{x+2}{x}}{\frac{x^2 - 4}{x}} = \frac{x+2}{x} \cdot \frac{x}{x^2 - 4} = \frac{x+2}{x} \cdot \frac{x}{(x+2)(x-2)} = \frac{1}{x-2}$$

56. $$\frac{\left(\frac{1}{x+1} + \frac{1}{2}\right)}{\left(\frac{3}{2x^2 + 4x + 2}\right)} = \frac{x+3}{2(x+1)} \cdot \frac{2(x+1)^2}{3} = \frac{(x+3)(x+1)}{3}$$

❑ Focus on Concepts

1. A solution of a system is an ordered pair that satisfies each equation in the system.

2. For a linear system the result will be contradictory equation such as $0 = N$, where N is a nonzero real number. For a nonlinear system there may be an equation with imaginary roots.

3. There will be a contradictory equation of the form $0 = N$, where N is a nonzero real number.

4. The algebraic methods yield exact solutions.

5. (a) One
(b) Two
(c) Four

6. The system has no solution.

7. The lines are distinct and parallel.

$x + 2y = 3$

$2x + 4y = 9$

8. (a) Interchange any two equations.
(b) Multiply an equation by a nonzero constant.
(c) Add a multiple of one equation to any other equation in the system.

9. No. When -2 times Equation 1 is added to Equation 2, the constant is -11.

10. The first system is inconsistent, because -4 times Equation 1 added to Equation 2 yields $0 = -4$.

11. d **12.** b **13.** c **14.** a

15. (a) The boundary would be included in the solution.
(b) The solution would be the half-plane on the opposite side of the boundary.

❑ Review Exercises for Chapter 7

Solutions to Even-Numbered Exercises

2. $2x = 3(y - 1)$

$y = x$

$2y = 3y - 2$

$3 = y$

$3 = x$

Answer: $(3, 3)$

4. $x^2 + y^2 = 169$

$3x + 2y = 39 \Rightarrow x = \frac{1}{3}(39 - 2y)$

$[\frac{1}{3}(39 - 2y)]^2 + y^2 = 169$

$\frac{1}{9}(1521 - 156y + 4y^2) + y^2 = 169$

$169 - \frac{52}{3y} + \frac{4}{9}y^2 + y^2 = 169$

$\frac{13}{9}y^2 - \frac{52}{3}y = 0$

$\frac{13}{3}y(\frac{1}{3}y - 4) = 0 \Rightarrow y = 0, 12$

$y = 0$: $x = \frac{1}{3}(39 - 2(0)) = 13$

$y = 12$: $x = \frac{1}{3}(30 - 2(12)) = 5$

Answer: $(13, 0), (5, 12)$

6. $x = y + 3$

$x = y^2 + 1$

$y + 3 = y^2 + 1$

$0 = y^2 - y - 2$

$0 = (y - 2)(y + 1) \Rightarrow y = 2, -1$

$y = 2$: $x = 2 + 3 = 5$

$y = -1$: $x = -1 + 3 = 2$

Answer: $(5, 2), (2, -1)$

8. $y = 2x^2 - 4x + 1$

$y = x^2 + 4x + 3$

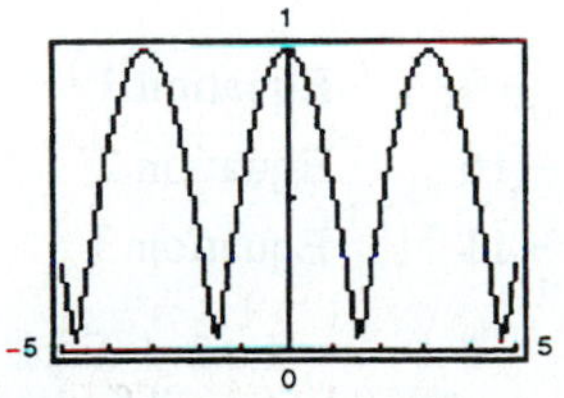

Points of intersection:

$(\sqrt{2}, 5 - 4\sqrt{2}), (-\sqrt{2}, 5 + 4\sqrt{2})$

or: $(1.41, -0.66), (-1.41, 10.66)$

10. $y = \ln(x - 1) - 3$

$y = 4 - \frac{1}{2}x$

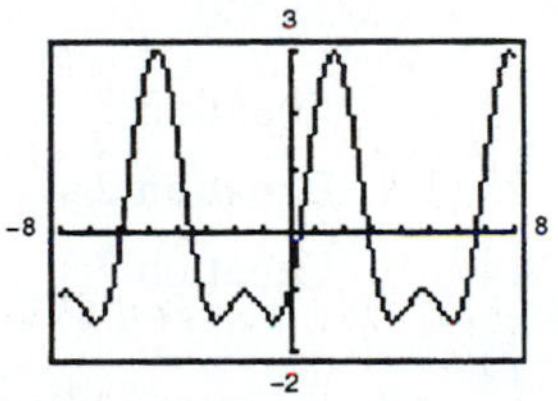

Points of intersection: $(0.68, -0.84)$

12. $40x + 30y = 24 \Rightarrow 40x + 30y = 24$

$20x - 50y = -14 \Rightarrow -40x + 100y = 28$

$130y = 52$

$y = \frac{2}{5}$

$x = \frac{3}{10}$

Answer: $(\frac{3}{10}, \frac{2}{5})$

14. $12x + 42y = -17 \Rightarrow 36x + 126y = -51$

$30x - 18y = 19 \Rightarrow 210x + 126y = 133$

$246x = 82$

$x = \frac{1}{3}$

$y = -\frac{1}{2}$

Answer: $(\frac{1}{3}, -\frac{1}{2})$

16. $7x + 12y = 63 \Rightarrow -7x - 12y = -63$

$2x + 3y = 15 \Rightarrow 8x + 12y = 60$

$x = -3$

$y = 7$

Answer: $(-3, 7)$

18. $1.5x + 1.5y = 8.5 \Rightarrow 3x + 5y = 17$

$6x + 10y = 24 \Rightarrow -3x - 5y = -12$

$0 = 5$

Incosistent; no solution

20. There are an infinite number of linear systems with the solution $(-6, 8)$. One possible solution is:

$$-6 + 8 = 2 \Rightarrow x + y = 2$$
$$-2(-6) + 3(8) = 12 \Rightarrow -2x + 3y = 12$$

22. $y = 22{,}500 + 0.015x$

$y = 20{,}000 + 0.02x$

$22{,}500 + 0.015x = 20{,}000 + 0.02x$

$2500 = 0.005x$

$\$500{,}000 = x$

24. x = number of \$9.95 cassette tapes

y = number of \$14.95 cassette tapes

$$x + y = 650 \Rightarrow y = 650 - x$$
$$9.95x + 14.90y = 7717.50$$
$$9.95x + 14.95(650 - x) = 7717.50$$
$$-5x = -2000$$
$$x = 400$$
$$y = 250$$

Answer: 400 at \$9.95 and 250 \$14.95

26. $2l + 2w = 480$

$l = 1.50w$

$2(1.50w) + 2w = 480$

$5w = 480$

$w = 96$

$l = 144$

The dimensions are 96×144 meters.

28. Supply = Demand

$45 + 0.0002x = 120 = 0.0001x$

$0.0003x = 75$

$x = 250{,}000$ units

$p = \$95.00$

Points of equilibrium: (250,000, 95)

30.

$$\begin{aligned} x + 3y - z &= 13 && \text{Equation 1}\\ 2x \quad\;\; - 5z &= 23 && \text{Equation 2}\\ 4x - y - 2z &= 14 && \text{Equation 3}\end{aligned}$$

$$\begin{aligned} x + 3y - z &= 13 \\ -6y - 2z &= -3 && -2\text{Eq.1} + \text{Eq.2}\\ -13y + 2z &= -38 && -4\text{Eq.1} + \text{Eq.3}\end{aligned}$$

$$\begin{aligned} x + 3y - z &= 13 \\ -6y - 3z &= -3 \\ \tfrac{17}{2}z &= -\tfrac{63}{2} && -\tfrac{13}{6}\text{ Eq.2} + \text{Eq.3}\end{aligned}$$

$$\tfrac{17}{2}z = -\tfrac{63}{2} \Rightarrow z = -\tfrac{63}{17}$$
$$-6y - 3\left(-\tfrac{63}{17}\right) = -3 \Rightarrow y = \tfrac{40}{17}$$
$$x + 3\left(\tfrac{40}{17}\right) - \left(-\tfrac{63}{17}\right) = 13 \Rightarrow x = \tfrac{38}{17}$$

Answer: $\left(\frac{38}{17}, \frac{40}{17}, -\frac{63}{17}\right)$

32.

$$\begin{aligned} 2x \quad\;\; + 6z &= -9 && \text{Equation 1}\\ 3x - 2y + 11z &= -16 && \text{Equation 2}\\ 3x - y + 7z &= -11 && \text{Equation 3}\end{aligned}$$

$$\begin{aligned} -x + 2y - 5z &= 7 && -\text{Eq.2} + \text{Eq.1}\\ 3x - 2y + 11z &= -16 \\ 3x - y + 7z &= -11 \end{aligned}$$

$$\begin{aligned} -x + 2y - 5z &= 7 \\ 4y - 4z &= 5 && 3\text{Eq.1} + \text{Eq.2}\\ 5y - 8z &= 10 && 3\text{Eq.1} + \text{Eq.2}\end{aligned}$$

$$\begin{aligned} -x + 2y - 5z &= 7 \\ 4y - 4z &= 5 \\ -3y \quad\quad &= 0 && -2\text{Eq.2} + \text{Eq.3}\end{aligned}$$

$$-3y = 0 \Rightarrow y = 0$$
$$4(0) - 4z = 5 \Rightarrow z = -\tfrac{5}{4}$$
$$-x + 2(0) - 5\left(-\tfrac{5}{4}\right) = 7 \Rightarrow x = -\tfrac{3}{4}$$

Answer: $\left(-\frac{3}{4}, 0, -\frac{5}{4}\right)$

34.
$$\begin{aligned}
-x + 3y + 2z + 2w &= 1 && \text{Equation 1}\\
2x + y + z + 2w &= -1 && \text{Equation 2}\\
5x - 2y + z - 3w &= 0 && \text{Equation 3}\\
3x + 2y + 3z - 5w &= 12 && \text{Equation 4}
\end{aligned}$$

$$\begin{aligned}
-x + 3y + 2z + 2w &= 1 &&\\
7y + 5z + 6w &= 1 && 2\text{Eq.}1 + \text{Eq.}2\\
13y + 11z + 7w &= 5 && 5\text{Eq.}1 + \text{Eq.}3\\
11y + 9z + w &= 15 && 3\text{Eq.}1 + \text{Eq.}4
\end{aligned}$$

$$\begin{aligned}
-x + 3y + 2z + 2w &= 1 &&\\
-y + z - 5w &= 3 && -2\text{Eq.}2 + \text{Eq.}3\\
7y + 5z + 6w &= 1 && \text{Eq.}2\\
11y + 9z + w &= 15 &&
\end{aligned}$$

$$\begin{aligned}
-x + 3y + 2z + 2w &= 1 &&\\
-y + z - 5w &= 3 &&\\
12z - 29w &= 22 && 7\text{Eq.}2 + \text{Eq.}3\\
20z - 54w &= 48 && 11\text{Eq.}2 + \text{Eq.}4
\end{aligned}$$

$$\begin{aligned}
-x + 3y + 2z + 2w &= 1 &&\\
-y + z - 5w &= 3 &&\\
60z - 145w &= 110 && 5\text{Eq.}3\\
60z - 162w &= 144 && 3\text{Eq.}4
\end{aligned}$$

$$\begin{aligned}
-x + 3y + 2z + 2w &= 1 &&\\
-y + z - 5w &= 3 &&\\
60z - 145w &= 110 &&\\
-17w &= 34 && -\text{Eq.}3 + \text{Eq.}4
\end{aligned}$$

$$\begin{aligned}
-17w &= 34 \Rightarrow w = -2\\
60z - 145(-2) &= 110 \Rightarrow z = -3\\
-y - 3 - 5(-2) &= 3 \Rightarrow y = 4\\
-x + 3(4) + 2(-3) + 2(-2) &= 1 \Rightarrow x = 1
\end{aligned}$$

Answer: $(1, 4, -3, -2)$

36. There are an infinite number of linear systems with the solution $\left(5, \frac{3}{2}, 2\right)$. One possible solution is:

$$\begin{aligned}
2(5) + 2\left(\tfrac{3}{2}\right) - 3(2) &= 7 \Rightarrow 2x + 2y - 3z = 7\\
5 - 2\left(\tfrac{3}{2}\right) + 2 &= 4 \Rightarrow 2x + 2y - 3z = 7\\
-5 + 4\left(\tfrac{3}{2}\right) - 2 &= -1 \Rightarrow -x + 4y - z = -1
\end{aligned}$$

38. $y = ax^2 + bx + c$ through $(-5, 6), (1, 0), (2, 20)$.

$(-5, 6)$: $6 = 25a - 5b + c \Rightarrow 24a - 6b = 6$

$(1, 0)$: $0 = a + b + c \Rightarrow c = -a - b$

$(2, 20)$: $20 = 4a + 2b + c \Rightarrow -8(3a + b = 20)$

$$\begin{aligned}
-14b &= -154\\
b &= 11\\
a &= 3\\
c = -11 - 3 &= -14
\end{aligned}$$

The equation of the parabola is $y = 3x^2 + 11x - 14$.

40. $x^2 + y^2 + Dx + Ey + F = 0$ through $(1, 4), (4, 3), (-2, -5)$.

$$\begin{array}{lll}
(1, 4): & 17 + D + 4E + F = 0 \Rightarrow D + 4D + F = -17 & \text{Equation 1} \\
(4, 3): & 25 + D + 4E + F = 0 \Rightarrow 4D + 3D + F = -25 & \text{Equation 2} \\
(-2, -5): & 29 + D + 4E + F = 0 \Rightarrow 2D + 5D - F = 29 & \text{Equation 3}
\end{array}$$

$$\begin{array}{rl}
D + 4E + F = -17 & \\
-13E - 3F = 43 & -4\text{Eq.}1 + \text{Eq.}2 \\
-3E - 3F = 63 & -2\text{Eq.}1 + \text{Eq.}3
\end{array}$$

$$\begin{array}{rl}
D + 4E + F = -17 & \\
-3E - 3F = 63 & \text{Interchange} \\
-13E - 3F = 43 & \text{Equations}
\end{array}$$

$$\begin{array}{rl}
D + 4E + F = -17 & \\
-3E - 3F = 63 & \\
10F = -230 & -\frac{13}{3}\text{Eq.}2 + \text{Eq.}3
\end{array}$$

$F = -23, E = 2, D = -2$

The equation of the circle is $x^2 + y^2 - 2x + 2y - 23 = 0$.

42. Let x = amount invested at 7%.

Let y = amount invested at 9%.

Let z = amount invested at 11%.

$$\begin{aligned}
x + y + z &= 20{,}000 \\
0.07x + 0.09y + 0.11z &= 1818 \\
y &= x - 3000
\end{aligned}$$

$$\begin{aligned}
x + x - 3000 + x - 1000 &= 20{,}000 \\
x &= 8000 \\
y &= 5000 \\
z &= 7000
\end{aligned}$$

Thus, \$8000 was invested at 7%, \$5000 was invested at 9%, and \$7000 was invested at 11%.

44.

$$\begin{aligned}
5c + 10a &= 9.1 \Rightarrow 10c - 20a = -18.2 \\
10b &= 8.0 \\
10c + 34a &= 19.8 \Rightarrow 10c - 34a = 19.8 \\
14a &= 1.6 \\
a &\approx 0.114 \\
c &= 1.591 \\
b &= 0.8
\end{aligned}$$

Least squares parabola: $y = 0.114x^2 + 0.800x + 1.591$

46.

$$\begin{aligned}
3x - 5y &= 8 \\
2x + k_1y &= k_2 \\
6x - 10y &= 16 \\
-6x - 3k_1y &= -3k_2 \\
\hline
(-10 - 3k_1)y &= 16 - 3k_2
\end{aligned}$$

For this system to have an infinite number of solutions, this last equation should be $0 = 0$. Thus,

$$\begin{array}{lll}
-10 - 3k_1 = 0 & \text{and} & 16 - 3k_2 = 0 \\
-10 = 3k_1 & & 16 = 3k_2 \\
-\frac{10}{3} = k_1 & & \frac{16}{3} = k_2
\end{array}$$

48. $2x + 3y \le 24$
$2x + \quad y \le 16$
$x \ge 0$
$y \ge 0$

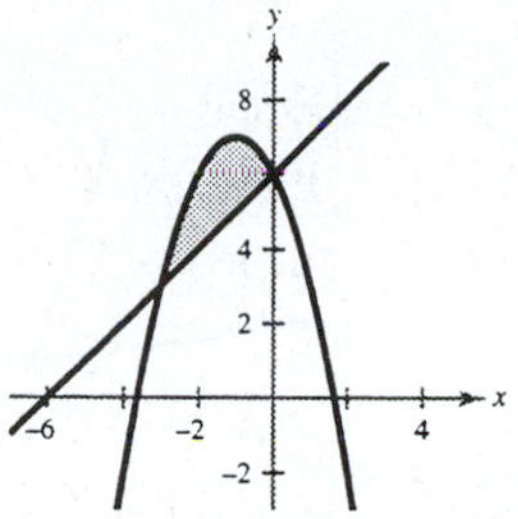

Vertex A	Vertex B	Vertex C
$2x + 3y = 24$	$2x + y = 16$	$x = 0$
$2x + \quad y = 16$	$x = 0$	$y = 0$
$(6, 4)$	$(0, 16)$	$(0, 0)$
	Outside the region	

Vertex D	Vertex E	Vertex F
$2x + 3y = 24$	$2x + 3y = 24$	$2x + y = 16$
$x = 0$	$y = 0$	$y = 0$
$(0, 8)$	$(12, 0)$	$(8, 0)$
	Outside the region	

50. $2x + \quad y \ge 16$
$x + 3y \ge 18$
$0 \le x \le 25$
$0 \le y \le 25$

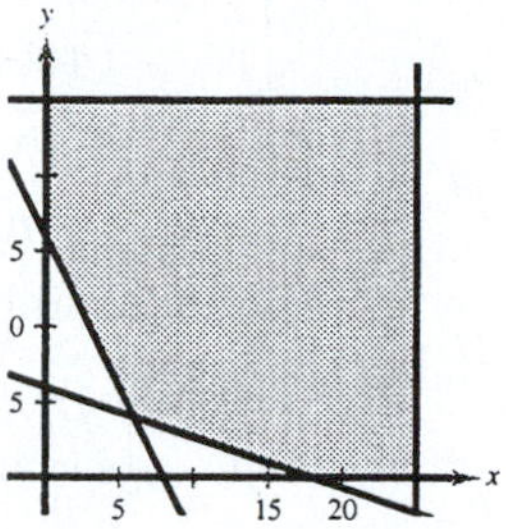

Vertex A	Vertex B	Vertex C	Vertex D	Vertex E
$2x + \quad y = 16$	$x + 3y = 18$	$x + 3y = 18$	$x = 0$	$x = 0$
$x + \quad 3y = 18$	$x = 0$	$x = 25$	$y = 0$	$y = 25$
$(6, 4)$	$(0, 6)$	$(0, 0)$	$(0, 0)$	$(0, 25)$
	Outside the region		Outside the region	

Vertex F	Vertex G	Vertex H	Vertex I	Vertex J
$x = 25$	$x = 25$	$2x + y = 16$	$2x + y = 16$	$2x + y = 16$
$y = 0$	$y = 25$	$x = 0$	$x = 25$	$y = 0$
$(25, 0)$	$(25, 25)$	$(0, 16)$	$(25, -34)$	$(8, 0)$
			Outside the region	Outside the region

Vertex K	Vertex L	Vertex M
$2x + y = 16$	$x + 3y = 18$	$x + 3y = 16$
$y = 25$	$y = 0$	$y = 25$
$(-4.5, 25)$	$(18, 0)$	$(-57, 25)$
Outside the region		Outside the region

52. $y \le 6 - 2 - x^2$
$y \ge x + 6$

Vertices: $x + 6 = 6 - 2x - x^2$
$x^2 + 3x = 0$
$x(x + 3) = 0 \Rightarrow x - 0, -3$

$(0, 6), (-3, 3)$

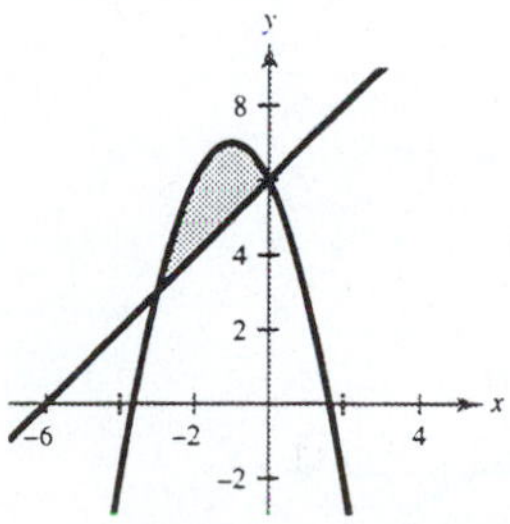

54. $x^2 + y^2 \le 9 \Rightarrow y^2 \le 9 - x^2$
$(x - 3)^2 + y^2 \le 9 \Rightarrow y^2 \le 9 - (x - 3)$

Vertices: $9 - x^2 = 9 - (x - 3)^2$
$(x - 3)^2 - x^2 = 0$
$x^2 - 6x + 9 - x^2 = 0$
$x = \frac{3}{2}$

$\left(\frac{3}{2}, 2.60\right), \left(\frac{3}{2}, -2.60\right)$

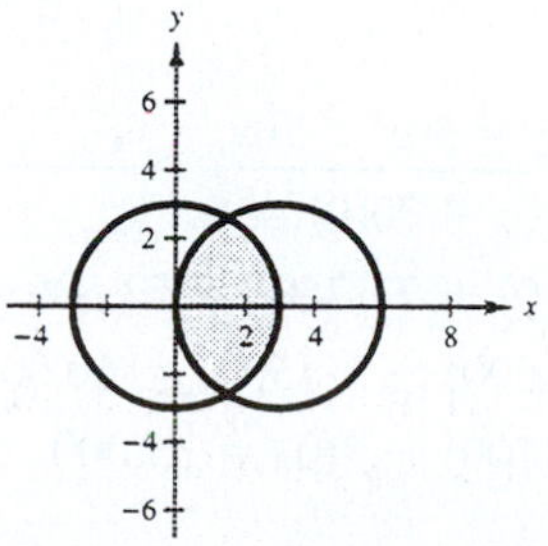

56. Line through (1, 2), (6, 7): $y = x + 1$

Line through (6, 7), (8, 1): $y = -3x + 25$

Line through (1, 2), (8, 1): $y = -\frac{1}{7} + \frac{15}{7} \Rightarrow -x + 15$

System of inequalities: $-x + y = \quad 1$

$3x + y = 25$

$x + 7y = 15$

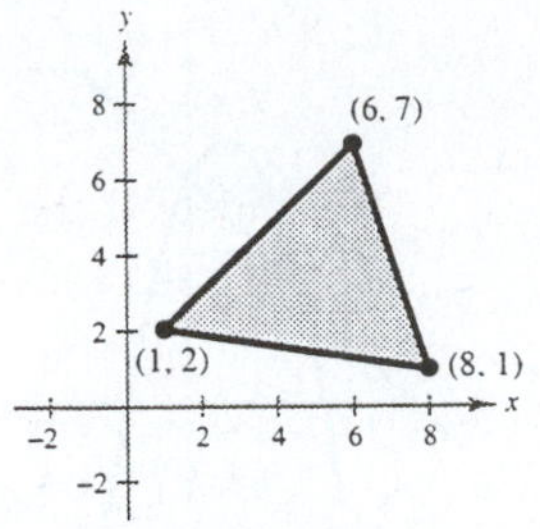

58. x = number of units of Product I

y = number of units of Product II

$20x + 30y \leq 24{,}000$

$12x + 8y \leq 12{,}400$

$x \geq 0$

$y \geq 0$

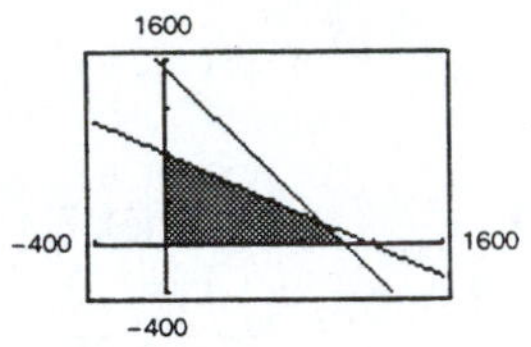

60. Demand = Supply

$130 - 0.0002x = 30 + 0.0003x$

$100 = 0.0005x$

$x = 200{,}000$ units

$p = \$90$

Point of equilibrium: (200,000, 90)

Consumer surplus: $\frac{1}{2}(200{,}000)(40) = \$4{,}000{,}000$

Producer surplus: $\frac{1}{2}(200{,}000)(60) = \$6{,}000{,}000$

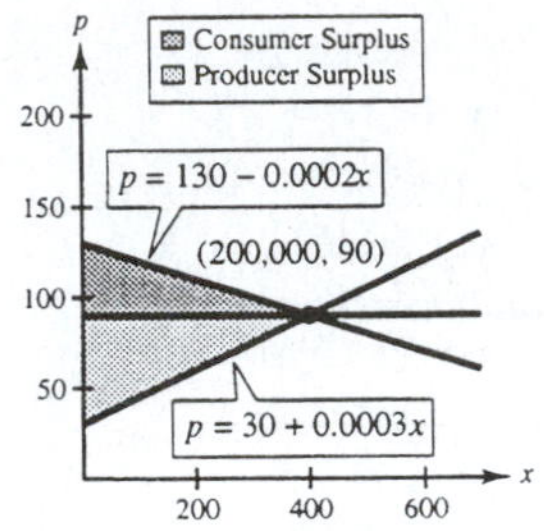

62. Minimize $z = 19x + 7y$ subject to the following constraints:

$x \geq \quad 0$

$y \geq \quad 0$

$2x + y \geq 100$

$x + y \geq \ 75$

Vertex	Value of $z = 10x + 7y$
At (0, 100):	$z = 10(0) + 7(100) = 700$
At (25, 50):	$z = 10(25) + 7(50) = 600$, minimum value
At (75, 0):	$z = 10(75) + 7(0) = 750$

64. Maximize $z = 50x + 70y$ subject to the following constraints:

$x \geq \quad 0$

$y \geq \quad 0$

$x + 2y \geq 1500$

$5x + 2y \geq 3500$

Vertex	Value of $z = 50x + 70y$
At (0, 0):	$z = 50(0) + 70(0) = 0$
At (0, 750):	$z = 50(0) + 77(750) = 52{,}500$
At (500, 500):	$z = 50(500) + 70(500) = 60{,}000$ maximum value
At (700, 0):	$z = 50(700) + 7(0) = 35{,}000$

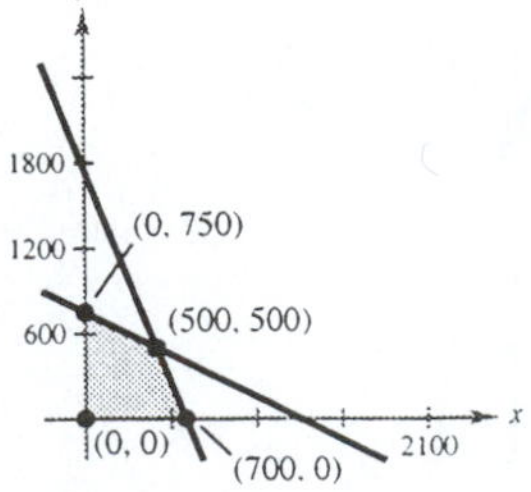

66. Let x = number of product A.

Let y = number of product B.

Maximize $P = 18x + 24y$ subject to the following constraints:

$4x + 2y \le 24$

$x + 2y \le 9$

$x + y \le 8$

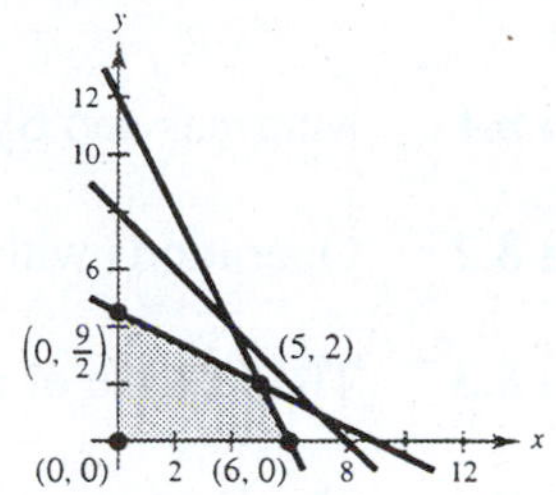

Vertex	Vertex of $P = 18x + 24y$
At (0, 0):	$P = 18(0) + 24(0) = 0$
At (6, 0):	$P = 18(6) + 24(0) = 108$
At (5, 2):	$P = 18(5) + 24(2) = 138$ maximum value
At $\left(0, \frac{9}{2}\right)$:	$P = 19(0) + 24\left(\frac{9}{2}\right) = 108$

To maximize profit, produce 5 units of product A and 2 units of product B.

68. x = fraction of Type A

y = fraction of Type B

Constraints: $80x + 92y \ge 88$

$x + y = 1$

$x \ge 0$

$y \ge 0$

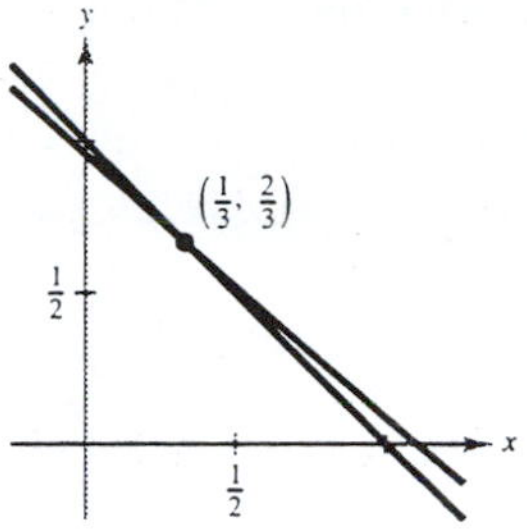

Objective function: $C = 1.25x + 1.55y$

At $\left(\frac{1}{3}, \frac{2}{3}\right)$: $C = 1.25\left(\frac{1}{3}\right) + 1.55\left(\frac{2}{3}\right) = 1.45$

The minimum cost is \$1.45 and occurs with a mixture of $\frac{1}{3}A$ and $\frac{2}{3}B$.

CHAPTER 8
Matrices and Determinants

CHAPTER 8
Matrices and Determinants

Section 8.1 Matrices and Systems of Equations

Solutions to Even-Numbered Exercises

2. Since the matrix has one row and four columns, its order is 1×4.

4. Since the matrix has three rows and four columns, its order is 3×4.

6. Since the matrix has one row and one column, its order is 1×1.

8. $7x + 4y = 22$

$5x - 9y = 15$

$$\begin{bmatrix} 7 & 4 & \vdots & 22 \\ 5 & -9 & \vdots & 15 \end{bmatrix}$$

10. $7x - 5y + z = 13$

$19x \quad -8z = 10$

$$\begin{bmatrix} 7 & -5 & 1 & \vdots & 13 \\ 19 & 0 & -8 & \vdots & 10 \end{bmatrix}$$

12. $$\begin{bmatrix} 7 & -5 & \vdots & 0 \\ 8 & 3 & \vdots & -2 \end{bmatrix}$$

$7x - 5y = 0$

$8x + 3y = -2$

14. $$\begin{bmatrix} 9 & 12 & 3 & 0 & \vdots & 0 \\ -2 & 18 & 5 & 2 & \vdots & 10 \\ 1 & 7 & -8 & 0 & \vdots & -4 \end{bmatrix}$$

$9x + 12y + 3z = 0$

$-2x + 18y + 5z + 2w = 10$

$x + 7y - 8z = -4$

16. $$\begin{bmatrix} 1 & 3 & 0 & 0 \\ 0 & 0 & 1 & 8 \\ 0 & 0 & 0 & 0 \end{bmatrix}$$

This matrix is in reduced row-echelon form.

18. $$\begin{bmatrix} 1 & 0 & 2 & 1 \\ 0 & 1 & -3 & 10 \\ 0 & 0 & 1 & 0 \end{bmatrix}$$

This matrix is in reduced row-echelon form.

20. $$\begin{bmatrix} 3 & 6 & 8 \\ 4 & -3 & 6 \end{bmatrix}$$

$$\tfrac{1}{3}R_1 \rightarrow \begin{bmatrix} 1 & \boxed{2} & \frac{8}{3} \\ 4 & -3 & 6 \end{bmatrix}$$

22. $$\begin{bmatrix} 2 & 4 & 8 & 3 \\ 1 & -1 & -3 & 2 \\ 2 & 6 & 4 & 9 \end{bmatrix}$$

$$\tfrac{1}{2}R_1 \rightarrow \begin{bmatrix} 1 & \boxed{2} & \boxed{4} & \boxed{\frac{3}{2}} \\ 1 & -1 & -3 & 2 \\ 2 & 6 & 4 & 9 \end{bmatrix}$$

$$\begin{matrix} \\ -R_1 + R_2 \rightarrow \\ -2R_1 + R_2 \rightarrow \end{matrix} \begin{bmatrix} 1 & 2 & 4 & \frac{3}{2} \\ 0 & \boxed{-3} & -7 & \frac{1}{2} \\ 0 & 2 & \boxed{-4} & \boxed{6} \end{bmatrix}$$

24. $\begin{bmatrix} 7 & 1 \\ 0 & 2 \\ -3 & 4 \\ 4 & 1 \end{bmatrix}$

(a) $\begin{bmatrix} 7 & 1 \\ 0 & 2 \\ -3 & 4 \\ 1 & 5 \end{bmatrix}$ (b) $\begin{bmatrix} 1 & 5 \\ 0 & 2 \\ -3 & 4 \\ 7 & 1 \end{bmatrix}$ (c) $\begin{bmatrix} 1 & 5 \\ 0 & 2 \\ 0 & 19 \\ 7 & 1 \end{bmatrix}$

(d) $\begin{bmatrix} 1 & 5 \\ 0 & 2 \\ 0 & 19 \\ 0 & -34 \end{bmatrix}$ (e) $\begin{bmatrix} 1 & 5 \\ 0 & 1 \\ 0 & 19 \\ 0 & -34 \end{bmatrix}$ (f) $\begin{bmatrix} 1 & 0 \\ 0 & 1 \\ 0 & 0 \\ 0 & 0 \end{bmatrix}$ This matrix is in reduced row-echelon form.

26. $\begin{bmatrix} 1 & 2 & -1 & 3 \\ 3 & 7 & -5 & 14 \\ -2 & -1 & -3 & 8 \end{bmatrix}$

$$\begin{matrix} \\ -3R_1 + R_2 \rightarrow \\ 2R_1 + R_3 \rightarrow \end{matrix} \begin{bmatrix} 1 & 2 & -1 & 3 \\ 0 & 1 & -2 & 5 \\ 0 & 3 & -5 & 14 \end{bmatrix}$$

$$\begin{matrix} \\ \\ -3R_2 + R_3 \rightarrow \end{matrix} \begin{bmatrix} 1 & 2 & -1 & 3 \\ 0 & 1 & -2 & 5 \\ 0 & 0 & 1 & -1 \end{bmatrix}$$

28. $\begin{bmatrix} 1 & -3 & 0 & -7 \\ -3 & 10 & 1 & 23 \\ 4 & -10 & 2 & -24 \end{bmatrix}$

$$\begin{matrix} \\ 3R_1 + R_2 \rightarrow \\ -4R_1 + R_3 \rightarrow \end{matrix} \begin{bmatrix} 1 & -3 & 0 & -7 \\ 0 & 1 & 1 & 2 \\ 0 & 2 & 2 & 4 \end{bmatrix}$$

$$\begin{matrix} \\ \\ -2R_2 + R_3 \rightarrow \end{matrix} \begin{bmatrix} 1 & -3 & 0 & -7 \\ 0 & 1 & 1 & 2 \\ 0 & 0 & 0 & 0 \end{bmatrix}$$

30. $\begin{bmatrix} 1 & 3 & 2 \\ 5 & 15 & 9 \\ 2 & 6 & 10 \end{bmatrix}$

$$\begin{matrix} \\ -5R_1 + R_2 \rightarrow \\ -2R_1 + R_3 \rightarrow \end{matrix} \begin{bmatrix} 1 & 3 & 2 \\ 0 & 0 & -1 \\ 0 & 0 & 6 \end{bmatrix}$$

$$\begin{matrix} \\ \\ 6R_2 + R_3 \rightarrow \end{matrix} \begin{bmatrix} 1 & 3 & 0 \\ 0 & 0 & -1 \\ 0 & 0 & 0 \end{bmatrix}$$

$$\begin{matrix} \\ -1R_2 \rightarrow \\ \\ \end{matrix} \begin{bmatrix} 1 & 3 & 0 \\ 0 & 0 & 1 \\ 0 & 0 & 0 \end{bmatrix}$$

32. $\begin{bmatrix} 1 & -3 \\ -1 & 8 \\ 0 & 4 \\ -2 & 10 \end{bmatrix}$

$$\begin{matrix} \\ R_1 + R_2 \rightarrow \\ \\ 2R_1 + R_4 \rightarrow \end{matrix} \begin{bmatrix} 1 & -3 \\ 0 & 5 \\ 0 & 4 \\ 0 & 4 \end{bmatrix}$$

$$\begin{matrix} \\ \frac{1}{5}R_2 \rightarrow \\ \\ \\ \end{matrix} \begin{bmatrix} 1 & -3 \\ 0 & 1 \\ 0 & 4 \\ 0 & 4 \end{bmatrix}$$

$$\begin{matrix} \\ \\ -4R_2 + R_3 \rightarrow \\ -4R_2 + R_4 \rightarrow \end{matrix} \begin{bmatrix} 1 & 0 \\ 0 & 1 \\ 0 & 0 \\ 0 & 0 \end{bmatrix}$$

34.
$$\begin{aligned} x + 5y &= 0 \\ y &= -1 \\ x + 5(-1) &= 0 \\ x &= 5 \end{aligned}$$

Answer: $(5, -1)$

36.
$$\begin{aligned} x + 2y - 2z &= -1 \\ y + z &= 9 \\ z &= -3 \\ y + (-3) &= 9 \\ y &= 12 \\ x + 2(12) - 2(-3) &= -1 \\ x &= -31 \end{aligned}$$

Answer: $(-31, 12, -3)$

38. $\begin{bmatrix} 1 & 0 & \vdots & -2 \\ 0 & 1 & \vdots & 4 \end{bmatrix}$

$x = -2$

$y = 4$

Answer: $(-2, 4)$

40. $\begin{bmatrix} 1 & 0 & 0 & \vdots & 3 \\ 0 & 1 & 0 & \vdots & -1 \\ 0 & 0 & 1 & \vdots & 0 \end{bmatrix}$

$x = 3$

$y = -1$

$x = 0$

Answer: $(3, -1, 0)$

42. $2x + 6y = 16$

$2x + 3y = 7$

$$\begin{bmatrix} 2 & 6 & \vdots & 16 \\ 2 & 3 & \vdots & 7 \end{bmatrix}$$

$$-R_1 + R_2 \rightarrow \begin{bmatrix} 2 & 6 & \vdots & 16 \\ 0 & -3 & \vdots & -9 \end{bmatrix}$$

$$\begin{matrix} \frac{1}{2}R_1 \rightarrow \\ -\frac{1}{3}R_2 \rightarrow \end{matrix} \begin{bmatrix} 1 & 3 & \vdots & 8 \\ 0 & 1 & \vdots & 3 \end{bmatrix}$$

$y = 3$

$x + 3(3) = 8 \Rightarrow x = -1$

Answer: $(-1, 3)$

44. $x + 2y = 0$

$x + y = 6$

$3x - 2y = 8$

$$\begin{bmatrix} 1 & 2 & \vdots & 0 \\ 1 & 1 & \vdots & 6 \\ 3 & -2 & \vdots & 8 \end{bmatrix}$$

$$\begin{matrix} \\ -R_1 + R_2 \rightarrow \\ -3R_1 + R_3 \rightarrow \end{matrix} \begin{bmatrix} 1 & 2 & \vdots & 0 \\ 0 & -1 & \vdots & 6 \\ 0 & -8 & \vdots & 8 \end{bmatrix}$$

$$\begin{matrix} \\ -R_2 \rightarrow \\ -8R_2 + R_3 \rightarrow \end{matrix} \begin{bmatrix} 1 & 2 & \vdots & 0 \\ 0 & 1 & \vdots & 6 \\ 0 & 0 & \vdots & -40 \end{bmatrix}$$

The system in inconsistent and there is no solution.

46. $2x - y = -0.1$

$3x + 2y = 1.6$

$$\begin{bmatrix} 2 & -1 & \vdots & -0.1 \\ 3 & 2 & \vdots & 1.6 \end{bmatrix}$$

$$-R_2 + R_1 \rightarrow \begin{bmatrix} -1 & -3 & \vdots & -1.7 \\ 3 & 2 & \vdots & 1.6 \end{bmatrix}$$

$$3R_1 + R_2 \rightarrow \begin{bmatrix} -1 & -3 & \vdots & -1.7 \\ 0 & -7 & \vdots & -3.5 \end{bmatrix}$$

$$\begin{matrix} -R_1 \rightarrow \\ -\frac{1}{7}R_2 \rightarrow \end{matrix} \begin{bmatrix} 1 & 3 & \vdots & 1.7 \\ 0 & 1 & \vdots & 0.5 \end{bmatrix}$$

$y = 0.5$

$x + 3(0.5) = 1.7 \Rightarrow x = 0.2$

Answer: $(0.2, 0.5)$

48. $x - 3y = 5$

$-2x + 6y = -10$

$$\begin{bmatrix} 1 & -3 & \vdots & 5 \\ -2 & 6 & \vdots & -10 \end{bmatrix}$$

$$2R_1 + R_2 \rightarrow \begin{bmatrix} 1 & -3 & \vdots & 5 \\ 0 & 0 & \vdots & 0 \end{bmatrix}$$

$y = a$

$x = 3a + 5$

Answer: $(3z + 5, a)$

50. $2x - y + 3z = 24$
$2y - z = 14$
$7x - 5y = 6$

$$\begin{bmatrix} 2 & -1 & 3 & \vdots & 24 \\ 0 & 2 & -1 & \vdots & 14 \\ 7 & -5 & 0 & \vdots & 6 \end{bmatrix}$$

$$R_3 - 3R_1 \rightarrow \begin{bmatrix} 1 & -2 & -9 & \vdots & -66 \\ 0 & 2 & -1 & \vdots & 14 \\ 7 & -5 & 0 & \vdots & 6 \end{bmatrix}$$

$$\begin{matrix} \\ \\ -7R_1 + R_3 \rightarrow \end{matrix} \begin{bmatrix} 1 & -2 & -9 & \vdots & -66 \\ 0 & 2 & -1 & \vdots & 14 \\ 0 & 9 & 63 & \vdots & 468 \end{bmatrix}$$

$$\begin{matrix} \\ 4R_2 \rightarrow \\ \\ \end{matrix} \begin{bmatrix} 1 & -2 & -9 & \vdots & -66 \\ 0 & 8 & -4 & \vdots & 56 \\ 0 & 9 & 63 & \vdots & 468 \end{bmatrix}$$

$$\begin{matrix} \\ -R_3 + R_2 \rightarrow \\ \\ \end{matrix} \begin{bmatrix} 1 & -2 & -9 & \vdots & -66 \\ 0 & -1 & -67 & \vdots & -412 \\ 0 & 9 & 63 & \vdots & 468 \end{bmatrix}$$

$$\begin{matrix} \\ \\ 9R_2 + R_3 \rightarrow \end{matrix} \begin{bmatrix} 1 & -2 & -9 & \vdots & -66 \\ 0 & -1 & -67 & \vdots & -412 \\ 0 & 0 & -540 & \vdots & -3240 \end{bmatrix}$$

$$\begin{matrix} \\ -R_2 \rightarrow \\ -\frac{1}{540}R_3 \rightarrow \end{matrix} \begin{bmatrix} 1 & -2 & -9 & \vdots & -66 \\ 0 & 1 & 67 & \vdots & 412 \\ 0 & 0 & 1 & \vdots & 6 \end{bmatrix}$$

$z = 6$
$y + 67(6) = 412 \Rightarrow y = 10$
$x - 2(10) - 9(6) = -66 \Rightarrow x = 8$
Answer: $(8, 10, 6)$

52. $2x + 3z = 3$
$4x - 3y + 7z = 5$
$8x - 9y + 15z = 9$

$$\begin{bmatrix} 2 & 0 & 3 & \vdots & 3 \\ 4 & -3 & 7 & \vdots & 5 \\ 8 & -9 & 15 & \vdots & 9 \end{bmatrix}$$

$$\begin{matrix} \\ -2R_1 + R_2 \rightarrow \\ -4R_1 + R_3 \rightarrow \end{matrix} \begin{bmatrix} 2 & 0 & 3 & \vdots & 3 \\ 0 & -3 & 1 & \vdots & -1 \\ 0 & -9 & 3 & \vdots & -3 \end{bmatrix}$$

$$\begin{matrix} \\ \\ -3R_2 + R_3 \rightarrow \end{matrix} \begin{bmatrix} 2 & 0 & 3 & \vdots & 3 \\ 0 & -3 & 1 & \vdots & -1 \\ 0 & 0 & 0 & \vdots & 0 \end{bmatrix}$$

$$\begin{matrix} \\ -\frac{1}{3}R_2 \rightarrow \\ \\ \end{matrix} \begin{bmatrix} 1 & 0 & \frac{3}{2} & \vdots & \frac{3}{2} \\ 0 & 1 & -\frac{1}{3} & \vdots & \frac{1}{3} \\ 0 & 0 & 0 & \vdots & 0 \end{bmatrix}$$

$z = a$
$y = \frac{1}{3}a + \frac{1}{3}$
$x = -\frac{3}{2}a + \frac{3}{2}$
Answer: $(-\frac{3}{2}a + \frac{3}{2}, \frac{1}{3}a + \frac{1}{3}, a)$

54. $4x + 12y - 7z - 20w = 22$
$3x + 9y - 5z - 28w = 30$

$$\begin{bmatrix} 4 & 12 & -7 & -20 & \vdots & 22 \\ 3 & 9 & -5 & -28 & \vdots & 30 \end{bmatrix}$$

$$\begin{matrix} -R_2 + R_1 \rightarrow \\ \\ \end{matrix} \begin{bmatrix} 1 & 3 & -2 & 8 & \vdots & -8 \\ 3 & 9 & -5 & -28 & \vdots & 30 \end{bmatrix}$$

$$\begin{matrix} \\ -3R_1 + R_2 \rightarrow \end{matrix} \begin{bmatrix} 1 & 3 & -2 & 8 & \vdots & -8 \\ 0 & 0 & 1 & -52 & \vdots & 54 \end{bmatrix}$$

$$\begin{matrix} 2R_2 + R_1 \rightarrow \\ \\ \end{matrix} \begin{bmatrix} 1 & 3 & 0 & -96 & \vdots & 100 \\ 0 & 0 & 1 & -52 & \vdots & 54 \end{bmatrix}$$

$w = a$
$z = 52a + 54$
$y = b$
$x = -3b + 96a + 100$
Answer: $(-3b + 96a + 100, b, 52a + 54, a)$

56. $x + 2y = 0$
$2x + 4y = 0$

$$\begin{bmatrix} 1 & 2 & \vdots & 0 \\ 2 & 4 & \vdots & 0 \end{bmatrix}$$

$$\begin{matrix} \\ -2R_1 + R_2 \rightarrow \end{matrix} \begin{bmatrix} 1 & 2 & \vdots & 0 \\ 0 & 0 & \vdots & 0 \end{bmatrix}$$

$y = a$
$x = -2a$
Answer: $(-2a, a)$

58. $$\begin{aligned} 2x + 10y + 2z &= 6 \\ x + 5y + 2z &= 6 \\ x + 5y + z &= 3 \\ -3x - 15y - 3z &= -9 \end{aligned}$$

$$\left[\begin{array}{rrr:r} 2 & 10 & 2 & 6 \\ 1 & 5 & 2 & 6 \\ 1 & 5 & 1 & 3 \\ -3 & -15 & -3 & -9 \end{array}\right]$$

$$\begin{array}{r} \frac{1}{2}R_1 \rightarrow \\ \\ \\ \\ \end{array} \left[\begin{array}{rrr:r} 1 & 5 & 1 & 3 \\ 1 & 5 & 2 & 6 \\ 1 & 5 & 1 & 3 \\ -3 & -15 & -3 & -9 \end{array}\right]$$

$$\begin{array}{r} \\ -R_1 + R_2 \rightarrow \\ -R_1 + R_3 \rightarrow \\ 3R_1 + R_4 \rightarrow \end{array} \left[\begin{array}{rrr:r} 1 & 5 & 1 & 3 \\ 0 & 0 & 1 & 3 \\ 0 & 0 & 0 & 0 \\ 0 & 0 & 0 & 0 \end{array}\right]$$

$$\begin{array}{r} -R_2 + R_1 \rightarrow \\ \\ \\ \\ \end{array} \left[\begin{array}{rrr:r} 1 & 5 & 0 & 0 \\ 0 & 0 & 1 & 3 \\ 0 & 0 & 0 & 0 \\ 0 & 0 & 0 & 0 \end{array}\right]$$

$z = 3, y = a, x = -5a$

Answer: $(-5a, a, 3)$

60. $$\begin{aligned} x + 2y + 2z + 4w &= 11 \\ 3x + 6y + 5z + 12w &= 30 \end{aligned}$$

$$\left[\begin{array}{rrrr:r} 1 & 2 & 2 & 4 & 11 \\ 3 & 6 & 5 & 12 & 30 \end{array}\right]$$

$$\begin{array}{r} \\ -3R_1 + R_2 \rightarrow \end{array} \left[\begin{array}{rrrr:r} 1 & 2 & 2 & 4 & 11 \\ 0 & 0 & -1 & 0 & -3 \end{array}\right]$$

$$\begin{array}{r} 2R_2 + R_1 \rightarrow \\ -R_2 \rightarrow \end{array} \left[\begin{array}{rrrr:r} 1 & 2 & 0 & 4 & 5 \\ 0 & 0 & 1 & 0 & 3 \end{array}\right]$$

$w = a, z = 3, y = b, x = -2b - 4a + 5$

Answer: $(-b - 4a + 5, b, 3, a)$

62. $$\begin{aligned} x + 2y + z + 3w &= 0 \\ x - y + w &= 0 \\ y - z + 2w &= 0 \end{aligned}$$

$$\left[\begin{array}{rrrr:r} 1 & 2 & 1 & 3 & 0 \\ 1 & -1 & 0 & 1 & 0 \\ 0 & 1 & -1 & 2 & 0 \end{array}\right]$$

$$\begin{array}{r} \\ -R_1 + R_2 \rightarrow \\ \\ \end{array} \left[\begin{array}{rrrr:r} 1 & 2 & 1 & 3 & 0 \\ 0 & -3 & -1 & -2 & 0 \\ 0 & 1 & -1 & 2 & 0 \end{array}\right]$$

$$\begin{array}{r} \\ 4R_3 + R_2 \rightarrow \\ \\ \end{array} \left[\begin{array}{rrrr:r} 1 & 2 & 1 & 3 & 0 \\ 0 & 1 & -5 & 6 & 0 \\ 0 & 1 & -1 & 2 & 0 \end{array}\right]$$

$$\begin{array}{r} -2R_2 + R_1 \rightarrow \\ \\ -R_2 + R_3 \rightarrow \end{array} \left[\begin{array}{rrrr:r} 1 & 0 & 11 & -9 & 0 \\ 0 & 1 & -5 & 6 & 0 \\ 0 & 0 & 4 & -4 & 0 \end{array}\right]$$

$$\begin{array}{r} \\ \\ \frac{1}{4}R_3 \rightarrow \end{array} \left[\begin{array}{rrrr:r} 1 & 0 & 11 & -9 & 0 \\ 0 & 1 & -5 & 6 & 0 \\ 0 & 0 & 1 & -1 & 0 \end{array}\right]$$

$$\begin{array}{r} -11R_3 + R_1 \rightarrow \\ 5R_3 + R_2 \rightarrow \\ \\ \end{array} \left[\begin{array}{rrrr:r} 1 & 0 & 0 & 2 & 0 \\ 0 & 1 & 0 & 1 & 0 \\ 0 & 0 & 1 & -1 & 0 \end{array}\right]$$

$w = a, z = a, y = -a, x = -2a$

Answer: $(-2a, -a, a, a)$

64. (a) In the row-echelon form of an augmented matrix that corresponds to an inconsistent system of linear equations, there exists a row consisting of all zeros except for the entry in the last column.

(b) In the row-echelon form of an augmented matrix that corresponds to a system with an infinite number of solutions, there are fewer rows with nonzero entries than there are variables.

66. x = amount at 9%, y = amount at 10%, z = amount at 12%

$$\begin{aligned} x + \quad y \quad z &= 500{,}000 \\ 0.09x + 0.010y + 0.12z &= 52{,}000 \\ 2.5x - \quad y \quad &= 0 \end{aligned}$$

$$\left[\begin{array}{ccc:c} 1 & 1 & 1 & 500{,}000 \\ 0.09 & 0.01 & 0.12 & 52{,}000 \\ 2.5 & -1 & 0 & 0 \end{array}\right]$$

$$\begin{array}{r} \\ -0.09R_1 + R_2 \rightarrow \\ -2.5R_1 + R_3 \rightarrow \end{array} \left[\begin{array}{ccc:c} 1 & 1 & 1 & 500{,}000 \\ 0 & 0.10 & 0.03 & 7{,}000 \\ 0 & -3.5 & -2.5 & -1{,}250{,}000 \end{array}\right]$$

$$\begin{array}{r} \\ 100R_2 \rightarrow \\ 2R_3 \rightarrow \end{array} \left[\begin{array}{ccc:c} 1 & 1 & 1 & 500{,}000 \\ 0 & 1 & 3 & 700{,}000 \\ 0 & -7 & -5 & -2{,}500{,}000 \end{array}\right]$$

$$\begin{array}{r} -R_2 + R_1 \rightarrow \\ \\ 7R_2 + R_3 \rightarrow \end{array} \left[\begin{array}{ccc:c} 1 & 0 & -2 & -200{,}000 \\ 0 & 1 & 3 & 700{,}000 \\ 0 & 0 & 16 & 2{,}400{,}000 \end{array}\right]$$

$$\begin{array}{r} \\ \\ \frac{1}{16}R_3 \rightarrow \end{array} \left[\begin{array}{ccc:c} 1 & 0 & -2 & -200{,}000 \\ 0 & 1 & 3 & 700{,}000 \\ 0 & 0 & 1 & 150{,}000 \end{array}\right]$$

$z = 150{,}000,\ y = 250{,}000,\ x = 100{,}000$

Answer: \$100,000 at 9%, \$250,000 at 10%, \$150,000 at 12%

68.

$$\begin{aligned} I_1 - I_2 + I_3 &= 0 \\ 2I_1 + 2I_2 \quad &= 7 \\ 2I_2 + 4I_3 &= 8 \end{aligned}$$

$$\left[\begin{array}{ccc:c} 1 & -1 & 1 & 0 \\ 2 & 2 & 0 & 7 \\ 0 & 2 & 4 & 8 \end{array}\right]$$

$$\begin{array}{r} \\ -2R_1 + R_2 \rightarrow \\ \\ \end{array} \left[\begin{array}{ccc:c} 1 & -1 & 1 & 0 \\ 0 & 4 & -2 & 7 \\ 0 & 2 & 4 & 8 \end{array}\right]$$

$$\begin{array}{r} \\ R_3 \rightarrow \\ R_2 \rightarrow \end{array} \left[\begin{array}{ccc:c} 1 & -1 & 1 & 0 \\ 0 & 2 & 4 & 8 \\ 0 & 4 & -2 & 7 \end{array}\right]$$

$$\begin{array}{r} \\ \frac{1}{2}R_2 \rightarrow \\ \\ \end{array} \left[\begin{array}{ccc:c} 1 & -1 & 1 & 0 \\ 0 & 1 & 2 & 4 \\ 0 & 4 & -2 & 7 \end{array}\right]$$

$$\begin{array}{r} \\ \\ -4R_2 + R_3 \rightarrow \end{array} \left[\begin{array}{ccc:c} 1 & -1 & 1 & 0 \\ 0 & 1 & 2 & 4 \\ 0 & 0 & -10 & -9 \end{array}\right]$$

$$\begin{array}{r} \\ \\ -\frac{1}{10}R_3 \rightarrow \end{array} \left[\begin{array}{ccc:c} 1 & -1 & 1 & 0 \\ 0 & 1 & 2 & 4 \\ 0 & 0 & 1 & \frac{9}{10} \end{array}\right]$$

$I_3 = \frac{9}{10}$ amperes

$I_2 + 2\left(\frac{9}{10}\right) = 4 \Rightarrow I_2 = \frac{11}{5}$ amperes

$I_1 - \frac{11}{5} + \frac{9}{10} = 0 \Rightarrow I_1 = \frac{13}{10}$ amperes

70. $f(x) = ax^2 + bx + c$

$f(1) = a + b + c = 9$

$f(2) = 4a + 2b + c = 8$

$f(3) = 9a + 3b + c = 5$

$$\left[\begin{array}{ccc:c} 1 & 1 & 1 & 9 \\ 4 & 2 & 1 & 8 \\ 9 & 3 & 1 & 5 \end{array}\right]$$

$$\begin{array}{r} \\ -4R_1 + R_2 \rightarrow \\ -9R_1 + R_3 \rightarrow \end{array} \left[\begin{array}{ccc:c} 1 & 1 & 1 & 9 \\ 0 & -2 & -3 & -28 \\ 0 & -6 & -8 & -76 \end{array}\right]$$

$$\begin{array}{r} \\ -\frac{1}{2}R_2 \rightarrow \\ -3R_2 + R_3 \rightarrow \end{array} \left[\begin{array}{ccc:c} 1 & 1 & 1 & 9 \\ 0 & 1 & \frac{3}{2} & 14 \\ 0 & 0 & 1 & 8 \end{array}\right]$$

$c = 8$

$b + \frac{3}{2}(8) = 14 \Rightarrow b = 2$

$a + 8 + 2 = 9 \Rightarrow a = -1$

Answer: $y = -x^2 + 2x + 8$

72. (a) (0, 670), (1, 280), (2, 231)

$f(x) = ax^2 + bx + c$

$f(0) = c = 670$

$f(1) = a + b + c = 280 \Rightarrow a + b = -390$

$f(2) = 4a + 2b + c = 231 \Rightarrow 4a + 2b = -439$

$$\begin{bmatrix} 1 & 1 & \vdots & -390 \\ 4 & 2 & \vdots & -439 \end{bmatrix}$$

$$\begin{matrix} \\ -4R_1 + R_2 \rightarrow \end{matrix} \begin{bmatrix} 1 & 1 & \vdots & -390 \\ 0 & -2 & \vdots & 1121 \end{bmatrix}$$

$$\begin{matrix} \\ -\frac{1}{2}R_2 \rightarrow \end{matrix} \begin{bmatrix} 1 & 1 & \vdots & -390 \\ 0 & 1 & \vdots & -560.5 \end{bmatrix}$$

$c = 670, b = -560.5$

$a - 560.5 = -390 \Rightarrow a = 170.5$

Answer: $y = 170.5t^2 - 560.5t + 670$

(b)

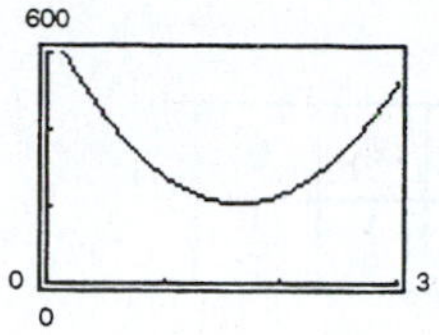

(c) $f(3) = 170.5(3)^2 - 560.5(3) + 670 = 523$

The value of new orders for civil jets in 1993 is estimated at \$523 million.

74. (a) $x_1 + x_2 = 300$

$x_1 + x_3 = 150 + x_4 \Rightarrow x_1 + x_3 - x_4 = 150$

$x_2 + 200 = x_3 + x_5 \Rightarrow x_2 - x_3 - x_5 = -200$

$x_4 + x_5 = 350$

$$\begin{bmatrix} 1 & 1 & 0 & 0 & 0 & \vdots & 300 \\ 1 & 0 & 1 & -1 & 0 & \vdots & 150 \\ 0 & 1 & -1 & 0 & -1 & \vdots & -200 \\ 0 & 0 & 0 & 1 & 1 & \vdots & 350 \end{bmatrix}$$

$$\begin{matrix} \\ -R_1 + R_2 \rightarrow \\ \\ \\ \end{matrix} \begin{bmatrix} 1 & 1 & 0 & 0 & 0 & \vdots & 300 \\ 0 & -1 & 1 & -1 & 0 & \vdots & -150 \\ 0 & 1 & -1 & 0 & -1 & \vdots & -200 \\ 0 & 0 & 0 & 1 & 1 & \vdots & 350 \end{bmatrix}$$

$$\begin{matrix} \\ \\ R_2 + R_3 \rightarrow \\ \\ \end{matrix} \begin{bmatrix} 1 & 1 & 0 & 0 & 0 & \vdots & 300 \\ 0 & -1 & 1 & -1 & 0 & \vdots & -150 \\ 0 & 0 & 0 & -1 & -1 & \vdots & -350 \\ 0 & 0 & 0 & 1 & 1 & \vdots & 350 \end{bmatrix}$$

$$\begin{matrix} \\ -R_2 \rightarrow \\ -R_3 \rightarrow \\ R_3 + R_4 \rightarrow \end{matrix} \begin{bmatrix} 1 & 1 & 0 & 0 & 0 & \vdots & 300 \\ 0 & 1 & -1 & 1 & 0 & \vdots & 150 \\ 0 & 0 & 0 & 1 & 1 & \vdots & 350 \\ 0 & 0 & 0 & 0 & 0 & \vdots & 0 \end{bmatrix}$$

Let $x_5 = t$.

$x_4 + t = 350 \Rightarrow x_4 = 350 - t$

Let $x_3 = s$.

$x_2 - s + 350 - t = 150 \Rightarrow x_2 = -200 + s + t$

$x_1 - 200 + s + t = 300 \Rightarrow x_1 = 500 - s - t$

(b) When $x_2 = 200$ and $x_3 = 50$,

$x_2 = -200 + s + t$

$200 = -200 + 50 + t \Rightarrow t = 350.$

$x_5 = 350, x_4 = 0, x_3 = 50, x_2 = 200,$

$x_1 = 100$

(c) When $x_2 = 150$ and $x_3 = 0$,

$x_2 = -200 + s + t$

$150 = -200 + 0 + t \Rightarrow t = 350.$

$x_5 = 350, x_4 = 0, x_3 = 0, x_2 = 150,$

$x_1 = 150$

76. $g(x) = 3^{-x/2}$

x	-2	0	2	4	6
y	3	1	$\frac{1}{3}$	$\frac{1}{9}$	$\frac{1}{27}$

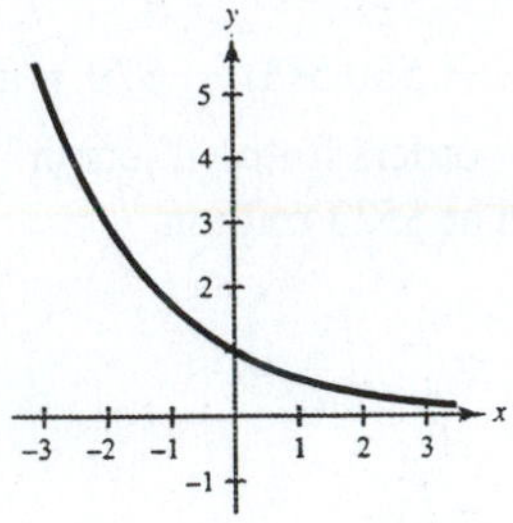

78. $f(x) = 3 + \ln x \Rightarrow y - 3 \ln x \Rightarrow e^{y-3} = x$

x	0.05	0.14	0.37	1	2.72
y	0	1	2	3	4

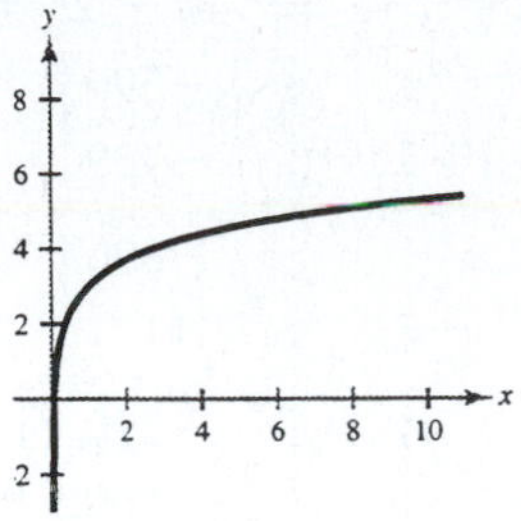

Section 8.2 Operations with Matrices

Solutions to Even-Numbered Exercises

2. $x = 13, y = 12$

4. $x + 2 = 2x + 6, \quad y + 2 = 11$

$-4 = x \qquad y = 9$

6. (a) $A + B = \begin{bmatrix} 1 & 2 \\ 2 & 1 \end{bmatrix} + \begin{bmatrix} -3 & -2 \\ 4 & 2 \end{bmatrix} = \begin{bmatrix} 1-3 & 2-2 \\ 2+4 & 1+2 \end{bmatrix} = \begin{bmatrix} -2 & 0 \\ 6 & 3 \end{bmatrix}$

(b) $A - B = \begin{bmatrix} 1 & 2 \\ 2 & 1 \end{bmatrix} - \begin{bmatrix} -3 & -2 \\ 4 & 2 \end{bmatrix} = \begin{bmatrix} 1+3 & 2+2 \\ 2-4 & 1-2 \end{bmatrix} = \begin{bmatrix} 4 & 4 \\ -2 & -1 \end{bmatrix}$

(c) $3A = 3\begin{bmatrix} 1 & 2 \\ 2 & 1 \end{bmatrix} = \begin{bmatrix} 3(1) & 3(2) \\ 3(2) & 3(1) \end{bmatrix} = \begin{bmatrix} 3 & 6 \\ 6 & 3 \end{bmatrix}$

(d) $3A - 2B = \begin{bmatrix} 3 & 6 \\ 6 & 3 \end{bmatrix} - 2\begin{bmatrix} -3 & -2 \\ 4 & 2 \end{bmatrix} = \begin{bmatrix} 3+6 & 6+4 \\ 6-8 & 3-4 \end{bmatrix} = \begin{bmatrix} 9 & 10 \\ -2 & -1 \end{bmatrix}$

8. (a) $A + B = \begin{bmatrix} 2 & 1 & 1 \\ -1 & -1 & 4 \end{bmatrix} + \begin{bmatrix} 2 & -3 & 4 \\ -3 & 1 & -2 \end{bmatrix} = \begin{bmatrix} 2+2 & 1-3 & 1+4 \\ -1-3 & -1+1 & 4-2 \end{bmatrix} = \begin{bmatrix} 4 & -2 & 5 \\ -4 & 0 & 2 \end{bmatrix}$

(b) $A - B = \begin{bmatrix} 2 & 1 & 1 \\ -1 & -1 & 4 \end{bmatrix} + \begin{bmatrix} 2 & -3 & 4 \\ -3 & 1 & -2 \end{bmatrix} = \begin{bmatrix} 2-2 & 1-(-3) & 1-4 \\ -1-(-3) & -1-1 & 4-(-2) \end{bmatrix} = \begin{bmatrix} 0 & 4 & -3 \\ 2 & -2 & 6 \end{bmatrix}$

(c) $3A = 3\begin{bmatrix} 2 & 1 & 1 \\ -1 & -1 & 4 \end{bmatrix} = \begin{bmatrix} 3(2) & 3(1) & 3(1) \\ 3(-1) & 3(-1) & 3(4) \end{bmatrix} = \begin{bmatrix} 6 & 3 & 3 \\ -3 & -3 & 12 \end{bmatrix}$

(d) $3A - 2B = \begin{bmatrix} 6 & 3 & 3 \\ -3 & -3 & 12 \end{bmatrix} - 2\begin{bmatrix} 2 & -3 & 4 \\ -3 & 1 & -2 \end{bmatrix} = \begin{bmatrix} 6 & 3 & 3 \\ -3 & -3 & 12 \end{bmatrix} + \begin{bmatrix} -4 & 6 & -8 \\ 6 & -2 & 4 \end{bmatrix}$

$= \begin{bmatrix} 2 & 9 & -5 \\ 3 & -5 & 16 \end{bmatrix}$

10. (a) $A + B = \begin{bmatrix} 3 \\ 2 \\ -1 \end{bmatrix} + \begin{bmatrix} -4 \\ 6 \\ 2 \end{bmatrix} = \begin{bmatrix} 3-4 \\ 2+6 \\ -1+2 \end{bmatrix} = \begin{bmatrix} -1 \\ 8 \\ 1 \end{bmatrix}$

(b) $A - B = \begin{bmatrix} 3 \\ 2 \\ -1 \end{bmatrix} - \begin{bmatrix} -4 \\ 6 \\ 2 \end{bmatrix} = \begin{bmatrix} 3+4 \\ 2-6 \\ -1-2 \end{bmatrix} = \begin{bmatrix} 7 \\ -4 \\ -3 \end{bmatrix}$

— CONTINUED —

10. — CONTINUED —

(c) $3A = 3\begin{bmatrix} 3 \\ 2 \\ -1 \end{bmatrix} = \begin{bmatrix} 3(3) \\ 3(2) \\ 3(-1) \end{bmatrix} = \begin{bmatrix} 9 \\ 6 \\ -3 \end{bmatrix}$

(d) $3A - 2B = \begin{bmatrix} 9 \\ 6 \\ -3 \end{bmatrix} - 2\begin{bmatrix} -4 \\ 6 \\ 2 \end{bmatrix} = \begin{bmatrix} 9 \\ 6 \\ -3 \end{bmatrix} + \begin{bmatrix} 8 \\ -12 \\ -4 \end{bmatrix} = \begin{bmatrix} 17 \\ -6 \\ -7 \end{bmatrix}$

12. $2X = 2A - B$

$$X = A - \tfrac{1}{2}B = \begin{bmatrix} -2 & -1 \\ 1 & 0 \\ 3 & -4 \end{bmatrix} - \tfrac{1}{2}\begin{bmatrix} 0 & 3 \\ 2 & 0 \\ -4 & -1 \end{bmatrix} = \begin{bmatrix} -2 & -1 \\ 1 & 0 \\ 3 & -4 \end{bmatrix} - \begin{bmatrix} 0 & \frac{3}{2} \\ 1 & 0 \\ -2 & -\frac{1}{2} \end{bmatrix} = \begin{bmatrix} -2 & -\frac{5}{2} \\ 0 & 0 \\ 5 & -\frac{7}{2} \end{bmatrix}$$

14. $2A + 4B = -2X$

$$X = -A - 2B = -1\begin{bmatrix} -2 & -1 \\ 1 & 0 \\ 3 & -4 \end{bmatrix} - 2\begin{bmatrix} 0 & 3 \\ 2 & 0 \\ -4 & -1 \end{bmatrix} = \begin{bmatrix} 2 & 1 \\ -1 & 0 \\ -3 & 4 \end{bmatrix} + \begin{bmatrix} 0 & -6 \\ -4 & 0 \\ 8 & 2 \end{bmatrix} = \begin{bmatrix} 2 & -5 \\ -5 & 0 \\ 5 & 6 \end{bmatrix}$$

16. (a) $AB = \begin{bmatrix} 2 & -1 \\ 1 & 4 \end{bmatrix}\begin{bmatrix} 0 & 0 \\ 3 & -3 \end{bmatrix} = \begin{bmatrix} 2(0) + (-1)3 & 2(0) + (-1)(-3) \\ 1(0) + 4(3) & 1(0) + 4(-3) \end{bmatrix} = \begin{bmatrix} -3 & 3 \\ 12 & -12 \end{bmatrix}$

(b) $BA = \begin{bmatrix} 0 & 0 \\ 3 & -3 \end{bmatrix}\begin{bmatrix} 2 & -1 \\ 1 & 4 \end{bmatrix} = \begin{bmatrix} 0(2) + (0)1 & 0(-1) + (0)(4) \\ 3(2) + (-3)(1) & 3(-1) + (-3)4 \end{bmatrix} = \begin{bmatrix} 0 & 0 \\ 3 & -15 \end{bmatrix}$

(c) $A^2 = \begin{bmatrix} 2 & -1 \\ 1 & 4 \end{bmatrix}\begin{bmatrix} 2 & -1 \\ 1 & 4 \end{bmatrix} = \begin{bmatrix} 2(2) + (-1)(1) & 2(-1) + (-1)4 \\ 1(2) + 4(1) & 1(-1) + 4(4) \end{bmatrix} = \begin{bmatrix} 3 & -6 \\ 6 & 15 \end{bmatrix}$

18. (a) $AB = \begin{bmatrix} 1 & -1 \\ 1 & 1 \end{bmatrix}\begin{bmatrix} 1 & 3 \\ -3 & 1 \end{bmatrix} = \begin{bmatrix} 1(1) + (-1)(-3) & 1(3) + (-1)(1) \\ 1(1) + 1(-3) & 1(3) + 1(1) \end{bmatrix} = \begin{bmatrix} 4 & 2 \\ -2 & 4 \end{bmatrix}$

(b) $BA = \begin{bmatrix} 1 & 3 \\ -3 & 1 \end{bmatrix}\begin{bmatrix} 1 & -1 \\ 1 & 1 \end{bmatrix} = \begin{bmatrix} 1(1) + (3)1 & 1(-1) + 3(1) \\ -3(1) + (1)(1) & -3(-1) + 1(1) \end{bmatrix} = \begin{bmatrix} 4 & 2 \\ -2 & 4 \end{bmatrix}$

(c) $A^2 = \begin{bmatrix} 1 & -1 \\ 1 & 1 \end{bmatrix}\begin{bmatrix} 1 & 1 \\ 1 & 1 \end{bmatrix} = \begin{bmatrix} 1(1) + (-1)(1) & 1(-1) + (-1)(1) \\ 1(1) + (1)(1) & 1(-1) + 1(1) \end{bmatrix} = \begin{bmatrix} 0 & -2 \\ 2 & 0 \end{bmatrix}$

20. (a) $AB = [3 \;\; 2 \;\; 1]\begin{bmatrix} 2 \\ 3 \\ 0 \end{bmatrix} = [3(2) + 2(3) + 1(0)] = [12]$

(b) $BA = \begin{bmatrix} 2 \\ 3 \\ 0 \end{bmatrix}[3 \;\; 2 \;\; 1] = \begin{bmatrix} 2(3) & 2(2) & 2(1) \\ 3(3) & 3(2) & 3(1) \\ 0(3) & 0(2) & 0(1) \end{bmatrix} = \begin{bmatrix} 6 & 4 & 2 \\ 9 & 6 & 3 \\ 0 & 0 & 0 \end{bmatrix}$

(c) The number of columns of A does not equal the number of rows of A; the multiplication is not possible.

22. A is 3×3, B is $3 \times 2 \Rightarrow AB$ is 3×2.

$$\begin{bmatrix} 0 & -1 & 0 \\ 4 & 0 & 2 \\ 8 & -1 & 7 \end{bmatrix}\begin{bmatrix} 2 & 1 \\ -3 & 4 \\ 1 & 6 \end{bmatrix} = \begin{bmatrix} 3 & -4 \\ 10 & 16 \\ 26 & 46 \end{bmatrix}$$

24. A is 3×3, B is $3 \times 3 \Rightarrow AB$ is 3×3.

$$\begin{bmatrix} 1 & 0 & 0 \\ 0 & 4 & 0 \\ 0 & 0 & -2 \end{bmatrix}\begin{bmatrix} 3 & 0 & 0 \\ 0 & -1 & 0 \\ 0 & 0 & 5 \end{bmatrix} = \begin{bmatrix} 3 & 0 & 0 \\ 0 & -4 & 0 \\ 0 & 0 & -10 \end{bmatrix}$$

26. A is 3×3, B is $3 \times 3 \Rightarrow AB$ is 3×3.

$$\begin{bmatrix} 0 & 0 & 5 \\ 0 & 0 & -3 \\ 0 & 0 & 4 \end{bmatrix}\begin{bmatrix} 6 & -11 & 4 \\ 8 & 16 & 4 \\ 0 & 0 & 0 \end{bmatrix} = \begin{bmatrix} 0 & 0 & 0 \\ 0 & 0 & 0 \\ 0 & 0 & 0 \end{bmatrix}$$

28. A is 2×4, B is $2 \times 2 \Rightarrow AB$ is not defined.

30. $\begin{bmatrix} 11 & -12 & 4 \\ 14 & 10 & 12 \\ 6 & -2 & 9 \end{bmatrix}\begin{bmatrix} 12 & 10 \\ -5 & 12 \\ 15 & 16 \end{bmatrix} = \begin{bmatrix} 252 & 30 \\ 298 & 452 \\ 217 & 180 \end{bmatrix}$

32. A is 3×3, B is $4 \times 2 \Rightarrow AB$ is not defined.

34. $\begin{bmatrix} 15 & -18 \\ -4 & 12 \\ -8 & 22 \end{bmatrix}\begin{bmatrix} -7 & 22 & 1 \\ 8 & 16 & 24 \end{bmatrix} = \begin{bmatrix} -249 & 42 & -417 \\ 124 & 104 & 284 \\ 232 & 176 & 520 \end{bmatrix}$

36. $A = \begin{bmatrix} 1 & -2 & 3 \\ -1 & 3 & -1 \\ 2 & -5 & 5 \end{bmatrix}, X = \begin{bmatrix} x \\ y \\ z \end{bmatrix}, B = \begin{bmatrix} 9 \\ -6 \\ 17 \end{bmatrix}$

$$\left[\begin{array}{ccc:c} 1 & -2 & 3 & 9 \\ -1 & 3 & -1 & -6 \\ 2 & -5 & 5 & 17 \end{array}\right]$$

$$\begin{array}{r} \\ R_1 + R_2 \rightarrow \\ -2R_1 + R_3 \rightarrow \end{array}\left[\begin{array}{ccc:c} 1 & -2 & 3 & 9 \\ 0 & 1 & 2 & 3 \\ 0 & -1 & -1 & -1 \end{array}\right]$$

$$\begin{array}{r} 2R_2 + R_1 \rightarrow \\ \\ R_2 + R_3 \rightarrow \end{array}\left[\begin{array}{ccc:c} 1 & 0 & 7 & 15 \\ 0 & 1 & 2 & 3 \\ 0 & 0 & 1 & 2 \end{array}\right]$$

$$\begin{array}{r} -7R_3 + R_1 \rightarrow \\ -2R_3 + R_2 \rightarrow \\ \\ \end{array}\left[\begin{array}{ccc:c} 1 & 0 & 0 & 1 \\ 0 & 1 & 0 & -1 \\ 0 & 0 & 1 & 2 \end{array}\right]$$

$x = 1, y = -1, z = 2$

38. $A = \begin{bmatrix} 1 & 1 & -3 \\ -1 & 2 & 0 \\ 0 & -1 & 1 \end{bmatrix}, X = \begin{bmatrix} x \\ y \\ z \end{bmatrix}, B = \begin{bmatrix} -1 \\ 1 \\ 0 \end{bmatrix}$

$$\left[\begin{array}{ccc:c} 1 & 1 & -3 & -1 \\ -1 & 2 & 0 & 1 \\ 0 & -1 & 1 & 0 \end{array}\right]$$

$$\begin{array}{r} \\ R_1 + R_2 \rightarrow \\ \\ \end{array}\left[\begin{array}{ccc:c} 1 & 1 & -3 & -1 \\ 0 & 3 & -3 & 0 \\ 0 & -1 & 1 & 0 \end{array}\right]$$

$$\begin{array}{r} \\ \frac{1}{3}R_2 \rightarrow \\ \frac{1}{3}R_2 + R_3 \rightarrow \end{array}\left[\begin{array}{ccc:c} 1 & 1 & -3 & -1 \\ 0 & 1 & -1 & 0 \\ 0 & 0 & 0 & 0 \end{array}\right]$$

$$\begin{array}{r} -R_2 + R_1 \rightarrow \\ \\ \\ \end{array}\left[\begin{array}{ccc:c} 1 & 0 & -2 & -1 \\ 0 & 1 & -1 & 0 \\ 0 & 0 & 0 & 0 \end{array}\right]$$

Let $z = a$, then $y = a, x = 2a - 1$.

40. $A = \begin{bmatrix} 5 & 4 \\ 1 & 2 \end{bmatrix}$

$f(A) = A^2 - 7A + 6 = \begin{bmatrix} 5 & 4 \\ 1 & 2 \end{bmatrix}\begin{bmatrix} 5 & 4 \\ 1 & 2 \end{bmatrix} - 7\begin{bmatrix} 5 & 4 \\ 1 & 2 \end{bmatrix} + 6\begin{bmatrix} 1 & 0 \\ 0 & 1 \end{bmatrix} = \begin{bmatrix} 0 & 0 \\ 0 & 0 \end{bmatrix}$

42. $A = \begin{bmatrix} 8 & -4 \\ 2 & 2 \end{bmatrix}$

$f(A) = A^2 - 10A + 24 = \begin{bmatrix} 8 & -4 \\ 2 & 2 \end{bmatrix}\begin{bmatrix} 8 & -4 \\ 2 & 2 \end{bmatrix} - 10\begin{bmatrix} 8 & -4 \\ 2 & 2 \end{bmatrix} + 24\begin{bmatrix} 1 & 0 \\ 0 & 1 \end{bmatrix} = \begin{bmatrix} 0 & 0 \\ 0 & 0 \end{bmatrix}$

44. $A = \begin{bmatrix} 3 & 3 \\ 4 & 4 \end{bmatrix}, B = \begin{bmatrix} 1 & -1 \\ -1 & 1 \end{bmatrix}$

$AB = \begin{bmatrix} 3 & 3 \\ 4 & 4 \end{bmatrix}\begin{bmatrix} 1 & -1 \\ -1 & 1 \end{bmatrix} = \begin{bmatrix} 0 & 0 \\ 0 & 0 \end{bmatrix}$

$AB = O$ but $A \neq O$ and $B \neq O$.

For 46–54, A is of order 2×3, B is of order 2×3, C is of order 3×2 and D is of order 2×2.

46. $B - 3C$ is not possible. B and C are not of the same order.

48. BC is possible. The resulting order is 2×2.

50. $CB - D$ is not possible. The order of CB is 3×3, but the order of D is 2×2.

52. $(BC)D$ is possible. The resulting order is 2×2.

54. $(BC - D)A$ is possible. The resulting order is 2×3.

56. $1.10\begin{bmatrix} 100 & 90 & 70 & 30 \\ 40 & 20 & 60 & 60 \end{bmatrix} = \begin{bmatrix} 110 & 99 & 77 & 33 \\ 44 & 22 & 66 & 66 \end{bmatrix}$

58. $BA = [\$20.50 \quad \$26.50 \quad \$29.50] = \begin{bmatrix} 5{,}000 & 4{,}000 \\ 6{,}000 & 10{,}000 \\ 8{,}000 & 5{,}000 \end{bmatrix} = [\$497{,}500 \quad \$494{,}500]$

The entries represent the costs of the three models of the product at the two warehouses.

60. $P^2 = \begin{bmatrix} 0.6 & 0.1 & 0.1 \\ 0.2 & 0.7 & 0.1 \\ 0.2 & 0.2 & 0.8 \end{bmatrix}\begin{bmatrix} 0.6 & 0.1 & 0.1 \\ 0.2 & 0.7 & 0.1 \\ 0.2 & 0.2 & 0.8 \end{bmatrix} = \begin{bmatrix} 0.4 & 0.15 & 0.15 \\ 0.28 & 0.53 & 0.17 \\ 0.32 & 0.32 & 0.68 \end{bmatrix}$

The P^2 matrix gives the transition probabilities from the first election to the third.

62. $P^3 = P^2P = \begin{bmatrix} 0.4 & 0.15 & 0.15 \\ 0.28 & 0.53 & 0.17 \\ 0.32 & 0.32 & 0.68 \end{bmatrix}\begin{bmatrix} 0.5 & 0.1 & 0.1 \\ 0.2 & 0.7 & 0.1 \\ 0.2 & 0.2 & 0.8 \end{bmatrix} = \begin{bmatrix} 0.300 & 0.175 & 0.175 \\ 0.308 & 0.433 & 0.217 \\ 0.392 & 0.392 & 0.608 \end{bmatrix}$

$P^4 = P^3P = \begin{bmatrix} 0.300 & 0.175 & 0.175 \\ 0.308 & 0.433 & 0.217 \\ 0.392 & 0.392 & 0.608 \end{bmatrix}\begin{bmatrix} 0.6 & 0.1 & 0.1 \\ 0.2 & 0.7 & 0.1 \\ 0.2 & 0.2 & 0.8 \end{bmatrix} = \begin{bmatrix} 0.250 & 0.188 & 0.188 \\ 0.315 & 0.377 & 0.248 \\ 0.435 & 0.435 & 0.565 \end{bmatrix}$

$P^5 = P^4P = \begin{bmatrix} 0.250 & 0.188 & 0.188 \\ 0.315 & 0.377 & 0.248 \\ 0.435 & 0.435 & 0.565 \end{bmatrix}\begin{bmatrix} 0.6 & 0.1 & 0.1 \\ 0.2 & 0.7 & 0.1 \\ 0.2 & 0.2 & 0.8 \end{bmatrix} = \begin{bmatrix} 0.225 & 0.194 & 0.194 \\ 0.314 & 0.345 & 0.267 \\ 0.461 & 0.461 & 0.539 \end{bmatrix}$

$P^6 = \begin{bmatrix} 0.213 & 0.197 & 0.197 \\ 0.311 & 0.326 & 0.280 \\ 0.477 & 0.477 & 0.523 \end{bmatrix}$

$P^7 = \begin{bmatrix} 0.206 & 0.198 & 0.198 \\ 0.308 & 0.316 & 0.288 \\ 0.486 & 0.486 & 0.514 \end{bmatrix}$

$P^8 = \begin{bmatrix} 0.203 & 0.199 & 0.199 \\ 0.305 & 0.309 & 0.292 \\ 0.492 & 0.492 & 0.508 \end{bmatrix}$

As P is raised to higher and higher powers, the resulting matrices appear to be approaching the matrix

$\begin{bmatrix} 0.2 & 0.2 & 0.2 \\ 0.3 & 0.3 & 0.3 \\ 0.5 & 0.5 & 0.5 \end{bmatrix}.$

64. $A = \begin{bmatrix} 0 & -i \\ i & 0 \end{bmatrix}$

$A^2 = \begin{bmatrix} 0 & -i \\ i & 0 \end{bmatrix}\begin{bmatrix} 0 & -i \\ i & 0 \end{bmatrix} = \begin{bmatrix} 1 & 0 \\ 0 & 1 \end{bmatrix} = I$, the identity matrix.

Section 8.3 The Inverse of a Square Matrix

Solutions to Even-Numbered Exercises

2. $AB = \begin{bmatrix} 1 & -1 \\ -1 & 2 \end{bmatrix}\begin{bmatrix} 2 & 1 \\ 1 & 1 \end{bmatrix} = \begin{bmatrix} 2-1 & 1-1 \\ -2+2 & -1=2 \end{bmatrix} = \begin{bmatrix} 1 & 0 \\ 0 & 1 \end{bmatrix}$

$BA = \begin{bmatrix} 2 & 1 \\ 1 & 1 \end{bmatrix}\begin{bmatrix} 1 & -1 \\ -1 & 2 \end{bmatrix} = \begin{bmatrix} 2-1 & -2+2 \\ 1-1 & -1+2 \end{bmatrix} = \begin{bmatrix} 1 & 0 \\ 0 & 1 \end{bmatrix}$

4. $AB = \begin{bmatrix} 1 & -1 \\ -1 & 2 \end{bmatrix}\begin{bmatrix} \frac{3}{5} & \frac{1}{5} \\ -\frac{2}{5} & \frac{1}{5} \end{bmatrix} = \begin{bmatrix} \frac{3}{5}+\frac{2}{5} & \frac{1}{5}-\frac{1}{5} \\ \frac{6}{5}-\frac{6}{5} & \frac{2}{5}+\frac{3}{5} \end{bmatrix} = \begin{bmatrix} 1 & 0 \\ 0 & 1 \end{bmatrix}$

$AB = \begin{bmatrix} \frac{3}{5} & \frac{1}{5} \\ -\frac{2}{5} & \frac{1}{5} \end{bmatrix}\begin{bmatrix} 1 & -1 \\ 2 & 3 \end{bmatrix} = \begin{bmatrix} \frac{3}{5}+\frac{2}{5} & -\frac{3}{5}+\frac{3}{5} \\ -\frac{2}{5}+\frac{2}{5} & \frac{2}{5}+\frac{3}{5} \end{bmatrix} = \begin{bmatrix} 1 & 0 \\ 0 & 1 \end{bmatrix}$

6. $AB = \begin{bmatrix} 2 & -17 & 11 \\ -1 & 11 & -7 \\ 0 & 3 & -2 \end{bmatrix}\begin{bmatrix} 1 & 1 & 2 \\ 2 & 4 & -3 \\ 3 & 6 & -5 \end{bmatrix}$

$= \begin{bmatrix} 2-34+33 & 2-68+66 & 4+51-55 \\ -1+22-21 & -1+44-42 & -2-33+35 \\ 6-6 & 12-12 & -9+10 \end{bmatrix} = \begin{bmatrix} 1 & 0 & 0 \\ 0 & 1 & 0 \\ 0 & 0 & 1 \end{bmatrix}$

$BA = \begin{bmatrix} 1 & 1 & 2 \\ 2 & 4 & -3 \\ 3 & 6 & -5 \end{bmatrix}\begin{bmatrix} 2 & -17 & 11 \\ -1 & 11 & -7 \\ 0 & 3 & -2 \end{bmatrix} = \begin{bmatrix} 2-1 & -17+11+6 & 11-7-4 \\ 4-4 & -34+44-9 & 22-28+6 \\ 6-6 & -51+66-15 & 33-42+10 \end{bmatrix}$

8. $AB = \frac{1}{3}\begin{bmatrix} -1 & 1 & 0 & -1 \\ 1 & -1 & 1 & 0 \\ -1 & 1 & 2 & 0 \\ 0 & -1 & 1 & 1 \end{bmatrix}\begin{bmatrix} -3 & 1 & 1 & -3 \\ -3 & -1 & 2 & -3 \\ 0 & 1 & 1 & 0 \\ -3 & -2 & 1 & 0 \end{bmatrix}$

$= \frac{1}{3}\begin{bmatrix} 3-3+0+3 & -1-1+0+2 & -1+2+0-1 & 3-3+0+0 \\ -3+3+0+0 & 1+1+1+0 & 1-2+1+0 & -3+3+0+0 \\ 3-3+0+0 & -1-1+2+0 & -1+2+2+0 & 3-3+0+0 \\ 0+3+0-3 & 0+1+1-2 & 0-2+1+1 & 0+3+0+0 \end{bmatrix}$

$= \frac{1}{3}\begin{bmatrix} 3 & 0 & 0 & 0 \\ 0 & 3 & 0 & 0 \\ 0 & 0 & 3 & 0 \\ 0 & 0 & 0 & 3 \end{bmatrix} = I_4$

$BA = \frac{1}{3}\begin{bmatrix} -3 & 1 & 1 & -3 \\ -3 & -1 & 2 & -3 \\ 0 & 1 & 1 & 0 \\ -3 & -2 & 1 & 0 \end{bmatrix}\begin{bmatrix} -1 & 1 & 0 & -1 \\ 1 & -1 & 1 & 0 \\ -1 & 1 & 2 & 0 \\ 0 & -1 & 1 & 1 \end{bmatrix}$

$= \frac{1}{3}\begin{bmatrix} 3+3-0+3 & -3-1+1+3 & 0+1+2-3 & 3+0+0-3 \\ 3-1-2+0 & -3+1+2+3 & 0-1+4-3 & 3+0+0-3 \\ 0+1-1+0 & 0-1+1+0 & 0+1+2+0 & 0+0+0+0 \\ 3-2-1+0 & -3+2+1+0 & 0-2+2+0 & 3+0+0+0 \end{bmatrix}$

$= \frac{1}{3}\begin{bmatrix} 3 & 0 & 0 & 0 \\ 0 & 3 & 0 & 0 \\ 0 & 0 & 3 & 0 \\ 0 & 0 & 0 & 3 \end{bmatrix} = I_4$

10. $[A \; \vdots \; I] = \left[\begin{array}{cc:cc} 1 & 2 & 1 & 0 \\ 3 & 7 & 0 & 1 \end{array}\right]$

$$-3R_1 + R_2 \rightarrow \left[\begin{array}{cc:cc} 1 & 2 & 1 & 0 \\ 0 & 1 & -3 & 1 \end{array}\right]$$

$$-2R_2 + R_1 \rightarrow \left[\begin{array}{cc:cc} 1 & 0 & 7 & -2 \\ 0 & 1 & -3 & 1 \end{array}\right] = [I \; \vdots \; A^{-1}]$$

$$A^{-1} = \begin{bmatrix} 7 & -2 \\ -3 & 1 \end{bmatrix}$$

12. $[A \; \vdots \; I] = \left[\begin{array}{cc:cc} -7 & 33 & 1 & 0 \\ 4 & -19 & 0 & 1 \end{array}\right]$

$$2R_2 + R_1 \rightarrow \left[\begin{array}{cc:cc} 1 & -5 & 1 & 2 \\ 4 & -19 & 0 & 1 \end{array}\right]$$

$$-4R_1 + R_2 \rightarrow \left[\begin{array}{cc:cc} 1 & -5 & 1 & 2 \\ 0 & 1 & -4 & -7 \end{array}\right]$$

$$5R_2 + R_1 \rightarrow \left[\begin{array}{cc:cc} 1 & 0 & -19 & -33 \\ 0 & 1 & -4 & -7 \end{array}\right] = [I \; \vdots \; A^{-1}]$$

$$A^{-1} = \begin{bmatrix} -19 & -33 \\ -4 & -7 \end{bmatrix}$$

14. $[A \; \vdots \; I] = \left[\begin{array}{cc:cc} 11 & 1 & 1 & 0 \\ -1 & 0 & 0 & 1 \end{array}\right]$

$$10R_2 + R_1 \rightarrow \left[\begin{array}{cc:cc} 1 & 1 & 1 & 10 \\ -1 & 0 & 0 & 1 \end{array}\right]$$

$$R_1 + R_2 \rightarrow \left[\begin{array}{cc:cc} 1 & 1 & 1 & 10 \\ 0 & 1 & 1 & 11 \end{array}\right]$$

$$-R_2 + R_1 \rightarrow \left[\begin{array}{cc:cc} 1 & 0 & 0 & -1 \\ 0 & 1 & 1 & 11 \end{array}\right] = [I \; \vdots \; A^{-1}]$$

$$A^{-1} = \begin{bmatrix} 0 & -1 \\ 1 & 11 \end{bmatrix}$$

16. $[A \; \vdots \; I] = \left[\begin{array}{cc:cc} 2 & 3 & 1 & 0 \\ 1 & 4 & 0 & 1 \end{array}\right]$

$$\begin{array}{r} R_2 \rightarrow \\ R_1 \rightarrow \end{array} \left[\begin{array}{cc:cc} 1 & 3 & 0 & 1 \\ 2 & 4 & 1 & 0 \end{array}\right]$$

$$-2R_1 + R_2 \rightarrow \left[\begin{array}{cc:cc} 1 & 4 & 0 & 1 \\ 0 & -5 & 1 & -2 \end{array}\right]$$

$$-\tfrac{1}{5}R_2 \rightarrow \left[\begin{array}{cc:cc} 1 & 4 & 0 & 1 \\ 0 & 1 & -\frac{1}{5} & \frac{2}{5} \end{array}\right]$$

$$-4R_2 + R_1 \rightarrow \left[\begin{array}{cc:cc} 1 & 0 & \frac{4}{5} & -\frac{3}{5} \\ 0 & 1 & -\frac{1}{5} & \frac{2}{5} \end{array}\right] = [I \; \vdots \; A^{-1}]$$

$$A^{-1} = \frac{1}{5}\begin{bmatrix} 4 & -3 \\ -1 & 2 \end{bmatrix}$$

18. $A = \begin{bmatrix} -2 & 5 \\ 6 & -15 \\ 0 & 1 \end{bmatrix}$ A has no inverse because it is not square.

20.

$$[A \;\vdots\; I] = \left[\begin{array}{rrr|rrr} 1 & 2 & 2 & 1 & 0 & 0 \\ 3 & 7 & 9 & 0 & 1 & 0 \\ -1 & -4 & -7 & 0 & 0 & 1 \end{array}\right]$$

$$\begin{array}{r} \\ -3R_1 + R_2 \rightarrow \\ R_1 + R_3 \rightarrow \end{array} \left[\begin{array}{rrr|rrr} 1 & 2 & 2 & 1 & 0 & 0 \\ 0 & 1 & 3 & -3 & 1 & 0 \\ 0 & -2 & -5 & 1 & 0 & 1 \end{array}\right]$$

$$\begin{array}{r} -2R_2 + R_1 \rightarrow \\ \\ 2R_2 + R_3 \rightarrow \end{array} \left[\begin{array}{rrr|rrr} 1 & 0 & -4 & 7 & -2 & 0 \\ 0 & 1 & 3 & -3 & 1 & 0 \\ 0 & 0 & 1 & -5 & 2 & 1 \end{array}\right]$$

$$\begin{array}{r} 4R_3 + R_1 \rightarrow \\ -3R_3 + R_2 \rightarrow \\ \\ \end{array} \left[\begin{array}{rrr|rrr} 1 & 0 & 0 & -13 & 6 & 4 \\ 0 & 1 & 0 & 12 & -5 & -3 \\ 0 & 0 & 1 & -5 & 2 & 1 \end{array}\right] = [I \;\vdots\; A^{-1}]$$

$$A^{-1} = \begin{bmatrix} -13 & 6 & 4 \\ 12 & -5 & -3 \\ -5 & 2 & 1 \end{bmatrix}$$

22. $$[A \;\vdots\; I] = \left[\begin{array}{rrr|rrr} 1 & 0 & 0 & 1 & 0 & 0 \\ 3 & 0 & 0 & 0 & 1 & 0 \\ 2 & 5 & 5 & 0 & 0 & 1 \end{array}\right] \begin{array}{r} \\ -3R_1 + R_2 \rightarrow \\ -2R_1 + R_3 \rightarrow \end{array} \left[\begin{array}{rrr|rrr} 1 & 0 & 0 & 1 & 0 & 0 \\ 0 & 0 & 0 & -3 & 1 & 0 \\ 0 & 5 & 5 & -2 & 0 & 1 \end{array}\right]$$

Since the first three entries of row 2 are all zeros, the inverse of A does not exist.

24.

$$[A \;\vdots\; I] = \left[\begin{array}{rrrr|rrrr} 1 & 3 & -2 & 0 & 1 & 0 & 0 & 0 \\ 0 & 2 & 4 & 6 & 0 & 1 & 0 & 0 \\ 0 & 0 & -2 & 1 & 0 & 0 & 1 & 0 \\ 0 & 0 & 0 & 5 & 0 & 0 & 0 & 1 \end{array}\right]$$

$$\begin{array}{r} \\ \frac{1}{2}R_2 \rightarrow \\ \\ \frac{1}{5}R_4 \rightarrow \end{array} \left[\begin{array}{rrrr|rrrr} 1 & 3 & -2 & 0 & 1 & 0 & 0 & 0 \\ 0 & 1 & 2 & 3 & 0 & \frac{1}{2} & 0 & 0 \\ 0 & 0 & -2 & 1 & 0 & 0 & 1 & 0 \\ 0 & 0 & 0 & 1 & 0 & 0 & 0 & \frac{1}{5} \end{array}\right]$$

$$\begin{array}{r} -3R_2 + R_1 \rightarrow \\ R_3 + R_2 \rightarrow \\ -R_4 + R_3 \rightarrow \\ \\ \end{array} \left[\begin{array}{rrrr|rrrr} 1 & 0 & -8 & -9 & 1 & -\frac{3}{2} & 0 & 0 \\ 0 & 1 & 0 & 4 & 0 & \frac{1}{2} & 1 & 0 \\ 0 & 0 & -2 & 0 & 0 & 0 & 1 & -\frac{1}{5} \\ 0 & 0 & 0 & 1 & 0 & 0 & 0 & \frac{1}{5} \end{array}\right]$$

$$\begin{array}{r} -4R_3 + R_1 \rightarrow \\ -4R_4 + R_2 \rightarrow \\ \frac{1}{2}R_3 \rightarrow \\ \\ \end{array} \left[\begin{array}{rrrr|rrrr} 1 & 0 & 0 & 0 & 1 & -\frac{3}{2} & -4 & \frac{4}{5} \\ 0 & 1 & 0 & 0 & 0 & \frac{1}{2} & 1 & -\frac{4}{5} \\ 0 & 0 & 1 & 0 & 0 & 0 & -\frac{1}{2} & \frac{1}{10} \\ 0 & 0 & 0 & 1 & 0 & 0 & 0 & \frac{1}{5} \end{array}\right]$$

$$\begin{array}{r} 9R_4 + R_1 \rightarrow \\ \\ \\ \\ \end{array} \left[\begin{array}{rrrr|rrrr} 1 & 0 & 0 & 0 & 1 & -\frac{3}{2} & -4 & \frac{13}{5} \\ 0 & 1 & 0 & 0 & 0 & \frac{1}{2} & 1 & -\frac{4}{5} \\ 0 & 0 & 1 & 0 & 0 & 0 & -\frac{1}{2} & \frac{1}{10} \\ 0 & 0 & 0 & 1 & 0 & 0 & 0 & \frac{1}{5} \end{array}\right] = [I \;\vdots\; A^{-1}]$$

$$A^{-1} = \frac{1}{10}\begin{bmatrix} 10 & -15 & -40 & 26 \\ 0 & 5 & 10 & -8 \\ 0 & 0 & -5 & 1 \\ 0 & 0 & 0 & 2 \end{bmatrix}$$

26. $A = \begin{bmatrix} 10 & 5 & -7 \\ -5 & 1 & 4 \\ 3 & 2 & -2 \end{bmatrix}$

$A^{-1} = \begin{bmatrix} -10 & -4 & 27 \\ 2 & 1 & -5 \\ -13 & -5 & 35 \end{bmatrix}$

28. $A = \begin{bmatrix} 3 & 2 & 2 \\ 2 & 2 & 2 \\ -4 & 4 & 3 \end{bmatrix}$

$A^{-1} = \frac{1}{2}\begin{bmatrix} 2 & -2 & 0 \\ 14 & -17 & 2 \\ -16 & 20 & -2 \end{bmatrix}$

30. $A = \begin{bmatrix} 2 & 0 & 0 \\ 0 & 3 & 0 \\ 0 & 0 & 5 \end{bmatrix}$

$A^{-1} = \frac{1}{30}\begin{bmatrix} 15 & 0 & 0 \\ 0 & 10 & 0 \\ 0 & 0 & 6 \end{bmatrix}$

32. $A = \begin{bmatrix} -1 & 0 & 1 & 0 \\ 0 & 2 & 0 & -1 \\ 2 & 0 & -1 & 0 \\ 0 & -1 & 0 & 1 \end{bmatrix}$

$A^{-1} = \begin{bmatrix} 1 & 0 & 1 & 0 \\ 0 & 1 & 0 & 1 \\ 2 & 0 & 1 & 0 \\ 0 & 1 & 0 & 2 \end{bmatrix}$

34. $A = \begin{bmatrix} 4 & 8 & -7 & 14 \\ 2 & 5 & -4 & 6 \\ 0 & 2 & 1 & -7 \\ 3 & 6 & -5 & 10 \end{bmatrix}$

$A^{-1} = \begin{bmatrix} 27 & -10 & 4 & -29 \\ -16 & 5 & -2 & 18 \\ -17 & 4 & -2 & 20 \\ -7 & 2 & -1 & 8 \end{bmatrix}$

36. $A = \begin{bmatrix} a & b \\ c & d \end{bmatrix}, A^{-1} = \frac{1}{ad - bc}\begin{bmatrix} d & -b \\ -c & a \end{bmatrix}$

(a) $A = \begin{bmatrix} 5 & -2 \\ 2 & 3 \end{bmatrix}$

$A^{-1} = \frac{1}{15 + 4}\begin{bmatrix} 3 & 2 \\ -2 & 5 \end{bmatrix} = \frac{1}{19}\begin{bmatrix} 3 & 2 \\ -2 & 5 \end{bmatrix}$

(b) $A = \begin{bmatrix} 7 & 12 \\ -8 & -5 \end{bmatrix}$

$A^{-1} = \frac{1}{-35 + 96}\begin{bmatrix} -5 & -12 \\ 8 & 7 \end{bmatrix} = \frac{1}{61}\begin{bmatrix} -5 & -12 \\ 8 & 75 \end{bmatrix}$

38. $\begin{bmatrix} x \\ y \end{bmatrix} = \begin{bmatrix} -3 & 2 \\ -2 & 1 \end{bmatrix}\begin{bmatrix} 0 \\ 3 \end{bmatrix} = \begin{bmatrix} 6 \\ 3 \end{bmatrix}$

Answer: (6, 3)

40. $\begin{bmatrix} x \\ y \end{bmatrix} = \begin{bmatrix} -3 & 2 \\ -2 & 1 \end{bmatrix}\begin{bmatrix} 1 \\ -2 \end{bmatrix} = \begin{bmatrix} -7 \\ -4 \end{bmatrix}$

Answer: (−7, −4)

42. $\begin{bmatrix} x \\ y \\ z \end{bmatrix} = \begin{bmatrix} 1 & 1 & -1 \\ -3 & 2 & -1 \\ 3 & -3 & 2 \end{bmatrix}\begin{bmatrix} -1 \\ 2 \\ 0 \end{bmatrix} = \begin{bmatrix} 1 \\ 7 \\ -9 \end{bmatrix}$

Answer: (1, 7, −9)

44. $\begin{bmatrix} x \\ y \\ z \\ w \end{bmatrix} = \begin{bmatrix} -24 & 7 & 1 & -2 \\ -10 & 3 & 0 & -1 \\ -29 & 7 & 3 & -2 \\ 12 & -3 & -1 & 1 \end{bmatrix}\begin{bmatrix} 1 \\ -2 \\ 0 \\ -3 \end{bmatrix} = \begin{bmatrix} -32 \\ -13 \\ -37 \\ 15 \end{bmatrix}$

Answer: (−32, −13, −37, 15)

46. $A = \begin{bmatrix} 18 & 12 \\ 30 & 24 \end{bmatrix}$

$A^{-1} = \frac{1}{423 - 360}\begin{bmatrix} 24 & -12 \\ -30 & 18 \end{bmatrix}$

Answer: $\left(\frac{1}{2}, \frac{1}{3}\right)$

48. $A = \begin{bmatrix} 13 & -6 \\ 26 & -12 \end{bmatrix}$

$A^{-1} = \frac{1}{-156 + 156}\begin{bmatrix} -12 & 6 \\ -26 & 13 \end{bmatrix} \Rightarrow A^{-1}$ does not exist.

No solution

50. $A = \begin{bmatrix} 3 & 2 \\ 2 & 10 \end{bmatrix}$

$A^{-1} = \dfrac{1}{30-4}\begin{bmatrix} 10 & -2 \\ -2 & 3 \end{bmatrix}$

$\begin{bmatrix} x \\ y \end{bmatrix} = \dfrac{1}{26}\begin{bmatrix} 10 & -2 \\ -2 & 3 \end{bmatrix}\begin{bmatrix} 1 \\ 6 \end{bmatrix} = \begin{bmatrix} -\frac{1}{13} \\ \frac{8}{13} \end{bmatrix}$

Answer: $\left(-\frac{1}{13}, \frac{8}{13}\right)$

52. $A = \begin{bmatrix} 4 & -2 & 3 \\ 2 & 2 & 5 \\ 8 & -5 & -2 \end{bmatrix}$

$A^{-1} = \dfrac{1}{82}\begin{bmatrix} -21 & 19 & 16 \\ -44 & 32 & 14 \\ 26 & -4 & -12 \end{bmatrix}$

$\begin{bmatrix} x \\ y \\ z \end{bmatrix} = \dfrac{1}{82}\begin{bmatrix} -21 & 19 & 16 \\ -44 & 32 & 14 \\ 26 & -4 & -12 \end{bmatrix}\begin{bmatrix} -2 \\ 16 \\ 4 \end{bmatrix} = \begin{bmatrix} 5 \\ 8 \\ -2 \end{bmatrix}$

Answer: $(5, 8, -2)$

54. $A = \begin{bmatrix} 2 & 3 & 5 \\ 3 & 5 & 9 \\ 5 & 9 & 17 \end{bmatrix}$ A^{-1} does not exist.

No solution

56. $A = \begin{bmatrix} 2 & 5 & 0 & 1 \\ 1 & 4 & 2 & -2 \\ 2 & -2 & 5 & 1 \\ 1 & 0 & 0 & -3 \end{bmatrix}$

$A^{-1} \approx \begin{bmatrix} 0.338 & -0.352 & 0.141 & 0.394 \\ 0.042 & 0.164 & -0.066 & -0.117 \\ -0.141 & 0.230 & 0.108 & -0.164 \\ 0.113 & -0.117 & 0.047 & -0.202 \end{bmatrix}$

$\begin{bmatrix} x \\ y \\ z \\ w \end{bmatrix} = \begin{bmatrix} 0.338 & -0.352 & 0.141 & 0.394 \\ 0.042 & 0.164 & -0.066 & -0.117 \\ -0.141 & 0.230 & 0.108 & -0.164 \\ 0.113 & -0.117 & 0.047 & -0.202 \end{bmatrix}\begin{bmatrix} 11 \\ -7 \\ 3 \\ -1 \end{bmatrix} = \begin{bmatrix} 6.21 \\ -0.77 \\ -2.67 \\ 2.40 \end{bmatrix}$

Answer: $(6.21, -0.77, -2.67, 2.40)$

For 58 and 60 use $A = \begin{bmatrix} 1 & 1 & 1 \\ 0.065 & 0.07 & 0.09 \\ 0 & 2 & -1 \end{bmatrix}$. Using the methods of this section, we have $A^{-1} = \frac{1}{11}\begin{bmatrix} 50 & -600 & -4 \\ -13 & 200 & 5 \\ -26 & 400 & -1 \end{bmatrix}$.

58. $X = A^{-1}B = \dfrac{1}{11}\begin{bmatrix} 50 & -600 & -4 \\ -13 & 200 & 5 \\ -26 & 400 & -1 \end{bmatrix}\begin{bmatrix} 45{,}000 \\ 3750 \\ 0 \end{bmatrix} = \begin{bmatrix} 0 \\ 15{,}000 \\ 30{,}000 \end{bmatrix}$

Answer: 0 in AAA bonds, \$15,000 in A bonds, and \$30,000 in B bonds.

60. $X = A^{-1}B = \dfrac{1}{11}\begin{bmatrix} 50 & -600 & -4 \\ -13 & 200 & 5 \\ -26 & 400 & -1 \end{bmatrix}\begin{bmatrix} 500{,}000 \\ 38{,}000 \\ 0 \end{bmatrix} = \begin{bmatrix} 200{,}000 \\ 100{,}000 \\ 200{,}000 \end{bmatrix}$

Answer: \$200,000 in AAA bonds, \$100,000 in A bonds, and \$200,000 in B bonds.

62. True. The definition of the inverse A^{-1} of an $n \times n$ matrix A is an $n \times n$ matrix such that $AA^{-1} = A = I$. Thus, the multiplication of an invertible matrix and its inverse is commutative. One example is:

$$\begin{bmatrix} 2 & 1 \\ 5 & 0 \end{bmatrix}\begin{bmatrix} 0 & \frac{1}{5} \\ 1 & -\frac{2}{5} \end{bmatrix} = \begin{bmatrix} 1 & 0 \\ 0 & 1 \end{bmatrix} = \begin{bmatrix} 0 & \frac{1}{5} \\ 1 & -\frac{2}{5} \end{bmatrix}\begin{bmatrix} 2 & 1 \\ 5 & 0 \end{bmatrix}$$

64. $A = \begin{bmatrix} 2 & 0 & 4 \\ 0 & 1 & 4 \\ 1 & 1 & -1 \end{bmatrix}$

$$A^{-1} = \frac{1}{14}\begin{bmatrix} 5 & -4 & 4 \\ -4 & 6 & 8 \\ 1 & 2 & -2 \end{bmatrix}$$

$$\begin{bmatrix} I_1 \\ I_2 \\ I_3 \end{bmatrix} = \frac{1}{14}\begin{bmatrix} 5 & -4 & 4 \\ -4 & 6 & 8 \\ 1 & 2 & -2 \end{bmatrix}\begin{bmatrix} 10 \\ 10 \\ 0 \end{bmatrix} = \begin{bmatrix} \frac{5}{7} \\ \frac{10}{7} \\ \frac{15}{7} \end{bmatrix}$$

Answer: $I_1 = \frac{5}{7}$ amps, $I_2 = \frac{10}{7}$ amps, $I_3 = \frac{15}{7}$ amps

66. $\begin{bmatrix} x_1' \\ x_2' \end{bmatrix} = \begin{bmatrix} \cos 30° & -\sin 30° \\ \sin 30° & \cos 30° \end{bmatrix} \cdot \begin{bmatrix} x_1 \\ y_1 \end{bmatrix}$

(a) $\begin{bmatrix} x_1' \\ x_2' \end{bmatrix} = \begin{bmatrix} \cos 30° & -\sin 30° \\ \sin 30° & \cos 30° \end{bmatrix} \cdot \begin{bmatrix} 3 \\ 1 \end{bmatrix} = \begin{bmatrix} \frac{\sqrt{3}}{2} & -\frac{1}{2} \\ \frac{1}{2} & \frac{\sqrt{3}}{2} \end{bmatrix}\begin{bmatrix} 3 \\ 1 \end{bmatrix} = \begin{bmatrix} \frac{3\sqrt{3}-1}{2} \\ \frac{3+\sqrt{3}}{2} \end{bmatrix}$

(b)

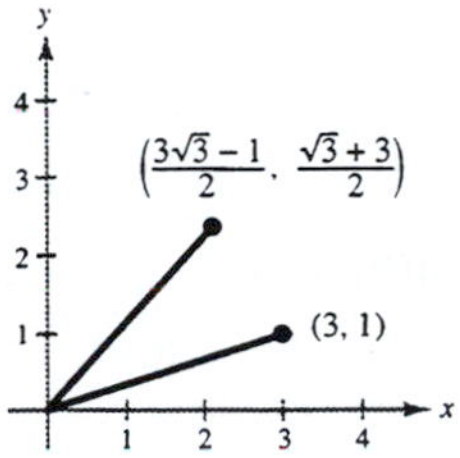

(c) The matrix product appears to produce a counterclockwise rotation of θ degrees ($\theta > 0$).

(d) Applying the inverse of the trigonometric matrix should produce a clockwise rotation of θ degrees ($\theta > 0$).

(e) $\frac{1}{ad - bc}\begin{bmatrix} d & -b \\ -c & a \end{bmatrix} = \frac{1}{\cos^2\theta - (-\sin^2\theta)}\begin{bmatrix} \cos\theta & \sin\theta \\ -\sin\theta & \cos\theta \end{bmatrix}$

$= \frac{1}{\cos^2\theta + \sin\theta}\begin{bmatrix} \cos\theta \ \sin\theta \\ -\sin\theta \ \cos\theta \end{bmatrix} = \begin{bmatrix} \cos\theta & \sin\theta \\ -\sin\theta & \cos\theta \end{bmatrix}$

(f) Another method for finding the inverse (to achieve the effect of producing a clockwise rotation of θ degrees) is to replace θ by $-\theta$ in the original matrix.

If $\theta = -\theta$:

$$\begin{bmatrix} \cos(-\theta) & \sin(-\theta) \\ \sin(-\theta) & \cos(-\theta) \end{bmatrix} = \begin{bmatrix} \cos\theta & \sin\theta \\ -\sin\theta & \cos\theta \end{bmatrix}$$

68. $s = 1.279 - 0.0049t$ $\quad\Rightarrow\quad$ $s + 0.0049t = 1.279$

$s = 1.411 - 0.0078t$ $\quad\quad\quad$ $s + 0.0078t = 1.411$

$$A = \begin{bmatrix} 1 & 0.0049 \\ 1 & 0.0078 \end{bmatrix}$$

$$A^{-1} \approx \begin{bmatrix} 2.69 & -1.69 \\ -344.83 & 344.83 \end{bmatrix}$$

$$\begin{bmatrix} s \\ t \end{bmatrix} = \begin{bmatrix} 2.69 & -1.69 \\ -344.83 & 344.83 \end{bmatrix}\begin{bmatrix} 1.279 \\ 1.411 \end{bmatrix} = \begin{bmatrix} 1.06 \\ 45.52 \end{bmatrix}$$

The year in which the women's time is less than the men's time is $1980 + 46 = 2026$.

70. $2000e^{-x/5} = 400$

$$e^{-x/5} = \frac{1}{5}$$

$$-\frac{x}{5} = \ln\frac{1}{5}$$

$$x = -5\ln\frac{1}{5} \approx 8.05$$

72. $\ln x + \ln(x-1) = 0$

$$\ln[x(x-1)] = 0$$

$$x(x-1) = 1$$

$$x^2 - x - 1 = 0$$

$$x = \frac{1 \pm \sqrt{1-4(-1)}}{2}$$

$$x = \frac{1 \pm \sqrt{5}}{2}$$

Choose the positive value only:

$$x = \frac{1+\sqrt{5}}{2} \approx 1.62$$

Section 8.4 The Determinant of a Square Matrix

Solutions to Even-Numbered Exercises

2. -8

4. $\begin{vmatrix} -3 & 1 \\ 5 & 2 \end{vmatrix} = (-3)(2) - (5)(1) = -11$

6. $\begin{vmatrix} 2 & -2 \\ 4 & 3 \end{vmatrix} = (2)(3) - (4)(-2) = 14$

8. $\begin{vmatrix} 4 & -3 \\ 0 & 0 \end{vmatrix} = (4)(0) - (0)(-3) = 0$

10. $\begin{vmatrix} 2 & -3 \\ -6 & 9 \end{vmatrix} = (2)(9) - (-6)(-3) = 0$

12. $\begin{vmatrix} -2 & 2 & 3 \\ 1 & -1 & 0 \\ 0 & 1 & 4 \end{vmatrix} = 0\begin{vmatrix} 2 & 3 \\ -1 & 0 \end{vmatrix} - 1\begin{vmatrix} -2 & 3 \\ 1 & 0 \end{vmatrix} + 4\begin{vmatrix} -2 & 2 \\ 1 & -1 \end{vmatrix} = 0(3) - 1(-3) + 4(0) = 3$

14. $\begin{vmatrix} 1 & 1 & 2 \\ 3 & 1 & 0 \\ -2 & 0 & 3 \end{vmatrix} = -2\begin{vmatrix} 1 & 2 \\ 1 & 0 \end{vmatrix} - 0\begin{vmatrix} 1 & 2 \\ 3 & 0 \end{vmatrix} + 3\begin{vmatrix} 1 & 1 \\ 3 & 1 \end{vmatrix} = -2$

16. $\begin{vmatrix} 1 & 0 & 0 \\ -4 & -1 & 0 \\ 5 & 1 & 5 \end{vmatrix} = (1)(-1)(5) = -5$ (Lower Triangular)

18. $\begin{vmatrix} 0.1 & 0.2 & 0.3 \\ -0.3 & 0.2 & 0.2 \\ 0.5 & 0.4 & 0.4 \end{vmatrix} = -0.022$

20. $\begin{vmatrix} 2 & 3 & 1 \\ 0 & 5 & -2 \\ 0 & 0 & -2 \end{vmatrix} = -20$

22. $\begin{bmatrix} 11 & 0 \\ -3 & 2 \end{bmatrix}$

(a) $M_{11} = 2$
$M_{12} = -3$
$M_{21} = 0$
$M_{22} = 11$

(b) $C_{11} = M_{11} = 2$
$C_{12} = M_{12} = 3$
$C_{21} = M_{21} = 0$
$C_{22} = M_{22} = 11$

24. $\begin{bmatrix} -2 & 9 & 4 \\ 7 & -6 & 0 \\ 6 & 7 & -6 \end{bmatrix}$

(a) $M_{11} = \begin{vmatrix} -6 & 0 \\ 7 & -6 \end{vmatrix} = 36$

$M_{12} = \begin{vmatrix} 7 & 0 \\ 6 & -6 \end{vmatrix} = -42$

$M_{13} = \begin{vmatrix} 7 & -6 \\ 6 & 7 \end{vmatrix} = 85$

$M_{21} = \begin{vmatrix} 9 & 4 \\ 7 & -6 \end{vmatrix} = -82$

$M_{22} = \begin{vmatrix} -2 & 4 \\ 6 & -6 \end{vmatrix} = -12$

$M_{23} = \begin{vmatrix} -2 & 9 \\ 6 & 7 \end{vmatrix} = -68$

$M_{31} = \begin{vmatrix} 9 & 4 \\ -6 & 0 \end{vmatrix} = 24$

$M_{32} = \begin{vmatrix} -2 & 4 \\ 7 & 0 \end{vmatrix} = -28$

$M_{33} = \begin{vmatrix} -2 & 9 \\ 7 & -6 \end{vmatrix} = -51$

(b) $C_{11} = (-1)^2 M_{11} = 36$

$C_{12} = (-1)^3 M_{12} = 42$

$C_{13} = (-1)^4 M_{13} = 85$

$C_{21} = (-1)^3 M_{21} = 82$

$C_{22} = (-1)^4 M_{22} = -12$

$C_{23} = (-1)^5 M_{23} = 68$

$C_{31} = (-1)^4 M_{31} = 24$

$C_{32} = (-1)^5 M_{32} = 28$

$C_{33} = (-1)^6 M_{33} = -51$

26. (a) $\begin{vmatrix} -3 & 4 & 2 \\ 6 & 3 & 1 \\ 4 & -7 & -8 \end{vmatrix} = -6\begin{vmatrix} 4 & 2 \\ -7 & -8 \end{vmatrix} + 3\begin{vmatrix} -3 & 2 \\ 4 & -8 \end{vmatrix} - 1\begin{vmatrix} -3 & 4 \\ 4 & -7 \end{vmatrix} = -6(-18) + 3(16) - (5) = 151$

(b) $\begin{vmatrix} -3 & 4 & 2 \\ 6 & 3 & 1 \\ 4 & -7 & -8 \end{vmatrix} = 2\begin{vmatrix} 6 & 3 \\ 4 & -7 \end{vmatrix} - \begin{vmatrix} -3 & 4 \\ 4 & -7 \end{vmatrix} - 8\begin{vmatrix} -3 & 4 \\ 6 & 3 \end{vmatrix} = 2(-54) - (5) - 8(-33) = 151$

28. (a) $\begin{vmatrix} 10 & -5 & 5 \\ 30 & 0 & 10 \\ 0 & 10 & 1 \end{vmatrix} = 0\begin{vmatrix} -5 & 5 \\ 0 & 10 \end{vmatrix} - 10\begin{vmatrix} 10 & 5 \\ 30 & 10 \end{vmatrix} + \begin{vmatrix} 10 & -5 \\ 30 & 0 \end{vmatrix} = 0(-50) - 10(-50) + 150 = 650$

(b) $\begin{vmatrix} 10 & -5 & 5 \\ 30 & 0 & 10 \\ 0 & 10 & 1 \end{vmatrix} = 10\begin{vmatrix} 0 & 10 \\ 10 & 1 \end{vmatrix} - 30\begin{vmatrix} -5 & 5 \\ 10 & 1 \end{vmatrix} + 0\begin{vmatrix} -5 & 5 \\ 0 & 10 \end{vmatrix} = 10(-100) - 30(-55) + 0(-50) = 650$

30. (a) $\begin{vmatrix} 10 & 8 & 3 & -7 \\ 4 & 0 & 5 & -6 \\ 0 & 3 & 2 & 7 \\ 1 & 0 & -3 & 2 \end{vmatrix} = 0\begin{vmatrix} 8 & 3 & -7 \\ 0 & 5 & -6 \\ 0 & -3 & 2 \end{vmatrix} - 3\begin{vmatrix} 10 & 3 & -7 \\ 4 & 5 & -6 \\ 1 & -3 & 2 \end{vmatrix} + 2\begin{vmatrix} 10 & 8 & -7 \\ 4 & 0 & -6 \\ 1 & 0 & 2 \end{vmatrix} - 7\begin{vmatrix} 10 & 8 & 3 \\ 4 & 0 & 5 \\ 1 & 0 & -3 \end{vmatrix}$

$= 0(-64) - 3(-3) + 2(-112) - 7(136) = -1167$

(b) $\begin{vmatrix} 10 & 8 & 3 & -7 \\ 4 & 0 & 5 & -6 \\ 0 & 3 & 2 & 7 \\ 1 & 0 & -3 & 2 \end{vmatrix} = 10\begin{vmatrix} 0 & 5 & -6 \\ 3 & 2 & 7 \\ 0 & -3 & 2 \end{vmatrix} - 4\begin{vmatrix} 8 & 3 & -7 \\ 3 & 2 & 7 \\ 0 & -3 & 2 \end{vmatrix} + 0\begin{vmatrix} 8 & 3 & -7 \\ 0 & 5 & -6 \\ 0 & -3 & 2 \end{vmatrix} - 1\begin{vmatrix} 8 & 3 & -7 \\ 0 & 5 & -6 \\ 3 & 2 & 7 \end{vmatrix}$

$= 10(24) - 4(245) + 0(-64) - 1(427) = -1167$

32. Expand by Row 3.

$$\begin{vmatrix} 2 & -1 & 3 \\ 1 & 4 & 4 \\ 1 & 0 & 2 \end{vmatrix} = 1\begin{vmatrix} -1 & 3 \\ 4 & 4 \end{vmatrix} + 2\begin{vmatrix} 2 & -1 \\ 1 & 4 \end{vmatrix} = 1(-16) + 2(9) = 2$$

34. $\begin{vmatrix} -3 & 0 & 0 \\ 7 & 11 & 4 \\ 1 & 2 & 2 \end{vmatrix} = (-3)(11)(2) = -66$ (Lower Triangular)

36. Expand by Row 2.

$$\begin{vmatrix} 3 & 6 & -5 & 4 \\ -2 & 0 & 6 & 0 \\ 1 & 1 & 2 & 2 \\ 0 & 3 & -1 & -1 \end{vmatrix} = -(-2)\begin{vmatrix} 6 & -5 & 4 \\ 1 & 2 & 2 \\ 3 & -1 & -1 \end{vmatrix} - 6\begin{vmatrix} 3 & 6 & 4 \\ 1 & 1 & 2 \\ 0 & 3 & -1 \end{vmatrix} = 2(-63) - 6(-3) = -108$$

38. Expand by Row 3.

$$\begin{vmatrix} 1 & 4 & 3 & 2 \\ -5 & 6 & 2 & 1 \\ 0 & 0 & 0 & 2 \\ 3 & -2 & 1 & 5 \end{vmatrix} = 0$$

40. Expand by Column 1.

$$\begin{vmatrix} 5 & 2 & 0 & 0 & -2 \\ 0 & 1 & 4 & 3 & 2 \\ 0 & 0 & 2 & 6 & 3 \\ 0 & 0 & 3 & 4 & 1 \\ 0 & 0 & 0 & 0 & 2 \end{vmatrix} = 5\begin{vmatrix} 1 & 4 & 3 & 2 \\ 0 & 2 & 6 & 3 \\ 0 & 3 & 4 & 1 \\ 0 & 0 & 0 & 2 \end{vmatrix} = 5 \cdot 1\begin{vmatrix} 2 & 6 & 3 \\ 3 & 4 & 1 \\ 0 & 0 & 2 \end{vmatrix} = 5(-20) = -100$$

42. $\begin{vmatrix} 5 & -8 & 0 \\ 9 & 7 & 4 \\ -8 & 7 & 1 \end{vmatrix} = 223$

44. $\begin{vmatrix} 3 & 0 & 0 \\ -2 & 5 & 0 \\ 12 & 5 & 7 \end{vmatrix} = 105$

46. $\begin{vmatrix} 0 & -3 & 8 & 2 \\ 8 & 1 & -1 & 6 \\ -4 & 6 & 0 & 9 \\ -7 & 0 & 0 & 14 \end{vmatrix} = 7441$

48. $\begin{vmatrix} -2 & 0 & 0 & 0 & 0 \\ 0 & 3 & 0 & 0 & 0 \\ 0 & 0 & -1 & 0 & 0 \\ 0 & 0 & 0 & 2 & 0 \\ 0 & 0 & 0 & 0 & -4 \end{vmatrix}$

50. $\begin{vmatrix} w & cx \\ y & cz \end{vmatrix} = cwz - cxy = c(wz - xy)$

$c\begin{vmatrix} w & x \\ y & z \end{vmatrix} = c(wz - xy)$

Thus, $\begin{vmatrix} w & cx \\ y & cz \end{vmatrix} = c\begin{vmatrix} w & x \\ y & z \end{vmatrix}$.

52. $\begin{vmatrix} w & x \\ cw & cx \end{vmatrix} = cxw - cxw = cxw = 0$

Thus, $\begin{vmatrix} w & x \\ cw & cx \end{vmatrix} = 0$.

54. $$\begin{vmatrix} a+b & a & a \\ a & a+b & a \\ a & a & a+b \end{vmatrix} = (a+b)\begin{vmatrix} a+b & a \\ a & a+b \end{vmatrix} - a\begin{vmatrix} a & a \\ a & a+b \end{vmatrix} + a\begin{vmatrix} a & a \\ a+b & a \end{vmatrix}$$

$$= (a+b)[(a+b)^2 - a^2] - a[a(a+b) - a^2] + a[a^2 - a(a+b)]$$

$$= (a+b)^3 - a^2(a+b) - a^2(a+b) + a^3 + a^3 - a^2(a+b)$$

$$= (a+b)^3 - 3a^2(a+b)2a^3$$

$$= a^3 + 3a^2b + 3ab^2 + b^3 - 3a^3 - 3a^2b + 2a^3$$

$$= 3ab^2 + b^3 = b^2(3a+b)$$

56. $$\begin{vmatrix} x-2 & -1 \\ -3 & x \end{vmatrix} = 0$$

$$x(x-2) - (-3)(-1) = 0$$
$$x^2 - 2x - 3 = 0$$
$$(x+1)(x-3) = 0$$
$$x = -1 \text{ or } x = 3$$

58. $$\begin{vmatrix} 3x^2 & -3y^2 \\ 1 & 1 \end{vmatrix} = 3x^2 - (-3y^2) = 3x^2 + 3y^2$$

60. $$\begin{vmatrix} e^{-x} & xe^{-x} \\ -e^{-x} & (1-x)e^{-x} \end{vmatrix} = (1-x)e^{-2x} - (-xe^{-2x}) = e^{-2x} - xe^{-2x} + xe^{-2x} = e^{-2x}$$

62. $$\begin{vmatrix} x & x\ln x \\ 1 & 1+\ln x \end{vmatrix} = x(1+\ln x) - x\ln x = x + x\ln x - x\ln x = x$$

64. (a) $|A| = \begin{vmatrix} -2 & 1 \\ 4 & -2 \end{vmatrix} = 0$ (b) $|B| = \begin{vmatrix} 1 & 2 \\ 0 & -2 \end{vmatrix} = -1$

(c) $AB = \begin{bmatrix} -2 & 1 \\ 4 & -2 \end{bmatrix}\begin{bmatrix} 1 & 2 \\ 0 & -1 \end{bmatrix} = \begin{bmatrix} -2 & -5 \\ 4 & 10 \end{bmatrix}$ (d) $|AB| = \begin{vmatrix} -2 & -5 \\ 4 & 10 \end{vmatrix} = 0$

66. (a) $|A| = \begin{vmatrix} 2 & 0 & 1 \\ 1 & -1 & 2 \\ 3 & 1 & 0 \end{vmatrix} = 0$ (b) $|B| = \begin{vmatrix} 2 & -1 & 4 \\ 0 & 1 & 3 \\ 3 & -2 & 1 \end{vmatrix} = -7$

(c) $AB = \begin{bmatrix} 2 & 0 & 1 \\ 1 & -1 & 2 \\ 3 & 1 & 0 \end{bmatrix}\begin{bmatrix} 2 & -1 & 4 \\ 0 & 1 & 3 \\ 3 & -2 & 1 \end{bmatrix} = \begin{bmatrix} 7 & -4 & 9 \\ 8 & -6 & 3 \\ 6 & -2 & 15 \end{bmatrix}$

(d) $|AB| = \begin{vmatrix} 7 & -4 & 9 \\ 8 & -6 & 3 \\ 6 & -2 & 15 \end{vmatrix} = 0$

68. (a) $\begin{vmatrix} 4 & 5 & 6 \\ 7 & 8 & 9 \\ 10 & 11 & 12 \end{vmatrix} = 0$ $\begin{vmatrix} 10 & 11 & 12 \\ 13 & 14 & 15 \\ 16 & 17 & 18 \end{vmatrix} = 0$

$\begin{vmatrix} 33 & 34 & 35 \\ 36 & 37 & 38 \\ 39 & 40 & 41 \end{vmatrix} = 0$ $\begin{vmatrix} -5 & -4 & -3 \\ -2 & -1 & 0 \\ 1 & 2 & 3 \end{vmatrix} = 0$

$\begin{vmatrix} 19 & 20 & 21 & 22 \\ 23 & 24 & 25 & 26 \\ 27 & 28 & 29 & 30 \\ 31 & 32 & 33 & 34 \end{vmatrix} = 0$ $\begin{vmatrix} 57 & 58 & 59 & 60 \\ 61 & 62 & 63 & 64 \\ 65 & 66 & 67 & 68 \\ 69 & 70 & 71 & 72 \end{vmatrix} = 0$

For an $n \times n$ matrix $(n > 2)$ with consecutive integer entries, the determinant appears to be 0.

(b) $$\begin{vmatrix} x & x+1 & x+2 \\ x+3 & x+4 & x+5 \\ x+6 & x+7 & x+8 \end{vmatrix} = x\begin{vmatrix} x+4 & x+5 \\ x+7 & x+8 \end{vmatrix} - (x+1)\begin{vmatrix} x+3 & x+5 \\ x+6 & x+8 \end{vmatrix} + (x+2)\begin{vmatrix} x+3 & x+4 \\ x+6 & x+7 \end{vmatrix}$$

$$= x[(x+4)(x+8) - (x+7)(x+5)] - (x+1)[(x+3)(x+8) - (x+6)(x+5)] + (x+2)[(x+3)(x+7) - (x+6)(x+4)]$$

$$= x[(x^2+12x+32) - (x^2+12x+35)] - (x+1)[(x^2+11x+24) - (x^2+11x+30)] + (x+2)[(x^2+10x+21) - (x^2+10x+24)]$$

$$= -3x - (x+1)(-6) + (x+2)(-3)$$

$$= -3x + 6x + 6 - 3x - 6 = 0$$

70. Let $A = \begin{bmatrix} x_{11} & x_{12} & x_{13} \\ x_{21} & x_{22} & x_{23} \\ x_{31} & x_{32} & x_{33} \end{bmatrix}$ and $|A| = 5$.

$$2A = \begin{bmatrix} 2x_{11} & 2x_{12} & 2x_{13} \\ 2x_{21} & 2x_{22} & 2x_{23} \\ 2x_{31} & 2x_{32} & 2x_{33} \end{bmatrix}$$

$$\begin{aligned} |2A| &= 2x_{11}\begin{vmatrix} 2x_{22} & 2x_{23} \\ 2x_{32} & 2x_{33} \end{vmatrix} - 2x_{12}\begin{vmatrix} 2x_{21} & 2x_{23} \\ 2x_{31} & 2x_{33} \end{vmatrix} + 2x_{13}\begin{vmatrix} 2x_{21} & 2x_{22} \\ 2x_{31} & 2x_{32} \end{vmatrix} \\ &= 2[x_{11}(4x_{22}x_{33} - 4x_{32}x_{23}) - x_{12}(4x_{21}x_{33} - 4x_{31}x_{23}) + x_{13}(4x_{21}x_{32} - 4x_{31}x_{22})] \\ &= 8[x_{11}(x_{22}x_{33} - x_{32}x_{23}) - x_{12}(x_{21}x_{33} - x_{31}x_{23}) + x_{13}(x_{21}x_{32} - x_{31}x_{22})] \\ &= 8|A| \end{aligned}$$

Thus, $|2A| = 8|A| = 8(5) = 40$.

72. Ellipse
Vertices: $(0, \pm 4)$
Foci: $(0, \pm 3)$
Vertical major axis
Center: $(0, 0)$
$a = 4, c = 3, b = \sqrt{16 - 9} = \sqrt{7}$

$$\frac{y^2}{a^2} + \frac{x^2}{b^2} = 1$$

$$\frac{y^2}{16} + \frac{x^2}{7} = 1$$

74. Hyperbola
Vertices: $(\pm 5, 0)$
Foci: $(\pm 6, 0)$
Horizontal transverse axis
Center: $(0, 0)$
$a = 5, c = 6, b = \sqrt{36 - 25} = \sqrt{11}$

$$\frac{x^2}{a^2} - \frac{x^2}{b^2} = 1$$

$$\frac{x^2}{25} - \frac{y^2}{11} = 1$$

Section 8.5 Applications of Matrices and Determinants

Solutions to Even-Numbered Exercises

2. $-0.4x + 0.8y = 1.6$
$0.2x + 0.3y = 2.2$

$$D = \begin{vmatrix} -0.4 & 0.8 \\ 0.2 & 0.3 \end{vmatrix} = -0.28$$

$$x = \frac{\begin{vmatrix} 1.6 & 0.8 \\ 2.2 & 0.3 \end{vmatrix}}{-0.28} = \frac{-1.18}{-0.28} = \frac{32}{7}$$

$$y = \frac{\begin{vmatrix} -0.4 & 1.6 \\ 0.2 & 2.2 \end{vmatrix}}{-0.28} = \frac{-1.20}{-0.28} = \frac{30}{7}$$

Answer: $\left(\frac{32}{7}, \frac{30}{7}\right)$

4. $4x - 2y + 3z = -2$
$2x + 2y + 5z = 16$
$8x - 5y - 2z = 4$

$$D = \begin{vmatrix} 4 & -2 & 3 \\ 2 & 2 & 5 \\ 8 & -5 & -2 \end{vmatrix} = -82$$

$$x = \frac{\begin{vmatrix} -2 & -2 & 3 \\ 16 & 2 & 5 \\ 4 & -5 & -2 \end{vmatrix}}{-82} = \frac{-401}{-82} = 5$$

$$y = \frac{\begin{vmatrix} 4 & -2 & 3 \\ 2 & 16 & 5 \\ 8 & 4 & -2 \end{vmatrix}}{-82} = \frac{-656}{-82} = 8$$

$$z = \frac{\begin{vmatrix} 4 & -2 & -2 \\ 2 & 2 & 16 \\ 8 & -5 & 4 \end{vmatrix}}{-82} = \frac{164}{-82} = -2$$

Answer: $(5, 8, -2)$

6. $2x + 3y + 5z = 4$
$3x + 5y + 9z = 7$
$5x + 9y + 17z = 13$

$$D = \begin{vmatrix} 2 & 3 & 5 \\ 3 & 5 & 9 \\ 5 & 9 & 17 \end{vmatrix} = 0$$

Cramer's Rule does not apply.

8. Vertices: $(0, 0), (4, 5), (5, -2)$

$$\text{Area} = -\frac{1}{2}\begin{vmatrix} 0 & 0 & 1 \\ 4 & 5 & 1 \\ 5 & -2 & 1 \end{vmatrix} = -\frac{1}{2}\begin{vmatrix} 4 & 5 \\ 5 & -2 \end{vmatrix} = \frac{33}{2} \text{ square units}$$

10. Vertices: $(-2, 1), (1, 6), (3, -1)$

$$\text{Area} = -\frac{1}{2}\begin{vmatrix} -2 & 1 & 1 \\ 1 & 6 & 1 \\ 3 & -1 & 1 \end{vmatrix} = -\frac{1}{2}(-19 + 1 - 13) = \frac{31}{2} \text{ square units}$$

12. Vertices: $(-4, -5), (6, 10), (6, -1)$

$$\text{Area} = -\frac{1}{2}\begin{vmatrix} -4 & -5 & 1 \\ 6 & 10 & 1 \\ 6 & -1 & 1 \end{vmatrix} = -\frac{1}{2}(-66 - 34 - 10) = 55 \text{ square units}$$

14. Vertices: $(0, -2), (-1, 4), (3, 5)$

$$\text{Area} = -\frac{1}{2}\begin{vmatrix} 0 & -2 & 1 \\ -1 & 4 & 1 \\ 3 & 5 & 1 \end{vmatrix} = -\frac{1}{2}(-17 - 6 - 2) = \frac{25}{2} \text{ square units}$$

16. Vertices: $(-2, 4), (1, 5), (3, -2)$

$$\text{Area} = -\frac{1}{2}\begin{vmatrix} -2 & 4 & 1 \\ 1 & 5 & 1 \\ 3 & -2 & 1 \end{vmatrix} = -\frac{1}{2}(-17 + 8 - 14) = \frac{23}{2} \text{ square units}$$

18. $$4 = \pm\frac{1}{2}\begin{vmatrix} -4 & 2 & 1 \\ -3 & 5 & 1 \\ -1 & x & 1 \end{vmatrix}$$

$$\pm 8 = \begin{vmatrix} -3 & 5 \\ -1 & x \end{vmatrix} - \begin{vmatrix} -4 & 2 \\ -1 & x \end{vmatrix} + \begin{vmatrix} -4 & 2 \\ -3 & 5 \end{vmatrix}$$

$\pm 8 = -3x + 5 - (-4x + 2) - 20 + 6$

$\pm 8 = -3x + 5 + 4x - 2 - 20 + 6$

$\pm 8 = x - 11$

$x = 11 \pm 8$

$x = 10$ or $x = 3$

20. Vertices: $(0, 30), (85, 0), (20, -50)$

$$\text{Area} = -\frac{1}{2}\begin{vmatrix} 0 & 30 & 1 \\ 85 & 0 & 1 \\ 20 & -50 & 1 \end{vmatrix} = 3100 \text{ square units}$$

22. Points: $(-3, -5), (6, 1), (10, 2)$

$$\begin{vmatrix} -3 & -5 & 1 \\ 6 & 1 & 1 \\ 10 & 2 & 1 \end{vmatrix} = \begin{vmatrix} 6 & 1 \\ 10 & 2 \end{vmatrix} - \begin{vmatrix} -3 & -5 \\ 10 & 2 \end{vmatrix} + \begin{vmatrix} -3 & -5 \\ 6 & 1 \end{vmatrix} = -15 \neq 0$$

The points are not collinear.

24. Points: $(0, 1), (4, -2), (-8, 7)$

$$\begin{vmatrix} 0 & 1 & 1 \\ 4 & -2 & 1 \\ -8 & 7 & 1 \end{vmatrix} = \begin{vmatrix} 4 & -2 \\ -8 & 7 \end{vmatrix} - \begin{vmatrix} 0 & 1 \\ -8 & 7 \end{vmatrix} + \begin{vmatrix} 0 & 1 \\ 4 & -2 \end{vmatrix} = 0$$

The points are collinear.

26. Points: $(2, 3), (3, 3.5), (-1, 2)$

$$\begin{vmatrix} 2 & 3 & 1 \\ 3 & 3.5 & 1 \\ -1 & 2 & 1 \end{vmatrix} = \begin{vmatrix} 3 & 3.5 \\ -1 & 2 \end{vmatrix} - \begin{vmatrix} 2 & 3 \\ -1 & 2 \end{vmatrix} + \begin{vmatrix} 2 & 3 \\ 3 & 3.5 \end{vmatrix} = \frac{1}{2} \neq 0$$

The points are not collinear.

28. Points: $(0, 0), (-2, 2)$

Equation: $\begin{vmatrix} x & y & 1 \\ 0 & 0 & 1 \\ -2 & 2 & 1 \end{vmatrix} = -(2x + 2y) = 0$ or $x + y = 0$

30. Points: $(10, 7), (-2, -7)$

Equation: $\begin{vmatrix} x & y & 1 \\ 10 & 7 & 1 \\ -2 & -7 & 1 \end{vmatrix} = -70 + 14 - (-7x + 2y) + 7x - 10y = 0$ or $7x - 6y = 28$

32. Points: $\left(\frac{2}{3}, 4\right), (6, 12)$

Equation: $\begin{vmatrix} x & y & 1 \\ \frac{2}{3} & 4 & 1 \\ 6 & 12 & 1 \end{vmatrix} = -16 - (12x - 6y) + 4x - \frac{2}{3}y = 0$ or $3x - 2y + 6 = 0$

34. $\begin{vmatrix} -6 & 2 & 1 \\ -5 & x & 1 \\ -3 & 5 & 1 \end{vmatrix} = 0$

$$\begin{vmatrix} -5 & x \\ -3 & 5 \end{vmatrix} - \begin{vmatrix} -6 & 2 \\ -3 & 5 \end{vmatrix} + \begin{vmatrix} -6 & 2 \\ -5 & x \end{vmatrix} = 0$$

$$-25 + 3x + 24 - 6x + 10 = 0$$

$$-3x = -9$$

$$x = 3$$

36. The uncoded row matrices are the rows of the 6 × 3 matrix on the left.

$$\begin{array}{ccc} P & L & E \\ A & S & E \\ & S & E \\ N & D & \\ M & O & N \\ E & Y & \end{array} \quad \begin{bmatrix}16&12&5\\1&19&5\\0&19&5\\14&4&0\\13&15&14\\5&25&0\end{bmatrix}\begin{bmatrix}4&2&1\\-3&-3&-1\\3&2&1\end{bmatrix}=\begin{bmatrix}43&6&9\\-38&-45&-13\\-42&-47&-14\\44&16&10\\49&9&12\\-55&-65&-20\end{bmatrix}$$

Answer: $[43\ \ 6\ \ 9], [-38\ \ -45\ \ -13], [-42\ \ -47\ \ -14], [44\ \ 16\ \ 10], [49\ \ 9\ \ 12], [-55\ \ -65\ \ -20]$

38. B E A M _ M E _ U P _ S C O T T Y _

$[2\ \ 5\ \ 1]\ [13\ \ 0\ \ 13]\ [5\ \ 0\ \ 21]\ [16\ \ 0\ \ 19]\ [3\ \ 15\ \ 20]\ [20\ \ 25\ \ 0]$

$$[2\ \ 5\ \ 1]A = [16\ \ 35\ \ 42]$$
$$[13\ \ 0\ \ 13]A = [0\ \ -26\ \ -65]$$
$$[5\ \ 0\ \ 21]A = [-16\ \ -74\ \ -137]$$
$$[16\ \ 0\ \ 19]A = [-3\ \ -44\ \ -101]$$
$$[3\ \ 15\ \ 20]A = [28\ \ 31\ \ 1]$$
$$[20\ \ 25\ \ 0]A = [95\ \ 215\ \ 265]$$

Cryptogram: 16 35 42 0 −26 −65 −16 −74 −137 −3 −44 −101 28 31 1 95 215 265

40. O P E R A T I O N _ O V E R L O A D

$[15\ \ 16\ \ 5]\ [18\ \ 1\ \ 20]\ [9\ \ 15\ \ 14]\ [0\ \ 15\ \ 22]\ [5\ \ 18\ \ 12]\ [15\ \ 1\ \ 4]$

$$[15\ \ 16\ \ 5]A = [58\ \ 122\ \ 139]$$
$$[18\ \ 1\ \ 20]A = [1\ \ -37\ \ -95]$$
$$[9\ \ 15\ \ 14]A = [40\ \ 67\ \ 55]$$
$$[0\ \ 15\ \ 22]A = [23\ \ 17\ \ -19]$$
$$[5\ \ 18\ \ 12]A = [47\ \ 88\ \ 88]$$
$$[15\ \ 1\ \ 4]A = [14\ \ 21\ \ 11]$$

Cryptogram: 58 122 139 1 −37 −95 40 67 55 23 17 −19 47 88 88 14 21 11

42. $A^{-1} = \begin{bmatrix}-2&-3&-1\\-3&-3&-1\\-2&-4&-1\end{bmatrix}$

$$\begin{bmatrix}9&-1&-9\\38&-19&-19\\28&-9&-19\\-80&25&41\\-64&21&31\\-7&-4&7\end{bmatrix}\begin{bmatrix}-2&-3&-1\\-3&-3&-1\\-2&-4&-1\end{bmatrix}=\begin{bmatrix}3&12&1\\19&19&0\\9&19&0\\3&1&14\\3&5&12\\12&5&4\end{bmatrix} \quad \begin{array}{ccc} C & L & A \\ S & S & \_ \\ I & S & \_ \\ C & A & N \\ C & E & L \\ L & E & D \end{array}$$

Message: CLASS IS CANCELLED

44. $A^{-1} = \begin{bmatrix} -13 & 6 & 4 \\ 12 & -5 & -3 \\ -5 & 2 & 1 \end{bmatrix}$

$$\begin{bmatrix} 13 & -9 & -59 \\ 61 & 112 & 106 \\ -17 & -73 & -131 \\ 11 & 24 & 29 \\ 65 & 144 & 172 \end{bmatrix} \begin{bmatrix} -13 & 6 & 4 \\ 12 & -5 & -3 \\ -5 & 2 & 1 \end{bmatrix} = \begin{bmatrix} 18 & 5 & 20 \\ 21 & 18 & 14 \\ 0 & 1 & 20 \\ 0 & 4 & 1 \\ 23 & 14 & 0 \end{bmatrix} \begin{matrix} \text{R} & \text{E} & \text{T} \\ \text{U} & \text{R} & \text{N} \\ \_ & \text{A} & \text{T} \\ \_ & \text{D} & \text{A} \\ \text{W} & \text{N} & \_ \end{matrix}$$

Message: RETURN AT DAWN

46. Let A be the 2×2 matrix needed to decode the message.

$$\begin{bmatrix} -19 & -19 \\ 37 & 16 \end{bmatrix} A = \begin{bmatrix} 0 & 19 \\ 21 & 5 \end{bmatrix} \begin{matrix} \_ & \text{S} \\ \text{U} & \text{E} \end{matrix}$$

$$A = \begin{bmatrix} -19 & -19 \\ 37 & 16 \end{bmatrix}^{-1} \begin{bmatrix} 0 & 19 \\ 21 & 5 \end{bmatrix} = \begin{bmatrix} \frac{16}{399} & \frac{19}{399} \\ -\frac{37}{399} & -\frac{19}{399} \end{bmatrix} \begin{bmatrix} 0 & 19 \\ 21 & 5 \end{bmatrix} = \begin{bmatrix} 1 & 1 \\ -1 & -2 \end{bmatrix}$$

$$\begin{bmatrix} 5 & 2 \\ 25 & 11 \\ -2 & -7 \\ -15 & -15 \\ 32 & 14 \\ -8 & -13 \\ 38 & 19 \\ -19 & -19 \\ 37 & 16 \end{bmatrix} \begin{bmatrix} 1 & 1 \\ -1 & -2 \end{bmatrix} = \begin{bmatrix} 3 & 1 \\ 14 & 3 \\ 5 & 12 \\ 0 & 15 \\ 18 & 4 \\ 5 & 18 \\ 19 & 0 \\ 0 & 19 \\ 21 & 5 \end{bmatrix} \begin{matrix} \text{C} & \text{A} \\ \text{N} & \text{C} \\ \text{E} & \text{L} \\ \_ & \text{O} \\ \text{R} & \text{D} \\ \text{E} & \text{R} \\ \text{S} & \_ \\ \_ & \text{S} \\ \text{U} & \text{E} \end{matrix}$$

Message: CANCEL ORDERS SUE

❑ Focus on Concepts

1. Interchange two rows.
Multiply a row by a nonzero constant.
Add a multiple of a row to another row.

2. They are the same.

3. A matrix in row-echelon form is in reduced row-echelon form if every column that has a leading 1 has zeros in every position above and below its leading 1.

4. Consistent—an infinite number of solutions

5. Inconsistent

6. Consistent—an unique solution

7. Consistent—an infinite number of solutions

8. (a) The operation can be performed.
(b) The operation can be performed.

9. (a) The operation can be performed.
(b) The operation cannot be performed. The number of rows in B must be the same as the number of columns in A.

10. (a) The operation cannot be performed. The orders A and B must be the same to perform the operations.
(b) The operation can be performed.

11. The matrix must be square and its determinant nonzero.

12. A square matrix is a square array of numbers, and a determinant is a real number associated with a square matrix.

13. No. The matrix must be square.

14. If A is a square matrix, the cofactor C_{ij} of the entry a_{ij} is $(-1)^{i+j} M_{ij}$, where M_{ij} is the determinant obtained by deleting the ith row and the jth column of A. The determinant of A is the sum of the entries of any row or column of A multiplied by their respective cofactors.

15. No. Each matrix is in row-echelon form, but the third matrix cannot be achieved from the first or second matrix with elementary row operations.

Review Exercises for Chapter 8

Solutions to Even-Numbered Exercises

2. $$\left[\begin{array}{rrr|r} 8 & -7 & 4 & 12 \\ 3 & -5 & 2 & 20 \\ 5 & 3 & -3 & 26 \end{array}\right]$$

4. $$\left[\begin{array}{rrrr|r} 13 & 16 & 7 & 3 & 2 \\ 1 & 21 & 8 & 5 & 12 \\ 4 & 10 & -4 & 3 & -1 \end{array}\right]$$

$$\begin{aligned} 13x + 16y + 7z + 3w &= 2 \\ x + 21y + 8z + 5w &= 12 \\ 4x + 10y - 4z + 3w &= -1 \end{aligned}$$

6. $$\begin{bmatrix} 1 & 1 & 1 & 0 \\ 1 & 1 & 0 & 1 \\ 1 & 0 & 1 & 1 \\ 0 & 1 & 1 & 1 \end{bmatrix}$$

$$\begin{matrix} \\ -R_1 + R_2 \rightarrow \\ -R_1 + R_3 \rightarrow \\ \\ \end{matrix} \begin{bmatrix} 1 & 1 & 1 & 0 \\ 0 & 0 & -1 & 1 \\ 0 & -1 & 0 & 1 \\ 0 & 1 & 1 & 1 \end{bmatrix}$$

$$\begin{matrix} R_3 + R_1 \rightarrow \\ \\ \\ R_3 + R_4 \rightarrow \end{matrix} \begin{bmatrix} 1 & 0 & 1 & 1 \\ 0 & 0 & -1 & 1 \\ 0 & -1 & 0 & 1 \\ 0 & 0 & 1 & 2 \end{bmatrix}$$

$$\begin{matrix} \\ -R_2 \rightarrow \\ -R_3 \rightarrow \\ R_2 + R_4 \rightarrow \end{matrix} \begin{bmatrix} 1 & 0 & 1 & 1 \\ 0 & 0 & 1 & -1 \\ 0 & 1 & 0 & -1 \\ 0 & 0 & 0 & 3 \end{bmatrix}$$

$$\begin{matrix} R_3 \rightarrow \\ R_2 \rightarrow \\ \frac{1}{3}R_4 \rightarrow \\ \\ \end{matrix} \begin{bmatrix} 1 & 0 & 1 & 1 \\ 0 & 1 & 0 & -1 \\ 0 & 0 & 1 & -1 \\ 0 & 0 & 0 & 1 \end{bmatrix}$$

$$\begin{matrix} -R_4 + R_1 \rightarrow \\ R_4 + R_2 \rightarrow \\ R_4 + R_3 \rightarrow \\ \\ \end{matrix} \begin{bmatrix} 1 & 0 & 1 & 0 \\ 0 & 1 & 0 & 0 \\ 0 & 0 & 1 & 0 \\ 0 & 0 & 0 & 1 \end{bmatrix}$$

$$\begin{matrix} -R_3 + R_1 \rightarrow \\ \\ \\ \\ \end{matrix} \begin{bmatrix} 1 & 0 & 0 & 0 \\ 0 & 1 & 0 & 0 \\ 0 & 0 & 1 & 0 \\ 0 & 0 & 0 & 1 \end{bmatrix}$$

8. $$\left[\begin{array}{rr|r} 2 & -5 & 2 \\ 3 & -7 & 1 \end{array}\right]$$

$$R_2 - R_1 \rightarrow \left[\begin{array}{rr|r} 1 & -2 & -1 \\ 3 & -7 & 1 \end{array}\right]$$

$$-3R_1 + R_2 \rightarrow \left[\begin{array}{rr|r} 1 & -2 & -1 \\ 0 & -1 & 4 \end{array}\right]$$

$$\begin{matrix} -2R_2 + R_1 \rightarrow \\ -R_2 \rightarrow \end{matrix} \left[\begin{array}{rr|r} 1 & 0 & -9 \\ 0 & 1 & -4 \end{array}\right]$$

$x = -9$

$y = -4$

Answer: $(-9, -4)$

10. $\left[\begin{array}{cc|c} 0.2 & -0.1 & 0.07 \\ 0.4 & -0.5 & -0.01 \end{array}\right]$

$\begin{array}{r} 5R_1 \rightarrow \\ -2R_1 + R_2 \rightarrow \end{array} \left[\begin{array}{cc|c} 1 & -0.5 & 0.35 \\ 0 & -0.3 & -0.15 \end{array}\right]$

$\begin{array}{r} -\frac{5}{3}R_2 + R_1 \rightarrow \\ -\frac{10}{3}R_2 \rightarrow \end{array} \left[\begin{array}{cc|c} 1 & 0 & 0.6 \\ 0 & 1 & 0.5 \end{array}\right]$

$x = 0.6$

$y = 0.5$

Answer: $(0.6, 0.5)$

12. $\left[\begin{array}{ccc|c} 2 & 3 & 1 & 10 \\ 2 & -3 & -3 & 22 \\ 4 & -2 & 3 & -2 \end{array}\right]$

$\begin{array}{r} R_3 \rightarrow \\ \\ R_1 \rightarrow \end{array} \left[\begin{array}{ccc|c} 4 & -2 & 3 & -2 \\ 2 & -3 & -3 & 22 \\ 2 & 3 & 1 & 10 \end{array}\right]$

$\begin{array}{r} \\ R_1 - 2R_2 \rightarrow \\ R_1 - 2R_3 \rightarrow \end{array} \left[\begin{array}{ccc|c} 4 & -2 & 3 & -2 \\ 0 & 4 & 9 & -46 \\ 0 & -8 & 1 & -22 \end{array}\right]$

$\begin{array}{r} \\ \\ 2R_2 + R_3 \rightarrow \end{array} \left[\begin{array}{ccc|c} 4 & -2 & 3 & -2 \\ 0 & 4 & 9 & -46 \\ 0 & 0 & 19 & -114 \end{array}\right]$

$19z = -114 \Rightarrow z = -6$

$4y + 9(-6) = -46 \Rightarrow y = 2$

$4x - 2(2) + 3(-6) = -2 \Rightarrow x = 5$

Answer: $(5, 2, -6)$

14. $\left[\begin{array}{ccc|c} 2 & 3 & 3 & 3 \\ 6 & 6 & 12 & 13 \\ 12 & 9 & -1 & 2 \end{array}\right]$

$\begin{array}{r} \\ -3R_1 + R_2 \rightarrow \\ -6R_1 + R_3 \rightarrow \end{array} \left[\begin{array}{ccc|c} 2 & 3 & 3 & 3 \\ 0 & -3 & 3 & 4 \\ 0 & -3 & -25 & -24 \end{array}\right]$

$\begin{array}{r} R_2 + R_1 \rightarrow \\ \\ -R_2 + R_3 \rightarrow \end{array} \left[\begin{array}{ccc|c} 2 & 0 & 6 & 7 \\ 0 & -3 & 3 & 4 \\ 0 & 0 & -28 & -28 \end{array}\right]$

$\begin{array}{r} \frac{1}{2}R_1 \rightarrow \\ -\frac{1}{3}R_2 \rightarrow \\ -\frac{1}{28}R_3 \rightarrow \end{array} \left[\begin{array}{ccc|c} 1 & 0 & 3 & \frac{7}{2} \\ 0 & 1 & -1 & -\frac{4}{3} \\ 0 & 0 & 1 & 1 \end{array}\right]$

$z = 1$

$y - 1 = -\frac{4}{3} \Rightarrow y = -\frac{1}{3}$

$x + 3(1) = \frac{7}{2} \Rightarrow x = \frac{1}{2}$

Answer: $\left(\frac{1}{2}, -\frac{1}{3}, 1\right)$

16. $\left[\begin{array}{ccc|c} 3 & 21 & -29 & -1 \\ 2 & 15 & -21 & 0 \end{array}\right]$

$\begin{array}{r} -R_2 + R_1 \rightarrow \\ \\ \end{array} \left[\begin{array}{ccc|c} 1 & 6 & -8 & -1 \\ 2 & 15 & -21 & 0 \end{array}\right]$

$\begin{array}{r} \\ -2R_1 + R_2 \rightarrow \end{array} \left[\begin{array}{ccc|c} 1 & 6 & -8 & -1 \\ 0 & 3 & -5 & 2 \end{array}\right]$

$\begin{array}{r} -2R_2 + R_1 \rightarrow \\ \\ \end{array} \left[\begin{array}{ccc|c} 1 & 0 & 2 & -5 \\ 0 & 3 & -5 & 2 \end{array}\right]$

Let $z = a$.

$3y - 5a = 2 \Rightarrow \frac{5}{3}a + \frac{2}{3}$

$x + 2a = -5 \Rightarrow -2a - 5$

Answer: $\left(-2a - 5, \frac{5}{3}a + \frac{2}{3}, a\right)$

18. $\begin{bmatrix} 1 & 2 & 0 & 1 & \vdots & 3 \\ 0 & -3 & 3 & 0 & \vdots & 0 \\ 4 & 4 & 1 & 2 & \vdots & 0 \\ 2 & 0 & 1 & 0 & \vdots & 3 \end{bmatrix}$

$$\begin{array}{r} \\ -\frac{1}{3}R_2 \rightarrow \\ -4R_1 + R_3 \rightarrow \\ -2R_1 + R_3 \rightarrow \end{array} \begin{bmatrix} 1 & 2 & 0 & 1 & \vdots & 3 \\ 0 & 1 & -1 & 0 & \vdots & 0 \\ 0 & -4 & 1 & -2 & \vdots & -12 \\ 0 & -4 & 1 & -2 & \vdots & -3 \end{bmatrix}$$

$$\begin{array}{r} \\ \\ \\ -R_3 + R_4 \rightarrow \end{array} \begin{bmatrix} 1 & 2 & 0 & 1 & \vdots & 3 \\ 0 & 1 & -1 & 0 & \vdots & 0 \\ 0 & -4 & 1 & -2 & \vdots & -12 \\ 0 & 0 & 0 & 0 & \vdots & 9 \end{bmatrix}$$

$$\begin{aligned} x + 2y \quad + w &= 3 \\ y - z \quad &= 0 \\ -4y + z - 2w &= -12 \\ 0 &= 9 \end{aligned}$$

Inconsistent, no solution

20. $x + 9 = A(x + 2)^2 + B(x + 1)(x + 2) + C(x + 1)$

$x + 9 = A(x^2 + 4x + 4) + B(x^2 + 3x + 2) + Cx + C$

$x + 9 = Ax^2 + 4Ax + 4A + Bx^2 + 3Bx + 2B + Cx + C$

$x + 9 = (A + B)x^2 + (4A + 3B + C)x + 4A + 2B + C$

Equating coefficients of corresponding terms:

$0 = A + B$

$1 = 4A + 3B + C$

$9 = 4A + 2B + C$

$$\begin{bmatrix} 1 & 1 & 0 & \vdots & 0 \\ 4 & 3 & 1 & \vdots & 1 \\ 4 & 2 & 1 & \vdots & 9 \end{bmatrix}$$

$$\begin{array}{r} \\ -4R_1 + R_2 \rightarrow \\ -4R_1 + R_3 \rightarrow \end{array} \begin{bmatrix} 1 & 1 & 0 & \vdots & 0 \\ 0 & -1 & 1 & \vdots & 1 \\ 0 & -2 & 1 & \vdots & 9 \end{bmatrix}$$

$$\begin{array}{r} \\ -R_2 \rightarrow \\ -2R_2 + R_3 \rightarrow \end{array} \begin{bmatrix} 1 & 1 & 0 & \vdots & 0 \\ 0 & 1 & -1 & \vdots & -1 \\ 0 & 0 & -1 & \vdots & 7 \end{bmatrix}$$

$-C = 7 \Rightarrow C = -7$

$B - (-7) = -1 \Rightarrow B = -8$

$A - 8 = 0 \Rightarrow A = 8$

$$\frac{x + 9}{(x + 1)(x + 2)^2} = \frac{8}{x + 1} - \frac{8}{x + 2} - \frac{7}{(x + 2)^2}$$

22. $-2\begin{bmatrix}1 & 2\\5 & -4\\6 & 0\end{bmatrix} + 8\begin{bmatrix}7 & 1\\1 & 2\\1 & 4\end{bmatrix} = \begin{bmatrix}-2 & -4\\-10 & 8\\-12 & 0\end{bmatrix} + \begin{bmatrix}56 & 8\\8 & 16\\8 & 32\end{bmatrix} = \begin{bmatrix}54 & 4\\-2 & 24\\-4 & 32\end{bmatrix}$

24. $\begin{bmatrix}1 & 5 & 6\\2 & -4 & 0\end{bmatrix}\begin{bmatrix}6 & -2 & 8\\4 & 0 & 0\end{bmatrix}$ is undefined.

26. $\begin{bmatrix}4\\6\end{bmatrix}\begin{bmatrix}6 & -2\end{bmatrix} = \begin{bmatrix}24 & -8\\36 & -12\end{bmatrix}$

28. $\begin{bmatrix}2 & 1\\6 & 0\end{bmatrix}\left(\begin{bmatrix}4 & 2\\-3 & 1\end{bmatrix} + \begin{bmatrix}-2 & 4\\0 & 4\end{bmatrix}\right) = \begin{bmatrix}2 & 1\\6 & 0\end{bmatrix}\begin{bmatrix}2 & 6\\-3 & 5\end{bmatrix}$

$$= \begin{bmatrix}2(2) + 1(-3) & 2(6) + 1(5)\\6(2) + 0 & 6(6) + 0\end{bmatrix}$$

$$= \begin{bmatrix}1 & 17\\12 & 36\end{bmatrix}$$

30. $-5\begin{bmatrix}2 & 0\\7 & -2\\8 & 2\end{bmatrix} + 4\begin{bmatrix}4 & -2\\6 & 11\\-1 & 3\end{bmatrix} = \begin{bmatrix}6 & -8\\-11 & 54\\-44 & 2\end{bmatrix}$

32. $\begin{bmatrix}-2 & 3 & 10\\4 & -2 & 2\end{bmatrix}\begin{bmatrix}1 & 1\\-5 & 2\\3 & 2\end{bmatrix} = \begin{bmatrix}13 & 24\\20 & 4\end{bmatrix}$

34. $X = \frac{1}{6}(4A + 3B) = \frac{1}{6}\left(4\begin{bmatrix}-4 & 0\\1 & -5\\-3 & 2\end{bmatrix} + 3\begin{bmatrix}1 & 2\\-2 & 1\\4 & 4\end{bmatrix}\right)$

$$= \frac{1}{6}\begin{bmatrix}-13 & 6\\-2 & -17\\0 & 20\end{bmatrix}$$

36. $X = \frac{1}{3}(2A - 5B) = \frac{1}{3}\left(2\begin{bmatrix}-4 & 0\\1 & -5\\-3 & 2\end{bmatrix} - 5\begin{bmatrix}1 & 2\\-2 & 1\\4 & 4\end{bmatrix}\right)$

$$= \frac{1}{3}\begin{bmatrix}-13 & -10\\12 & -15\\-26 & -16\end{bmatrix}$$

38. $2x + 3y + z = 10$
$2x - 3y - 3z = 22$
$4x - 2y + 3z = -2$

$$\begin{bmatrix}2 & 3 & 1\\2 & -3 & -3\\4 & -2 & 3\end{bmatrix}\begin{bmatrix}x\\y\\z\end{bmatrix} = \begin{bmatrix}10\\22\\-2\end{bmatrix}$$

40. $\begin{bmatrix}3 & -10\\4 & 2\end{bmatrix}^{-1} = \frac{1}{46}\begin{bmatrix}2 & 10\\-4 & 3\end{bmatrix}$

42. $A = \begin{bmatrix}1 & 4 & 6\\2 & -3 & 1\\-1 & 18 & 16\end{bmatrix}$

A^{-1} does not exist because $|A| = 0$.

44. $\begin{vmatrix}8 & 5\\2 & -4\end{vmatrix} = 8(-4) - 2(5) = -42$

46. $\begin{vmatrix} -5 & 6 & 0 & 0 \\ 0 & 1 & -1 & 2 \\ -3 & 4 & -5 & 1 \\ 1 & 6 & 0 & 3 \end{vmatrix} = -5\begin{vmatrix} 1 & -1 & 2 \\ 4 & -5 & 1 \\ 6 & 0 & 3 \end{vmatrix} - 6\begin{vmatrix} 0 & -1 & 2 \\ -3 & -5 & 1 \\ 1 & 0 & 3 \end{vmatrix}$ (Expansion along Row 1.)

$$= -5[6(-1 + 10) + 3(-5 + 4)] - 6[(-1 + 10) + 3(0 - 3)]$$
$$= -5[54 \quad -3] - 6[9 \quad -9]$$
$$= -255$$

48. $x + 3y = 23$

$-x + 2y = -18$

$$\begin{bmatrix} 1 & 3 \\ -6 & 2 \end{bmatrix}^{-1} \Rightarrow \begin{bmatrix} 0.1 & -0.15 \\ 0.3 & 0.05 \end{bmatrix}\begin{bmatrix} x \\ y \end{bmatrix} = \begin{bmatrix} 0.1 & -0.15 \\ 0.3 & 0.05 \end{bmatrix}\begin{bmatrix} 23 \\ -18 \end{bmatrix} = \begin{bmatrix} 5 \\ 6 \end{bmatrix}$$

$x = 5, y = 6$

Answer: $(5, 6)$

50. $x - 3y - 2z = 8$

$-2x + 7y + 3z = -19$

$x - y - 3z = 3$

$$\begin{bmatrix} 1 & -3 & -2 \\ -2 & 7 & 3 \\ 1 & -1 & -3 \end{bmatrix}^{-1} \Rightarrow \begin{bmatrix} -18 & -7 & 5 \\ -1 & -1 & 1 \\ -5 & -2 & 1 \end{bmatrix}\begin{bmatrix} x \\ y \\ z \end{bmatrix} = \begin{bmatrix} -18 & -7 & 5 \\ -1 & -1 & 1 \\ -5 & -2 & 1 \end{bmatrix}\begin{bmatrix} 8 \\ -19 \\ 3 \end{bmatrix} = \begin{bmatrix} 4 \\ -2 \\ 1 \end{bmatrix}$$

$x = 4, y = -2, z = 1$

Answer: $(4, -2, 1)$

52. $2x + 4y = -12$

$3x + 4y - 2z = -14$

$-x + y + 2z = -6$

$$\begin{bmatrix} 2 & 4 & 0 \\ 3 & 4 & -2 \\ -1 & 1 & 2 \end{bmatrix}^{-1} \Rightarrow \begin{bmatrix} 2.5 & -2 & -2 \\ -3 & 1 & 1 \\ 1.75 & -1.5 & -1 \end{bmatrix}\begin{bmatrix} x \\ y \\ z \end{bmatrix} = \begin{bmatrix} 2.5 & -2 & -2 \\ -3 & 1 & 1 \\ 1.75 & -1.5 & -1 \end{bmatrix}\begin{bmatrix} -12 \\ -14 \\ -6 \end{bmatrix} = \begin{bmatrix} 10 \\ -8 \\ 6 \end{bmatrix}$$

$x = 10, y = -8, z = 6$

Answer: $(10, -8, 6)$

54. $2x + 3y - 4z = 1$

$x - y + 2z = -4$

$3x + 7y - 10z = 0$

$$\begin{bmatrix} 2 & 3 & -4 \\ 1 & -1 & 2 \\ 3 & 7 & -10 \end{bmatrix}^{-1} \text{ does not exist.}$$

The system is inconsistent. No solution.

56. x = number of liters of 75% acid

y = number of liters of 50% acid

$x + y = 100$

$0.75x + 0.50y = 60$

$$\begin{bmatrix} 1 & 1 \\ 0.75 & 0.50 \end{bmatrix}\begin{bmatrix} x \\ y \end{bmatrix} = \begin{bmatrix} 100 \\ 60 \end{bmatrix}$$

$$D = \begin{vmatrix} 1 & 1 \\ 0.75 & 0.50 \end{vmatrix} = -0.25$$

$$x = \frac{\begin{vmatrix} 100 & 1 \\ 60 & 0.50 \end{vmatrix}}{-0.25} = \frac{-10}{-0.25} = 40$$

$$y = \frac{\begin{vmatrix} 1 & 100 \\ 0.75 & 60 \end{vmatrix}}{-0.25} = \frac{-15}{-0.25} = 60$$

Answer: 40 liters of 75% acid; 60 liters of 50% acid

58. x = number of units produced

y = number of units produced

$x - y = 0$

$-3.75x + 5.25y = 25{,}000$

$$\begin{bmatrix} 1 & -1 \\ -3.75 & 5.25 \end{bmatrix}\begin{bmatrix} x \\ y \end{bmatrix} = \begin{bmatrix} 0 \\ 25{,}000 \end{bmatrix}$$

$$D = \begin{vmatrix} 1 & -1 \\ -3.75 & 5.25 \end{vmatrix} = 1.5$$

$$y = \frac{\begin{vmatrix} 1 & 0 \\ -3.75 & 25{,}000 \end{vmatrix}}{1.5} \approx 16{,}667 \text{ units must be sold.}$$

60.
$$\begin{vmatrix} 2-\lambda & 5 \\ 3 & -8-\lambda \end{vmatrix} = 0$$

$$(2-\lambda)(-8-\lambda) - 15 = 0$$

$$-16 - 6\lambda + \lambda^2 - 15 = 0$$

$$\lambda^2 + 6\lambda - 31 = 0$$

$$\lambda = \frac{-6 \pm \sqrt{36 - 4)(-31)}}{2}$$

$$\lambda = -3 \pm 2\sqrt{10}$$

62. $(-4, 0), (4, 0), (0, 6)$

$$\text{Area} = \frac{1}{2}\begin{vmatrix} -4 & 0 & 1 \\ 4 & 0 & 1 \\ 0 & 6 & 1 \end{vmatrix} = \frac{1}{2}(48) = 24 \text{ square units}$$

64. $\left(\frac{3}{2}, 1\right), \left(4, -\frac{1}{2}\right), (4, 2)$

$$\text{Area} = \frac{1}{2}\begin{vmatrix} \frac{3}{2} & 1 & 1 \\ 4 & -\frac{1}{2} & 1 \\ 4 & 2 & 1 \end{vmatrix} = \frac{1}{2}\left(\frac{25}{4}\right) = \frac{25}{8} \text{ square units}$$

66. $(2, 5), (6, -1)$

$$\begin{vmatrix} x & y & 1 \\ 2 & 5 & 1 \\ 6 & -1 & 1 \end{vmatrix} = 0$$

$$6x + 4y - 32 = 0$$

$$2x + 2y - 16 = 0$$

68. $(-0.8, 0.2), (0.7, 3.2)$

$$\begin{vmatrix} x & y & 1 \\ -0.8 & 0.2 & 1 \\ 0.7 & 3.2 & 1 \end{vmatrix} = 0$$

$$-3x + 1.5y - 2.7 = 0$$

$$10x - 5y + 9 = 0$$

CHAPTER 9
Sequences and Probability

CHAPTER 9
Sequences and Probability

Section 9.1 Sequences and Summation Notation

Solutions to Even-Numbered Exercises

2. $a_n = 4n - 3$

$a_1 = 4(1) - 3 = 1$

$a_2 = 4(2) - 3 = 5$

$a_3 = 4(3) - 3 = 9$

$a_4 = 4(4) - 3 = 17$

$a_5 = 4(5) - 3 = 17$

4. $a_n = \left(\frac{1}{2}\right)^n$

$a_1 = \left(\frac{1}{2}\right)^1 = \frac{1}{2}$

$a_2 = \left(\frac{1}{2}\right)^2 = \frac{1}{4}$

$a_3 = \left(\frac{1}{2}\right)^3 = \frac{1}{8}$

$a_4 = \left(\frac{1}{2}\right)^4 = \frac{1}{16}$

$a_5 = \left(\frac{1}{2}\right)^5 = \frac{1}{32}$

6. $a_n = \left(-\frac{1}{2}\right)^n$

$a_1 = \left(-\frac{1}{2}\right)^1 = -\frac{1}{2}$

$a_2 = \left(-\frac{1}{2}\right)^2 = \frac{1}{4}$

$a_3 = \left(-\frac{1}{2}\right)^3 = -\frac{1}{8}$

$a_4 = \left(-\frac{1}{2}\right)^4 = \frac{1}{16}$

$a_5 = \left(-\frac{1}{2}\right)^5 = -\frac{1}{32}$

8. $a_n = \dfrac{n}{n+1}$

$a_1 = \dfrac{1}{1+1} = \dfrac{1}{2}$

$a_2 = \dfrac{2}{2+1} = \dfrac{2}{3}$

$a_3 = \dfrac{3}{3+1} = \dfrac{3}{4}$

$a_4 = \dfrac{4}{4+1} = \dfrac{4}{5}$

$a_5 = \dfrac{5}{5+1} = \dfrac{5}{6}$

10. $a_n = \dfrac{3n^2 - n + 4}{2n^2 + 1}$

$a_1 = \dfrac{3(1)^2 - 1 + 4}{2(1)^2 + 1} = 2$

$a_2 = \dfrac{3(2)^2 - 2 + 4}{2(2)^2 + 1} = \dfrac{14}{9}$

$a_3 = \dfrac{3(3) - 3 + 4}{2(3)^2 + 1} = \dfrac{28}{19}$

$a_4 = \dfrac{3(4) - 4 + 4}{2(4)^2 + 1} = \dfrac{16}{11}$

$a_5 = \dfrac{3(5)^2 - 5 + 4}{2(5)^2 + 1} = \dfrac{74}{51}$

12. $a_n = 1 + (-1)^n$

$a_1 = 1 + (-1)^1 = 0$

$a_2 = 1 + (-1)^2 = 2$

$a_3 = 1 + (-1)^3 = 0$

$a_4 = 1 + (-1)^4 = 2$

$a_5 = 1 + (-1)^5 = 0$

14. $a_n = \dfrac{3^n}{4^n}$

$a_1 = \dfrac{3^1}{4^1} = \dfrac{3}{4}$

$a_2 = \dfrac{3^2}{4^2} = \dfrac{9}{16}$

$a_3 = \dfrac{3^3}{4^3} = \dfrac{27}{64}$

$a_4 = \dfrac{3^4}{4^4} = \dfrac{81}{256}$

$a_5 = \dfrac{3^5}{4^5} = \dfrac{243}{1024}$

16. $a_n = \dfrac{10}{n^{2/3}} = \dfrac{10}{\sqrt[3]{n^2}}$

$a_1 = \dfrac{10}{1} = 10$

$a_2 = \dfrac{10}{\sqrt[3]{2^2}} = \dfrac{10}{\sqrt[3]{4}}$

$a_3 = \dfrac{10}{\sqrt[3]{3^2}} = \dfrac{10}{\sqrt[3]{9}}$

$a_4 = \dfrac{10}{\sqrt[3]{4^2}} = \dfrac{10}{\sqrt[3]{16}}$

$a_5 = \dfrac{10}{\sqrt[3]{3^2}} = \dfrac{10}{\sqrt[3]{9}}$

18. $a_n = \dfrac{n!}{n}$

$a_1 = \dfrac{1!}{1} = 1$

$a_2 = \dfrac{2!}{2} = 1$

$a_3 = \dfrac{3!}{3} = 2$

$a_4 = \dfrac{4!}{4} = 6$

$a_5 = \dfrac{5!}{5} = 24$

20. $a_n = (-1)^n\left(\frac{n}{n+1}\right)$

$a_1 = (-1)^1\frac{1}{1+1} = -\frac{1}{2}$

$a_2 = (-1)^2\frac{2}{1+2} = \frac{2}{3}$

$a_3 = (-1)^3\frac{3}{3+1} = -\frac{3}{4}$

$a_4 = (-1)^4\frac{4}{4+1} = \frac{4}{5}$

$a_5 = (-1)^5\frac{5}{5+1} = -\frac{5}{6}$

22. $a_n = n(n-1)(n-2)$

$a_1 = 1(1-1)(1-2) = 0$

$a_2 = 2(2-1)(2-2) = 0$

$a_3 = 3(3-1)(3-2) = 6$

$a_4 = 4(4-1)(4-2) = 24$

$a_5 = 5(5-1)(5-2) = 60$

24. $a_n = \frac{2^n}{n!}$

$a_{10} = \frac{2^{10}}{10!} = \frac{1024}{3{,}628{,}800}$

$= \frac{4}{14{,}175}$

26. $a_1 = 15,\quad a_{k+1} = a_k + 3$

$a_1 = 15$

$a_2 = a_1 + 3 = 15 + 3 = 18$

$a_3 = a_2 + 3 = 18 + 3 = 21$

$a_4 = a_3 + 3 = 21 + 3 = 24$

$a_5 = a_4 + 3 = 24 + 3 = 27$

28. $a_1 = 32,\quad a_{k+1} = \frac{1}{2}a_k$

$a_1 = 32$

$a_2 = \frac{1}{2}a_1 = \frac{1}{2}(32) = 16$

$a_3 = \frac{1}{2}a_2 = \frac{1}{2}(16) = 8$

$a_4 = \frac{1}{2}a_3 = \frac{1}{2}(8) = 4$

$a_5 = \frac{1}{2}a_4 = \frac{1}{2}(4) = 2$

30. $a_n = 2 - \frac{4}{n}$

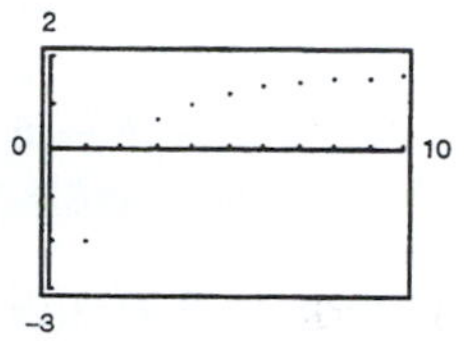

32. $a_n = 8(0.75)^{n-1}$

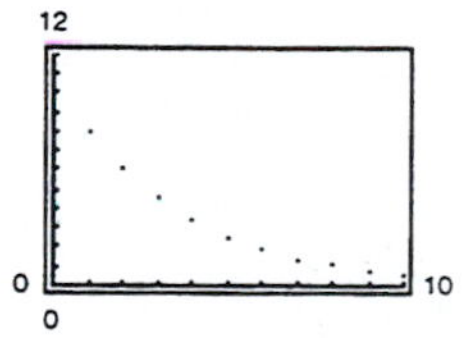

34. $a_n = \frac{3n^2}{n^2+1}$

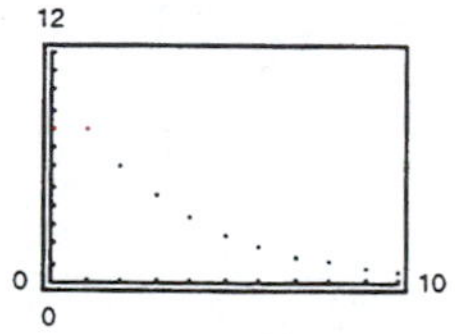

36. $a_n = \frac{8n}{n+1}$

$a_n \to 8$ as $n \to \infty$

$a_1 = 4,\quad a_4 = \frac{4^4}{4!} = \frac{256}{24} = 10\frac{2}{3}$

Matches graph (a).

38. $a_n = \frac{4^n}{n!}$

$a_n \to 0$ as $n \to \infty$

$a_1 = 4, a_4 = \frac{4^4}{4!} = \frac{256}{24} = 10\frac{2}{3}$

Matches graph (a).

40. $\frac{4!}{7!} = \frac{4!}{7 \cdot 6 \cdot 5 \cdot 4!} = \frac{1}{210}$

42. $\frac{25!}{23!} = \frac{25 \cdot 24 \cdot 23!}{23!} = 600$

44. $\frac{(n+2)!}{n!} = \frac{(n+2)(n+1)n!}{n!} = (n+2)(n+1)$

46. $\frac{(2n+2)!}{(2n)!} = \frac{(2n+2)(2n+1)(2n)!}{(2n)!}$

$= (2n+2)(2n+1)$

48. $3, 7, 11, 15, 19, \ldots$

$a_n = 4n - 1$

50. $1, \frac{1}{4}, \frac{1}{9}, \frac{1}{16}, \frac{1}{25}, \ldots$

$a_n = \frac{1}{n^2}$

52. $\frac{2}{1}, \frac{3}{3}, \frac{4}{5}, \frac{5}{7}, \frac{6}{9}, \ldots$

$a_n = \frac{n+1}{2n-1}$

54. $\frac{1}{3}, \frac{2}{9}, \frac{4}{27}, \frac{8}{81}, \ldots$

$a_n = \frac{2^{n-1}}{3^n}$

56. $1 + \frac{1}{2}, 1 + \frac{3}{4}, 1 + \frac{7}{8}, 1 + \frac{15}{16}, 1 + \frac{31}{32}, \ldots$

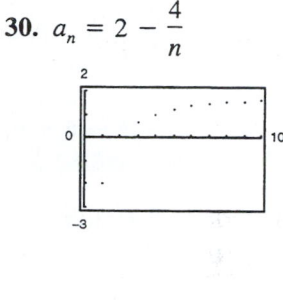

$a_n = 1 + \frac{2^n - 1}{2^n}$

58. $2, -4, 6, -8, 10, \ldots$

$a_n = (-1)^{n+1}(2n)$

60. $1, 2, \frac{2^2}{2}, \frac{2^3}{6}, \frac{2^4}{24}, \frac{2^5}{120}, \ldots$

$a_n = \frac{2^{n-1}}{(n-1)!}$

62. $a_1 = 25, \quad a_{k+1} = a_k - 5$

$a_1 = 25$

$a_2 = a_1 - 5 = 25 - 5 = 20$

$a_3 = a_2 - 5 = 20 - 5 = 15$

$a_4 = a_3 - 5 = 15 - 5 = 10$

$a_5 = a_4 - 5 = 10 - 5 = 5$

In general, $a_n = 30 - 5_n$.

64. $a_1 = 14, \quad a_{k+1} = (-2)a_k$

$a_1 = 14$

$a_2 = (-2)a_1 = (-2)(14) = -28$

$a_3 = (-2)a_2 = (-2)(-28) = 56$

$a_4 = (-2)a_3 = (-2)(56) = -112$

$a_5 = (-2)(a_4) = (-2)(-112) = 224$

In general, $a_n = 14(-2)^{n-1}$.

66. $\sum_{i=1}^{6}(3i-1) = (3 \cdot 1 - 1) + (3 \cdot 2 - 1) + (3 \cdot 3 - 1) + (3 \cdot 4 - 1) + (3 \cdot 5 - 1) + (3 \cdot 6 - 1) = 57$

68. $\sum_{k=1}^{5} 6 = 6 + 6 + 6 + 6 + 6 = 30$

70. $\sum_{k=0}^{5} 3i^2 = 3\sum_{i=0}^{5} i^2 = 3(0^2 + 1^2 + 2^2 + 3^2 + 4^2 + 5^2) = 165$

72. $\sum_{j=3}^{5} \frac{1}{j} = \frac{1}{3} + \frac{1}{4} + \frac{1}{5} = \frac{47}{60}$

74. $\sum_{k=2}^{5} (k+1)(k-3) = (2+1)(2-3) + (3+1)(3-3) + (4+1)(4-3) + (5+1)(5-3) = 14$

76. $\sum_{j=0}^{4}(-2)^j = (-2)^0 + (-2)^1 + (-2)^2 + (-2)^3 + (-2)^4 = 11$

78. $\sum_{j=1}^{10} \frac{3}{j+1} \approx 6.06$

80. $\sum_{k=0}^{4} \frac{(-1)^k}{k!} = \frac{3}{8} = 0.375$

82. $\frac{5}{1+1} + \frac{5}{1+2} + \frac{5}{1+3} + \cdots + \frac{5}{1+15} = \sum_{i=1}^{15} \frac{5}{1+i}$

84. $\left[1 - \left(\frac{1}{6}\right)^2\right] + \left[1 - \left(\frac{2}{6}\right)^2\right] + \cdots + \left[1 - \left(\frac{6}{6}\right)^2\right] = \sum_{k=1}^{6}\left[1 - \left(\frac{k}{6}\right)^2\right]$

86. $1 - \frac{1}{2} + \frac{1}{4} - \frac{1}{8} + \cdots - \frac{1}{128} = \frac{1}{2^0} - \frac{1}{2^1} + \frac{1}{2^2} - \frac{1}{2^3} + \cdots - \frac{1}{2^7} = \sum_{n=0}^{7}\left(-\frac{1}{2}\right)^2$

88. $\frac{1}{1 \cdot 3} + \frac{1}{2 \cdot 4} + \frac{1}{3 \cdot 5} + \cdots + \frac{1}{10 \cdot 12} = \sum_{k=1}^{10} \frac{1}{k(k+2)}$

90. $\frac{1}{2} + \frac{2}{4} + \frac{6}{8} + \frac{24}{16} + \frac{120}{32} + \frac{720}{64} = \sum_{k=1}^{6} \frac{k!}{2^k}$

92. (a) $A_1 = 100(101)[(1.01)^1 - 1] = \$101.00$

$A_2 = 100(101)[(1.01)^2 - 1] = \$203.01$

$A_3 = 100(101)[(1.01)^3 - 1] \approx \$306.04$

$A_4 = 100(101)[(1.01)^4 - 1] \approx \$410.10$

$A_5 = 100(101)[(1.01)^5 - 1] \approx \$515.20$

$A_6 = 100(101)[(1.01)^6 - 1] \approx \$621.35$

(b) $A_{60} = 100(101)[(1.01)^{60} - 1] \approx \$8248.64$

(c) $A_{240} = 100(101)[(1.01)^{240} - 1] \approx \$99{,}914.79$

94. $a_0 = 0.1\sqrt{82 + 9 \cdot 0^2} \approx 0.91$

$a_1 = 0.1\sqrt{82 + 9 \cdot 1^2} \approx 0.95$

$a_2 = 0.1\sqrt{82 + 9 \cdot 2^2} \approx 1.09$

$a_3 = 0.1\sqrt{82 + 9 \cdot 3^2} \approx 1.28$

$a_4 = 0.1\sqrt{82 + 9 \cdot 4^2} \approx 1.50$

$a_5 = 0.1\sqrt{82 + 9 \cdot 5^2} \approx 1.75$

$a_6 = 0.1\sqrt{82 + 9 \cdot 6^2} \approx 2.01$

$a_7 = 0.1\sqrt{82 + 9 \cdot 7^2} \approx 2.29$

$a_8 = 0.1\sqrt{82 + 9 \cdot 8^2} \approx 2.57$

$a_9 = 0.1\sqrt{82 + 9 \cdot 9^2} \approx 2.85$

$a_{10} = 0.1\sqrt{82 + 9 \cdot 10^2} \approx 3.13$

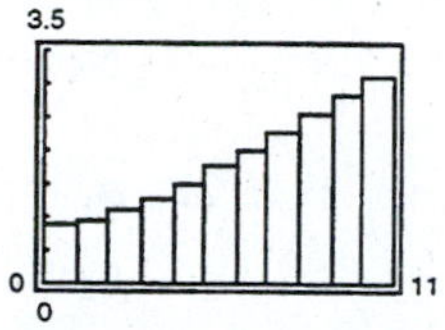

96. $\displaystyle\sum_{n=5}^{14} (1.39 + 0.18n - 1.02 \ln n) \approx \$8.55$

98. $b_n = \dfrac{a_{n+1}}{a_n};\ b_1 = 1,\ b_2 = 2,\ b_3 = \dfrac{3}{2},\ b_4 = \dfrac{5}{3}, \ldots$

$b_2 = 1 + \dfrac{1}{b_1} = 1 + \dfrac{1}{1} = 2$

$b_3 = 1 + \dfrac{1}{b_2} = 1 + \dfrac{1}{2} = \dfrac{3}{2}$

$b_4 = 1 + \dfrac{1}{b_3} = 1 + \dfrac{2}{3} = \dfrac{5}{3}$

$b_5 = 1 + \dfrac{1}{b_4} = 1 + \dfrac{3}{5} = \dfrac{8}{5}$

$b_n = 1 + \dfrac{1}{b_{n-1}}$

100. $\bar{x} = \dfrac{1}{n}\displaystyle\sum_{i=1}^{n} xi = \dfrac{1.279 + 1.259 + 1.289 + 1.329 + 1.349}{5} = \$1.301$

102.
$$\sum_{i=1}^{n}(x_i - \bar{x})^2 = \sum_{i=1}^{n}(x_i^2 - 2x_i\bar{x} + \bar{x}^2) = \sum_{i=1}^{n} x_i^2 - 2\bar{x}\sum_{i=1}^{n} x_i + n\bar{x}^2$$
$$= \sum_{i=1}^{n} x_i^2 - 2 \cdot \frac{1}{n}\sum_{i=1}^{n} x_i \sum_{i=1}^{n} x_i + n \cdot \frac{1}{n}\sum_{i=1}^{n} x_i \cdot \frac{1}{n}\sum_{i=1}^{n} x_i$$
$$= \sum_{i=1}^{n} x_i^2 + \sum_{i=1}^{n} x_i \sum_{i=1}^{n} x_i\left(-\frac{2}{n} + \frac{1}{n}\right) = \sum_{i=1}^{n} x_i^2 - \frac{1}{n}\left(\sum_{i=1}^{n} x_i\right)^2$$

104. $\displaystyle\sum_{j=1}^{4} 2^j = \sum_{j=3}^{6} 2^{j-2}$

True, because $2^1 + 2^2 + 2^3 + 2^4 = 2^{3-2} + 2^{4-2} + 2^{5-2} + 2^{6-2}$.

Section 9.2 Arithmetic Sequences

Solutions to Even-Numbered Exercises

2. $4, 7, 10, 13, 16, \ldots$

Arthimetic sequence, $d = 3$

4. $3, \frac{5}{2}, 2, \frac{3}{2}, 1, \ldots$

Arthimetic sequence, $d = -\frac{1}{2}$

6. $\frac{1}{3}, \frac{2}{3}, \frac{4}{3}, \frac{8}{3}, \frac{16}{3}, \ldots$

Not an arithmetic sequence

8. $\ln 1, \ln 2, \ln 3, \ln 4, \ln 5, \ldots$

Not an arithmetic sequence

10. $1^2, 2^2, 3^2, 4^2, 5^2, \ldots$

Not an arithmetic sequence

12. $a_n = (2^n)n$

$2, 8, 24, 64, 160$

Not an arithmetic sequence

14. $a_n = 1 + (n - 1)\,4$

$1, 5, 9, 13, 17$

Arthimetic sequence, $d = 4$

16. $a_n = 2^{n-1}$

$1, 2, 4, 8, 16$

Not an arithmetic sequence

18. $a_n = (-1)^n$

$-1, 1, -1, 1, -1$

Not an arithmetic sequence

20. $a_1 = 2, a_{k+1} = a_k + \frac{2}{3}$

$a_2 = 2 + \frac{2}{3} = \frac{8}{3}$

$a_3 = \frac{8}{3} + \frac{2}{3} = \frac{10}{3}$

$a_4 = \frac{10}{3} + \frac{2}{3} = \frac{12}{3} = 4$

$a_5 = 4 + \frac{2}{3} = \frac{14}{3}$

$a_n = \frac{4}{3} + \frac{2}{3}n$

22. $a_1 = 72, a_{k+1} = a_k - 6$

$a_2 = 72 - 6 = 66$

$a_3 = 66 - 6 = 60$

$a_4 = 60 - 6 = 54$

$a_5 = 54 - 6 = 48$

$a_n = 78 - 6n$

24. $a_1 = 0.375, a_{k+1} = a_k + 0.25$

$a_2 = 0.375 + 0.25 = 0.625$

$a_3 = 0.625 + 0.25 = 0.875$

$a_4 = 0.875 + 0.25 = 1.125$

$a_5 = 1.125 + 0.25 = 1.375$

$a_n = 0.125 + 0.25n$

26. $a_1 = 5, d = -\frac{3}{4}$

$a_1 = 5$

$a_2 = 5 - \frac{3}{4} = \frac{17}{4}$

$a_3 = \frac{17}{4} - \frac{3}{4} = \frac{14}{4} = \frac{7}{2}$

$a_4 = \frac{7}{2} - \frac{3}{4} = \frac{11}{4}$

$a_5 = \frac{11}{4} - \frac{3}{4} = \frac{8}{4} = 2$

28. $a_1 = 16.5, d = 0.25$

$a_1 = 16.5$

$a_2 = 16.5 + 0.25 = 16.75$

$a_3 = 16.75 + 0.25 = 17$

$a_4 = 17 + 0.25 = 17.25$

$a_5 = 17.25 + 0.25 = 17.5$

30. $a_4 = 16, a_{10} = 46$

$16 = a_4 = a_1 + (n - 1)d = a_1 + 3d$

$46 = a_{10} = a_1 + (n - 1)d = a_1 + 9d$

Answer: $a_1 = 1, d = 5$

$a_1 = 1$

$a_2 = 1 + 5 = 6$

$a_3 = 6 + 5 = 11$

$a_4 = 11 + 5 = 16$

$a_5 = 16 + 5 = 21$

32. $a_3 = 19, a_{15} = -1.7$

$19 = a_3 = a_1 + (n - 1)\,d = a_1 + 2d$

$-1.7 = a_{15} = a_1(n - 1)\,d = a_1 + 14d$

Answer: $a_1 = 22.45, d = -1.725$

$a_1 = 22.45$

$a_2 = 22.45 - 1.725 = 20.725$

$a_3 = 20.725 - 1.725 = 19$

$a_4 = 19 - 1.725 = 17.275$

$a_5 = 17.275 - 1.725 = 15.55$

34. $a_1 = 15, d = 4$

$a_n = a_1 + (n - 1)\,d = 15 + (n - 1)\,4$

36. $a_1 = 0, d = -\frac{2}{3}$

$a_n = a_1 + (n - 1)\,d = (n - 1)\left(-\frac{2}{3}\right)$

38. $a_1 = -y, d = 5y$

$a_n = a_1 + (n - 1)\,d = -y + (n - 1)\,(5y)$

40. $10, 5, 0, -5, -10, \ldots$

$d = -5$

$a_n = a_1 + (n - 1)\,d = 10 + (n - 1)\,(-5)$

42. $a_1 = -4, a_5 = 16$

$a_n = a_1 + (n-1)d$

$16 = -4 + 4d$

$d = 5$

$a_n = a_1 + (n-1)d = -4 + (n-1)5$

44. $a_5 = 190, a_{10} = 115$

$a_{10} = a_5 + 5d \Rightarrow 115 = 190 + 5d \Rightarrow d = -15$

$a_1 = a_5 - 4d \Rightarrow a_1 = 190 - 4(-15) = 250$

$a_n = a_1 + (n-1)d = 250 + (n-1)(-15)$

46. $a_n = 3n - 5$

$d = 3$ so the sequence is increasing and $a_1 = -2$.

Matches (d).

48. $a_n = 25 - 3n$

$d = -3$ so the sequence is decreasing and $a_1 = 22$.

Matches (a).

50. $a_n = -5 + 2n$

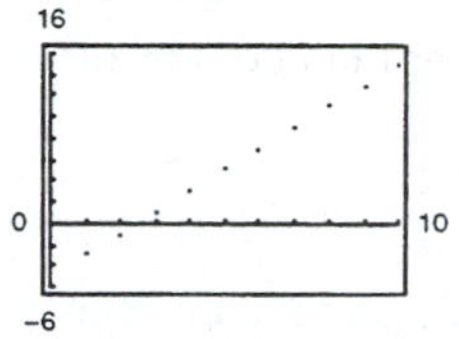

52. $a_n = -0.3n + 8$

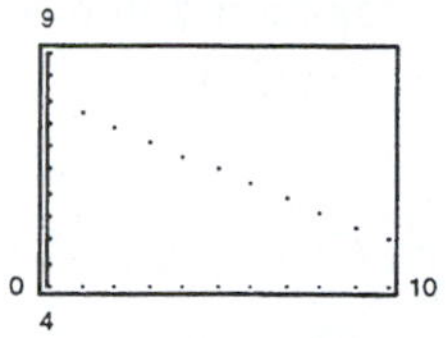

54. You can use the first two terms of an arithmetic sequence to find the nth term of the sequence by subtracting the first term from the second term to find the common difference and then find the nth term by adding $(n - 1)$ times the common difference to the first term.

56. $2, 8, 14, 20, \ldots, n = 25$

$a_n = 6n - 4$

$a_1 = 2$ and $a_{25} = 146$

$S_{25} = \frac{25}{2}(2 + 146) = 1850$

58. $0.5, 0.9, 1.3, 1.7, \ldots, n = 10$

$a_n = 0.4n + 0.1$

$a_1 = 0.5$ and $a_{10} = 4.1$

$S_{10} = \frac{10}{2}(0.5 + 4.1) = 23$

60. $1.50, 1.45, 1.40, 1.35, \ldots, n = 20$

$a_n = -0.05n + 1.55$

$a_1 = 1.50$ and $a_{20} = 0.55$

$S_{20} = \frac{20}{2}(1.50 + 0.55) = 20.5$

62. $a_1 = 15, a_{100} = 307, n = 100$

$S_{100} = \frac{100}{2}(15 + 307) = 16{,}100$

64. $a_n = 2n$

$a_1 = 2, a_{100} = 200, n = 100$

$$\sum_{n=1}^{100} 2n = \frac{100}{2}(2 + 200) = 10{,}100$$

66. $a_n = 7n$

$a_{51} = 357, a_{100} = 700$

$$\sum_{n=51}^{100} 7n = \frac{50}{2}(357 + 700) = 26{,}425$$

68. $$\sum_{n=51}^{100} n = \sum_{n=1}^{50} n = \frac{50}{2}(51 + 100) - \frac{50}{2}(1 + 50) = 3775 - 1275 = 2500$$

70. $a_n = 1000 - n$

$a_1 = 999, a_{250} = 750, n = 250$

$$\sum_{n=1}^{250}(1000 - n) = \frac{250}{2}(999 + 750) = 218{,}625$$

72. $a_1 = \frac{5}{2}, a_{100} = 52, n = 100$

$$\sum_{n=1}^{100} \frac{n+4}{2} = \frac{100}{2}\left(\frac{5}{2} + 52\right) = 2725$$

74. $a_0 = \frac{1}{2}, a_{100} = -18\frac{1}{4}, n = 101$

$$\sum_{n=0}^{100} \frac{8 - 3n}{16} = \frac{101}{2}\left(\frac{1}{2} - 18\frac{1}{4}\right) = -896.375$$

76. $a_1 = 4.525, a_{200} = 9.5, n = 200$

$$\sum_{j=1}^{200}(4.5 + 0.025j) = \frac{200}{2}(4.525 + 9.5) = 1402.5$$

78. $a_1 = -10, a_{61} = 50, n = 61$

$$\sum_{i=0}^{61}(i - 10) = \frac{61}{2}(-10 + 50) = 1220$$

80. (a) $a_1 = 36{,}800, d = 1750$

$a_6 = a_1 + 5d = 36{,}800 + 5(1750) = \$45{,}550$

(b) $S_6 = \frac{6}{2}[36{,}800 + 45{,}550] = \$247{,}050$

82. $a_1 = 15, d = 3, n = 36$

$a_{36} = 15 + 35(3) = 120$

$S_{36} = \frac{36}{2}(15 + 120) = 2430$ seats

84. $a_1 = 4.9, a_2 = 14.7, a_3 = 24.5,$

$a_4 = 34.3 \Rightarrow d = 9.8$

$a_1 = 4.9 = 9.8(1) + c \Rightarrow c = -4.9$

$a_n = 9.8n - 4.9$

$a_{10} = 9.8(10) - 4.9 = 93.1$

86. If an arithmetic sequence is defined by $a_1, a_2, a_3, a_4, \ldots$ and the common difference is $a_2 - a_1 = d.$

(a) A constant C is added to each term: $a_1 + C, a_2 + C, a_3 + C, a_4 + C, \ldots$

The resulting sequence is arithmetic, and the common difference is the original common difference:

$$a_2 + C - (a_1 + C) = a_2 + C - a_1 - C$$
$$= a_2 - a_1 = d.$$

(b) Each term is multiplied by a nonzero constant C: $Ca_1, Ca_2, Ca_3, Ca_4, \ldots$

The resulting sequence is arithmetic, and the common difference is C times the original common difference:

$Ca_2 - Ca_1 = C(a_2 - a_1) = Cd.$

(c) If each term is squared, the sequence is not arithmetic.

88. Let $S_n = \dfrac{n}{2}(a_1 + a_n)$ be the sum of the first n terms of the original sequence.

$$S_n' = \frac{n}{2}(a_1 + 5 + a_n + 5) = \frac{n}{2}(a_1 + a_n + 10) = \frac{n}{a}(a_1 + a_n) + \frac{n}{2}(10)$$
$$= \frac{n}{2}(a_1 + a_n) + 5n$$
$$= S_n + 5n$$

Section 9.3 Geometric Sequences

Solutions to Even-Numbered Exercises

2. $3, 12, 48, 192, \ldots$

Geometric sequence, $r = 4$

4. $1, -2, 4, -8, \ldots$

Geometric sequence, $r = -2$

6. $5, 1, 0.2, 0.04, \ldots$

Geometric sequence, $r = \frac{1}{5} = 0.2$

8. $9, -6, 4, -\frac{8}{3}, \ldots$

Geometric sequence, $r = -\frac{2}{3}$

10. $\frac{1}{5}, \frac{2}{7}, \frac{3}{9}, \frac{4}{11}, \ldots$

Not a geometric sequence

12. $a_1 = 6, r = 2$

$a_1 = 6$

$a_2 = 6(2)^1 = 12$

$a_3 = 6(2)^2 = 24$

$a_4 = 6(2)^3 = 48$

$a_5 = 6(2)^4 = 96$

14. $a_1 = 1, r = \frac{1}{3}$

$a_1 = 1$

$a_2 = 1\left(\frac{1}{3}\right)^1 = \frac{1}{3}$

$a_3 = 1\left(\frac{1}{3}\right)^2 = \frac{1}{9}$

$a_4 = 1\left(\frac{1}{3}\right)^3 = \frac{1}{27}$

$a_5 = 1\left(\frac{1}{3}\right)^4 = \frac{1}{81}$

16. $a_1 = 6, r = -\frac{1}{4}$

$a_1 = 6$

$a_2 = 6\left(-\frac{1}{4}\right)^1 = -\frac{3}{2}$

$a_3 = 6\left(-\frac{1}{4}\right)^2 = \frac{3}{8}$

18. $a_1 = 2, r = \sqrt{3}$

$a_1 = 2$

$a_2 = 2\left(\sqrt{3}\right)^1 = 2\sqrt{3}$

$a_3 = 2\left(\sqrt{3}\right)^2 = 6$

$a_4 = 2\left(\sqrt{3}\right)^3 = 6\sqrt{3}$

$a_5 = 2\left(\sqrt{3}\right)^4 = 18$

20. $a_1 = 5, r = 2x$

$a_1 = 5$

$a_2 = 5(2x)^1 = 10x$

$a_3 = 5(2x)^2 = 20x^2$

$a_4 = 5(2x)^3 = 40x^3$

$a_5 = 5(2x)^4 = 80x^4$

22. $a_1 = 81, a_{k+1} = \frac{1}{3}a_k$

$a_1 = 81$

$a_2 = \frac{1}{3}(81) = 27$

$a_3 = \frac{1}{3}(27) = 9$

$a_4 = \frac{1}{3}(9) = 3$

$a_5 = \frac{1}{3}(3) = 1$

$a_n = 243\left(\frac{1}{3}\right)^n$

24. $a_1 = 5, a_{k+1} = -2a_k$

$a_1 = 5$

$a_2 = -2(5) = -10$

$a_3 = -2(-10) = 20$

$a_4 = -2(20) = -40$

$a_5 = -2(-40) = 80$

$a_n = 5(-2)^{n-1}$

26. $a_1 = 36, \quad a_{k+1} = -\frac{2}{3}a_k$

$a_1 = 36$

$a_2 = -\frac{2}{3}(36) = -24$

$a_3 = -\frac{2}{3}(-24) = 16$

$a_4 = -\frac{2}{3}(16) = -\frac{32}{3}$

$a_5 = -\frac{2}{3}\left(-\frac{32}{3}\right) = \frac{64}{9}$

28. $a_1 = 5, \quad r = \frac{3}{2}, \quad n = 8$

$a_n = a_1 r^{n-1}$

$a_8 = 5\left(\frac{3}{2}\right)^7 = \frac{10,935}{128}$

30. $a_1 = 8, \quad r = \sqrt{5}, \quad n = 9$

$a_n = a_1 r^{n-1}$

$a_9 = 8(\sqrt{5})^8 = 5000$

32. $a_1 = 1, \quad r = -\frac{x}{3}, \quad n = 7$

$a_n = a_1 r^{n-1}$

$a_7 = 1\left(-\frac{x}{3}\right)^6 = \frac{x^6}{729}$

34. $a_1 = 1000, \quad r = 1.005, \quad n = 60$

$a_n = a_1 r^{n-1}$

$a_6 = 1000(1.005)^{59}$

36. $a_2 = 3, \quad a_5 = \frac{3}{64}, \quad n = 1$

$a_2 r^3 = a_5$

$3r^3 = \frac{3}{64}$

$r^3 = \frac{1}{64}$

$r = \frac{1}{4}$

$a_2 = a_1 r$

$3 = a_1 \frac{1}{4}$

$a_1 = 12$

38. $a_3 = \frac{16}{3}, \quad a_5 = \frac{64}{27}, \quad n = 7$

$a_3 r^2 = a_5$

$\frac{16}{3} r^2 = \frac{64}{27}$

$r^2 = \frac{4}{9}$

$r = \pm\frac{2}{3}$

$a_7 = a_5 r^2 = \frac{64}{27}\left(\pm\frac{2}{3}\right)^2 = \frac{256}{243}$

40. $a_n = 18\left(-\frac{2}{3}\right)^{n-1}$

$r = \left(-\frac{2}{3}\right) > -1$, so that the sequence alternates as it approaches 0. Matches (c).

42. $a_n = 18\left(-\frac{3}{2}\right)^{n-1}$

$r = \left(-\frac{3}{2}\right) < -1$, so the sequence alternates as it approaches ∞. Matches (d).

44. $a_n = 12(-0.4)^{n-1}$

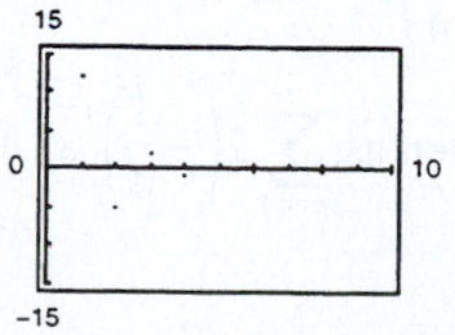

46. $a_n = 2(-1.4)^{n-1}$

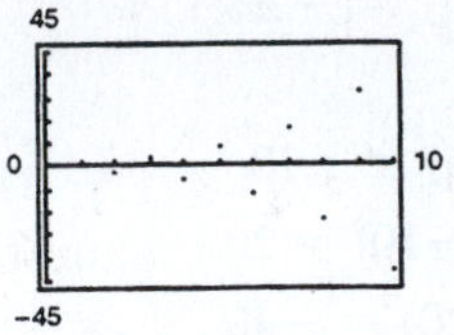

48. To use the first two terms of a geometric sequence to find the nth term, first divide the second term by the first term to obtain the common ratio. The nth term is the first term multiplied by the common ratio raised to the $(n - 1)$ power.

50. $A = P\left(1 + \frac{r}{n}\right)^{nt} = 2500\left(1 + \frac{0.12}{n}\right)^{n(20)}$

(a) $n = 1$, $A = 2500(1 + 0.12)^{20} \approx \$24{,}115.73$

(b) $n = 2$, $A = 2500\left(1 + \frac{0.12}{2}\right)^{2(20)} \approx \$25{,}714.29$

(c) $n = 4$, $A = 2500\left(1 + \frac{0.12}{4}\right)^{4(20)} \approx \$26{,}602.23$

(d) $n = 12$, $A = 2500\left(1 + \frac{0.12}{12}\right)^{12(20)} \approx \$27{,}231.38$

(e) $n = 365$, $A = 2500\left(1 + \frac{0.12}{365}\right)^{365(20)} \approx \$27{,}547.07$

52. P = population after n years

P_0 = initial population = 250,000

r = rate of increase = 1.3%

n = number of years = 30

$P = P_0(1 + r)^n = 250{,}000(1.013)^{30} \approx 368{,}318$

54. $8, 12, 18, 27, \frac{81}{2}, \ldots$

$S_1 = 8$

$S_2 = 8 + 12 = 20$

$S_3 = 8 + 12 + 18 = 38$

$S_4 = 8 + 12 + 18 + 27 = 65$

56. $\sum_{n=1}^{9} (-2)^{n-1} \Rightarrow a_1 = 1, r = -2$

$S_a = \frac{1(1 - (-2)^9)}{1 - (-2)} = 171$

58. $\sum_{i=1}^{6} 32\left(\frac{1}{4}\right)^{i-1} \Rightarrow a_1 = 32, r = \frac{1}{4}$

$S_6 = 32\frac{(1 - (1/4)^6)}{1 - (1/4)} = \frac{1365}{32}$

60. $\sum_{n=0}^{15} 2\left(\frac{4}{3}\right)^n = \sum_{n=1}^{16} 2\left(\frac{4}{3}\right)^{n-1} \Rightarrow a_1 = 2, r = \frac{4}{3}$

$S_{16} = 2\left(\frac{1 - (4/3)^{16}}{1 - (4/3)}\right) \approx 592.65$

62. $\sum_{i=1}^{10} 5\left(-\frac{1}{3}\right)^{i-1} \Rightarrow a_1 = 5, r = -\frac{1}{3}$

$S_{10} = 5\left(\frac{1 - (-1/3)^{10}}{1 - (-1/3)}\right) \approx 3.75$

64. $\sum_{n=0}^{6} 500(1.04)^n = \sum_{n=1}^{7} 500(1.04)^{n-1} \Rightarrow a_1 = 500, r = 1.04$

$S_7 = 500\left(\frac{1 - (1.04)^7}{1 - 1.04}\right) \approx 3949.15$

66. $2 - \frac{1}{2} + \frac{1}{8} - \cdots + \frac{1}{2048}$

$r = -\frac{1}{4}$ and $\frac{1}{2048} = 2\left(-\frac{1}{4}\right)^{n-1} \Rightarrow n = 7$

Thus, the sum can be written as $\sum_{n=1}^{7} 2\left(-\frac{1}{4}\right)^{n-1}$

68. $a = \sum_{n=1}^{60} 50\left(1 + \frac{0.12}{12}\right)^n$

$= 50(1.01) \cdot \frac{(1 - (1.01)^{60})}{(1 - 1.01)} \approx \$4124.32$

70. Let $N = 12t$ be the total number of deposits.

$$A = Pe^{r/12} + Pe^{2r/12} + \cdots + Pe^{Nr/12}$$

$$= \sum_{n=1}^{N} Pe^{r/12 \cdot n}$$

$$= Pe^{r/12}\frac{(1 - (e^{r/12})^N)}{(1 - e^{r/12})}$$

$$= Pe^{r/12}\frac{(1 - (e^{r/12})^{12t})}{1 - e^{r/12}}$$

$$= \frac{Pe^{r/12}(e^{rt} - 1)}{(e^{r/12} - 1)}$$

72. $P = \$75$, $r = 9\%$, $t = 25$ years

(a) Compounded monthly: $A = 75\left[\left(1 + \frac{0.09}{12}\right)^{12(25)} - 1\right]\left(1 + \frac{12}{0.09}\right) \approx \$84{,}714.78$

(b) Compounded continuously: $A = \frac{75e^{0.09/12}(e^{0.09(25)} - 1)}{e^{0.09/12} - 1} \approx \$85{,}196.05$

74. $P = \$20$, $r = 6\%$, $t = 50$ years

(a) Compounded monthly: $A = 20\left[\left(1 + \frac{0.06}{12}\right)^{12(50)} - 1\right]\left(1 + \frac{12}{0.06}\right) \approx \$76{,}122.54$

(b) Compounded continuously: $A = \frac{20e^{0.06/12}(e^{0.06(50)} - 1)}{e^{0.06/12} - 1} \approx \$76{,}533.16$

76. $W = \$2000$, $t = 20$, $r = 9\%$

$$P = W\left(\frac{12}{r}\right)\left[1 - \left(1 + \frac{r}{12}\right)^{-12t}\right]$$

$$P = 2000\left(\frac{12}{0.09}\right)\left[1 - \left(1 + \frac{0.09}{12}\right)^{-12(20)}\right] \approx \$222{,}289.91$$

78. $\sum_{n=5}^{14} 3.49e^{0.108n} = \sum_{n=6}^{15} 3.49(e^{0.108})^{n-1} = \sum_{n=1}^{15} 3.49(e^{0.108})^{n-1} - \sum_{n=1}^{5} 3.49(e^{0.108})^{n-1}$

$$= 3.49\left(\frac{1 - (e^{0.108})^{15}}{1 - e^{0.108}}\right) - 3.49\left(\frac{1 - (e^{0.108})^5}{1 - e^{0.108}}\right) \approx \$102.1 \text{ billion}$$

80. $a_n = 30{,}000(1.05)^{n-1}$

$$T = \sum_{n=1}^{40} 30{,}000(1.05)^{n-1} = 30{,}000\frac{(1 - 1.05^{40})}{(1 - 1.05)} \approx \$3{,}623{,}993.23$$

82. $a_1 = 2$, $r = \frac{2}{3}$

$$\sum_{n=0}^{\infty} 2\left(\frac{2}{3}\right)^n = \frac{a_1}{1 - r} = \frac{2}{1 - (2/3)} = 6$$

84. $a_1 = 2$, $r = -\frac{2}{3}$

$$\sum_{n=0}^{\infty} 2\left(-\frac{2}{3}\right)^n = \frac{a_1}{1 - r} = \frac{2}{1 - (-2/3)} = \frac{6}{5}$$

86. $a_1 = 1$, $r = \frac{1}{10}$

$$\sum_{n=0}^{\infty} \left(\frac{1}{10}\right)^n = \frac{a_1}{1 - r} = \frac{1}{1 - (1/10)} = \frac{10}{9}$$

88. $3 - 1 + \frac{1}{3} - \frac{1}{9} + \cdots = \sum_{n=0}^{\infty} 3\left(-\frac{1}{3}\right)^n = \frac{a_1}{1 - r} = \frac{3}{1 - (-1/3)} = \frac{9}{4}$

90. $0.\overline{297} = \sum_{n=0}^{\infty} 0.297(0.001)^n = \frac{0.297}{1 - 0.001} = \frac{0.297}{0.999} = \frac{297}{999} = \frac{11}{37}$

92. $1.3\overline{8} = 1.3 + \sum_{n=0}^{\infty} 0.08(0.1)^n = 1.3 + \frac{0.08}{1 - 0.1} = 1.3 + \frac{0.08}{0.9} = 1\frac{3}{10} + \frac{4}{45} = 1\frac{7}{18} = \frac{25}{18}$

94. $f(x) = 2\left[\frac{1 - (0.8)^x}{1 - (0.8)}\right], \sum_{n=0}^{\infty} 2\left(\frac{4}{5}\right)^n = \frac{2}{1 - (4/5)} = 10$

The horizontal asymptote of $f(x)$ is $y = 10$. This corresponds to the sum of the series.

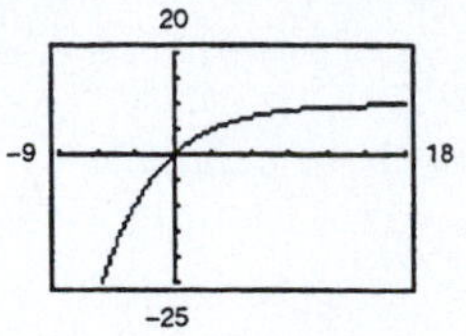

Section 9.4 Mathematical Induction

Solutions to Even-Numbered Exercises

2. $P_k = \frac{1}{(k + 1)(k + 3)}$

$P_{k+1} = \frac{1}{((k + 1) + 1)((k + 1) + 3)} = \frac{1}{(k + 2)(k + 4)}$

4. $P_k = \frac{k}{2}(3k - 1)$

$P_{k+1} = \frac{k + 1}{2}(3(k + 1) - 1) = \frac{k + 1}{2}(3k + 2)$

6. 1. When $n = 1$, $S_1 = 3 = 1(2 \cdot 1 + 1)$.

2. Assume that

$$S_k = 3 + 7 + 11 + 15 + \cdots + (4k - 1) = k(2k + 1).$$

Then,

$$\begin{aligned} S_{k+1} &= 3 + 7 + 11 + 15 + \cdots + (4k - 1) + [4(k + 1) - 1] \\ &= S_k + [4(k + 1) - 1] \\ &= k(2k + 1) + (4k + 3) \\ &= 2k^2 + 5k + 3 \\ &= (k + 1)(2k + 3) \\ &= (k + 1)[2(k + 1) + 1]. \end{aligned}$$

We conclude by mathematical induction that the formula is valid for all positive interger values of n.

8. 1. When $n = 1$,

$$S_1 = 1 = \frac{1}{2}(3 \cdot 1 - 1).$$

2. Assume that

$$S_k = 1 + 4 + 7 + 10 + \cdots + (3k - 2) = \frac{k}{2}(3k - 1).$$

Then,

$$\begin{aligned} S_{k+1} &= 1 + 4 + 7 + 10 + \cdots + (3k - 2) + (3(k + 1) - 2) \\ &= S_k + (3(k + 1) - 2) \\ &= \frac{k}{2}(3k - 1) + (3k + 1) \\ &= \frac{3k^2 - k + 6k + 2}{2} \\ &= \frac{3k^2 + 5k + 2}{2} \\ &= \frac{(k + 1)(3k + 2)}{2} \\ &= \frac{k + 1}{2}[3(k + 1) - 1]. \end{aligned}$$

Therefore, we conclude that this formula holds for all positive integer values of n.

10. 1. When $n = 1$, $S_1 = 2 = 3^1 - 1$.

2. Assume that

$$S_k = 2(1 + 3 + 3^2 + 3^3 + \cdots + 3^{k+1}) = 3^k - 1.$$

Then,

$$\begin{aligned} S_{k+1} &= 2(1 + 3 + 3^2 + 3^3 + \cdots + 3^{k-1}) + 2 \cdot 3^{k+1-1} \\ &= S_k + 2 \cdot 3^k \\ &= 3^k - 1 + 2 \cdot 3^k \\ &= 3 \cdot 3^k - 1 \\ &= 3^{k+1} - 1. \end{aligned}$$

Therefore, we conclude that this formula holds for all positive integer values of n.

12. 1. When $n = 1$,

$$S_1 = 1 = \frac{1(1 + 1)(2 \cdot 1 + 1)}{6}.$$

2. Assume that

$$S_k = 1^2 + 2^2 + 3^2 + 4^2 + \cdots + k^2 = \frac{k(k + 1)(2k + 1)}{6}.$$

Then,

$$\begin{aligned} S_{k+1} &= 1^2 + 2^2 + 3^2 + 4^2 + \cdots + k^2 + (k + 1)^2 \\ &= S_k + (k + 1)^2 \\ &= \frac{k(k + 1)(2k + 1)}{6} + (k + 1)^2 \\ &= \frac{k(k + 1)(2k + 1) + 6(k + 1)^2}{6} \\ &= \frac{(k + 1)[2k^2 + k + 6k + 6]}{6} \\ &= \frac{(k + 1)(k + 2)(2k + 3)}{6}. \end{aligned}$$

Therefore, we conclude that this formula holds for all positive integer values of n.

14. 1. When $n = 1$, $S_1 = 2 = 1 + 1$.

2. Assume that

$$S_k = \left(1 + \frac{1}{1}\right)\left(1 + \frac{1}{2}\right)\left(1 + \frac{1}{3}\right) \cdots \left(1 + \frac{1}{k}\right) = k + 1.$$

Then,

$$\begin{aligned} S_{k+1} &= \left(1 + \frac{1}{1}\right)\left(1 + \frac{1}{2}\right)\left(1 + \frac{1}{3}\right) \cdots \left(1 + \frac{1}{k}\right)\left(1 + \frac{1}{k + 1}\right) \\ &= (s_k)\left(1 + \frac{1}{k + 1}\right) \\ &= (k + 1)\left(1 + \frac{1}{k + 1}\right) \\ &= k + 1 + 1 \\ &= k + 2. \end{aligned}$$

Therefore, we conclude that this formula holds for all positive integer values of n.

16. 1. When $n = 1$,

$$S_1 = 1^4 = \frac{1(1+1)(2 \cdot 1 + 1)(3 \cdot 1^2 + 3 \cdot 1 - 1)}{30}.$$

2. Assume that

$$S_k = \sum_{i=1}^{k} i^4 = \frac{k(k+1)(2k+1)(3k^2+3k-1)}{30}.$$

Then,

$$\begin{aligned} S_{k+1} &= S_k + (k+1)^4 \\ &= \frac{k(k+1)(2k+1)(3k^2+3k-1)}{30} + (k+1)^4 \\ &= \frac{k(k+1)(2k+1)(3k^2+3k-1) + 30(k+1)^4}{30} \\ &= \frac{(k+1)[k(2k+1)(3k^2+3k-1) + 30(k+1)^3]}{30} \\ &= \frac{(k+1)(6k^4+39k^3+91k^2+89k+30)}{30} \\ &= \frac{(k+1)(k+2)(2k+3)(3k^2+9k+5)}{30} \\ &= \frac{(k+1)(k+2)(2(k+1)+1)(3(k+1)^2+3(k+1)-1)}{30}. \end{aligned}$$

Therefore, we conclude that this formula holds for all positive integer values of n.

18. 1. When $n = 1$,

$$S_1 = \frac{1}{2} = \frac{1}{2 \cdot 1 + 1}.$$

2. Assume that

$$S_k = \sum_{i=0}^{k} \frac{1}{(2i-1)(2i+1)} = \frac{k}{2k+1}.$$

Then,

$$\begin{aligned} S_{k+1} &= S_k + \frac{1}{(2(k+1)-1)(2(k+1)+1)} \\ &= \frac{k}{2k+1} + \frac{1}{(2k+1)(2k+3)} \\ &= \frac{k(2k+3)+1}{(2k+1)(2k+3)} \\ &= \frac{2k^2+3k+1}{(2k+1)(2k+3)} \\ &= \frac{(2k+1)(k+1)}{(2k+1)(2k+3)} \\ &= \frac{k+1}{2(k+1)+1}. \end{aligned}$$

Therefore, we conclude that this formula holds for all positive integer values of n.

20. $\sum_{n=1}^{50} n = \frac{50(50+1)}{2} = 1275$

22. $\sum_{n=1}^{10} n^3 = \frac{10^2(10+1)^2}{4} = 3025$

24. $\sum_{n=1}^{8} n^5 = \frac{8^2(8+1)^2(2(8)^2 + 2(8) - 1)}{12} = 61{,}776$

26. $\sum_{n=1}^{10} (n^3 - n^2) = \sum_{n=1}^{10} n^3 - \sum_{n=1}^{10} n^2 = \frac{10^2(10+1)^2}{4} - \frac{10(10+1)(2(10)+1)}{6}$

$= 3025 - 385 = 2640$

28. $\sum_{j=1}^{4}\left(2 + \frac{5}{2}j - \frac{3}{2}j^2\right) = \sum_{j=1}^{4} 2 + \frac{5}{2}\sum_{j=1}^{4} j - \frac{3}{2}\sum_{j=1}^{4} j^2$

$= 8 + \frac{5}{2}\left(\frac{4(4+1)}{2}\right) - \frac{3}{2}\left(\frac{4(4+1)(2(4)+1)}{6}\right)$

$= 8 + 25 - 45 = -12$

30. $25 + 22 + 19 + 16 + \cdots = \sum_{n=0}^{\infty}(25 - 3n)$

32. $3 - \frac{9}{2} + \frac{27}{4} - \frac{81}{8} + \cdots = \sum_{n=1}^{\infty} 3\left(-\frac{3}{2}\right)^{n-1}$

34. $\frac{1}{2 \cdot 3} + \frac{1}{3 \cdot 4} + \frac{1}{4 \cdot 5} + \frac{1}{5 \cdot 6} + \cdots + \frac{1}{(n+1)(n+2)} + \cdots = \sum_{n=1}^{\infty} \frac{1}{(n+1)(n+2)}$

36. 1. When $n = 7$, $\left(\frac{4}{3}\right)^7 \approx 7.4915 > 7$.

2. Assume that $\left(\frac{4}{3}\right)^k > k, k > 7$.

Then, $\left(\frac{4}{3}\right)^{k+1} = \left(\frac{4}{3}\right)^k\left(\frac{4}{3}\right) > k\left(\frac{4}{3}\right) = k + \frac{k}{3} > k + 1$ for $k > 7$.

Thus, $\left(\frac{4}{3}\right)^{k+1} > k + 1$.

Therefore, $\left(\frac{4}{3}\right)^n > n$.

38. 1. When $n = 1$, $\left(\frac{x}{y}\right)^2 < \left(\frac{x}{y}\right)$ and $(0 < x < y)$.

2. Assume that

$$\left(\frac{x}{y}\right)^{k+1} < \left(\frac{x}{y}\right)^k$$

$$\left(\frac{x}{y}\right)^{k+1} < \left(\frac{x}{y}\right)^k \Rightarrow \left(\frac{x}{y}\right)\left(\frac{x}{y}\right)^{k+1} < \left(\frac{x}{y}\right)\left(\frac{x}{y}\right)^k \Rightarrow \left(\frac{x}{y}\right)^{k+2} < \left(\frac{x}{y}\right)^{k+1}.$$

Therefore, $\left(\frac{x}{y}\right)^{n+1} < \left(\frac{x}{y}\right)^n$ for all integers $n \geq 1$.

40. 1. When $n = 1$, $\left(\frac{a}{b}\right)^1 = \frac{a^1}{b^1}$.

2. Assume that $\left(\frac{a}{b}\right)^k = \frac{a^k}{b^k}$.

Then, $\left(\frac{a}{b}\right)^{k+1} = \left(\frac{a}{b}\right)^k\left(\frac{a}{b}\right) = \frac{a^k}{b^k} \cdot \frac{a}{b} = \frac{a^{k+1}}{b^{k+1}}$.

Thus, $\left(\frac{a}{b}\right)^n = \frac{a^n}{b^n}$.

42. 1. When $n = 1$, $\ln x_1 = \ln x_1$.

2. Assume that

$$\ln(x_1 x_2 x_3 \dots x_k) = \ln x_1 + \ln x_2 + \ln x_3 + \cdots + \ln x_k.$$

Then,
$$\begin{aligned}\ln(x_1 x_2 x_3 \dots x_k x_{k+1}) &= \ln[(x_1 x_2 x_3 \dots x_k)x_{k+1}] \\ &= \ln(x_1 x_2 x_3 \dots x_k) + \ln x_{k+1} \\ &= \ln x_1 + \ln x_2 + \ln x_3 + \cdots + \ln x_k + \ln x_{k+1}.\end{aligned}$$

Thus, $\ln(x_1 x_2 x_3 \dots x_n) = \ln x_1 + \ln x_2 + \ln x_3 + \cdots \ln x_n$.

44. 1. When $n = 1$, $a + bi$ and $a - bi$ are complex conjugates by definition.

2. Assume that $(a + bi)^k$ and $(a - bi)^k$ are complex conjugates.

That is, if $(a + bi)^k = c + di$, then $(a - bi)^k = c - di$.

Then,
$$\begin{aligned}(a + bi)^{k+1} &= (a + bi)^k(a + bi) = (c + di)(a + bi) \\ &= (ac - bd) + i(bc + ad) \\ \text{and } (a - bi)^{k+1} &= (a - bi)^k(a - bi) = (c - di)(a - bi) \\ &= (ac - bd) - i(bc + ad).\end{aligned}$$

This implies that $(a + bi)^{k+1}$ and $(a - bi)^{k+1}$ are complex conjugates. Therefore, $(a + bi)^n$ and $(a - bi)^n$ are complex conjugates for $n \geq 1$.

46. 1. When $n = 1$, $\tan(x + \pi) = \frac{\tan x + \tan \pi}{1 - \tan x \tan \pi} = \frac{\tan x + 0}{1 - 0} = \tan x$.

2. Assume $\tan(x + k\pi) = \tan x$.

Then,
$$\begin{aligned}\tan(x + (k + 1)\pi) &= \tan[(x + \pi) + k\pi] \\ &= \frac{\tan(x + \pi) + \tan k\pi}{1 - \tan(x + \pi)\tan k\pi} \\ &= \frac{\tan x + \tan k\pi}{1 - \tan x \tan k\pi} \\ &= \frac{\tan x + 0}{1 - (\tan x)(0)} = \tan x.\end{aligned}$$

Thus, $\tan(x + n\pi) = \tan x$.

48. 1. When $n = 1$, $(2^{2(1)-1} + 3^{2(1)-1}) = 2 + 3 = 5$ and 5 is a factor.

2. Assume that 5 is a factor of $(2^{2k-1} + 3^{2k-1})$.

Then,
$$\begin{aligned}(2^{2(k+1)-1} + 3^{2(k+1)-1}) &= (2^{2k+2-1} + 3^{2k+2-1})\\ &= (2^{2k-1}2^2 + 3^{2k-1}3^2)\\ &= (4 \cdot 2^{2k-1} + 9 \cdot 3^{2k-1})\\ &= (2^{2k-1} + 3^{2k-1}) + (2^{2k-1} + 3^{2k-1})\\ &\quad + (2^{2k-1} + 3^{2k-1}) + (2^{2k-1} + 3^{2k-1}) + 5 \cdot 3^{2k-1}.\end{aligned}$$

Since 5 is a factor of each set of parenthesis and 5 is a factor of $5 \cdot 3^{2k-1}$, then 5 is a factor of the whole sum. Thus, 5 is a factor of $(2^{2n-1} + 3^{2n-1})$ for every positive integer n.

50. (a) If P_3 is true and P_k implies P_{k+1}, then P_n is true for integers $n \ge 3$.

(b) If $P_1, P_2, P_3, \ldots, P_{50}$ are all true, then P_n is true for integers $1 \le n \le 50$.

(c) If P_1, P_2, and P_3 are all true, but the truth of P_k does not imply that P_{k+1} is true, then you may only conclude that P_1, P_2, and P_3 are true.

(d) If P_2 is true and P_{2k} implies P_{2k+2}, then P_{2n} is true for any positive integer n.

52. $a_0 = 10, \quad a_n = 4a_{n-1}$

$a_0 = 10$

$a_1 = 4(10) = 40$

$a_2 = 4(40) = 160$

$a_3 = 4(160) = 640$

$a_4 = 4(640) = 2560$

54. $a_0 = 0, \quad a_1 = 2, \quad a_n = a_{n-1} + 2a_{n-2}$

$a_0 = 0$

$a_1 = 2$

$a_2 = 2 + 2(0) = 2$

$a_3 = 2 + 2(2) = 6$

$a_4 = 6 + 2(2) = 10$

56. $f(1) = 2, \quad a_n = n - a_{n-1}$

$a_1 = f(1) = 2$

$a_2 = n - a_1 = 2 - 2 = 0$

$a_3 = n - a_2 = 3 - 0 = 3$

$a_4 = n - a_3 = 4 - 3 = 1$

$a_5 = n - a_5 = 5 - 1 = 4$

a_n: 2 0 3 1 4

First differences: −2 3 −2 3

Second differences: 5 −5 5

Since neither the first differences nor the second differences are equal, the sequence does not have a linear or quadratic model.

58. $f(2) = -3, \quad a_n = -2a_{n-1}$

$a_2 = f(2) = -3 \Longrightarrow -3 = -2a_1$

$a_1 = \frac{3}{2}$

$a_3 = -2a_2 = -2(-3) = 6$

$a_4 = -2a_3 = -2(6) = -12$

$a_5 = -2a_4 = -2(-12) = 24$

a_n: $\frac{3}{2}$ −3 6 −12 24

First differences: $-\frac{9}{2}$ 9 −18 36

Second differences: $\frac{27}{2}$ −27 54

Since neither the first differences nor the second differences are equal, the sequence does not have a linear or quadratic model.

60. $a_0 = 2, \quad a_n = (a_{n-1})^2$

$a_0 = 2$

$a_1 = a_0^2 = 2^2 = 4$

$a_2 = a_1^2 = 4^2 = 16$

$a_3 = a_2^2 = 16^2 = 256$

$a_4 = a_3^2 = 256^2 = 65{,}536$

a_n: 2 4 16 256 65,536

First differences: 2 12 240 65,280

Second differences: 10 228 65,040

Since neither the first differences nor the second differences are equal, the sequence does not have a linear or quadratic model.

62. $f(1) = 0, \quad a_n = a_{n-1} + 2n$

$a_1 = 0$

$a_2 = a_1 + 2(2) = 0 + 4 = 4$

$a_3 = a_2 + 2(3) = 4 + 6 = 10$

$a_4 = a_3 + 2(4) = 10 + 8 = 18$

$a_5 = a_4 + 2(5) = 18 + 10 = 28$

a_n: 0 4 10 18 28

First differences: 4 6 8 10

Second differences: 2 2 2

Since the second differences are equal, the sequence has a quadratic model.

64. $a_0 = 0,\ a_n = a_{n-1} - 1$

$a_0 = 0$

$a_1 = a_0 - 1 = 0 - 1 = -1$

$a_2 = a_1 - 1 = -1 - 1 = -2$

$a_3 = a_2 - 1 = -2 - 1 = -3$

$a_4 = a_3 - 1 = -3 - 1 = -4$

a_n: 0 −1 −2 −3 −4

First differences: −1 −1 −1 −1

Second differences: 0 0 0

Since the first differences are equal, the sequence has a linear model.

66. $a_0 = 7,\ a_1 = 6,\ a_3 = 10$

Let $a_n = an^2 + bn + c$. Thus,

$$a_0 = a(0)^2 + b(0) + c = 7 \Rightarrow \qquad c = 7$$

$$a_1 = a(1)^2 + b(1) + c = 6 \Rightarrow a + b + c = 6$$

$$a + b = -1$$

$$a_3 = {}_a(3)^2 + b(3) + c = 10 \Rightarrow 9a + 3b + c = 10$$

$$9a + 3b = 3$$

$$3a + b = 1$$

By elimination: $-a - b = 1$

$3a + b = 1$

$2a = 2$

$a = 1 \Rightarrow b = -2$

Thus, $a_n = n^2 - 2n + 7$.

68. $a_0 = 3,\ a_2 = 0,\ a_6 = 36$

Let $a_n = an^2 + bn + c$. Thus,

$$a_0 = a(0)^2 + b(0) + c = 3 \Rightarrow \qquad c = 3$$

$$a_2 = a(2)^2 + b(2) + c = 0 \Rightarrow 4a + 2b + c = 0$$

$$4a + 2b = -3$$

$$a_6 = a(6)^2 + b(6) + c = 36 \Rightarrow 36a + 6b + c = 36$$

$$36a + 6b = 33$$

$$12a + 2b = 11$$

By elimination: $-4a - 2b = 3$

$12a + 2b = 11$

$8a = 14$

$a = \frac{7}{4} \Rightarrow b = -5$

Thus, $a_n = \frac{7}{4}n^2 - 5n + 3$.

70. $x - y^3 = 0 \Longrightarrow x = y^3$

$x - 2y^2 = 0$

$y^3 - 2y^2 = 0$

$y^2(y - 2) = 0 \Longrightarrow y = 0, 2$

When $y = 0$: $x = 0^3 = 0$.

When $y = 2$: $x = 2^3 = 8$.

Points of intersection: $(0, 0)$ and $(8, 2)$

72.
$$\begin{aligned} 2x + y - 2z &= 1 \\ x \quad\;\; - z &= 1 \\ 3x + 3y + z &= 12 \end{aligned}$$

$$A = \begin{bmatrix} 2 & 1 & -2 \\ 1 & 0 & -1 \\ 3 & 3 & 1 \end{bmatrix}, \quad A^{-1} = \frac{1}{4}\begin{bmatrix} -3 & 7 & 1 \\ 4 & -8 & 0 \\ -3 & 3 & 1 \end{bmatrix}$$

$$\begin{bmatrix} x \\ y \\ z \end{bmatrix} = \frac{1}{4}\begin{bmatrix} -3 & 7 & 1 \\ 4 & -8 & 0 \\ -3 & 3 & 1 \end{bmatrix}\begin{bmatrix} 1 \\ 1 \\ 12 \end{bmatrix} = \begin{bmatrix} 4 \\ -1 \\ 3 \end{bmatrix}$$

Thus, $x = 4$, $y = -1$, $z = 3$.

Answer: $(4, -1, 3)$

Section 9.5 The Binomial Theorem

Solutions to Even-Numbered Exercises

2. ${}_8C_6 = \dfrac{8!}{6!2!} = \dfrac{8 \cdot 7}{2 \cdot 1} = 28$

4. ${}_{20}C_{20} = \dfrac{20!}{20!0!} = 1$

6. ${}_{12}C_5 = \dfrac{12!}{5!7!} = \dfrac{12 \cdot 11 \cdot 10 \cdot 9 \cdot 8}{5 \cdot 4 \cdot 3 \cdot 2 \cdot 1} = 792$

8. ${}_{10}C_4 = \dfrac{10!}{6!4!} = \dfrac{10 \cdot 9 \cdot 8 \cdot 7}{4 \cdot 3 \cdot 2 \cdot 1} = 210$

10. ${}_{10}C_6 = \dfrac{10!}{6!4!} = 210$

12.

1
1 1
1 2 1
1 3 3 1
1 4 6 4 1
1 5 10 10 5 1
1 6 15 20 15 6 1
1 7 21 35 35 21 7 1
1 8 28 56 70 56 28 8 1

14.

1
1 1
1 2 1
1 3 3 1
1 4 6 4 1
1 5 10 10 5 1
1 6 15 (20) 15 6 1

${}_6C_3 = 20$, the 4th entry in the 7th row.

16.

1
1 1
1 2 1
1 3 3 1
1 4 6 4 1
1 5 10 10 5 1
1 6 15 20 15 6 1
1 7 21 35 35 21 7 1
1 8 28 56 70 56 28 (8) 1

${}_8C_7 = 8$, the 8th entry in the 9th row.

18. $(x + 1)^6 = {}_6C_0x^6 + {}_6C_1x^5(1) + {}_6C_2x^4(1)^2 + {}_6C_3x^3(1)^3 + {}_6C_4x^2(1)^4 + {}_6C_5x(1)^5 + {}_6C_6(1)^6$
$= x^6 + 6x^5 + 15x^4 + 20x^3 + 15x^2 + 6x + 1$

20. $(a + 3)^4 = {}_4C_0a^4 + {}_4C_1a^3(3) + {}_4C_2a^2(3)^2 + {}_4C_3a(3)^3 + {}_4C_4(3)^4$
$= a^4 + 12a^3 + 54a^2 + 108a + 81$

22. $(y - 2)^5 = {}_5C_0y^5 - {}_5C_1y^4(2) + {}_5C_2y^3(2)^2 - {}_5C_3y^2(2)^3 + {}_5C_4y(2)^4 - {}_5C_5(2)^5$
$= y^5 - 10y^4 + 40y^3 - 80y^2 + 80y - 32$

24. $(x + y)^6 = {}_6C_0x^6 + {}_6C_1x^5y + {}_6C_2x^4y^2 + {}_6C_3x^3y^3 + {}_6C_4x^2y^4 + {}_6C_5xy^5 + {}_6C_6y^6$
$= x^6 + 6x^5y + 15x^4y^2 + 20x^3y^3 + 15x^2y^4 + 6xy^5 + y^6$

26. $(x + 2y)^4 = {}_4C_0x^4 + {}_4C_1x^3(2y) + {}_4C_2x^2(2y)^2 + {}_4C_3x(2y)^3 + {}_4C_4(2y)^4$
$= x^4 + 4x^3(2y) + 6x^2(4y^2) + 4x(8y^3) + 16y^4$
$= x^4 + 8x^3y + 24x^2y^2 + 32xy^3 + 16y^4$

28. $(2x - y)^5 = {}_5C_0(2x)^5 - {}_5C_1(2x)^4y + {}_5C_2(2x)^3y^2 - {}_5C_3(2x)^2y^3 + {}_5C_4(2x)y^4 - {}_5C_5(2x)y^5$
$= 32x^5 - 5(16x^4)y + 10(8x^3)y^2 - 10(4x^2)y^3 + 5(2x)y^4 - y^5$
$= 32x^5 - 80x^4y + 80x^3y^2 - 40x^2y^3 + 10xy^4 - y^5$

30. $(5 - 3y)^3 = 5^3 - 3(5)^2 3y + 3(5)(3y)^2 - (3y)^3$
$= 125 - 225y + 135y^2 - 27y^3$

32. $(x^2 + y^2)^6 = {}_6C_0(x^2)^6 + {}_6C_1(x^2)^5(y^2) + {}_6C_2(x^2)^4(y^2)^2 + {}_6C_3(x^2)^3(y^2)^3 + {}_6C_4(x^2)^2(y^2)^4$
$+ {}_6C_5(x^2)(y^2)^5 + {}_6C_6(y^2)^6$
$= x^{12} + 6x^{10}y^2 + 15x^8y^4 + 20x^6y^6 + 15x^4y^8 + 6x^2y^{10} + y^{12}$

34. $\left(\frac{1}{x} + 2y\right)^6 = {}_6C_0\left(\frac{1}{x}\right)^6 + {}_6C_1\left(\frac{1}{x}\right)^5(2y) + {}_6C_2\left(\frac{1}{x}\right)^4(2y)^2 + {}_6C_3\left(\frac{1}{x}\right)^3(2y)^3 + {}_6C_4\left(\frac{1}{x}\right)^2(2y)^4 + {}_6C_5\left(\frac{1}{x}\right)(2y)^5 + {}_6C_6(2y)^6$
$= 1\left(\frac{1}{x}\right)^6 + 6(2)\left(\frac{1}{x}\right)^5 y + 15(4)\left(\frac{1}{x}\right)^4 y^2 + 20(8)\left(\frac{1}{x}\right)^3 y^3 + 15(16)\left(\frac{1}{x}\right)^2 + y^4 + 6(32)\left(\frac{1}{x}\right)y^5 + 1(64)y^6$
$= \frac{1}{x^6} + \frac{12y}{x^5} + \frac{60y^2}{x^4} + \frac{160y^3}{x^3} + \frac{240y^4}{x^2} + \frac{192y^5}{x} + 64y^6$

36. $3(x + 1)^5 - 4(x + 1)^3 = 3[{}_5C_0x^5 + {}_5C_1x^4(1) + {}_5C_2x^3(1)^2 + {}_5C_3x^2(1)^3 + {}_5C_4x(1)^4 + {}_5C_5(1)^5]$
$- 4[{}_3C_0x^3 + {}_3C_1x^2(1) + {}_3C_2x(1)^2 + {}_3C_3(1)^3]$
$= 3[(1)x^5 + 5x^4 + 10x^3 + 10x^2 + 5x + 1] - 4[(1)x^3 + 3x^2 + 3x + 1]$
$= 3x^5 + 15x^4 + 26x^3 + 18x^2 + 3x - 1$

38. 5th row of Pascal's Triangle: 1 5 10 5 1
$(x + 2y)^5 = (1)x^5 + 5x^4 2y + 10x^3(2y)^2 + 10x^2(2y)^3 + 5x(2y)^4 + (2y)^5$
$= x^5 + 10x^4y + 40x^3y^2 + 80x^2y^3 + 80xy^4 + 32y^5$

40. 5th row of Pascal's Triangle: 1 5 10 10 5 1
$(3y + 2)^5 = (3y)^5 + 5(3y)^4(2) + 10(3y)^3(2)^2 + 10(3y)^2(2)^3 + 5(3y)(2)^4 + (2)^5$
$= 243y^5 + 810y^4 + 1080y^3 + 720y^2 + 240y + 32$

42. The term involving x^8 in the expansion of $(x^2 + 3)^{12}$ is ${}_{12}C_8(x^2)^4(3)^8 = \frac{12!}{(12 - 8)!8!} \cdot 3^8x^8 = 3{,}247{,}695x^8$. The coefficient is 3,247,695.

44. The term involving x^2y^8 in the expansion of $(4x - y)^{10}$ is ${}_{10}C_8(4x)^2(-y)^8 = \frac{10!}{(10 - 8)!8!} \cdot 16x^2y^8 = 720x^2y^8$. The coefficient is 720.

46. The term involving x^6y^2 in the expansion of $(2x - 3y)^8$ is ${}_8C_2(2x)^6(-3y)^2 = \dfrac{8!}{(8-6)!2!}(64x^6)(9y^2) = 16{,}128x^6y^2$. The coefficient is 16,128.

48. The term involving z^6 in the expansion of $(z^2 - 1)^{12}$ is ${}_{12}C_9(z^2)^3(-1)^9 = \dfrac{12}{(12-9)!9!}z^6(-1) = -220z^6$. The coefficient is -220.

50. The expansions of $(x + y)^n$ and $(x - y)^n$ are almost the same except that the signs of the terms in the expansion of $(x - y)^n$ alternate from positive to negative.

52.
$$\begin{aligned}(2\sqrt{t} - 1)^3 &= (2\sqrt{t})^3 + 3(2\sqrt{t})^2(-1) + 3(2\sqrt{t})(-1) + (-1)^3\\ &= 8t^{3/2} - 12t + 6t^{1/2} - 1\end{aligned}$$

54.
$$\begin{aligned}(u^{3/5} + 2)^5 &= (u^{3/5})^5 + 5(u^{3/5})^4(2) + 10(u^{3/5})^3(2)^2 + 10(u^{3/5})^2(2)^3 + 5(u^{3/5})(2)^4 + 2^5\\ &= u^3 + 10u^{12/5} + 40u^{9/5} + 80u^{6/5} + 80u^{3/5} + 32\end{aligned}$$

56.
$$\begin{aligned}\frac{f(x+h) - f(x)}{h} &= \frac{(x+h)^4 - x^4}{h}\\ &= \frac{x^4 + 4x^3h + 6x^2h^2 + 4xh^3 + h^4 + x^4}{h}\\ &= \frac{h(4x^3 + 6x^2h + 4xh^2 + h^3)}{h}\\ &= 4x^3 + 6x^2h + 4xh^2 + h^3\end{aligned}$$

58.
$$\begin{aligned}\frac{f(x+h) - f(x)}{h} &= \frac{\dfrac{1}{x+h} - \dfrac{1}{x}}{h}\\ &= \frac{\dfrac{x - (x+h)}{x(x+h)}}{h}\\ &= \frac{\dfrac{-h}{x(x+h)}}{h}\\ &= -\frac{1}{x(x+h)}\end{aligned}$$

60.
$$\begin{aligned}(2 - i)^5 &= {}_5C_0 2^5 - {}_5C_1 2^4 i + {}_5C_2 2^3 i^2 - {}_5C_3 2^2 i^3 + {}_5C_4 2 i^4 - {}_5C_5 i^5\\ &= 32 - 80i - 80 + 40i + 10 - i\\ &= -38 - 41i\end{aligned}$$

62.
$$\begin{aligned}(5 + \sqrt{-9})^3 &= (5 + 3i)^3\\ &= 5^3 + 3 \cdot 5^2(3i) + 3 \cdot 5(3i)^2 + (3i)^3\\ &= 125 + 225i - 135 - 27i\\ &= -10 + 198i\end{aligned}$$

64.
$$\begin{aligned}(5 - \sqrt{3}i)^4 &= 5^4 - 4 \cdot 5^3(\sqrt{3}i) + 6 \cdot 5^2(\sqrt{3}i)^2 - 4 \cdot 5(\sqrt{3}i)^3 + (\sqrt{3}i)^4\\ &= 625 - 500\sqrt{3}i - 450 + 60\sqrt{3}i + 9\\ &= 184 - 440\sqrt{3}i\end{aligned}$$

66. ${}_{10}C_3\left(\frac{1}{4}\right)^3\left(\frac{3}{4}\right)^7 = 120\left(\frac{1}{64}\right)\left(\frac{2187}{16{,}384}\right) \approx 0.2503$

68. ${}_8C_4\left(\frac{1}{2}\right)^4\left(\frac{1}{2}\right)^4 = 70\left(\frac{1}{16}\right)\left(\frac{1}{16}\right) \approx 0.273$

70. $(2.005)^{10} = (2 + 0.005)^{10} = 2^{10} + 10(2)^9(0.005) + 45(2)^8(0.005)^2 + 120(2)^7(0.005)^3 + 210(2)^6(0.005)^4$

$+ 252(2)^5(0.005)^5 + 210(2)^4(0.005)^6 + 120(2)^3(0.005)^7 + 45(2)^8(0.005)^2$

$+ 10(2)(0.005)^9 + (0.005)^{10}$

$= 1024 + 25.6 + 0.288 + 0.00192 + 0.0000084 + \cdots$

≈ 1049.890

72. $(1.98)^9 = (2 - 0.02)^9 = 2^9 - 9(2)^8(0.02) + 36(2)^7(0.02)^2 - 84(2)^6(0.02)^3 + 126(2)^5(0.02)^4$

$- 126(2)^4(0.02)^5 + 84(2)^3(0.02)^6 - 36(2)^2(0.02)^7 + 9(2)(0.02)^8 - (0.02)^9$

$= 512 - 46.08 + 1.8432 - 0.043008 + 0.00064512$

$\approx 467{,}721$

74. $f(x) = -x^4 + 4x^2 - 1,\ g(x) = f(x - 3)$

$g(x) = f(x - 3)$

$= -(x - 3)^4 + 4(x - 3)^2 - 1$

$= -(x^4 + 4x^3(-3) + 6x^2(-3)^2 + 4x(-3)^3 + (-3)^4) + 4(x^2 - 6x + 9) - 1$

$= -x^4 + 12x^3 - 54x^2 + 108x - 81 + 4x^2 - 24x + 36 - 1$

$= -x^4 + 12x^3 - 50x^2 + 84x - 46$

The graph of g is the same as the graph of f shifted 3 units to the right.

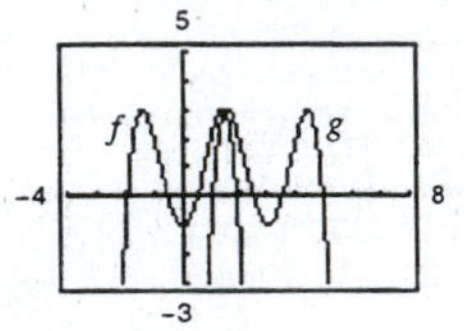

76. $0 = (1 - 1)^n = {}_nC_0 - {}_nC_1 + {}_nC_2 - {}_nC_3 + \cdots + (\pm {}_nC_n) = 0$

78. ${}_nC_0 + {}_nC_1 + {}_nC_2 + {}_nC_3 + \cdots + {}_nC_n = (1 + 1)^n - 2^n$

80. $f(t) = 0.1043t^2 + 0.7100t + 4.6852,\ 0 \le t \le 16$

(a) $g(t) = f(t + 4) \quad 0 \le t + 4 \le 16$

$= 0.1043(t + 4)^2 + 0.7100(t + 4) + 4.6852$

$= 0.1043(t^2 + 2t(4) + (4)^2) + 0.7100t + 2.8400 + 4.6852$

$= 0.1043(t^2 + 8t + 16) + 0.7100t + 7.5252$

$= 0.1043t^2 + 0.8344t + 1.6688 + 0.7100t + 7.5252$

$= 0.1043t^2 + 1.5444t + 9.1940, \quad -4 < t \le 12$

(b)

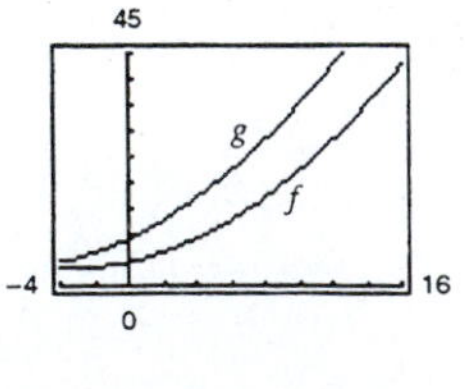

82. $g(x) = f(x) + 8$

$g(x)$ is shifted upward 8 units from $f(x)$.

84. $g(x) = f(-x)$

$g(x)$ is the reflection of $f(x)$ in the y-axis.

Section 9.6 Counting Principles

Solutions to Even-Numbered Exercises

2. Even integers: 2, 4, 6, 8, 10, 12

6 ways

4. Greater than 9: 10, 11, 12

3 ways

6. Divisible by 7: 7

1 way

8. Two *distinct* integers whose sum is 8:

$1 + 7, 2 + 6, 3 + 5$

3 ways

10. Monitors: 3

Keyboards: 2

Computers: 4

Total: $3 \cdot 2 \cdot 4 = 24$ ways

12. Math courses: 2

Science courses: 3

Social sciences and humanities courses: 5

Total: $2 \cdot 3 \cdot 5 = 30$ ways

14. $2^{10} = 1024$

16. 1st position: 4 choices

2nd position: 3 choices

3rd position: 2 choices

4th position: 1 choice

5th position: 6 choices

6th position: 5 choices

7th position: 4 choices

8th position: 3 choices

9th position: 2 choices

10th position: 1 choice

Total: $(4!)(6!) = 17{,}280$

18. $24 \cdot 24 \cdot 10 \cdot 10 \cdot 10 \cdot 10 = 5{,}760{,}000$

20. (a) $9 \cdot 10 \cdot 10 \cdot 10 = 9000$

(b) $9 \cdot 9 \cdot 8 \cdot 7 = 4536$

(c) $4 \cdot 10 \cdot 10 \cdot 10 = 4000$

(d) $9 \cdot 10 \cdot 10 \cdot 5 = 4500$

22. $50^3 = 125{,}000$

24. (a) $5! = 5 \cdot 4 \cdot 3 \cdot 2 \cdot 1 = 120$

(b) $(3!)(2!) = 3 \cdot 2 \cdot 1 \cdot 2 \cdot 1 = 12$

26. $_nP_r = \dfrac{n!}{(n-r)!}$

$$_5P_5 = \frac{5!}{(5-5)!} = \frac{5!}{0!} = 120$$

28. $_{20}P_2 = \dfrac{20!}{18!} = 20.19 = 380$

30. $_7P_4 = \dfrac{7!}{3!} = 7 \cdot 6 \cdot 5 \cdot 4 = 840$

32. $_nP_5 = 18 \cdot {}_{n-2}P_4$ Note: $n \geq 6$ for this to be defined.

$$\frac{n!}{(n-5)!} = 18\left(\frac{(n-2)!}{(n-6)!}\right)$$

$$n(n-1)(n-2)(n-3)(n-4) = 18(n-2)(n-3)(n-4)(n-5)$$

(We can divide by $(n-2)$, $(n-3)$, $(n-4)$ since $n \neq 2$, $n \neq 3$, and $n \neq 4$.)

$$n^2 - n = 18n - 90$$

$$n^2 - 19n + 90 = 0$$

$$(n-1)(n-10) = 0$$

$$n = 9 \text{ or } n = 10$$

34. ${}_{100}P_5 = 9{,}034{,}502{,}400$

36. ${}_{10}P_8 = 1{,}814{,}400$

38. ${}_{10}C_7 = 120$

40. The symbol ${}_nP_r$ means to choose and order r elements out of a collection of n elements.

42. A B C D
A C B D
D B C A
D C B A

44. $6! = 720$ ways

46. $4! = 24$ ways

48. $\dfrac{8!}{3!5!} = 56$

50. $\dfrac{11!}{1!4!4!2!} = \dfrac{11!}{4!4!2!} = 34{,}650$

52. ${}_6C_3 = \dfrac{6!}{3!3!} = 20$

ABC, ABD, ABE, ABF, ACD, ACE, ACF, ADE, ADF, AEF, BCD, BCE, BCF, BDE, BDF, BEF, CDE, CDF, CEF, DEF

54. ${}_{12}C_{10} = 66$ ways

56. ${}_{50}C_6 = 15{,}890{,}700$ ways

58. ${}_{80}C_5 = 24{,}040{,}016$ subsets

60. There are 9 good units and 3 defective units.

(a) ${}_9C_4 = 126$ ways

(b) ${}_9C_2 \cdot {}_3C_2 = 36 \cdot 3 = 108$ ways

(c) $$\begin{aligned} {}_9C_4 + {}_9C_3 \cdot {}_3C_1 + {}_9C_2 \cdot {}_3C_2 &= \frac{9!}{5!4!} + \frac{9!}{6!3!} \cdot \frac{3!}{2!1!} + \frac{9!}{7!2!} \cdot \frac{3!}{1!2!} \\ &= 126 + 84 \cdot 3 + 36 \cdot 3 \\ &= 486 \end{aligned}$$

62. Select type of card for three of a kind: ${}_{13}C_1$

Select three of four cards for three of a kind: ${}_4C_3$

Select type of card for pair: ${}_{12}C_1$

Select two of four cards for pair: ${}_4C_2$

${}_{13}C_1 \cdot {}_4C_3 \cdot {}_{12}C_1 \cdot {}_4C_2 = 13 \cdot 4 \cdot 12 \cdot 6 = 3744$ ways to get a full house

64. (a) ${}_3C_2 = \dfrac{3!}{2!1!} = 3$ relationships

(b) ${}_8C_2 = \dfrac{8!}{2!6!} = \dfrac{8 \cdot 7}{2} = 28$ relationships

(c) ${}_{12}C_2 = \dfrac{12!}{2!10!} = \dfrac{12 \cdot 11}{2} = 66$ relationships

(d) ${}_{20}C_2 = \dfrac{20!}{2!18!} = \dfrac{20 \cdot 19}{2} = 190$ relationships

66. ${}_6C_2 - 6 = 15 - 6 = 9$ diagonals

68. ${}_{10}C_2 - 10 = 45 - 10 = 35$ diagonals

70. ${}_nC_n = \dfrac{n!}{(n-n)!n!} = \dfrac{n!}{0!n!} = \dfrac{n!}{n!0!} = \dfrac{n!}{(n-0)!0!} = {}_nC_0$

72. $$\begin{aligned} {}_nC_r &= \frac{n!}{(n-r)!r!} \\ &= \frac{n(n-1)(n-2)\cdots(n-r+1)(n-r)!}{(n-r)!r!} \\ &= \frac{n(n-1)(n-2)\cdots(n-r+1)}{r!} \\ &= \frac{{}_nP_r}{r!} \end{aligned}$$

Section 9.7 Probability

Solutions to Even-Numbered Exercises

2. {2, 3, 4, 5, 6, 7, 8, 9, 10, 11, 12}

4. {(red, red), (red, blue), (red, black), (blue, blue), (blue, black)}

6. {SSS, SSF, SFS, FSS, SFF, FFS, FSF, FFF}

8. $E = \{\text{HHH, HHT, HTH, HTT}\}$

$$P(E) = \frac{n(E)}{n(s)} = \frac{4}{8} = \frac{1}{2}$$

10. $E = \{\text{HHH, HHT, HTH, THH}\}$

$$P(E) = \frac{n(E)}{n(s)} = \frac{4}{8} = \frac{1}{2}$$

12. The probability that the card is *not* a face card is the complement of getting a face card. (See Exercise 11.)

$$P(E') = 1 - P(E) = 1 - \frac{3}{13} = \frac{10}{13}$$

14. There are six possible cards in each of 4 suits: $6 \cdot 4 = 24$

$$P(E) = \frac{n(E)}{n(s)} = \frac{24}{52} = \frac{6}{13}$$

16. $E = \{(1, 6), (2, 5), (2, 6), (3, 4), (3, 5), (3, 6), (4, 3), (4, 4), (4, 5), (4, 6), (5, 2), (5, 3), (5, 4), (5, 5), (5, 6), (6, 1), (6, 2), (6, 3), (6, 4), (6, 5), (6, 6)\}$

$$P(E) = \frac{n(E)}{n(s)} = \frac{21}{36} = \frac{7}{12}$$

18. $E = \{(1, 1), (1, 2), (2, 1), (6, 6)\}$

$$P(E) = \frac{n(E)}{n(s)} = \frac{4}{36} = \frac{1}{9}$$

20. $E = \{(1, 1), (1, 2), (1, 4), (1, 6), (2, 1), (2, 3), (2, 5), (3, 2), (3, 4), (3, 6), (4, 1), (4, 3), (4, 5), (5, 2), (5, 4), (5, 6), (6, 1), (6, 3), (6, 5)\}$

$$P(E) = \frac{n(E)}{n(s)} = \frac{19}{36}$$

22. $$P(E) = \frac{{}_2C_2}{{}_6C_2} = \frac{1}{15}$$

24. $$P(E) = \frac{{}_1C_1 \cdot {}_2C_1 + {}_1C_1 \cdot {}_3C_1 + {}_2C_1 \cdot {}_3C_1}{{}_6C_2} = \frac{2 + 3 + 6}{15} = \frac{11}{15}$$

26. $1 - p = 1 - 0.36 = 0.64$

28. $1 - p = 1 - 0.84 = 0.16$

30. (a) $0.392(101) = 39.592$ million $= 39{,}592{,}000$

(b) 21.3%

(c) $21.3\% + 26.7\% = 48\%$

32. (a) $\frac{34}{100} = 0.34 = 34\%$

(b) $\frac{45}{100} = 0.45 = 45\%$

(c) $\frac{23}{100} = 0.23 = 23\%$

34. (a) $\frac{18 + 12}{72} = \frac{30}{72} = \frac{5}{12}$

(b) $1 - \frac{5}{12} = \frac{7}{12}$

(c) $\frac{10}{72} = \frac{5}{36}$

36. $1 - 0.37 - 0.44 = 0.19$

38. (a) $\frac{{}_6C_5}{{}_8C_5} = \frac{6}{56} = \frac{3}{28}$

(b) $\frac{{}_6C_4 \cdot {}_2C_1}{{}_8C_5} = \frac{15 \cdot 2}{56} = \frac{15}{28}$

(c) $\frac{3}{28} + \frac{15}{28} = \frac{18}{28} = \frac{9}{14}$

40. Total ways to insert paychecks: $5! = 120$ ways

5 correct: 1 way

4 correct: not possible

3 correct: 10 ways

2 correct: 20 ways

1 correct: 45 ways

0 correct: 44 ways

(a) $\dfrac{45}{120} = \dfrac{3}{8}$

(b) $\dfrac{45 + 20 + 10 + 1}{120} = \dfrac{19}{30}$

42. (a) $\dfrac{1}{{}_4P_4} = \dfrac{1}{24}$

(b) $\dfrac{1}{{}_3P_3} = \dfrac{1}{6}$

44. $\dfrac{{}_{13}C_1 \cdot {}_4C_3 \cdot {}_{12}C_1 \cdot {}_4C_2}{{}_{52}C_5} = \dfrac{13 \cdot 4 \cdot 12 \cdot 6}{2{,}598{,}960}$

$= \dfrac{3744}{2{,}598{,}960}$

$= \dfrac{6}{4165}$

46. (a) $\dfrac{{}_{16}C_5}{{}_{20}C_5} = \dfrac{4368}{15{,}504} = \dfrac{91}{323} \approx 0.282$ (5 good units)

(b) $\dfrac{{}_{16}C_4 \cdot {}_4C_1}{{}_{20}C_5} = \dfrac{1820 \cdot 4}{15{,}504} = \dfrac{455}{969} \approx 0.470$ (4 good units)

(c) Probability of at least one defective unit = 1 − (Probability of no defective units.)
Probability of no defective units = Probability of 5 good units

P (at least one defective unit) $= 1 - \dfrac{91}{323} = \dfrac{232}{323} \approx 0.718$

48. (a) $P(EE) = \frac{20}{40} \cdot \frac{20}{40} = \frac{1}{4}$

(b) $P(EO \text{ or } OE) = 2\left(\frac{20}{40}\right)\left(\frac{20}{40}\right) = \frac{1}{2}$

(c) $P(N_1 < 30, N_2 < 30) = \frac{29}{40} \cdot \frac{29}{40} = \frac{841}{1600}$

(d) $P(N_1N_1) = \frac{40}{40} \cdot \frac{1}{40} = \frac{1}{40}$

50. (a) $P(AA) = (0.90)^2 = 0.81$

(b) $P(NN) = (0.10)^2 = 0.01$

(c) $P(A) = 1 - P(NN) = 1 - 0.01 = 0.99$

52. (a) $P(BBBB) = \left(\frac{1}{2}\right)^4 = \frac{1}{16}$

(b) $P(BBBB) + P(GGGG) = \left(\frac{1}{2}\right)^4 + \left(\frac{1}{2}\right)^4 = \frac{1}{8}$

(c) $P(\text{at least one boy}) = 1 - P(\text{no boys})$

$= 1 - P(GGGG) = 1 - \frac{1}{16} = \frac{15}{16}$

54. $(0.78)^3 = 0.474552$

56. (a) If the *center* of the coin falls within the circle of radius $d/2$ around a vertex, the coin will cover the vertex.

$$P(\text{coin covers a vertex}) = \frac{\text{Area in which coin may fall so that it covers a vertex}}{\text{Total area}}$$

$$= \frac{n\left[\pi\left(\frac{d}{2}\right)^2\right]}{nd^2} = \frac{1}{4}\pi$$

(b) Experimental results will vary.

58. If a weather forecast indicates that the probability of rain is 40%, this means the meteorological records indicate that over an extended period of time with similar weather conditions it will rain 40% of the time.

60. $8.6\% + 11.4\% + 14.5\% + 9.2\% = 43.7\%$

❑ Focus on Concepts

1. Natural numbers

2. (a) Odd-numbered terms are negative.
(b) Even-numbered terms are negative.

3. True

4. True

5. True

6. True

7. (a) Each term is obtained by adding the same constant (common difference) to the previous term.
(b) Each term is obtained by multiplying the same constant (common ratio) by the previous term.

8. (a) Arithmetic. There is a constant difference between consecutive terms.
(b) Geometric. Each term is a constant multiple of the previous term. In this case the common ratio is greater than 1.

9. Each term of the sequence is defined in terms of the previous term.

10. Increased powers of real numbers between 0 and 1 approach zero.

11. d

12. a

13. b

14. c

15. The signs of the terms alternate in the expansion $(x - y)^n$.

16. They are the same because ${}_nC_r = \dfrac{n!}{(n-r)!r!} = \dfrac{n!}{r!(n-r)!} = {}_nC_{n-r}$.

17. ${}_{10}P_6 > {}_{10}C_6$. Changing the order of any of the six elements selected results in a different permutation but the same combination.

18. $0 \le p \le 1$, closed interval

19. $\frac{1}{3}$. The probability that an event does not occur is 1 minus the probability that it does occur.

20. Meteorological records indicate that over an extended period of time with similar weather conditions it will rain 60% of the time.

❑ Review Exercises for Chapter 9

Solutions to Even-Numbered Exercises

2. $a_n = \dfrac{5n}{2n - 1}$

$a_1 = \dfrac{5(1)}{2(1) - 1} = 5$

$a_2 = \dfrac{5(2)}{2(2) - 1} = \dfrac{10}{3}$

$a_3 = \dfrac{5(3)}{2(3) - 1} = 3$

$a_4 = \dfrac{5(4)}{2(4) - 1} = \dfrac{20}{7}$

$a_5 = \dfrac{5(5)}{2(5) - 1} = \dfrac{25}{9}$

4. $a_n = n(n - 1)$

$a_1 = 1(1 - 1) = 0$

$a_2 = 2(2 - 1) = 2$

$a_3 = 3(3 - 1) = 6$

$a_4 = 4(4 - 1) = 12$

$a_5 = 5(5 - 1) = 20$

6. $a_n = 4(0.4)^{n-1}$

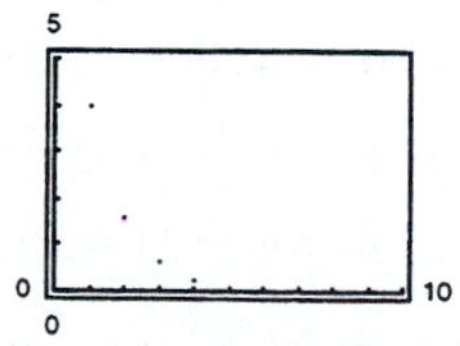

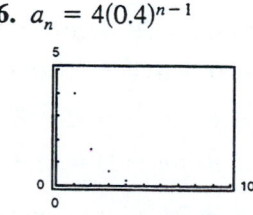

8. $a_n = 5 - \dfrac{3}{n}$

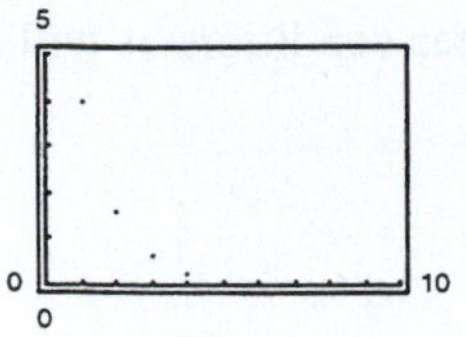

10. $2(1^2) + 2(2^2) + 2(3^2) + \cdots + 2(9^2) = \sum_{k=1}^{9} 2k^2$

12. $1 - \dfrac{1}{3} + \dfrac{1}{9} - \dfrac{1}{27} + \cdots = \sum_{k=0}^{\infty}\left(-\dfrac{1}{3}\right)^k$

14. $\sum_{k=2}^{5} 4k = 8 + 12 + 16 + 20 = 56$

16. $\sum_{i=1}^{8} \dfrac{i}{i+1} = \dfrac{1}{2} + \dfrac{2}{3} + \dfrac{3}{4} + \dfrac{4}{5} + \dfrac{5}{6} + \dfrac{6}{7} + \dfrac{7}{8} + \dfrac{8}{9} \approx 6.17$

18. $\sum_{j=0}^{4}(j^2 + 1) = \sum_{j=0}^{4} j^2 + \sum_{j=0}^{4} 1 = \dfrac{4(5)(9)}{6} + 5(1) = 35$

20. $\sum_{n=1}^{100}\left(\dfrac{1}{n} - \dfrac{1}{n+1}\right) = \left(\dfrac{1}{1} - \dfrac{1}{2}\right) + \left(\dfrac{1}{2} - \dfrac{1}{3}\right) + \left(\dfrac{1}{3} - \dfrac{1}{4}\right) + \cdots + \left(\dfrac{1}{99} - \dfrac{1}{100}\right) + \left(\dfrac{1}{100} - \dfrac{1}{101}\right)$

$= \dfrac{1}{1} - \dfrac{1}{101} = \dfrac{100}{101}$

22. $a_1 = 8, d = -2$

$a_1 = 8$

$a_2 = 8 - 2 = 6$

$a_3 = 6 - 2 = 4$

$a_4 = 4 - 2 = 2$

$a_5 = 2 - 2 = 0$

24. $a_2 = 14, a_6 = 22$

$a_6 = a_2 + 4d$

$22 = 14 + 4d$

$8 = 4d$

$2 = d$

$a_1 = a_2 - d$

$a_1 = 14 - 2 = 12$

$a_2 = 12 + 2 = 14$

$a_3 = 14 + 2 = 16$

$a_4 = 16 + 2 = 18$

$a_5 = 18 + 2 = 20$

26. $a_1 = 15, a_{k+1} = a_k + \frac{5}{2}$

$a_1 = 15$

$a_2 = 15 + \frac{5}{2} = \frac{35}{2}$

$a_3 = \frac{35}{2} + \frac{5}{2} = \frac{40}{2} = 20$

$a_4 = 20 + \frac{5}{2} = \frac{45}{2}$

$a_5 = \frac{45}{2} + \frac{5}{2} = \frac{50}{2} = 25$

$a_n = 15 + \frac{5}{2}(n - 1) = \frac{25}{2} + \frac{5}{2}n$

28. $a_1 = 100, a_{k+1} = a_k - 5$

$a_1 = 100$

$a_2 = 100 - 5 = 95$

$a_3 = 95 - 5 = 90$

$a_4 = 90 - 5 = 85$

$a_5 = 85 - 5 = 80$

$a_n = 100 - 5(n - 1) = 105 - 5n$

30. $a_3 = a_1 + 2d$

$28 = 10 + 2d$

$18 = 2d$

$9 = d$

$a_n = 10 + (n - 1)9 = 1 + 9n$

$\sum_{n=1}^{20}(1 + 9n) = \sum_{n=1}^{20} 1 + 9\sum_{n=1}^{20} n = 20(1) + 9\left[\dfrac{(20)(21)}{2}\right] = 1910$

32. $\sum_{j=1}^{8}(20 - 3j) = \sum_{j=1}^{8}20 - 3\sum_{j=1}^{8}j = 8(20) - 3\left[\frac{(8)(9)}{2}\right] = 52$

34. $\sum_{k=1}^{25}\left(\frac{3k+1}{4}\right) = \frac{3}{4}\sum_{k=1}^{25}k + \sum_{k=1}^{25}\frac{1}{4} = \frac{3}{4}\left[\frac{(25)(26)}{2}\right] + 25\left(\frac{1}{4}\right) = 250$

36. $\sum_{n=20}^{80}n = \sum_{n=1}^{80}n - \sum_{n=1}^{19}n = \frac{(80)(81)}{2} - \frac{(19)(20)}{2} = 3050$

38. $a_1 = 123, d = 112 - 123 = -11$

$n = 8$

$a_8 = 213 + 7(-11) = 46$

$S_8 = \frac{8}{2}(123 + 46) = 676$

40. $a_1 = 2, r = 2$

$a_1 = 2$

$a_2 = 2(2) = 4$

$a_3 = 4(2) = 8$

$a_4 = 8(2) = 16$

$a_5 = 16(2) = 32$

42. $a_1 = 2, a_3 = 12$

$a_3 = a_1r^2$

$12 = 2r^2$

$6 = r^2$

$\pm\sqrt{6} = r$

$a_1 = 2$		$a_1 = 2$
$a_2 = 2(\sqrt{6}) = 2\sqrt{6}$		$a_2 = 2(-\sqrt{6}) = -2\sqrt{6}$
$a_3 = 2\sqrt{6}(\sqrt{6}) = 12$	or	$a_3 = -2\sqrt{6}(-\sqrt{6}) = 12$
$a_4 = 12(\sqrt{6}) = 12\sqrt{6}$		$a_4 = 12(-\sqrt{6}) = -12\sqrt{6}$
$a_5 = 12\sqrt{6}(6) = 72$		$a_5 = -12\sqrt{6}(-\sqrt{6}) = 72$

44. $a_1 = 200, a_{k+1} = 0.1a_k$

$a_1 = 200$

$a_2 = 0.1(200) = 20$

$a_3 = 0.1(20) = 2$

$a_4 = 0.1(2) = 0.2$

$a_5 = 0.1(0.2) = 0.02$

$a_n = 200(0.1)^{n-1}$

46. $a_1 = 18, a_{k+1} = \frac{5}{3}a_k$

$a_1 = 18$

$a_2 = \frac{5}{3}(18) = 30$

$a_3 = \frac{5}{3}(30) = 50$

$a_4 = \frac{5}{3}(50) = \frac{250}{3}$

$a_5 = \frac{5}{3}\left(\frac{250}{3}\right) = \frac{1250}{9}$

$a_n = 18\left(\frac{5}{3}\right)^{n-1}$

48. $a_1 = 100, r = 1.05$

$a_n = 100(1.05)^{n-1}$

$\sum_{n=1}^{20}100(1.05)^{n-1} = 100\left[\frac{1 - (1.05)^{20}}{1 - 1.05}\right] \approx 3306.60$

50. $\sum_{i=1}^{5}3^{i-1} = \frac{1 - 3^5}{1 - 3} = 121$

52. $\sum_{i=1}^{\infty}\left(\frac{1}{3}\right)^{i-1} = \frac{1}{1 - (1/3)} = \frac{3}{2}$

54. $\sum_{k=1}^{\infty}1.3\left(\frac{1}{10}\right)^{k-1} = \frac{1.3}{1 - (1/10)} = \frac{13}{9}$

56. $\sum_{i=1}^{25}100(1.06)^{i-1} = 100\left[\frac{1 - (1.06)^{25}}{1 - 1.06}\right] \approx 5486.45$

58. $\sum_{i=1}^{40}32{,}000(1.055)^{i-1} = 32{,}000\left[\frac{1 - (1.055)^{40}}{1 - 1.055}\right]$

$= \$4{,}371{,}379.65$

60. $A = \sum_{i=1}^{120} 100\left(1 + \frac{0.065}{12}\right)^i = 100\left(1 + \frac{0.065}{12}\right)\left(\frac{1 - \left(1 + \frac{0.065}{12}\right)^{120}}{-\frac{0.065}{12}}\right) \approx \$16{,}931.53$

62. 1. When $n = 1$, $1 = \frac{1}{4}(1 + 3) = 1$.

2. Assume that $1 + \frac{3}{2} + 2 + \frac{5}{2} + \cdots + \frac{1}{2}(k + 1) = \frac{k}{4}(k + 3)$. Then,

$$\begin{aligned} 1 + \frac{3}{2} + 2 + \frac{5}{2} + \cdots + \frac{1}{2}(k + 1) + \frac{1}{2}(k + 2) &= \frac{k}{4}(k + 3) + \frac{1}{2}(k + 2) \\ &= \frac{k(k + 3) + 2(k + 2)}{4} \\ &= \frac{k^2 + 5k + 4}{4} \\ &= \frac{(k + 1)(k + 4)}{4} \\ &= \frac{k + 1}{4}[(k + 1) + 3]. \end{aligned}$$

Thus, the formula holds for all positive integers n.

64. 1. When $n = 1$, $a + 0 \cdot d = a = \frac{1}{2}[2a + (1 - 1)d] = a$.

2. Assume that $\sum_{k=0}^{i-1}(a + kd) = \frac{i}{2}[2a + (i - 1)d]$. Then,

$$\begin{aligned} \sum_{k=0}^{i+1-1}(a + kd) &= \frac{i}{2}[2a + (i - 1)d] + [a + id] \\ &= \frac{2ia + i(i - 1)d + 2a + 2id}{2} = \frac{2a(i + 1) + id(i + 1)}{2} = \left(\frac{i + 1}{2}\right)[2a + id]. \end{aligned}$$

Thus, the formula holds for all positive integers n.

66. ${}_{10}C_7 = \frac{10!}{3!7!} = 120$

68. ${}_{12}P_3 = \frac{12!}{9!} = 1320$

70. $(a - 3b)^5 = a^5 - 5a^4(3b) + 10a^3(3b)^2 - 10a^2(3b)^3 + 5a(3b)^4 - (3b)^5$
$= a^5 - 15a^4b + 90a^3b^2 - 270a^2b^3 + 405ab^4 - 243b^5$

72. $(3x + y^2)^7 = (3x)^7 + 7(3x)^6y^2 + 21(3x)^5(y^2)^2 + 35(3x)^4(y^2)^3 + 35(3x)^3(y^2)^4 + 21(3x)^2(y^2)^5 + 7(3x)(y^2)^6 + (y^2)^7$
$= 2187x^7 + 5103x^6y^2 + 5103x^5y^4 + 2835x^4y^6 + 945x^3y^8 + 189x^2y^{10} + 21xy^{12} + y^{14}$

74. $(4 - 5i)^3 = 4^3 - 3(4)^2(5i) + 3(4)(5i)^2 - (5i)^3$
$= 64 - 240i - 300 + 125i$
$= -236 - 115i$

76. $2^3 = 8$

78. $P(E) = \frac{n(E)}{n(S)} = \frac{1}{5!} = \frac{1}{120}$

80. $\left(\frac{6}{6}\right)\left(\frac{5}{6}\right)\left(\frac{4}{6}\right)\left(\frac{3}{6}\right)\left(\frac{2}{6}\right)\left(\frac{1}{6}\right) = \frac{6!}{6^6} = \frac{720}{46{,}656} = \frac{5}{324}$

82. $(0.8)^3 = 0.512$

84. (a) $\frac{208}{500} = 0.416$ or 41.6%
(b) $\frac{400}{500} = 0.8$ or 80%
(c) $\frac{37}{500} = 0.074$ or 7.4%

CHAPTER 10
Topics in Analytic Geometry

CHAPTER 10
Topics in Analytic Geometry

Section 10.1 Lines

Solutions to Even-Numbered Exercises

2. $m = \tan 45° = 1$

4. $m = \tan 124° \approx -1.4826$

6. $m = \tan 75.4° \approx 3.8391$

8. $m = \tan 145.5° \approx -0.6873$

10. $2 = \tan \theta$

$\theta = \tan^{-1} 2 \approx 63.4°$

12. $-\frac{5}{2} = \tan \theta$

$\theta = 180° + \tan^{-1}(-\frac{5}{2})$

$\approx 111.8°$

14. $m = \dfrac{-4 - 12}{-1 - 7} = 2$

$2 = \tan \theta$

$\theta = \tan^{-1} 2 \approx 63.4°$

16. $m = \dfrac{100 - 0}{0 - 50} = -2$

$-2 = \tan \theta$

$\theta = 180° + \tan^{-1}(-2) \approx 116.6°$

18. $4x + 5y - 9 = 0$

$y = -\frac{4}{5}x + \frac{9}{5} \Rightarrow m = -\frac{4}{5}$

$-\frac{4}{5} = \tan \theta$

$\theta = 180° + \tan^{-1}(-\frac{4}{5}) \approx 141.3°$

20. $x - y - 10 = 0$

$y = x - 10 \Rightarrow m = 1$

$1 = \tan \theta$

$\theta = \tan^{-1} 1 = 45°$

22. Slope: $m = \tan 4.5° \approx 0.0787$

Change in elevation: $\dfrac{x}{3(5280)} = \sin 4.5°$

$x = 15{,}840 \sin 4.5°$

$x \approx 1243$ feet

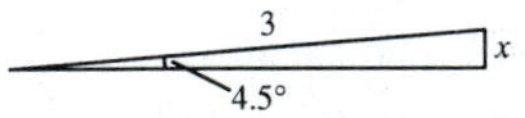

24. $\tan \theta = \dfrac{3}{5}$

$\theta = 31.0°$

26. $x + 3y = 2 \Rightarrow y = -\dfrac{1}{3}x + \dfrac{2}{3} \Rightarrow m_1 = -\dfrac{1}{3}$

$x - 2y = -3 \Rightarrow y = \dfrac{1}{2}x + \dfrac{3}{2} \Rightarrow m_2 = \dfrac{1}{2}$

$\tan \theta = \left|\dfrac{(1/2) - (-1/3)}{1 + (-1/3)(1/2)}\right| = 1$

$\theta = \tan^{-1} 1 = 45°$

28. $2x - y = 2 \Rightarrow y = 2x - 2 \Rightarrow m_1 = 2$

$4x + 3y = 24 \Rightarrow y = -\dfrac{4}{3}x + 8 \Rightarrow m_2 = -\dfrac{4}{3}$

$\tan \theta = \left|\dfrac{(-4/3) - 2}{1 + (2)(-4/3)}\right| = 2$

$\theta = \tan^{-1} 2 \approx 63.4°$

30. $5x + 3y = 18 \Rightarrow y = -\dfrac{5}{3}x + 6 \Rightarrow m_1 = -\dfrac{5}{3}$

$2x - 6y = -1 \Rightarrow y = \dfrac{1}{3}x + \dfrac{1}{6} \Rightarrow m_2 = \dfrac{1}{3}$

$\tan \theta = \left|\dfrac{(1/3) - (-5/3)}{1 + (-5/3)(1/3)}\right| = \dfrac{9}{2}$

$\theta = \tan^{-1}\left(\dfrac{9}{2}\right) \approx 77.5°$

32. $3x - 5y = 2 \Rightarrow y = \dfrac{3}{5}x - \dfrac{2}{5} \Rightarrow m_1 = \dfrac{3}{5}$

$2x + 5y = 13 \Rightarrow y = -\dfrac{2}{5}x + \dfrac{13}{5} \Rightarrow m_2 = -\dfrac{2}{5}$

$\tan \theta = \left|\dfrac{(-2/5) - (3/5)}{1 + (3/5)(-2/5)}\right| = \dfrac{25}{19}$

$\theta = \tan^{-1}\left(\dfrac{25}{19}\right) \approx 52.8°$

34. $0.02x - 0.05y = -0.19 \Rightarrow y = \frac{2}{5}x + \frac{19}{5} \Rightarrow m_1 = \frac{2}{5}$

$0.03x + 0.04y = 0.52 \Rightarrow y = -\frac{3}{4}x + 13 \Rightarrow m_2 = -\frac{3}{4}$

$$\tan\theta = \left|\frac{(-3/4) - (2/5)}{1 + (2/5)(-3/4)}\right| \approx \frac{23}{14}$$

$$\theta = \tan^{-1}\left(\frac{23}{14}\right) \approx 58.7°$$

36. Let $A = (-3, 2)$, $B = (1, 3)$, and $C = (2, 0)$.

Slope of AB: $m_1 = \frac{2 - 3}{-3 - 1} = \frac{1}{4}$

Slope of BC: $m_2 = \frac{3 - 0}{1 - 2} = -3$

Slope of AC: $m_3 = \frac{2 - 0}{-3 - 2} = -\frac{2}{5}$

$$\tan A = \left|\frac{(1/4) - (-2/5)}{1 + (-2/5)(1/4)}\right| = \frac{13/20}{18/20} = \frac{13}{18}$$

$$A = \tan^{-1}\left(\frac{13}{18}\right) \approx 35.8°$$

$$\tan C = \left|\frac{-3 - (-2/5)}{1 + (-3)(-2/5)}\right| \approx \frac{13/5}{11/5} = \frac{13}{11}$$

$$C = \tan^{-1}\left(\frac{13}{11}\right) \approx 49.8°$$

$$B = 180 - A - C \approx 180 - 35.8 - 49.8 = 94.4°$$

38. Let $A = (-3, 3)$, $B = (-1, 2)$, and $C = (3, 1)$.

Slope of AB: $m_1 = \frac{3 - 2}{-3 + 1} = -\frac{1}{2}$

Slope of BC: $m_2 = \frac{2 - 1}{-1 - 3} = -\frac{1}{4}$

Slope of AC: $m_3 = \frac{3 - 1}{-3 - 3} = -\frac{1}{3}$

$$\tan A = \left|\frac{(-1/3) - (-1/2)}{1 + (-1/2)(-1/3)}\right| = \frac{1/6}{7/6} = \frac{1}{7}$$

$$A = \tan^{-1}\left(\frac{1}{7}\right) \approx 8.1°$$

$$\tan C = \left|\frac{(-1/3) - (-1/4)}{1 + (-1/4)(-1/3)}\right| = \frac{1/12}{13/12} = \frac{1}{13}$$

$$C = \tan^{-1}\left(\frac{1}{13}\right) \approx 4.4°$$

$$B = 180 - A - C \approx 180 - 8.1 - 4.4 = 167.5°$$

40. $\tan\theta = \frac{6}{9}$

$\theta \approx 33.69°$

$B = 180° - 33.69° - 90° \approx 56.31°$

$\frac{6}{h} = \sin 56.31°$

$h = \frac{6}{\sin 56.31°}$

$\frac{6}{d} = \sin 33.69°$

$d = \frac{6}{\sin 33.69°}$

$\tan\alpha = \frac{h}{d} = \frac{6/\sin 56.31°}{6/\sin 33.69°}$

$\alpha \approx 33.69°$

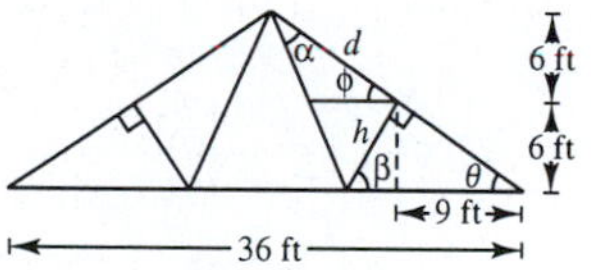

42. $(0, 0) \Rightarrow x_1 = 0$ and $y_1 = 0$

$2x - y = 4 \Rightarrow A = 2, B = -1$, and $C = -4$

$$d = \frac{|2(0) + (-1)(0) + (-4)|}{\sqrt{2^2 + (-1)^2}} = \frac{4}{\sqrt{5}} = \frac{4\sqrt{5}}{5} \approx 1.7889$$

44. $(-2, 1) \Rightarrow x_1 = -2$ and $y_1 = 1$

$x - y - 2 = 0 \Rightarrow A = 1, B = -1$, and $C = -2$

$$d = \frac{|1(-2) + (-1)(1) + (-2)|}{\sqrt{1^2 + (-1)^2}} = \frac{5}{\sqrt{2}} = \frac{5\sqrt{2}}{2} \approx 3.5355$$

46. $(10, 8) \Rightarrow x_1 = 10$ and $y_1 = 8$

$y - 4 = 0 \Rightarrow A = 0, B = 1$, and $C = -4$

$$d = \frac{|0(10) + 1(8) + (-4)|}{\sqrt{0^2 + 1^2}} = \frac{4}{1} = 4$$

48. $(4, 2) \Rightarrow x_1 = 4$ and $y_1 = 2$

$x - y - 20 = 0 \Rightarrow A = 1, B = -1$, and $C = -20$

$$d = \frac{|1(4) + (-1)(2) + (-20)|}{\sqrt{1^2 + (-1)^2}}$$

$$= \frac{18}{\sqrt{2}} = 9\sqrt{2} \approx 12.7279$$

50. (a) The slope of the line through AC is $m = \dfrac{0 + 2}{0 - 5} = -\dfrac{2}{5}$.

The equation of the line is $y - 0 = -\dfrac{2}{5}(x - 0) \Rightarrow 2x + 5y = 0$.

The distance between the line and $B = (4, 5)$ is $d = \dfrac{|2(4) + 5(5) + 0|}{\sqrt{2^2 + 5^2}} = \dfrac{33}{\sqrt{29}} = \dfrac{33\sqrt{29}}{29}$.

(b) The distance between A and C is $d = \sqrt{(0 - 5)^2 + (0 + 2)^2} = \sqrt{29}$.

$$A = \frac{1}{2}\sqrt{29}\left(\frac{33\sqrt{29}}{29}\right) = \frac{33}{2} \text{ square units}$$

52. (a) The slope of the line through AC is $m = \dfrac{10 + 5}{6 + 4} = \dfrac{3}{2}$.

The equation of the line through AC is $y - 10 = \dfrac{3}{2}(x - 6) \Rightarrow 3x - 2y + 2 = 0$.

The distance between the line and $B = (3, 10)$ is $d = \dfrac{|3(3) + (-2)(10) + 2|}{\sqrt{3^2 + (-2)^2}} = \dfrac{9}{\sqrt{13}} = \dfrac{9\sqrt{13}}{13}$.

(b) The distance between A and C is $d = \sqrt{(-4 - 6)^2 + (-5 - 10)^2} = 5\sqrt{13}$.

$$A = \frac{1}{2}(5\sqrt{13})\left(\frac{9\sqrt{13}}{13}\right) = \frac{45}{2} \text{ square units}$$

54. $3x - 4y = 1$

$3x - 4y = 10$

A point on $3x - 4y = 10$ is $\left(0, -\frac{5}{2}\right)$. The distance between $\left(0, -\frac{5}{2}\right)$ and $3x - 4y = 1$ is:

$A = 3, B = -4, C = -1, x_1 = 0, y_1 = -\dfrac{5}{2}$

$$d = \frac{|3(0) + (-4)(-5/2) - 1|}{\sqrt{3^2 + (-4)^2}} = \frac{9}{5}$$

56. Slope m and y-intercept $(0, 4)$

(a) $(x_1, y_1) = (3, 1)$ and line: $y = mx + 4$

$A = -m, B = 1, C = -4$

$$d = \frac{|(-m)(3) + (1)(1) + (-4)|}{\sqrt{(-m)^2 + 1^2}} = \frac{3|m + 1|}{\sqrt{m^2 + 1}}$$

(b)

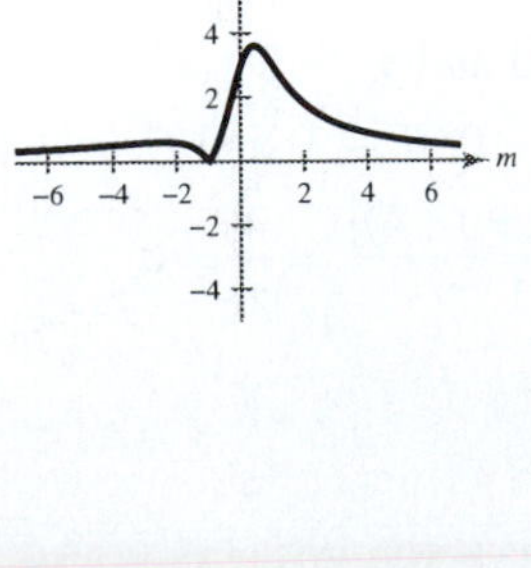

(c) From the graph it appears that the maximum distance is obtained when $m = 1$.

(d) From the graph it appears that the distance is 0 when $m = -1$.

(e) The asymptote of the graph in part (b) is $d = 3$. As the line approaches the vertical, the distance approaches 3.

Section 10.2 Introduction to Conics: Parabolas

Solutions to Even-Numbered Exercises

2. $x^2 = 2y$

Vertex: $(0, 0)$

$p = \frac{1}{2} > 0$

Opens upward.

Matches graph (b).

4. $y^2 = 12x$

Vertex: $(0, 0)$

$p = 3 > 0$

Opens to the right.

Matches graph (f).

6. $(x + 3)^2 = -2(y - 1)$

Vertex: $(-3, 1)$

$p = -\frac{1}{2} < 0$

Opens downward.

Matches graph (c).

8. $y = 2x^2 \Rightarrow x^2 = 4\left(\frac{1}{8}\right)y$

Vertex: $(0, 0)$

Focus: $\left(0, \frac{1}{8}\right)$

Directrix: $y = -\frac{1}{8}$

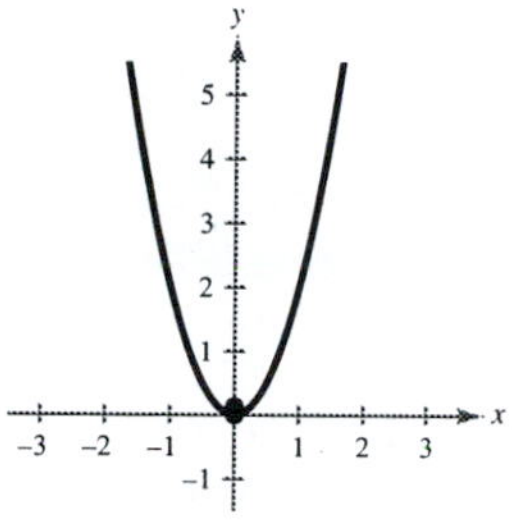

10. $y^2 = 3x \Rightarrow 4\left(\frac{3}{4}\right)x$

Vertex: $(0, 0)$

Focus: $\left(\frac{3}{4}, 0\right)$

Directrix: $x = -\frac{3}{4}$

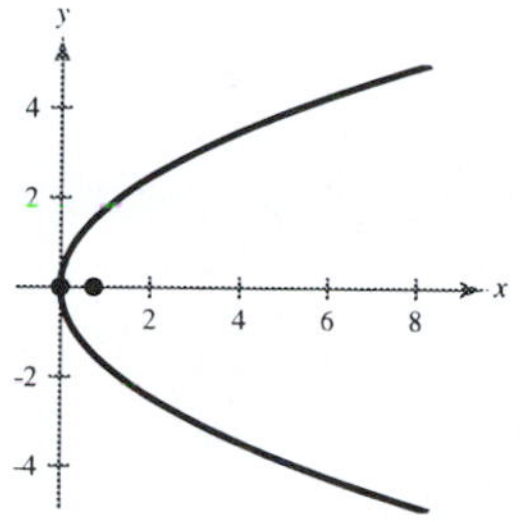

12. $x + y^2 = 0$

$$y^2 = -x = 4\left(-\frac{1}{4}\right)x$$

Vertex: $(0, 0)$

Focus: $\left(-\frac{1}{4}, 0\right)$

Directrix: $x = \frac{1}{4}$

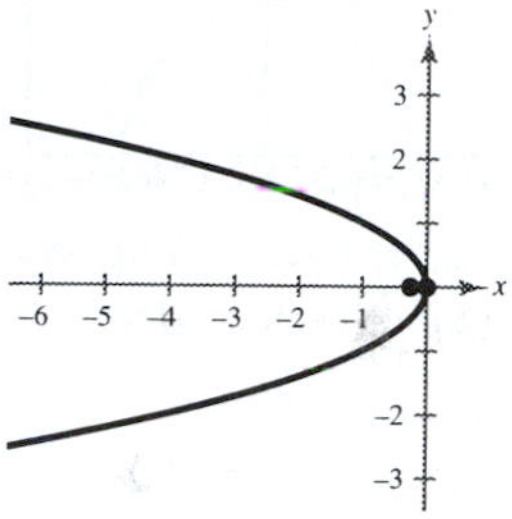

14. $(x + 3) + (y - 2)^2 = 0$

$$(y - 2)^2 = 4\left(-\frac{1}{4}\right)(x + 3)$$

Vertex: $(-3, 2)$

Focus: $\left(-3 + \left(-\frac{1}{4}\right), 2\right) \Rightarrow \left(-\frac{13}{4}, 2\right)$

Directrix: $x = -3 - \left(-\frac{1}{4}\right) = -\frac{11}{4}$

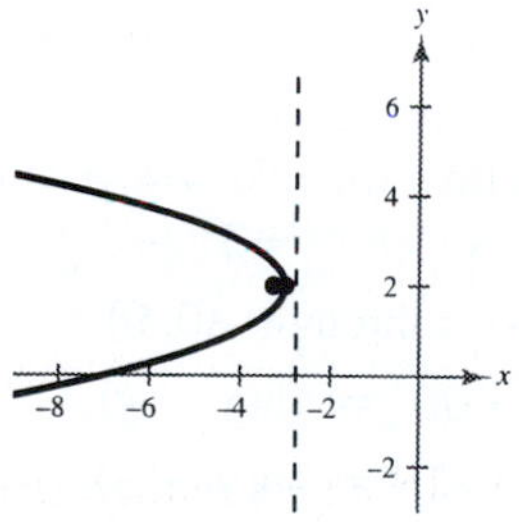

16. $\left(x + \frac{1}{2}\right)^2 = 4(y - 3) = 4(1)(y - 3)$

Vertex: $\left(-\frac{1}{2}, 3\right)$

Focus: $\left(-\frac{1}{2}, 3 + 1\right) \Rightarrow \left(-\frac{1}{2}, 4\right)$

Directrix: $y = 3 - 1 = 2$

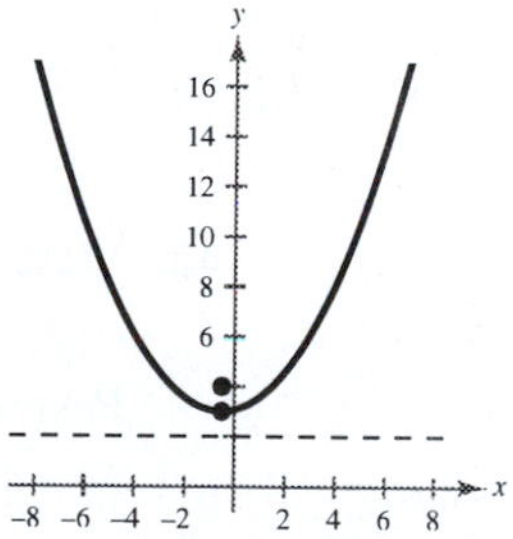

18. $4x - y^2 - 2y - 33 = 0$

$$y^2 + 2y + 1 = 4x - 33 + 1$$

$$(y + 1)^2 = 4(1)(x - 8)$$

Vertex: $(8, -1)$

Focus: $(9, -1)$

Directrix: $x = 7$

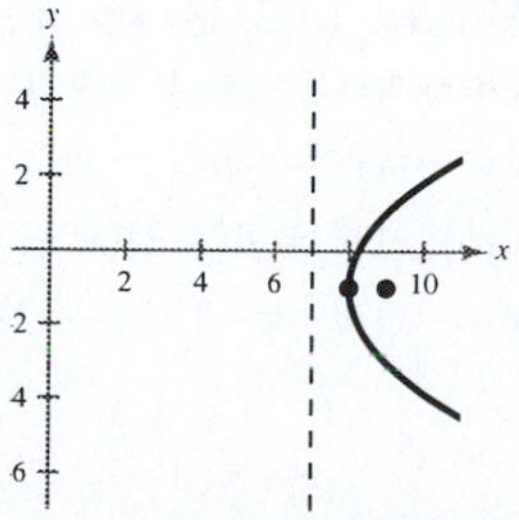

20. $y^2 - 4y - 4x = 0$

$$y^2 - 4y + 4 = 4x + 4$$

$$(y - 2)^2 = 4(1)(x + 1)$$

Vertex: $(-1, 2)$

Focus: $(0, 2)$

Directrix: $x = -2$

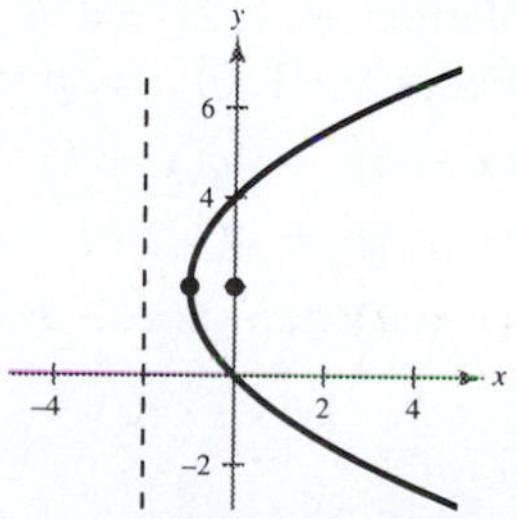

22. $x^2 - 2x + 8y + 9 = 0$

$x^2 - 2x + 1 = -8y - 9 + 1$

$(x - 1)^2 = -8(y + 1) = 4(-2)(y + 1)$

Vertex: $(1, -1)$
Focus: $(1, -3)$
Directrix: $y = 1$

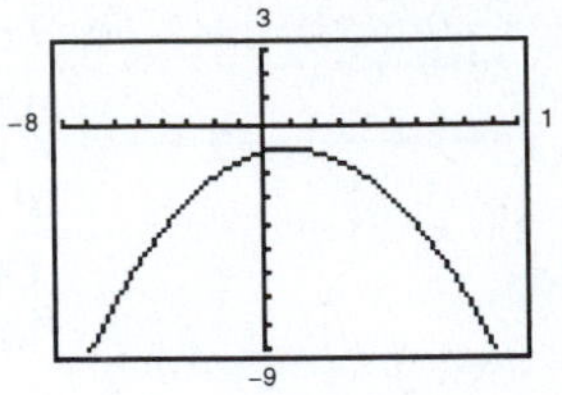

24. $y^2 - 4x - 4 = 0$

$y^2 = 4x + 4 = 4(1)(x + 1)$

Vertex: $(-1, 0)$
Focus: $(0, 0)$
Directrix: $x = -2$

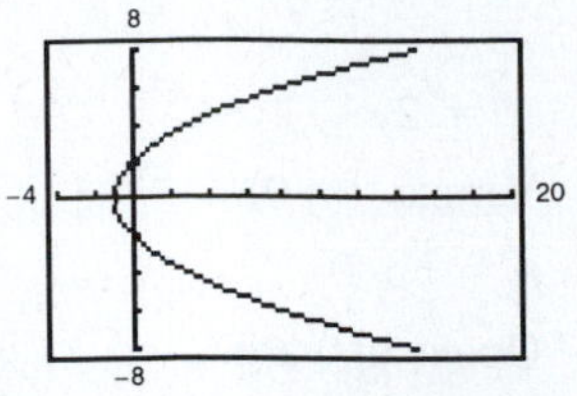

26. $x^2 + 12y = 0 \Rightarrow y_1 = -\frac{1}{12}x^2$

$x + y - 3 = 0 \Rightarrow y_2 = 3 - x$

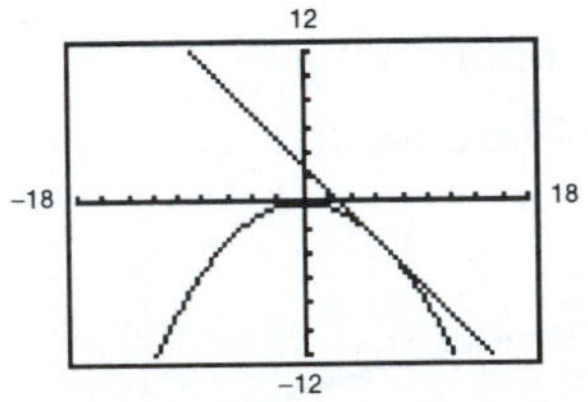

Using the trace or intersect feature, the point of tangency is $(6, -3)$.

28. No, it is not possible for a parabola to intersect its directrix. If the graph crossed the directrix there would exist points nearer the directrix than the focus.

30. Point: $(-2, 6)$

$x = ay^2$

$-2 = a(6)^2$

$-\frac{1}{18} = a$

$x = -\frac{1}{18}y^2$

32. Focus: $(2, 0) \Rightarrow p = 2$

$y^2 = 4px$

$y^2 = 8x$

34. Focus: $(0, -2) \Rightarrow p = -2$

$x^2 = 4py$

$x^2 = -8y$

36. Directrix: $x = 3 \Rightarrow p = -3$

$(y - k)^2 = 4p(x - h)$

$y^2 = 4(-3)x$

$y^2 = -12x$

38. Directrix: $x = -2 \Rightarrow p = 2$

$y^2 = 4px$

$y^2 = 8x$

40. Vertical axis
Passes through: $(-2, -2)$

$x^2 = 4py$

$(-2)^2 = 4p(-2)$

$4 = -8p$

$p = -\frac{1}{2}$

$x^2 = -2y$

42. Vertex: $(5, 3) \Rightarrow h - 5,$
$k = 3$

Passes through: $(4.5, 4)$

$(y - k)^2 = 4p(x - h)$

$(y - 3)^2 = 4p(x - 5)$

$1 = 4p(4.5 - 5)$

$p = -\frac{1}{2}$

$(y - 3)^2 = -2(x - 5)$

44. Vertex: $(3, -3) \Rightarrow h = 3,$
$k = -3$

Passes through: $(0, 0)$

$(x - h)^2 = 4p(y - k)$

$(x - 3)^3 = 4p(y + 3)$

$9 = 12p$

$p = \frac{3}{4}$

$(x - 3)^2 = 3(y + 3)$

46. Vertex: $(-1, 2) \Rightarrow h = -1, k = 2$
Focus: $(-1, 0) \Rightarrow p = -2$

$(x - h)^2 = 4p(y - k)$

$(x + 1)^2 = 4(-2)(y - 2)$

$(x + 1)^2 = -8(y - 2)$

48. Vertex: $(-2, 1) \Rightarrow h = -2, k = 1$
Directrix: $x = 1 \Rightarrow p = -3$

$(y - k)^2 = 4p(x - h)$

$(y - 1)^2 = 4(-3)(x - (-2))$

$(y - 1)^2 = -12(x + 2)$

50. Focus: $(0, 0)$

Directrix: $y = 4 \Rightarrow p = -2 \Rightarrow h = 0, k = 2$

$$(x - h)^2 = 4p(y - k)$$
$$x^2 = 4(-2)(y - 2)$$
$$x^2 = -8(y - 2)$$

52. $(y + 1)^2 = 2(x - 2)$

$$y + 1 = \pm\sqrt{2(x - 2)}$$
$$y = -1 \pm \sqrt{2(x - 2)}$$

Lower half of parabola: $y = -1 - \sqrt{2(x - 2)}$

54. (a)

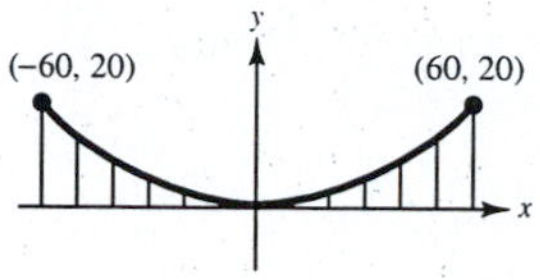

(b) $(x - 0)^2 = 4p(y - 0)$

$$x^2 = 4py$$

At $(60, 20)$: $60^2 = 4p(20) \Rightarrow p = 45$

$$x^2 = 4(45)y$$
$$y = \frac{x^2}{180}$$

(c)

x	0	20	40	60
y	0	$2\frac{2}{9}$	$8\frac{8}{9}$	20

56. Vertex: $(0, 0)$

$$(y - 0)^2 = 4p(x - 0)$$
$$y^2 = 4px$$

At $(1000, 800)$: $800^2 = 4p(1000) \Rightarrow p = 160$

$$y^2 = 4(160)x$$
$$y^2 = 640x$$

58. (a) $17{,}500\sqrt{2} \approx 24{,}749$ miles per hour

(b) Vertex: $(0, 4100)$

Focus: $(0, 0)$

$$(x - 0) = 4(-4100)(y - 4100)$$
$$x^2 = -16{,}400(y - 4100)$$

60. $0 = 30{,}000 - \dfrac{x^2}{39{,}204}$

$x \approx 34{,}294.6$ feet from the target or

$$34{,}294.6 \text{ feet} \cdot \frac{1 \text{ second}}{729 \text{ feet}} \approx 43.3 \text{ seconds}$$

before flying over the target.

62. $A = \dfrac{4}{3}pb^{3/2}$

(a) For $p = 2$ and $b = 4$: $A = \dfrac{4}{3}(2)(4)^{3/2} = \dfrac{64}{3}$

(b) The parabola becomes narrower for $0 \le y \le b$.

64.

$$2y = x^2$$
$$4\left(\frac{1}{2}\right)y = x^2$$
$$p = \frac{1}{2}$$

Focus: $\left(0, \dfrac{1}{2}\right)$

$$d_1 = \frac{1}{2} - b$$
$$d_2 = \sqrt{(-3 - 0)^2 + \left(\frac{9}{2} - \frac{1}{2}\right)^2} = 5$$
$$\frac{1}{2} - b = 5$$
$$b = -\frac{9}{2}$$
$$m = \frac{-(9/2) - (9/2)}{0 + 3} = -3$$

Tangent line: $y = -3x - \dfrac{9}{2} \Rightarrow 6x + 2y + 9 = 0$

x-intercept: $\left(-\dfrac{3}{2}, 0\right)$

66.

$$y = -2x^2$$
$$-\frac{1}{2}y = x^2$$
$$4\left(-\frac{1}{8}\right)y = x^2$$
$$p = -\frac{1}{8}$$

Focus: $\left(0, -\dfrac{1}{8}\right)$

$$d_1 = \frac{1}{8} + b$$
$$d_2 = \sqrt{(3 - 0)^2 + \left(-18 - \left(\frac{1}{8}\right)\right)^2} = \frac{145}{8}$$
$$\frac{1}{8} + b = \frac{145}{8}$$
$$b = \frac{144}{8} = 18$$
$$m = \frac{-18 - 18}{3 - 0} = -12$$

Tangent line: $y = -12x + 18 \Rightarrow 12x + y - 18 = 0$

x-intercept: $\left(\dfrac{3}{2}, 0\right)$

68. $f(x) = 2x^3 - 3x^2 + 50x - 75$

$\frac{3}{2}$	2	-3	50	-75
		3	0	75
	2	0	50	0

$$2x^3 - 3x^2 + 50x - 75 = \left(x - \tfrac{3}{2}\right)(2x^2 + 50)$$
$$= \left(x - \tfrac{3}{2}\right)(x - 5i)(x - 5i)$$

Zeros: $\frac{3}{2}, \pm 5i$

70. $h(x) = 2x^4 + x^3 - 19x^2 - 9x + 9$

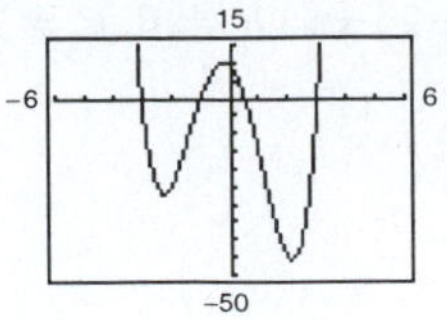

Zeros: $\pm 3, -1, \frac{1}{2}$

Section 10.3 Ellipses

Solutions to Even-Numbered Exercises

2. $\dfrac{x^2}{9} + \dfrac{y^2}{4} = 1$

Center: $(0, 0)$

$a = 3, b = 2$

Horizontal major axis

Matches graph (c).

4. $\dfrac{y^2}{4} + \dfrac{x^2}{4} = 1$

Center: $(0, 0)$

Circle of radius: 2

Matches graph (f).

6. $\dfrac{(x+2)^2}{4} + \dfrac{(y+2)^2}{16} = 1$

Center: $(-2, -2)$

$a = 4, b = 2$

Vertical major axis

Matches graph (e).

8. $\dfrac{x^2}{144} + \dfrac{y^2}{169} = 1$

$a^2 = 169, b^2 = 144, c^2 = 25$

Center: $(0, 0)$

Foci: $(0, \pm 5)$

Vertices: $(0, \pm 13)$

$e = \dfrac{5}{13}$

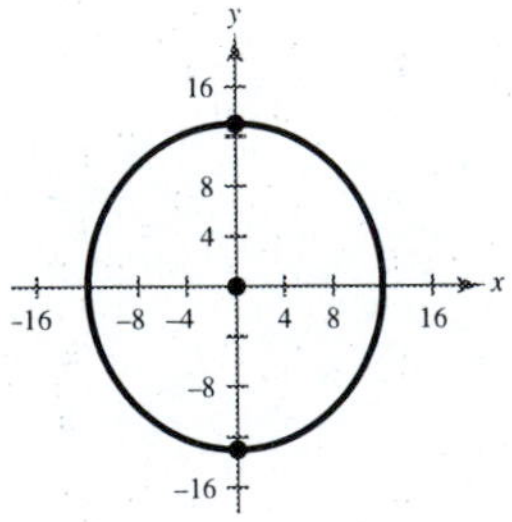

10. $\dfrac{x^2}{169} + \dfrac{y^2}{144} = 1$

$a^2 = 169, b^2 = 144, c^2 = 25$

Center: $(0, 0)$

Foci: $(\pm 5, 0)$

Vertices: $(\pm 13, 0)$

$e = \dfrac{c}{a} = \dfrac{5}{13}$

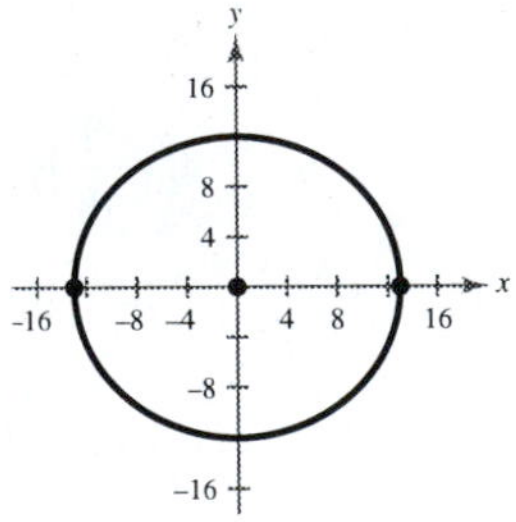

12. $\dfrac{x^2}{28} + \dfrac{y^2}{64} = 1$

$a^2 = 64, b^2 = 28, c^2 = 36$

Center: $(0, 0)$

Foci: $(0, \pm 6)$

Vertices: $(0, \pm 8)$

$e = \dfrac{c}{a} = \dfrac{6}{8} = \dfrac{3}{4}$

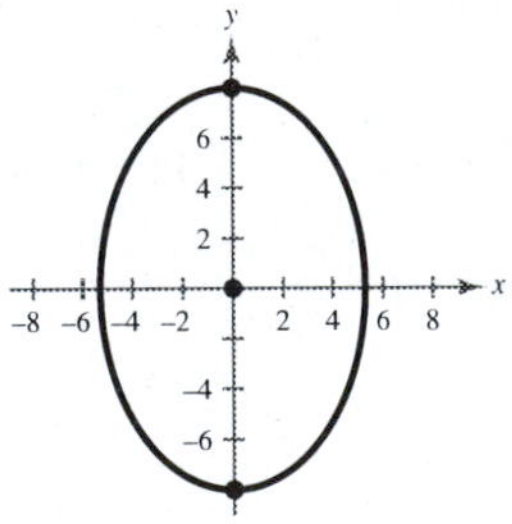

14. $\dfrac{(x+2)^2}{1} + \dfrac{(y+4)^2}{1/4} = 1$

$a^2 = 1, b^2 = \dfrac{1}{4}, c^2 = \dfrac{3}{4}$

Center: $(-2, -4)$

Foci: $\left(-2 \pm \dfrac{\sqrt{3}}{2}, -4\right)$

Vertices: $(-1, -4), (-3, -4)$

$e = \dfrac{\sqrt{3}}{2}$

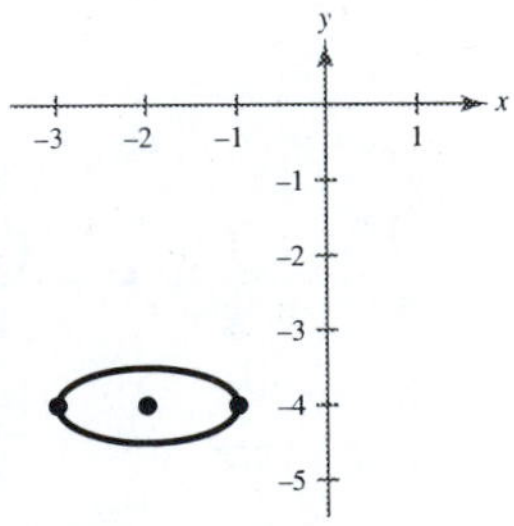

16. $9x^2 + 4y^2 - 36x + 8y + 31 = 0$

$$9(x^2 - 4x + 4) + 4(y^2 + 2y + 1) = -31 + 36 + 4$$

$$\frac{(x-2)^2}{1} + \frac{(y+1)^2}{9/4} = 1$$

$a^2 = \frac{9}{4}, b^2 = 1, c^2 = \frac{5}{4}$

Center: $(2, -1)$

Foci: $\left(2, -1 \pm \frac{\sqrt{5}}{2}\right)$

Vertices: $\left(2, \frac{1}{2}\right), \left(2, -\frac{5}{2}\right)$

$e = \frac{\sqrt{5}}{3}$

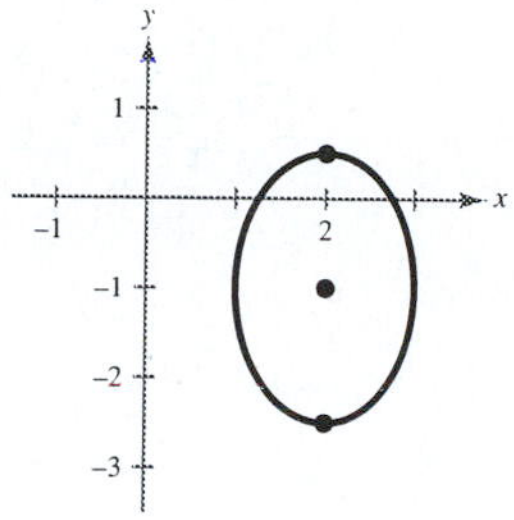

18. $9x^2 + 25y^2 - 36x - 50y + 61 = 0$

$$9(x^2 - 4x + 4) + 25(y^2 - 2y + 1) = -61 + 36 + 25$$

$$9(x-2)^2 + 25(y-1)^2 = 0$$

Degenerate ellipse with center $(2, 1)$ as the only point

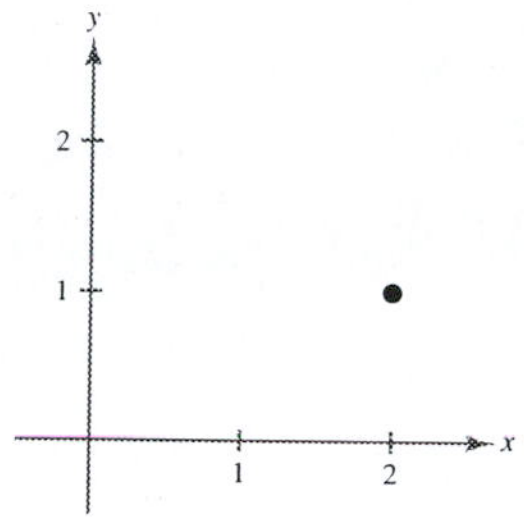

20. $x^2 + 4y^2 = 4$

$$\frac{x^2}{4} + \frac{y^2}{1} = 1$$

$a^2 = 4, b^2 = 1, c^2 = 3$

Center: $(0, 0)$

Foci: $(\pm\sqrt{3}, 0)$

Vertex: $(\pm 2, 0)$

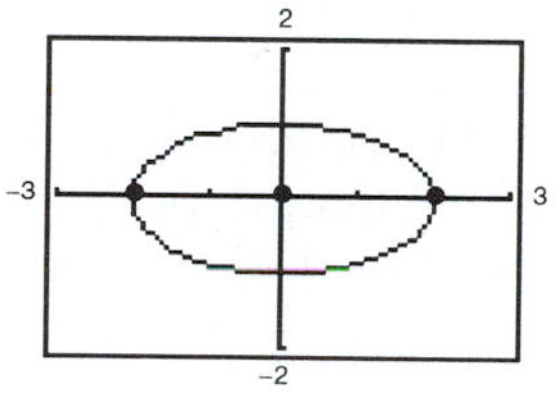

22. $36x^2 + 9y^2 + 48x - 36y + 43 = 0$

$$36\left(x^2 + \frac{4}{3}x + \frac{4}{9}\right) + 9(y^2 - 4y + 4) = -43 + 16 + 36$$

$$\frac{[x + (2/3)]^2}{1/4} + \frac{(y-2)^2}{1} = 1$$

$a^2 = 1, b^2 = \frac{1}{4}, c^2 = \frac{3}{4}$

Center: $\left(-\frac{2}{3}, 2\right)$

Foci: $\left(-\frac{2}{3}, 2 \pm \frac{\sqrt{3}}{2}\right)$

Vertices: $\left(-\frac{2}{3}, 3\right), \left(-\frac{2}{3}, 1\right)$

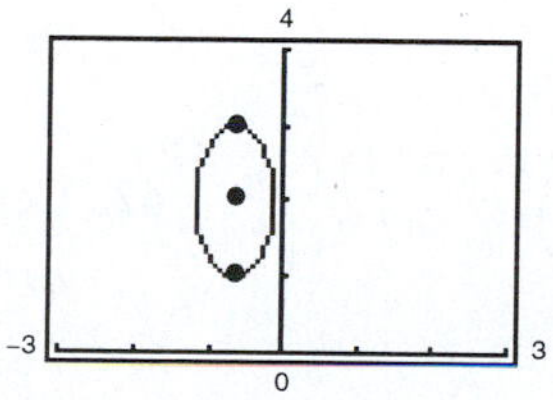

24. $\dfrac{(x+1)^2}{16} + \dfrac{(y-2)^2}{25} = 1$

$$\frac{y-2}{25} = 1 - \frac{(x+1)^2}{16}$$

$$\frac{y-2}{5} = \pm\sqrt{1 - \frac{(x+1)^2}{16}}$$

$$y - 2 = \pm 5\sqrt{\frac{16}{16} - \frac{(x+1)^2}{16}}$$

$$y - 2 = \pm\frac{5}{4}\sqrt{16 - (x+1)^2}$$

$$y = 2 \pm \frac{5}{4}\sqrt{16 - (x+1)^2}$$

Bottom half of ellipse: $y = 2 - \dfrac{5}{4}\sqrt{16 - (x+1)^2}$

26. Vertices: $(\pm 2, 0) \Rightarrow a = 2$

Endpoints of minor axis: $\left(0, \pm\dfrac{3}{2}\right) \Rightarrow b = \dfrac{3}{2}$

$$\frac{x^2}{a^2} + \frac{y^2}{b^2} = 1$$

$$\frac{x^2}{2^2} + \frac{y^2}{(3/2)^2} = 1$$

$$\frac{x^2}{4} + \frac{4y^2}{9} = 1$$

28. Vertices: $(0, \pm 8) \Rightarrow a = 8$

Foci: $(0, \pm 4) \Rightarrow c = 4$

$b^2 = a^2 - c^2 = 64 - 16 = 48$

Center: $(0, 0) = (h, k)$

$$\frac{(y-k)^2}{a^2} + \frac{(x-h)^2}{b^2} = 1$$

$$\frac{y^2}{64} + \frac{x^2}{48} = 1$$

30. Foci: $(\pm 2, 0) \Rightarrow c = 2$

Major axis length: $8 \Rightarrow a = 4$

$b^2 = a^2 - c^2 = 16 - 4 = 12$

$$\frac{x^2}{a^2} + \frac{y^2}{b^2} = 1$$

$$\frac{x^2}{16} + \frac{y^2}{12} = 1$$

32. Major axis vertical

Passes through: $(0, 4)$ and $(2, 0)$

$a = 4, b = 2$

$$\frac{x^2}{b^2} + \frac{y^2}{a^2} = 1$$

$$\frac{x^2}{4} + \frac{y^2}{16} = 1$$

34. Vertices: $(4, \pm 4) \Rightarrow a = 4$

Center: $(4, 0) \Rightarrow h = 4, k = 0$

Endpoints of minor axis: $(1, 0), (7, 0) \Rightarrow b = 3$

$$\frac{(x-h)^2}{b^2} + \frac{(y-k)^2}{a^2} = 1$$

$$\frac{(x-4)^2}{9} + \frac{y^2}{16} = 1$$

36. Vertices: $(0, -1), (4, -1) \Rightarrow a = 2$

Center: $(2, -1) \Rightarrow h = 2, k = -1$

Endpoints of minor axis: $(2, 0), (2, -2) \Rightarrow b = 1$

$$\frac{(x-h)^2}{a^2} + \frac{(y-k)^2}{b^2} = 1$$

$$\frac{(x-2)^2}{4} + \frac{(y+1)^2}{1} = 1$$

38. Foci: $(0, 0), (4, 0) \Rightarrow c = 2, h = 2, k = 0$

Major axis length: $8 \Rightarrow a = 4$

$b^2 = a^2 - c^2 = 16 - 4 = 12$

$$\frac{(x-h)^2}{a^2} + \frac{(y-k)^2}{b^2} = 1$$

$$\frac{(x-2)^2}{16} + \frac{y^2}{12} = 1$$

40. Center: $(2, -1) \Rightarrow h = 2, k = -1$

Vertex: $\left(2, \dfrac{1}{2}\right) \Rightarrow a = \dfrac{3}{2}$

Minor axis length: $2 \Rightarrow b = 1$

$$\frac{(x-h)}{b^2} + \frac{(y-k)^2}{a^2} = 1$$

$$\frac{(x-2)^2}{1} + \frac{(y+1)^2}{(3/2)^2} = 1$$

$$(x-2)^2 + \frac{4(y+1)^2}{9} = 1$$

42. Center: $(3, 2) = (h, k)$

$a = 3c$

Foci: $(1, 2), (5, 2) \Rightarrow c = 2, a = 6$

$b^2 = a^2 - c^2 = 36 - 4 = 32$

$$\frac{(x-h)^2}{a^2} + \frac{(y-k)^2}{b^2} = 1$$

$$\frac{(x-3)^2}{36} + \frac{(y-2)^2}{32} = 1$$

44. Vertices: $(5, 0), (5, 12) \Rightarrow a = 6$

Endpoints of the minor axis:

$(0, 6), (10, 6) \Rightarrow b = 5$

Center: $(5, 6) \Rightarrow h = 5, k = 6$

$$\frac{(x-h)^2}{b^2} + \frac{(y-k)^2}{a^2} = 1$$

$$\frac{(x-5)^2}{25} + \frac{(y-6)^2}{36} = 1$$

46. Vertices: $(\pm 3, 0) \Rightarrow a = 3$

Half of minor axis length: $2 \Rightarrow b = 2$

$c^2 = a^2 - b^2 = 9 - 4 = 5 \Rightarrow c = \sqrt{5}$

Place the tacks $\sqrt{5}$ feet from the center: $(\pm\sqrt{5}, 0)$

Length of string: $2a = 2(3) = 6$ feet

48. (a)

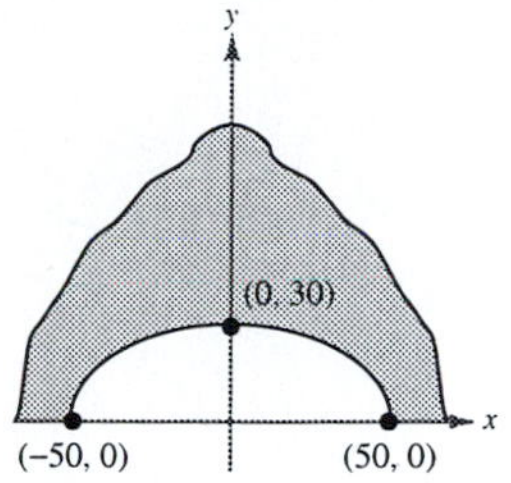

(b) $a = 50, b = 30$

$$\frac{x^2}{50^2} + \frac{y^2}{30^2} = 1$$

$$\frac{x^2}{2500} + \frac{y^2}{900} = 1$$

(c) When $x = \pm 45$, you are five feet from the edge of the tunnel.

$$\frac{45^2}{2500} + \frac{y^2}{900} = 1$$

$$\frac{2025}{2500} + \frac{y^2}{900} = 1$$

$$y^2 = \frac{475}{2500}(900)$$

$$y^2 = 171$$

$$y = \pm\sqrt{171} = \pm 3\sqrt{19}$$

The height of the tunnel five feet from its edge is $3\sqrt{19} \approx 13.08$ feet.

50. Area of ellipse = 2 (area of circle)

$$\pi ab = 2\pi r^2$$

$$\pi a(10) = 2\pi(10)^2$$

$$\pi a(10) = 200$$

$$a = 20$$

Length of major axis: $2a = 2(20) = 40$ units

52. Vertices: $(0, \pm 8) \Rightarrow a = 8, h = 0, k = 0$

Eccentricity: $e = \dfrac{1}{2} = \dfrac{c}{a}$

$$\frac{1}{2} = \frac{c}{8}$$

$$c = 4$$

$b^2 = c^2 - a^2 = 64 - 16 = 48$

$$\frac{x^2}{b^2} + \frac{y^2}{a^2} = 1$$

$$\frac{x^2}{48} + \frac{y^2}{64} = 1$$

54. Center: $(0, 0) \Rightarrow h = 0, k = 0$

$2a = 0.34 + 4.08 = 4.42$

$a = 2.21$

$c = 2.21 - 0.34 = 1.87$

$b^2 = a^2 - c^2 = 4.8841 - 3.4969 = 1.3872$

$$\frac{x^2}{a^2} + \frac{y^2}{b^2} = 1$$

$$\frac{x^2}{4.88} + \frac{y^2}{1.39} = 1$$

56. (a) $b^2 = a^2 - c^2 = a^2 - a^2\left(\dfrac{c^2}{a^2}\right)$

$$= a^2\left(1 - \frac{c^2}{a^2}\right) = a^2(1 - e^2)$$

$$\frac{(x-h)^2}{a^2} + \frac{(y-k)^2}{b^2} = 1$$

$$\frac{(x-h)^2}{a^2} + \frac{(y-k)^2}{a^2(1-e^2)} = 1$$

As $e \Rightarrow 0$, $1 - e^2 \Rightarrow 1$ and we have $\dfrac{x^2}{a^2} + \dfrac{y^2}{a^2} = 1$ or the circle $x^2 + y^2 = a^2$.

—CONTINUED—

56. —CONTINUED—

(b) $\dfrac{(x-2)^2}{4} + \dfrac{(y-3)^2}{4(1-e^2)} = 1$ for $e = 0.95, 0.75, 0.5, 0.25, 0.$

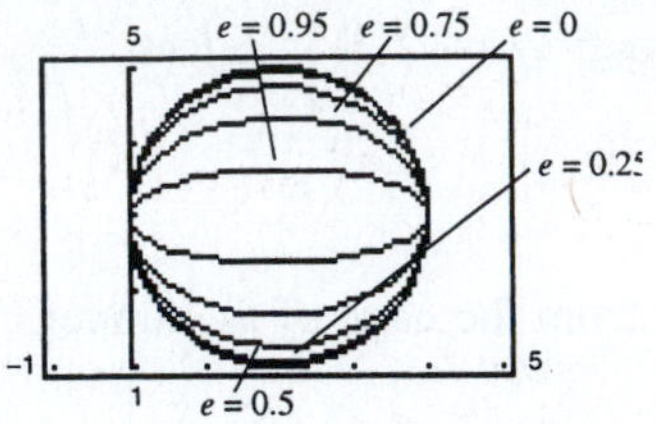

(c) As $e \to 0$, the shape of the ellipse becomes more nearly circular.

58. For $\dfrac{x^2}{a^2} + \dfrac{y^2}{b^2} = 1$, we have $c^2 = a^2 - b^2$.

When $x = c$:

$$\frac{c^2}{a^2} + \frac{y^2}{b^2} = 1 \Rightarrow y^2 = b^2\left(1 - \frac{a^2 - b^2}{a^2}\right) \Rightarrow y^2 = \frac{b^4}{a^2} \Rightarrow 2y = \frac{2b^2}{a}.$$

60. $\dfrac{x^2}{9} + \dfrac{y^2}{16} = 1$

$a = 4, b = 3, c = \sqrt{7}$

Points on the ellipse: $(\pm 3, 0), (0, \pm 4)$

Length of latus recta: $\dfrac{2b^2}{a} = \dfrac{2(3)^2}{4} = \dfrac{9}{2}$

Additional points: $\left(\pm\dfrac{9}{4}, -\sqrt{7}\right), \left(\pm\dfrac{9}{4}, \sqrt{7}\right)$

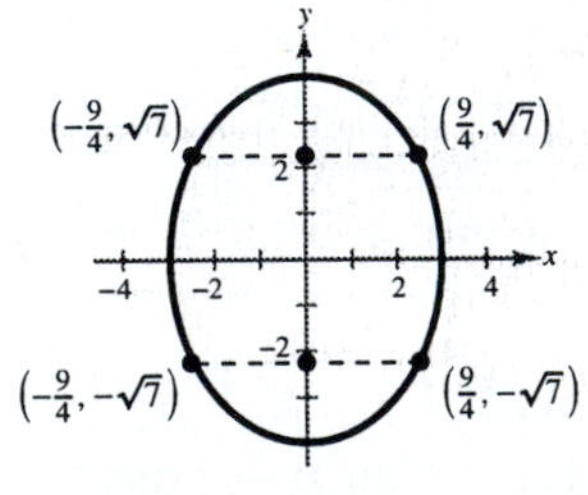

62. $5x^2 + 3y^2 = 15$

$\dfrac{x^3}{3} + \dfrac{y^2}{5} = 1$

$a = \sqrt{5}, b = \sqrt{3}, c = \sqrt{2}$

Points on the ellipse: $(\pm\sqrt{3}, 0), (0, \pm\sqrt{5})$

Length of latus recta: $\dfrac{2b^2}{a} = \dfrac{2 \cdot 3}{\sqrt{5}} = \dfrac{6\sqrt{5}}{5}$

Additional points: $\left(\pm\dfrac{3\sqrt{5}}{5}, -\sqrt{2}\right), \left(\pm\dfrac{3\sqrt{5}}{5}, \sqrt{2}\right)$

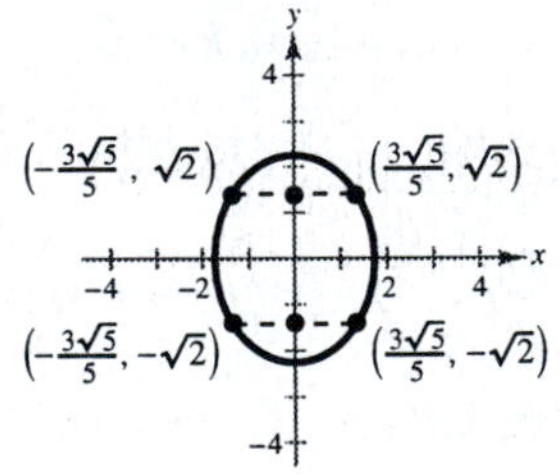

Section 10.4 Hyperbolas

Solutions to Even-Numbered Exercises

2. $\dfrac{y^2}{16} - \dfrac{x^2}{4} = 1$

Center: $(0, 0)$

$a = 4, b = 2$

Vertical transverse axis
Matches graph (c).

4. $\dfrac{y^2}{16} - \dfrac{x^2}{9} = 1$

Center: $(0, 0)$

$a = 4, b = 3$

Vertical transverse axis
Matches graph (d).

6. $\dfrac{(x+1)^2}{16} - \dfrac{(y-2)^2}{9} = 1$

Center: $(-1, 2)$

$a = 4, b = 3$

Horizontal transverse axis
Matches graph (f).

8. $\dfrac{x^2}{9} - \dfrac{y^2}{16} = 1$

$a = 3, b = 4,$

$c = \sqrt{4^2 + 3^2} = 5$

Center: $(0, 0)$

Vertices: $(\pm 3, 0)$

Foci: $(\pm 5, 0)$

Asymptotes: $y = \pm \dfrac{4}{3}x$

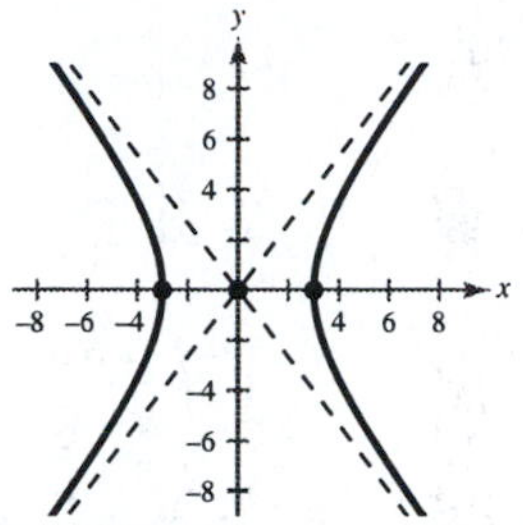

10. $\dfrac{y^2}{9} - \dfrac{x^2}{1} = 1$

$a = 3, b = 1,$

$c = \sqrt{3^2 + 1^2} = \sqrt{10}$

Center: $(0, 0)$

Vertices: $(0, \pm 3)$

Foci: $\left(0, \pm\sqrt{10}\right)$

Asymptotes: $y = \pm 3x$

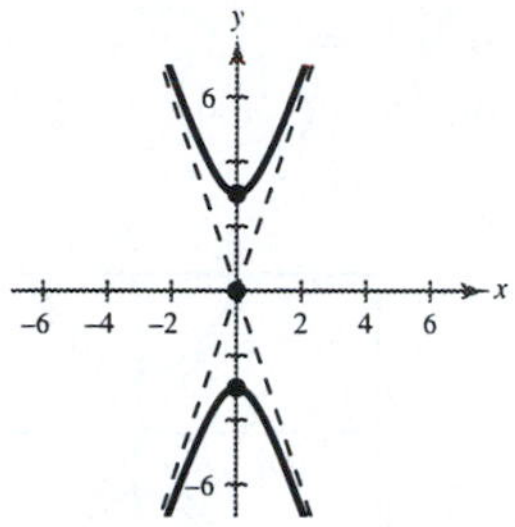

12. $\dfrac{x^2}{36} - \dfrac{y^2}{4} = 1$

$a = 6, b = 2,$

$c = \sqrt{36 + 4} = 2\sqrt{10}$

Center: $(0, 0)$

Vertices: $(\pm 6, 0)$

Foci: $\left(\pm 2\sqrt{10}, 0\right)$

Asymptotes: $y = \pm \dfrac{1}{3}x$

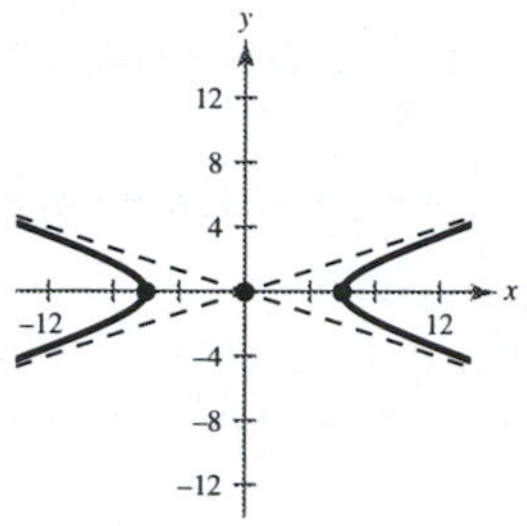

14. $\dfrac{(x + 1)^2}{144} - \dfrac{(y - 4)^2}{25} = 1$

$a = 12, b = 5, c = 13$

Center: $(-1, 4)$

Vertices: $(-13, 4), (11, 4)$

Foci: $(-14, 4), (12, 4)$

Asymptotes: $y = 4 \pm \dfrac{5}{12}(x + 1)$

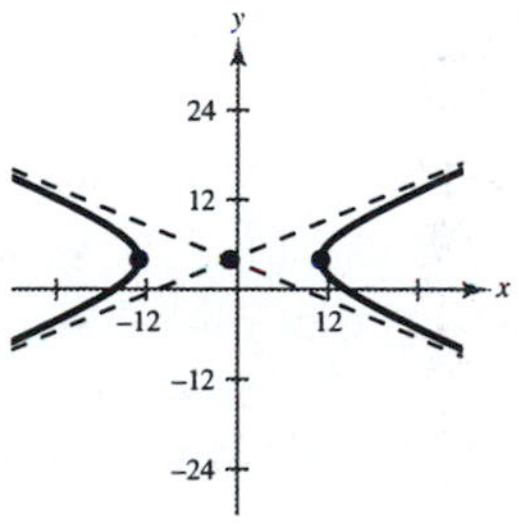

16. $\dfrac{(y - 1)^2}{1/4} - \dfrac{(x + 3)^2}{1/9} = 1$

$a = \dfrac{1}{2}, b = \dfrac{1}{3}, c = \dfrac{\sqrt{13}}{6}$

Center: $(-3, 1)$

Vertices: $\left(-3, \dfrac{1}{2}\right), \left(-3, \dfrac{3}{2}\right)$

Foci: $\left(-3, 1 \pm \dfrac{1}{6}\sqrt{13}\right)$

Asymptotes: $y = 1 \pm \dfrac{3}{2}(x + 3)$

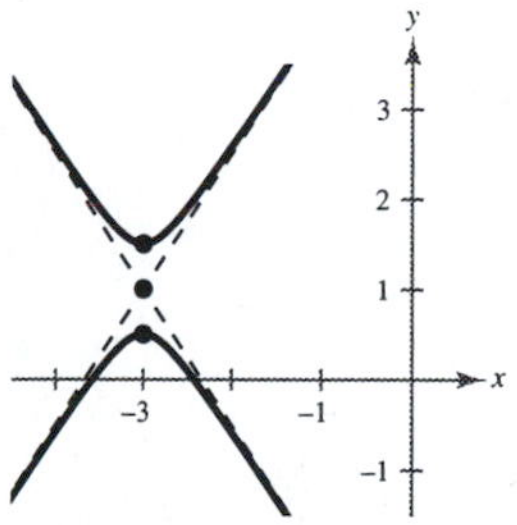

18. $x^2 - 9y^2 + 36y - 72 = 0$

$$x^2 - 9(y^2 - 4y + 4) = 72 - 36$$

$$x^2 - 9(y - 2)^2 = 36$$

$$\frac{x^2}{36} - \frac{(y - 2)^2}{4} = 1$$

$a = 6, b = 2, c = \sqrt{36 + 4} = 2\sqrt{10}$

Center: $(0, 2)$

Vertices: $(\pm 6, 2)$

Foci: $\left(\pm 2\sqrt{10}, 2\right)$

Asymptotes: $y = 2 \pm \dfrac{1}{3}x$

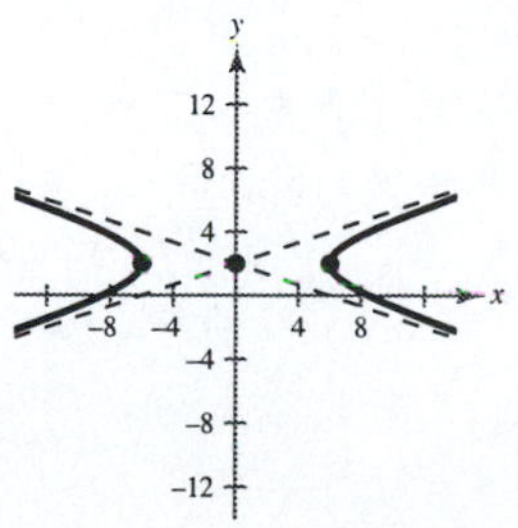

20.

$$16y^2 - x^2 + 2x + 64y + 63 = 0$$
$$16(y^2 + 4y + 4) - (x^2 - 2x + 1) = -63 + 64 - 1$$
$$16(y + 2)^2 - (x - 1) = 0$$
$$y + 2 = \pm\tfrac{1}{4}(x - 1)$$

Degenerate hyperbola is two intersecting lines.

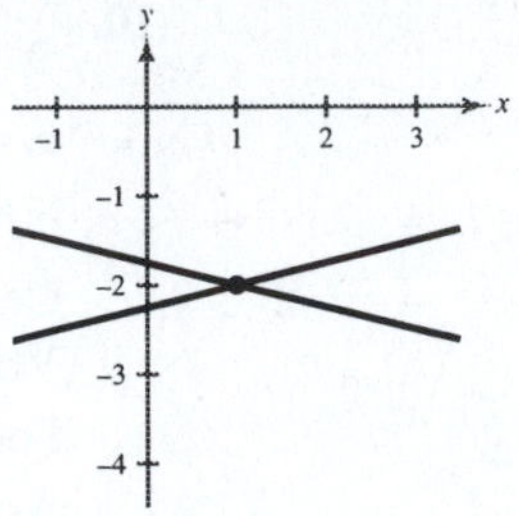

22. $3y^2 - 5x^2 = 15$

$$\frac{y^2}{5} - \frac{x^2}{3} = 1$$

$a = \sqrt{5}, b = \sqrt{3}, c = \sqrt{5 + 3} = 2\sqrt{2}$

Center: $(0, 0)$

Vertices: $\left(0, \pm\sqrt{5}\right)$

Foci: $\left(0, \pm 2\sqrt{2}\right)$

Asymptotes: $y = \pm\dfrac{\sqrt{5}}{\sqrt{3}}x$

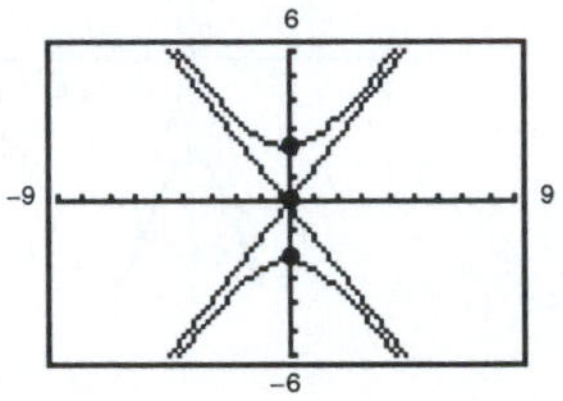

24.

$$9x^2 - y^2 + 54x + 10y + 55 = 0$$
$$9(x^2 + 6x + 9) - (y^2 - 10y + 25) = -55 + 81 - 25$$
$$\frac{(x + 3)^2}{1/9} - \frac{(y - 5)^2}{1} = 1$$

$a = \dfrac{1}{3}, b = 1, c = \dfrac{\sqrt{10}}{3}$

Center: $(-3, 5)$

Vertices: $\left(-3 \pm \dfrac{1}{3}, 5\right)$

Foci: $\left(-3 \pm \dfrac{\sqrt{10}}{3}, 5\right)$

Asymptotes: $y = 5 \pm 3(x + 3)$

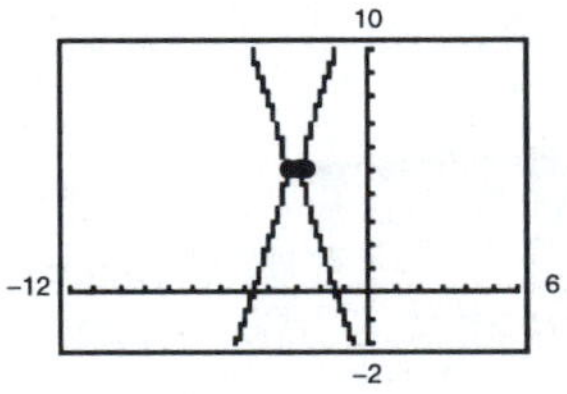

26. Vertices: $(\pm 2, 0) \implies a = 2$

Point on curve: $\left(3, \sqrt{3}\right)$

$$\frac{x^2}{a^2} - \frac{y^2}{b^2} = 1$$
$$\frac{3^2}{2^2} - \frac{\left(\sqrt{3}\right)^2}{b^2} = 1$$
$$b^2 = \frac{12}{5}$$
$$\frac{x^2}{4} - \frac{5y^2}{12} = 1$$

28. Vertices: $(\pm 3, 0) \implies a = 3$

Foci: $(\pm 5, 0) \implies c = 5$

$b^2 = c^2 - a^2 = 25 - 9 = 16 \implies b = 4$

$$\frac{x^2}{a^2} - \frac{y^2}{b^2} = 1$$
$$\frac{x^2}{9} - \frac{y^2}{16} = 1$$

30. Vertices: $(0, \pm 3) \implies a = 3$

Asymptotes: $y = \pm 3x \implies \dfrac{a}{b} = 3, b = 1$

Center: $(0, 0) = (h, k)$

$$\frac{(y - k)^2}{a^2} - \frac{(x - h)^2}{b^2} = 1$$
$$\frac{y^2}{9} - x^2 = 1$$

32. Foci: $(\pm 10, 0) \implies c = 10$

Asymptotes: $y = \pm\frac{3}{4}x \implies \frac{b}{a} = \frac{3m}{4m}$

$$c^2 = a^2 + b^2 \implies 100 = (3m)^2 + (4m)^2$$
$$100 = 25m^2$$
$$2 = m$$

$a = 3(2) = 6$

$b = 4(2) = 8$

$$\frac{x^2}{a^2} - \frac{y^2}{b^2} = 1$$

$$\frac{x^2}{36} - \frac{y^2}{64} = 1$$

34. Vertices: $(1, \pm 2) \implies a = 2$

Center: $(1, 0)$

Point on curve: $(0, \sqrt{5})$

$$\frac{(y-k)^2}{a^2} - \frac{(x-h)^2}{b^2} = 1$$

$$\frac{y^2}{4} - \frac{(x-1)^2}{b^2} = 1$$

$$\frac{5}{4} - \frac{1}{b^2} = 1$$

$$b^2 = 4$$

$$\frac{y^2}{4} - \frac{(x-1)^2}{4} = 1$$

36. Vertices: $(-8, 4), (0, 4) \implies a = 4$

Center: $(-4, 4)$

Point on curve: $(2, 0)$

$$\frac{(x-h)^2}{a^2} - \frac{(y-k)^2}{b^2} = 1$$

$$\frac{(x+4)^2}{16} - \frac{(y-4)^2}{b^2} = 1$$

$$\frac{36}{16} - \frac{16}{b^2} = 1$$

$$b^2 = \frac{64}{5}$$

$$\frac{(x+4)^2}{16} - \frac{(y-4)^2}{\frac{64}{5}} = 1$$

38. Vertices: $(2, 3), (2, -3) \implies a = 3$

Center: $(2, 0)$

Foci: $(2, 5), (2, -5) \implies c = 5$

$b^2 = c^2 - a^2 = 25 - 9 = 16$

$$\frac{(y-k)^2}{a^2} - \frac{(x-h)^2}{b^2} = 1$$

$$\frac{y^2}{9} - \frac{(x-2)^2}{16} = 1$$

40. Vertices: $(-2, 1), (2, 1) \implies a = 2$

Center: $(0, 1)$

Foci: $(-3, 1), (3, 1) \implies c = 3$

$b^2 = c^2 - a^2 = 9 - 4 = 5$

$$\frac{(x-h)^2}{a^2} - \frac{(y-k)^2}{b^2} = 1$$

$$\frac{x^2}{4} - \frac{(y-1)^2}{5} = 1$$

42. Vertices: $(-2, 1), (2, 1) \implies a = 2$

Solution point: $(4, 3)$

Center: $(0, 1) = (h, k)$

$$\frac{(x-h)^2}{a^2} - \frac{(y-k)^2}{b^2} = 1$$

$$\frac{x^2}{4} - \frac{(y-1)^2}{b^2} = 1 \implies b^2 = \frac{4(y-1)^2}{x^2-4} = \frac{4(2^2)}{16-4} = \frac{16}{12} = \frac{4}{3}$$

$$\frac{x^2}{4} - \frac{(y-1)^2}{4/3} = 1$$

44. Vertices: $(3, 0), (3, 4) \Rightarrow a = 2$

Asymptotes: $y = \frac{2}{3}x, y = 4 - \frac{2}{3}x$

$\frac{a}{b} = \frac{2}{3} \Rightarrow b = 3$

Center: $(3, 2) = (h, k)$

$$\frac{(y-k)^2}{a^2} - \frac{(x-h)^2}{b^2} = 1$$

$$\frac{(y-2)^2}{4} - \frac{(x-3)^2}{9} = 1$$

46. $\frac{(x-3)^2}{4} - \frac{(y-1)^2}{9} = 1$

$$\frac{(x-3)^2}{4} - 1 = \frac{(y-1)^2}{9}$$

$$\frac{(x-3)^2}{4} - \frac{4}{4} = \frac{(y-1)^2}{9}$$

$$\pm\frac{1}{2}\sqrt{(x-3)^2 - 4} = \frac{y-1}{3}$$

$$\pm\frac{3}{2}\sqrt{(x-3)^2 - 4} = y - 1$$

$$1 \pm \frac{3}{2}\sqrt{(x-3)^2 - 4} = y$$

$1 + \left(\frac{3}{2}\right)\sqrt{(x-3)^2 - 4} = y$ is the top half of the graph of the hyperbola.

48. Foci: $(\pm 150, 0) \Rightarrow c = 150$

Center: $(0, 0) = (h, k)$

$$\frac{d_2}{186{,}000} - \frac{d_1}{186{,}000} = 0.001 \Rightarrow 2a = 186, a = 93$$

$$b^2 = c^2 - a^2 = 150^2 - 93^2 = 13{,}851$$

$$\frac{x^2}{93^2} - \frac{y^2}{13{,}851} = 1$$

$$x^2 = 93^2\left(1 + \frac{75^2}{13{,}851}\right) \approx 12{,}161$$

$$x \approx 110.3 \text{ miles}$$

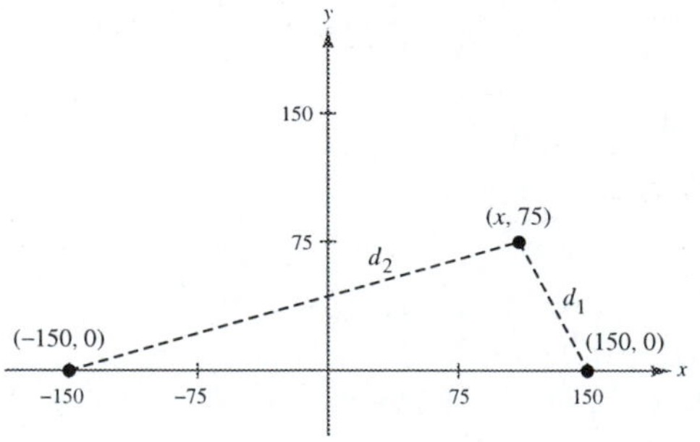

50. Let (x, y) be such that the difference of the distances from $(c, 0)$ and $(-c, 0)$ is $2a$ (again only deriving one of the forms).

$$2a = \left|\sqrt{(x+c)^2 + y^2} - \sqrt{(x-c) + y^2}\right|$$

$$2a + \sqrt{(x-c)^2 + y^2} = \sqrt{(x+c)^2 + y^2}$$

$$4a^2 + 4a\sqrt{(x-c)^2 + y^2} + (x-c)^2 + y^2 = (x+c)^2 + y^2$$

$$4a\sqrt{(x-c)^2 + y^2} = 4cx - 4a^2$$

$$a\sqrt{(x-c)^2 + y^2} = cx - a^2$$

$$a^2(x^2 - 2cx + c^2 + y^2) = c^2x^2 - 2a^2cx + a^4$$

$$a^2(c^2 - a^2) = (c^2 - a^2)x^2 - a^2y^2$$

Let $b^2 = c^2 - a^2$. Then $a^2b^2 = b^2x^2 - a^2y^2 \Rightarrow 1 = \frac{x^2}{a^2} - \frac{y^2}{b^2}$.

52. $x^2 + 4y^2 - 6x + 16y + 21 = 0$

$A = 1, C = 4$

$AC = 1(4) = 4 > 0 \Rightarrow$ Ellipse

54. $y^2 - 4y - 4x = 0$

$A = 0, C = 1$

$AC = 0(1) = 0 \Rightarrow$ Parabola

56. $4y^2 - 2x^2 - 4y - 8x - 15 = 0$

$A = -2, C = 4$

$AC = (-2)(4) = -8 < 0 \Rightarrow$ Hyperbola

58. $4x^2 + 4y^2 - 16y + 15 = 0$

$A = 4, C = 4$

$A = C \Rightarrow$ Circle

60. $\left(3x - \frac{1}{2}\right)(x + 4) = 3x^2 + 12x - \frac{1}{2}x - 2$

$= 3x^2 + \frac{23}{2}x - 2$

62. $[(x + y) + 3]^2 = (x + y)^2 + 2(3)(x + y) + 3^2$

$= x^2 + 2xy + y^2 + 6x + 6y + 9$

Section 10.5 Rotation of Conics

Solutions to Even-Numbered Exercises

2. $\theta = 45°$; Point: $(3, 3)$

$x' = x\cos\theta - y\sin\theta = 3\cos 45° - 3\sin 45° = 0$

$y' = x\sin\theta + y\cos\theta = 3\sin 45° + 3\cos 45° = 3\sqrt{2}$

Thus, $(x', y') = \left(0, 3\sqrt{2}\right)$.

4. $\theta = 60°$; Point: $(3, 1)$

$x' = x\cos\theta - y\sin\theta = 3\cos 60° - 1\sin 60° = \frac{3}{2} - \frac{\sqrt{3}}{2}$

$y' = x\sin\theta + y\cos\theta = 3\sin 60° + 1\cos 60° = \frac{3\sqrt{3}}{2} + \frac{1}{2}$

Thus, $(x', y') = \left(\frac{1}{2}(3 - \sqrt{3}), \frac{1}{2}(3\sqrt{3} + 1)\right)$.

6. $xy - 4 = 0$

$A = 0, B = 1, C = 0$

$\cot 2\theta = \frac{A - C}{B} = 0 \Rightarrow 2\theta = \frac{\pi}{2} \Rightarrow \theta = \frac{\pi}{4}$

$x = x'\cos\frac{\pi}{4} - y'\sin\frac{\pi}{4}$

$= x'\left(\frac{\sqrt{2}}{2}\right) - y'\left(\frac{\sqrt{2}}{2}\right)$

$= \frac{x' - y'}{\sqrt{2}}$

$y = x'\sin\frac{\pi}{4} + y'\cos\frac{\pi}{4}$

$= x'\left(\frac{\sqrt{2}}{2}\right) + y'\left(\frac{\sqrt{2}}{2}\right)$

$= \frac{x' + y'}{\sqrt{2}}$

$$xy - 4 = 0$$

$$\left(\frac{x' - y'}{\sqrt{2}}\right)\left(\frac{x' + y'}{\sqrt{2}}\right) - 4 = 0$$

$$\frac{(x')^2 - (y')^2}{2} = 4$$

$$\frac{(x')^2}{8} - \frac{(y')^2}{8} = 1$$

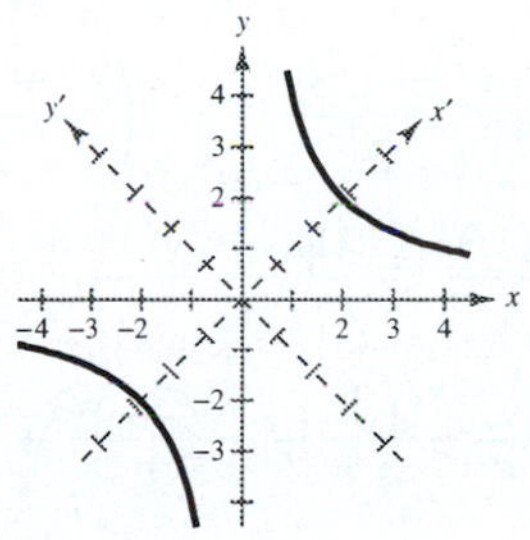

8. $xy + x - 2y + 3 = 0$

$A = 0, B = 1, C = 0$

$\cot 2\theta = \dfrac{A - C}{B} = 0 \Rightarrow 2\theta = \dfrac{\pi}{2} \Rightarrow \theta = \dfrac{\pi}{4}$

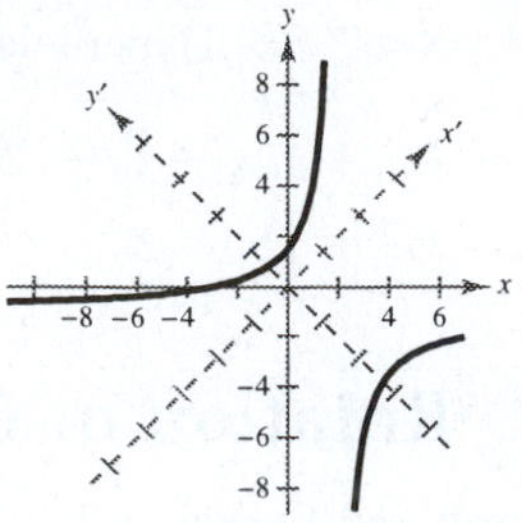

$$x = x'\cos\frac{\pi}{4} - y'\sin\frac{\pi}{4} = x'\left(\frac{\sqrt{2}}{2}\right) - y'\left(\frac{\sqrt{2}}{2}\right) = \frac{x' - y'}{\sqrt{2}}$$

$$y = x'\sin\frac{\pi}{4} + y'\cos\frac{\pi}{4} = x'\left(\frac{\sqrt{2}}{2}\right) + y'\left(\frac{\sqrt{2}}{2}\right) = \frac{x' + y'}{\sqrt{2}}$$

$$xy + x - 2y + 3 = 0$$

$$\left(\frac{x' - y'}{\sqrt{2}}\right)\left(\frac{x' + y'}{\sqrt{2}}\right) + \left(\frac{x' - y'}{\sqrt{2}}\right) - 2\left(\frac{x' + y'}{\sqrt{2}}\right) + 3 = 0$$

$$\frac{(x')^2}{2} - \frac{(y')^2}{2} + \frac{x'}{\sqrt{2}} - \frac{y'}{\sqrt{2}} - \frac{2x'}{\sqrt{2}} - \frac{2y'}{\sqrt{2}} + 3 = 0$$

$$\left[(x')^2 - \sqrt{2}x' + \left(\frac{\sqrt{2}}{2}\right)^2\right] - \left[(y')^2 + 3\sqrt{2}y' + \left(\frac{3\sqrt{2}}{2}\right)^2\right] = -6 + \left(\frac{\sqrt{2}}{2}\right)^2 - \left(\frac{3\sqrt{2}}{2}\right)^2$$

$$\left(x' - \frac{\sqrt{2}}{2}\right)^2 - \left(y' + \frac{3\sqrt{2}}{2}\right)^2 = -10$$

$$\frac{\left(y' + \frac{3\sqrt{2}}{2}\right)^2}{10} - \frac{\left(x' - \frac{\sqrt{2}}{2}\right)^2}{10} = 1$$

10. $13x^2 + 6\sqrt{3}xy + 7y^2 - 16 = 0$

$A = 13, B = 6\sqrt{3}, C = 7$

$\cot 2\theta = \dfrac{A - C}{B} = \dfrac{1}{\sqrt{3}} \Rightarrow 2\theta = \dfrac{\pi}{3} \Rightarrow \theta = \dfrac{\pi}{6}$

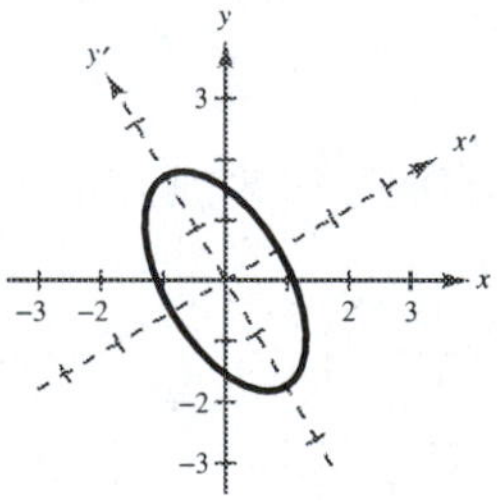

$$x = x'\cos\frac{\pi}{6} - y'\sin\frac{\pi}{6} = x'\left(\frac{\sqrt{3}}{2}\right) - y'\left(\frac{1}{2}\right) = \frac{\sqrt{3}x' - y'}{2}$$

$$y = x'\sin\frac{\pi}{6} + y'\cos\frac{\pi}{6} = x'\left(\frac{1}{2}\right) + y'\left(\frac{\sqrt{3}}{2}\right) = \frac{x' + \sqrt{3}y'}{2}$$

$$13x^2 + 6\sqrt{3}xy + 7y^2 - 16 = 0$$

$$13\left(\frac{\sqrt{3}x' - y'}{2}\right)^2 + 6\sqrt{3}\left(\frac{\sqrt{3}x' - y'}{2}\right)\left(\frac{x' + \sqrt{3}y'}{2}\right) + 7\left(\frac{x' + \sqrt{3}y'}{2}\right)^2 - 16 = 0$$

$$\frac{39(x')^2}{4} - \frac{13\sqrt{3}x'y'}{2} + \frac{13(y')^2}{4} + \frac{18(x')^2}{4} + \frac{18\sqrt{3}x'y'}{4} - \frac{6\sqrt{3}x'y'}{4}$$

$$- \frac{18(y')^2}{4} + \frac{7(x')^2}{4} + \frac{7\sqrt{3}x'y'}{2} + \frac{21(y')^2}{4} - 16 = 0$$

$$16(x')^2 + 4(y')^2 = 16$$

$$\frac{(x')^2}{1} + \frac{(y')^2}{4} = 1$$

12. $2x^2 - 3xy - 2y^2 + 10 = 0$

$A = 2, B = -3, C = -2$

$\cot 2\theta = \dfrac{A - C}{B} = -\dfrac{4}{3} \implies \theta \approx 71.57°$

$\cos 2\theta = -\dfrac{4}{5}$

$\sin\theta = \sqrt{\dfrac{1 - \cos 2\theta}{2}} = \sqrt{\dfrac{1 - (-4/5)}{2}} = \dfrac{3}{\sqrt{10}}$

$\cos\theta = \sqrt{\dfrac{1 + \cos 2\theta}{2}} = \sqrt{\dfrac{1 + (-4/5)}{2}} = \dfrac{1}{\sqrt{10}}$

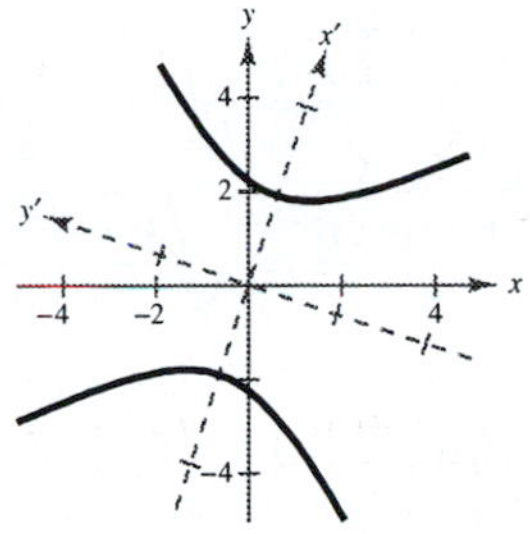

$x = x'\cos\theta - y'\sin\theta$

$= x'\left(\dfrac{1}{\sqrt{10}}\right) - y'\left(\dfrac{3}{\sqrt{10}}\right)$

$= \dfrac{x' - 3y'}{\sqrt{10}}$

$y = x'\sin\theta + y'\cos\theta$

$= x'\left(\dfrac{3}{\sqrt{10}}\right) + y'\left(\dfrac{1}{\sqrt{10}}\right)$

$= \dfrac{3x' + y'}{\sqrt{10}}$

$$2x^2 - 3xy - 2y^2 + 10 = 0$$

$$2\left(\frac{x' - 3y'}{\sqrt{10}}\right)^2 - 3\left(\frac{x' - 3y'}{\sqrt{10}}\right)\left(\frac{3x' + y'}{\sqrt{10}}\right) - 2\left(\frac{3x' + y'}{\sqrt{10}}\right)^2 + 10 = 0$$

$$\frac{(x')^2}{5} - \frac{6x'y'}{5} + \frac{9(y')^2}{5} - \frac{9(x')^2}{10} + \frac{24x'y'}{10} + \frac{9(y')^2}{10} - \frac{9(x')^2}{5} - \frac{6x'y'}{5} - \frac{(y')^2}{5} + 10 = 0$$

$$-\frac{5}{2}(x')^2 + \frac{5}{2}(y')^2 = -10$$

$$\frac{(x')^2}{4} - \frac{(y')^2}{4} = 1$$

14. $16x^2 - 24xy + 9y^2 - 60x - 80y + 100 = 0$

$A = 16, B = -24, C = 9$

$\cot 2\theta = \dfrac{A - C}{B} = -\dfrac{7}{24} \implies \theta \approx 53.13°$

$\cos 2\theta = -\dfrac{7}{25}$

$\sin\theta = \sqrt{\dfrac{1 - \cos 2\theta}{2}} = \sqrt{\dfrac{1 - (-7/25)}{2}} = \dfrac{4}{5}$

$\cos\theta = \sqrt{\dfrac{1 + \cos 2\theta}{2}} = \sqrt{\dfrac{1 + (-7/25)}{2}} = \dfrac{3}{5}$

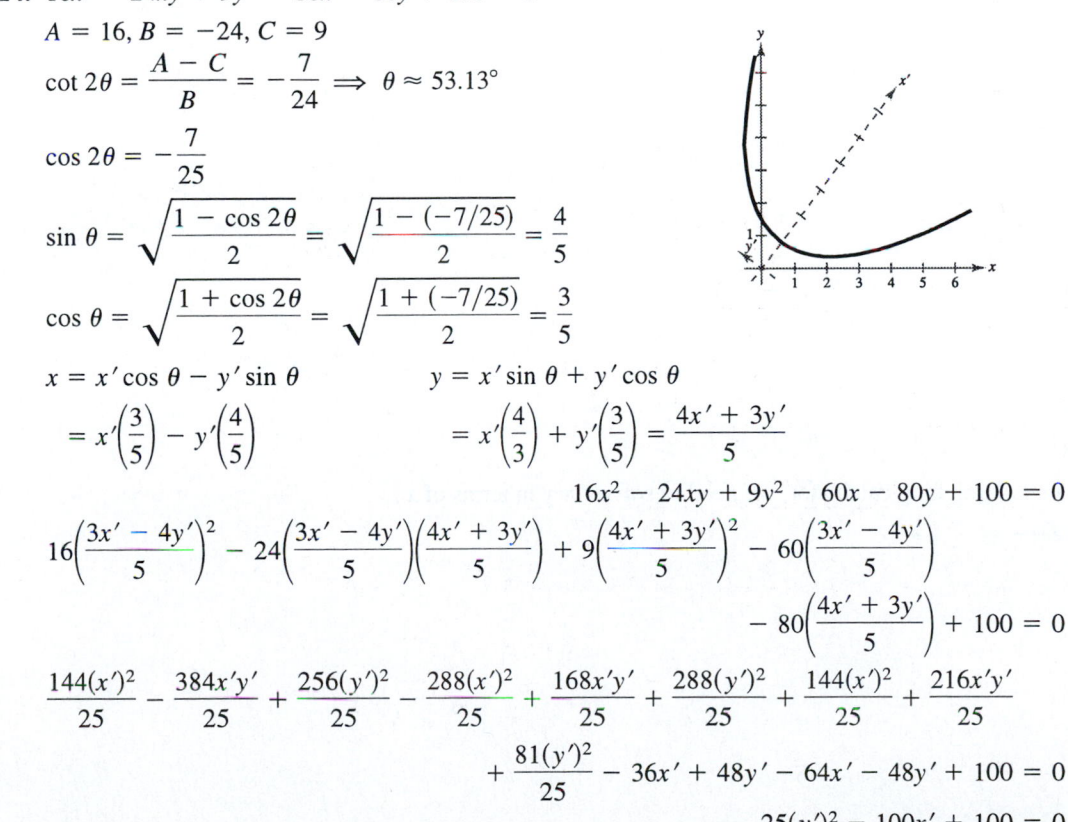

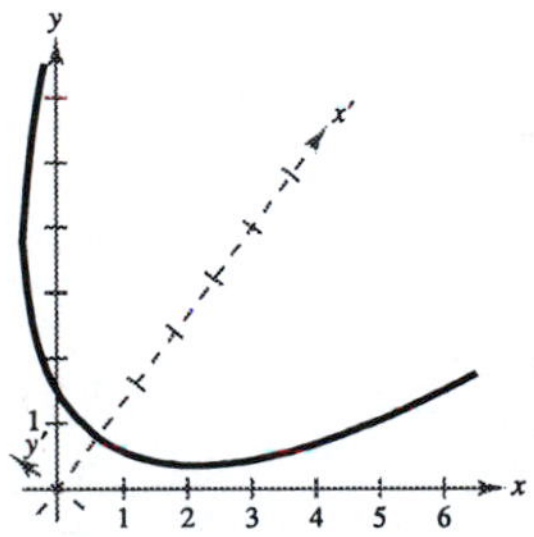

$x = x'\cos\theta - y'\sin\theta$

$= x'\left(\dfrac{3}{5}\right) - y'\left(\dfrac{4}{5}\right)$

$y = x'\sin\theta + y'\cos\theta$

$= x'\left(\dfrac{4}{3}\right) + y'\left(\dfrac{3}{5}\right) = \dfrac{4x' + 3y'}{5}$

$$16x^2 - 24xy + 9y^2 - 60x - 80y + 100 = 0$$

$$16\left(\frac{3x' - 4y'}{5}\right)^2 - 24\left(\frac{3x' - 4y'}{5}\right)\left(\frac{4x' + 3y'}{5}\right) + 9\left(\frac{4x' + 3y'}{5}\right)^2 - 60\left(\frac{3x' - 4y'}{5}\right)$$

$$- 80\left(\frac{4x' + 3y'}{5}\right) + 100 = 0$$

$$\frac{144(x')^2}{25} - \frac{384x'y'}{25} + \frac{256(y')^2}{25} - \frac{288(x')^2}{25} + \frac{168x'y'}{25} + \frac{288(y')^2}{25} + \frac{144(x')^2}{25} + \frac{216x'y'}{25}$$

$$+ \frac{81(y')^2}{25} - 36x' + 48y' - 64x' - 48y' + 100 = 0$$

$$25(y')^2 - 100x' + 100 = 0$$

$$(y')^2 = 4(x' - 1)$$

16. $9x^2 + 24xy + 16y^2 + 80x - 60y = 0$

$A = 9, B = 24, C = 16$

$\cot 2\theta = \dfrac{A - C}{B} = -\dfrac{7}{24} \Rightarrow \theta \approx 53.13°$

$\cos 2\theta = -\dfrac{7}{25}$

$\sin \theta = \sqrt{\dfrac{1 - \cos 2\theta}{2}} = \sqrt{\dfrac{1 - (-7/25)}{2}} = \dfrac{4}{5}$

$\cos \theta = \sqrt{\dfrac{1 + \cos 2\theta}{2}} = \sqrt{\dfrac{1 + (-7/25)}{2}} = \dfrac{3}{5}$

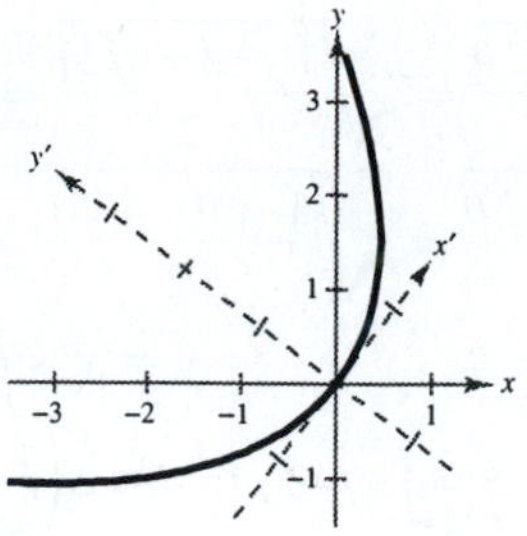

$x = x'\cos \theta - y'\sin \theta$
$= x'\left(\dfrac{3}{5}\right) - y'\left(\dfrac{4}{5}\right)$
$= \dfrac{3x' - 4y'}{5}$

$y = x'\sin \theta + y'\cos \theta$
$= x'\left(\dfrac{4}{5}\right) + y'\left(\dfrac{3}{5}\right)$
$= \dfrac{4x' + 3y}{5}$

$$9x^2 + 24xy + 16y^2 + 80x - 60y = 0$$

$$9\left(\frac{3x' - 4y'}{5}\right)^2 + 24\left(\frac{3xy - 4y'}{5}\right)\left(\frac{4x' + 3y'}{5}\right) + 16\left(\frac{4x' + 3y'}{5}\right)^2 + 80\left(\frac{3x' - 4y'}{5}\right) - 60\left(\frac{4x' + 3y'}{5}\right) = 0$$

$$\frac{81(x')^2}{25} - \frac{216x'y'}{25} + \frac{144(y')^2}{25} + \frac{288(x')^2}{25} - \frac{168x'y'}{25} - \frac{288(y')^2}{25} + \frac{256(x')^2}{25} + \frac{384x'y'}{25}$$

$$+ \frac{144(y')^2}{25} + 48x' - 64x' - 48x' - 36x' = 0$$

$$25(x')^2 - 100y' = 0$$

$$(x')^2 = 4y'$$

$$\frac{1}{4}(x')^2 = y'$$

18. $x^2 - 4xy + 2y^2 = 6$

$A = 1, B = -4, C = 2$

$\cot 2\theta = \dfrac{A - C}{B} = \dfrac{1 - 2}{-4} = \dfrac{1}{4}$

$\dfrac{1}{\tan 2\theta} = \dfrac{1}{4}$

$\tan 2\theta = 4$

$2\theta \approx 75.96$

$\theta \approx 37.98°$

To graph conic with a graphing calculator, we need to solve for y in terms of x.

—CONTINUED—

18. —CONTINUED—

$$x^2 - 4xy + 2y^2 = 6$$

$$y^2 - 2xy + x^2 = 3 - \frac{x^2}{2} + x^2$$

$$(y - x)^2 = 3 + \frac{x^2}{2}$$

$$y - x = \pm\sqrt{3 + \frac{x^2}{2}}$$

$$y = x \pm \sqrt{3 + \frac{x^2}{2}}$$

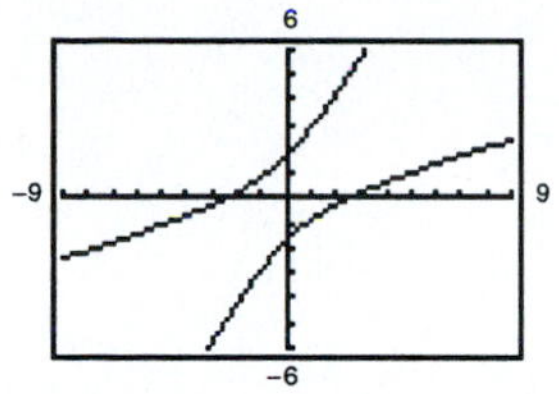

Enter $y_1 = x + \sqrt{3 + \frac{x^2}{2}}$ and $y_2 = x - \sqrt{3 + \frac{x^2}{2}}$.

20. $40x^2 + 36xy + 25y^2 = 52$

$A = 40, B = 36, C = 25$

$$\cot 2\theta = \frac{A - C}{B} = \frac{40 - 25}{36} = \frac{5}{12}$$

$$\frac{1}{\tan 2\theta} = \frac{5}{12}$$

$$\tan 2\theta = \frac{12}{5}$$

$$2\theta \approx 67.38°$$

$$\theta \approx 33.69°$$

Solve for y in terms of x by completing the square:

$$25y^2 + 36xy = 52 - 40x^2$$

$$y^2 + \frac{36}{25}xy = \frac{52}{25} - \frac{40}{25}x^2$$

$$y^2 + \frac{36}{25}xy + \frac{324}{625}x^2 = \frac{52}{25} - \frac{40}{25}x^2 + \frac{324}{625}x^2$$

$$\left(y + \frac{18}{25}x\right)^2 = \frac{1300 - 676x^2}{625}$$

$$y + \frac{18}{25}x = \pm\sqrt{\frac{1300 - 676x^2}{625}}$$

$$y = \frac{-18x \pm \sqrt{1300 - 676x^2}}{25}$$

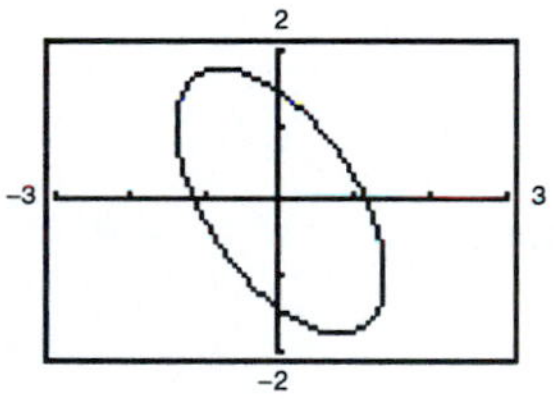

Enter $y_1 = \dfrac{-18x + \sqrt{1300 - 676x^2}}{25}$ and $y_2 = \dfrac{-18x - \sqrt{1300 - 676x^2}}{25}$.

22. $4x^2 - 12xy + 9y^2 + (4\sqrt{13} - 12)x - (6\sqrt{13} + 8)y = 91$

$A = 4, B = -12, C = 9$

$\cot 2\theta = \dfrac{A - C}{B} = \dfrac{4 - 9}{-12} = \dfrac{5}{12}$

$\dfrac{1}{\tan 2\theta} = \dfrac{5}{12}$

$\tan 2\theta = \dfrac{12}{5}$

$2\theta \approx 67.38°$

$\theta \approx 33.69°$

Solve for y in terms of x with the quadratic formula:

$$4x^2 - 12xy + 9y^2 + (4\sqrt{13} - 12)x - (6\sqrt{13} + 8)y = 91$$

$$9y^2 - (12x + 6\sqrt{13} + 8)y + (4x^2 + 4\sqrt{13}x - 12x - 91) = 0$$

$a = 9, b = -(12x + 6\sqrt{13} + 8), c = 4x^2 + 4\sqrt{13}x - 12x - 91$

$$y = \frac{-b \pm \sqrt{b^2 - 4ac}}{2a}$$

$$y = \frac{(12x + 6\sqrt{13} + 8) \pm \sqrt{(12x + 6\sqrt{13} + 8)^2 - 4(9)(4x^2 + 4\sqrt{13}x - 12x - 91)}}{18}$$

$$= \frac{(12x + 6\sqrt{13} + 8) \pm \sqrt{624x + 3808 + 96\sqrt{13}}}{18}$$

Enter $y_1 = \dfrac{12x + 6\sqrt{13} + 8 + \sqrt{624x + 3808 + 96\sqrt{13}}}{18}$

and $y_2 = \dfrac{12x + 6\sqrt{13} + 8 - \sqrt{624x + 3808 + 96\sqrt{13}}}{18}$.

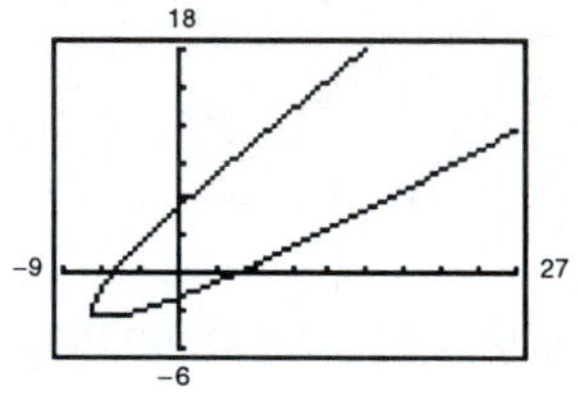

24. $x^2 + 2xy + y^2 = 0$

$(x + y)^2 = 0$

$x + y = 0$

$y = -x$

The graph is a line. Matches graph (f).

26. $x^2 - xy + 3y^2 - 5 = 0$

$A = 1, B = -1, C = 3$

$B^2 - 4AC = (-1)^2 - 4(1)(3) = -11$

The graph is an ellipse.

$\cot 2\theta = \dfrac{A - C}{B} = \dfrac{1 - 3}{-1} = 2 \implies \theta \approx 13.28°$

Matches graph (a).

28. $x^2 - 4xy + 4y^2 + 10x - 30 = 0$

$A = 1, B = -4, C = 4$

$B^2 - 4AC = (-4)^2 - 4(1)(4) = 0$

The graph is a parabola.

$\cot 2\theta = \dfrac{A - C}{B} = \dfrac{1 - 4}{-4} = \dfrac{3}{4} \implies \theta \approx 12.66°$

Matches graph (c).

30. $x^2 - 4xy - 2y^2 - 6 = 0$

$A = 1, B = -4, C = -2$

$B^2 - 4AC = (-4)^2 - 4(1)(-2) = 24 > 0$

Hyperbola

32. $2x^2 + 4xy + 5y^2 + 3x - 4y - 20 = 0$

$A = 2, B = 4, C = 5$

$B^2 - 4AC = 4^2 - 4(2)(5) = 16 - 40 = -24 < 0$

Ellipse

34. $36x^2 - 60xy + 25y^2 + 9y = 0$

$A = 36, B = -60, C = 25$

$B^2 - 4AC = (-60)^2 - 4(36)(25)$

$= 3600 - 3600 = 0$

Parabola

36. $x^2 + xy + 4y^2 + x + y - 4 = 0$

$A = 1, B = 1, C = 4$

$B^2 - 4AC = 1^2 - 4(1)(4) = -15 < 0$

Ellipse

38.
$$\begin{aligned} x^2 + y^2 - 2x + 6y + 10 &= 0 \\ (x^2 - 2x + 1) + (y^2 + 6y + 9) &= -10 + 1 + 9 \\ (x - 1)^2 + (y + 3)^2 &= 0 \end{aligned}$$

Point at $(1, -3)$

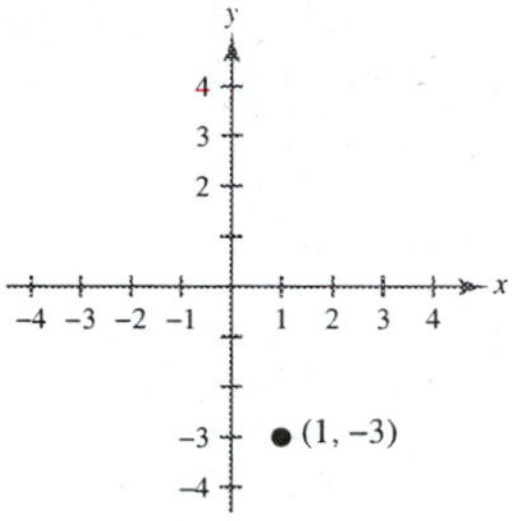

40.
$$\begin{aligned} x^2 + 10xy + y^2 &= 0 \\ y^2 - 10xy + 25x^2 &= 25x^2 - x^2 \\ (y - 5x)^2 &= 24x^2 \\ y - 5x &= \pm\sqrt{24x^2} \\ y &= 5x \pm 2\sqrt{6}x \\ y &= \left(5 \pm 2\sqrt{6}\right)x \end{aligned}$$

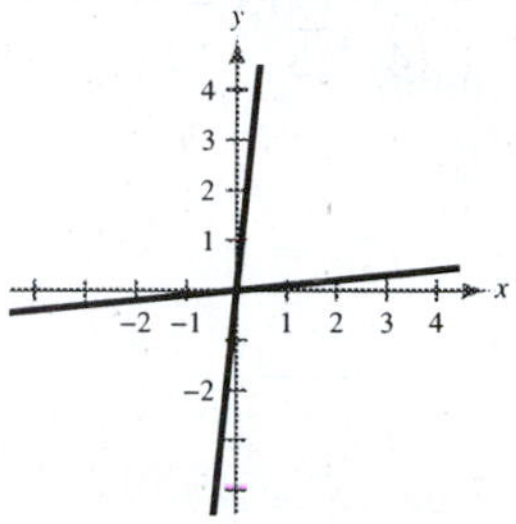

42.
$$\begin{aligned} -x^2 - y^2 - 8x + 20y - 7 &= 0 \Rightarrow (x + 4)^2 + (y - 10)^2 = 109 \\ x^2 + 9y^2 + 8x + 4y + 7 &= 0 \Rightarrow (x + 4)^2 + 9\left(y + \tfrac{2}{9}\right)^2 = \tfrac{85}{9} \\ \hline 8y^2 + 24y &= 0 \\ 8y(y + 3) &= 0 \\ y = 0 \text{ or } y &= -3 \end{aligned}$$

For $y = 0$: $x^2 + 9(0)^2 + 8x + 4(0) + 7 = 0$
$$(x + 7)(x + 1) = 0$$
$$x = -7, -1$$

For $y = -3$: $x^2 + 9(-3)^2 + 8x + 4(-3) + 7 = 0$
$$x^2 + 8x + 76 = 0$$

No real solution

Points of intersection: $(-7, 0), (-1, 0)$

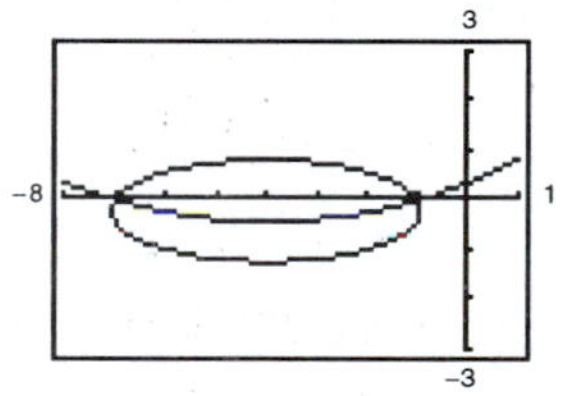

44.
$$\begin{aligned} x^2 - 4y^2 - 20x - 64y - 172 &= 0 \Rightarrow (x - 10)^2 - 4(y + 8)^2 = 16 \\ 16x^2 + 4y^2 - 320x + 64y - 1600 &= 0 \Rightarrow 16(x - 10)^2 + 4(y + 8)^2 = 256 \\ \hline 17x^2 - 340x + 1428 &= 0 \\ (17x - 238)(x - 6) &= 0 \\ x = 6 \text{ or } x &= 14 \end{aligned}$$

When $x = 6$: $6^2 - 4y^2 - 20(6) - 64y - 172 = 0$
$$\begin{aligned} -4y^2 - 64y - 256 &= 0 \\ y^2 + 16y + 64 &= 0 \\ (y + 8)^2 &= 0 \\ y &= -8 \end{aligned}$$

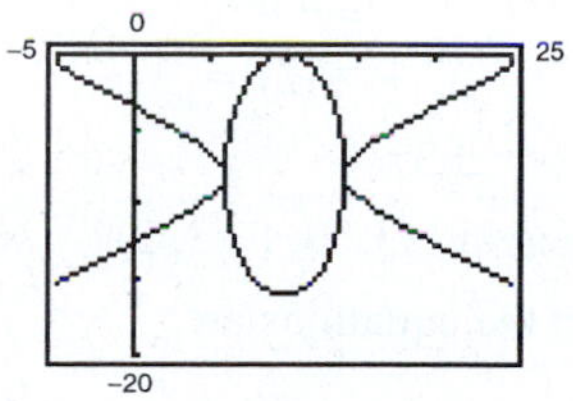

Points of intersection: $(6, -8), (14, -8)$

46. $x^2 + 4y^2 - 2x - 8y + 1 = 0 \Rightarrow (x-1)^2 + 4(y-1)^2 = 4$

$-x^2 \qquad + 2x - 4y - 1 = 0 \Rightarrow y = -\frac{1}{4}(x-1)^2$

$4y^2 \qquad -12y \qquad = 0$

$4y(y-3) = 0$

$y = 0$ or $y = 3$

When $y = 0$: $x^2 + 4(0)^2 - 2x - 8(0) + 1 = 0$

$x^2 - 2x + 1 = 0$

$(x-1)^2 = 0$

$x = 1$

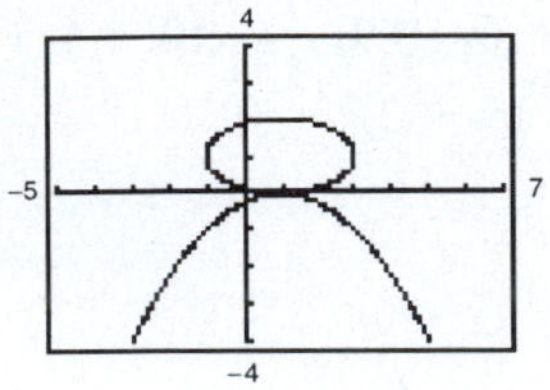

When $y = 3$: $-x^2 + 2x - 4(3) - 1 = 0$

$x^2 - 2x + 13 = 0$

No real solution

Point of intersection: $(1, 0)$

48. $16x^2 - y^2 \qquad + 16y - 128 = 0 \Rightarrow 16x^2 - (y-8)^2 = 64$

$y^2 - 48x - 16y - \ 32 = 0 \Rightarrow (y-8)^2 - 48x = 96$

$16x^2 \qquad - 48x \qquad - 160 = 0$

$16(x^2 - 3x - 10) = 0$

$(x-5)(x+2) = 0$

$x = 5$ or $x = -2$

When $x = 5$: $y^2 - 48(5) - 16y - 32 = 0$

$y^2 - 16y - 272 = 0$

$y = 8 \pm 4\sqrt{21}$

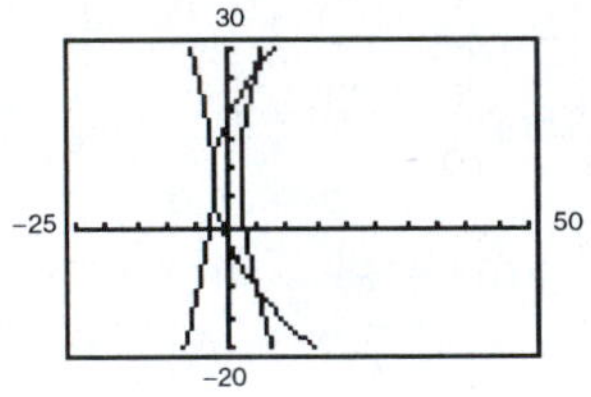

When $x = -2$: $y^2 - 48(-2) - 16y - 32 = 0$

$y^2 - 16y + 64 = 0$

$(y-8)^2 = 0$

$y = 8$

Points of intersection: $(5, 8 + 4\sqrt{21}), (5, 8 - 4\sqrt{21}), (-2, 8)$

50. $4x^2 + 9y^2 - 36y = 0$

$x^2 + 9y \ - \ 27 \ = 0 \Rightarrow x^2 = 27 - 9y$

$4(27 - 9y) + 9y^2 - 36y = 0$

$9y^2 - 72y + 108 = 0$

$9(y-6)(y-2) = 0$

$y = 6$ or $y = 2$

When $y = 6$: $x^2 = 27 - 9(6) = -27$

No real solution

When $y = 2$: $x^2 = 27 - 9(2) = 9$

$x = \pm 3$

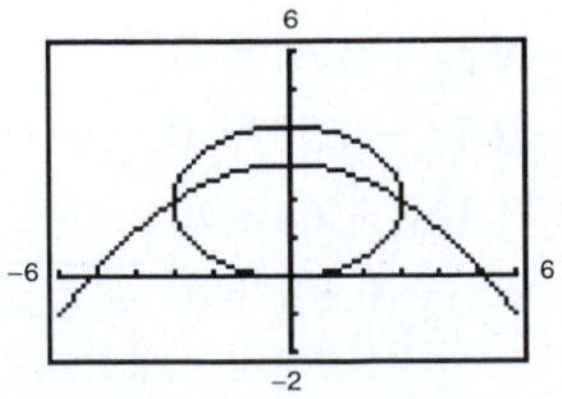

Points of intersection: $(3, 2), (-3, 2)$

In standard form the equations are:

$$\frac{x^2}{9} + \frac{(y-2)^2}{4} = 1$$

$$y = -\frac{x^2}{9} + 3$$

52. $x^2 + 2y^2 - 4x + 6y - 5 = 0$

$$x - 4x - y + 4 = 0 \Rightarrow x^2 - 4x = y - 4$$

$$y - 4 + 2y^2 + 6y - 5 = 0$$

$$2y^2 + 7y - 9 = 0$$

$$(2y + 9)(y - 1) = 0$$

$$y = -\frac{9}{2} \text{ or } y = 1$$

When $y = 1$: $x^2 - 4x - 1 + 4 = 0$

$$(x - 3)(x - 1) = 0$$

$$x = 1 \text{ or } x = 3$$

When $y = -\frac{9}{2}$: $x^2 - 4x - \left(-\frac{9}{2}\right) + 4 = 0$

$$x^2 - 4x + \frac{17}{2} = 0$$

No real solution

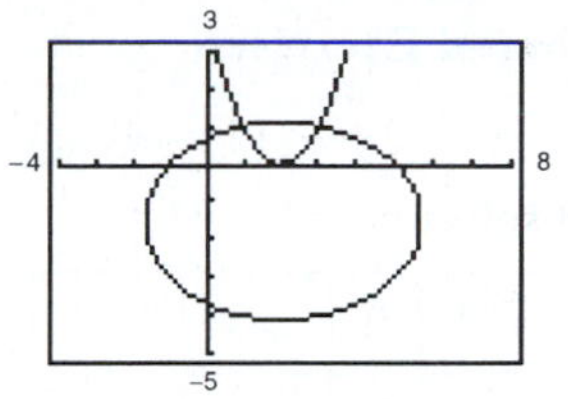

Points of intersection: (1, 1), (3, 1)

In standard form the equations are: $\dfrac{(x - 2)^2}{27/2} + \dfrac{2[y + (3/2)]^2}{27/2} = 1$

$$y = x^2 - 4x + 4$$

54. $5x^2 - 2xy + 5y^2 - 12 = 0$

$$x + y - 1 = 0 \Rightarrow y = 1 - x$$

$$5x^2 - 2x(1 - x) + 5(1 - x)^2 - 12 = 0$$

$$5x^2 - 2x + 2x^2 + 5(1 - 2x + x^2) - 12 = 0$$

$$5x^2 - 2x + 2x^2 + 5 - 10x + 5x^2 - 12 = 0$$

$$12x^2 - 12x - 7 = 0$$

$$x = \frac{3 \pm \sqrt{30}}{6}$$

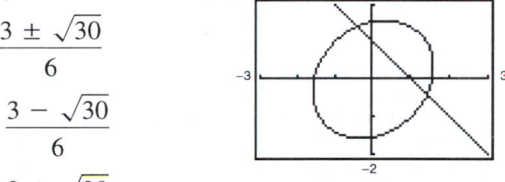

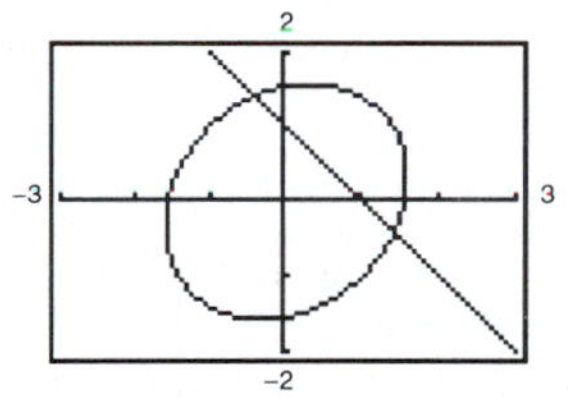

When $x = \dfrac{3 + \sqrt{30}}{6}$: $y = 1 - \dfrac{3 + \sqrt{30}}{6} = \dfrac{3 - \sqrt{30}}{6}$

When $x = \dfrac{3 - \sqrt{30}}{6}$: $y = 1 - \dfrac{3 - \sqrt{30}}{6} = \dfrac{3 + \sqrt{30}}{6}$

Points of intersection: $\left(\frac{1}{6}(3 + \sqrt{30}), \frac{1}{6}(3 - \sqrt{30})\right), \left(\frac{1}{6}(3 - \sqrt{30}), \frac{1}{6}(3 + \sqrt{30})\right)$

56. In Exercise 10, the equation of the rotated ellipse is:

$$\frac{(x')^2}{1} + \frac{(y')^2}{4} = 1$$

$a^2 = 4 \Rightarrow a = 2$

$b^2 = 1 \Rightarrow b = 1$

Length of major axis is $2a = 2(2) = 4$.

Length of minor axis is $2b = 2(1) = 2$.

58. $f(x) = \dfrac{2x}{2 - x}$

y-intercept: (0, 0)
Vertical asymptote: $x = 2$
Horizontal asymptote: $y = -2$

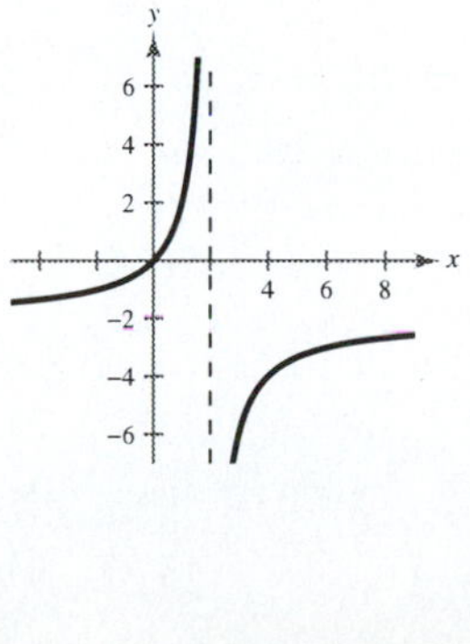

60. $g(s) = \dfrac{2}{4 - s^2}$

Intercept: $\left(0, \dfrac{1}{2}\right)$

Vertical asymptotes: $s = 2, s = -2$

Horizontal asymptote: $y = 0$

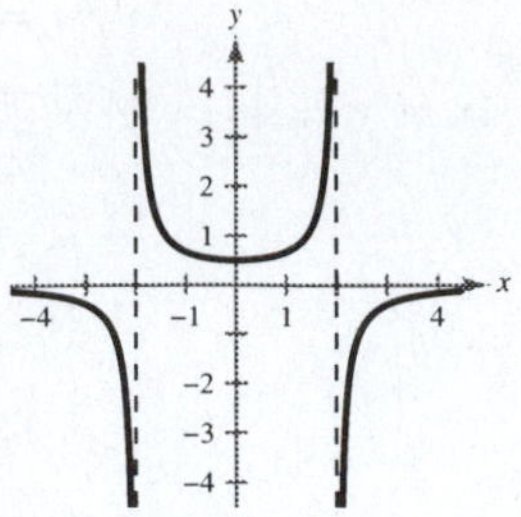

Section 10.6 Parametric Equations

Solutions to Even-Numbered Exercises

2. $x = 4\cos^2\theta, y = 2\sin\theta$

(a)

t	$-\pi/2$	$-\pi/4$	0	$\pi/4$	$\pi/2$
x	0	2	4	2	0
y	-2	$-\sqrt{2}$	0	$\sqrt{2}$	2

(b)

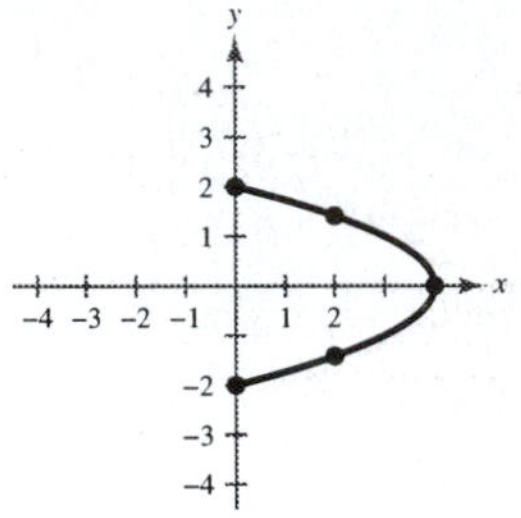

(c) $\dfrac{x}{4} = \cos^2\theta, \dfrac{y}{2} = \sin\theta$

$\cos^2\theta + \sin^2\theta = 1$

$\dfrac{x}{4} + \left(\dfrac{y}{2}\right)^2 = 1$

$\dfrac{x}{4} + \dfrac{y^2}{4} = 1$

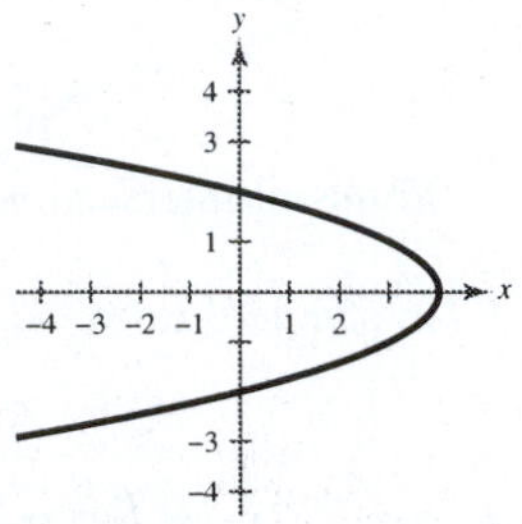

The rectangular version of the graph continues into the second and third quadrants.

4. $x = t$

$y = \frac{1}{2}t$

t	-2	-1	0	1	2
x	-2	-1	0	1	2
y	-1	$-\frac{1}{2}$	0	$\frac{1}{2}$	1

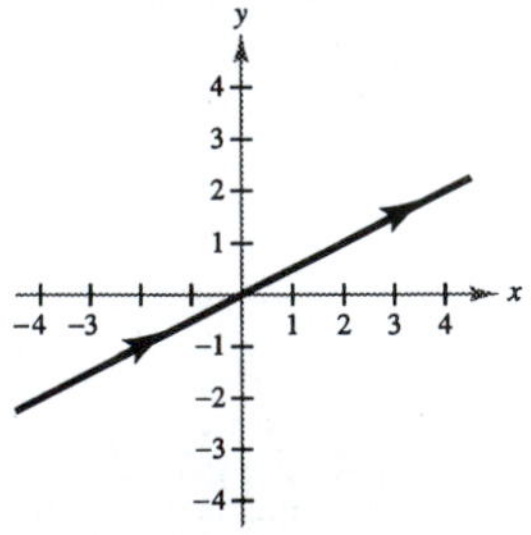

$y = \frac{1}{2}$

$2y = x$

$0 = x - 2y$

6. $x = 3 - 2t \Rightarrow t = -\frac{1}{2}x + \frac{3}{2}$

$y = 2 + 3t$

t	-3	-2	-1	0	1	2	3
x	9	7	5	3	1	-1	-3
y	-7	-4	-1	2	5	8	11

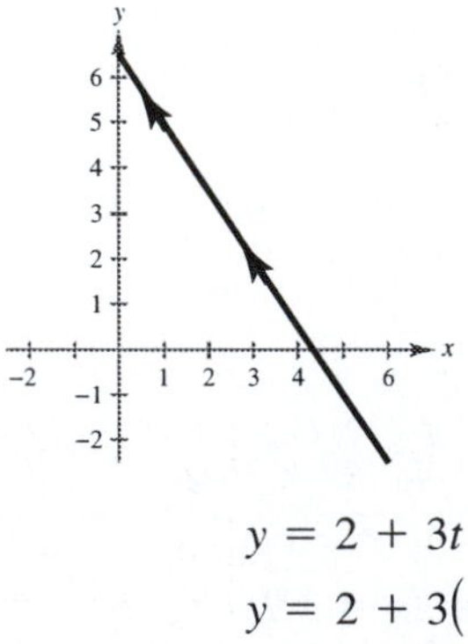

$y = 2 + 3t$

$y = 2 + 3\left(-\frac{1}{2}x + \frac{3}{2}\right)$

$y = 2 - \frac{3}{2}x + \frac{9}{2}$

$2y = 4 - 3x + 9$

$3x + 2y - 13 = 0$

8. $x = t$

$y = t^3$

t	-3	-2	-1	0	1	2	3
x	-3	-2	-1	0	1	2	3
y	-27	-8	-1	0	1	8	27

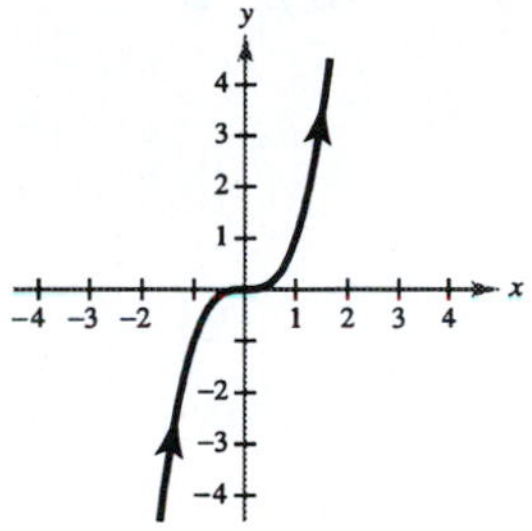

$y = t^3$

$y = x^3$

10. $x = \sqrt{t} \Rightarrow x^2 = t, t \geq 0$

$y = 1 - t$

t	0	1	2	3
x	0	1	$\sqrt{2}$	$\sqrt{3}$
y	1	0	-1	-2

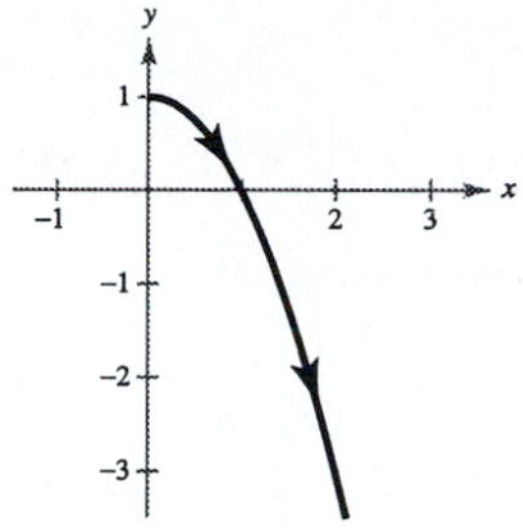

$y = 1 - t$

$y = 1 - x^2, x \geq 0$

12. $x = t - 1 \Rightarrow t = x + 1$

$y = \dfrac{t}{t-1}$

t	-3	-2	-1	0	2	3
x	-4	-3	-2	-1	1	2
y	$\frac{3}{4}$	$\frac{2}{3}$	$\frac{1}{2}$	0	2	$\frac{3}{2}$

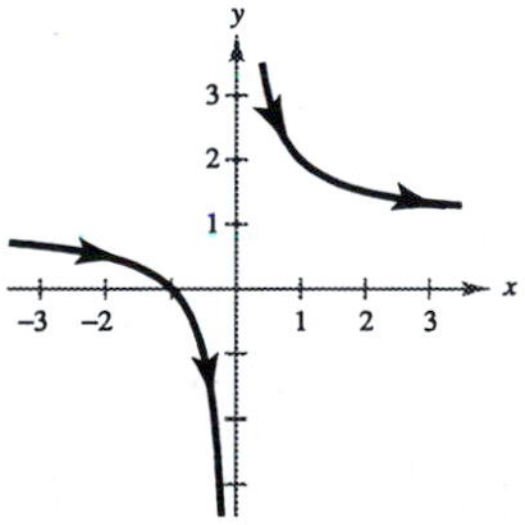

$y = \dfrac{t}{t-1}$

$y = \dfrac{x+1}{x+1-1} = \dfrac{x+1}{x}$

14. $x = |t - 1|$

$y = t + 2 \Rightarrow t = y - 2$

t	-3	-2	-1	0	1	2	3
x	4	3	2	1	0	1	2
y	-1	0	1	2	3	4	5

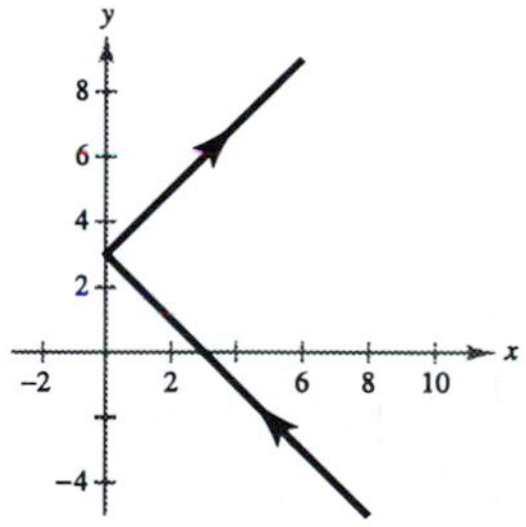

$x = |t - 1|$

$x = |y - 2 - 1| \Rightarrow x = |y - 3|$

16. $x = \cos\theta$

$y = 3\sin\theta \Rightarrow \sin\theta = \dfrac{y}{3}$

t	0	$\pi/4$	$\pi/2$	$3\pi/4$	π	$5\pi/4$	$3\pi/2$	$7\pi/4$	2π
x	1	$\sqrt{2}/2$	0	$-\sqrt{2}/2$	-1	$-\sqrt{2}/2$	0	$\sqrt{2}/2$	1
y	0	$3\sqrt{2}/2$	3	$3\sqrt{2}/2$	0	$-3\sqrt{2}/2$	-3	$-3\sqrt{2}/2$	0

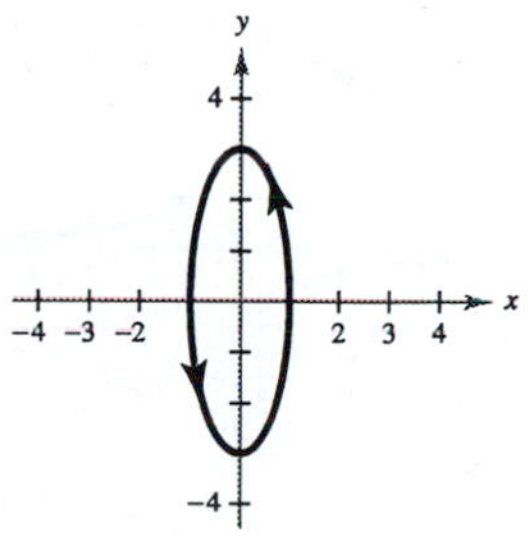

$\cos^2\theta + \sin^2\theta = 1$

$$x^2 + \left(\frac{y}{3}\right)^2 = 1$$

$$x^2 + \frac{y^2}{9} = 1$$

18. $x = \cos\theta$

$y = 2\sin 2\theta$

t	0	$\pi/4$	$\pi/2$	$3\pi/4$	π	$5\pi/4$	$3\pi/2$	$7\pi/4$	2π
x	1	$\sqrt{2}/2$	0	$-\sqrt{2}/2$	-1	$-\sqrt{2}/2$	0	$\sqrt{2}/2$	1
y	0	2	0	-2	0	2	0	-2	0

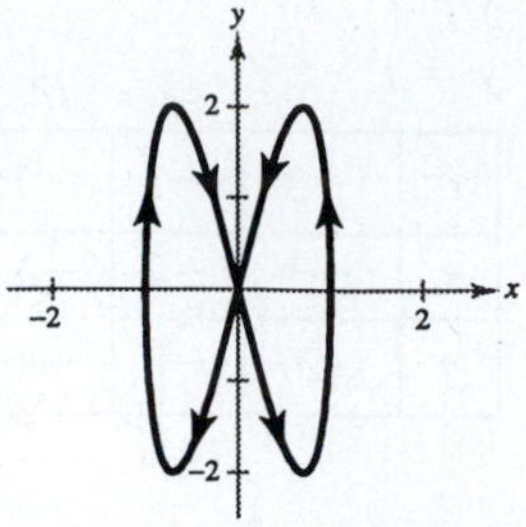

$y = 2\sin 2\theta$

$y = 2(2\sin\theta\cos\theta)$

$y^2 = 16\sin^2\theta\cos^2\theta$

$y^2 = 16(1 - x^2)x^2$

$y^2 = 16x^2(1 - x^2)$

20. $x = 4 + 2\cos\theta \implies \cos\theta = \dfrac{x-4}{2}$

$y = -1 + 2\sin\theta \implies \sin\theta = \dfrac{y+1}{2}$

t	0	$\pi/2$	π	$3\pi/2$	2π
x	6	4	2	4	6
y	-1	1	-1	-3	-1

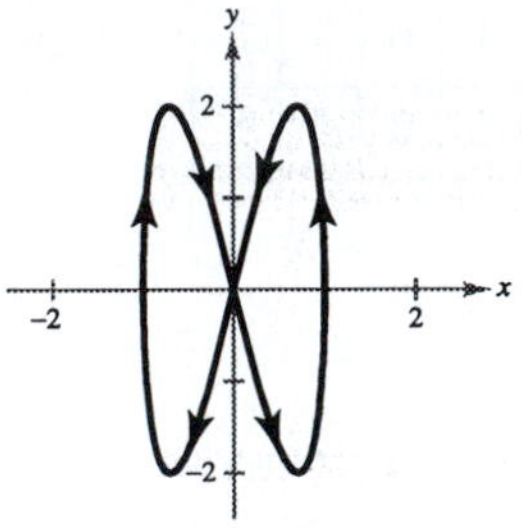

$\cos^2\theta + \sin^2\theta = 1$

$\left(\dfrac{x-4}{2}\right)^2 + \left(\dfrac{y+1}{2}\right)^2 = 1$

$(x-4)^2 + (y+1)^2 = 4$

22. $x = \sec\theta = \dfrac{1}{\cos\theta}$

$y = \tan\theta$

t	0	$\pi/4$	$3\pi/4$	π	$5\pi/4$	$7\pi/4$	2π
x	1	$\sqrt{2}$	$-\sqrt{2}$	-1	$-\sqrt{2}$	$\sqrt{2}$	1
y	0	1	-1	0	1	-1	0

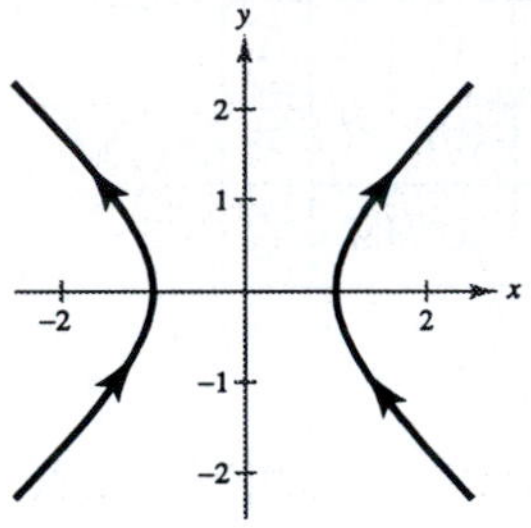

$\sec^2\theta - \tan^2\theta = 1$

$x^2 - y^2 = 1$

24. $x = \sec\theta = \dfrac{1}{\cos\theta} \implies \cos\theta = \dfrac{1}{x}$

$y = \cos\theta$

t	0	$\pi/4$	$3\pi/4$	π	$5\pi/4$	$7\pi/4$	2π
x	1	$\sqrt{2}$	$-\sqrt{2}$	-1	$-\sqrt{2}$	$\sqrt{2}$	1
y	1	$\sqrt{2}/2$	$-\sqrt{2}/2$	-1	$-\sqrt{2}/2$	$\sqrt{2}/2$	1

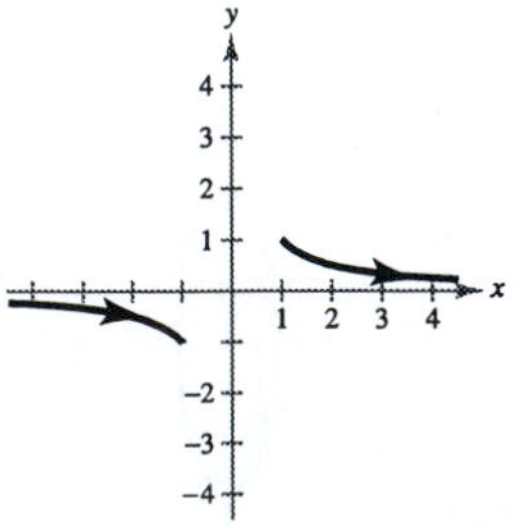

$y = \cos\theta = \dfrac{1}{x} \implies |x| \geq |y| \leq 1$

26. $x = e^{2t}$

$y = e^t \Rightarrow y^2 = e^{2t}$

t	-3	-2	-1	0	1	2
x	0.0025	0.0183	0.1353	1	7.3891	54.5982
y	0.0498	0.1353	0.3679	1	2.7183	7.3891

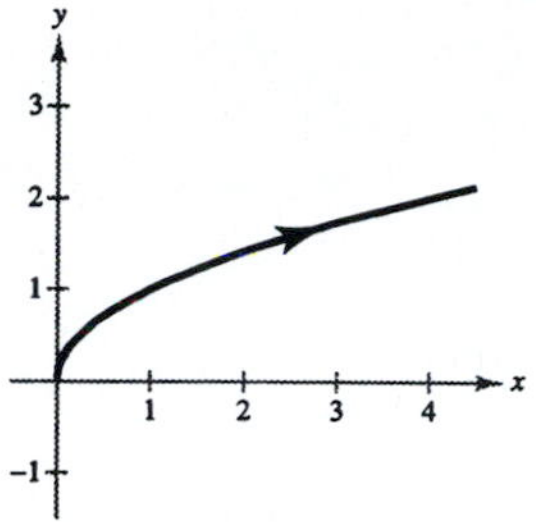

$x = e^{2t} = y^2$

$y^2 = x, y > 0$

28. $x = \ln 2t \Rightarrow t = \frac{1}{2}e^x$

$y = t^2$

t	1	2	3	4
x	0.6931	1.3863	1.7981	2.0794
y	1	4	9	16

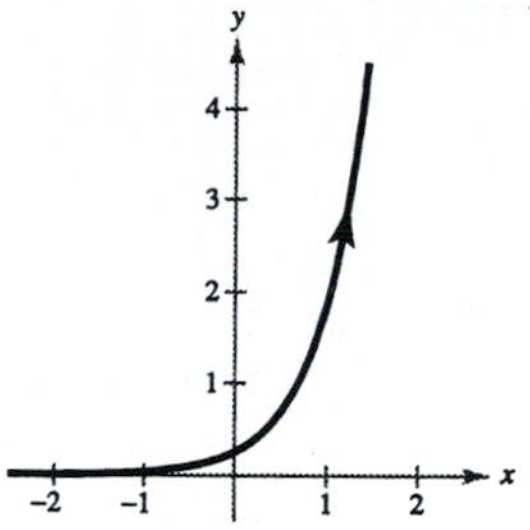

$y = t^2 = \left(\frac{1}{2}e^x\right)^2$

$y = \frac{1}{4}e^{2x}$

30. By eliminating the parameter, each curve represents a portion of $y = x^2 - 1$.

(a) $x = t$

$y = t^2 - 1$

There are no restrictions on x.

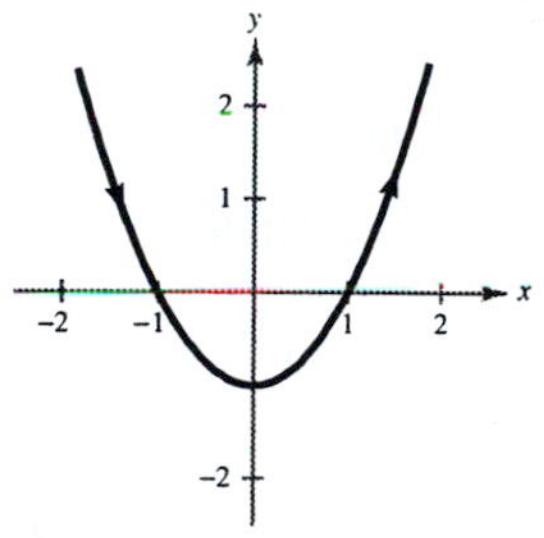

Domain: $(-\infty, \infty)$

Orientation: Left to right

(b) $x = t^2 \Rightarrow x \geq 0$

$y = t^4 - 1$

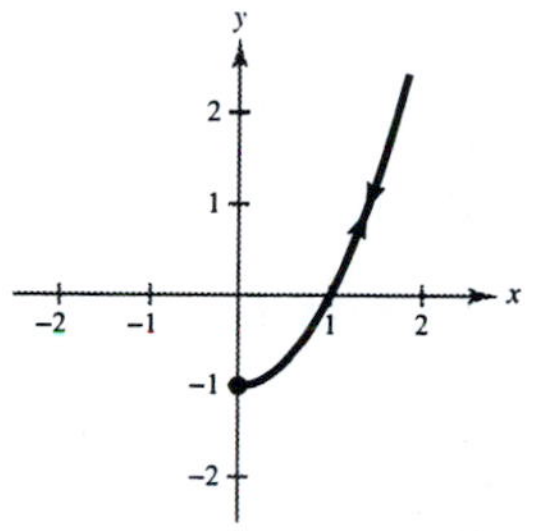

Domain: $[0, \infty)$

Orientation: Depends on t

(c) $x = \sin t \Rightarrow -1 \leq x \leq 1$

$y = \sin^2 t - 1$

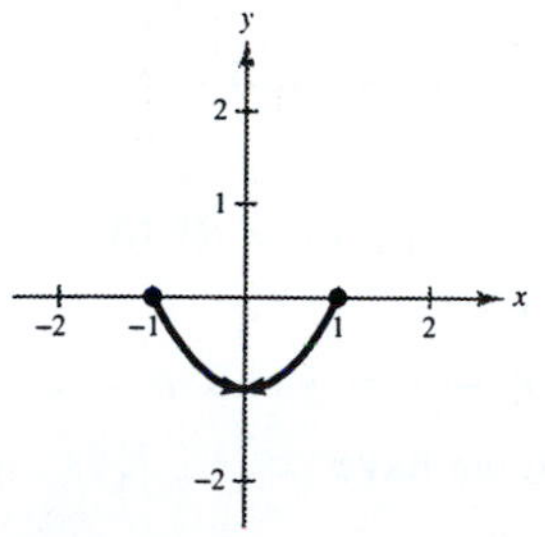

Domain: $[-1, 1]$

Orientation: Depends on t

(d) $x = e^t \Rightarrow x > 0$

$y = e^{2t} - 1$

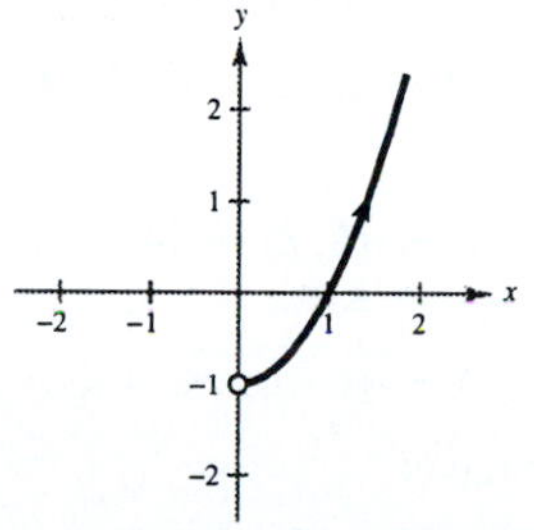

Domain: $(0, \infty)$

Orientation: Left to right

32. By eliminating the parameter, each curve represents a portion of $y = x$.

(a) $x = t, y = t$

There are no restrictions on x.

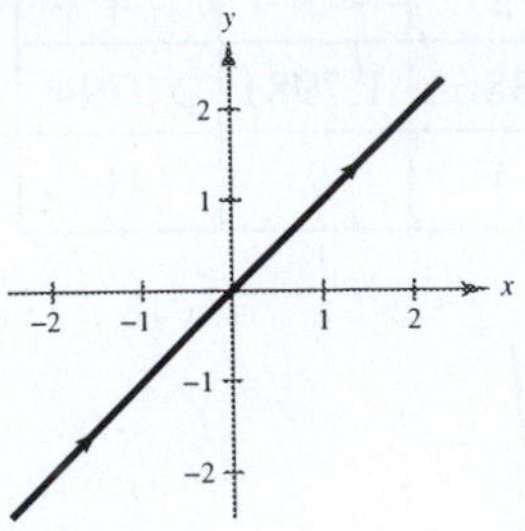

Domain: $(-\infty, \infty)$
Orientation: Left to right

(b) $x = t^2 \Rightarrow x \geq 0$

$y = t^2$

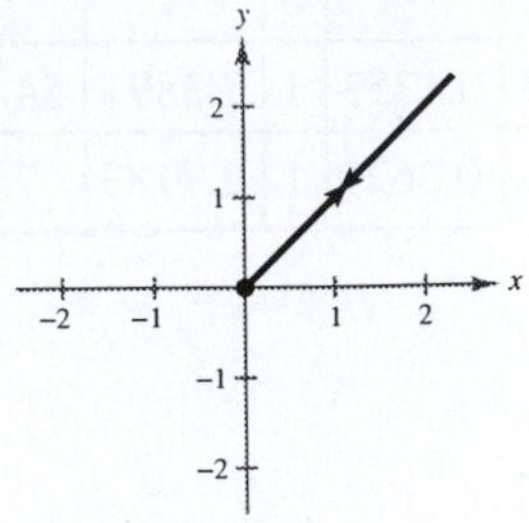

Domain: $[0, \infty)$
Orientation: Depends on t

(c) $x = -t$

$y = -t$

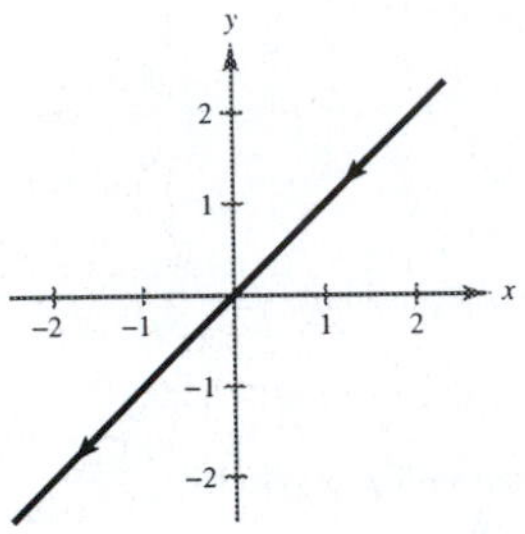

Domain: $(-\infty, \infty)$
Orientation: Right to left

(d) $x = t^3$

$y = t^3$

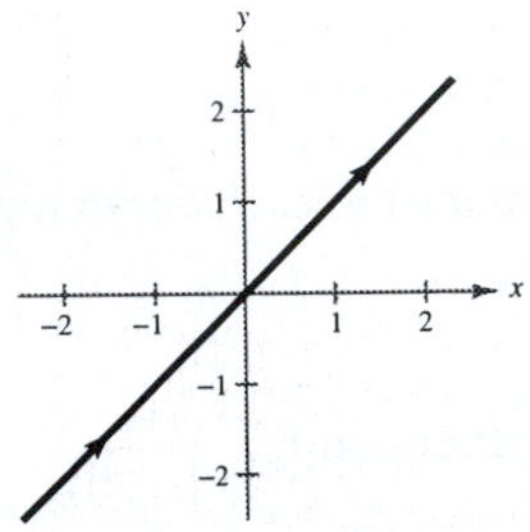

Domain: $(-\infty, \infty)$
Orientation: Left to right

34. $x = h + r\cos\theta, y = k + r\sin\theta$

$$\cos\theta = \frac{x-h}{r}, \sin\theta = \frac{y-k}{r}$$

$$\cos^2\theta + \sin^2\theta = \frac{(x-h)^2}{r^2} + \frac{(y-k)^2}{r^2} = 1$$

$$(x-h)^2 + (y-k)^2 = r^2$$

36. $x = h + a\sec\theta, y = k + b\tan\theta$

$$\frac{x-h}{a} = \sec\theta, \frac{y-k}{b} = \tan\theta$$

$$\frac{(x-h)^2}{a^2} - \frac{(y-k)^2}{b^2} = 1$$

38. Line through $(1, 4)$ and $(5, -2)$

From Exercise 33 we have:

$$x = x_1 + t(x_2 - x_1) = 1 + t(5 - 1) = 1 + 4t$$

$$y = y_1 + t(y_2 - y_1) = 4 + t(-2 - 4) = 4 - 6t$$

40. Circle with center $(-3, 1)$; radius: 3

From Exercise 34 we have:

$$x = h + r\cos\theta = -3 + 3\cos\theta$$

$$y = k + r\sin\theta = 1 + 3\cos\theta$$

42. Ellipse

Vertices: $(4, 7), (4, -3) \Rightarrow (h, k) = (4, 2), a = 5$

Foci: $(4, 5), (4, -1) \Rightarrow c = 3$

$$b^2 = a^2 - c^2 = 25 - 9 = 16 \Rightarrow b = 4$$

From Exercise 35 we have:

$$x = h + a\cos\theta = 4 + 5\cos\theta$$

$$y = k + b\sin\theta = 2 + 4\sin\theta$$

44. Hyperbola

Vertices: $(0, \pm 1) \Rightarrow (h, k) = (0, 0), a = 1$

Foci: $(0, \pm 2) \Rightarrow c = 2$

$$b^2 = c^2 - a^2 = 4 - 1 = 3 \Rightarrow b = \sqrt{3}$$

From Exercise 36 we have

$$x = h + a\sec\theta$$

$$y = k + b\tan\theta$$

for a horizontal transverse axis. However, this hyperbola has a vertical transverse axis. Thus,

$$x = h + b\tan\theta = \sqrt{3}\tan\theta$$

$$y = k + a\sec\theta = \sec\theta.$$

46. $y = \frac{1}{x}$

Examples

$x = t,\ y = \frac{1}{t}$

$x = \frac{t}{3},\ y = \frac{3}{t}$

$x = 2t,\ y = \frac{1}{2t}$

48. $y = x^2$

Examples

$x = t,\ y = t^2$

$x = \frac{1}{2}t,\ y = \frac{1}{4}t^2$

$x = \sin t,\ y = \sin^2 t$

50. $x = \theta + \sin \theta$

$y = 1 - \cos \theta$

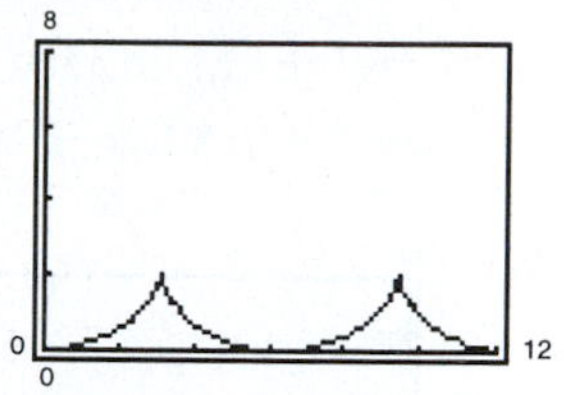

52. $x = 2\theta - 4 \sin \theta$

$y = 2 - 4 \cos \theta$

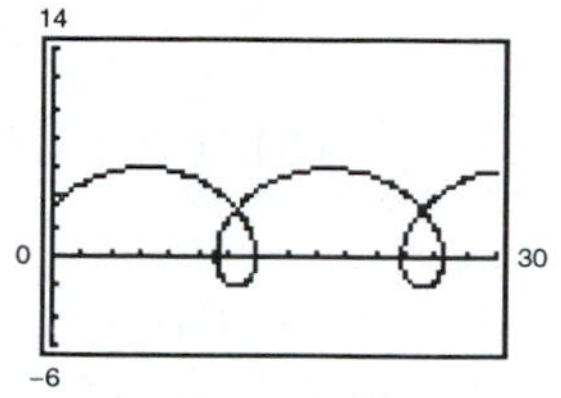

54. $x = 2\theta - \sin \theta$

$y = 2 - \cos \theta$

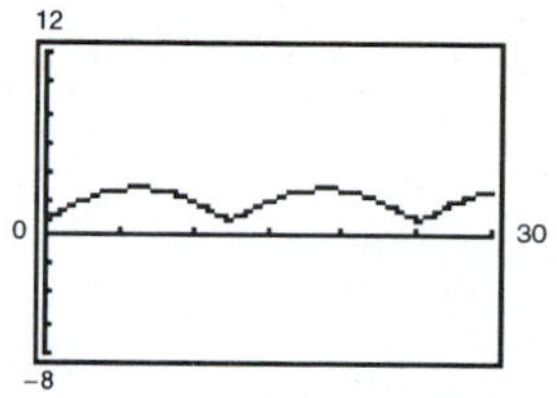

56. $x = \frac{3t}{1 + t^3}$

$y = \frac{3t^2}{1 + t^3}$

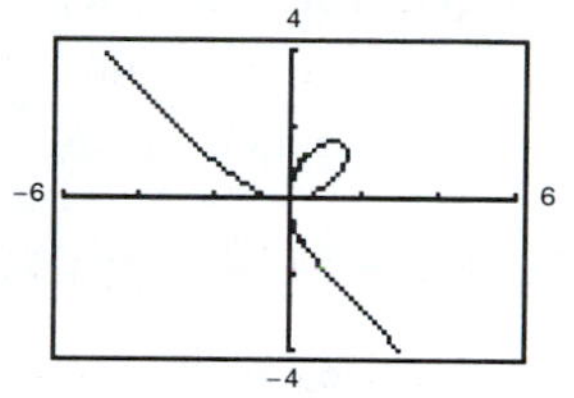

58. $x = 2 \cos^3 \theta \implies -2 \le x \le 2$

$y = 4 \sin^3 \theta \implies -4 \le y \le 4$

Matches graph (c).
Domain: $[-2, 2]$
Range: $[-4, 4]$

60. $x = \frac{1}{2} \cos \theta \implies -\infty < x < \infty$

$y = 4 \sin \theta \cos \theta \implies -2 \le y \le 2$

Matches graph (a).
Domain: $(-\infty, \infty)$
Range: $[-2, 2]$

62. $x = (v_0 \cos \theta)t$

$y = h + (v_0 \sin \theta)t - 16t^2$

(a) $\theta = 15°,\ v_0 = 60$ ft/sec

Maximum height: 3.8 feet
Range: 56.3 feet

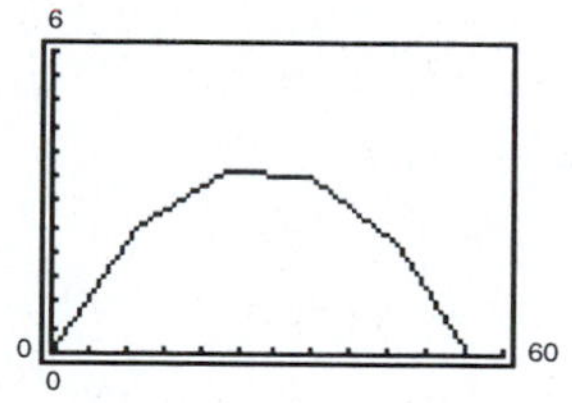

(b) $\theta = 15°,\ v_0 = 100$ ft/sec

Maximum height: 10.5 feet
Range: 156.3 feet

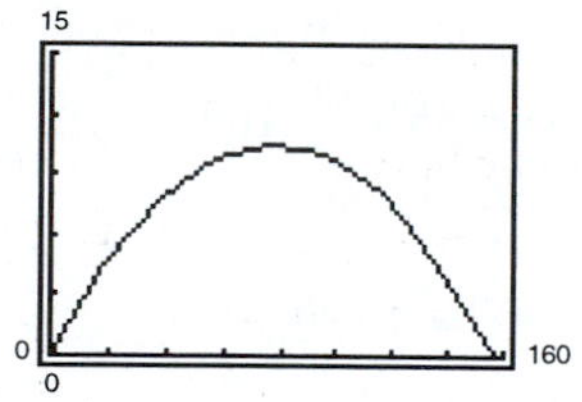

(c) $\theta = 30°,\ v_0 = 60$ ft/sec

Maximum height: 14.1 feet
Range: 97.4 feet

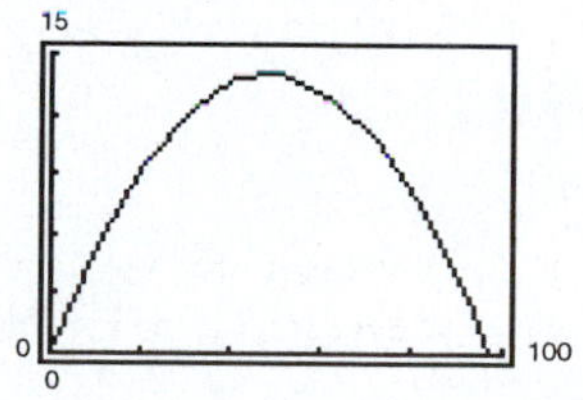

(d) $\theta = 30°,\ v_0 = 100$ ft/sec

Maximum height: 39.1 feet
Range: 270.6 feet

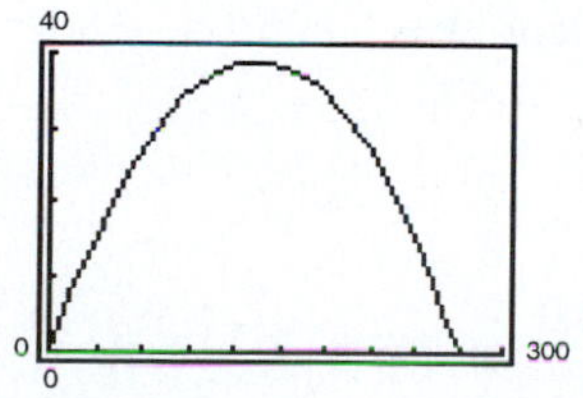

64. $\theta = 35°, h = 7$

(a) $x = (v_0 \cos \theta)t = (v_0 \cos 35°)t \approx 0.82v_0 t$

$$y = h + (v_0 \sin \theta)t - 16t^2$$
$$= 7 + (v_0 \sin 35°)t - 16t^2$$
$$\approx 7 + 0.57v_0 t - 16t^2$$

(b) Let $t = \dfrac{x}{0.82v_0}$.

$$y = 7 + (0.57v_0)\left(\frac{x}{0.82v_0}\right) - 16\left(\frac{x}{0.82v_0}\right)^2$$

When the ball is caught, $x = 30$ yards $= 90$ feet and $y = 4$ feet.

$$4 = 7 + (0.57v_0)\left(\frac{90}{0.82v_0}\right) - 16\left(\frac{90}{0.82v_0}\right)^2$$

$$\frac{1}{v_0^2} = \frac{4 - 7 - \dfrac{90(0.57)}{0.82}}{\dfrac{(-16)(90)^2}{(0.82)^2}}$$

$$v_0 = \sqrt{\frac{\dfrac{(-16)(90)^2}{(0.82)^2}}{4 - 7 - \dfrac{90(0.57)}{0.82}}} \approx 54.22 \text{ feet/second}$$

Thus, $x \approx 44.46t$

$y \approx 7 + 30.91t - 16t^2$.

(c)

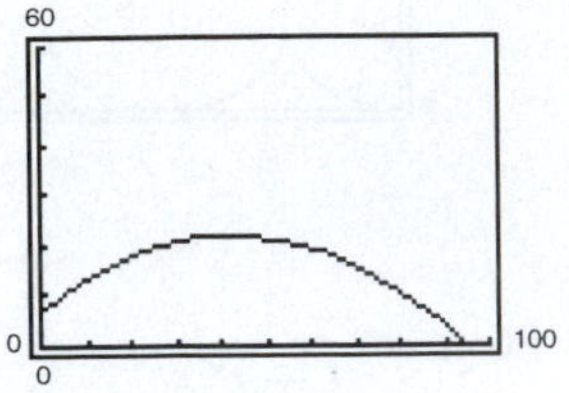

The maximum height of the football is approximately 21.9 feet.

(d) The ball is caught 90 feet downfield.

$$t = \frac{x}{44.46} = \frac{90}{44.46} \approx 2.02 \text{ seconds}$$

66. (a) $y = -0.005x^2 + x + 5$

$$y = -\frac{16 \sec^2 \theta}{v_0^2}x^2 + (\tan \theta)x + h$$

$h = 5$

$\tan \theta = 1 \implies \theta = 45°$

$$\frac{-16 \sec^2 \theta}{v_0^2} = -0.005$$

$$v_0^2 = \frac{16}{0.005 \cos^2 45°} = 6400$$

$$v_0 = 80$$

$$x = (v_0 \cos \theta)t = (80 \cos 45°)t = 40\sqrt{2}t$$
$$y = h + (v_0 \sin \theta)t - 16t^2$$
$$= 5 + (80 \sin 45°)t - 16t^2$$
$$= 5 + 40\sqrt{2}t - 16t^2$$

(b)

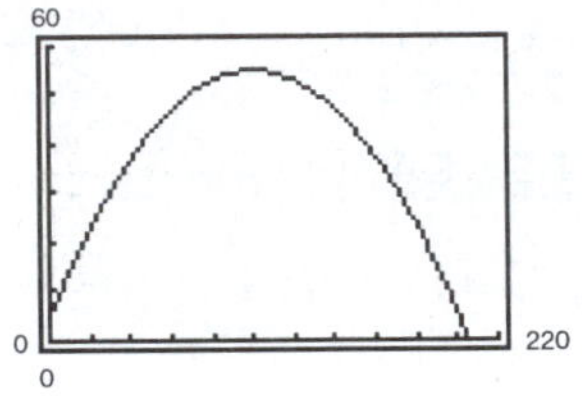

(c) Maximum height: 55 feet

Range: 205 feet

68. $2\theta = \text{arc } TS = \text{arc } TP = \phi$

$$y = \overline{OA} - \overline{OC}$$
$$\overline{OA} = 3 \sin \theta$$
$$\overline{OC} = \sin(180° - \theta - \phi)$$
$$= \sin(180° - 3\theta)$$
$$= \sin 180° \cos 3\theta - \sin 3\theta \cos 180°$$
$$= \sin 3\theta$$

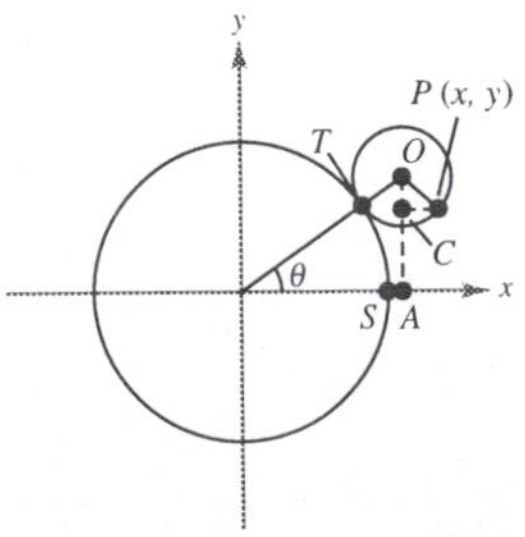

70. False

$x = t^2 \implies x \geq 0$

$y = t^2$

The graph of these parametric equations is the line $y = x$ for $x \geq 0$ *only*.

72.
$$\begin{aligned} 3x + 5y &= 9 \implies 6x + 10y = 18 \\ 4x - 2y &= -14 \implies \underline{20x - 10y = -70} \\ & \qquad\qquad\quad 26x = -52 \\ & \qquad\qquad\quad x = -2 \end{aligned}$$
$$3(-2) + 5y = 9 \implies y = 3$$

Solution: $(-2, 3)$

74.
$$5u + 7v + 9w = 4$$
$$u - 2v - 3w = 7$$
$$8u - 2v + w = 20$$

$$\begin{aligned} 5u + 7v + 9w &= 4 \\ \underline{3u - 6v - 9w} &\underline{= 21} \\ 8u + v &= 25 \end{aligned}$$

$$\begin{aligned} u - 2v - 3w &= 7 \\ \underline{24u - 6v + 3w} &\underline{= 60} \\ 25u - 8v &= 67 \end{aligned}$$

$$\begin{aligned} 8u + v = 25 &\implies 64u + 8v = 200 \\ 25u - 8v = 65 &\implies \underline{25u - 8v = 67} \\ & \qquad 89u = 267 \\ & \qquad u = 3 \end{aligned}$$
$$64(3) + 8v = 200 \implies v = 1$$
$$3 - 2(1) - 3w = 7 \implies w = -2$$

Solution: $(3, 1, -2)$

Section 10.7 Polar Coordinates

Solutions to Even-Numbered Exercises

2. Polar coordinates: $\left(4, \frac{3\pi}{2}\right) = (r, \theta)$

$x = r\cos\theta = 4\cos\left(\frac{3\pi}{2}\right) = 0$

$y = r\sin\theta = 4\sin\left(\frac{3\pi}{2}\right) = -4$

Rectangular coordinates: $(0, -4)$

4. Polar coordinates: $(0, -\pi) = (r, \theta)$

$x = r\cos\theta = 0\cos(-\pi) = 0$

$y = r\sin\theta = 0\sin(-\pi) = 0$

Rectangular coordinates: $(0, 0)$

6. Polar coordinates: $\left(-1, \frac{-3\pi}{4}\right)$

$x = -1\cos\left(\frac{-3\pi}{4}\right) = \frac{\sqrt{2}}{2}$

$y = -1\sin\left(\frac{-3\pi}{4}\right) = \frac{\sqrt{2}}{2}$

Rectangular coordinates: $\left(\frac{\sqrt{2}}{2}, \frac{\sqrt{2}}{2}\right)$

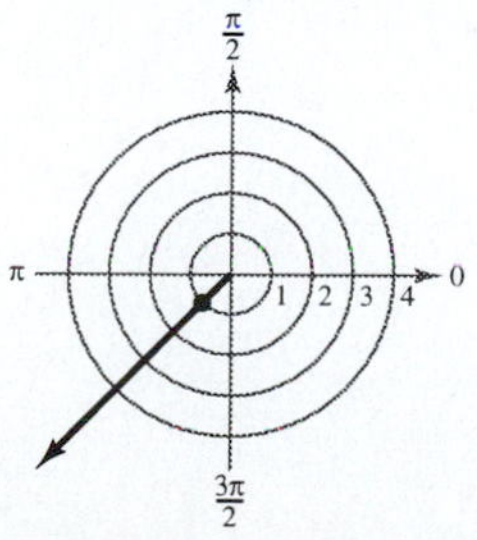

8. Polar coordinates: $\left(32, \frac{5\pi}{2}\right)$

$x = 32\cos\left(\frac{5\pi}{2}\right) = 0$

$y = 32\sin\left(\frac{5\pi}{2}\right) = 32$

Rectangular coordinates: $(0, 32)$

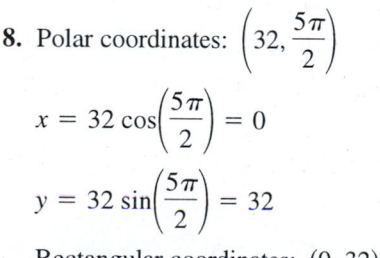

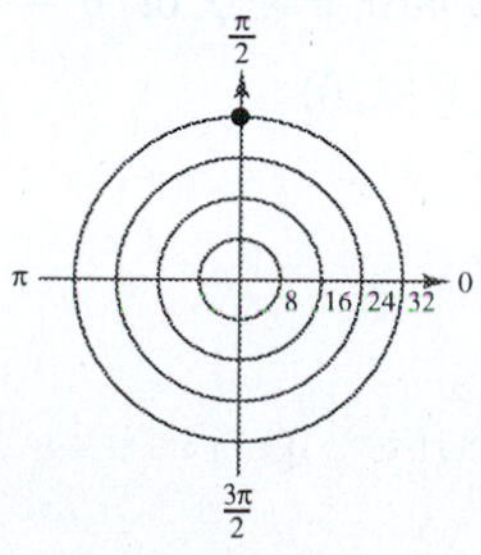

10. Polar coordinates: $(-3, -1.57)$

$x = -3\cos(-1.57) \approx -0.0024$

$y = -3\sin(-1.57) \approx 3.000$

Rectangular coordinates: $(-0.0024, 3)$

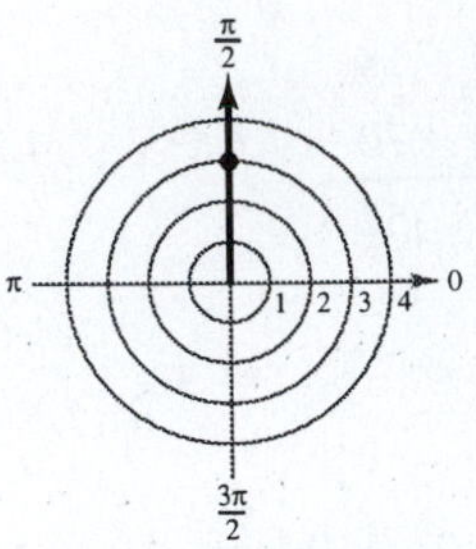

12. Polar coordinates: $\left(-2, \dfrac{7\pi}{6}\right)$

$P \blacktriangleright Rx\left(-2, \dfrac{7\pi}{6}\right) \approx 1.732$

$P \blacktriangleright Ry\left(-2, \dfrac{7\pi}{6}\right) = 1$

$= \left(\sqrt{2}, 1\right)$

14. Polar coordinates: $(8.25, 3.5)$

$P \blacktriangleright Rx(8.25, 3.5) \approx -7.726$

$P \blacktriangleright Ry(8.25, 3.5) \approx -2.894$

$\approx (-7.726, -2.894)$

16. Rectangular coordinates: $(0, -5)$

$r = \pm 5$, $\tan\theta$ undefined, $\theta = \dfrac{\pi}{2}$ or $\theta = \dfrac{3\pi}{2}$

Polar coordinates: $\left(5, \dfrac{3\pi}{2}\right), \left(-5, \dfrac{\pi}{2}\right)$

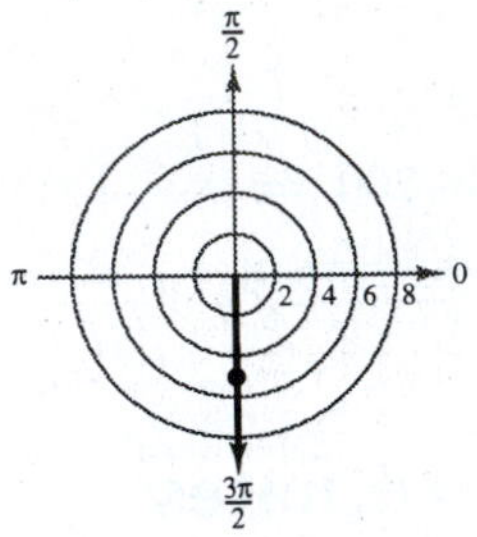

18. Rectangular coordinates: $(-3, -3)$

$r = \pm 3\sqrt{2}$, $\tan\theta = 1$, $\theta = \dfrac{\pi}{4}$ or $\theta = \dfrac{5\pi}{4}$

Polar coordinates: $\left(3\sqrt{2}, \dfrac{5\pi}{4}\right), \left(-3\sqrt{2}, \dfrac{\pi}{4}\right)$

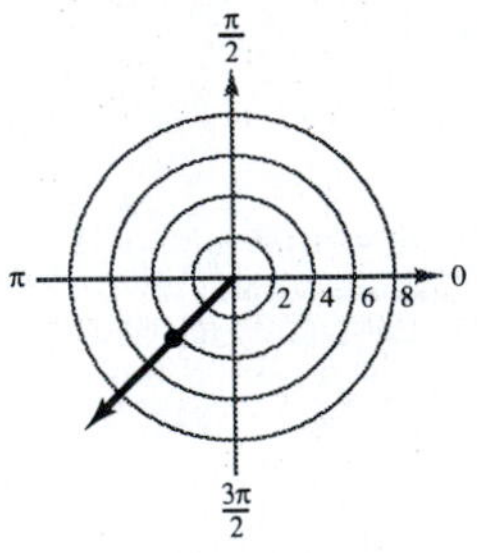

20. Rectangular coordinates: $(3, -1)$

$r = \sqrt{9+1} = \pm\sqrt{10}$, $\tan\theta = -\frac{1}{3}$, $\theta \approx -0.322 \approx 5.961$ or $\theta \approx 2.820$

Polar coordinates: $\left(-\sqrt{10}, 2.820\right), \left(\sqrt{10}, 5.961\right)$

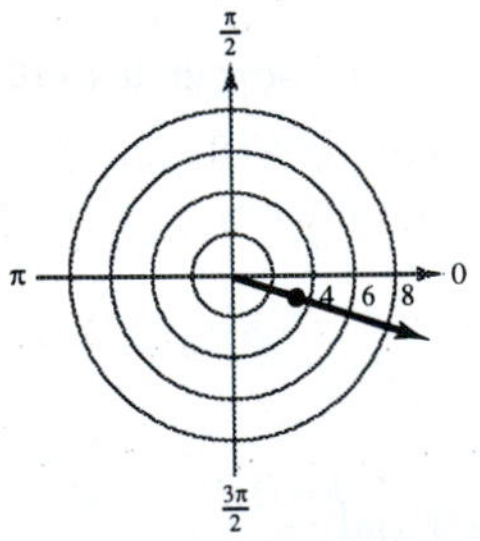

22. Rectangular coordinates: $(2, 0)$

$r = \sqrt{4+0} = \pm 2$, $\tan\theta = 0$, $\theta = 0$ or $\theta = \pi$

Polar coordinates: $(2, \pi), (-2, 0)$

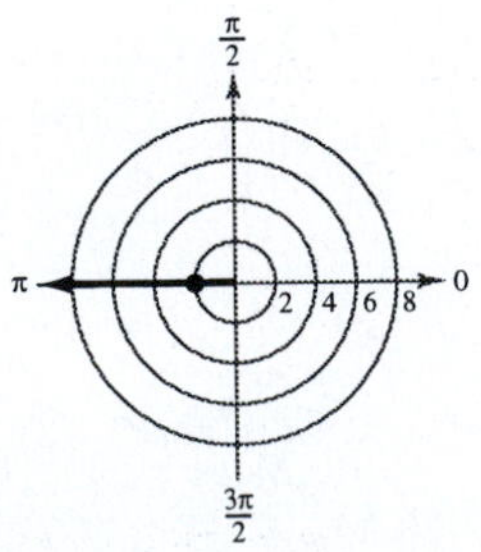

24. Rectangular coordinates: (5, 12)

$r = \sqrt{25 + 144} = \pm 13$, $\tan \theta = \frac{12}{5}$, $\theta \approx 1.176$ or $\theta \approx 4.318$

Polar coordinates: (13, 1.176), (−13, 4.318)

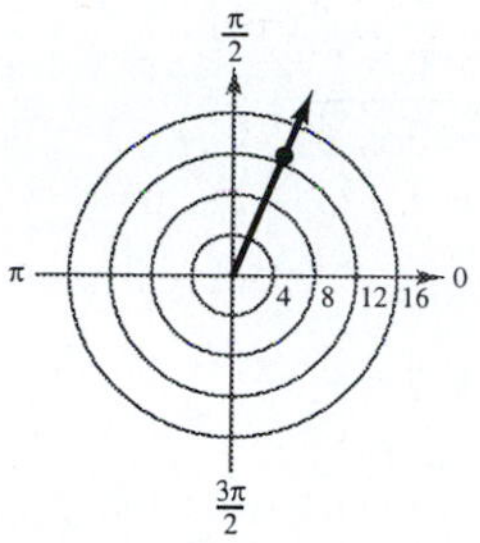

26. Rectangular coordinates: (−4, 1)

$R \blacktriangleright Pr(-4, 1) \approx 4.123$

$R \blacktriangleright P\theta(-4, 1) \approx 2.897$

$\approx \left(\sqrt{17}, 2.897\right)$

28. Rectangular coordinates: $\left(3\sqrt{2}, 3\sqrt{2}\right)$

$R \blacktriangleright Pr\left(3\sqrt{2}, 3\sqrt{2}\right) = 6$

$R \blacktriangleright P\theta\left(3\sqrt{2}, 3\sqrt{2}\right) \approx 0.785$

$= \left(6, \frac{\pi}{4}\right)$

30. Rectangular coordinates: (0, −5)

$R \blacktriangleright Pr(0, -5) = 5$

$R \blacktriangleright P\theta(0, -5) \approx -1.571$

$= \left(5, \frac{\pi}{2}\right)$

32. True.

$\theta_1 = \theta_2 + 2\pi n$, for n any integer.

34. $x^2 + y^2 = a^2$

$r = a$

36.
$$\begin{aligned} x^2 + y^2 - 2ay &= 0 \\ r^2 - 2ar \sin \theta &= 0 \\ r(r - 2a \sin \theta) &= 0 \\ r &= 2a \sin \theta \end{aligned}$$

38.
$$\begin{aligned} y &= b \\ r \sin \theta &= b \\ r &= b \csc \theta \end{aligned}$$

40.
$$\begin{aligned} x &= a \\ r \cos \theta &= a \\ r &= a \sec \theta \end{aligned}$$

42.
$$\begin{aligned} 4x + 7y - 2 &= 0 \\ 4r \cos \theta + 7r \sin \theta - 2 &= 0 \\ r(4 \cos \theta + 7 \sin \theta) &= 2 \\ r &= \frac{2}{4 \cos \theta + 7 \sin \theta} \end{aligned}$$

44.
$$\begin{aligned} y &= x \\ r \cos \theta &= r \sin \theta \\ 1 &= \tan \theta \\ \theta &= \frac{\pi}{4} \end{aligned}$$

46.
$$\begin{aligned} y^2 - 8x - 16 &= 0 \\ r^2 \sin^2 \theta - 8r \cos \theta &= 16 \\ r^2 - r^2 \cos^2 \theta - 8r \cos \theta - 16 &= 0 \\ r^2 \cos^2 \theta + 8r \cos \theta + 16 &= r^2 \\ (r \cos \theta + 4)^2 &= r^2 \\ r &= \pm(r \cos \theta + 4) \\ r &= \frac{4}{1 - \cos \theta} \\ \text{or } r &= \frac{-4}{1 + \cos \theta} \end{aligned}$$

48.
$$\begin{aligned} r &= 4 \cos \theta \\ r^2 &= 4r \cos \theta \\ x^2 + y^2 &= 4x \\ x^2 + y^2 - 4x &= 0 \end{aligned}$$

50.
$$\begin{aligned} r &= 4 \\ r^2 &= 16 \\ x^2 + y^2 &= 16 \end{aligned}$$

52.
$$\begin{aligned} r^2 &= \sin 2\theta = 2 \sin \theta \cos \theta \\ r^2 &= 2\left(\frac{y}{r}\right)\left(\frac{x}{r}\right) = \frac{2xy}{r^2} \\ r^4 &= 2xy \\ (x^2 + y^2)^2 &= 2xy \end{aligned}$$

54.
$$r = \frac{1}{1 - \cos\theta}$$
$$r - r\cos\theta = 1$$
$$\sqrt{x^2 + y^2} - x = 1$$
$$x^2 + y^2 = 1 + 2x + x^2$$
$$y^2 = 2x + 1$$

56.
$$r = \frac{6}{2\cos\theta - 3\sin\theta}$$
$$r = \frac{6}{2(x/r) - 3(y/r)}$$
$$r = \frac{6r}{2x - 3y}$$
$$1 = \frac{6}{2x - 3y}$$
$$2x - 3y = 6$$

58.
$$r = 8$$
$$r^2 = 64$$
$$x^2 + y^2 = 64$$

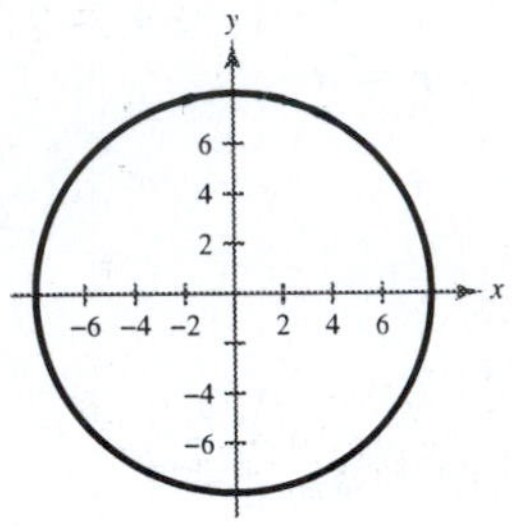

60.
$$\theta = \frac{5\pi}{6}$$
$$\tan\theta = \tan\frac{5\pi}{6}$$
$$\frac{y}{x} = -\frac{1}{\sqrt{3}}$$
$$\sqrt{3}y = -x$$
$$x + \sqrt{3}y = 0$$

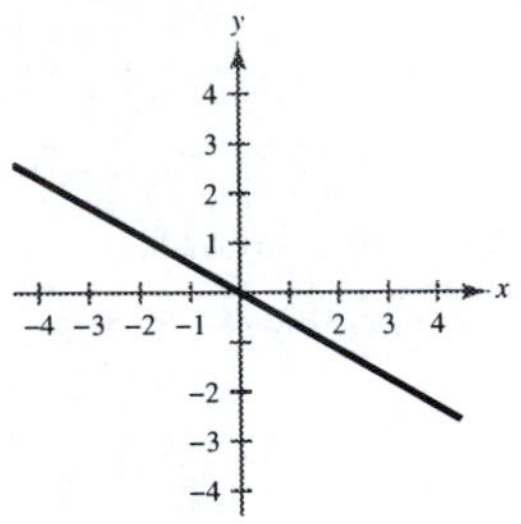

62.
$$r = 2\csc\theta$$
$$r\sin\theta = 2$$
$$y = 2$$
$$y - 2 = 0$$

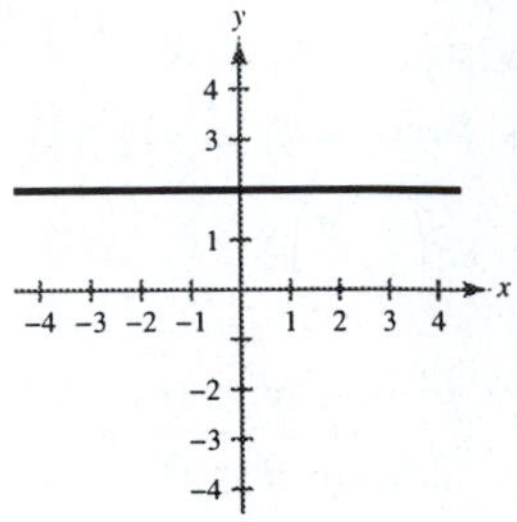

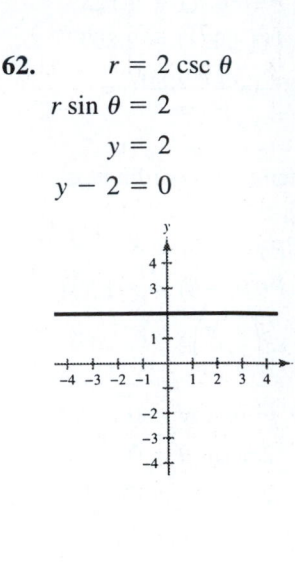

64.
$$r = \cos\theta + 3\sin\theta$$
$$r = \frac{x}{r} + \frac{3y}{r}$$
$$r^2 = x + 3y$$
$$x^2 + y^2 = x + 3y$$
$$x^2 - x + y^2 - 3y = 0$$
$$\left(x - \frac{1}{2}\right)^2 + \left(y - \frac{3}{2}\right)^2 = \frac{5}{2}$$

The graph is a circle.

66. (a) For horizontal moves, just the x-coordinate changes. For vertical moves, just the y-coordinate changes.

(b) For horizontal moves, both r and θ change. For vertical moves, both r and θ change.

(c) Unlike r and θ, x and y measure horizontal and vertical changes, respectively.

68. $3x + 5y = 10$

$4x - 2y = -5$

By Cramer's Rule we have: $x = \dfrac{\begin{vmatrix} 10 & 5 \\ -5 & -2 \end{vmatrix}}{\begin{vmatrix} 3 & 5 \\ 4 & -2 \end{vmatrix}} = \dfrac{5}{-26} = -\dfrac{5}{26}$

$$y = \frac{\begin{vmatrix} 3 & 10 \\ 4 & -5 \end{vmatrix}}{\begin{vmatrix} 3 & 5 \\ 4 & -2 \end{vmatrix}} = \frac{-55}{-26} = \frac{55}{26}$$

Solution: $\left(-\frac{5}{25}, \frac{55}{26}\right)$

70. $5u + 7v + 9w = 15$
$u - 2v - 3w = 7$
$8u - 2v + w = 0$

$$\begin{vmatrix} 5 & 7 & 9 \\ 1 & -2 & -3 \\ 8 & -2 & 1 \end{vmatrix} = -89$$

$$u = \frac{\begin{vmatrix} 15 & 7 & 9 \\ 7 & -2 & -3 \\ 0 & -2 & 1 \end{vmatrix}}{-89} = \frac{-295}{-89} = \frac{295}{89}$$

$$v = \frac{\begin{vmatrix} 5 & 15 & 9 \\ 1 & 7 & -3 \\ 8 & 0 & 1 \end{vmatrix}}{-89} = \frac{-844}{-89} = \frac{844}{89}$$

$$w = \frac{\begin{vmatrix} 5 & 7 & 15 \\ 1 & -2 & 7 \\ 8 & -2 & 0 \end{vmatrix}}{-89} = \frac{672}{-89} = -\frac{672}{89}$$

Solution: $\left(\frac{295}{89}, \frac{844}{89}, -\frac{672}{89}\right)$

72. $2y + 5z + 6w = 32$
$2x + 4y - 5z - w = -7$
$3x + 6y + z + 5w = 6$
$4x - 2y - z = -12$

$$\begin{vmatrix} 0 & 2 & 5 & 6 \\ 2 & 4 & -5 & -1 \\ 3 & -6 & 1 & 5 \\ 4 & -2 & -1 & 0 \end{vmatrix} = 852$$

$$x = \frac{\begin{vmatrix} 32 & 2 & 5 & 6 \\ -7 & 4 & -5 & -1 \\ 6 & -6 & 1 & 5 \\ -12 & -2 & -1 & 0 \end{vmatrix}}{852} = \frac{-1704}{852} = -2$$

$$y = \frac{\begin{vmatrix} 0 & 32 & 5 & 6 \\ 2 & -7 & -5 & -1 \\ 3 & 6 & 1 & 5 \\ 4 & -12 & -1 & 0 \end{vmatrix}}{852} = \frac{1278}{852} = \frac{3}{2}$$

$$z = \frac{\begin{vmatrix} 0 & 2 & 32 & 6 \\ 2 & 4 & -7 & -1 \\ 3 & -6 & 6 & 5 \\ 4 & -2 & -12 & 0 \end{vmatrix}}{852} = \frac{852}{852} = 1$$

$$w = \frac{\begin{vmatrix} 0 & 2 & 5 & 32 \\ 2 & 4 & -5 & -7 \\ 3 & -6 & 1 & 6 \\ 4 & -2 & -1 & -12 \end{vmatrix}}{852} = \frac{3408}{852} = 4$$

Solution: $\left(-2, \frac{3}{2}, 1, 4\right)$

Section 10.8 Graphs of Polar Equations

Solutions to Even-Numbered Exercises

2. $r = 5 - 5 \sin \theta$

Cardioid

4. $r^2 = 9 \cos \theta$

Lemniscate

6. $r = 1 + 4 \cos \theta$

Limaçon with inner loop

8. $r = 16 \cos 3\theta$

$\theta = \frac{\pi}{2}$: $-r = 16 \cos(3(-\theta))$

$-r = 16 \cos(-3\theta)$

$-r = 16 \cos 3\theta$

Not an equivalent equation

Polar axis: $r = 16 \cos(3(-\theta))$

$r = 16 \cos(-3\theta)$

$r = 16 \cos 3\theta$

Equivalent equation

Pole: $-r = 16 \cos 3\theta$

Not an equivalent equation

Answer: Symmetric with respect to polar axis

10. $r = 6 \sin \theta$

$\theta = \frac{\pi}{2}$: $-r = 6 \sin(-\theta)$

$-r = -6 \sin \theta$

$r = 6 \sin \theta$

Equivalent equation

Polar axis: $r = 6 \sin(-\theta)$

$r = -6 \sin \theta$

Not an equivalent equation

Pole: $-r = 6 \sin \theta$

Not an equivalent equation

Answer: Symmetric with respect to $\theta = \pi/2$

12. $r^2 = 25 \sin 2\theta$

$\theta = \frac{\pi}{2}$: $(-r)^2 = 25 \sin(2(-\theta))$

$r^2 = -25 \sin 2\theta$

Not an equivalent equation

Polar axis: $r^2 = 25 \sin(2(-\theta))$

$r^2 = -25 \sin 2\theta$

Not an equivalent equation

Pole: $(-r)^2 = 25 \sin(2(-\theta))$

$r^2 = 25 \sin 2\theta$

Equivalent equation

Answer: Symmetric with respect to pole

14. $|r| = |6 + 12 \cos \theta| \le |6| + |12 \cos \theta|$

$= 6 + 12|\cos \theta| \le 18$

$\cos \theta = 1$

$\theta = 0$

Maximum: $|r| = 18$ when $\theta = 0$

$0 = 6 + 12 \cos \theta$

$\cos \theta = -\frac{1}{2}$

$\theta = \frac{2\pi}{3}, \frac{4\pi}{3}$

Zero: $r = 0$ when $\theta = \frac{2\pi}{3}, \frac{4\pi}{3}$

16. $|r| = |5 \sin 2\theta| = 5|\sin 2\theta| \le 5$

$|\sin 2\theta| = 1$

$\sin 2\theta = \pm 1$

$\theta = \frac{\pi}{4}, \frac{3\pi}{4}, \frac{5\pi}{4}, \frac{7\pi}{4}$

Maximum: $|r| = 5$ when $\theta = \frac{\pi}{4}, \frac{3\pi}{4}, \frac{5\pi}{4}, \frac{7\pi}{4}$

$0 = 5 \sin 2\theta$

$\sin 2\theta = 0$

$\theta = 0, \frac{\pi}{2}, \pi, \frac{3\pi}{2}$

Zero: $r = 0$ when $\theta = 0, \frac{\pi}{2}, \pi, \frac{3\pi}{2}$

18. Circle: $r = 2$

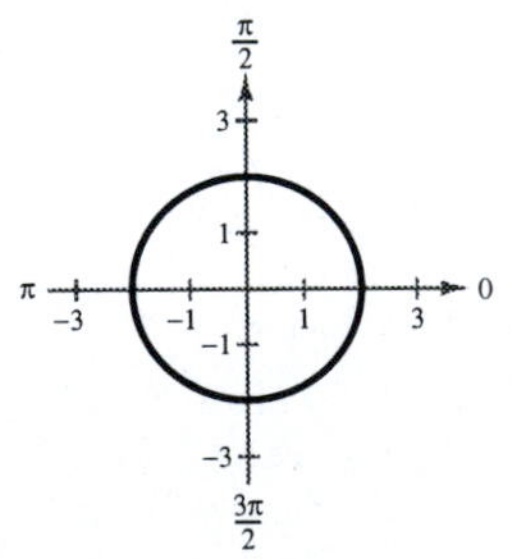

20. Circle: $r = -\frac{\pi}{4}$

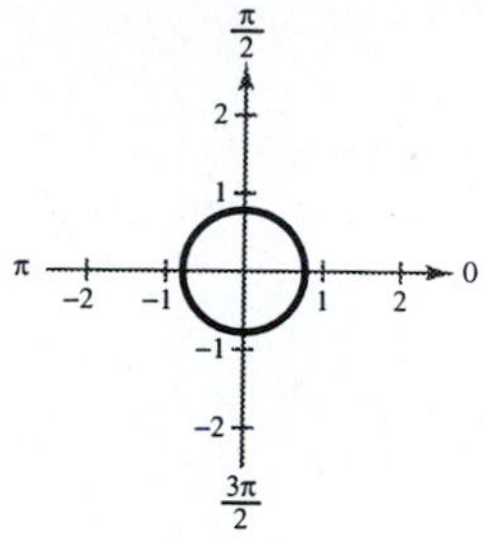

22. $r = 3 \cos \theta$

Symmetric with respect to the polar axis.
Circle with radius $3/2$

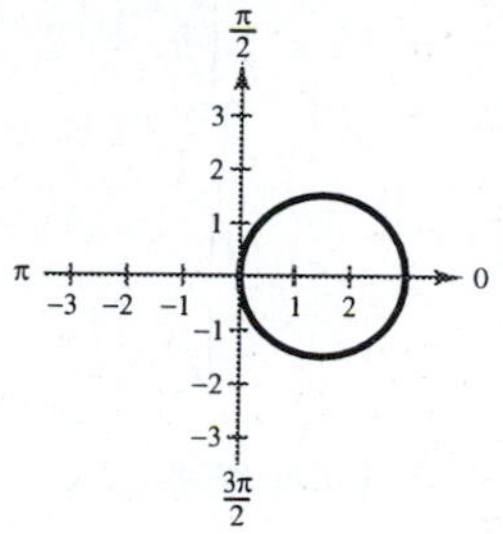

24. $r = 2(1 - \sin \theta)$

Symmetric with respect to $\theta = \pi/2$.

$\frac{a}{b} = \frac{2}{2} = 1 \Rightarrow$ Cardioid

$|r| = 4$ when $\theta = \frac{3\pi}{2}$.

$r = 0$ when $\theta = \frac{\pi}{2}$.

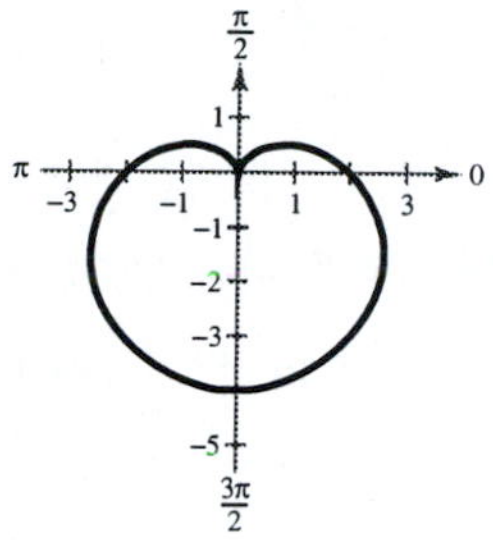

26. $r = 1 + \cos \theta$

Symmetric with respect to the polar axis.

$\frac{a}{b} = \frac{1}{1} = 1 \Rightarrow$ Cardioid

$|r| = 2$ when $\theta = 0$.

$r = 0$ when $\theta = \pi$.

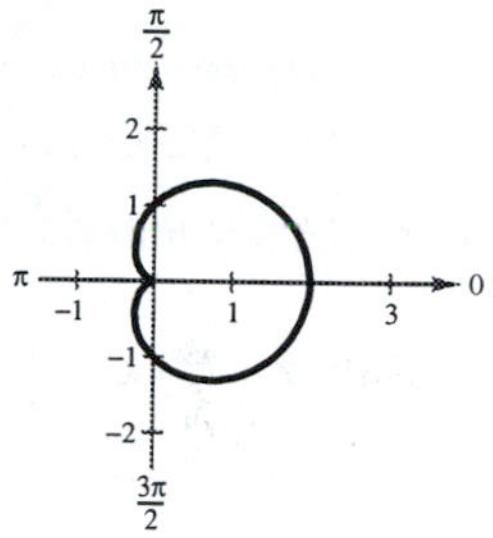

28. $r = 5 - 4 \sin \theta$

Symmetric with respect to $\theta = \pi/2$.

$a = 5, b = 4$

$\frac{a}{b} = \frac{5}{4} \Rightarrow$ Dimpled limaçon

$|r| = 9$ when $\theta = \frac{3\pi}{2}$.

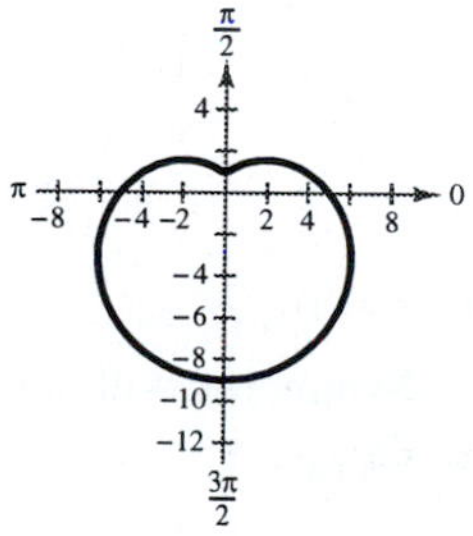

30. $r = 4 + 3 \cos \theta$

Symmetric with respect to the polar axis.

$\frac{a}{b} = \frac{4}{3} > 1 \Rightarrow$ Dimpled limaçon

$|r| = 7$ when $\theta = 0$.

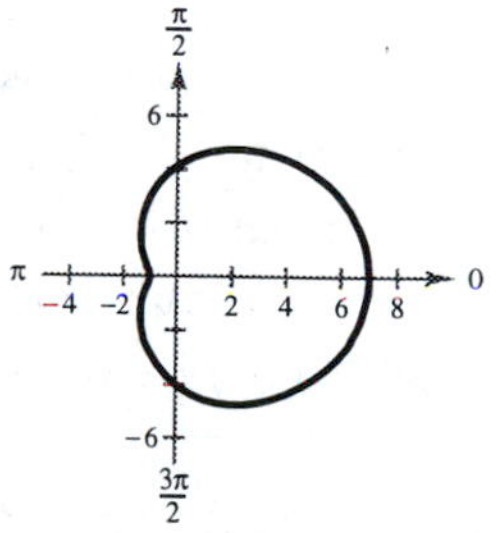

32. $r = 1 - 2 \cos \theta$

Symmetric with respect to the polar axis.

$\frac{a}{b} = \frac{1}{2} \Rightarrow$ Limaçon with inner loop

$|r| = 3$ when $\theta = \pi$.

$r = 0$ when $\theta = \frac{\pi}{3}, \frac{5\pi}{3}$.

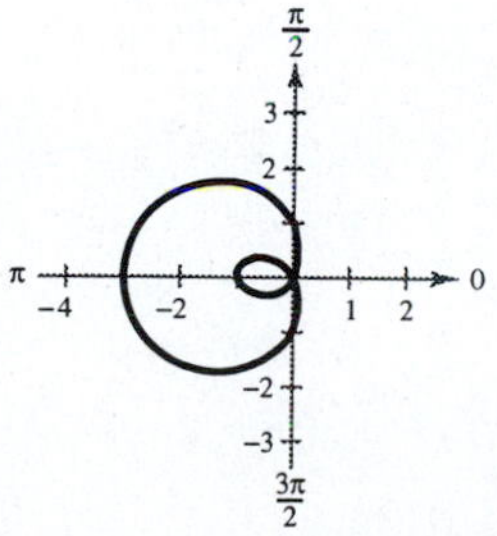

34. $r = 2(1 - 2 \sin \theta)$

$r = 2 - 4 \sin \theta$

Symmetric with respect to $\theta = \pi/2$.

$\frac{a}{b} = \frac{1}{2} \Rightarrow$ Limaçon with inner loop

$|r| = 6$ when $\theta = \frac{3\pi}{2}$.

$r = 0$ when $\theta = \frac{\pi}{6}, \frac{5\pi}{6}$.

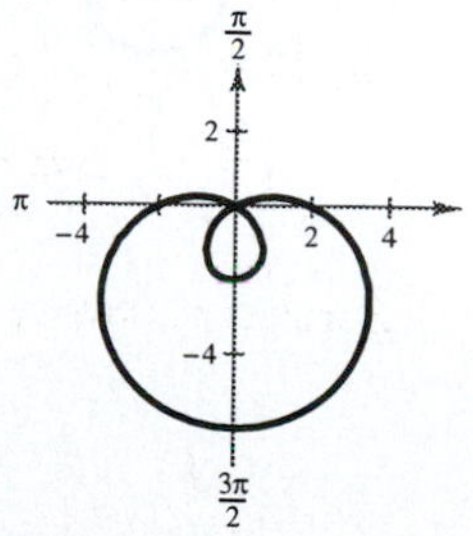

36. $r = -\sin 5\theta$

Symmetric with respect to $\theta = \pi/2$.

Rose curve five petals ($n = 5$)

$$|r| = 1 \text{ when } \theta = \frac{\pi}{12}, \frac{7\pi}{24}, \frac{\pi}{2}, \frac{17\pi}{24}, \frac{11\pi}{12}.$$

$$r = 0 \text{ when } \theta = 0, \frac{\pi}{5}, \frac{2\pi}{5}, \frac{3\pi}{5}, \frac{4\pi}{5}, \pi.$$

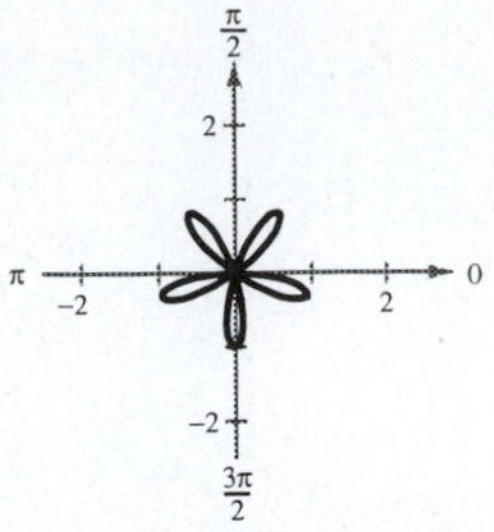

38. $r = 3\cos 2\theta$

Symmetric with respect to the polar axis.

Rose curve ($n = 2$) with four petals

$$|r| = 3 \text{ when } \theta = 0, \frac{\pi}{2}, \pi, \frac{3\pi}{2}.$$

$$r = 0 \text{ when } \theta = \frac{\pi}{4}, \frac{3\pi}{4}, \frac{5\pi}{4}, \frac{7\pi}{4}.$$

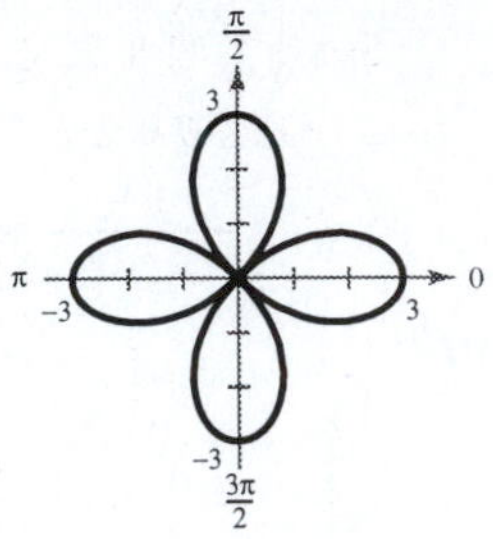

40. $r = 3\csc\theta = \dfrac{3}{\sin\theta}$

$r\sin\theta = 3$

$y = 3 \Rightarrow$ Line

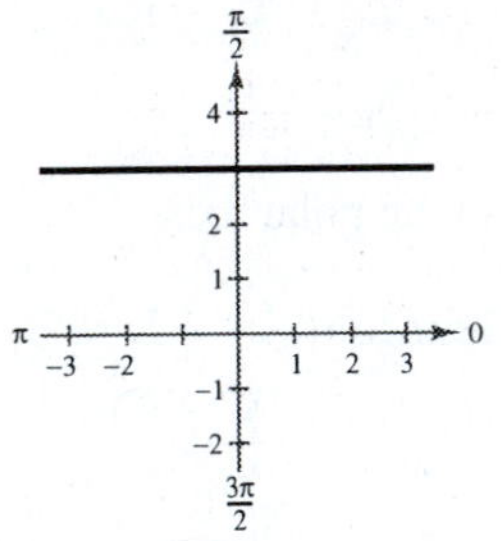

42.

$$r = \frac{6}{2\sin\theta - 3\cos\theta}$$

$$r(\sin\theta - 3\cos\theta) = 6$$

$$2y - 3x = 6$$

$$y = \frac{3}{2}x + 3 \Rightarrow \text{Line}$$

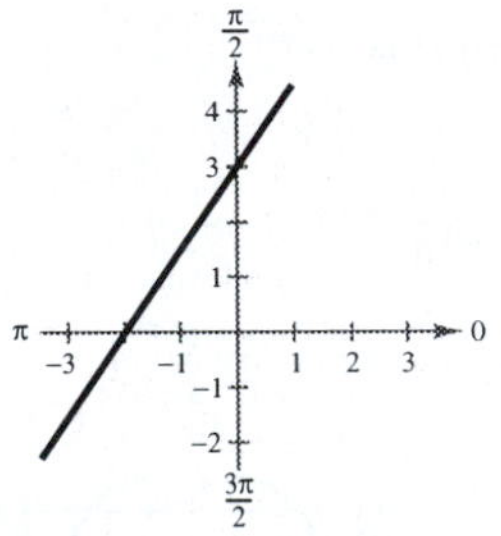

44. $r^2 = 4\sin\theta$

$r = 2\sqrt{\sin\theta}$

$r = -2\sqrt{\sin\theta}$

$0 \le \theta \le \pi$

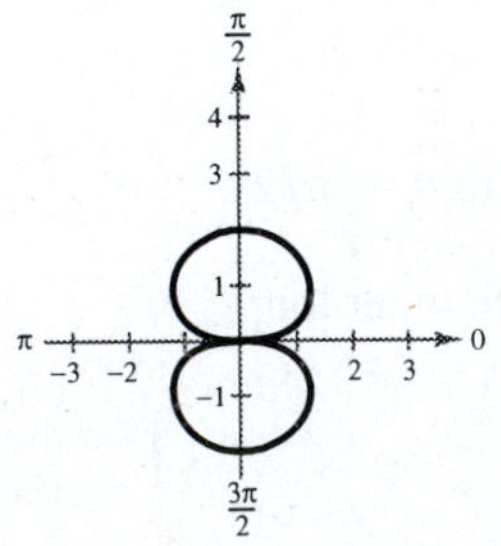

46. $r = 0$

Symmetric with respect to $\theta = \pi/2$.

Spiral

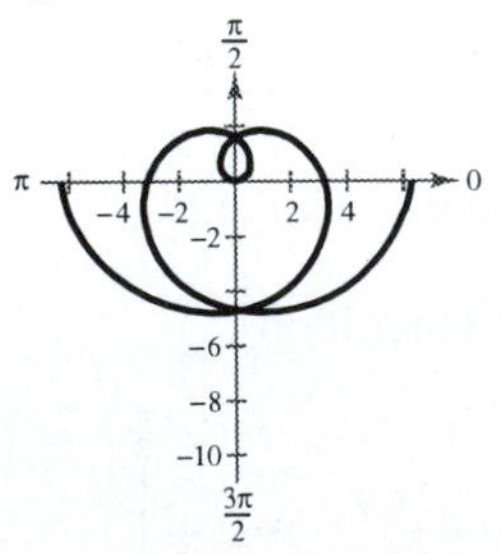

48. $r = \dfrac{\theta}{4}$

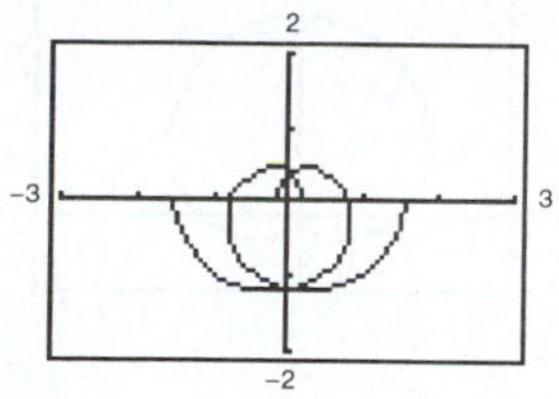

50. $r = \cos 2\theta$

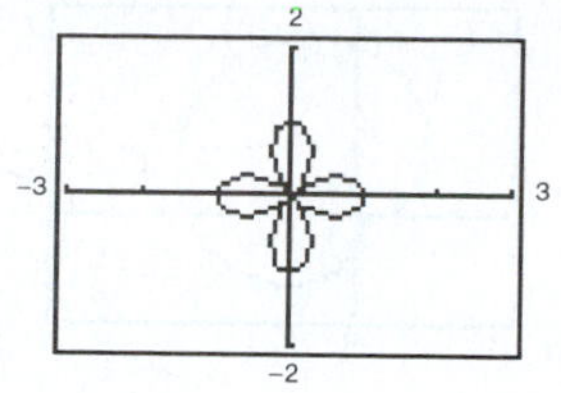

52. $r = 2\cos(3\theta - 2)$

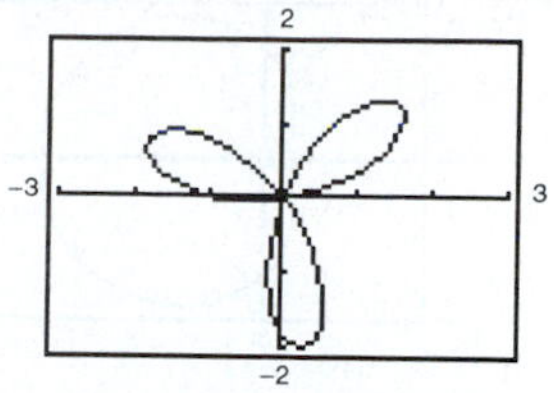

54. $r = 2 - \sec\theta$

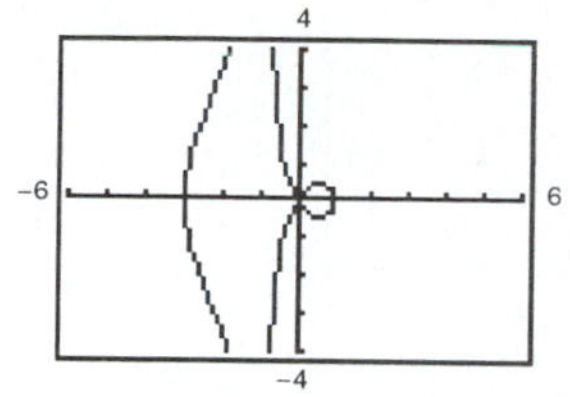

56. $r = 2(1 - 2\sin\theta)$

$0 \le \theta < 2\pi$

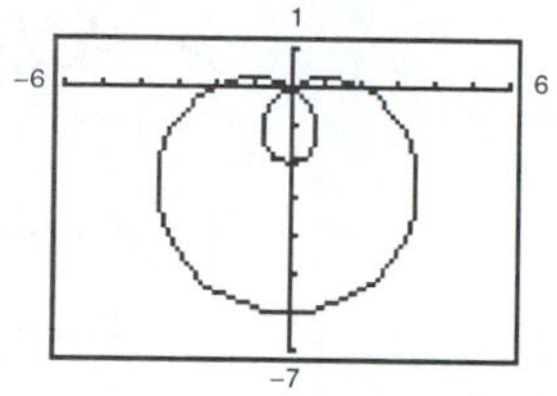

58. $r = 4 + 3\cos\theta$

$0 \le \theta < 2\pi$

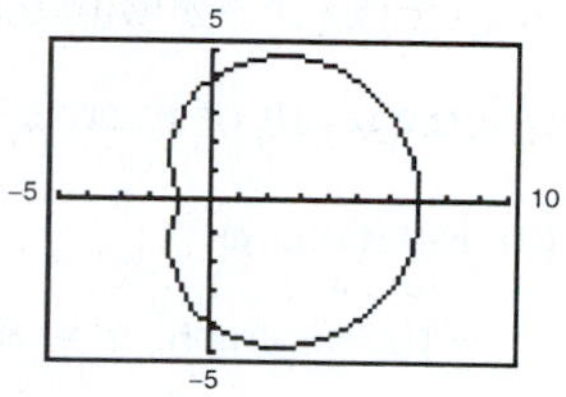

60. $r = 3\sin\left(\dfrac{5\theta}{2}\right)$

$0 \le \theta < 4\pi$

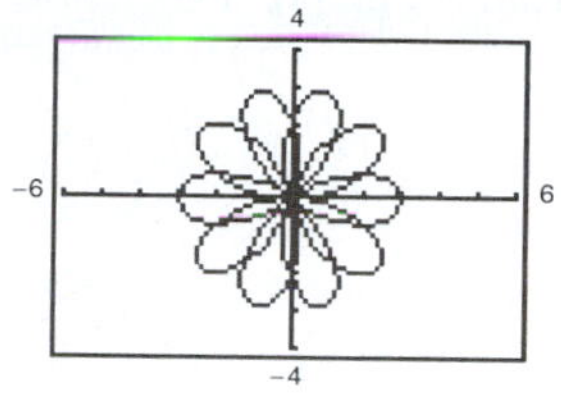

62. $r^2 = \dfrac{1}{\theta}$

$0 < \theta < \infty$

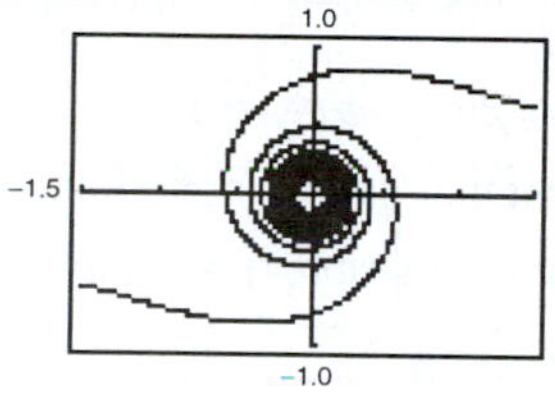

64.

$$r = 2 + \csc\theta = 2 + \frac{1}{\sin\theta}$$

$$r\sin\theta = 2\sin\theta + 1$$

$$r(r\sin\theta) = 2r\sin\theta + r$$

$$\left(\pm\sqrt{x^2 + y^2}\right)(y) = 2y + \left(\pm\sqrt{x^2 + y^2}\right)$$

$$\left(\pm\sqrt{x^2 + y^2}\right)(y - 1) = 2y$$

$$\left(\pm\sqrt{x^2 + y^2}\right) = \frac{2y}{y - 1}$$

$$x^2 + y^2 = \frac{4y^2}{(y - 1)^2}$$

$$x^2 = \frac{y^2(3 + 2y - y^2)}{(y - 1)^2}$$

$$x = \sqrt{\frac{y^2(3 + 2y - y^2)}{(y - 1)^2}} = \pm\left|\frac{y}{y - 1}\right|\sqrt{3 + 2y - y^2}$$

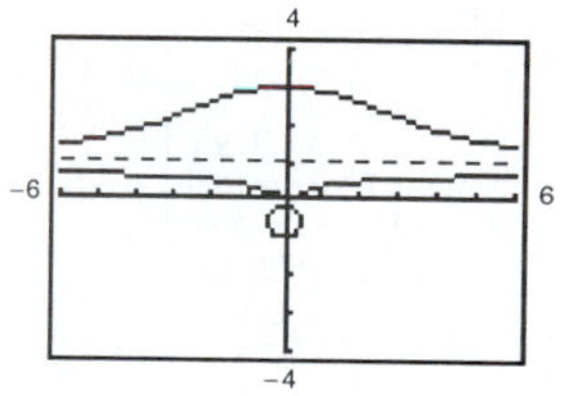

The graph has an asymptote at $y = 1$.

66.

$$r = 2\cos 2\theta\sec\theta = \frac{2\cos 2\theta}{\cos\theta}$$

$$r = \frac{2(\cos^2\theta - \sin^2\theta)}{\cos\theta}$$

$$r\cos\theta = 2(\cos^2\theta - \sin^2\theta)$$

$$x = 2(\cos^2\theta - \sin^2\theta)$$

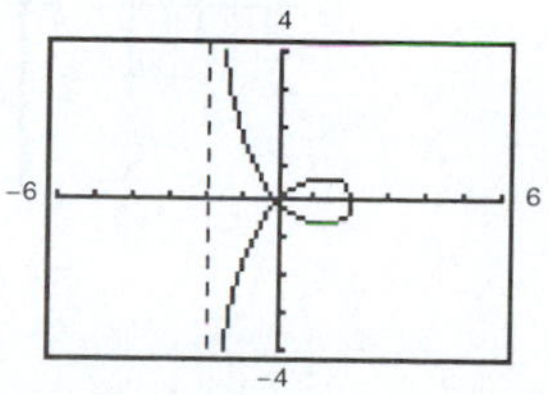

As $\theta \to \dfrac{\pi}{2}$, $x \to -2$.

68. $r = 6[1 + \cos(\theta - \phi)]$

(a) (b) (c)

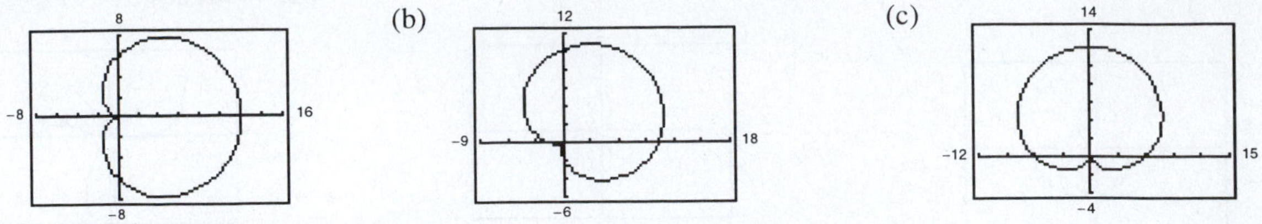

The angle ϕ has the effect of rotating the graph by the angle ϕ. For part (c),

$$r = 6\left[1 + \cos\left(\theta - \frac{\pi}{2}\right)\right]$$
$$= 6[1 + \sin\theta].$$

70. Use the result of Exercise 69.

(a) Rotation: $\phi = \dfrac{\pi}{2}$

Original graph: $r = f(\sin\theta)$

Rotated graph: $r = f\left(\sin\left(\theta - \dfrac{\pi}{2}\right)\right) = f(-\cos\theta)$

(b) Rotation: $\phi = \pi$

Original graph: $r = f(\sin\theta)$

Rotated graph: $r = f(\sin(\theta - \pi)) = f(-\sin\theta)$

(c) Rotation: $\phi = \dfrac{3\pi}{2}$

Original graph: $r = f(\sin\theta)$

Rotated graph: $r = f\left(\sin\left(\theta - \dfrac{3\pi}{2}\right)\right) = f(\cos\theta)$

72. $r = 2\sin 2\theta$

(a) $r = 2\sin\left[2\left(\theta - \dfrac{\pi}{6}\right)\right]$

$$= 2\left[2\sin\left(\theta - \frac{\pi}{6}\right)\cos\left(\theta - \frac{\pi}{6}\right)\right]$$
$$= 4\sin\left(\theta - \frac{\pi}{6}\right)\cos\left(\theta - \frac{\pi}{6}\right)$$

(b) $r = 2\sin\left[2\left(\theta - \dfrac{\pi}{2}\right)\right]$

$$= 2\sin(2\theta - \pi)$$
$$= -2\sin 2\theta$$
$$= -2(2\sin\theta\cos\theta)$$
$$= -4\sin\theta\cos\theta$$

(c) $r = 2\sin\left[2\left(\theta - \dfrac{2\pi}{3}\right)\right]$

$$= 2\left[2\sin\left(\theta - \frac{2\pi}{3}\right)\cos\left(\theta - \frac{2\pi}{3}\right)\right]$$
$$= 4\sin\left(\theta - \frac{2\pi}{3}\right)\cos\left(\theta - \frac{2\pi}{3}\right)$$

(d) $r = 2\sin[2(\theta - \pi)]$

$$= 2\sin(2\theta - 2\pi)$$
$$= 2\sin 2\theta$$
$$= 2[2\sin\theta\cos\theta]$$
$$= 4\sin\theta\cos\theta$$

74. (a) $r = 3\sec\theta$

$$r = \frac{3}{\cos\theta}$$

$$r\cos\theta = 3 \Rightarrow x = 3$$

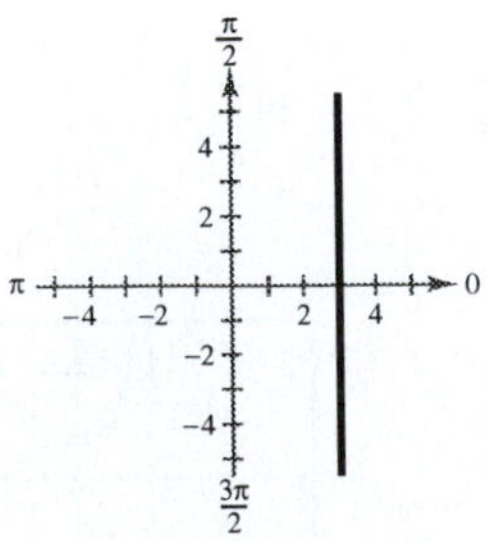

—CONTINUED—

74. —CONTINUED—

(b)

$$r = 3\sec\left(\theta - \frac{\pi}{4}\right)$$

$$r = \frac{3}{\cos(\theta - (\pi/4))}$$

$$r = \frac{3}{\cos\theta\cos(\pi/4) + \sin\theta\sin(\pi/4)}$$

$$\frac{\sqrt{2}}{2}r\cos\theta + \frac{\sqrt{2}}{2}r\sin\theta = 3$$

$$\frac{\sqrt{2}}{2}x + \frac{\sqrt{2}}{2}y = 3$$

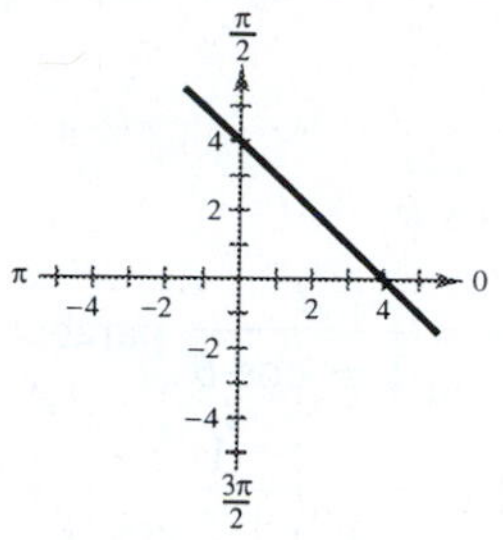

(c)

$$r = 3\sec\left(\theta + \frac{\pi}{3}\right)$$

$$r = \frac{3}{\cos(\theta + (\pi/3))}$$

$$r = \frac{3}{\cos\theta\cos(\pi/3) - \sin\theta\sin(\pi/3)}$$

$$\frac{1}{2}r\cos\theta - \frac{\sqrt{3}}{2}r\sin\theta = 3$$

$$\frac{1}{2}x - \frac{\sqrt{3}}{2}y = 3$$

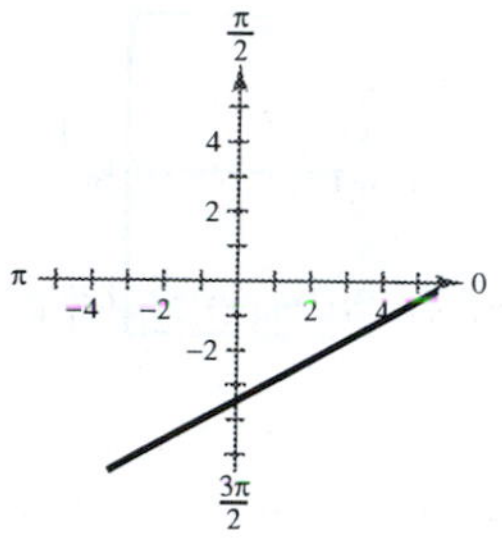

(d)

$$r = 3\sec\left(\theta - \frac{\pi}{2}\right)$$

$$r = \frac{3}{\cos(\theta - (\pi/2))}$$

$$r = \frac{3}{\cos\theta\cos(\pi/2) + \sin\theta\sin(\pi/2)}$$

$$r\sin\theta = 3 \Longrightarrow y = 3$$

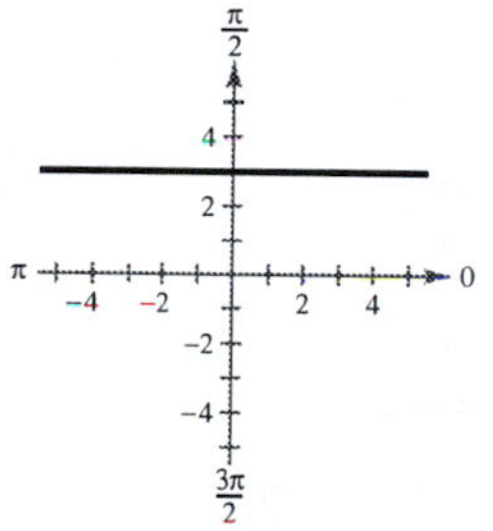

76. $r = 3\sin k\theta$

(a) $r = 3\sin 1.5\theta$

$0 \le \theta < 4\pi$

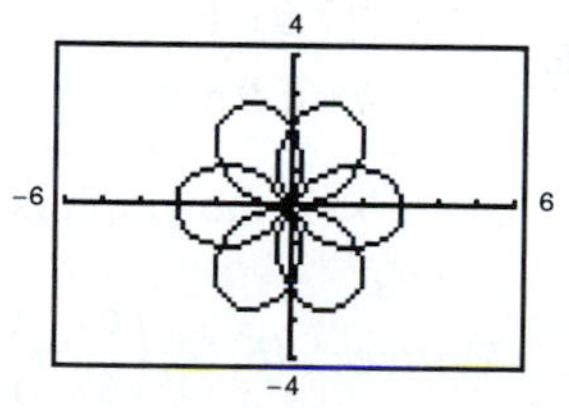

(b) $r = 3\sin 2.5\theta$

$0 \le \theta < 4\pi$

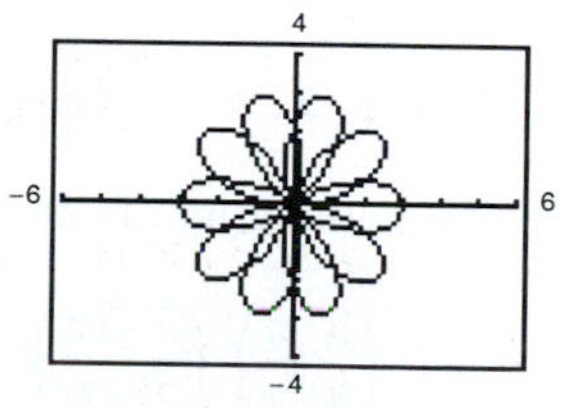

(c) Yes. $r = 3\sin(k\theta)$.

Find the minimum value of $\theta (\theta > 0)$ that is a multiple of 2π that makes $k\theta$ a multiple of 2π.

Section 10.9 Polar Equations of Conics

Solutions to Even-Numbered Exercises

2. $r = \dfrac{2e}{1 - e\cos\theta}$

(a) $e = 1, r = \dfrac{2}{1 - \cos\theta}$, parabola

(b) $e = 0.5, r = \dfrac{1}{1 - 0.5\cos\theta}$, ellipse

(c) $e = 1.5, r = \dfrac{3}{1 - 1.5\cos\theta}$, hyperbola

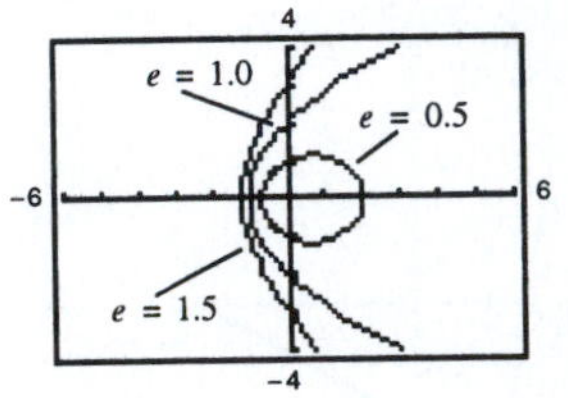

4. $r = \dfrac{2e}{1 + e\sin\theta}$

(a) $e = 1, r = \dfrac{2}{1 + \sin\theta}$, parabola

(b) $e = 0.5, r = \dfrac{1}{1 + 0.5\sin\theta}$, ellipse

(c) $e = 1.5, r = \dfrac{3}{1 + 1.5\sin\theta}$, hyperbola

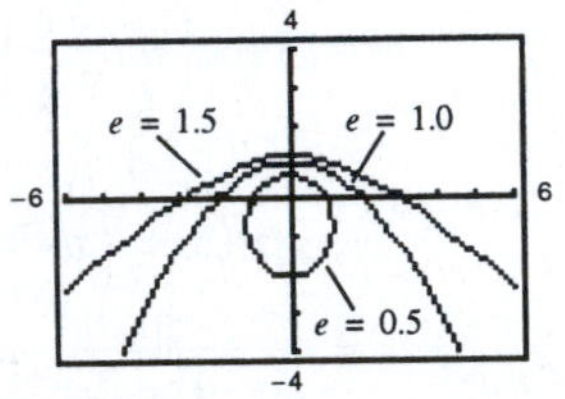

6. $r = \dfrac{3}{2 - \cos\theta}$

$e = \dfrac{1}{2} \Rightarrow$ Ellipse

Vertical directrix to the left of the pole
Matches graph (c).

8. $r = \dfrac{4}{1 + \sin\theta}$

$e = 1 \Rightarrow$ Parabola

Horizontal directrix above the pole
Matches graph (a).

10. $r = \dfrac{3}{1 + \sin\theta}$

$e = 1 \Rightarrow$ Parabola

Vertex: $\left(\dfrac{3}{2}, \dfrac{\pi}{2}\right)$

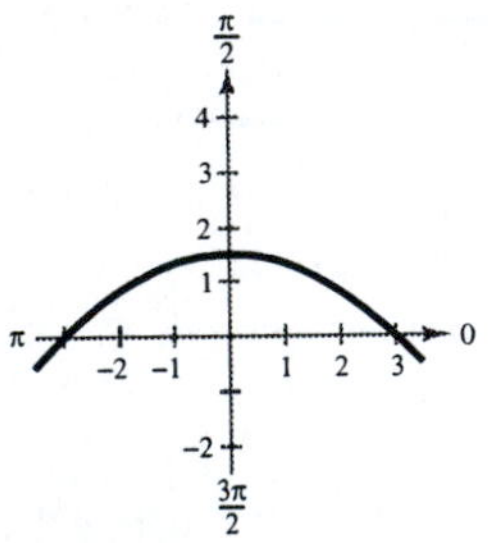

12. $r = \dfrac{6}{1 + \cos\theta}$

$e = 1 \Rightarrow$ Parabola
Vertex: $(3, 0)$

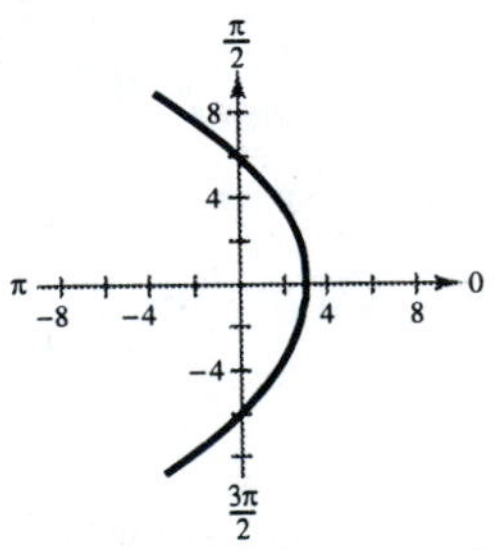

14. $r = \dfrac{3}{3 + \sin\theta} = \dfrac{1}{1 + (1/3)\sin\theta}$

$e = \dfrac{1}{3} < 1$, the graph is an ellipse.

Vertices: $\left(\dfrac{3}{4}, \dfrac{\pi}{2}\right), \left(\dfrac{3}{2}, \dfrac{3\pi}{2}\right)$

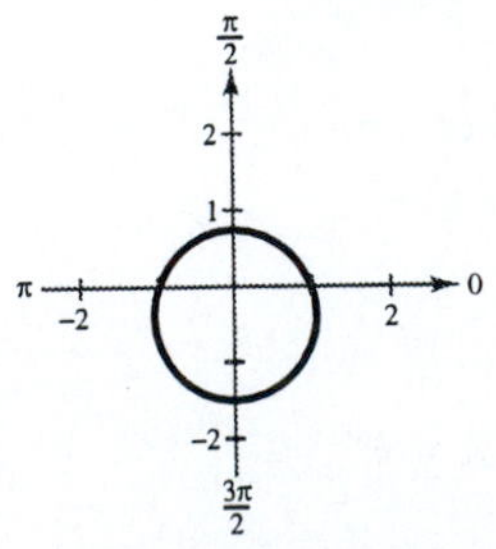

16. $r = \dfrac{6}{3 - 2\cos\theta} = \dfrac{2}{1 - (2/3)\cos\theta}$

$e = \dfrac{2}{3} < 1$, the graph is an ellipse.

Vertices: $(6, 0), \left(\dfrac{6}{5}, \pi\right)$

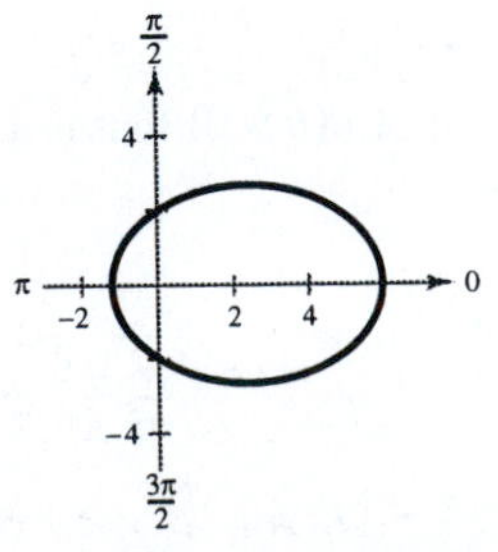

18. $r = \dfrac{5}{-1 + 2\cos\theta} = \dfrac{-5}{1 - 2\cos\theta}$

$e = 2 > 1$, the graph is a hyperbola.

Vertices: $(5, 0), \left(-\dfrac{5}{3}, \pi\right)$

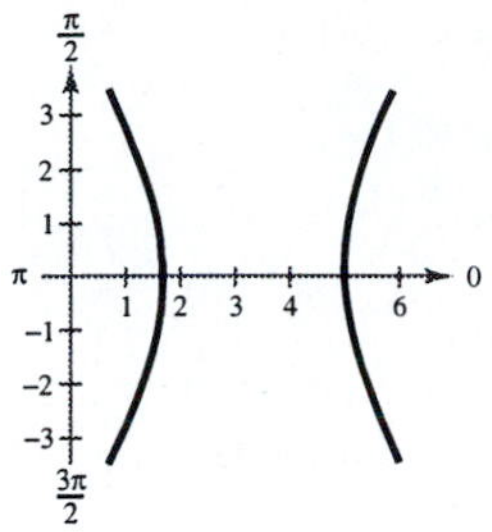

20. $r = \dfrac{3}{2 + 6\sin\theta} = \dfrac{3/2}{1 + 3\sin\theta}$

$e = 3 > 1$, the graph is a hyperbola.

Vertices: $\left(\dfrac{3}{8}, \dfrac{\pi}{2}\right), \left(-\dfrac{3}{4}, \dfrac{3\pi}{2}\right)$

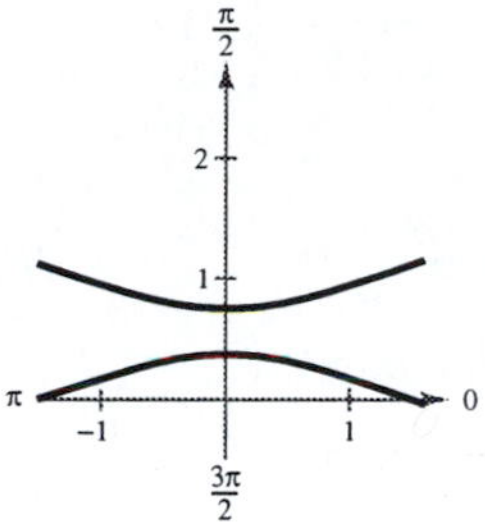

22. $r = \dfrac{2}{2 + 3\sin\theta} = \dfrac{1}{1 + (3/2)\sin\theta}$

$e = \dfrac{3}{2} > 1$, the graph is a hyperbola.

Vertices: $\left(\dfrac{2}{5}, \dfrac{\pi}{2}\right), \left(2, \dfrac{3\pi}{2}\right)$

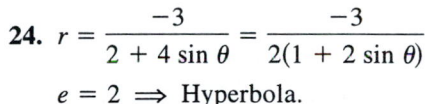

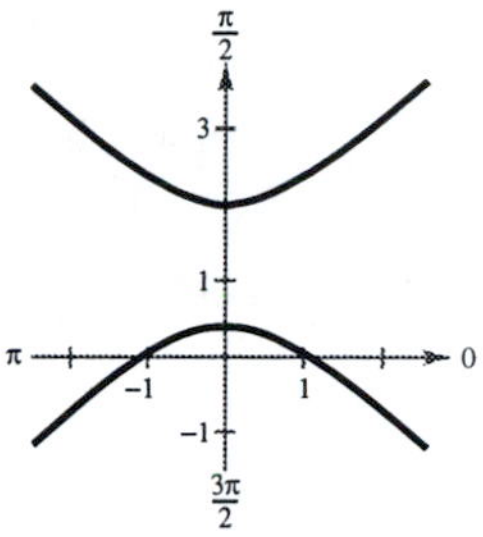

24. $r = \dfrac{-3}{2 + 4\sin\theta} = \dfrac{-3}{2(1 + 2\sin\theta)}$

$e = 2 \Rightarrow$ Hyperbola.

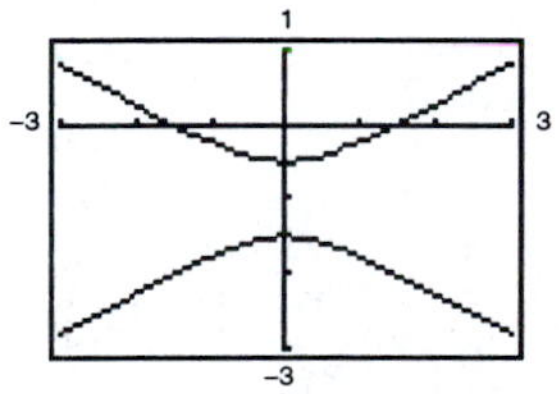

26. $r = \dfrac{4}{1 - 2\cos\theta}$

$e = 2 \Rightarrow$ Hyperbola.

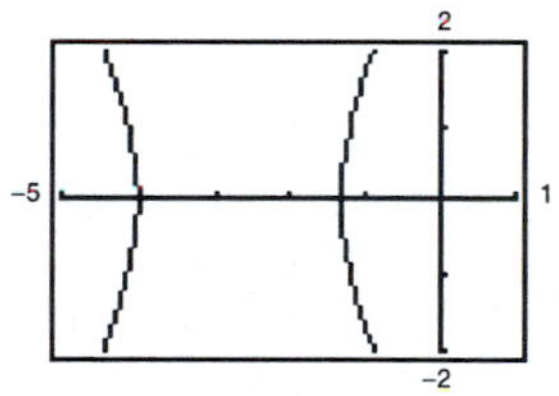

28. $r = \dfrac{3}{3 + \sin[\theta - (\pi/3)]}$

Rotate the graph in Exercise 14 through the angle $\pi/3$.

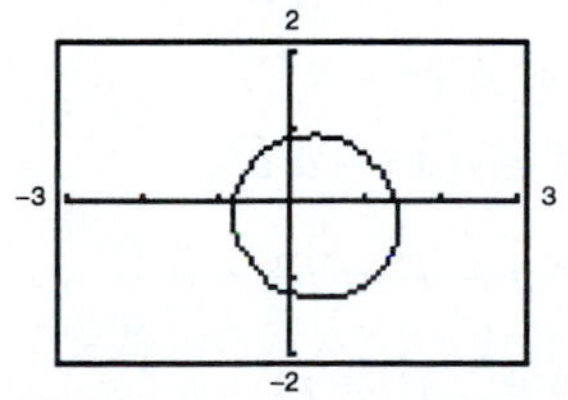

30. $r = \dfrac{5}{-1 + 2\cos[\theta + (2\pi/3)]}$

Rotate the graph in Exercise 18 through the angle $-2\pi/3$.

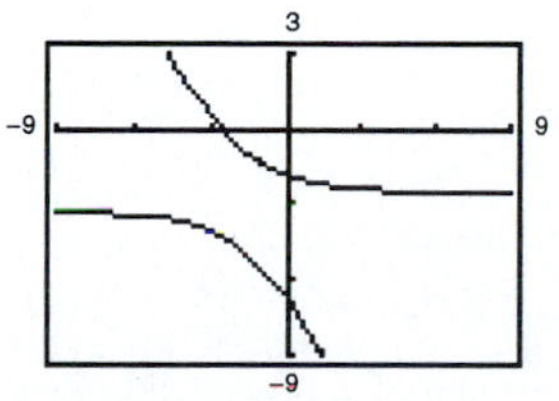

32. Parabola: $e = 1$

Directrix: $y = -2$

$p = 2$

Horizontal directrix below the pole

$r = \dfrac{1(2)}{1 - 1\sin\theta} = \dfrac{2}{1 - \sin\theta}$

34. Ellipse: $e = \dfrac{3}{4}$

Directrix: $y = -2$

$p = 2$

Horizontal directrix below the pole

$r = \dfrac{(3/4)(2)}{1 - (3/4)\sin\theta} = \dfrac{6}{4 - 3\sin\theta}$

36. Hyperbola: $e = \frac{3}{2}$

Directrix: $x = -1$

$p = 1$

Vertical directrix to the left of the pole

$$r = \frac{(3/2)(1)}{1 - (3/2)\cos\theta} = \frac{3}{2 - 3\cos\theta}$$

38. Parabola

Vertex: $(4, 0) \Rightarrow e = 1, p = 8$

Vertical directrix to the right of the pole

$$r = \frac{1(8)}{1 + 1\cos\theta} = \frac{8}{1 + \cos\theta}$$

40. Parabola

Vertex: $\left(10, \frac{\pi}{2}\right) \Rightarrow e = 1, p = 20$

Horizontal directrix above the pole

$$r = \frac{1(20)}{1 + 1\sin\theta} = \frac{20}{1 + \sin\theta}$$

42. Ellipse

Vertices: $\left(2, \frac{\pi}{2}\right), \left(4, \frac{3\pi}{2}\right)$

Center: $\left(1, \frac{3\pi}{2}\right)$; $c = 1, a = 3, e = \frac{1}{3}$

Horizontal directrix above the axis

$$r = \frac{1/3p}{1 + (1/3)\sin\theta} = \frac{p}{3 + \sin\theta}$$

$$2 = \frac{p}{3 + \sin(\pi/2)}$$

$$p = 8$$

$$r = \frac{8}{3 + \sin\theta}$$

44. Hyperbola

Vertices: $(2, 0), (10, 0)$

Center: $(6, 0)$: $c = 6, a = 4, e = \frac{3}{2}$

Vertical directrix to the right of the pole

$$r = \frac{3/2p}{1 + (3/2)\cos\theta} = \frac{3p}{2 + 3\cos\theta}$$

$$2 = \frac{3p}{2 + 3\cos 0}$$

$$p = \frac{10}{3}$$

$$r = \frac{3(10/3)}{2 + 3\cos\theta} = \frac{10}{2 + 3\cos\theta}$$

46. Hyperbola

Vertices: $\left(4, \frac{\pi}{2}\right), \left(-1, \frac{3\pi}{2}\right)$

Center: $\left(\frac{5}{2}, \frac{\pi}{2}\right)$; $c = \frac{5}{2}, a = \frac{3}{2}, e = \frac{5}{3}$

Horizontal directrix above the pole

$$r = \frac{5/3p}{1 + (5/3)\sin\theta} = \frac{5p}{3 + 5\sin\theta}$$

$$1 = \frac{5p}{3 + 5\sin(-3\pi/2)}$$

$$p = \frac{8}{5}$$

$$r = \frac{5(8/5)}{3 + 5\sin\theta} = \frac{8}{3 + 5\sin\theta}$$

48.

$$\frac{x^2}{a^2} - \frac{y^2}{b^2} = 1$$

$$\frac{r^2\cos^2\theta}{a^2} - \frac{r^2\sin^2\theta}{b^2} = 1$$

$$\frac{r^2\cos^2\theta}{a^2} - \frac{r^2(1 - \cos^2\theta)}{b^2} = 1$$

$$r^2b^2\cos^2\theta - r^2a^2 + r^2a^2\cos^2\theta = a^2b^2$$

$$r^2(b^2 + a^2)\cos^2\theta - r^2a^2 = a^2b^2$$

$$a^2 + b^2 = c^2$$

$$r^2c^2\cos^2\theta - r^2a^2 = a^2b^2$$

$$r^2\left(\frac{c}{a}\right)^2\cos^2\theta - r^2 = b^2, \quad e = \frac{c}{a}$$

$$r^2e^2\cos^2\theta - r^2 = b^2$$

$$r^2(e^2\cos^2\theta - 1) = b^2$$

$$r^2 = \frac{b^2}{e^2\cos^2\theta - 1}$$

$$= \frac{-b^2}{1 - e^2\cos^2\theta}$$

50. $\dfrac{x^2}{25} + \dfrac{y^2}{16} = 1$

$a = 5, b = 4, c = 3, e = \dfrac{3}{5}$

$r^2 = \dfrac{400}{25 - 9\cos^2\theta}$

52. $\dfrac{x^2}{36} - \dfrac{y^2}{4} = 1$

$a = 6, b = 2, c = 2\sqrt{10}, e = \dfrac{\sqrt{10}}{3}$

$r^2 = \dfrac{-4}{1 - (10/9)\cos^2\theta} = \dfrac{-36}{9 - 10\cos^2\theta}$

$= \dfrac{36}{10\cos^2\theta - 9}$

54. Ellipse

One focus: (4, 0)

Vertices: (5, 0), $(5, \pi)$

$a = 5, c = 4, b = 3, e = \dfrac{4}{5}$

$r^2 = \dfrac{9}{1 - (16/25)\cos^2\theta} = \dfrac{225}{25 - 16\cos^2\theta}$

56. Minimum distance occurs when $\theta = \pi$.

$r = \dfrac{(1 - e^2)a}{1 - e\cos\pi} = \dfrac{(1 - e)(1 + e)a}{1 + e} = a(1 - e)$

Maximum distance occurs when $\theta = 0$.

$r = \dfrac{(1 - e^2)a}{1 - e\cos 0} = \dfrac{(1 - e)(1 + e)a}{1 - e} = a(1 + e)$

58. $r = \dfrac{[1 - (0.0543)^2](1.427 \times 10^9)}{1 - 0.0543\cos\theta} \approx \dfrac{1.4228 \times 10^9}{1 - 0.0543\cos\theta}$

Perihelion distance: $r = 1.427 \times 10^9(1 - 0.0543) \approx 1.3495 \times 10^9$ kilometers

Aphelion distance: $r = 1.427 \times 10^9(1 + 0.0543) \approx 1.5045 \times 10^9$ kilometers

60. $r = \dfrac{[1 - (0.2056)^2](36.0 \times 10^6)}{1 - 0.2056\cos\theta} \approx \dfrac{3.4478 \times 10^7}{1 - 0.2056\cos\theta}$

Perihelion distance: $r = 36.0 \times 10^6(1 - 0.2056) \approx 2.8598 \times 10^7$ miles

Aphelion distance: $r = 36.0 \times 10^6(1 + 0.2056) \approx 4.3402 \times 10^7$ miles

62. Assume the earth's radius is 4000 miles.

$e = \dfrac{c}{a} = \dfrac{126{,}000 - 4119}{126{,}000 + 4119} \approx 0.937$

$r = \dfrac{ep}{1 - e\cos\theta}$

$(4119, \pi)$: $4119 = \dfrac{0.937}{1 - 0.937(-1)} \Longrightarrow p \approx 8516$

$r \approx \dfrac{7977.2}{1 - 0.937\cos\theta}$

When $\theta = 60°$: $r = \dfrac{7977.2}{1 - 0.937(1/2)} \approx 15{,}008$

Distance: 15,008 − 4000 = 11,008 miles

64. $g(x) = 4^{2-x}$

Exponential function

x	0	1	2	3
y	16	4	1	$\frac{1}{4}$

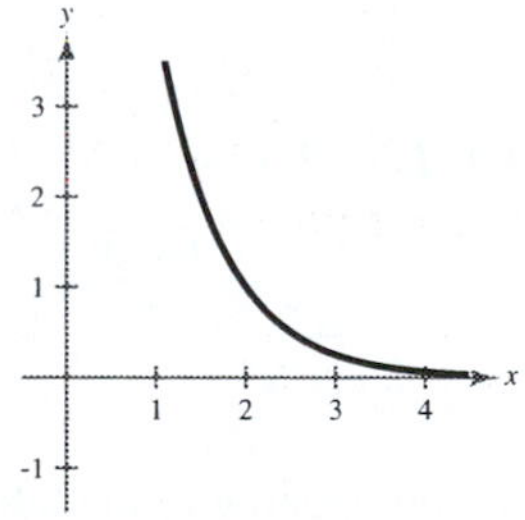

66. $h(s) = \log_4 s^2$

Logarithmic function

s	±1	±2	±4
y	0	1	2

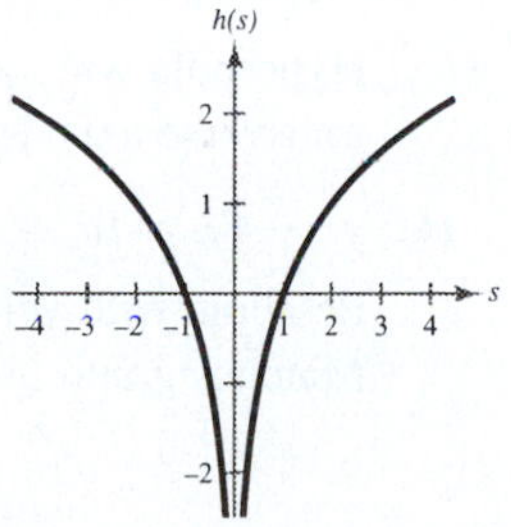

Focus on Concepts

1. (a) Vertical translation
   (b) Horizontal translation
   (c) Reflection in the y-axis
   (d) Parabola opens more slowly

2. (a) Major axis horizontal
   (b) Circle
   (c) Ellipse is flatter
   (d) Horizontal translation

3. The extended diagonals of the central rectangle are asymptotes of the hyperbola.

4. 5. The ellipse becomes more circular and approaches a circle of radius 5.

5. The orientation would be reversed.

6. (a) The speed would double.
   (b) The elliptical orbit would be flatter. The length of the major axis is greater.

7. False. The following are two sets of parametric equations for the line.
   $x = t,\ y = 3 - 2t$
   $x = 3t,\ y = 3 - 6t$

8. False. $(2, \pi/4)$, $(-2, 5\pi/4)$, and $(2, 9\pi/4)$ all represent the same point.

9. (a) Symmetric to the pole
   (b) Symmetric to the polar axis
   (c) Symmetric to $\pi/2$

10. Same graphs

Review Exercises for Chapter 10

Solutions to Even-Numbered Exercises

2. $m = \tan 55.8°$
   ≈ 1.4715

4. $y = \frac{3}{2}x - 2$
   $m = \frac{3}{2} = \tan\theta$
   $\theta = \arctan\left(\frac{3}{2}\right)$
   $\approx 56.3°$

6. $(0, 4) \Longrightarrow x_1 = 0, y_1 = 4$
   $x + 2y - 2 = 0 \Longrightarrow A = 1, B = 2, C = -2$
   $$d = \frac{|1(0) + (2)(4) + (-2)|}{\sqrt{1^2 + 2^2}} = \frac{6}{\sqrt{5}} = \frac{6\sqrt{5}}{5}$$

8. $x^2 = 4y$
   Parabola with vertex $(0, 0)$ and opening upward. Matches graph (i).

10. $y^2 = -4x$
    Parabola with vertex $(0, 0)$ and opening to the left. Matches graph (a).

12. $y^2 - 4x^2 = 4$
    $$\frac{y^2}{4} - \frac{x^2}{1} = 1$$
    Hyperbola with center $(0, 0)$ and a vertical transverse axis. Matches graph (j).

14. $x^2 + 5y^2 = 10$
    $$\frac{x^2}{10} + \frac{y^2}{2} = 1$$
    Ellipse with center $(0, 0)$ and a horizontal major axis. Matches graph (c).

16. $y^2 - 8x = 0$
    Parabola with vertex $(0, 0)$ and opening to the right. Matches graph (e).

18. $8y + x^2 = 0$

$-8y = x^2$

The graph is a parabola.
Vertex: $(0, 0)$

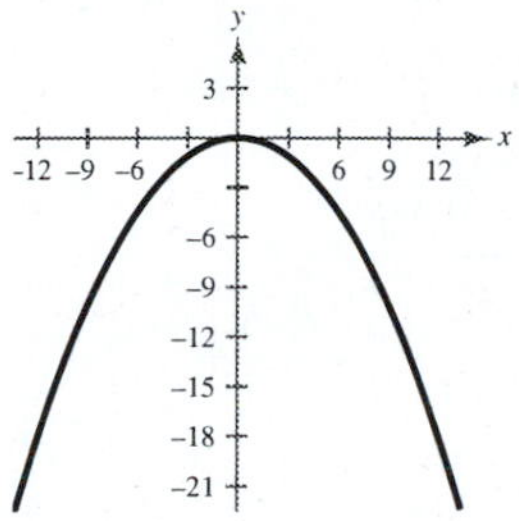

20. $y^2 - 12y - 8x + 20 = 0$

$AC = 0(1) = 0$

The graph is a parabola.

$y^2 - 12y - 8x + 20 = 0$

$(y - 6)^2 = 8(x + 2)$

Vertex: $(-2, 6)$

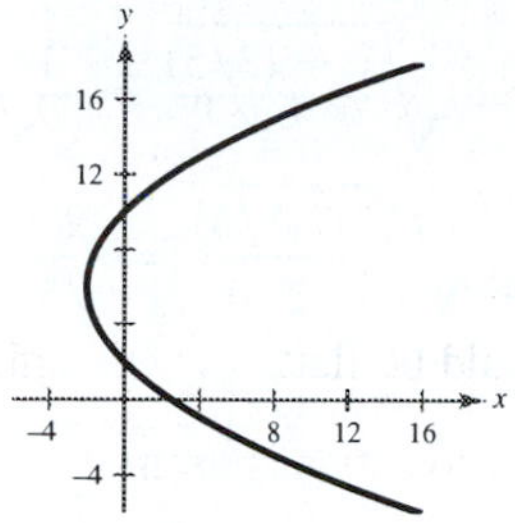

22. $16x^2 + 16y^2 - 16x + 24y - 3 = 0$

$A = C = 16$

The graph is a circle.

$16x^2 + 16y^2 - 16x + 24y - 3 = 0$

$$\left(x - \frac{1}{2}\right)^2 + \left(y + \frac{3}{4}\right)^2 = 1$$

Center: $\left(\frac{1}{2}, -\frac{3}{4}\right)$

Radius: 1

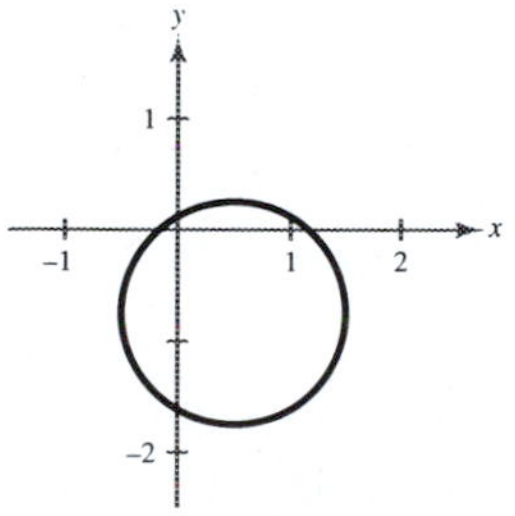

24. $2x^2 + 6y^2 = 18$

$AC = 2(6) = 12 > 0$

The graph is an ellipse.

$$\frac{2x^2}{18} + \frac{6y^2}{18} = \frac{18}{18}$$

$$\frac{x^2}{9} + \frac{y^2}{3} = 1$$

Center: $(0, 0)$

Vertex: $(\pm 3, 0)$

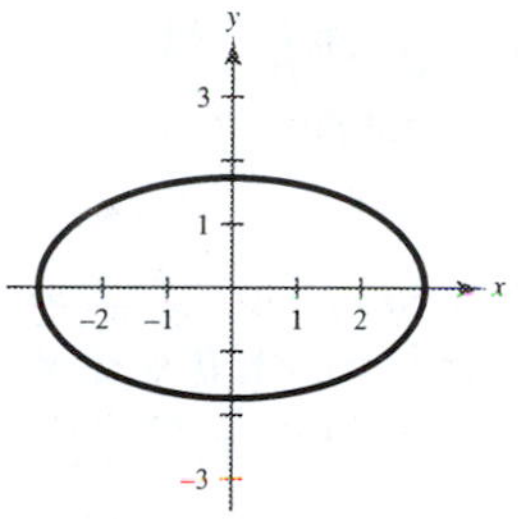

26. $4x^2 + y^2 - 16x + 15 = 0$

$AC = 4(1) > 0$

The graph is an ellipse.

$4x^2 + y^2 - 16x + 15 = 0$

$$\frac{(x - 2)^2}{1/4} + y^2 = 1$$

Center: $(2, 0)$

Vertices: $(2, \pm 1)$

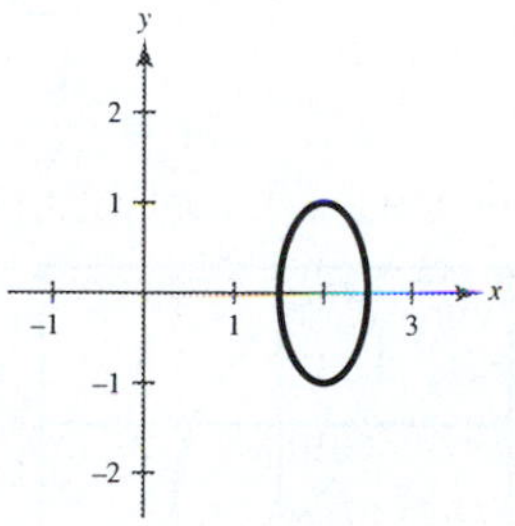

28. $x^2 - 9y^2 + 10x + 18y + 7 = 0$

$AC = 1(-9) = -9 < 0$

The graph is a hyperbola.

$x^2 + 10x - 9y^2 + 18y = -7$

$$\frac{(x + 5)^2}{9} - \frac{9(y - 1)^2}{9} = 1$$

Center: $(-5, 1)$

Vertices: $(-8, 1), (-2, 1)$

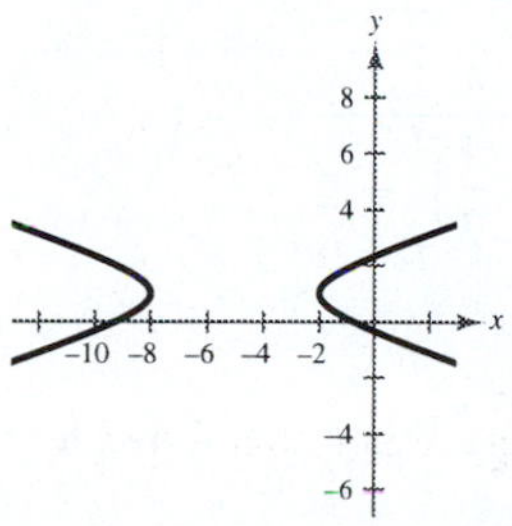

30. $9x^2 + 6y^2 + 4xy - 20 = 0$

$B^2 - 4AC = 4^2 - 4(9)(6) = -220 < 0$

The graph is an ellipse.

$$\cot 2\theta = \frac{A - C}{B} = \frac{9 - 6}{4} = \frac{3}{4} \Longrightarrow 2\theta \approx 53.1° \Longrightarrow \theta \approx 26.6°$$

$$\cot 2\theta = \frac{3}{5}$$

$$\sin \theta = \sqrt{\frac{1 - \cos 2\theta}{2}} = \sqrt{\frac{1 - (3/5)}{2}} = \frac{1}{\sqrt{5}}$$

$$\cos \theta = \sqrt{\frac{1 + \cos \theta}{2}} = \sqrt{\frac{1 + (3/5)}{2}} = \frac{2}{\sqrt{5}}$$

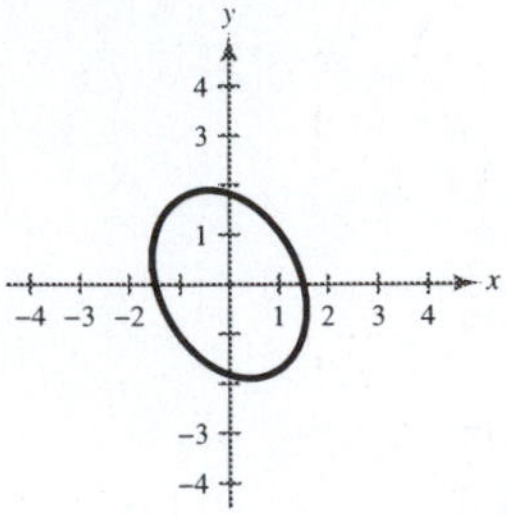

$$x = x'\cos \theta - y'\sin \theta = x'\left(\frac{2}{\sqrt{5}}\right) - y'\left(\frac{1}{\sqrt{5}}\right) = \frac{2x' - y'}{\sqrt{5}}$$

$$y = x'\sin \theta + y'\cos \theta = x'\left(\frac{1}{\sqrt{5}}\right) + y'\left(\frac{2}{\sqrt{5}}\right) = \frac{x' + 2y'}{\sqrt{5}}$$

$$9\left(\frac{2x' - y'}{\sqrt{5}}\right) + 6\left(\frac{x' + 2y'}{\sqrt{5}}\right) + 4\left(\frac{2x' - y'}{\sqrt{5}}\right)\left(\frac{x' + 2y'}{\sqrt{5}}\right) - 20 = 0$$

$$10(x')^2 + 5(y')^2 - 20 = 0$$

$$\frac{(x')^2}{2} + \frac{(y')^2}{4} = 1$$

Center: $(0, 0)$, $\theta \approx 26.6°$

32. $4x^2 - 4y^2 - 4x + 8y - 11 = 0$

$AC = 4(-4) = -16 < 0$

The graph is a hyperbola.

$4x^2 - 4y^2 - 4x + 8y - 11 = 0$

$$\frac{(x - 1/2)^2}{2} - \frac{(y - 1)^2}{2} = 1$$

Center: $\left(\frac{1}{2}, 1\right)$

Vertices: $\left(\frac{1}{2} \pm \sqrt{2}, 1\right)$

To use a graphing calculator, we need to solve for y in terms of x.

$$(y - 1)^2 = \left(x - \frac{1}{2}\right)^2 - 2$$

$$y = 1 \pm \sqrt{\left(x - \frac{1}{2}\right)^2 - 2}$$

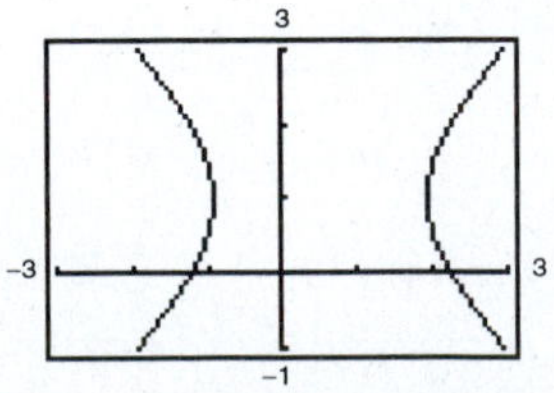

34. $40x^2 + 36xy + 25y^2 - 52 = 0$

$B^2 - 4AC = 36^2 - 4(40)(25) = -2704 < 0$

The graph is an ellipse.

To use a graphing calculator, we need to solve for y in terms of x.

$$25y^2 + 36xy = 52 - 40x^2$$

$$y^2 + \frac{36}{25}xy + \frac{324}{625} = \frac{1}{625}(1624 - 1000x^2)$$

$$\left(y + \frac{18}{25}x\right)^2 = \frac{1}{625}(1624 - 1000x^2)$$

$$y + \frac{18}{25}x = \pm\frac{1}{25}\sqrt{1624 - 1000x^2}$$

$$y = \frac{18}{25}x \pm \frac{1}{25}\sqrt{1624 - 1000x^2}$$

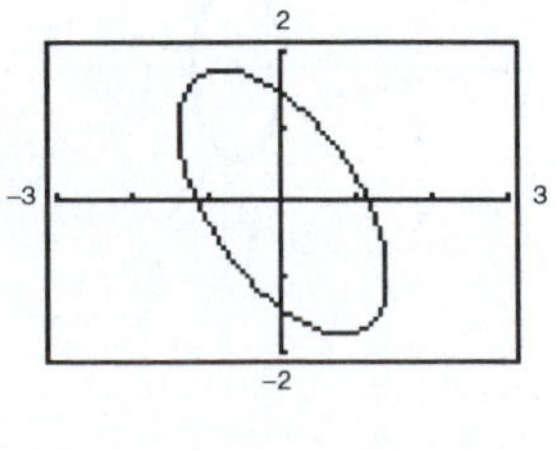

36. Vertex: $(2, 0) = (h, k)$
Focus: $(0, 0) \Longrightarrow p = -2$
$(y - k)^2 = 4p(x - h)$
$y^2 = -8(x - 2)$

38. Vertex: $(2, 2) = (h, k)$
Directrix: $y = 0 \Longrightarrow p = 2$
$(x - h)^2 = 4p(y - k)$
$(x - 2)^2 = 8(y - 2)$

40. Vertices: $(2, 0), (2, 4) \Longrightarrow a = 2, (h, k) = (2, 2)$
Foci: $(2, 1), (2, 3) \Longrightarrow c = 1$
$b^2 = a^2 - c^2 = 4 - 1 = 3$
$$\frac{(x - h)^2}{b^2} + \frac{(y - k)^2}{a^2} = 1$$
$$\frac{(x - 2)^2}{3} + \frac{(y - 2)^2}{4} = 1$$

42. Vertices: $(0, 1), (4, 1) \Longrightarrow a = 2, (h, k) = (2, 1)$
Endpoints of minor axis: $(2, 0), (2, 2) \Longrightarrow b = 1$
$$\frac{(x - h)^2}{a^2} + \frac{(y - k)^2}{b^2} = 1$$
$$\frac{(x - 2)^2}{4} + (y - 1)^2 = 1$$

44. Vertices: $(2, 2), (-2, 2) \Longrightarrow a = 2, (h, k) = (0, 2)$
Foci: $(4, 2), (-4, 2) \Longrightarrow c = 4$
$b^2 = c^2 - a^2 = 16 - 4 = 12$
$$\frac{(x - h)^2}{a^2} - \frac{(y - k)^2}{b^2} = 1$$
$$\frac{x^2}{4} - \frac{(y - 2)^2}{12} = 1$$

46. Foci: $(3, \pm 2) \Longrightarrow c = 2, (h, k) = (3, 0)$

Asymptotes: $y = \pm 2(x - 3) \Longrightarrow \frac{a}{b} = 2, a = 2b$

$b^2 = c^2 - a^2 = 4 - 4b^2 \Longrightarrow b^2 = \frac{4}{5}, a^2 = \frac{16}{5}$
$$\frac{(y - k)^2}{a^2} - \frac{(x - h)^2}{b^2} = 1$$
$$\frac{y^2}{16/5} - \frac{(x - 3)^2}{4/5} = 1 \Longrightarrow \frac{5y^2}{16} - \frac{5(x - 3)^2}{4} = 1$$

48. $a = 5, b = 4, c^2 = a^2 - b^2 = 25 - 16 = 9$
$c = 3$
The foci occur 3 feet from the center of the arch, on a line connecting the tops of the pillars.

50. False.
$\frac{x^2}{4} - y^4 = 1$ is a fourth-degree equation.
The equation of a hyperbola is a second degree equation.

52. $x = t^2, y = \sqrt{t}$
$t = y^2$
$x = (y^2)^2 = y^4, y \geq 0$
$\sqrt[4]{x} = y$

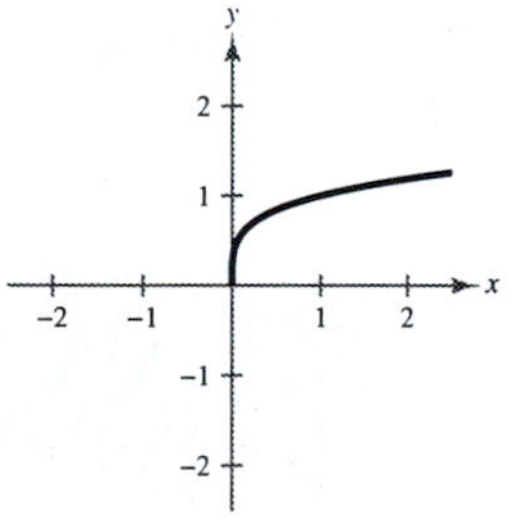

54. $x = t + 4, y = t^2$
$t = x - 4$
$y = (x - 4)^2$

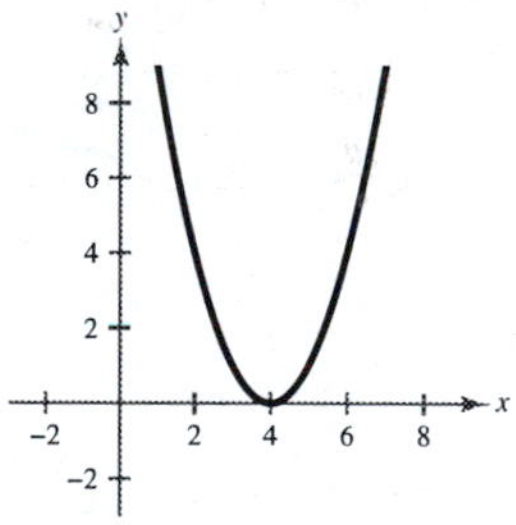

56. $x = \frac{1}{t}, y = 2t + 3$
$$t = \frac{1}{x}$$
$$y = \frac{2}{x} + 3$$
$$y = \frac{2 + 3x}{x}$$

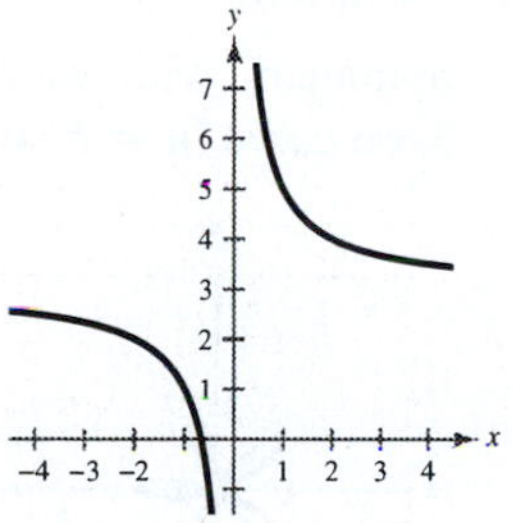

58. $x = 3 + 3 \cos \theta, y = 2 + 5 \sin \theta$
$$\cos \theta = \frac{x - 3}{3}, \sin \theta = \frac{y - 2}{5}$$
$$\frac{(x - 3)^2}{9} + \frac{(y - 2)^2}{25} = 1$$

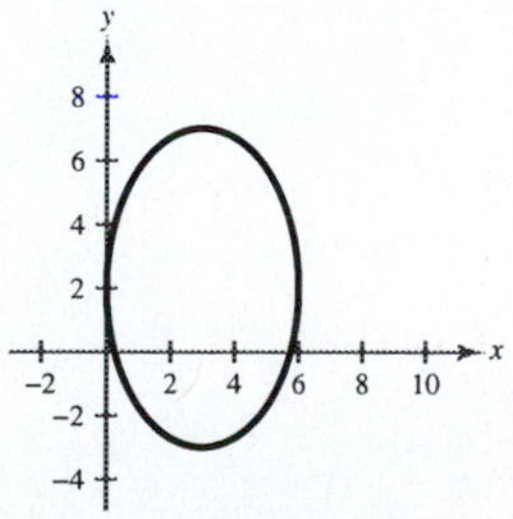

60. $\frac{y^2}{16} - \frac{x^2}{9} = 1$

$x = 3 \tan \theta$

$y = 4 \sec \theta$

This solution is not unique.

62. (a) $y = \overline{QB} - \overline{QA}$

$\overline{QP} = \text{arc } QC = r\theta$

$\overline{QA} = r\theta \sin(90° - \theta)$

$= r\theta \cos \theta$

$\overline{QB} = r \sin \theta$

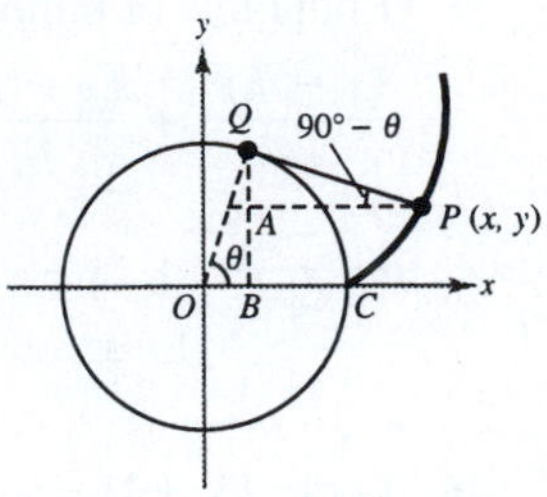

(b)

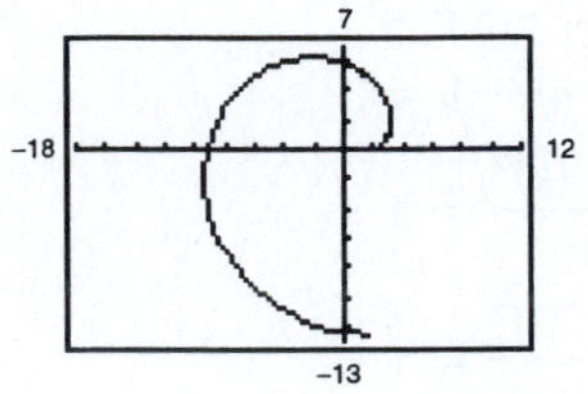

Therefore, $y = r \sin \theta - r\theta \cos \theta = r(\sin \theta - \theta \cos \theta)$.
Similarly, $x = \overline{OB} + \overline{AP}$. Therefore, $x = r \cos \theta + r\theta \sin \theta = r(\cos \theta + \theta \sin \theta)$.

64.

$$r = \frac{2}{1 + \sin \theta}$$
$$r + r \sin \theta = 2$$
$$r = 2 - r \sin \theta$$
$$r^2 = (2 - r \sin \theta)^2$$
$$x^2 + y^2 = (2 - y)^2$$
$$x^2 + y^2 = 4 - 4y + y^2$$
$$x^2 + 4y - 4 = 0$$

66.

$$r = 10$$
$$r^2 = 100$$
$$x^2 + y^2 = 100$$

68. $x^2 + y^2 - 4x = 0$

$$r^2 - 4r \cos \theta = 0$$
$$r = 4 \cos \theta$$

70. $\theta = \frac{\pi}{12}$

Line

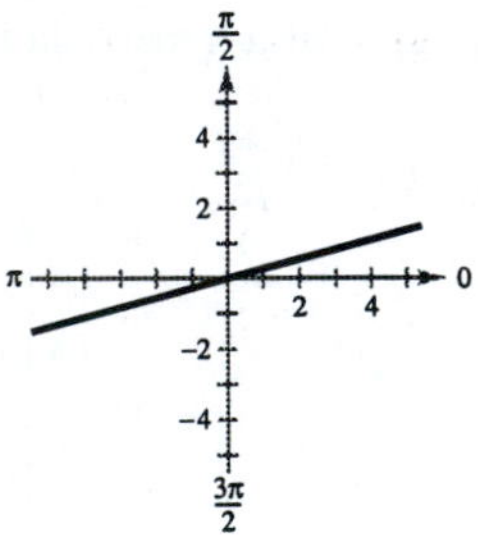

72. $r = 2\theta$

Symmetric with respect to $\theta = \pi/2$

Spiral

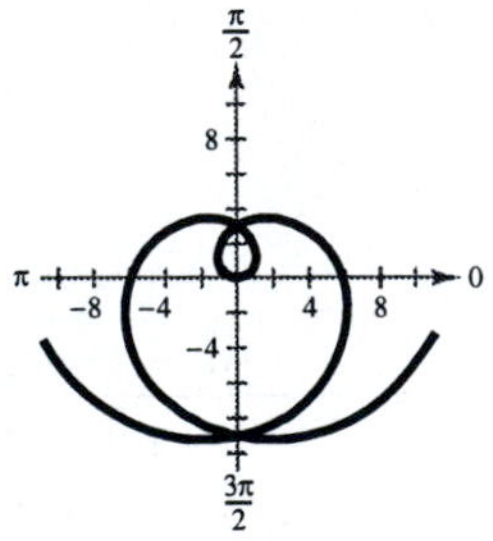

74. $r = 3 - 4 \cos \theta$

Symmetric with respect to polar axis

$\frac{a}{b} = \frac{3}{4} < 0 \Rightarrow$ Limaçon with inner loop

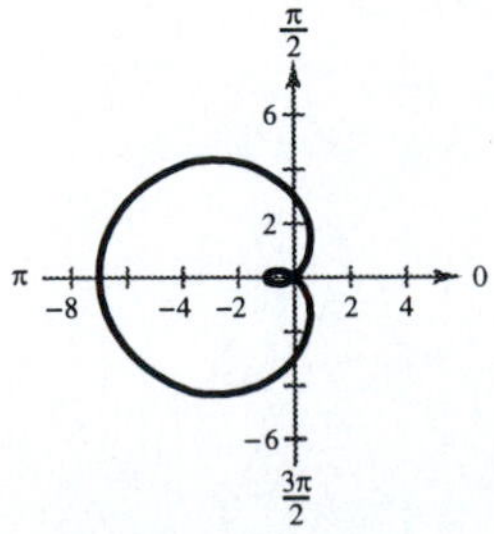

76. $r = \cos 5\theta$

Symmetric with respect to polar axis

Rose curve ($n = 5$) with five petals

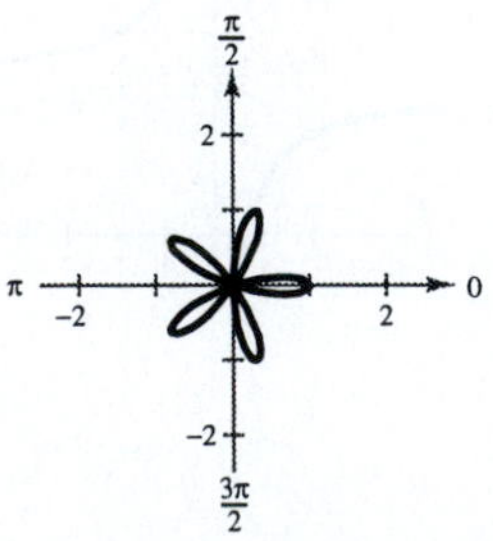

78. $r^2 = \cos\theta$

Symmetric with respect to polar axis

Leminscate

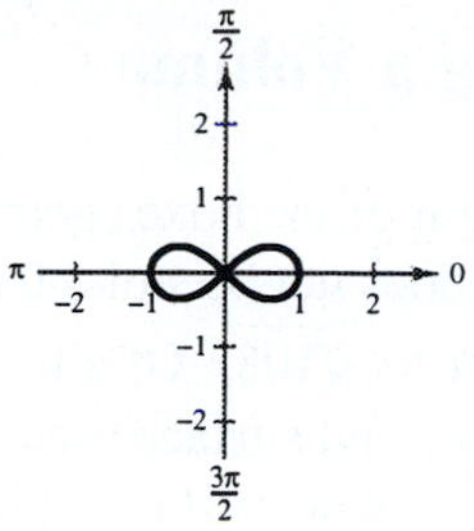

80. $r = \dfrac{1}{1 + 2\sin\theta}, e = 2$

Hyperbola symmetric with $\theta = \pi/2$ and having vertices at $(1/3, \pi/2)$ and $(-1, 3\pi/2)$.

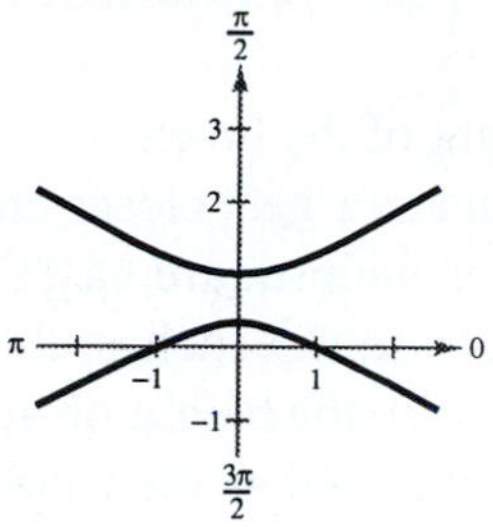

82.
$$r = 3\csc\theta$$
$$r\sin\theta = 3$$
$$y = 3$$

Line

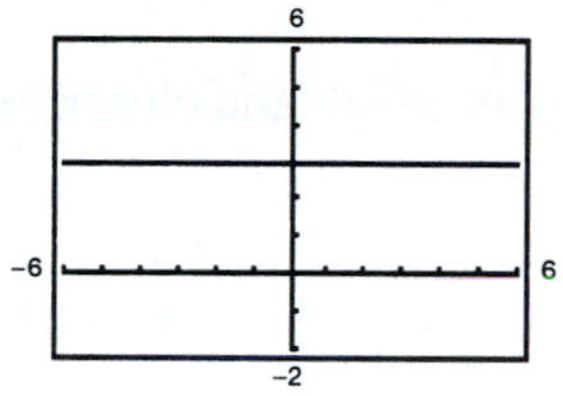

84. $r = \dfrac{4}{5 - 3\cos\theta}$

$r = \dfrac{4/5}{1 - (3/5)\cos\theta}, e = \dfrac{3}{5}$

Ellipse symmetric with polar axis and having vertices at $(2, 0)$ and $(1/2, \pi)$.

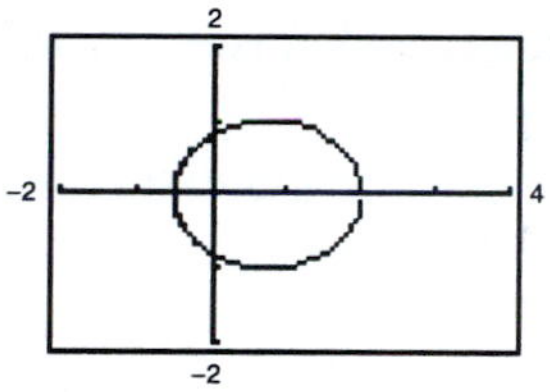

86.
$$m = \sqrt{3}$$
$$\tan\theta = \sqrt{3}$$
$$\theta = \frac{\pi}{3}$$

88. Parabola: $r = \dfrac{ep}{1 + e\sin\theta}, e = 1$

Vertex: $\left(2, \dfrac{\pi}{2}\right)$

Focus: $(0, 0) \implies p = 4$

$r = \dfrac{4}{1 + \sin\theta}$

90. Hyperbola: $r = \dfrac{ep}{1 + e\cos\theta}$

Vertices: $(1, 0), (7, 0) \implies a = 3$

One focus: $(0, 0) \implies c = 4$

$e = \dfrac{c}{a} = \dfrac{4}{3}, p = \dfrac{7}{4}$

$$r = \frac{(4/3)(7/4)}{1 + (4/3)\cos\theta} = \frac{7/3}{1 + (4/3)\cos\theta} = \frac{7}{3 + 4\cos\theta}$$

PART IV
Chapter Project Solutions

Chapter P Project A Numerical Approach to Maximizing a Volume

1. In Example 2, the height of the boxes increases without bound as x gets closer and closer to 0. Of all boxes with square bases and surface areas of 216 square inches, there is not a tallest box because the height of the boxes increases without bound as the length x of the base gets smaller and smaller.

2. In Example 2, the height of the boxes gets closer and closer (but does not become) 0 as x gets closer and closer to $\sqrt{108}$. Of all boxes with square bases and surface areas of 216 square inches, there is not a shortest box because the height of the boxes becomes infinitely small as the length x of the approaches $\sqrt{108}$.

3.

Base x	5.9	5.99	5.999	6.001	6.01	6.1
Volume V	215.9105	215.9991005	215.999991	215.999991	215.9990995	215.9095

The table lends further support to the conclusion of Example 2 because as x gets closer and closer to 6, the volume V gets closer and closer to 216.

4. Begin by writing a model for the volume of a rectangular box with a surface area of 216 square inches and base x inches by $2x$ inches.

Surface area = area of base + area of top + 2 · area of short side + 2 · area of long side

$$216 = 2x^2 + 2x^2 + 2(xh) + 2(2xh)$$

$$216 = 4x^2 + 6xh$$

$$216 - 4x^2 = 6xh$$

$$\frac{216 - 4x^2}{6x} = h$$

$$\frac{36}{x} - \frac{2x}{3} = h$$

Having written the height in terms of x, we can write the volume in terms of x.

Volume of box = length of box · width of box · height of box

$$V = 2x \cdot x \cdot \left(\frac{36}{x} - \frac{2x}{3}\right)$$

$$V = 72x - \frac{4x^3}{3}, \quad 0 \le x \le \sqrt{54}$$

Using this model for volume, we can make create a table like the one in Example 2 to find the maximum volume of this type of box.

Width of base, x	Length of base, $2x$	Height	Surface Area	Volume
1.0	2.0	35.333	216	70.667
1.5	3.0	23.000	216	103.500
2.0	4.0	16.667	216	133.333
2.5	5.0	12.733	216	159.167
3.0	6.0	10.000	216	180.000
3.5	7.0	7.952	216	194.833
4.0	8.0	6.333	216	202.667
4.5	9.0	5.000	216	202.500
5.0	10.0	3.867	216	193.333
5.5	11.0	2.879	216	174.167
6.0	12.0	2.000	216	144.000
6.5	13.0	1.205	216	101.833
7.0	14.0	0.476	216	46.667

From the results of the experiment, it appears that the box of greatest volume occurs when $3.5 < x < 4.5$. We can construct another table to narrow down the value of x giving the greatest volume.

Width of base, x	Length of base, $2x$	Height	Surface Area	Volume
3.6	7.2	7.600	216	196.992
3.7	7.4	7.263	216	198.863
3.8	7.6	6.940	216	200.44
3.9	7.9	6.631	216	201.708
4.0	8.0	6.333	216	202.667
4.1	8.2	6.047	216	203.305
4.2	8.4	5.771	216	203.616
4.3	8.6	5.505	216	203.591
4.4	8.8	5.248	216	203.221

From this table, it appears that the box of greatest volume occurs when $4.1 < x < 4.3$. We can construct yet another table to narrow down the value of x giving the greatest volume even more.

Width of base, x	Length of base, $2x$	Height	Surface Area	Volume
4.15	8.30	5.908	216	203.500
4.16	8.32	5.881	216	203.532
4.17	8.34	5.853	216	203.558
4.18	8.36	5.826	216	203.580
4.19	8.38	5.799	216	203.600
4.20	8.40	5.771	216	203.616
4.21	8.42	5.744	216	203.629
4.22	8.44	5.717	216	203.638
4.23	8.46	5.691	216	203.644
4.24	8.48	5.664	216	203.647
4.25	8.50	5.637	216	203.646
4.26	8.52	5.611	216	203.642

From the table, it appears that the box of the greatest volume occurs when $x \approx 4.24$ and $h \approx 5.66$. Better approximations may be obtained by creating additional tables (for example, creating a table covering values from 4.23 to 4.25 with an increment of 0.001).

Chapter 1 Project — A Graphical Approach to Maximization

1. (a) (0, 40): Because $(x, p) = (0, 40)$, no units are sold at a price of \$40. Thus, no one will buy the product at this price.
   (b) $(\sqrt{8}, 0)$: Because $(x, p) = (\sqrt{8}, 0)$, no more than $\sqrt{8}$ units will be accepted by the public at a price of \$0. Thus, you can't give more than $\sqrt{8}$ units away.

2. By tracing on the screen of the graphing utility, we find that at $x \approx 1.6329$, the revenue is maximized at \$43.546484 million. Yes, this improved accuracy is appropriate in the context of the problem. The new price is given by $p = 40 - 5(1.6329)^2 \approx \$26.67$.

3. By graphing $P = -5x^3 + 25x$ on a graphing utility, we find that a setting of $1.28 \le x \le 1.3$ and $21 \le y \le 22$ allows us to improve graphically the solution in Example 2.

4. In Example 2, if there is an initial cost of \$250,000 in addition to a cost of \$15 per unit, the new cost (in millions of dollars) is $C = 0.25 + 15x$. Thus, the new profit function is

$$\begin{aligned} P &= R - C \\ &= (40x - 5x^3) - (0.25 + 15x) \\ &= -5x^3 + 25x - 0.25 \end{aligned}$$

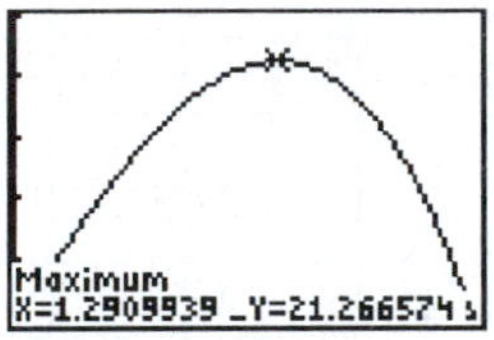

Using the maximum feature of a graphing utility, we find that the new maximum profit is approximately \$21.3 million, corresponding to $x \approx 1.29$ and the new price is $p \approx 40 - 5(1.29)^2 = \$31.68$.

5. Using the maximum feature of a graphing utility with the graph of $V = 54x - \frac{1}{2}x^3$, we find that the a box of maximum volume occurs when $x =$ 6. Thus the dimensions are 6 inches $\times$ 6 inches $\times$ 6 inches.

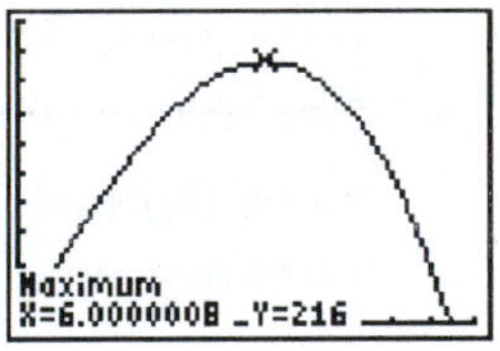

Chapter 2 Project — A Graphical Approach to Finding Zeros

1. From the graphing utility graph, we can see that the y-coordinate of the point of intersection is approximately 1.893.

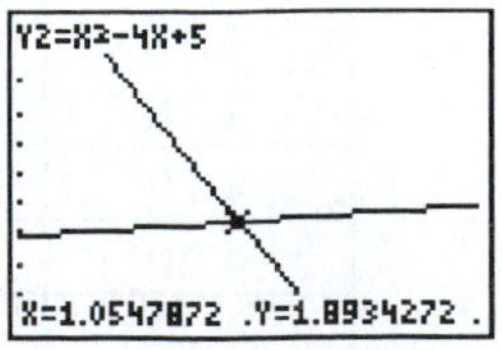

2. By graphing the polynomial $y = x^4 - 8x^3 + 32x^2 - 63x + 39$ with a graphing utility, we find the following.
   (a) Answers will vary. To approximate the first solution to five decimal places, one setting that can be used is demonstrated in the following figures.

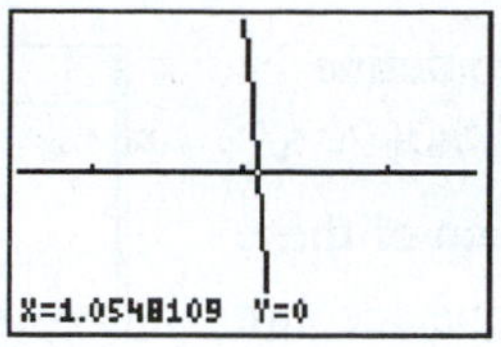

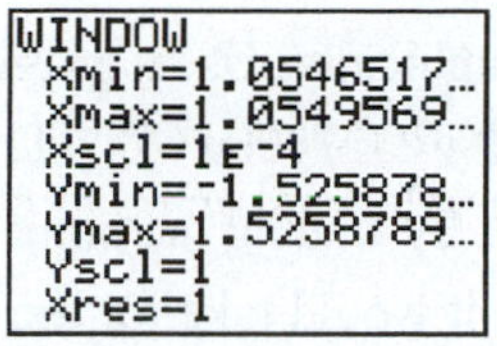

(b) Answers will vary. To approximate the second solution to five decimal places, one setting that can be used is demonstrated in the following figures.

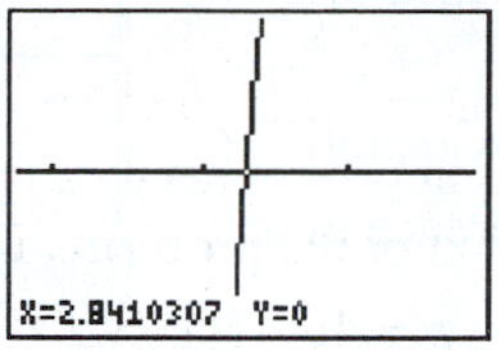

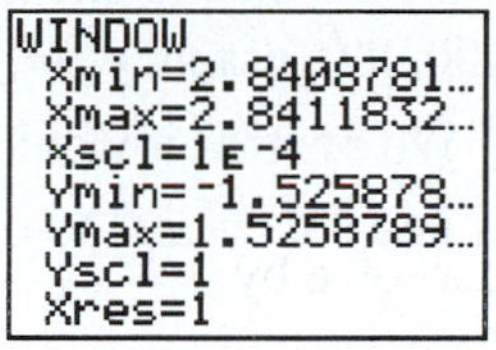

3. Using a graphing utility to graph $f(x) = x^2 + x - 1$, we obtain the graphs at the right. The zeros of the function are $x \approx -1.618$ and $x \approx 0.618$.

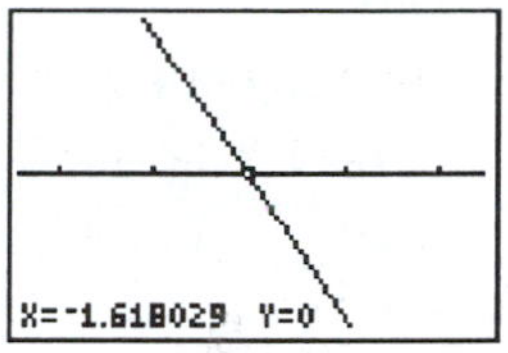

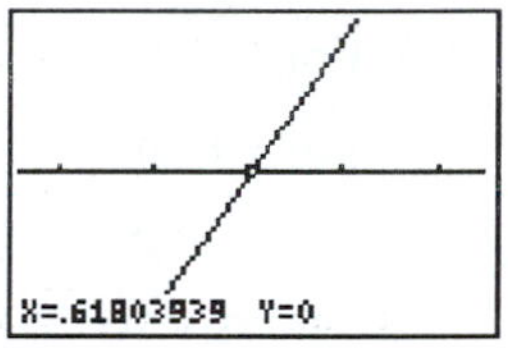

For the analytic solution, use the quadratic equation with $a = 1$, $b = 1$, and $c = -1$.

$$x = \frac{-1 \pm \sqrt{1^2 - 4(1)(-1)}}{2(1)}$$

$$= \frac{-1 \pm \sqrt{5}}{2}$$

$$\approx -1.618, \quad 0.618$$

4. Using a graphing utility to graph $y = x^2 - 3x + 2$ and $y = -2 + \sqrt{17 + 5x - x^2}$, we find the points of intersection to be approximately (– 0.034, 2.102) and (3.236, 2.765).

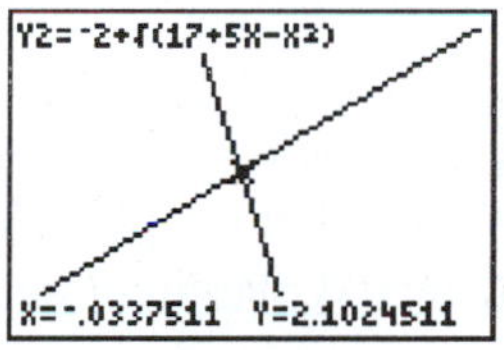

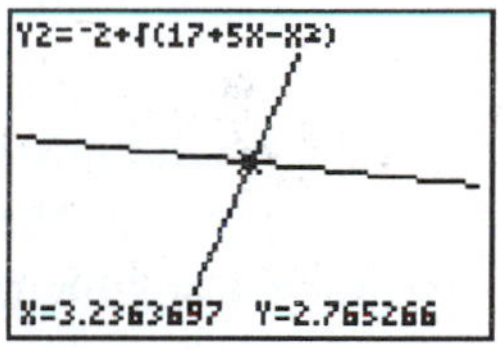

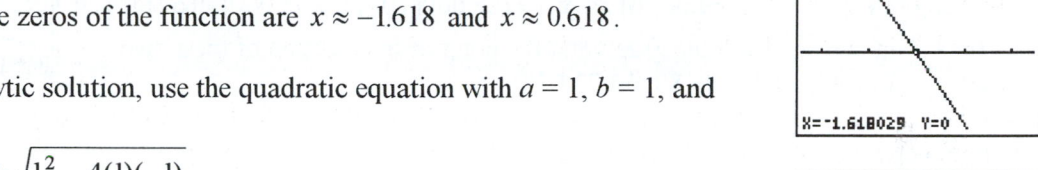

Chapter 3 Project A Graphical Approach to Compound Interest

1. To compare these two options, compute the effective yield for each.

8.05% compounded monthly:

$$\text{Effective yield} = \left(1 + \frac{0.0805}{12}\right)^{12(1)} \approx 1.0835$$

8% compounded continuously:

$$\text{Effective yield} = e^{0.08(1)} \approx 1.0833$$

Thus, an annual interest rate of 8.05% compounded monthly would produce a larger balance because this option has a slightly higher effective yield, approximately 8.35%, as compared to the option of 8% compounded continuously, having an effective yield of approximately 8.33%.

2. We can approach this problem graphically. Use a graphing utility to define $y_1 = 1000(1 + 0.0805/12)^{12x}$ and $y_2 = 1000e^{0.08x}$. Then graph $y_3 = y_1 - y_2$ and $y_4 = 100$ on the same screen. By finding the point of intersection of these two graphs, we can find how long it would take for the balance of one account to exceed the balance in the other account by \$100. We see that it would take approximately 32.4 years (or at least 33 years in whole years) to obtain a difference of at least \$100.

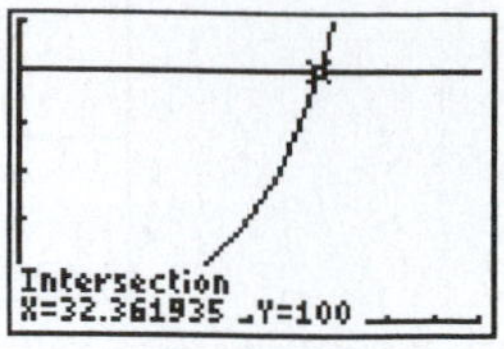

To find how long it would take for the balance of one account to exceed the balance in the other account by \$100,000, graph $y_3 = y_1 - y_2$ and $y_4 = 100{,}000$ on the same screen. We see that it would take approximately 104.01 years (or at least 105 years in whole years) to obtain a difference of at least \$100,000.

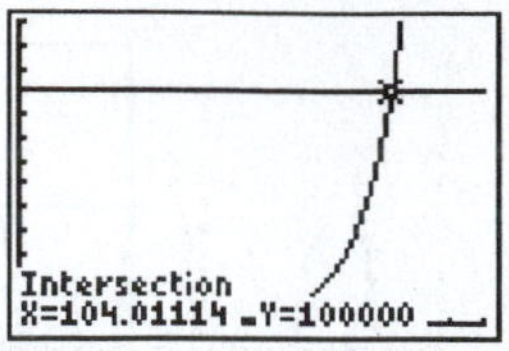

3. The answer will vary depending on how many years there are until retirement. This can be seen graphically by using a graphing utility to graph option (a) as $y_1 = 25{,}000(1.05)^x$ and option (b) as $y_2 = (25{,}000(1.07)^x - 25{,}000) \cdot 0.6 +$ 25,000. You can see from the graph that the point of intersection of these two graphs is approximately (16.80, 56,734.58).

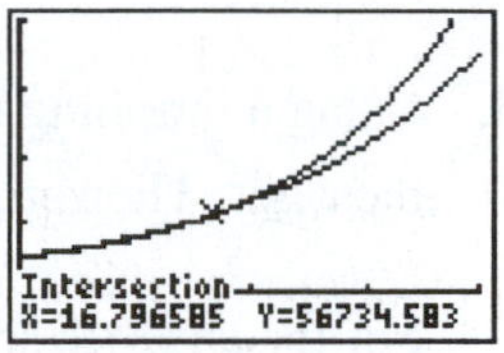

Let's compare each case at 12 years and at 40 years.

(a) For $x = 12$, $A = 25{,}000(1 + 0.05)^{12} \approx 44{,}896.41$.

For $x = 40$, $A = 25{,}000(1 + 0.05)^{40} \approx 175{,}999.72$.

(b) For $x = 12$, $A = 25{,}000(1 + 0.07)^{12} \approx 56{,}304.79$. After taxes on the earned interest, there would be $(56{,}304.79 - 25{,}000) \cdot 0.6 + 25{,}000 = 43{,}782.87$ left.

For $x = 40$, $A = 25{,}000(1 + 0.07)^{40} \approx 374{,}361.45$. After taxes on the earned interest, there would be $(374{,}361.45 - 25{,}000) \cdot 0.6 + 25{,}000 = 234{,}616.87$ left.

These calculations lead us to believe that if the length of the investment (i.e., time until retirement) is 16 years or less, the tax-free plan in option (a) is better. However, if the length of the investment is 17 years or more, the tax-deferred plan in option (b) is better.

Chapter 4 Project Analyzing a Graph

1.

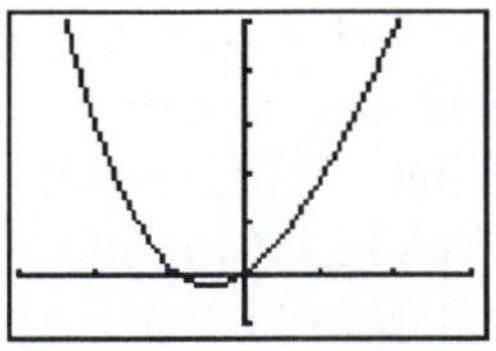

$-3 \le x \le 3$, $-1 \le y \le 5$

2.

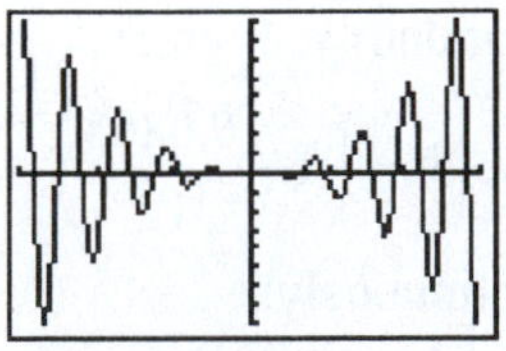

$-30 \le x \le 30$, $-720 \le y \le 720$

3.

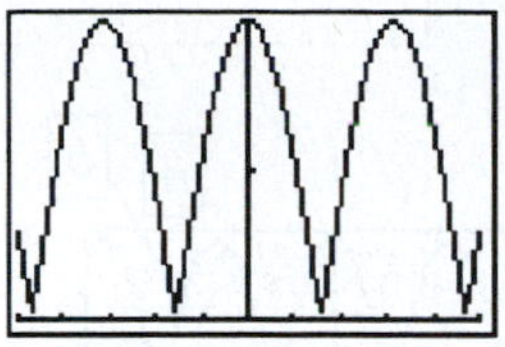

$-5 \le x \le 5,\ 0 \le y \le 1$

4.

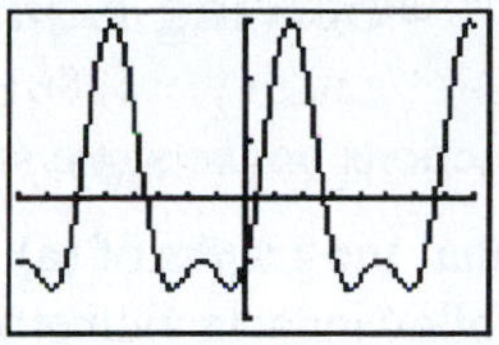

$-8 \le x \le 8,\ -2 \le y \le 3$

5.

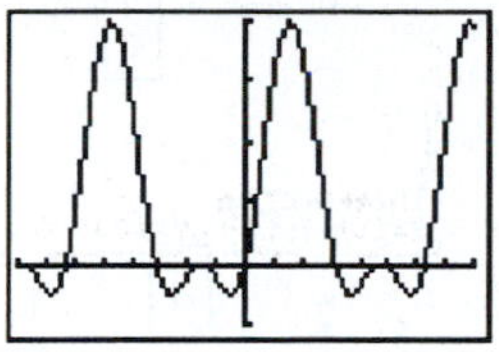

$-8 \le x \le 8,\ -0.5 \le y \le 2$

6.

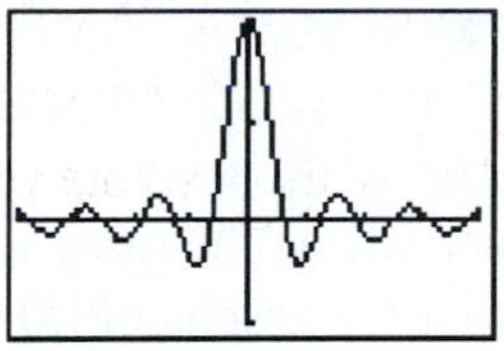

$-20 \le x \le 20,\ -0.5 \le y \le 1$

7. After sketching the model from Example 1 with a graphing utility, we can use either the trace or maximum/minimum feature of the graphing utility to find the highest and lowest levels of carbon dioxide between January 1988 and January 1990. (Each of the following graphs uses the ranges $28 \le x \le 30,\ 347 \le y \le 358$.)

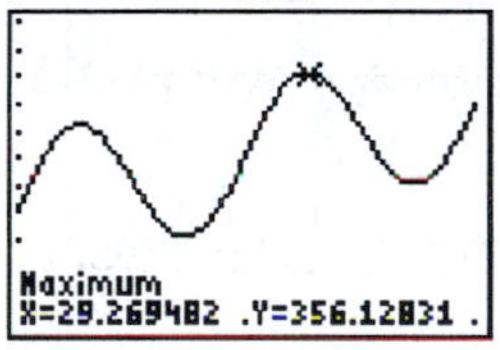

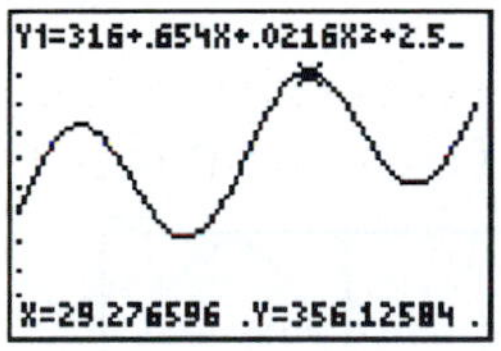

The highest level of carbon dioxide was approximately 356.13 parts per million in 1989.

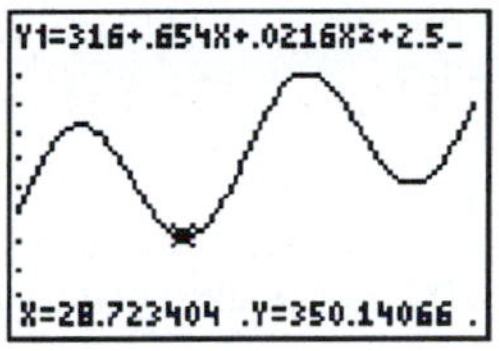

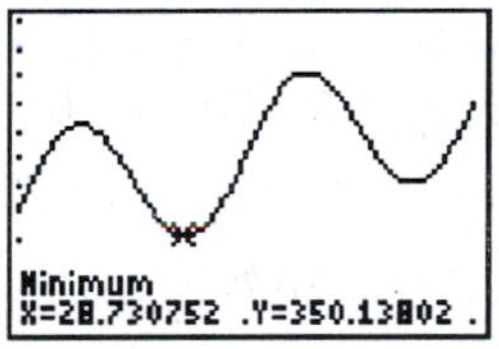

The lowest level of carbon dioxide was approximately 350.14 parts per million in 1988.

8. (a) Answers will vary. For instance, for $v = 45$ feet per second (or 66 miles per hour), $h = 5$ feet, and $\theta = 40°$, the sketch of the path of the shot put is as follows.

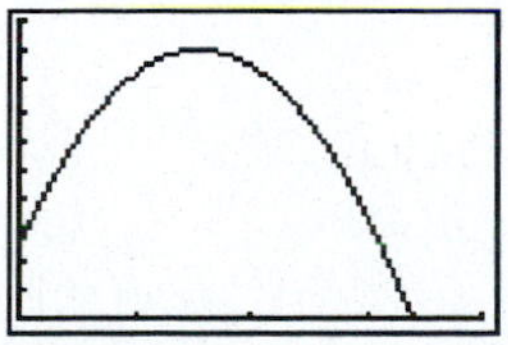

$0 \le x \le 80,\ 0 \le y \le 20$

Answers will vary depending on the graphs sketched. The values for v, h, and θ used in part (a) seem to be reasonable. For reference, the men's world record for shot put is 75 feet, 10.25 inches (set by Randy Barnes of the USA in 1990). The women's world record for shot put is 74 feet, 3 inches (set by Natalya Lisovskaya of the USSR in 1987). Notice from the graph in part (a) that the distance the shot put traveled was approximately 68 feet, which is reasonable in light of the world records.

Chapter 5 Project Solving Equations Graphically

1. By graphing all three functions on the same screen with a graphing utility, we see that the graphs of (a) and (b) are the same. This leads to the following trigonometric identity.

$$\sin^2 x = \tfrac{1}{2}(1-\cos 2x)$$
$$2\sin^2 x = 1-\cos 2x$$
$$\cos 2x = 1-2\sin^2 x$$

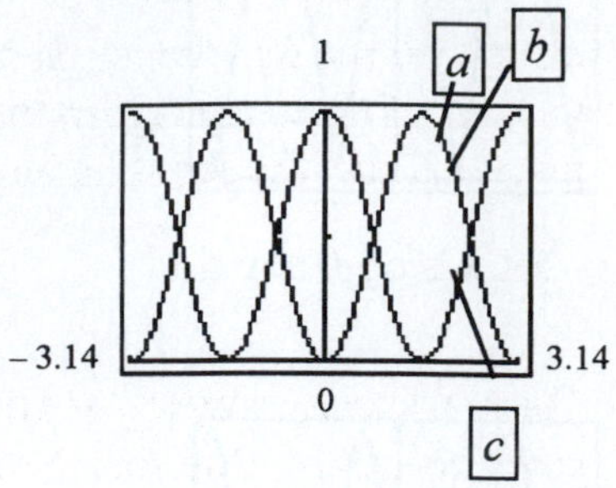

2. By graphing all three functions on the same screen with a graphing utility, we see that the graphs of (a) and (c) are the same. This leads to the following trigonometric identity.

$$2\cos^2 x = 1+\cos 2x$$
$$-\cos 2x = 1-2\cos^2 x$$
$$\cos 2x = 2\cos^2 x - 1$$

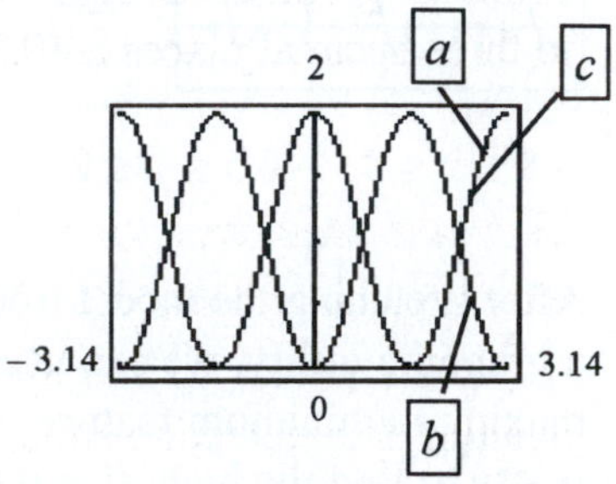

3. By graphing all three functions on the same screen with a graphing utility, we see that the graphs of (a) and (c) are the same. This leads to the trigonometric identity $2\sin x\cos 2x = \sin 3x - \sin x$.

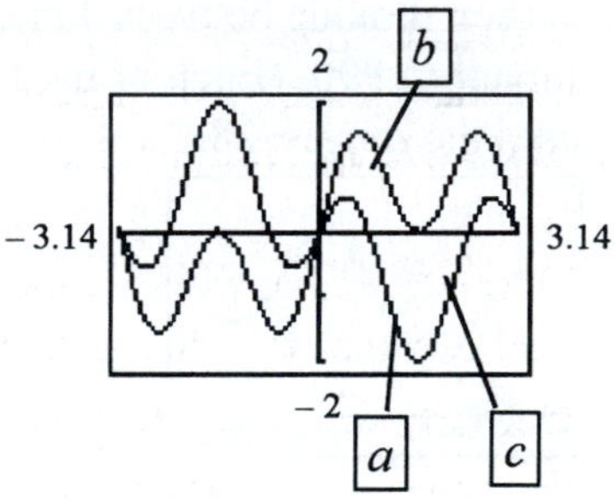

4. By graphing all three functions on the same screen with a graphing utility, we see that the graphs of (a) and (c) are the same. This leads to the trigonometric identity $\sin 2x = 2\sin x\cos x$.

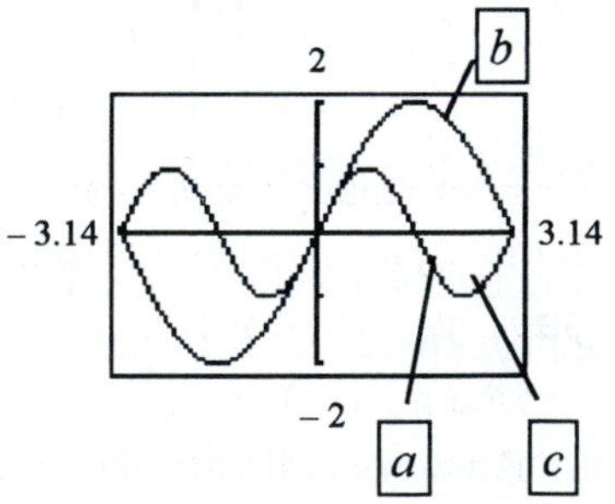

5. By graphing the equation $y = \sin 2x - x$ with a graphing utility, we see that the graph crosses the x-axis only three times. Thus, the equation has a finite number of solutions.

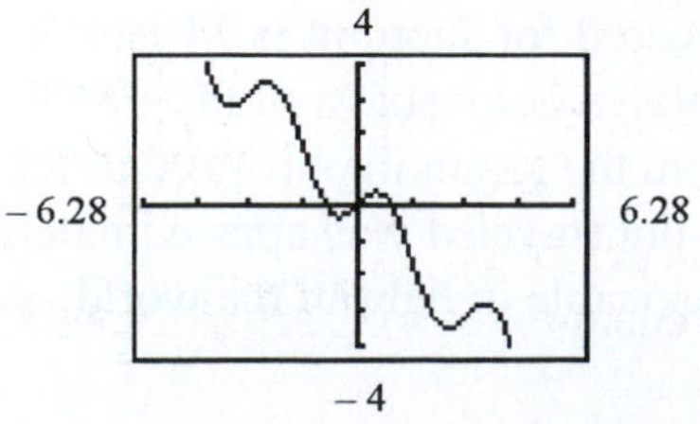

6. By graphing the equation $y = \tan 2x - x$ with a graphing utility, we see that the graph crosses the x-axis an infinite number of times (disregard the "connected" portions of the graph). Thus, the equation has an infinite number of solutions.

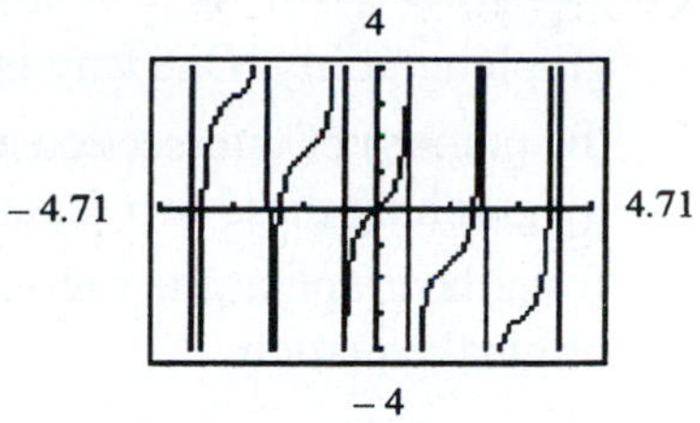

7. There are many ways to approximate the solution(s) of this equation with a graphing utility. One way is to graph $y_1 = x + \sin x$ and $y_2 = 1$, and find the point(s) of intersection either by tracing or using the intersect feature. From the graph at the right, we can see that the point of intersection correct to three decimal places is (0.511, 1). Thus, the solution of the equation is $x \approx 0.511$.

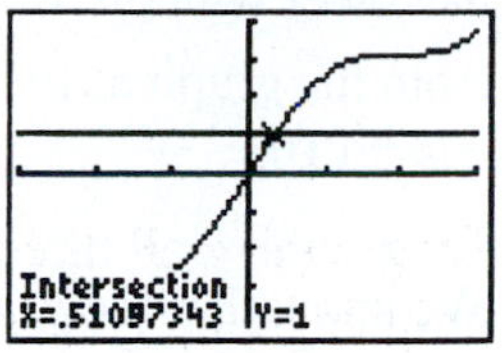

8. There are many ways to approximate the solution(s) of this equation with a graphing utility. One way is to graph $y_1 = x^2 + \cos x$ and $y_2 = 2$, and find the point(s) of intersection either by tracing or using the intersect feature. From the graphs at the right, we can see that the points of intersection correct to three decimal places are (– 1.326, 2) and (1.326, 2). Thus, the solutions of the equation are $x \approx -1.326,\ 1.326$.

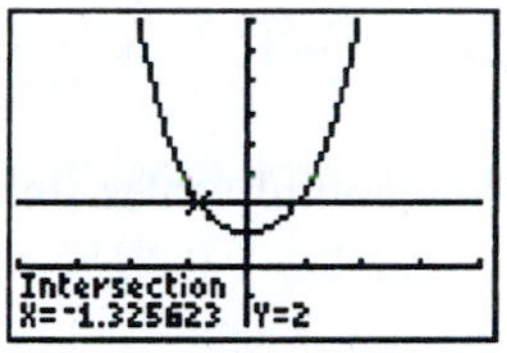

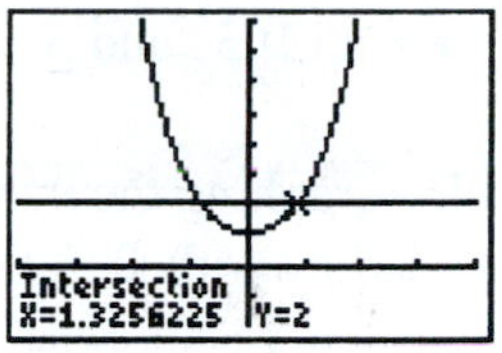

9. There are many ways to approximate the solution(s) of this equation with a graphing utility. One way is to graph $y_1 = 5\cos(1/(x^2+1))$ and $y_2 = 3$, and find the point(s) of intersection either by tracing or using the intersect feature. From the graphs at the right, we can see that the points of intersection correct to three decimal places are (– 0.280, 3) and (0.280, 3). Thus, the solutions of the equation are $x \approx -0.280,\ 0.280$.

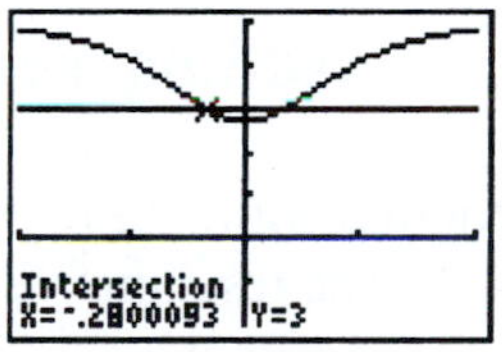

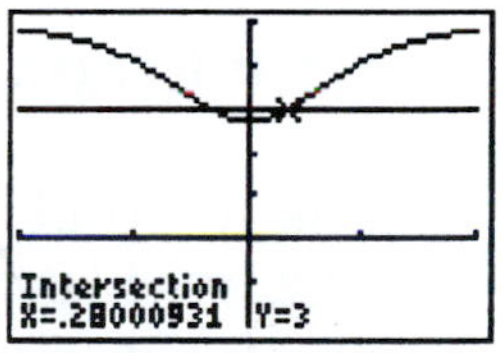

10. There are many ways to approximate the solution(s) of this equation with a graphing utility. One way is to graph $y_1 = |x| + \sec(1/(x^2+1))$ and $y_2 = 3$, and find the point(s) of intersection either by tracing or using the intersect feature. From the graphs at the right, we can see that the points of intersection correct to three decimal places are (– 1.979, 3) and (1.979, 3). Thus, the solutions of the equation are $x \approx -1.979,\ 1.979$.

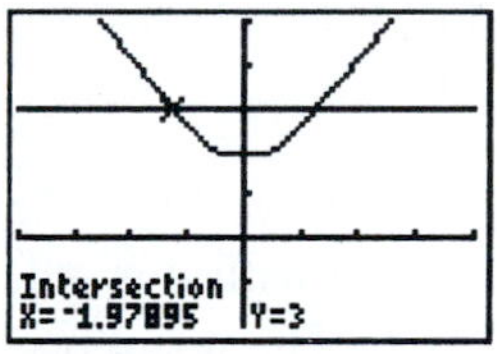

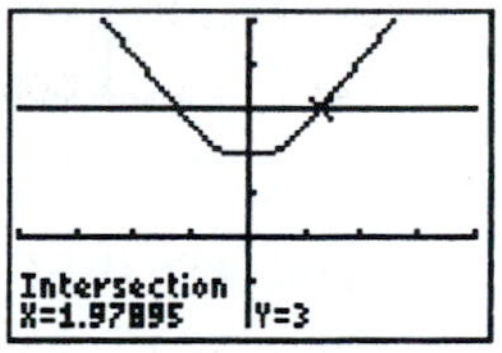

11. There are many ways to approximate the solution(s) of this equation with a graphing utility. One way is to graph $y_1 = x + 1.25$ and $y_2 = \arctan x$, and find the point(s) of intersection either by tracing or using the intersect feature. From the graph at the right, we can see that the point of intersection correct to three decimal places is (– 2.430, – 1.180). Thus, the solution of the equation is $x \approx -2.430$.

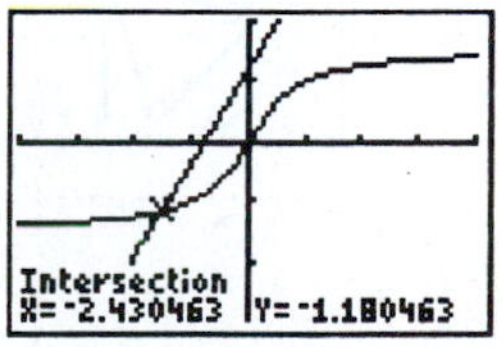

12. There are many ways to approximate the solution(s) of this equation with a graphing utility. One way is to graph $y_1 = \cos x$ and $y_2 = 1/(x^2+1)$, and find the point(s) of intersection either by tracing or using the intersect feature. Begin by graphing these two functions on $-3\pi \le x \le 3\pi$ and $-1 \le y \le 1$. We can see from the graph at the right that there are an infinite number of solutions to the original equation.

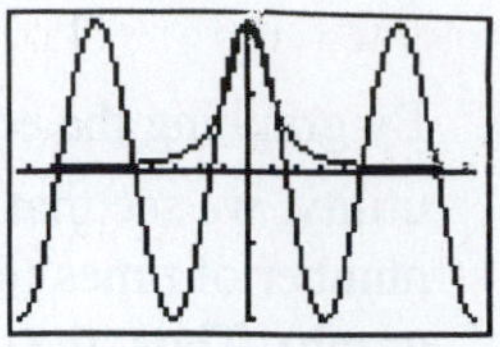

We can start estimating the solutions by graphing the two functions in a narrower window with $-\pi/2 \le x \le \pi/2$ and $-0.3 \le y \le 1.1$. As can be seen from the graph at the right, there are three solutions in this interval: $x = 0$ and $x \approx \pm 1.103$.

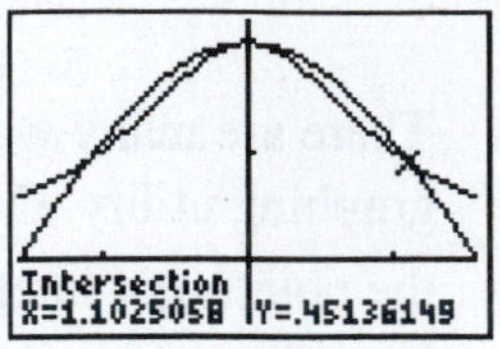

We know that the graph of $y_2 = 1/(x^2+1)$ approaches, but never crosses, $y = 0$ (the x-axis) as x approaches $\pm\infty$. Therefore, the solutions of the equation $\cos x = 1/(x^2+1)$ are approximated by the solutions to $\cos x = 0$ as x approaches $\pm\infty$. Thus, the solutions of the original equation are $x = 0$, $x \approx \pm 1.103$, and $x \approx \pm[(2n+1)\pi]/2$, for $n = 1, 2, 3, \ldots$

13. By graphing the model for monthly sales (in thousands) along with $y = 100$ in a window with $0 \le x \le 12$ and $0 \le y \le 125$, we see that sales exceeded 100,000 units in the months of February, March, April, and May.

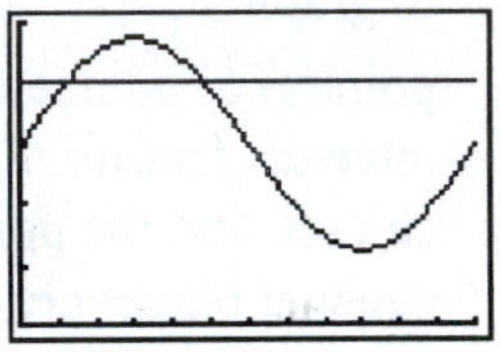

Chapter 6 Project Adding Vectors Graphically

1. Run the ADDVECT program with $A = 3$, $B = 4$, $C = -5$, and $D = 1$. The resulting graphing utility display is shown below.

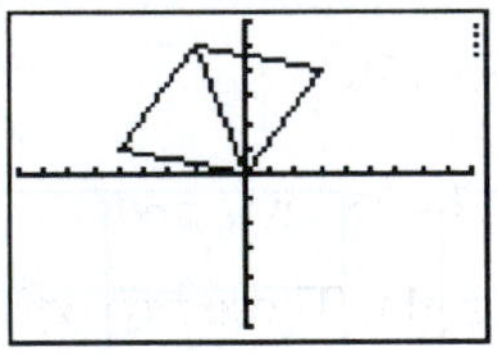

2. Run the ADDVECT program with $A = 5$, $B = -4$, $C = 3$, and $D = 2$. The resulting graphing utility display is shown below.

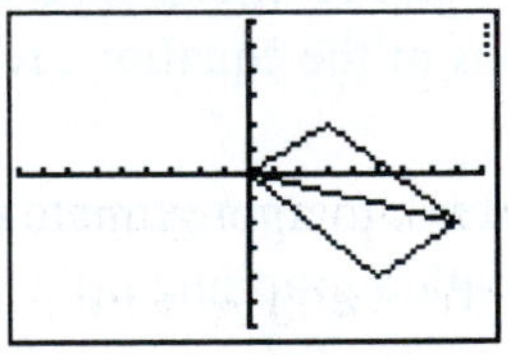

3. Run the ADDVECT program with $A = -4$, $B = 4$, $C = -2$, and $D = -6$. The resulting graphing utility display is shown below.

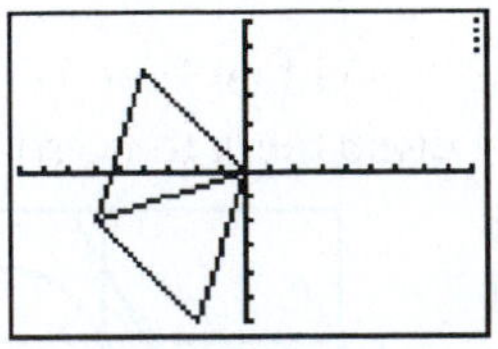

4. Run the ADDVECT program with $A = 7$, $B = 3$, $C = -2$, and $D = -6$. The resulting graphing utility display is shown below.

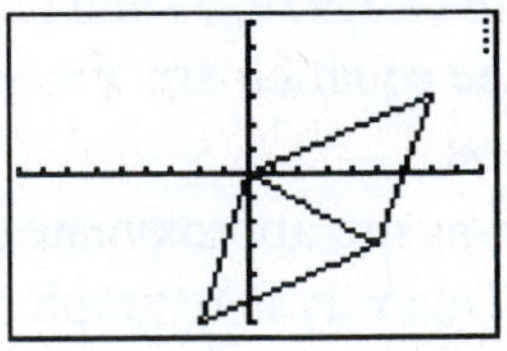

5. To find the speed of the airplane after encountering the wind, find the length of the resultant vector.

$$\|\mathbf{v}_1 + \mathbf{v}_2\| = \sqrt{(400\cos 150° + 75\cos 50°)^2 + (400\sin 150° + 75\sin 50°)^2}$$
$$\approx 394 \text{ miles per hour}$$

Because the original speed of the airplane was 400 miles per hour, the plane is now traveling slower.

6. To obtain the answer experimentally, try substituting different values for the wind's velocity into the program ADDVECT and approximate the components of the resultant vector. Then use the approximations to find the velocity. To obtain the answer analytically, solve the following equation for w, the wind's velocity.

$$\tan 140° = \frac{400\sin 150° + w\sin 50°}{400\cos 150° + w\cos 50°}$$
$$\tan 140°(400\cos 150° + w\cos 50°) = 400\sin 150° + w\sin 50°$$
$$400\tan 140° \cos 150° + w\tan 140° \cos 50° = 400\sin 150° + w\sin 50°$$
$$w(\tan 140° \cos 50° - \sin 50°) = 400\sin 150° - 400\tan 140° \cos 150°$$
$$w = \frac{400\sin 150° - 400\tan 140° \cos 150°}{\tan 140° \cos 50° - \sin 50°}$$
$$w \approx 69.5 \text{ miles per hour}$$

7. To obtain the answer experimentally, try substituting different values for the wind's direction into the program ADDVECT and approximate the components of the resultant vector. Then use the approximations to find the direction. To obtain the answer analytically, solve the following equation for θ, the wind's direction.

$$\tan 140° = \frac{400\sin 150° + 75\sin\theta}{400\cos 150° + 75\cos\theta}$$
$$\tan 140°(400\cos 150° + 75\cos\theta) = 400\sin 150° + 75\sin\theta$$
$$400\tan 140° \cos 150° + 75\tan 140° \cos\theta = 400\sin 150° + 75\sin\theta$$
$$75\tan 140° \cos\theta - 75\sin\theta = 400\sin 150° - 400\tan 140° \cos 150°$$

We can solve this equation graphically by graphing $y_1 = 75\tan 140° \cos\theta - 75\sin\theta$ and $y_2 = 400\sin 150° - 400\tan 140° \cos 150°$ in the same window, as at the right. These two equation intersect at approximately (27.8. – 90.7). Thus, the direction of the wind is about N 62.2 °E.

Chapter 7 Project Fitting Models to Data

1. Using the linear regression feature of a graphing utility, we find that the linear model that best represents the data is $y = 0.707t + 54.894$. To project the number of Sunday newspapers to be sold each week in 1998, evaluate the model at $t = 18$.

$$y = 0.707t + 54.894$$
$$= 0.707(18) + 54.894$$
$$= 67.62$$

In 1998, 67.62 million Sunday newspapers will be sold each week.

2. Using the linear regression feature of a graphing utility, we find that the linear model that best represents the data for morning newspaper companies is $y = 16.101t + 397.258$. The linear model for the evening newspaper company data is $y = -30.234t + 1376.939$. The linear model for the Sunday newspaper company data is $y = 12.448t + 736.924$.

3. In 1981, the average circulation per morning newspaper company was

$$\frac{30{,}600{,}000}{408} = 75{,}000 \text{ newspapers per day}$$

In 1992, the average circulation per morning newspaper company was

$$\frac{42{,}400{,}000}{596} \approx 71{,}141 \text{ newspapers per day}$$

Thus, the average circulation per morning newspaper company decreased from 1981 to 1992.

4. In 1981, the average circulation per evening newspaper company was

$$\frac{30{,}900{,}000}{1352} \approx 22{,}855 \text{ newspapers per day}$$

In 1992, the average circulation per evening newspaper company was

$$\frac{17{,}800{,}000}{996} \approx 17{,}871 \text{ newspapers per day}$$

Thus, the average circulation per evening newspaper company decreased from 1981 to 1992.

5. In 1981, the average circulation per Sunday newspaper company was

$$\frac{55{,}200{,}000}{755} \approx 73{,}113 \text{ newspapers per week}$$

In 1992, the average circulation per Sunday newspaper company was

$$\frac{62{,}200{,}000}{891} \approx 69{,}809 \text{ newspapers per week}$$

Thus, the average circulation per Sunday newspaper company decreased from 1981 to 1992.

6. The best newspaper investment choice would be a morning newspaper company because average circulation per morning newspaper company has not decreased as much as for evening newspaper or Sunday newspaper companies. Also, there are fewer morning newspaper companies to compete with.

7. Average number of newspapers of all types sold per day in 1981: $30.6 + 30.9 + 55.2/7 \approx 69.4$ papers.
Average number of newspapers of all types sold per day in 1982: $33.2 + 29.3 + 56.3/7 \approx 70.5$ papers
Average number of newspapers of all types sold per day in 1983: $33.8 + 28.8 + 56.7/7 \approx 70.7$ papers
Average number of newspapers of all types sold per day in 1984: $35.4 + 27.7 + 57.5/7 \approx 71.3$ papers
Average number of newspapers of all types sold per day in 1985: $36.4 + 26.4 + 58.8/7 \approx 71.2$ papers
Average number of newspapers of all types sold per day in 1986: $37.4 + 25.1 + 58.9/7 \approx 70.9$ papers
Average number of newspapers of all types sold per day in 1987: $39.1 + 23.7 + 60.1/7 \approx 71.4$ papers
Average number of newspapers of all types sold per day in 1988: $40.4 + 22.2 + 61.5/7 \approx 71.4$ papers
Average number of newspapers of all types sold per day in 1989: $40.7 + 21.8 + 62.0/7 \approx 71.4$ papers
Average number of newspapers of all types sold per day in 1990: $41.3 + 21.0 + 62.6/7 \approx 71.3$ papers
Average number of newspapers of all types sold per day in 1991: $41.5 + 19.2 + 62.2/7 \approx 69.6$ papers
Average number of newspapers of all types sold per day in 1992: $42.2 + 17.8 + 62.2/7 \approx 69.1$ papers.

We can see that the average number of newspapers of all types sold each day remained relatively constant from 1981 to 1992, while both the number of households and the total population increased.

Therefore, we can conclude that the percent of Americans who read newspapers was decreasing from 1981 through 1992.

Chapter 8 Project Solving Systems of Equations

1. $A=\begin{bmatrix}1&-1&10\\3&0&1\\7&2&1\end{bmatrix}$. Using a graphing utility, we find the inverse matrix $A^{-1}=\frac{1}{54}\begin{bmatrix}-2&21&-1\\4&-69&29\\6&-9&3\end{bmatrix}$.

We can use a graphing utility to find the matrix product $A^{-1}B=\frac{1}{54}\begin{bmatrix}-2&21&-1\\4&-69&29\\6&-9&3\end{bmatrix}\begin{bmatrix}2\\4\\0\end{bmatrix}\approx\begin{bmatrix}1.48\\-4.96\\-0.44\end{bmatrix}$.

2. $A=\begin{bmatrix}4&0&2\\-1&2&5\\3&1&-7\end{bmatrix}$. Using a graphing utility, we find the inverse matrix $A^{-1}=\frac{1}{90}\begin{bmatrix}19&-2&4\\-8&34&22\\7&4&-8\end{bmatrix}$.

We can use a graphing utility to find the matrix product $A^{-1}B=\frac{1}{90}\begin{bmatrix}19&-2&4\\-8&34&22\\7&4&-8\end{bmatrix}\begin{bmatrix}3\\-1\\10\end{bmatrix}=\begin{bmatrix}1.1\\1.8\\-0.7\end{bmatrix}$.

3. $A=\begin{bmatrix}2&6&-1\\3&1&2\\6&-1&3\end{bmatrix}$. Using a graphing utility, we find the inverse matrix $A^{-1}=\frac{1}{37}\begin{bmatrix}5&-17&13\\3&12&-7\\-9&38&-16\end{bmatrix}$.

We can use a graphing utility to find $A^{-1}B=\frac{1}{37}\begin{bmatrix}5&-17&13\\3&12&-7\\-9&38&-16\end{bmatrix}\begin{bmatrix}4\\-4\\1\end{bmatrix}\approx\begin{bmatrix}2.73\\-1.16\\-5.51\end{bmatrix}$.

4. $A=\begin{bmatrix}5&-3&4\\2&2&-1\\1&-1&2\end{bmatrix}$. Using a graphing utility, we find the inverse matrix $A^{-1}=\frac{1}{14}\begin{bmatrix}3&2&-5\\-5&6&13\\-4&2&16\end{bmatrix}$. We can use a graphing utility to find the matrix product $A^{-1}B=\frac{1}{14}\begin{bmatrix}3&2&-5\\-5&6&13\\-4&2&16\end{bmatrix}\begin{bmatrix}25\\-1\\9\end{bmatrix}=\begin{bmatrix}2\\-1\\3\end{bmatrix}$.

5. Use the information from the graph to form the following system of linear equations.

$$\begin{aligned}0.17x+0.09y+0.02z&=268.37\\0.05x+0.11y+0.03z&=160.04\\0.09x+0.19y+0.10z&=346.73\end{aligned}$$

We can use a graphing utility to solve this system of equations: $A^{-1}B=\begin{bmatrix}1113\\574\\1375\end{bmatrix}$. Therefore, \$1113 million dollars of gym shoes, \$574 million of jogging shoes, and \$1375 million of walking shoes were sold in 1992.

6. Using a graphing utility, we can solve the system of equations:

$$A^{-1}B=\begin{bmatrix}0.40&0.50\\0.60&0.50\end{bmatrix}^{-1}\begin{bmatrix}6.02\\8.24\end{bmatrix}=\begin{bmatrix}11.1\\3.16\end{bmatrix}.$$

Therefore, 11.1 million people participated in downhill skiing and 3.16 million people participated in cross-country skiing in 1992.

7. Using a graphing utility, we can solve the system of equations:

$$A^{-1}B = \begin{bmatrix} 0.94 & 0.92 & 0.80 \\ 0.04 & 0.02 & 0.04 \\ 0.02 & 0.06 & 0.16 \end{bmatrix}^{-1} \begin{bmatrix} 20{,}144 \\ 766 \\ 1990 \end{bmatrix} = \begin{bmatrix} 6600 \\ 7500 \\ 8800 \end{bmatrix}$$

Therefore, we can make 6600 grams of alloy X, 7500 grams of alloy Y, and 8800 grams of alloy Z.

Chapter 9 Project Recognizing Patterns in Data

1. Begin by constructing a table of the information given for the number of points n and the angle measure A_n. Add a row for the product nA_n.

n	5	6	7	8	9	10
A_n	36°	60°	$77\frac{1}{7}°$	90°	100°	108°
nA_n	180°	360°	540°	720°	900°	1080°

From this table, we can see that the products nA_n form an arithmetic sequence whose nth term is

$$nA_n = 180(n-4) \quad \Rightarrow \quad A_n = \frac{180(n-4)}{n}.$$

2. Again, begin by constructing a table of the information given for the number of points n and the angle measure A_n. Add a row for the product nA_n.

n	7	8	9	10	11	12
A_n	$25\frac{5}{7}°$	45°	60°	72°	$81\frac{9}{11}°$	90°
nA_n	180°	360°	540°	720°	900°	1080°

From this table, we can see that the products nA_n form an arithmetic sequence whose nth term is

$$nA_n = 180(n-6) \quad \Rightarrow \quad A_n = \frac{180(n-6)}{n}.$$

3. The models for the angle measures we found in Example 1, Question 1, and Question 2 are:

$$A_n = \frac{180(n-2)}{n}, \quad A_n = \frac{180(n-4)}{n}, \quad A_n = \frac{180(n-6)}{n}$$

These formulas form the pattern $A_n = \frac{180(n-2b)}{n}$, where the star is formed by connecting every bth point. Because the given star pattern connects every fourth point on a circle, b = 4, and the model for the angle measures of the star tips is given by

$$A_n = \frac{180(n-2(4))}{n} = \frac{180(n-8)}{n}.$$

Chapter 10 Project Graphing in Parametric and Polar Modes

1. Using the vertical line test, we can see that y is not a function of x.

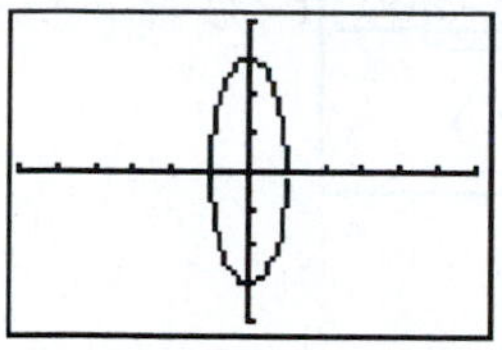

2. Using the vertical line test, we can see that y is not a function of x.

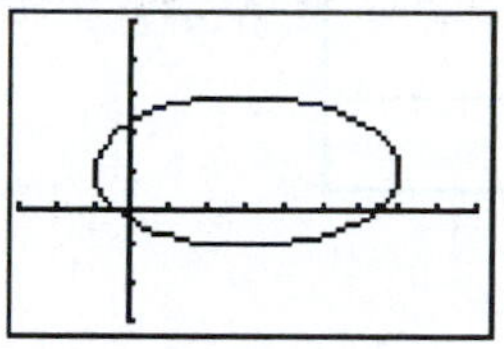

3. Using the vertical line test, we can see that y is not a function of x.

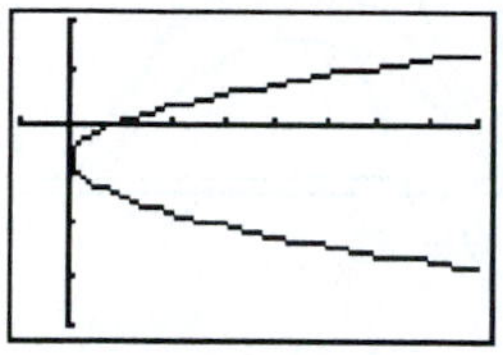

4. Using the vertical line test, we can see that y is a function of x.

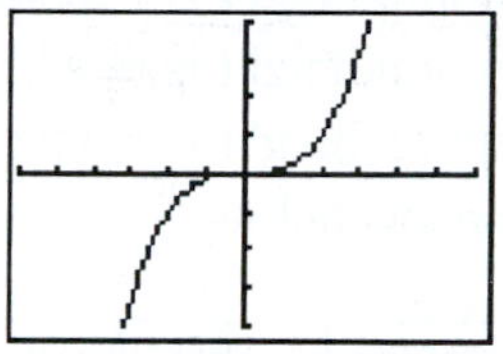

5. Using the vertical line test, we can see that y is not a function of x.

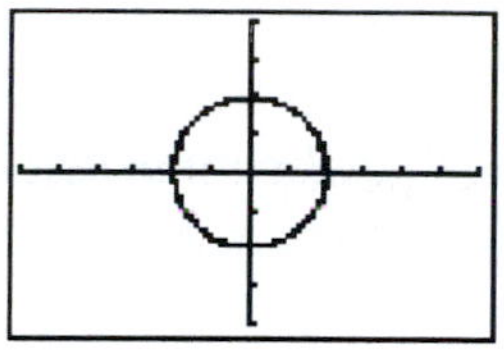

6. Using the vertical line test, we can see that y is not a function of x.

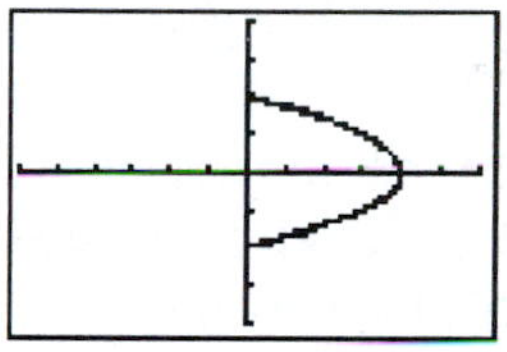

7. Dimpled limaçon

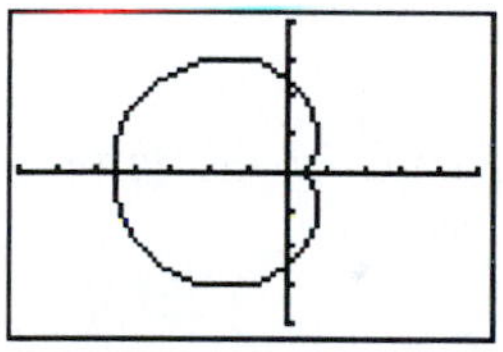

8. Vertical line

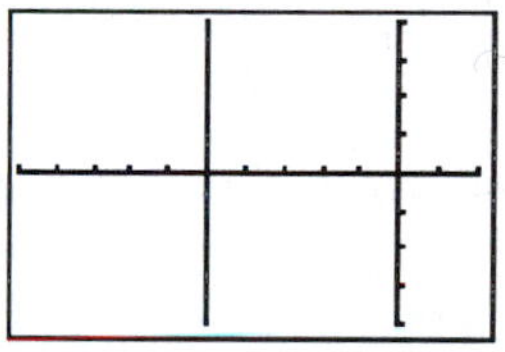

9. Hyperbola

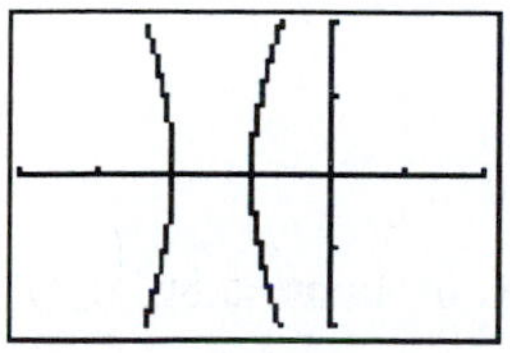

10. Ellipse

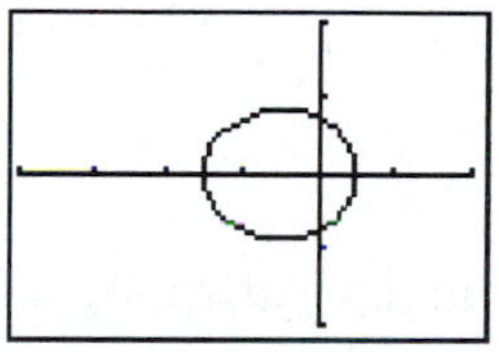

11. Parabola

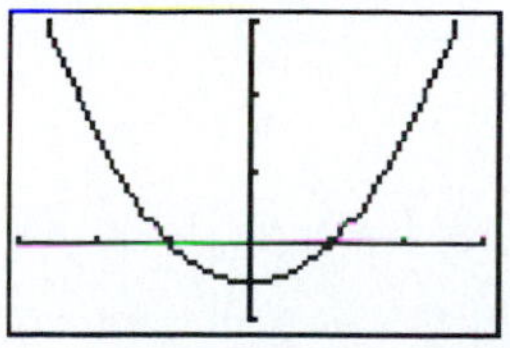

12. Rose curve

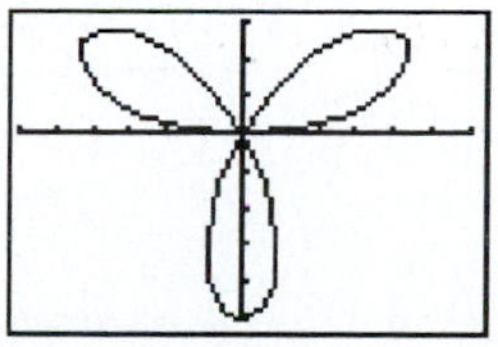

13. Horizontal line

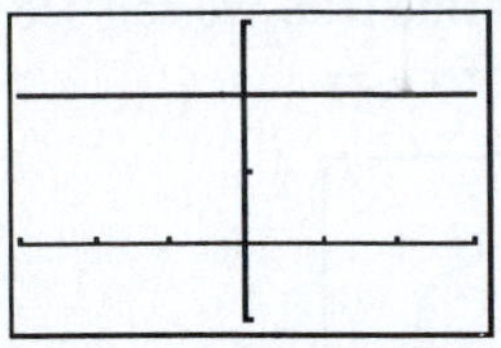

14. Rose curve

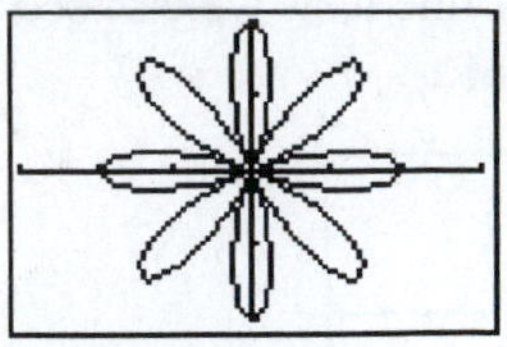

15. The graphs of all three baseball paths are shown at the right. We can use the graphing utility's trace feature to show that the baseballs do not reach their maximum heights at the same time. For instance, when $\theta = 35°$, the maximum height is reached at $t \approx 1.8$ seconds. When $\theta = 45°$, the maximum height is reached at $t \approx 2.2$ seconds. When $\theta = 55°$, the maximum height is reached at $t \approx 2.6$ seconds. Refer to the graphing utility screens below.

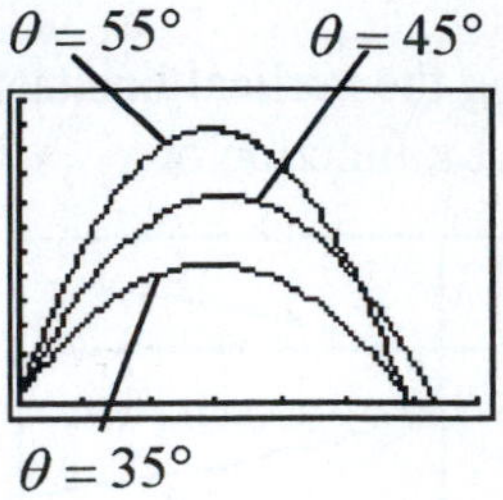

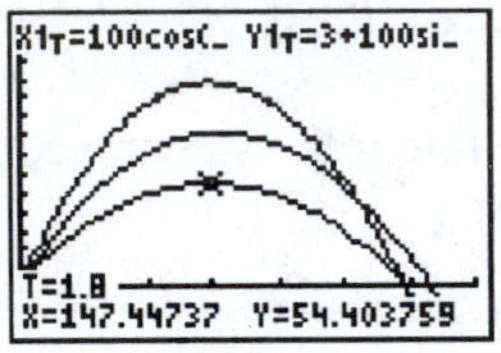

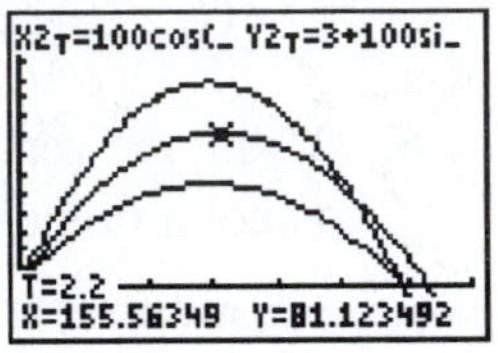

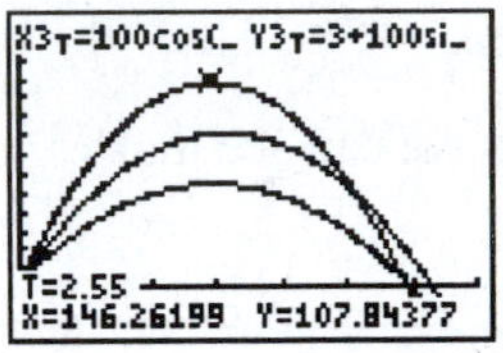

16. No, we can see from the graphs above that the baseballs do not reach the same maximum height. For instance when $\theta = 35°$, the maximum height is ≈ 54 feet. When $\theta = 45°$, the maximum height is ≈ 81 feet. When $\theta = 55°$, the maximum height is ≈ 108 feet.